# The Black Flies (Simuliidae) of North America

**Simulium vittatum Zett.**

Female of *Simulium vittatum* s. s. Image by Klaus Bolte, copyright Canadian Forest Service, Ottawa.

# The Black Flies (Simuliidae) of North America

PETER H. ADLER
DOUGLAS C. CURRIE
D. MONTY WOOD

*with a foreword*
*by Daniel H. Janzen*

&

*illustrations by Ralph M. Idema*
*& Lawrence W. Zettler*

COMSTOCK PUBLISHING ASSOCIATES,
*a division of*
*Cornell University Press*
*Ithaca & London*

*in association with the*
ROYAL ONTARIO MUSEUM

RŌM
A ROM PUBLICATION IN SCIENCE

**The Royal Ontario Museum gratefully acknowledges the Louise Hawley Stone Charitable Trust within the ROM Foundation for its generous support of this publication.**



First published 2004 by Cornell University Press in Association with the Royal Ontario Museum

A ROM PUBLICATION IN SCIENCE

Printed in the United States of America

Library of Congress Cataloging-in-Publication Data

Adler, Peter H. (Peter Holdridge), 1954–
The black flies (Simuliidae) of North America / Peter H. Adler, Douglas C. Currie, D. Montgomery Wood; with a foreword by Daniel H. Janzen; illustrations by Ralph M. Idema and Lawrence W. Zettler.
p. cm.
Includes bibliographical references and index.
ISBN 0-8014-2498-4 (cloth : alk. paper)
1. Simuliidae—North America. I. Currie, Douglas C. (Douglas Campbell), 1955– II. Wood, D. Montgomery (Donald Montgomery), 1933– III. Title.
QL537.S55A34 2003
595.77'2—dc21

2003004338

Cornell University Press strives to use environmentally responsible suppliers and materials to the fullest extent possible in the publishing of its books. Such materials include vegetable-based, low-VOC inks and acid-free papers that are recycled, totally chlorine-free, or partly composed of nonwood fibers. For further information, visit our website at www.cornellpress.cornell.edu.

Cloth printing 10 9 8 7 6 5 4 3 2 1

To my parents, Ammi H. Adler and Lorene T. Adler, role models par excellence
— P. H. A.

To my father, Robert MacNeill Currie, for his wisdom, humor, and unwavering support
— D. C. C.

To my wife, Grace C. Wood, for her patience and dedication
— D. M. W.

# Contents

# Foreword

Black flies revealed. We all know them. The pink welt with the droplet of blood in the center. And maybe you saw it, maybe you didn't. But if you look on the other side of your arm, there will be another. At their most in the north and the very south, tropical mountainsides with their share, where there are flowing waters there are black flies.

But this book emphasizes that black flies are far more than little precursors to an itchy welt. The tiny flies in this speciose group all seem maddeningly alike. If they were as big as grizzly bears, they would have been taxonomically understood centuries ago. Here their multitudinous diversity of larvae, pupae, genitalia, and chromosomes is revealed for all of North America with painstaking care. Their natural history and ecology, vital to both our appreciation and our many efforts at control of black flies, are here meticulously tied to their taxonomy, and vice versa. And their taxonomy has been cleaned and polished. Just a few moments turning these pages reveals that they are in fact not all maddeningly alike.

These black fly detectives have related who to whom by decades of painstaking phylogenetic inference based on morphologies and the happy accidents of a plethora of chromosomal inversions. These hundreds of species, their family tree so confidently put in place, are ripe for phylogenetic reanalysis through today's gene sequencing. This yet more micro look at morphology will be a rarely afforded test of the new by the old. And who knows. Maybe with a bit of genetic tweaking, the adults of these ubiquitous stream cleaners could even be caused to feed on some other pest than us, leaving us to welcome them with open arms rather than with covered arms.

This portal to everything we know about Simuliidae should be the envy of any major group of animals. We all stand at the exit from the cave. Behind is the darkness of hard-copy libraries and their precarious mission to sustain the growing stream of human knowledge. In front is the perpetual daylight of electronic information management. Team Black Fly has brought centuries of entomological toil to the cave exit. By creating this launch pad they have ensured that what we know already will not be left behind as we rocket into the brave new computerized world.

DAN JANZEN
*Philadelphia*
*15 April 2000*

# Preface

This work is a panoramic reference on the black flies of North America, here defined to include the continental United States, Canada, and Greenland. Our intent is to bring the black fly fauna of North America into the realm of familiarity held by butterflies, dragonflies, and other popular insects. In so doing, we endeavor to provide a model group of organisms for use in a wide range of biological investigations and to improve efforts to manage black flies as pests.

The underlying theme of the book is accurate identification, the key to understanding the biology of black flies and ensuring their successful management. Our focal point, therefore, is the species. We are keenly aware of the difficulties of identification inherent in the family Simuliidae and the rampant misidentifications in the literature. We advocate an integrated approach to identification that involves morphology, cytology, distribution, and ecology, realizing that molecular and other innovative techniques should join, but not supplant, this approach. We also emphasize the placement of species in a classification that serves as a storage and retrieval system and as a testable and predictive framework.

This book is geared to a general readership, with 10 chapters on various aspects of black flies. It should be of particular use to medical and veterinary entomologists, aquatic biologists, environmental consultants, systematists, natural-history enthusiasts, pest-management specialists, and students. Although the book has a North American focus, it is intended for a wider geographic audience. Our phylogenetic reconstruction includes all Palearctic and many Neotropical taxa above, and including, species groups. The book should be of special interest to students and workers in the Palearctic Region because 32 of the species that we treat in detail also occur in that zoogeographic area. We also treat 25 species that are found in Latin America.

We have attempted to summarize all that is known about North American black flies, believing that this approach provides the strongest foundation and the greatest opportunity to identify important questions that remain to be addressed. We have ensured that this information is well documented with citations to provide entry to the vast simuliid literature scattered over the past 240-plus years. Much of the gray literature, such as graduate theses, is included because it contains a wealth of unpublished information. We undoubtedly have not found all theses, and a few have been removed from the public domain. Abstracts, book reviews, unpublished technical reports, textbooks, and undergraduate theses are not included unless they contain significant novel information. In total, more than 2200 references, including those available through 2002, are provided (all on file in the Department of Entomology, Clemson University). Complementing the information gleaned from the literature are our own, heretofore unpublished observations, which appear throughout the text without attribution.

We have studied black flies, marveled at them, and cursed them for a combined total of 100 years. Yet we are humbled by all that remains to be learned of these remarkable creatures, each so small as to be expunged with the literal blink of an eye. We hope that this book, in some small measure, will inspire fascination for a group of organisms that has held companionship with humans throughout history yet has remained arcane to most.


PETER H. ADLER
*Clemson, South Carolina*
DOUGLAS C. CURRIE
*Toronto, Ontario*
D. MONTY WOOD
*Ottawa, Ontario*

*June 2002*

# Acknowledgments

Many friends and colleagues have shared their ideas, provided support, and contributed their energies in the preparation of this book. To all, we express a prodigious debt of gratitude. We especially thank our artists, Ralph M. Idema and Lawrence W. Zettler, whose incomparable illustrations were prepared over a period of 10 years and whose novel skills and techniques, some developed expressly for this book, extracted enormous detail from preserved specimens and breathed life into their subjects.

We have benefitted immeasurably from collectors who furnished a large amount of material and suffered requests for yet more. More than 100 individuals provided collections of black flies. In particular, we thank those who contributed numerous collections: C. R. L. Adler, G. H. Adler, N. H. Anderson, C. E. Beard, R. Beshear, D. S. Bidlack, G. S. Bidlack, R. Biggam, B. Bilyj, M. R. Bowen, D. E. Bowles, D. W. Boyd, A. R. Brigham, P. Bright, C. L. Brockhouse, J. F. Burger, S. G. Burgin, M. G. Butler, J. J. Ciborowski, M. G. Colacicco, G. W. Courtney, D. A. Craig, W. J. Cromartie, M. E. and R. W. Crosskey, R. M. Currie, J. M. Curro, D. M. Davies, S. Deisch, C. I. Dial, J. Doughman, W. R. English, J. H. Epler, C. L. Evans, L. C. Ferrington, D. Finn, G. Gaard, D. J. Giberson, G. M. Gíslason, E. W. Gray, R. D. Gray, A. I. Gryaznov, A. E. Hershey, F. F. Hunter, J. B. Keiper, B. C. Kondratieff, D. C. Kurtak, J. Lessard, G. T. Lester, J. D. Loerch, P. G. Mason, E. C. Masteller, J. W. McCreadie, M. J. Mendel, R. J. Mendel, R. W. Merritt, K. L. Minson, D. E. Mock, D. P. Molloy, J. K. Moulton, S. R. Moulton, B. A. Mullens, J. F. Napolitano, R. Noblet, R. T. Pachon, E. S. Paysen, B. V. Peterson, K. P. Pruess, L. Purcell, D. I. Rebuck, M. H. Reeves, W. K. Reeves, J. V. Robinson, D. Ruiter, R. Savignac, T. L. Schiefer, E. T. Schmidtmann, G. F. Shields, E. E. Simmons, M. L. Smith, S. M. Smith, J. C. Steven, C. A. Stoops, J. F. Sutcliffe, P. F. Wagner, M. E. Warfel, C. N. Watson, D. W. Webb, D. S. Werner, M. J. Wetzel, A. E. Woolwine, and L. W. Zettler.

We thank the following individuals who loaned museum material: N. E. Adams, H. Williams, A. L. Norrbom, and B. V. Peterson (National Museum of Natural History, Washington, D.C.); D. Azuma (Academy of Natural Sciences, Philadelphia); R. W. Brooks (University of Kansas, Lawrence); B. V. Brown (Los Angeles County Museum); P. J. Clausen and R. W. Holzenthal (University of Minnesota); R. Danielsson (Lund University, Sweden); C. Favret (Illinois Natural History Survey); E. R. Hoebeke (Cornell University), J. B. Keiper (Cleveland Museum of Natural History); B. C. Kondratieff (Colorado State University); R. E. Lewis (Iowa State University); F. W. Merickel (University of Idaho); P. D. Perkins (Harvard University); A. C. Pont (Hope Museum, Oxford, England); K. J. Ribardo (California Academy of Sciences); H. Takaoka (Oita Medical University, Japan); H. Wendt (Museum für Naturkunde, Berlin); and C. W. Young (Carnegie Museum of Natural History, Pittsburgh).

We are grateful to the following friends and colleagues who provided various bits of information or gave permission to cite their unpublished information: D. H. Arbegast, A. D. Kyle, and D. I. Rebuck (management of *Simulium jenningsi* in Pennsylvania); C. Back (site information for *Simulium* 'species O'); P. K. Basrur (history of cytotaxonomy); L. S. Bauer (natural enemies of *Twinnia tibblesi*); J. F. Burger (mating swarm of *Simulium tuberosum* species complex); G. Byrtus (status of *S. vampirum* in Alberta); T. R. Carrington (management of *S. jenningsi* in West Virginia); M. H. Colbo and J. W. McCreadie (management programs in Newfoundland); D. A. Craig (oviposition by *Gymnopais dichopticoides* and *Helodon susanae*); D. M. Davies (personal recollections of simuliid work); R. W. Dunbar (history of cytotaxonomy); D. J. Giberson (management programs on Prince Edward Island); E. W. Gray (number of generations for colony of *Simulium vittatum*); R. D. Gray (pest status of *Simulium meridionale* in Nevada); J. R. Harkrider (host records of possible *Tolypocladium* fungus); J. E. McPherson (nuisance outbreaks of *S. vittatum* complex in Illinois); R. W. Merritt (management programs in Michigan); K. Minson (management of *S. vittatum* in Utah); R. Noblet (leucocytozoonosis in domestic turkeys); J. Pasternak (history of cytotaxonomy); B. V. Peterson (personal recollections of simuliid work in Utah); K. P. Pruess (distribution of Nebraska species); W. S. Procunier (chromosomes of *Simulium arcticum* 'IIS-4,' *S. murmanum*, *S. defoliarti*, *S. malyschevi*); W. K. Reeves (parasitism by Nematomorpha and unknown pathogens from the Great Smoky Mountains); J. V. Robinson (economic impact of *Cnephia pecuarum* and *S. meridionale* in eastern Texas); K. H. Rothfels (chromosomes of *Simulium pugetense* 'Cypress Hills'); K. R. Simmons and J. P. Walz (management program in Minnesota); A. Stone (personal recollections of simuliid work; type locality of *Simulium verecundum*); F. C. Thompson (*S. jenningsi* as nuisance in Washington, D.C.); A. G. Wheeler (feeding at extrafloral nectaries); and R. S. Wotton (color of cocoon silk).

We extend special thanks to the following individuals for their valuable contributions: C. E. Beard identified a number of trichomycete fungi, obtained many obscure references, provided phenomenal computer expertise, and, through his multiple talents in the field and laboratory, solved numerous technical problems; E. C. Bernard identified the Collembola; M. R. Bowen, C. L. Evans, L. K. Klepfer, and C. N. Watson provided technical assistance; W. D. Bowman facilitated our access to the Boulder City, Colorado, watershed through the Mountain Research Station; S. G. Burgin provided superb assistance to D. C. C. and P. H. A. during a collecting expedition throughout the southern portion of the Northwest Territories; B. A. Caldwell identified the Chironomidae; J. E. Conn gave us chromosome maps of some members of the *Simulium virgatum* species group that she and K. H. Rothfels prepared in the early 1980s; F. A. Coyle identified the spiders; T. F. Gfeller served as an expert river guide for D. C. C. and P. H. A. on collecting expeditions by canoe down the Horton and Thelon Rivers in the Northwest Territories and Nunavut; D. J. Giberson assisted P. H. A. on a collecting expedition to the Magdalen Islands; E. W. Gray tracked down information for us on various management programs in the United States; H. S. Hill generated by computer the template from which our distribution maps were hand drawn; R. A. Humber identified the probable entomophthoraceous fungus in *Simulium vandalicum* and the oomycetes in other black flies; J. W. McCreadie enthusiastically shared his astute insights on simuliid ecology; J. K. Moulton, through his collecting efforts, helped solve several species problems and provided much distributional information; J. P. Norton illustrated Figs. 3.1 and 3.2; J. E. Raastad provided information about the types of *Stegopterna trigonium*, *Simulium annulus*, and *Simulium truncatum*; D. I. Rebuck, M. E. Warfel, and other members of the Pennsylvania Black Fly Suppression Program, especially S. L. Beers-Miller, A. J. Blascovich, H. M. Calaman, A. J. Dunkle, A. D. Kyle, B. D. Rush, B. J. Russell, and A. L. Shade, collected and reared important samples from Pennsylvania; P. Stephens-Bourgeault rendered a number of miscellaneous illustrations and painstakingly assembled the plates of all line illustrations; A. H. Undeen identified selected microsporidia; and J. W. Van Wagtendonk granted permission to collect in Yosemite National Park for the express purpose of this book.

We thank those individuals, some repeatedly imposed upon, who generously provided literature, photographs, and biographical information: E. C. Becker (information on C. R. Twinn and assistance in obtaining his photograph); J. L. Boisvert and M. P. Nadeau (photograph of an entomophthoraceous fungus); K. Bolte (imaging the frontispiece); G. W. Byers (information on W. T. Emery); W. E. Cox and B. Kirby (information in the Smithsonian Institution Archives); D. A. Craig, R. W. Crosskey, D. M. Davies, G. Gaard, A. I. Gryaznov, N. Hamada, S. Ibáñez Bernal, B. Malmqvist, R. J. Mendel, K. Minson, D. P. Molloy, W. K. Reeves, E. T. Schmidtmann, C. A. Stoops, F. C. Thompson, and D. S. Werner (pertinent literature and copies of various theses); J. Fitzpatrick (news articles on the New Jersey management program for *Simulium jenningsi*); N. K. Dawe (photograph of common loon); R. A. Fusco (photograph of *Bti* application); E. M. Galyean and C. C. Reading (photograph of E. C. O'Roke); E. W. Gray and J. Smink (photograph of *Bti* application); J. R. Harkrider (photograph of possible *Tolypocladium* fungus); R. W. Lichtwardt (photograph of zygospores of trichomycete fungus); S. A. Marshall (photograph of ovipositing flies); D. P. Molloy (information on H. Jamnback and photographs of DDT and nematode applications in New York State); G. Noblet (photograph of *Leucocytozoon* gametocyte); R. Noblet (photograph of turkey farm); P. W. Riegert (photographs of A. E. Cameron and C. R. Twinn); C. A. Stoops (photograph of G. S. Stains and accompanying military information); J. F. Sutcliffe (photograph of cattle under attack by *Simulium vampirum*); and D. S. Werner (some German translations and information on some Enderlein types).

We express our gratitude to the following people who reviewed various parts of the book and offered constructive comments: C. E. Beard (trichomycete fungi), J. J. Becnel (natural enemies), C. L. Brockhouse (Chapter 5), M. H. Colbo (Chapters 7, 8), G. W. Courtney (*Parasimulium*), D. A. Craig (Chapters 4, 6), D. M. Davies (Chapter 2), F. F. Hunter (Chapters 3, 5, 6), B. Malmqvist (Chapters 4, 6), J. W. McCreadie (Chapter 6), D. P. Molloy (Chapters 7, 8), J. K. Moulton (Chapter 9); W. S. Procunier (Chapter 5), F. C. Thompson (Chapter 2), A. G. Wheeler (Chapters 1, 2, 3), G. C. Wood (subgenus *Simulium*), and R. S. Wotton (*Simulium noelleri*). Several people deserve an exceptional thanks for their contributions to the text: C. R. L. Adler, with a keen eye and much patience, reviewed the entire manuscript—more than once—and tested the larval and pupal keys; R. W. Crosskey provided insightful suggestions on format and content early in the project and, by graciously reviewing the manuscript, afforded us the benefit of his unparalleled knowledge of the world Simuliidae; and A. G. Wheeler Jr. provided the spark to begin this project and served as a source of information and sagacious advice on many aspects of the text.

We gratefully acknowledge the financial support of Abbott Laboratories (now Valent Biosciences Corporation), the American Museum of Natural History (Theodore Roosevelt Memorial Fund to P. H. A.), Canadian National Defense, National Science Foundation (DEB-9629456 to P. H. A.), Royal Ontario Museum Foundation (to D. C. C.), S. C. Johnson & Sons (Racine, Wisconsin), the Louise Hawley Stone Charitable Trust of the Royal Ontario Museum, and Bohdan Bilyj (Weston, Ontario). We are particularly indebted to John F. Burger (University of New Hampshire), whose generous gift enabled a large number of the larval head capsules to be illustrated. In addition, funding for the production of many of the Canadian larval and pupal illustrations was provided by the Matching Investment Initiative Collaborative Research Agreement (MIICRA) between Agriculture and Agri-Food Canada and Biodiversity Incorporated. We thank R. A. Fusco, E. W. Gray, D. L. Lawson, R. W. Merritt, and D. C. Williams for facilitating the funding from industries; M. D. Engstrom for serving as liaison to the Louise Hawley Stone Charitable Trust of the Royal Ontario Museum Foundation; and J. M. Cumming for his unstinting support in mediating the MIICRA funding and supervising production of some larval and pupal drawings. Thanks also are extended to C. L. Brockhouse who hosted P. H. A. during two visits to Prince Edward Island, which were supported by a grant (CRG.CRG 972938) from the North Atlantic Treaty Organization International Collaborative Research Grants Programme. P. H. A. also was graciously hosted in 1993 at the Natural History Museum, London, by R. W. Crosskey. The following scientists in the Zoological Institute, St. Petersburg, Russia, offered splendid hospitality and assistance to P. H. A. during his 1997 study of the Rubtsov collection: A. A. Ilyina (deceased), V. A. Krivokhatsky, O. G. Ovtshinnikova, V. Richter, V. N. Tanasychuk, A. V. Yankovsky, and V. F. Zaitzev. We are indebted to the National Park Service, which granted collecting permits to us or those who collected for us.

The Department of Entomology at Clemson University provided P. H. A. an environment supportive of this project. The Royal Ontario Museum Foundation and National Sciences and Engineering Research Council (NSERC) provided field and laboratory support for D. C. C. Financial support to D. C. C. was also provided by the Biological Resources Program of Agriculture and Agri-Food Canada through an NSERC Postdoctoral Fellowship. The Diptera Unit of the Canadian National Collection of Insects, housed by the Biological Resources Program, provided facilities to D. C. C. and R. M. Idema.

We thank Peter J. Prescott, science editor of Cornell University Press, for his superb efforts in shepherding our manuscript along over the years and working closely with us to resolve many issues. We also thank manuscript editor Ange Romeo-Hall, designers Lou Robinson and Robert Tombs, and science coordinator Lena Bertone Rosenfield for their excellent technical assistance, and we extend our appreciation to freelance copyeditor Mary Babcock for her exceptionally careful work.

Finally, we acknowledge the titans that have passed. We recognize especially Klaus Herman Rothfels, friend, mentor, and phenomenon, who built the field of black fly cytogenetics, now an integral facet of simuliidology.

# PART I
# Background

# 1| Overview

Black flies are as much a part of the natural environment as the flowing waters in which they breed. Their abundance, wide distribution, nuisance habits, and economic impact have afforded them a high profile in the eyes of the public and as subjects of biologists. They constitute a group of organisms that bridge two radically different environments, the aquatic and the terrestrial, and consequently, they continue to be viewed paradoxically. As aquatic organisms, they are regarded as beneficial elements of the food web, and as terrestrial organisms, they are despised as intractable pests.

The family Simuliidae is rather small, ranking about 30th in number of species among the approximately 130 families of Diptera worldwide. At last count, 1772 valid, living species of black flies had been formally described from the planet (Crosskey 2002), representing less than 2% of the world's described Diptera. We recognize 254 living species in North America—roughly 15% of the total described simuliid fauna, about 1% of the Nearctic dipteran fauna, and a little more than 2% of all aquatic insect species recorded from the continent. Of the 254 species, 244 (96%) inhabit the United States and 161 (63%) reside in Canada.

These insects have acquired their own special assortment of vernacular names (Table 1.1), most of which are based on the females and many of which are local, obscure, sometimes inaccurate, and, though not provided here, often unacceptable in genteel company. "Black flies," often spelled as one word (*blackflies*), is the most universally applied common name for simuliids. We use the name *black flies* as two words to identify the simuliids as true flies distinct from "flies" in other orders (e.g., caddisflies, dragonflies, stoneflies) and to distinguish them from other insects of the same common name, such as the citrus blackfly (*Aleurocanthus woglumi*), a member of the homopteran family Aleyrodidae. Historical precedence also embraces a two-word moniker; the first unambiguous, printed use of the word *black flies*, dating from 1789, is in a two-word format (Simpson & Weiner 1989). Additional names for black flies also exist in many of the aboriginal languages of North America, particularly among the 53 native languages in Canada and the 20 in Alaska.

Black flies are small organisms that, without a microscope, would be virtually beyond the pale of study. Larvae, when mature, measure 3 to 15 mm in length, and adults exhibit a similar range in wingspan. The maddening structural uniformity of black flies has made them the bane of many who wish to study them. Much to the delight of taxonomists, the extent of variation within the family, nonetheless, can be striking, particularly in the color patterns of larvae, gill configurations of pupae, and terminalia of adults. The fundamental groundplan, however, is preserved and all black flies can be identified readily to the family level. Eggs are ovoid to bluntly triangular and remarkably smooth, with few taxonomically useful characters. Larvae have an elongate, posteriorly expanded body and a well-defined head capsule usually bearing a pair of labral fans for filtering food from the current. All pupae have a pair of spiracular gills on their thorax and a silk cocoon of varying complexity. Adults vary from black to orange and are humpbacked, with short legs, stout antennae, and strong anterior wing veins. Males of most species have eyes so large that they meet along the midline, contrasting markedly with the smaller, separated eyes of females.

Black flies, like all Diptera, have giant polytene chromosomes that reach their aesthetic zenith in the larval silk glands. These giant chromosomes are some of the most exceptionally developed among the Diptera, tempting speculation that if black flies were as easy to rear and mate in the laboratory as *Drosophila*, they might have become the workhorses of genetic studies. The demonstration that these giant chromosomes, with their hundreds of alternating dark and light transverse bands, could be used to infer reproductive isolation (Rothfels 1956) launched a new era of inquiry. By revealing sibling species, which are so similar in appearance as to defy easy separation, the polytene chromosomes have provided a wellspring of information. About one fourth of the 254 species treated in this book were first revealed through chromosomal study. Information on population structure, evolutionary relationships, and effects of toxicants also can be gleaned from the study of polytene chromosomes.

Flowing water is the lifeblood of all black flies. Within this environment, these flies have exploited every conceivable habitat from hot springs to glacial meltwaters and from trickles to roiling rivers. Wherever water flows, from the barren alpine regions to the coastal plains, black flies can be found. Females of most species release their eggs into the water as they fly low over its surface, or

**Table 1.1. Common Names Used for Members of the Family Simuliidae in North America**

| Common name | Species | Geographic area of usage |
|---|---|---|
| Adirondack black fly[1] | *Prosimulium mixtum*, *Simulium venustum* species complex | northeastern United States |
| Barren Lands black fly[2] | primarily *S. venustum* species complex | Northwest Territories, Nunavut |
| black fly[3] | general usage for all species | widespread |
| black gnat[4] | general usage for pestiferous species | widespread |
| blood-sucker[5] | *Simulium innoxium* | Ithaca (New York) |
| buckie fly[6] | probably *P. mixtum* | Rhode Island |
| buffalo gnat[7] | general usage, especially *Cnephia pecuarum* | widespread, especially southern United States |
| bull fly[8] | *Simulium meridionale* | Florida |
| cholera gnat[9] | possibly *Simulium parmatum* n. sp. and *Simulium jenningsi* | Virginia |
| gnat[10] | general usage, especially *S. jenningsi* | widespread |
| goose gnat[11] | *S. meridionale* | Mississippi River drainage |
| harmless black fly[12] | *S. innoxium* | Michigan |
| humpback gnat[13] | general usage for all species | widespread |
| innoxious black fly[14] | *S. innoxium* | New York |
| may fly[15] | *P. mixtum*, *S. venustum* | Maine, northeastern Pennsylvania |
| mouche noire[16] | general usage for all species | Quebec |
| northern black fly[17] | *Simulium arcticum* species complex | Canada |
| river leech[18] | *Simulium hippovorum* | Vancouver Island (British Columbia) |
| sand fly[19] | general usage; also *Simulium decimatum*, *S. venustum* species complex, *Simulium vampirum* | widespread, especially western Canada |
| simuliid[20] | general usage for all species | widespread |
| southern buffalo gnat[21] | *C. pecuarum* | Mississippi River drainage |
| striped black fly[22] | *Simulium vittatum* species complex | Canadian prairies |
| submarine spider[23] | general usage | western New York |
| swamp fly[24] | *S. vampirum* | central Saskatchewan |
| turkey gnat[25] | *S. meridionale* | central United States |
| western buffalo gnat[26] | *S. meridionale* | western United States |
| white-stockinged fly[27] | *S. venustum* species complex; also general usage | Alaska, Canadian Shield, New England |
| yellow gnat[28] | *Prosimulium fulvum* | western United States |

[1] Applied by guides and locals in the northeastern states, especially in the Adirondack Mountains of New York (Comstock 1924). The name was first used for *Simulium molestum* (Comstock & Comstock 1895) in general reference to members of the *Simulium venustum* species complex, but also has been applied to *Prosimulium mixtum* (Matheson 1938).

[2] Presumably coined by Waldron (1931), who described the namebearer as a "cruel parasite" of living things, particularly of caribou and humans.

[3] The most universal and preferred common name, evidently having originated in New England at least as early as 1789 (Simpson & Weiner 1989); often used as a single word ("blackfly") or sometimes hyphenated ("black-fly"), especially in European literature. In earlier years, it was used primarily to refer to northern simuliids, especially members of the *S. venustum* species complex (e.g., Osborn 1896). The name "black fly" also has been applied to the rhagionid fly *Symphoromyia atripes* (Ross 1940).

[4] Rather loosely applied, typically by the layperson, to pestiferous species of black flies (Anderson 1960, Stamps 1985); used in print at least as early as 20 May 1882 by the Inyo Independent newspaper, probably for *Simulium tescorum*, in the mining areas of California's Inyo Mountains. The name rarely appears in the scientific literature in reference to black flies but is sometimes used for members of the Ceratopogonidae.

[5] Although appropriate for the females of most species, the name in the mid-1800s actually was applied by young boys to larvae of *Simulium innoxium*, which aggregated in apparently menacing clumps (Howard 1888).

[6] Applied by locals of Rhode Island because the flies swarm and bite during the migration of the alewife (*Alosa pseudoharengus*), a fish locally known as the "buckie" (Dimond & Hart 1953).

[7] First use unknown, but the name appeared in print as early as 1822 in an account of life in southeastern Illinois, which noted only that "it was so dry, we had no buffalo gnats, and but few prairie flies or musquetoes [*sic*]" (p. 278 of Woods 1822). The name appeared again in 1826 in reference to *Cnephia pecuarum* (Fessenden 1826). The moniker presumably was applied because the humpbacked adult resembles a miniature buffalo, i.e., an American bison (*Bison bison*). It was used somewhat generally for black flies in the southern United States but was applied especially to *C. pecuarum* (Riley 1887). The name was familiar to those who crossed the prairies of the United States in the early to mid-1800s. Josiah Gregg (1844, p. 28), for example, found the buffalo gnat far more troubling and frequently encountered than the mosquito: "It not only attacks the face and hands, but even contrives to insinuate itself into those parts which one is most careful to guard against intrusion. Here it fastens itself and luxuriates, until completely satisfied. Its bite is so poisonous as to give the face, neck, and hands, or any other part of the person upon which its affectionate caresses have been bestowed, the appearance of a pustulated varioloid." Buffalo gnats were well known to the stagecoach and freight drivers of the Great Overland Stage Line that crossed the Central Plains in the 1860s, persecuting the cattle and turning the ears of the drivers raw (Root & Connelley 1901, p. 577). The primary species that ravaged these western pioneers probably was *Simulium meridionale*, which is a documented bane to humans in parts of the western

**Table 1.1. Continued**

United States (e.g., Townsend 1891, Mock & Adler 2002). (Most of the trouble on the plains was beyond the known range of *C. pecuarum*.) The name "buffalo gnat" (or "buffalo-gnat") is now essentially obsolete in general use but sometimes still is applied to *C. pecuarum*. A less frequently used variant is "buffalo fly."

[8] Origin obscure; mentioned in the literature only by Pinkovsky (1976).

[9] First used by Gilliam (in Riley 1888a) because females were believed to transmit the agent of "chicken cholera" (= *Leucocytozoon* disease?) to chickens and turkeys. Riley (1888a) erroneously linked the name to *S. meridionale*, a species that does not occur in the area where the name originally was applied.

[10] Used for any species that flies about the face, especially for *Simulium jenningsi* in the eastern United States (Adler & Kim 1986) and for the *Simulium vittatum* species complex in southern Idaho (Jessen 1977).

[11] A variant of "turkey gnat" used by locals because of problems inflicted on domestic geese (Riley 1887).

[12] Applied because the author (Boardman 1939) of the common name claimed (incorrectly) that *S. innoxium* (as *S. pictipes*) was "an exception to the rule that black flies bite." (Although *S. innoxium* does not bite humans, it does take blood from several species of mammals.)

[13] A general term applied infrequently by the general public (Stamps 1985) in reference to the arched thorax of the adult fly.

[14] Applied because the authors (Comstock & Comstock 1895) of the name had never known the species to bite, despite its abundance around their area of study in Ithaca, New York; also referred to as the "innoxious sand fly" (Garman 1912).

[15] Applied by locals because the pest species are active in May (Adler & Kim 1986); not to be confused with the nonbiting mayflies (Ephemeroptera).

[16] Used by francophones in Quebec, Canada; it is a direct translation of the words "black fly" to the French language.

[17] Although rarely used in print, the name was listed as an accepted common name in a list of French names of insects in Canada (Benoit 1975).

[18] Used by swimmers and rafters in reference to the copious silk extruded by the larvae, which sticks well to human bodies.

[19] Applied in historical times (sometimes as "sand-fly" or "sandfly") in three separate contexts: (1) to black flies as a group, from about 1910 to 1915 when these flies were thought to be vectors of some causal agent of pellagra (e.g., Hunter 1912b, 1913a, 1913b); (2) to pestiferous black flies, probably *Simulium decimatum* and the *S. venustum* species complex, especially by British explorers in the Yukon Territory and Nunavut (formerly Northwest Territories) (Currie 1997a); and (3) to females of *Simulium vampirum* (formerly *S. arcticum*) in Alberta and Saskatchewan that seemingly originated in great swarms from sand dunes but actually were carried by winds from the breeding areas (Fredeen 1973). The name also is applied to members of the family Ceratopogonidae and the subfamily Phlebotominae (Psychodidae).

[20] A semiscientific name used especially among entomologists and specialists of the Simuliidae. The family name Simuliidae is derived from the generic name *Simulium*, which means "little snub-nosed being" (Crosskey 1990).

[21] Used by Riley (1887) in his report of severe problems caused by females in the southern United States.

[22] Applied by locals in areas of Manitoba and Saskatchewan where the flies can be a pest of livestock and humans (Riegert 1999b).

[23] Evidently coined by Mr. Myron Pardee, a "wealthy gentleman of Oswego [New York], who propagates trout for his amusement and scientific purposes, he being a great naturalist" (McBride in Riley 1870c), and applied to larvae, which at the time were believed to spin "death webs" that ensnared fish fry (Green in Riley 1870a); "web-worm" was another variant that spun off from this misconception (Riley 1870c).

[24] Used by locals who believed that the flies originated from swamps in the affected area (Fredeen 1988).

[25] Used by Riley (1887) to refer to one of the hosts (turkey) of the female but evidently also used by locals in areas where the species was a pest.

[26] Used by Osborn (1896) to refer to *S. meridionale*, at that time recorded (as *S. occidentale*) only from New Mexico.

[27] Refers to the banded legs of the female; variants include "white socks," "whitesox," "white sox," and "white-sox." Reference to flies with "white stockings" appeared in the literature at least as early as the beginning of the 20th century (Needham & Betten 1901). In the Adirondack Mountains of New York, locals have an adage that the flies "put on their gray stockings," believing that the flies turn gray with age as the pest species changes from *P. mixtum*, with brown legs, to *S. venustum*, with white or "gray" legs (Jamnback 1969a). In Alaska, pestiferous black flies in general are referred to as "whitesox" (Gjullin et al. 1949a) or "white socks."

[28] Used by Essig (1928, 1942), whose travels in the western states acquainted him with this pestiferous, large, yellowish to orange species.

deposit them in masses on trailing vegetation and other wetted substrates. Larvae hatch in a few days to many months and pupate within a week to several months, the durations of each stage depending largely on temperature. Adults exit the submerged pupae and are buoyed to terrestrial life in jackets of air. The search for mates and nutriment—sugar for energy and bird or mammal blood for egg production—become prominent activities of adult life.

The female's requirement for blood has driven the role of black flies in history. North America is spared the human diseases caused by pathogens that are transmitted by black flies. Among the most serious diseases is river blindness (onchocerciasis), which afflicts nearly 18 million people in tropical Africa and parts of Central and South America (Crosskey 1990). Black flies in North America, nonetheless, have inflicted extraordinary misery and economic losses on many aspects of human life, from agriculture and forestry to tourism and entertainment. They appeared in North America's written record as early as 1604, when Samuel de Champlain wrote of bites so bad that they impaired vision (Davies et al. 1962). From that point forward black flies have been an emblem of the hardships offered by Nature. They were, for example, with

Lewis and Clark on the trek across the uncharted western United States, causing Lewis to note in June of 1805 that gnats, prickly pear cactuses, and mosquitoes were the "great trio of pests equal to any three curses that ever poor Egypt labored under" (Duncan & Burns 1997). Even today, black flies continue to exact a toll, their pest status intensifying as water quality improves. At the close of the 20th century, about 20 management programs were operating in the United States at an annual cost of more than $6 million.

# 2| History of Research

The history of research on black flies in North America has not previously been written, although Crosskey (1990) presented a few vignettes in a summary of the world scene, and Riegert (1999a, 1999b) recounted control efforts on the Canadian prairies and the history of the Northern Biting Fly Project. The following treatment focuses largely on the history of taxonomic research—the bulk of the early efforts—and is presented chronologically, concentrating on the years prior to 1970. Some of the history of control is presented in Chapter 8, and historical information on the classification of the world Simuliidae is given in Chapter 9. Many important aspects of simuliidology, such as community ecology, laboratory colonization, hydrodynamics, and filter feeding, are largely developments of the last 30 years of the 20th century. Having not yet weathered the ravages of time, they are not treated here.

Much of the biographical information was located through the works of Mallis (1971) and Gilbert (1977) or came from our personal knowledge and experiences. Some information about J. R. Malloch and R. C. Shannon was gleaned from the C. P. Alexander files in the Smithsonian Institution Archives. The history of cytotaxonomy was drawn from the recollections of P. K. Basrur, D. M. Davies, J. J. Pasternak, and especially R. W. Dunbar. Reflections of D. M. Davies, B. V. Peterson, A. Stone, and P. D. Syme were provided through their personal communications.

### The Beginning (1800s)

Thomas Say (1787–1834) (Fig. 2.1), the oft-called Father of American Entomology, initiated the study of black flies in North America when he described the first species from the continent in 1823. While traveling the Ohio River en route to the Rocky Mountains, the Philadelphia-born naturalist was annoyed by the "pungent" bite of a black fly that he named *Simulium venustum*. Studies of black flies in North America began comparatively late, relative to Europe, for Say's description came 65 years after Carolus Linnaeus (1707–1778) officially described the world's first black flies (*Simulium reptans* and *S. equinum*) in 1758. Although *S. venustum* was the first simuliid actually described from the continent, it was not the first North American black fly to receive a name. That distinction went to *Simulium maculatum*, a species described in 1804 from Germany by Johann Wilhelm Meigen (1764–1845), who shortly afterward bestowed two additional names (*pungens* [1806] and *subfasciatum* 1838) on this same species.

Thaddeus William Harris (1795–1856) (Fig. 2.2), a librarian at Harvard College (now Harvard University) and widely acclaimed founder of American economic entomology, was the second American to name black flies from the continent, adding *Simulium calceatum* (*nomen nudum*) in 1835—the name actually was coined by Say—and *Simulium molestum* in 1841. Harris (1841) also created the name *Simulium nocivum*, but it referred to a member of the family Ceratopogonidae. Harris's collection of insects, housed at Harvard University, contains 12 black flies, representing the oldest-known extant collection of North American simuliids, with the earliest specimen, a female of *Prosimulium mixtum*, mounted on a small sewing needle and dating from 1826. Our examination of the material, most in reasonably good condition, reveals that it includes three females and one male of *P. mixtum*, one female of *Simulium* possibly *anatinum*, two females of *Simulium bracteatum*, one female of *Simulium parnassum*, and four females of the *S. venustum* species complex. One of the females of *P. mixtum*—the specimen mounted on a sewing needle—is the only surviving black fly known definitely to have been examined by Thomas Say. Harris had sent the specimen to Say in November 1833 (Johnson 1925).

The next five names for North American black flies were provided by Europeans. *Simulium vittatum* was described from Greenland in 1838 by the Swedish worker Johann Wilhelm Zetterstedt (1785–1874). In 1848, Johannes Gistel (1803–1873) of Germany applied the name *Simulium nasale*, a *nomen nudum*, to *S. vittatum*. That same year, Francis Walker (1809–1874), an employee of the British Museum (now the Natural History Museum) known for having described more species (ca. 20,000) than any other human, authored the names *decorum* and *invenusta*. Walker's material for these descriptions was gleaned from a collection given to the British Museum by the Scotsman George Barnston (ca. 1800–1883), an employee of the Hudson's Bay Company and an avid insect collector who spent about six years at Martin Falls on the Albany River of northern Ontario, the type locality of the two Walker species. In 1859, Luigi Bellardi (1818–1889) of Italy provided the name *cinereum*, though preoccupied, for what would become *Simulium virgatum*.

Fig. 2.1. Thomas Say (possibly ca. 1833). National Archives.

Fig. 2.3. Charles V. Riley. Smithsonian Institution Archives.

Fig. 2.2. Thaddeus W. Harris. National Archives.

Fig. 2.4. William S. Barnard (ca. 1887). National Archives.

More than a decade passed before another name was applied to a North American black fly. One of history's most renowned entomologists, London-born Charles Valentine Riley (1843–1895) (Fig. 2.3), at the age of 17, entered the United States, where he sporadically directed his attention to black flies. In 1870, while serving as state entomologist of Missouri, he described *Simulium piscicidium* from material sent to him by Miss Sara J. McBride, an accomplished natural history enthusiast in Mumford, New York, who subsequently fell into obscurity. McBride's brush with simuliid fame came when Riley (1870d) published her careful observations refuting the myth (Green in Riley 1870a) that black flies (identified as *Simulium* by Riley 1870b) kill fish fry by ensnaring them in silken death webs, a claim that persisted in some quarters nearly to the end of the century (Howard 1894). With the year 1870 also came the first description and illustration of a larva and pupa of a North American black fly (*S. decorum*), again the work of Riley (1870c). The first description of the egg of a North American black fly (*Simulium innoxium*) appeared in 1880, published by William Stebbins Barnard (1849–1887) (Fig. 2.4), originally from Canton, Illinois.

Fig. 2.5. Hermann A. Hagen. National Archives.

Fig. 2.6. Francis M. Webster (ca. 1891). National Archives.

Barnard made his only contribution to simuliidology while he was an instructor of entomology and invertebrate zoology at Cornell University. The German emigrant Hermann August Hagen (1817–1893) (Fig. 2.5), professor of entomology at Harvard University, added *Simulium pictipes* to the list of North American species in 1880, and Jean Joseph Alexandre Laboulbène (1825–1898) of France contributed the name *Simulium hematophilum* in 1882.

Two of the continent's most arrant black flies, *Cnephia pecuarum* and *Simulium meridionale*, were described in 1887 by Riley, then entomologist to the U.S. Department of Agriculture (USDA) in Washington, D.C., and the first curator of insects at the National Museum of Natural History, then known as the United States National Museum (USNM). Riley's descriptions were based on material sent in by two of his devoted field agents, the self-made entomologist Francis Marion Webster (1849–1916) (Fig. 2.6), stationed at Vicksburg, Mississippi, and German-born Otto Lugger (1844–1901) (Fig. 2.7), who was posted at Memphis, Tennessee. The 1887 report by Riley was based largely on the fieldwork of four of his agents, most notably Special Agent Webster who worked on the buffalo-gnat problem in the Lower Mississippi River Valley from February 1886 to 1888 and who actually wrote much of the report and transmitted it to Riley (Webster 1887). The report was far more than a description of species, and it ranks among the most significant simuliid studies of all time. Notwithstanding some errors, it was replete with information on the natural history of black flies and was among the first to discuss larval diet, natural enemies, the cause of pest outbreaks, and larval control. Riley's magnificent career, which included the publication of 2896 papers, was terminated en route to work in mid-September 1895 when the bicycle he was riding—faster than usual, according to witnesses—careened into a chunk of granite, sending him violently to the pavement.

During the 1890s, nine new simuliid names appeared on the North American scene. Charles Henry Tyler

Fig. 2.7. Otto Lugger (ca. 1894). Smithsonian Institution Archives.

Townsend (1863–1944) (Fig. 2.8), a controversial entomologist who held various jobs including a stint as assistant to Riley, added two names (*occidentale* 1891 and *tamaulipense* 1897), although both have proved to be synonyms of *S. meridionale*. The prominent dipterologist and distinguished paleontologist Samuel Wendell Williston (1852–1918) (Fig. 2.9), professor of historical geology and paleontology at the University of Kansas, described *Simulium argus* in 1893 from material sent by Riley. The husband and wife team of John Henry Comstock

Fig. 2.8. Charles H. T. Townsend (1893). Smithsonian Institution Archives.

Fig. 2.10. John H. Comstock (1881). Smithsonian Institution Archives.

Fig. 2.9. Samuel W. Williston (ca. 1888). National Archives.

Fig. 2.11. Anna B. Comstock (ca. 1881). National Archives.

(1849–1931) (Fig. 2.10) and Anna Botsford Comstock (1854–1930) (Fig. 2.11) added the name *S. innoxium* in 1895. The name, however, was first used—and attributed to Williston—in a Bachelor of Science thesis written in 1890 by Rosina Olive Phillips (1872–19??), a student of John Comstock from Naples, New York. The thesis, perhaps the first on North American black flies, included perceptive observations, albeit not without errors, on morphology and natural history such as cocoon spinning, oviposition, overwintering, and larval food. Parts of her thesis were excerpted by Coquillett (1898) and Johannsen (1903b). Despite the thoroughness of her work, Miss Phillips, the first woman to write about the North American Simuliidae, abandoned the study of black flies and became a science examiner with the New York State Education Department in Albany, New York. John Comstock, a student of Hermann Hagen and the first entomologist at Cornell University, where he was stationed for about 60 years, became the father of entomological teaching in America. With his wife, an accomplished naturalist and illustrator, he published *A Manual for the Study of Insects* in which *S. innoxium* was officially described. The book was so successful that it led to the establishment of Comstock Publishing Company, which became the basis for

Fig. 2.12. Daniel W. Coquillett (ca. 1894). National Archives.

Cornell University Press. In 1897, Lugger, by then the state entomologist of Minnesota, described three species (*Simulium irritatum*, *S. minutum*, *S. tribulatum*) that, until now, have remained in the dustbin of synonymy.

At the close of the 19th century, 25 names had been proposed for 18 valid simuliid species now known to occur in North America. Thus, by 1900, about 7% of today's North American species had been named, and one genus (*Simulium*) was recognized. These early taxonomic efforts were summarized in the first key (males and females) to North American black flies, written in 1898 by Daniel William Coquillett (1856–1911) (Fig. 2.12), a USDA employee of Riley and honorary custodian of Diptera in the USNM. Although a bashful man who had been a tuberculosis patient in his early years, the Illinois-born Coquillett was a well-respected dipterologist among his peers. His 1898 paper also included descriptions of two new species, *Simulium griseum* and *Simulium bracteatum*.

### Establishment of a Framework (1900–1929)

The first 30 years of the 19th century witnessed a flurry of taxonomic activity on the Simuliidae, both national and foreign, that affected the North American situation. Keys were constructed, higher categories were established, new taxonomic characters were used, and bionomic studies were expanded. In 1903, Oskar Augustus Johannsen (1870–1961) (Fig. 2.13), born of Danish parents in Davenport, Iowa, recognized 15 described species north of Mexico, summarized most of the information about black flies on the continent, and produced the first keys to larvae and pupae of North American species. The work was based on his master's thesis, the first on black flies in North America, which was completed in 1902 under the direction of John Comstock at Cornell University. Ten years after Johannsen's work was published, status quo prevailed: Walter Titus Emery (1888–1957) from Wetmore, Kansas,

Fig. 2.13. Oskar A. Johannsen (1905). National Archives.

also recognized 15 described species north of Mexico and provided keys to larvae, pupae, and adults in his master's thesis, which was prepared at the University of Kansas and published in 1913.

Most generic and subgeneric names of North American species appeared during the first 30 years of the century, although all names but *Parasimulium* (Malloch 1914) were contributed by Emile Roubaud (1882–1962) (1906) in France and Günther Enderlein (1872–1968) (1921a, 1925, 1930) in Germany—and all without the benefit of genitalic information. The use of genitalia was introduced to the taxonomy of black flies in 1911 when Carl August Lundström (1844–1914) of Finland illustrated the male genitalia of 17 species, including those of 5 new species whose names would apply to North American black flies. Simuliid taxonomy previously had been dominated by the colors and patterns of the thorax, legs, and abdomen. The first use of simuliid genitalia in North America appeared in a 1916 government publication by Arthur W. Jobbins-Pomeroy (1891–1946), an entomological assistant with the USDA who worked on black flies in Spartanburg, South Carolina, and before that in Havana, Illinois, during the summer of 1912.

Jobbins-Pomeroy contributed insight into the anatomy and bionomics of black flies while investigating the idea that these insects were vectors of an agent that caused pellagra. This devastating disease—a blight of humans in southern Europe for nearly 200 years and first diagnosed in the United States (Georgia) in 1902—eventually was recognized as a nutritional malady related to a deficiency of nicotinic acid. Jobbins-Pomeroy resigned from the USDA on 20 November 1914, became government entomologist in Nigeria exactly one month later, and moved to the Gold Coast in 1925 to become medical entomologist.

Fig. 2.14. Harrison W. Garman (ca. 1900). National Archives.

Fig. 2.15. Samuel J. Hunter (ca. 1912). From Emery (1913).

The idea that black flies were vectors of some causal agent of pellagra was the brainchild of the European parasitologist Louis Westenra Sambon (1865–1931), who first suggested the possibility at a 1905 meeting of the British Medical Association. Five years later, Sambon (1910) published his idea and furnished the details to the United States' assistant surgeon, who summarized the so-called Sambon theory (Lavinder 1910), quickly bringing it to the attention of workers in the United States. Pellagra at this time was a serious health issue in much of the United States, particularly the South, killing at rates as high as 40%, ostracizing the afflicted who bore the hideous skin rashes and scaling, and creating panic. In 1915, as many as 165,000 people in the South were estimated to have pellagra (Etheridge 1972).

The U.S. government sprang into action, requesting that the surgeon general investigate pellagra. In light of the Sambon theory, entomologists were asked to do their part. Major studies of simuliids and their possible relation to pellagra were commissioned in Illinois, Kansas, Kentucky, and South Carolina. Small-scale investigations were conducted in states such as Georgia (Roberts 1911) and Texas (Jennings 1914). Heavy oils were suggested as a means of killing the larval black flies if Sambon's idea proved correct (Hewitt 1910). Most entomologists, however, were not satisfied with the implication that simuliids were responsible for pellagra, and they mustered considerable circumstantial evidence against the idea (Garman 1912; Jennings & King 1913a, 1913b; Jennings 1914). Among those who doubted the link between black flies and pellagra was Illinois-born Harrison W. Garman (1858–1944) (Fig. 2.14), who worked at the University of Kentucky from 1889 to 1929. Another was Allan Hinson Jennings (1866–1918), an assistant with the USDA's Bureau of Entomology who directed the work of Jobbins-Pomeroy and for whom one of the major simuliid scourges, *Simulium jenningsi*, was named. In early 1912, Jennings began working on the pellagra issue in Spartanburg, South Carolina, in cooperation with the Thompson-McFadden Pellagra Commission of the New York Post-Graduate Medical School, which established field headquarters in Spartanburg in June that same year. Sambon boosted the commission's morale and eminence when he spoke at a conference in Spartanburg in the summer of 1913. In the end it mattered little. The experimental work of Samuel John Hunter (1866–1946) (Fig. 2.15), initiated at the University of Kansas on 1 August 1911 and published in a series of papers (Hunter 1912a, 1912b, 1913a, 1913b, 1914), failed to confirm the Sambon theory, dealing a strong blow to the idea.

In 1915, the dedicated physician Joseph Goldberger (1874–1929) demonstrated that pellagra was a deficiency disease, essentially crushing the idea that simuliids (and other insects) were vectors, although many people refused to believe that substandard social conditions (i.e., poor diet) could cause disease. As for Sambon, the fallacious notion damaged his reputation, but it had fueled a great deal of solid research on various aspects of black flies, including life history, morphology, and taxonomy, and in many respects it brought the study of the Simuliidae to a higher level of sophistication—perhaps Sambon's real legacy in the pellagra saga.

The odyssey involving black flies and pellagra also spawned the first patronym for a North American black fly. The name resulted from a study that the Pellagra Commission of Illinois requested of Stephen Alfred Forbes (1844–1930) (Fig. 2.16), who served as state entomologist of Illinois from 1882 to 1927. The governor of Illinois appointed the commission to investigate the pellagra problem, which was particularly rampant in orphanages and the almshouses and insane asylums, as they were called in those days. Forbes was well suited to provide the entomological expertise. His interests in insects, as well as his characteristic heavy moustache, were largely products of his Civil War years as a soldier in the Seventh Illinois Cavalry. Forbes (1912, 1913) summarized the knowledge of black flies in Illinois and included descriptions of new species by the frail but energetic Charles Arthur Hart (1859–1918) (Fig. 2.17), one of his employees with the Illinois Natural History Survey. Hart described the larva, pupa,

Fig. 2.16. Stephen A. Forbes (1890). National Archives.

Fig. 2.17. Charles A. Hart (ca. 1892). National Archives.

male, and female of *Simulium johannseni* in 1912, honoring Johannsen who, by that point a professor and extension entomologist at the University of Maine, first pronounced the species new and undescribed. One hundred four patronyms now find their place among the 367 species names applied to North American black flies.

Three additional entomologists who worked, at least in part, with black flies were associates or students of Forbes at the Illinois Natural History Survey. Harrison Garman was a research associate with Forbes for six years, and Clarence M. Weed (1864–1946) was an associate for two years. Weed, while working in New Hampshire, became the first person in the 20th century to explore chemical means of controlling black flies (Weed 1904a, 1904b). Robert D. Glasgow (1879–1964), who received both a bachelor's and a doctoral degree under Forbes, went on to combat black flies with chemical pesticides in the Adirondack Mountains of New York while serving as the state entomologist of New York from 1928 to 1949.

Fig. 2.18. Frederick W. Knab (1906). National Archives.

From 1900 through 1929, 63 species names were proposed for 42 valid species of North American black flies. Coquillett (1902) was the first worker to provide new names (*fulvum*, *glaucum*, *virgatum*) in the 20th century. He was followed in 1904 by Charles Frederick Adams (1877–1950), then at the University of Chicago and later a public-health official in Indiana and Missouri, who added *Simulium notatum* to the list. In 1904, the German worker Paul Gustav Eduard Speiser (1877–1945) furnished a replacement name (*tephrodes*) for Bellardi's *cinereum*, but the species to which it would apply (*Simulium virgatum*) had been described two years earlier. Two names (*rubicundulum* and *mediovittatum*) were added in 1915 by Frederick Knab (1865–1918) (Fig. 2.18), a German who moved to Massachusetts in 1873 and succeeded Coquillett as custodian of Diptera at the USNM, where he worked from 1911 until he succumbed to leishmaniasis in 1918.

Through a nomenclatural faux pas, Alfred Ernest Cameron (1887–1952) (Fig. 2.19) from Aberdeen, Scotland, became the first worker in Canada to name a black fly, when in 1918 he applied the name *Simulium simile* to the species now known as *Simulium vampirum*. John Malloch (1919), however, coined the name *simile* and provided a formal description. But Cameron, who had sent specimens to Malloch for identification, unwittingly published the name a year before the appearance of Malloch's (1919) formal description, which actually applied to *Simulium decimatum*. Cameron's work, however, was solid. He established the Dominion Entomological Laboratory in Saskatoon, Saskatchewan, and there published the first papers (Cameron 1918, 1922) on *S. vampirum* (as *S. simile*), one of the most diabolical pests on the continent.

Fig. 2.19. Alfred E. Cameron. Photograph courtesy of P. W. Riegert.

Fig. 2.21. Harrison G. Dyar. Smithsonian Institution Archives.

Fig. 2.20. John R. Malloch (ca. 1920). Smithsonian Institution Archives.

Two German workers, Karl Friederichs (1879–1969) in 1920 and Enderlein (1921b, 1922, 1925, 1929a), contributed eight names (representing two valid species) that would apply to North American species. The British worker Frederick Wallace Edwards (1888–1940) in 1920 added the name *subornatum* that later would fall as a synonym of *noelleri*.

Nearly 62% of the species names authored from 1900 through 1929 were contributed by Malloch (1913, 1914, 1919) and Dyar and Shannon (1927). John Russell Malloch (1875–1963) (Fig. 2.20), renowned for his remarkable memory, was born in Scotland, came to the United States in 1909, and began working on black flies in 1912 as entomological assistant in the U.S. Bureau of Entomology. Although he was asked specifically to work on black flies, largely because of their possible connection with pellagra, he was not allowed to conduct fieldwork and had to confine his efforts to material in the USNM collection. His employment with the bureau was terminated on 30 June 1913, the official reason being lack of funds. From October 1914 to 31 May 1921, he worked as entomologist in the Illinois Natural History Survey. Malloch (1914) constructed comprehensive keys to females and males and recognized 26 species north of Mexico, including 14 that he described as new (11 valid). His 1914 description of the genus *Parasimulium* initiated a search for the immature stages that would last more than 70 years.

The sole simuliid enterprise by Harrison Gray Dyar (1866–1929) (Fig. 2.21) and Raymond Corbett Shannon (1894–1945) (Fig. 2.22), published in 1927, was one of the most influential taxonomic works on the North American Simuliidae. It contained more newly described species from the continent than any previous single work and any work done in the subsequent 75 years, and established for the first time the use of female terminalia in the taxonomy of the family. At the young age of 24, the brilliant but cantankerous Dyar secured his place in entomological history by demonstrating that head-capsule widths of insects reflect geometric progression in growth. This simple relationship, indelibly etched in the entomological lexicon as Dyar's rule, has been used many times in determining the number of instars in black flies (and other insects). In 1897, Dyar became the unpaid honorary custodian of Lepidoptera in the USNM, a post lasting until his death from a stroke in 1929. At the urging of Leland Ossian Howard (1857–1950), C. V. Riley's successor as chief of the

Fig. 2.22. Raymond C. Shannon (1928). Smithsonian Institution Archives.

Fig. 2.23. Thomas J. Headlee (1911). National Archives.

Bureau of Entomology at the USDA, Dyar began working on mosquitoes at about the turn of the 20th century. It was this work that would lead him eventually to black flies.

Dyar's coauthor on the simuliid monograph was Ray Shannon, an orphan whose foster mother provided lodgings for Malloch and Knab, thus affording young Shannon early exposure to Diptera. Shannon was born in Washington, D.C., and received his formal entomological training at Cornell University. He was employed as junior entomologist by the U.S. Bureau of Entomology from the time of his graduation from Cornell in 1923 to 1925, during which time he worked on black flies with Dyar. In later years, his work with medically important insects in South America subjected him to a number of health problems, including an attack of dengue that precipitated his demise from an intentional drug overdose on 8 March 1945 in a hotel room in Port of Spain, Trinidad. Together, Dyar and Shannon identified all North American simuliid material housed in the USNM—more than 3000 specimens—and described 25 new species (19 valid) from the continent north of Mexico. Four of their names commemorated individuals, including the ardent collectors Annie Trumbull Slosson (1838–1926) and Charles Vancouver Piper (1867–1926), both of whom had provided specimens. Dyar himself had collected a good portion of the material, mostly from western North America.

Other important contributions in the young century complemented the taxonomic work. Morphology, for example, received some attention. Thomas Jefferson Headlee (1877–1946) (Fig. 2.23), while a graduate student under John Comstock, weighed in (eventually at more than 200 pounds) with an excellent treatment of the rectal papillae in 1906, the same year he obtained his doctorate. This work would be his only effort on the Simuliidae, but as a professor of entomology at Rutgers University from 1912 to the last day of 1943, he would make substantial contributions to the field of mosquito control. Herbert Barker Hungerford (1885–1963), while working with Professor Samuel Hunter on the pellagra issue in Kansas, produced a master's thesis, published in 1913, on the internal and external anatomy of adult black flies. This was the first broad morphological study of black flies in North America. Master's theses on larval anatomy by Harold Coleman Hallock (1891–1976), who was born in Northville, South Dakota, and the Japanese-born Takayoshi Tanaka (1897–1927) were produced in Professor O. A. Johannsen's laboratory at Cornell University in 1922 and 1924, respectively. Lydia Ann Gambrell (née Jahn, 1904–1992), whose birthplace was Cleveland, Ohio, began her doctoral program at Ohio State University in the late 1920s and completed it in 1932, with a dissertation (Jahn 1932) on the embryology of *Simulium innoxium* (as *S. pictipes*). It was the first doctoral degree on the Simuliidae to be earned by a North American woman. Gambrell became a technical assistant at the New York Experiment Station in Geneva.

The first serious work on natural enemies of black flies can be traced to Edgar Harold "Strick" Strickland (1889–1962) (Fig. 2.24). Fresh from his homeland of England and working as a master's student on a Carnegie Scholarship at Harvard University, Strickland, in 1911, became the first person to describe parasites of black flies from the continent. He discussed and illustrated a chytrid fungus (as a gregarine protozoan, later described as *Coelomycidium simulii*), an unidentified mermithid nematode, and six species of microsporidia of which he named and described three (Strickland 1911, 1913a, 1913b). In 1913, Strickland became the first official entomologist in Alberta, establishing the Dominion Entomological Laboratory in Lethbridge. His entomological service was inter-

Fig. 2.24. Edgar A. Strickland (late 1950s). Photograph courtesy of D. A. Craig.

rupted for a stint in World War I with the Canadian Machine Gun Corps. Wounded in action, he returned to Canada and continued his entomological work. In 1922, he established the Department of Entomology at the University of Alberta. In later years, his legacy lived on—subsequent personnel at the Lethbridge station and the University of Alberta featured prominently in the study of black flies.

### The Expansion Years (1930s and 1940s)

The 1930s and 1940s marked a geographic expansion of effort that included the vigorous participation of workers in Canada and the beginning of research on the fauna of western North America. Taxonomic activity continued at a rapid pace, with the proposal of 77 species names, representing 39 valid species. High points of the two decades included the discovery of polytene chromosomes in black flies and the demonstration that the females of North American species could transmit disease agents to birds. Another landmark of this era was the first symposium on black flies in North America, which was held at the Higby Club in Big Moose, New York—in the heart of the Adirondack Mountains—on 29 September 1948.

Taxonomic activity in the 1930s by North American, British, German, Mexican, Russian, and Serbian workers produced 40 species names for only 18 valid species. The decade was dominated by personnel in Canada, working at a time when the Canadian simuliid fauna was poorly known. Only about 9 of the 160 species now known from Canada had been recorded from that vast country before 1930. Of the 40 names proposed during the 1930s that were

Fig. 2.25. Eric Hearle. Photograph courtesy of P. W. Riegert.

applicable to North American species, 21 were products of employees of the Canadian government. Eric Hearle (1893–1934) (Fig. 2.25), although born in India, was the first worker in Canada knowingly to describe North American black flies. (Cameron [1918] while working in Saskatchewan formally, albeit unknowingly, bestowed the name *Simulium simile* on the species now known as *S. vampirum*.) As a livestock entomologist with the Dominion Entomological Laboratory in Kamloops, British Columbia, from 1928 to 1934, Hearle (1932) described *Simulium canadense* and *Simulium kamloopsi* (= *S. argus*) from the area around the laboratory. Had he not died in the prime of life, undermined in health by brutal wounds received in World War I, he might have described more species.

Cecil Raymond Twinn (1897–1989) (Fig. 2.26), another veteran of World War I, was the most prolific of the two workers in Canada during the 1930s, contributing 19 species names for eight valid species. Twinn was born in London, England, and came to Canada via the United States in September 1914. He graduated from the University of Toronto in 1922 and that same year began his entomological career as junior entomologist in Canada's Department of Agriculture in Ottawa. Twinn (1936a, 1936b) treated 23 species in eastern Canada and produced the first identification keys (females, males, and pupae) to Canadian species.

The sole taxonomic effort in the United States during the 1930s was made by Utah native George Franklin Knowlton (1901–1987) (Fig. 2.27) and his graduate student John Allen Rowe (1907–?). Knowlton and Rowe (1934a) authored four names that applied to two species: *brevicercum* (and its synonym *nigresceum*) and *pilosum* (and its synonym *utahense*). From 1925 to 1967, Knowlton was an extension entomologist on the faculty of Utah State Agricultural College (now Utah State University). He was an

Fig. 2.26. Cecil R. Twinn (early 1950s). Photograph courtesy of P. W. Riegert.

Fig. 2.27. George F. Knowlton (ca. 1925). Smithsonian Institution Archives.

avid collector who prospected primarily in southern Idaho and the northern part of Utah, especially in Logan Canyon. Rowe obtained a doctorate on the bionomics of mosquitoes in 1942 from Iowa State College (now Iowa State University), and in the 1950s, he worked as senior scientist for the Communicable Disease Center (now the Centers for Disease Control) in Salt Lake City and Logan, Utah.

The year 1935 heralded the first, albeit unknowing, contribution to the North American fauna by the Russian

Fig. 2.28. Earl C. O'Roke (ca. 1955). Photograph courtesy of C. C. Reading.

worker Ivan Antonovich Rubtsov (1902–1993), who that year coauthored the names of five Holarctic species that he and his colleagues (Dorogostaisky et al. 1935) discovered in Russia. By the time of his death, Rubtsov had authored or coauthored 365 specific, subspecific, or varietal names for simuliids, of which 32 unwittingly would apply to 11 valid North American species. His life's work on the Simuliidae was summarized by Crosskey (1999b), and his large collection in the Zoological Institute in St. Petersburg, Russia, was discussed by Adler and Crosskey (1998).

During the 1930s, giant polytene chromosomes were discovered in black flies in Europe (Geitler 1934). A few years later, they were studied in the United States by the fulgent geneticist Theophilus Shickel Painter (1889–1969), who was born in Salem, Virginia, and his doctoral student, Allen Beattie Griffen (1914–?), from Corsicana, Texas. These workers investigated the structure of polytene chromosomes of *Simulium solarii* (as *S. virgatum*) from the area around Austin, Texas. Griffen (1939), in his doctoral thesis, actually prepared a detailed, hand-drawn map of the entire complement. This early work, conducted at the University of Texas (Austin) and published by Painter and Griffen (1937a, 1937b), presaged the use of polytene chromosomes in simuliid systematics that began 14 years later.

The 1930s also marked the first demonstration that North American black flies are vectors of disease agents. Earl Cleveland O'Roke (1887–1958) (Fig. 2.28), a native of Capioma, Kansas, who served until 1957 in the Department of Zoology at the University of Michigan, linked *Simulium rugglesi* (as *S. venustum*) to the transmission of the parasitic protozoan *Leucocytozoon simondi* (as *L. anatis*) in 1930. He followed in 1934 with a treatise on the morphology, pathogenicity, transmission, and control of the parasite. O'Roke's work with black flies dates to 1911–1912, when, as a student at the University of Kansas and a participant in the Kansas Biological Survey, he scouted for breeding sites of simuliids in an effort to correlate their distributions with pellagra cases. Louis Vallieres Skidmore (1889–1963), born in New York City and a faculty member of the University of Nebraska from 1920 to 1958, followed closely on the heels of O'Roke. He established *Simulium meridionale* (as *S. occidentale*) as a vector of *Leucocytozoon smithi* among turkeys in 1931 and provided an expanded treatment in 1932.

Taxonomic work continued in the 1940s with the addition of 37 species names (representing 21 valid species), including 13 during the heart of the war years. Twenty-six of these names were proposed by British, German, Japanese, Russian, and Central American taxonomists. Among the Central American workers was the Mexican entomologist Daniel Luis Vargas (1908–1994), who, with various colleagues, contributed some of the first Neotropical species names that would apply to North American species. The remaining names were products of workers in the United States (Stains & Knowlton 1940, 1943; Stone 1948, 1949b) and Canada (Davies 1949b).

One of the premiere simuliid workers in the United States was Brooklyn-born Alan Stone (1904–1999), who began employment as an entomologist with the USDA in Washington, D.C., on 21 October 1931, two years after receiving his doctorate on tabanids from Cornell University. Stone's position initially was offered to Raymond Shannon, who declined the offer, choosing instead the adventures of medical entomology in the tropics. Stone spent the next 40 years, his entire professional life, working on black flies, as well as culicids, tabanids, and tephritids. He also would earn the distinction of being one of the two longest-lived North American simuliid workers to date (the other being A. M. Fallis), reaching the age of 95. In the late 1940s, he contributed three new species names, plus the generic name *Gymnopais* (Stone 1948, 1949b). Although he enjoyed collecting black flies and was fascinated by their pupae, he retired earlier than he might have done otherwise, in part because of the taxonomic difficulties presented by the group. Despite his feeling that the species he described were not always legitimate, 25 (almost 90%) of the 28 species names that he authored or coauthored for North American black flies refer to valid species.

During the war years, George Savage Stains (1917–1997) (Fig. 2.29) and his graduate advisor George Knowlton at Utah State Agricultural College described a number of western species, mostly from material taken in Logan Canyon just above the University. In 1943, they treated 45 species, including 4 described as new, and furnished the first keys (females and males) to black flies west of the Mississippi River. The work was based on Stains's (1941) master's thesis. Stains was born in Delta, Utah, and after graduation worked as an active-duty entomologist with the U.S. Navy, eventually rising to the rank of captain. He worked at the Naval Medical Field Research Laboratory at Camp Lejeune, North Carolina, and later (1963–1969) was officer in charge of the Navy Disease Vector Ecology and Control Center in Bangor, Washington. He was instrumental in developing aerial pesticide-spray systems for helicopters, as well as the first ground-based ultralow-volume dispersal apparatus.

Fig. 2.29. George S. Stains (1960s). Photograph courtesy of United States Navy.

In Canada, Douglas Mackenzie Davies (1919–) described *Simulium euryadminiculum* (= *S. annulus*) in 1949, thus beginning a distinguished career at McMaster University, Ontario, that would last more than 50 years and involve the training of 14 graduate students and postdoctoral fellows in various areas of simuliid study. Davies was the first North American to devote the majority of his career to the Simuliidae. He was introduced to black flies in 1941 when Frederick Palmer Ide (1904–1996) of the University of Toronto asked him to conduct a preliminary study on the morphological differences in larval simuliids. In 1946, Ide encouraged Davies to examine the black flies from an eight-year collection of insects that he and his students had made with emergence traps over Costello Creek in Algonquin Park, Ontario. Through his work in Algonquin Park and the publication of numerous landmark papers on the biology of black flies, Davies helped turn the park into a mecca for simuliid investigators. The area around the research station in Algonquin Park is probably the most intensively studied plot of land for black flies on the planet.

### The Heydays (1950s and 1960s)

The most significant event of the 1950s, and probably in the history of simuliid taxonomy, was the advent of cytotaxonomy in the summer of 1951 at the University of Toronto, Ontario. This field of study was initiated in the Cytology Section of the Department of Botany and was directed for 35 years by Klaus Hermann Rothfels (1919–1986) (Fig. 2.30), whose work fundamentally changed the character of simuliid study.

Born in Berlin, Rothfels developed an early interest and talent in biology. He attended the University of Hamburg, but because of his Jewish ancestry, his enrollment was terminated by a Nazi decree. After leaving Germany, he entered the University of Aberdeen in Scotland, but in 1940 he was interned as an alien in the United Kingdom and

Fig. 2.30. Klaus H. Rothfels (ca. 1980).

later shipped to Sherbrooke, Quebec, where the greeting was less than cordial. Granted official refugee status a year later, he entered the University of Toronto, receiving his undergraduate degree and later a doctorate in 1948 on the cytogenetics of grasshoppers. He spent the remainder of his life at the University of Toronto, teaching, conducting research, and overseeing the theses of 25 master's and 28 doctoral students on various aspects of plant and animal genetics; 19 theses on black flies were produced in his laboratory. He is perhaps the only simuliid worker for whom a poem has been published (Golini 2000). By the early 1960s, Rothfels was regarded as one of the most famous cytologists in Canada. His research was held in such high esteem that the National Research Council of Canada (predecessor of the Natural Sciences and Engineering Research Council of Canada) routinely invited him to apply for funds. Rothfels died of a probable heart attack while still active in the chromosomal study of black flies. He was as unconventional in his approach to life as he was influential in science. A kindhearted man with a quick and often acerbic wit, he had little tolerance for those who interfered with his program or his students, and it was not uncommon for him to begin memoranda to administrators with the salutation "Dear Stuffed Shirt." The image of the father of simuliid cytotaxonomy seated at his microscope in undershorts and a laboratory coat, his hair disheveled and an old towel draped across his lap, is not easily forgotten.

Rothfels became intrigued with the chromosomes of black flies after having read a paper (Geitler 1934) on the giant chromosomes of a European black fly and after having seen drawings of the giant chromosomes of a member of the *Simulium vittatum* species complex that Douglas Davies had made in April 1942 while taking a cytology course with Leslie C. Coleman (1876–1950). Coleman later became Rothfels's doctoral advisor and passed along his technique (Coleman 1938) for the preparation of Feulgen stain, which Rothfels used extensively throughout his own career and which we routinely have used in our work. The cytotaxonomic work on black flies began in earnest when Robert Wilfrid Dunbar (1930–) arrived as a student in early 1951. Later that spring, Rothfels sent Dunbar to the Dominion Experimental Farm in Ottawa to learn about black flies from Guy Shewell.

In the early 1950s, Dunbar and Rothfels collected a mass of material from streams around Toronto, mainly along the Niagara escarpment. Among the first specimens examined were larvae of the *S. vittatum* species complex taken beneath the ice of the Don River behind Sunnybrook Hospital in Toronto. The material was, in part, the basis for the first paper (Rothfels & Dunbar 1953) on simuliid cytotaxonomy to come out of Rothfels's laboratory. This work presented a scene-setting description of chromosomal methodology and the first complete, semimodern chromosome map, now obsolete, for a black fly (*Simulium tribulatum* as *S. vittatum*). Chromosomes of larvae from the Don River indicated that colder water produced better preparations, leading Rothfels to choose the early-spring black flies, *Prosimulium*, as his first group to study in detail. His classic 1956 work on *Prosimulium* demonstrated the utility of chromosomes in distinguishing morphologically similar species and set forth the criteria for recognizing sibling species. At least 62 species of black flies in North America, nearly one fourth of the fauna, have been discovered as a result of the approach presented in that paper.

Dunbar remained with Rothfels for about 12 years, completing a master's thesis (1958a) and two papers on the *Simulium aureum* species complex in 1958 and 1959 and a doctoral dissertation on various basal species of *Simulium* in 1962. He left Rothfels's laboratory in 1962 to accept a research fellowship at Durham University, England, where he initiated the cytotaxonomic work on what would become the enormous *Simulium damnosum* species complex of tropical and southern Africa.

The cytotaxonomic group in the 1950s and early 1960s that assisted Rothfels in launching the field of simuliid cytotaxonomy included several other graduate students. Ruth Landau (née Zimring) joined the group in the spring of 1952, completed a master's degree in 1953 on the polytene chromosomes of five species of *Simulium*, worked as an assistant in the lab, and published a treatment of the *Simulium tuberosum* species complex in 1962. She eventually emigrated to Tel Aviv, Israel. In October 1955, Parvathi Koodathil Basrur arrived from India and took over a large portion of the *Prosimulium* work, completing her doctorate in 1958, along with papers in 1959 and 1962 on the polytene chromosomes of *Prosimulium*. She became a professor in the Ontario Veterinary College at the University of Guelph. Parvathi's husband, Vasanth Rao Basrur (1931–), completed a doctoral degree in 1957 and coauthored a 1959 paper with Rothfels on triploidy in *Stegopterna mutata*, following Douglas Davies's (1950) discovery of populations with female-biased sex ratios in Algonquin Park, Ontario. He subsequently accepted a faculty position at Sunnybrook Hospital in Toronto. John

Jacob "Jack" Pasternak (1938–) joined the group in 1959, having become interested in simuliid chromosomes after seeing a colossal set (ca. 0.9 × 1.2 m) from the *Simulium aureum* species complex photographed by Dunbar and displayed during an open house in the Department of Zoology at the University of Toronto. Pasternak completed his master's degree on *Simulium vittatum* s. l. in 1961 and published the work in 1964.

An important companion event to the cytotaxonomic work of the 1950s was the morphological validation of polytene chromosomes as tools for revealing sibling species of black flies. Following his 1956 discovery of three sibling species in what previously had been known in North America as *Prosimulium hirtipes*, Rothfels suggested that master's student Paul Denness Syme (1932–) and his advisor, Douglas Davies, search for morphological characters that would allow the siblings to be discriminated. The result was a publication in which Syme and Davies (1958) named the three sibling species (*Prosimulium fontanum, P. fuscum, P. mixtum*) and described them on the basis of structural characters. The specific name *mixtum* was born from the authors' frustration in morphologically identifying the specimens; they often "mixed 'em up" with those of the other two species. Further validation of cytotaxonomy came when Davies and Syme (1958) demonstrated ecological differences among the three species.

While cytotaxonomy was opening a new frontier of simuliid investigation, morphotaxonomy continued to thrive in the 1950s. Fourteen North Americans and the Russian worker Rubtsov contributed a total of 48 species names for 31 valid species. The decade began with a publication (1950) by graduate student Harry Page Nicholson (1913–2000) and his advisor Clarence Eugene Mickel (1892–1982), a mutillid wasp specialist who served on the faculty of the University of Minnesota from 1927 to 1960. They provided keys to males and females of Minnesota black flies and contributed three new names (*croxtoni, luggeri, rugglesi*). Nicholson was born in Crosby, Minnesota, and was the first North American to earn both a master's (1941) and doctorate (1949) specifically on black flies, receiving both degrees from the University of Minnesota. He later worked in pollution control in the southeastern United States, retiring in 1976. In 1951, Richard Walter Coleman (1922–) completed a doctoral degree on the black flies of California at the University of California, Berkeley, having traveled more than 80,000 km throughout the state to collect material. He followed in 1953 with the description of a new species (*Tlalocomyia* [as *Cnephia*] *stewarti*).

The northern regions of the continent were explored with great fervor during the 1950s, motivated largely by national defense interests. The Alaska Insect Project, with representatives from the Department of the Army and the USDA, conducted seasonal fieldwork from 4 May 1947 to 27 October 1948, with headquarters at Fort Richardson (Anchorage), Alaska. In 1948, 43 people, including 15 entomologists, were assigned to the project (Anonymous 1949). Bernard Valentine Travis (1907–1980), then with the USDA in Orlando, Florida, and eventually with Cornell University, was the project leader. Al Stone identified the majority of black flies for the project.

Kathryn Martha Sommerman (1915–2000), one of the most important morphotaxonomic workers of the 1950s, got her start with black flies in 1948 as a member of the Alaskan Insect Project. She was with the Army Medical Service Graduate School at Walter Reed Army Medical Center in Washington, D.C., and later the Department of Health, Education, and Welfare in Anchorage, Alaska. Sommerman's publications included a key to Alaskan larvae that established the standard format for illustration of the head capsule (Sommerman 1953) and a detailed study of the ecology of Alaskan black flies (Sommerman et al. 1955). Her larval key was based largely on structure, rather than colors, because she was color blind. In 1958, she became the first and only North American woman in the 20th century to contribute simuliid names (*frohnei, perspicuus*) to the North American fauna. The Brazilian Maria Aparecida Vulcano (1921–), junior author of the name *longithallum* in 1961, was the only other woman to contribute names to the North American fauna. Prior to Sommerman, only about five women had written theses (Phillips 1890), dissertations (Jahn [subsequently Gambrell] 1932), or publications (Anna Comstock [with her husband] 1895, Reeves 1910, Wu 1931, Gambrell 1933) that dealt in some fashion with North American black flies.

Al Stone's productivity was reaching its azimuth in the 1950s, with the description of one genus (*Twinnia*) and 14 species, plus significant works on the fauna of Alaska (Stone 1952), New York (Stone & Jamnback 1955), and California (Wirth & Stone 1956). His 1955 effort on the New York fauna was based partly on the doctoral work of Hugo Andrew Jamnback (1926–), who spent most of his professional career at the New York State Museum in Albany. Jamnback became one of the most prominent workers in the management of black flies and was the only North American worker who actively published from the beginning of the DDT era through the inception of the *Bti* era. He had been introduced to black flies by Gene Ray DeFoliart (1925–) from Stillwater, Oklahoma, who completed his doctoral dissertation at Cornell University in 1951 on the black flies of New York. As a professor at the University of Wyoming and later at the University of Wisconsin, DeFoliart coauthored nine names for western species in the late 1950s and early 1960s. In later years, DeFoliart devoted much of his effort to promoting the consumption of insects by humans. Stone's work with his colleague Willis Wagner Wirth (1916–1994) at the USDA in Washington, D.C., on the black flies of California was based, in part, on Richard Coleman's 1951 dissertation.

Workers in Canada also gave considerable attention to the black flies of the Far North, especially during the years of the Northern Biting Fly Project (1947–1955). Under this project, directed by Cecil Twinn (1950, 1954), the bionomics and control of northern black flies were investigated by several workers, particularly Brian Hocking (1914–1974), a native of London, England. Most of the northern work by Hocking and his colleagues was conducted in Labrador (Goose Bay), northern Manitoba (Churchill), and the Yukon Territory (Whitehorse). Hocking was on the faculty of the

University of Alberta from 1946 until his death 28 years later. He completed a doctorate on insect flight in 1953 under the supervision of Edgar Strickland and contributed important works on the bionomics, control, and physiology of black flies (e.g., Hocking 1950, 1952a, 1953a, 1953b; Hocking & Richards 1952; Hocking & Pickering 1954).

Guy Eaden Shewell (1913–1996), an entomologist with Agriculture Canada in Ottawa, became an important force in studies of the northern fauna, describing 11 species (8 valid) in a series of four papers (Shewell 1952, 1959a, 1959b; Shewell & Fredeen 1958). The English-born Shewell began working for the Entomological Branch of the federal Department of Agriculture in 1937, although his career was interrupted from 1939 to 1945 when he served in the Fifty-first Anti-tank Battalion of the Royal Canadian Artillery. He was a key player in the Northern Insect Survey of Canada that ran from 1947 to 1962. Shewell identified all the black flies that had been collected in the survey, which included nearly 7000 pinned adults and more than 1200 vials of larvae and pupae from 58 arctic and subarctic sites. Much of the early material had been sorted and temporary letters assigned to new taxa by Douglas Davies during a few weeks in each of the winters of 1948–1950. Although Shewell's simuliid work was nearly impeccable, he did not like working with black flies.

Another influential Canadian, and a junior author with Shewell of the names *duplex* (= *johannseni*) and *saskatchewana*, was Frederick John Hartley Fredeen (1920–2003). The intolerable deaths of livestock from black fly attacks in Saskatchewan in the 1940s prompted Cecil Twinn to create a position at the government research station (Dominion Entomological Laboratory) in Saskatoon that was filled by Fredeen on 25 April 1947. While an employee at the research station, Fredeen worked toward a master's degree on the black flies of Saskatchewan, receiving his diploma in 1951. His thesis provides excellent historical data on the distribution of larvae of *Simulium vampirum* (as *S. arcticum*) in the Saskatchewan River drainage. Fredeen spent nearly 40 years battling *S. vampirum* and *Simulium luggeri* on the Canadian prairies, publishing numerous papers on their bionomics and control. He was a pioneer in the control of black flies in large rivers. Fredeen's campaign against black flies has been told in detail by Riegert (1999b).

Near the end of the decade (1 November 1958), Shewell, Fredeen, and 11 other Canadians, plus one American (J. R. Anderson), attended the first international conference on black flies, which was held in the Entomology Laboratory of Canada's Department of Agriculture in Guelph, Ontario. A 32-page informal document ("Proceedings of Black-Fly [Simuliidae] Conference") was compiled from notes that were submitted by contributors to the conference.

Along with the previous decade, the 1960s were probably the most active years in the taxonomy of North American black flies, with research occurring on many fronts. The labors of 17 workers in the 1960s produced 41 species names for 29 valid species. The days of foreign contributions to North American simuliidology had nearly passed by the time the 1960s arrived. One of the last foreign taxonomic contributions dealing directly with North American black flies came in 1969 when the British simuliid specialist Roger Ward Crosskey (1930–) of the British Museum of Natural History established the genus *Metacnephia*, thus sorting out some of the species that previously had been lumped in the genus *Cnephia*. In the United States, Sommerman (1962a, 1964) added two new species of *Prosimulium*, and Stone authored or coauthored 11 names. Nine of Stone's names pertained to species from the poorly prospected southeastern United States, especially Alabama, and were presented in a faunistic work (Stone & Snoddy 1969) based largely on the doctoral dissertation of Edward Lewis Snoddy (1933–1997), a student at Auburn University, plus material from Stone's own expedition through the Southeast in April 1941. Stone also produced a list of 111 species in North America (Stone 1965a), an annotated world list of genus-group names (Stone 1963a), and a fully illustrated guide to the black flies of Connecticut (Stone 1964).

The illustrations in the Connecticut and southeastern treatments, as well as some that had been published earlier (e.g., Stone & DeFoliart 1959, Stone & Boreham 1965), were the work of Japanese artists in the 406th Medical General Laboratory of the U.S. Army in Tokyo, Japan. A conference at Walter Reed Army Medical Center on 31 January 1955 had considered how this group of artists could best be used to meet military needs. Conference participants recommended Stone's proposal, "Black Flies of the Boreal Regions," among a group of "second priority projects." Stone had intended these illustrations as part of a complete revision of the Nearctic Simuliidae. The revision, however, never came to fruition, in part because of his frustration over the growing number of sibling species that were being discovered by Rothfels and his students. Many of the remaining illustrations eventually were published by Peterson (1993, 1996) and Peterson and Kondratieff (1995). Most of the slide mounts from which the illustrations were made had been prepared in the Tokyo laboratory and were drawn as they appeared, which explains the sometimes distorted orientations.

Research efforts on black flies in Canada probably exceeded those in the United States during the 1960s. While the Canadian morphotaxonomists described new species, Rothfels and his students continued to discover sibling species and present hypotheses of their evolutionary relationships. For example, Pentti Olavi Ottonen (1939–) presented the cytotaxonomy of members of the *Prosimulium magnum* species group in 1966, Dunbar elucidated the cytotaxonomy of the subgenus *Hellichiella* in 1967, and Dharam Paul Madahar (1935–) revealed sibling species of *Stegopterna mutata* in 1969.

The most important taxonomic work in Canada during the 1960s was a triauthored treatment of Ontario black flies, including descriptions of six new species and keys to adults and pupae, by Davies et al. (1962). One of Davies's coauthors was Bobbie (Robert) Vern Peterson (1928–), who obtained his doctorate (1958a) (on black flies) from the University of Utah and during the 1960s published a number of key papers (e.g., Peterson 1960a, 1960b, 1962a, 1962b), both taxonomic and bionomic, on the simuliids of Utah and Ontario. Peterson was introduced to black flies as a master's student at the University of Utah, having accepted a project given to him by his advisor and respected mosquito worker, Don Merrill Rees (1901–1976). His work with black flies spanned five decades. He began working with Agriculture Canada's Medical and Veterinary Research Laboratory in Guelph, Ontario, in 1958. When the laboratory closed, he was transferred to the Biosystematics Research Institute in Ottawa, Ontario, where he worked for 20 years before moving to the National Museum of Natural History in Washington, D.C. He retired on 29 September 1994. During his career, he authored or coauthored a total of 37 species names for North American black flies, more than any other worker in the 20th century; the names represent 24 valid species. The third author of the Ontario paper was Donald Montgomery Wood (1933–), at that time a student of Davies, and the senior author of the complementary larval treatment of Ontario black flies (Wood et al. 1963), which included the first key to Canadian larvae. For both Ontario papers, Davies enlisted the artistic talents of Ralph Marten Idema (1944–), then a 15-year-old high-school student and president of the Hamilton Naturalists' Junior Club in Ontario, where Davies, the club's advisor, first met Idema. As a summer and seasonal technician, Idema prepared the illustrations that established the standard format and orientation for male and female terminalia, reflected in his drawings herein.

The first significant study of simuliid hosts was published in 1960 by Gordon Fraser Bennett (1930–1995). Born of missionary parents in India, Bennett came to Canada in 1949 and spent most of his professional life at the Ontario Research Foundation (1957–1966) and Memorial University in Newfoundland (1969–1995). With his doctoral advisor and later colleague, Albert Murray Fallis (1907–2003) of the Ontario Research Foundation and University of Toronto, Bennett contributed important papers (Bennett & Fallis 1960; Bennett 1961; Fallis & Bennett 1961, 1962, 1966) on blood parasites transmitted by black flies. The work of Bennett and Fallis was paralleled in the United States by that of John Richard Anderson (1931–), who completed his doctoral research in 1960 under the direction of Gene DeFoliart at the University of Wisconsin and became a faculty member at the University of California, Berkeley. Anderson's work involved host and leucocytozoon studies, as well as the taxonomy and bionomics of Wisconsin black flies (e.g., Anderson & Dicke 1960, Anderson & DeFoliart 1961, Anderson et al. 1962).

The intensity of interest in the Simuliidae, particularly in Canada, during the 1960s was reflected in two international conferences on black flies. The first one was held 23–24 September 1960 at Chaffey's Locks, Ontario (Queen's University Biological Station), attracting 25 workers, all but 7 from Canada (Fig. 2.31). A 117-page informal document ("Proceedings of the Second Conference on Black Flies") was edited by B. V. Peterson, and 39 copies were produced exclusively for participants of the conference. The second conference was held 14–15 September 1962 at the Wildlife Research Station in Algonquin Park, Ontario. Thirty-two participants attended—21 from Canada, 6 from the United States, and 5 from overseas. The conference was recorded on audio tape, and a 176-page informal typescript ("Proceedings of the Third Conference on Black Flies [Diptera:

Fig. 2.31. International conference on black flies, Chaffey's Locks (Queen's University Biological Station), Ontario, 23–24 September 1960. Standing (left to right): Unknown, Parvathi K. Basrur, Robert W. Dunbar, John J. Pasternak, Hugo A. Jamnback, Arne P. Arnason, R. John Phelps, D. Montgomery Wood, Alan Stone, John R. Anderson. Sitting (left to right): Gene R. DeFoliart, Kathryn M. Sommerman, F. J. Hartley Fredeen, Douglas M. Davies, Harold E. Welch, Hedley G. James, Bobbie V. Peterson, A. Murray Fallis, Gordon F. Bennett, Lewis Davies, Alan S. West, Klaus H. Rothfels, Guy E. Shewell, Douglas G. Peterson.

Simuliidae]") was produced by B. V. Peterson and Douglas Gordon Peterson (1921–).

### The Contemporary Era (1970s–Present)

Activity on black flies remained high during the 1970s. Twenty-nine names for 17 valid species were added; revisions were produced for *Prosimulium* (Peterson 1970b), *Parasimulium* (Peterson 1977a), and *Gymnopais* and *Twinnia* (Wood 1978); and a new genus (*Tlalocomyia*) that later would accommodate North American species was described from Central America (Wygodzinsky & Díaz Nájera 1970). Rothfels (1979) reviewed all published and unpublished information on black fly cytotaxonomy worldwide, the same year that the Russian black fly cytogeneticists Chubareva and Petrova (1979) independently presented their world review. The 1970s also ushered in the molecular age for black flies (Teshima 1972, Sohn et al. 1975, May et al. 1977).

An important simuliid conference was organized in the United States by John Frederick Burger (1940–) of the University of New Hampshire and was held at Dixville Notch, New Hampshire, from 31 January to 2 February 1977. The papers presented at the conference were reproduced in an informal document (Burger 1977) that was circulated after the meeting. The conference led to the official organiza-

Fig. 2.32. International conference on black flies, Pennsylvania State University, 28–31 May 1985. (Not all attendees are represented in the photograph.) Key: 1. Robert W. Lake; 2. Bernard W. Sweeney; 3. Kenneth R. Simmons; 4. Klaus H. Rothfels; 5. George L. Parks; 6. Joyce M. Nyhof; 7. Gail E. O'Grady; 8. Charles L. Brockhouse; 9. Ke Chung Kim; 10. James W. Amrine; 11. Unknown; 12. Mir S. Mulla; 13. Milan Trpis; 14. Terence L. Schiefer; 15. Omar M. Alamari; 16. Lise Molloy; 17. Daniel P. Molloy; 18. Douglas C. Currie; 19. Murray H. Colbo; 20. Peter H. Adler; 21. Bobbie V. Peterson; 22. Robin L. Vannote; 23. Eddie W. Cupp; 24. Douglas M. Davies; 25. Michael J. Roberts; 26. Paul Deepan; 27. John R. Anderson; 28. Victor I. Golini; 29. Paula J. Martin; 30. Richard W. Merritt; 31. Gisli M. Gíslasson; 32. Jean O. Lacoursière; 33. Jacques Boisvert; 34. Stefanie E. O. Meredith; 35. Margaret E. Crosskey; 36. Kathy E. Doisy; 37. Ann E. Gordon; 38. William S. Ettinger; 39. Unknown; 40. Charles W. Rutschky; 41. Keith Schuyler; 42. John F. Burger; 43. Markus Eymann; 44. Joseph A. Shemanchuk; 45. Unknown; 46. Mary M. Chance-Galloway; 47. Hiroyuki Takaoka; 48. Hans Knutti; 49. Fiona F. Hunter; 50. Harold Townson; 51. John N. Raybould; 52. John D. Edman; 53. Marcelo Jacobs-Lorena; 54. Lawrence A. Lacey; 55. K. Elizabeth Gibbs; 56. Lynda D. Corkum; 57. Bertha T. A. Maegga; 58. Roger W. Crosskey; 59. Unknown; 60. John L. Petersen; 61. David E. Leonard; 62. William G. Deutsch; 63. Unknown; 64. John W. McCreadie; 65. Unknown; 66. F. J. Hartley Fredeen; 67–68. Unknown; 69. David Courtemanch; 70. James F. Sutcliffe; 71. Steven B. Jacobs; 72. Unknown; 73. Susan B. McIver; 74. Erwin A. Elsner; 75. Marshall Laird; 76. D. Montgomery Wood; 77. William R. Beck; 78. Douglas A. Craig; 79. Peter Wenk; 80. Les J. Shipp; 81. David D. Hart; 82. Ralph Garms; 83. Christian Back; 84. Antoine Morin; 85. Roger S. Wotton; 86. Robert D. Sjogren; 87. Jan Conn; 88. Jan J. H. Ciborowski; 89. Bruce Wahle; 90. J. Bruce Wallace; 91. John B. Davies; 92. Michael W. Service; 93. Gregory W. Courtney; 94. Alice L. Millest; 95. Daniel C. Kurtak; 96. Angella Phillips; 97. Amin Muhammad; 98. Karin G. Leonhardt.

tion of Northeast Regional Project NE-118 (Black Fly Damage Thresholds, Biology and Control) that was active from 1 October 1977 to 30 September 1996. The regional project, under the auspices of the USDA, officially involved 15 states, two provinces, and the District of Columbia, as well as unofficial participants from throughout the world (Adler et al. 2001). During its lifetime, it ranked as one of the most productive regional projects in the nation and did more to unite simuliid workers than any other forum.

The 1980s were taxonomically lean. Only 10 new species (5 valid) were described, fewer than in any decade since the first of the century, although discovery of the immature stages of *Parasimulium* (Borkent & Wood 1986, Courtney 1986) brought closure to a search that had lasted more than 70 years. By contrast, studies in cytotaxonomy, ecology, morphology, laboratory colonization, natural enemies, and management flourished, and seven papers and theses were published on molecular aspects of black flies, particularly isozyme analysis. The decade was punctuated by the fourth international conference on black flies, held 28–31 May 1985 at Pennsylvania State University (Fig. 2.32). The conference attracted about 130 attendees from around the world and resulted in a coedited book (Kim & Merritt 1988). This gathering probably represented the high-water mark for the study of North American black flies.

In the 1990s, the pace of descriptive taxonomy increased over that of the previous decade, and 25 new species names for 18 valid species were added. The decade witnessed more molecular work than ever before, with the number of theses and published papers on molecular aspects more than doubling over that of all previous years, a trend that can be expected to increase in tempo.

In the preceding two and a half centuries, 324 formal species names were proposed—120 (37.0%) of them synonyms. We add 43 new names, bringing the grand total of formal species names for North American black flies to 367. In the past 200 years, more than 2000 papers and about 185 theses and dissertations have been written on North American black flies. At the close of the 20th century, however, fewer individuals were working on black flies and fewer students were using black flies as thesis subjects than at any time during the previous 50 years. Near the beginning of the new millennium—17–21 June 2000—the fifth international conference on black flies was held at Brock University in St. Catharine's, Ontario. Forty-two delegates attended, 35 of them from North America.

# 3| Techniques for Collection, Preparation, and Curation

Multiple methodologies often are required to obtain relevant biological information and accurate identifications of black flies. This chapter emphasizes the various methods for collecting black flies and preparing them for study and long-term storage. In all studies involving black flies, representative specimens (i.e., vouchers) of each taxon must be deposited in accessible collections. Voucher specimens help guarantee the validity of a study over time and provide an important source of historical material for future investigations.

## Collection and Fixation

### larvae and pupae

One of the simplest and most profitable means of obtaining material is to use forceps to collect by hand larvae and pupae from all available substrates, including trailing vegetation, debris, stones, and refuse such as plastic and glass. To obtain larvae and pupae from large inaccessible rivers, monofilament line or plastic tubing can be placed in the current, secured from a bank or bridge, and removed several days later. If larvae are scarce or pupate in bottom sediments, as do those of headwater species of *Greniera* and *Stegopterna*, sheets of plastic can be placed in the stream bottom and removed 4–6 days later (Hunter et al. 1994). Alternatively, pupae of these genera can be collected by disturbing the fine sediment while holding a net downstream of the area.

The required level of identification and intent of the investigator will determine the type of collecting fluid and procedure to be used. All samples, however, should be labeled with full collecting information, minimally the specific location, date, and name of the collector. The label, printed on high-rag-content paper, should be placed inside each sample vial. For identifications based strictly on morphology, specimens can be placed in 80%–95% ethanol. However, to extract maximum morphological information, we recommend acetic ethanol, known among simuliid workers as Carnoy's fixative, which consists of one part glacial acetic acid and three parts 95%–100% ethanol. Specimens for cytological analyses must be collected into acetic ethanol. The larvae should be blotted to remove excess water and immediately plunged into acetic ethanol. To ensure adequate fixation, a ratio of about one specimen per milliliter or more of solution should be used. The acetic ethanol should be decanted and replaced with fresh solution immediately after the specimens are collected and once or twice again within a few hours. Even with great care in fixation, chromosomal quality can vary from hopelessly poor to excellent. Higher water temperatures (e.g., during the summer months) and poor food quality significantly reduce chromosomal quality. For molecular analyses, specimens should be killed in 100% ethanol, which should be refreshed at least once within 1 hour (Post et al. 1993, Koch et al. 1998). Specimens fixed in absolute ethanol or acetic ethanol should be stored in darkness at temperatures below 0°C, pending analysis. Fixed larvae, if stored properly, can yield workable chromosomes more than 15 years later.

No single fixative is ideal for providing morphological, cytological, and molecular information. If acetic ethanol is used, the larval chromosomes are fixed for subsequent analysis, the labral fans are opened so that primary rays can be counted easily, and pigmentation patterns are retained, although some colors, especially reds, rapidly fade to shades of gray and brown. The major disadvantage of acetic ethanol is that it degrades DNA, yielding short fragments and fewer successful amplifications by polymerase chain reaction (Koch et al. 1998). Additionally, the rectal papillae of larvae often fail to extrude in acetic ethanol, although they can be dissected out if necessary. Live larvae can be prepared for multiple analyses by cutting them into three parts. The head is put in 80% ethanol or acetic ethanol for morphological study, the thorax and first four abdominal segments are placed in absolute ethanol for molecular analysis, and the posterior portion of the body is plunged into acetic ethanol for chromosomal analysis.

### adults

Despite their minute size, adults can be collected while they are engaged in natural behaviors. Ovipositing females can be netted during the waning hours of late afternoon as they fly upstream near the water surface, hover and dip to the surface, or crawl about on trailing vegetation and other wetted objects. Yellow flagging tape secured and allowed to trail in the current at the air-water interface serves as an excellent artificial substrate for attracting ovipositing females of some species (Golini

& Davies 1975). Adults rarely are seen at flowers, and mating swarms are frustratingly difficult to find. Both sexes can be collected by sweeping vegetation, especially near streams, where the adults sometimes rest. Females can be collected from hosts, using handheld nets or aspirators, although without special trapping techniques and permits, this type of collecting is often feasible only for more accessible hosts such as domestic animals and humans. Once adults are captured, they can be dispatched by freezing, by submersing them in 95% ethanol, or by using any of the standard killing agents (e.g., ethyl acetate); each of these techniques will provide specimens amenable to molecular and morphological study.

### Rearing and Associating Life Stages

To associate life stages and facilitate identification, larvae or pupae can be reared to adults and the preimaginal exuviae retained. Rearing adults from larvae requires more effort than rearing them from pupae and is not always feasible. Nonetheless, if larvae are placed in a moist dish and held on crushed ice, they can be returned to the laboratory, placed in aerated containers of cool, nonchlorinated water, and fed pulverized fish food until they pupate. Final instars with dark gill histoblasts sometimes pupate if they are refrigerated in a shallow dish with a thin film of water. More elaborate rearing techniques are available (e.g., Edman & Simmons 1985a, 1985b) but should not be necessary if the intent is simply to associate adults with antecedent life stages.

Rearing pupae to adults is simple and yields high success. Pupae can be removed from the substrate, using fine forceps to grasp the anterior margin of the cocoon, or they can be collected with a bit of their substrate. Individual pupae are then placed on moist filter paper in a petri dish or on packed cotton in a vial. Alternative techniques for pupal rearing have been developed, using multiple rearing tubes (Golini 1981) and sphagnum moss (*Sphagnum*) as a substrate for retaining moisture (Hunter et al. 1994) (Fig. 3.1). During rearing, the pupae should be held in darkness or dim light at room temperature or slightly cooler and kept moist but not in a film of water. Freshly emerged adults should be held in darkness for about 24 hours to allow tanning and hardening before being frozen, pinned, or placed in 95% ethanol. The legs of adults that have not fully tanned typically do not provide accurate color information.

The pupal stage often provides a link for associating larvae and adults. The cocoon sometimes retains the larval exuviae, including the head capsule, especially in species with a cocoon that encloses the entire pupa (e.g., some *Prosimulium* species) or is boot shaped (e.g., *Simulium malyschevi* species group). Association with the adult stage is possible if the pupa is reared to an adult or if it houses a developed (i.e., pharate) adult, which can be dissected from the pupa. The holotypes of a few species (e.g., *Prosimulium doveri*, *Simulium fionae*) include the adult with its associated larval and pupal exuviae and cocoon.

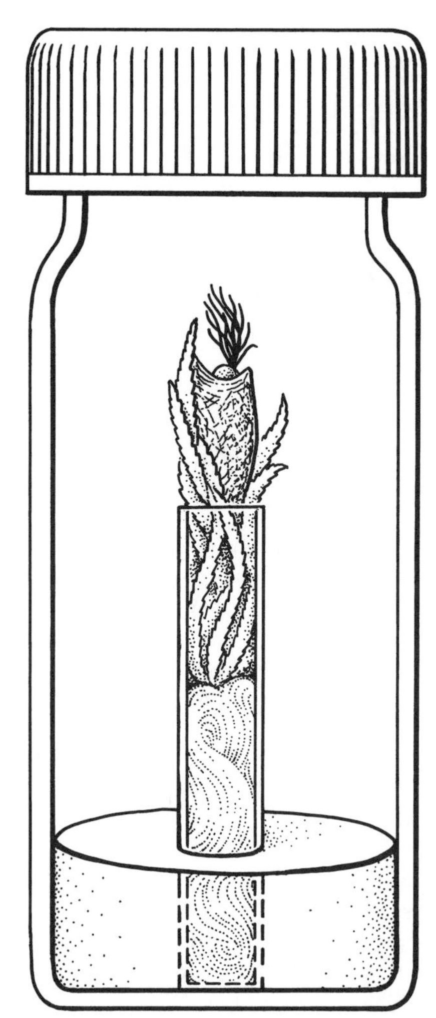

Fig. 3.1. Rearing vial for obtaining pristine adults from pupae. The pupa is placed on sphagnum moss that draws water from wet cotton in a smaller vial held in place with a Plastazote foam insert; the screwcap lid is loose to prevent condensation. Redrawn and modified from Hunter et al. (1994).

### Specialized Sampling and Trapping Techniques

#### Eggs

Eggs can be acquired directly from females or from the habitat. Oviposition can be induced by gently shaking gravid females in a vial with a small amount of water (Mokry et al. 1981). Exposure to carbon dioxide also can induce oviposition, although less reliably (Dalmat 1950). Eggs of species that oviposit on substrates such as trailing vegetation can be collected, often in large numbers, by gathering the substrate to which they are attached or by scraping them from rocks. Eggs of species that oviposit directly into the water must be recovered using special techniques (Shipp & Byrtus 1984) or dredged from the sediments and extracted using brine flotation (Fredeen

1959b). Eggs can be shipped long distances in wet cotton on ice or at room temperature (Tarshis 1965b).

#### LARVAE AND PUPAE

For specific ecological objectives, such as assessments of densities, spacing patterns, and drift rhythms, special sampling techniques typically are required. Artificial substrates, such as ceramic tiles and plastic strips, have been used to sample black flies in North America at least since 1958 (Wolfe & Peterson 1958). When placed in streams and rivers, these substrates can facilitate collection, provide many specimens, give estimates of relative abundance, and elucidate spacing patterns (e.g., Tarshis 1965a, 1968a; Johnson & Pengelly 1966; Curtis 1968; Lewis & Bennett 1974a; Boobar & Granett 1978; Fredeen & Spurr 1978; Noblet & Alverson 1978; Walsh et al. 1981; Morin 1987b; Colbo 1988). Artificial substrates, however, have limitations, and the variation among different types of samplers can be great (Colbo 1988). Clumped distributions, which are typical of larval and pupal black flies, can influence the efficiency of sampling as well as the estimation of densities (McCreadie & Colbo 1991c). Drift nets are useful for sampling black flies that have released their hold and are moving downstream in the water column (Adler et al. 1983b). Kick nets, Surber samplers, and similar devices tend to be inefficient for sampling, in part because the immature black flies often are clumped on trailing vegetation and are not readily dislodged.

#### ADULTS

Numerous sampling and trapping techniques have been used to collect both sexes (Tarshis 1978; Service 1981, 1988). Emergence traps, in use for black flies since the early 20th century (Fig. 2.15), can collect material and provide estimates of adult production. However, care must be taken not to bias the trapping toward females that might be attracted to the trap as an oviposition substrate or might enter it through a gap while flying low over the surface (Singh & Smith 1985, McCreadie et al. 1994b). Light traps often capture limited numbers of adults. Malaise traps, if deployed in appropriate habitats such as over streams, also are an effective collecting method and can be designed to segregate upstream and downstream catches. Vehicle-mounted traps capture males and females but have been used primarily as a means of collecting blood-fed females, with varying levels of success, perhaps depending on habitat (Davies & Roberts 1973, Morris 1978, Barnard 1979, Simmons et al. 1989). Additional trapping techniques include sticky panels (Bellec 1976, Walsh 1980, Adler et al. 1983a), airplane tow nets (Choe et al. 1984), suction traps (Johnson et al. 1982), and modified Manitoba fly traps (Peschken 1960, Peschken & Thorsteinson 1965). Resting adults can be obtained by fogging the canopy with knock-down insecticides such as resmethrin (Simmons et al. 1989).

The attraction of females to hosts and their odors, especially carbon dioxide, has been exploited in a variety of trap designs. Actual hosts have been used as bait in traps on the ground (Roberts 1965, Shemanchuk 1978b, McCreadie et al. 1984, Fletcher et al. 1988) and in the air (Bennett 1960). Carbon dioxide is an effective attractant alone (Snoddy & Hays 1966; Frommer et al. 1974, 1976) or sometimes in combination with octenol (Atwood & Meisch 1993), acetone, or crude animal fractions (Sutcliffe et al. 1994). Sham-host traps, often used with carbon dioxide, include silhouette traps (Fredeen 1961; Shipp 1983, 1985b, 1985c; Mason 1986; Mason & Kusters 1993, Sutcliffe et al. 1995), simulated-ear traps (Schmidtmann 1987), and entire host models coated with adhesive (Anderson & Yee 1995). The various traps for attracting females, however, can be biased in the proportions of species that they capture (e.g., Sutcliffe & Shemanchuk 1993).

For mark-recapture studies, larvae and pupae have been labeled with phosphorus 32, and the radioactive adults later captured and detected by autoradiographic techniques (Baldwin et al. 1966). Some of the best mark-recapture techniques involve self-marking devices that use fluorescent pigments to mark the adults (Dosdall et al. 1992). Aniline dyes are also effective for marking flies (Hunter & Jain 2000).

### Specimen Preparation and Morphological Identification

#### LARVAE AND PUPAE

For identification, larvae and pupae should be submerged in a shallow dish of ethanol or acetic ethanol and examined under a stereomicroscope. The fluid in the dish should be refreshed frequently to compensate for evaporation.

With few exceptions, larval identification does not require dissection or slide mounting. The hypostomal teeth generally can be seen adequately with a stereomicroscope; however, identification of some larvae of *Prosimulium* and *Hellichiella* is facilitated if the hypostoma is removed and slide mounted in Canada balsam, Euparal, or similar medium. To determine if abdominal setae are simple or branched (e.g., in some members of the subgenus *Nevermannia*), a small piece of cuticle (free of adherent tissue) can be removed posterodorsally from the abdomen and slide mounted temporarily in a drop of 50% acetic acid. If the rectal papillae are not extruded, the rectum can be dissected from the larva, slide mounted in a drop of ethanol or glycerin, and examined using phase-contrast microscopy. To distinguish larvae of the genus *Helodon* from those of *Prosimulium*, the lateral sclerite on either side of the apical article of the prothoracic proleg sometimes requires viewing. If the apical article is withdrawn, dissection with fine pins is necessary.

A structure critical for identifying larvae is the mature (i.e., dark) gill histoblast. The number, branching pattern, and surface sculpture of the gill filaments provide diagnostic information that is not always available without cutting the overlying transparent cuticle with fine pins and either teasing out the filaments while leaving the histoblast in situ or lifting the histoblast out entirely. Once the histoblast has been removed, the filaments must be uncurled. If the larva is in ethanol, the histoblast can be placed on a slide in a drop of polyvinyl lactophenol, and after a few minutes, the filaments can be spread with fine pins. If the larva is in acetic ethanol, uncurling can be effected by placing the histoblast on a slide in a drop of 50% acetic acid.

Pupae seldom require slide mounting or dissection. The surface sculpture of the cephalic and thoracic dorsum is sometimes more easily interpreted if a portion of the cuticle is slide mounted, especially for specimens containing pharate adults. In these specimens, the adult within the pupal cuticle can make the background too dark to allow the surface sculpture to be interpreted. The surface sculpture of the gill is sometimes more easily viewed on a slide. Pleurites on abdominal segments IV and V of *Helodon* and *Prosimulium* can be difficult to view if fixation has caused the pleural membrane to become inflected or if the pharate adult is too dark. In these cases, clearing the pupa and extracting the pharate adult can aid interpretation.

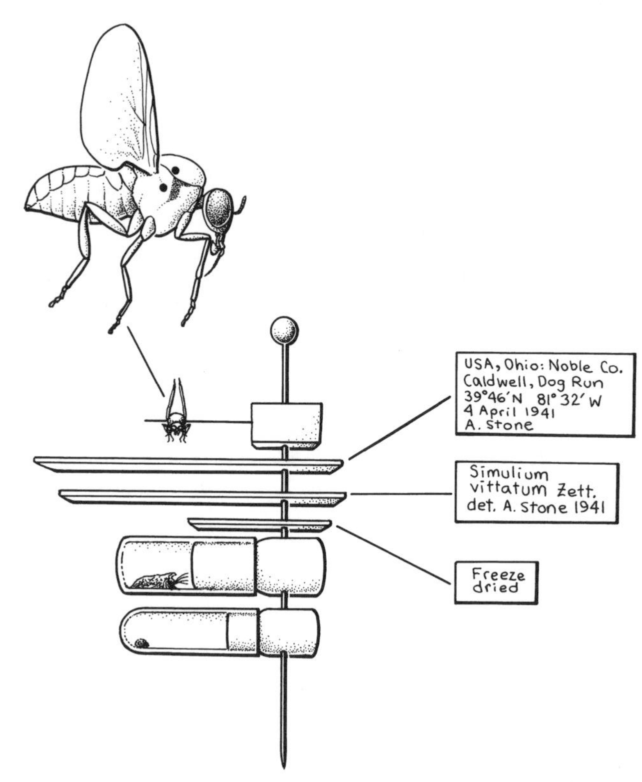


Fig. 3.2. Recommended technique for preserving and labeling adult black flies. The two black dots on the fly's thorax represent alternative locations for inserting the minuten pin. Labels include collection information, identification, and method of preservation. Microvials are placed below the labels to facilitate their removal for study of the contents. The larger stoppered vial holds the pupal exuviae and cocoon; the smaller stoppered vial houses the terminalia or other cleared structures.

**ADULTS**

Most adults can be identified in ethanol, but when colors and pollinosity patterns are required for identification, dried specimens are best. Colors of the adults eventually fade in material stored in ethanol. For some species (e.g., of the subgenus *Psilopelmia*), colors and patterns in alcohol differ from those of dried specimens and often do not permit accurate identification.

Freshly captured adults allowed to air dry will collapse and shrivel, obscuring many external features. Although air-dried specimens can be used to extract and analyze cuticular hydrocarbons (Carlson & Walsh 1981) and DNA (Koch et al. 1998), special drying techniques are recommended for morphological purposes. Fresh adults can be pinned and allowed to dry slowly for about 5 weeks in a freezer (ca. –20°C) (Wood & Davies 1966). Adults in ethanol can be dried by transferring them to ethyl acetate for an hour or more before allowing them to air dry (Vockeroth 1966). A simple recommended alternative is to place fresh specimens in 80% ethanol, dehydrate them to 100% ethanol, and transfer them to a thin film of hexamethyldisilazane (HMDS), which is then allowed to evaporate under a fume hood (Brown 1993). With this method, colors are retained and the specimens remain supple, allowing them to be pinned. Specimens treated in this manner also are suitable for molecular analysis. Adults can be mounted on a minuten pin, or they can be affixed to a paper point or directly to the pin with a bit of lacquer, clear nail polish, or glue. The preferred method is micropinning with a stainless-steel minuten pin through the side of the thorax just posterior to the anepisternal membrane or through the scutum above the membrane; the opposite end of the minuten is anchored in a small block of cork, polyurethane (Plastazote) foam, or similar material (Fig. 3.2).

If a series of specimens is available, we recommend drying some individuals by freezing and some with HMDS and keeping the remainder in 80% ethanol. For each pinned adult, a small label indicating the preservation method should be placed below the label bearing the collection data. Pupal exuviae can be associated with the adult by drying them and attaching them with a bit of lacquer to the pin or a piece of stiff paper (Crosskey 1993) or by dehydrating them through an ethanol series and storing them with glycerin in a microvial pinned through the stopper. Even if the glycerin evaporates over time, the specimen is protected in the microvial and can be reinfused with glycerin.

Males and females often require special preparation to interpret structures such as mouthparts and terminalia. Specimens or their parts, such as the terminalia, can be cleared with any of several agents. We recommend heating the material for about 5 minutes in 85% lactic acid. This agent does not overclear the material as do some of the classic clearing agents such as potassium hydroxide and sodium hydroxide that eventually produce ghosts of the material. In addition, the material can be transferred directly from lactic acid into ethanol or glycerin and, if necessary, returned to lactic acid for further clearing. The spermatheca of an unmated female typically collapses when the specimen is heated in lactic acid; that of a mated female, filled with sperm, does not collapse. Once material has been cleared, it can be placed in a drop of glycerin on a depression slide, and the relevant parts can be dissected under a stereomicroscope, properly oriented in the relevant aspect, and viewed with a compound microscope.

Cleared material should be placed in microvials with glycerin and associated with the dried adult by pinning the microvial through its stopper. Legs and mouthparts, including the maxillary palps, can be slide mounted in Canada balsam, Euparal, or other medium. The terminalia, however, should not be permanently slide mounted because orientation cannot be controlled and the benefit of viewing the terminalia in multiple aspects is lost.

### Storage of Specimens

All material in liquid media that is destined for long-term storage eventually should be transferred to 80% ethanol and housed in darkness to minimize fading. Two methods of storing alcohol specimens are used by museums. Specimens can be placed in individual 15-ml vials, with either ethanol-resistant stoppers or screw-on caps with conical polyethylene inserts. Alternatively, individual vials (5–15 ml) can be stoppered with cotton, inverted, and placed on a bed of cotton in jars of ethanol with tight-sealing caps. The storage vials and jars must be monitored over time and any evaporated ethanol replaced.

Pinned material can be placed in unit trays with bottoms of polyethylene or polyurethane (Plastazote) foam and stored in pest-proof drawers and cabinets with insecticides and fungicides. All of these materials are available from various biological supply houses.

### Mailing Specimens

Exchange of specimens, identifications by specialists, and deposition of vouchers in museums often are necessary. These processes frequently require that material be sent through the mail. Care, however, must be taken to package the material properly. Material in fluid should be shipped in plastic or glass vials with screw caps wrapped with a sealing film (e.g., Parafilm). Air bubbles should be eliminated from the vials to avoid rough jostling that can damage the specimens. Vials should be wrapped individually with packing material and placed in well-padded boxes.

Pinned material should be placed in a box with a Plastazote foam or comparable bottom. If a microvial is attached to the pin, it should be anchored by two pins fixed crosswise and driven into the foam bottom. The pinning box, in turn, should be placed in a much larger box and provided with packing material sufficient to protect the specimens even when a swift kick is administered to the outer box. Packages crossing country borders should carry customs labels indicating their contents (e.g., "dead, preserved insects for scientific study; no commercial value").

### Cytotaxonomic Procedures

To observe and analyze information in the polytene chromosomes of black flies, the banding patterns, as well as the larval gonads, first must be rendered visible by any of various staining procedures. Most cytotaxonomic needs are served by the Feulgen method and the lacto-propionic orcein or aceto-orcein methods. The Feulgen method allows multiple larvae to be stained simultaneously and permits staining of chromosomes and gonads in one step. The orcein method often gives better band resolution but requires that larvae be stained one at a time and that the gonads be poststained using the Feulgen method or other technique (e.g., Ballard 1988). Larval gender can be determined either morphologically or cytologically. Male gonads are spherical and, when stained and squashed, show clusters of meiotic figures. Female gonads are elongate with, at most, scattered meiotic activity. In the larvae of some taxa, such as the *Simulium vittatum* complex, the gonads are ensheathed at least partially with pigment, allowing their shape to be determined without staining.

We prefer the Feulgen method of Rothfels and Dunbar (1953). To prepare material for staining, we routinely sever the fixed larva at the fifth abdominal segment and stain only the posterior portion, keeping the anterior portion of the larva intact; alternatively, the entire larva can be stained. The staining procedure involves first splitting the abdomen along its ventral midline and placing it in distilled water for 20 minutes. The jelly-like contents of the silk glands can be removed by gently rolling the abdomen on a paper towel. The carcass is then placed sequentially into 1 N hydrochloric acid preheated to and maintained at 62°C–65°C (10 minutes), leucobasic fuchsin solution (ca. 1 hour), sulfur water (10 minutes; stock solution: 200 ml of distilled water, 1 gram of potassium metabisulfite, 10 ml of 1 N hydrochloric acid), and two rapid changes of cold tap water. The carcass is refrigerated in tap water until analysis, which should be conducted within 4 days, lest the chromosomes deteriorate. Silk-gland tissue and one gonad are removed with fine pins and placed in a drop of 50% acetic acid on a microscope slide. Extraneous tissue is then removed and the salivary-gland tissue is macerated with fine pins. The chromosomes are then squashed under a coverslip by applying a piece of absorbent paper and firm thumb pressure. We generally extract and mount the chromosomes of only one silk gland and place the stained abdomen with the intact silk gland back into acetic ethanol for refrigeration. Chromosomes in the intact silk gland are available for examination years later simply by placing them in a drop of 50% acetic acid and squashing them.

In the orcein method (Bedo 1975b), the silk glands from larvae fixed in acetic ethanol are dissected in 95% ethanol and the sheets of cells are transferred to a drop of lacto-propionic orcein, with final concentrations of 25% lactic acid, 25% propionic acid, and 0.5% orcein. A coverslip is applied and lightly tapped to spread the chromosomes, which then are squashed with thumb pressure. Additional, more specialized staining techniques, such as fluorescent staining and C banding, also have been applied, albeit infrequently, to the polytene chromosomes of black flies (Bedo 1975a, 1975b).

The coverslip can be ringed with 2% acetocarmine to prevent rapid evaporation and to mark the position of the coverslip for later reference if the mount is to be made permanent. Temporary preparations begin to degrade within a few hours, but they can be stored indefinitely in an ultralow (−80°C) freezer, from which they can be removed ad libitum, thawed, viewed, and returned. A more permanent chromosome mount can be made by placing the slide, coverslip down, on dry ice for an hour or more, popping the coverslip off with a razor blade, immersing the slide in a dish of absolute ethanol for about 30 seconds, removing the slide and blotting its edge, and quickly applying a small drop of mounting medium (e.g., Euparal) and a new coverslip. Slides should be stored in darkness. Chromosomes, nonetheless, suffer some shrinkage and fading with time. Photographs, therefore, should be made from temporary, rather than permanent, preparations.

Chromosomal analysis typically requires viewing with an oil immersion lens, especially when comparing banding patterns with those of standard maps. Interpreting banding sequences is largely a matter of pattern recognition in the context of variability that results from stretching, condensing, differential polytenization and staining, and other factors. Practice and patience are essential.

# PART II
# Biology

# 4 | Structure and Function

The long history of morphological study of Diptera in general, and black flies specifically, has spawned a voluminous lexicon of terms. In this book, we have adopted the morphological terminology of the lingua franca of modern dipterology, the *Manual of Nearctic Diptera* (McAlpine et al. 1981), except where subsequent research has suggested new interpretations and homologies. Structures in boldface type represent the adopted terminology and are generally labeled on our figures.

Our treatment of structure and function is applicable to the world scene, although our focus is on the Nearctic fauna. We have brought together information, whether hypothesized or demonstrated, on the functional significance of structures and at times have added our own interpretations, hoping that these ideas will provide springboards for future inquiries.

## ADULT (Fig. 4.1)

The characteristic form of the adult black fly revolves around its compact body, arched thorax, and broad wings with heavy anterior venation. The simuliid gestalt is preserved in the earliest-known Nearctic fossil, a female in New Jersey amber, estimated to be 90–94 million years old (Currie & Grimaldi 2000). Explanations for this homogeneity of body form across time and taxa are speculative but probably relate to the adaptations for emergence from the aquatic environment, flight, and movement through the feathers and hair of homeothermic hosts.

### HEAD (Figs. 4.2–4.4)

As highly visual organisms, black flies, particularly males, have prominent **compound eyes**. The eyes are reddish in all Holarctic species but can be iridescent blue in some Neotropical species. Female eyes are separated (head dichoptic) by a distinct **frons** with a width-length ratio of some taxonomic utility. The frons of the female bulges outward slightly above, or adjacent to, each antennal base, forming a **frontal dilation**. At the apex of each frontal dilation is the **nudiocular area** (*sensu* Crosskey 1990), or fronto-ocular triangle, a variously sized area of the female compound eye that lacks corneal facets. It sometimes is used as a taxonomic feature, especially in Latin American species. The frons of the male is greatly reduced and the eyes are dorsally contiguous (head holoptic) in all Nearctic species, except those of the subgenus *Parasimulium* and four of the six species known to couple on the ground. A nudiocular area is absent from the eye of holoptic males. An unusual situation exists in males of the Neotropical *Simulium* (*Notolepria*) *gonzalezi* in which both a dichoptic form (predominant) and a holoptic form occur (Shelley et al. 1989).

Male simuliids typically have more ommatidia than do females; for example, males of the *Simulium vittatum* species complex have 1300–1700 per eye, whereas females have 900–1200 (O'Grady & McIver 1987). Ornithophilic females generally have more ommatidia and a narrower frons than do mammalophilic females (Gryaznov 1984a, 1989), perhaps to detect smaller host targets. The **upper corneal facets** of the male eye are about twice as large as the **lower corneal facets**, except in the dichoptic species, whose facets are about the same size as those of females. Correlated with each large dorsal facet is an absence of the R7 cell (one of the eight retinular cells) and elongation of the six peripheral retinular cells (O'Grady & McIver 1987). The enlarged dorsal facets are adapted for detecting small objects (i.e., females) flying rapidly overhead. The pigment system has maximum sensitivity in ultraviolet at 340 nm, which increases the probability of detecting a female because resolution is greater at shorter wavelengths (Wenk 1988). Absence of the R7 cell, which is presumably a blue receptor, might mean that the male would see the female against a more strongly contrasting background (McIver & O'Grady 1987).

Ocelli are absent in black flies, but a shiny tubercle called the **stemmatic bulla** is positioned near the posterior margin of each compound eye in *Parasimulium*, *Gymnopais*, *Twinnia*, and those *Prosimulium* with reduced eyes such as *Prosimulium ursinum* (Fig. 4.52). This structure is believed to be a remnant of the larval eye (Wood 1978). The underlying black pigment can be seen through the less sclerotized bulla of *Twinnia* and even in species such as *Prosimulium fulvum* that lack a stemmatic bulla.

The antennae are diagnostic at the family level, appearing as slender inverted cones or strings of beads covered with microtrichia. A well-developed **scape** (first antennal segment), **pedicel** (second segment), and **flagellum** (third segment) are present in all taxa. Most Nearctic species have nine flagellomeres per antenna, but *Parasimulium*, *Helodon decemarticulatus*, *Prosimulium unicum*, and *Greniera denaria* have eight, and *Gymnopais*, *Twinnia*,

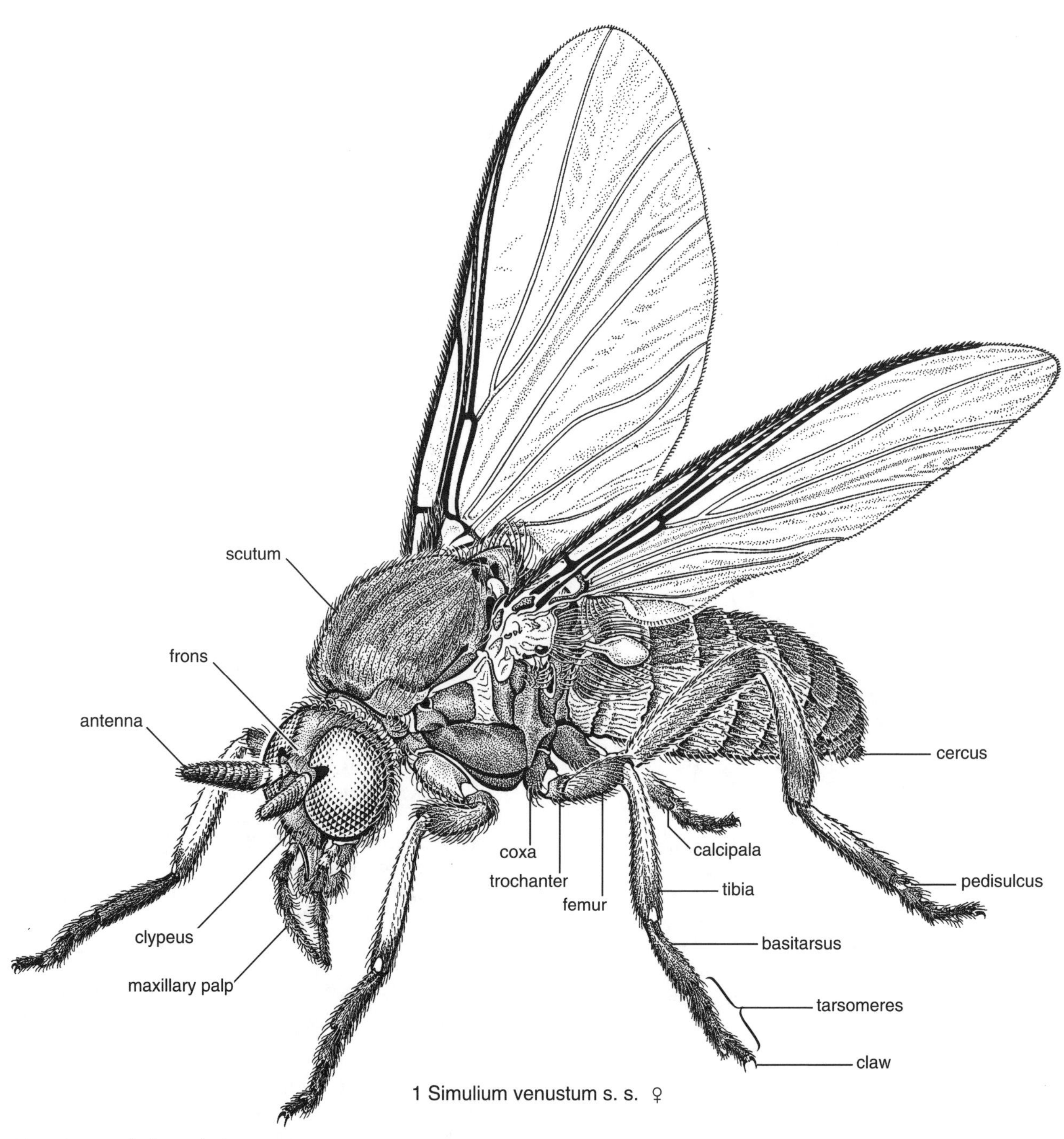


Fig. 4.1. Female, lateral view.

and *Helodon gibsoni* have seven. Each antenna is endowed with up to seven kinds of sensilla, numbering about 800 in nonbloodsucking species to nearly 1800 in some ornithophilic species (Mercer & McIver 1973a, Shipp et al. 1988b). These sensilla probably serve primarily in olfaction, but also in mechanoreception, contact chemoreception, and hygrothermoreception (McIver & Sutcliffe 1988, Shipp et al. 1988b, Sutcliffe et al. 1990). Each pedicel contains a small Johnston's organ possibly, in part, for monitoring air currents (Boo & Davies 1980).

**MOUTHPARTS** (Figs. 4.5–4.8)

The mouthparts, which constitute the proboscis, arise below the **clypeus**, a convex setose plate (sparsely haired in *Gymnopais*), larger in females than in males, to which the cibarial muscles attach. The length of the proboscis differs among taxa, its two extremes demonstrated by the stubby arrangement in species such as *Simulium transiens* and the elongate format in *Prosimulium longirostrum* n. sp. and its sister species *Prosimulium uinta*. The evolutionary significance of these variations in length, whether related to hosts, floral corollas, or other factors, is

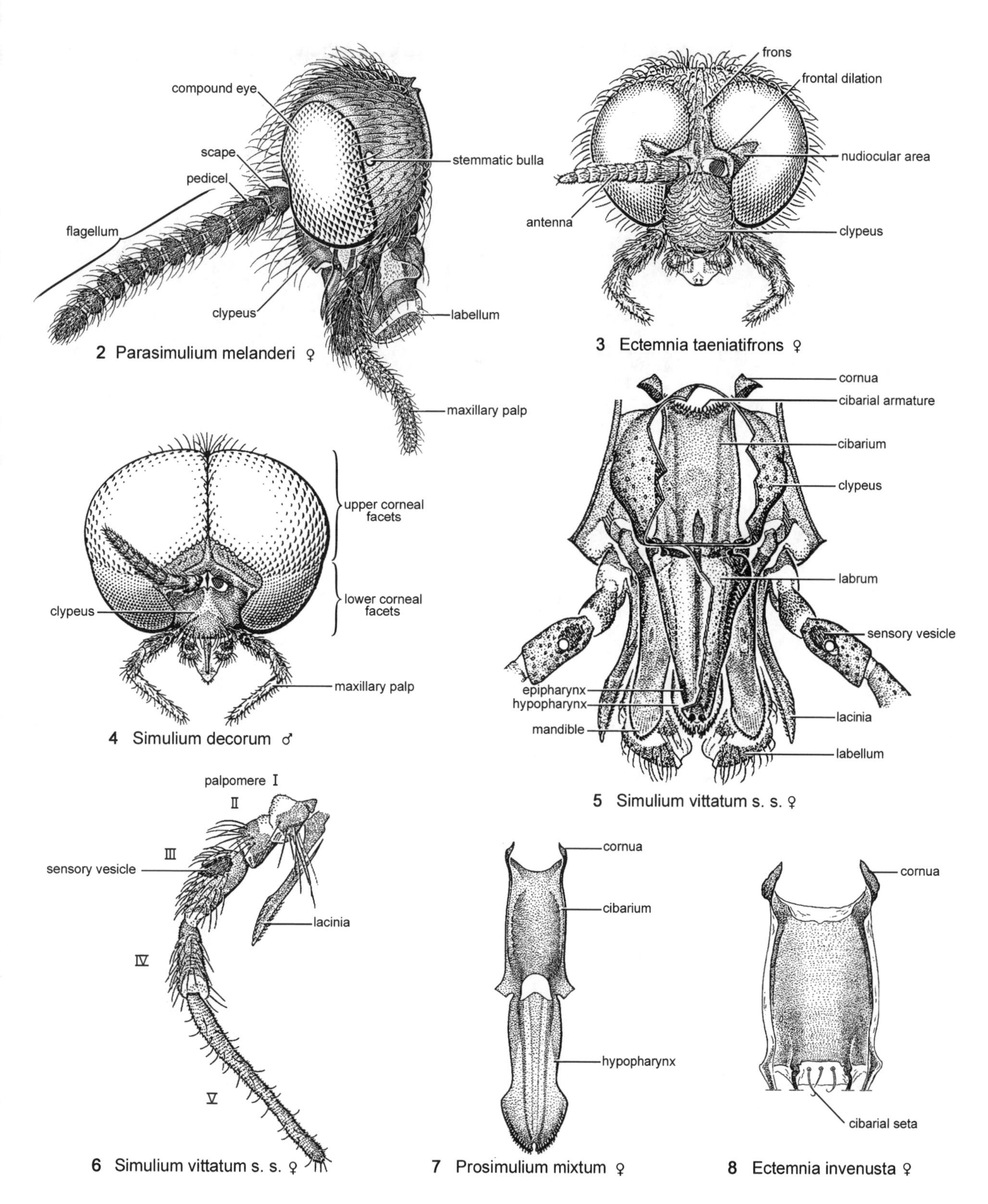


Figs. 4.2–4.8. Heads and mouthparts. (2) Lateral view. (3, 4) Anterior view. (5) Anterior cut-away view, with terminal segments of maxillary palps omitted. (6) Maxillary palp and lacinia. (7) Cibarium and hypopharynx. (8) Cibarium. (3–7) From Peterson (1981) by permission.

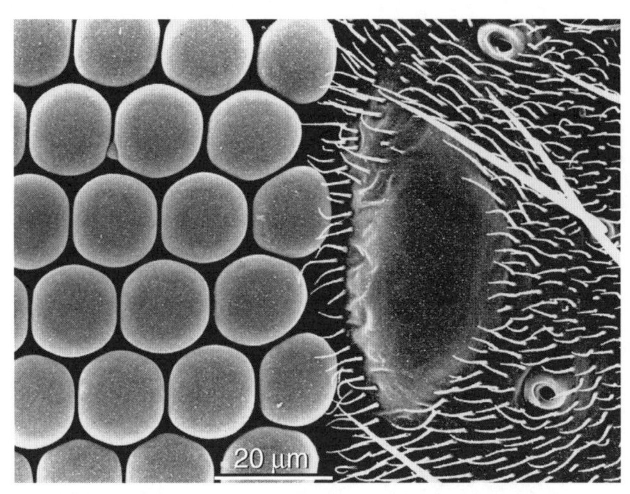

Fig. 4.52. Posterior margin of compound eye of male *Parasimulium crosskeyi*, showing corneal facets and stemmatic bulla.

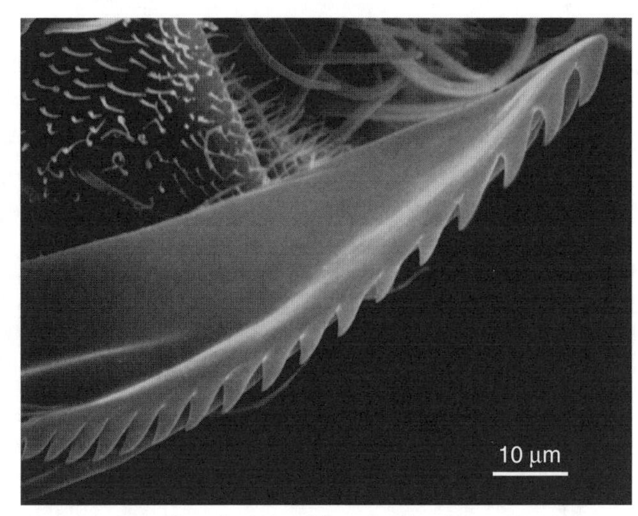

Fig. 4.54. Maxillary lacinia of female *Simulium vittatum* s. s.

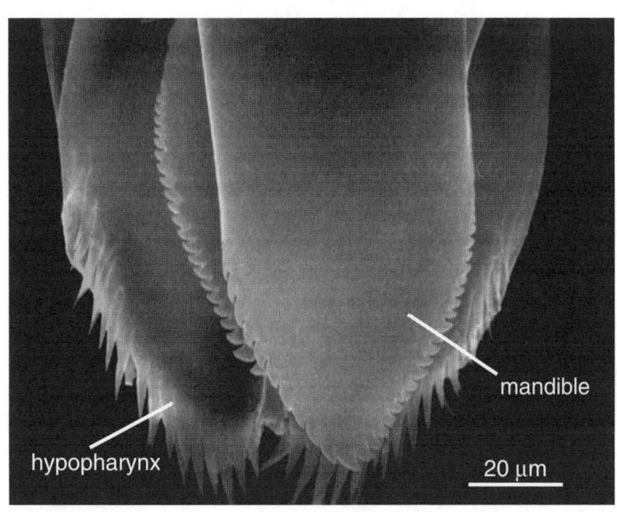

Fig. 4.53. Mandibles and hypopharynx of female *Simulium vittatum* s. s.

unknown. Articulated with the clypeus and forming the front of the proboscis is the somewhat triangular **labrum**, which bears a deep food channel on its inner surface and an apical pair of stout, heavily sclerotized, bifid or trifid labral teeth (absent from males and nonbiting females). The labrum of each sex has about 30–90 labral sensilla for chemoreception and mechanoreception, the number varying largely as a function of body size (Colbo et al. 1979). Some of these receptors probably detect mechanical stresses and stimulatory substances during feeding, and then determine if the food should be directed to the midgut (blood) or shunted into the crop (sugar) (Yang 1968, Colbo et al. 1979, Jefferies 1987).

The broad flat **mandibles** are minutely serrated, each having about 25–60 teeth in females of Nearctic biting species (Fig. 4.53). In males and nonbiting females, the mandibles have only ragged or hairlike margins and the mandibular musculature is correspondingly reduced (Krafchick 1942). Posterolateral to the mandibles are the **maxillae**, reduced except for the well-developed **laciniae** (Fig. 4.54) and prominent **maxillary palps**, each composed of five segments or palpomeres. The laciniae of biting females are armed with about 18–45 recurved (retrorse) teeth each, but they have only hairlike projections in males and nonbiting females.

The maxillary palps are long, slender, and covered with microtrichia, which on the two distalmost segments (IV and V) are arranged in rings. Each palp has up to several hundred sensilla of four structural types that are used in chemoreception and mechanoreception (Mercer & McIver 1973b); some of the chemoreceptors are sensitive to water, salts, and sugars (Angioy et al. 1982, Crnjar et al. 1983). The somewhat swollen third segment houses the **sensory vesicle** (Lutz's organ), its size and shape often of taxonomic value. The sensory vesicle opens to the outside by way of a short neck and is lined with bulblike chemosensilla for detecting odors such as carbon dioxide (Mercer & McIver 1973b). Ornithophilic females have a larger sensory vesicle with more sensilla than in males and mammalophilic females (Mercer & McIver 1973b, Gryaznov 1984a), perhaps to detect smaller, more concealed hosts. Although the sensory vesicle is smaller in males, its presence suggests some ability to respond to carbon dioxide or other odors. For example, males might use carbon dioxide to locate hosts of the females and then intercept the host-seeking females, as shown for males of the *Simulium venustum*/*verecundum* supercomplex (Mokry et al. 1981), or they might locate nectar sources by following carbon dioxide released by plants (Sutcliffe et al. 1987).

The **hypopharynx** is an elongate, sensilla-free, sclerotized plate continuous with the floor of the cibarium, armed distally with minute spines and furnished with a longitudinal salivary furrow on its anterior surface. Its base forms the inner surface of the prementum of the labium. The **labium** forms the back of the proboscis and ensheaths the apical portion of the other mouthparts with a pair of large fleshy lobes termed the **labella**. A pair of sclerotized basal plates constitutes part of the posterior portion of the prementum. The labium bears more than 130 chemoreceptors and mechanoreceptors of four structural types (Sutcliffe & McIver 1982). The chemoreceptors on the labella are capable of detecting water, salts, and sugars (Angioy et al. 1982).

The **cibarium** has a well-sclerotized venter, a proximal pair of arms (**cornuae**) for muscle attachment, and two pairs of chemoreceptive cibarial sensilla that probably evaluate the acceptability of food (Colbo et al. 1979). The area between the arms can be taxonomically important in shape and armature. This area is usually unarmed (i.e., smooth) or slightly tuberculate, but some taxa, such as the subgenera *Psilopelmia*, *Psilozia*, the Neotropical *Psaroniocompsa*, and some species of *Hemicnetha*, have an elaborate **cibarial armature**, typically consisting of spines and spicules that project into the lumen of the food channel. Various functions of this armature have been suggested: the destruction of ingested microfilariae that could establish a parasitic infection in the fly (Lehmann et al. 1994b), a valve mechanism preventing backflow of blood (Reid 1994), and the break up of ingested solids such as pollen (Crosskey 1990). The distal margin of the cibarium at the junction with the hypopharynx is rather uninformative taxonomically, but bears 1–11 **cibarial setae** of unknown function in females of *Ectemnia* (Moulton & Adler 1997).

In preparation for blood feeding, the labella are retracted and the labral teeth and hypopharyngeal spines stretch the skin taut as the head of the female bears down on the host. The mandibles then penetrate the skin, opening a wound that allows the hypopharynx and labrum to enter along with the maxillary laciniae, which anchor the mouthparts while the mandibles continue to snip the host flesh (Sutcliffe & McIver 1984). A subdermal hematoma is thus formed, and the pooled blood is ingested via the food canal that is formed when the mandibles fold over the labral channel. A minute tubercle on the anterior surface of each mandible abuts with the edge of the labral channel. Uptake of blood is facilitated by the tight seal formed by the membranous portions of the mouthparts; these membranes prevent both loss of blood from the wound and entry of air (Sutcliffe 1985). The actual uptake is effected by two muscular pumps, one in the cibarium and the other in the pharynx, in conjunction with the oral and postpharyngeal constrictors (McIver & Sutcliffe 1988). During feeding, saliva is released from the salivary glands, which are claimed to have species-specific structure (Bennett 1963b), and flows down the salivary gutter of the hypopharynx. The salivary secretions promote vasodilation, prevent clotting, and inhibit aggregation of platelets while affording local anesthesia (Cupp & Cupp 1997).

### THORAX (Figs. 4.9–4.16)

The thorax is dorsally convex, especially in males, reaching its most extreme development in *Simulium labellei* and *Simulium robynae*. It is least convex in flightless species (e.g., most *Gymnopais*) with degenerate flight muscles. The prothorax is represented on each side by a well-developed **postpronotal lobe** that is rather intimately associated with the mesothorax, plus a small, dorsolateral **antepronotal lobe** that is connected with its partner on the opposite side by a thin transverse strip. The membranous portion of the prothorax, the **cervix** (neck), joins the head to the thorax. The metathorax is reduced, leaving the mesothorax as the most prominent of the three segments, with its powerful set of indirect flight muscles to operate the single (front) pair of wings.

The **scutum** is the most obvious region of the mesothorax. It is bounded on each side by a narrow flange, the **paratergite**. The color and pattern of the scutum are important taxonomically, especially in the genus *Simulium*, and often belie the common familial name. Other than black or dark brown, shades of orange and yellow and various patterns of iridescence are most common, the patterns often changing with the angle of incident light. Some color patterns, for example in the subgenus *Psilopelmia*, are variable within species, especially with season, perhaps in response to temperature. Unstudied are possible ultraviolet reflection patterns of the scutum, as well as other parts of the body. The scutum is covered wholly or partly with microtrichia, which typically differ among females and males (Figs. 4.55, 4.56), and setae (hairs) of various types that impart the colors, especially brown, gold, and silver (Hannay & Bond 1971b, Lowry & Shelley 1990). Scalelike setae are present on the scutum of some Neotropical taxa (e.g., the subgenus *Notolepria*). Males are almost always darker than females and the microtrichia of males often impart a velvety appearance. The **scutellum** is usually subtriangular and setose, whereas the **postnotum** is typically bulbous and bare in Nearctic taxa, except in the subgenus *Eusimulium* and some females of the *Simulium vernum* species group, which have variably sized patches of golden hair. In *Gymnopais*, the postnotum has a longitudinal ridge (Fig. 4.14).

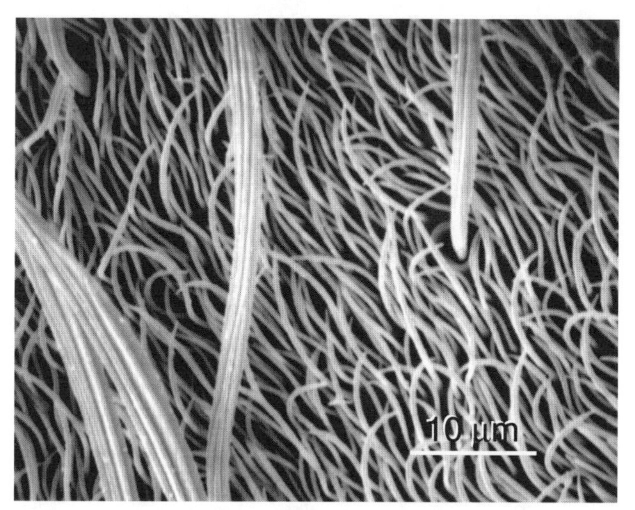


Fig. 4.55. Vestiture of posterolateral scutum of female *Simulium vittatum* s. s.

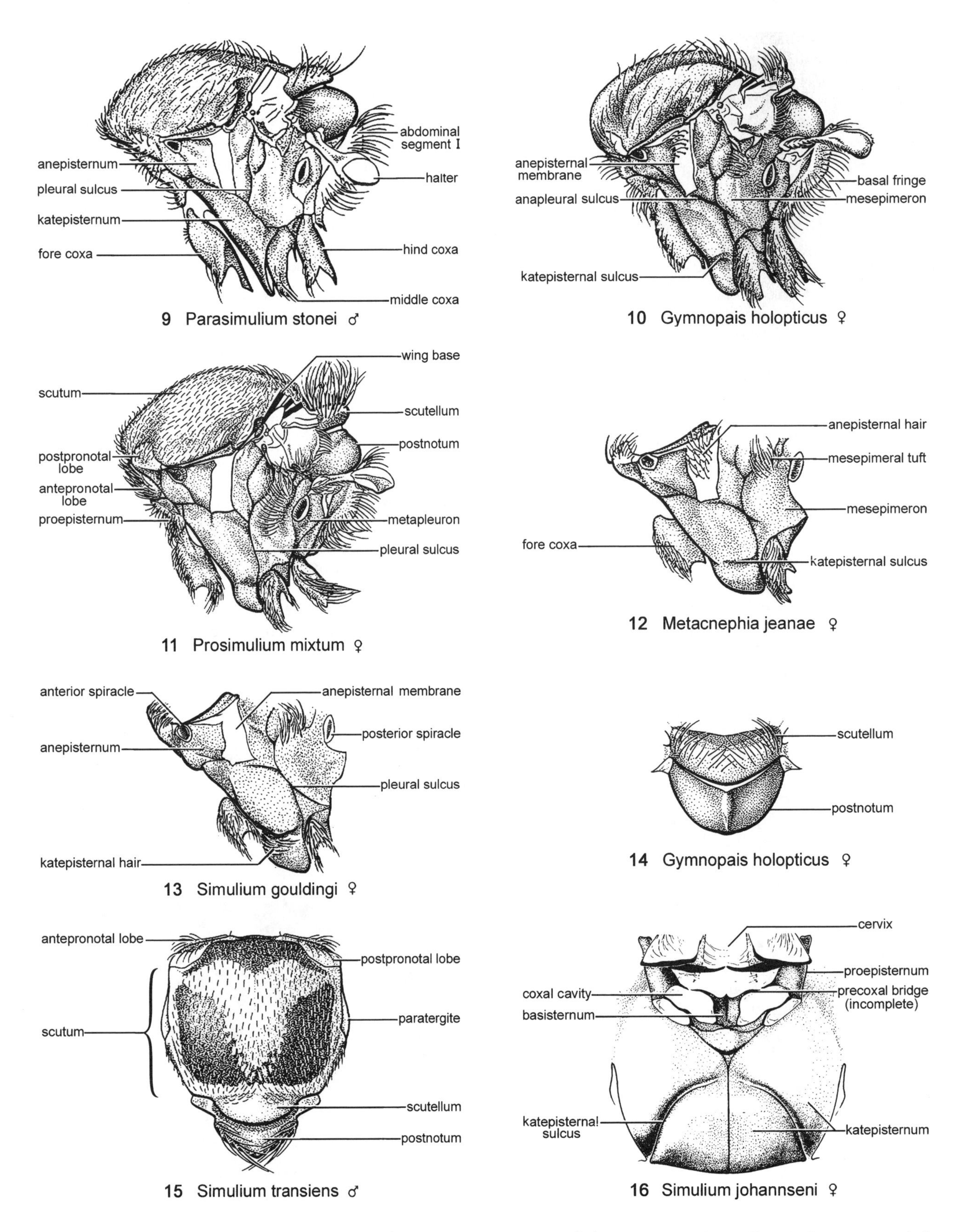


Figs. 4.9–4.16. Thoraces. (9–11) Lateral view. (12, 13) Lateral view of pleuron. (14) Dorsal view of scutellum and postnotum. (15) Dorsal view. (16) Ventral view. (9–12, 14, 15) From Peterson (1981) by permission.

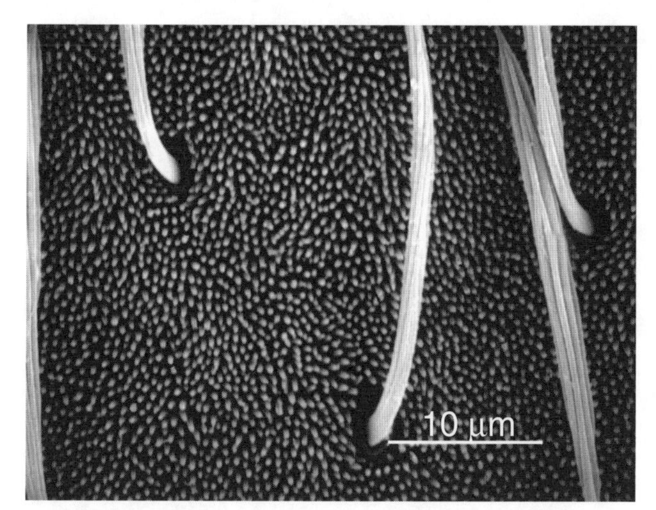


Fig. 4.56. Vestiture on middle of scutum of male *Simulium vittatum* s. s.

Laterally, the thorax consists chiefly of the large, mostly sclerotized mesothorax. The prothorax contributes the **proepisternum**, which sits above the base of each fore coxa, and the metathorax provides a small **metapleuron**. On each side of the fly, the mesothorax is divided into an anterior episternum and posterior epimeron by a nearly perpendicular line of inflected cuticle, the **pleural sulcus**, which strengthens the thorax for the stresses of locomotion. An **anterior spiracle** (mesothoracic spiracle) is located near the anterodorsal margin of the mesothorax. The episternum is subdivided by the **anapleural sulcus** into a dorsal and ventral region. The dorsal region is the **anepisternum**, which is bisected by the large **anepisternal membrane** (pleural membrane). Why it should be so large in the Simuliidae remains a mystery. It is usually bare but for unknown reasons has hair (**anepisternal hair**) in some Nearctic genera, namely, *Gymnopais* (except *G. fimbriatus*) and *Metacnephia* (except *M. saskatchewana*), and in some Palearctic subgenera, several Afrotropical species, the Mexican species *Tlalocomyia revelata*, and the *Simulium ornatum* species group. The lower portion of the episternum is the **katepisternum**, to which the vertical flight muscles are attached. It is exceptionally large, wrapping ventrally with its partner from the opposite side to occupy most of the ventral portion of the thorax. The katepisternum is divided by a transverse **katepisternal sulcus** in all taxa except *Parasimulium*. When present, the katepisternal sulcus is shallow, except in *Metacnephia* and *Simulium*, in which it is deep. In two Nearctic species (*Simulium croxtoni*, *S. gouldingi*), the katepisternum bears a diagnostic dorsal patch of hair (**katepisternal hair**, Fig. 4.13). The epimeron (**mesepimeron**), located anterior to the halter and **posterior spiracle** (metathoracic spiracle), has a tuft of setae (**mesepimeral tuft**), except in *Parasimulium*.

Ventral features of the thorax, including the internal apodemes, are underexploited in the taxonomy of the family. An exception is the **precoxal bridge**, which forms the anterior margin of the fore coxal cavity. It provides useful taxonomic information related to the extent to which it connects the mesothoracic **basisternum** to the proepisternum. It is complete in females of taxa such as the subgenus *Nevermannia* and in males of most taxa but is incomplete in most other taxa (Fig. 4.16). The shape of the invaginated **furcasternum**, which is located between the hind coxae, is of some taxonomic value.

**WINGS** (Figs. 4.17–4.20)

The wings are hyaline or smokey, never patterned, and are clothed with microtrichia. The opacity of the wings of taxa such as *Gymnopais* is evidently the result of microtrichia, which are particularly small and numerous on the wing membrane (Wood 1978). The wing membrane of all examined black flies has densely and regularly spaced, submicroscopic alary nipples (Hannay & Bond 1971a) that possibly waterproof the wing by trapping air. Each wing is broadened basally—an adaptation of rheophilic insects whose pupae are bound to the substrate beneath the water. The broad wing presumably allows the newly emerged adult to unfold its wings rapidly on arrival at the water surface (Hennig 1973). The **halteres**, although conspicuous as white, yellow, or brownish knobs endowed with various sensilla, rarely have been used for taxonomic purposes. As in other Diptera, the halteres serve as balancing organs to maintain stability during flight.

Wing venation has featured prominently in simuliid taxonomy since Roubaud (1906) used it to define some of the first modern genera. The leading area of the wing includes three strongly expressed veins: the **costa (C)**, **subcosta (Sc)**, and **radius (R)**. The anterior branch of the radius is referred to as $\mathbf{R_1}$. The posterior branch of the radius is the **radial sector (Rs)**. It is conspicuously branched ($\mathbf{R_{2+3}}$ and $\mathbf{R_{4+5}}$) in *Parasimulium* and the prosimuliines, but weakly so in the *Tlalocomyia osborni* species group, and scarcely so, if at all, in *Greniera*, *Cnephia*, *Ectemnia*, and some *Metacnephia*. In all other simuliids, the radial sector is unbranched. The branched **media** ($\mathbf{M_1}$ and $\mathbf{M_2}$), two **anterior cubital veins** ($\mathbf{CuA_1}$ and $\mathbf{CuA_2}$), the **posterior cubitus (CuP)**, and two **anal veins** ($\mathbf{A_1}$ and $\mathbf{A_2}$) are weakly expressed, especially CuP. Except for a straight rather than sinuous $CuA_2$ in *Gigantodax*, they are rather invariant throughout the family. A **false vein** (medial-cubital fold) lies between $M_2$ and $CuA_1$. Although unbranched in *Parasimulium*, this false vein is forked in all other taxa. A short **humeral cross vein** runs basally between the costa and subcosta. A small **basal medial cell** (bm), its apical margin formed by the **medial-cubital cross vein** (m-cu), is found in all Nearctic genera except *Parasimulium* and *Gigantodax* and the subgenus *Hellichiella*. A **basal radial cell** (br), its apical margin formed by the **radial-medial cross vein** (r-m), is also present in the Simuliidae, ranging in length from less than one fourth (e.g., *Parasimulium*, subgenus *Simulium*) to about one third the length of the wing (e.g., *Gymnopais*, *Cnephia*), as measured from its base. The relative lengths of the various veins and cells differ subtly among taxa and, from an engineering perspective, probably reflect differences in maneuverability and power of flight.

One of the most important taxonomic attributes of the wing veins is the macrotrichia (Fig. 4.57). *Parasimulium*,

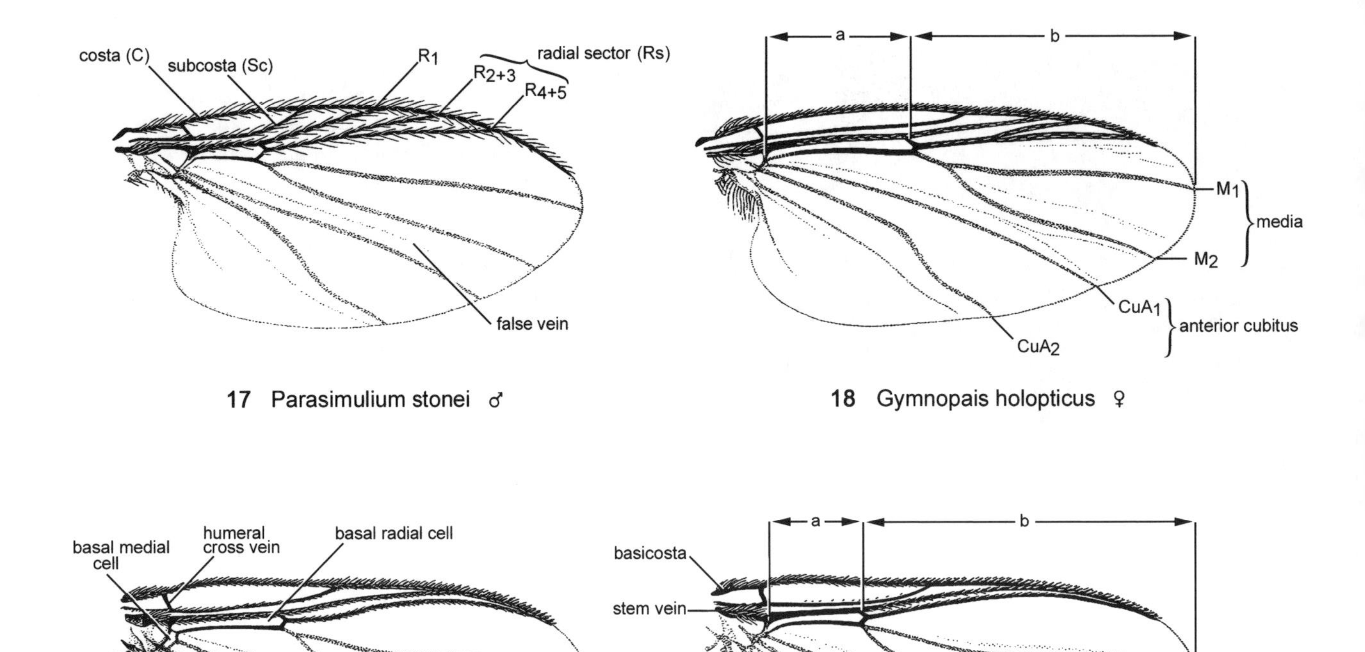


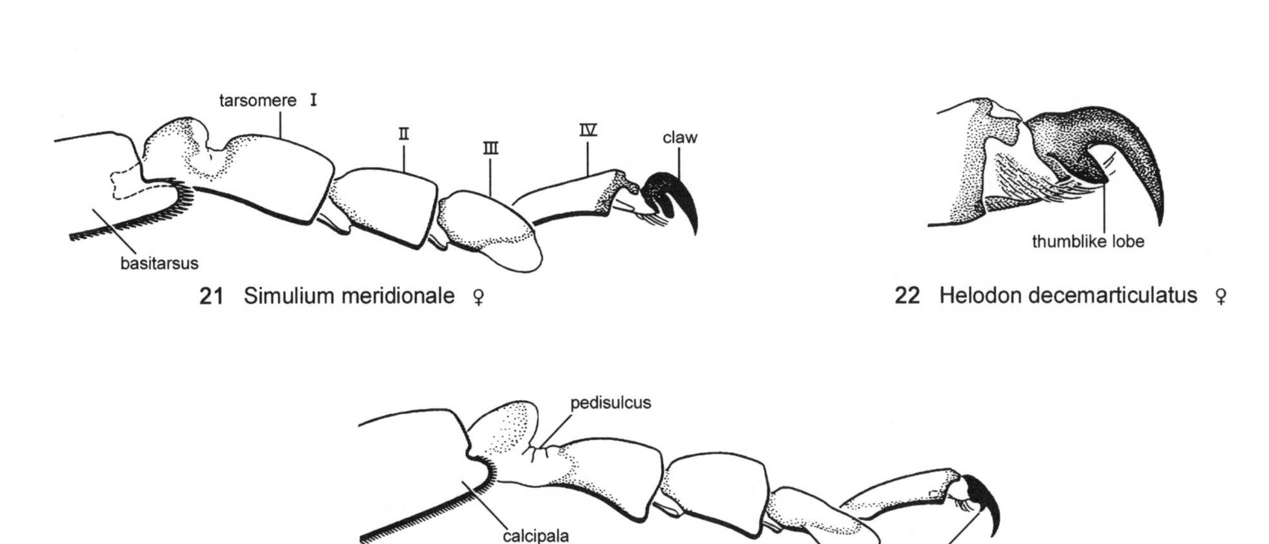


Figs. 4.17–4.23. Wings and legs. (17–20) Wings; a = basal section of radius; b = base of radial sector to apex of wing. (21, 23) Hind basitarsi and tarsi. (22) Claw and apex of tarsus. (17–23) From Peterson (1981) by permission.

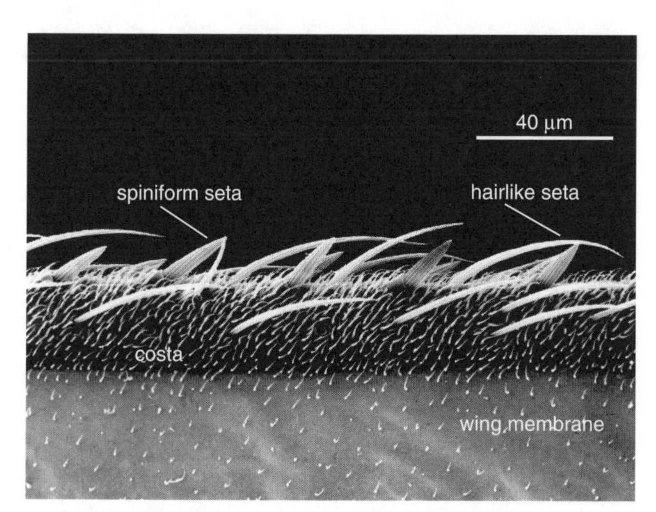

Fig. 4.57. Portion of costa from wing of female *Simulium vittatum* s. s.

the prosimuliines, and the fossil genus *Archicnephia* have only thin **hairlike setae** on the costa, whereas all other Holarctic taxa have short, usually dark **spiniform setae** (thinner in the *Greniera abdita* species group) interspersed among the hairlike setae. The presence of spiniform setae is correlated with stronger flight abilities (Crosskey 1990). Abundant microtrichia also cover the wing veins. All Nearctic taxa, except the *Simulium meridionale* species group and the subgenera *Psilozia* through *Simulium*, have fine setae dorsally on the base of the radius between the humeral cross vein and the beginning of the radial sector. These fine setae are readily rubbed off in several taxa (e.g., *Simulium johannseni* species group). The presence or absence of setae on the ventral surface of the subcosta sometimes can be of taxonomic value. Campaniform sensilla are located on some veins. The color of the setae on the **basicosta** (a plate at the costal base) and on the **stem vein** (the portion of the radius proximal to the humeral cross vein) can be important as a species-level character.

**LEGS** (Figs. 4.21–4.23)

The legs, especially of females, provide an abundant, yet weakly tapped source of taxonomic and phylogenetic information. Lengths of the legs differ among taxa, although the reasons for these differences are unknown (Sutcliffe 1975). The legs of prosimuliines and basal simuliines are generally unicolorous, whereas those of the clade containing the subgenera *Wilhelmia* through *Simulium* are banded, with the extent and affected segments often providing specific diagnostic information. The vestiture of the legs has not been exploited adequately as a character system for phylogenetic reconstruction. Legs of *Parasimulium*, the prosimuliines, and the basal simuliines, for example, generally have longer setae than do those of the most derived simuliines such as in the subgenus *Simulium*. The adaptive significance of the many variations in leg structures and color patterns has been scantily investigated, although grooming and sexual communication are probably important.

Leg structure is rather similar among the three thoracic segments. The **coxae** are short and subconical, the **trochanters** are small, and the **femora** are spindle shaped. The **tibiae** are elongate, cylindrical or flattened, and enlarged distally, especially on the hind legs. The middle and hind tibiae each bear a pair of variously sized apical spurs, whereas each fore tibia has but a single spur or none at all. These spurs contact the substrate when the fly is at rest, and because they have campaniform sensilla capable of detecting pressure, they probably provide the fly with positional information (Sutcliffe & McIver 1976). The apices of the fore and hind tibiae have dense patches of setae on their inner surfaces that are used in grooming (Sutcliffe & McIver 1974).

Variable in length and shape among species, the **basitarsus** lies distal to the tibia. Historically, it has been considered the first tarsomere of the tarsus (e.g., McAlpine 1981, Peterson 1981), but more recent interpretations suggest that in all insects it is a distinct, true segment (Kukalová-Peck 1992). The hind basitarsus is especially important taxonomically. It is outfitted along its ventral edge with a comb of short teeth for grooming. Its distal inner surface has a flattened flange or lobe, the **calcipala**, that is present in both sexes of Nearctic simuliines, except *Metacnephia*, although it differs in shape and degree of development among taxa. Its edge is serrated, suggesting a grooming function perhaps for the abdominal vestiture or the costal macrotrichia of the wing (Crosskey 1990).

Each **tarsus** (eutarsus) consists of four tarsomeres. The first tarsomere of each hind leg of males and females in the genus *Simulium* has a variously incised area, the **pedisulcus**, across its upper surface. Within the genus *Simulium*, the pedisulcus varies from long and shallow (e.g., subgenus *Hellichiella*) to short and deep (e.g., subgenus *Nevermannia*). The pedisulcus in some basal simuliines, such as *Stegopterna* and *Ectemnia*, is represented solely by faint wrinkling. The function of the pedisulcus might be to impart additional flexibility to the tarsus and to the leg generally, allowing it to be withdrawn more readily from the pupal leg sheaths during eclosion. Both the calcipala and pedisulcus were named by Enderlein (1930) but appeared earlier in illustrations by Johannsen (1903b). The second through fourth tarsomeres are short and have been little used taxonomically. The penultimate tarsomere differs from the second and fourth by being bilobed with a ventral pad of microtrichia. This pad aids movement on smooth surfaces and is used by ornithophilic females to adhere to mammalian hair (Gryaznov 1984a). The ventral surface of all tarsomeres, except the fourth, is endowed with extensive secretory epithelium that produces a wax, possibly for waterproofing, preventing entrapment in surface films, or inhibiting uptake of undesirable water-soluble compounds (Sutcliffe & McIver 1987).

The distalmost leg segment, the **acropod**, consists of a pair of **claws** and a claw-bearing assembly. A setiform **empodium** is present between the claws and is tiny in all simuliids except *Parasimulium*, in which it is nearly as long as the claws (Fig. 4.58). The claws of the female are replete with taxonomic and biological information. The

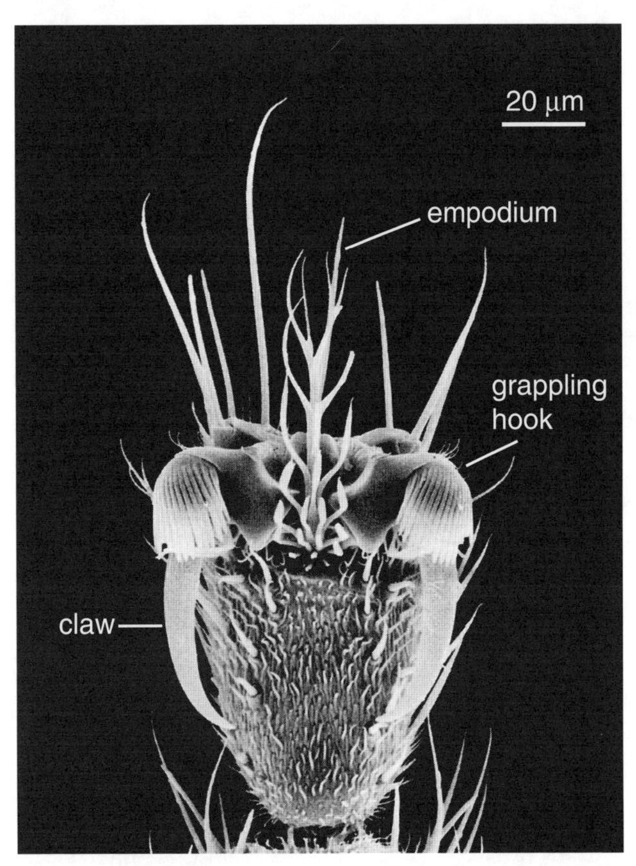


Fig. 4.58. Fifth tarsomere and acropod of male *Parasimulium crosskeyi*, ventral view.

simplest form is a gently curved talon, with the two members of a pair rather widely splayed. This design is typical of mammalophilic species and presumably is adapted for grasping host hair, although it is also present in nonbloodsucking species. Poorly understood variations on this theme include short claws (e.g., *Simulium jenningsi* species group) versus long claws (e.g., *Simulium parnassum*), the degree of curvature, and the presence of a minute **subbasal tooth** (e.g., *Simulium malyschevi* species group). Ornithophilic species have bifid claws, each of which includes a well-curved talon plus a variously developed, basal **thumblike lobe** (Fig. 4.22). This bifid arrangement helps the fly grasp feather barbules and also might prevent the claw from becoming entangled in the plumage. The inner margin of each bifid claw has oblique subparallel ridges, unlike the claw of mammalophilic species, which bears only superficial longitudinal ridges, a feature also seen in bifid claws (Sasaki 1988). The two claws of a bifid pair are generally less splayed than in mammalophilic species. Much variation exists within the ornithophilic design, suggesting that bifid claws have evolved more than once. The presence of two different claw designs within the same species is uncommon. Two Palearctic and some Neotropical species have forelegs with bifid claws but mid and hind legs with minutely toothed claws (Sasaki 1988). The claws of males are crowned with large, grooved lobes of cuticle—the **grappling hooks** (Fig. 4.58)—that grasp the female vestiture during coupling (D. A. Craig & R. E. G. Craig 1986). Male claws are regarded as rather uniform across taxa, but actually they are poorly studied from a taxonomic perspective.

The legs are bristling with sensilla. Each fly has from about 7500 to more than 16,000 external receptors of 11 different types, plus at least two internal chordotonal organs associated with the tarsal tendon (Sutcliffe & McIver 1976, 1987; McIver & Sutcliffe 1988). The majority (>80%) of the sensilla are trichoid-type mechanoreceptors generally distributed on the legs. Also scattered rather generally on the legs are about 180 campaniform and hair-plate sensilla for detecting stresses in the cuticle and relative positions of the legs and body (Sutcliffe & McIver 1976). The only leg sensilla that have been used taxonomically are the mechanoreceptive scalelike hairs on the femora and tibiae that are found in diverse taxa such as *Simulium annulus*, *Simulium rugglesi*, and various members of the *Simulium jenningsi* species group (Sutcliffe & McIver 1976 [*annulus* as *euryadminiculum*], Moulton & Adler 1995). Most chemoreceptors, both contact and olfactory types, are on the tarsi, especially the ventral surface (McIver & Sutcliffe 1988). Thus, the patting behavior of some females (e.g., of the *Simulium venustum* species complex) that crawl over the skin before probing is probably to facilitate the detection of contact stimuli (Sutcliffe & McIver 1976). The tarsal sensilla are sensitive to water, salts, and sugars (Angioy et al. 1982). The unique bifurcate chemoreceptors on the venter of the mesothoracic basitarsus are the only sensilla on the legs that show any sexual dimorphism; females have more than males (McIver et al. 1980).

## ABDOMEN

The elongate abdomen consists of 11 segments, with a pair of functional spiracles on the third through seventh segments. The first segment ("basal scale," a term first used by Malloch 1914) is a short ring that sports the **basal fringe** (Fig. 4.10), which is composed of long fine hair, the color of which can be taxonomically useful. The first segment also bears a vestigial pair of nonfunctional spiracles. The tergites of the first nine segments are usually well sclerotized, but are variously reduced in biting females. In males, the tergite and sternite of the ninth segment form a complete sclerite, the basal ring, anterior to the terminalia. The sternites of males and nonbiting females are usually developed on the first eight segments, but those of biting females are small and restricted to the first, eighth, and sometimes seventh segments. The remainder of the pregenital segments in both sexes consists of striate membrane. Abdominal color is usually similar to that of the scutum. The sclerotized plates and membranous areas are variously setose. Males of the subgenera *Wilhelmia* through *Simulium* have silvery spots laterally on the second and fifth through seventh terga. The adaptive significance of these spots is not known, but in mating swarms they might provide signals to females.

**FEMALE TERMINALIA** (Figs. 4.24, 4.25)

The terminalia of the female begin with the eighth segment. The sternite of the eighth segment (**sternite VIII**) is modified posteriorly as a pair of **hypogynial valves** (ovipositor lobes) that form the functional ovipositor. The size and shape of these valves are taxonomically important, varying from short truncate lobes (many species) to elongate processes (e.g., *Prosimulium*). The relation of valve shape to oviposition habits and to male terminalia during copulation has not been considered. However, the correlation between the long hypogynial valves and the typically long lip of the male ventral plate in both *Prosimulium* and the *Simulium paynei* species group suggests a possible functional complex. The ninth tergite (**tergite IX**) is well developed in all females and narrowly connected to the lateral arms of the highly modified, Y-shaped ninth sternite, or **genital fork**, except in *Gymnopais* and *Twinnia*, in which the ninth tergite and sternite are separated by membrane. The anteriorly directed, heavily sclerotized stem of the genital fork is internal and supports the dorsal wall of the genital chamber. The shape of the genital fork and the area between the two lateral arms provide specific diagnostic information, as does the degree of development of each **lateral plate** (terminal plate) and its anteriorly directed **apodeme**.

The tenth abdominal segment bears a small tergite plus a sternum that consists of a pair of setose sclerites, the **anal lobes** (paraprocts). These lobes are separate sclerites in all taxa except *Parasimulium*, in which they are narrowly connected (Wood & Borkent 1982, Currie 1988). The shape of the anal lobes has been used as a taxonomic character, but their real taxonomic value, the pattern of sclerotization and distribution of microtrichia, has remained essentially unappreciated until now. We suspect that these patterns partly reflect the way in which the anal lobes are contorted to accommodate the male genitalia during coupling and perhaps subsequently to clasp the spermatophore. The spermatophore of the male is held between the anal lobes and hypogynial valves and is enclosed dorsally in a cavity formed by the membrane that supports the genital fork (Wood 1978, Wenk 1988). The one-segmented setose **cerci** lie posterior to the anal lobes. The shape of the cerci provides some taxonomic information but is often quite variable within a species. As in males, the sensilla of the terminalia are unstudied but probably consist largely of trichoid mechanoreceptors.

The sperm-storage receptacle, or **spermatheca**, is best considered a part of the internal reproductive system, but because it usually is evaluated taxonomically along with the terminalia, we treat it here. The spermatheca is a sclerotized, pigmented structure in all taxa except the subgenus *Distosimulium*, in which it is a large, unpigmented, membranous sac. A single spermatheca is characteristic of the family Simuliidae. It connects with the genital chamber by a sclerotized **spermathecal duct** that is highly fluted externally. The duct is joined near its entry into the genital chamber by a pair of lateral ducts whose blind ends have not been investigated. The three ducts represent the probable ancestral condition of three spermathecae in Diptera (Wood & Borkent 1982). Only one species, a member of the *Simulium multistriatum* species group from Thailand, is known to have three spermathecae (Takaoka & Kuvangkadilok 1999), whereas the single known female of *Simulium* (*Schoenbaueria*) 'species Z' has two spermathecae. Aberrant, colony-reared individuals of *Simulium vittatum* have been found with two spermathecae (F. F. Hunter & P. H. Adler, unpublished data). The spermatheca is a rich source of taxonomic information, and its shape, degree of sclerotization at the junction with the duct, and the inner and outer surface sculpture provide useful characters (Evans & Adler 2000). Most prosimuliines and some simuliines (e.g., *Cnephia*, *Ectemnia*) have a wrinkled surface, whereas most other taxa have an unwrinkled but variously pitted, granulate, or minutely tuberculate surface, often with a faint to distinct polygonal pattern, representing the cellular imprints of epithelial cells that formed the spermathecal cuticle (Evans & Adler 2000). Internal spicules directed toward the duct are a feature in many simuliines. Adaptive explanations for these structural variations are obscure.

**MALE TERMINALIA** (Figs. 4.26–4.28)

The male terminalia consist of the genitalia, the tenth segment with its small tergite, and the **cerci**, which arise from the highly reduced eleventh segment. The male cerci are usually minute and little used in taxonomy, although they can be huge in some Polynesian species of the subgenus *Inseliellum* (Craig et al. 1995) and rather large in the subgenus *Eusimulium*.

The genitalia have assumed a key role in the taxonomy and phylogenetic reconstruction of the family. They are composed of the gonopods and the aedeagus with its associated parameres. The great diversity of male genitalic structure suggests that parts of the genitalia might be involved in internal courtship and that female selection might have played an important role in generating this diversity. The relations of the component parts of the male and female genitalia during mating require detailed study.

The **gonopods** (claspers) are a pair of two-segmented appendages, each composed of a basal gonocoxite and a distal gonostylus. The **gonocoxites** are large subconical structures varying among taxa most notably in their length-width ratios. Their more subtle features, such as shape and setal patterns, have not been scrutinized for taxonomic content. The **gonostyli** are far more variable in structure among taxa than are the gonocoxites, probably because they have more intimate contact with the female terminalia. In *Parasimulium* and the subgenus *Distosimulium*, the clasping action of the gonostyli is in a dorsoventral plane, whereas in other simuliids it is in a lateromedial plane (Wood & Borkent 1982). The female (at least of *Cnephia dacotensis*) is grasped between the ninth and tenth abdominal segments (Wood 1963a). The gonostyli vary in shape from finely tapered (e.g., subgenus *Hellichiella*) to paddle shaped (e.g., subgenus *Hemicnetha*),

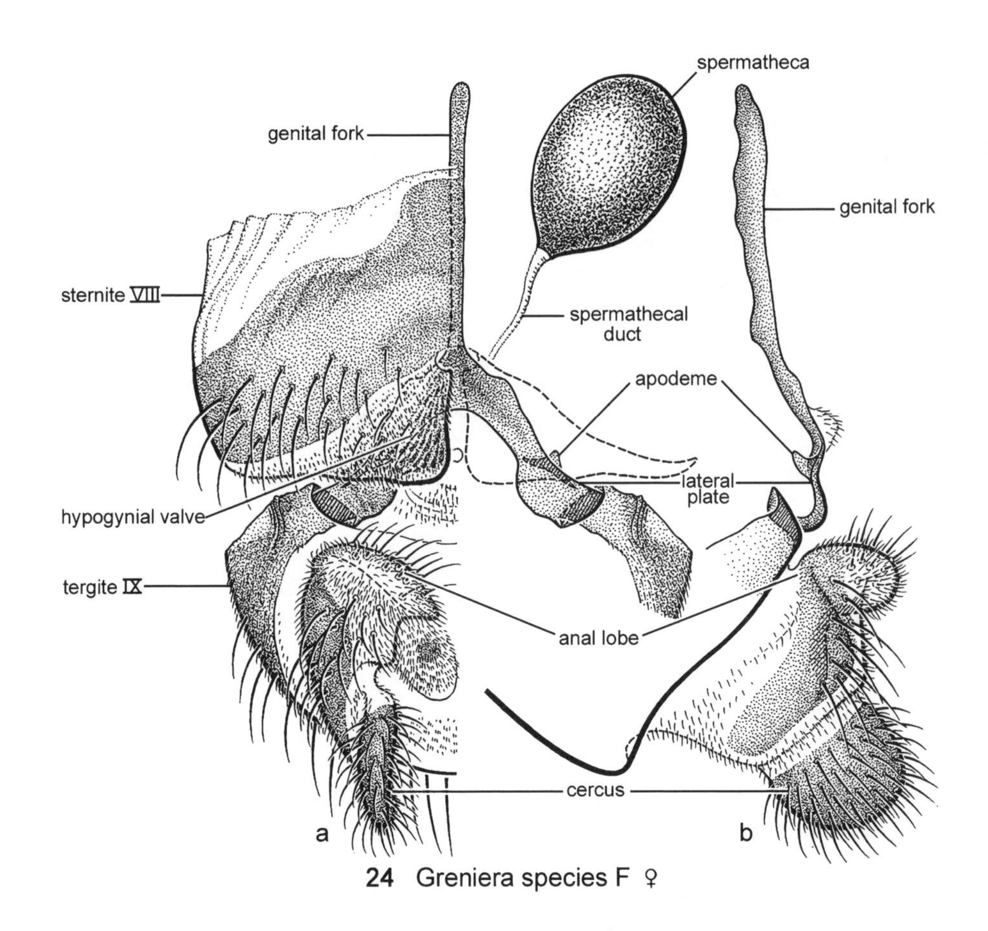


24 Greniera species F ♀

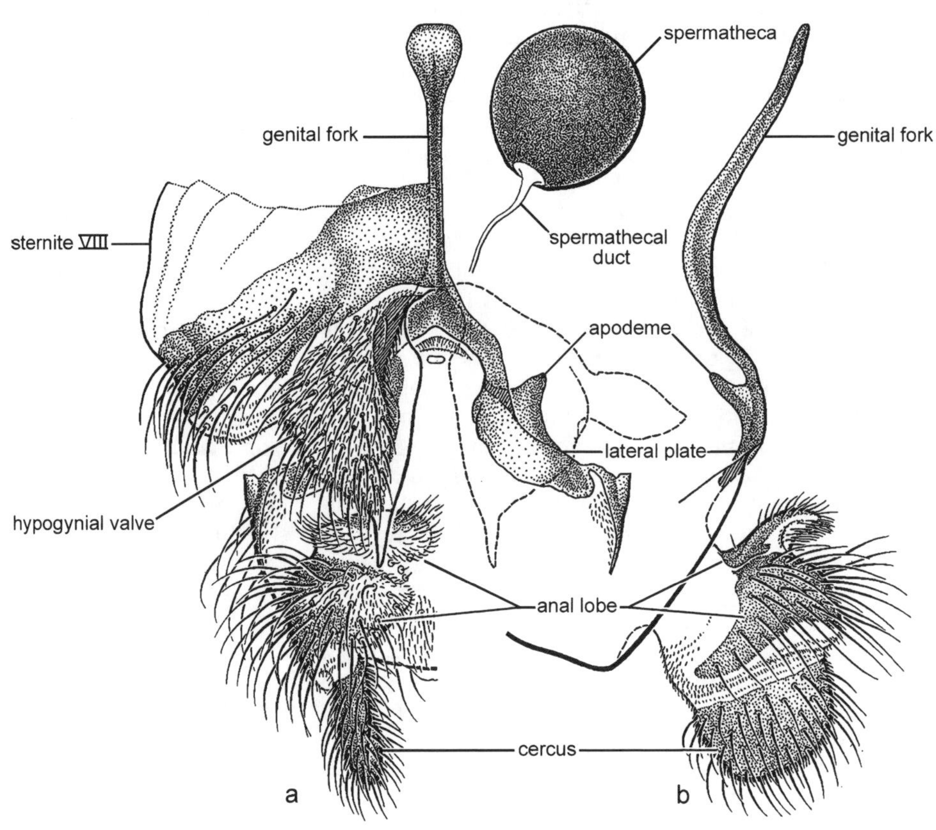


25 Simulium solarii ♀

Figs. 4.24, 4.25. Female terminalia; a, ventral view with left side of sternite VIII and left hypogynial valve, cercus, and anal lobe removed; b, right lateral view of genital fork, anal lobe, and cercus.

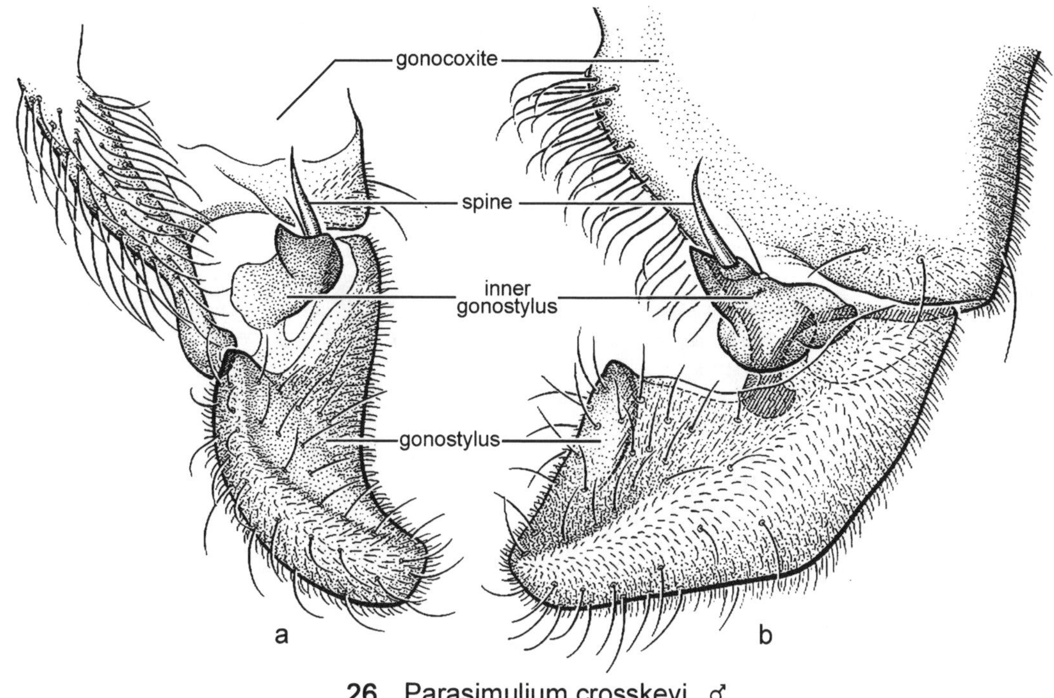


**26** Parasimulium crosskeyi ♂

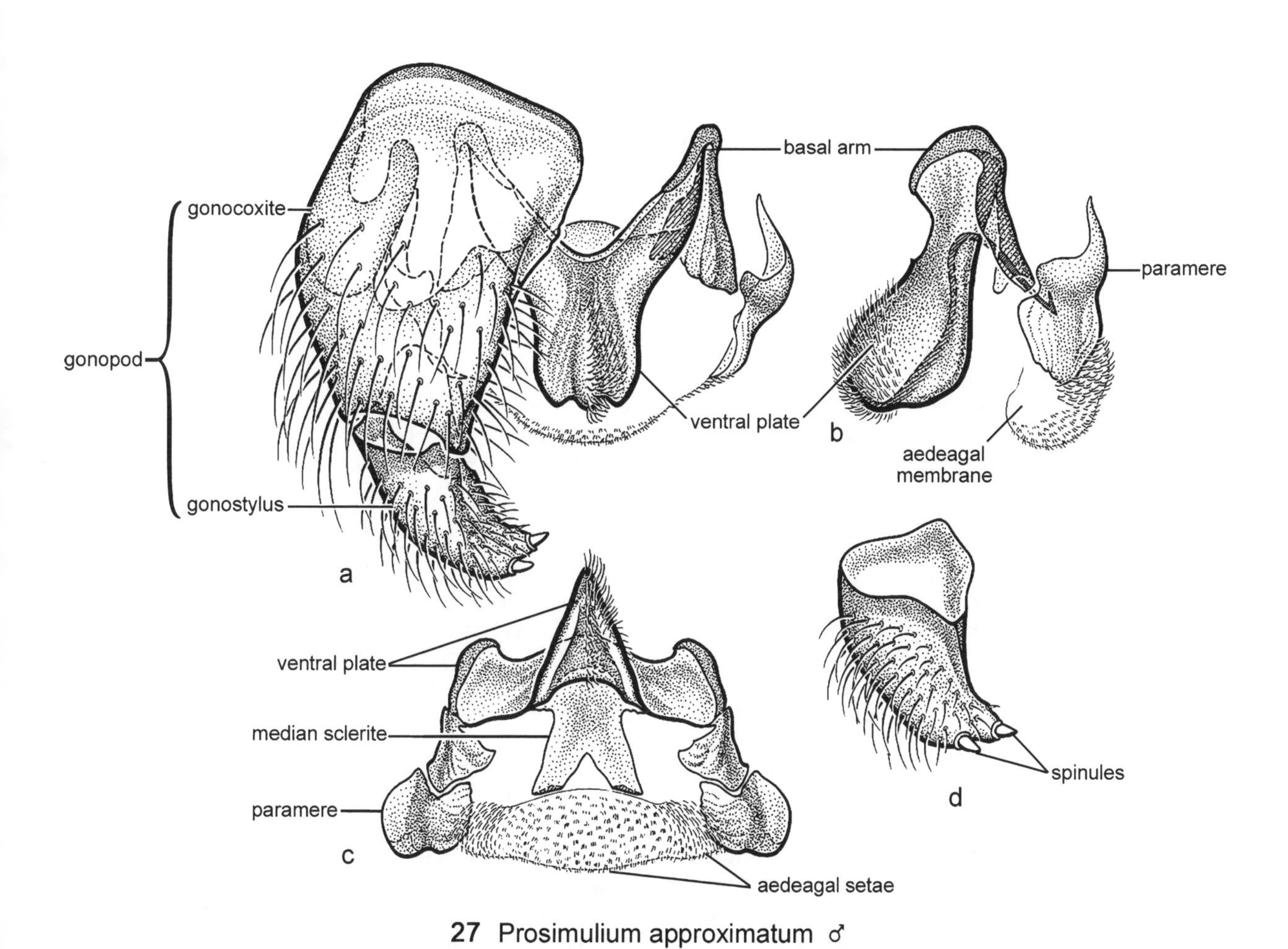


**27** Prosimulium approximatum ♂

Figs. 4.26, 4.27. Male genitalia. (26) Apex of gonocoxite, gonostylus, and inner gonostylus; a, dorsal view; b, medial view. (27) a, Ventral view with left gonopod removed; b, left lateral view (without gonopod); c, terminal (end) view (without gonopods); d, inner lateral view of right gonostylus.

and they often have subterminal flanges (e.g., subgenus *Nevermannia*), basal tubercles (e.g., *Simulium tuberosum* species group), or basal prongs (e.g., subgenus *Aspathia*). Most taxa have one or more apical or subapical **spinules** (i.e., peglike setae). Two, occasionally up to seven, apical spinules are characteristic of most prosimuliines, some basal simuliines (e.g., *Greniera*, *Stegopterna*, *Tlalocomyia*, *Gigantodax*, *Ectemnia*), some higher subgenera such as *Psilozia*, and scattered species of *Hemicnetha*. A single apical spinule characterizes *Twinnia*, the subgenus *Parahelodon*, and most simuliines. Apical spinules are lacking entirely in the genus *Parasimulium*, the *Simulium pictipes* species group, and *Simulium petersoni*. The genus *Parasimulium*, however, has a **spine** near the base of the gonostylus. In the subgenus *Parasimulium*, this spine arises from the medial surface of the **inner gonostylus**, a heavily sclerotized, freely movable structure situated between the gonocoxite and gonostylus (Fig. 4.26).

The **aedeagus** is a tubular muscular structure with two or three sclerites (plates) associated with its walls. The most conspicuous of the sclerites and most variable among species is the ventral plate. The heavily sclerotized **ventral plate** consists of a body and a pair of anterolateral apodemes (**basal arms**) that articulate with the gonocoxite and usually with the parameres. The body of the ventral plate varies from a slender, keel-shaped structure (e.g., subgenus *Eusimulium*) to a broad subrectangular plate, sometimes variously incised and crenulated or scalloped along its posterior margin, and often with a ventrally directed, variously sized, setose lip. The ventral plate serves, at least in part, to lift the anal lobes of the female away from the genital opening, presumably to enlarge the genital chamber to receive the spermatophore (Wood 1978). In some species, such as those of the subgenus *Eusimulium*, the ventral plate is probably too narrow to perform this function and evidently is inserted between the anal lobes (Wood 1963a). The **median sclerite** is a straplike extension of the dorsal surface of the ventral plate. It is articulated mediobasally with the ventral plate and projects posterodorsally as a midventral support for a conelike structure that bears at its apex the gonopore. The **dorsal plate**, when present, represents a third sclerite and is situated in the dorsal wall of the aedeagus. Its occurrence is scattered widely in the genus *Simulium*, appearing first in the subgenus *Hellichiella*. It varies from a thin strip, as in the *Simulium tuberosum* species group, to a large subrectangular or circular plate with a basal flange, reaching its most extreme expression in *Simulium loerchae*. The **aedeagal membrane** is bare or beset with variously sized setae or spines.

Each of the two **parameres**, one on either side of the aedeagal base, articulates with the apex of a basal arm of the ventral plate and with the dorsomedial base of the gonocoxite, except in the subgenera *Distosimulium* and *Parahelodon* (and the Palearctic *Levitinia*); in these taxa the connection with the ventral plate is lacking. In its simplest form, as seen in *Parasimulium*, the prosimuliines, and some basal simuliines (e.g., *Greniera*), each paramere consists of a subquadrate to subtriangular plate. In the remaining simuliines, a posterior extension gives rise to one to many **parameral spines** of various sizes. The spines typically originate from the apex of the paramere, although in some species (e.g., members of the *Simulium annulus* species group, Fig. 4.28), a single spine also arises near the parameral midpoint, and in other species (e.g., *Simulium baffinense*, Fig. 10.317), the parameral spines appear to be isolated in the aedeagal membrane adjacent to the paramere. In certain species of the subgenus *Psilopelmia* (e.g., *Simulium venator*, Fig. 10.347), each paramere has a finger-like anteromedial extension, of unknown significance, beyond the parameral spines. During coupling, the parameral spines push the hypogynial valves ventrally (Davies 1965a) and grip the sides and floor of the female genital chamber (Wood 1963a). The spines, therefore, help widen the female cavity to facilitate insertion of the aedeagus.

### SEXUAL MOSAICS

Aberrant individuals that display both male and female characters represent a small percentage (<0.05%) of adults in collections (Fredeen 1970a, Davies 1989). These sexual mosaics fall into two categories: gynandromorphs and intersexes. Gynandromorphs have a mix of male and female cells and phenotypically have male and female parts distributed in regular or random patterns. Most striking are the bilateral gynandromorphs, half male and half female, which provide opportunities for homologizing male and female structures. Intersexes have cells of a single genotype. Although usually bilaterally symmetrical, they typically show progressive anterior-posterior trends toward the opposite sex. Parasitism by mermithid nematodes is a common cause of intersexes.

### INTERNAL ANATOMY

Perhaps the most neglected morphological aspect of black flies is their internal anatomy. The internal anatomy of adults has been described and illustrated for one Nearctic species, a member of the *Simulium vittatum* species complex (Hungerford 1913), and several Palearctic species (e.g., Smart 1935). Among the finest illustrations of the internal anatomy and histology of adults are those of the Palearctic subgenus *Wilhelmia* (Jobling 1987). More focused treatments of Nearctic species include the musculature of the mouthparts and foregut of *Cnephia dacotensis* (Krafchick 1942); the gut of *Simulium innoxium*, including the transitional changes that occur during metamorphosis from the larva (Tanaka 1934; as *S. pictipes*); the female reproductive system of *Simulium verecundum* (Jobbins-Pomerory 1916); the gut of *Simulium jenningsi* (Cox 1938); the salivary glands of many species (Bennett 1963b, Gosbee et al. 1969); the hemocytes of *S. vittatum* (Luckhart et al. 1992); the peritrophic matrix (formerly termed "peritrophic membrane") of eight species (Yang 1968, Yang & Davies 1977, Ramos et al. 1994); and the male reproductive system of *S. innoxium* (Raminani & Cupp 1978; as *S. pictipes*). A miniature review of the relation of some internal systems to simuliid physiology has been presented (Cupp 1981), but more work is needed. Future studies should include comparative analyses of internal anatomy, particularly as a source of phylogenetic information.

### Pupa (Figs. 4.29, 4.30)

The body of the pupa is essentially uniform in shape for all black flies and reflects the shape of the adult, including most notably the arched thorax. The structures of the developing adult that are reflected in the overlying pupal cuticle have offered little taxonomic service; however, the elongate mouthparts of *Prosimulium longirostrum* n. sp. and *Prosimulium uinta* have correspondingly elongate, diagnostic labral sheaths. The **leg sheaths** of the first and second pair of legs are conspicuous laterally and ventrally. The third pair of legs, however, develop beneath the wing pads and are manifested as leg sheaths only distally where they wrap around the posterior margin of the **wing sheaths**, presumably to permit abdominal mobility (Wood & Borkent 1989). The eyes of the developing female are associated with a short broad **cephalic plate** and **antennal sheaths** that reach or exceed the posterior margin of the head. The larger eyes of the developing male are obvious as soon as pupation is complete, and are evidenced by a longer, narrower cephalic plate and antennal sheaths that extend only one half to three fourths of the distance to the posterior margin of the head.

The pupal abdomen has nine visible segments. Anterior to the first abdominal segment lies a small dorsal strip of cuticle, sometimes divided medially, that can cause confusion when counting abdominal segments. We have not found a suitable term for this strip of cuticle, and here apply the positional name **postscutellar bridge**. A survey of its shape throughout the family would be of potential phylogenetic value. Each side of the pupal abdomen has a striate **pleural membrane**, which on the fourth and fifth segments of *Prosimulium* and *Helodon* contains large diagnostic **pleurites**. The pleural membrane is absent from the third abdominal segment of *Parasimulium*, *Gymnopais*, and *Twinnia*, yielding a complete sclerotized ring. Ventrally, the sternites are rather uniform, and in all simuliids except *Parasimulium*, the sixth and seventh sternites (and sometimes fifth and eighth) are divided by a longitudinal striate **sternal membrane** (Fig. 4.59). The features of the pupa that are of greatest taxonomic importance are the gills, surface sculpture, armature (onchotaxy), and cocoon.

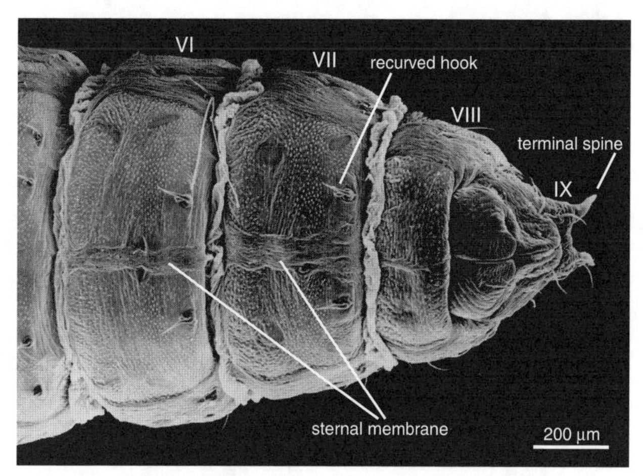


Fig. 4.59. Ventral surface of pupal segments VI–IX of *Prosimulium formosum*.

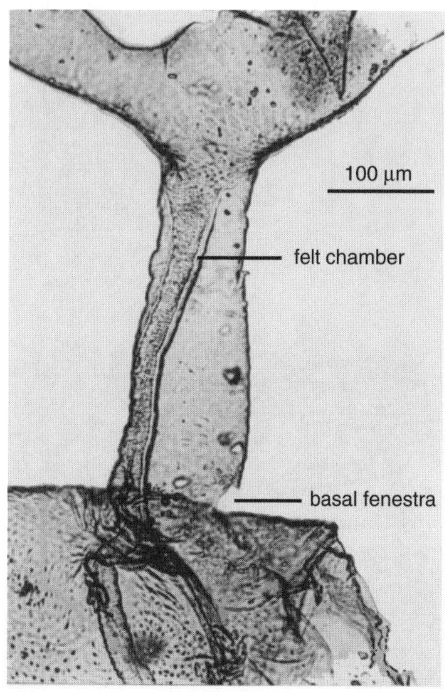


Fig. 4.60. Basal area of right pupal gill of *Parasimulium crosskeyi*, showing felt chamber.

#### Gills

The spiracular **gills** (respiratory organs) are the most characteristic feature of the pupa and the most structurally diverse attribute of the family. They arise as a bilateral pair of cuticular projections from the anterolateral corners of the thorax and generally extend forward or upward. Near the lateral base of each gill is a suboval area of thin cuticle, the **basal fenestra**, that bursts at the larva-to-pupa ecdysis, allowing water to enter the lumen of the gill. The shape of the gill thus is rendered independent of variation in hydrostatic pressure (Hinton 1957). The gill is connected dorsobasally with the anterior spiracle of the developing adult. Only the pupa of *Parasimulium* has a **felt chamber** (Fig. 4.60), a tubular continuation of the mesothoracic spiracle lined with dense fine setae, probably to reduce water loss during times of drought (Currie 1988). The felt chamber is formed by the dorsolongitudinal invagination of the gill, as suggested by the presence of an external longitudinal scar along the dorsal surface of the gill base. The gills serve in respiration and are designed to extract oxygen in and out of water, allowing for the possibility that the pupa, fixed by its cocoon to a substrate, might be left dry when water levels recede. The gills have long been believed to bear a plastron (Hinton 1964, 1976), but recent evidence casts some doubt, particularly because

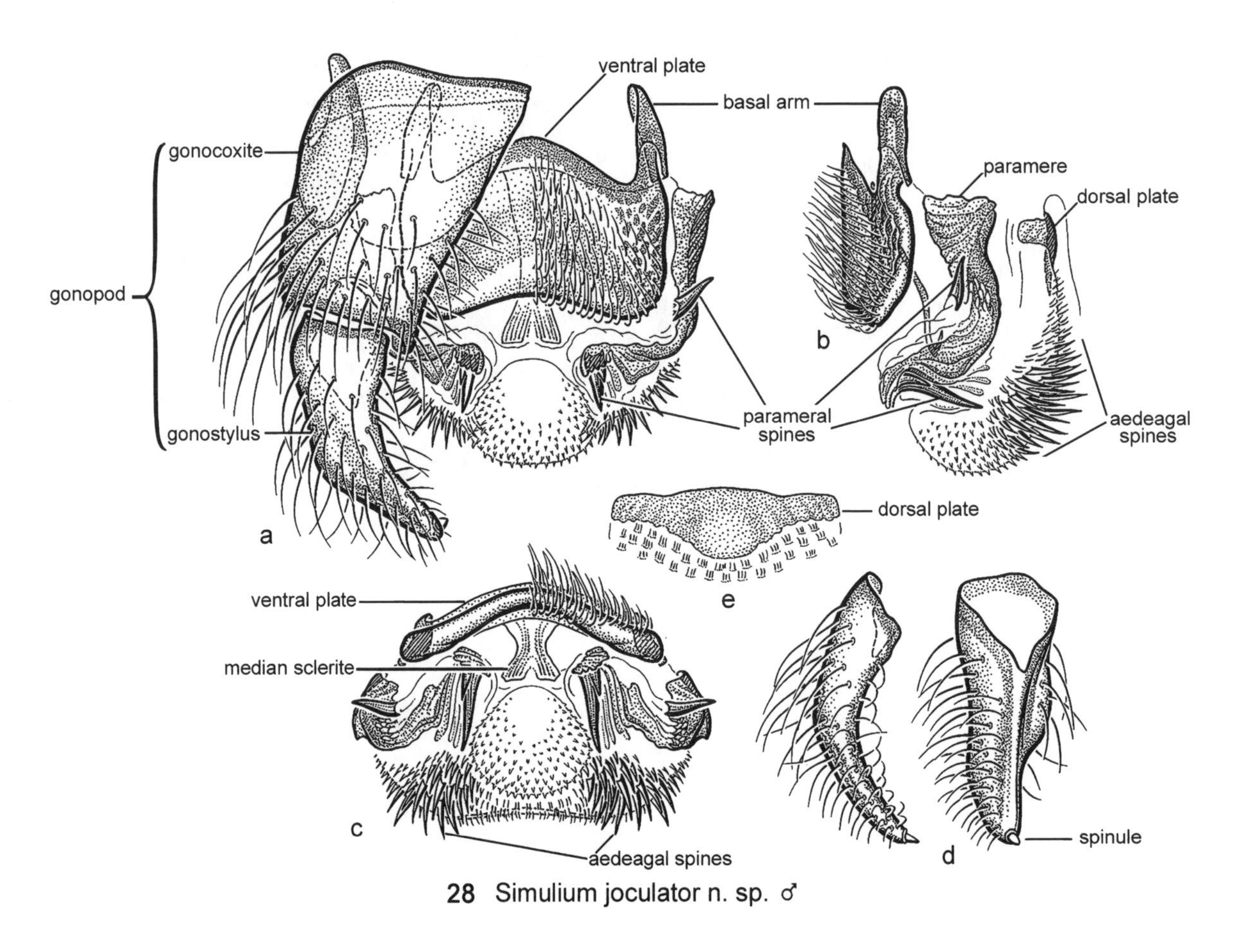


**28** Simulium joculator n. sp. ♂

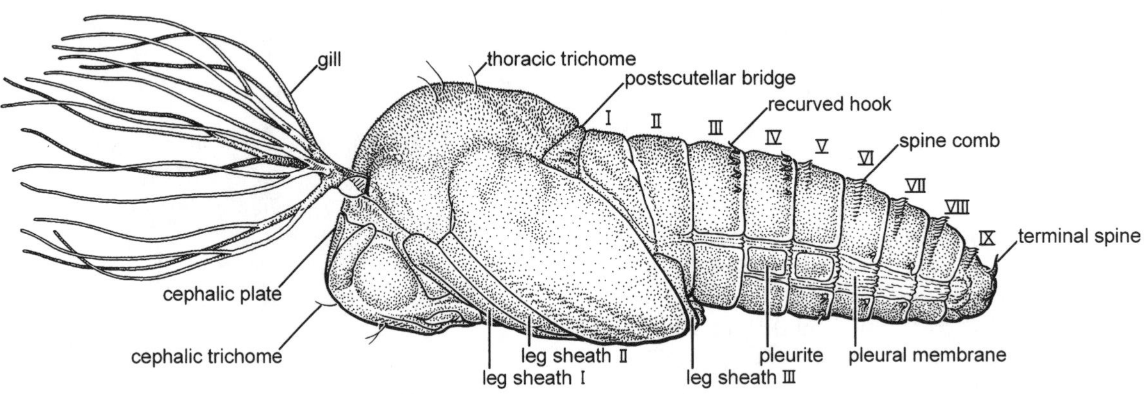


**29** Prosimulium clandestinum n. sp.

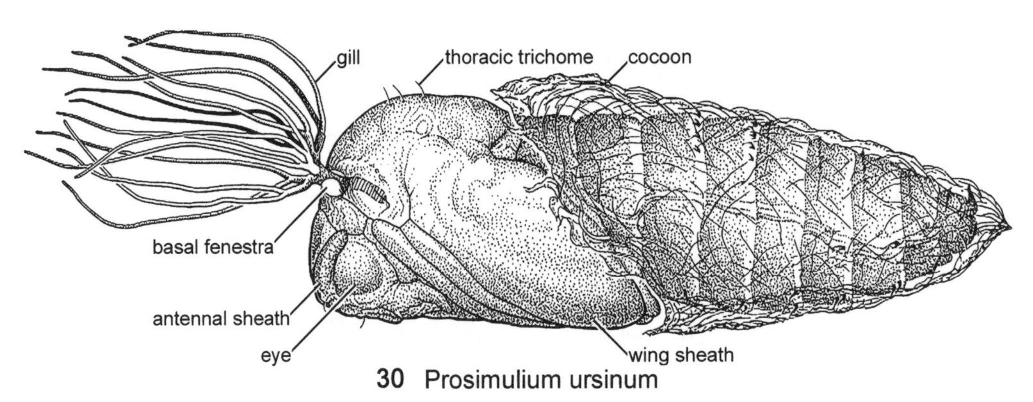


**30** Prosimulium ursinum

Figs. 4.28–4.30. Male genitalia and pupae. (28) Male genitalia; a, ventral view with left gonopod removed; b, left lateral view (without gonopod); c, terminal (end) view (without gonopods); d, right gonostylus, inner ventrolateral view (left), inner lateral view (right); e, dorsal plate. (29) Pupa, lateral view, cocoon omitted. (30) Pupa, lateral view. (30) From Peterson (1981) by permission.

of an apparent lack of a typical plastron surface with pores in the outer membrane (Williams et al. 1993). Nonetheless, a relation has been claimed between stream width and resistance of the plastron to wetting at excess pressure, the argument being that pupae in wider streams would be exposed to greater depths and, therefore, greater pressure during flooding (Hinton 1976).

Each gill consists of branches of various numbers, lengths, and thicknesses, typically filamentous in nature, but often tubular, clublike, or even spherical, and frequently ornamented with gracile secondary filaments. We generally use the term "filaments" in referring to the branches, even though they often are more tubular than filamentous. In the simuliid groundplan, a basal stalk gives rise to three main trunks (Currie 1988); in many species the number of trunks is secondarily reduced. The term "petiole" refers to secondary, bifurcated branches that arise from the primary trunks. The branching pattern can be presented as a formula. For example, a branching formula of (2 + 1) + (1 + 2) + (2) characterizes the gill of *Simulium bivittatum* (Fig. 10.456). Each set of parentheses indicates the filaments that arise from one of the three major trunks. In this example, the short, dorsalmost trunk gives rise to two filaments that share a dorsal petiole and a ventral filament that arises singly, whereas the middle trunk has the opposite arrangement, and the ventralmost trunk branches into two filaments.

The number of filaments per gill in Nearctic species varies from two in species such as *Gymnopais holopticus* to more than 100 in species such as *Prosimulium exigens* and *Simulium hunteri*. The gill of *Metacnephia villosa* can have more than 200 fine secondary filaments. Intraspecific variation in the number of filaments is common in many species, although species with 4–16 filaments typically have a more constant number than those with more filaments. An even number of filaments is typical of most species. For example, the number of North American species with 2, 3, 4, 5, 6, 7, 8, 9, 10, 11, 12, 13, and 14 filaments per gill as their typical number is, respectively, 1, 6, 36, 1, 37, 2, 32, 6, 18, 0, 27, 0, and 3. Individual pupae sometimes differ in filament number on opposite sides. The number of filaments, nonetheless, provides one of the simplest and most accurate characters for identification, particularly when combined with the branching pattern. The surface sculpture of the gill varies from smooth (e.g., *Simulium fibrinflatum*) to variously furrowed (e.g., *Simulium tuberosum*), reticulated (e.g., *Simulium jenningsi*), or tuberculate (e.g., *Simulium innoxium*) (Adler & Kim 1986; *innoxium* as *pictipes*), providing additional taxonomic characters.

The stunning variation in gill morphology begs questions of adaptive value, yet few hypotheses have been offered. Explanations probably cannot be sought by invoking a single selection force, for a gamut of forces is undoubtedly responsible. The limited range of behavioral options available to the stationary pupa might have necessitated greater structural changes to adapt the pupa to its environment (Eymann 1991b). Gills swollen and arranged so as to occlude the cocoon opening probably minimize access to the pupa by water mites, judging from lower rates of infestation (Gledhill et al. 1982), and also might lessen attacks by certain predators. Gills of species, such as *Metacnephia villosa*, with a plethora of fine filaments accumulate much sediment, but whether the accumulation is adaptive or artifactual remains unknown. Filament length generally decreases with stream size and velocity. Gaseous exchange could be facilitated by an increase in surface area, for example, by an increase in number of filaments. The surface area of gills varies from about 2 $mm^2$ to at least 12 $mm^2$ (Hinton 1965). An inverse relation between filament number and temperature (Bodrova 1980) and a positive relation between altitude and filament number (Konurbaev 1977) have been proposed, but so many exceptions can be found that the relations are questionable.

#### SURFACE SCULPTURE AND ARMATURE

The surface of the pupa is endowed with a diversity of textures and processes that contribute to identification. The cuticle of the head and thorax dorsally can be smooth and shiny (e.g., *Simulium aranti*) or covered with **microtubercles** varying from domes (e.g., *Simulium mysterium* n. sp., Fig. 10.825) and rounded granules (e.g., *Simulium carbunculum* n. sp., Fig. 10.826) to inverted cones (e.g., *Simulium dixiense*) and thin spines (e.g., *Simulium conicum* n. sp., Fig. 10.827). The density and arrangement of these microtubercles, whether regular or scattered, are of taxonomic importance and are generally consistent within a species; however, some species, such as several in the *Simulium venustum* complex, vary from having no microtubercles to a dense covering (Adler & Mason 1997). The cuticle also can be finely wrinkled (e.g., *Helodon alpestris*) or strongly rugose (e.g., *Prosimulium travisi*, *Simulium parnassum*, Figs. 10.525, 10.530). At least one species (*S. parnassum*) is dimorphic for the presence of rugosity, with the smooth form being far less common than the rugose form (Paysen & Adler 2000). The cuticle is sometimes overlain with abundant **microgranules**, so small that they typically cannot be detected without scanning electron microscopy; even species with cuticle that appears smooth and shiny under the dissecting microscope can be covered with microgranules (Fig. 10.828). In some species, especially those with dense microtubercles, a layer of organic matter often accumulates on the exposed thoracic cuticle (Adler & Currie 1986). Whether this adventitious layer is incidental or related to the adaptive significance of the surface sculpture is not known.

Arising from the head and thorax are unbranched or multiply branched, sometimes pigmented, sensory hairs, the **cephalic trichomes** and **thoracic trichomes**, of unknown function. The thorax typically has 4–7 pairs of trichomes, but in some species, especially in the *Simulium metallicum* species group, it has a dense covering of trichomes. Thoracic trichomes are absent in a few species such as *Greniera humeralis* n. sp.

The abdomen is outfitted with an array of hooks, spines, setae, and combs, most directed anteriorly and serving to secure the pupa within its cocoon. Most conspicuous are eight **recurved hooks** along the posterior

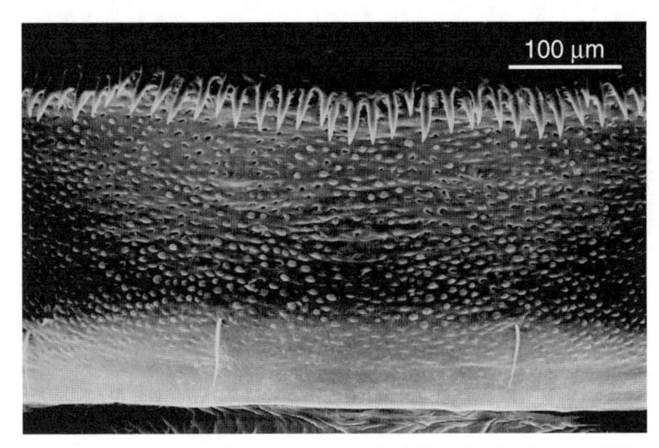


Fig. 4.61. Spine comb on anterior of abdominal segment VII of pupa of *Prosimulium formosum*, dorsal view.

margin of each of the third and fourth tergites, as first noted by Osten Sacken (1870). The ninth abdominal segment bears a pair of **terminal spines** that are long and slender in most prosimuliines and some basal simuliines (e.g., *Greniera*, *Stegopterna*, *Tlalocomyia*, *Cnephia*), rather short and stout in taxa such as *Ectemnia*, and small to apparently absent in *Parasimulium*, *Gymnopais*, and most remaining Nearctic simuliines. The sternites of the third to seventh segments are furnished variously with small recurved hooks. Pupae of *Gymnopais* have more sternal hooks than most taxa, probably to facilitate their attachment to the small ventral pad of silk, which is all that remains of their pupal cocoon. The pleural region of the sixth and seventh segments has one or more recurved hooks per side, lacking in *Parasimulium* and some *Greniera*. Members of the genus *Metacnephia* (and the Australian *Austrosimulium*) bear a set of anchor- or grapnel-shaped hooks laterally on the eighth and ninth segments. Small, usually unbranched setae are conservatively scattered over the abdomen, typically 1–6 pairs per tergite, 1–3 per sternite, and 0–3 in the pleural region. The complement of pupal armature also includes dorsal transverse rows of posteriorly directed **spine combs** on the anterior margins of, at most, the fourth through ninth tergites (Fig. 4.61). The Nearctic taxa without spine combs are *Parasimulium*, *Gymnopais*, *Twinnia*, and most *Greniera*. In these taxa, lack of spine combs is correlated with a reduced cocoon. Although the spine combs provide generic-level taxonomic information, their presence on the anteriormost segments varies within species such as *Cnephia ornithophilia*. Additional hooks and spines can be found, as a derived feature, in species such as the Mexican *Tlalocomyia revelata*.

#### COCOON

All known species of black flies produce a silk **cocoon**, variable in shape, complexity, thickness of individual strands, overall density of weave, and color. The color of the silk, which is typically brownish, might be related to sorption of fulvic and humic acids (R. S. Wotton, pers. comm.) in a manner similar to the sorption of pesticides to silk (Brereton et al. 1999). The sorption of pesticides and other chemicals to silk suggests that cocoons might serve as indicators of water quality; perhaps old cocoons in museum collections could provide historical information about water quality. Differences in physical properties of the silk suggest that silk composition varies among species, as shown for larval silk (Kiel 1997). Biotechnological applications of silk, such as its adhesive nature in water, have not been investigated but offer rich promise.

Cocoons are categorized as one of two major types. In *Parasimulium*, the prosimuliines, and the basal simuliines (e.g., *Greniera*, *Stegopterna*, *Tlalocomyia*, *Cnephia*), the cocoon is a shapeless saclike sleeve covering all or part of the pupa and sometimes the gills. In the most derived simuliines (*Ectemnia*, *Metacnephia*, *Sulcicnephia*, *Simulium*, and the Australian *Austrosimulium*), it is a well-formed housing with a specific architecture and often a fine weave that produces a smoother appearance, which at its most extreme resembles parchment, as in *Metacnephia jeanae*. Evolution of the well-shaped cocoon is associated, perhaps spuriously, with radiation into a broader range of aquatic habitats.

The shapeless cocoon can be as simple as a small ventral pad of silk that persists after the initial cocoon enclosing the pupa quickly disintegrates (*Gymnopais*). Or it can be as unusual as a transparent gelatinous envelope composed of slimy strands (*Twinnia*), a design that might protect the pupa if the flow in which it lives runs dry before development is completed (Wood 1978). More commonly, the shapeless cocoon is a sparsely to thickly woven covering of silk strands (e.g., Fig. 4.30). The density of the weave has been related to water velocity (Davies & Syme 1958). Taxa (e.g., *Prosimulium fontanum*, *Greniera*, *Stegopterna*) that pupate in slow waters, often buried in the sediments or ensconced in mosses and between leaves, have sparse cocoons or sometimes only a few strands of enveloping silk. Those taxa (e.g., *Prosimulium magnum* species group) that pupate in larger faster streams have thicker cocoons, either for anchorage or protection, and often form thick mats of silk during communal pupation.

Cocoons with a definite structure are either slipper shaped if the anterior margin lacks a collar and is flush with the substrate, or boot shaped if the anterior margin is raised as a collar (shoe shaped if the collar is short). The structure of these well-formed cocoons influences the dynamics of water flow around the pupa, promoting the formation of vortices that enhance aeration of the gills (Eymann 1991b). The evolutionary significance of boot-shaped cocoons might be to protect the gills from abrasion by suspended particles (Eymann 1991b). Species with cocoons of this ilk generally are found in swift water. Gills of species that make slipper-shaped cocoons gain protection from abrasion by orienting the anterior opening of the cocoon downstream. Boot-shaped cocoons also might limit access by water mites (Gledhill et al. 1982). Slipper-shaped cocoons of some basal members of *Simulium* (e.g., the subgenera *Hellichiella*, *Boreosimulium*, *Nevermannia*) have an "inner cocoon" that is the same shape as the pupa and is inside the exterior wall of the cocoon (Stuart & Hunter 1998b); its adaptive nature is unknown.

The slipper-shaped and boot-shaped cocoons have been subdivided into seven structural types (Crosskey 1990). Pedunculate cocoons (e.g., *Ectemnia*, Fig. 10.425) are borne subterminally on silk stalks up to 30 mm long, fastened to the substrate. The adaptive significance of a stalk is obscure, but it possibly serves to discourage predation or to elevate the larva and pupa above periphyton or ice that might form on the substrate. Truncate cocoons (e.g., *Simulium fontinale*, Fig. 10.522) have a shortened or scooped-out anterior margin, revealing much of the thorax. Hooded cocoons bear an anterodorsal projection or horn of various lengths (e.g., *Simulium innocens*, Fig. 10.429). The projection is generally a constant feature of a species, but in Nearctic populations of *Simulium bicorne*, it can be absent or well developed and bifurcate (Fig. 10.520). A prominent projection has arisen three or four times in the North American fauna (subgenus *Hellichiella*, *Simulium baffinense* species group, subgenus *Nevermannia*, *Simulium metallicum* species group). Cocoons with a projection are associated with species that inhabit small temporary streams. The projection might increase the flow of water over the gill filaments. Alternatively, we speculate that the function of the projection is to stem water loss at the ecdysial line when receding waters leave the pupa exposed to air; the projection often collapses onto the dorsum of the pupa during drying. Cribriform cocoons (e.g., *Simulium pictipes* species group, Fig. 10.480) are usually boot shaped and coarsely spun, producing a sieve-like mesh. Corbicular cocoons (e.g., *Simulium malyschevi* species group, Figs. 10.499–10.502) have loops of silk that arise from the anterior margin, yielding a patchwork of variously sized windows or fenestrae. Similar to the corbicular cocoons are the fenestrate cocoons (e.g., *Simulium jenningsi* species group, Fig. 10.489), also with anterior windows on either side but with a smooth margin at the cocoon opening. Cocoons with holes or windows form vortices that aerate the gills (Eymann 1991b). Patellate cocoons, with no examples in the Nearctic Region, are circular from a dorsal perspective.

### LARVA (Fig. 4.31)

The larval habitus consists of a well-sclerotized, external head capsule typically supporting a pair of labral fans, and an elongate, posteriorly expanded body with one prothoracic proleg and one posterior proleg. This basic design has remained unchanged over at least the past 120 million years, as indicated by the earliest-known larval fossils from Lower Cretaceous rocks in Australia (Jell & Duncan 1986). Changes in behavior and physiology, although poorly studied, might have played greater roles than structural changes in adapting larval black flies to their lotic environment (Eymann 1991b). The posteriorly expanded abdomen of all simuliid larvae might be designed partly to accommodate the enormous paired silk glands that run from the head into the abdomen, where they enlarge and double back on themselves. The consistency of larval structure across taxa also might be attributable to the uniformity of the fluid medium in which the larvae live. Only a limited number of body shapes permits efficient filter feeding (Eymann 1991b). The posterior portion of the abdomen probably reduces drag and determines the origin of vortices that aid filter feeding, whereas the narrow regions of the body facilitate bending and rotation (Chance & Craig 1986).

Although filter feeding might constrain structural variation among species that use this feeding mode, it does not explain the conservation of body form that extends even to the nonfiltering (i.e., fanless) taxa such as *Gymnopais* and *Twinnia*. In fanless larvae, the body generally is of the same design as in the filtering species, with slight modifications. Specifically, segments I–IV are narrow and ventrally corrugated, segment V expands abruptly, and the posterior proleg is small. This type of abdomen, which provides the flexibility necessary for the body to bend in a tight U shape, is an adaptation for a lifestyle that largely involves grazing food from the substrate (Currie & Craig 1988). It allows a broad C-shaped, rather than a narrow U-shaped, area to be grazed around the point of larval attachment. Although most fanned larvae occasionally graze, they do so less efficiently than the fanless species.

#### HEAD (Figs. 4.32, 4.33)

The importance of the head in simuliid taxonomy is demonstrated by its consistent illustration in descriptions of black flies worldwide. Although superficially similar in nearly all simuliids with labral fans, the shape and proportions of the head and its appendages are remarkably different among species and afford quantifiable taxonomic and phylogenetic information not yet investigated. The variety of shapes undoubtedly reflects biological differences in microhabitat choices and activities such as feeding. The posterior margin of the head is defined by a thin, well-sclerotized rim, the **postocciput**. In *Parasimulium*, the prosimuliines, and some *Greniera*, *Tlalocomyia*, *Gigantodax*, and a few other simuliines (e.g., some species of the Western Pacific subgenus *Inseliellum*), it nearly meets at the dorsal midline and encloses a pair of **cervical sclerites**. In most simuliines, however, the postocciput has a well-defined middorsal gap, leaving the cervical sclerites free.

One of the most taxonomically useful features of the head capsule is the pattern of pigmentation, usually in the form of spots on the **frontoclypeal apotome** (cephalic apotome), the region demarcated by the roughly U-shaped **ecdysial line**. These **head spots** typically consist of an **anteromedial group**, a **posteromedial group**, and one pair each of an **anterolateral group** and a **posterolateral group**. The pigmentation pattern results from deposition of pigment in the cuticle, particularly where muscles attach, and can be either a positive pattern in which dark head spots contrast with a paler background or a negative pattern in which the reverse arrangement is expressed. Sexual dimorphism in pigmentation patterns, especially the extent of infuscation around the head spots, is characteristic of taxa such as *Simulium definitum* and western members of the *Simulium annulus* species group, with females typically having more extensive

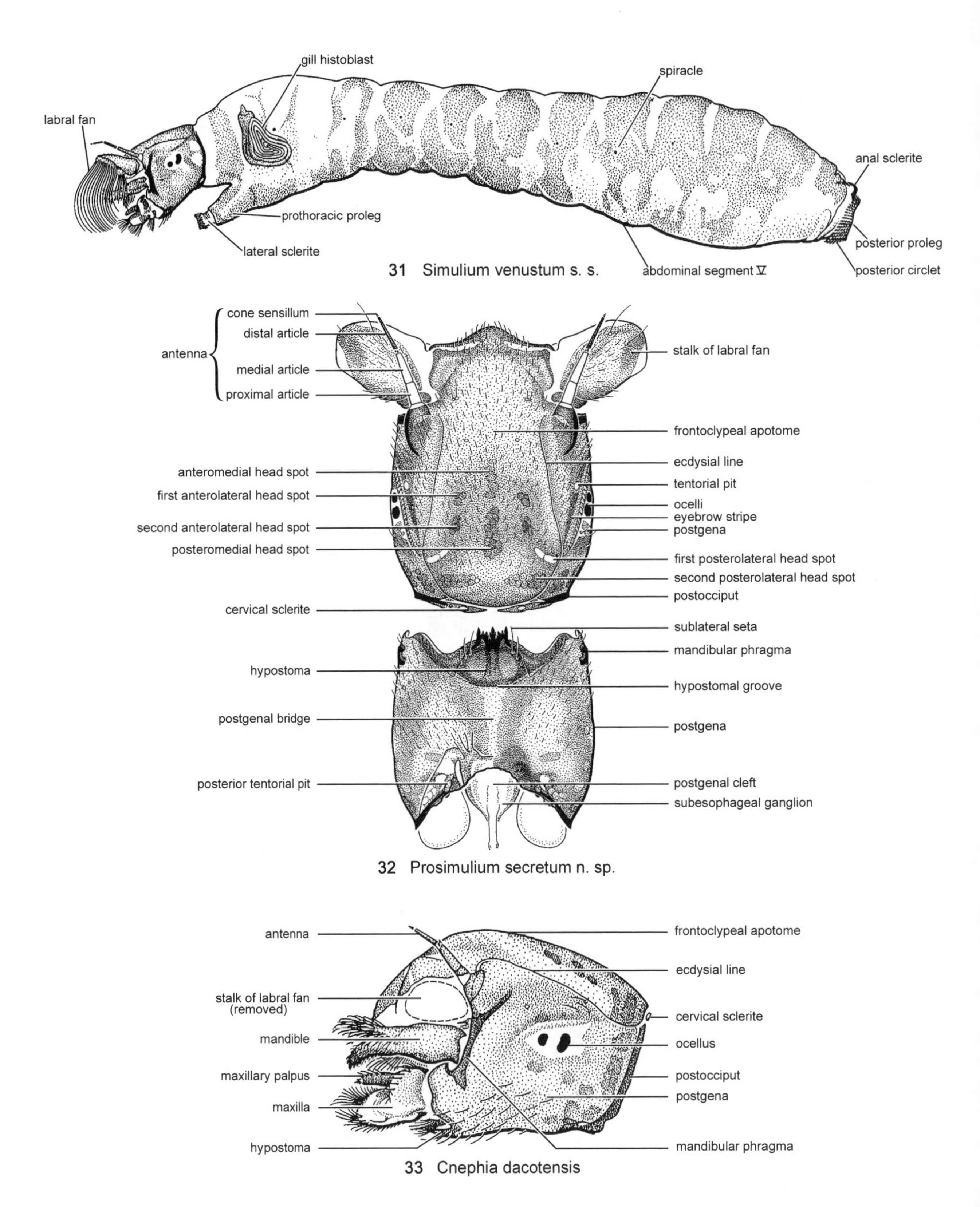


Figs. 4.31–4.33. Larval body and heads. (31) Larva, lateral view. (32) Head capsule, dorsal (top) and ventral (bottom) views. (33) Head capsule, lateral view. (31) From Peterson (1981); (33) redrawn and modified from Peterson (1996).

pigmentation than males. The basis for this sexual dimorphism is unclear but might relate to a differential need to protect certain internal attributes from ultraviolet radiation.

The entire head capsule is festooned with minute setae, as first shown by Puri (1925), which perhaps are useful in detecting water currents. In all North American species, the setae are unbranched and hairlike. The information contained in the setal patterns of late-instar larvae has not been tapped. First instars, however, have a small number of setae that, for the fanless simuliids, have been homologized with those of other nematocerous larvae; setae added at each successive molt are termed **secondary setae** (Craig 1974). A cephalic feature unique to first instars is the **egg burster**, a dorsomedial, darkly pigmented tubercle that helps free the pharate larva from its egg and subsequently is shed with the first molt.

Antennae arise anterolaterally from the dorsum of the head. Each **antenna** consists of a **proximal**, **medial**, and **distal article**, except in first instars and larvae of *Parasimulium*, which have a single article. The distal article has two basal multiporous sensilla and a terminal, uniporous **cone sensillum** (once considered a fourth antennal "segment"), all of which are probably chemosensory; five additional sensilla are located in the region surrounding the base of the proximal article (Craig & Batz 1982). Pigmentation patterns of the antennae are of taxonomic importance. For instance, a dark distal article contrasting with two colorless basal articles is a defining character for the prosimuliines (independently derived in *Metacnephia*), and annulated antennae characterize the subgenus *Boreosimulium*, most species of the subgenus *Hellichiella*, and some species of the subgenus *Hemicnetha*. Antennal length is inversely correlated with water velocity, the longest antennae occurring in groups such as *Greniera* that live in very slow flows and the shortest antennae occurring in fast-water taxa such as the subgenus *Hemicnetha*. Although antennal length in relation to the length of the labral-fan stalk can vary within a species (e.g., *Simulium hunteri*), it generally is a useful taxonomic character.

The portion of the head capsule lateral and ventral to the ecdysial line is composed largely of the **postgenae**, which generally appear as a single sclerite but are separate in taxa such as *Metacnephia*. Anterolaterally, each postgena is defined by a dark band, the **mandibular phragma**, that wraps ventrally a variable distance. Each side of the head capsule bears a **tentorial pit** and an adjacent set of three **ocelli** (eyespots), often with an overarching line of pigment, the **eyebrow stripe**. Two of the ocelli are heavily pigmented, except in *Parasimulium*. The third is unpigmented and externally invisible (Nyhof & McIver 1987). The ocelli can sense variation in light intensity, aiding in spatial orientation of the larva, and are perhaps sensitive to polarized light (McIver & Sutcliffe 1988). The ventral portion of the head capsule bears two important taxonomic features, the hypostoma and the postgenal cleft.

The **hypostoma** is an anteriorly toothed, trapezoidal plate delimited posteriorly by the **hypostomal groove**. It is beset with about 1–15 prominent **sublateral setae** per side (>30 in *Gymnopais fimbriatus*), each sometimes bifid or tufted apically, plus a variable number of minute scattered setae that are typically unbranched but can be bifid in taxa such as *Metacnephia*. The strongly sclerotized teeth of the dorsal wall and the serrations of the ventral wall have been homologized among genera (Currie 1986) (Figs. 4.34–4.43). They provide important diagnostic aids for generic and specific identification (Currie & Walker 1992). A single **median tooth**, variable in size and shape, is either simple or fitted with lateral denticles. It is bounded on each side by three sequential sets of teeth: the sublateral, lateral, and paralateral teeth. The **sublateral teeth** range in number from one to three per side in Nearctic species, with three being most typical. Small intercalated **intermediate teeth** on either side of one or more of the sublateral teeth can be present; they are best developed in the prosimuliines. The pair of **lateral teeth** (one tooth per side) can be as prominent as the median tooth or larger. In some groups (*Parasimulium*, *Greniera*, *Stegopterna*, *Tlalocomyia*) each lateral tooth is borne on a raised lobe in association with various numbers of sublateral and paralateral teeth. The small lateralmost teeth are the **paralateral teeth**, with one to four per side, except in the prosimuliines, which typically have none. The **lateral serrations** are distinct from the hypostomal teeth and are borne on the ventral, rather than dorsal, wall of the hypostoma. The evolutionary significance of interspecific variation in the hypostoma is poorly understood, although in many taxa it is used in grazing and also forms part of a functional complex with the mandibles to cut the silk lines (Craig 1977, Crosskey 1990). In the genus *Ectemnia*, the minutely toothed, concave margin probably helps shape the silk stalk that the larva constructs.

The **postgenal cleft**, its shape a rich source of taxonomic information, is an area of unpigmented, weakly sclerotized cuticle at the posterior margin of the head capsule anteromedial to the **posterior tentorial pits** and posterior to the **postgenal bridge**. The cuticle of the postgenal cleft is so transparent that some previous workers interpreted the cleft as an anterior extension of the occipital foramen. A correlation has been drawn between the depth of the postgenal cleft and the complexity of the cocoon (Shewell 1958). Taxa such as *Parasimulium* and the prosimuliines that lack a cleft or have small clefts build weak or simplistic cocoons, whereas taxa such as *Metacnephia* and members of the *Simulium jenningsi* and *Simulium malyschevi* species groups that have larger postgenal clefts construct more elaborate cocoons, often with raised anterior margins and fenestrae. This generalization carries exceptions, drawing into question cocoon spinning as the sole adaptive explanation. The cleft also might confer flexibility to the head capsule and permit it to withstand the forces of water that pass through the labral fans (Currie 1988). The **subesophageal ganglion**, if ensheathed with pigment, is visible through the postgenal cleft. Whether or not it is ensheathed with pigment offers a useful means of identifying some species, although within certain species the pigment can be present or absent (Adler & Kuusela 1994).

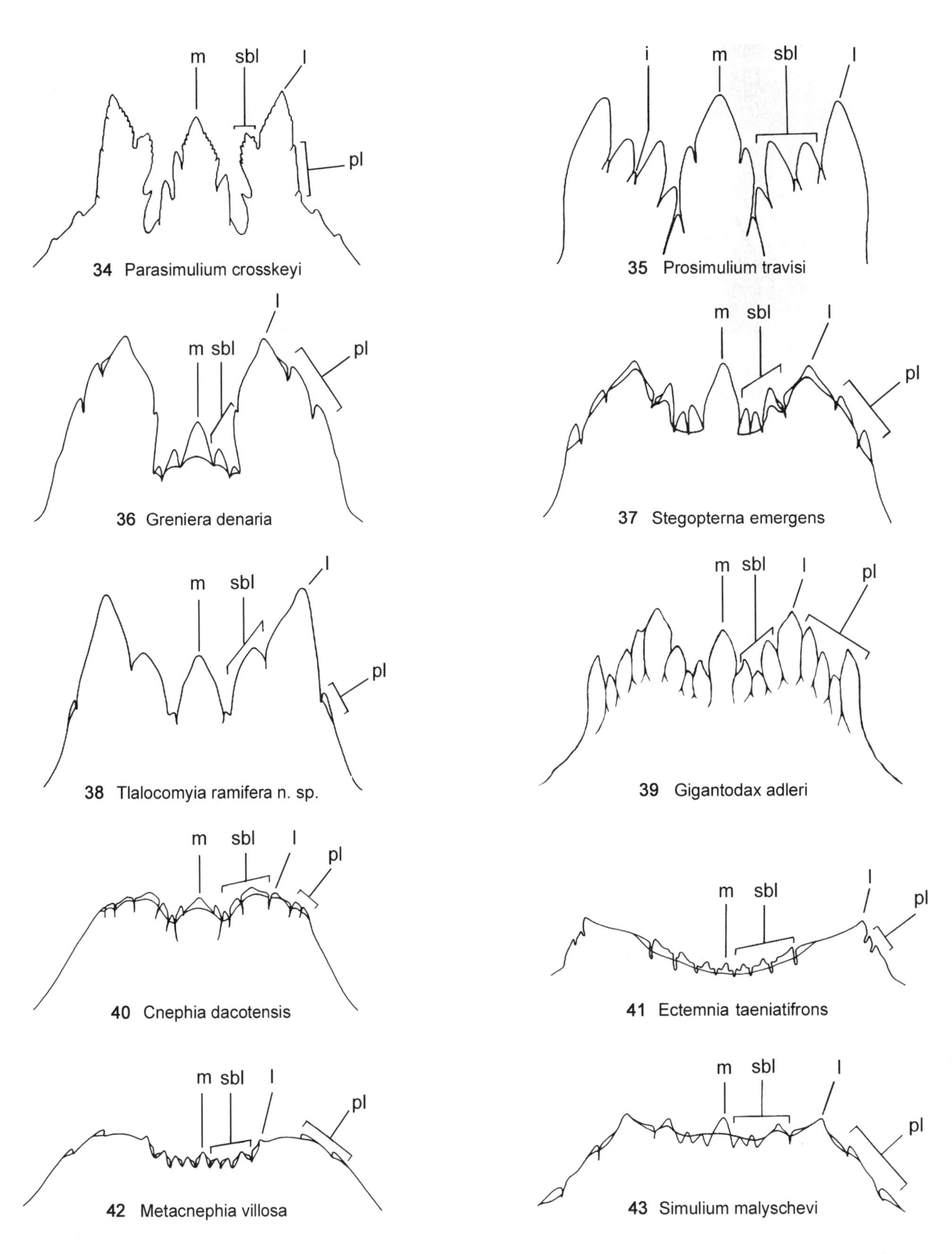


Figs. 4.34–4.43. Larval hypostomal teeth, anterior margin. i = intermediate tooth; l = lateral tooth; m = median tooth; pl = paralateral teeth; sbl = sublateral teeth.

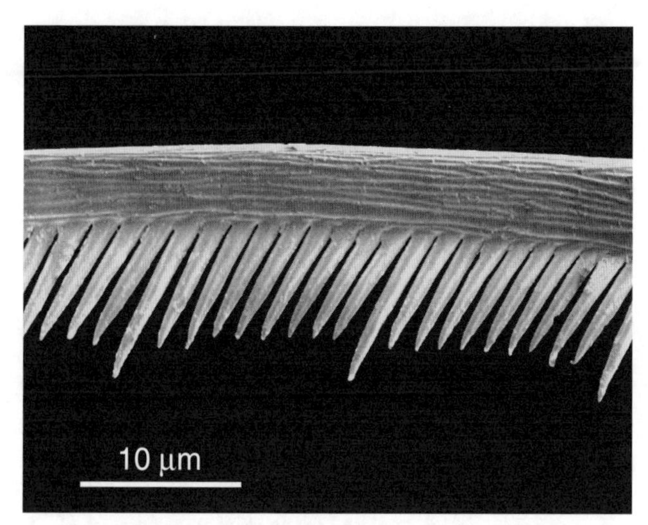


Fig. 4.62. Microtrichia on primary fan ray of larva of *Simulium vittatum* s. s.

**MOUTHPARTS**

The most conspicuous and probably most intensively studied larval feature is the pair of **labral fans** (cephalic fans), which embryologically are of labral origin (Craig 1974). They once were misunderstood as cilia that formed "a vortex by the rotary motion of the head . . . and the animalcules, thus engulfed in this miniature maelstrom, [were] irresistibly drawn towards the mouth" (Riley 1870b). Designed largely for filter feeding, as first deduced by Strickland (1911, 1913a), who described them functionally as "strainers," they are present in about 97% of North American species. Each labral fan consists of a stalk endowed with various sclerites such as the torma that serve as points for muscle attachment in the operation of the fan (Chance 1970a, Craig 1974, Wood & Borkent 1989). The evolution of the stalk perhaps is related to the need to have the fan situated far enough forward to allow its retraction into the cibarium during feeding and to provide enough separation between fans on opposite sides of the head (Craig 1974).

The fan actually consists of three sets of rays: the primary, secondary, and median. Forming the greatest portion of the functional filter are the long, curved, hollow, laterally flattened **primary rays**. Along the inner margin of each ray is a row of microtrichia (Fig. 4.62)—first reported by Osten Sacken (1870)—arranged in various patterns among species (Guttman 1967a, Chance 1970a). The structure of the primary fan is related to habitat. Fast-water species have small fans with short stout rays, whereas slow-water species have larger fans with elongate delicate rays (Zhang & Malmqvist 1996, Palmer & Craig 2000, Zhang 2000). The number of primary rays increases with instar. Final instars of North American species have an average of about 20 (e.g., *Tlalocomyia stewarti*) to 80 rays (*Simulium mysterium* n. sp.). Intraspecific variation in species such as *Simulium tribulatum*, which has about 30–60 primary rays, reflects developmental plasticity, and perhaps adaptation to local conditions, and dictates caution when using the fan for taxonomic purposes. Larvae living in slow water develop larger primary fans and longer stalks than do conspecifics in faster water (Zhang & Malmqvist 1997). And larvae living under low-food regimes develop more primary rays than do conspecifics that receive more food (Lucas & Hunter 1999). The **secondary rays** are smaller, fewer in number, and more basal than the primary rays. In most genera they form a flat triangular fan, whereas among some genera in the Nearctic Region (*Cnephia*, *Ectemnia*, *Metacnephia*, *Simulium*), they form a cupped fan with more rays. The **median rays** are straight, parallel, usually feathery (bare in species such as *Simulium decorum*), and located on the medial side of the stalk. A small set of 15 or fewer scalelike processes make up the **scale-fan** (*sensu* Crosskey 1990), located basally between the primary and secondary rays. The scale-fan holds the primary rays in the closed position as the stem of the fan moves outwardly, eventually releasing the rays and allowing them to whirl open rapidly; the blades thus act as a catch mechanism that allows energy to be stored until the fans click open (Kurtak 1973).

The fully fanned condition is considered ancestral (Wood 1978, Wood & Borkent 1989). Labral fans are secondarily absent in *Gymnopais* and *Twinnia* (and the Palearctic *Levitinia*), and the head, tapered anteriorly, does not require the musculature necessary to operate the fans nor a large volume to store the developing rays. The fan is reduced in first instars of all prosimuliines studied to date, as first discovered probably by Strickland (1913a) and later elucidated by L. Davies (1961). It consists of three rays and no stalk and presumably cannot filter or rake food effectively (Craig 1974, Ross & Craig 1979). The small size of the first-instar larva and the thickness of the boundary layer around the larval body make filter feeding in this early instar unlikely (Currie & Craig 1988).

In addition to the labral fans, the standard set of mouthparts consists of mandibles, maxillae, a labrum, and a fused labium and hypopharynx. Each **mandible** has about nine sets of setal brushes and bristles, several trichoid sensilla, and a series of variously sized, comblike **mandibular teeth** along its inner margin. The apicalmost teeth interdigitate with the teeth of the hypostoma to cut the ribbons of extruded silk (Craig 1977, Barr 1982). The basalmost tooth in most species, long referred to as a mandibular serration, is actually a sensillum (D. A. Craig & R. E. G. Craig 1986). The mandibles are rather uniform across filter-feeding species, but those of *Gymnopais* and *Twinnia* (and the Palearctic *Levitinia*) have flattened teeth (Figs. 4.44, 4.45). In *Gymnopais* and *Levitinia*, the outer apical margin of each mandible is beset with up to 12 rows of **comblike scales** (the modified apical brush) that might help transfer food to the cibarium. With their brushes and bristles, the mandibles play the most important role in removing filtered matter from the labral fans and directing it into the cibarium (Craig 1977), analogous to a pan and broom, with the mandibular brushes acting as the broom and the hypostoma as the pan (Currie & Craig 1988). The mandibles also play a role in scraping the substrate, especially in the fanless larvae, which often show well-worn teeth.

The **maxillae** are quite uniform within the family, although in grazing taxa (e.g., *Gymnopais*, *Twinnia*), they are rounded rather than tapered. Each maxilla is endowed with five brushes (Chance 1970a, Barr 1982) and a one-segmented **maxillary palpus** bearing 12 apical sensilla of six structural types that have been homologized within the family and with the sensilla of other nematoceran families (Craig & Borkent 1980). The palpal sensilla, which do not contact filtered food, might sense the substrate or dissolved substances (Craig 1977). The **labiohypopharynx** is fitted with various sclerites, sensilla, and setae (Chance 1970a, Barr 1982). Some of the setae might aid in cleaning the mandibular brushes (Currie & Craig 1988). Each labial palp bears five sensilla that probably assess the food as it is moved into the cibarium (Craig 1977). Silk, appearing in the form of flat filaments, is extruded through the common opening of the silk glands, which lies between the lower labial and upper hypopharyngeal lobes.

The **labrum** appears as a continuation of the frontoclypeal apotome and overhangs the opening to the cibarium. The bristly ventral portion of the labrum is the **labropalatum**, also called the epipharynx or palatum. It is supported by a spade-shaped labral sclerite, or **intertorma**, which is toothed at its apex (Chance 1970a, Craig 1974). Much of the labrum is covered with various setae, typically arranged in brushes. Marked differences are found in the labra of filtering versus fanless species. For example, the median apical portion of the labrum (i.e., the **palatal brush**), as well as the apical teeth of the intertorma, are uniquely adapted for grazing in the fanless genera such as *Gymnopais* and *Twinnia* (Craig 1974, Wood 1978). In *Gymnopais*, the ventral portion of the labrum bears a unique arrangement of setae, collectively referred to as the **epipharyngeal apparatus** (Wood 1978).

### THORAX AND ABDOMEN (Figs. 4.46–4.51)

The narrow, cylindrical body consists of three thoracic segments and nine apparent abdominal segments. Embryos have nine abdominal segments (Craig 1969), and larvae putatively have an intersegmental line between the eighth and ninth abdominal segments (Barr 1982). The larval body is rather uniform across taxa, particularly in the shape of the thorax and first four abdominal segments, but it can be categorized as having two distinct forms.

The most common body design is one in which the abdomen expands rather abruptly at the fifth segment and tapers at the eighth segment. Larvae of this form occupy a wide variety of habitats. This body shape is particularly exaggerated in the fanless species, presumably to give greater flexibility and permit grazing of a larger area around the point of attachment (Currie & Craig 1988). The nearly prehensile abdomen of *Ectemnia* exhibits the most abrupt expansion and is used to grasp the silk stalk while it is being constructed (Stuart & Hunter 1998a). The other body form is one in which the abdomen expands gradually toward the posterior. This body form, which often includes expansion of the posteroventral portion of the abdomen, characterizes larvae that live in swift currents and pack closely together in regularly spaced patterns. The ventrally swollen area of the abdomen might facilitate the rather upright posture that is assumed when the larvae are filter feeding (Eymann 1991c). In North American black flies, the gradually expanded abdomen has evolved independently in the *Prosimulium magnum* species group, *Cnephia*, *Metacnephia*, subgenus *Hemicnetha*, and the *Simulium malyschevi* species group.

In some species, such as those of the *Simulium meridionale* species group, the first five or more abdominal segments bear circlets of tubercles. These tubercles might increase the surface area for gas exchange; most species that have them are warm-water inhabitants. We have seen summer larvae of *Simulium lakei*, a species normally without tubercles, from southern Florida that have a pair of minute dorsal tubercles on the first through fifth abdominal segments, and summer larvae of *Simulium tuberosum* (cytotype AB) from Isle Royale, Michigan, with a pair of dorsal tubercles on the first through sixth abdominal segments.

The thoracic and abdominal cuticle is a thin, unpigmented, cellophane-like covering with nine pairs of nonfunctional **spiracles** represented by minute dark rings. (Oxygen diffuses through the integument and is distributed by the tracheal system.) In most Nearctic species, the cuticle is undifferentiated or, at most, sculpted into minute, irregular, transverse ridges (Eymann 1985). In the *S. meridionale* species group, the cuticle is thrown into minute, parallel and concentric ridges, as viewed with a compound microscope or scanning electron microscopy. In other taxa, such as African species phoretic on crabs and other aquatic invertebrates, the cuticle can be variously sculptured and might reduce the risk of desiccation (Crosskey 1990). The cuticle of all species is bestowed with minute unbranched setae (trichoid sensilla), with or without pigment, that are best viewed using a compound microscope, sometimes with phase contrast. Some species also have numerous multibranched, arborescent, or scale-like setae, often giving the larvae a hairy appearance. *Simulium meridionale* has slender, teardrop-shaped setae with minute apical papillae. In the prosimuliines and other basal taxa, the setae are typically unbranched, translucent, and widely scattered, but in the simuliines they are more varied. The setae on the body possibly help larvae sense water currents (Eymann 1985), although the adaptive significance of the various setal forms is not known.

Beneath the cuticle are chromatocytes containing pigment granules that impart the various colors and patterns (Hinton 1959) of taxonomic importance. Pigmentation, nonetheless, can vary within a species, often as a function of sex, diet, background, and exposure to ultraviolet radiation (Zettler et al. 1998). Only larvae of *Parasimulium* lack pigmentation. Larval color often differs between males and females; at least 16% of North American species of *Simulium* are sexually dimorphic with respect to color. Physiological and evolutionary reasons underlying sex-related color differences have not been explored and could offer a rich area of study. Internal organs are often ensheathed with pigment, perhaps for

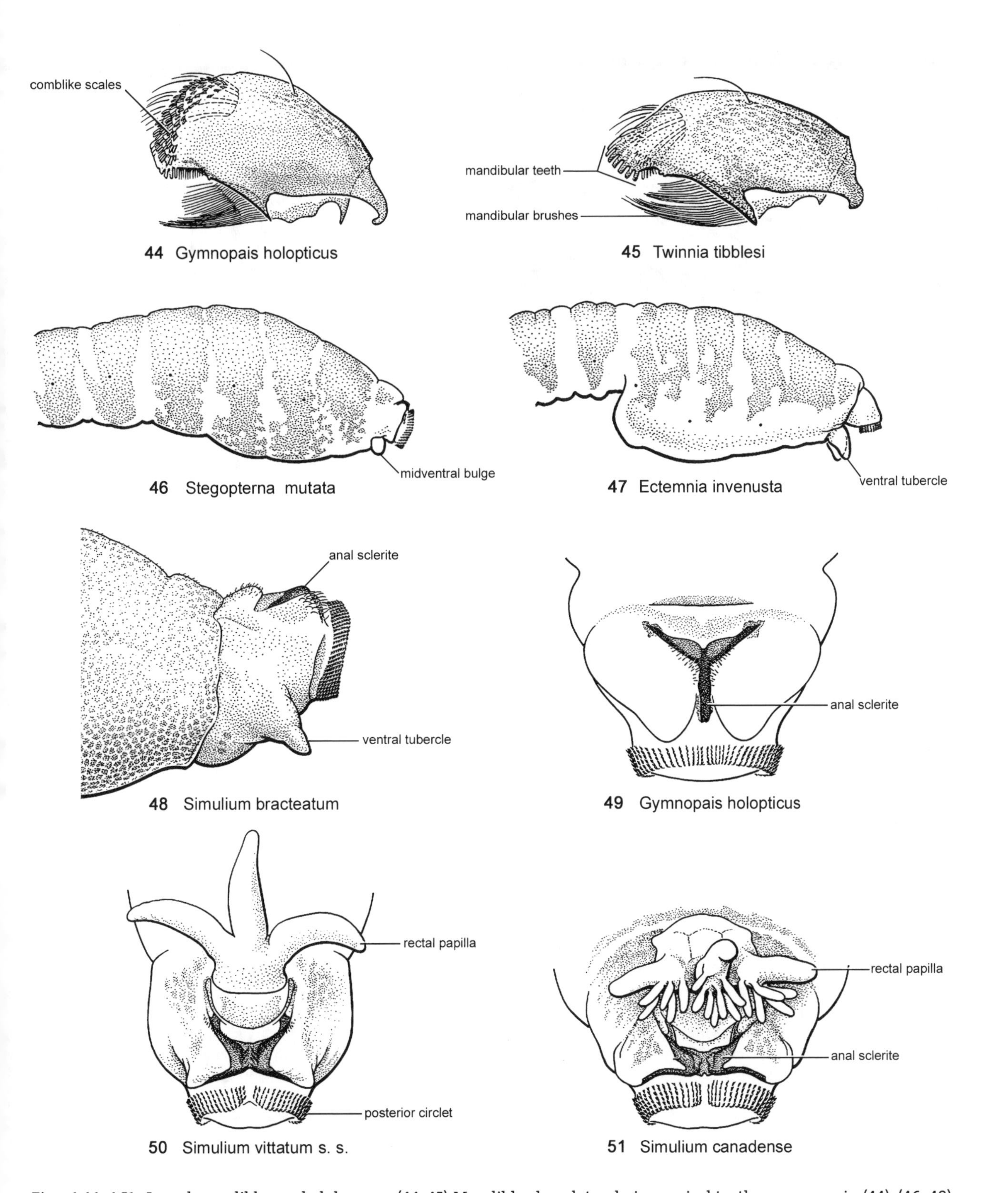


Figs. 4.44–4.51. Larval mandibles and abdomens. (44, 45) Mandible, dorsolateral view, apical teeth worn away in (44). (46–48) Abdomen, lateral view of posterior portion. (49–51) Abdomen, segment IX, dorsal view. (44–49) From Peterson (1981) by permission; (50, 51) redrawn and modified from Peterson (1996).

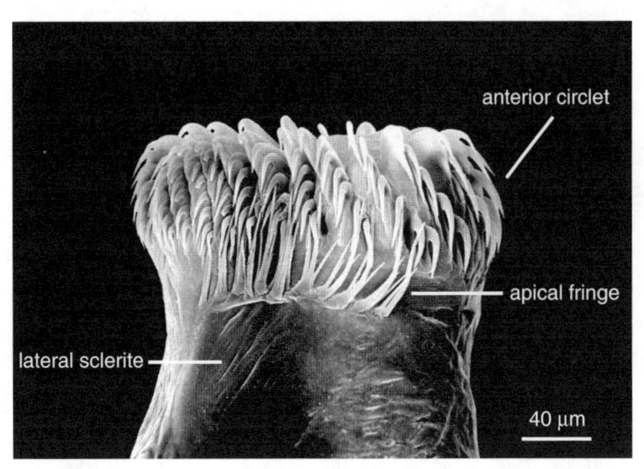

Fig. 4.63. Prothoracic proleg of larva of *Simulium decorum*, lateral view.

protection from ultraviolet radiation. They include the ventral nerve cord and gonads, the latter being conspicuous, particularly in males (e.g., in subgenus *Psilozia*), through the dorsal integument of the sixth abdominal segment. When a filter-feeding larva rotates its body 90–180 degrees, the anterior portion of the ventral nerve cord can be exposed to ultraviolet radiation. Larvae of the Polynesian black fly *Simulium cataractarum* do not twist their body; thus, their ventral nerve cord is not ordinarily exposed to direct sunlight and is unpigmented (Craig 1997).

The thorax has six histoblasts per side, the most prominent of which is that of the pupal gill, located on the prothorax. In the final instar, the dark **gill histoblast**, first noted as the future gill of the pupa by Riley (1887), can be dissected out to provide extralarval taxonomic information. The five additional histoblasts are those of the adult legs, halteres, and wings, as first described and illustrated by Strickland (1911). The extent of histoblast development has been used as a rapid and convenient means of sorting larvae into broad age classes (Tessler 1991) but is not always consistent (Hyder 1998).

The thorax supports a single ventromedial **prothoracic proleg** consisting of two articles. The distal article of the prothoracic proleg (Fig. 4.63) bears a pair of **lateral sclerites** (except in *Parasimulium*) of variable size and shape (Podszhun 1967), a pair of campaniform sensilla, and a ring of minute hooks (**anterior circlet**) that the larva uses to pull or hold threads of silk and to grasp the silk pad it spins on the substrate. An **apical fringe** of bristles on each lateral sclerite possibly helps extricate the hooks from the silk pad (Barr 1984).

The apparent ninth abdominal segment bears the single **posterior proleg** (Barr 1984), which appears as a continuation of the abdomen and, with its armature, serves to anchor the larva to its silk pad. The earlier idea of the posterior proleg functioning as a sucker (e.g., Frost 1932) now seems quaint. The armature consists of a ring of minute hooks (**posterior circlet**) that girdles the end of the posterior proleg and, when enmeshed with the silk pad, firmly secures the larva to the substrate. These hooks are arranged in rows and are similar to those of the prothoracic proleg except some have minute tubercles that probably prevent the hooks from loosening from the silk pad (Barr 1984). The number of hook rows in mature larvae of Nearctic species varies from about 54 (e.g., *Twinnia tibblesi*, *Simulium definitum*) to more than 300 (*Simulium paynei* species group), with about 7–30 hooks per row. Species of the Oriental subgenus *Daviesellum* have a phenomenal number of hooks—about 460 rows, with up to 53 per row (Takaoka & Adler 1997). Both the number of hook rows and the number of hooks per row are roughly correlated with water velocity of the larval habitat (Palmer & Craig 2000) and probably with the stickiness of the silk. Even within a species, individuals living in fast currents have more hooks than do those in slower currents (Konurbaev 1973). The posterior proleg is supplied with many sensilla, primarily translucent and setiform, including a minute pair at the very end, surrounded by the circlet of hooks. Anterior to the circlet of hooks is the dorsally situated **anal sclerite**, usually X shaped (Fig. 4.51) but subrectangular in some *Helodon*, Y shaped in fanless prosimuliines (Fig. 4.49) (star shaped in the Palearctic *Levitinia*), and absent in *Ectemnia* and first instars. The posteroventral arms of the anal sclerite form a continuous pigmented ring around the ninth segment in taxa such as *Gigantodax* and some members of the subgenus *Inseliellum* of Polynesia. In *Parasimulium*, they do not connect with the anterodorsal arms, but their apices nearly meet ventromedially (Currie 1988). Accessory anal sclerites are present in some taxa, such as *Parasimulium* and some members of the *Simulium metallicum* species group. Muscles that work the posterior proleg attach to the anal sclerite (Barr 1984). The anus is situated dorsally, just anterior to the anal sclerite. It has assumed a dorsal position because the posteroventral region of the abdomen is closely appressed to the substrate and is rendered unsuitable as a site for defecation.

The ninth segment is provided with additional features that disrupt its contour. Immediately anterior to the anal sclerite are the unpigmented eversible **rectal papillae** (rectal organ) (Figs. 4.50, 4.51), which arise from the rectal wall and function in osmoregulation (Komnick 1977). Their description by Headlee (1906), although functionally flawed, has not been bettered. In preserved larvae, they often are retracted into the rectum and cannot be viewed without dissection. They consist of an inflated base with either three simple digitiform lobes, sometimes with a few smaller lobules (*Parasimulium*, prosimuliines, some simuliines), or three compound lobes, with numerous smaller lobules. The adaptive significance of the two forms is unknown. Some species (e.g., *Simulium piperi*, *S. conicum* n. sp.) have both simple and compound rectal papillae, although the possibility that multiple species are involved has not been excluded. A patch of posteriorly directed, comblike rectal scales (rectal setulae) is present in many species just anterior to the opening for the rectal papillae. A large, conical, posteriorly directed lobe arises from the venter of the ninth segment of *Parasimulium*. Paired **ventral tubercles** arising from the ninth segment characterize genera such as *Greniera* and *Ectemnia*, as well as most ornithophilic *Simulium* (e.g., subgenera *Hellichiella* and *Nevermannia*) and some species of the mammalophilic subgenus *Aspathia* and the *Simulium jenningsi* species group (Figs. 4.47, 4.48). A transverse **midventral bulge** (considered fused ventral tubercles by Barr [1982]) is found on the ninth abdominal segment in the genus *Stegopterna* and *Tlalocomyia andersoni* n. sp. (Fig. 4.46). The function of these bulges and tubercles has not been studied but might be related to the interaction of hydrodynamics and feeding.

#### INTERNAL ANATOMY

The only comprehensive treatment of the internal anatomy of Nearctic larvae is that for *Simulium innoxium* (Hallock 1922; as *S. pictipes*). More specific larval treatments include the respiratory system of *Simulium verecundum* (Jobbins-Pomeroy 1916), the gut of *S. innoxium* (Tanaka 1924, 1934; as *S. pictipes*) and *Simulium bracteatum* (Strickland 1913a), the rectal papillae and associated musculature of *S. innoxium* (Headlee 1906; as *S. pictipes*), and the internal anatomy of the heads of five Nearctic genera (Craig 1974, Fry & Craig 1995). The embryology of larval simuliids has been studied in three Nearctic species, beginning with early perfunctory work on *S. verecundum* (Jobbins-Pomeroy 1916) and continuing with more comprehensive investigations of *S. innoxium* (Gambrell 1933; as *S. pictipes*), the *S. verecundum* species complex (Craig 1969, 1972b), the *Simulium vittatum* species complex (Goldie 1982), and *Simulium vampirum* (Shipp 1988; as *S. arcticum* IIS-10.11). One of the most comprehensive studies on the internal anatomy of a larval black fly is that by Puri (1925) for *Simulium noelleri* from England. The tracheal system of the larva, as well as that of the pupa and adult, was covered by Taylor (1902), presumably for a European species. The ultrastructure of the larval silk glands has been treated in detail for a European species (Macgregor & Mackie 1967), and the glands have been measured for various North American species (Rao 1966). Attempts to establish cell lines from neonate larvae and characterize their morphology have met with limited success (Ducros et al. 1992).

### Egg

The eggs of black flies are rather homogeneous throughout the family and do not provide the wealth of characters afforded by the other life stages. They are roughly oval in outline, especially when viewed dorsally, but from a lateral perspective they often appear subtriangular with rounded angles. This subtriangular shape is a defining synapomorphy for the family (Borkent & Wood 1986). The color of eggs is a function of age, changing from ivory white when freshly laid to brown shortly before the larvae hatch. Egg size within the family varies about fivefold. The largest eggs, those of autogenous northern species that produce relatively few ova, are slightly more than half a millimeter in length. Few species can be identified by their eggs because of the paucity of characters, but about one fourth of the approximately 20 North American species that lay their eggs in strings or masses can be distinguished on the basis of egg size and proportions and the manner of deposition. Five North American species, including four that deposit their eggs in masses, have been included in a key to simuliid eggs (Imhof & Smith 1979).

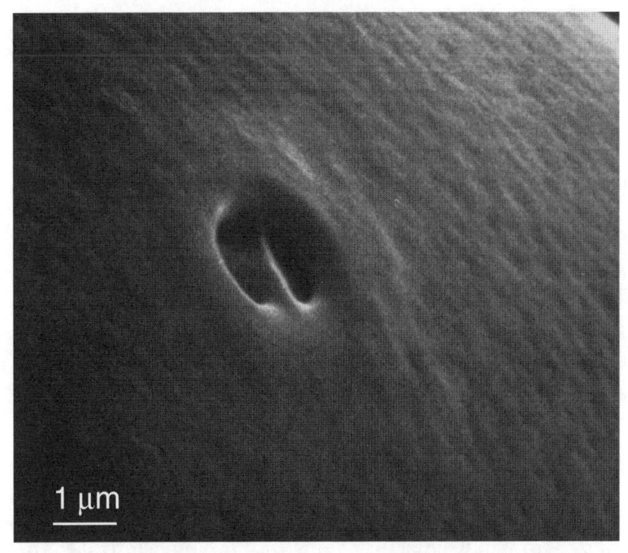


Fig. 4.64. Micropylar area of egg of *Simulium vittatum* s. s. Photograph and permission by P. Goldie.

The egg has an exochorion—an adhesive, unpigmented outer membrane of multiple layers—that evidently is homologous with the exochorion of the mosquito egg (Williams 1974). The outer double lamella of the exochorion has regularly spaced filaments, variable in length among species, that probably enhance adherence to a substrate by increasing the adhesive surface area (Williams 1974, Goldie 1982). The gelatinous matrix that constitutes the exochorion—probably an acidic mucopolysaccharide—often has bacteria and protozoa embedded in it, and might provide nutrition for the newly hatched larva, which will feed on the matrix (Goldie 1982). Beneath the outer lamella are three uniform layers and a transitional, possibly waxy layer between the exochorion and endochorion (Goldie 1982). When the exochorion is removed with a clearing agent, such as potassium hydroxide, the pigmented shell or endochorion is revealed. The endochorion has been examined at high magnification for only one North American species, a member of the *Simulium vittatum* complex. It has virtually no surface sculpture other than slight dimpling and an undifferentiated micropylar area (Fig. 4.64), which is located at the blunt pole of the egg (Goldie 1982, Adler & Kim 1986). A micropyle also has been observed in the egg of *Simulium innoxium* (Phillips 1890; as *S. pictipes*). Not all black fly eggs, however, have a micropyle, none having been found in the European *Simulium posticatum*, in which sperm possibly enter the egg by enzymatic activity (Williams in Crosskey 1990). The phylogenetic utility of the micropylar area has yet to be explored.

# 5| Cytology

The giant polytene chromosomes of black flies have played a pivotal role in the study of black flies worldwide, driving many of the investigations of the last half of the 20th century. Their study, often referred to as "cytotaxonomy," has become a subdiscipline of simuliidology central to the taxonomy and systematics of the family, with implications for evolutionary processes in general.

## Polytene Chromosomes

Among the hexapods, only the Collembola and Diptera have polytene chromosomes, first discovered in 1881 in the salivary-gland nuclei of larval chironomid midges. Polytene chromosomes were first reported in larval simuliids by Geitler in 1934. By virtue of their enormous size and transverse series of light and dark bands, polytene chromosomes are replete with micromorphological characters seen most easily in stained preparations.

Polytene chromosomes are formed in tissues that grow by cellular enlargement rather than by an increase in cell number. The chromosome duplication of polytenes differs from that of a typical mitotic cycle in a number of respects. The homologues remain paired, the chromosomes do not participate in the mitotic cycle of coiling and uncoiling, sister chromatids remain intimately paired at the end of each replication cycle, and the nuclear membrane remains intact throughout the cycles. As a result, each nucleus appears to exhibit a haploid number of chromosomes, with each of the paired homologues consisting of parallel interphase strands (chromatids). The word "polytene"—from the Greek word meaning "many sinew"—describes the multistranded nature of these chromosomes. The number of strands follows a $2^n$ series where n is the number of replication cycles. In simuliids, the level of polyteny varies from $2^9$ to $2^{11}$ (i.e., 512–2048 strands), depending on factors such as instar (Procunier & Smith 1993).

Polytene chromosomes are best developed in certain larval tissues, especially those with secretory activity, as in the midgut, rectal papillae, silk glands, and Malpighian tubules. In black flies, polytenes typically are amenable to detailed resolution of banding patterns only in the silk glands. Pupal and adult simuliids also have polytene chromosomes, but their potential has been underexploited, largely because they are unworkable or typically do not offer adequate detail. Nonetheless, some success has been managed with polytenes from the Malpighian tubules of pupae and females of sundry Australian species (Bedo 1976) and from females of some members of the African *Simulium damnosum* species complex (Procunier & Post 1986, Procunier 1989). These studies demonstrate that the banding pattern of the polytene chromosomes is conserved between larval and adult tissues. The degree of polytenization in adults is influenced by the physiological state and possibly by the source of the blood meal (Procunier & Post 1986). The conditions under which the larvae develop also influence the extent of polytenization in adults. Well-nourished larvae, for example, produce adults that, if examined shortly after emergence, can yield good polytene chromosomes (Adler 1983). Even polytenes of substandard quality sometimes offer gross karyotypical characters, such as presence of a chromocenter, that permit species identification (Procunier 1984, Procunier & Muro 1994).

## Chromosomal Complement

Nearly all black flies have a haploid number of three chromosomes (Fig. 5.1). The known exceptions, all of which have a haploid number of two chromosomes, are *Cnephia pallipes* (historically misidentified as *C. lapponica*) of the northern Palearctic Region (Procunier 1982b), *Simulium manense* of the Afrotropical Region (Leonhardt in Rothfels 1988), and all members of the widespread subgenus *Eusimulium* (Leonhardt 1985) (Fig. 5.2), of which five species occur in North America. In all of these exceptional cases, two of the three chromosomes fused, presumably facilitated by achiasmate meiosis in males that permitted a shift of centromeres to subterminal positions (Dunbar 1958b, Rothfels & Mason 1975).

Claims of more than three chromosomes are based on the presence of supernumerary chromosomes called "B chromosomes" (Fig. 5.3), which appear in the polytene complement as short, densely staining, usually banded chromosomes, often with a nucleolar organizer. Although absent in most species, B chromosomes are present in varying frequencies in some species. Preparations from larval gonads or neuroblasts are usually necessary to count the B chromosomes, which occur most often in even numbers (Procunier 1975b). In some species of *Cnephia*, B chromosomes are believed to enhance the maturation process by increasing the efficiency of ribosomal RNA synthesis between the sexes (Procunier 1982c). The presence

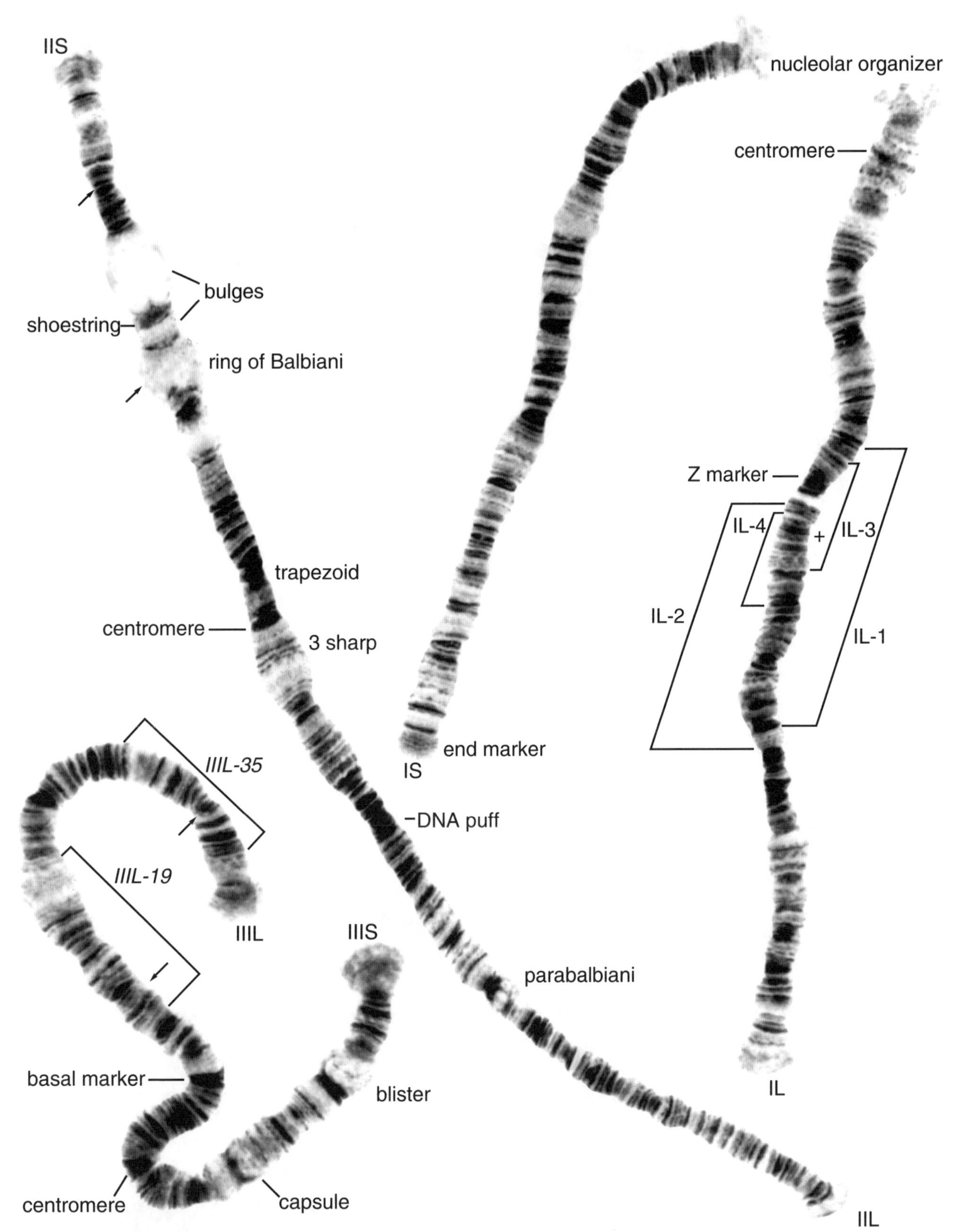


Fig. 5.1. Polytene chromosomes (haploid number = 3) of a female larva of *Simulium moultoni* n. sp.; Magdalen Islands, Quebec. The *IIIL-19* inversion is present and indicated by brackets; the limits of inversions IL-1pu, IL-2pu, IL-3pu, IL-4pu, and *IIIL-35* are shown, although the inversions themselves are not present. Arrows on the IIS and IIIL arms indicate the breakpoints of the IIS-1 and IIIL-1 inversions, respectively, although the inversions themselves are not present. + = location of the thick band (thin condition figured).

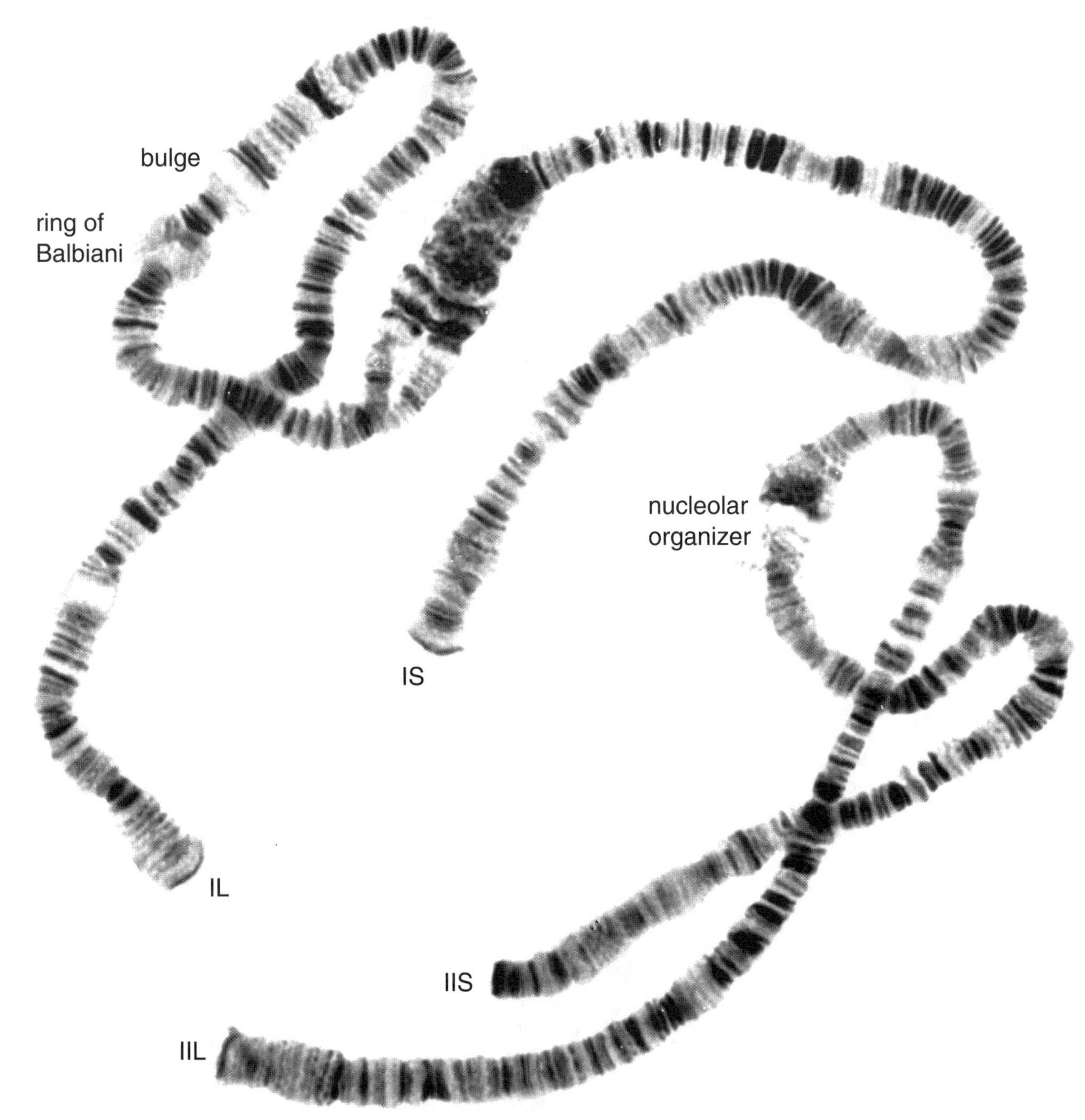


Fig. 5.2. Polytene chromosomes (haploid number = 2) of a male larva of *Simulium pilosum*; Horton River, Northwest Territories.

of B chromosomes has been correlated with water pollution in one Palearctic species (Kachvoryan et al. 1996).

Most of the world's black flies are diploid, the rare exception being about eight triploid species, four of which are found in North America. For the diploid majority, the two chromosomal homologues, one from the male and one from the female parent, are more or less intimately paired. The extent of pairing is often species specific but also has a seasonal component, at least for species of the temperate and boreal regions, with the greatest degree of unpairing in the colder months (Rothfels & Featherston

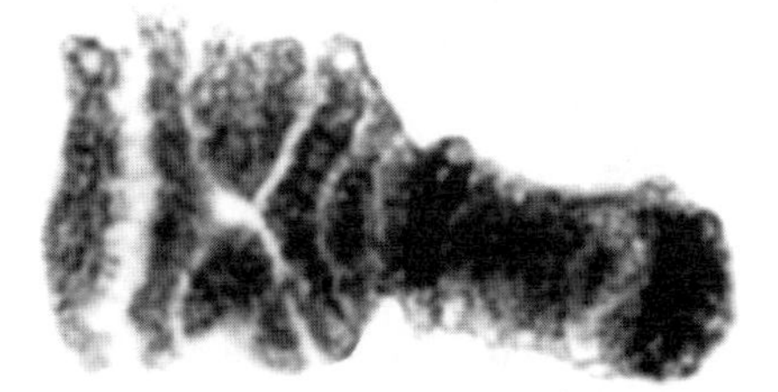

Fig. 5.3. B chromosome of a female larva of *Simulium verecundum* s. s.; Horton River, Northwest Territories.

Fig. 5.4. Polytene chromosomes of an autotriploid female larva of *Simulium rostratum*, showing three homologues in a section of chromosome II (top) and chromosome arm IIIL (bottom); Uranium City, Saskatchewan.

1981). Hybrids between species typically exhibit loose pairing of homologues (Rothfels & Nambiar 1981). Triploid species, which have three homologues, are parthenogenetic and can be of either autotriploid or allotriploid origin. *Stegopterna mutata* is probably of autotriploid origin (Basrur & Rothfels 1959), whereas the other North American triploids (*Gymnopais dichopticoides*, *G. holopticoides*, *Prosimulium ursinum*) are allotriploids, derived as hybrids from two different parental species (Rothfels 1979). Individuals that are autotriploid (Fig. 5.4) can be found in typical diploid species, usually at a frequency of 1 in more than 1500 larvae. Most autotriploids are probably products of fertilized unreduced eggs.

### Chromosomal Rearrangements

Much of the cytotaxonomy of black flies centers around chromosomal rearrangements, which are abundant throughout the family and possibly have been a driving force in speciation. Inversions (i.e., 180-degree reversals of banding sequences) are the most common type of rearrangement and serve to suppress crossing-over (i.e., genetic recombination). They are of two types, namely, interspecific or fixed inversions, so called because they do not occur heterozygously, and intraspecific or floating inversions (i.e., polymorphisms), which can occur heterozygously. Some species (e.g., *Simulium loerchae*, *S. parnassum*, *S. venustum* s. s.) lack floating inversions and are termed "monomorphic," whereas other species are quite polymorphic (e.g., *Prosimulium fuscum*, *Simulium vittatum* s. s., *S. rostratum*). The ecological correlates of chromosomal monomorphy and polymorphy in black flies are poorly known. A single inversion can have diverse fates over evolutionary time, becoming fixed in one line of descent, lost in another, polymorphic in a third, and sex-linked in still another (Rothfels et al. 1978). Most inversions are paracentric, occurring within a chromosomal arm. Fewer than 3% of inversions are pericentric, that is, involving the centromere (Bedo 1977). We have recognized, for example, more than 80 fixed and floating inversions, all paracentric, in the short arm of the second chromosome of the *Simulium jenningsi* species group.

Additional chromosomal rearrangements include deletions, insertions, duplications, band dimorphisms (heterobands, i.e., heteromorphic bands), interchanges (translocations), and chromocenters. Deletions and insertions of individual bands or sections are infrequent. The deletion of a small section from the base of the short arm of the second chromosome and its reinsertion in a similar position in the short arm of the third chromosome of *Simulium innoxium* provides one example (Bedo 1975a; as *S. pictipes* "B"). A heterozygous deletion of the nucleolar organizer from the centromere region of the first chromosome and its subsequent reinsertion in the short arm of the same chromosome has been observed once in *Twinnia hirticornis* (Rothfels & Freeman 1966; as *T. nova*). Duplication of bands has not been documented in the Simuliidae. Band dimorphisms, on the other hand, are quite common and are most conspicuous when the enhanced (thickened) bands occur in the heterozygous state (e.g., Bedo 1975a, Rothfels & Featherston 1981).

Translocations are infrequent. A single mid-arm translocation has been discovered for *Simulium woodi* in Africa (Procunier & Muro 1994). Whole-arm interchanges have been documented for 28 North American taxa: *Twinnia hirticornis*, *Twinnia nova*, *Helodon vernalis*, the 12 species of the subgenus *Helodon*, *Prosimulium transbrachium*, the four eastern species of the *Prosimulium magnum* species group, the seven nominal species in the genus *Metacnephia*, and *Simulium decimatum* (Ottonen 1966, Rothfels 1979, Procunier 1982a, Rothfels & Freeman 1983, Shields 1990; *decimatum* as *nigricoxum*). We have found an additional, and unique, whole-arm translocation in *Prosimulium clandestinum* n. sp. in which the base of the IIIL arm unites with the end of the IIIS arm.

Chromocenters, in which the chromosomal arms radiate from a central heterochromatic mass of varying size, are common and can provide a rapid, species-specific diagnostic aid (Fig. 5.5). They generally are a constant feature of the polytene complement, occurring infrequently as polymorphisms (Brockhouse et al. 1989). Several North American species (e.g., *Prosimulium fulvum*, *P. shewelli*) are polymorphic for a chromocenter. Ectopic pairing, or tight association, of centromeres forms a pseudochromocenter that can resemble a true chromocenter but typically does not occur in all nuclei of the larval silk glands or in all individuals in a population (Rothfels & Freeman 1977). Of the 217 species in North America that have been screened for a chromocenter, 39 (18%) are positive for this feature. The chromocenters of species such as *Simulium chromatinum* n. sp. and *Simulium chromocentrum* n. sp. are enormous and can be observed even in

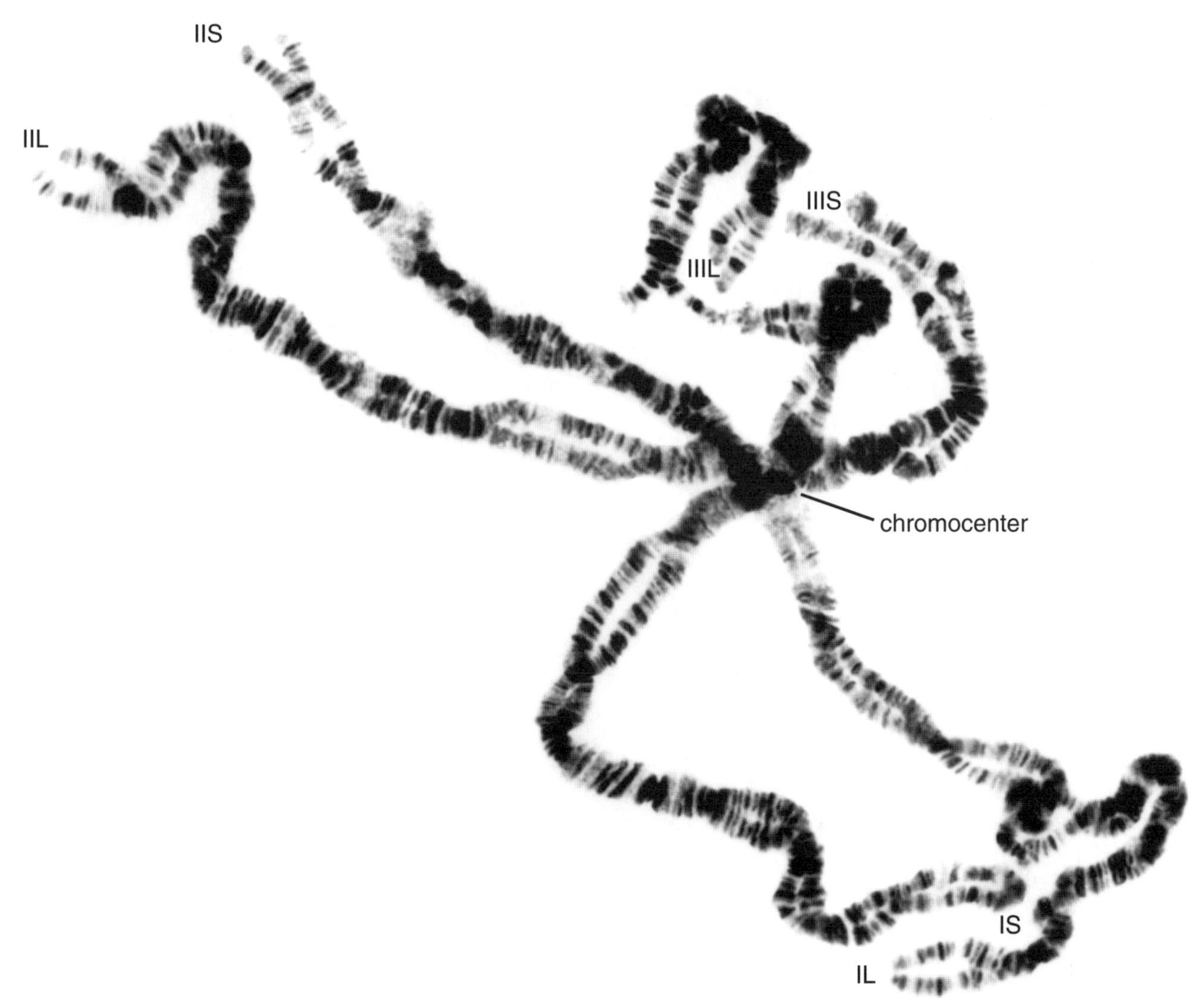


Fig. 5.5. Polytene chromosomes of a male larva of *Simulium arcticum* s. s., showing the chromocenter; Granite County, Montana.

normal somatic nuclei. A partial chromocenter, involving only two of the three chromosomes, is known in two Nearctic species: *Simulium croxtoni* (Hunter & Connolly 1986) and *Prosimulium impostor* (some Arizona populations only).

An important category of rearrangements includes those associated with the sex chromosomes. As in humans, male black flies are the heterogametic sex (XY) and females the homogametic sex (XX), except in a few taxa such as the Polynesian subgenus *Inseliellum* (Rothfels 1989) and one species each in Ecuador (Procunier et al. 1987) and Japan (Hadi et al. 1995), in which females are heterogametic. Any one of the three chromosomes can serve as the sex chromosome. The simplest condition is one in which the X and Y chromosomes cannot be distinguished morphologically in typical microscopic preparations. Undifferentiated sex chromosomes, presumed to be the ancestral condition ($X_0Y_0$) within a lineage, are present in at least some populations of approximately half of the 155 North American species that have been studied sufficiently. Many species, however, have cytologically differentiated sex chromosomes (e.g., $X_1Y_1$, $X_2Y_2$) that can be recognized by any of various rearrangements, especially inversions, but also band dimorphisms, supernumerary bands, heterochromatization of certain regions, and differential expression of the nucleolar organizer. In the heterogametic sex, the sex-differential segment exists in the heterozygous condition, which restricts recombination between the sex chromosomes. A single species can have multiple sex-chromosome rearrangements, as well as undifferentiated sex chromosomes (e.g., Procunier 1982a, McCreadie et al. 1995), although what often is presumed to be a single species with multiple sex chromosomes actually could be a composite of two or more species, each with a unique sex-chromosome sequence. Not all sex linkage, however, is complete. Sex-related rearrangements can be partially linked, and sex exceptions, possibly ancestral relicts or a result of crossing-over, can be found (Rothfels et al. 1978, Rothfels 1980). A special case of partial sex linkage is known as pseudo-partial sex linkage, which involves the coexistence of two structurally identical, inversion-bearing chromosomes, only one of which actually carries the sex locus (Rothfels 1980, Brockhouse & Adler 2002).

Most species differ in their sex chromosomes. Nonhomologous sex chromosomes (i.e., sex-linked rearrangements on different chromosomes) typically are associated with different species. In the *Simulium pictipes* species group, for example, the sex-linked rearrangement in each of the three member species is on a different chromosome (Bedo 1975a), possibly because the sex locus is mobile (Rothfels 1980, Procunier 1989). Sex-linked rearrangements of different species, however, need not be on different chromosomes. Often within a group of closely related species, different sex-related rearrangements on the same chromosomal arm are associated with separate species, as in the *Simulium tuberosum* species complex (Landau 1962) and the *Simulium arcticum* species complex (Fig. 5.6).

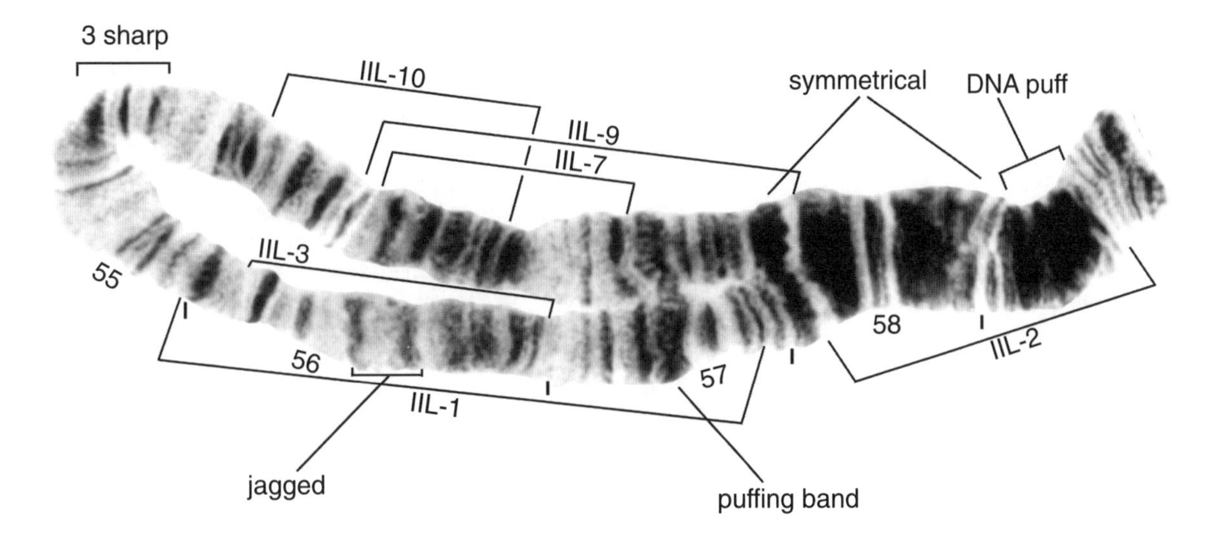


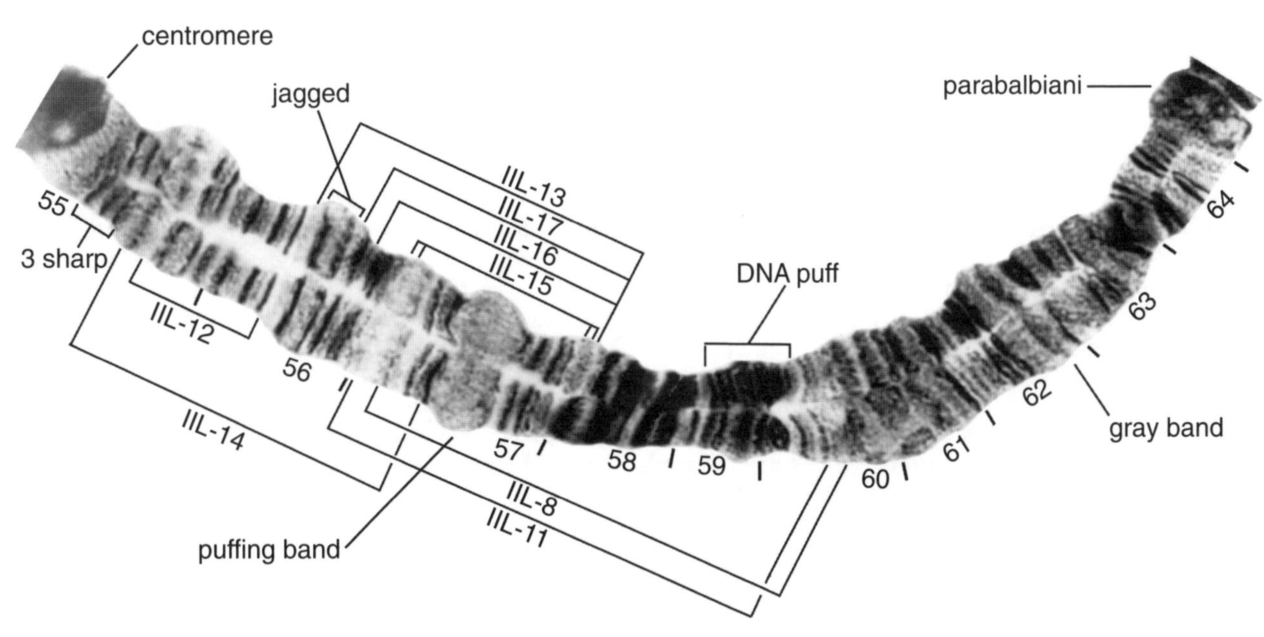


Fig. 5.6. Standard banding sequence for the basal half of the IIL arm of a female larva of the *Simulium arcticum* species complex, showing major landmarks, limits of sections 55–64, and various inversions important in the cytotaxonomy of the complex. Top, Josephine County, Oregon; bottom, Jefferson County, Montana. Numbering of sections follows that of Shields and Procunier (1982). All inversions are typically Y linked except IIL-2, which is X linked; all are indicated as solid (rather than dashed and dotted) brackets for ease of viewing.

Conversely, different sex chromosomes do not necessarily imply different species. A number of species have sex-chromosome polymorphisms in which some individuals carry a particular sex-linked rearrangement, and other individuals carry a different one, although typically on the same chromosomal arm. As many as five sex-chromosome polymorphisms have been found in *Simulium conundrum* n. sp. in an area less than 3500 km$^2$ on Newfoundland's Avalon Peninsula, although the possibility that some of these polymorphisms correspond to different species cannot be excluded (McCreadie et al. 1995, as *tuberosum* 'FGH'). Among the few species that share the same differentiated sex chromosomes are *Prosimulium mixtum* and *Prosimulium transbrachium* (Rothfels & Freeman 1983).

### Nomenclature

Associated with the cytotaxonomy of black flies is a specialized, somewhat universal argot distinct from that used in the cytotaxonomy of mosquitoes, drosophilids, and other Diptera. The current system was first proposed by Rothfels and Dunbar (1953) and further refined and expanded by Basrur (1959), Bedo (1977), Rothfels et al. (1978), and Rothfels (1988). Familiarity with this nomenclature is a prerequisite to understanding the cytotaxonomic literature on black flies.

The three chromosomes of black flies are numbered I, II, and III, from longest to shortest. They account for approximately 42%, 30%, and 28%, respectively, of the total complement (Dunbar 1966). Each is divided into two arms by a submedian centromere (C), which often lies within an expanded region. In the absence of an expanded region, the centromere can be indistinguishable from the remainder of the chromosome bands, as in *Simulium pugetense*. The location of the centromere then would need to be inferred by comparison with the banding sequence of a species whose centromere-band location is known, or by the presence of ectopic pairing of centromere bands. In some species of *Prosimulium*, the expanded centromere region is even more exaggerated than in closely related species and is referred to as a transformed centromere ($C_t$) (Basrur 1959).

The two chromosome arms are referred to as short (S) and long (L). In chromosomes II and III, the difference is readily apparent, whereas in chromosome I, the short arm is negligibly (ca. 2%) shorter. Application of short and long arms is consistent across species, except in the *Simulium vittatum* species complex, in which the designations for the short and long arms of chromosome I are reversed (Rothfels & Featherston 1981) and in species (e.g., members of the subgenus *Eusimulium*) in which fusion of chromosomes (typically II and III) has resulted in a haploid number of two. In species that have experienced whole-arm interchanges, the arm designations are retained (except in the subgenus *Helodon*), allowing recognition that interchanges exist (e.g., IS + IIIL and IL + IIIS in *Simulium decimatum*).

Arm recognition is most easily accomplished through landmarks (e.g., Rothfels et al. 1978, Rothfels 1988), which also provide a useful means of homologizing sections of chromosomes, particularly when banding patterns have been scrambled by one or more inversions. Arms IIS, IIL, and IIIS have universal landmarks such as, *inter alia*, the "ring of Balbiani" (a large permanent puff of distinctive texture) in IIS (Fig. 5.7), the "parabalbiani" and "3 sharp" in IIL, and the "blister" in IIIS. These arms, plus IS, IL, and IIIL, have numerous markers that are more taxonomically restricted, for example, the "basal 3" and "end marker" (i.e., terminal fine bands) in IS; the "Z marker" and "neck" in IL; the "trapezoid" and nearly universal "bulges" (= "double bubble") separated by the "shoestring" in IIS; the "DNA puff," "gray band," "jagged," "puffing band," "sawtooth," and "symmetrical" in IIL; the "cup and saucer" and "basal marker" in IIIL; and the "capsule" in IIIS (Figs. 5.1, 5.6). The nucleolar organizer is a landmark found in all black flies. It forms a single large nucleolus and is a site of heavy ribosomal RNA synthesis that appears as an exploded area somewhere in the stained complement. Its location is species or group specific and, therefore, can provide a simple means of identification. Although nearly always constant within a species, it is highly mobile throughout the family, not being tied to any one arm or location. A secondary nucleolar organizer, in addition to the primary nucleolar organizer, is present in varying frequencies in some species (e.g., *Simulium anatinum*; Rothfels & Golini 1983).

Inversions usually are designated by a number plus the arm in which they are found; for example, IIIL-1 would be the first inversion, arbitrarily named (though often in order of discovery), in the long arm (L) of chromosome III. Some authors also have used letter designations (e.g., IIS-B). Pericentric inversions are designated with the letter P (e.g., IIIP-1), rather than with S or L. Fixed inversions typically are underlined (e.g., IIS-1) or italicized (e.g., *IIS-1*) to distinguish them from floating inversions (e.g., IIS-1), which are sometimes designated by an abbreviated species name (e.g., IIS-1im in *Simulium impar*). Overlapping inversions are designated by commas, indicating order of occurrence (e.g., IIL-5, 8), whereas when one inversion is included within another or when two inversions are in tandem, a period is used (e.g., IL-1.2). Exact breakpoints of inversions are denoted by brackets on chromosomal maps (Figs. 5.6, 5.7). Sex-related inversions often are indicated on the maps by dotted brackets for X-linked inversions or by dashed brackets for Y-linked inversions. The three possible configurations for a particular banding sequence are designated "ss" (homozygous standard sequence), "si" (heterozygous condition), and "ii" (homozygous inverted sequence). Thus, "IL-1 ss" represents the standard homozygous sequence for the section of the chromosome in which the IL-1 inversion could occur. An alternative scheme can be written as St/St, IL-1/St, and IL-1/IL-1 for the same three respective configurations, or, in short form, St/St, 1/St, and 1/1.

A chromosomal map, which can be a drawing or a photograph, provides details of all chromosomal features and rearrangements. It generally is based on a standardized banding sequence—"the standard"—that ideally represents the most central or common sequence for each arm in the taxon under study. Consequently, the standard can

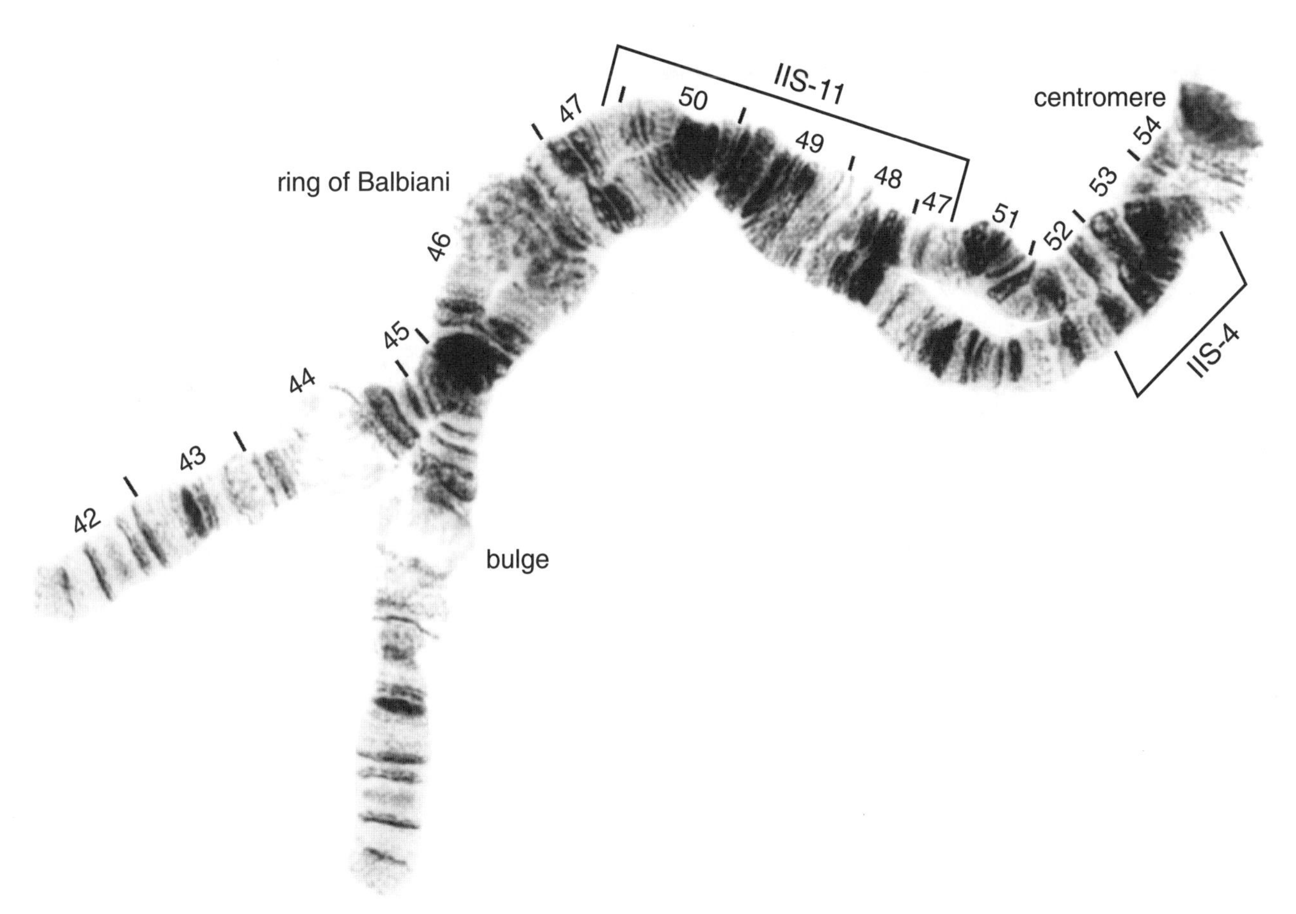


Fig. 5.7. IIS arm of a male larva of *Simulium apricarium* n. sp., showing the IIS-11 sequence and limits of IIS-4; Granite County, Montana. Numbering of sections follows that of Shields and Procunier (1982).

be tentative, pending additional taxon sampling, and does not always represent a real species. The entire complement is divided into 100 approximately equal sections, beginning with the end of IS and continuing through the end of IIIL. Sections can be subdivided such that bands are designated individually. For example, 96A3 is the third band in subsection A of section 96, which would lie in chromosome arm IIIL. Locations within a section are sometimes specified by the letters p, c, and d to indicate proximal, central, and distal, with respect to the direction of section numbers. Chromosomal maps often are complemented by idiograms, or stylized representations of the chromosomes, that show major landmarks and rearrangements (Fig. 5.8).

Because polytene chromosomes permit a high degree of taxonomic resolution, even to the level of population, a number of terms have arisen to describe the various levels of cytologically differentiated taxa. Ideally, we would recognize only species (i.e., reproductively isolated entities) and cytologically differentiated, but nonreproductively isolated, populations. The real world, however, is not so orderly, particularly because we are viewing it in a slice of evolutionary time. The speciation process, therefore, might not yet be complete in some cases, making taxonomic assignment of some populations problematic. At the finest level of differentiation are the cytotypes, populations that are not reproductively isolated but are nonetheless cytologically distinct. Often it is not possible to determine from cytological evidence alone whether a population is reproductively isolated, and so the term "cytotype" should be applied liberally and provisionally, pending further evidence. The term "cytoform" has been used whenever a cytological entity, whether reproductively isolated or not, does not carry a formal name (Crosskey & Howard 1997).

When reproductive isolation of morphologically similar or identical taxonomic entities can be demonstrated cytologically, the term "cytospecies" often is used. Cytospecies are special (i.e., chromosomally demonstrated) cases of "sibling species," or cryptic species, which are defined as structurally similar or indistinguishable species (i.e., isomorphic species). If species are morphologically inseparable and have the same chromosomal banding sequence, they are referred to as "homosequential sibling species," the logical extension of "homosequential species," which are morphologically distinguishable but carry the same chromosomal banding sequence. Two or more sibling species constitute a species complex. Once morphological discriminators are discovered and formal names are applied, sibling species are generally referred to simply as

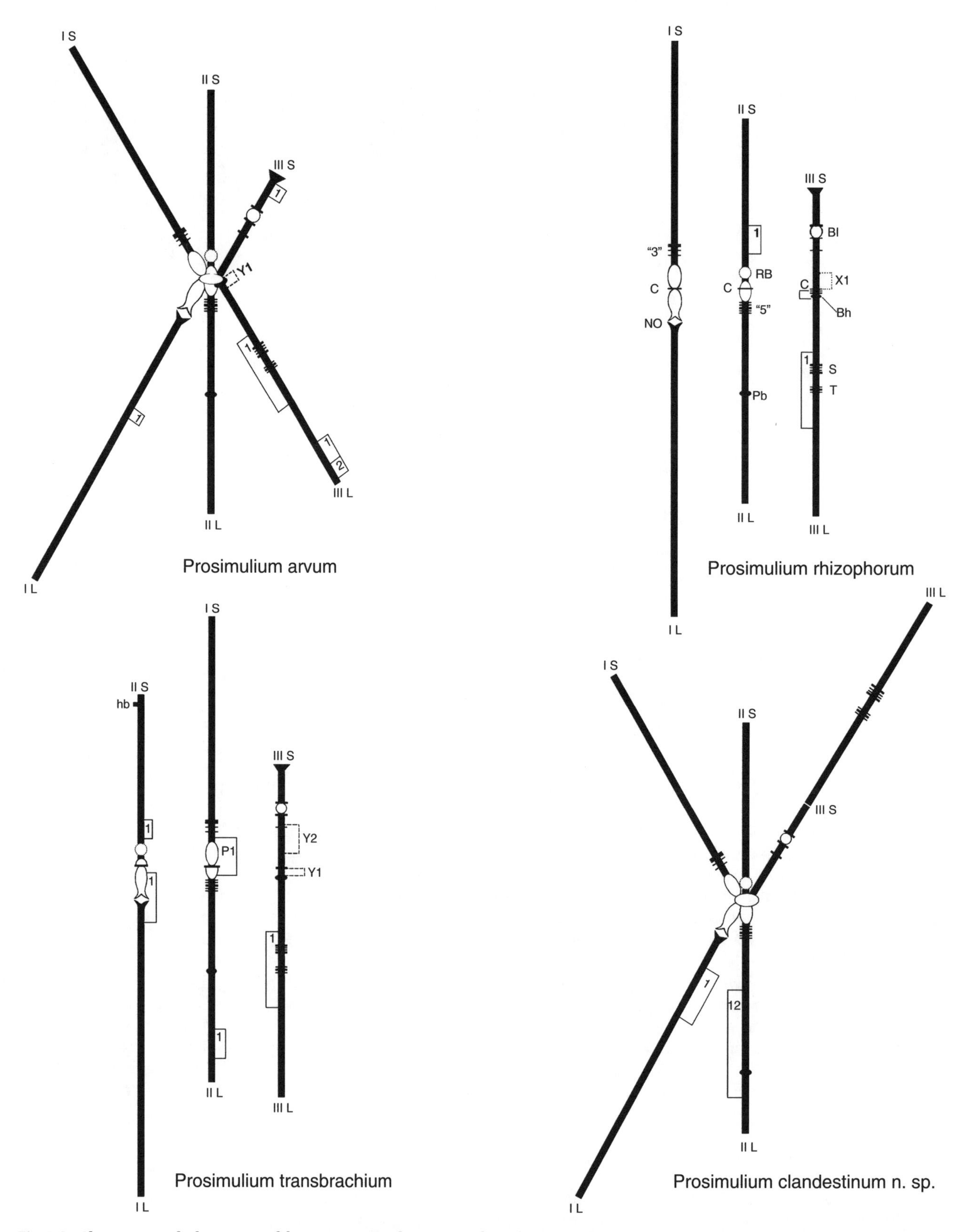


Fig. 5.8. Chromosomal idiograms of four species in the *Prosimulium hirtipes* group. *Prosimulium arvum* and *P. clandestinum* n. sp. are chromocentric; *P. clandestinum* n. sp. has a translocated IIIL arm; *P. rhizophorum* has standard arm associations; *P. transbrachium* has a whole-arm interchange. Bh = basal heterochromatin; Bl = blister; C = centromere; hb = heteroband; NO = nucleolar organizer; Pb = parabalbiani; P1 = pericentric inversion; RB = ring of Balbiani; S = shield; T = triad; "3" = group of 3 bands; "5" = group of 5 bands. Brackets to the left and right indicate fixed and floating inversions, respectively; dotted and dashed brackets denote inversions linked to the X and Y chromosomes, respectively.

"species," and they become formal members of "species groups," which are monophyletic assemblages of morphologically distinct species and any morphologically unresolved sibling species. The term "species complex" also is sometimes applied, and can be useful, as a label in routine identification for morphologically similar, chromosomally distinct, species that cannot be identified easily to a particular species, given the material at hand (e.g., *Simulium vittatum* species complex). An infrequently used term is "supercomplex," which refers to two or more similar species complexes (e.g., *Simulium venustum/verecundum* supercomplex).

### Utility

One of the most potent applications of cytotaxonomy has been the elucidation of sibling species through fixed-inversion differences, often additionally supported by unique sex chromosomes and autosomal polymorphism spectra (Rothfels 1956). At the heart of the recognition of these sibling species is the idea of reproductive isolation, typically inferred by an absence of inversion heterozygotes when two opposite banding sequences are present sympatrically in a significant number of individuals. By the same token, the presence of heterozygotes in a population of two valid species can be taken as evidence of hybridization (e.g., Rothfels & Nambiar 1981, Boakye et al. 2000).

As with any methodology, cytotaxonomy has its limitations and, whenever possible, should be used in conjunction with other approaches, including morphotaxonomy, molecular taxonomy, and ecology. One of the primary limitations, suffered equally by all approaches, is that imposed by allopatry. Even fixed-rearrangement differences do not signify different species if the populations under evaluation are allopatric. Ideally, reproductive isolation is best demonstrated when fixed rearrangement differences are found at the same site or in the same stream, the level of confidence in conferring species status dwindling with distance between sites. We argue that if allopatric populations cannot be differentiated cytologically, morphologically, and ecologically, they are best considered conspecific. This approach emphasizes similarities, not differences, among populations, including those in different zoogeographic regions. Ultimately, pragmatism drives species recognition in allopatric situations. Restraint should be exercised in formally naming allopatric populations with slight differences.

A potential limitation of cytotaxonomy involves populations that differ only in their Y chromosomes, or the more extreme cases in which populations differ only in the frequency of autosomal polymorphisms or do not differ at all in their chromosomes but, yet, might be different species. In these cases, reproductive isolation cannot be demonstrated by a lack of heterozygotes. Nonetheless, linkage disequilibria and heterozygote deficiencies of shared polymorphisms (typically inversions) have been used to argue compellingly for species status (Bedo 1979, Henderson 1986b). Heterozygote deficiencies, representing what is sometimes called a "Wahlund effect," are often recognized by significant deviations from Hardy-Weinberg equilibrium. The demonstration of homosequential sibling species indicates that speciation and evolution can occur without apparent chromosomal change, plausibly by gene mutations. Discovery and recognition of these species are the challenges of the taxonomist. We anticipate that many species remain to be discovered in the realm of sex-chromosome polymorphisms and homosequential sibling species.

Phylogenetic reconstruction of black flies has benefitted enormously from information gleaned from polytene chromosomes. Because chromosomal rearrangements, especially inversions, are assumed to be unique events, the existence of shared rearrangements, relative to the standard, is taken as evidence of common ancestry (Rothfels 1979). The more closely related the taxa, the more likely that chromosomal information will be useful in inferring relationships. The more distantly related the taxa, the greater the number of inversions separating them, making correct resolution of the scrambled complement extraordinarily difficult. Chromosomal resolution of distant taxa, however, is sometimes possible through intermediates.

Several caveats should be mentioned in the context of using chromosomal information in phylogenetic reconstruction. First, most reconstructions of evolutionary relationships are not truly cytophylogenies, but rather cytodendrograms, because they are not rooted; instead they are based on the most common arm sequences within the group under study. This problem can be overcome by using morphological or molecular criteria to root the phylogeny. Second, relationships should not be inferred simply on the basis of isolated similarities, such as the same location of the nucleolar organizer or the same banding sequence in one or a few arms, without regard to the entire complement. Third, relationships are not necessarily a function of the number of rearrangement differences; only shared rearrangements can demonstrate relationship. Finally, mimic inversions that differ by only one or a few bands, especially in areas of breakage hot spots (Landau 1962), can produce erroneous relationships if not recognized.

Polytene chromosomes also provide a potential tool for evaluating biological phenomena. Site-specific profiles of autosomal polymorphisms, first noted for *Simulium tuberosum* (Landau 1962), suggest that females return to natal waterways to breed (Rothfels 1981b) and that movement by ovipositing females occurs more readily along the same stream than between neighboring streams (Bedo 1975a). At least for the *Simulium venustum/verecundum* supercomplex, however, females do not tend to oviposit predominantly at natal sites (Hunter & Jain 2000). The phenomenon, nonetheless, is probably species specific and should be investigated in additional species. Differences in inversion profiles also might indicate some regional or localized selective value of the sequences. Nonetheless, a link between an inversion and a biological phenomenon has rarely been demonstrated. One exception is the Palearctic *Simulium erythrocephalum* in which a sex-linked inversion is correlated with faster larval development of males (Post 1985).

The impacts of environmental insults and the resultant changes in populations over time can be detected and monitored by chromosomal studies. The observation that heavy metals influence chromosomal features (e.g., Sanderson et al. 1982) suggests the utility of polytene chromosomes in studies of water quality. Different polymorphism spectra between populations separated by more than 10 years in Ontario's Chalk River have been attributed to insecticidal eradication of the original population followed by repopulation by individuals with different chromosomal features (Golini & Rothfels 1984). The presence of a particular inversion in an African species has been used as a marker to monitor the spread of insecticide resistance (Meredith et al. 1986).

An important conceptual matter that has surfaced through the study of polytene chromosomes concerns the mode(s) of speciation responsible for the plethora of black fly species. These chromosomal studies have demonstrated a consistent pattern, namely, that the most closely related species are nearly always sympatric (although sympatry facilitated their discovery), related species share polymorphisms, and different sex-chromosome systems characterize most species and are often the only chromosomal differences (Rothfels 1989). These patterns suggest a common process: sympatric speciation through coadaptation of polymorphic sex chromosomes in pairs reinforced by assortative mating (Rothfels 1989). Aside from the demonstrated sympatric origin of allotriploid species, this model provides evidence that other forms of sympatric speciation are at least as reasonable as allopatric speciation. Which mode of speciation is correct in which situations remains a vexing question. Regardless of the mode of speciation, considerable evidence suggests that sex-chromosome systems play a role in the speciation process (Feraday et al. 1989, Leonhardt & Feraday 1989, Rothfels 1989). Additional evidence suggests that hybridization, in some cases, also might lead to speciation or at least to the formation of new cytotypes (Boakye et al. 2000).

# 6 | Behavior and Ecology

Perhaps more has been written about the ecology and natural history than any other aspect of black flies. Our treatment, therefore, is not intended to be an exhaustive review. Rather, we provide salient information and general patterns while indicating areas that require further study. We emphasize the North American literature but draw amply on extralimital works. We refer the reader to the comprehensive global treatment of natural history by Crosskey (1990) for additional information on general biology.

## Distribution and Habitat

Black flies are intimately bound to their ancestral habitat—flowing freshwater—although reasons for this virtually exceptionless association are still poorly understood. Possibly the tie to running water is for respiratory reasons or because the primary larval feeding mode, filtering, requires delivery of food. Alone, neither explanation is satisfactory, for even the fanless, nonfiltering genera such as *Twinnia* and *Gymnopais* are restricted to flowing water, and larvae of many species can live in standing water for long periods of time. For all of North America, we have only one collection of black flies from a habitat other than running water. Larvae and pupae of the hardy *Simulium vittatum* species complex were collected once from the wave-washed shore of Lake Ontario, New York.

Black flies have exploited nearly all flowing freshwater habitats, from the ordinary to the more novel, the latter including sulfur springs, colossal rivers, hot springs, glacial meltwaters, swamp flows, seepages, impoundment outflows, crashing waterfalls, intermittent streams, and subterranean flows. Most flowing waters harbor black flies at some time during the year (Adler & McCreadie 1997). A stream site at any one time typically has about 3 or 4 species, rarely more than 9 or 10.

Each species of black fly is associated with a particular habitat and geographic area, and each has a distribution that generally is predictable on the basis of stream and landscape features (Corkum & Currie 1987, Burger 1988, Ciborowski & Adler 1990, McCreadie & Adler 1998). Predictors of distribution depend on a spatial scale ranging from a specific substrate to a zoogeographic region (Adler & McCreadie 1997, McCreadie & Adler 1998). Some of the factors, both abiotic and biotic, that influence simuliid distributions also affect their population dynamics (Ross & Merritt 1988).

Distributions at the finest scale—microhabitat (e.g., a rock)—are often represented by conspicuous aggregation patterns such as bands, clumps, and evenly spaced distributions that are characteristic of particular taxa but also vary with larval age and environmental factors. These distribution patterns (as well as densities) are influenced by microhydraulic factors, size and surface features of the substrate, food availability (Cressa 1977; Eymann 1985, 1991a, 1991c; Chance & Craig 1986; Ciborowski & Craig 1989; Beckett 1992; Lacoursière 1992), settlement rates from the drift (Fonseca 1996, Fonseca & Hart 2001), disturbance events (Love & Bailey 1992), and the presence of other organisms (including conspecifics) that might act as predators, facilitators, or competitors. Larvae of some species, for example, are territorial, with food typically the contested resource, whereas larvae of other species, especially those living at the outflows of impoundments where food is abundant, are not territorial (Harding & Colbo 1981; Hart 1986, 1987, 1988). Competition is apparently common with other benthic macroinvertebrates (McAuliffe 1984; Hemphill 1988, 1991; Hershey & Hiltner 1988; Dudley et al. 1990; Morin 1991; Kohler 1992), as well as with macroalgae (Dudley et al. 1986). Most studies indicate that fewer larvae occur on substrates with accumulated periphyton and debris than on cleaner surfaces, possibly because of the greater difficulty in attaching the silk pads (Barr 1982). A few North American studies, however, have shown a positive relation between the density of larval black flies and the amount of periphyton on the substrates (Pachón 1999, Ellsworth 2000). The interactions between larval simuliids and periphyton deserve closer study because the latter can serve as both a food source and a competitor (Morin 1987a). Remnants of other organisms, especially silk pads of larval black flies, also can influence microdistributions (Kiel et al. 1998a, 1998b).

At larger scales, such as within a stretch of flow up to 20 m, distributions are often correlated with substrate type (e.g., vegetation or rocks), velocity, and depth (Morin et al. 1986, McCreadie & Colbo 1993b, Adler & McCreadie 1997). Additionally, the presence of other organisms can influence distributions, as well as population levels within sections of a stream. For example, the spawning activities of salmon (e.g., redd excavation) and the subsequent death and decay of the fish can significantly influence simuliid distributions and biomass (Minakawa 1997). In longer reaches, physicochemical gradients and the presence of impoundment outflows are good predictors of distributions (Adler & Kim 1984, McCreadie & Colbo 1992). Densities of many species decrease markedly with distance downstream from an outlet (Sheldon & Oswood 1977, Wotton 1988a, Adler & McCreadie 1997), unless the impounded water is highly acidified (Curry & Powles 1991). The greatest estimates of secondary production for any stream macroinvertebrate are those for black flies below lake outlets (Wotton 1988b). Stream size, particularly width, and impoundment outfalls are excellent predictors of species distributions among streams in the Nearctic Region (McCreadie & Colbo 1991b, McCreadie & Adler 1998) as well as in other zoogeographic regions (Malmqvist et al. 1999, Hamada et al. 2002), suggesting some universality to the factors underlying distributions of the aquatic stages. At a still larger scale, species distributions often correspond with boundaries of ecoregions (i.e., areas of similar abiotic and biotic landscape). For example, the Fall Line, which represents the transition between the hard piedmont bedrock and the softer sediments of the Sandhills Region of the southeastern United States, sharply defines the distributional boundary of a number of species (e.g., *Simulium slossonae*, Map 231), even though no barrier to dispersal exists (McCreadie & Adler 1998).

Geographic areas with significant altitudinal variation offer a wide diversity of habitats and are richest in species, whereas areas with minimal relief have far fewer species. The Sierra Nevada (80,255 $km^2$) of California, for example, has about 63 species, whereas the five central states (North Dakota to Oklahoma) of the Great Plains, although 12 times larger in area, have only about 28. Species richness in the Nearctic Region varies with latitude: 89 species between 60° and 70° north, 111 between 50° and 60° north, 181 between 40° and 50° north, and 166 (plus no more than one or two additional species from the included area of Mexico) between 30° and 40° north. Further taxonomic effort is needed in other zoogeographic regions before global trends can be identified. Species richness, however, appears to be no greater in the tropics than in the temperate regions, perhaps because of the rather homogeneous nature of lotic habitats, relative to terrestrial habitats, and of the constraints imposed by flowing water. Investigations of the factors that influence species richness, both at local and at regional levels, are needed.

About 78% of the North American fauna north of Mexico is endemic. Twenty-five species (10%) in our fauna also have been found in Mexico and more are expected, but only five (2%) are shared with the Neotropical Region (i.e., tropical Mexico and the Carribean southward). Our fauna has a more intimate connection with the Palearctic Region (Currie 1997a, Adler et al. 1999, Currie et al. 2002). About 32 species (13%) are Holarctic, occurring in both regions. The number of species shared with the Palearctic Region increases with latitude in North America: 3 species between 30° and 40° north, 9 between 40° and 50° north, 19 between 50° and 60° north, and 32 between 60° and 70° north.

The causal factors that underlie the macrodistributions of black flies remain generally unknown. The testing of hypotheses that seek to explain the underlying reasons for these distributions represents an important frontier of investigation. Distributions of preimaginal black flies depend ultimately on the choice of oviposition sites by females, and both stream size and impoundment outfalls afford prominent visual features that might be used as cues by ovipositing females. At ecoregion boundaries, stream types (e.g., rocky versus sandy) and vegetation often change, and ovipositing females presumably perceive and respond to these changes. Mechanisms of large-scale site selection, for example, between streams of different sizes or between different ecoregions, nonetheless, remain virtually unexplored. Females presumably choose habitats in which structure and function of the immature stages are optimized. The characteristics of labral fans of larvae, for example, are strongly correlated with habitat characteristics (Malmqvist et al. 1999) and have been used to predict the distribution of simuliid species within a catchment (Palmer & Craig 2000). Studies of individual fitness under differing stream conditions are much needed and could provide insight into the causal factors that determine species distributions.

The distribution of a species is influenced by both contemporary and historical factors. Human influences probably have had a greater impact on simuliid distributions than any other factor since the end of the Wisconsinan glacial period about 10,000 years ago, and they continue to exert powerful effects, expanding and compressing distributions.

Logging, farming, urbanization, road construction, industrialization, groundwater withdrawals, impoundments, and other forms of habitat alteration affect distributions at spatial scales ranging from sections of streams to ecoregions. Many rivers and streams have been vastly altered by channelization, pollution, clearance of debris jams, and disturbance of riparian plant communities. Free-running rivers no longer exist in most of the contiguous United States. Extirpations and subsequent recoveries of beavers (*Castor canadensis*), the great landscape engineers of the natural world, undoubtedly have influenced simuliid distributions, with the outflows of their impoundments providing food-rich habitats for black flies. Human-built impoundments that now dot the landscape probably have expanded the ranges and abundances of a few species, namely, *Simulium decorum* and *Simulium tribulatum*. The impact of human disturbances, such as impoundments, siltation, and various forms of pollution, is reflected in the tolerances of different species. Black flies, therefore, can be useful indicators of water quality (Hilsenhoff 1982, 1987), a point perhaps first articulated by Paine and Gaufin (1956). Only one North American species (*S. tribulatum*) is capable of living in flows with heavy organic pollution (Adler & Kim 1984; as *S. vittatum* 'IIIL-1').

The impact of introduced and extirpated host species in recent times is unknown but they have undoubtedly influenced simuliid distributions. Once-abundant hosts such as the passenger pigeon (*Ectopistes migratorius*) have been driven to extinction, and others such as the American bison (*Bison bison*) have been decimated and their distributions reduced to a fraction of their former ranges. New hosts, such as cattle, have been introduced throughout much of the continent and now serve as blood sources for many species.

We are aware of no simuliid species that have become extinct or endangered in modern times. Some species of *Parasimulium*, which have specialized habitats and limited distributions, however, might be susceptible (Courtney 1993). Lacking biting mouthparts, these species are reasonable candidates for conservation. Even some of the apparently rare and localized biting species, such as *Simulium infernale* n. sp., might warrant conservation, certainly a novel concept for black flies. Despite the excellent colonizing abilities of black flies at local levels, no evidence exists that any species currently residing in North America has been introduced from extralimital areas, which is consistent with the lack of adventives anywhere in the world, except perhaps *Simulium bipunctatum* in the Galapagos Islands (Abedraabo et al. 1993).

Fig. 6.1. Oviposition on trailing vegetation by females of *Simulium vittatum*; Grand River, Ontario. Photograph by S. A. Marshall.

### Oviposition Behavior

Most black flies practice one of two oviposition strategies, depositing eggs either while in flight or while landed. Some species incorporate both strategies in their repertoire (e.g., *Simulium vittatum*; Davies & Peterson 1956), sometimes changing their strategy with season (e.g., *Simulium verecundum*; McCreadie & Colbo 1991b). Oviposition in nature, however, has been reported for only about 44 species, or 17%, of the North American fauna. Oviposition by species of the most ancestral genera, *Parasimulium* and *Gymnopais*, has not been observed in nature but is probably unique in both lineages. By virtue of the subterranean habitat of their larvae, females of *Parasimulium* probably crawl into seepages to oviposit. Females of *Gymnopais* either are unable to fly or show little inclination to do so and probably, by default, oviposit while standing on a substrate. A more bizarre strategy occurs in females in some far-northern populations of parthenogenetic species such as *Gymnopais dichopticoides* and *Prosimulium ursinum*. These species, the ultimate arctic insects, have been emancipated not only from mating and blood feeding but also, at times, from oviposition, their bodies deteriorating within the pupae and releasing viable eggs directly into the stream (Carlsson 1962, Currie 1997a).

The most common practice, though recorded in only about 27 North American species, is probably the deposition of eggs during flight. While the female flies upstream or drops repeatedly from a hovering position, it expels one to several eggs into the water or dabs them onto a wetted substrate. Eggs typically descend to the sediments, but in the case of the egg dabbers (*sensu* Crosskey 1990), such as certain members of the subgenus *Hemicnetha*, thousands of flies sometimes assemble to deposit eggs in a tightly circumscribed area, forming enormous egg masses on rocks and other wetted substrates. *Prosimulium caudatum* and *Prosimulium dicum*, which deposit eggs in tight clusters on stones in the splash zone (Speir 1975), also might be egg dabbers, but whether this habit is typical and whether the females land or repeatedly descend from a hovering flight to oviposit has not been reported.

The habit of laying eggs while stationed on a substrate, particularly trailing vegetation or rocks and sticks in flowing water, has been confirmed for about 17 North American species, mostly members of the genus *Simulium* (Fig. 6.1). Oviposition while situated on a substrate has evolved independently as the primary strategy at least five times in North American black flies (subgenus *Eusimulium*, subgenus *Psilozia*, some species of subgenus *Hemicnetha*, *Simulium noelleri* species group, *Simulium verecundum* species complex). The frequency of oviposition while landed is unknown for certain species of the genera *Prosimulium* and *Cnephia* and the subgenus *Aspathia*. Females of species in the subgenus *Eusimulium* lay their eggs on end in regular masses (DeFoliart 1951a), whereas those of the other species tend to place the eggs on their sides in long looping strings or irregular masses.

Most species that lay their eggs in masses typically do so communally. Freshly laid masses of eggs of some European and African species act as a stimulus to oviposition by emitting an oviposition pheromone (Coupland 1992, McCall et al. 1997). However, eggs that become covered by the sticky matrix of other eggs often fail to hatch (Imhof & Smith 1979), probably suffering oxygen deprivation. The advantage of mass oviposition might be that an individual female reduces the probability of its eggs (or neonates) succumbing to predation or desiccation, although in some cases, suitable substrate simply might be limiting.

Eggs generally cannot withstand desiccation, and those of some species succumb to drying in a few hours (Abdelnur 1968, Imhof & Smith 1979). All North American species for which oviposition is known deposit their eggs directly in water or in the wash and splash zones. The eggs of many species, nonetheless, experience periods during which the streambeds are dry. They presumably are protected in moist recesses of the stream bottom, although a portion of the eggs might succumb to desiccation. Terrestrial oviposition has been observed for several Palearctic species. Four Palearctic species of *Prosimulium* oviposit while crawling in moist streamside mosses, producing communal clumps of up to 56 million eggs, or about 20,000 eggs/cm$^2$ (Zwick & Zwick 1990, Baba & Takaoka 1991). The European *Simulium posticatum*, a member of the *Simulium venustum* species group, oviposits by crawling into semidry but humid cracks of riverbanks (Ladle et al. 1985). *Simulium noguerai*, a Brazilian species in the subgenus *Inaequalium*, deposits its eggs in communal masses on riparian vegetation of waterfalls up to 2.1 m above the stream surface (Moreira & Sato 1996). The closest approximation to terrestrial oviposition in a North American species is found in a member of the *Prosimulium hirtipes* species group in which many females were observed ovipositing among fine roots moistened by spray from a stream (Stone & Jamnback 1955).

The time of oviposition is mediated by light intensity and generally occurs toward the end of daylight (Imhof & Smith 1979) or sometimes in the morning (Corbet 1967) or under overcast skies. Cues used in finding a suitable site for oviposition, other than color (Peschken & Thorsteinson 1965, Golini & Davies 1988) and perhaps reflectance (Bellec 1976), are poorly studied but probably are arranged in a hierarchical sequence involving vision and possibly olfaction, as in host-finding behavior. Chemical cues are little studied, and the possibility of acoustical cues (e.g., purling water) has been ignored completely.

Actual fecundity (i.e., the number of mature eggs) in a single ovarian cycle varies from as few as 20 eggs in far northern species such as *Gymnopais holopticoides* (Downes 1965; as *Gymnopais* sp.) to more than 800 in some temperate species (Davies & Peterson 1956). Most species produce an average of 150–600 eggs/cycle (Davies & Peterson 1956, Abdelnur 1968, Pascuzzo 1976). Fecundity within a species typically is correlated positively with larval nutrition, water temperature, female size (Colbo & Porter 1981, Gíslason & Jóhannsson 1991), and size of the blood meal (Klowden & Lea 1979, Simmons 1985). For most species, the number of ovarian cycles is unknown. Females of some species, however, might undergo up to six cycles (Mokry in Wenk 1981), feeding on blood before each batch of eggs is matured and potentially producing more than 1500 eggs in a lifetime. Parous flies, those that have laid their eggs, can be distinguished from nulliparous flies, which have not yet oviposited, by ovariolar structure (Fredeen 1964a, Gryaznov 1995a), as well as factors such as the appearance of the Malpighian tubules and the amount of fat body (Smith & Hayton 1995).

### Development of Immature Stages

The most important factor influencing development is undoubtedly temperature (Merritt et al. 1982, Ross & Merritt 1988). Species that pass the warm summer months as eggs and develop during the cooler months often spend nearly a year from the time the eggs are laid until the adults emerge; those that develop in midsummer can complete their life cycle in less than 2 weeks. Eggs of some species hatch in about 4 days at 25°C, but when stored at temperatures slightly above freezing, at least some eggs remain viable for 2.5 years (Fredeen 1959b). Perhaps even longer periods of egg viability can be expected for species adapted to extreme northern or arid regions. Factors such as oxygen tension and photoperiod also might affect developmental rates in the egg and influence initiation of hatching.

Each species has a range of water temperature that maximizes individual fitness. Increasing the temperature above this optimum increases mortality, although it also reduces developmental time. A temperature increase from 17°C to 27°C, for example, can reduce developmental time by more than half in the *Simulium vittatum* species complex (Becker 1973). The trade-off, however, is a decrease in size, which carries with it a reduction in fecundity (Colbo & Porter 1981). At 25°C, for example, the *S. vittatum* species complex requires 10 times as much food to produce a fly as large and fecund as one at 20°C (Colbo & Porter 1981). Decreasing the water temperature below the optimum reduces survival and eventually arrests development, the lower limit varying with species. The *Simulium verecundum* species complex, for example, begins to suffer severe mortality below 10°C (Mokry 1976b [as *venustum*], McCreadie & Colbo 1991a), whereas development of *Prosimulium approximatum* is not curtailed until the temperature drops below 4°C (Mansingh & Steele 1973; as *P. mysticum*).

Species that have overcome the problems associated with increased temperatures, both as immatures and as adults, have the potential to develop year-round, producing multiple generations (multivoltinism). As many as seven generations have been reported for species in the southern United States (Stone & Snoddy 1969). Multivoltinism is confined to the genus *Simulium* and is most prominent in the more derived lineages of the genus. Even some single-generation, or univoltine, species of *Simulium*, such as *S. truncatum*, that rarely experience elevated temperatures have the ability to develop at temperatures as high as 30°C, albeit with reduced survival (McCreadie & Colbo 1991a; *truncatum* as 'EFG/C' cytotype). Approximately 37% of the species in North America are typically multivoltine. The remainder are univoltine and are represented in higher proportions at more northern latitudes and higher elevations where the flight season for adults is abbreviated. Between latitudes 60° and 70° north, nearly two thirds of the species are univoltine, whereas between 30° and 40° north, about half are univoltine. Multivoltine species generally have broader geographic ranges, perhaps owing to more time (i.e., more generations) historically for dispersal, greater temperature tolerances, and better means of dealing with problems of adult desiccation.

Black flies have a modal number of seven instars (e.g., Ross & Craig 1979, Fredeen 1981b), although even within a species, the number can vary from 7 (or fewer) to 11 (Colbo 1989), depending on environmental factors such as food availability and parasitism. Determinations of the number of instars frequently are made using histograms and Dyar's rule, Brooks's rule, or derivations of these (Craig 1975), but the assumptions often are not met because, for example, growth sometimes is not geometric through all instars and sexual dimorphism occurs (Fredeen 1976, Brenner et al. 1981, Bidlack 1997). Survival estimates for each instar are lacking, although about 10% of the larvae of a species of *Prosimulium* were estimated to survive to pupation in a stream in southern Quebec (Morin et al. 1987).

The final instar physiologically begins its molt to the pupa when the gill histoblast is still white (Hinton 1958). From the time that the larval cuticle separates from the epidermis, a process called "apolysis," until the old larval cuticle is shed (ecdysis), the pupa actually remains hidden in the larval skin, and is referred to as a "pharate pupa." The pharate pupa retains the larval appearance, continues to feed, locates a suitable pupation site, cleans the pupation area by scraping, and eventually spins the cocoon in a series of three to six distinct stages, after which it sheds the larval cuticle (Stuart & Hunter 1995, 1998a, 1998b). Factors influencing the choice of pupation sites are insufficiently known, but the dynamics of flow are probably important. Species that make loose, saclike cocoons generally spin where two surfaces form an angle (Stuart 1995). Pupal development typically requires a few days to a few weeks, depending largely on temperature. During this time, the larval tissues are lysed and the adult begins to form from tissue buds, or histoblasts, some of which (e.g., the leg histoblasts) are visible on the sides of the developing larva. Once development is complete, the pharate adult releases air from its respiratory system until the pressure splits the pupal cuticle along the ecdysial line, through which the adult will emerge.

### Larval Feeding Behavior

In addition to temperature, food strongly influences development, and the majority of larval life probably is spent feeding. After larvae hatch, they spin their silk attachment pads on a substrate and begin to feed; however, behavioral specifics and durations of these early events are poorly known. Larvae have been observed to feed on the gelatinous matrix of their eggs after eclosion (Goldie 1982). The dispersal rate of neonates from eggs deposited in masses is density dependent and possibly is related to reduced feeding rates through competition with neighboring neonates (Fonseca & Hart 1996).

Most of the large body of literature on feeding behavior has concentrated on the ancestral mode, filtering. By virtue of their filter-feeding mode, larvae have been termed "ecosystem engineers," removing fine particulate and dissolved matter from the water column and egesting larger pellets that can be ingested (coprophagy) and used by the benthic microbial and invertebrate community, including other black flies (Wotton 1980, Wotton et al. 1998). In some northern Palearctic rivers, the daily transport of larval fecal pellets past a site can reach an astonishing 429 tons of dry mass (Malmqvist et al. 2001). In other stream ecosystems, including those in North America, black flies likely do not regulate the downstream availability of particulate organic matter, probably accounting for no more than 11% of its deposition (Monaghan et al. 2001).

Far less attention has been given to grazing (or scraping), predation, deposit feeding, and engulfing algal filaments, each of which can be substantial even in species with fully developed fans (Burton 1973, Currie & Craig 1988, Miller et al. 1998). Inadequate information is available on the amount of time spent in the various modes of feeding and how these proportions might vary as a function of species, instar, density, microhabitat, and other environmental factors. Much needed are behavioral time budgets that quantify these different feeding behaviors in the context of other larval behaviors, such as molting, relocating, body grooming (if it occurs), and silk spinning.

Larval diets have been analyzed for many species, beginning with the early observations of Riley (1887), Phillips (1890), Kellogg (1901), and Muttkowski and Smith (1929). The diet includes a variety of materials—virtually anything of relevant size, such as bacteria, diatoms, leaf fragments, pollen, fecal pellets, protozoa, and minute arthropods, as well as inorganic matter. Larvae can digest diatoms, with the degree of digestion differing among simuliid species as a function of temperature (Kurtak 1979, Thompson 1989). Cellulase activity has not been detected in larvae (Kessler 1982); however, tannins and phenolic compounds in dietary leaf litter can exert a toxic effect on the midgut epithelium (David et al. 2000). Some ingested materials might serve simply as substrates for microbiota that is used as the actual food. Larvae, in fact, can develop to adults solely on a diet of bacteria, especially gram-negative species (Fredeen 1964b), and in the field bacterial carbon can account for up to 67% of daily larval growth (Edwards & Meyer 1987). In some systems, protozoa can account for a significant portion of the larval diet (Carlough 1990). The size of ingested particles ranges from 0.09 to 350 μm in diameter (Wallace & Merritt 1980) and, therefore, includes colloids (Wotton 1976). Larvae also are able to use surface films (Wotton 1982b) and dissolved organic matter (DOM) that has flocculated (Hershey et al. 1996, Ciborowski et al. 1997). They do not, however, drink water and, therefore, do not ingest nonflocculated DOM (Martin 1989, Martin & Edman 1991). Selectivity is generally minimal, and the ingested particles are largely a function of availability, the nature of the flow, and physical constraints of the labral fans (Braimah 1987a, 1987b; Thompson 1987a).

Filtering efficiency is influenced by factors such as temperature, current velocity, particle size and concentration, labral-fan morphology, and parasitism (Kurtak 1978, Wotton 1978a, Thompson 1987b). Efficiencies of particle capture typically have a range of 0.01%–1.6% and decrease with increasing velocity (Kurtak 1978, McCullough et al. 1979 [the value of 1% given by McCullough et al. is in error and should be 0.01%], Hart & Latta 1986). Larvae are capable of removing 60% of suspended algae in a stream (Maciolek & Tunzi 1968), and when their densities are high, they have the potential to regulate bacterial transport (Hall et al. 1996). Assimilation rates and nutritive values vary widely with diet (Wotton 1978b, Thompson 1987c, Martin & Edman 1993, Couch et al. 1996). About 20% of consumed protist carbon is retained by larvae in some southeastern blackwater rivers (Carlough 1990). The time required for material to pass through the gut typically ranges from less than 20 minutes to about 2 hours (Colbo & Wotton 1981).

Filter feeding is inextricably linked with hydrodynamics, and these two areas have been forged into a subdiscipline of simuliid study (Hart & Latta 1986, Craig & Galloway 1988, Lacoursière & Craig 1993). During filter feeding, larvae twist their bodies 90–180 degrees. In this position, one fan receives particulate matter resuspended from the substrate while the other fan filters water from the mainstream (Craig 1985, Chance & Craig 1986). The height at which larvae hold their fans represents a balance between the conflicting demands to minimize drag and maximize feeding (Hart et al. 1991). The amount of available food and the dynamics of flow influence the capture efficiency of the labral fans not only physically but also ontogenetically, by altering the size of the primary fan and the number of its rays (Zhang & Malmqvist 1997, Lucas & Hunter 1999). Larvae can optimize water velocity for bringing food to the labral fans by changing the spacing between themselves and their neighbors (Chance & Craig 1986, Eymann 1991a). During unfavorable conditions, such as spates, larvae cease filtering and burrow into the hyporheic zone (Eymann & Friend 1986).

### Larval Movement

Larvae of all instars relocate to new sites by looping (Reidelbach & Kiel 1990) or by drifting downstream, often on silk strands, particularly around dusk and during the night (Pearson & Franklin 1968, Cowell & Carew 1976, Reisen 1977, Adler et al. 1983b). Larvae that drift without the use of silk strands sink at rates positively related to body length (Fonseca 1999). Causal factors of simuliid drift include predators (Simmons 1982), ultraviolet radiation (Kiffney et al. 1997a, Donahue & Schindler 1998), changes in discharge (Jarvis 1987, Poff & Ward 1991), and anchor ice lifting larvae from the substrate as water temperature increases (Martin et al. 2000). Explanations of simuliid drift also might depend on factors such as instar, habitat, and parasitism. An absence of drift has been recorded when stream temperatures are near 0°C (Ross & Merritt 1978).

The actual distances moved in a single release and during the larval life of different species are poorly known. Experimentally induced releases by *Simulium noelleri*, an outlet-inhabiting species, indicate that most distances traveled by this species are about 3–4 cm each, but the larvae often successfully climb the silk thread, resulting in little net movement (Wotton 1986). Other species, especially those in large rivers, undoubtedly drift greater distances downstream, but just how far remains one of the mysteries of larval life. Estimates of several hundred meters to hundreds of kilometers have been offered (Rubtsov 1964d), but they remain within the realm of fantasy.

Although drift can be adaptive, particularly in avoiding predators and locating suitable microhabitats, it also can increase mortality, reduce pupation, and decrease the degree of gut filling (Kiel et al. 1998c). If frequent, perhaps drift also can prolong development, a possibility that has not yet been tested.

### Adult Emergence, Flight, and Longevity

Adult life begins with emergence from the pupal integument, typically a morning phenomenon (Davies 1950). Partially enveloped in a bubble of air (Hannay & Bond 1971a), the fly breaks the water-air interface either explosively "as if shot out of a gun" (Howard 1901) or more placidly, sometimes resting on the surface of the water (Adler 1988b). Substantial collections of adults in drift nets (Reisen & Fox 1970, Adler et al. 1983b) suggest that flies might ride the water before taking flight or experience significant mortality during the emergence process. Sex ratios at emergence are generally near unity, except in the parthenogenetic species, which lack males. Skewed sex ratios, however, often are obtained as a result of sampling bias because male larvae generally develop faster than female larvae and pupate and emerge from one to several days earlier, a condition known as "protandry." Females probably remain in the larval stage longer to acquire more fat reserves for egg development, at least in autogenous situations. A longer larval life could increase exposure of females to adverse environmental conditions, resulting in greater cumulative mortality (Brenner 1980).

Adults are considered diurnal, with activity often peaking just before the fading hours of daylight (Mason & Kusters 1990). The rather routine attraction of both sexes to lights at night (e.g., Frost 1949, 1964; Fredeen 1961; Lacey & Mulla 1977b), nonetheless, requires explanation. Optimal temperatures for most adult activities are largely unknown. Sustained flight for some Nearctic species, however, is possible at temperatures as low as 12.8°C and intermittent flight can occur at 7.8°C–8.9°C (Edmund 1952, Hocking 1953b). Other factors such as humidity and wind speed also influence flight activity (Underhill 1940, Grace & Shipp 1988). Female black flies are generally more prone to bite at low atmospheric pressures and following rapid drops in pressure (Underhill 1940). Resting places during both the day and the night and how they might vary with physiological state and gender are poorly known. Both males and females in various physiological states have been found during the day in the tree canopy (Simmons et al. 1989).

During their lives, females can disperse widely. Distances up to 225 km have been reported for *Simulium vampirum* (Fredeen 1969; as *S. arcticum*) and more than 500 km for some members of the African *Simulium damnosum* species complex (Garms & Walsh 1988). Dispersal distances of less than 15 km are perhaps more typical (Bennett 1963a, Moore & Noblet 1974, Baldwin et al. 1975), although *Simulium jenningsi* often might disperse at least 55 km from the breeding grounds (Amrine 1982). Nonetheless, good quantitative data and the ecological correlates of different dispersal distances are lacking for nearly all species; little is known of dispersal distances of males. Infrequent sightings of swarms of mated females laden with sugar meals but not in a blood-feeding mode perhaps represent the prelude to dispersal flights (Hocking 1953b). The tendency of females to return to their natal waters to oviposit varies among species (Rothfels 1981b, Hunter & Jain 2000).

Longevity in nature is ordinarily 10–35 days (Crosskey 1990), with females outliving males. The potential, however, is greater, as evidenced by a field record of 85 days for a tropical species (Dalmat 1952) and a laboratory lifespan of 63 days (at 11°C) for a Nearctic species (Davies 1953). Factors such as body size that might influence longevity in nature are not well studied.

### Laboratory Colonization

Most species are reared easily from eggs or larvae to adults, minimally requiring food and aeration (West 1964, Edman & Simmons 1985a). The first successful rearing of a North American black fly from larvae to adults involved *Simulium vampirum* in a constant flow of tap water through a galvanized iron tank (Cameron 1922; as *S. simile*). Earlier attempts to rear larvae to adults had been unsuccessful (McBride in Riley 1870c, Emery 1913). A member of the *Simulium verecundum* species complex was the first identified Nearctic species reared from eggs to adults (Hartley 1955; as *S. venustum*), although unidentified Nearctic species had been reared in elaborate systems about a decade previously (Thomas 1946), and egg-to-adult rearings had been accomplished in Europe even earlier (Puri 1925).

Continuous colonies, however, are more difficult to establish and are labor intensive to maintain (Cupp & Ramberg 1997, Gray & Noblet 1999). Colonization is simplified if autogenous species are used because blood feeding is obviated. The primary biological obstacles to colonization are egg diapause and aerial coupling. Only four North American species (*Stegopterna mutata, Simulium vittatum, Simulium innoxium, Simulium decorum*) have been colonized for multiple generations (Mokry 1978; Brenner et al. 1980; Bernardo et al. 1986a, 1986b; *S. innoxium* as *S. pictipes*). Of these, only the parthenogenetic *St. mutata* has an egg diapause. Along with *S. decorum*, which mates on the ground, *St. mutata* was the first North American species colonized for multiple, albeit only two, generations (Mokry 1978, Simmons & Edman 1978). The need for aerial coupling in *S. innoxium* and *S. vittatum* was overcome by confining virgin flies in a small tube (Bernardo et al. 1986a; *S. innoxium* as *S. pictipes*). *Simulium vittatum* has been in continuous colonization for about two decades (Cupp & Ramberg 1997, Brockhouse & Adler 2002), and as of 2002 had passed through more than 180 generations (E. W. Gray, pers. comm.). The effects of long-term colonization, particularly without infusion of field material, are poorly known. Cytogenetic changes, however, have occurred in the colony of *S. vittatum* (Brockhouse & Adler 2002).

Associated with colonization efforts has been an attempt to establish long-term storage techniques for Nearctic simuliid eggs through cryopreservation. Various cryoprotectants (e.g., ethylene glycol) have been evaluated, but surface barriers and lipid inclusions of the eggs have impeded these efforts, although some embryos have survived at 30°C for up to 66 hours in 3.0–5.0 M methanol (Goldie 1982).

### Mating Behavior

One of the least-observed facets of simuliid natural history is mating and its associated behaviors. Only four North American species (*Gymnopais dichopticoides, G. holopticoides, Prosimulium ursinum, Stegopterna mutata*) have abandoned mating altogether. These species are obligately parthenogenetic, lack males, and are strictly northern, except *St. mutata*, which extends from northern Canada deep into the southeastern states. Mating behavior has been recorded for only 38 (about 15%) of the 250 sexual species.

Much of the information about mating behavior concerns the species that couple on the ground at their emergence sites. The six North American species known to couple on the ground are *Gymnopais dichopticus, Gymnopais fimbriatus, Cnephia dacotensis, Cnephia eremites, Metacnephia coloradensis*, and *Simulium decorum*. These species are associated with northern or high-elevation environments, except *S. decorum*, which occurs over most of the continent. The only conspicuous structural adaptation to ground coupling involves the male eyes, which lack the prominent division into small lower and large upper facets characteristic of males that couple in the air. Of the ground-coupling species, only *C. eremites* and *S. decorum*, for unknown reasons, have retained the typical divided eyes.

Species of the most primitive genus, *Parasimulium*, are unique in their mating behavior. Males of the three species of the subgenus *Parasimulium* search for females that sit beneath the apices of leaves near the habitat of the aquatic stages; coupling has been observed in only one of the species (Wood & Borkent 1982). Males of these species, like those of most species that mate on the ground, have eyes with facets of uniform size. In the subgenus *Astoneomyia*, the male of *Parasimulium* 'species A' and possibly that of *P. melanderi*, for which only one headless male is known, have divided eyes, suggesting a different style of mating behavior.

Aerial coupling is probably in the simuliid groundplan, being predominant in the simuliid sister group (Ceratopogonidae + Chironomidae). We suspect that more than 96% of North America's sexual species include aerial coupling in their mating behavior. However, aerial swarming by males, a prelude to actual coupling, has been observed for only about 28 North American simuliid species. Swarms consist predominantly of males in loose aggregations of a few to thousands of individuals over or beside a landmark such as a path, a waterfall, a host of the female, or riparian vegetation. These swarm markers tend to have sharp boundaries, conspicuous angles, or distinct contrast against the sky or ground (Downes 1969). Females entering a swarm are engaged quickly, and the coupled pair exits the swarm by flying out or dropping to the ground. Coupling in nature has been observed for only 10 of these swarm-forming species. Hilltopping behavior, so common in many Diptera, rarely has been observed in the Simuliidae, although we have seen small swarms of males of the *Simulium aureum* species complex and large swarms (hundreds of males) of *Metacnephia saskatchewana* on hilly prominences.

The rarity with which mating swarms are observed in nature suggests that swarming is not always a prerequisite to mating and that black flies might use alternative strategies to locate mates. Unknown, however, is the frequency of canopy-level swarming beyond the range of unaided human vision, as observed for *Cnephia ornithophilia* (Mokry et al. 1981). A sit-and-wait strategy whereby males perch on streamside vegetation and dart out to intercept passing females has been suggested for some Palearctic species (Wenk 1988).

Virtually unknown and much sought are differences in mating behavior among closely related species, especially sibling species. Closely related species presumably differ in aspects such as timing, location, height of swarming, and choice of swarm markers (Bedo 1979, Adler 1983), but no data have been forthcoming, other than the obvious differences between ground- and aerial-coupling species, as in the genus *Cnephia*.

Attempts to understand the role of pheromones and possible auditory signals, male competition, and mate choice have been hampered by the small size of black flies, the rapidity of mating events, and the difficulty of duplicating swarms in the laboratory. Conventional wisdom holds that visual cues are of paramount importance in the mating behavior of swarming black flies (Wenk 1988). Males, for instance, might be able to recognize females by their flight patterns and speeds or structural features. Anecdotal evidence for some ground-coupling species suggests that contact pheromones might play a role (Edman & Simmons 1985a). Differences in sound production, especially wing-beat frequencies and abdominal vibrations, both among species and between sexes, although perhaps largely a function of body size, might provide auditory cues (Werner 1998). The possibility of assortative mating, on the basis of size or other factors, in both ground- and aerial-coupling species is unstudied. In at least one ground-mating species, larger males are more successful than smaller males in coupling with refractory females (Edman & Simmons 1988). Presumably there are optimal positions for males in a swarm. Males of *Simulium innoxium* are withered at the bottom of a swarm but robust toward the top; those near the bottom are said to be the ones that mate (Stone & Snoddy 1969; as *S. pictipes*), but the positioning dynamics of males and the entry routes of females into swarms are poorly known.

Actual coupling typically begins with the male above the female, both facing the same direction, and segues to an end-to-end position, with the male on its back. Coupling typically lasts from a few seconds to about 16 minutes (Sommerman 1958, Peterson 1977b) but can persist up to 2 hours in *Gymnopais* (Currie 1997a). Single matings are believed to be the rule in black flies (Wenk 1988), but double matings and use of sperm from both males have been reported for members of the African *Simulium damnosum* species complex (Boakye et al. 2000). Males of *Simulium vittatum* in the laboratory are capable of mating with two females in rapid succession and passing enough sperm to fill the spermatheca of each one (F. F. Hunter & P. H. Adler, unpublished data).

Males of most, if not all, sexual species of black flies transfer sperm in a small, opaque, two-chambered capsule, the spermatophore, which forms within the ejaculatory duct (Wood & Borkent 1989) or perhaps within the genital chamber of the female. Production of spermatophores, however, has been confirmed for a limited number of species and was not documented in the Simuliidae until the 1960s (Wood 1963a, Davies 1965a), though it was alluded to earlier (Rubtsov 1956). As sperm are released from the spermatophore, perhaps by enzymatic action, they are conveyed to the spermatheca for storage, probably for the life of the female. Within the spermatheca, the threadlike sperm appear as a small, fibrous or cottony mass. An average of 4048 spermatozoa are contained in a spermatophore (Linley & Simmons 1981, 1983)—approximately eight times the mean number of eggs produced per ovarian cycle. No information exists on possible sperm competition in multiply mated females. Limited evidence, however, suggests that within a species smaller males produce fewer sperm (Simmons 1985).

The storage of sperm in the spermatheca provides the opportunity for artificial insemination, one of the least-exploited aspects of reproductive study in black flies. In vitro fertilization can be achieved by mixing the spermathecal contents of a mated female with the eggs of a virgin female (Wood 1963a) in a drop of water or physiological saline solution. Inseminated females are readily collected in nature, and the in vitro procedure could help circumvent the difficulties of laboratory matings while also allowing the study of postmating isolating mechanisms of closely related species.

## Adult Feeding Behavior

### Sugar and Water

Probably all adult black flies are capable of imbibing water, a critical activity if they are to live more than a few days. Sugars, required as an energy source primarily for flight (Cooter 1982, 1983) but also of use in egg maturation (Anderson 1988), are probably taken by both sexes of most species. Some of the flightless *Gymnopais*, however, might not take sugars in the wild, although lack of sugar feeding for these species has not been documented.

Sugar sources remain poorly known for most species. Feeding on floral nectar is a little-observed behavior, having been recorded for only about 13 (5%) of the North American species. The flowers visited for nectar are generally small, white, and arranged in clusters. The paucity of observations on floral visits initially seems anomalous because most adults captured in the wild show evidence of having fed on sugars (McCreadie et al. 1994c). The dearth of records might be, in part, because nectar feeding occurs primarily late in the day. Nectar feeding, however, might be less frequent than often believed. Some species of black flies, for example, will feed on sap from plant wounds (Fredeen 1981a). Extrafloral nectaries of peonies (*Paeonia* sp.) also have been exploited by some North American black flies (A. G. Wheeler, pers. comm.). Honeydew from aphids and related insects provides a frequent additional source of sugar (Burgin & Hunter 1997a, 1997b). Sugar feeding in black flies, therefore, is probably opportunistic. We have, for example, seen scores of females of the *Simulium vittatum* species complex feeding on the sugary contents of exploded crops of tabanids and other Diptera on the windshields of vehicles. The effects on fitness of different sugars in honeydew and nectar are unstudied.

Coupled with observations of feeding on floral nectar is the idea that black flies serve as pollinators of the plants they visit. Virtually all such claims are anecdotal, however, and the leap is often made that floral visits equal pollination (e.g., Henderson et al. 1979). Popular literature and folklore often maintain that black flies are one of the most important pollinators of blueberries (*Vaccinium* spp.) (Bennet & Tiner 1993), yet experimental evidence indicates that they do not increase fruit set in these plants (Hunter et al. 2000). Nonetheless, because black flies are able to pick up pollen from the flowers they visit, their role as pollinators certainly requires further study.

### BLOOD

Female black flies take blood exclusively from warm-blooded vertebrates. Early reports of black flies feeding on insects (Hagen 1883, Williston 1908) were based on misidentifications of ceratopogonids. Host location and subsequent feeding involve a series of steps in which habitat features and host attributes, such as size, shape, color, odor (especially carbon dioxide), temperature, and various phagostimulants (e.g., adenosines), are sequentially evaluated (Simmons 1985; Sutcliffe 1986, 1987; S. A. Allan et al. 1987). Several host products (e.g., sweat) are attractive to black flies, particularly in the presence of carbon dioxide (Schofield 1994). Auditory cues such as bird songs seem not to play a role in host location, but they have not been evaluated adequately.

Different feeding and host-locating strategies might operate, depending on the simuliid species, host, and habitat. For example, some species might actively search for hosts, whereas other species might sit and wait for hosts to come within range of detection (Simmons 1985). The reasons that black flies are attracted to certain hosts and particular body regions have been addressed primarily at the proximate level (e.g., visual and olfactory cues) rather than at the evolutionary level (e.g., avoidance of host defenses). Areas on the host where feeding frequently occurs, such as in the ears and on the face and belly, tend to be thinly haired, rich in capillaries, and often difficult for the host to defend or groom. Animals such as caribou are particularly vulnerable when they are shedding their winter coats and have areas rather devoid of hair (Banfield 1954). Moose calves, which have long, dense natal hair, are less annoyed by black flies than older moose (Pledger 1978). Host defenses such as grooming are likely to have played an important role as agents of selection in determining feeding locations on the body (Simmons 1985). Nonetheless, host behavioral defenses in general are poorly studied, although several examples suggest that the defenses are diverse. Defense responses in cattle include tail switches, head shakes, ear flicks, and foot stomps (Kampani 1986). To minimize black fly attacks, moose sometimes seek refuge in water (Flook 1959), caribou move to the tops of windy ridges (Knap 1969) or form herds (Helle et al. 1992), cattle exhibit bunching behavior (Shemanchuk 1980a), and great horned owls shift the height at which they roost (Rohner et al. 2000).

Approximately 90% of North American black flies are capable of taking a blood meal. The 10% (26 species) with mouthparts unarmed for cutting flesh are obligatorily autogenous, maturing their eggs without a blood meal (Table 6.1). They are found at high elevations and particularly at northern latitudes (Currie 1997a). Between 60° and 70° north, 19% of the species in that area have untoothed mouthparts, whereas between 30° and 40° north, only one

**TABLE 6.1. NORTH AMERICAN BLACK FLIES WITH OBLIGATE AUTOGENY AND MOUTHPARTS TOO WEAK TO OBTAIN BLOOD**

| Species | Southernmost known latitude |
|---|---|
| *Parasimulium* 'species A' | 44°13′N |
| *Parasimulium melanderi* | 48°54′N[1] |
| *Parasimulium crosskeyi* | 45°34′N[2] |
| *Parasimulium furcatum* | 40°57′N |
| *Parasimulium stonei* | 40°16′N |
| *Gymnopais dichopticoides* | 50°37′N |
| *Gymnopais dichopticus* | 62°17′N |
| *Gymnopais fimbriatus* | 63°40′N |
| *Gymnopais holopticoides* | 53°15′N |
| *Gymnopais holopticus* | 62°35′N |
| *Twinnia tibblesi* | 41°13′N |
| *Helodon gibsoni* | 42°47′N |
| *Helodon alpestris* | 57°29′N |
| *Helodon clavatus* | 44°43′N |
| *Helodon irkutensis* | 60°49′N |
| *Prosimulium neomacropyga* | 40°03′N[3] |
| *Prosimulium ursinum* | 52°57′N |
| *Stegopterna decafilis* | 60°36′N |
| *Stegopterna emergens* | 45°42′N |
| *Cnephia dacotensis* | 40°41′N |
| *Cnephia eremites* | 58°43′N |
| *Metacnephia borealis* | 53°42′N |
| *Metacnephia coloradensis* | 39°17′N[3] |
| *Metacnephia sommermanae* | 61°35′N |
| *Simulium baffinense* | 40°34′N |
| *Simulium* 'species Z' | 63°10′N |

[1] Approximate latitude; exact southern location is unknown (Courtney 1993).
[2] The more southern record of Corvallis, Oregon, is believed to be in error (Courtney 1993).
[3] In the contiguous United States, found at elevations above 3000 m only.

species, which lives at high altitudes, has nonbiting mouthparts. An unknown number of species are facultatively autogenous, maturing at least an initial batch of eggs without a blood meal (Davies & Györkös 1990), a phenomenon first discovered in a member of the *Simulium vittatum* species complex (Wu 1931). These facultatively autogenous species are primarily northern in distribution, or their larvae develop in nutrient-rich habitats or early in the season when cooler water temperatures allow more assimilated energy to be shunted into larval fat reserves than into maintenance metabolism. The majority of simuliid species are probably anautogenous for at least one ovarian cycle, requiring a blood meal to produce a batch of eggs.

The design of the female claws has been used to infer principal hosts (birds or mammals) since Shewell (1955) suggested that a thumblike basal lobe on the claw indicates ornithophilic feeding habits. The lobe presumably aids purchase and movement of the flies through the feathers. The only known North American exception to this rule of thumb is *Cnephia pecuarum*, which has the lobe but is principally mammalophilic. The design of the tarsal claw, in conjunction with known feeding habits, indicates that 37% of the North American species that can take blood are primarily ornithophilic and 63% are chiefly mammalophilic. The distinction between the two groups, however, is sometimes blurred because some mammalophilic species (e.g., *Simulium jenningsi* and *S. venustum*) also feed on birds, and some ornithophilic species (e.g., *Simulium johannseni* and *S. meridionale*) sometimes feed on mammals, including humans. Of the blood-feeding species, those that attack birds increase in proportion with latitude, from 30% between 30° and 40° north to 46% between 50° and 60° north. About 81% of ornithophilic species are univoltine, whereas approximately 45% of mammalophilic species are univoltine. Underlying reasons for these trends are not immediately apparent.

The hosts for most species of black flies are incompletely known. We do not have a single specific host record for approximately 60% of the North American blood-feeding species, and even for many of those with host records, the information is meager. At least 32 species of mammals and 50 species of birds have been recorded as hosts of North American simuliids (Tables 6.2 and 6.3). Some large groups of mammals—bats, for instance—have never been recorded as hosts for black flies anywhere in the world. Either they are not fed on or they simply have not been adequately sampled. For most black flies, the known hosts are biased artificially toward those that are of economic importance and easily sampled, namely, humans and domestic animals. Available records indicate that most simuliid species feed on a suite of hosts that are similar in size or are found in a particular habitat such as a lakeshore or the canopy. A few species, such as *Simulium annulus*, feed on only one or a few host species, whereas other species, such as *S. venustum*, feed on many host species. At least seven records indicate that a single fly will feed on two different host species, with one host sometimes a bird and the other a mammal (Simmons et al. 1989). Female black flies often can be induced to feed on a host, even an atypical one, by holding them against the skin under an inverted vial. Exposure to carbon dioxide can induce blood-feeding behavior (Dalmat 1950). Much of our knowledge on the hosts of black flies comes from the landmark works of Bennett (1960) and Anderson and DeFoliart (1961).

More work on the hosts of black flies is needed, especially in the context of taxonomic rigor and appreciation of sibling species of simuliids (e.g., Hunter et al. 1993). Fitness effects, as a function of feeding on the blood of different host species, have been demonstrated (Klowden & Lea 1979, Mokry 1980a) but are insufficiently studied. Most species of black flies feed at particular sites on the host, especially in areas where the hair or feathers are sparse or thin such as around the eyes, in the ears, and on the venter. Possible differences in feeding sites among closely related black flies (e.g., members of a species complex) on the same host species are poorly studied. Additional studies of the relative attractiveness of different host attributes and host products (e.g., sweat) and their components are needed for more simuliid species. The study of natural host repellents, a fruitful area of investigation in plant-herbivore interactions, is virtually untouched for biting flies and their hosts. Limited data suggest that cattle produce odors repellent to black flies (Sutcliffe & Shemanchuk 1993).

## Natural Enemies and Symbionts

All black flies are subject to the selective forces of parasitism and predation. The eggs of black flies, however, are virtually free of parasites, other than the transovarial pathogens that generally express their effects in later host stages. Larval symbionts include mermithid nematodes, fungi (chytrids, trichomycetes, hyphomycetes), stramenopiles, microsporidia, helicosporidia, ichthyosporeans, protists (ciliates, haplosporidia), bacteria, viruses, and nematomorphs. Most of these symbionts also occur in pupae and adults, the latter life stage providing the primary means of parasite dispersal and recolonization. Parasites typically specific to adults are filarial nematodes, *Leucocytozoon* and trypanosomal protists, entomophthoraceous fungi, and water mites. Dipteran and hymenopteran parasitoids are virtually absent from black flies. The only records worldwide are one case of dipteran (Phoridae) parasitism (Baranov 1939) and four cases of hymenopteran parasitism (Enderlein 1921a, Peterson 1960b, Williams & Cory 1991); all cases are undoubtedly accidental. The relation between predators and black flies is largely one of opportunism, although at times some predacious species specialize on black flies (Snoddy 1968, Muotka 1993, Malmqvist 1994, Yoerg 1994).

Most taxa parasitic on black flies are encountered frequently throughout the world and nearly all, except nematomorphs and some entomophthoraceous and trichomycete fungi, attack only species of the family Simuliidae. Prevalence of infection varies from less than 1% to 100% in a population, although reported estimates of prevalence should be viewed cautiously because they

**Table 6.2. Mammalian Hosts of North American Black Flies, Based on Confirmed Biting Records in the Field or Serological Evidence**

| Host | Black fly |
|---|---|
| human (*Homo sapiens*) | *T. nova, P. doveri, P. esselbaughi, P. fontanum, P. formosum, P. fulvum, P. fuscum, P. mixtum, P. caudatum, P. magnum* complex, *St. acra* n. sp., *St. diplomutata/mutata, St. permutata, C. ornithophilia, C. pecuarum, E. taeniatifrons, M. saileri,*[1] *S. balteatum* n. sp./*canonicolum/quadratum, S. johannseni, S. parmatum* n. sp., *S. maculatum,*[1] *S. meridionale, S. bivittatum, S. clarum, S. griseum, S. mediovittatum, S. trivittatum, S. venator, S. argus, S. vittatum* complex, *S. vittatum* s. s., *S. hunteri, S. piperi, S. tescorum, S. innoxium, S. pictipes, S. infenestrum, S. jenningsi, S. luggeri, S. penobscotense, S. decimatum, S. malyschevi,*[1] *S. murmanum, S. arcticum* complex, *S. vampirum, S. decorum, S. noelleri,*[1] *S. parnassum, S. slossonae, S. transiens,*[1] *S. tuberosum* complex(?),[2] *S. venustum* complex, *S. hematophilum, S. irritatum, S. truncatum, S. venustum* s. s., *S. rostratum* |
| gray wolf (domestic dog) (*Canis lupus*) | *P. mixtum, C. pecuarum, S. venator, S. virgatum,*[1] *S. luggeri, S. penobscotense, S. arcticum* complex, *S. venustum* complex, *S. truncatum, S. venustum* s. s. |
| red fox (*Vulpes vulpes*) | *St. diplomutata/mutata, S. tuberosum* complex, *S. venustum* complex, *S. hematophilum, S. truncatum, S. venustum* s. s. |
| wild cat (domestic cat) (*Felis silvestris*) | *C. pecuarum, S. virgatum*[1] |
| Canadian lynx (*Lynx canadensis*) | *P. mixtum, S. venustum* complex, *S. hematophilum* |
| Canadian otter (*Lontra canadensis*) | *S. venustum* complex |
| American mink (domestic) (*Mustela vison*) | *S. venustum* complex, *S. truncatum, S. venustum* s. s. |
| raccoon (*Procyon lotor*) | *P. mixtum, P. magnum* complex, *St. diplomutata/mutata, S. croxtoni, S. gouldingi, S. jenningsi*(?), *S. tuberosum* complex, *S. venustum* complex, *S. verecundum* complex |
| American black bear (*Ursus americanus*) | *P. fontanum, P. fuscum, P. mixtum, S. decorum, S. parnassum, S. rugglesi, S. venustum* complex, *S. truncatum, S. venustum* s. s., *S. verecundum* complex, *S. verecundum* s. s. |
| donkey (burro) (*Equus asinus*) | *S. mediovittatum, S. virgatum*[1] |
| horse (*Equus caballus*) | *T. nova, P. fulvum, P. fuscum, P. mixtum, P. magnum* complex, *St. diplomutata/mutata, St. permutata*(?), *C. pecuarum, S. johannseni, S. meridionale, S. croxtoni, S. bivittatum, S. clarum, S. griseum, S. mediovittatum, S. robynae, S. venator, S. argus, S. vittatum* complex, *S. tribulatum, S. vittatum* s. s., *S. piperi, S. hippovorum, S. paynei,*[1] *S. solarii, S. virgatum,*[1] *S. innoxium, S. confusum, S. jenningsi, S. luggeri, S. murmanum, S. arcticum* complex, *S. vampirum, S. decorum, S. noelleri,*[1] *S. parnassum, S. tuberosum* complex(?), *S. venustum* complex, *S. venustum* s. s., *S. verecundum* complex, *S. rostratum* |
| mule (*Equus asinus–E. caballus* hybrid) | *P. magnum* complex, *C. pecuarum, S. mediovittatum, S. vittatum* complex, *S. virgatum,*[1] *S. innoxium, S. jenningsi, S. tuberosum* complex(?), *S. venustum* complex |
| wild boar (domestic pig) (*Sus scrofa*) | *C. pecuarum, S. bivittatum, S. griseum, S. vittatum* complex, *S. luggeri, S. vampirum, S. venustum* complex |
| elk (European red deer) (*Cervus elaphus*) | *P. fuscum, P. mixtum, St. diplomutata/mutata, S. vittatum* complex, *S. luggeri* |
| moose (European elk) (*Alces alces*) | *P. formosum, S. aureum* complex, *S. vittatum* complex, *S. pictipes, S. arcticum* complex, *S. decorum, S. venustum* complex |
| mule deer (*Odocoileus hemionus*) | *P. impostor* |
| white-tailed deer (*Odocoileus virginianus*) | *P. fuscum, P. mixtum, St. diplomutata/mutata, C. pecuarum, S. vittatum* complex, *S. vampirum, S. decorum, S. venustum* complex |
| caribou (reindeer) (*Rangifer tarandus*) | *M. saileri,*[1] *S. venustum* complex, *S. verecundum* complex |
| cow (cattle, ox) (*Bos taurus*) | *H. pleuralis,*[3] *P. fuscum, P. mixtum, P. magnum* complex, *St. diplomutata/mutata, C. pecuarum, M. saileri,*[1] *S. johannseni, S. maculatum,*[1,3] *S. meridionale, S. aureum* complex, *S. bivittatum, S. clarum, S. griseum, S. mediovittatum, S. argus, S. vittatum* complex, *S. tribulatum, S. vittatum* s. s., *S. hunteri, S. jacumbae, S. piperi, S. paynei,*[1] *S. virgatum,*[1] *S. innoxium, S. jenningsi, S. luggeri, S. penobscotense, S. defoliarti, S. murmanum, S. arcticum* complex, *S. vampirum, S. decorum, S. noelleri,*[1] *S. parnassum, S. tuberosum* complex(?), *S. venustum* complex, *S. verecundum* complex, *S. rostratum*[1] |

**Table 6.2. Continued**

| Host | Black fly |
|---|---|
| goat (domestic) (*Capra hircus*) | *S. virgatum*[1] |
| sheep (domestic) (*Ovis aries*) | *P. esselbaughi*, *P. fulvum*, *P. magnum* complex, *C. pecuarum*, *S. bivittatum*, *S. griseum*, *S. vittatum* complex, *S. piperi*, *S. virgatum*,[1] *S. luggeri*, *S. vampirum*, *S. venustum* complex |
| bighorn sheep (*Ovis canadensis*) | *S. vittatum* complex |
| woodchuck (*Marmota monax*) | *S. parnassum* |
| eastern gray squirrel (*Sciurus carolinensis*) | *S. tuberosum* complex |
| Uinta ground squirrel (*Spermophilus armatus*) | *S. tuberosum* complex, *S. venustum* complex |
| golden mantled ground squirrel (*Spermophilus lateralis*) | *S. tuberosum* complex |
| eastern chipmunk (*Tamias striatus*) | *S. tuberosum* complex |
| red squirrel (*Tamiasciurus hudsonicus*) | *S. tuberosum* complex, *S. venustum* complex, *S. truncatum*, *S. venustum* s. s. |
| beaver (*Castor canadensis*) | *S. venustum* complex |
| rat (*Rattus* sp.) | *S. noelleri*[1] |
| snowshoe hare (*Lepus americanus*) | *S. tuberosum* complex, *S. venustum* complex, *S. hematophilum* |
| black-tailed jackrabbit (*Lepus californicus*) | *S. griseum*, *S. mediovittatum*, *S. robynae* |
| European rabbit (domestic) (*Oryctolagus cuniculus*) | *S. tuberosum* complex, *S. venustum* complex, *S. truncatum*, *S. venustum* s. s. |
| "rabbit" | *S. clarum*, *S. vittatum* complex |

*Note*: Hosts are arranged according to the checklist of Wilson and Reeder (1993). For each host, black flies are listed in the order of their appearance in the taxonomic accounts (Chapter 10). Additional information, including frequency of host use and references, can be found in the text under individual taxonomic accounts of black flies.

[1] Based on records outside North America north of Mexico.

[2] (?) = Identification of black fly uncertain.

[3] Reported as attacking "livestock," here interpreted as cows.

**Table 6.3. Avian Hosts of North American Black Flies, Based on Confirmed Biting Records in the Field**

| Host | Black fly |
|---|---|
| emu (*Dromiceius novaehollandiae*) | *S. meridionale*, *S. lakei* |
| ostrich (*Struthio camelus*) | *S. meridionale* |
| common loon (great northern diver) (*Gavia immer*) | *S. annulus* |
| great blue heron (*Ardea herodias*) | *S. aureum* complex, *S. rugglesi*, *S. venustum* complex |
| black-crowned night heron (*Nycticorax nycticorax*) | *S. rugglesi* |
| Canada goose (*Branta canadensis*) | *E. invenusta*, *S. anatinum*, *S. rugglesi* |
| graylag (domestic goose) (*Anser anser*) | *H. pleuralis*, *M. saileri*, *M. saskatchewana*, *S. vittatum* complex, *S. rugglesi* |
| muscovy duck (*Cairina moschata*) | *S. rugglesi* |
| wood duck (*Aix sponsa*) | *S. rugglesi* |
| American black duck (*Anas rubripes*) | *H. decemarticulatus* |
| mallard (including domestic duck[1]) (*Anas platyrhynchos*) | *H. decemarticulatus*, *St. diplomutata/mutata*, *C. ornithophilia*, *E. invenusta*, *S. anatinum*, *S. rendalense*,[2] *S. usovae*,[2] *S. annulus*(?),[3] *S. emarginatum*, *S. aureum* complex, *S. decorum*, *S. rugglesi*, *S. venustum* complex, *S. truncatum*, *S. venustum* s. s. |
| blue-winged teal (*Anas discors*) | *S. rugglesi* |
| northern pintail (*Anas acuta*) | *S. rugglesi* |
| redhead (*Aythya americana*) | *S. anatinum*, *S. rugglesi* |
| ring-necked duck (*Aythya collaris*) | *S. rugglesi* |
| "duck" | *S. congareenarum*, *S. johannseni*, *S. vittatum* complex(?) |
| sharp-shinned hawk (*Accipiter striatus*) | *H. decemarticulatus*, *C. ornithophilia*, *S. aureum* complex |
| red-tailed hawk (*Buteo jamaicensis*) | *S. balteatum* n. sp./*canonicolum*/*quadratum* |
| prairie falcon (*Falco mexicanus*) | *S. balteatum* n. sp./*canonicolum*/*quadratum* |
| chicken (*Gallus gallus*) | *H. pleuralis*, *H. decemarticulatus*, *C. ornithophilia*, *E. taeniatifrons*, *S. congareenarum*, *S. annulus*(?), *S. johannseni*, *S. meridionale*, *S. aureum* complex, *S. vittatum* complex, *S. vampirum*, *S. decorum*, *S. rugglesi*, *S. slossonae*, *S. venustum* complex |
| ring-necked pheasant (*Phasianus colchicus*) | *E. taeniatifrons*, *S. johannseni*, *S. meridionale*, *S. aureum* complex, *S. rugglesi* |
| common peafowl (*Pavo cristatus*) | *S. meridionale* |

**Table 6.3. Continued**

| Host | Black fly |
|---|---|
| ruffed grouse (*Bonasa umbellus*) | *H. decemarticulatus*, *C. ornithophilia*, *E. invenusta*, *E. taeniatifrons*, *S. anatinum*, *S. annulus*(?), *S. aureum* complex, *S. croxtoni*, *S. quebecense*, *S. rugglesi*, *S. venustum* complex |
| blue grouse (*Dendragapus obscurus*) | *S. nebulosum*(?), *S. aureum* complex, *S. craigi*, *S. hunteri* |
| black grouse (*Tetrao tetrix*) | *S. bicorne*,[2] *S. transiens*[2] |
| "grouse" | *T. osborni*, *S. venustum* complex |
| wild (and domestic) turkey (*Meleagris gallopavo*) | *P. mixtum*, *C. ornithophilia*, *E. taeniatifrons*, *S. anatinum*, *S. congareenarum*, *S. johannseni*, *S. meridionale*, *S. aureum* complex, *S. jenningsi*, *S. rugglesi*, *S. slossonae* |
| sandhill crane (*Grus canadensis*) | *S. rugglesi* |
| spotted sandpiper (*Actitis macularia*) | *S. rugglesi* |
| herring gull (*Larus argentatus*) | *S. rugglesi* |
| black tern (*Chlidonias niger*) | *S. rugglesi* |
| "pigeon" (*Columba* sp.) | *S. meridionale* |
| ringed turtle-dove (*Streptopelia risoria*) | *S. aureum* complex, *S. rugglesi* |
| mourning dove (*Zenaida macroura*) | *S. meridionale*, *S. aureum* complex |
| yellow-headed Amazon parrot (*Amazona ochrocephala*) | *S. meridionale* |
| great horned owl (*Bubo virginianus*) | *H. pleuralis*, *H. decemarticulatus*, *S. canonicolum*, *S. aureum* complex |
| long-eared owl (*Asio otus*) | *S. balteatum* n. sp.(?) |
| northern saw-whet owl (*Aegolius acadicus*) | *H. decemarticulatus*, *S. aureum* complex, *S. croxtoni*, *S. rugglesi*, *S. venustum* complex |
| northern flicker (*Colaptes auratus*) | *H. decemarticulatus*, *C. ornithophilia* |
| gray jay (*Perisoreus canadensis*) | *H. decemarticulatus*, *C. ornithophilia*, *S. anatinum*, *S. aureum* complex, *S. croxtoni* |
| blue jay (*Cyanocitta cristata*) | *H. decemarticulatus*, *C. ornithophilia*, *S. aureum* complex, *S. croxtoni* |
| American crow (*Corvus brachyrhynchos*) | *H. decemarticulatus*, *C. ornithophilia*, *S. anatinum*, *S. emarginatum*, *S. aureum* complex, *S. croxtoni*, *S. decorum*, *S. rugglesi*, *S. venustum* complex |
| common raven (*Corvus corax*) | *H. decemarticulatus*, *C. ornithophilia*, *S. anatinum*, *S. aureum* complex, *S. croxtoni*, *S. rugglesi*, *S. venustum* complex |
| purple martin (*Progne subis*) | *S. meridionale* |
| tree swallow (*Tachycineta bicolor*) | *S. meridionale* |
| eastern bluebird (*Sialia sialis*) | *S. meridionale* |
| American robin (*Turdus migratorius*) | *H. decemarticulatus*, *C. ornithophilia*, *S. anatinum*, *S. aureum* complex, *S. croxtoni*, *S. quebecense*, *S. rugglesi*, *S. venustum* complex |
| European starling (*Sturnus vulgaris*) | *S. meridionale* |
| white-throated sparrow (*Zonotrichia albicollis*) | *H. decemarticulatus*, *C. ornithophilia*, *S. anatinum*, *S. aureum* complex, *S. croxtoni* |
| "sparrow" | *H. decemarticulatus*, *S. venustum* complex |
| red-winged blackbird (*Agelaius phoeniceus*) | *S. rugglesi* |
| common grackle (*Quiscalus quiscula*) | *H. decemarticulatus*, *C. ornithophilia*, *S. aureum* complex, *S. croxtoni*, *S. rugglesi*, *S. venustum* complex |
| purple finch (*Carpodacus purpureus*) | *H. decemarticulatus*, *S. aureum* complex, *S. croxtoni* |

*Note*: Host names follow the checklist of the American Ornithologists' Union (1998). For each host, black flies are listed in the order of their appearance in the taxonomic accounts (Chapter 10). Additional information, including frequency of host use and references, can be found in the text under individual taxonomic accounts of black flies.

[1] "Domestic duck" in the literature is assumed to be a derivative of the mallard, unless otherwise stated.

[2] Based on records outside North America north of Mexico.

[3] (?) = Identification of black fly uncertain.

often are based only on patent infections and vary with age of the host population. Infected individuals typically become more prominent and relatively more prevalent as a population ages, in part because infected individuals—at least those with microsporidia—generally have a longer larval stage (Maurand 1975).

Patterns of host-parasite relations are derived from a sound taxonomy of both the hosts and the parasites. Taxonomic work, however, is much needed for most parasites of black flies. Apparent lack of host specificity, morphological variation, and wide geographic distributions suggest that sibling species exist among some of the parasites that attack black flies (Federici et al. 1977, Vávra & Undeen 1981, Adler et al. 1996). Probable sibling species have been demonstrated in mermithid nematodes of black flies. They occur in the same streams but are asynchronous, attack different hosts, and do not cross breed in the laboratory (Ebsary & Bennett 1975b, Bailey et al. 1977, Colbo & Porter 1980).

### NEMATODES

Three families of nematodes have been found in black flies: Mermithidae, Onchocercidae, and Robertdollfusidae. Only the first two families are represented in North American black flies. Mermithid nematodes alternate between parasitic and free-living stages, whereas the onchocercids are obligatorily parasitic, developing in both adult black flies and the vertebrate hosts to which they are transmitted.

Mermithid nematodes were the first parasites found in black flies, their discovery dating to 1848 in Germany (Siebold 1848) and to 1888 in North America (Webster 1914). They are responsible for some of the highest levels of infection of all black fly parasites (Molloy 1981). The recently hatched preparasitic worms enter the host by piercing the integument (Molloy & Jamnback 1975). Multiple worms can infect an individual host, with the proportion of male worms increasing as the parasite load increases (Ezenwa & Carter 1975). The worms grow until they occupy most of the larval hemocoel. Infections cause major histological changes in the host, particularly in the fat body (von Ahlefeldt 1968). Juveniles exit the larvae or persist through the pupal stage into the abdomen of the adult hosts and exit shortly thereafter as the adults fly over a stream. Infections in adults typically impede gonadal development and often produce intersexes, resulting in altered behavior (Molloy 1981) such as oviposition flights by feminized males. Parasitism also inhibits host seeking and blood feeding by some simuliid species (Simmons 1985) but not by others (Anderson & Shemanchuk 1987b). Whether these effects on behavior are related to the species of simuliid host, the species of mermithid, or both is unknown. Postparasitic juveniles molt to adults and then mate and deposit eggs in the streambed (Ebsary & Bennett 1973, Poinar 1981). Distributions of nematodes among streams are associated with factors such as stream size and water chemistry (McCreadie & Adler 1999).

The taxonomy of mermithid nematodes that attack black flies is poorly developed and is based primarily on the free-living adult stage, even though only the juvenile stage is encountered in black flies. The taxonomic state of affairs is exacerbated by the controversial efforts of the Russian worker I. A. Rubtsov, who described many new species of mermithids from the Palearctic Region based only on questionably useful characters of the juvenile stage (R. Gordon 1984). Molecular tools could be of great benefit in sorting out mermithid taxonomy. Procedures are available for mass collecting and culturing the free-living adult stage (Bailey et al. 1974, 1977). Seventeen described species of mermithid nematodes have been found in about 22 species of North American black flies (although some of these records are only from the Palearctic Region), and unidentified mermithids have been found in 124 simuliid species on the continent (Table 6.4). Worldwide, nearly 70 mermithid species have been found in black flies (Crosskey 1990).

Filarial nematodes are obligate parasites that complete their life cycle in two hosts. Those associated with black flies are in the family Onchocercidae and include *Onchocerca volvulus*, the causal agent for human onchocerciasis (river blindness) that is transmitted by members of the (non-Nearctic) *Simulium damnosum* complex. Female filarial nematodes produce live young, the microfilariae, that circulate in the host blood and are acquired by blood-feeding black flies. The microfilariae develop in female black flies and are transmitted, during a subsequent blood meal, to a vertebrate host, where maturation and reproduction occur. Of the world's 12 described species of filarial nematodes for which black flies serve as intermediate hosts (summarized by Adler & McCreadie 2002), 4 are found in North America. One of these (*Splendidofilaria fallisensis*) is a parasite of ducks, and the remainder parasitize large mammals (Table 6.5).

Among the more recently discovered symbionts of black flies are the tiny nematodes of the family Robertdollfusidae, which are parasites of mammals. They were first found in black flies in the early 1990s during gut dissections of females of the *S. damnosum* species complex from Cameroon, Africa (Bain & Renz 1993). They have not been found in black flies from North America, but perhaps only because no one has looked for them.

### FUNGI

Four major groups of fungi have been found in black flies. A single common species (*Coelomycidium simulii*) in the class Chytridiomycetes attacks larval black flies throughout the world. Members of the class Trichomycetes are abundant in black flies worldwide but are not detectable without dissection of the gut. Species in the class Zygomycetes typically attack adult black flies, although larvae evidently can become infected as well; all species attacking black flies are members of the order Entomophthorales. Members of the class Hyphomycetes are rarely observed in black flies.

The chytrid fungus *C. simulii* was first reported (as a gregarine) from the larva of an unnamed species of *Simulium* around Boston, Massachusetts (Strickland 1913b), and later was named from a European black fly (Debaisieux 1919). It is one of the most widespread parasites of black flies, but patent infections typically are found in less than 1% of any larval population (Jamnback 1973a, McCreadie & Adler 1999). Patently infected larvae, recognized by the small spherical sporangia that fill the body cavity (Fig. 6.2), eventually assume a C shape, whereupon the cuticle ruptures, releasing the fungal zoospores (Tarrant 1984). Modes of transmission have not been resolved unequivocally, but evidence suggests transovarial (vertical) transmission (Tarrant 1984, Tarrant & Soper 1986). Larva-to-larva (horizontal) transmission is assumed,

**Table 6.4. Mermithid Nematodes (Family Mermithidae) of North American Black Flies**

| Mermithid nematode | Simuliid host |
|---|---|
| *Gastromermis bobrovae* | *H. alpestris*[1] |
| *Gastromermis rosalba* | *S. rostratum*[1] |
| *Gastromermis viridis* | *P. mixtum*, *S. vittatum* complex, *S. murmanum*, *S. venustum* complex |
| *Gastromermis* sp. | *P. caudatum*, *P. dicum*, *S. canadense*, *S. arcticum* complex |
| *Hydromermis* sp. | *S. venustum* complex |
| *Isomermis rossica* | *S. rostratum*[1] |
| *Isomermis wisconsinensis* | *P. mixtum*,[2] *S. vittatum* complex, *S. venustum* complex |
| *Isomermis* sp. | *S. vampirum* |
| *Mesomermis albicans* | *S. noelleri*[1] |
| *Mesomermis arctica* | *S. giganteum*[1] |
| *Mesomermis baicalensis* | *H. alpestris*[1] |
| *Mesomermis camdenensis* | *S. tuberosum* complex (probably *S. appalachiense* n. sp.), *S. venustum* complex |
| *Mesomermis canescens* | *S. noelleri*[1] |
| *Mesomermis flumenalis*[3] | *P. mixtum*, *P. magnum* complex, *St. mutata*, *S. aureum* complex, *S. donovani*, *S. argus*,[2] *S. vittatum* complex, *S. piperi*, *S. hippovorum*, *S. murmanum*, *S. noelleri*, *S. tuberosum* complex, *S. venustum* complex, *S. venustum* s. s. |
| *Mesomermis melusinae* | *S. noelleri*,[1] *S. rostratum*[1] |
| *Mesomermis paradisus* | *P. exigens* |
| *Mesomermis parallela* | *H. alpestris*[1] |
| *Mesomermis pivaniensis* | *S. vulgare*[1] |
| *Mesomermis sibirica* | *S. murmanum*[1] |
| *Mesomermis* sp. | *P. magnum* complex |
| Unidentified species[4] | *Gymnopais* (2 species), *Helodon* (7 species), *Prosimulium* (19 species), *Stegopterna* (3 species), *Tlalocomyia* (2 species), *Cnephia* (4 species), *Ectemnia* (1 species), *Metacnephia* (1 species), *Simulium* (85 species) |

*Note*: For each mermithid, black flies are listed in the order of their appearance in the taxonomic accounts (Chapter 10). Additional information and references can be found in the text under individual species accounts of black flies.

[1] Based on records outside North America north of Mexico.

[2] Identification of mermithid species uncertain.

[3] Often placed unnecessarily in the genus *Neomesomermis* (R. Gordon 1984).

[4] Includes our unpublished records, as well as records of mermithids cited in older literature as *Hydromermis* (a genus usually associated with Chironomidae), *Limnomermis*, and unidentified. Information on individual species of black flies infected with unidentified mermithids is given in the section on species accounts (Chapter 10).

**Table 6.5. Filarial Nematodes (Family Onchocercidae) of North American Black Flies**

| Filarial nematode | Simuliid (intermediate) host | Vertebrate host[1] |
|---|---|---|
| *Dirofilaria ursi* | *S. venustum* complex[2] | American black bear |
| *Onchocerca cervipedis* | *P. impostor*, *S. decorum*, *S. venustum* complex | mule deer, moose |
| *Onchocerca lienalis*[3] | *S. jenningsi* | cow |
| *Splendidofilaria fallisensis*[4,5] | *S. anatinum*, *S. rugglesi* | domestic duck, American black duck |

*Note*: References and additional information can be found in the text under individual species accounts of black flies (Chapter 10).

[1] Scientific names of vertebrate hosts are given in Tables 6.2 and 6.3.

[2] Includes *S. venustum* s. s. and perhaps *S. truncatum* (Hunter 1990).

[3] *Simulium vittatum*, *S. innoxium* (as *S. pictipes*), and *S. decorum* support development of filaria in the laboratory but have not been incriminated as vectors (Lok et al. 1983a).

[4] Formerly in the genus *Ornithofilaria*.

[5] *Simulium parnassum* and the *S. venustum* species complex support development to the infective stage in the laboratory, but are unlikely to be natural vectors (Anderson 1956).

possibly through an alternate host (Lacey & Undeen 1988). *Coelomycidium simulii* has been found in 95 (37%) of the North American black flies (Table 6.6). It might, however, be a complex of multiple species. For example, we have larvae of *Simulium malyschevi* from the Horton River, Northwest Territories, with infections of a fungus, presumably *C. simulii*, that have either large or small sporangia.

Fungi of the order Entomophthorales were first reported from black flies more than 100 years ago (Thaxter 1888). These fungi often kill the adult flies, usually through penetration of the cuticle. Infected flies typically bear a

**Table 6.6. Fungi and Stramenopiles of North American Black Flies**

| Fungus or stramenopile | Simuliid host |
|---|---|
| class Chytridiomycetes (fungi) | |
| order Chytridiales | |
| *Coelomycidium simulii* | *T. tibblesi, H. alpestris, H. irkutensis, H. onychodactylus* complex, *H. onychodactylus* s. s., *H. susanae, P. daviesi, P. mixtum, P. secretum* n. sp., *P. travisi, P. neomacropyga, P. exigens, P. impostor, P. magnum* complex, *P. albionense, St. acra* n. sp., *St. diplomutata, St. emergens, St. mutata, St. permutata, C. eremites, C. ornithophilia, E. primaeva, E. reclusa, M. borealis, M. saileri, M. villosa, Simulium* species (70 species) |
| class Hyphomycetes[1] (fungi) | |
| *Tolypocladium*? sp. | *S. piperi, S. arcticum* complex |
| class Trichomycetes (fungi) | |
| order Harpellales | |
| *Genistellospora homothallica* | *P. exigens, C. dacotensis, S. argus, S. vittatum* complex, *S. tribulatum, S. hippovorum, S. claricentrum, S. innoxium, S. arcticum* complex, *S. tuberosum* complex, *S. venustum* complex, *S. verecundum* complex, *S. verecundum* s. s. |
| *Harpella leptosa* | *G. adleri, S. canonicolum, S. argus, S. hunteri, S. piperi, S. virgatum, S. apricarium* n. sp., *S. venustum* complex |
| *Harpella melusinae* | *H. onychodactylus* complex, *P. fuscum, P. mixtum, P. saltus, P. neomacropyga, P. ursinum, P. constrictistylum, P. dicum, P. exigens, P. magnum* s. s., *G. humeralis* n. sp., *St. mutata, C. dacotensis, C. eremites, E. reclusa, S. congareenarum, S. annulus, S. aestivum, S. argus, S. vittatum* complex, *S. tribulatum, S. vittatum* s. s., *S. hippovorum, S. claricentrum, S. innoxium, S. dixiense, S. fibrinflatum, S. arcticum* complex, *S. decorum, S. noelleri, S. parnassum, S. tuberosum* complex, *S. tuberosum* s. s., *S. ubiquitum* n. sp., *S. vandalicum, S. venustum* complex, *S. venustum* s. s., *S. verecundum* complex, *S. verecundum* s. s. |
| *Pennella angustispora* | *S. argus, S. vittatum* complex, *S. hippovorum, S. arcticum* complex |
| *Pennella arctica* | *P. exigens, S. canonicolum, S. arcticum* complex |
| *Pennella hovassi* | *P. mixtum,*[2] *S. vittatum* complex |
| *Pennella* near *hovassi* | *S. vittatum* complex, *S. tribulatum, S. innoxium* |
| *Pennella simulii* | *P. mixtum, St. mutata, S. vittatum* complex, *S. tribulatum, S. venustum* complex, *S. verecundum* s. s. |
| *Pennella* sp. | *C. dacotensis, E. primaeva, S. vittatum* complex |
| *Simuliomyces microsporus* | *P. mixtum, C. dacotensis, S. argus, S. vittatum* complex, *S. tribulatum, S. innoxium, S. arcticum* complex, *S. tuberosum* complex |
| *Smittium coloradense* | *Prosimulium* sp.[3] |
| *Smittium culicis* | *C. dacotensis,*[2] *S. vittatum* complex |
| *Smittium culisetae* | *S. vittatum* complex, *S. tribulatum, S. verecundum* s. s. |
| *Smittium megazygosporum* | *S. vittatum* complex, *S. tribulatum, S. vittatum* s. s.,[4] *S. innoxium*[4] |
| *Smittium pennelli* | *H. onychodactylus* complex, *P. exigens, S. defoliarti* |
| *Smittium simulii* | *S. argus, S. hippovorum, S. noelleri,*[5] *S. tuberosum* complex |
| *Smittium* sp. | *P. mixtum, St. mutata, S. vittatum* s. s. |
| *Stachylina* sp. | *S. vittatum* complex |
| unidentified species | *St. mutata, M. coloradensis* |
| class Zygomycetes (fungi) | |
| order Entomophthorales | |
| *Entomophaga* nr. *limoniae* | *S. verecundum* complex |
| *Entomophthora culicis* | *S. vittatum* complex, *S. venustum* complex, *S. rostratum* |
| *Erynia conica* | *S. vittatum* complex, *S. noelleri,*[5] *S. venustum* complex, *S. verecundum* complex, *S. rostratum* |
| *Erynia curvispora* | *S. vittatum* complex, *S. vittatum* s. s.,[4] *S. innoxium,*[4] *S. decorum* |
| unidentified species | *S. vandalicum* |
| class Oomycetes (stramenopiles) | |
| order Saprolegniales | |
| *Pythiopsis cymosa* | *P. mixtum* |
| *Saprolegnia ferax*[6] | *S. vittatum* complex |
| unknown | *P. magnum* s. s., *S. subpusillum, S. malyschevi, S. tuberosum* s. s. |

*Note*: For each fungus and stramenopile, black flies are listed in the order of their appearance in the taxonomic accounts (Chapter 10). Additional information and references can be found in the text under individual species accounts of black flies.

[1] Within the Hyphomycetes, no orders (or families) are recognized in current classifications.

[2] Identification of fungus species uncertain.

[3] Williams and Lichtwardt (1987).

[4] Infection induced in laboratory (Kramer 1983 for *Erynia curvispora*, Beard & Adler 2000 for *Smittium megazygosporum*).

[5] Based on records outside North America north of Mexico.

[6] Apparently nonpathogenic (Nolan 1976).

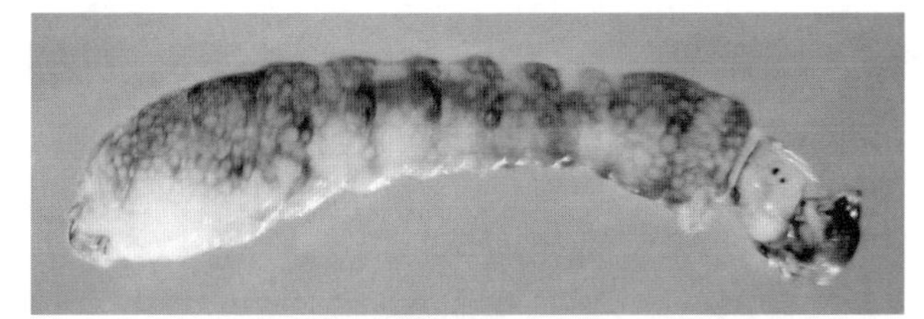

Fig. 6.2. Larva of *Simulium tuberosum* s. s. infected with the chytrid fungus *Coelomycidium simulii*; Horton River, Northwest Territories. Spherical sporangia are distributed throughout the body.

Fig. 6.3. Female of a member of the *Simulium vittatum* complex infected with the entomophthoraceous fungus *Erynia conica*; Saint-Maurice Reserve, Quebec. Eggs surround the female black fly. Photograph by J. L. Boisvert and M. P. Nadeau.

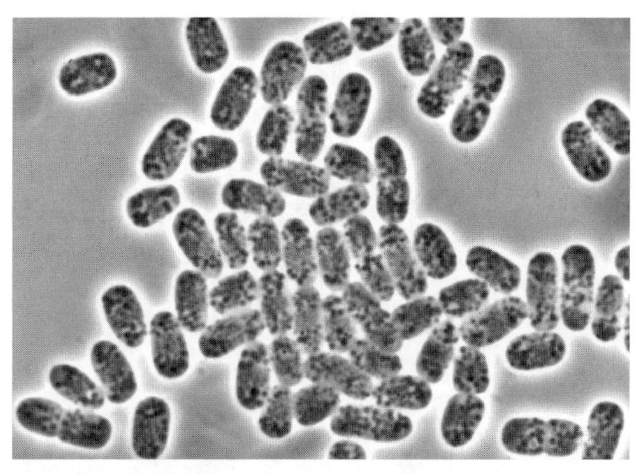

Fig. 6.4. Probable entomophthoraceous fungus (phase-contrast) from a larva of *Simulium vandalicum*; Pickens County, South Carolina.

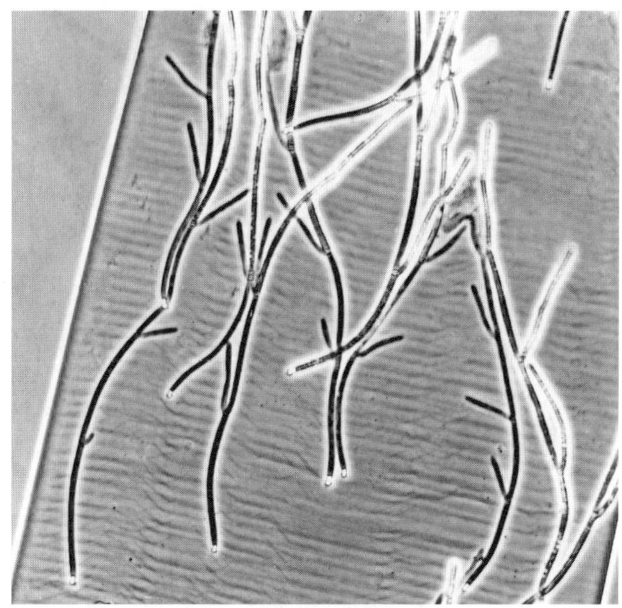

Fig. 6.5. The trichomycete fungus *Harpella melusinae* (phase-contrast) attached to the peritrophic matrix of a larva of *Simulum innoxium*; Pickens County, South Carolina. The peritrophic matrix is pleated in this simuliid species.

white coat of conidia and can be found, often in groups, plastered to a substrate at the oviposition site (Shemanchuk & Humber 1978) (Fig. 6.3). Certain species of black flies are apparently immune to some of these fungi, possibly because of the chemical composition of the fly's cuticle (Nadeau et al. 1994, 1996). A probable entomophthoraceous fungus with rodlike hyphal bodies (Fig. 6.4) has been found in the larval hemocoel of one North American black fly. Five species of Entomophthorales have been found in about six species of North American black flies (Table 6.6).

Trichomycete fungi are common inhabitants of the guts of larval black flies. Species of the genera *Harpella* (Fig. 6.5) and *Stachylina* attach their holdfasts to the cellophane-like peritrophic matrix of the midgut. All other species affix themselves to the cuticular wall of the hindgut. Trichomycetes are detrimental to females, replacing their eggs with fungal cysts (Lichtwardt 1996). In larvae, which are colonized when free trichospores released from other larvae are ingested, the relationship is described as commensalistic, although experimental demonstration of the nature of the relationship is wanting. Prevalence of trichomycetes in larval populations is typically seasonal and widely variable, sometimes reaching 100% (Taylor et al. 1996, Beard & Adler 2002). Worldwide, at least 35 species of trichomycetes have been found in black flies. They can be identified by the characteristics of their zygospores (e.g., Fig. 6.6), trichospores, and thalli. At least 15 species, identifiable with the keys of Lichtwardt (1986), have been found in North American black flies (Table 6.6). All but six species (*Smittium coloradense*, *S. culicis*, *S. culisetae*, *S. megazygosporum*, *S. simulii*, *Stachylina* sp.) restrict their reproduction to simuliid hosts, but host specificity within the North American Simuliidae is poorly developed; host preferences, however, have not been investigated. *Harpella*

Fig. 6.6. Tip of two conjugated thalli of the trichomycete fungus *Pennella simulii*, producing zygospores (length = 85 $\mu$m) (phase-contrast), from the hindgut of a larva of *Prosimulium mixtum*; Avalon Peninsula, Newfoundland. Photograph by R. W. Lichtwardt.

*melusinae*, the first trichomycete discovered in black flies (Léger & Duboscq 1929), is the most common and widespread species and probably occurs in nearly all species of black flies on the continent.

Additional fungi are associated with black flies, albeit infrequently. A species, perhaps of the genus *Tolypocladium* (Class Hyphomycetes), has been recovered from eight larvae in southern California (J. R. Harkrider, pers. comm.) (Table 6.6; Fig. 6.7).

#### STRAMENOPILES

At least two members of the class Oomycetes, both in the order Saprolegniales, have been found in association with black flies (Table 6.6). *Pythiopsis cymosa* has been isolated from a pupa (Nolan & Lewis 1974), and *Saprolegnia ferax* has been found in larval guts (Nolan 1976). Little is known of the frequency or nature of these associations, although *P. cymosa* was considered a parasite, whereas *S. ferax* was deemed nonpathogenic. We have found additional oomycete infections in which the larvae and pupae were coated with hyphal growth of an unknown species, but whether or not the infections were primary or secondary is not known.

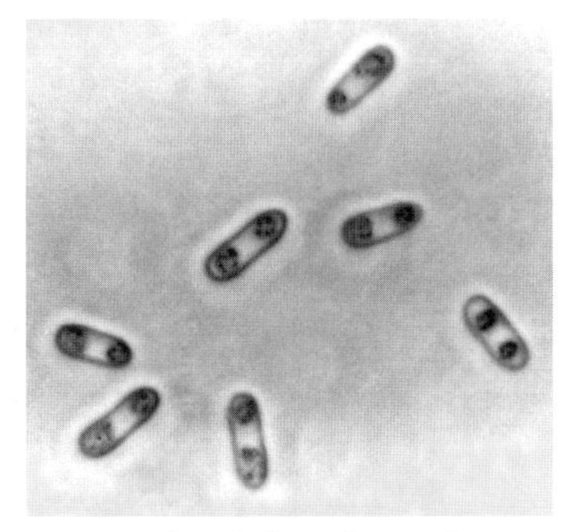

Fig. 6.7. *Tolypocladium*? fungus from a larva of *Simulium piperi*; Los Angeles County, California. Photograph by J. R. Harkrider.

#### MICROSPORIDIA

In the current molecular age, the higher classification of organisms such as microsporidia is undergoing substantial reassessment. Long classified as protozoans, the microsporidia are now considered by many workers to have phylogenetic affinities with the true fungi (Mathis 2000).

Microsporidia were first discovered in black flies by Léger (1897), who described *Amblyospora varians* (as *Thelohania varians*) from France. Patently infected larvae have conspicuous lobate cysts (Fig. 6.8). These cysts are typically white in North American black flies, although cysts produced by *Amblyospora bracteata* and *Janacekia debaisieuxi* are sometimes reddish in certain hosts. The cysts are located in the fat body, the only tissue known definitely to support development in black flies. Patent infections typically are found in less than 1% of a larval population (Crosskey 1990). Individual larvae can be infected simultaneously with two species of microsporidia.

Vertical transmission of at least one species of microsporidium (*Janacekia debaisieuxi*) has been demonstrated in black flies (Tarrant 1984), but horizontal transmission, although assumed, has not been confirmed and might require an alternate host (Lacey & Undeen 1988). In one set of experiments, 1%–4% of larvae reared from surface-sterilized eggs were infected with microsporidia, about the same level found in larvae from the same stream, suggesting that transovarial transmission alone can account for typical infection levels (Jamnback 1973a). On the other hand, vertical transmission is claimed not to occur in *Polydispyrenia simulii* (Castello Branco 1999).

Thirteen described species of microsporidia plus about seven additional, possibly undescribed species (Figs. 6.9–6.14) are known from North American black flies (Table 6.7). In addition, sibling species of microsporidia

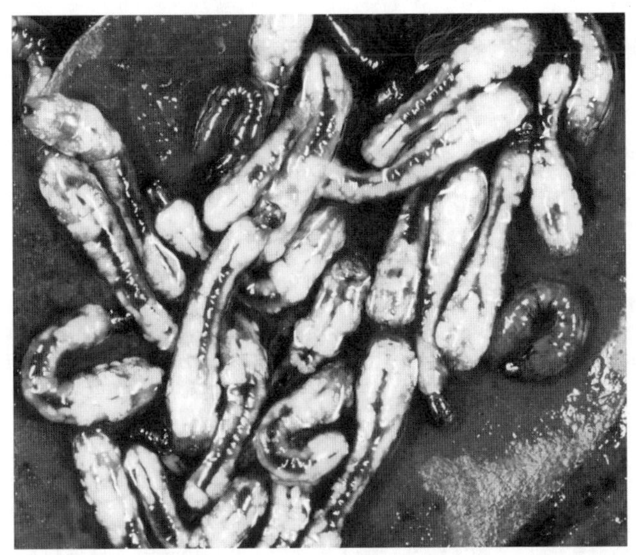

Fig. 6.8. Larvae of *Cnephia ornithophilia* infected with the microsporidium *Caudospora palustris*; Sumter County, South Carolina.

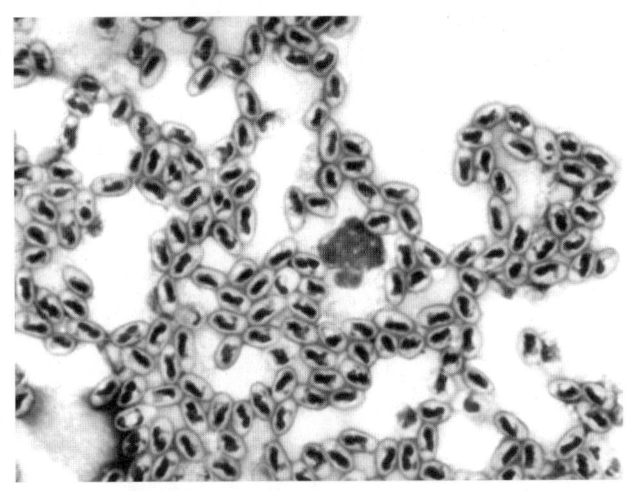

Fig. 6.10. Giemsa-stained spores, each approximately 6 μm long, of the microsporidium *Caudospora palustris* from a larva of *Cnephia ornithophilia*; Calhoun County, Georgia.

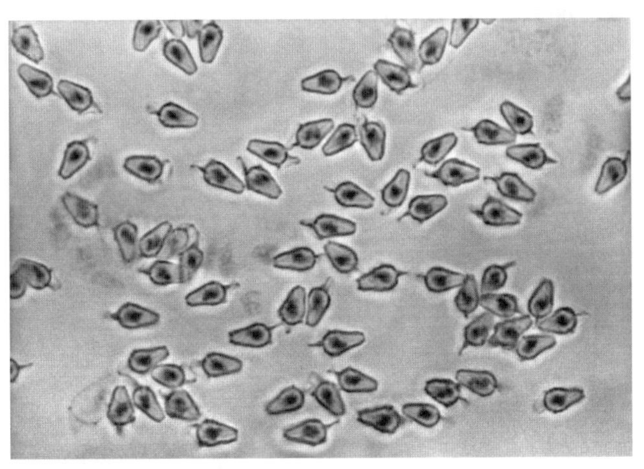

Fig. 6.9. Feulgen-stained spores, each approximately 5 μm long, of the microsporidium *Caudospora alaskensis* from a larva of *Prosimulium neomacropyga*; Steese Highway, mile 78, Alaska.

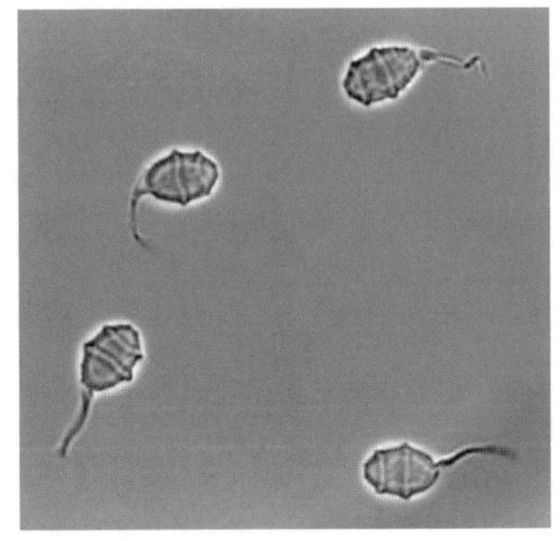

Fig. 6.11. Ethanol-fixed spores, each approximately 5 μm long, of the microsporidium *Caudospora polymorpha* from a larva of *Stegopterna diplomutata*; Burlington County, New Jersey.

probably exist. For example, ultrastructural studies suggest that *A. bracteata* is a complex of species (Hazard & Oldacre 1975). The 13 described species of microsporidia that attack black flies can be identified with the descriptions of Vávra and Undeen (1981), Weiser and Undeen (1981), and Adler et al. (2000). About 35 described species of microsporidia have been found in black flies worldwide.

Host specificity is greatest in the family Caudosporidae. The members of this family typically attack the more basal simuliid lineages such as *Gymnopais*, *Helodon*, *Prosimulium*, *Stegopterna*, and *Cnephia* (Adler et al. 2000). Most of the remaining microsporidia have broad host ranges but infect primarily members of the genus *Simulium*. *Polydispyrenia simulii* (Fig. 6.13) and *J. debaisieuxi* (Fig. 6.12), for example, have been found in 65 (26%) and 59 (23%), respectively, of the North American simuliid species. These broad host ranges, however, might reflect the presence of sibling species of microsporidia. Populations of *J. debaisieuxi* from different hosts often vary in spore size and ultrastructural details, lending support to the idea that more than one species is involved (Vávra & Undeen 1981).

#### HELICOSPORIDIA

An unidentified species (*Helicosporidium* sp.) of the order Helicosporidia represents one of the most recently discovered pathogens of black flies. It was found in the larvae of *Simulium jonesi* in Florida in September 1998 (Boucias et al. 2001). Larvae are infected by ingesting the cysts, which contain three ovoid cells and one filamentous cell. They subsequently manifest a cloudy appearance in the posterior portion of the abdomen. The detailed developmental and morphological analyses of Boucias et al.

**Table 6.7. Microsporidia of North American Black Flies**

| Microsporidium | Simuliid host |
|---|---|
| family Duboscqiidae | |
| *Pegmatheca simulii* | *S. jonesi*, *S. tuberosum* complex, *S. ubiquitum* n. sp. |
| *Polydispyrenia simulii*[1] | *P. arvum*, *St. acra* n. sp., *C. dacotensis*, *M. borealis*, *M. jeanae*, *M. saileri*, *S. congareenarum*, *S. innocens*, *S. annulus*, *S. canonicolum*, *S. joculator* n. sp., *S. johannseni*, *S. maculatum*,[2] *S. aureum* complex, *S. donovani*, *S. pilosum*, *S. violator* n. sp., *S. bicorne*, *S. carbunculum* n. sp., *S. conicum* n. sp., *S. craigi*, *S. fontinale*, *S. moultoni* n. sp., *S. silvestre*, *S. bivittatum*, *S. mediovittatum*, *S. trivittatum*, *S. venator*, *S. vittatum* complex, *S. tribulatum*, *S. vittatum* s. s., *S. hunteri*, *S. tescorum*, *S. paynei*, *S. solarii*, *S. claricentrum*, *S. innoxium*, *S. confusum*, *S. definitum*, *S. dixiense*, *S. fibrinflatum*, *S. jenningsi*, *S. jonesi*, *S. krebsorum*, *S. lakei*, *S. notiale*, *S. taxodium*, *S. murmanum*, *S. apricarium* n. sp., *S. brevicercum*, *S. negativum* n. sp., *S. saxosum* n. sp., *S. decorum*, *S. noelleri*, *S. parnassum*, *S. tuberosum* complex, *S. appalachiense* n. sp., *S. conundrum* n. sp., *S. perissum*, *S. tuberosum* s. s., *S. twinni*, *S. ubiquitum* n. sp., *S. vandalicum*, *S. vulgare*, *S. venustum* complex, *S. irritatum*, *S. venustum* s. s., *S. verecundum* complex, *S. rostratum*, *S. verecundum* s. s. |
| family Janacekiidae | |
| *Janacekia debaisieuxi*[3] | *H. onychodactylus* complex,[4] *H. diadelphus* n. sp.,[4] *H. susanae*,[4] *P. arvum*, *P. clandestinum* n. sp., *P. fuscum*, *P. mixtum*, *P. rhizophorum*, *P. saltus*, *St. diplomutata*, *St. mutata*, *C. ornithophilia*, *C. pecuarum*, *M. borealis*, *M. saileri*, *S. congareenarum*, *S. rendalense*, *S. curriei*,[4] *S. annulus*, *S. maculatum*,[2] *S. bracteatum*, *S. exulatum* n. sp., *S. violator* n. sp., *S. craigi*, *S. croxtoni*, *S. fionae*, *S. fontinale*, *S. modicum* n. sp., *S. moultoni* n. sp., *S. silvestre*, *S. vittatum* complex, *S. tribulatum*, *S. vittatum* s. s., *S. hunteri*, *S. piperi*, *S. confusum*, *S. dixiense*, *S. fibrinflatum*, *S. jenningsi*, *S. jonesi*, *S. lakei*, *S. defoliarti*, *S. murmanum*, *S. apricarium* n. sp., *S. chromatinum* n. sp., *S. saxosum* n. sp., *S. decorum*, *S. noelleri*, *S. slossonae*, *S. tuberosum* complex, *S. appalachiense* n. sp., *S. tuberosum* s. s., *S. twinni*, *S. ubiquitum* n. sp., *S. vandalicum*, *S. vulgare*, *S. venustum* complex, *S. irritatum*, *S. tormentor* n. sp., *S. truncatum*,[2] *S. venustum* s. s., *S. verecundum* complex, *S. rostratum*, *S. verecundum* s. s. |
| family Amblyosporidae[5] | |
| *Amblyospora bracteata* | *M. borealis*, *M. saileri*, *S. aureum* complex, *S. silvestre*, *S. furculatum*, *S. argus*, *S. vittatum* complex, *S. tribulatum*, *S. vittatum* s. s., *S. hunteri*, *S. piperi*, *S. paynei*,[2] *S. murmanum*, *S. brevicercum*, *S. chromatinum* n. sp., *S. decorum*, *S. noelleri*, *S. tuberosum* s. s., *S. venustum* complex, *S. tormentor* n. sp., *S. venustum* s. s., *S. verecundum* complex, *S. rostratum* |
| *Amblyospora fibrata* | *P. dicum*, *S. aureum* complex, *S. vittatum* complex, *S. hunteri*, *S. iriartei*, *S. piperi*, *S. canadense*, *S. paynei*,[2] *S. jenningsi*, *S. luggeri*, *S. noelleri*,[2] *S. petersoni*, *S. tuberosum* complex, *S. vandalicum*, *S. venustum* complex |
| *Amblyospora varians* | *S. maculatum*,[2] *S. bicorne*, *S. impar*, *S. loerchae*, *S. argus*, *S. vittatum* complex, *S. tribulatum*, *S. vittatum* s. s., *S. saxosum* n. sp., *S. decorum*, *S. noelleri*,[2] *S. petersoni*, *S. tuberosum* complex, *S. tuberosum* s. s., *S. twinni*, *S. ubiquitum* n. sp., *S. vandalicum*, *S. venustum* complex, *S. verecundum* complex, *S. rostratum*, *S. verecundum* s. s. |
| *Amblyospora bracteata/varians* | *S. jonesi*, *S. appalachiense* n. sp. |
| *Amblyospora* spp.[6] | *M. saileri*, *S. loerchae*, *S. modicum* n. sp., *S. anchistinum*, *S. apricarium* n. sp., *S. negativum* n. sp., *S. vampirum* |
| family Caudosporidae | |
| *Caudospora alaskensis*[7] | *T. nova*, *H. alpestris*, *H. onychodactylus* complex, *H. beardi* n. sp., *H. chaos* n. sp., *H. onychodactylus* s. s., *P. fulvithorax*, *P. shewelli*, *P. travisi*, *P. neomacropyga*, *P. exigens*, *St. diplomutata/mutata*, *S. craigi* |
| *Caudospora palustris* | *St. diplomutata*, *St. mutata*, *C. ornithophilia* |
| *Caudospora pennsylvanica* | *P. magnum* complex |
| *Caudospora polymorpha*[8] | *St. diplomutata*, *St. mutata* |
| *Caudospora simulii* | *P. fuscum*, *P. mixtum*, *P. rhizophorum*, *P. multidentatum*, *P. magnum* complex, *P. canutum* n. sp., *P. magnum* s. s. |
| *Caudospora stricklandi* | *St. emergens* |
| *Weiseria sommermanae* | *Gymnopais* sp. |
| unplaced to family[9] | |
| species 1 | *S. pictipes* |
| species 2 | *S. griseum*, *S. notatum* |
| species 3 | *S. apricarium* n. sp., *S. negativum* n. sp., *S. vampirum*, *S. petersoni* |
| species 4 | *S. negativum* n. sp. |
| species 5 | *P. exigens* |
| species 6 | *P. doveri*, *P. frohnei* |
| species 7 | *P. travisi* |

**TABLE 6.7. CONTINUED**

| Microsporidium | Simuliid host |
|---|---|
| unidentified species | *G. dichopticus, G. holopticus, T. tibblesi, H. decemarticulatus, H. gibsoni, H. vernalis, P. formosum, P. impostor, G. humeralis* n. sp., *St. trigonium, M. sommermanae, S. loerchae, S. pugetense, S. wyomingense, S. furculatum, S. jenningsi, S. rugglesi, S. rubtzovi, S. hematophilum* |

*Note*: Classification of microsporidia follows that of Sprague et al. (1992). For each microsporidium, black flies are listed in the order of their appearance in the taxonomic accounts (Chapter 10). Additional information and references can be found in the text under individual taxonomic accounts of black flies.

[1] Appeared in earlier North American literature as *Pleistophora multispora* and *Thelohania multispora*.

[2] Based on records outside North America north of Mexico.

[3] Appeared in earlier North American literature as *Tuzetia debaisieuxi*.

[4] Identification of microsporidium species uncertain.

[5] The three nominal species of *Amblyospora* infecting North American black flies formerly were placed in the genus *Thelohania*.

[6] Probably *A. bracteata* and/or *A. varians*.

[7] Probably a synonym of *Weiseria laurenti* (Adler et al. 2000).

[8] Appeared in earlier North American literature as *Caudospora brevicauda*.

[9] All seven "species" are distinct from any of the known named microsporidia in North America, based on spore shape; at least some probably represent undescribed species.

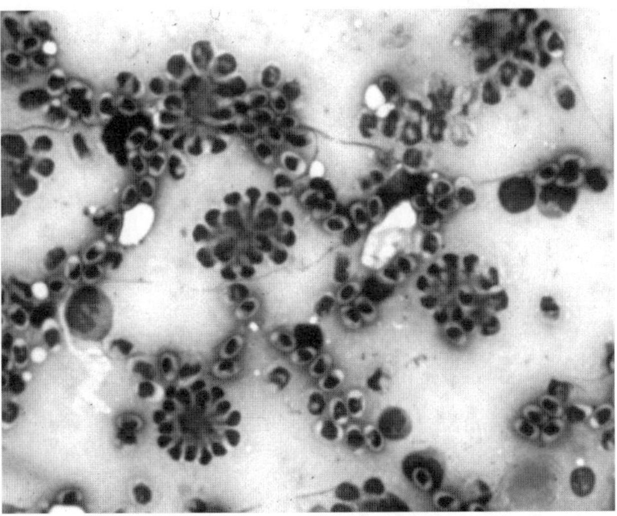

Fig. 6.12. Giemsa-stained spores, each approximately 6 $\mu$m long, and multiple fission of sporogonial plasmodia of the microsporidium *Janacekia debaisieuxi* from a larva of *Simulium tuberosum* s. s.; Pickens County, South Carolina.

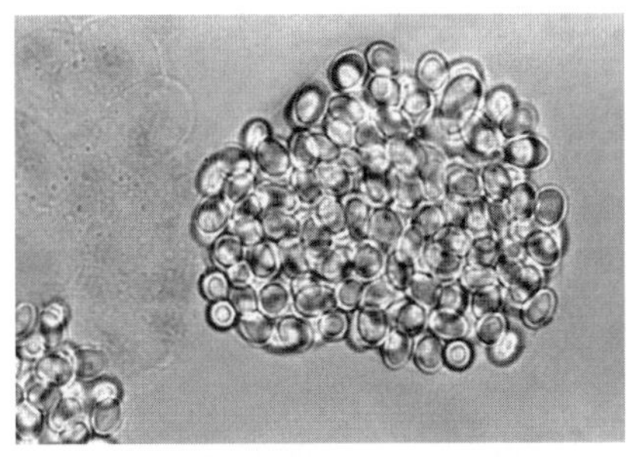

Fig. 6.13. Ethanol-fixed spores, each approximately 4 $\mu$m long, of the microsporidium *Polydispyrenia simulii* from a larva of *Simulium verecundum*; Gilchrist County, Florida.

(2001) indicate that this pathogen is capable of infecting a wide range of aquatic and terrestrial insects. Molecular analyses indicate that the helicosporidians are actually green algae (Chlorophyta) (Tartar et al. 2002).

**ICHTHYOSPOREANS**

Species of the genera *Amoebidium* and *Paramoebidium* historically were treated as trichomycete fungi (e.g., Lichtwardt 1986). Molecular evidence, however, suggests that *Amoebidium* is not a fungus but a member of the class Ichthyosporea (Benny & O'Donnell 2000). *Paramoebidium* is presumably close to *Amoebidium* but has not been investigated at the molecular level and here is placed only tentatively with the ichthyosporeans. Species of *Paramoe-*

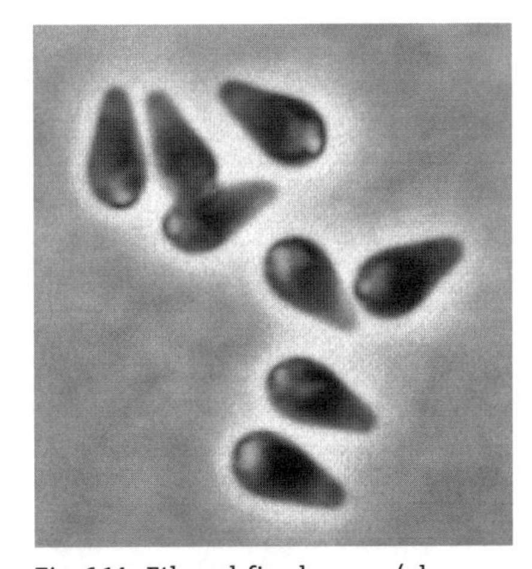

Fig. 6.14. Ethanol-fixed spores (phase contrast), each approximately 5 $\mu$m long, of a probable undescribed species of microsporidium from a larva of *Simulium petersoni*; Summit County, Utah.

*bidium* are common hindgut inhabitants of larval black flies, attaching their thallus-like structures to the cuticular lining by means of a holdfast. These thallus-like structures eventually release amoeboid cells that settle on the substrate, encyst, and probably are ingested by larval black flies (Lichtwardt 1976). *Paramoebidium curvum* was the first ichthyosporean recorded from black flies (Chatton & Roubaud 1909; as *Amoebidium* sp.). Four species of *Paramoebidium* are known from black flies, and at least two of these occur in North America. One described species of *Amoebidium* has been found on the rectal papillae of larval black flies in Costa Rica (Lichtwardt 1997).

### PROTISTS

The protists that attack black flies constitute a disparate assemblage of organisms that fall into four groups. Ciliates, members of the phylum Ciliophora, are found sporadically in black flies in various parts of the world. One species of the phylum Haplosporidia is known from the larvae of a single species of North American black fly. Species of the phyla Apicomplexa (*Leucocytozoon*) and Sarcomastigophora (*Trypanosoma*) exist in a parasitic relationship that involves the female black fly and the vertebrate hosts to which the parasites are transmitted.

Ciliates are poorly known in black flies, although they are probably more common than generally appreciated. They occur in the hemocoel of larvae, pupae, and adults, but their mode of transmission is unknown. Only two species have been described from black flies in the world. In North America, one described species (*Tetrahymena rotunda*) and at least one unidentified species have been found in nine species of black flies (Table 6.8). The prevalence of infection by *T. rotunda* is typically less than 2% (Lynn et al. 1981). In black flies, infections with *Tetrahymena* generally are not conspicuous but at times can become pathogenic and kill the host (Batson 1983). In South Carolina, noticeably infected larvae have been found with large numbers of ciliates throughout the hemocoel (C. E. Beard and P. H. Adler, unpublished data). Infections by *Tetrahymena* in alcohol-fixed larvae superficially resemble those caused by the fungus *Coelomycidium simulii*.

A single haplosporidian species (*Haplosporidium simulii*), the only one known from black flies, has been found in larvae of the *Simulium venustum* species complex in Pennsylvania (Beaudoin & Wills 1968). Its pathogenicity and mode of transmission are unknown.

*Leucocytozoon* protozoa are blood parasites of birds, producing a malaria-like disease known as "leucocytozoonosis." They have a complex life cycle involving both avian and simuliid hosts (Long et al. 1987, Steele & Noblet 2001). Female black flies acquire *Leucocytozoon* gametocytes (Fig. 6.15) from the blood of infected birds and then

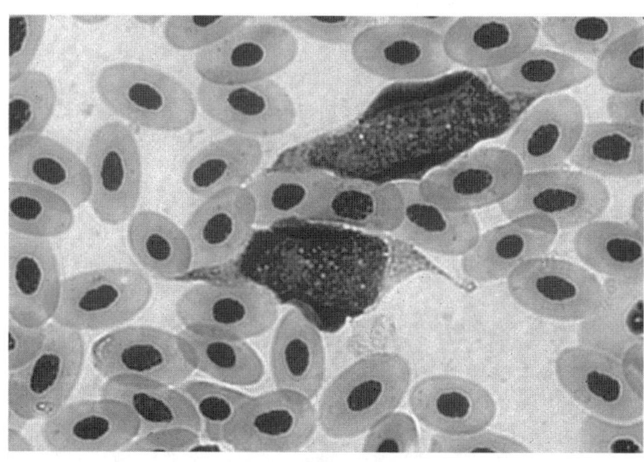

Fig. 6.15. Two gametocytes of *Leucocytozoon smithi*, each approximately 12 μm long, among red blood cells of a turkey; South Carolina. Photograph by G. P. Noblet.

**Table 6.8. Parasitic Protists, Helicosporidia, and Ichthyosporeans of North American Black Flies**

| Symbiont | Simuliid host |
|---|---|
| phylum Ciliophora | |
| *Tetrahymena rotunda* | *S. tuberosum* complex, *S. venustum* complex |
| *Tetrahymena* sp. | *S. congareenarum*, *S. anchistinum*, *S. fibrinflatum*, *S. jonesi*, *S. lakei*, *S. verecundum* complex |
| unidentified species | *S. jenningsi* |
| phylum Haplosporidia | |
| *Haplosporidium simulii* | *S. venustum* complex (probably *S. venustum* s. s.) |
| order Helicosporidia | |
| *Helicosporidium* sp. | *S. jonesi* |
| class Ichthyosporea | |
| *Paramoebidium chattoni*[1,2] | *S. vittatum* complex, *S. tribulatum*, *S. innoxium* |
| *Paramoebidium curvum*[1] | *S. vittatum* complex, *S. tribulatum*, *S. innoxium*, *S. verecundum* complex, *Simulium* sp.[3] |
| *Paramoebidium* sp.[1,4] | *S. vittatum* complex, *S. tribulatum*, *S. innoxium*, *S. fibrinflatum* |

*Note*: For each symbiont, black flies are listed in the order of their appearance in the taxonomic accounts (Chapter 10). Additional information and references can be found in the text under individual species accounts of black flies.

[1] Previously considered trichomycete fungi.

[2] Sensu Moss (1970); species name technically a *nomen nudum*.

[3] Dang and Lichtwardt (1979).

[4] Includes *Paramoebidium* "B" (Beard & Adler 2002).

support both asexual and sexual development of the parasite. During a subsequent blood meal, sporozoites of the parasite are transmitted to birds, which then serve as hosts for an asexual stage. *Leucocytozoon* protozoa are prevalent in simuliids; for example, 90%–100% of ornithophilic species in Algonquin Park, Ontario, have sporozoites in their salivary glands during the summer (Bennett & Squires-Parsons 1992). The taxonomy of the parasites is in some confusion, but each species of parasite appears to be specific to an avian family (Bennett et al. 1991, p. 1408). The parasites are widespread in birds, but the vectors, although generally assumed to be simuliids, are largely unknown. Worldwide, at least 12 described species of *Leucocytozoon* are known to be transmitted by black flies. About 18 species of black flies have been implicated as natural vectors of 9 *Leucocytozoon* species in North America (Table 6.9), but considerably more species undoubtedly serve as vectors.

**Table 6.9. Species of *Leucocytozoon* and *Trypanosoma* of North American Black Flies and Examples of Their Avian Hosts**

| Protist | Simuliid host[1] | Avian host[2,3] (family) |
|---|---|---|
| phylum Apicomplexa | | |
| *Leucocytozoon cambournaci*[4] | *H. decemarticulatus*, *C. ornithophilia*, *S. aureum* complex, *S. craigi*(?),[5] *S. quebecense*(?), *S. silvestre*(?) | white-throated sparrow (Emberizidae) |
| *Leucocytozoon dubreuili*[6] | *H. decemarticulatus*, *C. ornithophilia*, *S. aureum* complex, *S. craigi*(?), *S. croxtoni*, *S. quebecense*(?), *S. silvestre*(?) | American robin (Turdidae) |
| *Leucocytozoon icteris*[4] | *H. decemarticulatus*, *C. ornithophilia*, *S. anatinum*, *S. annulus*, *S. aureum* complex, *S. craigi*(?), *S. croxtoni*, *S. quebecense*(?), *S. silvestre*(?), *S. venustum* complex | common grackle (Icteridae) |
| *Leucocytozoon lovati*[7] | *S. aureum* complex, *S. craigi*(?), *S. croxtoni*, *S. quebecense*(?), *S. silvestre*(?) | blue grouse, ruffed grouse (Phasianidae) |
| *Leucocytozoon sakharoffi*[8] | *H. decemarticulatus*, *S. aureum* complex | common raven, American crow, blue jay (Corvidae) |
| *Leucocytozoon simondi*[9] | *C. ornithophilia*, *S. anatinum*, *S. rendalense*,[10] *S. usovae*,[10] *S. rugglesi*, *S. venustum* complex | domestic duck, American black duck, mallard, wood duck, redhead, northern pintail, Canada goose, snow goose (Anatidae) |
| *Leucocytozoon smithi* | *P. mixtum*(?), *S. congareenarum*, *S. meridionale*, *S. aureum* complex, *S. jenningsi* group, *S. jenningsi*(?), *S. slossonae* | turkey (domestic and wild) (Phasianidae) |
| *Leucocytozoon toddi* | *H. decemarticulatus*, *S. aureum* complex, *S. quebecense*(?) | sharp-shinned hawk (Accipitridae) |
| *Leucocytozoon ziemanni*[11] | *H. decemarticulatus*, *S. aureum* complex, *S. craigi*(?), *S. silvestre*(?) | northern saw-whet owl (Strigidae) |
| phylum Sarcomastigophora | | |
| *Trypanosoma confusum*[12] | *H. decemarticulatus*, *S. aureum* complex, *S. craigi*(?), *S. croxtoni*, *S. quebecense*(?), *S. silvestre*(?), *S. rugglesi* | domestic duck, ruffed grouse, blue grouse, blue jay, gray jay, American robin, purple finch, white-throated sparrow (numerous families) |

*Note*: For each protist, black flies are listed in the order of their appearance in the taxonomic accounts (Chapter 10). Additional information and references can be found in the text under individual species accounts of black flies.

[1] The species listed can serve as hosts (i.e., support development) of the parasite but have not necessarily been demonstrated to be vectors in the wild.

[2] Scientific names of avian hosts are given in Table 6.3.

[3] Black flies have not been observed feeding on some hosts for which *Leucocytozoon* infections have been reported (e.g., Laird & Bennett 1970). As of 1975, *Leucocytozoon* infections had been found in 114 (17.7%) of 645 species of birds surveyed in North America (Greiner et al. 1975).

[4] Previously known as *L. fringillinarum* (Bennett & Squires-Parsons 1992).

[5] (?) = Identification of simuliid species uncertain.

[6] Previously known as *L. mirandae* (Khan & Fallis 1970).

[7] Previously known as *L. bonasae* (Bennett et al. 1991).

[8] Previously known as *L. berestneffi* (Bennett & Peirce 1992).

[9] Previously known as *L. anatis*.

[10] Based on records outside North America north of Mexico.

[11] Previously known as *L. danilewskyi* (Bennett et al. 1993).

[12] Previously known as *T. avium*.

Trypanosomes also are blood parasites of birds, requiring a female black fly to complete part of their life cycle and to transfer them to their avian host. Transmission evidently occurs when contaminated fecal droplets from the fly enter the bite (Bennett 1961). In North America, one described species, *Trypanosoma confusum* (formerly *T. avium*), possibly a species complex, is transmitted by at least seven species of black flies (Table 6.9).

#### BACTERIA

Reports of bacteria causing pathogenic effects in black flies are rare, although the few known examples suggest inadequate investigation. An unidentified bacterium that produces white cysts has been found in a larva of the *Simulium venustum* complex (Weiser & Undeen 1981). A similar pathology was recorded for a larva of the *Simulium tuberosum* complex that was infected with the bacterium *Bacillus amyloliquefaciens* (Reeves & Nayduch 2002). A gram-negative bacterium in the wings of *Cnephia ornithophilia* had unknown effects on the host (Weiser & Undeen 1981).

#### VIRUSES

Various viruses have been found in larval black flies since the first report by Weiser (1968) of an iridescent virus in Czechoslovakian larvae. Their mode of transmission is largely unknown, although laboratory infections have been produced by feeding homogenates of diseased larvae to first instars (Federici & Lacey 1987). Probably the most common virus is a cytoplasmic polyhedrosis virus, which attacks the midgut epithelium and gastric caeca (Weiser & Undeen 1981). A patent infection in a live larva can be seen through the integument as an opaque whitish band around the midgut. This virus has been reported from at least 10 species of North American black flies. An iridescent virus that produces bluish larvae (Weiser & Undeen 1981) has been recorded from four species of North American black flies. Other virus-like particles have been recorded from three North American species (Charpentier et al. 1986, Federici & Lacey 1987) (Table 6.10).

#### OTHER PARASITES

Black flies are hosts of additional pathogens and parasites, some accidental, some rare, and some simply overlooked because they are not readily apparent in the host. Laboratory experiments demonstrate that larval black flies can serve as second intermediate hosts of trematode cercariae (Bušta & Našincová 1986, Jacobs et al. 1993), suggesting that larvae in the wild also might carry infections. An immature gordian worm, *Paragordius varius* (Nematomorpha), has been found in a single larva of *Simulium impar* in North Carolina (W. K. Reeves, pers. comm.). Claims of immature gordian worms in larval black flies in Wisconsin (White 1969) very well might apply to mermithid nematodes. Whether parasitism of simuliids by gordian worms is accidental, or common but overlooked because of their small size, remains unknown. Several unidentified organisms use black flies as hosts. For example, a cystlike cluster of an unknown organism (Fig. 6.16) was discovered by W. K. Reeves in the abdomen of a female of *Prosimulium mixtum* from the Great Smoky Mountains of North Carolina.

#### ECTOPARASITES

Water mites (Hydracarina or Hydrachnellae) are the only known ectoparasites of black flies, other than an occasional terrestrial trombidioid mite (Davies 1959). Larval water mites enter the pupal cocoon where they presumably rest, eventually gaining access to the adult flies as they emerge. Once they have boarded the flies, water mites assume a parasitic existence, penetrating the cuticle with their chelicerae. Female flies typically have two to

**TABLE 6.10. VIRUSES OF NORTH AMERICAN BLACK FLIES**

| Virus | Simuliid host |
|---|---|
| larval black fly as host | |
| cytoplasmic polyhedrosis virus | *P. mixtum*, *P. rhizophorum*, *St. mutata*, *S. aureum* complex, *S. donovani*, *S. croxtoni*, *S. vittatum* complex, *S. tribulatum*, *S. vittatum* s. s., *S. noelleri*,[1] *S. tuberosum* complex, *S. venustum* complex |
| intranuclear virus-like particle | *S. vittatum* complex |
| iridescent virus | *P. mixtum*,[2] *S. vittatum* complex, *S. paynei*,[1] *S. luggeri* |
| virus-like particle | *S. donovani*,[3] *S. argus*,[3] *S. vittatum* complex |
| adult black fly as host | |
| bunyavirus | *S. bivittatum* |
| eastern equine encephalitis virus | *S. johannseni*, *S. meridionale*[4] |
| snowshoe hare virus | *S. arcticum* complex |
| vesicular stomatitis virus | *S. notatum*, *S. vittatum* complex, *S. vittatum* s. s. |
| unidentified arbovirus | *S. meridionale* |

*Note*: For each virus, black flies are listed in the order of their appearance in the taxonomic accounts (Chapter 10). Additional information and references can be found in the text under individual species accounts of black flies.

[1] Based on records outside North America north of Mexico.

[2] Identification of simuliid species questionable.

[3] Infection induced in laboratory (Federici & Lacey 1987).

[4] Suspected host (Anderson et al. 1961, DeFoliart & Rao 1965).

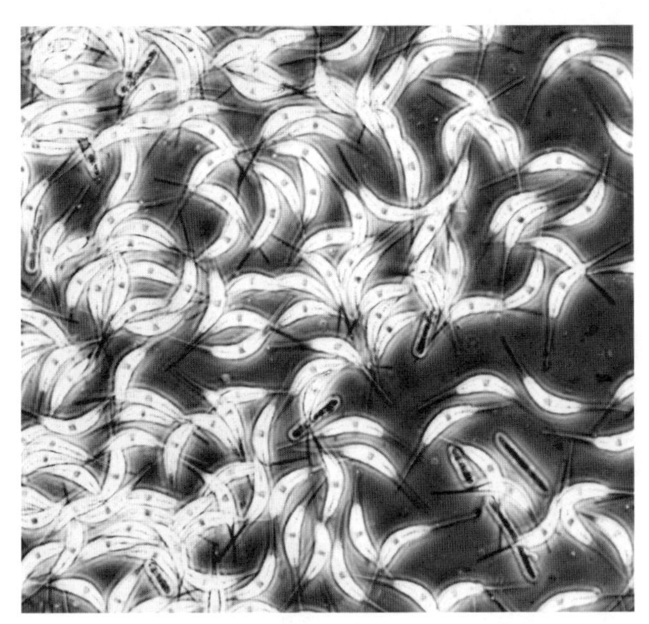

Fig. 6.16. Unidentified symbiont (phase-contrast) from the abdomen of a female of *Prosimulium mixtum*; Great Smoky Mountains National Park, North Carolina.

five times more water mites per fly than do males; more than 30 water mites have been found on a single female fly (Davies 1959). Attachment sites are primarily at the base of the abdomen (Davies 1959, Pascuzzo 1976). Infestations reduce fecundity (Pascuzzo 1976), but further investigations of the effects on fitness are needed, especially in relation to different levels of infestation. Larval water mites eventually return to running water where they will transform into predaceous adults. This transport is accomplished, at least in part, by females returning to oviposit (Davies 1959). Only two species of water mites are known from black flies in North America. *Sperchon ?jasperensis* has been recorded from about 10 species of black flies in Canada (Davies 1959), and *Sperchon texana* from one species in Texas (Davis & Cook 1985).

**PREDATORS**

Reports of predation on black flies are extensive (Davies 1981, 1991; Crosskey 1990) and include cannibalism (Burton 1971) and even routine predation on larvae by the Yanomami Indians of Brazil (Shelley & Luna Dias 1989). At least one record of predation by a carnivorous pitcher plant (*Sarracenia purpurea*) has been reported in North America; adult black flies constituted about 0.8% of the plants' total prey (Cresswell 1991). A few cases have been documented where black flies are so abundant that they provide a major portion of the diet of certain predators and influence their population dynamics and reproduction. In Iceland, for example, the enormous populations of *Simulium vittatum* in the River Laxá drive production of animals such as brown trout (*Salmo trutta*), harlequin duck (*Histrionicus histrionicus*), and Barrow's goldeneye (*Bucephala islandica*) (Gíslason 1985, Gardarsson & Einarsson 1994). Black flies also constitute a major portion of the diet of some North American ducks (e.g., harlequin duck) (Rodway 1998, Robert & Cloutier 2001) and numerous fish species (Power 1969, Greger & Deacon 1987, Gutowski & Stauffer 1993). More than 1600 adult simuliids and about 600 larvae have been found in the stomachs of individual brook trout (*Salvelinus fontinalis*) in Algonquin Park, Ontario (Ide 1942). The effects of predation on black fly populations, however, have rarely been documented. Mortality attributed to predation was estimated at 80.3%–86.5% for four species of black flies in Oregon (Speir 1975). A Swedish study determined that 0.2%–10.0% of larval black flies were consumed per day by 11 species of predators (Malmqvist 1994).

Studies of behavioral interactions between predators and their simuliid prey and the evolutionary impacts on the prey are scant, as are data on antipredator defenses, particularly of eggs, pupae, and adults. The preimaginal stages of some simuliid species might minimize predation by adjusting their phenologies (Malmqvist 1994) or occupying predator-free space such as fast currents (Fuller & DeStaffan 1988, Hart & Merz 1998) and glacial meltwater streams. Those of other species might overwhelm predators numerically (predator satiation) by massing in enormous densities (Wotton & Merritt 1988). Predation also might be avoided if size selection is practiced, as it is in some fish (Moore & Moore 1974). One of the most common active defense mechanisms of larvae is downstream drifting, typically a few centimeters, with or without a silk lifeline; this drift is triggered especially when the posterior portion of the abdomen is touched (Simmons 1982). Larvae also respond to predators by curling the anterior of their body against the posterior portion, thus forming a C shape, or they relocate by moving in a looping fashion. Larvae might be able to detect some aquatic predators chemically and relocate prior to contact. Limited evidence suggests that larvae can distinguish certain predatory from nonpredatory Trichoptera (Wiley & Kohler 1981). Retaliation might be a common means of defense, for larvae often spar with other simuliid larvae, probably as a means of maintaining space. Large larvae of the *Helodon onychodactylus* species complex repel small stoneflies by repeatedly biting the predators about the head (J. D. Allan et al. 1987). Other larvae, under laboratory conditions, have bound and killed mayfly nymphs with silk threads by lashing at them (Bradt 1932). Larvae of many simuliid species are assumed to be cryptic; for example, brown larvae of various *Prosimulium* species are difficult to see against brown leaves. The adaptive value of crypsis, however, has been tested in only one case. Larvae of the *Simulium vittatum* species complex, which generally match the color of their substrate (Adler & Kim 1984), as first observed by McBride (in Riley 1870c), are conferred a selective advantage against mosquitofish (*Gambusia holbrooki*) when attached to an appropriately colored substrate (Zettler et al. 1998).

# PART III
# Economic Aspects

# 7| Social and Economic Impact

The ravages of black flies undoubtedly span human history on the continent, although most information has been lost to the pantheon of time. Little is known of the problems experienced by the natives or how they coped with black flies before the first explorers and their domestic animals arrived. The written history of troubles inflicted by black flies in North America is about 400 years old, dating to the writings of Samuel de Champlain, who journeyed through eastern Canada from 1599 to 1613. Writings of other early explorers, especially in Canada, provide ample documentation of maddening encounters with black flies (Agassiz 1850, Davies et al. 1962, Laird et al. 1982, Currie 1997a).

A mere 16% of the North American species of black flies have caused economic losses and suffering of humans and their animals (Table 7.1). Perhaps only a third of these species are consistent major pests; the remainder are sporadic, minor, or localized pests. Together these pests have provoked enormous efforts to control their ravages (see Chapter 8). Few readers will disagree that black flies have been devastating not only to the composure of humanity but also to the economy. The following treatment illustrates both the problems caused by black flies and the dearth of knowledge about their impact, especially in dollar figures.

## Factors Promoting Pest Problems

The factors that drive pest status are poorly known. Certain species are habitual biters in some geographic areas but not in others. *Simulium venustum* s. s., for example, is one of the most relentless human biters in northeastern North America, but for unknown reasons it rarely attacks humans south of Pennsylvania, despite its abundance. Some Alaskan species are aggressive human biters in late September and October, but not earlier in the year (Sailer 1954). Within a species, larger females—those produced under more optimal larval conditions—are more prone to bite than are smaller females (Mokry 1980b, Simmons 1985). Thus, water quality, or the quantity and quality of food in a stream, sometimes could influence pest status in localized areas. Many of the continent's worst pests, past and present, have been big-river species, reflecting the enormous populations that large waterways often produce.

Pest problems can appear without warning, depending on numerous factors (Cupp 1988) that often are poorly understood. Completion of hydroelectric dams on the North and South Saskatchewan Rivers of the Canadian prairies changed the river conditions so dramatically that the historical pest *Simulium vampirum* (originally known as *S. arcticum*) was replaced by an even more hostile pest, *Simulium luggeri* (Fredeen 1985b). Large numbers of flies sometimes invade previously untroubled areas, brought in by prevailing winds (Webster 1902) or traveling frontal systems associated with fluctuations in atmospheric pressure (Wellington 1974). Hurricane Agnes, for example, was suggested as an explanation for the record numbers of black flies in the town of Deep River, Ontario, in 1972 (Baldwin et al. 1977). El Niño of 1998 was blamed for the outbreak of *Simulium slossonae* as a nuisance of humans in parts of Florida, presumably because it brought rains that created excess breeding habitats. Much credence has been given to improved water quality, stemming primarily from the federal Clean Water Act of 1972, as a factor promoting pest problems, particularly those caused by the big-river species such as *Simulium jenningsi* in eastern North America.

## Biting and Nuisance Problems

### Humans

About 33 species—13% of the North American simuliid fauna—cause discomfort for humans by biting, ceaselessly swarming around the head and entering the orifices, or furtively working their way into openings in the clothing (Table 7.1). At least 54 North American species have been recorded to bite humans at least once (Table 6.2), but only about half of these can be considered pests. No known species in North America, or anywhere in the world, is exclusively anthropophilic.

Even when not biting, a few flies darting about the face and entering the eyes can make outdoor activities unpleasant. As few as five flies captured around a person in 10 overhead sweeps of a standard insect net is deemed unacceptable on a golf course, and 10 flies per sweep session is unacceptable for the general public of Pennsylvania (Gray et al. 1996). With a bit of imagination, one might begin to appreciate the situation in some northern

**Table 7.1. Pests of Humans, Livestock, and Poultry in North America**

| Pest species | Geographic area of problem | Economic impact[1] |
|---|---|---|
| *Twinnia nova* | Rocky Mountains of Alberta and British Columbia | sporadic, minor biting and nuisance pest of humans and horses |
| *Prosimulium fontanum* | northeastern North America | sporadic, minor nuisance to humans |
| *Prosimulium fulvum* | mountains from Alaska to Utah | sporadic biting and nuisance pest of humans and livestock |
| *Prosimulium mixtum* | eastern North America | major biting and nuisance pest of humans; perhaps the worst simuliid pest of humans during springtime |
| *Prosimulium magnum* species complex | eastern United States | sporadic, minor biting pest of humans and livestock |
| *Stegopterna diplomutata/ mutata* | eastern North America | sporadic, minor nuisance to humans |
| *Stegopterna acra* n. sp. | mountains of western North America | sporadic, minor nuisance to humans |
| *Cnephia pecuarum* | southern Mississippi River Valley | historically one of the continent's worst biting and nuisance pests of livestock; responsible for untold millions of dollars in economic losses and thousands of animal deaths since at least 1818; only a fraction of actual livestock deaths (about 12,950) was recorded; single counties lost up to $0.5 million in 1882; putatively responsible for several human deaths; current problems affecting livestock and pets are highly localized |
| *Simulium anatinum* | Canada | vector of *Leucocytozoon simondi* among ducks and geese |
| *Simulium congareenarum* | southeastern United States | vector of *Leucocytozoon smithi* among turkeys; historically played a role in damaging turkey production |
| *Simulium johannseni* | midwestern North America | localized biting and nuisance pest of humans and livestock |
| *Simulium meridionale* | midwestern North America | biting pest of poultry, exotic birds, and native birds such as purple martins, tree swallows, and bluebirds; deaths of chickens and turkeys sometimes caused by exsanguination; $1.5 million lost to emu and ostrich industry in Texas in 1993; vector of *L. smithi* among turkeys, historically causing enormous economic losses; sporadic biting and nuisance pest of livestock and humans, often causing great discomfort |
| *Simulium aureum* species complex | North America | biting pest of poultry; vector of *L. smithi* among turkeys |
| *Simulium bivittatum* | western North America | biting pest of livestock and sporadic biting pest of humans |
| *Simulium clarum* | southern California | biting pest of livestock and sporadic biting pest of humans; responsible for $818 in lost milk production on three dairy farms in Merced County in 1962 |
| *Simulium griseum* | western North America | biting pest of livestock and sporadic biting pest of humans |
| *Simulium mediovittatum* | Texas | sporadic biting pest of livestock and humans; attacks have been severe in some years |
| *Simulium notatum* | southwestern United States | vector of vesicular stomatitis virus among livestock |
| *Simulium venator* | western United States | sporadic, minor biting pest of horses and humans |
| *Simulium argus* | western United States | biting pest of livestock |
| *Simulium tribulatum* | North America | biting pest of livestock (especially horses and sheep) and nuisance to humans[2]; $34,000/month lost in green fees by five golf courses in Los Angeles County, California, during the 1994–1995 season |
| *Simulium vittatum* s. s. | North America | biting pest of livestock (especially horses and sheep) and nuisance to humans[2]; vector of vesicular stomatitis virus among livestock |
| *Simulium hunteri* | western North America | sporadic, minor biting and nuisance pest of humans and cattle |
| *Simulium tescorum* | southwestern United States | biting pest of humans; frequently requires management |
| *Simulium hippovorum* | southern California | biting pest of horses |
| *Simulium infenestrum* | western South Carolina | biting and nuisance pest of humans |
| *Simulium jenningsi* | eastern North America | major nuisance to humans; perhaps the greatest black fly pest of humans at the end of the 20th century; subject of largest North American management program for black flies; biting pest of cattle; vector of *Onchocerca lienalis* among cattle |
| *Simulium jenningsi* species group[3] | southeastern United States | sporadic nuisance to humans; occasional biting pest of livestock; $27,000 lost to economy of South Carolina in 1 year (early 1990s) as result of problems on one golf course |

**Table 7.1. Continued**

| Pest species | Geographic area of problem | Economic impact[1] |
|---|---|---|
| *Simulium luggeri* | Manitoba, Saskatchewan, Minnesota | major biting pest of cattle and sporadic biting pest of humans since early 1970s; worst outbreak was in Saskatchewan in 1978 when $3,058,000 was lost to the beef and dairy industries, and at least 14 livestock deaths were reported; major problems continued into the 21st century |
| *Simulium penobscotense* | Maine | biting pest of humans and cattle |
| *Simulium decimatum* | Alaska, Yukon Territory, Northwest Territories | major biting and nuisance pest of humans |
| *Simulium defoliarti* | southern British Columbia | sporadic biting pest of cattle; more than $24,000 in beef cattle lost in Cherryville District in 1952 |
| *Simulium vampirum* | prairies of Alberta and Saskatchewan | historically (as *S. arcticum*) one of the continent's worst biting and nuisance pests of livestock; responsible for enormous economic losses and livestock deaths since 1886; more than 3500 deaths and losses of about $1.6 million recorded (most losses not recorded); up to 16% reduction in egg production in poultry industry; humans attacked in severe outbreaks; some problems remain in Alberta but not in Saskatchewan |
| *Simulium decorum* | North America | sporadic, minor biting pest of humans; probable mechanical vector of causal agent of tularemia |
| *Simulium parnassum* | mountains of eastern North America | biting pest of humans and sporadic biting pest of livestock |
| *Simulium rugglesi* | Canada and northcentral United States | vector of *L. simondi* among ducks and geese |
| *Simulium slossonae* | southeastern United States | vector of *L. smithi* among turkeys, often producing high mortality levels; historically played a major role in damaging turkey production; for example, $11,250 lost to turkey industry in Jasper County, South Carolina, in 1952; occasional biting and sporadic nuisance pest of humans in Florida (severe in winter of 1997–1998) |
| *Simulium hematophilum* | northeastern North America | biting pest of humans |
| *Simulium irritatum* | northern North America | biting pest of humans |
| *Simulium truncatum* | northern North America | biting pest of humans |
| *Simulium venustum* s. s.[4] | northeastern North America | major biting and nuisance pest of humans |

*Note*: Black flies are listed in the taxonomic order in which they appear in Chapter 10. References and details of the impacts of each pest can be found under the relevant taxonomic accounts in Chapter 10.

[1] Dollars have not been adjusted for inflation; Canadian dollars are typically given for Canadian species.

[2] See also detailed treatment under *Simulium vittatum* species complex.

[3] Includes species in the *Simulium jenningsi* group (other than *S. infenestrum*, *S. jenningsi* s. s., *S. luggeri*, and *S. penobscotense*), especially *S. anchistinum*, *S. confusum*, *S. fibrinflatum*, *S. jonesi*, *S. lakei*, and *S. notiale*.

[4] Other members of the *Simulium venustum* species complex (in addition to *S. hematophilum*, *S. irritatum*, *S. truncatum*, and *S. venustum* s. s.) are probably biting and nuisance pests of humans, but inability to distinguish females of all species in the complex precludes determination of their hosts and economic impacts.

areas where thousands of black flies can descend on a person within minutes (Fig. 7.1). Such astronomical numbers of nuisance black flies, however, are not unique to the remote north woods. In the early 1980s, three overhead passes of an insect net yielded as many as 1800 black flies in the metropolitan area of Minnesota's Twin Cities below the Coon Rapids Dam (Stamps 1985).

People differ in their attractiveness to black flies, perhaps because of differing production rates of exhaled carbon dioxide, and some people are bitten more readily than others (Schofield & Sutcliffe 1996, 1997). Fresh bites seem to prompt additional flies, especially those of the *Simulium venustum* species complex, to lacerate nearby flesh, often producing concentrations of bites (Stokes 1914, Simmons 1985). In fact, an "invitation effect" has been demonstrated in a member of the African *Simulium damnosum* complex in which the proportion of flies feeding increases with the number of conspecifics present, suggesting the influence of pheromones or host odors (McCall & Lemoh 1997).

The pathology of the bites inflicted on humans has been well described (Stokes 1914, Gudgel & Grauer 1954). Externally, the bite appears as a small reddish or purplish hemorrhage in the center of a variously raised area. The bites themselves are often painless, and their presence is sometimes first detected by streaks of blood oozing from the wound, facilitated by anticoagulant properties of the saliva. Anecdotal statements that the bites of species preferring human blood seem less painful (Scoles in Luoma 1984) call for more studies of the saliva, particularly its

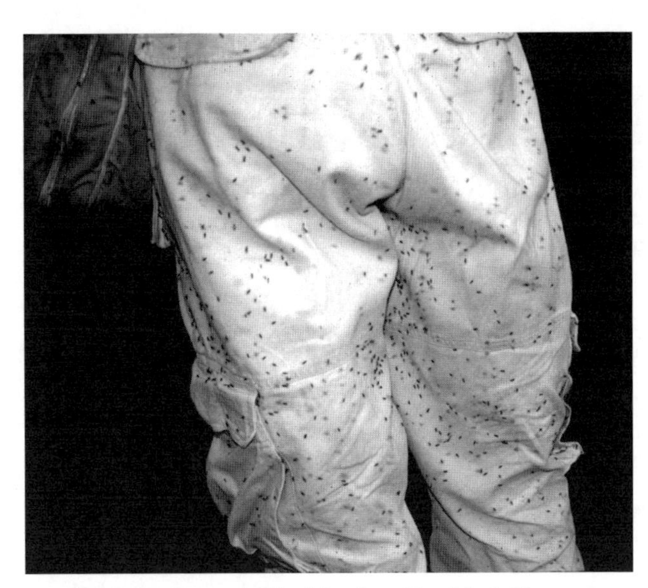

Fig. 7.1. Trousers covered by blood-seeking black flies, mainly members of the *Simulium venustum* complex, in July along the Horton River, Northwest Territories.

anesthetic properties. Individual reactions subsequently vary, with a fortunate portion of the population experiencing little anguish. For many people, however, the bites can induce an aggravating itch that persists for several days or more; excessive scratching can precipitate secondary infections. Numerous bites can cause more general reactions. Bites around the eyes can impair vision by causing the surrounding tissue to swell (Jamnback & Collins 1955), a condition infrequently referred to as "bung-eye" (Smart 1952). In eastern Canada, painful swelling of lymph nodes at the base of the head and ears is called "black fly stiff neck" (Peterson & Wolfe 1958). "Black fly fever," as it is known especially in the northeastern United States, involves headaches, nausea, fever, and swollen lymph nodes in the neck (Jamnback 1969a, Molloy 1984).

About 10% of people experience allergic reactions to bites (Reiling et al. 1988). Severe allergic reactions occur (Fredeen 1969, Brown & Bernton 1970, Frazier 1973, Grande 1997) but infrequently find their way into the literature. Children generally are more susceptible than adults (Craig 1972c), and in areas such as Baie Comeau, Quebec, they have been unable to attend school because of infections caused by bites (Prevost 1946). If repeatedly bitten, children can develop hypochronic anemia and experience a dramatic increase in the number of white blood cells; complete recovery often requires more than 6 months (Knap 1969). Some immunity to bites evidently can develop (Stokes 1914, Davies in Odum 1973), but immunity to bites of *Prosimulium* seems not to confer immunity to those of *Simulium* (Wood 1985).

The consequences of simuliid attacks can be unpleasant or even life threatening, but human mortality from bites is exceedingly rare. All human deaths attributed directly to the bites of black flies were recorded before the 20th century. Most cases might be apocryphal. Certainly, the citizens of severely afflicted areas often were whipped into a lather of hyperbole when the black flies attacked: "Having once tasted human gore, may we not expect, in future, to hear of man-eating Buffalo gnats?" (Anonymous in Doran 1887). Nonetheless, as many as four deaths could be legitimate, including one from Newfoundland (Noble 1861), perhaps caused by members of the *S. venustum* species complex, and three from Arkansas and Louisiana, putatively caused by *Cnephia pecuarum* (Riley 1887, Buck in Riley 1888b, Webster 1904).

Psychological reactions to attacks by black flies are very real and can lower individual efficiency (Knap 1969), with susceptible individuals rapidly reaching an "emotional state bordering on dementia" (Hocking 1952a). Luoma's (1984) account of a trip to the Minnesota north country paints a vivid image: "The swarm stays attached to you, like a huge, living helmet . . . [and] this chaotic mass around your head can plunge you into a metaphysical abyss, destroy all perspective." Perhaps the early Canadian accounts of humans driven insane by black flies stem from fierce cases of black fly fever; whether the madness was temporary or permanent seems not to have been mentioned.

#### DOMESTIC AND WILD ANIMALS

Livestock pests represent about 9% of the simuliid fauna on the continent and include some of history's most damaging species (Table 7.1, Fig. 7.2). Massive attacks by several species, notably *Cnephia pecuarum*, *Simulium luggeri*, and *Simulium vampirum*, have caused mortality of horses, mules, pigs, cattle, and sheep (Riley 1887, Bradley 1935a, Fredeen 1977b; *vampirum* as *arcticum*); details of these attacks are provided under the respective species accounts. Deaths of livestock usually are attributed to toxic shock (simuliotoxicosis) from the salivary injections of many females. They also occur as a result of accidents from frenzied behavior (Fredeen 1977b) or as a consequence of asphyxiation from fly-clogged respiratory passages (Tucker 1918) and infections from inhalation (Atwood 1996). Irritation from many bites around the nostrils is said to cause cattle to snort and take deep breaths, resulting in the inhalation of large numbers of flies that irritate the respiratory passages and cause coughing (Hadwen 1923). Immunological responses of cattle to the bites have been described (Beck 1980). Damage to horses in the mountains of Alberta, manifested as swellings under the jaws and belly, bleeding from the mouth and nostrils, and rapid death, once was known locally as "swamp fever" (Fredeen 1969). Attacks on livestock also cause reduced milk production, hyperactivity, malnutrition, impotence, weight loss, delayed conception, and possibly eye inflammations and stress-related diseases such as pneumonia (Jessen 1977; Fredeen 1984, 1985b; Kampani 1986). The tendency of animals under attack to bunch together day after day in the same spot might enhance transmission of pink eye (conjunctivitis) through contact and might lead to foot rot by creating a low-lying wet area (Kampani 1986).

Many livestock pests (e.g., *Simulium vittatum*) feed inconspicuously in the ears of their hosts, often causing

Fig. 7.2. Cattle under attack by a cloud of *Simulium vampirum*; Athabasca River Valley, Alberta. Photograph by J. F. Sutcliffe.

severe scabbing and irritation that leads to unruly behavior. In horses, a temporary condition known as "lop ear" results from inflammation of the bites (Merritt 1979). Eight flies per ear produce sufficient scabbing to warrant prophylactic measures, and 14 flies are enough to cause adverse behavior by the horses (Townsend 1975, Townsend et al. 1977). In areas, such as eastern Ontario, where a cloud of at least 10,000 black flies was reported swarming around a single horse (MacNay 1959a), the potential for damage is great. Some simuliid species feed principally on the thinly haired ventral regions of the body such as the udder, impairing nursing and milking (Rempel & Arnason 1947, Anderson & Voskuil 1963).

Poultry pests account for about 3% of the total simuliid species in North America, and they afflict their hosts primarily through the transmission of disease agents (see below). Poultry under attack experience reduced egg production, inflammation, egg desertion, loss of appetite, and sometimes death (Swenk & Mussehl 1928, Edgar 1953). Feeding on poultry is often concentrated around the eyes, head, and neck. Exotic birds such as cockatoos and parrots also have been killed by excessive attacks (Mock & Adler 2002).

The impact of biting and swarming on wildlife is poorly known. For many wild hosts, the species of black flies that feed on them are not even known. Nonetheless, available evidence suggests that the impact of black flies on wild hosts can be severe. Members of the *Simulium venustum* species complex, for example, can be so vexatious to caribou that, along with other flies, they might influence the extent and timing of their migrations (Harper 1955). We have paddled canoes through a choking smoke of black flies that arose from a herd of caribou crossing the Horton River in the Northwest Territories. Similarly, we have seen furious swarms of pestilential black flies around bison in the Northwest Territories. Birds sitting on their nests are likely targets (Fig. 7.3), and a few reports indicate that nestlings can be killed by massive biting attacks. One or more species of the *Simulium annulus* group have killed nestlings of red-tailed hawks by causing them to dehydrate or jump from their nest (Fitch et al. 1946, Smith et al. 1998). A suite of species, including members of the *S. annulus* species group, similarly have killed great horned owls less than 10 weeks old (Hunter et al. 1997a). Females of *Simulium meridionale* have caused nestling mortality of purple martins (Hill 1994), tree swallows, and bluebirds (Gaard 2001, 2002; as *Cnephia taeniatifrons*).

Despite the many zoos and wild animal parks in North America that maintain a variety of animals in outdoor arenas, not a single black fly has been recorded as harassing or feeding on the animals. We expect, however, that incidents often occur but either go unnoticed or are not reported in the technical literature.

Fig. 7.3. Common loon beseiged by a probable member of the *Simulium annulus* species group; Fawn Lake, British Columbia. Photograph and permission by Neil K. Dawe, Canadian Wildlife Service.

## Economic Losses

### HUMAN ACTIVITIES

Actual dollar losses in the entertainment and work sectors are largely undocumented, although many reports (e.g., Metcalf 1932, Glasgow 1936, D. Peterson 1955, Jamnback 1973b, Laird et al. 1982) consider black flies an integral part of the economy because they discourage tourism and interfere with timbering, mining, and construction. The value of the Adirondack Mountains of New York in the early 1930s was reduced by about 40%–50% each season (mid-May to mid-July) by black flies that wreaked havoc not only on the tourists and workers (e.g., road construction crews) but also on the patients of the large tuberculosis sanitariums whose therapeutic outdoor routines were severely hampered (Metcalf 1932). Some conservationists, however, were not happy with these statistics, suggesting that control efforts would destroy the natural conditions (Vogt 1936). But the crusade against black flies moved forward. A program to control black flies in the Adirondack town of North Elba purportedly bolstered the economy through increased tourism in the 1970s by $367,500 per year (Elliott 1983). By the late 1990s, North Elba was spending more than $63,000 per year for a suppression program. The town of Waterville Valley, New Hampshire, witnessed an 85% decline in tourism during the black fly season (May–July) of 1974, resulting in a loss of $244,800 (Martin 1981).

Costly losses to the economy have been documented outside the northeastern United States. During outbreaks of *Cnephia pecuarum* in the late 20th century, livestock producers in Texas and Arkansas worked an average of 20 additional hours per week, costing an average of up to $1200 in extra labor each year (Atwood 1996). A paper company in Texas lost $10,000 for each roll of product in which a single adult of *C. pecuarum* became mashed after having been attracted to the factory lights (Guyette 1999). Employees driving to work at the paper company were exposed to hazardous conditions in heavy swarms of this species (Guyette 1994). Residents in the Lower Colorado River Basin of the southwestern United States overwhelmingly believed that black flies were a detriment to outdoor living and pleasure in their area, with hundreds of residents seeking medical attention for bites (Mulla & Lacey 1976b). In the pulpwood-cutting regions of Canada, labor turnover was high, with three men often being hired in succession for one summer job along the North Shore of Quebec (West 1961). Industrial and recreational development of northern Canada, in general, has been delayed by simuliids and other biting flies (Laird 1980b, Laird et al. 1982). Even in southwestern Canada, forest-tree planters refused to work until they could be outfitted with head nets and veils, which then impeded their planting operations (Wyeth in Watts 1976).

Black flies are considered second only to mosquitoes as major arthropod pests in recreation areas of North America (Newson 1977), disrupting outdoor sports, picnics, vacations, and other activities. More than $27,000 was lost to the economy of South Carolina in 1994 as a result of nuisance black flies at a single golf course (Gray et al. 1996). Five golf courses along California's Los Angeles River lost approximately $34,000 per month in green fees during the 1994–1995 season (Shaw in O'Connor et al. 2001). Even Major League Baseball games have been disrupted by pestiferous black flies (O'Connor et al. 2001). Before the DDT era, black flies typically delayed the opening of summer hotels and resort camps in New York by as much as three to six weeks (Glasgow 1942, 1948). Additional problems can be masked because the public often perceives a mosquito, rather than a simuliid, problem (Crans & McCuiston 1970b). In 1927 in Yellowstone National Park, for example, harsh control measures were applied against mosquitoes for biting tourists when the real culprits were black flies. Flies have been so numerous in some areas that they gum up the windshields of passing automobiles (Jones 1934) and create a hazard for bicyclists. One can only imagine the effect that black flies have on the fishing, hunting, hiking, and camping industries. The misery often has been painted in colorful prose (e.g., Hafele 1997).

Almost no figures exist for the direct impact of black fly attacks on humans through reduced efficiency and loss of job time. However, an infantry division of the U.S. Reserve and National Guard at Camp Drum, New York, lost approximately 700 human hours of training time during a nine-day exercise (Frommer et al. 1975). In the 1960s and 1970s, about 100 people per year sought treatment for black fly bites at the naval station's dispensary in Kodiak, Alaska (Fussel 1970). Fort Leonard Wood in Missouri and the Holston Army Ammunition Plant in Kingsport, Tennessee, are among other military installations in the United States that have experienced severe problems with black flies (Frommer 1981). Personnel at the Royal Canadian Airforce Station at Mount Apica, Quebec, sustained serious problems from black flies, including hospitalizations and treatments with penicillin (Snider 1958). Three to five bites per minute at Canadian forces bases usually prompted adulticiding measures (Laird et al. 1982).

Somewhat ironically, the economic losses caused by black flies gain a certain balance with the economic returns to the management, protection, medical, and veterinary industries, not to mention the creation of employment opportunities. About 88% of people in the simuliid-problem zone along Maine's Penobscot River use some kind of personal protection, and in 1986–1987, about 20% of households spent an average of $6.00 per year for medical services, including anti-itch creams (Reiling et al. 1988, 1989). The two million-plus citizens in the problem areas of Pennsylvania paid about $1.58 per person in 1993 to suppress the nuisance pest *Simulium jenningsi* (Arbegast 1994). In the black fly season of 2002, nine biologists were employed full time in Pennsylvania's management program for *S. jenningsi*. Parts of southern Illinois (e.g., Carbondale) sometimes experience massive outbreaks of *C. pecuarum* (J. E. McPherson, pers. comm.); during these outbreaks some car-wash owners have claimed a temporary business boom from fly-splattered automobiles. Even the entertainment sector has capitalized on problems caused by black flies. For example, the community of Inlet in the Adirondack Mountains of New York calls itself the Black Fly Capital of the World and has held a festival to celebrate black flies (Lidz 1981). A Black Fly Open Golf Tournament is held in Copper Harbor, Michigan, every year near the end of May. The World Wide Web now has sites that provide weekly reports of the severity of the black fly problem in various afflicted areas, helping people plan their outdoor activities.

### LIVESTOCK AND POULTRY

Despite the enormous and pervasive impact of black flies, actual figures on economic losses are scarce (Table 7.1). Some of the best records, although representing an iota of the actual problem, come from the livestock industry. Deaths of livestock from withering attacks have been mentioned in the literature at least since the beginning of the 19th century (Cornelius 1818), but detailed assessments rarely have been made. About 16,450 deaths of domestic mammals, probably a fraction of the total number, actually have been tallied for the continent. Published figures on lost dollars are similarly meager, with only about $12.7 million in losses to the livestock industry having been reported in the published literature since the initial settlement of North America. The obvious inadequacy of this figure is illustrated by the estimation of nearly $6 million in losses to cattle producers in a single outbreak year in just two counties along the Arkansas-Texas border (Atwood 1996). Similarly, black flies caused an estimated loss of 2.0%–2.5% of total cash receipts from livestock farming in Saskatchewan in 1985 (Kampani 1986).

The poultry industry also has experienced enormous losses. About two thirds of lost dollars reported by the industry can be attributed to *Leucocytozoon* disease (discussed later), the causal agents of which are transmitted by black flies. Deaths from massive attacks also can occur, presumably as a result of toxemic shock or withdrawal of excessive blood. The deaths of more than 240 chickens and 20 turkeys in southwestern Manitoba in a single day were attributed to black fly attacks (Burgess in Riegert 1999b). One poultryman in Iowa lost 600 chickens and turkeys to black flies in a few weeks (Ainslie 1929). The ostrich and emu industries lost many birds and about $400,000 along the Trinity River in Texas during 1993 as a result of attacks by black flies (J. V. Robinson, pers. comm.).

We stress emphatically that the reported figures for deaths and lost dollars are inadequate vis-à-vis actual losses. They reflect a woeful failure, and suggest a challenging task, to document the economic impact of black flies. Grossly underreported are the effects of black flies on pets, although deaths and hospitalizations have been recorded (Atwood 1996). In addition, the indirect effects that pest outbreaks cause on the economy too often go unrecognized. The historical attacks of *Cnephia pecuarum*, for example, not only affected the livestock directly but also hampered the planting of crops, a task borne by horses and mules in the days before mechanization. Similarly, harassment of sheep has led to destruction of rangeland vegetation because the besieged animals stand in one spot and trample the vegetation (Jessen 1977).

## Vector-borne Diseases

Blood feeding is a prerequisite for the transmission of most disease agents. However, at least two blood meals are necessary for transmission, one meal to acquire the causal agent and one to transmit it. Anautogenous black flies, those that must take a blood meal, therefore, are among the most important vectors of disease agents. Black flies that are facultatively autogenous, maturing at least the first batch of eggs without a blood meal, probably play a lesser role in the transmission of disease agents. Changes in the appearance of the Malpighian tubules can be used to determine the proportion of a population returning for at least a second blood meal (Smith & Hayton 1995).

At least 10 species of protozoa, 4 filarial nematodes, an unknown number of arboviruses, and possibly a bacterium are transmitted by black flies in North America. Simuliid-borne pathogens of humans are restricted to Africa and Central and South America; native cases of human disease have not been documented in North America. Some potential, however, exists for black flies to transmit eastern equine encephalitis virus (DeFoliart & Rao 1965) and perhaps, in rare or aberrant situations, other pathogens to humans. However, of more than 3000 simuliids screened for arboviruses in southern California, not a single specimen tested positive (Reeves & Milby 1990).

### Avian Diseases

The most pernicious disease linked to North American black flies is leucocytozoonosis, an avian malaria-like affliction caused by nine described species of *Leucocytozoon* protozoa that are transmitted by simuliids (Table 6.9). Two species of parasite, *Leucocytozoon simondi* and *Leucocytozoon smithi*, have created the greatest economic problems. *Leucocytozoon simondi* is borne primarily by *Simulium anatinum* and *Simulium rugglesi* and is pathogenic to ducks and geese. *Leucocytozoon smithi* is transmitted to turkeys by *Simulium congareenarum*, *Simulium meridionale*, and *Simulium slossonae*. The disease is known colloquially as "duck malaria" if caused by *L. simondi* (Swales 1936), and "gnat fever," "turkey malaria," or "gnat disease" if caused by *L. smithi* (Bierer 1954).

Prevalence of infection approaches 100% in some areas (O'Roke 1934, Herman et al. 1975, Noblet & Moore 1975, Alverson & Noblet 1977). The degree of pathogenicity, however, appears to vary with host species and age, geographic location (e.g., Skidmore 1932, Byrd 1959, Fallis & Bennett 1966, Desser & Ryckman 1976), abundance of the simuliid vectors (Fallis et al. 1951), and other poorly understood factors such as concurrent infections with other diseases (Simpson et al. 1956). Pathogenicity is typically greatest in young domestic birds (Khan & Fallis 1968). Infected hosts that appear healthy and exhibit little or no adverse affects, nonetheless, might be more susceptible to predation, secondary infections, and other stresses. Chronically infected hosts experience reduced reproduction and depressed immune systems, and they serve as reservoirs for perpetuating the disease. Severely infected birds display reduced appetite, lethargy, emaciation, and convulsions that quickly lead to death. Subsequent necropsies reveal dehydration, enlarged liver and spleen, pale heart muscle, and congested lungs (Skidmore 1932, O'Roke 1934).

Economic losses from leucocytozoon disease can be great. Mortality levels in some turkey and duck operations range from 5% to 100% (e.g., Skidmore 1932, O'Roke 1934, Fallis et al. 1951, Jones & Richey 1956, Simpson et al. 1956). Mortality of domestic turkeys attributed to *L. smithi* in the United States from 1942 to 1951 was estimated at 0.3%, with an annual average economic loss of $708,000 (Agricultural Research Service 1954). An epizootic in one county in South Carolina in the early 1950s, plus a second in two counties in the early 1970s, resulted in losses that ran into hundreds of thousands of dollars in the abandonment of physical facilities alone (Barnett 1977). In those years, turkey production was largely an outdoor operation, often with large concentrations of birds that provided ideal conditions for the transmission of *L. smithi* (Fig. 7.4). A single farm near Manitoba's Assiniboine River suffered about 63% mortality of its 8000 turkeys in 1944 as a result of an unidentified species of *Leucocytozoon* presumably transmitted by black flies (Burgess in Riegert 1999b). High levels of leucocytozoonosis contributed to the discontinuation of Agriculture Canada's program to establish geese (from Montreal) around Fort Chimo, Quebec, as a new source of fresh food for the Inuit population (Laird & Bennett 1970).

*Leucocytozoon* infections are common in wild birds (e.g., Fallis & Bennett 1958, 1961, 1962), except in the arctic where vectors are lacking (Laird 1961). Reports of mortal-

Fig. 7.4. Turkey farm in South Carolina in the 1970s, providing ideal conditions for transmission of the agent of leucocytozoon disease. Photograph by R. Noblet.

ity and clinical disease, however, are rare, perhaps in part because of the greater difficulty in studying wild populations and the lesser attention paid to economically less important species. *Leucocytozoon simondi* is believed to be the prime cause of gosling deaths in populations of Canada geese in Seney National Wildlife Refuge, Michigan (Herman et al. 1975). *Leucocytozoon* infections were involved with mortality in great horned owls 10–14 weeks old in the Yukon Territory (Hunter et al. 1997a) and in red-tailed hawks about 1–6 weeks old in Wyoming (Smith et al. 1998). However, the relative effects of bites and blood loss, independent of the *Leucocytozoon* infections, were not clarified. The deleterious effects of both the bites and the disease are probably greater in years of lean food supply (Hunter et al. 1997a) and can vary, depending on weather (Smith et al. 1998). Of potentially great importance, but virtually unknown, are the effects of subclinical *Leucocytozoon* infections on host fitness (Hunter et al. 1997b).

The impacts of other North American simuliid-borne diseases on their avian hosts are virtually unknown. The protozoan *Trypanosoma confusum* occurs in a number of taxonomically unrelated host families (Bennett 1961; as *T. avium*) (Table 6.9). *Splendidofilaria fallisensis*, the only known avian filarial nematode transmitted by black flies, attacks the subcutaneous connective tissues of ducks in Ontario, Canada (Anderson 1956, 1968) (Table 6.5). Eastern equine encephalitis virus is believed to be transmitted, perhaps mechanically, to turkeys by *S. meridionale* and has been isolated from females of *Simulium johannseni* (Anderson et al. 1961, Spalatin et al. 1961). An unidentified arbovirus has been isolated from *S. meridionale* in Wisconsin (DeFoliart et al. 1969).

**MAMMALIAN DISEASES**

Three species of filarial nematodes are transmitted to mammals in North America. *Dirofilaria ursi*, a parasite of American black bears, is transmitted by members of the *Simulium venustum* species complex in Algonquin Park, Ontario (Addison 1980), and probably other areas such as Massachusetts (Simmons et al. 1989). The parasite develops in the Malpighian tubules of the simuliid host. In the bear, subcutaneous tissues house the microfilariae, whereas peritracheal tissues might be the preferred definitive site of the adult parasites (Addison 1980). The impact of the parasite on black bears is not known.

*Onchocerca lienalis* infects cattle in North America and Europe, causing bovine onchocerciasis. In New York State, it is transmitted to cattle by *Simulium jenningsi*, which feeds primarily in the umbilical region (Lok et al. 1983b). Infections typically do not cause overt symptoms in cattle, and the economic impact of the parasite has not been assessed. This parasite-vector system, however, serves as a laboratory model for studying the debilitating human filarial nematode *Onchocerca volvulus* and its vectors (members of the *Simulium damnosum* species complex) (Lok et al. 1980, 1983a).

The third filarial nematode transmitted to mammals by North American black flies is *Onchocerca cervipedis*. The adult worms are found primarily in subcutaneous connective tissues of the legs, but their impact on the vertebrate hosts is inadequately known. *Prosimulium impostor* has been implicated in the transmission of this nematode to mule deer in California, where the parasite is known as "footworm" (Weinmann et al. 1973). *Simulium decorum* and the *S. venustum* species complex have been incriminated as the vectors in Alberta, where the nematode infects moose and is called "legworm" (Pledger et al. 1980).

Black flies are capable of transmitting various viruses to mammals. Considerable evidence suggests that *Simulium vittatum* s. s. and members of the subgenus *Psilopelmia*, especially *Simulium notatum*, transmit both the New Jersey and Indiana serotypes of vesicular stomatitis virus to livestock (Schnitzlein & Reichmann 1985; Francy et al. 1988; Cupp et al. 1992; Mead et al. 1997, 1999, 2000a; Mead 1999; Schmidtmann et al. 1999). This virus causes epithelial lesions, primarily in horses and cattle, although humans, wildlife, and other domestic animals are also susceptible. A noninfected black fly feeding on a nonviremic host can become infected with this virus merely by cofeeding with an infected fly (Mead et al. 2000b). If black flies are shown to be significant vectors of this virus in some of the outbreaks, then their economic impact could be great. The state of New Mexico alone, for example, lost about $15 million in the outbreak of 1995 (Bridges et al. 1997). Other viruses isolated from black flies include snowshoe hare virus from females of a member of the *Simulium malyschevi* species group (Ritter & Feltz 1974, Sommerman 1977), a bunyavirus from females of *Simulium bivittatum* (Kramer et al. 1990), and eastern equine encephalitis virus from females of *Simulium johannseni* (Anderson et al. 1961). Although female black flies feed on sheep infected with bluetongue disease, they have not been shown to transmit the causal virus (Jones 1981).

Bacterial transmission by black flies is poorly studied. *Simulium decorum* probably contributes to the natural incidence of tularemia among rabbits and other susceptible hosts by mechanical transfer of the causal bacterium *Francisella tularensis* during biting (Philip & Jellison 1986). This same causal agent has been isolated from females of *Simulium maculatum* in Russia (Yakuba 1963), but populations of this black fly in North America have not been screened for *F. tularensis*. Black flies at one time were suspected of transmitting the causal agent (rickettsia) of Potomac horse fever (Fletcher 1987) but later were shown experimentally to be unlikely vectors (Hahn et al. 1989, Hahn 1990).

# 8| Management

The management of black flies in North America is largely a 20th-century phenomenon (Table 8.1), born from the increasing devastation wrought by these insects on the economy and citizenry of Canada and the United States (see Chapter 7). Personal measures of protection predate all other means of managing black flies and continue to provide relief. But the first large-scale triumphs over black flies were achieved with chemicals aimed at the larvae, including the early use of oils and the heady but fleeting success of DDT (dichlorodiphenyltrichloroethane) and its replacement compounds. By the end of the 20th century, however, chemical insecticides as a means of managing black flies in North America had virtually disappeared. The successes, woes, and controversies stemming from the use of chemical insecticides against black flies are well illustrated in the history of New York's Adirondack Mountains, where black flies have long been a problem (Lintner 1884, p. 50) and control has been waged annually since 1948 (Elliott 1983). Many additional means for controlling black flies have been suggested, investigated, and implemented, from habitat alteration to natural enemies. The current phenomenon in the war against black flies is the naturally occurring bacterium *Bacillus thuringiensis* variety *israelensis* (*Bti*). How long it will remain the ideal weapon is a matter of useful speculation.

## Chemical Control

### OILS AND OTHER EARLY COMPOUNDS

The first control efforts seem to have sprung, in 1884, from the desire to suppress the horrific outbreaks of the southern buffalo gnat (*Cnephia pecuarum*) in the lower Mississippi River Valley. Even before the immature stages had been discovered, Riley (1884) speculated that "it ought not to be difficult to kill the insect in its earlier stages on a large scale by the introduction of some poisonous substance, even at the expense of the food-fishes." Riley's field assistant, Francis Webster (1887), tried to kill larvae by exposing them to currents of various substances, such as salt water or kerosene, in glass tubes. Otto Lugger, another of Riley's field men, made the first field tests in streams in the early spring of 1886, applying freshly burned lime "in pieces the size of an Irish potato," emulsion of kerosene, powdered pyrethrum, carbon-bisulphide, powdered cocculus indicus (the berry of the Indomalaysian vine *Anamirta cocculus*), and tobacco soap (Riley 1887). Large amounts of material killed simuliid larvae, as well as other aquatic insects and fish, but too much toxicant was required to make larval control a worthy venture.

Not until the early 20th century did efforts to control black flies begin in earnest. From that point forward, attempts to control black flies often paralleled, or were conducted in tandem with, efforts to control mosquitoes. Clarence M. Weed (1904a, 1904b), a professor at New Hampshire State College (now University of New Hampshire), became impressed by the use of Phinotas oil (a proprietary compound of the erstwhile Phinotas Chemical Co., New York) against mosquitoes. Unlike oils such as kerosene that floated on water, Phinotas oil sank to the bottom. Weed sent his field assistant, Albert F. Conradi, to Dixville Notch, New Hampshire, to test the compound's efficacy against black flies. At the headwaters of Mohawk Creek (ca. 1.5 m wide), Conradi killed the immature stages by throwing about a liter of insoluble oil into the stream. By applying about 19 liters of oil to a stream 3 m wide, Conradi (1905) killed nearly all larvae for more than 5 km downstream, thereby affording relief to guests in the neighboring resort. A knapsack sprayer and nozzle later were recommended for more economical applications (Sanderson 1910). Further work demonstrated that if satisfactory levels of larval mortality were to be achieved, the oil would have to be applied in a soluble form in amounts great enough "to make the water look as white as milk" for at least 1.5 minutes (O'Kane 1926). Crude petroleum also was poured into streams as a means of control around the turn of the century (Webster 1914).

The effects of oiling were often disastrous for fish and other aquatic life (O'Kane 1926), but the application of miscible oils to streams was recommended at least through the 1930s in the United States and Canada (e.g., Swenk & Mussehl 1928, Twinn 1933b, Hearle 1938) despite laws in some areas that prohibited the use of any poisonous substance to kill fish (Weed 1904b). The use of pyrethrum extract in petroleum oil—another tactic inspired by mosquito-larvicide technology—was tried in New York (Glasgow 1939, 1942) and recommended as a larvicide in small streams (Bishopp 1942) but was not widely adopted. Pyrethrum also found use in one of the earliest efforts to control adult black flies with chemicals. As an aerosol, it was effective in Canadian tests against adult black flies in

**Table 8.1. Control Programs[1] for North American Black Flies, Listed Chronologically**

| Dates | Location | Target species | Control method[2] | Selected references |
|---|---|---|---|---|
| 1948–present | New York, Adirondack Mountains | *Prosimulium mixtum*, *Simulium venustum* complex | DDT (aerial and ground) against larvae and adults (1948–1965); methoxychlor (1966–1981); naled (aerial) against adults (ca. 1978–1983); *Bti* (1984–present); 1–10 treatments/site/year | Jamnback & Collins 1955; Elliott 1983; Rutley 2000, 2001; D. P. Molloy, pers. comm. |
| 1948–present | central Saskatchewan | *Simulium luggeri*, *Simulium vampirum* (formerly *S. arcticum*) | DDT (aerial and ground) against larvae of *S. vampirum* (1948–1967); methoxychlor against *S. luggeri* and *S. vampirum* (1968–1986); *Bti* against *S. luggeri* (1987–present) | Fredeen 1988, Riegert 1999b, Lipsit 2001 |
| 1951–1953, 1955 | Manitoba, Souris River drainage | *S. venustum* complex | DDT (ground) against larvae | Riegert 1999b |
| 1952–1965 (longer?) | Quebec, North Shore | *Prosimulium hirtipes* group, *S. venustum* complex | DDT (aerial and ground) against larvae and adults | West 1961, 1973 |
| 1952–1953 (longer?) | Quebec, Halet | *P. hirtipes* group, *S. venustum* complex | DDT (ground) against larvae | Roach 1954 |
| 1953 | British Columbia, Cherryville | *Simulium defoliarti* | DDT (ground) against larvae | Curtis 1954 |
| 1950s–present | Labrador, Goose Bay | *P. hirtipes* group, *S. venustum* complex | DDT (aerial and ground) (early years); *Bti* (aerial and ground) (1980s–present) | based on program developed by Hocking & Richards (1952); M. H. Colbo, pers. comm. |
| 1955 | South Carolina, Jasper County | *Simulium congareenarum*, *Simulium slossonae* | DDT (aerial) against larvae | Anthony & Richey 1958 |
| 1957–? | Quebec, Mont Apica | unspecified | DDT (aerial and ground) against larvae and adults | Snider 1958 |
| early 1960s–present | Labrador, Labrador City–Wabush area | *P. hirtipes* group, *S. venustum* complex | DDT (1960s); methoxychlor and Abate (1970–1983); *Bti* (aerial and ground) (1984–present); 3 treatments/site/year | M. H. Colbo & J. W. McCreadie, pers. comm. |
| mid-1960s–present | Labrador, Churchill Falls | *P. hirtipes* group, *S. venustum* complex | Abate and methoxychlor (1960s–1980s); *Bti* (aerial and ground) (mid-1980s–present) | M. H. Colbo, pers. comm. |
| 1969–1970, 1996–present | California, Los Angeles County, San Gabriel Valley | *Simulium vittatum* complex, *Simulium tescorum* | Abate (ground) (1969–1970); *Bti* (ground) in small creeks and channels; larger water-management channel shut off for 48 hours every 7–10 days (1996–present) | Pelsue et al. 1970; Pelsue 1971; E. W. Gray, pers. comm. |
| 1970 | Alaska, Kodiak | unspecified | naled (aerial) against adults | Fussel 1970 |
| 1972 | South Carolina, Chesterfield County | *S. congareenarum*, *S. slossonae* | Abate (ground) | Kissam et al. 1975 |
| 1972–1975 (longer?) | Ontario, Chalk River and Deep River (towns) | *S. venustum* complex | Abate (aerial) | Baldwin et al. 1977 |
| 1973, 1990–present | Idaho, Twin Falls County | *S. vittatum* complex | methoxychlor (ground) (1973, 1990–1995); *Bti* (ground) (1996–present); ca. 7 or 8 treatments/site/year | Jessen 1977; E. W. Gray, pers. comm. |
| 1979–1987 | Alberta, Athabasca River, downstream of Athabasca (town) | *S. vampirum* (formerly *S. arcticum*) | methoxychlor (ground); 2 treatments/site/year | Byrtus & Jackson 1988 |

**Table 8.1. Continued**

| Dates | Location | Target species | Control method[2] | Selected references |
|---|---|---|---|---|
| 1979–ca. 1992 | New Hampshire, Waterville Valley | *S. venustum* complex | adulticiding (1979–1984); *Bti* (ground) (1985–ca. 1992) | Martin 1981, Koontz 1992 |
| 1979–present | Utah, southern Salt Lake County | *Simulium tribulatum* | methoxychlor (ground) (1979–1984); *Bti* (ground) (1985–present); ca. 1 treatment/site/2 weeks | K. L. Minson, pers. comm. |
| 1980s–present | Quebec, nearly 30 towns | unspecified | *Bti* | Mason et al. 2002 |
| 1980, 1983, 1984, 1986, 1989, 1993–present | Arkansas-Texas, Sulphur River drainage | *Cnephia pecuarum* | Abate (ground) (1980, 1983, 1984); *Bti* (ground) (1986, 1989, 1993–present); 1 treatment/site/year | Atwood 1996 |
| 1981–1982 | South Carolina, Marlboro County | *S. slossonae* | *Bti* (ground) | Horosko & Noblet 1986a |
| 1983, 1985–present | Pennsylvania, 2570 km of flowing water in 33 counties by 1999 | *Simulium jenningsi* | *Bti* (mostly aerial [helicopter]; some treatments of smaller streams by backpack sprayers); 6–10 treatments/site/year | Arbegast 1994; A. D. Kyle & D. I. Rebuck, pers. comm. |
| 1984–present | Minnesota, 7 counties around Minneapolis –St. Paul | *Simulium johannseni*, *Simulium meridionale*, *S. luggeri*, *S. venustum* complex | *Bti* (ground); 1–17 treatments/site/year | Sanzone 1995; Sanzone in Gray et al. 1999; K. R. Simmons & J. P. Walz, pers. comm. |
| 1984–present | Arizona-Nevada border, Colorado River below Davis Dam, Bullhead City | *S. tribulatum* | *Bti* (ground); treatments every 2 or 3 weeks | E. W. Gray, pers. comm. |
| 1985–present | northern Nevada, Humboldt River | *S. meridionale*, *Simulium venator* | methoxychlor (ground) (1985–1991, 1993–1998); *Bti* (ground) (1997–present ); up to 9 treatments/site/year | R. D. Gray, pers. comm. |
| 1985–present | New Hampshire, Dixville Notch | *S. venustum* complex | *Bti* (ground) | Koontz 1992 |
| 1986–present | West Virginia, Bluestone River, Greenbrier River, and New River | *S. jenningsi* | *Bti* (aerial [helicopter]); ca. 14 treatments/site/year | Smithson in Gray et al. 1999; T. R. Carrington, pers. comm. |
| mid-1980s–present | western Kentucky, Tradewater River | *C. pecuarum* | *Bti* (ground); 1 treatment (mid-January)/site/year | M. Buhlig in Gray et al. 1999; E. W. Gray, pers. comm. |
| 1988–present | Michigan, Benzie County, Thompsonville | *P. hirtipes* group, *S. venustum* complex | *Bti* (ground); 6 treatments/site/year | R. W. Merritt, pers. comm. |
| 1990–1992 | Tennessee, Holston River at Kingsport | *S. tribulatum* | *Bti* (ground) | E. W. Gray, pers. comm. |
| 1990–present | California, Los Angeles County, Santa Monica Mountains, especially Malibu Creek | *S. tescorum* | *Bti* (ground); year-round treatments | E. W. Gray, pers. comm. |
| ca. 1992–present | Prince Edward Island | *P. mixtum*, *S. venustum* complex | *Bti* (ground) | D. J. Giberson, pers. comm. |
| 1992–1997 | Michigan, Keweenaw County, Copper Harbor | *P. hirtipes* group | *Bti* (ground); 1 treatment/site/year | R. W. Merritt, pers. comm. |
| 1994–present | California, Los Angeles County, Los Angeles River | *S. tribulatum* | *Bti* (ground); 7–10 treatments/site/year | O'Connor et al. 2001; E. W. Gray, pers. comm. |
| 1994–present | South Carolina, Enoree and Tyger River drainages | *S. jenningsi* group | *Bti* (ground); 12–14 treatments/site/year | Gray et al. 1996; E. W. Gray, pers. comm. |

**Table 8.1. Continued**

| Dates | Location | Target species | Control method[2] | Selected references |
|---|---|---|---|---|
| 1996–present | New Jersey, Delaware and Raritan River drainages | *S. jenningsi* | *Bti* (aerial [helicopter] and ground); 6–10 treatments/site/year | Grande 1997; McNelly & Crans 1999; Crans et al. 2001; A. D. Kyle & D. I. Rebuck, pers. comm. |
| 1998, 2000 | Florida, Pasco County | *S. slossonae* | *Bti* (ground) | E. W. Gray, pers. comm. |
| 2000–present | Labrador, Red Bay to near Forteau | *P. hirtipes* group, *S. venustum* complex | *Bti* (aerial and ground) | M. H. Colbo, pers. comm. |

*Note*: The many experimental field trials and personal protection schemes are not included. Additional details and references can be found in the text of the current chapter and under the relevant taxonomic accounts of the targeted species in Chapter 10.

[1] A distinction is sometimes unclear between actual control programs and studies designed to determine the feasibility for control, the latter often bringing relief to an infested area and forming the basis for subsequent control programs. In general, if the authors indicated that their study was experimental in nature and intended to determine feasibility, methodology, and potential for success, the study is not included in the table. Information on some control programs was never published, unlike the results of experimental and exploratory studies for which many examples appear in the text of this chapter. Therefore, the tabulation of control programs is markedly incomplete.

[2] All treatments with *Bti*, methoxychlor, and Abate (temephos) have been aimed at larvae.

confined spaces (Monro et al. 1943) but generally was not pursued as a means of control.

**DDT**

No satisfactory chemical means of controlling black flies appeared until DDT, probably the most famous and economical chemical insecticide of all time, was rediscovered in the 1930s and registered as patent no. 226,180 with the Swiss Patent Office in 1940. (DDT was first synthesized in 1874 in Germany.) The earliest recorded test of DDT against black flies was in 1943, when F. C. Bishopp, assistant to the chief of the U.S. Bureau of Entomology and Plant Quarantine, applied an emulsion of kerosene, DDT, soap, and water to Sligo Creek in Montgomery County, Maryland, killing larvae of the *Simulium venustum* species complex (Jamnback 1981). In Canada, DDT was first tested on 21 July 1944 in Costello Creek, Algonquin Park, Ontario (Davies 1950). From 1945 well into the 1960s, DDT was considered the solution to the black fly problem. Gustave Prevost's (1949) eradication of black flies from streams that he treated with DDT in 1946 and 1947 in Mont Tremblant, Quebec, elicited virtual euphoria: "Black flies, just about the most tormenting pest in northern latitudes, have at last met their Waterloo—and DDT was the weapon."

Most of the research on DDT was focused in Alaska, northern Canada, the northeastern United States, and the agricultural areas of Saskatchewan. Efforts to control black flies in Alaska and northern Canada stemmed primarily from military interests, including establishment of the Distant Early Warning (DEW) Line of radar stations, and from development of natural resources, especially in the mining and pulpwood industries. DDT consistently provided excellent larval control, even when it was applied to the snow and ice covering the breeding sites (Anonymous 1947). A number of other synthetic organic insecticides were tested in these areas, including aldrin; chlordane; chlorinated camphene; dieldrin; gamma-benzene hexachloride; heptachlor; methoxychlor; parathion; a pyrethrum-piperonyl butoxide preparation; schradan; TDE (= DDD; 1,1-dichloro-2,2-bis(*p*-chlorophenyl)ethane); toxaphene; and 1,2,4-trichlorobenzene (Fig. 8.1). Applications of DDT, usually in fuel oil, both by aircraft and by hand (Fig. 8.2), produced the greatest larval mortality but often raised concerns for nontarget organisms (Anonymous 1949; Cope et al. 1949; Gjullin et al. 1949a, 1949b, 1950; Hocking et al. 1949; Travis 1949; Hocking 1950, 1953c; Twinn 1950, 1952; Hocking & Richards 1952; Smith et al. 1952; Roach 1954; D. Peterson 1955; Snider 1958; West et al. 1960).

Fig. 8.1. Insecticide dispenser releasing aldrin emulsion for black fly control near Whitehorse, Yukon Territory, 1949. Photograph courtesy of D. A. Craig.

Many of these same chemicals were tested against black flies in New Hampshire, New York, and Pennsylva-

Fig. 8.2. Brian Hocking using a hand sprayer to apply 10% DDT in fuel oil for control of immature black flies at Churchill, Manitoba, July 1947. Photograph courtesy of D. A. Craig.

nia, where recreational activities were hampered by black flies. Results again indicated that DDT was the superior killing agent, both in aerial applications and in impregnated blocks of plaster placed in streams (Kindler & Regan 1949; Goulding & Deonier 1950; DeFoliart 1951a; Jamnback 1951, 1952, 1953; Travis et al. 1951a; Collins et al. 1952; Jamnback & Collins 1955). In New York, applications by aircraft were recommended, with planes flying a swath across each stream every 0.4 km (Travis 1967). DDT was applied even in the sluggish swamp streams of the southeastern United States, where a series of tests from 1954 to 1956 in Florida and South Carolina provided 100% control for up to 4.5 km (Davis et al. 1957).

In Saskatchewan, efforts to control black flies were driven by the devastating outbreaks of *Simulium vampirum* (historically known as *S. arcticum*) that threw the livestock industry into near chaos. The history of these efforts was told in detail by Riegert (1999b). DDT was first applied to the large rivers that provided larval habitat for this pest in the spring of 1948; application of nearly 280 kg of DDT ridded the South Saskatchewan River of larvae for at least 27 km (Arnason et al. 1949, Riegert 1999b). DDT treatments continued in the branches of the Saskatchewan River, once or twice per year, through 1967 (Fredeen 1977c), each treatment often eliminating larvae for 185 km of the river (Fredeen et al. 1953a). Adsorption of the insecticide to silt particles contributed to the long-distance transport (Fredeen et al. 1953b). The pest problem on cattle in southern British Columbia, caused by *Simulium defoliarti*, was virtually eliminated with applications of DDT in the early 1950s (Curtis 1954). Similarly, black flies (primarily members of the *Simulium venustum* species complex) troublesome to livestock owners in southern Manitoba were controlled with DDT applications to the Souris River and nearby flows from 1951 through at least 1955 (Riegert 1999b).

The majority of control efforts with DDT were aimed at larvae, although DDT was applied by helicopter and ground equipment against adults as early as 1945 when the communities of Webb and Blue Mountain Lake in the Adirondack Mountains of New York were fogged to assess the feasibility of control (Glasgow & Collins 1946). The portable Todd smoke-screen generators used by the U.S. Navy ashore and on landing crafts in World War II were adapted—and dubbed the Todd Insecticide Fog Applicator—to create the DDT fog in these field trials (Glasgow & Collins 1946, Collins & Jamnback 1958). They also were used in trials around Whitehorse in the Yukon Territory (Twinn 1950). The success of the fogging trials in the Adirondack Mountains led to one of the first actual simuliid control programs in North America (Figs. 8.3, 8.4). The town of Webb, seeking to extend its tourist season, sponsored DDT fogging by helicopter during 14 days of June 1948 (Elliott 1983). The first use of DDT against adults in Canada was probably in Quebec in 1946, when it was applied by airplane, with xylol and furnace oil as solvents (Prevost 1946). Along Quebec's North Shore and in Labrador, the control of both larvae and adults was common, but when funds were limited, adulticidal air sprays were the method of choice from about 1957 to 1965 (West 1973). Detailed protocols were developed for the use of ground-based aerosol generators and aircraft applications (Twinn & Peterson 1955).

Fig. 8.3. DDT fogging by helicopter to control adult black flies in Old Forge, town of Webb, in the Adirondack Mountains of New York, late 1940s. Photograph courtesy of D. P. Molloy.

Fogging with DDT provided mixed results. It was deemed ineffective or gave ambiguous results in northern

Fig. 8.4. DDT fogging by truck-mounted aerosol generator (Todd Insecticide Fog Applicator) to control adult black flies in Old Forge, town of Webb, in the Adirondack Mountains of New York, late 1940s. Photograph courtesy of D. P. Molloy.

Manitoba (Anonymous 1947), Alaska (Goldsmith et al. 1949, Travis 1949, Wilson et al. 1949), and Labrador (Brown et al. 1951, Brown 1952), largely because extensive areas required treatment to stem the rapid influx of black flies from adjacent areas. Adulticiding, nonetheless, was considered efficacious in Quebec (Peterson & West 1960, West 1961) and the Adirondacks (Jamnback 1952, Jamnback & Collins 1955). In the Adirondacks, however, adulticiding with DDT was relatively expensive. An area 228 $km^2$ required \$6000 in 1950 to treat for adults versus \$2500 for larvae (Jamnback 1951). The drawbacks to DDT treatments for adults established insecticidal fogging as an infrequent means of management, although fogging was used through the end of the 20th century on an ad hoc basis, especially with permethrin products (Guyette 1999).

By 1960, problems inherent with DDT, particularly resistance and effects on nontarget organisms, had risen to prominence (e.g., Jamnback & Eabry 1962) and were being documented routinely (e.g., Hatfield 1969). As early as 1954, the possibility had been raised that larval black flies in Quebec might be developing resistance to DDT (Roach 1954). Resistance to DDT in North American black flies was first documented in 1967 for larvae of the *S. venustum* species complex from the Petit Bras River of Quebec, Canada, where aerial spraying had been conducted for 10 years (West in Brown & Pal 1971, pp. 251–252). In 1968, black flies exhibited resistance to DDT in the Adirondacks, where the compound had been used since the mid-1940s (Jamnback & West 1970). DDT eventually was banned in Canada in 1970 and in the United States at the beginning of 1973, although in some areas, such as New York State, it had been banned as early as 1965.

#### DDT REPLACEMENTS

The search for substitute compounds began in earnest in the early 1960s and continued for nearly 25 years. Most efforts to find effective and safe replacements for DDT were directed at larvae. The search for adulticide replacements proved more difficult, although the carbamate Baygon provided rapid knockdown and high mortality (West 1973). From the early 1960s to the early 1980s, about 80 alternative compounds, mostly synthetic insecticides, were tested against larval black flies in North American laboratories and, in some cases, in the field (Jamnback 1962, 1969b, 1969c; Travis & Wilton 1965; Frempong-Boadu 1966a, 1966b; Jamnback & Frempong-Boadu 1966; Travis 1966; Travis & Guttman 1966; Travis et al. 1967, 1970; Travis & Schuchman 1968; Chance 1970b; Wallace 1971; Kurtak et al. 1972; Lacey & Mulla 1977d; Mohsen & Mulla 1981b, 1982b; Mohsen 1982; Rodrigues 1982; Frommer et al. 1983; Rodrigues et al. 1983). Various simulated stream systems were devised for testing these compounds (e.g., Jamnback 1964, Wilton & Travis 1965, Guttman et al. 1966, Travis 1968, Gaugler et al. 1980, Rodrigues & Kaushik 1984a), and constant-rate dispensers were designed to provide injection of the chemicals at a uniform rate (e.g., Jamnback & Means 1966, Fredeen 1970b). Nearly all of the evaluations considered only larval mortality rates; few studies (e.g., Frempong-Boadu 1966a) looked at the effects of the insecticides on larval behavior. Some of the more promising compounds, such as methoxychlor and temephos (typically Abate), also were tested in the field.

Methoxychlor, a chlorinated hydrocarbon, was probably the most touted chemical insecticide to fill the void left by DDT. It was considered effective and more environmentally safe because it was readily broken down into water-soluble molecules, eliminating the problem of biomagnification. Methoxychlor had been tested against black flies as early as 1947 in Alaska (Travis 1949). By 1966, it had largely replaced DDT in control programs in New York State (Jamnback & Means 1966). It was registered for use as a larvicide against black flies in the United States in 1968 and in Canada in 1970. Methoxychlor was used in the Saskatchewan River system primarily to control *Simulium luggeri*, but also *Simulium vampirum*, from July 1968 to at least 1986 (Fredeen 1983, 1988; *vampirum* as *arcticum*). Like DDT, methoxychlor adsorbed to particles suspended in the water, making it particularly effective against filter-feeding larval black flies (Fredeen et al. 1975). Single injections killed the majority of larvae for distances of more than 160 km (Fredeen 1974, 1975a, 1977c). Methoxychlor also was applied against larvae of the *Simulium vittatum* species complex in canals of southern Idaho in 1973 (Jessen 1977) and against *S. vampirum* in Alberta's Athabasca River, beginning in 1974 (Depner 1978, 1979; Charnetski & Haufe 1981; *vampirum* as *arcticum*). It was effective against all species of black flies tested, although the level of susceptibility differed among some species (Wallace et al. 1976, Dosdall & Lehmkuhl 1989). Methoxychlor usually was used as a larvicide, but it also was registered for use against adult simuliids in Canada (Blake et al. 1981).

The major drawback of methoxychlor (and similar compounds), both as particulate and as liquid formulations, was the rather nonspecific action against aquatic invertebrates and fish (Burdick et al. 1968, Wallace 1971, Wallace et al. 1973, Helson & West 1978, Sebastien & Lockhart 1981, Wallace & Hynes 1981). A single injection into Alberta's Athabasca River, for example, caused catastrophic drift and a reduction in standing crop for distances of more than 400 km (Flannagan et al. 1979). Methoxychlor, nonetheless, was still being used to suppress black flies in northern Nevada as late as 1998 (Gray et al. 1999).

Abate, an organophosphate, became for a short while the larvicide of choice because of its effectiveness, low persistence in the environment, low mammalian toxicity, and somewhat lower toxicity to nontarget organisms. It proved effective in field tests as early as 1964 in New York State (Jamnback & Frempong-Boadu 1966, Jamnback & Means 1968) and 1966 in Ontario (Swabey et al. 1967). The compound was applied in eastern Canada (Baldwin et al. 1977, Back et al. 1979), and by the early 1980s it was the only product registered for use against larval black flies in Canada (Blake et al. 1981). Abate was used effectively against *Simulium slossonae* in South Carolina (Kissam et al. 1973, 1975) and against *Simulium tescorum* and a member of the *S. vittatum* species complex in southern California (Pelsue et al. 1970, Pelsue 1971), but particulate formulations typically did not provide adequate larval control when applied by air (Helson 1972). As with most chemical insecticides used against black flies, Abate negatively impacted the nontarget organisms (Wallace 1971). The last major use of the product against North American black flies was in a program to manage *Cnephia pecuarum* along the Arkansas-Texas border from 1980 through 1984, under an emergency permit granted by the U.S. Environmental Protection Agency (Atwood 1996).

The problem of nonspecificity coupled with resistance and low toxicity of compounds such as temephos in cold waters (Rodrigues & Kaushik 1984b) stimulated the search for new toxic agents. Heptachlor, an organochlorine, was investigated as a potential larvicide in the Canadian prairies (Fredeen 1962) but did not gain wide use. Naled (Dibrom), an organophosphate, was effective against adult simuliids at Kodiak, Alaska, providing a protective swath of 640 m when applied at a concentration of 0.36 kg/liter (Fussel 1970). It also was used for at least 5 years against adult black flies in New York's Adirondack Mountains (Elliott 1983). A few additional synthetic adulticides, such as malathion, were tested in the field but were ineffective (Carestia et al. 1974, Elliott 1983).

#### NONTRADITIONAL COMPOUNDS

Evaluation of nontraditional chemical larvicides against black flies in North America began in 1972. Premiere among these alternative compounds were the insect-growth regulators (IGRs), such as juvenile-hormone analogues, that disrupted insect hormonal action. Methoprene (typically Altosid) was among the first compounds tested. By adding as little as 0.001 ppm of the compound to the rearing water of larvae, adult emergence could be reduced or completely inhibited (Cumming & McKague 1973, McKague & Wood 1974, Dove & McKague 1975). The compound often produced larval-pupal abnormalities (Garris & Adkins 1974). Because mortality generally occurred in the pupal stage, the timing of field applications needed to be more precise than if conventional insecticides were used (Thompson & Adams 1979). Field tests with formulated blocks that released Altosid over a 48-hour period were conducted in streams in British Columbia, with complete suppression of adult emergence for more than 3 weeks (McKague et al. 1978). Field trials on Newfoundland's Avalon Peninsula also demonstrated up to 99% mortality (Thompson & Adams 1979).

Diflubenzuron (typically Dimilin), an IGR that inhibits chitin synthesis, also was evaluated. Laboratory bioassays with various species of black flies showed that the compound had larvicidal properties and killed eggs if concentrations and exposure times were great enough (Lacey & Mulla 1977c, 1978a; Rodrigues & Kaushik 1986). Field tests in streams in New York's Adirondack Mountains, British Columbia, and southern California significantly reduced adult emergence (McKague et al. 1978, Lacey & Mulla 1979a, McKague & Pridmore 1979, Rodrigues 1982). Even when applied as an insecticide against forest-insect pests, diflubenzuron reduced the populations of black flies in streams running through the treated areas (Martin 1993). At least three additional IGRs produced larval mortality in laboratory tests against black flies (Lacey & Mulla 1978b).

The IGRs had very low toxicity to vertebrates and broke down rapidly in the field, but they were nonselective and impacted the nontarget benthic insects. This same problem plagued the use of the plant-derived phototoxin alpha-terthienyl, which killed larvae in laboratory trials (Philogène et al. 1985) and in field tests in southeastern Ontario (Dosdall et al. 1991).

Compounds aimed at other pest organisms typically have had few adverse effects on larval black flies. For example, the use of 3-trifluoromethyl-4-nitrophenol (TFM) against the sea lamprey (*Petromyzon marinus*) in streams of the Great Lakes watershed does not negatively impact the simuliid populations (Dermott & Spence 1984). Nor do the herbicides hexazinone and triclopyr ester pose a risk to simuliids when used at typical concentrations (Kreutzweiser et al. 1992). Nonetheless, certain forest insecticides (e.g., mexacarbate) have the potential to affect larval black flies (Poirier & Surgeoner 1987).

### Physical Control

As with chemical control, Riley (1887) seems to have been the first to suggest habitat manipulation as a means of reducing pest populations of black flies in North America. He recommended the removal of logs and debris to control *Cnephia pecuarum*, and mentioned that repair of the broken levees along the Mississippi River might solve the pest problem. The relation between broken levees and outbreaks of the southern buffalo gnat was generally recognized by inhabitants of the infested area and championed by Francis Webster (1889). As history demonstrates, local knowledge is often correct, and eventual repair of the levees along the Mississippi River reduced pest problems significantly.

The removal of twigs, leaves, and other debris that could serve as larval attachment sites was recognized as a viable method of reducing larval populations. Most workers (Reeves 1910, Swenk & Mussehl 1928) suggested that this technique would function only in streams where suitable substrate was limited, for example, in sandy or silty streams; the procedure was deemed worthless in stony streams. In reality, the procedure rarely has been practiced in any stream. Herbicides that were used to kill aquatic plants in irrigation canals also reduced available attachment sites, negatively impacting the larvae of pestiferous black flies (*Simulium vittatum* species complex) (Jessen 1977).

The use of wire brooms and iron rakes to remove larvae from streambeds dates from the 1890s, when they were used in the resort area of Dixville Notch, New Hampshire (Weed 1904b). The method was recommended especially as a means of avoiding fish kills. More recent incarnations of this technique were tested, also at Dixville Notch, to reduce highly localized larval populations of *Simulium decorum* at lake and pond outlets (LaScala and Burger 1981). Streams in areas with pulpwood operations often were flooded each spring to move the logs downstream, which had the unintended but beneficial consequence of scouring the streams and eliminating the larval black flies (Peterson & Wolfe 1958).

Altering water levels, especially by damming streams, has been recommended since the beginning of the 20th century (Conradi 1905). Dams, both artificial and natural, however, can create prime breeding sites for pest species. Therefore, their design, especially with regard to the amount of surface area, should be fashioned or manipulated to reduce populations of pest species that breed on the face of these structures and immediately downstream. Recommended designs include dams with sheer drops into deep pools and with multiple spillways that alternately could be allowed to run dry (Metcalf & Sanderson 1931, 1932). Management of water levels in the Tennessee River Basin proved effective in reducing pest populations of the *S. vittatum* species complex (Snow et al. 1958b). Lowering water levels by regulation of dams or erection of current barriers to strand larvae above water long enough to kill them was recommended, but not implemented, as a means of reducing pest populations of *Simulium penobscotense* in Maine (Granett & Boobar 1979). The ability of larvae to survive physical manipulations of the habitat and the effects of habitat alteration on nontarget organisms rarely have been evaluated.

Few other means of physically or mechanically reducing populations of pest black flies have been employed. The use of traps or baits to reduce adult populations has not been explored, probably because it has been considered ineffectual. Interest in the sterile-male technique led to preliminary testing, in the early 1970s, of the effects of gamma radiation on pupation, emergence, and morphology of black flies (Gross et al. 1972), but the technique was not pursued.

## Biological Control

### POTENTIAL BIOCONTROL AGENTS

Riley (1887) foreshadowed biological control of simuliids when he documented six predators that consumed larvae and adults of *Cnephia pecuarum*. The notion that natural enemies could control black flies, however, was not unique to Riley. An 1885 newspaper in Washington, D.C., for example, reported L. O. Howard's portrayal of a larval hydropsychid caddisfly as "a philanthropical being which prevents the increase of this pest" (putatively *Simulium venustum*). Strickland (1913b) was probably the first to suggest the potential biological control value of parasites when he titled one of his papers "Some Parasites of *Simulium* Larvae and Their Possible Economic Value." But no actual discussion of biological control was provided—only the evocative title and his statement (which for most parasites still defies demonstration) that "transferring these parasites from one species of *Simulium* to another . . . should be no great difficulty." Most pre-1970 efforts in biological control were simply surveys to determine what parasitized and preyed on black flies (Twinn 1952, Jenkins 1964, Strand et al. 1977).

The heyday of studies on natural enemies and their potential use in control programs for simuliids in North America began in the 1970s (Laird 1972, 1978, 1980a, 1980b; Maser 1973; Anonymous 1977; Diarrassouba 1977) and culminated with the publication of a book (Laird 1981) entitled *Blackflies: The Future for Biological Methods in Integrated Control*. Most parasitic taxa known from black flies, as well as many nonsimuliid parasites, were investigated and screened as biological control agents during the 1970s and early 1980s. Predators, however, were little studied.

Despite excellent studies during those years, many parasites remain too poorly understood taxonomically and biologically to allow their effective use in biological control programs. Life cycles, for example, are unresolved for most simuliid parasites. Suggestions have been made that spores of microsporidia could be mixed with water or dusts and sprayed into streams (Jamnback 1973a). Larval black flies to date, however, are refractory to *per os* infection with microsporidian spores, as well as with zoospores of the common, simuliid-specific fungus *Coelomycidium simulii* (Undeen [p. 14] in Anonymous 1977, Lacey & Undeen 1988). *Coelomycidium simulii*, however, can be cultivated in vitro, and intrahemocoelic injections can produce infections in larval black flies (Bauer, Soper & Roberts in Nolan 1981). Some of the entomophthoraceous fungi can cause epizootics (Shemanchuk & Humber 1978) and they can be cultured in the laboratory (Nolan 1981), but their biological control potential is largely unstudied. Most bacteria isolated from black flies have little apparent consequence to the larvae (Lacey & Undeen 1988), and no bacterium has yet been transmitted back to its host. Cytoplasmic polyhedrosis virus has been transmitted experimentally in the laboratory (Bailey 1977), but cultivation has not been achieved and the biology of this disease agent remains inadequately known.

Mermithid nematodes have been among the most promising biological control agents of black flies (Gordon et al. 1973, Poinar 1981, R. Gordon 1984), although taxonomic problems have hindered their use in simuliid control (Colbo & Porter 1980, Colbo 1990). Laboratory studies with *Mesomermis flumenalis* demonstrated average transmission levels of 80% in first instars, suggesting the potential of mermithids for managing simuli-

Fig. 8.5. Ernie Thibodeau, technician with the New York State Museum, treating small stream in Upstate New York in the first and only field trial conducted in North America, using mermithid nematodes against black flies, 1975. Photograph by D. P. Molloy.

ids (Molloy & Jamnback 1975). Field releases of this mermithid in New York State produced infections in more than 70% of first instars of the *Simulium verecundum* species complex immediately downstream from the treatment point (Fig. 8.5); however, production costs, using in vivo cultivation, were prohibitively expensive (Molloy & Jamnback 1977; as *S. venustum*). Laboratory rearing techniques were developed for the free-living stages of this mermithid species (Bailey et al. 1977), and some progress was made with in vitro production of the infective juveniles (Finney 1976, 1981a, 1981b). But until mass cultivation can be achieved economically, further development of mermithids as biological control agents will remain stymied (Molloy 1976, 1981).

A number of pathogens and parasites that do not naturally infect black flies have been bioassayed for their use against larval simuliids. The mosquito fungi *Culicinomyces clavosporus* and *Tolypocladium cylindrosporum* were tested but were not sufficiently virulent to be biological control agents of black flies (Knight 1980, Gaugler & Jaronski 1983, Sweeney & Roberts 1983, Nadeau & Boisvert 1994). Various fungal toxins also have been tested against larval simuliids (LePage et al. 1992). The trematode *Plagiorchis noblei* can parasitize larvae under experimental conditions and was believed to have some control utility under certain circumstances (Jacobs 1991, Jacobs et al. 1993). Attempts to use various entomogenous mermithids against black flies have met with limited success. *Romanomermis culicivorax*, a mermithid nematode of mosquitoes, was tested in the laboratory against larvae of *Simulium vittatum* and the *S. verecundum* species complex (as *S. venustum*) but proved unsuitable as a control agent (Poinar et al. 1979, Finney & Mokry 1980). *Steinernema feltiae* (formerly known as *Neoaplectana carpocapsae*), a nematode of terrestrial insects, produced up to 100% mortality in laboratory tests and averaged about 50% in field tests against late instars of the *S. verecundum* and *S. vittatum* species complexes but was ineffectual against early instars (Molloy et al. 1980; Gaugler & Molloy 1981a, 1981b). *Steinernema carpocapsae*, another nematode of terrestrial insects, also significantly reduced simuliid populations in the field and had a negligible effect on nontarget benthic insects but did not persist in the stream environment (Georgis et al. 1991).

#### *BACILLUS THURINGIENSIS* VARIETY *ISRAELENSIS*

The foremost biological control agent for black flies does not naturally occur in these insects. The bacterium *Bacillus thuringiensis* variety *israelensis* (*Bti*), initially referred to as *B. thuringiensis* H-14 serotype, was discovered in larval mosquitoes (*Culex pipiens*) in the Negev Desert of Israel in 1976. Before the widespread use of *Bti*, various strains of *B. thuringiensis* that were pathogenic especially to lepidopteran larvae were tested against black flies, but they produced mortality only at high dosages and prolonged exposures (Lacey & Mulla 1977a, Lacey 1978, Lacey et al. 1978, Finney & Harding 1982). These strains lacked *Bti*'s highly toxic component, a delta-endotoxin in the parasporal inclusions produced during sporulation. Particles the size of parasporal inclusions (0.2–0.8 μm) are easily filtered by larval black flies. Once ingested, the inclusions are solubilized by the proteolytic enzymes in the alkaline midgut (pH 8.2–11.4), and the released toxins rapidly disrupt the midgut cells (Lacey & Federici 1979).

Two years after its discovery in mosquitoes, *Bti* was shown to be highly toxic to larval black flies in laboratory trials. The world's first laboratory trial was conducted at Memorial University of Newfoundland in Canada (Undeen & Nagel 1978). Various bioassay apparatuses were designed to facilitate laboratory testing (Hembree et al. 1980). Further laboratory tests on the efficacy of *Bti* demonstrated the influence of factors such as species, instar, concentration and type of suspended food particles, formulation and concentration of the *Bti*, and water temperature (Frommer et al. 1980; Gaugler & Molloy 1980; Molloy et al. 1981, 1984; Lacoursière & Charpentier 1988; Nixon 1988; Atwood et al. 1992; Atwood 1996). Only one study has examined the influence of *Bti* on adults. The solubilized parasporal crystals of *Bti*, administered via enemas to the midgut of adult females, produced up to 100% mortality (Klowden et al. 1985).

The first field trials with *Bti* in North America were conducted in July 1978 in small streams on Newfoundland's Avalon Peninsula (Undeen 1980). These tests demonstrated up to 100% larval mortality (Undeen & Colbo 1980). Unofficial field testing of *Bti* against black flies in Pennsylvania and West Virginia in 1980 also demonstrated the agent's killing power (Anonymous 1982). Field trials in streams in eastern Tennessee further confirmed the efficacy of *Bti* (Frommer et al. 1981a, 1981b, 1981c, 1981d; Lacey & Undeen 1984; Undeen et al. 1984a; Lacey & Heitzman 1985), as did field tests in southern Idaho (Stoltz 1982), New York State (Molloy & Jamnback 1981), and South Carolina (Horosko & Noblet 1983). These field trials demonstrated the influence of vegetation, pools, discharge, and formulation on the rate and distance that the *Bti* was carried. The effective distance (i.e., carry) was identified as the greatest limiting factor to the efficacy of *Bti* (Gaugler & Finney 1982). The hyporheic zone was shown to be a major source of *Bti* loss, with higher discharges reducing the losses and providing greater carry of the product (Boisvert et al. 2001c, 2002). Proteolytic bacteria that occur naturally in rivers do not interfere with the efficacy of *Bti* (Khachatourians 1990). Adsorption onto periphyton, however, affects its carry (Boisvert et al. 2001b). Conceptual models estimating the sensitivity of larvae to *Bti* indicated that larval size, water temperature, seston concentration, length of contact, and variability of larval feeding rates were important in determining mortality rates (Morin et al. 1989).

Nearly all field trials demonstrated a lack of adverse direct effects on nontarget insects and fish (Colbo & Undeen 1980; Molloy & Jamnback 1981; Burton 1984; Pistrang & Burger 1984; Duckitt 1986; Gibbs et al. 1986; Merritt et al. 1989; Jackson et al. 1994, 2002; Brancato 1996). One Canadian study (Back et al. 1985), conducted at a dosage about 3–15 times greater than operational levels (Molloy 1990), documented severe effects on larval blephariceird midges and some toxemia to larvae of three genera of chironomid midges, presumably because the *Bti* attached to periphyton on rocks. Filter-feeding chironomids in the genus *Rheotanytarsus* also have shown susceptibility to *Bti* (Molloy 1992). Mortality occurred in larvae of one tipulid species and a heptageniid mayfly at dosages more than 50 times higher than recommended field rates (Wipfli & Merritt 1994b), and fish were killed in the laboratory when recommended dosage rates were exceeded by a factor of 12,000 (Wipfli et al. 1994). Fish mortality also was linked to xylene in early formulations (Fortin et al. 1986). Consumption of *Bti*-killed larval black flies by predators and detritivores is generally harmless (Merritt et al. 1991, Wipfli 1992, Wipfli & Merritt 1994b). *Bti*, however, like any management tool, can exert indirect effects by reducing food resources and forcing predation on less preferred prey (Wipfli & Merritt 1994a).

Fig. 8.6. Application of *Bti* by helicopter, upstream of the Clark's Ferry Bridge on the Susquehanna River, Pennsylvania, 2000. Photograph by R. A. Fusco.

The promise of *Bti* as a management tool for black flies led to a workshop with 28 participants in 1981, with the purpose of reporting progress and recommending standardized protocols for laboratory and field evaluations (Lacey et al. 1982, Undeen & Lacey 1982). The same year that the proceedings of the workshop were published (Molloy 1982), *Bti* was registered (30 June 1982) by the U.S. Environmental Protection Agency for use against black flies. Additional field trials and pilot control programs using *Bti* produced larval mortalities of 90%–100% and reduced the biting and nuisance levels in areas such as Maine (Gibbs et al. 1988), New Hampshire (Pistrang & Burger 1984), the Adirondack Mountains of New York (Molloy & Struble 1989), Labrador (Colbo 1984), New Brunswick (Riley & Fusco 1990), and Newfoundland's Avalon Peninsula (Colbo & O'Brien 1984). *Bti* rapidly became the agent of choice to manage black fly pests in these and other areas such as Minnesota (Simmons & Sjogren 1984), Pennsylvania (Arbegast 1994), and South Carolina (Horosko & Noblet 1986a, 1986b; Gray et al. 1996). By 2000, nearly 30 towns in Quebec were using *Bti* (Mason et al. 2002). Even areas such as West Virginia's New River Valley came to rely on *Bti*, despite vehement opposition from fishermen and other environmental groups (Cawthon 1981, Byrd 1984).

By the end of the 20th century, nearly all management programs for black flies in the United States and Canada were using *Bti* (e.g., Lehmkuhl 1990; Koontz 1992; Gray et al. 1999; Rutley 2000, 2001; Lipsit 2001). The *Bti* program in the Adirondack Mountains began in Indian Lake (Hamilton County) in 1984 (D. P. Molloy, pers. comm.). By 2001, the program included 30 communities in 11 counties (Rutley 2001). Not all afflicted towns participated, in part because some residents viewed black flies as an ingredient of life and adventure in the rugged mountains (McKibben 1999), just as some fishermen look on black flies as part of the experience of being outdoors (Finogle 2001). The largest-ever North American management program for black flies (aimed at *Simulium jenningsi*) has been conducted in the state of Pennsylvania, where helicopters are used to apply *Bti* to an extensive system of streams and rivers (Fig. 8.6). The Pennsylvania program began in 1983 and reached an annual operating budget of $5 million by the end of the

Fig. 8.7. Application of *Bti* from carboy by Elmer W. Gray, Enoree River, South Carolina, 2002. Photograph by J. Smink.

century (Arbegast 1994; D. H. Arbegast, pers. comm.). In smaller programs, *Bti* can be applied effectively by hand (Fig. 8.7). The effectiveness of a management program using *Bti* often is limited by recolonization from untreated areas. The problem of recolonization has been addressed to some extent by treating entire streams (McCracken & Matthews 1997) or enormous geographic areas, as in the Pennsylvania program.

*Bti* is now used worldwide, and new formulations are frequently tested with various field and laboratory bioassays (Barton et al. 1991; Lacey 1997; Boisvert et al. 2001a, 2001b). The use of *Bti* in a wider geographic context has been reviewed (Lacey & Undeen 1986, 1988; Cibulsky & Fusco 1988; Knutti & Beck 1988; Molloy 1990; M. Boisvert & J. Boisvert 2000), as have the costs of using the product in area-wide management (Gray et al. 1999).

The success of *Bti* as a biological control agent can be attributed to its host specificity, tremendous efficacy, safety for humans, and inexpensive mass production. Current formulations provide carry of up to 9 km, with 90% or greater mortality (Anonymous 1999). Ironically, the success of *Bti* has dampened the study of other natural control agents for simuliids. *Bti* is such an effective control agent that it often is viewed as a panacea. But if the history of pest control has provided contemporary workers a message, it is surely that resistance will develop, whether physiological or behavioral, particularly under the intensive selection of repeated applications that often produce 99% mortality. Monitoring for resistance, therefore, should be an integral part of any management program.

## Protection

### NATURAL PREVENTIVES

The futility of controlling black flies through the ages has spawned a long history of protective strategies for people and animals, liberally sprinkled with dubious folk remedies and fraudulent products. One of the time-honored means of protection has been the use of smoke. Smudges were particularly recommended. These smoldering masses were designed to produce a dense stifling smoke by placing materials such as damp moss, leather, cloth, green grass, or dried dung on a fire with an ample bed of embers. Smudges were the sine qua non during the historical outbreaks of the southern buffalo gnat and were used to protect livestock when farms and cities came under siege (Riley 1887). Entire fields often had many smudges smoking simultaneously (Robertson 1937). Livestock were outfitted with smudge pails tied around their necks (Webster 1889, Tucker 1918). As late as the 1990s, livestock producers in Arkansas and Texas burned wood piles and hay to deter the southern buffalo gnat (Atwood 1996).

Smudges also were used to protect livestock during outbreaks of *Simulium vampirum* in central Saskatchewan (Cameron 1918, 1922; as *S. arcticum* and *S. similis*), and their use against this pest continued well into the 1980s, although burning permits often were required (Fredeen 1988).

Smudges also found service with people. They were used, for example, by the Hudson's Bay Company in Canada. Company personnel also burned pyrethrum powder on pieces of bark to drive black flies from their tents and houses (Lugger 1897a, 1897b). Fires ignited by the lumbermen's smudges in northern Canada were a great hazard (Asselin 1935). Smudges were used in the 1890s at resorts in northern New Hampshire (Weed 1904b) and well into the 20th century in the Adirondack Mountains (Metcalf & Sanderson 1932). They were recommended even for use in tents, although they were considered hard on the eyes (Dunn 1925). Some outdoorsmen swear by the use of cigars, with their nicotine-laden puffs of smoke, to gain personal protection, but the effectiveness dwindles with increasing numbers of flies.

Bodily applications of repellent substances have been popular presumably since humans first encountered black flies. The earliest nostrums probably included various herbal brews, mud, and fatty substances. The men of Louis Agassiz's 1848 expedition along the north shore of Lake Superior anointed themselves with a thick coat of camphorated oil, which was too thick for the mouthparts to penetrate and entangled the flies (Agassiz 1850). Oil of tar was provided for employees of the Hudson's Bay Company (Riley 1887), and a now-mysterious "Black-fly Cream" (made in Portland, Maine) was recommended for people in the northeastern United States (Lugger 1897a, 1897b). Minnesota fishermen greased their faces with a mixture of mutton tallow and kerosene (Lugger 1897a, 1897b), while woodsmen of the Adirondack Mountains used blends of oil of tar and lard (Metcalf & Sanderson 1932) or mutton tallow, beeswax, and camphor that went by names such as Pflueger's Perfumed Shoo Fly Cream and Wood's Improved Lollacapop (Lidz 1981). The hardy woodsmen often conditioned their skin over a period of time, beginning with the liberal use of lard followed by gradual increases in the amount of oil of tar as their skin hardened. The procedure was continued until a fine glaze formed on their skin, which was reinforced daily, without washing off the previous days' accumulation. Pulpwood cutters on Quebec's northern shore of the Gulf of St. Lawrence used a mixture of spruce gum, turpentine, and olive oil to provide protection from the black flies that "could bite through galvanized iron" (Asselin 1935). A mixture of bacon grease and balsam gum was the solution for the survey crew on the Little Abitibi hydroelectric project in northern Ontario, as recorded in 1955 in "The Black Fly Song" by Wade Hemsworth. More pleasing means of protection included the use of compounds derived from plants such as pennyroyal (*Hedeoma pulegioides*) and citronella (*Cymbopogon nardus*), although these products, too, typically were applied in a grease (Metcalf & Sanderson 1932). *Harper's Weekly* recommended "a mixture of two and a half ounces of olive-oil, one ounce of oil of pennyroyal, and half an ounce of tar" (Anonymous 1875).

Many of the elixirs and balms touted today, such as various skin softeners and bath oils, are too volatile to provide extended protection or have not been tested critically. Nonetheless, even if found wanting in protection, they might have placebo effects. In some simuliid-problem areas, satisfaction with personal-protection measures probably plays an important role in driving the lack of homeowner interest in contributing funds to control black flies (Reiling et al. 1989).

Protective potions similar to those used on humans, as well as measures too harsh or unpleasant for most people, have been used on livestock. Oil of tar was favored in the days of punishing attacks by the southern buffalo gnat because it was easy to make and apply. A "small quantity" of oil of tar or oil of turpentine was mixed with coal tar, water was added, the concoction was left for several days to become impregnated with the odor, and stock were then slathered with the brew (Riley 1887). A coat of mud or syrup (molasses) also was useful for livestock (Riley 1887). Various plant concoctions using alder (*Alnus*), pennyroyal, or other herbs were ineffective, and applications with insecticidal properties such as tobacco soap and pyrethrum powder in water offered less than 2 hours of solace (Riley 1887). Oil of citronella in light mineral oil was recommended as a spray for the outside feathers of poultry (Benbrook 1943). Crushed leaves of walnut (*Juglans*) were rubbed on the ears of workstock in Tennessee (Snow et al. 1958b).

The most effective preventives used on livestock were various oils and greases, such as fish oil mixed with pine tar and sulfur (Skipwith 1888), tallow and pine tar (Pearce 1888), lard or cottonseed oil mixed with tar, gnat oil (i.e., any kind of stinking oil) mixed with kerosene, or axle grease and kerosene (Riley 1887, Cameron 1918). Alligator grease was said to be most efficacious (Webster 1889), as was train oil (Frierson 1889), often with a small amount of sulfur (Marlatt 1889). During the 1882 outbreak of the southern buffalo gnat in the Mississippi River Valley, more than 11,300 liters of gnat oil were sold by a single druggist in Vicksburg, Mississippi (Webster 1889). Though effective, many of these oil- and grease-based applications were "very apt to remove the hair" and impart their own unpleasant consequences (Riley 1887). Efforts, therefore, were made to devise more benign oil-based repellents such as lubricating oil and fish-oil soap diluted with water (Schwardt 1935b).

The application of grease and facsimiles has been carried into the modern era. Farmers in the Tennessee River Basin applied crankcase oil to the ears of horses and mules (Snow et al. 1958b). White petroleum jelly (Vaseline)—a more benevolent material than axle grease and kerosene or lard and turpentine (Knowlton & Rowe 1934c)—is recommended as a preventive in the ears of horses and provides three days of protection (Townsend & Turner 1976). Motor oil and fish oil were applied to newborn calves through the end of the 20th century to stem the assault of the southern buffalo gnat (Atwood 1996).

#### SYNTHETIC REPELLENTS

The first synthetic repellents were tested against black flies in Mississippi in 1942 and in the Adirondack Mountains in 1943. They were effective for more than 7 hours on humans and 4 hours on livestock—about the same amount of time afforded by citronella (Travis et al. 1951b). Tests in the Adirondacks, using nine different compounds and mixtures, indicated that certain species (e.g., members of the *Prosimulium hirtipes* species group) were repelled for shorter periods of time than other species (DeFoliart 1951b). One of the most effective synthetic repellents for use on humans has been *N,N*-diethyl-meta-toluamide (DEET). The familiar repellent OFF, with DEET as the active ingredient, was developed in 1957. Clothing, especially jackets, treated with compounds such as pyrethroids (e.g., permethrin) and DEET have proved effective (Frommer et al. 1975, Schiefer et al. 1976, Lindsay & McAndless 1978). Although DEET remains the most effective repellent, its deleterious effects on paints and plastics, as well as concerns over its safety, have led to the testing of additional compounds (e.g., piperidines), usually on live hosts (Schreck et al. 1979, Robert et al. 1992, Debboun et al. 2000), but also with ersatz membrane and blood systems (Mokry 1980c, Bernardo & Cupp 1986).

A number of pyrethroids and other synthetic compounds, generally unsuitable for humans, have been tested and recommended for livestock protection in the form of sprays, dusts, pour-ons, and ear tags (Shemanchuk 1978a, 1980b, 1981, 1982b, 1985; Shemanchuk & Taylor 1983, 1984; Expert Committee on Arthropod Pests of Animals 1993). One such compound, phosmet, provided good protection as a pour-on formulation for cattle in Alberta's Athabasca River Valley (Khan 1978, 1980, 1981). Other compounds such as dichlorvos (Vapona) and permethrin have found service on cattle and ponies, and residual sprays such as dimethoate (Cygon), naled, and permethrins have been used for poultry (Sanford et al. 1993, Schmidtmann et al. 2001). Pyrethrum sprays applied to the exterior of the houses of purple martins and bluebirds, after plugging the entrances with rags, have been recommended when simuliid problems are severe (Chambers & Hill 1996, Gaard 2002). Ear tags (Bovaid) impregnated with fenvalerate protected cattle from *Simulium luggeri* in central Saskatchewan for 6–7 weeks, permitting weight gains that justified expenditures for the ear tags (Fredeen 1988). Earlier studies with ear tags in the same area, however, indicated a lack of weight gains (Kampani 1986). Malathion and various formulations of permethrin, when applied with back rubbers, proved effective in reducing black flies (mainly *S. luggeri*) around cattle in Saskatchewan (Kampani 1986, Mason & Kusters 1991a). Using back rubbers to apply insecticides is simple and convenient and, under forced-use conditions, can treat an entire herd with little wastage of the pesticide. Sprayers that dispense insecticides as the livestock walk through the unit are also effective, particularly when they are modified to treat the undersides of the animals (Shemanchuk 1982a). The deployment of self-application devices, such as dust bags and back rubbers with repellents or insecticides, along paths and at mineral licks and watering holes could provide protection not only for cattle but also for wild ungulates (e.g., moose) that are being farmed (Pledger 1978).

#### ALTERNATIVES TO REPELLENTS

Nearly all repellents used against black flies have been external applications to the skin. Scant investigation has been aimed at internal or systemic means of conferring protection against simuliid attacks. An interest in developing vaccines has been discussed (e.g., McDaniel 1971, Wood 1985), but nothing practical has been developed. Folk remedies of limited or unknown value include hanging fabric softeners on hats and ingesting vitamin $B_1$ (Lidz 1981, Currie 1995). Once a person has been bitten, various salves are recommended to relieve the pain and itching; we find camphor-based products to be among the most effective. Electronic devices purported to emit repellent, high-frequency noises are not effective against black flies.

An appropriate style of dress can afford relative comfort while braving infested areas. Light-colored clothing is usually recommended; dark colors should be avoided, especially dark blue (Davies 1951). We, nonetheless, have found that white is attractive to black flies in parts of the Canadian arctic where the number of black flies can boggle the mind (Fig. 7.1). Clothing with zippers rather than buttons is encouraged, and pants should be tucked into boots or socks (Metcalf & Sanderson 1932, Hocking 1952a) or secured tightly at the cuffs with wide rubber bands. Head nets of fine mesh protect the head but impair vision and maneuverability. Fuel oil rubbed over an aluminum hard hat has been proclaimed effective, with the shiny hat attracting the flies and the oil trapping them (Hilton 1970).

Mules and horses have been dressed with bags or sleeves over their ears for protection against ear-infesting species (Hearle 1938), and workstock in Tennessee have been fitted with leafy branches on their bridles (Snow et al. 1958b). The udders of milk cows under attack by *Simulium mediovittatum* have been covered with an unspecified material (Bishopp 1935).

Provisioning infested areas with shelters, which many species of black flies typically will not enter, can offer significant protection. In areas that experience heavy attacks, dark shelters for livestock have been recommended for many years (Riley 1887) and are sometimes coupled with self-application devices for delivering repellents as the animals exit the shelters (Khan 1980, Mason & Shemanchuk 1990). The move by the turkey industry from outdoor to indoor production probably reduced the problem of leucocytozoonosis more than any other factor in the southeastern United States (R. Noblet, pers. comm.). Shelters combined with space sprays (e.g., dichlorvos or pyrethrins with piperonyl butoxide) have been used to protect poultry in Texas (Sanford et al. 1993).

Although black flies generally are considered not to enter structures, some species (e.g., *Simulium meridionale*) routinely enter nest boxes of native birds such as purple martins, bluebirds, and tree swallows. The problem can be curtailed by various nonchemical methods such as by covering the vent holes and slots with tape, placing an adhesive substance (e.g., Tanglefoot) around the vents, or using a house design that does not have vents (Gaard 2002).

Selection of more tolerant animal breeds can be important in establishing livestock and poultry operations in areas infested with pestiferous black flies. Cattle breeds such as Highland with longer, denser hair are apparently more tolerant of attacks by *Simulium vampirum* than are shorter-haired breeds such as Angus and Hereford (Khan 1980, Khan & Kozub 1985, Shemanchuk 1988; *S. vampirum* as *S. arcticum*). Long-haired animals, however, were considered at greater risk to damage by *Cnephia pecuarum* during the 1800s, and clipping the hair was recommended (Riley 1887). Some Texan farmers recently have truncated the calving season and restricted it to the fall to avoid attacks by *C. pecuarum*.

# PART IV
# Systematics and Taxonomy

# 9| Phylogeny and Classification of Holarctic Black Flies

In this chapter, we provide the logical basis for the classification used in this book. An understanding of phylogenetic relationships is needed to create a more natural classification of black flies and is fundamental to interpretations about simuliid evolution and biogeography. Knowledge of relationships is far from complete, but sufficient information has accumulated that we can offer a system based on our best estimate of simuliid evolutionary history. North American simuliids cannot be considered in isolation because of their close relationship with black flies from other geographic regions. Accordingly, we have included representatives of all Holarctic taxa at the species-group level and above. For the purposes of phylogenetic analysis, we have assumed the validity of the supraspecific taxa recognized by Crosskey and Howard (1997) but rejected paraphyletic groupings once they were identified. Much of the reconstruction of the basal lineages (Parasimuliinae and Prosimuliini) is based on the work of Currie (1988). Here we present the first modern evolutionary reconstruction for all supraspecific taxa with representatives in the Holarctic Region. Readers of this chapter are assumed to have some familiarity with the theory and practice of phylogenetic systematics. Those unfamiliar with terminology are referred to Schuh (2000).

Morphological, cytological, and molecular data have contributed to our knowledge of simuliid relationships. Behavioral data recently have provided additional insight into simuliid relationships (Stuart & Hunter 1998a, 1998b; Stuart et al. 2002). Taxon sampling, however, has been uneven among these data sets. Relatively few taxa have received behavioral, cytological, or molecular scrutiny, and even morphological character states are incompletely surveyed, particularly among the Palearctic taxa. It is, therefore, impractical to attempt a comprehensive numerical analysis at this time. Instead, we offer a series of trees derived from Hennigian argumentation of morphological, cytological, and molecular evidence. Presumed synapomorphic character states (i.e., shared derived character states) or appropriate literature citations are optimized on the trees to show support for each node (i.e., branching point).

The system outlined in this chapter should not be construed as the azimuth of phylogeny reconstruction for black flies. Rather, it is an attempt to derive an explicit phylogenetic framework on which to base future work. We hope that problem areas such as weakly supported nodes and polytomies (i.e., three or more branches arising from a single node) will stimulate further research.

Deriving a classification from a phylogeny entails a certain degree of subjectivity. Other than the requirement of monophyly, considerable latitude exists for the ranking and recognition of supraspecific taxa in a cladistic system. We generally have endeavored to be as nomenclaturally conservative as possible, bearing in mind the small size and structural homogeneity of the Simuliidae. We believe that such a system, which recognizes fewer genera, with numerous subgenera and species groups, carries substantially more information than systems consisting of myriad small genera. In this sense, we have inclined more toward the approach to classification of Crosskey (e.g., 1988) than that of Rubtsov (e.g., 1974a), workers who have contributed substantially to classification in modern times.

### Historical Context

The family Simuliidae as a whole is easily recognized, and its relationships with other Nematocera are reasonably well established. Yet, relationships within the family have remained somewhat obscure. The difficulty lies partly with an almost uninterrupted sequence in form between plesiomorphic (i.e., primitive) and more apomorphic (i.e., derived) members. Another problem is the incomplete knowledge of the life-history stages of some plesiomorphic taxa, which has frustrated attempts to polarize certain key character states, particularly those of the immature stages. Black flies are few in the fossil record, and those that have been described shed little light on phylogenetic relationships. In the absence of a complete fossil record, the evolutionary history of the Simuliidae must be inferred from the extant fauna.

The taxonomic history of black flies began with the monumental publication of *Systema Naturae* (Linnaeus 1758). Smart (1945) gave a detailed review of works to 1945. Reviews of later works were provided by Rubtsov (1974a), Crosskey (1988), and Crosskey and Howard (1997). The following taxa are central to understanding the early evolutionary pathways of Simuliidae, and so in this section special emphasis is placed on the timing of their discovery and the role that their discovery played in shaping present-day notions about classification: *Parasimulium*, *Prosimulium* s. l., *Gymnopais*, *Twinnia*, *Paracnephia*, and *Crozetia*. We present a brief account of some classificatory schemes proposed to date, followed by summaries of other contributions or discoveries that have shaped current notions about phylogenetic relationships.

### HIGHER CLASSIFICATION OF BLACK FLIES

Linnaeus (1758) placed the first described simuliids in the mosquito genus *Culex*. The distinctiveness of black flies went unrecognized until 1802, when Latreille established the genus *Simulium* to accommodate all the species known at that time. Thirty-two years later, the taxon "Simuliites" was created by Newman (1834), providing the modern-day basis of the family name. Apart from scattered species descriptions, the state of simuliid classification remained unchanged from that point until the 20th century.

Until the early 1900s, all described simuliids (30–40 species) were relegated to the genus *Simulium*. The first step toward recognizing simuliids as a polytypic group was taken by Roubaud (1906), who distinguished two subgenera in *Simulium* ("*Pro-Simulium*" and "*Eu-Simulium*") based primarily on details of the hind leg. Within 5 years, these taxa were ranked, along with *Simulium* s. s., as full genera (Surcouf & Gonzalez-Rincones 1911).

The taxonomic significance of the wing was recognized first by Malloch (1914), who distinguished, on features of the radial sector, *Prosimulium* (radial sector forked) from *Simulium* (radial sector unforked). Another of Malloch's (1914) contributions was the description of the genus *Parasimulium*, which was characterized by several unique features of the eyes (widely separated at the vertex, with larger facets near the middle of the eye) and wings (absence of a basal medial cell). Although the original description was based on a single specimen, its features seemed distinctive enough to warrant generic recognition. The distinctiveness of *Parasimulium* was magnified by the discovery that the specimen was a male (Knab 1915a), not a female as originally described by Malloch (1914). Malloch (1914) ironically failed to note one of the most conspicuous features of *Parasimulium*, the widely separated branches of the radial sector.

The status quo in simuliid classification prevailed until the 1920s, when a multitude of supraspecific names was introduced. This proliferation of names was due almost entirely to the efforts of the German worker Günther Enderlein. From 1914 to 1943, he published 25 papers on simuliids, in which about 50 genera were recognized, most described by Enderlein himself. Moreover, Enderlein (1921a) was the first worker to erect a suprageneric classification of the Simuliidae, and over the course of his career he recognized a total of seven subfamilies and five tribes: Prosimuliinae, Hellichiinae, Ectemniinae, Cnesiinae, Stegopterninae, Nevermanniinae (Nevermanniini, Friesiini, and Wilhelmiini), and Simuliinae (Simuliini and Odagmiini).

Frederick W. Edwards, a contemporary of Enderlein, forwarded a radically different classificatory scheme. In his final attempt at a classification, Edwards (1931b) recognized only two genera of black flies: *Parasimulium* and *Simulium*. Seven subgenera were delimited in the latter genus: *Prosimulium*, *Cnephia*, *Gigantodax*, *Austrosimulium*, *Simulium*, *Morops*, and *Eusimulium*. He made no attempt at a suprageneric classification of the family.

Smart (1945), in recognizing the need for a clear monographic account of the family, provided a critical analysis of both Enderlein's and Edwards's works. He criticized Enderlein for disregarding the treatments of other taxonomists, for his use of spurious structural characters, and for his failure to consider both male and female characters. These deficiencies contributed to an overestimation of supraspecific taxa. Of Edwards's work, Smart (1945) cited a failure to define phylogenetically significant characters, with too few genera being recognized as a result. As an alternative, Smart (1945) proposed that simuliids be organized into a system of six genera and two subfamilies. *Parasimulium* was isolated in its own subfamily (Parasimuliinae), and *Prosimulium*, *Cnephia*, *Gigantodax*, *Austrosimulium*, and *Simulium* were united in another (Simuliinae). No tribes or subgenera were recognized. Smart (1945) was the first worker to interpret the widely branched radial sector of *Parasimulium* as plesiomorphic within the Simuliidae, although Stone (1941) had first described the condition. This wing character provides the foundation on which most subsequent classifications are based, for the degree to which the radial sector is branched is one of the primary criteria by which "primitiveness" in the Simuliidae is judged (i.e., the wider the branching, the more primitive a black fly is presumed to be). Contemporaneous with the classification of Smart (1945) was the minimalist approach of Vargas (1945c), who recognized only two genera: *Parasimulium* and *Simulium*.

Stone (1949b) described a new genus, *Gymnopais*, with a larva that lacked the typical feeding apparatus (labral fans) of black flies. A similarly modified but distinct genus, *Twinnia*, was subsequently discovered in the mountains of eastern and western North America (Stone & Jamnback 1955). The fanless condition in black flies generally was assumed to be primitive (based on the assumption of a close relationship between the Simuliidae and Chironomidae), and most workers believed that a substantial phylogenetic gap existed between *Gymnopais* and *Twinnia* on the one hand and *Prosimulium* s. l. on the other (Rubtsov 1956, Grenier & Rageau 1960, Dumbleton 1963). Rubtsov (1955) was so impressed with the fanless condition that he erected the subfamily Gymnopaidinae. Other workers were concerned about the manifest similarity of adults of *Twinnia* and *Prosimulium* s. l. and regarded these two forms as congeneric (Shewell 1958). However, most workers during the 1960s relegated *Gymnopais* and *Twinnia* to a separate subfamily or tribe.

In a classification that reflected notions about simuliid relationships during the 1960s, Grenier and Rageau (1960) accepted a three-subfamily system of Simuliidae: Parasimuliinae, Prosimuliinae, and Simuliinae. As in most previous classifications, *Parasimulium* was segregated as a separate subfamily. Two tribes were recognized in Prosimuliinae: Gymnopaidini (*Gymnopais* and *Twinnia*) and Prosimuliini (*Prosimulium*). The subfamily Simuliinae included three tribes: Cnephiini (*Cnephia* s. l. ), Austrosimuliini (*Austrosimulium*, *Gigantodax*), and Simuliini

(*Simulium* s. l.). This system was essentially the same as that adopted by Dumbleton (1963), except that *Gigantodax* and *Cnephia* s. l. were relegated to Prosimuliinae.

Although restricted in geographic scope, Stone's (1965a) classification of the North American Simuliidae represented a considerable departure from other classifications proposed during the 1960s, particularly with respect to the taxonomic rank accorded certain primitive-grade taxa. Stone was not impressed enough with the features of *Parasimulium* to accord subfamilial status to members of that genus alone. Instead, he united the tribes Parasimuliini (*Parasimulium*), Prosimuliini (*Prosimulium* s. l.), and Gymnopaidini (*Gymnopais*, *Twinnia*) in the single subfamily Prosimuliinae. The genera *Cnephia* s. l. and *Simulium* s. l. constituted the only other recognized subfamily, Simuliinae. No tribal-level segregates were recognized in this latter subfamily.

Most present-day notions about simuliid classification have been drawn from the efforts of two workers, I. A. Rubtsov and R. W. Crosskey. The former produced numerous works on the supraspecific classification of the family from 1937 to 1988. The most comprehensive account of Rubtsov's classification is embodied in his 1974 treatise about the evolution, phylogeny, and classification of the Simuliidae. There he recognized a total of 59 genera in four subfamilies, as follows: Parasimuliinae (*Parasimulium*), Gymnopaidinae (*Gymnopais*, *Twinnia*), Prosimuliinae (*Gigantodax*, *Paracnephia*, *Procnephia*, *Prosimulium* s. l.), and Simuliinae (all other genera). A total of five tribes was recognized in the Simuliinae: Austrosimuliini, Cnephiini, Eusimuliini, Simuliini, and Wilhelmiini. This system was followed by most eastern European workers. Rubtsov is one of the few workers to have attempted to unravel the phylogenetic relationships of black flies and to incorporate this information in a classification. Yankovsky (1992), in an effort to provide a compromise between the classifications of Rubtsov (1974a) and Crosskey (1981, 1988), produced a classification in which he recognized nine tribes in two subfamilies (Prosimuliinae and Simuliinae).

In the classification followed by most western European and North American workers, Crosskey and Howard (1997) recognized only two subfamilies of black flies, Parasimuliinae (*Parasimulium*) and Simuliinae (all other simuliids). The subfamily Simuliinae was divided into two tribes as follows: Simuliini (*Austrosimulium*, *Simulium*) and Prosimuliini (all other simuliids). As opposed to the large number of genera recognized by Rubtsov (1974a), Crosskey and Howard (1997) adopted a rather conservative approach, recognizing 24 extant genera. The intent of their classification was not necessarily to reflect phylogenetic relationships among black flies, but rather to find some kind of "phyletic cleavage" (*sensu* Crosskey 1981) between plesiomorphic and more apomorphic members.

#### OTHER TAXONOMIC CONTRIBUTIONS

A number of African forms are characterized by the lack of a pedisulcus, a weakly developed calcipala, absence of spiniform setae on the costa and other wing veins, an ill-formed pupal cocoon, and long terminal spines on the pupal abdomen. All of these character states, which are shared to some degree by members of *Parasimulium* and *Prosimulium* s. l., have been considered plesiomorphic in the Simuliidae. The wing venation of these African forms, however, is evidently derived in lacking a definite fork in the radial sector. As a result of this combination of plesiomorphic and derived traits, the placement of these black flies has remained problematic. Most of the species were originally described as either *Prosimulium* (De Meillon & Hardy 1951) or *Cnephia* (e.g., Freeman & De Meillon 1953). A separate genus, *Paracnephia*, was not recognized until 1962 (Rubtsov 1962f). Crosskey (1969) accepted this name as a subgenus of *Prosimulium* s. l. and erected another subgenus, *Procnephia*, for a small group of species that appeared to be intermediate in form between *Prosimulium* s. s. and *Paracnephia* (e.g., costa of female with hairlike macrotrichia only; male with some macrotrichia thickened, but not fully spiniform). The phylogenetic position of *Procnephia* and *Paracnephia* has remained uncertain to the present day. Crosskey (1988) retained these taxa as subgenera of *Prosimulium* s. l., whereas Rubtsov (1974a) regarded them as full genera in the Prosimuliinae. In addition to the forms living in southern Africa, Rubtsov (1962f) referred the Australian species of "*Cnephia* s. l." to *Paracnephia*. Crosskey and Howard (1997) recognized *Paracnephia* as a full genus, with two subgenera, *Paracnephia* s. s. and *Procnephia*. These authors also assigned the Australian "*Cnephia* s. l." to *Paracnephia* s. l. (though unplaced at the subgeneric level).

Davies (1965b, 1974) studied the genus *Crozetia*, which has larvae with rudimentary, rakelike labral fans. Instead of filter feeding in the typical simuliid fashion, the larvae use their fans to rake filamentous algae from the stones on which they live. Davies (1965b, 1974) compared this arrangement with the fans of typical simuliids, together with the minute, bristle-like appendages that arise from the labra of first-instar larvae of *Gymnopais*, *Twinnia*, and *Prosimulium* s. l. He concluded that bristle-like, labral appendages are "primitive" for Simuliidae (an example of "ontogeny recapitulating phylogeny"), and that evolution has proceeded in two directions: to complete loss in *Gymnopais* and *Twinnia*, whose ancestors never before possessed fans; and to elaboration of the bristles into a raking device, such as in larvae of *Crozetia*. Davies (1965b, 1974) assumed that this latter type of fan gave rise to the elegant filtering devices of typical simuliid larvae. No attempt was made to reclassify the family, but the assumption was implicit that *Gymnopais* and *Twinnia* would have to be considered the sister group of all other simuliids.

Wood (1978) took the opposite view. He argued that fans are a fundamental and homologous feature not only of the Simuliidae but also of related families (Culicidae, Dixidae, Ptychopteridae, Tanyderidae). Thus, absence of fully developed fans in larval Simuliidae (and by extension in larvae of other families of Culicomorpha, viz., Ceratopogonidae, Chaoboridae, Chironomidae, Thaumaleidae) must be the result of loss. The discovery of fully developed labral fans in larval *Parasimulium* confirms that filter feeding is in the groundplan of the Simuliidae (Currie 1988).

Wygodzinsky and Coscarón (1973) surveyed the primitive-grade simuliids (Prosimuliini *sensu* Crosskey) of the Neotropical Region. Although they did not concern themselves directly with the higher classification of the family, Wygodzinsky and Coscarón (1973) proposed sister-group relationships among a number of the genera. Moreover, they maintained that most definitions of Prosimuliini were based on symplesiomorphic characters (i.e., shared primitive characters) and, hence, could not be maintained in a cladistic system. Although they did not develop an explicit phylogeny, Wygodzinsky and Coscarón (1973) were the first authors to view simuliid relationships in a cladistic sense.

Special mention is required about the role of cytology in resolving simuliid relationships. The pattern of chromosomal inversions, the breakpoints of which can occur at any of several thousand sites, can be analyzed in a sequential fashion to generate a cytological transformation series, the rationale being that the probability of the same inversion occurring independently more than once is remote. The transformation series is unrooted because the point of origin, the so-called standard sequence, is chosen on the basis of its centrality and not on the basis of out-group comparison. In other words, for any given arm of the chromosome, the standard sequence occurs in a number of related taxa and gives rise to the largest number of individual derivatives. Thus, although Rothfels and Freeman (1966) demonstrated the transition *Prosimulium* ↔ *Helodon* ↔ *Twinnia* ↔ *Gymnopais*, they could not determine the direction of evolution (i.e., it is unclear whether *Prosimulium* or *Gymnopais* is the most plesiomorphic member of the sequence).

As the world simuliid fauna becomes better known chromosomally, the many small cytological transformation series eventually might be joined together into a comprehensive system of hypothesized relationships. However, it will remain to the morphotaxonomist (or molecular systematist) to determine the direction of evolution, by using external structural (or molecular) information to root the system of relationships suggested by cytology. This one weakness not withstanding, cytotaxonomy will continue to be an important component of phylogenetic investigation because it provides an independent means by which morphologically and molecularly derived hypotheses can be tested.

### PHYLOGENETIC STUDIES OF THE SIMULIIDAE

The first phylogenies based on an explicit phylogenetic (i.e., cladistic) framework were developed during the early 1980s, spurred by the rediscovery of *Parasimulium*. The primitive status of *Parasimulium* has been inferred largely from the wing venation, which has $R_1$, $R_{2+3}$, and $R_{4+5}$ widely separated at the costa; these veins are more crowded in other simuliids. This condition, and several unique features of the head and terminalia, have led most workers to recognize a separate subfamily for members of the genus (e.g., Smart 1945, Shewell 1958, Grenier & Rageau 1960, Crosskey 1969, Rubtsov 1974a, Peterson 1977a). Interpretations prior to 1981 were based on the adult males because only they were known. The only modern worker not according subfamilial rank to *Parasimulium* was Stone (1965a), who instead recognized the tribe Parasimuliini in the subfamily Prosimuliinae.

Because of the apparent primitive status of *Parasimulium*, many efforts were directed at collecting the female and immature stages. General agreement held that the phylogenetic position of *Parasimulium* could be established only in view of all life-history stages. But despite repeated attempts to obtain additional material, a period of 46 years elapsed between the last collection of *Parasimulium* in 1935 and the recollection of *P. crosskeyi* in 1981 (Wood & Borkent 1982). This latter discovery was particularly important because as well as significantly adding to the number of known male specimens, the female of *Parasimulium* was collected for the first time.

Using both male and female characters, Wood and Borkent (1982) reassessed the phylogenetic position of *Parasimulium*, and with cladistic techniques they proposed a sister-group relationship between that genus and all other simuliids. Their work supported the notions of most previous workers whose conclusions were based on a noncladistic or dialectical approach to systematics. Characters given in support of the monophyly of *Parasimulium* included reduction of the katepisternum, loss of the mesepimeral tuft, and loss of a peglike seta at the apex of the gonostylus. Synapomorphic characters given for the Simuliidae, exclusive of *Parasimulium*, included the following: male eye with a line of discontinuity between the large upper and small lower facets, crowding of the branches of the radial sector ($R_{2+3}$ and $R_{4+5}$), and sternum X of the female divided into anal lobes. This evidence was given in support of a two-subfamily arrangement for the Simuliidae, in which Parasimuliinae comprised *Parasimulium* and Simuliinae comprised all other simuliids (Wood & Borkent 1982).

These same authors (Borkent & Wood 1986) later obtained first- and second-instar larvae of *Parasimulium stonei* from eggs laid in vitro. This material provided additional information by which their original hypothesis, based on adult characters alone, could be tested. But instead of corroborating their earlier hypothesis, larval characters suggested a close relationship between *Parasimulium* and some typical members of the Prosimuliini *sensu* Crosskey (1969). This revised interpretation lent support to the classification of Stone (1965a), in which *Parasimulium* was considered merely an aberrant side branch of the tribe Prosimuliini. The primary character given in support of this relationship was loss of the labral fan in first-instar larvae. On this basis, *Parasimulium* was presumed to have shared an immediate common ancestor with *Crozetia*, *Gymnopais*, *Prosimulium*, and *Twinnia*.

Courtney (1986) discovered later-instar larvae and pupae of *Parasimulium crosskeyi* and provided brief descriptions. Currie (1988) redescribed this material, evaluated its character states, and provided for the first time a phylogenetic analysis based on all life-history stages of the Simuliidae. His analysis supported the hypothesis of a sister-group relationship of *Parasimulium* (Parasimuliinae) and all other black flies (Simuliinae), and provided evidence for two monophyletic tribes within the latter subfamily: Prosimuliini and Simuliini. Although nominally similar to Crosskey's (1988) suprageneric classification, Currie's (1988) concept of the Prosimuliini was much more restricted, consisting only of the genera *Gymnopais*, *Helodon* s. l., *Levitinia*, *Prosimulium*, *Urosimulium*, and *Twinnia*. All other genera were relegated to the tribe Simuliini. In addition to his interpretation of the initial two dichotomies in simuliid phylogeny, Currie (1988) executed individual cladistic analyses of the Parasimuliinae and the Prosimuliini (this latter at the genus-group level). No attempt was made to resolve relationships within the Simuliini. Later, however, Currie and Grimaldi (2000) modified Currie's (1988) concept of the Simuliini to accommodate the newly described fossil genus *Archicnephia*.

The only other morphologically based cladistic analysis of the Simuliidae is that of Py-Daniel (1990). His analysis of 20 genera of black flies in the New World resulted in a two-subfamily arrangement, with the Gymnopaidinae consisting of *Gymnopais* and *Twinnia*, and the Simuliinae housing the other 18 genera. His analysis did not include reference to *Parasimulium*. The proposed sister-group relationship between the Gymnopaidinae and all other simuliids evidently stemmed from the belief that the fanless condition is plesiomorphic within the Simuliidae. However, Wood (1978) and Currie (1988) demonstrated that absence of labral fans is secondary in the family. Other anomalous relationships are suggested by Py-Daniel's (1990) analysis, owing largely to his choice of exemplar taxa and his consideration of characters of only the immature stages.

#### MOLECULAR SYSTEMATICS OF THE SIMULIIDAE

The 1990s heralded the first modern attempts to infer simuliid relationships using molecular techniques (Xiong & Kocher 1991, Pruess et al. 1992, Tang et al. 1995b), although cruder attempts were made as early as the 1970s (e.g., Teshima 1972, Sohn et al. 1975). Each of these studies was of limited taxonomic scope, shedding little light on outstanding problems in simuliid phylogeny. They all suffered from poor choice of genes, small data sets, and inadequate taxon sampling. More recently, Pruess et al. (2000) assessed the utility of the mitochondrial cytochrome oxidase II gene in resolving relationships among 13 species and four genera of black flies. And Smith (2002), using one nuclear and two mitochondrial genes, achieved good resolution of relationships of selected members of the subgenus *Simulium*. The most detailed and comprehensive molecular phylogeny of the family was that of Moulton (1997, 2000), who examined exemplars of 20 genera, using molecular sequences from four nuclear and two mitochondrial loci. The phylogenetic trees, believed to be the best estimates of relationships, reflected Currie's (1988) interpretations at the subfamilial and tribal levels. Moulton's (1997) analyses also included an interpretation about relationships in the tribe Simuliini. Although his preferred phylogeny (Fig. 3.20 of Moulton 1997) contained a number of expected relationships (e.g., *Cnesia* + *Gigantodax*, *Metacnephia* + *Simulium* s. l.), it also contained some unexpected results, such as the placement of *Austrosimulium* at the base of the Simuliini. Crosskey (1969, 1990 [p. 52]), in contrast, considered *Austrosimulium* to be more closely related to *Simulium* s. l. than to any other genus, and included these two genera as the only members of the tribe Simuliini. Morphological data strongly support this latter interpretation. No single source of data is likely to resolve relationships adequately throughout the entire family.

## PHYLOGENY OF THE HOLARCTIC SIMULIIDAE

In this section, we provide evidence for the monophyly and relationships of black flies in the Holarctic Region. Phylogenetic trees at the species-group level and above are presented using Hennigian argumentation. Character numbers appear on the accompanying trees (Figs. 9.1–9.8).

#### MONOPHYLY OF THE FAMILY SIMULIIDAE (Fig. 9.1)

*Adult Groundplan Apomorphies*

1. *Pedicel reduced, not much wider than first flagellomere, cylindrical, similar in both sexes.* The plesiomorphic state is for the pedicel to be large relative to the flagellomeres (Wood & Borkent 1982).

2. *Eye of male with line of discontinuity between large upper facets and small lower facets.* The plesiomorphic state, which is for no such line of discontinuity to be present, is typical of most Diptera (Wood & Borkent 1982). The derived state is evident in several other families of Nematocera, including the Blephariceridae (in part), Axymyiidae, Bibionidae, and Cecidomyiidae (*Trisopsis*). However, the condition is unknown in any other family of Culicomorpha and, therefore, must be considered an independently derived trait of the Simuliidae. The dichoptic males of *Gymnopais* and the subgenus *Parasimulium* have no evident line of discontinuity between the upper and lower eye facets but were undoubtedly derived from an ancestor whose eyes were so divided. Indeed, the holoptic males of certain plesiomorphic species of *Gymnopais* (e.g., *G. holopticus*) have a divided eye. The condition found in dichoptic simuliid males, therefore, is considered a reversal toward the plesiomorphic form.

3. *Wing greatly broadened at base.* Black flies are unique among the Culicomorpha in having the wing greatly broadened at its base. In this respect, the wings resemble those of the distantly related Blephariceridae and Deuterophlebiidae (Hennig 1973).

4. *A small but distinct basal medial (bm) cell.* We do not agree with the interpretation of Hennig (1973) that the medial-cubital cross vein (m-cu) is absent from the Simuliidae. This cross vein, which forms the apical margin of the bm cell, arises close to the wing base of most black flies. The consequence of such an arrangement is a small but distinct bm cell in most genera. The groundplan condition for the Culicomorpha is for the bm cell to be much larger, with cross vein m-cu arising more apically on the wing at or near the level of the radial-medial cross vein (r-m). This latter condition is basic to Chaoboridae, Chironomidae, Culicidae, Dixidae, and Thaumaleidae, although secondar-

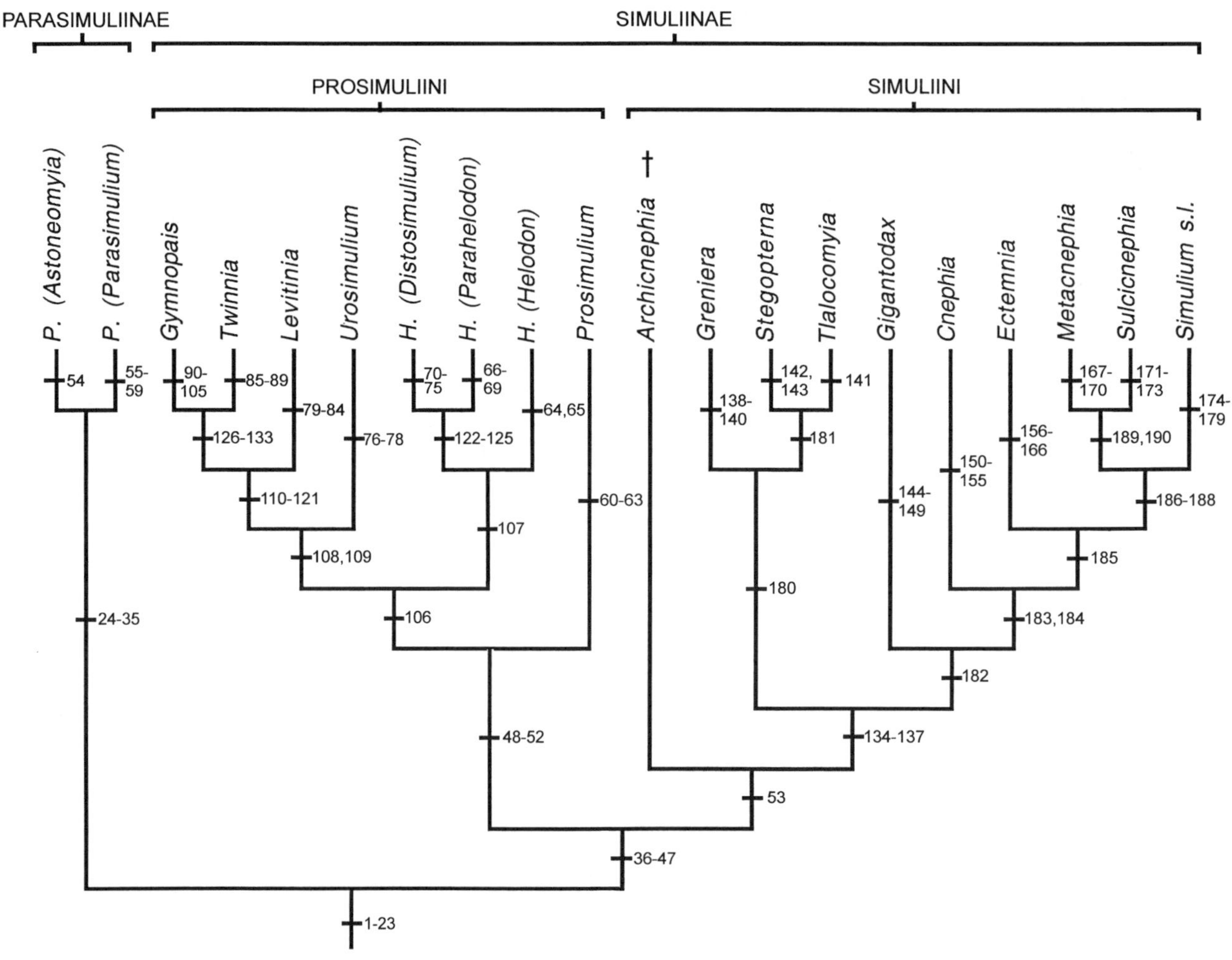


Fig. 9.1. Phylogeny of the genera and subgenera of black files in the Holarctic Region (excluding subgenera of *Simulium* s. l.).

ily lost from some lineages. The small bm cell in Simuliidae (absent from some genera) is derived with respect to these families. The condition in Ceratopogonidae is further derived in lacking cross vein m-cu altogether.

5. *Claw of male with grooved lobes of cuticle on dorsal side at base.* The claw of other Culicomorpha males is simple basally (Wood & Borkent 1982).

6. *Hind basitarsus laterally flattened and ventrally keeled.* The plesiomorphic condition is for the hind basitarsus to be more cylindrical (Wood & Borkent 1982).

7. *Tergite I of abdomen with posterior fringe of long hairs (= basal fringe) on each side.* A basal fringe is not evident in any other Nematocera (Wood & Borkent 1982).

8. *A single, large, median spermatheca, with retention of the two lateral ducts.* Three spermathecae are present in the groundplan of Diptera (Wood & Borkent 1982). In the Simuliidae, a central spermatheca is flanked by two lateral ducts.

9. *Feeding behavior of adult female.* At least two major differences distinguish the blood-feeding behavior of black flies and that of other hematophagous Nematocera. First, blood feeding in the simuliid groundplan is strictly diurnal, whereas in Ceratopogonidae, Culicidae, and Phlebotominae, feeding also occurs nocturnally. Second, black flies spend a relatively long period of time crawling and probing before biting a host, whereas females of Ceratopogonidae, Culicidae, and other hematophagous Diptera bite rapidly upon settling. Both of these characteristics can be interpreted as derived in the Simuliidae. We disagree with the interpretation of Rubtsov (1974a) that simuliids are primitively nonhematophagous, a view held by early as well as modern workers (e.g., Townsend 1895, Crosskey 1990) who considered them to have originally fed on plants. Relatively few species have females that are incapable of taking a blood meal, and their mouthparts are similar to those of biting species, except that their mandibles and maxillae are weakly developed (Nicholson 1945).

### *Pupal Groundplan Apomorphies*

10. *Gill with a basal fenestra laterally.* Most simuliids have a basal fenestra associated with the gill. The gill of *Crozetia crozetensis* (Womersley) lacks a basal fenestra, but the greatly prostrate form of the gill suggests that the fenestra might have been lost from this species.

11. *Gill multibranched.* In its basic form, the pupal gill of Culicomorpha consists of a simple cuticular tube. A branched gill is found only among members of the Simuliidae and some relatively derived lineages of the Chironomidae (e.g., Chironomini and some Pseudochironomini). However, the condition in these two families has been derived independently. In the simuliid groundplan, a common gill base gives rise to three main branches, although secondary filaments can arise from these primary trunks. Such a condition is found in at least some members of the following plesiomorphic genera: *Gymnopais, Paracnephia, Parasimulium, Prosimulium* s. l., and *Twinnia*.

12. *Pupal spiracle with regulatory apparatus operated by pharate adult.* The pharate adult of simuliids can regulate the opening and closing of the pupal spiracle by moving an apodeme that bridges the pupal and adult cuticles (Hinton 1957). Among the Diptera, only the Psychodidae possess a similar mechanism, presumably as the result of convergence.

13. *Abdominal segments III– or IV–VIII with tergites and sternites widely separated by pleural membrane.* The abdominal tergites and sternites of most Culicomorpha are evidently fused to each other laterally (i.e., they are apparently undivided by pleural membrane). The groundplan for the Simuliidae is to have a wide area of striate pleural membrane dividing abdominal segments III– or IV–VIII. Such a condition is found in nearly all simuliid genera, including *Parasimulium*. The tergites and sternites of *Crozetia* are only narrowly separated by pleural membrane, but this arrangement is probably a secondary development, as inferred from other characters. Similarly, the ringlike appearance of abdominal segment III in *Gymnopais* and *Twinnia* can be interpreted as a secondarily derived feature.

14. *Pharate pupa feeds and spins its cocoon.* Black flies are unique among Diptera in that the pharate pupa is able to feed and spin its own cocoon (Hinton 1958). In nearly all other endopterygotes, feeding ceases after the larval-pupal apolysis, and the larva spins the cocoon.

### *Larval Groundplan Apomorphies*

15. *Supernumerary and unfixed number of larval instars.* The number of larval instars reported for the Simuliidae varies from 4 to 11, depending on the species and environmental conditions (e.g., Yakuba 1960, Craig 1975, Fredeen 1976, Colbo 1989). However, the report (Yakuba 1960) of four instars is open to question. Most species so far investigated have either six or seven larval instars, and the groundplan for the Simuliidae is probably in this range. Most other Nematocera have a fixed number of larval instars.

16. *Form of labrum and labropalatum.* As indicated by Craig (1974) and Wood (1978), the labral fan of simuliids is fundamentally the same structure as the lateral palatal brush of other nematoceran families (e.g., Culicidae, Dixidae, and Ptychopteridae). Wood (1978) argued that the labral fans of these families are homologous, providing evidence of their common ancestry. Nevertheless, several independently derived features of the simuliid labrum are evident:

a. *Reduced number of rows of labral-fan rays.* The labral fans of Ptychopteridae and most Culicomorpha consist of many rows of numerous hairs. In the Simuliidae, the number of rows is reduced to three (i.e., the primary fan, the secondary fan, and the median fan).
b. *Intertorma with anteroventrally directed stem.* The intertormal stem of simuliids does not have a homologue in other Nematocera (Craig 1974).
c. *Interdigitation of ventral fascicles of the posterior frontolabral muscles.* Interdigitation of the posterior frontolabral muscles of the Simuliidae is apomorphic in relation to the noninterdigitated condition of most Nematocera (Craig 1974). This latter condition also is evident in larvae of *Gymnopais* and *Twinnia* but probably as a reversal.

17. *Silk glands large, folded.* The silk glands of simuliid larvae are perhaps the largest of any nematoceran relative to the size of the body. From the common opening of the silk glands, each gland extends posteriorly to segment VII, whereon it doubles back on itself.

18. *Bacteria-covered multiporous sensillum.* This sensillum is situated ventral to the membranous antennal base of simuliid larvae (Craig & Batz 1982); it is also present in the larvae of *Parasimulium stonei* (Borkent & Wood 1986) and *Parasimulium crosskeyi*. Craig and Batz (1982) suggested that the multiporous sensillum is a possible homologue of Lauterborn's organ, a compound antennal sensory organ of chironomid larvae, but the position and form of the former sensillum render this interpretation unlikely. No homologue of the multiporous sensillum has been identified in any other Culicomorpha.

19. *Dorsal anus.* Hennig (1973) remarked on the presence of a dorsal anus in larval simuliids. The most common condition in Diptera is for the anus to be ventrally situated, although it is terminally situated in some Nematocera (Teskey 1981). The anus of larval Thaumaleidae occupies a similar dorsal position, so a dorsal anus is possibly part of the groundplan of the Chironomoidea. The condition in simuliids, however, is exaggerated because of the intervention of the anal sclerite between the anus and the posterior proleg. We interpret this latter state as synapomorphic for the Simuliidae.

20. *Rectal papillae of three lobes.* Simuliids are unique among the Culicomorpha in possessing an odd number of rectal papillae. The papillae either are simple, as in most simuliid genera, or bear multiple secondary lobules, as in most species of the genus *Simulium*. All other culicomorph larvae, as far as we are aware, possess an even number (two or four) of simple rectal papillae. Four is evidently the groundplan number for the Culicomorpha.

21. *Anal sclerite.* The anal sclerite is visible as an X-, Y-, or subrectangular-shaped structure on the dorsum of abdominal segment IX. Inserted on the sclerite are muscles that operate the posterior proleg (Barr 1984). No other culicomorph larvae possess such a structure. The anal sclerite is absent from *Ectemnia*, but this condition is probably the result of loss.

22. *Hook rows of prothoracic and posterior prolegs.* Simuliidae is the only family of Chironomoidea in which the hooks of the prothoracic and posterior prolegs are arranged in longitudinal rows. The hooks of the prolegs of the Ceratopogonidae, Chironomidae, and Thaumaleidae are variously and irregularly arranged, but not in the linear fashion of simuliid larvae.

*Egg Groundplan Apomorphies*

23. *Egg subtriangular, with a dorsal transverse bulge.* The subtriangular egg shape of the Simuliidae is synapomorphic for the family (Borkent & Wood 1986). The typical condition in Nematocera is for the egg to be oval or nearly cylindrical in outline.

**MONOPHYLY OF THE SUBFAMILY PARASIMULIINAE** (Fig. 9.1)

*Adult Groundplan Apomorphies*

24. *Stemmatic bulla.* Species of the genus *Parasimulium* share with those of *Gymnopais* and *Twinnia* a stemmatic bulla near the posterior margin of the compound eye. All other simuliids lack a prominent bulla, although one is suggested in some species of the genera *Prosimulium* and *Helodon* (Wood 1978). The only other flies with a similar structure are the Nymphomyiidae and some aberrant Chironomidae. The presence of a bulla is correlated with reduction of the compound eye (McAlpine 1981) and has undoubtedly arisen several times independently in the Diptera. In the Simuliidae, a prominent stemmatic bulla has evolved at least twice: once in *Parasimulium* and once in the *Gymnopais-Twinnia* clade.

25. *Katepisternum markedly reduced, almost pointed ventrally in profile.* The plesiomorphic state is for the katepisternum to be much more convex ventrally (Wood & Borkent 1982).

26. *Mesepimeral tuft lost.* The plesiomorphic state is for a tuft of setae to be present on the dorsal part of the mesepimeron (Wood & Borkent 1982).

27. *Basal medial (bm) cell lost.* A small bm cell is widely distributed among simuliid genera and is interpreted as part of the simuliid groundplan. Absence of the cell, therefore, is considered derivative and is evidence of the monophyly of *Parasimulium* s. s. + *Astoneomyia*. However, a regressive bm cell also is found in other lineages of black flies (*Gigantodax*, *Simulium* s. l.), and by itself does not provide strong evidence of the monophyly of the genus *Parasimulium*.

28. *Basal radial (br) cell reduced in length.* A relatively long br cell is evidently part of the culicomorph groundplan, at least as inferred from the distribution of this character throughout the infraorder. In Culicoidea, the br cell is equal to about half the length of the wing, as measured from the base of the cell. In the Chironomoidea, the cell is rather varied in length and evidently has become reduced through convergence a number of times. Two length categories of the br cell are recognized in the Simuliidae: one, in which the br cell is equal to about one third the length of the wing as measured from the base of the cell; and another, in which the cell is less than one fourth that length. Because the former condition is widespread among simuliids and most closely approximates the condition found in Culicoidea, we regard it as plesiomorphic. A derived (shortened) br cell is characteristic of the subgenera *Astoneomyia* and *Parasimulium* and at least two other simuliid lineages (*Austrosimulium*, subgenus *Simulium*). Based on other characters, the condition in the genus *Parasimulium* is independently derived from the others.

29. *Gonostylus with apical spinule (i.e., peglike seta) lost.* The plesiomorphic state is a gonostylus with one or more spinules (Wood & Borkent 1982). The seta evidently has been lost from various other lineages of Culicomorpha, including some Simuliidae (e.g., *Simulium pictipes* species group).

30. *Autogenous females.* Female simuliids are presumed here to be primitively anautogenous because their mouthparts agree in many important features with those of the following Diptera, which also have representatives with blood-sucking mouthparts: Blephariceridae, Ceratopogonidae, Chaoboridae (*Corethrella*), Chironomidae (*Archaeochlus*), Culicidae, Psychodidae, Rhagionidae, and Tabanidae. Females of the genus *Parasimulium* are autogenous. The apomorphic state evidently evolved a number of times independently in the Simuliidae. All species of *Gymnopais* are autogenous, as are isolated species of other genera.

31. *Mating behavior.* Females rest motionless on the undersides of leaves while males search underneath such leaves for prospective mates. The typical culicomorph condition, as in most Diptera, is for the males to form aerial swarms into which the females fly to become mated. The mating behavior of *Parasimulium* is probably apomorphic for the Simuliidae; however, the majority of genera have yet to be surveyed for this character.

*Preimaginal Groundplan Apomorphies*

32. *Maxillary palpus of first-instar larva with subapically arranged sensilla.* Instead of the typical culicomorph condition, in which the sensilla are positioned apically and in a circular configuration (Craig and Borkent 1980), the sensilla of the first instar of the subgenus *Parasimulium* are situated subapically on the palpus, although still in the typical circular arrangement (Borkent & Wood 1986). The condition in first-instar *Parasimulium* was interpreted by Borkent and Wood (1986) as an intermediate stage of a transformation series between the usual culicomorph condition (as in most simuliid genera examined) and the laterally positioned but linear arrangement of sensilla in first instars of *Gymnopais*, *Prosimulium*, and *Twinnia*. In the present analysis, we regard the condition in the subgenus *Parasimulium* as independently derived. The arrangement of palpal sensilla in first instars of *Astoneomyia* is unknown. The majority of simuliid genera has yet to be surveyed for this character.

33. *Abdominal segment IX of larva with single, large, conical lobe ventrally.* In the groundplan of the Chironomoidea, the venter of abdominal segment IX is simple

(i.e., without obvious projections or extensions of the abdominal cuticle). A simple segment IX is probably also part of the groundplan for the Simuliidae, as inferred from the occurrence of such a condition in the following plesiomorphic taxa: *Crozetia*, *Gymnopais*, *Levitinia*, *Paracnephia*, *Prosimulium*, *Twinnia*, and most Neotropical "Prosimuliini" *sensu* Wygodzinsky and Coscarón (1973). The larva of *Parasimulium crosskeyi* is derived with respect to other simuliids in having a single, large tubercle or lobe on the venter of abdominal segment IX. The only other simuliids having a similar structure are in the genus *Stegopterna* and *Tlalocomyia*. However, the single transverse midventral bulge of these genera is probably not homologous with the structure in *Parasimulium*, as judged from other characters. Similarly, the paired conical ventral tubercles of certain other genera (*Austrosimulium*, *Ectemnia*, *Greniera*, *Lutzsimulium*, and some *Simulium* s. l.) probably are derived independently.

34. *Hypogean habitat*. Courtney (1986) described the habitat of larval *Parasimulium* as hyporheic; however, we consider the seepages in which they occur to be hypogean or phreatic in origin. Although the differences between these two types of habitats might appear trivial and perhaps difficult to distinguish where they occur together, marked biological differences exist between them (i.e., there are elements that occur in the seep that do not occur in the main stream, and vice versa). If the proposition is accepted that the immature stages of *Parasimulium* are hypogean, then the habitat is clearly unique for the Simuliidae and perhaps even for Diptera. If the immature stages do prove to be hyporheic, the distinction still must be drawn between the apparent obligatory existence of *Parasimulium* in that habitat, and the facultative occurrence there of other simuliids (Currie & Craig 1988). The groundplan condition for black flies is for the immature stages to be attached to exposed substrates in surface-flowing streams.

35. *Lack of pigmentation*. Larvae of *Parasimulium crosskeyi* and *Parasimulium stonei* are apparently blind and completely unpigmented, except for the heavily sclerotized anterior margin of the hypostoma. The pupal integument of *P. crosskeyi* is nearly transparent, revealing the developing adult underneath. These adaptations are associated with the subsurface environment of the immature stages. Among the Culicomorpha, only larvae of the Chaoboridae (in part) are translucent or transparent. The immature stages of all other simuliids have at least some pigmentation, and two of the three pairs of larval ocelli are darkly pigmented.

### *Systematics of the Subfamily Parasimuliinae*

The monophyly of the subfamily seems reasonably well supported by the 12 character states just listed. However, only character states 24–30 are confirmed for both *Astoneomyia* and *Parasimulium* s. s. The five other character states have been confirmed only for *Parasimulium* s. s. Discovery of the immature stages of *Astoneomyia* is needed to firmly establish the monophyly of the subfamily Parasimuliinae.

One of the most curious features of *Parasimulium* s. l. is the arrangement of wing veins. The primitive status of the genus has been inferred largely by the widely separated branches of the radial sector. However, several derived features of the wing are also evident, such as loss of the bm cell (character 27) and reduction of the br cell (character 28). Such discord raises questions about the direction of character evolution in the simuliid wing. One possibility is that a widely branched radial sector in simuliids is a reversal toward the culicoid condition and, thus, is evidence of the common ancestry of *Parasimulium* + Prosimuliini, as defined in the present work. This hypothesis is based on the presence of a rather short-branched radial sector in the groundplan of the Ceratopogonidae, Chironomidae, and Thaumaleidae. Evidence for a close relationship between *Parasimulium* and some typical members of the Prosimuliini *sensu* Crosskey already has been forwarded based on larval characters (Borkent & Wood 1986), and although this hypothesis cannot be dismissed out of hand, we believe that other evidence indicates that *Parasimulium* is only distantly related to other black flies. We, therefore, have retained the traditional view that a widely branched radial sector is plesiomorphic for the Simuliidae, and that a short-branched or unbranched radial sector represents increasingly derived states.

The phylogenetic significance of the anal sclerite of larval *Parasimulium* remains uncertain. Of particular interest are the ventral arms, which are not articulated dorsally with the dorsomedian sclerite. Ventrally, the ventral arms almost meet along the midline and are separated by only a small spindle-shaped sclerite. No other simuliid has such an arrangement. In larvae of *Crozetia*, the apices of the ventral arms almost meet midventrally, but like all other simuliids they are articulated dorsally with the dorsomedian sclerite. Also, the ventromedian sclerite is lacking in *Crozetia*. In most other simuliids, the ventral arms are shorter than in *Parasimulium* and *Crozetia*, and they project ventrally to a point no farther than about half the distance around abdominal segment IX. Whether the short- or long-arm state represents the plesiomorphic condition for the Simuliidae cannot be determined using out-group comparison, and the distribution of these states within the family sheds little light. The condition in *Parasimulium* and *Crozetia* should not be confused with that in certain other simuliid genera (*Austrosimulium* in part, *Gigantodax*, and *Simulium* in part), in which a semicircular sclerite contours the ventral half of abdominal segment IX. The semicircular sclerite is a separate structure, although probably serving as an accessory to the anal sclerite, and probably has developed along its own lines several times independently.

## MONOPHYLY OF THE SUBFAMILY SIMULIINAE (Fig. 9.1)

### *Adult Groundplan Apomorphies*

36. *Radial sector with branches* ($R_{2+3}$ *and* $R_{4+5}$) *closely approximated*. The plesiomorphic state is for the branches of the radial sector to be more widely separated (Wood &

Borkent 1982). This character is considered in greater detail under character 53.

37. *False vein (medial-cubital fold) forked apically.* Simuliidae, Ceratopogonidae, and Chironomidae are the only Culicomorpha that have a false vein between the posterior branch of M and $CuA_1$. This arrangement can be taken as further evidence of the common ancestry of these three families. A simple or unbranched false vein is characteristic of Ceratopogonidae and Chironomidae, but among simuliids, only the genus *Parasimulium* has such a condition (which must be considered basic to these three families). The derived condition, a false vein that is forked apically, is characteristic of all other simuliids and can be taken as evidence of the monophyly of the Simuliinae.

38. *Katepisternal sulcus (mesepisternal groove).* Most simuliids have the katepisternum divided into a dorsal and a ventral part by a horizontal katepisternal sulcus. The only simuliids lacking this sulcus are species of the genus *Parasimulium*. All other Diptera share this condition with *Parasimulium*. However, as indicated under character 25, the katepisternum of *Parasimulium* is reduced ventrally, so the sulcus possibly has become secondarily lost in that genus. If this were so, the katepisternal sulcus would have to be considered a synapomorphy of the entire family. Until further evidence is brought to bear on the problem, the presence of a katepisternal sulcus only tentatively can be considered a synapomorphy of the Simuliidae, exclusive of the genus *Parasimulium*.

39. *Sternum X of female divided medially.* The plesiomorphic state is for sternum X to be incompletely divided in the female. Wood and Borkent (1982) compared the completely divided sternum X (anal lobes) of most simuliids with the undivided sternum of other Culicomorpha. They tentatively interpreted the condition in *Parasimulium crosskeyi* (which has sternum X deeply notched but narrowly continuous medially) as forming an intermediate stage of a transformation between the condition in other Culicomorpha and that in all other simuliids. Examination of the female of *Parasimulium stonei* corroborates this conclusion, which is accepted here. Whether the condition in *Parasimulium* s. s. is truly plesiomorphic for the Simuliidae, or whether it represents a secondary fusion of the anal lobes is still not clear. If a secondary fusion, then the completely divided state would have to be considered a synapomorphy of the entire family.

### *Pupal Groundplan Apomorphies*

Hinton (1957, 1968, 1976) provided detailed accounts of the pupal gill of the Simuliidae. Although varied in external appearance, the gill has the same basic structure and function throughout the family. At least three derived features of the gill, discussed below, are evident in the Simuliidae, exclusive of *Parasimulium*.

40. *Plastron network covering entire gill.* The Simuliidae and Chironomidae are the only Culicomorpha in which a plastron is held on the pupal gill (Hinton 1976). Examination of the pupa of *Thaumalea americana* reveals, however, that a spiracular gill is also present in the Thaumaleidae. Hinton (1968) considered the plastron to be derived independently in most families of Diptera, but the argument also can be made that a plastron is part of the groundplan of the Chironomoidea. In the Thaumaleidae and plesiomorphic Chironomidae, the plastron (= plastron plate) is restricted to the apex of the gill. A similar condition is found in the pupa of *Parasimulium crosskeyi*, in which the plastron is held on the three branches of the gill but not on the elongate gill base. In all other simuliids, at least as far as we are aware, the plastron covers the entire gill, including the base. Such an arrangement is unique among the Culicomorpha and can be taken as strong evidence of the monophyly of the Simuliinae. We interpret the condition in *Parasimulium* as forming an intermediate stage of a transformation series between the plastron plates of other Chironomoidea (which are considered plesiomorphic) and the extensive plastron network of all other simuliids.

41. *Felt chamber of gill lost.* The felt chamber (i.e., the internal component of the culicomorph pupal gill) is visible in the gill of the Ceratopogonidae, Chaoboridae, Chironomidae, Culicidae, Dixidae, and Thaumaleidae. To date, *Parasimulium crosskeyi* and *Parasimulium stonei* are the only simuliids known to have such a structure. A felt chamber eventually might prove to be characteristic of *Parasimulium* as a whole, but this will be confirmed only through additional discoveries. The condition in *P. crosskeyi* is plesiomorphic, as judged from out-group comparison. Absence of a felt chamber, therefore, must be derived and can be considered a convincing synapomorphy of the Simuliidae, exclusive of *Parasimulium*.

42. *Pupal spiracle communicating with lumen of plastron.* In the culicomorph groundplan (as in most other aquatic Nematocera), the mesothoracic spiracle communicates directly with the base of the gill. Such a condition is characteristic of the Ceratopogonidae, Chaoboridae, Culicidae, Dixidae, and Thaumaleidae. The same pattern is found in several plesiomorphic lineages of the Chironomidae. Among black flies, only *P. crosskeyi* has the mesothoracic spiracle communicating directly with the lumen of the gill. In other simuliids, the spiracle is situated somewhat more dorsally, communicating directly with the thickened area of plastron on the dorsal base of the gill (Hinton 1957). This latter condition appears to be unique and is here considered a synapomorphy of the Simuliinae.

43. *Sternites VI and VII (at least) divided medially by a semimembranous longitudinal striate area.* In almost all simuliid pupae, sternites VI and VII are divided along their midline by a longitudinal tract of striate membrane. This division is, in some specimens, also evident on sternites V and VIII, although not as conspicuously. Even in members of the Simuliini *sensu* Crosskey (1969) (i.e., *Austrosimulium*, *Metacnephia*, *Simulium* s. l.), which are widely reputed to have sternites VI and VII undivided (e.g., p. 142 of Wygodzinsky & Coscarón 1973), the division can be resolved with the aid of a compound microscope. The condition is obscured in such taxa because the pupal abdomen is weakly sclerotized, making the distinction between sternite and membrane difficult. The pupa of *P. crosskeyi* lacks any suggestion of a mediolongitudinal divi-

sion of sternites VI and VII, which is the condition found in all other Culicomorpha. The divided state, therefore, can be considered a derived feature of the Simuliidae, exclusive of *Parasimulium*.

44. *Abdominal tergites V–IX (sometimes also IV) each with an anterior row of spine combs.* In most simuliid genera, the anterior margin of abdominal tergites V–IX (sometimes also IV) has a single row of spine combs. The condition in the pupa of *P. crosskeyi* is similar to that of other Culicomorpha in lacking such spine combs. (Spines occur on the abdominal tergites of many culicomorph pupae, but their position and form render homology with the simuliid spine combs unlikely.) Because spine combs have evidently evolved early in black flies—as inferred by their presence in such plesiomorphic taxa as *Paracnephia*, *Prosimulium*, and most Neotropical "Prosimuliini" *sensu* Wygodzinsky and Coscarón (1973)—their absence in *Parasimulium* might be considered plesiomorphic. One difficulty with this interpretation, however, is that the spine combs are subject to reduction or loss. For example, pupae of *Gymnopais* and *Twinnia* lack spine combs but are derived from a *Prosimulium*-like ancestor that had them (Wood 1978). Thus, the question arises as to whether spine combs should be considered a groundplan feature of the entire family. In some taxa, the loss is correlated with the reduction of the pupal cocoon (*Crozetia*, *Gymnopais*, *Twinnia*); in others, it might be part of a trend toward desclerotization of the pupal abdomen (*Austrosimulium* and some species of *Simulium* s. l., especially the subgenus *Afrosimulium*). *Parasimulium* has neither a greatly reduced cocoon nor an especially desclerotized abdomen. The condition in *Parasimulium*, therefore, is possibly plesiomorphic and not the result of loss.

### *Larval Groundplan Apomorphies*

45. *Antenna of three articles.* In the Simuliidae, the antenna of first-instar larvae consists of a single article; at first molt, a basal article is added. In subsequent instars, the basal article becomes annulated to give the appearance of a three-articled antenna. In all larval instars of *Parasimulium* so far examined, the antenna consists of a single antennal article (Borkent & Wood 1986). Because a single antennomere is present in most other culicomorph larvae (Ceratopogonidae, Chaoboridae, Culicidae, Dixidae, and Thaumaleidae), the condition is taken as plesiomorphic for the Simuliidae. The three-articled antenna of later instars, therefore, can be considered a synapomorphy of the family, exclusive of *Parasimulium*. The only other culicomorph family with a multiarticled larval antenna is the Chironomidae; however, the basic structure of the antenna is much more complex than in the Simuliidae (e.g., presence of antennal blade, Lauterborn's organ). Because the condition in the Simuliinae is not manifest in any other Culicomorpha, it is considered a synapomorphy of the subfamily.

46. *Postgenal cleft.* As far as we are aware, a postgenal cleft is not evident in any other Nematocera. Larvae of *Parasimulium* lack any suggestion of a postgenal cleft, a condition that must be considered plesiomorphic. The larvae of *Gigantodax* (in part), some *Simulium* (e.g., *Montisimulium*), and the fanless *Gymnopais*, *Levitinia*, and *Twinnia* have only a small area of slightly wrinkled cuticle in the area of the cleft, giving the impression of a small or rudimentary postgenal cleft, if any at all. We have interpreted this latter condition as a reversal toward the plesiomorphic condition. The irregularity of the cuticle in the region of the cleft lends credence to this hypothesis, as no such wrinkling is evident in *Parasimulium* or in any other culicomorph larvae of which we are aware. In *Gymnopais*, *Levitinia*, and *Twinnia*, the loss of the cleft is probably related to the loss of the labral fans.

47. *Lateral sclerite of prothoracic proleg.* A pair of apically fringed lateral sclerites is situated immediately proximal to the apical circlet of hooks on the prothoracic proleg of most simuliid larvae. A lateral sclerite is present in the larva of all genera examined except *Parasimulium*. Because no homologue of the lateral sclerite has been identified in any other Chironomoidea, we regard the structure as a synapomorphy of the Simuliinae.

### *Systematics of the Subfamily Simuliinae*

The monophyly of the Simuliinae, as defined here, is strongly supported by the 12 synapomorphies just listed. Particularly convincing are the pupal characters, which until now have received little attention in phylogenetic reconstruction. Future work should center on structural details of the pupal gill, as our current knowledge about this important character system is inadequate. The larva, too, has an array of presumed autapomorphies, but additional studies will be needed before all characters can be analyzed. At present, we are unsure how to interpret certain features of the hypostoma and anal sclerite.

## MONOPHYLY OF THE TRIBE PROSIMULIINI (Fig. 9.1)

48. *Pupal abdomen with large pleural plates (pleurites) on segments IV and V.* Pupae of *Distosimulium*, *Gymnopais*, *Helodon*, *Levitinia*, *Parahelodon*, *Prosimulium*, *Twinnia*, and *Urosimulium* are unique among the Simuliidae in having large pleural plates on each of abdominal segments IV and V (present only on segment V in *Levitinia*). The plates are so large that they are separated from the tergites and sternites by only a narrow, longitudinally striated band. Segments IV and V of other simuliids lack large plates, although small rounded pleurites such as those on segments VI–VIII can be present in some taxa (e.g., some *Paracnephia*). Abdominal segment IV of *Gymnopais* appears ringlike, with the pleurite of the segment evidently being fused to the adjacent sternite and tergite. This condition possibly represents a further derived state.

49. *Larval antenna with proximal and medial articles unpigmented, contrasting with black distal article.* Among the most distinctive features of larval prosimuliines is the coloration of the antenna. The proximal and medial articles typically are devoid of pigment, rendering the basal two thirds of the antenna nearly transparent and sharply contrasting with the dark pigmentation of the distal article. The only other simuliids with similarly colored antennae are the distantly related *Metacnephia*, but this

situation is undoubtedly an independent development. No other culicomorph larva has the basal portion of the antenna completely unpigmented. Species of *Urosimulium* have an irregular fuscous mottling on the basal two articles, but this trait is probably secondarily derived, as inferred from other characters. Similarly, the pigmentation on the basal two antennal articles of *Levitinia* probably is derived independently. The plesiomorphic state is for the basal two articles to be pigmented.

50. *Larval hypostoma lacking paralateral teeth.* The larval hypostoma of Prosimuliini differs from that of most other simuliids in lacking paralateral teeth. As a result of this arrangement, the lateral margins of the toothed portion of the hypostoma appear to be rather parallel. Absence of paralateral teeth is considered here a synapomorphic feature of Prosimuliini. A similar, but probably nonhomologous, arrangement of the hypostoma is found in the larvae of *Crozetia*. The plesiomorphic state is for the paralateral teeth to be present.

51. *Anteromedian palatal brush of first-instar larva consisting of scoop-shaped, fringed hairs.* First-instar larvae of all examined prosimuliines possess an anteromedian palatal brush consisting of a series of paired, scoop-shaped, apically fringed, anteromedian palatal brushes (Craig 1974). Such a condition is probably synapomorphic for the Prosimuliini, although the majority of simuliid genera have yet to be surveyed for this character. The presumed plesiomorphic condition is for the palatal brush to consist of small simple hairs, as in first instars of *Parasimulium*, *Cnephia*, *Crozetia*, *Metacnephia*, and *Simulium* (Craig 1974, Borkent & Wood 1986). Simple hairs also make up the anteromedian palatal brush of first instars of the mosquito *Aedes aegypti*.

52. *Maxillary palpal sensilla of first-instar larva subapically situated and in a linear configuration.* First-instar larvae of *Gymnopais*, *Helodon*, *Prosimulium*, and *Twinnia* are unique in having the sensilla laterally situated on the palpus in a more or less linear fashion (Craig & Borkent 1980). This condition contrasts with the presumed plesiomorphic condition for Nematocera in which the sensilla of first instars are apically situated on the palpus in a circular configuration (Craig & Borkent 1980). The condition in the first instar of *Parasimulium stonei* approaches that of the Prosimuliini in having the sensilla laterally positioned on the palpus; however, it differs in having the sensilla arranged in the plesiomorphic circular configuration (Borkent & Wood 1986). Although this stage has been interpreted as intermediate in a transformation series between the arrangements found in the Prosimuliini and other Nematocera (Borkent & Wood 1986), we here regard the condition in *Parasimulium* as independently derived.

### *Systematics of the Tribe Prosimuliini*

The following questions must be addressed for a satisfactory definition of the Prosimuliini: (a) Do other simuliids share an immediate common ancestor with *Prosimulium* that is not shared by any other simuliid? (b) Are any simuliids currently included in *Prosimulium* that can be shown to be paraphyletic?

The first question already has been answered, as it now seems well established that *Gymnopais* and *Twinnia* are derivative lineages in the Prosimuliini (Craig 1974, Wood 1978, Craig & Borkent 1980, Borkent & Wood 1986, Currie 1988). The first-instar larva of *Levitinia* has yet to be discovered; nonetheless, the members of *Levitinia* belong to a monophyletic lineage including *Gymnopais* and *Twinnia*. The second question is more difficult to answer because of a lack of information about key Afrotropical forms (viz., *Paracnephia*, *Procnephia*). Crosskey (1969) formerly treated these taxa as subgenera of *Prosimulium* s. l., but more recently Crosskey and Howard (1997) recognized *Paracnephia* as a valid genus with two subgenera. Rubtsov (1974a) placed *Procnephia* and *Paracnephia* as a "monophyletic" assemblage including *Gigantodax* and *Prosimulium* s. l. However, on the basis of available information, no close relationship exists between these African forms and *Prosimulium*. Rubtsov's (1974a) inclusion of *Gigantodax* in the Prosimuliini does not seem justified, as any similarities among these taxa are exclusively symplesiomorphic.

Individually, none of the five synapomorphies just listed provide convincing evidence about the monophyly of the Prosimuliini. Characters 49 and 50 seem promising as constitutive features but require further investigation. Collectively, however, the monophyly of the tribe seems reasonably well established by the five characters listed. Molecular sequence data also support the monophyly of the Prosimuliini as defined here (Moulton 1997, 2000).

### MONOPHYLY OF THE TRIBE SIMULIINI, INCLUDING THE FOSSIL GENUS *ARCHICNEPHIA* (Fig. 9.1)

53. *Radial sector unbranched, or with an obscure apical fork that is conspicuously shorter than its petiole.* This condition represents the third stage of a transformation series that begins with the conspicuously branched radial sector of Parasimuliinae. The second or intermediate stage is exemplified by the radial sector of Prosimuliini, in which the branches are more closely approximated. In its most highly derived state, the radial sector either is unbranched or is represented by an obscure apical fork. The differences between these patterns might be argued as merely matters of degree; however, each state is immediately recognizable, and no forms have yet been found that could serve as a link between them. Differences in the radial sector, therefore, possibly signify a substantial phylogenetic gap between the Parasimuliinae and the Simuliinae on the one hand, and the Prosimuliini and the Simuliini on the other. The plesiomorphic state is for the radial sector to have a long distinct fork.

Currie's (1988) concept of the tribe Simuliini was expanded by Currie and Grimaldi (2000) to include the new genus *Archicnephia*, which was established for an adult female preserved in Turonian-aged amber of central New Jersey about 90–94 million years ago. Some of the character states originally ascribed to the simuliine groundplan are not present in *Archicnephia*, and other character states cannot be evaluated because they belong to other life-history stages. Accordingly, the newly defined tribe currently is supported by only one character state, an unbranched or obscurely branched radial sector. However, one or two character states listed in the section "Monophyly of the Extant Holarctic Simuliini" (characters 136, 137) could apply to the Simuliini as a whole.

### RELATIONSHIPS OF THE PRIMARY LINEAGES OF THE FAMILY SIMULIIDAE

Our model is conservative in that it places the genus *Parasimulium* as the sister taxon of all other simuliids, a point on which most modern workers agree. Nevertheless, the hypothesis of a close relationship between *Parasimulium* and some other primitive-grade simuliids also deserves consideration. Enderlein (1921a) was the first worker to suggest such a relationship by including *Cnephia*, *Helodon*, *Parasimulium*, and *Prosimulium* in the subfamily Prosimuliinae. Stone (1963a, 1965a) adopted a similar system by including *Parasimulium*, *Gymnopais*, *Prosimulium*, and *Twinnia* in the subfamily Prosimuliinae. Neither classification, however, was prefaced with a discussion of characters, and they have received little support over the years. Borkent and Wood (1986) provided a model in which *Crozetia* was placed as the sister taxon of the above-mentioned genera. This entire assemblage was held together by a single synapomorphy, namely, reduction of the labral fan in the first-instar larva. A reduced fan is apomorphic for the Simuliidae (Wood 1978), but questions remain about the reliability of this character as a phylogenetic indicator. First, it is a regressive feature and cannot be taken as strong phylogenetic evidence. Second, the character is homoplastic in the Culicomorpha. If well-developed labral fans are in the groundplan of the infraorder, as suggested by Wood (1978), then they must have been lost in the Ceratopogonidae, Chaoboridae, Chironomidae, and Thaumaleidae. The character is even homoplastic within the Simuliidae, because members of the *Simulium oviceps* species group also have reduced fans (Craig 1977); *S. oviceps* Edwards is only remotely related to the fanless Prosimuliini. Two other characters, both features of the first-instar larva, were provided in support of the monophyly of *Parasimulium* + *Prosimulium* s. l. + *Gymnopais* + *Twinnia* (viz., maxillary palpus with apical spicules; maxillary palpal sensilla that are subapical). Neither character has been widely surveyed in the Culicomorpha, and we are inclined to accept the larger suite of characters (36–47) used to support the monophyly of Simuliinae, as defined in the present work.

The initial dichotomy in the Simuliinae remains somewhat more problematic. A sister-group relationship is suggested between *Prosimulium* + *Urosimulium* + *Helodon* + *Levitinia* + *Twinnia* + *Gymnopais* (= Prosimuliini), and all other Simuliinae (= Simuliini). However, monophyly of this latter tribe, as recently redefined by Currie and Grimaldi (2000) in view of the fossil genus *Archicnephia*, is supported by only one synapomorphy. The possibility remains that certain primitive-grade simuliines (e.g., *Paracnephia*) really belong to the Prosimuliini, although this possibility assumes independent evolution of several apomorphic character states ascribed to the groundplan of the extant Simuliini. (See characters 134–137.) In the absence of complete information about the first-instar larva (in which two key prosimuliine autapomorphies are manifest [i.e., characters 51 and 52]), the hypothesis of a close relationship between *Paracnephia* and the Simuliini is preferred.

### MONOPHYLY OF GENUS-GROUP TAXA OF THE SUBFAMILY PARASIMULIINAE

We concur with Wood and Borkent (1982) that *Astoneomyia* and *Parasimulium* s. s. are probably sister taxa; hence, *Parasimulium melanderi* and *Parasimulium* 'species A' together are placed as the sister group of all other *Parasimulium*. In recognition of the close relationship between *Astoneomyia* and *Parasimulium* (Borkent & Currie 2001), and in keeping with the classifications of other specialists, we recognize one genus and two subgenera. Apomorphic character states for *Parasimulium* s. s. were discussed by Wood and Borkent (1982).

#### *Subgenus* Astoneomyia

54. *Ventral plate with pommel-shaped apicomedial projection.* Plesiomorphic: ventral plate typically truncate or broadly rounded apically, without pommel-shaped projection.

#### *Subgenus* Parasimulium

55. *Supra-alar notch narrow and deep.* Plesiomorphic: other simuliids have a shallower notch.

56. *Gonocoxite with apicolateral finger-like extension and row of setae along adjacent edge.* Plesiomorphic: gonocoxite not so arranged in other simuliids.

57. *Gonostylus with subapical cusp (inner gonostylus) on dorsal side.* Plesiomorphic: gonostylus without such a cusp in other simuliids.

58. *Ventral plate with forked apex, the apicolateral prongs immobilized by an "inner gonostylus."* Plesiomorphic: ventral plate with truncate or rounded apex.

59. *Median sclerite long and straplike, with apically widened portion.* Plesiomorphic: median sclerite shorter and more triangular, narrowed apically in most other simuliids.

### MONOPHYLY OF GENUS-GROUP TAXA OF THE TRIBE PROSIMULIINI (Fig. 9.1)

#### *Genus* Prosimulium

The generic concept used here is narrower than the concepts favored by Rubtsov (1956, 1974a) and Crosskey and Howard (1997) (as subgenus), and corresponds more closely with the one used by Peterson (1970b) for the North American species of *Prosimulium* s. s. The first-mentioned authors refer to *Prosimulium* a number of species that have a well-developed basal or subbasal tooth on the female tarsal claw, which is considered a groundplan apomorphy of the Prosimuliini, exclusive of *Prosimulium*. Further, prosimuliines with a toothed claw lack any of the four synapomorphies used to define *Prosimulium*. Hence, most previous concepts of *Prosimulium* are nonmonophyletic and cannot be accepted in a cladistic system.

Cytologically, five major species groups of *Prosimulium* exist (Rothfels 1979): the Palaearctic *Prosimulium hirtipes* group ($C_t$), the Nearctic *Prosimulium mixtum* group (IIIL-1), the western Nearctic *Prosimulium esselbaughi* group (IIIL-2), the Nearctic *Prosimulium magnum* group (IIIS-1), and the northern Holarctic *Prosimulium macropyga/ ursinum* group. The last-mentioned group is the only one that does not have a distinctive cytological marker. The phylogenetic relationships among these five groups have not yet been resolved satisfactorily, either cytologically or cladistically. An acceptable arrangement of named species groups within *Prosimulium*, therefore, must await further study.

The assignment of species to *Prosimulium* based on purely cytological grounds (Rothfels 1979) closely reflects the arrangement proposed here, based on external structural characters. The only exception is the placement of *Urosimulium aculeatum* (Rivosecchi), which Rothfels (1979) suggested (as *U. stefanii* Contini) might belong to the Palearctic *P. hirtipes* species group. Species of *Urosimulium* share none of the synapomorphies listed below for *Prosimulium*, but instead have the single constitutive feature listed for the Prosimuliini, exclusive of *Prosimulium* (i.e., presence of a distinct basal or subbasal tooth on the female tarsal claw). Rothfels (1979) did not provide any justification for his placement of *U. aculeatum* in *Prosimulium*, although perhaps his decision was based on the chromosome map of Frizzi et al. (1970).

60. *Hypogynial valves long, with anteromedial corner produced nipple-like.* Females of *Prosimulium* are distinguished from those of all other Prosimuliini by the form of the hypogynial valves. These paired processes might be homologous with the anterior gonapophyses of the orthopteroid ovipositor (McAlpine 1981). In *Prosimulium*, these valves are relatively long and narrowly rounded or pointed apically and project posteriorly to the level of the anal lobe or beyond. This arrangement gives the female abdomen a rather pointed appearance terminally. Another distinctive feature is that the anteromedial corner of each valve is produced nipple-like. In other Prosimuliini, the valves are markedly shorter and more truncated posteriorly, and they do not project posteriorly as far as the anal lobes. The consequence of this arrangement is an abdomen that is more blunt apically. Further, the anteromedial corner of each valve is not produced nipple-like and instead appears rather truncated. The condition in Prosimuliini, exclusive of *Prosimulium*, is also characteristic of *Parasimulium* and most Simuliini and can be considered plesiomorphic. Examination of out-group families (Ceratopogonidae, Chironomidae, Thaumaleidae) corroborates this conclusion. The derived condition, as found in *Prosimulium*, is also evident in some distantly related Simuliini (e.g., subgenus *Hemicnetha*), but this condition undoubtedly is derived independently, as judged from other characters.

61. *Spermatheca with large differentiated area at junction with spermathecal duct.* The spermatheca is a highly varied structure in the Simuliidae and has been used as a definitive character at various taxonomic levels (e.g., Peterson 1970b, Wygodzinsky & Coscarón 1973). However, the pattern of variation is complex, undoubtedly owing to parallelisms or reversals. Even in the Prosimuliini, no arrangement that completely eliminates homoplasy can be constructed. Conclusions about the direction of character evolution must be made with caution. Based on out-group comparison with other Culicomorpha, the groundplan for the Simuliidae is for the spermatheca either to be complete basally (i.e., with no differentiated membranous area at the junction with the spermathecal duct) or to have only a small rounded area of differentiated membrane at the junction of the spermathecal duct (Evans & Adler 2000). Such a condition is found in *Parasimulium* and many Simuliini. In Prosimuliini, the plesiomorphic condition is characteristic of *Gymnopais*, *Helodon*, *Parahelodon*, and *Urosimulium*. A spermatheca with a large differentiated circular area at the junction with the spermathecal duct, therefore, must be derived and can be taken as evidence of the monophyly of *Prosimulium*. The only other Prosimuliini with a similar condition are *Levitinia* and *Twinnia*; however, this condition is probably an independently derived feature, as judged from other characters.

62. *Ventral plate of aedeagus not flattened dorsoventrally, but with a prominent lip or emargination apicoventrally.* Wood and Borkent (1982) suggested homologies between the ventral plate of male Simuliidae and variously named structures of other male Culicomorpha. The Ceratopogonidae and the chironomid genus *Buchonomyia* are the only Culicomorpha that have structures readily recognizable as ventral plates, and so conclusions about the basic plan in the Simuliidae are based on out-group comparison with these taxa. Primitively, the ventral plate is inferred to be rather flattened dorsoventrally, such as in *Buchonomyia*, Ceratopogonidae, and *Parasimulium*. In *Prosimulium*, a pronounced lip is present apicoventrally, which gives the ventral plate a distinctive triangular appearance in terminal view. No other prosimuliine has such a pronounced lip on the ventral plate. Certain species of the subgenus *Helodon* have a short lip, but we have interpreted this as an independent modification of the basic plan. In the Simuliini, the form of the ventral plate is varied, and any similarity with the ventral plate of *Prosimulium* must be attributed to parallelism. A flattened ventral plate is probably also in the groundplan of the Simuliini, as inferred from the occurrence of that state in some of the most plesiomorphic lineages of that tribe (e.g., *Paracnephia*, *Tlalocomyia osborni* species group). Another distinctive feature of *Prosimulium* is a dorsal concavity near the base of the ventral plate (Peterson 1970b). It, too, is a possible synapomorphy of *Prosimulium*, but we are unable to draw any definite conclusions about the polarity of this character from out-group families. No such concavity is evident in *Parasimulium* or any other prosimuliines.

63. *Lateral sclerite of prothoracic proleg broad, with vertical portion well developed.* The lateral sclerite of the prothoracic proleg is interpreted as a groundplan apomorphy of the Simuliinae. Two character states of the lateral sclerite exist: one in which the sclerite is in the form of a

narrow, sclerotized horizontal bar with little or no indication of vertical development, and another in which the sclerite is a broader structure with pronounced vertical development. Out-group comparison with members of the Simuliini indicates that the first-mentioned condition is plesiomorphic for the Prosimuliini. This condition is found in the Prosimuliini, exclusive of the genus *Prosimulium*. A broad lateral sclerite with pronounced vertical development, therefore, must be derived and can be taken as evidence of the monophyly of the genus *Prosimulium*.

*Subgenus* Helodon

Enderlein (1921a) described the genus *Helodon* by distinguishing it in a key to genera, based largely on the toothed claw of the female, and by designating the genotype *Simulia ferruginea* Wahlberg. Enderlein did not expand on the description in any of his subsequent works and referred only one additional species (*Helodon pleuralis*) to the genus. Rubtsov (1940b, 1956) followed Enderlein's lead by accepting *Helodon* as a valid genus, although in a somewhat different sense.

*Helodon*, as defined by Rubtsov, is considerably more restricted than originally conceived by Enderlein (1921a). If Rubtsov's (1956) generic limits were to be followed, only several species would be referable to *Helodon* worldwide (e.g., *H. ferrugineus*, *H. onychodactylus*). Rubtsov (1956) evidently did not put much weight on the toothed condition of the female tarsal claw, and instead was impressed more with both the orange color of the adult and the clublike form of the pupal gill. However, as Stone (1963a) indicated, most of the characters used in Rubtsov's concept of *Helodon* are found in various combinations in other prosimuliine species. For example, both males and females of *Prosimulium fulvum* and *Prosimulium secretum* n. sp. are orange, as are females of *Prosimulium fulvithorax*. The lack of clear-cut characters has caused most western workers to regard *Helodon* as either "a weakly defined subgenus of *Prosimulium*" (Stone 1963a), or "at most a species-group within *Prosimulium* s. s." (Crosskey 1969).

Peterson (1970b), in his revision of *Prosimulium* s. l. of Canada and Alaska, defined the subgenus *Helodon* in a sense that was less inclusive than the one proposed by Enderlein (i.e., not including all prosimuliine species whose females possess a toothed claw), but more inclusive than the one advocated by Rubtsov (i.e., including species whose adults are black and whose pupal gills are not in the form of clublike structures supporting many fine filaments). Most of the characters used to define this segregate are symplesiomorphic, but at least two evidently are derived and can be taken as evidence of the monophyly of the subgenus *Helodon*, as defined in the present work.

Rothfels (1979) provided cytological evidence for the monophyly of the subgenus *Helodon*, as defined in the present work. Chromosomal rearrangements of this segregate differ profoundly from those of other Prosimuliini and include a whole-arm interchange (IIIS + IIL), a nucleolar organizer in IIIL, and at least four autapomorphic inversions.

64. *Spermatheca elongate, delicate, and rather lightly pigmented*. Females of the subgenus *Helodon* have a spermatheca typically in the form of an elongate, delicate, rather lightly pigmented sac. In the groundplan of the Prosimuliini, the spermatheca is inferred to be a more rounded and densely sclerotized structure, such as in females of *Parasimulium*, *Gymnopais*, *Parahelodon*, *Urosimulium*, and many Simuliini. The spermathecae of females of subgenus *Prosimulium* are also rather delicate; however, the apex is more acuminate in forms that are elongate, and a large area of differentiated membrane occurs at the junction with the spermathecal duct (see character 61). Similarly, the delicate, lightly pigmented spermathecae of *Levitinia* and *Twinnia* are distinguished by their mushroom-like appearance and by a large area of differentiated membrane basally. Females of *Distosimulium* have a spermatheca in the form of a greatly inflated, unpigmented sac (see character 71).

65. *Postgenal cleft biarctate or subrectangular, with faint anterior margin*. Larvae of the subgenus *Helodon* are distinguished from those of most other Prosimuliini by the form of the postgenal cleft, particularly if the comparison is restricted to the Prosimuliini, exclusive of *Prosimulium* s. s. The postgenal cleft of this latter group is of greatly varied shape, with forms approaching the types found in the subgenus *Helodon*. Typically, the cleft of *Helodon* is broader than deep and has along its anterior margin a short, posteriorly directed, sclerotized process. The effect is a postgenal cleft that is biarctate anteriorly. This form is clearly unique among the Prosimuliini and is considered derivative. In certain species of *Helodon* (e.g., *H. perspicuus*), however, the posteriorly directed process is not clearly manifest, and the postgenal cleft appears as a rather subrectangular structure with an indefinite anterior margin.

This latter form approaches the condition found in certain species of *Prosimulium* s. s., and a subrectangular cleft possibly represents the progenitor of the biarctate form. If this were so, the biarctate condition could be used to define only groups of species within the subgenus *Helodon* and could not be construed as a synapomorphy of the entire subgenus. Cytological evidence, however, demonstrates that *H. perspicuus* is derived from an ancestor that had a biarctate postgenal cleft (Rothfels 1979), and so in at least one species, a subrectangular postgenal cleft with a faint anterior margin is derived from the biarctate condition. Because the postgenal cleft of larvae of *Prosimulium* s. s. is highly varied, conclusions about character-state polarity in other Prosimuliini are difficult to establish. Character states in the Prosimuliini, exclusive of *Prosimulium* s. s., are as follows: inverted U shape (*Distosimulium*; *Parahelodon*, two species; *Urosimulium*, two species), biarctate (*Helodon* in part), subrectangular with faint anterior margin (*Helodon* in part), inverted V shape (*Parahelodon*, one species), subrectangular with distinct anterior margin (*Urosimulium*), and nearly absent (*Gymnopais*, *Levitinia*, *Twinnia*). Because the inverted U shape of the postgenal cleft is widespread among the in group, and is also common in *Prosimulium* s. s. and the plesiomorphic Simuliini, we have interpreted it as plesiomorphic for members of the Prosimuliini, exclusive of *Prosimulium* s. s. A cleft that is biarctate or subrectangular with a faint anterior margin, therefore, must be derived and can be taken as evidence of the monophyly of *Helodon* s. s.

### *Subgenus* Parahelodon

The subgenus *Parahelodon* was erected for a group of three moderately distinctive species (Peterson 1970b). Most of the characters used to define this segregate are symplesiomorphic and cannot be used to demonstrate phylogenetic relationships. However, the four apomorphic characters enumerated below provide strong evidence of the monophyly of the subgenus.

Rothfels (1979), in his cytological transformation series of species derived from *Helodon* s. l., indicated that an inversion in the IIS arm is a constitutive feature of the three nominal species of *Parahelodon*. Cytological information also provided evidence of a sister-group relationship between *Helodon vernalis* and *Helodon decemarticulatus* + *Helodon gibsoni*. These three species together were placed as the sister group of *Helodon* (*Distosimulium*) *pleuralis*, and this entire assemblage was placed as the sister group of *Helodon* s. s. (Fig. 3 of Rothfels 1979).

66. *Arm of genital fork (sternite IX) bearing a typically wrinkled or denticulate, pronounced lateral plate.* Not only are the lateral plates relatively large in comparison with those of most other Prosimuliini, but also they are typically wrinkled or denticulate. No such wrinkling or denticulation is evident on the lateral plates of other Prosimuliini. The lateral plate of *Prosimulium* s. s. is rather varied in form, but any similarity with that of *Parahelodon* is due to convergence. The lateral plate of *Distosimulium* is larger than that of *Parahelodon*; however, the cuticle of that plate is neither wrinkled nor denticulate. A large lateral plate is interpreted as a groundplan apomorphy of *Distosimulium* + *Parahelodon*.

67. *Ventral plate of aedeagus with digitiform dorsomedial projection basally, to which base of median sclerite is fused.* Males of *Parahelodon* are unique among the Simuliidae in having the anterodorsal surface of the ventral plate produced finger-like well beyond the apices of the basal arms. This projection gives rise to the median sclerite, which projects posterodorsally to serve as a midventral support for the aedeagus. In other simuliids, the anterodorsal surface of the ventral plate is not produced finger-like beyond the apices of the basal arms, and the median sclerite has, at most, a short rounded base.

68. *Ventral plate of aedeagus with anterolateral apodeme (basal arm) short.* Typically, the basal arm is relatively long and broad and accounts for a substantial proportion of the total length of the ventral plate, as in *Distosimulium*, *Gymnopais*, *Helodon*, *Prosimulium*, *Twinnia*, and *Urosimulium*. Males of *Levitinia* have a basal arm, although broad and conspicuous, that is comparatively short relative to the total length of the ventral plate. In males of *Parahelodon*, the basal arm is shorter yet and rather slender. Out-group comparison with *Parasimulium* and the Simuliini indicates that the long-armed condition is probably in the groundplan of the Prosimuliini, and that a short slender arm is a secondary development. The short-armed condition in *Levitinia* and *Parahelodon* probably is derived independently, as judged from other characters.

69. *Gonostylus with single spinule (i.e., peglike seta) apically.* Males of *Parahelodon* are distinguished from those of most other Prosimuliini by a single peglike seta near the apex of the gonostylus. The number of apical spinules in other genus groups is as follows: one to three in *Gymnopais*; one, rarely two, in *Distosimulium*, *Levitinia*, *Twinnia*, and *Urosimulium*; two to five in *Helodon* s. s.; and two to six, rarely one, in *Prosimulium* s. s. Although the number of apical spinules is varied in *Prosimulium* and *Helodon*, two is the most commonly encountered condition. We interpret two apical spinules as the groundplan number for Prosimuliini. The presence of one spinule, therefore, must be a loss and is evidence of the monophyly of *Parahelodon*. However, the gonostyli of males of *Gymnopais* (in part), *Prosimulium unispinum* Rubtsov, and *Twinnia* also have a single spinule apically. Because of homoplasy, the derived state can be considered only a weak phylogenetic indicator.

### *Subgenus* Distosimulium

The phylogenetic relationships of *Distosimulium* are reasonably well established. In erecting the subgenus, Peterson (1970b) indicated that *Helodon pleuralis* is more closely related to species of *Helodon* s. s. and *Parahelodon* than it is to species of *Prosimulium* s. s. However, the structural attributes of this species appeared distinctive enough to warrant subgeneric recognition. Uemoto et al. (1976) indicated that *Helodon daisetsensis* is phylogenetically intermediate between *Distosimulium* and *Parahelodon* but relegated the species to the former subgenus based mainly on similarities of the male and female terminalia. Rothfels (1979), in his chromosomal transformation series of species derived from the *Helodon* standard, chose not to recognize the subgenus *Distosimulium* and instead relegated *H. pleuralis*, along with species of *Parahelodon*, to "*Helodon* s. l." *Helodon pleuralis* was considered the sister taxon of *vernalis* + *decemarticulatus* + *gibsoni* (= *Parahelodon sensu* Peterson 1970b), and this entire assemblage was placed as the sister group of all other *Helodon* (= *Helodon* s. s.).

70. *Arm of genital fork (sternite IX) with markedly pronounced lateral plate bearing patch of setae.* Each lateral plate of *Distosimulium* is so large that its posterior margin projects conspicuously beyond the posterior margin of the hypogynial valves. The condition in other Prosimuliini is for the posterior margin of the lateral plate to project slightly beyond the posterior margin of the hypogynial valves (*Parahelodon*), or to be overlain completely by the valves (all remaining genus groups). Although we have interpreted a large lateral plate as a groundplan apomorphy of *Distosimulium* + *Parahelodon* (character 122), the exceptionally large size of that structure in females of *Distosimulium* is a further derived state. Another distinctive feature of the lateral plate of *Distosimulium* is the

central patch of long setae. No other prosimuliine has such pronounced setation on the lateral plate, although a few scattered setae are present on the lateral plates of some females of *Gymnopais fimbriatus* and *Twinnia*. The combination of a markedly pronounced lateral plate with a central patch of setae is taken as evidence of the monophyly of *Distosimulium*.

71. *Spermatheca a greatly enlarged, delicate, thin, unpigmented sac*. The spermathecae of females of *Distosimulium* are distinguished from those of all other females of the Simuliidae by their large size and lack of pigmentation. The spermathecae of some taxa have a variously sized area of unsclerotized differentiated membrane at the junction with the spermathecal duct, but these are invariably pigmented or patterned apically.

72. *Ventral plate of aedeagus deeply cleft apically*. The ventral plate of *Distosimulium* males is unique among the Prosimuliini in having the posterior margin deeply cleft. Coupled with the relatively long anterolateral arms of the ventral plate, the entire structure appears H shaped in ventral view. The ventral plates of other Prosimuliini are neither deeply cleft nor H shaped in ventral view and are, at most, only shallowly emarginate apically. The only other simuliids with an H-shaped ventral plate are *Parasimulium furcatum* and *Simulium pictipes*; however, based on other characters, we suggest that the H-shaped condition in these taxa is derived independently.

73. *Gonostylus with apical half markedly thinner than proximal half, tapered posteriorly to an acute point*. In the prosimuliine groundplan, the gonostylus appears to be a rather evenly tapered structure in ventral view. Further, the gonostylus is typically narrowly to broadly rounded apically. This condition is found in members of *Helodon*, *Levitinia*, *Parahelodon*, *Twinnia*, *Urosimulium*, and most *Prosimulium*. The gonostyli of males of *Distosimulium* differ from those of most other Prosimuliini in that the apical half is markedly thinner than the proximal half in ventral view and the apex is tapered to an acute point. Males of *Gymnopais* are the only other prosimuliines with the gonostylus markedly tapered near the apex, although this would have to be considered an independent development (see also character 102).

74. *Gonostylus flexed in dorsoventral plane, and its apex opposed to spiniform, dorsoventrally curved paramere*. In the nematoceran groundplan, the gonocoxites diverge laterally, and the gonostyli oppose each other apically in a lateromedial plane (Wood & Borkent 1982). Unlike the condition of most male simuliids, the gonostyli of males of *Distosimulium* do not oppose each other medially and, instead, are flexed in a dorsoventral plane. When fully flexed, the gonostylus is opposed to the apex of the paramere, which is in the form of a dorsoventrally curved, spinelike structure fused to the gonocoxal apodeme. In effect, the gonostylus and paramere act as a set of pincers, which evidently impinge on the enlarged lateral plate of the female genital fork (see also character 70). Both the mode of action of the gonostylus and the form of the paramere can be taken as strong evidence of the monophyly of *Distosimulium*. The only other simuliids with a similar arrangement are males of *Parasimulium*; however, the dorsoventrally flexed gonostyli are opposed to the apicolateral angle of the ventral plate and not to the paramere (Wood & Borkent 1982). The condition in *Parasimulium* is, therefore, an independent development.

75. *Postgenal cleft of larva deep and broad*. Larvae of *Distosimulium* have the most extensive postgenal cleft of any larval prosimuliine. It is a relatively broad, inverted U, and is extended from one third to one half the distance from the posterior tentorial pits to the hypostomal groove. The form of the postgenal cleft is rather varied in other larval Prosimuliini but is extended in few taxa farther than about one fourth the distance from the posterior tentorial pits to the hypostomal groove. A shallow postgenal cleft is probably in the groundplan of the Prosimuliini, as judged from the distribution of this character state throughout the tribe. A deep and broad cleft, therefore, must be derived and is evidence of the monophyly of *Distosimulium*.

### *Genus* Urosimulium

The genus *Urosimulium* consists of three species with a western Mediterranean distribution. The relationship of *Urosimulium* to other prosimuliines has not been clearly established. Some authors (e.g., Rivosecchi 1978) regard this aggregate as a distinct genus, whereas other workers (Crosskey & Howard 1997) regard it as a moderately distinct species group of *Prosimulium* s. s. (i.e., the *aculeatum* species group). Rothfels (1979) provided support for this latter view by suggesting that *Urosimulium aculeatum* (as *U. stefanii*) might belong cytologically to the *hirtipes* species group of *Prosimulium* s. s. However, members of *Urosimulium* lack any of the synapomorphies used to define *Prosimulium* and, instead, have the single synapomorphy (bifid female claw) used to define the sister group of *Prosimulium*. Until the relationships of *Urosimulium* are better understood, we prefer to recognize this segregate as a full genus.

76. *Cercus of female elongate, produced apically to fine point*. The cercus of *Urosimulium* females is much longer than wide and is tapered apically to a fine, posterodorsally directed point. In lateral view, the cercus is rather scimitar shaped. This condition is evidently unique to females of *Urosimulium* and provides strong evidence of the monophyly of the three included species. Females of all other Prosimuliini have a cercus that is broader than long and appears subrectangular or subquadrate in lateral view. Out-group comparison with *Parasimulium* and the Simuliini indicates that a subrectangular female cercus is probably basic to the Prosimuliini.

77. *Gonostylus with an accessory lobe laterally at base*. This condition must not be confused with the inner gonostylus of many other Chironomoidea (Wood & Borkent 1982). The inner gonostylus is an articulated appendage situated on the dorsomedial surface of the gonopod between the gonostylus and the gonocoxite. Members of the subgenus *Parasimulium* are the only Simuliidae with such a structure. A similarly positioned (but probably nonhomologous) structure is present in certain members of the subgenus *Simulium*, which have a nonarticulated

prong or spinose lobe lateromedially on the base of the gonostylus. The situation in *Urosimulium* males differs from the conditions just described in that the accessory lobe is a nonarticulated appendage positioned laterally at the base of the gonostylus. No other simuliid has such a condition, which must be considered apomorphic. The groundplan condition for the Prosimuliini is a simple gonostylus.

78. *Larval antenna with proximal and medial articles pigmented.* The basal two articles of the larval antenna are presumed to be colorless in the groundplan of the Prosimuliini. This condition is found in members of the following genera: *Gymnopais*, *Helodon*, *Prosimulium*, and *Twinnia*. The three included species of *Urosimulium* have a larval antenna that is pigmented basally, which approaches the presumed plesiomorphic condition for the Simuliidae. The distal antennal article, however, is markedly darker than the basal two articles, and so the pigmentation is possibly a reversal toward the plesiomorphic form. The larva of *Levitinia* has a similarly colored antenna, but other characters suggest that this is an independent derivation from the plesiomorphic form.

*Genus* Levitinia

The genus *Levitinia* was recognized for a new species (*L. tacobi*) of fanless black flies and placed, along with *Gymnopais* and *Twinnia*, in the subfamily Gymnopaidinae (Chubareva & Petrova 1981). The description was based on a series of larvae collected from the mountains of Tadzhikistan. The features listed as diagnostic for the genus included the five-armed anal sclerite (as compared to three armed in *Gymnopais* and *Twinnia*) and the 10-branched respiratory organ (as compared to 2–6 branches in *Gymnopais* and 14 or 16 branches in *Twinnia*). Chromosomal features of the larva confirmed that *Levitinia* shares a close relationship with the other two genera (Chubareva & Petrova 1981), but evidently did not reveal how the three taxa are related to each other. On the basis of external structural characters (especially the form of the larval mandible), *Levitinia* was said to occupy a position phylogenetically intermediate between *Gymnopais* and *Twinnia*.

Beaucournu-Saguez and Braverman (1987) described larvae, pupae, and adults of a second species of *Levitinia* (*L. freidbergi*) from the Golan Heights of Syria. In recognizing the similarity between adults of *Levitinia* and *Prosimulium*, and the overall close relationship between the first-mentioned genus and *Gymnopais* and *Twinnia*, these authors preferred not to recognize the subfamily Gymnopaidinae and instead relegated the fanless simuliids to the tribe Prosimuliini *sensu* Crosskey (1988). They compared the various life-history stages of *Levitinia* with those of *Gymnopais*, *Prosimulium* s. l., and *Twinnia* but did not arrive at any definite conclusions about interrelationships among these genera.

79. *Tergite IX of female elongate posteriorly, projected shieldlike over cercus.* The female of *Levitinia* can be distinguished from females of all other Prosimuliini by the form of tergite IX. The posterior margin is produced posteriorly as a dartlike structure that projects well beyond the apex of the cercus. The apex of the tergite is tapered to a fine point, and the overall appearance is that of a shield. Tergite IX of females of all other Prosimuliini is a triangular or subtriangular sclerite that does not project shieldlike over the cerci; the apex is narrowly to broadly pointed. This latter condition is plesiomorphic.

80. *Lateral plate of genital fork (sternite IX) not connected directly to tergite IX.* The lateral plate of the genital fork of *Levitinia* females is not directly connected with tergite IX. Instead, the tergite and sternite of that segment are separated by membrane. The only other simuliid with a similar condition is *Gymnopais*. A membranous connection between the tergite and sternite of segment IX is derived with respect to other simuliids, which typically have a distinct sclerotized connection between these two sclerites. Although it might be argued that the apomorphic state is evidence of the immediate common ancestry of *Gymnopais* and *Levitinia*, evidence presented later suggests that *Levitinia* is the sister taxon of *Gymnopais* + *Twinnia*. Two interpretations are possible: either the lack of a connection is a groundplan feature of the fanless prosimuliines, with subsequent reversal to the plesiomorphic form in *Twinnia*, or two independent derivations of the apomorphic state have occurred, one in *Gymnopais* and one in *Levitinia*. The latter hypothesis is favored in this instance because loss of a connection is an easier step developmentally.

81. *Ventral plate of aedeagus with short anterolateral apodemes (basal arms).* Males of *Levitinia* have basal arms that are comparatively short relative to the total length of the ventral plate, although they are broad and conspicuous. In males of *Parahelodon* the arms are shorter yet and rather slender. Out-group comparison with *Parasimulium* and the Simuliini indicates that the long-armed condition is in the groundplan of the Prosimuliini, and that short slender arms are a secondary development. The short-armed condition in *Parahelodon* and *Levitinia* probably is derived independently, as judged from other characters.

82. *Paramere rudimentary, not connected to anterolateral apodeme (basal arm) of ventral plate.* In males of *Distosimulium*, *Levitinia*, and *Parahelodon*, the straplike connection between the paramere and the basal arm evidently has been lost, and the paramere remains articulated only with the gonocoxal apodeme. This condition, which is derivative, evidently has evolved a number of times independently in the Simuliidae and at least twice in the Prosimuliini: once in *Levitinia* and once in *Distosimulium* + *Parahelodon*. The paramere of *Levitinia* is further derived in that it is in the form of a fine fingerlike projection fused to the gonocoxal apodeme. No clear separation exists where the paramere and the gonocoxal apodeme are joined together. No other Prosimuliini has such an ill-formed paramere, and the connection with the gonocoxal apodeme is typically rather tenuous. The paramere of *Distosimulium* is fused to the gonocoxal apodeme, but the distal portion of that structure is produced into a long, strongly sclerotized spine that curves ventrally to oppose the apex of the gonostylus (see character 74).

83. *Hypostoma of larva with lateral serrations produced as spiniform setae.* Larvae of *Levitinia* are evidently unique among the Simuliidae in that the lateral serrations of the hypostoma are in the form of elongate spiniform setae. The typical condition in the Simuliidae is for the lateral serrations to be in the form of short, broadly to narrowly rounded denticles.

84. *Anal sclerite with a prominent supernumerary arm projected posteromedially, resulting in an inverted, star-shaped structure.* In the groundplan of the Prosimuliini, the anal sclerite is hypothesized to consist of a median plate and four radiating arms, two anterodorsal and two posteroventral. This condition is characteristic of the following genus groups: *Distosimulium*, *Helodon* s. s., *Parahelodon*, and *Prosimulium*. The same basic pattern also can be seen in the anal sclerites of other genus groups (*Gymnopais*, *Levitinia*, *Twinnia*, *Urosimulium*), except that there is an additional projection arising posteromedially from the area between the posteroventral arms. We suggest in the discussion of character 109 that a short posteromedially directed projection, such as in *Urosimulium*, might represent the groundplan condition for *Urosimulium* + the fanless Prosimuliini. The anal sclerite of *Levitinia* seems further derived in that the projection is in the form of a prominent arm, subequal in proportion to the anterodorsal and posteroventral arms. The resulting inverted star shape of the anal sclerite is unique among the Simuliidae and provides evidence of the monophyly of *Levitinia*. The plesiomorphic condition is for the anal sclerite to be X shaped, sometimes with only a short, posteromedially directed projection.

### *Genus* Twinnia

The nominal taxon *Twinnia* was recognized originally for two species (*T. nova*, *T. tibblesi*) whose adults appeared to be closely related to *Prosimulium* but whose fanless larvae suggested a close relationship with *Gymnopais* (Stone & Jamnback 1955). *Twinnia* was considered distinctive enough to rank at the generic level and was placed along with *Cnephia*, *Gymnopais*, and *Prosimulium* in the subfamily Prosimuliinae.

Shewell (1958) accepted *Gymnopais* as a separate genus of the Prosimuliinae but preferred to rank *Twinnia* as a subgenus of *Prosimulium*, pointing to the apparent lack of differences between the adults and pupae of these latter two segregates as justification for this action. Rubtsov (1974a), on the other hand, rejected the hypothesis of a close relationship between *Prosimulium* and *Twinnia* and relegated the latter genus to the subfamily Gymnopaidinae.

Craig (1974) and Wood (1978) provided convincing evidence that the fanless condition is secondary, and that *Gymnopais* and *Twinnia* share an immediate common ancestor with *Prosimulium* s. l. This interpretation is supported by cytological evidence, which shows that *Gymnopais* and *Twinnia* are derived from the *Prosimulium* standard sequence (Rothfels & Freeman 1966, Rothfels 1979).

85. *Body of ventral plate of aedeagus strongly emarginated laterally near base of anterolateral apodeme (basal arm).* Males of *Twinnia* are distinguished from those of most other Prosimuliini by the form of the ventral plate. In ventral view, the body of the plate is rather broad posteriorly and is excavated laterally near the base of each basal arm. This lateral emargination is magnified by the divergence of the basal arms from each other apically. The form of the ventral plate is rather varied in other Prosimuliini, but it typically is not excavated laterally near the base of the basal arms. In species that have a laterally emarginated ventral plate, the basal arms are not markedly divergent apically. Out-group comparison with *Parasimulium* and the plesiomorphic Simuliini indicates that a laterally emarginated ventral plate is apomorphic and should be taken as evidence of the monophyly of *Twinnia*. The presumed plesiomorphic condition in the Prosimuliini is for the ventral plate to have more or less parallel sides laterally and for the basal arms to project parallel to each other.

86. *Gonostylus with a single spinule (i.e., peglike seta) apically.* The groundplan number of apical peglike setae on the gonostylus is presumed to be two. The gonostylus of *Twinnia* is characterized by the presence of a single, large, peglike seta near the apex, which must be considered derivative. The only exception to this rule is the occasional rare specimen of *Twinnia hirticornis*, which has two apical spinules (Wood 1978). *Twinnia hirticornis*, however, is a relatively derived species of *Twinnia*, and the majority of specimens have one apical spinule. The occasional extra spinule is possibly a reversal toward the plesiomorphic condition. The derived state also defines members of *Parahelodon*. Because the apomorphic state is regressive and subject to homoplasy, it cannot be taken as strong evidence of the monophyly of *Twinnia*.

87. *Mandible of larva not markedly curved medially near apex.* The simuliid larval mandible typically is curved distally, with the apical mandibular teeth directed toward the larval midline. The larval mandible of *Twinnia* differs from this general plan in that its apex is not curved markedly toward the midline but instead is more in line with the longitudinal axis of the mandible. The consequence of this arrangement is that the anteroventral surface of the mandible (and hence the apical brush) is not brought into contact with the substrate during grazing. Therefore, the apical mandibular teeth evidently act alone to transfer food into the cibarium.

88. *Mandible of larva with comblike scales on anteroventral surface of mandible reduced in number and size.* The apical mandibular brush of *Twinnia* larvae consists of three or four rows of slender, short, bristle-like setae. This condition is similar to that of most other prosimuliine larvae, except that the brush typically consists of longer finer hairs. In addition to these hairlike setae, three or four additional spinelike setae are situated immediately basal to the apical mandibular teeth (Fig. 46 of Chance 1970a). These spines are similar to the comblike scales (= apical mandibular brush) on the anteroventral surface of the larval mandibles of *Gymnopais* and *Levitinia*, except that the scales in these two segregates are organized in 7–12 rows. Other evidence suggests that *Levitinia* is the sister taxon of *Gymnopais* + *Twinnia*. If the hypothesized set of

relationships is accepted, then the form of the apical mandibular brush of *Twinnia* must be interpreted as a reduction or loss. Because the anteroventral surface of the mandible of *Twinnia* is probably not brought into contact with the substrate during grazing, the rudimentary nature of the apical brush is possibly the result of loss. This interpretation is followed here. Character states are as follows: apical brush consisting of hairlike or bristle-like setae; apical brush consisting of 7–12 rows of comblike scales; apical brush with comblike scales reduced to a single row of three or four spines.

89. *Pupal cocoon a transparent gelatinous envelope that encloses all of pupa, retaining larval exuviae.* In the prosimuliine groundplan, the cocoon is a thin to thickly woven silk sac covering the pupal abdomen and typically most of the thorax. The larval exuviae are held temporarily between the pupal gills, but the anterior portion of the cocoon sloughs off shortly after pupation, and the larval exuviae typically are swept away by the current. The pupal cocoon of *Twinnia* differs from this general plan in two major respects. First, it is a transparent gelatinous envelope that covers all of the pupa, including the gill. The envelope consists of thick slimy threads that become manifest only after fixation in preservative. This condition contrasts with the thin, discretely formed, darkened threads of the cocoons of most other simuliids. Second, the larval exuviae remain trapped between the gills by the surrounding gelatinous matrix. Both of these features are unique and are evidence of the monophyly of *Twinnia*. Evidence presented later will show that *Twinnia* belongs to a monophyletic group that includes *Gymnopais* and *Levitinia*. The cocoon of these latter two genera is in the form of a small ventral pad by which the pupa remains attached to the substrate. A reduced cocoon is possibly in the groundplan of *Levitinia* + *Twinnia* + *Gymnopais*, and the condition in *Twinnia* might be a further modification of this general plan. The pharate pupa of *Gymnopais* is enveloped in the same slimy matrix as the pharate pupa of *Twinnia*, except that all but the ventralmost portion of this matrix disappears at pupation (Wood 1978). If this condition is taken as the groundplan of *Levitinia* + *Twinnia* + *Gymnopais*, then the condition in *Twinnia* can be interpreted as a neotenic retention of the transparent envelope throughout the pupal stage. Character states are as follows: a well-developed cocoon consisting of darkened threads; an ill-formed cocoon that is in the form of a small ventral pad; a cocoon that is in the form of a gelatinous envelope enclosing all of the pupa, retaining the larval exuviae.

### *Genus* Gymnopais

The genus *Gymnopais* was described for two new species discovered in Alaska (Stone 1949b). Larvae of this segregate appeared different from all others known at the time by the complete absence of labral fans. Other life stages of *Gymnopais* were thought to agree with *Prosimulium* s. l. in most respects, except as follows: absence of fine recumbent hairs on the body, elongate petiole of $M_{1+2}$, a bulla near the posterior margin of the compound eye, and virtual absence of recurved hooks on the dorsum of the pupal abdomen.

Rubtsov (1955) was so impressed with the fanless condition of larval *Gymnopais* that he recognized a separate subfamily, the Gymnopaidinae. This action presumably was based on the assumption that black flies are derived from a chironomid ancestor—chironomid larvae also lack labral fans—and that the fanless condition in the Simuliidae, therefore, must be plesiomorphic. However, as indicated under the section on *Twinnia*, the fanless condition in the Simuliidae is probably derivative, and *Gymnopais* and its fanless relatives probably are derived from a *Prosimulium*-like ancestor.

Many of the features used to characterize *Gymnopais* are no more than specializations to life in wind-blown localities, and various combinations of these same specializations are found in other groups of Diptera. Rubtsov (1974a) used one of these characteristics (a flattened thorax) to help justify his supposition that *Gymnopais* is a link between the Chironomidae and Simuliidae. Caution should be exercised, however, when evaluating characters that are subject to parallel selection.

90. *General vestiture of adult sparse, short, erect.* The general vestiture of adults of *Gymnopais* consists of a sparse covering of short, coarse, erect (or semierect) setae. Absent is the dense covering of fine recumbent setae (pile) characteristic of the adults of nearly all other simuliids. The apomorphic state is unique to *Gymnopais* and is strong evidence of the monophyly of the genus. The plesiomorphic state is vestiture mainly of pile, with only a few coarse, erect setae interspersed.

91. *Clypeus nearly devoid of setae.* The clypeus of the adult of *Gymnopais* is devoid of setae centrally and has only a few irregularly situated setae laterally. In adults of all other prosimuliines, like those of other simuliids, the clypeus is entirely covered with setae. Out-group comparison with other Chironomoidea indicates that this latter condition is plesiomorphic.

92. *Mandible and lacinia of female without teeth.* Simuliid females are presumed in the present work to be primitively anautogenous (character 30). Females of *Gymnopais* lack serrated mandibles and laciniae, which must be considered a secondary development. The apomorphic state is a regressive feature that is subject to homoplasy and does not provide convincing evidence of phylogenetic relationships.

93. *Thorax of adult slightly arched.* One of the most characteristic features of adult simuliids is the humpbacked appearance. Adults are typically strong fliers, and the markedly arched (and usually high) thorax contains well-developed flight muscles. Nearctic species of *Gymnopais* are either poor fliers or incapable of flight altogether. The weakly arched thorax is probably related to the degenerate condition of the flight muscles.

94. *Postnotum relatively small and markedly arched, with variously distinct median longitudinal ridge.* The apomorphic state is present only in members of *Gymnopais*. The condition in other prosimuliines is for the postnotum to be larger and rather evenly arched dorsally, with no suggestion of a dorsomedian longitudinal ridge. Out-group

comparison with other simuliids indicates that these latter two conditions are plesiomorphic.

95. *Anepisternal (pleural) membrane with a small group of hairs.* The anepisternal membrane is bare in *Parasimulium* and most members of the tribes Prosimuliini and Simuliini. Adults of *Gymnopais* are distinguished from those of most other simuliids by the presence of a few setae on the anepisternal membrane. Adults of *Gymnopais fimbriatus* apparently lack such setae (Wood 1978); however, other characters suggest that this condition is a reversal toward the plesiomorphic condition. The apomorphic condition apparently has evolved independently in other taxa, such as *Tlalocomyia revelata*, *Metacnephia*, the subgenera *Anasolen* and *Wilhelmia*, and the *Simulium ornatum* species group.

96. *Mesepimeron with vestiture (mesepimeral tuft) confined to dorsal part of sclerite above level of metathoracic spiracle.* In *Gymnopais* adults the mesepimeral tuft is confined to the dorsalmost portion of the sclerite above the level of the ventral margin of the metathoracic spiracle. This condition differs from that in other adult prosimuliines, in which the mesepimeral tuft is more extensive ventrally and extends into the area below the level of the metathoracic spiracle. We have interpreted this latter condition as a groundplan feature of the Prosimuliini.

97. *Wing membrane fumose and slightly wrinkled.* The wing of *Gymnopais* is somewhat wrinkled and has a pale smokey brown appearance (nearly opaque in some specimens). In all other prosimuliines, the wing is less wrinkled and the membrane is hyaline. Out-group comparison with other simuliids indicates that these latter two conditions are plesiomorphic.

98. *Petiole of $M_{1+2}$ elongate, about half as long as petiole of radial sector.* The petiole or stalk of veins $M_{1+2}$ is exceptionally long in *Gymnopais*. It is about half as long as the petiole of the radial sector ($R_{2+3}$ and $R_{4+5}$), as compared to about one fifth that length in other Prosimuliini. A short petiole is probably in the groundplan of the Prosimuliini, as inferred from the occurrence of this condition in the Parasimuliinae and Simuliini.

99. *Lateral plate of genital fork (sternite IX) not connected directly to tergite IX.* The lateral plate of the genital fork of *Gymnopais* females is not directly connected with tergite IX. Instead, the tergite and sternite of that segment are separated by membrane. The only other simuliid with a similar condition is *Levitinia*. A membranous connection between the tergite and sternite of segment IX is derived with respect to other simuliids, which typically have a distinct sclerotized connection between these two sclerites.

100. *Spermatheca basally with a short or long neck to which the spermathecal duct is connected.* In the groundplan of the Prosimuliini, the apex of the spermathecal duct is inferred to be connected with a variously sized membranous ring at the base of the spermatheca. This condition is found in *Helodon* s. l., *Levitinia*, *Prosimulium*, *Twinnia*, and *Urosimulium*. The spermatheca of *Gymnopais* differs from the general plan in that the base is produced into a variously sized sclerotized neck. This condition is unique among prosimuliines.

101. *Anal lobe and cercus of female fused into a single solid sclerite.* Females of *Gymnopais* are unique among the Prosimuliini in that the anal lobe and cercus are fused into a solid sclerite. The typical condition is for the anal lobe and cercus to be separated by membrane. Out-group comparison with the Parasimuliinae and Simuliini indicates that this latter condition is plesiomorphic.

102. *Gonostylus slender apically, with apical spinule (i.e., peglike seta) minute.* The gonostylus of *Gymnopais* males is relatively narrow, compared with that of other prosimuliines, and is tapered apically to an acute point. The apical peglike setae are so small that they are barely visible under magnification with a dissecting microscope. The plesiomorphic condition for the Prosimuliini is for the gonostylus to be relatively broad throughout with a narrowly to broadly rounded apex, and for the apical peglike setae to be readily visible under magnification with a dissecting microscope. The only other prosimuliine with a narrow gonostylus is the male of *Distosimulium*; however, its apical peglike setae are conspicuous. Other characters, such as the unique clasping mechanism of the gonostylus of *Distosimulium*, suggest that any similarity with the gonostylus of *Gymnopais* is due to convergence (see characters 73 and 74).

103. *Sternites IV–VII of pupal abdomen with up to five pairs of recurved hooks each.* Pupae of *Gymnopais* have a relatively large number of sternal hooks compared to those of other prosimuliines. The exact arrangement of hooks varies among species and even among individuals of the same species. However, the overall number of sternal hooks per pupa of *Gymnopais* far exceeds the number in most other simuliid pupae. The typical prosimuliine pupa has no more than two pairs of hooks per sternite.

104. *Pleurite IV of pupal abdomen hardly differentiated from (or fused to) adjacent sternite and tergite.* Pupae of *Gymnopais* are unique in that abdominal pleurite IV is slightly differentiated from, or fused to, the adjacent sternite and tergite. In this respect, segment IV is similar in appearance to segment III, which also has the pleurite fused to the sternite and tergite. Presence of a discrete pleurite on each of abdominal segments IV and V is interpreted as a groundplan feature of the Prosimuliini. Hence, any connection between the pleurite and the adjacent sternite and tergite must be a secondary development. All other prosimuliine pupae have a narrow but distinct band of striate pleural membrane between the pleurite and the adjacent sternite and tergite.

105. *Form of labrum and labropalatum of larva.* Larvae of *Gymnopais* are distinguished from those of other fanless prosimuliines by several features of the labrum and labropalatum. For the purposes of the present analysis, we have combined these features under a single character state. Synapomorphic features of the labrum and labropalatum of *Gymnopais* are as follows:

a. *Labrum conspicuously longer than wide.* The labrum of *Gymnopais* is proportionally longer and narrower than that of both *Levitinia* and *Twinnia*. Because the condition in fanned prosimuliines is for the labrum to be short and broadly rounded apically, we have interpreted the exceptionally long and narrow labrum of the larvae of *Gymnopais* as derivative.

b. *Labrum with a transparent tubercle on each side.* Another distinctive (and apparently autapomorphic) feature of *Gymnopais* larvae is the conspicuous tubercle on each side of the labrum. No such tubercle is present on the labrum of any other simuliid.
c. *Epipharyngeal apparatus with V-shaped brush anteriorly.* Larvae of *Gymnopais* have a unique arrangement of hairs on the ventral surface of the labrum. The anteriormost setae of the epipharyngeal apparatus are organized into a distinct V-shaped structure (Wood 1978). The anteriormost setae of the labra of *Levitinia* and *Twinnia* are more irregularly situated and are not organized into a V-shaped structure. We have interpreted this latter condition as plesiomorphic.

#### RELATIONSHIPS OF THE PROSIMULIINE GENUS-GROUP TAXA

##### *Monophyly of the Tribe Prosimuliini, Exclusive of the Genus* Prosimulium (Fig. 9.1)

106. *Claw of female with a distinct basal or subbasal tooth.* The taxonomic value of the female tarsal claw has long been recognized. Differences in the claw have been used to distinguish not only species and groups of species but also higher taxa. Rubtsov (1940b) listed the tarsal claw as the single most important diagnostic feature of the Simuliidae. The relative merits of this character system might be open to debate, but little doubt exists as to its importance as a phylogenetic indicator. An extreme amount of homoplasy, however, exists within the family, and so hypotheses about relationship must be based on careful reference to the defined out groups.

Rubtsov (1940b), in addressing the character polarity of the female tarsal claw, concluded that claws with a conspicuous basal tooth and those that are simple are plesiomorphic. The presumed derived condition is one of claws with a small basal or subbasal tooth. In support of this hypothesis, Rubtsov (1940b) suggested that the two "primitive" types of claw are formed early in pupal development, an example of "ontogeny recapitulating phylogeny." Later, Rubtsov (1974a) realized that the distribution of character states was more complex than originally supposed and that a conspicuous basal tooth was best regarded as a secondary development in certain taxa.

How two completely different character states (i.e., a simple versus a complex tarsal claw) can simultaneously be construed as "primitive" is difficult to comprehend. Out-group comparison with other Chironomoidea reveals that a simple claw is the only type in females of Chironomidae and Thaumaleidae. In the Ceratopogonidae, the pattern is more complex, with the tarsal claw either simple or bearing a variously sized basal or subbasal tooth. However, a simple claw is evident in two plesiomorphic clades of the Ceratopogonidae (viz., Austroconopinae and most Leptoconopinae), and so a simple female claw is probably basic to that family as well. Within the Simuliidae, a simple claw (or one with a minute basal or subbasal tooth) is characteristic of *Parasimulium*, *Prosimulium* s. s., and many simuliines, and so we conclude that a simple claw represents the plesiomorphic condition.

Although a large basal tooth can be assigned to the groundplan of *Helodon* s. l., it has been lost secondarily in some species. For example, on overall balance of characters, *Prosimulium aridum* Rubtsov (known only from the female) is best assigned to *Helodon* s. l.; however, the female of *P. aridum* has only a minutely toothed claw. The mandible and maxillary lacinia of *P. aridum* are devoid of teeth and, therefore, are rendered incapable of piercing the skin. As discussed under character 30, blood feeding is presumed to be a plesiomorphic feature of the Simuliidae, and so autogeny must be considered a secondary, or derivative, feature. Having lost the ability to blood feed, the female of *P. aridum* no longer has the need for a large-toothed claw, and so the tooth has presumably become reduced over time. The same argument can be applied to the autogenous female of *Prosimulium phytofagum* Rubtsov, which also lacks a conspicuous basal tooth on the tarsal claw. This species is best assigned to the subgenus *Helodon* because it lacks the autapomorphies attributable to other Prosimuliini genus groups.

In summary, the initial dichotomy is between prosimuliines with females that are mammalophilic (*Prosimulium*) and those with females that are primitively ornithophilic (Prosimuliini, exclusive of *Prosimulium*). Several members of this latter group have reverted back to mammalophily or have lost the ability to blood feed altogether. In such instances, the basal or subbasal tooth of the tarsal claw has become reduced or lost. The derived state also occurs widely within the Simuliini, including members of the most plesiomorphic lineages. If the polarity arguments presented here are accepted, then a bifid claw (and by extension, ornithophily) evolved independently in the common ancestor of the Simuliini.

##### *Monophyly of the Genus* Helodon (Distosimulium, Helodon *s. s.*, Parahelodon) (Fig. 9.1)

107. *Chromosomes with fixed inversions in IIIL (IIIL-1) and IIS.* The overall close relationship of *Distosimulium*, *Helodon*, and *Parahelodon* has not been questioned, yet we have not been able to find convincing structural characters to unite them. The monophyly of the group is supported solely on cytological grounds. In his chromosomal transformation series of the "Prosimuliinae," Rothfels (1979) characterized the members of *Helodon* s. l. as having fixed inversion *IIIL-1* plus a fixed inversion in the IIS arm. These inversions are evidently unique and are taken as evidence of the common ancestry of *Helodon* s. l. Two cytological subgroups were recognized within this group: one including the nominate subgenera *Distosimulium* and *Parahelodon* and another including members of *Helodon* s. s.

##### *Monophyly of the Clade* Urosimulium + *Fanless Prosimuliini* (Gymnopais, Levitinia, Twinnia) (Fig. 9.1)

108. *Mandible of larva with apical brush of bristle-like setae.* Primitively, the apical brush consists of a series of

rows of fine hairlike setae, such as in larvae of *Parasimulium*, *Helodon* s. l., and *Prosimulium* s. s. The condition in larvae of *Urosimulium* differs in that the apical brush is composed of shorter, thicker bristle-like setae. In this respect, the apical brush of *Urosimulium* represents an intermediate stage of a transformation series between the presumed plesiomorphic configuration, as characteristic of the majority of simuliids, and the stout comblike scales that constitute the apical brush of larvae of *Gymnopais* and *Levitinia*. The more extensive apical brush of fanless prosimuliines probably is related to the obligate grazing habit of the larvae. The apical mandibular brush of *Twinnia* consists of two types of setae: an anterior row of three or four spinelike setae, and several more posterior rows of bristle-like setae. We suggested in the discussion of character 88 that the condition in *Twinnia* might be a reversal from the type of mandible of *Gymnopais* and *Levitinia*. Character states are as follows: apical brush of fine setae; apical brush of bristle-like setae; apical brush of comblike scales; apical brush with comblike scales reduced in number and size.

109. *Anal sclerite with a short, posteromedially directed projection between posteroventral arms.* The anal sclerite has been interpreted as a groundplan apomorphy of the Simuliidae. Primitively, the anal sclerite is an X-shaped structure that consists of a plate and four radiating arms, two anterodorsal and two posteroventral. This pattern is basic to all major lineages of the Simuliidae, including the Prosimuliini. An X-shaped anal sclerite characterizes all members of *Prosimulium* s. s., all members of *Helodon* s. s., both species of *Distosimulium*, and one species of *Parahelodon* (subrectangular in the other two species). The anal sclerite of *Urosimulium* larvae differs from the basic plan by having a small posteromedial projection arising from the plate between the posteroventral arms. Other character states of the prosimuliines include an inverted star-shaped anal sclerite (*Levitinia*) and a Y-shaped anal sclerite (*Gymnopais*, *Twinnia*). Developmentally, the type of anal sclerite in larvae of *Urosimulium* is the most easily derivable from the plesiomorphic form. All that would be required is development of a short, posteriorly directed sclerotized projection between the posteroventral arms. If this scenario can be accepted as a logical first step, then the inverted star-shaped anal sclerite of *Levitinia* could be derived through lengthening and widening of this posteromedially directed process (see also character 84). The Y-shaped anal sclerite of *Gymnopais* and *Twinnia* could be derived from the type of anal sclerite found in *Urosimulium* through reduction or loss of the posteroventral arms and development of an elongate common base for the anterodorsal arms. These interpretations are followed in the present work. Character states are as follows: anal sclerite X shaped; anal sclerite X shaped with a short, supernumerary, posteromedially directed process; anal sclerite with supernumerary process in the form of a prominent arm subequal in length with anterodorsal and posteroventral arms; anal sclerite with posteroventral arms reduced and with anterodorsal arms borne on an elongate common base.

*Monophyly of the Fanless Prosimuliini (*Gymnopais, Levitinia, Twinnia*)* (Fig. 9.1)

110. *Cocoon rudimentary.* In the simuliid groundplan, the cocoon is an irregular, thin to thickly woven silk sac enclosing all of the pupal abdomen and typically most of the thorax. This condition, which is considered plesiomorphic, is characteristic of *Parasimulium*, *Helodon* s. l., *Prosimulium*, and most plesiomorphic genera of the Simuliini. The cocoons of *Gymnopais* and *Levitinia* are rudimentary in comparison with those of most other simuliids because most of the cocoon sloughs off at the time of pupation. All that remains is a small ventral pad to which the naked pupa remains attached to the substrate. The only other simuliids with a similar arrangement are the two species of *Crozetia* and *Tlalocomyia revelata*; however, other characters suggest that this modification must be independently derived from the simuliid groundplan. The cocoon of *Twinnia* is in the form of thick gelatinous transparent strands that cover the entire pupa in a slimy envelope. As discussed under character 89, this condition is considered to be a further modification of the condition in *Gymnopais* and *Levitinia*.

111. *Abdominal tergites IV and V of pupa without an anterior row of spine combs.* Spine combs are interpreted as a groundplan feature of the Simuliinae. Primitively, spine combs are situated on each of abdominal tergites IV–IX. This condition is found in *Distosimulium*, *Helodon*, *Parahelodon*, *Prosimulium*, *Urosimulium*, and many Simuliini (although spine combs sometimes are lacking on tergites IV and V in certain species within these taxa). In the only examined pupa of *Levitinia* (*L. freidbergi*) spine combs are lacking from each of abdominal tergites IV and V but are present on tergites VI–IX. Pupae of *Gymnopais* and *Twinnia* lack spine combs altogether. We have interpreted the condition in *Levitinia* as forming part of a transformation series between the complete onchotaxy typical of pupae of *Helodon* s. l., *Prosimulium*, and *Urosimulium*, and the total absence of spine combs from pupae of *Gymnopais* and *Twinnia*.

112. *Labral fans absent from second through final instars.* The form of the larval head is one of the most characteristic features of the fanless prosimuliines. A substantial phylogenetic gap appears to exist between fanless and fully fanned simuliids, but Craig (1974) showed that the two types of head are markedly similar in the first instar. For example, first instars of *Crozetia*, *Gymnopais*, *Helodon* s. s., *Parahelodon*, *Prosimulium*, and *Twinnia* have simple labral fans (i.e., the stalk of the fan is rudimentary or lacking and only zero to four fan rays are present), and the head is rather ovoid in dorsal view. This same type of arrangement occurs in first instars of *Parasimulium stonei* (Borkent & Wood 1986). The fan of other first-instar simuliids, although somewhat more complex than in the taxa just listed, is still less complex than those of later instars. Wood (1978) suggested that reduction of fans in the first instar is characteristic of the entire family. For reasons yet unknown, the fans are not fully expressed until the later instars.

Larvae of *Gymnopais*, *Levitinia*, and *Twinnia* are derived with respect to most other simuliids in that their fans are suppressed throughout the entire larval stage, not just in the first instar. Later-instar larvae of *Crozetia crozetensis*, *Crozetia seguyi* Beaucournu-Saguez and Vernon, *Simulium neoviceps* Craig, and *Simulium oviceps* Edwards have heads that are structured similarly, but at least a few fan rays are present in each of these species. Other characters indicate no close relationship between any of these species and the fanless prosimuliines. Suppression of the labral fans in all larval instars has undoubtedly been derived several times independently in simuliids.

In addition to the complete absence of labral fans in second- through final-instar larvae, the following related modifications occur in the larval head of the fanless prosimuliines and provide evidence of their common ancestry:

a. *Head widest near base, ovoid in dorsal view*. The head in other simuliids has rather parallel sides.
b. *Labrum elongate, terminated in greatly pronounced palatal brush*. The labrum is shorter, with an inconspicuous palatal brush in other simuliids.
c. *Ecdysial line broadly V shaped posteriorly*. The ecdysial line is broadly U shaped posteriorly in most other simuliids but broadly V shaped in *Crozetia*, *Simulium neoviceps*, *Simulium oviceps*, and *Simulium infernale* n. sp.

113. *Head of larva with posterior frontolabral muscles divided into two fascicles*. In the Simuliidae, as in other Culicomorpha, the posterior frontolabral muscles, which control the labral fans, insert anteriorly on the messors and originate posteriorly near the posterior margin of the frontoclypeal apotome. Anteriorly, the muscles are in the form of a single fascicle; posteriorly, they are divided into two or more fascicles. Craig (1974) recognized two basic arrangements in his study of the labrum of larval simuliids. In fully fanned simuliids, the posterior frontolabral muscles are divided into three fascicles posteriorly. The fanless larvae of *Gymnopais*, *Levitinia*, and *Twinnia* have only two fascicles posteriorly. The frontolabral muscles of the Dixidae are similar to those in fanned simuliids in that there are three fascicles posteriorly; we, therefore, have interpreted this condition as plesiomorphic. Presence of only two fascicles must be derived (a loss) and can be taken as a synapomorphy of the fanless Prosimuliini.

114. *Head of larva with ventral fascicles of posterior frontolabral muscles not interdigitated*. In fanned simuliids, the posterior fascicles of the right and left muscles are fully interdigitated (Craig 1974). This condition contrasts with that in *Gymnopais* and *Twinnia*, in which the posterior fascicles do not interdigitate posteriorly. We have confirmed this latter condition for *Levitinia* as well. As discussed under character 16, we have interpreted the interdigitated condition as a synapomorphy of the Simuliidae. The condition in *Gymnopais*, *Levitinia*, and *Twinnia*, therefore, is considered a reversal toward the plesiomorphic condition.

115. *Mandible of larva broad apically, with flattened, evenly sized apical teeth*. The basic form of the larval mandible is similar in *Gymnopais*, *Levitinia*, and *Twinnia*. It is relatively broad apically and bears a number of flattened, more or less equally sized bladelike teeth. In most other simuliid larvae, the mandible is more slender apically and the teeth are more conical and pointed. Further, the teeth are of rather unequal length, with one tooth (the apical mandibular tooth) being the most prominent. Outgroup comparison with other Culicomorpha reveals that the latter condition is plesiomorphic.

116. *Mandible of larva with apical brush extensive, consisting of numerous rows of comblike scales*. The larval mandible of simuliids has a series of rows of flattened setae or bristles (apical mandibular brush) on the aboral or ventral surface (Craig 1977). Typically, the apical brush consists of slender setae arranged in several rows immediately basal to the apical mandibular teeth. The type of mandible in *Gymnopais* and *Levitinia* differs from this basic plan in that the apical brush is more extensive basally, consisting of 7–12 rows of scales. Another major difference is that the scales are in the form of stout, conical, comblike outgrowths of cuticle. Both of these features are adaptations to grazing (Currie & Craig 1988) and are considered derivative.

The apical brush in *Twinnia* larvae appears to be intermediate in form between the two types just described. It has a distal row of three or four spinelike setae (similar to the scales in *Gymnopais* and *Levitinia*) and an additional three or four rows of slender bristle-like setae (similar to the setae of *Urosimulium*; see character 88). We suggest later that *Levitinia* is the sister taxon of *Gymnopais* + *Twinnia*, and so the apparent intermediate condition in *Twinnia* is possibly the reversal from the form found in *Gymnopais* and *Levitinia*. This interpretation is followed in the present work. Character states are as follows: apical brush consisting of long fine setae; apical brush consisting of bristle-like setae; apical brush consisting of comblike scales; apical brush with comblike scales reduced in number and size.

117. *Mandible of larva with covering brush, first external brush, and second external brush reduced or lost*. The mandible of larval simuliids typically possesses a series of brushes that comb the labral fan. In the fanless Prosimuliini, four of the mandibular brushes are either modified or reduced. For example, the larval mandible of *Twinnia* has apparently lost the covering and second external brushes, and the first external brush has shorter smaller bristles than that in filter-feeding larvae (Chance 1970a). The larval mandibles of *Gymnopais* and *Levitinia* are similarly modified. The reduced or modified condition of mandibular brushes is undoubtedly related to the loss of the labral fans.

118. *Hypostoma of larva with teeth dorsoventrally flattened, bladelike, and slightly sclerotized, not inclined dorsally*. The larval hypostomata of *Gymnopais*, *Levitinia*, and *Twinnia* bear an apical series of flattened, bladelike, slightly sclerotized teeth. Another distinctive feature is that the teeth are all perpendicular to the ventral wall of the hypostoma. This type of hypostoma serves as a pan into which algae and other organic matter is swept by the broomlike mandibles (Currie & Craig 1988). Although

other larval simuliids are capable of feeding in a similar fashion, their teeth are more conical and pointed apically, and the median tooth is inclined dorsally. As a consequence of this arrangement, the typical fanned larva is not able to graze as efficiently as fanless larvae. The pan type of hypostoma is adapted for grazing and can be considered a synapomorphy of *Gymnopais*, *Levitinia*, and *Twinnia*. The only other prosimuliine with a similar arrangement is *Helodon pleuralis*, but its median tooth is inclined dorsally, and its nominal sister species, *Helodon daisetsensis*, has the typical hypostomal configuration for the Prosimuliini. The similarity probably is due to convergence. The pan type of hypostoma evidently has evolved several times independently in members of the Simuliini with markedly reduced labral fans (e.g., *Crozetia*).

119. *Hypostoma of larva with median tooth relatively short, its apex not extended anteriorly beyond apex of shortest lateral or sublateral tooth.* In larvae of the fanless Prosimuliini, the median hypostomal tooth is relatively short compared with the lateral and sublateral teeth, giving the hypostoma a concave appearance medially. In larvae of the fanned Prosimuliini, the median tooth is typically longer than the one just described, with its apex extended anteriorly beyond the apex of the shortest lateral or sublateral tooth. Out-group comparison with *Parasimulium* and the Simuliini indicates that this latter state is plesiomorphic. The only other prosimuliines with a short median tooth are members of the subgenera *Distosimulium* and *Parahelodon*; however, other differences in the hypostoma suggest that this modification might be independently derived. We, therefore, interpret the apomorphic state as being derived twice: once in *Distosimulium* + *Parahelodon* (see character 125) and once in the fanless prosimuliines.

120. *Postgenal cleft rudimentary or absent.* We argued previously that the postgenal cleft is synapomorphic in members of the Simuliinae (see character 46). This feature of the larval head is present in most simuliines and all fanned prosimuliines. In this latter group, the cleft is in the shape of an inverted U, an inverted V, or a shallow subrectangular notch. In fanless members of the Prosimuliini, the postgenae are completely sclerotized medially and posteromedially, and the cuticle of the posteroventral margin of the head capsule appears wrinkled. We suggested that the cleft became lost from members of the fanless Prosimuliini and that the loss is correlated with loss of the labral fans. The postgenal cleft evidently was lost independently in members of the simuliine genus *Gigantodax* (in part).

121. *Abdomen of larva with segments I–IV narrow and ventrally corrugated, expanded abruptly at segment V, and tapered posteriorly to a small posterior proleg.* The fanless larvae of *Gymnopais*, *Levitinia*, and *Twinnia* have this type of body form. The larval body of fanned prosimuliines is typically more evenly tapered posteriorly, and the abdomen is not corrugated ventrally. The posterior proleg is proportionally larger than in members of the fanless Prosimuliini. Evidently, the derived body form has evolved independently in members of the Simuliini with markedly reduced labral fans (*Crozetia*, *Simulium oviceps* species group).

*Monophyly of the Clade* Parahelodon + Distosimulium (Fig. 9.1)

122. *Arm of genital fork (sternite IX) slender, with pronounced lateral plate.* The genital fork of females of *Distosimulium* and *Parahelodon* is distinguished by slender arms and a markedly pronounced lateral plate that arises from the apex of each arm. In females of other Prosimuliini, the arm of the genital fork is typically rather broad, and the lateral plate is comparatively small (i.e., not projected posteriorly beyond the posterior margin of the hypogynial valves). Out-group comparison with the genus *Parasimulium* and the simuliines indicates that this latter condition is plesiomorphic. The combination of a slender arm and a greatly pronounced lateral plate, therefore, must be derived and is taken as evidence of the common ancestry of *Distosimulium* and *Parahelodon*.

123. *Median sclerite of aedeagus with arms fused together apically.* Primitively, the median sclerite is a Y-shaped structure, with its simple end articulated with the ventral plate. In the Parasimuliinae, the sclerite is only shallowly notched apically, possibly representing the groundplan condition for the Simuliidae. In the Simuliinae, the arms typically are distinctly separated. The median sclerites of *Distosimulium* and *Parahelodon* differ from those of most other simuliids in that the arms are fused together apically. However, a variably distinct space or groove remains between the two arms. This condition is unique among prosimuliines and is taken as evidence of the common ancestry of *Distosimulium* and *Parahelodon*.

124. *Paramere not connected to anterolateral apodeme (basal arm) of ventral plate.* In the simuliid groundplan, the paramere is connected to both the basal arm of the ventral plate and the anterodorsal margin of the gonocoxite (the gonocoxal apodeme). As discussed under character 136, the paramere is typically connected to the basal arm by a variously sized sclerotized strap. A long, distinct, straplike connection with the basal arm is characteristic of most male prosimuliines except those in *Distosimulium*, *Levitinia*, and *Parahelodon*. The parameres in these three groups lack any suggestion of a straplike connection; instead, they simply are fused to the gonocoxal apodeme. Other characters suggest that the apomorphic condition has been derived twice in the Prosimuliini: once in *Levitinia* (character 82) and once in *Distosimulium* + *Parahelodon*. The parameral connection also has been lost independently from certain members of the Simuliini.

125. *Hypostoma with median tooth relatively short, its apex not extended anteriorly beyond apex of shortest lateral or sublateral tooth.* The form of the larval hypostoma in *Distosimulium* and *Parahelodon* is distinctive among the fanned Prosimuliini. The median tooth is relatively short compared to the lateral and sublateral teeth, giving the hypostoma a concave appearance medially. In other fanned prosimuliines, the median tooth is typically longer than the one just described, with its apex extended anteri-

orly beyond the apex of the shortest lateral or sublateral tooth. The median tooth of *Helodon* s. s. is typically the longest. In comparison with other fanned prosimuliines, therefore, a short median tooth is derived. A short median tooth is also evident in the fanless prosimuliines, and so the apomorphic state is possibly indicative of a more inclusive monophyletic group. However, other differences in the hypostoma might suggest independent development of the derived condition (see character 119). For reasons described more fully later, we interpret the apomorphic state as being derived twice: once in *Distosimulium* + *Parahelodon* and once in the fanless prosimuliines.

*Monophyly of the Clade* Gymnopais + Twinnia (Fig. 9.1)

126. *Antennal flagellum with seven articles.* Adults of *Gymnopais* and *Twinnia* are distinguished from those of most other Prosimuliini by their seven-articled flagellum. A nine-articled flagellum is the most common condition in the Prosimuliini, although eight articles characterize isolated species (*Helodon decemarticulatus*, *Prosimulium unicum*). A nine-articled flagellum is interpreted as the groundplan condition for the Simuliidae, and so the lesser number can be considered derivative in *Gymnopais* and *Twinnia*. The only other prosimuliine with a seven-articled flagellum is *Helodon gibsoni;* however, other characters do not indicate a close relationship between this species and *Gymnopais* and *Twinnia*.

127. *Stemmatic bulla near posterior margin of compound eye.* A prominent bulla apparently has evolved twice in the Simuliidae: once in *Parasimulium* and once in *Gymnopais* and *Twinnia* (see character 24). The plesiomorphic condition in the Simuliidae is for the stemmatic bulla to be absent.

128. *Tarsal claw of female without basal or subbasal tooth.* A tarsal claw with a distinct basal or subbasal tooth is presumed to be the groundplan condition for the Prosimuliini, exclusive of the genus *Prosimulium* (see character 106). Such a condition characterizes the female of *Levitinia*, which has a tooth about one fourth the length of the claw. If *Gymnopais* and *Twinnia* share an immediate common ancestor with *Levitinia* (and all available information points to this conclusion), then absence of a subbasal tooth must be interpreted as a loss.

129. *Segment III of pupal abdomen ringlike, the tergite and sternite of that segment fused together laterally.* Pupae of *Gymnopais* and *Twinnia* are unique among the Prosimuliini in that the tergite and sternite of abdominal segment III are fused together laterally. In all other prosimuliine pupae, abdominal segment III is divided laterally by pleural membrane. This latter condition has been interpreted as part of the prosimuliine groundplan and is considered plesiomorphic.

130. *Tergites III and IV of pupal abdomen with recurved hooks reduced in number or lost, occupying a position between middle of tergite and its posterior margin.* A characteristic feature of simuliid pupae is the presence of four pairs of anteriorly directed hooks along the posterior margin on each of abdominal tergites III and IV. This pattern is probably basic to the Prosimuliini, as suggested by its presence in the following lineages: *Helodon* s. l., *Levitinia*, *Prosimulium*, and *Urosimulium*. Pupae of *Gymnopais* and *Twinnia* can be distinguished from those of other Prosimuliini by the presence of a maximum of three pairs of hooks per tergite; instead of the hooks occupying the posteriormost margin of the tergite, as in most other simuliids, they are situated between the middle and posterior edge of the tergite. This arrangement is unique and is evidence of the monophyly of *Gymnopais* + *Twinnia*. The tergal hooks of pupae of *Gymnopais* are less well developed than they are in *Twinnia*.

131. *Tergites V–IX of pupal abdomen without anterior row of spine combs.* This condition represents the third stage of a transformation series that begins with the complete pupal onchotaxy of *Prosimulium* s. s. and *Helodon* s. l. (spine combs on each of abdominal tergites IV–IX) (see also character 111). An intermediate condition characterizes the pupa of *Levitinia*, which has spine combs only on tergites VI–IX. If this latter state is interpreted as the groundplan condition for the fanless prosimuliines, then complete absence of spine combs from *Gymnopais* and *Twinnia* must be considered a further derived state.

132. *Anal sclerite with posteroventral arms rudimentary or absent.* The Y-shaped anal sclerites of the larvae of *Gymnopais* and *Twinnia* are distinguished by the rudimentary appearance of the posteroventral arms. They are lacking entirely or are represented by a short protuberance on either side of the stem near the base. The posteroventral arms are presumed to be well developed in the groundplan of the Prosimuliini. The rudimentary form of the posteroventral arms, therefore, must be the result of reduction or loss and is evidence of the common ancestry of *Gymnopais* and *Twinnia*. Larvae of *Helodon decemarticulatus* and *Helodon gibsoni* evidently have lost both the anterodorsal and the posteroventral arms of the anal sclerite (resulting in a subrectangular sclerite), but there is no evidence to suggest that this condition is in any way homologous with the arrangement in *Gymnopais* and *Twinnia*. The most plesiomorphic species of subgenus *Parahelodon* has the hypothesized plesiomorphic type of anal sclerite for the Prosimuliini.

133. *Anal sclerite with anterodorsal arms borne on elongate common stalk, resulting in a Y-shaped structure.* The anal sclerite of most simuliids typically is in the form of a rectangular or subrectangular plate that gives rise to a pair of anterodorsal and of posteroventral arms. In *Gymnopais* and *Twinnia*, the plate is considerably narrower and longer than the ones found in other Prosimuliini, with the effect of displacing the anterodorsal arms anteriorly. No other simuliid has the anterodorsal arms situated so far anteriorly. Combined with the rudimentary form of the posterodorsal arms, the anal sclerite appears Y shaped in dorsal view. This arrangement is unique and is strong evidence of the monophyly of *Gymnopais* + *Twinnia*.

**MONOPHYLY OF THE EXTANT HOLARCTIC SIMULIINI** (Fig. 9.1)

134. *Calcipala.* We are unaware of any homologue of the calcipala among out-group families, and no sugges-

tion of a calcipala is evident in any member of the Parasimuliinae or the Prosimuliini in the sense of the present work; however, a variously developed calcipala is widely and commonly distributed among members of the Simuliini. The calcipala probably evolved early, as evidenced by its presence in such plesiomorphic simuliines as *Paracnephia* and the Neotropical genera formerly assigned to the tribe Prosimuliini (cf. Wygodzinsky & Coscarón 1973). Two of the three species assigned to *Paracnephia* (*Procnephia*) have only a weakly developed calcipala, but the third, *P. rhodesiana* (Crosskey), lacks one altogether. Whether the apparent absence of a calcipala in *P. rhodesiana* is truly primitive or represents a reversal toward the primitive form cannot be evaluated at present. Similarly, we are unable to interpret the apparent absence of a calcipala in *Crozetia*. *Sulcicnephia*, *Metacnephia* (in part), and a few *Simulium* species also lack a calcipala, but other characters suggest that this absence is the result of loss. Because of homoplasy, the origin of the calcipala is only tentatively considered a synapomorphy of the extant Simuliini.

135. *Costa with dimorphic setae*. Hackman and Väisänen (1985) concluded that a relatively simple type of costal chaetotaxy, in which slender, irregularly situated, hairlike macrotrichia are the only setal type present, is the plesiomorphic condition for Diptera. This condition also occurs in the Mecoptera (*Panorpa*), nearly all the nematocerous superfamilies, and many other groups of Diptera. Among the Culicomorpha, the plesiomorphic condition is evident in the Ceratopogonidae, Chironomidae, Dixidae, Thaumaleidae, Parasimuliinae, Prosimuliini, and *Archicnephia*. The Culicidae and Chaoboridae have setae that are scalelike, but Hackman and Väisänen (1985) considered this condition to be a minor modification of the plesiomorphic condition. Similarly, the long thin setae of Psychodidae are considered an independent development from the general plan. Most simuliine wings differ from those of the Parasimuliinae and Prosimuliini in having a second setal type interspersed among the characteristic hairlike macrotrichia. Hackman and Väisänen (1985) suggested that the presence in Simuliidae of sparsely arranged costal spinules could be considered a synapomorphy of a group of genera, but failed to elaborate. We concur with this conclusion, although they examined only a limited number of genera (*Cnephia*, *Metacnephia*, *Simulium*), and the distinction between the two types of setation is not as clear as they suggested. Primitively, the differentiated setae of the Simuliini appear as a somewhat thickened version of the hairlike macrotrichia. This condition is found in *Greniera* (in part), the Australian *Paracnephia* (in part), and the males of *Paracnephia* (*Procnephia*). Females of *Paracnephia* (*Procnephia*) lack differentiated costal setae, but whether this feature represents the plesiomorphic condition or a reversal to the plesiomorphic form is unclear. In its most highly derived form, the differentiated setal type appears as a conspicuous blackened spinule (spiniform seta). However, the distinction between the two types of setae is rather arbitrary, as an almost uninterrupted transformation series can be formed between the hairlike and spinelike setae in members of the Simuliini. Possibly, lineages within the tribe can be defined by the form of the costal setation, but this area requires further study.

136. *Straplike connection between paramere and ventral plate of male arising subapically on anterolateral apodeme (basal arm) of ventral plate*. Males of the Parasimuliinae share with those of the Prosimuliini a straplike connection between the paramere and the basal arm of the ventral plate. The strap arises from near the apex of the arm and projects anterodorsally to a point where it connects with the paramere. Wood and Borkent (1982) compared this arrangement with the condition found in most other simuliids, in which the paramere connects with an angular point on the side of the ventral plate. The angular point is probably the remnant of the straplike connection, which simply has become fused to the lateral margin of the ventral plate. The basal arm in this latter arrangement appears as a solid finger-like projection, free of any association with the paramere. An intermediate condition is evident in males of certain genera (e.g., *Cnesiamima*, *Crozetia*, *Paracnephia*, *Stegopterna* in part) in which the straplike connection arises subapically on the basal arm. We have interpreted this intermediate stage as the groundplan condition for the Simuliini. The plesiomorphic condition is for the straplike connection of the paramere to originate from the apex of the basal arm. Out-group comparison is not possible at present because we are unable to identify homologous structures among the out groups.

137. *Genital fork (sternite IX) of female with lateral plate bearing a variously developed dorsal apodeme*. The genital fork of simuliine females possesses a variously distinct apodeme dorsally on the lateral plate. Primitively (e.g., *Stegopterna*), the apodeme appears as a narrow transverse ridge on the dorsal surface of the plate. The apodeme is better developed in most other simuliines, and in some taxa the apex is produced far beyond the anterior margin of the lateral plate (e.g., subgenus *Boreosimulium*). No suggestion of a dorsal apodeme is present in females of the Parasimuliinae or Prosimuliini. The genital fork is not visible in the only known specimen of *Archicnephia*, so the derived state is possibly a synapomorphy of the entire tribe.

Monophyly of the extant Simuliini is reasonably well supported based on morphological evidence, although two of the four putative synapomorphies (characters 134, 135) exhibit various degrees of homoplasy. However, no arrangement can be constructed in which instances of homoplasy can be eliminated completely. The most problematic species belong to *Paracnephia* (*Procnephia*), which evidently is one of the earliest lineages of the tribe. Several of the features just discussed are developed only incipiently in this taxon. Furthermore, the subgenus as a whole is inadequately collected, and not all character states have been firmly established. For example, the male of *Paracnephia* (*Procnephia*) *damarensis* (De Meillon & Hardy) is known only from parts dissected from pharate adults, and so the apparent absence of parameral spines might be the result of incomplete development. Additional support for the monophyly of the extant Simuliini, as defined here, comes from molecular sequence data (Moulton 1997, 2000).

**MONOPHYLY OF THE EXTANT GENERA OF HOLARCTIC SIMULIINI** (Fig. 9.1)

*Genus* Greniera

138. *Labral fan relatively large (as long as or longer than head capsule), bearing large number of primary fan rays (typically >60).* The labral fan of *Greniera* can be distinguished from that of most other simuliids by its relatively large size and large number of primary rays. Larvae of *Simulium* (*Hellichiella*) have fans that approach, or even exceed, those of *Greniera* in terms of relative size and number of primary rays; however, we infer that the derived state has evolved independently in these two lineages. A large fan with numerous rays is evidently an adaptation to life in slowly flowing water.

139. *Pupal gill with numerous slender, weakly supported filaments.* The pupal gill of *Greniera* consists of numerous (12+) slender filaments that readily collapse when removed from water. The gill filaments of most other simuliids are thicker and more rigid, thus maintaining their basic form when removed from water. The derived condition has evolved several times independently in simuliids that have gills with a large number of filaments (e.g., *Prosimulium dicum, P. exigens, Metacnephia villosa, Simulium hunteri*).

140. *Larval antenna markedly elongate, far exceeding length of labral-fan stalk.* The plesiomorphic state is for the antenna to be subequal in length to or only moderately longer than the labral-fan stalk. Larvae of *Greniera* can be distinguished from those of most other simuliids by the extraordinary length of their antennae, which far exceeds the length of the fan stalk. Only in certain species of *Simulium* (*Hellichiella*) do the antennae approach the length characteristic of *Greniera*. Other characters suggest that the elongate antenna has evolved independently in these two lineages.

*Genus* Tlalocomyia

141. *Form of pupal gill.* The pupal gill of *Tlalocomyia*, although varied, exhibits a characteristic form that distinguishes it from that of most other simuliids. In species that have eight or more filaments (e.g., *T. pachecolunai, T. ramifera* n. sp.), the gill base is short and the filaments branch fanlike horizontally from inflated prostrate trunks. A trend in certain species of *Tlalocomyia*, however, is for the filaments to coalesce into fewer tubular or inflated saclike filaments (e.g., *T. aguirrei, T. grenieri, T. salasi, T. stewarti*). In all species, except those with the most inflated gills (e.g., *T. salasi*), the filaments have superficial, but conspicuous, transverse annulations. The general form of the gill, as exemplified by species with eight or more filaments, is inferred to be in the groundplan of *Tlalocomyia*. The short common gill base with inflated annulated trunks and filaments radiating fanlike horizontally is apomorphic and can be taken as evidence of the common ancestry of *Tlalocomyia*, as defined here. The presence of markedly inflated gill filaments represents a further derived state that might define monophyletic groupings within *Tlalocomyia*.

*Genus* Stegopterna

142. *Female tarsal claw simple.* As discussed under character 106 and argued by Currie and Grimaldi (2000), a claw with a distinct basal or subbasal tooth is inferred to be in the groundplan of the Simuliini. This condition, which is associated with ornithophily, is characteristic of most simuliine genera, including plesiomorphic genera such as *Archicnephia, Crozetia, Gigantodax, Greniera, Tlalocomyia*, and *Paracnephia*. A simple claw, therefore, must be derived within the Simuliini and can be used to define monophyletic groups. *Stegopterna* is unique among the primitive-grade simuliines in having no suggestion of a basal or subbasal tooth on the tarsal claw. A simple claw has evolved independently in the more derivative simuliine genera *Simulium* s. l. (in part) and *Austrosimulium* (in part).

143. *Tibial spurs of hind leg markedly elongate, with hyaline apex.* The hind tibial spurs of *Stegopterna* can be distinguished from those of other simuliids by a combination of their length and hyaline apex. The length of each spur far exceeds the width of the tibia at the spur's point of attachment. The typical simuliid condition is for each spur to be subequal in length to or shorter than the width of the tibia. Furthermore, the dark scalelike setae that typically adorn the tibial spur are not present apically, giving the spur a distinct bicolored appearance (i.e., black base and white apex). These features are derived and constitute convincing evidence of the monophyly of *Stegopterna*.

*Genus* Gigantodax

144. *$CuA_2$ and $A_1$ straight.* The derived condition is unique to *Gigantodax*. In all other simuliids, $CuA_2$ has a strong sigmoidal curvature at about its midpoint, and $A_1$ is gently curved apically toward $CuA_2$.

145. *Genital fork with anteriorly directed apodemes markedly elongate.* As discussed under character 137, an anteriorly directed apodeme on the lateral plate of the genital fork is inferred to be basic to the Simuliini, exclusive of *Archicnephia*. The apodeme in *Gigantodax* is much longer than is typical, and thus can be considered a further derived state. The only other simuliid with such an elongate apodeme is the South American genus *Cnesia*, indicating that the derived state belongs to a more inclusive monophyletic group.

146. *Paramere apparently represented only by apical spines (apex of parameral arm rudimentary or not discernible).* The presence of well-developed parameral arms is plesiomorphic. We infer that the arms have been reduced through desclerotization in most *Gigantodax* species, giving the appearance that the parameral spines are isolated in the aedeagal membrane.

147. *Mandible of larva with three outer (one long and two short) teeth and one apical tooth.* The plesiomorphic condition is for the larval mandible to consist of two or four outer teeth (Wygodzinsky & Coscarón 1973).

148. *Anal sclerite with posterior arms extended ventrally around abdomen, forming a complete ring around base of posterior circlet of hooks.* The plesiomorphic state in the Simuliini is for the anal sclerite to be X shaped, without the posterior arms extended ventrally in a ringlike fashion. The condition in *Gigantodax* is similar to that in *Crozetia crozetensis*, in which the posterior arms of the anal sclerite in certain specimens is joined together midventrally. However, the sclerite is noticeably thinner at the point of union, and the arms are extended as little as two thirds around the abdomen in other specimens (L. Davies 1974). Larvae of *Parasimulium crosskeyi* also have markedly elongate posterior arms, but they are separated midventrally by a spindle-shaped sclerite. The complete ring formed by the posterior arms in *Gigantodax*, therefore, is unique and can be considered a synapomorphy of the genus. The derived state should not be confused with the condition in *Austrosimulium*, in which a semicircular sclerite contours the ventral half of the abdomen at the base of the circlet of hooks. The semicircular sclerite is a separate structure that is not joined to the ventral arms. Accordingly, we conclude that the condition in *Austrosimulium* is independently derived.

149. *Anal sclerite with interarm strut projected posteriorly from apex of anterior arm.* The plesiomorphic state is for no such strut to be present. The interarm strut on the anal sclerite of *Paraustrosimulium* has evolved independently (Wygodzinsky & Coscarón 1973), as has that in *Austrosimulium*.

*Genus* Cnephia

150. *Spermatheca in the form of a greatly enlarged, wrinkled spherical sac.* In *Cnephia*, the length of the spermatheca is markedly longer than the stem of the genital fork. Only females of *Helodon* (*Distosimulium*) have a spermatheca that approaches the same relative proportions; however, the spermatheca of this distantly related taxon is neither pigmented nor patterned. Similarly, the elongate spermathecae (about as long as the stem of the genital fork) of *Ectemnia* species can be distinguished from those of *Cnephia* by their narrower width and presence of a large differentiated area near the junction with the spermathecal duct. We have concluded that similarities in the spermathecae of *Cnephia*, *Distosimulium*, and *Ectemnia* are due to convergence. The plesiomorphic condition is for the spermatheca to be markedly shorter than the stem of the genital fork.

151. *Pupal gill with filaments arising singly from bulbous central knob.* To our knowledge, the derived state is confined to *Cnephia*. The gill filaments of most other simuliids arise variously from two or, more typically, three primary trunks.

152. *Pupal abdomen with hooklike setae in pleural membrane of segments VIII and IX.* The pleural setae of most other simuliid genera are rather varied in form (straight, twisted, anchor-like, or grapnel-like), but not a simple hook as in *Cnephia*. *Lutzsimulium* has similar setae in the pleural membrane of segments VIII and IX, but they are so completely looped that the apex is typically doubled back on itself. The derived condition, therefore, probably has evolved independently in *Lutzsimulium*.

153. *Larval head capsule with frontoclypeal apotome strongly convex in lateral view.* The derived state has evidently evolved independently in *Metacnephia*. The plesiomorphic condition is for the frontoclypeal apotome not to be noticeably arched or strongly convex.

154. *Larval antenna conspicuously shorter than labral-fan stalk.* The larval antenna of *Cnephia* is among the shortest of any simuliid, being markedly shorter than the stalk of the labral fan. The plesiomorphic condition is for the antenna to be subequal in length to the stalk of the labral fan.

155. *Labral-fan rays with markedly elongate microtrichia interspersed among normal-sized microtrichia.* Larvae of *Cnephia* have highly modified microtrichia borne on delicate primary fan rays. The normal elongate microtrichia are produced to an extraordinary degree and alternately extend laterally toward the other fan rays. The intervening microtrichia are variously reduced but are much shorter. As a result of this arrangement, the fan serves as a true sieve (= direct interception) (Currie & Craig 1988). In other fanned simuliids, the elongate microtrichia are only moderately longer than the intervening microtrichia, and food capture is primarily through "inertial impaction" rather than direct interception.

*Genus* Ectemnia

156. *Cibarium with 1–11 setae on anteroventral margin.* The derived state is otherwise known only from *Simulium mie* Ogata & Sasa, a distantly related species in the subgenus *Nevermannia* (Moulton & Adler 1997).

157. *Ventral plate in ventral view at least 2.5 times wider than long.* The derived state is for the ventral plate to be relatively narrower (Moulton & Adler 1997).

158. *Gill filaments convergent apically.* The filaments are randomly arrayed or divergent in other simuliid pupae (Moulton & Adler 1997).

159. *Gill filaments with surface sculpture imbricate.* The plesiomorphic condition is for the surface sculpture to be smooth or furrowed (Moulton & Adler 1997).

160. *Pupal tergite V with stout hooks.* Stout hooks are absent from pupal tergite V in other simuliids (Moulton & Adler 1997).

161. *Terminal spines of pupal abdomen inflated, hollow.* The plesiomorphic condition is for the terminal spines to be slender and solid. No other simuliid exhibits the derived condition.

162. *Hypostoma with anterior margin concave.* The anterior margin of the hypostoma is straight or convex in other simuliids (Moulton & Adler 1997).

163. *Larval mandible with inner subapical margin smooth.* The subapical margin is dentate in other simuliids (Moulton & Adler 1997).

164. *Larval abdominal segments V–VIII expanded ventrally and laterally.* The plesiomorphic condition is for segments V–VIII to be expanded more gradually (Moulton & Adler 1997).

165. *Anal sclerite absent.* All other simuliid larvae have a well-formed anal sclerite (Moulton & Adler 1997).

166. *Larval construction of an elongate stalk on which to feed and pupate.* Stalk building is otherwise known only from two or three species of *Paracnephia* in southeastern Australia. However, the form of the stalk differs from that of *Ectemnia* species, and pupation occurs on the substrate (Moulton & Adler 1997).

*Genus* Metacnephia

167. *Anepisternal (pleural) membrane with tuft of setae dorsally.* As discussed under character 95, the anepisternal membrane is bare in most simuliids. The derived state has evolved independently in taxa such as *Gymnopais*, *Tlalocomyia revelata*, *Metacnephia*, and the *Simulium ornatum* species group. Within *Metacnephia*, only *M. saskatchewana* lacks the characteristic tuft of setae.

168. *Pupal abdomen with numerous triramous (grapnel-like) setae in pleural membrane of segments VIII and IX.* These modified setae are evidently unique to *Metacnephia*. The common base of each seta gives rise to three equal-length branches with apices curved backward toward the base. The sporadic occurrence of biramous (anchor-like) setae in *Austrosimulium*, *Gigantodax*, and *Greniera* probably represents independently derived states in those genera. Similarly, the triramous setae of certain species of *Tlalocomyia* (e.g., *T. pachecolunai*) and *Paraustrosimulium anthracinum* (Bigot) are not as numerous as in *Metacnephia*, nor are the branches curved backward toward the base.

169. *Larval antenna with proximal and medial articles unpigmented, contrasting with dark distal article.* The plesiomorphic condition is for the basal two articles to be pigmented and the distal article to be more lightly colored. The derived state is shared by members of the Prosimuliini and *Metacnephia*; however, as discussed under character 49, the condition is believed to be independently derived in those two lineages.

170. *Chromosomes with whole-arm interchange between chromosomes I and II.* The derived state, as described by Procunier (1982a), is not known in any other simuliid.

*Genus* Sulcicnephia

171. *Gonostylus subtruncate, with an internal triangular lobe bearing one apical spinule.* The plesiomorphic state is for the gonostylus to be subconical and evenly tapered to a variously pointed apex, with the spinule(s) either apical or, at most, only slightly subapical. A condition similar to the derived state is also present in males of *Simulium* (*Nevermannia*); however, other characters indicate that the condition is derived independently in this subgenus.

172. *Hypostoma triangular in shape anteriorly, with median tooth most prominent and with sublateral and lateral teeth progressively decreasing in length laterally.* Among simuliine genera that are characterized by uniformly small teeth (i.e., *Cnephia*, *Ectemnia*, *Metacnephia*, *Sulcicnephia*, and *Simulium*), the median and lateral teeth are typically most prominent and the sublateral teeth are variously shorter. The condition in *Sulcicnephia* differs from the typical arrangement in that the median tooth projects farthest anteriorly, with the submedian and lateral teeth progressively decreasing in length laterally. We interpret the resulting triangular-shaped hypostoma to be derived. Similar arrangements of hypostomal teeth in certain lineages of *Simulium* s. l. are probably due to convergence.

173. *Pedisulcus present, typically deep and distinct.* The plesiomorphic state is for the pedisulcus to be absent. Among simuliids, the derived state is present only in the genera *Austrosimulium*, *Simulium*, and *Sulcicnephia*. A pedisulcus can be ascribed to the common ancestor of *Simulium* and *Austrosimulium*, but the pedisulcus of *Sulcicnephia* might not be a homologous structure. Evidence presented later suggests a sister-group relationship between *Sulcicnephia* and *Metacnephia*, the latter of which lacks any suggestion of a pedisulcus. Because *Sulcicnephia* and *Metacnephia* together form the probable sister group of *Austrosimulium* + *Simulium*, two equally parsimonious explanations are possible: (a) a pedisulcus is in the groundplan of all four genera, with a reversal toward the primitive form in *Metacnephia*; and (b) a well-formed pedisulcus evolved twice in the Simuliidae—once in *Sulcicnephia* and once in *Austrosimulium* + *Simulium*. The latter interpretation is followed for the purposes of the present analysis.

*Genus* Simulium

174. *Basal radial cell short, equal to much less than one third the distance from base of radial sector to apex of wing.* The plesiomorphic state is for the basal radial cell to be relatively longer, rarely less than one third the distance from the base of the radial sector to the apex of the wing.

175. *Basal medial cell absent or markedly reduced.* The plesiomorphic state is for the basal medial cell to be present and distinguishable. The derived state evidently has evolved several times independently in the Simuliidae and, therefore, does not provide convincing evidence of common ancestry in the genus *Simulium*.

176. *Furcasternum with internal dorsal arms each bearing a ventrally directed apodeme.* The plesiomorphic state is for the ventrally directed apodemes to be absent. The derived state occurs in most species of the genus *Simulium*, with an apparent reversal toward the primitive form in isolated species (e.g., *S. quebecense*, *S. slossonae*). Wygodzinsky and Coscarón (1973) found ventrally directed apodemes in *Paraustrosimulium* and *Lutzsimulium*, and the condition also occurs in *Austrosimulium*, perhaps indicating that the derived state is synapomorphic for a more inclusive monophyletic group.

177. *Pedisulcus present, typically deep and distinct.* The plesiomorphic state is for the pedisulcus to be absent. (See discussion of character 173 for further explanation.)

178. *Cocoon with thickened rim anteriorly (secondarily reduced in some species).* The plesiomorphic state is for the cocoon to lack a thickened rim anteriorly. The derived state is present only in the genus *Simulium* (in part),

*Austrosimulium* (in part), and *Paraustrosimulium*, perhaps constituting a synapomorphy of these three genera.

179. *Rectal papillae compound (secondarily simple in certain species).* The derived state is known only among species of the genus *Simulium*. Larvae of all other simuliid genera have simple rectal papillae. Isolated species of *Simulium* (e.g., *S. vittatum*) have simple rectal papillae, but other characters suggest that this condition represents a reversal toward the primitive form.

### RELATIONSHIPS OF THE EXTANT GENERA OF HOLARCTIC SIMULIINI

#### *Monophyly of the Clade* Greniera *to* Tlalocomyia (Fig. 9.1)

180. *Molecular sequence data. Greniera, Stegopterna,* and *Tlalocomyia* evidently diverged early in simuliine phylogeny, and we are unable to identify structural features that convincingly demonstrate their relationships. Molecular sequence data support the monophyly of *Tlalocomyia* + *Stegopterna*, with one species of *Greniera* (*G. denaria*) resolved as the sister group of those two genera together (Fig. 3.20 of Moulton 1997). Another exemplar of *Greniera* included in Moulton's (1997) analysis, *G. fabri*, was placed as the sister group of a lineage informally referred to as the "higher simuliines." Morphological evidence presented earlier (characters 138–140) suggests that *Greniera* is monophyletic, and we tentatively place that genus as the sister group of *Tlalocomyia* and *Stegopterna* together. Evidence for the monophyly of this clade is weak, and we acknowledge the possibility that *Greniera* might prove to nest elsewhere among the primitive-grade simuliines.

#### *Monophyly of the Clade* Stegopterna + Tlalocomyia (Fig. 9.1)

181. *Larval abdominal segment VIII with transverse midventral bulge.* The derived state is expressed variably in all exemplars of *Tlalocomyia* that we have examined and most members of *Stegopterna* (except *St. emergens*). In certain members of the *T. pachecolunai* species group, the transverse bulge is somewhat excised medially, giving the appearance of a pair of ventral tubercles. This condition should not, however, be confused with the ventral tubercles of the genera *Greniera, Ectemnia,* and *Simulium* (in part) in which the tubercles are more conical and do not form part of a transverse ridge. The plesiomorphic condition is for the venter of abdominal segment VIII to be simple.

#### *Monophyly of the Clade* Gigantodax *to* Simulium *s. l.* (Fig. 9.1)

182. *Paramere with well-developed apical spines.* The presence of parameral spines, regardless of their form, is a widely and commonly encountered feature in the Simuliini. The typical condition in Diptera is for the paramere to be simple apically (i.e., without spines), a feature shared by the Parasimuliinae, Prosimuliini, and certain plesiomorphic genera of the Simuliini. The presence of parameral spines, therefore, must be derived and can be taken as a synapomorphy of simuliine genera that exhibit the condition.

#### *Monophyly of the Clade* Cnephia *to* Simulium *s. l.* (Fig. 9.1)

Monophyly of this clade is supported by molecular sequence data (Moulton 1997) and by the two synapomorphic character states described here.

183. *Hypostomal teeth uniformly small, not arranged on prominent lobes.* The plesiomorphic condition is for the hypostomal teeth to be larger and arranged on two or three prominent lobes.

184. *Postgenal cleft deep, extended anteriorly about one third or more distance to hypostomal groove.* The plesiomorphic condition is for the postgenal cleft to be shallow, extending anteriorly no more than about one fourth the distance to the hypostomal groove.

#### *Monophyly of the Clade* Ectemnia *to* Simulium *s. l.* (Fig. 9.1)

Evidence for the monophyly of this clade is provided by molecular sequence data (Moulton 1997) and by the synapomorphy described here.

185. *Cocoon with definitely formed rigid walls.* The plesiomorphic condition is for the cocoon to be an ill-formed sac. Stuart and Hunter (1998a), who conducted a cladistic analysis of cocoon-spinning behavior, identified seven behavioral character states that support the monophyly of *Ectemnia* and other simuliids that construct a well-formed cocoon.

#### *Monophyly of the Clade* Metacnephia *to* Simulium *s. l.* (Fig. 9.1)

Monophyly of this clade is supported by molecular sequence data (Moulton 1997) and by the three synapomorphic character states described here.

186. *Katepisternal sulcus narrow and deep throughout.* The plesiomorphic condition is for the katepisternal sulcus to be broad, shallow, and evanescent anteriorly.

187. *Aedeagus with dorsal plate.* The derived state is present in *Metacnephia* (in part), *Sulcicnephia*, and *Simulium* s. l. (in part). The plesiomorphic condition is for the dorsal plate to be absent.

188. *Pupal abdomen weakly sclerotized, typically collapsed after emergence of adult.* The plesiomorphic condition is for the pupal abdomen to be well sclerotized, typically maintaining its form after emergence of the adult.

#### *Monophyly of the Clade* Metacnephia + Sulcicnephia (Fig. 9.1)

189. *Cocoon boot shaped, constructed of finely spun silk.* The plesiomorphic condition is for the cocoon to be slipper shaped and constructed of more coarsely spun silk.

190. *Postgenal cleft wide and deep throughout, extended anteriorly to level of or beyond hypostomal groove.* The plesiomorphic condition is for the postgenal cleft to be shorter, narrower, and typically not extended anteriorly to the level of the hypostomal groove (except in

the subgenus *Byssodon* and members of the *Simulium slossonae* species group).

**RELATIONSHIPS OF THE SUBGENERA OF HOLARCTIC *SIMULIUM* S. L.** (Fig. 9.2)

For the purposes of the present study we have accepted most of the subgeneric concepts circumscribed by Crosskey and Howard (1997). However, we recognize two additional subgenera (*Boreosimulium* and *Aspathia*) and adopt an expanded concept of *Hemicnetha*, which subsumes the genus-group names *Hearlea*, *Obuchovia*, and *Shewellomyia*. We consider the *S. argenteostriatum* species group of the subgenus *Simulium* to be related only distantly to that subgenus. Evidence presented here suggests that the *argenteostriatum* species group shares a common ancestry with three closely related subgenera of *Simulium* (viz., *Crosskeyellum*, *Hemicnetha*, *Himalayum*). Until relationships of the *S. argenteostriatum* species group are established more firmly, we prefer to leave it unassigned at the subgeneric level. Our knowledge of certain Palearctic and Pacific subgenera is based mainly on the literature, and we defer to the opinion of specialists on those faunas about issues of monophyly. Accordingly, only synapomorphic character states are provided in this section. Evidence for the monophyly of selected subgenera (*Boreosimulium*, *Aspathia*, *Simulium*) is provided in the section "Monophyly and Relationships of Holarctic Species Groups."

*Monophyly of the Clade* Boreosimulium *to* Wallacellum (Fig. 9.2)

191. *Parameral spines differentiated, with up to nine large spines and various numbers of markedly smaller spines.* Plesiomorphic: parameral spines not markedly differentiated, all spines similar in size and shape.

*Monophyly of the Clade* Boreosimulium + Byssodon (Fig. 9.2)

192. *Precoxal bridge incomplete, or if complete, markedly narrow medially.* Plesiomorphic: precoxal bridge complete, broad throughout.

193. *Ventral plate broad, lamellate, without a well-developed apical lip.* Plesiomorphic: ventral plate narrower, more convex, typically bearing a prominent apical lip.

*Monophyly of the Clade* Eusimulium *to* Schoenbaueria (Fig. 9.2)

194. *Paramere with one or two large spines, and up to two smaller accessory spines.* Plesiomorphic: paramere with greater number of large and small spines.

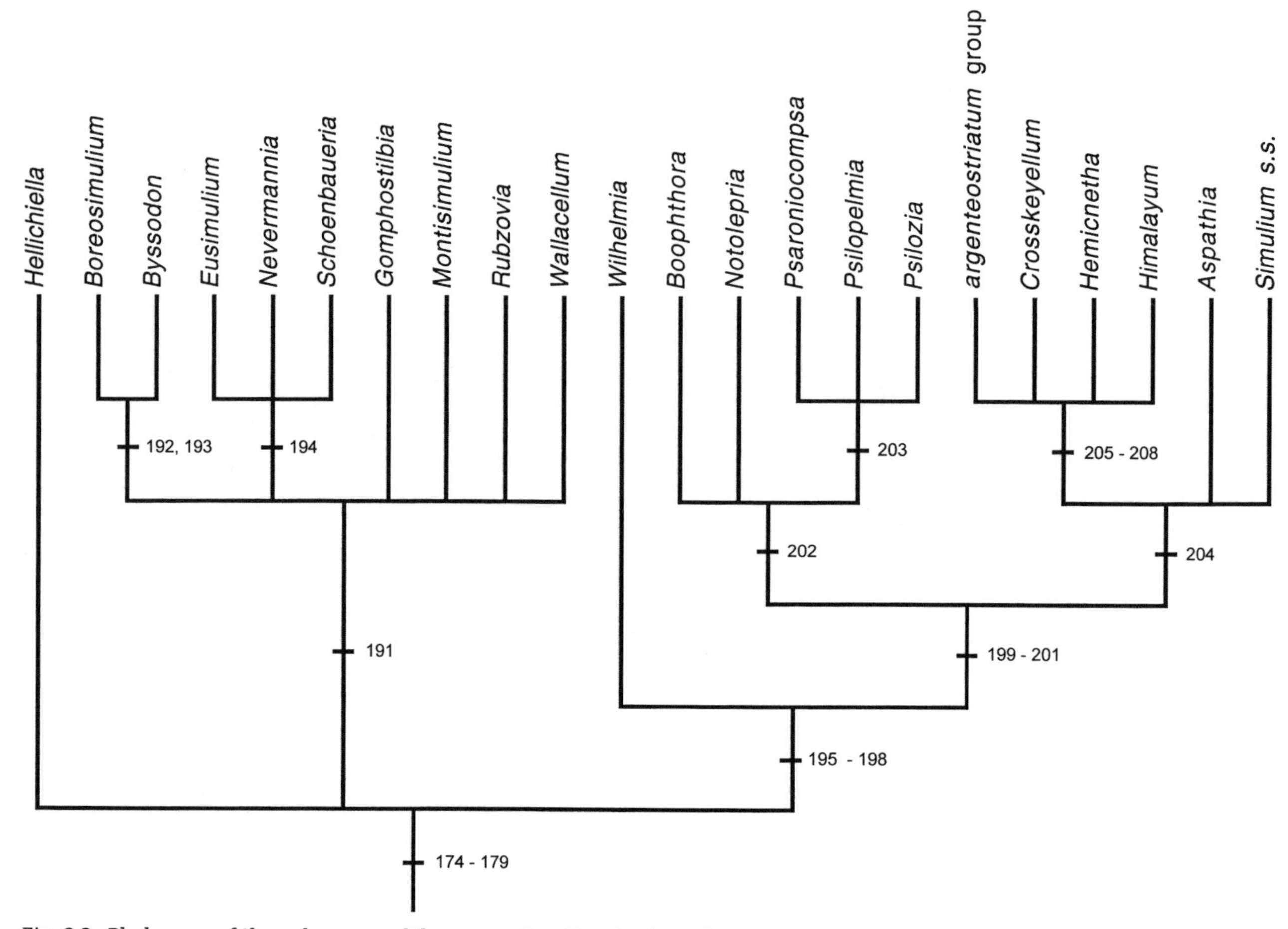


Fig. 9.2. Phylogeny of the subgenera of the genus *Simulium* in the Holarctic Region.

*Monophyly of the Clade* Wilhelmia *to* Simulium *s. s.* (Fig. 9.2)

195. *Scutum adorned with silvery spots and/or stripes (secondarily lost in some species).* Plesiomorphic: scutum without such ornamentation.

196. *Legs variously banded.* Plesiomorphic: legs uniformly colored. The derived state has evolved independently in members of the subgenus *Eusimulium.*

197. *Males with silvery spots laterally on abdomen (secondarily lost in some species).* Plesiomorphic: males without silvery spots on abdomen.

198. *Female claw simple, or bearing a small subbasal tooth that is typically much less than one sixth length of claw.* Plesiomorphic: female claw distinctly bifid. The claws of females in the *Simulium slossonae* species group evidently reacquired the bifid state.

*Monophyly of the Clade* Boophthora *to* Simulium *s. s.* (Fig. 9.2)

199. *Frons of female shiny (pollinose in some species, but underlying cuticle shiny).* Plesiomorphic: frons of female dull, not markedly different in luster from adjacent cranium.

200. *Basal section of radius bare dorsally.* Plesiomorphic: basal section of radius with setae dorsally.

201. *Gonostylus flattened dorsoventrally.* Plesiomorphic: gonostylus cylindrical or subcylindrical.

*Monophyly of the Clade* Boophthora *to* Psilozia (Fig. 9.2)

202. *Gonostylus subquadrate apically.* Plesiomorphic: gonostylus pointed or narrowly rounded apically.

*Monophyly of the Clade* Psaroniocompsa *to* Psilozia (Fig. 9.2)

203. *Cibarial armature present.* Plesiomorphic: cibarial armature absent. The derived state evidently has evolved independently in several taxa such as a few members of *Aspathia* and *Hemicnetha.* This character has not been surveyed adequately so the derived state might be ascribed to a more inclusive group.

*Monophyly of the Clade* argenteostriatum *Species Group to* Simulium *s. s.* (Fig. 9.2)

204. *Gonostylus markedly elongate, much longer than gonocoxite.* Plesiomorphic: gonostylus subequal in length to or shorter than gonocoxite.

*Monophyly of the Clade* argenteostriatum *Species Group to* Himalayum (Fig. 9.2)

Taxa assigned to this clade share a large number of character states, none of which alone constitute convincing evidence of monophyly. However, the similarity in all life stages is so striking that it is difficult to escape the conclusion that they are derived from an immediate common ancestor. We retain three of the four genus-group names referable to this clade, subsuming *Obuchovia* (as the *auricoma* species group) in *Hemicnetha.* To this clade we also refer the *argenteostriatum* species group, which hitherto had been assigned to the subgenus *Simulium.* The relationships of these four taxa need further study, and additional synonymies might be warranted.

205. *Cocoon boot shaped, with silk either loosely woven or adorned with various openings anteriorly.* Plesiomorphic: cocoon slipper shaped, typically without openings anteriorly. The boot-shaped condition evolved independently in the clade *Metacnephia* + *Sulcicnephia,* and in certain lineages of *Simulium* s. s. Certain species in the latter subgenus also have openings, which presumably represent independent evolutionary events. Among members of the clade *argenteostriatum* species group to *Himalayum,* the derived condition is present widely except in the *canadense* species group of *Hemicnetha,* in which the cocoon is typically slipper shaped. Because of homoplasy, the derived state does not provide convincing evidence of common ancestry.

206. *Larva elongate and fusiform, adapted to life in rapidly flowing streams and rivers.* Plesiomorphic: larva neither elongate nor fusiform; habitat various.

207. *Larval body color typically blackish dorsally and whitish ventrally.* Plesiomorphic: body variously colored but not black dorsally and white ventrally.

208. *Postgenal cleft deep, extended anteriorly to a level approaching the hypostomal groove.* Plesiomorphic: postgenal cleft not as deep. A deep cleft has evolved several times independently in *Simulium* s. l. and in *Metacnephia* + *Sulcicnephia.*

## MONOPHYLY AND RELATIONSHIPS OF HOLARCTIC SPECIES GROUPS

Prosimulium *s. s.* (Fig. 9.3)

Monophyly of *Prosimulium* s. s. is supported by four synapomorphies (characters 60–63). We are unable to

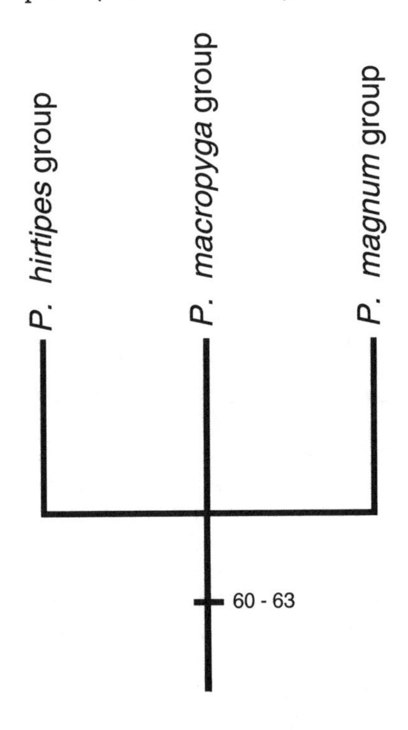


Fig. 9.3. Phylogeny of *Prosimulium* species groups.

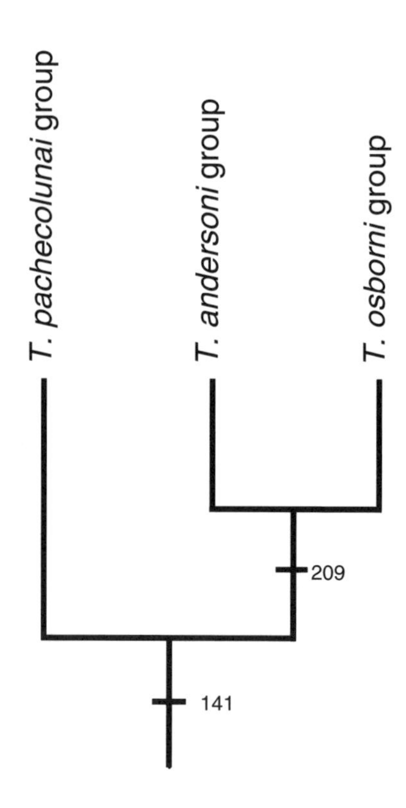


Fig. 9.4. Phylogeny of *Tlalocomyia* species groups.

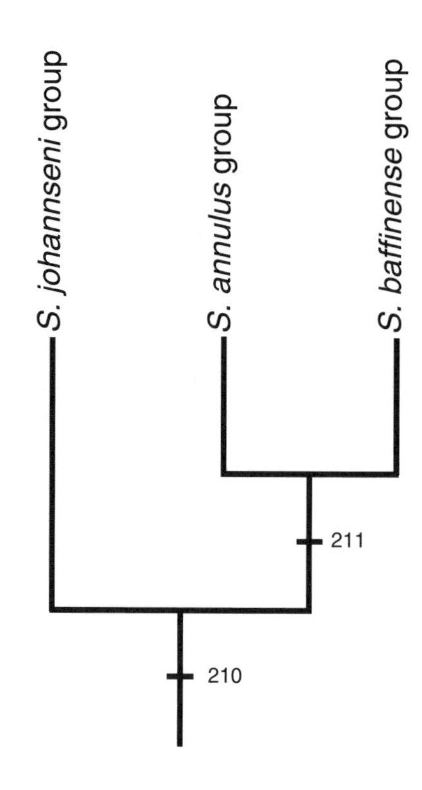


Fig. 9.5. Phylogeny of *Simulium* (*Boreosimulium*) species groups.

identify character states that convincingly resolve relationships among the three included species groups.

Tlalocomyia (Fig. 9.4)

Monophyly of *Tlalocomyia* is supported by only one putative synapomorphy (see character 141). We recognize the possibility that two species groups assigned to this genus (viz., *osborni* species group, *andersoni* species group) might have closer affinity to *Stegopterna*. Nonetheless, members of the *osborni* and *andersoni* species groups are phenetically more similar to *Tlalocomyia* than they are to *Stegopterna*, and we believe that most specialists will prefer to retain them in *Tlalocomyia*, pending convincing evidence to the contrary.

Monophyly of the Clade *osborni* Species Group + andersoni Species Group (Fig. 9.4)

209. *Radial sector with a distinct apical fork, with branches narrowly separated by membrane*. In the discussion of character 53, we argued that an unbranched, or obscurely branched, radial sector represents the groundplan condition of the Simuliini. *Archicnephia*, the most plesiomorphic simuliine, has an unbranched radial sector, as do the majority of genera assigned to the tribe. The radial sector of certain other simuliine genera (e.g., *Cnephia*) has an indistinct apical fork with the branches closely appressed and scarcely separated, if at all, by membrane. Members of the *osborni* and *andersoni* species groups can be distinguished from those of all other simuliines in having a distinct apical fork in which the branches are separated by membrane. This fork is markedly shorter than the petiole and cannot be confused with the long fork of *Parasimulium* and the Prosimuliini. Given the unforked radial sector in closely related taxa (members of the *pachecolunai* species group, *Stegopterna*), we interpret the condition in the *osborni* and *andersoni* species groups to be an atavism.

Simulium (Boreosimulium)

Monophyly of *Boreosimulium* (Fig. 9.5)

210. *Larval antenna annulated*. Plesiomorphic: larval antenna not annulated. The derived state has evolved independently several times in the Simuliidae, including the subgenus *Hellichiella*.

Monophyly of the Clade *annulus* Species Group + *baffinense* Species Group (Fig. 9.5)

211. *Aedeagal membrane beset with series of densely packed stout spines*. Plesiomorphic: aedeagal membrane bare or with fine setae only.

Simulium (Aspathia)

Monophyly of *Aspathia* (Fig. 9.6)

212. *Ventral plate with markedly produced, sparsely haired lip*. Plesiomorphic: ventral plate not markedly produced apically, and lip normally haired.

Monophyly of the Clade *griseifrons* Species Group to *multistriatum* Species Group (Fig. 9.6)

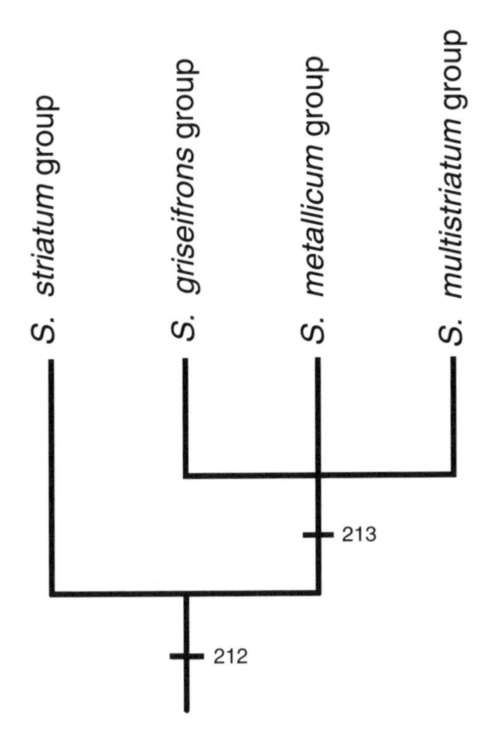


Fig. 9.6. Phylogeny of *Simulium* (*Aspathia*) species groups in the Holarctic Region.

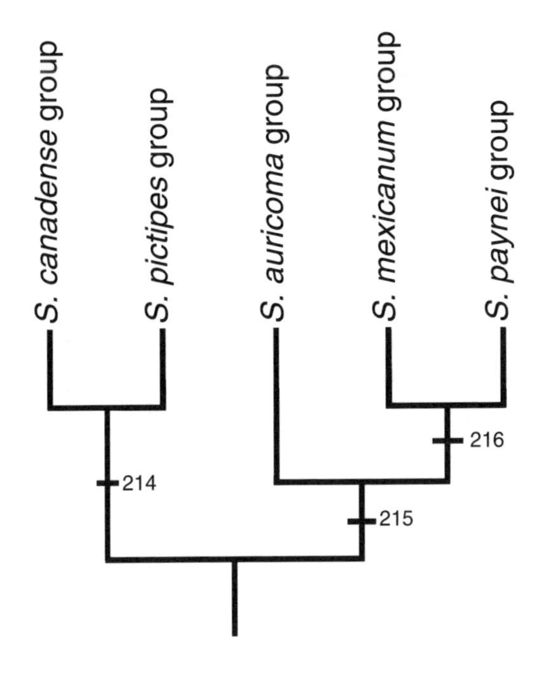


Fig. 9.7. Phylogeny of *Simulium* (*Hemicnetha*) species groups.

213. *Gonostylus with markedly elongate, pointed basal process.* Plesiomorphic: gonostylus either lacking a basal process, or process short and broadly rounded.

Simulium (Hemicnetha)

Monophyly of *Hemicnetha* (Fig. 9.7)

*Hemicnetha* is closely related to several other taxa in the genus *Simulium* (viz., *Crosskeyellum*, *Himalayum*, and the *argenteostriatum* species group of the subgenus *Simulium*). Relationships among these groups are not resolved, and one or more of these names eventually might fall into synonymy. We subsume the genus-group taxon *Obuchovia* in *Hemicnetha* based on similarities in the female terminalia (see character 215). However, we are unable to identify any convincing synapomorphies for the subgenus *Hemicnetha* as a whole.

Monophyly of the Clade *canadense* Species Group + *pictipes* Species Group (Fig. 9.7)

214. *Ventral plate notched apicomedially.* Plesiomorphic: ventral plate not notched apicomedially.

Monophyly of the Clade *auricoma* Species Group to *paynei* Species Group (Fig. 9.7)

215. *Hypogynial valves markedly elongate, typically extended to anterior margin of, or beyond, cercus.* Plesiomorphic: hypogynial valves of normal shape and length.

Monophyly of the Clade *mexicanum* Species Group + *paynei* Species Group (Fig. 9.7)

216. *Lateral plate of genital fork with ventrally directed spine.* Plesiomorphic: plate of genital fork without a ventrally directed spine.

Simulium *s. s.*

Monophyly of the Subgenus Simulium (Fig. 9.8)

217. *Gonostylus with apex broadly rounded and flattened paddle-like.* Plesiomorphic: gonostylus pointed or narrowly rounded apically. The derived state here differs from that ascribed to the clade *Boophthora* to *Psilozia*, in which the gonostylus is truncate apically.

218. *Scutal stripes absent.* Plesiomorphic: scutal stripes present. Members of the *Simulium ornatum* species group evidently have reacquired scutal stripes.

Monophyly of the Clade *bezzii* Species Group to *venustum* Species Group (Fig. 9.8)

219. *Ventral plate dentate apicolaterally.* Plesiomorphic: ventral plate not dentate apicolaterally.

Monophyly of the Clade *bezzii* Species Group to *reptans* Species Group (Fig. 9.8)

220. *Body of ventral plate laterally compressed.* Plesiomorphic: body of ventral plate not laterally compressed.

221. *Cocoon loosely woven anteriorly, with anterior rim absent or rudimentary.* Plesiomorphic: cocoon tightly woven anteriorly, with distinct anterior rim.

Monophyly of the Clade *bezzii* Species Group to *variegatum* Species Group (Fig. 9.8)

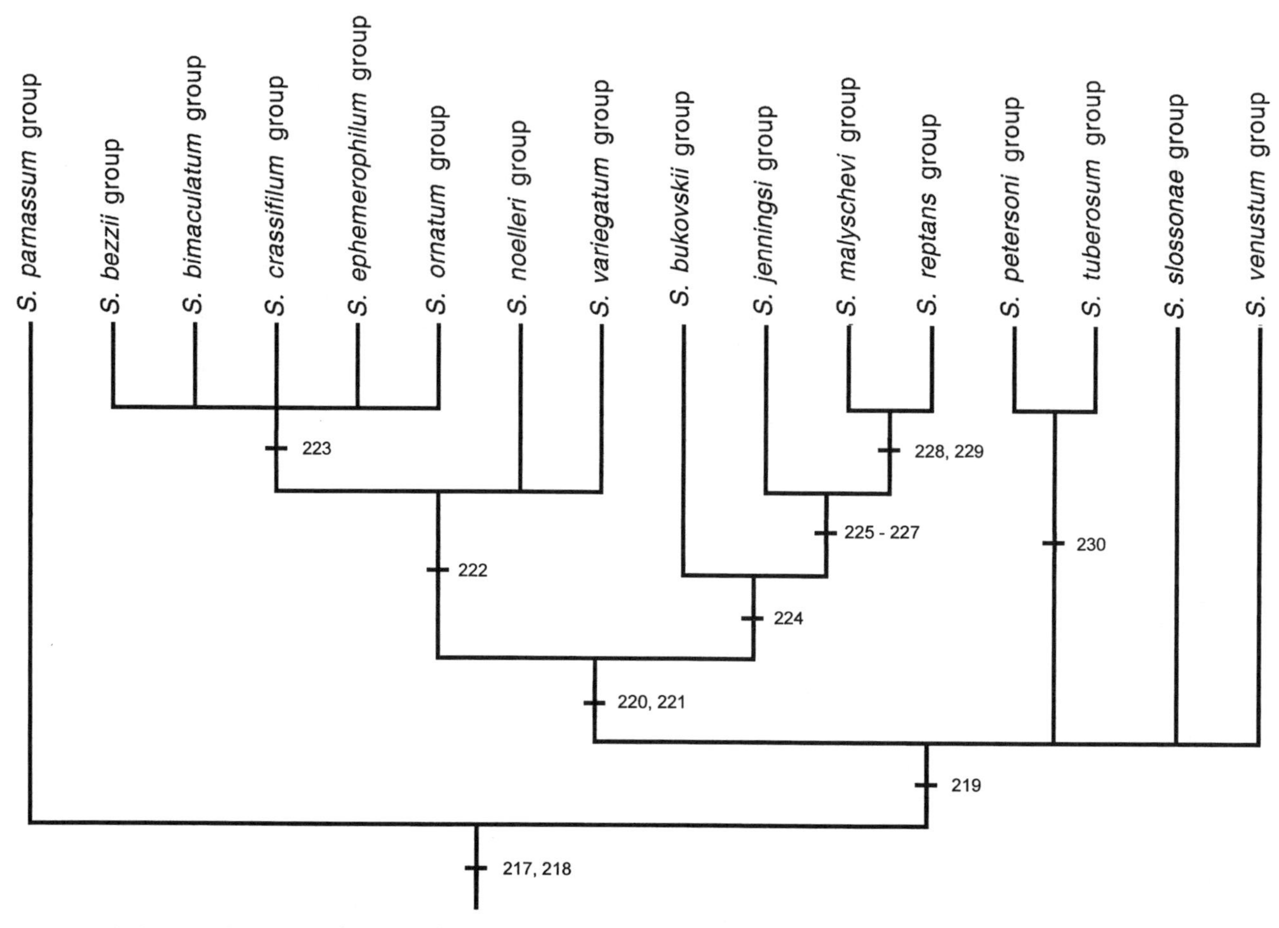


Fig. 9.8. Phylogeny of *Simulium* (*Simulium*) species groups in the Holarctic Region.

222. *Chromosomal inversion IIIL-1 present.* Plesiomorphic: IIIL arm lacking this inversion. The derived state has been confirmed only for certain members of the *Simulium bezzii, S. ornatum, S. noelleri,* and *S. variegatum* species groups (Adler & Kachvorian 2001).

Monophyly of the Clade *bezzii* Species Group to *ornatum* Species Group (Fig. 9.8)

223. *Anepisternal membrane with patch of hair.* Plesiomorphic: pleural membrane without hair.

Monophyly of the Clade *bukovskii* Species Group to *reptans* Species Group (Fig. 9.8)

224. *Cocoon shoe shaped or boot shaped.* Plesiomorphic: cocoon slipper shaped.

Monophyly of the Clade *jenningsi* Species Group to *reptans* Species Group (Fig. 9.8)

225. *Cocoon with variously sized fenestrae anteriorly.* Plesiomorphic: cocoon without fenestrae anteriorly.

226. *Chromosomal inversion present in base of IIL arm.* Plesiomorphic: chromosomal inversion absent in base of IIL arm.

227. *Chromosomal inversion present near middle of IIIL arm.* Plesiomorphic: chromosomal inversion absent near middle of IIIL arm.

Monophyly of the Clade *malyschevi* Species Group + *reptans* Species Group (Fig. 9.8)

228. *Chromosomal inversion present in middle of IIIL arm.* Plesiomorphic: chromosomal inversion absent in middle of IIIL arm.

229. *Chromosomal inversion present in base of IIIS arm (middle of section 78 to end of section 81 on standard map of Rothfels et al. [1978]).* Plesiomorphic: IIIS arm lacking this inversion.

Monophyly of the Clade *petersoni* Species Group + *tuberosum* Species Group (Fig. 9.8)

230. *Molecular sequence data.* A combined analysis of sequences from the cytochrome-*b*, cytochrome oxidase II, and elongation factor-1∝ genes provides evidence of a sister-group relationship between the *petersoni* species group and the *tuberosum* species group (Smith 2002).

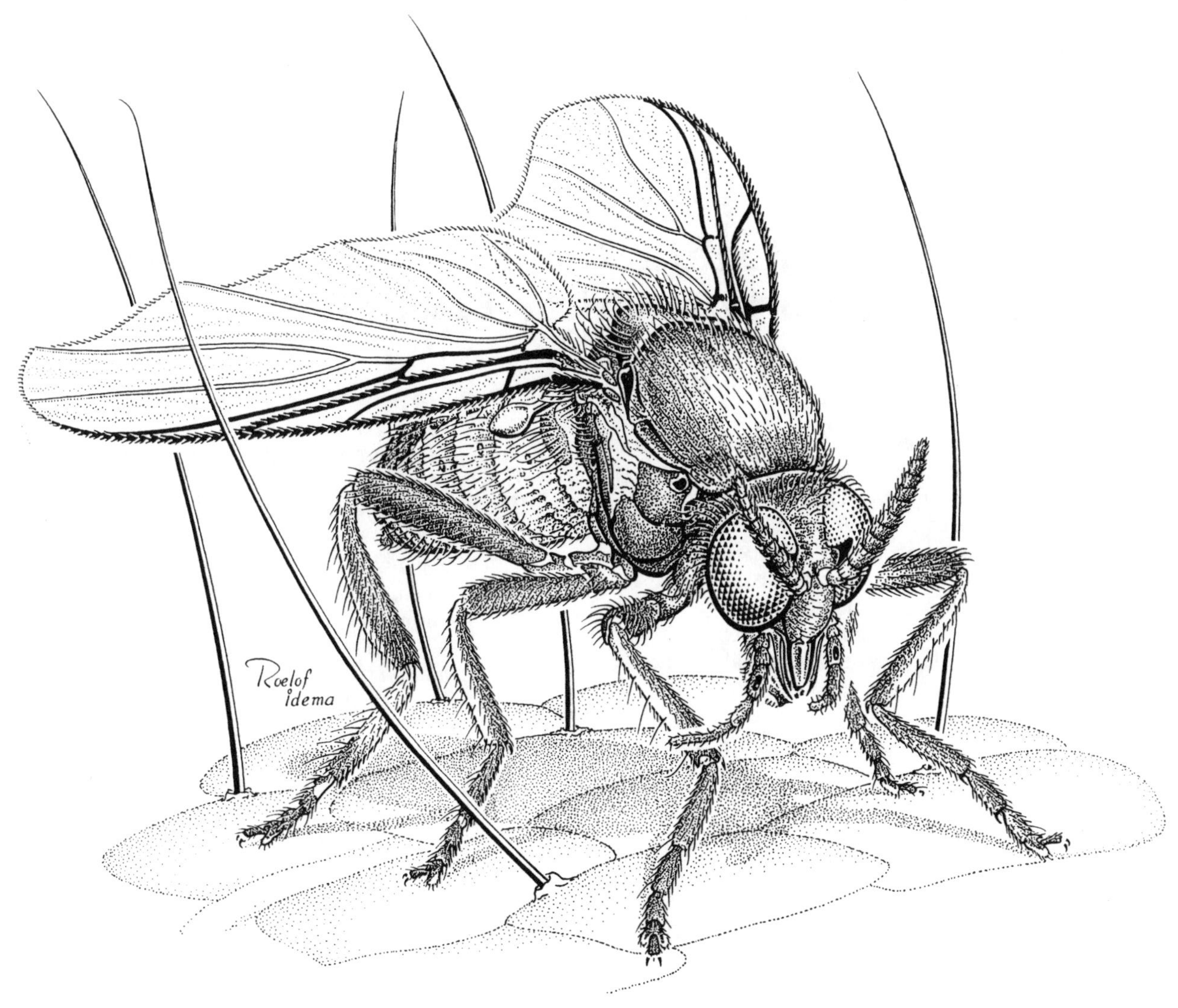


Simulium tuberosum ♀

# 10| Synoptic List, Identification Keys, and Taxonomic Accounts of North American Black Flies

## Synoptic List

The following list of extant taxa and formal synonyms is arranged phylogenetically by subfamily, tribe, genus, and subgenus. Species groups and species are arranged alphabetically within each genus or subgenus. Species complexes and their component species appear at the end of the relevant subgenus or species group. New nomenclatural acts are shown in boldface type. Vernacular names, misidentifications, erroneous spellings, and further details are provided under individual taxonomic accounts.

Family Simuliidae Newman 1834
Subfamily Parasimuliinae Smart 1945
Parasimuliinae Smart 1944a *nomen nudum*
Genus *Parasimulium* Malloch 1914
Subgenus *Astoneomyia* Peterson 1977a
*melanderi* Stone 1963b
'species A'
Subgenus *Parasimulium* Malloch 1914
*crosskeyi* Peterson 1977a
*furcatum* Malloch 1914
*stonei* Peterson 1977a
Subfamily Simuliinae Newman 1834
Tribe Prosimuliini Enderlein 1921a
Prosimuliinae Enderlein 1921a
Hellichiini Enderlein 1925
Gymnopaidinae Rubtsov 1955
Helodoini [*sic*, correctly Helodontini] Ono 1982
Kovalevimyiinae Kalugina 1991
Genus *Gymnopais* Stone 1949b
*dichopticoides* Wood 1978
*dichopticus* Stone 1949b
*fimbriatus* Wood 1978
*holopticoides* Wood 1978
*holopticus* Stone 1949b
Genus *Twinnia* Stone & Jamnback 1955
*hirticornis* Wood 1978
*nova* (Dyar & Shannon 1927)
*biclavata* Shewell 1959b
*tibblesi* Stone & Jamnback 1955
Genus *Helodon* Enderlein 1921a
Subgenus *Distosimulium* Peterson 1970b
*pleuralis* (Malloch 1914)
*tenuicalx* (Enderlein 1925)
*pancerastes* (Dyar & Shannon 1927)
Subgenus *Parahelodon* Peterson 1970b
*decemarticulatus* (Twinn 1936a)
*gibsoni* (Twinn 1936a)
*vernalis* (Shewell 1952)
Subgenus *Helodon* Enderlein 1921a
*Haimophaga* Rubtsov 1977
*Ahaimophaga* Rubtsov & Chubareva in Rubtsov 1977 (unavailable)
*Ahaimophaga* Chubareva & Rubtsov in Chubareva 1978
*alpestris* (Dorogostaisky, Rubtsov & Vlasenko 1935)
*altaicus* (Rubtsov 1956)
*relensis* (Rubtsov 1956)
*komandorensis* (Rubtsov 1971b)
*clavatus* (Peterson 1970b)
*albertensis* (Peterson & Depner 1972) **new synonym**
*irkutensis* (Rubtsov 1956)
*perspicuus* (Sommerman 1958) **new synonym**
*martini* (Peterson 1970b) **new synonym**
*onychodactylus* Species Complex
*beardi* Adler, Currie & Wood **new species**
*chaos* Adler, Currie & Wood **new species**
*diadelphus* Adler, Currie & Wood **new species**
*mccreadiei* Adler, Currie & Wood **new species**
*newmani* Adler, Currie & Wood **new species**
*onychodactylus* (Dyar & Shannon 1927)
*protus* Adler, Currie & Wood **new species**
*susanae* (Peterson 1970b)
*trochus* Adler, Currie & Wood **new species**
Genus *Prosimulium* Roubaud 1906
*Hellichia* Enderlein 1925
*Taeniopterna* Enderlein 1925

*Mallochella* Enderlein 1930 (preoccupied)
*Mallochianella* Vargas & Díaz Nájera 1948a (substitute name)
*Piezosimulium* Peterson 1989 **new synonym**
*hirtipes* Species Group
*approximatum* Peterson 1970b
*arvum* Adler & Kim 1985
*clandestinum* Adler, Currie & Wood **new species**
*daviesi* Peterson & DeFoliart 1960
*doveri* Sommerman 1962a ("1961")
*esselbaughi* Sommerman 1964
*fontanum* Syme & Davies 1958
*formosum* Shewell 1959b
*frohnei* Sommerman 1958
*fulvithorax* Shewell 1959b
*fulvum* (Coquillett 1902)
*fuscum* Syme & Davies 1958
*mysticum* Peterson 1970b **new synonym**
*idemai* Adler, Currie & Wood **new species**
*minifulvum* Adler, Currie & Wood **new species**
*mixtum* Syme & Davies 1958
*calceatum* (Say in Harris 1835) ***nomen nudum***
*rhizophorum* Stone & Jamnback 1955
*rusticum* Adler, Currie & Wood **new species**
*saltus* Stone & Jamnback 1955
*secretum* Adler, Currie & Wood **new species**
*shewelli* Peterson & DeFoliart 1960
*woodorum* Peterson 1970b **new synonym**
*opleri* Peterson & Kondratieff 1995 ("1994") **new synonym**
*transbrachium* Adler & Kim 1985
*travisi* Stone 1952
*macropyga* Species Group
*neomacropyga* Peterson 1970b
*jeanninae* (Peterson 1989) new synonym
*wui* Peterson & Kondratieff 1995 ("1994") new synonym
*ursinum* (Edwards 1935b)
*ursinum* (Edwards 1935a) *nomen nudum*
*browni* (Twinn 1936a)
*magnum* Species Group
*caudatum* Shewell 1959b
*constrictistylum* Peterson 1970b
*dicentum* Dyar & Shannon 1927
*dicum* Dyar & Shannon 1927
*exigens* Dyar & Shannon 1927
*hardyi* (Stains & Knowlton 1940)
*longilobum* Peterson & DeFoliart 1960 **new synonym**
*flaviantennus* (Stains & Knowlton 1940)
*impostor* Peterson 1970b
*longirostrum* Currie, Adler & Wood **new species**
*multidentatum* (Twinn 1936a)
*uinta* Peterson & DeFoliart 1960
*unicum* (Twinn 1938)
*magnum* Species Complex
*albionense* Rothfels 1956 **revalidated**
*canutum* Adler, Currie & Wood **new species**
*magnum* Dyar & Shannon 1927
*frisoni* (Dyar & Shannon 1927)
Tribe Simuliini Newman 1834
Nevermanniini Enderlein 1921a
Wilhelmiini Baranov 1926
Ectemniinae Enderlein 1930
Stegopterninae Enderlein 1930
Cnesiinae Enderlein 1934a
Friesiini Enderlein 1936b (unavailable)
Odagmiini Enderlein 1936b (unavailable)
Austrosimuliini Smart 1945
Cnephiini Grenier & Rageau 1960
Eusimuliini Rubtsov 1974a
Genus *Greniera* Doby & David 1959
*abdita* Species Group
*abdita* (Peterson 1962a)
*abditoides* (Wood 1963b)
*humeralis* Currie, Adler & Wood **new species**
*fabri* Species Group
*denaria* (Davies, Peterson & Wood 1962)
*longicornis* Currie, Adler & Wood **new species**
'species F'
Genus *Stegopterna* Enderlein 1930
*acra* Currie, Adler & Wood **new species**
*decafilis* Rubtsov 1971a
*diplomutata* Currie & Hunter 2003
*emergens* (Stone 1952)
*tschukotensis* Rubtsov 1971a **new synonym**
*mutata* (Malloch 1914)
*permutata* (Dyar & Shannon 1927)
*trigonium* (Lundström 1911)
*freyi* (Enderlein 1929c)
*richteri* Enderlein 1930
*majalis* Rubtsov & Carlsson 1965
*dentata* Rubtsov & Carlsson 1965
*haematophaga* Rubtsov & Carlsson 1965
*longicoxa* Rubtsov 1971a
*xantha* Currie, Adler & Wood **new species**
Genus *Tlalocomyia* Wygodzinsky & Díaz Nájera 1970
*Stewartella* Coleman 1951 (**unavailable**)
*Mayacnephia* Wygodzinsky & Coscarón 1973 **new synonym**
*andersoni* Species Group **newly recognized**

*andersoni* Currie, Adler & Wood **new species**
*osborni* Species Group **newly recognized**
*osborni* (Stains & Knowlton 1943) **new combination**
*ramifera* Currie, Adler & Wood **new species**
*stewarti* (Coleman 1953) **new combination**
Genus *Gigantodax* Enderlein 1925
*Archicnesia* Enderlein 1934a
*wrighti* Species Group
*adleri* Moulton 1996
Genus *Cnephia* Enderlein 1921a
*Astega* Enderlein 1930
*dacotensis* (Dyar & Shannon 1927)
*lascivum* (Twinn 1936a)
*eremites* Shewell 1952
*arborescens* Rubtsov 1971a **new synonym**
*ornithophilia* Davies, Peterson & Wood 1962
*pecuarum* (Riley 1887)
Genus *Ectemnia* Enderlein 1930
*invenusta* (Walker 1848)
*loisae* (Stone & Jamnback 1955)
*primaeva* Moulton & Adler 1997
*reclusa* Moulton & Adler 1997
*taeniatifrons* (Enderlein 1925)
Genus *Metacnephia* Crosskey 1969
*borealis* (Malloch 1919)
*arctocanadensis* Yankovsky 1996 (unnecessary substitute name)
*coloradensis* Peterson & Kondratieff 1995 ("1994")
*jeanae* (DeFoliart & Peterson 1960)
*macrocerca* Peterson 1958a (**unavailable**)
*saileri* (Stone 1952)
*becherii* (Rivosecchi 1964a) *nomen nudum*
*saskatchewana* (Shewell & Fredeen 1958)
*sommermanae* (Stone 1952)
*crassifistula* (Rubtsov 1956) **new synonym**
*villosa* (DeFoliart & Peterson 1960)
*freytagi* (DeFoliart & Peterson 1960) **new synonym**
Genus *Simulium* Latreille 1802
Subgenus *Hellichiella* Rivosecchi & Cardinali 1975
*Parahellichiella* Golini 1982 (**unavailable**)
*congareenarum* Species Group
*anatinum* Wood 1963b
*congareenarum* (Dyar & Shannon 1927)
*innocens* (Shewell 1952)
*minus* (Dyar & Shannon 1927)
*nebulosum* Currie & Adler 1986
*rendalense* (Golini 1975)
*usovae* (Golini 1987)
'species O'
*rivuli* Species Group
*curriei* Adler & Wood 1991
*excisum* Davies, Peterson & Wood 1962
*mysterium* Adler, Currie & Wood **new species**
*rivuli* Twinn 1936a
Subgenus *Boreosimulium* Rubtsov & Yankovsky 1982
*annulus* Species Group
*annulus* (Lundström 1911)
*euryadminiculum* Davies 1949b **new synonym**
*balteatum* Adler, Currie & Wood **new species**
*canonicolum* (Dyar & Shannon 1927)
*clarkei* Stone & Snoddy 1969
*emarginatum* Davies, Peterson & Wood 1962
*joculator* Adler, Currie & Wood **new species**
*quadratum* (Stains & Knowlton 1943) **revalidated**
*zephyrus* Adler, Currie & Wood **new species**
*baffinense* Species Group
*baffinense* Twinn 1936a
*pallens* Twinn 1936a
*johannseni* Species Group **newly recognized**
*johannseni* Hart in Forbes 1912
*duplex* Shewell & Fredeen 1958 **new synonym**
*parmatum* Adler, Currie & Wood **new species**
*rothfelsi* Adler, Brockhouse & Currie 2003
Subgenus *Byssodon* Enderlein 1925
*Psilocnetha* Enderlein 1935
*Titanopteryx* Enderlein 1935
*Echinosimulium* Baranov 1938
*Gibbinsiellum* Rubtsov 1962f
*meridionale* Species Group
*maculatum* (Meigen 1804)
*pungens* (Meigen in Panzer [1806]) (unjustified substitute name)
*subfasciatum* Meigen 1838
*vigintiquaterni* (Enderlein 1929a)
*echinatum* (Baranov 1938)
*ussurianum* Rubtsov 1940b
*koidzumii* (Takahasi 1940)
*danubense* (Rubtsov 1956)
*uralense* (Rubtsov 1956)
*lenae* (Rubtsov 1956)
*gutsevitshi* (Yankovsky 1978)
*meridionale* Riley 1887
*occidentale* Townsend 1891
*tamaulipense* Townsend 1897

*forbesi* Malloch 1914
Subgenus *Eusimulium* Roubaud 1906
(= *S. aureum* Species Group or Species Complex)
*bracteatum* Coquillett 1898
*donovani* Vargas 1943b
*diazi* De León 1945 ("1944")
*exulatum* Adler, Currie & Wood **new species**
*pilosum* (Knowlton & Rowe 1934a)
*utahense* (Knowlton & Rowe 1934a)
*violator* Adler, Currie & Wood **new species**
Subgenus *Nevermannia* Enderlein 1921a
*Cnetha* Enderlein 1921a
*Stilboplax* Enderlein 1921a
*Pseudonevermannia* Baranov 1926
*Cryptectemnia* Enderlein 1936b
*Chelocnetha* Enderlein 1936b
*Dexomyia* Crosskey 1969
*vernum* Species Group
*aestivum* Davies, Peterson & Wood 1962
*bicorne* Dorogostaisky, Rubtsov & Vlasenko 1935
*paracorniferum* (Yankovsky 1979) **new synonym**
*corniferum* (Yankovsky 1979) **new synonym**
*burgeri* Adler, Currie & Wood **new species**
*carbunculum* Adler, Currie & Wood **new species**
*attenuatum* Peterson 1958a (**unavailable**)
*conicum* Adler, Currie & Wood **new species**
*craigi* Adler & Currie 1986
*croxtoni* Nicholson & Mickel 1950
*dendrofilum* (Patrusheva 1962)
*fionae* Adler 1990
*fontinale* Radzivilovskaya 1948
*decolletum* Adler & Currie 1986 **new synonym**
*gouldingi* Stone 1952
*impar* Davies, Peterson & Wood 1962
*loerchae* Adler 1987
*merritti* Adler, Currie & Wood **new species**
*modicum* Adler, Currie & Wood **new species**
*moultoni* Adler, Currie & Wood **new species**
*pugetense* (Dyar & Shannon 1927)
*quebecense* Twinn 1936a
*silvestre* (Rubtsov 1956)
*caledonense* Adler & Currie 1986 **new synonym**
*wyomingense* Stone & DeFoliart 1959
*alpinum* (Peterson 1958a) (**unavailable**)
Subgenus *Schoenbaueria* Enderlein 1921a
*Miodasia* Enderlein 1936a
*Gallipodus* Usova & Reva 2000
*furculatum* (Shewell 1952)
*giganteum* Rubtsov 1940b
*subpusillum* Rubtsov 1940b
'species Z'
Subgenus *Psilopelmia* Enderlein 1934a
*Lanea* Vargas, Martínez Palacios & Díaz Nájera 1946
*escomeli* Species Group
*bivittatum* Malloch 1914
*idahoense* Twinn 1938
*clarum* (Dyar & Shannon 1927)
*griseum* Coquillett 1898
*labellei* Peterson 1993
*longithallum* Díaz Nájera & Vulcano 1962 ("1961")
*mediovittatum* Knab 1915b
*meyerae* Moulton & Adler 2002a
*notatum* Adams 1904
*robynae* Peterson 1993
*trivittatum* Malloch 1914
*distinctum* Malloch 1913 (preoccupied)
*mazzottii* Díaz Nájera 1981 ("1979")
*venator* Dyar & Shannon 1927
*beameri* Stains & Knowlton 1943
Subgenus *Psilozia* Enderlein 1936b
*Neosimulium* Rubtsov 1940b *nomen nudum*
*Neosimulium* Vargas, Martínez Palacios & Díaz Nájera 1946
*argus* Williston 1893
*obtusum* (Dyar & Shannon 1927)
*kamloopsi* Hearle 1932
*hearlei* Twinn 1938
*encisoi* Vargas & Díaz Nájera 1949
*infernale* Adler, Currie & Wood **new species**
*vittatum* Species Complex
*tribulatum* Lugger 1897a ("1896") **revalidated**
*glaucum* Coquillett 1902
*venustoides* Hart in Forbes 1912
*vittatum* Zetterstedt 1838
*vittata* Zetterstedt 1837 *nomen nudum*
*nasale* Gistel 1848 *nomen nudum*
*groenlandicum* (Enderlein 1936b) (preoccupied)
*asakakae* Smart 1944b (substitute name)
Subgenus *Aspathia* Enderlein 1935
*Striatosimulium* Rubtsov & Yankovsky 1982
*Jalacingomyia* Py-Daniel in Py-Daniel & Moreira Sampaio 1994
*metallicum* Species Group
*anduzei* Vargas & Díaz Nájera 1948b

*patziciaense* Takaoka & Takahasi 1982 **new synonym**
*hechti* Vargas, Martínez Palacios & Díaz Nájera 1946
*hunteri* Malloch 1914
*lassmanni* Vargas, Martínez Palacios & Díaz Nájera 1946
*iriartei* Vargas, Martínez Palacios & Díaz Nájera 1946
*jacumbae* Dyar & Shannon 1927
*guatemalense* De León 1945 ("1944")
*piperi* Dyar & Shannon 1927
*sayi* Dyar & Shannon 1927
*knowltoni* Twinn 1938
*stonei* Stains & Knowlton 1943
*puigi* Vargas, Martínez Palacios & Díaz Nájera 1946
*tescorum* Stone & Boreham 1965
Subgenus *Hemicnetha* Enderlein 1934b
*Hearlea* Rubtsov 1940b *nomen nudum*
*Dyarella* Vargas, Martínez Palacios & Díaz Nájera 1946
*Hearlea* Vargas, Martínez Palacios & Díaz Nájera 1946 **new synonym**
*Obuchovia* Rubtsov 1947 **new synonym**
*Hagenomyia* Shewell 1959a (preoccupied) **new synonym**
*Shewellomyia* Peterson 1975 (substitute name) **new synonym**
*canadense* Species Group **newly recognized**
*canadense* Hearle 1932
*fraternum* Twinn 1938
*mexicanum* Species Group
*freemani* Vargas & Díaz Nájera 1949
*paynei* Species Group
*bricenoi* Vargas, Martínez Palacios & Díaz Nájera 1946
*wirthi* Peterson & Craig 1997 ("1996") **new synonym**
*hippovorum* Malloch 1914 **revalidated**
*paynei* Vargas 1942 (substitute name)
*mexicanum* (Enderlein 1934b) (preoccupied)
*mathesoni* Vargas 1943b
*bilimekae* Smart 1944b (unjustified substitute name)
*acatenangoense* Dalmat 1951
*conviti* Ramírez-Pérez & Vulcano 1973
*solarii* Stone 1948
*virgatum* Coquillett 1902
*cinereum* Bellardi 1859 (preoccupied)
*tephrodes* Speiser 1904 (substitute name)
*rubicundulum* Knab 1915a ("1914")
*chiapanense* Hoffman 1930b
*pictipes* Species Group **newly recognized**
*claricentrum* Adler 1990
*innoxium* Comstock & Comstock 1895 **revalidated**
*innoxium* Williston in Phillips 1890 (**unavailable**)
*aldrichianum* (Enderlein 1936b)
*pictipes* Hagen 1880
*longistylatum* Shewell 1959a **new synonym**
Subgenus *Simulium* Latreille 1802
*Odagmia* Enderlein 1921a
*Friesia* Enderlein 1922
*Discosphyria* Enderlein 1922
*Gynonychodon* Enderlein 1925
*Pseudodagmia* Baranov 1926
*Danubiosimulium* Baranov 1935
*Cleitosimulium* Séguy & Dorier 1936
*Gnus* Rubtsov 1940b
*Tetisimulium* Rubtsov 1960b (unavailable substitute name)
*Tetisimulium* Rubtsov 1963a
*Parabyssodon* Rubtsov 1964c
*Phosterodoros* Stone & Snoddy 1969
*Phoretodagmia* Rubtsov 1972a
*Paragnus* Rubtsov & Yankovsky 1982
*Archesimulium* Rubtsov & Yankovsky 1982
*Argentisimulium* Rubtsov & Yankovsky 1982
*jenningsi* Species Group
*anchistinum* Moulton & Adler 1995
*aranti* Stone & Snoddy 1969
*chlorum* Moulton & Adler 1995
*chlorosum* Moulton 1992 (unavailable)
*confusum* Moulton & Adler 1995
*definitum* Moulton & Adler 1995
*dixiense* Stone & Snoddy 1969
*fibrinflatum* Twinn 1936b
*underhilli* Stone & Snoddy 1969
*octobranchium* Moulton 1992 (unavailable)
*haysi* Stone & Snoddy 1969
*infenestrum* Moulton & Adler 1995
*jenningsi* Malloch 1914
*nigroparvum* Twinn 1936b
*jonesi* Stone & Snoddy 1969
*krebsorum* Moulton & Adler 1992
*lakei* Snoddy 1976
*luggeri* Nicholson & Mickel 1950
*notiale* Stone & Snoddy 1969
*nyssa* Stone & Snoddy 1969
*ozarkense* Moulton & Adler 1995
*penobscotense* Snoddy & Bauer 1978
*podostemi* Snoddy 1971
*remissum* Moulton & Adler 1995
*snowi* Stone & Snoddy 1969
*taxodium* Snoddy & Beshear 1968
*malyschevi* Species Group
*decimatum* Dorogostaisky, Rubtsov & Vlasenko 1935
*simile* Malloch 1919 (preoccupied)
*wagneri* Rubtsov 1940b

*nigricoxum* Stone 1952 (substitute name) **new synonym**
*xerophilum* (Rubtsov 1969)
*defoliarti* Stone & Peterson 1958
*curtisi* Riegert 1999b ***nomen nudum***
*malyschevi* Dorogostaisky, Rubtsov & Vlasenko 1935
*albipes* (Rubtsov 1956)
*lucidum* (Rubtsov 1956)
*murmanum* Enderlein 1935
*corbis* Twinn 1936b
*relictum* Rubtsov 1940b
*forsi* (Carlsson 1962)
*arcticum* Species Complex
*apricarium* Adler, Currie & Wood **new species**
*arcticum* Malloch 1914
*arcticum* Malloch 1914 'cytospecies IIL-1'
*arcticum* Malloch 1914 'cytospecies IIS-4'
*brevicercum* Knowlton & Rowe 1934a **revalidated**
*nigresceum* Knowlton & Rowe 1934a **new synonym**
*chromatinum* Adler, Currie & Wood **new species**
*negativum* Adler, Currie & Wood **new species**
*saxosum* Adler, Currie & Wood **new species**
*vampirum* Adler, Currie & Wood **new substitute name**
*simile* Cameron 1918 (preoccupied)
*noelleri* Species Group
*decorum* Walker 1848
*katmai* Dyar & Shannon 1927
*ottawaense* Twinn 1936b
*noelleri* Friederichs 1920
*subornatum* Edwards 1920
*tenuimanus* Enderlein 1921a *nomen nudum*
*tenuimanus* Enderlein 1921b
*septentrionale* Enderlein 1935
*lindneri* (Enderlein in Lindner 1943)
*avidum* Rubtsov 1963b (unavailable)
*bonomii* Rubtsov 1964b (unavailable)
*parnassum* Species Group
*parnassum* Malloch 1914
*rileyanum* (Enderlein 1922) **new synonym**
*hydationis* Dyar & Shannon 1927
*petersoni* Species Group **newly recognized**
*petersoni* Stone & DeFoliart 1959
*slossonae* Species Group
*rugglesi* Nicholson & Mickel 1950
*slossonae* Dyar & Shannon 1927
*transiens* Rubtsov 1940b
*tuberosum* Species Group
*tuberosum* Species Complex
*appalachiense* Adler, Currie & Wood **new species**
*chromocentrum* Adler, Currie & Wood **new species**
*conundrum* Adler, Currie & Wood **new species**
*perissum* Dyar & Shannon 1927 **revalidated**
*tuberosum* (Lundström 1911)
*twinni* Stains & Knowlton 1940 **revalidated**
*ubiquitum* Adler, Currie & Wood **new species**
*vandalicum* Dyar & Shannon 1927 **revalidated**
*turmale* Twinn 1938 **new synonym**
*vulgare* Dorogostaisky, Rubtsov & Vlasenko 1935
*venustum* Species Group
*rubtzovi* Smart 1945 (substitute name)
*simile* Rubtsov 1940a ("1939") (preoccupied)
*venustum* Species Complex
*hematophilum* Laboulbène 1882 **revalidated**
*incognitum* Adler & Mason 1997
*irritatum* Lugger 1897a ("1896") **revalidated**
*minutum* Lugger 1897a ("1896") **revalidated**
*molestum* Harris 1841 **revalidated**
*piscicidium* Riley 1870d **revalidated**
*tormentor* Adler, Currie & Wood **new species**
*truncatum* (Lundström 1911)
*venustum* Say 1823
*venustum* Say 1823 'cytospecies JJ'
*verecundum* Species Complex
*rostratum* (Lundström 1911)
*wilhelmii* Enderlein 1922
*groenlandicum* Enderlein 1935
*sublacustre* Davies 1966
*verecundum* Stone & Jamnback 1955

## Identification Keys

In the following keys, we have emphasized ease of use rather than phylogenetic arrangement, although the two are not mutually exclusive. All structural terms can be interpreted by reference to the labeled illustrations in Figs. 4.1–4.51. To aid the probability of correct identification, we have included some variable species more than once in the keys. Some species have defied our efforts to separate them and, therefore, are lumped together in a couplet. We have tended to be conservative in the separation of species in the keys. Certain species that can be distinguished in one geographic area but not in another are generally not separated in the keys. Identification sometimes rests on subtle characters and combinations of characters (gestalt) that are not easily articulated. A good reference collection, experience, attention to subtle details in the illustrations, and bionomic information (e.g., geographic distribution, habitat, and seasonality) often permit identification, even of species that we have grouped together in the keys. For larvae, we have used chromosomal characters when morphological characters are weak or unavailable. Space does not allow us to provide chromosome maps for all species; however, references to the relevant literature are given in the key (and under individual species accounts) for those who wish to use chromosomal characters. Despite our efforts, black flies remain challenging to identify.

### KEY TO ADULTS OF NORTH AMERICAN BLACK FLIES

Terminalia typically require slide mounting in glycerin, with subsequent viewing under a compound microscope; other structures require slide mounting only if indicated parenthetically.

#### Genera

1. Wing vein $R_1$ joined to costa near middle of wing (Fig. 4.17). Radial sector with posterior branch ($R_{4+5}$) terminated markedly before apex of costa. False vein (m-cu fold) unforked apically. Katepisternum pointed ventrally, not divided horizontally by katepisternal sulcus (Fig. 4.9). Pacific Northwest (subfamily Parasimuliinae)........***Parasimulium*** (p. 170)
– Wing vein $R_1$ joined to costa beyond middle of wing (Fig. 4.18). Radial sector with posterior branch ($R_{4+5}$) terminated near apex of costa. False vein (m-cu fold) forked apically. Katepisternum broadly rounded ventrally, divided horizontally by variously developed katepisternal sulcus (Figs. 4.10–4.12). Widespread (subfamily Simuliinae)........2

2. Radial sector with fork longer than its stem (Figs. 4.18, 10.77). Costa with hairlike setae only. Katepisternal sulcus wide, shallow (Figs. 4.10, 4.11). Basitarsus of hind leg without calcipala and pedisulcus (Fig. 10.34) (tribe Prosimuliini)........3
– Radial sector unforked, or with fork shorter than its stem (Figs. 10.78–10.81). Costa with spiniform setae interspersed among hairlike setae (Fig. 10.79) (spiniform setae absent from costa in certain species of *Greniera*). Katepisternal sulcus narrow, deep (Figs. 4.12, 4.13). Basitarsus of hind leg with or without calcipala and pedisulcus (Figs. 4.21, 10.36–10.43) (tribe Simuliini)........8

3. Antenna with 7 flagellomeres. Eye with shiny, dark, raised tubercle (stemmatic bulla) just behind posterior margin (Figs. 10.3–10.7). Claws of female toothless (Figs. 10.46, 10.47)........4
– Antenna with 8 or more flagellomeres (Fig. 10.14), or if with 7 flagellomeres, then posterior margin of eye without stemmatic bulla, and claws of female each with small subbasal tooth. Claws of female toothless (Figs. 10.52–10.56), or each with variously sized basal or subbasal tooth or lobe (Figs. 10.48–10.51)........5

4. Vestiture of head and body consisting of short, sparse erect hairs (Fig. 10.3). Wing brown or gray, opaque, slightly wrinkled. Clypeus bare except for few erect hairs laterally (Fig. 10.6). Gonostylus of male with 2 or more minute apical spinules (not readily visible under dissecting microscope) (Figs. 10.231–10.233). Cercus and anal lobe of female fused into single sclerite (Fig. 10.90)........***Gymnopais*** (p. 171)
– Vestiture of head and body consisting of long, dense recumbent hairs (Fig. 10.11). Wing pale brown, hyaline, not especially wrinkled. Clypeus entirely covered with hair (Fig. 10.12). Gonostylus of male with 1, rarely 2, large apical spinules (clearly visible under dissecting microscope) (Fig. 10.234). Cercus and anal lobe of female separated by membrane (Fig. 10.93)........***Twinnia*** (p. 171)

5. Male........6
– Female........7

6. Ventral plate in lateral view flattened, with lip absent, short, or slender (Figs. 10.237–10.244)........***Helodon*** (p. 172)

– Ventral plate in lateral view not markedly flattened, typically with prominent lip (Figs. 10.245–10.277) ........ ***Prosimulium*** (p. 174)

7. Claws (slide mounted if necessary) each with variously sized but distinct basal or subbasal tooth or lobe (Figs. 10.48–10.51). Hypogynial valve short, broadly rounded or truncated posteriorly, not extended to anal lobe (giving abdomen somewhat rounded appearance posteriorly). Spermatheca with small or no unpigmented area at junction with spermathecal duct (Figs. 10.95–10.102), or entirely unpigmented (Fig. 10.94) ........ ***Helodon*** (p. 172)

– Claws toothless, or each with minute subbasal tooth (Figs. 10.52–10.56). Hypogynial valve long, produced posteriorly, extended beyond anal lobe (giving abdomen pointed appearance posteriorly). Spermatheca with large unpigmented area at junction with spermathecal duct (Figs. 10.103-10.123) ........ ***Prosimulium*** (p. 173)

8. Wing vein $CuA_2$ straight. Mountains of Arizona and New Mexico (Map 86) ........ ***Gigantodax*** (***G. adleri***, p. 294)

– Wing vein $CuA_2$ sinuous (Fig. 4.19). Widespread ........ 9

9. Basitarsus of hind leg with pedisulcus deep and conspicuous (Figs. 4.21, 4.23); calcipala well developed (small in subgenus *Psilozia*, Fig. 10.43). Wing without basal medial cell (Fig. 10.81); radius with or without setae dorsobasally ........ ***Simulium*** (in part) (p. 180)

– Basitarsus of hind leg with pedisulcus absent or represented by shallow depression or wrinkles; calcipala poorly to well developed (Figs. 10.35–10.40). Wing with basal medial cell, although often minute (Fig. 4.19); radius with setae dorsobasally (Fig. 10.80) ........ 10

10. Costa with only pale setae, some of which may be short and stiff but neither dark nor fully spiniform ........ ***Greniera*** (in part) (p. 176)

– Costa with stout, black spiniform setae interspersed among longer, paler hairlike setae (the former more prevalent near apex of costa) (Figs. 10.78, 10.79) ........ 11

11. Radial sector bifurcated at apex, with branches separated by membrane (Fig. 10.79). Subcosta of male bare ventrally, or bearing at most 1 or 2 short setae. Rocky Mountains westward ........ ***Tlalocomyia*** (p. 177)

– Radial sector unbranched, or if bifurcated, then branches closely approximated and not separated by membrane (Figs. 10.78, 10.80). Subcosta of male setose ventrally (bare in males of certain specimens of *Ectemnia*). Widespread ........ 12

12. Male ........ 13

– Female ........ 18

13. $R_1$ dorsally with hairlike setae and scattered, black spiniform setae on distal two thirds or more; spiniform setae near apex more numerous than, and as stout as, those on costa ........ 14

– $R_1$ dorsally with hairlike setae only, or if spiniform setae present, these confined to apical half or less and not as stout as those on costa ........ 16

14. Gonostylus with 2 or 3 apical spinules. Ventral plate in ventral view about 2.5–3.0 times as broad as long (Figs. 10.295–10.297) ........ ***Ectemnia*** (p. 179)

– Gonostylus with 1 apical spinule. Ventral plate in ventral view less than 2 times as broad as long (Figs. 10.298–10.305) ........ 15

15. Anepisternal membrane typically with pale hair dorsally (as in Fig. 4.12) (absent in *M. saskatchewana*). Basitarsus of hind leg with pedisulcus represented by, at most, shallow wrinkles ........ ***Metacnephia*** (p. 179)

– Anepisternal membrane bare. Basitarsus of hind leg with pedisulcus represented by shallow depression (Fig. 10.41) ........ ***Simulium*** (in part) (subgenus *Hellichiella*) (p. 186)

16. Basitarsus of hind leg with calcipala typically prominent, lamellate, rounded apically; in medial view overlapping base of second tarsomere (as in Figs. 10.36–39); if calcipala rather small, then hind tibial spurs with hyaline apices and longer than width of tibia at point of attachment (as in Fig. 10.44) ........ ***Stegopterna*** (p. 177)

– Basitarsus of hind leg with calcipala either small and rather pointed apically, or apparently absent; in medial view typically not overlapping base of second tarsomere. Hind tibial spurs uniformly dark, subequal in length to, or shorter than, width of tibia (as in Fig. 10.45) ........ 17

17. Parameral spines present (Fig. 10.291). Widespread east of Rocky Mountains .................. ***Cnephia*** (p. 178)
   – Parameral spines absent (Fig. 10.279). Mountains from southern British Columbia to California (Map 70) .................. ***Greniera*** (in part) (***G. humeralis*** n. sp., p. 281)

18. $R_1$ dorsally with hairlike setae and scattered, black spiniform setae on distal half. Wing hyaline, with basal medial cell minute or absent .................. 19
   – $R_1$ dorsally with hairlike setae only, or if with few, spiniform setae, then wing rather smokey, and with basal medial cell distinct .................. 20

19. Anepisternal membrane typically with pale hair dorsally (Fig. 4.12) (absent in *M. saskatchewana*). Basitarsus of hind leg with pedisulcus represented by, at most, shallow wrinkles .................. ***Metacnephia*** (p. 179)
   – Anepisternal membrane bare. Basitarsus of hind leg with pedisulcus represented by shallow depression (as in Fig. 10.41) .................. ***Simulium*** (in part) (subgenus *Hellichiella*) (p. 180)

20. Claws toothless. Basitarsus of hind leg with calcipala typically prominent, lamellate, rounded apically; in medial view overlapping base of second tarsomere (Figs. 10.36–10.39); if calcipala rather small, then hind tibial spurs with hyaline apices and longer than width of tibia at point of attachment (Fig. 10.44) .................. ***Stegopterna*** (p. 177)
   – Claws each with small subbasal tooth or large basal thumblike lobe (Figs. 10.57, 10.60, 10.61). Basitarsus of hind leg with calcipala either small and rather pointed apically or apparently absent; in medial view not overlapping base of second tarsomere (Fig. 10.40); if calcipala slightly lamellate, then hind tibial spurs uniformly dark, subequal in length to or shorter than width of tibia (Fig. 10.45) .................. 21

21. Spermatheca elongate, with large unpigmented area at junction with spermathecal duct (Figs. 10.139, 10.140). Anteroventral margin of cibarium with 1–11 setae (Fig. 4.8) .................. ***Ectemnia*** (p. 178)
   – Spermatheca spherical or reniform, with small or no unpigmented area at junction with spermathecal duct (Figs. 10.125, 10.135–10.138). Anteroventral margin of cibarium without setae .................. 22

22. Spermatheca spherical, wrinkled, large (Figs. 10.135–10.138). Widespread east of Rocky Mountains .................. ***Cnephia*** (p. 178)
   – Spermatheca reniform, smooth, small (Fig. 10.125). Mountains from southern British Columbia to California (Map 70) .................. ***Greniera*** (in part) (***G. humeralis*** n. sp., p. 281)

## Species

### Females of Genus *Parasimulium*

The female of *Parasimulium* 'species A' is unknown.

1. Anal lobe and cercus in lateral view rounded (Fig. 10.82). Genital fork with arms broadly connected to tergite IX. Spermatheca with pigment extended slightly onto spermathecal duct .................. ***melanderi*** (p. 236)
   – Anal lobe and cercus in lateral view bilobed (Figs. 10.83–10.85). Genital fork with arms narrowly connected to tergite IX. Spermatheca with small unpigmented area at junction with spermathecal duct .................. 2

2. Scutum and postnotum dark brown, concolorous with head. Abdominal tergites I and II darkly pigmented .................. ***furcatum*** (p. 237)
   – Scutum and postnotum orange, contrasting with brown head. Abdominal tergites I and II unpigmented .................. 3

3. Abdominal tergite III mostly yellow, with brown spots laterally; tergites IV–VI mostly brown, with distinct yellow area dorsomedially. Anal lobe in lateral view nearly straight (Fig. 10.83) .................. ***crosskeyi*** (p. 237)
   – Abdominal tergites III–VI almost uniformly brown (some specimens with slightly paler area dorsomedially on tergites III and IV). Anal lobe in lateral view curved ventrally (Fig. 10.85) .................. ***stonei*** (p. 237)

### Males of Genus *Parasimulium*

1. Eyes touching dorsomedially, with transverse line of separation between large upper and small lower facets. Ventral plate in ventral view narrow or somewhat V shaped, with bulbous liplike projection (Figs. 10.226, 10.227) (subgenus *Astoneomyia*) .................. 2

– Eyes widely separated dorsally, without transverse line of separation between upper and lower facets, nearly all of which are of similar size (Fig. 10.2). Ventral plate in ventral view broad, somewhat H shaped (Figs. 10.228–10.230) (subgenus *Parasimulium*) ........................................3

2. Ventral plate in ventral view narrow, without basal arms (Fig. 10.226) ........................***melanderi*** (p. 236)
– Ventral plate in ventral view somewhat V shaped, with basal arms (Fig. 10.227) .......**'species A'** (p. 236)

3. Abdominal tergites I and II darkly pigmented. Gonostylus in ventral view with inner apical margin upturned (Fig. 10.229) ........................................***furcatum*** (p. 237)
– Abdominal tergites I and II unpigmented. Gonostylus in ventral view rounded or pointed apically, without inner apical margin upturned (Figs. 10.228, 10.230) ........................................4

4. Antenna with scape yellow, contrasting with color of frons. Gonostylus in ventral view abruptly enlarged at about one half to two thirds its length (Figs. 4.26, 10.228) ........................***crosskeyi*** (p. 237)
– Antenna with scape shiny brown, similar to color of frons. Gonostylus in ventral view slender, sinuous (Fig. 10.230) ........................................***stonei*** (p. 237)

### Females of Genus *Gymnopais*

1. Palpomeres IV and V not distinctly separated, usually appearing fused (Fig. 10.9). Genital fork with arms curved toward one another medially (Fig. 10.88). Anepisternal membrane bare ........................................***fimbriatus*** (p. 240)
– Palpomeres IV and V distinctly separated (Fig. 10.3). Genital fork with arms curved toward lateral extremities of sternite IX (Fig. 10.87). Anepisternal membrane typically with hair (Fig. 4.10) ............2

2. Fore femur longer than 2.5 times height of eye. Spermathecal diameter less than diameter of anal lobe (Figs. 10.86, 10.87) ........................................3
– Fore femur shorter than 2.5 times height of eye. Spermathecal diameter equal to or greater than diameter of anal lobe (Figs. 10.89, 10.90) ........................................4

3. Tergite IX in lateral view with lateral extension narrow, rounded apically, smaller than anal lobe ........................................***dichopticoides*** (p. 239)
– Tergite IX in lateral view with lateral extension broad, parallel sided, nearly truncated, and as wide laterally as anal lobe ........................................***dichopticus*** (p. 240)

4. Eye with about 23–25 horizontal rows of facets, and in profile not projected beyond frons. Spermatheca with pigmented portion of duct shorter than one third diameter of spermatheca (Fig. 10.89) ........................................***holopticoides*** (p. 240)
– Eye with about 30 horizontal rows of facets, and in profile projected anterior to and slightly above frons. Spermatheca with pigmented portion of duct longer than one third diameter of spermatheca (Fig. 10.90) ........................................***holopticus*** (p. 241)

### Males of Genus *Gymnopais*

Males of the parthenogenetic *Gymnopais dichopticoides* and *G. holopticoides* do not exist.

1. Eyes meeting medially above antennae (head holoptic); upper facets enlarged (Fig. 10.8) ........................................***holopticus*** (p. 241)
– Eyes not meeting above antennae, separated by frons (head dichoptic); all facets subequal in size (Figs. 10.5, 10.10) ........................................2

2. Ventral plate in ventral view with body wider than long, straight or concave posteriorly (Fig. 10.231) ........................................***dichopticus*** (p. 240)
– Ventral plate in ventral view with body longer than wide, convex posteriorly (Fig. 10.232) ........................................***fimbriatus*** (p. 240)

### Females of Genus *Twinnia*

1. Lacinia (slide mounted) without retrorse teeth; mandible (slide mounted) without serrations. Northeastern North America (Map 13) ........................................***tibblesi*** (p. 242)
– Lacinia (slide mounted) with retrorse teeth; mandible (slide mounted) with serrations. Rocky Mountains westward (Maps 11, 12) ........................................2

2. Genital fork with each lateral plate bearing prominent posteromedial angle; lateral plate with posterior margin evenly concave and with 5 or more setae (Fig. 10.91) ................................***hirticornis*** (p. 242)
– Genital fork with each lateral plate lacking distinct posteromedial angle; lateral plate typically with irregular jagged posterior margin and fewer than 5 setae (Fig. 10.92) ..................................***nova*** (p. 242)

### Males of Genus *Twinnia*

1. Clypeus sparsely haired. Northeastern North America (Map 13) .......... ***tibblesi*** (p. 242)
– Clypeus densely haired. Rocky Mountains westward (Maps 11, 12) .......... 2
2. Scutum with brown hair. Antennal scape with dense hair extended beyond first flagellomere (Fig. 10.20). Ventral plate in ventral view with body subrectangular, and lateral margins rather parallel (Fig. 10.234) .......... ***hirticornis*** (p. 242)
– Scutum typically with pale golden hair and some brown hairs. Antennal scape with sparse hair seldom extended beyond first flagellomere (Fig. 10.21). Ventral plate in ventral view narrower, with lateral margins bowed inward (Fig. 10.235) .......... ***nova*** (p. 242)

### Females of Genus *Helodon*

1. Spermatheca unpigmented, difficult to see (Fig. 10.94). Anepisternum with small patch of hair on either side of anepisternal membrane (in addition to mesepimeral tuft) (subgenus *Distosimulium*) .......... ***pleuralis*** (p. 243)
– Spermatheca pigmented (brown), conspicuous (Fig. 10.95). Anepisternum without patch of hair on either side of anepisternal membrane .......... 2
2. Genital fork with each lateral plate typically bearing medial, wrinkled or denticulate process (Figs. 10.95–10.97) (subgenus *Parahelodon*) .......... 3
– Genital fork with each lateral plate not bearing medial, wrinkled or denticulate process (Figs. 10.98–10.102) (subgenus *Helodon*) .......... 5
3. Antenna with 9 flagellomeres. Stem vein with dark hair .......... ***vernalis*** (p. 245)
– Antenna with 7 or 8 flagellomeres. Stem vein with pale hair .......... 4
4. Antenna with 8 flagellomeres. Antennae and legs yellowish orange, contrasting with brown body. Lacinia (slide mounted) with retrorse teeth .......... ***decemarticulatus*** (p. 244)
– Antenna with 7 flagellomeres. Antennae and legs brown, concolorous with body. Lacinia (slide mounted) with fine hair only .......... ***gibsoni*** (p. 244)
5. Frons narrow, at narrowest part not more than one third as wide as long. Lacinia (slide mounted) with retrorse teeth. Thoracic integument, especially of katepisternum, typically brownish orange (*onychodactylus* species complex, p. 248) .......... ***onychodactylus*** s. s. (p. 251), ***susanae*** (p. 252) (females of the following species are unknown but probably key out here: ***beardi*** n. sp. [p. 248], ***chaos*** n. sp. [p. 249], ***diadelphus*** n. sp. [p. 250], ***mccreadiei*** n. sp. [p. 250], ***newmani*** n. sp. [p. 251], ***protus*** n. sp. [p. 252], ***trochus*** n. sp. [p. 253])
– Frons broad, at narrowest part half or more as wide as long. Lacinia (slide mounted) with fine hair only. Thoracic integument, especially of katepisternum, typically brown .......... 6
6. Abdominal sternites II–VII membranous .......... ***alpestris*** (p. 246)
– Abdominal sternites IV–VII heavily sclerotized .......... 7
7. Hypogynial valves truncated (Fig. 10.99). Anal lobe in lateral view with anteroventral margin unsclerotized .......... ***clavatus*** (p. 246)
– Hypogynial valves broadly rounded (Fig. 10.100). Anal lobe in lateral view with anteroventral margin sclerotized .......... ***irkutensis*** (p. 247)

### Males of Genus *Helodon*

1. Ventral plate in ventral view deeply cleft posteriorly on either side of midline (Fig. 10.237). Anepisternum with small patch of hair on either side of anepisternal membrane (in addition to mesepimeral tuft) (subgenus *Distosimulium*) .......... ***pleuralis*** (p. 243)
– Ventral plate in ventral view at most shallowly concave posteriorly (Figs. 10.238–10.244). Anepisternum without patch of hair on either side of anepisternal membrane .......... 2
2. Ventral plate in ventral view with anterior margin extended beyond apices of basal arms as long, slender process (Figs. 10.238–10.240); gonostylus with 1 apical spinule (subgenus *Parahelodon*) .......... 3
– Ventral plate in ventral view with anterior margin at most slightly convex (Figs. 10.241–10.244); gonostylus with 2 or more apical spinules (subgenus *Helodon*) .......... 5
3. Antenna with 9 flagellomeres .......... ***vernalis*** (p. 245)
– Antenna with 7 or 8 flagellomeres .......... 4

4. Ventral plate in ventral view (excluding anterior process) wider than long; median sclerite blunt apically (Fig. 10.238) .......... ***decemarticulatus*** (p. 244)
– Ventral plate in ventral view (excluding anterior process) longer than wide; median sclerite pointed apically (Fig. 10.239) .......... ***gibsoni*** (p. 244)

5. Gonostylus with 4 or 5 apical spinules (Fig. 10.243) .......... ***irkutensis*** (p. 247)
– Gonostylus with 2 or 3 apical spinules (Figs. 10.241, 10.242, 10.244) .......... 6

6. Ventral plate in ventral view with lateral margins subparallel (Fig. 10.244) (*onychodactylus* species complex, p. 248) .......... ***onychodactylus*** s. s. (p. 251), ***susanae*** (p. 252) (males of the following species are unknown but probably key out here: ***beardi*** n. sp. [p. 248], ***chaos*** n. sp. [p. 249], ***diadelphus*** n. sp. [p. 250], ***mccreadiei*** n. sp. [p. 250], ***newmani*** n. sp. [p. 251], ***protus*** n. sp. [p. 252], ***trochus*** n. sp. [p. 253])
– Ventral plate in ventral view with lateral margins bowed outward (Figs. 10.241, 10.242) .......... 7

7. Ventral plate in ventral view with medial notch posteriorly; gonostylus in inner lateral view rounded apically, with rather straight lateral margins (Fig. 10.241) .......... ***alpestris*** (p. 246)
– Ventral plate in ventral view without medial notch posteriorly, although sometimes shallowly concave posteriorly; gonostylus in inner lateral view pointed apically, slightly curved (Fig. 10.242) .......... ***clavatus*** (p. 246)

### Females of Genus *Prosimulium*

The females of *Prosimulium canutum* n. sp. and *P. idemai* n. sp. are unknown.

1. Eastern North America (i.e., east of 100° longitude) .......... 2
– Western North America (i.e., west of 100° longitude) .......... 6

2. Lacinia (slide mounted) without retrorse teeth; mandible (slide mounted) without serrations. Spermatheca minute, about one third as long as stem of genital fork, not compressed apicobasally (Fig. 10.115). Northern Canada (Map 53) .......... ***ursinum*** (in part) (p. 271)
– Lacinia (slide mounted) with retrorse teeth; mandible (slide mounted) with serrations. Spermatheca more than one third as long as stem of genital fork (Fig. 10.110), or rather compressed apicobasally (Fig. 10.121). Widely distributed .......... 3

3. Antenna with first flagellomere typically longer than pedicel. Spermatheca wider than long (Figs. 10.121, 10.123) (*magnum* species group, in part) .......... ***albionense*** (p. 278), ***magnum*** s. s. (p. 279), ***multidentatum*** (p. 276) (female of ***canutum*** n. sp. [p. 278] is unknown but probably keys out here)
– Antenna with first flagellomere subequal in length to pedicel. Spermatheca longer than, or as long as, wide (Figs. 10.108, 10.110) (*hirtipes* species group, in part) .......... 4

4. Palpomere III in lateral view with sensory vesicle opening to outside directly or through short neck by means of wide mouth (Fig. 10.24) .......... ***fontanum*** (p. 258)
– Palpomere III in lateral view with sensory vesicle opening to outside through distinct neck that is not expanded into wide mouth (Fig. 10.25) .......... 5

5. Hypogynial valve with sclerotization of inner margin expanded anterolaterally (Fig. 10.108) .......... ***approximatum*** (p. 254), ***arvum*** (in part) (p. 255), ***fuscum*** (in part) (p. 260)
– Hypogynial valve with sclerotization of inner margin typically not expanded anterolaterally (Fig. 10.110) .......... ***arvum*** (in part) (p. 255), ***clandestinum*** n. sp. (p. 255), ***mixtum*** (p. 263), ***rhizophorum*** (p. 266), ***saltus*** (p. 267), ***transbrachium*** (p. 269)

6. Thoracic integument orange or yellowish .......... 7
– Thoracic integument brown (although hair can appear orange) .......... 8

7. Large species, wing length greater than 3.3 mm .......... ***daviesi*** (in part) (p. 257), ***fulvithorax*** (p. 260), ***fulvum*** (p. 260), ***secretum*** n. sp. (p. 267)
– Small species, wing length less than 3.6 mm .......... ***minifulvum*** n. sp. (p. 262)

8. Antenna yellowish brown .......... ***flaviantennus*** (p. 274)
– Antenna brownish, although base sometimes yellowish .......... 9

9. Lacinia (slide mounted) without retrorse teeth; mandible (slide mounted) without serrations. Spermatheca minute, about one third as long as stem of genital fork, not compressed apicobasally (Fig. 10.115). Alaska, northern Canada, and high elevations (>3000 m) in Rocky Mountains (Maps 52, 53) .......... ***neomacropyga*** (p. 270), ***ursinum*** (in part) (p. 271)

– Lacinia (slide mounted) with retrorse teeth; mandible (slide mounted) with serrations. Spermatheca more than one third as long as stem of genital fork (Fig. 10.114), or rather compressed apicobasally (Fig. 10.117). Widely distributed....10

10. Spermatheca wider than long (Figs. 10.116, 10.117) (*magnum* species group, in part)....11
   – Spermatheca longer than, or as long as, wide (Figs. 10.111–10.113) (*hirtipes* species group, in part)....16
11. Mouthparts longer than clypeus (as in Fig. 10.14)....12
   – Mouthparts as long as or shorter than clypeus....13
12. Spermatheca more than half as wide as length of stem of genital fork (Fig. 10.120). Known from Oregon (Map 61)....***longirostrum*** n. sp. (p. 275)
   – Spermatheca less than half as wide as length of stem of genital fork (Fig. 10.122). Known from Arizona, Utah, and Wyoming (Map 63)....***uinta*** (p. 276)
13. Antenna with 8 flagellomeres....***unicum*** (p. 276)
   – Antenna with 9 flagellomeres....14
14. Hypogynial valve long, slender, bowed outward only along proximal one third of outer margin (Fig. 10.118)....***dicentum*** (p. 272), ***dicum*** (p. 273), ***exigens*** (p. 273)
   – Hypogynial valve shorter, bowed outward along half or more of outer margin (Figs. 10.116, 10.117, 10.119)....15
15. Anal lobe in lateral view nearly straight along posterior margin (Figs. 10.116, 10.119)....***caudatum*** (p. 272), ***impostor*** (p. 275)
   – Anal lobe in lateral view curved along posterior margin (Fig. 10.117)....***constrictistylum*** (p. 272)
16. Anal lobe projected posteriorly, nearly reaching posterior margin of cercus. Hypogynial valve rather slender, long, nearly reaching posterior margin of anal lobe; inner margin rather straight (Fig. 10.108). East of Rocky Mountains (Map 41)....***fuscum*** (in part) (p. 260)
   – Anal lobe weakly projected posteriorly, not reaching posterior margin of cercus. Hypogynial valve rather broad, often short, not reaching posterior margin of anal lobe; inner margin often sinuous (Figs. 10.103–10.106). Rocky Mountains westward....17
17. Femora rather bright yellowish orange. Hypogynial valve short, about twice as long as wide (Figs. 10.103–10.105)....18
   – Femora rather dull yellowish or straw colored to brownish. Hypogynial valve long, more than twice as long as wide (Figs. 10.106, 10.111, 10.113)....19
18. Spermatheca less tapered and more bluntly rounded apically (Figs. 10.103, 10.105)....***daviesi*** (in part) (p. 257), ***formosum*** (p. 259)
   – Spermatheca typically rather tapered apically (Fig. 10.104)....***doveri*** (p. 257)
19. Spermatheca bluntly rounded apically (Figs. 10.111, 10.114)....***esselbaughi*** (p. 258), ***rusticum*** n. sp. (p. 266), ***travisi*** (p. 269)
   – Spermatheca typically rather tapered apically (Figs. 10.106, 10.113)....***frohnei*** (p. 259), ***shewelli*** (p. 268)

#### Males of Genus *Prosimulium*

The males of *Prosimulium canutum* n. sp. and *P. idemai* n. sp. are unknown. The male of the parthenogenetic *P. ursinum* does not exist.

1. Eastern North America (i.e., east of 100° longitude)....2
   – Western North America (i.e., west of 100° longitude)....9
2. Antenna with first flagellomere longer than pedicel. Ventral plate in ventral view evenly tapered and broadly rounded posteriorly (Figs. 10.274, 10.277) (*magnum* species group, in part)....3
   – Antenna with first flagellomere subequal in length to pedicel. Ventral plate in ventral view abruptly tapered (Fig. 10.245) or subrectangular, not evenly tapered or broadly rounded posteriorly (Figs. 10.246, 10.247, 10.255, 10.257, 10.258) (*hirtipes* species group, in part)....4
3. Gonostylus with 2 or 3 (rarely 4) apical spinules (Fig. 10.277) (*magnum* species complex, in part, p. 277)....***albionense*** (p. 278), ***magnum*** s. s. (p. 279)
   – Gonostylus with 4–7 apical spinules (Fig. 10.274)....***multidentatum*** (p. 276) (male of ***canutum*** n. sp. [p. 278] is unknown but possibly keys out here)

4. Ventral plate in ventral view with body abruptly narrowed, keel-like (Fig. 10.245) ........................................................................ ***approximatum*** (p. 254)
– Ventral plate in ventral view with body somewhat square or rectangular, not abruptly narrowed (Figs. 10.246, 10.247, 10.255, 10.257, 10.258) ........................................ 5
5. Ventral plate in lateral view with lip small, strongly upturned (Fig. 10.247) ..... ***clandestinum*** n. sp. (p. 255)
– Ventral plate in lateral view with lip large, broadly rounded, slightly or not at all upturned (Figs. 10.246, 10.255, 10.257, 10.258) ........................................ 6
6. Palpomere III in lateral view with sensory vesicle opening to outside directly or through short neck and wide mouth (as in Fig. 10.24) ........................................ ***fontanum*** (p. 258)
– Palpomere III in lateral view with sensory vesicle opening to outside through distinct neck that is not expanded into wide mouth (as in Fig. 10.25) ........................................ 7
7. Ventral plate in ventral view with body nearly rectangular (Fig. 10.246) ............ ***arvum*** (in part) (p. 255)
– Ventral plate in ventral view with body rather square (Figs. 10.255, 10.257, 10.258, 10.264) ........................ 8
8. Basal fringe typically brownish ............................ ***arvum*** (in part) (p. 255), ***mixtum*** (p. 263), ***saltus*** (p. 267)
– Basal fringe typically grayish, silvery, or pale golden ........................................ ***fuscum*** (in part) (p. 260), ***rhizophorum*** (p. 266), ***transbrachium*** (p. 269)
9. Scutum with integument orange; much of body with integument often orange ........................................ 10
– Scutum and remainder of body with integument brown, slightly orange, or orange tinged (hair can appear orange) ........................................ 11
10. Ventral plate in ventral view about as wide as long, with lateral margins subparallel; in lateral view with lip directed posteromedially (Fig. 10.254) ........................................ ***fulvum*** (in part) (p. 260)
– Ventral plate in ventral view about 1.5 or more times wider than long, with lateral margins bowed outward; in lateral view with lip directed medially (Figs. 10.260, 10.261) .............. ***secretum*** n. sp. (p. 267)
11. Mouthparts nearly 3 times longer than clypeus (Fig. 10.14) ........................................ 12
– Mouthparts less than 2.5 times as long as clypeus ........................................ 13
12. Ventral plate in ventral view with lateral margins convergent (Fig. 10.273). Known from Oregon (Map 61) ........................................ ***longirostrum*** n. sp. (p. 275)
– Ventral plate in ventral view with lateral margins subparallel or weakly convergent (Fig. 10.275). Known from Arizona, Utah, and Wyoming (Map 63) ........................................ ***uinta*** (p. 276)
13. Antenna yellowish brown. Gonostylus in inner lateral view with apical spinules rather flattened (Fig. 10.271) ........................................ ***flaviantennus*** (p. 274)
– Antenna brownish, although base sometimes yellowish. Gonostylus in inner lateral view with apical spinules conical (Fig. 10.270) ........................................ 14
14. Gonostylus in ventral and inner lateral views abruptly narrowed at midlength (Fig. 10.268) ........................................ ***constrictistylum*** (p. 272)
– Gonostylus in ventral and inner lateral views uniform in width or gradually tapered, but not abruptly narrowed at midlength (Figs. 10.267, 10.269) ........................................ 15
15. Ventral plate in terminal and ventral views with lip elongate, snoutlike (Figs. 10.267, 10.272) .............. 16
– Ventral plate in terminal and ventral views with lip shorter (Figs. 10.251, 10.269, 10.270, 10.276) .......... 17
16. Ventral plate in all views with lip enormous, at least twice as long as wide (Fig. 10.267). Occiput dorsally and laterally with brown hair ........................................ ***caudatum*** (p. 272)
– Ventral plate in all views with lip only slightly longer than wide (Fig. 10.272). Occiput dorsally and laterally with pale golden hair ........................................ ***impostor*** (p. 275)
17. Ventral plate in ventral view somewhat trilobed posteriorly, with lip protruded into concavity of posterior margin (Figs. 10.269, 10.276) ........................................ 18
– Ventral plate in ventral view bilobed, straight, or convex posteriorly, without lip protruded into concavity of posterior margin (Figs. 10.250–10.253) ........................................ 20
18. Ventral plate in ventral view with lateral margins subparallel; in lateral view with lip directed anteromedially (Fig. 10.276). Antenna with 8 flagellomeres ........................................ ***unicum*** (p. 276)
– Ventral plate in ventral view with lateral margins convergent posteriorly; in lateral view with lip directed posteromedially (Fig. 10.270). Antenna with 9 flagellomeres ........................................ 19
19. Ventral plate in ventral view with shallow concavity on either side of lip .................... ***dicentum*** (p. 272)

– Ventral plate in ventral view with rather deep concavity on either side of lip (Figs. 10.269, 10.270)........................................................................................................***dicum*** (p. 273), ***exigens*** (p. 273)

20. Gonostylus in ventral and inner lateral views with apex acutely pointed (Fig. 10.266) ...................................................................................................................***neomacropyga*** (p. 270)

– Gonostylus in ventral and inner lateral views with apex rounded (Fig. 10.259) (*hirtipes* species group, in part)........................................................................................................................................21

21. Ventral plate in ventral view with body about twice as wide as long (Fig. 10.252)...........***frohnei*** (p. 259)

– Ventral plate in ventral view with body less than twice as wide as long (Figs. 10.248–10.251) ..............22

22. Scutum with integument slightly orange or orange tinged.........................................................................23

– Scutum with integument brown....................................................................................................................24

23. Ventral plate in terminal view subtriangular; in ventral view rather straight posteriorly (Fig. 10.248) ...................................................................................................................***daviesi*** (in part) (p. 257)

– Ventral plate in terminal view broadly rounded; in ventral view convex posteriorly (Fig. 10.254) .......... ...................................................................................................................***fulvum*** (in part) (p. 260)

24. Ventral plate in ventral view with posterior margin deeply concave (Fig. 10.251). Legs yellowish; body hair often appearing orange ..........................................................................................***formosum*** (p. 259)

– Ventral plate in ventral view with posterior margin shallowly concave (Figs. 10.248–10.250). Legs typically brown, sometimes pale yellowish; body hair typically yellowish or brownish ..........................25

25. Ventral plate in terminal view narrow, somewhat pointed apically (Fig. 10.255). East of Rocky Mountains (Map 41).........................................................................................***fuscum*** (in part) (p. 260)

– Ventral plate in terminal view broad, often well rounded apically (Figs. 10.262, 10.263). Rocky Mountains westward ..........................................................................................................................26

26. Ventral plate in ventral view with basal arms short (Figs. 10.262, 10.263).........................***shewelli*** (p. 268)

– Ventral plate in ventral view with basal arms longer (Figs. 10.248–10.250, 10.253, 10.256, 10.259, 10.265).........................................................................................................................................27

27. Femora yellowish orange.....................................................................................***daviesi*** (in part) (p. 257)

– Femora brownish to pale yellowish ...........................................................***doveri*** (p. 257), ***esselbaughi*** (p. 258), ***fulvithorax*** (p. 260), ***minifulvum*** n. sp. (p. 262), ***rusticum*** n. sp. (p. 266), ***travisi*** (p. 269)

### Females of Genus *Greniera*

1. Costa with stout, black spiniform setae interspersed among longer, paler hairlike setae (the former more prevalent near apex of costa) (Fig. 10.78). Spermatheca reniform (Fig. 10.125). West of Rocky Mountains (Map 70)......................................................................................***humeralis*** n. sp. (p. 281)

– Costa with only pale setae, some of which may be shorter and stiffer, but neither dark nor fully spiniform. Spermatheca somewhat oblong, not reniform (Figs. 10.124, 10.126, 10.127). Widespread.............2

2. Antenna with 8 flagellomeres ...................................................................................................***denaria*** (p. 282)

– Antenna with 9 flagellomeres ........................................................................................................................3

3. Antenna extended beyond lateral margin of head by more than half its length (Fig. 10.16). Eastern coastal plain (Map 72)............................................................................................***longicornis*** n. sp. (p. 282)

– Antenna extended beyond lateral margin of head by less than half its length (Fig. 10.15). California, Canada, and northeastern United States.........................................................................................4

4. Clypeus large, somewhat bulbous, about half length of head (Fig. 10.15)...............................***abdita*** (p. 280)

– Clypeus small, not bulbous, less than half length of head ..........................................................................5

5. Spermatheca with small unpigmented area at junction with spermathecal duct. Canada and northern United States (Map 69) ......................................................................................................***abditoides*** (p. 281)

– Spermatheca with pigmentation extended onto spermathecal duct (Fig. 4.24). California (Map 73) ....... .....................................................................................................................................**'species F'** (p. 283)

### Males of Genus *Greniera*

The male of *Greniera* 'species F' is unknown.

1. Costa with stout, black spiniform setae interspersed among longer, paler hairlike setae (the former more prevalent near apex of costa) (Fig. 10.78). West of Rocky Mountains (Map 70).........................................................................................................***humeralis*** n. sp. (p. 281)

– Costa with only pale setae, some of which may be shorter and stiffer, but neither dark nor fully spiniform. Widespread ........ 2

2. Antenna with 8 flagellomeres. Gonostylus tapered to pointed apex, with 1 apical spinule (Fig. 10.280) ........ ***denaria*** (p. 282)

– Antenna with 9 flagellomeres. Gonostylus only slightly tapered to rounded apex, with 2 apical spinules (Figs. 10.278, 10.281) ........ 3

3. Antenna extended beyond lateral margin of head by more than half its length (Fig. 10.17). Ventral plate in ventral view tapered posteriorly (Fig. 10.281). Eastern coastal plain (Map 72) ........ ***longicornis*** n. sp. (p. 282)

– Antenna extended beyond lateral margin of head by less than half its length. Ventral plate in ventral view truncated posteriorly (Fig. 10.278). Canada and northeastern United States (Maps 68, 69) ........ 4

4. Scutum with whitish hair ........ ***abdita*** (p. 280)

– Scutum with brownish hair ........ ***abditoides*** (p. 281)

#### Females of Genus *Stegopterna*

The female of *Stegopterna trigonium* from North America is unknown.

1. Calcipala about half or more as wide as apex of basitarsus (Figs. 10.36, 10.38, 10.39). Spermatheca with small unpigmented area at junction with spermathecal duct (Fig. 10.128). Lacinia (slide mounted) with retrorse teeth; mandible (slide mounted) with serrations ........ 2

– Calcipala less than half as wide as apex of basitarsus (Fig. 10.37). Spermatheca with pigmentation extended slightly onto spermathecal duct (Figs. 10.129, 10.130). Lacinia (slide mounted) without retrorse teeth; mandible (slide mounted) without serrations ........ 4

2. Great Plains eastward (Maps 76, 78) ........ ***diplomutata*** (p. 285), ***mutata*** (p. 287)

– West of Great Plains ........ 3

3. Calcipala about half as wide as apex of basitarsus (Fig. 10.36) ........ ***acra*** n. sp. (p. 284)

– Calcipala about three fourths as wide as apex of basitarsus (Fig. 10.39) ........ ***permutata*** (p. 288), ***xantha*** n. sp. (p. 289)

4. Spermatheca with slightly raised rounded areas. Anal lobe in ventral view broad, unsclerotized anteriorly (Fig. 10.129) ........ ***decafilis*** (p. 285)

– Spermatheca smooth, without raised rounded areas. Anal lobe in ventral view narrow, sclerotized anteriorly (Fig. 10.130) ........ ***emergens*** (p. 286)

#### Males of Genus *Stegopterna*

The male of *Stegopterna trigonium* from North America is unknown. The male of the parthenogenetic *St. mutata* does not exist.

1. Calcipala about half or more as wide as apex of basitarsus (as in Figs. 10.36, 10.38, 10.39) ........ 2

– Calcipala less than half as wide as apex of basitarsus (as in Fig. 10.37) ........ 4

2. Great Plains eastward (Map 76) ........ ***diplomutata*** (p. 285)

– West of Great Plains ........ 3

3. Calcipala about half as wide as apex of basitarsus (as in Fig. 10.36) ........ ***acra*** n. sp. (p. 284)

– Calcipala about three fourths as wide as apex of basitarsus (as in Fig. 10.39) ........ ***permutata*** (p. 288), ***xantha*** n. sp. (p. 289)

4. Paramere with apical margin jagged; aedeagal membrane with hair only (Fig. 10.283) ........ ***decafilis*** (p. 285)

– Paramere with apical margin smooth or finely toothed; aedeagal membrane with fine spines and hair (Fig. 10.285) ........ ***emergens*** (p. 286)

#### Females of Genus *Tlalocomyia*

1. Spermatheca elongate, about half as long as stem of genital fork (Fig. 10.131) (*andersoni* species group) ........ ***andersoni*** n. sp. (p. 290)

– Spermatheca somewhat round, less than half as long as stem of genital fork (Figs. 10.132, 10.133) (*osborni* species group) ........ 2

2. Spermatheca with small unpigmented area at junction with spermathecal duct (Fig. 10.132) ........ ***osborni*** (p. 291)

– Spermatheca with large unpigmented area at junction with spermathecal duct (Fig. 10.133) ........................................................ ***ramifera*** n. sp. (p. 292), ***stewarti*** (p. 293)

### Males of Genus *Tlalocomyia*

1. Ventral plate in lateral view with distinct lip; paramere small, subtriangular, not twisted (Figs. 10.286, 10.287) ........................................................ 2
– Ventral plate in lateral view without distinct lip; paramere large, subrectangular, twisted (Figs. 10.288, 10.289) ........................................................ 3
2. Basal fringe with hairs typically dark basally, pale apically. Ventral plate in lateral view slender; aedeagal membrane with pale spines as well as hair (Fig. 10.286) (*andersoni* species group) ........................................................ ***andersoni*** n. sp. (p. 290)
– Basal fringe entirely pale. Ventral plate in lateral view more robust; aedeagal membrane with hair only (Fig. 10.287) ........................................................ ***osborni*** (p. 291)
3. Ventral plate in ventral view subrectangular (Fig. 10.288) ........................................................ ***ramifera*** n. sp. (p. 292)
– Ventral plate in ventral view subquadrate (Fig. 10.289) ........................................................ ***stewarti*** (p. 293)

### Females of Genus *Cnephia*

1. Posterior margin of abdominal tergite IX produced as narrow snoutlike projection (Fig. 10.135). Claws toothless, or each with minute subbasal tooth ........................................................ ***dacotensis*** (p. 294)
– Posterior margin of abdominal tergite IX evenly rounded (Figs. 10.136–10.138). Claws each with basal thumblike lobe (Fig. 10.60) ........................................................ 2
2. Basitarsus of hind leg without distinct calcipala, although with small acute projection. Alaska and northern Canada (Map 88) ........................................................ ***eremites*** (p. 296)
– Basitarsus of hind leg with distinct, apically rounded calcipala. Eastern North America (i.e., east of 105° longitude) (Maps 89, 90) ........................................................ 3
3. Abdominal tergites VIII and IX shiny, contrasting with dull preceding segments. Spermatheca with small but distinct unpigmented area at junction with spermathecal duct (Fig. 10.137). Eastern North America (Map 89) ........................................................ ***ornithophilia*** (p. 296)
– Abdominal tergites VIII and IX dull. Spermatheca without unpigmented area at junction with spermathecal duct (Fig. 10.138). Mississippi River area (Map 90) ........................................................ ***pecuarum*** (p. 297)

### Males of Genus *Cnephia*

1. Gonostylus in dry specimens shiny on apical one third or more ........................................................ 2
– Gonostylus in dry specimens dull throughout ........................................................ 3
2. Gonostylus in inner lateral view not markedly curved near apex (as in Fig. 10.294d). Ventral plate in ventral view broad, about as long as wide (Fig. 10.291). Widespread (Map 87) ........ ***dacotensis*** (p. 294)
– Gonostylus in inner lateral view markedly curved near apex (Fig. 10.292d). Ventral plate in ventral view narrow, longer than wide (Fig. 10.292). Alaska and northern Canada (Map 88) ......... ***eremites*** (p. 296)
3. Abdominal sternites III–VII with relatively few hairs, those on segments III and IV moderately long and extended laterally beyond margins of abdomen, but those on remaining segments decreasing in length and mostly short on sternites VI and VII. Ventral plate in ventral view truncated posteriorly (Fig. 10.293). Eastern North America (Map 89) ........................................................ ***ornithophilia*** (p. 296)
– Abdominal sternites III–VII with numerous long hairs, those on segments III–V extended laterally far beyond margins of abdomen. Ventral plate in ventral view with posterior margin broadly rounded (Fig. 10.294). Mississippi River area (Map 90) ........................................................ ***pecuarum*** (p. 297)

### Females of Genus *Ectemnia*

1. Frons about 3 times as long as narrowest width. Southeastern coastal plain and Sandhills (Map 93) ........................................................ ***reclusa*** (p. 301)
– Frons about 4 times as long as narrowest width. Widespread ........................................................ 2
2. Canada and northern United States south through Appalachian Mountains (Map 91) ........................................................ ***invenusta*** (p. 300)
– Widespread but not in mountains ........................................................ 3

3. Mesepimeral tuft covering upper one fourth of mesepimeron. Southern coastal plain and piedmont (Map 92) ........ ***primaeva*** (p. 301)

– Mesepimeral tuft typically covering upper half of mesepimeron. Central North America (Map 94) ........ ***taeniatifrons*** (p. 301)

### Males of Genus *Ectemnia*

1. Ventral plate in ventral view tapered posteriorly (Fig. 10.295). Canada and northern United States south through Appalachian Mountains (Map 91) ........ ***invenusta*** (p. 300)

– Ventral plate in ventral view subrectangular, not tapered posteriorly (Figs. 10.296, 10.297). Widespread but not in mountains ........ 2

2. Mesepimeral tuft, basal fringe, and scutal hair coppery brown. Southeastern coastal plain and Sandhills (Map 93) ........ ***reclusa*** (p. 301)

– Mesepimeral tuft and often basal fringe and scutal hair pale golden. Widespread ........ 3

3. Mesepimeral tuft covering upper one fourth of mesepimeron. Southern coastal plain and piedmont (Map 92) ........ ***primaeva*** (p. 301)

– Mesepimeral tuft typically covering upper half of mesepimeron. Central North America (Map 94) ........ ***taeniatifrons*** (p. 301)

### Females of Genus *Metacnephia*

1. Lacinia (slide mounted) without retrorse teeth; mandible (slide mounted) without serrations. Spermatheca with pigment extended onto spermathecal duct for distance greater than basal width of duct (Figs. 10.141, 10.144) ........ 2

– Lacinia (slide mounted) with retrorse teeth; mandible (slide mounted) with serrations. Spermatheca with small unpigmented area at junction with spermathecal duct (Figs. 10.143, 10.145), or with pigmentation extended onto duct for no more than distance about equal to basal width of duct (Fig. 10.142) ........ 3

2. Alaska and northern Canada (Maps 95, 100) ........ ***borealis*** (p. 302), ***sommermanae*** (p. 305)

– High elevations in Rocky Mountains of United States (Map 96) ........ ***coloradensis*** (p. 303)

3. Spermatheca with pigmentation extended at least slightly onto spermathecal duct (Fig. 10.142) ........ 4

– Spermatheca with small unpigmented area at junction with spermathecal duct (Figs. 10.143, 10.145) ........ 5

4. Western United States south of Canada (Map 97) ........ ***jeanae*** (p. 303)

– Alaska and Canada (Map 98) ........ ***saileri*** (p. 304)

5. Anepisternal membrane without hair. Anal lobe in ventral view with outer margin strongly emarginate (Fig. 10.143) ........ ***saskatchewana*** (p. 305)

– Anepisternal membrane with hair. Anal lobe in ventral view with outer margin rather straight (Fig. 10.145) ........ ***villosa*** (p. 305)

### Males of Genus *Metacnephia*

1. Eyes not meeting above antennae, separated by frons (head dichoptic); all facets subequal in size ........ ***coloradensis*** (p. 303)

– Eyes meeting above antennae (head holoptic); upper facets enlarged ........ 2

2. Gonostylus in inner lateral view with apex curved sharply inward (Fig. 10.300d). Anepisternal membrane without hair ........ ***saskatchewana*** (p. 305)

– Gonostylus in inner lateral view with apex rather straight or curved smoothly inward (Figs. 10.298d, 10.301d). Anepisternal membrane with hair (sometimes rubbed off) ........ 3

3. Mesepimeral tuft and hair of stem vein brown ........ 4

– Mesepimeral tuft and hair of stem vein typically silvery ........ 6

4. Ventral plate in terminal view triangular, with lateral margins only weakly bowed inward (Fig. 10.302) ........ ***villosa*** (p. 305)

– Ventral plate in terminal view with lateral margins bowed inward (Fig. 10.299) ........ 5

5. Western United States south of Canada (Map 97) ........ ***jeanae*** (p. 303)

– Alaska and Canada (Map 98) ........ ***saileri*** (p. 304)

6. Gonostylus in inner lateral view nearly twice as long as basal width (Fig. 10.298). Ventral plate in terminal and lateral views with lip long, well developed ........ ***borealis*** (p. 302)
– Gonostylus in inner lateral view only slightly longer than basal width (Fig. 10.301). Ventral plate in terminal and lateral views with lip short, weakly developed ........ ***sommermanae*** (p. 305)

#### Females of Genus *Simulium*

The females of *Simulium* 'species O' and *S. rothfelsi* are unknown.

1. Wing with basal section of radius bearing hair dorsally ........ 2
– Wing with basal section of radius without hair, or with few hairs, dorsally (Fig. 10.81) ........ 37

2. Claws toothless, or each with small subbasal tooth (Fig. 10.66) ........ 3
– Claws each with basal thumblike lobe (Figs. 10.64, 10.65, 10.67) ........ 7

3. Lacinia (slide mounted) without retrorse teeth; mandible (slide mounted) without serrations. Alaska and northern Canada southward along Rocky Mountains ........ 4
– Lacinia (slide mounted) with retrorse teeth; mandible (slide mounted) with serrations. Widespread ........ 5

4. Anal lobe in ventral view with medial region unsclerotized (Fig. 10.157). Spermatheca single ........ ***baffinense*** (p. 320)
– Anal lobe in ventral view with medial region sclerotized (Fig. 10.179). Spermatheca double (whether this latter feature is typical or aberrant must await discovery of more than the single known specimen) ........ **'species Z'** (p. 348)

5. Claws each with small subbasal tooth ........ ***giganteum*** (p. 347)
– Claws toothless ........ 6

6. Spermatheca with pigmentation typically extended onto spermathecal duct (Fig. 10.177) ........ ***furculatum*** (p. 347)
– Spermatheca with unpigmented area at junction with spermathecal duct (Fig. 10.178) ........ ***subpusillum*** (p. 348)

7. Tibia with integument and hair yellow on basal two thirds, contrasting with brown pigment on apical one third. Postscutellum with patches of hair posteriorly (sometimes rubbed off). Spermatheca pigmented at junction with spermathecal duct (Fig. 10.160) (subgenus *Eusimulium*, p. 326) ........ ***bracteatum*** (p. 328), ***donovani*** (p. 328), ***exulatum*** n. sp. (p. 329), ***pilosum*** (p. 330), ***violator*** n. sp. (p. 330)
– Tibia with integument typically uniformly brown (paler in teneral specimens) and hair silvery or pale golden. Postscutellum without patches of hair. Spermatheca lacking pigment at junction with spermathecal duct (Fig. 10.146) ........ 8

8. Scutum with 3 dark gray or brown longitudinal stripes (especially prominent in rubbed specimens) (as in Plate 1a) ........ ***parmatum*** n. sp. (in part) (p. 322)
– Scutum unicolorous, without stripes or with at most 3 faint, pale longitudinal stripes visible in rubbed specimens ........ 9

9. Precoxal bridge incomplete (Fig. 10.32). Genital fork with each arm expanded into triangular lateral plate, typically with its posterior margin perpendicular to longitudinal body axis; each lateral plate with (Figs. 10.154–10.156) or without (Fig. 10.158) prominent anteriorly directed apodeme (subgenus *Boreosimulium*, in part) ........ 10
– Precoxal bridge complete (Fig. 10.33). Genital fork with each arm variously expanded, but not as triangular lateral plate; each lateral plate without prominent anteriorly directed apodeme (Figs. 10.146–10.149, 10.162–10.165) ........ 15

10. Genital fork with each lateral plate bearing weakly developed, anteriorly directed apodeme (Fig. 10.158). Central North America and southeastern coastal plain (Map 123) ........ ***johannseni*** (in part) (p. 321)
– Genital fork with each lateral plate bearing prominent, anteriorly directed apodeme (Figs. 10.154–10.156). Widespread (*annulus* species group) ........ 11

11. East of Great Plains ........ 12
– West of Great Plains ........ 14

12. Anal lobe in ventral view not deeply incised medially; in lateral view with anterior margin sclerotized except distally (Fig. 10.155). Southeastern piedmont and coastal plain (Map 117) ........ ***clarkei*** (p. 317)

– Anal lobe in ventral view incised medially; in lateral view with anterior margin unsclerotized (Fig. 10.154). Widespread ........ 13

13. Western and eastern North America north of Pennsylvania (Map 114) ........ ***annulus*** (in part) (p. 314)
 – Eastern North America (Map 118) ........ ***emarginatum*** (p. 317)

14. Anal lobe in ventral view with anteromedial one third unsclerotized (Fig. 10.154) ........ ***annulus*** (in part) (p. 314)
 – Anal lobe in ventral view with anteromedial half unsclerotized (Fig. 10.156) ........ ***balteatum*** n. sp. (p. 315), ***canonicolum*** (p. 316), ***joculator*** n. sp. (p. 318), ***quadratum*** (p. 318), ***zephyrus*** n. sp. (p. 319)

15. Basitarsus of hind leg with pedisulcus typically long and shallow, its depth less than one third width of segment (Fig. 10.41). Antenna with pedicel larger than basal flagellomere (Fig. 10.22) (subgenus *Hellichiella*) ........ 16
 – Basitarsus of hind leg with pedisulcus short and deep, its depth at least one third to one half width of segment (as in Fig. 10.42). Antenna with pedicel smaller than basal flagellomere (Fig. 10.23) (subgenus *Nevermannia*) ........ 23

16. Fore coxa with integument yellowish, contrasting with color of adjacent thorax ........ 17
 – Fore coxa with integument gray or brown, not contrasting with color of adjacent thorax ........ 18

17. Claws (slide mounted) each about half or less length of last tarsomere. Spermatheca about three fourths as long as stem of genital fork (Fig. 10.147). Canada and eastern United States; present in coastal plain (Map 103) ........ ***congareenarum*** (p. 308)
 – Claws (slide mounted) each about two thirds length of last tarsomere. Spermatheca about half as long as stem of genital fork (Fig. 10.153). Northeastern North America; absent from coastal plain (Map 113) ........ ***rivuli*** (p. 313)

18. Frons at narrowest part one ninth to one twelfth width of head ........ ***innocens*** (p. 308), ***minus*** (p. 309), ***nebulosum*** (p. 309)
 – Frons at narrowest part one fifth to one eighth width of head ........ 19

19. Palpomere III (slide mounted) in lateral view rather bulbous, with sensory vesicle occupying about one third to one half its area (Fig. 10.27). West of Great Plains (Map 110) ........ ***curriei*** (p. 311)
 – Palpomere III (slide mounted) in lateral view slender (as in Fig. 10.26) or rather bulbous, with sensory vesicle occupying about one fourth or less its area. Widespread ........ 20

20. Anal lobe with about three fourths of medial area unsclerotized (Fig. 10.152). Northwestern United States (Map 112) ........ ***mysterium*** n. sp. (p. 312)
 – Anal lobe with about half of medial area unsclerotized (Figs. 10.146, 10.150, 10.151). Widespread ........ 21

21. Genital fork with lateral plate of each arm bearing dark sclerotized line along posterior margin (Fig. 10.151) ........ ***excisum*** (p. 312)
 – Genital fork with lateral plate of each arm weakly sclerotized along posterior margin (Figs. 10.146, 10.150) ........ 22

22. Legs uniformly dark brown (nonteneral specimens) ........ ***anatinum*** (p. 307), ***rendalense*** (p. 310)
 – Legs variegated, yellowish and dark brown (nonteneral specimens) ........ ***usovae*** (p. 310)

23. Katepisternum with patch of hair on dorsal margin (Fig. 4.13) (in rubbed specimens, sockets visible after clearing) ........ 24
 – Katepisternum bare ........ 25

24. Scutum with silvery hair; katepisternum with patch of hair typically central, small (<20 hairs) ........ ***croxtoni*** (p. 337)
 – Scutum with golden hair; katepisternum with patch of hair typically extended across dorsal margin (Fig. 4.13) ........ ***gouldingi*** (p. 340)

25. Anal lobe in ventral view entirely sclerotized and covered with microtrichia (Fig. 10.167). Far northern North America (Map 140) ........ ***dendrofilum*** (p. 338)
 – Anal lobe in ventral view with at least anterior one third unsclerotized and lacking microtrichia (Figs. 10.162–10.165). Widespread ........ 26

26. Genital fork with each arm nearly uniform in width along its entire length, almost as narrow at junction as at point of attachment to lateral plate (Figs. 10.162, 10.169, 10.171) ........ 27
 – Genital fork with each arm narrowed distally, about twice or more as wide at junction as at point of attachment to lateral plate (Figs. 10.172–10.176) ........ 28

27. East of Great Plains (Map 144) ........................................ ***impar*** (p. 340)
– West of Great Plains (Maps 134, 142) ........................................ ***bicorne*** (p. 332), ***fontinale*** (p. 338)

28. Scutum with hair pale brassy. Anal lobe in lateral and ventral views with half or more unsclerotized (Fig. 10.174). East of Great Plains (Map 150) ........................................ ***quebecense*** (p. 344)
– Scutum with hair golden, pale golden, or silvery. Anal lobe in ventral view with less than half unsclerotized (Figs. 10.163–10.165). Widespread ........................................ 29

29. Scutum with hair predominantly silvery ........................................ 30
– Scutum with hair predominantly pale golden to golden ........................................ 31

30. Anal lobe in ventral view with anteromedial one third unsclerotized; in lateral view with small ventral spot unsclerotized (Fig. 10.163). Canada and northeastern United States (Map 135) ........................................ ***burgeri*** n. sp. (p. 333)
– Anal lobe in ventral view with strip of unsclerotized cuticle just posterior to anterior margin; in lateral view with anteroventral margin unsclerotized (Fig. 10.176). Rocky Mountains westward (Map 152) ........................................ ***wyomingense*** (p. 346)

31. Anal lobe in lateral and ventral views almost bisected transversely by unsclerotized area (Figs. 10.164, 10.165, 10.168) ........................................ 32
– Anal lobe in lateral and ventral views unsclerotized in anterior half or less (Figs. 10.161, 10.172, 10.173, 10.175) ........................................ 34

32. Genital fork with lateral plate of each arm expanded into large, posteromedially directed thumblike lobe (Fig. 10.165). Anal lobe in lateral view with anterior margin weakly sclerotized ......... ***craigi*** (p. 336)
– Genital fork with lateral plate of each arm truncated posteromedially, or if expanded into thumblike lobe, then anal lobe in lateral view with anterior margin well sclerotized (Figs. 10.164, 10.168) ............ 33

33. East of Rocky Mountains (Map 141) ........................................ ***fionae*** (p. 338)
– West of Rocky Mountains (Map 137) ........................................ ***conicum*** n. sp. (p. 335)

34. Stem vein with golden hair. Northeastern North America (Map 133) ........................................ ***aestivum*** (p. 332)
– Stem vein typically with shiny brown hair. Widespread ........................................ 35

35. Genital fork with space enclosed by arms as long as wide or longer, often subcircular (Fig. 10.173). Abdomen often pale gray or brownish orange ........................................ ***carbunculum*** n. sp. (p. 334), ***merritti*** n. sp. (p. 341), ***modicum*** n. sp. (p. 342), ***moultoni*** n. sp. (p. 343), ***pugetense*** (p. 343)
– Genital fork with space enclosed by arms wider than long, often subrectangular or suboval (Figs. 10.172, 10.175). Abdomen typically gray ........................................ 36

36. Widespread (Map 151) ........................................ ***silvestre*** (p. 344)
– Eastern United States, especially southeastern portion (Map 145) ........................................ ***loerchae*** (p. 341)

37. Claws each with basal thumblike lobe (Fig. 10.42) ........................................ 38
– Claws either toothless or each with small subbasal tooth (Figs. 10.68–10.74) ........................................ 44

38. Scutum with 3 dark gray or brown longitudinal stripes (especially prominent in rubbed specimens) (Plate 1a) ........................................ 39
– Scutum unicolorous, without stripes, or with at most 3 faint, pale longitudinal stripes visible in rubbed specimens ........................................ 41

39. Genital fork with lateral plate of each arm subtriangular and without conspicuous anteriorly directed apodeme ........................................ ***parmatum*** n. sp. (in part) (p. 322)
– Genital fork with lateral plate of each arm slender and bearing anteriorly directed apodeme (Fig. 10.159) (subgenus *Byssodon*) ........................................ 40

40. Alaska (possibly Yukon Territory) (Map 126) ........................................ ***maculatum*** (p. 323)
– Canada and United States, excluding Alaska (Map 127) ........................................ ***meridionale*** (p. 324)

41. Frons gray, pollinose. Scutum not shiny. Central North America and southeastern coastal plain (Map 123) ........................................ ***johannseni*** (in part) (p. 321)
– Frons bluish gray to brownish, shiny. Scutum subshiny to shiny (*slossonae* species group). Widespread ........................................ 42

42. Wing with subcosta bearing row of hairs on ventral surface. Antenna uniformly brown, or with basal 2 segments only slightly paler than remainder ........................................ ***transiens*** (p. 411)

- Wing with subcosta bare on ventral surface. Antenna with basal 2 segments distinctly paler than remainder .....43

43. Scutum strongly shiny, not pollinose. Eastern coastal plain (Map 231) ..... ***slossonae*** (p. 410)
- Scutum subshiny, with thin gray pollinosity. Canada and northern United States; absent from coastal plain (Map 230) ..... ***rugglesi*** (p. 409)

44. Claws (slide mounted if necessary) each with small subbasal tooth (Figs. 10.68, 10.70–10.73) .....45
- Claws (slide mounted if necessary) toothless (Figs. 10.69, 10.74) .....63

45. Hypogynial valve elongate, forming abrupt outer angle with sternite VIII (Figs. 10.199–10.204). Large, grayish, nonshiny species, sometimes with reddish orange tinges on scutum. Frons pollinose, nonshiny, not contrasting with pollinose clypeus (subgenus *Hemicnetha*, in part) .....46
- Hypogynial valve short, its outer margin smoothly divergent from sternite VIII (Figs. 10.197, 10.198, 10.210, 10.211). Small to large shiny species (except *canadense*, which is dull and grayish), without reddish orange tinges on scutum. Frons typically shiny, contrasting with pollinose clypeus (except in *canadense*) .....50

46. Hypogynial valve extended posteriorly as fine, weakly sclerotized point (Fig. 10.203). Genital fork with lateral plate of each arm not bearing strongly sclerotized, ventrally directed tubercle ..... ***solarii*** (p. 373)
- Hypogynial valve rounded posteriorly. Genital fork with lateral plate of each arm bearing strongly sclerotized, ventrally directed tubercle (Figs. 10.199–10.202, 10.204) .....47

47. Scutum at least partially reddish orange, or with reddish cast ..... ***paynei*** (p. 372), ***virgatum*** (p. 374)
- Scutum black and gray, not reddish orange (Plate 2f) .....48

48. British Columbia and Pacific Coastal states (Map 184) ..... ***hippovorum*** (p. 371)
- Interior of western United States (Maps 182, 183) .....49

49. Hypogynial valve with inner margin straight, weakly sclerotized (Fig. 10.200). Anal lobe in lateral view with 1 small sclerotized patch ..... ***bricenoi*** (p. 371)
- Hypogynial valve with inner margin bulged inward, well sclerotized, free of microtrichia (Fig. 10.199). Anal lobe in lateral view with more than 1 small sclerotized patch ..... ***freemani*** (p. 370)

50. Frons and most of body nonshiny. Anal lobe in lateral view subrectangular, with polished, rather bare area along anterior margin (Fig. 10.198) ..... ***canadense*** (p. 369)
- Frons and scutum shiny. Anal lobe in lateral view subtriangular to subquadrate, with well sclerotized but unpolished setose area (Figs. 10.209–10.214, 10.216) .....51

51. Claws long, slightly sigmoidal (Fig. 10.73). Eastern North America (i.e., east of 95° longitude) (Map 228) ..... ***parnassum*** (p. 407)
- Claws small, smoothly curved (Figs. 10.70–10.72). Widespread .....52

52. Hypogynial valves in ventral view swollen, with inner margins divergent (Figs. 10.209, 10.211) .....53
- Hypogynial valves in ventral view not swollen, with inner margins subparallel (Figs. 10.190–10.197, 10.210, 10.212–10.214) .....54

53. Fore coxa brown ..... ***decimatum*** (p. 390)
- Fore coxa yellowish ..... ***malyschevi*** (p. 392)

54. Anal lobe in ventral view large, ovoid to nearly circular, well sclerotized; in lateral view subquadrate (Figs. 10.210, 10.212–10.214) (*malyschevi* species group, in part) .....55
- Anal lobe in ventral view smaller, variously shaped, partially sclerotized; in lateral view subtriangular (Figs. 10.190–10.197) (subgenus *Aspathia*) .....57

55. Femora mostly yellowish ..... ***defoliarti*** (p. 391)
- Femora mostly brown .....56

56. Western North America (i.e., west of 100° longitude) (*arcticum* species complex, p. 394) ..... ***apricarium*** n. sp. (p. 395), ***arcticum*** s. s. (p. 396), ***brevicercum*** (p. 398), ***chromatinum*** n. sp. (p. 398), ***negativum*** n. sp. (p. 399), ***saxosum*** n. sp. (p. 399), ***vampirum*** (p. 400) (females of ***arcticum* 'cytospecies IIL-1'** [p. 397] and ***arcticum* 'cytospecies IIS-4'** [p. 397] are unknown but probably key out here)
- Canada and northern United States (Map 216) ..... ***murmanum*** (p. 393)

57. Fore coxa pale yellowish .....58
- Fore coxa brownish .....60

58. Scutum in anterior view without distinct longitudinal stripes but with pair of bluish green iridescent spots. Anal lobe in lateral view entirely sclerotized, not polished or bicolored (Fig. 10.196). Known from southwestern New Mexico (Map 179) ........ ***puigi*** (p. 368)
- Scutum in anterior view with distinct longitudinal stripes (as in Plate 2d). Anal lobe in lateral view either entirely sclerotized, polished, and bicolored (Figs. 10.192, 10.193), or with ventral portion unsclerotized (Fig. 10.190). Widespread ........ 59

59. Legs, except tarsi, mostly yellow. Stem vein and basicosta with brown hair. Anal lobe in lateral view not polished or bicolored; ventral portion unsclerotized (Fig. 10.190) ........ ***anduzei*** (p. 365)
- Legs mostly brownish. Stem vein and basicosta with golden hair. Anal lobe in lateral view well sclerotized, polished, bicolored; ventral portion well sclerotized (Figs. 10.192, 10.193) ........ ***hunteri*** (p. 366), ***iriartei*** (p. 366)

60. Scutum without longitudinal stripes ........ ***hechti*** (p. 366)
- Scutum with longitudinal stripes, those on either side of midline strongest (Plate 2d, e) ........ 61

61. Anal lobe in lateral and ventral views with sclerotized portion distinctly bicolored, that is, with hyaline band separating darker areas (Fig. 10.195). Stem vein and basicosta with brown or golden hair ........ ***piperi*** (p. 367)
- Anal lobe in ventral view with sclerotized portion not noticeably bicolored (Figs. 10.194, 10.197). Stem vein and basicosta with golden hair ........ 62

62. Anal lobe in lateral view with sclerotized portion about twice as wide as long (Fig. 10.194). Scutum in anterodorsal view without pinkish iridescence ........ ***jacumbae*** (p. 367)
- Anal lobe in lateral view with sclerotized portion somewhat squarish (Fig. 10.197). Scutum in anterodorsal view typically with slight pinkish iridescence (Plate 2e) ........ ***tescorum*** (in part) (p. 368)

63. Scutum in lateral view strongly arched, its anterior face nearly vertical. Counties along southern Rio Grande (Maps 160, 165) ........ 64
- Scutum in lateral view not strongly arched, its anterior face smoothly curved posteriorly. Widespread ........ 65

64. Scutum brownish black ........ ***labellei*** (p. 351)
- Scutum yellowish orange to brownish orange, or if brownish black, then with yellowish or orange patches ........ ***robynae*** (p. 352)

65. Scutum in anterodorsal view with distinct longitudinal stripe(s) (Plates 1b–d, f; 2b, e) ........ 66
- Scutum in anterodorsal view without longitudinal stripe(s), although pruinose spots can be present anteriorly and extended slightly posteriorly (*aranti* has short, pale, vague stripes) ........ 78

66. Scutum grayish pruinose, with 1 medial longitudinal black or brown stripe (Plate 1d, h, i) ........ 67
- Scutum variously colored, with 2 or more longitudinal stripes (Plates 1b, c, f; 2b, e, g) ........ 68

67. Anal lobe in lateral view with short, blunt ventral extension (Fig. 10.182). Texas (Map 162) ........ ***mediovittatum*** (p. 352)
- Anal lobe in lateral view with elongate, pointed ventral extension (Fig. 10.185). North and west of Texas (Map 167) ........ ***venator*** (p. 353)

68. Frons and scutum shiny ........ ***tescorum*** (in part) (p. 368)
- Frons and scutum pollinose, not shiny ........ 69

69. Scutum orange to black, with 2 silvery white or silvery blue longitudinal stripes (Plate 1b, c, f) ........ 70
- Scutum grayish, with dark gray or brown longitudinal stripes (Plate 2b, g) ........ 73

70. Anal lobe in lateral view with elongate, pointed ventral extension about 3 times longer than its basal width (Fig. 10.184) ........ 71
- Anal lobe in lateral view with blunt or pointed ventral extension subequal in width and length (Fig. 10.180) ........ 72

71. Southeastern Arizona (Map 161) ........ ***longithallum*** (p. 351)
- Oklahoma and Texas (Map 166) ........ ***trivittatum*** (p. 353)

72. Scutum orange (rarely brownish black), with 2 silvery white longitudinal stripes (Plate 1b). Widespread (Map 157) ........ ***bivittatum*** (p. 349)
- Scutum brownish black, with 2 silvery blue longitudinal stripes (Plate 1c). California (Map 158) ........ ***clarum*** (p. 350)

73. Fore tibia anteriorly without white patch. Abdomen with single row of black patches dorsally. Subcosta typically with row of hairs on ventral surface. Sternum VII with fringe of long hair (Figs. 10.205, 10.206) (*pictipes* species group) ........................................ 74
– Fore tibia anteriorly with broad white patch. Abdomen with black patches dorsally and laterally. Subcosta bare on ventral surface, or with at most 3 hairs. Sternum VII without fringe of long hair (subgenus *Psilozia*) ........................................ 75

74. Anal lobe in lateral view with posteroventral margin extended as acute point (Fig. 10.206) ........................................ ***pictipes*** (p. 377)
– Anal lobe in lateral view with posteroventral margin blunt, rounded (Fig. 10.205) .............. ***claricentrum*** (p. 374), ***innoxium*** (p. 375)

75. Anal lobe in lateral view with outer margin mostly unpigmented, weakly sclerotized (Fig. 10.189). Widespread (*vittatum* species complex, p. 356) ...................... ***tribulatum*** (p. 361), ***vittatum*** s. s. (p. 363)
– Anal lobe in lateral view with outer margin mostly pigmented, sclerotized (Figs. 10.186–10.188). West of Mississippi River ........................................ 76

76. Anal lobe in lateral view short, blunt apically (Fig. 10.188). Central California (Map 170) ........................................ ***infernale*** n. sp. (p. 355)
– Anal lobe in lateral view elongate, rather pointed apically (Figs. 10.186, 10.187). Widespread ................ 77

77. Anal lobe in lateral view rather bluntly pointed ventrally, well sclerotized along anterior margin (Fig. 10.186) ........................................ ***argus*** (p. 354)
– Anal lobe in lateral view rather sharply pointed ventrally, weakly sclerotized along anterior margin (Fig. 10.187) ........................................ ***encisoi*** (p. 355)

78. Scutum not shiny (Plate 1e). Anal lobe in lateral view with pointed ventral extension (Fig. 10.181). West of Mississippi River (Maps 159, 163, 164) ................. ***griseum*** (p. 350), ***meyerae*** (p. 352), ***notatum*** (p. 352)
– Scutum subshiny to shiny. Anal lobe in lateral view blunt, without pointed ventral extension (Fig. 10.207). Widespread ........................................ 79

79. Subcosta bare on ventral surface, or with at most 4 hairs. Hypogynial valves with inner margins divergent, each valve divided obliquely into anterior, weakly sclerotized setose region, and posterior, well-sclerotized bare region (Figs. 10.207, 10.208) (*jenningsi* species group) .................................. 80
– Subcosta with row of hairs on ventral surface. Hypogynial valves with inner margins subparallel, each valve rather uniformly sclerotized, covered with microtrichia (Fig. 10.221) .................................. 84

80. Stem vein with yellow or golden hair ........................................ 81
– Stem vein with golden brown to brown hair ........................................ 82

81. Scutum in anterodorsal view with pair of short, pale, vague stripes (dark stripes in posterodorsal view) ........................................ ***aranti*** (p. 379)
– Scutum in anterodorsal view without stripes .......................................... ***luggeri*** (p. 386), ***ozarkense*** (p. 388)

82. Hind femur (slide mounted) with scalelike setae 3–7 times longer than greatest width ............ ***dixiense*** (p. 381), ***haysi*** (p. 382), ***jonesi*** (p. 384), ***krebsorum*** (p. 385)
– Hind femur (slide mounted) with scalelike setae 7–10 times longer than greatest width ...................... 83

83. Genital fork with small patch(es) of sclerotization adjacent to inner margin of each arm (Fig. 10.208) ........................................ ***anchistinum*** (p. 378), ***infenestrum*** (p. 382), ***jenningsi*** (p. 382), ***nyssa*** (p. 388), ***podostemi*** (p. 389), ***remissum*** (p. 389)
– Genital fork with, at most, faint trace of sclerotization adjacent to inner margin of each arm (Fig. 10.207) ............................... ***chlorum*** (p. 379), ***confusum*** (p. 379), ***definitum*** (p. 380), ***fibrinflatum*** (p. 381), ***lakei*** (p. 385), ***notiale*** (p. 387), ***penobscotense*** (p. 388), ***snowi*** (p. 390), ***taxodium*** (p. 390)

84. Fore coxa brown ........................................ 85
– Fore coxa yellowish to yellowish brown ........................................ 86

85. Stem vein with pale golden hair. Western United States (Map 229) ................................ ***petersoni*** (p. 408)
– Stem vein with brown hair. Alaska and northern Canada (Maps 237a, b; 241) .................. ***tuberosum*** s. s. (in part) (p. 417), ***vulgare*** (p. 420)

86. Anal lobe in lateral view subquadrate, about as long as wide (Fig. 10.215). Frons grayish, dull, thinly pollinose (*noelleri* species group) ........................................ 87
– Anal lobe in lateral view subtriangular, wider than long (Figs. 10.221–10.225). Frons brownish black, shiny, not pollinose ........................................ 88

87. Widespread, including Alaska and northern Canada (Map 226) ........................................ ***decorum*** (p. 403)
– Alaska and northern Canada (Map 227) ........................................ ***noelleri*** (p. 405)

88. Anal lobe in lateral and ventral views unsclerotized along anterior margin, although dark line of setae can be present anteriorly (Fig. 10.224) (*venustum* species complex, p. 421) ........................ ***hematophilum*** (p. 425), ***incognitum*** (p. 425), ***irritatum*** (p. 426), ***minutum*** (p. 426), ***molestum*** (p. 427), ***piscicidium*** (p. 427), ***truncatum*** (p. 428), ***venustum*** s. s. (p. 429) (females of ***tormentor*** n. sp. [p. 427] and ***venustum*** **'cytospecies JJ'** [p. 430] are unknown but probably key out here)
– Anal lobe in lateral and ventral views sclerotized along anterior margin (Figs. 10.221–10.223, 10.225) ........................................ 89

89. Anal lobe in ventral view with thin, smoothly curved sclerotized line along anterior margin; posterior area more weakly sclerotized but not in form of distinct hyaline band (Figs. 10.221, 10.222) (*tuberosum* species complex, in part, p. 411) ........................ ***appalachiense*** n. sp. (p. 413), ***chromocentrum*** n. sp. (p. 414), ***conundrum*** n. sp. (p. 415), ***perissum*** (p. 416), ***tuberosum*** s. s. (in part) (p. 417), ***twinni*** (p. 418), ***ubiquitum*** n. sp. (p. 418), ***vandalicum*** (p. 419)
– Anal lobe in ventral view with variously wide, sclerotized band along anterior margin, separated from larger posterior sclerotized area by thin unsclerotized hyaline band (Figs. 10.223, 10.225) ................ 90

90. Anal lobe in ventral view with anterior sclerotized band wide; outer anterolateral angle evenly rounded, not produced (Fig. 10.225). Widespread (*verecundum* species complex, p. 430) (Maps 253, 254) ........................................ ***rostratum*** (p. 432), ***verecundum*** s. s. (p. 433)
– Anal lobe in ventral view with anterior sclerotized band thin; outer anterolateral angle produced as small rounded process (Fig. 10.223). Alaska and northwestern Canada (Map 242) .......... ***rubtzovi*** (p. 421)

### Males of Genus *Simulium*

The males of *Simulium minus*, *S.* 'species O,' *S. rothfelsi*, *S. dendrofilum*, and *S.* 'species Z' are unknown or insufficiently known to include in the following key.

1. Wing with basal section of radius bearing hair dorsally ........................................ 2
– Wing with basal section of radius without hair, or with few hairs, dorsally ........................................ 35

2. Ventral plate in ventral view laterally compressed, keel like, with basal arms extended laterally (Figs. 10.321, 10.322) (subgenus *Eusimulium*, p. 326) ........................................ 3
– Ventral plate in ventral view not laterally compressed, but can be tapered posteriorly; basal arms extended anteriorly (Figs. 10.323–10.325) ........................................ 5

3. Ventral plate in lateral view with body somewhat pointed apically (Fig. 10.322). Gonostylus in dorsal view with base forming 90-degree angle with narrowed portion ................... ***exulatum*** n. sp. (p. 329), ***pilosum*** (p. 330), ***violator*** n. sp. (p. 330)
– Ventral plate in lateral view with body broadly rounded apically (Fig. 10.321). Gonostylus in dorsal view with base forming 45-degree angle with narrowed portion ........................................ 4

4. Alaska, Canada, eastern United States (Map 128) ........................................ ***bracteatum*** (p. 328)
– Western United States (Map 129) ........................................ ***donovani*** (p. 328)

5. Dorsal plate anchor shaped (Fig. 10.339). Far North (Map 154) ........................................ ***giganteum*** (p. 347)
– Dorsal plate absent or variously shaped, but not anchor shaped (Figs. 10.306, 10.317, 10.323, 10.334, 10.336). Widespread ........................................ 6

6. Gonostylus in inner lateral view bulged well beyond spinule (Figs. 10.338d, 10.340d). Dorsal plate with lip well developed, and with flange (collar) about 3 times wider than lip at greatest width. Alaska, Canada, northern United States (Maps 153, 155) ........................................ 7
– Gonostylus in inner lateral view not bulged, or if so, then not markedly beyond spinule (Figs. 10.310–10.316, 10.334–10.337). Dorsal plate absent, or with lip poorly developed (Fig. 10.312e), or with flange (collar) less than 3 times wider than lip at greatest width (Figs. 10.324e, 10.334e). Widespread ........................................ 8

7. Ventral plate in ventral view with lateral margins convergent posteriorly (Fig. 10.338) ........................................ ***furculatum*** (p. 347)
– Ventral plate in ventral view with lateral margins subparallel (Fig. 10.340) ............ ***subpusillum*** (p. 348)

8. Gonostylus in ventral view narrow, uniformly tapered to pointed apex (Figs. 10.310–10.319) ................. 9

– Gonostylus in ventral view with medially directed, subtriangular flange (Figs. 10.323–10.337) (subgenus *Nevermannia*) (male of *dendrofilum* [Patrusheva 1962] is too poorly known to include in this key, but probably would key to this subgenus)........22

9. Paramere bearing 1 stout spine at midlength and 1–3 stout spines distally (Figs. 10.311–10.316). Aedeagal membrane with long fine spines (*annulus* species group)........10
– Paramere without stout spine at midlength, but with 1 stout detached spine (Fig. 10.317) or more than 4 stout spines distally (Figs. 10.303–10.310). Aedeagal membrane without long fine spines, although cluster of numerous short stout spines can be present........14

10. Ventral plate in ventral view with posterolateral finger-like lobe on each side (Fig. 10.311)........***annulus*** (p. 314)
– Ventral plate in ventral view without posterolateral finger-like lobe on each side (Figs. 10.312–10.316)........11

11. Gonostylus in inner lateral view with margin not bulged beyond apical spinule (Figs. 10.314, 10.315). East of Great Plains........12
– Gonostylus in inner lateral view with margin bulged beyond apical spinule (Figs. 10.312, 10.313, 10.316). West of Great Plains........13

12. Ventral plate in ventral view enormous, shallowly and narrowly concave posteriorly (Fig. 10.314)........***clarkei*** (p. 317)
– Ventral plate in ventral view broad and short, deeply and broadly concave posteriorly (Fig. 10.315)........***emarginatum*** (p. 317)

13. Paramere with 3 distal spines (Fig. 10.312)........***balteatum*** n. sp. (p. 315)
– Paramere with 1 or 2 distal spines (Figs. 10.313, 10.316)........***canonicolum*** (p. 316), ***joculator*** n. sp. (p. 318), ***quadratum*** (p. 318), ***zephyrus*** n. sp. (p. 319)

14. Ventral plate in lateral view with lip truncated (Fig. 10.317). Paramere with 1 detached stout spine; aedeagal membrane with cluster of many small spines........***baffinense*** (p. 320)
– Ventral plate in lateral view with lip pointed or well rounded (Figs. 10.303–10.310, 10.318, 10.319). Paramere with numerous stout spines; aedeagal membrane bearing hair only........15

15. Basitarsus of hind leg with pedisulcus short and deep, its depth at least one third to one half width of segment (as in Fig. 10.42). Antenna with pedicel distinctly shorter than basal flagellomere (*johannseni* species group)........16
– Basitarsus of hind leg with pedisulcus typically long and shallow, its depth less than one third width of segment (as in Fig. 10.41). Antenna with pedicel as long as or only slightly shorter than basal flagellomere (subgenus *Hellichiella*)........17

16. Ventral plate in ventral view broadly rounded posteriorly (Fig. 10.318). Scutum unicolorous, without stripes. Central North America and southeastern coastal plain (Map 123)........***johannseni*** (p. 321)
– Ventral plate in ventral view acutely tapered posteriorly (Fig. 10.319). Scutum with 3 dark gray or brown longitudinal stripes (most prominent in rubbed specimens). Southeastern United States, west into Texas (Map 124)........***parmatum*** n. sp. (p. 322)

17. Paramere with spines small, the longest about 2–3 times longer than its basal width (Fig. 10.308)........***excisum*** (p. 312)
– Paramere with spines long, the longest more than 3 times longer than its basal width (Figs. 10.303–10.307, 10.309, 10.310)........18

18. Paramere with longest spine at least 5 times longer than its basal width (Figs. 10.303, 10.306)........19
– Paramere with longest spine about 3–5 times longer than its basal width (Figs. 10.304, 10.305, 10.307, 10.309, 10.310)........21

19. Legs variegated, yellowish and dark brown (nonteneral specimens)........***usovae*** (p. 310)
– Legs uniformly dark brown (nonteneral specimens)........20

20. Mesepimeral tuft brown. Canada and northern United States (Maps 102, 107)........***anatinum*** (p. 307), ***rendalense*** (p. 310)
– Mesepimeral tuft silvery. Canada and eastern United States (Map 103)........***congareenarum*** (p. 308)

21. Ventral plate in lateral view well rounded apically (Figs. 10.307, 10.309). Rocky Mountains westward (Maps 110, 112)........***curriei*** (p. 311), ***mysterium*** n. sp. (p. 312)

– Ventral plate in lateral view pointed apically (Figs. 10.304, 10.305, 10.310). Widespread ............***innocens*** (p. 308), ***nebulosum*** (p. 309), ***rivuli*** (p. 313)

22. Katepisternum with patch of golden hair dorsally (as in Fig. 4.13) ..................................***gouldingi*** (p. 340)
 – Katepisternum bare ..........................................................................................................................23

23. Dorsal plate large, subrectangular (Fig. 10.334) .........................................................................***loerchae*** (p. 341)
 – Dorsal plate oblong to circular, with flanged collar (Figs. 10.335–10.337) ........................................................24

24. Gonostylus in ventral view with medially directed flange markedly produced, concave posteriorly (Fig. 10.327). West of Great Plains (Map 137)..................................................................***conicum*** n. sp. (p. 335)
 – Gonostylus in ventral view with medially directed flange variously produced, convex or only slightly concave posteriorly (Figs. 10.335–10.337). Widespread..........................................................................25

25. Dorsal plate 3 or more times longer than width at midpoint (Fig. 10.323). Northeastern North America (Map 133)...........................................................................................................................***aestivum*** (p. 332)
 – Dorsal plate no more than 2 times as long as width at midpoint (Figs. 10.328, 10.329). Widespread ...................................................................................................................................26

26. Ventral plate in ventral view convex or straight posteriorly (Figs. 10.325, 10.328–10.330, 10.336, 10.337) ..........................................................................................................................................27
 – Ventral plate in ventral view at least slightly concave posteriorly, sometimes with small tubercle extended into concavity (Figs. 10.324, 10.326, 10.331, 10.333, 10.335).............................................................32

27. Body hair coppery brown.........................................................................................................***burgeri*** n. sp. (p. 333)
 – Body hair, especially on scutum, at least partially golden, pale yellow, or silvery....................................28

28. Gonostylus in inner lateral view with spinule flush with posterior margin (Figs. 10.328, 10.337) ............29
 – Gonostylus in inner lateral view with spinule anterior to posterior margin (Figs. 10.329, 10.330, 10.336) .............................................................................................................................................30

29. Ventral plate in ventral view subrectangular, about twice as wide as long (Fig. 10.328). Mesepimeral tuft pale yellow to golden. Widespread (Map 138).........................................................................***craigi*** (p. 336)
 – Ventral plate in ventral view less than twice as wide as long (Fig. 10.337). Mesepimeral tuft typically brown. West of Great Plains (Map 152) ...............................................................***wyomingense*** (p. 346)

30. Gonostylus in ventral and inner lateral views with outer margin bulged (Fig. 10.336). Dorsal plate typically subcircular beyond flanged collar.............................................................................***silvestre*** (p. 344)
 – Gonostylus in ventral and inner lateral views with outer margin straight (Figs. 10.329, 10.330). Dorsal plate rather oval beyond flanged collar ..................................................................................................31

31. Ventral plate in ventral view rounded posteriorly (Fig. 10.329) ..............................................***croxtoni*** (p. 337)
 – Ventral plate in ventral view rather truncated posteriorly (Fig. 10.330)................................***fionae*** (p. 338)

32. Ventral plate in terminal view with lip swollen (Fig. 10.335). East of Great Plains (Map 150) ................................................................................................................................***quebecense*** (p. 344)
 – Ventral plate in terminal view with lip not appreciably swollen (Figs. 10.324, 10.326, 10.331, 10.333). Widespread ...........................................................................................................................................33

33. Gonostylus in ventral view with anterior margin of medial flange directed posteriorly (Figs. 10.324, 10.331). Ventral plate with posterior margin weakly scalloped. Northwestern North America (Maps 134, 142)..................................................................................***bicorne*** (p. 332), ***fontinale*** (p. 338)
 – Gonostylus in ventral view with anterior margin of medial flange directed medially (Figs. 10.326, 10.333). Ventral plate with posterior margin not scalloped. Widespread..................................................34

34. Ventral plate in ventral view with central tubercle of posterior margin prominent; basal arms straight or slightly bowed (Fig. 10.326). Widespread.....................................................***carbunculum*** n. sp. (p. 334), ***merritti*** n. sp. (p. 341), ***modicum*** n. sp. (p. 342), ***moultoni*** n. sp. (p. 343), ***pugetense*** (p. 343)
 – Ventral plate in ventral view with central tubercle of posterior margin absent or weakly defined; basal arms strongly bowed, with apices directed medially (Fig. 10.333). East of Great Plains (Map 144) ...........................................................................................................................................***impar*** (p. 340)

35. Gonostylus about as long as gonocoxa, with 1 apical spinule; in inner lateral view conical, uniformly tapered to pointed apex (Figs. 10.318–10.320) .........................................................................................36
 – Gonostylus shorter to longer than gonocoxa, with 0–3 apical spinules; in inner lateral view neither conical nor uniformly tapered to pointed apex (Figs. 10.343, 10.349, 10.353, 10.360).............................38

36. Ventral plate in ventral view broadly U or V shaped, less than twice as wide as long (Figs. 10.318, 10.319)....go back to 16
- Ventral plate in ventral view subrectangular, about twice as wide as long (Fig. 10.320) (subgenus *Byssodon*)....37

37. Alaska (possibly Yukon Territory) (Map 126)....***maculatum*** (p. 323)
- Canada and United States, excluding Alaska (Map 127)....***meridionale*** (p. 324)

38. Gonostylus shorter than gonocoxite (Figs. 10.341–10.351)....39
- Gonostylus longer than gonocoxite (Figs. 10.352–10.409)....53

39. Gonostylus with 2–5 apical spinules (Figs. 10.348–10.351). Fore leg mostly brown or dark brown, with dull silvery patch on tibia. Widespread (subgenus *Psilozia*)....40
- Gonostylus with 1 apical spinule (Figs. 10.341–10.347). Fore leg typically pale yellow (brown in few species) from coxa through tibia. West of Mississippi River (subgenus *Psilopelmia*)....43

40. Scutum in anterior view grayish, without paler stripes. Central California (Map 170)....***infernale*** n. sp. (p. 355)
- Scutum in anterior view velvety black, with 2 variably distinct, grayish to silver patches or stripes (Plate 2a, c). Widespread....41

41. Scutum in anterior view with 2 variably distinct, grayish pruinose triangular patches (Plate 2c). Widespread (*vittatum* species complex, p. 356)....***tribulatum*** (p. 361), ***vittatum*** s. s. (p. 363)
- Scutum in anterior view with 2 shiny silver stripes (Plate 2a). West of Mississippi River....42

42. Gonostylus with 3–5 (rarely 2) apical spinules (Fig. 10.348). Ventral plate in ventral view nearly truncate posteriorly....***argus*** (p. 354)
- Gonostylus with 2 apical spinules (Fig. 10.349). Ventral plate in ventral view tapered posteriorly....***encisoi*** (p. 355)

43. Scutum in lateral view strongly arched, its anterior face nearly vertical. Ventral plate in ventral view less than 2 times wider than long, truncated posteriorly (Fig. 10.345). Counties along southern Rio Grande (Maps 160, 165)....44
- Scutum in lateral view not strongly arched, its anterior face smoothly curved posteriorly. Ventral plate in ventral view either about 3 times wider than long or tapered posteriorly (Figs. 10.341–10.344, 10.346, 10.347). Widespread....45

44. Legs brown....***labellei*** (p. 351)
- Fore and middle legs pale yellow or yellowish brown, except parts of tarsi....***robynae*** (p. 352)

45. Ventral plate in ventral view with anterior margin convex, broadly rounded; in lateral view with distinct, anteriorly directed flange (Fig. 10.346)....46
- Ventral plate in ventral view with anterior margin slightly concave to slightly convex; in lateral view without anteriorly directed flange (Figs. 10.342, 10.343)....47

46. Scutum with silvery stripes subparallel, not tapered posteriorly. Known from Arizona (Map 161)....***longithallum*** (p. 351)
- Scutum with silvery stripes broadest anteriorly, tapered posteriorly (Plate 1g). Known from Oklahoma and Texas (Map 166)....***trivittatum*** (p. 353)

47. Ventral plate in ventral view about 2 times or less wider than long, tapered or broadly rounded posteriorly (Figs. 10.343, 10.344, 10.347)....48
- Ventral plate in ventral view about 3 times wider than long, truncated or slightly convex posteriorly (Figs. 10.341, 10.342)....50

48. Ventral plate in ventral view broadly rounded posteriorly (Fig. 10.344). Rio Grande of New Mexico (Map 163)....***meyerae*** (p. 352)
- Ventral plate in ventral view tapered and rather pointed posteriorly (Figs. 10.343, 10.347). Widespread....49

49. Gonostylus in inner lateral view pointed apically (Fig. 10.343). Texas (Map 162)....***mediovittatum*** (p. 352)
- Gonostylus in inner lateral view truncated apically (Fig. 10.347). North and west of Texas (Map 167)....***venator*** (p. 353)

50. Scutum rather uniformly grayish pruinose, especially in anterior view....***griseum*** (p. 350)
- Scutum dark brown or black, often with gray, orange, or silvery patches or stripes....51

51. Legs mostly brown. California (Map 158) .......... ***clarum*** (p. 350)
– Fore legs and middle legs mostly yellowish, except parts of tarsi. Widespread .......... 52

52. Scutum in anterior view black, sometimes orange, with pair of gray or silvery stripes or variably sized orange patches. Widespread (Map 157) .......... ***bivittatum*** (p. 349)
– Scutum in anterior view dark brown, with large coppery green to bluish patch on each side. Arizona and possibly southern California (Map 164) .......... ***notatum*** (p. 352)

53. Gonostylus in inner lateral view with distinct basal or subbasal, rounded or spinelike lobe (Figs. 10.352–10.359, 10.386, 10.393, 10.395–10.398) (minute in *petersoni*, Fig. 10.394) .......... 54
– Gonostylus in inner lateral view without distinct basal or subbasal lobe (Figs. 10.362, 10.373, 10.385, 10.387, 10.409) .......... 72

54. Gonostylus in ventral view about 2 times as long as greatest width, bulged anteromedially and posterolaterally (Fig. 10.386). Ventral plate in ventral view longer than wide, slightly concave posteriorly. Alaska and northern Canada (Map 213) .......... ***decimatum*** (p. 390)
– Gonostylus in ventral view about 3 times or more as long as greatest width, or if shorter, then bulged along only 1 margin (Figs. 10.354, 10.395–10.398). Ventral plate in ventral view wider than long, or if longer than wide, convex posteriorly. Widespread .......... 55

55. Gonostylus in inner lateral view with minute subbasal lobe; apical spinule absent (Fig. 10.394). Great Basin of western United States (Map 229) .......... ***petersoni*** (p. 408)
– Gonostylus in inner lateral view with large basal or subbasal lobe or prong; apical spinule present (Figs. 10.352–10.359, 10.395–10.398). Widespread .......... 56

56. Gonostylus with basal lobe bearing numerous short stout spines on anterior surface (Figs. 10.397, 10.398) .......... 57
– Gonostylus with basal or subbasal lobe bearing, at most, fine hair on anterior surface (Figs. 10.354, 10.393, 10.395) .......... 63

57. Ventral plate in ventral view about 3 times wider than long, with posterolateral corners flared outward (Fig. 10.397). Scutum in anterior view with shiny, steel-gray, V-shaped area (Fig. 4.15) .......... ***transiens*** (p. 411)
– Ventral plate in ventral view 2.5 times or less as wide as long (Figs. 10.398–10.406). Scutum in anterior view with pair of pruinose patches anterolaterally (*tuberosum* species complex, p. 411) .......... 58

58. Ventral plate in ventral view deeply concave posteriorly, somewhat W shaped (Fig. 10.404). West of Great Plains (Map 238) .......... ***twinni*** (p. 418)
– Ventral plate in ventral view convex or slightly concave posteriorly (Figs. 10.398–10.403). Widespread .......... 59

59. Ventral plate in lateral view with lip variably bulged posteriorly (Figs. 10.398, 10.399). East of Mississippi River (Map 233) .......... ***appalachiense*** n. sp. (p. 413)
– Ventral plate in lateral view with lip directed medially or anteriorly (Figs. 10.400–10.403). Widespread .......... 60

60. Ventral plate in lateral view with lip short, small (Fig. 10.405). Eastern, mainly southeastern, United States (Map 239) .......... ***ubiquitum*** n. sp. (p. 418)
– Ventral plate in lateral view with lip longer, larger (Figs. 10.400–10.403). Widespread .......... 61

61. Ventral plate in lateral view with posterior margin directed medially (Figs. 10.402, 10.403, 10.406) .......... ***perissum*** (p. 416), ***tuberosum*** s. s. (p. 417), ***vandalicum*** (p. 419), ***vulgare*** (p. 420)
– Ventral plate in lateral view with posterior margin directed anteriorly (Figs. 10.400, 10.401) .......... 62

62. Gonostylus in ventral view bulged on inner margin about three fourths distance to its apex (Fig. 10.401). Canada and eastern mountains (Map 235) .......... ***conundrum*** n. sp. (p. 415)
– Gonostylus in ventral view bulged basally on inner margin (Fig. 10.400). Sierra Nevada of California (Map 234) .......... ***chromocentrum*** n. sp. (p. 414)

63. Gonostylus about 2 times as long as greatest width (Figs. 10.395, 10.396). Ventral plate in ventral and terminal views with lateral margins serrated .......... 64
– Gonostylus about 3 times or more as long as greatest width (Figs. 10.352–10.359, 10.393). Ventral plate in ventral and terminal views with lateral margins smooth .......... 65

64. Gonostylus in inner lateral and ventral views with subbasal lobe narrow (Fig. 10.395). Transcontinental north of 40° latitude (Map 230) .......... ***rugglesi*** (p. 409)

– Gonostylus in inner lateral and ventral views with subbasal lobe broad (Fig. 10.396). Eastern coastal plain (Map 231) .......... ***slossonae*** (p. 410)

65. Gonostylus in ventral view with prominent, rounded, medially directed bend near midlength (Fig. 10.393). East of Great Plains (Map 228) .......... ***parnassum*** (p. 407)

– Gonostylus in ventral view rather straight, without prominent, medially directed bend near midlength (Figs. 10.352–10.359). West of Mississippi River (subgenus *Aspathia*) .......... 66

66. Gonostylus in ventral view not tapered toward apex, about as wide apically as basally (Figs. 10.358, 10.359) .......... 67

– Gonostylus in ventral view abruptly or gradually tapered toward apex, wider at base than at apex (Figs. 10.352–10.357) .......... 68

67. Gonostylus in ventral view with outer margin nearly straight (Fig. 10.358) .......... ***puigi*** (p. 368)

– Gonostylus in ventral view sinuous (Fig. 10.359) .......... ***tescorum*** (p. 368)

68. Gonostylus in inner lateral and ventral views with massive basal, non-spinelike lobe (Fig. 10.352). Southern Arizona and New Mexico (Map 173) .......... ***anduzei*** (p. 365)

– Gonostylus in inner lateral and ventral views with thinner basal, blunt or spinelike lobe (Figs. 10.353–10.357). Widespread .......... 69

69. Gonostylus in inner lateral view with basal lobe blunt, not spinelike (Fig. 10.356) .......... ***jacumbae*** (p. 367)

– Gonostylus in inner lateral view with basal lobe pointed, strongly spinelike (Figs. 10.353–10.355, 10.357) .......... 70

70. Ventral plate in ventral view with cleft on either side of lip; in lateral view with narrow heel-like projection (Fig. 10.357) .......... ***piperi*** (p. 367)

– Ventral plate in ventral view convex or straight posteriorly, without cleft on either side of lip; in lateral view without heel-like projection (Figs. 10.353–10.355) .......... 71

71. Ventral plate in terminal and tilted ventral views with lip broad (Fig. 10.353). Southern Arizona and New Mexico (Map 174) .......... ***hechti*** (p. 366)

– Ventral plate in terminal and tilted ventral views with lip narrow, triangular (Figs. 10.354, 10.355). Widespread (Maps 175, 176) .......... ***hunteri*** (p. 366), ***iriartei*** (p. 366)

72. Ventral plate in ventral view with deep medial cleft posteriorly (Figs. 10.360, 10.369–10.371). Gonostylus in ventral view about 4 times as long as basal width .......... 73

– Ventral plate in ventral view convex, concave, or laterally compressed posteriorly, but without deep medial cleft (Figs. 10.362, 10.366, 10.374, 10.387, 10.408). Gonostylus in ventral view less than 4 times as long as basal width .......... 76

73. Ventral plate in ventral view with medial cleft extended about one third to one half distance to anterior margin (Fig. 10.360). Gonostylus with apical spinule .......... ***canadense*** (p. 369)

– Ventral plate in ventral view with medial cleft extended more than half distance to anterior margin (Figs. 10.369–10.371). Gonostylus without apical spinule (*pictipes* species group) .......... 74

74. Ventral plate in ventral view with outer lateral margins bowed outward (Fig. 10.370) .......... ***innoxium*** (p. 375)

– Ventral plate in ventral view with outer lateral margins straight or convergent posteriorly (Figs. 10.369, 10.371) .......... 75

75. Ventral plate in ventral view with margins of medial cleft divergent (Fig. 10.369) .......... ***claricentrum*** (p. 374)

– Ventral plate in ventral view with margins of medial cleft subparallel (Fig. 10.371) .......... ***pictipes*** (p. 377)

76. Ventral plate in lateral view large, nearly circular (Fig. 10.361). Arizona, New Mexico, and Utah (Map 182) .......... ***freemani*** (p. 370)

– Ventral plate in lateral view smaller, noncircular (Figs. 10.362, 10.373, 10.391, 10.409). Widespread .......... 77

77. Ventral plate in ventral view with snoutlike lip extended well beyond posterior margin of body of plate (Figs. 10.362–10.368). West of Mississippi River (*paynei* species group) .......... 78

– Ventral plate in ventral view without snoutlike lip, although entire body of plate posterior to basal arms can be laterally compressed (Figs. 10.373, 10.389–10.392, 10.409). Widespread .......... 82

78. Ventral plate in ventral view rather deeply concave on either side of lip (Figs. 10.366, 10.367)........................................................................................................................................***solarii*** (p. 373)

– Ventral plate in ventral view convex to weakly concave on either side of lip (Figs. 10.362–10.365, 10.368) ..........................................................................................................................................................79

79. Gonostylus in ventral view nearly parallel sided (Figs. 10.364, 10.365) ..................................***paynei*** (p. 372)

– Gonostylus in ventral view with at least 1 lateral margin bowed (Figs. 10.362, 10.363, 10.368) .............80

80. Gonostylus in ventral view with medial margin bowed near midpoint (Fig. 10.363) ...................................................................................................................................................***hippovorum*** (p. 371)

– Gonostylus in ventral view with medial margin straight from base to at least midpoint (Figs. 10.362, 10.368)...........................................................................................................................................................81

81. Ventral plate in ventral view with snoutlike lip extended beyond posterior margin of plate by about 2 times its basal width; posterior margin not concave on either side of snoutlike lip (Fig. 10.362). Arizona and New Mexico (Map 183) ...........................................................................................***bricenoi*** (p. 371)

– Ventral plate in ventral view with snoutlike lip extended beyond posterior margin of plate by about 3–4 times its basal width; posterior margin slightly concave on either side of snoutlike lip (Fig. 10.368). Widespread (Map 187)........................................................................................***virgatum*** (p. 374)

82. Ventral plate in ventral view with basal arms each bearing distinct lateral projection (Fig. 10.375). Scutum with posterior one third shiny and bearing fine indistinct hair (Plate 2 h) (*jenningsi* species group)...................................................................................................................................................83

– Ventral plate in ventral view with basal arms each not bearing distinct lateral projection (Figs. 10.389–10.392, 10.407–10.409), or if these present, then scutum with posterior one fourth shiny and clothed with coarse hair.........................................................................................................................................95

83. Gonostylus in ventral view with outer margin sinuous (Fig. 10.385)..............................***podostemi*** (p. 389)

– Gonostylus in ventral view with outer margin nearly straight (Fig. 10.383).................................................84

84. Ventral plate in lateral and terminal views flat, without lip (Fig. 10.381) .....................***krebsorum*** (p. 385)

– Ventral plate in lateral and terminal views, with prominent lip (Figs. 10.374–10.377)..............................85

85. Ventral plate in ventral view with body about 3 times as long as distal width, slightly narrowed posteriorly or with margins subparallel (Fig. 10.372). Southeastern piedmont to mountains (Map 192) ......................................................................................................................................***aranti*** (p. 379)

– Ventral plate in ventral view with body less than 3 times as long as distal width, expanded or narrowed posteriorly or with margins subparallel (Figs. 10.373–10.377). Widespread in various ecoregions ..........................................................................................................................................................86

86. Ventral plate in ventral view narrowed posteriorly (Figs. 10.375, 10.384)......................................................87

– Ventral plate in ventral view expanded posteriorly or with sides subparallel (Figs. 10.379, 10.380, 10.382) ..........................................................................................................................................................88

87. Ventral plate in lateral view with minute lip (Fig. 10.375). Southeastern coastal plain and Sandhills (Map 196) ..................................................................................................................................***dixiense*** (p. 381)

– Ventral plate in lateral view with well-developed lip (Fig. 10.384). Interior lowlands of central United States (Map 207) ......................................................................................................................***ozarkense*** (p. 388)

88. Ventral plate in ventral view with lateral margins of body subparallel (Figs. 10.379, 10.382, 10.383) ..........................................................................................................................................................89

– Ventral plate in ventral view with lateral margins of body divergent (Figs. 10.374, 10.376–10.378, 10.380) ..........................................................................................................................................................91

89. Ventral plate in lateral view with lip rather thin (Fig. 10.382) ..............................................***luggeri*** (p. 386)

– Ventral plate in lateral view with lip thicker (Figs. 10.379, 10.383) ..................................................................90

90. Ventral plate in ventral view with space between basal arms somewhat V shaped (Fig. 10.379) ...................................................................................................................................................***jenningsi*** (p. 382)

– Ventral plate in ventral view with space between basal arms broadly U shaped (Fig. 10.383)...........................***anchistinum*** (p. 378), ***nyssa*** (p. 388), ***penobscotense*** (p. 388), ***remissum*** (p. 389)

91. Ventral plate in terminal view twice as wide as long (excluding apical projection) (Fig. 10.374) ...................................................................................................................................................***definitum*** (p. 380)

– Ventral plate in terminal view less than twice as wide as long (excluding apical projection) (Figs. 10.373, 10.376–10.378) ..........................................................................................................................................92

92. Ventral plate in ventral view with distal width of body greater than length along midline (Fig. 10.380) ........ ***jonesi*** (p. 384)

– Ventral plate in ventral view with distal width of body equal to or less than length along midline (Figs. 10.373, 10.376–10.378 ) ........ 93

93. Ventral plate in terminal view with apical projection broadly joined to body (Fig. 10.377). Coastal plain of Alabama and western Florida to Texas (Map 198) ........ ***haysi*** (p. 382)

– Ventral plate in terminal view with apical projection more abruptly and narrowly joined to body (Figs. 10.373, 10.376, 10.378). Widespread in various ecoregions ........ 94

94. Ventral plate in terminal view with lateral margins bowed outward (Fig. 10.376) ........ ***fibrinflatum*** (p. 381), ***notiale*** (p. 387), ***snowi*** (p. 390)

– Ventral plate in terminal view with lateral margins straighter, not bowed outward (Figs. 10.373, 10.378) ...... ***chlorum*** (p. 379), ***confusum*** (p. 379), ***infenestrum*** (p. 382), ***lakei*** (p. 385), ***taxodium*** (p. 390)

95. Ventral plate in ventral view with body rather broad, and lateral margins slightly divergent; in terminal view somewhat trilobed, about as long as wide (Fig. 10.408) (*venustum* species complex, p. 421) ........ ***hematophilum*** (p. 425), ***incognitum*** (p. 425), ***irritatum*** (p. 426), ***minutum*** (p. 426), ***molestum*** (p. 427), ***piscicidium*** (p. 427), ***truncatum*** (p. 428), ***venustum*** s. s. (p. 429) (males of ***tormentor*** n. sp. [p. 427] and ***venustum* 'cytospecies JJ'** [p. 430] are unknown but probably key out here)

– Ventral plate in ventral view with body laterally compressed, somewhat V or Y shaped; in terminal view slender, longer than wide (Figs. 10.387–10.392, 10.407, 10.409) ........ 96

96. Ventral plate in ventral view broadly V shaped; in lateral view large, with lip bluntly rounded (Fig. 10.392) (*noelleri* species group) ........ 97

– Ventral plate in ventral view narrowly V or Y shaped; in lateral view small, with lip rather pointed (Figs. 10.387–10.391, 10.407, 10.409) ........ 98

97. Widespread, including Alaska and northern Canada (Map 226) ........ ***decorum*** (p. 403)

– Alaska and northern Canada (Map 227) ........ ***noelleri*** (p. 405)

98. Ventral plate in ventral view strongly Y shaped; in lateral view with inner margin nearly straight (Figs. 10.387–10.391) (*malyschevi* species group, in part) ........ 99

– Ventral plate in ventral view rather V shaped; in lateral view with inner margin produced as prominent lip (Figs. 10.407, 10.409) ........ 102

99. Femora mostly yellowish ........ ***defoliarti*** (p. 391)

– Femora mostly brown ........ 100

100. Ventral plate in ventral view with body wider than each arm (Fig. 10.388). Alaska, central and western Canada (Map 215) ........ ***malyschevi*** (p. 392)

– Ventral plate in ventral view with body as wide as or narrower than each arm (Figs. 10.389–10.391). Widespread ........ 101

101. Ventral plate in ventral view slightly expanded posteriorly or with lateral margins subparallel (Figs. 10.390, 10.391) (*arcticum* species complex, p. 394) ........ ***apricarium*** n. sp. (p. 395), ***arcticum*** s. s. (p. 396), ***brevicercum*** (p. 398), ***chromatinum*** n. sp. (p. 398), ***negativum*** n. sp. (p. 399), ***saxosum*** n. sp. (p. 399), ***vampirum*** (p. 400) (males of ***arcticum* 'cytospecies IIL-1'** [p. 397] and ***arcticum*** **'cytospecies IIS-4'** [p. 397] are unknown but probably key out here)

– Ventral plate in ventral view slightly tapered posteriorly or strongly compressed (almost flattened) laterally (Fig. 10.389) ........ ***murmanum*** (p. 393)

102. Ventral plate in ventral and terminal views gradually tapered (Fig. 10.409). Widespread (Maps 253, 254) (*verecundum* species complex, p. 430) ........ ***rostratum*** (p. 432), ***verecundum*** s. s. (p. 433)

– Ventral plate in ventral and terminal views slightly expanded at midpoint of body (Fig. 10.407). Alaska and northwestern Canada (Map 242) ........ ***rubtzovi*** (p. 421)

### KEY TO PUPAE OF NORTH AMERICAN BLACK FLIES

For species identification of pupae, users should proceed directly to the Summary Key to Species, which follows the key to Genera.

#### Genera

1. Cocoon rudimentary or shapeless, saclike, covering various portions of abdomen and thorax (Figs. 10.411, 10.412, 10.416, 10.422) ........ 2

– Cocoon shaped like slipper, shoe, or boot, with definitely formed, rigid walls (sometimes coarsely woven), typically covering abdomen and thorax (Figs. 10.427, 10.428, 10.435, 10.480, 10.499)............10

2. Gill of 3 moderately inflated filaments arising from elongate base (Fig. 10.410); felt chamber visible dorsally within gill base (Fig. 4.60). Abdominal sternites VI and VII complete, not divided medially by longitudinal striate membrane. Pacific Northwest............................................................***Parasimulium***

– Gill of various numbers of filaments arising from short base; felt chamber absent from gill base. Abdominal sternites VI and VII divided by longitudinal striate membrane (Fig. 4.59). Widespread...................3

3. Abdominal sternites IV–VII each with 2–5 pairs of strong recurved hooks (Figs. 10.512–10.516). Gill with 2–4 (rarely 5) slightly inflated filaments (Figs. 10.536–10.538). Cocoon a small ventral pad (typically not collected with pupa)..........................................................................................................***Gymnopais***

– Abdominal sternites IV–VII each with no more than 2 pairs of recurved hooks. Gill with more than 4 filaments. Cocoon covering at least part of pupal dorsum...............................................................................4

4. Gill of 5 inflated tubes each bearing 0–10 tiny secondary filaments (Fig. 10.423). Southwestern United States (Map 86)..........................................................................................***Gigantodax*** (***G. adleri***, p. 294)

– Gill of 6 or more slender filaments or inflated tubes with or without tiny secondary filaments. Widespread...........................................................................................................................................5

5. Abdominal segments IV and V each with large pleurite set in striate pleural membrane of each side (Fig. 4.29); abdominal segment III either ringlike and lacking striate membrane, or with pleurite fused with sternite and separated from tergite by narrow longitudinal band of striate membrane..........................................................................................................................................6

– Abdominal segments III–V with longitudinal band of striate pleural membrane, without pleurites or with, at most, minute pleurites (Figs. 10.414, 10.420)...........................................................................7

6. Abdominal tergites VI–IX each with transverse row of spine combs along anterior margin (Figs. 4.29, 4.61); tergites III and IV each with 4 pairs of recurved hooks. Cocoon of fine, nongelatinous, readily discernible threads variously enclosing abdomen and sometimes thorax (Figs. 10.411, 10.412), but often sufficiently open anteriorly to expose gill and allow loss of larval head capsule; if head capsule retained, then not bullet shaped .......................................................................***Prosimulium***, ***Helodon***

– Abdominal tergites devoid of spine combs; tergites III and IV each with 3 pairs of recurved hooks. Cocoon of thick, gelatinous, transparent strands not readily discernible except when covered with debris, enclosing pupa and gill, and retaining bullet-shaped larval head capsule (Fig. 10.685)....................................................................................................................***Twinnia***

7. Abdominal segments VIII and IX laterally with hook-shaped setae (Fig. 10.600). Gill of 17–50 filaments arising in more than 3 groups from short, slightly inflated base (Fig. 10.424)............................***Cnephia***

– Abdominal segments VIII and IX laterally either with straight or slightly curved setae, or without setae (Fig. 10.599). Gill of various numbers of filaments arising from base in 2 or 3 main groups (Figs. 10.414–10.422) ........................................................................................................................8

8. Gill of 10 or 12 filaments arising from base on 2 or 3 slender trunks (Figs. 10.419, 10.574, 10.575)..............................................................................................................................***Stegopterna***

– Gill of fewer than 10 filaments, or with 15–30 filaments; if gill of 12 filaments, then base rather inflated and filaments arising in 4 or 5 main groups ..........................................................................................9

9. Gill of 6–12 moderately inflated, rigid filaments, typically radiated laterally from base (Figs. 10.421, 10.422, 10.576, 10.577) ..........................................................................................................***Tlalocomyia***

– Gill of 15–30 slender, delicate filaments, typically projected forward (Figs. 10.414–10.418) .........***Greniera***

10. Gill of 8 or 10 stout filaments arising close to base and converging anteriorly toward common point (Figs. 10.578–10.580). Cocoon slipper shaped, with somewhat irregular anterior margin, and attached to silk stalk with basal holdfast (Fig. 10.425). Terminal spines relatively prominent, stout (Fig. 10.601)........................................................................................................................***Ectemnia***

– Gill with various numbers of filaments, but if 8 or 10, then filaments neither inflated nor converging anteriorly toward common point. Cocoon variously shaped, but not attached to silk stalk. Terminal spines typically short or apparently absent (Fig. 10.602), or elongate and slender in a few species...........................................................................................................................................11

11. Cocoon boot shaped (sometimes short in *M. borealis*), loosely or tightly woven, but without definitely formed anterior apertures or loops (Fig. 10.427). Pleural region of abdominal segments VIII and IX with numerous anchor- or grapnel-shaped setae (Fig. 10.602) ..............................................***Metacnephia***

– Cocoon variously shaped (Figs. 10.428–10.442), but if boot shaped, then with definitely formed apertures, loops, or perforations anteriorly (Figs. 10.480, 10.499–10.502), except in *S. freemani* and *S. solarii* (Figs. 10.477, 10.479). Pleural region of abdominal segments VIII and IX at most with unbranched setae ........ ***Simulium***

### Summary Key to Species

Pupae of *Parasimulium* 'species A,' *Parasimulium melanderi*, *Parasimulium furcatum*, *Greniera* 'species F,' *Simulium* 'species O,' and *Simulium* 'species Z' are unknown. Other species for which the pupa is unknown are keyed on the basis of the gill histoblast of the mature larva.

In the following summary key, the term "filaments" refers to all terminal branches of the gill regardless of their thickness. Filaments sometimes differ in number and branching pattern on opposite sides of the same individual. Rare variants (<5% of individuals) in filament number are not included in the key. *Metacnephia villosa*, which has 4 or 5 inflated trunks from which arise 150 or more short threadlike filaments (Fig. 10.585) that often are broken off in older specimens, is considered to have more than 17 filaments. *Gigantodax adleri*, which has 5 inflated tubes from each of which arise 0–10 tiny filaments (Fig. 10.423), is considered to have 5 or 6 filaments.

| Number of filaments per gill | Starting couplet |
|---|---|
| 1 or 2 | 1 |
| 3 or 4 | 5 |
| 5 or 6 | 43 |
| 7 | 66 |
| 8 | 68 |
| 9–12 | 101 |
| 13–16 | 138 |
| 17 or more | 162 |

### Species

1. Gill of 1 annulate club, with 1 dorsal and 1 ventral projection basally (Fig. 10.476) ........ ***Simulium canadense*** (p. 369)
– Gill of 2 or more filaments, sometimes arising from swollen club ........ 2

2. Gill of 2 filaments subequal in length and thickness ........ 3
– Gill of 3 or more filaments; if 2 filaments, then these unequal in length and thickness ........ 5

3. Cocoon a small ventral pad (typically not collected with pupa) ........ ***Gymnopais holopticus*** (p. 241)
– Cocoon slipper shaped, covering most of pupal body ........ 4

4. Gill filaments thickened basally (Fig. 10.586). Central North America south of Northwest Territories (Map 123) ........ ***Simulium johannseni*** (in part) (p. 321)
– Gill filaments not markedly thickened. Alaska and northern Canada (Map 155) ........ ***Simulium subpusillum*** (in part) (p. 348)

5. Cocoon a small ventral pad (typically not collected with pupa). Gill of 3 or 4 (rarely 2) filaments (Figs. 10.536–10.538). Abdominal tergites without spine combs (Figs. 10.512–10.516) ........ 6
– Cocoon covering at least part of pupal dorsum (Figs. 10.411–10.413, 10.425–10.428). Gill of 3–200 filaments. Abdominal tergites either with spine combs on at least segments VIII and IX or without spine combs ........ 9

6. Abdominal sternites IV and V each with 2 pairs of recurved hooks (Fig. 10.514) ........ ***Gymnopais fimbriatus*** (p. 240)
– Abdominal sternites IV and V each with 4 or 5 pairs of recurved hooks (Figs. 10.512, 10.513, 10.515, 10.516) ........ 7

7. Terminal spines present, although small (Fig. 10.515). Gill filaments not particularly swollen at base (Fig. 10.538) ........ ***Gymnopais holopticoides*** (p. 240)
– Terminal spines absent (Figs. 10.512, 10.513). Gill filaments swollen at base (Figs. 10.536, 10.537) ........ 8

8. Gill typically of 3 (sometimes 4) branches arranged as 2 short, stout dorsal branches and usually 1 long, undivided ventral branch, but if divided, the 2 filaments unequal in length (Figs. 10.512, 10.536)........ ***Gymnopais dichopticoides*** (p. 239)
– Gill of 4 branches arranged as 2 short, stout dorsal branches and 2 longer, slender ventral branches nearly equal in length (Figs. 10.513, 10.537)........ ***Gymnopais dichopticus*** (p. 240)

9. Gill of 3 filaments, sometimes with rudimentary fourth filament (Figs. 10.410, 10.438)........ 10
– Gill of 4 or more filaments........ 13

10. Cocoon shapeless, saclike. Pacific Northwest (Maps 3, 5)........ ***Parasimulium crosskeyi, P. stonei*** (p. 237)
– Cocoon well formed, slipper shaped (Figs. 10.438, 10.439). Widespread........ 11

11. Cocoon with anterodorsal projection (Fig. 10.438)........ ***Simulium baffinense*** (p. 320)
– Cocoon without anterodorsal projection (Fig. 10.439)........ 12

12. Gill with filaments thickened basally (Figs. 10.439, 10.586). Central North America south of Northwest Territories (Map 123)........ ***Simulium johannseni*** (in part) (p. 321)
– Gill with filaments not markedly thickened basally. Alaska and northern Canada (Map 155)........ ***Simulium subpusillum*** (in part) (p. 348)

13. Gill of 4 filaments........ 14
– Gill of 5 or more filaments........ 42

14. Cocoon with long anterodorsal projection (Figs. 10.433, 10.444, 10.450, 10.520b)........ 15
– Cocoon without anterodorsal projection (Figs. 10.434–10.437), although anterodorsal margin may be ragged (Fig. 10.520a)........ 19

15. Gill in lateral view with dorsalmost filament divergent basally from ventralmost filament at angle of 50 degrees or greater (Figs. 10.433, 10.450)........ 16
– Gill in lateral view with dorsalmost filament divergent basally from ventralmost filament at angle of 50 degrees or less (Fig. 10.444)........ 18

16. Head and thorax with flat, uniformly distributed microtubercles (as in Fig. 10.836). Eastern United States, especially southeastern portion (Map 145)........ ***Simulium loerchae*** (p. 341)
– Head and thorax with rounded, irregularly distributed microtubercles (as in Figs. 10.825, 10.834). Widespread........ 17

17. Gill with filaments of dorsal pair branched more in horizontal plane (Fig. 10.433). Northeastern North America (Map 113)........ ***Simulium rivuli*** (p. 313)
– Gill with filaments of dorsal pair branched more in vertical plane (as in Fig. 10.450). Widespread (Map 151)........ ***Simulium silvestre*** (p. 344)

18. Cocoon with anterodorsal projection bifurcate, or if not bifurcate, short and broad (Fig. 10.520)........ ***Simulium bicorne*** (in part) (p. 332)
– Cocoon with anterodorsal projection long, slender, not bifurcate (Fig. 10.444)........ ***Simulium craigi*** (p. 336)

19. Cocoon with 1 or more large anterior apertures on each side (Fig. 10.497)........ ***Simulium snowi*** (p. 390)
– Cocoon without large apertures........ 20

20. Gill with ventral trunk at least 4 times length of dorsal trunk (Fig. 10.449)........ ***Simulium impar*** (p. 340)
– Gill with ventral trunk less than 3 times length of dorsal trunk (Figs. 10.434, 10.436, 10.437, 10.440)........ 21

21. Gill with filaments markedly downturned and running along substrate (Fig. 10.440); surface sculpture minutely reticulated, especially basally. Southeastern United States, west into Texas (Map 124)........ ***Simulium parmatum*** n. sp. (p. 322)
– Gill with filaments projected anteriorly, not markedly downturned (Figs. 10.439, 10.443, 10.451); surface sculpture ridged, furrowed, or reticulated (Figs. 10.452, 10.586, as in 10.591, 10.837). Widespread........ 22

22. Gill with at least 2 of the 4 filaments swollen basally (Figs. 10.439, 10.441, 10.586)........ 23
– Gill with none or 1 of the 4 filaments swollen basally (Figs. 10.434–10.437)........ 24

23. Gill with dorsal 2 filaments markedly swollen basally, about 2–3 times thicker than ventralmost filament (Fig. 10.441). Northeastern North America (Map 125)........ ***Simulium rothfelsi*** (p. 323)

– Gill with all filaments swollen basally (Figs. 10.439, 10.586). Central North America and southeastern coastal plain (Map 123)........................................................................***Simulium johannseni*** (in part) (p. 321)

24. Gill with either dorsal or ventral trunk at least 5 times longer than wide (Figs. 10.434, 10.436)...........25
   – Gill with dorsal and ventral trunks each less than 4 times longer than wide (Figs. 10.435, 10.437, 10.443)..........................................................................................................................27

25. Gill with dorsal trunk more than 1.25 times as thick as ventral trunk (Fig. 10.436). Southeastern United States (Map 117)........................................................................................***Simulium clarkei*** (in part) (p. 317)
   – Gill with dorsal and ventral trunks subequal in thickness (Fig. 10.434). Western North America (i.e., west of 90° longitude)...........................................................................................................26

26. Head and thorax smooth, or with sparsely distributed microtubercles. Widespread (Map 115) ........................................................................................................***Simulium balteatum*** n. sp. (p. 315)
   – Head and thorax covered with fine microtubercles. Alaska and northern Canada (Map 155).........***Simulium subpusillum*** (in part) (p. 348)

27. Gill in lateral view with dorsalmost filament strongly divergent from other 3 filaments (Fig. 10.443) (subgenus *Eusimulium*, p. 326).................................***Simulium bracteatum*** (p. 328), ***S. donovani*** (p. 328), ***S. exulatum*** n. sp. (p. 329), ***S. pilosum*** (p. 330), ***S. violator*** n. sp. (p. 330)
   – Gill in lateral view with dorsalmost filament not strongly divergent from other 3 filaments (Figs. 10.435, 10.451, 10.452).........................................................................................................28

28. Head and thorax smooth, virtually without microtubercles (microgranules can be present and visible with scanning electron microscopy, Fig. 10.828)............................................................................29
   – Head and thorax with numerous microtubercles (Figs. 10.826, 10.827, 10.829)...........................................30

29. Gill in lateral view with filaments dorsoventrally divergent (as in Fig. 10.435)..........................***Simulium joculator*** n. sp. (p. 318)
   – Gill in lateral view with filaments subparallel, in compact bundle, not divergent....................***Simulium merritti*** n. sp. (p. 341)

30. Head and thorax with densely distributed, thin spinelike microtubercles (Fig. 10.827). Western United States and southwestern Canada (Map 137)............................................***Simulium conicum*** n. sp. (p. 335)
   – Head and thorax with numerous rounded, sometimes very fine microtubercles (Figs. 10.826, 10.829). Widespread.......................................................................................................................31

31. Gill filaments (slide mounted) with reticulate surface pattern (Fig. 10.452). Eastern North America (i.e., east of 100° longitude) (Map 150).........................................................***Simulium quebecense*** (p. 344)
   – Gill filaments (slide mounted, if necessary) with transverse furrows (as in Figs. 10.591, 10.837). Widespread.......................................................................................................................32

32. Gill in lateral view with filaments divergent dorsoventrally at base (Figs. 10.435, 10.455, 10.507).........33
   – Gill in lateral view with filaments in compact bundle at base, not strongly divergent dorsoventrally (Figs. 10.451, 10.837)........................................................................................................37

33. Gill with ventral pair of filaments branched in horizontal plane (Fig. 10.455)............................***Simulium subpusillum*** (in part) (p. 348)
   – Gill with ventral pair of filaments branched in vertical plane (Figs. 10.435, 10.507).................................34

34. Gill with filaments nearly sessile, trunks no longer than wide; both trunks subequal in thickness, and bases of all filaments subequal in thickness (Fig. 10.507). Alaska to Manitoba (Map 232).........................................................................................***Simulium transiens*** (p. 411)
   – Gill typically with at least 1 trunk 2 or more times longer than wide; dorsal trunk 1.5–2.0 times thicker than ventral trunk, and bases of dorsal filaments up to 2 times thicker than bases of ventral filaments (Figs. 10.435, 10.437). Widespread (*S. annulus* species group, in part).....................................35

35. Eastern North America (i.e., east of 100° longitude)..............................................................................36
   – Western North America (i.e., west of 100° longitude)..........................***Simulium annulus*** (in part) (p. 314), ***S. canonicolum*** (p. 316), ***S. quadratum*** (p. 318), ***S. zephyrus*** n. sp. (p. 319)

36. Rocky or sandy streams in eastern North America (Maps 114, 118)................................***Simulium annulus*** (in part) (p. 314), ***S. emarginatum*** (p. 317)
   – Sandy or swampy streams in southeastern United States (Map 117)................................***Simulium clarkei*** (in part) (p. 317)

37. Gill with filaments of each pair branched in vertical plane (Fig. 10.451) ........................................ 38
– Gill with filaments of each pair branched in horizontal plane (Figs. 10.520a, 10.522) ........................................ 40

38. Head and thorax with scattered, small microtubercles (Fig. 10.829) ........................................ ***Simulium modicum*** n. sp. (in part) (p. 342), ***S. pugetense*** (p. 343)
– Head and thorax with densely distributed, large microtubercles (Fig. 10.826) ........................................ 39

39. Eastern North America (i.e., east of 100° longitude) ........................................ ***Simulium modicum*** n. sp. (in part) (p. 342), ***S. moultoni*** n. sp. (p. 343)
– Western North America (i.e., west of 100° longitude) ........................................ ***Simulium carbunculum*** n. sp. (p. 334)

40. Northeastern North America (Map 133) ........................................ ***Simulium aestivum*** (p. 332)
– Alaska and northwestern Canada (Maps 134, 142) ........................................ 41

41. Cocoon in dorsal view more elongate than circular, not exposing most of thorax; anterior margin weakly reinforced and often ragged, at least medially (Fig. 10.520a) ........................................ ***Simulium bicorne*** (in part) (p. 332)
– Cocoon in dorsal view nearly circular, exposing most of thorax; anterior margin smooth and strongly reinforced (Fig. 10.522) ........................................ ***Simulium fontinale*** (p. 338)

42. Gill of 5 or 6 filaments ........................................ 43
– Gill of 7 or more filaments ........................................ 65

43. Gill of swollen tubular filaments, with at least 2 filaments projected posteriorly (Figs. 10.423, 10.577). Cocoon a loosely woven, shapeless sleeve ........................................ 44
– Gill of thin unswollen filaments, or if swollen, all filaments projected anteriorly. Cocoon well formed, shaped like slipper, shoe, or boot ........................................ 45

44. Gill of 5 tubular filaments, each with 0–10 tiny secondary filaments (Fig. 10.423). Southwestern United States (Map 86) ........................................ ***Gigantodax adleri*** (p. 294)
– Gill of 6 tubular filaments lacking secondary filaments (Fig. 10.577). California mountains (Map 85) ........................................ ***Tlalocomyia stewarti*** (p. 293)

45. Cocoon with anterodorsal projection (sometimes short) (Figs. 10.431, 10.432, 10.448, 10.496, 10.518, 10.519) ........................................ 46
– Cocoon without anterodorsal projection (Figs. 10.486, 10.504, 10.508–10.510) ........................................ 52

46. Gill filaments arising from elongate swollen base (Figs. 10.432, 10.519) ........................................ ***Simulium mysterium*** n. sp. (p. 312)
– Gill filaments arising from short unswollen or only slightly swollen base (Figs. 10.431, 10.448, 10.496, 10.506, 10.518) ........................................ 47

47. Cocoon with anterodorsal projection short (Figs. 10.496, 10.506) ........................................ 48
– Cocoon with anterodorsal projection long (Figs. 10.431, 10.448, 10.518) ........................................ 49

48. Gill with each trunk 3 times or less longer than wide (Fig. 10.496). Southeastern mountains (Map 210) ........................................ ***Simulium remissum*** (p. 389)
– Gill with at least 1 trunk 5 times or more longer than wide (Fig. 10.506). Eastern coastal plain (Map 231) ........................................ ***Simulium slossonae*** (p. 410)

49. Gill of 5 or 6 filaments, with 2 ventral trunks arising from common stem (Fig. 10.518). Cocoon in dorsal view appearing top heavy, with anterodorsal projection broad basally. West of Rocky Mountains (Maps 105, 106) ........................................ ***Simulium minus*** (p. 309), ***S. nebulosum*** (p. 309)
– Gill of 6 filaments, with 2 ventral trunks typically arising independently from base (Figs. 10.431, 10.448). Cocoon in dorsal view with anterodorsal projection narrower basally. Widespread ........................................ 50

50. Gill with at least 1 pair of filaments branched in vertical plane (Fig. 10.448). Eastern North America (i.e., east of 95° longitude) (Map 143) ........................................ ***Simulium gouldingi*** (p. 340)
–410 ll with each pair of filaments branched in horizontal plane (Fig. 10.431). Widespread ........................................ 51

51. Head and thorax with microtubercles raised, rounded, irregularly spaced (as in Fig. 10.825). Rocky Mountains westward (Map 110) ........................................ ***Simulium curriei*** (in part) (p. 311)
– Head and thorax with microtubercles rather flattened, uniformly spaced (as in Figs. 10.835, 10.836). Widespread (Map 111) ........................................ ***Simulium excisum*** (p. 312)

52. Cocoon with 1 or more anterior apertures per side (Figs. 10.486, 10.493) ........................................ 53

- Cocoon without apertures (Figs. 10.460, 10.462, 10.504, 10.508–10.510)....54

53. Gill with middle pair of filaments on short petiole arising from base of ventral pair; filaments often, but not always, swollen their entire length (Fig. 10.486)........***Simulium fibrinflatum*** (in part) (p. 381)
- Gill with middle pair of filaments nearly sessile, and arising from base of gill or from base of dorsal pair; filaments never swollen their entire length (Fig. 10.493)....***Simulium notiale*** (p. 387)

54. Head and thorax with reticulate rugosity (Fig. 10.530). Eastern North America (i.e., east of 95° longitude) (Map 228)....***Simulium parnassum*** (in part) (p. 407)
- Head and thorax smooth or with microtubercles, but without rugosity (Figs. 10.830, 10.835). Widespread....55

55. Cocoon shoe shaped, with prominent raised anteroventral collar (Fig. 10.504)....***Simulium petersoni*** (p. 408)
- Cocoon slipper shaped, without raised anteroventral collar (Figs. 10.474, 10.475, 10.508–10.510) or with small raised anteroventral collar in some specimens (Fig. 10.460)....56

56. Thoracic trichomes multiply branched (Fig. 10.475)....***Simulium tescorum*** (p. 368)
- Thoracic trichomes unbranched or bifid (Figs. 10.462, 10.474)....57

57. Gill with 2 trunks arising from base; at least 1 trunk or 1 petiole more than 3 times longer than wide (Figs. 10.460, 10.462)....58
- Gill with 3 trunks arising from base (Figs. 10.474, 10.509, 10.510, 10.597, 10.598), or if 2 arising from base, then with trunks less than 3 times longer than wide (Fig. 10.508)....59

58. Thorax dorsally with densely distributed microtubercles. Gill of 5 or 6 filaments (Fig. 10.460). Rio Grande of New Mexico (Map 163)....***Simulium meyerae*** (p. 352)
- Thorax dorsally with scattered microtubercles. Gill typically of 6 filaments (Fig. 10.462). Oklahoma and Texas (Map 166)....***Simulium trivittatum*** (p. 353)

59. Gill with at least 2 trunks longer than wide (Figs. 10.474, 10.509)....60
- Gill with at least 2 trunks subequal in length and width (Figs. 10.597, 10.598)....61

60. Head and thorax nonshiny, with densely and uniformly distributed microtubercles (Figs. 10.835, 10.836). Eastern United States (Map 239)....***Simulium ubiquitum*** n. sp. (p. 418)
- Head and thorax shiny, with scattered, irregularly distributed microtubercles. Southwestern United States (Map 179)....***Simulium puigi*** (p. 368)

61. Gill filaments in lateral view compact, narrowly spread (Fig. 10.508). Sierra Nevada of California; high elevations (Map 234)....***Simulium chromocentrum*** n. sp. (p. 414)
- Gill filaments in lateral view moderately to widely spread dorsally to ventrally (Figs. 10.510, 10.597, 10.598). Widespread....62

62. Gill filaments in dorsal view divergent laterally (Fig. 10.524). Head and thorax with uniformly arranged microtubercles (as in Figs. 10.835, 10.836) (*Simulium verecundum* species complex, p. 430)....***Simulium rostratum*** (p. 432), ***S. verecundum*** s. s. (p. 433)
- Gill filaments in dorsal view aligned more or less in vertical plane (Fig. 10.523). Head and thorax smooth or with variously arranged microtubercles....63

63. Gill typically more than half as long as pupal body, and with all filaments subequal in thickness (Fig. 10.597). Abdomen of young pupa typically other than dark gray or blackish (*Simulium venustum* species complex, p. 421)....***Simulium hematophilum*** (p. 425), ***S. incognitum*** (p. 425), ***S. irritatum*** (p. 426), ***S. minutum*** (p. 426), ***S. molestum*** (p. 427), ***S. piscicidium*** [p. 427], ***S. truncatum*** (p. 428), ***S. venustum*** s. s. (p. 429) (pupae of ***S. tormentor*** n. sp. [p. 427] and ***S. venustum* 'cytospecies JJ'** [p. 430] are unknown but presumably key out here)
- Gill often about half as long as pupal body, and with filaments decreasing in thickness from dorsalmost to ventralmost (Fig. 10.510). Abdomen of young pupa dark gray or blackish....64

64. Head and thorax with numerous microtubercles (Figs. 10.831, 10.834). Widespread; abundant (*Simulium tuberosum* species complex, in part, p. 411)....***Simulium appalachiense*** n. sp. (p. 413), ***S. conundrum*** n. sp. (p. 415), ***S. perissum*** (p. 416), ***S. tuberosum*** s. s. (p. 417), ***S. twinni*** (p. 418), ***S. vandalicum*** (p. 419), ***S. vulgare*** (p. 420)
- Head and thorax smooth, without microtubercles. Eastern North America; rare....***Simulium parnassum*** (in part) (p. 407)

65. Gill of 7 filaments. Cocoon with 1 or more large anterior apertures per side (Figs. 10.487, 10.491) .......66
– Gill of 8 or more filaments. Cocoon with or without large anterior apertures .......67

66. Gill with dorsal trunk giving rise to 1 pair of filaments (Fig. 10.487). Large streams along Gulf Coast (Map 198) ....... ***Simulium haysi*** (p. 382)
– Gill with dorsal trunk giving rise to 2 pairs of filaments (Fig. 10.491). Small streams in Sandhills of North Carolina and South Carolina (Map 202) ....... ***Simulium krebsorum*** (p. 385)

67. Gill of 8 filaments .......68
– Gill of 9 or more filaments .......100

68. Cocoon with anterodorsal projection (sometimes short) (Figs. 10.430, 10.445–10.447) .......69
– Cocoon without anterodorsal projection (Figs. 10.453, 10.456–10.459, 10.477, 10.478) .......74

69. Gill of 4 pairs of filaments (Fig. 10.430) .......70
– Gill of 1 pair of filaments and 2 triads, arranged 2 + 3 + 3 (Figs. 10.445–10.447, 10.587) .......71

70. Cocoon with anterodorsal projection long. Gill with middle trunk giving rise to 2 pairs of filaments (Fig. 10.430). Western United States, southwestern Canada (Map 110) ....... ***Simulium curriei*** (in part) (p. 311)
– Cocoon with anterodorsal projection short. Gill with filaments arising singly or in pairs directly from base. Northern Canada, possibly Alaska (Map 154) ....... ***Simulium giganteum*** (in part) (p. 347)

71. Gill in lateral view with dorsalmost filament not strongly divergent from other filaments; filaments not widely splayed (Fig. 10.446). Filaments in dorsal view subparallel .......72
– Gill in lateral view with dorsalmost filament divergent from other filaments, or with filaments widely splayed (Figs. 10.445, 10.447, 10.587). Filaments in dorsal view divergent laterally .......73

72. Cocoon with anterodorsal projection long (Fig. 10.446) ....... ***Simulium dendrofilum*** (p. 338)
– Cocoon with anterodorsal projection very short ....... ***Simulium furculatum*** (in part) (p. 347)

73. Gill with dorsalmost filament typically divergent at nearly 90-degree angle (Figs. 10.447, 10.587) ....... ***Simulium burgeri*** n. sp. (p. 333), ***S. fionae*** (p. 338)
– Gill with dorsalmost filament typically divergent at less than 90-degree angle (Fig. 10.445) ....... ***Simulium croxtoni*** (p. 337), ***S. giganteum*** (in part) (p. 347)

74. Pupa on silk stalk (Fig. 10.425). Gill filaments all converging distally toward common point (Fig. 10.578) .......75
– Pupa not on silk stalk. Gill filaments not all converging distally toward common point (Figs. 10.422, 10.477, 10.589) .......76

75. Gill with third pair of filaments from dorsum nearly sessile (Fig. 10.578). Widespread (Map 91) ....... ***Ectemnia invenusta*** (p. 300)
– Gill with third pair of filaments from dorsum petiolate (Fig. 10.426). Southeastern United States (Map 93) ....... ***Ectemnia reclusa*** (p. 301)

76. Cocoon saclike, loosely woven, covering abdomen only. Gill an open rosette of slightly inflated filaments (Fig. 10.422) ....... ***Tlalocomyia ramifera*** n. sp. (p. 292)
– Cocoon well formed, shaped like slipper or boot, covering most of pupa (Figs. 10.477, 10.495). Gill not an open rosette; filaments inflated or not .......77

77. Cocoon boot shaped, with raised anteroventral collar (Figs. 10.477, 10.478) .......78
– Cocoon slipper shaped, without raised anteroventral collar (Figs. 10.453, 10.498) .......82

78. Thorax with reticulate rugosity (Fig. 10.529) ....... ***Simulium bricenoi*** (p. 371)
– Thorax smooth or with microtubercles, but without rugosity .......79

79. Cocoon complete, without anterior apertures (Fig. 10.477) ....... ***Simulium freemani*** (p. 370)
– Cocoon with large loops of silk forming anterior apertures (Fig. 10.478) .......80

80. Alaska and northern Canada (Map 213) ....... ***Simulium decimatum*** (in part) (p. 390)
– Western United States and southwestern Canada .......81

81. Gill in frontal view open, palmate (Fig. 10.589), contiguous with gill of opposite side. Southern British Columbia to southern California (Map 184) ....... ***Simulium hippovorum*** (p. 371)
– Gill in frontal view clumped, not palmate, distinctly separated from gill of opposite side. East of California (Maps 185, 187) ....... ***Simulium paynei*** (p. 372), ***S. virgatum*** (p. 374)

82. Cocoon with 1 or more large anterior apertures per side (Fig. 10.498)............83
- Cocoon without large apertures (Fig. 10.453), although numerous small gaps may be present (Fig. 10.503)............88

83. Head and thorax smooth, without microtubercles............***Simulium fibrinflatum*** (in part) (p. 381)
- Head and thorax with sparsely to densely distributed microtubercles............84

84. Gill with ventral trunk and ventralmost petiole somewhat swollen (Fig. 10.484)............***Simulium definitum*** (in part) (p. 380)
- Gill typically without swelling of either ventral trunk or ventralmost petiole (Figs. 10.483, 10.498)............85

85. Gill with ventralmost petiole of ventral trunk typically more than one sixth length of gill (Fig. 10.483). Piedmont (Maps 193, 194)............86
- Gill with ventralmost petiole of ventral trunk typically less than one eighth length of gill (Fig. 10.498). Coastal plain and interior of Pennsylvania (Maps 203, 212)............87

86. Abdomen of freshly formed specimens yellowish green. Pennsylvania to South Carolina (Map 193)............***Simulium chlorum*** (p. 379)
- Abdomen of freshly formed specimens banded with red or gray. Southeastern United States, west to Nebraska and Texas (Map 194)............***Simulium confusum*** (p. 379)

87. Gill with dorsal petiole of ventral trunk typically about 2 times or less length of ventralmost petiole, or if more than 2 times length of ventralmost petiole, then occurring north of North Carolina. Pennsylvania to Florida (Map 203)............***Simulium lakei*** (in part) (p. 385)
- Gill with dorsal petiole of ventral trunk about 3 or more times length of ventralmost petiole (Fig. 10.498). North Carolina to Alabama and Florida (Map 212)............***Simulium taxodium*** (p. 390)

88. Gill with at least 1 filament arising independently from base, or with at least 1 trio of filaments on common trunk or petiole (Figs. 10.453, 10.458, 10.503)............89
- Gill with 4 pairs of petiolate filaments (Figs. 10.488, 10.495, 10.505, 10.511)............96

89. Gill in lateral view with dorsalmost filament strongly divergent from other filaments; dorsalmost 2 filaments typically thicker than other filaments (Fig. 10.453)............***Simulium wyomingense*** (p. 346)
- Gill in lateral view with dorsalmost filament not strongly divergent from other filaments; all filaments subequal in thickness (Figs. 10.454, 10.456–10.459, 10.503)............90

90. Cocoon loosely woven, with many small gaps, especially anteriorly. Gill with at least 1 filament arising independently from base (Fig. 10.503). Widespread (*Simulium noelleri* species group)............91
- Cocoon tightly woven, without gaps. Gill typically without filaments arising singly from base (Figs. 10.454, 10.456–10.459)............92

91. Widespread, including Alaska and northern Canada (Map 226)............***Simulium decorum*** (p. 403)
- Alaska and northern Canada (Map 227)............***Simulium noelleri*** (p. 405)

92. Gill with dorsal trunk giving rise to 2 filaments (Fig. 10.454). Canada and northeastern United States (Map 153)............***Simulium furculatum*** (in part) (p. 347)
- Gill with dorsal trunk giving rise to 3 or more filaments (Figs. 10.456–10.459). Western North America (i.e., west of 95° longitude) (subgenus *Psilopelmia*, in part)............93

93. Gill with 2 elongate trunks arising from base (Fig. 10.458). Arizona (Map 161)............***Simulium longithallum*** (p. 351)
- Gill with 3 short trunks arising from base (Fig. 10.456). Widespread............94

94. Gill with middle group of filaments branching 2 + 1 (Fig. 10.459). Texas (Map 162)............***Simulium mediovittatum*** (p. 352)
- Gill with middle group of filaments branching 1 + 2 (Fig. 10.456). Widespread............95

95. Head and thorax in lateral view forming rather smooth dorsal arch (Figs. 10.456, 10.457). Thoracic trichomes unbranched. Widespread............***Simulium bivittatum*** (p. 349), ***S. clarum*** (in part) (p. 350), ***S. griseum*** (p. 350), ***S. notatum*** (p. 352), ***S. venator*** (p. 353)
- Head in lateral view set off dorsally from thorax, such that anterior face of thorax is nearly perpendicular to dorsum of head (Fig. 10.461). Thoracic trichomes multiply branched. Southern New Mexico and Texas (Maps 160, 165)............***Simulium labellei*** (p. 351), ***S. robynae*** (p. 352)

96. Alaska and northwestern Canada (Map 242). Typically in small streams............***Simulium rubtzovi*** (p. 421)

– Widespread. Typically in rivers ....................97

97. Gill in lateral view with filaments tightly grouped, typically about half length of pupal body (Fig. 10.488). Known from North Carolina and South Carolina (Map 199).................... ***Simulium infenestrum*** (in part) (p. 382)

– Gill in lateral view with filaments not as tightly grouped, typically more than half length of pupal body (Figs. 10.495, 10.505). Widespread ....................98

98. Pennsylvania southward (Map 209) .................... ***Simulium podostemi*** (p. 389)

– North of Pennsylvania (Maps 208, 230) ....................99

99. Gill in dorsal view with filaments subparallel (Fig. 10.495). Head dorsally with microtubercles uniformly distributed (as in Figs. 10.835, 10.836). Northeastern North America (Map 208)........ ***Simulium penobscotense*** (in part) (p. 388)

– Gill in dorsal view with filaments laterally divergent (Fig. 10.505). Head dorsally with microtubercles irregularly distributed (as in Fig. 10.834). Canada and northern United States (Map 230).................... ***Simulium rugglesi*** (p. 409)

100. Gill of 9–12 filaments ....................101

– Gill of 13 or more filaments ....................137

101. Cocoon with anterodorsal projection (Figs. 10.428, 10.429, 10.473) ....................102

– Cocoon without anterodorsal projection (Figs. 10.480–10.483, 10.501, 10.502) ....................104

102. Thoracic trichomes 20 or more in number (Fig. 10.473).................... ***Simulium piperi*** (p. 367)

– Thoracic trichomes 10 or fewer in number (Figs. 10.428, 10.429)....................103

103. Gill of 10 filaments (Fig. 10.429).................... ***Simulium innocens*** (p. 308)

– Gill of 12 filaments (Fig. 10.428).................... ***Simulium anatinum*** (p. 307), ***S. congareenarum*** (p. 308), ***S. rendalense*** (p. 310), ***S. usovae*** (p. 310)

104. Pupa on silk stalk. Gill filaments all converging distally toward common point; filaments with surface sculpture scalelike (Figs. 10.579, 10.580) ....................105

– Pupa not on silk stalk. Gill filaments not all converging distally toward common point (Figs. 10.574, 10.591); filaments with surface sculpture furrowed or granular (Figs. 10.591, 10.592)....................106

105. Gill with fourth pair of filaments from dorsum petiolate (Fig. 10.579).......... ***Ectemnia primaeva*** (p. 301)

– Gill with fourth pair of filaments from dorsum nearly sessile (Fig. 10.580).................... ***Ectemnia taeniatifrons*** (p. 301)

106. Cocoon shapeless, saclike (as in Figs. 10.411, 10.412)....................107

– Cocoon well formed, shaped like slipper, shoe, or boot (Figs. 10.492, 10.501) ....................115

107. Gill of 9 filaments (Fig. 10.543).................... ***Helodon decemarticulatus*** (p. 244)

– Gill of 10–12 filaments ....................108

108. Gill of 10 or 11, rarely 12, filaments (Figs. 10.412, 10.574) ....................109

– Gill of 12, rarely 10 or 11, filaments (Figs. 10.575, 10.576) ....................110

109. Gill with ventral group of 4 filaments (Fig. 10.574). Alaska and northwestern Canada (Map 75) .................... ***Stegopterna decafilis*** (p. 285)

– Gill with ventral group of 3 filaments (Fig. 10.412). Southwestern United States (Map 46).................... ***Prosimulium rusticum*** n. sp. (p. 266)

110. Gill with ventral group of 5 filaments (Figs. 10.419, 10.575, 10.576)....................111

– Gill with ventral group of 3 or 4 filaments (Figs. 10.413, 10.421, 10.561) ....................113

111. Gill in anterior view opened cuplike (Figs. 10.420, 10.576).............. ***Tlalocomyia andersoni*** n. sp. (p. 290)

– Gill in anterior view not opened cuplike (Figs. 10.419, 10.575)....................112

112. Gill typically of 3 main groups of filaments, with branching pattern 4 + 3 + 5 (Fig. 10.575). Northern United States and Canada (Maps 77, 80)............ ***Stegopterna emergens*** (p. 286), ***St. trigonium*** (p. 289)

– Gill typically of 2 main groups of filaments, with branching pattern 7 + 5 (Fig. 10.419). Widespread.................... ***Stegopterna acra*** n. sp. (p. 284), ***St. diplomutata*** (p. 285), ***St. mutata*** (p. 287), ***St. permutata*** (p. 288), ***St. xantha*** n. sp. (p. 289)

113. Gill with dorsal group of 6 filaments (Fig. 10.413).................... ***Prosimulium unicum*** (p. 276)

– Gill with dorsal group of 4 filaments (Figs. 10.421, 10.561)....................114

114. Abdomen with pleurites on segments IV and V (as in Fig. 4.29)................***Prosimulium shewelli*** (p. 268)
– Abdomen without pleurites (Fig. 10.421)................................................................***Tlalocomyia osborni*** (p. 291)

115. Gill of 11 or 12 filaments..........................................................................................................................116
– Gill of 9 or 10 filaments ............................................................................................................................119

116. Gill swollen basally; filaments in anterior view forming open rosette (Fig. 10.535) ...................***Simulium defoliarti*** (p. 391)
– Gill not swollen basally; filaments in anterior view not forming open rosette (Figs. 10.492, 10.581, 10.596) ..........................................................................................................................................117

117. Cocoon without anterior apertures. Pleural region of abdominal segments VIII and IX with numerous anchor- or grapnel-shaped setae (Fig. 10.602). Northern Canada north of 50° latitude (Map 95)..........................................................................................***Metacnephia borealis*** (in part) (p. 302)
– Cocoon with 1 or more anterior apertures per side. Pleural region of abdominal segments VIII and IX with unbranched setae. Widespread......................................................................................................118

118. Cocoon boot shaped, with anterior loops of silk forming multiple apertures (Fig. 10.502). Gill with ventral 6 filaments in pairs (Fig. 10.596). Western North America (i.e., west of 95° longitude) (*Simulium arcticum* species complex, p. 394) ....................................***Simulium apricarium*** n. sp. (p. 395), ***S. arcticum*** s. s. (p. 396), ***S. arcticum*** **'cytospecies IIL-1'** (p. 397), ***S. arcticum*** **'cytospecies IIS-4'** (p. 397), ***S. brevicercum*** (p. 398), ***S. chromatinum*** n. sp. (p. 398), ***S. negativum*** n. sp. (p. 399), ***S. saxosum*** n. sp. (p. 399), ***S. vampirum*** (p. 400)
– Cocoon slipper shaped or slightly shoe shaped, with 1 large anterior aperture per side. Gill with ventral 6 filaments in triads (Fig. 10.492). Widespread (Map 204)................................***Simulium luggeri*** (p. 386)

119. Cocoon boot shaped (Figs. 10.480, 10.499, 10.501).....................................................................................120
– Cocoon slipper or shoe shaped (Figs. 10.481, 10.483, 10.488) .......................................................................124

120. Gill of 9 filaments, arranged in an open rosette in anterior view (Fig. 10.534). Cocoon with small gaps throughout (Fig. 10.480) (*Simulium pictipes* species group) ..........................................................................121
– Gill of 10 filaments, not arranged in open rosette. Cocoon with large anterior loops of silk forming apertures (Figs. 10.499, 10.501) .................................................................................................................123

121. Gill filaments nongranulate, but with transverse furrows (Fig. 10.591)....***Simulium claricentrum*** (p. 374)
– Gill filaments with granular surface sculpture (Fig. 10.592) ..........................................................................122

122. Gill with all filaments inflated basally (weakly inflated in westernmost populations) (Fig. 10.593) ..........................................................................................................................***Simulium pictipes*** (p. 377)
– Gill with, at most, only some filaments inflated basally (Fig. 10.592) ...........***Simulium innoxium*** (p. 375)

123. Gill with all filaments paired (Figs. 10.501, 10.595) ........................................***Simulium murmanum*** (p. 393)
– Gill with some filaments unpaired and arranged in triads (Fig. 10.499)........................................***Simulium decimatum*** (in part) (p. 390)

124. Cocoon without apertures (Figs. 10.457, 10.463), but sometimes loosely woven with numerous small holes ......................................................................................................................................................125
– Cocoon with 1 or more large anterior apertures per side (Figs. 10.482, 10.485, 10.489) (*Simulium jenningsi* species group, in part) ...........................................................................................................................128

125. East of Mississippi River...............................................................................................................................126
– West of Mississippi River...............................................................................................................................127

126. Gill with filaments tightly grouped, typically less than half length of pupal body (Fig. 10.488). Southeastern United States (Map 199) ....................................................***Simulium infenestrum*** (in part) (p. 382)
– Gill with filaments loosely grouped, typically more than half length of pupal body. Northeastern North America (Map 208) ...........................................................................***Simulium penobscotense*** (in part) (p. 388)

127. Gill in lateral view compact (Fig. 10.457). California (Map 158)...........***Simulium clarum*** (in part) (p. 350)
– Gill in lateral view loosely arranged (Fig. 10.463). Widespread in western United States and southwestern Canada (Map 168)..................................................................................................***Simulium argus*** (p. 354)

128. Gill strongly annulated, one fourth or less length of pupal body (Fig. 10.482). Head and thorax smooth, without microtubercles ..........................................................................................***Simulium aranti*** (p. 379)
– Gill shallowly furrowed or with reticulate surface sculpture, one third or more length of pupal body (Figs. 10.485, 10.489, 10.490). Head and thorax with microtubercles.........................................................129

129. Gill with 5 petiolate pairs of filaments (Fig. 10.485). Head and thorax with minute conical microtubercles ........ ***Simulium dixiense*** (p. 381)

– Gill with at least 1 trio of filaments (Figs. 10.489, 10.490). Head and thorax with rounded microtubercles (as in Fig. 10.834) ........ 130

130. Gill filaments with reticulate surface sculpture (Fig. 10.489) ........ ***Simulium jenningsi*** (p. 382)

– Gill filaments with transverse furrows ........ 131

131. Gill with ventral trunk swollen (Fig. 10.490), or with ventralmost petiole swollen for more than one half its length (Fig. 10.484) ........ 132

– Gill with ventralmost petiole typically not swollen for more than one half its length (Figs. 10.481, 10.494) ........ 133

132. Gill of 9 filaments, with filaments 5 and 6 from dorsum typically petiolate (Fig. 10.484) ........ ***Simulium definitum*** (in part) (p. 380)

– Gill of 10 filaments, with filaments 5 and 6 from dorsum typically sessile (Fig. 10.490) ........ ***Simulium jonesi*** (p. 384)

133. Gill with filament 7 from dorsum arising basally (Fig. 10.494) ........ ***Simulium ozarkense*** (p. 388)

– Gill with filament 7 from dorsum arising remote from base (Figs. 10.481, 10.483) ........ 134

134. Gill three fifths or less length of pupal body, typically slightly swollen basally (Fig. 10.481) ........ 135

– Gill two thirds or more length of pupal body, typically not swollen basally (Fig. 10.483), except in some populations of *S. confusum* ........ 136

135. Head dorsally with densely and uniformly distributed microtubercles (as in Figs. 10.835, 10.836) ........ ***Simulium anchistinum*** (p. 378)

– Head dorsally with sparse, irregularly distributed microtubercles (as in Fig. 10.834) ........ ***Simulium nyssa*** (p. 388)

136. Gill with ventralmost petiole of ventral trunk typically less than one eighth length of gill. Coastal Plain and interior of Pennsylvania (Map 203) ........ ***Simulium lakei*** (in part) (p. 385)

– Gill with ventralmost petiole of ventral trunk typically more than one sixth length of gill (Fig. 10.483). Southern piedmont, rarely Pennsylvania ........ go back to 86

137. Gill of 13–16 filaments ........ 138

– Gill of 17 or more filaments ........ 162

138. Cocoon well formed, slipper, shoe, or boot shaped (Figs. 10.466, 10.479, 10.500) ........ 139

– Cocoon shapeless, saclike (Figs. 10.411–10.413, 10.416) ........ 144

139. Pleural region of abdominal segments VIII and IX with anchor- or grapnel-shaped setae (Fig. 10.602). Northern Canada north of 50° latitude (Map 95) ........ ***Metacnephia borealis*** (in part) (p. 302)

– Pleural region of abdominal segments VIII and IX without anchor- or grapnel-shaped setae. Widespread ........ 140

140. Cocoon slipper shaped (Figs. 10.465, 10.466, 10.469) ........ 141

– Cocoon boot shaped, with anteroventral margin raised above substrate (Figs. 10.479, 10.500) ........ 143

141. Gill of 13 or 14 filaments, with 2 or 3 elongate filaments extended far beyond basal cluster of fine filaments (Fig. 10.469). Thorax with numerous slender, spinelike microtubercles. Southwestern United States (Map 174) ........ ***Simulium hechti*** (p. 366)

– Gill of 14–16 filaments, with no filaments extended far beyond remainder of cluster (Figs. 10.465, 10.466). Thorax with minute flattened microtubercles. Widespread ........ 142

142. Cocoon loosely woven, with numerous gaps (Fig. 10.465). Abdomen with ventral recurved hooks and terminal spines larger than dorsal recurved hooks. California (Map 170); small, sulfur springs ........ ***Simulium infernale*** n. sp. (p. 355)

– Cocoon densely woven, without gaps (Fig. 10.466). Abdomen with ventral recurved hooks and terminal spines no larger than dorsal recurved hooks. Widespread (Maps 171, 172); many habitats (*Simulium vittatum* species complex, p. 356) ........ ***Simulium tribulatum*** (p. 361), ***S. vittatum*** s. s. (p. 363)

143. Cocoon with large loops of silk forming anterior apertures (Fig. 10.500). Alaska and Canada (Map 215) ........ ***Simulium malyschevi*** (p. 392)

– Cocoon without apertures (Fig. 10.479). Texas (Map 186) ........ ***Simulium solarii*** (p. 373)

144. Gill of 2 long, inflated, nonannulated clubs, each with 8 slender filaments (Figs. 10.539, 10.540) .......145
– Gill of various forms, but not of 2 long inflated clubs (Figs. 10.541, 10.558, 10.559, 10.562) ......................146

145. Gill with at least 1 pair of terminal filaments arising from common stem on each club (Fig. 10.539)........................................................................................................................***Twinnia hirticornis*** (p. 242)
– Gill with terminal filaments arising independently from each club (Fig. 10.540) .......***Twinnia nova*** (p. 242)

146. Gill of 1 inflated club with 16 slender filaments (Fig. 10.558)..............***Prosimulium rhizophorum*** (p. 266)
– Gill of 13–16 slender filaments (Figs. 10.414, 10.541, 10.554–10.557), sometimes arising from 3 inflated trunks (Fig. 10.559).......................................................................................................................................147

147. Thoracic dorsum rugose (Fig. 10.525). Gill filaments in compact bundle (Fig. 10.562)...........***Prosimulium travisi*** (p. 269)
– Thoracic dorsum smooth or with microtubercles, but not rugose. Gill filaments typically in loose bundle or in 2 or more bundles (Figs. 10.414, 10.541, 10.550–10.557)..........................................................148

148. Abdominal tergites each without anterior row of spine combs (Fig. 10.414) ...............................................149
– Abdominal tergites VII–IX each with anterior row of spine combs (Figs. 4.29, 4.61)...............................150

149. Gill with filaments arising from 2 short trunks subequal in thickness (Fig. 10.414) .....................***Greniera abdita*** (in part) (p. 280)
– Gill with filaments arising from 3 distinct, somewhat thickened trunks, at least 1 of which is longer than wide (Fig. 10.541).......................................................................................................***Twinnia tibblesi*** (p. 242)

150. Gill of 16 slender filaments arising from 3 swollen trunks (Fig. 10.559) ........***Prosimulium saltus*** (p. 267)
– Gill of 13–16 slender filaments arising from unswollen to moderately swollen trunks (Figs. 10.550–10.557) ..............................................................................................................................................151

151. Gill with 3 elongate trunks (at least 2 trunks 4 times longer than wide) arising from base and giving rise to 14 or 15, rarely 16, filaments (Fig. 10.554) ....................................***Prosimulium fulvithorax*** (p. 260)
– Gill with at least 2 basal trunks less than 3 times longer than wide and giving rise to 13–16 filaments (Figs. 10.544, 10.553, 10.564) ..........................................................................................................................152

152. Gill of 13 or 14, rarely 15 or 16, filaments ..........................................................................................................153
– Gill of 16, rarely 14 or 15, filaments ......................................................................................................................156

153. Gill with basal trunks moderately swollen, and dorsal trunk in lateral view strongly divergent from lateral and ventral trunks (Fig. 10.553)................................................................***Prosimulium frohnei*** (p. 259)
– Gill with basal trunks not swollen; in lateral view with trunks rather evenly spaced (Figs. 10.544, 10.564)..........................................................................................................................................................154

154. Pacific Coast (Map 42) ..........................................................................................***Prosimulium idemai*** n. sp. (p. 262)
– Northern North America and high elevations in Rocky Mountains ..............................................................155

155. Gill with dorsal group of filaments on trunk longer than wide (Fig. 10.544)..........***Helodon gibsoni*** (p. 244)
– Gill with dorsal group of filaments on trunk subequal in length and width (Fig. 10.564) ..................................................................................***Prosimulium neomacropyga*** (p. 270), ***P. ursinum*** (p. 271)

156. Gill filaments in lateral view in 2 distinct groups (Fig. 10.551) ..............***Prosimulium esselbaughi*** (p. 258)
– Gill filaments in lateral view in 1 or 3 groups (Figs. 10.555–10.557)....................................................................157

157. Gill shorter than half length of pupal body (Fig. 10.548). Alaska and northern Canada (Map 20) ..............................................................................................................***Helodon irkutensis*** (in part) (p. 247)
– Gill longer than half length of pupal body (Fig. 10.411). Widespread.............................................................158

158. Eastern North America (i.e., east of 100° longitude), plus Saskatchewan ......................................................159
– Western North America (i.e., west of 100° longitude).........................................................................................160

159. Gill with dorsal trunk bearing 2 main branches as follows: an outer branch with 3 filaments and an inner branch that divides into an innermost branch with 2 filaments and a mesal branch with 3 filaments (Fig. 10.545) ..................................................................................................***Helodon vernalis*** (p. 245)
– Gill with dorsal trunk bearing 3 main branches, or if appearing as 2, then arranged as follows: an inner branch with 3 filaments and an outer branch that divides into an inner branch with 2 filaments and another branch with 3 filaments (Figs. 4.29, 10.411, 10.550, 10.557) ........................................***Prosimulium approximatum*** (p. 254), ***P. arvum*** (p. 255), ***P. clandestinum*** n. sp. (p. 255), ***P. fontanum*** (p. 258), ***P. fuscum*** (p. 260), ***P. mixtum*** (p. 263), ***P. transbrachium*** (p. 269)

160. Gill in lateral view with filaments typically in 3 distinct clusters (Fig. 10.556)....................... ***Prosimulium minifulvum*** n. sp. (p. 262)

– Gill in lateral view with all filaments rather evenly spaced in 1 cluster (Figs. 10.552, 10.555, 10.560)........................................................................................................................................161

161. Pupa in ventral view markedly tapered anteriorly (Fig. 10.532)......................... ***Prosimulium daviesi*** (p. 257), ***P. doveri*** (p. 257), ***P. formosum*** (p. 259), ***P. fulvum*** (in part, female pupa) (p. 260), ***P. secretum*** n. sp. (p. 267)

– Pupa in ventral view only moderately tapered anteriorly (Fig. 10.531)...................................... ***Prosimulium fulvum*** (in part, male pupa) (p. 260)

162. Cocoon well formed, shaped like slipper or boot (Figs. 10.442, 10.464, 10.471, 10.472) ..............................163

– Cocoon shapeless, saclike (as in Figs. 10.411–10.413)..........................................................................................175

163. Cocoon with anterodorsal projection (Figs. 10.471, 10.472)...................................................................................164

– Cocoon without anterodorsal projection (Figs. 10.442, 10.464, 10.467, 10.468).............................................165

164. Gill of 20–26 slender filaments (Fig. 10.472)........................................................... ***Simulium jacumbae*** (p. 367)

– Gill antler-like, of 55–80 slender filaments (often broken) arising from swollen trunks (Fig. 10.471)........................................................................................................................ ***Simulium iriartei*** (p. 366)

165. Gill of 17–21 filaments (Figs. 10.467, 10.468, 10.584) ..............................................................................................166

– Gill of more than 21 filaments ........................................................................................................................................167

166. Cocoon boot shaped. Northern North America (Map 99) ................ ***Metacnephia saskatchewana*** (p. 305)

– Cocoon slipper shaped (Figs. 10.467, 10.468). Southwestern United States (Map 173) ................. ***Simulium anduzei*** (p. 365)

167. Cocoon slipper shaped (Figs. 10.464, 10.470) or shoe shaped, sometimes with anteroventral collar raised slightly above substrate (Fig. 10.442) ..........................................................................................................168

– Cocoon boot shaped, with anteroventral collar raised high above substrate (Figs. 10.427, 10.517).............171

168. Gill of about 60–130 filaments. Thoracic trichomes 10 or more in number, typically forming dense covering (Fig. 10.470)............................................................................................................ ***Simulium hunteri*** (p. 366)

– Gill of 22–26 filaments. Thoracic trichomes fewer than 10 in number (Figs. 10.442, 10.464)................169

169. Gill of 23 or 24 filaments, more than half length of pupal body. Cocoon slipper shaped, without anterior collar (Fig. 10.464) ...................................................................................................... ***Simulium encisoi*** (p. 355)

– Gill of 22–26 filaments, half or less length of pupal body. Cocoon typically shoe shaped, with anterior collar (Fig. 10.442) (subgenus *Byssodon*)..........................................................................................................170

170. Alaska (possibly Yukon Territory) (Map 126) ........................................................ ***Simulium maculatum*** (p. 323)

– Canada and United States, excluding Alaska (Map 127)............................... ***Simulium meridionale*** (p. 324)

171. Gill antler-like, of 30–70 filaments arising from swollen base (Fig. 10.517)...................................................172

– Gill not antler-like, of 25 to more than 150 slender filaments (Figs. 10.582, 10.583, 10.585)......................173

172. Colorado (Map 96)........................................................................................... ***Metacnephia coloradensis*** (p. 303)

– Alaska and western Canada (Map 100)................................................... ***Metacnephia sommermanae*** (p. 305)

173. Gill of about 150 or more short threadlike filaments arising from 4 or 5 inflated trunks (Figs. 10.427, 10.585) ......................................................................................................................... ***Metacnephia villosa*** (p. 305)

– Gill of 25–110 long filaments arising from 4 or 5 slender trunks (Figs. 10.582, 10.583) .............................174

174. Cocoon smooth, parchment-like, translucent, iridescent. Western United States south of Canada (Map 97)............................................................................................................................ ***Metacnephia jeanae*** (p. 303)

– Cocoon coarser, brown, nontranslucent, noniridescent. Alaska and Canada (Map 98)........... ***Metacnephia saileri*** (p. 304)

175. Gill of 30–50 slender filaments arising from 1 or 2 elongate swollen clubs (Figs. 10.547, 10.549) ...............176

– Gill of 17 or more slender filaments not arising from swollen clubs (Figs. 10.414–10.418, 10.424, 10.546, 10.565–10.573).................................................................................................................................................177

176. Gill of 1 club, with 30–50 slender filaments (Fig. 10.547) ......................................... ***Helodon clavatus*** (p. 246)

– Gill of 2 divergent clubs, with 15–20 filaments on dorsal club and 20–30 filaments on ventral club (Fig. 10.549) (*Helodon onychodactylus* species complex, p. 248)............. ***Helodon onychodactylus*** s. s. (p. 251), ***H. susanae*** (p. 252) (pupae of the following species are unknown but are keyed here on the basis of the larval gill histoblast: ***H. beardi*** n. sp. [p. 248], ***H. chaos*** n. sp. [p. 249], ***H. diadelphus*** n. sp. [p. 250], ***H. mccreadiei*** n. sp. [p. 250], ***H. newmani*** n. sp. [p. 251], ***H. protus*** n. sp. [p. 252], ***H. trochus*** n. sp. [p. 253])

177. Abdomen without spine combs on any tergites (Figs. 10.414, 10.418) ............ 178
- Abdomen with spine combs on at least tergites VII and VIII (as in Fig. 4.29) (typically minute and pale in *Greniera denaria*, Fig. 10.417) ............ 181

178. Thorax with protuberance posterodorsal to each gill base (Fig. 10.416) ........ ***Greniera humeralis*** n. sp. (p. 281)
- Thorax without protuberance posterodorsal to each gill base (Fig. 10.414) ............ 179

179. Abdomen with terminal spines directed posterolaterally (Fig. 10.418) ............ ***Greniera longicornis*** n. sp. (p. 282)
- Abdomen with terminal spines directed dorsally (Fig. 10.414) ............ 180

180. Gill of 17–23 filaments; ventralmost trunk giving rise to 4–7 filaments (Fig. 10.414) ............ ***Greniera abdita*** (in part) (p. 280)
- Gill of 21–26 filaments; ventralmost trunk giving rise to 9 or 10 filaments (Fig. 10.415) ............ ***Greniera abditoides*** (p. 281)

181. Gill typically of 22 filaments arising from 3 main trunks and branching 10 + 4 + 8 (Fig. 10.417) ............ ***Greniera denaria*** (p. 282)
- Gill of 17 or more filaments arising from various numbers of trunks and with various branching patterns ............ 182

182. Abdominal segments VIII and IX with strongly curled setae laterally (Fig. 10.600). Abdomen without pleurites (Fig. 10.424). Gill of 17–50 filaments arising from a short, thickened knoblike base (genus *Cnephia*) ............ ***Cnephia dacotensis*** (p. 294), ***C. eremites*** (p. 296), ***C. ornithophilia*** (p. 296), ***C. pecuarum*** (p. 297)
- Abdominal segments VIII and IX with straight or slightly curved setae laterally (as in Fig. 10.599). Abdomen with large pleurites on segments IV and V (as in Fig. 4.29). Gill of 17 to more than 100 filaments variously arising from base ............ 183

183. Eastern North America (i.e., east of 100° longitude), excluding Nunavut ............ 184
- Western North America (i.e., west of 100° longitude), plus Nunavut ............ 186

184. Gill with ventral primary trunk of 8 filaments typically in 4 petiolate pairs (Fig. 10.542). Northeastern North America (Map 14) ............ ***Helodon pleuralis*** (in part) (p. 243)
- Gill with ventral primary trunk typically of fewer or more than 8 filaments not all in petiolate pairs (Figs. 10.571, 10.573). Southern Canada southward ............ 185

185. Gill of 24–75 filaments (Fig. 10.573) (*Prosimulium magnum* species complex, in part, p. 277) ............ ***Prosimulium albionense*** (p. 278), ***P. magnum*** s. s. (p. 279)
- Gill of 20–28 filaments (Fig. 10.571) ............ ***Prosimulium canutum*** n. sp. (pupa is unknown but is keyed here on the basis of the larval gill histoblast, p. 278), ***P. multidentatum*** (p. 276)

186. Gill of 29 to more than 100 filaments (Figs. 10.546, 10.567) ............ 187
- Gill of fewer than 29 filaments (Figs. 10.548, 10.568–10.570) ............ 190

187. Gill of 29–46 filaments (Fig. 10.546). Northern British Columbia, Alaska, Yukon Territory (Map 18) ............ ***Helodon alpestris*** (p. 246)
- Gill of 40 to more than 100 filaments (Fig. 10.567). Southernmost Alaska southward ............ 188

188. Thoracic dorsum strongly rugose (Fig. 10.526) ............ ***Prosimulium dicentum*** (p. 272)
- Thoracic dorsum smooth or with microtubercles, but not strongly rugose (Figs. 10.527, 10.528) ............ 189

189. Thoracic dorsum not shiny, but with raised or flattened microtubercles (Fig. 10.527) ............ ***Prosimulium dicum*** (p. 273)
- Thoracic dorsum shiny, entirely smooth, without microtubercles (Fig. 10.528) ............ ***Prosimulium exigens*** (p. 273)

190. Labral sheath longer than wide (Fig. 10.533) ............ 191
- Labral sheath subequal in width and length (as in Fig. 10.532) ............ 192

191. Oregon (Map 61) ............ ***Prosimulium longirostrum*** n. sp. (p. 275)
- Wyoming to southern Arizona (Map 63) ............ ***Prosimulium uinta*** (p. 276)

192. Thoracic dorsum smooth or with only faint reticulate pattern, but without microtubercles ............ 193
- Thoracic dorsum with minute, flattened or raised microtubercles ............ 194

193. Gill of 24–28 filaments (Fig. 10.542). Alaska to southern Idaho (Map 14)....................***Helodon pleuralis*** (in part) (p. 243)

– Gill of 19–25 filaments (Figs. 10.566, 10.568). Great Basin of British Columbia southward (Maps 55, 59)....................***Prosimulium constrictistylum*** (p. 272), ***P. flaviantennus*** (p. 274)

194. Alaska and northern Canada (Map 20)....................***Helodon irkutensis*** (in part) (p. 247)

– British Columbia southward....................195

195. Gill of 18–23 filaments, loosely arranged, typically with at least 2 petioles more than 4 times as long as wide (Fig. 10.569)....................***Prosimulium impostor*** (p. 275)

– Gill of 22–27 filaments, compactly arranged, typically with fewer than 2 petioles more than 3 times as long as wide (Fig. 10.565)....................***Prosimulium caudatum*** (p. 272)

## KEY TO LARVAE OF NORTH AMERICAN BLACK FLIES

The following key is based on mature (i.e., final-instar) larvae fixed in acetic ethanol. In many cases, it can be used for younger larvae as well as larvae fixed in ethanol. All characters, except those of chromosomes, can be assessed with a dissecting microscope unless otherwise indicated. Gill histoblasts should be dissected out and uncurled in a drop of 50% acetic acid (if the larva is in acetic ethanol) or a drop of polyvinyl lactophenol (if the larva is in ethanol). Mild clearing of the head with a caustic agent (e.g., potassium hydroxide) might be required if the postgenal cleft is difficult to see in specimens with a well-pigmented subesophageal ganglion.

### Genera

1. Labral fans absent (Plate 3b–d). Head tapered anteriorly to conical labrum that terminates in pronounced palatal brush (Figs. 10.684, 10.685). Hypostomal teeth markedly flattened, rounded apically; in doubtful specimens, paralateral teeth absent (Figs. 10.622, 10.623). Postgenal cleft absent. Anal sclerite Y shaped (Fig. 10.618)....................2

– Labral fans present (often closed, but with stalks obvious) (Plate 3e, f). Head with sides subparallel (Figs. 10.686–10.688). Hypostomal teeth neither markedly flattened nor rounded apically, each produced to acute point; in doubtful specimens, paralateral teeth present (Figs. 10.625–10.628). Postgenal cleft present or absent. Anal sclerite X shaped (Figs. 10.619, 10.620), subrectangular, or absent....................3

2. Frontoclypeal apotome with posterolateral head spots present; medial spots typically extended anteriorly to no more than level of posteriormost eyespot (Fig. 10.684). Mandible (slide mounted) with rows of comblike scales on outer apical surface (Fig. 4.44). Streams often associated with permanent ice in western mountains and high arctic....................***Gymnopais*** (p. 210)

– Frontoclypeal apotome with posterolateral head spots absent; medial spots extended anteriorly to level of anteriormost eyespot (Fig. 10.685). Mandible (slide mounted) without rows of comblike scales on outer apical surface (Fig. 4.45). Headwater streams in mountains and foothills of eastern and western North America....................***Twinnia*** (p. 210)

3. Anal sclerite with posterior arms nearly or completely encircling base of posterior proleg, or if posterior arms difficult to resolve, then body of larva completely unpigmented. Postgenal cleft absent (Figs. 10.683, 10.732). Western mountains....................4

– Anal sclerite, if present, with posterior arms absent or extended ventrally no farther than about half distance around base of posterior proleg (Figs. 10.619, 10.620), or if posterior arms difficult to resolve, then body of larva variously pigmented. Postgenal cleft typically present (in some species, small or apparently absent) (Figs. 10.686, 10.700, 10.722, 10.738, 10.741, 10.761). Widely distributed....................5

4. Body unpigmented (Plate 3a). Head pale yellowish, without head spots (Fig. 10.683). Ocelli unpigmented. Antenna of single slender (distal) article. Hypostoma with long median tooth bearing 3 or 4 irregularly situated lateral denticles; median tooth and apicomedial margins of lateral and sublateral teeth minutely serrated (Fig. 10.621). Seepages in southern British Columbia to California....................***Parasimulium*** (p. 235) (*P. crosskeyi* and *P. stonei* cannot be distinguished morphologically; larvae of *P.* 'species A,' *P. furcatum*, and *P. melanderi* are unknown)

– Body pigmented (Plate 8f). Head yellowish brown to dark brown, with prominent head spots (Fig. 10.732). Ocelli darkly pigmented. Antenna of 3 articles. Hypostoma with short median tooth not bearing lateral denticles; all teeth smooth, without serrations (Fig. 10.635). High mountain streams in Arizona and New Mexico (Map 86)....................***Gigantodax*** (***G. adleri***, p. 294)

5. Antenna with proximal and medial articles transparent, colorless, contrasting with dark brown distal article (Fig. 10.686). Postocciput completely enclosing cervical sclerites, leaving narrow medial gap (Figs. 10.687–10.690). Hypostoma with 7 primary teeth as follows: 1 median tooth, 2 pairs sublateral teeth, and 1 pair lateral teeth; median tooth with pair of lateral denticles typically arising distinctly anterior to base (giving tooth tridentate appearance); intermediate teeth present between primary teeth; paralateral teeth absent (Fig. 4.35) ........ 6

– Antenna with proximal, medial, or both articles typically opaque, pigmented (Figs. 10.719–10.724), or if colorless, then postgenal cleft extended anteriorly to or beyond hypostomal groove (Fig. 10.738). Postocciput not enclosing cervical sclerites, leaving wide medial gap (Fig. 10.724) (except in some species of *Greniera*, *Tlalocomyia*, and *Gigantodax*; Figs. 10.720, 10.729–10.732). Hypostoma either with 9 or more primary teeth, or various numbers of teeth arranged in 3 distinct groups (1 median, 2 lateral) (Figs. 4.36–4.38, 4.40–4.43); median tooth typically single, or appearing so, with pair of lateral denticles typically arising close to base (except in *Tlalocomyia*, in part, Figs. 10.631, 10.632); intermediate teeth absent; 1 or more paralateral teeth typically present ........ 7

6. Prothoracic proleg with lateral sclerite a narrow bar lying parallel to base of hooks, extended at most one third distance to base of apical article (Figs. 10.603, 10.604) (requires dissection if apical article is withdrawn) ........ ***Helodon*** (p. 210)

– Prothoracic proleg with lateral sclerite better developed than above, extended half or more distance to base of apical article (Fig. 10.605) (requires dissection if apical article is withdrawn) ........ ***Prosimulium*** (p. 212)

7. Hypostoma with lateral and sublateral teeth not clustered on prominent common lobes (Figs. 4.40–4.43, 10.636–10.650) (median and lateral teeth can be extended beyond sublateral teeth, but not on prominent lobes); heavily sclerotized anterior portion occupying at most one sixth total length of hypostoma; in doubtful specimens (i.e., subgenus *Hellichiella*, in part), hypostoma (slide mounted) with anterolateral margin of ventral wall typically bearing longitudinal sulcus (Figs. 10.640, 10.641). Postgenal cleft of various sizes and shapes, but in many species extended anteriorly more than one third distance to hypostomal groove ........ 8

– Hypostoma with 1 or more sublateral teeth and 1 or more paralateral teeth clustered on prominent common lobes, giving hypostoma trilobed appearance (consisting of median tooth and 2 lateral lobes); heavily sclerotized anterior portion occupying at least one third total length of hypostoma (Figs. 4.36–4.38, 10.626–10.634). Postgenal cleft in shape of inverted U or V, typically extended anteriorly no farther than one third distance to hypostomal groove (longer in some specimens of *Greniera*) (Figs. 10.719–10.731) ........ 11

8. Hypostoma with apex of median tooth extended anteriorly to about same level as or beyond apices of lateral teeth; sublateral teeth variously but distinctly posterior to median and lateral teeth (Figs. 4.43, 10.640–10.650). Rectal papillae of 3 simple or compound lobes (Figs. 10.619, 10.620) ........ ***Simulium*** (p. 217)

– Hypostoma with apex of median tooth posterior to apices of lateral teeth, or all teeth uniformly small; sublateral teeth with apices extended to various levels (Figs. 4.40–4.42, 10.636–10.639). Rectal papillae of 3 simple lobes without secondary lobules (as in Fig. 10.619) ........ 9

9. Postgenal cleft extended anteriorly to or slightly beyond hypostomal groove (Figs. 10.738–10.740). Antenna with transparent, colorless proximal and medial articles contrasting with dark brown distal article (Figs. 10.738–10.740). Hypostoma with lateral or paralateral teeth directed anteromedially (Figs. 4.42, 10.639) ........ ***Metacnephia*** (p. 216)

– Postgenal cleft typically extended anteriorly half or less distance to hypostomal groove, broadly rounded or pointed anteriorly (Figs. 10.733–10.737). Antenna with proximal and medial articles variously pigmented, but not entirely transparent. Hypostoma with lateral and paralateral teeth directed anteriorly or anterolaterally (Figs. 4.40, 4.41, 10.636–10.638) ........ 10

10. Abdomen without abrupt lateral and ventral expansion at segment V (Plate 9a, b), and without pair of ventral tubercles on segment IX. Antenna shorter than stalk of labral fan (Figs. 10.733, 10.734). Hypostoma with anterior margin in form of 3 short convex lobes (Figs. 4.40, 10.636). Anal sclerite present ........ ***Cnephia*** (p. 216)

– Abdomen with abrupt lateral and ventral expansion at segment V (Plate 9c–e), and with pair of ventral tubercles on segment IX (Fig. 4.47). Antenna longer than stalk of labral fan (Figs. 10.735–10.737). Hypostoma with anterior margin distinctly concave (Figs. 4.41, 10.637, 10.638). Anal sclerite absent ........ ***Ectemnia*** (p. 216)

11. Abdominal segment IX with pair of prominent ventral tubercles (as in Fig. 4.48) ........ ***Greniera*** (p. 215)
– Abdominal segment IX without pair of ventral tubercles, but with at most 1 transverse midventral bulge (Fig. 4.46) ........ 12

12. Antenna longer than stalk of labral fan by about half length of distal article (Figs. 10.723–10.727). Hypostoma with outer margin of lateral cluster of teeth typically sloped inward (Figs. 4.37, 10.630). Frontoclypeal apotome with second posterolateral head spots typically parallel to postocciput and not closely approximated to base of posteromedial head spot (Figs. 10.723–10.727) ........ ***Stegopterna*** (p. 215)
– Antenna equal in length to or marginally extended beyond stalk of labral fan (Figs. 10.728–10.731). Hypostoma with outer margin of lateral cluster of teeth either approximately straight or sloped somewhat outward (Figs. 4.38, 10.631–10.634). Frontoclypeal apotome with second posterolateral head spots typically angled toward, and closely approximated to or confluent with, base of posteromedial head spot (Figs. 10.728–10.731) ........ ***Tlalocomyia*** (p. 216)

## Species

### Larvae of *Gymnopais*

1. Hypostomal teeth obscured laterally by dense covering of setae (Fig. 10.622) ........ ***fimbriatus*** (p. 240)
– Hypostomal teeth not obscured by setae, which are few in number and well spaced (Fig. 10.623) ........ 2

2. Gill histoblast of 2 filaments equal in length (as in Fig. 10.516) ........ ***holopticus*** (p. 241)
– Gill histoblast of 3 or 4 branches, some of which might be short (if 2 filaments, then these unequal in length and thickness) (Figs. 10.536–10.538) ........ 3

3. Gill histoblast not particularly swollen at base (Fig. 10.538). Chromosomes triploid (as in Fig. 5.4) ........ ***holopticoides*** (p. 240)
– Gill histoblast swollen at base (Figs. 10.536, 10.537). Chromosomes diploid or triploid ........ 4

4. Gill histoblast of 2 short stout branches, and 1 long single or branched filament (Fig. 10.536). Chromosomes triploid (as in Fig. 5.4) ........ ***dichopticoides*** (p. 239)
– Gill histoblast of 2 short stout branches and 2 long slender filaments subequal in length (Fig. 10.537). Chromosomes diploid ........ ***dichopticus*** (p. 240)

### Larvae of *Twinnia*

1. East of Great Plains (Map 13) ........ ***tibblesi*** (p. 242)
– West of Great Plains ........ 2

2. Gill histoblast with at least 1 pair of terminal filaments on each basal trunk arising from common stem (Fig. 10.539). Nucleolar organizer in distal third of chromosome arm IL ........ ***hirticornis*** (p. 242)
– Gill histoblast with terminal filament arising independently on basal trunks (Fig. 10.540). Nucleolar organizer in center of chromosome I ........ ***nova*** (p. 242)

### Larvae of *Helodon*

1. Postgenal cleft extended about half distance to hypostomal groove, in shape of broad inverted U (Fig. 10.686). Hypostoma with teeth weakly sclerotized; sublateral teeth broad (Fig. 10.651) (subgenus *Distosimulium*) ........ ***pleuralis*** (p. 243)
– Postgenal cleft extended less than half distance to hypostomal groove, variously shaped, but not in shape of broad inverted U (Figs. 10.687–10.692). Hypostoma with teeth strongly sclerotized; sublateral teeth narrow (Figs. 10.652–10.658 ) ........ 2

2. Hypostoma with median tooth extended anteriorly to same level as or beyond lateral teeth (Figs. 10.655–10.658). Postgenal cleft subrectangular, biarctate in most species (Figs. 10.690–10.692) (subgenus *Helodon*) ........ 3
– Hypostoma with lateral teeth extended anteriorly beyond median tooth (Figs. 10.652–10.654). Postgenal cleft typically narrowed anteriorly, not biarctate (Figs. 10.687–10.689) (subgenus *Parahelodon*) ........ 14

3. Postgenal cleft with anterior margin weakly defined, truncated or slightly rounded (Fig. 10.691). Head pale yellowish, typically contrasting with dark posterolateral head spots (Plate 4c). Alaska and northern Canada (Map 20) ........ ***irkutensis*** (p. 247)

– Postgenal cleft with anterior margin well defined, typically biarctate (Figs. 10.690, 10.692). Head yellowish brown to brown, typically not contrasting markedly with posterolateral head spots (Plate 4b, d). Widespread in western North America ....................4

4. Hypostoma with median, sublateral, and lateral teeth extended anteriorly to about same level (Fig. 10.656). Body pale grayish (Plate 4b). Gill histoblast of single club with numerous fine filaments (Fig. 10.547) .................... ***clavatus*** (p. 246)

– Hypostoma with median, sublateral, and lateral teeth extended anteriorly to different levels (Figs. 10.655, 10.658). Body brownish (Plate 4d). Gill histoblast of slender filaments only (Fig. 10.546), or of 2 clubs bearing numerous fine filaments (Fig. 10.549) ....................5

5. Hypostoma typically convex anteriorly, with teeth (from median tooth outward) decreasing in level to which they are extended anteriorly (Fig. 10.655), although lateral teeth sometimes extended anteriorly farther than sublateral teeth. Gill histoblast of 29–46 slender filaments (Fig. 10.546) .................... ***alpestris*** (p. 246)

– Hypostoma straight or concave anteriorly, with lateral teeth extended anteriorly beyond sublateral teeth, which are extended anteriorly to same level (Fig. 10.658). Gill histoblast of 2 clubs bearing numerous fine filaments (Fig. 10.549) (*onychodactylus* species complex, p. 248) ....................6

6. Chromosomes with chromocenter (as in Fig. 5.5) .................... ***trochus*** n. sp. (p. 253)

– Chromosomes without chromocenter ....................7

7. Chromosomes with IIL-E homozygously inverted or heterozygous (Fig. 3 of Henderson 1986a for breakpoints) ....................8

– Chromosomes without IIL-E (Fig. 3 of Henderson 1986a) ....................10

8. Chromosomes without IIL-F (Fig. 3 of Henderson 1986a) .................... ***chaos*** n. sp. (p. 249)

– Chromosomes with IIL-F homozygously inverted or heterozygous (Fig. 3 of Henderson 1986a for breakpoints) ....................9

9. Chromosomes typically with IL-1 homozygously inverted or heterozygous, IIS-1 absent, and IIIL-1 homozygously inverted or heterozygous (Figs. 2, 3, 4 of Henderson 1986a) .................... ***diadelphus*** n. sp. (p. 250)

– Chromosomes typically with IL-1 absent, IIS-1 homozygously inverted or heterozygous, and IIIL-1 absent (Figs. 2, 3, 4 of Henderson 1986a) .................... ***susanae*** (p. 252)

10. Chromosomes typically with IIIS-B homozygously inverted or heterozygous (Fig. 4 of Henderson 1986a for breakpoints) ....................11

– Chromosomes without IIIS-B (Fig. 4 of Henderson 1986a) ....................12

11. Chromosomes of male typically with IIIS-D heterozygous; chromosomes of female and male without IIIL-3 (Fig. 4 of Henderson 1986a) .................... ***mccreadiei*** n. sp. (p. 250)

– Chromosomes of male without IIIS-D; chromosomes of female and male sometimes with IIIL-3 homozygously inverted or heterozygous (Fig. 4 of Henderson 1986a) .................... ***onychodactylus*** s. s. (p. 251)

12. Chromosomes typically with IIIS-D homozygously inverted or heterozygous (Fig. 4 of Henderson 1986a for breakpoints) .................... ***newmani*** n. sp. (p. 251)

– Chromosomes typically without IIIS-D (Fig. 4 of Henderson 1986a) ....................13

13. Chromosomes with IIS-1 homozygously inverted (Fig. 3 of Henderson 1986a for breakpoints) .................... ***beardi*** n. sp. (p. 248)

– Chromosomes with IIS-1 typically absent (Fig. 3 of Henderson 1986a) .................... ***protus*** n. sp. (p. 252)

14. Anal sclerite X shaped, with well-developed anterodorsal and posteroventral arms (as in Fig. 10.619; Plate 4a). Postgenal cleft in shape of inverted U or V, extended about one fourth or less distance to hypostomal groove (Fig. 10.689) .................... ***vernalis*** (p. 245)

– Anal sclerite subrectangular, with anterodorsal and posteroventral arms weakly, if at all, developed (Plate 3f). Postgenal cleft in shape of inverted U or square notch, extended more than one fourth distance to hypostomal groove (Figs. 10.687, 10.688) ....................15

15. Hypostoma with median tooth extended nearly to same level as sublateral teeth, which are posterior to lateral teeth (Fig. 10.652). Gill histoblast of 9 filaments .................... ***decemarticulatus*** (p. 244)

– Hypostoma with median tooth posterior to sublateral teeth, which are extended anteriorly to same level as lateral teeth (Fig. 10.653). Gill histoblast of 14 filaments .................... ***gibsoni*** (p. 244)

## Larvae of *Prosimulium*

1. Eastern North America (i.e., east of 100° longitude) ....... 2
– Western North America (i.e., west of 100° longitude) ....... 15

2. Hypostoma with sublateral teeth extended anteriorly about half as far as lateral teeth, and decreasing inward in height (Fig. 10.625). Chromosomes triploid (as in Fig. 5.4). Northern Canada (Map 53) ....... ***ursinum*** (in part) (p. 271)
– Hypostoma with sublateral and lateral teeth extended anteriorly to same level, or if lateral teeth extended farther anterior, then sublateral teeth not decreasing inward in height (Figs. 10.624, 10.659, 10.666, 10.669, 10.680, 10.682). Chromosomes diploid. Widely distributed ....... 3

3. Antenna shorter than stalk of labral fan (Figs. 10.716, 10.718). Abdomen gradually expanded posteriorly (Plate 6e, f). Gill histoblast of more than 20 filaments (*magnum* species group, in part) ....... 4
– Antenna as long as or longer than stalk of labral fan (Figs. 10.693–10.695, 10.701, 10.704, 10.705). Abdomen rather abruptly expanded at segment V (Plates 4e, f, 5a–c). Gill histoblast of 16 or fewer filaments (*hirtipes* species group, in part) ....... 7

4. Body grayish, banded (i.e., with conspicuous intersegmental lines) (Plate 6e) ....... ***multidentatum*** (p. 276)
– Body pale grayish to dark brown, not banded (i.e., with thin intersegmental lines) (Plate 6f) (*magnum* species complex, p. 277) ....... 5

5. Body grayish. Head yellowish. Chromosomes with IIS-25 homozygously inverted (Fig. 1 of Ottonen & Nambiar 1969) ....... ***canutum*** n. sp. (p. 278)
– Body brownish (Plate 6f). Head yellowish brown to dark brown. Chromosomes without IIS-25 (Fig. 28 of Ottonen 1966) ....... 6

6. Sex chromosomes undifferentiated (Fig. 8 of Ottonen 1966) ....... ***magnum*** s. s. (p. 279)
– Sex chromosomes differentiated in base of IIIS arm and sometimes IIIL arm (e.g., Figs. 9–13 of Ottonen 1966) ....... ***albionense*** (p. 278)

7. Postgenal cleft narrowed anteriorly, in shape of inverted U or V (Fig. 10.695). Body grayish, typically strongly banded. Chromosomes with parabalbiani marker inverted and in distal half of IIL arm; IIIL arm attached to end of IIIS arm (Fig. 5.8) ....... ***clandestinum*** n. sp. (p. 255)
– Postgenal cleft subrectangular, with truncated or slightly rounded anterior margin (Figs. 10.693, 10.694, 10.701, 10.704, 10.705). Body brownish to grayish, typically not strongly banded (Plates 4f, 5a–c). Chromosomes with parabalbiani marker in standard orientation and in proximal half of IIL arm; IIIL arm attached to IIIS arm at centromere (Fig. 5.8) ....... 8

8. Frontoclypeal apotome yellowish white, contrasting strongly with dark brown head spots (Fig. 10.694, Plate 4f) ....... ***arvum*** (p. 255)
– Frontoclypeal apotome yellowish brown to brown, variously contrasting with brown head spots (Figs. 10.693, 10.701, 10.704, 10.705; Plate 5a–c) ....... 9

9. Frontoclypeal apotome with anterolateral head spots yellowish or apparently absent (Fig. 10.705). Gill histoblast of 1 inflated club with 16 filaments (Fig. 10.558) ....... ***rhizophorum*** (p. 266)
– Frontoclypeal apotome with anterolateral head spots typically brownish (Figs. 10.693, 10.701, 10.704). Gill histoblast of 3 primary trunks with 14–16 total filaments (Figs. 10.557, 10.559) ....... 10

10. Frontoclypeal apotome with half of first posterolateral head spot brown (Fig. 10.701). Body typically grayish, slightly banded (Plate 5a). Chromosome arm IIIS with "blister" marker inverted in at least 1 homologue (Figs. 10, 11 of Rothfels & Freeman 1977) ....... ***fuscum*** (in part) (p. 260)
– Frontoclypeal apotome with first posterolateral head spot typically yellowish or apparently absent (Figs. 10.693, 10.704). Body brownish, not banded (Plates 4e, 5b). Chromosome arm IIIS without "blister" inversion (Fig. 8 of Basrur 1962) ....... 11

11. Gill histoblast with filaments arising from 3 inflated trunks (Fig. 10.559) ....... ***saltus*** (p. 267)
– Gill histoblast with filaments arising from 3 unswollen to slightly swollen trunks (Figs. 10.550, 10.557) ....... 12

12. Hypostoma (slide mounted) with each of 2 outer sublateral teeth on each side asymmetrical, extended anteriorly to same level as lateral teeth, giving anterior margin straight or convex appearance (Fig. 10.659) ....... ***approximatum*** (p. 254)

– Hypostoma (slide mounted) with each of 2 outer sublateral teeth on each side symmetrical, slightly to distinctly posterior to lateral teeth, giving anterior margin concave appearance (Figs. 10.624, 10.669) ....13

13. Hypostoma (slide mounted) with lateral teeth extended anteriorly well beyond sublateral teeth (Fig. 10.624).... ***fontanum*** (p. 258)

– Hypostoma (slide mounted) with lateral teeth extended anteriorly slightly beyond sublateral teeth (Fig. 10.669)....14

14. Chromosomes with standard arm associations (IS + IL, IIS + IIL; Fig. 1a, d of Rothfels & Freeman 1983). Widespread (Map 44) .... ***mixtum*** (p. 263)

– Chromosomes with whole-arm interchange (IS + IIL, IL + IIS; Fig. 1b, c of Rothfels & Freeman 1983). Known only from valleys of Pennsylvania and Virginia (Map 50).... ***transbrachium*** (p. 269)

15. Hypostoma (slide mounted) with lateral teeth slightly posterior to, or extended anteriorly to about same level as, sublateral teeth (Figs. 10.666, 10.675, 10.676, 10.678, 10.679) ....16

– Hypostoma (slide mounted) with lateral teeth extended anteriorly well beyond sublateral teeth (Figs. 10.660–10.665, 10.667, 10.668, 10.677)....25

16. Hypostoma (slide mounted) with median and lateral teeth extended anteriorly to about same level (Fig. 10.666). Abdomen expanded rather abruptly at segment V (Plate 5a). Gill histoblast of 16 filaments. East of Rocky Mountains (Map 41).... ***fuscum*** (in part) (p. 260)

– Hypostoma (slide mounted) with median tooth extended anteriorly beyond lateral teeth (Figs. 10.675, 10.676, 10.678, 10.681). Abdomen gradually expanded posteriorly (Plate 6a–d). Gill histoblast of fewer than 13 or more than 20 filaments. Widespread (*magnum* species group, in part) ....17

17. Hypostoma (slide mounted) with median tooth (including its denticles) broad, at widest part, nearly one fourth as wide as distance across all teeth (Fig. 10.675). Postgenal cleft wide, one third or more distance across head, with anterior margin straight (Fig. 10.711). Gill histoblast of 22–27 filaments (Fig. 10.565). Pacific Coast (Map 54) .... ***caudatum*** (p. 272)

– Hypostoma (slide mounted) with median tooth (including its denticles) slender, at widest part, less than one fourth as wide as distance across all teeth (Figs. 10.676, 10.678, 10.679, 10.681). Postgenal cleft of various widths, with anterior margin typically somewhat rounded or slightly irregular (Figs. 10.712, 10.714, 10.715). Gill histoblast of various numbers of filaments. Widespread ....18

18. Head pale yellow, contrasting strongly with brown head spots (Fig. 10.715, Plate 6d). Gill histoblast of 20–24 filaments (Fig. 10.570). Known only from Oregon (Map 61).... ***longirostrum*** n. sp. (p. 275)

– Head brownish orange to dark brown, variously contrasting with brown head spots (Figs. 10.712, 10.714, 10.717; Plate 6c). Gill histoblast of various numbers of filaments. Widespread ....19

19. Hypostoma (slide mounted) with sublateral teeth slightly posterior to lateral teeth (Fig. 10.678) .... ***impostor*** (p. 275)

– Hypostoma (slide mounted) with sublateral teeth extended anteriorly to same level as or beyond lateral teeth (Figs. 10.676, 10.681)....20

20. Gill histoblast of fewer than 30 filaments....21

– Gill histoblast of more than 40 filaments ....22

21. Gill histoblast of 21–24 filaments (Fig. 10.572) .... ***uinta*** (p. 276)

– Gill histoblast of 10–12 filaments (as in Fig. 10.413).... ***unicum*** (p. 276)

22. Chromosome I with centromere region tightly paired....23

– Chromosome I with centromere region typically unpaired ....24

23. Nucleolar organizer heterozygously expressed .... ***dicentum*** (in part, male larva) (p. 272)

– Nucleolar organizer homozygously expressed .... ***dicentum*** (in part, female larva) (p. 272), ***dicum*** (in part, female larva) (p. 273), ***exigens*** (in part, female larva) (p. 273)

24. Chromosome I with centromere band present in both homologues (Fig. 25 of Ottonen 1966) .... ***dicum*** (in part, male larva) (p. 273)

– Chromosome I with centromere band typically absent in 1 homologue (Fig. 26 of Ottonen 1966) .... ***exigens*** (in part, male larva) (p. 273)

25. Mandibular phragma connected with or closely approaching posterolateral margin of hypostoma (Figs. 10.698, 10.710)....26

– Mandibular phragma typically not connected with posterolateral margin of hypostoma, leaving wide gap (Figs. 10.696, 10.697, 10.700) ....... 28

26. Chromosomes triploid (as in Fig. 5.4) ....... ***ursinum*** (in part) (p. 271)

– Chromosomes diploid ....... 27

27. Chromosomes with chromocenter (as in Fig. 5.5) ....... ***frohnei*** (p. 259)

– Chromosomes without chromocenter ....... ***neomacropyga*** (p. 270)

28. Hypostoma (slide mounted) with median tooth posterior to lateral teeth, and/or median tooth with its denticles posterior to most anteriorly projected sublateral teeth (Figs. 10.660, 10.662, 10.664, 10.670, 10.673, 10.677) ....... 29

– Hypostoma (slide mounted) with median tooth extended anteriorly to same level as or beyond lateral teeth; median tooth with its denticles extended anteriorly to same level as most anteriorly projected sublateral teeth (Figs. 10.661, 10.665, 10.667, 10.668, 10.672, 10.674) ....... 37

29. Antenna as long as or longer than stalk of labral fan (Figs. 10.696, 10.699, 10.700). Gill histoblast of 14–16 filaments ....... 30

– Antenna about three fourths as long as stalk of labral fan (Figs. 10.697, 10.706, 10.708, 10.713). Gill histoblast of 10–25 filaments ....... 33

30. Frontoclypeal apotome with head spots absent or paler than surrounding pigment (Fig. 10.700) ....... ***fulvum*** (in part) (p. 260)

– Frontoclypeal apotome with at least anterior head spots darker than surrounding pigment (Figs. 10.696, 10.699) ....... 31

31. Frontoclypeal apotome with posteromedial head spot paler than anteromedial head spot (Fig. 10.699). Gill histoblast with at least 2 trunks 4 times longer than wide; 14 or 15, rarely 16, total filaments (Fig. 10.554) ....... ***fulvithorax*** (in part) (p. 260)

– Frontoclypeal apotome with posteromedial head spot as dark and distinct as anteromedial head spot (Fig. 10.696). Gill histoblast with at least 2 basal trunks less than 3 times longer than wide; 16 total filaments ....... 32

32. Chromosomes with IIL-1 inversion of Basrur (1962) homozygously inverted, heterozygous, or absent. Pacific Coast and Rocky Mountains (Map 33) ....... ***daviesi*** (p. 257)

– Chromosomes with IIL-1 homozygously inverted (Fig. 7 of Basrur 1962). Alaska and Pacific Coast (Map 34) ....... ***doveri*** (p. 257)

33. Gill histoblast of 16 filaments (Fig. 10.552) ....... ***formosum*** (p. 259)

– Gill histoblast of 10–12 or 19–25 filaments ....... 34

34. Gill histoblast of 10–12 filaments. Postgenal cleft truncated anteriorly, or in shape of inverted U or V (Figs. 10.706, 10.708) ....... 35

– Gill histoblast of 19–25 filaments. Postgenal cleft truncated, or only slightly rounded anteriorly (Fig. 10.713) ....... 36

35. Frontoclypeal apotome with all but anteriormost head spot typically absent or paler than surrounding pigment (Fig. 10.706) ....... ***rusticum*** n. sp. (p. 266)

– Frontoclypeal apotome with posteromedian and second posterolateral head spots darker than surrounding pigment (Fig. 10.708) ....... ***shewelli*** (in part) (p. 268)

36. Chromosomes with chromocenter (as in Fig. 5.5) ....... ***constrictistylum*** (p. 272)

– Chromosomes without chromocenter ....... ***flaviantennus*** (p. 274)

37. Hypostoma with median and lateral teeth typically extended anteriorly to same level (Figs. 10.665, 10.671, 10.672) ....... 38

– Hypostoma with median tooth extended anteriorly well beyond lateral teeth (Figs. 10.661, 10.667, 10.668, 10.674) ....... 40

38. Frontoclypeal apotome with all head spots, except first posterolateral, typically darker than surrounding pigment (Figs. 4.32, 10.707) ....... ***secretum*** n. sp. (p. 267)

– Frontoclypeal apotome with most posterior head spots absent or paler than surrounding pigment (Figs. 10.699, 10.700) ....... 39

39. Frontoclypeal apotome with at least anterior head spots darker than surrounding pigment (Fig. 10.699). Gill histoblast with at least 2 trunks 4 times longer than wide; 14 or 15, rarely 16, total filaments (Fig. 10.554)........................................................................................***fulvithorax*** (in part) (p. 260)
- Frontoclypeal apotome with head spots absent or paler than surrounding pigment (Fig. 10.700). Gill histoblast with at least 2 basal trunks less than 3 times longer than wide; 16 total filaments (Fig. 10.555)........................................................................................***fulvum*** (in part) (p. 260)

40. Postgenal cleft in shape of inverted U or rounded V (Fig. 10.709). Body grayish brown (Plate 5e) ........................................................................................***travisi*** (p. 269)
- Postgenal cleft with straight or slightly rounded anterior margin (Figs. 10.702, 10.703). Body brownish ........................................................................................41

41. Gill histoblast of 14 filaments........................................................................................***idemai*** n. sp. (p. 262)
- Gill histoblast of 16 filaments........................................................................................42

42. Chromosome arm IIIL with "shield" marker closer than "triad" marker to centromere (Fig. 12 of Basrur 1962)........................................................................................***esselbaughi*** (p. 258)
- Chromosome arm IIIL with "triad" marker closer than "shield" marker to centromere............***minifulvum*** n. sp. (p. 262)

**Larvae of *Greniera***

The larva of *Greniera* 'species F' is unknown.

1. Antenna with proximal article 2 times or more longer than medial article; distal article (slide mounted) with dark irregular or spiral bands or mottling (Figs. 10.719, 10.720) (*abdita* species group)........................................................................................2
- Antenna with proximal article as long as or shorter than medial article; distal article entirely dark (Figs. 10.721, 10.722) (*fabri* species group)........................................................................................4

2. Hypostoma with lateral teeth acutely pointed, subequal in width to, and extended anteriorly to same level as or slightly beyond median tooth (Fig. 10.627)............................................***humeralis*** n. sp. (p. 281)
- Hypostoma with lateral teeth broadly pointed, broader than, and extended anteriorly beyond median tooth (Fig. 10.626)........................................................................................3

3. Antenna with proximal article 2–3 times as long as medial article. Head spots pale, indistinct (Fig. 10.719). Gill histoblast of 15–23 filaments........................................................................................***abdita*** (p. 280)
- Antenna with proximal article 3 times as long as medial article. Head spots dark, distinct. Gill histoblast of 21–26 filaments........................................................................................***abditoides*** (p. 281)

4. Antenna (slide mounted) with medial article bearing minute spicules (Fig. 10.721). Postgenal cleft extended one third to one half distance to hypostomal groove. Canada, northern United States, and western mountains (Map 71)........................................................................................***denaria*** (p. 282)
- Antenna (slide mounted) with medial article smooth, without spicules. Postgenal cleft extended less than one third distance to hypostomal groove (Fig. 10.722). Coastal Plain of eastern United States (Map 72)........................................................................................***longicornis*** n. sp. (p. 282)

**Larvae of *Stegopterna***

1. Postgenal cleft typically extended more than one fourth distance to hypostomal groove (Fig. 10.725). Canada and northern United States (Map 77)........................................................................................***emergens*** (p. 286)
- Postgenal cleft extended one fourth or less distance to hypostomal groove (Fig. 10.726). Widespread..........2

2. Eastern North America (i.e., east of 100° longitude)........................................................................................3
- Western North America (i.e., west of 100° longitude) and Nunavut........................................................................................4

3. Chromosomes triploid (e.g., Fig. 5.4). Female larvae only (determined by staining gonads; males do not exist)........................................................................................***mutata*** (p. 287)
- Chromosomes diploid. Female and male larvae (determined by staining gonads).........***diplomutata*** (p. 285)

4. Head brownish orange to brown. Body grayish brown to brown, typically not banded (Plate 7e)..........5
- Head pale yellowish to yellowish brown. Body grayish, sometimes banded (Plates 7f, 8b)......................6

5. Chromosomes with parabalbiani marker in proximal half of IIL arm (as in Figs. 9, 10 of Madahar 1969); males with complex heterozygous inversion in chromosome II..................................***acra*** n. sp. (p. 284)
- Chromosomes with parabalbiani marker in distal half of IIL arm (Figs. 11, 12 of Madahar 1969); males without complex heterozygous inversion in chromosome II........................................***permutata*** (p. 288)

6. Gill histoblast of 10 filaments (Fig. 10.574). Alaska and northwestern Canada (Map 75) .... ***decafilis*** (p. 285)
– Gill histoblast of 12 filaments. Widespread .......... 7
7. Body typically banded. Alaska and northern Canada (Map 80) .......... ***trigonium*** (p. 289)
– Body not banded (Plate 8b). Southwestern Canada southward (Map 81) .......... ***xantha*** n. sp. (p. 289)

### Larvae of *Tlalocomyia*

1. Hypostoma with median and lateral teeth extended anteriorly to about same level (Fig. 10.631) (*andersoni* species group) .......... ***andersoni*** n. sp. (p. 290)
– Hypostoma with median tooth extended either anteriorly beyond or distinctly posterior to lateral teeth (Figs. 10.632–10.634) (*osborni* species group) .......... 2
2. Hypostoma with median tooth extended anteriorly beyond lateral teeth (Fig. 10.632). Body pale brownish (Plate 8d). Gill histoblast of 10–12 filaments .......... ***osborni*** (p. 291)
– Hypostoma with median tooth posterior to lateral teeth (Figs. 10.633, 10.634). Body dark gray or brown (Plate 8e). Gill histoblast of 6–8 filaments .......... 3
3. Gill histoblast of 8 rather slender or slightly inflated filaments, each decreasing in diameter along its length (as in Fig. 10.422). Northwestern North America (Map 84) .......... ***ramifera*** n. sp. (p. 292)
– Gill histoblast of 6 inflated filaments, each subequal in diameter along its length (Fig. 10.577). Sierra Nevada of California (Map 85) .......... ***stewarti*** (p. 293)

### Larvae of *Cnephia*

1. Postgenal cleft triangular (Fig. 10.734). Alaska and Canadian arctic (Map 88) .......... ***eremites*** (p. 296)
– Postgenal cleft subrectangular (Fig. 10.733). Widely distributed east of Rocky Mountains .......... 2
2. Body pale (Plate 9b). Mississippi River area (Map 90) .......... ***pecuarum*** (p. 297)
– Body typically dark (Plate 9a; some larvae of *C. ornithophilia* are pale). Widely distributed .......... 3
3. Chromosome arm IIIL standard (Fig. 7 of Procunier 1982b) .......... ***dacotensis*** (p. 294)
– Chromosome arm IIIL with fixed inversions *IIIL-4* and *IIIL-5* (Fig. 7 of Procunier 1982b for breakpoints) .......... ***ornithophilia*** (p. 296)

### Larvae of *Ectemnia*

1. Frontoclypeal apotome with head spots absent or restricted to faint anteromedial and/or anterolateral groups (Fig. 10.737). Abdomen weakly banded (Plate 9e). Gill histoblast of 8 filaments. Southeastern coastal plain and Sandhills (Map 93) .......... ***reclusa*** (p. 301)
– Frontoclypeal apotome typically with pale head spots surrounded by dark pigment (Figs. 10.735, 10.736). Abdomen typically banded (Plate 9d). Gill histoblast of 8 or 10 filaments. Widely distributed .......... 2
2. Hypostoma (slide mounted) with anterior margin moderately concave (Fig. 10.638). Nucleolar organizer in middle of chromosome arm IS. Gill histoblast of 10 filaments. Central Canada and north-central United States (Map 94) .......... ***taeniatifrons*** (p. 301)
– Hypostoma (slide mounted) with anterior margin slightly concave (Fig. 10.637). Nucleolar organizer in base of chromosome arm IS. Gill histoblast of 8 or 10 filaments. Widely distributed .......... 3
3. Gill histoblast of 8 filaments. Rocky rivers in Canada and northern United States southward along Appalachian Mountains (Map 91) .......... ***invenusta*** (p. 300)
– Gill histoblast of 10 filaments. Sandy rivers in southern coastal plain (Map 92) .......... ***primaeva*** (p. 301)

### Larvae of *Metacnephia*

1. Gill histoblast of 10–14 filaments (Fig. 10.581). Far northern Canada (and possibly Alaska) (Map 95) .......... ***borealis*** (p. 302)
– Gill histoblast of 16 to more than 200 filaments. Widely distributed .......... 2
2. Postgenal cleft with narrowest width subequal to or greater than length of hypostoma .......... ***saskatchewana*** (p. 305)
– Postgenal cleft with narrowest width less than length of hypostoma (Figs. 10.738–10.740) .......... 3
3. Gill histoblast of more than 150 short fine filaments arising from 4 or 5 inflated trunks (Fig. 10.585) .......... ***villosa*** (p. 305)

– Gill histoblast of 25–110 long filaments arising from 4–6 trunks (Figs. 10.582, 10.583); histoblast antler-like in some species (as in Fig. 10.517)....4
4. Gill histoblast antler-like, of 30–70 short filaments arising from swollen base (as in Fig. 10.517)....5
– Gill histoblast not antler-like, of 25–110 long slender filaments arising from mildly swollen base (Figs. 10.582, 10.583)....6
5. Chromosome arm IL with standard sequence (females; Fig. 5 of Procunier 1982a) or with large heterozygous loop (males). Colorado at high elevations (Map 96)....***coloradensis*** (p. 303)
– Chromosome arm IL homozygous for inversion with breakpoints in sections 25 and 35 relative to standard sequence in Fig. 5 of Procunier (1982a). Alaska and northwestern Canada (Map 100)....***sommermanae*** (p. 305)
6. Western United States south of Canada (Map 97)....***jeanae*** (p. 303)
– Alaska and Canada (Map 98)....***saileri*** (p. 304)

### Larvae of *Simulium*

Larvae of *Simulium* 'species O,' *S. dendrofilum*, *S. giganteum*, and *S.* 'species Z' are unknown or insufficiently known to include in the following key.

1. Postgenal cleft extended less than half distance from posterior tentorial pits to hypostomal groove (Figs. 10.741–10.757)....2
– Postgenal cleft extended half or more distance from posterior tentorial pits to hypostomal groove (Figs. 10.758–10.761, 10.791–10.824); in doubtful cases, abdomen gradually expanded posteriorly (Plates 17e; 18e, f), or frontoclypeal apotome with negative head-spot pattern or with head spots faint, obscure, or apparently absent (Figs. 10.780, 10.781)....55
2. Frontoclypeal apotome narrowed posteriorly (Fig. 10.783, Plate 16c). Hypostoma with all teeth on either side of median tooth nearly uniform in size (Fig. 10.645). California (Map 170); warm sulfur springs....***infernale*** n. sp. (p. 355)
– Frontoclypeal apotome broadened posteriorly (Fig. 10.784). Hypostoma with sublateral teeth smaller than and posterior to lateral teeth (Fig. 10.646). Widespread; many habitats....3
3. Antenna shorter than stalk of labral fan, or extended beyond apex of stalk by no more than half length of distal article (Figs. 10.779, 10.782, 10.784, 10.786, 10.787). Abdominal segment IX in lateral view with ventral tubercles typically inconspicuous (Fig. 10.610)....4
– Antenna extended beyond apex of stalk of labral fan by more than half length of distal article (Figs. 10.741–10.757). Abdominal segment IX in lateral view with ventral tubercles conspicuous (Figs. 10.606, 10.608)....11
4. Postgenal cleft broadly rounded (Fig. 10.779). Labral fan with more than 60 primary rays. Body brownish or grayish, strongly banded or mottled. Gill histoblast of 2–4 filaments. Northern Alaska and northern Canada (Map 155)....***subpusillum*** (in part) (p. 348)
– Postgenal cleft subquadrate or subtriangular (Figs. 10.782, 10.784, 10.786, 10.787). Labral fan with 60 or fewer primary rays. Body variously pigmented. Gill histoblast of 10 or more filaments. Widespread....5
5. Postgenal cleft subquadrate, with anterior margin straight or rounded (Figs. 10.782, 10.784). Antenna concolorous or with 1 or 2 pale spots on medial article. Widespread (subgenus *Psilozia*, in part)....6
– Postgenal cleft subtriangular (Fig. 10.787). Antenna with medial article paler than proximal and distal articles. Western North America (i.e., west of 100° longitude)....9
6. Gena with spots, when visible, paler than surrounding pigment. Frontoclypeal apotome sometimes with dark pigment posteriorly or sometimes with lateral head spots weak relative to medial head spots (Fig. 10.782). Abdomen often with dorsocentral and lateral, longitudinal, unpigmented stripes (Plate 16b). Gill histoblast of 10 filaments. Western United States and southwestern Canada (west of 95° longitude) (Map 168)....***argus*** (p. 354)
– Gena with spots, when visible, darker than surrounding pigment, although 1 spot sometimes paler. Frontoclypeal apotome without dark pigment posteriorly, or if with dark pigment, then typically rather equally distributed around head spots; anterolateral head spots weak or strong relative to medial head spots (Fig. 10.784). Abdomen typically without dorsocentral and lateral, longitudinal, unpigmented stripes, or if dorsocentral stripe present, then not continuous (Plate 16d). Gill histoblast of 14–24 filaments. Widespread....7

7. Frontoclypeal apotome with anterolateral head spots weak relative to medial head spots. Gill histoblast of 23 or 24 filaments. Western United States (west of 100° longitude) (Map 169) ........................................ ***encisoi*** (p. 355)
   – Frontoclypeal apotome with anterolateral head spots typically strong relative to medial head spots (Fig. 10.784). Gill histoblast of 14–16 filaments. Widespread (*vittatum* species complex, p. 356) .........8

8. Chromosomes with standard sequence in distal half of IS (Fig. 2c of Rothfels & Featherston 1981); males typically with IIIL-1 heterozygous (Fig. 12c of Rothfels & Featherston 1981)................***tribulatum*** (p. 361)
   – Chromosomes typically with IS-7 homozygously inverted or heterozygous (Fig. 2a, b of Rothfels & Featherston 1981; exceptional individuals can be common in some eastern areas); males without IIIL-1 inversion (Fig. 12b of Rothfels & Featherston 1981)............................................***vittatum*** s. s. (p. 363)

9. Gena with spots paler than surrounding pigment. Frontoclypeal apotome with head spots typically weak or obscured by surrounding pigment, sometimes absent (Plate 16f). Postgenal cleft extended less than one third distance to hypostomal groove (Fig. 10.786). Gill histoblast of fewer than 20 slender filaments ........................................***hechti*** (p. 366)
   – Gena with spots darker than surrounding pigment. Frontoclypeal apotome with head spots typically strong (Plate 17a). Postgenal cleft extended one third to nearly one half distance to hypostomal groove (Fig. 10.787). Gill histoblast of more than 50 filaments ........................................10

10. Postgenal cleft in shape of rounded inverted V; subesophageal ganglion with or without pigmented sheath (Fig. 10.787). Gill histoblast of about 60–130 fine slender filaments (as in Fig. 10.470). Widespread in western North America (Map 175) ........................................***hunteri*** (p. 366)
   – Postgenal cleft in shape of rather acute inverted V; subesophageal ganglion typically without pigmented sheath. Gill histoblast antler-like, of 55–80 slender filaments arising from swollen trunks (as in Fig. 10.471). Southwestern United States (Map 176) ........................................***iriartei*** (p. 366)

11. Frontoclypeal apotome with head spots absent or paler than surrounding pigment (Fig. 10.790). Gill histoblast of 6 filaments. Southwestern United States ........................................12
   – Frontoclypeal apotome with head spots darker than surrounding pigment (Figs. 10.741, 10.789). Gill histoblast of various numbers of filaments. Widespread........................................13

12. Postgenal cleft rather truncated anteriorly, extended less than one third distance to hypostomal groove. Abdomen dorsally with segments VII–IX typically darker than, and contrasting with, preceding segments........................................***puigi*** (p. 368)
   – Postgenal cleft rounded anteriorly, extended about one third distance to hypostomal groove (Fig. 10.790). Abdomen with segments VII–IX concolorous with or paler than preceding segments (Plate 17d)........................................***tescorum*** (p. 368)

13. Postgenal cleft in shape of acute or rounded inverted V (Figs. 10.785, 10.788, 10.789). Antenna with medial article paler than proximal and distal articles. West of Mississippi River ........................................14
   – Postgenal cleft subquadrate, subrectangular, or broadly U shaped (Figs. 10.741–10.757). Antenna with medial article variously pigmented. Widespread........................................17

14. Postgenal cleft extended at least one third distance to hypostomal groove (Fig. 10.787). Gill histoblast of more than 40 filaments ........................................go back to 10
   – Postgenal cleft typically a minute, variously broadened notch, extended less than one third distance to hypostomal groove (Figs. 10.785, 10.788, 10.789). Gill histoblast of fewer than 30 filaments....................15

15. Frontoclypeal apotome with anteromedial head spot typically weak or absent (Plate 16e). Postgenal cleft almost absent (Fig. 10.785). Gill histoblast of 18 or 19 filaments. Southern Arizona and New Mexico (Map 173)........................................***anduzei*** (p. 365)
   – Frontoclypeal apotome with anteromedial head spot typically as strong as other head spots, often bold. Postgenal cleft small but typically broader or more deeply incised (Figs. 10.788, 10.789). Gill histoblast of 9–13 or 20–26 filaments. Widespread ........................................16

16. Frontoclypeal apotome with head spots small, distinct, but not bold (Fig. 10.788). Body brownish, mottled or strongly banded (Plate 17b). Gill histoblast of 20–26 filaments................***jacumbae*** (p. 367)
   – Frontoclypeal apotome with head spots large, often bold (Fig. 10.789). Body grayish, with pigment typically uniformly distributed (Plate 17c) (banded in some populations). Gill histoblast of 9–13 filaments ........................................***piperi*** (p. 367)

17. Antenna with medial article bearing 3 or more pale rings (can be difficult to see in pale specimens; slide mount if in doubt) (Figs. 10.741–10.745, 10.749–10.757). Gill histoblast of 3–12 filaments ..........................18

– Antenna with medial article not bearing pale rings (Figs. 10.746, 10.748, 10.762–10.764). Gill histoblast of 4 or 6 filaments ........................................................................................................ 35

18. Hypostoma with lateral teeth largest, extended anteriorly beyond median tooth (Figs. 10.640, 10.642, 10.745). Gill histoblast of 3 or 5–12 filaments. Typically small streams ........................................ 19

– Hypostoma with median and lateral teeth subequal in prominence, extended anteriorly to about same level (Figs. 10.749–10.756). Gill histoblast of 4 slender filaments. Typically wide streams and rivers (*annulus* species group) ........................................................................................................ 28

19. Hypostoma (slide mounted) with anterolateral margin of ventral wall not bearing longitudinal sulcus (Fig. 10.642). Antenna with medial article typically bearing 6 or more pale rings (Fig. 10.757). Gill histoblast of 3 swollen filaments ........................................................ ***baffinense*** (p. 320)

– Hypostoma (slide mounted) with anterolateral margin of ventral wall typically bearing longitudinal sulcus (Fig. 10.640). Antenna with medial article typically bearing fewer than 6 pale rings. Gill histoblast of 5–12 filaments (subgenus *Hellichiella*, in part) ........................................................ 20

20. Postgenal cleft slightly wider than long, markedly biarctate (Figs. 10.742–10.744). Gill histoblast of 5, 6, or 10 filaments ........................................................................................................ 21

– Postgenal cleft about as wide as long, rounded or truncated anteriorly, at most weakly biarctate (Figs. 10.741, 10.745, 10.747). Gill histoblast of 6–8 or 12 filaments ........................................ 23

21. Head typically brownish orange, with brown flecks (Fig. 10.744, Plate 10d). Gill histoblast of 5 or 6 filaments. Chromosomes with nucleolar organizer in base of IIS ........................ ***nebulosum*** (p. 309)

– Head typically pale yellowish, without brown flecks (Figs. 10.742, 10.743; Plate 10b, c). Gill histoblast of 5, 6, or 10 filaments. Chromosomes with nucleolar organizer in base of IL or IIIS ........................ 22

22. Gill histoblast with 10 filaments. Chromosomes with nucleolar organizer in base of IL (within expanded centromere region). Widespread (Map 104) ........................................ ***innocens*** (p. 308)

– Gill histoblast with 5 or 6 filaments. Chromosomes with nucleolar organizer in base of IIIS. California (Map 105) ........................................................................................................ ***minus*** (p. 309)

23. Abdomen with reddish pigment arranged either in bands (Plate 10a) or in 2 dorsal longitudinal rows of spots. Gill histoblast of 12 filaments ........................................................................ 24

– Abdomen with grayish or brownish pigment (Plates 10e, 11a), typically rather uniformly distributed. Gill histoblast of 6–8 or 12 filaments ........................................................................ 25

24. Chromosomes with IIS-1 homozygously inverted, so that ring of Balbiani is closer than "bulge" marker to centromere (Fig. 41 of Dunbar 1967). Canada and northern United States (Map 102); absent from coastal plain. ('**Species O**' [p. 311] possibly keys out here; it can be distinguished from *S. anatinum* by having IIS-2 homozygously inverted; Fig. 17 of Rothfels & Golini 1983) ........................ ***anatinum*** (p. 307)

– Chromosome arm IIS with standard sequence, so that "bulge" marker is closer than ring of Balbiani to centromere (Fig. 43 of Dunbar 1967). Canada and eastern United States (Map 103); present in coastal plain ........................................................................ ***congareenarum*** (p. 308)

25. Antenna dark brown, contrasting strongly with pale rings. Head brownish yellow to dark brown (Figs. 10.745, 10.747; Plates 10e, 11a). Gill histoblast of 6–8 filaments. Southern Canada and Rocky Mountains westward ........................................................................................................ 26

– Antenna pale brown or yellowish, contrasting rather weakly with pale rings. Head pale yellowish to brownish yellow. Gill histoblast of 12 filaments. Northern Canada ........................................ 27

26. Gill histoblast of 6–8 filaments arising from short base (as in Fig. 10.430) ........................ ***curriei*** (p. 311)

– Gill histoblast of 6 filaments arising from elongate swollen base (as in Fig. 10.432) ................ ***mysterium*** n. sp. (p. 312)

27. Chromosomes with IIS-5 present in at least 1 homologue (Fig. 18 of Rothfels & Golini 1983); IIS-3 absent ........................................................................................................ ***rendalense*** (p. 310)

– Chromosomes with IIS-3 homozygously inverted (Fig. 20 of Rothfels & Golini 1983); IIS-5 absent ........................................................................................................ ***usovae*** (p. 310)

28. Postgenal cleft strongly biarctate (Fig. 10.749) ........................................................ ***annulus*** (p. 314)

– Postgenal cleft with anterior margin straight or rounded (Figs. 10.750–10.756); if biarctate, then abdomen spotted, typically not banded (Plate 12a) ........................................................ 29

29. Eastern North America (i.e., east of 100° longitude) ........................................................ 30

– Western North America (i.e., west of 100° longitude) ........................................................ 31

30. Abdomen banded (Plate 11f) or spotted. Gill histoblast with trunks of unequal length (as in Fig. 10.436) or sometimes equal length (as in Fig. 10.437). Eastern piedmont and coastal plain (Map 117) .......... ***clarkei*** (p. 317)

– Abdomen spotted (Plate 12a). Gill histoblast typically with short trunks of approximately equal length (trunks of variable length in piedmont). Eastern mountains, infrequent in piedmont (Map 118) .......... ***emarginatum*** (p. 317)

31. Abdomen with pigment of first segment dark, contrasting strongly with remaining segments (Plate 11d) .......... ***balteatum*** n. sp. (p. 315)

– Abdomen with pigment of first segment subequal in intensity to that of remaining segments (Plates 11e, 12b) .......... 32

32. Subesophageal ganglion lacking pigmented sheath. Postgenal cleft with anterior margin clearly defined (Figs. 10.751, 10.752) .......... ***canonicolum*** (p. 316)

– Subesophageal ganglion with pigmented sheath. Postgenal cleft with anterior margin clearly defined or obscure (Figs. 10.755, 10.756) .......... 33

33. Body pale, weakly banded (Plate 12b) .......... ***joculator*** n. sp. (p. 318)

– Body darker, strongly banded (as in Plate 11e) .......... 34

34. Chromosome arm IIIL with homozygous mimic inversion of IIIL-1, such that first band in section 89 is lacking (Fig. 13 of Golini & Rothfels 1984 for standard sequence showing breakpoints of true IIIL-1); IIS of male larva with large heterozygous inversion from section 47d to 53d (Fig. 15 of Golini & Rothfels 1984 for standard sequence) .......... ***quadratum*** (p. 318)

– Chromosomes with IIIL-1 homozygously inverted (breakpoints shown in Fig. 13 of Golini & Rothfels 1984); IIS of male larva without large heterozygous inversion (Fig. 15 of Golini & Rothfels 1984) .......... ***zephyrus*** n. sp. (p. 319)

35. Anterolateral head spots on each side not closely approximated to one another (Figs. 10.746, 10.748). Hypostoma (slide mounted) with anterolateral margin of ventral wall bearing longitudinal sulcus (Fig. 10.641) .......... 36

– Anterolateral head spots on each side closely approximated to one another (Figs. 10.762–10.764). Hypostoma (slide mounted) with anterolateral margin of ventral wall not bearing longitudinal sulcus (Figs. 10.643, 10.644) .......... 37

36. Hypostoma with teeth not conspicuously arranged in 3 groups; median and lateral teeth extended anteriorly to about same level (Figs. 10.641, 10.746). Gill histoblast of 6 filaments .......... ***excisum*** (p. 312)

– Hypostoma with teeth arranged in 3 groups; median tooth extended anteriorly beyond lateral teeth (Fig. 10.748). Gill histoblast of 4 filaments .......... ***rivuli*** (p. 313)

37. Frontoclypeal apotome with all head spots except anteromedial and sometimes anterolateral head spots enclosed in nearly solid dark pigment (Fig. 10.764, Plate 13c) .......... ***aestivum*** (p. 332)

– Frontoclypeal apotome with head spots typically not enclosed in dark pigment except sometimes centrally (Fig. 10.766), or if enclosed, then completely and often faintly so (Figs. 10.768, 10.773, 10.777), or with pale yellowish area posterior and slightly lateral to posteromedial head spot (Fig. 10.763) .......... 38

38. Postgenal cleft parallel sided, with rather straight anterior margin (Figs. 10.762, 10.763). Rectal papillae of 3 simple lobes (as in Fig. 10.619). Chromosomes with haploid number of 2 (Fig. 5.2) (subgenus *Eusimulium*) .......... 39

– Postgenal cleft in shape of small notch or inverted U, with broadly rounded or pointed anterior margin (Figs. 10.766, 10.768, 10.771–10.773). Rectal papillae of 3 simple or compound lobes. Chromosomes with haploid number of 3 (subgenus *Nevermannia*, in part) .......... 43

39. Frontoclypeal apotome with head spots typically enclosed in dark pigment except for pale yellowish area posterior and slightly lateral to posteromedial head spot (Fig. 10.763). Chromosomes with IL-10 homozygously inverted, often with IL-11 homozygously inverted or heterozygous (Fig. 2 of Leonhardt 1985 for breakpoints) .......... ***violator*** n. sp. (p. 330)

– Frontoclypeal apotome with head spots not enclosed in pigment, or if so, then only centrally (Fig. 10.762; Plate 13a, b). Chromosomes without IL-10 or IL-11 inversions (Fig. 2 of Leonhardt 1985) .......... 40

40. Head pale yellowish or whitish, with little or no trace of darker pigment surrounding head spots (Plate 13b). Chromosomes with IL-18 homozygously inverted (Fig. 2 of Leonhardt 1985 for breakpoints); "bulge" marker closer than ring of Balbiani to centromere. Western United States (west of 100° longitude), typically at low elevations (Map 129) .......... ***donovani*** (p. 328)

– Head variously pigmented. Chromosomes without IL-18 inversion (Fig. 2 of Leonhardt 1985); "bulge" marker variously located relative to ring of Balbiani. Widespread ....................41

41. Head pale yellowish, typically with little or no trace of darker pigment surrounding head spots. Chromosomes with ring of Balbiani closer than "bulge" marker to centromere (as in Fig. 5.2), without IS-3 inversion (as in Fig. 11 of Dunbar 1958b). Alaska, Canada, and high elevations in Rocky Mountains (Map 130) .................... ***exulatum*** n. sp. (p. 329)

– Head brownish yellow or brownish orange, with or without some darker pigment surrounding head spots (Fig. 10.762, Plate 13a). Chromosomes either with "bulge" marker closer than ring of Balbiani to centromere (Fig. 15 of Dunbar 1958b), or with IS-3 homozygously inverted (Fig. 10 of Dunbar 1958b). Widespread ....................42

42. Postgenal cleft rather straight anteriorly, not biarctate, typically extended slightly less than one half distance to hypostomal groove. Frontoclypeal apotome sometimes with dark pigment surrounding more than part of posteromedial head spot. Chromosomes with "bulge" marker closer than ring of Balbiani to centromere. Canada south through Appalachian Mountains (Map 128) .................... ***bracteatum*** (p. 328)

– Postgenal cleft rather straight or biarctate anteriorly, extended about one third distance to hypostomal groove. Frontoclypeal apotome with, at most, small amount of dark pigment surrounding part of posteromedial head spot (Fig. 10.762). Chromosomes with ring of Balbiani closer than "bulge" marker to centromere (Fig. 5.2). Widespread (Map 131) .................... ***pilosum*** (p. 330)

43. Postgenal cleft surrounded by faint brownish pigment in shape of H, contrasting with pale yellowish pigment of ventral head (Fig. 10.777). Abdominal cuticle posterodorsally with numerous dark, unbranched setae (sometimes broken off, but if so, then visible by viewing slide-mounted piece of cuticle under compound microscope) (Fig. 10.840) .................... ***silvestre*** (in part) (p. 344)

– Postgenal cleft typically not surrounded by brownish pigment in shape of H (Figs. 10.768, 10.772, 10.773). Abdominal cuticle posterodorsally without numerous dark setae (if setae present, these sparse or pale) ....................44

44. Gill histoblast of 6 filaments. Eastern North America (i.e., east of 95° longitude) (Map 143) .................... ***gouldingi*** (in part) (p. 340)

– Gill histoblast of 4 filaments. Widespread ....................45

45. Abdomen with reddish patch posterordorsally (Plate 14f), or (for larvae in Ozark Mountains) with reddish bands (in either case, color not obvious in older stored specimens). Gill histoblast with ventral trunk at least 4 times length of dorsal trunk (as in Fig. 10.449). Eastern North America (i.e., east of 95° longitude) (Map 144) .................... ***impar*** (p. 340)

– Abdomen variously pigmented, but without reddish patch posterodorsally (Plates 13e, f; 14a). Gill histoblast with ventral trunk no more than 3 times length of dorsal trunk (as in Fig. 10.444). Widespread ....................46

46. Frontoclypeal apotome with head spots typically encompassed by faint pigment in form of inverted U (Plate 14a). Postgenal cleft rounded anteriorly (Fig. 10.768) .................... ***craigi*** (in part) (p. 336)

– Frontoclypeal apotome with head spots typically free of surrounding pigment in form of inverted U (Plates 13e, f; 14d). Postgenal cleft truncated, rounded, or tapered anteriorly, or a small squarish notch (Figs. 10.766, 10.767, 10.771) ....................47

47. Postgenal cleft typically truncated anteriorly, with widest point just posterior to apex (Fig. 10.771). Alaska and western Canada (Map 142) .................... ***fontinale*** (p. 338)

– Postgenal cleft typically tapered or rounded anteriorly, with widest point near midlength (Fig. 10.775); if truncated anteriorly, then cleft in form of a small squarish notch (Figs. 10.767). Widespread ....................48

48. Chromosomes without IIIS-1 inversion (as in Fig. 10 of Hunter & Connolly 1986). Alaska and northern Canada (Map 134) .................... ***bicorne*** (p. 332)

– Chromosomes with IIIS-1 homozygously inverted (Fig. 10 of Hunter & Connolly 1986 for breakpoints; Fig. 5.1). Widespread ....................49

49. Head dark brownish orange (Plate 13e). Postgenal cleft a small squarish notch (Fig. 10.766). Rocky Mountains westward (Map 136) .................... ***carbunculum*** n. sp. (in part) (p. 334)

– Head yellowish brown to brownish orange (Plates 13f, 15b). Postgenal cleft variously shaped. Widespread ....................50

50. Postgenal cleft a small squarish notch (Fig. 10.767), or tapered or broadly rounded anteriorly. Chromosomes with homozygous, subterminal inversion *IIIL-35* (Fig. 5.1 for inversion limits). Western United States and southwestern Canada (west of 110° longitude) ........................................ ***conicum*** n. sp. (p. 335)
 – Postgenal cleft tapered or broadly rounded anteriorly (Fig. 10.775). Chromosomes without subterminal inversion *IIIL-35* (Fig. 5.1). Widespread........................................................................................................51

51. Chromosomes with IL-1 homozygously inverted (Fig. 5.1 for inversion limits) .............................................52
 – Chromosomes with IL-1 heterozygous or absent (Fig. 5.1).......................................................................................53

52. Chromosomes without IL-4 (Fig. 5.1). Widespread (Map 136)........................... ***carbunculum*** n. sp. (in part) (p. 334), ***merritti*** n. sp. (in part, female larva) (p. 341)
 – Chromosomes with IL-4 heterozygous (Fig. 5.1 for inversion limits). Known only from northern Utah (Map 146) ..................................................................................... ***merritti*** n. sp. (in part, male larva) (p. 341)

53. Chromosomes with IL-2 and IL-3 heterozygous (Fig. 5.1 for inversion limits). Northeastern North America (Map 148) ........................................................................ ***moultoni*** n. sp. (in part, male larva) (p. 343)
 – Chromosomes with IL-2 homozygously inverted, heterozygous, or absent; IL-3 absent (Fig. 5.1 for inversion limits). Widespread......................................................................................................................54

54. Chromosomes without IL-2, IIS-1, or IIIL-1 (Fig. 5.1). Widespread............................. ***modicum*** n. sp. (p. 342), ***moultoni*** n. sp. (in part, female larva) (p. 343), ***pugetense*** (in part, some female larvae) (p. 343)
 – Chromosomes with at least 1 of the following inversions heterozygous or homozygous: IL-2, IIS-1, or IIIL-1 (Fig. 5.1 for inversion limits). Pacific Northwest (Map 149).................. ***pugetense*** (in part) (p. 343)

55. Postgenal cleft typically broad anteriorly, and reaching or nearly reaching hypostomal groove (Figs. 10.761, 10.814–10.816). Abdominal cuticle (slide mounted) posterodorsally with numerous tubular or multiply branched, pale or dark setae (Figs. 10.838, 10.841, 10.842), often giving abdomen minutely hairy appearance ...............................................................................................................................56
 – Postgenal cleft not reaching hypostomal groove (Figs. 10.791–10.803, 10.805–10.812), or if so, then only by narrowed or pointed extension (Figs. 10.804, 10.813). Abdominal cuticle (slide mounted) posterodorsally with or without numerous tubular or multiply branched, pale or dark setae ..............60

56. Abdominal segments I–V with circlet of tubercles (Plate 12f). Gill histoblast of 20 or more filaments (subgenus *Byssodon*) ..............................................................................................................................57
 – Abdominal segments I–V without circlet of tubercles (Plates 22e, f; 23a). Gill histoblast of 4–8 filaments (*slossonae* species group).......................................................................................................................58

57. Alaska (possibly Yukon Territory) (Map 126) ........................................................................ ***maculatum*** (p. 323)
 – Canada and United States, excluding Alaska (Map 127).................................................... ***meridionale*** (p. 324)

58. Abdominal cuticle (slide mounted) with numerous dark, apically branched setae (Fig. 10.841). Gill histoblast of 6 filaments. Eastern coastal plain (Map 231)...................................................... ***slossonae*** (p. 410)
 – Abdominal cuticle (slide mounted) with numerous pale or dark, basally branched setae (Fig. 10.842). Gill histoblast of 4 or 8 filaments. Canada and northern United States....................................................59

59. Frontoclypeal apotome with head spots typically not surrounded by dark pigment (Fig. 10.814, Plate 22e). Abdominal cuticle (slide mounted) with numerous pale, multiply branched setae. Gill histoblast of 8 filaments............................................................................................................... ***rugglesi*** (p. 409)
 – Frontoclypeal apotome with head spots typically surrounded by dark pigment (Fig. 10.816, Plate 23a). Abdominal cuticle (slide mounted) with numerous dark, arborescent setae (Fig. 10.842). Gill histoblast of 4 filaments......................................................................................................... ***transiens*** (p. 411)

60. Body pale green (Plate 19f). Head in dorsal view slender, less than three fourths as wide as long, with positive head-spot pattern. Postgenal cleft elongate (Fig. 10.799). Gill histoblast of 10 filaments. Southeastern coastal plain and Sandhills (Map 196)........................................................... ***dixiense*** (p. 381)
 – Body variously pigmented. Head in dorsal view three fourths or more as wide as long, with various head-spot patterns. Postgenal cleft variously shaped (Figs. 10.791–10.798, 10.800–10.808). Gill histoblast of various numbers of filaments. Widespread .........................................................................................61

61. Postgenal cleft large, bulbous, often nearly circular in outline, about as long as wide, not gradually tapered (although sometimes slightly pointed) anteriorly (Figs. 10.758–10.760, 10.798, 10.800–10.803). Frontoclypeal apotome with positive head-spot pattern.................................................................................62

– Postgenal cleft of various sizes and shapes, but typically not bulbous or circular in outline; typically longer than wide, pointed, rounded, or gradually tapered anteriorly (Figs. 10.804–10.813, 10.817–10.824). Frontoclypeal apotome with positive or negative head-spot pattern, or with head spots apparently absent....90

62. Abdominal segment IX, in lateral view, without ventral tubercles, although wrinkle of cuticle often present (Fig. 10.615) (*jenningsi* species group, in part)....63

– Abdominal segment IX, in lateral view, with ventral tubercles minute and often slender (Figs. 10.614, 10.616, 10.617) or conspicuous and conical (Fig. 10.607)....69

63. Head typically dark brown, with head spots not strongly differentiated from background pigment (Fig. 10.798). Antenna short, subequal in length to stalk of labral fan. Abdomen gradually expanded posteriorly (Plate 19b). Gill histoblast of 10 short, strongly annulated filaments (as in Fig. 10.482)....***aranti*** (p. 379)

– Head typically yellowish brown, with distinct head spots (Figs. 10.800, 10.801). Antenna extended beyond apex of stalk of labral fan. Abdomen expanded rather abruptly at segment V (Plate 20a, b, f). Gill histoblast of various numbers of short to long filaments, but not strongly annulated....64

64. Gill histoblast of 4–8 filaments....65

– Gill histoblast of 10–12 filaments....67

65. Gill histoblast of 4 filaments....***snowi*** (p. 390)

– Gill histoblast of 6–8 filaments....66

66. Gill histoblast of 6–8 filaments, with either middle pair of filaments or third pair from dorsum on short petiole that arises from base of ventral pair; filaments often swollen their entire length (as in Fig. 10.486)....***fibrinflatum*** (p. 381)

– Gill histoblast of 6 filaments, with middle pair of filaments lacking petiole and arising from base of gill or from base of dorsal pair; filaments not swollen their entire length (as in Fig. 10.493)....***notiale*** (p. 387)

67. Gill histoblast of 12 filaments....***luggeri*** (in part) (p. 386)

– Gill histoblast of 10 filaments....68

68. Gill histoblast (slide mounted) with filaments bearing reticulate surface sculpture (as in Fig. 10.489)....***jenningsi*** (in part) (p. 382)

– Gill histoblast (slide mounted) with filaments bearing transverse furrows (as in Fig. 10.591)....***ozarkense*** (in part) (p. 388)

69. Gill histoblast of 2–4 filaments (*johannseni* species group)....70

– Gill histoblast of 6–12 filaments (*jenningsi* species group, in part)....72

70. Gill histoblast of 4 filaments, with dorsal 2 filaments swollen basally, about 2–3 times thicker than ventralmost filament (as in Fig. 10.441). Northeastern North America (Map 125)....***rothfelsi*** (in part) (p. 323)

– Gill histoblast of 2–4 filaments, with all filaments mildly to strongly swollen basally (as in Figs. 10.439, 10.440). Widespread....71

71. Body with brownish bands of subequal intensity (Plate 12d). Frontoclypeal apotome with head spots free of dark surrounding pigment, or enclosed and sometimes obscured with mottled brown pigment (Fig. 10.758). Central North America and southeastern coastal plain (Map 123)....***johannseni*** (p. 321)

– Body with brownish, reddish, or dark gray bands, typically with pigment of abdominal segments I and V heaviest (Plate 12e). Frontoclypeal apotome with head spots free of surrounding pigment (male larva, Fig. 10.759), or obscured by heavy, almost black pigment (female larva, Plate 12e). Southeastern United States west into Texas (Map 124)....***parmatum*** n. sp. (p. 322)

72. Gill histoblast of 6 filaments. Southern Appalachian Mountains (Map 210)....***remissum*** (p. 389)

– Gill histoblast of 7–12 filaments. Widespread....73

73. Gill histoblast of 7 filaments....74

– Gill histoblast of 8–12 filaments....75

74. Frontoclypeal apotome pale yellowish brown. Body variously colored, often with first abdominal segment more heavily pigmented than remaining abdominal segments. Gill histoblast with dorsal trunk giving rise to 1 pair of filaments (as in Fig. 10.487). Large streams along Gulf Coast (Map 198) .......... ***haysi*** (p. 382)

– Frontoclypeal apotome fulvous. Body reddish or dark green, with pigment of all abdominal segments subequal in intensity (Plate 20d). Gill histoblast with dorsal trunk giving rise to 2 pairs of filaments (as in Fig. 10.491). Small streams in Sandhills of North and South Carolina (Map 202) .......... ***krebsorum*** (p. 385)

75. Gill histoblast of 12 filaments .......... ***luggeri*** (in part) (p. 386)

– Gill histoblast of 8–10 filaments .......... 76

76. Frontoclypeal apotome darkened posteriorly, encompassing all head spots (Plate 19e). Gill histoblast of 9 (rarely 8) filaments .......... ***definitum*** (in part, female larva) (p. 380)

– Frontoclypeal apotome without dark pigment surrounding head spots, or if present, then typically not encompassing all head spots (Figs. 10.801, 10.803). Gill histoblast of 8–10 filaments .......... 77

77. Gill histoblast (uncurled) typically short, about 2 times or less length of head. Body pigment typically greenish or grayish, with pigment of all abdominal segments typically subequal in intensity (Plate 19a). Abdominal segment IX with ventral tubercles typically minute and rather slender (Figs. 10.614, 10.616, 10.617) .......... 78

– Gill histoblast (uncurled) typically long, more than 2 times length of head. Body variously pigmented, with abdominal segment I (sometimes V) typically more heavily pigmented than remaining abdominal segments (Plates 19c, d; 20c). Abdominal segment IX with ventral tubercles typically larger and more conical .......... 85

78. Gill histoblast (slide mounted) of 10 filaments with reticulate surface sculpture (as in Fig. 10.489) .......... ***jenningsi*** (in part) (p. 382)

– Gill histoblast (slide mounted) of 8–10 filaments with transverse furrows (as in Fig. 10.591) .......... 79

79. Gill histoblast with 10 filaments, branching 2, 2, 2, 1, (2 + 1) (as in Fig. 10.494). Kentucky to Texas (Map 207) .......... ***ozarkense*** (in part) (p. 388)

– Gill histoblast with 8–10 filaments; if 10 filaments, then branching 2, 2, (1 + 2), (1 + 2) or 2, 2, (1 + 2), (2 + 1) (as in Figs. 10.481, 10.488). Widespread .......... 80

80. Gill histoblast of 8 filaments. Posterior proleg (slide mounted) typically with more than 80 rows of hooks .......... ***podostemi*** (p. 389)

– Gill histoblast of 8–10 filaments. Posterior proleg (slide mounted) typically with 80 or fewer rows of hooks .......... 81

81. Gill histoblast of 8 or 9 filaments .......... 82

– Gill histoblast of 10 filaments .......... 83

82. Southern Appalachian Mountains and foothills (Map 199) .......... ***infenestrum*** (p. 382)

– Northeastern North America (Map 208) .......... ***penobscotense*** (p. 388)

83. Gill histoblast with filaments branching 2, 2, (1 + 2), (2 + 1) (as in Fig. 10.488) .......... go back to 82

– Gill histoblast with filaments branching 2, 2, (1 + 2), (1 + 2) (as in Fig. 10.481) .......... 84

84. Chromosome arm IIS with ring of Balbiani closer than "bulge" marker to centromere .......... ***anchistinum*** (p. 378)

– Chromosome arm IIS with "bulge" marker closer than ring of Balbiani to centromere .......... ***nyssa*** (p. 388)

85. Gill histoblast of 10 filaments arising from 1 long, basally swollen, gradually tapered trunk; filaments 5 and 6 (dorsal to ventral) typically not petiolate (as in Fig. 10.490) .......... ***jonesi*** (p. 384)

– Gill histoblast of 8–10 filaments arising from variously thickened trunk; filaments 5 and 6 (dorsal to ventral) petiolate (as in Figs. 10.483, 10.484, 10.498) .......... 86

86. Gill histoblast typically of 9 (rarely 8) filaments, with ventralmost petiole swollen (as in Fig. 10.484) .......... ***definitum*** (in part, male larva) (p. 380)

– Gill histoblast of 8–10 filaments, with ventralmost petiole typically not swollen for more than half its length, or if swollen for more than half its length, then gill histoblast of 10 filaments (as in Figs. 10.483, 10.498) .......... 87

87. Gill histoblast with ventralmost petiole typically one tenth or less length of uncurled gill histoblast (as in Fig. 10.498). Coastal plain .......... 88

– Gill histoblast with ventralmost petiole typically one seventh or more length of uncurled gill histoblast (as in Fig. 10.483). Primarily piedmont, also coastal plain along Gulf Coast ................................89

88. Gill histoblast of 8–10 filaments; petiole of third pair of filaments (from dorsum) typically about 2 times longer than petiole of fourth pair. Chromosome arm IIS with "bulge" marker closer than ring of Balbiani to centromere ................................ ***lakei*** (p. 385)

– Gill histoblast typically of 8 filaments (9 in some specimens); petiole of third pair of filaments (from dorsum) typically 3 or more times longer than petiole of fourth pair (as in Fig. 10.498). Chromosome arm IIS with ring of Balbiani closer than "bulge" marker to centromere ................... ***taxodium*** (p. 390)

89. Body greenish, with heavy band on first abdominal segment green or pale reddish (Plate 19c). Gill histoblast typically of 8, less frequently 9 or 10, filaments. Pennsylvania to South Carolina (Map 193) ................................ ***chlorum*** (p. 379)

– Body brownish, grayish, or reddish, with heavy band on first abdominal segment, if present, brownish, grayish, or reddish (Plate 19d). Gill histoblast of 9 or 10, less frequently 8, filaments. Widespread in southeastern United States west to Nebraska and Texas (Map 194) ....................... ***confusum*** (p. 379)

90. Postgenal cleft rounded anteriorly, extended slightly more than half distance to hypostomal groove (Fig. 10.779). Body brownish or grayish, strongly banded or mottled. Gill histoblast of 2–4 filaments. Northern Alaska and northern Canada (Map 155) ........................................ ***subpusillum*** (in part) (p. 348)

– Postgenal cleft variously shaped anteriorly, extended various distances to hypostomal groove (Figs. 10.769, 10.770, 10.791–10.796, 10.805–10.813). Body variously pigmented. Gill histoblast of various numbers of filaments. Widespread ................................91

91. Abdominal segment IX with pair of conspicuous conical ventral tubercles (Fig. 10.609). Frontoclypeal apotome with positive head-spot pattern (Figs. 10.769, 10.774) ................................92

– Abdominal segment IX without pair of ventral tubercles (Fig. 10.613) or, at most, with pair of small rounded tubercles. Frontoclypeal apotome with positive (Figs. 10.818–10.820) or negative head-spot pattern (Fig. 10.823), or with spots apparently absent (Fig. 10.780) ................................102

92. Frontoclypeal apotome with medial brown stripe (sometimes faint) extended anteriorly beyond antennal bases (Fig. 10.769, Plate 14b). Abdominal cuticle (slide mounted) posterodorsally with numerous minute, unbranched, dark setae (as in Fig. 10.840). Gill histoblast of 8 filaments ................................ ***croxtoni*** (p. 337)

– Frontoclypeal apotome typically without medial brown stripe extended anteriorly (Figs. 10.774, 10.776). Abdominal cuticle (slide mounted) posterodorsally with numerous or scattered, transparent or dark, unbranched or multiply branched setae. Gill histoblast of 4–8 filaments ................................93

93. Abdomen posterodorsally with dark reddish or burgundy patch (Plate 15a). Postgenal cleft narrow, about twice as long as wide (Fig. 10.774). Gill histoblast of 4 filaments ......................... ***loerchae*** (p. 341)

– Abdomen posterodorsally variously colored, but without dark reddish or burgundy patch (Plates 13d; 14c, e; 15c). Postgenal cleft variously shaped, typically less than twice as long as wide (Figs. 10.765, 10.770, 10.776). Gill histoblast of 4–8 filaments ................................94

94. Body greenish gray to reddish, with pigment on first segment darker and denser than on subsequent 3 segments (Plate 15c). Gill histoblast of 4 filaments with reticulate surface sculpture (as in Fig. 10.452) ................................ ***quebecense*** (p. 344)

– Body variously pigmented, with pigment of first segment subequal in intensity to that on subsequent 3 segments (Plates 14c, 15d). Gill histoblast of 4–8 filaments; if 4 filaments, not with reticulate surface sculpture ................................95

95. Abdominal cuticle (slide mounted) dorsally with numerous dark, multiply branched setae (Fig. 10.839). Gill histoblast of 8 filaments ................................ ***fionae*** (p. 338)

– Abdominal cuticle (slide mounted) dorsally with dark or pale unbranched setae (Figs. 10.840), or if with multiply branched setae, these pale, typically visible only with phase contrast. Gill histoblast of 4–8 filaments ................................96

96. Abdominal cuticle (slide mounted, phase contrast) dorsally with numerous unpigmented, multiply branched setae. Gill histoblast of 8 filaments ................................ ***burgeri*** n. sp. (p. 333)

– Abdominal cuticle (slide mounted) dorsally with numerous or scattered, dark or pale unbranched setae (Fig. 10.840). Gill histoblast of 4–8 filaments ................................97

97. Abdominal cuticle (slide mounted) posterodorsally with scattered pale setae. Gill histoblast of 4 or 6 filaments ................................98

– Abdominal cuticle (slide mounted) posterodorsally with numerous dark setae (Fig. 10.840). Gill histoblast of 4 or 8 filaments ........................................ 100

98. Frontoclypeal apotome with head spots encompassed by faint pigment in form of inverted U (Fig. 10.768, Plate 14a). Gill histoblast of 4 filaments. Widespread (Map 138) ............ ***craigi*** (in part) (p. 336)

– Frontoclypeal apotome with head spots typically free of surrounding pigment in form of inverted U, although anterior spots sometimes surrounded by pigment (Figs. 10.760, 10.772). Gill histoblast of 4 or 6 filaments. Eastern North America (i.e., east of 95° longitude) ........................................ 99

99. Postgenal cleft extended only slightly more than half distance to hypostomal groove (Fig. 10.772). Gill histoblast of 6 filaments, each subequal in thickness. Widespread (Map 143) ........................................ ***gouldingi*** (in part) (p. 340)

– Postgenal cleft extended three fourths or more distance to hypostomal groove (Fig. 10.760). Gill histoblast of 4 filaments, with 2 thick and 2 thin (as in Fig. 10.441). Northeastern North America (Map 125) ........................................ ***rothfelsi*** (in part) (p. 323)

100. Postgenal cleft surrounded by faint brownish pigment in shape of H, contrasting with pale yellowish pigment of ventral head (Fig. 10.777). Gill histoblast of 4 filaments ................ ***silvestre*** (in part) (p. 344)

– Postgenal cleft not surrounded by pale brownish pigment in shape of H (Fig. 10.778). Gill histoblast of 8 filaments ........................................ 101

101. Antenna extended beyond apex of stalk of labral fan by half or more length of distal article. Head brownish, with head spots somewhat distinct, typically not surrounded by heavy dark pigment. Body brownish. Rocky Mountains westward (Map 152) ........................................ ***wyomingense*** (p. 346)

– Antenna shorter than apex of labral fan or extended beyond apex of labral fan by no more than half length of distal article (Fig. 10.778). Head yellowish to brownish yellow, with head spots bold, sometimes surrounded, at least centrally, by dark pigment. Body pale grayish to reddish (Plate 15e). Alaska, Canada, and northeastern United States (Map 153) ........................................ ***furculatum*** (in part) (p. 347)

102. Head pale yellowish to brownish yellow, with head spots bold, sometimes surrounded, at least centrally, by dark pigment (Fig. 10.778). Abdominal cuticle (slide mounted) posterodorsally with numerous dark setae. Body pale grayish to reddish (Plate 15e). Gill histoblast of 8 filaments. Alaska, Canada, and northeastern United States (Map 153) ........................................ ***furculatum*** (in part) (p. 347)

– Head variously pigmented, with head spots variously expressed (Figs. 10.780, 10.781, 10.791–10.797, 10.805–10.812). Abdominal cuticle (slide mounted) posterodorsally with scattered pale or dark setae. Body variously pigmented. Gill histoblast of 6–16 filaments. Widespread ........................................ 103

103. Postgenal cleft a broadly rounded inverted U, about as wide as long, extended about half distance to hypostomal groove (Figs. 10.780, 10.781). Frontoclypeal apotome with negative head-spot pattern, or with faint, obscure, or apparently absent head spots (Plates 15f, 16a). Rectal papillae of 3 simple lobes (sometimes with few secondary lobules) (as in Fig. 10.619). Gill histoblast of 6–10 filaments. West of Mississippi River (subgenus *Psilopelmia*) ........................................ 104

– Postgenal cleft subtriangular and tapered apically, or broadly rounded, typically longer than wide, extended half or more distance to hypostomal groove (Figs. 10.791–10.797, 10.804–10.813). Frontoclypeal apotome with various head-spot patterns. Rectal papillae of 3 compound lobes (as in Fig. 10.620). Gill histoblast of 6–16 filaments. Widespread ........................................ 110

104. Antenna brown (Fig. 10.781, Plate 16a). Gill histoblast of 6 filaments. Oklahoma and Texas (Map 166) ........................................ ***trivittatum*** (p. 353)

– Antenna colorless or pale yellowish brown to brown (Fig. 10.780, Plate 15f). Gill histoblast of 4–6 or 8–10 filaments. Widespread ........................................ 105

105. Gill histoblast of 4–6 filaments. Labral fan with more than 50 primary rays. Rio Grande of New Mexico (Map 163) ........................................ ***meyerae*** (p. 352)

– Gill histoblast of 8–10 filaments. Labral fan with various numbers of primary rays. Widespread ............ 106

106. Gill histoblast with 2 elongate trunks arising from base (as in Fig. 10.458). Southern Arizona (Map 161) ........................................ ***longithallum*** (p. 351)

– Gill histoblast with 3 short trunks arising from base. Widespread ........................................ 107

107. Gill histoblast with middle group of filaments branching 2 + 1 (as in Fig. 10.459). Texas (Map 162) ........................................ ***mediovittatum*** (p. 352)

– Gill histoblast with middle group of filaments branching 1 + 2 (as in Fig. 10.456). Widespread ......... 108

108. Gill histoblast of 9 or 10 filaments. California (Map 158) ........................................ ***clarum*** (in part) (p. 350)
– Gill histoblast of 8 filaments. Widespread ........................................ 109

109. Labral fan with more than 50 primary rays. Rio Grande of Texas and southern New Mexico (Maps 160, 165) ........................................ ***labellei*** (p. 351), ***robynae*** (p. 352)
– Labral fan with fewer than 50 primary rays. Widespread ........................................ ***bivittatum*** (p. 349), ***clarum*** (in part) (p. 350), ***griseum*** (p. 350), ***notatum*** (p. 352), ***venator*** (p. 353)

110. Abdomen dark gray to blackish, with brownish patches dorsolaterally on segments III–V (Plate 17e). Frontoclypeal apotome pale yellowish, with indistinct head spots; gena brownish orange, contrastingly darker than apotome (Fig. 10.791). Gill histoblast of 1 annulate club, with 1 dorsal and 1 ventral projection basally (as in Fig. 10.476). Western North America (i.e., west of 100° longitude), with 1 record from southern Florida (Map 181) ........................................ ***canadense*** (p. 369)
– Abdomen variously pigmented, but if dark gray to blackish, then not with brownish patches dorsolaterally on segments III–V. Frontoclypeal apotome variously colored, with head spots negative or positive, variously expressed; gena variously pigmented. Gill histoblast of 6–16 slender filaments. Widespread ........................................ 111

111. Hypostoma with median tooth broader than and extended anteriorly beyond lateral teeth; anterolateral margin typically somewhat rounded (Figs. 10.648, 10.649). Postgenal cleft subtriangular (Figs. 10.792–10.797). Large species, typically 8.0 mm or longer. Gill histoblast of 8, 9, or 15 filaments (subgenus *Hemicnetha*, in part) ........................................ 112
– Hypostoma with median and lateral teeth subequal in width and extended anteriorly to same level; anterolateral margin abruptly angled (Fig. 10.650). Postgenal cleft of various shapes (Figs. 10.804–10.813). Smaller species, typically shorter than 8.5 mm. Gill histoblast of various numbers of filaments (subgenus *Simulium*, in part) ........................................ 119

112. Abdomen posterodorsally with large, blackish, anteriorly pointed patch of pigment contrasting with paler adjacent pigment (Plate 18e). Gill histoblast of 9 filaments. Eastern North America (i.e., east of 100° longitude) (Map 188) ........................................ ***claricentrum*** (p. 374)
– Abdomen dorsally with pigment uniformly distributed, or if patchy, then without large, blackish, anteriorly pointed, posterodorsal patch (Plates 17f; 18a–d, f). Gill histoblast of 8, 9, or 15 filaments. Widespread ........................................ 113

113. Body pigment blackish and uniformly distributed dorsally (Plate 18f). Gill histoblast of 9 filaments. Eastern North America (i.e., east of 100° longitude) and northern Canada (Maps 189, 190) ........................................ 114
– Body pigment brownish to grayish, or if blackish, then distributed patchily (Plates 17f, 18a–d); if uniformly distributed, then posterior proleg (slide mounted) with more than 250 rows of hooks. Gill histoblast of 6–16 filaments. Western United States (i.e., west of 100° longitude) and British Columbia ........................................ 115

114. Gill histoblast with, at most, only some filaments inflated basally (Fig. 10.592) ........................................ ***innoxium*** (p. 375)
– Gill histoblast with all filaments inflated basally (weakly inflated in westernmost populations) (Fig. 10.593) ........................................ ***pictipes*** (p. 377)

115. Frontoclypeal apotome with negative head-spot pattern (Fig. 10.794, Plate 18b), or with rather faint positive head spots against pale yellowish to yellowish brown background (Fig. 10.795); if head spots positive, then body with pale brownish or grayish pigment (Plate 18c). Gill histoblast of 15 filaments. Southern Texas and New Mexico (Map 186) ........................................ ***solarii*** (p. 373)
– Frontoclypeal apotome with positive head-spot pattern (Figs. 10.792, 10.793). Gill histoblast of 8 filaments. Widespread west of Mississippi River ........................................ 116

116. Head (excluding head spots) rather uniformly brownish orange. Frontoclypeal apotome with anteromedial head spot nearly absent, anterolateral head spots large and dark, and posteromedial head spot slender anteriorly and bulbous posteriorly (Fig. 10.792). Abdomen rather uniformly brownish or grayish (Plate 17f) ........................................ ***freemani*** (p. 370)
– Head (excluding head spots) typically with contrasting dark and pale areas. Frontoclypeal apotome with head spots variously expressed; posteromedial head spot gradually tapered anteriorly (Fig. 10.793). Abdomen typically dark gray with small whitish patches (Plate 18a, d) (*paynei* species group, in part) ........................................ 117

117. Southern British Columbia to southern California (Map 184) ........................................ ***hippovorum*** (p. 371)
– East of California ........................................ 118

118. Chromosomes with chromocenter (as in Fig. 5.5) ........ ***bricenoi*** (p. 371)

– Chromosomes without chromocenter ........ ***paynei*** (p. 372), ***virgatum*** (p. 374)

119. Postgenal cleft triangular, acutely pointed anteriorly; subesophageal ganglion not ensheathed with pigment so that area encompassed by postgenal cleft is white and contrasting with brown head (Fig. 10.812). Body rather uniformly brown (Plate 22c) or sometimes brown banded. Gill histoblast of 6 filaments. Eastern North America (i.e., east of 95° longitude) ........ ***parnassum*** (p. 407)

– Postgenal cleft variously shaped, but typically not acutely pointed and triangular; subesophageal ganglion with or without ensheathing pigment (Figs. 10.804–10.811). Body variously pigmented. Gill histoblast of 6–16 filaments. Widespread ........ 120

120. Head with negative head-spot pattern (Figs. 10.811, 10.822–10.824) or without spots. Gill histoblast of 6 or 8 filaments ........ 121

– Head with positive head-spot pattern, although some head spots can be weak, absent, or negative (Figs. 10.804–10.810, 10.818–10.820); if with complete negative head-spot pattern, then body black (Plate 23c) or gill histoblast of 10 or 12 filaments. Gill histoblast of 6–16 filaments ........ 134

121. Frontoclypeal apotome with dark pigment typically restricted to central, elongate or H-shaped pattern (Plates 22b, 24b). Subesophageal ganglion typically not ensheathed with pigment (Figs. 10.811, 10.822). Gill histoblast of 8 filaments ........ 122

– Frontoclypeal apotome with dark pigment either absent or typically not forming central, elongate or H-shaped pattern (Figs. 10.823, 10.824; Plate 24c–e). Subesophageal ganglion with or without ensheathing pigment. Gill histoblast of 6 filaments (*venustum* species complex, p. 421; *verecundum* species complex, p. 430) ........ 124

122. Body pale greenish (Plate 24b). Head typically pale yellowish, except for dark pigment surrounding head spots (Fig. 10.822). Alaska and northwestern Canada (Map 242) ........ ***rubtzovi*** (p. 421)

– Body brownish (Plate 22b). Head typically brownish laterally and ventrally (Fig. 10.811). Widespread (*noelleri* species group) ........ 123

123. Chromosome arm IL with heavy band basally. Widespread (Map 226) ........ ***decorum*** (p. 403)

– Chromosome arm IL without heavy band basally. Alaska and northern Canada (Map 227) ........ ***noelleri*** (p. 405)

124. Chromosome arm IIS heterozygous or homozygous for inversion (= H sequence) that places ring of Balbiani nearly at end of arm (much closer than does A inversion in Fig. 11 of Rothfels et al. 1978). Northern Canada (Map 249) ........ ***tormentor*** n. sp. (p. 427)

– Chromosome arm IIS with various sequences, but without H sequence (Figs. 11, 12 of Rothfels et al. 1978). Widespread ........ 125

125. Chromosome arm IIS with homozygous inversion (= J sequence) from second band in section 45 to first band in section 51 (Fig. 11 of Rothfels et al. 1978 for standard sequence). Pacific Northwest (Map 252) ........ ***venustum* 'cytospecies JJ'** (p. 430)

– Chromosome arm IIS with various sequences, but without J sequence (Figs. 11, 12 of Rothfels et al. 1978). Widespread ........ 126

126. Chromosome arm IIS with EFG sequence homozygously inverted or heterozygous (Fig. 12 of Rothfels et al. 1978). Canada and northern United States (Map 250) ........ ***truncatum*** (p. 428)

– Chromosome arm IIS with various sequences, but without EFG sequence (Fig. 11 of Rothfels et al. 1978). Widespread ........ 127

127. Chromosomes with IIL-4 homozygously inverted or heterozygous, changing orientation of parabalbiani marker (Figs. 19, 20 of Rothfels et al. 1978) ........ ***rostratum*** (p. 432)

– Chromosomes without IIL-4 inversion (Fig. 15 of Rothfels et al. 1978) ........ 128

128. Chromosomes typically without IIIL-5 inversion (Fig. 23 of Rothfels et al. 1978, Fig. 2 of Adler & Mason 1997) ........ 129

– Chromosomes with IIIL-5 homozygously inverted or heterozygous (Fig. 22 of Rothfels et al. 1978) ........ 130

129. Chromosomes with IIL arm typically standard (Fig. 15 of Rothfels et al. 1978). Central Canadian provinces and adjacent states (Map 244) ........ ***incognitum*** (p. 425) (small percentage of individuals carry IIIL-5 inversion)

– Chromosomes minimally with IIL-1, 2 homozygously inverted (Fig. 17 of Rothfels et al. 1978). Widespread (Map 254) ........ ***verecundum*** s. s. (p. 433)

130. Chromosome arm IIS with A sequence homozygously inverted or heterozygous (Fig. 11 of Rothfels et al. 1978 for breakpoints) .......................... ***molestum*** (p. 427), ***piscicidium*** (p. 427) (population analysis is necessary for species separation [Rothfels et al. 1978, McCreadie et al. 1994a])

– Chromosome arm IIS with CC sequence (Fig. 11 of Rothfels et al. 1978) .......................... 131

131. Chromosomes with IIIL-5 heterozygous .......................... 132

– Chromosomes with IIIL-5 homozygously inverted (Fig. 22 of Rothfels et al. 1978) .......................... 133

132. Chromosomes without IIIL-6 .......................... ***irritatum*** (in part, some male larvae) (p. 426)

– Chromosomes with IIIL-6 heterozygous (Fig. 21 of Rothfels et al. 1978 for breakpoints) .......... ***minutum*** (in part, male larva, p. 426)

133. Chromosomes with IIL-1 homozygously inverted or heterozygous (Fig. 16 of Rothfels et al. 1978) .......................... ***hematophilum*** (in part) (p. 425), ***irritatum*** (in part) (p. 426) (only *irritatum* occurs west of 90° longitude; east of 90° longitude, population analysis is necessary for species separation [McCreadie et al. 1994a])

– Chromosomes without IIL-1 inversion (Fig. 15 of Rothfels et al. 1978) .................. ***hematophilum*** (in part) (p. 425), ***irritatum*** (in part) (p. 426), ***minutum*** (in part, female larva, p. 426), ***venustum*** s. s. (p. 429)

134. Body brown, with pigment uniformly distributed, not in distinct bands (Plate 22d). Head dark brownish orange. Antenna with medial article bearing pale band (Fig. 10.813). Gill histoblast of 6 filaments. Great Basin of western United States (Map 229) .......................... ***petersoni*** (p. 408)

– Body variously pigmented, often banded. Head variously pigmented. Antenna with or without pale band. Gill histoblast of 6–16 filaments. Widespread .......................... 135

135. Abdomen rather gradually expanded posteriorly (Plates 21a–f, 22a). Large species, typically more than 6 mm long. Gill histoblast of 10–16 (rarely 8) filaments (*malyschevi* species group) .......................... 136

– Abdomen rather abruptly expanded at segment V (Plates 23b–f, 24a). Small species, typically less than 6 mm long. Gill histoblast of 6 filaments (*tuberosum* species complex, p. 411) .......................... 149

136. Frontoclypeal apotome with posterior head spots typically encompassed by brown, subtriangular patch of pigment (Fig. 10.807). Gill histoblast of 10 filaments. Canada and northern United States (Map 216) .......................... ***murmanum*** (p. 393)

– Frontoclypeal apotome with posterior head spots not encompassed by brown pigment, or if encompassed, then not by subtriangular patch (Figs. 10.804–10.806, 10.808–10.810). Gill histoblast of 10–16 (rarely 8) filaments. Widespread .......................... 137

137. Postgenal cleft extended about three fourths distance to hypostomal groove (Fig. 10.805). Head brownish orange, with diffuse head spots typically encompassed by dark pigment (Plate 21b). Gill histoblast of 12 basally swollen filaments (as in Fig. 10.535) .......................... ***defoliarti*** (p. 391)

– Postgenal cleft typically extended to, or nearly to, hypostomal groove, often as narrowed nipple-like extension (Figs. 10.804, 10.806, 10.808–10.810). Head variously pigmented, often yellowish brown, with head spots variously distinct. Gill histoblast of 10–16 (rarely 8) slender filaments .................. 138

138. Head spots typically diffuse (Fig. 10.806, Plate 21c). Gill histoblast of 13–16 (rarely 12) filaments. Alaska and Canada (Map 215) .......................... ***malyschevi*** (p. 392)

– Head spots variously expressed. Gill histoblast of 10–12 (rarely 8) filaments. Widespread .................. 139

139. Postgenal cleft extended to hypostomal groove as narrowed projection (Fig. 10.804). Gill histoblast of 10 (rarely 8) filaments. Chromosomes with whole-arm interchange (IS + IIIL and IL + IIIS). Alaska and northern Canada (Map 213) .......................... ***decimatum*** (p. 390)

– Postgenal cleft not extended to hypostomal groove (Figs. 10.808–10.810). Gill histoblast of 12 filaments. Chromosomes with standard arm associations (IS + IL and IIIS + IIIL). Widespread (*arcticum* species complex, p. 394; numerous cytotypes currently unassigned to a species will not key out in the subsequent couplets; see Cytology section under *S. arcticum* species complex and Figs. 5.6, 5.7) .......... 140

140. Body uniformly brown, not banded. Head typically dark brown (Fig. 10.808). Chromosomes with enormous, darkly staining chromocenter; IIL-11 homozygously inverted or heterozygous (Fig. 5.6 for breakpoints). Rocky Mountains of United States (Map 222) .......................... ***chromatinum*** n. sp. (p. 398)

– Body variously pigmented, often banded. Head variously pigmented, often yellowish brown or brownish orange. Chromosomes with small (Fig. 5.5) or no chromocenter; IIL-11 absent (Fig. 5.6). Widespread .......................... 141

141. Subesophageal ganglion not ensheathed with pigment (Fig. 10.809) ....................142
– Subesophageal ganglion typically ensheathed with pigment (Fig. 10.810) ....................143

142. Head brownish, with pale yellowish head spots (female larva, Plate 22a), or pale yellowish, with faint brown head spots (male larva, Fig. 10.809). Body well pigmented. Chromosomes with chromocenter (as in Fig. 5.5), but without IIL-8 (Fig. 5.6). Widespread (Map 223) ....................***negativum*** n. sp. (p. 399)
– Head pale yellowish, almost white, with faint brownish head spots. Body typically pale, weakly pigmented. Chromosomes without chromocenter, but with IIL-8 homozygously inverted (Fig. 5.6 for breakpoints). Canadian prairie provinces and southern Northwest Territories (Map 225) ....................***vampirum*** (p. 400)

143. Chromosomes with IIL-2 homozygously inverted or heterozygous (Fig. 5.6 for breakpoints). Alaska and Pacific Coast, rare inland (Map 224) ....................***saxosum*** n. sp. (p. 399)
– Chromosomes typically without IIL-2 (Fig. 5.6). Widespread ....................144

144. Chromosomes with IIS-11 homozygously inverted (Fig. 5.7) or heterozygous; IIL-7 typically homozygously inverted or heterozygous (Fig. 5.6 for breakpoints) ....................***apricarium*** n. sp. (p. 395)
– Chromosomes without IIS-11 or IIL-7 (Fig. 5.6) ....................145

145. Chromosomes with IIL-1 heterozygous (Fig. 5.6 for breakpoints) ....................***arcticum* 'cytospecies IIL-1'** (in part, male larva) (p. 397)
– Chromosomes without IIL-1 inversion (Fig. 5.6) ....................146

146. Chromosomes with IIL-3 heterozygous (Fig. 5.6 for breakpoints) ....................***arcticum*** s. s. (in part, male larva) (p. 396)
– Chromosomes without IIL-3 inversion (Fig. 5.6) ....................147

147. Chromosomes with IS-1 homozygously inverted (Fig. 3 of Shields & Procunier 1982 for breakpoints) ....................***arcticum* 'cytospecies IIS-4'** (p. 397)
– Chromosomes with or without IS-1 homozygously inverted (Fig. 3 of Shields & Procuier 1982 for breakpoints) ....................148

148. Chromosomes with sections 55–57 in base of IIL arm failing to pair, although no inversion present ....................***brevicercum*** (in part, some male larvae) (p. 398)
– Chromosomes with sections 55–57 paired in base of IIL arm ....................***arcticum*** s. s. (in part, female larva, p. 396), ***arcticum* 'cytospecies IIL-1'** (in part, female larva, p. 397), ***brevicercum*** (in part, female larva, some male larvae, p. 398)

149. Body black, with pigment distributed uniformly (Plate 23c). Head darkly pigmented, with small pale head spots (negative pattern) (Fig. 10.817). Sierra Nevada of California; high elevations (Map 234) ....................***chromocentrum*** n. sp. (p. 414)
– Body grayish, blackish, or brownish, with pigment distributed uniformly or in bands (Plates 23b, d–f; 24a). Head pale yellowish brown to brownish orange, with small dark head spots (positive pattern) (Figs. 10.818–10.821). Widespread; low to high elevations ....................150

150. Postgenal cleft with margins slightly or not at all bowed outward; subesophageal ganglion not fully ensheathed with pigment so that area encompassed by postgenal cleft is white and contrasting with brownish head (Fig. 10.821). Alaska and northern Canada (Map 241) ....................***vulgare*** (p. 420)
– Postgenal cleft with margins slightly to strongly bowed outward; subesophageal ganglion ensheathed with pigment (Figs. 10.818–10.820). Widespread ....................151

151. Frontoclypeal apotome with anteromedial and posteromedial head spots closely approximated (Fig. 10.819). Body brownish (male larva, Plate 23f) or grayish (female larva). Gill histoblast with all 3 trunks longer than wide (as in Fig. 10.509). Eastern, mainly southeastern, United States (Map 239) ....................***ubiquitum*** n. sp. (p. 418)
– Frontoclypeal apotome with anteromedial and posteromedial head spots typically separated by distinct gap (Figs. 10.818, 10.820). Body variously pigmented. Gill histoblast with no more than 1 trunk longer than wide (as in Fig. 10.510). Widespread ....................152

152. Body dark gray or blackish, with pigment rather uniformly distributed, or if banded, then intersegmental areas mottled with grayish pigment (Plate 23d). Head brownish orange, with head spots often obscure, or with anterolateral head spots darkest. Chromosome arm IIS with "bulge" marker

closer than ring of Balbiani to centromere. Canada and east of Mississippi River in United States (Map 235) ........ ***conundrum*** n. sp. (p. 415)

– Body gray to brown, with pigment distributed in bands (Plates 23b, e; 24a). Head yellowish brown to brownish orange, with head spots variously distinct (Figs. 10.818, 10.820); if head brownish orange, then chromosome arm IIS with ring of Balbiani closer than "bulge" marker to centromere. Widespread ........ 153

153. Body with large whitish area on posterior half of thorax, contrasting with grayish pigment anteriorly and posteriorly (visible even with naked eye) (Plate 23b). Chromosome arm IIS with "bulge" marker often closer than ring of Balbiani to centromere. East of Mississippi River (Map 233) ........ ***appalachiense*** n. sp. (p. 413)

– Body with smaller whitish area on thorax (Plates 23e, 24a). Chromosome arm IIS with ring of Balbiani closer than "bulge" marker to centromere. Widespread ........ 154

154. Body strongly banded. Head pale yellowish brown, typically with head spots obscure and not surrounded by brown pigment. Rocky Mountains westward (Map 238) ........ ***twinni*** (p. 418)

– Body variously banded. Head pale yellowish brown to brownish orange, with head spots variously distinct and with or without brown surrounding pigment (Figs. 10.818, 10.820). Widespread ........ 155

155. Head brownish orange, typically with head spots surrounded by brown pigment, at least posteriorly (Fig. 10.820). Body dark gray (Plate 24a). Chromosome arm IIS with FG sequence heterozygous or homozygous; A and AB sequences absent (Figs. 1, 21 of Landau 1962) ........ ***vandalicum*** (p. 419)

– Head typically yellowish or yellowish brown, with head spots not surrounded by brown pigment (Fig. 10.818), except in some northern populations. Body pale gray (Plate 23e). Chromosome arm IIS without FG sequence, but with A or AB sequence heterozygous or homozygous (Figs. 1, 22, 23 of Landau 1962) ........ 156

156. Chromosome arm IIS with ring of Balbiani and "bulge" marker closely approximated, without any intervening bands; IIS without heavy subterminal band (Fig. 1 of Landau 1962). United States and southernmost Canada east of Rocky Mountains (Map 236) ........ ***perissum*** (p. 416)

– Chromosome arm IIS with ring of Balbiani and "bulge" marker separated by intervening bands; IIS with heavy, subterminal band present heterozygously or homozygously (Figs. 1, 23 of Landau 1962). Widespread (Maps 237a, b) ........ ***tuberosum*** s. s. (p. 417)

## Taxonomic Accounts

### EXPLANATORY INFORMATION

*Taxonomic Accounts*

This chapter contains accounts of 254 extant species, plus 70 higher taxa known from North America north of Mexico. Subfamilies, tribes, genera, and subgenera are arranged phylogenetically. Species in each genus are arranged alphabetically according to subgenus, species group, and species complex; species complexes appear at the end of the relevant subgenus or species group. A diagnosis and an overview are provided for each taxon above the species level. Detailed treatments focus on species, although we also treat eight species complexes that, believed for many years to be single species, have acquired a large literature. Most North American species are now known as females (94%), males (93%), pupae (93%), and larvae (98%).

We emphasize that much additional taxonomic work is needed for the North American fauna. In a number of cases, we have indicated populations that are probably valid species, but we have deferred formal description until more material can be studied. The localized distributions of many species suggest that additional species are yet to be discovered in the countless kilometers of unprospected flowing water and unsampled habitats. Taxa such as the subgenera *Psilopelmia* and *Aspathia* have not yet been studied chromosomally in North America and probably contain sibling species. Many of the known chromosomal variants, after closer scrutiny, might deserve species status. Homosequential sibling species are probably fairly common but rarely have been considered. On the other hand, the number of synonyms is unlikely to increase by more than a few names. Thus, with increased chromosomal study, and perhaps molecular analysis, as well as further prospecting in poorly sampled areas, the total number of species recognized in North America is certain to increase, perhaps to nearly 300 species.

Each account of a species or species complex includes a summary of all pertinent North American information known to us on taxonomy, morphology, cytology, physiology, molecular systematics, and bionomics. To the fullest extent, this information is based on accurate identifications. Whenever possible, we coupled chromosomal and morphological characters to ensure accuracy. To verify literature records and accounts, we examined museum material and visited many of the original study sites. Our choice of Palearctic and Neotropical literature is eclectic and includes primarily the major works for which identifications are reliable. Large lacunae characterize the information base for individual species, and we often specifically indicate these voids.

Synonymies and Types

All synonyms for tribal, generic, and subgeneric names are provided. A synonymy, including all vernacular names and known misidentifications, also is provided for each species. Approximately 367 formal names and more than 390 vernacular names have been applied to North American species. The first entry in the species synonymy is the current name, followed by all other names, both formal and vernacular, arranged chronologically. Vernacular names are indicated by single quotation marks. We were unable to associate 9 vernacular names with the relevant species; these 9 names are treated at the end of the taxonomic accounts. Misidentifications follow the formal and vernacular names and are presented chronologically in brackets; more than 615 misidentifications are listed for the North American fauna, but many undoubtedly have been missed. Information in the species accounts is sometimes drawn from these corrected identifications, and familiarity with the synonymy will ensure full benefit from the cited literature. Misspellings of names are listed after the correct names.

Data for type specimens include author and full pagination for the original description (plus figure numbers that do not fall within that pagination), life stages described, curatorial status (i.e., whether pinned, on a slide, or in an alcohol or glycerin vial), accession number (when given), depository, and type locality and associated information gleaned from the type labels and published descriptions. To facilitate location of type localities, we provide the state and county for all species described from the United States, and give miles and kilometers according to which was used in the original publication; all elevations are in metric units. A question mark before a name indicates that the suggested synonymy requires further study; we used query marks especially when we were unable to examine actual material. Types that we examined are marked with an asterisk (*). We examined 242 (about 70%) of the 347 primary types and collected fresh material at more than one third of the type localities. When material was available, we designated lectotypes and neotypes to establish the current concept of each relevant name, especially because sibling species are common in the Simuliidae. Most primary types (70%) are housed in the Canadian National Collection of Insects, Ottawa, Ontario (CNC) (Cooper 1991) and the National Museum of Natural History, Washington, D.C. (USNM).

New Species

Forty-three new species are described herein. Thirty-six are authored by Adler, Currie, and Wood, and seven (*Prosimulium longirostrum*, *Greniera humeralis*, *G. longicornis*, *Tlalocomyia andersoni*, *T. ramifera*, *Stegopterna acra*, *St. xantha*) are authored by Currie, Adler, and Wood. We have provided concise descriptions and diagnoses of new species; additional characters can be obtained from the diagnoses of inclusive higher taxa. Larval descriptions are based on final instars fixed in acetic ethanol. Adult colors are taken from pinned specimens. Holotypes are pinned or stored in 80% ethanol. Pinned holotypes either were frozen until dry or were taken through an ethanol series and dried with Peldri II (Brown 1990) or hexamethyldisilazane (Brown 1993). Holotypes are deposited in the CNC or USNM. Paratypes are deposited in the Natural History Museum in London (BMNH), CNC, Clemson University Arthropod Collection (CUAC, Clemson, South Carolina), and USNM. Paratypes for some specimens include photographic negatives of chromosomes, which are deposited in the CNC. Additional specimens that we examined are not listed in the text but can be found in lists deposited in the CNC, CUAC, Royal Ontario Museum (ROM, Toronto, Ontario), and USNM.

Whenever possible, we have conserved names, especially for use with members of species complexes. For example, about four formal names previously were held in synonymy with *Simulium venustum*, now known to be an aggregate of species; we applied these formal names pragmatically to the valid species formerly known as *S. venustum*. We have avoided formal description of species known from single specimens and have avoided formally naming disjunct populations with minor differences.

Taxonomy

The taxonomic literature for each species and species complex is summarized chronologically as citations, with parenthetical indication of the life stage or gender treated. These taxonomic references include descriptions, keys, illustrations, and photographs. New taxonomic information, such as our rationale for synonymies and notes on variation, is provided as narrative.

Morphology

Morphological works are those dealing with the interpretation of structures beyond a purely taxonomic perspective. Morphological literature is summarized as citations, each with a short parenthetical description of the subject and life stage or gender treated (unless the structure is restricted to one stage or gender). The *Simulium vittatum* species complex has been featured in more than a third of the references that contain morphological information on North American black flies. By consulting the references under this species complex, one can expediently enter the morphological literature on black flies.

Physiology

Citations of the literature on physiology include short parenthetical descriptions of the subject and life stage or gender to which the work pertains. The *S. vittatum* species complex has been the workhorse in physiological research on black flies, having been represented in more than 60 percent of the roughly 54 references to the physiology of North American species.

Cytology

All cytological papers are cited for each species and species complex. The vast majority of cytological references include descriptions, photographs, illustrations, keys, and idiograms of polytene chromosomes. Cytological information other than these aspects of polytene chromosomes is indicated parenthetically. New information is provided as narrative, with reference to published standard maps when available. Breakpoints of inversions, when given, are sometimes expressed as c (central), d (distal), and p (proximal). Fixed inversions are in italic type. Photographic negatives and maps of the polytene chromosomes of many species are housed in the CUAC.

Molecular Systematics

Citations of papers dealing with molecular systematics include both biochemical (e.g., allozyme) and molecular works.

Bionomics

Bionomic information is presented for each species and species complex and includes distribution and habitat, oviposition and related aspects (e.g., autogeny), development (including larval feeding behavior), mating, natural enemies, and hosts and economic importance. General information that applies to all species of a higher group is included in the treatment of the higher taxon. For example, the fact that all species of *Prosimulium* are univoltine is presented under the generic treatment. The section on natural enemies provides records of predators and parasites, including trichomycete fungi, which are believed to be commensalistic in the larval gut but pathogenic in the adult female.

Many organisms other than black flies are mentioned in the bionomics sections. Common names of mammals and birds are given in the text and listed with their scientific names in Tables 6.2 and 6.3; scientific names for birds and mammals are given in the text only if they do not appear in these tables. For all other organisms, we give both scientific and, when available, common names. We have eschewed the use of author names for all non-simuliid organisms because they can be found in the specialized literature and are not necessary for the comprehension of simuliid bionomics.

### *Illustrations*

More than 820 line drawings, 150 color illustrations, and about 20 photographs are provided as identification aids. Not all species are illustrated in each life stage, either because we could not distinguish the particular life stage or sex from that of a closely related species or, infrequently, because of lack of material. We, therefore, caution that accurate identification begins with use of the keys, supplemented by reference to the figures and other data; exclusive use of pictures for identification could result in errors.

Color Illustrations

Color illustrations were prepared with colored pencils and an inked outline of the body. The length of each larva was adjusted to one size by enlargement or reduction with a photocopier. Larvae were drawn within a month after collection in acetic ethanol. Some colors, however, fade over time, particularly when larvae are transferred to ethanol; for example, red colors eventually become grayish green or brown. The numbers of primary rays illustrated for the labral fans were based on counts of rays for 1–50 (usually 10–15) final-instar larvae. For *Parasimulium crosskeyi*, the largest available larva (possibly an antepenultimate instar) was illustrated because final instars have never been collected. The gut, which is often seen through the larval integument, was not illustrated for any species. Antennae and labral fans sometimes were drawn slightly darker so that they could be seen adequately. Color illustrations of adult scutal patterns were based on pinned specimens illuminated from multiple angles to show the complete pattern. All specimens used for color illustrations are housed in the CUAC.

Line Illustrations

Head capsules of final-instar larvae were placed in ethanol, positioned in cotton batting, and sketched, along with the subesophageal ganglion when ensheathed with pigment, at 100× magnification, using a stereomicroscope fitted with a camera lucida. The capsules subsequently were cleared in hot lactic acid, and the mouthparts and internal structures were removed. Each head capsule then was placed on a spot of shellac glue in a depression slide and flooded with glycerin. Details of the head were drawn on the original sketch at 100× magnification, using a compound microscope and camera lucida. Within a few days, capillary action of the glycerin separated the specimen from the glue, allowing the specimen to be flipped, reglued, and illustrated in ventral view. As a novel character system, all setae on the head were drawn. These setae generally are not easily viewed on the specimens until the head capsules are cleared, slide mounted in glycerin or a permanent medium, and viewed with a compound microscope (with phase contrast if necessary). Hypostomata were drawn while the head capsules were positioned ventrally under the compound microscope at 200× magnification.

Pupae were placed in cotton batting or white sand, flooded with ethanol, and sketched using a stereomicroscope (40×) with a camera lucida. If the gill was markedly pale, it was highlighted by slipping a piece of black sandpaper behind it.

Adult terminalia were cleared in hot lactic acid, mounted on a spot of shellac glue in a depression slide, and flooded with glycerin. The aedeagal membrane of some taxa (*Simulium annulus* species group and *S. furculatum*) first was inflated by inserting a microsyringe in the abdomen of a fresh specimen and gently injecting ethanol. All terminalia were illustrated intact with the aid of a camera lucida on a compound microscope (250–400×). They then were dissected as necessary to reveal greater detail and to position component parts such as gonostyli. Male genitalia were illustrated in ventral, lateral, and terminal views; gonostyli often were illustrated in one or two additional views. Because orientation of the ventral plate in situ can be quite different from that in isolation (e.g., in the subgenus *Aspathia*), both views sometimes were provided. Female terminalia were drawn in ventral and lateral views. Great care was taken to show patterns of sclerotization and microtrichia of the female anal lobes. These patterns provide excellent, but previously underused, diagnostic information.

A standard size was determined for each type of illustration (e.g., head capsule). All subsequent illustrations of that type were adjusted to the same size before the final inking. For each line drawing, up to 10 (typically 3–5) specimens from the same locality or geographic area were drawn in pencil. Scale bars were added to the rough sketches, preserving the range of sizes drawn for each species. These rough sketches were equalized in size by enlargement or reduction with a photocopier. They then were overlaid on a light table, and a composite sketch was produced. The composite sketch was overlaid with a mylar acetate sheet with drawing film embedded on the surface, and a final illustration was produced in ink. Highlights and contrasts were created by scraping the ink and surface film off the mylar. All specimens that contributed to the composite were labeled and deposited in the CNC. Each specimen was linked, by a number, with its original pencil drawing. Original drawings making up the composite are archived in the CNC, providing a record of species variability.

### *Distribution Maps*

Distributional information is drawn from literature records that we believe to be accurate, museum collections, and collections that we made or that were provided by colleagues and friends and identified by us. We used distribution records that span the period from 1823 to 2003. The distribution of each species is summarized in a statement under the Bionomics section and on maps prepared expressly for this book. These maps show states, provinces, counties, and 1980-census districts for western Canadian provinces; they were prepared before Nunavut was split from the Northwest Territories. Greenland is excluded from the maps because the region is sparsely populated by only three species (*Prosimulium ursinum*, *Simulium vittatum*, *S. rostratrum*). The maps not only provide a rough approximation of species distributions but also reflect collecting effort. Large areas of the continent, particularly Nunavut, and remote areas of western Alaska, including the Aleutian Islands, remain virtually unsampled. Even readily accessible areas such as Montana and the Dakotas are poorly sampled.

Each dot on a map represents one or more collections. Information pertaining to a collection is recorded in a database, arranged by species. The information includes location, date, collector, number of specimens of each life stage or gender, depository, and often data on stream characteristics, natural enemies, chromosomes, hosts, and bionomics. Hard copies and CD ROMs of this database are deposited in the CNC, CUAC, ROM, and USNM.

### *Commonly Used Abbreviations*

| | |
|---|---|
| C | chromosomes |
| ca. | circa |
| cDNA | complementary deoxyribonucleic acid |
| cm | centimeter(s) |
| Co. | county |
| E | egg |
| elev. | elevation |
| F | female |
| Hwy. | highway |
| km | kilometer(s) |
| L | larva |
| m | meter(s) |
| M | male |
| mi. | mile(s) |

| | |
|---|---|
| min | minute(s) |
| misident. | misidentification |
| ml | milliliter(s) |
| mm | millimeter(s) |
| mtDNA | mitochondrial deoxyribonucleic acid |
| n. sp. | new species |
| p. | page (pp., pages) |
| P | pupa |
| pers. comm. | personal communication |
| rDNA | ribosomal deoxyribonucleic acid |
| Rt. | route |
| sec | second(s) |
| s. l. | *sensu lato* |
| s. s. | *sensu stricto* |
| * | type examined by us |

*Institutional Abbreviations*

| | |
|---|---|
| BBM | Bernice P. Bishop Museum, Honolulu, Hawaii, USA |
| BMNH | Natural History Museum, London, England |
| CAS | California Academy of Sciences, San Francisco, USA |
| CNC | Canadian National Collection, Ottawa, Ontario, Canada |
| CU | Cornell University, Ithaca, New York, USA |
| HNHM | Hungarian Natural History Museum, Budapest, Hungary |
| HUS | Hokkaido University, Sapporo, Japan |
| IDSV | Instituto de Dermatología Sanitaria, Villa de Cura, Venezuela |
| IMZ | Istituto e Museo di Zoologia, University of Turin, Turin, Italy |
| INDRE | Instituto Nacional de Diagnóstico y Referencia Entomológicos, Mexico City, Mexico |
| INHS | Illinois Natural History Survey, Urbana, Illinois, USA |
| MCZ | Museum of Comparative Zoology, Harvard University, Cambridge, Massachusetts, USA |
| MNHNP | Muséum National d'Histoire Naturelle, Paris, France |
| SMNL | Staatliches Museum für Naturkunde, Ludwigsburg, Germany |
| UKAL | University of Kansas, Lawrence, Kansas, USA |
| UMSP | University of Minnesota, St. Paul, Minnesota, USA |
| USNM | National Museum of Natural History, Washington, D.C., USA |
| UZMH | Zoological Museum, Helsinki, Finland |
| ZIL | Zoological Institute, University of Lund, Lund, Sweden |
| ZISP | Zoological Institute of St. Petersburg, St. Petersburg, Russia |
| ZMHU | Museum für Naturkunde der Humboldt Universität, Berlin, Germany |

### Family Simuliidae Newman

Simuliites Newman 1834: 379, 387–388 (as 'Natural Order'). Type genus: *Simulium* Latreille 1802: 426

The latest world inventories (Crosskey & Howard 1997; Crosskey 1999a, 2002) list 1772 formally named, extant species, to which, after eliminating synonyms, we add a net of 21 formally described species, bringing the world's total number of extant nominal species to 1793. In North America, we recognize 247 formally described, extant species, plus another 7 for which insufficient material is available to permit formal description, bringing the total number of extant species in North America to 254.

### Subfamily Parasimuliinae Smart

Parasimuliinae Smart 1945: 479. Type genus: *Parasimulium* Malloch 1914: 24

Parasimuliinae Smart 1944a: 24. *Nomen nudum*

**Diagnosis.** Adults: Stemmatic bulla present near posterior margin of compound eye. Antenna with 8 flagellomeres. Radius with basal section less than one fourth distance from base of radial sector to apex of wing; radial sector with branches widely separated by membrane; $R_1$ connected to costa near middle of wing; $R_{4+5}$ connected before apex of costa; costa, subcosta, and radial sector with long setae dorsally and ventrally; false vein unforked apically; $CuA_2$ slightly sinuous; basal medial cell absent; basal radial cell one fourth or less length of wing (measured from apex of stem vein). Katepisternum in lateral view pointed ventrally; sulcus absent. Mesepimeral tuft absent. Calcipala and pedisulcus absent. Empodium elongate, spiculose. Female: Lacinia without retrorse teeth; mandible without serrations. Claws toothless. Anal lobes (sternum X) undivided medially. Spermatheca spherical, with or without unsclerotized ring around apex of spermathecal duct. Male: Head dichoptic or holoptic. Gonostylus without apical spinule. Ventral plate with straplike connection between apex of each basal arm and paramere. Pupa: Gill of 3 filaments arising from elongate base; plastron only on branches; felt chamber present in base of gill; mesothoracic spiracle communicating with lumen of gill. Abdominal segment III undivided by pleural membrane. Sternites VI and VII undivided medially by longitudinal striate membrane. Pleuron without recurved hooks. Tergites without spine combs. Larva: Antenna of 1 article. Postgenal cleft absent. Prothoracic proleg without lateral sclerites. Abdominal segment IX with large conical lobe ventrally. Anal sclerite with posteroventral arms not articulated dorsally; apices nearly in contact midventrally, separated by small spindle-shaped sclerite.

**Overview.** This tiny monotypic subfamily is endemic to the Nearctic Region and includes 5 species—less than 2% of the total simuliid fauna in North America.

### Genus *Parasimulium* Malloch

*Parasimulium* Malloch 1914: 24 (as genus). Type species: *Parasimulium furcatum* Malloch 1914: 24–25, by original designation

**Diagnosis.** Same as that given for the subfamily.

**Overview.** *Parasimulium* is the most plesiomorphic genus of black flies. It includes 5 univoltine species and is endemic to forested regions of the Pacific Northwest, with its distributional heart in the Cascade Range of Oregon and Washington. Most sites where *Parasimulium* has been found are in mesic coniferous forests dominated by western hemlock (*Tsuga heterophylla*) and Douglas fir

(*Pseudotsuga menziesii*). Some species are associated with ancient forests and might be ecologically sensitive indicators of the health of these forests (Courtney 1993). Fossil fragments of larvae about 11,000 years old have been recovered from lake sediments in southwestern British Columbia (Currie & Walker 1992). Present-day members of the genus probably occupy a reduced version of their original range, but they could not have occupied their current range until the early Eocene Epoch (Currie 1986).

Discovery of eggs, larvae, and pupae of *Parasimulium* (Borkent & Wood 1986, Courtney 1986) came nearly three fourths of a century after the first adult was described by Malloch (1914). Still, larvae and pupae are known for only 2 species. They are obligate inhabitants of subterranean waters that flow through coarse, poorly sorted substrates. These flows have been described as part of the hyporheic zone, that is, a region of flowing water beneath and lateral to the streambed (Courtney 1993), and as hypogean or phreatic, referring to underground aquatic habitats such as subterranean springs (Currie 1988). Larvae have well-developed labral fans for filter feeding, but they exhibit structural characteristics of cavernicolous organisms, such as blindness and unpigmented cuticle. Final-instar larvae have never been found, suggesting that they move elsewhere, possibly into the sediments, to pupate. No natural parasites of the species in this genus have been recorded. Larval chromosomes have not been studied, although our preliminary examination indicates that they are workable.

Adults are readily collected from streamside vegetation where resting and mating occur (Wood & Borkent 1982). Nothing is known of egg deposition, although females can crawl about in water at the bottom of vials, suggesting that they are adapted to entering springs and seepages. Females are not fitted with biting mouthparts and therefore cannot take blood.

### Subgenus *Astoneomyia* Peterson

*Astoneomyia* Peterson 1977a: 105 (as subgenus of *Parasimulium*). Type species: *Parasimulium melanderi* Stone 1963b: 127, by original designation

**Diagnosis.** Adults: Body uniformly dark brown, except halter in some specimens partially pale. Radial sector with stem as long as, or distinctly longer than, posterior branch ($R_{4+5}$) of fork; false vein weak but extended nearly to wing margin; $A_2$ distinct. Male: Head holoptic. Gonostylus broad, with broadly rounded apex. Ventral plate with short, bulbous, liplike projection. Median sclerite short, slightly longer than wide.

**Overview.** This subgenus contains 2 species represented by only 4 known adults from the forests of the Pacific Northwest. Larvae and pupae are unknown for this subgenus.

#### *Parasimulium* (*Astoneomyia*) *melanderi* Stone

(Figs. 4.2, 10.1, 10.75, 10.76, 10.82, 10.226; Map 1)

*Parasimulium melanderi* Stone 1963b: 127, Figs. 1–4 (M). Holotype* male (pinned, 1 wing and terminalia in glycerin vial below, 1 wing and abdominal fragments on slide [head missing]; USNM). Washington, Whatcom Co., Nooksack River, Mt. Baker, 11 August 1925 (A. L. Melander)

**Taxonomy**

Stone 1963b (M); Corredor 1975 (M); Peterson 1977a, 1981, 1996 (M); Borkent & Currie 2001 (FM).

**Bionomics**

**Habitat.** This black fly is one of the rarest in North America, known only from the headless type specimen (male) taken in the Nooksack Valley of Washington's North Cascade Range, and from two females collected in a cave on Vancouver Island, British Columbia (Borkent & Currie 2001). Despite considerable prospecting around the type locality, additional specimens have not been found, leading to speculation that the species might have been extirpated from the area, since disturbed by logging (Courtney 1993). The elongate hairs on the legs and front of the head suggest that adults can move about in darkness, providing further evidence that the adults are associated with subterranean environments (Borkent & Currie 2001).

**Oviposition.** Unknown.

**Development.** The 3 known specimens were taken in August.

**Mating.** Unknown.

**Natural Enemies.** No records.

**Hosts and Economic Importance.** The mouthparts are not developed for biting.

#### *Parasimulium* (*Astoneomyia*) 'species A'

(Fig. 10.227, Map 2)

*Parasimulium* 'n. sp.' Courtney 1993: 368 (Oregon record) (also p. 368 as 'undescribed species')

'undescribed species from Oregon' Borkent & Currie 2001: 550 (Oregon record)

**Bionomics**

**Habitat.** This rare fly is known from a single male collected beside Mack Creek in the H. J. Andrews Experimental Forest, Lane Co., Oregon (44°13′N, 122°09′W, 800 m elev.) (Courtney 1993). It was captured with males of *P. stonei*.

**Oviposition.** Unknown.

**Development.** The singular specimen was taken on 14 July 1987.

**Mating.** Unknown.

**Natural Enemies.** No records.

**Hosts and Economic Importance.** Unknown.

### Subgenus *Parasimulium* Malloch

*Parasimulium* Malloch 1914: 24 (as genus). Type species: *Parasimulium furcatum* Malloch 1914: 24–25, by original designation

**Diagnosis.** Adults: Body at least with some lightly pigmented parts, including antennae, palps, much of thorax, and abdominal tergite I. Radial sector with stem as long as, or shorter than, posterior branch ($R_{4+5}$) of fork; false vein evanescent at about three fourths its length; $A_2$ faint or absent. Male: Head dichoptic. Gonocoxite with apicolateral, finger-like extension; gonostylus narrowed, somewhat rounded or pointed distally, with subapical cusp; inner gonostylus present. Ventral plate broad, flat, forked apically, without bulbous apical lip. Median sclerite long, straplike.

**Overview.** This subgenus contains 3 species in the Pacific Northwest.

### *Parasimulium* (*Parasimulium*) *crosskeyi* Peterson

(Figs. 4.26, 4.34, 4.52, 4.58, 4.60, 10.83, 10.228, 10.410, 10.621, 10.683; Plate 3a; Map 3)

*Parasimulium* (*Parasimulium*) *crosskeyi* Peterson 1977a: 104–105, Figs. 12–14, 19 (M). Holotype* male (pinned, terminalia in glycerin vial below; USNM). Oregon, Multnomah Co., Benson Park, 24 June 1935 (A. L. Melander)

[*Parasimulium furcatum*: Stone (1963b) not Malloch 1914 (misident. in part, M from Eagle Creek Park and Corvallis)]

**Taxonomy**

Peterson 1977a (M), 1996 (L); Wood & Borkent 1982 (FM); Peterson & Courtney 1985 (F); Courtney 1986 (PL); Currie 1988 (MPL).

**Morphology**

Currie 1988 (gill), Evans & Adler 2000 (spermatheca), Palmer & Craig 2000 (labral fan).

**Molecular Systematics**

Moulton 1997, 2000 (DNA sequences).

**Bionomics**

**Habitat.** This species is known from elevations below 250 m in the Columbia River Gorge of Oregon. The first larvae and pupae of this species—and the first for the genus—were found in the expansive subterranean flow beside Wahkeena Creek. This flow harbors one of the largest known populations of any species in the genus (Courtney 1986). Water velocity of the subterranean channel ranges from 20 to 40 cm/sec (Currie 1988), and water temperature ranges from 3°C to 10°C (Courtney 1993). Larvae have been found up to 60 cm into the seep, with pupae occurring nearer the mouth of the seep (Currie 1988).

**Oviposition.** Unknown.

**Development.** The majority of larvae and pupae are found in May and June, with a single record (1 larva) from early January (Courtney 1986). Larvae are primarily filter feeders (Currie 1988), although gut contents indicate that they also might be deposit feeders (Courtney 1986). Adults fly from late May into July, with peak abundance during the first 2 weeks of June (Courtney 1986).

**Mating.** From morning to dusk, males hover, usually singly beneath leaf tips, especially those of broadleaf maple (*Acer macrophyllum*), investigating small dark objects such as leaf blemishes, resting females, and spiders (Wood & Borkent 1982). Initial coupling might occur in the air near the leaves; however, the only observed copulation, lasting about 20 minutes, was on the underside of a leaf (Wood & Borkent 1982).

**Natural Enemies.** Males searching for females often become trapped in spider webs (Wood & Borkent 1982).

**Hosts and Economic Importance.** Females are incapable of biting.

### *Parasimulium* (*Parasimulium*) *furcatum* Malloch

(Figs. 10.84, 10.229; Map 4)

*Parasimulium furcatum* Malloch 1914: 24–25, Plate I (Fig. 4) (M, incorrectly given as F). Holotype* male (pinned #15405 in 2 glycerin vials; USNM). California, Humboldt Co., Redwood Creek, Bair's Ranch (later became Merillon's Ranch [Coleman 1951]; actual site possibly Redwood Creek Ranch, 40°57′N, 123°50′W), 9 June 1903 (H. S. Barber)

**Taxonomy**

Malloch 1914 (M); Knab 1915a (M); Riley & Johannsen 1915 (M); Dyar & Shannon 1927 (M); Stains 1941 (M); Stone 1941 (M); Stains & Knowlton 1943 (M); Coleman 1951 (M); Wirth & Stone 1956 (M); Peterson 1977a, 1981, 1996 (M).

The immature stages have not been discovered.

**Bionomics**

**Habitat.** Although the most widespread member of the genus, *P. furcatum* is known from just 7 sites in Oregon and California and on Vancouver Island, British Columbia. Adults can be swept from riparian vegetation (Borkent 1992).

**Oviposition.** Unknown.

**Development.** Adults have been collected from July to mid-August.

**Mating.** Males swarm beneath dead limbs, lichens, and conifers near streams (Courtney 1993).

**Natural Enemies.** No records.

**Hosts and Economic Importance.** Females are incapable of biting.

### *Parasimulium* (*Parasimulium*) *stonei* Peterson

(Figs. 4.9, 4.17, 10.2, 10.85, 10.230; Map 5)

*Parasimulium* (*Parasimulium*) *stonei* Peterson 1977a: 102–104, Figs. 2, 9–11 (M). Holotype* male (pinned, terminalia in glycerin vial below; USNM). California, Humboldt Co., Bolling Park (probably Bolling Grove, Humboldt Redwoods [Stone 1963b]), 19 June 1935 (A. L. Melander)

[*Parasimulium furcatum*: Stone (1962b) not Malloch 1914 (misident.)]

[*Parasimulium furcatum*: Stone (1963b) not Malloch 1914 (misident. in part, M from Bolling Park and Viento)]

**Taxonomy**

Stone 1962b (M); Peterson 1977a, 1981, 1996 (M); Peterson & Courtney 1985 (F); Borkent & Wood 1986 (LE).

**Morphology**

Borkent & Wood 1986 (L head).

**Bionomics**

**Habitat.** *Parasimulium stonei* is most common in the Cascade Range of Oregon and Washington, with additional records from the Coast Range of Oregon and northern California. Larvae and pupae have been found by digging beneath debris dams and large boulders that form stairsteps in the stream; typically the surface flow disappears just upstream of the stairsteps (Courtney 1993).

**Oviposition.** Eggs have been obtained by placing field-caught females in stoppered vials, each with a leaf and a small amount of moisture (Borkent & Wood 1986). Eggs collected in this manner were refrigerated in distilled water, and the first larvae of the genus ever observed hatched about 6 months later (Borkent & Wood 1986).

**Development.** Larvae and pupae have been found from late May to late June. Adults fly from mid-June to late August, with peak abundance during July.

**Mating.** Males swarm beneath leaves of deciduous trees, searching for resting females (Currie 1988, Courtney 1993).

**Natural Enemies.** No records.

**Hosts and Economic Importance.** Females are incapable of biting.

### Subfamily Simuliinae Newman

Simuliites Newman 1834: 379, 387–388 (as 'Natural Order' = family)

**Diagnosis.** Adults: $R_1$ connected to costa well beyond middle of wing; radial sector unbranched, or with branches closely approximated; $R_{4+5}$ (or $R_{2+3}$ and $R_{4+5}$ if radial sector unbranched) connected near apex of costa; costa, subcosta, and radial sector with short setae; false vein forked apically; $CuA_2$ sinuous. Katepisternum in lateral view rounded ventrally; sulcus present. Mesepimeral tuft present. Calcipala and pedisulcus present or absent; claws toothless or each with small subbasal tooth or basal thumblike lobe. Female: Sternum X divided medially. Male: Head typically holoptic, with upper facets larger than lower facets. Gonostylus typically with 1 or more apical spinules. Ventral plate with or without straplike connection between apex of each basal arm and paramere. Pupa: Gill with plastron over entire surface, including base; base of gill without felt chamber internally; mesothoracic spiracle in contact with lumen of thickened portion of plastron on dorsal side of base. Abdominal segments III (or IV in *Gymnopais* and *Twinnia*)–VIII divided by pleural membrane. Sternites VI and VII (sometimes V and VIII) divided medially by longitudinal striate membrane. Pleuron typically with recurved hooks. Abdominal tergites VI–IX (often IV and V) each typically with spine combs (absent in taxa such as *Gymnopais*, *Twinnia*, and some *Greniera*). Larva: Antenna of 3 articles. Postgenal cleft typically present (rudimentary in some taxa). Prothoracic proleg with lateral sclerites. Abdominal segment IX without ventral protuberances, or with pair of ventral tubercles or 1 transverse midventral bulge. Anal sclerite with posteroventral arms articulated dorsally; apices widely separated ventrally, or if apices of arms closely approximated ventrally, then not separated by spindle-shaped sclerite.

**Overview.** This subfamily has a worldwide distribution and includes all simuliids of the world except the 5 species of *Parasimulium*. Its members are worldwide in distribution, excluding Antarctica and some isolated oceanic islands.

#### Tribe Prosimuliini Enderlein

Prosimuliinae Enderlein 1921a: 199. Type genus: *Prosimulium* Roubaud 1906: 519–521

Hellichiini Enderlein 1925: 203. Type genus: *Hellichia* Enderlein 1925: 203–204

Gymnopaidinae Rubtsov 1955: 329–330. Type genus: *Gymnopais* Stone 1949b: 260–261

Helodoini [*sic*, correctly Helodontini] Ono 1982: 280, 282. Type genus: *Helodon* Enderlein 1921a: 199

Kovalevimyiinae Kalugina 1991: 71–72. Type genus: *Kovalevimyia* Kalugina 1991: 72 (fossil)

**Diagnosis.** Adults: Radial sector distinctly forked, with forked portion longer than its petiole; costa with uniformly sized, hairlike setae only. Calcipala and pedisulcus absent. Female: Maxillary palpomere V typically, at most, slightly longer than palpomere III. Claws toothless or each with basal thumblike lobe. Genital fork without anteriorly directed apodeme on lateral plate of each arm. Male: Ventral plate with straplike connection between apex of each basal arm and paramere (except in *Distosimulium*, *Levitinia*, and *Parahelodon*). Paramere without apical spines. Pupa: Abdominal segments IV and V each with large pleurites set in striate membrane (pleurites absent from segment IV of *Levitinia*); segment III either without pleural membrane or with membrane in form of narrow longitudinal band. Cocoon shapeless. Larva: Antenna with proximal and medial articles unpigmented, contrasting with dark brown distal article. Hypostoma lacking paralateral teeth. Anteromedian palatal brush of scoop-shaped, fringed plates in first instars. Maxillary palpal sensilla of first instar subapically situated, linearly arranged. Postocciput completely enclosing cervical sclerites, leaving narrow medial gap. Abdominal segment IX without ventral tubercles or protuberances. Rectal papillae of 3 simple lobes.

**Overview.** This tribe consists of 5 extant genera (*Gymnopais*, *Helodon*, *Levitinia*, *Prosimulium*, and *Twinnia*). All but *Levitinia* are found in North America. Members of the tribe account for less than 8% of the world's simuliid fauna. All are univoltine cold-water species found only in the Nearctic and Palearctic Regions, from as far north as Bjørnøya (Bear Island), Norway, to as far south as North Africa (Morocco). In North America, the tribe has 62 known species, or about 24% of the fauna.

#### Genus *Gymnopais* Stone

*Gymnopais* Stone 1949b: 260–261 (as genus). Type species: *Gymnopais dichopticus* Stone 1949b: 261–265, by original designation; *Cymnopais* (subsequent misspelling)

**Diagnosis.** Adults: Antenna with 7 flagellomeres. Clypeus nearly devoid of setae. Lacinia without retrorse teeth; mandible without serrations. Vestiture of sparse, short, coarse, erect setae. Thorax only slightly arched. Postnotum small, strongly arched, with medial longitudinal ridge. Anepisternal membrane with small cluster of hair. Mesepimeral tuft confined to dorsalmost portion of mesepimeron above ventral margin of metathoracic spiracle. Wing fumose, slightly wrinkled; basal radial cell about one third length of wing (measured from apex of stem vein); petiole of $M_{1+2}$ about half as long as petiole of radial sector. Female: Claws toothless. Genital fork with lateral plates not connected directly to tergum IX, but separated by membrane. Anal lobe and cercus fused into 1 sclerite. Spermatheca with pigmentation extended onto spermathecal duct. Male: Head dichoptic or holoptic. Gonostylus slender distally, with 1–3 minute apical spinules. Pupa: Gill of 2–6, typically swollen filaments. Abdominal tergites without spine combs; tergites III and IV (and sometimes V) with 1–3 pairs of fine recurved hooks; sternites IV–VII each with up to 5 pairs of recurved hooks;

pleurite IV slightly differentiated from or fused to adjacent tergum and sternum. Cocoon a small ventral pad (although initially covering pupa before disintegrating). Larva: Cephalic apotome with medial head spots extended anteriorly to level below posteriormost eyespot. Mandible with 7–12 rows of comblike scales on outer apical surface. Labral fans absent (greatly reduced in first instar). Labrum longer than wide, with lateral transparent tubercle on each side; epipharyngeal apparatus with V-shaped brush anteriorly. Hypostoma weakly sclerotized, with broad, apically rounded teeth. Postgenal cleft absent. Anal sclerite Y shaped. Chromosomes: Fixed inversions *IIS-1* and *IIIS-4* present (Rothfels & Freeman 1966, Rothfels 1979).

**Overview.** The genus *Gymnopais* consists of 12 species distributed in mountainous areas and high latitudes of Alaska, Canada, and the Palearctic Region. In North America, the genus contains 5 univoltine species, all adapted to life in small, fast, icy (<6°C), low-productivity streams where, although often abundant, they eke out an existence under the most austere conditions. Most habitats of the immature stages are headwater streams originating from glaciers, springs, and permafrost. One of the assets of life in these harsh habitats is often freedom from predators.

Two species of *Gymnopais* are parthenogenetic triploids that probably originated in Beringia as recently as late Wisconsinan times (Wood 1978, Rothfels 1989) and dispersed widely, perhaps as a result of having been liberated from the necessity to mate. The three sexual diploid species never dispersed far from Beringia. The parthenogenetic species are less synchronous than their sexual counterparts, often being the first and last to pupate. Several species of the genus often coexist in the same stream.

Larvae hatch after 8 months from eggs held at 1°C (Wood 1978), suggesting an obligatory egg diapause (Wood & Davies 1965). Lacking labral fans, they obtain food primarily by using their highly modified mouthparts to graze organic matter from stones. They generally pupate under stones and form a cocoon of unorganized silk, but the meshwork soon disintegrates, leaving only a small ventral pad to secure the pupa to its substrate. The exposed nature of the pupa inspired the generic name, which means "naked child." During dry periods, larvae burrow into the substrate, aided by their tapered fanless heads, and pupate in the streambed (Currie & Craig 1988). The pupal cuticle is thick and leathery, often allowing the exuviae to persist to the subsequent year in protected recesses of the substrate. The tough pupal integument, stout gills with reduced filaments, and deterioration of the cocoon are probably adaptations to habitats that often run dry (Wood 1978). The microsporidium *Weiseria sommermanae* was described from larvae of an unidentified member of the genus (Jamnback 1970).

Adults are classic examples of arctic-adapted insects, so much so that the casual observer might not recognize them as black flies. With their slight bodies and vestiture of sparse short hairs, they can be found resting beneath streamside stones or walking about on long slender legs (Currie 1997a). Females are autogenous and emerge with a full complement of 18–90 comparatively large eggs (Wood 1978, Davies & Györkös 1990, Currie 1997a), which they deposit freely when placed in vials. The low number of eggs suggests that members of the genus live in a stable and predictable environment (Currie 1997b). Mating takes place on the ground. Females of the parthenogenetic species resist mating with diploid males by curling their abdomens ventrally (Wood 1978). Adults live less than a week. Females have untoothed mouthparts, rendering them incapable of feeding on animals.

### *Gymnopais dichopticoides* Wood

(Figs. 10.3, 10.86, 10.512, 10.536; Map 6)

*Gymnopais dichopticoides* Wood 1978: 1315–1316, Figs. 18, 29, 37, 58, 63 (FPL). Holotype* female with pupal exuviae (pinned #15765; CNC). Yukon, Alaska Hwy., mi. 1057, small stream flowing into south end of Kluane Lake, 30 July 1963 (D. M. & G. C. Wood)

'undescribed species of *Gymnopais*' L. Davies 1960: 81–84 (L)

*Gymnopais* 'undescribed parthenogenetic species close to *dichopticus*' Davies 1965b: 160–167 (FL)

*Gymnopais* 'sp.' Craig 1972a: 61–62 (embryo)

*Gymnopais* 'sp. (near *dichopticus*)' Craig 1974: 135 (L)

*Gymnopais* 'sp. near *dichopticus*' Davies 1974: 213, 224 (L)

*Gymnopais* 'sp.' Peterson 1996: 600 (P, atypical gill)

**Taxonomy**

Davies 1965b (FL); Wood 1978 (FPL); Currie 1986, 1988 (PL); Peterson 1996 (P).

**Morphology**

L. Davies 1960 (L labrum), 1965b (L head), 1974 (L labrum); Craig 1972a (embryology), 1974 (L head); Craig & Borkent 1980 (L maxillary palpal sensilla); Fry 1994 (L internal head and thorax); Fry & Craig 1995 (L internal head and thorax).

**Cytology**

Rothfels 1979, 1989.

This species consists of only parthenogenetic triploid females, with 2 chromosomal homologues derived from the female of *G. dichopticus* and 1 from the male of *G. holopticus* (Rothfels 1979). Morphological variation suggests that hybridization occurred more than once (Wood 1978).

**Bionomics**

**Habitat.** *Gymnopais dichopticoides*, probably the most common species of the genus, is known from Alaska and the Yukon southward at high elevations into Alberta and British Columbia. It is less closely tied to the mountains than other members of the genus (Currie 1997a). Immature forms reside in streams originating from alpine glaciers, cold mountain springs, or permafrost areas such as *Sphagnum* bogs.

**Oviposition.** In the laboratory females from Alberta deposit 22–53 eggs within 2 days after emergence and then die in a few hours (D. A. Craig, pers. comm.). Females trapped in ice might be able to release viable eggs after disintegration of the body wall (Currie 1997a).

**Development.** Eggs overwinter, and larvae begin to hatch after the ice breaks up in June. Five or 6 instars precede pupation (Craig in Currie 1986). Emergence is highly asynchronous, occurring from late July to mid-October. Females are fully winged but show no inclination to fly (Currie 1997a). They rest under streamside stones for up to 5 days before laying their eggs (Currie 1986).

**Mating.** Females are parthenogenetic; males do not exist.

**Natural Enemies.** Unknown.

**Hosts and Economic Importance.** Females are incapable of biting and possibly do not even take sugar.

### *Gymnopais dichopticus* Stone

(Figs. 10.4, 10.5, 10.46, 10.87, 10.231, 10.513, 10.537; Map 7)

*Gymnopais dichopticus* Stone 1949b: 261–265 (FMPL). Holotype* male with pupal exuviae (pinned #51229; USNM). Alaska, Steese Hwy., mi. 19.1 (now probably mi. 19.3), north of Fairbanks, 14 September 1948 (Alaska Insect Project, collector unknown)

[*Prosimulium novum*: Jenkins (1948) not Dyar & Shannon 1927 (misident. in part)]

**Taxonomy**

Stone 1949b (FMPL), 1952 (FMP); Sommerman 1953 (L); Davies & Peterson 1956 (E); Smith 1970 (FMP); Wood 1978 (FMPL); Currie 1988 (FM).

**Morphology**

Craig 1974 (L head).

**Cytology**

Rothfels & Freeman 1966; Rothfels 1979, 1989.

**Bionomics**

**Habitat.** This species occurs in the unglaciated hills and mountains of Alaska and the Yukon. Its immature stages dwell in flows less than a meter wide.

**Oviposition.** Females emerge with an average of 43 eggs (Davies & Györkös 1990).

**Development.** Larvae begin hatching from overwintered eggs in early June when water temperatures are 3°C–4°C. Larval and pupal development requires about 8 and 3 weeks, respectively (Sommerman et al. 1955). Adults emerge from mid-July well into September. Males emerge slightly earlier than females (Currie 1997a).

**Mating.** Both sexes show little inclination to fly, although they can skim the ground for short distances when disturbed. This trait, plus the undivided eyes of the male, adapts the species for mating on the ground (Wood 1978, Currie 1997a). Coupling lasts about 27–46 minutes at ambient temperature (Currie 1997a) when pairs are confined in petri dishes. Males attempt to mate with any black fly that they contact.

**Natural Enemies.** Larvae are attacked by an unidentified microsporidium and mermithid nematode (Sommerman et al. 1955).

**Hosts and Economic Importance.** Females are incapable of biting.

### *Gymnopais fimbriatus* Wood

(Figs. 10.9, 10.10, 10.88, 10.232, 10.514, 10.622; Plate 3b; Map 8)

*Gymnopais fimbriatus* Wood 1978: 1316–1317, Figs. 9, 21, 22, 25, 43, 46, 53 (FMPL). Holotype* male (pinned #15766; CNC). Yukon, Dempster Hwy. (Ogilvie Mountain Road), small stream, mi. 47.5, ca. 3.5 mi. south of North Fork Pass (highest point on southern part of road), 4 September 1963 (D. M. & G. C. Wood)

**Taxonomy**

Wood 1978 (FMPL).

**Cytology**

Rothfels 1979.

**Molecular Systematics**

Moulton 1997, 2000 (DNA sequences).

**Bionomics**

**Habitat.** This least-collected species of the genus is known from the Yukon and southeastern Alaska. Along with *G. dichopticus*, it has the most stringent habitat requirements, being confined to very small flows in areas that were ice free during the Wisconsinan glaciation.

**Oviposition.** Females emerge with 32–35 eggs and oviposit within a day (Wood 1978). Some individuals lack flight muscles, and the eggs protrude into the thorax (Currie 1988).

**Development.** Eggs overwinter and larvae hatch in June. Adults emerge from mid-July into September but are unable to fly (Wood 1978).

**Mating.** Mating occurs on the ground and on the surface of pools and slow flows where males, with their markedly reduced eyes, sometimes grapple with one another (Currie 1997a). Males attempt to mate with any black fly that they contact, abandoning nonconspecific attempts only after the terminalia touch; adults mate readily in small vials shortly after emergence (Wood 1978). Coupling lasts 64–119 minutes at ambient temperature (Currie 1997a).

**Natural Enemies.** No records.

**Hosts and Economic Importance.** Females cannot bite.

### *Gymnopais holopticoides* Wood

(Figs. 10.6, 10.89, 10.515, 10.538; Plate 3c; Map 9)

*Gymnopais holopticoides* Wood 1978: 1313, Figs. 17, 28, 36, 59, 62 (FPL). Holotype* female with pupal exuviae (pinned #15764; CNC). Northwest Territories, Victoria Island, Kuujjua River Valley, small stream flowing down northwest side of hill, 71°17′N, 114°0′W, 23–28 July 1975 (D. M. & G. C. Wood)

'new species of *Gymnopais*' Basrur & Rothfels 1959: 571 (triploidy) (also p. 588 as *Gymnopais* 'sp.')

*Gymnopais* 'n. sp.' Downes 1962: 154 (bionomics); Downes 1964: 300 (bionomics)

*Gymnopais* 'sp.' Wood 1963a: 33, Fig. 58a (L)

*Gymnopais* 'sp.' Downes 1964: 294 (development)

'parthenogenetic *Gymnopais*' Downes 1965: 265 (development) (also p. 265 as *Gymnopais* 'sp.')

[*Gymnopais holopticus*: Stone (1949b) (misident. in part, many paratypes)]

[*Gymnopais holopticus*: Hocking & Richards (1952) not Stone 1949b (misident.)]

[*Gymnopais holopticus*: Shewell (1957) not Stone 1949b (misident.)]

[*Gymnopais holopticus*: Shewell (1958) not Stone 1949b (misident. in part, material from Baffin and Southampton Islands)]

**Taxonomy**

Wood 1978 (FPL).

Populations east of the Yukon are morphologically homogeneous but in the Yukon and Alaska are highly variable, suggesting that hybridization might have taken place more than once (Wood 1978). *Gymnopais holopticoides* is considered the most highly modified black fly (Currie 1997b).

**Morphology**

Wood 1963a (L labrum).

**Cytology**

Rothfels 1979, 1989.

This species is a parthenogenetic allotriploid composed of females only, with 2 chromosomal homologues derived from the female of *G. holopticus* and 1 from the male of *G. dichopticus* (Rothfels 1979).

**Bionomics**

**Habitat.** This trans-Canadian species of the high arctic ranges farther north than other species of *Gymnopais* and is the only member of the genus that extends east of the Rocky Mountains. Its type locality (Victoria Island) is near the northern limit for North American black flies. Larvae and pupae are found in small to moderate-sized streams with rapid flows.

**Oviposition.** Females produce about 20 eggs that they deposit within a day or two after emergence (Downes 1965).

**Development.** Larvae probably begin to hatch from overwintered eggs in June. Pupae and adults can be found during July and August. At the northern limit of distribution, emergence is restricted to a small temporal window in mid-August (Downes 1964). Females cannot fly (Currie 1986).

**Mating.** Females are parthenogenetic; males do not exist.

**Natural Enemies.** No records.

**Hosts and Economic Importance.** Females are incapable of biting.

### *Gymnopais holopticus* Stone

(Figs. 4.10, 4.14, 4.18, 4.44, 4.49, 10.7, 10.8, 10.47, 10.90, 10.233, 10.516, 10.623, 10.684; Map 10)

*Gymnopais holopticus* Stone 1949b: 265–267, Figs. 4, 5, 7, 9, 10 (FMPL). Holotype* male with pupal exuviae (pinned #59230; USNM). Alaska, Steese Hwy., mi. 16.2, north of Fairbanks, 17 August 1948 (Alaska Insect Project, collector unknown)

*Gymnopais* 'sp.' Peterson 1996: 620 (M)

[*Prosimulium novum*: Jenkins (1948) not Dyar & Shannon 1927 (misident. in part)]

**Taxonomy**

Stone 1949b (FMPL), 1952 (FMP); Sommerman 1953 (L); Davies & Peterson 1956 (E); Smith 1970 (FMP); Peterson 1981 (F), 1996 (FMP); Currie 1988 (F).

**Morphology**

Colbo et al. 1979 (F labral and cibarial sensilla), Currie & Craig 1988 (L head and mouthparts).

**Cytology**

Rothfels & Freeman 1966; Rothfels 1979, 1989.

**Bionomics**

**Habitat.** This species occupies the glacial refugia of Alaska and the Yukon. Its immatures have the broadest habitat range of any species in the genus, occupying very small flows as well as large streams that sometimes support fish.

**Oviposition.** Females emerge with an average of 46 eggs (Davies & Györkös 1990).

**Development.** Larvae begin hatching from overwintered eggs in early June when streams reach about 3°C (Sommerman et al. 1955). Pupae and adults first appear in late July and can be found into September.

**Mating.** *Gymnopais holopticus* is the only Nearctic member of the genus with holoptic males and with both sexes capable of sustained flight, suggesting the possibility of aerial coupling. Mating behavior has not been observed in nature, but coupling lasts about 10–18 minutes at ambient temperature (Currie 1997a) for pairs confined in petri dishes.

**Natural Enemies.** Larvae are attacked by an unidentified microsporidium and mermithid nematode (Sommerman et al. 1955). They have fallen prey to larvae of *Prosimulium ursinum* (Currie & Craig 1988).

**Hosts and Economic Importance.** Females are incapable of biting.

### Genus *Twinnia* Stone & Jamnback

*Twinnia* Stone & Jamnback 1955: 18–19 (as genus). Type species: *Twinnia tibblesi* Stone & Jamnback 1955: 19–21, by original designation

**Diagnosis.** Adults: Antenna with 7 flagellomeres. Female: Claws toothless. Spermatheca wider than long, with large unpigmented area basally. Male: Gonostylus with 1, rarely 2, apical spinules. Ventral plate strongly emarginated laterally near attachment of basal arms. Pupa: Gill of 14–16 filaments, arising from 2 swollen clubs or 3 slightly swollen trunks. Abdominal segment III ringlike, pleurite fused with adjacent tergite and sternite; all tergites without spine combs; tergites III and IV with 3 pairs of well-developed, recurved hooks. Cocoon transparent, gelatinous, enclosing entire pupa, retaining larval exuviae. Larva: Cephalic apotome with medial head spots extended anteriorly to level of anteriormost eyespot. Mandible not markedly curved medially near apex, with 3 or 4 spinelike setae on outer apical surface. Labral fans absent (greatly reduced in first instar). Hypostoma weakly sclerotized, with broad, apically rounded teeth. Postgenal cleft absent. Anal sclerite Y shaped. Chromosomes: Fixed inversions *IL-1* and *IIIS-3* present (Rothfels & Freeman 1966, Rothfels 1979).

**Overview.** The genus *Twinnia* consists of 10 species distributed throughout the temperate forests of the Nearctic and Palearctic Regions. The 3 North American species are univoltine, northern, and confined to mountainous regions. The immature stages nearly always are found near the source of small streams or in streams that simulate this type of habitat, such as outflows from impoundments. Larvae lack labral fans and use their modified mouthparts to graze food from stones and submerged forest litter and to engulf strands of algae. Pupation often takes place between stones and the sandy stream bottom, where desiccation is less likely when the streams run dry (Wood 1978). Females probably fly upstream to oviposit near the origins of flow. The species that seek blood are mammalophilic.

### *Twinnia hirticornis* Wood

(Figs. 10.11, 10.20, 10.91, 10.234, 10.539; Map 11)

*Twinnia hirticornis* Wood 1978: 1305–1307, Figs. 3, 5, 8, 12, 49, 52, 55 (FMPL). Holotype* male with pupal exuviae (pinned #15763; CNC). British Columbia, near Kamloops, mi. 2.1 on Cold Creek Road, pond seepage, 13–16 May 1964 (D. M. & G. C. Wood)

[*Twinnia nova*: Rothfels & Freeman (1966) not Dyar & Shannon 1927 (misident.)]

**Taxonomy**

Wood 1978 (FMPL).

**Cytology**

Rothfels & Freeman 1966; Rothfels 1979, 1980.

**Bionomics**

**Habitat.** This species, the least common of the genus, is known from southern British Columbia and adjacent Montana to central California. Larvae and pupae are found at the headwaters of small cool streams and seepages where they attach to stones and debris.

**Oviposition.** Unknown.

**Development.** Winter is passed as eggs, and larvae are present from May through June. When this species is found with *T. nova*, it is slightly later in development.

**Mating.** Unknown.

**Natural Enemies.** No records.

**Hosts and Economic Importance.** Unknown; presumably mammalophilic.

### *Twinnia nova* (Dyar & Shannon)

(Figs. 10.12, 10.21, 10.92, 10.235, 10.540; Map 12)

*Prosimulium novum* Dyar & Shannon 1927: 5–6, Figs. 14, 15 (F). Lectotype* female (designated by Wood 1978: 1308; pinned #28325, head and terminalia in glycerin vial below; USNM). Montana, Glacier Co., Glacier National Park, Two Medicine Lake, 4 July 1921 (H. G. Dyar)

*Twinnia biclavata* Shewell 1959b: 686–688 (FMPL). Holotype* male abdomen in glycerin vial pinned #6925 (remaining parts of body on slide #GES-5902-05F, larval head capsule and pupal exuviae on separate slide #GES5902-10A; CNC). British Columbia, Vancouver Island, Bowser, small stream near railroad track, 3 June (pupa), 5 June (adult) 1955 (G. E. Shewell)

[*Gymnopais*: Crosskey (1990) not Stone 1949b (misident. in part, Fig. 6.5, larval head)]

**Taxonomy**

Dyar & Shannon 1927 (F); Hearle 1932 (F); Stains 1941 (F); Stains & Knowlton 1943 (F); Coleman 1951 (F); Peterson 1958a (F), 1996 (P); Shewell 1959b (FMPL); Corredor 1975 (F); Wood 1978 (FMPL); Currie 1986, 1988 (PL).

**Morphology**

Chance 1969, 1970a (L mouthparts); Craig 1974 (L head); Craig & Borkent 1980 (L maxillary palpal sensilla).

**Cytology**

Rothfels & Freeman 1966, Rothfels 1979.

**Molecular Systematics**

Moulton 1997, 2000 (DNA sequences).

**Bionomics**

**Habitat.** *Twinnia nova* ranges from the Rocky Mountains of Alberta west onto Vancouver Island, British Columbia, and south to Utah. Larvae and pupae are found within 100 m of the sources of cold spring-fed streams and seepages, where they fasten to stones and floating mats of filamentous algae. Less often, they affix themselves to grasses, with older larvae occupying more proximal portions of the blades.

**Oviposition.** Unknown.

**Development.** Larvae of inland populations hatch in spring, often during May, and complete 6 instars before pupating in June (Currie 1986). On Vancouver Island, pupae have been found as early as late March. *Twinnia nova* is slightly earlier in development when it occurs in the same streams with *T. hirticornis*. Adults are on the wing especially during July.

**Mating.** Unknown.

**Natural Enemies.** One larva infected with the microsporidium *Caudospora alaskensis* has been found (Adler et al. 2000).

**Hosts and Economic Importance.** Females can be pestiferous to horses at high altitudes (Hearle 1932). They are attracted to sheep but have not been reported to bite (Jessen 1977). They can deliver fierce bites to humans but usually are attracted in numbers too few to be severe pests.

### *Twinnia tibblesi* Stone & Jamnback

(Figs. 4.45, 10.13, 10.93, 10.236, 10.541, 10.685; Plate 3d; Map 13)

*Twinnia tibblesi* Stone & Jamnback 1955: 19–21, Figs. 17, 33, 55, 76, 79, 81, 89, 98, 115 (FMPL). Holotype* female (slide #6525; CNC). Labrador, Goose Bay, 30 August 1950 (J. J. Tibbles)

*Gymnopais* 'sp.' Jamnback 1953: 24 (L)

**Taxonomy**

Jamnback 1953 (L); Stone & Jamnback 1955 (FMPL); Davies et al. 1962 (FMP); Wood et al. 1963 (L); Stone 1964 (FMPL); Holbrook 1967 (L); Pinkovsky 1970 (MPL); Wood 1978 (FMPL); Peterson 1981 (FML), 1996 (FMPL); Adler & Kim 1986 (PL); Currie 1988 (FM).

**Morphology**

L. Davies 1960 (L labrum), 1965b (L head); Craig 1974 (L head).

**Cytology**

Rothfels & Freeman 1966, Rothfels 1979.

**Bionomics**

**Habitat.** *Twinnia tibblesi* is fairly common from northeastern Canada south to the Pocono Mountains of Pennsylvania. The immature stages occupy the sources of spring-fed streams as well as seepages below beaver dams and earthen embankments. These flows are often temporary, remain below about 11°C during larval and pupal development, and generally run for only a few meters before vanishing underground or feeding a larger stream. Larvae pupate in silt and sand on the sides and undersurfaces of stones, leaves, and wood, or among strands of algae.

**Oviposition.** Females are autogenous, emerging with fully developed eggs that are probably deposited within a few hours after emergence (Davies et al. 1962). Total fecundity is unknown.

**Development.** Eggs overwinter, and larvae hatch as early as March in New York or April in Canada (Stone & Jamnback 1955, Davies et al. 1962). From southern Canada southward, pupae and adults are found from early May through June, but at more northern latitudes and at high elevations they can be found as late as September (Stone & Jamnback 1955, Wood 1978).

**Mating.** Adults probably mate in aerial swarms within a few hours after emergence (Davies et al. 1962, Wood 1978).

**Natural Enemies.** Larvae sometimes are infected with the chytrid fungus *Coelomycidium simulii* and an unidentified microsporidium (L. S. Bauer, pers. comm.).

**Hosts and Economic Importance.** The untoothed mouthparts of the female are incapable of piercing flesh.

### Genus *Helodon* Enderlein

*Helodon* Enderlein 1921a: 199 (as genus). Type species: *Simulia ferruginea* Wahlberg 1844: 110, by original designation

**Diagnosis.** Female: Hypogynial valves short, truncated or broadly rounded distally, giving abdomen rounded or truncated appearance; anteromedial corner not produced nipple-like. Claws each with small basal tooth or thumb-like lobe. Spermatheca with or without small unpigmented ring basally. Male: Ventral plate rather flat; lip in lateral view typically small or absent. Larva: Postgenal cleft of various sizes and shapes. Prothoracic proleg with each lateral sclerite a narrow horizontal strip; vertical portion weakly indicated at most. Chromosomes: Fixed inversions in IIIS and IIIL (Rothfels 1979).

**Overview.** This Holarctic genus consists of 3 subgenera and 43 species. In North America, we recognize 3 subgenera and 16 species. Members of the genus are mostly northern and represent a range of female feeding habits.

### Subgenus *Distosimulium* Peterson

*Distosimulium* Peterson 1970b: 30 (as subgenus of *Prosimulium*). Type species: *Prosimulium pleurale* Malloch 1914: 17, by original designation

**Diagnosis.** Female: Genital fork with each lateral plate bearing patch of setae. Spermatheca large, thin walled, unpigmented. Male: Gonostylus with distal half markedly thinner than proximal half, tapered to acute point, flexed in dorsoventral plane, with apex opposed to spiniform, dorsoventrally curved paramere. Ventral plate deeply cleft posteriorly. Pupa: Gill of 24–29 filaments. Larva: Hypostoma with lateral and sublateral teeth broad, weakly sclerotized. Postgenal cleft broadly rounded, extended nearly one half distance to hypostomal groove. Chromosomes (only *H. pleuralis* has been examined): Fixed inversions in IS, IL, IIS, IIL, and IIIL (Rothfels 1979).

**Overview.** This subgenus contains 2 or 3 species, one of which lives in North America, another in Japan (*H. daisetsensis* [Uemoto et al. 1976]), and a possible third (*H. mesenevi* [Patrusheva 1975]) in Siberia (Taymyr). Little is known about the natural history of *Distosimulium* species. The immature stages are encountered in boreal rivers, mountain streams, and bog drainages with rocky bottoms. All species are presumably ornithophilic.

#### *Helodon* (*Distosimulium*) *pleuralis* (Malloch)

(Figs. 10.94, 10.237, 10.542, 10.651, 10.686; Plate 3e; Map 14)

*Prosimulium pleurale* Malloch 1914: 17, Fig. 1 (F). Holotype* female (pinned #15403, terminalia on slide; USNM). British Columbia, Kaslo, 18 June 1903 (R. P. Currie)

*Prosimulium tenuicalx* Enderlein 1925: 203 (F). Lectotype female (designated by Stone 1962a: 209; pinned, terminalia on slide; ZMHU). Idaho, Latah Co., Moscow, date and collector unknown

*Prosimulium pancerastes* Dyar & Shannon 1927: 10–11 (M). Lectotype* male (designated but not labeled as such by Stone 1952: 76; pinned #28330, abdomen and hind legs on slide; USNM). Idaho, Nez Perce Co., Peck, 8 April 1900 (J. M. Aldrich)

[*Prosimulium* (*Prosimulium*) *multidentatum*: Shewell (1957) not Twinn 1936a (misident. in part, Labrador record)]

**Taxonomy**

Malloch 1914 (F); Riley & Johannsen 1915 (F); Enderlein 1925 (F); Dyar & Shannon 1927 (FM); Hearle 1932 (F); Stains 1941 (FM); Stains & Knowlton 1943 (FM); Coleman 1951 (FM); Stone 1952 (FMP); Sommerman 1953 (L); Peterson & DeFoliart 1960 (FP); Abdelnur 1966, 1968 (FMPL); Peterson 1970b (FMPL), 1996 (FM); Lewis 1973 (FMPL); Corredor 1975 (FM); Rubtsov & Yankovsky 1984 (FML); Currie 1986 (PL).

**Cytology**

Rothfels 1979, 1981a. Fig. 5 of Carlsson (1966) is not the IIIS arm of *H. pleuralis*.

On the basis of 2 fixed-inversion differences in chromosome arm IIIL, Rothfels (1979) suggested that distinct species existed, 1 eastern and 1 western. Because eastern and western populations are allopatric, we conservatively consider them a single species.

**Bionomics**

**Habitat.** This species can be found from Alaska and the Yukon to Washington and Idaho and from Labrador and Quebec to the Maritime Provinces. The immensely disjunct distribution of *H. pleuralis*, with more than 2500 km between eastern and western populations, is virtually unique among North American black flies. Northwestern populations breed primarily in the great crashing rivers of the mountains and foothills. Northeastern populations occupy large flows in the rolling relief of the boreal forest. All of these rivers are swift and rocky, with temperatures less than 10°C. A fossil fragment of a larva about 3300 years old was found on Vancouver Island, British Columbia (Currie & Walker 1992).

**Oviposition.** Females are anautogenous (Currie 1997a). Oviposition behavior is unknown.

**Development.** In Alaska and much of Canada, larvae hatch from mid-August to October, develop through the winter, and mature in May (Sommerman et al. 1955). Pupae are present in May and June, and adults fly from June to August. Development is up to a month earlier south of Canada.

**Mating.** Males clasp females between the gonostyli and a process of each paramere (Wood 1991), as inferred from genitalic structure.

**Natural Enemies.** We have larvae that were recovered from the stomachs of young chinook salmon (*Oncorhynchus tshawytscha*).

**Hosts and Economic Importance.** Females feed on birds such as great horned owls (Hunter et al. 1997a). We have feeding records from domestic geese and bantam chickens. Females are sometimes attracted to humans and occasionally attack livestock (Jenkins 1948, Shewell 1957).

### Subgenus *Parahelodon* Peterson

*Parahelodon* Peterson 1970b: 36–37 (as subgenus of *Prosimulium*). Type species: *Simulium decemarticulatum* Twinn 1936a: 110–112, by original designation

**Diagnosis.** Adults: Antenna with 7–9 flagellomeres. Female: Genital fork with lateral plates each typically bearing medial, wrinkled or denticulate process. Male: Gonostylus with 1 apical spinule. Ventral plate with digitiform dorsomedial projection extended anteriorly and fused with base of median sclerite; basal arms short. Pupa: Gill of 9–16 filaments. Larva: Postgenal cleft in shape of inverted U or V. Anal sclerite X shaped or subrectangular. Chromosomes: Fixed inversion in IIS (Rothfels 1979).

**Overview.** This taxon, containing 3 species, is endemic to North America. Each species develops in small cold streams, including springs, bog seeps, drainage ditches, and outflows from beaver ponds. Adult behavior is little known. Females of *H. gibsoni* are incapable of taking blood. The other 2 species are ornithophilic.

### *Helodon* (*Parahelodon*) *decemarticulatus* (Twinn)

(Figs. 4.22, 10.48, 10.95, 10.238, 10.543, 10.603, 10.652, 10.687; Map 15)

*Simulium* (*Prosimulium*) *decemarticulatum* Twinn 1936a: 110–112, Fig. 1D (FMP). Holotype* female (pinned #4122, terminalia on slide #P.32; CNC). Ontario, Carleton Co., Goulbourn Township, near Carleton Place (1.5 mi. west of Stanley Corners), roadside ditch (Fig. 4B of Twinn 1936a), 10 May 1935 (C. R. Twinn); *decemartilulatum* (subsequent misspelling)

**Taxonomy**

Twinn 1936a (FMP); Stone 1952 (FMP), 1964 (FMPL); Sommerman 1953 (L); Anderson 1960 (FMPL); Davies et al. 1962 (FMP); Wood et al. 1963 (L); Abdelnur 1966, 1968 (FMPL); Peterson 1970b (FMPL), 1981 (FM), 1996 (FMPL); Rubtsov & Yankovsky 1984 (FMPL); Fredeen 1985a (FMP); Currie 1986 (PL); Peterson & Kondratieff 1995 (FMPL).

**Morphology**

Bennett 1963b (F salivary gland), Guttman 1967a (labral fan), Yang 1968 (F peritrophic matrix, crop duct, labral and cibarial sensilla), Yang & Davies 1977 (F peritrophic matrix).

**Physiology**

Yang 1968 (F blood digestion, FM trypsin); Yang & Davies 1968a (M amylase), 1968b (FM trypsin).

**Cytology**

Rothfels 1956, 1979; Rothfels & Mason 1975 (M achiasmate meiosis).

Rothfels (1979) stated that *H. decemarticulatus* includes at least 2 sibling species corresponding to the 2 different Y chromosomes referred to in his Fig. 3. We feel, however, that additional evidence is required before species status can be conferred on these 2 chromosomal entities.

**Bionomics**

**Habitat.** This species is fairly common across Canada, Alaska, and the northern United States. The immature stages live in streams less than 2 m wide that are often temporary and drain coniferous forests or emanate from beaver dams, bogs, and springs.

**Oviposition.** Unknown.

**Development.** Larvae begin hatching in April and become final instars from May to mid-June, depending on latitude (Anderson & Dicke 1960, Davies et al. 1962, Currie 1986). The pupal stage lasts as little as 4–6 days (Anderson & Dicke 1960). Adults can be found from late May well into July but are most common in June.

**Mating.** Unknown.

**Natural Enemies.** Larvae are attacked by unidentified mermithid nematodes (Anderson & DeFoliart 1962). An unidentified microsporidium has been reported from the fat body of females (Yang & Davies 1977).

**Hosts and Economic Importance.** Females feed in the late evening 1.5 m or more above the ground (Bennett 1960, Golini 1970). Hosts include a variety of forest birds, such as sharp-shinned hawks, ruffed grouse, northern saw-whet owls, northern flickers, gray jays, blue jays, American crows, common ravens, American robins, white-throated sparrows, common grackles, and purple finches (Bennett 1960), as well as American black ducks, mallards (domestic and wild), and bantam chickens that are offered in a woodland habitat (Bennett 1960, Smith 1966). Great horned owlets also are attacked (Hunter et al. 1997a, 1997b), as are nestlings of woodland birds such as sparrows (Abdelnur 1968). Females have been taken from traps baited with live moose (Pledger et al. 1980), but these captures might have been incidental.

Females transmit the blood protozoans *Leucocytozoon cambournaci* (as *L. fringillinarum* in part), *L. dubreuili* (as *L. mirandae*), *L. icteris* (as *L. fringillinarum* in part), *L. sakharoffi*, and *L. ziemanni* to woodland birds (Fallis & Bennett 1961; Khan & Fallis 1970, 1971; Khan 1975). They are probable vectors of the protozoa *Trypanosoma confusum* (as *T. avium*) and *L. toddi* in numerous birds (Bennett 1961, Bennett et al. 1993) but are unsuitable vectors of *L. lovati* (as *L. bonasae*) (Fallis & Bennett 1962) and *L. simondi* (Fallis & Bennett 1966).

### *Helodon* (*Parahelodon*) *gibsoni* (Twinn)

(Figs. 10.96, 10.239, 10.544, 10.653, 10.688; Plate 3f; Map 16)

*Simulium* (*Prosimulium*) *gibsoni* Twinn 1936a: 108, 110, Figs. 1C, 4, 7A (FMP). Holotype* female (pinned #4121, terminalia and hind legs on slide #P.42; CNC). Ontario, Carleton Co., Goulbourn Township, near Carleton Place (1.5 mi. west of Stanley Corners), roadside ditch (Fig. 4B of Twinn 1936a), 8 May (pupa), 10 May (adult) 1935 (C. R. Twinn)

**Taxonomy**

Twinn 1936a (FMP), Nicholson 1949 (FMP), Nicholson & Mickel 1950 (FMP), Anderson 1960 (FMPL), Davies et al. 1962 (FMP), Wood et al. 1963 (L), Stone 1964 (FMPL), Peterson 1970b (FMPL), Amrine 1971 (FMPL), Merritt et al. 1978b (PL), Fredeen 1985a (FMP), Currie 1986 (PL).

**Morphology**

Guttman 1967a (labral fan), Smith 1970 (ovarian follicles), Wood 1978 (first-instar labral fan), Colbo et al. 1979 (F labral and cibarial sensilla).

**Cytology**

Rothfels 1956, 1979; Rothfels & Freeman 1966; Rothfels & Mason 1975 (absence of M achiasmate meiosis).

The 2 sibling species mentioned by Rothfels (1979), corresponding to the 2 different Y chromosomes referred to in his Fig. 3, are best considered Y-chromosome polymorphisms, rather than distinct species, until more evidence is adduced.

**Bionomics**

**Habitat.** This northern species ranges from the Atlantic seaboard to eastern Alberta. Larvae and pupae develop in cold, clear, often evanescent headwater streams less than 1.5 m wide, with gravel, sand, or stone bottoms. Turbid stream conditions are not tolerated (Anderson & Dicke 1960). The weakly silked pupae can be found in protected nooks of the streambed.

**Oviposition.** Females are obligately autogenous and often emerge with mature eggs (Davies & Györkös 1990). Nothing is known of their oviposition habits.

**Development.** Larvae hatch in late winter or early spring and pupate within about 6 weeks (Anderson & Dicke 1960, Merritt et al. 1978b); they might, however, overwinter in northern Manitoba (Hocking & Pickering 1954). Adults begin emerging in late April or early May at the southern limit of distribution, but as late as the end of June at the northern limit (Davies et al. 1962). Adults have been reared from first-instar larvae (Wood & Davies 1966).

**Mating.** Unknown.

**Natural Enemies.** Larvae are occasional hosts of unidentified mermithid nematodes (Anderson & DeFoliart 1962) and microsporidia (Twinn 1939).

**Hosts and Economic Importance.** The weakly developed mouthparts do not permit biting.

### *Helodon* (*Parahelodon*) *vernalis* (Shewell)

(Figs. 10.97, 10.240, 10.545, 10.654, 10.689; Plate 4a; Map 17)

*Prosimulium vernale* Shewell 1952: 33–36 (FMP). Holotype* female (pinned #5987; CNC). Ontario, Bell's Corners, south of Ottawa, small stream draining swamp and wooded area, 9 May 1950 (G. E. Shewell)

[*Prosimulium gibsoni*: Tarshis & Stuht (1970) not Twinn 1936a (misident.)]

**Taxonomy**

Shewell 1952 (FMP), Davies et al. 1962 (FMP), Wood et al. 1963 (L), Stone 1964 (FMP), Peterson 1970b (FMPL).

**Cytology**

Rothfels & Freeman 1966, Rothfels 1979.

**Bionomics**

**Habitat.** This lowland species is known from southern Ontario to southern Virginia, with an isolated record in Arkansas. It is probably more common and widely distributed but is likely overlooked because of its early development and its resemblance to *Prosimulium arvum*. Streams inhabited by the immature stages are typically less than 1–2 m wide, cold, swift, and temporary. We also have found larvae in large, swampy blackwater flows. Larvae attach themselves to leaves and the bases of submerged vegetation, and pupate in loosely woven silk on bottom debris and stones (Davies et al. 1962).

**Oviposition.** Unknown.

**Development.** Larvae begin hatching in autumn and often develop under the ice, maturing from March in Maryland and Delaware to April in Ontario (Davies et al. 1962, Tarshis & Stuht 1970). This species is one of the earliest to fly in Ontario. Adults emerge in early May, with males emerging a few days earlier than females (Shewell 1952). Larvae can be reared to adults in aquaria, even at a water temperature of 22°C (Tarshis 1971).

**Mating.** Unknown.

**Natural Enemies.** We have one larva from the Blackwater River in Virginia that is infected with an unidentified microsporidium.

**Hosts and Economic Importance.** Unknown; presumably ornithophilic.

### Subgenus *Helodon* Enderlein

*Helodon* Enderlein 1921a: 199 (as genus). Type species: *Simulia ferruginea* Wahlberg 1844: 110, by original designation

*Haimophaga* Rubtsov 1977: 49 (as subgenus of *Ahaimophaga*). Type species: *Prosimulium multicaulis* Popov 1968: 444–447, by original designation

*Ahaimophaga* Rubtsov & Chubareva in Rubtsov 1977: 47–49. Unavailable (no designated type species)

*Ahaimophaga* Chubareva & Rubtsov in Chubareva 1978: 42 (as genus). Type species: *Prosimulium alpestre* Dorogostaisky, Rubtsov & Vlasenko 1935: 136–139, by original designation

**Diagnosis.** Female: Spermatheca typically elongate. Male: Gonostylus with 2–6 apical spinules. Larva: Postgenal cleft biarctate or subrectangular with faint anterior margin. Chromosomes: Whole-arm interchange (IIL + IIIS, IIS + IIIL). Nucleolar organizer in middle of IIL arm (i.e., the new IIIL arm).

**Overview.** Worldwide, the subgenus *Helodon* includes 38 described species, of which 12 occur in North America. The immature stages develop in swift mountain streams and rivers. First instars are fitted with labral fans of only 3 rays that are probably ineffectual in filter feeding (Craig 1974). Adult behavior is poorly known. Females of a number of species have feebly developed mouthparts incapable of taking blood. The blood feeders are primarily ornithophilic.

**Helodon (Helodon) alpestris (Dorogostaisky, Rubtsov & Vlasenko)**

(Figs. 10.49, 10.98, 10.241, 10.546, 10.655; Map 18)

*Prosimulium alpestre* Dorogostaisky, Rubtsov & Vlasenko 1935: 136–139 (FMPL). Lectotype* female (designated by Yankovsky 1995: 8; slide #3043, terminalia [without spermatheca], one pair of legs, mouthparts, anterior part of one wing only; ZISP). Russia, Irkutsk Region, Slyudyanka River (Yankovsky 1995 says Medlyaika River), 14 August 1934 (V. C. Dorogostaisky & I. A. Rubtsov)

*Prosimulium* 'J' Hocking 1950: 497 (Yukon record)

*Prosimulium alpestre altaicum* Rubtsov 1956: 215–216 (ML). Holotype* larva (slide #2664; ZISP). Russia, Altay, Katon-Karagay, Katonka River, 23 July 1915 (Shvanvich)

*Prosimulium alpestre relense* Rubtsov 1956: 215–217 (L). Lectotype* larva (designated by Yankovsky 1995: 45; slide #7291; ZISP). Russia, Irkutsk Region, Rel' River, entering northwestern side of Lake Baikal, 3 August 1952 (Gilev & Romanov)

*Prosimulium alpestre komandorense* Rubtsov 1971b (FM). Holotype* male (pinned #12263, terminalia on celluloid mount below; ZISP). Russia, Commander Islands, Preobrajenskovo District, 5 September 1959 (K. B. Gorodkov)

**Taxonomy**

Dorogostaisky et al. 1935 (FMPL); Rubtsov 1940b, 1956, 1960b (FMPL), 1971b (FM); Stone 1952 (FMP); Sommerman 1953 (L), 1958 (M); Peterson & DeFoliart 1960 (F); Peterson 1970b (FMPL); Patrusheva 1982 (FM); Rubtsov & Yankovsky 1984 (FML); Bodrova 1988 (PL); Yankovsky 1999 (PL).

The possibility that *H. alpestris* is a species complex has been raised (Sommerman et al. 1955, Rubtsov 1956, Stone 1965a) but not investigated. We have examined the types of *H. alpestris*, *H. altaicus*, *H. relensis*, and *H. komandorensis*, and confirmed morphologically that they are conspecific.

**Morphology**

Prokofyeva 1959 (ovariole), Rubtsov 1960a (P fat body, hemolymph, ovarioles, M reproductive system), Yankovsky 1977 (labral fan), Colbo et al. 1979 (FM labral and cibarial sensilla), Currie 1988 (L antenna).

**Cytology**

Rothfels & Freeman 1966, Chubareva 1978, Rothfels 1979.

**Bionomics**

**Habitat.** This Holarctic species is found in the eastern Palearctic Region and from Alaska and the western Northwest Territories southward to northern British Columbia. Larvae and pupae are common on stones and submerged branches in rocky mountain streams less than 12°C and from less than a meter to 25 m wide. Larvae often pupate in masses.

**Oviposition.** Females are obligately autogenous. In Russia, they mature 150–200 eggs (Prokofyeva 1959). Females oviposit over smooth swift water while in upstream flight and will deposit eggs in small containers of water (Sommerman 1958).

**Development.** Larvae hatch from overwintered eggs and can be found from May to August, requiring 7–12 weeks to reach maturity (Sommerman et al. 1955). Adults fly in August and September.

**Mating.** Adults mate soon after emergence (Rubtsov 1960a).

**Natural Enemies.** Larvae in North America are attacked by the chytrid fungus *Coelomycidium simulii* (Jamnback 1973a), the microsporidium *Caudospora alaskensis* (Jamnback 1970), and unidentified mermithid nematodes (Sommerman et al. 1955). Three species of mermithid nematodes (*Gastromermis bobrovae*, *Mesomermis baicalensis*, *M. parallela*) have been described from *H. alpestris* in the Palearctic Region (Rubtsov 1972b, 1974b).

**Hosts and Economic Importance.** Males and females feed on nectar and water (Rubtsov 1960a). The mouthparts are not adapted for biting.

**Helodon (Helodon) clavatus (Peterson)**

(Figs. 10.99, 10.242, 10.547, 10.656, 10.690; Plate 4b; Map 19)

*Prosimulium* (*Helodon*) *clavatum* Peterson 1970b: 57–60, Figs. 3, 10, 41, 65, 102, 135 (FMPL). Holotype* male with pupal exuviae (pinned #9596; CNC). Yukon, Alaska Hwy., mi. 907, Wolf Creek, 25 August 1963 (the year [1953] given in the original description is incorrect) (D. M. & G. C. Wood)

*Prosimulium* (*Helodon*) *albertense* Peterson & Depner 1972: 289–294 (FMPL). Holotype* male with pupal exuviae (alcohol vial #11819; CNC). Alberta, Crowsnest Pass, Gold Creek, 5 mi. north of Frank, 3–4 September 1970 (reared from larva collected 24 August 1970) (K. R. Depner). New Synonym

**Taxonomy**

Peterson 1970b (FMPL), Peterson & Depner 1972 (FMPL), Currie 1986 (PL).

Currie (1997a) suggested that the Palearctic *H. rhizomorphus* Rubtsov (1971a) might be conspecific with *H. clavatus*. However, our study of the types indicates otherwise. In the former species, the gill filaments arise more apically from the club, and the median tooth of the hypostoma extends farther anteriorly.

The intervening distance between the type localities of *H. clavatus* and *H. albertensis* might account for morphological differences between these entities. In light of this allopatry, as well as variation that we have found in the diagnostic characters, we consider *H. albertensis* a synonym of *H. clavatus*.

**Morphology**

Palmer & Craig 2000 (labral fan).

**Bionomics**

**Habitat.** *Helodon clavatus* is found from the Yukon Territory southward at high elevations to Washington and Wyoming. The immatures develop in cold rocky streams 3 m or more in width. Larvae pupate under rocks (Peterson & Depner 1972).

**Oviposition.** Females are obligately autogenous. Oviposition behavior is unknown.

**Development.** This species is one of the latest members of the genus to appear each year. Eggs overwinter, larvae appear in July or August, and adults emerge primarily in September.

**Mating.** Possible mating swarms have been observed within 3 m above ground, with males and females cruising in and out of the shade of large streamside spruce trees (*Picea*) (Peterson & Depner 1972).

**Natural Enemies.** No records.

**Hosts and Economic Importance.** The mouthparts are weakly sclerotized, rendering females incapable of taking blood.

### *Helodon* (*Helodon*) *irkutensis* (Rubtsov)

(Figs. 10.50, 10.100, 10.243, 10.548, 10.657, 10.691; Plate 4c; Map 20)

*Prosimulium irkutense* Rubtsov 1956: 267–268 (L). Lectotype* larva (designated by Yankovsky 1995: 28; slide #7474; ZISP). Russia, Irkutsk Region, village of Mond, Irkut River, 1000 m elev., 12 August 1953 (Luzina); *ircutense* (subsequent misspelling)

?*Prosimulium* 'sp. X' Sommerman 1953: 260, 263, 266 (L)

*Prosimulium* 'sp. #' Sommerman 1953: 260, 265, 268 (L)

*Prosimulium perspicuum* Sommerman 1958: 199–202, Figs. 11, 25–28 (FMPL). Holotype* female with exuviae (alcohol vial; USNM). Alaska, 13.5 mi. north of Anchorage, Eagle River, emerged 28 August 1957 (K. M. Sommerman). New Synonym

*Prosimulium* (*Helodon*) *martini* Peterson 1970b: 61–62, Fig. 42 (M). Holotype* male (pinned #9597; CNC). Nunavut (as Northwest Territories), Baffin Island, head of Clyde Inlet, 7 August 1958 (J. E. H. Martin). New Synonym

*Prosimulium* 'sp.' Peterson et al. 1985: 1385 (L response to phosphorus enrichment)

**Taxonomy**

Sommerman 1953 (L), 1958 (FMPL); Rubtsov 1956, 1961a (L), 1971b (FP); Peterson & DeFoliart 1960 (F); Peterson 1970b (FMPL); Smith 1970 (FPL); Patrusheva 1973 (FM), 1982 (FMP); Bodrova 1980 (P).

Currie (1997a) suggested that the name *perspicuus* might be synonymous with the names of 1 or 2 Palearctic species (*H. buturlini* and *H. irkutensis*) described by Rubtsov (1956). Our study of the holotype larva of *H. irkutensis* suggests that it is conspecific with *H. perspicuus*, although the head spots of *H. irkutensis* are slightly paler. Mature larvae from Kular, Yakutia (ZISP), labeled by Rubtsov as *irkutensis*, are identical to those of *H. perspicuus*. Similarly, Rubtsov's (1971b) concept of the pupa and female of *H. irkutensis* agrees with topotypical material of *H. perspicuus*.

We prefer to apply the name *buturlini* (type locality: Russia, Kolyma River basin) to the species whose larvae Rubtsov (1956) tentatively associated with the type. Although the holotype female of *H. buturlini* conforms to the female of *H. irkutensis*, the associated larvae are different. In the larva of *H. buturlini*, the median tooth of the hypostoma extends anteriorly beyond all other teeth, which in turn extend progressively less anteriorly, giving the entire anterior margin of the hypostoma a pointed appearance. Similar larvae (labeled by Rubtsov as "*Ahaimophaga* aff. *irkutense*"; ZISP) have gills of 3 stout branches, each with 10 or more filaments.

The holotype male of *H. czekanowskii* Rubtsov 1956 (type locality: Russia, Yakutia, Golimer River) is similar to the male of *H. irkutensis* from North America; however, the female has uniquely shaped hypogynial valves. Further study of properly associated life stages and chromosomes is needed for the Palearctic species related to *H. irkutensis*.

We consider the diagnostic features given by Peterson (1970b) for the sole known specimen of *H. martini* to be within the range of intraspecific variation for *H. irkutensis*. We, therefore, synonymize *H. martini* with *H. irkutensis*.

The species known as *Prosimulium* 'sp. X' Sommerman (1953) is tentatively placed under *H. irkutensis*, but might represent an undescribed species. Sommerman (1953) noted that *P.* 'sp. X' had 16 filaments in each gill histoblast. Peterson (1970b), however, found that at least some of Sommerman's material had 12 filaments per gill histoblast.

**Cytology**

Rothfels & Freeman 1966, Rothfels 1979.

**Bionomics**

**Habitat.** Reaching a latitude of 72° north, this Holarctic species shares top honors with 2 other species (*Gymnopais holopticoides* and *Metacnephia borealis*) as the most northern black fly on the North American continent. It is known from Banks Island and Baffin Island to southern Yukon and Alaska, as well as from Siberia. The immatures are particularly common in cold, clear to milky glacial streams from 2 to more than 20 m wide, flowing through tundra. Larvae are sometimes difficult to see in these streams because they can be nearly transparent, a trait that inspired the name *perspicuus*. On Victoria Island, larvae often pupate inside empty cocoons of *M. borealis*. Phosphorus enrichment of the larval habitat indirectly increases the size of larvae and pupae by augmenting algal and microbial biomass (Hiltner & Hershey 1992). Larval densities decrease, however, probably because concomitant increases in caddisfly populations cause dislodgement (Hershey & Hiltner 1988).

**Oviposition.** Females are obligately autogenous and produce up to 45 well-developed eggs in about a week, usually after taking a sugar meal (Sommerman 1958, Smith 1970). Oviposition behavior has not been reported.

**Development.** Larvae probably hatch in June from overwintered eggs. Pupae appear in July, and adults are most prevalent from August to early September (Sommerman 1958). Larvae feed on detritus, diatoms, filamentous algae, and early-instar black flies and chironomids (Hershey et al. 1997). Sex ratios range from near unity to about 3 : 1 in favor of males (Sommerman 1958, Smith 1970). Adults have been kept alive on sugar and water for up to 20 days (Sommerman 1958).

**Mating.** In nature, one mating has been observed on the leaf of a cottonwood tree (*Populus*) while hordes of males swarmed around the trees and crawled on the leaves, often attempting to mate with one another (Sommerman 1958). Caged flies mate on the substrate and remain coupled for an average of 11.5 minutes at an unspecified temperature; males seem unable to recognize females except by contact, at which time they grab the female, curl their abdomen under her, and move backward until the genitalia are coupled (Sommerman 1958). A virgin female will mate on the day of emergence and offer no resistance; however, a mated female counters male advances by tucking its head and abdomen under its thorax, often pushing the males with their hind legs (Sommerman 1958).

**Natural Enemies.** Larvae are attacked by the chytrid fungus *Coelomycidium simulii* and an unidentified mermithid nematode.

**Hosts and Economic Importance.** The mouthparts are too weakly developed to pierce skin.

### *Helodon* (*Helodon*) *onychodactylus* Species Complex

**Diagnosis.** Female: Thoracic integument orange to brownish orange. Spermatheca typically elongate, sometimes rather round. Male: Ventral plate with lateral margins subparallel. Pupa: Gill of 2 divergent clubs, each with 15–30 slender filaments. Larva: Hypostoma with median tooth extended anterior to or to about the same level as lateral teeth. Chromosomes: Standard banding sequence as given by Henderson (1986a).

**Overview.** Fourteen cytological entities have been recognized in this species complex (Henderson 1986a). Based on evidence of reproductive isolation (e.g., Newman 1983; Henderson 1986a, 1986b), we recognize 9 of these entities as valid species (with cytological designations in parentheses): *H. beardi* n. sp. ('3'), *H. chaos* n. sp. ('1a,' '1b'), *H. diadelphus* n. sp. ('2b'), *H. mccreadiei* n. sp. ('7a,' '7b'), *H. newmani* n. sp. ('5'), *H. onychodactylus* s. s. ('6,' '10'), *H. protus* n. sp. ('4a,' '4b'), *H. susanae* ('2a,' '9'), *H. trochus* n. sp. ('8'). All members of the complex are excruciatingly similar in their known life stages but can be identified by features of their larval polytene chromosomes. We have found numerous populations that do not fit the chromosomal criteria established by Newman (1983) and Henderson (1986a, 1986b). Rather than designate new species or cytotypes, we have included these populations under the species that they most resemble, realizing that further work is required to clarify their status. Information on the complex as a whole is presented below, followed by accounts of each of the 9 member species.

**Taxonomy**

Dyar & Shannon 1927 (F); Hearle 1932 (PL); Stains 1941 (F); Stains & Knowlton 1943 (F); Coleman 1951 (FM); Stone 1952 (FMP); Sommerman 1953 (L); Wirth & Stone 1956 (FMP); Peterson 1958a (FMPL), 1960a (FMP), 1970b (FMPL), 1981 (FM), 1996 (FMP); Rivosecchi 1964b (FMPL); Abdelnur 1966, 1968 (FMPL); Corredor 1975 (FM); Currie 1986 (PL), 1988 (FM); Peterson & Kondratieff 1995 (FMPL).

**Morphology**

Craig 1974 (L head), Craig & Borkent 1980 (L maxillary palpal sensilla), Evans & Adler 2000 (spermatheca), Palmer & Craig 2000 (labral fan).

**Cytology**

Rothfels & Freeman 1966; Rothfels 1979; Newman 1983; Henderson 1985, 1986a, 1986b.

**Molecular Systematics**

Zhu 1990 (mtDNA); Pruess et al. 1992, 2000 (mtDNA); Tang et al. 1996b (mtDNA); Moulton 1997, 2000 (DNA sequences).

**Bionomics**

**Habitat.** This species complex is widespread and abundant in mountainous areas from Alaska to New Mexico, with most species and chromosomal forms inhabiting a strip from Washington to central California. It has been found as far south as the mountains of southern California (Mohsen & Mulla 1981a) and southern Arizona. Swift rocky streams and rivers are typical breeding sites. These watercourses are clear and cold, generally of low productivity, and from less than a meter to more than 15 m wide. In many of these streams, granitic boulders constitute the substrate. Larvae are easily found in large numbers on trailing vegetation, sticks, and rocks. One report noted a preference for darker vegetation (Peterson 1959b). Larvae have been reported from microhabitats with high turbulence (average Froude number of 1.7) (Eymann 1993). Pupae usually are ensconced cryptically in sand at the base of large stones. Less often, they can be found on rocks and gravel or on branches trailing in the current (Hearle 1932, Jenkins 1948). Viable pupae can be obtained by holding fully mature larvae in cold (4°C–13°C) shallow water in a petri dish (Wood & Davies 1966).

**Oviposition.** Unknown.

**Development.** Species in this complex are univoltine and show marked seasonal succession. In Oregon, for example, larvae can be collected nearly year-round, reflecting a sequence of up to 10 cytological entities in a single stream (Newman 1983). In areas such as Utah and Alaska, eggs overwinter and hatch from late April to mid-June (Sommerman et al. 1955, Peterson 1959b), reflecting the presence of fewer cytologically distinct taxa.

**Mating.** Unknown.

**Natural Enemies.** The microsporidia *Caudospora alaskensis* and possibly *Janacekia debaisieuxi* and the chytrid fungus *Coelomycidium simulii* infect larvae. The trichomycete fungi *Harpella melusinae* and *Smittium pennelli* have been found in larval guts (Lichtwardt 1984, Lichtwardt & Williams 1988). Predators of larvae include rhyacophilid caddisfly larvae and larvae of the empidid fly *Oreogeton* (Peterson 1960b, Sommerman 1962b). Larvae have been fed to stoneflies (*Hesperoperla pacifica*) in studies of prey preference (Allan & Flecker 1988). By biting at the heads of smaller stoneflies, larvae are capable of warding off attacks (J. D. Allan et al. 1987).

**Hosts and Economic Importance.** Surprisingly, no feeding records exist for a group of species so common and widely distributed. The bifid claws of the females suggest that birds are hosts. A few individuals have been attracted to a human (Abdelnur 1968).

### *Helodon* (*Helodon*) *beardi* Adler, Currie & Wood, New Species

(Map 21)

*Prosimulium onychodactylum* 'Sibling species 3' Newman 1983: 2822–2823 (C)

[*Helodon onychodactylus*: Moulton (1997, 2000) not Dyar & Shannon 1927 (misident.)]

**Taxonomy**

**Female.** Unknown.

**Male.** Unknown.

**Pupa.** Unknown.

**Larva.** Length 7.3–9.3 mm. Head capsule dark orange-brown, with central portion of frontoclypeal apotome often darker than surrounding pigment; head spots diffuse or absent. Antenna extended three fourths length of stalk of labral fan. Hypostoma with median tooth extended anteriorly beyond lateral teeth; sublateral teeth and denticles of median tooth extended anteriorly to about same level (as in Fig. 10.658). Postgenal cleft slightly rounded anteriorly, weakly or not at all biarctate, about twice as wide as long, extended about one fourth to one third distance to hypostomal groove; subesophageal ganglion sometimes ensheathed with pigment. Body pigment brownish gray, uniformly distributed (as in Plate 4d). Labral fan with 30–33 primary rays. Posterior proleg with 9–14 hooks in each of 67–74 rows; anal sclerite with posterior arms typically not extended around posterior proleg as lightly pigmented band.

**Diagnosis.** This species can be differentiated from other members of the *H. onychodactylus* complex by fixation of inversion *IIS-1* and sex chromosomes that are either undifferentiated or based on loose pairing of the homologues in the IIL arm (Newman 1983, Henderson 1986a).

**Holotype.** Larva (mature, ethanol vial; originally collected in acetic ethanol). Arizona, Graham Co., Big Creek at Hospital Flat, Pinaleno Mountains, 32°40.06′N, 109°52.40′W, 20 May 1994, J. K. Moulton (USNM).

**Paratypes.** Arizona, Apache Co., Apache National Forest, Greer Recreation Area, unnamed tributary of West Fork Little Colorado River, Forest Service Road 1120, 33°59′N, 109°28′W, 2590 m elev., 22 May 1994, J. K. Moulton (18 larvae); Graham Co., same data as holotype (24 larvae).

**Etymology.** We name this species in honor of Charles Edward Beard who has been of great assistance in the preparation of this book and who has contributed significantly to our knowledge of the trichomycete fungi that colonize black flies.

**Cytology**

Newman 1983, Henderson 1986a.

Arizona populations have a slightly expanded centromere region, a distinct centromere band in chromosome I, and occasional ectopic pairing of centromeres. Populations in Arizona and some in California lack differentiated sex chromosomes.

**Molecular Systematics**

Moulton 1997, 2000 (DNA sequences).

**Bionomics**

**Habitat.** This species is known from Arizona, California, and Oregon. Larvae have been taken from cold streams 1–6 m wide.

**Oviposition.** Unknown.

**Development.** Mature larvae can be found from April into June.

**Mating.** Unknown.

**Natural Enemies.** Larvae are infected by the microsporidium *Caudospora alaskensis* (Adler et al. 2000) and unidentified mermithid nematodes.

**Hosts and Economic Importance.** Unknown; presumably ornithophilic.

### *Helodon (Helodon) chaos* Adler, Currie & Wood, New Species

(Map 22)

*Prosimulium onychodactylum* 'Sibling species 1a and 1b' Newman 1983: 2820 (C)

**Taxonomy**

**Female.** Unknown.

**Male.** Unknown.

**Pupa.** Unknown.

**Larva.** Not differing from that of *H. beardi* n. sp. except as follows: Length 7.3–8.7 mm. Head capsule (as in Fig. 10.692) pale yellowish to brownish orange; head spots brown to dark brown, variably distinct. Postgenal cleft typically biarctate. Labral fan with 23–27 primary rays. Posterior proleg with 9–13 hooks in each of 67–75 rows.

**Diagnosis.** This species can be distinguished from other members of the *H. onychodactylus* complex by the unique X chromosome, which is defined by the E sequence in chromosome arm IIL (Newman 1983).

**Holotype.** Larva (mature, ethanol vial; originally collected in acetic ethanol). Oregon, Lincoln Co., Big Elk Creek drainage, near Harlan, 44°33′N, 123°41′W, 12 March 1993, D. C. Currie & D. M. Wood (USNM).

**Paratypes.** Same data as holotype (10 larvae).

**Etymology.** The specific name is from Greek and is used as a noun in apposition, alluding to the taxonomic confusion often created by sibling species.

**Cytology**

Newman 1983, Henderson 1986a.

Newman (1983) recognized 2 synchronic sibling species, '1a' and '1b,' on the basis of differences in polymorphism frequencies. Some of our samples indicate that the issue is more muddled. Populations in central California (e.g., Tuolumne Co., Rt. 108, 9.3 km east of Kennedy Meadow, 11 June 1990) are polymorphic for the IIS-1 inversion. Most samples from central California are small and often contain larvae with cytological characteristics of *H. beardi* n. sp. Some populations in Oregon (e.g., type locality, 12 March 1993) have a high proportion of the IIIL-1 inversion. Whether or not 2 or more species are involved awaits further study. One hybrid larva, a cross between *H. chaos* (= '1a') and *H. mccreadiei* n. sp. (= '7a/b'), has been found (Newman 1983).

**Bionomics**

**Habitat.** This member of the *H. onychodactylus* species complex is distributed in the coastal states from Washington to central California. Larvae have been found in cold rocky streams 3 m or less in width.

**Oviposition.** Unknown.

**Development.** This species is one of the earliest members of the complex to appear each year. Larvae mature from early December through June.

**Mating.** Unknown.

**Natural Enemies.** Larvae are hosts of the microsporidium *Caudospora alaskensis* (Adler et al. 2000).

**Hosts and Economic Importance.** Unknown; presumably ornithophilic.

### *Helodon* (*Helodon*) *diadelphus* Adler, Currie & Wood, New Species

(Map 23)

*Prosimulium onychodactylum* 'Sibling species 2b' Newman 1983: 2820–2822, Fig. 5c, d (C)

**Taxonomy**

**Female.** Unknown.

**Male.** Unknown.

**Pupa.** Unknown.

**Larva.** Not differing from that of *H. beardi* n. sp. except as follows: Length 7.4–9.5 mm. Head capsule (as in Fig. 10.692) pale yellowish to brownish orange; head spots brown to dark brown, variably distinct, often bold. Postgenal cleft typically biarctate. Labral fan with 21–30 primary rays. Posterior proleg with 10–12 hooks in each of 72–80 rows; anal sclerite with posterior arms often extended variable distance around posterior proleg as lightly pigmented band.

**Diagnosis.** This species can be distinguished from other members of the *H. onychodactylus* complex by the following features of its larval polytene chromosomes: inversions IL-1, IIIL-1, IIIL-2, and IIIL-7 common; inversion IIS-1 uncommon; IIIS standard or infrequently with the D inversion; and the X chromosome defined by IIL–E + F (Newman 1983; Henderson 1986a, 1986b).

**Holotype.** Larva (mature, ethanol vial; originally collected in acetic ethanol). Wyoming, Lincoln Co., Little White Creek, Rt. 89, 12.3 mi. south of Smoot, Allred Flat Recreation Area, 42°29.1′N, 110°57.7′W, 28 May 1989, P. H. Adler (USNM).

**Paratypes.** Washington, Jefferson Co., Olympic National Park, Dosewallips River, Dose Forks Campground, 47°44.5′N, 123°11.4′W, 27 June 1999, D. S. Bidlack (4 larvae). British Columbia, Alaska Hwy., beaver pond, specific location not given, July 1989, D. M. Wood (41 larvae).

**Etymology.** The specific name is from Greek, meaning "two brothers," an allusion to the cytological status of this species as a homosequential sibling of *H. susanae*.

**Cytology**

Newman 1983; Henderson 1985, 1986a, 1986b.

One hybrid larva between this species and *H. mccreadiei* n. sp. (= '7a/b') has been found (Newman 1983).

**Bionomics**

**Habitat.** This species is broadly sympatric with its homosequential sibling species *H. susanae*, although it does not range as far north or south. The immature stages are found in a variety of cold, swift rocky streams typically less than 5 m, but up to 40 m, in width.

**Oviposition.** Unknown.

**Development.** Larvae probably hatch during the winter or early spring. Mature larvae are most prevalent in June but can be taken from May into August.

**Mating.** Unknown.

**Natural Enemies.** We have found 3 larvae infected with a microsporidium (possibly *Janacekia debaisieuxi*) and 1 with an unidentified mermithid nematode.

**Hosts and Economic Importance.** Unknown; presumably ornithophilic.

### *Helodon* (*Helodon*) *mccreadiei* Adler, Currie & Wood, New Species

(Map 24)

*Prosimulium onychodactylum* 'Sibling species 7a' Newman 1983: 2824–2825 (C)

*Prosimulium onychodactylum* 'Sibling species 7b' Newman 1983: 2825, Fig. 3 (C)

**Taxonomy**

**Female.** Unknown.

**Male.** Unknown.

**Pupa.** Unknown.

**Larva.** Not differing from that of *H. beardi* n. sp. except as follows: Head capsule (as in Fig. 10.692) brownish orange; head spots brown to dark brown, variably distinct. Postgenal cleft typically biarctate. Labral fan with 26–29 primary rays. Posterior proleg with 9–12 hooks in each of 72–80 rows.

**Diagnosis.** This species can be distinguished from other members of the *H. onychodactylus* complex by the following features of its larval chromosomes: inversion *IIIL-2* fixed; the standard sequence in chromosomes I and II; an X chromosome with the B, C, or standard sequence; and a Y chromosome with either the D or standard sequence (Newman 1983).

**Holotype.** Larva (mature, ethanol vial; originally collected in acetic ethanol). Washington, Whatcom Co., Ruth Creek, about 1 km above confluence with North Fork of Nooksack River, 48°53′N, 121°39′W, 28 August 1986, G. W. Courtney (USNM).

**Paratypes.** Washington, Lewis Co., Smith Creek, Rt. 12, 46°34′N, 121°42′W, 19 June 1986, G. W. Courtney (1 larva).

**Etymology.** We name this species in honor of John William McCreadie, University of South Alabama, for his insightful contributions to the ecology of black flies.

**Cytology**

Newman 1983, Henderson 1986a.

Populations of '7a' and '7b' (*sensu* Newman 1983) are chromosomally almost identical and differ only slightly in maturation dates. We, therefore, consider them a single species. Nonetheless, the demonstration that *H. susanae* (= '2a') and *H. diadelphus* n. sp. (= '2b') are valid species (Newman 1983, Henderson 1986b) serves as a caveat that '7a' and '7b' also might be distinct species.

This member of the *H. onychodactylus* species complex is the most frequent as a parent of hybrid larvae. Hybrids have been found with *H. mccreadiei* as one parent and *H. chaos* n. sp. (= '1a'), *H. diadelphus* n. sp. (= '2b'), *H. newmani* n. sp. (= '5'), *H. protus* n. sp. (= '4a/b'), or *H. clavatus* (as *Prosimulium albertense*) as the other parent (Newman 1983).

We have 6 larvae from the headwaters of Shepherd Creek (Inyo Co., California, 13 August 2001) in which the sex chromosomes were classic for this species—males were B/D heterozygotes and females were homozygous for the B inversion (Newman 1983)—but the IIIL-3 inversion was consistently homozygous and IIIL-2 appeared in only 1 larva. We tentatively consider these larvae under the name *H. mccreadiei*, although they might represent a distinct species.

**Bionomics**

**Habitat.** This species has been found in the mountains from northern Washington to central California. No information is available on its breeding sites.

**Oviposition.** Unknown.

**Development.** Mature larvae can be found from June to September (Newman 1983), making this species the last member of the *H. onychodactylus* complex to develop each year.

**Mating.** Unknown.

**Natural Enemies.** No records.

**Hosts and Economic Importance.** Unknown; presumably ornithophilic.

### *Helodon (Helodon) newmani* Adler, Currie & Wood, New Species

(Map 25)

*Prosimulium onychodactylum* 'Sibling species 5' Newman 1983: 2823–2824 (C)

**Taxonomy**

**Female.** Unknown.

**Male.** Unknown.

**Pupa.** Unknown.

**Larva.** Not differing from that of *H. beardi* n. sp. except as follows: Head capsule (as in Fig. 10.692) brownish orange; head spots brown to dark brown, variably distinct. Postgenal cleft typically biarctate. Labral fan with 31–37 primary rays. Posterior proleg with 9–14 hooks in each of about 85 rows.

**Diagnosis.** This species is separable from other members of the *H. onychodactylus* complex by the banding patterns of its larval chromosomes, which are standard in all arms, but with IIS-1 a common polymorphism and the X chromosome typically carrying the D sequence (Newman 1983).

**Holotype.** Larva (final instar [white gill histoblasts], ethanol vial; originally collected in acetic ethanol). Oregon, Lane Co., Roaring River, 43°57'N, 122°05'W, 790 m elev., 22 June 1996, G. W. Courtney (USNM).

**Paratypes.** Same data as holotype (6 larvae).

**Etymology.** *Helodon newmani* is named in honor of Lester J. Newman of Portland State University, Oregon, whose detailed chromosomal study initially resolved the sibling species of the *H. onychodactylus* complex.

**Cytology**

Newman 1983; Henderson 1985, 1986a.

This taxon is similar to *H. beardi* n. sp. but is not chromosomally fixed for IIS-1 and typically has the IIIS-D inversion as its X-chromosome sequence. Newman's (1983) data suggest that the 2 taxa (as '3' and '5') are reproductively isolated.

We have larvae from 3 sites in California (e.g., Mono Co., Rt. 89, Mountaineer Creek), all taken in mid-July, that we tentatively include under *H. newmani*. The chromosomes of these larvae are fixed for *IIIS-D*, are nearly fixed for IIIL-3, and do not carry IIS-1. They are similar to the larvae of *H. newmani* to the extent that the IIIS-D sequence is present, although in *H. newmani*, IIIS-D serves as the X sequence and in the California populations the sex chromosomes are undifferentiated. We suspect that these larvae represent a separate species. The late maturation date of these larvae, among the last in the species complex to be found in California streams, also suggests that they are a separate species. Hybrids are known, with *H. newmani* as one parent and *H. beardi* n. sp. (= '3'), *H. mccreadiei* n. sp. (= '7a/b'), or true *H. onychodactylus* (= '6') as the other parent (Newman 1983).

**Bionomics**

**Habitat.** This species is known from Oregon to central California. Larvae and pupae inhabit cold streams and rivers up to 8 m wide that originate from lake ouflows and springs and have substrates of boulders and rubble.

**Oviposition.** Unknown.

**Development.** Larvae begin to hatch in late fall and mature from May into July.

**Mating.** Unknown.

**Natural Enemies.** No records.

**Hosts and Economic Importance.** Unknown; presumably ornithophilic.

### *Helodon (Helodon) onychodactylus* (Dyar & Shannon)

(Figs. 10.101, 10.244, 10.549, 10.604, 10.658, 10.692; Map 26)

*Prosimulium onychodactylum* Dyar & Shannon 1927: 4 (F). Holotype* female (pinned #28324 [terminalia missing], 2 slides, each with 1 leg; USNM). Colorado, Boulder Co., Long's Peak Trail, 3414–3444 m elev. (timberline), 28 August, year unknown (T. D. A. Cockerell); *onycodactylum* (subsequent misspelling)

?*Helodon onychodactylum* 'B' Rothfels 1979: 515 (C)

*Prosimulium onychodactylum* 'Sibling species 6' Newman 1983: 2824 (C)

*Prosimulium onychodactylum* 'sibling species 10a, 10b, 10c, and 10d' Henderson 1985: 45–48 (C)

*Prosimulium onychodactylum* 'form 10' Henderson 1986a: 40 (C)

**Taxonomy**

Dyar & Shannon 1927 (F), Stains 1941 (F), Stains & Knowlton 1943 (F).

We consider the species known cytologically as 'form 10' (*sensu* Henderson 1986a) to be true *H. onychodactylus*, based on its distribution, late collection date, and abundance around the type locality.

**Cytology**

Newman 1983; Henderson 1985, 1986a.

We regard 'sibling species 6' of Newman (1983) and 'form 10' of Henderson (1986a) as the same species, given small differences in polymorphism frequencies and allopatry. Single hybrids are known between this species and *H. newmani* n. sp. (= '5') and between this species and *H. susanae* (= '2a') (Newman 1983, Henderson 1986a).

**Molecular Systematics**

Zhu 1990 (mtDNA); Pruess et al. 1992, 2000 (mtDNA); Tang et al. 1996b (mtDNA). (We cytologically determined that material used for these publications was true *H. onychodactylus*.)

**Bionomics**

**Habitat.** This species is common from Alaska southward through the Rocky Mountains to New Mexico and into the Sierra Nevada of California. Streams inhabited by the immature stages are cold and rocky, often spring fed, and generally from less than a meter to about 6 m wide. Larvae affix themselves to trailing vegetation and sticks.

**Oviposition.** Unknown.

**Development.** This species is one of the latest members of the complex to develop. Larvae are found from early June into August, with the bulk of collections having been made in July (Newman 1983, Currie & Adler 1986, Henderson 1986a). Larvae probably begin hatching as early as May.

**Mating.** Unknown.

**Natural Enemies.** Larvae are hosts of the microsporidium *Caudospora alaskensis* (Adler et al. 2000), the chytrid fungus *Coelomycidium simulii*, and an unidentified mermithid nematode.

**Hosts and Economic Importance.** Unknown; presumably ornithophilic.

### *Helodon* (*Helodon*) *protus* Adler, Currie & Wood, New Species

(Map 27)

*Prosimulium onychodactylum* 'Sibling species 4a and 4b' Newman 1983: 2823, Fig. 6a, b (C)

**Taxonomy**

**Female.** Unknown.

**Male.** Unknown.

**Pupa.** Unknown.

**Larva.** Not differing from that of *H. beardi* n. sp. except as follows: Head capsule (as in Fig. 10.692) pale yellowish to brownish orange; head spots brown to dark brown, variably distinct. Postgenal cleft typically biarctate. Labral fan with 23 or 24 primary rays. Posterior proleg with 9–12 hooks in each of about 80 rows.

**Diagnosis.** This species can be separated from other members of the *H. onychodactylus* complex by the banding patterns of its larval chromosomes. All arms carry the standard banding sequence, and the sex chromosomes are weakly differentiated with, at most, loose pairing of the homologues in the IIIS arm (Newman 1983; Henderson 1986a, 1986b).

**Holotype.** Larva (mature, ethanol vial; originally collected in acetic ethanol). Idaho, Teton Co., Targhee National Forest, Rt. 33, Moose Creek, 43°33.78′N, 111°04.11′W, 11 March 1994, D. S. Bidlack (USNM).

**Paratypes.** Same data as holotype (1 larva).

**Etymology.** The species name is from Greek, meaning "first," in reference to the early seasonal occurrence of mature larvae.

**Cytology**

Newman 1983; Henderson 1985, 1986a.

Newman (1983) treated '4a' and '4b' as homosequential sibling species on the basis of slight differences in inversion frequencies and larval maturation times, with '4a' maturing from January to March and '4b' from April to late May. Differences in maturation times might reflect cohorts that hatch at different times. Until additional evidence comes to light, we consider '4a' and '4b' conspecific. One hybrid with *H. mccreadiei* n. sp. (= '7a/b') is known (Newman 1983).

**Bionomics**

**Habitat.** Mountain streams and small rivers from Alaska to Oregon and Idaho are breeding sites of this species.

**Oviposition.** Unknown.

**Development.** *Helodon protus* is one of the earliest members of the *H. onychodactylus* species complex to develop, with larvae maturing from late January to the end of May (Newman 1983).

**Mating.** Unknown.

**Natural Enemies.** No records.

**Hosts and Economic Importance.** Unknown; presumably ornithophilic.

### *Helodon* (*Helodon*) *susanae* (Peterson)

(Figs. 10.51, 10.102; Plate 4d; Map 28)

*Prosimulium* (*Helodon*) *susanae* Peterson 1970b: 71–73, Fig. 13 (F). Holotype* female (pinned #9598, head, abdomen, and left wing in glycerin vial below, slide with tibia and tarsus of right hind leg; CNC). British Columbia, near Muncho Lake, Alaska Hwy., mi. 451, 23 August 1948 (W. R. M. Mason)

?'new pupa with unusual breathing organs and the corresponding larva' Hearle 1932: 18–19 (PL)

*Helodon onychodactylum* 'A' Rothfels 1979: 515 (C)

*Prosimulium onychodactylum* 'Sibling species 2a' Newman 1983: 2820–2822 (C)

*Prosimulium onychodactylum* 'form 9' Henderson 1986a: 39–40 (C)

**Taxonomy**

Hearle 1932 (PL), Peterson 1970b (F), Currie 1995 (F).

We believe the name *H. susanae* applies to one of the cytological entities recognized by Newman (1983) and Henderson (1986a, 1986b). The most common cytological segregates in the area encompassing the type locality of *H. susanae* are '2a' and '10' (i.e., *H. onychodactylus* s. s.). Given the late collection date of the type specimen of *H. susanae*, *H. onychodactylus* s. s. might be the most likely candidate, but to conserve names, we match the name *H. susanae* with cytospecies '2a.'

*Helodon susanae* was described from a single female that differs from *H. onychodactylus* s. l. principally in having a round rather than elongate, spermatheca and browner body color. The terminalia of the holotype are illustrated (Fig. 10.102). The darker color might represent seasonal variation, as it does in females of species such as *Simulium bivittatum*. We have seen populations of the *H. onychodactylus* species complex in which the spermatheca varied from round to elongate. A population from the outlet of Cavell Lake in Jasper National Park, Alberta, has been referred to in the past as *H. susanae*, largely because the females are brownish. We tentatively accept these determinations but note that material from this site has not been examined cytologically and that the females fly later in the season than might be expected, given the known larval phenology of the '2a' entity.

All life stages of *H. susanae*, as currently recognized, are similar to those of *H. onychodactylus* s. s.

**Morphology**

Craig 1974 (L head), Craig & Borkent 1980 (L maxillary palpal sensilla), Evans & Adler 2000 (spermatheca).

## Cytology

Newman 1983; Henderson 1985, 1986a, 1986b.

Most populations of this species conform chromosomally to the criteria of Henderson (1986a, 1986b). However, we found populations in Utah in which IIS-1 is fixed and IIL-F no longer functions as the X sequence. Of 24 larvae from the Upper Provo River, Summit Co., Utah (29 May 1989), 14 females and 7 males were homozygous inverted for IIL-F, 2 females were heterozygous, and 1 female was homozygous standard. IIL-E still functioned as the X sequence and IIS-1 was homozygous inverted in all larvae. No other polymorphisms were observed. A population from Upper Beaver River in Beaver Co. (12 June 1996) was similar. Populations in California (Mono and Tuolumne Cos.) have the classic sex chromosomes but are otherwise nearly always standard for other sequences. Of 17 larvae from California, only 1 carried autosomal polymorphisms; it was heterozygous for IIS-1. Further study is needed to determine the status of these populations.

We place form '9,' which differs only in its Y chromosome, under *H. susanae* until evidence of separate species status can be found. Because form '9' has a high frequency of IIS-1, we consider form '9' under *H. susanae* rather than under *H. diadelphus* n. sp.

## Bionomics

**Habitat.** This species is one of the most widespread and abundant members of the *H. onychodactylus* species complex, ranging over most of the mountainous terrain of western North America. The immature stages inhabit cold swift streams and small rivers less than 5 m wide; however, we found a sizable population of larvae in a productive shallow stream (Champion Creek, Custer Co., Idaho) of 24°C—unusually warm for any species of the Prosimuliini. In the laboratory, larvae pupate in bottom sediments of the rearing chambers (D. A. Craig, pers. comm.) and probably do likewise in stream gravel.

**Oviposition.** Field notes made by D. A. Craig on 1 and 2 October 1970 describe thousands of females flying over the surface of the water at the outlet of Cavell Lake, Alberta, dipping the tips of their abdomens into the water to dispense single eggs brightly visible in the sunlight. Using a plankton net, Craig retrieved more than 220,000 drifting eggs. A small peak of oviposition occurred between 0900 and 1200. A second enormous peak, with thousands of females participating, took place from 1600 to 1900. By dark, only a few flies remained active. Air temperature during oviposition ranged from 10°C to 15°C.

**Development.** Hatching probably takes place in late fall or winter because larvae large enough to cytotype have been found as early as March in Alaska (Henderson 1986a). When eggs from Cavell Lake are held in the dark at 2°C, larvae hatch in 7 months (D. A. Craig, pers. comm.). Mature larvae are found from April through July, depending on latitude and elevation (Newman 1983, Henderson 1986a). Females from the type locality and Cavell Lake have been collected from late August to early October.

**Mating.** Previous reports that *H. susanae* might be parthenogenetic (Currie 1986) are erroneous.

**Natural Enemies.** Larvae are infrequent hosts of the chytrid fungus *Coelomycidium simulii* and a microsporidium resembling *Janacekia debaisieuxi* in spore shape.

**Hosts and Economic Importance.** Unknown; presumably ornithophilic.

### *Helodon (Helodon) trochus* Adler, Currie & Wood, New Species

(Map 29)

*Prosimulium onychodactylum* 'form 8' Henderson 1986a: 41–43 (C)

## Taxonomy

**Female.** Unknown.

**Male.** Unknown.

**Pupa.** Unknown.

**Larva.** Not differing from that of *H. beardi* n. sp. except as follows: Head capsule (as in Fig. 10.692) pale yellowish to brownish orange; head spots brown to dark brown, variably distinct. Postgenal cleft typically biarctate. Labral fan with 26–29 primary rays. Posterior proleg with 9–12 hooks in each of about 75 rows.

**Diagnosis.** This species can be distinguished from all other members of the *H. onychodactylus* species complex by the conspicuous chromocenter of its larval polytene chromosomes (Henderson 1986a).

**Holotype.** Larva (final instar [white gill histoblasts], ethanol vial; originally collected in acetic ethanol). British Columbia, New Denver, Carpenter Creek, 49°59′N, 117°23′W, 11 July 1994, D. C. Currie (CNC).

**Paratypes.** Same data as holotype (2 larvae).

**Etymology.** The species name is from the Greek noun meaning "wheel," in reference to the arms of the polytene chromosomes that radiate from the chromocenter like the spokes of a wheel.

## Cytology

Henderson 1985, 1986a.

Under this species, we include 3 female larvae from Nevada (White Pine Co., Snake Range, Lehman Creek, 2987 m elev., 25 August 1996, R. D. Gray) that had a conspicuous chromocenter and were homozygous for the B and IIL-2 inversions of Henderson (1986a). They differed from Henderson's (1986a) chromosomal description by lacking IIS-1, IIIL-3, and IIIL-7.

## Bionomics

**Habitat.** Three sites in southern British Columbia and 1 in eastern Nevada are known for this species, the least-collected member of the *H. onychodactylus* complex. Little is known of the habitat of the immature stages. The river at 1 collection site was 25 m wide.

**Oviposition.** Unknown.

**Development.** Larvae are present from at least June through August.

**Mating.** Unknown.

**Natural Enemies.** No records.

**Hosts and Economic Importance.** Unknown; presumably ornithophilic.

### Genus *Prosimulium* Roubaud

*Prosimulium* Roubaud 1906: 519–521 (as subgenus of *Simulium*). Type species: *Simulia hirtipes* Fries 1824: 17–18, designated by Malloch 1914: 16

*Hellichia* Enderlein 1925: 203–204 (as genus). Type species: *Hellichia latifrons* Enderlein 1925: 204 (= *Melusina macropyga* Lundström 1911: 20–21), by original designation
*Taeniopterna* Enderlein 1925: 203 (as genus). Type species: *Melusina macropyga* Lundström 1911: 20–21, by original designation
*Mallochella* Enderlein 1930: 84 (preoccupied) (as genus). Type species: *Mallochella sibirica* Enderlein 1930: 84 (= *Simulia hirtipes* Fries 1824: 17–18), by original designation
*Mallochianella* Vargas & Díaz Nájera 1948a: 67 (substitute name for *Mallochella* Enderlein 1930; same type species)
*Piezosimulium* Peterson 1989: 317–318 (as genus). Type species: *Piezosimulium jeanninae* Peterson 1989: 317–330 (= *Prosimulium neomacropyga* Peterson 1970b: 134–139), by original designation. New Synonym

**Diagnosis.** Female: Claws toothless, or each with minute subbasal tooth. Hypogynial valves long (typically extended to, or beyond, middle of anal lobe), giving abdomen somewhat pointed appearance; anteromedial corner produced nipple-like. Spermatheca with large unpigmented area proximally; external surface with loose reticulate pattern. Male: Gonostylus in inner lateral view typically conical. Ventral plate not flattened; lip in lateral view typically large, prominent. Paramere narrow anteriorly, broadened into subquadrangular plate posteriorly. Median sclerite short, bifid apically. Pupa: Gill of 9 or more filaments. Thoracic trichomes unbranched. Larva: Postgenal cleft variable, typically truncated or rounded apically. Prothoracic proleg with each lateral sclerite broad; vertical portion well developed.

**Overview.** The genus *Prosimulium* is the largest segregate of the tribe Prosimuliini, consisting of 68 described species worldwide. Its members are distributed throughout the Holarctic Region. In North America, we recognize 3 species groups (*hirtipes*, *macropyga*, and *magnum* species groups) and 38 species. Ecologically, *Prosimulium* is the most diverse genus of the Prosimuliini. All members of the genus are univoltine. The immature stages are cold-water life forms that occur in a wide variety of running-water habitats from tiny headwater springs to large rivers, and from food-rich lake outflows to oligotrophic alpine streams. Females of most species feed primarily on mammalian blood, although several northern species possess weakly developed mouthparts.

Our rationale for synonymizing *Piezosimulium* with *Prosimulium* is explained in the Taxonomy section for *Prosimulium neomacropyga*.

### *Prosimulium hirtipes* Species Group

**Diagnosis.** Female: Lacinia with retrorse teeth; mandible with serrations. Spermatheca about as long as, or longer than, wide; about half as long as stem of genital fork. Male: Gonostylus with 2 (rarely 3) apical spinules. Pupa: Gill of 9–16 filaments. Larva: Mandibular phragma typically well separated from posterolateral corner of hypostoma (except in *P. frohnei*). Abdomen rather abruptly expanded at segment V. Chromosomes: Standard banding sequence as given by Basrur (1959, 1962).

**Overview.** Worldwide, 45 described species are recognized in this group, with 22 of these in North America. The group has been referred to more parochially as the *P. mixtum* species group (Peterson 1970b). North American species have been subdivided into eastern (IIIL-1) and western (IIIL-2) groups that are monophyletic and geographically nonoverlapping (Rothfels 1979). Many of the species in this group are difficult to identify, and caution should be exercised in making identifications of isolated individuals. The colors of the adult integument and hair often vary considerably in some species.

Additional species undoubtedly remain to be discovered, especially in the mountains of western North America. We have, for example, a single mature larva from Mariposa Co., California (Yosemite National Park, Rt. 120, 3.5 km west of Old Big Oak Flat Creek Trail, 8 May 1997) that probably represents a new species. The gill histoblast consists of a short thick base that diverges into 2 short primary trunks, with the dorsal trunk giving rise to 4 petiolate pairs of filaments and the ventral trunk giving rise to 2 petiolate pairs. The hypostomal teeth resemble those of *P. esselbaughi*.

Members of this univoltine species group develop in the cool waters of mountains and high latitudes, where they are a major component of the early-season simuliid fauna. Oviposition is not well documented for the group. Most observations (Davies & Syme 1958, Colbo 1979) indicate that females typically oviposit during flights over the water. At least 1 species, however, oviposits among fine roots embedded in loose streamside soil (Stone & Jamnback 1955), and several European members of the group lay masses of eggs on moist terrestrial mosses along streams, resulting in clusters of up to 56 million eggs (Zwick & Zwick 1990). All known Nearctic species are capable of taking a blood meal, although some are autogenous under favorable larval conditions. Mammals are the principal hosts. No more than 5 North American species are pests of humans and livestock. They usually become pestiferous when the buds of the forest trees begin to burst.

### *Prosimulium approximatum* Peterson
(Figs. 4.27, 10.245, 10.550, 10.659, 10.693; Plate 4e; Map 30)

*Prosimulium* (*Prosimulium*) *approximatum* Peterson 1970b: 74–75, Fig. 44 (M). Holotype* male (pinned #9600, terminalia in glycerin vial below; CNC). Quebec, Missisquoi River, South Bolton, taken from swarm, 7 May 1953 (G. E. Shewell)
*Prosimulium* 'another species of the *hirtipes* complex' Syme & Davies 1958: 717 (Ontario record) (also p. 697 as 'one other cytologically distinct species')
'species . . . related to *Prosimulium fuscum* and *P. mixtum*' Davies et al. 1962: 81 (in part; Ontario record)
[*Prosimulium* (*Prosimulium*) *mysticum*: Peterson (1970b) (misident. in part, L)]
[*Prosimulium mysticum*: Mansingh et al. (1972) not Peterson 1970b (misident.)]
[*Prosimulium mysticum*: Mansingh & Steele (1973) not Peterson 1970b (misident.)]
[*Prosimulium mysticum*: Rothfels & Freeman (1977) not Peterson 1970b (misident.)]

[*Prosimulium mysticum*: Merritt et al. (1978b) not Peterson 1970b (misident. in part, L)]
[*Prosimulium mysticum*: Rothfels (1979) not Peterson 1970b (misident.)]
[*Prosimulium gibsoni*: Cupp & Gordon (1983) not Twinn 1936a (misident. in part, pupa from Manistique River, Michigan)]

**Taxonomy**

Peterson 1970b (ML), Merritt et al. 1978b (L).

All previous references to *P. mysticum* (a synonym of *P. fuscum*), with the exception of the pupa, male, and female in the original description and the pupa in Merritt et al. (1978b), refer to *P. approximatum*. (See discussion under *P. fuscum*.)

The female is very similar to that of *P. fuscum*. Variation in the male genitalia is illustrated in Fig. 4.27 (Ontario, Nipissing District, Amable du Fond River) and Fig. 10.245 (type locality).

**Physiology**

Mansingh & Steele 1973 (L respiration and metabolism).

**Cytology**

Rothfels & Freeman 1977, Rothfels 1979.

**Bionomics**

**Habitat.** This northern species breeds in medium-sized to large, cold rocky rivers from Newfoundland and the Maritime Provinces of Canada to Massachusetts and Wisconsin.

**Oviposition.** Unknown.

**Development.** Larvae probably hatch in the fall and pass through about 7 instars before completing development as early as April (Mansingh et al. 1972) or as late as June. Larval development is optimal at about 9°C but is curtailed below 4°C, inducing a kind of dormancy, termed "oligopause," during which feeding is greatly reduced (Mansingh & Steele 1973). Larvae can survive in unaerated water for about a day, provided the temperature is near freezing; they can withstand entrapment in ice for about a day (Mansingh & Steele 1973).

**Mating.** We have observed males swarming about 3 m above ground by riverside trees.

**Natural Enemies.** We have 1 larva infected with an unidentified mermithid nematode.

**Hosts and Economic Importance.** Unknown; presumably mammalophilic.

### *Prosimulium arvum* Adler & Kim

(Figs. 5.8, 10.246, 10.694; Plate 4f; Map 31)

*Prosimulium arvum* Adler & Kim 1985: 41–45 (FMPLC). Holotype* final-instar female larva (alcohol vial, slide and photographic negatives of chromosomes; USNM). Pennsylvania, Centre Co., State College, Slab Cabin Run, 40°46′25″N, 77°50′02″W, 14 April 1983 (P. H. Adler)
*Prosimulium hirtipes* '4' Syme 1957: 51–52 (New York and Ontario records [Trenton, Ontario, is actually Trenton, New York])
'species . . . related to *Prosimulium fuscum* and *P. mixtum*' Davies et al. 1962: 81 (in part; Ontario record)
[*Simulium pecuarum*: Garman (1912) not Riley 1887 (misident.)]
[*Prosimulium approximatum*: Rothfels & Freeman (1977) not Peterson 1970b (misident.)]
[*Prosimulium approximatum*: Rothfels (1979) not Peterson 1970b (misident.)]
[*Prosimulium* (*Prosimulium*) *approximatum*: Adler (1983) not Peterson 1970b (misident.)]
[*Prosimulium approximatum*: Adler et al. (1983b) not Peterson 1970b (misident.)]
[*Prosimulium* (*Prosimulium*) *approximatum*: Cupp & Gordon (1983) not Peterson 1970b (misident.)]

**Taxonomy**

Garman 1912 (FL); Adler 1983 (PL); Adler & Kim 1985 (FMPL), 1986 (PL).

In typical populations, larvae have a subrectangular postgenal cleft and bold head spots contrasting markedly with the pale yellowish frontoclypeal apotome (Fig. 10.694); males have a ventral plate shaped much like that of *P. fuscum* and *P. mixtum*. However, in some South Carolina populations (e.g., Byrd Creek and Turkey Creek, McCormick Co.), the larval head spots are weak (especially the lateral spots), the postgenal cleft is narrowed anteriorly, and the ventral plate is considerably wider than long (Fig. 10.246). These South Carolina populations might represent a separate species.

**Cytology**

Rothfels & Freeman 1977, Rothfels 1979, Adler & Kim 1985.

Populations from Pennsylvania northward have differentiated sex chromosomes and a glassy chromocenter (Rothfels & Freeman 1977). At least some southern populations, including those from South Carolina (discussed under the Taxonomy section above), have undifferentiated sex chromosomes and lack a glassy chromocenter, although the centromere regions are often associated, mainly as a result of extra heterochromatin in the centromere region of chromosome II.

**Bionomics**

**Habitat.** *Prosimulium arvum* is distributed widely in eastern North America, infiltrating the Southeast deeper than other members of the *P. hirtipes* species group. The immature stages are found most often in moderately productive, sparsely shaded rocky streams less than 20°C and from about 2 to more than 10 m in width.

**Oviposition.** Unknown.

**Development.** Larvae begin hatching during the winter and mature from February to April south of Pennsylvania and from April to June northward.

**Mating.** Unknown.

**Natural Enemies.** Larvae are hosts of unidentified mermithid nematodes (Adler & Kim 1986) and the microsporidia *Janacekia debaisieuxi* and *Polydispyrenia simulii* (Ledin 1994).

**Hosts and Economic Importance.** Unknown; presumably mammalophilic.

### *Prosimulium clandestinum* Adler, Currie & Wood, New Species

(Figs. 4.29, 5.8, 10.247, 10.695; Map 32)

**Taxonomy**

**Female.** Thorax dark brown; colors of antennae, frons, legs, abdomen, and hair unavailable. Mandible with 38 serrations; lacinia with 23 retrorse teeth. Sensory vesicle in lateral view about one third as long as, and occupying about one fourth to one third of, palpomere III, opening to outside via short neck with small mouth. Claws toothless. Terminalia: Anal lobe in lateral view slender, curved. Genital fork with each arm short, slender, and expanded into small lateral plate. Hypogynial valve extended beyond middle of anal lobe; sclerotization of inner margin strong, minimally extended anterolaterally. Spermatheca broadest proximally.

**Male.** Wing length 3.1–3.2 mm. Scutum, abdomen, and antennae brownish black; hair coppery, with golden reflections, to brown. Legs brown; hair brown, with golden reflections. Genitalia (Fig. 10.247): Ventral plate in ventral view with body about 1.5 times wider than long, slightly tapered posteriorly; body in terminal view short, triangular; lip in lateral view small, strongly upturned.

**Pupa.** Length 3.6–4.3 mm. Gill of 16 filaments, about one third to one half as long as pupa (Fig. 4.29); base about as long as wide, giving rise to 3 short trunks, branching (2 + 3 + 3) + (2 + 2) + (2 + 2). Head and thorax dorsally with densely distributed, minute, dome-shaped microtubercles. Cocoon variably covering pupa.

**Larva.** Length 6.3–8.7 mm. Head capsule (Fig. 10.695) pale yellowish or yellowish brown to brown, faintly flecked with brown pigment in some specimens; head spots brown; second posterolateral head spot faint. Antenna equal in length to stalk of labral fan, or extended beyond by about one fourth length of distal article. Hypostoma with median and lateral teeth extended anteriorly to about same level. Postgenal cleft narrowed, rounded or straight anteriorly, extended about one third distance to hypostomal groove; subesophageal ganglion faintly ensheathed with pigment. Labral fan with 34–48 primary rays. Prothoracic proleg with vertical portion of each lateral sclerite rather weakly developed. Body pigment grayish, arranged in distinct bands. Posterior proleg with 9–12 hooks in each of 61–67 rows.

**Diagnosis.** Females are similar to other eastern members of the *P. hirtipes* group. Males are distinguished from those of other species of *Prosimulium* by the small, markedly upturned lip of the ventral plate in lateral view. Pupae are indistinguishable from those of other eastern members of the *P. hirtipes* species group. Larvae are distinguished from those of all other *Prosimulium* species by the combination of an anteriorly narrowed postgenal cleft and a gray-banded abdomen. Unique chromosomal features include the fixed inversion in the long arm of chromosome II and the translocation of the IIIL arm to the end of the IIIS arm.

**Holotype.** Larva (mature, ethanol vial; originally collected in acetic ethanol). New Jersey, Warren Co., Delaware Water Gap National Recreation Area, Delaware River, Poxono Island, 41°02.9′N, 75°01.1′W, 26 March 1998, D. S. Bidlack (USNM).

**Paratypes.** New Jersey, Sussex Co., Delaware Water Gap National Recreation Area, Little Flat Brook, upstream of confluence with Big Flat Brook, 41°11.4′N, 74°51′W, 26 March 1998, D. S. Bidlack (3 larvae); same data as holotype (10 larvae); North Carolina, Halifax Co., Fishing Creek, Bellamy Lake Road, 36°09′N, 77°45′W, 20 February 2002, P. H. Adler & J. K. Moulton (1 pupa, 47 larvae); Nash Co., Tar River, Rt. 581, 35°52.93′N, 78°05.38′W, 20 February 2002, P. H. Adler & J. K. Moulton (2♂ + exuviae, 8 pupae [1 pharate ♀, 3 pharate ♂], 53 larvae); Northampton Co., Meherrin River, Branchs Bridge Road, 36°31.92′N, 77°15.68′W, 20 February 2002, P. H. Adler & J. K. Moulton (103 larvae). Pennsylvania, Pike Co., Rattlesnake Creek, Spring Brook Road (T424), 41°21.82′N, 74°57.89′W, 17 December 2001, M. E. Warfel (2 larvae), 6 February 2002, M. E. Warfel (36 larvae); Shohola Creek, Rt. 739, Lords Valley, 41°21.60′N, 75°03.37′W, 17 January 2002, M. E. Warfel (1 larva), 8 February 2002, M. E. Warfel (21 larvae); York Creek, Interstate 84, 41°21.51′N, 75°04.15′W, 6 February 2002, M. E. Warfel (205 larvae), 12 February 2002, M. E. Warfel (67 larvae). Virginia, Brunswick Co., Nottoway River, Rt. 609, 36°54.07′N, 77°40.38′W, 21 February 2002, P. H. Adler & J. K. Moulton (96 larvae); Meherrin River, Rt. 639, 36°46.26′N, 77°58.89′W, 21 February 2002, P. H. Adler & J. K. Moulton (34 larvae); Dinwiddie Co., Nottoway River, Rt. 612, 36°59.01′N, 77°48.00′W, 21 February 2002, P. H. Adler & J. K. Moulton (104 larvae).

**Etymology.** The specific name is from the Latin adjective meaning "secret" or "hidden" and alludes to the clandestine nature of this long-hidden species in the heavily prospected eastern United States.

**Remarks.** The New Jersey material was submitted to us by D. S. Bidlack in unsorted collections that included 4 other species of *Prosimulium* (*P. arvum*, *P. fuscum*, *P. magnum* s. s., *P. mixtum*). The Pennsylvania material was submitted by M. E. Warfel and D. I. Rebuck in collections that also included *P. fuscum* and *P. mixtum*. In collections from North Carolina and Virginia, *P. clandestinum* is typically the sole species of *Prosimulium*.

**Cytology**

Our analysis of the chromosomes of 8 larvae from New Jersey, Pennsylvania, and Virginia revealed the following features, relative to the standard sequence of Basrur (1962). A large, pale, glassy chromocenter is present. The standard banding sequence occurs in IS, IL, IIS, IIIS, and IIIL (i.e., IIIL does not have the *IIIL-1* inversion), and the centromere regions of chromosomes I and II are standard, with heterochromatin on either side of the chromocenter in CII. III has a novel homozygous inversion, *IIL-12* (limits 64c–69d), that inverts the parabalbiani marker. The most unusual feature is the translocation of the entire IIIL arm to the end of the IIIS arm, such that the base of IIIL is continuous with the end of IIIS. Thus, only 5 arms radiate from the chromocenter. IL carries a unique polymorphism (limits 28–30 inclusive), with the inverted sequence predominating. Our limited analysis precludes a statement about sex chromosomes.

**Bionomics**

**Habitat.** This species is quite common in the piedmont region of North Carolina and Virginia, and reasonably so in the Delaware River system that drains the northern mountains of New Jersey and Pennsylvania. Larvae and pupae inhabit sandy to rocky rivers and streams typically more than 3 m wide.

**Oviposition.** Unknown.

**Development.** Larvae hatch in late fall. Middle instars appear as early as mid-December in Pennsylvania. Final instars are present in February in North Carolina and Virginia and by late March in New Jersey.

**Mating.** Unknown.

**Natural Enemies.** Larvae are hosts of the microsporidium *Janacekia debaisieuxi*.

**Hosts and Economic Importance.** Unknown; presumably mammalophilic.

### *Prosimulium daviesi* Peterson & DeFoliart

(Figs. 10.103, 10.248, 10.660, 10.696; Map 33)

*Prosimulium daviesi* Peterson & DeFoliart 1960: 85–91 (FMPL). Holotype* female with pupal exuviae (alcohol vial, terminalia on slide; USNM). Utah, Cache Co., small stream 19.3 mi. up Logan Canyon, 1890 m elev., 26 May 1957 (B. V. Peterson)

*Prosimulium* 'sp. 199' Peterson 1959a: 147 (M in swarm)

[*Prosimulium doveri*: Rothfels (1979) not Sommerman 1962a (misident. in part, Utah and Colorado segregates, p. 516)]

**Taxonomy**

Peterson 1958a (FMPL), 1960a (FMP); Peterson & DeFoliart 1960 (FMPL).

The morphological distinction between *P. daviesi* and *P. doveri* is not fully resolved. We illustrate topotypical material of the 2 species, showing subtle differences in genitalia that require further study to determine consistency. Females of *P. daviesi* from the type locality are orange.

**Cytology**

Rothfels 1979.

Populations from southern Wyoming southward, including those at and around the type locality, match the description of Rothfels (1979, Fig. 4) for *P. doveri* in Utah and Colorado. Relative to the standard banding sequence of Basrur (1962), the centromere region of chromosome I is transformed; IS-1, IS-2, IIIL-2, and IIIL-3 are fixed; IIL is standard; and sex chromosomes are undifferentiated. The IIL-1 inversion is polymorphic in Montana and northern Wyoming.

Populations in Oregon and Washington cloud the chromosomal distinction between *P. daviesi* and topotypical *P. doveri* of Basrur (1962). Two separate breeding populations with morphologically similar larvae were present in a small stream in the Dose Forks Campground of Olympic National Park, Washington (28 June 1999). Of 23 larvae in the collection, 17 were homozygous for IIL-1, whereas 6 had the standard banding sequence in the IIL arm; no IIL heterozygotes were present. Both the IIL-1 and IIL-standard groups were fixed for IS-1 and IIIL-2 and polymorphic for IIIL-3. The IS-2 inversion was floating in the IIL-1 larvae and absent in the IIL-standard larvae. The centromere region of chromosome I was unpaired in males of the IIL-1 group; no males in the IIL-standard group were available for study. Neither of these groups fits all chromosomal criteria for *P. daviesi* or *P. doveri* at their respective type localities. However, we consider the IIL-1 population to be *P. doveri* and the IIL-standard population to be *P. daviesi*, based on the IIL-1 sequence, which is fixed at the type locality of *P. doveri* and absent at that of *P. daviesi*. These assignments conserve the 2 names, although we recognize that additional species might exist.

Under *P. daviesi*, we also include larvae from a seep at 1100 m on Marys Peak, Benton Co., Oregon (14 June 1996, 18 July 1997). The chromosomes of these larvae were fixed for inversions IS-1 and IIIL-2, 3 but did not carry IS-2 or IIL-1. They had a common floating inversion with breakpoints in sections 7 and 16 of chromosome arm IS.

Overall, we consider the chromosomes of *P. daviesi* to be fixed for *IS-1* and *IIIL-2* and polymorphic for IS-2, IIL-1, and IIIL-3. These chromosomal criteria, and those for *P. doveri* (below), might require revision if additional species are involved. The centromere region in chromosome I is transformed in all populations of *P. daviesi*.

**Bionomics**

**Habitat.** *Prosimulium daviesi* is rather common but spotty in the mountains from southern British Columbia and Montana to southern New Mexico. The immature stages are restricted to cold, sandy and rocky streams less than 2 m wide, often with patches of trailing or submerged grasses. Larvae adhere to live vegetation and the undersides of stones. Final instars often work their way between embedded stones and the sandy stream bottom, making the pupae difficult to find in their sand-covered cocoons.

**Oviposition.** Unknown.

**Development.** Larvae hatch in late winter. Pupae first appear from about mid to late May and can be found through July at higher elevations.

**Mating.** Unknown.

**Natural Enemies.** Larvae are hosts of an unidentified mermithid nematode and the chytrid fungus *Coelomycidium simulii*.

**Hosts and Economic Importance.** Females identified as *P. daviesi* fly around humans and sheep in southern Idaho but evidently do not feed on them (Jessen 1977).

### *Prosimulium doveri* Sommerman

(Figs. 10.52, 10.104, 10.249; Map 34)

*Prosimulium doveri* Sommerman 1962a ('1961'): 225–235 (FMPL). Holotype* female with pupal exuviae and larval head capsule (alcohol vial; USNM). Alaska, 40 mi. northeast of Anchorage, trickle 11-B near margin of Eklutna Lake, 267 m elev., 21 July (pupa), 29 July (adult) 1958 (K. M. Sommerman)

*Prosimulium hirtipes* 'E' Sommerman 1958: 193 (overwintering); Basrur 1962: 1019–1033 (C)

*Prosimulium* 'n. sp.' Rothfels 1979: 516 (C)

[*Prosimulium travisi*: Sommerman (1953) not Stone 1952 (misident.)]

[*Prosimulium travisi*: Sommerman et al. (1955) not Stone 1952 (misident.)]

**Taxonomy**

Sommerman 1953 (L), 1962a (FMPL); Peterson 1970b (FMPL).

*Prosimulium doveri* is perhaps the species referred to as *P.* n. sp. in Fig. 4 of Rothfels (1979), but other than a cursory chromosomal description, no information was provided, nor has the species ever been mentioned again. Alternatively, it might refer to our incompletely analyzed populations of *P. secretum* n. sp. on Vancouver Island, British Columbia.

**Cytology**

Basrur 1962, Rothfels 1979.

We consider the chromosomes of *P. doveri* to be fixed for the *IS-1*, *IIL-1*, and *IIIL-2* inversions and polymorphic for IS-2 and IIIL-3, relative to the standard banding sequence of Basrur (1962). All populations have a transformed centromere region in chromosome I. These criteria differ from those established for topotypical material (Basrur 1962) by having IS-2 and IIIL-3 as polymorphisms. (See also the entry under *P. daviesi.*)

**Bionomics**

**Habitat.** *Prosimulium doveri* is found from the foothills to above the treeline along the Pacific Coast from Anchorage, Alaska, to Washington. The immature stages reside in cold spring-fed streams less than a meter wide, 5 cm deep or less, and often intermittent. Pupation takes place on stones and debris, usually at the interface with the substrate.

**Oviposition.** Females are anautogenous (Sommerman 1962a). Nothing is known of oviposition behavior.

**Development.** Hatching probably begins in late April, with larval development requiring about 8 weeks (Sommerman 1962a). Pupae first appear in late June or early July and can be found into October. Most adults emerge from mid-July to mid-August, with males emerging slightly before females, and live up to 3 weeks on a sugar diet (Sommerman 1962a).

**Mating.** Unknown.

**Natural Enemies.** We have a topotypical larva that harbors a possible new species of microsporidium with elliptical spores. Larvae of empidid flies (*Oreogeton*) prey on the immature stages (Sommerman 1962b).

**Hosts and Economic Importance.** Females have been described as "vicious human biters" capable of delivering wounds accompanied by maddening burning and itching (Sommerman 1962a). They will take up to 7 blood meals in the laboratory (Sommerman 1962a). The report of facial feeding on humans in the Queen Charlotte Islands suggests that the identification of those flies as *P. doveri* is legitimate (Currie & Adler 1986).

### ***Prosimulium esselbaughi* Sommerman**

(Figs. 10.250, 10.551, 10.661; Map 35)

*Prosimulium esselbaughi* Sommerman 1964: 141–145 (FMPL). Holotype* female with pupal exuviae (alcohol vial; USNM). Alaska, Fort Richardson, Station 272 (see Sommerman et al. 1955), 14 July 1959 (K. M. Sommerman)

*Prosimulium hirtipes* '2' Sommerman 1958: 193–195 (FM)

*Prosimulium* 'Alaskan *mixtum*' Syme & Davies 1958: 714–716 (FML)

*Prosimulium hirtipes* '2 (Alaska)' Basrur 1962: 1030 (C)

[*Prosimulium hirtipes*: Sommerman (1953) not Fries 1824 (misident.)]

[*Prosimulium hirtipes*: L. Davies (1957) not Fries 1824 (misident. in part, material from Alaska)]

**Taxonomy**

Sommerman 1953 (L), 1958 (FM), 1962a, 1964 (FMPL); L. Davies 1957 (L); Syme & Davies 1958 (FML); Peterson 1970b (FMPL); Corredor 1975 (FM); Currie 1986 (PL).

**Cytology**

Basrur 1962, Rothfels 1979.

We have tentatively identified larvae from Mariposa Co., California, as *P. esselbaughi*. The chromosomes of these larvae are like those of classic *P. esselbaughi* (*sensu* Basrur 1962), except males have a heterozygous subbasal inversion in IIIL that is larger than the one in males of classic *P. esselbaughi*; the inversion resembles IIIL-3. Our record of *P. esselbaughi* from Nevada is based on the chromosomes of 3 female larvae.

**Bionomics**

**Habitat.** This species is known from Alaska southward into California. Larvae and pupae inhabit small, cold, permanent spring-fed streams in forested areas. They also have been found in outlets of the hyporheic zone (Courtney 1986). Larvae pupate, often in patches, beneath stones or in nooks and crannies of rocks and mosses.

**Oviposition.** Females are autogenous. Some eggs have been laid in the laboratory (Sommerman 1964), but oviposition in nature has not been recorded.

**Development.** Larvae hatch as early as late August, develop through the winter, and pupate from late May into July (Sommerman 1964). Males emerge first, and both sexes can live at least 2 weeks on sugar and water (Sommerman 1964).

**Mating.** Unknown.

**Natural Enemies.** Larvae are infrequent targets of unidentified mermithid nematodes. They are preyed on by larvae of empidid flies (*Oreogeton*) (Sommerman 1962b).

**Hosts and Economic Importance.** Females sometimes attack domestic sheep in alpine pastures (Mason & Shemanchuk 1990). They are attracted to blue grouse but do not feed on them (Williams et al. 1980). Females also are attracted to humans and can administer an effective puncture, despite their tendency not to bite in the field (Sommerman 1964, Currie & Adler 1986).

### ***Prosimulium fontanum* Syme & Davies**

(Figs. 10.24, 10.624; Map 36)

*Prosimulium fontanum* Syme & Davies 1958: 708–711, Figs. 7, 9, 10, 13 (FMPL). Holotype* female (pinned #6984, pupal exuviae in glycerin vial below; CNC). Ontario, Algonquin Provincial Park, 1 mi. from Wildlife Research Station, first small stream on Tote Road, east side of Lake Sasajewun, 22 June (pupa), 26 June (adult) 1956 (P. D. Syme & D. M. Davies); *fontanatum* (subsequent misspelling)

*Prosimulium* '16' Rothfels 1956: 120 (C)

*Prosimulium hirtipes* '3' Syme & Davies 1958: 700–701 (F)

**Taxonomy**

Syme 1957 (FMPL), Syme & Davies 1958 (FMPL), Anderson 1960 (FMPL), Peterson & DeFoliart 1960 (FM), Davies et al. 1962 (FMP), Wood et al. 1963 (L), Stone 1964 (FMPL), Holbrook 1967 (L), Peterson 1970b (FMPL), Adler 1983 (PL), Adler & Kim 1986 (PL).

**Morphology**

Guttman 1967a (labral fan); Chance 1969, 1970a (labral fan).

**Cytology**

Rothfels 1956, 1979; Basrur 1958, 1959, 1962; Rothfels & Freeman 1977; Adler & Kim 1986.

**Bionomics**

**Habitat.** This northeastern species has not been recorded south of Pennsylvania. The immature stages dwell in small forest streams that are fed by springs or bogs and remain below 17°C the entire year. Larvae pupate in mosses and on sticks and upper surfaces of rocks, constructing feeble cocoons in small slow streams, but forming more heavily silked cocoons in larger, faster currents where dense pupal aggregations often accumulate much sediment (Peterson 1970b, Tessler 1991).

**Oviposition.** Females are facultatively autogenous for the first batch of eggs, requiring about 2 weeks for oogenesis at 8.5°C when fed sucrose and water (Davies & Györkös 1990). Oviposition habits are unknown.

**Development.** Larvae hatch after March and linger into August or September in some streams, making this species the last of the eastern *P. hirtipes* species group to appear and the last to disappear (Davies & Syme 1958, Adler & Kim 1986, Tessler 1991). Males emerge a few days before females. Both sexes are on the wing from May to August or September but are most prevalent during June.

**Mating.** Unknown.

**Natural Enemies.** Larvae are hosts of an unidentified mermithid nematode.

**Hosts and Economic Importance.** Females feed on American black bears (Addison 1980). They swarm about humans (Davies & Syme 1958, L. Davies 1961, White & Morris 1985b, Adler & Kim 1986) but have been recorded biting only once (LaScala 1979).

### *Prosimulium formosum* Shewell

(Figs. 4.59, 4.61, 10.105, 10.251, 10.662, 10.697; Map 37)

*Prosimulium formosum* Shewell 1959b: 692–694 (FMPL). Holotype* female (pinned #6927, pupal exuviae in glycerin vial below; CNC). British Columbia, Vancouver Island, Bowser, small stream near railroad track, 13 June (pupa), 16 June (adult) 1955 (G. E. Shewell)

**Taxonomy**

Shewell 1959b (FMPL), Peterson 1970b (FMPL), Corredor 1975 (FM), Currie 1986 (PL).

The female of *P. formosum* typically has a brownish scutum, but we have seen a few specimens with an orange scutum. The color of the scutal hair of both sexes varies from dull to bright gold and often has an orange cast.

**Physiology**

McKague et al. 1978 (response to growth regulators).

**Cytology**

Basrur 1962, Rothfels 1979.

Inversion *IS-5* of Basrur (1962) is diagnostic, although the breakpoint nearest the centromere should be between that shown for IS-2 and IS-4. The centromere region of chromosome I is transformed (*sensu* Basrur 1962), as emended by Rothfels (1979).

**Molecular Systematics**

Moulton 1997, 2000 (DNA sequences).

**Bionomics**

**Habitat.** *Prosimulium formosum* is a species of lower montane habitats in western North America from southern Yukon to southern Arizona and is particularly common along the coast. The immature stages inhabit cool woodland streams that are less than 5 m wide, sluggish or rapidly tumbling, and usually with rocky bottoms, although sometimes coursing through bogs. A 6500-year-old fossil hypostoma, possibly of this species, has been recovered from lake sediments on the Queen Charlotte Islands of British Columbia (Currie & Walker 1992).

**Oviposition.** Unknown.

**Development.** Larvae probably overwinter at lower elevations, such as along the California coast where larvae, pupae, and adults can be found in March. At higher elevations in California and from British Columbia northward, eggs overwinter and larvae are present into June. In Alberta, larvae are found from May to August, with peak adult activity in the first half of July (Pledger et al. 1980).

**Mating.** Unknown.

**Natural Enemies.** Larvae are sometimes infected with unidentified mermithid nematodes and microsporidia.

**Hosts and Economic Importance.** Females feed on moose (Pledger et al. 1980) and probably other mammals. They sometimes fly around humans, but we have only 1 biting record.

### *Prosimulium frohnei* Sommerman

(Figs. 10.106, 10.252, 10.553, 10.663, 10.698; Map 38)

*Prosimulium frohnei* Sommerman 1958: 196–197, Figs. 1, 4, 6–8, 12–15 (FMPL). Holotype* female with pupal exuviae (alcohol vial; USNM). Alaska, Eklutna Lake, 267 m elev., stream paralleling road, 26 July (pupa), 31 July (adult) 1956 (K. M. Sommerman)

**Taxonomy**

Sommerman 1958 (FMPL), Peterson & DeFoliart 1960 (FML), Carlsson 1968 (L), Currie 1986 (PL), Peterson & Kondratieff 1995 (FMPL), Peterson 1996 (P).

**Morphology**

Chance 1969, 1970a (L mouthparts); Craig & Chance 1982 (labral fan).

**Cytology**

Basrur 1962, Carlsson 1966, Rothfels 1979.

Larvae in central California do not have a chromocenter but otherwise are chromosomally similar to larvae from Alaska, as described by Basrur (1962).

**Bionomics**

**Habitat.** This species is distributed from Alaska southward, typically above the timberline into California and Colorado. Larvae and pupae are found in cold headwater flows less than a meter wide, originating from springs, snowfields, and glaciers.

**Oviposition.** Unknown.

**Development.** Overwintered eggs begin to yield larvae in May or later at higher elevations. Adults fly from late June well into August (Sommerman 1958, Currie 1986). Larvae are often predaceous and include other black fly larvae among their prey.

**Mating.** Unknown.

**Natural Enemies.** We have 1 larva from a tributary of Mosquito Creek in Park Co., Colorado, that harbors a probable new species of microsporidium with rather elliptical spores.

**Hosts and Economic Importance.** Unknown; presumably mammalophilic.

### *Prosimulium fulvithorax* Shewell

(Figs. 10.107, 10.253, 10.554, 10.664, 10.699; Map 39)

*Prosimulium fulvithorax* Shewell 1959b: 694–696 (FP). Holotype* female (pinned #6928, pupal exuviae on slide #GES5902-05A; CNC). British Columbia, Vancouver Island, Bowser, stream near railroad track, 4 June (pupa), 7 June (adult) 1955 (G. E. Shewell)

*Prosimulium* 'species 1' Basrur 1962: 1030–1031 (C)

**Taxonomy**

Shewell 1959b (FP), Peterson 1970b (FMPL).

**Cytology**

Basrur 1962, Rothfels 1979.

At least 3 X-linked inversions are known in the base of IS (Basrur 1962), and we have found females with the standard sequence in IS (Montana), as well as occasional males (California) homozygous for IS sp-1 and IS sp-2 of Basrur (1962).

**Bionomics**

**Habitat.** This species is sporadic from southern British Columbia and western Montana southward into the high Sierra Nevada of California. Its immature stages have been found in cool, sometimes tumbling, often temporary, stony streams less than 4 m wide.

**Oviposition.** Unknown.

**Development.** Larvae and pupae are present from March or earlier in British Columbia to July in the high Sierra Nevada.

**Mating.** Unknown.

**Natural Enemies.** Larvae are hosts of the microsporidium *Caudospora alaskensis* (Adler et al. 2000) and unidentified mermithid nematodes.

**Hosts and Economic Importance.** Unknown; presumably mammalophilic.

### *Prosimulium fulvum* (Coquillett)

(Figs. 10.254, 10.531, 10.555, 10.665, 10.700; Map 40)

*Simulium fulvum* Coquillett 1902: 96 (FM). Holotype* male (slide #6182; USNM). Montana, county unknown, Bear Paw Mountains, 3 September 1891 (H. G. Hubbard)

[*Simulium ochraceum*: Baker (1897) not Walker 1861 (misident.)]

[*Simulium ochraceum*: Coquillett (1898, 1900) not Walker 1861 (misident.)]

**Taxonomy**

Coquillett 1898, 1902 (FM); Johannsen 1902, 1903b (F); Emery 1913 (F); Malloch 1914 (FM); Riley & Johannsen 1915 (F); Cole & Lovett 1921 (F); Dyar & Shannon 1927 (FM); Stains 1941 (FM); Stains & Knowlton 1943 (FM); Coleman 1951 (F); Fredeen 1951 (FM); Stone 1952 (FMP); Sommerman 1953 (L), 1958 (M), 1962a (FMPL); Wirth & Stone 1956 (FMP); Peterson 1958a (FMPL), 1960a (FMP), 1970b (FMPL); Shewell 1959b (F); Abdelnur 1966, 1968 (FMPL); Corredor 1975 (FM); Currie 1986 (P only); Peterson & Kondratieff 1995 (FMPL).

The name *P. fulvum* probably has been applied over the years to several species with orange females. Shewell (1959b) recognized *P. fulvithorax* as a distinct species, and we describe *P. minifulvum* n. sp. and *P. secretum* n. sp., both of which previously might have been identified as *P. fulvum*. All of these species are sympatric in the high Sierra Nevada of California. Other species that previously might have been included under the name *P. fulvum* are *P. daviesi* and *P. formosum*, both of which sometimes have orange females. *Prosimulium fulvum*, as currently recognized, might be a complex of species. For example, pupae at high elevations in the Sierra Nevada have gills resembling those of *P. esselbaughi*.

**Cytology**

Basrur 1962, Rothfels 1979.

Populations of *P. fulvum* in Alaska differ chromosomally from more southern populations (Rothfels 1979), providing further evidence that *P. fulvum* might be a complex of species.

**Bionomics**

**Habitat.** The large orange females are among the most conspicuous black flies in the western mountains, from Alaska southward to California and Colorado. The immature stages, which are less frequently collected than females, are typically found on mosses and on or beneath stones in cold mountain streams less than a meter wide. Females are strong fliers and occasionally enter the Canadian prairie east of the Rocky Mountains, presumably aided by prevailing winds (Currie 1986).

**Oviposition.** Unknown.

**Development.** In Alaska, larvae hatch from late August to mid-October and mature the following May; pupal development requires about 2–3 weeks (Sommerman et al. 1955). In the high Sierra Nevada, mature larvae and pupae can be found in June. Adults fly predominantly from June to August but as early as May and as late as October.

**Mating.** Unknown.

**Natural Enemies.** Larvae are preyed on by larval empidid flies (*Oreogeton*) (Sommerman 1962b).

**Hosts and Economic Importance.** Females feed on horses (Baker 1897, McAtee 1922, Hearle 1932) and sheep (Mason & Shemanchuk 1990). They are readily attracted to humans and will bite (Essig 1928, Jenkins 1948, Travis 1949, Sailer 1953).

### *Prosimulium fuscum* Syme & Davies

(Figs. 10.108, 10.255, 10.411, 10.666, 10.701;
Plate 5a; Map 41)

*Prosimulium fuscum* Syme & Davies 1958: 702–706, Figs. 1, 3, 11, 26, 27 (FMPL). Holotype* female (pinned #6982; CNC). Ontario, Muskoka District, Morrison Township, Kahshe River, south of Gravenhurst, Hwy. 11, 27 April (pupa), 3 May (adult) 1956 (P. D. Syme); *fuseum* (subsequent misspelling)

*Prosimulium hirtipes* '1' Rothfels 1956: 114–121 (C)

?*Prosimulium* 'species #2' Anderson & Dicke 1960: 398 (Wisconsin record)

*Prosimulium (Prosimulium) mysticum* Peterson 1970b: 122–125, Figs. 23, 55, 85 (FMP). Holotype* male with pupal exuviae (pinned #9602; CNC). Wisconsin, Shewano Co., Pulcifer, Oconto River, 19 April 1964 (D. M. & G. C. Wood). New Synonym

*Prosimulium*, Merritt & Wallace 2001: 201 (utility in forensic investigation)

[*Prosimulium hirtipes*: MacNay (1954) not Fries 1824 (misident.)]

[*Prosimulium hirtipes*: Davies & Peterson (1956) not Fries 1824 (misident. in part, see Davies & Syme 1958, p. 744)]

[*Prosimulium (Prosimulium) multidentatum*: Shewell (1957) not Twinn 1936a (misident. in part, Manitoba record)]

[*Prosimulium fulvum*: Travis et al. (1974) not Coquillett 1902 (*lapsus calami* in part, footnote p. 195)]

**Taxonomy**

Syme 1957 (FMPL), Syme & Davies 1958 (FMPL), Anderson 1960 (FMPL), Peterson & DeFoliart 1960 (FML), Davies et al. 1962 (FMP), Sommerman 1962a (P), Wood et al. 1963 (L), Stone 1964 (FMPL), Holbrook 1967 (L), Peterson 1970b (FMPL), Pinkovsky 1970 (FMPL), Amrine 1971 (FMPL), Lewis 1973 (FMPL), Merritt et al. 1978b (PL), Adler 1983 (PL), Fredeen 1985a (FMP), Adler & Kim 1986 (PL).

Our study of the type series of *P. mysticum* and of material that we collected (23 April 1995) from the type locality on the Oconto River at Pulcifer, Wisconsin, shows that Peterson (1970b) unknowingly chose a male of *P. fuscum* as the holotype of *P. mysticum*. The pupae and females that he described are also those of *P. fuscum*, but the majority of larvae in the type series of *P. mysticum* are actually those of *P. approximatum*. Rothfels, in 1964, chromosomally established that the larvae he examined from the original series were those of a new species, but he did not at that time publish his observations. *Prosimulium fuscum* is slightly earlier in development when it occurs with *P. approximatum*, resulting in the mixture of the 2 species in the type series.

**Morphology**

L. Davies 1960 (early-instar head); Guttman 1967a (labral fan); Yang 1968 (F cibarial sensilla); Chance 1969, 1970a (labral fan); Kurtak 1973 (labral fan); Colbo et al. 1979 (FM labral and cibarial sensilla); Hayton 1979 (instars); Ross & Craig 1979 (instars); Craig & Borkent 1980 (L maxillary palpal sensilla); Linton 1982 (L morphometrics); Snyder & Linton 1983 (L morphometrics).

**Physiology**

Guttman 1967b (L anesthetic); Yang 1968 (F amylase and invertase, FM pepsin and trypsin); Yang & Davies 1968a (F amylase), 1968b (FM trypsin), 1968c (FM invertase); Martin et al. 1985 (L digestive enzymes).

**Cytology**

Rothfels 1956, 1979, 1980, 1981a; Basrur 1958, 1959, 1962; Rothfels & Freeman 1977; Adler & Kim 1986.

**Molecular Systematics**

Sohn 1973 (DNA hybridization); Sohn et al. 1975 (DNA hybridization); Linton 1982 (allozymes); Snyder & Linton 1983, 1984 (allozymes); Feraday & Leonhardt 1989 (allozymes); Xiong & Kocher 1991 (mtDNA); Xiong 1992 (mtDNA).

**Bionomics**

**Habitat.** *Prosimulium fuscum* is common in northeastern North America west to Saskatchewan and south into Pennsylvania. It breeds in productive streams and rivers more than 2 m wide, often below lake outlets. Larval densities decrease exponentially with increasing distance downstream from lake outlets (Morin et al. 1986). Larvae do not tolerate periphyton on substrates (Morin & Peters 1988). They often pupate in groups (Davies & Syme 1958).

**Oviposition.** Most females are autogenous for the first batch of eggs (L. Davies 1961, Davies & Györkös 1990). After emergence, they rest in nearby trees, and the majority probably return to their birth streams to oviposit (L. Davies 1961), which tends to build site-specific profiles of chromosomal polymorphisms (Rothfels 1981a). Peak oviposition occurs about 6 or 7 days after emergence (L. Davies 1961), which is the approximate time required to mature the eggs at 16°C–17°C (Davies & Györkös 1990). The potential fecundity (i.e., number of ovarioles) in the first ovarian cycle is about 200–500 and is positively related to female size (Pascuzzo 1976). Eggs are released as females dip their abdomens to the water surface while in flight (Davies & Syme 1958). Females and males parasitized by mermithid nematodes exhibit the same behavior (L. Davies 1961, Hayton 1979).

**Development.** Larvae hatch over a period of about 2–5 months, beginning in the fall. They develop during the winter at temperatures just above freezing, and pupate synchronously once a critical temperature is reached (L. Davies 1961). Larvae undergo 6 or 7 instars and require about 240 degree-days above 0°C to complete their development in Michigan (Ross & Merritt 1978, Ross & Craig 1979), although higher values (ca. 2000 degree-days above 0°C) have been reported for southern Ontario (Davies & Syme 1958). They pupate from March well into June, depending on latitude, which is earlier than other members of the eastern *P. hirtipes* species group. Emergence takes place during daylight hours, with a sharp peak about 2–3 hours after sunrise and a lower broader peak between midafternoon and sunset (Hayton 1979). Adults have been reared from first instars in the laboratory (Wood & Davies 1966). Filter-feeding efficiency decreases as temperature increases (Kurtak 1979).

*Prosimulium fuscum* was once used in a court of law to help convict a murderer—the only known use of black flies in forensic entomology. Cocoons and pupal exuviae found on a submerged car containing a human body correctly placed the time of death, countering the testimony of the felon (Wolkomir & Wolkomir 1992, Merritt & Wallace 2001).

**Mating.** Males swarm in the lee of large trees and buildings. Swarms can be 10 m high and cover an area 60 m × 8 m; copulating pairs drop to the ground and remain attached for up to 6 minutes (White 1983). Matings have been observed in the laboratory (Edman & Simmons 1985a).

**Natural Enemies.** Larvae, pupae, and adults often are infected with unidentified mermithid nematodes (Davies et al. 1962). Records of *Gastromermis* and *Isomermis* from larvae in Wisconsin (Anderson & DeFoliart 1962) are possibly based on erroneous identifications (R. Gordon 1984) and are here considered unidentified mermithid nematodes. Larvae are hosts of the microsporidia *Caudospora simulii* (Jamnback 1970) and *Janacekia debaisieuxi* and the trichomycete fungus *Harpella melusinae*. They are consumed by perlid stoneflies and corydalids in the laboratory (Merritt et al. 1991, Wotton et al. 1993). Adults are preyed on by the ants *Formica subnuda* and *F. podzolica* (White 1983).

**Hosts and Economic Importance.** Hosts include American black bears, horses, elks, white-tailed deer, and cattle (Anderson & DeFoliart 1961, Merritt et al. 1978b, Addison 1980). Females are attracted to humans but are not considered serious pests because most are autogenous for the first ovarian cycle and only about 10% of parous individuals survive to take blood (L. Davies 1961). On the other hand, females have been cited as the major nuisance species in areas such as northern New Hampshire (Veit 1986), suggesting either problems with identification or that the species is not universally autogenous.

### *Prosimulium idemai* Adler, Currie & Wood, New Species

(Figs. 10.667, 10.702; Map 42)

**Taxonomy**

**Female.** Unknown.

**Male.** Unknown.

**Pupa.** Unknown.

**Larva.** Length 6.4–6.7 mm. Head capsule (Fig. 10.702) pale brownish orange; head spots brown, diffuse, variably distinct; anteromedial and posteromedial head spots darkest, often in nearly continuous line; first posterolateral head spot yellow. Antenna subequal in length to stalk of labral fan. Hypostoma with median tooth largest and extended farthest anteriorly; denticles of median tooth extended anteriorly to about same level as inner sublateral teeth; outer sublateral teeth extended slightly posterior to inner sublateral teeth; lateral teeth extended anteriorly beyond sublateral teeth (Fig. 10.667). Postgenal cleft straight anteriorly, extended about one fourth to one third distance to hypostomal groove; subesophageal ganglion not ensheathed with pigment. Labral fan with 28–32 primary rays. Gill histoblast of 13 or 14 filaments. Body pigment pale brown, weakly arranged in bands. Posterior proleg with 8–11 hooks in each of 64–69 rows.

**Diagnosis.** The gill of 13 or 14 filaments, without any swollen trunks, distinguishes the pupae (and mature larvae) from those of all other species of *Prosimulium* in the lower 48 United States.

**Holotype.** Larva (mature, ethanol vial; originally collected in acetic ethanol). California, Marin Co., Redwood Creek, 37°53′N, 122°34′W, 21 March 1990, P. H. & C. R. L. Adler (USNM).

**Paratypes.** California, Marin Co., Mt. Tamalpais State Park, Panorama Parkway, 37°54.5′N, 122°36.1′W, Bootjack Picnic Area, 21 March 1990, P. H. Adler (1 larva); same data as holotype (9 larvae).

**Etymology.** This species is named in honor of Ralph Idema, who set the standard for simuliid illustrations during his more than 40 years of experience.

**Remarks.** Two pupae and 2 exuviae have been lost.

**Cytology**

In our preliminary chromosomal examination, the IIIL arm was not analyzed, although the 3 males we examined had a heterozygous inversion in the base. All other arms have the standard banding sequence of Basrur (1962). A chromocenter is lacking, and the centromere region of chromosome I is not transformed.

**Bionomics**

**Habitat.** This species is known from the forests of northwestern California. Its larvae and pupae have been found in rocky streams less than 5 m wide.

**Oviposition.** Unknown.

**Development.** Larvae probably hatch during the winter. Mature larvae appear in March, and pupae can be found into early May.

**Mating.** Unknown.

**Natural Enemies.** No records.

**Hosts and Economic Importance.** Unknown; presumably mammalophilic.

### *Prosimulium minifulvum* Adler, Currie & Wood, New Species

(Figs. 10.109, 10.256, 10.556, 10.668, 10.703; Map 43)

**Taxonomy**

**Female.** Wing length 3.1–3.6 mm. Thorax orange. All hair golden. Antenna brownish orange. Frons dark brown. Mandible with 33–36 serrations; lacinia with 24 or 25 retrorse teeth. Sensory vesicle in lateral view about one fourth to one third as long as, and occupying about one fifth to one fourth of, palpomere III. Legs pale yellow, except tarsi brown; claws toothless. Abdomen grayish brown. Terminalia (Fig. 10.109): Anal lobe in ventral view unsclerotized anteriorly. Genital fork with each arm slender and expanded into large subtriangular lateral plate directed posteromedially. Hypogynial valve extended beyond middle of anal lobe; inner and outer margins subparallel basally. Spermatheca broadest proximally.

**Male.** Wing length 2.8–3.5 mm. Scutum brownish black. All hair pale golden to orange, except base of basal fringe coppery brown. Legs pale brown. Genitalia (Fig. 10.256): Ventral plate in ventral view with body nearly as long as wide, slightly concave posteriorly; lip in lateral view short, broad, slightly upturned.

**Pupa.** Length 3.5–4.3 mm. Gill of 16 filaments, about one third to one half as long as pupa (Fig. 10.556); in lateral view typically in 3 distinct clusters (8 + 4 + 4), although sometimes in 2 distinct clusters. Head and thorax dorsally with densely distributed, dome-shaped microtubercles. Orange integument of pharate female (not male) readily visible through pupal cuticle. Cocoon variably covering pupa.

**Larva.** Length 5.7–7.3 mm. Head capsule (Fig. 10.703) brownish orange; head spots brown, variably distinct, sometimes surrounded by darker pigment; first posterolateral head spot pale yellow; remaining posterior head spots sometimes absent. Antenna shorter than stalk of labral fan by about one third length of distal article. Hypostoma with median tooth largest and extended farthest anteriorly; denticles of median tooth and inner sublateral teeth extended anteriorly to about same level; outer sublateral teeth extended slightly posterior to inner sublateral teeth; lateral teeth extended anteriorly beyond sublateral teeth (Fig. 10.668). Postgenal cleft straight or slightly rounded anteriorly, extended one third or less distance to hypostomal groove; subesophageal ganglion sometimes ensheathed with pigment. Labral fan with 22–29 primary rays. Body pigment brown, uniformly distributed. Posterior proleg with 9–13 hooks in each of 68–81 rows.

**Diagnosis.** The small orange females are not easily distinguished from the orange females of other North American *Prosimulium*, and the males are similar to those of many western members of the *P. hirtipes* species group. The gill of 3 clusters (in lateral view) is generally diagnostic among the pupae of North American black flies. Larvae are similar to those of other species of *Prosimulium* that have the median hypostomal tooth extended anteriorly well beyond all other teeth. The standard orientation of the "shield" and "triad" markers in chromosome arm IIIL is diagnostic among the western species of *Prosimulium*.

**Holotype.** Female (Peldri II, pinned) with pupal exuviae (glycerin vial below). California, Mono Co., Rt. 108, Leavitt Meadow Campground, Leavitt Creek, 38°19′N, 119°33′W, 10 June 1990, P. H. Adler (USNM).

**Paratypes.** California, Alpine Co., Rt. 4, Upper Cascade Creek, 24 June 1992, P. H. Adler (2♀ + exuviae, 4 pupae, 5 larvae); Mariposa Co., Yosemite National Park, Bridalveil Creek, Glacier Point Road, 37°39.95′N, 119°37.0′W, 10 May 1992, P. H. Adler (4♀ & 2♂ + exuviae, 6 pupae, 22 larvae); same data as holotype (15 larvae); Tuolumne Co., Rt. 108, Deadman Creek, east of Kennedy Meadow turnoff, 38°18′N, 119°44′W, 25 June 1991, P. H. Adler (11♀ & 2♂ + exuviae, 14 pupae, 51 larvae). Idaho, Blaine Co., Rt. 75, near Alexander Ross Historic Site, 43°N, 114°W, 10 June 1994, P. H. Adler (30 larvae, photographic negatives of larval polytene chromosomes).

**Etymology.** The specific name is from Latin and refers to the orange females, which superficially resemble small versions of *P. fulvum* females.

**Remarks.** In Yosemite National Park, *P. minifulvum* is often found in the same streams with larvae of a species that we have identified as *P. esselbaughi*. (See Cytology section under *P. esselbaughi*.) At these sites, larvae of the 2 species appear morphologically indistinguishable, and some pupae also might be difficult to separate.

**Cytology**

IS, IL, IIS, IIL, and IIIS have the standard banding sequence for the *Prosimulium* IIIL-2 group of Basrur (1962). A chromocenter is lacking and the centromere region of chromosome I is not transformed. The IIIL arm is characterized by a large fixed inversion that runs from the center of section 85 to the center of section 94 and is overlain on inversions IIIL-2 and IIIL-3. This large inversion moves the "shield" and "triad" markers into standard orientation. The Y chromosome of these larvae is characterized by a large inversion in IIIL between sections 84 and 95.

**Bionomics**

**Habitat.** This species is known from the central Sierra Nevada of California and the Sawtooth Range of Idaho. Larvae and pupae are found in cool, forested rocky streams less than 8 m wide that are often tumbling. Larvae frequently pupate on the tops and sides of dark smooth stones.

**Oviposition.** Unknown.

**Development.** Eggs probably overwinter. Larvae and pupae can be found from at least May to the end of June.

**Mating.** Unknown.

**Natural Enemies.** Larvae are parasitized by unidentified mermithid nematodes.

**Hosts and Economic Importance.** Unknown; presumably mammalophilic.

***Prosimulium mixtum* Syme & Davies**

(Figs. 4.7, 4.11, 10.25, 10.31, 10.110, 10.257, 10.605, 10.669, 10.704; Plate 5b; Map 44)

*Prosimulium mixtum* Syme & Davies 1958: 706–709, Figs. 2, 4, 5, 12, 15, 20, 22 (FMPL). Holotype* female (pinned #6983; CNC). Ontario, Peel Co., Chinguacousy Township, 2 mi. northwest of Terra Cotta, 13th sideroad, Concession 6, 1 May (pupa), 6 May (adult) 1956 (P. D. Syme)

*Simulium calceatum* Say in Harris 1835: 595. *Nomen nudum*

*Prosimulium hirtipes* '2' Rothfels 1956: 114–121 (C)

*Prosimulium* 'spp. complex (*fuscum*/*mixtum*)?' Frost 1970: 890–891 (parasites)

*Prosimulium* 'species 1' Pinkovsky 1970: 23, 31, 80–82, 177, Figs. 8, 29, 50, 64, 72 (L; aberrant gill histoblasts)

*Prosimulium* 'sp.' Frost & Manier 1971: 777 (parasites)

*Prosimulium mixtum*/*fuscum*, Gersabeck 1978: 3, 5–8, 12–18, 25–29, 32–39; Gersabeck & Merritt 1979: 34–38 (bionomics)

*Prosimulium* 'sp.' Merritt et al. 1978b: 199 (L)

[*Simulium hirtipes*: Johannsen (1912) not Fries 1824 (misident.)]

[*Prosimulium hirtipes*: Johnson (1925) not Fries 1824 (misident. in part, No. 36, No. 321, and specimen marked Franconia)]

[*Prosimulium hirtipes*: Matheson (1938) not Fries 1824 (misident.)]

[*Prosimulium hirtipes*: Stone & Jamnback (1955) not Fries 1824 (misident.)]

[*Prosimulium hirtipes*: Davies & Peterson (1956) not Fries 1824 (misident. in part, see Davies & Syme 1958, p. 744)]

[*Prosimulium hirtipes*: Shaw (1959) not Fries 1824 (misident.)]

[*Prosimulium fuscum*: Pickavance et al. (1970) not Syme & Davies 1958 (misident.)]

[*Prosimulium fuscum*: Frost & Nolan (1972) not Syme & Davies 1958 (misident. in part, sites A–C)]

[*Prosimulium fuscum*: Lewis & Bennett (1974b, 1975) not Syme & Davies 1958 (misident.)]

[*Prosimulium fuscum*: Nolan & Lewis (1974) not Syme & Davies 1958 (misident.)]

[*Prosimulium fuscum*: Reilly (1975) not Syme & Davies 1958 (misident.)]

[*Prosimulium fuscum*: Bruder & Crans (1979) not Syme & Davies 1958 (misident.)]

[*Prosimulium fuscum*: Adler et al. (1982) not Syme & Davies 1958 (misident.)]

[*Twinnia tibblesi*: Cupp & Gordon (1983) not Stone & Jamnback 1955 (misident. in part, material from Michigan)]

[*Prosimulium* (*Parahelodon*) *decemarticulatum*: Cupp & Gordon (1983) not Twinn 1936a (misident. in part, material from Walsh Creek, Michigan)]

[*Prosimulium* (*Parahelodon*) *gibsoni*: Cupp & Gordon (1983) not Twinn 1936a (misident. in part, larvae from Manistique River and Pine Creek, Michigan)]

[*Prosimulium* (*Parahelodon*) *vernale*: Cupp & Gordon (1983) not Shewell 1952 (misident in part, material from Michigan)]

[*Prosimulium* (*Prosimulium*) *fontanum*: Cupp & Gordon (1983) not Syme & Davies 1958 (misident. in part, material from Manistique River, Michigan)]

[*Prosimulium* (*Prosimulium*) *saltus*: Cupp & Gordon (1983) not Stone & Jamnback 1955 (misident. in part, material from Michigan)]

[*Prosimulium* (*Prosimulium*) *multidentatum*: Cupp & Gordon (1983) not Twinn 1936a (misident. in part, material from Schoolcraft Co., Michigan)]

**Taxonomy**

Stone & Jamnback 1955 (FMPL); Syme 1957 (FMPL); Syme & Davies 1958 (FMPL); Anderson 1960 (FMPL); Peterson & DeFoliart 1960 (FML); Davies et al. 1962 (FMP); Sommerman 1962a (P), 1964 (FMPL); Wood et al. 1963 (L); Stone 1964 (FMPL); Snoddy 1966 (FMPL); Holbrook 1967 (L); Stone & Snoddy 1969 (FMPL); Peterson 1970b (FMPL), 1981, 1996 (FM); Pinkovsky 1970 (FMPL); Amrine 1971 (FMPL); Lewis 1973 (FMPL); Snoddy & Noblet 1976 (PL); Merritt et al. 1978b (PL); Webb & Brigham 1982 (L); Adler 1983 (PL); Adler & Kim 1985 (L), 1986 (PL); Stuart & Hunter 1998b (cocoon).

Most of the early (pre-1960) North American literature under the name *hirtipes* probably refers to *P. mixtum*. The name *hirtipes* was first applied to North American material by Johannsen (1903a) and was used until the chromosomal demonstration (Rothfels 1956, Basrur 1959) that *P. hirtipes* is a Palearctic species.

Several species of the eastern *P. hirtipes* species group, chiefly *P. fuscum*, have been confused with *P. mixtum*, and the label *P. mixtum/fuscum* has often been applied when either or both species are present. We have assigned certain of these references to either or both species based on examination of the original material, recollections, or information given about the location and habitat. Nearly all references to *P. mixtum/fuscum* on Newfoundland's Avalon Peninsula, for example, refer to *P. mixtum*, the only member of the *P. hirtipes* species group, other than *P. approximatum*, known from the peninsula.

The name *calceatum* was first used, without elaboration, by Harris (1835) in a list of Massachusetts insects. The name, however, actually was applied to Harris's material by Thomas Say, who examined one of Harris's specimens labeled 'No. 36 u,' the 'u' indicating that Say had examined the specimen (Johnson 1925). No. 36 u is a female of *P. mixtum* collected in April 1826, presumably by Harris in the vicinity of Boston, Massachusetts. The name *calceatum*, which clearly applies to *P. mixtum*, means "wearing shoes," suggesting that in this species, the banded "feet" mentioned in Say's (1823) description of *Simulium venustum* were covered by shoes, rendering the legs a uniform color.

*Prosimulium mixtum* might be a complex of sibling species, as suggested by sex-chromosome variation (Rothfels & Freeman 1977), alloyzme data (Snyder & Linton 1984), diversity of larval habitats (Adler 1988a), and wide distribution. The legs of the female vary in color from brown to yellow. Populations in southern Illinois and Indiana have pupal gills with swollen trunks somewhat intermediate between the gills of typical *P. mixtum* and those of *P. saltus*. The chromosomes of these populations, however, are identical to those of typical *P. mixtum* with a $Y_1$ chromosome (*sensu* Rothfels & Freeman 1977).

**Morphology**

L. Davies 1960 (first-instar head), 1961 (ovariole), 1965a (spermatophore); Guttman 1967a (labral fan); Hannay & Bond 1971a (FM wing); Condon 1975 (L neuroendocrine system); Condon et al. 1976 (L neuroendocrine system); Colbo et al. 1979 (FM labral and cibarial sensilla); Ross & Craig 1979 (instars), 1980 (labral fan); Craig & Batz 1982 (L antennal sensilla); Linton 1982 (L morphometrics); Snyder & Linton 1983 (L morphometrics); Thompson 1989 (labral-fan microtrichia); Palmer & Craig 2000 (labral fan).

**Physiology**

Condon 1975 (mermithid effects on L); Gordon & Bailey 1976 (L hemolymph); Condon & Gordon 1977 (mermithid effects on L); Gordon et al. 1978 (mermithid effects on L hemolymph), 1979 (L hemolymph lipids); Hall et al. 1988 (pH effects on L cation concentrations).

**Cytology**

Rothfels 1956, 1979, 1980, 1981a; Basrur 1958, 1959, 1962; Ottonen 1964; Rothfels & Freeman 1966 (Fig. 8), 1977, 1983; Adler & Kim 1986.

**Molecular Systematics**

Linton 1982 (allozymes); Snyder & Linton 1983, 1984 (allozymes); Feraday & Leonhardt 1989 (allozymes).

**Bionomics**

**Habitat.** *Prosimulium mixtum* is probably the most abundant black fly in wooded tracts of eastern North America from midwinter to spring. The immature stages are found in a variety of cold streams and rivers but are most common in small shaded rocky streams. When attached to rocks, larvae are found almost exclusively on the upper surfaces (Tessler 1991). Larvae are aggressive toward one another, resulting in evenly spaced microdistribution patterns (Colbo 1979, Gersabeck & Merrit 1979). However, they are sensitive to contact and are readily displaced by those of species such as *Cnephia ornithophilia* (Harding & Colbo 1981). Larvae drift primarily at night (Adler et al. 1983b), except during the winter when drift is negligible (Ross & Merritt 1978). Females are active mainly in forested areas (Martin 1987).

**Oviposition.** Females can be autogenous (Colbo 1982) or anautogenous (L. Davies 1961), depending on environmental factors such as food and temperature during larval development. Autogenous females can mature an average of about 280–300 eggs in their first ovarian cycle (Mokry 1980b, Colbo 1982). Fecundity of anautogenous females is influenced by blood source, with flies fed on avian blood having lower fecundity than those fed on human blood (Mokry 1980b). Fecundity is reduced in the second ovarian cycle (Mokry 1980b). Females deposit eggs by tapping their abdomens to the water as they fly above the surface (Davies & Syme 1958). Great numbers of ovipositing females have been observed at outlets of impoundments (Colbo 1979). *Prosimulium mixtum* (as *P. hirtipes*) might be the species that Stone and Jamnback (1955) observed crawling among fine streamside roots to deposit their eggs.

**Development.** Within the same stream, 1 cohort can hatch in the fall, while a second hatches in late winter when eggs in exposed sediments are inundated (Merritt et al. 1982). Larvae pass through 6 or 7 instars, requiring about 240 degree-days above 0°C to complete development in Michigan (Ross & Merritt 1978, Ross & Craig 1979), although 1196 degree-days (22 weeks) have been reported in New Jersey (Bruder & Crans 1979) and 2000 in southern Ontario (Davies & Syme 1958). Development proceeds even in frigid waters that flow beneath ice and snow for 3 months (Lewis & Bennett 1974b). Larvae can be maintained for 6 weeks in standing water at 1.5°C–7.0°C (Davies & Syme 1958).

Larvae spend about 75% of their time feeding (Hart & Latta 1986). Ingestion rates increase by a factor of 2.5 when water velocity increases from 25 to 100 cm/sec (Charpentier & Morin 1994). A diet of diatoms produces larger larvae than do diets of bacteria, leaf litter, or green algae (Thompson 1987c). Digestion of diatoms is more complete at lower temperatures (Thompson 1987b). First instars, which have highly reduced labral fans, acquire food by grazing periphyton (Thompson 1987a).

Adults are present in Canada from mid-April into June or even August at higher latitudes (Davies & Syme 1958, Peterson 1970b), in Pennsylvania from late March to early June (McCreadie et al. 1994b), and in Georgia and South Carolina from February into April. The sex ratio is near unity (Singh & Smith 1985). Females can live at least 10 days in the field (White & Morris 1985b).

**Mating.** Mating behavior is probably similar to that of *P. fuscum* but has not been documented with certainty.

**Natural Enemies.** *Prosimulium mixtum* is a common target for parasites. Larvae are infected during the first and second instars by the mermithid nematode *Mesomermis flumenalis* (Ezenwa & Carter 1975, Bailey & Gordon 1977, Bailey et al. 1977, Bruder & Crans 1979). Multiple infections are common, and as many as 9 nematodes can pass to a single female fly (Ebsary & Bennett 1975b, Colbo & Porter 1980). The mermithids *Gastromermis viridis* and *Isomermis* possibly *wisconsinensis* also have been recovered from female flies (Mokry & Finney 1977). *Caudospora simulii* is the most common microsporidium that attacks *P. mixtum* (Ebsary 1973, Vávra & Undeen 1981, Undeen et al. 1984b, Adler & Kim 1986), but it also is host to *Janacekia debaisieuxi*. A cytoplasmic polyhedrosis virus attacks larvae (Adler & Kim 1986), although susceptibility is low (Bailey 1977). A report (Avery & Bauer 1984) of iridescent virus from the larva of a species of *Prosimulium* probably refers to *P. mixtum*. Larvae are hosts of the chytrid fungus *Coelomycidium simulii* (Adler & Kim 1986) and the trichomycete fungi *Harpella melusinae*, *Pennella simulii*, *Pennella* possibly *hovassi*, *Simuliomyces microsporus*, and *Smittium* sp. (Frost & Manier 1971, Lichtwardt et al. 2001). Female ovaries are attacked by a trichomycete fungus, originally identified as a phycomycete (Undeen & Nolan 1977, Yeboah et al. 1984, Moss & Descals 1986). Some of the many bacteria that have been isolated from the larval gut might be pathogenic (Malone & Nolan 1978). Cercariae of the trematode *Plagiorchis noblei* can infect larvae under experimental conditions (Jacobs 1991, Jacobs et al. 1993). The oomycete *Pythiopsis cymosa* has been isolated from a pupa (Nolan & Lewis 1974). An unknown pathogen (Fig. 6.16) was discovered in the abdomen of an adult female from the Great Smoky Mountains of Tennessee (W. K. Reeves, pers. comm.).

Predators of larvae include crayfish (*Cambarus bartonii*) (Tessler in Davies 1991) and brook trout (*Salvelinus fontinalis*) (Light 1983). Perlid stoneflies and corydalids consume larvae in the laboratory (Merritt et al. 1991). Females have been taken from the clutches of empidid flies of the *Hilara femorata* species group.

**Hosts and Economic Importance.** Both sexes visit flowers of plants such as pussy willow (*Salix discolor*) (Ezenwa 1974a). Females are most active from midmorning to late afternoon, with low levels of activity throughout the night (McCreadie et al. 1985). They are capable of dispersing at least 7 km (White & Morris 1985a) but probably do most of their host seeking in forested areas (Martin 1987, Martin et al. 1994). The greatest proportion of blood-fed flies is collected in the evening (McCreadie et al. 1986). Host-seeking females are especially attracted to red or black nonglossy targets (Bradbury 1972, Bradbury & Bennett 1974a, Browne & Bennett 1980) and are repelled by ultraviolet-reflecting pigments (Simmons 1985). Hosts include humans, dogs, Canadian lynxes, raccoons, American black bears, horses, elks, white-tailed deer, and cattle (Anderson & DeFoliart 1961, Merritt et al. 1978b, Addison 1980, Simmons 1985, Simmons et al. 1989, McCreadie et al. 1994c). Females readily accept blood of cows, pigs, humans, ducks, and, to a lesser extent, dogs and geese when offered with an artificial membrane system; however, in the field, pigs are not attractive (Mokry 1980a).

*Prosimulium mixtum* is the principal human biter during the spring in many parts of northeastern North America (e.g., Matheson 1938, Schreck et al. 1980, White & Morris 1985b). Ten-minute sweep samples around a human can capture an average of 790 females during mid-May in the Adirondack Mountains of New York (White 1984). The pest problem is exacerbated because about 20% of females can take a second blood meal (L. Davies 1961). Bites can be "stingingly painful" and bleed freely after females have engorged for up to 10 minutes (Anderson & DeFoliart 1961). South of Pennsylvania, however, females are only a nuisance as they swarm about the face, biting infrequently. In Delaware, the species was first recorded as a pest in 1951 (Milliron 1958).

*Prosimulium mixtum* (as *P. hirtipes*) was cited as a probable vector of *Leucocytozoon* to turkeys in Virginia (Byrd 1959). However, in New York *P. mixtum* evidently carries no infections of *Leucocytozoon smithi* and is not attracted to turkeys (Kiszewski & Cupp 1986).

**_Prosimulium rhizophorum_ Stone & Jamnback**
(Figs. 5.8, 10.258, 10.558, 10.705; Plate 5c; Map 45)

*Prosimulium rhizophorum* Stone & Jamnback 1955: 28–29, Figs. 56, 82, 109 (FMPL). Holotype* male with pupal exuviae (pinned #62358, terminalia on slide; USNM). Pennsylvania, Luzerne Co., Bear Creek Township, waterfalls, 13 May 1948 (R. L. Goulding); *rhizophum*, *rhyzophorum* (subsequent misspellings)

*Prosimulium* 'sp.' O'Kane 1926: 21 (PL)

'undescribed species of *Prosimulium*' Dimond & Hart 1953: 239 (Rhode Island record)

*Prosimulium* 'sp. 14' Jamnback 1953: 30 (L)

**Taxonomy**

O'Kane 1926 (PL); Jamnback 1953, 1956 (L); Stone & Jamnback 1955 (FMPL); Peterson & DeFoliart 1960 (F); Stone 1964 (FMPL); Snoddy 1966 (FMPL); Holbrook 1967 (L); Stone & Snoddy 1969 (FMPL); Pinkovsky 1970 (PL); Amrine 1971 (FMPL); Snoddy & Noblet 1976 (PL); Webb & Brigham 1982 (L); Adler 1983 (PL); Adler & Kim 1986 (PL).

**Morphology**

Guttman 1967a (labral fan).

**Cytology**

Rothfels & Freeman 1977, Rothfels 1979.

**Bionomics**

**Habitat.** A common species in the eastern United States, *P. rhizophorum* also should be present in southern Quebec. The immatures inhabit shallow woodland streams, often temporary and near a spring, less than 3 m in width, and ranging in temperature from about 0°C to 17°C. Pupae in their flimsy cocoons are conspicuous when attached to live vegetation and fallen leaves, but can be well concealed in mosses, leaf packs, and pits and fissures of stones. Adults rest on streamside vegetation in sunny openings (Stone & Snoddy 1969).

**Oviposition.** Unknown.

**Development.** Larvae hatch from November to February, and pupate in March and early April from Pennsylvania southward or from April through June farther north (Stone & Snoddy 1969, Adler & Kim 1986, Adler 1988a, Tessler 1991). In addition to filter feeding, larvae prey on small invertebrates such as isopods (Stone & Snoddy 1969).

**Mating.** Unknown.

**Natural Enemies.** Larvae are sometimes infected with a cytoplasmic polyhedrosis virus (Adler & Kim 1986) and the microsporidia *Caudospora simulii* (Adler et al. 2000) and *Janacekia debaisieuxi*.

**Hosts and Economic Importance.** Unknown; presumably mammalophilic.

**_Prosimulium rusticum_ Adler, Currie & Wood, New Species**
(Figs. 10.111, 10.259, 10.412, 10.670, 10.706; Map 46)

?*Prosimulium* 'new species from northwestern New Mexico and northeastern Arizona' Peterson & Kondratieff 1995: 25 (P)

**Taxonomy**

**Female.** Wing length 3.4–3.8 mm. Scutum brownish black. All hair pale golden. Mandible with 46–49 serrations; laciniae with 26–28 retrorse teeth. Sensory vesicle in lateral view about 3 times as long as wide, one half as long as and occupying about one fourth to one third of palpomere III; neck short. Legs pale yellow, except coxae and tarsi pale brown to brown; claws toothless. Terminalia (Fig. 10.111): Anal lobe in ventral view unsclerotized anteriorly. Genital fork with each arm slender and expanded into large subtriangular lateral plate directed posteromedially. Hypogynial valve elongate, somewhat pointed distally, extended to about middle of anal lobe; outer margin smoothly curved posteromedially. Spermatheca broadest proximally.

**Male.** Wing length 3.3 mm. Scutum brownish black. All hair pale golden. Genitalia (Fig. 10.259): Ventral plate in ventral view with body subrectangular, about two thirds as long as wide, slightly concave posteriorly; lip in lateral view short, broad.

**Pupa.** Length 3.9–4.5 mm. Gill of 10 (infrequently 9, 11, or 12) widely splayed filaments arising from 3 short trunks, branching 3 (4 or 5) + 4 + 3 (Fig. 10.412). Head and thorax dorsally with densely distributed, dome-shaped microtubercles. Cocoon often restricted to abdomen.

**Larva.** Length 6.6–7.3 mm. Head capsule (Fig. 10.706) pale brownish orange to brown; head spots brown (except first posterolateral head spot pale yellow), variably distinct, all but anterior group and first posterolateral head spot often absent; <10% of larvae with head spots paler than background (negative pattern). Antenna subequal in length to stalk of labral fan. Hypostoma with lateral teeth extended farthest anteriorly; sublateral teeth all extended anteriorly to about same level, but posterior to apex of median tooth (Fig. 10.670). Postgenal cleft rounded anteriorly, often subtriangular, extended about one third or less distance to hypostomal groove; subesophageal ganglion lacking pigmented sheath. Labral fan with 33–36 primary rays. Body pigment pale brown, rather uniformly distributed. Posterior proleg with 7–9 hooks in each of 66–68 rows.

**Diagnosis.** Adults are similar to those of other western species of *Prosimulium*. The number of filaments in the pupal gill varies from 9 to 12, but the typical number of 10 is nearly unique among North American species of *Prosimulium*. (The pupa of *P. unicum* sometimes has 10 filaments.) Larvae can be distinguished from those of other species of *Prosimulium* by the dark anteriormost head spots that contrast with pale, often absent posterior head spots.

**Holotype.** Female (hexamethyldisilazane, pinned) with pupal exuviae (glycerin vial below). Arizona, Coconino Co., Coconino National Forest, Oak Creek, Rt. 89, north of Sedona and Slide Rock, 34°59′N, 111°44′W, 18 April 1996, P. H. Adler & J. K. Moulton (USNM).

**Paratypes.** Arizona, Cochise Co., Dragoon Mountains, 25 April 1992, J. K. Moulton (6 pupae, 4 larvae); same data as holotype, 19 April 1997, J. K. Moulton (1♀ + exuviae, 6 pupae, 100 larvae); Pima Co., Marshall Gulch, Santa Catalina Mountains, Mt. Lemmon Hwy., 3 May 1996, J. K. Moulton (1♀ + exuviae).

**Etymology.** The species name is from the Latin adjective meaning "rural" or "of the country," referring to the rural mountainous habitat of this species.

**Cytology**

The chromosomes of larvae (4♀, 3♂) from the type locality (19 April 1997) are fixed for inversions *IIIL-2, 3*. They are chromocentric and the centromere region of chromosome I is not transformed. Otherwise, they are standard for the *Prosimulium* sequence of Basrur (1962). Males have a small band differential in the base of the IIIS arm (section 80).

**Bionomics**

**Habitat.** This new species is confirmed only from the mountains of Arizona. The immature stages are found in rocky streams less than 5 m wide.

**Oviposition.** Unknown.

**Development.** Mature larvae and pupae have been taken from early April to early May.

**Mating.** Unknown.

**Natural Enemies.** No records.

**Hosts and Economic Importance.** Unknown; presumably mammalophilic.

### *Prosimulium saltus* Stone & Jamnback

(Fig. 10.559, Map 47)

*Prosimulium saltus* Stone & Jamnback 1955: 29–30, Fig. 57 (FMP). Holotype* female with pupal exuviae (separate slides #62359; USNM). New York, Schuyler Co., cascade flowing into outlet of Cayuta Lake, ca. 1 mi. from lake on east side, 17 May 1950 (A. Stone)

[*Prosimulium* (*Prosimulium*) *rhizophorum*: Cupp & Gordon (1983) not Stone & Jamnback 1955 (misident. in part, material from Summers Co., West Virginia)]

**Taxonomy**

Stone & Jamnback 1955 (FMP), Sommerman 1958 (PL), Peterson & DeFoliart 1960 (F), Stone 1964 (FMPL).

**Morphology**

Guttman 1967a (labral fan).

**Cytology**

Rothfels & Freeman 1977, Rothfels 1979.

**Bionomics**

**Habitat.** This species is rather common in the toe of the Appalachian Mountains but scarce in the mountains northward. Small shallow mountainside streams with sandstone cobble and small cascades provide typical habitat for the immature stages. In northwestern Georgia, larvae were taken from a small subterranean stream in complete darkness, suggesting that females had flown upstream and entered a cave to oviposit (Reeves & Paysen 1999).

**Oviposition.** Unknown.

**Development.** Eggs overwinter. Pupae are present from late March to mid-May.

**Mating.** Unknown.

**Natural Enemies.** Larvae are hosts of the trichomycete fungus *Harpella melusinae* and the microsporidium *Janacekia debaisieuxi*. Larval hydropsychid caddisflies prey on larvae in the laboratory (Reeves & Paysen 1999).

**Hosts and Economic Importance.** Unknown; presumably mammalophilic.

### *Prosimulium secretum* Adler, Currie & Wood, New Species

(Figs. 4.32, 10.112, 10.260, 10.261, 10.532, 10.560, 10.671, 10.672, 10.707; Plate 5d; Map 48)

[*Prosimulium fulvum*: Currie (1986) not Coquillett 1902 (misident. in part, L; Fig. 59 based on material from unknown location but not Alberta)]

**Taxonomy**

**Female.** Wing length 4.3 mm. Body orange. All hair golden. Mandible with 40–43 serrations; lacinia with 26 or 27 retrorse teeth. Sensory vesicle about one third as long as and occupying about one fifth to one fourth of palpomere III. Legs orange, except tarsi brownish orange; claws toothless. Terminalia (Fig. 10.112): Anal lobe in ventral view unsclerotized anteriorly. Genital fork with each arm slender and expanded into large subtriangular lateral plate directed posteromedially. Hypogynial valve extended to about middle of anal lobe; outer margin smoothly curved posteromedially. Spermatheca broadest proximally.

**Male.** Wing length 3.9 mm. Body orange; tarsi brownish orange; abdomen grayish brown. All hair golden. Head small; compound eye dorsally about one fourth wider than long. Genitalia (Figs. 10.260, 10.261): Ventral plate in ventral view with body about two thirds to three fourths as long as wide, concave posteriorly, convex posterolaterally; lip in lateral view short, broad, upturned.

**Pupa.** Length 3.6–5.2 mm. Gill of 16 thin filaments, less than half as long as pupa, arising from 3 short petioles (8 + 4 + 4) (Fig. 10.560). Head small, about 35% narrower than greatest width of thorax (Fig. 10.532). Head and thorax dorsally with densely distributed, dome-shaped microtubercles; thoracic cuticle finely rippled. Orange integument of pharate adult visible through pupal cuticle. Cocoon covering most of pupa.

**Larva.** Length 6.5–8.2 mm. Head capsule (Figs. 4.32, 10.707) brownish orange, variably dark; head spots brown (except first posterolateral head spot pale yellow or absent), variably distinct, often surrounded by brown pigment. Antenna subequal in length to stalk of labral fan. Hypostoma with median and lateral teeth extended anteriorly to about same level; sublateral teeth and denticles of median tooth extended anteriorly to about same level (Figs. 10.671, 10.672). Postgenal cleft straight or slightly rounded and sometimes narrowed anteriorly, shallow, extended about one fourth distance to hypostomal groove; subesophageal ganglion sometimes lightly ensheathed with pigment. Labral fan with 21–28 primary rays. Body pigment pale brown to brown, uniformly distributed (Plate 5d). Posterior proleg with 10–13 hooks in each of 73 or 74 rows.

**Diagnosis.** Females are similar to other orange females of *Prosimulium*. Only one other North American species (*P. fulvum*) has bright orange males, but the head of *P. secretum* n. sp. is considerably smaller. Pupae are not easily distinguished from those of other western *Prosimulium* species. The pale body contrasting with the dark head capsule generally distinguishes the larvae from those of other North American species of *Prosimulium*.

**Holotype.** Pupa with pharate male (ethanol vial). California, Mono Co., Rt. 108, near Sonora Pass, small stream on north side, 38°19′N, 119°37′W, 23 June 1991, P. H. Adler (USNM).

**Paratypes.** California, Amador Co., Rt. 88, Oyster Creek Rest Area, tributary of Oyster Creek, 38°40.4′N, 120°07.0′W, 13 May 1997, P. H. Adler (52 larvae); Mariposa Co., Yosemite National Park, Rt. 41/140, Fern Spring (tributary of Merced River), 37°42.8′N, 119°39.9′W, 12 May 1997, P. H. Adler (44 larvae); same data as holotype (1♀ & 1♂ + exuviae, 4 pupae, 92 larvae), 9 June 1990 (3 pupae, 35 larvae).

**Etymology.** The species name is from the Latin noun meaning "something hidden," in reference to the small, sometimes obscure streams in which the immature stages develop.

**Remarks.** We provide illustrations of the larval head capsule and hypostoma, pupal gill, and male genitalia for material from both the type locality (Figs. 4.32, 10.260, 10.532, 10.560a, 10.671) and Vancouver Island, British Columbia (Figs. 10.261, 10.560b, 10.672, 10.707). The illustrations of the female terminalia (Fig. 10.112) and the larva (Plate 5d) were made from specimens collected at the type locality. Populations in these 2 areas might represent distinct species. (See Cytology section below.)

**Cytology**

Our chromosomal material suggests that the species morphologically characterized as *P. secretum* n. sp. consists of 2 species. Populations in California are homogeneous. All arms carry the standard sequence of Basrur (1962) except IIIL, which has *IIIL-2, 3*. A chromocenter is absent and the centromere region of chromosome I is large and transformed. Males have a rearrangement causing a lack of pairing in the centromere region of chromosome I, and are typically heterozygous for expression of the nucleolar organizer. A possible second species is represented by 2 populations on Vancouver Island, British Columbia. Our partial analysis of the chromosomes of one of these populations (Englishman River Falls Provincial Park, 4 June 1991) shows that the centromere region of chromosome I is transformed and a chromocenter is absent. IIL has a homozygous rearrangement that inverts the parabalbiani and is very similar, if not identical, to IIL-1 of *P. doveri*. IS carries a floating inversion similar to IS-1 of Basrur (1962), but poor chromosomal quality does not permit a more definitive resolution of IS or other arms. Until material from intervening sites can be examined, we prefer to recognize a single species, particularly in light of the morphological similarity between populations in California and British Columbia.

**Bionomics**

**Habitat.** The Sierra Nevada of California and British Columbia's Vancouver Island are home to this species. The immature stages are found in cool flows less than a meter wide, often with those of *P. fulvum*.

**Oviposition.** Unknown.

**Development.** Eggs overwinter, and larvae and pupae have been taken from mid-May to mid-July. When *P. secretum* is present in the same streams with *P. fulvum*, it is about a week later in development.

**Mating.** Unknown.

**Natural Enemies.** We have found unidentified mermithid nematodes and the chytrid fungus *Coelomycidium simulii* in larvae.

**Hosts and Economic Importance.** Unknown; presumably mammalophilic.

### *Prosimulium shewelli* Peterson & DeFoliart

(Figs. 10.113, 10.262, 10.263, 10.561, 10.673, 10.708; Map 49)

*Prosimulium shewelli* Peterson & DeFoliart 1960: 96–100 (FMPL). Holotype* female with pupal exuviae (alcohol vial, slides: 1 with head and mouthparts, 1 with terminalia; USNM). Wyoming, Teton Co., 7 mi. north of Leeks Lodge, small stream crossing Rt. 89–287, 16 June (pupa), 20 June (adult) 1958 (G. R. DeFoliart)

*Prosimulium* (*Prosimulium*) *woodorum* Peterson 1970b: 130–133 (FMPL). Holotype* male (pinned #9604, larval head capsule, pupal exuviae, and terminalia in glycerin vial below; CNC). British Columbia, 12.8 mi. west of Kinnaird, 14 June 1964 (D. M. & G. C. Wood). New Synonym

*Prosimulium* (*Prosimulium*) *opleri* Peterson & Kondratieff 1995 ("1994"): 27–28, Fig. 118 (M). Holotype male (alcohol vial; putatively USNM but presumably lost). Colorado, Grand Co., Rocky Mountain National Park, Lake Irene, 3231 m elev., mercury-vapor light trap, 3 July 1991 (P. A. Opler). New Synonym

**Taxonomy**

Peterson 1960a (FMP), 1970b (FMPL); Peterson & DeFoliart 1960 (FMPL); Peterson & Kondratieff 1995 (M).

The structure, habitat, and chromosomes of *P. woodorum* and *P. shewelli* are so similar that we prefer to recognize a single species, particularly given the distributional gap of more than 800 km that separates *P. woodorum* from the nearest known population of *P. shewelli*. Figure 10.263 presents the genitalia of a male from the type series of *P. woodorum*, whereas Fig. 10.262 shows the genitalia of a specimen taken in the high Sierra Nevada of California.

The only known male of *P. opleri* is so similar to that of *P. shewelli* that synonymy seems warranted. However, we were unable to locate the single known specimen (holotype) of *P. opleri* either in the reported depository (USNM) or through communication with both authors of the name.

**Cytology**

Rothfels 1979.

Chromosomes of larvae (1♀, 2♂) that we examined from Summit Co., Utah (29 May 1989), have the following characteristics. The centromere region of chromosome I (CI) is transformed, a chromocenter is present, IIIL has multiple rearrangements that place the "shield" marker in reverse orientation and distal to the "triad" marker, sex chromosomes are undifferentiated, and all other arms carry the standard banding sequence of Basrur (1962). Larvae from the Sierra Nevada of California are also chromocentric but their chromosomes have not been examined in detail. Chromosomes of larvae from the type locality of *P. woodorum* in British Columbia have CI transformed, *IIIL-2* and *IIIL-3* fixed, 2 additional fixed inversions (breakpoints unspecified) in IIIL, no chromocenter, and undifferentiated sex chromosomes (Rothfels 1979). Whether or not the presence of a chromocenter justifies recognition of a separate species cannot be decided without studying intervening populations; future study also should include a comparison of the IIIL arm.

**Bionomics**

**Habitat.** *Prosimulium shewelli* occupies high elevations in the central Rocky Mountains and Sierra Nevada. Larvae and pupae are found in cool shallow streams less than a meter wide, running through forests and along roadsides.

**Oviposition.** Unknown.

**Development.** Eggs overwinter. Pupae are present from late May through June. We have observed larvae feeding on ceratopogonid larvae.

**Mating.** Unknown.

**Natural Enemies.** The microsporidium *Caudospora alaskensis* has been found in larvae (Adler et al. 2000). We have recovered unidentified mermithid nematodes from larvae.

**Hosts and Economic Importance.** Unknown; presumably mammalophilic.

### *Prosimulium transbrachium* Adler & Kim

(Figs. 5.8, 10.264; Map 50)

*Prosimulium transbrachium* Adler & Kim 1985: 45–47 (FMPLC). Holotype* final-instar female larva (alcohol vial, slide and photographic negatives of chromosomes; USNM). Pennsylvania, Centre Co., State College, Slab Cabin Run, 40°46′25″N, 77°50′02″W, 14 April 1983 (P. H. Adler)

*Prosimulium* (*Prosimulium*) 'sp. X' Adler 1983: 183–185 (PLC)

**Taxonomy**

Adler 1983 (PL); Adler & Kim 1985 (FMPL), 1986 (PL).

**Cytology**

Rothfels in Adler 1983; Rothfels & Freeman 1983; Adler & Kim 1985, 1986.

**Bionomics**

**Habitat.** Rare and extremely local, *P. transbrachium* has been found in only 5 streams in Pennsylvania and Virginia. These are productive unshaded streams flowing over the valley floors of the Appalachian Mountains. The streams are underlain by limestone or dolomite, warm rapidly in the spring, and are 1–12 m wide. Most flow year-round, though some are impermanent, the dry streambeds harboring eggs in diapause for up to 6 months, awaiting the winter flooding (Adler & Kim 1986). Larvae and pupae have been taken at water temperatures from just above freezing to nearly 20°C. Larvae cling to leaves, rocks, and trailing vegetation and often pupate in the axils of submerged grasses.

**Oviposition.** Unknown.

**Development.** Larvae begin hatching during the winter and mature in early March in Virginia. Larvae and pupae disappear by about mid-May in Pennsylvania (Adler & Kim 1986).

**Mating.** Unknown.

**Natural Enemies.** No records.

**Hosts and Economic Importance.** Unknown; presumably mammalophilic.

### *Prosimulium travisi* Stone

(Figs. 4.35, 10.114, 10.265, 10.525, 10.562, 10.674, 10.709; Plate 5e; Map 51)

*Prosimulium travisi* Stone 1952: 76–77 (FMP). Holotype* female (pinned #61188, pupal exuviae on slide; USNM). Alaska, Anchorage, second stream on Ski Run Road, 30 September 1948 (K. M. Sommerman & L. H. Dover)

[*Prosimulium esselbaughi*: Elgmork & Saether (1970) not Sommerman 1964 (misident.)]

[*Prosimulium esselbaughi*: Saether (1970) not Sommerman 1964 (misident.)]

[*Prosimulium hirtipes*: Bushnell et al. (1987) not Fries 1824 (misident.)]

**Taxonomy**

Stone 1952 (FMP); Wirth & Stone 1956 (FMP); Syme 1957 (F palp); Sommerman 1958, 1962a (FMPL); Syme & Davies 1958 (F palp); Peterson 1960a (FMP), 1970b (FMPL); Peterson & DeFoliart 1960 (FMP); Abdelnur 1966, 1968 (FMPL); Corredor 1975 (FM); Currie 1986 (PL); Peterson & Kondratieff 1995 (FMPL).

**Morphology**

Chance 1969, 1970a (L mouthparts); Craig 1974 (L head); Craig & Chance 1982 (L head); Currie & Craig 1988 (L head).

**Cytology**

Basrur 1962, Rothfels 1979.

The chromosomes of larvae from Rocky Mountain National Park, Colorado, differ from those of Alaskan larvae described by Basrur (1962) by lacking inversion IS-3 and having instead a similar fixed inversion with breakpoints in sections 2c and 10c. Further study might show that *P. travisi* consists of more than 1 species.

**Bionomics**

**Habitat.** *Prosimulium travisi* is common in the mountains from Alaska through the Sierra Nevada of California and the Rocky Mountains to southern Arizona and New Mexico. Immature forms inhabit cool rocky, often low-productivity streams, generally less than a meter wide, that flow through forested areas or above timberline.

**Oviposition.** Unknown.

**Development.** Larvae in Alberta begin hatching in May and adults emerge from late July through September (Currie 1986). In the Sierra Nevada, adults are on the wing by early June, but at high elevations in the Sierra Nevada and Rocky Mountains, larvae and pupae can be found into August and early September. Larvae are filter feeders but also prey on chironomid larvae (Currie & Craig 1988).

**Mating.** Unknown.

**Natural Enemies.** Larvae are attacked by the microsporidium *Caudospora alaskensis* (Adler et al. 2000), the chytrid fungus *Coelomycidium simulii*, an unidentified mermithid nematode, and a possible new species of microsporidium with subspherical spores from near Fremont Pass, Lake Co., Colorado. Larval empidid flies (*Oreogeton*) prey on larvae (Sommerman 1962b). We have observed linyphiid spiders (*Erigone* probably *dentosa*), common in dry pockets beneath rocks in streams of the Sierra Nevada, preying on teneral adults.

**Hosts and Economic Importance.** Specific feeding habits are unknown, but females probably feed mainly on mammals. We cannot confirm records (Williams et al. 1980) of females attracted to blue grouse.

### *Prosimulium macropyga* Species Group

**Diagnosis.** Female: Lacinia typically without retrorse teeth; mandible typically without serrations. Spermatheca small, about one third as long as stem of genital fork. Male: Gonostylus with 2 (rarely 3) apical spinules. Pupa: Gill of 13 or 14 (sometimes 15 or 16) filaments. Larva: Mandibular phragma reaching or nearly reaching posterolateral corner of hypostoma. Abdomen rather abruptly expanded at segment V. Chromosomes: Diploid or triploid.

**Overview.** This species group, first recognized by Rubtsov (1956), also has been referred to as the *P. ursinum* species group (Peterson 1970b). Two of the world's 6 nominal species in the group live in North America. Several of the species are allotriploids (Rothfels 1979) and often live in the same streams as their diploid progenitors. All species are high-latitude life forms infrequently found south of the 60th parallel. Consequently, they have a number of features, such as elimination of blood feeding, that reflect adaptation to far northern environments.

### *Prosimulium neomacropyga* Peterson

(Figs. 10.266, 10.563, 10.710; Plate 5f; Map 52)

*Prosimulium* (*Prosimulium*) *neomacropyga* Peterson 1970b: 134–139, Figs. 26, 57, 75, 123, 150 (FMPL). Holotype* male (pinned #9601, pupal exuviae dry in microvial below; CNC). Alaska, Cape Thompson, emerged 30 July 1961 (R. Madge)

*Prosimulium* 'n. sp.' Bushnell et al. 1987: 506, 508 (Colorado record)

*Prosimulium* 'n. sp. B' Bushnell et al. 1987: 508 (Colorado record)

*Prosimulium* (*Prosimulium*) 'sp. near *frohnei*' Bushnell et al. 1987: 506 (Colorado record)

*Piezosimulium jeanninae* Peterson 1989: 318–330 (FMP). Holotype* male with fragments of pupal exuviae (alcohol vial; USNM). Colorado, Boulder Co., North Boulder Creek (originating from glacier on Niwot Ridge), 3600 m elev., 11 September 1981 (J. Bushnell). New Synonym

'near *Prosimulium neomacropyga*' Peterson 1989: 327 (Colorado record)

*Prosimulium* (*Prosimulium*) *wui* Peterson & Kondratieff 1995 ("1994"): 28–32, Figs. 134–143 (FMPL). Holotype male (pinned, associated with pupal exuviae; putatively USNM but presumably lost). Colorado, Boulder Co., North Boulder Creek between Green Lakes 5 and 4, 3566 m elev., 23 August 1988 (B. V. Peterson & S. K. Wu). New Synonym

[*Prosimulium ursinum*: Stone (1952) not Edwards 1935b (misident. in part, M)]

[*Prosimulium ursinum*: Sommerman (1953) not Edwards 1935b (misident.)]

[*Prosimulium ursinum*: Syme (1957) not Edwards 1935b (misident.)]

[*Prosimulium ursinum*: Syme & Davies (1958) not Edwards 1935b (misident.)]

[*Prosimulium ursinum*: Peterson & DeFoliart (1960) not Edwards 1935b (misident. in part, M)]

[*Prosimulium ursinum*: Madahar (1967) not Edwards 1935b (misident. in part, diploids)]

[*Prosimulium ursinum*: Elgmork & Saether (1970) not Edwards 1935b (misident.)]

[*Prosimulium ursinum*: Saether (1970) not Edwards 1935b (misident.)]

[*Prosimulium ursinum*: Madahar (1973) not Edwards 1935b (misident. in part, diploids)]

[*Prosimulium* (*Prosimulium*) *ursinum*: Bushnell et al. (1987) not Edwards 1935b (misident.)]

**Taxonomy**

Sommerman 1953 (L); Syme 1957 (FM); Syme & Davies 1958 (FM); Peterson 1970b (FMPL), 1989 (FMP), 1996 (MP); Peterson & Kondratieff 1995 (FMPL).

Our study of the holotype of *P. jeanninae*, plus inspection of material that we collected at the type locality in 1997, confirms our original supposition (p. 92 in Crosskey & Howard 1997) that *P. jeanninae* is not only a typical member of the *P. macropyga* species group but also conspecific with *P. neomacropyga*. We, therefore, formally synonymize *jeanninae* with *neomacropyga*. The unusual reproductive structures of the holotype of *P. jeanninae*, on which Peterson's (1989) description is based and which are represented accurately in the illustrations that accompany the description, are the result of a developmental abnormality, probably caused by the species of mermithid nematode that is common at the type locality. The so-called sperm pump appears to be a grossly disfigured spermatheca. The "sclerotized, setose plate" includes a portion of the genital fork.

The name *wui* also is synonymous with *neomacropyga*, based on our collection of a large series of material from the type locality of *P. wui*, and the lack of convincing structural and chromosomal differences between populations from or near the respective type localities. The holotype, however, could not be located in its reported depository (USNM) or through communication with either of the 2 authors of the name.

*Prosimulium neomacropyga* probably also occurs in the Palearctic Region and, therefore, might have an older applicable name (Currie 1997a). In particular, the relationship of *P. neomacropyga* to *P. macropyga* (type locality: Finland) needs to be resolved.

**Cytology**

Madahar 1967, 1973; Rothfels 1979.

The chromosomal banding sequence of larvae (10♀, 10♂) that we examined from the type locality (22 August 1997) of *P. jeanninae* and *P. wui* is identical to that of *P. neomacropyga*. All male larvae, however, have an inversion in the base of IL, coupled with heterozygous expression of the nucleolar organizer, compared with undifferentiated sex chromosomes of populations of *P. neomacropyga* from Alaska and the Yukon. We consider larvae from Colorado to represent a Y-chromosome polymorphism. Two male larvae from the Beartooth Mountains (Wyoming, Park Co., 15.1 km northeast of Beartooth Lake, 18 July 1994) have chromosomes identical to those of *P. neomacropyga*, including undifferentiated sex chromosomes.

**Bionomics**

**Habitat.** The distribution of *P. neomacropyga* is disjunct, although more collecting along the Rocky Mountains might reveal intervening populations. Northern populations are found little beyond the ice-free margins of Alaska and the Yukon. Southern populations occur in treeless alpine meadows above 3000 m in Wyoming and Colorado. The immature forms are confined largely to the headwaters of cold clear streams about a meter wide and originating from tundra, although at 1 site in Colorado, they have been found at a productive lake outlet. Larvae and pupae often affix themselves to the undersides of stones.

**Oviposition.** Females are obligately autogenous. Oviposition behavior is unknown.

**Development.** The winter is spent as eggs. Larvae can be found from June to September and pupae from July to September. Larval guts sometimes contain small chironomid larvae, suggesting that the larvae might be facultative predators.

**Mating.** Unknown.

**Natural Enemies.** Larvae occasionally harbor the microsporidium *Caudospora alaskensis* (Adler et al. 2000), the chytrid fungus *Coelomycidium simulii*, and the trichomycete fungus *Harpella melusinae*. At the type locality of *P. jeanninae*, 29 (23%) of a sample of 128 larvae (22 August 1997) were patently infected with an unidentified mermithid nematode, presumably the same species that caused the abnormalities in the holotype.

**Host and Economic Importance.** The weakly sclerotized mandibles and laciniae are powerless to pierce flesh.

### *Prosimulium ursinum* (Edwards)

(Figs. 4.30, 10.34, 10.53, 10.77, 10.115, 10.564, 10.625; Map 53)

*Simulium* (*Prosimulium*) *ursinum* Edwards 1935b: 535–536 (PL). Lectotype* larva (specimen figured by Edwards [1935b] designated as lectotype by Peterson 1970b: 143; 2 dissected heads and mouthparts that cannot be associated are on a slide; BMNH). Norway, Bjørnøya (Bear Island), Salmon River, 9 August 1932 (G. C. L. Bertram)

*Simulium* (*Prosimulium*) *ursinum* Edwards 1935a: 473. *Nomen nudum*

*Simulium* (*Prosimulium*) *browni* Twinn 1936a: 113 (FP). Holotype* female (pinned #4123, terminalia on slide #P.25; CNC). Nunavut (as Northwest Territories), Baffin Island, Lake Harbour, 10 August 1935 (W. J. Brown)

*Prosimulium* 'O' Rothfels 1956: 113–122 (C)

*Prosimulium* 'sp.' Merritt et al. 1978b: 204 (P)

[*Prosimulium macropyga*: Basrur & Rothfels (1959) not Lundström 1911 (misident.)]

**Taxonomy**

Edwards 1935b (PL); Twinn 1936a (FP); Davies 1954 (E); Rubtsov 1956, 1961a (FPL); Peterson & DeFoliart 1960 (FP); Carlsson 1962 (FPL); Peterson 1970b, 1977b, 1981, 1996 (FPL); Smith 1970 (FPL); Jensen 1997 (PL); Merritt et al. 1978b (P); Yankovsky 1999 (PL).

**Cytology**

Rothfels 1956, 1979; Carlsson 1962; Madahar 1967, 1973.

*Prosimulium ursinum* is an allotriploid species, with populations in Alaska and Nunavut apparently having different origins but both with *P. neomacropyga* as 1 parent (Madahar 1973, Rothfels 1979), suggesting a possible species complex of allotriploids. Additional study is required to clarify the status of these populations, as well as others throughout the range.

**Bionomics**

**Habitat.** The distribution of this large Holarctic species stretches across northern Alaska and Canada to Greenland, Iceland, and Fennoscandia. The species is the world's most northern black fly, reaching a latitude of 74°30′N (Bjørnøya [Bear Island], Norway). The immature stages occur in cold upland streams less than 3 m wide.

**Oviposition.** Females exhibit obligate autogeny, with about 30–150 large eggs maturing as early as the pupal stage (Davies 1954, Carlsson 1962, Downes 1965). Oviposition in nature has not been observed, but females will lay eggs on moist filter paper in vials (Wood & Davies 1966). Pharate females in the harsh environment of northern Norway sometimes deteriorate within the pupae, releasing fully developed, viable eggs (Carlsson 1962).

**Development.** Eggs overwinter. They can survive desiccation (Carlsson 1962). Throughout much of Canada, pupae begin to appear in late July, and adults emerge from late July into September. Adults in Labrador are active from late June to late August (Hocking & Richards 1952). Suspended food is sometimes so meager in the larval habitat that, in addition to filtering their food from the current, larvae also feed on deposited organic matter and ingest prey items a third or more their own size (Currie & Craig 1988).

**Mating.** Females are parthenogenetic; males do not exist.

**Natural Enemies.** Larvae are hosts of unidentified mermithid nematodes and the trichomycete fungus *Harpella melusinae*.

**Hosts and Economic Importance.** The mouthparts are not adapted for biting.

### *Prosimulium magnum* Species Group

**Diagnosis.** Female: Spermatheca wider than long. Male: Gonostylus with 3–7 (rarely 2) apical spinules. Pupa: Gill of 10 to more than 100 filaments. Larva: Abdomen gradually expanded posteriorly. Chromosomes: Fixed inversion *IIIS-1* present (Ottonen 1966) (except in *P. constrictistylum* and *P. flaviantennus*).

**Overview.** The *P. magnum* species group was recognized first by Peterson (1970b). The 14 species of this monophyletic group are endemic to North America. The 10 western species are isolated from the 4 eastern species by the monotonous relief of the Great Plains and the Prairie Provinces.

The large streamlined larvae are associated with swift waters where they pack tightly together on rocks, grasses, and twigs. Larvae of most species vary in color, with either dark brown bodies and heads or grayish bodies and brownish orange heads. All species are capable of taking blood, but only 1 or 2 are occasional pests.

### *Prosimulium caudatum* Shewell

(Figs. 10.54, 10.116, 10.267, 10.565, 10.675, 10.711; Plate 6a; Map 54)

*Prosimulium caudatum* Shewell 1959b: 688–692 (FMPL). Holotype* female (pinned #6926, pupal exuviae in glycerin vial below; CNC). British Columbia, Vancouver Island, Qualicum, Nile Creek, 12 June (pupa), 13 June (adult) 1955 (G. E. Shewell)

[*Prosimulium dicum*: Dyar & Shannon (1927) (misident. in part, 1 F from Hoodsport, Washington)]

**Taxonomy**

Shewell 1959b (FMPL), Peterson 1970b (FMPL), Corredor 1975 (FM), Speir 1975 (E).

**Cytology**

Ottonen 1964, 1966; Rothfels 1979.

**Bionomics**

**Habitat.** This species is common along the western coastal tract from southern British Columbia to southern California. The immature stages dwell in cool, shallow woodland streams, often temporary, usually less than 3 m wide, and with swift riffles and numerous pools. Larvae adhere principally to dark rocks with minimal periphyton, and pupae cluster on the upper sides of rocks.

**Oviposition.** Females emerge with immature ovaries. They deposit flat, tightly packed egg masses (2–7 mm in diameter) in the splash zone on the downstream side of dark fine-grained stones near pools (Speir 1975). Whether these masses are formed while stationed on the rocks or by repeatedly dipping to the rock from a hovering flight has not been reported.

**Development.** Eggs are laid in the spring and spend 6–10 months in diapause above water. Larvae hatch in February after the eggs become covered by rising waters of about 3°C (Speir 1975). Six larval instars complete development in slightly more than 10 weeks at a mean stream temperature of about 9°C (Speir 1975). Larvae construct a cocoon in an average of 80 minutes, and the subsequent pupal stage lasts about 5–7 days at a mean water temperature of 12°C–13°C (Speir 1975). Larval gut contents generally reflect the composition of suspended materials in the water column; approximately 50 minutes is required to clear the gut at water temperatures of 7°C–14°C (Speir 1975). Emerging adults have a sex ratio near unity or slightly skewed in favor of females (Speir 1975). Adults weigh about 25% less than final-instar larvae (Speir & Anderson 1974), and males are longer and heavier than females (Speir 1975).

**Mating.** Unknown.

**Natural Enemies.** Larvae are parasitized by mermithid nematodes of the genus *Gastromermis* (Speir 1975).

**Hosts and Economic Importance.** Females are probably mammalophilic. They will bite humans but are more prone simply to fly about the face.

### *Prosimulium constrictistylum* Peterson

(Figs. 10.117, 10.268, 10.566; Map 55)

*Prosimulium* (*Prosimulium*) *constrictistylum* Peterson 1970b: 149–153, Figs. 28, 58, 90, 116, 149 (FMPL). Holotype* male (pinned #9603, pupal exuviae in glycerin vial below; CNC). British Columbia, 1 mi. east of Osoyoos, 11 June 1964 (D. M. & G. C. Wood)

*Prosimulium* 'X' Ottonen 1964: 14, Figs. 4, 32; Ottonen 1966: 680, Fig. 3 (not Sommerman 1953 or Sommerman et al. 1955 or Adler 1983) (C)

**Taxonomy**

Peterson 1970b (FMPL).

**Cytology**

Ottonen 1964, 1966; Rothfels 1979.

**Bionomics**

**Habitat.** This species is recorded from southern British Columbia and northern Nevada. It colonizes unshaded intermittent streams and headwater flows running through canyons and deep gullies in sagebrush country of the Great Basin.

**Oviposition.** Unknown.

**Development.** At the type locality in British Columbia, all life stages are present in mid-June, suggesting that eggs overwinter. In Nevada, larvae hatch at various times during the winter, and at some sites, pupae are present as early as mid-April, while some larvae and pupae linger through June. In some Nevada populations, we have found the larval guts stuffed with purple isotomid Collembola (*Proisotoma titusi*), which stain the peritrophic matrix purple.

**Mating.** Unknown.

**Natural Enemies.** Larvae are hosts of the trichomycete fungus *Harpella melusinae*.

**Hosts and Economic Importance.** Unknown; presumably mammalophilic.

### *Prosimulium dicentum* Dyar & Shannon

(Fig. 10.526, Map 56)

*Prosimulium dicentum* Dyar & Shannon 1927: 7, Figs. 7, 8 (F). Holotype* female (slide #28328, wings on point; USNM). California, Nevada Co., Truckee, 28 April 1921 (published date: 22 April 1921) (H. G. Dyar)

[*Prosimulium dicum*: Wirth & Stone (1956) not Dyar & Shannon 1927 (misident. in part, P)]

[*Prosimulium dicum*: Peterson & DeFoliart (1960) not Dyar & Shannon 1927 (misident.)]

**Taxonomy**

Dyar & Shannon 1927 (F), Stains 1941 (F), Stains & Knowlton 1943 (F), Wirth & Stone 1956 (P), Peterson & DeFoliart 1960 (FMP), Peterson 1970a (FMPL).

**Cytology**

The chromosomes of 2 male larvae that we examined from Trinity Co., California, are similar to those of *P. dicum* but differ by having the nucleolar organizer heterozygously expressed and the centromere region of chromosome I tightly paired.

**Molecular Systematics**

Tang et al. 1996b (mtDNA), Pruess et al. 2000 (mtDNA).

**Bionomics**

**Habitat.** Although rather common along the Pacific Coast and in sparse localities in the Sierra Nevada of California, this large species remains little known biologically. The immature stages live in small woodland streams and roadside ditches strewn with rocks.

**Oviposition.** Females might be autogenous for the first cycle of egg production (Peterson 1970a). Nothing is known of their oviposition habits.

**Development.** This species spends at least a portion of the winter as larvae. Pupation occurs from about January or earlier in southern California to late April and early May in the northern half of the range.

**Mating.** Unknown.

**Natural Enemies.** No records.

**Hosts and Economic Importance.** Unknown; presumably mammalophilic.

### *Prosimulium dicum* Dyar & Shannon

(Figs. 10.55, 10.118, 10.269, 10.527; Map 57)

*Prosimulium dicum* Dyar & Shannon 1927: 7, Figs. 5, 6 (F). Lectotype* female (designated by Peterson 1970a: 120; slide #28327; USNM). Washington, Mason Co., Hoodsport, 11 May 1924 (H. G. Dyar)

[*Prosimulium dicentum*: Hearle (1932) not Dyar & Shannon 1927 (misident. in part, few specimens)]

**Taxonomy**

Dyar & Shannon 1927 (F); Stains 1941 (F); Stains & Knowlton 1943 (F); Wirth & Stone 1956 (F); Shewell 1959b (FPL); Peterson 1970a, 1970b (FMPL); Corredor 1975 (FM); Speir 1975 (E); Currie 1986 (PL).

The thoracic cuticle of the pupa typically bears well-raised microtubercles that provide a handy diagnostic aid. However, pupae in some populations have flattened microtubercles.

**Cytology**

Ottonen 1964, 1966; Rothfels 1979.

Most males are characterized by the $Y_1$ sequence, although in some males the Y chromosome is undifferentiated (Ottonen 1966). (The $Y_2$ and $Y_3$ sequences of Ottonen [1966] pertain to *P. exigens* only.) Females are apparently indistinguishable chromosomally from those of *P. dicentum*, *P. exigens*, and *P. uinta*.

**Bionomics**

**Habitat.** This species is most common along the Pacific Coast and in the Sierra Nevada of California but also penetrates the dry interior of British Columbia, Arizona, and New Mexico. The immature stages occupy shallow woodland streams usually less than 5 m wide, with stony bottoms and numerous pools. In California, *P. dicum* is collected infrequently from the same streams as *P. exigens*. Larvae and pupae attach themselves, often in great numbers, to rocks and sticks. Fossil hypostomata, possibly of *P. dicum* and dating from about 12,500 years past, have been discovered in lake beds of southwestern British Columbia (Currie & Walker 1992).

**Oviposition.** Females emerge with immature ovaries, later ovipositing in masses on the downstream side of dark fine-grained stones in the splash zone near pools (Speir 1975).

**Development.** Eggs are deposited in the spring and spend 6–10 months in diapause, often above water (Speir 1975). In Oregon, hatching occurs in mid-February when the water temperature is about 4°C (Speir 1975). Larvae pass through 6 instars, completing development after about 10 weeks at a mean water temperature near 9°C; the pupal stage lasts 4–6 days at a mean water temperature of 12°C–13°C (Speir 1975). Larvae feed primarily on particulate matter in the water column, and they are able to clear the gut within 50 minutes at water temperatures of 7°C–14°C (Speir 1975). A weight loss of nearly 28% takes place between the final-instar larva and the adult (Speir & Anderson 1974). Emerging adults have a sex ratio near unity (Speir 1975).

**Mating.** Unknown.

**Natural Enemies.** Larvae are parasitized by mermithid nematodes of the genus *Gastromermis* (Speir 1975) and by the microsporidium *Amblyospora fibrata*. We have found the trichomycete fungus *Harpella melusinae* in the larval midgut.

**Hosts and Economic Importance.** Females are probably mammalophilic. In one case, they were collected while probing the warm hood of a brown vehicle, suggesting that large mammals serve as hosts (Currie & Adler 1986). Small numbers have been taken from around blue grouse (Williams et al. 1980).

### *Prosimulium exigens* Dyar & Shannon

(Figs. 10.270, 10.528, 10.567, 10.676, 10.712; Map 58)

*Prosimulium exigens* Dyar & Shannon 1927: 10, Figs. 3, 4, 30, 31 (FM). Lectotype* male (designated by Stone 1962a: 209; slide #28329; USNM). Idaho, Latah Co., Moscow, date and collector unknown

*Simulium* (*Eusimulium*) *hardyi* Stains & Knowlton 1940: 78–79 (M). Holotype* male (pinned, terminalia on slide; USNM). Utah, Millard Co., Kanosh Canyon, 27 May 1939 (G. F. Knowlton & F. C. Harmston). (Stains and Knowlton [1940] gave Mill Creek Canyon as the type locality and 24 June 1938 as the date of collection; these data pertain to one of the paratypes)

*Prosimulium longilobum* Peterson & DeFoliart 1960: 100–102 (F). Holotype* female (2 slides: 1 with terminalia, 1 with remainder of body; USNM). Utah, Duchesne Co., Mirror Lake, 3063 m elev., 26 July 1952 (L. T. Neilsen). New Synonym

'provisionally . . . *Prosimulium exigens*' Peterson & Kondratieff 1995: 26 (MP)

[*Prosimulium tenuicalx*: Enderlein (1925) (misident. in part, M)]

[*Prosimulium dicentum*: Hearle (1932) not Dyar & Shannon 1927 (misident. in part, most material)]

[*Prosimulium dicum*: Ottonen (1964) not Dyar & Shannon 1927 (misident. in part, material from Colorado, Utah, and Bowser, Kamloops, and Osoyoos, British Columbia; Figs. 20, 42)]

[*Prosimulium dicum*: Ottonen (1966) not Dyar & Shannon 1927 (misident. in part, material from Colorado, Utah, and Bowser, Kamloops, and Osoyoos, British Columbia; Fig. 26)]

[*Prosimulium dicum*: Rothfels (1979) not Dyar & Shannon 1927 (misident. in part, $Y_2$ and $Y_3$; p. 516)]

**Taxonomy**

Enderlein 1925 (M); Dyar & Shannon 1927 (FM); Twinn 1938 (F); Stains & Knowlton 1940 (M), 1943 (FM); Stains 1941 (FM); Wirth & Stone 1956 (FMP); Peterson 1958a (FMPL), 1959a (E), 1960a (FMP), 1970a, 1970b (FMPL), 1996 (P); Shewell 1959b (M); Peterson & DeFoliart 1960 (FMPL); Corredor 1975 (FM); Currie 1986 (PL); Peterson & Kondratieff 1995 (FMPL).

We can find no evidence that *P. longilobum* is a separate species, and consider it a small summer representative of *P. exigens*.

One of the most diagnostic characters for *P. exigens* is the smooth thoracic cuticle of the pupa (Peterson 1970b). However, the thoracic cuticle ranges from entirely smooth and shiny, especially in populations of the Rocky Mountains, to slightly wrinkled with very minute, flattened, uniformly distributed microtubercles in some of the westernmost populations. Whitish gill filaments, mentioned by Peterson and Kondratieff (1995), seem to be associated with older exuviae.

**Cytology**

Ottonen 1964, 1966; Rothfels 1979.

Most males are characterized by the $Y_2$ or $Y_3$ sequence in chromosome I, although some males apparently have an undifferentiated Y chromosome (Ottonen 1966). (The $Y_1$ sequence of Ottonen [1966] refers to *P. dicum*.) Males in some California populations (e.g., Bridalveil Creek, Yosemite National Park) have the $Y_2$ sequence plus an inversion with limits slightly broader than IS-3 of Ottonen (1966). Females are apparently indistinguishable chromosomally from those of *P. dicentum*, *P. dicum*, and *P. uinta*.

Under the name *P. exigens*, we tentatively include populations (Alder Creek and Grouse Creek, Yosemite National Park, California, 12 May 1997), probably representing a separate species, that feature a large chromocenter and, in males, heterozygous expression of the nucleolar organizer and lack of pairing on both sides of the centromere that runs through section 18 of chromosome I. Larvae, pupae, males, and females are morphologically similar to those of typical *P. exigens*.

**Bionomics**

**Habitat.** With a range extending from the Pacific Coast to the eastern edge of the Rocky Mountains and Black Hills of South Dakota, this species occupies both the wet coastal areas and the dry belt of the interior. It is one of the most common species of *Prosimulium* in the Rocky Mountains. The immature stages are found in a variety of cool streams, from those tumbling over boulders to those with slower flows, scant rocks, and ample trailing vegetation. All of these streams typically are shaded and less than 5 m wide. Larvae form clumps on sticks, rocks, and corrugations of culvert pipes. They pupate in masses, sometimes working their way under existing pupae, and form mats so thick that pupae on the bottom fail to yield adults (Peterson 1959b). Adults rest during the day in cool locations such as under bridges and in culvert pipes.

**Oviposition.** Females, which can carry at least 336 eggs, oviposit by dipping to the surface of the water while they are in upstream flight (Peterson 1959a).

**Development.** Over most of the range, larvae hatch in the fall and adults emerge in the spring, although at higher elevations, where adults fly well into the summer, larvae probably hatch in the spring. In laboratory rearings, small groups of pupae yield more successful emergence of adults than do singletons, suggesting the role of a positive thigmotaxis (Peterson 1958a).

**Mating.** A possible mating swarm has been observed about 15–30 cm above a small stream (Peterson 1959a).

**Natural Enemies.** Larvae are hosts of the microsporidium *Caudospora alaskensis* (Adler et al. 2000). We also have found larvae infected with the chytrid fungus *Coelomycidium simulii*, as well as a possible new species of microsporidium from Water Creek, Humboldt Co., Nevada, with elongate spores that are slightly narrowed in the middle. The mermithid nematode *Mesomermis paradisus* was described from larvae of *P. exigens* from California (Poinar & Hess 1979) and might be the same species that Peterson (1960b) mentioned under the name *Hydromermis* in Utah. The trichomycete fungi *Genistellospora homothallica*, *Harpella melusinae*, *Pennella arctica*, and *Smittium pennelli* inhabit the larval gut (Lichtwardt 1967, 1972, 1984). Larvae of hydropsychid and rhyacophilid caddisflies prey on the larval flies (Peterson 1960b).

**Hosts and Economic Importance.** Swarming females can be annoying to humans, sometimes entering the facial orifices but apparently not biting (Peterson 1959a).

### *Prosimulium flaviantennus* (Stains & Knowlton)

(Figs. 10.271, 10.568, 10.677, 10.713; Plate 6b; Map 59)

*Simulium* (*Eusimulium*) *flaviantennus* Stains & Knowlton 1940: 79–80 (F). Holotype* female (pinned, terminalia on slide; USNM). Utah, Cache Co., 10 July 1938 (A. T. & D. E. Hardy)

**Taxonomy**

Stains & Knowlton 1940 (F), 1943 (FM); Stains 1941 (F); Peterson 1958a, 1958b (FMPL), 1959a (E), 1960a (FMP), 1970b (FMPL); Peterson & DeFoliart 1960 (FMP); Peterson & Kondratieff 1995 (FMPL).

**Cytology**

Based on 2 larvae from each of 3 sites (Mono Co., California; Sandoval Co., New Mexico; Cache Co., Utah), chromosomes are identical with those of *P. constrictistylum* but lack a chromocenter. Sex chromosomes were undifferentiated in the small samples we examined.

**Bionomics**

**Habitat.** This species inhabits the Sierra Nevada of California and the Rocky Mountains of the United States. The immature stages are found in streams less than a meter wide, with open canopies and plenty of trailing vegetation. Pupation occurs communally, especially on rocks. Adults rest in cool shaded situations, such as culvert pipes, during the heat of the day.

**Oviposition.** Gravid females carry at least 205 eggs (Peterson 1959a). Oviposition behavior has not been reported.

**Development.** Larvae overwinter (Peterson 1958b) and reach maturity from early May to mid-July, depending on elevation. The flight period lasts from the end of May to late July. In addition to filter feeding, larvae often ingest large volumes of algal filaments.

**Mating.** Unknown.

**Natural Enemies.** No records.

**Hosts and Economic Importance.** Females fly about humans, cows, and sheep, but have not been observed biting (Peterson 1958b).

### *Prosimulium impostor* Peterson

(Figs. 10.119, 10.272, 10.569, 10.678, 10.714; Plate 6c; Map 60)

*Prosimulium* (*Prosimulium*) *impostor* Peterson 1970b: 162–165, Figs. 30, 60, 89, 118, 154 (FMPL). Holotype* male with pupal exuviae (pinned #9599; CNC). British Columbia, 11.5 mi. north of Boston Bar, 10 May 1964 (D. M. & G. C. Wood)

*Prosimulium* 'Y' Ottonen 1964: 17–19, Figs. 4, 5; 1966: 681–682, Figs. 5, 21 (C)

**Taxonomy**

Peterson 1970b (FMPL).

Antennal length can be shorter or longer than the stalk of the labral fan. *Prosimulium impostor* might be a species complex, based on chromosomal and morphological variation as well as the variety of larval habitats.

**Cytology**

Ottonen 1966, Rothfels 1979.

Ottonen (1966) analyzed samples with undifferentiated sex chromosomes and with distinct X and Y sequences. Most of our samples contain males with undifferentiated Y chromosomes, although some samples also contain males with various rearrangements, particularly in IS and IL. In our samples, the banding sequence in all arms is exactly as described by Ottonen (1966). However, populations from southern Arizona lack a chromocenter, although the centromere bands of chromosomes II and III associate, weakly in some populations (e.g., Ramsey Canyon, Cochise Co.), but strongly (partial chromocenter) in others (Madera Canyon, Cochise Co.); the centromere of chromosome I does not participate in either case. In other populations (e.g., Tanque Verde Canyon, Pima Co.), all 3 centromeres associate weakly. These Arizona populations have undifferentiated sex chromosomes.

**Molecular Systematics**

Moulton 1997, 2000 (DNA sequences).

**Bionomics**

**Habitat.** The range of *P. impostor* stretches in a thin band from southern British Columbia to the Arizona border with Mexico. The species has the most southern distribution of any *Prosimulium* in North America. It breeds in a variety of cool flows, from mountain streams of the high Sierra Nevada of California to semipermanent streams of arid lowlands and open desert. These streams are typically less than a few meters wide, although some as wide as 8 m support significant populations.

**Oviposition.** Unknown.

**Development.** In lowland areas, larvae probably begin hatching in the fall, based on the presence of pupae in March. At higher elevations, where larvae pupate primarily in June, they might hatch in the spring.

**Mating.** Unknown.

**Natural Enemies.** Larvae are sometimes infected with the chytrid fungus *Coelomycidium simulii* and an unidentified microsporidium and mermithid nematode.

**Hosts and Economic Importance.** Females have been incriminated as vectors of footworm (*Onchocerca cervipedis*) in mule deer (Weinmann et al. 1973).

### *Prosimulium longirostrum* Currie, Adler & Wood, New Species

(Figs. 10.14, 10.120, 10.273, 10.533, 10.570, 10.679, 10.715; Plate 6d; Map 61)

'newly described species of black fly found in Oregon' Currie 1995: 24–25 (F)

**Taxonomy**

**Female.** Wing length 3.5–3.7 mm. Scutum dark brown. All hair pale golden. Mouthparts markedly elongate, about twice as long as clypeus. Mandible with 46–50 serrations; lacinia with 32 or 33 retrorse teeth. Sensory vesicle in lateral view about one third as long as palpomere III; neck short. Legs brown; claws toothless. Terminalia (Fig. 10.120): Genital fork with each arm short, slender, and expanded into subtriangular lateral plate directed posteromedially. Hypogynial valve elongate, extended nearly to posterior margin of anal lobe.

**Male.** Wing length 3.4 mm. Scutum dark brown. All hair pale golden, except basal fringe coppery brown proximally. Mouthparts markedly elongate, more than twice as long as clypeus (Fig. 10.14). Genitalia (Fig. 10.273): Gonostylus with 3 apical spinules. Ventral plate in ventral view broadly U shaped, with lip protruded slightly beyond posterior margin; arms thick, extended anterolaterally.

**Pupa.** Length 4.6–5.0 mm. Gill of 20–24 filaments, about half as long as pupa, arising from 3 short trunks; dorsal group of 8–10 filaments, medial and ventral groups each of 6 or 7 filaments (Fig. 10.570). Head and thorax dorsally with densely distributed, dome-shaped microtubercles. Labral sheath longer than wide (Fig. 10.533). Cocoon covering most of pupa.

**Larva.** Length 7.2–8.2 mm. Head capsule (Fig. 10.715) pale yellow; head spots brown, distinct. Antenna subequal in length to stalk of labral fan. Hypostoma with median tooth extended slightly anterior to all other teeth, which are extended to about same level (Fig. 10.679). Postgenal cleft straight anteriorly, wider than long, extended about one third distance to hypostomal groove; subesophageal ganglion sometimes ensheathed with pigment. Labral fan with 50–60 primary rays. Body pigment pale gray to gray, arranged in bands (Plate 6d). Posterior proleg with 9–13 hooks in each of 71–78 rows.

**Diagnosis.** The elongate mouthparts of the adults and the elongate labral sheath of the pupa distinguish these life stages from those of all other simuliid species except *P. uinta*. The pale grayish body and pale head capsule with dark head spots distinguish the larvae from those of other species in the *P. magnum* species group.

**Holotype.** Male (hexamethyldisilazane, pinned) with pupal exuviae (glycerin vial below). Oregon, Benton Co., ditch (76 m long) into tributary of Beaver Creek, junction of Bellfountain and Greenberry Roads, 44°27′N, 123°20′W, 12 March 1993, D. C. Currie & D. M. Wood (CNC).

**Paratypes.** Same data as holotype (15♀ & 6♂ + pupal exuviae, 34 larvae).

**Etymology.** The specific name is from Latin and refers to the strikingly elongate proboscis of both sexes.

**Cytology**

The chromosomes of larvae (10♀, 5♂) from the type locality (12 March 1993) differ from the standard sequence of Ottonen (1966) by having a chromocenter and *IIIS-1* as a fixed inversion. The centromere region of chromosome I is transformed. The Y chromosome is characterized by a minimum of 1 inversion in the basal half of IL.

**Bionomics**

**Habitat.** This species is known only from the type locality in western Oregon, where it breeds in a swift cold stream less than a meter wide with large stones. The stream runs through farmland.

**Oviposition.** Unknown.

**Development.** Larvae, pupae, and adults have been collected in mid-March.

**Mating.** Unknown.

**Natural Enemies.** No records.

**Hosts and Economic Importance.** Hosts are unknown but are probably mammals. The function of the elongate proboscis of males and females is a mystery, although it might be related to lengths of corollas of flowers visited for nectar. Recently emerged, unmated females that are confined under a vial will feed on humans and can raise a nasty itching welt.

***Prosimulium multidentatum* (Twinn)**

(Figs. 10.121, 10.274, 10.571, 10.680, 10.716; Plate 6e; Map 62)

*Simulium* (*Prosimulium*) *multidentatum* Twinn 1936a: 106, 108, Figs. 1B, 3 (FMP). Holotype* male (slide #4120; CNC). Quebec, near Hull, 3 May 1935 (C. R. Twinn)

**Taxonomy**

Twinn 1936a (FMP), Peterson & DeFoliart 1960 (P), Davies et al. 1962 (FMP), Wood et al. 1963 (L), Ottonen 1964 (L), Stone 1964 (FMPL), Holbrook 1967 (L), Peterson 1970b (FMPL), Amrine 1971 (FMPL), Merritt et al. 1978b (PL).

**Morphology**

Chance 1969, 1970a (labral fan).

**Cytology**

Rothfels 1956, 1979; Ottonen 1964, 1966; Carlsson 1966; Ottonen & Nambiar 1969; Rothfels & Nambiar 1975, 1981.

Occasional hybridization with *P. canutum* n. sp. (as 'Form 2') and with *P. albionense* (as 'Form 3') occurs (Rothfels & Nambiar 1981).

**Molecular Systematics**

Teshima 1970, 1972 (DNA hybridization); Sohn 1973 (DNA hybridization); Sohn et al. 1975 (DNA hybridization); Feraday & Leonhardt 1989 (allozymes); Brockhouse 1991 (rDNA).

**Bionomics**

**Habitat.** Populations of this large species are clustered from southeastern Canada, Maine, and New York west to Wisconsin and south to Kentucky. The immature stages are found in small streams flowing over granitic rock, as well as in productive streams and rivers running through agricultural lands. Larvae often pupate in masses.

**Oviposition.** Females can mature approximately 230 eggs autogenously in about a week at 20°C when fed sugar and water (Davies & Györkös 1990).

**Development.** In Michigan, larvae develop throughout the winter and pupate in mid-March unless the stream freezes solid, in which case they hatch later and pupate in early April (Merritt et al. 1978b). In Canada and Wisconsin, pupae appear later in April and adults emerge primarily in May. *Prosimulium multidentatum* is about 3 weeks ahead of *P. albionense* in development when the 2 species occupy the same streams (Rothfels & Nambiar 1981).

**Mating.** Unknown.

**Natural Enemies.** Larvae are parasitized by unidentified mermithid nematodes (Twinn 1936a, 1939) and the microsporidium *Caudospora simulii* (Jamnback 1970).

**Hosts and Economic Importance.** Unknown; presumably mammalophilic.

***Prosimulium uinta* Peterson & DeFoliart**

(Figs. 10.122, 10.275, 10.572, 10.681, 10.717; Map 63)

*Prosimulium uinta* Peterson & DeFoliart 1960: 91–96 (FMPL). Holotype* male with pupal exuviae (alcohol vial, terminalia on slide; USNM). Wyoming, Sublette Co., Sweeny Creek, Skyline Drive, mi. 8.4, Pinedale, 26 June (pupa), 27 June (adult) 1957 (G. R. DeFoliart)

**Taxonomy**

Peterson 1958a (FMPL), 1960a (FMP); Peterson & DeFoliart 1960 (FMPL).

**Cytology**

Based on larvae (9♀, 4♂) from Benny Creek, Apache Co., Arizona (15 May 1995), the centromere region of chromosome I is transformed, *IIIS-1* is fixed, and a chromocenter is absent; chromosomes otherwise have the standard banding sequence of Ottonen (1966). Males show a failure of pairing in sections 18 and 19 of IS, similar to that in males of *P. dicum*.

**Bionomics**

**Habitat.** Poorly known and sparsely distributed, *P. uinta* inhabits the mountains of Arizona, Utah, and Wyoming. Its immature stages have been found in shallow streams 1–2 m wide, with scattered stones and trailing grasses. Pupation often takes place in masses.

**Oviposition.** Unknown.

**Development.** The winter probably is spent as eggs. Adults emerge from mid-May through late June.

**Mating.** Unknown.

**Natural Enemies.** No records.

**Hosts and Economic Importance.** Hosts are unknown, as is the significance of the elongate proboscis of both sexes. Females are probably mammalophilic.

***Prosimulium unicum* (Twinn)**

(Figs. 10.56, 10.276, 10.413; Map 64)

*Simulium* (*Prosimulium*) *unicum* Twinn 1938: 49, Fig. 1 (F). Holotype* female (slide #4447; CNC). Utah, Morgan Co., Morgan, 6 May 1937 (G. F. Knowlton)

**Taxonomy**

Twinn 1938 (F); Stains 1941 (F); Stains & Knowlton 1943 (F); Peterson 1958a, 1960a (F).

The identity of the Arizona population as *P. unicum* was drawn to our attention by J. K. Moulton. The male (Fig. 10.276) and pupa (Fig. 10.413) are illustrated here for the first time. The pupal gill consists of 12 (infrequently 10 or 11) filaments, with a branching pattern of (2 + 2 + 2) + 3 + 3. The larva is similar to that of *P. dicum*.

**Cytology**

Chromosomes of larvae (4♀, 2♂) from Oak Creek Canyon, Arizona (19 April 1997), differ from the standard banding sequence of Ottonen (1966) by having a chromocenter, *IIIS-1* fixed, and a Y-chromosome sequence that involves unpairing in section 19 followed by an inversion that extends to about section 15; the centromere region of chromosome I is transformed.

**Bionomics**

**Habitat.** Prior to our collections in Oak Creek Canyon, Arizona, *P. unicum* was known only from the type locality in Utah, about 650 km to the north. Oak Creek is a rocky productive stream 5 m or more in width.

**Oviposition.** Unknown.

**Development.** Larvae, pupae, and adults have been collected in April. The holotype was collected in early May.

**Mating.** Unknown.

**Natural Enemies.** No records.

**Hosts and Economic Importance.** Unknown; presumably mammalophilic.

### *Prosimulium magnum* Species Complex

**Diagnosis.** Male: Ventral plate tapered, broadly rounded posteriorly. Chromosomes: Whole-arm interchange (IS + IIS, IL + IIL) and fixed inversions *IIL-12* and *IIIS-1* present (all shared with *P. multidentatum*), but lacking chromocenter (Ottonen 1966).

**Overview.** This complex consists of at least 3 species that were first recognized cytologically as forms 1, 2, and 3 (Ottonen & Nambiar 1969) but are now known as *P. magnum* s. s., *P. canutum* n. sp., and *P. albionense*, respectively. All 3 species are morphologically similar to one another. Information on the complex as a whole is presented below, followed by information on each of the 3 member species. We do not know to which members the following vernacular names and misidentifications apply:

*Prosimulium* 'sp.' Jamnback 1953: 36 (L)

*Prosimulium* 'species #1' Anderson & Dicke 1960: 398 (bionomics)

[*Prosimulium hirtipes*: Johannsen (1902, 1903b, 1934) not Fries 1824 (misident. in part, PL)]

[*Prosimulium hirtipes*: Emery (1913) not Fries 1824 (misident. in part, P)]

[*Prosimulium hirtipes*: Malloch (1914) not Fries 1824 (misident. in part, PL)]

[*Prosimulium multidentatum*: Jamnback (1953) not Twinn 1936a (misident.)]

[*Prosimulium multidentatum*: Davies & Peterson (1956) not Twinn 1936a (misident.)]

?[*Prosimulium multidentatum*: Stone (1964) not Twinn 1936a (misident. in part, M)]

**Taxonomy**

Johannsen 1902, 1903b, (PL), 1934 (P); Emery 1913 (P); Malloch 1914 (PL); Dyar & Shannon 1927 (FMP); DeFoliart 1951a (L); Jamnback 1953, 1956 (L); Stone & Jamnback 1955 (FMPL); Davies & Peterson 1956 (E); Anderson 1960 (FMPL); Peterson & DeFoliart 1960 (FMP); Davies et al. 1962 (FMP); Wood et al. 1963 (L); Ottonen 1964 (L); Stone 1964 (FMPL); Snoddy 1966 (FMPL); Holbrook 1967 (L); Stone & Snoddy 1969 (FMPL); Peterson 1970b (FMPL); Pinkovsky 1970 (FMPL); Amrine 1971 (FMPL); Snoddy & Noblet 1976 (PL); Webb & Brigham 1982 (L); Adler 1983 (PL); Adler & Kim 1986 (PL).

**Morphology**

Krafchick 1941, 1942 (F mandibular muscles); Snodgrass 1944 (F mandibular muscles); Guttman 1967a (labral fan); von Ahlefeldt 1968 (histology of parasitized and nonparasitized L); Kurtak 1973 (labral fan); Craig 1974 (L head); Evans & Adler 2000 (spermatheca).

**Cytology**

Rothfels 1956, 1979, 1980; Ottonen 1964, 1966; Carlsson 1966; Ottonen & Nambiar 1969; Rothfels & Nambiar 1981.

**Molecular Systematics**

Sohn 1973 (DNA hybridization), Sohn et al. 1975 (DNA hybridization), Feraday & Leonhardt 1989 (allozymes), Xiong & Kocher 1991 (mtDNA), Xiong 1992 (mtDNA).

**Bionomics**

**Habitat.** This species complex fairly blankets eastern North America south of the 46th parallel. The immature stages occupy diverse habitats but are especially prevalent in swift rocky rivers and forest streams, both permanent and temporary. They also are found in pastureland streams and on spillways of impoundments and lips of waterfalls. Larvae and pupae aggregate in dense clusters on rocks and twigs, at times attaining enormous densities.

**Oviposition.** Females are sometimes autogenous, producing the first batch of mature eggs without a blood meal (Moobola 1981, Davies & Györkös 1990). They oviposit by tapping their abdomens to the water while in flight, often upstream of slack water or crests of waterfalls (Davies et al. 1962, Stone & Snoddy 1969). Gravid females carry at least 250 eggs (Davies et al. 1962).

**Development.** Larvae begin hatching from early December to mid-February (Stone & Snoddy 1969, Peterson 1970b, Bruder & Crans 1979, Doisy et al. 1986) or as late as March and April in Wisconsin (Anderson & Dicke 1960). They begin pupating from February in the southern states to April and early May farther north. The average period for larval development is about 4–5 weeks in Wisconsin (Anderson & Dicke 1960) to almost 19 weeks (1393 degree-days above 0°C) in New Jersey (Bruder & Crans 1979). Members of this species complex are among the last of the eastern *Prosimulium* to occupy the streams, and from Pennsylvania northward they often are found well into June and early July. Early-instar larvae have been reared to adults (Wood & Davies 1966). Larvae can ingest large quantities of leaf fragments (Kurtak 1979). They are more efficient at ingesting particles 5–250 µm in diameter at 30 cm/sec than at higher velocities (Kurtak 1978).

**Mating.** Unknown.

**Natural Enemies.** Mermithid nematodes, mostly unidentified though including *Mesomermis flumenalis*, are common in larvae (Anderson & DeFoliart 1962 [footnote, p. 832 of R. Gordon 1984], Phelps & DeFoliart 1964, Reilly 1975, Bruder & Crans 1979). As many as 14 small juveniles of *Mesomermis* sp. have been found in a single larva (von Ahlefeldt 1968). Larvae infected with the chytrid fungus *Coelomycidium simulii* are less frequent (Jamnback 1973a, Adler & Kim 1986). Larvae are also hosts of the microsporidia *Caudospora simulii* and *Caudospora pennsylvanica*, of which the latter was first described from an unknown member of the *P. magnum* species complex (Beaudoin & Wills 1965, Jamnback 1970). Of these 2 microsporidia, we have found only *Caudospora simulii* in members of the *P. magnum* species complex. The absence of alae was described by Beaudoin and Wills (1965) as a diagnostic feature of *C. pennsylvanica*; however, our preparations typically had spores with and without lateral alae. The specific distinctness of *C. pennsylvanica*, vis-à-vis *Caudospora simulii*, requires further study. Males have been found in spider webs (Davies & Peterson 1956).

**Hosts and Economic Importance.** Host-seeking females are capable of dispersing at least 3 km from their natal waters (White & Morris 1985a). They feed on humans, horses, mules, cows, and sheep, sometimes reaching pest status on cattle as they feed inside the ears (Snow et al. 1958b, Anderson & DeFoliart 1961, Stone & Snoddy 1969). Serological evidence suggests that they also feed on raccoons and birds (Simmons et al. 1989). They have been netted around turkeys (Kiszewski & Cupp 1986). Females are attracted to humans (White & Morris 1985b) and the bites can raise welts, but the entire species complex is best considered no more than a minor pest of humans.

### *Prosimulium albionense* Rothfels
(Map 65)

*Prosimulium albionense* Rothfels 1956: 119–121 (C). Revalidated

'species . . . related to *Prosimulium fuscum* and *P. mixtum*' Davies et al. 1962: 81 (in part, Ontario record)

*Prosimulium* 'sp. (nr. *magnum*)' Phelps & DeFoliart 1964: 71 (Wisconsin record)

*Prosimulium magnum* 'Form 3' Ottonen & Nambiar 1969: 944 (C)

*Prosimulium magnum* '$Y_cX_o$ hybrid' Ottonen & Nambiar 1969: 945 (C)

*Prosimulium magnum* 'intermediate' Ottonen & Nambiar 1969: 946 (C); Rothfels & Nambiar 1981: 675 (C)

[*Prosimulium magnum*: Phelps (1962) not Dyar & Shannon 1927 (misident.)]

[*Prosimulium magnum*: Ottonen (1964, 1966) not Dyar & Shannon 1927 (misident. in part, populations with differentiated sex chromosomes)]

[*Prosimulium magnum*: Phelps & DeFoliart (1964) not Dyar & Shannon 1927 (misident.)]

[*Prosimulium magnum*: Rothfels (1979) not Dyar & Shannon 1927 (misident. in part, populations with differentiated sex chromosomes)]

**Taxonomy**

The name *albionense* was used originally by Rothfels (1956) in his classic paper on sibling species. Peterson (1965b) considered the name a *nomen nudum*. We maintain, however, that the name is available from Rothfels (1956) because (1) the drawing in Rothfels's Fig. 9 (even though composite) provides a visual characterization of *P. albionense*; (2) the 2 right-hand photographs in his Fig. 8 differentiate the taxon *P. albionense* from *P. multidentatum* by virtue of the centromeric differences (the centromere of *P. multidentatum* being thicker because it is united in a chromocenter, as illustrated in Rothfels's Fig. 6); and (3) descriptive words are provided about *P. albionense*. Our decision agrees with the implicit listing of the name as valid by Crosskey (1988) and Crosskey and Howard (1997). The species can be distinguished only on the basis of its chromosomes, as originally presented by Rothfels (1956). Our nomenclatural decision provides the simplest course of action and gives fitting attribution to the individual who first recognized the species. No type specimen exists.

**Cytology**

Rothfels 1956, 1979; Ottonen 1964, 1966; Ottonen & Nambiar 1969; Rothfels & Nambiar 1981.

Under this taxon, we include all populations with sex determination based on chromosome III. Four Y chromosomes have been described (Ottonen & Nambiar 1969), and we have seen an additional Y in material from Arkansas that involves a small rearrangement in the centromere region of chromosome III. Under the name *P. albionense*, we tentatively include populations described by Ottonen and Nambiar (1969) as "intermediate" or "exceptional hybrids," which have an undifferentiated X chromosome coupled with the $Y_c$ chromosome. The $X_oY_c$ segregate might represent a good species, as suggested by its wide distribution, occurrence in nearly pure populations, and tendency for ectopic pairing. *Prosimulium albionense* has hybridized with *P. magnum* and especially with *P. multidentatum* (Rothfels & Nambiar 1981).

**Bionomics**

**Habitat.** This species is common in forests from southern Ontario and Wisconsin south to Alabama, Arkansas, and Georgia. Its immature stages reside in rocky streams and rivers that are perhaps less productive than those in which other members of the complex are found. Adults often rest during the day beneath overhanging rocks near streams inhabited by the immature stages.

**Oviposition.** Unknown.

**Development.** Larvae are present during March and April in the southeastern states and during April and May farther north.

**Mating.** Unknown.

**Natural Enemies.** Larvae are hosts of the chytrid fungus *Coelomycidium simulii* and an unidentified mermithid nematode.

**Hosts and Economic Importance.** Unknown; presumably mammalophilic.

### *Prosimulium canutum* Adler, Currie & Wood, New Species
(Map 66)

*Prosimulium magnum* 'Form 2' Ottonen & Nambiar 1969: 944 (C)

[*Prosimulium magnum*: Ottonen (1964) not Dyar & Shannon 1927 (misident. in part, collections from Mecklenburg [in part], Taughannock [in part], Bad River, Pulcifer)]
[*Prosimulium magnum*: Ottonen (1966) not Dyar & Shannon 1927 (misident. in part, sites 9, 10, 13, 39, 40)]
[*Prosimulium multidentatum*: Crans & McCuiston (1970a) not Twinn 1936a (misident.)]
[*Prosimulium multidentatum*: Crans (1996) not Twinn 1936a (misident.)]

**Taxonomy**

**Female.** Unknown.

**Male.** Unknown.

**Pupa.** Unknown.

**Larva.** Length 6.3–9.5 mm. Head capsule pale yellowish brown to brown; head spots brown, variably distinct. Antenna shorter than stalk of labral fan by about one half length of distal article. Hypostoma with median tooth extended farthest anteriorly; remaining teeth extended progressively less anteriorly from median tooth outward. Postgenal cleft straight anteriorly, wider than long, extended about one fourth or less distance to hypostomal groove; subesophageal ganglion lacking pigmented sheath. Labral fan with 39–53 primary rays. Gill histoblast of 24–28 filaments. Body pigment pale brownish gray, rather uniformly distributed or arranged in bands. Posterior proleg with 13–15 hooks in each of 90–95 rows.

**Diagnosis.** The larval body and head capsule of *P. canutum* are paler than those of other members of the eastern *P. magnum* species group, except *P. multidentatum*. The fixed chromosomal inversion *IIS-25* (Ottonen & Nabmiar 1969) affords the most reliable means of identification.

**Holotype.** Larva (mature, ethanol vial; originally collected in acetic ethanol). Wisconsin, Shawano Co., Pulcifer, Oconto River, 44°50.2′N, 88°22.3′W, 23 April 1995, D. M. Wood (USNM).

**Paratypes.** New Jersey, Sussex Co., Delaware Water Gap National Recreation Area, Flat Brook, Old Mine Road, 41°06.4′N, 74°57.0′W, 26 March 1998, D. S. Bidlack (11 larvae). Wisconsin, Marinette Co., creek west of Nathan, 45°35′N, 87°52′W, 4 May 1996, D. M. Wood (43 larvae); same data as holotype (74 larvae).

**Etymology.** The specific name is from the Latin adjective meaning "gray", in reference to the color of the larval body.

**Remarks.** Two males in collections with the larvae of this new species resemble those of *P. multidentatum* in having 4–6 spinules on each gonostylus. Because we are uncertain of the identity of these males, we do not describe them or include them in the type series.

**Cytology**

Ottonen 1964, 1966; Ottonen & Nambiar 1969; Rothfels & Nambiar 1981.

We interpret each breakpoint of inversion IIS-25 as being 2 bands closer to the centromere than the breakpoints shown by Ottonen (1966) and Ottonen and Nambiar (1969). Infrequent hybridization occurs with *P. magnum* and *P. multidentatum* (Rothfels & Nambiar 1981).

**Bionomics**

**Habitat.** This species has been found from New York and Wisconsin south to West Virginia. The immature stages occur in wide streams and rocky productive rivers.

**Oviposition.** Unknown.

**Development.** Middle- to late-instar larvae are present from at least early April to mid-May.

**Mating.** Unknown.

**Natural Enemies.** We have larvae infected with the microsporidium *Caudospora simulii* and unidentified mermithid nematodes.

**Hosts and Economic Importance.** Unknown; presumably mammalophilic.

### *Prosimulium magnum* Dyar & Shannon

(Figs. 10.123, 10.277, 10.573, 10.682, 10.718; Plate 6f; Map 67)

*Prosimulium magnum* Dyar & Shannon 1927: 6, Figs. 1, 2, 22, 23 (FMP). Holotype* male (terminalia and pupal exuviae on slide #28326, all else missing; USNM). Virginia, Fairfax Co., Dead Run, bank of Potomac River, 12 April 1925 (R. C. Shannon)
*Eusimulium frisoni* Dyar & Shannon 1927: 18, Fig. 1E (F). Holotype* female (pinned #28725, terminalia on slide; USNM). Illinois, Union Co., Alto Pass, 8 May 1917 (collector unknown)
*Prosimulium magnum* 'Form 1' Ottonen & Nambiar 1969: 944 (C)
*Prosimulium magnum* 'B' Traoré 1992: 21 (*lapsus calami*)
[*Prosimulium hirtipes*: Emery (1913) not Fries 1824 (misident. P)]
[*Prosimulium hirtipes*: Malloch (1914) not Fries 1824 (misident. in part, PL)]

**Taxonomy**

Dyar & Shannon 1927 (FMP).

Chromosomal study of larvae collected near the type locality of *P. magnum* suggests that 'Form 1' represents true *P. magnum* (Ottonen & Nambiar 1969). Our chromosomal analysis of larvae from the type locality of *P. frisoni* indicates that the name *frisoni* applies to 'Form 1' and is, therefore, a synonym of *P. magnum*.

**Morphology**

Evans & Adler 2000 (spermatheca).

**Cytology**

Ottonen 1964, 1966 ($X_0X_0$ and $X_0Y_0$ only); Ottonen & Nambiar 1969; Rothfels 1979 ($X_0Y_0$ only); Rothfels & Nambiar 1981.

We have found occasional males heterozygous for expression of the nucleolar organizer. Hybrids with *P. albionense* and *P. canutum* n. sp. (as *P. magnum* 'Form 2') have been found (Ottonen & Nambiar 1969, Rothfels & Nambiar 1981).

**Bionomics**

**Habitat.** This species ranges from New Hampshire south to Georgia and west to Arkansas. Its immature stages are found in rocky streams and shallow rivers that are often rather productive.

**Oviposition.** Females drop their eggs into the water while in flight, often upstream of pools.

**Development.** In the southernmost states, larvae are quite large by February and begin to pupate in March, suggesting that they hatch in early winter. From Pennsylvania northward, larvae appear in March and, along with pupae, are conspicuously common in April and May.

**Mating.** Unknown.

**Natural Enemies.** The microsporidium *Caudospora simulii* and an unidentified mermithid nematode frequently infect the larvae. We have found the trichomycete fungus *Harpella melusinae* in the larval midgut. We also have larvae from Pope Co., Illinois, that are covered with a probable oomycete. They were dead but attached to the substrate when collected.

**Hosts and Economic Importance.** Specific hosts are unknown, but females probably feed primarily on mammals.

### Tribe Simuliini Newman

Simuliites Newman 1834: 379, 387–388 (as 'Natural Order' = family). Type genus: *Simulium* Latreille 1802: 426.

Nevermanniini Enderlein 1921a: 199. Type genus: *Nevermannia* Enderlein 1921a: 199

Wilhelmiini Baranov 1926: 164. Type genus: *Wilhelmia* Enderlein 1921a: 199

Ectemniinae Enderlein 1930: 81. Type genus: *Ectemnia* Enderlein 1930: 88

Stegopterninae Enderlein 1930: 81. Type genus: *Stegopterna* Enderlein 1930: 89–90

Cnesiinae Enderlein 1934a: 273. Type genus: *Cnesia* Enderlein 1934a: 273–274

Friesiini Enderlein 1936b: 117. Unavailable (post-1930 uncharacterized family-group name)

Odagmiini Enderlein 1936b: 127. Unavailable (post-1930 uncharacterized family-group name)

Austrosimuliini Smart 1945: 472. Type genus: *Austrosimulium* Tonnoir 1925: 215–230, 239–255

Cnephiini Grenier & Rageau 1960: 739. Type genus: *Cnephia* Enderlein 1921a: 199

Eusimuliini Rubtsov 1974a: 256, 275. Type genus: *Eusimulium* Roubaud 1906: 521

**Diagnosis.** Adults: Radial sector unforked, or with apical fork obscure and shorter than its petiole; costa with spiniform setae interspersed with fine setae; $R_1$ typically with stout dark setae interspersed with fine setae on apical half (except in most *Greniera*). Calcipala typically present; pedisulcus well developed or represented, at most, by shallow wrinkles. Female: Genital fork with anteriorly directed apodeme on lateral plate of each arm weakly to strongly developed. Male: Ventral plate with straplike connection to paramere arising subapically on basal arm (represented in some species by angular point on side of ventral plate). Paramere typically with 1 to numerous spines (absent from taxa such as *Greniera*). Pupa: Abdominal segment III widely divided by pleural membrane; segments IV and V without large pleurites (some species with small, rounded pleurites). Larva: Antenna with distal article variously pigmented; proximal and medial articles typically pigmented (unpigmented in *Metacnephia*). Hypostoma with paralateral teeth. Anteromedian palatal brush of simple hairs in first instar. Maxillary palpal sensilla of first instar apically situated, circularly arranged. Postocciput typically not completely enclosing cervical sclerites, leaving wide medial gap (except in some members of taxa such as *Gigantodax*, *Greniera*, and *Tlalocomyia*).

**Overview.** The tribe Simuliini includes 19 valid extant genera, of which 7 occur in North America. About 1652 described species are referred to the tribe at present, representing more than 90% of the world's total simuliid fauna. They are found in all zoogeographic regions of the world. North America has 187 species in the tribe.

### Genus *Greniera* Doby & David

*Greniera* Doby & David 1959: 763–765 (as genus). Type species: *Greniera fabri* Doby & David 1959: 763–765, by original designation

*Cnephia* 'subgenus uncertain' Stone 1964: 12–14

**Diagnosis.** Adults: Wing with basal medial cell; costa with fine pale setae, without dark spinules (except in *G. humeralis* n. sp.); radius with hair dorsobasally. Calcipala small, pointed apically; pedisulcus represented, at most, by shallow wrinkles. Female: Claws each with basal thumb-like lobe. Male: Gonostylus typically not strongly tapered, with 1 or 2 apical spinules. Paramere and aedeagal membrane without spines. Pupa: Gill of 12 or more slender filaments, typically lacking rigidity. Abdominal tergites with spine combs absent or restricted to various number of segments. Terminal spines long, slender. Cocoon shapeless, loosely woven, typically covering only abdomen. Larva: Antenna extended well beyond stalk of labral fan. Hypostoma with teeth arranged in 3 distinct groups. Labral fan large, as long as or longer than head capsule, with 40 or more primary rays. Abdominal segment IX with prominent ventral tubercles. Rectal papillae of 3 simple lobes.

**Overview.** Thirteen species are included in this Holarctic genus; 6 are found in the Nearctic Region. The phylogenetic relationships of and within the genus require deeper study. All species are univoltine. The immature stages of most species are difficult to find, often living in the tiniest, most cryptic flows. Adults are rarely encountered and little is known of their behavior. Their feeding habits are unknown, although the toothed claws of the females suggest that birds are hosts.

### *Greniera abdita* Species Group

**Diagnosis.** Adult: Radial sector unbranched apically. Antenna with 9 flagellomeres. Female: Anal lobe with anterior margin unsclerotized. Male: Pleural membrane of abdominal segments III and IV with tufts of long erect setae. Median sclerite weakly cleft apically. Pupa: Gill of 15–30 gracile threadlike filaments. Abdominal tergites without spine combs. Larva: Antenna with proximal article 2 or more times longer than medial article; distal article mottled, or with spiral or irregular, dark bands.

**Overview.** This species group, first recognized by Peterson (1981), has 3 members, all in the Nearctic Region.

### *Greniera abdita* (Peterson)

(Figs. 10.15, 10.124, 10.278, 10.414, 10.626, 10.719; Plate 7a; Map 68)

*Cnephia abdita* Peterson 1962a: 96–102 (FMPL). Holotype*

female (pinned #7525, terminalia and pupal exuviae in separate glycerin vials below; CNC). Ontario, Algonquin Provincial Park, ca. 0.25 mi. north of Hwy. 60, mi. 20 from west gate, small stream crossing Lake of Two Rivers Nature Trail, 3 June 1959 (B. V. Peterson)

**Taxonomy**

Davies et al. 1962 (FMP, Fig. 59 is *G. abditoides*); Peterson 1962a (FMPL), 1981 (FL), 1996 (FPL); Wood 1963a (FMPL); Wood et al. 1963 (L); Stone 1964 (FMPL).

The number of filaments in the pupal gill varies from 15 to 23, with 19 being most typical.

**Bionomics**

**Habitat.** This minuscule, often overlooked species is found in thin populations in northeastern North America as far south as Pennsylvania. The immature stages are usually found near sources of temporary, spring-fed streamlets often only several centimeters in width. Larvae typically are on the undersides of grasses, leaves, and debris. Pupae, weakly enveloped with silk and possessing pale gills, tend to be concealed in rolled up leaves and cracks of sticks in pools (Peterson 1962a).

**Oviposition.** Females are anautogenous (Davies et al. 1962). Oviposition habits are unknown.

**Development.** Larvae hatch from late February or early March in Pennsylvania (Tessler 1991) to April or early May in Ontario when the water is just above freezing (Davies et al. 1962). Adults are on the wing from early May well into June.

**Mating.** Unknown.

**Natural Enemies.** No records.

**Hosts and Economic Importance.** Unknown; presumably ornithophilic.

### *Greniera abditoides* (Wood)

(Fig. 10.415, Map 69)

*Cnephia abditoides* Wood 1963b: 95–96, Figs. 1–3 (FMP). Holotype* female (pinned #8195, pupal exuviae in glycerin vial below; CNC). Ontario, Muskoka District, 3 mi. south of Huntsville, roadside ditch at Gryffin Sideroad and Hwy. 11, 9 May 1962 (D. M. Wood)

[*Cnephia abdita*: Davies et al. (1962) not Peterson 1962a (misident. in part, Fig. 59 and occurrence in warmer streams)]

**Taxonomy**

Davies et al. 1962 (M); Wood 1963a (FMPL), 1963b (FMP); Wood et al. 1963 (L).

*Greniera abditoides* is structurally similar to its sister species *G. abdita* and can be difficult to separate in all life stages except the female, which has a large clypeus in the latter species (Fig. 10.15).

**Bionomics**

**Habitat.** This species is sparsely distributed across much of northern North America. Its apparent rarity is perhaps spurious, a result of its inconspicuous breeding habitats. The type locality, where it has not been seen in recent years, is a warm roadside streamlet emanating from a partially wooded swamp and running through an open field. Larvae have been found in slow-flowing areas of the stream and like those of *G. abdita*, they probably pupate in debris on the stream bottom.

**Oviposition.** Females are autogenous and emerge with mature eggs (Wood 1963b, Davies & Györkös 1990). Oviposition habits are unknown.

**Development.** Eggs probably overwinter. Larvae are present in April and May, and the adults fly until at least mid-June.

**Mating.** Unknown.

**Natural Enemies.** No records.

**Hosts and Economic Importance.** Unknown; presumably ornithophilic.

### *Greniera humeralis* Currie, Adler & Wood, New Species

(Figs. 4.19, 10.57, 10.78, 10.125, 10.279, 10.416, 10.627, 10.720; Plate 7b; Map 70)

*Greniera* 'sp.' Peterson 1981: 365, 366, 389 (FML)

*Greniera* 'undescribed' Peterson 1981: 370 (FM)

'undescribed species from western North America' Currie & Grimaldi 2000: 479 (membership in *G. abdita* species group)

**Taxonomy**

**Female.** Wing length 2.6–3.0 mm. Costa with short, stout, black spiniform setae interspersed among longer, paler hairlike setae (Fig. 10.78). Scutum dark brown. All hair pale golden, except that of basicosta, stem vein, and scutellum golden brown. Mandible with 35–40 serrations; lacinia with 25–27 retrorse teeth. Sensory vesicle in lateral view about one fourth as long as and occupying about one sixth of palpomere III. Legs brown; calcipala minute; claws each with basal thumblike lobe (Fig. 10.57). Halter brown. Terminalia (Fig. 10.125): Anal lobe in ventral view with medial three fourths unsclerotized. Genital fork with each arm slender and expanded into subquadrate lateral plate. Spermatheca reniform, smooth.

**Male.** Wing length 2.5–2.6 mm. Costa with short, stout, black spiniform setae interspersed among longer, paler hairlike setae. Scutum rather velvety, brownish black. All hair coppery brown. Legs brown. Halter brown. Genitalia (Fig. 10.279): Gonostylus in ventral view with 2 apical spinules directed medially. Ventral plate in ventral view with body nearly as wide as long; lip short. Median sclerite elongate, slightly bifurcate. Paramere subtriangular, without spines.

**Pupa.** Length 2.7–3.6 mm. Gill of 18–30 pale slender filaments arising from 2 tiny petioles (Fig. 10.416). Thorax at humeral angle projected over base of gill. Head and thorax dorsally with densely distributed clusters of dome-shaped microtubercles bearing secondary granules; thoracic trichomes absent. Ventral onchotaxy more developed than dorsal onchotaxy. Cocoon feeble, covering abdomen.

**Larva.** Length 4.6–5.9 mm. Head capsule (Fig. 10.720) pale brownish orange; head spots pale brown, often indistinct; posteromedial head spot heaviest. Antenna extended beyond stalk of labral fan by about one half length of distal article; medial article about as long as wide; proximal and medial articles dark brown; distal article with spiral brown annulations. Hypostoma with teeth in 3 prominent groups; lateral groups extended anteriorly to same level as, or beyond, median group (Fig.

10.627). Postgenal cleft a small notch, about as long as wide; subesophageal ganglion without pigmented sheath. Labral fan with 40–44 primary rays. Body pigment pale brown, uniformly distributed or weakly arranged in bands (Plate 7b); ventral tubercles about one third depth of abdomen at attachment points. Posterior proleg with 11–14 hooks in each of 77–89 rows.

**Diagnosis.** Females are unique among North American black flies in having a reniform spermatheca. The male can be recognized by the following combination of characters: costa with short stout setae interspersed among longer hairlike setae; halter entirely brown; ventral plate in ventral view narrow; and paramere without spines. The tuberculate humeral projection above each gill base of the pupa is unique among North American black flies. Larvae are distinguished from those of all other North American species by the spiral annulations of the distal antennal article and the form of the hypostomal teeth.

**Holotype.** Male (Peldri II, pinned) with pupal exuviae (glycerin vial below). California, Lake Co., Mendocino National Forest, small creek, CR 301 (Elk Mountain Road, Forest Route M1), 2.9 km north of Middle Creek Campground, 39°15′N, 122°56′W, 23 March 1990, P. H. & C. R. L. Adler (CNC).

**Paratypes.** California, same data as holotype (1♀ & 2♂ + exuviae, 4 exuviae, 82 larvae). British Columbia, 15 km south of junction Rt. 1 and Rt. 8, Spence's Bridge, 7 June 1964, G. C. & D. M. Wood (7♀ & 3♂ + exuviae).

**Etymology.** The specific name is from Latin, meaning "pertaining to the shoulder," and refers to the unique humeral projection at the base of each pupal gill.

**Cytology**

Based on larvae (3♀, 7♂) from Lake Co., California (23 March 1990), arm associations are standard, the nucleolar organizer is in the base of IIIL, and the Y chromosome is anucleolate and carries a rearrangement in IIIL near the nucleolar organizer.

**Bionomics**

**Habitat.** This new species is the most common member of the genus and is encountered frequently in wooded areas of dry sagebrush country from southern British Columbia to southern California. It is less common in forested areas of the high Sierra Nevada of California. Larvae and pupae typically are found on the undersurfaces of rocks and leaves in cool flows less than a meter wide.

**Oviposition.** Unknown.

**Development.** Larvae probably hatch from late winter to spring. In California, larvae, pupae, and adults can be found from mid-March to mid-April, and in British Columbia from April to mid-May.

**Mating.** Unknown.

**Natural Enemies.** We have an unidentified species of microsporidium with short elliptical spores from a larva collected in Lake Co., California. The trichomycete fungus *Harpella melusinae* has been found in the larval midgut.

**Hosts and Economic Importance.** Unknown; presumably ornithophilic.

### *Greniera fabri* Species Group

**Diagnosis.** Adult: Radial sector obscurely bifurcate apically. Antenna with 8 or 9 flagellomeres. Female: Anal lobe with anterior third unsclerotized or sclerotized. Male: Median sclerite deeply cleft apically. Pupa: Abdominal tergites with spine combs absent, or restricted to various number of segments. Larva: Antenna with medial article as long as or longer than proximal article; distal article typically dark (that of *G. fabri* with dark spiral bands).

**Overview.** Ten species constitute this Holarctic species group, first recognized by Currie and Grimaldi (2000). Three species are found exclusively in North America. *Greniera fabri* Doby & David (1959), the type species of the genus and namesake of the species group, occurs from France and Italy to Algeria and Morocco.

### *Greniera denaria* (Davies, Peterson & Wood)

(Figs. 4.36, 10.35, 10.58, 10.126, 10.280, 10.417, 10.628, 10.721; Plate 7c; Map 71)

*Cnephia denaria* Davies, Peterson & Wood 1962: 97–100, Figs. 19, 60 (FMP). Holotype* female (slide #7995, pupal exuviae in alcohol vial; CNC). Ontario, Renfrew Co., 1 mi. west of Renfrew, 3 May 1961 (B. V. Peterson & E. Bond)

**Taxonomy**

Davies et al. 1962 (FMP), Wood 1963a (FMPL), Wood et al. 1963 (L), Currie 1986 (PL).

*Greniera denaria* is perhaps the sister species of *G. brachiata* (Rubtsov 1961b), which is known only as pupae from Karelia. The pupae of both species have 22 filaments in each gill but differ in the branching patterns of the filaments.

**Molecular Systematics**

Moulton 1997, 2000 (DNA sequences).

**Bionomics**

**Habitat.** A little known species, *G. denaria* is sparsely distributed across Canada southward into the Rocky Mountains of Utah. The immature forms are found in shallow streams and ditches half a meter or less in width, often sporting trailing vegetation and flowing through open swampy land. Larvae cling to trailing vegetation, but pupae are difficult to find in the mud and sand of the stream bottom or attached to the undersurfaces of stones.

**Oviposition.** Unknown.

**Development.** The winter is passed as eggs. Mature larvae appear from April to mid-May. Adults are on the wing primarily during May.

**Mating.** Unknown.

**Natural Enemies.** No records.

**Hosts and Economic Importance.** Unknown; presumably ornithophilic.

### *Greniera longicornis* Currie, Adler & Wood, New Species

(Figs. 10.16, 10.17, 10.127, 10.281, 10.418, 10.629, 10.722; Plate 7d; Map 72)

*Cnephia* 'species undetermined No. 1' Pinkovsky 1976: 268–270; Pinkovsky & Butler 1978: 261 (L)

**Taxonomy**

**Female.** Wing length 3.5–3.6 mm. Scutum dull, brownish black, pollinose, with pale golden hair; scutellum with golden brown hair. Mandible with 45–53 serrations; lacinia with 33–36 retrorse teeth. Sensory vesicle in lateral view about half as long as and occupying about one third to one half of palpomere III; neck short. Antenna extended beyond lateral margin of compound eye by about two thirds its length (Fig. 10.16). Wing with radial sector bifurcate near apex; hair of basicosta and stem vein golden brown. Basal fringe and mesepimeral tuft silvery to pale golden. Legs pale golden brown; calcipala minute, pointed; claws each with basal thumblike lobe. Terminalia (Fig. 10.127): Anal lobe in ventral view unsclerotized except posterolaterally. Genital fork with each arm short, slender, and expanded into subquadrate to subrectangular lateral plate. Spermatheca slightly longer than wide, about half or less as long as stem of genital fork, with small unsclerotized area at junction with spermathecal duct.

**Male.** Wing length 3.0–3.1 mm. Scutum brownish black. All hair coppery brown. Antenna extended beyond lateral margin of compound eye by about two thirds its length (Fig. 10.17). Wing with radial sector bifurcate at apex. Legs pale golden brown. Genitalia (Fig. 10.281): Gonostylus in ventral view nearly 4 times longer than basal width, with 2 apical spinules directed anteromedially. Ventral plate in ventral view small, tapered posteriorly. Median sclerite short, broad, bifurcate apically. Paramere elongate, without spines.

**Pupa.** Length 3.8–4.1 mm. Gill of 22 or 23 slender filaments arising from 3 primary trunks (Fig. 10.418). Terminal spines divergent, directed posteriorly. Head and thorax dorsally with densely distributed, barely perceptible, dome-shaped microtubercles; trichomes unbranched. Abdomen without spine combs. Cocoon feeble.

**Larva.** Length 6.1–6.9 mm. Head capsule pale brown; head spots brown, variably distinct; medial head spots heaviest (Fig. 10.722). Antenna dark brown, extended beyond stalk of labral fan by about one third length of distal article. Hypostoma with lateral teeth broad, extended anteriorly slightly beyond slender median tooth; sublateral teeth extended anteriorly about half as far as median tooth (Fig. 10.629). Postgenal cleft a tiny notch, nearly absent; subesophageal ganglion with pigmented sheath. Labral fan with 62–72 primary rays. Body pigment brown, arranged in bands (Plate 7d); ventral tubercles about one third depth of abdomen at attachment points. Posterior proleg with 9–11 hooks in each of 65–68 rows.

**Diagnosis.** The long antennae of the adults and the long, posteriorly directed terminal spines of the pupa are unique among all North American black flies. Larvae can be distinguished from those of all other North American species by the form of the hypostomal teeth; the long, smooth brown antennae; and the shallow postgenal cleft.

**Holotype.** Male with pupal exuviae (frozen dried, pinned). Delaware, Sussex Co., Ellendale State Forest, 1 km east and 4.3 km south of junction Rt. 16 and Rt. 113, 38°46.25′N, 75°25.73′W, 20 April 1992, D. M. Wood (CNC).

**Paratypes.** Same data as holotype (1♀ + exuviae, 1 pupa, 9 larvae), 26 March 1992 (53 larvae).

**Etymology.** The specific name is from Latin and refers to the elongate antennae of both sexes.

**Bionomics**

**Habitat.** This rare species is known only from the type locality in the sandy coastal plain of southern Delaware and from a small stream in Gainesville, Florida. The stream at the type locality was 1–2 m wide and strewn with leaves and sticks. Larvae were on submerged grasses, and pupae were in leaf mats on the bottom of the stream.

**Oviposition.** Unknown.

**Development.** Mature larvae at the type locality have been found from early March to late April, suggesting staggered hatching. The earliest pupae have been found in mid-March.

**Mating.** Unknown.

**Natural Enemies.** No records.

**Hosts and Economic Importance.** Unknown; presumably ornithophilic.

### *Greniera* 'species F'
(Fig. 4.24, Map 73)

**Taxonomy**

The sole representative (female) of this species somewhat resembles the female of *G. denaria*. However, it has 9 rather than 8 flagellomeres, the anal lobe is large and unsclerotized anteroventrally, and the pigmentation of the spermatheca extends onto the spermathecal duct.

**Bionomics**

**Habitat.** The only known specimen was collected by M. J. and R. J. Mendel at an elevation of 1500 m in California, El Dorado Co., near Wolf Creek at the Union Valley Reservoir.

**Oviposition.** Unknown.

**Development.** The singular specimen was collected on 7 July 1998, which is rather late for members of the genus *Greniera*.

**Mating.** Unknown.

**Natural Enemies.** No records.

**Hosts and Economic Importance.** The single known female was collected while crawling on the collectors. Its hosts, although presumably birds, remain unknown.

### Genus *Stegopterna* Enderlein

*Stegopterna* Enderlein 1930: 89–90 (as genus). Type species: *Stegopterna richteri* Enderlein 1930: 90 (= *Melusina trigonium* Lundström 1911: 18–19), by original designation; *Stegoptora* (subsequent misspelling)

**Diagnosis.** Adults: Wing with basal medial cell; radius with length of basal section one third or more remaining distance to apex of wing, with hair dorsobasally; radial sector unbranched. Calcipala typically large and lamellate (smaller in a few species); pedisulcus represented, at most, by weak wrinkles; hind tibial spurs long, each with hyaline apex. Female: Claws toothless. Spermatheca typically with strongly raised polygonal pattern. Male: Subcosta setose ventrally. Gonostylus not strongly tapered, with 2 apical spinules. Paramere without stout spines, although distal margin can be jagged or wrinkled. Aedeagal membrane with fine spines or hair. Pupa: Gill of 10 or 12 slender filaments arranged in 2 divergent groups, with dorsal trunk bearing 6 or 7 filaments and ventral trunk bearing 4 or 5 filaments. Terminal spines long, slender.

Cocoon loosely woven, shapeless, sometimes covering only part of abdomen. Larva: Hypostoma with teeth arranged in 3 distinct groups; outside margin of lateral cluster of teeth typically sloped inward. Postgenal cleft shallow, pointed or rounded anteriorly, shaped like inverted V. Labral fan with 30–60 primary rays. Abdominal segment IX with transverse midventral bulge. Rectal papillae of 3 simple lobes. Chromosomes: Standard banding sequence as given by Madahar (1969).

**Overview.** Thirteen nominal species in the Holarctic Region constitute the genus *Stegopterna*; 8 species occur in North America. A fossil head capsule, about 6600 years old, is known from Vancouver Island (Currie & Walker 1992). Most members of the genus breed in small streams of the mountains and tundra, although *St. diplomutata* and *St. mutata* also have invaded the coastal plain. Pupae are difficult to find in the sediment next to stones, in tiny crevices of rocks, among sand grains trapped in leaf mats, under stones, or occasionally between blades of trailing grasses. The species that take blood are primarily mammalophilic. Two species do not have biting mouthparts.

***Stegopterna acra* Currie, Adler & Wood, New Species**
(Figs. 10.36, 10.44, 10.128, 10.282, 10.723;
Plate 7e; Map 74)

*Cnephia* 'W' Madahar 1969: 117, Figs. 1, 19 (C)
[*Cnephia mutata*: Rees & Peterson (1953) not Malloch 1914 (misident.)]
[*Cnephia mutata*: Peterson (1955, 1956, 1959a) not Malloch 1914 (misident.)]
[*Cnephia mutata*: Peterson (1958a) not Malloch 1914 (misident. in part, material from Utah)]
[*Cnephia* (*Stegopterna*) *mutata*: Peterson (1960a) not Malloch 1914 (misident.)]
[*Stegopterna mutata*: Peterson & Kondratieff (1995) not Malloch 1914 (misident. in part, material from Colorado)]

**Taxonomy**

**Female.** Wing length 2.7–3.1 mm. Scutum brownish black. All hair pale golden, except that of basicosta, stem vein, and scutellum golden brown. Mandible with 29–33 serrations; lacinia with 18–20 retrorse teeth. Sensory vesicle in lateral view about one fourth as long as and occupying about one sixth of palpomere III. Legs brown; calcipala about half as wide as apex of basitarsus; claws toothless (Fig. 10.36); hind tibial spurs with hyaline apex, and longer than width of tibia at point of attachment (Fig. 10.44). Terminalia (Fig. 10.128): Anal lobe in ventral view with anterior one third to one half unsclerotized. Genital fork with each arm gradually expanded into subrectangular lateral plate. Spermatheca slightly longer than wide, with small unpigmented area proximally; surface with polygonal pattern.

**Male.** Wing length 2.4–3.1 mm. Scutum velvety, black. All hair coppery brown. Legs brown; calcipala small, in lateral view about half as wide as apex of basitarsus; hind tibial spurs with hyaline apex, and longer than width of tibia at point of attachment. Genitalia (Fig. 10.282): Gonostylus in ventral view about 3 times longer than wide, with 2 apical spinules directed medially. Ventral plate in ventral view with body about two thirds as long as wide, rounded posterolaterally. Paramere elongate, closely associated with fine pale spines in aedeagal membrane. Median sclerite bifurcate apically.

**Pupa.** Length 3.6–4.1 mm. Gill of 12 thin filaments, about as long as pupa; dorsal group of 1 triplet and 2 pairs; ventral group of 1 triplet and 1 pair. Head and thorax dorsally with densely distributed, rounded microtubercles, giving surface roughened appearance; trichomes unbranched. Cocoon covering most of pupa, often with adherent debris.

**Larva.** Length 6.2–8.5 mm. Head capsule (Fig. 10.723) brownish orange to brown, often with darker flecks of brown; ecdysial line often bordered by dark pigment; head spots dark brown. Antenna brown, extended beyond stalk of labral fan by about half length of distal article. Hypostoma with teeth in 3 prominent groups, extended anteriorly to about same level; lateral groups broadest. Postgenal cleft small, subtriangular, broadly rounded or acute anteriorly; subesophageal ganglion lacking pigmented sheath. Labral fan with 49–53 primary rays. Body pigment grayish brown to brown, rather uniformly distributed (Plate 7e). Posterior proleg with 9–17 hooks in each of 67–73 rows.

**Diagnosis.** Females and males are best distinguished from other members of the genus by the form of the calcipala, which is about half as wide as the apex of the basitarsus. Pupae are not separable from those of most other species of *Stegopterna*. The brown head and body generally distinguish the larvae of this species from those of other group members. The *IILW-1* inversion (Madahar 1969) is unique.

**Holotype.** Male with pupal exuviae (frozen dried, pinned). Arizona, Greenlee Co., 40 km south of Alpine, Rt. 191, K. P. Cienega Campground, 2804 m elev., 33°34.5′N, 109°21.4′W, 10 May 1967, D. M. Wood (CNC).

**Paratypes.** Arizona, same data as holotype (12♀ & 1♂ + exuviae). California, Alpine Co., Rt. 4, 0.2 km east of Ebbetts Pass, seepage, ca. 2660 m elev., 38°32.75′N, 119°48.67′W, 24 June 1991, P. H. Adler (1 exuviae, 30 larvae); Rt. 4, Ebbetts Pass, 2600 m elev., 38°32.67′N, 119°48.73′W, 24 June 1991, P. H. Adler (1♂ + exuviae, 50 larvae); Mono Co., Saddlebag Lake Road, 1.3 km off Rt. 120, outside eastern boundary of Yosemite National Park, 37°56.86′N, 119°15.43′W, 25 June 1991, P. H. Adler (61 larvae); Tuolumne Co., Rt. 108, 0.5 km west of St. Mary's Pass Trail head, 2400–2700 m elev., 11 June 1990, P. H. Adler (23 larvae). Colorado, Grand Co., Rabbit Ears Pass, Dumont Lake, 2926 m elev., 40°23′N, 106°37′W, 30 June 1963, D. M. & G. C. Wood (51♀ & 14♂ + exuviae, 16 pupae, 36 larvae). Nevada, Humboldt Co., Bloody Run Range, Paradise Creek, headwaters, 41°13′N, 117°45′W, 1920 m elev., 23 May 1995, R. D. Gray (3 pupae, 10 larvae).

**Etymology.** The species name is from Greek, meaning "summit," in reference to the montane habitat of this species.

**Morphology**

Evans & Adler 2000 (spermatheca).

**Cytology**

Madahar 1969.

In addition to the Y chromosome described by Madahar (1969), several other Y chromosomes are found. In California, for example, males carry a complex rearrangement in chromosome II, whereas in Utah the Y chromosome is characterized by an inversion that includes the parabalbiani.

**Bionomics**

**Habitat.** This species inhabits the mountains of the western United States. Larvae and pupae develop in shallow streams from a few centimeters to a meter in width. These flows are usually cool but can reach 18°C in sunlight.

**Oviposition.** Unknown.

**Development.** Larvae begin to hatch at various times during the winter and early spring. In northern Nevada, larvae can be found by March, and in the Sierra Nevada of California, larvae generally complete development by the end of June. In Arizona, adults begin to emerge by early May.

**Mating.** Unknown.

**Natural Enemies.** Larvae are hosts of the chytrid fungus *Coelomycidium simulii* and unidentified mermithid nematodes. We have 1 larva from Tulare Co., California, that is infected with the microsporidium *Polydispyrenia simulii*. We have observed a larval dytiscid beetle (*Agabus*) consuming larvae.

**Hosts and Economic Importance.** Females bite humans infrequently (Peterson 1956, 1959a). We have encountered minor biting and nuisance problems with this species in the mountains of Colorado.

### *Stegopterna decafilis* Rubtsov

(Figs. 10.129, 10.283, 10.574, 10.724; Plate 7f; Map 75)

*Stegopterna sibirica decafilis* Rubtsov 1971a: 177–179, Fig. 6 (PL). Holotype* (mature) larva (slide #19716; ZISP). Russia, Yakutia, near Kular, Ilisty stream, 15 August 1967 (E. I. Borobets)

'Undescribed species' Corkum & Currie 1987: 207 (in part, site 101) (Alaska record)

**Taxonomy**

Rubtsov 1971a (PL), Usova & Bodrova 1979 (MP), Patrusheva 1982 (MPL).

**Cytology**

Our analyses of 2 male larvae (Alaska, Richardson Hwy., mi. 200.4, 17 July 1994) and 3 female larvae (Northwest Territories, Deline, 30 June 2001) indicate that *IS-1* and *IIIS-1* of Madahar (1969) are present; the polarity of the IS-2 sequence was not determined. Other arms have the standard banding sequence; the nucleolar organizer is in the base of IS; 1 male and all females lacked heterozygous inversions and 1 male had a subterminal heterozygous inversion in IL.

**Bionomics**

**Habitat.** This species is known from Alaska, the Northwest Territories, and the Yukon, as well as Siberia. Larvae and pupae are found mainly on stones in cold headwater streams 1–4 m wide that often are associated with mountains.

**Oviposition.** Females are obligately autogenous. Oviposition behavior is unknown.

**Development.** Winter is passed in the egg stage. Larvae hatch in early June and begin to pupate in July (Currie 1997a).

**Mating.** Unknown.

**Natural Enemies.** No records.

**Hosts and Economic Importance.** The weakly developed mouthparts of the females cannot pierce flesh.

### *Stegopterna diplomutata* Currie & Hunter

(Figs. 10.284, 10.419; Plate 8a; Map 76)

*Stegopterna diplomutata* Currie & Hunter 2003: 1–11 (FMPLC). Holotype male (pinned, pupal exuviae in glycerin vial below; CNC). Ontario, Nipissing District, Algonquin Provincial Park, Booth Lake, Booth's Rock Trail, 18 May 1992 (F. F. Hunter)

*Cnephia mutata* '2n' Basrur 1957: 4–28 (PLC)

*Cnephia mutata* 'diploid' Basrur & Rothfels 1959: 571–589 (C)

[*Simulium* (*Eusimulium*) *mutatum*: Twinn (1936a) not Malloch 1914 (misident.)]

[*Simulium venustum*: Prevost (1946) not Say 1823 (misident. in part, Fig. 16)]

[*Cnephia mutatum*: Nicholson (1949) not Malloch 1914 (misident. in part, M)]

[*Cnephia mutatum*: Davies (1950) not Malloch 1914 (misident. in part, M)]

[*Cnephia mutatum*: Nicholson & Mickel (1950) not Malloch 1914 (misident. in part, M)]

[*Cnephia* (*Mallochianella*) *mutata*: Stone & Jamnback (1955) not Malloch 1914 (misident. in part, M)]

[*Cnephia mutata*: Peterson (1958a, 1960a) not Malloch 1914 (misident. in part, M)]

[*Cnephia* (*Stegopterna*) *mutata*: Anderson (1960) not Malloch 1914 (misident. in part, M)]

[*Cnephia mutata*: Davies et al. (1962) not Malloch 1914 (misident. in part, M)]

[*Cnephia mutata*: Wood (1963a) not Malloch 1914 (misident. in part, M)]

[*Cnephia* (*Stegopterna*) *mutata*: Stone (1964) not Malloch 1914 (misident. in part, M)]

[*Cnephia* (*Stegopterna*) *mutata*: Snoddy (1966) not Malloch 1914 (misident. in part, M)]

[*Cnephia* (*Stegopterna*) *mutata*: Stone & Snoddy (1969) not Malloch 1914 (misident. in part, M)]

[*Cnephia mutata*: Pinkovsky (1970) not Malloch 1914 (misident. in part, M)]

[*Cnephia* (*Stegopterna*) *mutata*: Amrine (1971) not Malloch 1914 (misident. in part, M)]

[*Stegopterna mutata*: Wygodzinsky & Coscarón (1973) not Malloch 1914 (misident. in part, M)]

[*Cnephia* (*Stegopterna*) *mutata*: Back & Harper (1978, 1979) not Malloch 1914 (misident. in part, M and some F)]

[*Stegopterna mutata*: Adler et al. (1982) not Malloch 1914 (misident. in part, material from Sixmile Creek)]

[*Stegopterna mutata*: Singh & Smith (1985) not Malloch 1914 (misident. in part, M)]

[*Stegopterna mutata*: Peterson & Kondratieff (1995) not Malloch 1914 (misident. in part, M)]

[*Stegopterna* 'W': Moulton (2000) not Madahar 1969 (misident.)]

**Taxonomy**

Twinn 1936a (FMP); Nicholson 1949 (M); Nicholson & Mickel 1950 (M); Stone & Jamnback 1955 (M); Basrur 1957

(PL); Peterson 1958a, 1960a (M); Basrur & Rothfels 1959 (P); Anderson 1960 (M); Davies et al. 1962 (M); Wood 1963a (M); Stone 1964 (M); Snoddy 1966 (M); Stone & Snoddy 1969 (M); Amrine 1971 (M); Wygodzinsky & Coscarón 1973 (M); Peterson & Kondratieff 1995 (M); Currie & Hunter 2003 (FMPL).

**Cytology**

Basrur 1957; Basrur & Rothfels 1959; Madahar 1967, 1969; Adler & Kim 1986; Currie & Hunter 2003.

Two or more species might be included under this taxon. Y chromosomes, in addition to the one described by Basrur and Rothfels (1959), include an anucleolate Y alone or coupled with unpairing in the centromere region of chromosome I. On the Magdalen Islands of Quebec, a complex Y chromosome is characterized by lack of nucleolar expression and as many as 4 inversions in chromosome I: a somewhat small and subterminal inversion plus a large basal inversion involving the expanded centromere region in chromosome arm IS, and a large basal inversion plus a smaller midarm inversion in IL.

**Molecular Systematics**

Moulton 1997, 2000 (DNA sequences).

**Bionomics**

**Habitat.** This eastern species is fairly common from northern Quebec to Arkansas and southern Mississippi. Larvae and pupae occupy 2 rather distinct habitats: low-productivity, temporary streams less than a meter wide in forested regions, and sandy streams up to 10 m wide in the piedmont and coastal plain.

**Oviposition.** Unknown.

**Development.** Larvae begin hatching in late fall or early winter (Adler & Kim 1986). This species overlaps in development with *St. mutata* but is the earlier of the 2, with larvae maturing in late March and early April (Basrur & Rothfels 1959). Most adult emergence is completed by June (Davies et al. 1962, Back & Harper 1979, Adler & Kim 1986).

**Mating.** Unknown.

**Natural Enemies.** Larvae are hosts of the chytrid fungus *Coelomycidium simulii*, an unidentified mermithid nematode (Adler & Kim 1986), and the microsporidia *Caudospora palustris*, *C. polymorpha* (Adler et al. 2000), and *Janacekia debaisieuxi*.

**Hosts and Economic Importance.** See entry under *St. mutata*.

### *Stegopterna emergens* (Stone)

(Figs. 4.37, 10.37, 10.130, 10.285, 10.575, 10.725; Map 77)

*Cnephia emergens* Stone 1952: 80–81 (FM). Holotype* female (pinned #61189; USNM). Alaska, Fairbanks, Station 526 (see Sommerman et al. 1955), emergence trap, 19 June 1948 (Alaska Insect Project, collector unknown)

*Cnephia* 'species # 2' Anderson & Dicke 1960: 395 (Wisconsin record)

*Stegopterna tschukotensis* Rubtsov 1971a: 174–177 (FM). Holotype* male (pinned #19112, terminalia on celluloid mount below; ZISP). Russia, Chukotka, Chaun River, Krasnoarmeyskiy, 8 July 1963 (K. B. Gorodkov). New Synonym

[*Cnephia mutata*: Stone (1952) not Malloch 1914 (misident. in part, record from College)]

[*Cnephia mutata*: Hopla (1965) not Malloch 1914 (misident.)]

[*Cnephia* (*Stegopterna*) *mutata*: Abdelnur (1966, 1968) not Malloch 1914 (misident.)]

[*Stegopterna richteri dentata*: Bodrova (1977) not Rubtsov & Carlsson 1965 (misident.)]

[*Stegopterna mutata* complex: Fredeen (1985a) not Malloch 1914 (misident.)]

[*Stegopterna mutata*: Shipp (1985a) not Malloch 1914 (misident.)]

[*Stegopterna mutata*: Currie (1986) not Malloch 1914 (misident.)]

[*Stegopterna mutata*: Hershey & Hiltner (1988) not Malloch 1914 (misident.)]

[*Stegopterna mutata*: Ciborowski & Adler (1990) not Malloch 1914 (misident.)]

[*Stegopterna mutata*: Peterson et al. (1993) not Malloch 1914 (misident.)]

**Taxonomy**

Stone 1952 (FM); Sommerman 1953 (L); Anderson 1960 (FMPL); Davies et al. 1962 (FMP); Wood 1963a (FM); Wood et al. 1963 (L); Abdelnur 1966, 1968 (FMPLE); Smith 1970 (FMPL); Rubtsov 1971a (FM); Bodrova 1977 (FPL); Patrusheva 1982 (FM); Fredeen 1985a (FMP); Currie 1986 (PL).

We conclude that the name *tschukotensis* is synonymous with *emergens*. The holotype and allotype of *St. tschukotensis* match specimens of *St. emergens* from North America in the details of the terminalia. Additionally, the females have laciniae entirely without teeth and mandibles with only weak apical teeth. Females of all other described species of the genus *Stegopterna*, except *St. decafilis*, have mouthparts adapted for blood feeding.

**Cytology**

Basrur 1957; Basrur & Rothfels 1959 (footnote, p. 572); Madahar 1967, 1969.

**Bionomics**

**Habitat.** *Stegopterna emergens* is common across the arctic tundra from Alaska to Labrador, southward in the coniferous forests to Wisconsin. It also inhabits the Russian Far East (Currie 1997a). The immature stages are found most frequently in cool, sometimes evanescent rocky streams less than 4 m wide. They also have been taken from enriched streams draining cattle feedlots.

**Oviposition.** Females are obligately autogenous. Fecundity is variable, with females in Nunavut producing 1–66 eggs in an ovarian cycle (Smith 1970) and those in Alberta producing 125–211 (Abdelnur 1968). Oviposition behavior is unknown.

**Development.** In Alaska, eggs overwinter, and larvae hatch from late May at low elevations to late June at higher elevations; larval development requires 3–7 weeks (Sommerman et al. 1955). In Alberta, Manitoba, and Wisconsin, pupae are found during May (Anderson 1960, Abdelnur 1968, Crosskey 1994a). Streams that run dry while harboring larvae still produce adults, suggesting that larvae tunnel into the streambed and pupate in the absence of running water. Adults fly from May into August, depending on latitude and elevation. They have a sex ratio near unity and can live on honey in captivity for up to 28 days (Smith 1970).

**Mating.** Unknown.

**Natural Enemies.** Larvae are attacked by unidentified mermithid nematodes (Anderson & DeFoliart 1962 [footnote, p. 832 of R. Gordon 1984]), the microsporidium *Caudospora stricklandi* (Maurand 1975), and the chytrid fungus *Coelomycidium simulii*.

**Hosts and Economic Importance.** Mouthparts of the female are not adapted for biting.

### *Stegopterna mutata* (Malloch)

(Figs. 4.46, 10.38, 10.599, 10.630, 10.726; Map 78)

*Prosimulium mutatum* Malloch 1914: 20–21 (F). Holotype* female (pinned #15404; USNM). New Jersey, Gloucester Co., Glassboro, 28 March 1910 (C. T. Greene)

*Simulium* 'species undescribed' Strickland 1911: 321 (in part, larvae bearing "bi-annulated ovoid" microsporidia and possibly larvae bearing "ovoid bodies having the 'flagellum' replaced by a transparent disc") (possibly *St. diplomutata*)

*Simulium* 'Undescribed Species No. 2' Ritcher 1931: 242–246, Plate VI (PL) (possibly *St. diplomutata*)

*Simulium* 'sp. A' Johannsen 1934: 61 (PL) (possibly *St. diplomutata*)

*Cnephia mutata* '3n' Basrur 1957: 4–28 (PLC)

*Cnephia mutata* 'triploid' Basrur & Rothfels 1959: 571–589 (PC)

*Stegopterna mutata* '(IIIL-1 sibling)' Crosskey 1990: 288 (*lapsus calami*)

[*Greniera abditoides*: Cupp & Gordon (1983) not Wood 1963b (misident. in part, material from Michigan)] (possibly *St. diplomutata*)

[*Simulium* (*Hellichiella*) *rivuli*: Cupp & Gordon (1983) not Twinn 1936a (misident. in part, material from Pine Creek and some from Commencement Creek, Michigan)] (possibly *St. diplomutata*)

[*Simulium* (*Simulium*) *luggeri*: Cupp & Gordon (1983) not Nicholson & Mickel 1950 (misident. in part, material from Schoolcraft Co., Michigan)] (possibly *St. diplomutata*)

**Taxonomy**

Malloch 1914 (F); Riley & Johannsen 1915 (F); Dyar & Shannon 1927 (F); Ritcher 1931 (PL); Johannsen 1934 (PL); Nicholson 1949 (FP); Nicholson & Mickel 1950 (FP); DeFoliart 1951a (L); Jamnback 1953, 1956 (L); Stone & Jamnback 1955 (FPL); Davies & Peterson 1956 (E); Basrur 1957 (PL); Young 1958 (F); Basrur & Rothfels 1959 (P); Anderson 1960 (FPL); Davies et al. 1962 (FPL); Wood 1963a (FL); Wood et al. 1963 (L); Stone 1964 (FPL); Snoddy 1966 (FPL); Holbrook 1967 (L); Stone & Snoddy 1969 (FPL); Pinkovsky 1970 (FPL); Amrine 1971 (FPL); Wygodzinsky & Coscarón 1973 (L); Snoddy & Noblet 1976 (PL); Merritt et al. 1978b (PL); Peterson 1981 (FL), 1996 (FPL); Webb & Brigham 1982 (L); Adler 1983 (PL); Adler & Kim 1986 (PL); Peterson & Kondratieff 1995 (FPL, illustrations only); Stuart 1995 (cocoon); Stuart & Hunter 1998b (cocoon).

**Morphology**

Guttman 1967a (labral fan), Kurtak 1973 (labral fan), Colbo et al. 1979 (F labral and cibarial sensilla), Ross 1979 (instars), Craig & Borkent 1980 (L palpal sensilla), Yeboah et al. 1984 (FPL ovary), Thompson 1989 (labral-fan microtrichia), Palmer & Craig 2000 (labral fan).

**Physiology**

Madahar 1967 (oogenesis).

**Cytology**

Basrur 1957; Basrur & Rothfels 1959; Madahar 1967, 1969; Adler & Kim 1986.

**Molecular Systematics**

Xiong & Kocher 1991 (mtDNA), Xiong 1992 (mtDNA).

**Bionomics**

**Habitat.** This species is distributed from Labrador and northern Quebec southward to Georgia and Louisiana. Records from British Columbia (Basrur 1957, Basrur & Rothfels 1959) are based on labeling errors. The immature stages are common in slower areas of small, temporary or permanent streams less than a meter or 2 wide in forests of mountains and hills. They occur in many of the same streams with *St. diplomutata*, but also in more productive habitats such as impoundment outflows. Their densities decrease with distance from the outlets of impoundments (Morin et al. 1986). Larvae respond negatively to light and seek the undersides of stones (Kurtak 1973), especially when they are in pure populations; however, when they occur with larvae of *Prosimulium*, they are more common on the tops of the substrates (Tessler 1991). Larvae drift primarily during the night (Adler et al. 1983b). Preliminary work suggests that the larvae are not adept at selecting appropriate temperatures (Thomas 1967). Females are most likely to be collected in forest habitats (Martin et al. 1994).

**Oviposition.** Females emerge with as many as 337 oocytes and, even starved, can mature more than 200 of these in 36 hours (Chutter 1970). Actual fecundity ranges from 82 to 301 eggs per ovarian cycle and is positively related to female size (Pascuzzo 1976). Females from some populations require 5 days at either 8.5°C or 16.5°C to mature their eggs autogenously (Davies & Györkös 1990). They oviposit in flight by tapping their abdomens to the surface of the stream (Davies & Peterson 1956).

**Development.** Eggs undergo a prolonged diapause. Larvae eventually hatch from eggs stored below 15°C, but they do not hatch from eggs stored above this temperature (Colbo 1979). Alternatively, diapause can be broken by holding eggs for 6–8 weeks at 17°C–22°C and then transferring them to lower temperatures (8°C–12°C); larvae begin to hatch in an additional 8–11 weeks (Mokry [p. 5] in Anonymous 1977). Embryonated eggs can survive freezing temperatures of at least –15°C (Colbo & Wotton 1981). Hatching times at a single stream site can vary, producing several cohorts (Ross & Merritt 1978, Morin et al. 1988c). In Pennsylvania, larvae begin hatching in late fall and early winter, and adults emerge from mid-March to mid-June (Adler & Kim 1986, McCreadie et al. 1994b). On Newfoundland's Avalon Peninsula, larvae begin hatching in October and can be found well into May (Colbo 1979). They can spend 3 months beneath snow and ice yet still show significant growth (Lewis 1973, Lewis & Bennett 1974b). In Michigan, larvae hatch in January or, if the streams freeze solid, in March (Merritt et al. 1978b). Temperature appears to play a greater role than does food in the growth and production of overwintering larvae (Merritt et al. 1982). Larval growth requires about 250–275 degree-days above 0°C (Ross & Merritt 1978). Development from egg to adult includes 6 larval instars (Ross 1979) and averages 12 weeks at 9°C–10°C (Mokry 1978). In Algonquin Park, Ontario, most adults emerge during the morning from early May to early June (Davies 1950). This species has been reared for 2 complete generations in the laboratory (Mokry 1978).

The efficiency of filter feeding is low compared with that for other species, probably because the microtrichia are short (Kurtak 1978) and because the larvae can spend considerable time grazing food from the substrate; the gut can contain more than 50% diatoms (Kurtak 1973). The extent of grazing, however, apparently varies among populations. Larvae in some populations acquire nearly all of their food from the seston (Thompson 1989).

**Mating.** Females are parthenogenetic; males do not exist.

**Natural Enemies.** Larvae are hosts of numerous parasites. The chytrid fungus *Coelomycidium simulii* attacks larvae (Adler & Kim 1986), and the trichomycete fungi *Harpella melusinae*, *Pennella simulii*, and *Smittium* sp. have been found in the larval gut (Frost & Manier 1971, Lichtwardt et al. 2001). Larvae are infected by the microsporidia *Caudospora palustris*, *C. polymorpha* (formerly *C. brevicauda*), and *Janacekia debaisieuxi* (Frost & Nolan 1972, Vávra & Undeen 1981, Adler et al. 2000). We have a single larva of either *St. diplomutata* or *St. mutata* from Lyon Co., Kentucky, that is infected with a microsporidium resembling *Caudospora alaskensis* (Adler et al. 2000). Similarly, Strickland (1911, Fig. 16) illustrated a microsporidium that appears to be *C. alaskensis*, which probably was from a larva of either *St. diplomutata* or *St. mutata*, and Frost and Nolan (1972) found a microsporidium, *Caudospora* sp., tentatively recognized as *C. alaskensis* (Adler et al. 2000). *Caudospora polymorpha*, discovered early in the 20th century (Strickland 1911), was formally described from larvae of *St. mutata* (Jamnback 1970) and is probably host specific for this species and *St. diplomutata*. Prevalence of this microsporidium in a larval population can reach 20% (Merritt et al. 1978b). The mermithid nematode *Mesomermis flumenalis* has been found in larvae, but infections typically are rare, possibly a result of larval behavior or inability of preparasites to recognize the host (Bruder 1974, Bailey & Gordon 1977, Colbo & Porter 1980). A cytoplasmic polyhedrosis virus is common in larvae at some sites and is probably transmitted transovarially (Bailey et al. 1975, Bailey 1977, Mokry 1978). In the laboratory, cercariae of the trematode *Plagiorchis noblei* infect larvae (Jacobs 1991, Jacobs et al. 1993).

Females are infected by a trichomycete, originally identified as a phycomycete, that undergoes at least part of its development in the adult ovary and replaces the eggs with spores (Undeen 1979a, Yeboah 1980, Yeboah et al. 1984, Moss & Descals 1986). Female size is not affected by this trichomycete (Colbo 1982).

**Hosts and Economic Importance.** Host-seeking females orient primarily to the color black (somewhat to red) and to points of convergence on silhouettes (Bradbury & Bennett 1974a, 1974b; Browne & Bennett 1980). Most host-seeking females have taken a sugar meal (McCreadie et al. 1994c). Nectar of sour-top blueberry (*Vaccinium myrtilloides*) serves as 1 source of sugar (Davies & Peterson 1956).

Because females of the 2 eastern species of *Stegopterna* are currently indistinguishable, the following information might pertain to either or both species. Females typically feed on large mammals such as horses, elks, white-tailed deer, and cows (Davies & Peterson 1956, Downe & Morrison 1957, Anderson & DeFoliart 1961, Merritt et al. 1978b, McCreadie et al. 1985). One or 2 feedings on a red fox (Hunter et al. 1993), a raccoon, domestic ducks (Pickavance et al. 1970, Hunter et al. 1993), and a passerine bird (Simmons et al. 1989) also have been recorded. Females have been attracted in small numbers to American minks and domestic rabbits (Hunter et al. 1993). They fly around humans and can be a nuisance, crawling about the neck and ears (Stone & Jamnback 1955, Davies 1963, Jamnback 1969a, Adler & Kim 1986). They are rarely biting pests of humans, although attacks do occur (Mokry 1978, Lewis & Bennett 1979).

### *Stegopterna permutata* (Dyar & Shannon)
(Map 79)

*Eusimulium mutatum permutatum* Dyar & Shannon 1927: 17–18, Fig. 36 (F). Holotype* female (pinned #28332, abdomen, 1 hind leg, 1 middle leg on slide; USNM). British Columbia, Prince Rupert, 17 June 1919 (H. G. Dyar)

*Cnephia* 'X' Madahar 1969: 117, Figs. 1, 5–8, 11–13, 15, 16 (C)

**Taxonomy**

Dyar & Shannon 1927 (F), Stains 1941 (F), Stains & Knowlton 1943 (F).

The larva is similar to that of *St. acra* n. sp. The other life stages are similar to those of *St. diplomutata*.

**Cytology**

Madahar 1967, 1969.

**Bionomics**

**Habitat.** This species is fairly common in forested areas along the Pacific Coast but less common in the Sierra Nevada of California. Cool, shallow, brownwater flows less than a meter wide and often temporary are home to the immature stages, although shallow streams as wide as 5 m sometimes have larvae.

**Oviposition.** Unknown.

**Development.** Larvae are present by March and adults are produced by May at lower elevations along the coast. In more mountainous areas, larvae are present in May and June. Above 3500 m in the Sierra Nevada, larvae and pupae are found into September.

**Mating.** Unknown.

**Natural Enemies.** Larvae are hosts of the chytrid fungus *Coelomycidium simulii*.

**Hosts and Economic Importance.** Females are attracted to humans (Currie & Adler 1986), but we have only 1 record of a human being bitten. A female taken (presumably feeding) from a horse in British Columbia (Hearle 1932) might be of this species.

**_Stegopterna trigonium_ (Lundström)**
(Map 80)

*Melusina trigonium* Lundström 1911: 18–19, Fig. 16 (M). Lectotype male (designated by Rubtsov 1961b: 222; microvial; UZMH). Finland, Kittilä, date unknown (R. Frey)

*Cnetha Freyi* [*sic*] Enderlein 1929c: 73–74 (M [F misassociated; Zwick 1995]). Lectotype male (designated by Zwick 1995: 145–147; pinned, terminalia on slide below; ZMHU). Russia, Murmansk Region, high swamp Alexandrowsk, July 1926 (W. Richter)

*Stegopterna Richteri* [*sic*] Enderlein 1930: 90 (F). Lectotype female (designated by Zwick 1995: 146–147; pinned, terminalia on side below, hind leg on separate slide; ZMHU). Russia, Murmansk Region, high swamp Alexandrowsk, 20 July 1926 (W. Richter); *richerti* (subsequent misspelling)

*Stegopterna richteri majalis* Rubtsov & Carlsson 1965: 12–15 (FMPL). Holotype* male (pinned, terminalia on slide #14036; ZISP). Russia, Leningrad (St. Petersburg) Region, tributary of Sitenka River, Luzhski District, 10 May 1961 (I. A. Rubtsov)

*Stegopterna richteri dentata* Rubtsov & Carlsson 1965: 15–17 (FMPL). Holotype* male (slide #17319; ZISP). Russia, Sverdlovsk Region, Sjalinsky area, rill, 4 June 1962 (G. I. Kotelnikova)

*Stegopterna richteri haematophaga* Rubtsov & Carlsson 1965: 17 (F). Holotype female (slide #17257; ZISP). Russia, Perm Region, Lysvensk area, 17 June 1960 (K. N. Beltukova)

*Stegopterna richteri longicoxa* Rubtsov 1971a: 172–174 (FM). Holotype* male (slide #19379; ZISP). Russia, region unknown, Noril'skie Mountains, 13–28 July 1967 (K. Y. Grunin)

**Taxonomy**

Lundström 1911 (M); Enderlein 1929c (M only), 1930 (F); Rubtsov 1956, 1961b, 1971a (FMPL); Usova 1961 (FMPL); Rubtsov & Carlsson 1965 (FMPL); Patrusheva 1982 (FMPL); Gryaznov 1984b (F); Rubtsov & Yankovsky 1984 (MPL); Zwick 1995 (FM).

We provisionally apply the name *St. trigonium* to a species for which we have immatures from the North Slope of Alaska, the Northwest Territories, and western Nunavut. These larvae are not conspecific with any North American member of the genus. They have a shallow postgenal cleft, a brown- or gray-banded abdomen, and a brownish yellow head capsule with dark brown head spots; the anteromedian and posteromedian head spots of some larvae are nearly contiguous. The larvae are similar to those of *St. decafilis* but have a gill of 12, rather than 10, filaments configured in the same fashion as the filaments of *St. emergens*. We have not seen North American adults and, therefore, do not include the species in the keys to females and males.

We suspect that this species is Holarctic and have seen morphologically similar larvae from Sweden, suggesting that an existing Palearctic name (e.g., *trigonium*) would be appropriate for the North American material. The Palearctic species of *Stegopterna*, however, are in need of revision. In particular, the status of the synonyms of *trigonium* requires clarification in light of the cytological evidence that northern Sweden has at least 2 species of *Stegopterna* (Adler et al. 1999). We, therefore, do not delve deeply into the literature on *St. trigonium*, nearly all of which has appeared under the names *freyi* and *richteri*, as discussed by Zwick (1995).

**Cytology**

We have examined the chromosomes of 3 female larvae from the Northwest Territories: one from north of Fort Simpson, one from Deline, and a third from a tributary of the Thelon River. These larvae were homozygous for the *IS-1*, *IIIS-1*, and *IIIL-1* inversions but otherwise had the standard banding sequence of *Stegopterna*, as given by Madahar (1967); the polarity of the IS-2 sequence, however, was not determined in our material. Chromosomally, these larvae are very similar to those of *St. emergens*. They differ from those of *St. decafilis* by the presence of *IIIL-1*, and from morphologically similar larvae in Sweden described cytologically by Adler et al. (1999) in having *IIIL-1* and lacking the IIIL inversion with section limits 89p–93d.

**Bionomics**

**Habitat.** Our larvae of *St. trigonium* were collected from the North Slope of Alaska, the Northwest Territories, and western Nunavut, often with the larvae of *St. emergens*. The streams were cold and rocky. In the Palearctic Region, *St. trigonium* is known from Russia and Fennoscandia.

**Oviposition.** Unknown.

**Development.** Final and mid-instar larvae have been taken from late June to mid-July.

**Mating.** Unknown.

**Natural Enemies.** We have 1 Alaskan larva infected with an unidentified, possibly new, microsporidium.

**Hosts and Economic Importance.** Unknown; presumably mammalophilic.

**_Stegopterna xantha_ Currie, Adler & Wood, New Species**
(Figs. 10.39, 10.727; Plate 8b; Map 81)

*Cnephia* 'O' Madahar 1969: 117, Fig. 1 (C)

*Cnephia* 'Y' Madahar 1969: 117, Fig. 1 (C)

**Taxonomy**

**Female.** Not differing from that of *St. acra* n. sp. except as follows: Wing length 2.9–3.3 mm. Calcipala large, about three fourths as wide as apex of basitarsus (Fig. 10.39).

**Male.** Not differing from that of *St. acra* n. sp. except as follows: Calcipala large, about three fourths as wide as apex of basitarsus.

**Pupa.** Not differing from that of *St. acra* n. sp. except as follows: Length 3.1–3.4 mm.

**Larva.** Not differing from that of *St. acra* n. sp. except as follows: Length 6.0–7.3 mm. Head capsule (Fig. 10.727) and antenna yellowish to brownish yellow. Postgenal cleft a tiny rounded notch, often nearly absent; subesophageal ganglion sometimes ensheathed with pigment. Labral fan with 40–47 primary rays. Body pigment pale gray to brownish gray (Plate 8b). Posterior proleg with 10–13 hooks in each of 60–74 rows.

**Diagnosis.** The female, male, and pupa are not easily distinguished from those of most other western members of the genus *Stegopterna*. The pale body and head capsule of the larva are unique among western members of the genus.

**Holotype.** Female with pupal exuviae (frozen dried, pinned). British Columbia, Bowser, railroad track stream, 7 May 1964, D. M. & G. C. Wood (CNC).

**Paratypes.** California, Marin Co., Rt. 1, 1.3 km south of Olema, 38°01′N, 122°46′W, 21 March 1990, P. H. & C. R. L. Adler (17 larvae); Trinity Co., Weaverville, Sidney Gulch at west end of town, Rt. 299, 40°44′N, 122°56′W, 24 April 1991, D. C. Currie (3 pupae, 15 larvae). Oregon, Benton Co., Corvallis, 60th Street, Oak Burn, 1–30 April 1994, N. H. Anderson (6♀, 6♂); Clatsop Co., 3.2 km northeast of Jewell, Rt. 202, 45°56′N, 123°32′W, 21 March 1992, D. C. Currie (21 larvae). British Columbia, Vancouver Island, Miracle Beach Provincial Park, tributary of Black Creek, 29 March 1991, D. C. & R. M. Currie (35 larvae); 2 km southeast of Osoyoos, Rt. 3, 5 June 1990, D. C. & R. M. Currie (16 larvae); Cultus Lake (town), 28 April 1964, D. M. & G. C. Wood (33 larvae).

**Etymology.** The specific name is from the Greek adjective meaning "yellow," in reference to the color of the larval head capsule.

**Cytology**

Madahar 1967, 1969.

Chromosome arm IIL carries 2 inversions that reinvert the parabalbiani, giving the appearance that the arm has the standard sequence, as claimed by Madahar (1969).

We never rediscovered larvae of Madahar's (1969) *Stegopterna* 'O' despite searches around Cultus Lake, British Columbia, the area of its original discovery. We tentatively consider it a Y-chromosome variant.

**Bionomics**

**Habitat.** *Stegopterna xantha* is most common along the Pacific Coast from southern British Columbia to northern California, but it also inhabits the Sierra Nevada of California and reaches as far east as Montana. The immature stages inhabit cool, clear shallow streams 3 m or less in width.

**Oviposition.** Unknown.

**Development.** In coastal streams, larval development is well underway in March, but larvae can be found in May and June in the Sierra Nevada.

**Mating.** Unknown.

**Natural Enemies.** No records.

**Hosts and Economic Importance.** Unknown; presumably mammalophilic.

### Genus *Tlalocomyia* Wygodzinsky & Díaz Nájera

*Tlalocomyia* Wygodzinsky & Díaz Nájera 1970: 83–88 (as genus). Type species: *Tlalocomyia revelata* Wygodzinsky & Díaz Nájera 1970: 88–109, by original designation

*Stewartella* Coleman 1951: 132. Unavailable (unpublished thesis name)

*Mayacnephia* Wygodzinsky & Coscarón 1973: 144–148 (as genus). Type species: *Simulium pachecolunai* De León 1945: 67–68, by original designation. New Synonym

**Diagnosis.** Adults: Wing with basal medial cell; radius with hair dorsobasally; radial sector unbranched (species south of United States) or with short apical fork. Calcipala typically large and lamellate; pedisulcus represented, at most, by weak wrinkles. Female: Claws with basal thumblike lobe. Spermatheca typically with raised pattern of serpentine wrinkles. Male: Gonostylus not strongly tapered, with 2 (rarely 1) apical spinules. Paramere without stout spines, although distal margin can be wrinkled. Aedeagal membrane with fine spines or hair, and with or without small pale flattened spines. Pupa: Gill of 2–15 variously inflated filaments, arising from short base, and radiated fanlike (one basally inflated filament and 4 slender filaments in *T. revelata*). Terminal spines long, slender (absent in *T. revelata*). Cocoon loosely woven, shapeless, covering all or part of abdomen. Larva: Hypostoma with teeth arranged in 3 distinct groups; outside margin of lateral cluster of teeth typically straight or sloped slightly outward. Postgenal cleft shallow, pointed or rounded apically, shaped like inverted V. Abdominal segment IX with or without transverse midventral bulge. Rectal papillae of 3 simple lobes.

**Overview.** *Tlalocomyia revelata* was described by Wygodzinsky and Díaz Nájera (1970) and until now has remained the only member of the genus. All life stages of *T. revelata* are similar to those of the genus *Mayacnephia* described by Wygodzinsky and Coscarón (1973). Among the few distinguishing characters of *T. revelata* are the well-developed onchotaxy and lack of terminal spines of the pupa, and presence of hairs on the anepisternal membrane of the adults. *Mayacnephia* species and *T. revelata* are so similar that we synonymize the more euphonius name *Mayacnephia* with the older name *Tlalocomyia*.

The genus *Tlalocomyia*, as currently recognized, comprises 3 species groups and 16 species in the Nearctic and Neotropical Regions. Two species groups—the *T. andersoni* group with 1 species and the *T. osborni* group with 3 species—occur in North America north of Mexico. The newly recognized *T. pachecolunai* species group, which includes 12 species, is found from Mexico south to Panama. The following names, which apply to species in the *T. pachecolunai* group, represent new combinations: *T. aguirrei*, *T. alticola*, *T. atzompensis*, *T. fortunensis*, *T. grenieri*, *T. mixensis*, *T. muzquicensis*, *T. pachecolunai*, *T. roblesi*, *T. salasi*, and *T. tadai*; *T. revelata* is also a member of this group.

The immature stages of all species of *Tlalocomyia* live in small cool streams. All species are univoltine. The females are ornithophilic.

#### *Tlalocomyia andersoni* Species Group, Newly Recognized

**Diagnosis.** Same as that given below for the single representative species (*Tlalocomyia andersoni*).

**Overview.** A single species, endemic to the Pacific Coast of the United States, constitutes this group.

#### *Tlalocomyia andersoni* Currie, Adler & Wood, New Species

(Figs. 10.131, 10.286, 10.420, 10.576, 10.631, 10.728; Plate 8c; Map 82)

**Taxonomy**

**Female.** Wing length 3.2–3.6 mm. Scutum brownish black, pollinose. All hair pale golden, except that of basicosta, stem vein, and scutellum golden brown. Mandible with 36–40 serrations; lacinia with 20–22 retrorse teeth. Sensory vesicle in lateral view about half as long as and occupying about one third of palpomere III. Radial sector with short apical fork. Legs golden brown; calcipala well developed, rounded; claws each with basal thumblike lobe. Terminalia (Fig. 10.131): Anal lobe in ventral view with anterior half unsclerotized. Genital fork with each arm slender and expanded into subrectangular lateral plate. Spermatheca nearly twice as long as wide, unpigmented proximally; surface with raised pattern of serpentine wrinkles.

**Male.** Wing length 2.8–3.0 mm. Scutum blackish brown. All hair coppery brown. Radial sector with short apical fork; subcosta bare ventrally. Legs brown. Genitalia (Fig. 10.286): Gonostylus in ventral view about 3 times as long as basal width, with 2 apical spinules directed anteromedially. Ventral plate in ventral view with body subrectangular; lip in lateral view slender. Paramere subtriangular, without spines; aedeagal membrane with rather flat, weakly sclerotized, pale spines. Median sclerite bifurcate apically.

**Pupa.** Length 2.9–3.3 mm. Gill of 12 filaments arising from rosette of 4 primary trunks (Figs. 10.576). Head and thorax dorsally with densely distributed, barely perceptible, dome-shaped microtubercles; trichomes unbranched. Abdomen sometimes with minute pleural sclerites in membrane (Fig. 10.420). Cocoon feeble, usually restricted to abdomen.

**Larva.** Length 5.3–5.6 mm. Head capsule (Fig. 10.728) pale brownish orange; head spots brown, variably distinct; medial head spots heaviest. Antenna subequal in length to stalk of labral fan; distal article pale brown; other articles unpigmented. Hypostoma with teeth in 3 prominent groups; lateral and median groups extended anteriorly to about same level (Fig. 10.631). Postgenal cleft a tiny notch or absent; subesophageal ganglion without pigmented sheath. Labral fan with 27–31 primary rays. Body pigment pale brown, arranged in bands (Plate 8c); abdominal segment IX with small, transverse midventral bulge. Posterior proleg with 10–13 hooks in each of 60–62 rows.

**Diagnosis.** The elongate spermatheca and the pale spines in the aedeagal membrane distinguish the female and male, respectively, from those of all other North American species with a weakly bifurcate radial sector. The cuplike arrangement of the 12 gill filaments, with 5 filaments in the ventral group, is unique among the pupae of North American simuliids. Larvae are similar to those of the genus *Stegopterna* but have a broader head anteriorly, more parallel-sided lateral hypostomal teeth, and longer labral-fan stalks with fewer (27–31) primary rays.

**Holotype.** Male (frozen dried, pinned) with pupal exuviae (glycerin vial below). Oregon, Benton Co., Corvallis, 60th Street, Dimple Hill, Outgate Beck, 44°35.9′N, 123°19.0′W, emergence trap, 9 March 1993, D. C. Currie & D. M. Wood (CNC).

**Paratypes.** California, Santa Barbara Co., Santa Ynez Mountains, Nojoqui Falls, 34°31.8′N, 120°10.57′W, 3 February 1993, J. R. Vockeroth (1♀). Oregon, same data as holotype (11♀ & 8♂ + exuviae), 27 January–23 February 1993, N. H. Anderson (3♀, 2♂), 12 February 1993, N. H. Anderson (28 larvae), 21 February–20 March 1994, N. H. Anderson (4♀, 4♂), 9–10 March 1993, D. C. Currie & D. M. Wood (7 larvae), 10–18 March 1993, N. H. Anderson (2♀, 1♂), 29 March–4 April 1993, N. H. Anderson (2♀, 1♂), 17–31 May 1993, N. H. Anderson (1♀); Corvallis, 60th Street, Oak Burn, emergence trap, 27 January–23 February 1993, N. H. Anderson (1♀); Corvallis, 60th Street, Studio Run, emergence trap, 20–28 March 1993, N. H. Anderson (2♀); 12.8 km north of Corvallis, Rt. 99, Burry Creek (temporary), 44°42′N, 123°13′W, 31 March 1984, A. Carratti (2 pupae).

**Etymology.** This species is named in honor of Norman H. Anderson, an entomologist formerly of Oregon State University, on whose property a large population was discovered.

**Bionomics**

**Habitat.** This rare species is known from 3 counties, 1 in Oregon and 2 in California. The immature stages reside in shallow, temporary or permanent headwater streams less than half a meter wide that flow through oak (*Quercus*) forests.

**Oviposition.** Unknown.

**Development.** *Tlalocomyia andersoni* is probably often overlooked because of its early maturation. Adults fly by February in southern California and by the end of March farther north.

**Mating.** Unknown.

**Natural Enemies.** No records.

**Hosts and Economic Importance.** Unknown; presumably birds.

### *Tlalocomyia osborni* Species Group, Newly Recognized

**Diagnosis.** Adults: Radial sector with short apical fork. Male: Subcosta ventrally bare or with 1 or 2 setae. Ventral plate in lateral view robust; aedeagal membrane with hair only. Pupa: Gill of 6–12 variously thickened filaments, typically arranged in rosette. Larva: Labral fan with 18–25 primary rays. Abdominal segment IX with transverse midventral bulge minute or absent.

**Overview.** The United States and Canada are home to the 3 infrequently collected species of this group. The immature stages live in small headwater flows in mountainous regions. Nearly nothing is known of adult behavior, although females have bifid claws adapted for feeding on birds.

### *Tlalocomyia osborni* (Stains & Knowlton), New Combination

(Figs. 10.132, 10.287, 10.421, 10.632, 10.729; Plate 8d; Map 83)

*Eusimulium osborni* Stains & Knowlton 1943: 271, Figs. 132, 133 (F). Holotype* female (pinned #6348; CAS [terminalia, hind leg, wing on slide; USNM]). California, Tuolumne Co., 4.5 mi. north northwest of Brightman Flat, 1859 m elev., on grouse, June 1941 (R. T. Orr)

*Mayacnephia* 'sp. nr. *osborni*' Moulton 1997: 152; 2000: 99 (DNA sequences)

**Taxonomy**

Stains & Knowlton 1943 (F), Coleman 1951 (F), Wirth & Stone 1956 (F).

The pinned holotype has a red label with the name "Eusimulium intermedius" but also carries a small scrap of paper stating that Knowlton wrote a letter to the California Academy of Sciences (CAS) in 1946 to indicate that the species was published under the name *Eusimulium osborni*. The USNM contains a single slide with 1 hind leg, 1 wing, and the terminalia of this same species. This slide is evidently the partner to the pinned holotype. It bears a red star, which was affixed to all the type slides prepared by Stains and Knowlton, and is labeled "Eusimulium intermedius." The specific name was changed because Alan Stone informed the authors that it was already in use for a European species (letter from G. F. Knowlton to the CAS, 3 April 1946, on file at CAS). The change from "intermedius" to *osborni* occurred after the types were sent to the depositories but before the manuscript went to press.

Two or more species might exist under the name *Tlalocomyia osborni*. Populations in Arizona and British Columbia require further study to determine if they are conspecific with the type.

**Molecular Systematics**

Moulton 1997, 2000 (DNA sequences).

**Bionomics**

**Habitat.** This taxon is known from small, cool, often temporary streams and roadside ditches from southern British Columbia southward through the high Sierra Nevada of California to southern Arizona. Larvae and pupae often are found under rocks.

**Oviposition.** Unknown.

**Development.** Larvae pupate from about late May to mid-June in the interior and from early January to March along the coast and in Arizona.

**Mating.** Unknown.

**Natural Enemies.** No records.

**Hosts and Economic Importance.** Females in the type series were collected (presumably feeding) from an unspecified species of grouse.

### *Tlalocomyia ramifera* Currie, Adler & Wood, New Species

(Figs. 4.38, 10.79, 10.133, 10.288, 10.422, 10.633, 10.730; Map 84)

*Cnephia* 'G' Hocking 1950: 497 (Yukon record)

*Cnephia* 'n. sp. near *stewarti*' Madahar 1967: 12, 18, 64, 75, 79, Figs. 1, 6, 31–36 (C)

*Mayacnephia* 'sp.' Peterson 1981: 365, 389 (FL); 1996: 599, 607 (FL)

*Mayacnephia* 'species X' Currie 1986: 20–21, Figs. 9, 16, 27 (PL)

*Mayacnephia* 'sp.' Currie 1988: 122 (FP)

*Mayacnephia* 'unnamed species X' Ramírez-Pérez et al. 1988: 74 (P)

**Taxonomy**

**Female.** Wing length 3.9–4.7 mm. Scutum dark brown. All hair pale golden. Mandible with 42–45 serrations; lacinia with 34–36 retrorse teeth. Sensory vesicle in lateral view about half as long as and occupying about one third to one fourth of palpomere III. Legs brown; calcipala well developed; claws each with basal thumblike lobe. Terminalia (Fig. 10.133): Anal lobe in ventral view with anterior half unsclerotized. Genital fork with each arm expanded into large subrectangular lateral plate. Spermatheca small, less than half as long as stem of genital fork, about two thirds as wide as long, unpigmented proximally; surface externally with raised pattern of serpentine wrinkles.

**Male.** Wing length 3.1–3.2 mm. Scutum brownish black. All hair pale golden, except that of basicosta and stem vein golden brown. Legs pale golden brown. Genitalia (Fig. 10.288): Gonostylus in ventral view about 3 times longer than basal width, with 2 apical spinules directed medially. Ventral plate in ventral view with body subrectangular; lip weak. Median sclerite bifurcate apically. Paramere large, subtriangular, without spines.

**Pupa.** Length 4.5–4.7 mm. Gill of 8 inflated, widely splayed filaments (Fig. 10.422). Head and thorax dorsally with densely distributed, dome-shaped microtubercles; trichomes unbranched. Cocoon usually covering only abdomen.

**Larva.** Length 6.6–7.3 mm. Head capsule (Fig. 10.730) brownish orange; head spots brown, variably distinct; anteromedial head spot elongate. Antennna subequal in length to stalk of labral fan, with proximal and medial articles pale brown or unpigmented, and distal article brown. Hypostoma with teeth in 3 groups; lateral teeth extended well beyond other teeth; median tooth and outer sublateral teeth extended anteriorly to about same level (Fig. 4.38), or outer sublateral teeth extended farther anteriorly (Fig. 10.633). Postgenal cleft a small, broadly rounded or triangular notch; subesophageal ganglion not ensheathed with pigment. Labral fan with 21–23 primary rays. Body pigment pale brown to grayish brown; abdominal segment IX with minute midventral bulge. Posterior proleg with 8–10 hooks in each of 77–80 rows.

**Diagnosis.** The adults and immature larvae are not easily separated from those of *T. stewarti*. Pupae with their small saclike cocoon and gill of 8, slightly inflated filaments arranged in a rosette are unique among North American black flies.

**Holotype.** Female with pupal exuviae (frozen dried, pinned). British Columbia, Kamloops, Cold Creek Road, mi. 2.1 (km 3.4), pond outlet, 13–16 May 1964, D. M. & G. C. Wood (CNC).

**Paratypes.** Washington, Chelan Co., Wenatchee National Forest, Riverbend Campground, Malaise trap, 13–14 June 1989, E. Fuller (1♂). Alberta, Banff National Park, seepage behind Lake Minnewanka, 27 May 1981, D. C. Currie (2 larvae), 31 May 1982, D. C. Currie (10 larvae), 20 June 1982, D. C. Currie (1 larva). British Columbia, same data as holotype (25♀ & 32♂ + exuviae); Kamloops, near Cold Creek Road, 16 May 1964, D. M. & G. C. Wood (1 pupa, 7 larvae), 14 May 1964, D. M. & G. C. Wood (1 exuviae, 3 larvae); 11.9 km west of Kinnaird, 25 May 1964, D. M. & G. C. Wood (34♀ & 6♂ + exuviae); 20.6 km west of Kinnaird, 49°17′N, 117°39′W, 2 May 1964, D. M. & G. C. Wood (12 ♀ & 1 ♂+ exuviae, 1 pupa, 2 larvae), 14 June 1964, D. M. & G. C. Wood (4♀ + exuviae). Yukon Territory, Alaska Hwy., km 1683, small stream flowing through alpine meadow, 60°58′N, 139°07′W, 7 June 1979, R. Jaagumagi (3 larvae).

**Etymology.** The specific name is from Latin meaning "branch bearing," and refers to the radiating nature of the gills.

**Cytology**

Madahar 1967.

Based on Madahar's (1967) chromosome maps and preliminary analysis, the nucleolar organizer is in the base of IS, and IIS is identical to the *Stegopterna* standard banding sequence of Madahar (1969).

**Bionomics**

**Habitat.** This species is sporadic from southern Yukon to Idaho. Larvae and pupae have been taken from shallow headwater streams and seepages less than a meter wide in forested areas.

**Oviposition.** Unknown.

**Development.** Eggs overwinter and produce larvae in the spring after ice breakup (Currie 1986). Pupae appear from May to mid-July, depending on elevation.

**Mating.** Unknown.

**Natural Enemies.** Larvae are attacked by an unidentified mermithid nematode.

**Hosts and Economic Importance.** Unknown; presumably ornithophilic.

### *Tlalocomyia stewarti* (Coleman), New Combination

(Figs. 10.59, 10.289, 10.577, 10.634, 10.731; Plate 8e; Map 85)

*Cnephia stewarti* Coleman 1953: 45 (FMP). Holotype unknown (putatively USNM but presumably lost). California, Plumas Co., near Spring Garden, 19 May 1948 (R. W. Coleman & W. W. Wirth)

**Taxonomy**

Coleman 1951, 1953 (FMP); Wirth & Stone 1956 (FMP); Ramírez-Pérez et al. 1988 (P); Peterson 1996 (P).

We were unable to locate the holotype of this species in the USNM, the putative depository. We did, however, find 3 pinned paratypes (with small yellow paratype labels) among the undetermined simuliid material. In addition, we found 8 slides and 14 vials of alcohol material without type status. In light of these findings, the type specimen might exist somewhere in the USNM collection. We, therefore, refrain from designating a neotype.

**Morphology**

Evans & Adler 2000 (spermatheca).

**Cytology**

A perfunctory analysis indicates that arm associations are standard, the nucleolar organizer is in the base of IS, and polymorphisms are common.

**Bionomics**

**Habitat.** This large species hails from high elevations in the Sierra Nevada of California. The immature forms dwell in cool rivulets derived from springs, seeps, and snow melt. These tiny flows are often temporary and less than a meter wide. Larvae and pupae adhere to debris and rocks, often on the undersurfaces.

**Oviposition.** Unknown.

**Development.** The winter is probably passed as eggs, with pupae appearing from May to late June.

**Mating.** Unknown.

**Natural Enemies.** Larvae are hosts of an unidentified mermithid nematode.

**Hosts and Economic Importance.** Unknown; presumably ornithophilic.

## Genus *Gigantodax* Enderlein

*Gigantodax* Enderlein 1925: 205 (as genus). Type species: *Gigantodax bolivianus* Enderlein 1925: 205–206, by original designation; *Gygantodax* (subsequent misspelling)

*Archicnesia* Enderlein 1934a: 273 (as genus). Type species: *Simulium femineum* Edwards 1931b: 135–137, by original designation; *Archinesia* (subsequent misspelling)

**Diagnosis.** Adults: Wing without basal medial cell; radius with hair dorsobasally; $R_1$ dorsally with black spiniform setae and fine hair; $CuA_2$ straight. Calcipala present or absent; pedisulcus represented, at most, by shallow wrinkles. Female: Claws each with variously sized and shaped basal thumblike lobe or tooth. Genital fork with each arm bearing long, slender, anteriorly directed process. Male: Gonostylus typically tapered, with 1–3 apical spinules. Paramere rudimentary apically, with parameral spines apparently isolated in aedeagal membrane. Pupa: Gill of various configurations, arborescent or globose. Terminal spines thin or stout. Cocoon shapeless, covering all or part of pupa. Larva: Mandible with 3 outer teeth and 1 apical tooth. Postgenal cleft a small notch or absent. Abdominal segment IX with ventral tubercles small to prominent; anal sclerite ringlike, with posterior arms encircling posterior proleg. Rectal papillae of 3 simple or compound lobes.

**Overview.** This genus contains 7 species groups and 65 nominal species, 60 of which were treated in a revision by Wygodzinsky and Coscarón (1989). The *cortesi* species group, formerly in the genus *Gigantodax*, was assigned to a new genus, *Pedrowygomyia*, by Coscarón and Miranda-Esquivel (1998). Chromosomes of only 3 species of *Gigantodax*, all from Argentina, have been studied in detail (Coscarón Arias 1998). All but 1 species of *Gigantodax* occur south of the United States, and most are South American. Members of the genus are found from sea level to 4700 m. Until its 1994 discovery in Arizona, this genus had not been found within 1200 km of the United States (Moulton 1996).

The immature stages usually occupy small, clear to muddy streams and sometimes must tolerate water temperatures of 0°C–25°C in a single day (Wygodzinsky & Coscarón 1989). Adult behavior is not well known, although females are presumably ornithophilic. We are unaware of any host records for this genus.

### *Gigantodax wrighti* Species Group

**Diagnosis.** Female: Body color yellowish to reddish brown. Claws each with subtriangular, basal thumblike lobe. Male: Ventral plate without conspicuous concavity posteriorly. Aedeagal membrane with hair. Pupa: Gill of 4–14 tubular or globose filaments bearing short slender processes or wrinkled surface. Cocoon weakly developed. Larva: Hypostoma with median and lateral teeth typically extended anteriorly to about same level.

**Overview.** The *G. wrighti* group contains 18 species distributed from the southwestern United States to Tierra del Fuego. It has been subdivided further into 2 subgroups, based in part on whether or not the branches of the pupal gill have fine processes.

### *Gigantodax adleri* Moulton

(Figs. 4.39, 10.134, 10.290, 10.423, 10.635, 10.732; Plate 8f; Map 86)

*Gigantodax adleri* Moulton 1996: 741–751 (FMPL). Holotype* female (pinned; USNM). Arizona, Apache Co., Government Spring, 50 m south of Forest Service Road 1120, 33°59′34″N, 109°27′54″W, 2591 m elev., 21 May 1994 (J. K. Moulton)

**Taxonomy**

Moulton 1996 (FMPL).

**Cytology**

Two male larvae that we examined from Catron Co., New Mexico, had standard arm associations, the nucleolar organizer in the base of IS, and no heterozygous rearrangements.

**Molecular Systematics**

Moulton 1997, 2000 (DNA sequences).

**Bionomics**

**Habitat.** As the sole representative of the genus north of Mexico, *G. adleri* is known only from southwestern New Mexico and southeastern Arizona. The immature stages are found in springs 0.1–1.0 m wide and 8°C–10°C that percolate through the rubble of high-elevation slopes (Moulton 1996). Larvae are found on the undersides of stones and leaves of watercress (*Nasturtium officinale*). Pupae in their flimsy cocoons covered with fine debris are ensconced in the substrate or in rock crevices, under leaves, and in algal clumps (Moulton 1996).

**Oviposition.** Unknown.

**Development.** Larvae have been taken from March into October, suggesting that the species is multivoltine; the overwintering stage is not known (Moulton 1996). Pupae have been taken as early as April. Larval guts contain a preponderance of diatoms.

**Mating.** Unknown.

**Natural Enemies.** Larvae are hosts of the trichomycete fungus *Harpella* near *leptosa* (C. Beard, pers. comm.).

**Hosts and Economic Importance.** Unknown; presumably ornithophilic.

### Genus *Cnephia* Enderlein

*Cnephia* Enderlein 1921a: 199 (as genus). Type species: *Simulium pecuarum* Riley 1887: 512–513, by original designation

*Astega* Enderlein 1930: 83 (as genus). Type species: *Cnetha lapponica* Enderlein 1921b: 213 (= *Simulia pallipes* Fries 1824: 19), by original designation

**Diagnosis.** Adults: Wing with basal medial cell; basal radial cell about one third length of wing (measured from apex of stem vein); radius with hair dorsobasally; costa and apical half of $R_1$ dorsally with scattered, black spiniform setae. Calcipala minute or absent; pedisulcus represented, at most, by shallow wrinkles. Female: Claws each with minute basal tooth or thumblike lobe. Spermatheca large, spherical, wrinkled. Male: Gonostylus curved, tapered, bearing 1 apical spinule. Ventral plate in ventral view typically wider than long. Paramere with numerous short spines. Pupa: Gill of 17–50 slender filaments arising from somewhat swollen knob. Segments VIII and IX with unbranched, twisted, hooklike setae laterally. Terminal spines elongate. Cocoon shapeless, typically covering most of pupa. Larva: Head in lateral view strongly convex anterodorsally. Antenna shorter than stalk of labral fan. Hypostoma with anterior margin in form of 3 convex lobes bearing minute teeth of uniform size; lateral teeth with apices not noticeably displaced laterally from sublateral teeth. Postgenal cleft subquadrate or bluntly pointed anteriorly, extended less than one half distance to hypostomal groove. Labral fan with primary rays bearing elongate and typical-sized microtrichia. Abdomen gradually expanded posteriorly; segment IX without ventral tubercles. Rectal papillae of 3 simple lobes. Chromosomes: Standard banding sequence as given by Procunier (1982b).

**Overview.** The genus *Cnephia* is distributed throughout much of the Holarctic Region. Of the world's 8 nominal species, 4 are found in North America. Members of the genus are associated with lake and pond outlets and other productive flows, where they often achieve impressive populations. The elongate streamlined larvae pack tightly on the substrate and often pupate en masse. All species are univoltine. About half of the world's species are anautogenous, including one of history's most notorious blood feeders, *Cnephia pecuarum*. Hosts of the blood feeders are mammals and birds.

### *Cnephia dacotensis* (Dyar & Shannon)

(Figs. 4.33, 4.40, 10.18, 10.80, 10.135, 10.291, 10.424, 10.600, 10.636, 10.733; Plate 9a; Map 87)

*Eusimulium dacotense* Dyar & Shannon 1927: 20–21, Figs. 48–51 (FM). Lectotype* male (designated by Stone 1964: 25; pinned #28334; USNM). South Dakota, Brookings Co., Brookings, date unknown (J. M. Aldrich); *dakotense* (subsequent misspelling)

*Simulium* (*Eusimulium*) *lascivum* Twinn 1936a: 127–130, Plate I (Fig. 4), Fig. 8 (FMP). Holotype* female (pinned #4127; CNC). Ontario, Rockcliffe, McKay Lake, 13 May 1935 (C. R. Twinn)

'species . . . close to *C. dacotensis*' Davies et al. 1962: 81 (in part; Ontario record)

*Cnephia* 'sp. A' Pruess et al. 2000: 287, 289–291 (mtDNA)

**Taxonomy**

Dyar & Shannon 1927 (FM); Twinn 1936a (FMP); Stains 1941 (FM); Stains & Knowlton 1943 (FM); Davies 1949a (M), 1949c (ME); Nicholson 1949 (FMP); Nicholson & Mickel 1950 (FMP); DeFoliart 1951a (PL); Jamnback 1953, 1956 (L); Stone & Jamnback 1955 (FMPL); Davies & Peterson 1956 (E); Anderson 1960 (FMPL); Davies et al. 1962 (FMP); Wood et al. 1963 (L); Stone 1964 (FMPL); Abdelnur 1966, 1968 (FMPLE); Holbrook 1967 (L); Pinkovsky 1970 (MPL); Amrine 1971 (FMPL); Lewis 1973 (FMPL); Merritt et al. 1978b (PL); Westwood 1979 (FMPL); Peterson 1981 (FML), 1996 (FMPL); Adler 1983 (PL); Fredeen 1985a (FMP); Adler & Kim 1986 (PL); Currie 1986 (PL); Stuart 1995 (cocoon); Stuart & Hunter 1998b (cocoon).

**Morphology**

Krafchick 1941, 1942 (F mouthparts); Nicholson 1941, 1945 (FM mouthparts); Davies & Peterson 1956 (M ommatidial number); Wood 1963a (spermatophore, FM terminalia *in copula*); Davies 1965b (F wing and metepisternum); Guttman 1967a (labral fan); Chance 1969, 1970a (L mouthparts); Smith 1970 (ovarian follicles); Hannay & Bond 1971a (FM wing); Craig 1972a (L internal head and recurrent nerve), 1974 (labral fan); Kurtak 1973 (labral fan); Colbo et al. 1979 (FM labral and cibarial sensilla); Craig & Borkent 1980 (L palpal sensilla); Ross & Craig 1980 (L internal head); Craig & Batz 1982 (L antennal sensilla); McIver & O'Grady 1987 (M ommatidia); Currie & Craig 1988 (labral-fan microtrichia); Wood 1991 (M terminalia); Palmer & Craig 2000 (labral fan).

**Cytology**

Basrur 1957; Madahar 1967; Procunier 1974, 1975a, 1975b, 1980, 1982b, 1982c; Rothfels 1979, 1980; Adler 1986.

**Physiology**

Yang 1968 (FM trypsin), Yang & Davies 1968b (FM trypsin).

**Molecular Systematics**

Teshima 1970, 1972 (DNA hybridization); Sohn 1973 (DNA hybridization); Sohn et al. 1975 (DNA hybridization); Feraday & Leonhardt 1989 (allozymes); Zhu 1990 (mtDNA); Xiong & Kocher 1991 (mtDNA); Pruess et al. 1992, 2000 (mtDNA); Xiong 1992 (mtDNA); Tang et al. 1995b, 1996b, 1998 (mtDNA and rDNA); Zhu et al. 1998 (mtDNA).

**Bionomics**

**Habitat.** *Cnephia dacotensis* is common in a wide band across much of the continent from Quebec and Pennsylvania to western Alberta, Kansas, and Nebraska, with the Rocky Mountains serving as a natural barrier at the western fringe. Enormous masses of larvae and pupae often blanket the substrate below small impoundments, especially those enriched by pastures and feedlots. Population densities, however, can fluctuate markedly at individual sites from year to year (Chmielewski & Hall 1993). Larvae are clumped when water velocity is 50–60 cm/sec, but they lie in parallel bands perpendicular to the direction of flow when the velocity is about 30 cm/sec (Brenner & Cupp 1980a; Eymann 1991a, 1991c). These bands force the water to accelerate between adjacent larvae, thereby creating velocities at which filter feeding is more efficient (Eymann 1991a). Optimal velocities for filter feeding have been reported to be around 50 cm/sec (Kurtak 1978), although larval relocation studies suggest a preferred velocity of about 5–10 cm/sec for all instars (Gersabeck & Merritt 1979). Some of these discrepancies might be attributable to the difficulty of measuring relevant microvelocities. Larvae are rather tolerant of pH depressions to 3.5, although pupation is reduced (Chmielewski & Hall 1992).

**Oviposition.** Females are autogenous and emerge with an average of about 280–710 oocytes, depending on female size and larval nutrition (Abdelnur 1968, Chutter 1970). They can oviposit within a few hours after emergence and mating (Davies & Peterson 1956, Fredeen 1959a), but they usually require a day or longer to complete ovarian development and lay down the chorion (Abdelnur 1968, Davies & Györkös 1990). Females release eggs by tapping their abdomens to the water during short flights over the surface (Davies & Peterson 1956). Large numbers of eggs have been recovered from stream sediments (Fredeen 1959b). Females sometimes oviposit on moist logs, rocks, and grasses (Anderson 1960, Martin 1989) and can be induced to oviposit in vials with a little water (Davies & Peterson 1956).

**Development.** Larvae hatch either in the late fall or from overwintered eggs in March or April and pupate from early May into June. Larval development is strongly dependent on temperature, requiring about 5–6 weeks at 9°C–13°C (Anderson 1960), about 3 weeks at an average of 23°C (Fredeen 1959a), or 475 degree-days above 0°C, although this latter estimate is possibly too high (Ross & Merritt 1978). Early instars can develop to the pupal stage in less than 1 week at an unspecified laboratory temperature (Gersabeck & Merritt 1979). At least 6 instars are produced, but instar determination above the third is problematic (Ross & Merritt 1978). Larvae require an angle of 2 surfaces to spin their cocoons (Stuart 1995). *Cnephia dacotensis* was the second North American black fly reared from egg to adult in the laboratory, although an obligatory egg diapause precluded colonization (Fredeen 1959a).

Larvae are highly efficient at filtering a wide variety of particle sizes, probably an adaptation to the autogenous nature of the females (Kurtak 1973, 1978). Larval feeding rates are low compared with those of non-outlet-inhabiting species, possibly reflecting the nutritive quality of the seston (Morin et al. 1988b). Larvae do not drink water (Martin & Edman 1991).

Adults emerge throughout the day during May and June; the long daily emergence period might be related to the short duration of seasonal emergence (Davies 1950, Back & Harper 1979). Emergence is synchronous within a population (Davies & Peterson 1956), resulting in huge aggregations of flies crawling over every conceivable object. Males outnumber females by about 3:1 or 4:1 (Davies 1950, Back & Harper 1979). Longevity is short, although reports of adults living only a few days in captivity (Nicholson 1945) are probably underestimates. Adults apparently are not attracted to lights (Twinn 1936a).

**Mating.** This species is one of North America's few black flies that does not form aerial mating swarms. Consequently, the eyes of the male are not divided into large and small facets. Mating takes place throughout the day in massive streamside aggregations, often within minutes after emergence. Coupling lasts 3 minutes or less (Abdelnur 1968). Layers of flies sometimes cover the water, and coupling occurs as these rafts of flies float downstream. Such high densities promote mating but increase early mortality (Fredeen 1959a) because balls of flies composed of many males and a female sometimes tumble into the water (Davies & Peterson 1956). Males attempt to couple with other males and various insects but do not persist (Nicholson & Mickel 1950, Downes 1958). Contact pheromones might be involved in the mating behavior (Edman & Simmons 1985a).

**Natural Enemies.** Larvae are prone to parasitism by unidentified mermithid nematodes in Ontario (Davies et al. 1962). The microsporidium *Polydispyrenia simulii* has been recorded from larvae (D. Davies 1957). Trichomycete fungi recorded from larval guts include *Genistellospora homothallica*, *Harpella melusinae*, *Pennella* sp., *Simuliomyces microsporus*, and *Smittium* possibly *culicis*

(Labeyrie et al. 1996). Fungal hyphae found on pupae might be those of a saprophytic species (Twinn 1939). Adults are sometimes infested with larval water mites (*Sperchon ?jasperensis*) (Davies 1959).

**Hosts and Economic Importance.** Females are not fitted with biting mouthparts. They can, however, take sugar solutions, which markedly increase longevity (Smith 1970).

### *Cnephia eremites* Shewell

(Figs. 10.60, 10.136, 10.292, 10.734; Map 88)

*Cnephia eremites* Shewell 1952: 36–38 (FMP). Holotype* female (pinned #5988, pupal exuviae in alcohol vial; CNC). Nunavut (as Northwest Territories), Southampton Island, Coral Harbour, ca. 1 mi. east of Kathleen Falls on Kirchoffer River, small stream draining south end of large shallow lake, 18 July (pupa), 25 July (published date: 22 July) (adult) 1948 (G. E. Shewell); *erimites* (subsequent misspelling)

*Eusimulium* 'species A' Twinn et al. 1948: 352 (bionomics)

*Cnephia arborescens* Rubtsov 1971a: 169–172 (FPL). Holotype* female with pupal exuviae (slide #19889; ZISP). Russia, Taymyr, stream in the flooded region of the Popigaya River, 13 July 1967 (Mezenev). New Synonym

**Taxonomy**

Shewell 1952 (FMP), Stone 1952 (FMP), Sommerman 1953 (L), Davies & Peterson 1956 (E), Rubtsov 1971a (FPL), Bodrova 1977 (MPL), Patrusheva 1982 (FMPL).

Our study of the holotype of *C. arborescens*, plus chromosomal confirmation that *C. eremites* occurs in the Palearctic Region (Adler et al. 1999), substantiates the suggestion of Currie (1997a) that *arborescens* is a synonym of *eremites*. *Cnephia angarensis* Rubtsov (1956), described only from immature larvae collected in the Angara River of Yakutia, Russia, might be conspecific with *C. eremites*. All of its features, particularly the triangular postgenal cleft, match those of *C. eremites*. We have not, however, seen type material. (Curiously, Yankovsky [1995] designated a pupal lectotype and mentioned a pupal paralectotype even though the original description states that the pupa is unknown.)

**Morphology**

Davies & Peterson 1956 (M ommatidial facets, F mouthparts), Colbo et al. 1979 (F labral and cibarial sensilla).

**Cytology**

Basrur 1957; Rothfels 1979; Procunier 1980, 1982b; Adler et al. 1999.

**Bionomics**

**Habitat.** This Holarctic species breeds at lake outlets across the tundra from northern Manitoba and Nunavut to Alaska. We also have seen material from Sweden, the Taymyr Peninsula, and Chukotka. Larvae and pupae experience water temperatures of 6°C–21°C (Sommerman et al. 1955). Larvae often pupate in dense mats several layers thick.

**Oviposition.** Females are obligately autogenous and emerge with an average of about 90–180 eggs (maximum 211) half to fully mature (Davies & Peterson 1956, Smith 1970). Oviposition probably occurs shortly after emergence but has not been recorded.

**Development.** Larvae probably hatch during May and even well into June when water temperatures are about 6°C–9°C; pupae can be found from early June to mid-July (Twinn et al. 1948, Hocking & Pickering 1954, Sommerman et al. 1955). Adults are active from June to August (Currie 1997a). Females outnumber males by about 1.5 : 1.0 and can live on honey up to 21 days in the laboratory (Smith 1970).

**Mating.** Both sexes can be found on streamside vegetation, where males seek females soon after emergence; adults readily mate in captivity (Smith 1970).

**Natural Enemies.** Larvae are hosts of the chytrid fungus *Coelomycidium simulii*, the trichomycete fungus *Harpella melusinae*, and an unidentified mermithid nematode.

**Hosts and Economic Importance.** A single female has been recorded as a visitor to the flowers of Labrador lousewort (*Pedicularis labradorica*), although its activity at the flower is unknown (MacInnes 1973). The weak, nearly untoothed mouthparts render females incapable of biting.

### *Cnephia ornithophilia* Davies, Peterson & Wood

(Figs. 10.137, 10.293; Map 89)

*Cnephia ornithophilia* Davies, Peterson & Wood 1962: 102–104, Fig. 23 (F). Holotype* female (slide #7996; CNC). Ontario, Nipissing District, Algonquin Provincial Park, Wildlife Research Station, on blue jay, 27 May 1959 (G. F. Bennett); *ornitophilia* (subsequent misspelling)

*Cnephia* '(near *pecuarum*)' Hocking & Richards 1952: 241 (Labrador record) (also p. 242 as 'undescribed species of *Cnephia*')

*Cnephia* (*Cnephia*) 'U' Bennett 1960: 380 (F feeding habits)

[*Cnephia dacotensis*: Hocking & Richards (1952) not Dyar & Shannon 1927 (misident.)]

[*Cnephia pecuarum*: Jones & Richey (1956) not Riley 1887 (misident.)]

[*Cnephia pecuarum*: Davis et al. (1957) not Riley 1887 (misident.)]

[*Cnephia pecuarum*: Anthony & Richey (1958) not Riley 1887 (misident.)]

[*Cnephia* (*Cnephia*) *pecuarum*: Snoddy (1966) not Riley 1887 (misident.)]

[*Cnephia pecuarum*: Snoddy & Hays (1966) not Riley 1887 (misident.)]

[*Cnephia* (*Cnephia*) *pecuarum*: Stone & Snoddy (1969) not Riley 1887 (misident. in part, material from Alabama and Florida)]

[*Cnephia dacotensis*: Crans & McCuiston (1970a) not Dyar & Shannon 1927 (misident.)]

[*Cnephia dacotensis*: Ebsary (1973) not Dyar & Shannon 1927 (misident.)]

[*Cnephia* (*Cnephia*) *pecuarum*: Garris (1973) not Riley 1887 (misident.)]

[*Cnephia* (*Cnephia*) *dacotensis*: Lewis (1973) not Dyar & Shannon 1927 (misident. except keys)]

[*Cnephia* (*Cnephia*) *dacotensis*: Lewis & Bennett (1973) not Dyar & Shannon 1927 (misident.)]

[*Cnephia dacotensis*: Ezenwa (1974a, 1974b) not Dyar & Shannon 1927 (misident.)]

[*Cnephia* (*Cnephia*) *pecuarum*: Garris & Noblet (1976) not Riley 1887 (misident.)]

[*Cnephia* (*Cnephia*) *pecuarum*: Noblet et al. (1978) not Riley 1887 (misident.)]
[*Cnephia pecuarum*: Snoddy in Davies (1981) not Riley 1887 (misident.)]
[*Cnephia pecuarum*: Smock et al. (1985) not Riley 1887 (misident.)]

**Taxonomy**

Davies et al. 1962 (F), Snoddy 1966 (FPL), Stone & Snoddy 1969 (FPL), Amrine 1971 (FPL), Pinkovsky 1976 (FMPLE), Snoddy & Noblet 1976 (PL), Merritt et al. 1978b (PL), Webb & Brigham 1982 (L).

**Morphology**

Bennett 1963b (F salivary gland); Pinkovsky 1976 (F salivary gland); Okaeme 1983 (instars); Thompson 1987a, 1989 (labral fan); Colbo & Okaeme 1988 (instars); Evans & Adler 2000 (spermatheca).

**Physiology**

Undeen 1979b (L midgut pH).

**Cytology**

Madahar 1967; Procunier 1974, 1975a, 1975b, 1980, 1982b, 1982c; Rothfels 1979.

**Molecular Systematics**

Feraday & Leonhardt 1989 (allozymes); Moulton 1997, 2000 (DNA sequences).

**Bionomics**

**Habitat.** This species is common over an impressive range of climatic conditions from Labrador to Florida and southern Texas. In Canada, it breeds especially at outlets of impoundments. Southward it occupies the coastal plain, surviving in a wide variety of running waters, from muddy bayous to swamp streams and blackwater rivers. During larval development, most of these flows have a pH of about 4.0 to slightly above neutral and temperatures from 0°C to 15°C (as high as 24°C in Florida) (Pinkovsky 1976, Colbo 1979, Okaeme 1983). The broad geographical and habitat distribution might be related to the B chromosomes, which give this species an additional measure of variability (Procunier 1982b). Larvae sometimes pack themselves densely on substrates and can displace or exclude other species (Harding & Colbo 1981). At some sites, they occur predominantly under stones during the winter, a phenomenon perhaps driven by temperature (Okaeme 1983).

**Oviposition.** Eggs have been recovered from stream sediments, suggesting that airborne females drop their eggs into the water (Tarshis & Stuht 1970).

**Development.** Eggs are deposited in the spring and larvae begin to hatch during the fall. However, in streams that freeze solid during the winter, hatch is delayed until late winter or early spring (Merritt et al. 1978b). Larvae of this species hatch before those of *C. dacotensis* when the 2 species occupy the same streams (Merritt et al. 1978b, Back & Harper 1979). Larvae progress through about 9 instars, although instars above the third or fourth are difficult to distinguish (Colbo & Okaeme 1988). South of Pennsylvania, larvae are present from late October into April, with peak abundance from December into March. In Canada, larvae usually disappear by June and pupae appear by May. Canadian larvae generally develop at water temperatures of 0°C–5°C (Colbo & Okaeme 1988); however, adults can be reared from first instars at temperatures above 20°C (Colbo & Wotton 1981). Bacteria and diatoms provide relatively high larval growth rates compared with food sources such as leaf litter (Thompson 1987c). Late-instar larvae feed selectively on very small particles (5–15 μm), capturing a narrower size range of particles than most other species, probably because the microtrichia of the primary fans form a mesh of about 5 μm (Thompson 1987a).

Males emerge only slightly ahead of females, the latter living at least a month at 21°C–25°C (Tarshis 1973). Females fly during late May and early June in Canada (Bennett 1960, Back & Harper 1979) and from January into April south of Pennsylvania (Tarshis & Stuht 1970, Pinkovsky 1976).

**Mating.** Males swarm at dusk near treetop level, probably awaiting host-seeking females (Mokry et al. 1981).

**Natural Enemies.** Larvae are attacked by the chytrid fungus *Coelomycidium simulii*, an unidentified mermithid nematode (McCreadie & Adler 1999), and the microsporidia *Caudospora palustris* (Adler et al. 2000) and *Janacekia debaisieuxi* (Vávra & Undeen 1981). The largest microsporidian infestation ever recorded in a population of black flies involved *C. palustris* and *Cnephia ornithophilia*, the type host; densities of patently infected larvae reached 10,600/m² (Adler et al. 2000). A gram-negative bacterium has been found in the wings (Weiser & Undeen 1981). Adults have been preyed on by the cattle egret (*Bubulcus ibis*) (Snoddy in Davies 1981).

**Hosts and Economic Importance.** Females search for hosts above the forest floor and feed chiefly during sunset on a variety of birds including mallards (domestic ducks), sharp-shinned hawks, ruffed grouse, gray jays, blue jays, American crows, common ravens, American robins, white-throated sparrows, common grackles (Bennett 1960), and northern flickers (Stone & Snoddy 1969). Chickens and turkeys also are attacked (Pinkovsky 1976). For example, we have females replete with blood that were collected from chickens in early January along the South Carolina coast. Feeding is often centered around the anus and heels and tends to cause lacerations that remain inflamed for up to 3 days (Bennett 1960). Females become engorged in 5–8 minutes and can digest all blood in 72–96 hours at 21°C–25°C (Tarshis 1972). Small numbers of females have been taken in cattle-baited traps (McCreadie et al. 1985), perhaps incidentally. We have experienced minor nuisance problems when populations were large, and we have 1 specimen that was taken as it was biting a human.

Females have been implicated in the transmission of the following *Leucocytozoon* protozoa to woodland birds: *L. cambournaci* (as *L. fringillinarum* in part), *L. dubreuili* (as *L. mirandae*), *L. icteris* (as *L. fringillinarum* in part) (Fallis & Bennett 1961, 1962; Khan & Fallis 1970). Under laboratory conditions, females can transmit *L. simondi* to mallard ducklings but not to domestic goslings (Tarshis 1976). They probably do not transmit *L. smithi* to turkeys (Pinkovsky et al. 1981).

### Cnephia *pecuarum* (Riley)

(Figs. 10.138, 10.294; Plate 9b; Map 90)

*Simulium pecuarum* Riley 1887: 512–513, Plates VI–VIII (FMPL). Lectotype* female (here designated; pinned

#772; USNM). Louisiana, Tensas Parish, Somerset Landing (ca. 32°09′09″N, 91°10′36″W), 5 mi. from Mississippi River (as it existed in 1886), 10 April 1886 (F. M. Webster); *pecuarium* (subsequent misspelling)

'destructive Insect' Cornelius 1818: 328–330 (pest of horses)

*Simulium* 'sp.' Packard 1886: 650 (bionomics)

'probably an undescribed species' Doran 1887: 239–242 (pest status)

*Cnephia* 'sp. B' Pruess et al. 2000: 287, 289–291 (mtDNA)

*Cnephia* 'sp. C' Pruess et al. 2000: 287, 289–291 (mtDNA)

[*Simulium meridionale*: Lugger (1897a, 1897b) not Riley 1887 (misident.)]

[*Simulium invenustum*: Coquillett (1898) not Walker 1848 (misident. in part, material from Mississippi and Louisiana)]

[*Simulium invenustum*: Howard (1901) not Walker 1848 (misident.)]

[*Simulium invenustum*: Webster (1904) not Walker 1848 (misident.)]

**Taxonomy**

Riley 1887 (FMPL); Osborn 1896 (FMPL); Johannsen 1902, 1903b (FMPL), 1934 (PL); Webster 1904 (FL); Emery 1913 (FMPL); Malloch 1914 (FMPL); Riley & Johannsen 1915 (F); Dyar & Shannon 1927 (F); Bradley 1935a (ME); Stains 1941 (F); Stains & Knowlton 1943 (F); Davies et al. 1962 (F); Snoddy 1966 (FMPL); Stone & Snoddy 1969 (FMPL); Amrine 1971 (FMPL); Snoddy & Noblet 1976 (PL); Rubtsov & Petrova 1977 (FMPL); Webb & Brigham 1982 (L).

Riley (1887) examined an unknown number of specimens when he described *C. pecuarum* but did not designate a holotype. We designate, as lectotype, the single female in the USNM type collection to establish the current concept of the name.

**Morphology**

Smith 1890 (F mouthparts), King 1991 (adult cardia).

**Cytology**

Rothfels 1979; Procunier in Rothfels 1980; Procunier 1980, 1982b.

**Molecular Systematics**

Pruess et al. 2000 (mtDNA).

**Bionomics**

**Habitat.** The infamous southern buffalo gnat breeds in the low country of the Mississippi River Valley from Illinois and Indiana to the Gulf Coast. Larvae and pupae inhabit swollen turbid bayous and rivers broken with debris jams and submerged vegetation. Pupae can be found at depths up to 3 m (Riley 1887). Females have a flight range of about 24 km (Robinson in Atwood 1996).

**Oviposition.** Females carry 500–750 eggs (Webster 1914). They oviposit just before dusk, flying low and upstream near the middle of rivers, where they periodically dip to the surface to release their eggs; these females show evidence of blood meals and will oviposit in jars (Bradley 1935a).

**Development.** Eggs are laid in the spring and undergo obligatory diapause. Larvae hatch from October to December and develop at water temperatures from 2°C to 20°C (Bradley 1935b, Atwood et al. 1992). When this species occurs in the same streams with *C. ornithophilia*, it is about 2 larval instars ahead in development (Procunier 1982b). Adults fly primarily from November through February in Louisiana and Texas and as late as April in Arkansas and May in Illinois and Indiana.

**Mating.** Males form large loose swarms 2–3 m above open fields during sunny afternoons and fly with their legs hanging down and abdomens curled upward (Bradley 1935a). One swarm of males extended to a height 6–7 m above a vehicle late in the afternoon (Atwood 1996). Coupled pairs fly to the ground or to nearby vegetation and buildings (Bradley 1935a) and remain joined for up to 2 minutes or more (Atwood 1996).

**Natural Enemies.** Larvae are attacked by the microsporidium *Janacekia debaisieuxi* and an unidentified mermithid nematode. Predators of larvae include cyprinid fish; adults are consumed by northern mockingbirds (*Mimus polyglottos*), winter wrens (*Troglodytes troglodytes*), chickens, dragonflies, and asilid flies (Riley 1887).

**Hosts and Economic Importance.** Females attack humans, dogs, cats, horses, mules, pigs, white-tailed deer, cattle, and sheep. They also might attack birds, although many of the bird-feeding records (e.g., Riley 1887) probably pertain to *C. ornithophilia*, *Simulium meridionale*, and perhaps *S. parmatum* n. sp. *Cnephia pecuarum* is the only major mammal feeder in North America that is fitted with a basal thumblike lobe on the tarsal claw, a characteristic of bird feeders. Host-seeking females are most active during early morning and evening but also fly during moonlit nights and cloudy days (Riley 1887). Octenol sometimes enhances the attractiveness of carbon dioxide in bait traps (Atwood & Meisch 1993). Adults visit the flowers of spreading chervil (*Chaerophyllum procumbens*), sassafras (*Sassafras albidum* as *S. variifolium*), sandbar willow (*Salix exigua* as *S. longifolia*), and golden alexanders (*Zizia aurea*) (Robertson 1928).

The menace of this species dates by oral record to the earliest settlers of the Mississippi River Valley. The first known written account of what is surely this species comes from 1818 and recounts the destruction of horses in Choctaw country (now Mississippi) (Cornelius 1818). Shortly thereafter about 40 deaths of horses were recorded in Shelby County, Tennessee (Fessenden 1826). Other early records of its ravages include 1843 in Illinois and Indiana and 1846 in Louisiana (Riley 1887, Webster 1891), but it was the years during and after the Civil War that produced the legendary attacks of the dreaded buffalo gnats along the lower Mississippi River (Webster 1887). In the words of one who experienced the buffalo gnats during these years, they were "the greatest pest that has ever afflicted this country" (Gunby in Riley 1884). An experienced observer in Tennessee described the buffalo gnat as "a blood-thirsty monster, and more reckless of his own life than Herr Most, the bloody anarchist" (Anonymous in Doran 1887). From the perspective of a Union soldier during the Civil War, "to be besieged by Arkansas gnats is absolutely beyond endurance" (Browne 1865).

The deterioration of the great levees on the Mississippi River during and after the Civil War caused massive overflows into the brushy alluvial zone, creating vast breeding grounds (Webster 1904). Gargantuan debris rafts in some bayous caused flooding that provided additional breeding sites, and cries for government removal of the rafts often were made (Marlatt 1889, Riley & Howard 1889). Invaded areas were afflicted for a few days to 6 weeks. Huge clouds of females—compared to swarms of bees by one observer (Boardman in Webster 1891)—would suddenly issue from dense vegetation or low-lying wet areas to overtake animals. These immense swarms sometimes were swept in by prevailing winds blowing from the breeding areas up to 19 km away (Webster 1902). At times, a single sweep of the hand through the air gleaned hundreds of flies (Riley 1887).

Farm animals reacted frantically to attacks, bellowing with pain, running, rolling, and dashing through dense thickets and into streams. As the armies of the North and South battered one another during the Civil War, many cavalry horses and artillery mules of both armies fell dead under the onslaught of the southern buffalo gnat (Webster 1904). The gnats wreaked havoc on the horses, mules, and cattle during the Vicksburg campaign of 1863, killing a number of stock (Maury 1894, p. 177). Plantation owners frequently lost every horse and mule in their charge. Disastrous problems continued after the Civil War. More than 4000 mules and horses were killed during a few days in 1866 in 3 parishes in Louisiana. In a single week in 1882, 3200 head of stock died in Franklin Parish, Louisiana, and 1 county in western Tennessee suffered half a million dollars in losses (Webster 1889). Livestock kills also occurred in southern Illinois and Indiana in 1882–1884 (Webster 1891). In the following 2 years, more than 2200 mules were killed in 3 counties in Mississippi (Anonymous 1883, 1884). More than 300 mules, worth about $37,500, died in 3 or 4 Louisiana parishes in 1889 (Webster 1889). Mules and horses on the streets and in stables were killed in Vicksburg, Mississippi, and in Memphis, Tennessee (Webster 1904). An estimated 1000 horses died from attacks in western Kentucky in 1892 (Riley & Howard 1892a). The flies were said to have killed horses and mules within a few hours (Thompson in Riley 1884). Pigs were killed by the attacks, and tormented sheep often burned to death when they approached too close to the smoldering fires set to provide protective smoke. White-tailed deer were sometimes driven to the fires for protection and would allow people to rub the flies from their bodies; some were said to lie down in the burning embers or hot ashes (Riley 1887). The deer were claimed to have been virtually exterminated in parts of Louisiana during the attacks of 1882 (Webster 1904). During invasions, enormous numbers of flies often covered the windows of buildings (Riley 1887).

Most livestock deaths probably were attributable to toxic shock from the introduction of saliva, or to exsanguination, so much blood being lost that it became too thick to transport oxygen efficiently. Some livestock deaths were rumored to have been caused by suffocation from flies entering the nostrils (Tucker 1918) or from subsequent respiratory infection (Atwood 1996). The collection of the USNM contains females that were removed from the respiratory tracts of cattle. Moribund livestock putatively were revived by immersing them in streams (Bishopp 1913); administering repeated doses of whiskey, sometimes mixed with carbonate of ammonia (Herrick 1899) or camphor (Barnett in Harned 1927); or using nitrates and similar remedies that caused copious urination (Marten in Webster 1891). More hideous measures called for oral dosages of liquid ammonia in warm lard oil every 2 hours until the animals recovered (Webster 1889).

From the late 1800s through the end of the 20th century, the southern buffalo gnat resurged many times, often with fatal consequences. Reprieves were granted in years when water levels were low (Frierson 1892, Cockerell 1895, Howard 1895), and gnat problems along the Mississippi River eased somewhat after 1887 when levees were built or reconstructed to control the immense overflows. Yet in 1905, the deaths of more than 200 stock were reported in Louisiana (Tucker 1918), and in 1917 some mules were killed in western Kentucky (Garman 1917). From 1927 to at least 1934, another series of outbreaks resulted in the deaths of more than 1800 horses and mules in Mississippi and eastern Arkansas (Harned 1927, 1929, 1931; Bradley & McNeel 1928; Bradley 1932, 1934, 1935a; State Plant Board of Mississippi 1933; Anonymous 1934; Jones 1934). February 1927 was particularly dreadful in Yazoo County, Mississippi. Livestock under attack rolled and squatted, swelling, running against objects, and soon falling dead; nearly 75 animals died and the udders of cows were badly bloodied (Barnett in Harned 1927). The Red Cross was called on for assistance during the 1931 outbreaks in Mississippi (Harned 1931), and the 1934 outbreak was so bad that the windshields of passing automobiles became plastered (Jones 1934). In subsequent years, the flies continued to be troublesome in Mississippi and eastern Arkansas, but no livestock deaths were recorded in those states (e.g., Lyle & Assistants 1935, Schwardt 1935a, Muldrow 1937). About 100 livestock were killed in 1936 in western Kentucky (Price 1936). In 1961, flies were blamed for the deaths of 15 horses and mules in northeastern Louisiana (Anonymous 1961). These outbreaks of the early to mid-20th century probably resulted from flooding of tributaries of the Mississippi River.

The year 1979 brought a severe outbreak to the area around the Sulphur River in southwestern Arkansas and eastern Texas (Atwood et al. 1992), where problems had occurred for much of the century (Robertson 1937, Schwardt 1938). That year in Bowie Co., Texas, the flies killed 42 calves, 2 cows, and 12 horses (Robinson in Atwood 1996), and in eastern Texas, cattle ranchers lost about $700,000 (Guyette 1994). From 1980 to 1989, 17 additional calf deaths were reported in Bowie Co., Texas, and Miller Co., Arkansas, and the total economic impact on cattle producers in the area from 1979 to 1989 was estimated to run as high as $5.4 million per year (Atwood 1996). Individual horse breeders estimated losses of up to $11,250 in stud fees during the 1988–1989 season, and pet owners reported some deaths and frequent hospitalization of their animals (Atwood 1996). The flies seldom bit people during these outbreaks but created great annoyance by swarming around the face. Problems in eastern Texas and southwestern Arkansas continued through the end of the

20th century, with not only livestock deaths but also losses to the paper industry from flies that were squashed in rolls of freshly milled paper (Guyette 1999). In 1992, calves were killed by the bites, and horses were injured as they attempted to escape the flies (J. V. Robinson, pers. comm.). In the winter of 2002, heavy rains made management of larval populations difficult and the adults wreaked havoc.

Females bite humans, sometimes with grievous consequences, although most reports of this nature date from the 1800s. Francis Webster (1914), as C. V. Riley's point man in the field, studied the buffalo gnat during its heyday and recounted being bitten so many times about the face and neck that shaving became impossible for weeks. He also told of civil engineers working on levees of the St. Francis River at Madison, Arkansas, who were overwhelmed by flies crawling under their clothes and down their boots, biting with such ferocity that months later their calves and ankles appeared as though they had been severely beaten. Webster (1904) compared the bite to a puncture "by a blunt, hot pin or awl, leaving behind a dull aching pain." Human deaths, said to have been the result of relentless attacks by buffalo gnats, were reported in Louisiana and Arkansas during the early 1880s (Doran 1887, Riley 1887, Buck in Riley 1888b, Webster 1904). Some of these cases are probably authentic. Crosskey (1990), however, has speculated that some of the deaths might have been caused by bites from rattlesnakes (*Crotalus*).

### Genus *Ectemnia* Enderlein

*Ectemnia* Enderlein 1930: 88 (as genus). Type species: *Cnetha taeniatifrons* Enderlein 1925: 206–207, by original designation

**Diagnosis.** Adults: Wing with basal medial cell; costa with spinules and fine setae; radius with hair dorsobasally. Anepisternal membrane bare. Calcipala small; pedisulcus represented, at most, by shallow wrinkles. Female: Cibarium with distal margin bearing 1–11 setae. Subcosta setose ventrally. Claws each with large, basal thumblike lobe. Spermatheca elongate, wrinkled, unpigmented basally. Male: Ventral plate in ventral view with body about 2.5 times broader than long; gonostylus with 2 or 3 apical spinules; parameral spines short, numerous. Pupa: Gill of 8–10 apically convergent filaments, with imbricate surface sculpture. Abdominal segment V with stout hooks; segments VI–VIII with spine combs. Terminal spines short, stout, conical, hollow. Cocoon slipper shaped, with irregular anterior margin, situated subapically on stalk. Larva: Hypostoma concave anteriorly, with all teeth small; 2 or 3 paralateral teeth per side; lateral serrations absent. Mandible with inner subapical margin smooth. Postgenal cleft extended about one third to one half distance to hypostomal groove, rounded anteriorly. Abdomen with ventrolateral flanges on segments V–VIII; segment IX with prominent ventral tubercles; anal sclerite absent. Rectal papillae of 3 simple lobes. Chromosomes: Standard banding sequence as given by Moulton and Adler (1997).

**Overview.** The genus *Ectemnia* and its 4 constituent species are endemic to Canada and the eastern United States. A single fossil, however, is known from Baltic amber (Crosskey 1994b). The genus was revised by Moulton and Adler (1997). The larvae and pupae of these univoltine species are often difficult to collect in the fast, deep, and frigid waters of winter and early spring. Larvae fashion a silk stalk up to 30 mm long that is impregnated with adherent detritus. This stalk-building habit was first observed in 1947 (Fredeen 1951). The only other black flies in which stalk building occurs are a few Australian species, which feed while situated on the stalk but abandon it before pupation (Moulton & Adler 1997). The stalk of *Ectemnia* is attached to the substrate by a splayed silk holdfast. During construction, the larva sits subapically on the stalk and applies silk the entire length, from the apex onto the substrate, both perpendicular and parallel to the stalk (Stuart & Hunter 1998a). We have seen second and subsequent instars on stalks, but we do not know if first instars construct stalks. Larvae engage in filter feeding and pupate while positioned subapically on the stalk. Parasites and pathogens rarely have been recorded in the larvae of this genus. We do not know if this is an artifact of sampling or a real trend, possibly related to life on the stalk. Little is known of adult behavior. Females are ornithophilic but have not been implicated as pests.

#### *Ectemnia invenusta* (Walker)

(Figs. 4.8, 4.47, 10.139, 10.295, 10.425, 10.578, 10.637, 10.735; Plate 9c; Map 91)

*Simulium invenustum* Walker 1848: 112 (FM, sexes not differentiated in description). Lectotype* male (designated by Crosskey in Crosskey & Lowry 1990: 217; pinned, terminalia on celluloid mount below; BMNH). Ontario, Hudson Bay, Martin Falls (as St. Martin's Falls), Albany River, date unknown but probably between 1834 and 1843 (1844 is the year given, but is actually the date of registration in the collection [see Arthur 1985]) (G. Barnston)

'? Nov. sp.' Jamnback 1951: 3 (New York record)

*Cnephia* 'sp. 1' Jamnback 1953: 43 (L)

*Cnephia* (*Ectemnia*) *loisae* Stone & Jamnback 1955: 35–38, Figs. 22, 39, 65, 72, 85, 102, 111 (FMPL). Holotype* male with pupal exuviae (separate slides; USNM). New York, Oneida Co., Forestport, Pine Creek, 10 April 1952 (H. A. Jamnback)

[*Cnephia taeniatifrons*: MacNay (1956) not Enderlein 1925 (misident.)]

**Taxonomy**

Walker 1848 (FM); Johannsen 1902, 1903b (F); Emery 1913 (F); Jamnback 1953 (L); Stone & Jamnback 1955 (FMPL); Wolfe & Peterson 1959 (P); Davies et al. 1962 (FMP); Wood et al. 1963 (L); Stone 1964 (FMPL); Peterson 1981 (L), 1996 (PL); Currie 1986 (PL); Moulton & Adler 1997 (FMPL).

**Morphology**

Guttman 1967a (labral fan), Colbo et al. 1979 (FM labral and cibarial sensilla), Stuart & Hunter 1998a (cocoon construction).

**Cytology**

Madahar 1967, Rothfels 1979, Moulton & Adler 1997.

Larvae from Pennsylvania have undifferentiated sex chromosomes, although the IIIL-6 inversion of

Moulton and Adler (1997) is present as an autosomal polymorphism.

**Bionomics**

**Habitat.** *Ectemnia invenusta* ranges from Alberta to eastern Canada, southward along the Appalachian Mountains to Georgia. Larvae and pupae dwell in swift, bitterly cold streams and rivers, where they attach their stalks to rocks, logs, and submerged mosses at depths up to 1.5 m.

**Oviposition.** Unknown.

**Development.** Larvae hatch in the fall, usually in October, and develop during the winter. Construction of the stalk is interspersed with filter feeding (Stuart & Hunter 1998a). This species is one of the earliest black flies on the wing, emerging from early April through May in Canada (Wolfe & Peterson 1959, Davies et al. 1962). At the southern extreme of the range, adults emerge in late January and February.

**Mating.** Unknown.

**Natural Enemies.** No records.

**Hosts and Economic Importance.** Females feed on Canada geese, mallards, and ruffed grouse, but they do not transmit the agents of *Leucocytozoon* disease (Bennett 1960, Tarshis & Herman 1965, Tarshis 1972). One study reported females flying about humans (White & Morris 1985b).

### *Ectemnia primaeva* Moulton & Adler

(Figs. 10.579, 10.736; Plate 9d; Map 92)

*Ectemnia primaeva* Moulton & Adler 1997: 1899–1914 (FMPLC). Holotype* male (pupal exuviae in glycerin vial below; USNM). South Carolina, Williamsburg Co., Black River, Rt. 377, 33°35.2′N, 79°49.1′W, 31 December 1991 (J. K. Moulton)

*Ectemnia* 'p' Crosskey 1994b: 279 (F calcipala and pedisulcus)

**Taxonomy**

Moulton & Adler 1997 (FMPL).

**Cytology**

Moulton & Adler 1997.

**Bionomics**

**Habitat.** *Ectemnia primaeva* resides in the coastal plain and piedmont from Virginia to Texas. The immature stages are found in blackwater to clear sandy rivers where they are often inaccessible without a boat. This species is probably quite common but seldom encountered because of its development in the dead of winter. Larvae and pupae attach their stalks to submerged branches, twigs, debris, stalks of conspecifics, and the posterior ends of cases of brachycentrid caddisflies such as *Brachycentrus numerosus*.

**Oviposition.** Unknown.

**Development.** Larvae hatch in the fall, begin pupating in late December, and usually disappear, along with pupae, by the end of January, although in some years large populations persist well into February. Adults fly from January to mid-March.

**Mating.** Unknown.

**Natural Enemies.** We have recovered unidentified mermithid nematodes from 3 larvae and the chytrid fungus *Coleomycidium simulii* from 1 larva. Larval guts are often bristling with trichomycete fungi, especially of the genus *Pennella*.

**Hosts and Economic Importance.** Unknown; presumably ornithophilic.

### *Ectemnia reclusa* Moulton & Adler

(Figs. 10.296, 10.426, 10.737; Plate 9e; Map 93)

*Ectemnia reclusa* Moulton & Adler 1997: 1901–1914 (FMPLC). Holotype* male (pupal exuviae in glycerin vial below; USNM). South Carolina, Kershaw Co., Spears Creek, Porter's Crossroads (Co. Road 47), 34°07.6′N, 80°42.4′W, 2 January 1993 (J. K. Moulton)

*Ectemnia* 'r' Crosskey 1994b: 279 (F calcipala and pedisulcus)

[*Cnephia invenusta*: Garris (1973) not Walker 1848 (misident.)]

[*Cnephia invenusta*: Arnold (1974) not Walker 1848 (misident.)]

[*Cnephia invenusta*: Garris & Noblet (1976) not Walker 1848 (misident.)]

[*Cnephia* (*Ectemnia*) *invenusta*: Noblet et al. (1978) not Walker 1848 (misident.)]

[*Ectemnia invenusta*: Lake (1983) not Walker 1848 (misident.)]

**Taxonomy**

Moulton & Adler 1997 (FMPL).

**Cytology**

Moulton & Adler 1997.

**Molecular Systematics**

Moulton 1997, 2000 (DNA sequences).

**Bionomics**

**Habitat.** *Ectemnia reclusa* occurs in small populations in the coastal plain and the Sandhills Region from North Carolina to southern Georgia. The immatures of this species are found in smaller watercourses than are those of other species in the genus, but they can be difficult to collect in the deep brownwater streams bordered with thickets of greenbriar (*Smilax*). Larvae trail on stalks attached to the tips of small twigs, submerged forbs, and even to the posterior ends of cases of the caddisfly *Brachycentrus chelatus*. Stalks can be arborescent when small larvae construct their stalks on those of larger larvae. On contact, larvae often release from their stalks.

**Oviposition.** Unknown.

**Development.** Larvae begin to hatch in October. Most larvae pupate in January and February, although some pupate as early as the first of December and as late as early March.

**Mating.** Unknown.

**Natural Enemies.** We have found 1 mid-instar larva with the chytrid fungus *Coelomycidium simulii* and numerous larvae with the trichomycete fungus *Harpella melusinae* in their midguts.

**Hosts and Economic Importance.** Unknown; presumably ornithophilic.

### *Ectemnia taeniatifrons* (Enderlein)

(Figs. 4.3, 4.41, 10.19, 10.40, 10.45, 10.61, 10.140, 10.297, 10.580, 10.601, 10.638; Map 94)

*Cnetha taeniatifrons* Enderlein 1925: 206–207 (F). Holotype* female (pinned; ZMHU). Illinois, county unknown, north of St. Louis, Mississippi (presumably River), date and collector unknown

'Simuliid 1' Fredeen 1951: 12, 95–97, Plate 18 (Figs. H–K) (L)

[*Cnephia invenustum*: Nicholson (1949) not Walker 1848 (misident. in part, all material except 1 F from Pine Co.)]

[*Cnephia invenustum*: Nicholson & Mickel (1950) not Walker 1848 (misident. in part, all material except 1 F from Pine Co.)]

[*Cnephia invenustum*: Fredeen (1951) not Walker 1848 (misident.)]

**Taxonomy**

Enderlein 1925 (F); Nicholson 1949 (FM); Nicholson & Mickel 1950 (FM); Fredeen 1951 (FML), 1981a (FMPL), 1985a (FMP); Anderson 1960 (FMPL); Peterson 1981 (F), 1996 (FP); Currie 1986 (PL); Mason & Kusters 1991b (FMPL); Moulton & Adler 1997 (FMPL).

**Cytology**

Madahar 1967, Rothfels 1979, Moulton & Adler 1997.

**Bionomics**

**Habitat.** The distribution of this species is centered in the north-central states and provinces. Larvae and pupae are found in swift rivers at depths up to a meter. Larvae often attach their stalks to the undersurfaces of rocks, whereas pupae are more likely to be found on the tops and sides (Anderson & Dicke 1960).

**Oviposition.** Unknown.

**Development.** Larvae hatch in late summer and fall and often develop under ice. Pupae appear as early as March in water slightly above freezing. This species is one of the first to emerge each year, with adults taking to the air in late March and April before the first spring plants begin to flower. Adults disappear by the end of May, although at higher latitudes such as Churchill, Manitoba, they fly into July.

**Mating.** Mating takes place during the evening in sparse swarms 3 m or more above the ground in clearings among trees (Fredeen 1951, 1981a).

**Natural Enemies.** No records.

**Hosts and Economic Importance.** Both sexes feed on sap seeping from wounds in the bark of birch trees (Fredeen 1951, 1981a). Females feed on birds such as chickens, ring-necked pheasants, ruffed grouse, and especially domestic turkeys, with the flies landing on the backs or wings and crawling beneath the feathers (Anderson & DeFoliart 1961). A small sample of females has been screened for eastern encephalitis virus, with negative results (Anderson et al. 1961). Females are attracted to humans, and in areas such as Saskatchewan, they sometimes bite (Anderson & DeFoliart 1961, Fredeen 1985a).

## Genus *Metacnephia* Crosskey

*Metacnephia* Crosskey 1969: 26–30 (as genus). Type species: *Cnephia saileri* Stone 1952: 82–84, by original designation

*Cnephia pallipes* 'group' (by virtue of misidentification of *saileri* as *pallipes*) Rubtsov 1956: 295

*Cnephia* 'Group B' Madahar 1967: 14–15, 20–24

**Diagnosis.** Adults: Wing with basal medial cell; apical one half to two thirds of $R_1$ dorsally with numerous black spiniform setae and fine hair; radius with hair dorsobasally. Calcipala minute or absent; pedisulcus represented, at most, by shallow wrinkles. Katepisternal sulcus typically narrow, deep, nearly complete anteriorly. Anepisternal membrane typically haired. Female: Maxillary palpomere V elongate, typically much longer than palpomere III. Claws each with basal thumblike lobe. Spermatheca with small unpigmented ring basally, or with pigmentation extended onto spermathecal duct. Male: Gonostylus tapered, slightly curved, of various diameters, with 1 apical spinule. Ventral plate subrectangular or subtriangular. Paramere with numerous short stout spines. Pupa: Gill of 12 to more than 150 filaments, in some species arising from swollen trunks. Abdominal segments VII and VIII (sometimes V and VI) with spine combs; segments VIII and IX with anchor- and grapnel-shaped setae laterally. Terminal spines minute. Cocoon typically boot shaped (shoe shaped in *M. borealis*, pedunculate in the Palearctic *M. pedipupalis*). Larva: Antenna with proximal and medial articles unpigmented, contrasting with dark brown distal article. Hypostoma with all teeth minute; lateral and/or paralateral teeth directed anteromedially; setae of central area often bifid. Postgenal cleft extended to or beyond hypostomal groove, with lateral margins subparallel or convergent. Abdomen gradually expanded posteriorly; segment IX without ventral tubercles. Rectal papillae of 3 simple lobes. Chromosomes: Whole-arm interchange (IS + IIL, IL + IIS). Standard banding sequence as given by Procunier (1982a).

**Overview.** The world *Metacnephia* fauna consists of about 55 nominal species distributed throughout much of the Holarctic Region. Seven species occupy North America. Most members of the genus occur at northern latitudes or high altitudes and breed in a variety of habitats ranging from lake outlets and small temporary streams to large rivers. A few Palearctic species are adapted for life in low-elevation streams in hot areas such as Andalusia, Spain. All species are univoltine, although some populations of *M. saileri* are possibly bivoltine. Several species have nonbiting mouthparts. Biting species have claws designed to feed on birds, but hosts and adult behavior in general are poorly known.

### *Metacnephia borealis* (Malloch)

(Figs. 10.62, 10.141, 10.298, 10.581, 10.602; Map 95)

*Prosimulium borealis* Malloch 1919: 41–42 (F, incorrectly given as M). Holotype* female (pinned #1148, terminalia on slide #E66; CNC). Northwest Territories, Victoria Island, Wollaston Peninsula, summer 1915 (D. Jenness)

*Simulium* 'sp. 3' Malloch 1919: 43, Fig. 13 (P)

*Simulium* (*Cnephia*) 'sp. indet.' Edwards 1933: 614 (3 F from Akpatok Island)

*Cnephia* 'sp. X' Davies 1950: 358 (infestation by water mites) (also p. 358 as 'species of *Cnephia*')

*Metacnephia arctocanadensis* Yankovsky 1996: 115 (unnecessary substitute name)

[*Metacnephia tredecimata*: Bodrova (1980) not Edwards 1920 (misident.)]

**Taxonomy**

Malloch 1919 (FP), Twinn 1936a (FM), Stains 1941 (FM), Stains & Knowlton 1943 (FM), Smith 1970 (FMPL), Bodrova 1980 (FP).

The number of filaments in the pupal gill varies from 10 to 14 but is typically 12 or 13.

**Morphology**

Smith 1970 (ovarian follicles), Davies 1989 (gynandromorph), Crosskey 1990 (gynandromorph).

**Cytology**

Procunier 1980, 1982a.

**Bionomics**

**Habitat.** This species, 1 of the 3 most northern black flies in North America, is indigenous to the arctic. In the Nearctic Region it occurs from Baffin Island and northern Quebec west to the Mackenzie River Valley. Its absence from Alaska and the Yukon is puzzling, given the large tracts of suitable habitat and its existence in the Russian Far East (Currie 1997a). We predict that it eventually will be found on the North Slope of Alaska and the Yukon. Larvae and pupae are often abundant in large swift rocky streams and rivers but also can be found in streams a few meters wide.

**Oviposition.** Females are autogenous, emerging with follicles in an advanced stage and producing up to 158 eggs, with an average of about 50; egg production depends on larval nutrition (Smith 1970).

**Development.** Larvae probably begin to hatch from late spring to early summer. Most adults fly during July and August. Adults can live up to 26 days in captivity on honey; sex ratios are skewed in favor of females (Smith 1970).

**Mating.** Adults mate readily in vials under some conditions (Downes 1962) but not others (Smith 1970). Large numbers of adults sometimes can be found on streamside vegetation; however, males often form swarms from which a mating pair has been recovered (Smith 1970). Thus, *M. borealis* apparently couples in the air and is an exception to the rule that highly precocious ovarian development in arctic species is correlated with ground-mating habits.

**Natural Enemies.** We have found larvae infected with the chytrid fungus *Coelomycidium simulii* and the microsporidia *Amblyospora bracteata*, *Janacekia debaisieuxi*, and *Polydispyrenia simulii*. Water mites (*Sperchon ?jasperensis*) sometimes infest adults (Davies 1959).

**Hosts and Economic Importance.** The mouthparts are not designed to cut flesh.

### *Metacnephia coloradensis* Peterson & Kondratieff
(Map 96)

*Metacnephia coloradensis* Peterson & Kondratieff 1995 ("1994"): 19–24, Figs. 19–28 (FMPL). Holotype male and pupal exuviae (alcohol vial; putatively USNM but presumably lost). Colorado, Boulder Co., North Boulder Creek between Green Lakes 5 and 4, 3566 m elev., 23 August 1988 (B. V. Peterson & S. K. Wu)

Simuliidae 'sp.' Saether 1970: 105 (P)

*Metacnephia* 'sp. near *jeanae*' Bushnell et al. 1987: 506–508 (Colorado record)

'new species of *Metacnephia*' Peterson 1989: 327 (Colorado record)

**Taxonomy**

Saether 1970 (P), Peterson & Kondratieff 1995 (FMPL).

The holotype could not be located in the putative depository (USNM) or through communication with either author of the name.

**Cytology**

*Metacnephia coloradensis* from the type locality (22 August 1997) has a chromosomal banding sequence identical to that of its sister species *M. sommermanae*, except that IL is standard. The Y sequence of *M. coloradensis*, however, is equivalent to the IL inversion that is fixed in *M. sommermanae*, such that males carry a large loop in IL.

**Bionomics**

**Habitat.** This alpine species is possibly endemic to lake and stream chains in the glacially formed valleys of Colorado at elevations above 3500 m. The lake outlets where the immatures live are rocky, cold (<10°C), and remarkably productive, with a pH of 6–7. At these outlets, this species is the numerically dominant organism (Bushnell et al. 1987). Larvae and pupae cluster in great numbers on the sides and upper surfaces of rocks. We recorded densities of more than 350,000 penultimate and final instars per square meter at the type locality, with silk production so great that webs 20 cm wide and more than half a meter long trailed in the current. A single sweep of the hand across a rock yielded a handful of larvae. Pupae in their pale cocoons render the streambed gray, interrupted by bands of brown larvae.

**Oviposition.** Females emerge with 80–100 large eggs (Peterson & Kondratieff 1995).

**Development.** Larvae and pupae of this late-maturing species have been collected from July to September. Larvae nearest the lake outlet develop slightly earlier and grow larger than those downstream. Adults are produced from late August into September. Larval guts are packed with bacteria.

**Mating.** The undifferentiated eyes of the male and weak flight abilities are associated with mating on the ground. Just after emergence, adults mate on rocks in and beside the stream in groups of a few to several hundred flies (Peterson & Kondratieff 1995).

**Natural Enemies.** Larval hindguts typically contain unidentified trichomycete fungi.

**Hosts and Economic Importance.** The mouthparts lack serrations, rendering females incapable of acquiring blood.

### *Metacnephia jeanae* (DeFoliart & Peterson)
(Figs. 4.12, 10.582; Map 97)

*Cnephia jeanae* DeFoliart & Peterson 1960: 218–219, Figs. 15–25 (FMPLE). Holotype* male with pupal exuviae (alcohol vial; USNM). Utah, Summit Co., Chalk Creek Canyon, 15 June 1958 (B. V. Peterson)

*Cnephia macrocerca* Peterson 1958a: 164–168. Unavailable (unpublished thesis name)

**Taxonomy**

Peterson 1958a (FMPL), 1960a (FMP), 1981 (F), 1996 (FP); DeFoliart & Peterson 1960 (FMPLE).

Only the female of *M. macrocerca*, which was described in detail in a doctoral thesis (Peterson 1958a), is known. It resembles that of *M. jeanae* but is said to lack anepisternal hair, which is likely to be lost in some field-collected material.

**Cytology**

Larvae from the outlet of Tioga Lake in Mono Co., California (9 June 1990), have the standard *Metacnephia* banding sequence of Procunier (1982a). The X chromosome is undifferentiated and the Y is either undifferentiated (13♂ larvae) or carries a small inversion (section limits 86p–89c) in IIIL (9♂ larvae).

**Bionomics**

**Habitat.** This species is known from the central Sierra Nevada of California and the central Rocky Mountains. The immature stages are found at high-altitude lake outlets and in rocky streams up to 10 m wide.

**Oviposition.** Females carry at least 365 mature eggs (DeFoliart & Peterson 1960). Oviposition has not been observed, although there is 1 report of a gravid female that landed on a grass blade, purportedly to lay eggs (Peterson 1958a).

**Development.** Larvae hatch in the spring and can be found into July. Pupae first appear in May or June.

**Mating.** Unknown.

**Natural Enemies.** Larvae are hosts of the microsporidium *Polydispyrenia simulii*. They are preyed on by larvae of hydropsychid caddisflies (Peterson 1960b).

**Hosts and Economic Importance.** Unknown; presumably ornithophilic.

### *Metacnephia saileri* (Stone)

(Figs. 10.142, 10.299, 10.583, 10.639, 10.738; Plate 9f; Map 98)

*Cnephia saileri* Stone 1952: 82–84 (FMP). Holotype* female with pupal exuviae (pinned #61190; USNM). Alaska, Anchorage, Glenn Hwy., mi. 16, outlet of Lower Fire Lake, 29 May 1948 (Alaska Insect Project, collector unknown)

*Simulium* 'sp. III' Puri 1926: 164 (P)

*Eusimulium* 'species C' Twinn et al. 1948: 353 (Manitoba record)

*Cnephia* 'D' Hocking 1950: 498 (Yukon record)

'species . . . belonging to the *Cnephia saileri* Stone group' Davies et al. 1962: 81 (in part; Ontario record)

*Simulium becherii* Rivosecchi 1964a: 139. *Nomen nudum*

*Cnephia* 'H' Madahar 1967: 13, 21–23, 27, 77, 78, 80–84, Figs. 1, 3, 4, 43–47 (C)

*Metacnephia* 'sp. (near *saileri*)' Craig 1974: 147 (L)

*Metacnephia* 'unidentified species from Tuktoyaktuk' Rothfels 1979: 520 (C)

*Metacnephia* 'sp. H' Rothfels 1979: 521 (C)

*Metacnephia* 'Tuk' Procunier 1980: 37 (C)

*Metacnephia* 'H' Procunier 1980: 41 (C)

*Metacnephia* 'IIIS-2 sibling (*M.* H.)' Procunier 1982a: 2857 (C)

*Metacnephia* 'IIL-3 + 4, IIL-3, 5 sibling' Procunier 1982a: 2857–2858 (C)

*Metacnephia* 'sp.' Peterson 1996: 595, 608 (FL)

[*Simulium pallipes*: Dorogostaisky et al. (1935) not Fries 1824 (misident.)]

[*Cnephia* (and *Metacnephia*) *pallipes*: Rubtsov (1940b) and subsequent authors not Fries 1824 (misident.)]

**Taxonomy**

Puri 1926 (P); Dorogostaisky et al. 1935 (FMPL); Rubtsov 1940b, 1956, 1961b, 1962d (FMPL); Stone 1952 (FMP); Sommerman 1953 (L); Usova 1961 (FMPL); Carlsson 1962 (FMPL); Abdelnur 1966, 1968 (FMPL); Smith 1970 (FMPL); Peterson 1981, 1996 (FL); Rubtsov & Yankovsky 1984 (FMPL); Currie 1986 (PL); Jensen 1997 (PL).

From about 1935 until Zwick (1995) corrected the taxonomy, this species was recognized under the name *pallipes*. The name *fuscipes* (Fries) could apply to *M. saileri* (Adler et al. 1999) and would become the legitimate name; however, we have not studied type material. *Metacnephia saileri* as currently recognized is a Holarctic species, and some Palearctic names in the genus *Metacnephia* (e.g., some authored by Usova and Bazarova in Bazarova 1990) might be synonyms. The diversity of sex-chromosome systems (Procunier 1982a), variability in number of gill filaments (25–110), variation in habitat of the immature stages, and wide geographic distribution suggest that *M. saileri* is a species complex.

**Morphology**

Craig 1974 (labral fan), Yankovsky 1977 (labral fan), Craig & Borkent 1980 (L palpal sensilla), Evans & Adler 2000 (spermatheca), Palmer & Craig 2000 (labral fan).

**Cytology**

Madahar 1967; Rothfels 1979; Procunier 1980, 1982a.

**Bionomics**

**Habitat.** This species is Holarctic. In North America it ranges across the northern portion of the continent from Alaska to Quebec. Larvae and pupae occupy cool, shallow rocky habitats ranging from crackling headwater springs to swift rivers more than 100 m wide. They usually affix themselves to rocks.

**Oviposition.** Females are anautogenous (Smith 1970) and mature at least 110 large eggs. Oviposition behavior is unknown.

**Development.** Larvae hatch from overwintered eggs shortly after ice breakup or when water temperatures reach about 7°C, but these events vary with altitude and latitude to the extent that larvae can be found from May through September (Sommerman et al. 1955, Currie 1986). In Alaska, where some populations are presumably bivoltine, larvae can complete development in about a month and pupae in a week (Sommerman et al. 1955). Adults have a nearly even sex ratio and live up to 3 weeks when fed honey in the laboratory (Smith 1970).

**Mating.** Unknown.

**Natural Enemies.** Larvae are hosts of the chytrid fungus *Coelomycidium simulii*; the microsporidia *Amblyospora bracteata*, *Janacekia debaisieuxi*, and *Polydispyrenia simulii*; and an unidentified mermithid nematode. Larvae and adults are preyed on by Atlantic salmon (*Salmo salar*) (Back in Davies 1981).

**Hosts and Economic Importance.** Immature domestic geese are the only host records for North America; 15

females were taken from these birds in Fort Chimo, Quebec. Despite the bifid claws that adapt the females for ornithophily, humans, reindeer (caribou), and cattle have been recorded as hosts in Russia and Scandinavia (Rubtsov 1956, Carlsson 1962, Golini et al. 1976).

### *Metacnephia saskatchewana* (Shewell & Fredeen)

(Figs. 10.143, 10.300, 10.584; Map 99)

*Cnephia saskatchewana* Shewell & Fredeen 1958: 733–735 (FMPL). Holotype* female (pinned #6644, terminalia and pupal exuviae on separate slides; CNC). Saskatchewan, Prince Albert, Shell River, 12 May 1949 (F. J. H. Fredeen)

*Simulium* 'sp.' Twinn et al. 1948: 351 (Manitoba record)

*Cnephia* 'N' Fredeen 1951: 11, 16, 20–23, Plate 3 (Figs. J–N) (FP); 1956: 4 (Saskatchewan record)

*Cnephia* 'species N' Hocking & Pickering 1954: 100 (Manitoba record)

**Taxonomy**

Fredeen 1951 (FP); Shewell & Fredeen 1958 (FMPL); Abdelnur 1966, 1968 (FMPL); Fredeen 1981a (FMPL), 1985a (FMP); Currie 1986 (PL); Mason & Kusters 1991b (FMPL).

The number of filaments in each pupal gill is typically 17–19, although some pupae have 16.

**Morphology**

Smith 1970 (F abdominal musculature, ovarian follicle).

**Cytology**

Madahar 1967; Procunier 1980, 1982a.

**Bionomics**

**Habitat.** *Metacnephia saskatchewana* is distributed from Quebec to the Yukon, and we expect that it will be found in northern Minnesota. The immature stages live in slow-moving streams and small to large rivers at low elevations.

**Oviposition.** Unknown.

**Development.** Larvae probably overwinter under the ice of permanent flows (Fredeen 1981a). Adults are on the wing from May to mid-June in Alberta and Saskatchewan and into early August farther north.

**Mating.** We have seen swarms containing hundreds of males 2–3 m above hilly prominences on the Canadian tundra.

**Natural Enemies.** No records.

**Hosts and Economic Importance.** Males have been taken from the flowers of three-bristle saxifrage (*Saxifraga tricuspidata*), and females feed on immature domestic geese (Currie 1997a). Females are attracted to cattle (Fredeen 1969) and humans but do not bite them.

### *Metacnephia sommermanae* (Stone)

(Figs. 10.63, 10.144, 10.301, 10.517, 10.739; Map 100)

*Cnephia sommermanae* Stone 1952: 84–86 (FMP). Holotype* female with pupal exuviae (pinned #61191; USNM). Alaska, Steese Hwy., mi. 32, 3 September 1948 (Alaska Insect Project, collector unknown)

*Cnephia crassifistula* Rubtsov 1956: 305–307 (FMPL). Holotype* larva (slide #2777; ZISP). Russia, Altay, Altay State Nature Reserve, Karlagash stream, 23 July 1947 (B. G. Iogansen); *crassifictula* (subsequent misspelling). New Synonym

[*Cnephia borealis*: Jenkins (1948) not Malloch 1919 (misident.)]

**Taxonomy**

Stone 1952 (FMP); Sommerman 1953 (L); Rubtsov 1956, 1961b (FMPL); Bodrova 1980 (P); Yankovsky 1999 (PL).

Currie (1997a) stated that the name *M. sommermanae* should be applied to populations of *M. crassifistula* in the Russian Far East. We extend the synonymy to all populations under the name *crassifistula*, recognizing, however, that further study could reveal sibling species.

**Cytology**

Madahar 1967; Procunier 1980, 1982a.

Procunier's (1982a) analysis of *M. sommermanae* was incomplete. Our analysis of material from Alaska's Steese Hwy. (mi. 57, 25 July 1994) indicates that inversions *IIS-2* and *IIIS-6* of Procunier (1982a) are fixed. Chromosome arms IS, IIL, and IIIL are standard. IL has a fixed inversion from the middle of section 25 to about the middle of section 35 but is otherwise standard. Sex chromosomes are undifferentiated or males are heterozygous for expression of the nucleolar organizer. Autosomal polymorphisms include a subterminal inversion in IL.

**Molecular Systematics**

Moulton 1997, 2000 (DNA sequences).

**Bionomics**

**Habitat.** This species is found in Alaska, the Yukon, Siberia, and the Russian Far East. The immature stages live in small, cold, shallow rocky streams associated especially with alpine tundra.

**Oviposition.** Females are obligately autogenous. Oviposition behavior is unknown.

**Development.** Larvae hatch from late May to mid-June (Sommerman et al. 1955). Adults fly from July to September.

**Mating.** Unknown.

**Natural Enemies.** Larvae are hosts of an unidentified microsporidium.

**Hosts and Economic Importance.** Females have nonbiting mouthparts.

### *Metacnephia villosa* (DeFoliart & Peterson)

(Figs. 4.42, 10.145, 10.302, 10.427, 10.585, 10.740; Map 101)

*Cnephia villosa* DeFoliart & Peterson 1960: 213–216 (FMPL). Holotype* male with pupal exuviae (alcohol vial; USNM). Wyoming, Sublette Co., Sweeney Creek, Skyline Drive, 10 mi. north of Pinedale, ca. 2438 m elev., 26 June 1957 (G. R. DeFoliart)

*Cnephia freytagi* DeFoliart & Peterson 1960: 216–218 (FMPL). Holotype* male (alcohol vial; USNM). Wyoming, Lincoln Co., Snake River Canyon, ca. 10 mi. east of Alpine, small stream crossing Rt. 89, 15 June (pupa), 16 June (adult) 1958 (G. R. DeFoliart). New Synonym

[*Eusimulium boreale*: Dyar & Shannon (1927) not Malloch 1919 (misident.)]

[*Eusimulium boreale*: Hearle (1932) not Malloch 1919 (misident.)]

[*Simulium* (*Eusimulium*) *boreale*: Twinn (1936a) not Malloch 1919 (misident. in part, P)]

[*Eusimulium borealis*: Stains (1941) not Malloch 1919 (misident. in part, material from Utah)]
[*Eusimulium borealis*: Stains & Knowlton (1943) not Malloch 1919 (misident. in part, material from Utah)]
[*Cnephia saileri*: Stone (1952) (misident. in part, material from Montana)]
[*Eusimulium borealis*: Newell (1970) not Malloch 1919 (misident.)]

**Taxonomy**

Dyar & Shannon 1927 (F); Hearle 1932 (FP); Twinn 1936a (P); Peterson 1958a (FMPL), 1960a (FMP); DeFoliart & Peterson 1960 (FMPL); Currie 1986 (PL).

The larger size of *M. villosa*, compared with *M. freytagi*, was a major factor contributing to the original recognition of 2 separate species (DeFoliart & Peterson 1960). Our collections from both type localities produced striking, but exactly opposite, size differences. Other minor differences given by DeFoliart and Peterson (1960) seem to represent intraspecific variation, justifying synonymy.

**Cytology**

Madahar 1967; Procunier 1980, 1982a.

Larvae from the type locality of *M. villosa* have undifferentiated sex chromosomes, in contrast to populations from Utah, which Procunier (1982a) showed to have a differentiated Y chromosome. Additional chromosomal work is required, particularly an examination of material from the type locality of *M. freytagi*.

**Bionomics**

**Habitat.** This species is found from the Yukon south into the mountains of Utah and central California. The immature forms are most common in cool streams several meters or less in width that often drain beaver ponds and other impoundments.

**Oviposition.** Unknown.

**Development.** Eggs overwinter. Pupae are present from May through June.

**Mating.** Unknown.

**Natural Enemies.** Larvae are hosts of the chytrid fungus *Coelomycidium simulii*.

**Hosts and Economic Importance.** Unknown; presumably ornithophilic.

### Genus *Simulium* Latreille

*Simulium* Latreille 1802: 426 (as genus). Type species: *Oestrus columbacensis* Scopoli 1780: 133, by monotypy; *Simulia* (subsequent misspelling) (Thompson [2001] discusses historical details of authorship, date, and spelling of the type species)
[*Melusina*: authors (since 1908) (misident.)]
[*Atractocera*: Meigen (1803, 1804) (misident.)] (Crosskey in Crosskey and Howard 1997 [pp. 84–85] discusses the history of the error)

**Diagnosis.** Adults: Wing with basal medial cell small or absent; basal radial cell about one fourth length of wing (measured from apex of stem vein); radius with or without setae dorsally; radial sector unbranched; $R_1$ dorsally with numerous dark spiniform setae on apical two thirds. Calcipala typically present; pedisulcus typically deep (shallow in taxa such as subgenus *Hellichiella*). Katepisternal sulcus narrow, deep, complete anteriorly. Furcasternum with internal dorsal arms bearing ventrally directed apodeme. Anepisternal membrane with or without hair. Female: Maxillary palpomere V typically much longer than palpomere III. Claws each with or without subbasal tooth or basal, thumblike lobe. Male: Gonostylus typically with 1 apical spinule (absent or 2 or more in several taxa). Paramere with 1 or more spines. Dorsal plate of aedeagus typically present. Pupa: Spine combs present or absent. Terminal spines typically short. Cocoon typically well formed, slipper, shoe, or boot shaped; anterior rim well formed; some taxa with anterodorsal projection, lateral apertures, and other modifications. Larva: Hypostoma with median and lateral teeth typically extended anteriorly to about same level; sublateral teeth with apices posterior to those of median and lateral teeth. Abdominal segment IX with or without prominent ventral tubercles. Rectal papillae of 3 simple or compound lobes.

**Overview.** This large genus contains about 40 subgenera worldwide. In North America, we recognize 11 subgenera and 153 species. Members of the genus can be found in nearly all habitats throughout the world.

#### Subgenus *Hellichiella* Rivosecchi & Cardinali

*Hellichiella* Rivosecchi & Cardinali 1975: 69 (as genus). Type species: *Eusimulium saccai* Rivosecchi 1967: 63–70, by original designation
*Eusimulium* 'Group 1' Dunbar 1962: 23, 25
*Simulium* (*Eusimulium*) 'Group 1—The *rivuli* Group' Wood 1963a: 134–136
*Parahellichiella* Golini 1982: 40–44. Unavailable (unpublished thesis name)
[*Hellichia*: Rubtsov (1956) not Enderlein 1925 (misident. in part)]

**Diagnosis.** Adults: Radius with hair dorsobasally. Antenna with pedicel larger than basal flagellomere. Scutum unpatterned. Female: Sensory vesicle typically occupying about one fourth or less of palpomere III. Pedisulcus shallow; claws each with basal thumblike lobe. Precoxal bridge complete. Anal lobe in ventral view with medial half or more unsclerotized. Genital fork with lateral plates subrectangular, broadly joined to corresponding tergite. Spermatheca with unpigmented ring proximally; wall with polygonal pattern. Male: Gonostylus slender, evenly tapered, moderately curved, bearing 1 apical spinule. Ventral plate rather flat, tapered posteriorly. Paramere closely associated with multiple spines, although not all of these directly connected to it. Dorsal plate a weakly sclerotized, horizontal strip. Pupa: Gill of 4–12 slender filaments. Cocoon with anterodorsal projection typically thickened along margin. Larva: Antenna with 3 or more hyaline bands or spots (except in *S. excisum* and *S. rivuli*, which have none). Hypostoma with teeth arranged in 3 distinct groups; anterolateral margin of ventral wall bearing longitudinal sulcus. Abdominal segment IX with prominent ventral tubercles. Rectal papillae of 3 simple lobes.

**Overview.** The Holarctic Region is home to the world's 22 known species of *Hellichiella*. Twelve of these species, including one still undescribed, inhabit North America. The subgenus consists of 2 monophyletic species groups (*S.*

congareenarum group and S. rivuli group) that are differentiated by 15 fixed chromosomal inversions (Dunbar 1967) but are nearly indistinguishable morphologically.

The immature stages typically inhabit small, open-canopy, eurythermal streams, rivulets, and seepages, often with abundant trailing and emergent vegetation. All species are univoltine and develop early in the season. Females are ornithophilic, and those of several species are vectors of various avian pathogens.

### *Simulium (Hellichiella) congareenarum* Species Group

**Diagnosis.** Female: Spermatheca more than half as long as stem of genital fork. Pupa: Gill of 5–12 slender filaments. Larva: Antenna with 3 or more hyaline bands or spots. Chromosomes: Nucleolar organizer typically in IL beside centromere (IIIS in *S. minus*, IIS in *S. nebulosum*). Standard banding sequence as given by Dunbar (1967).

**Overview.** This species group was first recognized by Dunbar (1962, 1967), who characterized it cytologically and referred to it as *Eusimulium* 'Subgroup A.' Wood (1963a) independently characterized it morphologically as a subgroup of what we now call subgenus *Hellichiella*. At the time that Wood (1963) recognized this subgroup, all species in the subgenus *Hellichiella* were not known, and his sole diagnostic character (i.e., presence of hyaline bands on the larval antennae) now also is known in members of the *S. rivuli* species group. Eighteen species comprise the Holarctic *S. congareenarum* group, of which 8 are found in the Nearctic Region.

#### *Simulium (Hellichiella) anatinum* Wood

(Figs. 10.41, 10.146, 10.303; Map 102)

*Simulium anatinum* Wood 1963b: 96–98 (FMP). Holotype* female (pinned #8196, pupal exuviae in glycerin vial below; CNC). Ontario, Muskoka District, 3 mi. south of Huntsville, roadside ditch, 15 May 1962 (D. M. Wood)

*Simulium (Eusimulium)* 'H' Bennett 1960: 379–388 (host records)

*Cnephia* 'sp. T' Anderson & DeFoliart 1961: 719–720 (host records)

?[*Simulium venustum*: Johnson (1925) not Say 1823 (misident. in part, No. 212)]

[*Eusimulium croxtoni*: Anderson (1955) not Nicholson & Mickel 1950 (misident.)]

[*Eusimulium euryadminiculum*: Anderson (1955) not Davies 1949b (misident.)]

[*Eusimulium latipes*: Anderson (1955) not Meigen 1804 (misident.)]

[*Simulium (Eusimulium) croxtoni*: Anderson (1956) not Nicholson & Mickel 1950 (misident.)]

[*Simulium (Eusimulium) euryadminiculum*: Anderson (1956) not Davies 1949b (misident.)]

[*Simulium (Eusimulium) latipes*: Anderson (1956) not Meigen 1804 (misident.)]

[*Simulium croxtoni*: Davies & Peterson (1956) not Nicholson & Mickel 1950 (misident. in part, p. 624)]

[*Simulium euryadminiculum*: Davies & Peterson (1956) not Davies 1949b (misident. in part, p. 624)]

[*Simulium latipes*: Davies & Peterson (1956) not Meigen 1804 (misident. in part, p. 624)]

[*Simulium croxtoni*: Fallis et al. (1956) not Nicholson & Mickel 1950 (misident.)]

[*Simulium euryadminiculum*: Fallis et al. (1956) not Davies 1949b (misident.)]

[*Simulium latipes*: Fallis et al. (1956) not Meigen 1804 (misident.)]

[*Simulium (Eusimulium) subexcisum*: Bennett (1960) not Edwards 1915 (misident.)]

[*Simulium congareenarum*: Davies et al. (1962) not Dyar & Shannon 1927 (misident. in part, bionomics, couplet 9 on p. 86, couplet 10 on p. 89)]

[*Simulium innocens*: Tarshis (1972) not Shewell 1952 (misident.)]

[*Simulium innocens*: Herman et al. (1975) not Shewell 1952 (misident.)]

[*Simulium innocens*: Desser et al. (1978) not Shewell 1952 (misident.)]

[*Simulium (Hellichiella) excisum*: Cupp & Gordon (1983) not Davies et al. 1962 (misident. in part, material from Schoolcraft Co., Michigan)]

?[*Simulium congareenarum*: Perez (1999) not Dyar & Shannon 1927 (misident.)]

**Taxonomy**

Wood 1963a (FMPL), 1963b (FMP); Wood et al. 1963 (L); Golini 1982 (FMPL); Currie 1986 (PL); Stuart 1995 (cocoon).

Males, larvae, and pupae are indistinguishable from those of *S. congareenarum*.

**Morphology**

Bennett 1963b (F salivary gland).

**Cytology**

Dunbar 1962, 1967; Golini 1982; Rothfels & Golini 1983.

**Bionomics**

**Habitat.** *Simulium anatinum* is distributed across Canada and part of the northern United States. The immature stages develop in small streams draining swamps and bogs or coursing through open fields.

**Oviposition.** Unknown.

**Development.** Eggs overwinter and larvae begin to mature in May (Wood 1963a). *Simulium anatinum* occupies the same streams with its morphologically similar relative, *S. congareenarum*, in areas such as Algonquin Park, Ontario, where it is at least a week earlier in development. The former pupates before leaves appear on the trees, whereas the latter pupates after leaf development. Adults are on the wing in late May.

**Mating.** Unknown.

**Natural Enemies.** No records.

**Hosts and Economic Importance.** Females can be found along lake shores where they feed on Canada geese and ducks such as redheads and mallards (domestic and wild) during early evening (Bennett 1960, Anderson & DeFoliart 1961, Fallis & Smith 1964a, Tarshis 1972). Limited feeding also has been recorded on ruffed grouse, domestic turkeys, gray jays, American crows, common ravens, American robins, and white-throated sparrows (Bennett 1960, Anderson & DeFoliart 1961). Females have been collected from common loons, but no statement of feeding was given by the authors (Fallis & Smith 1964a). Extracts of the uropygial glands of domestic ducklings evidently do not attract this species (Smith 1966). Black models against a white background are particularly attractive (Mercer 1972).

Females transmit the protozoan *Leucocytozoon simondi* to ducks and Canada geese in Ontario and Michigan (Fallis & Bennett 1966, Tarshis 1972, Herman et al. 1975, Desser et al. 1978) and subarctic Quebec (Laird & Bennett 1970) and possibly to blue geese (= snow geese) (*Chen caerulescens*) and Canada geese in Nunavut (Bennett & MacInnes 1972). They can transmit *Leucocytozoon icteris* to common grackles under laboratory conditions (Fallis & Bennett 1962; as *L. fringillinarum*). *Simulium anatinum* also transmits the filarial nematode *Splendidofilaria fallisensis* to ducks (Anderson 1968).

### *Simulium* (*Hellichiella*) *congareenarum* (Dyar & Shannon)

(Figs. 10.22, 10.26, 10.147, 10.428, 10.741; Plate 10a; Map 103)

*Eusimulium congareenarum* Dyar & Shannon 1927: 20, Fig. 45 (F). Holotype* female (slide #28333, terminalia, 3 legs, and all claws missing; USNM). South Carolina, Richland Co., Congaree, 22 April 1912 (A. H. Jennings & W. V. King)

*Eusimulium congareenarum* 'b' Dunbar 1967: 387 (C)

[*Simulium meridionale*: Malloch (1914) not Riley 1887 (misident. in part, material from Congaree, South Carolina)]

[*Simulium meridionale*: Stone & Snoddy (1969) not Riley 1887 (misident. in part, Fig. 126)]

**Taxonomy**

Dyar & Shannon 1927 (F), Jamnback & Stone 1957 (FMPL), Davies et al. 1962 (FMP), Wood 1963a (FMPL), Wood et al. 1963 (L), Stone 1964 (FMPL), Snoddy 1966 (FMPL), Stone & Snoddy 1969 (FMPL), Amrine 1971 (FMPL), Pinkovsky 1976 (FMPL), Snoddy & Noblet 1976 (PL), Golini 1982 (FMPL), Webb & Brigham 1982 (L), Stuart 1995 (cocoon).

**Morphology**

Evans & Adler 2000 (spermatheca).

**Cytology**

Dunbar 1962, 1967; Golini 1982; Rothfels & Golini 1983.

Larvae from the Northwest Territories are chromosomally similar to those characterized by Dunbar (1967) from Ontario's Bruce Peninsula in that the 4 larvae (3♀, 1♂) we examined had the fixed inversions *IIIL-6.7*. These western larvae were unique in that 7 of 8 constituents had the same IS-1 inversion that is fixed in *S.* 'species O' ('Opinaca'), as illustrated in Fig. 7 of Rothfels and Golini (1983). One of the 3 female larvae was heterozygous for a large inversion in IL, and the male was heterozygous for a simple subterminal inversion in each of IS and IIIL.

**Molecular Systematics**

Moulton 1997, 2000 (DNA sequences).

**Bionomics**

**Habitat.** *Simulium congareenarum* occurs farther south than other members of the subgenus *Hellichiella* and has one of the most unusual distributions of all North American black flies. It is a common resident of the eastern coastal plain and Sandhills Region, with its western limit bounded abruptly by the Fall Line. Records beyond this range come from Ontario's Bruce Peninsula and Algonquin Park and from the Northwest Territories around Great Slave Lake, suggesting that the species probably occurs across much of Canada. Larvae and pupae are found on trailing vegetation, fallen leaves, and twigs in sluggish, sandy or swampy, brownwater and blackwater streams that are generally quite acidic and from less than a meter to more than 15 m in width. In Canada, they apparently are restricted to flows less than 0.5 m wide.

**Oviposition.** At least some females are autogenous for the first cycle of egg production (Pinkovsky 1976). Oviposition habits are unknown.

**Development.** A single generation per year is probably typical. From South Carolina to Florida, larvae begin to hatch in early fall and pupate as early as November; however, larvae and pupae are most common from February to April. Northward, overwintered eggs produce larvae no later than April and pupae appear in May. Larvae and pupae in some northern populations linger into July, leading to speculation that the species at times might be multivoltine (Jamnback & Stone 1957, Davies et al. 1962).

**Mating.** Unknown.

**Natural Enemies.** Our records indicate that larvae are hosts of the chytrid fungus *Coelomycidium simulii*, the trichomycete fungus *Harpella melusinae*, the microsporidia *Janacekia debaisieuxi* and *Polydispyrenia simulii*, and an unidentified mermithid nematode. A ciliate protozoan of the genus *Tetrahymena* has been found in a larva in southeastern Georgia (C. E. Beard, pers. comm.).

**Hosts and Economic Importance.** Females feed on ducks, chickens, and domestic and wild turkeys, especially on the head and neck (Jones & Richey 1956, Noblet et al. 1972). Approximately 3–5 minutes is required to complete a blood meal (Pinkovsky et al. 1981). This species transmits the blood protozoan *Leucocytozoon smithi* to turkeys (Noblet et al. 1972, Pinkovsky et al. 1981), probably serving as the principal vector during the winter and early spring (Noblet et al. 1975, Garris & Noblet 1976). As a vector, *S. congareenarum* played a secondary role in damaging the southeastern turkey industry in the early 1970s where its distribution overlapped major production areas (Arnold & Noblet 1975, Noblet et al. 1976).

### *Simulium* (*Hellichiella*) *innocens* (Shewell)

(Figs. 10.148, 10.304, 10.429, 10.606, 10.742; Plate 10b; Map 104)

*Eusimulium innocens* Shewell 1952: 38–39 (FMP). Holotype* female (pinned #5989, pupal exuviae in alcohol vial; CNC). Ontario, Bell's Corners, 2 June (pupa), 6 June (adult) 1950 (G. E. Shewell)

**Taxonomy**

Shewell 1952 (FMP), Davies et al. 1962 (FMP), Wood 1963a (FMPL), Wood et al. 1963 (L), Stone 1964 (FMP), Golini 1982 (FMPL).

**Cytology**

Dunbar 1962, 1967; Golini 1982; Rothfels & Golini 1983.

**Bionomics**

**Habitat.** This trans-Canadian species ranges south into the Rocky Mountains of Utah. Its immature stages are found in small, shallow, often temporary streams slowly passing through grassy swales and boggy areas.

**Oviposition.** Unknown.

**Development.** Eggs overwinter and larvae and pupae are found from April to June, with adults emerging in May and June (Davies et al. 1962).

**Mating.** Unknown.

**Natural Enemies.** We have 1 larva infected with the microsporidium *Polydispyrenia simulii*.

**Hosts and Economic Importance.** Females have been implicated as vectors of the protozoan *Leucocytozoon simondi* among waterfowl (Tarshis 1972, Herman et al. 1975, Desser et al. 1978), but our examination of representative specimens indicates that these records pertain to *S. anatinum*. Hosts remain unknown.

### *Simulium (Hellichiella) minus* (Dyar & Shannon)

(Fig. 10.743, Plate 10c, Map 105)

*Eusimulium minus* Dyar & Shannon 1927: 21, Fig. 39 (F). Holotype* female (pinned #28335, head in glycerin vial below, terminalia missing; USNM). California, Mariposa Co., Yosemite National Park, 14 May 1916 (H. G. Dyar). (A slide of a female [also #28335, but 17 May 1916] bears a type label in the handwriting of A. Stone. We assume that Stone inadvertently failed to indicate that this specimen is a paratype.)

**Taxonomy**

Dyar & Shannon 1927 (F), Stains 1941 (F), Stains & Knowlton 1943 (F), Coleman 1951 (F), Stone 1952 (F), Wirth & Stone 1956 (F), Corredor 1975 (F).

The name of this formerly enigmatic species has been applied to a number of species in the subgenus *Hellichiella* and other groups. Nearly all previous references to *S. minus*, other than references to the type, are probably misidentifications. Previous confusion over the identity of this species centered around the choice of an unassociated female as the name bearer.

The larva of *S. minus* is illustrated here for the first time (Fig. 10.743, Plate 10c). Pupae have either 5 or 6 gill filaments per side and are indistinguishable from those of *S. nebulosum*. *Simulium minus* is one of the few North American species for which the male is unknown.

**Cytology**

Chromosomes of larvae that we examined from the type locality (9 May 1992) are identical to those of *S. innocens* except as follows: The nucleolar organizer is in the extreme base of chromosome arm IIIS, IL has 1 fixed inversion (limits 39c-41d), IIIL is reorganized between sections 89 and 96, and in the Y chromosome the centromere region of chromosome I is unpaired and the base of IS is rearranged through section 15. One male was heterozygous for an inversion in sections 62–72. The chromosomes indicate that *S. minus* is the sister species of *S. nebulosum* and that together these 2 species are the sister group of *S. innocens*.

**Bionomics**

**Habitat.** *Simulium minus* is known only from the type locality, Yosemite National Park, California, where the immature stages inhabit streams up to 2 m wide in grassy meadows bordered by forest. The morphologically similar *S. nebulosum* occurs in some of the same streams but is far more common and widespread.

**Oviposition.** Unknown.

**Development.** Eggs overwinter. Larvae, pupae, and adults have been found in May.

**Mating.** Unknown.

**Natural Enemies.** Larvae are hosts of the chytrid fungus *Coelomycidium simulii*.

**Hosts and Economic Importance.** Unknown; presumably ornithophilic. (Also see the entry under *S. nebulosum*.)

### *Simulium (Hellichiella) nebulosum* Currie & Adler

(Figs. 10.149, 10.305, 10.518, 10.640, 10.744; Plate 10d; Map 106)

*Simulium (Hellichiella) nebulosum* Currie & Adler 1986: 221–224 (FMPL). Holotype* female with pupal exuviae (pinned #19459 [published as #18819]; CNC). British Columbia, 1.1 mi. south of Kaslo, 49°55′N, 116°55′W, 22 May 1964 (D. M. & G. C. Wood)

?[*Cnephia minus*: Woo (1964) not Dyar & Shannon 1927 (misident.)]

?[*Cnephia minus*: Gibson (1965) not Dyar & Shannon 1927 (misident.)]

?[*Cnephia minus*: Williams et al. (1980) not Dyar & Shannon 1927 (misident.)]

[*Hellichiella minus*: Golini (1982) not Dyar & Shannon 1927 (misident. in part, 6-filamented P)]

[*Simulium anatinum*: Mahrt (1982) not Wood 1963b (misident.)]

[*Simulium gouldingi*: Mahrt (1982) not Stone 1952 (misident.)]

**Taxonomy**

Currie & Adler 1986 (FMPL).

Pupae have gills of 5 or 6 filaments, sometimes in the same individual.

**Cytology**

Chromosomes of material from British Columbia and California differ from those of *S. minus* as follows: The nucleolar organizer is in the base of IIS virtually at the centromere, a secondary nucleolar organizer is common in section 36, and the sex chromosomes generally are undifferentiated, although in some males the centromere region of chromosome I is unpaired. Floating inversions have not been found. A sample, mixed with *S. minus*, from near the type locality (California, Mariposa Co., Yosemite National Park, Glacier Point Road, Summit Meadow, 9 May 1992) confirmed reproductive isolation of these 2 species.

**Bionomics**

**Habitat.** This species ranges from British Columbia and southern Alaska to southern Idaho and central California. It is especially common in the Pacific Coastal and Columbian forests. Larvae and pupae are found on stones and trailing vegetation in small, sluggish brownwater streams and seepages arising from swampy and boggy ground.

**Oviposition.** Unknown.

**Development.** Pupae are present from early March to mid-June in southern British Columbia and Oregon and as late as July in the high Sierra Nevada of California and in northern British Columbia.

**Mating.** Unknown.

**Natural Enemies.** Larvae are sporadically infected with the chytrid fungus *Coelomycidium simulii* (Currie & Adler 1986).

**Hosts and Economic Importance.** *Simulium nebulosum* might be the species, at least in part, collected from blue grouse and found with microfilariae and oocysts of avian blood protozoa (Woo 1964, Gibson 1965, Williams et al. 1980). However, our examination of a portion of the original Gibson (1965) collection of black flies from Trout Creek, British Columbia, revealed only *S. craigi* and an unidentified member of the subgenus *Eusimulium*, suggesting that the original identification as *S. minus* is incorrect.

### *Simulium* (*Hellichiella*) *rendalense* (Golini)

(Figs. 10.150, 10.306; Map 107)

*Eusimulium rendalense* Golini 1975: 229–238 (FE). Holotype* female (alcohol vial #13913, CNC). Norway, Hedmark, Ytre Rendal (Rendalen), Rena River, 61°43′N, 11°24′E, carbon dioxide trap at water level, 16 July 1968 (V. I. Golini); *randalense* (subsequent misspelling)

*Eusimulium* 'sp. 1' Golini 1970: 117, 175–176, 219–224 (F) (in part)

*Simulium* 'close to *Simulium dogieli*' Eide & Fallis 1972: 414 (vector of *Leucocytozoon simondi*) (in part)

**Taxonomy**

Golini 1970 (F), 1975 (FE), 1982 (FMPL), 1987 (MPL).

The specific distinctness of *S. rendalense*, with respect to several other Palearctic species of the subgenus *Hellichiella*, especially *S. dogieli* and *S. fallisi*, is not entirely clear (Golini 1987). Some authors (e.g., Raastad 1979) consider *rendalense* a synonym of *dogieli*. Nonetheless, the species that we call *S. rendalense* is distinct, especially chromosomally and as larvae, from all other members of *Hellichiella* in North America. The chromosomal match between populations in North America and the type locality confirms its Holarctic distribution. Pupae, males, and females resemble those of *S. anatinum*; the adults might be distinguishable on the basis of subtle differences, as yet untested, in the terminalia.

**Cytology**

Golini 1982, Rothfels & Golini 1983.

Nearctic material (8♀, 7♂) that we examined is chromosomally similar to that, characterized by Rothfels and Golini (1983), from the type locality in Norway. Specifically, the banding sequence of all chromosome arms, including the complexly rearranged IIIL arm, is identical between populations in the Nearctic and Palearctic Regions. Two X chromosomes, $X_1$ (IIS-5) and $X_2$ (IIS-5,6), occur in both regions, with $X_2$ predominating. The 2 Y chromosomes found at the type locality have not been seen in our material. Rather, the Y chromosome of Nearctic material apparently has the standard sequence for the IIS arm, although a small inversion in section 49 might be present. In the Nearctic Region, the chromosomes are loosely paired, and a secondary nucleolar organizer is typically, but not always, present in section 36. Autosomal polymorphisms are common among Nearctic larvae but are not shared with topotypical specimens. The most common polymorphisms that we found are a subterminal inversion in IS, a basal inversion in IL, and an inversion each in IIL (sections 59p–65d) and IIIL (sections 97c–98c in Fig. 34 of Rothfels & Golini [1983]). A conspicuous heteroband was found for the heaviest band in section 41 of 1 larva.

**Bionomics**

**Habitat.** Until we found *S. rendalense* in the Northwest Territories, it was known only from the type locality in southern Norway and from a few larvae from Sweden identified chromosomally by Adler et al. (1999). It is undoubtedly more widespread in both the Old and the New World. In late June and early July, it is one of the most common members of the subgenus *Hellichiella* in the southern portion of the Northwest Territories. Larvae and pupae are found in tiny flows slowly coursing through open bogs and along roadsides. In Norway, they are found more frequently at the edges, rather than the center, of bogs (Golini 1982).

**Oviposition.** Blood-fed females mature their eggs in 5–7 days at 13°C–14°C and produce an average of 121 eggs in each ovarian cycle (Golini 1975). Oviposition behavior is unknown.

**Development.** Eggs overwinter. We have found larvae and pupae from late June to early July. In Norway, pupation begins in mid-June and adults emerge from about the third week of June to mid-July (Golini 1975). Females can live on sugar water in the laboratory for up to 18 days (Golini 1975).

**Mating.** Unknown.

**Natural Enemies.** Larvae in the Nearctic Region are hosts of the microsporidium *Janacekia debaisieuxi* and an unidentified mermithid nematode.

**Hosts and Economic Importance.** Host-seeking females fly within 10 m of rivers and 1 m above the water (Golini 1975). They are attracted to carbon dioxide, especially when it is combined with extract from the uropygial gland of the domestic ducks on which they feed. Females are natural vectors of the protozoan *Leucocytozoon simondi* among ducks (Eide & Fallis 1972, Golini 1975).

### *Simulium* (*Hellichiella*) *usovae* (Golini)

(Map 108)

*Eusimulium* (*Hellichiella*) *usovae* Golini 1987: 708–713 (FE). Holotype female with pupal exuviae (pinned #18692, CNC). Norway, Rendalen, Åsmyrtjörna bog-fen, 61°43′N, 11°06′E, 2 July 1980 (V. I. Golini)

*Eusimulium* 'sp. 1' Golini 1970: 117, 175–176 (F) (in part)

*Simulium* 'close to *Simulium dogieli*' Eide & Fallis 1972: 414 (vector of *Leucocytozoon simondi*) (in part)

*Eusimulium* 'near *dogieli*' Golini 1975: 237 (F) (also p. 234 as 'close to *E. dogieli*')

**Taxonomy**

Golini 1982, 1987 (FMPL).

Larvae and pupae are not morphologically separable from those of *S. rendalense*. Females and males have variegated, yellow and dark brown legs (Golini 1987), compared with the uniformly dark brown legs of the adults of closely related species such as *S. anatinum* and *S. rendalense*. Comparisons, however, must be made using well-tanned and sclerotized adults, not freshly emerged specimens.

**Cytology**

Golini 1982, Rothfels & Golini 1983.

The remarkable chromosomal similarity between the female larvae that we examined from western Nunavut and those from the type locality in Norway leaves little doubt that *S. usovae* is Holarctic. Nearctic larvae, like those from the type locality, are standard in all chromosomal arms except IIS, which is fixed for *IIS-3* of Rothfels and Golini (1983). Unlike Palearctic larvae, however, our material had a conspicuous secondary nucleolar organizer in section 36 of chromosome arm IL. Not having seen the chromosomes of male larvae, we cannot comment on the sex chromosomes.

**Bionomics**

**Habitat.** Prior to our collection of larvae from a 3-m wide tributary of the Thelon River in Nunavut, *S. usovae* was known only from a single bog in Rendalen, Norway.

**Oviposition.** Females are anautogenous (Golini 1987), but oviposition behavior is unknown.

**Development.** Our collection of 6 immature larvae was made on 8 July. In Norway, larvae are found from early June to early July, and adults emerge from mid-June to mid-July (Golini 1987).

**Mating.** Unknown.

**Natural Enemies.** No records.

**Hosts and Economic Importance.** Females are attracted to carbon dioxide, feed on domestic ducks, and are natural vectors of the protozoan *Leucocytozoon simondi* (Golini 1987).

### *Simulium* (*Hellichiella*) 'species O'

(Map 109)

*Hellichiella* 'Opinaga [*sic*] Cytotype' Golini 1982: 14–15, Figs. 2, 17, 35 (C)

*Hellichiella* 'sp. Opinaca' (also 'Opinaca sp.' and 'Opinaga') Rothfels & Golini 1983: 1221–1230, Figs. 2, 17, 35 (C)

**Taxonomy**

No known material of this species exists in collections.

**Cytology**

Golini 1982, Rothfels & Golini 1983.

**Bionomics**

**Habitat.** The sole known location for this species lies in the *Sphagnum*-bog habitat of Eastmain-Opinaca along the Eastmain River, which flows into the southern portion of James Bay, Quebec (Rothfels & Golini 1983; C. Back, pers. comm.). The stream from which larvae were collected was slow (ca. 30 cm/sec), with low discharge (ca. 50 liters/sec), much trailing vegetation and dead branches, and a bottom covered with organic debris (C. Back, pers. comm.).

**Oviposition.** Unknown.

**Development.** Six immature larvae were taken in late July, about 2 months later than a larval collection of *S. anatinum* from the same stream (Rothfels & Golini 1983).

**Mating.** Unknown.

**Natural Enemies.** No records.

**Hosts and Economic Importance.** Unknown; presumably ornithophilic.

### *Simulium* (*Hellichiella*) *rivuli* Species Group

**Diagnosis.** Female: Spermatheca typically one half or less as long as stem of genital fork. Pupa: Gill of 4–8 slender filaments. Larva: Antenna with or without hyaline bands or spots. Chromosomes: Nucleolar organizer in IS, two thirds distance from end. Standard banding sequence as given by Dunbar (1967).

**Overview.** The *S. rivuli* species group was first recognized by Dunbar (1962, 1967), who referred to it as *Eusimulium* 'Subgroup B' and characterized it cytologically. Wood (1963a) independently recognized it on weak morphological grounds as a subgroup of what we now call subgenus *Hellichiella*. Wood's (1963a) diagnostic criterion (i.e., lack of hyaline bands on the larval antennae) is now known not to hold for all members in the subgroup. Nonetheless, monophyly of the group is supported by at least 15 fixed chromosomal inversions (Dunbar 1967). The *S. rivuli* species group with its 4 members might be endemic to North America, although several Palearctic species of the subgenus *Hellichiella* have not yet been studied chromosomally to determine their placement.

### *Simulium* (*Hellichiella*) *curriei* Adler & Wood

(Figs. 10.27, 10.307, 10.430, 10.745; Plate 10e; Map 110)

*Simulium* (*Hellichiella*) *curriei* Adler & Wood 1991: 2867–2872 (FMPLEC). Holotype* female (pinned, pupal exuviae in glycerin vial below; USNM). California, Mono Co., Rt. 120, trickle beside Tioga Lake outlet, 37°55′45″N, 119°15′01″W, 9 June 1990 (P. H. Adler)

[*Hellichiella minus*: Golini (1982) not Dyar & Shannon 1927 (misident. in part, 8-filamented P)]

**Taxonomy**

Adler & Wood 1991 (FMPLE).

This species is composed predominantly of individuals with 8 gill filaments per side. However, we have samples of individuals with 6 filaments, for which we were unable to find additional morphological differences. A sample from a tributary of Beaver Creek at Decker Road in Benton Co., Oregon (12 March 1993), was composed of 2 individuals with 8 filaments, 1 with 7 and 8 filaments per side, 1 with 6 and 7 filaments, and 20 with 6 filaments per side.

**Cytology**

Adler & Wood 1991.

A preliminary chromosomal analysis of 2 individuals with 6 filaments revealed no differences from individuals with 8 filaments.

**Molecular Systematics**

Moulton 1997, 2000 (DNA sequences).

**Bionomics**

**Habitat.** *Simulium curriei* is a mountain resident of western North America. Larvae and pupae are found on trailing grasses or the undersides of stones in tiny unshaded rills flowing through grassy roadsides and meadows or trickling from impounded waters.

**Oviposition.** Females oviposit near the sources of rivulets and release eggs as they repeatedly dip to the water's surface from a hovering height of about 6 cm (Adler & Wood 1991). This behavior might be characteristic of the subgenus *Hellichiella*, for we also have seen the European *S. latipes* ovipositing in this fashion in southern England.

**Development.** Eggs overwinter, and mature larvae and pupae are found from April and May into June or even through July at higher elevations.

**Mating.** Unknown.

**Natural Enemies.** Larvae are hosts of an unidentified mermithid nematode (Adler & Wood 1991) and a microsporidium resembling *Janacekia debaisieuxi*.

**Hosts and Economic Importance.** Unknown; presumably ornithophilic.

### *Simulium* (*Hellichiella*) *excisum* Davies, Peterson & Wood

(Figs. 10.64, 10.151, 10.308, 10.431, 10.641, 10.746; Plate 10f; Map 111)

*Simulium excisum* Davies, Peterson & Wood 1962: 113–114, Figs. 3B, 7A, 29, 68 (FMP). Holotype* female with pupal exuviae (pinned #7993; CNC). Ontario, Carleton Co., Goulbourn Township, 1.5 mi. west of Stanley Corners, roadside ditch (Fig. 4B of Twinn 1936a), 5 May 1961 (D. M. Wood & E. Bond)

*Simulium* 'sp. near *subexcisum*' Shewell 1957: 2, Map 31 (Manitoba record)

*Simulium* (*Hellichiella*) 'sp.' Currie 1997a: 567, 592 (Alaska record)

[*Simulium* (*Eusimulium*) *subexcisum*: Twinn (1936a) not Edwards 1915 (misident.)]

[*Cnephia subexcisum*: Davies (1950) not Edwards 1915 (misident.)]

[*Cnephia minus*: Stone (1952) not Dyar & Shannon 1927 (misident.)]

[*Simulium* (*Eusimulium*) *gouldingi*: Stone (1952) (misident. in part, material from Alaska; labled as paratypes of *S. gouldingi*)]

[*Simulium* (*Eusimulium*) *gouldingi*: Sommerman (1953) not Stone 1952 (misident.)]

[*Simulium gouldingi*: Sommerman et al. (1955) not Stone 1952 (misident.)]

[*Simulium* (*Eusimulium*) *subexcisum*: Shewell & Fredeen (1958) not Edwards 1915 (misident.)]

[*Simulium* (*Eusimulium*) *subexcisum*: Bennett (1960) not Edwards 1915 (misident.)]

[*Simulium gouldingi*: Hopla (1965) not Stone 1952 (misident.)]

**Taxonomy**

Twinn 1936a (FMP), Stone 1952 (MP), Sommerman 1953 (L), Davies et al. 1962 (FMP), Wood 1963a (FMPL), Wood et al. 1963 (L), Amrine 1971 (FMPL), Lewis 1973 (FMPL), Merritt et al. 1978b (PL), Hilsenhoff 1982 (L).

**Morphology**

Smith 1970 (ovarian follicles), Colbo et al. 1979 (F labral and cibarial sensilla).

**Cytology**

Dunbar 1962, 1967; Rothfels & Golini 1983 (Fig. 37).

**Bionomics**

**Habitat.** *Simulium excisum* occurs in Alaska and across Canada, south into the northernmost states. The immature stages are confined to temporary or permanent streams and ditches less than a meter wide that are exposed to the sun and often drain bogs or small impoundments.

**Oviposition.** Females of some populations emerge with an average of 200–250 yolkless oocytes and small fat bodies (Chutter 1970). If given only water, they undergo some egg maturation but tend to die prematurely (Smith 1970). In other populations, females emerge with much stored nutrient and eggs about half mature, suggesting that these females are autogenous for their first ovarian cycle (Davies et al. 1962). Oviposition behavior is unknown.

**Development.** Winter is passed as eggs, with larvae appearing in early March or April and pupation beginning as early as mid-April and lasting into June at more northern latitudes (Hocking & Pickering 1954, Sommerman et al. 1955, Davies et al. 1962, Merritt et al. 1978b, Back et al. 1983).

**Mating.** We have observed males during midday in a loose swarm about a meter above a small stream in an open area. Coupled pairs fell from the swarm onto streamside grasses.

**Natural Enemies.** No records.

**Hosts and Economic Importance.** Females are ornithophilic, but specific host records given by Bennett (1960) require confirmation (Davies et al. 1962).

### *Simulium* (*Hellichiella*) *mysterium* Adler, Currie & Wood, New Species

(Figs. 10.152, 10.309, 10.432, 10.519, 10.747, 10.825; Plate 11a; Map 112)

**Taxonomy**

**Female.** Wing length 2.3 mm. Colors not available. Frons at narrowest point about one sixth width of head. Mandible with about 50 serrations; lacinia with 25–29 retrorse teeth. Sensory vesicle in lateral view occupying about one sixth of palpomere III. Claws each with basal thumblike lobe. Terminalia (Fig. 10.152): Anal lobe in ventral view with medial three fourths or more unsclerotized. Genital fork with arms slender throughout. Spermatheca subspherical.

**Male.** Wing length 2.1–2.5 mm. Colors not available. Genitalia (Fig. 10.309): Ventral plate in ventral view roughly V shaped, rounded or slightly concave posteriorly. Paramere with about 20 spines, the longest spine about 5 times its width. Median sclerite elongate. Dorsal plate weakly sclerotized, wider than long.

**Pupa.** Length 2.9–3.4 mm. Gill of 6 filaments in 3 pairs, as long as or longer than pupa (Figs. 10.432, 10.519); base swollen, slightly wrinkled, giving rise to 2 thin petiolate ventral pairs and 1 distal pair; filaments of each pair branched in horizontal plane and with numerous transverse furrows. Head and thorax with numerous irregularly distributed, dark rounded microtubercles (Fig. 10.825); trichomes unbranched. Cocoon densely woven, with anterodorsal projection.

**Larva.** Length 4.7–5.7 mm. Head capsule yellowish brown; head spots brown, typically weak (Fig. 10.747). Antenna brown, extended beyond stalk of labral fan by more than length of distal article; medial article with 4 hyaline bands or spots; proximal article with 1 or 2 hyaline spots. Postgenal cleft straight or weakly biarctate anteriorly, slightly wider than long, extended about one fourth distance to hypostomal groove; subesophageal ganglion ensheathed with pigment. Labral fan with 75–84 primary rays. Body pigment grayish brown, rather uniformly distributed (Plate 11a); abdominal setae unbranched, translucent, sparse; ventral tubercles about one third depth of abdomen at attachment points. Posterior proleg with 8–12 hooks in each of 61–63 rows.

**Diagnosis.** The largely unsclerotized anal lobe of the female is diagnostic among members of the subgenus *Hellichiella*. Males resemble those of several other species in the subgenus. The gill of 6 filaments that arise from a swollen base is unique among the simuliids of North America and immediately distinguishes mature larvae and pupae.

**Holotype.** Pupa with pharate female (ethanol vial). California, Trinity Co., Weaverville, Sidney Gulch, west end of town, Rt. 299, 40°44′N, 122°56′W, 24 April 1991, D. C. Currie (CNC).

**Paratypes.** Same data as holotype (2♀ & 3♂ [in glycerin vials] + exuviae, 4 pupae, 33 larvae).

**Etymology.** The species name is from the Latin noun meaning "mystery" or "secret," in allusion to the small, rather cryptic preimaginal habitat, the infrequency with which the species has been collected, and the lack of biological information about the species.

**Cytology**

Our preliminary analysis suggests that topotypical larvae (4♀, 2♂) are similar to those of *S. curriei* but have an enormous chromocenter and a possible Y chromosome characterized by a subbasal rearrangement in IIIL. *Simulium mysterium* is the only North American species of the subgenus *Hellichiella* with a chromocenter.

**Bionomics**

**Habitat.** This inadequately collected species is known from 2 sites in northwestern California. Streams inhabited by the immature stages are small, warm, shallow, and rather productive, with stone and pebble substrates.

**Oviposition.** Unknown.

**Development.** Mature larvae and pupae have been taken from late March to late April.

**Mating.** Unknown.

**Natural Enemies.** No records.

**Hosts and Economic Importance.** Unknown; presumably ornithophilic.

### *Simulium* (*Hellichiella*) *rivuli* Twinn

(Figs. 10.153, 10.310, 10.433, 10.748; Plate 11b; Map 113)

*Simulium* (*Eusimulium*) *rivuli* Twinn 1936a: 120–121, Figs. 4B, 6D (MP). Holotype* male (slide #4125; CNC). Ontario, Carleton Co., Goulbourn Township, near Carleton Place (1.5 mi. west of Stanley Corners), roadside ditch (Fig. 4B of Twinn 1936a), 8 May (pupa), 13 May (adult) 1935 (C. R. Twinn)

*Simulium* (*Eusimulium*) 'sp. nr. *innocens*' Adler et al. 1982: 254 (Pennsylvania record)

**Taxonomy**

Twinn 1936a (MP), Davies et al. 1962 (FPL), Wood 1963a (FMPL), Wood et al. 1963 (L), Stone 1964 (FM), Holbrook 1967 (L), Adler & Kim 1986 (PL), Stuart 1995 (cocoon), Stuart & Hunter 1998b (cocoon).

**Morphology**

Bennett 1963b (F salivary gland), Guttman 1967a (labral fan).

**Cytology**

Dunbar 1962, 1967.

**Bionomics**

**Habitat.** *Simulium rivuli* is confined to northeastern North America. The immature stages inhabit small streams, often ephemeral, in grassy swales with ample trailing vegetation and in open woodlands with fallen leaves and pebbled beds.

**Oviposition.** Females are anautogenous and emerge with an average of 220 immature oocytes and minimal fat body (Chutter 1970, Davies & Györkös 1990). Oviposition habits are unknown.

**Development.** Eggs overwinter, and larvae and pupae are found primarily during April and May (Davies et al. 1962, Adler & Kim 1986).

**Mating.** Unknown.

**Natural Enemies.** We have found larvae infected with unidentified mermithid nematodes.

**Hosts and Economic Importance.** Unknown; presumably ornithophilic.

### Subgenus *Boreosimulium* Rubtsov & Yankovsky

*Boreosimulium* Rubtsov & Yankovsky 1982: 183–184 (as genus). Type species: *Melusina annulus* Lundström 1911: 17–18, by original designation

[*Hellichia*: Rubtsov (1956) not Enderlein 1925 (misident. in part)]

**Diagnosis.** Adults: Radius with hair dorsobasally (sparse or easily rubbed off in *S. johannseni* species group). Scutum typically unpatterned (except in *S. parmatum* n. sp.), although 3 faint longitudinal stripes can be present in underlying cuticle. Female: Claws each with basal thumb-like lobe (except in *S. baffinense* species group). Spermatheca unpigmented proximally; wall with polygonal pattern. Male: Gonostylus slender, tapered to pointed apex, with 1 apical spinule. Ventral plate with median keel typically weak. Paramere with 0, 1, or multiple apical spines. Aedeagal membrane with or without small spines. Dorsal plate present or absent. Pupa: Gill of 4 filaments (3 in *S. baffinense*, 2 in some specimens of *S. johannseni*) in 2 petiolate pairs. Cocoon slipper shaped, without anterodorsal projection (except in *S. baffinense* species group). Larva: Antenna typically with 3 or more hyaline bands or spots (often obscure or as few as 1 in some specimens of *S. johannseni* species group). Abdominal segment IX with prominent ventral tubercles; abdominal setae unbranched, translucent, sparse. Rectal papillae of 3 compound lobes.

**Overview.** Sixteen known species, all in the Holarctic Region, are members of this subgenus. North America has 12 species, which we assign to the following 3 species groups: *S. annulus* group, *S. baffinense* group, and *S. johannseni* group.

The immature stages of all but the *S. baffinense* species group inhabit cold streams and rivers more than 5 m wide with smooth placid sections interspersed with riffles. The immatures of the *S. baffinense* species group, like those of the subgenus *Hellichiella*, are found in small warm streams. All species are univoltine, and most develop in winter or early spring. Females of most species feed on birds.

### *Simulium* (*Boreosimulium*) *annulus* Species Group

**Diagnosis.** Female: Sensory vesicle occupying about one fourth to one half of palpomere III. Genital fork with each arm expanded into large triangular lateral plate bearing prominent, anteriorly directed apodeme. Precoxal bridge incomplete. Male: Gonostylus of most species bulged on dorsomedial margin. Ventral plate broad, flat, nearly lacking median keel. Paramere weakly sclerotized distally, with 1 stout spine near midlength, and 1–3 stout apical spines. Aedeagal membrane with numerous small spines. Dorsal plate broad, subtriangular. Median sclerite elongate. Pupa: Gill of 4 slender filaments. Larva: Hypostoma with teeth of rather uniform size. Postgenal cleft rounded or biarctate anteriorly, extended about one third or less distance to hypostomal groove. Chromosomes: Nucleolar organizer in base of IIIL. Centromere region of chromosome I expanded. Standard banding sequence as given by Golini and Rothfels (1984).

**Overview.** The *S. annulus* species group was recognized cytologically as *Eusimulium* 'Group 3' by Dunbar (1962) and morphologically as *Simulium* 'Group 3—The *euryadminiculum* Group' by Wood (1963a). It also was referred to as the *S. canonicolum* species group by Golini and Rothfels (1984), who characterized it chromosomally. The group includes 10 species, of which 8 are found in North America. In addition to the North American species, we include 2 Palearctic representatives, *S. annuliforme* Rubtsov and *S. olonicum* (Usova), of which the latter might be synonymous with *S. annulus*. The remaining species placed in the group by Crosskey and Howard (1997) are members of other taxa. *Simulium arctium*, described by Rubtsov (1956) from Murmansk, Russia, is a member of the *S. baffinense* species group. *Simulium tsheburovae*, also described by Rubtsov (1956) from Murmansk, is probably in the subgenus *Hellichiella* (*congareenarum* species group) based on the numerous parameral spines. *Simulium kariyai*, described by Takahasi (1940) from Mongolia, might be in the subgenus *Simulium*, based on the patterned scutum, bare base of the radius, and bicolored legs.

Larvae of all North American species except *S. clarkei*, *S. emarginatum*, and *S. joculator* n. sp. are sexually dimorphic. Male larvae have small distinct head spots (e.g., Fig. 10.752), whereas female larvae have head spots that are larger or enclosed in a darkly pigmented area (e.g., Fig. 10.751) that becomes progressively darker in later instars. Adult females feed on birds.

### *Simulium* (*Boreosimulium*) *annulus* (Lundström)

(Figs. 10.154, 10.311, 10.749; Plate 11c; Map 114)

*Melusina annulus* Lundström 1911: 17–18, Fig. 15 (M). Syntype males (7 pinned, each with terminalia on mount below; UZMH). Finland, Kittilä (2 males) (R. Frey), Lapland (3 males) (E. Palmén), Muonio (1 male) (R. Frey), dates unknown for all; *annula*, *annulum* (subsequent misspellings)

*Simulium euryadminiculum* Davies 1949b: 45–49 (FMP). Holotype* male (slide #5867 with head, 1 wing, 1 leg, and terminalia [all else missing]; CNC). Ontario, Algonquin Provincial Park, Costello Creek, 20 May 1940 (F. P. Ide). New Synonym

[*Eusimulium canonicolum*: MacNay (1954) not Dyar & Shannon 1927 (misident.)]

[*Simulium canonicolum*: Fredeen (1956, 1958) not Dyar & Shannon 1927 (misident.)]

?[*Simulium* (*Eusimulium*) *canonicolum*: Shewell (1957) not Dyar & Shannon 1927 (misident.)]

[*Simulium canonicolum*: Shewell & Fredeen (1958) not Dyar & Shannon 1927 (misident.)]

[*Simulium* (*Hellichia*) *canonicolum*: Bennett (1960) not Dyar & Shannon 1927 (misident.)]

[*Simulium* (*Eusimulium*) *emarginatum*: Cupp & Gordon (1983) not Davies et al. 1962 (misident.)]

**Taxonomy**

Lundström 1911 (M); Rubtsov 1940b (M), 1956, 1962a (FMPL); Davies 1949b, 1949c (FMP); Fredeen 1951, 1981a (FMPL), 1985a (FMP); Anderson 1960 (FM); Wood 1963a (FMPL); Wood et al. 1963 (L); Stone 1964 (FMPL); Lewis 1973 (FMPL); Hilsenhoff 1982 (L); Currie 1986 (PL); Mason & Kusters 1991b (FMPL); Stuart 1995 (cocoon); Jensen 1997 (PL); Stuart & Hunter 1998b (cocoon); Yankovsky 1999 (PL).

Larvae, pupae, males, and females in North America are virtually identical to those in Sweden. This morphological correspondence, coupled with the chromosomal similarity shown by Adler et al. (1999), demonstrates that populations in the Nearctic and Palearctic Regions are conspecific. We, therefore, synonymize *euryadminiculum* with *annulus*, as foreshadowed by Adler et al. (1999).

**Morphology**

Bennett 1963b (F salivary gland); Guttman 1967a (labral fan); Mercer 1972 (F antennal and palpal sensilla); Mercer & McIver 1973a (F antennal sensilla), 1973b (F palpal sensilla); Sutcliffe 1975 (F legs); Sutcliffe & McIver 1976 (F leg sensilla); Hayton 1979 (instars).

**Cytology**

Dunbar 1962, Golini & Rothfels 1984, Adler et al. 1999.

**Molecular Systematics**

Brockhouse 1991 (rDNA).

**Bionomics**

**Habitat.** *Simulium annulus* populates much of Canada, part of the northern United States, Fennoscandia, and western Russia. We have no records west of the Rocky Mountains, but circumstantial evidence suggests that it has a trans-Canadian distribution (Fig. 7.3). The immature stages inhabit streams and rivers typically more than 10 m wide, but as small as 2 m in width, often downstream of a lake outlet. The positive association with stream size and proximity to lake outlets might reflect a need to be near hosts (loons) of the females (Malmqvist & Hoffsten 2000). Females are more likely to be collected in fens than in forested and shrubby habitats (Graham 1992).

**Oviposition.** Females are anautogenous (Davies & Györkös 1990) and have an average of about 550 ovarioles (Pascuzzo 1976). Eggs possibly are dropped freely into the water during flight (Davies et al. 1962).

**Development.** Larvae develop during the winter. They pass through 7 instars (Hayton 1979). Pupae begin to appear in late April on Newfoundland's Avalon Peninsula and in Ontario, central Saskatchewan, and New York. They can be found into June in Alberta and Quebec (Wolfe & Peterson 1959, Currie 1986). Adults begin emerging in May and can be found through June or even into early July. Emergence is greatest about 2 hours after sunrise (Hayton 1979). Females live about 2–3 weeks and have a typical flight range of 2.5 km, although distances of at least 8 km are possible (Bennett & Fallis 1971).

**Mating.** The highly specific association between blood-seeking females and the common loon suggests that males might intercept host-seeking females downwind of the host (Mokry et al. 1981).

**Natural Enemies.** The trichomycete *Harpella melusinae* has been recorded from the larval midgut (Frost & Manier 1971). We have found larvae infected with the chytrid fungus *Coelomycidium simulii*, the microsporidium *Polydispyrenia simulii*, and unidentified mermithid nematodes. Palearctic larvae are attacked by the microsporidium *Janacekia debaisieuxi*, as well as *P. simulii* (Adler et al. 1999). Females flying over lakes have been preyed on by empidid flies (Peterson & Davies 1960).

**Hosts and Economic Importance.** This species is the most host-specific black fly yet known, having been confirmed feeding only on the common loon (Fallis & Smith 1964a). It is perhaps the only simuliid that regularly feeds on the common loon. A collection of more than 1000 females from a single, freshly shot common loon, and continued collections from the same specimen after it had been stuffed and stored in paradichlorobenzene, originally suggested this extraordinary host specificity (Lowther & Wood 1964, Smith 1966). Experiments confirmed a strong attraction to extracts of the common loon's uropygial gland (Fallis & Smith 1964a, 1964b). Carbon dioxide is a poor attractant but has a synergistic effect in combination with uropygial gland extract from the common loon (Bennett et al. 1972) but not from the domestic duck (Mercer 1972). Although the loon extract attracts females—hundreds sometimes gathering within minutes of exposure—it does not evoke a landing response (Smith 1966). Rather, visual stimuli, such as color, are important in near-range orientation. Black is attractive when used with three-dimensional models, but white is attractive with two-dimensional models; attraction to white might be associated with a resting response (Bennett et al. 1972, Bradbury & Bennett 1974b). Most host-seeking females land on the leeward side of the head and neck, the most prominent body regions, as the loon swims low in the water (Lowther & Wood 1964, Bennett et al. 1972). Females with immature eggs and nectar-laden crops fly over lakes throughout the day, especially during early evening, probably in search of loons (Davies & Peterson 1957, Bennett & Fallis 1971). Females also swarm around the nests of the common loon (Bennett & Fallis 1971).

Collections of females (unfed) from penned moose (Pledger et al. 1980), attacks on chickens (Fredeen 1981a), engorgement on domestic ducks and ruffed grouse (Bennett 1960), induced feedings on the defeathered abdomens of domestic ducks (Tarshis 1972), and swarms around humans have been reported, but at least some of these cases might represent misidentifications. Females have transmitted *Leucocytozoon icteris* to common grackles under laboratory conditions (Fallis & Bennett 1962; as *L. fringillinarum*). Flowers of willow (*Salix* sp.) provide a source of sugar (Davies & Peterson 1956).

### *Simulium* (*Boreosimulium*) *balteatum* Adler, Currie & Wood, New Species

(Figs. 10.28, 10.312, 10.434, 10.750; Plate 11d; Map 115)

*Simulium* 'Undescribed species' Corkum & Currie 1987: 207 (in part, sites 46 and 47) (records from British Columbia)

*Simulium* 'n. sp. near *canonicolum*' Currie 1997a: 567, 578 (Yukon record)

*Simulium* 'near *canonicolum*' McCreadie et al. 1997: 764 (Wyoming record)

[*Eusimulium clarum*: Hearle (1932) not Dyar & Shannon 1927 (misident.)]

**Taxonomy**

**Female.** Wing length 2.5–2.7 mm. Scutum grayish black; thoracic hair silvery, with golden reflections. Mandible with 28–36 serrations; lacinia with 22–25 retrorse teeth. Sensory vesicle in lateral view occupying about one third of palpomere III (Fig. 10.28). Hair of stem vein and basicosta silvery to pale golden. Mesepimeral tuft, basal fringe, and abdominal hair silvery. Legs brown, with silvery hair; pedisulcus deeply incised. Terminalia: Anal lobe in ventral view with unsclerotized medial area deeply and broadly concave. Spermatheca longer than wide.

**Male.** Wing length 2.3–2.6 mm. Scutum velvety, brownish black; thoracic hair pale golden, although that of scutellum sometimes coppery. Hair of stem vein and basicosta coppery. Mesepimeral tuft pale golden. Basal fringe and abdominal hair coppery. Legs brown, with pale golden hair. Genitalia (Fig. 10.312): Gonostylus in inner lateral view with mesal margin bulged. Ventral plate in ventral view with body subrectangular, broadly concave posteriorly, convex anteriorly. Paramere with 1 medial and 3 distal spines.

**Pupa.** Length 2.9–3.5 mm. Gill of 4 filaments, as long as or longer than pupa (Fig. 10.434); base slender, 2 or more times its width, giving rise to 2 petiolate pairs; dorsal petiole about twice as long as ventral petiole, branching near middle of gill; filaments with numerous transverse furrows. Head and thorax smooth dorsally or with scattered, rounded microtubercles; trichomes unbranched. Cocoon densely woven, heavily reinforced anterodorsally often with short anterodorsal stub.

**Larva.** Length 4.9–5.5 mm. Head capsule (Fig. 10.750) pale yellow to yellowish brown; head spots dark brown, enclosed in dark brown pigment (female), or slightly infuscated, or free of surrounding pigment (male); apex of frontoclypeal apotome suffused with brown pigment. Antenna yellowish brown to dark brown, extended beyond stalk of labral fan by length of distal article; medial article with 3 or 4 hyaline bands. Hypostoma with lateral and median teeth small, typically extended anteriorly to about same level. Postgenal cleft straight or biarctate anteriorly, as wide as or slightly wider than long, extended less than one third distance to hypostomal groove; subesophageal ganglion ensheathed with pigment. Labral fan with 43–52 primary rays. Body slender; pigment gray to brown, arranged in bands, heaviest on first abdominal segment (Plate 11d). Posterior proleg with 8–12 hooks in each of 56–60 rows.

**Diagnosis.** Females are indistinguishable from those of other western members of the subgenus *Boreosimulium*. The male is the only member of the subgenus with 3 stout spines at the distal end of the paramere. The 4-filamented pupal gill with its dorsal petiole branching near the middle of the gill is unique among North American species. Larvae can be distinguished from those of all other members of the subgenus by their small size and conspicuous band on the first abdominal segment.

**Holotype.** Male (Peldri II, pinned) with pupal exuviae (glycerin vial below). California, Shasta Co., East Fork Hat Creek, Dersch Meadow, 40°30.03′N, 121°26.32′W, 11 July 1988, D. C. Currie & P. H. Adler (USNM).

**Paratypes.** California, same data as holotype (12♀ & 16♂ + exuviae, 11 pupae, 179 larvae, photographic negatives of larval polytene chromosomes). Wyoming, Park Co., Yellowstone National Park, Cascade Meadows, Canyon-Norris Road, west of Canyon Junction, Cascade Creek, 44°44.15′N, 110°30.08′W, 2438 m elev., 2 July 1993, J. F. Burger (13 larvae). British Columbia, south of Silver Creek (town), Silverhope Creek, inflow Silver Lake, 49°18′N, 121°24′W, 24 June 1991, D. C. Currie (64 larvae); Silver Creek (town), Silverhope Creek, Flood-Hope Road, 49°22′N, 121°27′W, 24 June 1991, D. C. Currie (1♀ & 5♂ + exuviae, 2 pupae, 39 larvae); McLeese Lake, inflow south of town, Rt. 97, 52°25′N, 122°18′W, 12 June 1991, D. C. Currie (2 pupae, 12 larvae).

**Etymology.** The specific name is from the Latin adjective meaning "belted," in reference to the conspicuous dark band on the first abdominal segment of the larva.

**Cytology**

The basic banding sequence is identical to the standard sequence of Golini and Rothfels (1984) in all chromosomal arms except IIL, which carries the *IIL-1* inversion. The Y chromosome is consistently anucleolate. Floating inversions are common in the distal portions of IS and IIL; IIIL carries a common autosomal polymorphism deceptively close to IIIL-1 but differing in its proximal breakpoint, which is 2 bands more distal than that of IIIL-1. In the central Rocky Mountains (Yellowstone National Park), this inversion appears to function as an X chromosome, with the Y usually being standard but anucleolate. Populations in the central Rocky Mountains have fewer polymorphisms than do more western populations.

**Bionomics**

**Habitat.** This species, the smallest of the *S. annulus* group, ranges from southern Yukon southward into the Rocky Mountains and Sierra Nevada of California, with heavy representation in British Columbia. The immature stages are found in cool swift streams and rivers up to 20 m wide, with a cobble and boulder substrate.

**Oviposition.** Unknown.

**Development.** Eggs overwinter and larvae are found from May into July. All life stages are most common during June, making this species the latest member of the *S. annulus* species group to develop.

**Mating.** Unknown.

**Natural Enemies.** Larvae sometimes bear infections of the chytrid fungus *Coelomycidium simulii*.

**Hosts and Economic Importance.** One female, probably of *S. balteatum*, has been recorded feeding on the head of a long-eared owl (Hearle 1932). Also see the entry under *S. canonicolum*.

### *Simulium* (*Boreosimulium*) *canonicolum* (Dyar & Shannon)

(Figs. 10.313, 10.435, 10.751, 10.752; Plate 11e; Map 116)

*Eusimulium canonicolum* Dyar & Shannon 1927: 22, Fig. 40 (F). Holotype* female (pinned #28337, terminalia and leg on slide; USNM). Wyoming, Park Co., Yellowstone National Park, Yellowstone Canyon, 3 July 1922 (H. G. Dyar)

*Simulium canonicola* Peterson & Kondratieff 1995: 34 (unjustified emendation)

[*Simulium pugetense*: Peterson (1955) not Dyar & Shannon 1927 (misident.)]

[*Simulium* (*Eusimulium*) *baffinense*: Lichtwardt (1984) not Twinn 1936a (misident.)]

**Taxonomy**

Dyar & Shannon 1927 (F); Stains 1941 (F); Stains & Knowlton 1943 (F); Peterson 1958a (FMPL), 1960a (FMP), 1996 (FM); Currie 1986 (PL); Peterson & Kondratieff 1995 (FMPL).

Many past identifications of *S. canonicolum* are dubious, owing to our recent discoveries of 3 new western species in the *S. annulus* species group (*S. balteatum*, *S. joculator*, *S. zephyrus*), revalidation of *S. quadratum* as a legitimate species, and morphological homogeneity of the females.

Freshly collected larvae typically have reddish bands on the body (Plate 11e) that soon fade to greenish gray (female) or brown (male).

**Morphology**

A series of intersexes from Robbins Creek, near Kamloops, British Columbia (10 June 1964), resides in the CNC.

**Cytology**

Golini & Rothfels 1984.

This species typically features homozygous expression of the nucleolar organizer in both sexes. IIIL-1 of Golini and Rothfels (1984) is present in nearly all populations and is virtually fixed in some (Logan River, Logan, Utah). Populations from Alberta to northwestern Wyoming have the typical Y chromosome of Golini and Rothfels (1984), whereas those from Utah southward generally have undifferentiated sex chromosomes.

**Molecular Systematics**

Moulton 1997, 2000 (DNA sequences).

**Bionomics**

**Habitat.** Ranging from the Yukon Territory to southern Arizona, this rather common species is concentrated in the Rocky Mountains. Larvae and pupae live in cool clear streams and rivers, especially those 5–30 m wide.

**Oviposition.** Females, presumably of this species, release eggs during flight as they dip to the surface of the water (Peterson 1959a).

**Development.** This early-season species passes at least part of the winter as eggs. Larvae have been found as early as March (Shipp & Procunier 1986), but they are most common, as are pupae, in May and early June.

**Mating.** A possible mating swarm containing males and females was reported flying less than 1 m above smooth water (Peterson 1959a).

**Natural Enemies.** Larvae are hosts of the trichomycete fungi *Harpella leptosa* (Moss & Lichtwardt 1980) and *Pennella arctica* (Lichtwardt 1984), the chytrid fungus *Coelomycidium simulii*, and the microsporidium *Polydispyrenia simulii.*

**Hosts and Economic Importance.** Males and females feed on the nectar of silver buffalo-berry (*Shepherdia argentea*). Deaths of fledgling great horned owls have been attributed to anemia from feeding by the females of *S. canonicolum*, in concert with infection by *Leucocytozoon* sp. (Hunter et al. 1997a, 1997b). The following information pertains to *S. balteatum* n. sp., *S. canonicolum*, or *S. quadratum*: Females are severe pests of red-tailed hawks, biting nestlings and causing them to dehydrate or jump to their death; about 12% of nestlings died in northern Wyoming as a result of black fly attacks (Smith et al. 1998). The USNM collection contains 3 females taken from prairie falcons in northern Utah. One bite on a human has been reported (Peterson 1959a).

### *Simulium (Boreosimulium) clarkei* Stone & Snoddy

(Figs. 10.155, 10.314, 10.436, 10.437, 10.753;
Plate 11f; Map 117)

*Simulium* (*Eusimulium*) *clarkei* Stone & Snoddy 1969: 26–27, Figs. 100, 127, 146, 191, 218, 246, 268, 290 (MPL). Holotype* male (pinned #68977, pupal exuviae, terminalia and 1 leg on slide; USNM). Virginia, Prince George Co., near Prince George, Blackwater Swamp, 19 April 1942 (J. F. G. Clarke)

*Simulium* (*Eusimulium*) 'sp. No. 4' Snow et al. 1958b: 19 (Tennessee record)

*Simulium* (*Eusimulium*) '#4 new species' Snoddy 1966: 47–48 (also pp. 26, 31, 36 as 'sp. #4')

*Simulium* 'near *clarkei*' McCreadie & Adler 1998: 82 (South Carolina record)

[*Simulium slossonae*: Stone & Snoddy (1969) not Dyar & Shannon 1927 (misident. in part, Fig. 127)]

**Taxonomy**

Snoddy 1966 (MPL), Stone & Snoddy 1969 (MPL), Snoddy & Noblet 1976 (PL), Webb & Brigham 1982 (L), Moulton & Adler 2002a (FMPL).

The pigmentation patterns of larvae vary from spotted, as in *S. emarginatum* (Plate 12a), to banded (Plate 11f). The lengths of the dorsal and ventral trunks of the pupal gill vary but are nearly always considerably longer than wide (Figs. 10.436, 10.437). Where *S. clarkei* and *S. emarginatum* occur together, the larvae can be difficult to distinguish. The ventral plate of the male—the largest of all North American species—is diagnostic.

**Cytology**

Moulton & Adler 2002a.

**Bionomics**

**Habitat.** *Simulium clarkei* is known from eastern Virginia south to Georgia. A record from a cobble-bottomed stream in eastern Tennessee (Stone & Snoddy 1969) should be reevaluated chromosomally and with reared adults. Larvae and pupae are found on leaf mats in wide sandy rivers of the piedmont and less often in swampy blackwater streams of the coastal plain.

**Oviposition.** Unknown.

**Development.** This species is exceptionally early. Larvae hatch in the winter, and pupae can be collected as early as mid-January. Pupae are particularly common in February, but stragglers, including the holotype, have been found as late as April.

**Mating.** Unknown.

**Natural Enemies.** No records.

**Hosts and Economic Importance.** Unknown; presumably ornithophilic.

### *Simulium (Boreosimulium) emarginatum* Davies, Peterson & Wood

(Figs. 10.65, 10.315, 10.754; Plate 12a; Map 118)

*Simulium emarginatum* Davies, Peterson & Wood 1962: 110–112, Figs. 6a & b, 32, 64 (FMP). Holotype* female with pupal exuviae (pinned #7994; CNC). Ontario, Nipissing District, Bonfield Township, 0.5 mi. west of Rutherglen, Sharpes Creek (= Sparks Creek), Hwy. 17, 5 May 1959 (D. M. Davies & D. M. Wood)

*Simulium* 'n. sp.' Bennett 1960: 380 (host feeding) (also p. 380 as 'third undescribed species of *Simulium* (*Eusimulium*)' and p. 385 as *Eusimulium* 'n. sp.')

*Simulium* (*Eusimulium*) 'sp. near *emarginatum*' Adler 1983: 205 (PL)

[*Simulium euryadminiculum*: Davies & Peterson (1956) not Davies 1949b (misident. in part, p. 640)]

[*Simulium euryadminiculum*: Anderson (1960) not Davies 1949b (misident.)]

[*Simulium* (*Hellichiella*) *rivuli*: Cupp & Gordon (1983) not Twinn 1936a (misident. in part, some material from Commencement Creek, Michigan)]

**Taxonomy**

Anderson 1960 (PL), Davies et al. 1962 (FMP), Wood 1963a (FMPL), Wood et al. 1963 (L), Amrine 1971 (FMPL), Adler 1983 (PL), Adler & Kim 1986 (PL).

The dorsal and ventral trunks of the pupal gill are more variable in length than previously appreciated, causing some difficulty in distinguishing the pupae from those of *S. clarkei* where the 2 species occur together. The trunks of *S. emarginatum* tend to be shorter and more equal in length than those of *S. clarkei.*

**Cytology**

Dunbar 1962, Golini & Rothfels 1984.

**Bionomics**

**Habitat.** *Simulium emarginatum* ranges from Ontario southward along the mountains to Georgia and Arkansas, developing in cold rocky streams and rivers more than 10 m wide. We also have found it in wide sandy streams near the eastern limit of the piedmont region of Virginia.

**Oviposition.** Females, presumably ovipositing, fly within a few centimeters above fine sand gently bathed by stream water in protected areas (Davies et al. 1962).

**Development.** Larvae hatch in late fall and develop during the winter. Adults emerge from April through May in Canada (Davies et al. 1962) and during February and March in the southern states.

**Mating.** Unknown.

**Natural Enemies.** We have found an unidentified mermithid nematode in a larva from South Carolina.

**Hosts and Economic Importance.** A few females have been recorded engorging on domestic ducks and American crows (Bennett 1960).

### *Simulium* (*Boreosimulium*) *joculator* Adler, Currie & Wood, New Species

(Figs. 4.28, 10.29, 10.316, 10.755; Plate 12b; Map 119)

**Taxonomy**

**Female.** Not differing from that of *S. balteatum* n. sp. except as follows: Wing length 2.8–3.1 mm. Sensory vesicle in lateral view occupying about one fourth to one half of palpomere III (Fig. 10.29).

**Male.** Not differing from that of *S. balteatum* n. sp. except as follows: Wing length 2.9–3.3 mm. Genitalia (Figs. 4.28, 10.316): Paramere with 1 medial and 1 or 2 distal spines.

**Pupa.** Length 3.3–3.7 mm. Gill of 4 filaments, slightly longer than pupa; base short, giving rise to 2 petiolate pairs; ventral petiole thinner and usually slightly longer than dorsal petiole; filaments with numerous transverse furrows. Head and thorax smooth dorsally, without microtubercles; trichomes unbranched. Cocoon densely woven.

**Larva.** Length 5.9–7.1 mm. Head capsule (Fig. 10.755) yellowish brown; area posterior to ocelli suffused with brown, continuous with eyebrow stripe; head spots dark brown, with anteromedial group sometimes weak. Antenna yellowish brown to dark brown, extended beyond stalk of labral fan by length of distal article; medial article with 4 hyaline bands. Hypostoma with lateral and median teeth small, typically extended anteriorly to about same level. Postgenal cleft straight or gently rounded anteriorly, slightly wider than long, extended more than one third distance to hypostomal groove; subesophageal ganglion ensheathed with pigment, although sometimes weakly. Labral fan with 44–49 primary rays. Body pigment pale gray, rather uniformly distributed or weakly banded, suffused with pink posteriorly in fresh specimens (Plate 12b). Posterior proleg with 10–13 hooks in each of 70–72 rows.

**Diagnosis.** Females resemble those of other western members of the subgenus *Boreosimulium*. Males can be distinguished from those of other members of the subgenus by the combination of a strong dorsomedial bulge on each gonostylus and 1 or 2 apical spines on each paramere. Variation in the male genitalia is illustrated in Figs. 4.28 and 10.316. The smooth cuticle of the thorax, in combination with the near-basal branching of the 4-filamented gill, is distinct among pupae of the subgenus. The pigmentation pattern of the head, the pale body, and dark antennae with 4 hyaline bands distinguish the larvae from those of all other species on the continent.

**Holotype.** Male (Peldri II, pinned) with pupal exuviae (glycerin vial below). California, Mariposa Co., Yosemite National Park, Merced River, Rt. 120, in front of El Capitan, 37°43.3′N, 119°38.40′W, 10 May 1992, P. H. Adler (USNM).

**Paratypes.** California, Alpine Co., Rt. 4, Upper Cascade Creek, 24 June 1991, P. H. Adler (2♀ + exuviae, 30 larvae); Mariposa Co., same data as holotype (12♀ & 10♂ + exuviae, 10 pupae, 6 exuviae, 159 larvae); Yosemite National Park, Bridalveil Creek, Glacier Point Road, 37°39.95′N, 119°37.0′W, 10 May 1992, P. H. Adler (3♂ + exuviae, 6 pupae, 46 larvae); Mono Co., Rock Creek, Rt. 395, 37°33.5′N, 118°39.6′W, 13 May 2001, P. H. Adler (3♂ + exuviae, 12 pupae, 31 larvae, photographic negatives of larval polytene chromosomes).

**Etymology.** The specific name is from the Latin noun meaning "jester," in reference to the sportive markings of the larval head capsule.

**Cytology**

Larvae (2♀, 3♂) from California, Inyo Co., Rock Creek (13 May 2001), had the standard banding sequence of Golini and Rothfels (1984) in all arms except IIL, which carried *IIL-1* homozygously. Centromeres were associated weakly in some nuclei. No sex chromosomes or autosomal polymorphisms were found. The chromosomes, therefore, are similar to those of *Simulium zephyrus* n. sp.

**Bionomics**

**Habitat.** This species is encountered in the central Sierra Nevada of California and the Sawtooth Mountains of Idaho. Larvae and pupae are found on trailing vegetation especially in smooth areas of cool, medium-sized rocky streams and rivers.

**Oviposition.** Unknown.

**Development.** Eggs overwinter. Subsequent life stages can be found from early May to late June, depending on elevation.

**Mating.** Unknown.

**Natural Enemies.** Larvae are attacked by the chytrid fungus *Coelomycidium simulii* and the microsporidium *Polydispyrenia simulii*.

**Hosts and Economic Importance.** Unknown; presumably ornithophilic.

### *Simulium* (*Boreosimulium*) *quadratum* (Stains & Knowlton)

(Fig. 10.756, Map 120)

*Eusimulium quadratus* Stains & Knowlton 1943: 271, Figs. 43–45 (M). Holotype* male (slide, head and terminalia only; USNM). Utah, Cache Co., Logan Canyon, 28 June 1940 (G. F. Knowlton). Revalidated

*Simulium* 'near Hinton' McCreadie et al. 1997: 764 (Wyoming record)

?[*Eusimulium clarum*: Dyar & Shannon (1927) (misident. in part, F)]

?[*Eusimulium clarum*: Stains (1941) not Dyar & Shannon 1927 (misident. in part, F)]
?[*Eusimulium clarum*: Stains & Knowlton (1943) not Dyar & Shannon 1927 (misident. in part, F)]
?[*Eusimulium clarum*: Fitch et al. (1946) not Dyar & Shannon 1927 (misident.)]

**Taxonomy**

Stains 1941 (M), Stains & Knowlton 1943 (M).

Larvae of *S. quadratum* can be distinguished from those of *S. canonicolum* by the presence of a pigmented sheath around the subesophageal ganglion and a poorly defined anterior margin of the postgenal cleft. Adults and pupae of these 2 species are indistinguishable. In the interest of conserving names, we apply the name *S. quadratum* to the species whose chromosomes are described below.

**Cytology**

The *IIL-1* inversion of Golini and Rothfels (1984) is fixed. Males are homozygous for expression of the nucleolar organizer and heterozygous for a large diagnostic inversion in IIS (limits 47d–53d). The IIIL arm commonly carries the same mimic inversion of IIIL-1 as *S. balteatum* n. sp.; true IIIL-1 of Golini and Rothfels (1984) is absent. IIIL eu-5 is a common autosomal polymorphism; additional inversions overlapping it are less frequent, as are inversions basally in IL and subterminally in IIL.

**Bionomics**

**Habitat.** This species is fairly common in the central Rocky Mountains and in the mountains from Washington to southern California. In the Rocky Mountains, it frequently occupies the same streams and rivers as *S. canonicolum*. Both species occur together at each of their type localities.

**Oviposition.** Unknown.

**Development.** Larvae probably hatch during late fall or winter along the coast of California, judging from the presence of pupae by the third week of March. At higher elevations and farther inland, larvae and pupae are most common during May and June, suggesting that hatching is delayed until late winter or early spring.

**Mating.** Unknown.

**Natural Enemies.** Larvae are sometimes infected with the chytrid fungus *Coelomycidium simulii* and unidentified mermithid nematodes.

**Hosts and Economic Importance.** See entry under *S. canonicolum*. This species might be the one, referred to as *Eusimulium clarum* by Fitch et al. (1946), that killed nestling red-tailed hawks in California.

### *Simulium* (*Boreosimulium*) *zephyrus* Adler, Currie & Wood, New Species

(Fig. 10.156, Map 121)

*Simulium* 'sp. (Undescribed Species)' Speir 1969: 25 (bionomics)
?*Simulium* 'sp. #1' Speir 1969: 51 (bionomics)
*Eusimulium* 'Hinton' Golini & Rothfels 1984: 2097–2108 (C)

**Taxonomy**

**Female.** Not differing from that of *S. balteatum* n. sp. except as follows: Wing length 2.6–3.0 mm. Terminalia (Fig. 10.156): Anal lobe in lateral view with slightly less sclerotization.

**Male.** Not differing from that of *S. balteatum* n. sp. except as follows: Wing length 2.7–2.9 mm.

**Pupa.** Not differing from that of *S. joculator* n. sp. except as follows: Length 3.3- 4.0 mm. Head and thorax dorsally with numerous, irregularly distributed, rounded microtubercles.

**Larva.** Length 6.3–6.6 mm. Head capsule pale yellowish; head spots brown, free of surrounding brown pigment (male), or enclosed in dark pigment, or with posteromedial and second posterolateral head spots enlarged (female). Antenna brown, extended beyond stalk of labral fan by more than length of distal article; medial article with 4–6 hyaline bands. Hypostoma with teeth small, typically extended anteriorly to about same level. Postgenal cleft straight or weakly rounded anteriorly; subequal in width and length, extended about one third distance to hypostomal groove; subesophageal ganglion ensheathed with pigment. Labral fan with 39–47 primary rays. Body pigment reddish in freshly fixed specimens (as in Plate 11e), fading to greenish gray (female) or brownish (male), arranged in bands. Posterior proleg with 9–13 hooks in each of 66–71 rows.

**Diagnosis.** None of the life stages can be distinguished reliably from those of *S. quadratum*. We, nonetheless, recognize this species as distinct from *S. quadratum* on the basis of its more western distribution and its chromosomes, specifically the anucleolate Y chromosome (or sometimes undifferentiated sex chromosomes) and low frequency of polymorphisms.

**Holotype.** Male (hexamethyldisilazane, pinned) with pupal exuviae (glycerin vial below). Oregon, Benton Co., Beaver Creek, near junction of Bellfountain and Greenberry Roads, 44°27′N, 123°20′W, 17 March 1986, G. W. Courtney (USNM).

**Paratypes.** California, Marin Co., Samuel P. Taylor State Park, Lagunitas (Papermill) Creek, 38°01.34′N, 122°44.08′W, 21 March 1990, P. H. & C. R. L. Adler (photographic negatives of larval polytene chromosomes [IIIL arm only]); Mt. Tamalpais State Park, Redwood Creek, 37°53′N, 122°34′W, 21 March 1990, P. H. & C. R. L. Adler (photographic negatives of larval polytene chromosomes). Montana, Powell Co., Washington Creek, Rt. 41, north of Avon, 46°45.76′N, 112°42.23′W, 23 May 1993, K. P. Pruess (6♀ & 3♂ + exuviae, 6 pupae, 11 larvae). Oregon, same data as holotype (1♀ + exuviae, 2 pupae, 73 larvae); Clatsop Co., North Fork Nehalem River, Rt. 53, 45°42.5′N, 123°51.5′W, 13 March 1993, D. C. Currie & D. M. Wood (25 larvae); Nehalem River, Fishhawk Road, 6 km west of Mist, 46°00′N, 123°20′W, 21 March 1992, D. C. Currie (20 larvae); Tillamook Co., Tillamook River, Rt. 101, 45°24′N, 123°48′W, 21 March 1992, D. C. Currie (1♀ & 1♂ + exuviae, 2 pupae, 18 larvae), 12 March 1993, D. C. Currie & D. M. Wood (22 larvae).

**Etymology.** The specific name is from the Latin noun meaning "West wind," in allusion to the western distribution of the species.

**Cytology**

Golini & Rothfels 1984.

The chromosomes are consistent with the original description (as "Hinton") by Golini and Rothfels (1984). The IIIL-1 inversion is polymorphic in nearly all populations.

IIL-1 is fixed in most populations but is polymorphic in a few Oregon populations (e.g., Clatsop Co., Nehalem River). Other polymorphisms are present but infrequent. The Y chromosome is often anucleolate.

**Bionomics**

**Habitat.** This species is most common in the Pacific Coastal states but ranges eastward into the Rocky Mountains of Alberta and Montana. Larvae and pupae are found on trailing vegetation in medium-sized rocky streams and rivers with generous stretches of smooth flow.

**Oviposition.** Unknown.

**Development.** Larvae hatch from late fall near the coast to late winter and early spring in the Rocky Mountains. Coastward, adults begin emerging as early as mid-March. Farther inland, larvae, pupae, and adults can be found through May, with stragglers into June.

**Mating.** Unknown.

**Natural Enemies.** Larvae are hosts of the chytrid fungus *Coelomycidium simulii*.

**Hosts and Economic Importance.** Unknown; presumably ornithophilic.

### *Simulium* (*Boreosimulium*) *baffinense* Species Group

**Diagnosis.** Female: Mandible without serrations; lacinia without retrorse teeth. Palpomere III with minute sensory vesicle. Claws toothless. Genital fork with anteriorly directed apodeme of each lateral plate weakly developed. Male: Ventral plate broad, with median keel. Paramere elongate. Aedeagal membrane with pair of single stout spines not connected to parameres, and with cluster of numerous short stout spines in addition to many fine smaller spines. Dorsal plate subtriangular. Pupa: Gill of 3 or 4 basally inflated filaments; dorsalmost filament strongly divergent from other filaments. Cocoon with anterodorsal projection. Larva: Hypostoma with teeth in 3 prominent groups. Postgenal cleft minute.

**Overview.** The *S. baffinense* species group was first recognized cytologically as *Eusimulium* 'Group 2' (Dunbar 1962) and morphologically as 'Group 2—The *baffinense* Group' (Wood 1963a). *Simulium baffinense* later was considered a member of the subgenus *Hellichiella* (Crosskey 1988, Crosskey & Howard 1997). However, on the basis of both morphological and chromosomal evidence, we reassign *S. baffinense* to the *S. baffinense* species group in the subgenus *Boreosimulium*. We also include the Palearctic species *S. arctium* (Rubtsov 1956) and *S. crassum* (Rubtsov 1956) in this group. The species of this group differ ecologically from other members of *Boreosimulium* in their preimaginal habitat; they occupy small warm streams. The females do not have biting mouthparts.

### *Simulium* (*Boreosimulium*) *baffinense* Twinn

(Figs. 10.66, 10.157, 10.317, 10.438, 10.642, 10.757; Plate 12c; Map 122)

*Simulium* (*Eusimulium*) *baffinense* Twinn 1936a: 121–123 (FM). Holotype* female (pinned #4126, left hind leg and terminalia on slide #B.9.3; CNC). Nunavut (as Northwest Territories), Baffin Island, Lake Harbour, 10 August 1935 (W. J. Brown)

*Simulium* (*Eusimulium*) *baffinense* form *pallens* Twinn 1936a: 123 (F). Syntype* females (3 each on separate slide #17218; CNC). Nunavut (as Northwest Territories), Baffin Island, Lake Harbour, 10 August 1935 (W. J. Brown); *palens* (subsequent misspelling)

*Eusimulium* 'new species 1' Jenkins 1948: 149–150 (Alaska record)

*Simulium* (*Hellichiella*) 'sp.' Hershey et al. 1995: 284 (Alaska record)

**Taxonomy**

Twinn 1936a (FM); Stone 1952 (FMP); Sommerman 1953 (L); Rubtsov 1956, 1962a (FMPL); Peterson 1958a (FMPL), 1960a (FMP); Davies et al. 1962 (FMP); Wood 1963a (FMPL); Wood et al. 1963 (L); Smith 1970 (FMPL); Rubtsov & Yankovsky 1984 (FMPL); Currie 1986 (PL); Jensen 1997 (PL); Yankovsky 1999 (PL).

Closer study is required to determine if *S. baffinense* is a species complex and if it is specifically distinct from *S. arctium* and *S. crassum*, both of which were described by Rubtsov (1956). The major diagnostic feature of *S. crassum* is its gill of 4, rather than 3, filaments. We have found Nearctic pupae of *S. baffinense* with a fourth, albeit rudimentary, filament (Fig. 10.438).

**Morphology**

Mercer 1972 (FM antennal and palpal sensilla); Mercer & McIver 1973a (FM antennal sensilla), 1973b (FM palpal sensilla); Sutcliffe 1975 (FM legs); Sutcliffe & McIver 1976 (FM leg sensilla).

**Cytology**

Dunbar (1962), on the basis of material possibly from Churchill, Manitoba, presented a low-magnification photograph of the entire complement; homologized IIS with the standard banding sequence for the *S. vernum* species group; and declared the nucleolar organizer to be in the base of IL and associated with a heavy band. Our examination of larvae from Alaska's Taylor Highway and from the Horton River drainage in the Northwest Territories shows that IIS and IIIS have the same banding sequence as the standard maps for the *S. vernum* and *S. annulus* species groups, the centromere region of chromosome I is greatly expanded, the nucleolar organizer is in the base of IS (not IL) and is often associated with a heavy band, and males minimally are heterozygous for nucleolar expression. Chromosome arms IL, IS, IIL, and IIIL are complexly rearranged relative to the standard banding sequences for the subgenus *Hellichiella* as given by Dunbar (1967), the *S. annulus* species group of Golini and Rothfels (1984, as *S. canonicolum* group), and the *S. vernum* species group of Brockhouse (1985).

**Bionomics**

**Habitat.** This Holarctic species is found across the tundra from Alaska to Quebec, south through the boreal forests into the mountains of northern Utah. In the Palearctic Region, it occurs in Scandinavia and Russia. It breeds in seepages and small flows that drain bogs, tarns, and swampy areas. Abundant emergent vegetation characterizes these streams and serves as the principal substrate for larvae and pupae.

**Oviposition.** Females produce up to 218 eggs autogenously (Smith 1970). Ovarian development is precocious in the arctic (Smith 1970) but less advanced farther south (Davies et al. 1962). Egg-laying behavior is unknown.

**Development.** Winter is passed as eggs. Larvae are present in Ontario during June (Davies et al. 1962) and in Alaska and the Northwest Territories from June to September. Adults have an even sex ratio and live up to 15 days in the laboratory if fed honey (Smith 1970).

**Mating.** Mating has not been observed in nature, but it occurs readily in vials, suggesting that the species mates on the ground, probably near emergence sites (Downes 1962).

**Natural Enemies.** Larvae are hosts of the chytrid fungus *Coelomycidium simulii* and an unidentified mermithid nematode.

**Hosts and Economic Importance.** Mouthparts of females cannot pierce skin to take blood, but both sexes take liquid carbohydrates.

### *Simulium* (*Boreosimulium*) *johannseni* Species Group, Newly Recognized

**Diagnosis.** Adults: Radius with hair dorsobasally, although often rubbed off or sparse. Scutum with or without 3 dark longitudinal stripes. Female: Precoxal bridge incomplete. Genital fork with each arm slender and expanded into triangular lateral plate. Male: Ventral plate tapered posteriorly. Paramere closely associated with multiple spines, although not all of these directly connected to it. Dorsal plate absent. Pupa: Gill of 2–4 filaments, at least 2 of which are inflated basally. Cocoon without anterodorsal projection. Larva: Antenna with 1–3 hyaline bands. Postgenal cleft rounded, as wide as long, extended about half or more distance to hypostomal groove.

**Overview.** We recognize *S. johannseni* and 2 closely related species as members of this new species group. Previously, *S. johannseni* was placed in the subgenus *Nevermannia* (Crosskey & Howard 1997). The 3 species in the group are endemic to North America and share unique similarities in the shape of the pupal gill filaments and larval postgenal cleft. The male of *S. johannseni* has 1 stout spine located near the midlength of the paramere, an arrangement similar to that in the *S. annulus* species group.

The immature stages typically are found in large streams and rivers and develop early in the year. Females are primarily ornithophilic, but those of at least 1 species (*S. johannseni*) also feed on mammals and can be pests of humans and domestic animals.

### *Simulium* (*Boreosimulium*) *johannseni* Hart

(Figs. 4.16, 10.32, 10.158, 10.318, 10.439, 10.586, 10.607, 10.758; Plate 12d; Map 123)

*Simulium johannseni* Hart in Forbes 1912: 31–37 (FMPL). Lectotype* female (designated by Frison 1927: 181; pinned; INHS). Illinois, Mason Co., Havana, shore of Illinois River, on house boat, 26 April 1912 (collector unknown)

*Eusimulium* 'sp. near *johannseni*' MacNay 1954: 156 (Saskatchewan record)

*Simulium* '2' Fredeen 1956: 4 (Saskatchewan record)

*Simulium* (*Hellichia*) *johannseni duplex* Shewell & Fredeen 1958: 734–738 (FMP). Holotype* female (pinned #6645, cocoon in glycerin vial below, pupal exuviae on slide #GES-5705-14B; CNC). Saskatchewan, Shell River, Hwy. 55, 5 mi. west of Prince Albert, 2.5 mi. above confluence with North Saskatchewan River, 26 May 1955 (F. J. H. Fredeen). New Synonym

*Simulium* (*Hellichia*) 'sp.?' Shewell & Fredeen 1958: 737 (MP)

*Simulium* (*Eusimulium*) *johannseni johannseni*, Stone 1965a: 186 (distribution)

*Simulium* 'sp. (prob. new sp.) near *johannseni*' Edwards & Meyer 1987: 243 (larval growth); Hauer & Benke 1987: 252 (larval growth)

*Simulium* 'near *johannseni*' Benke & Parsons 1990: 171 (production)

**Taxonomy**

Hart in Forbes 1912 (FMPL); Malloch 1914 (FMP); Riley & Johannsen 1915 (F); Dyar & Shannon 1927 (FM); Johannsen 1934 (PL); Stains 1941 (FM); Stains & Knowlton 1943 (FM); Nicholson 1949 (FMP); Nicholson & Mickel 1950 (FMP); Fredeen 1951 (FMP), 1981a (FMPL), 1985a (FMP); Shewell & Fredeen 1958 (FMP); Anderson 1960 (FMPL); Wood 1963a (FM); Stone 1964 (FMPL); Amrine 1971 (FMPL); Westwood 1979 (FMPL); Currie 1986 (PL); Mason & Kusters 1991b (FMPL).

The pupal gill consists of 2 or 4 (infrequently 3) filaments. The form with 2 filaments is most common in the northern part of the range and originally was considered a subspecies of *S. johannseni*, and later a full species (*S. duplex*) (e.g., Fredeen 1981a, Currie 1986). Additional structural differences between the 2 pupal forms have not been found. We, therefore, consider the differences in filament number to represent intraspecific variation.

Populations in the southern coastal plain might represent a separate species. However, until supporting evidence for species status is found, we retain them under *S. johannseni*.

**Cytology**

Rothfels & Golini 1984 (p. 2097).

**Molecular Systematics**

Zhu 1990 (mtDNA), Tang et al. 1996b (mtDNA), Zhu et al. 1998 (mtDNA).

**Bionomics**

**Habitat.** Primarily a species of midwestern North America, *S. johannseni* ranges from southern Alberta south to Texas and southern Alabama, with a few records from the southern Atlantic coastal plain. It breeds in large, turbid, meandering streams and rivers with beds of sand and silt. It was once abundant in some of the large rivers, such as the Illinois River (Forbes 1912), which eventually became too polluted to support populations. Larvae and pupae anchor themselves to trailing and submerged vegetation and driftwood.

**Oviposition.** Unknown.

**Development.** In Canada, eggs overwinter and larvae hatch just after ice breakup, requiring about 20–30 days to mature (Westwood & Brust 1981, Currie 1986). In Nebraska, larvae begin hatching as early as November (Pruess & Peterson 1987). Throughout much of the distribution, pupae appear from March through May. At the southern limit of distribution pupae are found in January and February. Adults first appear in March in Nebraska, in April or

May in Wisconsin, and in May in Canada (Anderson & Dicke 1960, DeFoliart et al. 1967, Westwood & Brust 1981, Currie 1986, Pruess & Peterson 1987). Females have been maintained at 8°C for 42 days on sugar water (Anderson & DeFoliart 1961). Adults are most active in the morning and early evening (Van Deveire 1981).

**Mating.** Unknown.

**Natural Enemies.** Larvae are hosts of unidentified mermithid nematodes and the microsporidium *Polydispyrenia simulii.*

**Hosts and Economic Importance.** Females feed on gallinaceous birds, such as chickens, ring-necked pheasants, and domestic turkeys, particularly when these hosts are exposed in the lower canopy of woodlands; they also enter poultry sheds and brooder houses to feed (Anderson & DeFoliart 1961, Spalatin et al. 1961). They feed especially on the unfeathered heads and wattles of chickens and turkeys, but also crawl beneath the body feathers (Van Deveire 1981). They also feed on ducks but do not transmit the blood protozoan *Leucocytozoon simondi* (Anderson et al. 1962). Eastern equine encephalitis virus has been isolated from females (Anderson et al. 1961).

The claws are designed for ornithophily, but females also engorge on horses and cattle, sometimes attaining pest status (Westwood & Brust 1981). They are attracted to other mammals, such as raccoons and woodchucks, but rarely feed on them (Wright & DeFoliart 1970). They bite and crawl on humans and can be a genuine pest in some areas, such as Manitoba (Westwood & Brust 1981), Minnesota, and North Dakota. At one time, they might have been a biting problem in Illinois (Jobbins-Pomeroy 1916).

### *Simulium* (*Boreosimulium*) *parmatum* Adler, Currie & Wood, New Species

(Figs. 10.319, 10.440, 10.759; Plate 12e; Map 124)

*Simulium* 'near *johannseni*' McCreadie & Adler 1998: 82 (South Carolina record)

[*Simulium meridionale*: Malloch (1914) not Riley 1887 (misident. in part, material from Georgia, South Carolina [except Congaree], and Tennessee)]

[*Simulium occidentale*: Dyar & Shannon (1927) not Townsend 1891 (misident. in part, material from Florida, Georgia, South Carolina, and Tennessee)]

[*Simulium occidentale*: Brimley (1938) not Townsend 1891 (misident.)]

[*Simulium meridionale*: Pinkovsky (1976) not Riley 1887 (misident. in part, material from Duval Co.)]

[*Simulium meridionale*: Noblet et al. (1978) not Riley 1887 (misident.)]

[*Simulium meridionale*: Pinkovsky & Butler (1978) not Riley 1887 (misident. in part, material from Duval Co.)]

**Taxonomy**

**Female.** Wing length 2.4–3.0 mm. Scutum gray, with central longitudinal, thin, dark gray stripe and 2 lateral, posteriorly divergent, longitudinal dark gray to brown stripes. All hair silvery. Mandible with 33–35 serrations; lacinia with 19 or 20 retrorse teeth. Sensory vesicle in lateral view occupying about one fourth of palpomere III. Legs brown; pedisulcus deeply incised. Terminalia: Anal lobe in ventral view with anteromedial area unsclerotized. Genital fork with each arm slender and expanded into triangular, posteromedially projected lateral plate. Spermatheca subspherical or slightly longer than wide.

**Male.** Wing length 2.5–2.7 mm. Scutal pattern similar to that of female. All hair silvery, with faint, pale golden reflections. Legs brown. Genitalia (Fig. 10.319): Ventral plate in ventral view with body subtriangular, acutely tapered posteriorly. Paramere with about 15 long and short spines. Median sclerite elongate. Dorsal plate absent.

**Pupa.** Length 3.1–3.7 mm. Gill of 4 filaments, about as long as pupa (Fig. 10.440); base short, giving rise to 2 petiolate pairs of filaments; dorsal and ventral petioles short, subequal in length and width; filaments directed anteroventrally, slightly swollen proximally, tapered distally, with reticulate pattern. Head and thorax dorsally with numerous irregularly distributed, dome-shaped microtubercles; trichomes unbranched. Cocoon densely woven.

**Larva.** Length 5.9–6.3 mm. Head capsule pale yellowish; head spots brown, enclosed in black pigment (female, Plate 12e) or pale brown pigment, or free of surrounding pigment (male, Fig. 10.759). Antenna pale yellowish, extended beyond stalk of labral fan by about half length of distal article; medial article with 1–3 hyaline bands. Hypostoma with lateral and median teeth small, typically extended anteriorly to about same level. Postgenal cleft rounded, as wide as long, extended about half or more distance to hypostomal groove; subesophageal ganglion ensheathed with pigment. Labral fan with 33–42 primary rays. Body pigment dark gray (female, Plate 12e) to brown (male) or occasionally reddish, arranged in distinct bands, heaviest on first and fifth abdominal segments. Posterior proleg with 9–12 hooks in each of 59–63 rows.

**Holotype.** Male (frozen dried, pinned) with pupal exuviae (glycerin vial below). South Carolina, Abbeville Co., Little River, Rt. 71, 34°11.90′N, 82°28.98′W, 18 March 1993, P. H. Adler (USNM).

**Paratypes.** Arkansas, Clark Co., L'Eau Frais Creek, Rt. 7, 8.7 km southeast of Arkadelphia, 34°04.11′N, 92°58.98′W, 10 March 1993, D. S. Bidlack (1 pupa, 24 larvae). North Carolina, Northampton Co., Meherrin River, Branchs Bridge Road, 36°31.92′N, 77°15.68′W, 20 February 2002, P. H. Adler (4 pupae, 19 larvae). South Carolina, same data as holotype (1♀ & 4♂ + exuviae, 5 pupae, 101 larvae); Anderson Co., Hen Coop Creek, Co. Road S-4-244, 34°22.76′N, 82°33.62′W, 21 March 1993, J. W. McCreadie (12 pupae, 40 larvae).

**Diagnosis.** Females resemble those of *S. meridionale* externally, although the terminalia are nearly identical to those of *S. johannseni.* The V-shaped ventral plate (in ventral view) and faintly striped grayish scutum are unique among males of all North American species. The pupal gill with its basally swollen, downward-curved filaments is diagnostic among all species in the southeastern portion of the continent. Within the range of this species, the black pigment on the head capsule of the female larva, in combination with the 4-filamented gill histoblast, is the best diagnostic aid for larvae.

**Etymology.** The species name is from the Latin adjective meaning "armed with a shield," in reference to the dark patch on the head of the female larva.

**Bionomics**

**Habitat.** This species lives primarily in the piedmont and Sandhills Region from North Carolina to eastern Texas. The immature stages are found in turbid streams and rivers with sandy bottoms, averaging about 8 m wide. Larvae affix themselves to leaves and other debris.

**Oviposition.** Unknown.

**Development.** Larvae begin hatching in late fall and early winter. Pupae are present during a small window of early spring. In the piedmont of Georgia and South Carolina, for instance, pupae are found almost exclusively in March, although we have collected pupae from the Meherrin River in northern North Carolina as early as the third week of February.

**Mating.** Unknown.

**Natural Enemies.** Larvae are hosts of unidentified mermithid nematodes (McCreadie & Adler 1999).

**Hosts and Economic Importance.** This species previously has been misidentified as *S. meridionale*. It might be a pest of chickens and turkeys and possibly transmits the agent of *Leucocytozoon* disease. The specimen recorded by Malloch (1914) as biting a human hand represents the only confirmed host record, although humans evidently are not typical hosts.

### *Simulium* (*Boreosimulium*) *rothfelsi* Adler, Brockhouse & Currie

(Figs. 10.441, 10.760; Map 125)

*Simulium rothfelsi* Adler, Brockhouse & Currie 2003: 9–12 (PL). Holotype pupal exuviae (ethanol vial; USNM). Nova Scotia, Lunenberg Co., near New Germany, LaHave River, 44°33′N, 64°43′W, 21 May 1999 (C. L. Brockhouse)

**Taxonomy**

Adler et al. 2003 (PL).

Adults are unknown. We have, however, seen females from Orono, Maine (May), and Axton, New York, in the Adirondack Mountains (12–22 June 1901) (CU collection) that were identified as *S. johannseni* by Alan Stone. These specimens, in fact, do resemble *S. johannseni* and, given their distribution, might well be the females of *S. rothfelsi*. Published reports of *S. johannseni* from northeastern North America (e.g., Stone 1964) might refer, at least in part, to *S. rothfelsi*.

**Bionomics**

**Habitat.** Our only confirmed record for this species is from the LaHave River near New Germany, Nova Scotia. The river at the collection site is more than 30 m wide, rocky, and forested on either side. Larvae have been collected from submerged grasses.

**Oviposition.** Unknown.

**Development.** Eggs hatch in early May and larvae and pupae develop rapidly, disappearing before the end of May (Adler et al. 2003).

**Mating.** Unknown.

**Natural Enemies.** No records.

**Hosts and Economic Importance.** Unknown; presumably ornithophilic.

### Subgenus *Byssodon* Enderlein

*Byssodon* Enderlein 1925: 209 (as genus). Type species: *Simulium forbesi* Malloch 1914: 63–65 (= *Simulium meridionale* Riley 1887: 513–514), by original designation

*Psilocnetha* Enderlein 1935: 359 (as genus). Type species: *Psilocnetha scapulata* Enderlein 1935: 359 (= *Simulium griseicollis* Becker 1903: 78–79), by original designation

*Titanopteryx* Enderlein 1935: 360–361 (as genus). Type species: *Atractocera maculata* Meigen 1804: 95–96, by original designation; *Titanopterix* (subsequent misspelling)

*Echinosimulium* Baranov 1938: 317, 322 (as genus). Type species: *Echinosimulium echinatum* Baranov 1938: 313, 323 (= *Atractocera maculata* Meigen 1804: 95–96), by original designation

*Gibbinsiellum* Rubtsov 1962f: 1494–1496 (as genus). Type species: *Simulium griseicollis* Becker 1903: 78–79, by original designation

**Diagnosis.** Adults: Radius without hair dorsobasally. Female: Scutum typically with dark, lyre-shaped pattern. Precoxal bridge incomplete. Claws each with basal thumb-like lobe. Genital fork with slender arms; each arm with lateral plate bearing prominent, anteriorly directed apodeme. Male: Gonostylus evenly tapered, moderately curved, bearing 1 apical spinule. Ventral plate broad, rather flat. Paramere with 3 or more stout spines. Pupa: Gill shorter than pupal body, of 2–6 or 22–26 branches. Cocoon slipper or shoe shaped. Larva: Head spots negative or weakly positive. Antenna weakly pigmented. Postgenal cleft large, extended nearly to or beyond hypostomal groove. Integument of body typically with dark, unbranched, tubular or fan-shaped setae (except in Oriental species). Abdominal segments I–V with or without circlet of tubercles; segment IX with or without ventral tubercles. Rectal papillae typically of 3 compound lobes (simple in Sri Lankan species).

**Overview.** The subgenus *Byssodon* consists of 12 species in 2 species groups. Members of the *S. griseicolle* species group are found in the Afrotropical, Oriental, and Palearctic Regions. Species in the *S. meridionale* group occur in the Nearctic, Oriental, and Palearctic Regions.

### *Simulium* (*Byssodon*) *meridionale* Species Group

**Diagnosis.** Pupa: Gill of 6 or 22–26 slender filaments. Larva: Postgenal cleft convex laterally, extended to or beyond hypostomal groove. Integument of body typically with minute parallel and concentric ridges (visible under compound microscope), and tubular setae darkly pigmented basally and unpigmented apically. Abdominal segments I–V (sometimes also all thoracic segments and abdominal segments VI and VII) with circlet of tubercles; segment IX with or without ventral tubercles.

**Overview.** Two of the 5 species in the *S. meridionale* species group are found in North America. They breed in big rivers and are capable of multiple generations and large populations. The 2 species treated here are often high-profile pests of humans and animals.

### *Simulium* (*Byssodon*) *maculatum* (Meigen)

(Map 126)

*Atractocera maculata* Meigen 1804: 95–96 (F). Lectotype female (designated by Zwick & Crosskey 1981: 230–231;

pinned, terminalia in glycerin vial below; MNHNP). Germany, probably Stolberg area, near ponds and ditches, on Meigen's hat, date unknown (J. W. Meigen)

*Atractocera pungens* Meigen in Panzer [1806]: 8, color plate on reverse (text as *pungens*, plate as *maculata*) (F) (unnecessary substitute name for *maculata* Meigen 1804; Zwick & Crosskey 1981)

*Simulia subfasciata* Meigen 1838: 54 (F). Type material lost. Germany, Stolberg, date unknown (Förster)

*Prosimulium vigintiquaterni* Enderlein 1929a: 222–223 (FP). Lectotype female (designated by Zwick 1974: 75–76; pinned with pupal exuviae [gills damaged]; ZMHU). Germany, Brandenburg, Güntersberg, Oder River, 31 August 1922 (M. Hering)

*Echinosimulium echinatum* Baranov 1938: 313 (PL). Type pupal material lost. Serbia, Danube River, date and collector unknown

*Simulium (Nevermannia) maculatum ussurianum* Rubtsov 1940b: 353–354 (English version p. 501), Figs. 16N, 69B, 74H, 75T (F). Type female (curatorial status unknown to us; ZISP). Russia, Ussuri area, De Castri Bay (near Vladivostok), 20 July 1909 (Dörbeck) (Lectotype female designated by Yankovsky [1995] is invalid if the type mentioned by Rubtsov [1940b] exists)

*Titanopteryx koidzumii* Takahasi 1940: 67–70 (FM). Holotype female (pinned?; HUS). China, Inner Mongolia Autonomous Region, Hailar, 3–4 August 1940 (H. Takahasi); *coidzumii* (subsequent misspelling)

*Titanopteryx maculata danubensis* Rubtsov 1956: 370–371 (FMPL). Type material unknown to us (depository unknown). Serbia/Romania, Danube River, date unknown (I. A. Rubtsov)

*Titanopteryx maculata uralensis* Rubtsov 1956: 371–373 (FMPL). Lectotype male (designated by Yankovsky 1995: 55; slide #2952; ZISP). Kazakhstan, western area, tributary of Ural River above Yanvartsevo, 17 June 1950 (Akatova)

*Titanopteryx maculata lenae* Rubtsov 1956: 373 (FM). Lectotype female (designated by Yankovsky 1995: 33; slide #3868; ZISP). Russia, Yakutia Region, Yakutsk, Vilyuisk, Vilyui River, 1 August 1927 (Bianki)

*Byssodon gutsevitshi* Yankovsky 1978: 175–178 (FMPL). Holotype female (slide #21330; ZISP). Russia, Jewish Autonomous Region, Babstovo village, 1 August 1935 (Gutsevich)

[*Simulium (Simulium) meridionale*: Stone (1952) not Riley 1887 (misident. in part, F from Alaska)]

[*Simulium meridionale*: Hopla (1965) not Riley 1887 (misident.)]

**Taxonomy**

Meigen 1804, 1838 (F); Meigen in Panzer 1806 (F); Friederichs 1922 (FMPL); Enderlein 1929a (FP); Baranov 1938 (PL); Rubtsov 1940b (FM), 1956, 1961, 1962a, 1962d, 1964b (FMPL); Takahasi 1940 (FM); Dinulescu 1966 (FMPL); Patrusheva 1971 (FMPL), 1982 (FMPL); Rivosecchi 1978 (FMPL); Yankovsky 1978 (FMPL); Rubtsov & Yankovsky 1984 (FMPL); Jensen 1997 (PL); Saether 2000 (L).

We are unable to separate the life stages of this species from those of *S. meridionale*. Both species probably exist in North America, separated by the vast tract of boreal forest between Alaska and the southern Northwest Territories. This 1100-km gap is probably not merely a collecting artifact (Currie 1997a). The taxonomy of *S. maculatum* is complicated by the existence of closely related taxa in the Palearctic Region (Yankovsky 1978). The wide distribution of *S. maculatum* suggests that it is a complex of species.

**Morphology**

Yankovsky 1977 (labral fan).

**Bionomics**

**Habitat.** *Simulium maculatum* has been reported from an enormous expanse of the Palearctic Region, breeding in large rivers such as the Danube, the Kolyma, the Ob, the Lena, and the Volga. It has become scarce, if not extinct, in much of Europe because of pollution in the big rivers (Zwick & Crosskey 1981, Jensen 1997). In the Nearctic Region, it is known only from adults collected at scattered sites in Alaska, where it probably breeds in large piedmont flows such as the Yukon River.

**Oviposition.** Females are autogenous for the first ovarian cycle (Shipitsina 1962). Oviposition takes place primarily in the evening (Crosskey 1990).

**Development.** Eggs overwinter. Females in Alaska have been taken from early June through the end of July. In the Palearctic Region, 1 generation is produced in the summer, possibly 2 at the southern limit of distribution (Rubtsov 1956). Adults in the laboratory can live nearly 3 weeks on sugar water (Friederichs 1922).

**Mating.** Mating occurs readily in the laboratory and lasts as long as 40 minutes (Friederichs 1922).

**Natural Enemies.** Larvae in the Palearctic Region are hosts of the microsporidia *Amblyospora varians*, *Janacekia debaisieuxi*, and *Polydispyrenia simulii* (Mitrokhin 1979) and an unidentified mermithid nematode (Shipitsina 1963).

**Hosts and Economic Importance.** *Simulium maculatum* has been described as a "nefarious bloodsucker" of humans and animals and of great economic concern in the forest zone of the Palearctic Region (Rubtsov 1956). It is one of the most significant pests of humans and livestock in the floodplains of Russia (Yankovsky 1978). The causal agent of tularemia has been isolated from females in Yakutia, Russia (Yakuba 1963).

### *Simulium* (*Byssodon*) *meridionale* Riley

(Figs. 4.21, 10.42, 10.159, 10.320, 10.442, 10.761, 10.838; Plates 1a, 12f; Map 127)

*Simulium meridionale* Riley 1887: 513–514, Fig. 6 (F; MPL probably *S. venustum*). Lectotype* female (here designated; pinned #773, 1 hind leg and abdomen on slide #3892; USNM). Mississippi, Desoto Co., probably Lake View (Dyar & Shannon 1927, p. 32), 16 March 1886, "bred" (probably O. Lugger); *meribionale*, *meridianole*, *meriodionale* (subsequent misspellings)

*Simulium occidentale* Townsend 1891: 107 (F). Lectotype* female (here designated; pinned #55; UKAL). New Mexico, Dona Ana Co., Las Cruces, 14 May 1891 (C. H. T. Townsend)

*Simulium tamaulipense* Townsend 1897: 171–172 (F). Syntype females (4, curatorial status and depository unknown). Mexico, Tamaulipas, Reynosa, window of railway train, 10 May 1897 (C. H. T. Townsend)

*Simulium forbesi* Malloch 1914: 63–65 (FMP). Holotype* female (pinned; INHS). Illinois, Mason Co., Havana, White Oak Run, 7 June 1912 (A. W. Jobbins-Pomeroy)

?*Eusimulium* 'sp.' Ainslie 1929: 208 (pest problems)

?[*Prosimulium pecuarum*: Gibson (1930) not Riley 1887 (misident.)]

[*Simulium congareenarum*: Stone & Snoddy (1969) not Dyar & Shannon 1927 (misident. in part, Fig. 125)]

[*Cnephia taeniatifrons*: Gaard (2002) not Enderlein 1925 (misident.)]

**Taxonomy**

Riley 1887 (F only); Townsend 1891, 1897 (F); Coquillett 1898 (F only); Johannsen 1902, 1903b (F only), 1934 (P, as *occidentale* only); Forbes 1912 (F); Emery 1913 (F only); Malloch 1914 (FMP); Dyar & Shannon 1927 (FM); Stains 1941 (FM); Stains & Knowlton 1943 (FM); Nicholson 1949 (FM); Nicholson & Mickel 1950 (FM); Fredeen 1951, 1981a (FMPL), 1985a (FMP); Stone 1952 (FM); Wirth & Stone 1956 (FMP); Peterson 1958a, 1960a (FMP), 1981 (FM), 1996 (FMP); Young 1958 (F); Anderson 1960 (FMPL); Abdelnur 1966, 1968 (FMP); Snoddy 1966 (FMPL); Stone & Snoddy 1969 (FMPL); Amrine 1971 (FMPL); Pinkovsky 1976 (FMPL); Snoddy & Noblet 1976 (PL); Yankovsky 1978 (FM); Westwood 1979 (FMPL); Webb & Brigham 1982 (L); Currie 1986 (PL); Mason & Kusters 1991b (FMPL); Peterson & Kondratieff 1995 (FMPL).

Riley (1887) had many specimens before him when he described *S. meridionale*, but he did not designate a holotype for this important pest species. We, therefore, designate the female specimen in the type collection of the USNM as lectotype to establish the current concept of the name. Four cotypes of *S. occidentale* exist in the UKAL collection. To fix the current concept of the name, we designate and label, as lectotype, the specimen in best condition. The remaining 3 females, now with paralectotype status, bear identical labels evidently placed on the pins by Townsend.

Most pre-1950 reports of the immature stages of *S. meridionale* actually refer to *S. venustum*. In 1914, Malloch correctly associated the pupa under the name *forbesi*, which Stone (1952) later synonymized with *meridionale*.

**Morphology**

Palmer & Craig 2000 (labral fan).

**Molecular Systematics**

Zhu 1990 (mtDNA); Tang et al. 1995b, 1996a, 1996b (mtDNA).

**Bionomics**

**Habitat.** *Simulium meridionale* is common from the southern Northwest Territories and the agricultural areas of Alberta and Saskatchewan southward through the Great Basin and the Great Plains to the Mexican border and eastward into the Apalachicola River basin of Florida. The immature stages are found in swift areas of bayous and meandering rivers with sandy bottoms or a rocky base overlain with alluvial matter. Larvae and pupae can be found on trailing vegetation, debris, and sometimes the upper surfaces of stones. Larval drift is greatest during the night (Jarvis 1987). Females will travel at least 30 km from their natal sites (Fredeen 1956) and have been collected at heights up to 1500 m above ground (Glick 1939).

**Oviposition.** Females are anautogenous (Van Deveire 1981). Eggs have been dredged from bottom sediments, suggesting that females oviposit into the water while in flight (Fredeen 1959b). Eggs remain viable for at least 2 months after storage at 0.5°C–1.5°C (Fredeen 1959b).

**Development.** This species is typically multivoltine. In Alabama and Florida, however, it has a single generation in the spring (Pinkovsky & Butler 1978). Eggs overwinter from Nebraska to Canada. In Wisconsin and Canada, larvae hatch within a month after the ice breaks up, usually by early May. Maturation of first-generation larvae in Saskatchewan requires about 369 degree-days above 0°C (Jarvis 1987). Development from hatch to pupation requires 20–25 days at 15°C or 18–23 days at 22°C (Westwood & Brust 1981). Adults in Wisconsin are on the wing from May to October, representing at least 4 generations (Anderson & DeFoliart 1961). In the Gulf Coast states, they fly by late March. Adults have been reared in the laboratory from eggs and first instars (Fredeen 1959a, 1959b) and maintained on water and raisins for up to 10 days (Pinkovsky 1976).

**Mating.** Unknown.

**Natural Enemies.** No records.

**Hosts and Economic Importance.** In earlier days, the feeding habits of the females of *S. meridionale* often were not distinguished from those of the southern buffalo gnat, *Cnephia pecuarum* (e.g., Riley 1887), or they referred, at least in part, to other species such as *S. venustum* (e.g., Forbes 1912). Females feed on both birds and mammals. Confirmed avian hosts include chickens, ring-necked pheasants, domestic turkeys, mourning doves, and European starlings, all of which are particularly attractive when located in the lower canopy rather than on the ground (Anderson & DeFoliart 1961). Females also feed on the soft featherless skin around the eyes of purple martin nestlings, and in 1984 they caused mortality and colony abandonment in several states of the Central Plains (Hill 1994). Females, probably of *S. meridionale*, presumably caused parent purple martins to abandon their young in Arkansas (Chambers & Hill 1996). In some areas of southern Wisconsin, the flies have killed as many as 96 nestling bluebirds in 68 nest boxes during June and July (Gaard 2002). They land on the nest box and gain entrance by crawling through the vent holes, main opening, and slots; as many as 500 females have been trapped in a small ring of adhesive around the vent holes of a single bluebird house (Gaard 2002). Tree swallows in Wisconsin also suffer nestling mortality from attacks by *S. meridionale* (Gaard 2002). Feeding is most conspicuous on the unfeathered regions of birds, but the flies also crawl beneath the feathers to take blood (Anderson & DeFoliart 1961).

Females have wreaked havoc on poultry from Canada to Florida, often feeding in large numbers, particularly in evenings, and causing swelling, inflammation, and decreased egg production (Edgar 1953, Anonymous 1976, Riegert 1999b). They have caused fatalities among chickens in Alabama, Florida, Iowa, Kansas, Missouri, Nebraska, and Saskatchewan, with deaths often attributed to exsanguination (Burrill 1922; Swenk 1922; Swenk & Mussehl 1928; Drake 1933, 1935; Edgar 1953; Pinkovsky 1976; Fredeen 1981a; Mock & Adler 2002). Severe anemia that caused heavy losses in chicks in western Iowa in 1928 probably was caused by *S. meridionale*, the chicks ingesting so many flies during the attacks that their crops became distended (Benbrook 1943). Old chickens, as well as young chicks, are killed (Drake 1933). In contrast to the females of most simuliids, those of *S. meridionale* readily enter poultry shelters (Anderson & DeFoliart 1961). Females also have killed pigeon squabs and attacked common peafowls in Kansas (Mock & Adler 2002). In March and April 1993, this species caused economic losses of about $1.5 million to the emu and ostrich industry along the Trinity River in eastern Texas, with birds exhibiting signs of anaphylactic shock and dying within 24 hours (Sanford et al. 1993). The female flies fed especially on the head and neck (J. V. Robinson pers. comm.). Deaths of ostriches in the Platte River Valley of Colorado in 1995 (Cranshaw 1995) probably can be ascribed to the females of *S. meridionale*. Females have killed other exotic birds, including a double yellow-headed Amazon parrot and probably cockatoos (Mock & Adler 2002).

Females also plague mammals. They are attracted to moose (Pledger et al. 1980) and can be annoying to horses and cattle, which they sometimes bite (Westwood & Brust 1981, Peterson & Kondratieff 1995). They also can be an aggravating nuisance to humans and will bite, often resulting in inflammation, pain, and itching for several days (Townsend 1891, Riley & Howard 1892b, Jobbins-Pomeroy 1916, DeFoliart & Rao 1965, Wright & DeFoliart 1970, Pinkovsky 1976, Van Deveire 1981, Mock & Adler 2002). The bites have caused high fever and swelling, sometimes leading to hospitalization (Atwood 1996). Females around Winnemucca, Nevada, have been such a nuisance to humans that management has been required since 1985 (R. D. Gray, pers. comm.). Movement, such as swinging a net around a person's head, is said to attract females (Nicholson & Mickel 1950). Carbon dioxide is a suitable attractant, but octenol alone and in combination with carbon dioxide is not (Atwood & Meisch 1993).

Females are vectors of *Leucocytozoon smithi* among turkeys (Skidmore 1931, 1932; Anderson & DeFoliart 1961; Fredeen 1981a; Pinkovsky et al. 1981). About 285 birds (74%) perished from infections on a single farm in Nebraska in 1930 (Skidmore 1932). Females also are suspected vectors of equine encephalitis virus among turkeys (Anderson et al. 1961) and have been suggested as vectors of this virus to humans (DeFoliart & Rao 1965). An unidentified arbovirus has been isolated from females in Wisconsin (Rao 1966, DeFoliart et al. 1969).

### Subgenus *Eusimulium* (= *S. aureum* Species Group or *S. aureum* Species Complex)

*Eusimulium* Roubaud 1906: 521 (as subgenus of *Simulium*).
Type species: *Simulia aurea* Fries 1824: 16, by monotypy
*Eusimulium* 'Group *aureum*' Rubtsov 1956: 502
*Eusimulium* 'Group 5' Dunbar 1962: 23, 52
*Simulium* (*Eusimulium*) 'Group 8—The *aureum* Group' Wood 1963a: 149–151

**Diagnosis.** Adults: Radius with hair dorsobasally. Scutum unpatterned. Postnotum with patch of decumbent, golden hair. Female: Precoxal bridge complete. Legs brown and yellow; claws each with basal thumblike lobe. Spermatheca with pigmentation typically reaching junction with spermathecal duct or extended onto duct. Male: Cercus large. Gonostylus disproportionately smaller than gonocoxa, broad basally, abruptly narrowed to slender apex, with 1 apical spinule. Ventral plate in ventral view with body laterally compressed, keel-like, and more than twice as long as wide. Paramere with 1 stout spine, sometimes with 1 additional, variously sized spine. Aedeagal membrane with rows of minute spinose combs. Median sclerite long, slender. Dorsal plate a narrow horizontal strip. Pupa: Gill of 4 slender filaments; dorsalmost filament strongly divergent from other filaments. Cocoon slipper shaped, without anterodorsal projection. Larva: Antenna without hyaline bands. Hypostoma with lateral and median teeth rather prominent, extended anteriorly to about same level. Postgenal cleft extended about one third or less distance to hypostomal groove, with anterior margin rounded, squared, or biarctate. Abdominal segment IX with prominent ventral tubercles; abdominal setae unbranched, translucent, sparse. Rectal papillae of 3 simple lobes. Chromosomes: Haploid number = 2. Nucleolar organizer typically in base of IIS. Standard banding sequence as given by Leonhardt (1985).

**Overview.** This morphologically homogeneous subgenus, often referred to as the *S. aureum* species complex, consists of 34 formally named species centered in the Holarctic Region, with 1 species in the Oriental Region and another (*S. donovani*) extending into the Neotropical Region. All known members of this subgenus have 2 pairs of chromosomes, strongly supporting the monophyly of the group. Chromosome I is the product of a fusion between chromosomes II and III of related species (Leonhardt 1985).

At least 13 of the world's 34 known species in this subgenus have been studied cytologically and designated with the letters A through N, excluding H (Leonhardt 1985, 1987). Most of these cytological designates have been associated with formal names. In North America, 5 species (with their cytological epithets) are recognized: *S. bracteatum* ('A'), *S. donovani* ('G'), *S. exulatum* n. sp. ('C'), *S. pilosum* ('B'), and *S. violator* n. sp. ('D'). True *S. aureum* ('F') is strictly a Palearctic species. A few of the Nearctic species can be distinguished in 1 or 2 life stages, but in general, the Nearctic members of this subgenus are so similar morphologically that they rarely have been identified individually. Rather, they have been treated as the *S. aureum* species complex, or simply as *S. aureum*. Up to 4 of these species can be found in a single stream at the same time of year. Because much of the literature has not distinguished among the 5 Nearctic species, we first treat them together and then follow with individual treatments.

**Taxonomy**

Coquillett 1898 (FM); Johannsen 1902, 1903b (FM), 1934 (PL); Emery 1913 (FM); Strickland 1913a (PL); Malloch 1914 (FMP); Jobbins-Pomeroy 1916 (MPL); Dyar & Shannon 1927 (FM); Hearle 1932 (F); Knowlton & Rowe 1934a (F); Twinn

1936a (FMP), 1938 (F); Stains 1941 (FM); De León 1943, 1945 (P); Stains & Knowlton 1943 (FM); Vargas 1943b (F); Vargas et al. 1946 (L); Nicholson 1949 (FMP); Vargas & Díaz Nájera 1949 (P), 1957 (FMP); Nicholson & Mickel 1950 (FMP); Coleman 1951 (FMP); DeFoliart 1951a (PLE); Fredeen 1951 (FMPL), 1985a (FMP); Stone 1952 (FMP), 1964 (FMPL); Jamnback 1953, 1956 (L); Sommerman 1953 (L); Dalmat 1955 (FMPL); Stone & Jamnback 1955 (FMPL); Davies & Peterson 1956 (E); Wirth & Stone 1956 (FMP); Dunbar 1958b (L); Peterson 1958a (FMPL), 1960a (FMP), 1977b (FMPL = *S. aureum* s. s., except Fig. 26), 1981 (FM = *S. aureum* s. s.), 1996 (P; FM = *S. aureum* s. s.); Young 1958 (F); Anderson 1960 (FMPL); Davies et al. 1962 (FMP); Wood 1963a (FMPL); Wood et al. 1963 (L); Abdelnur 1966, 1968 (FMPLE); Holbrook 1967 (L); Pinkovsky 1970 (FMPL); Amrine 1971 (FMPL); Hall 1973 (FMPL), 1974 (L); Lewis 1973 (FMPL); Corredor 1975 (FM); Merritt et al. 1978b (PL; figures = *S. aureum* s. s., except Fig. 33); Hilsenhoff 1982 (L); Adler 1983 (PL); Adler & Kim 1986 (PL); Currie 1986 (PL); Peterson & Kondratieff 1995 (FMPL).

**Cytology**

Dunbar 1958a, 1958b, 1959, 1962, 1965; Rothfels & Mason 1975; Leonhardt 1985, 1987; Adler & Kim 1986; Currie & Adler 1986; Leonhardt & Feraday 1989.

**Morphology**

Bennett 1963b (F salivary gland), Guttman 1967a (labral fan), Yang 1968 (F gut, peritrophic matrix), Yang & Davies 1977 (F peritrophic matrix), Hayton 1979 (F age-related changes in Malpighian tubules), Smith & Hayton 1995 (F age-related changes in Malpighian tubules), Evans & Adler 2000 (spermatheca).

**Physiology**

Yang 1968 (F blood digestion).

**Molecular Systematics**

Zhu 1990 (mtDNA); Pruess et al. 1992, 2000 (mtDNA); Tang et al. 1996b (mtDNA).

**Bionomics**

**Habitat.** This subgenus blankets the continent. A member of the subgenus is 1 of only 2 simuliid species known from the Aleutian Islands of Alaska (Shewell 1957). The immature stages are found in small, fairly warm streams, usually with much vegetation and often flowing from ponds and lakes or through open areas such as bogs and swales.

**Oviposition.** Females are anautogenous and can produce more than 800 eggs per ovarian cycle; fecundity varies positively with female size (Davies & Peterson 1956, Abdelnur 1968, Pascuzzo 1976). They are capable of entering a second ovarian cycle within 2 weeks after the first (Wood & Davies 1966). Females oviposit on trailing vegetation and possibly while in flight (Davies et al. 1962). They deposit their eggs on end in masses typically 1 layer deep (DeFoliart 1951a). Eggs have been obtained in the laboratory by confining females in petri dishes with a blade of grass (Wood & Davies 1966).

**Development.** All members of this subgenus are multivoltine and spend a portion of the winter in the egg stage. Eggs remain viable for at least 9 months at 0.5°C–1.5°C (Fredeen 1959b) or up to 14 months at 4°C (Tarshis 1968b). Egg diapause can be broken by maintaining eggs at 1.5°C–4.5°C for about 2 months and then returning them to room temperature (Wood 1963a; Wood & Davies 1965, 1966). Eggs lose their viability if frozen or allowed to desiccate (Tarshis 1968b). Development from egg to adult requires about 17–25 days at 21°C–23°C (DeFoliart 1951a, Fredeen 1959a). Larvae can be found from at least March into November, depending on latitude and elevation. In Alberta and Wisconsin, where 2 or 3 generations are produced annually, larvae hatch in late May and adults appear in mid-June (Anderson & Dicke 1960, Abdelnur 1968). Larvae hatch from late April to early June in New Hampshire where there are 2 generations (Lake & Burger 1983). Hatching takes place in late March in central Michigan (Merritt et al. 1978b) and from mid-May into June in Alaska (Sommerman et al. 1955). Adults fly by early June in southern Quebec and in Algonquin Park, Ontario (Davies 1950, Back & Harper 1979). In Pennsylvania, the flight period is from May into November (McCreadie et al. 1994b). Adults fly from late April to mid-September in Utah where there are 3 or 4 generations yearly (Peterson 1958a, 1959b).

**Mating.** We have observed males swarming on hilltops about 3 m above the ground in southern Quebec.

**Natural Enemies.** Species of the subgenus *Eusimulium* are rather infrequent hosts of mermithid nematodes (Anderson 1960, Peterson 1960b, Abdelnur 1968) such as *Mesomermis flumenalis* (Bruder & Crans 1979). The microsporidia *Amblyospora bracteata*, *A. fibrata*, and *Polydispyrenia simulii* originally were described by Strickland (1913a) from larvae of the subgenus *Eusimulium*. A cytoplasmic polyhedrosis virus attacks larvae (Takaoka 1980). Adults, particularly females, are parasitized by larval water mites (*Sperchon ?jasperensis*) (Davies 1959). Larvae have been recovered from guts of larval hydropsychid and rhyacophilid caddisflies and from the clutches of a dolichopodid fly (Peterson 1960b).

**Hosts and Economic Importance.** Adults feed on the nectar of wild parsnip (*Pastinaca sativa*) (Adler & Kim 1986). Females feed in late evening, about 2–8 m above the ground, on birds such as great blue herons, mallards (domestic and wild), sharp-shinned hawks, chickens, ring-necked pheasants, ruffed grouse, domestic turkeys, ringed turtle-doves, mourning doves, northern saw-whet owls, gray jays, blue jays, American crows, common ravens, American robins, white-throated sparrows, common grackles, and purple finches (Bennett 1960, Anderson & DeFoliart 1961, Greiner 1975). They can be pests of chickens (Shemanchuk & Depner 1971) and great horned owls (Hunter et al. 1997a, 1997b), and we have specimens that fed on blue grouse. Females in small numbers have been reported to feed on cattle (Abdelnur 1968) and moose (Pledger et al. 1980), to fly around sheep (Jessen 1977), and to probe the face of a dead black bear (Davies & Peterson 1956). They rarely are attracted to humans (Sailer 1953, White & Morris 1985b). Despite the many feeding records for the subgenus *Eusimulium*, none have been associated unambiguously with any of the component species.

Females transmit *Leucocytozoon lovati* (as *L. bonasae*) to ruffed grouse (Fallis & Bennett 1958) and probably to blue grouse (Woo 1964, Williams et al. 1980). They have been incriminated as vectors of *L. smithi* among turkeys

(Anderson & DeFoliart 1961, Kiszewski 1984, Kiszewski & Cupp 1986). They are also hosts of *L. cambournaci* (as *L. fringillinarum* in part), *L. dubreuili* (as *L. mirandae*), *L. icteris* (as *L. fringillinarum* in part), *L. sakharoffi*, *L. ziemanni* (as *L. danilewskyi*), and unidentified *Leucocytozoon* species that infect woodland birds such as northern saw-whet owls, blue jays, American crows, common ravens, American robins, and common grackles (Fallis & Bennett 1961, 1962, 1966; Bennett et al. 1965; Khan & Fallis 1970, 1971; Bennett & Coombs 1975; Khan 1975). Females have been taken from corpses of great horned owls that died from anemia and *Leucocytozoon* infections (Hunter et al. 1997a, 1997b). They probably transmit *L. toddi* to sharp-shinned hawks (Bennett et al. 1993). They are also probable vectors of *Trypanosoma confusum* in numerous birds (Bennett 1961; as *T. avium*). Females have tested negative for eastern encephalitis virus in Wisconsin (Anderson et al. 1961) and vesicular stomatitis virus in Colorado and Utah (Kramer et al. 1990).

### *Simulium (Eusimulium) bracteatum* Coquillett

(Figs. 4.48, 10.321, 10.443; Plate 13a; Map 128)

*Simulium bracteatum* Coquillett 1898: 69 (FM). Lectotype* male (here designated; slide #10380; USNM). Massachusetts, Middlesex Co., Cambridge, 31 May 1889 (published date), 31 May 1869 (slide label) (H. A. Hagen)

*Eusimulium aureum bracteatum*, Dyar & Shannon 1927: 14–15, Figs. 24–26 (M)

*Eusimulium aureum* 'Sibling A' Dunbar 1958b: 23–44 (C)

[*Simulium venustum*: Johnson (1925) not Say 1823 (misident. in part, 2 of No. 206)]

[*Eusimulium euryadminiculum*: Rothfels & Dunbar (1953) not Davies 1949b (misident. in part)]

[*Simulium (Nevermannia) aestivum*: Cupp & Gordon (1983) not Davies et al. 1962 (misident. in part, material from Michigan)]

**Taxonomy**

Coquillett 1898 (FM); Johannsen 1902, 1903b (FM); Strickland 1913a (PL); Jobbins-Pomeroy 1916 (MPL); Dyar & Shannon 1927 (M); Dunbar 1958b (L); Davies et al. 1962 (FM); Wood 1963a (FML); Wood et al. 1963 (L).

Coquillett (1898) had 2 females and 2 males before him when he described *S. bracteatum*, but did not designate a holotype. We designate the slide-mounted male in the USNM type collection, which bears a red type label, as the lectotype to establish the current concept of the name.

Intraspecific variability makes separation of this species and others of *Eusimulium* difficult. The body of the male ventral plate in lateral view is more rounded and the larval postgenal cleft tends to be longer in *S. bracteatum* than in other North American members of the subgenus.

**Cytology**

Dunbar 1958a, 1958b, 1959, 1962, 1965; Rothfels & Mason 1975; Leonhardt 1985, 1987; Adler & Kim 1986.

**Bionomics**

**Habitat.** The range of this species spans Canada and Alaska, dipping south along the Appalachian Mountains to Georgia. The immature stages occupy fairly cool streams less than a few meters wide, with plentiful aquatic macrophytes. Throughout most of southern Canada and the northeastern United States, *S. bracteatum* is often found in the same streams with *S. pilosum*.

**Oviposition.** Females deposit egg masses on trailing vegetation.

**Development.** Diapause of overwintering eggs can be broken by storing the eggs at 1.7°C–4.5°C for at least 3 months (Wood & Davies 1966). Larvae hatch in the spring in Canada and the northeastern United States but in late winter in the southeastern states. They can be found through October (Leonhardt 1985, Adler & Kim 1986).

**Mating.** Mating behavior is unknown. However, *S. bracteatum* has been involved in one of the only attempts to interbreed 2 species of black flies. By mixing mature eggs from a female of *S. bracteatum* with sperm from the spermatheca of a female of *S. pilosum*, fertilization was obtained and embryos developed to the "eye-spot" stage before the experiment was terminated (Wood 1963a).

**Natural Enemies.** Larvae are hosts of the chytrid fungus *Coelomycidium simulii*, the microsporidium *Janacekia debaisieuxi*, and an unidentified mermithid nematode.

**Hosts and Economic Importance.** Unknown; presumably ornithophilic.

### *Simulium (Eusimulium) donovani* Vargas

(Plate 13b, Map 129)

*Simulium (Eusimulium) donovani* Vargas 1943b: 359–370, Figs. 34–36 (F). Holotype female (alcohol vial; INDRE). Mexico, Chiapas, Pueblo de Chamula, 2350 m elev., 17 November 1940 (J. Parra)

*Simulium* 'sp. E' De León 1943: 99–100 (P)

*Simulium diazi* De León 1945 ("1944"): 70, Fig. 7 (P). Type material unknown to us (depository unknown). Guatemala, other collection information unknown

*Eusimulium aureum* 'material received from . . . California' Dunbar 1958b: 39 (C) (also p. 24 as 'fifth sibling')

*Eusimulium aureum* 'Form G' Dunbar 1959: 499–518 (C)

*Simulium arrum* Reeves & Milby 1990: 139 (*lapsus calami*)

*Simulium (Eusimulium) donovani* 'G1' Pruess et al. 2000: 287, 289 (mtDNA)

*Simulium (Eusimulium) donovani* 'G2' Pruess et al. 2000: 287, 289 (mtDNA)

*Simulium donovani* 'A' Pruess et al. 2000: 290–291 (mtDNA)

*Simulium donovani* 'B' Pruess et al. 2000: 290–291 (mtDNA)

[*Simulium bracteatum*: Coquillett (1898) (misident. in part, F from California)]

[*Eusimulium aureum*: Dyar & Shannon (1927) not Fries 1824 (misident. in part, F from California)]

[*Eusimulium obtusum*: Dyar & Shannon (1927) (misident. in part, paralectotype M)]

[*Simulium (Eusimulium) aureum*: authors of all Guatemalan and Mexican literature, not Fries 1824 (misident.)]

[*Simulium aureum*: Boreham (1962) not Fries 1824 (misident.)]

[*Simulium aureum*: Anderson & Voskuil (1963) not Fries 1824 (misident.)]

[*Simulium aureum*: Pelsue (1971) not Fries 1824 (misident.)]

[*Simulium aureum*: Hall (1972, 1973, 1974) not Fries 1824 (misident.)]
[*Simulium aureum*: Lacey (1978) not Fries 1824 (misident.)]
[*Simulium aureum*: Lacey & Mulla (1978a, 1980) not Fries 1824 (misident.)]
[*Simulium aureum*: Lacey et al. (1978) not Fries 1824 (misident.)]
[*Simulium aureum*: Mohsen (1981) not Fries 1824 (misident.)]
[*Simulium aureum*: Mohsen & Mulla (1982a) not Fries 1824 (misident.)]
[*Simulium aureum*: Federici & Lacey (1987) not Fries 1824 (misident.)]
[*Simulium aureum*: Harkrider (1988) not Fries 1824 (misident.)]
[*Simulium aureum*: Hart (1988) not Fries 1824 (misident.)]
[*Simulium aureum*: Tietze & Mulla (1989) not Fries 1824 (misident.)]

**Taxonomy**

De León 1943, 1945 (P); Vargas 1943b (F); Vargas et al. 1946 (L); Vargas & Díaz Nájera 1949 (P), 1957 (FMP); Dalmat 1955 (FMPL); Hall 1973 (FMPL), 1974 (L).

The suggestion that cytospecies 'G' corresponds to *S. donovani* and *S. diazi* (Crosskey 1988) is correct; this cytospecies is the only one known from Central America, the region of the type localities of *S. donovani* and *S. diazi*.

**Cytology**

Dunbar 1958a, 1959, 1965; Leonhardt 1985, 1987; Leonhardt & Feraday 1989.

**Molecular Systematics**

Pruess et al. 2000 (mtDNA).

**Bionomics**

**Habitat.** *Simulium donovani* ranges from Oregon south to Guatemala, east into Colorado. It is the exclusive member of the subgenus *Eusimulium* in its most common habitat, the open desert and drier areas of the western states. It also can be found, albeit less frequently, with *S. pilosum* in lusher areas of the mountains. The immatures live in small warm streams, where they can be found on trailing vegetation. Larvae show even dispersion patterns and are aggressive toward one another (Hart 1988).

**Oviposition.** Females deposit eggs on trailing vegetation.

**Development.** Larvae typically are found from at least March through October. In areas, such as southern California, however, they can be found from May through January (Harkrider 1988). We have found all life stages in mid-March in northern California, suggesting that larvae hatch during the winter in some areas. Larvae are most common in late spring and summer.

**Mating.** Unknown.

**Natural Enemies.** Larvae are hosts of the mermithid nematode *Mesomermis flumenalis* (Harkrider 1988), a cytoplasmic polyhedrosis virus (Takaoka 1980), the chytrid fungus *Coelomycidium simulii*, and the microsporidium *Polydispyrenia simulii*. An intranuclear disease of uncertain etiology has been induced in larvae in the laboratory (Federici & Lacey 1987).

**Hosts and Economic Importance.** Unknown; presumably ornithophilic.

### *Simulium* (*Eusimulium*) *exulatum* Adler, Currie & Wood, New Species

(Map 130)

*Eusimulium aureum* 'Sibling C' Dunbar 1958b: 23, 38–39 (C)
[*Eusimulium excisum*: Perez (1999) not Davies et al. 1962 (misident. in part, material from site R11)]

**Taxonomy**

**Female.** Wing length 3.0–3.4 mm. Scutum grayish black. All hair pale golden to golden yellow. Mandible with 23–26 serrations; lacinia with 28–31 retrorse teeth. Sensory vesicle in lateral view occupying about one third to one half of palpomere III; neck short. Legs brown, except whitish on anterior face of front tibia, yellow on basal one half to three fourths of all femora and tibiae and on basal one fourth of hind basitarsus; pedisulcus about one fourth depth of segment. Abdomen largely unsclerotized. Terminalia (cf. Fig. 10.160): Hypogynial valves with medial margins slightly divergent. Anal lobe in ventral view slender, shaped like comma, with medial three fourths lacking microtrichia. Genital fork with arms slender; lateral plate of each arm bearing short, anteriorly directed apodeme. Spermatheca slightly longer than wide.

**Male.** Wing length 2.7–3.0 mm. Scutum velvety, black. All hair pale golden to golden yellow, except that of postocciput, basicosta, and base of basal fringe coppery brown. Legs brown, except whitish on anterior face of front tibia, yellow on basal one fourth to one third of all tibiae. Genitalia (cf. Fig. 10.322): As for subgenus.

**Pupa.** Length 2.7–3.6 mm. Gill of 4 filaments, as long as or longer than pupa; dorsal pair with petiole about as long as wide; ventral pair typically without petiole; filaments with numerous transverse furrows. Head and thorax dorsally with densely and irregularly distributed, rounded microtubercles; trichomes unbranched.

**Larva.** Length 5.7–7.3 mm. Head capsule pale yellow, slightly darker laterally and ventrally; head spots dark brown; posteromedial head spot elongate, rarely surrounded by darker pigment. Antenna brown (medial article pale), extended beyond stalk of labral fan by about one third antennal length. Postgenal cleft straight anteriorly (infrequently biarctate), about as long as or slightly longer than wide, extended about one third distance to hypostomal groove; subesophageal ganglion ensheathed with pigment. Labral fan with 47–57 primary rays. Body pigment gray, grayish brown, or pale reddish brown, arranged in bands. Posterior proleg with 10–14 hooks in each of 68–71 rows.

**Diagnosis.** Females, males, and pupae are not distinguishable from those of most other North American members in the subgenus *Eusimulium*. The pale head capsule is useful for distinguishing the larvae from those of other northern or high-altitude species in the subgenus. The chromosomes, specifically the presence of *IL-1* and the absence of IS-3 and IL-10, provide the most reliable means of identification.

**Holotype.** Male (air dried, pinned). Manitoba, Churchill, first large coastal stream ca. 180 m on road past rocket-launching site, 2 August 1957, R. W. Dunbar (CNC).

**Paratypes.** Colorado, Park Co., Forest Road 438, ca. 1 km east of Oliver Twist Lake, roadside trickle, 39°17′N,

106°09′W, 23 August 1997, P. H. Adler & D. C. Currie (5 pupae, 14 larvae). British Columbia, Vancouver Island, west of Courtenay, 3 km east of Wood Mountain Provincial Ski Park, 5 June 1991, D. C. Currie (5 larvae); tributary of Cranberry Creek, Rt. 23, ca. 20 km north of Shelter Bay, 12 July 1994, D. C. Currie (2 pupae, 5 larvae). Manitoba, same data as holotype (5♀, 17♂).

**Etymology.** The species name is from the Latin adjective meaning "banished" or "exiled," in reference to the far northern and high-altitude distribution.

**Remarks.** The polytene chromosomes of larvae collected concomitantly with the holotype were studied in detail by Dunbar (1959).

**Cytology**

Dunbar 1958a, 1958b, 1959, 1962, 1965; Rothfels & Mason 1975; Leonhardt 1985, 1987; Leonhardt & Feraday 1989.

**Bionomics**

**Habitat.** *Simulium exulatum* is a transcontinental species of the far north, ranging southward at high elevations at least to southern British Columbia and Colorado. Larvae are found in small streams, often with sedges (*Carex*), running through tundra meadows or draining beaver ponds.

**Oviposition.** Unknown.

**Development.** Eggs overwinter. Larvae have been found from June through August.

**Mating.** Unknown.

**Natural Enemies.** We have found larvae with infections of the microsporidium *Janacekia debaisieuxi*.

**Hosts and Economic Importance.** Unknown; presumably ornithophilic.

### *Simulium* (*Eusimulium*) *pilosum* (Knowlton & Rowe)

(Figs. 5.2, 10.160, 10.322, 10.643, 10.762; Map 131)

*Eusimulium pilosum* Knowlton & Rowe 1934a: 580–582 (F). Holotype* female (pinned #51407; USNM). Utah, Davis Co., Bountiful, 10 May 1929 (H. J. Pack)

*Eusimulium utahense* Knowlton & Rowe 1934a: 582, Figs. 5, 6 (F). Holotype* female (pinned #51408, genital fork on slide; USNM). Utah, Cache Co., Logan, 5 May 1934 (published date), 15 May 1923 (pin label), 3 June 1934 (slide label) (G. F. Knowlton)

*Eusimulium aureum* 'Sibling B' Dunbar 1958b: 23–44 (C)

*Simulium* (*Eusimulium*) *aureum* 'Complex' Pruess & Peterson 1987: 529 (Nebraska records)

[*Eusimulium aureum*: Stains (1941) not Fries 1824 (misident.)]

[*Eusimulium aureum*: Stains & Knowlton (1943) not Fries 1824 (misident.)]

[*Eusimulium euryadminiculum*: Rothfels & Dunbar (1953) not Davies 1949b (misident. in part)]

[*Simulium* (*Eusimulium*) *aureum*: Stone (1964) not Fries 1824 (misident.)]

[*Simulium* (*Hellichiella*) *baffinense*: Cupp & Gordon (1983) not Twinn 1936a (misident.)]

[*Simulium aureum*: Smith & Rapp (1985) not Fries 1824 (misident.)]

[*Simulium aureum*: Zhu (1990) not Fries 1824 (misident.)]

[*Simulium aureum*: Pruess et al. (1992) not Fries 1824 (misident.)]

[*Simulium* (*Eusimulium*) *aureum*: Peterson & Kondratieff (1995) not Fries 1824 (misident.)]

**Taxonomy**

Knowlton & Rowe 1934a (F), Stains 1941 (FM), Stains & Knowlton 1943 (FM), Dunbar 1958b (L), Wood 1963a (ML), Stone 1964 (FMPL), Peterson & Kondratieff 1995 (FMPL).

Our collections from the type localities of *S. pilosum* and *S. utahense*, and subsequent chromosomal study, confirm Crosskey's (1988) association of the names *pilosum* and *utahense* with cytospecies 'B.'

**Morphology**

Evans & Adler 2000 (spermatheca).

**Cytology**

Dunbar 1958a, 1958b, 1959, 1965; Leonhardt 1985, 1987; Adler & Kim 1986; Currie & Adler 1986.

The diagnostic IS-3 inversion, which previously has been reported as fixed in all populations of this species, is polymorphic on Quebec's Magdalen Islands, where the easternmost known population occurs.

**Molecular Systematics**

Zhu 1990 (mtDNA); Pruess et al. 1992, 2000 (mtDNA); Tang et al. 1996b (mtDNA).

**Bionomics**

**Habitat.** The distribution of this species stretches from coast to coast and from northern Canada southward into the Great Plains and along the major mountain chains. The immature forms occupy small warm streams, often below beaver dams, with open canopies and plenty of vegetation for attachment.

**Oviposition.** Females deposit their eggs on trailing vegetation.

**Development.** Eggs overwinter and larvae can be found from April through October (Leonhardt 1985, Adler & Kim 1986, Pruess & Peterson 1987).

**Mating.** Unknown.

**Natural Enemies.** Larvae are sporadically attacked by the chytrid fungus *Coelomycidium simulii* (Adler & Kim 1986), the microsporidium *Polydispyrenia simulii*, and unidentified mermithid nematodes.

**Hosts and Economic Importance.** Unknown; presumably ornithophilic.

### *Simulium* (*Eusimulium*) *violator* Adler, Currie & Wood, New Species

(Fig. 10.763, Map 132)

*Eusimulium aureum* 'Sibling D' Dunbar 1958b: 23, 38–39 (C)

[*Simulium aureum*: Currie (1986) not Fries 1824 (misident. in part, Figs. 3, 110)]

?[*Eusimulium aestivum*: Perez (1999) not Davies et al. 1962 (misident.)]

**Taxonomy**

**Female.** Not differing from that of *S. exulatum* n. sp.

**Male.** Not differing from that of *S. exulatum* n. sp.

**Pupa.** Not differing from that of *S. exulatum* n. sp.

**Larva.** Not differing from that of *S. exulatum* n. sp. except as follows: Head capsule (Fig. 10.763) brownish yellow dorsally, grading to brown ventrally; head spots dark brown, typically enclosed in brown pigment of varying extent; subesophageal ganglion with or without ensheathing pigment. Labral fan with 57–70 primary rays.

**Diagnosis.** Females, males, and pupae are not distinguishable from those of most other North American members in the subgenus *Eusimulium*. The pattern of brown pigmentation enclosing the head spots is useful for distinguishing most larvae from those of other species in the subgenus. The chromosomes, specifically the presence of *IL-1* and *IL-10* and often IL-11, provide the most reliable means of identification.

**Holotype.** Female (hexamethyldisilazane, pinned) with pupal exuviae (glycerin vial below). Northwest Territories, ca. 60 km east of Yellowknife, Ingraham Trail (Rt. 4), 62°29.49′N, 113°26.80′W, 30 June 2001, P. H. Adler & D. C. Currie (USNM).

**Paratypes.** Northwest Territories, same data as holotype (3♀ & 1♂ + exuviae); Swede Creek, Rt. 1, 40 km north of Alberta, 60°16.31′N, 116°33.95′W, 26 June 2001, P. H. Adler, D. C. Currie & S. G. Burgin (27 larvae, 1 pupa); Escarpment Creek, Rt. 1, 77 km north of Alberta, 60°31.60′N, 116°12.93′W, 27 June 2001, P. H. Adler, D. C. Currie & S. G. Burgin (38 larvae); Ingraham Trail (Rt. 4), 41.5 km east of Yellowknife, 62°31.16′N, 113°47.50′W, 1 July 2001, P. H. Adler & D. C. Currie (32 larvae); Vee Lake Road off Rt. 4, near Yellowknife, 62°30.85′N, 114°21.71′W, 29 June 2001, P. H. Adler & D. C. Currie (21 larvae). Saskatchewan, Ace Creek, Ace Lake Road, 9.1 km east of Uranium City, 59°33.48′N, 108°27.32′W, 30 June 1986, J. J. Ciborowski & E. Whiting (10 larvae); Ace Creek, upstream of Ace lake, 9.9 km east of Uranium City, 59°34.70′N, 108°26.35′W, 27 June 1986, J. J. Ciborowski & E. Whiting (1 pupa, 6 larvae); Fredette Lake outflow, 3.6 km north of Uranium City, 29 June 1986, J. J. Ciborowski (1 pupa); Melville Lake outflow, 3.7 km east of Uranium City, 59°33.53′N, 108°33.22′W, 26 June 1986, J. J. Ciborowski & E. Whiting; creek into east side of Jean Lake, 1.5 km northwest of Uranium City, 50°34.70′N, 108°38.05′W, 30 June 1986, J. J. Ciborowski & E. Whiting (16 larvae); Strike Lake outflow, 9.6 km east of Uranium City, 59°34.35′N, 108°25.60′W, 27 June 1986, J. J. Ciborowski & E. Whiting (18 larvae, photographic negatives of larval polytene chromosomes [mitotic figures from testis and IL arm only]).

**Etymology.** The species name is from the Latin noun meaning "injurer," suggesting that the females impact birds through their blood feeding and perhaps by transmission of the agents of leucocytozoonosis.

**Cytology**

Dunbar 1958a, 1958b, 1959, 1962, 1965; Rothfels & Mason 1975; Leonhardt 1985, 1987; Leonhardt & Feraday 1989; Adler & McCreadie 1997.

**Bionomics.**

**Habitat.** This common species occupies Canada and the extreme northern United States. Larvae and pupae can be found in a variety of streams in forested areas (Ciborowski & Adler 1990) but are most prevalent in small streams emanating from boggy meadows, ponds, and lakes.

**Oviposition.** Unknown.

**Development.** Eggs overwinter and larvae are present from May through August.

**Mating.** Unknown.

**Natural Enemies.** Larvae are hosts of the chytrid fungus *Coelomycidium simulii*, the microsporidia *Janacekia debaisieuxi* and *Polydispyrenia simulii*, and an unidentified mermithid nematode.

**Hosts and Economic Importance.** Unknown; presumably ornithophilic.

### Subgenus *Nevermannia* Enderlein

*Nevermannia* Enderlein 1921a: 199 (as genus). Type species: *Simulium annulipes* Becker 1908: 72–73 (= *Simulium ruficorne* Macquart 1838: 88), by original designation

*Cnetha* Enderlein 1921a: 199 (as genus). Type species: *Simulium vernum* Macquart 1826: 79, by designation of International Commission on Zoological Nomenclature (1986), Opinion 1416

*Stilboplax* Enderlein 1921a: 199 (as genus). Type species: *Simulium speculiventre* Enderlein 1914: 374–375, by original designation

*Pseudonevermannia* Baranov 1926: 164 (as subgenus of *Nevermannia*). Type species: *Simulium vernum* Macquart 1826: 79, by designation of International Commission on Zoological Nomenclature (1986), Opinion 1416

*Cryptectemnia* Enderlein 1936b: 114 (as genus). Type species: *Cryptectemnia laticalx* Enderlein 1936b: 114 (= *Simulium orsovae* Smart 1944b: 131), by original designation

*Chelocnetha* Enderlein 1936b: 117 (as genus). Type species: *Chelocnetha biroi* Enderlein 1936b: 117 (= *Simulium ornatipes* Skuse 1890: 632–633), by original designation

*Dexomyia* Crosskey 1969: 49–52 (as subgenus of *Simulium*). Type species: *Simulium atlanticum* Crosskey 1969: 52–56, by original designation

**Diagnosis.** Adults: Radius with hair dorsobasally. Scutum unpatterned. Female: Claws each typically with basal thumblike lobe. Spermatheca with polygonal pattern, without internal spicules. Male: Gonostylus typically truncated (sometimes tapered), with 1 apical spinule. Ventral plate typically broad, with variously developed median keel. Paramere with 1–6 stout spines. Pupa: Gill of 3–14, typically slender filaments, as long as, or longer than, pupa. Cocoon typically slipper shaped, with or without anterodorsal projection. Larva: Head spots positive, typically well defined. Postgenal cleft variously developed, from minute to large and miter shaped. Abdominal segment IX with prominent ventral tubercles. Rectal papillae of 3 simple or compound lobes.

**Overview.** This large subgenus consists of 5 species groups and more than 185 described species in the Holarctic, Afrotropical, and Oriental Regions. Only the *S. vernum* species group is represented in North America.

### *Simulium* (*Nevermannia*) *vernum* Species Group

**Diagnosis.** Adults: Pedisulcus typically deep. Female: Precoxal bridge complete. Male: Gonostylus distally with medially directed, subtriangular flange. Paramere with 1 or 2 large distal spines. Median sclerite elongate, Y shaped. Dorsal plate well developed, variously shaped. Pupa: Gill of 4–8 slender filaments. Cocoon slipper shaped, with or without anterodorsal projection. Larva: Antenna without hyaline bands. Hypostomal teeth with lateral teeth only moderately enlarged. Abdominal integument with unbranched or multiply branched, pigmented or unpigmented setae. Chromosomes: Standard banding sequence as given by Hunter and Connolly (1986).

**Overview.** Rubtsov (1956) first referred to the *S. vernum* species group (as the *Eusimulium latipes* Group), followed by Dunbar (1962), who characterized it chromosomally and referred to it as *Eusimulium* 'Group 4.' Dunbar (1962) further subdivided the group into 2 subgroups: 'Subgroup C' (with 4 gill filaments) and 'Subgroup D' (with >4 gill filaments). Wood (1963a) characterized the group morphologically and referred to it as *Simulium* 'Group 5—The *latipes* Group.' The 108 described members of this species group are distributed widely in the Holarctic and Oriental Regions. In North America, we recognize 20 species. Many past identifications of Nearctic species in this group are erroneous, in part because they were made before the recognition of several new species. In addition, females are especially difficult to identify. Because of the history of misidentification, we report only biological information that we are confident applies to each species.

The members of this group are typically univoltine. The immature stages of most species inhabit small forest streams. Females are ornithophilic, and those of a number of species transmit the agents of various *Leucocytozoon* diseases.

### *Simulium (Nevermannia) aestivum* Davies, Peterson & Wood

(Figs. 10.33, 10.161, 10.323, 10.764; Plate 13c; Map 133)

*Simulium aestivum* Davies, Peterson & Wood 1962: 104–106, Figs. 36, 71 (FMP). Holotype* female with pupal exuviae (pinned #7992; CNC). Ontario, Renfrew Co., Rolph Township, Point Alexander, small stream flowing into Ottawa River at North Star Lodge, 25 June 1959 (D. M. Davies & D. M. Wood)

*Eusimulium* 'Labrador' Brockhouse 1984: 42; 1985: 2153 (C)

**Taxonomy**

Davies et al. 1962 (FMP), Wood 1963a (FMPL), Wood et al. 1963 (L).

**Cytology**

Dunbar 1962; Brockhouse 1984, 1985; Hunter & Connolly 1986; Hunter 1987a.

Errors in the original chromosomal interpretation by Hunter and Connolly (1986) were corrected by Hunter (1987a, 1987b, 2002).

**Bionomics**

**Habitat.** *Simulium aestivum* is spottily distributed in northeastern North America. The immatures inhabit small cool forest streams often arising from springs or bogs. The largest population has been found at the type locality, a sandy stream that emerges from the pine-covered hills constituting the ancient banks of the Ottawa River.

**Oviposition.** Unknown.

**Development.** Eggs of this univoltine species overwinter, and larvae pupate in May and June (Davies et al. 1962). In some years, pupation persists well into July.

**Mating.** Unknown.

**Natural Enemies.** The trichomycete fungus *Harpella melusinae* inhabits the larval midgut.

**Hosts and Economic Importance.** Unknown; presumably ornithophilic.

### *Simulium (Nevermannia) bicorne* Dorogostaisky, Rubtsov & Vlasenko

(Figs. 10.67, 10.162, 10.324, 10.520; Map 134)

*Simulium bicornis* Dorogostaisky, Rubtsov & Vlasenko 1935: 178–180, Fig. 20 (FMPL). Lectotype* larva (designated by Yankovsky 1995: 13; slide #3061; ZISP). Russia, Irkutsk Region, Mol'ka River, June 1932 (I. A. Rubtsov). (The paralectotype designated by Yankovsky [1995] is not valid because it was collected in 1936, which was after the publication of Dorogostaisky et al. [1935].)

*Cnetha paracornifera* Yankovsky 1979: 100–103 (FMPL). Holotype* female (pharate) and pupal exuviae (slide #21326; ZISP). Russia, Khabarovsk Territory, Pivan Station, tributary of Pivan stream, 15 July 1977 (A. V. Yankovsky). New Synonym

*Cnetha cornifera* Yankovsky 1979: 103–104 (FMPL). Holotype* female (pharate) and pupal exuviae (slide #21331; ZISP). Russia, Murmansk Region, Olen'ya Station, 20 July 1951 (Z. V. Usova). New Synonym

?*Eusimulium* 'species 1' Jenkins 1948: 149–150 (Alaska record)

?*Eusimulium vernum* 'Alaska #23' Brockhouse 1984: 42 (C)

**Taxonomy**

Dorogostaisky et al. 1935 (FMPL); Rubtsov 1940b, 1956, 1962b, 1962d (FMPL); Stone 1952 (FMP); Sommerman 1953 (L); Peterson 1958a (FP), 1960a (FMP); Yankovsky 1979 (FMPL); Patrusheva 1982 (FMPL); Jensen 1997 (PL); Yankovsky 1999 (PL).

The dubious nature of the characters given as diagnostic for *S. corniferum* and *S. paracorniferum*, first noted by Currie (1997a), and the similarity of the type specimens to individuals of *S. bicorne* support our decision to synonymize the names.

In its most distinctive form, the cocoon bears a long, deeply bifurcate anterodorsal projection. Considerable variation in cocoon structure exists in Nearctic populations. The projection can be quite long and bifurcate, short (with or without a shallow bifurcation), or lacking altogether except for a bit of roughness at the anterodorsal margin (Fig. 10.520). Possibly more than 1 species is involved. The entire range of variation is sometimes present in a single population and can create confusion in distinguishing some pupae of *S. bicorne* from those of *S. fontinale* and *S. craigi*, which tend to show little variation in their cocoons. Nonetheless, *S. bicorne*, *S. craigi*, and *S. fontinale* are valid species, based on consistent morphological differences, notwithstanding the ambiguous chromosomal evidence presented below.

**Morphology**

Rubtsov 1956 (L fat body), 1960a (L gonad); Zhang & Malmqvist 1996 (labral fan).

**Cytology**

Brockhouse 1984, 1985; Adler et al. 1999.

*Simulium bicorne* is not well resolved chromosomally. We have examined the chromosomes of fewer than 15 North American larvae (Horton River drainage, Northwest Territories), all of which were homozygous for inversion IL-

2 of Hunter (1987b) and most often carried the IIIL-F sequence of Hunter (1987a, 2002). We also found the IIIL-C and IIIL-E sequences, as well as the following 3 sequences, numbered according to the scheme of Hunter (1987a, 2002): 1 4.5 6.7 2.3 8, 1 3.2 7.6 4.5 8, and 1 3.2 7.6 5.4 8. In Sweden, all females that we examined were homozygous for the IIIL-F sequence, whereas males were heterozygous for IIIL-C/F. In the Horton River drainage, larvae were polymorphic for the same subterminal inversion in IL that is described for *S. fontinale* (below). We have not seen larvae that match the chromosomal description of *S. bicorne* given by Brockhouse (1985). A more comprehensive study of the chromosomes of *S. bicorne* is needed.

**Bionomics**

**Habitat.** In North America, this Holarctic species is found in Alaska, the Northwest Territories, and the Yukon. The record from Utah (Peterson 1958a, 1959c) requires confirmation. In the Palearctic Region, *S. bicorne* has been recorded from China, Mongolia, Russia, and Scandinavia. Larvae and pupae are found on stones and trailing vegetation in streams less than 22°C and narrower than 3 m in areas of tundra and forest. In the taiga streams of Russia, they are found principally on the undersides of stones (Dorogostaisky et al. 1935).

**Oviposition.** Unknown.

**Development.** One generation is produced per year, with larvae hatching in late May or early June and persisting through August (Sommerman et al. 1955). In the Palearctic Region, where the egg stage lasts up to 10 months (Jensen 1997), the life history is similar (Rubtsov 1956).

**Mating.** Unknown.

**Natural Enemies.** Larvae are hosts of mermithid nematodes (Sommerman et al. 1955), the microsporidia *Amblyospora varians* and *Polydispyrenia simulii*, and the chytrid fungus *Coelomycidium simulii*.

**Hosts and Economic Importance.** One engorged female has been taken from a black grouse in Finland (Ojanen et al. 2002). No host records exist for the Nearctic Region.

### *Simulium* (*Nevermannia*) *burgeri* Adler, Currie & Wood, New Species

(Figs. 10.163, 10.325, 10.587, 10.765; Plate 13d; Map 135)

*Simulium* 'sp. nr. *fionae*' Adler & Mason 1997: 86 (Saskatchewan record)

'undescribed species near *S. fionae*' Currie 1997a: 579 (comparison with *S. croxtoni*)

**Taxonomy**

**Female.** Wing length 2.8–3.0 mm. Scutum grayish black, with silvery hair. Mandible with 36–38 serrations; lacinia with 26–28 retrorse teeth. Sensory vesicle in lateral view occupying about one third to one half of palpomere III. Stem vein with golden brown hair. Basal fringe of silvery hair, with faint pale golden reflections. All other hair silvery. Legs brown; pedisulcus short, about one fourth to one third depth of segment. Terminalia (Fig. 10.163): Anal lobe in ventral view with medial one third unsclerotized and lacking microtrichia. Genital fork with each arm broadened at junction with stem, thin distally, and with posteromedial projection. Spermatheca slightly longer than wide, unpigmented proximally.

**Male.** Wing length 2.6–2.9 mm. Scutum velvety, black. All hair brown to coppery. Genitalia (Fig. 10.325): Gonostylus in inner lateral view truncate distally. Ventral plate in ventral view with body subrectangular, slightly convex posteriorly. Paramere with 1 large distal spine, and 2–4 minute spines in membrane near base of distal spine. Dorsal plate subcircular, with flanged collar.

**Pupa.** Length 3.8–4.2 mm. Gill of 8 filaments (Fig. 10.587), about as long as pupa; base as long as wide, giving rise to 3 short trunks (2 + 3 + 3); dorsalmost group of 2 widely divergent filaments, its ventral filament thicker and longer than all others; filaments with numerous transverse furrows. Head and thorax with numerous, irregularly distributed, rounded microtubercles; trichomes unbranched. Cocoon with anterodorsal projection.

**Larva.** Length 5.9–7.0 mm. Head capsule (Fig. 10.765) pale yellowish brown, darker laterally and ventrally; head spots brown, defining infuscated area. Antenna yellowish brown, extended beyond stalk of labral fan by about one half length of distal article. Hypostoma with lateral and median teeth small; median tooth extended slightly anterior to lateral teeth. Postgenal cleft rounded anteriorly, often ragged, widest at midpoint, extended about one half to two thirds distance to hypostomal groove; subesophageal ganglion with or without pigmented sheath. Labral fan with 35–41 primary rays. Body pigment brown, arranged in bands (Plate 13d); abdominal setae multiply branched, unpigmented. Posterior proleg with 8–10 hooks in each of 57–61 rows. Rectal papillae of 3 compound lobes.

**Diagnosis.** The silvery hair on the scutum and the rather uniformly sclerotized anal lobe (in lateral view) distinguish the females from those of most other members of the *S. vernum* species group. For males, the coppery brown hair of the scutum is diagnostic among species in this group. The pupa is not readily distinguished from that of *S. croxtoni*. The unpigmented, multiply branched abdominal setae of the larvae are unique among North American species.

**Holotype.** Male (frozen dried, pinned) with pupal exuviae (dry). New Brunswick, Restigouche Co., junction Hwy. 17 and Restigouche River Road, 23 May 1995, D. M. Wood (CNC).

**Paratypes.** New Hampshire, Coos Co., Dixville Notch, Two-Towns Pond outlet, 44°52′N, 71°18′W, 27–29 April 1987, J. F. Burger (3 larvae, photographic negatives of larval polytene chromosomes). New Brunswick, same data as holotype (7♀ & 9♂ + exuviae); Restigouche Co., Dawsonville, 23 May 1995, D. M. Wood (32 larvae). Ontario, Hwy. 71, 2 km south of junction Hwy. 17, trickle between marshy ponds, 8 May 1986, F. F. Hunter (17 larvae). Saskatchewan, trickle into North Armit River, Hwy. 3 at bridge, 16 May 1991, S. G. Burgin (1 pupa [pharate ♂]).

**Etymology.** We name this species in honor of John F. Burger, University of New Hampshire, for his contributions to the study of North American black flies.

**Cytology**

We examined the chromosomes of 1 female larva from Algonquin Park, Ontario. A conspicuous chromocenter was present. The IIS and IIIS arms had the standard banding sequence of Hunter and Connolly (1986). The sequence for IL was similar to that of *S. gouldingi* but was rearranged between sections 29a and 34a and did not carry inversion IL-4. The IS and IIIL sequences were identical to those of *S. gouldingi*. IIL was rearranged but was not studied in detail.

**Bionomics**

**Habitat.** The distribution of this species is patchy from the northeastern United States and the Maritime Provinces of Canada westward to Saskatchewan. Small productive flows from beaver dams and marshy areas afford suitable habitat for the immature stages.

**Oviposition.** Unknown.

**Development.** Eggs overwinter. Most pupae are found in May. *Simulium burgeri* usually occurs in the same streams with its closest relatives, *S. croxtoni* and *S. fionae*, and is the earliest member of the trio to pupate, followed by *S. fionae*.

**Mating.** Unknown.

**Natural Enemies.** Larvae are hosts of an unidentified mermithid nematode.

**Hosts and Economic Importance.** Unknown; presumably ornithophilic.

### *Simulium* (*Nevermannia*) *carbunculum* Adler, Currie & Wood, New Species

(Figs. 10.326, 10.521, 10.766, 10.826, 10.837; Plate 13e; Map 136)

*Eusimulium attenuatum* Peterson 1958a: 214–221. Unavailable (unpublished thesis name)

*Simulium* (*Eusimulium*) 'sp. A' Peterson 1960b: 266–267 (prey of trichopteran larvae)

*Eusimulium pugetense* 'Cypress' Dunbar 1962: 30, Figs. 14, 40 (C)

*Eusimulium pugetense* 'Cypress Hills population' Hunter & Connolly 1986: 301–302 (C)

*Simulium pugetense* 'Cypress Hills' McCreadie et al. 1997: 764 (Wyoming record)

*Simulium pugetense* 'near Cypress Hills' McCreadie et al. 1997: 764 (Wyoming record)

*Simulium* (*Nevermannia*) *pugetense* 'complex' Moulton 1997: 152 (DNA sequences)

*Simulium* (*Nevermannia*) *pugetense* 'cpx' Moulton 2000: 99 (DNA sequences)

[*Simulium costatum*: Fredeen (1951) not Friederichs 1920 (misident.)]

[*Eusimulium costatum*: MacNay (1954) not Friederichs 1920 (misident.)]

[*Simulium* (*Eusimulium*) *pugetense*: Vargas & Díaz Nájera (1957) not Dyar & Shannon 1927 (misident.)]

[*Simulium pugetense*: Fredeen (1958) not Dyar & Shannon 1927 (misident.)]

[*Simulium* (*Eusimulium*) *pugetense*: Peterson (1960a) not Dyar & Shannon 1927 (misident. in part, Utah material)]

[*Simulium* (*Eusimulium*) *pugetense*: Fredeen (1985a) not Dyar & Shannon 1927 (misident.)]

**Taxonomy**

**Female.** Wing length 3.2–3.8 mm. Scutum grayish brown. All hair pale golden. Mandible with 36–39 serrations; lacinia with 25–28 retrorse teeth. Sensory vesicle in lateral view occupying about half of palpomere III. Legs yellowish brown, darker at apices of tibiae and femora; coxae brown. Abdomen pale brownish gray. Terminalia as in Fig. 10.173. Anal lobe in ventral view with anteromedial area unsclerotized. Genital fork with arms broadened at junction with stem. Spermatheca longer than wide, unpigmented proximally.

**Male.** Wing length 3.2–3.5 mm. Scutum velvety, brownish black. All hair pale golden to golden, except that of stem vein golden brown, and basal fringe pale golden to pale coppery. Legs golden brown to brown, with apices of femora and tibiae darker; front and middle tarsi dark brown. Genitalia (Fig. 10.326): Gonostylus in inner lateral view acutely V shaped distally. Ventral plate in ventral view with body subrectangular, concave posteriorly except for setose central tubercle. Paramere with 1 large distal spine and usually 1 minute spine in membrane near base of distal spine. Dorsal plate subcircular, with flanged collar.

**Pupa.** Length 3.2–4.0 mm. Gill of 4 tightly clustered filaments as long as or longer than pupa (Fig. 10.521), directed ventrally and running parallel to substrate; base about as long as wide, giving rise to 2 short petioles; ventral petiole thinner than and 1–2 times longer than dorsal petiole; base and both petioles densely and finely granulate; filaments with numerous, moderately deep transverse furrows (Fig. 10.837). Head and thorax dorsally with densely and irregularly distributed, large dark rounded microtubercles and densely distributed microgranules (Fig. 10.826); trichomes unbranched. Thorax at base of each gill with slight prominence. Cocoon excavated anteriorly.

**Larva.** Length 6.7–7.4 mm. Head capsule (Fig. 10.766) pale yellowish brown to brownish orange; head spots brown; posteromedial head spot usually subtriangular. Antenna brown, extended beyond stalk of labral fan by about length of distal article. Hypostoma with lateral and median teeth small, typically extended anteriorly to about same level. Postgenal cleft subquadrate, often small, irregular, and asymmetrical or with rounded or pointed anterior margin, smooth sided, and symmetrical, extended about one fourth to one third distance to hypostomal groove; subesophageal ganglion without pigmented sheath. Labral fan with 32–43 primary rays. Body pigment pale gray or grayish brown, rather uniformly distributed (Plate 13e); abdominal setae minute, unpigmented. Posterior proleg with 10–13 hooks in each of 75–84 rows. Rectal papillae of 3 simple or compound lobes.

**Diagnosis.** Females and males are not distinguishable from those of *S. pugetense* and related species. Pupae are separable from those of other western members of the *S. vernum* species group by the large granulate microtubercles on the head and thorax. Larvae with small irregular postgenal clefts and darker heads generally can be distinguished from those of *S. pugetense* and related species; otherwise chromosomal identification is necessary.

**Holotype.** Male (Peldri II) with pupal exuviae (glycerin vial below). California, Tuolumne Co., Rt. 108, tributary of Deadman Creek, 9.3 km east of Kennedy Meadow turnoff, 2.1 km east of Chipmunk Flat, 38°19.3′N, 119°40.8′W, 2400–2700 m elev., 11 June 1990, P. H. Adler (USNM).

**Paratypes.** California, Tuolumne Co., same data as holotype (1♀ + exuviae, 7 pupae, 31 larvae); Rt. 108, Deadman Creek, 3.7 km east of Kennedy Meadow turnoff, 38°18′N, 119°44′W, 11 June 1990, P. H. Adler (1 pupa, 3 larvae), 25 June 1991, P. H. Adler (5♂ + exuviae, 3 pupae, 39 larvae). Nevada, Humboldt Co., Santa Rosa Mountains, Morey Creek, 2255 m elev., 28 June 1997, R. D. Gray (23 pupae, 6 larvae). Utah, Cache Co., Logan Canyon, Rt. 89, Guinavah-Malibu Campground, 41°45.7′N, 111°42.0′W, 28 May 1989, P. H. Adler (1♀ + exuviae, 2 exuviae, 4 larvae, photographic negatives of larval polytene chromosomes), 6 June 1992, P. H. Adler (9 pupae, 104 larvae); Logan Canyon, Rt. 89, Temple Fork, 41°50.01′N, 111°35.37′W, 28 May 1989, P. H. Adler (1 exuviae, 1 pupa, 5 larvae, photographic negatives of larval polytene chromosomes).

**Etymology.** The specific name is from the Latin adjective meaning "carbuncular" or "with severe pustules," in reference to the large rounded microtubercles on the pupal head and thorax.

**Remarks.** We consider the fixation of chromosomal inversion *IL-1pu* of Hunter and Connolly (1986) and the large, densely distributed microtubercles of the pupa sufficient evidence to warrant species status. Two species possibly are included under the name *S. carbunculum*. (See Cytology section.) Larvae of the classic 'Cypress Hills' chromosomal type (*sensu* Hunter & Connolly 1986) have paler heads and larger, more regularly defined postgenal clefts than do the larvae of the second chromosomal type (referred to as 'near Cypress Hills' by McCreadie et al. 1997). In his master's thesis, Fredeen (1951) described, under the name *S. costatum*, the female, male, pupa, and larva of the classic chromosomal type. In his doctoral dissertation, Peterson (1958a) described, under the name *S. attenuatum* (unavailable), the female, male, and pupa of what probably was the second chromosomal type. Vargas and Díaz Nájera (1957) illustrated the same life stages (chromosomal type unknown). Our type series and illustrations are strictly of the second chromosomal type.

**Cytology**

Dunbar 1962, Hunter & Connolly 1986, Hunter 1987a.

Populations fall into 1 of 2 chromosomal types, each possibly representing a separate species. The classic type, known informally as the 'Cypress Hills' cytotype of *S. pugetense*, has the fixed inversion *IL-1pu* (Fig. 5.1) but is otherwise monomorphic (Hunter & Connolly 1986). The interpretation by Dunbar (1962) of a whole arm interchange in the population from Maple Creek, Cypress Hills, Alberta, is a misinterpretation; no interchange exists (K. H. Rothfels, pers. comm.). We have found larvae of the classic 'Cypress Hills' with a weak chromocenter. The second chromosomal type is similar to the classic type and is usually chromocentric, but females are homozygous and males are heterozygous for a thick band slightly distal to the "Z marker" (Fig. 5.1) of Hunter and Connolly (1986). This band plays an important role in the species that formerly were treated under the name *S. pugetense*. The band is autosomal and predominant in *S. conicum* n. sp., linked to the X chromosome in *S. merritti* n. sp., present in the inverted constituent in *S. pugetense*, and absent in *S. modicum* n. sp. and *S. moultoni* n. sp.

**Molecular Systematics**

Moulton 1997, 2000 (DNA sequences).

**Bionomics**

**Habitat.** *Simulium carbunculum* is fairly common in the mountains from southwestern Canada southward through Arizona and New Mexico into Durango, Mexico. Larvae and pupae inhabit cold streams usually less than a meter wide.

**Oviposition.** Unknown.

**Development.** This species is multivoltine. Larvae can be found throughout the year, often beneath the ice and snow that covers their streams in the winter.

**Mating.** Unknown.

**Natural Enemies.** Larvae are hosts of the chytrid fungus *Coelomycidium simulii* and the microsporidium *Polydispyrenia simulii*. Predators of larvae include the larvae of hydropsychid, rhyacophilid, and limnephilid caddisflies (Peterson 1960b).

**Hosts and Economic Importance.** Unknown; presumably ornithophilic.

### *Simulium* (*Nevermannia*) *conicum* Adler, Currie & Wood, New Species

(Figs. 10.164, 10.327, 10.767, 10.827; Plate 13f; Map 137)

[*Simulium pugetense*: Courtney (1986) not Dyar & Shannon 1927 (misident.)]

**Taxonomy**

**Female.** Not differing from that of *S. carbunculum* n. sp. except as follows: Wing length 3.7–3.8 mm. Hair of basicosta golden brown. Mandible with 31–34 serrations; lacinia with 23–25 retrorse teeth. Terminalia (Fig. 10.164): Anal lobe with about three fourths of anterior area unsclerotized. Genital fork with arms slightly broadened at junction with stem. Spermatheca slightly longer than wide.

**Male.** Not differing from that of *S. carbunculum* n. sp. except as follows: Wing length 2.9–3.2 mm. Scutum with golden hair; legs with mixed golden and brown hair; all remaining hair coppery brown. Genitalia (Fig. 10.327): Gonostylus in ventral view with medially directed, subtriangular flange strongly produced, concave on posteromedial margin. Ventral plate in ventral view straight or slightly concave posteriorly, with central tubercle extended beyond posterior margin.

**Pupa.** Length 3.3–3.9 mm. Gill of 4 tightly clustered filaments as long as or longer than pupa, directed ventrally and running parallel to substrate; base as long as wide, giving rise to 2 short, slightly swollen petioles subequal in length; base and both petioles densely granulate; filaments with reticulate pattern of numerous, moderately deep transverse furrows that impart rough texture. Head and thorax dorsally with densely distributed, minute, thin spinelike microtubercles (Fig. 10.827); trichomes unbranched. Cocoon excavated anteriorly.

**Larva.** Length 6.1–7.6 mm. Head capsule (Fig. 10.767) pale brownish orange; head spots brown, diffuse; posteromedial head spot subtriangular. Antenna pale brownish orange, extended beyond stalk of labral fan by about one third antennal length. Hypostoma with median and lateral teeth rather prominent, extended anteriorly to about same level. Postgenal cleft rounded or straight, often ragged anteriorly, as long as to slightly longer than wide, extended about one third distance to hypostomal groove; subesophageal ganglion without pigmented sheath. Labral fan with 21–33 primary rays. Body pigment pale grayish brown, uniformly distributed (Plate 13f); abdominal setae minute, sparse, unpigmented. Posterior proleg with 9–14 hooks in each of 77–81 rows. Rectal papillae of 3 compound lobes, although each lobe with as few as 2 lobules.

**Diagnosis.** The shape of the genital fork and the sclerotization pattern of the anal lobe distinguish the females from those of other species in the *S. vernum* group. The male is readily distinguished by the pronounced, posteriorly concave flange on the gonostylus. The slender spinose microtubercles of the pupal thorax are unique among North American species. Larvae are identified most reliably by the unique banding pattern near the end of chromosome arm IIIL.

**Holotype.** Male (Peldri II, pinned) with pupal exuviae (glycerin vial below). California, Marin Co., Mount Tamalpais State Park, Panorama Parkway, 37°53.0′N, 122°33.2′W, 21 March 1990, P. H. & C. R. L. Adler (USNM).

**Paratypes.** California, same data as holotype (10 larvae). British Columbia, Goldstream Provincial Park, Langsford, seepages along trail, 48°28′N, 123°33′W, 28 March 1991, D. C. Currie (1♂ + exuviae, 5 exuviae, 4 pupae, 7 larvae), 31 March 1991 (4♀ & 1♂ + exuviae, 27 larvae), 2 April 1992 (11♀ & 8♂ + exuviae, 26 exuviae, 12 pupae, 57 larvae); Englishman River Provincial Park, seepages along trail to falls, 49°15′N, 124°21′W, 30 March 1991, D. C. & R. M. Currie (2 exuviae, 5 pupae, 9 larvae), 1 June 1991, D. C. & R. M. Currie (3♂ + exuviae), 4 June 1991, D. C. Currie (2 exuviae, 1 pupa, 9 larvae).

**Etymology.** The specific name is from the Latin adjective meaning "conical," in allusion to the thin spine-like or conical microtubercles of the pupal head and thorax.

**Cytology**

Larvae from Marin Co. (5♀, 8♂) and Del Norte Co. (1♀, 4♂), California, and Clallam Co., Washington (5♀, 13♂), were fixed for the subterminal inversion *IIIL-35* (limits shown in Fig. 5.1) and were homozygous (94%) or heterozygous (6%) for IL-1pu of Hunter and Connolly (1986). More than 58% had a novel polymorphism (limits 31A–39C3) superimposed on IL-1pu. Larvae from Clallam and Del Norte Cos. were homozygous (87%) or heterozygous (9%) for a thick band (lacking in 1 larva) slightly distal to the "Z marker." All males from Marin Co. had a small heterozygous inversion in IIIS between the centromere and the "capsule" marker. The IIIL-19 inversion was polymorphic in the population from Del Norte Co. Ectopic pairing was occasional in all populations.

**Bionomics**

**Habitat.** This species is distributed in a strip from southern British Columbia to southern California, with an isolated record from Idaho. The immature stages are most prevalent in cool streamlets and seepages, often no more than 25 cm wide, in forested areas. Larvae and pupae have been found with those of *Parasimulium crosskeyi* (Courtney 1986).

**Oviposition.** Unknown.

**Development.** Larvae and pupae have been taken from January through July and adults have been found as late as September, suggesting that 2 or more generations occur annually in some areas.

**Mating.** Unknown.

**Natural Enemies.** Larvae are attacked by the chytrid fungus *Coelomycidium simulii* and the microsporidium *Polydispyrenia simulii*.

**Hosts and Economic Importance.** Unknown; presumably ornithophilic.

### *Simulium (Nevermannia) craigi* Adler & Currie

(Figs. 10.23, 10.30, 10.165, 10.328, 10.444, 10.608, 10.644, 10.768; Plate 14a; Map 138)

*Simulium craigi* Adler & Currie 1986: 1208–1212, Figs. 6, 9, 13, 17, 19 (FMPLC). Holotype* male (pinned #19483, pupal exuviae in glycerin vial below; CNC). Alberta, 3.1 km west of Obed, Hwy. 16, 53°33′N, 117°14′W, 11 June 1984 (P. H. Adler & D. C. Currie)

*Simulium* (*Eusimulium*) *vernum* 'complex' Adler 1983: 203–204, Fig. 41 (P)

*Simulium* (*Eusimulium*) *vernum* 'cytotype A' Rothfels & Brockhouse in Okaeme 1983: 57

*Eusimulium vernum* 'Cypress Hills' Brockhouse 1984: 28–29, Fig. 17; 1985: 2150 (C)

*Eusimulium vernum* 'Gothic' Brockhouse 1984: 29–32, Figs. 18–19; 1985: 2150, 2152, Fig. 15 (C)

*Eusimulium vernum* 'Nipigon' Brockhouse 1984: 33–34, Fig. 20; 1985: 2150, Fig. 14 (C)

*Simulium* 'X' Hunter 1987a: 64–68, Figs. 52–57 (C)

[*Simulium latipes*: Fredeen (1951) not Meigen 1804 (misident.)]

[*Eusimulium latipes*: Peterson (1958a, 1959c, 1960a) not Meigen 1804 (misident.)]

[*Eusimulium latipes*: Dunbar (1962) not Meigen 1804 (misident.)]

[*Simulium* (*Eusimulium*) *latipes*: Stone (1964) not Meigen 1804 (misident. in part, Figs. 53, 54 except 54f [left illustration] and g)]

[*Simulium (Nevermannia) impar*: Cupp & Gordon (1983) not Davies et al. 1962 (misident. in part, some material from Commencement Creek, Michigan)]

[*Simulium* (*Eusimulium*) *vernum*: Okaeme (1983) not Macquart 1826 (misident. in part)]

[*Simulium vernum*: Colbo & O'Brien (1984) not Macquart 1826 (misident. in part)]

[*Simulium vernum*: Currie (1986) not Macquart 1826 (misident. in part, Fig. 109)]

[*Simulium* (*Nevermannia*) *vernum*: Peterson & Kondratieff (1995) not Macquart 1826 (misident. in part, Figs. 260–272, 273b)]

**Taxonomy**

Fredeen 1951 (FMPL); Peterson 1958a (FMPL), 1960a (FMP); Stone 1964 (FMPL); Adler 1983 (P); Adler & Currie 1986 (FMPL); Currie 1986 (L); Peterson & Kondratieff 1995 (FMPL); Stuart 1995 (cocoon); Stuart & Hunter 1998b (cocoon).

See *S. silvestre* for discussion of the names *latipes* and *vernum* as they have been applied to *S. craigi*.

**Cytology**

Dunbar 1962; Brockhouse 1984, 1985, 1991; Brockhouse in Adler & Currie 1986; Hunter 1987a, 2002.

Hunter (1987a, 2002) reinterpreted the analyses of Brockhouse (1985), who analyzed *S. craigi* under the vernacular names 'Cypress Hills,' 'Gothic,' and 'Nipigon;' she demonstrated that *S. craigi* is fixed for inversions *IL-2*, *IIIL-1*, and *IIIL-9*. A number of polymorphisms exist in the IIIL arm (Hunter 1987a, 2002), suggesting that *S. craigi* might be a complex of species. Our analysis of larvae from the western half of the Northwest Territories supports Hunter's (1987a, 2002) reinterpretation and indicates that the IIIL-F sequence is predominant, followed by the IIIL-C and IIIL-E sequences, with IIIL-G being rare. We have seen 7 zygotic combinations of these sequences (CC, CF, EE, EF, FF, FG, GG) in the Northwest Territories, but the most common is FF in females and CF in males, suggesting at least some linkage of the C sequence to the Y chromosome.

**Bionomics**

**Habitat.** This species is common across Canada and Alaska southward along the major mountain chains to Arizona, central California, New Mexico, and Pennsylvania. The immatures reside in cool, well-shaded streams less than 2 m wide, most commonly in those less than a meter wide. The oxygen content of the water is generally high (McCreadie et al. 1995), and the streambed is typically composed of rocks or pebbles. Breeding sites tend not to be associated with impoundments (Ciborowski & Adler 1990). Larvae and pupae often are found on the undersurfaces of stones.

**Oviposition.** Unknown.

**Development.** Eggs overwinter, and larvae and pupae are collected in May and June or later at higher latitudes and elevations. A single generation per year is typical.

**Mating.** Unknown.

**Natural Enemies.** Larvae are hosts of an unidentified mermithid nematode (Adler & Currie 1986); the chytrid fungus *Coelomycidium simulii*; and the microsporidia *Caudospora alaskensis*, *Janacekia debaisieuxi*, and *Polydispyrenia simulii*. Four larvae with pale orange to pink abdominal cysts packed with spores of *C. alaskensis* were taken in the Northwest Territories at 2 sites around Great Slave Lake. The infections represent the only record of this microsporidium infecting a species of the genus *Simulium* in North America.

**Hosts and Economic Importance.** Taxonomic confusion in past literature prevents us from providing a list of specific hosts, although the list of Bennett (1960), under the name *S.* "*latipes*," is probably representative. We have females, plus a bilateral gynandromorph, taken from blue grouse in British Columbia. Females, probably of this species and of *S. silvestre* (both under the name *S. latipes*), are vectors of *Leucocytozoon cambournaci* (as *L. fringillinarum* in part), *L. dubreuili* (as *L. mirandae*), *L. icteris* (as *L. fringillinarum* in part), *L. lovati* (as *L. bonasae*), *L. ziemanni* (as *L. danilewskyi*) (Fallis & Bennett 1958, 1962, 1966; Bennett & Fallis 1960; Khan 1975; Greiner 1991), unidentified *Leucocytozoon* species (Bennett & Coombs 1975), and probably *Trypanosoma confusum* (as *T. avium*) (Bennett 1961).

### *Simulium (Nevermannia) croxtoni* Nicholson & Mickel

(Figs. 10.166, 10.329, 10.445, 10.769; Plate 14b; Map 139)

*Simulium croxtoni* Nicholson & Mickel 1950: 41–42, Fig. 20A, B (FP). Holotype* female (pinned; UMSP). Minnesota, Koochiching Co., west of International Falls, reared, 2 June 1941 (H. P. Nicholson)

*Simulium (Eusimulium)* 'sp. E' Shewell 1957: 2–3, Map 26 (Labrador record)

**Taxonomy**

Nicholson 1949 (FP), Nicholson & Mickel 1950 (FP), Stone & Jamnback 1955 (FP), Anderson 1960 (FPL), Davies et al. 1962 (FMP), Wood 1963a (FMPL), Wood et al. 1963 (L), Stone 1964 (FMPL), Holbrook 1967 (L), Pinkovsky 1970 (MPL), Lewis 1973 (FMPL), Hilsenhoff 1982 (L), Fredeen 1985a (FMP), Currie 1986 (PL), Stuart 1995 (cocoon), Stuart & Hunter 1998b (cocoon).

**Morphology**

Bennett 1963b (F salivary gland), Guttman 1967a (labral fan), Yang 1968 (F midgut), Yang & Davies 1977 (F peritrophic matrix).

**Cytology**

Zimring 1953, Dunbar 1962, Hunter & Connolly 1986, Hunter 1987a.

**Bionomics**

**Habitat.** This fairly common trans-Canadian species ranges stateside into the northern tier of states and the Rocky Mountains. Larvae and pupae are found on grasses, stones, and sticks in temporary and permanent streams less than 5 m wide, usually below the outflows of beaver ponds and other small impoundments.

**Oviposition.** Unknown.

**Development.** *Simulium croxtoni* overwinters as eggs and is typically univoltine, although some populations in eastern Canada might be bivoltine (Davies et al. 1962). Mature larvae and pupae are present from May into July.

**Mating.** Unknown.

**Natural Enemies.** Larvae are attacked by the chytrid fungus *Coelomycidium simulii*, the microsporidium *Janacekia debaisieuxi*, a cytoplasmic polyhedrosis virus, and an unidentified mermithid nematode.

**Hosts and Economic Importance.** Females feed primarily during late afternoon and evening on woodland birds that are active above the forest floor (Bennett 1960, Fallis & Smith 1964a, Smith 1966, Golini 1970). Host species include ruffed grouse, northern saw-whet owl, gray jay, blue jay, American crow, common raven, American robin, white-throated sparrow, common grackle, and purple finch (Bennett 1960). Small numbers of females have been taken in a trap enclosing a moose (Pledger et al. 1980). One record implicates a mixed blood meal from a raccoon and a horse (Simmons et al. 1989). A blood meal averages about 3.3 $mm^3$ (Bennett 1963a). Females have been recorded as probable vectors of *Leucocytozoon dubreuili* (as *L. mirandae*), *L. icteris* (as *L. fringillinarum*), *L. lovati* (as *L. bonasae*)

(Fallis & Bennett 1961), and *Trypanosoma confusum* (as *T. avium*) (Bennett 1961).

### *Simulium* (*Nevermannia*) *dendrofilum* (Patrusheva)

(Figs. 10.167, 10.446; Map 140)

*Eusimulium dendrofila* Patrusheva 1962: 100–102 (MPL). Holotype male with pupal exuviae(?) (curatorial status and depository unknown to us). Russia (Siberia), Tomsk District, near Asino, Yuksa River (tributary of Chulym River), day and month unknown, 1960–1961 (possibly V. D. Patrusheva)

*Simulium* (*Eusimulium*) 'sp. near *croxtoni*' Smith 1970: 100, 113, 115, 117, 121, 123, 125, 126, 129, 136, 147, 247 (FPL)

**Taxonomy**

Patrusheva 1962 (MPL), 1982 (FMPL); Rubtsov 1964a (MPL); Smith 1970 (FPL); Bodrova 1977 (PL).

The strong medially directed flange and single parameral spine of the male, illustrated in the original description (Patrusheva 1962), suggest that this species belongs in the subgenus *Nevermannia*.

**Bionomics**

**Habitat.** Described from Siberia, *S. dendrofilum* is known in North America only from warm streams around Baker Lake in Nunavut and from the outfall of a productive lake at the Toolik Lake Field Station on the North Slope of Alaska. It also has been collected in northern Sweden (Adler et al. 1999).

**Oviposition.** Unknown.

**Development.** The pupae known from North America were collected in late June and July. In the Russian Far East, a single generation is produced annually, with emergence ending by early August (Bodrova 1977).

**Mating.** Unknown.

**Natural Enemies.** No records.

**Hosts and Economic Importance.** The mouthparts of the single studied female (pharate) from North America have been described previously as too weakly developed for biting (Currie 1997a). A reassessment, however, indicates that the mouthparts are fully armed for cutting flesh. Each lacinia bears 25 retrorse teeth, and the mandibles are well serrated. Females have bifid claws and, therefore, are presumably ornithophilic.

### *Simulium* (*Nevermannia*) *fionae* Adler

(Figs. 10.168, 10.330, 10.447, 10.770, 10.839; Plate 14c; Map 141)

*Simulium fionae* Adler 1990: 431–437 (FMPLC). Holotype* male (pinned, larval and pupal exuviae in glycerin vial below; USNM). New Hampshire, Coos Co., Dixville Notch (The Balsams Resort), Two-Towns Pond outlet, 44°52′N, 71°18′W, 24 May 1988 (P. H. Adler)

*Eusimulium vernum* 'A-Type' Brockhouse 1984: 40–41, Figs. 3, 22 (C)

*Eusimulium vernum* 'Eastmain' Brockhouse 1985: 2148, 2153 (C)

*Simulium* (*Nevermannia*) 'species near *furculatum/croxtoni*' Adler & Kim 1986: 29, Figs. 9, 41, 58 (PL)

*Simulium* 'sp.' Hunter & Connolly 1986: 300–301, Fig. 18 (C)

*Simulium* 'sp. near *croxtoni-furculatum*' Hunter 1987a: 52–54, Figs. 17, 19, 31 (C)

[*Simulium* (*Eusimulium*) *furculatum*: Adler (1983) not Shewell 1952 (misident.)]

**Taxonomy**

Adler 1983 (P), 1990 (FMPL); Adler & Kim 1986 (PL).

**Cytology**

Brockhouse 1984, 1985; Hunter & Connolly 1986; Hunter 1987a; Adler 1990.

Having examined 4 female larvae from Algonquin Park, Ontario, we conclude that *S. vernum* 'Eastmain' of Brockhouse (1985) (= 'A-Type' of Brockhouse [1984]) is synonymous with *S. fionae*, based on possible misinterpretations (e.g., *IS-1* of Hunter & Connolly [1986] overlooked) and heretofore unknown polymorphisms, such as IIIS-2, previously recognized as fixed by Hunter and Connolly (1986).

**Bionomics**

**Habitat.** *Simulium fionae* is known from New Brunswick to eastern Saskatchewan, south to Michigan, New Jersey, and Pennsylvania. Larvae and pupae inhabit outflows of beaver ponds and other impoundments, where they can be found on trailing vegetation, sticks, and the undersides of stones. They usually are collected with larvae of *Cnephia dacotensis*, which they resemble in coloration.

**Oviposition.** Unknown.

**Development.** A single generation is produced annually, with eggs overwintering and mature larvae and pupae appearing in May or as late as June in New Brunswick.

**Mating.** Unknown.

**Natural Enemies.** Larvae are attacked by the chytrid fungus *Coelomycidium simulii*, an unidentified mermithid nematode (Adler 1990), and the microsporidium *Janacekia debaisieuxi*.

**Hosts and Economic Importance.** Unknown; presumably ornithophilic.

### *Simulium* (*Nevermannia*) *fontinale* Radzivilovskaya

(Figs. 10.169, 10.331, 10.522, 10.771; Plate 14d; Map 142)

*Simulium fontinale* Radzivilovskaya 1948: 137–139 (FPL). Lectotype* pupa (here designated; slide #4832; ZISP). Russia, Primor'ye Territory, Kedrovaya Pad Reserve, just east of Vladivostok, day and month unknown, 1939–1940 (Z. A. Radzivilovskaya)

*Eusimulium* 'X' Choate 1984: 52–60 (C)

*Eusimulium* 'Yukon #9' Brockhouse 1984: 43–44 (C)

*Eusimulium* 'Yukon' Brockhouse 1985: 2154 (C)

*Simulium decolletum* Adler & Currie 1986: 1216–1219, Figs. 3, 8, 11, 12, 15 (FMPLC). Holotype* male (pinned #19460, pupal exuviae in glycerin vial below; CNC). Alberta, Swan Hills, 1 km northwest of Edith Lake, 54°48′N, 115°23′W, 2 August 1984 (P. H. Adler & D. C. Currie). New Synonym

?[*Simulium costatum*: Frohne & Sleeper (1951) not Friederichs 1920 (misident.)]

[*Simulium pugetense*: Sommerman et al. (1955) not Dyar & Shannon 1927 (misident. in part, populations with compound rectal papillae ["gills"])]

[*Simulium longipile*: Rubtsov (1956) not Radzivilovskaya 1948 (misident. in part, Lake Baikal form)]
?[*Eusimulium pugetense*: Rubtsov & Violovich (1965) not Dyar & Shannon 1927 (misident.)]
[*Stegopterna tschukotensis*: Rubtsov (1971a) (misident. in part, 1M paratype #19114, as listed under *tschukotensis* by Yankovsky [1995])]

**Taxonomy**

Radzivilovskaya 1948 (FPL); Rubtsov 1956, 1962c (FMPL); Rubtsov & Violovich 1965 (FMP); Choate 1984 (PL); Adler & Currie 1986 (FMPL); Jensen 1997 (PL).

Currie (1997a) concluded, on the basis of Fig. 201 of Rubtsov 1956, that *S. decolletum*, previously considered a Nearctic species, also occurs in the Palearctic Region under the name *longipile* Rubtsov 1956. He noted, however, that only the Lake Baikal form (Zhilishche spring, 14 km north of the mouth of the Angara River) was conspecific with Nearctic populations, and he regarded as problematic the relationship of *S. longipile* and *S. decolletum* to *S. fontinale*, a species described from the Russian Far East by Radzivilovskaya (1948). Our subsequent examination of the black fly collection in the ZISP, St. Petersburg, revealed an original series (now with lectotype and paralectotype status) of *S. fontinale*, consisting of 9 slides with numbers written in purple wax pencil by Radzivilovskaya and bearing a female (#4833), 2 pupae (#4832, 4928), and 6 larvae (#4834, 4836, 4840, 4867 4868, 4895). This material appears to be conspecific with Nearctic *S. decolletum*, and the latter name falls as a synonym of *fontinale*. We also examined females, males, pupae, and larvae in the USNM that were collected in Zhilishche spring, Irkutsk Region, Russia (27 July 1953), and that are probably from the same site as the specimens depicted as *S. longipile* in Rubtsov's (1956) Fig. 201. The specimens in the USNM collection, however, were identified by Rubtsov as *S. fontinale* rather than *S. longipile*. They, in any event, are nearly identical with those of Nearctic *S. decolletum*.

In her 1948 paper, Radzivilovskaya used the name *longipile*, based on Rubtsov's unpublished work and his manuscript name "longipile." In comparing *fontinale* with "longipile," she inadvertently described the latter. The name *longipile* thus dates to 1948, with Radzivilovskaya as the author but without a type specimen. We do not know if any of the material that Radzivilovskaya identified as *longipile* is in the ZISP. Rubtsov (1956), evidently not realizing that Radzivilovskaya had formalized the name *longipile*, described *longipile* as a new species and suggested that Radzivilovskaya's material of *longipile* was probably *fontinale*. The variability in the Russian material, such as in the thickness and lengths of the gill petioles, suggests that more than 1 species could be involved. Until the issue can be resolved, we decline to synonymize *longipile* Radzivilovskaya with *fontinale*. Rubtsov (1965) considered the Lake Baikal form of *S. longipile* conspecific with *S. pugetense*; however, *S. pugetense* has not been found beyond the Nearctic Region.

**Cytology**

Brockhouse 1984, 1985; Choate 1984; Adler & Currie 1986; Allison & Shields 1989.

Our examination of material from a number of sites in the southwestern corner of the Northwest Territories and from British Columbia (Sikanni River area) showed that the chromosomes of *S. fontinale* are fixed for the *IL-2* inversion of Hunter (1987b) and the *IIIL-1* inversion of Hunter (1987a, 2002). As in the chromosomally similar *S. craigi*, the most common sequence in the IIIL arm is the IIIL-F sequence, followed by the IIIL-C sequence. In addition, however, we have found at least 4 novel sequences. The arrangement of these 4 sequences (based on the numbering scheme of Hunter 1987a, 2002), listed from most to least common, is as follows: 1 4.5 6.7 2.3 8, 1 3.2 7.6 4.5 8, 1 3.2 7.6 5.4 8, and 1 6.7 2.3 5.4 8. We have found these sequences, plus IIIL-C and IIIL-F, arranged in 10 different zygotic combinations. Too few specimens at any 1 site have been analyzed to determine if 1 or more of the sequences is sex linked, although males tend to be heterozygous and carry the IIIL-C sequence more often than females. In *S. fontinale*, unlike in *S. craigi*, an inversion in IIIL often has one of its breaks between sections 90A and 87A2, suggesting that the IIIL-9 inversion, which is fixed in *S. craigi*, might be polymorphic in *S. fontinale*. A few females that we analyzed from northern Alaska (Upper Oksrukuyik River, Dalton Hwy., 14 July 1992) were homozygous for the 1 3.2 7.6 4.5 8 sequence and homozygous for the *IL-2* inversion plus an included inversion that reinverts the IL-2 section into the standard configuration except for section 42b, which remains in place against section 40b. A central puff in section 85 of IIIL is present as a polymorphism, as it also is in *S. bicorne* and *S. craigi*. Many individuals (those with only the C and/or F sequences) cannot be differentiated chromosomally from those of *S. craigi*; only the individuals that carry 1 of the IIIL sequences unique to each species can be identified chromosomally. The chromosomal distinction between *S. fontinale* and *S. bicorne* is not clear.

Choate (1984) and Allison and Shields (1989) reported that the chromosomes of populations in Alaska and the Yukon are fixed for the IIIL-19 inversion and have a white puff in section 68 of the IIL arm, a central puff in section 85 of IIIL, a variety of floating inversions, and a Y-linked inversion in IIIL. An actual photograph of the IL arm in Choate's (1984) thesis (Fig. 12), although labeled as standard for the *S. vernum* sequence, clearly shows the IL-2 inversion of Hunter (1987b). Similarly, Choate's (1984) photograph of the IIIL arm (Fig. 16), although labeled as standard for the *S. vernum* sequence, shows the IIIL-F sequence of Hunter (1987a, 2002). Allison and Shields (1989) erroneously claimed that the IIIL arm was fixed for the IIIL-19 inversion of Brockhouse (1985), and presented a drawing of the arm based on the photograph (Fig. 16) in Choate's (1984) thesis. We conclude that both Choate (1984) and Allison and Shields (1989) had material of true *S. fontinale* but misinterpreted the chromosomes.

Brockhouse (1985) characterized the chromosomes of populations (under the name 'Yukon') in Alberta (including the type locality of *S. decolletum*) and the Yukon as fixed for only 1 inversion (*IIIL-29*), relative to his standard sequence, and as highly polymorphic but with undifferentiated sex chromosomes. Photographs were not provided, and we have not seen populations with these chromosomal characteristics. The specimens on which the characterization is based, however, are morphologically identical to those that we have examined chromosomally. A reexamination of the chromosomes from topotypical material is needed.

**Bionomics**

**Habitat.** *Simulium fontinale* is a Holarctic species. In North America, it is found from northern Alaska to the Swan Hills of Alberta. In the Palearctic Region, it inhabits Siberia and possibly northern Europe. The immature stages are found in small cold streams flowing from springs and bogs. In northern Alaska, it also has been found in rocky rivers up to 15 m wide.

**Oviposition.** Unknown.

**Development.** This species overwinters in the egg stage and passes through 1 generation annually, with mature larvae appearing from June to August (Allison & Shields 1989).

**Mating.** Unknown.

**Natural Enemies.** Larvae are hosts of the chytrid fungus *Coelomycidium simulii* (Adler & Currie 1986), the microsporidia *Janacekia debaisieuxi* and *Polydispyrenia simulii*, and an unidentified mermithid nematode.

**Hosts and Economic Importance.** Unknown; presumably ornithophilic.

### *Simulium* (*Nevermannia*) *gouldingi* Stone

(Figs. 4.13, 10.170, 10.332, 10.448, 10.772; Plate 14e; Map 143)

*Simulium* (*Eusimulium*) *gouldingi* Stone 1952: 90–91 (FMP). Holotype* female with pupal exuviae (pinned #61192; USNM). Pennsylvania, Luzerne Co., Rt. 115, 14.5 mi. west of Wilkes Barre, 5 June 1948 (A. Stone)

*Simulium* (*Eusimulium*) 'species 1' DeFoliart 1951a: 67 (PL)

'Sp. 27' Jamnback 1951: 3 (New York record)

[*Simulium costatum*: Goulding & Deonier (1950) not Friederichs 1920 (misident. in part)]

**Taxonomy**

DeFoliart 1951a (PL); Stone 1952 (FMP), 1964 (FMPL); Jamnback 1953, 1956 (L); Stone & Jamnback 1955 (FMPL); Anderson 1960 (FMPL); Davies et al. 1962 (FMP); Wood 1963a (FMPL); Wood et al. 1963 (L); Holbrook 1967 (L); Amrine 1971 (FMPL); Lewis 1973 (FMPL); Hilsenhoff 1982 (L); Adler 1983 (PL); Adler & Kim 1986 (PL); Stuart 1995 (cocoon); Stuart & Hunter 1998b (cocoon).

**Morphology**

Guttman 1967a (labral fan).

**Cytology**

Dunbar 1962, Hunter & Connolly 1986, Hunter 1987a.

**Bionomics**

**Habitat.** This black fly is fairly common in eastern North America, reaching the far borders of Arkansas, Kentucky, and Wisconsin. Shallow woodland streams arising from springs, swamps, ponds, and bogs provide typical habitat. These streams flow year-round or intermittently and are usually less than 2 m wide. Larvae and pupae are found on sticks, trailing vegetation, and the undersides of stones.

**Oviposition.** Females emerge with immature eggs and little stored fat (Davies et al. 1962). A few females have been observed dipping from a hovering flight to deposit eggs on the surface of the water (Imhof 1977).

**Development.** *Simulium gouldingi* overwinters as eggs and is generally univoltine. Mature larvae and pupae can be found from April to June. At some sites, however, larvae have been collected into July and August, and adults are on the wing as late as August and September (Davies et al. 1962, Adler & Kim 1986), suggesting the possibility of more than 1 generation per year.

**Mating.** Unknown.

**Natural Enemies.** Larvae are infrequently infected with the chytrid fungus *Coelomycidium simulii*.

**Hosts and Economic Importance.** Females probably feed chiefly on birds, although a blood meal from a raccoon has been reported (Simmons et al. 1989).

### *Simulium* (*Nevermannia*) *impar* Davies, Peterson & Wood

(Figs. 10.171, 10.333, 10.449, 10.773; Plate 14f; Map 144)

*Simulium impar* Davies, Peterson & Wood 1962: 116–117, Figs. 40, 76 (FMP). Holotype* female with pupal exuviae (pinned #8012; CNC). Ontario, Renfrew Co., Rolph Township, small stream crossing Laurentian Point Road and flowing into Ottawa River ca. 3 mi. west of Point Alexander, 25 June 1959 (D. M. Davies & D. M. Wood)

[*Simulium costatum*: Goulding & Deonier (1950) not Friederichs 1920 (misident. in part)]

**Taxonomy**

Davies et al. 1962 (FMP), Wood 1963a (FMPL), Wood et al. 1963 (L), Adler 1983 (PL), Adler & Kim 1986 (PL).

Populations in the Ozark Mountains might represent a different species. The larvae are distinctly banded with red pigment, unlike those eastward, which have a characteristic, deep red splotch posterodorsally on the abdomen (Plate 14f).

**Cytology**

Hunter & Connolly 1986, Hunter 1987a.

**Bionomics**

**Habitat.** *Simulium impar* is quite common from southeastern Canada through the Appalachian Mountains, with populations in the Ozark and Ouachita Mountains of Arkansas. The immatures are denizens of cool woodland streams typically flowing over sandstone and less than 2 m wide. Larvae and pupae are found singly or in small groups in protected microhabitats where the current is slow (Tessler 1991). Larval drift is greatest during the night (Adler & Kim 1986).

**Oviposition.** Unknown.

**Development.** In Canada, larvae hatch in May (Davies et al. 1962), while in the southernmost states larvae first appear in March. *Simulium impar* is typically univoltine, completing larval development by mid-spring to early summer, although occasional populations persist throughout the summer (Adler & Kim 1986).

**Mating.** Unknown.

**Natural Enemies.** Larval pathogens include the chytrid fungus *Coelomycidium simulii*, the microsporidium *Amblyospora varians*, and an unidentified mermithid nematode. A larval gordian worm (Nematomorpha) was found in a larva from a small stream in the Great Smoky

Mountains National Park, North Carolina (W. K. Reeves, pers. comm.). Brook trout (*Salvelinus fontinalis*) feed on the larvae (Adler & Kim 1986).

**Hosts and Economic Importance.** Unknown; presumably ornithophilic.

### *Simulium* (*Nevermannia*) *loerchae* Adler

(Figs. 10.172, 10.334, 10.450, 10.609, 10.774; Plate 15a; Map 145)

*Simulium loerchae* Adler 1987: 673–681 (FMPLC). Holotype* male (pinned, pupal exuviae in glycerin vial below; USNM). South Carolina, Pickens Co., tributary of Indian Creek, 1.0 km from east entrance of Clemson University Experimental Forest, 34°44′40″N, 82°50′51″W, 22 February 1986 (C. R. Loerch Adler)

*Simulium* 'species undetermined No. 1' Pinkovsky 1976: 271–276; Pinkovsky & Butler 1978: 261 (PL)

*Simulium* 'new species' Hunter 1987a: 63 (C)

**Taxonomy**

Pinkovsky 1976 (PL), Pinkovsky & Butler 1978 (PL), Adler 1987 (FMPL).

**Cytology**

Adler 1987, Hunter 1987a.

**Bionomics**

**Habitat.** *Simulium loerchae*, one of the most southern members of the subgenus *Nevermannia*, ranges from Pennsylvania to the Florida panhandle westward into Arkansas. Populations have been found at elevations up to 900 m in the mountains but are most common in the foothills and piedmont. This morphologically distinct species was probably overlooked in early studies of eastern black flies because of its restricted, rather inconspicuous habitat: slow, shallow, spring-fed streams narrower than 1.5 m and less than 20°C. These woodland streams have sandy bottoms scattered with deciduous leaves on which the larvae and pupae occur; pupae are found most easily by gently washing adherent sediment from the leaves. The species has been cited as an occasional habitat associate of the rare monkey-face orchid, *Platanthera integrilabia* (Zettler & Fairey 1990).

**Oviposition.** Unknown.

**Development.** In Pennsylvania, larvae of this multivoltine species can be found year-round but are most common in the summer (Tessler 1991). Southward, larvae are most common in the winter and early spring and are rarely found during the summer months.

**Mating.** Unknown.

**Natural Enemies.** Larval pathogens include the chytrid fungus *Coelomycidium simulii* (Adler 1987), the microsporidium *Amblyospora varians*, and an unidentified species of microsporidium.

**Hosts and Economic Importance.** Unknown; presumably ornithophilic.

### *Simulium* (*Nevermannia*) *merritti* Adler, Currie & Wood, New Species

(Fig. 10.828, Map 146)

**Taxonomy**

**Female.** Not differing from that of *S. carbunculum* n. sp. except as follows: Wing length 3.2–3.3 mm. Mandible with 37–39 serrations; lacinia with 24 retrorse teeth. Abdomen brownish orange.

**Male.** Not differing from that of *S. carbunculum* n. sp. except as follows: Wing length 3.0–3.1 mm. Scutum with golden hair; legs with mixed golden and brown hair; all remaining hair coppery brown.

**Pupa.** Length 3.1–3.2 mm. Gill of 4 thin, tightly clustered filaments as long as or longer than pupa, directed ventrally and running parallel to substrate; base about as long as wide, giving rise to 2 short petioles; ventral petiole as long to twice as long as dorsal petiole; base and both petioles densely granulate; filaments with numerous shallow transverse furrows. Head and thorax appearing smooth and shiny dorsally but with densely distributed microgranules visible at more than 1000× (Fig. 10.828); 2 of 19 pupae with large, dark, rounded microtubercles; trichomes unbranched. Cocoon excavated anteriorly.

**Larva.** Length 6.2–6.6 mm. Head capsule brownish orange; head spots brown; posteromedial head spot subtriangular. Antenna brownish orange, extended beyond stalk of labral fan by about length of distal article. Hypostoma with lateral and median teeth small, extended anteriorly to about same level. Postgenal cleft rounded, or straight but ragged anteriorly, as long as to slightly longer than wide, extended about one third distance to hypostomal groove; subesophageal ganglion without pigmented sheath. Labral fan with 38–42 primary rays. Body pigment pale gray, uniformly distributed; abdominal setae minute, unpigmented. Posterior proleg with 9–14 hooks in each of 74–83 rows. Rectal papillae of 3 simple lobes.

**Diagnosis.** We recognize this species as distinct from *S. pugetense* and related species on the basis of the smooth cuticle of the pupal head and thorax, thin gill filaments, orange-brown pleurites and tergites of the female, and unique Y-chromosome inversion. The chromosomally related species *S. carbunculum* n. sp. has pupae with consistently prominent surface sculpture and females with gray tergites and pleurites.

**Holotype.** Female (frozen dried, pinned) with pupal exuviae (glycerin vial below). Utah, Wasatch Co., Wasatch Mountain State Park, spring-fed stream running under road before Wasatch Park Cafe, 40°32.11′N, 111°28.98′W, 30 May 1989, P. H. Adler (USNM).

**Paratypes.** Same data as holotype (8♀ & 3♂ + exuviae, 6 pupae, 85 larvae).

**Etymology.** This species is named in honor of Richard W. Merritt of Michigan State University for his contributions to the study of North American black flies.

**Cytology**

The chromosomes of larvae (7♀, 4♂) from the type locality (30 May 1989) are fixed for *IL-1pu* of Hunter and Connolly (1986) and have a weak chromocenter. The X chromosome is characterized by a thick band slightly distal to the "Z marker," whereas the Y chromosome lacks the thick band but carries inversion IL-4pu. Autosomal polymorphisms have not been found. Aside from its Y chromosome, this species is chromosomally similar to its nearest relative, *S. carbunculum* n. sp.

**Bionomics**

**Habitat.** We have collected this species at only 1 site: the headwaters of a cool, shaded, spring-fed runnel choked with watercress (*Nasturtium officinale*) at the entrance to Wasatch Mountain State Park in Utah. The watercress served as the predominant substrate for the larvae and pupae.

**Oviposition.** Unknown.

**Development.** The single collection of larvae and pupae was made on 30 May.

**Mating.** Unknown.

**Natural Enemies.** No records.

**Hosts and Economic Importance.** Unknown; presumably ornithophilic.

### *Simulium (Nevermannia) modicum* Adler, Currie & Wood, New Species

(Map 147)

*Eusimulium pugetense* 'IL-st' Choate 1984: 60–64 (C)

*Eusimulium pugetense* 'cytotype A' Hunter & Connolly 1986: 302 (C)

[*Simulium pugetense*: Davies et al. (1962) not Dyar & Shannon 1927 (misident.)]

[*Simulium pugetense*: Dunbar (1962) not Dyar & Shannon 1927 (misident. in part, material from Ontario)]

[*Simulium pugetense*: Wood (1963a) not Dyar & Shannon 1927 (misident.)]

[*Simulium pugetense*: Wood et al. (1963) not Dyar & Shannon 1927 (misident.)]

[*Simulium pugetense*: Wood & Davies (1966) not Dyar & Shannon 1927 (misident.)]

[*Simulium pugetense*: Ross (1977) not Dyar & Shannon 1927 (misident.)]

[*Simulium pugetense*: Merritt et al. (1978b) not Dyar & Shannon 1927 (misident.)]

[*Simulium pugetense*: Davies & Györkös (1990) not Dyar & Shannon 1927 (misident.)]

**Taxonomy**

Davies et al. 1962 (FMP), Wood 1963a (FMPL), Wood et al. 1963 (L), Merritt et al. 1978b (PL), Choate 1984 (PL).

**Female.** Wing length 3.2–3.9 mm. Scutum grayish brown. All hair pale golden, except that of basicosta and stem vein brassy in western populations. Mandible with 32–39 serrations; lacinia with 26–30 retrorse teeth. Sensory vesicle in lateral view occupying half or more of palpomere III. Legs yellowish brown, with apices of femora and tibiae darker; coxae and front and middle tarsi brown. Abdomen pale brownish gray. Terminalia: Not differing from those of *S. carbunculum* n. sp.

**Male.** Wing length 2.9–3.5 mm. Scutum velvety, brownish black, with golden hair. All remaining hair coppery. Legs brownish, with apices of femora and tibiae darker; front and middle tarsi dark brown. Genitalia: Not differing from those of *S. carbunculum* n. sp.

**Pupa.** Not differing from that of *S. carbunculum* n. sp. except as follows: Gill with base and both petioles less granulate. Head and thorax with irregularly distributed, rounded microtubercles (sparse in some western populations to densely distributed in eastern populations). Thorax at base of gill typically not forming slight prominence.

**Larva.** Not differing from that of *S. carbunculum* n. sp. except as follows: Length 6.0–6.8 mm. Postgenal cleft subquadrate, with rounded or pointed anterior margin. Labral fan with 29–39 primary rays. Posterior proleg with 10–12 hooks in each of 69–75 rows.

**Diagnosis.** All life stages of this species are nearly indistinguishable from those of *S. moultoni* n. sp. and *S. pugetense*. We consider *S. modicum* a distinct species largely on the basis of its monomorphic chromosomes (e.g., Hunter & Connolly 1986) and its tendency in western North America to have sparser microtubercles on the dorsum of the pupal head and thorax than in closely related species.

**Holotype.** Male with pupal exuviae (frozen dried, pinned). Ontario, Peel Co., Caledon Township, Cataract, outlet of pond, grid reference 785519, 7 April 1958, D. M. Davies & D. M. Wood (CNC).

**Paratypes.** California, Marin Co., Mt. Tamalpais State Park, Redwood Creek, Redwood Creek Trail at Heather Cutoff Trail, Muir Woods Road, 37°53.3′N, 122°34.0′W, 21 March 1990, C. R. L. & P. H. Adler (3♀ & 5♂ + exuviae [alcohol], 6♀ & 5♂ + exuviae [pinned], 9 pupae, 129 larvae); Shasta Co., Dersch Meadow, trickle, 40°30′N, 121°26′W, 11 July 1988, P. H. Adler & D. C. Currie (4♀ & 6♂ + exuviae, 19 pupae, 121 larvae, photographic negatives of larval polytene chromosomes [IL arm only]). Michigan, Antrim Co., Jordan River, Section 21, T31N, R5W, 2 May 1992, G. S. Bidlack (1 pupa, 10 larvae). Wyoming, Teton Co., Yellowstone National Park, Arnica Creek, West Thumb to Lake Road, 44°28.65′N, 110°32.51′W, 2370 m elev., 14 July 1994, J. F. Burger (10 pupae, 19 larvae). Ontario, same data as holotype (8♀ & 11♂ + exuviae).

**Etymology.** The specific name is from the Latin adjective meaning "average" or "moderate," in reference to the monomorphic chromosomes.

**Cytology**

Dunbar 1962, Choate 1984, Hunter & Connolly 1986, Hunter 1987a, Allison & Shields 1989.

This species is chromosomally monomorphic. However, we also include populations (e.g., Redwood Cr., Marin Co., California, 21 March 1990) with tightly associated centromeres (i.e., chromocenter).

**Bionomics**

**Habitat.** *Simulium modicum* ranges from southern Alaska eastward across Canada and the northern United States southward into California and the Rocky Mountains. It is common in forested areas of California and the central Rockies but is patchy eastward. Cool, rocky, spring-fed streams less than 2 m wide provide standard habitat. Larvae and pupae are found most often on live vegetation.

**Oviposition.** Females are presumed to be anautogenous (Davies & Györkös 1990). Oviposition behavior is unknown.

**Development.** In Ontario, larvae overwinter and begin pupating in late March when the water temperature reaches about 4°C; they disappear by late July (Davies et al. 1962). Some populations are univoltine, but most western populations appear to be multivoltine. In California, for example, larvae and pupae are common from March to August.

**Mating.** Unknown.

**Natural Enemies.** Larvae are attacked by the chytrid fungus *Coelomycidium simulii*, the microsporidia *Amblyospora* sp. and *Janacekia debaisieuxi*, and unidentified mermithid nematodes.

**Hosts and Economic Importance.** Unknown; presumably ornithophilic.

### *Simulium* (*Nevermannia*) *moultoni* Adler, Currie & Wood, New Species

(Fig. 5.1, Map 148)

*Eusimulium pugetense* 'cytotype C' Hunter & Connolly 1986: 302 (C)

?[*Simulium* (*Eusimulium*) *pugetense*: Stone & Jamnback (1955) not Dyar & Shannon 1927 (misident.)]

**Taxonomy**

**Female.** Not differing from that of *S. carbunculum* n. sp. (Wing length and colors not available; pharate individuals only.)

**Male.** Not differing from that of *S. carbunculum* n. sp. except as follows: Wing length and colors not available (pharate individuals only). Ventral plate with central tubercle more prominent and bearing more setae.

**Pupa.** Not differing from that of *S. carbunculum* n. sp. except as follows: Length 3.0–3.9 mm. Gill with base and both petioles less granulate; base 1–3 times as long as wide. Thorax at base of gill typically without slight prominence.

**Larva.** Not differing from that of *S. carbunculum* n. sp. except as follows: Length 5.8–6.9 mm. Head capsule pale yellowish to pale brownish orange. Postgenal cleft in shape of smooth-sided, inverted U (sometimes pointed), extended slightly more than one third distance to hypostomal groove. Labral fan with 41–46 primary rays. Posterior proleg with 10–12 hooks in each of 60–70 rows. Rectal papillae of 3 simple lobes.

**Diagnosis.** *Simulium moultoni* is cytologically consistent throughout its wide distribution. We, therefore, recognize it as a valid species, although no unique morphological features have been found to distinguish it from *S. modicum* n. sp. or *S. pugetense*. The chromosomes of the male larva provide the only reliable means of separating this species from *S. modicum* n. sp. and *S. pugetense*. Males are double heterozygotes for the IL-2 and IL-3 inversions (Fig. 5.1).

**Holotype.** Pupa with pharate male (in ethanol). New Hampshire, Grafton Co., Waterville Valley, Flume Brook, 43°59′N, 71°28′W, 25 May 1988, P. H. Adler (USNM).

**Paratypes.** New Hampshire, same data as holotype, 10 August 1974, E. White (18 pupae, 3 exuviae), 25 May 1988, P. H. Adler (2 pupae, 8 larvae). Prince Edward Island, Launching, 30 April 1994, D. J. Giberson (10 pupae, 40 larvae); near Stanhope, Winter Creek, 28 July 1998, M. L. Smith (5 pupae, 1 exuviae, 123 larvae); Prince Edward Island National Park, Cavendish, Green Gables Golf Course, Balsam Hollow, 28 July 1998, M. L. Smith (1 pupa, 67 larvae). Quebec, Magdalen Islands, Ile du Cap aux Meules, Le Ruisseau, Buck Parc, 47°22.84′N, 61°53.12′W, 5 May 2001, P. H. Adler & D. J. Giberson (26 larvae, photographic negatives of larval polytene chromosomes), 12 August 2001, L. Purcell (1 pupa, 81 larvae).

**Etymology.** This species is named in honor of the simuliid specialist John Kevin Moulton, in recognition of his contributions to black fly systematics.

**Cytology**

Hunter & Connolly 1986, Hunter 1987a.

We present the entire chromosomal complement of a female larva in Fig. 5.1.

**Bionomics**

**Habitat.** This species is known from northeastern North America. The immature stages are found in small, cold, spring-fed forest streams with bottoms of gravel, sand, and small stones.

**Oviposition.** Unknown.

**Development.** The overwintering stage is unknown. Two or more generations are produced in some areas. On Prince Edward Island, for example, pupae first appear in April and then again from late July into October.

**Mating.** Unknown.

**Natural Enemies.** Larvae are hosts of the chytrid fungus *Coelomycidium simulii*, the microsporidia *Janacekia debaisieuxi* and *Polydispyrenia simulii*, and an unidentified mermithid nematode.

**Hosts and Economic Importance.** Unknown; presumably ornithophilic.

### *Simulium* (*Nevermannia*) *pugetense* (Dyar & Shannon)

(Figs. 10.173, 10.451, 10.775, 10.829; Plate 15b; Map 149)

*Eusimulium pugetense* Dyar & Shannon 1927: 23, Figs. 121–123 (M). Holotype* male (pinned #28338, slide with terminalia; USNM). Washington, King Co., Seattle, date unknown (C. V. Piper); *prugetens* (subsequent misspelling)

*Eusimulium pugetense* 'IL-1' Choate 1984: 60–72 (C)

*Eusimulium pugetense* 'cytotype B' Hunter & Connolly 1986: 301–302 (C)

*Eusimulium pugetense* 'cytotype D' Hunter & Connolly 1986: 301–302 (C)

*Eusimulium pugetense* 'cytotype E' Hunter & Connolly 1986: 301–302 (C)

**Taxonomy**

Dyar & Shannon 1927 (M).

'Cytotype D' of Hunter and Connolly (1986) is the most common cytological segregate in the area around the type locality. We, therefore, consider it to represent true *S. pugetense*.

**Cytology**

Choate 1984, Hunter & Connolly 1986, Hunter 1987a, Allison & Shields 1989.

Of 128 larvae that we cytotyped from 32 sites, 79 (62%) had sex chromosomes (X = IL-2pu) corresponding to those of 'cytotype D' of Hunter and Connolly (1986), whereas 39 (30%) had the undifferentiated sex chromosomes of 'cytotype E.' Of the remaining larvae, 1 male (Inyo Co., near Lone Pine, Lone Pine Creek, 22 March 1992) was homozygous inverted for IL-2pu, and 9 females were heterozygous for IL-2pu. Unlike populations from the Queen Charlotte Islands of British Columbia scored by Hunter and Connolly (1986), all of our larvae with IL-2pu had a thick band located slightly distal to the "Z marker" (Fig. 5.1) that was associated with the inverted constituent. IIS-1pu and IIIL-1pu were fixed autosomal inversions in both 'cytotypes D' and 'E' that Hunter and Connolly (1986) analyzed from the Queen Charlotte Islands, British Columbia, but they were often polymorphic in the mainland populations that we examined. Our pooled data suggest that 'cytotypes D' and 'E' are each valid species. However, we examined the chromosomes of too few larvae at any 1 site to draw a definitive conclusion regarding species status. We, therefore, consider them a single species until larger samples from sites where both cytotypes are present can be studied.

Under *S. pugetense*, we tentatively include 'cytotype B' of Hunter and Connolly (1986) (= cytotype 'IL-1' of Choate [1984]). This decision is not entirely satisfactory because 'cytotype B' might represent a distinct species. However, we include it here on the basis of its differentiated Y chromosome, albeit a unique one (IL-1pu), and the presence of IIS-1pu and IIIL-1pu. Further work is required to determine its status.

**Bionomics**

**Habitat.** *Simulium pugetense* is a species of the Pacific Coast, ranging from Alaska into California. The immature stages are most common in cool streams less than about 3 m wide, but small populations can be found in larger flows (e.g., Currie & Adler 1986).

**Oviposition.** Unknown.

**Development.** Over much of the range, pupae can be found from at least February into September, suggesting that larvae overwinter and that multiple generations occur at some sites. In Alaska, larvae of 'cytotype B' probably overwinter because pupae can be found from the beginning of April to the middle of May (Allison & Shields 1989).

**Mating.** Unknown.

**Natural Enemies.** Larvae are attacked by the chytrid fungus *Coelomycidium simulii* (Weiser & Žižka 1974a, 1974b) and unidentified microsporidia and mermithid nematodes.

**Hosts and Economic Importance.** Unknown; presumably ornithophilic.

### *Simulium* (*Nevermannia*) *quebecense* Twinn

(Figs. 10.174, 10.335, 10.452, 10.776; Plate 15c; Map 150)

*Simulium quebecense* Twinn 1936a: 117–118, Figs. 6B, 7A (FMP). Holotype* female (slide #4124; CNC). Quebec, ca. 5 mi. south of Perkins Mills, Blanche River, downstream of falls, 22 May (pupa), 26 May (adult) 1935 (C. R. Twinn); *quebencences* (subsequent misspelling)

[*Eusimulium pugetense*: Stains (1941) not Dyar & Shannon 1927 (misident.)]

[*Eusimulium pugetense*: Stains & Knowlton (1943) not Dyar & Shannon 1927 (misident.)]

[*Simulium croxtoni*: Davies & Peterson (1956) not Nicholson & Mickel 1950 (misident. in part, ovipositing female)]

[*Simulium* (*Eusimulium*) *latipes*: Pinkovsky (1970) not Meigen 1804 (misident. in part, material from Newfield)]

[*Simulium* (*Nevermannia*) *gouldingi*: Cupp & Gordon (1983) not Stone 1952 (misident. in part, material from Driggs River, Michigan)]

[*Simulium* (*Nevermannia*) *impar*: Cupp & Gordon (1983) not Davies et al. 1962 (misident. in part, material from East Branch Fox River, Michigan)]

**Taxonomy**

Twinn 1936a (FMP), Stains 1941 (FM), Stains & Knowlton 1943 (FM), Davies et al. 1962 (FMP), Wood 1963a (FMPL), Wood et al. 1963 (L), Lewis 1973 (FMPL), Adler 1983 (PL), Adler & Kim 1986 (PL).

The larval head capsule ranges from pale yellowish with well-defined brown head spots to dark brown with obscure brown head spots.

**Morphology**

Bennett 1963b (F salivary gland), Yang 1968 (F peritrophic matrix, labral sensilla), Yang & Davies 1977 (F peritrophic matrix), Colbo et al. 1979 (F labral and cibarial sensilla).

**Physiology**

Yang 1968 (F blood digestion, trypsin).

**Cytology**

Hunter & Connolly 1986, Hunter 1987a.

Three separate X chromosomes have been identified (Hunter & Connolly 1986), possibly representing more than 1 species.

**Bionomics**

**Habitat.** *Simulium quebecense* ranges throughout northeastern North America southward along the Appalachian Mountains. It is the only North American member of the subgenus *Nevermannia* that consistently is found in watercourses more than 5 m wide. These flows are shallow, rocky, and clear.

**Oviposition.** Females oviposit in flight, releasing eggs into gentle streamside waves (Davies et al. 1962).

**Development.** One generation is produced per year, with larvae hatching from early March in the southernmost states to late April and May in southern Canada.

**Mating.** Unknown.

**Natural Enemies.** No records.

**Hosts and Economic Importance.** Females feed on birds such as ruffed grouse and American robins (Davies et al. 1962, Fallis & Bennett 1962). If the flies were identified correctly, they are probable vectors of the blood protozoans *Leucocytozoon cambournaci* (as *L. fringillinarum* in part), *L. dubreuili* (as *L. mirandae*), *L. icteris* (as *L. fringillinarum* in part), *L. lovati* (as *L. bonasae*) (Fallis & Bennett 1962, Khan & Fallis 1970), *L. toddi* (Bennett et al. 1993), and *Trypanosoma confusum* (as *T. avium*) (Bennett 1961).

### *Simulium* (*Nevermannia*) *silvestre* (Rubtsov)

(Figs. 10.175, 10.336, 10.777, 10.840; Plate 15d; Map 151)

*Eusimulium silvestre* Rubtsov 1956: 433–434 (FMPL). Holotype* male (pinned #7237, terminalia on celluloid mount below; ZISP). Russia, Irkutsk Region, near Karolog River, near village of Pashka, 9 July 1953 (I. A. Rubtsov)

*Eusimulium latipes* 'St. Rose' Landau in Dunbar 1962: 31, Figs. 17, 43, 51 (C)

*Simulium (Eusimulium) vernum* 'complex' Lake 1980: 37–39; Lake & Burger 1983: 2522–2525, 2528–2532 (bionomics)
*Simulium (Eusimulium) vernum* 'cytotype B' Rothfels & Brockhouse in Okaeme 1983: 57 (bionomics)
*Eusimulium vernum* 'Caledon' Brockhouse 1984: 35–39, Figs. 7, 11, 14, 21; 1985: 2152, Figs. 8, 12 (C)
*Simulium caledonense* Adler & Currie 1986: 1212–1216, Figs. 2, 18 (FMPLC). Holotype* male (pinned #19461, pupal exuviae in glycerin vial below; CNC). Alberta, 3.1 km west of Obed, Hwy. 16, 53°33′N, 117°14′W, 15 June 1984 (P. H. Adler & D. C. Currie). New Synonym
[*Simulium (Eusimulium) latipes*: Sommerman (1953) not Meigen 1804 (misident.)]
[*Eusimulium latipes*: Zimring (1953) not Meigen 1804 (misident.)]
[*Simulium latipes*: Davies et al. (1962) not Meigen 1804 (misident. in part, Figs. 37, 72)]
[*Simulium latipes*: Wood (1963a) not Meigen 1804 (misident. in part, Figs. 25, 50, 75)]
[*Simulium latipes*: Wood et al. (1963) not Meigen 1804 (misident. in part, Fig. 45)]
[*Simulium (Eusimulium) latipes*: Stone (1964) not Meigen 1804 (misident. in part, Fig. 54f [left illustration only], g)]
[*Simulium vernum*: Imhof (1977) not Macquart 1826 (misident.)]
[*Simulium vernum*: Imhof & Smith (1979) not Macquart 1826 (misident.)]
[*Simulium (Nevermannia) impar*: Cupp & Gordon (1983) not Davies et al. 1962 (misident. in part, some material from Commencement Creek, Michigan)]
[*Simulium (Eusimulium) vernum*: Okaeme (1983) not Macquart 1826 (misident. in part)]
[*Eusimulium vernum*: Choate (1984) not Macquart 1826 (misident.)]
[*Simulium vernum*: Colbo & O'Brien (1984) not Macquart 1826 (misident. in part)]
[*Eusimulium vernum* 'Knebworth': Brockhouse (1985) (misident. in part, Alaskan population)]
[*Eusimulium vernum* 'Knebworth': Allison & Shields (1989) not Brockhouse 1985 (misident.)]
[*Eusimulium canonicolum*: Perez (1999) not Dyar & Shannon 1927 (misident.)]
[*Eusimulium excisum*: Perez (1999) not Davies et al. 1962 (misident. in part, material from site S22)]

**Taxonomy**

Sommerman 1953 (L); Rubtsov 1956, 1962b (FMPL); Davies et al. 1962 (FM); Wood 1963a (FML); Wood et al. 1963 (L); Stone 1964 (P); Imhof 1977 (E); Imhof & Smith 1979 (E); Choate 1984 (PL); Adler & Currie 1986 (FMPL); Adler & Kim 1986 (PL); Stuart 1995 (cocoon); Jensen 1997 (PL); Stuart & Hunter 1998b (cocoon); Yankovsky 1999 (PL).

The similarity of the holotypes and other life stages of *S. caledonense* and *S. silvestre*, as well as identical chromosomal patterns between western Nearctic and Palearctic populations (see Cytology section below) provides the basis for our decision to synonymize the name *caledonense* with *silvestre*. The illustrations that we provide are based on male and female paratypes of *S. caledonense* and larvae from Coos Co., New Hampshire.

*Simulium silvestre* and *S. craigi* were referred to in the North American literature prior to the early 1970s as *S. latipes* (Meigen), itself a misidentification of *S. vernum* (Crosskey & Davies 1972). *Simulium vernum*, however, is strictly a Palearctic species, and references to *S. vernum* in North America are morphological misidentifications largely of *S. silvestre* and *S. craigi*.

**Morphology**

Zhang & Malmqvist 1996 (labral fan).

**Cytology**

Zimring 1953; Landau in Dunbar 1962; Brockhouse 1984, 1985; Choate 1984; Adler & Currie 1986; Allison & Shields 1989.

Over its wide range, *S. silvestre* is chromosomally variable. Most populations east of the Rocky Mountains differ from western populations by as many as 10 fixed inversions. From the Rocky Mountains westward, nearly all inversions described by Brockhouse (1985) as fixed in populations east of the Canadian Rocky Mountains are lacking or polymorphic. For example, populations in the Tumbler Ridge area of British Columbia and the Horton River drainage and Inuvik area of the Northwest Territories are polymorphic for inversions IIL-1, IIL-2, IIIL-1, and IIIL-2. Populations in the Horton River drainage and the Inuvik area are predominantly homozygous inverted for IIIL-1 and IIIL-2, have B chromosomes, and frequently carry IIIL-19 (absent in most eastern populations) as an autosomal polymorphism. In Wyoming (Sublette Co., Pinedale), all inversions that are present in populations of eastern North America are lacking, except IIIL-1, which characterizes the X chromosome; the Y chromosome exists in 1 of 3 forms: standard, characterized by IIIL-19 (and standard for IIIL-1), or identical to the X chromosome. Along the Thelon River of Nunavut, IIIL-1 is nearly fixed, IIIL-19 is a common autosomal polymorphism, and IIIS-1 is present but infrequent. Alaskan and Yukon populations have IIIL-19 as a fixed inversion and often have B chromosomes, but otherwise have the standard banding sequence for the *S. vernum* species group of Brockhouse (1985). Populations on Vancouver Island and in Oregon and Sweden are similar but carry IIIL-19 as an autosomal polymorphism. Populations in which IIIL-19 is floating are chromosomally identical to those of nominotypical *S. vernum* (= "Knebworth") in England, as defined by Brockhouse (1985). *Simulium vernum* and *S. silvestre* in part are, therefore, homosequential species.

The chromosomal data suggest that *S. silvestre* consists of sibling species. Most populations east of the Rocky Mountains, including the foothills, are chromosomally cohesive and if shown to represent a distinct species, would take the name *S. caledonense*. Populations in Alaska, the Yukon Territory, and the Pacific Northwest are characterized by the IIIL-19 inversion and are similar to populations in the Palearctic Region, where the name *S. silvestre* applies. Even within the Nearctic Region, we cannot disregard the possibility that 2 sibling species exist among these westernmost populations, with the northernmost populations having IIIL-19 as a fixed inversion and the more southern populations being polymorphic for this inversion. Populations from the Northwest Territories and western Nunavut through the Rocky Mountains are variously intermediate between the eastern and western populations, providing our rationale for recognizing a single widespread polymorphic species. These intermediate populations conceivably could represent 1 or more sibling species or even a broad hybrid zone between eastern and western populations.

**Bionomics**

**Habitat.** This Holarctic species is common across Alaska and Canada southward along the major mountain chains. In the Palearctic Region, it is found from Scandinavia to Siberia, Mongolia, and China. The immature stages dwell in streams less than a meter wide that issue from bogs and beaver ponds or flow through swales in forests and tundra. In Newfoundland, these streams tend to be well shaded (McCreadie et al. 1995), but elsewhere they are typically exposed. Larvae and pupae can be found on trailing vegetation or the undersides of stones.

**Oviposition.** During the late afternoon and evening, females hover 5–20 cm above the water and dip to the surface to release their eggs, one at a time, in areas with negligible current, often with decaying organic matter and vegetation such as cattail (*Typha*); these oviposition sites are typically along the channel margins or sometimes in small isolated pools (Imhof & Smith 1979).

**Development.** Eggs overwinter. Major larval populations are present during May and June, although in Alaska and western Canada they are prevalent into August. Pupation begins as early as April in areas such as Vancouver Island, British Columbia, but more typically it commences in May or June. Most populations are univoltine, but some might be bivoltine.

**Mating.** Unknown.

**Natural Enemies.** Larvae are attacked by an unidentified mermithid nematode (Adler & Currie 1986) and the microsporidia *Amblyospora bracteata*, *Janacekia debaisieuxi*, and *Polydispyrenia simulii*.

**Hosts and Economic Importance.** Past taxonomic confusion precludes a listing of the hosts of *S. silvestre*, although the list by Bennett (1960), under the name *S. "latipes,"* is probably reasonable. We do know that females will engorge on chickens in the laboratory, and we have a few females taken from swarms around humans. Rubtsov (1956) claimed that females sometimes attack humans.

*Simulium silvestre* and *S. craigi* are possibly the species (as *S. latipes*) implicated by Fallis and Bennett (1958, 1962, 1966), Bennett and Fallis (1960), Bennett (1961), Bennett and Coombs (1975), Khan (1975), and Greiner (1991) as vectors of various blood protozoans among woodland birds: *Leucocytozoon cambournaci* (as *L. fringillinarum* in part), *L. dubreuili* (as *L. mirandae*), *L. icteris* (as *L. fringillinarum* in part), *L. lovati* (as *L. bonasae*), *L. ziemanni* (as *L. danilewskyi*), and *Trypanosoma confusum* (as *T. avium*).

### *Simulium* (*Nevermannia*) *wyomingense* Stone & DeFoliart

(Figs. 10.176, 10.337, 10.453; Map 152)

*Simulium* (*Eusimulium*) *wyomingensis* Stone & DeFoliart 1959: 395–400 (FMPL). Holotype* male (pinned, pupal exuviae on slide; USNM). Wyoming, Albany Co., Little Laramie River Valley, McGill Ranch, irrigation ditch, 10 June 1957 (G. R. DeFoliart)

*Eusimulium alpinum* Peterson 1958a: 206–213. Unavailable (unpublished thesis name)

[*Eusimulium minus*: Hearle (1932) not Dyar & Shannon 1927 (misident.)]

[*Cnephia minus*: Coleman (1951) not Dyar & Shannon 1927 (misident. in part, P)]

**Taxonomy**

Hearle 1932 (P); Coleman 1951 (P); Peterson 1958a (FMPL), 1960a (FMP); Stone and DeFoliart 1959 (FMPL).

**Cytology**

Our cursory examination of the chromosomes indicates the presence of a chromocenter and the standard banding sequence of Hunter and Connolly (1986) in chromosome arms IIS and IIIS.

**Bionomics**

**Habitat.** This little-known species inhabits mountainous areas of western North America from the Rocky Mountains to the Pacific Coast. The immatures dwell in small streams along roadsides and in meadows and open woodlands.

**Oviposition.** Unknown.

**Development.** Eggs overwinter, and larvae and pupae can be collected in May and June.

**Mating.** Unknown.

**Natural Enemies.** An unidentified microsporidium infects larvae.

**Hosts and Economic Importance.** Unknown; presumably ornithophilic.

### Subgenus *Schoenbaueria* Enderlein

*Schoenbaueria* Enderlein 1921a: 199 (as genus). Type species: *Schoenbaueria matthiesseni* Enderlein 1921a: 199 (= *Atractocera nigra* Meigen 1804: 96), by original designation; *Schonbaueria* (subsequent misspelling)

*Miodasia* Enderlein 1936a: 39 (as genus). Type species: *Miodasia opalinipennis* Enderlein 1936a: 39 (= *Atractocera nigra* Meigen 1804: 96), by original designation

*Simulium* (*Eusimulium*) 'Group 4—The *furculatum* Group' Wood 1963a: 141–143

*Gallipodus* Usova & Reva 2000: 109–112 (as genus). Type species *Gallipodus raastadi* Usova & Reva 2000: 110–112, by original designation.

**Diagnosis.** Adults: Radius with hair dorsobasally. Scutum unpatterned. Female: Precoxal bridge complete. Pedisulcus deep; claws toothless or with minute subbasal tooth. Genital fork with each lateral plate bearing short, broad, anteriorly directed apodeme. Male: Gonostylus in inner lateral view with bulge or flange, and 1 apical spinule typically directed medially. Paramere narrowed distally before attachment to 1 or more stout spines. Dorsal plate somewhat T shaped. Pupa: Gill with 4–8 (rarely 2 or 3) slender filaments. Cocoon slipper shaped, sometimes with minute anterodorsal process. Larva: Postgenal cleft extended about one half or more distance to hypostomal groove. Abdominal segment IX with small ventral tubercles. Rectal papillae of 3 compound lobes.

**Overview.** This Holarctic subgenus has about 21 known species; 4 inhabit North America. The immature stages typically develop in large streams and rivers in northern areas. Adult behaviors, including biting habits, are inadequately known. Several members of the subgenus in Russia are blood-sucking pests of humans and animals (Rubtsov 1956).

***Simulium* (*Schoenbaueria*) *furculatum* (Shewell)**

(Figs. 10.177, 10.338, 10.454, 10.778; Plate 15e; Map 153)

*Eusimulium furculatum* Shewell 1952: 40–42 (FMP). Holotype* male (pinned #5990; CNC). Manitoba, Churchill, Goose River, 9 July 1947 (C. R. Twinn)

*Eusimulium* 'species B' Twinn et al. 1948: 353 (bionomics)

**Taxonomy**

Shewell 1952 (FMP); Stone 1952 (FMP), 1964 (FMPL); Sommerman 1953 (L); Davies et al. 1962 (FMP); Wood 1963a (FMPL); Wood et al. 1963 (L); Smith 1970 (FMPL); Lewis 1973 (FMPL); Fredeen 1985a (FMP); Currie 1986 (PL).

*Simulium furculatum* might occur in the Palearctic Region, possibly under the name *tsharae* Yankovsky 1982 (Currie 1997a). Whether or not *S. furculatum* and *S. tsharae* are specifically distinct requires further study, preferably using fresh material of *S. tsharae* from the type locality (Russia, Chitinskaya Region, Charskaya Hollow, Ingur, flow from Lake Gryaznoye). *Simulium furculatum* is distinct from the Palearctic *S. pusillum* described by Fries (1824).

Larvae vary considerably in head-spot pattern and body color. The head spots can be dark and heavily infuscated or somewhat weak and completely free of surrounding dark pigment. Body color varies from brick red (Plate 15e) to pale gray.

**Cytology**

Hunter 1987a, 1989.

**Bionomics**

**Habitat.** This species ranges across Alaska and Canada into the northeastern United States. It is especially common on the tundra and in areas with sparse trees. Larvae and pupae inhabit swift, medium-sized streams and rivers, as well as wide swift lake outfalls and smaller streams draining partially wooded muskeg.

**Oviposition.** Females emerge with about 350 yolkless oocytes that mature within 36 hours, suggesting that they are autogenous for the first ovarian cycle (Hocking & Pickering 1954, Chutter 1970). Nothing is known of oviposition habits.

**Development.** Larvae of this univoltine species hatch in the spring. Throughout southern Canada, they are present from late April to mid-June, and pupae are found from late May through June (Davies et al. 1962, Currie 1986). Pupae begin to appear in mid-May in Maine. At higher latitudes, such as Churchill, Manitoba, development is a few weeks later (Twinn et al. 1948, Hocking & Pickering 1954).

**Mating.** Mating swarms have been seen facing into the wind in the early evening above tamaracks (*Larix laricina*) (ca. 3.5 m high) and streamside willows (*Salix*) (Hocking & Pickering 1954).

**Natural Enemies.** Larvae are hosts of unidentified microsporidia and mermithid nematodes (Hocking & Pickering 1954). We have larvae infected with the microsporidium *Amblyospora bracteata* that were taken from the mouth of the Yellowknife River in the Northwest Territories. Adults are attacked by water mites (*Sperchon ?jasperensis*) (Davies 1959).

**Hosts and Economic Importance.** Both sexes visit the flowers of lousewort (*Pedicularis* spp.), gray-leaf willow (*Salix glauca* as *S. cordifolia*), bramble (*Rubus acaulis*), and yarrow (*Achillea millefolium*) (Hocking & Pickering 1954, MacInnes 1973). A few females have been taken in traps baited with moose (Pledger et al. 1980), and we have females swept from nuisance swarms around humans. No biting records exist.

***Simulium* (*Schoenbaueria*) *giganteum* Rubtsov**

(Fig. 10.339, Map 154)

*Simulium* (*Schoenbaueria*) *giganteum* Rubtsov 1940b: 398–399 (English version pp. 503–504), Fig. 64 I, K (F). Holotype* female (2 slides #3560: 1 with head, 1 with terminalia [spermatheca missing] and leg; ZISP). Russia, western Siberia, Big Ob' River, Yuganovo, 20 June 1928 (Samko)

**Taxonomy**

Rubtsov 1940b (F), 1956, 1962e (FM); Petrova et al. 1971 (PL); Chubareva et al. 1976 (F); Patrusheva 1982 (PL); Yankovsky 1982 (FPL); Rubtsov & Yankovsky 1984 (FM).

We have not seen larvae or pupae of this species, and they remain poorly characterized. We include *S. giganteum* in our key to females on the basis of our study of the type specimen, most of which is missing, and the description by Rubtsov (1956). The male genitalia, especially the anchor-shaped dorsal plate (Fig. 10.339), are unique among the world's black flies. The species is included in our pupal key on the basis of the illustrations and description of Petrova et al. (1971). The larva is insufficiently known for it to be placed confidently in a key. Placement of *S. giganteum* in the subgenus *Schoenbaueria* is tentative until all life stages and the chromosomes can be studied in greater detail.

**Cytology**

Petrova et al. 1971, Chubareva et al. 1976.

The claim that this species is related chromosomally to members of the subgenus *Hemicnetha* (Petrova et al. 1971, Chubareva et al. 1976) requires further study.

**Bionomics**

**Habitat.** Only 1 specimen of this primarily Siberian species has been taken in North America. A male was collected in Nunavut at the mouth of the McConnell River.

**Oviposition.** Unknown.

**Development.** The sole North American specimen was collected on 3 August (MacInnes 1973). In Siberia, adults emerge in June and July (Rubtsov 1956).

**Mating.** Unknown.

**Natural Enemies.** A mermithid nematode (*Mesomermis arctica*) was described from *S. giganteum* in the Palearctic Region (Rubtsov 1972b).

**Hosts and Economic Importance.** The single male known from North America was collected from a flower of Labrador lousewort (*Pedicularis labradorica*); its activity at the flower is unknown (MacInnes 1973). Animal hosts of the female, although unknown, are probably mammals.

**Simulium (Schoenbaueria) subpusillum Rubtsov**
(Figs. 10.178, 10.340, 10.455, 10.779; Map 155)

*Simulium* (*Schoenbaueria*) *subpusillum* Rubtsov 1940b: 404–405 (English version pp. 505–506), Fig. 74 N, S (F). Neotype* male with pupal exuviae (here designated; slide #17764, abdomen and thorax of male missing; ZISP). Russia, Irkutsk Region, Ust'-Kut, Lena River, sample 58, 8 July 1960 (S. Grebel'sky)

*Simulium* 'sp. 2' Malloch 1919: 43, Fig. 4 (P)

*Simulium* (*Eusimulium*) 'sp. near *subpusilla*' Smith 1970: 100, 112, 115, 117, 120, 123–125, 129, 131, 136, 146, 147, 175, 244, 247, 249 (FMPL, bionomics)

*Simulium* (*Eusimulium*) 'sp. near *subpusilla* (#2)' Smith 1970: 129 (F)

*Simulium* (*Schoenbaueria*) 'sp.' Hershey et al. 1995: 282 (Alaska record)

[*Schoenbaueria annulitarsis*: Carlsson (1962) not Zetterstedt 1838 (misident. in part, MPL)]

[*Schönbaueria pusilla*: Bodrova (1977) not Fries 1824 (misident.)]

[*Schönbaueria annulitarsis*: Carlsson et al. (1977) not Zetterstedt 1838 (misident.)]

[*Simulium* (*Schoenbaueria*) *annulitarse*: Jensen (1997) not Zetterstedt 1838 (misident.)]

**Taxonomy**

Malloch 1919 (P); Rubtsov 1940b (F), 1956, 1962e (FMPL); Usova 1961 (FMPL); Carlsson 1962 (MPL, not F); Smith 1970 (FMPL); Bodrova 1980 (P); Jensen 1997 (PL).

The lectotype designated by Yankovsky (1995) has no status because the specimen was collected (8 June 1955) after the date of publication of the original description by Rubtsov (1940b). We, therefore, designate a neotype to fix the current concept of the name.

Several reports of *S. subpusillum* in the Palearctic Region have appeared under the name *annulitarse*. This latter name, however, correctly applies to a member of the *S. tuberosum* species group (Adler et al. 1999).

Smith (1970) included 2 species under the name *S.* 'sp. near *subpusilla*' but did not differentiate them. His material was not available to us for study, but his descriptions of the males and larvae are much like those of topotypical *S. subpusillum*. The female claws are said to have 2 subbasal teeth each (Smith 1970), unlike the claws of *S. subpusillum*, which are toothless. The pupae are described as having a flimsy, nearly transparent cocoon and 2–4 gill filaments that arise from petioles of variable length (often quite long), as illustrated by Bodrova (1980) for material from the Russian Far East. The gill of *S. subpusillum* at the type locality and in most North American populations that we have examined typically branches near the base (Fig. 10.455). The gills of pupae from the Thelon River in Nunavut, however, occasionally branch at varying distances from the base. We consider the variation in the branching pattern of the gill to represent a single species, but encourage future workers to test this hypothesis.

**Bionomics**

**Habitat.** *Simulium subpusillum* inhabits the far northern realms of Alaska, the Northwest Territories, and Nunavut, as well as Fennoscandia and Russia. Larvae and pupae have been taken below a lake outlet in northern Alaska and from the slow marginal flows of rivers such as the Horton, Prince, and Thelon in the Northwest Territories and Nunavut. In Russia, larvae and pupae are found in small tundra streams (Bodrova 1977) and medium to large silty rivers with ample vegetation (Rubtsov 1956). In Sweden, the immatures typically inhabit large rivers and lake outlets (Carlsson et al. 1977, Adler et al. 1999).

**Oviposition.** Females from the Prince River in Nunavut are autogenous, producing 16–43 eggs (Smith 1970). Oviposition behavior is unknown.

**Development.** Mature larvae have been collected from early July through August in the Nearctic Region. In northern Sweden, larvae hatch in the middle of June (Carlsson et al. 1977). Two generations are produced annually in Russia (Rubtsov 1956).

**Mating.** Unknown.

**Natural Enemies.** We have larvae from the Horton River that are infected with an unidentified mermithid nematode and an oomycete probably of the order Saprolegniales. Adults from this river often have water mites of the genus *Sperchon* attached to them.

**Hosts and Economic Importance.** Unknown; presumably mammalophilic.

**Simulium (Schoenbaueria) 'species Z'**
(Fig. 10.179, Map 156)

*Simulium* 'n. sp. near *giganteum*' Currie 1997a: 567, 590 (Alaska record)

**Taxonomy**

Currie 1997a (F).

This distinctive species is known from a single female. The eyes are reduced, the frons is broad (nearly one fourth the width of the head at its narrowest point), and the maxillae and mandibles have only fine hairs. The hair has been rubbed from the scutum, but that on the legs and remainder of the body is pale golden. The claws are toothless. The anal lobe with its sinuous outer margin and anterior hyaline band is unique (Fig. 10.179). Particularly distinctive is the condition of 2 spermathecae, each on a short duct that arises from a common duct. Because we have only 1 female, we do not know if the presence of 2 spermathecae is aberrant or typical.

**Bionomics**

**Habitat.** The single known female was taken by sweeping vegetation in a subalpine meadow (Isabel Pass) at mile 206 (63°10'N, 145°35'W) on the Richardson Highway, Alaska.

**Oviposition.** Females are obligately autogenous. Oviposition behavior is unknown.

**Development.** The only known specimen was collected on 16 July 1987.

**Mating.** Unknown.

**Natural Enemies.** No records.

**Hosts and Economic Importance.** This species is the only known member of the subgenus *Schoenbaueria* that has mouthparts not adapted for biting.

**Subgenus *Psilopelmia* Enderlein**

*Psilopelmia* Enderlein 1934a: 283 (as genus). Type species: *Psilopelmia rufidorsum* Enderlein 1934a: 283 (=

*Simulium escomeli* Roubaud 1909: 428–429), by original designation; *Pselopelmia* (subsequent misspelling)

*Lanea* Vargas, Martínez Palacios & Díaz Nájera 1946: 103, 107, 160 (as subgenus of *Simulium*). Type species: *Simulium haematopotum* Malloch 1914: 62–63, by original designation

**Diagnosis.** Adults: Radius with or without hair dorsobasally. Legs typically bicolored. Female: Body pollinose. Cibarium typically with armature between cornuae. Claws each with or without small subbasal tooth. Scutum typically with some yellow or orange areas and/or variously colored longitudinal stripe(s). Genital fork with anteriorly directed apodemes typically well developed. Anal lobe with ventral margin acutely or bluntly produced. Male: Gonostylus half or more as long as gonocoxite; in inner lateral view typically subquadrate, with distal angle produced and typically bearing 1 apical spinule. Pupa: Gill of 4–250 filaments. Cephalic trichomes unbranched or multiply branched. Cocoon slipper or slightly shoe shaped, loosely to tightly woven. Larva: Frontoclypeal apotome with negative or faintly positive head-spot pattern (sometimes no pattern). Postgenal cleft rounded or pointed anteriorly, extended one third or more distance to hypostomal groove. Abdominal segment IX without ventral tubercles. Rectal papillae of 3 simple or compound lobes.

**Overview.** The subgenus *Psilopelmia* is largely Neotropical. The systematics of the subgenus are controversial, especially with regard to the subgeneric limits vis-à-vis the closely related subgenus *Ectemnaspis* (Miranda Esquivel & Muñoz de Hoyos 1995). Crosskey and Howard (1997) recognized 5 species groups in the subgenus *Psilopelmia*, whereas Coscarón et al. (1996) restricted the subgenus *Psilopelmia* to the *S. escomeli* species group of Crosskey and Howard (1997). Coscarón (1990) recognized 4 species groups in *Ectemnaspis*, whereas Crosskey and Howard (1997) restricted *Ectemnaspis* to the *S. bicoloratum* species group of Coscarón (1990). We follow the arrangement of Crosskey and Howard (1997) but recognize that additional effort is required to resolve the issue. Of the approximately 50 species recognized in the subgenus *Psilopelmia* by Crosskey and Howard (1997), 11 of them, all in the *S. escomeli* species group, occur north of Mexico.

### *Simulium (Psilopelmia) escomeli* Species Group

**Diagnosis.** Adults: Radius typically without hair dorsobasally. Pupa: Gill of 4, 6, 8, or 20 filaments. Cephalic trichomes typically unbranched or bifid. Cocoon tightly woven. Larva: Frontoclypeal apotome with negative head pattern (sometimes no pattern). Postgenal cleft rounded anteriorly, extended about half or more distance to hypostomal groove. Rectal papillae typically of 3 simple lobes.

**Overview.** This group ranges from western Canada to southern Peru and has its greatest concentration of species in Mexico. It consists of 29 species, of which 11 occur north of Mexico. The group requires revision, including comprehensive chromosomal work, to determine if it is monophyletic. The North American species were treated by Peterson (1993) and Moulton (1998).

The immature stages of the *S. escomeli* species group develop in warm, open, productive, and usually rather wide watercourses. Larvae attach themselves to trailing vegetation and are pale yellowish green when alive but become purple, aquamarine, or turquoise when fixed in acetic ethanol. All species are multivoltine. Females are among the most colorful of North American black flies, although their scutal patterns in alcohol are markedly different from when they are dry. Females are mammalophilic and sometimes pestiferous to humans and livestock. Those of some species have been implicated as vectors of vesicular stomatitis virus.

### *Simulium (Psilopelmia) bivittatum* Malloch

(Figs. 10.180, 10.341, 10.456; Plate 1b; Map 157)

*Simulium bivittatum* Malloch 1914: 31–32 (F). Holotype* female (pinned #15415; USNM). New Mexico, San Miguel Co., East Las Vegas, 1 June 1901 (T. D. A. Cockerell)

*Simulium (Simulium) idahoense* Twinn 1938: 50–51, Fig. 3 (M). Holotype* male (slide #4449; CNC). Idaho, Franklin Co., Riverdale, at light, 11 September 1934 (C. F. Smith)

*Simulium (Psilopelmia) bivittatum* 'A' Pruess et al. 2000: 287, 289–292 (mtDNA)

*Simulium (Psilopelmia) bivittatum* 'B' Pruess et al. 2000: 287, 289–292 (mtDNA)

[*Simulium (Psilopelmia) griseum*: Pruess & Peterson (1987) not Coquillett 1898 (misident. in part, F from Loup Co.)]

**Taxonomy**

Malloch 1914 (F); Dyar & Shannon 1927 (F); Twinn 1938 (FM); Stains 1941 (FM); Stains & Knowlton 1943 (FM); Coleman 1951 (FM); Fredeen 1951 (FM), 1981a (FMPL), 1985a (FMP); Peterson 1958a (FMP), 1959a (E), 1960a (FMP), 1981 (FM), 1993 (FMPL), 1996 (FM); Vargas & Díaz Nájera 1958 (FMPL); Abdelnur 1966, 1968 (FMPL); Hall 1973 (FMP); Corredor 1975 (FM); Currie 1986 (PL); Mason & Kusters 1991b (FMPL); Mead 1995 (F); Peterson & Kondratieff 1995 (FMPL); Coscarón et al. 1996 (FMPL); Moulton 1998 (FMP).

The longitudinal stripes on the female scutum vary from bright orange to dark brown, and occasionally dark blue, possibly depending on the developmental temperature. The wide distribution of *S. bivittatum* suggests that it is a species complex.

**Morphology**

Braimah 1985, 1987b (labral-fan microtrichia); Currie & Craig 1988 (labral-fan microtrichia); Palmer & Craig 2000 (labral fan).

**Physiology**

Cupp et al. 1994 (F salivary vasodilators), 1995 (F salivary apyrase).

**Molecular Systematics**

Zhu 1990 (mtDNA); Pruess et al. 1992 (mtDNA); Tang et al. 1995b, 1996a, 1996b, 1998 (mtDNA and rDNA); Zhu et al. 1998 (mtDNA).

**Bionomics**

**Habitat.** *Simulium bivittatum*, the most common and widely distributed member of the subgenus *Psilopelmia*, occupies the lowlands from Nebraska to Oregon and from the prairies and low foothills of southern Alberta and Saskatchewan south to the states of San Luis Potosí and Zacatecas, Mexico. Warm shallow rivers with bottoms of sand, gravel, or rock provide habitat for the immature stages. Adults have been taken in large numbers in light traps baited with carbon dioxide (Kramer et al. 1990).

**Oviposition.** An egg load of 191 has been recorded for 1 female (Peterson 1959a). Nothing is known of oviposition behavior.

**Development.** Eggs overwinter in the northern portion of the range (Currie 1986, Pruess & Peterson 1987), but larvae can be found in small numbers throughout the year in southern Arizona. Larvae and pupae can be found from late May to early September at the northern distributional limit (Currie 1986). Three or more generations a year might be produced in Nebraska (Pruess & Peterson 1987).

Larvae consume a maximum number of particles at velocities of 10–25 cm/sec and are most efficient in capturing small particles (0.5–5.7 μm) (Braimah 1987a, 1987c). The microtrichia on the labral fans function as a nearly impervious wall with no flow between adjacent microtrichia (Braimah 1987b).

**Mating.** We observed a presumed mating swarm at midafternoon (11 June) along southern Idaho's Bruneau River. A loose swarm of about 50 males faced into the wind while hovering about 1 m above the water adjacent to streamside willows (*Salix*). The males hung their legs downward and pointed their abdomens skyward. Both sexes rested on the willows, but we saw no mated pairs.

**Natural Enemies.** Larvae are attacked by the chytrid fungus *Coelomycidium simulii*, the microsporidium *Polydispyrenia simulii*, and unidentified mermithid nematodes. Larval hydropsychid caddisflies prey on the larvae (Peterson 1960b). We have found water mites (*Sperchon* sp.) attached to adults.

**Hosts and Economic Importance.** Females attack horses, pigs, cattle, sheep, and humans (Schroeder 1939, Fredeen 1981a). Feeding occurs especially during the morning and evening and is centered around the belly, brisket, and jaw (Knowlton 1935). Females have been described as "wicked little biters" of humans in the area around Oregon's Hood River during July (Cole 1969). Historically, severe outbreaks forced people indoors in some areas of southern Saskatchewan and Alberta (Fredeen 1981a).

A bunyavirus has been isolated from females (Kramer et al. 1990). Vesicular stomatitis virus has been isolated from a pool of black flies that included *S. bivittatum* and *S. vittatum* s. l. (Francy et al. 1988). Replication of vesicular stomatitis virus can be sustained in *S. bivittatum* for at least 10 days; however, *S. bivittatum* is an unlikely vector of the virus because of midgut barriers to dissemination (Mead 1995, Mead et al. 1997).

### *Simulium* (*Psilopelmia*) *clarum* Dyar & Shannon

(Fig. 10.457, Plate 1c, Map 158)

*Eusimulium clarum* Dyar & Shannon 1927: 21–22, Figs. 52, 53 (M). Lectotype* male (designated by Peterson 1993: 309; slide #28336; USNM). California, Fresno Co., Fresno, 12 May 1923 (M. E. Phillips); *clarium* (misspelling in caption for Fig. 52 of original description)

[*Simulium trivittatum*: Coleman (1951) not Malloch 1914 (misident. in part, F)]

[*Simulium* (*Lanea*) *bivittatum*: Wirth & Stone (1956) not Malloch 1914 (misident. in part, California records)]

[*Simulium* (*Lanea*) *trivittatum*: Wirth & Stone (1956) not Malloch 1914 (misident. in part, California records)]

[*Simulium* (*Psilopelmia*) *bivittatum*: Anderson & Voskuil (1963) not Malloch 1914 (misident.)]

[*Simulium* (*Psilopelmia*) *trivittatum*: Anderson & Voskuil (1963) not Malloch 1914 (misident.)]

[*Simulium trivittatum*: Peters & Womeldorf (1966) not Malloch 1914 (misident.)]

[*Simulium* (*Psilopelmia*) *trivittatum*: Cole (1969) not Malloch 1914 (misident. in part, California records)]

[*Simulium* (*Psilopelmia*) *bivittatum*: Hall (1974) not Malloch 1914 (misident.)]

[*Simulium* (*Psilopelmia*) *trivittatum*: Hall (1974) not Malloch 1914 (misident.)]

[*Simulium bivittatum*: Reeves & Milby (1990) not Malloch 1914 (misident.)]

[*Simulium* (*Psilopelmia*) *bivittatum*: Peterson (1993) not Malloch 1914 (misident. in part, California records)]

[*Simulium* (*Psilopelmia*) *trivittatum*: Peterson (1993) not Malloch 1914 (misident. in part, California records)]

[*Simulium trivittatum*: Anderson & Yee (1995) not Malloch 1914 (misident.)]

[*Simulium trivittatum*: Yee & Anderson (1995) not Malloch 1914 (misident.)]

**Taxonomy**

Dyar & Shannon 1927 (M), Stains 1941 (M only), Stains & Knowlton 1943 (M only), Moulton 1998 (FMPL).

The validity of this species was established by Moulton (1998), primarily on the basis of the pupal gill.

**Molecular Systematics**

Tang et al. 1998 (mtDNA and rDNA).

**Bionomics**

**Habitat.** *Simulium clarum* is known only from the Central Valley of California. Its immature stages inhabit medium-sized to large rivers.

**Oviposition.** Unknown.

**Development.** Adults are present from at least March into October.

**Mating.** Unknown.

**Natural Enemies.** Larvae are attacked by unidentified mermithid nematodes.

**Hosts and Economic Importance.** Females feed on horses, cattle, and humans (Anderson & Voskuil 1963). Blood-meal analyses indicate that they also feed on birds and rabbits (Reeves & Milby 1990). On cattle and horses, they work primarily the underside of the body (Anderson & Voskuil 1963, Anderson & Yee 1995). Outbreaks have been severe enough to cause $818 in lost milk production at 3 dairies in less than 1 month (Anderson & Voskuil 1963). The females can be annoying by flying into the eyes and hair (Michelbacher 1938). Bites on humans can produce a red wheal up to 2 cm in diameter (Peters & Womeldorf 1966).

### *Simulium* (*Psilopelmia*) *griseum* Coquillett

(Figs. 10.181, 10.342; Map 159)

*Simulium griseum* Coquillett 1898: 69 (FM). Lectotype* male (here designated; pinned #10381, terminalia on slide; USNM). Colorado, specific location and date unknown (C. F. Baker); *grisem* (subsequent misspelling)

*Simulium* 'sp. undet.' Cameron 1922: 4, 24 (biting cattle and horses in Saskatchewan)

[*Simulium* (*Psilopelmia*) *venator*: Peterson (1993) not Dyar & Shannon 1927 (misident. in part, Fig. 76a, c)]

[*Simulium* (*Psilopelmia*) *venator*: Peterson & Kondratieff (1995) not Dyar & Shannon 1927 (misident. in part, Fig. 315a, c)]

**Taxonomy**

Coquillett 1898 (FM); Johannsen 1902, 1903b (FM); Emery 1913 (FM); Malloch 1914 (FM); Dyar & Shannon 1927 (FM); Stains 1941 (FM); Stains & Knowlton 1943 (FM); Coleman 1951 (FM); Fredeen 1951 (FMP), 1981a (FMPL), 1985a (FMP); Wirth & Stone 1956 (FM); Peterson 1958a (FM), 1960a (FMP), 1993 (FMPL); Abdelnur 1966, 1968 (FMPL); Hall 1973 (FM); Currie 1986 (PL); Mason & Kusters 1991b (FMPL); Peterson & Kondratieff 1995 (FMPL); Moulton 1998 (FMP).

Coquillett (1898) had 3 females and 1 male before him when he described *S. griseum*, but he did not designate a holotype. Peterson (1993) called the single male a holotype, although it should have been designated a lectotype. We formally designate and label the single male in the USNM type collection as lectotype to establish the current concept of the name.

**Morphology**

Hannay & Bond 1971a (FM wing).

**Molecular Systematics**

Tang et al. 1998 (mtDNA and rDNA).

**Bionomics**

**Habitat.** *Simulium griseum* is distributed at low elevations from southern Alberta and Saskatchewan southward into Arizona, California, Nebraska, and New Mexico. Its immature stages can be found with those of *S. bivittatum*, especially in shallow rivers. However, *S. griseum* is a less common species and is more likely than *S. bivittatum* to be found in irrigation systems and slower smaller flows.

**Oviposition.** Unknown.

**Development.** Eggs overwinter throughout most of the range, and larvae and pupae are present from spring to fall. Adults are most common during the summer.

**Mating.** Unknown.

**Natural Enemies.** Larvae are attacked by unidentified mermithid nematodes. In the Rio Grande of New Mexico at Las Cruces they are hosts of a new species of microsporidium with elongate spores in clusters of 8. Females are attacked by water mites (*Sperchon* sp.).

**Hosts and Economic Importance.** Females are pests of horses, feeding principally on the venter (MacNay 1958, 1959b; Anderson & Yee 1995) and around the eyes (Lord & Tabachnick 2002). They also can be pests of cattle, feeding especially in the ears (Edmunds 1954) and on the muzzle (Lord & Tabachnick 2002). Black-tailed jackrabbits suffer attacks in the ears (Ryckman 1961). Females have been taken in traps baited with sheep (Jones 1961) and will feed on both sheep and hogs (Fredeen 1981a). They have been trapped around chicken flocks but do not appear to be ornithophilic (Shemanchuk & Depner 1971). At times, they annoy humans by biting the ankles (Moulton 1998). Outbreaks were intense enough to drive people indoors in southern Saskatchewan until 1968, when a hydroelectric dam was constructed on the South Saskatchewan River; the dam was believed to have prohibited drift of eggs and larvae from near the Alberta border (Fredeen 1977a, 1979, 1981a).

### *Simulium* (*Psilopelmia*) *labellei* Peterson
(Map 160)

*Simulium* (*Psilopelmia*) *labellei* Peterson 1993: 360–368, Figs. 110, 111, 114, 115, 118, 120, 144–152 (FMPL). Holotype* female (pinned; USNM). Texas, Presidio Co., Presidio, 12 January 1949 (collector unknown)

**Taxonomy**

Peterson 1993 (FMPL), Moulton 1998 (FMPL).

Further study is needed to determine if *S. labellei* is actually a species distinct from *S. robynae* or merely a darker, cold-season variant of *S. robynae*.

**Bionomics**

**Habitat.** This poorly collected species is sympatric with the structurally similar *S. robynae* along the Texas-Mexico border. The immature stages have been collected too infrequently to permit a description of their habitat. Adults have been taken in light traps.

**Oviposition.** Unknown.

**Development.** Adults have been collected from December to February, with 1 collection in mid-April.

**Mating.** Unknown.

**Natural Enemies.** No records.

**Hosts and Economic Importance.** Unknown; presumably mammalophilic.

### *Simulium* (*Psilopelmia*) *longithallum* Díaz Nájera & Vulcano
(Fig. 10.458, Map 161)

*Simulium* (*Psilopelmia*) *longithallum* Díaz Nájera & Vulcano 1962a ("1961"): 221–235 (FMPL). Holotype male (pinned [in part] and on slide [in part] #6516; INDRE). Mexico, Morelos State, Tepoztlan, 5 March 1962 (A. Barrera & M. A. Vulcano)

**Taxonomy**

Díaz Nájera & Vulcano 1962a (FMPL), Coscarón et al. 1996 (FMPL), Moulton 1998 (FMPL).

This species is morphologically very similar to *S. gonzalezherrejoni* Díaz Nájera from Mexico, differing slightly only in the configuration of its pupal gill. Further work is needed to confirm their specific distinctness.

**Molecular Systematics**

Tang et al. 1998 (mtDNA and rDNA).

**Bionomics**

**Habitat.** This species is found from southern Arizona to southern Mexico. It breeds in shallow lowland streams and rivers up to 5 m wide.

**Oviposition.** Unknown.

**Development.** Larvae and pupae have been taken in Arizona from December into May. They nearly disappear during the heat of the summer (Moulton 1998).

**Mating.** Unknown.

**Natural Enemies.** Larvae are hosts of unidentified mermithid nematodes.

**Hosts and Economic Importance.** Females presumably feed on mammals. They can support viral replication of vesicular stomatitis virus in the laboratory for at least 10 days but do not serve as vectors because the midgut acts as a barrier to dissemination (Mead 1995, Mead et al. 1997).

### *Simulium* (*Psilopelmia*) *mediovittatum* Knab

(Figs. 10.182, 10.343, 10.459; Plate 1d; Map 162)

*Simulium mediovittatum* Knab 1915b: 77–78 (F). Holotype* female (pinned #19635; USNM). Texas, Tarrant Co., Arlington, 28 October 1914 (F. C. Bishopp)

**Taxonomy**

Knab 1915b (F); Dyar & Shannon 1927 (F); Stains 1941 (FM); Stains & Knowlton 1943 (FM); Coleman 1951 (FMP); Vargas & Díaz Nájera 1954 (FPL), 1957, 1958 (FMP); Peterson 1958a (F), 1993 (FMPL); Coscarón et al. 1996 (FMPL); Moulton 1998 (FMP).

**Molecular Systematics**

Tang et al. 1998 (mtDNA and rDNA).

**Bionomics**

**Habitat.** This species ranges from Texas southward to the state of San Luis Potosí in Mexico. The immature stages are found on trailing vegetation in ditches and shallow streams and rivers.

**Oviposition.** Unknown.

**Development.** All life stages are probably present throughout the year over at least part of the range.

**Mating.** Unknown.

**Natural Enemies.** Larvae are attacked by the microsporidium *Polydispyrenia simulii* and unidentified mermithid nematodes.

**Hosts and Economic Importance.** Females feed on burros (donkeys), horses, mules, cattle, and black-tailed jackrabbits (Bishopp 1935, Jones et al. 1977, Peterson 1993). As many as 5000 flies have been found on 1 mule, especially around the ears and along the belly (Bishopp 1935). At times, they swarm about humans, especially around the ankles, inflicting painful bites that can cause swelling and allergic reactions (Wiseman & Eads 1960).

### *Simulium* (*Psilopelmia*) *meyerae* Moulton & Adler

(Figs. 10.344, 10.460; Map 163)

*Simulium* (*Psilopelmia*) *meyerae* Moulton & Adler 2002b: 213–218 (FMPL). Holotype* male (pinned, pupal exuviae in glycerin vial below; USNM). New Mexico, Dona Ana Co., Rio Grande, Rt. 70 (Picacho Avenue), Riverside Park, 1 July 2001 (J. K. Moulton & S. K. Meyer)

**Taxonomy**

Moulton & Adler 2002b (FMPL).

**Bionomics**

**Habitat.** *Simulium meyerae* is known only from the type locality, a limited section of the Rio Grande in the city limits of Las Cruces, New Mexico. Larvae and pupae were collected in small numbers, relative to the large populations of *S. griseum* and *S. meridionale*.

**Oviposition.** Unknown.

**Development.** Larvae, pupae, and reared adults have been collected from late June to early July but are probably present throughout most of the year.

**Mating.** Unknown.

**Natural Enemies.** No records.

**Hosts and Economic Importance.** Unknown; presumably mammalophilic.

### *Simulium* (*Psilopelmia*) *notatum* Adams

(Plate 1e, Map 164)

*Simulium notatum* Adams 1904: 434 (F). Lectotype* female (designated by Peterson 1993: 338; pinned #54, head and abdomen in glycerin vial below; UKAL). Arizona, Mohave Co. or La Paz Co., Bill Williams River (as Bill Williams Fork), July, day and year unknown (F. H. Snow)

[*Simulium* (*Psilopelmia*) *griseum*: Peterson (1993) not Coquillett 1898 (misident. in part, material from southern Arizona)]

**Taxonomy**

Adams 1904 (F), Stains 1941 (F), Stains & Knowlton 1943 (F), Peterson 1993 (F), Mead 1995 (F), Moulton 1998 (FMPL).

*Simulium notatum* was an enigmatic species for most of the 20th century, known only from 2 females collected in Arizona. Moulton (1998) associated its immature stages and male, and considered it distinct from *S. griseum*. Further work is required to resolve the taxonomic limits of *S. notatum* vis-à-vis *S. griseum*.

**Molecular Systematics**

Tang et al. 1998 (mtDNA and rDNA).

**Bionomics**

**Habitat.** This tiny species is known only from the southern half of Arizona, where its immature forms live in swift, small to large streams and rivers.

**Oviposition.** Unknown.

**Development.** The immatures and adults have been collected from July to December (Moulton 1998) and are probably present throughout most of the year.

**Mating.** Unknown.

**Natural Enemies.** Two larvae from the Gila River at Winkleman, Pinal Co., Arizona, harbored a new species of microsporidium with elongate spores in clusters of 8.

**Hosts and Economic Importance.** Females are presumably mammalophilic. They are competent laboratory vectors of the Indiana and New Jersey serotypes of vesicular stomatitis virus and might be involved in epizootics (Mead 1995; Mead et al. 1997, 2000a).

### *Simulium* (*Psilopelmia*) *robynae* Peterson

(Figs. 10.183, 10.345, 10.461, 10.780; Plate 15f; Map 165)

*Simulium* (*Psilopelmia*) *robynae* Peterson 1993: 368–382 (FMPL). Holotype* female (pinned; USNM). Texas, Val Verde Co., 10 mi. west of Del Rio, 13 October 1953 (H. M. Brundrette)

*Simulium* 'sp. 1' Jones et al. 1977: 444 (feeding on horses in Texas)

[*Simulium ochraceum*: Cockerell (1897) not Walker 1861 (misident.)]

[*Simulium notatum*: Malloch (1914) not Adams 1904 (misident. in part, F except type)]

[*Simulium notatum*: Dyar & Shannon (1927) not Adams 1904 (misident. in part, F except type)]

**Taxonomy**

Malloch 1914 (F), Dyar & Shannon 1927 (F), Peterson 1993 (FMPL), Moulton 1998 (FMP).

**Molecular Systematics**

Tang et al. 1998 (mtDNA and rDNA).

**Bionomics**

**Habitat.** This species hugs the New Mexico and Texas border with Mexico. The immature stages are found on trailing vegetation in shallow flows of the Rio Grande drainage. Adults are attracted to lights (Peterson 1993).

**Oviposition.** Unknown.

**Development.** Adults have been collected from March into October. All life stages might occur throughout the year.

**Mating.** Unknown.

**Natural Enemies.** Larvae are hosts of unidentified mermithid nematodes.

**Hosts and Economic Importance.** Females feed on horses and black-tailed jackrabbits (Peterson 1993).

### *Simulium (Psilopelmia) trivittatum* Malloch
(Figs. 10.184, 10.346, 10.462, 10.781; Plates 1f, g, 16a; Map 166)

*Simulium trivittatum* Malloch 1914: 30 (F). Holotype* female (pinned #15608; USNM). Mexico, Tampico, 17 December, year unknown (E. A. Schwarz)

*Simulium distinctum* Malloch 1913: 133–134 (FM). Holotype* male (slide #15958; USNM). Texas, Val Verde Co., Devil's River, at light, 5 May 1907 (F. C. Bishopp & H. S. Pratt). Junior primary homonym of *S. distinctum* Lutz 1910: 241–243

*Simulium (Psilopelmia) mazzottii* Díaz Nájera 1981 ("1979"): 554–561 (FMPL). Holotype male (slide; INDRE). Mexico, Coahuila, Múzquiz, Arroyo Las Salinas, 7–13 November 1970 (A. Díaz Nájera)

**Taxonomy**

Malloch 1913, 1914 (FM); Knab 1915a (F); Dyar & Shannon 1927 (FM); Coleman 1951 (MP); Wirth & Stone 1956 (FMP); Peterson 1958a, 1960a (FMP), 1993 (FMPL, intersex); Hall 1973 (FM); Díaz Nájera 1981 (FMPL); Coscarón et al. 1996 (FMP); Moulton 1998 (FMPL).

Much of the Mexican and other Neotropical material previously identified as *S. trivittatum* actually represents a similar species, *S. bobpetersoni*, which has a pupal gill of 8, rather than 6, filaments (Coscarón et al. 1996). We have seen occasional Texan specimens with 8 filaments, suggesting either that variation exists within *S. trivittatum* or that *S. bobpetersoni* extends northward into the United States. The validity of *S. bobpetersoni* as a distinct species requires confirmation.

**Molecular Systematics**

Tang et al. 1998 (mtDNA and rDNA).

**Bionomics**

**Habitat.** *Simulium trivittatum* is most common in Texas but extends into Oklahoma and Mexico. Larvae and pupae can be found on trailing vegetation in small grassy ditches and shallow rivers. Larvae drift primarily during the night and early morning (Reisen 1977).

**Oviposition.** Unknown.

**Development.** Larvae and pupae are present year-round.

**Mating.** Unknown.

**Natural Enemies.** Larvae are hosts of the microsporidium *Polydispyrenia simulii* and unidentified mermithid nematodes.

**Hosts and Economic Importance.** Hosts are not known but are presumably mammals. In the United States, a single fly was reported to have bitten a human (Reisen 1974a). Nearly all published biting records under the name of this species pertain to *S. clarum*.

### *Simulium (Psilopelmia) venator* Dyar & Shannon
(Figs. 10.185, 10.347; Plate 1h, i; Map 167)

*Simulium venator* Dyar & Shannon 1927: 36, Figs. 92, 93 (FM). Holotype* female (pinned #28343, terminalia on slide; USNM). Nevada, Washoe Co., Reno, 7 July 1916 (H. G. Dyar)

*Simulium beameri* Stains & Knowlton 1943: 279–280, Figs. 70, 80 (FM). Holotype* female (pinned #3283, terminalia on slide [poor condition]; UKAL). California, Inyo Co., Lone Pine, 28 July 1940 (R. H. Beamer)

[*Simulium mediovittatum*: Stains (1941) not Knab 1915b (misident.)]

[*Simulium mediovittatum*: Stains & Knowlton (1943) not Knab 1915b (misident.)]

[*Simulium mediovittatum*: Knowlton & Fronk (1950) not Knab 1915b (misident.)]

[*Simulium (Psilopelmia) mediovittatum*: Peterson (1960a) not Knab 1915b (misident. in part, material from Utah)]

[*Simulium mediovittatum*: Newell (1970) not Knab 1915b (misident.)]

[*Simulium (Psilopelmia) griseum*: Peterson (1993) not Coquillett 1898 (misident. in part, Fig. 25a, c)]

[*Simulium (Psilopelmia) griseum*: Peterson & Kondratieff (1995) not Coquillett 1898 (misident. in part, Fig. 299a, c)]

**Taxonomy**

Dyar & Shannon 1927 (FM); Stains 1941 (FM): Stains & Knowlton 1943 (FM); Coleman 1951 (FM); Wirth & Stone 1956 (FM); Peterson 1958a, 1960a (FM), 1993 (FMPL); Peterson & Kondratieff 1995 (FMPL); Moulton 1998 (FMP).

The female scutum has either a black or a brown medial stripe (Plate 1h, i), with black predominating in California and brown in the Great Basin. We consider these forms variants of the same species, although this hypothesis requires testing.

**Bionomics**

**Habitat.** *Simulium venator* is a lowland species that occurs from the upper Great Basin of the United States to southern California. The immature stages have been found in moderately wide rivers.

**Oviposition.** Unknown.

**Development.** Larvae are present from at least April into October.

**Mating.** Unknown.

**Natural Enemies.** Larvae are sometimes infected by the microsporidium *Polydispyrenia simulii*.

**Hosts and Economic Importance.** Females feed on dogs, horses, and humans (Twinn 1938, Peterson 1993). On horses, feeding usually occurs in the ears or around the eyes and nostrils (Knowlton 1935).

## Subgenus *Psilozia* Enderlein

*Psilozia* Enderlein 1936b: 113–114 (as genus). Type species: *Psilozia groenlandica* Enderlein 1936b: 114 (= *Simulia vittata* Zetterstedt 1838: 803), by original designation; *Psilosia* (subsequent misspelling)

*Neosimulium* Rubtsov 1940b: 116, 121, 124, 130–132, 138. *Nomen nudum*

*Neosimulium* Vargas, Martínez Palacios & Díaz Nájera 1946: 103, 108, 160 (as subgenus of *Simulium*). Type species: *Simulia vittata* Zetterstedt 1838: 803, by original designation

**Diagnosis.** Adults: Radius without hair dorsobasally. Subcosta bare ventrally (or with at most 2 or 3 hairs). Legs bicolored; calcipala small. Female: Grayish, pollinose. Cibarium with spinose armature between cornuae. Fore tibia anteriorly with broad white patch. Scutum with dark stripes. Claws toothless. Abdomen gray, with black patches dorsally and laterally. Anal lobe subtriangular, produced ventrally. Genital fork with anteriorly directed apodemes well developed. Spermatheca with internal spicules. Male: Scutum velvety black, with grayish, whitish, or silvery patches anteriorly. Fore leg mostly brown, with dull, silvery or yellowish white patch on tibia. Gonostylus about half as long as gonocoxite, in inner lateral view with distal angle at least slightly produced and bearing 2–5 apical spinules. Ventral plate in ventral view typically subtriangular, rounded posteriorly. Median sclerite bifurcate. Pupa: Gill of 10–24 filaments. Cephalic and thoracic trichomes unbranched. Cocoon slipper shaped, loosely to tightly woven. Larva: Frontoclypeal apotome with positive head-spot pattern. Antenna with 1–3 hyaline patches. Postgenal cleft rounded or squared anteriorly, extended less than one fourth distance to hypostomal groove; subesophageal ganglion ensheathed with pigment. Abdominal segment IX without ventral tubercles; abdominal setae unbranched, translucent. Rectal papillae simple, although each lobe occasionally with few secondary lobules.

**Overview.** The 5 described species in this subgenus are centered in the Nearctic Region. Three of them also occur in the Neotropical Region, and 1 (*S. vittatum* s. s.) barely slips into the Palearctic Region. In addition, we have 5 small larvae collected by D. C. Kurtak on 8 January 2001 from the Owens River, about 13 km northwest of Bishop in Inyo Co., California, that do not match any of the currently known species. The larvae resemble those of the *S. vittatum* complex, except that the apical article of the proleg is almost entirely sclerotized (i.e., the lateral sclerites are large and confluent), only the central head spots are manifested and these are diffuse, and the ecdysial lines converge posteriorly. These larvae might represent a sixth species, although their small size precludes a definitive judgment.

All species of *Psilozia* are associated with productive habitats, and most have done well in the face of massive habitat alterations. They all are probably multivoltine. Eggs generally are laid on wetted substrates. Females are mammalophilic, and those of several species are pests of livestock and occasional nuisances to humans.

### *Simulium* (*Psilozia*) *argus* Williston

(Figs. 10.186, 10.348, 10.463, 10.782; Plate 16b; Map 168)

*Simulium argus* Williston 1893: 253–254 (F). Holotype* female (pinned #53, terminalia missing; UKAL). California, county unknown, Argus Mountains, May 1891 (through C. V. Riley, collector unknown); *arguo* (subsequent misspelling)

*Eusimulium obtusum* Dyar & Shannon 1927: 15, Figs. 27–29 (M). Lectotype* male (designated by Stone 1949a: 138; slide #28331 [indicated on slide by penciled asterisk]; USNM). California, San Bernardino Co., Redlands, day and month unknown, 1914 (F. R. Cole)

'species *Simulium*' Brues 1932: 244 (PL in hot springs)

*Simulium kamloopsi* Hearle 1932: 12–13 (FMP). Holotype* male (pinned #3444; CNC). British Columbia, Kamloops, Peterson Creek, 2 September 1918 (E. Hearle)

*Simulium hearlei* Twinn 1938: 50, Fig. 2 (M). Holotype* male (slide #4448; CNC). Utah, Uintah Co., Ft. Duchesne, swept from grass, 11 July 1937 (published date), 11 August 1937 (slide label) (G. F. Knowlton & F. C. Harmston); *haerlei* (subsequent misspelling)

[*Simulium trivittatum*: Knowlton & Fronk (1950) not Malloch 1914 (misident.)]

[*Simulium trivittatum*: Peterson (1958c) not Malloch 1914 (misident.)]

[*Simulium virgatum*: El-Buni & Lichtwardt (1976) not Coquillett 1902 (misident.)]

[*Simulium* (*Psilopelmia*) *trivittatum*: Peterson (1993) not Malloch 1914 (misident. in part, material from Arizona and New Mexico)]

**Taxonomy**

Williston 1893 (F); Johannsen 1902, 1903b (F); Emery 1913 (F); Dyar & Shannon 1927 (M); Hearle 1932 (FMP); Twinn 1938 (M); Stains 1941 (M); Stains & Knowlton 1943 (M); Vargas et al. 1946 (L); Vargas & Díaz Nájera 1948a (F), 1948b (M), 1949 (P), 1957 (FMP); Stone 1949a (FM); Coleman 1951 (FMP); Wirth & Stone 1956 (FMP); Peterson 1958a, 1960a (FMP); Hidalgo Escalante 1959 (L); Hall 1973 (FMPL), 1974 (L); Corredor 1975 (FM); Peterson & Kondratieff 1995 (FMPL).

The abundance and wide distribution of *S. argus* are suggestive of a species complex.

**Physiology**

Cross et al. 1993a (F salivary antigens); Abebe 1994 (F salivary anticoagulants); Abebe et al. 1994 (F salivary anticoagulant); Cupp et al. 1994 (F salivary vasodilators), 1995 (F salivary apyrase).

**Cytology**

Pasternak 1964 (footnote, p. 135), Rothfels 1979.

**Molecular Systematics**

Zhu 1990 (mtDNA).

**Bionomics**

**Habitat.** *Simulium argus* is abundant throughout western North America from southern British Columbia to southern Mexico. It breeds in productive farm streams, irrigation ditches, spring-fed desert streams, and watercourses issuing from impoundments. These flows are typically less than 5 m wide, though large populations often develop in shallow rivers up to 25 m wide. The immature stages often inhabit the same streams as those of the *S. vit-*

*tatum* species complex. Larvae and pupae have been found in hot springs at 31°C (Brues 1932). In fact, *S. argus* is one of the most heat-tolerant black flies, occurring at temperatures up to 37°C, although populations decline as temperatures exceed 30°C (Mohsen & Mulla 1982a). Like the larvae of the *S. vittatum* species complex, those of *S. argus* tend to match the color of the substrate to which they are attached. Larvae colonize colored polyethylene strips more frequently than they colonize clear polyethylene strips (Mohsen 1981).

**Oviposition.** Females oviposit on trailing vegetation.

**Development.** Larvae and pupae are present year-round in many areas of the United States (Mohsen & Mulla 1982a). In more northern areas, such as Oregon, some streams support only 1 generation annually (Speir 1969). Larvae feed indiscriminately on diatoms and detritus and also ingest filamentous green algae (Koslucher & Minshall 1973). Their assimilation efficiency on diatoms averages about 55% (McCullough et al. 1979).

**Mating.** Unknown.

**Natural Enemies.** Larvae are hosts of various microsporidia (Mohsen & Mulla 1984), including *Amblyospora varians* (Peterson 1960b) and *A. bracteata*. They also are infected by the chytrid fungus *Coelomycidium simulii* and mermithid nematodes such as *Mesomermis* probably *flumenalis* (Peterson 1960b, Harkrider 1988). Infections with a virus-like particle (Lacey in Molloy 1988), originally believed to be a densonucleosis virus, can be induced in larvae (Federici & Lacey 1987). The larval gut is colonized by the trichomycetes *Genistellospora homothallica, Harpella melusinae, Pennella angustispora, Simuliomyces microsporus, Smittium simulii* (Lichtwardt 1964, 1967, 1972), and *Harpella leptosa*. Hydropsychid and rhyacophilid caddisflies and odonates are predators of larvae (Peterson 1960b, Koslucher & Minshall 1973). Pupae have been found with unidentified mites attached to them (Peterson 1960b).

**Hosts and Economic Importance.** Females feed on horses and cattle, especially in the ears (Anderson & Voskuil 1963, Hall 1972, Jones et al. 1977, Yee & Anderson 1995). Horses are possibly not a preferred host and might even be repellent (Anderson & Yee 1995). Females rarely bite humans (Hall 1972).

### *Simulium* (*Psilozia*) *encisoi* Vargas & Díaz Nájera

(Figs. 10.187, 10.349, 10.464; Plate 2a; Map 169)

*Simulium* (*Neosimulium*) *encisoi* Vargas & Díaz Nájera 1949: 292–295, Figs. 29–37 (FMP). Holotype male (pinned #3965, terminalia on slide; INDRE). Mexico, Hidalgo State, San Miguel Regulatory District, 25 May 1949 (L. Vargas)

**Taxonomy**

Vargas & Díaz Nájera 1949, 1957 (FMP); Hall 1973 (FM).

**Molecular Systematics**

Moulton 1997, 2000 (DNA sequences); Tang et al. 1998 (mtDNA and rDNA); Pruess et al. 2000 (mtDNA).

**Bionomics**

**Habitat.** This western species has been collected in the arid regions from southern Idaho to central Mexico. Although it has a much greater range in the United States than indicated by its initial discovery in this country (Boreham 1962), it is common only in southern Arizona and New Mexico. Larvae and pupae are found in streams less than 15 m wide. Larvae on green vegetation are paler than those on sticks, dead leaves, and rocks.

**Oviposition.** Females oviposit on trailing vegetation.

**Development.** Pupae are present by April or earlier in Idaho. Southward, all life stages can be found throughout the year.

**Mating.** Unknown.

**Natural Enemies.** Unidentified mermithid nematodes infect larvae.

**Hosts and Economic Importance.** Unknown; presumably mammalophilic.

### *Simulium* (*Psilozia*) *infernale* Adler, Currie & Wood, New Species

(Figs. 10.188, 10.350, 10.465, 10.645, 10.783; Plate 16c; Map 170)

**Taxonomy**

**Female.** Wing length 2.6–3.3 mm. Scutum pale grayish, pollinose, with anterior pair of whitish pruinose spots (appearing black at some angles) and 3 brown stripes. All hair silvery. Mandible with 31–35 serrations; lacinia with 23–26 retrorse teeth. Sensory vesicle in lateral view occupying about half or more of palpomere III. Legs brown except as follows: fore leg with femur pale yellowish brown, and tibia with anterior, whitish patch; middle and hind legs with proximal three fourths of femora and tibiae and proximal half of basitarsi pale yellowish brown. Terminalia (Fig. 10.188): Anal lobe in ventral view bilobed, well sclerotized, heavily setose, especially anteriorly. Genital fork with space between arms a broadly inverted U. Spermatheca slightly longer than wide.

**Male.** Wing length 2.5–2.9 mm. Scutum grayish, pollinose, with pair of whitish pruinose spots anteriorly and 1 faint brown medial stripe and 1 faint brownish patch on either side of midline; anterior face of scutum nearly vertical in some specimens. Legs brown except as follows: fore tibia with yellowish white patch on anterior face; middle and hind legs with pale brown patches centrally on femora and tibiae, and yellowish brown patch on proximal half of basitarsi. Genitalia (Fig. 10.350): Gonostylus with 3 or 4 apical spinules. Ventral plate in ventral view with body about 3 times as wide as long.

**Pupa.** Length 3.8–4.1 mm. Gill of 16 filaments in 8 pairs, about half or more as long as pupa (Fig. 10.465). Head and thorax with densely and regularly distributed, barely raised microtubercles. Abdominal onchotaxy well developed, with sternal hooks longest. Terminal spines long, dark. Cocoon loosely woven.

**Larva.** Length 6.6–7.5 mm. Head capsule (Fig. 10.783) brown, except frontoclypeal apotome yellowish brown; ecdysial line directed posteromedially; head spots dark brown; anterior group often surrounded by brown pigment. Antenna brown, with 2 or 3 hyaline patches, subequal in length to stalk of labral fan. Hypostoma convex, with median tooth broadest, extended farthest anteriorly; remaining teeth similar to one another, 8 in number per side (Fig. 10.645). Postgenal cleft rounded anteriorly. Labral fan with 26–30 primary rays. Body pigment gray, mottled, arranged in distinct bands; medial area of each segment variously unpigmented, especially on posteriormost segments (Plate 16c). Posterior proleg with 13–21 hooks in each of 108–111 rows.

**Diagnosis.** Females resemble those of other members of the subgenus *Psilozia* but can be differentiated by the more rounded anal lobe (in lateral view). Males are distinguished from those of other members of the subgenus *Psilozia* by the grayish unstriped scutum. The well-developed onchotaxy, especially the large sternal hooks and terminal spines, characterizes the pupa. A number of larval features are unique to this species, including the form of the hypostomal teeth and shape of the ecdysial line.

**Holotype.** Female (frozen dried, pinned) with pupal exuviae (glycerin vial below). California, Santa Clara Co., Alum Rock Park, 37°24′N, 121°47′W, 22 March 1993, D. M. Wood (USNM).

**Paratypes.** Same data as holotype (45♀ & 3♂ + exuviae, 16 larvae), 6 December 1991 (2 ethanol vials: 106 larvae, 536 larvae).

**Etymology.** The specific name is from the Latin adjective meaning "hellish," in reference to the hot sulfurous habitat of the immature stages.

### Cytology

The chromosomal complement is highly rearranged relative to the *S. vittatum* standard banding sequence of Pasternak (1964), with only the IIS arm retaining the standard sequence.

### Bionomics

**Habitat.** *Simulium infernale* is known from only 1 site near San Jose, California, where it occurs in a pure population, apparently specialized for life in tiny thermal springs. These milky seepages, stinking of sulfur and holding at about 27°C year-round, issue from shrubby banks and form trickles less than 10 cm wide that flow over the edges of slime-covered rocks, grasses, and fallen leaves. Larvae and pupae can be found beneath these substrates.

**Oviposition.** Unknown.

**Development.** Larvae and pupae have been collected in December and March, although the species is probably present throughout most of the year. In some years, flow ceases and the immatures cannot be found.

**Mating.** Unknown.

**Natural Enemies.** No records.

**Hosts and Economic Importance.** Unknown; presumably mammalophilic.

## *Simulium (Psilozia) vittatum* Species Complex

**Diagnosis.** Female: Anal lobe unsclerotized along entire ventral margin. Male: Gonostylus smoothly curved, with 3–5 (typically 3) apical spinules. Pupa: Gill of 14–16 filaments. Chromosomes: Standard banding sequence as given by Pasternak (1964) and Rothfels and Featherston (1981).

This species complex contains at least 2 morphologically indistinguishable cytological segregates, formerly known as 'IIIL-1' and 'IS-7' (Rothfels & Featherston 1981), to which we assign the names *S. tribulatum* and *S. vittatum*, respectively. Evidence from cytology, distribution, and ecology (Rothfels & Featherston 1981, Adler & Kim 1984) favors formal recognition of these entities as valid species, although molecular work has yielded ambiguous results (Zhu et al. 1998). The complex has appeared in more biological studies than has any other comparable group of black flies (Adler & McCreadie 1997). We present most of the information under the species complex as a whole but also treat the 2 component species in detail.

### Taxonomy

Zetterstedt 1838 (F); Osten Sacken 1870 (PL, except Figs. 145, 146); Riley 1870b (L only, Fig. 143); Lugger 1897a, 1897b (FMPL); Coquillett 1898 (F), 1902 (M); Johannsen 1902, 1903b (FMPL), 1934 (PL); Forbes 1912 (FL); Hart in Forbes 1912 (M); Emery 1913 (FMPLE); Hungerford 1913 (FMPL); Malloch 1914 (FMPL); Riley & Johannsen 1915 (FM); Jobbins-Pomeroy 1916 (MPL); Dyar & Shannon 1927 (FM); Wu 1931 (E); Hearle 1932 (FMP); Knowlton & Rowe 1934a (F); Enderlein 1936b (F); Twinn 1936b (FMPLE); Stains 1941 (FM); Smith et al. 1943 (FM); Stains & Knowlton 1943 (FM); Vargas & Díaz Nájera 1948a (F), 1948b (M), 1949 (P); Davies 1949a (MP); Nicholson 1949 (FMP); Nicholson & Mickel 1950 (FMP); Coleman 1951 (FMP); DeFoliart 1951a (PL); Fredeen 1951 (FMPLE), 1981a (FMPL), 1985a (FMP); Stone 1952 (FMP), 1964 (FMPL); Jamnback 1953, 1956 (L); Sommerman 1953 (L); Gill & West 1955 (FPL); Stone & Jamnback 1955 (FMPL); Davies & Peterson 1956 (E); Wirth & Stone 1956 (FMP); Peterson 1958a, 1960a (FMP), 1965a (F), 1977b, 1981, 1996 (FMPL); Young 1958 (F); Anderson 1960 (FMPL); Davies et al. 1962 (FMP); Wood et al. 1963 (L); Abdelnur 1966, 1968 (FMPLE); Snoddy 1966 (FMPL); Holbrook 1967 (L); Stone & Snoddy 1969 (FMPL); Pinkovsky 1970 (FMPL), 1976 (FMPL); Smith 1970 (PL); Amrine 1971 (FMPL); Hall 1973 (FMPL), 1974 (L); Lewis 1973 (FMPL); Corredor 1975 (FM); Pascuzzo 1976 (E); Snoddy & Noblet 1976 (PL); Imhof 1977 (E); Merritt et al. 1978b (PL); Imhof & Smith 1979 (E); Westwood 1979 (FMPL); Hilsenhoff 1982 (L); Webb & Brigham 1982 (L); Adler 1983 (PLE); Adler & Kim 1984 (L), 1986 (FPLE); Rubtsov & Yankovsky 1984 (FMPL); Currie 1986 (PL), 1988 (F); Coscarón 1987 (FM); Golini & Davies 1988 (E); Mason & Kusters 1991b (FMPL); Peterson & Kondratieff 1995 (FMPL); Stuart 1995 (cocoon); Jensen 1997 (PL); Stuart & Hunter 1998b (cocoon).

### Morphology

Hungerford 1913 (F internal anatomy); Buerger 1967 (FM labrum); Guttman 1967a (labral fan); Craig 1968 (L maxillary musculature); Yang 1968 (F gut, peritrophic matrix); Chance 1969, 1970a (L mouthparts); Chen 1969 (oocytes, ovarioles); Gosbee et al. 1969 (FM salivary gland); Fredeen 1970a (gynandromorph); Smith 1970 (ovarian follicles); Hannay & Bond 1971a (FM wing, F cornea), 1971b (FM thoracic pollinosity); Liu & Davies 1971, 1972a (F flight muscle), 1972b, 1972c (L fat body), 1972d, 1972e (F fat body), 1972f, 1973 (oocytes); Liu 1972 (L fat body), 1973, 1974 (oocytes); Kurtak 1973 (labral fan); Craig 1974 (L head); Liu et al. 1975 (follicular cells); Pascuzzo 1976 (oocytes, ovarioles); Yang & Davies 1977 (F peritrophic matrix); Colbo et al. 1979 (FM labral and cibarial sensilla and armature); Hayton 1979 (F age-related changes in Malpighian tubules); Lacey & Federici 1979 (L midgut epithelium); Mokry 1979 (ovarian development); Ross 1979 (instars); Boo & Davies 1980 (FM Johnston's organ); Craig & Borkent 1980 (L maxillary palpal sensilla); Ross & Craig 1980 (L dorsal gland of labrum, labral-fan microtrichia); Watts 1981b (F salivary gland removal); Barr 1982 (L mouthparts, silk, prolegs), 1984 (L labial gland, mouthparts, prolegs, silk); Craig & Batz 1982 (L antennal sensilla); Goldie 1982 (embryology, E ultrastructure); Eymann 1985 (L abdominal sensilla and cuticle); Craig & Craig 1986 (L mandibular sensillum); O'Grady 1986 (FM compound eye); Nyhof & McIver 1987 (L ocelli); O'Grady & McIver 1987 (FM compound eye); Sutcliffe & McIver 1987 (FM tarsal sensilla); Thompson 1987a, 1989 (labral-fan microtrichia); Currie 1988 (F costa); Currie & Craig 1988 (L mandibular-hypostomal functional complex); Colbo 1989 (instars); Davies 1989 (gynandromorph); Ramos 1991 (F peritrophic matrix); Luckhart et al. 1992 (F hemocytes); Lacoursière & Craig 1993 (labral fan); Fry 1994 (L internal head and thorax); Hudson 1994 (instars); Ramos et al. 1994 (F peritrophic matrix); Fry & Craig 1995 (L internal head and thorax); Gryaznov 1995a (ovarioles); Smith & Hayton 1995 (F age-related changes in Malpighian tubules); Stuart & Hunter 1995 (cocoon construction); Evans & Adler 2000 (spermatheca); Palmer & Craig 2000 (labral fan).

**Physiology**

Hocking 1953a, 1953b (F crop capacity, flight, metabolic rate); Hutcheon & Chivers-Wilson 1953 (F salivary anticoagulants and histamine); Guttman 1967b (L anesthetic); Madahar 1967 (oogenesis); Yang 1968 (F salivary agglutinin and anticoagulant, amylase, trypsin); Yang & Davies 1968a (F amylase), 1968b (F trypsin), 1974 (F salivary agglutinin and anticoagulant); Chen 1969 (oogenesis); Gordon & Bailey 1976 (L osmotic pressure, composition of hemolymph); Lacey et al. 1978 (L gut pH); Lacey & Federici 1979 (L gut pH); Undeen 1979b (L gut pH); Watts 1981a (FPL content and molecular weights of salivary gland proteins); Barr 1982, 1984 (L silk composition); Jacobs et al. 1990 (F salivary anticoagulant); Grant 1991 (F glutathione *S*-transferase); Ramos 1991 (F peritrophic matrix proteins); Cross et al. 1993a, 1993b (F salivary antigens), 1994 (F salivary gland extract); Cupp et al. 1993 (FM salivary apyrase), 1994 (F salivary vasodilator), 1997 (F hemolymph response to microfilariae injections); Abebe 1994 (F salivary anticoagulants); Abebe et al. 1994 (F salivary anticoagulant), 1995 (F salivary protein with antithrombin activity), 1996 (F salivary anticoagulant); Ramos et al. 1994 (F peritrophic matrix proteins); Cupp & Cupp 1997 (F salivary secretions); Ribeiro et al. 2000 (F salivary hyaluronidase); Noriega et al. 2002 (oogenesis).

**Cytology**

Rothfels & Dunbar 1953; Zimring 1953; Pasternak 1961, 1964; Barley 1964; Rothfels 1979, 1980, 1981a, 1981b; Rothfels & Featherston 1981; Featherston 1982; Sanderson et al. 1982 (effects of aluminum); Adler 1983, 1986; Adler & Kim 1986; Brockhouse & Adler 2002.

**Molecular Systematics**

Teshima 1970, 1972 (DNA hybridization); Snyder 1981 (allozymes, p. 98); Jacobs-Lorena et al. 1988 (DNA sequences); Feraday & Leonhardt 1989 (allozymes); Zhu 1990 (mtDNA); Brockhouse 1991 (rDNA); Ramos 1991 (gut-specific protease genes); Xiong & Kocher 1991, 1993b (mtDNA); Pruess et al. 1992, 2000 (mtDNA); Xiong 1992 (mtDNA); Brockhouse et al. 1993 (FPL silk proteins); Procunier & Smith 1993 (p. 171, in situ hybridization of rDNA probe to polytene chromosomes); Ramos et al. 1993 (gut-specific protease genes); Tang et al. 1995b, 1996a, 1996b, 1998 (mtDNA and rDNA); Xiong & Jacobs-Lorena 1995a (trypsin gene), 1995b (carboxypeptidase gene); Miller et al. 1997 (rDNA); Cupp et al. 1998 (cDNA and recombinant protein for F vasoactive protein in saliva); Koch et al. 1998 (effects of preservation methods, gut contents, and parasites on DNA); Zhu et al. 1998 (mtDNA); Pruess et al. 2000 (mtDNA).

**Bionomics**

**Habitat.** This species complex is abundant throughout continental North America into the Federal District of Mexico, southern Greenland, Iceland, and the Faeroe Islands. A member of the complex is 1 of only 2 simuliid species recorded from the Aleutian Islands of Alaska (Gorham 1975). The immature stages inhabit a wide range of flowing waters, from tiny seeps to rivers more than 300 m wide. They are particularly prevalent in streams coursing through agricultural lands and below outfalls of impoundments, but are generally absent from heavily forested streams unless these are secondarily enriched. Dense populations often form below outfalls (Stone & Snoddy 1969, Gíslason & Gardarsson 1988), with biomass decreasing as distance from the outfall increases (Morin & Peters 1988, Gíslason et al. 1994). Instances of black flies inhabiting lake shores are rare, but we have larvae of this species complex that were collected beneath rocks on the shore of Lake Ontario, near Kuckville, New York (19 June 1987).

Larvae are tolerant of extreme temperatures (0°C–33°C), low oxygen tensions, a wide range of current velocities, and organic and industrial pollution (Stone & Snoddy 1969, Ali et al. 1974). Some populations, however, are unable to tolerate decreases in pH to 4.0 (Chmielewski & Hall 1992). Macrocurrent velocities experienced by larvae generally fall within a range of 0.09–1.86 m/sec (Westwood & Brust 1981, Shipp 1985a). However, the velocities in the larval microhabitats are considerably lower than those in the water column (Osborne et al. 1985). Larvae stranded in stagnant pools or placed in still water can survive for at least several days (Edmunds 1954, Bacon & McCauley 1959, Anderson & Shemanchuk 1975). Larvae avoid scouring during floods by entering fissures in bedrock or the hyporheic zone to a depth of 30 cm (Colbo 1985, Eymann & Friend 1986).

When larvae hatch, they usually produce silk pads for attachment. Subsequent dispersal of neonate larvae is density dependent (Fonseca & Hart 1996). Most larval drift, at least for later instars, occurs at night (Adler et al. 1983b). Larvae sink at a rate that is positively related to body length (Fonseca 1999). They attach themselves to virtually every available substrate and tend to match the color of the substrate to which they are attached (Adler & Kim 1984). This correspondence in color between larvae and background is influenced by ultraviolet radiation, diet, and substrate color (Zettler et al. 1998).

Some studies have shown that larvae select locations with high turbulence (i.e., Froude numbers = ca. 0.7–1.6) (Wetmore et al. 1990, Eymann 1993), whereas other studies report that larval densities are highest in areas with low turbulence (i.e., Froude number = ca. 0.25) (Beckett 1992). The abundance of larvae at a particular spot in a stream can be predicted from current speeds 2 mm above the substrate (Hart et al. 1996). Larval abundances are greater when the levels of dissolved organic matter are higher, putatively because ultraviolet radiation is attenuated (Kelly et al. 2001). Surface features of the substrate, such as texture and the amount of periphyton, also influence larval colonization and microdistributions (Beckett 1992), as do settlement rates from the drift (Fonseca 1996).

Larvae are agonistic when contacted and, therefore, good competitors for space (Harding & Colbo 1981, Eymann et al. 1987). Although actual spacing is the result of larval interactions, spacing patterns are a function of habitat; for example, larvae tend to clump when near impoundment outflows (Eymann & Friend 1988). Larval positions within aggregations vary with the interaction of flow and food concentration (Ciborowski & Craig 1989). Larvae, for example, often space themselves to avoid low velocities created by the presence of upstream conspecifics (Fonseca & Hart 2001). Aggregation behavior appears unrelated to the presence of predaceous stoneflies (Ciborowski & Craig 1991).

The activity of adults peaks in the morning and early evening (Mason and Kusters 1990). Adults rest in riparian vegetation but move to moist rocks as daytime temperatures increase (Lacey & Mulla 1977b). Lawns, patios, buildings, and boat docks also serve as resting sites during the heat of the day (Mulla & Lacey 1976b). Maximum flight speeds for short distances approach 300 cm/sec, and unfed females are capable of flying distances of 104 km in still air (Hocking 1953b). Adults have been taken in light traps (Frost 1949, Bacon 1953, Bacon & McCauley 1959).

**Oviposition.** Females are often autogenous for the first ovarian cycle, with the degree of expression positively correlated with larval nutrition (Mokry 1980b). Females that emerge from overwintered larvae are autogenous (Davies & Peterson 1956). Some females apparently are able to produce a second, albeit smaller, batch of eggs autogenously (Jóhannsson 1988). The degree of autogeny decreases with subsequent generations as water temperature increases (Chutter 1970). An increase in the rearing temperature from 19°C to 20°C approximately halves autogenous egg production (Colbo & Wotton 1981). Autogenous production of mature eggs requires about 2–6 days and depends largely on the temperature at which the females are held (Davies & Györkös 1990). If deprived of sugar, females resorb some follicles (Moobola 1981). Autogenous females will feed on blood in the laboratory (Moobola 1981), which significantly increases the number of eggs they mature; feeding on sugar may or may not increase actual fecundity (Pascuzzo 1976, Mokry 1980b). Females will feed on blood even when they are gravid (Moobola 1981).

The potential number of eggs produced per ovarian cycle ranges from 200 to more than 600 (Davies & Peterson 1956, Abdelnur 1968, Chutter 1970, Pascuzzo 1976). Female size greatly influences fecundity, with larger flies producing more and larger eggs (Pascuzzo 1976, Mokry 1980b, Gíslason & Jóhannsson 1991). Females derived from overwintered larvae produce a mean of 562 eggs in the first cycle, whereas females from the summer generation produce an average of 221 eggs (Colbo & Porter 1981).

Oviposition begins about 24 hours after emergence (Wolfe & Peterson 1959) and occurs primarily around sunset (Imhof & Smith 1979), although indirect evidence suggests a smaller postdawn period of oviposition (Corbet 1967). Gravid females fly upstream near the surface of the water in search of oviposition sites (Adler et al. 1983a). On landing, they oviposit on any substrate bathed or sprayed by water, and produce long strings of eggs that loop about and form irregular masses. The density of an egg mass is highly correlated with current speed (Fonseca 1999). White or green-yellow substrates floating over a streambed are preferred for egg laying (Golini 1974). Consequently, trailing vegetation is a common oviposition substrate. Females also oviposit underwater and, perhaps more rarely, in still water (Davies & Peterson 1956). And they have been reported to oviposit by tapping their abdomens on the water surface as they fly (e.g., Davies & Peterson 1956). Gravid females can be induced to oviposit by placing them in containers with a small amount of water (Fredeen 1959a, Tarrant et al. 1987) or by shaking them in vials half-filled with water (Mokry et al. 1981). Crushing the female's head also induces oviposition (Pascuzzo 1976).

**Development.** Larvae hatch within 3–5 days at 24°C and within 4 days at 15°C, but if massed together, only the outermost eggs produce larvae (Imhof & Smith 1979). No hatching occurs if the eggs are frozen (Tarshis 1968b, Colbo & Wotton 1981). Less than 10% of larvae hatch if the eggs are stored for 2–3 months at temperatures just above freezing (Fredeen 1959b). Eggs held at 4°C in moist cotton for more than a year will yield larvae when placed in water (Tarshis 1968b). The time of year when eggs are collected is apparently important in hatching success after storage, with eggs collected in October producing a better yield than those collected in August (Waters 1969). Eggs of the *S. vittatum* species complex are vulnerable to desiccation, particularly when young (Imhof & Smith 1979). Members of the complex, nonetheless, are premier colonizers of temporary streams.

The duration of larval and pupal development varies, especially with water temperature (Wu 1931). The larval stage lasts about 13 days at 17°C and 8 days at 27°C; the pupal stage lasts 6 days at 17°C and 3 days at 27°C (Becker 1973). Food quality and quantity, however, can significantly modify developmental rates (Colbo & Porter 1979, Fuller et al. 1988). For example, at 25°C, a 10-fold increase in larval food is required to produce a female as large as one reared from larvae at 20°C (Colbo & Porter 1981). Adults can develop from larvae fed only bacterial suspensions (Fredeen 1960, 1964b). Algae, free-living bacteria, and conditioned leaf-particulate matter are equally good sources of energy (Fuller & Fry 1991), although larvae in some studies grow larger on a diet of diatoms (Thompson 1987c). The number of larval instars varies from 7 to 11, perhaps depending on nutritional state (Ross 1979, Colbo 1989). The pharate pupa cleans the substrate and then constructs its cocoon in 6 distinct stages (Stuart & Hunter 1995). The slipper-shaped cocoon generates small vortices that flow upward through the gill filaments (Eymann 1991b).

Larvae generally overwinter, developing even beneath ice. In some northern areas, however, eggs overwinter (Sommerman et al. 1955, Lewis & Bennett 1973). In the northern United States, eggs and larvae sometimes overwinter in the same streams, a situation that prompted early speculation about the existence of sibling species (Anderson & Dicke 1960). In some parts of the country, such as Florida and southern Oklahoma, larvae and pupae are seldom collected during the summer (Reisen 1975b, Pinkovsky & Butler 1978). The annual number of population peaks varies with latitude: 1 in Iceland, but with 2 cohorts (Gíslason 1985, Gíslason & Jóhannsson 1985), sometimes called generations (Jónsson et al. 1986, Einarsson 1988); 2 or 3 in Alaska, Oregon, Ontario, Quebec, the Maritime Provinces, and insular Newfoundland (Davies 1950; Sommerman et al. 1955; Speir 1969; Lewis & Bennett 1973, 1979; Back & Harper 1979); 4 or 5 in Michigan, New York, Pennsylvania, Rhode Island, Wisconsin, Manitoba, and Saskatchewan (Dimond & Hart 1953, Anderson 1960, Pinkovsky 1970, Merritt et al. 1978b, Westwood & Brust 1981, Adler et al. 1982, Jarvis 1987); and perhaps 7 in Idaho (Jessen 1977) and Alabama, where emergence frequently continues through the winter (Stone & Snoddy 1969). Multiple cohorts often are produced (Morin et al. 1988c). Seasonal abundance of larvae has been positively correlated with water temperature (Shipp & Procunier 1986).

At northern latitudes, adults are among the first black flies of the year to emerge and among the last to disappear. Over much of the United States, they often emerge on warm winter days. They emerge primarily during the morning (Abdelnur 1968) and exhibit an even sex ratio (Wu 1931, Davies 1950), with males emerging slightly earlier than females (Wolfe & Peterson 1959, Colbo & Porter 1979). During emergence, the fly is covered in a film of gas that is concentrated in a belt around the neck (Hannay & Bond 1971a). On breaking the water-air interface, adults pause on the surface of the water or take flight immediately, only to rest soon thereafter (Adler 1988b). Females live longer than males (Emery 1913), with a maximum life span of 2 months on dry sucrose and water at 11°C (Davies 1953). Females that feed on sugar water live significantly longer than those that feed only on water (Moobola 1981).

Rearing *S. vittatum* s. l. from eggs to adults can be accomplished in any container of aerated water (Fredeen 1959a; Tarshis 1968b, 1971) or in more elaborate systems of running water (Wood & Davies 1966). Aerated water also has been used to transport and ship larvae for long distances (Tarshis 1966, Tarshis & Adkins 1971). Mass production has been achieved in systems of recirculated water (Brenner & Cupp 1980a). Members of the species complex mate in swarms and for many years were intractable to laboratory mating. Attempts to hand-pair adults have met with limited success (Field et al. 1967, Mokry et al. 1981). Large-scale laboratory rearings through multiple generations, nonetheless, have proved successful, with the percentage of individuals that mate increasing in subsequent generations (Tarrant et al. 1983, Bernardo & Cupp 1986, Bernardo et al. 1986a, Cupp & Ramberg 1997, Gray & Noblet 1999). A colony that was started in December 1982 by E. W. Cupp's group at Cornell University and continued by E. W. Gray and R. Noblet at the University of Georgia (Brockhouse & Adler 2002) was still in operation as of 2003, never having had any additional flies introduced.

*Simulium vittatum* s. l. has played a central role in studies of larval hydrodynamics and filter feeding (Chance 1977, Craig & Chance 1982, Cheer & Koehl 1987, Craig & Galloway 1988, Eymann 1988, Lacoursière 1990). At one time, larvae were believed to maintain their bodies and labral fans within the boundary layer (Chance & Craig 1986), but microhydraulic studies suggest that most larvae hold their labral fans above the boundary layer (Lacoursière 1992). In the standard feeding position, 1 labral fan filters water from the mainstream, while the other fan filters water arising from the surface of the substrate (Merritt et al. 1996). The posture that filter-feeding larvae assume, including the height at which they hold their labral fans, is believed to be a compromise between reducing drag and maximizing food intake (Hart et al. 1991). As water velocity increases, larvae open their fans wider (Lacoursière & Craig 1993). Larvae can filter 2.8 ml of water in 80 minutes at 20°C (Webster 1973).

Larvae feed on a wide range of particulate matter. They show little qualitative selection of algae in the water column (Jóhannsson 1984, Thompson 1987a), although most particles ingested are 45 μm or less in diameter (Chance 1970a). At velocities greater than 0.3 m/sec, the efficiency of large-particle capture decreases rapidly, probably due to the short microtrichia on the primary fan rays (Kurtak 1978). Earlier instars retain more coarse mineral particles, relative to fine particles, than do later instars (Merritt et al. 1978a), and they ingest a greater proportion of small algae than is present in the seston (Thompson 1987a). Material on the microtrichia of the labral fans, once believed to be a sticky mucosubstance facilitating fine-particle capture (Ross & Craig 1980), is actually flocculated dissolved organic matter, which contributes to larval growth (Fry & Craig 1995, Ciborowski et al. 1997). Some studies have shown that temperature and particulate concentration influence larval feeding rates and efficiencies (Lacey & Mulla 1979b, Thompson 1987b), whereas other studies have demonstrated that food concentration has no effect on feeding behavior (Finelli et al. 2002). Diel periodicity in feeding rates has not been found (Mulla & Lacey 1976a). Assimilation efficiencies of larvae feeding on diatoms average about 57% (McCullough et al. 1979). Larvae in some populations graze periphyton from the substrate (Kurtak 1973), whereas larvae in other populations do not (McCullough 1975). They do not recycle their old silk pads (Barr 1982). The rather rapid rate at which material moves through the gut suggests that some microencapsulated insecticides might not release their toxins fast enough to cause high larval mortality (Sibley & Kaushik 1991).

**Mating.** Mating generally occurs at about sunset (Peterson 1962b) or sometimes in the morning (Lacey & Mulla 1977b). Swarms begin as diffuse aggregations oriented into the wind, and they extend about 1–9 m above the ground; eventually these loose aggregations condense into smaller swarms over markers such as people, bushes, and vehicles, with moving markers accumulating more flies (Lacey & Mulla 1977b). Mating swarms also have been observed above the lip of a lake outlet (Peterson 1962b), 3 m above the lip of a waterfall (Davies & Peterson 1956), within a few meters of a cow (Mokry in Wenk 1981), and in open streamside areas above sun-warmed rocks (Stone & Snoddy 1969). At times, swarming males were so abundant over paved areas of Ottawa's Central Experimental Farm that they became a nuisance to bicyclists.

Swarms have a tripartite pattern of movement; individuals oscillate within larger vertical movements, while the entire swarm moves horizontally (Peterson 1962b). Males account for 95% or more of the individuals in a swarm, and their interactions result in instantaneous repulsion, although they sometimes couple with one another and drop from the swarm (Lacey & Mulla 1977b). Mating pairs leave the swarm either by dropping to the ground, where passive males are dragged along by females, or by flying out vertically (Peterson 1962b, Lacey & Mulla 1977b). During copulation, the male uses the grappling hooks on its claws to grip the abdominal vestiture of the female (Craig & Craig 1986). Coupling lasts less than 30 seconds (Lacey & Mulla 1977b, Peterson 1977b).

**Natural Enemies.** Larvae are parasitized by several species of mermithid nematodes, including *Gastromermis viridis*, *Isomermis wisconsinensis* (Anderson & DeFoliart 1962, Phelps & DeFoliart 1964, Jessen 1977), *Mesomermis flumenalis* (Ebsary & Bennett 1975a, Bailey & Gordon 1977, Harkrider 1988), and unidentified species (Fredeen & Shemanchuk 1960, Peterson 1960b). A member of the *S. vittatum* complex is the type host for *G. viridis* and *I. wisconsinensis* (Welch 1962). Larvae in some extensively sampled areas, such as New Jersey (Bruder & Crans 1979), Pennsylvania (Adler & Kim 1984), and South Carolina, are apparently free of mermithids. Larvae are capable of grasping and killing host-searching preparasites of *M. flumenalis* (Molloy 1981). In laboratory tests, larvae are minimally susceptible to a mermithid nematode of mosquitoes (*Romanomermis culicivorax*) (Finney & Mokry 1980) and to cercariae of the trematode *Plagiorchis noblei* (Jacobs 1991, Jacobs et al. 1993).

A cytoplasmic polyhedrosis virus commonly infects the larval midgut (Federici & Lacey 1976, Bailey 1977), and an intranuclear virus-like particle invades the gut epithelium (Charpentier et al. 1986). Larvae also are infected by an iridescent virus (Erlandson & Mason 1990) and by a virus-like particle (Lacey in Molloy 1988) originally labeled as a densonucleosis virus (Federici & Lacey 1976, 1987).

A number of fungi have been recorded from the various life stages. The chytrid fungus *Coelomycidium simulii* infects larvae over a broad geographic range, though infection rates are generally less than 1% (Jamnback 1973a, Ezenwa 1974b, Federici et al. 1977). The first evidence for vertical transmission of this fungus came from work on the *S. vittatum* complex (Tarrant 1984, Tarrant & Soper 1986). The trichomycete fungi *Genistellospora homothallica*, *Harpella melusinae*, *Pennella angustispora*, *P. simulii*, *P. hovassi*, *P.* near *hovassi*, *Pennella* sp., *Simuliomyces microsporus*, *Smittium culicis*, *Smittium culisetae*, *Smittium megazygosporum*, and *Stachylina* sp. have been recovered from the larval gut (Lichtwardt 1967, 1972, 1986; Frost & Manier 1971; Williams & Lichtwardt 1971; Labeyrie et al. 1996; Beard & Adler 2000, 2002; Lichtwardt et al. 2001). The entomophthoraceous fungus *Entomophthora culicis* kills young adults (Shemanchuk & Humber 1978, Shemanchuk 1980c). Experimental and natural infections have been obtained with the entomophthoraceous fungus *Erynia curvispora* (Kramer 1983, Nadeau et al. 1994). Both *E. curvispora* and *E. conica* might reduce fly populations by killing females before they have completed oviposition (Nadeau et al. 1994). *Tolypocladium cylindrosporum*, a fungal pathogen of mosquitoes, kills larvae of the *S. vittatum* complex in the laboratory (Nadeau & Boisvert 1994). *Saprolegnia ferax*, an oomycete, has been isolated from the gut of larvae in Newfoundland but is apparently nonpathogenic (Nolan 1976).

Among the microsporidia that infect the larvae are *Amblyospora bracteata*, *A. fibrata*, *A. varians*, *Janacekia debaisieuxi*, and *Polydispyrenia simulii* (Strickland 1913a, Twinn 1939, Frost 1970, Ebsary & Bennett 1975a, Vávra & Undeen 1981). Both larval sexes are infected (Undeen et al. 1984b). Some of the few reports of microsporidia in adult black flies involve the *S. vittatum* complex (Undeen 1981, Tarrant 1984). The ichthyosporean protists *Paramoebidium chattoni*, *P. curvum*, and *Paramoebidium* sp. occur in the larval hind gut (Lichtwardt 1976, Beard & Adler 2002).

Larvae are preyed on by caddisfly larvae of the families Hydropsychidae, Rhyacophilidae, Brachycentridae, and Limnephilidae (Peterson 1960b, Peterson & Davies 1960); hydrophilid beetle larvae (Peterson 1960b); leeches (Fredeen & Shemanchuk 1960); triclad flatworms (Waters 1969, Hansen et al. 1991, Hart & Merz 1998); hydra (Hannay & Bond 1971a); and the slimy sculpin *Cottus cognatus* (Adler & Kim 1986). The crayfish *Cambarus bartonii* consumes larvae in the laboratory (Zettler 1996). Because large populations are often attained, the larvae can play a key role in the population dynamics and reproduction of vertebrate predators such as birds (e.g., wood ducks and harlequin ducks) and fish (Twinn 1939, James 1968, Lindsay & Dimmick 1983, Gíslason 1985, Einarsson 1988, Gíslason & Gardarsson 1988, Gardarsson & Einarsson 1994, Steingrímsson & Gíslason 2002). Invertebrate predators of adults include spiders (Longstaff 1932, Adler et al. 1983a), water mites (*Sperchon ?jasperensis*) (Davies 1959), and asilid flies (Lavigne et al. 1993).

**Hosts and Economic Importance.** Males and females feed on floral nectar of alfalfa (*Medicago sativa*), Chinese tamarisk (*Tamarix chinensis* as *T. pentandra*), wild parsnip (*Pastinaca sativa*), and white heath aster (*Aster pilosus*) (Stone & Snoddy 1969, Lacey & Mulla 1977b, Adler & Kim 1986). Females have been collected as they fed on the extrafloral nectaries of peonies (*Paeonia* sp.) in Pennsylvania (A. G. Wheeler, pers. comm.). A curious report of females drawing sap from the leaves and stems of gray-leaf willow (*Salix glauca*) in Greenland (Longstaff 1932) might refer to feeding on homopteran honeydew, although Crosskey (1990) cited evidence that females of some species actually penetrate plant tissues to obtain sap.

Host searching by blood-seeking females is influenced primarily by light intensity during the summer and by temperature in other seasons (Lacey & Mulla 1977b). Females are attracted to black, blue, red, and white silhouettes but less so to those that are yellow (Bradbury 1972; Bradbury & Bennett 1974a, 1974b). A shaded area on the host has been suggested as important in the selection of a place to bite (Jessen 1977). Ground beef, particularly when putrefied, has served on 1 occasion as an unusual attractant to both sexes (Davis & James 1957). Failure to mate does not deter females from blood feeding (Mokry 1976a). Some females are capable of taking blood through a membrane system 1–2 hours after emergence, but do so less readily than those 3–5 days old (Mokry 1976a). A single blood meal has an average volume of 2.5 $mm^3$ (Townsend et al. 1977).

Females readily attack horses, mules, cattle, and sheep and feed principally in the ears and on the underparts (Knowlton & Rowe 1934b; Knowlton 1935; Cottral 1938; Snow et al. 1958b; Teskey 1960; Anderson & DeFoliart 1961; Jones 1961, 1981; Anderson & Voskuil 1963; Jessen 1977; Townsend et al. 1977; Shemanchuk 1978b; Anderson & Yee 1995; Mock & Adler 2002). Females are common on equine farms (Burg et al. 1991) where feeding in the horses' ears can cause intense scabbing and make the animals unruly (Swenk 1938, Townsend 1975, Townsend et al. 1977); surgical treatment of repeatedly bitten areas is sometimes necessary (Jessen 1977). Cattle under heavy attack will toss their heads, crowd together, and kick up dust (Jessen 1977). Females in southern Idaho can be serious pests of sheep. Harassed sheep bunch together, trample the vegetation beneath them, and refuse to graze, or they stick their heads into bushes or lie down with their heads and ears closely appressed to the ground (Jessen 1977). In years of severe attacks, weight reductions of up to 4.5 kg/lamb have been reported (Noh in Jessen 1977). Elks, moose, and white-tailed deer also serve as hosts (Merritt et al. 1978b, Pledger et al. 1980, Mason & Shemanchuk 1990), and we have specimens provided by B. A. Mullens from southern California that were biting the ears of bighorn sheep in the early morning.

Other animals have been reported less commonly as hosts. Some specimens in the CNC were collected from domestic geese in Fort Chimo, Quebec. Females have been induced to feed for a short while on domestic ducks and turkeys (Davies & Peterson 1956), and females, putatively of the *S. vittatum* complex, have attacked ducks and chickens in Michigan (Pettit 1929). Engorged females have been taken in traps baited with chickens and, to a lesser extent, rabbits (Shemanchuk 1988). Under experimental conditions, females have fed on guinea pigs (Downe 1957). They feed on pig blood that is offered artificially (Mokry 1980a, Bernardo & Cupp 1986) and have been taken from swarms around pigs (Fredeen 1969). Serological evidence indicates that pigs are hosts (Downe & Morrison 1957), but actual pig-feeding records are sparse (Bryson 1931). The *S. vittatum* species complex holds the only confirmed record of a black fly having fed on an invertebrate. A single dead female was found with its mouthparts embedded in the abdomen of the ant *Formica fusca* (Peterson 1956); however, because the ant includes sugars in its diet, the fly might actually have been seeking a sugar meal (Crosskey 1990).

Females rarely bite humans. When they do bite, however, severe allergic reactions can result (Douglass 1940, Peterson 1959a, Abdelnur 1968, Mulla & Lacey 1976b). In some regions, dense swarms cause alarm even though bites are not incurred (Peterson 1977b, Fredeen 1985a, Carstens 2001). Females are notorious for swarming about the head, darting into body openings, sometimes partially clogging the nostrils, and crawling about on the skin, into the clothes, and through the hair (Sailer 1953, Peterson 1956, Judd 1957, Eckhart & Snetsinger 1969, O'Connor et al. 2001). Adults are attracted to lights (Frost 1949, Bacon & McCauley 1959, Jessen 1977). Large numbers at incandescent lights in the southwestern states can be a nuisance (Lacey & Mulla 1977b).

Females have been infected experimentally with the parasites *Leucocytozoon smithi* and *Onchocerca lienalis* but are unlikely to be natural vectors (Lok et al. 1983a, Kiszewski & Cupp 1986). Nonetheless, the saliva of females contains orientation factors that promote directed movement of the microfilarial stage of *O. lienalis* to the site of blood feeding (Stallings et al. 2002). *Leucocytozoon simondi* can develop to the ookinete stage in females under experimental conditions (Fallis et al. 1951). The *S. vittatum* complex is capable of transmitting vesicular stomatitis virus (Maré et al. 1991; Cupp et al. 1992; Mead et al. 1999, 2000a, 2000b), as originally suggested by Francy et al. (1988). Studies in the early 20th century focused on these flies as possible vectors of the causal agent of pellagra (Hunter 1912a, 1912b, 1913a, 1913b), since known to be a nutritional malady.

### *Simulium* (*Psilozia*) *tribulatum* Lugger
(Fig. 10.784, Map 171)

*Simulium tribulatum* Lugger 1897a ("1896"): 205–207 (FMPL). Type material lost. Minnesota, other information unknown. (pp. 179–181 of Lugger 1897b include an exact replica of the entry for *S. tribulatum* that appeared on pp. 205–207 of Lugger 1897a). Revalidated

*Simulium*, Osten Sacken 1870: 229–231 (PL from Washington, DC)

*Simulium glaucum* Coquillett 1902: 97 (M). Holotype* male (slide #6184; USNM). Missouri, probably Jackson Co., Kansas City, 8 April 1898 (C. F. Adams); *glacum* (subsequent misspelling)

*Simulium venustoides* Hart in Forbes 1912: 42–43 (M only). Lectotype* male (designated by Frison 1927: 181; pinned; INHS). Illinois, McHenry Co., Algonquin, 8 July 1896 (W. A. Nason); *venustoide* (subsequent misspelling)

*Simulium vittatum* 'IIIL-1' Rothfels & Featherston 1981: 1858–1882 (C)

*Simulium* 'sp.' Osborne et al. 1985: 158–159 (larval microhabitats)

*Simulium vittatum* 'cytotype IIIL-2' Sutcliffe & McIver 1987: 13 (*lapsus calami*)
?*Simulium* 'spp.' Mullens & Dada 1992: 478–479 (in part) (attraction to bighorn sheep)
[*Simulium piscicidium*: Malloch (1914) not Riley 1870d (misident. in part, M)]
[*Simulium decorum*: Dyar & Shannon (1927) not Walker 1848 (misident. in part, M)]
[*Simulium pictipes*: Strickland (1938) not Hagen 1880 (misident.)]
[*Simulium clarkei*: Stone & Snoddy (1969) (misident. in part, Fig. 124)]

**Taxonomy**

Lugger 1897a, 1897b (FMPL); Coquillett 1902 (M); Hart in Forbes 1912 (M); Malloch 1914 (M, as *piscicidium*); Dyar & Shannon 1927 (M, as *decorum*); Stone & Snoddy 1969 (M); Adler 1983 (LE); Adler & Kim 1984 (L), 1986 (PLE).

Lugger (1897a, 1897b), through his illustrations, associated the name *tribulatum* with a member of the *S. vittatum* complex, although he described the species as a "great tormentor to humanity" with a "very severe" bite. He clearly confused the adult behavior of this species with that of the *S. venustum* complex. Likewise, several figure numbers that he cited in the text do not apply to the correct figures. We associate the name *tribulatum* and the above synonyms with cytospecies 'IIIL-1,' the predominant member of the *S. vittatum* species complex around the type localities. We have not attempted to sort out the many past misidentifications of *S. tribulatum* and *S. vittatum* s. s.

**Cytology**

Rothfels & Dunbar 1953; Pasternak 1961, 1964; Barley 1964; Rothfels 1979, 1980, 1981a, 1981b; Rothfels & Featherston 1981; Featherston 1982; Adler 1983, 1986; Adler & Kim 1986.

**Molecular Systematics**

Feraday & Leonhardt 1989 (allozymes); Zhu 1990 (mtDNA); Brockhouse 1991 (rDNA); Brockhouse et al. 1993 (F silk proteins); Procunier & Smith 1993 (p. 171, in situ hybridization of rDNA probe to polytene chromosomes); Tang et al. 1996a, 1996b (mtDNA); Zhu et al. 1998 (mtDNA); Pruess et al. 2000 (mtDNA).

**Bionomics**

**Habitat.** Perhaps the most abundant species of black fly east of the Rocky Mountains, *S. tribulatum* cuts a vast swath across North America, although it is rather scarce along the Pacific Coast and in the Far North. Its presence in Alaska is based on a single record (Rothfels & Featherston 1981) that requires confirmation. It also has been reported from River Laxa in Iceland (Rothfels in Gíslason & Jóhannsson 1985), but in many samples from this river we were unable to confirm the presence of anything other than a population (included under *S. vittatum* s. s.) with undifferentiated sex chromosomes and all major autosomal polymorphisms in Hardy-Weinberg equilibrium.

The immature stages are adapted to a wide range of habitats, from small trickles to enormous rivers, but they are especially common in warm enriched streams and rivers flowing through agricultural lands or draining impoundments (Adler & Kim 1984). This species is probably the most pollution-tolerant black fly in North America and is often among the few aquatic insects in organically fouled watercourses. Consequently, it has figured prominently in biotic indexes for evaluating organic pollution (Hilsenhoff 1987). Larvae are found at temperatures from 0°C to more than 31°C, although population levels decrease at the higher temperatures (Adler & Kim 1986). Not surprisingly, human influence, particularly agricultural practices, sewage disposal, and creation of impoundments, has favored the distribution and abundance of this species more so than for any other North American black fly.

Larvae and pupae are found on almost any substrate at depths up to 4 m. At some sites, especially those that are nutrient rich, they are found almost exclusively on the undersides of stones (Adler & Kim 1984, Adler 1986). Larvae vary in color from dark gray to pale yellowish green and typically match the color of the substrate to which they are attached (Adler & Kim 1984, Zettler et al. 1998). Interactions among larvae generate characteristic spacing patterns (Eymann & Friend 1988). Larvae drift primarily during the night, beginning around sunset (Adler et al. 1983b).

**Oviposition.** Gravid females fly predominantly upstream (Adler et al. 1983a), landing to oviposit on trailing grasses, rocks, and sticks bathed by a film of water (Adler & Kim 1986). Oviposition occurs primarily in the late afternoon and evening (Adler & Kim 1986). Potential fecundity in the first ovarian (autogenous) cycle averages 314–351 eggs (Tarrant 1984).

**Development.** This species might pass at least part of the winter as eggs at northern latitudes, such as upper Michigan, Alberta, and Saskatchewan (Rothfels & Featherston 1981, Adler 1986, Adler & Mason 1997), but in many parts of the continent, larvae can be found throughout the year. Along northern streams artificially warmed during the winter, we have found moribund adults littering the snow-covered streambanks. Techniques for inducing oviposition, rearing the larvae, and obtaining matings in the laboratory pertain to *S. tribulatum*, material of which was collected originally from the outflow of Dryden Lake, Dryden, New York (Tarrant 1984; Tarrant et al. 1983, 1987).

**Mating.** The mating behavior of *S. tribulatum*, as separate from that of *S. vittatum* s. s., is not known, but it undoubtedly involves many of the components presented for the *S. vittatum* species complex. Laboratory matings of *S. tribulatum* (as *S. vittatum*) have been obtained by confining adults to sections of clear flexible tubing 2 cm in diameter (Tarrant 1984; Tarrant et al. 1983).

**Natural Enemies.** Larvae are hosts of the chytrid fungus *Coelomycidium simulii*; a cytoplasmic polyhedrosis virus; unidentified mermithid nematodes; and the microsporidia *Amblyospora bracteata*, *A. varians*, *Janacekia debaisieuxi* (Adler 1983, 1986; Tarrant 1984; Adler & Kim 1986; Ledin 1994; Adler & Mason 1997), and *Polydispyrenia simulii*. Among the trichomycete fungi recovered from larval guts are *Harpella melusinae*, *Genistellospora homothallica*, *Pennella* near *hovassi*, *Pennella simulii*, *Simuliomyces microsporus*, *Smittium culisetae*, and *Smittium megazygosporum* (Slaymaker 1998; Beard & Adler

2000, 2002; Beard 2002). The ichthyosporean protists *Paramoebidium chattoni* and *P. curvum*, as well as an unidentified species of *Paramoebidium*, occur in the larval hind gut (Beard & Adler 2002). Adults frequently are captured in the webs of spiders (*Epeira* sp. and *Tetragnatha elongata*) that are constructed over streams, with females having a higher probability of escape than males (Adler et al. 1983a).

**Hosts and Economic Importance.** Females are biting pests of horses and cattle. Females, either of this species or of *S. vittatum* s. s., are attracted to bighorn sheep (Mullens & Dada 1992) and feed in the ears. No bona fide biting records of humans are known, but the species can be a terrific nuisance to people. Salt Lake Co., Utah, has had a management program in place since 1979 to control nuisance populations of this species that boil out of the Jordan River and its irrigation canals (K. L. Minson, pers. comm.). Similarly, the city of Kingsport, Tennessee, initiated a management program in 1991 for problems originating in the Holston River. Nuisance problems on 5 golf courses in Los Angeles Co., California, resulted in losses of about $34,000 each month in green fees during the 1994–1995 season (O'Connor et al. 2001). In Bullhead City, Arizona, and surrounding areas, the nuisance problems caused by females that originate from the Colorado River have been so severe at times that locals refer to waving away the flies as the "Bullhead City Salute." Spectators and players at baseball games in Dodger Stadium, Los Angeles, California, perform the "Black Fly Wave" as they brush the pestiferous flies away from their faces (O'Connor et al. 2001).

### *Simulium (Psilozia) vittatum* Zetterstedt

(Figs. 4.5, 4.6, 4.50, 4.53–57, 4.62, 4.64, 10.43, 10.189, 10.351, 10.466, 10.610, 10.619, 10.646; Plates 2b, c, 16d; Map 172)

*Simulia vittata* Zetterstedt 1838: 803 (F). Lectotype female (designated by Peterson 1965a: 231; pinned, terminalia in glycerin vial below; ZIL). Greenland, other information unknown

*Simulia vittata* Zetterstedt 1837: 58. *Nomen nudum*

*Simulium nasale* Gistel 1848: 151. *Nomen nudum*. (P. 497 of Gistel & Bromme 1850 includes an exact replica of the entry for *S. nasale* that appeared on p. 151 of Gistel 1848)

*Psilozia groenlandica* Enderlein 1936b: 114 (F). Holotype female (pinned, left front and hind legs and terminalia on slide; ZMHU). West Greenland, Asakak, 15 August 1893 (Vanhöffen). Junior secondary homonym of *S. groenlandicum* Enderlein 1935: 363–364

*Simulium asakakae* Smart 1944b: 131 (substitute name for *groenlandica* Enderlein 1936b)

'simuliid' Craig 1968: 31 (L)

*Simulium vittatum* 'IS-7' Rothfels & Featherston 1981: 1858–1882 (C)

?*Simulium* 'sp.' Grant 1991: 436 (glutathione *S*-transferase)

*Simulium vittatum* '/?' Tang et al. 1996a: 229 (material from Iceland)

*Simulium (Psilozia) vittatum* 'Iceland' Pruess et al. 2000: 287, 289–291 (mtDNA)

[*Simulium canadensis*: Loomis et al. (1975) not Hearle 1932 (misident.)]

[*Psilozia dahlgruni*: Rubtsov & Yankovsky (1988) not Enderlein 1921b (misident.)]

**Taxonomy**

Zetterstedt 1838 (F); Enderlein 1936b (F); Peterson 1965a (F), 1977b (FMPL); Adler 1983 (L).

Pending further study, we lump under the name *S. vittatum* s. s. all populations with sex chromosomes not based on the IIIL-1 inversion, thereby providing nomenclatural stability, preserving the name *S. vittatum* over a wide geographic area, and avoiding the establishment of new names. In addition to populations with the classic IS-7 X-chromosome inversion, we include under the name *S. vittatum* s. s., populations on Newfoundland's Avalon Peninsula and in southwestern Iceland that have unique differentiated sex chromosomes not based on either IIIL-1 or IS-7, and populations in River Laxa (Iceland) and on Prince Edward Island with undifferentiated sex chromosomes that represent the hypothesized ancestral form (Rothfels & Featherston 1981). Also included are populations on the Magdalen Islands of Quebec, which we discovered do not have differentiated sex chromosomes. Material from Greenland (type locality) and the Faeroe Islands has not been examined chromosomally but, like that from Iceland, it probably does not have sex chromosomes based on either the IIIL-1 or the IS-7 inversion. We have not attempted to untangle the many misidentifications of *S. vittatum* s. s. and *S. tribulatum*.

**Morphology**

Craig 1968 (L maxillary musculature), Mokry 1979 (ovarian development), Barr 1984 (L labial gland, mouthparts, prolegs, silk), Ramos 1991 (F peritrophic matrix), Luckhart et al. 1992 (F hemocytes), Hudson 1994 (instars), Ramos et al. 1994 (F peritrophic matrix), Evans & Adler 2000 (spermatheca).

**Physiology**

Jacobs et al. 1990 (F salivary anticoagulant); Grant 1991 (F glutathione *S*-transferase); Ramos 1991 (F peritrophic matrix proteins); Cross et al. 1993a, 1993b (F salivary antigens and extracts), 1994 (F salivary gland extract); Cupp et al. 1993 (FM salivary apyrase), 1994 (F salivary vasodilator), 1997 (F hemolymph response to microfilariae injections); Abebe 1994 (F salivary anticoagulants); Abebe et al. 1994 (F salivary anticoagulant), 1995 (F salivary protein with antithrombin activity), 1996 (F salivary anticoagulant); Ramos et al. 1994 (F peritrophic matrix proteins); E. W. Cupp & M. S. Cupp 1997 (F salivary secretions); Ribeiro et al. 2000 (F salivary hyaluronidase); Noriega et al. 2002 (oogenesis).

**Cytology**

Pasternak 1961, 1964; Rothfels 1979, 1980, 1981a, 1981b; Rothfels & Featherston 1981; Featherston 1982; Adler 1983, 1986; Adler & Kim 1986; Brockhouse & Adler 2002.

Populations with modified Y chromosomes in northwestern North America might represent additional species but have been retained under *S. vittatum* s. s. (Rothfels & Featherston 1981, Rothfels 1989).

**Molecular Systematics**

Zhu 1990 (mtDNA); Brockhouse 1991 (rDNA); Ramos 1991 (gut-specific protease genes); Brockhouse et al. 1993 (FPL silk proteins); Ramos et al. 1993 (gut-specific protease genes); Xiong & Jacobs-Lorena 1995a (trypsin gene), 1995b (carboxypeptidase gene); Tang et al. 1996a, 1996b (mtDNA); Miller et al. 1997 (rDNA); Zhu et al. 1998 (mtDNA); Pruess et al. 2000 (mtDNA).

**Bionomics**

**Habitat.** *Simulium vittatum* s. s. is transcontinental, ranging from north of the Arctic Circle south at least to the 34th parallel. At higher latitudes, it is more common than its sister species, *S. tribulatum*, and on the West Coast, it virtually replaces *S. tribulatum*. Throughout eastern North America, the immature stages of *S. vittatum* s. s. inhabit cooler cleaner waters than those of *S. tribulatum* (Adler & Kim 1984). Consequently, *S. vittatum* s. s. is underrepresented in habitats such as warm rivers and outflows from the surface of impoundments. The proportions of *S. tribulatum* and *S. vittatum* s. s. at a particular stream site, therefore, are good indicators of water quality (Adler & Kim 1983). In parts of western North America, however, it inhabits lake outflows as well as large warm watercourses (Adler 1986, Ciborowski & Adler 1990). In 1 case, larvae and pupae were found in the stock tanks and pipe outlets of a Canadian trout hatchery (Crosskey 1994a). Large numbers of larvae and pupae can be found in streams flowing across the sandy beaches of the Magdalen Islands (Quebec) and Prince Edward Island. At high tide, the immatures are subjected to innundation with salt water and the scouring effects of the waves. In these habitats, they can experience salinities of 10,000 mg/liter (Giberson et al. 2002). Larvae and pupae adhere to nearly all substrates, and larvae show the same color variation with respect to substrate as those of *S. tribulatum* (Adler & Kim 1984, Zettler et al. 1998). Larval drift patterns are similar to those of *S. tribulatum* (Adler et al. 1983b).

**Oviposition.** Females are autogenous for the first ovarian cycle when larval conditions are optimal; mean fecundity per cycle ranges from 180 to 445 eggs (Bernardo et al 1986a). Oviposition habits are similar to those of *S. tribulatum*.

**Development.** The best life-history data come from studies of the first 24 generations of a laboratory colony (Bernardo 1986, Bernardo et al. 1986a) maintained continuously for more than 2 decades (Cupp & Ramberg 1997, Gray & Noblet 1999). Eggs held at 21°C hatched about 72 hours after oviposition. The time from hatching to 50% adult emergence required 460 degree-days above 0°C, although larval density and nutrition affected developmental rates. Males emerged about 1 day before females. Nearly all emergence occurred during light hours, with subsequent behavior being highly phototactic. Larvae have 7 instars, with females being significantly larger than males in the final instar (Hudson 1994). Colony larvae, which are typically robust on a high-quality diet, show low growth rates when fed a diet of natural seston (Dey 1992).

Larvae are found throughout the year in most areas (Rothfels & Featherston 1981, Larson & Colbo 1983). In Iceland, a single generation with 2 cohorts occurs each year (Gíslason 1985, Gíslason & Jóhannsson 1985). Egg mortality in the laboratory increases significantly after 2 weeks at 2°C–3°C, suggesting that overwintering as eggs would be unlikely for this species (Bernardo et al. 1986a). For 1–2 days after hatching, larvae graze on the substrate, after which filter feeding predominates (Bernardo et al. 1986a). In addition to ingesting particulate matter (Jóhannsson 1984), larvae also consume small invertebrates and strands of algae (Adler & Kim 1986).

**Mating.** Although numerous field observations have been made on the mating behavior of the *S. vittatum* species complex, these observations never have been associated specifically with *S. vittatum* s. s. The natural mating behavior of *S. vittatum* s. s., separate from that of *S. tribulatum*, therefore, remains unknown. However, laboratory matings have been achieved by confining adults in sections of flexible tubing (Bernardo et al. 1986a).

**Natural Enemies.** Larvae are infected by the same parasites and pathogens that attack *S. tribulatum* (Adler 1983, 1986; Adler & Kim 1986; Adler & Mason 1997), although fewer species of trichomycete fungi (only *Harpella melusinae* and *Smittium* sp.) and no ichthyosporean protists have been found in the larval guts (Lichtwardt et al. 2001), a reflection of less intensive sampling. Colonization of larvae by the trichomycete fungus *Smittium megazygosporum* has been obtained in the laboratory (Beard & Adler 2000). According to Bernardo et al. (1986a), *S. vittatum* s. s. is the species that Kramer (1983) experimentally infected with the entomophthoraceous fungus *Erynia curvispora*. We have seen larvae preyed on by larval dytiscid beetles. Ducks and brown trout (*Salmo trutta*) are major predators of larvae, pupae, and adults in Iceland (Gardarsson & Einarsson 1994, Steingrímsson & Gíslason 2002).

**Hosts and Economic Importance.** In the laboratory, females feed on extracted blood from cows, horses, pigs, sheep, and turkeys; females are typically mammalophilic, but feeding rates on turkey blood are significantly higher than those on mammalian blood (Bernardo & Cupp 1986). This species is a biting pest of horses and cattle. Large swarms of flies are attracted to humans and can be a severe nuisance, although only a single bite has been documented (Peterson 1977b).

Females of *S. vittatum* s. s. have been shown experimentally to be competent vectors of the New Jersey and the Indiana serotypes of vesicular stomatitis virus, suggesting that they are involved in epizootic transmission of the virus (Maré et al. 1991; Cupp et al. 1992; Mead et al. 1999, 2000a, 2000b). They are also good surrogate hosts for *Onchocerca lienalis*, although their cibarial armature destroys most ingested microfilariae (Lehmann 1994; Lehmann et al. 1994a, 1994b, 1995a, 1995b). According to Bernardo et al. (1986a), *S. vittatum* s. s. was used by Kiszewski and Cupp (1986) as a surrogate host for the blood protozoan *Leucocytozoon smithi*.

### Subgenus *Aspathia* Enderlein

*Aspathia* Enderlein 1935: 359 (as genus). Type species: *Simulium hunteri* Malloch 1914: 59, by original designation

*Striatosimulium* Rubtsov & Yankovsky 1982: 186 (as subgenus of *Simulium*). Type species: *Simulium japonicum* Matsumura 1931: 407, by original designation

*Jalacingomyia* Py-Daniel in Py-Daniel & Moreira Sampaio 1994: 127, 129–134 (as genus). Type species: *Simulium anduzei* Vargas & Díaz Nájera 1948b: 328–330, by original designation; *Jalacongomyia* (subsequent misspelling)

**Diagnosis.** Female: Scutum typically striped. Cibarium between cornuae smooth, with minute nodules or with stout central projection. Male: Abdomen with segments II and V-VII with pearlaceous spots laterally. Gonostylus elongate, slender, typically with basal tubercle or prong. Ventral plate with posterior margin typically smooth (crenulated in a few species); lip typically slender, apically rounded and hyaline. Paramere with multiple spines of various lengths. Pupa: Gill of 3 to about 130 filaments. Cocoon shaped like shoe or slipper, with or without anterodorsal projection, with or without anterior perforations or windows. Larva: Antenna with medial article bearing minute, transverse, pigmented striations. Postgenal cleft of various shapes and sizes. Abdominal segment IX with or without prominent ventral tubercles. Rectal papillae of 3 simple or compound lobes.

**Overview.** We recognize the subgenus *Aspathia* on the basis of the characters discussed in Chapter 9, and we include 4 species groups: the *S. griseifrons*, *S. metallicum*, *S. multistriatum*, and *S. striatum* species groups. The *S. eximium* species group, as recognized by Takaoka and Davies (1996), also might be a member of this subgenus, but it is strictly Oriental and, therefore, not considered further here. The subgenus includes approximately 80 described species distributed in the Nearctic, Oriental, and Palearctic Regions. Only the *S. metallicum* species group occurs in North America.

### *Simulium* (*Aspathia*) *metallicum* Species Group

**Diagnosis.** Female: Anal lobe well sclerotized anteroventrally. Male: Ventral plate with posterior margin smooth. Pupa: Dorsum of thorax often with supernumerary trichomes. Cocoon slipper shaped, with or without anterodorsal projection. Larva: Postgenal cleft typically subtriangular, extended less than one half distance to hypostomal groove. Abdominal segment IX typically with prominent ventral tubercles.

**Overview.** Members of this group previously were placed in 2 separate groups: the *S. hunteri* species group and the *S. metallicum* species group (Crosskey & Howard 1997), both of which were revised by Coscarón et al. (1999). Because of similarities among all species and the few, simple-character differences between the 2 previously recognized groups (Coscarón et al. 1999), we prefer to include all species in the *S. metallicum* species group. As now recognized, the group contains 25 formally described species distributed from northern Canada to Venezuela, with 8 of these in western North America north of Mexico. We expect the number of species known in North America to increase when chromosomal work is conducted. *Simulium metallicum*, which is strictly Neotropical, is a complex of at least 7 cytospecies (Conn et al. 1989).

Members of the group are typically multivoltine. The immature forms generally develop in small cool flows associated with springs. The females of a few species can be pests of humans and livestock.

### *Simulium* (*Aspathia*) *anduzei* Vargas & Díaz Nájera

(Figs. 10.190, 10.352, 10.467, 10.468, 10.785; Plate 16e; Map 173)

*Simulium anduzei* Vargas & Díaz Nájera 1948b: 328–330, Figs. 25–31 (MPL). Holotype pharate male (body and pupal exuviae in alcohol vial #3920, terminalia on slide; INDRE). Mexico, State of Veracruz, Jalacingo, 1980 m elev., August 1946 (J. Parra Sevilla)

*Simulium* (*Simulium*) *patziciaense* Takaoka & Takahasi 1982: 63–67 (FMPL). Holotype female with pupal exuviae (slide; BBM). Guatemala, Department of Chimaltenango, Patzicía, stream crossing road between Patzicía and Acatenango, ca. 2 km from Patzicía Town, ca. 2200 m elev., 15 December 1979 (H. Takaoka); *patzicianense* (subsequent misspelling). New Synonym

**Taxonomy**

Vargas & Díaz Nájera 1948b (MPL), 1949 (FM), 1957 (FMP); Takaoka & Takahasi 1982 (FMPL); Coscarón et al. 1999 (FMPL).

The lengths of the basal stalk and trunks of the pupal gill constitute one of the few features purported to distinguish *S. anduzei* and *S. patziciaense* (Takaoka & Takahasi 1982). Our material from the United States is rather intermediate between that illustrated by Vargas and Díaz Nájera (1957) for *S. anduzei* and by Takaoka and Takahasi (1982) for *S. patziciaense*. The basal section of the female radius is said to be haired in *S. anduzei* but bare in *S. patziciaense* (Takaoka & Takahasi 1982); it is bare in our material. The striking similarity between *S. anduzei* and *S. patziciaense* leads us to synonymize the latter with the former.

Cocoon structure differs among populations in the United States. At some sites (e.g., Apache Co., near Greer), it covers most of the thorax and has much adherent debris (Fig. 10.467), as also illustrated for *S. patziciaense* by Takaoka and Takahasi (1982) from Guatemala. At other sites (e.g., Ramsey Canyon), the cocoon covers only the abdomen and lacks adherent debris (Fig. 10.468). These populations, however, are chromosomally similar, suggesting that they represent the same species.

**Cytology**

Chromosomes of populations in Arizona (Apache Co., near Greer, 19 April 1996; Cochise Co., Ramsey Canyon, 23 September 1998) have a Y-linked inversion near the base of IIIL. The nucleolus is in the base of IIS. IIIS carries an inversion that reverses the positions of the "capsule" and "blister" markers relative to the subgeneric standard sequence of Rothfels et al. (1978).

**Bionomics**

**Habitat.** This species is known from southern Arizona and New Mexico southward to the mountains of Guatemala. It breeds in spring-fed streams and seepages 9°C–14°C on steep rocky slopes above 2200 m, where snow melt contributes to the flow of water. Larvae are found on fallen leaves and the undersurfaces of rocks.

**Oviposition.** Unknown.

**Development.** In the United States, larvae and pupae have been taken from April through September. In Guatemala, they are present year-round (Takaoka & Takahasi 1982). The presence of larvae and pupae is probably limited by the availability of flowing water.

**Mating.** Unknown.

**Natural Enemies.** No records.

**Hosts and Economic Importance.** Unknown; presumably mammalophilic.

### *Simulium (Aspathia) hechti* Vargas, Martínez Palacios & Díaz Nájera

(Figs. 10.191, 10.353, 10.469, 10.786; Plate 16f; Map 174)

*Simulium (Simulium) hechti* Vargas, Martínez Palacios & Díaz Nájera 1946: 142–144, 173, Figs. 79–82 (FMPL). Holotype male (slide #3805; INDRE). Mexico, Federal District, Contreras, Las Dínamos, 7 April 1944 (A. Díaz Nájera); *hetchi* (subsequent misspelling)

[*Simulium (Simulium) sayi*: Vargas et al. (1943) not Dyar & Shannon 1927 (misident.)]

**Taxonomy**

Vargas et al. 1943 (M), 1946 (FMPL); Vargas & Díaz Nájera 1948a (F), 1948b (M), 1949 (P), 1957 (FMP); Hidalgo Escalante 1959 (L); Coscarón et al. 1999 (FMPL).

**Bionomics**

**Habitat.** This chiefly Mexican species extends into the mountains of southern Arizona and southwestern New Mexico. Larvae and pupae reside in small cool, rocky streams.

**Oviposition.** Unknown.

**Development.** Larvae and pupae have been collected from March through December in the United States.

**Mating.** Unknown.

**Natural Enemies.** No records.

**Hosts and Economic Importance.** Unknown; presumably mammalophilic.

### *Simulium (Aspathia) hunteri* Malloch

(Figs. 10.192, 10.354, 10.470, 10.611, 10.787; Plate 17a; Map 175)

*Simulium hunteri* Malloch 1914: 59 (F). Holotype* female (pinned #15413; terminalia missing, USNM). Colorado, Larimer Co., Virginia Dale, 31 [sic] September 1912 (F. C. Bishopp)

*Simulium (Simulium) lassmanni* Vargas, Martínez Palacios & Díaz Nájera 1946: 149–151, 174–175, Figs. 91–97, 123 (FMPL). Holotype male (pinned and slide #3793; INDRE). Mexico, State of Mexico, Los Remedios, 19 October 1944 (A. Díaz Nájera)

[*Simulium virgatum*: Hearle (1929) not Coquillett 1902 (misident. in part, specimen from Alberta)]

**Taxonomy**

Malloch 1914 (F); Dyar & Shannon 1927 (F); Hearle 1932 (F); Stains 1941 (F); Stains & Knowlton 1943 (F); Vargas et al. 1946 (FMPL); Vargas & Díaz Nájera 1948a (F), 1948b (M), 1949 (P), 1957 (FMP); Stone 1952 (FMP); Sommerman 1953 (L); Wirth & Stone 1956 (FMP); Peterson 1958a, 1960a, 1996 (FMP); Abdelnur 1966, 1968 (FMPL); Corredor 1975 (FM); Fredeen 1985a (FMP); Currie 1986 (PL); Peterson & Kondratieff 1995 (FMPL); Coscarón et al. 1999 (FMPL).

*Simulium hunteri* might be a complex of sibling species, given its broad distribution in western North America and variation in the number of pupal gill filaments and density of thoracic trichomes. If it does consist of 2 or more species, the synonymy of *S. lassmani* with *S. hunteri*, as established by Coscarón et al. (1999), will need to be reevaluated.

**Cytology**

The IIIS inversion and the location of the nucleolus are the same as in *S. anduzei*.

**Molecular Systematics**

Smith 2002 (nuclear and mtDNA).

**Bionomics**

**Habitat.** *Simulium hunteri* ranges from southern Alaska and the Yukon south to the Federal District of Mexico. It is one of the most common black flies on the coastal islands of British Columbia. The immature stages live in cool, sometimes temporary streams usually less than a few meters wide. Larvae occupy microhabitats with a laminar or turbulent boundary layer (Eymann 1993).

**Oviposition.** Unknown.

**Development.** Near the southern border of the United States, *S. hunteri* can be found year-round. Farther north, the immature stages are most common from May into August. In Alberta, larvae appear in June and adults are on the wing into September (Currie 1986).

**Mating.** Unknown.

**Natural Enemies.** Larvae are attacked by the chytrid fungus *Coelomycidium simulii* (Currie & Adler 1986); the microsporidia *Amblyospora bracteata*, *A. fibrata*, *Janacekia debaisieuxi*, and *Polydispyrenia simulii*; and an unidentified mermithid nematode. The trichomycete fungus *Harpella leptosa* colonizes the larval midgut. We have found first-instar larvae of the chironomid genus *Corynoneura* and larvae and pupae of a chironomid species in the *Eukiefferiella claripennis* group within the cocoons of living pupae, probably using the cocoons as substrate. Predators of larvae include larval rhyacophilid caddisflies (Peterson 1960b).

**Hosts and Economic Importance.** Females feed on cattle (Malloch 1914), blue grouse (Williams et al. 1980), and humans (Peterson 1959a, Currie 1997a), and are attracted to sheep (Jessen 1977).

### *Simulium (Aspathia) iriartei* Vargas, Martínez Palacios & Díaz Nájera

(Figs. 10.68, 10.193, 10.355, 10.471; Map 176)

*Simulium iriartei* Vargas, Martínez Palacios & Díaz Nájera 1946: 144–146, 173–174, Figs. 74–78, 148 (FMPL). Holotype male (slide #3852; INDRE). Mexico, Federal District, Contreras, Las Dínamos, 7 April 1944 (A. Díaz Nájera)

[*Simulium (Prosimulium) hirtipes*: Vargas (1943a) not Fries 1824 (misident.)]

**Taxonomy**

Vargas et al. 1946 (FMPL); Vargas & Díaz Nájera 1948a (F), 1948b (M), 1949 (P), 1957 (FMP); Coscarón et al. 1999 (FMPL).

**Bionomics**

**Habitat.** This species lives in the mountains from central Arizona and New Mexico southward through Mexico. Cool rocky streams up to 8 m wide offer habitat for the immature stages.

**Oviposition.** Unknown.

**Development.** In Arizona, larvae have been collected from March to August. They might be present throughout the year.

**Mating.** Unknown.

**Natural Enemies.** Larvae are hosts of the microsporidium *Amblyospora fibrata* and unidentified mermithid nematodes.

**Hosts and Economic Importance.** Unknown; presumably mammalophilic.

### *Simulium (Aspathia) jacumbae* Dyar & Shannon

(Figs. 10.194, 10.356, 10.472, 10.788; Plate 17b; Map 177)

*Simulium jacumbae* Dyar & Shannon 1927: 44–45, Figs. 113, 114 (M). Holotype* male (slide #28348; USNM). California, San Diego Co., Jacumba Springs, date unknown (E. A. McGregor)

*Simulium guatemalensis* De León 1945 ("1944"): 75–76 (P). Type material unknown to us (depository unknown). Guatemala, Chimaltenango, Yepocapa, date and collector unknown

*Cnephia (Cnephia)* 'nr. *abditoides*' Snyder & Huggins 1980: 31 (Kansas record)

**Taxonomy**

Dyar & Shannon 1927 (M); Stains 1941 (M); Stains & Knowlton 1943 (M); De León 1945 (P); Vargas et al. 1946 (FMPL); Vargas & Díaz Nájera 1948a (F), 1948b (M), 1949 (P), 1957 (FMP); Coleman 1951 (FMP); Dalmat 1955 (FMPL); Wirth & Stone 1956 (FM); Peterson 1958a (FMP), 1960a (FM); Hidalgo Escalante 1959 (PL); Stone & Boreham 1965 (FMPL); Hall 1973 (FMPL), 1974 (L); Onishi et al. 1977 (PL); Peterson & Kondratieff 1995 (FMPL); Coscarón et al. 1999 (FMPL).

**Morphology**

Matsuo & Ochoa 1979 (L abdominal setae).

**Cytology**

The IIIS inversion and the location of the nucleolus are the same as in *S. anduzei*.

**Molecular Systematics**

Pruess et al. 2000 (mtDNA).

**Bionomics**

**Habitat.** A species of the lowlands, *S. jacumbae* ranges from central California east into Nebraska, north into Wyoming, and south into Guatemala. Larvae and pupae live in small, cool, spring-fed streams that sometimes run dry part of the year.

**Oviposition.** One report claims that females oviposit in flight by dipping their abdomens to the surface of the water and that they carry at least 130 rather spherical eggs (Peterson 1959a).

**Development.** The seasonality of *S. jacumbae* in Nebraska is probably typical, with pupae of the first generation appearing in late March and early April, and those of the final generation persisting into October and November. Larvae and pupae can be found during the winter near the Mexican border.

**Mating.** Unknown.

**Natural Enemies.** No records.

**Hosts and Economic Importance.** We have females collected by E. T. Schmidtmann that had fed on calves in Wyoming.

### *Simulium (Aspathia) piperi* Dyar & Shannon

(Figs. 10.195, 10.357, 10.473, 10.588, 10.612, 10.789; Plates 2d, 17c; Map 178)

*Simulium piperi* Dyar & Shannon 1927: 38 (M). Holotype* male (pinned #28347, terminalia on slide; USNM). Washington, King Co., specific location and date unknown (C. V. Piper)

*Simulium sayi* Dyar & Shannon 1927: 40 (F). Holotype* female (pinned #15415, terminalia on slide; USNM). Colorado, Larimer Co., Virginia Dale, on cow, 31 [sic] September 1912 (F. C. Bishopp)

*Simulium knowltoni* Twinn 1938: 53 (M). Holotype* male (on slide #4451; CNC). Utah, Cache Co., Logan Canyon, at light, 23 July 1935 (G. F. Knowlton & C. F. Smith)

*Simulium stonei* Stains & Knowlton 1943: 277 (M). Holotype* male (pinned, terminalia on slide; USNM). Utah, Cache Co., Logan Canyon, at light, 16 September 1939 (G. F. Knowlton & G. S. Stains)

[*Simulium hunteri*: Malloch (1914) (misident. in part, 3 F in type series)]

**Taxonomy**

Dyar & Shannon 1927 (FM); Hearle 1932 (FMP); Twinn 1938 (M); Stains 1941 (FM); Stains & Knowlton 1943 (FM); Coleman 1951 (FMP); Wirth & Stone 1956 (FMP); Vargas & Díaz Nájera 1957 (FMP); Peterson 1958a, 1960a (FMP), 1996 (P); Abdelnur 1966, 1968 (FMPL); Hall 1973 (FMPL), 1974 (L); Corredor 1975 (FM); Fredeen 1985a (FMP); Currie 1986 (PL); Peterson & Kondratieff 1995 (FMPL); Coscarón et al. 1999 (FMPL).

Given its wide distribution (Map 178), abundance, and variation in the configuration and number (9–13) of gill filaments, *S. piperi* is probably a species complex. Most populations along the Rocky Mountains and eastward have a pupal gill with the filaments of the dorsal trunk diverging markedly from a tight cluster of ventral filaments (Fig. 10.473). If these populations represent a separate species, the name *S. sayi* would probably apply, with *knowltoni* and *stonei* as synonyms. Pupae with divergent filaments are also found west of the Rockies; however, pupae with 3 clusters of filaments (Fig. 10.588) are common west of the Rockies and also can be found in the Rockies (e.g., Jasper National Park, Alberta). In addition, some populations have an intermediate gill arrangement. Females have either dark brown or golden hair on the stem vein and costa, again suggesting that multiple species are involved. At the moment, we consider it reckless to recognize separate species, particularly without chromosomal studies.

**Morphology**

Evans & Adler 2000 (spermatheca).

**Cytology**

Golini in Fredeen 1985a (location of nucleolar organizer); Conn et al. 1989 (cytological intermediate linking *S. metallicum* species complex and subgenus *Simulium*).

The IIIS inversion and the location of the nucleolus are the same as in *S. anduzei*. IS has the standard banding sequence of the subgenus *Simulium* of Rothfels et al. (1978).

**Molecular Systematics**

Zhu 1990 (mtDNA), Tang et al. 1996b (mtDNA), Zhu et al. 1998 (mtDNA), Pruess et al. 2000 (mtDNA), Smith 2002 (nuclear and mtDNA).

**Bionomics**

**Habitat.** *Simulium piperi* is common and widely distributed in western North America from southern Canada to southern Mexico. Larvae and pupae cling to vegetation such as watercress (*Nasturtium officinale*) in small, cool, spring-fed streams. Larvae can be found in turbulent and laminar boundary layers (Eymann 1993). They are considered territorial but generally attack only upstream neighbors, presumably to drive them from the path of food delivery (Hart 1986).

**Oviposition.** Little is known of oviposition behavior. One report claims that eggs are deposited in masses on trailing vegetation (Speir 1969), but this assertion requires verification.

**Development.** Several generations are produced yearly, and larvae are present from March to September in Alberta (Currie 1986) and from February to October in Nebraska (Pruess & Peterson 1987). In southern California, larvae can be found during much of the winter (Hart 1986, Harkrider 1988), as well as the summer and fall (Hall 1972, Mohsen & Mulla 1982a, Tietze & Mulla 1989). Larvae can grow in water that contains only dissolved organic matter (Ciborowski et al. 1997).

**Mating.** Unknown.

**Natural Enemies.** Larvae are parasitized by mermithid nematodes such as *Mesomermis flumenalis* (Peterson 1960b, Harkrider 1988). We routinely find larvae infected with the chytrid fungus *Coelomycidium simulii* and the microsporidia *Amblyospora bracteata*, *A. fibrata*, and *Janacekia debaisieuxi*, the last-mentioned parasite forming orange cysts. The trichomycete fungus *Harpella leptosa* inhabits the larval midgut. A possible fungus similar to *Tolypocladium* has been found in 7 larvae from San Dimas Canyon in southern California in December 1983 and January 1984 (J. R. Harkrider, pers. comm.) (Fig. 6.7). Water mites have been found on pupae; larvae are preyed on by larval dytiscid beetles, hydrophilid beetles, hydropsychid and brachycentrid caddisflies, and probably empidid flies (Peterson 1960b).

**Hosts and Economic Importance.** Horses, cattle, and sheep are hosts (Hearle 1932, Jones 1961, Jessen 1977). Humans have been bitten in California (Coleman 1951).

### *Simulium* (*Aspathia*) *puigi* Vargas, Martínez Palacios & Díaz Nájera

(Figs. 10.196, 10.358, 10.474; Map 179)

*Simulium* (*Simulium*) *puigi* Vargas, Martínez Palacios & Díaz Nájera 1946: 153–155, 176, Figs. 106–109 (FMPL). Holotype male (pinned and slide #3816; INDRE). Mexico, Aldea El Oriente, Mariscal, Chiapas, 2350 m elev., 19 November 1944 (J. Parra)

**Taxonomy**

Vargas et al. 1946 (FMPL); Vargas & Díaz Nájera 1948a (F), 1948b (M), 1949 (P), 1957 (FMP); Coscarón et al. 1999 (FMPL).

**Bionomics**

**Habitat.** *Simulium puigi* is known from Costa Rica, Guatemala, Mexico, and a single site in Catron Co., New Mexico, where Moulton (1996) discovered it breeding at 2744 m above sea level in a steep, rocky, woodland, spring-fed stream less than 0.3 m wide.

**Oviposition.** Unknown.

**Development.** Larvae and pupae have been taken from April into July in New Mexico.

**Mating.** Unknown.

**Natural Enemies.** No records.

**Hosts and Economic Importance.** Unknown; presumably mammalophilic.

### *Simulium* (*Aspathia*) *tescorum* Stone & Boreham

(Figs. 10.197, 10.359, 10.475, 10.790; Plates 2e, 17d; Map 180)

*Simulium* (*Simulium*) *tescorum* Stone & Boreham 1965: 164–170 (FMPL). Holotype* male (pinned #68193, pupal exuviae on slide; USNM). California, San Bernardino Co., 5 mi. north of Earp, 16 February 1962 (M. M. Boreham)

*Simulium* 'spp.' Mullens & Dada 1992 (in part) (attraction to bighorn sheep)

**Taxonomy**

Stone & Boreham 1965 (FMPL); Hall 1973 (FMPL), 1974 (L); Lacey & Mulla 1980 (E); Coscarón et al. 1999 (FMPL).

**Morphology**

Lacey & Mulla 1980 (instars).

**Molecular Systematics**

Smith 2002 (nuclear and mtDNA).

**Bionomics**

**Habitat.** This species breeds in small, spring-fed streams of the southwestern desert and in larger streams of the coastal canyons of southern California. Larvae are found in alkaline waters of 10°C–32°C; distribution is possibly restricted by winter water temperatures (Lacey & Mulla 1980). They also occur in streams heavily influenced by effluent from fish farms (Pachón 1999). Adults can be found even in the heart of Death Valley, California.

**Oviposition.** Females deposit batches of 96–275 eggs on trailing vegetation, each egg being placed on its large end (Lacey & Mulla 1980).

**Development.** *Simulium tescorum* has 7 larval instars (Lacey & Mulla 1980). Larvae, pupae, and adults can be collected throughout the year. During the summer, larvae and pupae practically disappear at low elevations (Lacey & Mulla 1980) but can be quite prevalent in cooler canyon streams at higher elevations (Tietze & Mulla 1989). Larvae show no diurnal feeding rhythms; they can clear their guts within 30 minutes at 30°C (Mulla & Lacey 1976a).

**Mating.** Males form mating swarms in the late afternoon or before a storm, about 1–3 m above ground over small streamside trees or toward their leeward side (Stone & Boreham 1965, Lacey & Mulla 1980).

**Natural Enemies.** We have found larvae infected with the microsporidium *Polydispyrenia simulii* and females infested with unidentified mites.

**Hosts and Economic Importance.** Females are attracted to bighorn sheep (Mullens & Dada 1992) and might feed on them. They are also fierce biters of humans. They are attracted initially to the head but feed primarily on the arms and hands, requiring 3–8 minutes to engorge (Lacey & Mulla 1980). Biting occurs predominately in the morning and around dusk and is especially intense during the spring (Lacey & Mulla 1980), sometimes prompting management efforts (Pelsue et al. 1970, Pelsue 1971). An integrated management program was in effect in some areas of the Santa Monica Mountains of southern California during the 1990s and into the 21st century.

### Subgenus *Hemicnetha* Enderlein

*Hemicnetha* Enderlein 1934b: 190 (as genus). Type species: *Hemicnetha mexicana* Enderlein 1934b: 190–191 (= *Simulium paynei* Vargas 1942: 246), by original designation

*Hearlea* Rubtsov 1940b: 116, 121, 124, 126, 135, 138. *Nomen nudum*

*Dyarella* Vargas, Martínez Palacios & Díaz Nájera 1946: 104–106 (as subgenus of *Simulium*). Type species: *Simulium mexicanum* Bellardi 1862: 6, by original designation

*Hearlea* Vargas, Martínez Palacios & Díaz Nájera 1946: 104, 106, 159 (as subgenus of *Simulium*). Type species: *Simulium virgatum canadensis* Hearle 1932: 14–15, by original designation. New Synonym

*Obuchovia* Rubtsov 1947: 90, 105 (as subgenus of *Simulium*). Type species: *Simulium albellum* Rubtsov 1947: 116, by monotypy. New Synonym

*Hagenomyia* Shewell 1959a: 83–84 (as subgenus of *Simulium*). Type species: *Simulium pictipes* Hagen 1880: 305–307, by original designation. Junior homonym of *Hagenomyia* Banks 1911: 8 (created for *H. tristis*, an African neuropteran). New Synonym

*Shewellomyia* Peterson 1975: 111 (substitute name for *Hagenomyia* Shewell 1959a; same type species); *Shewellomia* (subsequent misspelling). New Synonym

**Diagnosis.** Adults: Radius typically without hair dorsobasally. Legs bicolored. Female: Body pollinose, typically grayish in part. Cibarium with or without armature between cornuae. Claws each with or without subbasal tooth. Scutum typically with longitudinal stripes. Genital fork with anteriorly directed apodemes typically well developed; lateral plate with or without ventrally directed tubercle. Male: Gonostylus elongate, with 0–3 apical spinules. Pupa: Gill of 6–16 filaments. Cocoon slipper, shoe, or boot shaped, with or without apertures. Larva: Hypostoma with anterolateral angles rather smoothly rounded, and median tooth typically extended farthest anteriorly; 8 or more setae sublaterally. Postgenal cleft subtriangular, extended one half or more distance to hypostomal groove; subesophageal ganglion typically ensheathed with pigment. Abdomen elongate, typically dark, gradually expanded posteriorly and posteroventrally; segment IX without ventral tubercles. Posterior proleg with more than 100 rows of hooks. Rectal papillae of 3 compound lobes.

**Overview.** The species that we include in the subgenus *Hemicnetha* previously were partitioned among 4 subgenera (*Hearlea*, *Hemicnetha*, *Obuchovia*, *Shewellomyia*). The former subgenera *Hearlea* and *Shewellomyia* correspond to our *S. canadense* and *S. pictipes* species groups, respectively. They are separated by as few as 3 fixed chromosomal inversions (Gibson in Rothfels 1979). The subgenus *Hemicnetha*, as previously recognized, corresponds to our *S. mexicanum* and *S. paynei* species groups. All 14 species formerly placed in the subgenus *Obuchovia* occur in the Palearctic Region; they are retained in the previously recognized *S. auricoma* species group (which now includes members of the previously recognized *S. albellum* species group). As now defined, the subgenus *Hemicnetha* includes 6 species groups and 60 described species in the Nearctic, Neotropical, and Palearctic Regions. We recognize 4 species groups and 10 species in North America north of Mexico.

The species of *Hemicnetha* are among the largest black flies on the planet. The elongate larvae with their sticky salivary-gland secretions are ideally suited for life in the swiftest waters, where they usually cling, often in large clusters, to rocks. Most species are probably multivoltine. Females are mammalophilic, feeding primarily on large hosts, but they are not particularly pestiferous.

#### *Simulium* (*Hemicnetha*) *canadense* Species Group, New Status

**Diagnosis.** Female: Cibarium typically without armature between cornuae. Claws each with small subbasal tooth. Hypogynial valve short, with posterior and inner margins slightly emarginate. Genital fork with lateral plate of each arm lacking ventrally directed tubercle. Anal lobe typically elongate, subrectangular, about half sclerotized and polished. Male: Gonostylus long, slender, with small basal lobe and 1 apical spinule. Pupa: Gill typically of 2 or 3 (occasionally more) swollen trunks, sometimes with short pointed basal projections. Cocoon slipper or shoe shaped, without apertures.

**Overview.** Only 1 of the 20 described members of this species group is found north of Mexico.

#### *Simulium* (*Hemicnetha*) *canadense* Hearle

(Figs. 4.51, 10.198, 10.360, 10.476, 10.613, 10.620, 10.647, 10.791; Plate 17e; Map 181)

*Simulium virgatum canadensis* Hearle 1932: 14–15 (FM). Holotype* male (pinned #3454; CNC). British Columbia, Kamloops, Lanes Creek, 6 August 1931 (T. K. Moilliett & R. T. Turner); *canidense* (subsequent misspelling)

*Simulium* (*Simulium*) *fraternum* Twinn 1938: 53–54, Fig. 6 (M). Holotype* male (pinned #4452, terminalia on slide; CNC). Utah, Davis Co., Farmington, 4 September 1934 (G. F. Knowlton & C. F. Smith)

*Simulium* 'sp.' Merritt et al. 1978b: 203 (L)

[*Simulium virgatum*: Hearle (1929) not Coquillett 1902 (misident.)]

**Taxonomy**

Hearle 1929 (FMPL), 1932 (FM), 1935 (FMP); Twinn 1938 (M); Stains 1941 (FM); Stains & Knowlton 1943 (FM); Vargas 1945a (MP); Vargas et al. 1946 (ML); Vargas & Díaz Nájera 1948a (F), 1948b (M), 1949 (P), 1957 (FMP); Coleman 1951 (FMP); Wirth & Stone 1956 (FMP); Peterson 1958a (FMP), 1959a (E), 1960a (FMP), 1981 (FM), 1996 (FMPL); Díaz Nájera & Vulcano 1962b (L); Hall 1973 (FMPL), 1974 (L); Corredor 1975 (FM); Speir 1975 (E); Merritt et al. 1978b (L); Peterson & Kondratieff 1995 (FMPL).

We have 2 females of *S. canadense* collected by J. K. Moulton in a light trap at Lake Tohopekaliga, Rt. 525A, Kissimmee, Florida (17 December 1991). The occurrence of

*Hemicnetha* in central Florida is odd and begs additional prospecting in the area.

**Cytology**

Gibson in Rothfels 1979.

Three fixed inversions separate the chromosomes of *S. canadense* from the standard banding sequence given by Bedo (1975a) for the *S. pictipes* group (Gibson in Rothfels 1979). Our cursory analysis indicates that the inversion differences are in chromosome arms IL and IIL; the nucleolar organizer is located near the base of the IS arm. Given its broad geographic range, *S. canadense* might be a species complex.

**Molecular Systematics**

Brockhouse & Tanguay 1996 (salivary-gland proteins).

**Bionomics**

**Habitat.** *Simulium canadense* is distributed from British Columbia to southern Mexico, east into the Rocky Mountains and Black Hills of South Dakota, with an unusual record from central Florida. The immatures inhabit rocky streams several meters in width. Larvae can be found at water temperatures from 0°C to more than 20°C. Water velocity significantly affects larval and pupal density (Mohsen & Mulla 1982a). Larvae occupy microhabitats with high turbulence (i.e., average Froude number = 1.6) (Eymann 1993). In some streams, larvae and pupae are found almost exclusively on vegetation (Mohsen & Mulla 1982a), but in other flows they are found primarily on rocks (Hall 1972). Water flowing around the cocoon forms a pair of vortices, one between the gill filaments and the other anterior to the filaments (Eymann 1991b).

**Oviposition.** Females generally emerge with immature eggs. For the first generation, they are probably often autogenous. Oviposition begins in the afternoon and lasts until after sunset. Females, each carrying at least 336 eggs, repeatedly dart from the air to moist rocks or the surface of smooth water to release 1 or more eggs, blunt end first; eventually, loose irregular egg masses are formed on rocks (Peterson 1959a). In the wash zone of a single rock, we have seen masses up to 60 cm long, 8 cm high, and 1–2 mm deep, containing nearly 3.5 million eggs.

**Development.** This species is generally multivoltine. In some streams, however, it passes through 1 generation annually (Speir 1969). Larvae overwinter throughout most of the range, sometimes spending the winter under the cover of ice (Hearle 1929, Mohsen & Mulla 1982a). In some Oregon streams, however, eggs overwinter and larvae begin hatching in April (Speir 1975). Six instars have been reported, with mean larval development requiring 10.5 weeks at 7°C–21°C; the pupal stage lasts 3–5 days at 15°C–16°C (Speir 1975). Larvae require about an hour to clear the gut at water temperatures of 7°C–14°C; gut contents reflect the proportions of materials suspended in the water column (Speir 1975).

Adults have been kept alive in the laboratory for 13 days but probably can live longer; they exhibit a sex ratio that is slightly female biased (Speir 1975). About 21% of body weight is lost between the final-instar larva and the adult (Speir & Anderson 1974). Females are longer and heavier than males (Speir 1975).

**Mating.** Unknown.

**Natural Enemies.** Larvae are attacked by mermithid nematodes of the genus *Gastromermis* (Speir 1975, Mohsen & Mulla 1984), the microsporidium *Amblyospora fibrata*, and the chytrid fungus *Coelomycidium simulii*. They are preyed on by larvae of hydropsychid and rhyacophilid caddisflies (Peterson 1960b).

**Hosts and Economic Importance.** Females attack unspecified "larger domestic animals" (Hearle 1932). A few blood-fed specimens have been collected near blue grouse (Williams et al. 1980).

### *Simulium* (*Hemicnetha*) *mexicanum* Species Group

**Diagnosis.** Female: Cibarium with or without armature between cornuae. Claws each with small subbasal tooth. Hypogynial valve elongate. Anal lobe large, mostly membranous. Male: Gonostylus broad, flat, with at least 1 lateral margin sinuous and 1 apical spinule. Ventral plate without enlarged lip extended beyond posterior margin. Pupa: Gill of 8–16 filaments. Cocoon shoe or boot shaped, typically tightly woven, not corbicular.

**Overview.** This species group consists of 13 described species, only 1 of which is found north of Mexico.

#### *Simulium* (*Hemicnetha*) *freemani* Vargas & Díaz Nájera

(Figs. 10.199, 10.361, 10.477, 10.792; Plate 17f; Map 182)

*Simulium* (*Dyarella*) *freemani* Vargas & Díaz Nájera 1949: 289–292, Figs. 19–28 (FMP). Holotype male (pinned #3964, legs, wing, terminalia on slide, pupal exuviae in alcohol vial; INDRE). Mexico, State of Oaxaca, San Pablo, Etla, 21 January 1949 (F. Reyes Salgado)

**Taxonomy**

Vargas & Díaz Nájera 1949 (FMP), 1957 (FMP); Peterson et al. 1988 (P); Ibáñez Bernal 1992 (FMP).

The distinctive larva is illustrated here for the first time (Fig. 10.792, Plate 17f).

**Cytology**

The chromosomes of larvae that we examined from Box Elder Co., Utah, were nonchromocentric and had a subterminal nucleolar organizer in IS, leaving a small terminal piece of the chromosome detached from the remainder of the arm.

**Bionomics**

**Habitat.** This poorly known species is found from northern Utah through Arizona and New Mexico southward into Costa Rica. The immature stages occupy swift rocky streams several meters wide.

**Oviposition.** Unknown.

**Development.** In the United States, larvae and pupae have been collected from March through August and are probably present throughout much of the year.

**Mating.** Unknown.

**Natural Enemies.** No records.

**Hosts and Economic Importance.** Unknown; presumably mammalophilic.

### *Simulium* (*Hemicnetha*) *paynei* Species Group

**Diagnosis.** Female: Cibarium with or without armature between cornuae. Claws each with small subbasal tooth. Hypogynial valve elongate. Genital fork with or without lateral plate of each arm bearing ventrally directed tubercle. Anal lobe large, mostly membranous. Male: Gonostylus broad, flat, with at least 1 lateral margin sinuous and 1–3 apical spinules. Ventral plate with lip enlarged and extended well beyond posterior margin. Pupa: Gill of 6–15 filaments. Cocoon boot shaped, tightly woven or corbicular.

**Overview.** This group, as currently recognized, has 10 species distributed from southern Canada to Argentina. It is common and widespread west of the Mississippi River. We recognize 5 species north of Mexico. The group is in great need of revision. We have combined the *S. brachycladum* and *S. paynei* species groups of Coscarón (1987), having found no compelling characters that define the 2 as separate groups.

### *Simulium* (*Hemicnetha*) *bricenoi* Vargas, Martínez Palacios & Díaz Nájera

(Figs. 10.200, 10.362, 10.529, 10.648, 10.793; Plate 18a; Map 183)

*Simulium* (*Dyarella*) *bricenoi* Vargas, Martínez Palacios & Díaz Nájera 1946: 115–118, 177, Figs. 8–15 (FMPL). Holotype male (slide #3846 with wing, legs, terminalia; INDRE). Mexico, State of Mexico, Dos Ríos, 2645 m elev., 12 March 1944 (A. Díaz Nájera); *briseñoi* (subsequent misspelling)

?*Simulium virgatum* 'C & D' Muhammad 1988: 66–70, Figs. 6, 16, 18 (C)

*Simulium* (*Hemicnetha*) *wirthi* Peterson & Craig 1997 ("1996"): 212–220 (FMPL). Holotype* male (pinned, pupal exuviae in glycerin vial below; USNM). New Mexico, Grant Co., Gallinas Creek, Railroad Canyon, Rt. 152, 1890 m elev., 15 May 1992 (B. V. Peterson & M. E. Craig). New Synonym

**Taxonomy**

Vargas et al. 1946 (FMPL); Vargas & Díaz Nájera 1948a (F), 1948b (M), 1949 (P), 1957 (FMP); Peterson et al. 1988 (P); Ibáñez Bernal 1992 (FMPL); Coscarón & Ibáñez Bernal 1993 (PL); Peterson & Craig 1997 (FMPL).

We can find no convincing differences to separate *S. wirthi* from *S. bricenoi*. Peterson and Craig (1997) stated that the branching patterns of the pupal gills and the terminalia differed, but did not explain these differences. They did, however, point out that the windows of the cocoon were poorly developed in *S. bricenoi*. We attribute these minor differences to intraspecific variation.

**Cytology**

Muhammad 1988.

Larvae from Tanque Verde Canyon, Pima Co., Arizona, are chromocentric and have a Y-linked inversion in the IIL arm, but we have not studied the chromosomes in detail. Because *S. bricenoi* is the only known chromocentric member of the subgenus *Hemicnetha* north of Mexico and because the sex differential segment is in chromosome arm IIL, we tentatively associate the name *S. bricenoi* with *S. virgatum* 'C & D' of Muhammad (1988). We have not, however, seen material from western Texas, where 'C & D' were collected and, therefore, cannot confirm this association.

**Bionomics**

**Habitat.** *Simulium bricenoi* ranges from southern Arizona and New Mexico southward to Morelos, Mexico. It breeds in cool, swift mountain streams about 1–3 m wide, with sand and rock bottoms and intermittent or perennial flow.

**Oviposition.** Unknown.

**Development.** Larvae, pupae, and adults have been collected from December through May in the United States. One pupa was taken in October (Peterson & Craig 1997).

**Mating.** Unknown.

**Natural Enemies.** No records.

**Hosts and Economic Importance.** Hosts are unknown, although females are presumably mammalophilic. One female was recorded feeding on a human in the laboratory (Vargas et al. 1946).

### *Simulium* (*Hemicnetha*) *hippovorum* Malloch

(Figs. 10.201, 10.363, 10.478, 10.589; Plate 2f; Map 184)

*Simulium hippovorum* Malloch 1914: 28–29, Plate 2 (Fig. 12) (F). Holotype* female (pinned #15407, abdomen and right hind leg on separate slides; USNM). Mexico, Sierra Madre, Chihuahua, head of Río Piedras Verdes, ca. 2225 m elev., in ear of horse, 27 July, year unknown (C. H. T. Townsend). Revalidated

*Simulium virgatum* 'complex' Eymann 1991c: 82; 1993: 62–64 (L microhabitat)

[*Simulium virgatum*: Dyar & Shannon (1927) not Coquillett 1902 (misident. in part, material from California)]

[*Simulium* (*Dyarella*) *virgatum*: Stone (1948) not Coquillett 1902 (misident. in part, material from California and Washington)]

[*Simulium virgatum*: Coleman (1951) not Coquillett 1902 (misident.)]

[*Simulium* (*Dyarella*) *virgatum*: Wirth & Stone (1956) not Coquillett 1902 (misident.)]

[*Simulium virgatum*: Lichtwardt (1967, 1972) not Coquillett 1902 (misident.)]

[*Simulium virgatum*: Hall (1972, 1973, 1974) not Coquillett 1902 (misident.)]

[*Simulium virgatum*: Kramer & Mulla (1980) not Coquillett 1902 (misident.)]

[*Simulium* (*Hemicnetha*) *virgatum*: Mohsen (1981) not Coquillett 1902 (misident.)]

[*Simulium* (*Hemicnetha*) *virgatum*: Peterson (1981, 1996) not Coquillett 1902 (misident.)]

[*Simulium* (*Hemicnetha*) *virgatum*: Mohsen & Mulla (1982a) not Coquillett 1902 (misident.)]

[*Simulium virgatum*: Hemphill & Cooper (1983) not Coquillett 1902 (misident.)]

[*Simulium virgatum*: Dudley et al. (1986, 1990) not Coquillett 1902 (misident.)]

[*Simulium virgatum*: Peterson & Lichtwardt (1987) not Coquillett 1902 (misident.)]

[*Simulium virgatum*: Harkrider (1988) not Coquillett 1902 (misident.)]

[*Simulium virgatum*: Hemphill (1988, 1989, 1991) not Coquillett 1902 (misident.)]

[*Simulium virgatum*: Reeves & Milby (1990) not Coquillett 1902 (misident.)]

[*Simulium virgatum*: Eymann (1991b) not Coquillett 1902 (misident.)]
[*Simulium* (*Hemicnetha*) *virgatum*: Brockhouse et al. (1993) not Coquillett 1902 (misident.)]
[*Simulium virgatum*: Anderson & Yee (1995) not Coquillett 1902 (misident.)]
[*Simulium virgatum*: Yee & Anderson (1995) not Coquillett 1902 (misident.)]
[*Simulium* (*Hemicnetha*) *virgatum*: Brockhouse & Tanguay (1996) not Coquillett 1902 (misident.)]
[*Simulium virgatum*: Pachón (1999) not Coquillett 1902 (misident.)]

**Taxonomy**

Malloch 1914 (F); Coleman 1951 (FMP); Wirth & Stone 1956 (FMP); Hall 1973 (FMPL), 1974 (L); Peterson 1981, 1996 (FM).

We apply the name *S. hippovorum* to the species found from Vancouver Island south through the Pacific states. This decision conserves names in the *S. paynei* species group, pending a full revision of the group complete with chromosomal study. Assignment of the name is reasonable because the scutum of the holotype, like that of the material from the Pacific region, is gray and black rather than fuscous or reddish, as in *S. paynei* and *S. virgatum*.

**Molecular Systematics**

Brockhouse & Tanguay 1996 (salivary-gland proteins).

**Bionomics**

**Habitat.** This species ranges from Vancouver Island, British Columbia, southward through the Pacific Coastal states into the Sierra Madre of Mexico. Larvae and pupae are found in swift rocky flows. They readily colonize natural substrates, including those coated with periphyton (Pachón 1999), but anchor themselves especially to rocks. Microhabitat velocities are about 50–210 cm/sec (Eymann 1993). Larvae are often found with those of *Hydropsyche* caddisflies but are weaker competitors for space and can be killed by the caddisflies (Hemphill & Cooper 1983, Hemphill 1988). Competition with caddisfly larvae is highly seasonal and is mediated by disturbances such as floods (Hemphill 1991). Larvae also compete with macroalgae for space (Dudley et al. 1986) and are involved in interference competition with blephariceridlarvae (Dudley et al. 1990). Pupation often occurs in clumps. The fenestrated cocoons form vortices that bathe the gill filaments (Eymann 1991b).

**Oviposition.** Females deposit masses of eggs in the splash zone, an area within 4 cm of the water's edge (Hemphill 1989).

**Development.** All life stages are present year-round in southern California, where larval development, including possibly 8 instars, requires 45–76 days in winter and 29–49 days in summer at mean water temperatures of 12.5°C and 19.5°C, respectively (Hemphill 1989). Northward, adults are on the wing at least from March through October.

**Mating.** Unknown.

**Natural Enemies.** Larvae are infected by the mermithid nematode *Mesomermis flumenalis* but are not highly susceptible hosts (Harkrider 1988). The trichomycete fungi *Genistellospora homothallica*, *Harpella melusinae*, *Pennella angustispora*, and *Smittium simulii* have been found in the larval guts (Lichtwardt 1967, 1972; Peterson & Lichtwardt 1987). We have found pupae destroyed by larvae of the chironomid genus *Cardiocladius*, which subsequently pupated in the host cocoon.

**Hosts and Economic Importance.** Females feed on horses, especially during the morning, causing discomfort to the animals (Hall 1972). Nearly all feeding occurs in the ears, and landing rates are greatest on ears oriented horizontally (Anderson & Yee 1995, Yee & Anderson 1995). Females are not pests of humans (Kramer & Mulla 1980).

### ***Simulium* (*Hemicnetha*) *paynei* Vargas**

(Figs. 10.202, 10.364, 10.365; Map 185)

*Simulium* (*Simulium*) *paynei* Vargas 1942: 246 (substitute name for *mexicanum* Enderlein 1934b); *peynei* (subsequent misspelling)
*Hemicnetha mexicanum* Enderlein 1934b: 190–191 (FM). Lectotype female (pinned; presumably ZMHU, but not listed as such by Werner 1996). Mexico, specific location and date unknown, 1883 (D. Bilimek). Junior secondary homonym of *S. mexicanum* Bellardi 1862: 6
*Simulium mathesoni* Vargas 1943b: 360–363, Figs. 39–43 (MP). Holotype male (curatorial status unknown to us; INDRE). Mexico, State of Morelos, Temixco, 1400–1500 m elev., 21 November 1943 (A. Martínez Palacios)
*Simulium bilimekae* Smart 1944b: 132 (unjustified substitute name for *mexicanum* Enderlein 1934b)
*Simulium* (*Dyarella*) *acatenangoensis* Dalmat 1951: 31–38 (FMP). Holotype* male (3 slides, pupal exuviae in alcohol vial; USNM). Guatemala, Department of Chimaltenango, Acatenango, Río Ladrillera, Finca La Esperanza Pérez, 25 November 1948 (A. Calí & J. Marroquín)
*Simulium* (*Hemicnetha*) *conviti* Ramírez-Pérez & Vulcano 1973: 376–379 (FMP). Holotype male (pinned; IDSV). Venezuela, State of Bolívar, Gran Sabana, date and collector unknown
*Hemicnetha* 'undescribed related species . . . from Turner Falls' Rothfels 1979: 525 (Oklahoma record)
*Simulium virgatum* 'B' Muhammad 1988: 62–64, Figs. 2, 7–9 (C)
[*Simulium virgatum*: Bedo (1973, p. 6) not Coquillett 1902 (misident.)]
[*Simulium virgatum*: Reisen (1974a, 1974b, 1975b) not Coquillett 1902 (misident.)]
[*Simulium* (*Hemicnetha*) *virgatum*: Reisen (1975a, 1977) not Coquillett 1902 (misident.)]
[*Simulium* (*Hemicnetha*) *virgatum*: Peterson & Kondratieff (1995) not Coquillett 1902 (misident. in part, illustrations)]

**Taxonomy**

Enderlein 1934b (FM); Vargas 1943b (MP); Vargas et al. 1946 (ML); Vargas & Díaz Nájera 1948a (F), 1948b (M), 1949 (P), 1957 (FMP); Dalmat 1951 (FMP), 1955 (FMPL); Ramírez-Pérez & Vulcano 1973 (FMP); Chubareva et al. 1976 (FM); Ramírez-Pérez 1983 (FMP); Ibáñez Bernal 1992 (FMPL); Peterson & Kondratieff 1995 (FMPL).

Given its ecological and morphological variation and wide geographic distribution, *S. paynei* is probably a complex of 2 or more species. Within the United States, for example, differences occur in the male terminalia. Figure

10.364 depicts the genitalia of a male from Murray Co., Oklahoma, whereas Fig. 10.365 shows the genitalia of a male from southern Arizona. A tandem chromosomal-morphological study throughout the range of *S. paynei* is needed to resolve this issue.

Coscarón (1987) indicated that *S. falculatum* (Enderlein 1929b) from Mexico (Veracruz, Bora del Monte) is possibly a synonym of *S. paynei*, based on his examination of the holotype female (pinned; ZMHU) in 1985. Our examination of the holotype suggests that *S. falculatum* is not conspecific with *S. paynei* nor with any nominal member of the subgenus *Hemicnetha* north of Mexico. The long claws, shiny frons and abdomen, and dark forelegs are not characteristic of any of the species in our treatment. We, therefore, remove the name *falculatum* and its unnecessary substitute name, *coffeae* Vargas 1945b, from synonymy with *paynei*. The absence of terminalia on the type of *S. falculatum*, however, impedes further diagnosis. We suggest that the name *falculatum*—among the older names in the subgenus *Hemicnetha*—represents a valid species.

**Cytology**

Chubareva et al. 1976, Muhammad 1988 (number of gill filaments incorrectly given as 15).

*Simulium paynei* and *S. virgatum* 'B,' both described cytologically by Muhammad (1988), differ by about 3 fixed inversions, again suggesting the likelihood that *S. paynei*, as treated here, is a species complex.

**Morphology**

Reisen 1974a, 1975a (instars).

**Bionomics**

**Habitat.** This species is distributed from Arizona, New Mexico, and Oklahoma southward through Central America to Peru. No other North American black fly extends so far south. In the United States, the immature stages are found in swift rocky sections of lowland rivers typically with stretches of flat sedimentary rock or travertine. Most larvae are found at depths of less than 10 cm (Reisen 1977). In Guatemala, they occupy smaller streams, from less than a meter to about 2 m wide, that flow on volcanic slopes (Dalmat 1951). Larval drift rates are highest during the afternoon (Reisen 1977).

**Oviposition.** Egg laying is often communal, with swarms of females forming in the evening over water-splashed rocks near riffles and waterfalls; females drop from the swarm to oviposit on the substrate, eventually forming dense masses of eggs (Reisen 1974a). Field-collected females carry as many as 314 eggs (Reisen 1974a).

**Development.** Larvae hatch within 4 days at 25°C–27°C, pass through 6 instars, and pupate in as few as 21 days; the pupal stage lasts 4–5 days (Reisen 1974a, 1975a). Emergence from the pupa occurs during daylight (Reisen 1977). Larvae can be found year-round, although oviposition does not occur from November to March (Reisen 1974a).

**Mating.** In Oklahoma, males form swarms over riffles and seize entering females; pairs leave the swarm and mate on the ground for a matter of seconds (Reisen 1974a).

**Natural Enemies.** We have found Texan larvae infected with the microsporidium *Polydispyrenia simulii*. Guatemalan larvae are hosts of the microsporidia *Amblyospora bracteata* and *A. fibrata*, a chytrid fungus in the genus *Coelomycidium*, and unidentified mermithid nematodes (Takaoka 1980). Larvae in Mexico are hosts of an iridescent virus (Hernández et al. 2000).

**Hosts and Economic Importance.** A few females have been collected from cattle and horses in Guatemala (Dalmat 1955).

### *Simulium* (*Hemicnetha*) *solarii* Stone

(Figs. 4.25, 10.203, 10.366, 10.367, 10.479, 10.590, 10.794, 10.795; Plate 18b, c; Map 186)

*Simulium* (*Dyarella*) *solarii* Stone 1948: 402–404, Figs. 2, 4, 6, 8, 10 (FMP). Holotype* male (slide #58956; USNM). Texas, Menard Co., San Saba River, 23 April 1941 (A. Stone)

*Simulium*, Davis & Cook 1985: 113, 119–120 (hosts of water mites)

[*Simulium virgatum*: Painter & Griffen (1937a, 1937b) not Coquillett 1902 (misident.)]

[*Simulium virgatum*: Griffen (1939) not Coquillett 1902 (misident.)]

[*Simulium virgatum*: Painter (1939) not Coquillett 1902 (misident.)]

[*Simulium virgatum*: Stains (1941) not Coquillett 1902 (misident. in part, Figs. 90, 91)]

[*Simulium virgatum*: Stains & Knowlton (1943) not Coquillett 1902 (misident. in part, Figs. 90, 91)]

[*Simulium* (*Dyarella*) *virgatum*: Vargas et al. (1946) not Coquillett 1902 (misident.)]

[*Simulium* (*Dyarella*) *virgatum*: Vargas & Díaz Nájera (1948a, 1948b, 1949) not Coquillett 1902 (misident.)]

[*Prosimulium* 'sp.': Sites & Nichols (1990) not Roubaud 1906 (misident.)]

**Taxonomy**

Stains 1941 (M); Stains & Knowlton 1943 (M); Vargas et al. 1946 (ML); Stone 1948 (FMP); Vargas & Díaz Nájera 1948a (F), 1948b (M), 1949 (P), 1957 (FMP); Peterson et al. 1988 (P); Ibáñez Bernal 1992 (FMPL).

The larval head pattern varies from dark brown with yellowish head spots (Fig. 10.794) to yellowish with brown head spots (Fig. 10.795). Dark brown and pale brown bodies, respectively, are associated with these head patterns (Plate 18b, c). Both larval forms occur in the same streams throughout the known range of *S. solarii*. Slight variation in the female terminalia is illustrated in Figs. 4.25 and 10.203, and in the male genitalia in Figs. 10.366 and 10.367.

**Cytology**

Painter & Griffen 1937a, 1937b; Griffen 1939; Painter 1939; Muhammad 1988.

*Simulium solarii* was the first North American species to have its polytene chromosomes studied (Painter & Griffen 1937a, 1937b). Originally reported as *S. virgatum*, the Painter and Griffen specimens, now in the USNM, later were identified by Stone (1948) as *S. solarii*.

Larvae with both types of head pattern have a Y-linked inversion in the base of chromosome arm IIIL. A careful band-by-band chromosomal comparison between larvae of the 2 color forms is needed to corroborate our opinion that they represent a single species.

**Bionomics**

**Habitat.** This species is known from Texas, where it is particularly common in the Balcones Escarpment, west to southeastern New Mexico and south to San Luis Potosí and Mexico State, Mexico. The immature stages are found in rocky streams and rivers up to 30 m wide.

**Oviposition.** Unknown.

**Development.** Larvae have been collected throughout the year, and adults have been captured from February into August.

**Mating.** Unknown.

**Natural Enemies.** Larvae are hosts of the microsporidium *Polydispyrenia simulii*. They have been used as prey to rear the naucorid bug *Ambrysus lunatus* (Sites & Nichols 1990). Adults are attacked by parasitic larvae of the water mite *Sperchon texana* (Davis & Cook 1985).

**Hosts and Economic Importance.** A female has been taken from the underside of a horse (Stone 1948).

### *Simulium* (*Hemicnetha*) *virgatum* Coquillett

(Figs. 10.204, 10.368; Plate 18d; Map 187)

*Simulium virgatum* Coquillett 1902: 97 (FM). Holotype* male (pinned #6183, terminalia on slide; USNM). New Mexico, San Miguel Co., Las Vegas Hot Springs, 4 August 1901 (H. S. Barber)

*Simulium cinereum* Bellardi 1859: 13–14 (FM). Type material unknown to us (curatorial status unknown; IMZ). Mexico, Morelia, date unknown (H. L. F. de Saussure). Junior primary homonym of *S. cinereum* Macquart 1834: 174–175

'species of *Simulium*' Townsend 1893: 45–48 (PL)

*Simulium tephrodes* Speiser 1904: 148 (substitute name for *cinereum* Bellardi 1859)

*Simulium rubicundulum* Knab 1915a ("1914"): 178–179 (F). Holotype* female (pinned #19112, abdomen and hind leg on slide; USNM). Mexico, Córdoba, 17 December 1907 (F. Knab)

*Simulium virgatum chiapanense* Hoffman 1930b: 293–297 (F). Type material unknown to us (depository unknown). Mexico, Chiapas, Santa Anita District, 650 m elev., July 1930 (C. C. Hoffman); *chiapense* (subsequent misspelling)

?[*Simulium mexicanum*: Muhammad (1988) not Bellardi 1862 (misident.)]

**Taxonomy**

Bellardi 1859 (FM); Townsend 1893 (PL); Coquillett 1902 (FM); Johannsen 1902, 1903b (FMPL); Emery 1913 (FMPL); Malloch 1914 (FM); Knab 1915a (F); Dyar & Shannon 1927 (FM); Hoffman 1930b (F), 1931 (L); Bequaert 1934 (FP); Hearle 1935 (FM); Fairchild 1940 (FP); Stains 1941 (F); Stains & Knowlton 1943 (F); Vargas et al. 1946 (L); Stone 1948 (FMP); Vargas & Díaz Nájera 1948a (F), 1948b (M), 1949 (P), 1957 (FMP); Dalmat 1951 (FMP), 1955 (FMPL); Peterson 1958a, 1960a (FMP); Corredor 1975 (FM); Peterson et al. 1988 (P); Ibáñez Bernal 1992 (FMPL); Coscarón & Ibáñez Bernal 1993 (PL).

*Simulium virgatum* is probably a complex of species, requiring morphological and chromosomal study throughout its range. The color of the female scutum is highly variable, ranging from reddish orange, as in the type specimen of *S. rubicundulum*, to chestnut. This variation might be related to temperature and therefore might differ with season and latitude, or it might be associated with sibling species.

**Cytology**

Muhammad 1988.

**Bionomics**

**Habitat.** This monstrous species occupies a longitudinal strip from South Dakota through Guatemala and Panama. The immature stages live in rocky streams 1–8 m wide, with temperatures sometimes exceeding 25°C. In North America, these streams usually flow through low, somewhat arid regions. The larvae and pupae attach themselves to stones and bedrock.

**Oviposition.** Unknown.

**Development.** In the northern portion of the range, large larvae can be found at least from March through August. In Arizona, larvae might be present during most of the year. We have adults from southern Arizona that emerged in early February.

**Mating.** Unknown.

**Natural Enemies.** The trichomycete fungus *Harpella leptosa* colonizes the larval midgut.

**Hosts and Economic Importance.** Females in Guatemala attack dogs, cats, donkeys, mules, cattle, goats, and sheep, but they are especially common on horses, particularly in the ears and on the belly (Dalmat 1955). No feeding records are available for the continent north of Mexico.

### *Simulium* (*Hemicnetha*) *pictipes* Species Group, New Status

**Diagnosis.** Female: Cibarium without armature between cornuae. Tibiae without bright white patches; claws toothless. Sternite VII with fringe of long hairs posteriorly. Hypogynial valve short, with inner margins concave. Genital fork with lateral plate of each arm lacking ventrally directed tubercle. Anal lobe subquadrate, with large polished sclerotized area. Male: Gonostylus nearly 4 times as long as wide, without apical spinule. Ventral plate in ventral view with deep medial cleft posteriorly. Pupa: Gill of 9 rather swollen filaments. Cocoon boot shaped, cribriform. Chromosomes: Standard banding sequence as given by Bedo (1975a).

**Overview.** The *S. pictipes* group contains 3 large species endemic to the United States and Canada. Cytological and molecular data indicate that *S. innoxium* and *S. pictipes* are more closely related to one another than either is to *S. claricentrum* (Bedo 1975a, Brockhouse 1991). The larvae and pupae form clusters in the rapids of large streams and rivers. The females feed on large mammals.

### *Simulium* (*Hemicnetha*) *claricentrum* Adler

(Figs. 10.369, 10.591, 10.796; Plate 2g, 18e; Map 188)

*Simulium claricentrum* Adler 1990: 437–442 (FMPLC). Holotype* male (pinned, pupal exuviae and larval head capsule in glycerin vial below; USNM). Pennsylvania, Erie Co., Northeast (town), Sixteenmile Creek, junction Washington Street and Shadduck Road, 42°11′N, 79°50′W, 10 August 1988 (P. H. & C. R. L. Adler)

*Simulium pictipes* 'A' Bedo 1973: 11–39, 46–49, Figs. 1, 3, 5–9, 14–16, 20–22, 25, 27–30, 42 (C); 1975a: 1147–1163 (C)
(*Hagenomyia*) 'species A' Reisen 1974a: 72 (bionomics)
*Simulium* 'species A' Reisen 1974b: 275–276 (bionomics)
*Simulium* 'sp. a' Reisen 1975b: 27 (bionomics)
[*Simulium pictipes*: Bell (1960) not Hagen 1880 (misident. in part, material from Arkansas)]
[*Simulium* (*Hagenomyia*) *pictipes*: Stone (1964) not Hagen 1880 (misident. in part, PL)]

**Taxonomy**

Stone 1964 (PL), Adler & Kim 1986 (L), Adler 1990 (FMPL).

**Morphology**

Bell 1960 (instars), Reisen 1975a (instars).

**Cytology**

Bedo 1973, 1975a; Adler 1990.

**Molecular Systematics**

Brockhouse 1991 (rDNA).

**Bionomics**

**Habitat.** *Simulium claricentrum* ranges from the southern shore of Lake Erie southwest into Oklahoma. Larvae and pupae aggregate in mosslike clumps in swift, smooth-bottomed streams of shale, siltstone, or travertine, with small cascades. Larvae have been collected less than 10 cm beneath the water in flows 10–20 m wide and 10°C–29°C. Larval drift, particularly of early instars, peaks during the night (Reisen 1977).

**Oviposition.** Females are autogenous to some degree (Bell 1960). While standing on a substrate or darting into the water, typically near the lip of a cascade, they deposit their eggs, placing 1 or 2 on the substrate (Bell 1960). Both methods of oviposition often result in the females, each carrying about 200–450 eggs, forming dense communal masses of up to 750,000 eggs (Reisen 1974a). Females typically oviposit at dusk (Reisen 1974a) and can become stuck in the gelatinous matrix of the egg masses (Bell 1960).

**Development.** Larvae hatch in about 4 days at 25°C–27°C, and the larval and pupal stages last 21–35 and 4–5 days, respectively (Reisen 1974a, 1975a). At about 11°C, larvae hatch in 6–14 days, with those at the top of the egg mass hatching first; larvae require about 5 weeks to mature and pupae 4–6 days at an average stream temperature of about 13°C (Bell 1960). Six to 7 larval instars are produced (Bell 1960, Reisen 1975a). The species overwinters as eggs in Pennsylvania, with the first larvae appearing by May (Adler 1990). In the southern half of the range, larvae overwinter. Four generations are produced annually in Arkansas (Bell 1960). Larvae and pupae can be found year-round in Oklahoma, but although adults emerge throughout the year, the females evidently do not oviposit from about November to March (Reisen 1977). Many pupae die during the winter when they become overgrown by algae and cemented into the substrate by travertine deposition (Reisen 1977). Larval abundance is highest in the spring and is correlated with periphyton abundance (Reisen 1977). Adults emerge during the day, especially in the morning (Reisen 1977).

**Mating.** Mating swarms form over riffles. Females enter these swarms from downstream, and coupled pairs fall to the ground, where copulation lasts only seconds (Reisen 1974a). Males in the swarms fly in an up-and-down manner (Bell 1960).

**Natural Enemies.** Larvae are attacked by the microsporidium *Polydispyrenia simulii* (Adler 1990). The trichomycete fungi *Genistellospora homothallica* and *Harpella melusinae* have been found in the guts of larvae.

**Hosts and Economic Importance.** *Simulium claricentrum* is probably mammalophilic, although no host records exist. It does not pester humans.

### *Simulium* (*Hemicnetha*) *innoxium* Comstock & Comstock

(Figs. 10.205, 10.370, 10.480, 10.534, 10.592, 10.649, 10.797; Plate 18f; Map 189)

*Simulium innoxium* Comstock & Comstock 1895: 452–453 (PL). Neotype* male (here designated; pinned; CU). New York, Tompkins Co., Ithaca, Fall Creek, 1 July 1886 (collector unknown). Revalidated
*Simulium*, Barnard 1880: 191 (MPLE, bionomics)
*Simulium* 'sp.' Riley 1887: 592, Fig. 7 (E)
*Simulium*, Howard 1888: 99 (bionomics)
*Simulium innoxium* Williston in Phillips 1890: 21–23. Unavailable (unpublished thesis name)
'Another species' Garman 1893: 215 (silk spinning)
*Schoenbaueria aldrichiana* Enderlein 1936b: 120 (F). Holotype* female (pinned, left wing on slide; ZMHU). New York, Tompkins Co., Ithaca, 18 August 1928 (G. Enderlein)
*Simulium pictipes* 'B' Bedo 1975a: 1147–1163 (C)
[*Simulium pictipes*: authors (except Hagen 1880, 1881; Riley 1881; Nicholson & Mickel 1950 [in part, Figs. 27A, B, material from Cook Co.]; Fredeen 1951 [in part, Plate 10 (Figs. F, G)]; Hocking & Richards 1952; Davies & Peterson 1956) not Hagen 1880 (misident.)]

**Taxonomy**

Barnard 1880 (MPLE); Riley 1887 (E); Phillips 1890 (FMPLE); J. H. Comstock & A. B. Comstock 1895 (PL); Coquillett 1898 (FM); Johannsen 1902, 1903b (FMPL), 1934 (PL); Garman 1912 (L); Emery 1913 (FMPL); Malloch 1914 (FMPL); Herrick 1916 (PL); Jobbins-Pomeroy 1916 (MPLE); Dyar & Shannon 1927 (FM); Jahn 1932 (E); Gambrell 1933 (E); Smart 1934 (PLE); Enderlein 1936b (F); Twinn 1936b (FMP); Nicholson 1949 (MP); Nicholson & Mickel 1950 (MP); DeFoliart 1951a (PL); Fredeen 1951 (MPL); Jamnback 1953, 1956 (L); Gill & West 1955 (FPL); Stone & Jamnback 1955 (FMPL); Shewell 1959a (FP); Anderson 1960 (FMPL); Davies et al. 1962 (FMP); Wood et al. 1963 (L); Stone 1964 (FMPL; Fig. 62 = *S. claricentrum*); Abdelnur 1966, 1968 (FMPL); Snoddy 1966 (FMPL); Stone & Snoddy 1969 (FMPL); Pinkovsky 1970 (FMPL); Amrine 1971 (FMPL); Snoddy & Noblet 1976 (PL); Peterson 1981, 1996 (FM); Hilsenhoff 1982 (L); Webb & Brigham 1982 (L); Adler 1983 (PL); Adler & Kim 1986 (P); Stuart 1995 (cocoon); Stuart & Hunter 1998b (cocoon).

Our study of the syntypes of *S. pictipes* shows that the species known by this name for more than 100 years has been misidentified. All specimens in the type series of *S. pictipes* are conspecific. They agree in every respect with the species described as *S. longistylatum* by Shewell (1959a), who evidently did not examine the types of *S. pictipes*. The name *longistylatum*, therefore, becomes a synonym of *pictipes*, and another name is needed for the species previously misidentified as *S. pictipes*. The oldest available name for the species is *S. innoxium* Comstock & Comstock (1895). To establish the current concept of the name of this common species, we designate a neotype for *S. innoxium*, selected from material collected in the area where the Comstocks and their students did much of their work.

**Morphology**

Headlee 1906 (rectal papillae); Hallock 1922 (L external and internal anatomy); Tanaka 1924 (L gut), 1934 (FPL gut); Jahn 1932 (embryology); Gambrell 1933 (embryology); Wood 1963a (labral fan); Guttman 1967a (labral fan); Kurtak 1973 (labral fan); Raminani & Cupp 1978 (M reproductive system), Bidlack 1997 (instars); Evans & Adler 2000 (spermatheca); Palmer & Craig 2000 (labral fan).

**Physiology**

Guttman 1967b (L anesthetic), Moobola 1981 (oogenesis).

**Cytology**

Bedo 1973, 1975a; Dechene 1973.

**Molecular Systematics**

Sohn 1973 (DNA hybridization), Sohn et al. 1975 (DNA hybridization), Jacobs-Lorena et al. 1988 (DNA sequences), Brockhouse 1991 (rDNA).

**Bionomics**

**Habitat.** The distribution of this species wraps around the Great Lakes, thence southward to Alabama and Missouri. *Simulium innoxium* is most common in the Appalachian highlands. Larvae and pupae form mosslike clusters and carpets on the crests and faces of waterfalls and dams and on boulders or flat sedimentary rocks in swift rivers. The larvae often arrange themselves in bands perpendicular to the current in areas where the flow passes over irregularities in the substrate (Kurtak 1973). Larvae have been found at water temperatures up to 31°C (Pinkovsky 1970).

**Oviposition.** Females are anautogenous (Bernardo et al. 1986b). Their average fecundity ranges from 118 to more than 200 eggs per ovarian cycle (Bernardo & Cupp 1986). The eggs of this species were probably the first to be described for a North American black fly (Barnard 1880). Females stand in a film of water, facing upstream, and deposit their eggs on bedrock or, less frequently, sticks and vegetation (Smart 1934, Stone & Snoddy 1969). Where the oviposition site is lapped or sprayed by water, the females hover about 30 cm above the site and repeatedly dip to the substrate to attach 5–8 eggs (Phillips in Coquillett 1898, Gambrell 1933, Smart 1934, Stone & Snoddy 1969). Females often oviposit in groups of up to several hundred individuals. Their eggs become massed on top of one another, sometimes to a depth of about 1 cm, and the ovipositing females can become entrapped in the sticky matrix (Gambrell 1933, Stone & Snoddy 1969). A mass of eggs can cover an area about 0.2 $m^2$ (Smart 1934). In the late fall, northern females oviposit in moist crevices, where the eggs will overwinter (Kurtak 1974). Oviposition occurs in early morning and late afternoon or throughout cool overcast days. Less commonly, females oviposit after dark (Jahn 1932, Gambrell 1933).

**Development.** Eggs develop in as few as 2.5 days at 25°C (Smart 1934). They cannot withstand desiccation but can be frozen and still produce larvae (Smart 1934, Kurtak 1974). The larval stage includes 8 instars (Bidlack 1997) and lasts 4–6 weeks in the summer, whereas the pupal stage lasts about 4.5 days at 25°C (Smart 1934). Larvae overwinter in the southern states, but from Pennsylvania northward the eggs overwinter. Four or 5 generations are produced annually in the northern part of the range (Smart 1934, Pinkovsky 1970), and 3–7 are produced in the southern part (Jobbins-Pomeroy 1916, Stone & Snoddy 1969). In New York, the first generation is on the wing by early May (Phillips in Howard 1901). The sex ratio is near unity, with males emerging slightly before females (Smart 1934). Females can live at least 10 days on water at 22°C–24°C (Moobola 1981).

Filter-feeding behavior has been studied in both the field and the laboratory (Kurtak 1978, Bidlack 1997). Larvae ingest a large number of diatoms and can digest them, at least partially, during normal gut-retention times (Kurtak 1979). The efficiency of filter feeding is highly variable, depending on particle size and concentration, water velocity, and temperature, but the species is one of the most efficient in capturing large particles at high velocities (Kurtak 1978). At velocities of 0.5–1.0 m/sec, larvae feed with their bodies almost perpendicular to the substrate, but at velocities greater than 1.3 m/sec, they assume a nearly horizontal position (Kurtak 1973). Larvae are also capable of ingesting dissolved organic matter but have not been observed grazing (Hershey et al. 1996).

Adults can be reared from eggs or first instars in systems of continuously replaced water (Wood & Davies 1966) or recirculated water (Brenner & Cupp 1980a). Vigorous females can be obtained, blood fed with a membrane apparatus, and induced to oviposit in the laboratory through multiple generations; however, the percentage of females that mate successfully is typically less than 10% (Bernardo & Cupp 1986, Bernardo et al. 1986b).

**Mating.** Swarms of males form about 2 m above the water but later settle close to the surface above the pupal beds (Smart 1934, Snow et al. 1958b). The males bounce up and down while facing into the wind, with their abdomens curved upward, ready to couple with females flying over them (Snow et al. 1958b). Swarms are often composed of clusters of 5–8 males, but they can become enormous, forming cones 4.5 m wide that taper to a point 3.5 m above the water (Snow et al. 1958b, Stone & Snoddy 1969). Individuals at the base are often emaciated, but those near the top are plump and move downward as those on the bottom die or couple and leave (Stone & Snoddy 1969). Swarms continue throughout the night if temperatures are above 15°C (Stone & Snoddy 1969).

**Natural Enemies.** Larvae occasionally sport infections of the chytrid fungus *Coelomycidium simulii* and the microsporidium *Polydispyrenia simulii* (McCreadie & Adler 1999). Prevalence of the trichomycete fungus *Harpella melusinae* in the larval midgut often reaches 100% in a population (Beard & Adler 2002). The trichomycetes *Genistellospora homothallica*, *Pennella* near *hovassi*, and *Simuliomyces microsporus* have been found in the larval hindgut, and laboratory colonization by the trichomycete *Smittium megazygosporum* has been induced (Beard & Adler 2000, Beard 2002). The ichthyosporean protists *Paramoebidium chattoni, P. curvum*, and an unidentified species of *Paramoebidium* have been found in the hindguts of larvae (Beard & Adler 2002). Fungal hyphae recorded from pupae might be those of a saprophytic species (Twinn 1939). Adults are susceptible to experimental infections with the entomophthoraceous fungus *Erynia curvispora* (Kramer 1983). Water mites infesting adults have been recorded once (Twinn 1939). Larval *Rhyacophila* (as Phryganid [*sic*]) caddisflies prey on larvae (Howard 1888), and common grackles feed voraciously on all immature stages (Snoddy 1967).

**Hosts and Economic Importance.** Females can be nasty pests of horses and cattle as they feed in the ears, often shoving aside other species (Stone & Snoddy 1969). They also have been taken from the ears of mules (Jobbins-Pomeroy 1916, Snow et al. 1958b). In the laboratory, within 3–5 hours after emergence, they will drink distilled water and feed on blood of cows, horses, pigs, sheep, and (less so) turkeys that is offered through a membrane system (Bernardo & Cupp 1986). *Simulium innoxium* is sometimes attracted to humans (White & Morris 1985b), but only 2 bites have been recorded (Snoddy 1966, Stone & Snoddy 1969). It can be induced to feed on humans in the laboratory (Smart 1934).

Females serve as surrogate vectors of the bovine parasite *Onchocerca lienalis* and the human parasite *O. volvulus* (Lok et al. 1980, 1983a). In contrast, they are poor experimental hosts for the avian parasite *Leucocytozoon smithi* (Kiszewski & Cupp 1986).

### *Simulium* (*Hemicnetha*) *pictipes* Hagen

(Figs. 10.206, 10.371, 10.593; Map 190)

*Simulium pictipes* Hagen 1880: 305–307 (FMPL). Syntype* pupae (27 pupae with cocoons [1 pharate F, 4 pharate M], 24 exuviae with cocoons, 8 empty cocoons, 6 exuviae fragments in ethanol vial #12532; MCZ). New York, Clinton Co., Adirondack Mountains, Ausable River, attached to stones in rapids, August 1879 (R. P. Edes & H. P. Bowditch)

*Simulium* 'n. sp. nr. *pictipes*' Shewell 1957: 2, Map 47 (Canada records)

*Simulium* (*Hagenomyia*) *longistylatum* Shewell 1959a: 84–87 (FMPL). Holotype* male (pinned #6695; CNC). Quebec, Baie Comeau, Outardes River, 21 July 1955 (L. S. Wolfe). New Synonym

*Simulium* 'sp. M near *pictipes*' Wolfe & Peterson 1959: 148 (bionomics)

**Taxonomy**

Hagen 1880 (FMPL), 1881 (P); Nicholson & Mickel 1950 (F only); Fredeen 1951 (F only), 1985a (FMP); Davies & Peterson 1956 (E); Shewell 1959a (FMPL); Davies et al. 1962 (FMP); Wood et al. 1963 (L); Stone 1964 (FMP); Imhof 1977 (E); Imhof & Smith 1979 (E); Currie 1986 (PL).

After examining the type series of *S. pictipes*, we apply the name *pictipes* to the species formerly known as *S. longistylatum*. (See the discussion under the Taxonomy section of *S. innoxium*.)

**Cytology**

Bedo 1973, 1975a.

**Molecular Systematics**

Feraday & Leonhardt 1989 (allozymes), Brockhouse 1991 (rDNA).

**Bionomics**

**Habitat.** This behemoth occupies northeastern North America and tracks the Precambrian Shield west into Saskatchewan, northeastern Alberta, and the Northwest Territories. Larvae and pupae form mosslike aggregations in the clear torrential currents of waterfalls and swift rocky rivers and lake outfalls. Early instars tolerate slower currents than later instars, but nearly all have a low-velocity threshold near 10 cm/sec (Wong 1976).

**Oviposition.** Females emerge with immature eggs, but at least some females are autogenous and at 16°C–17°C they require about 12 days to complete oogenesis when fed sucrose and water (Imhof & Smith 1979, Davies & Györkös 1990). Eggs are deposited in dense masses during the evening on water-splashed rocks and mosses near the fastest currents of waterfalls (Davies & Peterson 1956, Imhof & Smith 1979). A single mass, produced by multiple females, can contain more than 100,000 eggs (Imhof & Smith 1979).

**Development.** Eggs laid in autumn undergo diapause, and the larvae from these eggs hatch in about 5 months when incubated at 20°C (Imhof & Smith 1979). At least 2 generations are produced annually, with larvae of the first generation appearing by May (Wolfe & Peterson 1959, Davies et al. 1962). Adults are on the wing from June to October.

**Mating.** Males form swarms facing into the wind, within 1–3 m over waterfalls (Wolfe & Peterson 1959, Davies et al. 1962). Constant misting from the falls often permits swarming from morning to evening in full sunlight (Davies & Peterson 1956).

**Natural Enemies.** A larva from Newfoundland was reported as a host of a potentially new species of microsporidium (Vávra & Undeen 1981). Larvae are hosts of an unidentified mermithid nematode. We have taken many male flies from streamside webs of the araneid spider *Larinioides patagiata*.

**Hosts and Economic Importance.** Females feed on floral nectar of blueberry (*Vaccinium*) (Wolfe & Peterson 1959). A single female has been taken from the ear of a moose (Nicholson & Mickel 1950). We occasionally have netted this species in swarms around humans. However, we are aware of only 2 biting records (Davies & Peterson 1956, Pistrang 1983).

### Subgenus *Simulium* Latreille

*Simulium* Latreille 1802: 426 (as genus). Type species: *Oestrus columbacensis* Scopoli 1780: 133, by original designation (Thompson [2001] discusses historical details of authorship, date, and spelling of the type species)

*Odagmia* Enderlein 1921a: 199 (as genus). Type species: *Simulia ornata* Meigen 1818: 290, by original designation

*Friesia* Enderlein 1922: 69 (as genus). Type species: *Nevermannia tristrigata* Enderlein 1921b: 213 (= *Melusina bezzii* Corti 1914: 197–198), by original designation

*Discosphyria* Enderlein 1922: 72 (as genus). Type species: *Discosphyria odagmiina* Enderlein 1922: 72–73 (= *Simulium trifasciatum* Curtis 1839: 765 [2 pages + color plate]), by original designation

*Gynonychodon* Enderlein 1925: 208 (as genus). Type species: *Simulium nobile* De Meijere 1907: 206–207, by original designation

*Pseudodagmia* Baranov 1926: 161, 164 (as subgenus of *Odagmia*). Type species: *Simulia variegata* Meigen 1818: 292, by original designation

*Danubiosimulium* Baranov 1935: 158 (as subgenus of *Simulium*). Type species: *Culex columbaczensis* Schönbauer 1795: 24–26 (= *Oestrus columbacensis* Scopoli 1780: 133), by monotypy

*Cleitosimulium* Séguy & Dorier 1936: 141–142 (as subgenus of *Simulium*). Type species: *Simulium rupicolum* Séguy & Dorier 1936: 133–146 (= *Simulia argentiostriata* Strobl 1898: 594–595), by original designation

*Gnus* Rubtsov 1940b: 363–364 (as subgenus of *Simulium*). Type species: *Simulium decimatum* Dorogostaisky, Rubtsov & Vlasenko 1935: 145–148, by original designation

*Tetisimulium* Rubtsov 1960b: 117–122 (unavailable substitute name for *Friesia* Enderlein 1922; not expressly proposed as substitute name)

*Tetisimulium* Rubtsov 1963a: 497 (as genus). Type species: *Melusina bezzii* Corti 1914: 197–198, by original designation

*Parabyssodon* Rubtsov 1964c: 623 (as genus). Type species: *Simulium transiens* Rubtsov 1940b: 361–363, by original designation

*Phosterodoros* Stone & Snoddy 1969: 32–33 (as subgenus of *Simulium*). Type species: *Simulium jenningsi* Malloch 1914: 41–43, by original designation; *Phosterodorus* (subsequent misspelling)

*Phoretodagmia* Rubtsov 1972a: 406 (as genus). Type species: *Simulium ephemerophilum* Rubtsov 1947: 90, 115, by original designation; *Foretodagmia* (subsequent misspelling)

*Paragnus* Rubtsov & Yankovsky 1982: 184–185 (as genus). Type species: *Simulium bukovskii* Rubtsov 1940b: 418–419, by original designation

*Archesimulium* Rubtsov & Yankovsky 1982: 185 (as subgenus of *Simulium*). Type species: *Melusina tuberosa* Lundström 1911: 14–15, by original designation

*Argentisimulium* Rubtsov & Yankovsky 1982: 185–186 (as subgenus of *Simulium*). Type species: *Simulium nölleri* Friederichs 1920: 567–569, by original designation

**Diagnosis.** Adults: Radius typically without hair dorsobasally. Legs typically bicolored. Female: Cibarium smooth or with minute nodules between cornuae. Claws each with or without small subbasal tooth (*S. slossonae* species group with basal thumblike lobe). Scutum typically subshiny to shiny (pollinose in some taxa). Spermatheca typically with internal spicules. Male: Abdomen with segments II and V–VII with pearlaceous spots laterally. Gonostylus longer than gonocoxite. Ventral plate typically with crenulated posterior margin. Paramere with multiple spines of various sizes. Pupa: Gill of 4–30 filaments. Spine combs typically present. Cocoon shoe or slipper shaped. Larva: Postgenal cleft typically well developed, but reduced to minute notch in some species. Abdominal segment IX with or without prominent ventral tubercles. Rectal papillae of 3 simple or compound lobes. Chromosomes: Standard banding sequence for IS, IL, IIL, and IIIS as given by Rothfels et al. (1978).

**Overview.** This subgenus is the largest in the family, containing 21 species groups worldwide, of which 8 are represented in North America north of Mexico.

#### *Simulium (Simulium) jenningsi* Species Group

**Diagnosis.** Adults: Subcosta without hairs ventrally. Female: Scutum shiny. Legs with scalelike hairs; claws toothless. Hypogynial valve with well-sclerotized, nonsetose posterior portion and oblique anterior portion bearing long hairs. Anal lobe in ventral view bluntly pointed anteriorly and with unsclerotized medial notch. Male: Scutum with anterolateral pair of iridescent spots, and posterior third shiny, bearing fine indistinct hair. Ventral plate with each arm bearing distinct lateral projection. Pupa: Gill of 4–12 filaments. Cocoon slipper or shoe shaped, typically with lateral apertures (absent in 4 species). Larva: Postgenal cleft large, broad, rounded. Abdominal segment IX with or without ventral tubercles. Rectal papillae of 3 compound lobes. Chromosomes: Standard banding sequence as given by A. E. Gordon (1984b).

**Overview.** This species group is strictly Nearctic and has 22 members concentrated in the southeastern United States, with about 5 reaching Canada. Phylogenetic relationships within the group indicate an origin in cooler habitats, with subsequent radiation in the warmer southeastern region (Moulton & Adler 1995).

All members of the group are multivoltine, and their immature stages live in warm streams and rivers. The females probably oviposit while in flight by dipping to the surface of the water. They are principally mammalophilic, and some have been implicated as vectors of filarial nematodes in cattle. Most members of the group swarm around humans and can be miserable pests, although only a few commonly bite. They also can be pests of cattle and horses in areas such as Georgia (Sutton 1929) and South Carolina. The poor ability to identify the females of most species in this group is at odds with their economic importance. Molecular markers eventually might permit identification.

#### *Simulium (Simulium) anchistinum* Moulton & Adler

(Figs. 10.481, 10.614; Plate 19a; Map 191)

*Simulium (Simulium) anchistinum* Moulton & Adler 1995: 7–8, Figs. 10, 24, 48, 68, 86, 103E (FMPLC). Holotype* male (pinned, pupal exuviae in glycerin vial below; USNM). South Carolina, Pickens Co., Twelve Mile Creek, Rt. 137, 34°46′29″N, 82°46′18″W, 17 July 1990 (J. K. Moulton)

[*Simulium* (*Phosterodoros*) *nyssa*: Stone & Snoddy (1969) (misident. in part, paratypes from Bedford and Montgomery Cos., Virginia)]

**Taxonomy**

Moulton 1992 (FMPL), Moulton & Adler 1995 (FMPL).

**Cytology**

Moulton & Adler 1995.

**Bionomics**

**Habitat.** This species is rather common in the piedmont from southern New York to Georgia, penetrating slightly into the mountains. The immature stages are most prevalent in swift sandy rivers more than 15 m wide and interspersed with rocky riffles.

**Oviposition.** Unknown.

**Development.** Eggs overwinter. Larvae hatch in early spring and disappear in November.

**Mating.** Unknown.

**Natural Enemies.** Larvae are hosts of unidentified mermithid nematodes and a species of the microsporidium *Amblyospora*. A ciliate protozoan of the genus *Tetrahymena* has been found in larvae in South Carolina (C. E. Beard & P. H. Adler, unpublished data).

**Hosts and Economic Importance.** Females sometimes swarm around humans and can be a nuisance on golf courses (Gray et al. 1996).

### *Simulium* (*Simulium*) *aranti* Stone & Snoddy

(Figs. 10.372, 10.482, 10.798; Plate 19b; Map 192)

*Simulium* (*Phosterodoros*) *aranti* Stone & Snoddy 1969: 33–34, Figs. 20, 45, 106, 174, 197, 214, 225, 251, 274 (FMPL; the larva was said to be unknown, but Figs. 251 & 274 are of this species). Holotype* female (pinned #68978, terminalia on slide; USNM). Alabama, Coosa Co., Rockford, Swamp Creek, 12 June 1964 (E. L. Snoddy)

*Simulium* '(Subgenus Novum) #116' Snoddy 1966: 56–58 (FPL)

**Taxonomy**

Snoddy 1966 (FPL), Stone & Snoddy 1969 (FMPL), Snoddy & Noblet 1976 (P), Moulton 1992 (FMPL), Moulton & Adler 1995 (FMPL).

**Cytology**

Moulton & Adler 1995.

**Bionomics**

**Habitat.** *Simulium aranti* is an uncommon resident of the southeastern piedmont and the mountain rivers of Tennessee. Its immature stages occupy shallow rivers where water flows swiftly over large flat outcroppings of sedimentary rock. Larvae and pupae adhere to aquatic vegetation such as hornleaf riverweed (*Podostemum ceratophyllum*).

**Oviposition.** Unknown.

**Development.** Eggs overwinter and larvae are present from March to November.

**Mating.** Unknown.

**Natural Enemies.** Larvae are subject to parasitism by an unidentified mermithid nematode (McCreadie & Adler 1999).

**Hosts and Economic Importance.** Females infrequently swarm about humans and enter the eyes and nostrils (Snoddy 1966).

### *Simulium* (*Simulium*) *chlorum* Moulton & Adler

(Plate 19c, Map 193)

*Simulium* (*Simulium*) *chlorum* Moulton & Adler 1995: 9–10, Fig. 103D (FMPLC). Holotype* male (pinned, pupal exuviae in glycerin vial below; USNM). South Carolina, Pickens Co., Six Mile Creek, Issaqueena Experimental Forest, 2.5 km from west gate, near Lake Issaqueena, 34°45′29″N, 82°51′33″W, 23 April 1992 (J. K. Moulton)

*Simulium* (*Simulium*) *chlorosum* Moulton 1992: 31, 238, 239. Unavailable (unpublished thesis name)

**Taxonomy**

Moulton 1992 (FMPL), Moulton & Adler 1995 (FMPL).

Each pupal gill has 8–10 filaments, with 8 being most common. In *S. confusum*, the most similar species, pupae most often have 10 filaments. The original larval description indicated that each labral fan has 38–44 primary rays. Pennsylvania larvae have as many as 55 primary rays. They, however, have the green body color characteristic of the species.

**Cytology**

Moulton & Adler 1995.

**Bionomics**

**Habitat.** This species is common in the piedmont of South Carolina, beyond which it is known only from Clark's Creek in Dauphin Co., Pennsylvania. The immature stages are found in slow sandy streams, usually less than 5 m wide, with scattered rocks.

**Oviposition.** Unknown.

**Development.** Larvae begin hatching from late winter to early spring. Pupae can be found from April to November.

**Mating.** Unknown.

**Natural Enemies.** No records.

**Hosts and Economic Importance.** Unknown; presumably mammalophilic.

### *Simulium* (*Simulium*) *confusum* Moulton & Adler

(Figs. 10.373, 10.483; Plate 19d; Map 194)

*Simulium* (*Simulium*) *confusum* Moulton & Adler 1995: 10–12, Figs. 11, 26, 50, 70, 88, 103A, 106 (FMPLC). Holotype* male (pinned, pupal exuviae in glycerin vial below; USNM). South Carolina, Laurens Co., Duncan Creek, Rt. 56, 2.5 km north of junction with Interstate 26, 34°32′02″N, 81°50′51″W, 16 November 1991 (J. K. Moulton)

[*Simulium jenningsi*: Malloch (1914) (misident. in part, material from Frierson's Mill, Louisiana)]

[*Simulium jenningsi*: Jobbins-Pomeroy (1916) not Malloch 1914 (misident.)]

[*Simulium jenningsi*: Johannsen (1934) not Malloch 1914 (misident.)]

[*Simulium* (*Simulium*) *jenningsi*: Stone (1949a) not Malloch 1914 (misident. in part, material from Frierson's Mill, Louisiana)]

[*Simulium* (*Phosterodoros*) *nyssa*: Stone & Snoddy (1969) (misident. in part, material from New Kent Co., Virginia; Leon Co., Texas; Bienville and DeSoto Parishes, Louisiana; Fig. 207)]

[*Simulium nyssa*: Arnold (1974) not Stone & Snoddy 1969 (misident. in part, material from Chesterfield Co.)]

[*Simulium taxodium*: Arnold (1974) not Snoddy & Beshear 1968 (misident. in part, material from Chesterfield Co.)]
[*Simulium nyssa*: Garris & Noblet (1976) not Stone & Snoddy 1969 (misident.)]
[*Simulium jenningsi*: Noblet et al. (1976) not Malloch 1914 (misident.)]
[*Simulium nyssa*: Snoddy & Noblet (1976) not Stone & Snoddy 1969 (misident. in part, Fig. 22)]
[*Simulium jenningsi*: Noblet et al. (1978) not Malloch 1914 (misident. in part, material from Fairfield, Greenville, Greenwood, Lancaster, Lee, Newberry, Orangeburg, Richland Cos.)]
[*Simulium lakei*: Noblet et al. (1978) not Snoddy 1976 (misident. in part, material from Abbeville, Chester, Fairfield, Greenville, Greenwood, Laurens, McCormick, Newberry, York Cos.)]
[*Simulium nyssa*: Noblet et al. (1978) not Stone & Snoddy 1969 (misident. in part, material from Abbeville, Fairfield, Lancaster, Newberry Cos.)]
[*Simulium taxodium*: Noblet et al. (1978) not Snoddy & Beshear 1968 (misident. in part, material from Abbeville, Anderson, Cherokee, Chester, Chesterfield, Fairfield, Greenville, Greenwood, Laurens, McCormick, Newberry, Oconee, Pickens, Spartanburg Cos.)]
[*Simulium jenningsi*: Stoneburner & Smock (1979) not Malloch 1914 (misident.)]
[*Simulium taxodium*: Horosko (1982) not Snoddy & Beshear 1968 (misident.)]
[*Simulium (Simulium) nyssa*: Adler & Kim (1986) not Stone & Snoddy 1969 (misident. in part, material from North Carolina)]
[*Simulium (Simulium) podostemi*: Adler & Kim (1986) not Snoddy 1971 (misident. in part, material from North Carolina)]
[*Simulium (Phosterodoros) nyssa*: Pruess & Peterson (1987) not Stone & Snoddy 1969 (misident.)]
[*Simulium (Simulium) anchistinum* Moulton (1992) (misident. in part, material from Nebraska)]
[*Simulium (Simulium) nyssa*: Pruess et al. (1992) not Stone & Snoddy 1969 (misident.)]
[*Simulium nyssa*: Tang et al. (1995a) not Stone & Snoddy 1969 (misident.)]
[*Simulium nyssa*: Tang et al. (1996b) not Stone & Snoddy 1969 (misident. in part)]

**Taxonomy**

Jobbins-Pomeroy 1916 (MPL), Johannsen 1934 (PL), Stone & Snoddy 1969 (P, Fig. 207 only), Moulton 1992 (FMPL), Moulton & Adler 1995 (FMPL).

Chromosomal, morphological, and biological information suggests that *S. confusum* is a species complex (Moulton & Adler 1995). The number of filaments in each gill varies from 8 to 10 and has been the source of numerous misidentifications. Western populations (e.g., in Texas) generally have 10 gill filaments, whereas other populations have a mixture of 8, 9, or 10 filaments.

**Cytology**

Moulton & Adler 1995.

**Molecular Systematics**

Pruess et al. 1992, 2000 (mtDNA); Tang et al. 1995a, 1996b (mtDNA).

**Bionomics**

**Habitat.** This species is one of the most common members of the *S. jenningsi* group in sandy streams of the southern piedmont westward into Nebraska, Oklahoma, and Texas. It rarely enters the Sandhills Region or coastal plain, but we have found the immatures in blackwater bayous of Louisiana. Inhabited streams are usually of circumneutral pH and about 5–10 m wide. Larvae and pupae cling to fallen leaves and debris in sandy streams where aquatic plants and trailing vegetation are sparse.

**Oviposition.** Unknown.

**Development.** Larvae begin hatching in late winter and, along with pupae, can be found from late March to late November, attaining their greatest numbers in the summer.

**Mating.** Unknown.

**Natural Enemies.** Larvae are hosts of unidentified mermithid nematodes, the chytrid fungus *Coelomycidium simulii*, and the microsporidia *Janacekia debaisieuxi* and *Polydispyrenia simulii* (McCreadie & Adler 1999).

**Hosts and Economic Importance.** Females feed on horses (Moulton & Adler 1995). They also swarm around humans (Gray et al. 1996).

### *Simulium (Simulium) definitum* Moulton & Adler

(Figs. 10.374, 10.484; Plate 19e; Map 195)

*Simulium (Simulium) definitum* Moulton & Adler 1995: 13–14, Figs. 12, 13, 27, 51, 71, 89, 103A, 108 (FMPLC). Holotype* male (pinned, pupal exuviae in glycerin vial below; USNM). South Carolina, Pickens Co., Six Mile Creek, Issaqueena Experimental Forest, near Lake Issaqueena, 34°45′29″N, 82°51′33″W, 6 October 1991 (J. K. Moulton)
[*Simulium (Phosterodoros) lakei*: Snoddy (1976) (misident. in part)]
[*Simulium (Simulium) lakei*: Cupp & Gordon (1983) not Snoddy 1976 (misident. in part)]
[*Simulium (Simulium) lakei*: Adler & Kim (1986) not Snoddy 1976 (misident.)]

**Taxonomy**

Adler & Kim 1986 (PL), Moulton 1992 (FMPL), Moulton & Adler 1995 (FMPL).

Pupae typically have 9 filaments per gill, although about 5%–10% have 8 filaments.

**Cytology**

Moulton & Adler 1995.

**Bionomics**

**Habitat.** *Simulium definitum* is found in small scattered populations from Pennsylvania to South Carolina. The immature stages inhabit shallow streams 1–12 m wide. These streams are typically sandy with trailing vegetation, although some are rocky with sparse aquatic vegetation.

**Oviposition.** Unknown.

**Development.** This species overwinters as eggs and is the only member of the *S. jenningsi* species group that declines in numbers during the hot summer months. It is most prevalent from April to mid-June and from September to November.

**Mating.** Unknown.

**Natural Enemies.** Larvae are hosts of the microsporidium *Polydispyrenia simulii* (McCreadie & Adler 1999) and an unidentified mermithid nematode.

**Hosts and Economic Importance.** Unknown; presumably mammalophilic.

### *Simulium (Simulium) dixiense* Stone & Snoddy

(Figs. 10.375, 10.485, 10.799; Plate 19f; Map 196)

*Simulium (Phosterodoros) dixiense* Stone & Snoddy 1969: 34–36, Figs. 46, 200, 226 (FMP). Holotype* female with pupal exuviae (pinned #68979; USNM). Alabama, Washington Co., Lewis Creek, 6 November 1965 (E. L. Snoddy)

*Simulium* '(Subgenus Novum) #17' Snoddy 1966: 58–60 (FMP only)

[*Simulium lakei*: Moulton (1992) not Snoddy 1976 (misident. in part, Figs. 53A and B reversed)]

**Taxonomy**

Snoddy 1966 (FMP), Stone & Snoddy 1969 (FMP), Pinkovsky 1976 (FMPL), Snoddy & Noblet 1976 (P), Webb & Brigham 1982 (L), Moulton 1992 (FMPL), Moulton & Adler 1995 (FMPL), Stuart 1995 (cocoon), Stuart & Hunter 1998b (cocoon).

**Cytology**

Moulton & Adler 1995.

**Molecular Systematics**

Smith 2002 (nuclear and mtDNA).

**Bionomics**

**Habitat.** This species is common in a thin strip from the Sandhills Region of North Carolina to the coastal plain of Mississippi. Its distribution is sharply limited by the Fall Line, which marks the boundary between the piedmont and the Sandhills (McCreadie & Adler 1998). The pale green larvae and pale pupae are most numerous on trailing macrophytes, such as bur-reed (*Sparganium americanum*), in swift sections of sandy swamp streams that average about 6 m in width. These streams are nearly always acidic, with pH as low as 3.75 (Arnold 1974, Pinkovsky 1976).

**Oviposition.** Unknown.

**Development.** Larvae and pupae of *S. dixiense* often are present year-round. South Carolina might have 5–7 generations annually, with the largest populations in the spring and summer (Garris & Noblet 1976).

**Mating.** Unknown.

**Natural Enemies.** Larvae are hosts of unidentified mermithid nematodes (Pinkovsky 1976); the chytrid fungus *Coelomycidium simulii*; the microsporidia *Janacekia debaisieuxi*, which often forms pink, red, or purple cysts in the abdomen, and *Polydispyrenia simulii* (McCreadie & Adler 1999); and the trichomycete fungus *Harpella melusinae*.

**Hosts and Economic Importance.** Females are presumably mammalophilic, but nothing is known of their specific feeding habits. One report implicates a few females flying about the head of a human (Moulton 1992).

### *Simulium (Simulium) fibrinflatum* Twinn

(Figs. 10.207, 10.376, 10.486, 10.615, 10.800; Plate 20a; Map 197)

*Simulium (Simulium) fibrinflatum* Twinn 1936b: 141–142 (FMP). Holotype* male (pinned #4128, terminalia on slide; CNC). Ontario, Ottawa, Ottawa River, Remic Rapids, Cunningham Island, 17 August 1935 (C. R. Twinn)

*Simulium (Simulium)* 'sp. No. 7' Snow et al. 1958b: 19 (Alabama record)

*Simulium* '(Subgenus Novum) #7' Snoddy 1966: 77–80 (FMPL)

*Simulium (Phosterodoros) underhilli* Stone & Snoddy 1969: 45–47, Figs. 28, 54, 113, 181, 208 (FMPL). Holotype* female with pupal exuviae (pinned #68985; USNM). Alabama, Lee Co., Meadows Mill, 25 August 1963 (E. L. Snoddy)

*Simulium (Simulium) octobranchium* Moulton 1992: 113. Unavailable (unpublished thesis name)

[*Simulium notiale*: Stone & Snoddy (1969) (misident. in part, 1 F from Lee Co., Alabama)]

**Taxonomy**

Twinn 1936b (FMP), Stone & Jamnback 1955 (FMPL), Davies et al. 1962 (FMP), Wood et al. 1963 (L), Stone 1964 (FMPL), Snoddy 1966 (FMPL), Stone & Snoddy 1969 (FMPL), Snoddy & Noblet 1976 (PL), Webb & Brigham 1982 (L), Adler 1983 (PL), Adler & Kim 1986 (PL), Al-Amari 1988 (P), Moulton 1992 (FMPL), Moulton & Adler 1995 (FMPL).

The pupal filaments vary in number from 6 to 8 per gill and in structure from thin to markedly inflated, suggesting that *S. fibrinflatum* is a species complex (Moulton & Adler 1995). Populations in the southeastern United States (especially Georgia and South Carolina) are the most variable. In the piedmont of Georgia, pupae of *S. fibrinflatum* sometimes can be difficult to distinguish from those of *S. notiale*. Additional research is required to understand the limits of these 2 species.

**Morphology**

Guttman 1967a (labral fan).

**Cytology**

Rothfels 1979 (IIS misinterpreted, not based on 'B' inversion), A. E. Gordon 1984b, Al-Amari 1988, Moulton & Adler 1995.

**Bionomics**

**Habitat.** This common species ranges from southern Ontario and Nova Scotia to the Gulf Coast of Alabama. In the Appalachian Mountains, larvae and pupae are found most often on submerged vegetation in swift rocky flows, sometimes downstream of impoundments. They also live in sandy streams of the southeastern piedmont and Alabama's coastal plain. Streams inhabited by *S. fibrinflatum* in the Adirondack Mountains of New York are mildly acidic to circumneutral (A. E. Gordon 1984a). Larvae are found at depths up to at least 1 m (Granett 1979a).

**Oviposition.** Females are anautogenous. When given enemas of rat blood, they can produce up to 226 eggs; enemas of human blood result in higher egg production (Klowden & Lea 1979). Oviposition behavior is unknown.

**Development.** Eggs overwinter and pupae are present from May to October in the northeastern states and from March to early December in the southeastern states. At least 2 generations per year are suspected in Canada (Davies et al. 1962). Four or more generations are possible at the southern end of the range. Isolation of bacteria from the larvae led to the suggestion that the gut could serve as an assay of water quality (Snoddy & Chipley 1971).

**Mating.** Mating has taken place in confined tubes after flies were aspirated from clusters on the wall of a dam spillway (Snow et al. 1958b).

**Natural Enemies.** Larvae are hosts of unidentified mermithid nematodes, the chytrid fungus *Coelomycidium simulii*, the trichomycete fungus *Harpella melusinae*, the ichthyosporean protist *Paramoebidium* sp., and the microsporidia *Janacekia debaisieuxi* and *Polydispyrenia simulii*. Infections with *J. debaisieuxi* often turn the fat body bright orange or red. A ciliate protozoan of the genus *Tetrahymena* has been found in larvae in South Carolina (C. E. Beard & P. H. Adler, unpublished data).

**Hosts and Economic Importance.** Females do not bite humans (Granett 1979b), although they frequently swarm around people and can be a nuisance (Boyer 1989). More than $27,000 was lost to the economy of South Carolina in 1 year as a result of nuisance problems caused by this species and several other members of the group at a single golf course (Gray et al. 1996).

### *Simulium* (*Simulium*) *haysi* Stone & Snoddy

(Figs. 10.377, 10.487; Map 198)

*Simulium* (*Phosterodoros*) *haysi* Stone & Snoddy 1969: 36–37, Figs. 22, 52, 202, 228, 252, 298 (FMPL). Holotype* female with pupal exuviae (slide #68980; USNM). Alabama, Escambia Co., Brewton, Burnt Corn Creek, 20 September 1964 (E. L. Snoddy)

*Simulium* '(Subgenus Novum) #114' Snoddy 1966: 62–63 (FMPL)

**Taxonomy**

Snoddy 1966 (FMPL), Stone & Snoddy 1969 (FMPL), Pinkovsky 1976 (FMPL), Snoddy & Noblet 1976 (PL), Webb & Brigham 1982 (L), Moulton 1992 (FMPL), Moulton & Adler 1995 (FMPL).

**Cytology**

Moulton & Adler 1995.

**Bionomics**

**Habitat.** *Simulium haysi* is found along the Gulf Coast from northwestern Florida to eastern Texas and is most common in the Escambia and Perdido River drainages of Florida and Alabama. It breeds in wide shallow streams with sandy bottoms and sparse aquatic macrophytes. A pH as low as 3.75 has been recorded in these streams (Pinkovsky 1976).

**Oviposition.** Unknown.

**Development.** Larvae and pupae are present nearly year-round (Moulton & Adler 1995), although populations are small in winter.

**Mating.** Unknown.

**Natural Enemies.** Larvae are hosts of unidentified mermithid nematodes.

**Hosts and Economic Importance.** Unknown; presumably mammalophilic.

### *Simulium* (*Simulium*) *infenestrum* Moulton & Adler

(Figs. 10.378, 10.488; Map 199)

*Simulium* (*Simulium*) *infenestrum* Moulton & Adler 1995: 17–18, Figs. 7, 34, 56, 75, 93 (FMPL). Holotype* male (pinned, pupal exuviae in glycerin vial below; USNM). South Carolina, Pickens Co., Rocky Bottom Creek, Co. Road 100, 34°58′N, 82°51′W, 21 July 1991 (J. K. Moulton)

[*Simulium rugglesi*: Noblet et al. (1976) not Nicholson & Mickel 1950 (misident. in part, material from Pickens and Oconee Cos.)]

[*Simulium rugglesi*: Barnett (1977) not Nicholson & Mickel 1950 (misident. in part, material from mountains of South Carolina)]

[*Simulium* (*Byssodon*) *rugglesi*: Noblet et al. (1978) not Nicholson & Mickel 1950 (misident. in part, all material except that from Berkeley Co.)]

[*Simulium rugglesi*: Horosko (1982) not Nicholson & Mickel 1950 (misident.)]

[*Simulium podostemi*: Moulton & Adler (1995) not Snoddy 1971 (misident. in part, Figs. 7 and 8 reversed)]

**Taxonomy**

Moulton 1992 (FMPL), Moulton & Adler 1995 (FMPL).

The number of filaments in each pupal gill varies from 8 to 10.

**Bionomics**

**Habitat.** This tiny black fly, the smallest in North America, hails from the western Carolinas. The immature stages usually are found in rocky streams and rivers 5–40 m wide, but they also occupy silt-bottomed streams with trailing vegetation. Mature larvae, often barely 3 mm long, and pupae are sometimes so small that they escape the notice of collectors.

**Oviposition.** Unknown.

**Development.** Eggs overwinter and mature larvae are present from early April to mid-November. In the summer, *S. infenestrum* can be locally abundant.

**Mating.** Unknown.

**Natural Enemies.** Larvae are attacked by the chytrid fungus *Coelomycidium simulii* and an unidentified mermithid nematode (McCreadie & Adler 1999).

**Hosts and Economic Importance.** Females swarm about humans and will bite the face, ears, and arms (Moulton & Adler 1995).

### *Simulium* (*Simulium*) *jenningsi* Malloch

(Figs. 10.208, 10.379, 10.489, 10.616, 10.801; Plates 2h, 20b; Map 200)

*Simulium jenningsi* Malloch 1914: 41–43, Plates 3 (Fig. 1), 5 (Fig. 12) (FMPL). Holotype* female (pinned #15412, terminalia on slide; USNM). Maryland, Montgomery Co., Plummer's Island, 8 July 1904 (R. P. Currie)

*Simulium venustum* 'var. a' Johannsen 1902: 47, Plate 5 (Figs. 8–14), Plate 6 (Figs. 4–6); 1903b: 381, Plate 37 (Figs. 8–14), Plate 38 (Figs. 4–6) (FPL)

*Simulium* (*Simulium*) *nigroparvum* Twinn 1936b: 142–145, Fig. 13B (FMP). Holotype* female (pinned #4129, terminalia on slide; CNC). Ontario, Ottawa, Ottawa River, Remic Rapids, 10 June 1935 (C. R. Twinn); *nigrovum* (subsequent misspelling)

*Simulium jenningsi* 'subsp. *jenningsi*' Nicholson 1949: 97–101, Fig. 29 (FMP); Nicholson & Mickel 1950: 52–54, Fig. 29 (FMP)

**Taxonomy**

Johannsen 1902, 1903b (FPL), 1934 (PL, as *venustum* 'var. a' only); Emery 1913 (FML); Malloch 1914 (FMPL); Riley &

Johannsen 1915 (F); Twinn 1936b (FMP); Underhill 1944 (FMPL); Nicholson 1949 (FMP); Stone 1949a (F), 1964 (FMPL); Nicholson & Mickel 1950 (FMP); DeFoliart 1951a (PL); Jamnback 1953, 1956 (L); Stone & Jamnback 1955 (FMPL); Young 1958 (F); Anderson 1960 (FMPL); Davies et al. 1962 (FMP); Wood et al. 1963 (L); Snoddy 1966 (FMPL); Stone & Snoddy 1969 (FMPL); Pinkovsky 1970 (FMPL); Amrine 1971 (FMPL), 1982 (PE); Snoddy & Noblet 1976 (PL); Merritt et al. 1978b (PL); Peterson 1981, 1996 (M); Hilsenhoff 1982 (L); Webb & Brigham 1982 (L); Adler 1983 (PL); Adler & Kim 1986 (PL); Currie 1986 (PL); Al-Amari 1988 (P); Moulton 1992 (FMPL); Moulton & Adler 1995 (FMPL).

*Simulium jenningsi* is possibly a complex of 2 or more sibling species (Moulton & Adler 1995). The number of filaments in each pupal gill is typically 10; however, we have seen rare individuals with 9, 11, and 12 filaments.

**Morphology**

Cox 1938 (F mouthparts, gut), Davies 1965a (spermatophore), Hannay & Bond 1971a (FM wing), Kurtak 1973 (labral fan).

**Cytology**

Rothfels 1979 (IIS misinterpreted, not based on 'B' inversion), A. E. Gordon 1984b, Al-Amari 1988, Moulton & Adler 1995.

**Molecular Systematics**

May et al. 1977 (allozymes), Zhu 1990 (mtDNA), Pruess et al. 1992 (mtDNA), Smith 2002 (nuclear and mtDNA).

**Bionomics**

**Habitat.** *Simulium jenningsi* is common throughout the eastern United States and southeastern Canada, where it breeds in rocky rivers more than 6 m wide. Some of the biggest rivers can churn out nearly a billion flies per kilometer per day (Amrine 1982). Larvae and pupae occur in water with a pH of about 6–9 (A. E. Gordon 1984a) and tolerate water temperatures above 30°C. They can be found on rocks and aquatic macrophytes, such as water-willow (*Justicia americana*) and hornleaf riverweed (*Podostemum ceratophyllum*), but they do not tolerate substrates with silt and algal growth (Adler & Kim 1986). Consequently, pest outbreaks sometimes follow heavy rains and floods that scour and clean the riverbeds, probably opening habitat for larval colonization. Female-flight activity is greatest in the late afternoon and early evening before sunset and occurs primarily within the boundary layer (Choe et al. 1984). Females can disperse more than 55 km from their natal watercourses (Amrine 1982).

**Oviposition.** Females are anautogenous (Brenner & Cupp 1980b). Those fed human blood develop 112–127 eggs within 5 days (Amrine 1982). Females probably fly above the current and drop their eggs directly into the water. Reports of females ovipositing on substrates are possibly erroneous.

**Development.** Eggs overwinter throughout the range. Larvae pass through 7 instars (Voshell 1991). Pupal development requires 4–5 days at 25°C (Amrine 1982). In Pennsylvania, Wisconsin, and southern Ontario, larvae of the first generation hatch in April and pupate in May (Anderson & Dicke 1960, Davies et al. 1962, Adler & Kim 1986). In Missouri, larvae can reach maturity by mid-April (Doisy et al. 1986), and adults emerge from late April to October. In South Carolina and the Virginias, larvae are present from February or March into October or November, with peak adult emergence in mid-July (Amrine 1982, Voshell 1991, Moulton & Adler 1995). Adults fly from the end of May through October in southern New York (Kiszewski 1984). Three generations are produced annually in Ontario (Twinn 1936b) and 5 or more in Alabama and Virginia (Stone & Snoddy 1969, Voshell 1991). Larval feeding efficiency over a range of particle sizes is low relative to other species (Kurtak 1978).

**Mating.** Mating occurs in small swarms near rivers harboring the immature stages (Anderson 1960).

**Natural Enemies.** Larvae are attacked by mermithid nematodes (Anderson & DeFoliart 1962) of questionable identity (footnote, p. 832 of R. Gordon 1984); the chytrid fungus *Coelomycidium simulii*; and the microsporidia *Amblyospora fibrata*, *Janacekia debaisieuxi*, and *Polydispyrenia simulii*. Unidentified microsporidia and ciliate protozoa have been found in females (Kiszewski & Cupp 1986). A putative fungus found on pupae might be a saprophytic species (Twinn 1939).

Invertebrate predators of larvae include the hellgrammite *Corydalus cornutus*, hydropsychid caddisflies, and the American rubyspot damselfly (*Hetaerina americana*) (Kondratieff & Voshell 1983). Minnows such as the silver shiner (*Notropis photogenis*) include larvae as a minor part of their diet (Amrine 1982), as do other fish such as banded darters (*Etheostoma zonale*) (Brancato 1996). The sphecid wasp *Ectemnius stirpicola* includes adults among the prey with which it provisions its nests (Krombein 1960).

**Hosts and Economic Importance.** Females frequently imbibe sugars (Brenner & Cupp 1980b). They take blood from horses, mules, cattle, and turkeys (Underhill 1944, Anderson & DeFoliart 1961) and swarm around dogs (Adler & Kim 1986). An unknown member of the *S. jenningsi* species group, possibly *S. jenningsi*, has taken blood from raccoons (Simmons et al. 1989). On cattle, feeding occurs around the umbilical region (Lok et al. 1983b) and in the ears (Schmidtmann 1987). On turkeys, most feeding is around the head and neck and is greatest when the air temperature is 24°C–29°C, relative humidity is high, and barometric pressure is low (Underhill 1939, 1944). In laboratory trials, the percentage of females that feed is higher on horse blood than on cow or sheep blood (Bernardo & Cupp 1986).

Humans are bitten, though infrequently given the many females often in attendant swarms (Amrine 1982, Adler & Kim 1986), indicating that humans are not the principal hosts. Feeding to repletion on humans requires 2–4 minutes and results in redness and edema for up to 3 days, with itching for up to 3 weeks (Amrine 1982). Allergic reactions, including allergic asthma, have been reported as responses to the bites, especially among children (Brown & Bernton 1970, Grande 1997). Sixty-three cases of allergic reactions to suspected bites of *S. jenningsi* were reported in Raleigh Co., West Virginia, in 1983 (Anonymous 1983). The real misery inflicted by these "gnats," as they often are called, comes from the persistent swarms of females entering the eyes, ears, nose, and mouth, making outdoor activities insufferable (McComb & Bickley 1959, Amrine 1982).

*Simulium jenningsi* in historical times might have been unbearable; the name "Punxsutawney," the town in Pennsylvania famous for its tradition of having woodchucks foretell the weather, actually means "land of gnats" in the language of the Native Americans who once lived there. In recent times, these gnats have become legendary in areas such as the New River around Hinton, West Virginia, and the Susquehanna River around Harrisburg, Pennsylvania. Many of the big rivers that now produce *S. jenningsi* had been polluted—some since the 19th century—by industrial wastes, acid-mine drainage, and agricultural pesticides that precluded or restricted development of the immature stages. After the Federal Water Pollution Control Amendments ("Clean Water Act"; PL 92–500) of 1972, the rivers eventually became clean enough to support massive populations of *S. jenningsi*.

*Simulium jenningsi* is a pest of increasing importance throughout much of eastern North America, especially as water quality continues to improve in the large rivers. Flies emerging from the Potomac River in Washington, D.C., have irritated presidents (e.g., Jimmy Carter) and their families to the degree that both the Secret Service and the Federal Bureau of Investigation have submitted specimens to the U.S. National Museum (F. C. Thompson, pers. comm.). Since the late 1990s, *S. jenningsi* has become a pest in parts of eastern Canada, such as the Montreal area.

The Susquehanna River and its major tributaries produced the first recorded major outbreak in Pennsylvania in 1982. In 1983, citizens of Harrisburg, Pennsylvania, formed a group called "NAG," or Neighbors Against Gnats, which launched a public awareness campaign to raise money and pressure legislators for a solution. As a result, a limited management program was initiated in 1983 and included treatment of 2 large streams in the Harrisburg area plus about 32 km of the Susquehanna River (Arbegast 1994). A suppression program was not conducted in 1984, but in 1985, half a million dollars was allocated for treatments with *Bacillus thuringiensis israelensis* (*Bti*). The Pennsylvania program continued to expand (Arbegast 1996), and by the end of 1999, it had become a $5 million operation, using *Bti* to treat more than 2570 km of waterways in 33 Pennsylvania counties, with treatments approximately every 2 weeks from May to the first week of September (D. H. Arbegast, pers. comm.).

Similar programs have been formed in other areas troubled by *S. jenningsi*. The citizens of Hunterdon Co., New Jersey, for example, formed a NAG group in 1995 to address the *S. jenningsi* problem that began to arise primarily from the Delaware River in about 1991 (Cole 1996, Page 1996). As a result, about 80 km of the Delaware River was treated with *Bti* in the spring of 1996 at a cost of approximately $400,000; the state subsequently earmarked more than one quarter of a million dollars annually to manage the pest (Grande 1997), and increased the extent of treatment in the Delaware River to about 113 km (McNelly & Crans 1999). New Jersey's management program is compatible with the objectives of the state's conservation foundation (Devito 1996). West Virginia has had a program to manage *S. jenningsi* in the New River since 1986.

Females have been incriminated as vectors of the filarial nematode *Onchocerca lienalis* in cattle (Lok 1981, Lok et al. 1983b) but have a rather effective physiological defense response to the microfilariae (Lehmann et al. 1994b). Females are susceptible to laboratory infections of *Leucocytozoon smithi* but do not seem to be natural vectors to turkeys, at least in New York (Kiszewski & Cupp 1986) and the piedmont of Virginia (Byrd 1959), although the latter record might pertain to another member of the *S. jenningsi* species group. Conversely, females have been reported (possibly based on misidentifications) as vectors of this blood parasite among turkeys in the western part of Virginia (Johnson et al. 1938, Johnson 1942), and an unknown species of the *S. jenningsi* species group is a vector in the mountains of South Carolina (Noblet & Moore 1975). Females once were suspected as vectors of the causal agent of Potomac horse fever (Fletcher 1987), although this possibility was discredited later (Hahn et al. 1989, Hahn 1990).

### *Simulium (Simulium) jonesi* Stone & Snoddy

(Figs. 10.380, 10.490; Plate 20c; Map 201)

*Simulium* (*Phosterodoros*) *jonesi* Stone & Snoddy 1969: 38–40, Figs. 24, 49, 86, 109, 132, 154, 177, 203, 230, 254, 278, 299 (FMPL). Holotype* female with pupal exuviae (pinned #68981; USNM). Alabama, Baldwin Co., Fish River, 28 August 1963 (E. L. Snoddy)

*Simulium* 'species No. 58' Simpson et al. 1956: 574 (Florida record)

*Simulium* '(Subgenus Novum) #58' Snoddy 1966: 65–68 (FMPL)

[*Simulium jenningsi*: Lake (1983) not Malloch 1914 (misident.)]

**Taxonomy**

Snoddy 1966 (FMPL), Stone & Snoddy 1969 (FMPL), Pinkovsky 1976 (FMPL), Snoddy & Noblet 1976 (PL), Webb & Brigham 1982 (L), Moulton 1992 (FMPL), Moulton & Adler 1995 (FMPL).

**Cytology**

Moulton & Adler 1995.

**Molecular Systematics**

Smith 2002 (nuclear and mtDNA).

**Bionomics**

**Habitat.** This species is very common in the southeastern piedmont and coastal plain from Delaware to Texas. The immature stages are found in shallow, small to medium-sized sandy streams with limited vegetation, as well as in permanent swamp streams with acidic pH and thick growths of trailing aquatic vegetation. Larvae have been found at temperatures up to 30°C and at a pH as low as 3.5 (Stone & Snoddy 1969, Pinkovsky 1976).

**Oviposition.** The claim that females attach their eggs to grasses (Stone & Snoddy 1969) is probably erroneous. Females, instead, probably deposit their eggs freely into the water.

**Development.** Eggs usually overwinter in all but the southernmost portion of the range (Moulton & Adler 1995). In Florida, where all life stages can be found throughout the year, 7 generations are possible (Pinkovsky

1976). Four or 5 generations might be produced in South Carolina (Arnold 1974), where small populations of larvae and pupae occasionally are found in some streams during the winter.

**Mating.** Unknown.

**Natural Enemies.** Larvae are hosts of unidentified mermithid nematodes; the chytrid fungus *Coelomycidium simulii*; the microsporidia *Amblyospora bracteata/varians, Janacekia debaisieuxi, Polydispyrenia simulii* (McCreadie & Adler 1999), and *Pegmatheca simulii* (Ledin 1994); a parasitic green alga of the genus *Helicosporidium* (Boucias et al. 2001, Tartar et al. 2002); and a ciliate protozoan of the genus *Tetrahymena*.

**Hosts and Economic Importance.** Females presumably feed chiefly on mammals. Only a few have been induced to feed on turkeys in the laboratory, suggesting that *S. jonesi* is an unlikely vector of the protozoan parasite *Leucocytozoon smithi* (Pinkovsky et al. 1981). Females probably form a small component of nuisance swarms around humans (Gray et al. 1996).

### *Simulium (Simulium) krebsorum* Moulton & Adler

(Figs. 10.381, 10.491, 10.802; Plate 20d; Map 202)

*Simulium (Simulium) krebsorum* Moulton & Adler 1992: 393–397 (FMPLC). Holotype* male (pinned with pupal exuviae in glycerin vial below; USNM). South Carolina, Richland Co., Co. Road 86, outflow of Harmon's Mill Pond, 33°58′31″N, 80°49′40″W, 27 October 1990 (J. K. Moulton)

[*Simulium haysi*: Garris (1973) not Stone & Snoddy 1969 (misident.)]

[*Simulium haysi*: Arnold (1974) not Stone & Snoddy 1969 (misident.)]

[*Simulium haysi*: Garris & Noblet (1976) not Stone & Snoddy 1969 (misident.)]

[*Simulium (Phosterodoros) haysi*: Noblet et al. (1978) not Stone & Snoddy 1969 (misident.)]

[*Simulium haysi*: Horosko & Noblet (1986a) not Stone & Snoddy 1969 (misident.)]

**Taxonomy**

Moulton 1992 (FMPL); Moulton & Adler 1992, 1995 (FMPL); Stuart 1995 (cocoon); Stuart & Hunter 1998b (cocoon).

**Cytology**

Moulton & Adler 1992, 1995.

**Bionomics**

**Habitat.** *Simulium krebsorum* is known only from the Sandhills Region and inner fringe of the coastal plain of North and South Carolina. Streams inhabited by this species are acidic, sluggish, sandy, clear to blackwater, and less than 5 m wide. Larvae and pupae are especially prevalent on trailing vegetation such as bur-reed (*Sparganium americanum*).

**Oviposition.** Unknown.

**Development.** Larvae and pupae are rare or absent from November to March, indicating that the winter is passed as eggs (Moulton & Adler 1992).

**Mating.** Unknown.

**Natural Enemies.** Larvae are hosts of the microsporidium *Polydispyrenia simulii* (McCreadie & Adler 1999).

**Hosts and Economic Importance.** Unknown; presumably mammalophilic.

### *Simulium (Simulium) lakei* Snoddy

(Fig. 10.69, Plate 20e, Map 203)

*Simulium (Phosterodoros) lakei* Snoddy 1976: 173–176 (FMPL). Neotype* male (designated by Moulton & Adler 1995: 20–21; pinned, pupal exuviae in glycerin vial below; USNM). Delaware, Kent Co., roadside ditch near Rt. 10, 39°04′N, 75°38′W, 17 June 1991 (J. K. Moulton)

*Simulium jenningsi* 'group' Frost 1964: 156 (light trap catches)

*Simulium* 'Species 120a' Arnold 1974: 115–117 (South Carolina record)

*Simulium* 'Species 120b' Arnold 1974: 117–118 (South Carolina record)

*Simulium* 'sp.' Benke et al. 1984: 61 (Georgia record)

*Simulium* 'sp. (prob. new sp.) near *jenningsi*' Edwards & Meyer 1987: 243; Hauer & Benke 1987: 252 (larval growth)

*Simulium* 'near *jenningsi*' Benke & Parsons 1990: 171 (production dynamics)

*Simulium* 'sp.' Benke & Parsons 1990: 171 (production dynamics)

[*Simulium (Phosterodoros) nyssa*: Stone & Snoddy (1969) (misident. in part, material from Florida and South Carolina)]

[*Simulium jenningsi*: Arnold (1974) not Malloch 1914 (misident. in part, material from Lee and Orangeburg Cos.)]

[*Simulium rugglesi*: Arnold (1974) not Nicholson & Mickel 1950 (misident.)]

[*Simulium jenningsi*: Arnold & Noblet (1975) not Malloch 1914 (misident. in part, material from Lee and Orangeburg Cos.)]

[*Simulium jenningsi*: Noblet et al. (1976) not Malloch 1914 (misident. in part, material from Lee and Orangeburg Cos.)]

[*Simulium rugglesi*: Noblet et al. (1976) not Nicholson & Mickel 1950 (misident. in part, material from coastal plain of South Carolina)]

[*Simulium (Phosterodoros) jenningsi*: Pinkovsky (1976) not Malloch 1914 (misident.)]

[*Simulium (Phosterodoros) nyssa*: Pinkovsky (1976) not Stone & Snoddy 1969 (misident. in part, material from Orange Co.)]

[*Simulium (Phosterodoros) taxodium*: Pinkovsky (1976) not Snoddy & Beshear 1968 (misident. in part)]

[*Simulium rugglesi*: Barnett (1977) not Nicholson & Mickel 1950 (misident. in part, material from coastal plain of South Carolina)]

[*Simulium (Phosterodoros) jenningsi*: Noblet et al. (1978) not Malloch 1914 (misident. in part, material from Hampton, Lee, and Orangeburg Cos.)]

[*Simulium (Byssodon) rugglesi*: Noblet et al. (1978) not Nicholson & Mickel 1950 (misident. in part, material from Berkeley Co.)]

[*Simulium (Phosterodoros) taxodium*: Noblet et al. (1978) not Snoddy & Beshear 1968 (misident. in part, material from Jasper Co.)]

[*Simulium* (*Phosterodoros*) *jenningsi*: Pinkovsky & Butler (1978) not Malloch 1914 (misident.)]
[*Simulium* (*Phosterodoros*) *nyssa*: Pinkovsky & Butler (1978) not Stone & Snoddy 1969 (misident. in part, material from Orange Co.)]
[*Simulium* (*Phosterodoros*) *taxodium*: Pinkovsky & Butler (1978) not Snoddy & Beshear 1968 (misident.)]
[*Simulium jenningsi*: Pinkovsky et al. (1981) not Malloch 1914 (misident.)]
[*Simulium taxodium*: Pinkovsky et al. (1981) not Snoddy & Beshear 1968 (misident.)]
[*Simulium taxodium*: Benke et al. (1984) not Snoddy & Beshear 1968 (misident.)]
[*Simulium taxodium*: Smock et al. (1985) not Snoddy & Beshear 1968 (misident.)]
[*Simulium taxodium*: Smock & Roeding (1986) not Snoddy & Beshear 1968 (misident.)]
[*Simulium taxodium*: Edwards & Meyer (1987) not Snoddy & Beshear 1968 (misident.)]
[*Simulium taxodium*: Hauer & Benke (1987) not Snoddy & Beshear 1968 (misident.)]
[*Simulium taxodium*: Smock (1988) not Snoddy & Beshear 1968 (misident.)]
[*Simulium taxodium*: Benke & Parsons (1990) not Snoddy & Beshear 1968 (misident.)]
[*Simulium dixiense*: Moulton (1992) not Stone & Snoddy 1969 (misident. in part, Figs. 53A and B reversed)]

**Taxonomy**

Pinkovsky 1976 (FMPL), Snoddy 1976 (FMPL), Snoddy & Noblet 1976 (PL), Webb & Brigham 1982 (L), Moulton 1992 (FMPL), Moulton & Adler 1995 (FMPL).

The variability in the number of gill filaments (8–10) is responsible for many previous misidentifications. Larval color patterns are highly variable.

**Cytology**

Moulton & Adler 1995.

Chromosomal information suggests that *S. lakei* is a complex of 2 species, a northern one from Pennsylvania to at least North Carolina and a southern one from South Carolina to Florida (Moulton & Adler 1995).

**Bionomics**

**Habitat.** This black fly inhabits the coastal plain of the eastern United States, although it also has been found in the interior of northern Pennsylvania. It breeds in shallow streams about 1–12 m wide in the northern portion of its range and in blackwater rivers from South Carolina to Florida. It is the most common black fly in large streams and rivers in Florida.

**Oviposition.** Unknown.

**Development.** From Pennsylvania to South Carolina, *S. lakei* overwinters primarily as eggs. Nonetheless, larvae and pupae can be found throughout the year in some streams in South Carolina, completing at least 6 generations per year and reaching peak densities in the late fall (Smock et al. 1985, Smock 1988). In Florida, larvae and pupae can be found throughout the year.

**Mating.** Unknown.

**Natural Enemies.** We have larvae infected with unidentified mermithid nematodes, the microsporidia *Janacekia debaisieuxi* and *Polydispyrenia simulii*, and a ciliate protozoan of the genus *Tetrahymena*.

**Hosts and Economic Importance.** A few females have fed on turkeys in the laboratory (Pinkovsky 1976). Females can be biting pests of emus on farms in South Carolina's coastal plain. They fly around the heads of humans, and we suspect that they also feed on mammals.

### ***Simulium* (*Simulium*) *luggeri* Nicholson & Mickel**

(Figs. 10.382, 10.492; Plate 20f; Map 204)

*Simulium jenningsi luggeri* Nicholson & Mickel 1950: 54–55 (FMP). Holotype* male (pinned; UMSP). Minnesota, Anoka Co., Coon Rapids, reared, 2 July 1939 (H. P. Nicholson); *luggari* (subsequent misspelling)
*Simulium* (*Simulium*) 'L' Fredeen 1951: 12, 25, 27, 74–79, Plate 14 (FMP)
'species . . . resembling *Simulium luggeri*' Davies et al. 1962: 81 (in part; Ontario record)
*Simulium* 'n. sp. aff. *luggeri*' Pistrang 1983: 33 (swarming about humans)
[*Simulium* (*Phosterodoros*) *jenningsi*: Currie (1986) not Malloch 1914 (misident. in part, Alberta material)]
?[*Simulium nyssa*: Tang et al. (1996b) not Stone & Snoddy 1969 (misident. in part)]

**Taxonomy**

Nicholson 1949 (FMP); Nicholson & Mickel 1950 (FMP); Fredeen 1951 (FMP), 1981a (FMPL), 1985a (FMP); Anderson 1960 (FPL); Stone 1964 (FMPL); Abdelnur 1966, 1968 (FMPLE); Snoddy 1966 (FMPL); Stone & Snoddy 1969 (FMPL); Pinkovsky 1970 (FMPL); Amrine 1971 (FMPL); Snoddy & Noblet 1976 (PL); Merritt et al. 1978b (PL); Westwood 1979 (FMPL); Hilsenhoff 1982 (L); Webb & Brigham 1982 (L); Adler 1983 (PL); Adler & Kim 1986 (PL); Currie 1986 (PL); Al-Amari 1988 (P); Mason & Kusters 1991b (FMPL); Moulton 1992 (FMPL); Moulton & Adler 1995 (FMPL).

*Simulium luggeri* is probably a complex of at least 1 eastern and 1 western species (Moulton & Adler 1995, Adler & Mason 1997).

**Morphology**

Yang 1968 (F cibarium), Colbo et al. 1979 (F labral and cibarial sensilla and armature), Fredeen 1981b (instars), Palmer & Craig 2000 (labral fan).

**Cytology**

A. E. Gordon 1984b, Procunier in Adler & Kim 1986, Al-Amari 1988, Moulton & Adler 1995, Adler & Mason 1997.

Eastern populations have undifferentiated sex chromosomes, whereas western populations, probably representing a separate species, have differentiated sex chromosomes (Adler & Mason 1997). Western populations might consist of additional species, based on the diversity of Y chromosomes. Populations in Nebraska, southeastern Manitoba, central Saskatchewan, and southern Alberta each have a different Y chromosome. Two larvae, 1 from Oklahoma and 1 from Arkansas, differed from typical *S. luggeri* by having the same sequence in IIS as *S. fibrinflatum*; no other larvae from these areas have been examined, but the evidence suggests that these larvae represent a separate species.

**Molecular Systematics**

Zhu 1990 (mtDNA); Pruess et al. 1992, 2000 (mtDNA); Tang et al. 1996b (mtDNA); Zhu et al. 1998 (mtDNA).

**Bionomics**

**Habitat.** As the most northern and western member of the *S. jenningsi* species group, *S. luggeri* occurs from the Mackenzie River in the Northwest Territories southeastward through New Hampshire, Oklahoma, and Georgia. It is most common in the central and northern portions of its range. Larvae and pupae are found on vegetation in warm, wide, rocky or sandy rivers. In New York, these rivers have basic pH and are nutrient rich (A. E. Gordon 1984a). Larval drift is typically greatest between sunset and sunrise (Jarvis 1987).

**Oviposition.** Females are anautogenous (Davies & Györkös 1990). They produce 135–174 eggs (Abdelnur 1968) and drop them in the water during flight (Fredeen 1977a). They also have been reported to attach the eggs in masses to emergent vegetation (Fredeen 1981a).

**Development.** Eggs overwinter. They remain viable when stored at 0.5°C–1.5°C for 2–9 months (Fredeen 1959b). Hatching in the large rivers of Saskatchewan occurs for a period of about 1 month (Jarvis 1987). Larvae pass through 7 instars (Fredeen 1981b) and require about 255–321 degree days above 0°C to complete development in Saskatchewan (Jarvis 1987). The pupal stage lasts about 4 days in summer (Pruess 1989). Development from first instar to adult can be completed in 25 days at an average temperature of 22°C (Fredeen 1959a). Larvae and pupae are present from June through September in Alberta (Depner 1971, Currie 1986); May to October or November in Nebraska, Manitoba, and Saskatchewan (Westwood & Brust 1981, Jarvis 1987, Pruess & Peterson 1987); and late April through September in Pennsylvania and Missouri (Adler & Kim 1986, Doisy et al. 1986). Three to 5 overlapping generations are produced annually (Pinkovsky 1970, Fredeen 1981a, Westwood & Brust 1981).

**Mating.** Mating occurs in small swarms near the breeding sites (Abdelnur 1968).

**Natural Enemies.** Larvae are hosts of unidentified mermithid nematodes (Anderson & DeFoliart 1962 [p. 832, footnote of R. Gordon 1984]), an iridescent virus (Erlandson & Mason 1990), the microsporidium *Amblyospora fibrata* (Adler & Mason 1997), and the chytrid fungus *Coelomycidium simulii.*

**Hosts and Economic Importance.** Host-seeking females can travel at least 42 km from their natal waters (Fredeen et al. 1953c) and perhaps more than 300 km with the aid of favorable winds (Fredeen 1985b). They exhibit a small morning peak of activity and a much larger peak in the afternoon and evening (Fredeen & Mason 1991). Meteorological conditions influence their activity, although no one variable regulates host-seeking behavior; parous females might be more tolerant of dry conditions than nulliparous females (Fredeen & Mason 1991).

Hosts include humans, dogs, horses, pigs, elks, cattle, sheep (Anderson & DeFoliart 1961, Fredeen 1985b, Mason & Kusters 1990), and probably moose (Flook 1959). Females are attracted to both the head and the body (Mason 1986), feeding especially inside the ears and around the eyes and, to some extent, on the underbelly and inner thighs (Van Deveire 1981). Nearly all females attracted to cattle contain sugar and have mated; less than 1% are parasitized with mermithid nematodes (Fredeen & Mason 1991). Females feed on humans in Minnesota, Manitoba, and Saskatchewan, causing swelling and redness (Westwood & Brust 1981, Fredeen 1985b, Sanzone in Gray et al. 1999). Host records are almost wholly lacking from eastern North America, perhaps because eastern *S. luggeri* has been confused with other members of the *S. jenningsi* group and because it might be a distinct sibling species with different feeding habits.

*Simulium luggeri* has had a colossal economic impact in east-central Saskatchewan since the 1970s. It was first detected in the South Saskatchewan River in 1968, the same year that a hydroelectric dam was completed. The dam modified the river, creating ideal larval habitat—warm, shallow, nutrient-rich flow with extensive beds of aquatic macrophytes (Fredeen 1979). The first harassment of cattle came in 1971 (Riegert 1999b), and by 1976 the population had exploded (Fredeen 1985b). Weather conditions during the outbreak years were particularly favorable for adult activity (Fredeen 1977a), and females were able to disperse up to 170 km (Fredeen 1985b). Massive numbers of females emerged throughout the summer and attacked livestock, causing hyperactivity, stampedes, reduced milk production, malnutrition in calves, impotence in bulls, delayed conception, weight loss, and possibly bovine keratitis, pink eye, mastitis, and stress-related diseases such as hoof rot and pneumonia (Fredeen 1984, 1985b). One rancher lost 0.8 km of fencing when his cattle stampeded under the onslaught of the flies (Fredeen 1977a). In the most severe year, 1978, actual losses to the beef and dairy industries were estimated at nearly $3 million and $58 thousand, respectively (Fredeen 1985b). A single breeding herd of 1570 animals experienced losses of $111,590, with 14 deaths that year (Fredeen in Laird et al. 1982). Humans also experienced great anguish during outbreaks, and even with repellent, they often could not work outdoors (Fredeen 1985b). Medical attention was often required (Fredeen 1977b). The outbreak of 1985 also brought major losses to livestock producers in Saskatchewan's simuliid problem zone. For example, farmers experienced about 2.5%–3.0% annual milk loss per cow (Kampani 1986). Management of *S. luggeri*, using *Bti*, was conducted in central Saskatchewan from the 1980s (Khachatourians 1990) through the 1990s, with an annual budget in Canadian currency of $100,000–$400,000 (Kampani 1986, Lipsit 2001).

### *Simulium* (*Simulium*) *notiale* Stone & Snoddy

(Fig. 10.493, Map 205)

*Simulium* (*Phosterodoros*) *notiale* Stone & Snoddy 1969: 40–42, Figs. 26, 51, 88, 111, 134, 156, 179, 205, 232 (FMP). Holotype* male with pupal exuviae (pinned #68982; USNM). Alabama, Lee Co., Meadows Mill, 27 March 1963 (E. L. Snoddy); *notaliale* (subsequent misspelling)

*Simulium* '(Subgenus Novum) #52' Snoddy 1966: 69–72 (FMP)

*Simulium (Phosterodoros)* 'sp. near *notiale-underhilli*' Adler 1983: 220–221, Fig. 47 (PL)

[*Simulium (Phosterodoros) underhilli*: Stone & Snoddy (1969) (misident. in part, 1 M, 2 F from Meadows Mill, Alabama; pupal fragments from Big Otter Creek, Virginia)]

**Taxonomy**

Snoddy 1966 (FMP), Stone & Snoddy 1969 (FMP), Pinkovsky 1976 (FMPL), Snoddy & Noblet 1976 (P), Adler 1983 (PL), Adler & Kim 1986 (PL), Moulton 1992 (FMPL), Moulton & Adler 1995 (FMPL).

**Cytology**

Moulton & Adler 1995.

**Bionomics**

**Habitat.** This species is common on the eastern side of the Appalachian Mountains from Pennsylvania to northern Florida, but it has been found only once (Obed River, Cumberland Co., Tennessee) west of the Appalachians. The immature stages are collected most often in shallow rocky flows more than 3 m wide, especially those with exposed bedrock. They are present in smaller numbers in sandy streams toward the coast.

**Development.** Eggs overwinter and larvae and pupae are present from March to November (Moulton & Adler 1995) or through the winter in Florida (Pinkovsky & Butler 1978).

**Oviposition.** Unknown.

**Mating.** Unknown.

**Natural Enemies.** Larvae are hosts of the chytrid fungus *Coelomycidium simulii* (Garris & Noblet 1975), unidentified mermithid nematodes (McCreadie & Adler 1999), and the microsporidium *Polydispyrenia simulii*.

**Hosts and Economic Importance.** Females probably are chiefly mammalophilic. A few have fed on turkeys in the laboratory (Pinkovsky et al. 1981). They swarm around humans (Gray et al. 1996).

### *Simulium (Simulium) nyssa* Stone & Snoddy

(Fig. 10.383, Map 206)

*Simulium (Phosterodoros) nyssa* Stone & Snoddy 1969: 42–44, Figs. 207, 216 (FMPL). Holotype* male with pupal exuviae (slide #68983; USNM). Alabama, Lee Co., Meadows Mill, 2 September 1964 (E. L. Snoddy)

*Simulium* '(Subgenus Novum) #117' Snoddy 1966: 72–74 (FMPL)

**Taxonomy**

Snoddy 1966 (FMPL), Stone & Snoddy 1969 (FMPL), Amrine 1971 (FMP), Snoddy & Noblet 1976 (L), Webb & Brigham 1982 (L), Adler & Kim 1986 (PL), Al-Amari 1988 (P), Moulton 1992 (FMPL), Moulton & Adler 1995 (FMPL).

The original material examined by Stone and Snoddy (1969), including the paratype series, contained at least 5 separate species (*S. anchistinum*, *S. confusum*, *S. lakei*, *S. nyssa*, *S. ozarkense*) (Moulton & Adler 1995).

The sparse, irregularly distributed microtubercles of the pupal head provide the only reliable morphological means of distinguishing this species from *S. anchistinum*.

**Cytology**

Al-Amari 1988 (IIS misinterpreted), Moulton & Adler 1995.

**Molecular Systematics**

May et al. 1977 (allozymes).

**Bionomics**

**Habitat.** *Simulium nyssa* is fairly common along the eastern interior from Nova Scotia to Arkansas. It breeds in the shallow rapids of rocky streams and rivers with circumneutral or slightly basic pH and temperatures up to 28°C. Larvae and pupae attach themselves chiefly to vegetation.

**Oviposition.** Females oviposit as they oscillate up and down while tapping their abdomens to water trickling over the face of a dam (Stone & Snoddy 1969).

**Development.** Eggs overwinter. Three or more generations are produced annually in Maine (Cupp & Gordon 1983) and up to 5 occur in Alabama (Stone & Snoddy 1969). Larvae begin to mature in March at the southern end of the distribution but do not appear until May in Maine. Larvae disappear between September and November, depending on latitude.

**Mating.** Unknown.

**Natural Enemies.** Unidentified mermithid nematodes attack larvae.

**Hosts and Economic Importance.** The status as a biting pest of livestock (Stone & Snoddy 1969) is unconfirmed. Females might be a nuisance to humans (Cupp & Gordon 1983), but biting, although suspected, has not been confirmed (May et al. 1977).

### *Simulium (Simulium) ozarkense* Moulton & Adler

(Figs. 10.384, 10.494; Map 207)

*Simulium (Simulium) ozarkense* Moulton & Adler 1995: 23–24, Figs. 42, 63, 81, 99, 103A (FMPLC). Holotype* male (pinned, pupal exuviae in glycerin vial below; USNM). Missouri, Morgan Co., Lamine River, Rt. 50, 38°44′N, 92°57′W, 13 May 1992 (J. K. Moulton)

[*Simulium (Phosterodoros) nyssa*: Stone & Snoddy (1969) (misident. in part, material from Oklahoma and Dardenne Creek, Missouri)]

**Taxonomy**

Moulton 1992 (FMPL), Moulton & Adler 1995 (FMPL).

**Cytology**

Moulton & Adler 1995.

**Bionomics**

**Habitat.** This species is fairly common from eastern Kentucky to eastern Texas. Larvae and pupae usually cling to trailing vegetation in small to medium-sized, stony rivers.

**Oviposition.** Unknown.

**Development.** Larvae probably hatch in late winter, judging from the presence of pupae by early March. Larvae and pupae are present throughout the spring and summer.

**Mating.** Unknown.

**Natural Enemies.** No records.

**Hosts and Economic Importance.** Unknown; presumably mammalophilic.

### *Simulium (Simulium) penobscotense* Snoddy & Bauer

(Fig. 10.495, Map 208)

*Simulium (Phosterodoros) penobscotensis* Snoddy & Bauer 1978: 579–581 (FMPL). Neotype* pupa (designated by

Moulton & Adler 1995: 25; in alcohol vial; USNM). Maine, Penobscot Co., Penobscot River, Winn, 18 August 1990 (R. J. Mendel)

'species that is a serious nuisance' Granett 1977: 17 (pest status) (also p. 18 as 'newly described biting species')

*Simulium* 'n. sp. P' May et al. 1977: 637–640 (allozymes)

'species first described from the Penobscot River' LaPointe & Martin 1990: 32 (Massachusetts record)

[*Simulium rugglesi*: Stone (1964) not Nicholson & Mickel 1950 (misident. in part, Fig. 57g, j)]

[*Simulium* 'species no. 117': Waters (1969) not Snoddy 1966 (misident. in part)]

[*Simulium nyssa*: Sleeper (1975) not Stone & Snoddy 1969 (misident.)]

[*Simulium podostemi*: Moulton & Adler (1995) not Snoddy 1971 (misident. in part, Figs. 64 and 65 reversed)]

**Taxonomy**

Stone 1964 (M terminalia), Bauer 1977 (FMPL), Snoddy & Bauer 1978 (FMPL), Moulton 1992 (FMPL), Moulton & Adler 1995 (FMPL).

The apparently restricted northern distribution of this species aids its identification. The adults resemble those of *S. nyssa*; the larva is similar to but larger than that of *S. infenestrum*; and the pupa resembles that of *S. podostemi*.

**Morphology**

Granett 1979a (instars).

**Molecular Systematics**

May et al. 1977 (allozymes).

**Bionomics**

**Habitat.** *Simulium penobscotense*, known from Maine and Massachusetts, is the only member of the *S. jenningsi* species group that is strictly northern in distribution. Larvae and pupae are found on aquatic macrophytes, especially pondweed (*Potamogeton*), in large streams and rivers up to 0.5 km wide (Boobar 1979, Granett & Boobar 1979, Boobar & Granett 1980). This species has capitalized on the improved water quality in urban rivers in New England by recolonizing previously polluted waters (LaPointe & Martin 1990).

**Oviposition.** Females are anautogenous and drop their eggs into flowing water during flight (Cupp & Gordon 1983).

**Development.** Eggs overwinter. Photoperiod has been suggested as the principal factor controlling egg diapause (Granett 1979b). Three generations are produced per year, with the later 2 consisting of several cohorts; larvae can be found from June through October, though they become scarce after August (Bauer & Granett 1979, Granett 1979b). Larvae pass through 7 instars in less than 2 weeks during July and August (Granett 1979a, 1979b).

**Mating.** Unknown.

**Natural Enemies.** The chytrid fungus *Coelomycidium simulii* has been isolated from a larva (Bauer & Soper in Tarrant 1984). Larvae are attacked by unidentified mermithid nematodes.

**Hosts and Economic Importance.** This species bites humans and is a formidable pest in Maine during the summer and fall (Sleeper 1975, May et al. 1977, Bauer & Granett 1979). Dogs and cattle also are attacked (Cupp & Gordon 1983). *Simulium penobscotense* probably has increased in pest status since the Clean Water Act of 1972 led to improved water quality in the major breeding areas (Boyer 1989).

### *Simulium (Simulium) podostemi* Snoddy

(Figs. 10.385, 10.617, 10.803; Map 209)

*Simulium (Phosterodoros) podostemi* Snoddy 1971: 196–199 (FMPL). Neotype* male (designated by Moulton & Adler 1995: 25; pinned, pupal exuviae in glycerin vial below; USNM). Mississippi, Tishomingo Co., Tishomingo State Park, Bear Creek, 34°35′N, 88°11′W, 2 November 1990 (J. K. Moulton)

[*Simulium infenestrum*: Moulton & Adler (1995) (misident. in part, Figs. 7 and 8 reversed)]

[*Simulium penobscotense*: Moulton & Adler (1995) not Snoddy & Bauer 1978 (misident. in part, Figs. 64 and 65 reversed)]

**Taxonomy**

Snoddy 1971 (FMPL), Snoddy & Noblet 1976 (PL), Webb & Brigham 1982 (L), Adler & Kim 1986 (P), Moulton 1992 (FMPL), Moulton & Adler 1995 (FMPL).

**Cytology**

Moulton 1992, Moulton & Adler 1995.

We have a single antepenultimate larva that is a hybrid between *S. podostemi* and *S. luggeri*. It was collected in the Caddo River, Arkansas, and morphologically resembles *S. podostemi*.

**Bionomics**

**Habitat.** A species of the eastern piedmont, *S. podostemi* occurs in patchy populations from southern Pennsylvania to Georgia and Arkansas. Larvae and pupae adhere to vegetation in large streams and broad rivers. Water temperature at some sites can exceed 30°C during the summer, producing minuscule adults.

**Oviposition.** Unknown.

**Development.** Larvae and pupae can be found nearly year-round in the southern part of the range. The largest populations are present during the summer.

**Mating.** Unknown.

**Natural Enemies.** Larvae are attacked by unidentified mermithid nematodes (McCreadie & Adler 1999).

**Hosts and Economic Importance.** Unknown; presumably mammalophilic.

### *Simulium (Simulium) remissum* Moulton & Adler

(Fig. 10.496, Map 210)

*Simulium (Simulium) remissum* Moulton & Adler 1995: 25–26, Figs. 45, 103A (FMPLC). Holotype* male (pinned, pupal exuviae and terminalia in glycerin vial below; USNM). North Carolina, Ashe Co., North Fork New River, Rt. 16, near Crumpler, 36°32′N, 81°25′W, 18 September 1992 (P. H. Adler)

**Taxonomy**

Moulton & Adler 1995 (FMPL).

The pupal cocoon, with its short, broad anterodorsal projection, is unique among North American black flies, but the adults resemble those of *S. anchistinum* and *S. nyssa*.

**Cytology**

Moulton & Adler 1995.

**Bionomics**

**Habitat.** This scarce species is a resident at lower elevations (320–850 m) in the mountains of western North Carolina. It is one of only 2 known black flies endemic to the southern Appalachian Mountains. Larvae and pupae have been found on leaves trapped in debris jams of medium-sized rivers with rocky and sandy bottoms.

**Oviposition.** Unknown.

**Development.** Eggs overwinter. Larvae appear in the spring and decline to negligible numbers by October.

**Mating.** Unknown.

**Natural Enemies.** No records.

**Hosts and Economic Importance.** Unknown; presumably mammalophilic.

### *Simulium* (*Simulium*) *snowi* Stone & Snoddy
(Fig. 10.497, Map 211)

*Simulium* (*Phosterodoros*) *snowi* Stone & Snoddy 1969: 44–45, Figs. 27, 53, 89, 112, 135, 157, 180, 206, 233, 257, 280, 301 (FMPL). Holotype* female with pupal exuviae (pinned #68984; USNM). Alabama, Lee Co., Chewacla Creek, 17 August 1963 (E. L. Snoddy)

*Simulium* (*Simulium*) 'sp. No. 16' Snow et al. 1958b: 19 (Alabama record)

*Simulium* '(Subgenus Novum) #16' Snoddy 1966: 75–77 (FMPL)

**Taxonomy**

Snoddy 1966 (FMPL), Stone & Snoddy 1969 (FMPL), Snoddy & Noblet 1976 (PL), Webb & Brigham 1982 (L), Moulton 1992 (FMPL), Moulton & Adler 1995 (FMPL).

The 4-filamented pupal gill provides the only means of distinguishing this species from *S. fibrinflatum* and *S. notiale*.

**Cytology**

Moulton & Adler 1995.

**Bionomics**

**Habitat.** This little-collected species is known from small populations in Alabama and Tennessee. The immature stages can be found on aquatic vegetation such as hornleaf riverweed (*Podostemum ceratophyllum*) in riffles of clear rocky streams up to 20 m wide.

**Oviposition.** Unknown.

**Development.** Part of the winter is passed as eggs. Mature larvae and pupae are present by late March.

**Mating.** Unknown.

**Natural Enemies.** No records.

**Hosts and Economic Importance.** Unknown; presumably mammalophilic.

### *Simulium* (*Simulium*) *taxodium* Snoddy & Beshear
(Fig. 10.498, Map 212)

*Simulium* (*Simulium*) *taxodium* Snoddy & Beshear 1968: 123–125 (FMPL). Neotype* male (designated by Moulton & Adler 1995: 27; pinned, pupal exuviae in glycerin vial below; USNM). Georgia, Baker Co., Chickasawhatchee Creek, Rts. 37 & 216, near Elmodel, 31°25′N, 84°29′W, 16 March 1991 (J. K. Moulton)

[*Simulium* (*Simulium*) *confusum*: Moulton (1992) not Moulton & Adler 1995 (misident. in part, larva from Johnston Co., North Carolina)]

**Taxonomy**

Snoddy & Beshear 1968 (FMPL), Snoddy & Noblet 1976 (PL), Webb & Brigham 1982 (L), Moulton 1992 (FMPL), Moulton & Adler 1995 (FMPL).

**Cytology**

Moulton & Adler 1995.

**Bionomics**

**Habitat.** The immature stages of this species have been found in large numbers in medium-sized lowland rivers from North Carolina to southern Alabama. The type locality is a blackwater river, swift and murky in the winter, with stands of bald cypress, *Taxodium distichum*, the namesake of the fly. In Florida, the single collection site is a greenwater river flowing through limestone country (Moulton 1992).

**Oviposition.** Early-season females at the type locality emerge with a wealth of stored nutrient, suggesting that they are autogenous for the first generation. Oviposition behavior is unknown.

**Development.** Larvae and pupae have been collected year-round.

**Mating.** Unknown.

**Natural Enemies.** We have collected topotypical larvae harboring the chytrid fungus *Coelomycidium simulii*, the microsporidium *Polydispyrenia simulii*, and an unidentified mermithid nematode.

**Hosts and Economic Importance.** Unknown; presumably mammalophilic.

### *Simulium* (*Simulium*) *malyschevi* Species Group

**Diagnosis.** Female: Claws each with minute subbasal tooth. Anal lobe large, subquadrate in lateral view. Male: Gonostylus typically rather broad. Ventral plate typically compressed laterally and keel-like. Pupa: Gill of 10–16 (rarely 8) short filaments. Cocoon shoe or boot shaped, corbicular. Larva: Postgenal cleft extended three fourths or more distance to hypostomal groove, widest at midpoint. Abdomen gradually expanded posteriorly; segment IX without ventral tubercles; abdominal setae unbranched, translucent, short. Rectal papillae of 3 compound lobes.

**Overview.** About 35 nominal species constitute this Holarctic group. In the Nearctic Region, we recognize 11 described species and 2 cytospecies. Nearly all group members develop in the cold racing streams and rivers of northern areas and higher elevations. Voltinism varies among species. Females deposit their eggs during flight. All species are mammalophilic, and several are notorious pests of humans and livestock.

### *Simulium* (*Simulium*) *decimatum* Dorogostaisky, Rubtsov & Vlasenko
(Figs. 10.70, 10.209, 10.386, 10.499, 10.804; Plate 21a; Map 213)

*Simulium decimatum* Dorogostaisky, Rubtsov & Vlasenko 1935: 145–148 (FMPL). Lectotype* male (designated by Yankovsky 1995: 19; slide #3713; ZISP). Russia, Irkutsk Region, Ushakovka River, August 1932 (I. A. Rubtsov)

*Simulium similis* Malloch 1919: 42–43 (F). Holotype* female (pinned #1147; CNC). Nunavut (as Northwest Territories), Arctic Sound, Coronation Gulf, Hood River, 28 August 1915 (R. M. Anderson). Junior primary homonym of *S. simile* Silva Figueroa 1917: 33–35, and *S. similis* Cameron 1918: 558–559

*Simulium wagneri* Rubtsov 1940b: 450 (English version p. 518), Fig. 68A, B (M). Neotype* male (here designated; slide #3545, partial abdomen and terminalia only; ZISP). Russia, mouth of Matur River, trib. of Abakan River, 30 June 1897 (Wagner)

*Simulium (Simulium) nigricoxum* Stone 1952: 94–95 (F) (substitute name for *similis* Malloch 1919); *nigrocoxum* (subsequent misspelling). New Synonym

*Gnus decimatum xerophilus* Rubtsov 1969: 119–121 (FM). Holotype male (pinned #528, terminalia on slide #19322; HNHM). Mongolia, Central Aimak, 2–7 km west of Somon Lun, 1200 m elev., 17 June 1966 (Z. Kaszab)

**Taxonomy**

Malloch 1919 (F); Dorogostaisky et al. 1935 (FMPL); Rubtsov 1940b, 1956, 1962d, 1962h (FMPL), 1969 (FM); Stone 1952 (F); Peterson 1958a, 1960a (F); Rubtsov & Violovich 1965 (P); Yankovsky 1999 (PL).

Conspecificity of Nearctic *S. nigricoxum* and Palearctic *S. decimatum* was suggested by Currie (1997a). Our study of the type specimens, plus series of all life stages (except the egg) from Alaska and Russia, supports the notion of conspecificity. The vast geographic range of this species, however, suggests that sibling species might be involved. We have seen populations from various sites on the Horton River, Northwest Territories, that consist either of individuals with 8 filaments in their pupal gills or of a mix of individuals with 8 or 10 filaments. Larvae in western North America have a positive head pattern (Fig. 10.804), whereas those of the central Canadian arctic (e.g., Thelon River) have a distinct negative pattern.

The slide-mounted male abdomen of *S. wagneri* that we designate as neotype evidently was intended by Rubtsov, at one time, to be the type. The slide bears the name "*Gnus ?decimatum*" and has the words "*wagneri* typus" crossed out. No other material of *S. wagneri* could be located in the ZISP, including the specimen indicated as type by Rubtsov (1940b). By designating a neotype, we establish the current concept for the name of this economically important Holarctic black fly.

We accept the synonymy suggested by Crosskey (1988) of *S. xerophilum* with *S. decimatum*, having examined Rubtsov's (ZISP) material. According to Rubtsov (1969), the type of *S. xerophilum* is in Budapest (HNHM). Yankovsky (1995), however, reported that the type is in St. Petersburg (ZISP), although we were unable to find it.

*Simulium mongolicum* (Rubtsov 1969) is similar to *S. decimatum* in all life stages. We have examined larvae, possibly associated with adults that resemble *S. decimatum*, from the Orhon Gol River in Mongolia. These larvae have negative head patterns, like those from the central Canadian arctic. *Simulium mongolicum* might be conspecific with *S. decimatum*.

**Morphology**

Yankovsky 1977 (labral fan).

**Cytology**

Shields (1990).

**Molecular Systematics**

Smith 2002 (nuclear and mtDNA).

**Bionomics**

**Habitat.** This species inhabits the arctic and sparsely treed tundra from Alaska to the western shore of Hudson Bay. In the Palearctic Region, it ranges across the taiga and steppes of Eurasia into China and Mongolia. A record from Utah (Peterson 1959c) is questionable. The immature forms are poorly known, in part because of possible confusion with those of the earlier developing *S. murmanum*. Most records are based on females taken from swarms around humans. Larvae and pupae cling to stones in rocky streams and rivers about 15–300 m wide.

**Oviposition.** Blood-fed females can be induced to oviposit by holding them beneath a drop of water (Wood & Davies 1966). Oviposition in nature has not been reported.

**Development.** Eggs overwinter. A single generation is probably produced each year in North America (Currie 1997a), although 2 have been reported in the Palearctic Region (Rubtsov 1956). Females have been collected from July to early September.

**Mating.** Unknown.

**Natural Enemies.** Larvae are attacked by unidentified mermithid nematodes.

**Hosts and Economic Importance.** Females swarm around humans and will take their blood. They also feed on humans in the laboratory (Wood & Davies 1966). *Simulium decimatum* is probably the principal species that inflicted grief on members of Captain John Franklin's second expedition to the hinterlands of northwestern North America in 1825–1827 (Currie 1997a). In Russia, females are "one of the most nefarious bloodsuckers" of humans (Rubtsov 1956).

### *Simulium (Simulium) defoliarti* Stone & Peterson

(Figs. 10.210, 10.387, 10.535, 10.805;
Plate 21b; Map 214)

*Simulium defoliarti* Stone & Peterson 1958: 1–6 (FMPL). Holotype* female (pinned, pupal exuviae on slide; USNM). Wyoming, Lincoln Co., Smith's Fork Creek at Lander Trail, 8.5 mi. from Smoot entrance, 11 August 1956 (published date), 10 August 1956 (slide label), 12 August 1956 (pin label) (G. R. DeFoliart)

*Simulium* 'sp. #73' Rees & Peterson 1953: 58 (Utah record)

'undescribed species' Curtis 1954: 4 (cattle pest)

*Simulium* 'sp. near *arcticum*' MacNay 1955: 201 (cattle pest)

*Simulium defoliarti* '(IIS-14.15)' Shipp & Procunier 1986: 1493 (life history)

*Simulium curtisi* Riegert 1999b: 36. *Nomen nudum*

**Taxonomy**

Peterson 1958a (FMPL), 1960a (FMP); Stone & Peterson 1958 (FMPL); Corredor 1975 (FM); Currie 1986 (PL); Currie & Walker 1992 (L); Peterson & Kondratieff 1995 (FMPL).

Larval body color in some populations is linked to sex, with males being brownish (Plate 21b) and females grayish.

**Cytology**

Relative to the standard banding sequence for the *S. arcticum* species complex, as given by Shields and Procunier (1982), the chromosomes of *S. defoliarti* differ by 2 fixed inversions in IIS (*IIS-14.15*) and by fixation of enhanced centromere bands in all chromosomes; 2 Y chromosomes occur, 1 in IS and 1 in IL (W. S. Procunier, pers. comm.). *Simulium defoliarti* has not been evaluated adequately for the presence of sibling species over its wide range. We do not know if the different sex chromosomes represent different species.

**Bionomics**

**Habitat.** This species is common in forested areas of the western mountains from southern British Columbia and southwestern Alberta into New Mexico and southern California. The immature stages occupy cool, clear, turbulent streams, from 1 to more than 30 m in width, with rocky bottoms. Larval abundance is positively correlated with river discharge and turbidity (Shipp & Procunier 1986).

**Oviposition.** Unknown.

**Development.** Eggs overwinter and larvae are present from April at least into August. One or possibly 2 generations are produced annually (Stone & Peterson 1958, Shipp & Procunier 1986). Mean developmental time for the 7 larval instars in Alberta is about 300 degree-days above 0°C (Shipp & Procunier 1986). Peak emergence occurs in June and July.

**Mating.** Unknown.

**Natural Enemies.** Larvae are occasional hosts of unidentified mermithid nematodes, the chytrid fungus *Coelomycidium simulii*, and the microsporidium *Janacekia debaisieuxi*, which often produces reddish cysts in the abdomen. The trichomycete fungus *Smittium pennelli* inhabits the larval hindgut (Lichtwardt 1984).

**Hosts and Economic Importance.** Females are sometimes pests of cattle, attacking the eyelids and soft underbelly, particularly the udders so that cows will not tolerate suckling calves. These attacks can lead to considerable weight loss, and in 1952 in a small district (Cherryville) in southern British Columbia, they caused losses of more than $24,000 in beef cattle, almost driving some ranchers out of business (Curtis 1954). Females fly about humans but do not bite (Stone & Peterson 1958).

### *Simulium* (*Simulium*) *malyschevi* Dorogostaisky, Rubtsov & Vlasenko

(Figs. 4.43, 10.71, 10.211, 10.388, 10.500, 10.594, 10.806; Plate 21c; Map 215)

*Simulium malyschevi* Dorogostaisky, Rubtsov & Vlasenko 1935: 142–144 (FMPL). Lectotype* male (designated by Yankovsky 1995: 36; slide #3717; ZISP). Russia, Irkutsk Region, near Irkutsk, Ushakovka, day and month unknown, 1934 (I. A. Rubtsov); *malyshevi* (subsequent misspelling)

*Simulium* 'sp. 4' Malloch 1919: 43–44, Figs. 1–3, 7, 12 (PL)

*Simulium* 'sp.' Twinn et al. 1948: 353 (Manitoba record)

*Gnus malyshevi* [sic] *albipes* Rubtsov 1956: 615–616 (FMPL). Lectotype* female (designated by Yankovsky 1995: 7; pinned, terminalia on celluloid mount below; ZISP). Russia, Primor'ye Territory, Barabash, Kedrovaya Pad Reserve, 1940 (Z. A. Radzivilovskaya)

*Gnus malyshevi* [sic] *lucidum* Rubtsov 1956: 616–617 (FMPL). Lectotype* male (head, terminalia, and legs only) with pupal exuviae (cocoon absent) (designated by Yankovsky 1995: 35; slide #7435; ZISP). Russia, Angara River, Pashki, 11 August 1953 (I. A. Rubtsov)

**Taxonomy**

Malloch 1919 (PL); Dorogostaisky et al. 1935 (FMPL); Rubtsov 1940b, 1956, 1962d, 1962h (FMPL); Fredeen 1951, 1985a (FMP); Stone 1952 (FMP); Sommerman 1953 (L); Rubtsov & Violovich 1965 (F); Currie 1986 (PL); Yoon & Song 1989 (PL); Yankovsky 1999 (PL).

The type slide (#3717) of *S. malyschevi* has no specimen beneath the coverslip, nor could a corresponding pinned specimen be found. If further search fails to reveal a pinned specimen that corresponds with the barren slide, a neotype will need to be established.

We examined the types of *S. albipes* and *S. lucidum* and confirmed that they are morphologically identical with our concept of *S. malyschevi*. The wide distribution of *S. malyschevi*, however, indicates a possible complex of species. The names of at least 3 other species from the eastern Palearctic Region might apply to Nearctic populations (Currie 1997a).

The number of filaments in the pupal gill is typically 16, but varies from 13 to 16 (rarely 12) in Nearctic populations.

**Morphology**

Yankovsky 1977 (labral fan).

**Cytology**

The chromosomes of Nearctic material differ from the standard banding sequence given by Shields and Procunier (1982) for the *S. arcticum* species complex as follows: IS and IIIL are complexly rearranged, IL and IIL have 2 inversions each, IIS has *IIS-10.11* of Procunier (1984) plus 1 additional inversion, IIIS is standard, a chromocenter is absent, and the sex chromosomes are undifferentiated (W. S. Procunier, pers. comm.).

**Molecular Systematics**

Smith 2002 (nuclear and mtDNA).

**Bionomics**

**Habitat.** This species is Holarctic and boreal. In North America, it is known from Alaska to Manitoba. In the Palearctic Region, it has been recorded from Siberia south into Japan, Korea, and China. Larvae and pupae are found in large cool streams and rivers up to 150 m wide. The pupal gills are bathed by anterior vortices flowing over the top of the cocoon and from lateral vortices flowing through the fenestrae (Eymann 1991b).

**Oviposition.** Unknown.

**Development.** Eggs of this multivoltine species overwinter, and pupae are present from late May into September.

**Mating.** Unknown.

**Natural Enemies.** Larvae are hosts of mermithid nematodes (Sommerman et al. 1955) and the chytrid fungus *Coelomycidium simulii*. We have found many dead larvae on rocks in the Horton River, Northwest Territories, that were infected with an oomycete, probably of the order Saprolegniales.

**Hosts and Economic Importance.** Hosts probably include moose (Flook 1959). Humans are among the hosts in Siberia (Rubtsov 1962d).

**Simulium (Simulium) murmanum Enderlein**
(Figs. 10.212, 10.389, 10.501, 10.595, 10.807;
Plate 21d; Map 216)

*Simulium murmanum* Enderlein 1935: 363 (F). Lectotype* female (designated by Raastad & Adler 2001; pinned #T348, terminalia and left front and hind legs on slide mount below; ZMHU). Russia, Murmansk, Renntierinsel (Reindeer Island) near Alexandrowsk, 17 July 1925 (published date), 17 July 1926 (pin date) (W. Richter)

*Simulium* 'sp.' O'Kane 1926: 21–22, Figs. 3-6 (PL)

*Simulium* 'sp. II' Puri 1926: 164 (P)

*Simulium (Simulium) corbis* Twinn 1936b: 147–148, Fig. 15B (FMP). Holotype* female (pinned #4131, terminalia and 1 leg on slide; CNC). Quebec, ca. 5 mi. south of Perkins Mills, Blanche River, 22 May (pupa), 26 May (adult) 1935 (C. R. Twinn)

*Simulium* (s. str.) *relictum* Rubtsov 1940b: 425–428 (English version pp. 516–517), Figs. 5F, 15E (FMPL). Lectotype* female (designated by Raastad, Usova & Kuusela in Raastad & Adler 2001: 184; on slide; ZISP). Russia, Irkutsk Region, Malaya Iret River, 29 June 1935 (I. A. Rubtsov)

*Gnus forsi* Carlsson 1962: 90–93, Figs. 6F, 7H, 8H, 9K, 10A, 11, 17B, 19B, 21A, 22G (FMPL). Holotype male with pupal exuviae (pinned?; presumably in G. Carlsson private collection). Sweden, Jämtland, Bispfors, Edset, 20 June 1958 (G. Carlsson)

[*Simulium tenuimanus*: Rubtsov (1956) not Enderlein 1921b (misident.)]

?[*Simulium hunteri*: Peschken (1960) not Malloch 1914 (misident.)]

[*Gnus rostratum*: Rubtsov (1962h) not Lundström 1911 (misident.)]

[*Gnus rostratum*: Golini (1970) not Lundström 1911 (misident.)]

[*Simulium rostratum*: Golini et al. (1976) not Lundström 1911 (misident.)]

[*Gnus rostratum*: Patrusheva (1982) not Lundström 1911 (misident.)]

[*Gnus rostratum*: Rubtsov & Yankovsky (1988) not Lundström 1911 (misident.)]

**Taxonomy**

O'Kane 1926 (PL); Puri 1926 (P); Enderlein 1935 (F); Twinn 1936b (FMP); Rubtsov 1940b, 1956, 1962d, 1962h (FMPL), 1964a (FMP); Nicholson 1949 (FMP); Nicholson & Mickel 1950 (FMP); DeFoliart 1951a (PL); Stone 1952 (FMP), 1964 (FMPL); Sommerman 1953 (L); Stone & Jamnback 1955 (FMPL); Peterson 1958a, 1960a (FMP); Anderson 1960 (FMPL); Carlsson 1962 (FMPL); Davies et al. 1962 (FMP); Wood et al. 1963 (L); Abdelnur 1966, 1968 (FMPL); Holbrook 1967 (L); Lewis 1973 (FMPL); Hilsenhoff 1982 (L); Currie 1986 (PL); Peterson & Kondratieff 1995 (FMPL); Zwick 1995 (F); Jensen 1997 (PL); Yankovsky 1999 (PL); Raastad & Adler 2001 (F).

*Simulium murmanum* might be a complex of species. If so, the name *S. murmanum* would apply to at least some populations in western North America, whereas the name *S. corbis* would apply to populations in eastern North America. (See also Cytology section below.) We have a sample of larvae, pupae, and a pharate male from the Northwest Territories (Dempster Hwy., south of Inuvik, near Lake Campbell, 21 July 1982) in which the pupal cocoons are delicately spun, as compared with the more typical, robust cocoons. This variation might be a result of slower water velocity or could represent a distinct species.

**Morphology**

Zhang & Malmqvist 1996 (labral fan).

**Cytology**

The IS arm has 1 fixed inversion, IIS carries fixed inversions *IIS-10.11* of Procunier (1984) plus 1 additional fixed inversion, IIL has 2 fixed inversions, and all other arms have the standard banding sequence described by Shields and Procunier (1982) for the *S. arcticum* species complex (W. S. Procunier, pers. comm.). A chromocenter is absent, and the nucleolar organizer is transposed into the distal third of the IIIL arm (section 95). Eastern populations (Labrador, Quebec, Ontario) have undifferentiated sex chromosomes, whereas western populations (Alaska, Yukon, Alberta) have at least 3 Y-chromosome sequences based on IIL; some autosomal polymorphisms are shared between eastern and western populations (Shields & Procunier in Rothfels 1988; W. S. Procunier, pers. comm.). More work is needed to determine if the different sex-chromosome systems represent distinct species.

**Bionomics**

**Habitat.** This Holarctic species is common in the boreal forests of Alaska and northeastern North America but is sporadic in the north-central region of the continent. In the Palearctic Region, it has been recorded from Fennoscandia and Russia. Larvae and pupae are most prevalent in rocky rivers 1–15 m wide and less than 20°C, where they affix themselves firmly to sticks and rocks.

**Oviposition.** Eggs are dropped singly into the water as females fly over rapids (Wolfe & Peterson 1959).

**Development.** *Simulium murmanum* is typically univoltine. It might, however, complete 2 generations in some areas (Martin 1981). Winter is passed in the egg stage. Larvae usually hatch in late April and May, and pupae are present from May into July (Sommerman et al. 1955, Lewis & Bennett 1973, Lake & Burger 1983, Currie 1986, Crosskey 1994a).

**Mating.** Unknown.

**Natural Enemies.** North American larvae are attacked by the mermithid nematodes *Gastromermis viridis* and *Mesomermis flumenalis*; the chytrid fungus *Coelomycidium simulii*; and the microsporidia *Amblyospora bracteata*, *Polydispyrenia simulii* (Ezenwa 1973, 1974b; Ebsary & Bennett 1975a); and *Janacekia debaisieuxi*. In the Palearctic Region, larvae are hosts of the mermithid nematode *Mesomermis sibirica* (Rubtsov 1972b).

**Hosts and Economic Importance.** Serological evidence indicates that females feed on horses and cattle (Downe & Morrison 1957). In Norway, they can be significant pests of cattle (Golini et al. 1976). A few females have been taken from traps baited with caribou and snowshoe hares (McCreadie et al. 1994c). Females east of Ontario can be aggravating to humans as they swarm about the face (Hocking & Richards 1952, White & Morris 1985b, Gibbs et al. 1986), but we are aware of only 1 record of a human being bitten in North America (Pistrang 1983). Females in western North America are rarely attracted to humans.

Records implicating them as biters in Alaska (Smith et al. 1952) probably are based on misidentifications. Females are said to attack humans in the Palearctic Region (Rubtsov 1956), although the frequency of biting is low (Golini et al. 1976).

### *Simulium* (*Simulium*) *arcticum* Species Complex

**Diagnosis.** Male: Ventral plate in ventral view Y shaped, slightly expanded posteriorly, or with lateral margins subparallel. Pupa: Gill of 12 short filaments. Chromosomes: Standard banding sequence as given by Shields and Procunier (1982).

**Overview.** We recognize 9 species in this strictly Nearctic complex. We apply formal names to 7 of these, 5 of which were known previously by cytological designations (in parentheses): *S. apricarium* n. sp., *S. arcticum* s. s. ('IIL-3'), *S. brevicercum* ('IIL-st'), *S. chromatinum* n. sp., *S. negativum* n. sp. ('IL-3.4'), *S. saxosum* n. sp. ('IIL-2'), and *S. vampirum* ('IIS-10.11'). We have insufficient material to describe 2 additional species that are recognized only on cytological grounds: 'cytospecies IIL-1' and 'cytospecies IIS-4.' The extent of cytological differentiation in the *S. arcticum* species complex is spectacular. Under the Cytology section, we briefly describe 10 cytotypes, which cannot be assigned to any of the 9 recognized species. Some or all of these cytotypes might represent valid species.

Information of a general nature or that cannot be assigned to 1 of the 9 species in the complex is presented below, followed by accounts of each of the 9 species that we recognize as valid.

**Taxonomy**

The following references cannot presently be assigned to the relevant members of the *S. arcticum* species complex: Johannsen 1902, 1903b (PL); Emery 1913 (PL); Dyar & Shannon 1927 (FM); Hearle 1932 (FP); Stains 1941 (FM); Stains & Knowlton 1943 (FM); Coleman 1951 (FMP); Stone 1952 (FMP); Sommerman 1953 (L); Wirth & Stone 1956 (FMP); Peterson 1958a (FMP), 1959a (E), 1960a (FMP), 1981 (F), 1996 (FM); Abdelnur 1966, 1968 (FMPLE); Corredor 1975 (FM); Speir 1975 (E); Currie 1986 (PL); Mason & Kusters 1991b (FMPL); Peterson & Kondratieff 1995 (FMPL).

Good taxonomic characters that will distinguish the adults and pupae of the species in this complex have not been found.

The name *S. arcticum*, long known as the label for the notorious cattle pest of Canada, applies to a species in the complex that is of little economic consequence. We apply the name *S. arcticum* to the 'IIL-3' cytospecies, the most likely candidate based on distribution, and use the replacement name *S. vampirum* for the cattle pest. Most literature on the *S. arcticum* species complex actually pertains to *S. vampirum* and is presented under this species.

A possible new species is known only from females taken in Alaska and the Yukon. It was noted first by Ritter and Feltz (1974), who misidentified it as *S. malyschevi*. The misidentification was pointed out by Sommerman (1977), who referred to it as *Simulium* 'n. sp?' (also as *Simulium malyschevi*?). Females are attracted to dogs, humans, and automobiles (Sommerman 1977). Females (presumably of this species) that we have seen along the Yukon River north of Whitehorse are like those of other members of the *S. arcticum* species complex but are smaller and more yellowish. We note the species here so that future workers might associate the female with the male and immature stages.

**Morphology**

The following references cannot presently be assigned to the relevant members of the *S. arcticum* species complex: Sutcliffe & McIver 1974 (F leg), 1976 (FM leg sensilla); Speir 1975 (instars); Sutcliffe 1975 (FM legs); Palmer & Craig 2000 (labral fan).

**Cytology**

Shields & Procunier 1982, Procunier & Shemanchuk 1983, Procunier 1984, Procunier et al. 1984.

Based on cytological evidence, we believe that the *S. arcticum* complex comprises more than 9 species. We have seen populations, especially in the northwestern United States, that chromosomally do not fit our 9 recognized species. Most of these are geographically restricted. Each might represent a distinct species, although we recognize them here as cytotypes pending further study. Populations in Montana are being studied by G. F. Shields.

**Cytotype 1**: Populations in northern Nevada (e.g., Humboldt Co., Water Creek, 21 April–3 July 1995) are chromocentric, and male larvae are heteromorphic for the centromere band of chromosome II (CII) and heterozygous for an inversion in the base of IIS from about the middle of section 49 to near the end of section 52 (Fig. 5.7; inversion not shown).

**Cytotype 2**: Male larvae from the Metolius River, Oregon (9 July 1988), are heteromorphic for the CII band and heterozygous for an inversion in the base of IIL from sections 57 to 58 inclusive (Fig. 5.6; inversion not shown).

**Cytotypes 3, 4, and 9**: Certain populations in western Montana (e.g., Missoula Co., Clearwater Creek and Blackfoot River; Granite Co., Rock Creek) tend to be chromocentric, with male larvae heteromorphic for the CII band and heterozygous for 1 of 3 unique inversions in the base of IIL: IIL-13 (cytotype 3), IIL-9 (cytotype 4), and IIL-17 (cytotype 9) (Fig. 5.6). We do not know if these inversions represent separate breeding populations. We do know, however, that these 3 cytotypes are reproductively isolated from *S. apricarium* n. sp. because no hybrids have been detected when these taxa occupy the same streams.

**Cytotype 5**: An additional, novel, Y-linked inversion, IIL-10, occurs in Montana (e.g., Fergus Co., Upper Spring Creek, 2 September 2000) (Fig. 5.6).

**Cytotype 6**: In the Great Bear River at its entry into Great Bear Lake, Northwest Territories (1 July 2001), all male larvae are heterozygous for inversion IIL-14 (Fig. 5.6). Female larvae have the standard sequence, and both males and females are chromocentric.

**Cytotype 7**: Male larvae in the Yellowstone and Gallatin Rivers of southern Montana (24 June 2001) are heteromorphic for the CII band and heterozygous for inversion IIL-15 (Fig. 5.6); some males carry the IIS-4 inversion (Fig. 5.7).

**Cytotype 8**: Larvae from River Between Two Mountains (38 km south of Wrigley) in the Northwest Territories (9 July 2001) are fixed chromosomally for the IS-1 inversion of Shields and Procunier (1982), and male larvae are heteromorphic for the CII band and heterozygous for inversion

IIL-16 (Fig. 5.6). Some male larvae also have heterozygous inversions in the base of IIS (e.g., IIS-4). We found a single male larva bearing the same heterozygous inversion in IIL, but heterozygous for IS-1, in the Buckinghorse River of northern British Columbia. These larvae might belong to 'cytospecies IIS-4.'

**Cytotype 9**: See under Cytotypes 3, 4, and 9 above.

**Cytotype 10**: Male larvae from the Smith River, Del Norte Co., California (31 May 2002); the Rogue River, Curry Co., Oregon (29 May 2002); and the Illinois River, Josephine Co., Oregon (8 May 1986), are heterozygous for inversion IIL-12 (Fig. 5.6). This inversion also characterizes male larvae from the Thelon River (early July 2002) in the Northwest Territories and Nunavut; however, the chromosomes of both female and male larvae in the Thelon River (unlike those of larvae in California and Oregon) are fixed for the IS-1 inversion of Shields and Procunier (1982).

**Bionomics**

**Habitat.** This species complex is abundant throughout western North America, breeding in a majority of the rocky streams and rivers. Larvae are found at higher macrovelocities than most species (Fredeen & Shemanchuk 1960, Craig & Galloway 1988, Eymann 1993). Larvae held in standing water at 19°C can survive for 3 days or more, although mortality is high (Anderson & Shemanchuk 1975). The pupal gills are bathed by 2 pairs of vortices that form as water passes around the boot-shaped, fenestrated cocoon (Eymann 1991b).

**Oviposition.** Females carry 116 to more than 200 eggs per ovarian cycle and typically drop them into smooth water while flying upstream; occasionally females crawl beneath the water or on a rock to deposit eggs (Peterson 1956, 1959a; Abdelnur 1968).

**Development.** Eggs and larvae overwinter (Sommerman et al. 1955, Speir & Anderson 1974, Harkrider 1988). Eggs can remain viable after storage at 0.5°C–1.5°C for 2–9 months (Fredeen 1959b). Some species in the complex are multivoltine. Others consist of a single generation with multiple cohorts that can be found from spring to fall (Anderson & Shemanchuk 1987b). At least 6 larval instars are produced (Speir 1975). In the Canadian subarctic, about 5%–13% of larval gut contents consist of algae, although not all algal species are digested (Moore 1977a, 1977b). Cocoon construction requires about 50 minutes at 14.5°C, and the pupal stage lasts 2.5–4.3 days at about 14°C–16°C (Speir 1975). Males emerge before females, often by several days (Speir 1969).

**Mating.** Swarms of males, females, and mated pairs have been observed flying up and down a stream about a meter above the water (Peterson 1959a).

**Natural Enemies.** Larvae and adults often are infected with mermithid nematodes, such as *Gastromermis* (Peterson 1960b, Speir 1975). Trichomycete fungi recorded from larval guts include *Genistellospora homothallica*, *Harpella melusinae*, *Pennella angustispora*, *P. arctica*, and *Simuliomyces microsporus* (Lichtwardt 1967, 1972, 1984, 1986). A single larva infected with a possible fungus resembling *Tolypocladium* has been taken from San Dimas Canyon in southern California (J. R. Harkrider, pers. comm.) (as in Fig. 6.7).

Predators of larvae include trichopteran larvae (Peterson 1960b) and the crayfish *Pacifastacus leniusculus* (Speir 1969). Empidid flies prey on adults (Peterson 1960b). An individual parasitoid of the family Scelionidae that emerged from a reared female of the *S. arcticum* complex (Peterson 1960b) represents the only North American record of Hymenoptera parasitizing a black fly.

**Hosts and Economic Importance.** Most host records and accounts of destructive feeding by *S. arcticum* on the Canadian prairies (e.g., Fredeen 1977b) pertain to *S. vampirum* (treated below). Records not attributable to *S. vampirum* include blood feeding in the ears of horses (Knowlton & Maddock 1944, Peterson 1959a), on cattle, on a dog (Peterson 1959a), on moose (Pledger et al. 1980), and on humans (Sommerman et al. 1955, Peterson 1959a). These limited records suggest that members of the *S. arcticum* species complex, other than *S. vampirum*, are not pests. Snowshoe hare virus has been isolated from females of an unknown member of the complex (Sommerman 1977).

***Simulium* (*Simulium*) *apricarium* Adler, Currie & Wood, New Species**

(Fig. 5.7, Map 217)

?*Simulium* 'species' Johannsen 1902: 53, Plate 3 (Figs. 4–7); 1903b: 387–388, Plate 35 (Figs. 4–7) (PL)

[*Simulium* (*Simulium*) *arcticum*: Pruess & Peterson (1987) not Malloch 1914 (misident.)]

[*Simulium arcticum*: Zhu (1990) not Malloch 1914 (misident.)]

?[*Simulium articum* [*sic*]: Kiffney et al. (1997b) not Malloch 1914 (misident.)]

**Taxonomy**

**Female.** Wing length 3.0–3.7 mm. Scutum brassy brown, subshining, with pair of variably distinct, anterior, silvery pruinose spots. All hair pale golden. Mandible with 40–50 serrations; lacinia with 24–31 retrorse teeth. Sensory vesicle in lateral view occupying about one fourth to one third of palpomere III. Frons shiny. Fore leg pale yellow from coxa through basal three fourths of tibia, although femur sometimes suffused with brown; remainder of leg dark brown. Middle and hind legs pale yellow, with brown on each coxa, apical two thirds of femur, apical one third of tibia and basitarsus, and apical one third of second tarsomere through last tarsomere. Claws each with minute subbasal tooth. Terminalia as in Fig. 10.213. Anal lobe in ventral view large, rounded, well sclerotized, with minute unsclerotized spot on mesal margin. Genital fork with arms slender, usually with anteriorly directed apodeme arising from each lateral plate. Spermatheca subspherical; wall rather smooth, with internal spicules.

**Male.** Wing length 3.1–3.7 mm. Scutum velvety black, with pair of anterior, oblique silvery patches; lateral and posterior areas silvery, pruinose; hair golden. All remaining hair coppery brown. Legs mostly brown, with paler patches, especially on fore tibia and bases of segments. Genitalia as in Fig. 10.390. Gonostylus in ventral view as wide at apex as at base. Ventral plate in ventral view with body as narrow as arms, toothed apically. Median sclerite elongate, expanded distally.

**Pupa.** Length 3.0–4.1 mm. Gill of 12 short filaments in 3 groups (as in Fig. 10.596): dorsal group of 2 petiolate pairs arising from common trunk, medial group of 2 petiolate pairs each arising independently from base or from short trunk, ventral group of 2 petiolate pairs arising from common trunk; filaments with numerous transverse furrows. Head and thorax dorsally with densely and irregularly distributed, dome-shaped microtubercles; trichomes unbranched. Cocoon boot shaped, corbicular, densely woven.

**Larva.** Length 6.6–8.0 mm. Head capsule yellowish brown or brownish orange to brown; head spots brown, diffuse, variably distinct, sometimes obscured in surrounding brown pigment; anteromedial group weakest. Antenna yellowish brown to brown, extended beyond stalk of labral fan by about one half length of distal article; medial article with hyaline spot. Hypostoma with all teeth small; lateral and median teeth extended anteriorly to about same level. Postgenal cleft subtriangular, bowed outward at midpoint, extended nearly to hypostomal groove; epidermis in cleft sometimes slightly pigmented; subesophageal ganglion at least partially ensheathed with pigment. Labral fan with 32–46 primary rays. Body pigment grayish brown to brown, arranged in bands. Posterior proleg with 10–16 hooks in each of 79–90 rows.

**Diagnosis.** Adults, pupae, and larvae cannot be distinguished from those of other members of the *S. arcticum* species complex. Chromosome inversion IIS-11 provides the most reliable means of identifying this species (see Cytology section below).

**Holotype.** Female (hexamethyldisilazane, pinned)with pupal exuviae (glycerin vial below). Wyoming, Lincoln Co., Green River, Rt. 189, south of LaBarge, 42°14.80′N, 110°10.90′W, 8 June 1994, P. H. Adler (USNM).

**Paratypes.** Arizona, Coconino Co., Coconino National Forest, Oak Creek, Rt. 89, north of Sedona and Slide Rock, 34°59′N, 111°44′W, 18 April 1996, P. H. Adler & J. K. Moulton (33 larvae). Nebraska, Sioux Co., White River, 1.6 km west of Glen, 27 March 1989, K. P. Pruess (36 larvae). Utah, Uintah Co., Jones Hole Creek, 1.6 km from confluence with Green River, Dinosaur National Monument, 40.33°N, 109.03°W, 19 June 1998, W. K. Reeves (21 pupae, 59 larvae). Wyoming, same data as holotype (11♀ & 8♂ + exuviae [pinned], 13♀ & 4♂ + exuviae [ethanol], 22 pupae, 79 larvae), 7 June 1992 (8 pupae, 19 larvae, photographic negatives of larval polytene chromosomes [chromosome II only]).

**Etymology.** The specific name is from the Latin adjective meaning "of the open," in reference to the typical habitat in which the species is found.

**Cytology**

Under this species we include populations in which inversion IIS-11 of Procunier (1984) is present. This same inversion appears to characterize the first of 2 inversions in the IIS arm of *S. vampirum*, and when imposed on the *S. arcticum* standard banding sequence of Shields and Procunier (1982), it mimics the IIS-B sequence of Rothfels et al. (1978). Although deceptively close to the IIS-B sequence, the IIS arm of *S. apricarium* actually differs by 2 inversions between the beginning of section 45 and the ring of Balbiani on the standard map of Shields and Procunier (1982); these 2 inversions remove a few thin bands from section 45 and place them next to the ring of Balbiani. Most populations of *S. apricarium* are fixed for IIS-11 (Fig. 5.7), but in all other arms they have the standard banding sequence for the *S. arcticum* complex. The fixation of IIS-11 in most populations of *S. apricarium* provides evidence of reproductive isolation from *S. arcticum* s. s. and *S. chromatinum* n. sp. when these species occur in the same streams. An additional inversion, here designated IIL-7 (Fig. 5.6), is present in most populations of *S. apricarium*, with males predominantly heterozygous. Males also tend to be heteromorphic for the centromere band of chromosome II (females being homozygous for the enhanced band) and, in some populations, heterozygous for IIS-4. A chromocenter can be present or absent.

In some peripheral western populations (e.g., California's Kern and Mokelumne Rivers), IIS-11 tends to be polymorphic and IIL-7 absent. Some California populations have an additional inversion in IIL (limits 58–60 inclusive), which has been found in males (and 1 female) only in the heterozygous state.

**Molecular Systematics**

Zhu 1990 (mtDNA).

**Bionomics**

**Habitat.** This common species inhabits the lowlands of the western United States. It is the only member of the *S. arcticum* complex consistently found in dry open areas, extending even into the Great Plains of Nebraska and the southern desert of Arizona. Streams harboring the immatures are swift, more than 4 m wide, and sometimes turbid, with at least some reaches of rock or gravel substrate. Larvae and pupae are found predominantly on trailing vegetation and debris.

**Oviposition.** Unknown.

**Development.** Larvae are often present throughout the year, and pupae are found from March to November or year-round at the southern limit of distribution. At least 4 generations are produced annually in Nebraska (Pruess & Peterson 1987).

**Mating.** Unknown.

**Natural Enemies.** Larvae are attacked by the chytrid fungus *Coelomycidium simulii*; the microsporidia *Amblyospora* sp., *Janacekia debaisieuxi*, and *Polydispyrenia simulii*; and unidentified mermithid nematodes. We have 1 larva from Oak Creek Canyon in Coconino Co., Arizona, that harbors a possible new species of microsporidium with pyriform spores in clusters of 8 (as in Fig. 6.14). Larvae are hosts of the trichomycete fungus *Harpella leptosa*.

**Hosts and Economic Importance.** Unknown; presumably mammalophilic.

### *Simulium (Simulium) arcticum* Malloch

(Figs. 4.23, 5.5, 10.390, 10.502, 10.596; Plate 21e; Map 218)

*Simulium arcticum* Malloch 1914: 37, Plate II (Fig. 4) (F). Holotype* female (pinned #15410; USNM). British Columbia, Kaslo, 4 July (probably 1903) (R. P. Currie) (Malloch 1914 gives H. G. Dyar as collector); *articum* (subsequent misspelling)

*Simulium* 'IIL-3 *arcticum*' Shields & Procunier 1982: 185–187, Fig. 9c, d (C)

**Taxonomy**

Malloch 1914 (F).

This cytological segregate is the best candidate for true *S. arcticum*, based on distribution and habitat. For purposes of nomenclatural stability, therefore, we link the name *S. arcticum* to the 'IIL-3' cytospecies. The pupae and adults of *S. arcticum* s. s. are essentially indistinguishable from those of other members of the *S. arcticum* species complex. Reliable identifications rest on chromosomal examination.

**Morphology**

Shipp & Kokko 1987 (F mouthparts); Sutcliffe et al. 1987 (FM palpal bulb sensilla), 1990 (F antennal sensilla); Shipp et al. 1988b (F antennal sensilla).

**Cytology**

Shields & Procunier 1982, Procunier 1984 (adult polytene chromosomes), Procunier et al. 1984 (adult polytene chromosomes).

The limits of the IIL-3 inversion given by Shields and Procunier (1982) in their Fig. 6 include 1 band more than in their Fig. 9c. Additional confusion arises because the IIL-3 inversion is actually mapped on top of the IIL-1 inversion in their Fig. 6. (Although the figure caption states that Fig. 6 is the standard IIL arm, it is really heterozygous for the IIL-1 inversion.) Fig. 9c of Shields and Procunier (1982) shows the actual IIL-3 inversion (heterozygously), although the section limits of 56 in their figure include 1 less band—the last band in section 55—than in their standard map shown in Fig. 6. The breakpoints of the IIL-3 inversion also can be seen in our Fig. 5.6.

**Bionomics**

**Habitat.** This species is common from northern Saskatchewan into the central Rocky Mountains. Clear cool streams and rivers up to 30 m or more in width, with cobble, boulders, and swift flow provide typical habitat for larvae and pupae. Ambient temperature and vapor pressure significantly influence the flight activity of females (Shipp & Grace 1988, Shipp et al. 1988a). Flight activity is more sustained at higher humidities (Grace & Shipp 1988).

**Oviposition.** Females are obligatorily anautogenous (Anderson & Shemanchuk 1987a). Oviposition behavior is unknown.

**Development.** Larvae are present throughout the year. Up to 3 generations are completed annually, and mean developmental time for the 7 larval instars requires 246 and 309 degree-days above 0°C for the summer and overwintered generations, respectively (Shipp & Procunier 1986). Larval abundance is positively correlated with water temperature, and peak abundance coincides with maximum discharge in the spring (Shipp & Procunier 1986).

**Mating.** Unknown.

**Natural Enemies.** Larvae are attacked by the chytrid fungus *Coelomycidium simulii* and unidentified mermithid nematodes.

**Hosts and Economic Importance.** Unknown; presumably mammalophilic.

## *Simulium* (*Simulium*) *arcticum* Malloch 'cytospecies IIL-1'

(Map 219)

'Two siblings . . .' Shields & Procunier 1981: 151 (in part, Alaskan record)

*Simulium* 'IIL-1 *arcticum*' Shields & Procunier 1982: 184–186, Figs. 6, 9a, b (C)

**Cytology**

Shields & Procunier 1982, Procunier & Shemanchuk 1983.

We have not seen material of this cytological segregate; however, its consistent chromosomal composition across Alaska and half of Canada supports its species status. Figure 6 in Shields and Procunier (1982) was labeled as the standard configuration for the IIL arm, but it is actually heterozygous for the IIL-1 inversion; the top homologue carries the inversion. Based on the limits given by Shields and Procunier (1982), we show the breakpoints in our Fig. 5.6.

**Bionomics**

**Habitat.** This taxon is known from Alaska, Alberta, and Manitoba. The immature stages live in cool rocky rivers.

**Oviposition.** Unknown.

**Development.** Larvae are present from at least April through September, suggesting that 2 or more generations are produced yearly (Shields & Procunier 1982, Procunier 1984).

**Mating.** Unknown.

**Natural Enemies.** No records.

**Hosts and Economic Importance.** Unknown; presumably mammalophilic.

## *Simulium* (*Simulium*) *arcticum* Malloch 'cytospecies IIS-4'

(Map 220)

*Simulium* 'IIS-4 *arcticum*' Procunier 1984: 29 (Alberta record)

**Cytology**

This cytological segregate was mentioned as a new sibling species by Procunier (1984) but was not described cytologically. Procunier (pers. comm.) later characterized it as fixed for *IS-1*, with 2 X and 7 Y chromosomes, including IIS-4 as a Y chromosome (limits as in Shields & Procunier 1982). Because it has the *IS-1* inversion fixed and is distinct when sympatric with at least 4 other members of the *S. arcticum* complex (Procunier 1984), we consider it a valid species, although we have not seen material.

**Bionomics**

**Habitat.** This species was described as ubiquitous in central Alberta (Procunier 1984). It breeds in swift cool rocky rivers ranging in width from about 6 to more than 100 m.

**Oviposition.** Unknown.

**Development.** Larvae have been collected from May through August (Procunier 1984).

**Mating.** Unknown.

**Natural Enemies.** No records.

**Hosts and Economic Importance.** Unknown; presumably mammalophilic.

**_Simulium (Simulium) brevicercum_ Knowlton & Rowe**
(Fig. 10.213, Plate 21f, Map 221)

*Simulium brevicercum* Knowlton & Rowe 1934a: 583, Figs. 7, 8 (F). Holotype* female (pinned #51405; USNM). Utah, Cache Co., Logan Canyon, 4 May 1933 (published date), 14 May 1933 (pin label) (J. S. Stanford); *brevicerum* (subsequent misspelling). Revalidated

*Simulium nigresceum* Knowlton & Rowe 1934a: 583–584, Figs. 9, 10 (M). Holotype* male (pinned, terminalia on slide; USNM). Utah, Cache Co., Logan, 10 April 1909 (collector unknown). New Synonym

'Two siblings . . .' Shields & Procunier 1981: 151 (in part, Alaskan record)

*Simulium* 'IIL-st *arcticum*' Shields & Procunier 1982: 184–186, Fig. 2 (C)

[*Simulium corbis*: Knowlton et al. (1938) not Twinn 1936b (misident.)]

[*Simulium corbis*: Peterson (1958c) not Twinn 1936b (misident.)]

**Taxonomy**

Knowlton & Rowe 1934a (FM).

We apply the name *S. brevicercum* (and its synonym *nigresceum*) to the cytological segregate described below, which is the only member of the *S. arcticum* complex that we have found in collections at the type locality.

In Utah, larval males are brownish with rather distinct head spots, whereas larval females are grayish with head spots that are typically obscured by surrounding pigment. Pupae and adults are virtually indistinguishable from those of other members of the *S. arcticum* species complex.

**Cytology**

Shields & Procunier 1982.

The standard banding sequence for the *S. arcticum* species complex is preserved in all arms, and a chromocenter is generally present. In most populations, the sex chromosomes are undifferentiated. At the type locality, however, males are typically heteromorphic for the CII band, exhibit nonpairing in the base of IIL from section 55 into 57 even though the region is not inverted, and often carry various repatternings and differential band expressions in section 56 and in the base of IIS, especially at the junction of sections 53 and 54 and in sections 50 and 51. Populations with undifferentiated sex chromosomes might represent a species distinct from topotypical material, but until additional evidence materializes, we prefer to recognize both chromosomal types under a single species.

**Bionomics**

**Habitat.** Populations of *S. brevicercum* have been found from southern Alaska into the mountains of central Utah, as well as in the high Sierra Nevada of California. The immature stages occupy cool rocky flows more than 4 m wide that typically tumble through forested areas.

**Oviposition.** Unknown.

**Development.** Most populations overwinter as larvae. In Alaska, the larvae appear earlier than those of other members of the *S. arcticum* species complex and are claimed to have 1 generation annually (Shields & Procunier 1982). Most of our collections were made from March through June, although we also have material from July and August.

**Mating.** Unknown.

**Natural Enemies.** Larvae often are infected with unidentified mermithid nematodes, the chytrid fungus *Coelomycidium simulii*, and the microsporidia *Polydispyrenia simulii* and *Amblyospora bracteata*. *Rhyacophila* caddisflies prey on larvae.

**Hosts and Economic Importance.** Unknown; presumably mammalophilic.

**_Simulium (Simulium) chromatinum_ Adler, Currie & Wood, New Species**
(Fig. 10.808, Map 222)

[*Simulium arcticum*: Tang et al. (1996b) not Malloch 1914 (misident.)]

**Taxonomy**

**Female.** Not differing from that of *S. apricarium* n. sp.

**Male.** Not differing from that of *S. apricarium* n. sp.

**Pupa.** Not differing from that of *S. apricarium* n. sp.

**Larva.** Not differing from that of *S. apricarium* n. sp. except as follows: Length 6.2–8.2 mm. Head capsule (Fig. 10.808) dark brown with faint pale head spots at most, or brownish orange with either no head spots or diffuse brown head spots partially obscured by surrounding pigment on posterior half of frontoclypeal apotome. Labral fan with 41–50 primary rays. Body pigment grayish brown to dark brown, uniformly distributed or weakly arranged in bands. Posterior proleg with 9–16 hooks in each of 94–96 rows.

**Diagnosis.** Adults and pupae are not distinguishable from those of other members of the *S. arcticum* species complex. Larvae tend to be darker than those of other members of the complex. The large chromocenter and the X chromosome provide the best means of identification. (See Cytology section below.)

**Holotype.** Male (hexamethyldisilazane, pinned) with pupal exuviae (glycerin vial below). Arizona, Apache Co., West Fork Little Colorado River, State Road 1120, Greer, 33°59.6′N, 109°27.9′W, 2590 m elev., 21 May 1994, J. K. Moulton (USNM).

**Paratypes.** Arizona, same data as holotype (5♀ & 8♂ + exuviae [pinned], 3♀ & 13♂ + exuviae [in ethanol], 7 pupae, 131 larvae). Colorado, Jackson Co., North Michigan River, Ranger Lakes, side channel from beaver pond, 40°33′N, 106°00′W, 3 August 1989, K. P. Pruess (102 larvae); Park Co., outflow of Oliver Twist Lake, Forest Road 438, 39°17.65′N, 106°09.83′W, 3749 m elev., 24 August 1997, P. H. Adler & D. C. Currie (1 pupa, 42 larvae). Wyoming, Lincoln Co., Rt. 89, Allred Flat Recreation Area, Little White Creek, 42°29.20′N, 110°57.66′W, 28 May 1989, P. H. Adler (7 pupae, 37 larvae).

**Etymology.** The specific name is from Greek, meaning "color," in reference to the large, darkly staining chromocenter of this species.

**Cytology**

The large heterochromatic chromocenter, visible even in interphase nuclei of somatic tissues, is diagnostic. The Y chromosome is typically standard, but the X chromosome is characterized by inversion IIL-11 (Fig. 5.6). The X-linked inversion is strikingly close to *IIL-8*, which occurs in *S. vampirum*, and might prove, on closer inspection, to be identical to it.

**Molecular Systematics**

Tang et al. 1996b (mtDNA).

**Bionomics**

**Habitat.** A denizen of the southern Rocky Mountains, this species has been taken from Wyoming to southern Arizona. Larvae and pupae live in cool, clear rocky streams about 1–4 m wide, flowing through mixed hardwood-conifer forests of mountainous terrain.

**Oviposition.** Unknown.

**Development.** Mature larvae have been found from May through August.

**Mating.** Unknown.

**Natural Enemies.** Larvae often are attacked by the chytrid fungus *Coelomycidium simulii*, the microsporidia *Amblyospora bracteata* and *Janacekia debaisieuxi*, and an unidentified mermithid nematode.

**Hosts and Economic Importance.** Unknown; presumably mammalophilic.

### *Simulium* (*Simulium*) *negativum* Adler, Currie & Wood, New Species

(Figs. 10.72, 10.809; Plate 22a; Map 223)

*Simulium* 'IL-3.4 *arcticum*' Shields & Procunier 1982: 187 (C)

**Taxonomy**

**Female.** Not differing from that of *S. apricarium* n. sp. except as follows: Wing length 2.6–2.9 mm. All femora more extensively brown.

**Male.** Not differing from that of *S. apricarium* n. sp. except as follows: Wing length 2.5–2.8 mm.

**Pupa.** Not differing from that of *S. apricarium* n. sp. except as follows: Length 3.1–3.8 mm.

**Larva.** Not differing from that of *S. apricarium* n. sp. except as follows: Head capsule of female brown, often pale yellowish at anterior and lateral margins of frontoclypeal apotome, typically with pale yellow head spots (Plate 22a); that of male (Fig. 10.809) pale yellowish, with faint brown head spots. Subesophageal ganglion not ensheathed with pigment. Labral fan with 33–43 primary rays. Body pigment gray to brown, arranged in distinct bands. Posterior proleg with 12–17 hooks in each of 83–86 rows.

**Diagnosis.** Adults and pupae are not easily separated from those of other members of the *S. arcticum* species complex. The negative head-spot pattern of the female and the unpigmented subesophageal ganglion provide a useful diagnostic aid for distinguishing the larvae from those of other members of the *S. arcticum* species complex.

**Holotype.** Male (hexamethyldisilazane, pinned) with pupal exuviae (glycerin vial below). Wyoming, Uinta Co., Bear River, Evanston, Interstate 80, exit 6, 41°16.08′N, 110°56.74′W, 8 June 1994, P. H. Adler (USNM).

**Paratypes.** Colorado, Moffat Co., Yampa River, Anderson Hole, Dinosaur National Monument, 40.28°N, 108.34°W, 16 June 1998, W. K. Reeves (19 pupae, 3 larvae). Nevada, Lander Co., Argenta, Humboldt River, 40°40.8′N, 116°42.3′W, 23 April 1995, R. D. Gray (1 pupa, 27 larvae). Wyoming, Lincoln Co., Rt. 89, 9.2 km south of Allred Flat Recreation Area, Dipper Creek, 42°25′N, 110°59′W, 28 May 1989, P. H. Adler (55 pupae, 59 larvae); same data as holotype (9♀ & 5♂ + exuviae [pinned], 51♀ & 14♂ + exuviae [ethanol], 70 pupae, 125 larvae), 29 May 1989 (10 pupae, 49 larvae). Yukon, Yukon River, Carmacks, 30 July 1994, D. C. Currie & D. M. Wood (51 larvae).

**Etymology.** The specific name is from Latin, in reference to the negative head-spot pattern of the female larva.

**Morphology**

Evans & Adler 2000 (spermatheca).

**Cytology**

Shields & Procunier 1982.

We have examined the chromosomes of populations from the Yukon Territory to New Mexico that are like those described by Shields and Procunier (1982) except that they are fixed for *IIS-10.11* of Procunier et al. (1984), rather than for IIS-5 of Shields and Procunier (1982) that to our eyes is equivalent to IIS-10 of Procunier et al. (1984).

**Bionomics**

**Habitat.** This rather small species is distributed from Alaska to New Mexico. It breeds in a variety of flows ranging from shallow rocky streams as little as a meter in width to the monstrous Yukon River half a kilometer wide.

**Oviposition.** Unknown.

**Development.** Two or more generations are produced annually (Shields & Procunier 1982). Larvae and pupae are present from at least June into August in Canada and Alaska. Pupae can be found in April farther south, such as in Nevada.

**Mating.** Unknown.

**Natural Enemies.** Larvae are hosts of an unidentified mermithid nematode, the microsporidia *Amblyospora* sp. and *Polydispyrenia simulii*, and 2 possibly new microsporidia, one with narrow ovate spores in clusters of 6 (Yampa River, Moffat Co., Colorado) and another with pyriform spores in clusters of 8 (Yukon River, Carmacks, Yukon) (as in Fig. 6.14).

**Hosts and Economic Importance.** Unknown; presumably mammalophilic.

### *Simulium* (*Simulium*) *saxosum* Adler, Currie & Wood, New Species

(Figs. 10.391, 10.810; Map 224)

*Simulium* 'IIL-2 *arcticum*' Shields & Procunier 1982: 184–185 (C)

**Taxonomy**

**Female.** Not differing from that of *S. apricarium* n. sp.

**Male.** Not differing from that of *S. apricarium* n. sp. Genitalia as in Fig. 10.391.

**Pupa.** Not differing from that of *S. apricarium* n. sp.

**Larva.** Not differing from that of *S. apricarium* n. sp. except as follows: Length 5.9–7.6 mm. Head capsule (Fig. 10.810) pale yellowish to brown; head spots variably expressed and infuscated. Labral fan with 40–47 primary rays. Body pigment gray, arranged in distinct bands. Posterior proleg with 12–22 hooks in each of 94–106 rows.

**Diagnosis.** Adults, pupae, and larvae resemble those of most other species in the *S. arcticum* species complex. The X chromosome, marked by inversion IIL-2 (Shields & Procunier 1982), provides the best means of identifying this species (Fig. 5.6).

**Holotype.** Female (Peldri II, pinned) with pupal exuviae (glycerin vial below). California, Marin Co., Samuel P. Taylor State Park, Lagunitas (Papermill) Creek, 38°01.34′N, 122°44.08′W, 21 March 1990, P. H. & C. R. L. Adler (USNM).

**Paratypes.** Alaska, George Parks Hwy., mi. 88, Sheep Creek, 61°59.7′N, 150°03.0′W, 18 July 1989, D. M. Wood (4 pupae, 169 larvae). California, same data as holotype (1♀ + exuviae, 57 larvae); Sonoma Co., Russian River, Healdsburg, Healdsburg Memorial Beach Co. Park, below dam, 38°35′N, 122°51′W, 22 March 1990, P. H. & C. R. L. Adler (3♀ & 3♂ + exuviae, 1 pupa, 22 larvae). Oregon, Benton Co., Beaver Creek, south of Philomath, 44°27′N, 123°22′W, 21 March 1992, D. C. Currie (5 pupae, 63 larvae); Clatsop Co., Nehalem River, Fishhawk Road, 6 km west of Mist, 46°00′N, 123°20′W, 21 March 1992, D. C. Currie (2 pupae, 57 larvae).

**Etymology.** The specific name is from the Latin adjective meaning "rocky," in reference to the habitat of the immature stages.

**Cytology**

Shields & Procunier 1982, Currie & Adler 1986.

**Bionomics**

**Habitat.** *Simulium saxosum* is well represented along the Pacific Coast from northern Alaska to central California but is uncommon inland to central Alberta. Larvae and pupae inhabit rocky flows from about 4 m to more than 100 m wide.

**Oviposition.** Unknown.

**Development.** Larvae begin hatching in the winter, and pupae begin to appear in March along the coast of the Pacific Northwest. Development is delayed at higher elevations and latitudes. Two generations possibly are produced in Alaska (Shields & Procunier 1982).

**Mating.** Unknown.

**Natural Enemies.** Larvae are hosts of the chytrid fungus *Coelomycidium simulii*; unidentified mermithid nematodes (Currie & Adler 1986); and the microsporidia *Amblyospora varians*, *Janacekia debaisieuxi*, and *Polydispyrenia simulii*.

**Hosts and Economic Importance.** Unknown; presumably mammalophilic.

### *Simulium (Simulium) vampirum* Adler, Currie & Wood

(Fig. 10.214, Map 225)

*Simulium vampirum* Adler, Currie & Wood (New Substitute Name for *similis* Cameron 1918)

*Simulium similis* Cameron (attributed to Malloch) 1918: 558–559 (FL). Neotype* female (here designated; pinned; USNM). Saskatchewan (probably near Saskatoon, ca. 52°N, 106°W], cow farm, 27 June 1917 (A. E. Cameron). Junior primary homonym of *S. simile* Silva Figueroa 1917: 33–35.

'black flies' Anonymous 1930: 260 (attacks on livestock)

*Simulium* 'IIS-8.9 *arcticum*' Procunier & Shemanchuk 1983: 35 (C)

*Simulium* 'IIS-10 *arcticum*' Procunier & Shemanchuk 1983: 36 (C)

*Simulium* 'IIS-10 (IL-8) *arcticum*' Procunier & Shemanchuk 1983: 36 (C)

*Simulium* '(IIS-10.11) *arcticum*' Procunier et al. 1984: 37–40 (C)

[*Simulium simile*: Cameron (1922) not Malloch 1919 (misident.)]

[*Simulium arcticum*: Millar & Rempel (1944) not Malloch 1914 (misident.)]

[*Simulium arcticum*: Rempel & Arnason (1947) not Malloch 1914 (misident.)]

[*Simulium arcticum*: Arnason et al. (1949) not Malloch 1914 (misident.)]

[*Simulium arcticum*: Fredeen et al. (1951, 1953a, 1953b) not Malloch 1914 (misident.)]

[*Simulium arcticum*: Fredeen (1951, 1956, 1958, 1969, 1981a, 1985a) not Malloch 1914 (misident. in part, illustrations and pest populations)]

[*Simulium arcticum*: MacNay (1954) not Malloch 1914 (misident.)]

[*Simulium arcticum*: Fredeen (1959a, 1960, 1961, 1963, 1964b, 1970a, 1973, 1974, 1975a, 1975b, 1976, 1977a, 1977b, 1979, 1983) not Malloch 1914 (misident.)]

[*Simulium arcticum*: Sutcliffe & McIver (1976) not Malloch 1914 (misident.)]

[*Simulium arcticum*: Depner (1978, 1979) not Malloch 1914 (misident.)]

[*Simulium arcticum*: Fredeen & Spurr (1978) not Malloch 1914 (misident.)]

[*Simulium arcticum*: Khan (1978, 1981) not Malloch 1914 (misident.)]

[*Simulium arcticum*: Shemanchuk (1978b, 1980a, 1980b, 1982a, 1987) not Malloch 1914 (misident.)]

[*Simulium arcticum*: Haufe & Croome (1980) not Malloch 1914 (misident.)]

[*Simulium arcticum*: Shemanchuk & Anderson (1980) not Malloch 1914 (misident.)]

[*Simulium arcticum*: Charnetski & Haufe (1981) not Malloch 1914 (misident.)]

[*Simulium arcticum*: Ryan & Hilchie (1982) not Malloch 1914 (misident.)]

[*Simulium arcticum*: Shemanchuk & Taylor (1983) not Malloch 1914 (misident.)]

[*Simulium arcticum*: Shipp (1983, 1985b, 1985c) not Malloch 1914 (misident.)]

[*Simulium arcticum*: Schaefer & Bent (1984) not Malloch 1914 (misident.)]

[*Simulium arcticum*: Shipp & Byrtus (1984) not Malloch 1914 (misident.)]

[*Simulium arcticum*: Khan & Kozub (1985) not Malloch 1914 (misident.)]

[*Simulium arcticum*: Anderson & Shemanchuk (1987a) not Malloch 1914 (misident. in part, material from Athabasca River)]

[*Simulium arcticum*: Anderson & Shemanchuk (1987b) not Malloch 1914 (misident.)]

[*Simulium arcticum*: Shipp et al. (1987) not Malloch 1914 (misident.)]

[*Simulium arcticum*: Shipp & Grace (1988) not Malloch 1914 (misident.)]

[*Simulium arcticum*: Mason & Shemanchuk (1990) not Malloch 1914 (misident.)]

[*Simulium arcticum*: Sutcliffe & Shemanchuk (1993) not Malloch 1914 (misident.)]
[*Simulium arcticum*: Riegert (1999a, 1999b) not Malloch 1914 (misident.)]

**Taxonomy**

Cameron 1918 (FL), 1922 (FMPLE); Fredeen 1951 (FMPL), 1981a (FMPL), 1985a (FMP).

Cameron sent females of this species to Malloch (probably in 1917 or 1918), who identified the material as *S. similis* and noted that he recently had described the species (Cameron 1922). But Cameron (1918) published the name, along with illustrations of larvae and a female plus a description of host-attacking behavior, thus validating the name a year before Malloch's (1919) description actually appeared. The name authored by Malloch, however, pertains to the species now known as *S. decimatum*. No type specimen of *S. simile* Cameron exists; we, therefore, designate a neotype to establish the concept of the name of this important pest species. The specimen chosen as neotype is a female collected by Cameron and probably examined by him during the preparation of his 1918 paper. Because the name *simile* Cameron (1918) is a junior primary homonym of the name *simile* Silva Figueroa (1917), we provide the replacement name *vampirum*, which is the Latinized form of the word "vampire," in reference to the legendary blood-sucking habits of the females of this species.

We treat as *S. vampirum* all literature referring to *S. arcticum* from the problem zones of Alberta and Saskatchewan. Adults and pupae of *S. vampirum* resemble those of most other species in the *S. arcticum* species complex. The larvae are the palest of any species in the complex. The chromosomes, specifically the combination of fixed inversions *IIL-8* and *IIS-10.11* (Procunier 1984), provide the most reliable means of identifying this species.

**Morphology**

Fredeen 1970a (gynandromorphs, intersexes), 1976 (instars); Sutcliffe & McIver 1976 (FM leg sensilla); Shipp & Kokko 1987 (F mouthparts); Shipp 1988 (embryology); Shipp et al. 1988b (F antennal sensilla); Sutcliffe et al. 1990 (F antennal sensilla).

**Cytology**

Procunier & Shemanchuk 1983, Procunier 1984, Procunier et al. 1984.

**Bionomics**

**Habitat.** This species is restricted to the lowlands of Alberta, Saskatchewan, and the southern portion of the Northwest Territories. It breeds especially in the rapids of big, nutrient-rich silty rivers such as the mighty Athabasca, Slave, and North Saskatchewan that are subjected to severe ice scouring during the spring. Smaller populations inhabit the irrigation canals of southern Alberta (Procunier & Shemanchuk 1983). After hatching, first-instar larvae drift downstream, attaching themselves to gravel on the flooded beaches; later instars form masses on boulders in swift rapids (Fredeen et al. 1951). More than 7 billion pupae were estimated once to exist in a rocky weir across the North Saskatchewan River (Fredeen 1958).

**Oviposition.** Females are anautogenous and can produce mature eggs within 4.5 days at 18°C–21°C after feeding on cattle blood (Anderson & Shemanchuk 1987a, 1987b). Earlier findings (Fredeen 1963) suggesting that females are autogenous for the first ovarian cycle have not been supported (Anderson & Shemanchuk 1987a). Females carry as many as 294 eggs and complete at least 2 ovarian cycles (Anderson & Shemanchuk 1987b). Oviposition begins around sunset, when the females leave their resting places beneath leaves and on shaded areas of boulders to fly over the river and drop their eggs singly into the water (Fredeen et al. 1951).

**Development.** The first diapausing eggs are laid in the latter half of June, and by mid-July all deposited eggs are in diapause (Shipp 1987). Diapausing eggs overwinter in the riverbeds, where they can be covered with at least 8 cm of sediment and still produce larvae (Fredeen et al. 1951). Diapause can be terminated by exposure to 0.5°C for 71 days (Shipp & Whitfield 1987a, 1987b). Larvae begin hatching when water temperatures reach a daily mean of 8.3°C–9.4°C (Fredeen et al. 1951). They can be found from April to at least September (Procunier 1984, Procunier et al. 1984), representing a single generation with 3 cohorts (Anderson & Shemanchuk 1987b). The 7 larval instars (Fredeen 1976) can develop to maturity on a bacterial diet (Fredeen 1960, 1964b), although the water must be kept cool and well aerated (Fredeen 1959a). The sex ratio of adults is near unity (Cameron 1922), with males emerging about a day before females (Fredeen 1958). Adults first appear in late May. Teneral adults are transported downstream as they rest on the surface of the water before taking flight (Shemanchuk 1988).

**Mating.** Mating occurs in swarms at distances up to 64 km from the breeding site (Fredeen 1956). Swarms can be 2–3 m in height or larger and extend up to 15 m above the ground; they have been found in the shelter of roadside trees as well as in direct wind (Cameron 1922, Fredeen 1951). Four possible mating swarms of 30–100 males each have been observed during early evening in shaded areas about 30–50 cm over a river (Shemanchuk 1988). Swarms are seen most commonly in the evenings of June and July, especially before advancing thunderstorms (Fredeen 1951).

**Natural Enemies.** Information on natural enemies of *S. vampirum* is sparse. Mermithid nematodes (*Isomermis* sp.) in the adults inhibit gonadal development, but not mating or feeding (Anderson & Shemanchuk 1987b). We found 2 species of microsporidia infecting larvae from the Slave River in the Northwest Territories, one a species of *Amblyospora* and the other a probable new species with pyriform spores in clusters of 8 (as in Fig. 6.14). Predators of larvae and pupae include larval mayflies, stoneflies, dragonflies, and the white sucker *Catostomus commersoni* (Cameron 1922).

**Hosts and Economic Importance.** *Simulium vampirum* historically has been the major cattle pest in the prairies of Alberta and Saskatchewan, inflicting hideous damage during massive outbreaks. In irrigation areas of southwestern Alberta, it has not been a pest, possibly because it manages only 1 cohort annually before the irrigation water is terminated (Procunier 1984). Females take blood from horses, pigs, white-tailed deer, cattle, sheep, and probably elks, moose, and caribou; during severe outbreaks, females attack humans and other animals such as chickens (Rempel & Arnason 1947; Fredeen 1969, 1977b; Shemanchuk 1988). We have swept females from bison, implicating them as a likely host. Most feeding occurs on the forequarters, underbelly, and neck (Cameron 1922, Rempel & Arnason 1947), with the flies also entering the eyes, ears, and nostrils (Cameron 1918). Breeds of cattle differ in their tolerance to attacks (Khan & Kozub 1985, Shemanchuk 1987).

Females seek hosts only during the day but come to light traps at night (Fredeen 1961). Their activity around cattle during the summer occurs from about 0600 to 2230 hours, with peak activity at about 1300 and 1830 hours (Shemanchuk & Anderson 1980). Their flight activity is influenced by weather variables, especially vapor pressure, but also air temperature, light intensity, and wind velocity (Shipp et al. 1987). Host-seeking females tend to fly higher above the ground than do non-host-seeking females (Sutcliffe et al. 1995). They are strongly attracted to carbon dioxide and congregate around the heads of cattle; the response, however, is visually mediated (Sutcliffe & Shemanchuk 1993, Sutcliffe et al. 1995). Attractiveness increases if acetone or crude bovine odor is combined with carbon dioxide (Sutcliffe et al. 1994).

The first documented report of economic losses attributable to this species dates from 1886, when massive attacks caused the loss of 6 cattle, 2 horses, and an ox south of Saskatoon (Fredeen 1977b, 1979). Prior to that time, the area was too sparsely settled to have been plagued with economic losses. Livestock again were killed in 1896, this time near the present site of Rosthern (Fredeen 1951). In 1913, about 100–300 cattle died from attacks in Saskatchewan; one witness claimed that the flies had a "pungent, mustard-like taste" (Cameron 1918, Rempel & Arnason 1947). Losses in the province were severe in 1919 and 1930, and from 1944 through 1946, the flies killed more than 500 cattle and at least 38 horses, 17 sheep, and 3 pigs, all worth nearly $100,000 (Rempel & Arnason 1947). In 1947, at least 210 animals died from attacks (Fredeen 1977b). From 1948 through 1972, a period during which larviciding was often intense, about 125 animals died in Saskatchewan, with 18 succumbing in 1972 alone (Fredeen 1977b, 1981a). Wild deer also died from the attacks (Fredeen 1977b). Additional but untallied losses included sterilized bulls, weight losses, unrealized weight gains, reduced milk production, delays in calving, and damage during stampedes (Rempel & Arnason 1947, Fredeen 1977b, Shemanchuk & Taylor 1983). Entire livestock enterprises shut down as a result of damage (Fredeen 1969). Egg production in chicken flocks was reduced up to 16% (Fredeen 1977b). Though not usually human biters, flies could be intolerable during severe outbreaks, driving many people indoors or to seek medical attention for bites, a single one of which could impair circulation in a limb (Fredeen 1969, 1977b).

Outbreaks were unpredictable but usually struck from mid-May through June, with minor outbreaks in August and September (Fredeen 1977b). At times, they occurred as far as 150–225 km from the nearest breeding site (Fredeen 1969, Charnetski & Haufe 1981). Massive clouds of flies seethed from rivers such as the north and south branches of the Saskatchewan River. So many flies were produced at times that estimates suggest they could have withdrawn 2 liters of blood from every head of cattle in Saskatchewan's 75,000-km$^2$ problem zone (Fredeen 1977b). Livestock deaths—usually attributed to toxemia and shock—were most severe in older animals, probably because they had shorter hair and more bare areas (Rempel & Arnason 1947). Death could occur within minutes to hours after an attack (Millar & Rempel 1944).

Larviciding and construction of hydroelectric dams (e.g., the Big Horn Dam on the North Saskatchewan River and the Diefenbaker Dam on the South Saskatchewan River), followed by eutrophication, greatly reduced populations of *S. vampirum* in Saskatchewan; by 1975 larvae were scarce (Fredeen 1977a, 1977b; Riegert 1999b). The last substantial outbreak in Saskatchewan occurred in 1972 (Fredeen 1975a). The species is no longer targeted for management in Saskatchewan.

Losses near Alberta's Athabasca River occurred sporadically from 1955 to the early 1970s (Fredeen 1969, 1977b). The outbreaks along the Athabasca River often lasted up to a month—considerably longer than those in Saskatchewan, which typically lasted a few days (Fredeen 1969). Livestock were killed in 1955 (or 1956), 1963 (6 deaths), and 1964 (Fredeen 1969). The flies were so thick that they created a loud buzzing fog as far and high as the eye could see (Depner in Riegert 1999b). In 1971, 973 livestock—7.5% of Athabasca Co.'s entire cattle population—were killed and losses of about $600,000 were recorded (Charnetski & Haufe 1981). In 1972, 449 animals were killed and approximately $600,000 was lost (Ryan & Hilchie 1982). From 1977 to 1979, more than $283,000 worth of animals was lost, with at least 255 cattle killed in 1978 alone (Ryan & Hilchie 1982). Smaller outbreaks that killed cattle were reported in 1956 from areas such as Manville and Minburn in eastern Alberta (Fredeen 1969). Even mild attacks caused irritability among cattle and induced vigorous tail switching, licking of the body, and constant walking—behaviors that detracted from feeding (Shemanchuk 1980a).

Livestock producers eventually petitioned the Alberta provincial government for assistance, leading to a research program to investigate management efforts, which included the first treatment of the river, with methoxychlor, from 1974 to 1976 (Haufe & Croome 1980, Ryan & Hilchie 1982). The 1974–1976 treatments were part of a pilot program to evaluate the efficacy and environmental impacts of a management program in the Athabasca River. As a result of the pilot program's success, an effective management program operated from 1979 to 1987. After 1987, management was limited to oilers and electrostatic sprayers, using the synthetic pyrethroid Ectiban. As of 2002, the problem was restricted to the northern edge of the farming area around the Wandering River (G. Byrtus, pers. comm.).

### *Simulium (Simulium) noelleri* Species Group

**Diagnosis.** Female: Frons and terminal abdominal tergites pollinose. Claws toothless. Anal lobe large, subquadrate. Male: Ventral plate V shaped. Pupa: Gill of 8 filaments. Cocoon slipper shaped, loosely woven, especially anteriorly. Larva: Frontoclypeal apotome with pale head spots surrounded by dark pigment (negative pattern) often in shape of H. Postgenal cleft pointed anteriorly, broadest at midpoint, extended about three fourths distance to hypostomal groove; subesophageal ganglion without pigmented sheath. Abdominal segment IX without ventral tubercles; abdominal setae unbranched, translucent. Rectal papillae of 3 compound lobes. Chromosomes: Standard banding sequence as given by Adler and Kachvorian (2001).

**Overview.** The members of this group represent one of the most morphologically uniform species groups in the family. Nine species are recognized from the Holarctic Region; 2 inhabit North America. Effort is needed to determine if all the Palearctic species are legitimate. The aquatic stages generally are found at nutrient-rich sites, especially the outflows of impoundments.

### *Simulium (Simulium) decorum* Walker

(Figs. 4.4, 4.63, 10.215, 10.392, 10.503, 10.811; Plate 22b; Map 226)

*Simulium decorum* Walker 1848: 112 (F). Lectotype* female (designated by Crosskey in Crosskey & Lowry 1990: 210; pinned; BMNH). Ontario, Hudson Bay, Martin Falls (as St. Martin's Falls), Albany River, date unknown but probably between 1834 and 1843 (1844 is the year given but is actually the date of registration in the collection [Arthur 1985]) (G. Barnston)

*Simulium venustum* 'var. *piscicidium*' Johannsen 1902: 48–49, Plate 5 (Figs. 2, 5, 7), Plate 6 (Figs. 1–3, 19); 1903b: 381–383, Plate 37 (Figs. 2, 5, 7), Plate 38 (Figs. 1–3, 19) (FPL)

*Simulium decorum katmai* Dyar & Shannon 1927: 31 (F). Holotype* female (pinned #28341, terminalia on slide; USNM). Alaska, Katmai, August 1917 (H. S. Hine)

*Simulium* (*Simulium*) *ottawaense* Twinn 1936b: 146–147 (FMP). Holotype* male (slide #4130; CNC). Quebec, near Hull, Leamy Creek, Pinks Lake, 26 July 1935 (C. R. Twinn)

*Simulium* 'C' Fredeen 1956: 4 (Alberta record)

*Simulium* 'sp. No. 1' Anderson & Dicke 1960: 397 (host of mermithid nematode)

[*Simulium venustum*: Needham & Betten (1901) not Say 1823 (misident.)]

[*Simulium venustoides*: Hart in Forbes (1912) (misident. in part, F)]

[*Simulium piscicidium*: O'Kane (1926) not Riley 1870d (misident.)]

[*Simulium jacumbae*: Peterson & Kondratieff (1995) not Dyar & Shannon 1927 (misident. in part, material from Gunnison Co.)]

**Taxonomy**

Walker 1848 (F); Riley 1870c (P only, Fig. 143); Lugger 1897a, 1897b (F); Johannsen 1902, 1903b (FPL), 1934 (PL); Hart in Forbes 1912 (F); Emery 1913 (FPL); Malloch 1914 (FP); Dyar & Shannon 1927 (F); Hearle 1932 (F); Twinn 1936b (FMP); Stains & Knowlton 1943 (F); Nicholson 1949 (FMP); Nicholson & Mickel 1950 (FMP); DeFoliart 1951a (PL); Fredeen 1951, 1981a (FMPL), 1985a (FMP); Stone 1952 (FMP), 1964 (FMPL); Jamnback 1953, 1956 (L); Sommerman 1953 (L); Stone & Jamnback 1955 (FMPL); Davies & Peterson 1956 (E); Peterson 1958a, 1960a (FMP), 1981, 1996 (M); Shewell 1958 (F); Anderson 1960 (FMPL); Davies et al. 1962 (FMP); Wood et al. 1963 (L); Abdelnur 1966, 1968 (FMPLE); Snoddy 1966 (FMPL); Holbrook 1967 (L); Stone & Snoddy 1969 (FMPL); Pinkovsky 1970, 1976 (FMPL); Amrine 1971 (FMPL); Lewis 1973 (FMPL); Corredor 1975 (FM); Snoddy & Noblet 1976 (PL); Imhof 1977 (E); Merritt et al. 1978b (PL); Imhof & Smith 1979 (E); Westwood 1979 (FMPL); Hilsenhoff 1982 (L); Webb & Brigham 1982 (L); Adler 1983 (PL); Adler & Kim 1986 (PL); Currie 1986 (PL), 1988 (L); Mason & Kusters 1991b (FMPL); Peterson & Kondratieff 1995 (FMPL); Stuart 1995 (cocoon); Stuart & Hunter 1998b (cocoon).

**Morphology**

Bennett 1963b (F salivary gland); Guttman 1967a (labral fan); Yang 1968 (F peritrophic matrix); Chance 1969, 1970a (L mouthparts); Hannay & Bond 1971a (FM wing); Yang & Davies 1977 (FM peritrophic matrix); Colbo et al. 1979 (F labral and cibarial sensilla and armature); Brenner et al. 1981 (L sexual dimorphism); Watts 1981b (F salivary gland removal); Crnjar et al. 1983 (F chemosensilla of labella, palps, and tarsi); Currie 1988 (L antenna and proleg); Davies 1989 (gynandromorph).

**Physiology**

Yang 1968 (F salivary agglutinin and anticoagulant), Cumming & McKague 1973 (response to juvenile-hormone analogues), McKague & Wood 1974 (response to development inhibitors), Yang & Davies 1974 (F salivary agglutinin and anticoagulant), McKague et al. 1978 (response to growth regulators), Watts 1981a (FPL content and molecular weights of salivary gland proteins), Crnjar et al. 1983 (F taste chemosensation).

**Cytology**

Zimring 1953; Rothfels et al. 1978 (IIS only); Rothfels 1979, 1981a, 1989; Feraday 1984; Feraday & Leonhardt 1989 (pp. 531, 535); Adler & Kachvorian 2001.

Our examination of material from several sites on and near Alaska's Kenai Peninsula reveals a heavy band in the base of IL and a perinucleolar inversion characterizing the Y chromosome. These chromosomal features characterize typical *S. decorum*, as described by Rothfels (1981a). We, therefore, conclude that the name *katmai*, which is based on material from Katmai (near the Kenai Peninsula), is synonymous with *decorum*.

**Molecular Systematics**

Feraday & Leonhardt 1989 (allozymes); Zhu 1990 (mtDNA); Brockhouse 1991 (rDNA); Xiong & Kocher 1991, 1993b (mtDNA); Xiong 1992 (mtDNA); Tang et al. 1995b, 1996a, 1996b (mtDNA); Brockhouse & Tanguay 1996 (salivary-gland proteins); Smith 2002 (nuclear and mtDNA).

**Bionomics**

**Habitat.** This abundant species is distributed across North America except the Southwest. It is a habitat specialist occupying food-rich outlets of impounded waters, where it often forms extensive larval and pupal mats. The abundance and distribution of this species probably have been augmented by the proliferation of artificial impoundments. By virtue of its habitat, *S. decorum* is exposed to a broad spectrum of temperature and flow and is one of the most heat-tolerant black flies. Larvae are found at temperatures of 0°C–33°C, macrocurrent speeds from less than 0.1 to about 2.0 m/sec, and an average Froude number of about 2.6 (Stone & Snoddy 1969, Westwood & Brust 1981, Eymann 1993). Small larvae tolerate lower current speeds than do larger larvae (Wong 1976). Some larvae can live without current at 20°C for as long as 8 days (Simmons & Edman 1981). When flow ceases, larvae in dense populations sometimes form communal cables of silk more than 7 m long as they relocate (Tarshis & Neil 1970). Whether this behavior is adaptive or incidental is unclear, although it might minimize larval desiccation. Larvae are somewhat tolerant of pH reductions to 3.5 (Chmielewski & Hall 1992). Larvae attach themselves to any available substrate but prefer certain species of aquatic vegetation over others (Hudson & Hays 1975). They will tolerate some accumulation of periphyton. Sometimes they can be found on the undersides of stones, particularly those in direct sunlight (Sommerman et al. 1955). Contact among larvae results in clumping or spacing, depending on the microhabitat (Eymann & Friend 1988).

**Oviposition.** Sugar is important for autogenous development of eggs (Moobola 1981), and oogenesis can be completed in 12 days at 16°C–17°C on a diet of sucrose and water (Davies & Györkös 1990). Some females, however, can produce mature eggs and oviposit within 2 days of emergence (Wolfe & Peterson 1959). A greater degree of anautogeny is expressed during the summer, when females emerge with less stored nutrient (Abdelnur 1968, Chutter 1970). Blood feeding does not increase fecundity, and mating is not essential for oviposition (Simmons & Edman 1981). Females can produce nearly 700 eggs in an ovarian cycle, although the number of eggs per female averages about 400–580 and is positively correlated with female size (Davies & Peterson 1956, Abdelnur 1968, Chutter 1970, Pascuzzo 1976, Brenner et al. 1980).

Gravid females fly upstream and hover while facing an outfall (Imhof & Smith 1979). On landing, they oviposit at the crest or foot of the outfall in areas covered with a thin film of water or in areas wetted by spray. They will oviposit throughout the day but primarily just before dusk, forming irregular masses or strings of eggs on almost any object, including pupae and other females trapped in the gelatinous matrix (Davies & Peterson 1956, Imhof & Smith 1979). More than 11,000 eggs/cm$^2$ can accumulate (Davies & Peterson 1956). Before ovipositing, females probe the moist substrate with their fore tarsi and maxillary palps (Imhof & Smith 1979) and will not oviposit if the water is fouled (Simmons & Edman 1981). Each deposition of an egg is preceded by raising the abdomen, then lowering and arching it toward the substrate (Imhof & Smith 1979, Brenner et al. 1980). Oviposition is followed by grooming (Imhof & Smith 1979). Less commonly, females oviposit in flight, releasing 1 to several eggs as they dip their abdomens into the water, usually above an outfall (Davies & Peterson 1956, Anderson 1960). Eggs found in stream sediments (Fredeen 1959b, 1981a) might be a result of this behavior or of the break up of egg masses on fixed substrates. Females can be induced to oviposit in moist vials (Imhof & Smith 1979).

**Development.** Eggs remain viable for up to 14 months at 4°C but cannot withstand desiccation or freezing (Abdelnur 1968, Tarshis 1968b). Larvae hatch in less than 4 days and complete development in as little as 12 days (pupae in 3 days) at about 21°C (DeFoliart 1951a). Larvae pass through 7 instars (Brenner et al. 1981). Populations from different geographic areas vary in their developmental rates, with males of some populations developing from eggs in as little as 14 days at a temperature of 21°C (ca. 203 degree-days above 6.5°C); females require about 2 days longer (Brenner et al. 1981). Adults emerge during daylight hours (Brenner et al. 1980). Females can live at least 2 months on dry sucrose and water at approximately 11°C (Davies 1953), and because they have greater caloric reserves, they live longer than males (Magnarelli & Burger 1984). A diet of sugar water increases longevity significantly over a diet of water alone (Moobola 1981).

Eggs are the principal overwintering stage, although in the southern states all immature stages are found throughout the winter. Overwintered eggs produce larvae shortly after ice breakup, and the first generation pupates from mid-April to late May (Sommerman et al. 1955, Davies et al. 1962, Merritt et al. 1978b, Lewis & Bennett 1979, Currie 1986, Adler & Mason 1997). Two to 4 generations are produced each year in Alaska and Canada (Wolfe & Peterson 1959, Abdelnur 1968, Back & Harper 1979, Lewis & Bennett 1979, Westwood & Brust 1981, Adler 1986). Some New England states might have up to 5 generations annually (Cupp & Gordon 1983). In Alabama, the first major adult emergence occurs in April, followed by 5 or 6 additional generations (Stone & Snoddy 1969).

Larvae select particles 25 μm in diameter (Chance 1970a). They feed at a relatively low rate, compared with non-outlet-inhabiting species, possibly reflecting the higher nutritive quality of the seston (Morin et al. 1988b). During development, larvae are able to adjust the number of rays in their primary fans, decreasing the number as food increases and increasing the number as food decreases (Lucas 1995, Lucas & Hunter 1999). For unknown reasons, gut contents are about threefold more densely packed in small larvae than in large larvae (Morin et al. 1988a).

Adults have been reared from eggs in various aquarium systems (Fredeen 1959a; Tarshis 1968b, 1971; Lacoursière & Boisvert 1987). *Simulium decorum* was one of the first 2 species in North America to be colonized for multiple generations (Simmons & Edman 1978). Large numbers of adults can be reared synchronously from eggs and induced to mate and oviposit through many generations (Simmons & Edman 1978, 1981; Brenner & Cupp 1980a; Brenner et al. 1980, 1981; Edman & Simmons 1985b).

**Mating.** Mating takes place on the ground at emergence and oviposition sites. Contact pheromones might be involved in mating (Edman & Simmons 1985a). Males become sexually aggressive after voiding the meconium (i.e., the first material excreted from the gut as an adult)—usually within several hours following emergence—and attempt to mate with both sexes, though they transfer spermatophores only to females (Brenner et al. 1980). Spermatophores are placed in a cavity beneath the eighth sternite and between the arms of the genital fork; they contain an average of 4048 spermatozoa (Linley & Simmons 1981, 1983). Coupling lasts 0.5–4.0 minutes (Brenner et al. 1980). Males mate with newly emerged females, as well as those that are gravid and have returned to oviposit (Davies & Peterson 1956). Larger males show greater sexual activity than smaller individuals (Simmons & Edman 1981). Males also dip their abdomens into egg masses, raising the intriguing possibility of external fertilization (Davies & Peterson 1956). Matings between individuals from Georgia and New York produce viable progeny (Brenner et al. 1981).

**Natural Enemies.** Larvae are hosts of the microsporidia *Amblyospora bracteata*, *A. varians*, *Janacekia debaisieuxi*, and *Polydispyrenia simulii* (D. Davies 1957, Adler 1986, Ledin 1994). They are infrequent hosts of mermithid nematodes of questionable identity (Anderson & DeFoliart 1962 [footnote, p. 832 of R. Gordon 1984], Adler & Mason 1997). Larvae are susceptible to laboratory infections by cercariae of the trematode *Plagiorchis noblei* (Jacobs 1991, Jacobs et al. 1993). Larval midguts often harbor the trichomycete fungus *Harpella melusinae*.

Males and females are attacked by the entomophthoraceous fungus *Erynia curvispora* (Kramer 1983) but are apparently immune from attack by *E. conica*, possibly because of the chemical composition of the fly's cuticle (Nadeau et al. 1994, 1996). Water mites (*Sperchon ?jasperensis*) attack adults, with parasite loads sometimes exceeding 30 mites/fly (Davies 1959). These mites commonly attach themselves to the base of the abdomen, possibly reducing fecundity (Pascuzzo 1976). Larval trombidioid mites also have been found on adults (Davies 1959).

Lake chub (*Couesius plumbeus*), dolichopodid flies, and larvae of various caddisflies prey on the immature stages (Peterson & Davies 1960, LaScala 1979). Spiders, empidid flies, and ants prey on adults (Peterson & Davies 1960, Pinkovsky 1976, Adler 1986).

**Hosts and Economic Importance.** Females imbibe the nectar of low-bush blueberry (*Vaccinium angustifolium*) and sour-top blueberry (*V. myrtilloides*) (Davies & Peterson 1956). They feed chiefly on large mammals such as black bears, horses, moose, white-tailed deer, and cattle (Downe & Morrison 1957; Anderson & DeFoliart 1961; Shemanchuk 1978b; Addison 1980; Pledger et al. 1980; McCreadie et al. 1984, 1985). Engorged females have been taken from a domestic duck, an American crow (Bennett 1960), and traps baited with chickens (Shemanchuk 1988). Females held under inverted vials will feed on rabbits (Mokry et al. 1981) and defeathered areas of domestic ducks, geese, ruffed grouse, turkeys, and American crows (Davies & Peterson 1956). Small numbers of females have been attracted to red foxes and American minks (Hunter et al. 1993). Females can be bothersome to humans (Eckhart & Snetsinger 1969, White & Morris 1985b), especially after a period of cool weather (DeFoliart & Rao 1965), as they fly into the eyes and ears and sometimes bite (Davies & Peterson 1956, Abdelnur 1968, Pinkovsky 1976, Currie & Adler 1986).

This species has been incriminated as a vector of legworm, *Onchocerca cervipedis*, in moose (Pledger 1978, Pledger et al. 1980). It also appears capable of transmitting the agent of tularemia among rabbits and other susceptible hosts by mechanical transfer during biting (Philip in Parker 1934, Philip & Jellison 1986). Though not a natural vector, it supports development of the blood protozoan *Leucocytozoon simondi* to the ookinete stage (Fallis et al. 1951) and development of the bovine parasite *O. lienalis* and the human parasite *O. volvulus* to the third larval stage (Lok et al. 1980, 1983a; Lok 1981).

### *Simulium* (*Simulium*) *noelleri* Friederichs
(Map 227)

*Simulium nölleri* Friederichs 1920: 567–569 (FMP). Lectotype male (designated by Zwick 1986: 140; pinned, terminalia on mount below; ZMHU). Germany, Thüringen, Paulinzella, 21 September 1920 (W. Nöller); *nolleri* (subsequent misspelling)

*Simulium subornatum* Edwards 1920: 227–228, Figs. 1c, 2b, 3b, 4d, 5b (FMPL). Holotype male (pinned, terminalia on celluloid mount below; BMNH). England, Nottinghamshire, Bulwell Hall Park, 21 June 1916 (J. W. Carr)

*Simulium tenuimanus* Enderlein 1921a: 200. *Nomen nudum*

*Simulium tenuimanus* Enderlein 1921b: 222–223 (F). Lectotype female (designated by Zwick 1995: 155; pinned, terminalia in glycerin vial below; ZMHU). Germany, Berlin, Straussberg, day and month unknown, 1900 (G. Enderlein)

*Simulium* 'sp. I' Puri 1926: 163–164 (P)

?*Simulium* '(s. str.) sp. indet.' Edwards 1933: 613–614 (45 F from Akpatok Island)

*Simulium septentrionale* Enderlein 1935: 362–363 (FM). Lectotype male (designated by Zwick 1995: 155; pinned, terminalia on slide mount below; ZMHU). Russia, Murmansk Region, along coast in high swamp Alexandrowsk, 20 July 1926 (W. Richter)

*Cryptectemnia lindneri* Enderlein in Lindner 1943: 245–246 (M). Holotype male (pinned [wing only], terminalia on slide mount below; SMNL). Austria, Lunz, Ybbstaler Alpen, lower Danube River, lower lake, 13 August 1940 (E. Lindner)

*Simulium* (*Simulium*) 'sp. nr. *decorum*' Shewell 1957: 2, 3, Map 36 (Canada records)

*Simulium argyreatum avidum* Rubtsov 1963b: 547 (M). Unavailable (varietal name published after 1960, and junior primary homonym of *S. avidum* Hoffman 1930a: 51–53; without status in nomenclature)

*Simulium nölleri bonomii* Rubtsov 1964b: 75–78 (FMPL). Unavailable (varietal name published after 1960, without status in nomenclature)

*Simulium argyreatum* 'separate sibling' Rothfels 1981a: 24–25 (C)

*Simulium decorum* 'cytospecies nr. churchill' Hershey et al. 1995: 284 (Alaska record)

[*Simulium ottawaense*: Twinn et al. (1948) not Twinn 1936b (misident.)]

[*Simulium decorum*: Hocking & Pickering (1954) not Walker 1848 (misident.)]
[*Simulium decorum*: Ezenwa (1973) not Walker 1848 (misident.)]
[*Simulium argyreatum*: authors (during the period 1963–1980) not Meigen 1838 (misident.)]
[*Simulium decorum*: authors (when applied to Palearctic populations) not Walker 1848 (misident.)]

**Taxonomy**

Taxonomic information on North American populations of *S. noelleri* has not been reported previously. The following list provides a sample of the taxonomic references for *S. noelleri* in the Palearctic Region:

Edwards 1920 (FMPL); Friederichs 1920, 1922 (FMP); Enderlein 1921b (F), 1935 (FM); Puri 1925 (PLE), 1926 (P); Rubtsov 1940b, 1956, 1962d, 1963b, 1964b (FMPL); Enderlein in Lindner 1943 (M); Grenier 1953 (FMPL); Usova 1961 (FMPL); Carlsson 1962 (FMPL); Knoz 1965 (FMPL); Davies 1966, 1968 (FMPL); Dinulescu 1966 (FMPL); Rivosecchi 1978 (FMPL); Gryaznov 1984b (F); Rubtsov & Yankovsky 1984 (FMPL); Jensen 1997 (PL); Bass 1998 (PL); Yankovksy 1999 (PL).

Rubtsov (1963b) incorrectly synonymized the name *noelleri* with *argyreatum*. The error was not corrected until 18 years later (Zwick & Crosskey 1981). True *S. argyreatum* is a member of the *S. variegatum* species group and is restricted to the Palearctic Region. Despite the recommendation by Stone (1965b) to synonymize the names *noelleri* and *decorum* with the name *argyreatum*, the names *argyreatum* and *noelleri* seem never to have been used, until now, for any North American population. Shewell (1957, 1958) noted that females of *S. decorum* on the tundra, from the Mackenzie Delta to Hudson Bay, had a less hump-backed thorax than those from the rest of the continent, and he suggested that they were at least subspecifically distinct. Presumably, at least some of these specimens with the putatively flatter thorax were *S. noelleri*. We have found, however, that the degree of thoracic curvature is not a useful character for distinguishing the females of *S. noelleri* from those of *S. decorum*; females of the 2 species currently cannot be distinguished morphologically. The other life stages also are indistinguishable between the 2 species, and identification must rely on chromosomal characters.

A revision of the *S. noelleri* species group is needed. Several Palearctic names might be synonymous with *noelleri*. Examples include *vershininae* by Yankovsky (1982) and *beringovi* by Bodrova (1988), which were applied to populations in eastern Siberia and on the Commander Islands of Russia, respectively. On the other hand, *S. palustre*, described by Rubtsov (1956) from the Karolog River, Irkutsk, Russia, is specifically distinct from *S. noelleri*, based on our study of the type series. In particular, the inner margin of the male gonostylus is more sinuous, and the ventral plate in ventral view is narrower than that of *S. noelleri*.

**Morphology**

Puri 1925 (L anatomy), Williams 1974 (E exochorion), Gryaznov 1989 (F ommatidial number), Zhang & Malmqvist 1996 (labral fan).

**Cytology**

Shcherbakov 1966; Rothfels 1979, 1981a (as *S. argyreatum*); Adler in Currie 1997a; Adler & Kachvorian 2001.

*Simulium noelleri* consists of at least 3 cytologically distinct populations conservatively recognized as cytotypes (Adler & Kachvorian 2001). Cytotype B is Holarctic and occurs in North America from Alaska to at least as far east as the Thelon River drainage in Nunavut; if it represents a distinct species, the name *S. septentrionale* Enderlein would apply.

**Molecular Systematics**

Tang et al. 1996a (mtDNA), Smith 2002 (nuclear and mtDNA).

**Bionomics**

**Habitat.** This black fly is found from northern Alaska to northern Manitoba and across much of the Palearctic Region. Specimens from Akpatok Island (Edwards 1933) possibly belong to this species, but we have seen only females. In the Palearctic Region, *S. noelleri* typically inhabits impoundment outfalls and sluices and is often associated with anthropogenic habitats. In North America, populations are concentrated at the outflows of tundra pools and small lakes. The bionomics of this species are poorly known in North America, but a rather impressive literature has accrued in the Palearctic Region. Remarkably high larval densities—more than 1 million larvae/$m^2$—have been recorded below lake outlets in England (Wotton 1987, 1992), yielding the highest secondary production estimates for any freshwater invertebrate (Wotton 1988b). Larvae are tolerant of short-term oxygen depletion (Kiel & Frutiger 1997). Small larvae tolerate slower current velocities than do larger larvae (Wotton 1985). The larval silk pads are quite durable, remaining on a substrate for up to a month (Kiel 1997). Larvae of *S. noelleri*, unlike those of certain other species, are tolerant of silk-pad remnants left by previous colonizers (Kiel et al. 1998a).

**Oviposition.** Females are autogenous for at least the first cycle of egg production (Wotton 1982a). They are capable of producing more than 650 eggs per cycle, with an average of about 300–420 eggs (Schütte 1990, Gryaznov 1995b). Females that emerge from the densest parts of aggregations are larger and presumably more fecund than those from areas of lower density (Wotton 1988a). Masses of eggs are deposited on moist substrates (Rühm 1975) in a fashion similar to that described for *S. decorum*.

**Development.** Eggs overwinter. At least 2 generations are produced annually in North America. Larvae of the first generation in northern Alaska are present from late June to late July, and those of the second generation appear in late July and early August. In Churchill, Manitoba, pupae have been found in June (Hocking & Pickering 1954). Up to 4 generations per year are produced in England (Wotton 1987) and 5 in Germany (Schütte 1990). Seven larval instars have been recorded for Czechoslovakian populations (Ptáček & Knoz 1971). Differences in particle-capture efficiency among larvae of different sizes might reduce intraspecific competition for food, thereby aiding the formation of dense larval aggregations (Wotton 1984). Larvae are capable of ingesting surface films in addition to particulate matter (Wotton 1982b). When reared at 20°C, males begin to pupate 23 days, and females 26 days, after the larvae hatch (Wotton 1987). Laboratory rearings show continuous production of pupae for up to 40 days from larval hatch, with early-emerging adults being smaller than those in the same cohort that emerge later (Wotton 1987). Males emerge before females, are smaller in size, and are produced in slightly higher numbers, with the sex ratio (males to females) being correlated inversely with population density (Wotton 1987).

**Mating.** Mating behavior is apparently unreported, suggesting that it might not be similar to that of *S. decorum*, which forms conspicuous mating aggregations on the ground. Small swarms of males, possibly representing mating swarms, have been observed hovering under a tree in England (Edwards 1920).

**Natural Enemies.** Larvae in the Nearctic Region are hosts of the mermithid nematode *Mesomermis flumenalis*; the microsporidia *Amblyospora bracteata*, *Polydispyrenia simulii* (Ezenwa 1973), and *Janacekia debaisieuxi*; and the trichomycete fungus *Harpella melusinae*. Additional parasites have been recorded from *S. noelleri* (usually as *S. argyreatum*) in the Palearctic Region, including the mermithid nematodes *Mesomermis albicans*, *M. melusinae* (Rubtsov 1966), and *M. canescens* (Rubtsov 1972b); the trichomycete fungus *Smittium simulii* (Moss in Crosskey 1990); a cytoplasmic polyhedrosis virus; and the microsporidia *A. fibrata* (Weiser 1978) and *A. varians* (Mitrokhin 1979). In Europe, larvae have been infected experimentally with cercariae of the trematode *Plagiorchis neomidis* (Bušta & Našincová 1986). Palearctic adults are attacked by the entomophthoraceous fungus *Erynia conica* (Crosskey 1990). Larvae and pupae in Europe are preyed on by larvae of the muscid fly *Limnophora riparia* (Merritt & Wotton 1988), although predation is reduced by aggregation of the prey (Wotton & Merritt 1988).

**Hosts and Economic Importance.** Humans, horses, and cows have been reported as hosts in Poland and Scandinavia (Zwolski 1956, Carlsson 1962), but females are believed not to take blood in England (Crosskey 1985). Blood-meal analyses suggest that females in Germany feed on rats (*Rattus*) (Schütte 1990).

### *Simulium* (*Simulium*) *parnassum* Species Group

**Diagnosis.** Female: Body shiny black. Claws long, slightly sigmoidal, each with small subbasal tooth. Male: Gonostylus with medially directed, subbasal bulge. Ventral plate with posterior margin smooth, not crenulated; dorsal plate a thin horizontal strip. Pupa: Gill of 6 filaments in 3 pairs. Head and thorax typically rugose. Cocoon slipper shaped, with anterior margin strongly reinforced. Larva: Postgenal cleft triangular; subesophageal ganglion not ensheathed with pigment. Abdominal segment IX without ventral tubercles. Rectal papillae of 3 compound lobes. Chromosomes: Standard banding sequence as given by Paysen and Adler (2000).

**Overview.** The single member of this species group is endemic to eastern North America. Because it is distinct in all life stages, as well as chromosomally, and does not fit satisfactorily into any species group, it was placed in its own species group (Paysen & Adler 2000).

### *Simulium* (*Simulium*) *parnassum* Malloch

(Figs. 10.73, 10.216, 10.393, 10.530, 10.812;
Plate 22c; Map 228)

*Simulium parnassum* Malloch 1914: 36–37, Plates II (Fig. 8) and V (Fig. 11) (F). Syntype* females (9 on 1 pinned card mount #15409; USNM). New Hampshire, Carroll Co., Red Hill, Moultonborough, 5 August 1902 (H. G. Dyar)

*Boophthora rileyana* Enderlein 1922: 75–76 (F). Lectotype* female (here designated; pinned; ZMHU). New York, Hamilton Co., Adirondack Mountains, Long Lake, day and month unknown, 1907 (Horváth). New Synonym

*Simulium hydationis* Dyar & Shannon 1927: 28 (M). Holotype* male (slide #28340; USNM). Virginia, Fairfax Co., Dead Run, 21 May 1914 (R. C. Shannon)

[*Simulium venustum*: Johnson (1925) not Say 1823 (misident. in part, 1 F from New Hampshire 20 August 1851)]

**Taxonomy**

Malloch 1914 (F); Enderlein 1922 (F); Dyar & Shannon 1927 (FM); DeFoliart 1951a (PL); Jamnback 1953, 1956 (L); Stone & Jamnback 1955 (FMPL); Davies & Peterson 1956 (E); Davies et al. 1962 (FMP); Wood et al. 1963 (L); Stone 1964 (FMPL); Snoddy 1966 (FMPL); Holbrook 1967 (L); Stone & Snoddy 1969 (FMPL); Pinkovsky 1970 (FMPL); Amrine 1971 (FMPL); Snoddy & Noblet 1976 (PL); Webb & Brigham 1982 (L); Adler 1983 (PL); Adler & Kim 1986 (PL); Paysen 1999 (FPL); Paysen & Adler 2000 (P).

Our examination of the 2 syntypes of *Boophthora rileyana* indicates that both are actually *Simulium parnassum*. The former name, therefore, becomes a synonym of the latter. We designate one of the specimens of *Boophthora rileyana* as lectotype (so labeled) to establish the current concept of the name; the second specimen now has paralectotype status. In addition to Enderlein's type labels, both specimens carry undated determination labels of Coquillett, who identified the females as *Simulium venustum*.

*Simulium parnassum* consists of a rugose pupal form, which is widespread and common, and a smooth pupal form that has been found only in 1 stream in Massachusetts, 1 in North Carolina, and 2 in South Carolina (Paysen & Adler 2000).

**Morphology**

Guttman 1967a (labral fan), Kurtak 1973 (labral fan), Hayton 1979 (F age-related changes in Malpighian tubules), Smith & Hayton 1995 (F age-related changes in Malpighian tubules).

**Cytology**

Rothfels 1979, Paysen 1999, Paysen & Adler 2000.

**Molecular Systematics**

Smith 2002 (nuclear and mtDNA).

**Bionomics**

**Habitat.** This species is found from Labrador southward along the spine of the Appalachian Mountains, with 1 record from the Ozark Mountains in Missouri. The immature stages occupy swift rocky streams 1–9 m wide that run year-round through woodlands. Larvae are usually collected at temperatures below 20°C. Larvae and pupae have been recovered from a stream up to 14 m into a cave in complete darkness, suggesting that females flying upstream entered the cave to oviposit (Reeves & Paysen 1999). Larvae do not tolerate contact, and when touched by another, they thrash about until one relocates (Kurtak 1973). The labral fans are robust and designed for filtering efficiency in swift water (Kurtak 1978). Adults have been swept from forest vegetation (Whittaker 1952).

**Oviposition.** Females produce mature eggs within a week after a blood meal (Davies & Peterson 1956). The potential fecundity (i.e., number of ovarioles) in the first ovarian cycle averages about 385 (Pascuzzo 1976). Egg-laying behavior is unknown.

**Development.** *Simulium parnassum* is typically a univoltine species. Some populations, however, have overlapping cohorts or might even be bivoltine (Tessler 1991). In central Pennsylvania, larvae hatch from overwintered eggs in early to mid-May when the water temperature reaches 9°C–10°C (Adler & Kim 1986). From Pennsylvania northward, larvae and pupae are most prevalent from June to August and adults from July through August (Stone 1964, Martin 1981, Gibbs et al. 1988). Southward, larvae hatch as early as March and adults are on the wing from April through August.

**Mating.** Unknown.

**Natural Enemies.** Larvae are hosts of the chytrid fungus *Coelomycidium simulii* (Jamnback 1973a), the microsporidium *Polydispyrenia simulii* (McCreadie & Adler 1999), and the trichomycete fungus *Harpella melusinae*. Large numbers of larvae are sometimes infected with unidentified mermithid nematodes. Larval hydropsychid caddisflies prey on larvae in the laboratory (Reeves & Paysen 1999). A female bearing a water mite (*Sperchon ?jasperensis*) has been reported (Davies 1959).

**Hosts and Economic Importance.** Females feed on horses and cattle and can be occasional pests of livestock in eastern Canada (Downe & Morrison 1957, Fredeen 1973). They also feed on American black bears (Addison 1980) and woodchucks (Fuller 1940) and have been induced to feed on the defeathered skin of domestic ducks (Anderson 1956, Davies & Peterson 1956). Under experimental conditions, they have fed on guinea pigs (*Cavia cobaya*) (Downe 1957). Females can be aggravating to humans, biting frequently (Jamnback & Collins 1955, Davies & Peterson 1956, Eckhart & Snetsinger 1969, Adler & Kim 1986, Gibbs et al. 1986). They are often the principal human biter in the late summer and are particularly annoying in the morning and evening as they bite the hands, arms, and face. Populations, however, are generally too small to create a major pest problem.

Females are not known to transmit parasites or pathogens naturally. However, the filarial nematode *Splendidofilaria fallisensis* can develop to the infective stage under experimental conditions (Anderson 1956).

### *Simulium* (*Simulium*) *petersoni* Species Group, Newly Recognized

**Diagnosis.** Female: Subcosta without hair ventrally. Cibarium with minute nodules between cornuae. Legs including fore coxa dark, except for pale areas on middle and hind legs; claws toothless. Male: Gonostylus flat, with small, slender subbasal tubercle; apical spinule absent. Pupa: Gill of 6 filaments in 3 pairs. Cocoon shoe shaped. Larva: Antenna dark, with 1 hyaline band. Postgenal cleft widest at midpoint, narrowed anteriorly, extended three fourths or more distance to hypostomal groove; subesophageal ganglion ensheathed with pigment. Abdominal segment IX without ventral tubercles. Rectal papillae of 3 compound lobes; in some populations, each lobe bears only 0–3 lobules.

**Overview.** *Simulium petersoni* previously was placed in the *S. tuberosum* species group (Crosskey & Howard 1997) based on a pupal gill with 6 filaments and a male gonostylus with a subbasal tubercle that is questionably homologous with the larger, more bulbous tubercle of members of the *S. tuberosum* species complex. *Simulium petersoni* is chromosomally and morphologically unique and distinct from members of the homogeneous *S. tuberosum* species group. Molecular evidence indicates that *S. petersoni* is the sister taxon of the *S. tuberosum* species group (Smith 2002). Pending further insight into its phylogenetic affinities, we prefer to place *S. petersoni* in its own species group.

### *Simulium* (*Simulium*) *petersoni* Stone & DeFoliart

(Figs. 10.217, 10.394, 10.504, 10.813; Plate 22d; Map 229)

*Simulium petersoni* Stone & DeFoliart 1959: 394–397 (FMPL). Holotype* male (pinned, pupal exuviae and terminalia on slide; USNM). Wyoming, Albany Co., confluence of School Creek and North Sybille Creek, 18 June 1956 (G. R. DeFoliart)

*Simulium* 'species #3' B. Peterson 1955: 114 (Utah record)

**Taxonomy**

Stone & DeFoliart 1959 (FMPL), Peterson 1960a (FMP), Corredor 1975 (FM).

**Cytology**

Our analyses of larvae from Kamas, Summit Co., Utah (8 June 1994), and Rock Creek (near Golconda), Humboldt Co., Nevada (17 April 1995), show that relative to the *Simulium* subgeneric standard of Rothfels et al. (1978), IIIS is standard, IS and IL each carry a single midarm inversion, IIS is based on the B inversion, IIL and IIIL are complexly rearranged, the nucleolar organizer is in the standard position, and the Y chromosome is demarcated by 2 inversions in the IIL arm.

**Molecular Systematics**

Smith 2002 (nuclear and mtDNA).

**Bionomics**

**Habitat.** This species is sporadic in sagebrush areas of the western United States. Larvae and pupae form tightly packed masses on rocks and debris in streams a few meters wide, often bordered with willows (*Salix*) and other riparian vegetation.

**Oviposition.** Unknown.

**Development.** Larvae hatch in March. Several generations are produced yearly.

**Mating.** Unknown.

**Natural Enemies.** Larvae are hosts of the microsporidia *Amblyospora fibrata* and *A. varians*, plus a possible new species with pyriform spores in clusters of 8 from

Kamas, Summit Co., Utah (Fig. 6.14). Ants prey on larvae, and empidid flies cause larvae to drift by probing them (Peterson 1960b). *Tetragnatha* spiders capture adults in webs built over streams.

**Hosts and Economic Importance.** Females are presumably mammalophilic. They have been reported to fly around sheep but not to bite (Jessen 1977).

### *Simulium (Simulium) slossonae* Species Group

**Diagnosis.** Female: Frons shiny. Scutum subshiny to shiny. Claws each with basal thumblike lobe. Male: Gonostylus with basal or subbasal lobe. Dorsal plate present or absent. Pupa: Gill of 4–8 filaments. Cocoon slipper shaped. Larva: Postgenal cleft large, extended to hypostomal groove, pointed to broadly rounded anteriorly. Abdominal integument with branched setae. Abdominal segment IX with well-developed ventral tubercles. Rectal papillae of 3 compound lobes.

**Overview.** Three of the 4 species in this group occur in North America, with 1 (*S. transiens*) also inhabiting the Palearctic Region. They were formerly members of the subgenus *Parabyssodon* until Currie (1997a) placed them as a species group in the subgenus *Simulium*. *Simulium rugglesi* and *S. slossonae* are sister species based on a high degree of morphological similarity. Molecular evidence indicates that *S. transiens* is a member of the *S. slossonae* species group (Smith 2002). A fourth member of the group recently was described from Japan (Takaoka & Saito 2002). The immature stages of the Nearctic group members are found in northern rivers and southeastern swamp streams. Females are ornithophilic.

### *Simulium (Simulium) rugglesi* Nicholson & Mickel

(Figs. 10.218, 10.395, 10.505, 10.814; Plate 22e; Map 230)

*Simulium rugglesi* Nicholson & Mickel 1950: 60–61, Fig. 23 (F). Holotype* female (pinned; UMSP). Minnesota, Todd Co., on goose, 24 June 1937 (collector unknown)

*Simulium* (*Simulium*) 'M' Fredeen 1951: 11, 25, 27, 80–84, Plate 15 (FMP)

[*Simulium decorum*: Walker (1848) (misident. in part, 2 F recognized as paralectotypes by Crosskey in Crosskey & Lowry 1990 and referred to as "species of the *Simulium venustum* Say group")]

[*Simulium bracteatum*: Walker (1927) not Coquillett 1898 (misident.)]

[*Simulium venustum*: O'Roke (1930, 1931, 1934) not Say 1823 (misident.)]

[*Simulium venustum*: Twinn (1933a) not Say 1823 (misident. in part, F collected 9 July 1925)]

[*Simulium venestum* [sic]: Twinn in Mail & Gregson (1940) not Say 1823 (misident.)]

[*Simulium venustum*: Fallis et al. (1951) not Say 1823 (misident.)]

[*Simulium (Simulium) fibrinflatum*: Cupp & Gordon (1983) not Twinn 1936b (misident. in part, material from Michigan)]

**Taxonomy**

Nicholson 1949 (F); Nicholson & Mickel 1950 (F); Fredeen 1951 (FMP), 1981a (FMPL), 1985a (FMP); Stone 1952 (FMP), 1964 (FMPL, not Fig. 57g or j); Davies & Peterson 1956 (E); Peterson 1958a, 1960a (FMP); Anderson 1960 (FMPL); Davies et al. 1962 (FMP); Wood et al. 1963 (L); Abdelnur 1966, 1968 (FMPL); Lewis 1973 (FMPL); Hilsenhoff 1982 (L); Currie 1986 (PL); Mason & Kusters 1991b (FMPL).

**Morphology**

Bennett 1963b (F salivary gland); Guttman 1967a (labral fan); Yang 1968 (F salivary gland ducts); Johnson & Pengelly 1969 (instars); Mercer 1972 (FM antennal and palpal sensilla); Mercer & McIver 1973a (FM antennal sensilla), 1973b (FM palpal sensilla); Sutcliffe 1975 (FM legs); Sutcliffe & McIver 1976 (FM leg sensilla); Yang & Davies 1977 (F peritrophic matrix); Koehler et al. 2002 (F midgut musculature).

**Physiology**

Yang 1968 (F blood digestion, trypsin), Yang & Davies 1968b (F trypsin).

**Molecular Systematics**

Smith 2002 (nuclear and mtDNA).

**Bionomics**

**Habitat.** *Simulium rugglesi* occurs across the northern United States and Canada. The immature stages inhabit clear, briskly flowing rivers and large streams more than 3 m wide. Larvae are most common in shallow riffles but can be found as deep as a meter or more. They cling to trailing vegetation where, alive, they often appear emerald green.

**Oviposition.** The number of eggs per ovarian cycle ranges from 187 to 348, but because a female potentially could undergo 4–6 cycles, lifetime production could reach 1000–1500 eggs (Davies & Peterson 1956, Bennett 1963a). Females probably drop their eggs into the water while in flight.

**Development.** This species is probably univoltine. Eggs overwinter in riverbed sediments. Larvae hatch in the spring about a month after ice breakup in Saskatchewan and as early as April in Algonquin Park, Ontario (Pilfrey 1963, Fredeen 1981a). Larvae pass through 7 instars (Johnson & Pengelly 1969). Pupation begins in early to late May and at higher latitudes lasts well into July; adults are on the wing from late May to August (Wolfe & Peterson 1959, Anderson & Dicke 1960, Bennett 1960, Barrow et al. 1968, Currie 1986). Females can live at least 28 days in the field (Bennett 1963a).

**Mating.** Swarms of mating flies have been observed in a forest about 1.6 km from the breeding site, but details are not available (Fredeen 1951).

**Natural Enemies.** An unidentified microsporidium has been reported from the fat body of females (Yang & Davies 1977). We have seen 1 larva infected with an unidentified mermithid nematode. Predators of adults include empidid flies (Peterson & Davies 1960).

**Hosts and Economic Importance.** Females congregate over marshes, lake shores, and open areas near lake shores, where they feed on birds, especially ducks, in the evening and occasionally just after dawn (Bennett 1960, Barrow et al. 1968). The attraction to ducks declines with decreasing host size, as follows: domestic duck, mallard (wild), redhead, wood duck, northern pintail, ring-necked duck, and blue-winged teal (Anderson & DeFoliart 1961). Additional though limited host-species records include great blue heron, black-crowned night heron, Canada goose, domestic goose, muscovy duck, chicken, ring-necked pheasant, ruffed grouse, domestic turkey, sandhill crane, spotted sandpiper, herring gull, black tern, ringed turtle-dove, northern saw-whet owl, American crow, common raven, American robin, red-winged blackbird, and common grackle (Bennett 1960, Anderson & DeFoliart 1961, Barrow et al. 1968, Tarshis 1972). Under natural conditions, the principal hosts—chiefly ducks and geese—are found in marshes and along lake shores. Female flies are attracted not only to the birds but also to holding cages containing droppings and the remains of sheath scales from young feathers (Barrow et al. 1968).

Only 1 mammalian host, the American black bear, has been reported (Addison 1980). Although females are sometimes attracted to white-tailed deer in marshes, they apparently do not feed on them (Anderson & DeFoliart 1961). Where populations are large, females infrequently annoy humans by crawling into the ears and nostrils (Leonard in Davies & Peterson 1956).

Females generally search for hosts within 0.3 m of the ground and water (Smith 1966). Host visibility is an important factor in attracting females, which approach the host from the shaded ventral surface and land primarily on the breast and bases of the legs (Anderson & DeFoliart 1961), sometimes gaining access by crawling up the legs (Shewell 1955). One experiment, however, indicated greatest attraction to the heads, sides, and tails of mallard decoys, with little attraction to the breast area (Mercer 1972). Hosts against a light background attract more flies than those against a dull backdrop, and some evidence indicates that groups of ducklings provide stronger attraction than singletons (Pilfrey 1963). Extracts of the uropygial glands of ducks are poor attractants unless combined with a carbon dioxide source, which alone is attractive (Fallis & Smith 1964a, 1964b; Mercer 1972). Engorgement on the host requires about 4–12 minutes at an unspecified temperature (Fallis & Smith in Fallis 1964). An average blood meal is about 1.9 $mm^3$ and is digested and eliminated in 4–6 days (Bennett 1963a). Females will disperse at least 3 km from the site at which they last obtained a blood meal (Bennett 1963a).

*Simulium rugglesi* is a confirmed vector of *Leucocytozoon simondi* among domestic and wild ducks and Canada geese, particularly during June and July (O'Roke 1934, Fallis et al. 1956, Anderson et al. 1962, Fallis & Bennett 1966, Tarshis 1972, Herman et al. 1975, Desser & Ryckman 1976, Desser et al. 1978). It might have been one of the vectors that inflicted heavy mortality among domestic geese in subarctic Quebec, interfering with Canada's program to establish the geese as a year-round food for the native people (Laird & Bennett 1970). It also has been incriminated as a vector of the filarial nematode *Splendidofilaria fallisensis* (Anderson 1955, 1956) and the protozoan *Trypanosoma confusum* (as *T. avium*) among ducks (Desser et al. 1975, Desser 1977).

### *Simulium (Simulium) slossonae* Dyar & Shannon

(Figs. 10.219, 10.396, 10.506, 10.815, 10.841; Plates 2i, 22f; Map 231)

*Simulium slossonae* Dyar & Shannon 1927: 34–35, Figs. 58, 59, 124, 125 (FM). Holotype* male (slide #28342; USNM). Florida, Dade Co., Biscayne Bay, date unknown (A. T. Slosson)

[*Simulium venustum*: Coquillett (1898) not Say 1823 (misident. in part, Florida material)]

[*Simulium jenningsi*: Malloch (1914) (misident. in part, Florida material)]

[*Simulium vittatum*: Stone & Snoddy (1969) not Zetterstedt 1838 (misident. in part, Fig. 128)]

**Taxonomy**

Dyar & Shannon 1927 (FM), Underhill 1944 (FMPL), Snoddy 1966 (FMPL), Stone & Snoddy 1969 (FMPL), Pinkovsky 1976 (FMPL), Snoddy & Noblet 1976 (PL), Webb & Brigham 1982 (L).

**Morphology**

Evans & Adler 2000 (spermatheca).

**Cytology**

Evans 2001.

The chromosomes are complexly rearranged relative to the *Simulium* subgeneric standard of Rothfels et al. (1978). The nucleolar organizer is in the base of IS and polymorphisms are common, especially in the IIIL arm. At least 3 different sex-chromosome systems exist, each possibly representing a separate sibling species (Evans 2001).

**Molecular Systematics**

Smith 2002 (nuclear and mtDNA).

**Bionomics**

**Habitat.** A southeastern species, *S. slossonae* occupies the coastal plain from Virginia to Texas. It is one of the most common black flies in Florida, although it probably has been extirpated from much of southern Florida, including the type locality. A record from Maine (Cupp & Gordon 1983) requires confirmation. The immatures live in both clear and darkwater streams with sand-mud bottoms, rather sluggish flows, and abundant vegetation. These streams are generally acidic (pH as low as 3.5; Pinkovsky 1976) and less than 4 m wide. Larvae have been collected at water temperatures up to 29°C and, along with pupae, are found most often on trailing vegetation.

**Oviposition.** Females are anautogenous and probably drop their eggs into the water during flight.

**Development.** This species is multivoltine. All life stages are found year-round in the southern portion of the range (Stone & Snoddy 1969, Pinkovsky & Butler 1978). Farther north, as in South Carolina and Virginia, the eggs overwinter. These northern populations begin to increase in early spring and peak in summer and early autumn, with occasional larvae or pupae appearing in the winter (Underhill 1944, Arnold 1974). As many as 6 generations occur each year in South Carolina (Garris & Noblet 1976).

**Mating.** Unknown.

**Natural Enemies.** We have seen larvae infected with the microsporidium *Janacekia debaisieuxi* and the chytrid fungus *Coelomycidium simulii*.

**Hosts and Economic Importance.** Females are well-known pests of turkeys. We also have records of feeding on chickens. Feeding usually occurs on exposed areas of the head and neck of turkeys (Underhill 1939, Jones & Richey 1956), though sometimes on the lower abdomen and legs and beneath feathers on the back of the neck (Garris et al. 1975, Pinkovsky et al. 1981). Mature male turkeys are particularly attractive to blood-seeking females, especially about the eyes (Moore & Noblet 1974, Garris et al. 1975). Females feed throughout the day, although peak activity occurs during midafternoon (Jones & Richey 1956, Alverson & Noblet 1976), when up to 700 females have fed on a single turkey in a 30-minute period (Kissam et al. 1975). Engorgement requires an average of about 4 minutes, with some flies becoming so replete that they drop to the ground (Garris et al. 1975, Pinkovsky et al. 1981). Females disperse up to 6.4 km after a blood meal, possibly following streams as they fly toward swampy areas to oviposit (Moore & Noblet 1974).

The literature includes records of a few females attracted to humans, plus 1 record of a human being bitten (Snow et al. 1958a, Pinkovsky 1976). However, we have numerous Florida records of females swarming around and sometimes biting humans. In fact, *S. slossonae* caused one of the continent's most southern black fly problems in the winter of 1997–1998, when large numbers swarmed around and bit humans in Florida, especially in Pasco, Hernando, and Volusia Cos. The unusually heavy winter rains created vast areas of suitable breeding habitat.

Females transmit the protozoan blood parasite *Leucocytozoon smithi* to domestic and wild turkeys, causing leucocytozoonosis, known colloquially as "gnat fever" or "turkey malaria" (Richey & Ware 1955, Jones & Richey 1956, Vatne 1972, Greiner & Forrester 1979, Pinkovsky et al. 1981). Infection rates as high as 100% have been reported for wild turkeys in the coastal plain of South Carolina, where *S. slossonae* is very common (Moore 1973). The highest levels of transmission coincide with peak abundance of adult flies during summer and early fall (Noblet et al. 1975). *Simulium slossonae* was largely responsible for crippling the southeastern turkey industry in the 1950s and 1970s (Jones & Richey 1956, Arnold & Noblet 1975, Noblet et al. 1976, Barnett 1977). Annual losses of turkeys in some counties of South Carolina were as high as 5%, and the industry lost approximately $11,250 in Jasper Co., South Carolina, in 1952 as a result of leucocytozoonosis (Jones & Richey 1956, Jones & Richey in Wehr 1962). Deaths of about 300 (20%) of 1500 turkeys on 1 farm in Florida were attributed to *Leucocytozoon smithi*, probably transmitted by *S. slossonae* (Simpson et al. 1956).

*Simulium slossonae* is still common in the southeastern coastal plain. However, turkey production now occurs primarily in conventional poultry houses, rather than outdoors, which has resulted in reduced levels of feeding activity by *S. slossonae* and consequently reduced levels of leucocytozoonosis (R. Noblet, pers. comm.). Early computer models suggested that housing turkey poults would reduce losses to leucocytozoonosis (Alverson et al. 1976). Flies typically do not venture into enclosed buildings (Noblet in Alverson 1976).

### *Simulium (Simulium) transiens* Rubtsov

(Figs. 4.15, 10.220, 10.397, 10.507, 10.816, 10.842; Plate 23a; Map 232)

*Simulium (Byssodon) transiens* Rubtsov 1940b: 361–363 (English version p. 502), Figs. 13N, 16A, 21P, 65M, N, 68N, O, 73B, 74U, 75H, I (FM). Lectotype* female with pupal exuviae (cocoon absent) (designated by Yankovsky 1995: 53; slide #3174; ZISP). Russia, Irkutsk Region, Belaya River, 12 July 1934 (I. A. Rubtsov)

*Simulium (Simulium)* 'R' Fredeen 1951: 11, 27, 93–95, Plate 18 (Figs. A–G) (MP)

**Taxonomy**

Rubtsov 1940b (FM), 1956, 1962d, 1962g (FMPL); Fredeen 1951 (MP), 1985a (FMP); Usova 1961 (FMPL); Abdelnur 1966, 1968 (FMPL); Peterson 1981, 1996 (FM); Rubtsov & Yankovsky 1984 (FMPL); Currie 1986 (PL); Jensen 1997 (PL).

**Morphology**

Yankovsky 1977 (labral fan).

**Molecular Systematics**

Smith 2002 (nuclear and mtDNA).

**Bionomics**

**Habitat.** This Holarctic species occurs primarily in the boreal forests of northwestern Canada and Alaska. In the Palearctic Region, it ranges from Fennoscandia across much of Russia, Mongolia, and China. Larvae and pupae live in medium to large piedmont rivers at temperatures up to 26°C. Larvae exhibit a spaced dispersion pattern (Eymann 1991c).

**Oviposition.** Unknown.

**Development.** Larvae and pupae are present from late May to mid-August. In Siberia, eggs overwinter and 2 generations are produced annually (Rubtsov 1956, 1962d).

**Mating.** Unknown.

**Natural Enemies.** Larvae are infrequently attacked by unidentified mermithid nematodes.

**Hosts and Economic Importance.** No host records exist for this species in North America. In Finland, engorged females have been taken from black grouse (Ojanen et al. 2002). Females are claimed to be fierce but sporadic human biters in Russia (Rubtsov 1956).

### *Simulium (Simulium) tuberosum* Species Group

**Diagnosis.** Female: Subcosta with row of hair ventrally. Claws toothless. Male: Gonostylus with subbasal lobe bearing numerous short stout spines on anterior face. Ventral plate wider than long. Pupa: Gill of 6 filaments, half or more as long as pupa. Cocoon slipper shaped. Larva: Hypostoma with all teeth small; lateral and median teeth extended anteriorly to about same level. Postgenal cleft subtriangular or bowed outward at midpoint, extended about three fourths distance to hypostomal groove. Abdominal segment IX without ventral tubercles; abdominal setae unbranched, translucent. Rectal papillae of 3 compound lobes.

**Overview.** Worldwide, 35 nominal species are members of this group. They occupy the Holarctic and Oriental Regions. At least 9 species are found in North America. Members of the group develop in a wide range of flowing waters, from the smallest streams to the largest rivers. Females of all species are presumably mammalophilic.

### *Simulium (Simulium) tuberosum* Species Complex

**Diagnosis.** Female: Anal lobe in ventral view sclerotized, except minute spot on mesal margin; heaviest sclerotization along smoothly curved anterior margin. Genital fork with arms slender, and each lateral plate bearing anteriorly directed apodeme. Spermatheca large, subspherical; wall rather smooth, with internal spicules. Chromosomes: Standard banding sequence as given by Landau (1962).

**Overview.** Once viewed as a widely distributed and ecologically versatile black fly, *S. tuberosum* is known now to be a conglomerate of at least 9 species in the Nearctic Region and an untold number in the Palearctic Region. Because the complex has been treated as a single species for the better part of a century, a muddled literature has resulted, most of which cannot be assigned to the proper species. References to *S. tuberosum* in North America also have appeared under the names *S. perissum* Dyar & Shannon, *S. turmale* Twinn, *S. twinni* Stains & Knowlton, and *S. vandalicum* Dyar & Shannon, all of which we have associated with the appropriate cytological segregates after recollecting at the type localities and studying the holotypes. The 9 Nearctic species (and their previous cytological designations) are *S. appalachiense* n. sp. ('CDE' and 'CDEM'), *S. chromocentrum* n. sp., *S. conundrum* n. sp. ('FGH'), *S. perissum* ('CKL'), *S. tuberosum* s. s. ('A' and 'AB'), *S. twinni*, *S. ubiquitum* n. sp. ('F'), *S. vandalicum* ('FG'), and *S. vulgare* ('FGI'). Information dealing with the species complex as a whole is presented below, followed by specific information on each of the 9 named species.

**Taxonomy**

Lundström 1911 (M); Forbes 1912 (PL); Garman 1912 (PL); Dyar & Shannon 1927 (FM); Dorogostaisky et al. 1935 (FMPL); Twinn 1936b (FMP), 1938 (M); Rubtsov 1940b, 1956, 1962d, 1963a (FMPL); Stains & Knowlton 1940, 1943 (M); Stains 1941 (M); Vargas et al. 1946 (ML); Vargas & Díaz Nájera 1948a (F), 1949 (P), 1957 (FMP); Coleman 1951 (FMP); DeFoliart 1951a (PL); Stone 1952 (FMP), 1964 (FMPL); Jamnback 1953, 1956 (L); Sommerman 1953 (L); Stone & Jamnback 1955 (FMPL); Davies & Peterson 1956 (E); Wirth & Stone 1956 (FMP); Peterson 1958a (FMP), 1959a (E), 1960a (FMP), 1981 (M), 1996 (MP); Anderson 1960 (FMPL); Davies et al. 1962 (FMP); Wood et al. 1963 (L); Abdelnur 1966, 1968 (FMPLE); Dinulescu 1966 (FMPL); Snoddy 1966 (FMPL); Holbrook 1967 (L); Stone & Snoddy 1969 (FMPL); Pinkovsky 1970 (FMPL), 1976 (FMPL); Smith 1970 (FMP); Amrine 1971 (FMPL); Lewis 1973 (FMPL); Zwick 1974 (FMPL); Corredor 1975 (FM); Snoddy & Noblet 1976 (MPL); Merritt et al. 1978b (PL); Westwood 1979 (FMPL); Fredeen 1981a (FMPL), 1985a (FMP); Hilsenhoff 1982 (L); Webb & Brigham 1982 (L); Adler 1983 (PL), 1986 (L); Pistrang 1983 (L); Rubtsov & Yankovsky 1984 (FMP); Adler & Kim 1986 (PL); Currie 1986 (PL); Pistrang & Burger 1988 (L); Hunter 1990 (FP); Mason & Kusters 1991b (FMPL); Adler & Kuusela 1994 (MPL); Colgan 1995 (F); Peterson & Kondratieff 1995 (FMPL); Stuart 1995 (cocoon); Jensen 1997 (PL); Stuart & Hunter 1998b (cocoon); Yankovsky 1999 (PL).

**Morphology**

Guttman 1967a (labral fan), Kurtak 1973 (labral fan), Yankovsky 1977 (labral fan), Simmons 1982 (instars), Shipp & Procunier 1986 (instars), Zhang & Malmqvist 1996 (labral fan).

**Physiology**

McKague & Wood 1974 (response to development inhibitors).

**Cytology**

Zimring 1953; Landau 1962; Rothfels et al. 1978; Rothfels 1979, 1981a; Mason 1982, 1984; Adler 1986; Adler & Kim 1986; Shipp & Procunier 1986; Adler & Kuusela 1994; McCreadie et al. 1995.

**Molecular Systematics**

Teshima 1970, 1972 (DNA hybridization); Zhu 1990 (mtDNA); Pruess et al. 1992 (mtDNA); Tang et al. 1996b (mtDNA).

**Bionomics**

**Habitat.** This species complex is abundant throughout the United States, Canada, Mexico, and the Palearctic Region. The immature stages are nearly omnipresent in the riffles of a great panoply of watercourses, from the high mountains to the coastal plains (Adler & McCreadie 1997). Macrocurrent velocities for larvae range from less than 0.2 to 2.5 m/sec; temperatures, from about 0°C to more than 29°C; and stream widths, from less than a meter to more than 500 m. Larvae and pupae attach to virtually any substrate, particularly sticks and rocks. Larvae show a tendency to attach to larger pebbles while avoiding smaller ones (Cressa 1977). Pupation on snails (*Oxytrema carinifera*) is incidental (Vinikour 1982).

**Oviposition.** Females are anautogenous and can complete more than 1 ovarian cycle (Smith 1970, Magnarelli & Cupp 1977, Davies & Györkös 1990). They have been observed on the surface of calm water, depositing 10–20 eggs at a time (Wolfe & Peterson 1959). Reports claiming that females oviposit on trailing vegetation (Stone & Snoddy 1969) are likely based on misidentifications. Fecundity per ovarian cycle varies from about 200 to 400 eggs (Davies & Peterson 1956, Peterson 1959a, Abdelnur 1968, Pascuzzo 1976).

**Development.** In the northern United States and Canada, eggs typically overwinter and larvae appear between March and June; adults are on the wing from late April to mid-November (Back & Harper 1979, McCreadie et al. 1994b). Larvae and pupae can be found year-round in the Ozark Mountains (Bowles & Pinkovsky 1993) and the Great Smoky Mountains (Traoré 1992). All life stages are present throughout the year in the southern United States. The number of population peaks per year varies with latitude: 1 or 2 in Alaska and Newfoundland (Sommerman et al. 1955, Lewis & Bennett 1973), 3 or 4 (sometimes 5) in the northern states and much of Canada (Stone & Jamnback 1955, Anderson & Dicke 1960, Davies et al. 1962, Abdelnur 1968, Eckhart 1968, Pinkovsky 1970, Back & Harper 1979, Lewis & Bennett 1979, Adler et al. 1982), and 5–7 in Alabama and Florida (Stone & Snoddy 1969, Pinkovsky 1976). Larvae pass through 6 or 7 instars (Simmons 1982, Shipp & Procunier 1986). Adults can live up to a month on honey (Smith 1970).

Interactions between larvae commonly result in local displacement (Kurtak 1973, Wiley & Kohler 1981). Larval drift rhythms are variable, showing no clear periodicity (Reisen 1977), increasing dramatically at night (Simmons 1982), or depending somewhat on size, with the smallest larvae drifting primarily during the afternoon and the largest larvae drifting mainly during the night (Wong 1976). Nocturnal increases in drift have been related to contact by foraging stoneflies (Simmons 1982).

The seston provides most of the larval food, suggesting high reliance on filter feeding (Thompson 1987a). Long strands of algae often found in larval guts (Reilly 1975), however, might not be acquired by filter feeding. Larvae feed efficiently at concentrations of approximately 50–200 dye particles/mm$^2$/sec (Kurtak 1978). Feeding rates are among the highest recorded for black flies (Morin et al. 1988b). Ingestion rates increase about tenfold between 25 and 100 cm/sec (Charpentier & Morin 1994). Larvae are able to digest diatoms at least partially during normal gut-retention times (Kurtak 1979). They also can assimilate bacterial extracellular polysaccharide (Couch et al. 1996). The density of the gut contents declines exponentially with increasing body length, though reasons for this relationship are unclear (Morin et al. 1988a).

**Mating.** Only 1 probable mating swarm has been reported. It was observed in the morning about 2.3 m above ground near a group of trees on the edge of a golf course (J. F. Burger, pers. comm.).

**Natural Enemies.** Larvae are attacked by the mermithid nematodes *Mesomermis flumenalis* (Ebsary & Bennett 1975a, Bailey & Gordon 1977, Bruder & Crans 1979) and *Mesomermis camdenensis* (Molloy 1979), as well as unidentified species (Sommerman et al. 1955, McComb & Bickley 1959, Peterson 1960b, Abdelnur 1968, Garris & Noblet 1975, Pinkovsky 1976, Adler & Kim 1986). A record of *Isomermis* sp. from larvae in Wisconsin (Anderson & DeFoliart 1962) is of questionable identity (R. Gordon 1984). Microsporidia infecting larvae include *Amblyospora fibrata*, *A. varians*, *Pegmatheca simulii*, *Polydispyrenia simulii* (Ezenwa 1974b, Ebsary & Bennett 1975a, Lewis & Bennett 1975, Hazard & Oldacre 1975, Vávra & Undeen 1981, Canning & Hazard 1982), and *Janacekia debaisieuxi*. The lobate cysts in larvae infected with *J. debaisieuxi* sometimes are partially enveloped with grayish pigment. The ciliate protozoan *Tetrahymena rotunda* was described from the larval hemolymph (Lynn et al. 1981). Larvae are also hosts of the chytrid fungus *Coelomycidium simulii* (Jamnback 1973a, Adler & Kim 1986); the trichomycete fungi *Genistellospora homothallica*, *Harpella melusinae*, *Simuliomyces microsporus*, and *Smittium simulii* (Lichtwardt 1967, 1972; Labeyrie et al. 1996); and a cytoplasmic polyhedrosis virus (Bailey 1977). One larva from the Great Smoky Mountains National Park, North Carolina, was severely infected with the bacterium *Bacillus amyloliquefaciens*, creating a large white mass in the abdomen (Reeves & Nayduch 2002).

Predators of larvae include ants, larval caddisflies, possibly empidid flies (Peterson 1960b), and perlid stoneflies; larvae are often capable of avoiding stonefly predation (Simmons 1982). Adults are parasitized by water mites (*Sperchon ?jasperensis*) (Davies 1959).

**Hosts and Economic Importance.** Adults feed on the nectar of alfalfa (*Medicago sativa*) (Stone & Snoddy 1969). We have observed females taking nectar from the flowers of wild parsnip (*Pastinaca sativa*) and bearded meadow parsnip (*Thaspium barbinode*).

The literature is rife with confusion over the hosts of the *S. tuberosum* complex. Serological evidence implicates cattle and horses (Downe & Morrison 1957, Simmons et al. 1989). Females have been branded the most persistent pests of livestock in Alabama, putatively feeding in the ears or on the underparts (Stone & Snoddy 1969). Similar problems have been reported in South Carolina (Snoddy & Noblet 1976). Elsewhere, females have been reported to feed in the ears of horses and mules (Snow et al. 1958b) and to swarm around cattle and pigs (Fredeen 1969). We suspect, however, that many of these records are based on misidentifications. We believe that females of the complex feed primarily on small mammals, with the host-species list for these including the Uinta ground squirrel, golden mantled ground squirrel (Peterson 1956), red fox, raccoon, eastern chipmunk, red squirrel, snowshoe hare, and domestic rabbit (Hunter 1990, Hunter et al. 1993, McCreadie et al. 1994c). To this list we add eastern gray squirrel. The arctic ground squirrel (*Spermophilus parryii*) is believed to be the host of females in the Baker Lake area of Nunavut (Smith 1970). Laboratory-reared females have fed on rabbits (Mokry et al. 1981). Attempts to blood feed females on turkeys have proved unsuccessful (Pinkovsky 1976).

Our information suggests that members of the *S. tuberosum* complex are rarely a nuisance to humans. Some reports claim that females can be a nuisance as they swarm about the face and enter the eyes and ears (Sailer 1953, Snow et al. 1958b, Peterson 1959a, Wolfe & Peterson 1959). Yet over much of the continent, biting records are sparse (Anderson & DeFoliart 1961, Abdelnur 1968, Adler & Kim 1986, Burger & Pistrang 1987) or restricted to altitudes above 2100 m (Peterson 1956) or to late in the season (Hayton 1979). Reports of females biting humans in the southeastern United States (Stone & Snoddy 1969, Snoddy & Noblet 1976) probably are based on misidentifications of members of the *S. jenningsi* species group and *S. venustum* species complex.

Females are not known to transmit parasites or pathogens. The avian protozoan *Leucocytozoon simondi*, however, can develop to the ookinete stage under experimental conditions (Fallis et al. 1951).

### *Simulium (Simulium) appalachiense* Adler, Currie & Wood, New Species

(Figs. 10.398, 10.399; Plate 23b; Map 233)

*Simulium tuberosum* 'CDE sibling' Landau 1962: 928, Figs. 1, 8, 13, 23, 25 (C)
'Simuliidae' Reisen & Fox 1970: 624–625 (in part) (L drift)
'Simuliidae' Reisen & Prins 1972: 878–881 (in part) (L drift)
*Simulium* 'sp.' Reisen 1977: 333 (in part) (L drift)
*Simulium tuberosum* 'CDE-2' Mason 1982: 293 (C)
*Simulium* 'CDEM *tuberosum*' Mason 1984: 654–655 (C)
?[*Simulium tuberosum*: Molloy (1976, 1979) not Lundström 1911 (misident.)]

**Taxonomy**

**Female.** Wing length 2.2–2.6 mm. Scutum gray with coppery to golden sheen. Mandible with 40–44 serrations; lacinia with 30–40 retrorse teeth. Sensory vesicle in lateral view occupying about one fourth of palpomere III. Frons shiny. Fore leg brown, except coxa yellowish, and tibia with anterior whitish yellow patch. Middle and hind legs brown except basal one fourth of tibia, half of basitarsus, and one fourth of second tarsomere pale yellowish brown. Claws toothless. All hair pale golden, except that of scutellum, basicosta, and stem vein coppery brown. Genitalia as for species complex.

**Male.** Wing length 2.1–2.5 mm. Scutum velvety black, with pair of anterior, variably sized, silvery pruinose patches; lateral and posterior areas silvery, pruinose. All hair coppery brown. Legs rather uniformly brown, slightly paler at base of some segments. Genitalia (Figs. 10.398, 10.399): Gonostylus in ventral view narrowed at midpoint; in inner lateral view with large, spinose, subbasal, somewhat slender tubercle. Ventral plate in ventral view with posterior margin truncated to strongly tapered, but always with bulbous lip extended beyond posterior margin; lip in lateral view directed posteriorly (downward). Paramere with numerous, variously sized spines. Dorsal plate a thin horizontal strip. Median sclerite expanded distally, slightly bifurcate.

**Pupa.** Length 2.5–3.1 mm. Gill of 6 filaments, three fourths or less length of pupa; filaments in 3 pairs, each trunk about as long as wide; dorsalmost filament thickest; each subsequent filament reduced in thickness; all filaments with transverse furrows. Head and thorax dorsally with densely, irregularly distributed, rounded microtubercles; trichomes unbranched.

**Larva.** Length 4.3–4.9 mm. Head capsule pale yellow to yellowish brown; head spots brown, variably distinct, infrequently surrounded by brown pigment; anteromedial group weakest. Antenna pale yellowish, extended beyond stalk of labral fan by about half length of distal article. Postgenal cleft somewhat horseshoe shaped; subesophageal ganglion ensheathed with pigment. Labral fan with 27–49 primary rays. Body pigment gray to brownish gray, arranged in distinct bands; last 2 thoracic segments usually lacking pigment dorsally and laterally, producing conspicuous white area (Plate 23b). Posterior proleg with 10–12 hooks in each of 61–67 rows.

**Diagnosis.** Females and pupae are not distinguishable from those of several other members of the *S. tuberosum* species complex. The bulbous lip of the ventral plate, although variably expressed (Figs. 10.398, 10.399), distinguishes the males from those of other species in the group. The large white patch on the thorax generally distinguishes the larvae from those of related species. The chromosomes are particularly useful for larval identification.

**Holotype.** Male (frozen dried, pinned) with pupal exuviae and abdomen (separate glycerin vials below). South Carolina, Oconee Co., Chauga River, Chau Ram Co. Park, 34°40.88′N, 83°08.80′W, 20 March 1988, P. H. & C. R. L. Adler (USNM).

**Paratypes.** New York, Washington Co., Camden Creek, 4.8 km northeast of Shushan, 43°07.34′N, 73°17.47′W, 11 August 1987, J. F. Burger (4♀ + exuviae, 50 pupae, 49 larvae). North Carolina, Buncombe Co., Bent Creek, 35°28.94′N, 82°37.85′W, 1 April 1988, C. R. L. Adler (2♂ + exuviae, 20 larvae). South Carolina, same data as holotype (7♀ & 6♂ + exuviae, 14 pupae, 30 larvae), 15 February 1986 (125 larvae), 6 March 1988 (3♀, 108 larvae), 4 April 1992 (photographic negatives of larval polytene chromosomes [IIS arm only]); Pickens Co., Eastatoe River, Co. Road S-39–143, 34°57.50′N, 82°51.17′W, 17 November 1991, P. H. Adler (10♀), 20 November 1991, P. H. Adler (65♀, 37 pupae, 22 larvae).

**Etymology.** This species is named for the Appalachian Mountains, which represent its distributional heartland; the name was suggested by our color artist L. W. Zettler.

**Remarks.** The holotype represents the 'CDEM' cytotype. The degree to which the lip of the ventral plate extends posteriorly is variable (Fig. 10.399a, d, e); the extended condition is more exaggerated in northern populations.

**Cytology**

Landau 1962; Mason 1982, 1984; Adler & Kim 1986.

The 'CDE' and 'CDEM' cytotypes are believed to represent a single species (Mason 1984); however, differences in seasonality (Adler & Kim 1986) suggest that closer study is warranted.

**Bionomics**

**Habitat.** This species is common from southeastern Canada and northern Michigan southward along the mountainous axis of the Appalachian chain. It breeds in shallow rocky streams and rivers with swift currents, temperatures below 25°C, and an average width of 8 m. The distribution of the immature stages is highly predictable based on factors such as stream size (McCreadie & Adler 1998).

**Oviposition.** We have observed thousands of females ovipositing singly and in groups of several hundred during overcast days at air temperatures above 11°C. They generally hover within 10 cm above fine sand of the shoreline that is bathed with gentle waves. When the waves are out, females dart down to the sand and deposit a few eggs, then quickly fly up again, sometimes being overpowered by an incoming wave but usually recovering.

**Development.** This multivoltine species is present throughout the year in southern states such as Georgia and South Carolina. Farther north, larvae of the 'CDEM' cytotype can be found from spring to midfall, whereas those of the 'CDE' cytotype do not appear until late spring or summer (Adler & Kim 1986, Pistrang & Burger 1988).

**Mating.** Unknown.

**Natural Enemies.** Larvae are hosts of the chytrid fungus *Coelomycidium simulii* and the microsporidia *Amblyospora bracteata/varians*, *Janacekia debaisieuxi*, and *Polydispyrenia simulii* (McCreadie & Adler 1999). *Simulium appalachiense* is probably the species (as *S. tuberosum*) that was heavily parasitized by the mermithid nematode *Mesomermis camdenensis* (Molloy 1979).

**Hosts and Economic Importance.** Unknown; presumably mammalophilic.

### *Simulium (Simulium) chromocentrum* Adler, Currie & Wood, New Species

(Figs. 10.400, 10.508, 10.817, 10.830; Plate 23c; Map 234)

**Taxonomy**

**Female.** Not differing from that of *S. appalachiense* n. sp.

**Male.** Not differing from that of *S. appalachiense* n. sp. except as follows: Genitalia (Fig. 10.400): Gonostylus in ventral view uniform in width beyond stout subbasal tubercle; outer lateral margin with slight crook. Ventral plate in ventral view with body subrectangular, concave posteriorly; in lateral view with lip directed medially.

**Pupa.** Not differing from that of *S. appalachiense* n. sp. except as follows: Length 2.6–3.2 mm. Gill (Fig. 10.508) with dorsal pair of filaments arising from thick trunk; middle pair usually arising directly from trunk of dorsal pair; ventral pair arising from thin trunk 1–3 times as long as wide. Head and thorax dorsally with microtubercles as in *S. appalachiense* n. sp., but larger (Fig. 10.830).

**Larva.** Not differing from that of *S. appalachiense* n. sp. except as follows: Length 5.0–5.7 mm. Head capsule (Fig. 10.817) dark brown; head spots absent or paler than head capsule (negative pattern). Antenna brown, subequal in length to stalk of labral fan. Postgenal cleft narrow, subtriangular. Labral fan with 45–50 primary rays. Body pigment nearly black, uniformly distributed (Plate 23c). Posterior proleg with 10–13 hooks in each of 62–66 rows.

**Diagnosis.** Females cannot be separated from those of most other members of the *S. tuberosum* species complex. Males are not easily distinguished from those of other group members, although the short wide ventral plate can be diagnostic. The compact gill of the pupa is unique among species in the group, as is the chromocenter of the polytene chromosomes. The nearly black larvae with their faint negative head-spot patterns are unique among North American black flies.

**Holotype.** Male (Peldri II, pinned) with pupal exuviae (glycerin vial below). California, Mono Co., Saddlebag Lake Road, 1.3 km off Rt. 120, outside eastern boundary of Yosemite National Park, 37°56.86′N, 119°15.43′W, 23 June (pupa), 30 June (adult) 1991, P. H. Adler (USNM).

**Paratypes.** California, Inyo Co., headwaters of Shepherd Creek, 3460 m elev., 36°41.2′N, 118°20.5′W, 13 August 2001, D. C. Kurtak (23 pupae, 6 larvae); Mariposa Co., Yosemite National Park, Rt. 120, small stream at Crane Flat General Store, 37°45.47′N, 119°48.29′W, 11 May 1997, P. H. Adler (1♀ + exuviae); Mono Co., inlet to Tioga Lake, Rt. 120, Tioga Pass, 3030 m elev., 37°55.17′N, 119°14.90′W, 16 July 1988, P. H. Adler & D. C. Currie (1♀ + exuviae, 2 pupae, 4 exuviae); same data as holotype (4♂ + exuviae, 289 larvae), 10 June 1990 (74 larvae), 25 June 1991 (3 pupae, 51 larvae, photographic negatives of larval polytene chromosomes).

**Etymology.** This species is named for the chromocenter that characterizes its polytene chromosomes.

**Cytology**

The chromosomes have the standard banding sequence of the *S. tuberosum* species complex in all arms except IIS, which has the FG sequence of Landau (1962). The large heterochromatic chromocenter is unique among all known members of the *S. tuberosum* complex and is conspicuous even in early instars. The Y chromosome is defined by a set of unique inversions in the IIS and IIL arms.

**Bionomics**

**Habitat.** This species has been found in 7 streams, all but one at elevations above 3000 m, in the Sierra Nevada of California. Larvae—the darkest of all North American species—are found in unshaded spring-fed streams about a third of a meter wide, with plentiful trailing vegetation. We discovered a pure population of this species developing in the inlet stream to Tioga Lake at the same time that a sizable pure population of the closely related, morphologically similar *S. vandalicum* was developing at the lake outlet.

**Oviposition.** Unknown.

**Development.** Larvae in Mono Co. hatch by May and pupae first appear from mid-May to mid-June, disappearing by July. In Fresno and Inyo Cos., larvae and pupae persist well into August. The number of generations per year is unknown.

**Mating.** Unknown.

**Natural Enemies.** Larvae are hosts of the chytrid fungus *Coelomycidium simulii*.

**Hosts and Economic Importance.** Unknown; presumably mammalophilic.

### *Simulium (Simulium) conundrum* Adler, Currie & Wood, New Species

(Figs. 10.401, 10.831–10.833; Plate 23d; Map 235)

*Simulium tuberosum* 'FGH sibling' Landau 1962: 928 (C)

[*Simulium tuberosum*: Zimring (1953) not Lundström 1911 (misident. in part, material from Rattlesnake Point)]

**Taxonomy**

**Female.** Not differing from that of *S. appalachiense* n. sp.

**Male.** Not differing from that of *S. appalachiense* n. sp. except as follows: Genitalia (Fig. 10.401): Gonostylus in ventral view slightly narrowed posterior to stout subbasal tubercle. Ventral plate in ventral view with body subrectangular, concave posteriorly; in lateral view with lip directed anteromedially (upward).

**Pupa.** Not differing from that of *S. appalachiense* n. sp. except as follows: Length 2.9–3.4 mm. Head and thorax dorsally with microtubercles as in *S. appalachiense* n. sp., but larger, with or without microgranules visible at more than 1000× magnification (Figs. 10.831–10.833).

**Larva.** Not differing from that of *S. appalachiense* n. sp. except as follows: Length 4.5–5.9 mm. Head capsule pale yellowish brown to dark brownish orange; head spots brown, indistinct, often apparently absent or obscured by darker surrounding pigment; anterolateral head spots heaviest. Antenna pale brown, subequal in length to stalk of labral fan. Postgenal cleft subtriangular, bowed slightly outward at midpoint. Labral fan with 28–48 primary rays; mean number of rays greater at higher latitudes. Body pigment gray to dark gray, uniformly distributed (Plate 23d) (infrequently arranged in distinct bands). Posterior proleg with 10–14 hooks in each of 65–73 rows.

**Diagnosis.** Females and pupae are indistinguishable from those of most other members of the *S. tuberosum* species complex. Males can be separated from those of other species of the complex by the short wide ventral plate and gonostylus with a bulge about three fourths of the distance from its apex. Larvae often can be distinguished by their dark, rather uniformly distributed body pigment and faint, nearly obscure head-spot pattern. The banding sequence of the IIS chromosome arm provides the best diagnostic tool.

**Holotype.** Male (hexamethyldisilazane, pinned) with pupal exuviae (glycerin vial below). West Virginia, Kanawha Co., tributary of Davis Creek, Kanawha State Forest, 38°14′N, 81°39′W, 2 May 1997, P. H. & C. R. L. Adler (USNM).

**Paratypes.** New Hampshire, Grafton Co., Waterville Valley, 43°58′N, 71°31′W, 26 May 1987, J. F. Burger (photographic negatives of larval polytene chromosomes). North Carolina, Haywood Co., East Fork Pigeon River, 1.5 km west of Yellowstone Falls, 35°19.24′N, 82°50.93′W, 19 May 1988, C. I. Dial (17 larvae), 31 May 1988, P. H. Adler (1♀, 9♂). West Virginia, same data as holotype (2♀, 7♂, 7 pupae, 96 larvae); Pocahontas Co., Cranberry Glades, Monongahela National Forest, creek at boundary before parking lot, 38°11′N, 80°16′W, 23 May 1987, P. H. & C. R. L. Adler (79 larvae), 4 June 1988, P. H. Adler & C. R. L. Adler (2♀ & 2♂ + exuviae, 26 larvae, photographic negatives of larval polytene chromosomes). Alberta, 3.2 km east of Obed Summit, Hwy. 16, 53°31′N, 117°17′W, 27 June 1984, P. H. Adler (2 exuviae, 66 larvae). British Columbia, unnamed creek, 32 km north of Tumbler Ridge on Hwy. 29, 55°17′N, 121°19′W, 900 m elev., 3 July 1987, D. C. Currie (5 pupae, 19 larvae, photographic negatives of larval polytene chromosomes).

**Etymology.** The specific name refers to the unknown significance of the numerous Y-linked inversions on the short arm of chromosome II.

**Cytology**

Zimring 1953; Landau 1962; Mason 1982, 1984; Adler & Kim 1986; McCreadie et al. 1995.

Undifferentiated sex chromosomes represent the most common and widespread condition in this species. However, at least 5 different Y-linked IIS sequences have been reported in localized populations (Mason 1984, McCreadie et al. 1995), and we have found additional Y-linked sequences on the IIS arm. Some of these Y chromosomes might represent separate species. The most compelling evidence supporting this suggestion comes from a newly discovered Y-chromosome sequence in IIS that superficially resembles the FG sequence of Landau (1962). Larvae with this Y chromosome occur in spring-fed streams (e.g., Grady Branch, Watershed 18) about a meter wide at the Coweeta Hydrologic Laboratory, Macon Co, North Carolina (April and May). These springs empty into Ball Creek, which is about 5 m wide, with a powerful discharge, and which supports a pure population of larvae with undifferentiated sex chromosomes. Either these 2 populations represent sex-chromosome polymorphisms that show remarkable site fidelity or they are actually separate species.

Two larval hybrids between *S. conundrum* and *S. vandalicum* have been found (Mason 1984).

**Bionomics**

**Habitat.** This common trans-Canadian species extends southward through the Appalachian highlands. It breeds predominantly in cool rocky forest streams less than a few meters wide. In some areas, such as New Hampshire and Newfoundland's Avalon Peninsula, its distribution is independent of stream size (Pistrang & Burger 1988, McCreadie et al. 1995).

**Oviposition.** Unknown.

**Development.** This species is the only eastern member of the *S. tuberosum* complex that is univoltine (Adler & Kim 1986, Pistrang & Burger 1988), although overlapping cohorts occur (Tessler 1991). Eggs overwinter and larvae are present from April through June in the eastern half of the range, being most common in May. In western Canada, larvae are present from June into August (Adler 1986).

**Mating.** Unknown.

**Natural Enemies.** Larvae are hosts of unidentified mermithid nematodes (Adler 1986), the chytrid fungus *Coelomycidium simulii*, and the microsporidium *Polydispyrenia simulii*. *Rhyacophila* caddisflies prey on larvae (Adler 1986).

**Hosts and Economic Importance.** Unknown; presumably mammalophilic.

### *Simulium (Simulium) perissum* Dyar & Shannon
(Fig. 10.402, Map 236)

*Simulium perissum* Dyar & Shannon 1927: 43–44, Figs. 84, 85, 119, 120 (FM). Holotype* male with pupal exuviae (slide #28346; USNM). Virginia, Fairfax Co., Dead Run, 12 April 1925 (R. C. Shannon). Revalidated

*Simulium tuberosum* 'CKL-type' Landau 1962: 928, Fig. 1 (C)

'black fly larvae' Belluck & Furnish 1981: 204 (prey of staphylinid beetles)

*Simulium (Simulium) tuberosum* 'Complex' Pruess & Peterson 1987: 532 (Nebraska records)

[*Simulium venustum*: Forbes (1912) not Say 1823 (misident. in part, Figs. 19–21)]

[*Simulium venustum*: Garman (1912) not Say 1823 (misident.)]

[*Simulium tuberosum*: Zhu (1990) not Lundström 1911 (misident.)]

[*Simulium tuberosum*: Pruess et al. (1992) not Lundström 1911 (misident.)]

**Taxonomy**

Forbes 1912 (PL), Garman 1912 (PL), Dyar & Shannon 1927 (FM).

Our visit (25 April 1988) to the now-polluted type locality of *S. perissum* yielded 1 pupa. The gill of this pupa had rather widely spaced filaments that match the gill of the pupal exuviae of the type specimen of *S. perissum* and the gill histoblast of mature larvae identified cytologically as 'CKL.' An adjacent stream, Turkey Run, also supported a population that we identified cytologically as 'CKL.' We, therefore, link the name *S. perissum* with cytospecies 'CKL.'

Larval body color is typically sex linked, with dark gray males and brown females.

**Cytology**

Landau 1962; Mason 1982, 1984; Adler & Kim 1986.

In addition to undifferentiated sex chromosomes, we have found a widespread differentiated Y chromosome characterized by a small inversion that reorients the "bulge" marker in the IIS arm.

Two hybrid larvae between *S. perissum* and *S. tuberosum* cytotype A are known, one from southern Ontario

(Landau 1962) and one (♂) from Mississippi (Wilkinson Co., Clark Creek Natural Area, 11 March 1989). A hybrid between *S. perissum* and *S. tuberosum* cytotype AB was found in Virginia (Mason 1984).

**Molecular Systematics**

Zhu 1990 (mtDNA), Pruess et al. 1992 (mtDNA).

**Bionomics**

**Habitat.** This species is found from the eastern coast of the United States to western Nebraska and is the most common member of the *S. tuberosum* species complex in the central states. In Canada, it has been recorded only around Toronto. The immature stages live in highly productive, small streams and shallow cobble-bottomed rivers with open canopies.

**Oviposition.** Unknown.

**Development.** Eggs of this multivoltine species overwinter, although larvae and pupae are present year-round in the southernmost states.

**Mating.** Unknown.

**Natural Enemies.** Larvae are attacked by the chytrid fungus *Coelomycidium simulii*, the microsporidium *Polydispyrenia simulii*, and unidentified mermithid nematodes. They have been consumed by staphylinid beetles (*Psephidonus brunneus*) (Belluck & Furnish 1981)

**Hosts and Economic Importance.** Unknown; presumably mammalophilic.

### *Simulium (Simulium) tuberosum* (Lundström)

(Figs. 10.221, 10.403, 10.818, 10.834;
Plate 23e; Map 237a, b)

*Melusina tuberosa* Lundström 1911: 14–15, Fig. 10 (M). Lectotype* male (designated by Zwick 1974: 103; pinned #4178, terminalia [in Canada Balsam] on plastic mount below; UZMH). Finland, Enontekiö, date unknown (R. Frey); *tuberozum* (subsequent misspelling)

*Simulium* '(s. str.) sp. indet.' Edwards 1933: 614 (1F from Akpatok Island)

?*Simulium* 'species 2' Jenkins 1948: 149–151 (Alaska record)

*Simulium tuberosum* 'cytotype AB' Landau 1962: 928, Figs. 1, 6, 12, 14, 15, 17, 18 (C)

*Simulium (Simulium) tuberosum* 'A' Wong 1976: 40–42 (L velocity tolerance)

*Simulium tuberosum* 'A sibling' Mason 1982: 293–300 (C)

*Simulium tuberosum* '$Y_2$' Mason 1982: 295–297 (C)

?[*Simulium meridionale*: O'Kane (1926) not Riley 1887 (misident.)]

**Taxonomy**

Lundström 1911 (M); Rubtsov 1940b, 1956, 1963a (FMPL); Zwick 1974 (M); Peterson 1981 (M), 1996 (MP); Rubtsov & Yankovsky 1984 (FMP); Adler 1986 (L); Adler & Kuusela 1994 (MPL); Stuart 1995 (cocoon); Stuart & Hunter 1998b (cocoon); Yankovsky 1999 (PL).

Larvae collected in August 1994 from Isle Royale National Park, Michigan, have tuberculate abdomens (similar to those of *S. meridionale*), allowing rapid separation from the larvae of *S. vandalicum*, which occupy most of the same streams. These tuberculate larvae were chromosomally classic for the 'AB' cytotype. Larvae collected in May 1997 were darkly pigmented and lacked tubercles.

Figures 10.221, 10.403, 10.818, and 10.834 represent the 'AB' cytotype, whereas the color illustration (Plate 23e) represents the 'A' cytotype.

**Morphology**

Zhang & Malmqvist 1996 (labral fan).

**Cytology**

Zimring 1953; Landau 1962; Rothfels 1981a; Mason 1982, 1984; Adler & Kim 1986; Adler & Kuusela 1994.

**Molecular Systematics**

Smith 2002 (nuclear and mtDNA).

This species is composed of 3 cytologically distinct entities that we recognize conservatively as cytotypes: 'A,' 'AB' (Nearctic), and 'AB $Y_2$' (Palearctic). Possibly 2 or all of these cytotypes are valid species. However, in light of the allopatric nature of Nearctic and Palearctic populations, we prefer to recognize a single species, following Adler and Kuusela (1994). The situation in North America, nonetheless, remains problematic. Throughout most of the sympatric portion of their range, 'A' and 'AB' show no evidence of hybridization. However, at 2 sites, A/AB females have been found (Mason 1984). At Fishing Brook, Long Lake, Franklin Co., New York (the actual location for the site given by Mason [1984] as near Newcomb), we scored 129 larvae collected from 6 June to 11 August 1988 as follows: 38 A/A females, 9 AB/AB females, 5 A/AB females, 60 St/A males, and 17 St/AB males. Fewer female heterozygotes were found than expected by chance. One inversion near the end of IS was shared between cytotypes. Whether 'A' and 'AB' are distinct species that occasionally hybridize or are merely X-chromosome polymorphisms remains unresolved. Because 'A' and 'AB' have distinct distributions and because they might later prove to be valid species, we have plotted their ranges separately (Maps 237a, 237b).

One or 2 hybrids between *S. tuberosum* ('AB' cytotype) and *S. vandalicum* and between *S. tuberosum* ('AB' and 'A' cytotypes) and *S. perissum* have been found (Landau 1962, Mason 1984, Adler 1986).

**Bionomics**

**Habitat.** *Simulium tuberosum* is Holarctic. In the Palearctic Region, it probably has a broad distribution but has not been confirmed chromosomally beyond Fennoscandia. In North America, it occupies most of Canada and Alaska and extends southward in sparse populations through the Rocky Mountains and in abundance east of the Mississippi River. Only the 'AB' cytotype is found in the western half of the continent, and only the 'A' cytotype occupies the area south of Virginia; elsewhere, both cytotypes are found. The immatures live in a wide variety of watercourses but are most frequent in larger streams and rivers (McCreadie et al. 1995). The smaller streams inhabited by the immatures are often organically enriched (Adler 1986). In Canada and the northern United States, both cytotypes are found in rocky productive streams and rivers. In the southeastern United States, the 'A' cytotype is one of the most abundant black flies and is particularly prevalent in wide sandy piedmont streams (Adler & McCreadie 1997).

**Oviposition.** Unknown.

**Development.** In Canada, eggs overwinter and mature larvae begin to appear by May. In Pennsylvania, larvae and pupae can be found from April to October (Adler & Kim 1986). South of Virginia, they can be found year-round. Adults in Missouri emerge from mid-February through mid-November (Doisy et al. 1986). The average yearly development time from hatch to pupation is about 44 days in Georgia (Freeman & Wallace 1984).

**Mating.** Unknown.

**Natural Enemies.** Larvae are hosts of unidentified mermithid nematodes; the chytrid fungus *Coelomycidium simulii*; the microsporidia *Amblyospora varians* (Adler 1986, Adler & Mason 1997), *A. bracteata*, *Janacekia debaisieuxi*, and *Polydispyrenia simulii* (McCreadie & Adler 1999); and the trichomycete fungus *Harpella melusinae*. We have found larvae and pupae bearing hyphal masses of an unidentified oomycete, possibly in the order Saprolegniales. We have taken males and females from streamside webs of the araneid spider *Larinioides patagiata*.

**Hosts and Economic Importance.** Unknown; presumably mammalophilic.

**_Simulium (Simulium) twinni_ Stains & Knowlton**
(Fig. 10.404, Map 238)

*Simulium* (*Simulium*) *twinni* Stains & Knowlton 1940: 77–78 (M). Holotype* male (slide, terminalia only; USNM). Utah, Cache Co., 4 mi. above mouth of Logan Canyon, light trap, 8 August 1939 (G. F. Knowlton). Revalidated

*Simulium tuberosum* 'FG/FG' McCreadie et al. 1997: 764 (Wyoming record)

[*Simulium vandalicum*: Stains (1941) not Dyar & Shannon 1927 (misident.)]

[*Simulium vandalicum*: Stains & Knowlton (1943) not Dyar & Shannon 1927 (misident.)]

**Taxonomy**

Stains & Knowlton 1940, 1943 (M); Stains 1941 (M).

We resurrect the name *twinni* from synonymy with *tuberosum* on the basis of the unique W-shaped ventral plate of the male (Fig. 10.404a); the pale larval head capsule that virtually lacks head spots; the fixed FG banding sequence (*sensu* Landau 1962) in the IIS chromosomal arm; and the wide, swift, and rocky preimaginal habitat.

**Cytology**

Under the name *S. twinni*, we include populations from the Rocky Mountains westward that occur in streams wider than 3 m and are fixed for the FG sequence in IIS. We have seen 1 larva from Boulder Co., Colorado, with B chromosomes.

**Bionomics**

**Habitat.** This species is found from southern British Columbia and Alberta into the Sierra Nevada of California and the Rocky Mountains of Colorado. The immature stages inhabit large rocky streams and rivers with swift current.

**Oviposition.** Unknown.

**Development.** This species is probably multivoltine. Mature larvae are present from at least May through August.

**Mating.** Unknown.

**Natural Enemies.** We have collected larvae infected with unidentified mermithid nematodes; the chytrid fungus *Coelomycidium simulii*; and the microsporidia *Amblyospora varians*, *Janacekia debaisieuxi*, and *Polydispyrenia simulii*.

**Hosts and Economic Importance.** Unknown; presumably mammalophilic.

**_Simulium (Simulium) ubiquitum_ Adler, Currie & Wood, New Species**
(Figs. 10.222, 10.405, 10.509, 10.819, 10.835, 10.836; Plate 23f; Map 239)

*Simulium tuberosum* 'complex' Ledin 1994: 37 (in part, records under *Pegmatheca simulii*)

*Simulium tuberosum* 'F' Adler & McCreadie 1997: 157–158 (ecology)

[*Simulium tuberosum*: Sutherland & Darsie (1960a, 1960b) not Lundström 1911 (misident.)]

[*Simulium venustum* 'complex': Arnold (1974) (misident. in part, p. 113, P)]

[*Simulium tuberosum*: Hazard & Oldacre (1975) not Lundström 1911 (misident.)]

[*Simulium tuberosum*: Pinkovsky (1976) not Lundström 1911 (misident. in part)]

[*Simulium tuberosum* 'FGH': Mason in Rothfels (1981a) (misident. in part, records from Arkansas)]

[*Simulium tuberosum* 'FGH': Mason (1982) (misident. in part, material from Arkansas and Delaware)]

[*Simulium tuberosum* 'CKL': Mason (1984) (misident. in part, material from South Carolina)]

[*Simulium tuberosum* 'cytospecies FGH': Adler (1987) (misident.)]

**Taxonomy**

**Female.** Not differing from that of *S. appalachiense* n. sp. Terminalia as in Fig. 10.222.

**Male.** Not differing from that of *S. appalachiense* n. sp. except as follows: Genitalia (Fig. 10.405): Gonostylus in ventral view slightly narrowed posterior to stout subbasal tubercle. Ventral plate in ventral view with body subrectangular, slightly concave posteriorly; lip in lateral view small, short, directed medially.

**Pupa.** Not differing from that of *S. appalachiense* n. sp. except as follows: Length 2.3–3.5 mm. Gill of 6 rather widely spaced filaments, about as long as pupa (Fig. 10.509); filaments subequal in thickness, grouped in 3 pairs; trunks typically 2–6 times longer than wide. Head and thorax with densely and regularly distributed, dome-shaped microtubercles (Figs. 10.835, 10.836).

**Larva.** Not differing from that of *S. appalachiense* n. sp. except as follows: Length 4.2–6.0 mm. Head capsule (Fig. 10.819) brownish orange; head spots brown, diffuse, often indistinct; anteromedial and posteromedial head spots often nearly contiguous. Labral fan with 26–45 primary rays. Body pigment grayish (female) or pale brown (male, Plate 23f), arranged in bands. Posterior proleg with 7–12 hooks in each of 64–68 rows.

**Diagnosis.** Females are indistinguishable from those of most other members of the *S. tuberosum* species group. The tiny lip of the ventral plate (in lateral view) is useful for distinguishing males from those of other species in the group. The densely and uniformly distributed microtubercles of the thorax, in combination with the elongate trunks of the 6-filamented gill, distinguish the pupae from those of all other North American species. This species is the only member of the *S. tuberosum* group with sexually dimorphic larvae in which males are brownish and females grayish.

**Holotype.** Male (frozen dried, pinned) with pupal exuviae (glycerin vial below). South Carolina, Oconee Co., Co. Road TU-37 (Cobbs Bridge Road), tributary of Chauga River, 34°43′N, 83°10′W, 19 April 1987, P. H. and C. R. L. Adler (USNM).

**Paratypes.** Delaware, Kent Co., junction of Rt. 384 and Rt. 385, north end of Killen Pond State Park, Spring Pond Creek, 38°59.5′N, 75°32.6′W, 26 October 1991, P. H. Adler & D. C. Currie (7 pupae, 39 larvae). Florida, Gadsden Co., Crooked Creek, Rt. 270, 30°34.9′N, 84°52.8′W, 27 February 1988, P. H. Adler (9♀ & 2♂ + exuviae); Marion Co., Ocala National Forest, Ninemile Creek, 29°05.65′N, 81°36.49′W, 20 February 1988, P. H. Adler (4♀ & 5♂ + exuviae, 17 pupae, 133 larvae); Juniper Springs, 29°11.12′N, 81°42.21′W, 20 February 1988, P. H. Adler (6♀ & 3♂ + exuviae, 12 pupae, 256 larvae); Washington Co., Falling Waters State Recreation Area, 30°43′N, 85°32′W, 27 February 1988, P. H. Adler (2♀ & 4♂ + exuviae, 7 pupae, 80 larvae). South Carolina, Aiken Co., Aiken State Park, Cypress Stump Picnic Area, 33°32′N, 81°29′W, 19 July 1987, P. H. Adler (6♀ & 1♂ + exuviae, 6 pupae, 107 larvae), 16 February 1990, J. K. Moulton (5 pupae, 61 larvae); Clarendon Co., Rt. 378, Rocky Bluff Swamp, 33°53.8′N, 80°10.2′W, 24 March 1988, P. H. Adler (1♀ + exuviae); same data as holotype (10♀ & 13♂ + exuviae, 6 pupae, 23 larvae, photographic negatives of larval polytene chromosomes [IIS arm only]), 4 February 1986 (19 larvae, photographic negatives of larval polytene chromosomes [IIS arm only]), 11 July 1987 (3♀ & 2♂ + exuviae, 4 pupae, 65 larvae); Pickens Co., Table Rock State Park, Green Creek, 35°02.17′N 82°42.14′W, 9 May 1986, P. H. Adler (1♂ + exuviae).

**Etymology.** The specific name is from the Latin adjective meaning "ubiquitous" or "existing everywhere," in reference to the occurrence of this species in most streams of the southeastern United States.

**Cytology**

All chromosome arms except IIS are standard for the *S. tuberosum* sequence of Landau (1962). The IIS sequence is deceptively close to the FGH sequence of *S. conundrum* n. sp. Sex chromosomes are undifferentiated.

**Bionomics**

**Habitat.** *Simulium ubiquitum* is one of the most common black flies in the southeastern United States. In the Sandhills Region and the coastal plain, the immature stages are found in clear and brownwater streams up to a meter or more deep and about 1–15 m wide, with plentiful stands of aquatic macrophytes such as *Sparganium*. A pH as low as 3.5 has been recorded for the streams of some populations in Florida (Pinkovsky 1976). In upland areas, larvae and pupae inhabit cool, shallow, sandy, spring-fed streams less than 2 m wide, sometimes with small stones and pebbles but usually devoid of instream vegetation.

**Oviposition.** Unknown.

**Development.** Larvae and pupae are found throughout the year but are least common during the hot summer months. Seven or more generations might occur in Florida (Pinkovsky 1976).

**Mating.** Unknown.

**Natural Enemies.** Larvae frequently are attacked by unidentified mermithid nematodes; the chytrid fungus *Coelomycidium simulii*; the microsporidia *Amblyospora varians*, *Janacekia debaisieuxi*, *Polydispyrenia simulii* (McCreadie & Adler 1999), and *Pegmatheca simulii* (Ledin 1994); and the trichomycete fungus *Harpella melusinae*.

**Hosts and Economic Importance.** Hosts are unknown but presumably are mammals. Females evidently do not feed on turkeys (Pinkovsky 1976).

### *Simulium (Simulium) vandalicum* Dyar & Shannon

(Figs. 10.510, 10.820; Plate 24a; Map 240)

*Simulium vandalicum* Dyar & Shannon 1927: 44, Figs. 111, 112 (FM). Holotype* male (pinned #28347, terminalia on slide; USNM). California, El Dorado Co., Lake Tahoe, Fallen Leaf, 28 June 1916 (H. G. Dyar). Revalidated

*Simulium (Simulium) turmale* Twinn 1938: 51–53 (M). Holotype* male (slide #4450; CNC). Utah, Cache Co., Logan, at light, 16 September 1937 (G. F. Knowlton & F. C. Harmston). New Synonym

*Eusimulium* 'species' Frost 1949: 129 (in part, specimen collected 28 May; member of the *S. tuberosum* species complex, probably *S. vandalicum*)

*Simulium tuberosum* 'FG sibling' Landau 1962: 928, Figs. 7, 8, 10, 11, 16, 19, 21 (C)

*Simulium (Simulium) tuberosum* 'B' Wong 1976: 41 (L velocity tolerance)

*Simulium tuberosum* 's. l.' Tang et al. 1996b: 40–44 (mtDNA)

[*Simulium tuberosum*: Zimring (1953) not Lundström 1911 (misident. in part, material from Credit Forks)]

[*Simulium tuberosum*: Wood et al. (1963) not Lundström 1911 (misident. in part, Fig. 54)]

[*Simulium tuberosum*: Merritt et al. (1978b) not Lundström 1911 (misident. in part, Fig. 16)]

[*Simulium (Simulium) parnassum*: Cupp & Gordon (1983) not Dyar & Shannon 1927 (misident. in part, all material from Michigan)]

[*Simulium tuberosum*: Bowles & Pinkovsky (1993) not Lundström 1911 (misident.)]

**Taxonomy**

Dyar & Shannon 1927 (FM); Twinn 1938 (M); Wood et al. 1963 (L); Merritt et al. 1978b (L); Pistrang 1983 (L); Adler 1983, 1986 (L); Adler & Kim 1986 (PL); Pistrang & Burger 1988 (L).

The female, male, and pupa of *S. vandalicum* resemble those of *S. tuberosum* s. s.

**Cytology**

Zimring 1953; Landau 1962; Mason 1982, 1984; Adler 1986; Adler & Kim 1986.

Our cytological examination of larvae from the type locality of *S. vandalicum*, along with study of the holotype, indicates that the name *S. vandalicum* should be applied to the cytological entity described by Landau (1962) as 'FG.' Similarly, our cytological examination of larvae from the type locality of *S. turmale* and study of the holotype indicate that the name *S. turmale* also is equivalent to cytospecies 'FG.' Male larvae at both type localities had the

FG/St sequence in the IIS arm; this sequence occurs across most of the continent.

However, additional, more localized Y-chromosome sequences also occur, some of which might represent distinct species. The FG/CDE sequence is common in the Appalachian Mountains (Mason 1982, 1984). Larvae that we examined from the Ozark Mountains (e.g., Devil's Den State Park, Arkansas, 23 April 1989) had undifferentiated sex chromosomes, as did larvae from several sites in Wyoming (e.g., Yellowstone National Park, 17.5 km east of Mammoth Hot Springs on Mammoth-Tower Road, 30 June 1993) and California (Mono Co., Tioga Lake area, 9 June 1990). These populations with undifferentiated sex chromosomes do not belong to *S. twinni* because the larvae are almost black, the associated males have a ventral plate characteristic of *S. vandalicum* rather than the W-shaped ventral plate of *S. twinni*, and the immature stages are found in streams less than a meter wide rather than the large swift streams of *S. twinni*. A population near Michigan Technological University, Houghton Co., Michigan (10 May 1990), was homozygous for the FG sequence in both sexes, but all 7 male larvae carried a small inversion in the base of IIS between section 54 and the ring of Balbiani, whereas the 3 females did not; on 11 August of the same year, all 3 male larvae analyzed from this site had the St/FG sequence. Seven male larvae examined from Silverhope Creek in the town of Silver Creek, British Columbia (24 June 1991), had the St/FG sequence plus an inversion on the St sequence of Landau (1962) from about the middle of section 44 to the middle of section 46. We tentatively consider all of the above populations to represent a single species.

Hybrids between *S. vandalicum* ('FG') and *S. tuberosum* ('AB') are known. One was from Alberta (Adler 1986) and the other from Manitoba (14.3 km south of Grand Rapids, 26 May 1986); both were female larvae. Two hybrids between *S. vandalicum* ('FG') and *S. conundrum* n. sp. ('FGH') are also known (Mason 1984).

**Molecular Systematics**

Tang et al. 1996b (mtDNA), Smith 2002 (nuclear and mtDNA).

**Bionomics**

**Habitat.** With a distribution spanning the entire continent, *S. vandalicum* is one of the most common black flies in North America. It is probably the species recorded by Vargas and Díaz Nájera (1957) as *S. tuberosum* from Mexico. Larvae and pupae inhabit rocky streams 1–15 m wide in forested areas of mountainous or hilly terrain. Larvae often form clusters on the top surfaces of rocks, leaves, and twigs in areas of swift current (Tessler 1991) with both laminar and turbulent boundary layers (Eymann 1993). Peak larval drift begins at sunset and lasts through the night (Adler et al. 1983b).

**Oviposition.** Females fly predominantly upstream (Adler & Kim 1986), presumably in search of oviposition sites. Oviposition behavior has not been reported.

**Development.** This multivoltine species can be found throughout the year as far north as Pennsylvania, where both eggs and larvae overwinter and adults fly from about late April into November (Adler & Kim 1986, McCreadie et al. 1994b). In Canada, mature larvae can be found at least by May (Adler 1986, Adler & Mason 1997). Development is delayed in streams that remain below 20°C year-round, with mature larvae first appearing in June (Pistrang & Burger 1988). Larvae of the summer generation in southern Alberta require about 475 degree-days above 0°C to complete development (Shipp & Procunier 1986).

**Mating.** Unknown.

**Natural Enemies.** Larvae often harbor unidentified mermithid nematodes (Adler & Kim 1986, Adler & Mason 1997); the chytrid fungus *Coelomycidium simulii*; the microsporidia *Amblyospora varians*, *Janacekia debaisieuxi*, *Polydispyrenia simulii* (McCreadie & Adler 1999), and *A. fibrata*; and the trichomycete fungus *Harpella melusinae*. An unidentified, probably entomophthoraceous fungus with uninucleate, rodlike hyphal bodies (Fig. 6.4) has been found in the hemocoel of 4 larvae from South Carolina and 1 from western Virginia.

**Hosts and Economic Importance.** Unknown; presumably mammalophilic.

### *Simulium* (*Simulium*) *vulgare* Dorogostaisky, Rubtsov & Vlasenko

(Figs. 10.406, 10.821; Map 241)

*Simulium vulgaris* Dorogostaisky, Rubtsov & Vlasenko 1935: 166–168 (FMPL). Lectotype* female (designated by Yankovsky 1995: 57; 1 wing, 1 foreleg, and femur and tibia of another leg on slide #3427, other parts missing; ZISP). Russia, near Irkutsk, Karolog River, 21 June 1931 (published date), 21 June 1932 (slide label) (V. C. Dorogostaisky)

*Simulium* (s. str.) *tuberosum* var. *vulgare*, Rubtsov 1940b: 421–422, Figs. 15G, 58J, 62B, I, 65L, 72K, 73O, 87C–E, I (FMPL)

*Simulium tuberosum* 'FGI' Mason 1982: 295–302 (C)

[*Simulium tuberosum*: Sommerman (1953) not Lundström 1911 (misident.)]

**Taxonomy**

Dorogostaisky et al. 1935 (FMPL); Rubtsov 1940b, 1956, 1962d, 1963a (FMPL); Sommerman 1953 (L); Adler & Kuusela 1994 (MPL); Yankovsky 1999 (PL).

This species and other Palearctic members of the *S. tuberosum* species complex frequently have been misidentified. We, therefore, have not included much of the Palearctic literature purporting to deal with *S. vulgare*. Much of the Alaskan treatment by Sommerman et al. (1955) probably pertains to *S. vulgare*; however, *S. tuberosum* cytotype ('AB') and *S. vandalicum* are also common in Alaska.

Mason (1984) reported sexual dimorphism in larval color, but we have noticed it infrequently. Larvae tend to be more darkly pigmented in smaller flows.

**Morphology**

Yankovsky 1977 (labral fan).

**Cytology**

Mason 1982, 1984; Adler & Kuusela 1994.

**Bionomics**

**Habitat.** This species is Holarctic. In North America, it is common from Alaska to the Mackenzie River in the Northwest Territories but is uncommon southward into Alberta and eastward into Nunavut. In the Palearctic

Region, we have seen bona fide material from Fennoscandia, Russia, and Mongolia. The immature stages reside in cool rocky flows as narrow as 10–20 cm and as wide as 100 m.

**Oviposition.** Unknown.

**Development.** Throughout the range of this multivoltine species, eggs overwinter and larvae and pupae develop from May through August.

**Mating.** Unknown.

**Natural Enemies.** Larvae frequently suffer infections with the chytrid fungus *Coelomycidium simulii*, the microsporidia *Janacekia debaisieuxi* and *Polydispyrenia simulii*, and unidentified mermithid nematodes. The mermithid *Mesomermis pivaniensis* was described from a larva in the Palearctic Region (Rubtsov 1980). Females are attacked by unidentified water mites.

**Hosts and Economic Importance.** Unknown; presumably mammalophilic.

### *Simulium* (*Simulium*) *venustum* Species Group

**Diagnosis.** Female: Subcosta with row of hair ventrally. Claws toothless. Male: Gonostylus in inner lateral view with slight basal bulge. Pupa: Gill of 6 or 8 filaments in pairs. Cocoon slipper shaped. Larva: Hypostoma with all teeth small; median tooth extended anteriorly as far as or slightly farther than lateral teeth. Frontoclypeal apotome with head spots pale and surrounded by darker pigment (negative pattern), or indistinct. Postgenal cleft typically broadly rounded, extended about three fourths distance to hypostomal groove. Abdominal segment IX without ventral tubercles; abdominal setae unbranched, translucent. Rectal papillae of 3 compound lobes.

**Overview.** This species group occurs throughout the Nearctic and Palearctic Regions and includes 37 species. North America has 13 recognized species. The group consists of univoltine and multivoltine species that occupy a wide variety of watercourses. All species are mammalophilic, and some are among the most noxious pests of humans.

### *Simulium* (*Simulium*) *rubtzovi* Smart

(Figs. 10.223, 10.407, 10.511, 10.822;
Plate 24b; Map 242)

*Simulium rubtzovi* Smart 1945: 528 (substitute name for *similis* Rubtsov 1940a; further information provided by Smart 1946)

*Simulium similis* Rubtsov 1940a ("1939"): 196 (FMPL). Lectotype* male with pupal exuviae (here designated; slide, pupal thorax and gills, male thorax, abdomen and one hind leg only; ZISP). Russia, Irkutsk Region, Onokhoika River (slide label reads Onokhoi, stream #1), date unknown but probably 20–29 June 1936 (I. A. Rubtsov). Junior primary homonym of *S. simile* Silva Figueroa 1917: 33–35, *S. similis* Cameron 1918: 558–559, and *S. similis* Malloch 1919: 42–43

**Taxonomy**

Rubtsov 1940a, 1956, 1963c (FMPL); Stone 1952 (FMP).

To establish the current concept of the name *rubtzovi*, we designate as lectotype the slide-mounted specimen that Rubtsov evidently intended as the type. The slide bears the word "Typus," although the word has been scratched out, along with a name that cannot be read, and the words "*Simulium rubtzovi* Smart" written over top. Also appearing on the slide but with lines drawn through them are the names "*S. morsitans*" and "near zabaicali," the latter perhaps a manuscript name. All handwriting on the slide is that of Rubtsov. Yankovsky (1995) overlooked this slide, and the name *similis* is omitted from his list of simuliid types in the ZISP.

The lengths of the 2 ventral petioles of the pupal gill of this species are variable; each petiole is about 2–18 times longer than its basal width.

**Morphology**

Zhang & Malmqvist 1996 (labral fan).

**Cytology**

Adler et al. 1999.

Based on larvae (10♀, 3♂) from Alaska (Richardson Highway, milepost 161, 19 July 1989), the chromosomes relative to the *Simulium* subgeneric standard sequence of Rothfels et al. (1978) differ by 7 fixed inversions. IS, IIL, and IIIS are standard. IL has 2 fixed inversions (sections 25p–31p and 32c–34d, both shared with the Palearctic *S. morsitans*). IIS has 3 fixed inversions (sequence = 42–44c.45p–44c.48c–50p.48d–45p.48c–48d.51d–54). IIIL has 2 overlapping fixed inversions yielding the sequence 83–91p.94p–96c.93d–91p.96c–100 (shared with *S. morsitans*). Sex chromosomes are undifferentiated and floating inversions have not been found. The chromosomes are identical with those of larvae from northern Sweden, although we have found B chromosomes in Swedish larvae.

**Bionomics**

**Habitat.** *Simulium rubtzovi* is primarily a Palearctic species. In North America, it is known from the lowlands of Alaska, Yukon, and the western Northwest Territories. Larvae attach themselves to sedges in sluggish brownwater flows about 6 m or less in width, coursing through boggy terrain of the boreal forest. In the Palearctic Region, larvae and pupae live in slower currents of stony streams densely bordered with vegetation (Rubtsov 1956).

**Oviposition.** Unknown.

**Development.** In the Nearctic Region, larvae and pupae are present from June to at least mid-August (Sommerman et al. 1955), suggesting the possibility of 2 generations per year, as in the Palearctic Region (Rubtsov 1956).

**Mating.** Unknown.

**Natural Enemies.** Alaskan larvae are attacked by an unidentified mermithid nematode and microsporidium.

**Hosts and Economic Importance.** Unknown; presumably mammalophilic.

### *Simulium* (*Simulium*) *venustum* Species Complex

**Diagnosis.** Female: Anal lobe unsclerotized along anterior margin. Male: Ventral plate in ventral view tooth shaped, with lateral margins slightly divergent posteriorly. Pupa: Gill of 6 filaments. Chromosomes: Standard banding sequence as given by Rothfels et al. (1978).

**Overview.** All members of the *S. venustum* complex, except the Holarctic *S. truncatum*, are endemic to North America. We recognize the following 10 species (with their previous cytological designations): *S. hematophilum* ('CC2'),

*S. incognitum* ('CC4'), *S. irritatum* ('CC3'), *S. minutum* ('CC1'), *S. molestum* ('A/C'), *S. piscicidium* ('AC(gB)'), *S. tormentor* n. sp. ('H/C'), *S. truncatum* ('EFG/C'), *S. venustum* s. s. ('CC'), and *S. venustum* 'cytospecies JJ.' Our use of names long held in synonymy with *S. venustum* s. l.—viz., *S. hematophilum*, *S. irritatum*, *S. minutum*, *S. molestum*, and *S. piscicidium*—provides formal names for 5 cytospecies while obviating the need to generate new names. The members of the *S. venustum* complex are structurally homogeneous; chromosomal identifications are typically necessary.

Members of this complex routinely have been confused with the 2 species in the *S. verecundum* complex, and the 2 species complexes often have been referred to jointly as the *S. venustum/verecundum* complex (or supercomplex *sensu* Crosskey 1994a). The abundance, wide distribution, and major pest status of several species in the complex have helped make *S. venustum* one of the most familiar names for black flies in North America. The name has appeared in nearly 400 papers (Adler & McCreadie 1997). Most of this literature cannot be assigned to the relevant species. We, therefore, present information for the complex as a whole, followed by individual accounts of the 10 component species.

**Taxonomy**

Say 1823 (FM); Harris 1841 (F); Riley 1870d (F); Laboulbène 1882 (F); Treille 1882 (F); Lugger 1897a, 1897b (FM); Coquillett 1898 (FM); Johannsen 1902, 1903b (FMPL), 1934 (PL); Washburn 1905 (FM); Forbes 1912 (FMPL, not Figs. 19, 20); Emery 1913 (FMP); Malloch 1914 (FMPL); Riley & Johannsen 1915 (F); Dyar & Shannon 1927 (FM); Hearle 1932 (F); Twinn 1936b (FMP); Rubtsov 1940b, 1956 (FMPL); Nicholson 1949 (FMP, not Fig. 30F); Nicholson & Mickel 1950 (FMP, not Fig. 30F); DeFoliart 1951a (PL); Fredeen 1951, 1981a (FMPL), 1985a (FMP); Chivers-Wilson 1952 (FM); Stone 1952 (FMP), 1964 (FMPL); Jamnback 1953, 1956 (L); Sommerman 1953 (L); Gill & West 1955 (FPL); Stone & Jamnback 1955 (FMPL); Davies & Peterson 1956 (E); Wirth & Stone 1956 (FMP); Peterson 1958a, 1960a (FMP), 1981, 1996 (FL); Young 1958 (F); Anderson 1960 (FMPL); Davies et al. 1962 (FMP); Wood et al. 1963 (L); Abdelnur 1966, 1968 (FMPL); Snoddy 1966 (FMPL); Holbrook 1967 (L); Stone & Snoddy 1969 (FMPL); Pinkovsky 1970 (FMPL); Smith 1970 (FMPL); Amrine 1971 (FMPL); Lewis 1973 (FMPL); Corredor 1975 (FM); Snoddy & Noblet 1976 (MPL); Merritt et al. 1978b (PL); Westwood 1979 (FMPL); Hilsenhoff 1982 (L); Webb & Brigham 1982 (L); Adler 1983 (PL); Adler & Kim 1986 (PL); Currie 1986 (PL); Hunter 1990 (FMPL); Mason & Kusters 1991b (FMPL); Peterson & Kondratieff 1995 (FMPL); Stuart 1995 (cocoon); Adler & Mason 1997 (FMPL); Stuart & Hunter 1998b (cocoon).

**Morphology**

Nicholson 1941, 1945 (F mouthparts); Snodgrass 1944 (F mouthparts); Leyon & Eklund 1950 (FM wing); Chivers-Wilson 1952 (F mouthparts); Wolfe & Peterson 1959 (gynandromorphs); Bennett 1963b (F salivary gland); Davies 1965a (spermatophore), 1965b (F mesepisternum); Buerger 1967 (FM labrum); Guttman 1967a (labral fan); Yang 1968 (F cibarium, peritrophic matrix); Chance 1969, 1970a (L mouthparts); Gosbee et al. 1969 (F salivary gland); Smith 1970 (ovarian follicles); Hannay & Bond 1971a (FM wing); Mercer 1972 (F antennal and palpal sensilla); Mercer & McIver 1973a (F antennal sensilla), 1973b (F palpal sensilla); Condon et al. 1976 (L neuroendocrine system); Yang & Davies 1977 (F peritrophic matrix); Sutcliffe 1978, 1985 (F mouthparts); Colbo et al. 1979 (FM labral and cibarial sensilla and armature); Hayton 1979 (F age-related changes in Malpighian tubules, ovarioles); McIver et al. 1980 (F tarsal sensilla); Sutcliffe & McIver 1982 (F mouthpart sensilla), 1984 (F mouthparts); Davies 1989 (gynandromorphs); Gryaznov 1995a (ovarioles); Smith & Hayton 1995 (F age-related changes in Malpighian tubules, ovarioles); Palmer & Craig 2000 (labral fan); Koehler et al. 2002 (F midgut musculature).

**Physiology**

Chivers-Wilson 1952 (FM salivary anticoagulant and histamines); Hocking 1953a, 1953b (F crop capacity, flight, metabolic rate); Hutcheon & Chivers-Wilson 1953 (FM salivary anticoagulant and histamines); Yang 1968 (F crop contents, salivary agglutinin and anticoagulant, invertase, FM amylase and trypsin); Yang & Davies 1968a (FM amylase), 1968b (FM trypsin), 1968c (FM invertase), 1974 (F salivary agglutinin and anticoagulant); Gordon & Bailey 1974 (L amino acids in hemolymph), 1976 (L hemolymph); Condon & Gordon 1977 (mermithid effects on L); Gordon et al. 1978 (mermithid effects on L hemolymph), 1979 (L hemolymph lipids), 1980 (L hemolymph sterols); Angioy et al. 1982 (F taste receptors).

**Cytology**

Rothfels et al. 1978; Rothfels 1979, 1980, 1981a; Adler 1983, 1986; Adler & Kim 1986; Ciborowski & Adler 1990; Hunter 1990; McCreadie 1991; McCreadie et al. 1994a; Adler & Mason 1997.

**Molecular Systematics**

Teshima 1970, 1972 (DNA hybridization); Sohn 1973 (DNA hybridization); Sohn et al. 1975 (DNA hybridization); Snyder 1981, 1982 (allozymes); Xiong & Kocher 1991, 1993a, 1993b (mtDNA); Xiong 1992 (mtDNA); Tang et al. 1995b (mtDNA).

**Bionomics**

**Habitat.** Nearly the entire continent is home to this abundant species complex. The immature stages are found in most clear streams and rivers. They are often the dominant macroinvertebrate at sites such as beaver dams (Clifford et al. 1993). Larvae can be found in both laminar and turbulent boundary layers (Eymann 1993). Larval survival in standing water decreases markedly with increasing temperature (Anderson & Shemanchuk 1975), and some evidence suggests that larvae in standing water will select temperatures of about 13°C (Thomas 1967). Larvae can tolerate pH depressions to 4.0 for 3 days (Chmielewski & Hall 1992). Females can travel an average of 9–13 km (maximum = 35 km) from their natal waterways after only 2 days (Baldwin et al. 1975). They are capable of speeds of about 2.4 m/sec over short distances and have a maximum flight range, if unfed, of 116 km in still air (Hocking 1953b).

**Oviposition.** Females are anautogenous (Anderson & Shemanchuk 1987a, Davies & Györkös 1990) and typically carry about 200–500 eggs, with fecundity increasing with female size (Abdelnur 1968, Pascuzzo 1976). A single female was estimated to produce 1050 eggs, the greatest number recorded for a North American black fly (Hocking &

Pickering 1954). Females can complete up to 5 or 6 ovarian cycles (Mokry in Wenk 1981), although in the field they might survive only long enough to initiate a third cycle (Magnarelli & Cupp 1977). All members of this species complex oviposit in flight by tapping the abdomen to the water's surface. References to females laying eggs in masses on substrates, such as trailing vegetation, pertain strictly to *S. rostratum* and *S. verecundum* s. s. Females tend to oviposit at any suitable site, rather than at the sites where they emerged (Hunter & Jain 2000).

**Development.** All members of this species complex are principally univoltine, although some occasionally have a second generation or cohort. Eggs remain viable when stored for a year or more at 0°C–0.5°C (Fredeen 1959a). Some eggs, if partially embryonated, can hatch if frozen at –15°C (Colbo & Wotton 1981). Eggs generally overwinter and larvae begin hatching in April in Wisconsin (Anderson & Dicke 1960) and in May in southern Ontario, eastern Quebec, and Alaska (Davies 1950, Sommerman et al. 1955, Wolfe & Peterson 1959, Davies et al. 1962). In the southern states, such as Arkansas, pupae can be found from March to May. Adults generally fly from April through June or July, depending on latitude (Twinn et al. 1948, Baldwin & Gross 1972, Back & Harper 1979).

Larvae include algae in their diet, although some species of algae are not digested (Moore 1977a, 1977b). The density of the gut contents decreases with increasing body length (Morin et al. 1988a). Larval feeding rates approximately double between 10°C and 20°C (Morin et al. 1988b), and ingestion rates increase about fourfold when water velocity increases from 25 to 100 cm/sec (Charpentier & Morin 1994). Adults can be reared from larvae fed on bacterial suspensions (Fredeen 1960, 1964b). They can be produced from eggs in as little as 13 days under unspecified conditions (Fredeen 1958). Females live longer than males, with a maximum of 37 days recorded for a female maintained on dry sucrose and water at ambient temperature (Davies 1953). The sex ratio is even (Smith 1970).

**Mating.** Mating swarms have been observed in the evening and on overcast days about 2–5 m above roadways (Peterson 1962b). The males bounce up and down as the swarm moves laterally; they couple with females and the pair flies up and out of the swarm (Peterson 1962b). Males are attracted to carbon dioxide and swarm downwind of the source; females attracted to the source are intercepted in flight, and the coupled pair eventually falls to the ground (Mokry et al. 1981).

**Natural Enemies.** Species in the *S. venustum* complex serve as hosts for numerous mermithid nematodes, including *Gastromermis viridis*, *Hydromermis* sp., *Isomermis wisconsinensis*, *Mesomermis camdenensis*, and *M. flumenalis* (Phelps & DeFoliart 1964, Ezenwa 1974b, Ezenwa & Carter 1975, Bailey & Gordon 1977, Bailey et al. 1977, Mokry & Finney 1977, Molloy 1979), as well as unidentified species (Anderson & DeFoliart 1962 [footnote, p. 832 of R. Gordon 1984]). A member of the *S. venustum* species complex is the type host for *M. flumenalis* (Welch 1962). As many as 4 species of mermithids might infect larvae at a single site (Finney in Colbo & Porter 1980). *Mesomermis flumenalis* is probably a complex of species because the worms that attack larvae of the *S. venustum* complex are reproductively isolated in time from those attacking the larvae of *Prosimulium mixtum* in the same stream (Ebsary & Bennett 1975b). The redescription of *M. flumenalis*, however, was based on material from both *P. mixtum* and the *S. venustum* species complex (Ebsary & Bennett 1974). Larvae can be infected as early as the first instar (Colbo & Porter 1980). Infections can persist into adult females, sometimes affecting their size (Colbo 1982). Adult males, however, seem not to be infected and might be killed as larvae by the mermithids (Mokry & Finney 1977, Colbo & Porter 1980).

The chytrid fungus *Coelomycidium simulii* attacks larvae (Ebsary & Bennett 1975a, Lewis & Bennett 1975), whereas the entomophthoraceous fungus *Entomophthora culicis* kills young adults (Nadeau et al. 1994). *Erynia conica* is another entomophthoraceous fungus that attacks adults (Nadeau et al. 1994). The trichomycete fungi *Genistellospora homothallica*, *Harpella leptosa*, *H. melusinae*, and *Pennella simulii* inhabit the guts of larvae (Lichtwardt 1967, 1972, 1986; Williams & Lichtwardt 1971; Moss & Lichtwardt 1980). A member of the *S. venustum* species complex is the host from which the type of *G. homothallica* was described (Lichtwardt 1972).

Larvae also serve as hosts for the microsporidia *Amblyospora bracteata*, *A. fibrata*, *A. varians*, *Janacekia debaisieuxi*, and *Polydispyrenia simulii* (Ezenwa 1973, Ebsary & Bennett 1975a, Lewis & Bennett 1975, Vávra & Undeen 1981). Both larval sexes are infected (Undeen et al. 1984b). Attempts to infect larvae in the laboratory have not succeeded, leaving unanswered the question of whether oral infections are possible (Vávra & Undeen 1980). One of the earliest, although unverified, reports of a microsporidium in an adult black fly is that of *A. bracteata* in an unknown member of the *S. venustum* species complex (Fantham et al. 1941).

A number of additional pathogens have been found in members of this species complex. The haplosporidian protist *Haplosporidium simulii* has been recorded from larvae (Beaudoin & Wills 1968), and the ciliate protozoan *Tetrahymena rotunda* is a parasite in the hemolymph of larvae, pupae, and adults (Lynn et al. 1981). A cytoplasmic polyhedrosis virus infects larvae (Bailey 1977). Unidentified bacteria, one of which produces microsporidium-like white cysts, have been recovered from larvae (Colbo [p. 14] in Anonymous 1977, Weiser & Undeen 1981).

Predators of larvae include larval rhyacophilid and hydropsychid caddisflies, dragonfly larvae (Peterson & Davies 1960, Ezenwa 1974a), chironomid larvae, conspecifics (Peterson & Davies 1960), the perlodid stonefly *Isoperla holochlora* (Adler & Kim 1986), the crayfish *Cambarus robustus*, and the mottled sculpin *Cottus bairdi* (Coffman 1967). Predation by trout occurs when larvae drift (Mokry 1975). Predators of adults include empidid and dolichopodid flies, ants (Peterson & Davies 1960), dragonflies (Twinn 1939), and white-crowned sparrows (Hocking & Pickering 1954). Unidentified simuliids captured by the yellowjacket *Vespula albida* (Hocking 1952b) probably were members of the *S. venustum* species complex. Water mites (*Sperchon ?jasperensis*) gather in the pupal cocoon, probably to await the emerging adults, which they parasitize (D. M. Davies 1959, 1960). Larval trombidioid mites also have been found on adults (D. M. Davies 1959).

**Hosts and Economic Importance.** Females feed on the nectar of low-bush blueberry (*Vaccinium angustifolium*), sour-top blueberry (*V. myrtilloides*) (Davies & Peterson 1956), bramble (*Rubus acaulis*), cow-parsnip (*Heracleum sphondylium* (as *H. maximum*), yarrow (*Achillea millefolium*), water-hemlock (*Cicuta mackenzieana*), and Labrador tea (*Ledum groenlandicum*) (Hocking & Pickering 1954). Most host-seeking females have taken a sugar meal (McCreadie et al. 1994c, Burgin & Hunter 1997b). Adults feed on homopteran honeydew and floral nectar in proportion to availability, which can change with habitat (Burgin 1996; Burgin & Hunter 1997a, 1997b, 1997c).

Females are found in a variety of habitats but are more prone to harass hosts in forests than in open areas (Martin et al. 1994). They tend to feed within 2 km of their natal waters (Fredeen et al. 1953c) but generally do not attempt to blood feed for 12–24 hours after emergence or oviposition (Moobola 1981). Temperature and wind speed influence host-seeking behavior. For example, activity drops sharply at temperatures below 11°C (McCreadie et al. 1986), with 7.8°C representing the lower threshold for flight, at least in northern populations (e.g., Churchill, Manitoba) (Edmund 1952). Rapid changes in atmospheric pressure, especially when falling, increase biting rates (Davies 1952). Consequently, increased biting is often associated with approaching thunderstorms (Edmund 1952) or even snow storms (Anonymous 1949). Host-seeking females are most active in midmorning and before sunset (McCreadie et al. 1985). Some studies indicate that parous flies are most frequent around hosts in late afternoon and evening (Davies 1963), whereas other studies show no overall relation between parity and time of activity (Moobola 1981). More than 90% of biting females carry sperm, although insemination is not a prerequisite for seeking blood (Magnarelli & Cupp 1977).

Both vision and odors, especially carbon dioxide, are important in the orientation and landing of females (Fallis et al. 1967). Human sweat and urine are attractive (Schofield 1994). Most human attractiveness comes from exhaled odors, with differences in attractiveness among humans probably stemming from different production rates of exhaled carbon dioxide (Schofield & Sutcliffe 1996). Females generally locate a source of carbon dioxide by flying upwind (Golini & Davies 1971). Blood-seeking females land more often on surfaces reflecting less light (Davies 1972). Blue is a preferred color (Davies 1951; Bradbury 1972; Bradbury & Bennett 1974a, 1974b; Browne & Bennett 1980), whereas green, yellow, and white—preferred colors for oviposition—are least attractive (D. M. Davies 1961, 1972). Gorging behavior is elicited by stimulants such as various adenosines (ATP, ADP, AMP, cAMP) and red blood cells (Sutcliffe & McIver 1975, 1979; Smith & Friend 1982). Humans differ in their ability to elicit biting responses (Schofield & Sutcliffe 1997).

Host records for the *S. venustum* species complex are difficult to tease apart from those for the *S. verecundum* complex; undoubtedly there is some overlap in hosts. We can state confidently, however, that only members of the *S. venustum* complex are pests of humans. We suspect that the following information pertains primarily to the *S. venustum* complex. Females are principally mammalophilic, with hosts including humans, dogs, red foxes, Canadian lynxes, Canadian otters, American minks, raccoons, American black bears, horses, mules, pigs, moose, white-tailed deer, caribou, cattle, sheep, Uinta ground squirrels, red squirrels, beavers, snowshoe hares, and domestic rabbits (Ruggles 1933, Olsen & Fenstermacher 1942, Smith et al. 1952, Davies & Peterson 1956, Downe & Morrison 1957, Snow et al. 1958b, Peterson 1959a, Anderson & DeFoliart 1961, Abdelnur 1968, Addison 1980, Pledger et al. 1980, Simmons et al. 1989, Hunter et al. 1993, McCreadie et al. 1994c). They also swarm around North American martens (*Martes americana*) (Davies & Peterson 1956) and rabbits (Shemanchuk 1988) and engorge on sheep and cow blood in the laboratory (Bernardo & Cupp 1986). Under experimental conditions, they have fed on guinea pigs (Downe 1957). Feeding on birds occurs occasionally, mainly on larger species, such as great blue heron, domestic duck, ruffed grouse, northern saw-whet owl, American crow, common raven, American robin, common grackle (Davies & Peterson 1956, Bennett 1960), and chicken (Shemanchuk 1988). Additional hosts include unspecified species of grouse (Gibson 1965) and sparrow (Abdelnur 1968). Females have been captured from a snapping turtle (*Chelydra serpentina*) but had not engorged (Smith 1969).

Females tend to bite areas of the body where they can penetrate the hair (Shemanchuk 1988). Moose, for example, are attacked in great numbers around the eyes (Smith in Anderson & Lankester 1974). Caribou suffer greatly from attacks on the lips, muzzle, and ears, around the eyes, and on areas from which the hair has been rubbed (Banfield 1954). In some areas of the continent, such as Wisconsin, females of the *S. venustum* complex are the principal pests of cattle and horses (Anderson & DeFoliart 1961), and in areas such as the Canadian prairie, they can reduce milk production by half (Fredeen 1958). Females have been implicated in dermatitis cases of cattle (Burghardt et al. 1951). So great are the numbers of flies at times that they probably can influence the migrations of caribou (Harper 1955).

Females of the *S. venustum* complex can be prodigious biters of humans. The black flies that tormented the early explorers, missionaries, and settlers of eastern Canada—recorded as early as the beginning of the seventeenth century (Agassiz 1850, Packard 1869, Davies et al. 1962)—undoubtedly included members of the *S. venustum* complex. They are the major simuliid pests of humans in areas such as Alaska (Frohne & Sleeper 1951), Maine (McDaniel 1975, Gibbs et al. 1988), New Hampshire (LaScala 1979), New York (Jamnback 1969a), Pennsylvania (Eckhart & Snetsinger 1969), Labrador (Hocking & Richards 1952), Manitoba (Twinn et al. 1948, Twinn 1950), Newfoundland (Lewis & Bennett 1973), New Brunswick, Nova Scotia, Prince Edward Island (Lewis & Bennett 1979), Northwest Territories (Currie & Adler 2000), Ontario (Davies et al. 1962), and Quebec (Wolfe & Peterson 1959). They are notorious for driving humans from the outdoors and making life miserable (Reeves 1910, Fredeen 1958). In parts of Canada, landing rates on a human can reach 78 flies/ 6.5 $mm^2$/min (Davies 1978), and biting rates (coupled with those of the *P. hirtipes* species group) can reach 60 bites/ 10 min on untreated arms in the Adirondack Mountains of New York (Travis et al. 1951b). Females usually bite the back of the neck and scalp or behind the ears (Wolfe & Peterson 1960). Human responses to the bites follow a pattern of pathological change over time (Stokes 1914). In some areas, such as Massachusetts and parts of Maine, the complex might be incriminated unjustly as a frequent biter, with far fewer flies biting than are attracted (Schreck et al. 1980, Simmons et al. 1989).

Females have been incriminated as vectors of legworm (*Onchocerca cervipedis*) in moose (Pledger 1978, Pledger et al. 1980) and the filarial nematode *Dirofilaria ursi* in American black bears (Addison 1980, Simmons et al. 1989). They have fed on dogs infected with heartworm (*Dirofilaria immitis*), resulting in destruction of their Malpighian tubules (Simmons et al. 1989). Females can support development of the filarial nematode *Splendidofilaria fallisensis* to the infective stage but are unlikely to be natural vectors (Anderson 1956). They also can support development of *Leucocytozoon icteris* (as *L. fringillinarum*) and *L. simondi* but probably play an insignificant role as vectors (Fallis & Bennett 1962, Desser & Yang 1973). Females are not suitable hosts of *L. lovati* (Fallis & Bennett 1958; as *L. bonasae*).

### *Simulium* (*Simulium*) *hematophilum* Laboulbène
(Map 243)

*Simulia hematophila* Laboulbène 1882: 222–223 (F). Type material lost (described from field notes and unpublished sketches made by Treille 1882). Eastern Newfoundland, other information unknown. Revalidated

*Simulium venustum* 'CC2 (Cal 4; CTC 4)' Rothfels et al. 1978: 1119–1121 (C)

[*Simulium venustum* 'CC': Colbo (1990) not Rothfels et al. 1978 (misident.)]

**Taxonomy**

Treille 1882 (F), Laboulbène 1882 (F).

Bequaert (1945) first linked the name *hematophilum* to a member of the family Simuliidae, although Stone (1964) believed that the description did not satisfactorily fit any *Simulium*. Lewis and Bennett (1973) suggested that the name applied to *P. fuscum* or *P. mixtum*, based primarily on seasonality. However, we believe that the name applies to a member of the *S. venustum* species complex. Treille (1882) stated that the thorax is brilliant and the wings have bluish reflections; these characters suggest that the name *hematophilum* applies to females of the *S. venustum* complex, members of which begin to appear in large numbers in June. We, therefore, associate the name *S. hematophilum* with cytospecies 'CC2,' a common species implicated by McCreadie et al. (1994c) as a biter of humans in Newfoundland. We have insufficient material to designate a neotype.

**Cytology**

Rothfels et al. 1978; Rothfels 1980, 1981a; Adler & Kim 1986; McCreadie 1991; McCreadie et al. 1994a.

*Simulium hematophilum* is reproductively isolated from *S. irritatum* ('CC3'), *S. piscicidium* ('AC(gB)'), true *S. venustum* ('CC') (Rothfels et al. 1978, McCreadie et al. 1994a), and *S. molestum* ('A/C'). Two or 3 species, distinguished primarily on the basis of sex chromosomes, might be included under *S. hematophilum* (McCreadie et al. 1994a), but more collections, particularly in areas between cytologically differentiated populations, are needed. The IIL-1 inversion is less often fixed in southern populations than indicated by McCreadie et al. (1994a). In addition to the sex-chromosome variation previously reported (McCreadie et al. 1994a), we have seen material from the Magdalen Islands of Quebec in which males are heterozygous for the IIL-1 inversion and females are homozygous standard.

**Bionomics**

**Habitat.** *Simulium hematophilum* is spotty from Newfoundland's Avalon Peninsula southward along the Appalachian Mountains to Alabama. In the United States, larvae and pupae are found most often in stony streams less than a few meters wide and are usually taken with the immatures of *S. venustum*. In Newfoundland, they are found in a wide variety of streams but are nearly always remote from the outlets of impoundments (McCreadie & Colbo 1991b, Adler & McCreadie 1997).

**Oviposition.** Unknown.

**Development.** Larvae hatch from late winter to early spring. Mature larvae are present during March and April from North Carolina to Alabama and from April to June northward (Rothfels et al. 1978, Adler & Kim 1986). This species is typically univoltine, although at some sites in Newfoundland, larvae are present into August, suggesting an occasional extra generation (McCreadie & Colbo 1993a).

**Mating.** Unknown.

**Natural Enemies.** Larvae are infected by an unidentified microsporidium.

**Hosts and Economic Importance.** Probable hosts include red foxes, Canadian lynxes, snowshoe hares, and humans (McCreadie et al. 1994c).

### *Simulium* (*Simulium*) *incognitum* Adler & Mason
(Map 244)

*Simulium* (*Simulium*) *incognitum* Adler & Mason 1997: 85–88 (FMPLC). Holotype* male (pinned, exuviae and terminalia in separate glycerin vials below; USNM). Saskatchewan, Torch River, northeast of Love, Elves Farm, 53°38′N, 104°15′W, 3 June (pupa), 6 June (adult) 1992 (P. Burgess)

*Simulium venustum* 'cytotype CC4' Adler 1986: 13 (C)

**Taxonomy**

Adler & Mason 1997 (FMPL).

**Cytology**

Adler 1986, Adler & Mason 1997.

**Molecular Systematics**

Xiong 1992 (mtDNA), Xiong & Kocher 1993b (mtDNA).

**Bionomics**

**Habitat.** This species occurs from Alberta to Manitoba, with 1 record from Idaho. It is characteristic of Canada's Parkland Ecoregion. The immature stages are found in productive streams and small rivers with open canopies, usually below beaver dams.

**Oviposition.** Unknown.

**Development.** *Simulium incognitum* overwinters as eggs and passes through 1 or 2 generations, or perhaps cohorts, with the first pupating in late May and early June and the second, when present, pupating in mid to late June (Adler 1986, Adler & Mason 1997).

**Mating.** Unknown.

**Natural Enemies.** Larvae are hosts of unidentified mermithid nematodes (Adler 1986, Adler & Mason 1997).

**Hosts and Economic Importance.** Unknown; presumably mammalophilic.

### *Simulium (Simulium) irritatum* Lugger

(Map 245)

*Simulium irritatum* Lugger 1897a ("1896"): 203–205 (FM). Neotype* male (here designated; pinned, pupal exuviae in glycerin vial below; USNM). Utah, Summit Co., Weber River, Rt. 32, 40°42.5′N, 111°17.5′W, 8 June 1994 (P. H. Adler). (Pp. 177–179 of Lugger 1897b include an exact replica of the entry for *S. irritatum* that appeared on pp. 203–205 of Lugger 1897a). Revalidated

*Simulium venustum* 'CC(gB)B' Rothfels 1979: 526 (Quebec record)

*Simulium venustum* 'CC3' Rothfels 1981a: 27 (C) (alluded to by Rothfels 1980, Fig. 13.10)

**Taxonomy**

Lugger 1897a, 1897b (FM).

In the interest of conserving names, we apply the name *irritatum* to cytospecies 'CC3.' The species is, in fact, one of the irritating members of the *S. venustum* complex. *Simulium irritatum* was described originally from an unspecified location in Minnesota (Lugger 1897a, 1897b). The type material has been lost. We, therefore, designate a neotype to establish the current concept of the name.

**Cytology**

Rothfels 1979, 1980, 1981a; Adler 1986; Ciborowski & Adler 1990; McCreadie 1991; McCreadie et al. 1994a.

**Molecular Systematics**

Xiong 1992 (mtDNA); Xiong & Kocher 1993a, 1993b (mtDNA).

**Bionomics**

**Habitat.** This species is found across Canada and Alaska southward along the Rocky Mountains into the forests and bordering sagebrush areas of Wyoming. It is the most common member of the *S. venustum* species complex throughout western North America. The immature forms are common in cool streams 1 to several meters wide, particularly in the western half of the continent, but they also inhabit the midreaches of rivers up to 50 m wide, especially those with coarse substrate and small amounts of fine particulate organic matter (Adler 1986, Ciborowski & Adler 1990, McCreadie & Colbo 1991b). We found the larvae so abundant below a beaver dam in Alberta that the slippery larva-coated rocks forced us to crawl back to shore.

**Oviposition.** Unknown.

**Development.** This species is probably univoltine, with eggs overwintering (McCreadie & Colbo 1993a). Larval development is underway in May, with larvae and pupae most abundant during June and July.

**Mating.** Unknown.

**Natural Enemies.** Larvae are parasitized by an unidentified mermithid nematode (Adler 1986), the chytrid fungus *Coelomycidium simulii*, and the microsporidia *Janacekia debaisieuxi* and *Polydispyrenia simulii*.

**Hosts and Economic Importance.** Females swarm around and bite humans in Wyoming. The contribution of this species to the pest problem in Canada is probably great.

### *Simulium (Simulium) minutum* Lugger

(Map 246)

*Simulium minutum* Lugger 1897a ("1896"): 201–203 (F, Fig. 143). Type material lost. Minnesota, county unknown, possibly near Minneapolis, between 15 May and 1 June, year and collector unknown. (Pp. 175–177 of Lugger 1897b include an exact replica of the entry for *S. minutum* that appeared on pp. 201–203 of Lugger 1897a). Revalidated

*Simulium venustum* 'CC1 (Cal 1 and CTC1)' Rothfels et al. 1978: 1118–1119 (C)

**Taxonomy**

Lugger 1897a, 1897b (F).

We assign the name *minutum* to cytospecies 'CC1,' which probably occurs in Minnesota, the area from which *S. minutum* was described by Lugger (1897a, 1897b). We make this decision primarily to conserve names and avoid cluttering the literature with additional names. Nicholson and Mickel (1950) discussed the possibility that the names *S. minutum* and *S. irritatum* apply to *S. jenningsi* and *S. luggeri*, although the restricted flight period (15 May–1 June) given by Lugger (1897a) is more suggestive of a member of the *S. venustum* species complex. The type material has been lost, however, so the true identity of the original *S. minutum* and *S. irritatum* probably never will be known. We have insufficient material to establish a neotype.

**Cytology**

Rothfels et al. 1978; Rothfels 1980, 1981a; Hunter 1990.

Rothfels et al. (1978) argued for the specific distinctness of *S. minutum* ('CC1') on the basis of its earlier development in sympatry with true *S. venustum* ('CC'). The consistent early seasonality and wide distribution (Adler & Mason 1997) support the idea that *S. minutum* is a distinct species.

**Molecular Systematics**

Snyder 1982 (allozymes).

**Bionomics**

**Habitat.** This taxon is patchy from Saskatchewan to southeastern Ontario and Michigan. Larvae have been taken from small streams.

**Oviposition.** Unknown.

**Development.** Late-instar larvae have been collected from late April to mid-May. *Simulium minutum* is earlier in development when it occupies the same streams with true *S. venustum* (Rothfels et al. 1978).

**Mating.** Unknown.

**Natural Enemies.** No records.

**Hosts and Economic Importance.** Unknown; presumably mammalophilic.

### *Simulium (Simulium) molestum* Harris
(Map 247)

*Simulium molestum* Harris 1841: 405 (F). Syntype females (2 pinned; presumably MCZ [Johnson 1925, Bequaert 1945]). New Hampshire, county unknown, White Mountains, date unknown (probably T. W. Harris). Revalidated

*Simulium* 'A/C *venustum*' Rothfels et al. 1978: 1119 (C)

**Taxonomy**

Harris 1841 (F).

We link the name *S. molestum* to cytospecies 'A/C' on the basis of distribution and in the interest of conserving names. If 1 or both of the syntypes still exist, our assignment can be tested, perhaps by molecular means.

**Cytology**

Rothfels et al. 1978; Rothfels 1980, 1981a.

Rothfels et al. (1978) provided evidence that this cytological segregate is reproductively isolated from true *S. venustum* ('CC'), and further suggested that it is specifically distinct from *S. piscicidium* ('AC(gB)'). A lack of hybrids in mixed populations of *S. molestum* and *S. hematophilum* ('CC2') on the Magdalen Islands of Quebec confirms the reproductive isolation of these 2 species.

**Molecular Systematics**

Snyder 1981, 1982 (allozymes).

**Bionomics**

**Habitat.** Records for this species are scattered, coming from northern New Hampshire, the Upper Peninsula of Michigan, Prince Edward Island, and Quebec's Magdalen Islands. The species has been reported from Alaska (Shields in Rothfels 1988), but we have no further knowledge of the record. The immature stages occupy cool streams in heavily forested areas.

**Oviposition.** Unknown.

**Development.** This species is univoltine. Winter is passed as eggs, with larvae hatching in early spring and maturing from about mid-May to early June in New Hampshire (Lake & Burger 1983) or late June in Canada. When *S. molestum* occurs in the same streams with *S. venustum* s. s., it is later in development (Rothfels et al. 1978).

**Mating.** Unknown.

**Natural Enemies.** We have found 1 larva infected with an unidentified mermithid nematode.

**Hosts and Economic Importance.** Unknown; presumably mammalophilic.

### *Simulium (Simulium) piscicidium* Riley
(Plate 24c, Map 248)

*Simulium piscicidium* Riley 1870d: 367 (F). Syntype* females (2 on 1 pin #771, terminalia of bottom female in glycerin vial below and of top female on slide; USNM). New York, Monroe Co., Mumford, Spring Creek, day and month unknown, 1870 (S. J. McBride); *piscioidium*, *piscidium* (subsequent misspellings). Revalidated

*Simulium* 'AC(gB) *venustum* (AC *venustum*)' Rothfels et al. 1978: 1119 (C)

*Simulium venustum* 'Ac(qb)' Colbo & Porter 1980: 1487 (*lapsus calami*)

**Taxonomy**

Riley 1870d (F).

Cytospecies 'AC(gB)' has not been confirmed cytologically from the type locality of *S. piscicidium*. However, in the interest of conserving names, we assign the name *piscicidium* to this cytospecies. Because 2 syntypes exist, our assignment of the name *piscicidium* to 'AC(gB)' eventually can be tested, perhaps by molecular means.

**Cytology**

Rothfels et al. 1978; Rothfels 1980, 1981a; McCreadie 1991; McCreadie et al. 1994a.

This taxon is reproductively isolated from true *S. venustum* ('CC') and *S. molestum* ('A/C') (Lake & Burger 1983), as well as *S. hematophilum* ('CC2') and *S. irritatum* ('CC3') (McCreadie et al. 1994a). It might consist of 2 sibling species (McCreadie et al. 1994a).

**Molecular Systematics**

Xiong 1992 (mtDNA); Xiong & Kocher 1993a, 1993b (mtDNA).

**Bionomics**

**Habitat.** *Simulium piscicidium* is known from Maine, New Hampshire, New York, Newfoundland, and Quebec. In Canada, the immature stages are most common in canopied streams remote from outlets, with temperatures around 13.5°C, pH of 4.8–7.2, and beds of small stones and rubble (McCreadie & Colbo 1991b, 1992). In New Hampshire, however, they have been taken below lake outlets (Lake & Burger 1983).

**Oviposition.** Unknown.

**Development.** One generation per year is typical, with mature larvae occurring from late May to late June; however, a late-summer generation (or cohort) has been reported for some streams (Lake & Burger 1983, McCreadie & Colbo 1993a).

**Mating.** Unknown.

**Natural Enemies.** No records.

**Hosts and Economic Importance.** Unknown; presumably mammalophilic.

### *Simulium (Simulium) tormentor* Adler, Currie & Wood, New Species
(Map 249)

*Simulium venustum* 'CH(gB)' Rothfels 1979: 526 (Quebec record)

*Simulium venustum* 'H/C sibling' Rothfels 1981a: 27 (C) (alluded to by Rothfels 1980, Fig. 13.10, as 'H/C')

**Taxonomy**

**Female.** Unknown.

**Male.** Unknown.

**Pupa.** Unknown.

**Larva.** Length 5.8–6.6 mm. Head capsule pale yellowish, variably suffused with brown pigment ventrally; head spots pale yellowish, variably surrounded by pale to dark brown pigment (negative head-spot pattern); posteromedian head spot typically largest, most distinct. Antenna brownish yellow, subequal in length to stalk of labral fan. Postgenal cleft horseshoe shaped, variably pointed apically, extended about three fourths distance to hypostomal groove; subesophageal ganglion with or without ensheathing pigment. Labral fan with 43–60 primary rays.

Gill histoblast of 6 filaments. Body pigment grayish brown to brown, arranged in bands. Posterior proleg with 9–14 hooks in each of 63–66 rows.

**Diagnosis.** The chromosomes provide the only reliable means of distinguishing the larvae of *S. tormentor* from those of other members of the *S. venustum* species complex. They are unique in having the H inversion (heterozygous or homozygous) in the IIS arm. We generally were able to distinguish the brown larvae of *S. tormentor* from the typically greenish larvae of *S. irritatum* with which they often were found in the Horton River drainage of the Northwest Territories. The color distinction is not general, however, for larvae of *S. irritatum* are brownish in many areas.

**Holotype.** Larva (mature female, ethanol vial) with photographic negatives of larval polytene chromosomes. Northwest Territories, tributary of Horton River, outflow of shallow pool, 67°59.41′N, 123°13.92′W, 20 July 2000, P. H. Adler & D. C. Currie (CNC).

**Paratypes.** Northwest Territories, outflow of Horton Lake, 67°32.50′N, 122°16.55′W, 17 July 2000, P. H. Adler & D. C. Currie (69 larvae); Stefansson Creek, near confluence with Horton River, 68°48.90′N, 125°17.08′W, 29 July 2000, P. H. Adler & D. C. Currie (11 larvae); braided tributary of Horton River, 69°14.89′N, 126°42.26′W, 3 August 2000, P. H. Adler & D. C. Currie (6 larvae).

**Etymology.** The specific name is from Latin and refers to the probable tendency of the females to inflict misery on their hosts.

**Remarks.** The validity of this species is supported by its broad distribution and the maintenance of its chromosomal integrity when sympatric with both *S. irritatum* and *S. truncatum*.

**Cytology**

Rothfels 1979, 1980, 1981a.

All larvae examined to date have the standard chromosomal banding sequence of Rothfels et al. (1978) in the IIIL arm, although inversion polymorphisms are common in this arm. The IIL-1 inversion predominates in all studied populations and is nearly fixed (inverted for 65 of 68 sequences) in the Northwest Territories. In Quebec, females are homozygous for the H sequence of Rothfels (1981a), whereas males are heterozygous for the H/C sequence. We have found sex-exceptional males (HH) and females (HC) in the Northwest Territories, where males also can have the EFG/H sequence, demonstrating a relationship with *S. truncatum*. For example, at the outlet of Horton Lake in the Northwest Territories, all 7 females were HH, whereas 1 male was H/C, 1 was HH, and 9 were EFG/H. The H inversion shares 1 breakpoint (nearest the centromere) with the A inversion of Rothfels et al. (1978), whereas the other breakpoint (reading from chromosomal end to centromere) is just after the first band of the "4 doublets" marker in section 42 of Rothfels et al. (1978).

**Molecular Systematics**

Smith 2002 (nuclear and mtDNA).

**Bionomics**

**Habitat.** This species was first discovered in a large pure sample from the Lac Pau area beside the Caniapiscau River, which flows north into Ungava Bay, Quebec (Rothfels 1981a; C. Back, pers. comm.). We subsequently found it in several drainages in the Northwest Territories and Nunavut. The immature stages occupy cool streams and rivers 2–20 m wide, especially those that are rocky.

**Oviposition.** Unknown.

**Development.** We have taken larvae and pupae from early July to early August in the Northwest Territories and Nunavut.

**Mating.** Unknown.

**Natural Enemies.** We have found larvae infected with the chytrid fungus *Coelomycidium simulii* and the microsporidia *Amblyospora bracteata* and *Janacekia debaisieuxi*.

**Hosts and Economic Importance.** This species is presumably one of the members of the *S. venustum* species complex that demonstrates a rapacious capacity to savage human flesh. Specific hosts, however, are unknown.

### *Simulium* (*Simulium*) *truncatum* (Lundström)
(Map 250)

*Melusina reptans* var. *truncata* Lundström 1911: 13, Fig. 6 (M). Syntype males (4 on separate slides; UZMH). Finland, Lapland, specific location, date, and collector unknown

*Simulium venustum* 'EFG/C (CEFG, CEF)' Rothfels et al. 1978: 117 (C)

'species near *Simulium truncatum*' Gordon & Cupp 1980: 975 (New York record)

[*Simulium* (s. str.) *venustum*: Rubtsov (1940b) not Say 1823 (misident.)]

[*Simulium venustum*: Addison (1980) not Say 1823 (misident., 1976 specimens in part)]

**Taxonomy**

Lundström 1911 (M); Rubtsov 1940b, 1956, 1963b (FMPL); Hunter 1990 (FMPL); Jensen 1997 (PL).

Many references to *S. truncatum* in the Palearctic Region are suspect because of past difficulties with identification. We have, for instance, received hundreds of larvae from the Palearctic Region labeled as *S. truncatum* but that chromosomally proved largely to be *S. rostratum*, *S. posticatum*, and other members of the *S. venustum* species group. *Simulium truncatum* has been known from the Nearctic Region since 1978 (Rothfels et al. 1978).

**Morphology**

McCreadie & Colbo 1990 (L allometry), McCreadie 1991 (L allometry), Zhang & Malmqvist 1996 (labral fan).

**Cytology**

Rothfels et al. 1978; Rothfels 1979, 1981a; Hunter 1990; McCreadie 1991; McCreadie et al. 1994a.

**Molecular Systematics**

Snyder 1981, 1982 (allozymes); Smith 2002 (nuclear and mtDNA).

**Bionomics**

**Habitat.** *Simulium truncatum* is Holarctic. In North America, it occurs across Canada and Alaska, dipping as far south as New York. Throughout its range, *S. truncatum* breeds below outfalls of impoundments, including small pools and ponds, and is rarely found more than 500 m from an outlet. Larval size decreases markedly with distance downstream from outlets, possibly reflecting decreasing food (McCreadie & Robertson 1998). Larvae are most common in slow (0.36 m/sec), shallow water (McCreadie & Colbo 1993b), with proportionally more larvae on rocks than on vegetation (McCreadie & Colbo 1993a).

**Oviposition.** Females drop their eggs singly into the water while flying near the surface (Colbo in Rothfels et al. 1978).

**Development.** This early-season univoltine species overwinters as eggs and pupates as early as late April. Larvae are present through July north of the Arctic Circle. Larval survival in the laboratory is greatest at 15°C and 20°C, but even at 30°C a third of the larvae survive (McCreadie & Colbo 1991a). The mean number of days to complete larval development ranges from about 9 at 30°C to 33 at 10°C (245 degree-days at 15°C, threshold = 0°C); female larvae require significantly more time to mature than do male larvae (McCreadie & Colbo 1991a).

**Mating.** Unknown.

**Natural Enemies.** Larvae are hosts of the chytrid fungus *Coelomycidium simulii* and unidentified mermithid nematodes. In the Palearctic Region, they are attacked by the microsporidium *Janacekia debaisieuxi* (Adler et al. 1999).

**Hosts and Economic Importance.** Hosts include humans, dogs, red foxes, American minks, American black bears, red squirrels, domestic rabbits, and sometimes young domestic ducks (Hunter 1990, Hunter & Bayly 1991, Hunter et al. 1993).

### *Simulium* (*Simulium*) *venustum* Say

(Figs. 4.1, 4.20, 4.31, 10.74, 10.81, 10.224, 10.408, 10.523, 10.597, 10.650, 10.823; Plate 24d; Map 251)

*Simulium venustum* Say 1823: 28–29 (FM). Neotype* female (here designated; pinned, pupal exuviae in glycerin vial below, USNM). North Carolina, Haywood Co., Great Smoky Mountains National Park, tributary of Palmer Creek, 35°38′N, 83°07′W, 16 April 1997 (P. H. Adler); *venestum, venustem, venustom, venustrum* (subsequent misspellings)

*Simulium* 'sp.' Forbes 1912: 23 (L)

*Simulium* 'CC *venustum*' Rothfels et al. 1978: 1118 (C)

[*Simulium meridionale*: Riley (1887) (misident. in part, PL)]

[*Simulium meridionale*: Johannsen (1902, 1903b) not Riley 1887 (misident. in part, PL)]

?[*Simulium meridionale*: Webster & Newell (1902) not Riley 1887 (misident.)]

[*Simulium invenustum*: Webster (1904) not Walker 1848 (misident. in part, Fig. 3)]

**Taxonomy**

Say 1823 (FM); Riley 1887 (PL); Johannsen 1902, 1903b (FMPL); Webster 1904 (P); Forbes 1912 (L); Adler 1983 (L); Adler & Kim 1986 (PL); Hunter 1990 (FMP); Stuart 1995 (cocoon); Stuart & Hunter 1998b (cocoon).

*Simulium venustum* was the first black fly described from North America by a North American worker. Thomas Say (1823) described a male and a female that he collected from Shippingsport in Jefferson Co., Kentucky, sometime around 5 May–9 June in either 1819 or 1820. Shippingsport is the old landing on the Kentucky side of the Ohio River below the falls, now covered by the city of Louisville (Coquillett 1900). The type material has long since been lost. We, therefore, designate a neotype to fix and stabilize the current concept of the name.

**Physiology**

Angioy et al. 1982 (F taste receptors).

**Cytology**

Rothfels et al. 1978; Rothfels 1980, 1981a; Adler 1983; Adler & Kim 1986; Hunter 1990; Adler & Mason 1997.

The chromosomes of *S. venustum* s. s. are generally monomorphic, although polymorphisms occur rarely. We characterize the chromosomes of true *S. venustum* as fixed for *IIIL-5* and *IIS-CC* and standard for the IIL-1 sequence, although the possibility that 1 or more of these sequences is polymorphic in some populations, particularly in the northern United States and Canada, cannot be discounted.

**Molecular Systematics**

Snyder 1981, 1982 (allozymes); Xiong 1992 (mtDNA); Xiong & Kocher 1993a, 1993b (mtDNA); Smith 2002 (nuclear and mtDNA).

**Bionomics**

**Habitat.** *Simulium venustum* is one of the most common black flies in North America. It is abundant, if not ubiquitous, throughout the east but spotty in the west. The immature stages are found in a wide variety of streams and small rivers from the mountains to the coastal plain, but are encountered most frequently in rocky woodland streams less than 10 m wide. Larvae and pupae tolerate a broad range of physical and chemical conditions (Gordon & Cupp 1980). Larvae can be found in slow and fast currents and on the tops and bottoms of stones (Tessler 1991).

**Oviposition.** Females are anautogenous. They require about 4–5 days after a blood meal to mature the eggs at 22°C and can mature an average of about 183 eggs in the first ovarian cycle (Simmons 1985). They release eggs as they fly above the water. Peak oviposition occurs in the early evening, but about 2 hours later for mermithid-parasitized females (Simmons 1985).

**Development.** This species is univoltine; however, some stenothermal streams produce 2 or more generations (or cohorts) per year, judging from the presence of larvae from April through August (Adler & Kim 1986). Eggs can overwinter in the sediments of dry streambeds (Adler & Kim 1986). In New Hampshire, larvae hatch from early March well into May (Lake & Burger 1983). In southern states such as Alabama and Arkansas, adults emerge from February through May. In Canada, emergence is generally from May into July, whereas at high elevations in Colorado, emergence lasts into late August. Females can live up to 2 weeks when held in the laboratory at 22°C and fed a 10% sucrose solution (Simmons 1985).

**Mating.** Unknown.

**Natural Enemies.** Larvae are hosts of mermithid nematodes (Adler & Mason 1997) including *Mesomermis flumenalis* (Simmons 1985); the chytrid fungus *Coelomycidium simulii* (Adler & Kim 1986); the trichomycete fungus *Harpella melusinae*; and the microsporidia *Amblyospora bracteata, Janacekia debaisieuxi,* and *Polydispyrenia simulii*. The haplosporidian protist *Haplosporidium simulii* was described from a member of the *S. venustum* complex (Beaudoin & Wills 1968), probably *S. venustum* s. s., the only member of the complex recorded from the type locality of this pathogen.

**Hosts and Economic Importance.** The tendency of females to feed on blood increases with age, with maximum feeding at 7 days, or perhaps older, at 21°C–22°C; most host-seeking females, therefore, are probably older than 5 days (Simmons 1985). Confirmed hosts include humans, dogs, red foxes, American minks, American black bears, horses, red squirrels, domestic rabbits, and domestic ducks (Simmons 1985, Hunter et al. 1993). Dogs are attacked primarily on the undersides (Simmons 1985). Females in areas such as Massachusetts are pests of horses (Simmons 1982).

*Simulium venustum* s. s. is probably one of the major biting pests of humans in North America. Human subjects, nonetheless, show marked differences in attractiveness to female flies, probably because of differences in production rates of carbon dioxide (Schofield & Sutcliffe 1996), which in combination with sweat is highly attractive (Schofield 1994). In some parts of the country, no more than 20% of females swarming around humans actually bite, with larger females being more likely to bite, usually after walking on the skin for 2–5 seconds (Simmons 1985). Over much of its range, particularly south of New York, *S. venustum* is not a pest. Why this species can be a horrendous pest in some areas but not others remains a mystery.

True *S. venustum* is probably the principal member of the *S. venustum* species complex implicated as a vector of the filarial nematode *Dirofilaria ursi* to American black bears in central Ontario, although *S. truncatum* also might be involved (Addison 1980, Hunter 1990).

### *Simulium (Simulium) venustum* Say 'cytospecies JJ' (Map 252)

*Simulium venustum* 'JJ, The Oregon Sibling' Rothfels 1981a: 27 (C)

#### Taxonomy

We tentatively consider this cytological segregate a valid species. However, we have not seen material and, to our knowledge, none exists in collections.

#### Cytology

Rothfels 1981a.

The chromosomal banding sequence of this species differs from the standard sequence (*sensu* Rothfels et al. 1978) by a single inversion, called J, in IIS (Rothfels 1981a). Old photographs given to us by Klaus Rothfels indicate that the J inversion in IIS runs from the beginning of the second band in section 45 to the first band in section 51. Sex chromosomes are undifferentiated (Rothfels 1981a).

#### Bionomics

**Habitat.** This species is known only from an unspecified site on the Nehalem River in the northwestern corner of Oregon.

**Oviposition.** Unknown.

**Development.** Unknown.

**Mating.** Unknown.

**Natural Enemies.** No records.

**Hosts and Economic Importance.** Unknown; presumably mammalophilic.

### *Simulium (Simulium) verecundum* Species Complex

**Diagnosis.** Female: Anal lobe sclerotized along anterior margin; in ventral view with sclerotized region divided by hyaline band. Male: Ventral plate in ventral view narrow, tapered posteriorly. Pupa: Gill of 6 filaments. Chromosomes: Banding sequences as given by Rothfels et al. (1978).

**Overview.** This species complex consists of at least 2 isomorphic species, *S. rostratum* and *S. verecundum* s. s., which overlap broadly across much of Canada and the northern United States. *Simulium rostratum*, which is well represented in the Palearctic Region, is more common in Canada, whereas *S. verecundum* s. s. is more common in the United States.

Information dealing with the species complex as a whole is presented below, followed by specific information on each of the 2 component species. The Taxonomy, Morphology, and Cytology sections for the species complex are compiled from references that could not be assigned to the appropriate member of the complex, as well as from references dealing specifically with *S. rostratum* and *S. verecundum* s. s.

#### Taxonomy

Lundström 1911 (M); Jobbins-Pomeroy 1916 (FMPLE); Malloch 1919 (F); Enderlein 1922, 1935 (F); Wu 1931 (E); Hearle 1938 (FE); Stains 1941 (M); Stains & Knowlton 1943 (M); Nicholson 1949 (M, Fig. 30F); Nicholson & Mickel 1950 (M, Fig. 30F); Jamnback 1953 (L); Stone & Jamnback 1955 (FMPL); Rubtsov 1956 (FMPL); Anderson 1960 (FMPL); Davies et al. 1962 (FMP); Wood et al. 1963 (L); Stone 1964 (FMPL); Abdelnur 1966, 1968 (FMPLE); Davies 1966, 1968 (FMPL); Snoddy 1966 (FMPL); Holbrook 1967 (L); Stone & Snoddy 1969 (FMPL); Pinkovsky 1970, 1976 (FMPL); Amrine 1971 (FMPL); Lewis 1973 (FMPL); Zwick & Rühm 1973 (F); Corredor 1975 (FM); Snoddy & Noblet 1976 (MPL); Imhof 1977 (E); Merritt et al. 1978b (PL); Imhof & Smith 1979 (E); Westwood 1979 (FMPL); Fredeen 1981a (FMPL), 1985a (FMP); Hilsenhoff 1982 (L); Webb & Brigham 1982 (L); Adler 1983 (PL); Gryaznov 1984b (F); Jensen 1984, 1997 (PL); Adler & Kim 1986 (PL); Zwick 1987 (M); Hunter 1990 (FMP); Mason & Kusters 1991b (FMPL); Peterson & Kondratieff 1995 (FMPL); Stuart 1995 (cocoon); Bass 1998 (PL); Stuart & Hunter 1998b (cocoon).

The members of this species complex historically have been confused with those of the *S. venustum* complex, even though females, males, and pupae of the 2 complexes can be separated with ease on the basis of the sclerotization pattern of the anal lobe, shape of the ventral plate, and configuration of the gill, respectively, as detailed in our identification keys. Larvae are best distinguished chromosomally.

#### Morphology

Jobbins-Pomeroy 1916 (FLE internal anatomy); Guttman 1967a (labral fan); Craig 1969, 1972b (embryology); Davis 1971 (embryonated E); Colbo et al. 1979 (F labral and cibarial sensilla and armature); Nübel 1984 (labral fan); Thompson 1989 (labral-fan microtrichia); McCreadie & Colbo 1990 (L allometry); McCreadie 1991 (L allometry); Evans & Adler 2000 (spermatheca).

**Physiology**

Dove & McKague 1975 (response to development inhibitors), McKague et al. 1978 (response to growth regulators), Undeen 1979b (L midgut pH), Philogène et al. 1985 (L response to phototoxin).

**Cytology**

Zimring 1953; Rothfels et al. 1978; Gordon & Cupp 1980; Rothfels 1980, 1981a; Adler & Kim 1986; Currie & Adler 1986; Hunter 1990; McCreadie 1991; McCreadie et al. 1994a.

**Molecular Systematics**

Snyder 1981, 1982 (allozymes); Xiong 1992 (mtDNA); Xiong & Kocher 1993a, 1993b (mtDNA).

**Bionomics**

**Habitat.** This species complex is found throughout the continent, except in the southwestern United States. The immature stages occupy a wide variety of streams and rivers, often with those of the *S. venustum* complex. They are particularly common in highly productive watercourses with trailing vegetation. Larvae occupy both laminar and turbulent boundary layers (Eymann 1993) and are tolerant of pH depressions to 4.0 for at least 3 days (Chmielewski & Hall 1992).

**Oviposition.** Females are anautogenous (Anderson & Shemanchuk 1987a, Davies & Györkös 1990) and carry an average of 417 eggs, although potential fecundity (i.e., number of ovarioles) can exceed 860 eggs per ovarian cycle (Golini & Davies 1975, Pascuzzo 1976). In the late afternoon, females begin skimming upstream over the water, eventually landing on trailing vegetation where they walk to the substrate-water interface and deposit masses of eggs; this behavior often occurs communally, producing multilayered egg masses (Imhof & Smith 1979). Green and yellow substrates are preferred for oviposition, although this preference can be modified by the background color of the streambed (Golini & Davies 1975). Oviposition also takes place on rocks, sticks, and debris (Cupp & Gordon 1983).

**Development.** Eggs overwinter throughout most of the range, although in parts of the southeastern states, larvae overwinter. Eggs remain viable if held at 4°C for 424 days (Tarshis 1968b), but they do not produce larvae after being held at –15°C (Colbo & Wotton 1981). Hatching success for eggs on the bottom and lower layers of multilayered masses is generally less than half that on the top layers (Imhof & Smith 1979). Eggs are highly susceptible to desiccation (Imhof & Smith 1979). Prior to hatching, the pharate larva imbibes fluids within the egg and begins to swell, bringing the egg burster in contact with the chorion and eventually rupturing it (Davis 1971). Larvae begin hatching in 7–9 days at 15°C or in 5 days at both 20°C and 24°C (Imhof & Smith 1979). Larval development is poor below 15°C and requires about 3 weeks at both 18°C–21°C and 24°C, during which time larvae pass through 6 instars (Mokry 1976b; Colbo & Porter 1979, 1981).

A member of the *S. verecundum* complex was one of the first North American species of black fly reared from egg to adult (Hartley 1955). Early-instar larvae have been reared to adults in the laboratory in as little as 21 days at 18.5°C–22.0°C, with males emerging 1–2 days before females (Hartley 1955, Colbo & Thompson 1978). Emergence occurs between sunrise and sunset, with a peak shortly after sunrise (Hayton 1979). Adults in Canada fly from about May into October (Bradbury & Bennett 1974a, Back & Harper 1979).

Larvae feed primarily by filtering rather than by grazing (Thompson 1987a). If particulate concentration in the water is low or if temperature is high, larvae spend more time filtering than if the concentration is higher or the temperature is lower (Thompson 1987b). The highest assimilation rates (27.8% efficiency) for larvae are achieved with a diet of bacteria, followed by diatoms (19.5%) and green algae (4.0%) (Martin & Edman 1993). A diet of diatoms, nonetheless, produces larger larvae than do diets of bacteria, green algae, or leaf litter (Thompson 1987c). Diatoms are digested more completely at 19.5°C than at 11.5°C (Thompson 1987b). Larvae ingest negligible amounts of water and are unable to ingest soluble glucose, indicating that they cannot use dissolved organic matter that is not flocculated (Martin 1988, Martin & Edman 1991). Adults can develop from first-instar larvae fed only a bacterial suspension (Fredeen 1964b).

**Mating.** Mating has been observed in a swarm (Downes 1958) and on vegetation where oviposition was occurring (Waters 1969).

**Natural Enemies.** Larvae are sometimes attacked by unidentified mermithid nematodes. In the laboratory, they are minimally susceptible to *Romanomermis culicivorax*, a mermithid nematode of mosquitoes (Finney & Mokry 1980). Larvae are hosts of the microsporidia *Amblyospora bracteata*, *Janacekia debaisieuxi*, *Polydispyrenia simulii* (Vàvra & Undeen 1981, Canning & Hazard 1982), and *A. varians*. A *Tetrahymena*-like protozoan has been recorded from females (Mokry in Weiser & Undeen 1981), and the ichthyosporean protist *Paramoebidium curvum* has been found in the larval hind gut (Lichtwardt et al. 2001). The trichomycete fungi *Genistellospora homothallica* and *Harpella melusinae* have been found in ovipositing females (Labeyrie et al. 1996). Spores of a fungus originally called an ovarian phycomycete have been found among egg masses on stream vegetation (Undeen 1979a, Yeboah 1980, Yeboah et al. 1984); this fungus is now believed to be a trichomycete (Moss & Descals 1986). The entomophthoraceous fungi *Entomophaga* near *limoniae* and *Erynia conica* attack adults (Nadeau et al. 1994).

**Hosts and Economic Importance.** Feeding records for this species complex have previously been confused with those for the *S. venustum* complex. Members of the *S. verecundum* complex seem to feed on larger hosts than do those of the *S. venustum* complex. Reliable records of hosts include raccoons, horses, cattle (Simmons et al. 1989), American black bears (Addison 1980, Hunter 1990), and caribou (McCreadie et al. 1994c). Females will feed on rabbits in the laboratory (Mokry et al. 1981). They do not feed well on human blood in the laboratory (Mokry 1976a), although most females will accept it and subsequently produce eggs (Wood & Davies 1966, Simmons 1985). They are not pests of humans (McCreadie et al. 1994c).

### *Simulium (Simulium) rostratum* (Lundström)

(Figs. 10.225, 10.409, 10.524, 10.598, 10.824; Plate 24e; Map 253)

*Melusina reptans* var. *rostrata* Lundström 1911: 13–14, Figs. 7, 9 (M). Lectotype male (designated by Zwick 1987: 26; pinned, terminalia in glycerin vial below; UZMH). Finland, Kangasala, date unknown (R. Frey); *restratum* (subsequent misspelling)

*Simulium* 'sp. 1' Malloch 1919: 42 (F)

*Simulium wilhelmii* Enderlein 1922: 74 (F). Lectotype female (designated by Zwick 1995: 164; pinned, terminalia on slide mount below; ZMHU). "Deutschland Kr[eis] Oststernberg Bechenfliess bei Langenpfuhl" (Poland, ca. 60 km east of Frankfurt), 20 August 1921 (J. Wilhelmi); *wilhelmi* (subsequent misspelling)

*Simulium groenlandicum* Enderlein 1935: 363–364 (F). Holotype female (pinned; ZMHU). West Greenland, Asakak (mid-August 1893) (Vanhöffen)

*Simulium (Simulium) sublacustre* Davies 1966: 491–493, Figs. 32g–k, 36, 37c, 42h, 44m, 46g, 48d (FMPL). Holotype* male with pupal exuviae (pinned; BMNH). England, Westmorland, Brathay River, outlet of Little Langdale Tarn, 19 September 1963 (R. W. Dunbar)

*Simulium* 'ACD *verecundum*' Rothfels et al. 1978: 1122–1123 (C)

[*Simulium reptans*: Johannsen (1902, 1903b) not Linnaeus 1758 (misident. in part, Greenland references)]

[*Simulium reptans*: Johannsen (1912) not Linnaeus 1758 (misident.)]

[*Simulium reptans*: Emery (1913) not Linnaeus 1758 (misident.)]

[*Simulium arcticum*: Edwards (1931a) not Malloch 1914 (misident.)]

[*Simulium (Simulium) venustum*: Twinn (1936b) not Say 1823 (misident. in part, Plate I [Figs. 2, 5, 6])]

[*Simulium venustum*: Hearle (1938) not Say 1823 (misident.)]

[*Simulium venustum*: Stains (1941) not Say 1823 (misident. in part, p. 23, Figs. 98–100)]

[*Simulium venustum*: Stains & Knowlton (1943) not Say 1823 (misident. in part, p. 273, Figs. 98–100)]

[*Simulium venustum*: Zimring (1953) not Say 1823 (misident.)]

[*Simulium venustum*: Hocking & Pickering (1954) not Say 1823 (misident. in part, eggs)]

[*Simulium venustum*: Hartley (1955) not Say 1823 (misident.)]

[*Simulium (Simulium) verecundum*: Stone & Jamnback (1955) (misident. in part, material from Alaska and Washington)]

[*Simulium argyreatum*: Rubtsov (1956) not Meigen 1838 (misident.)]

[*Simulium (Simulium) venustum*: Shewell (1957) not Say 1823 (misident. in part, material from Greenland)]

[*Simulium (Simulium) venustum*: Abdelnur (1966, 1968) not Say 1823 (misident. in part, oviposition and egg desiccation)]

[*Simulium venustum*: Craig (1969, 1972b) not Say 1823 (misident.)]

[*Simulium venustum*: Shemanchuk & Humber (1978) not Say 1823 (misident.)]

[*Simulium venustum*: Poinar et al. (1979) not Say 1823 (misident.)]

[*Simulium venustum*: Shemanchuk (1980c) not Say 1823 (misident.)]

[*Simulium verecundum*: authors (when applied to Palearctic and some Nearctic populations) not Stone & Jamnback 1955 (misident.)]

**Taxonomy**

Lundström 1911 (M); Malloch 1919 (F); Enderlein 1922, 1935 (F); Hearle 1938 (FE); Stains 1941 (M); Stains & Knowlton 1943 (M); Rubtsov 1956 (FMPL); Davies 1966, 1968 (FMPL); Zwick & Rühm 1973 (F); Gryaznov 1984b (F); Jensen 1984, 1997 (PL); Zwick 1987 (M); Hunter 1990 (FMP); Stuart 1995 (cocoon); Bass 1998 (PL); Stuart & Hunter 1998b (cocoon).

**Morphology**

Craig 1969, 1972b (embryology); Nübel 1984 (labral fan); McCreadie & Colbo 1990 (L allometry); McCreadie 1991 (L allometry); Evans & Adler 2000 (spermatheca).

**Cytology**

Zimring 1953; Rothfels et al. 1978; Gordon & Cupp 1980; Rothfels 1980, 1981a; Currie & Adler 1986; Hunter 1990; McCreadie 1991; McCreadie et al. 1994a.

**Molecular Systematics**

Snyder 1981, 1982 (allozymes); Xiong 1992 (mtDNA); Xiong & Kocher 1993a, 1993b (mtDNA).

**Bionomics**

**Habitat.** This species is abundant across Canada and Alaska south to Oregon, Wyoming, and New York. It also inhabits Greenland and a large portion of the Palearctic Region. In Canada and Alaska, it is one of the most common black flies. The immatures are prevalent in enriched flows, especially at the outfalls of lakes and beaver ponds (Davies 1966, Ciborowski & Adler 1990, McCreadie & Colbo 1991b) and in streams draining lowland bogs and marshes. They also can be found in less enriched, slightly acidic streams (Gordon & Cupp 1980). Larvae attach themselves proportionately more to trailing vegetation than to rocks (McCreadie & Colbo 1993a) and exhibit spaced dispersion patterns (Eymann 1991a). Optimal current velocity for the larvae is 0.69–0.73 m/sec (McCreadie & Colbo 1993b).

**Oviposition.** Females deposit eggs in masses on trailing vegetation (McCreadie & Colbo 1991a, 1991b), beginning when light intensity is decreasing at about 2%/min (Reuter & Rühm 1976). Females in the Palearctic Region have an average of about 590 ovarioles, with a range of about 470–690 (Gryaznov 1995b).

**Development.** Eggs of this multivoltine species overwinter. They can be stored for at least 14 months at 2°C, although viability decreases with storage time (Nadeau 1995). In New Hampshire, larvae hatch in late April (Lake and Burger 1983), and in Canada mature larvae can be found from May to mid-September (Adler 1986, McCreadie & Colbo 1993a). In Ontario, adults emerge until late October (Hunter 1990). Larval development requires an average of about 16 days at 15°C (ca. 235 degree-days, threshold = 0°C) (McCreadie & Colbo 1991a). During development, the number of rays in the primary labral fan decreases as food concentration increases (Lucas 1995, Lucas & Hunter 1999). This species has been reared in the laboratory from egg to adult (Nadeau 1995).

**Mating.** Unknown.

**Natural Enemies.** Larvae are hosts of unidentified mermithid nematodes; the microsporidia *Amblyospora varians*, *Janacekia debaisieuxi*, *Polydispyrenia simulii* (Adler & Mason 1997, McCreadie & Adler 1999), and *A. bracteata*; and the chytrid fungus *Coelomycidium simulii* (Adler 1986). Larvae in the Palearctic Region (under the name *S. verecundum*) are hosts of the mermithids *Gastromermis rosalba*, *Isomermis rossica*, and *Mesomermis melusinae* (Rubtsov 1966, 1967, 1968, 1974b). Adults are attacked by the entomophthoraceous fungi *Entomophthora culicis* (Shemanchuk & Humber 1978, Shemanchuk 1980c) and *Erynia conica* (Nadeau et al. 1995, 1996). We have collected males and females from streamside webs of the araneid spider *Larinioides patagiata*.

**Hosts and Economic Importance.** Few reliable feeding records exist for this species in the Nearctic Region other than 1 instance of blood feeding from the shoulder of a human (Currie 1997a) and 1 from a horse. We also have specimens swept from around cattle. In the Palearctic Region, females feed on cattle (Golini et al. 1976) and have been collected around reindeer (caribou) (Helle et al. 1992). Nowhere in the world are females pests of humans.

### *Simulium (Simulium) verecundum* Stone & Jamnback
(Fig. 5.3, Map 254)

*Simulium (Simulium) verecundum* Stone & Jamnback 1955: 83–84, Figs. 25, 41 (FMPL). Holotype* male with pupal exuviae (pinned #62631, terminalia on slide; USNM). Pennsylvania, Monroe Co. (exact site not specified but west of Mt. Pocono on Rt. 115 or Rt. 940, A. Stone, pers. comm.), 4 June 1948 (A. Stone)

*Simulium* '(probably *venustum*)' Davis 1971: 333–336 (hatching)

*Simulium* 'AA *verecundum* (Transposition Sibling)' Rothfels et al. 1978: 1121–1122 (C)

*Simulium* 'A/C *verecundum* (PS3)' Rothfels et al. 1978: 1122 (C)

*Simulium* 'AA-AC *verecundum*' Gordon & Cupp 1980: 975–980 (C, bionomics)

*Simulium* 'CC *ver.*' Rothfels 1981a: 26–27 (C)

*Simulium verecundum* 'AA-A/C-CC' Ciborowski & Adler 1990: 2115–2119 (bionomics)

[*Simulium venustum*: Jobbins-Pomeroy (1916) not Say 1823 (misident.)]

[*Simulium venustum*: Nicholson & Mickel (1950) not Say 1823 (misident. in part, Fig. 30F)]

[*Simulium venustum*: Mokry (1976b) not Say 1823 (misident.)]

[*Simulium venustum*: Molloy & Jamnback (1977) not Say 1823 (misident.)]

**Taxonomy**

Jobbins-Pomeroy 1916 (FMPLE), Nicholson 1949 (M, Fig. 30F), Nicholson & Mickel 1950 (M, Fig. 30F), Stone & Jamnback 1955 (FMPL), Pinkovsky 1976 (FMPL), Adler & Kim 1986 (P), Hunter 1990 (FMP).

All life stages are virtually indistinguishable from those of *S. rostratum*. Females of *S. verecundum* s. s. in southern populations tend to have dark hair on the stem vein, whereas those in northern populations generally are endowed with golden hair.

**Morphology**

Jobbins-Pomeroy 1916 (FLE internal anatomy), Davis 1971 (embryonated E).

**Cytology**

Rothfels et al. 1978; Gordon & Cupp 1980; Rothfels 1980, 1981a; Adler & Kim 1986; Hunter 1990; McCreadie 1991; McCreadie et al. 1994a.

The status of cytological segregates 'AA,' 'A/C,' and 'CC' of Rothfels et al. (1978) and Rothfels (1981a) remains unknown; all might be valid species. The northernmost populations are predominantly CC, although in the Northwest Territories, A/C males are present with CC males and females and all larvae have the IIL-1, 2 sequence, suggesting that, in this case, the A sequence in the IIS arm is a Y-chromosome polymorphism. B chromosomes are common in the Northwest Territories.

Populations from northern New Jersey and Pennsylvania to Florida might represent a new species. These populations are characterized by a new inversion on top of IIL-1,2.3. With reference to the IIL-1,2.3 sequence in Fig. 17 of Rothfels et al. (1978), this new inversion has breakpoints at the distal break for IIL-3 and the distal junction between 61 and 63, producing the sequence 58 59 60 62 61 60 61 63 62 63 64 65. This new inversion, on top of IIL-1,2.3, typically occurs in the homozygous condition, but the new inversion (on top of IIL-1,2.3) also can occur heterozygously with IIL-1,2. The DNA puff inversion is polymorphic and the IIS sequence is typically AA, although AC is occasional. Differentiated sex chromosomes are apparently lacking. We have examined 3 larvae from Luzerne Co., Pennsylvania (Nescopeck Creek), of which 2 were IIL-1,2.3 homozygotes and the third was homozygous for the new inversion on top of IIL-1,2.3. We refrain from naming a new species until more geographically intermediate material (e.g., from New York) can be evaluated.

**Molecular Systematics**

Snyder 1982 (allozymes); Xiong 1992 (mtDNA); Xiong & Kocher 1993a, 1993b (mtDNA).

**Bionomics**

**Habitat.** *Simulium verecundum* is found throughout eastern North America northwest through Saskatchewan deep into the Northwest Territories. In the southeastern states, it is nearly ubiquitous. The immature stages are often abundant in small to large streams, particularly when beds of trailing grasses and aquatic macrophytes such as *Sparganium* are present. They are most frequent in downstream areas remote from lake outlets, in contrast to the sublacustrine habits of their nearest relative *S. rostratum* (Ciborowski & Adler 1990; McCreadie & Colbo 1991b, 1992). Proportionately more larvae are found on vegetation than on rocks (Ciborowski & Adler 1990, McCreadie & Colbo 1991b).

**Oviposition.** Females generally oviposit in masses on trailing vegetation, especially at the tips of the leaves. However, during the fall in northern areas such as Newfoundland, females deposit their eggs while in flight over open water (McCreadie & Colbo 1991b).

**Development.** Several generations are produced annually, and in most of the southeastern states, larvae and pupae are present year-round. In New Hampshire, hatching occurs in late April (Lake & Burger 1983). In New York, larvae can be found from at least June into October (Gordon & Cupp 1980), whereas in Florida, larvae and pupae are present from about October to June (Pinkovsky & Butler 1978). Mature larvae can be found in Newfoundland from late May to mid-September (McCreadie & Colbo 1993a). Larvae are unable to complete development at 30°C and have low survival rates at 5°C; maximum survival occurs at 15°C–25°C (McCreadie & Colbo 1991a). The average time required for larval development is about 18 days at 15°C (265 degree-days, threshold = 0°C); pupal development requires about 5 days (McCreadie & Colbo 1991a). Adults fly from about March to November in the southeastern states.

**Mating.** Unknown.

**Natural Enemies.** Larvae are hosts of the trichomycete fungi *Genistellospora homothallica*, *Harpella melusinae*, *Pennella simulii*, and *Smittium culisetae* (Beard & Adler 2000); the chytrid fungus *Coelomycidium simulii*; unidentified mermithid nematodes; and the microsporidia *Amblyospora varians*, *Janacekia debaisieuxi*, and *Polydispyrenia simulii*. We have found individual larvae with double infections of *J. debaisieuxi* (which often creates reddish cysts in the abdomen) and *P. simulii*.

**Hosts and Economic Importance.** Hosts are poorly known, although American black bears are among the confirmed hosts (Hunter 1990). Humans are not attacked.

## Species Erroneously Recorded from North America

*Simulium* (*Ectemnaspis*) 'sp.' Wygodzinsky 1973: 10–12

Wygodzinsky (1973) recorded 2 females—members of what he called the *S. bicoloratum* species complex—from the foothills of the Catskill Mountains in Ulster Co., New York. The specimens were collected in a Malaise trap that had been deployed during the first 2 weeks of July 1971. Further prospecting by Wygodzinsky yielded no additional specimens. Aside from this curious New York record, *S. bicoloratum* and its relatives are strictly Neotropical. Wygodzinsky was well known for his meticulous efforts, but he also worked almost exclusively with Neotropical black flies, having collected in South America numerous times. In light of Wygodzinsky's South American connection, we attribute the New York record to contamination.

*Simulium irritans*, Reeves & Milby 1990: 139

This name appeared in a list of black flies collected in Kern Co., California (Reeves & Milby 1990). It is an error in listing and applies to the horn fly (*Haematobia irritans*). The only simuliid with the name *irritans* is in the genus *Prosimulium*, and it occurs in Siberia and China.

## Unassociated Names

We are unable to associate the following 9 vernacular names with any known North American species, primarily because the published records lack adequate information and the original material has long since disappeared.

*Prosimulium* 'undescribed species' Peterson & Kondratieff 1995: 32

One male of this species was taken on 13 August 1993 from Roaring Fork River, Rt. 83 (mi. 52), Pitkin Co., Colorado. The specimen could not be located in the Colorado State University collection or in the USNM.

'new triploid species of *Prosimulium*' Basrur & Rothfels 1959: 583, as footnote

Basrur and Rothfels (1959) reported "numbers" of triploid males of a *Prosimulium* species from the North Bay area of Ontario. The 4 known North American triploid species (*Gymnopais dichopticoides*, *G. holopticoides*, *Prosimulium ursinum*, *Stegopterna mutata*) are composed of females only. Rothfels discovered the population just south of the city of North Bay, but it was never found again, its habitat presumably wrecked by urban sprawl before enough material could be collected to characterize it (R. W. Dunbar, pers. comm.).

*Cnephia* 'L' Wolfe & Peterson 1959: 142

This species, said at the time to be undescribed, is known from only 4 females collected in 1954–1955 near Lac la Loutre (10 July) and on the Manicouagan Peninsula (30 July) of Quebec. The informal letter designation was assigned by G. E. Shewell, but we could find no reference to it in Shewell's correspondence files housed in the CNC.

*Cnephia* 'sp.' Nicholson 1949: 49–51, Fig. 21; Nicholson & Mickel 1950: 29–30, Fig. 21

Five females of this species, said to represent a probable new species, were collected at 5 separate localities in Minnesota (1–11 May 1941). They most likely belong to 1 or more members of the subgenus *Hellichiella*, probably *S. anatinum* and perhaps *S. excisum*.

*Cnephia* 'sp. #1' Anderson & Dicke 1960: 395

This species was mentioned as an associate of the larvae of *Helodon decemarticulatus* in a tiny stream draining a forest bog in Wisconsin in early May 1958. The record might refer to a species of *Greniera*.

*Cnephia* 'unidentified species' Peterson & Kondratieff 1995: 18

The 2 females and 1 male on which this putatively new species is based were from Jefferson, Park Co. (1 August 1938), and Rabbit Ears Pass, Grand Co.-Jackson Co. (18–26 June, year not given), Colorado. The material could not be located in the Colorado State University collection or the USNM.

*Metacnephia* 'unidentified species' Peterson & Kondratieff 1995: 24

The material on which this putatively new species is based came from South Fork (20 June 1972) and Beaver Creek (21 June 1972), both in Rio Grande Co., Colorado. The material could not be found in the Colorado State University collection or the USNM.

*Simulium* (*Simulium*) 'sp. (probably new)' Watson, Davis & Hanson 1966: 580

This species appeared in a list of invertebrates collected during a survey of Cape Thompson, Alaska, during the summers of 1959–1961. No additional information was provided.

*Simulium* 'new species' Jessen 1977: 22

Females of this species were collected from caged sheep, though apparently not biting, in forested areas of southern Idaho. No further information was provided.

**ADDENDUM:** As this book was going to press, we discovered the larvae and pupae of ***Simulium annulitarse*** Zetterstedt in streams 0.5–2.0 m wide in the Nunavut communities of Baker Lake and Rankin Inlet during mid-July. The species previously was known from Finland (Adler & Kuusela 1994; as *tuberosum* 'FGIO') and Sweden (Adler et al. 1999). It is the tenth member of the *Simulium tuberosum* species complex and the 255th species known from North America.

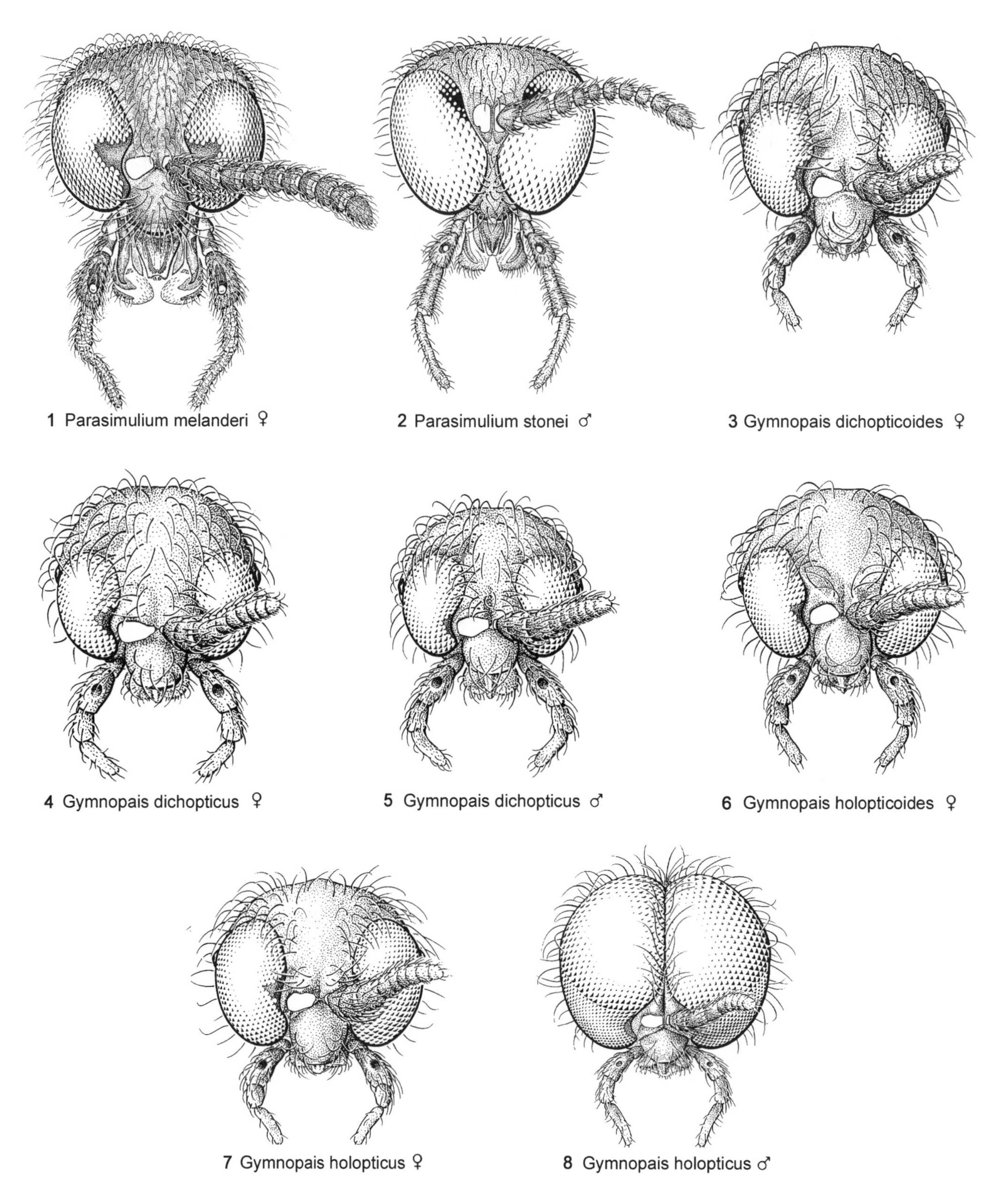


Figs. 10.1–10.8. Adult heads, anterior view; right antennae omitted.

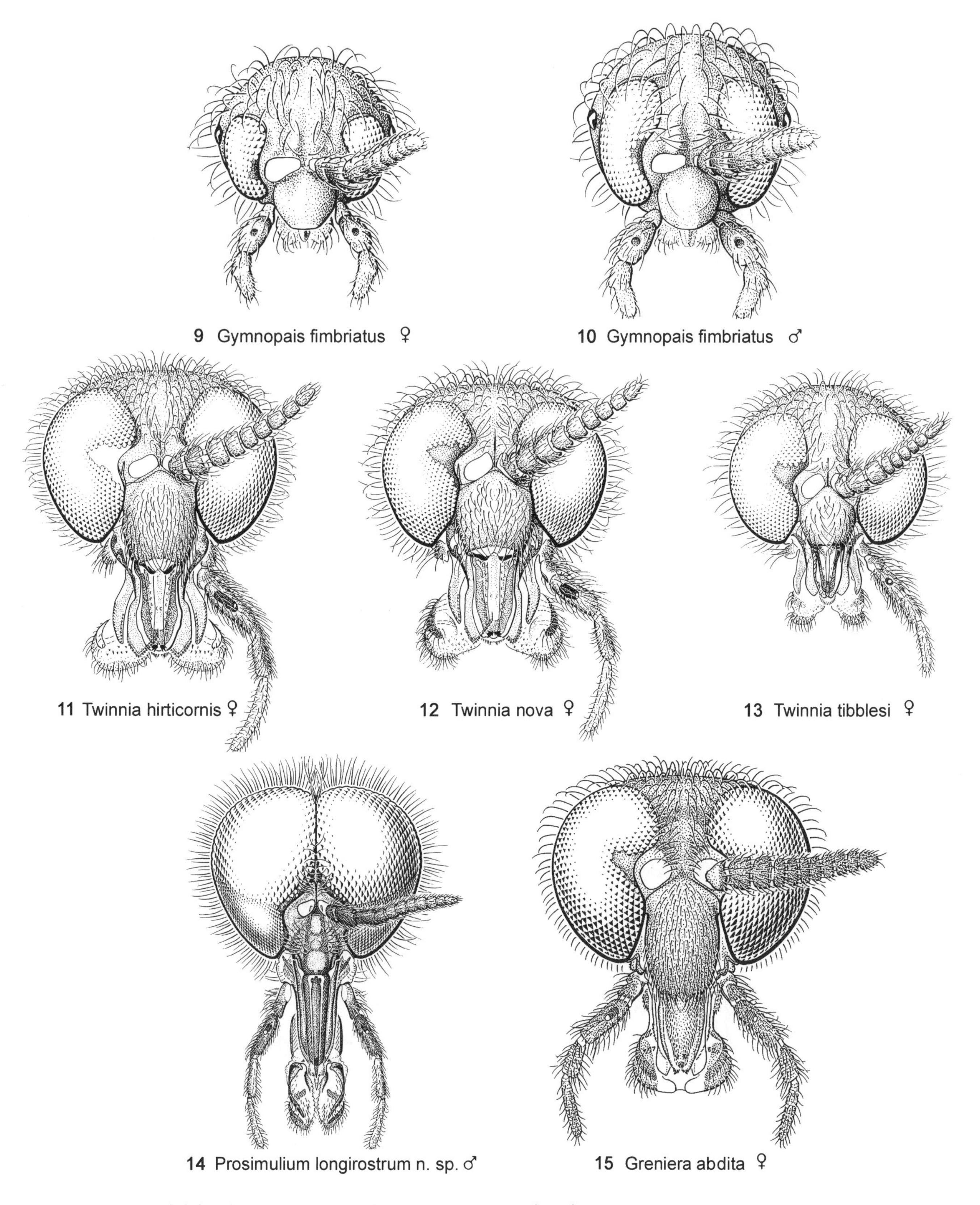


Figs. 10.9–10.15. Adult heads, anterior view; right antennae omitted; (11–13) right palps omitted.

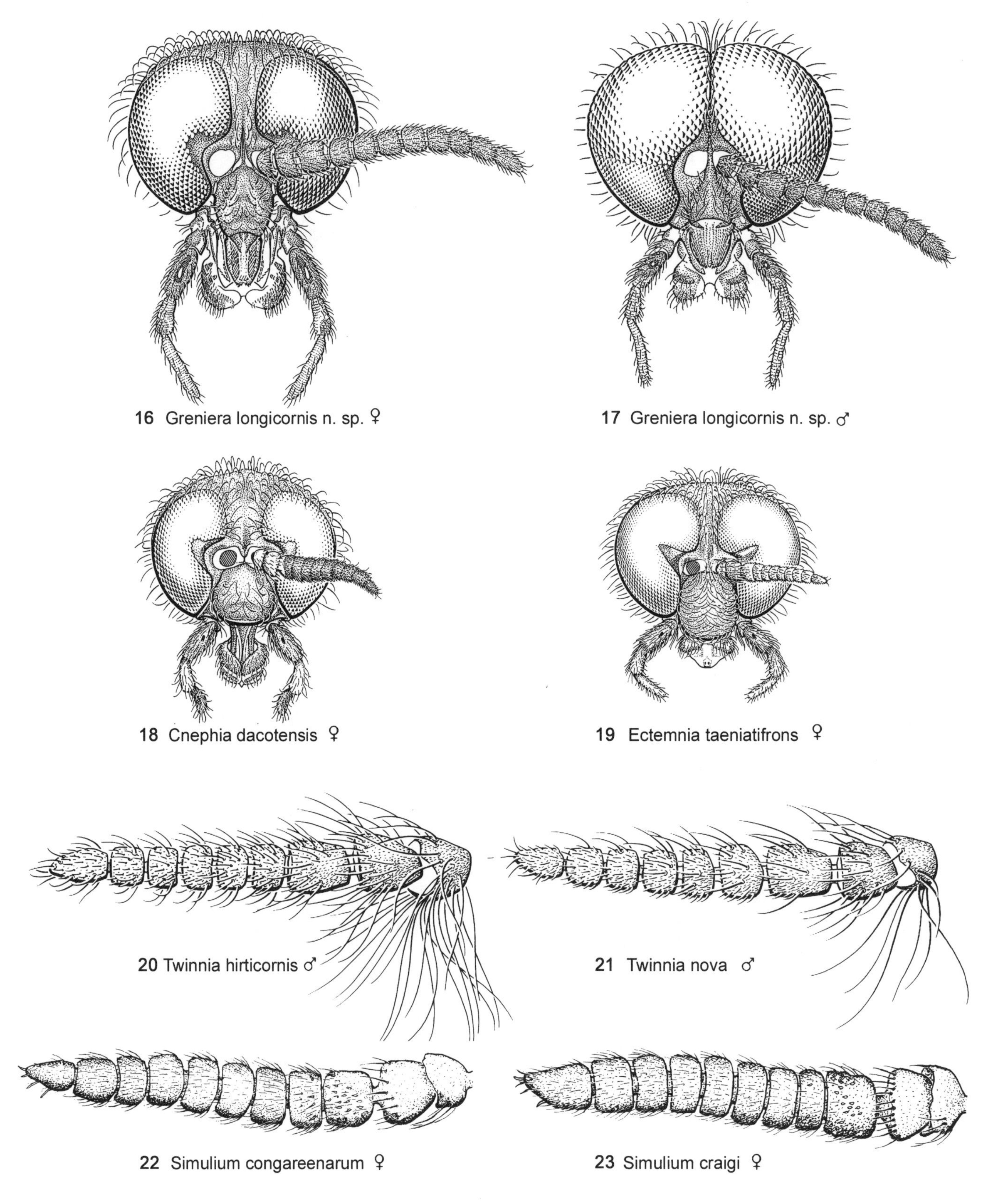


Figs. 10.16–10.23. Adult heads and antennae. (16–19) Heads; right antennae omitted. (20–23) Antennae. (18, 19) From Peterson (1981) by permission.

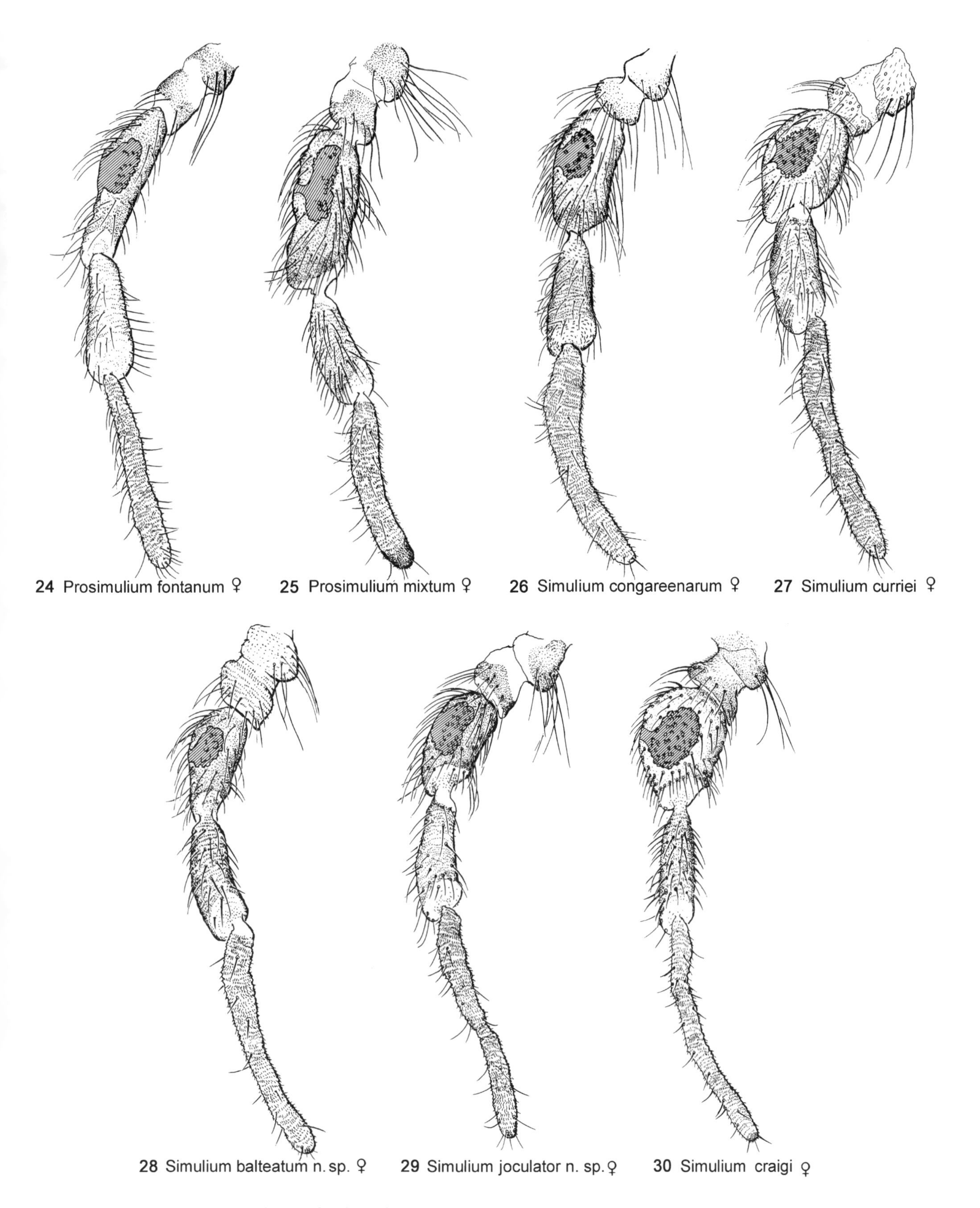


Figs. 10.24–10.30. Female maxillary palps, lateral view.

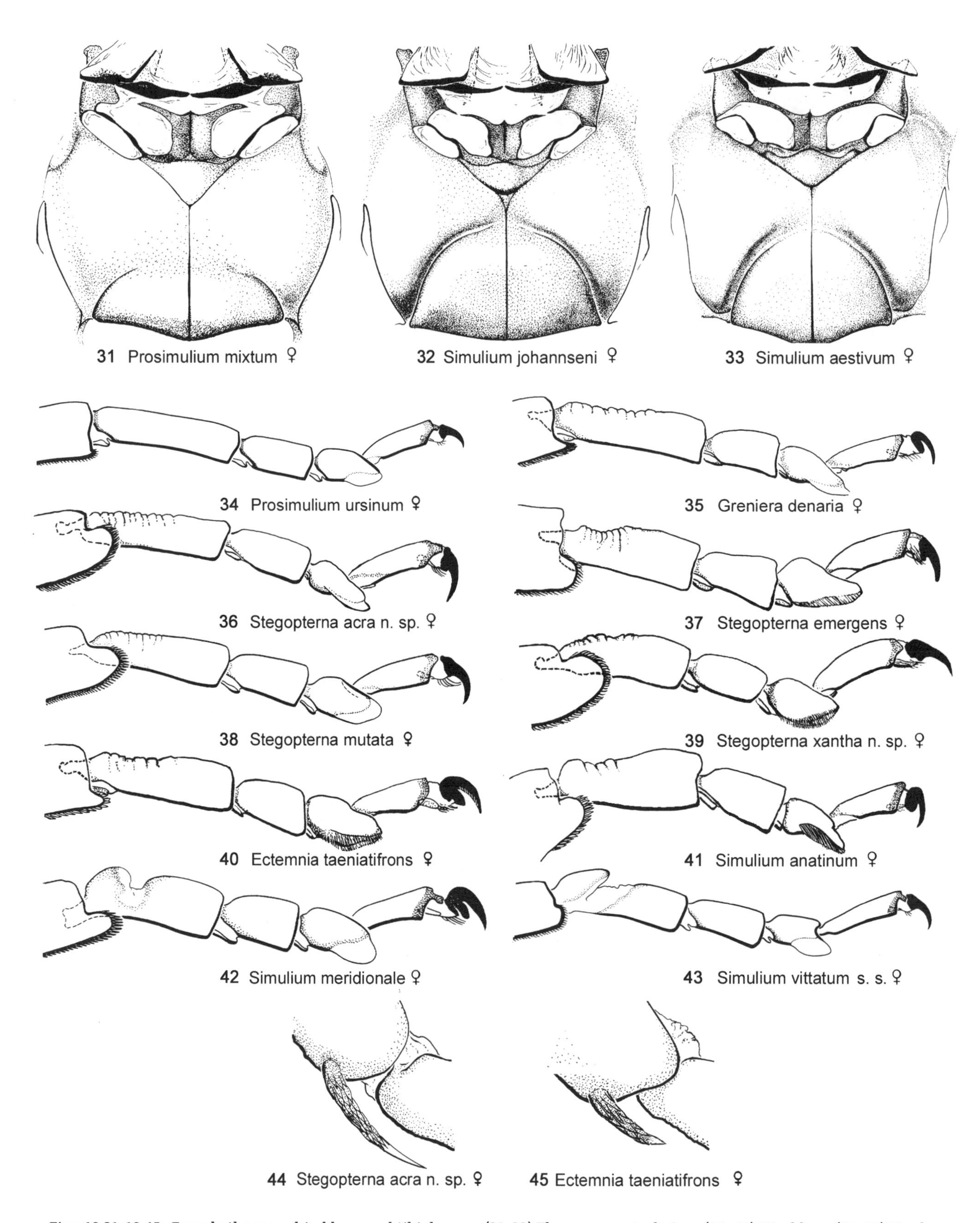


Figs. 10.31–10.45. Female thoraces, hind legs, and tibial spurs. (31–33) Thoraces, ventral view. (34–43) Hind legs. (44, 45) Hind tibial spurs. (34, 38, 42, 43) From Peterson (1981) by permission.

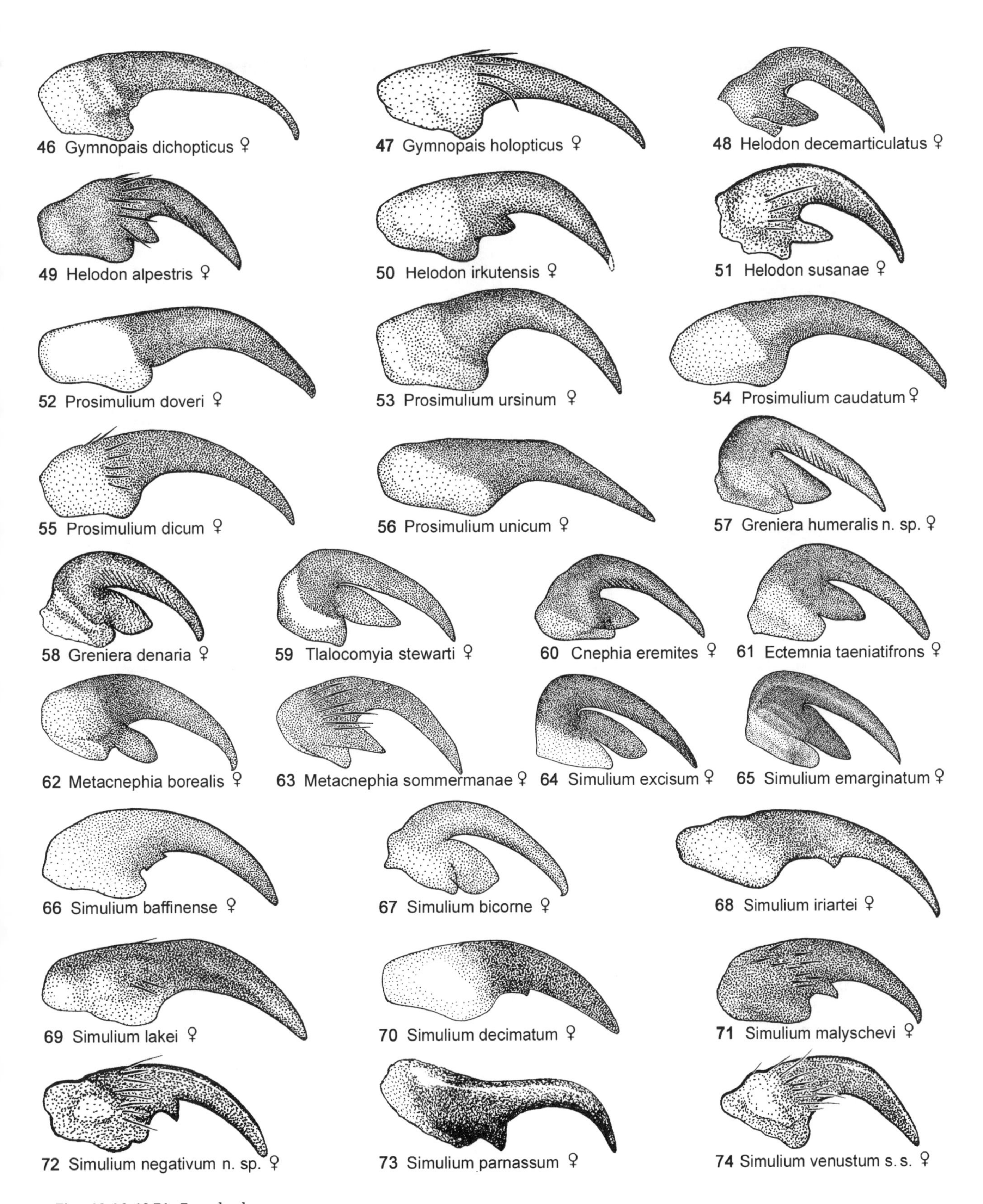


Figs. 10.46–10.74. Female claws.

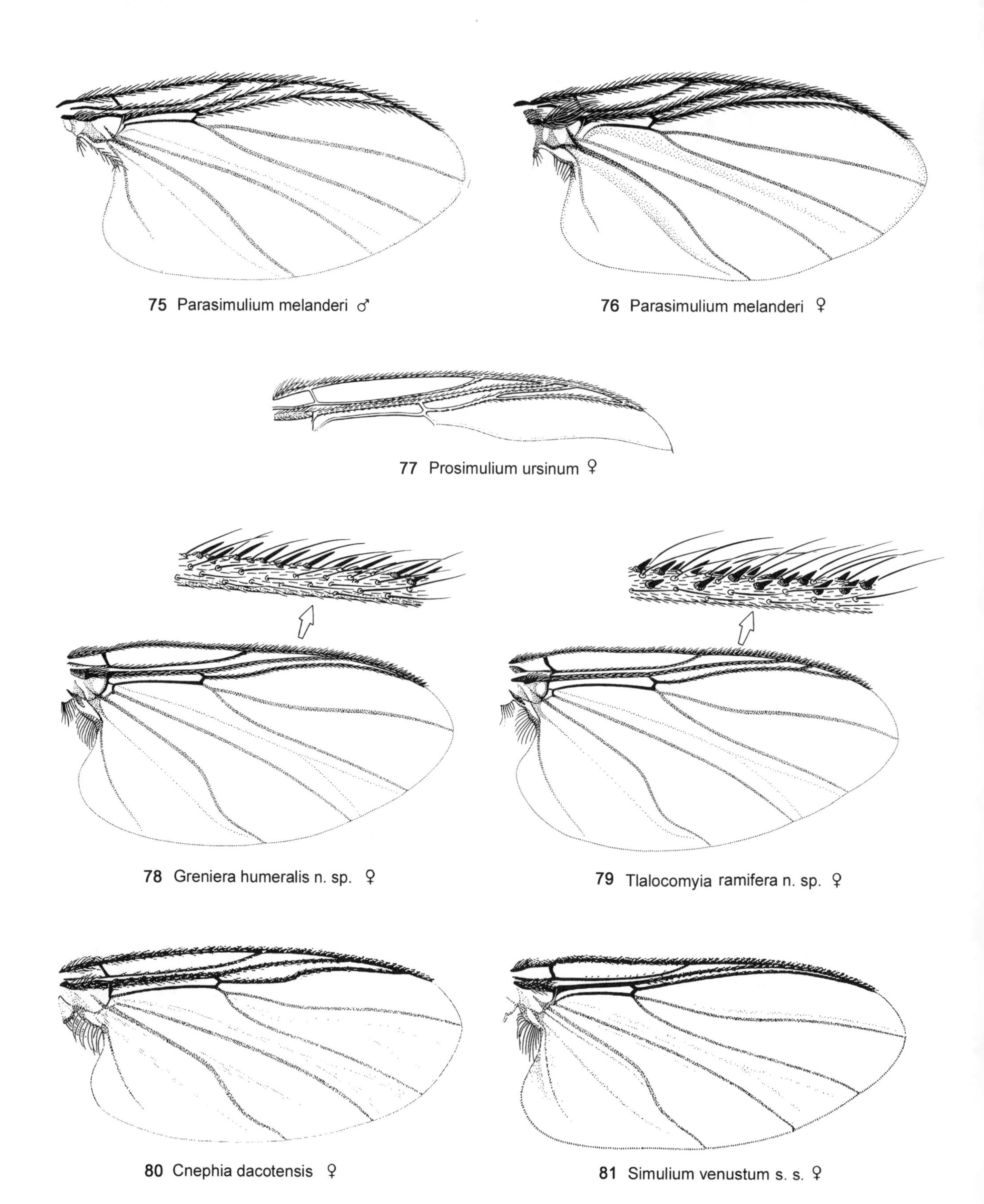


Figs. 10.75–10.81. Wings. (75, 77–81) From Peterson (1981) by permission.

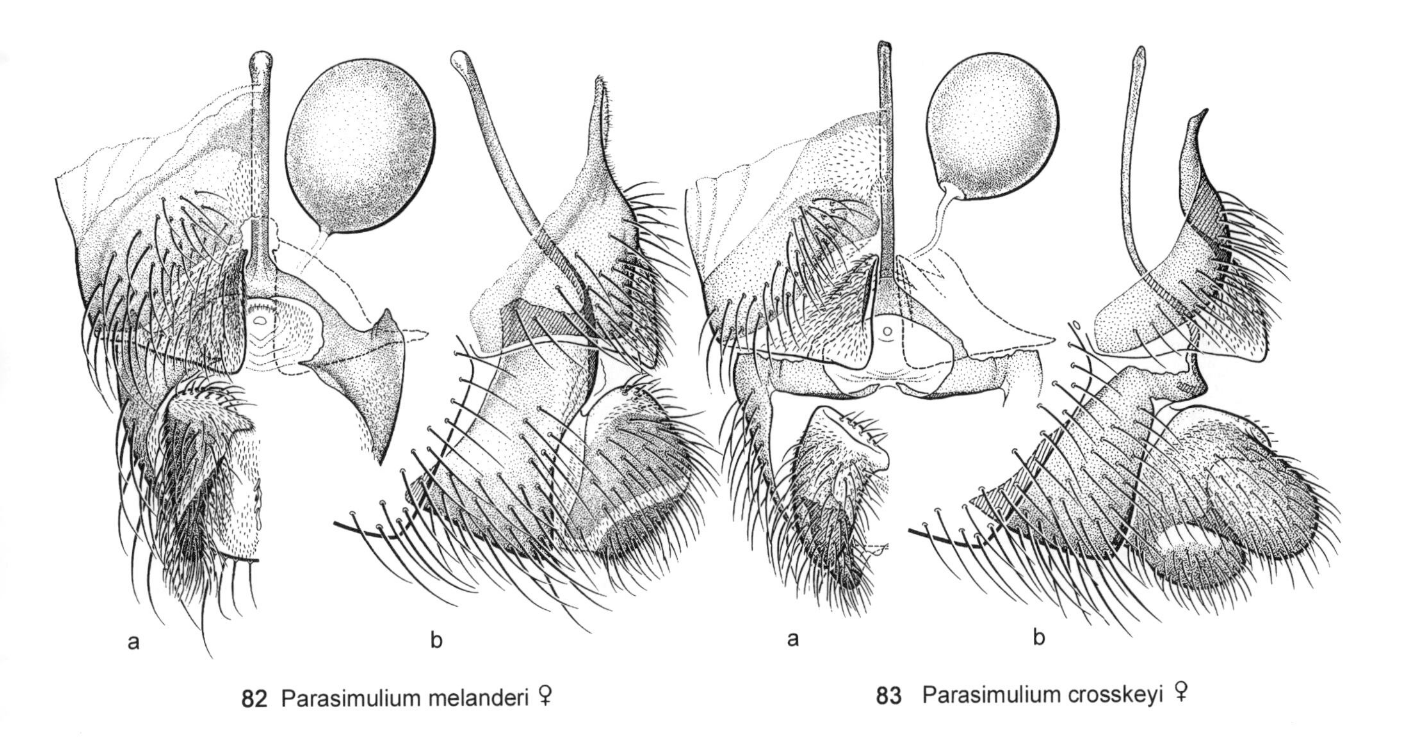


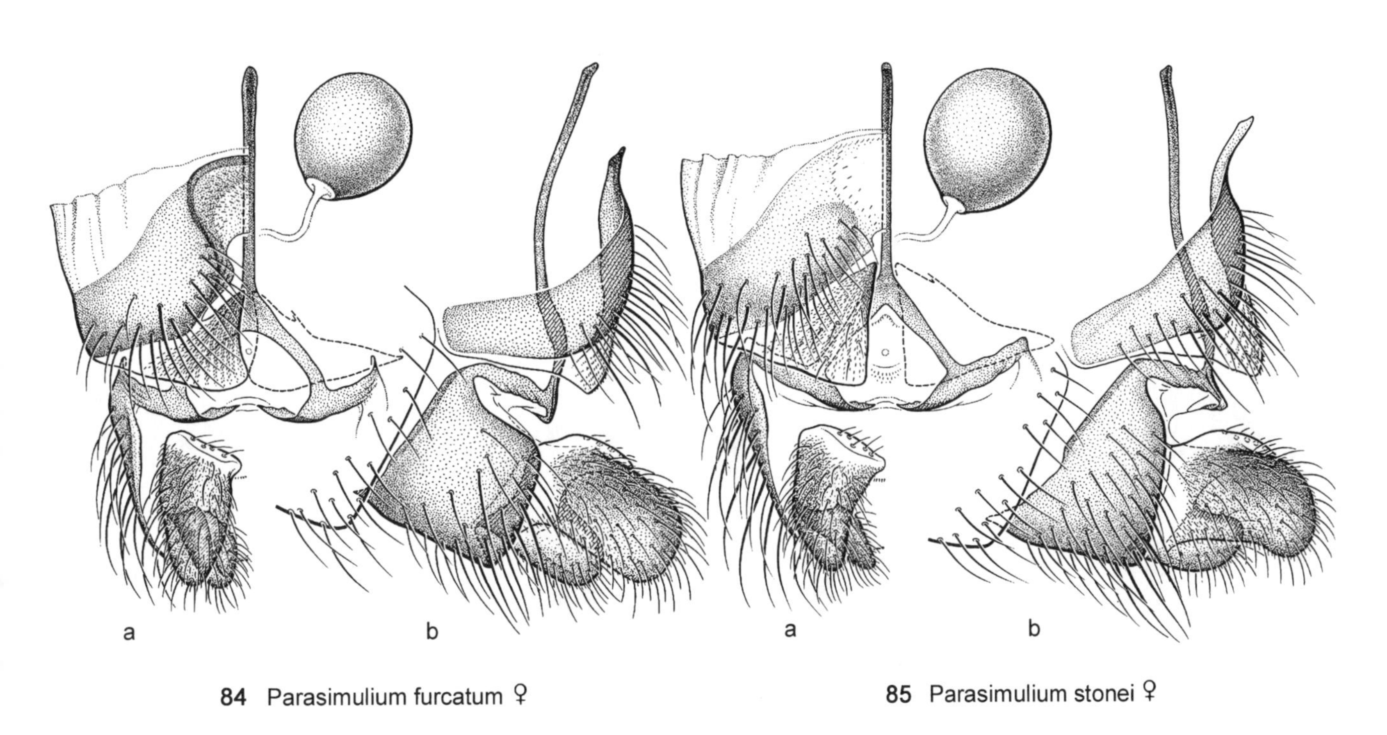


Figs. 10.82–10.85. Female terminalia; a, ventral view; b, right lateral view.

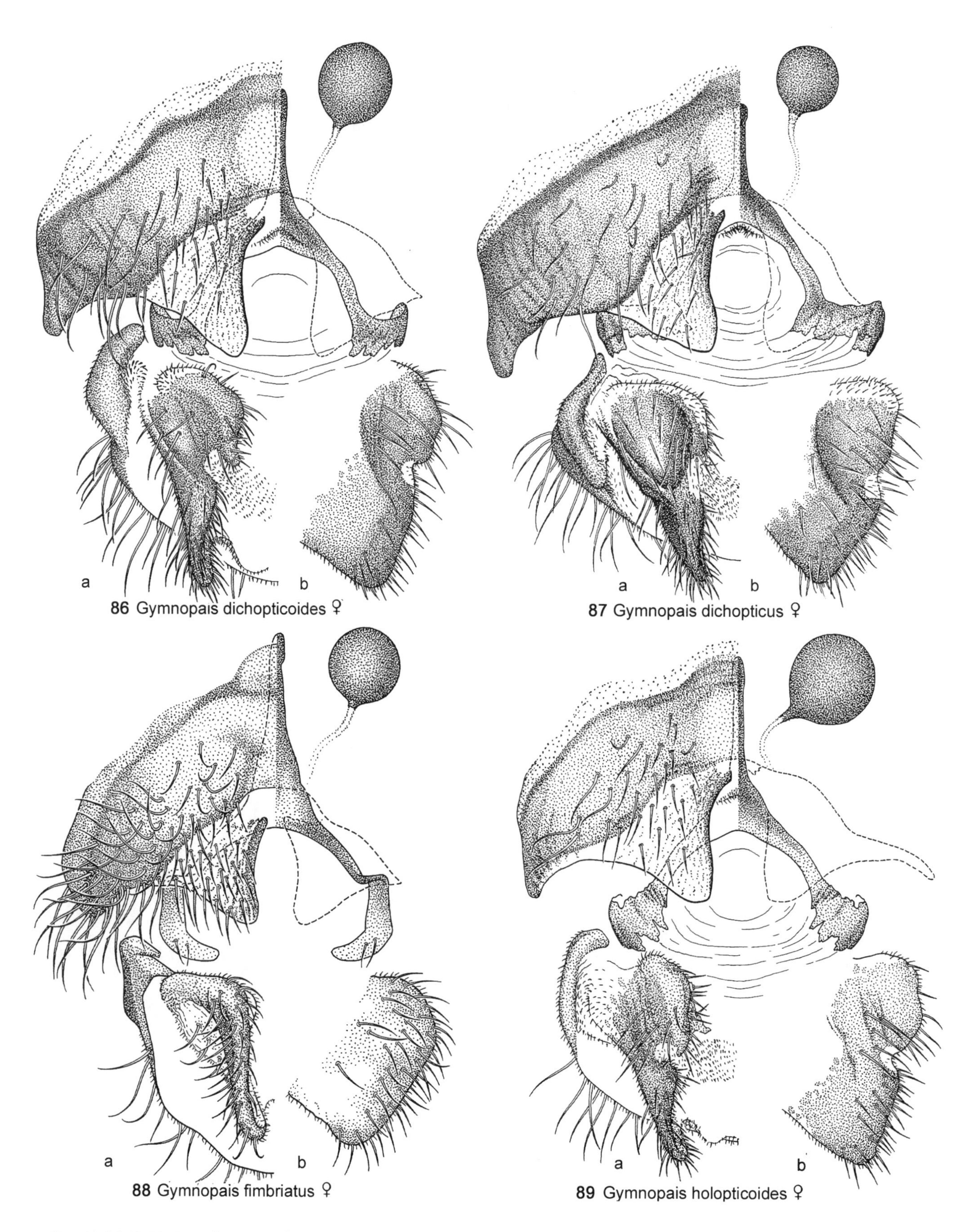


Figs. 10.86–10.89. Female terminalia; a, ventral view; b, right lateral view.

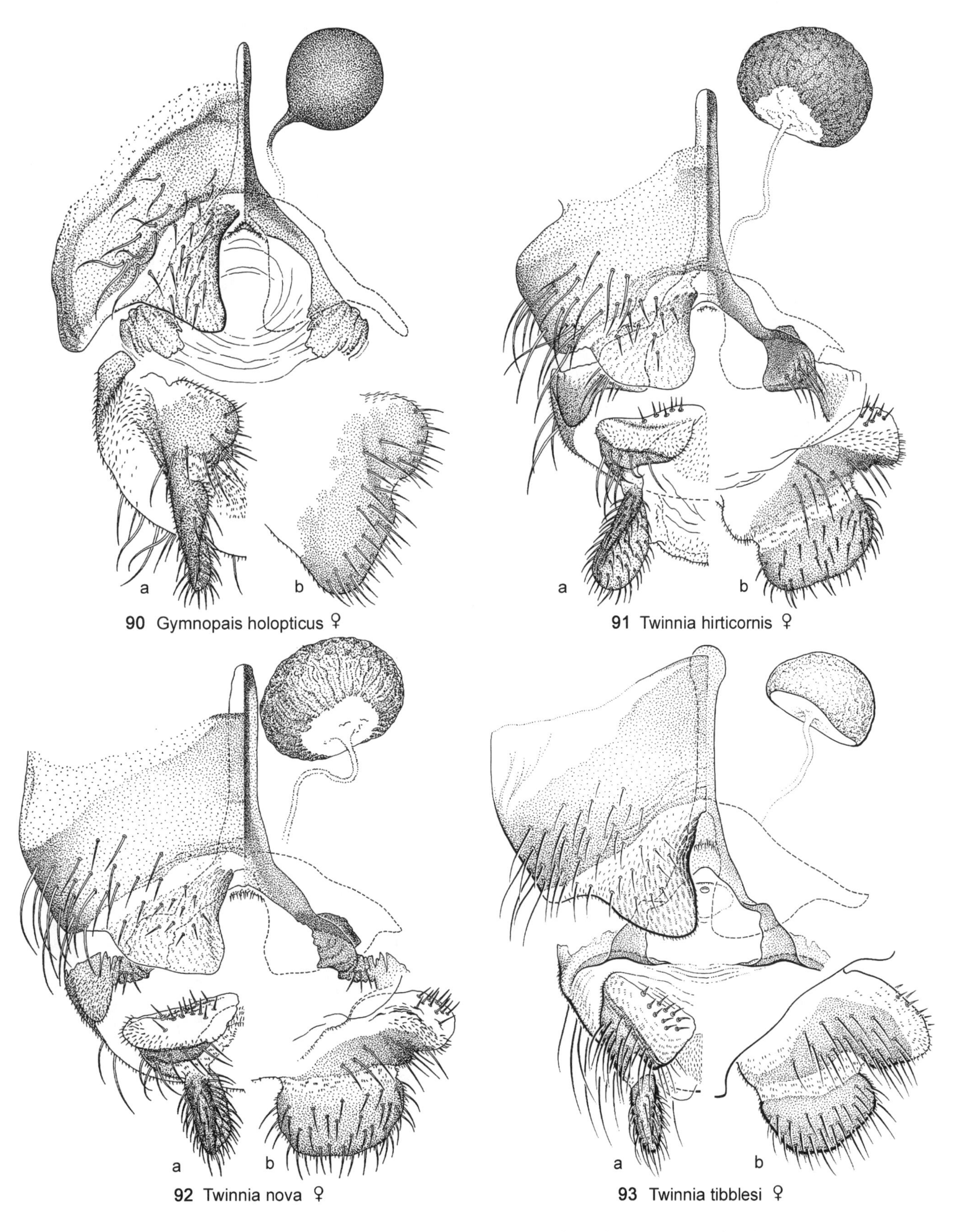


Figs. 10.90–10.93. Female terminalia; a, ventral view; b, right lateral view. (90, 93) From Peterson (1981) by permission.

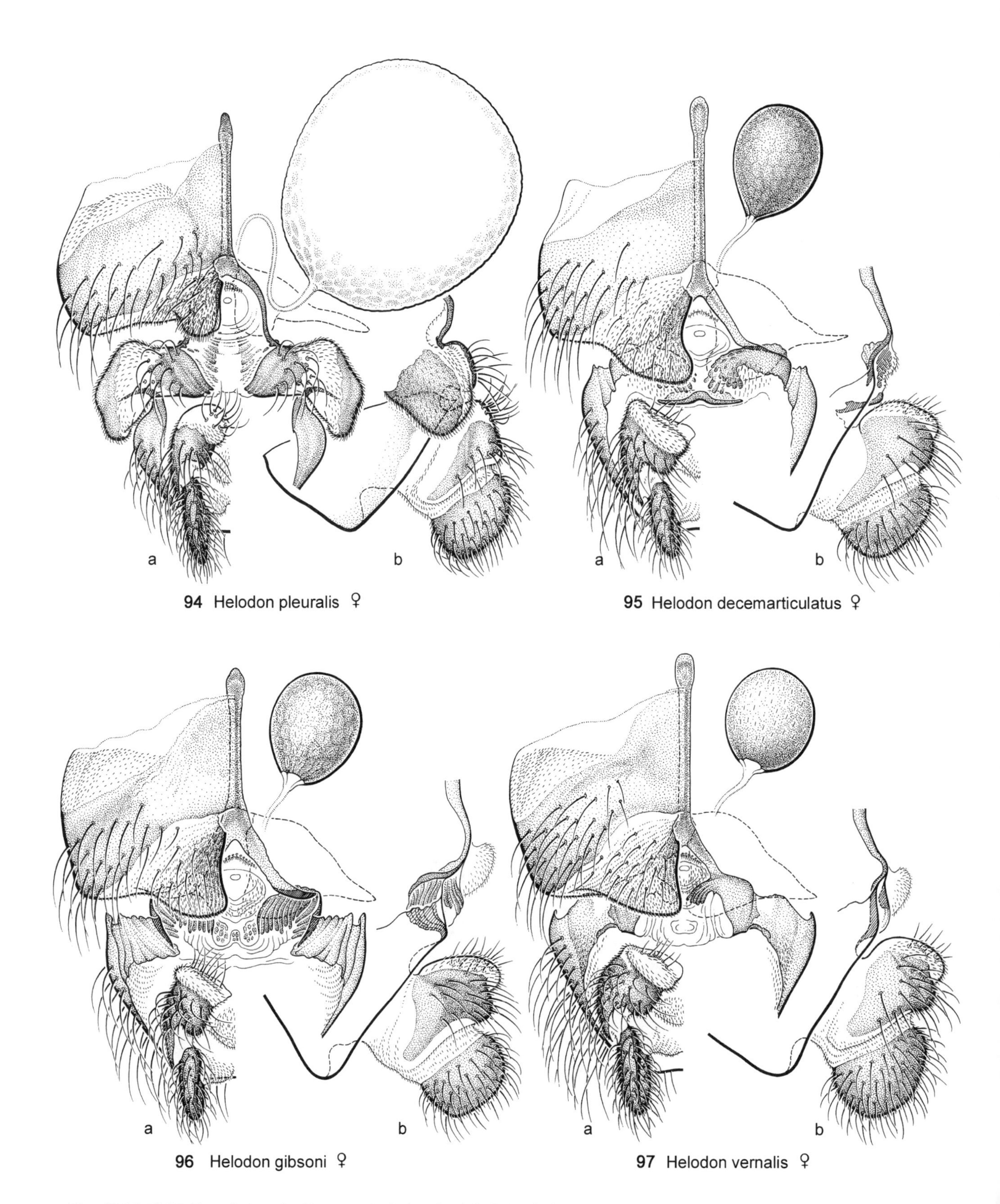


Figs. 10.94–10.97. Female terminalia; a, ventral view; b, right lateral view.

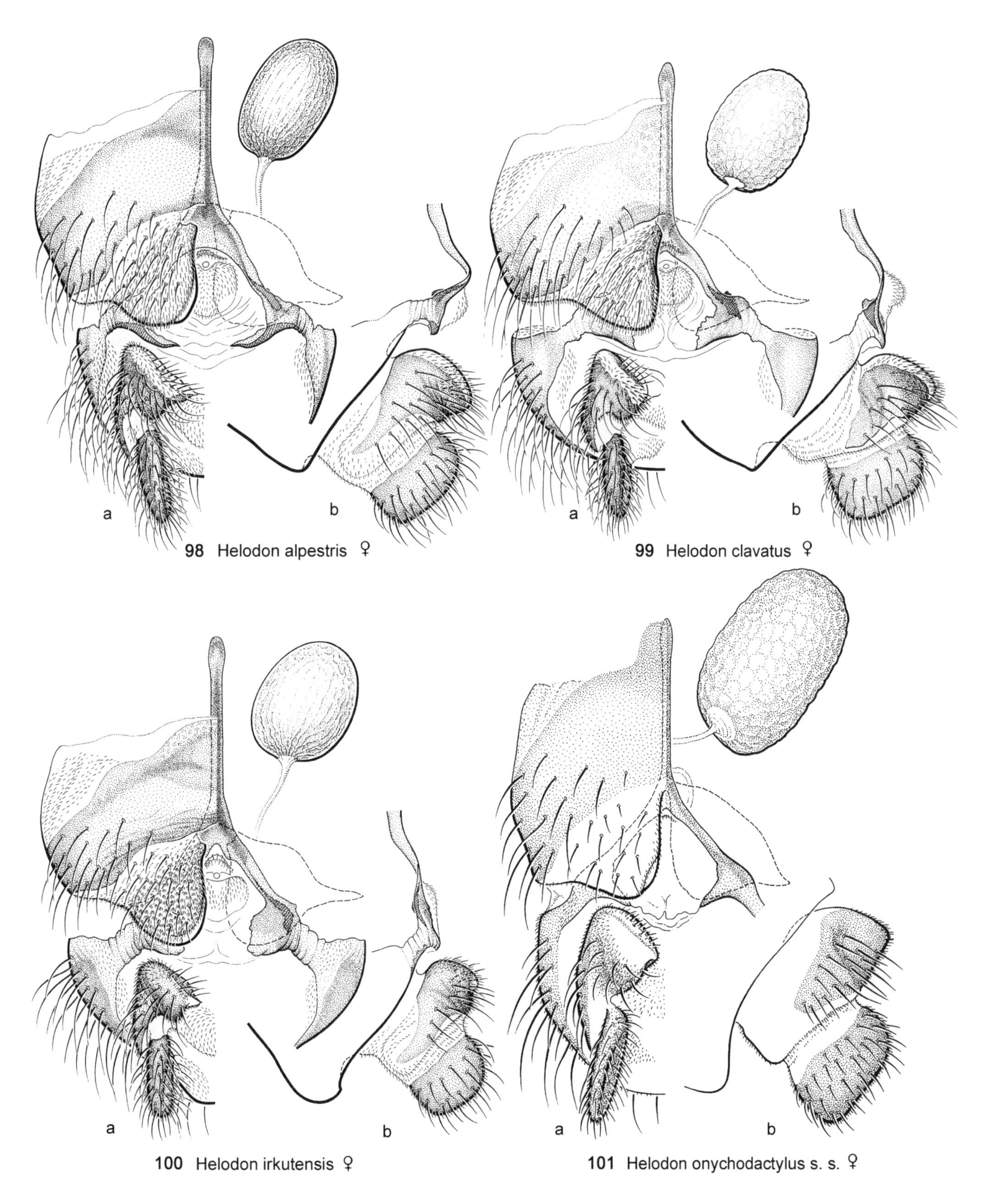


Figs. 10.98–10.101. Female terminalia; a, ventral view; b, right lateral view. (101) From Peterson (1981) by permission.

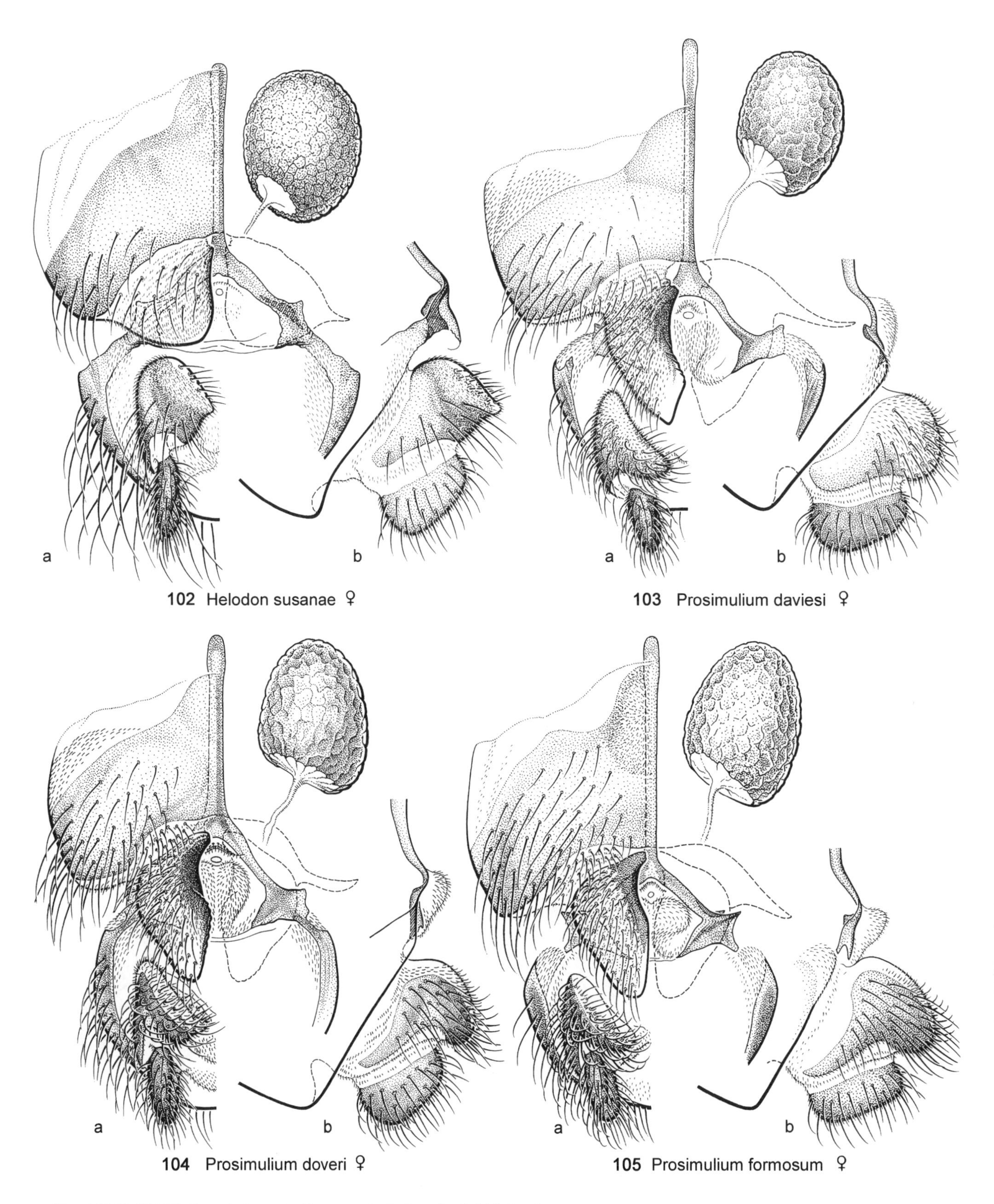


Figs. 10.102–10.105. Female terminalia; a, ventral view; b, right lateral view.

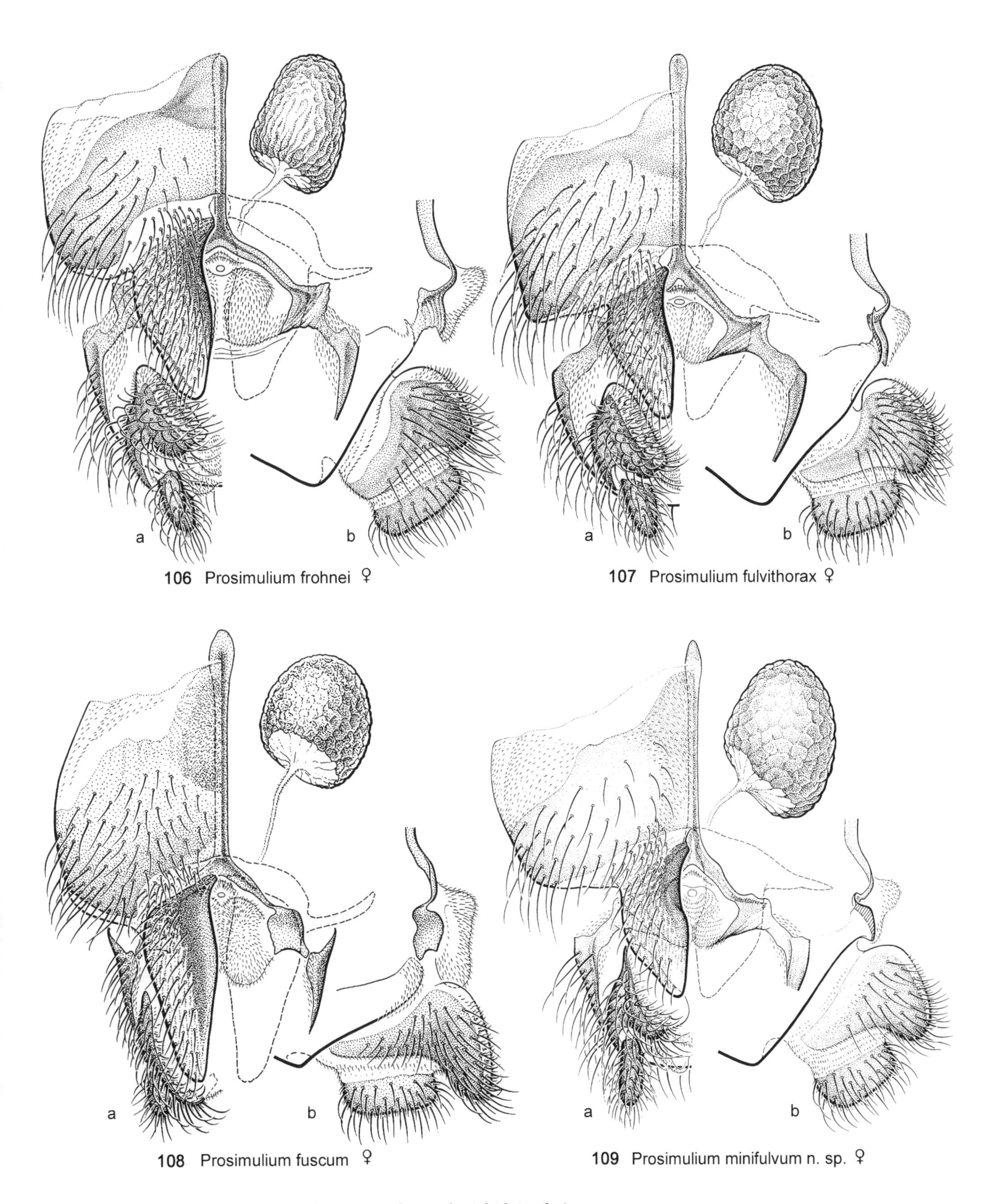


Figs. 10.106–10.109. Female terminalia; a, ventral view; b, right lateral view.

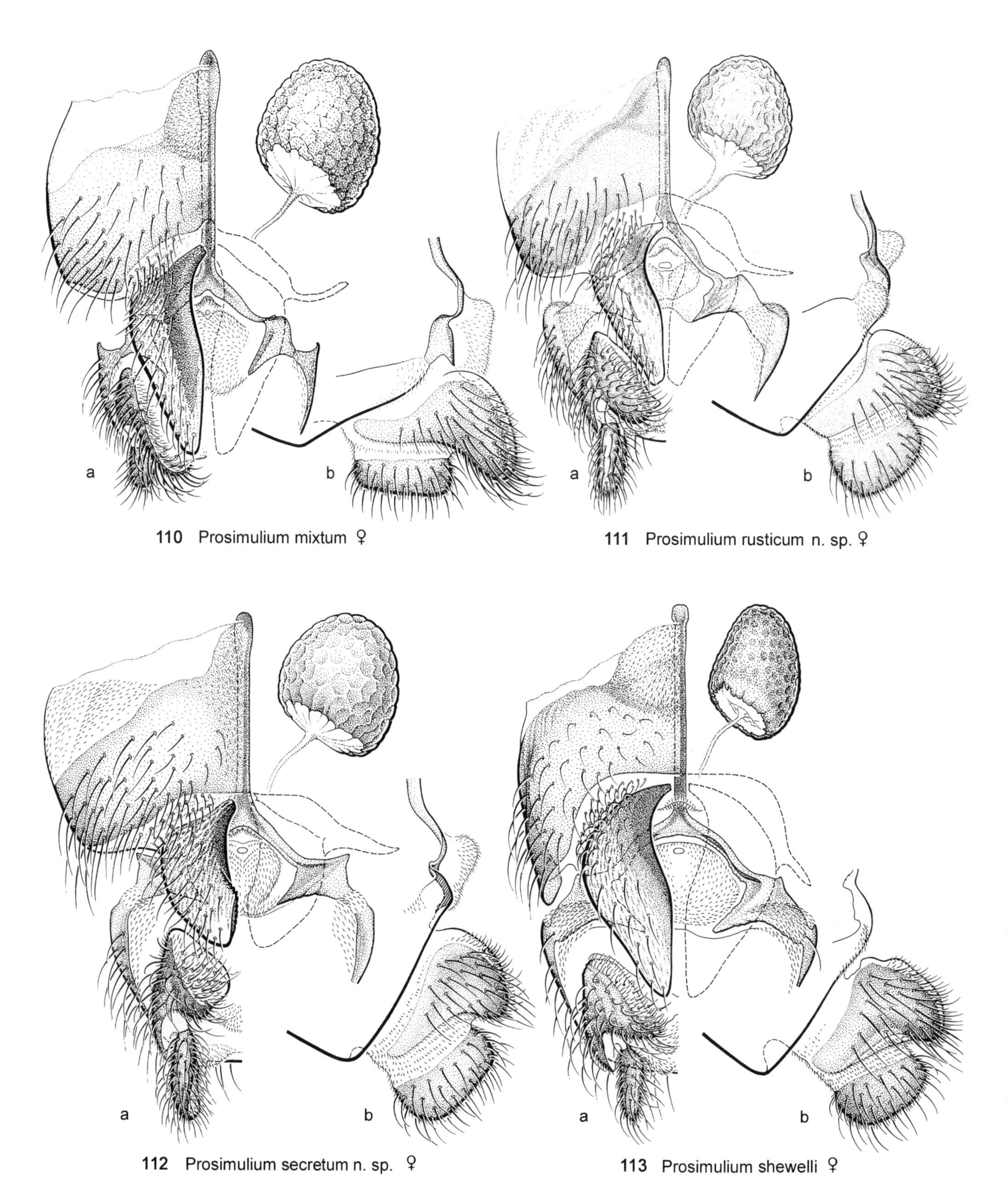


Figs. 10.110–10.113. Female terminalia; a, ventral view; b, right lateral view.

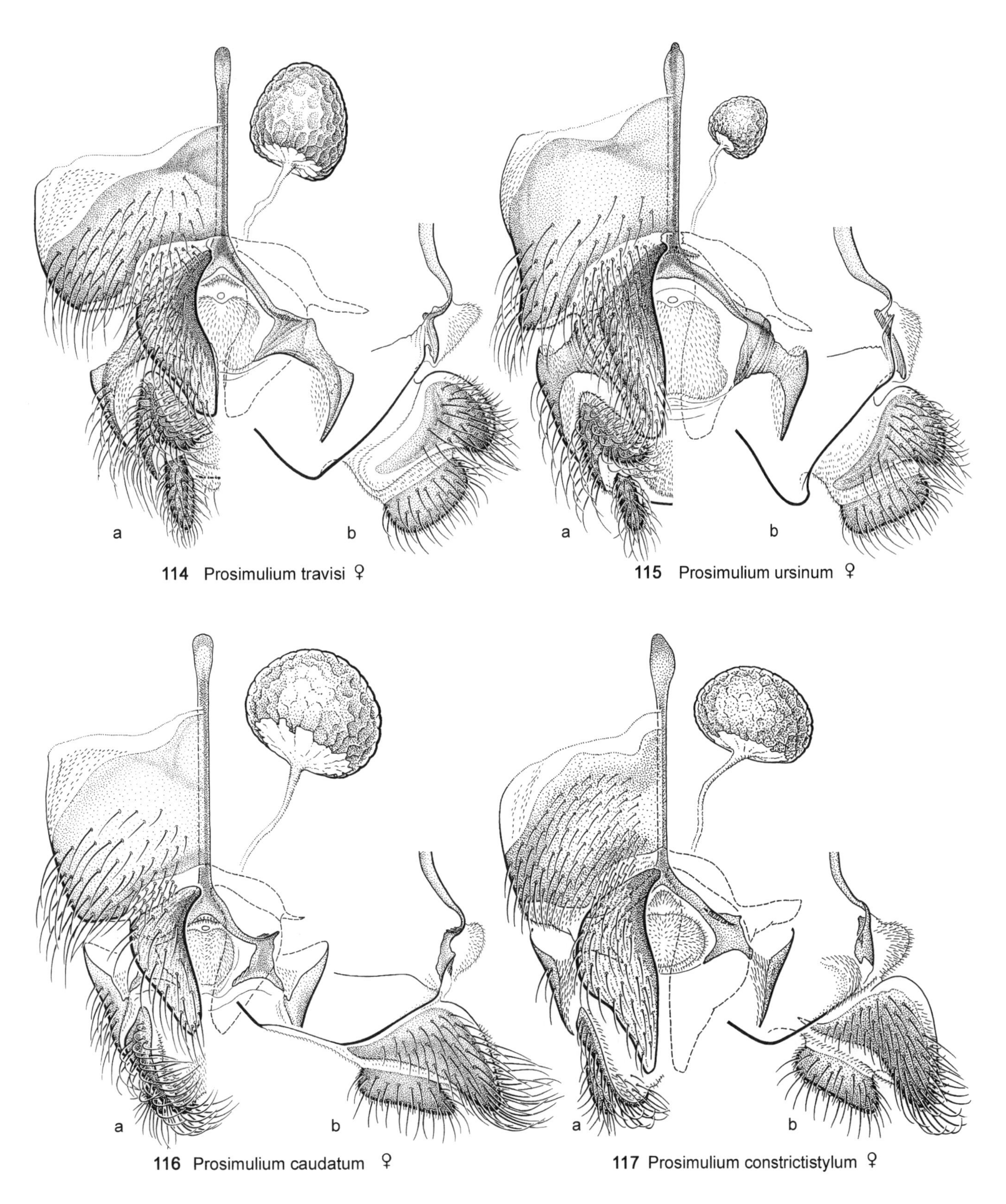


Figs. 10.114–10.117. Female terminalia; a, ventral view; b, right lateral view.

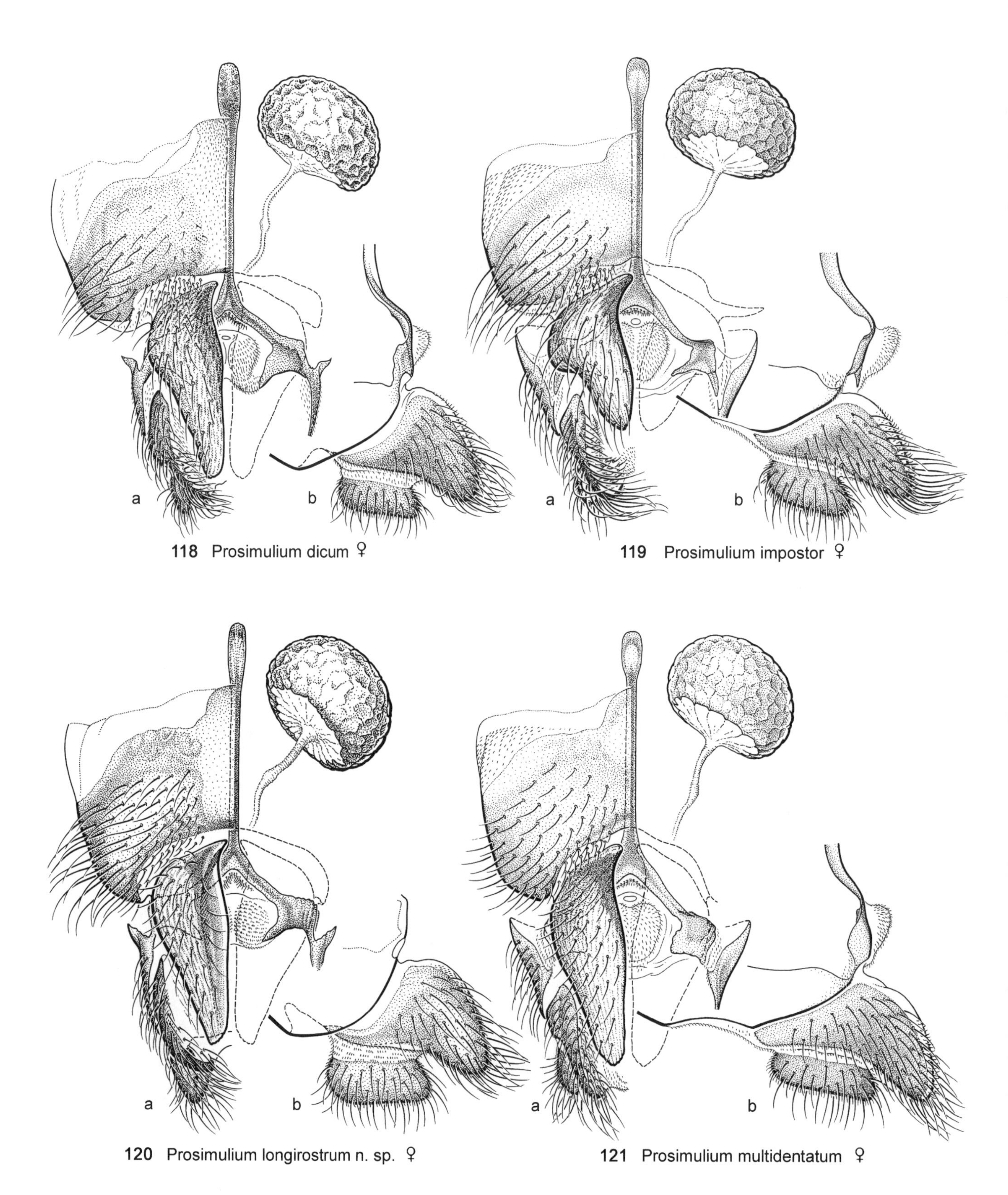


Figs. 10.118–10.121. Female terminalia; a, ventral view; b, right lateral view.

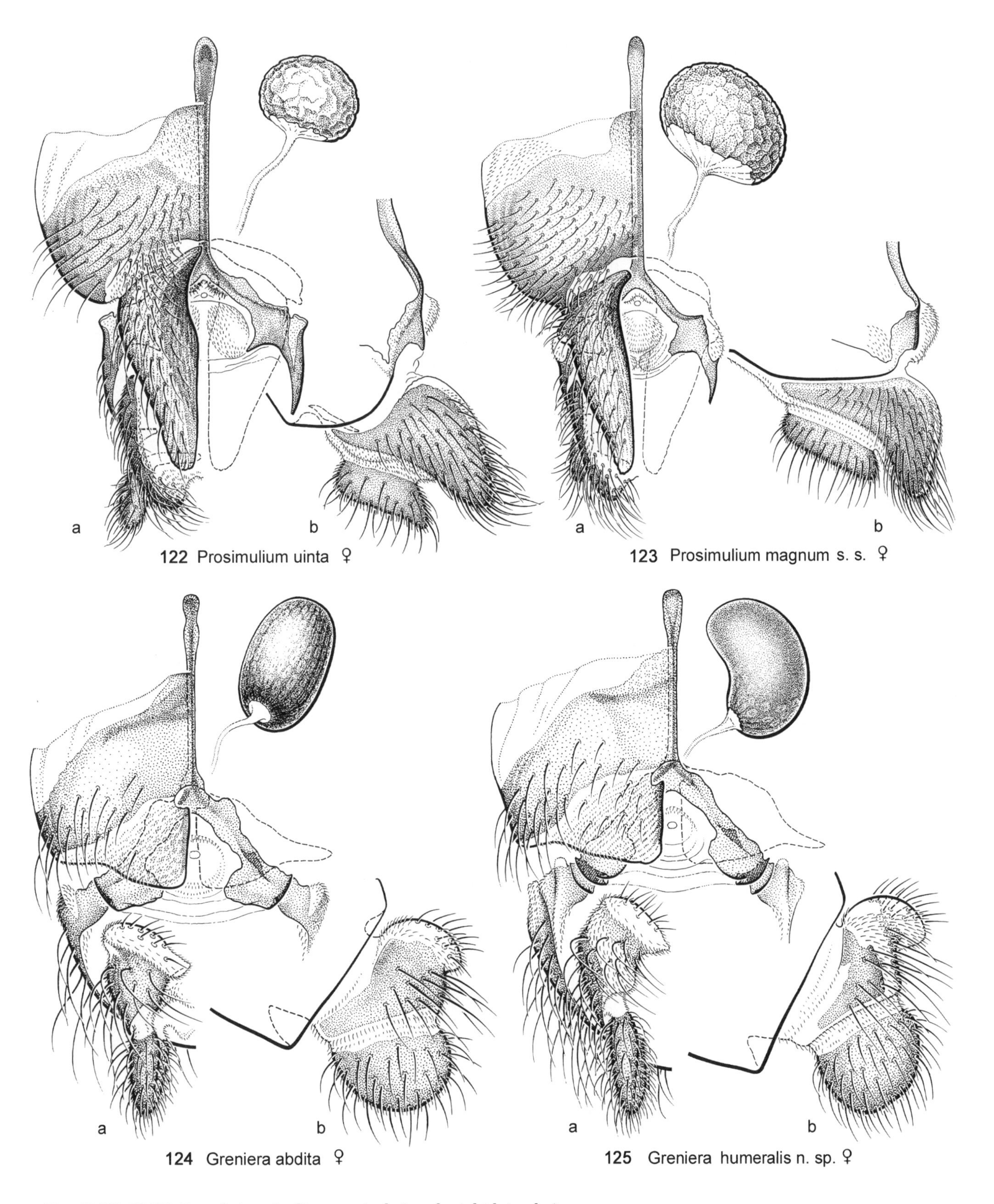


Figs. 10.122–10.125. Female terminalia; a, ventral view; b, right lateral view.

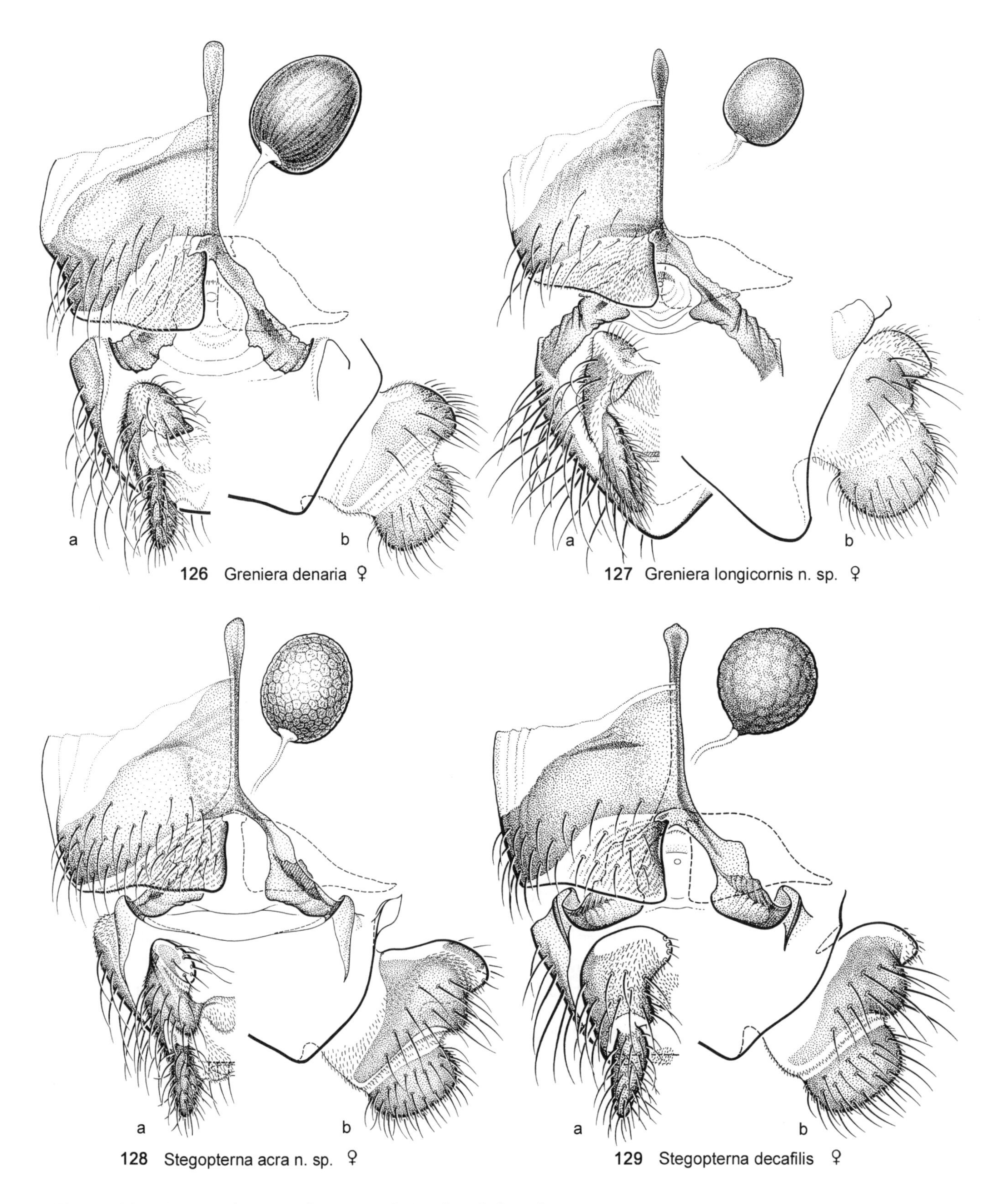


Figs. 10.126–10.129. Female terminalia; a, ventral view; b, right lateral view.

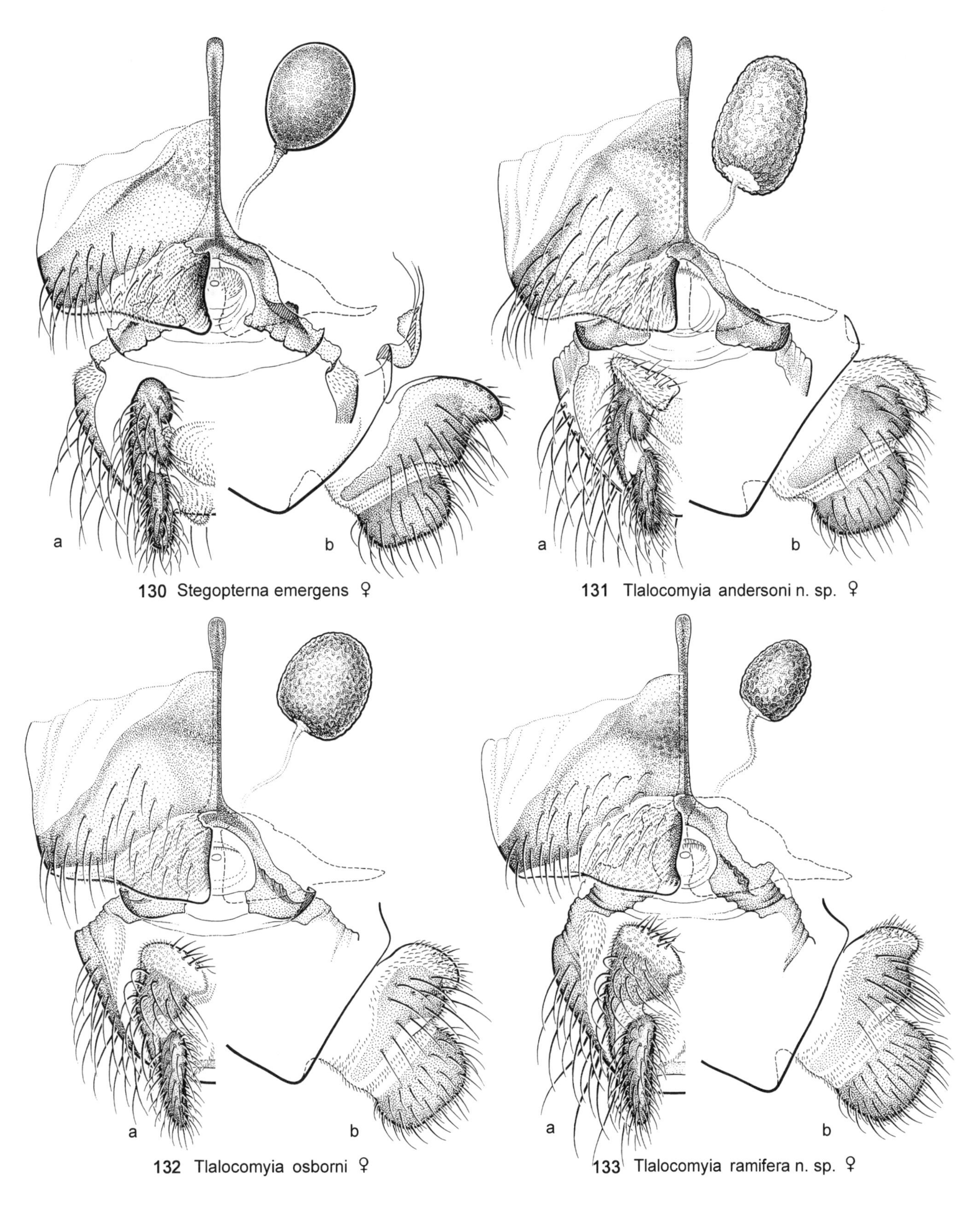


Figs. 10.130–10.133. Female terminalia; a, ventral view; b, right lateral view.

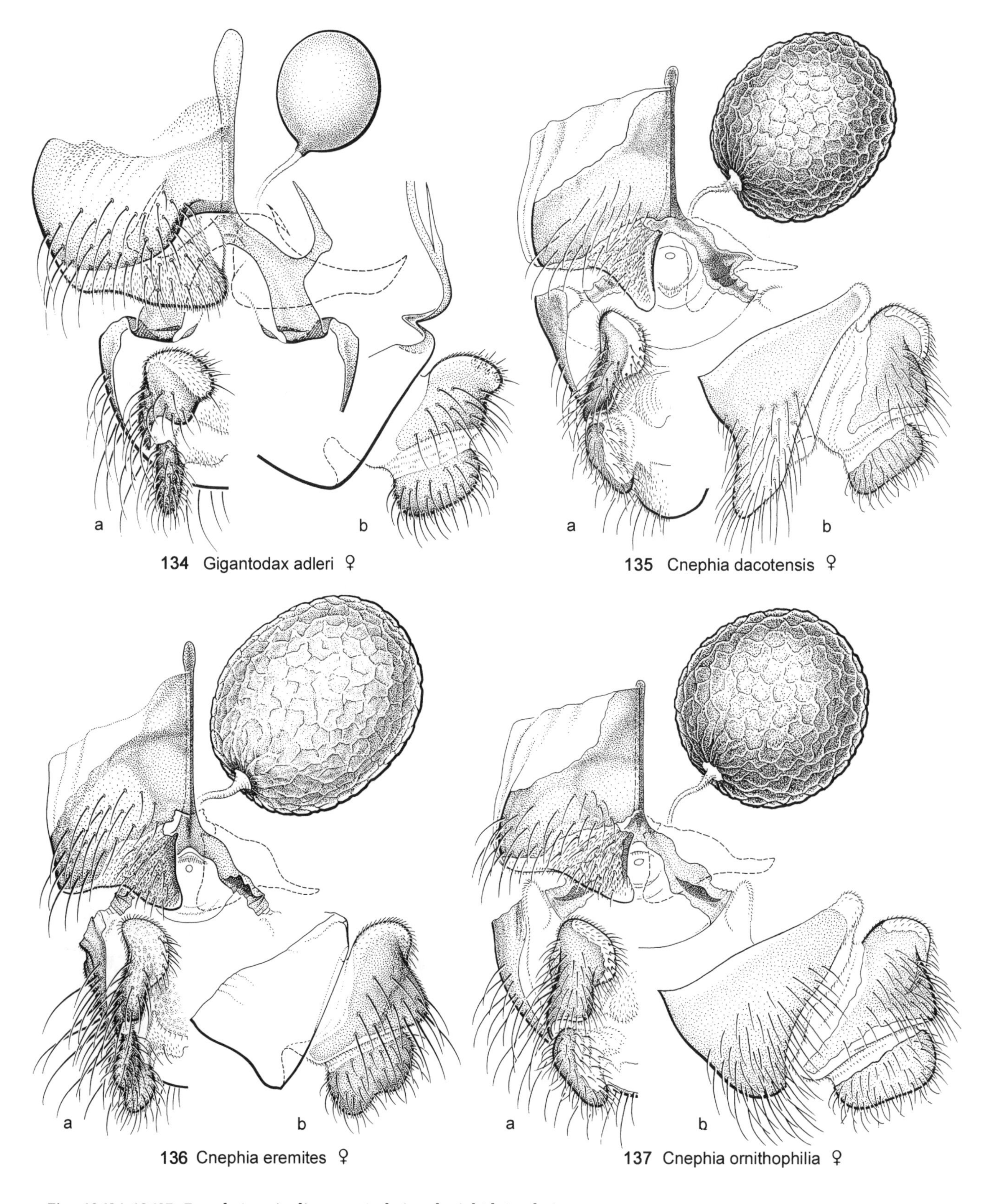


Figs. 10.134–10.137. Female terminalia; a, ventral view; b, right lateral view.

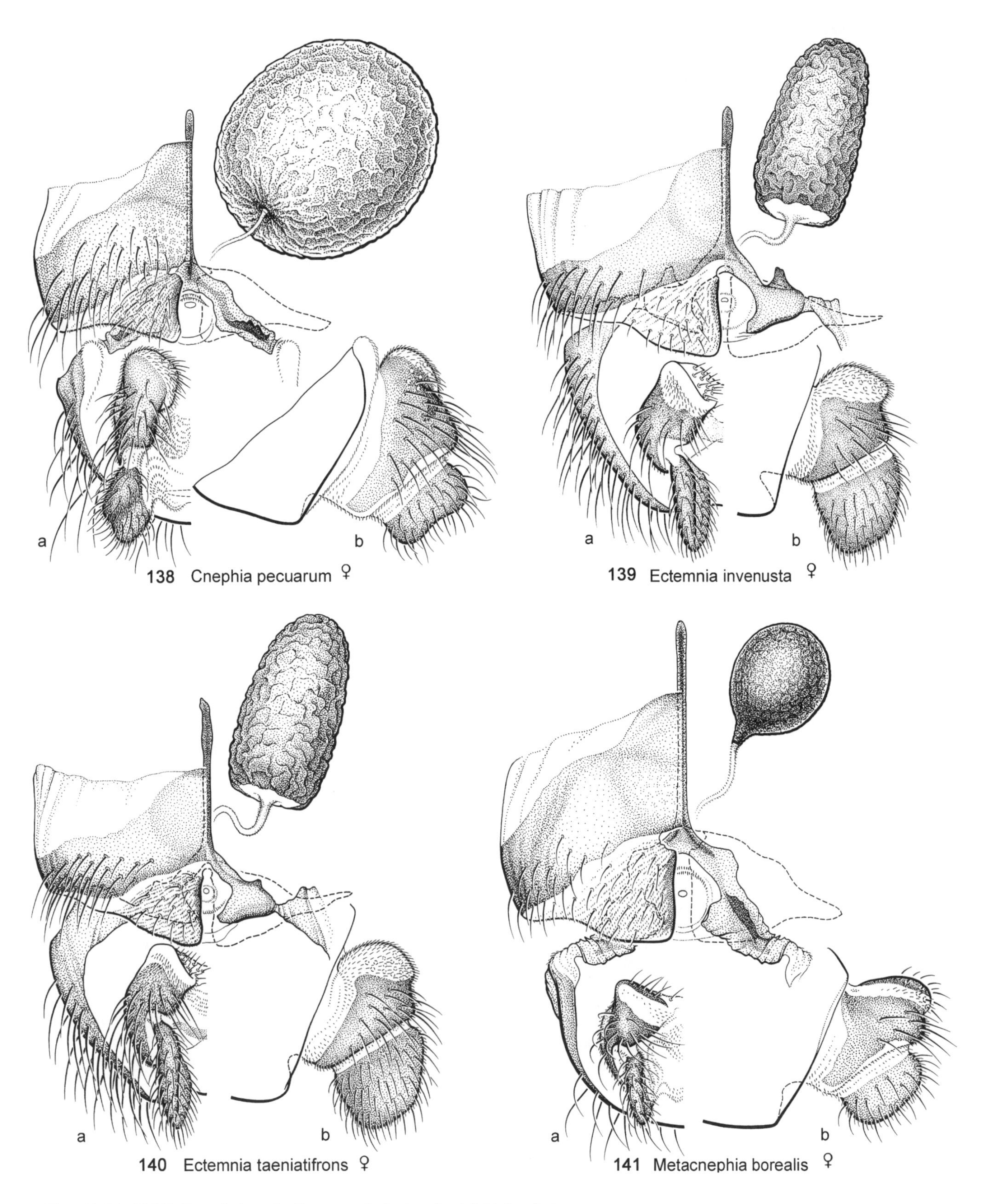


Figs. 10.138–10.141. Female terminalia; a, ventral view; b, right lateral view.

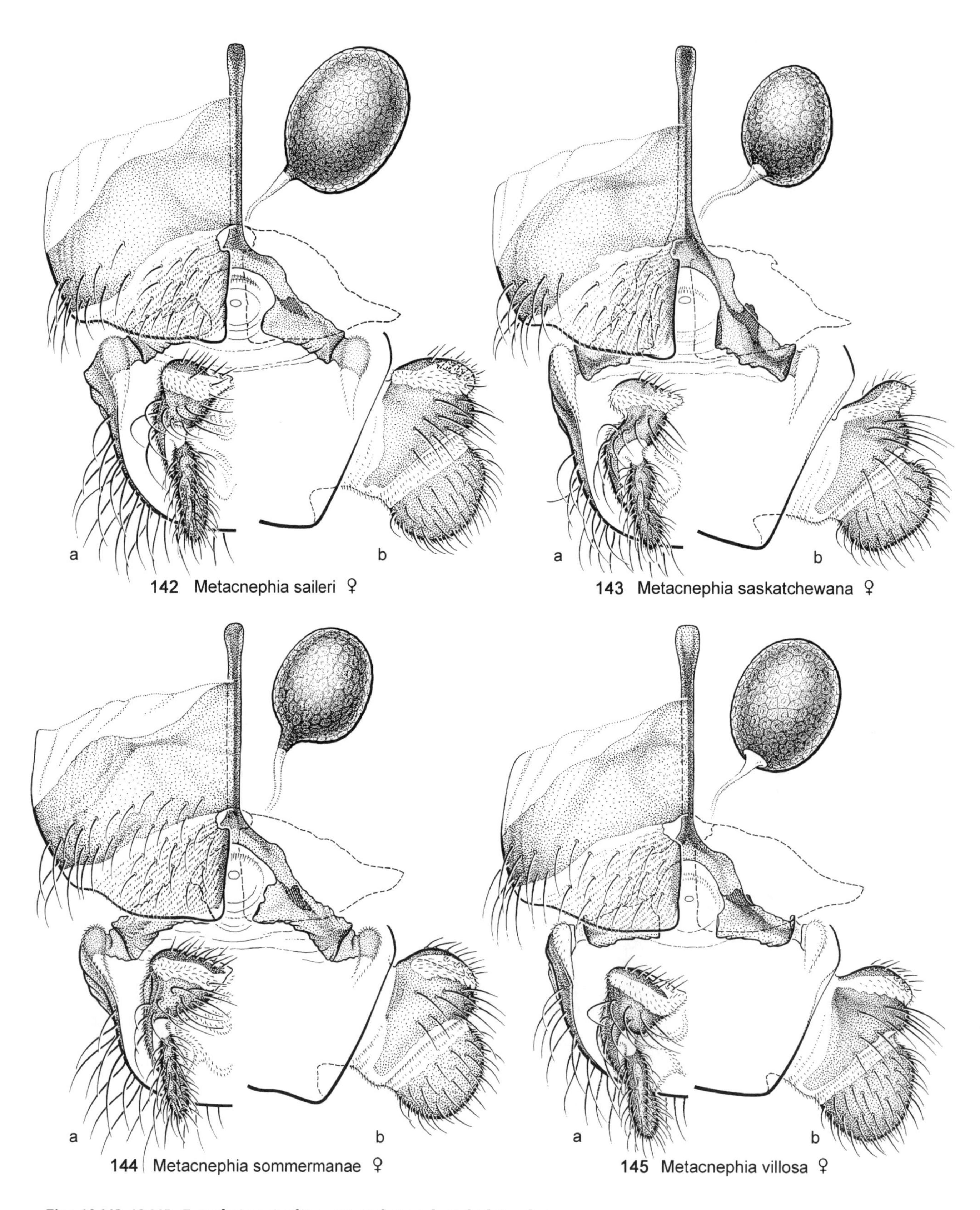


Figs. 10.142–10.145. Female terminalia; a, ventral view; b, right lateral view.

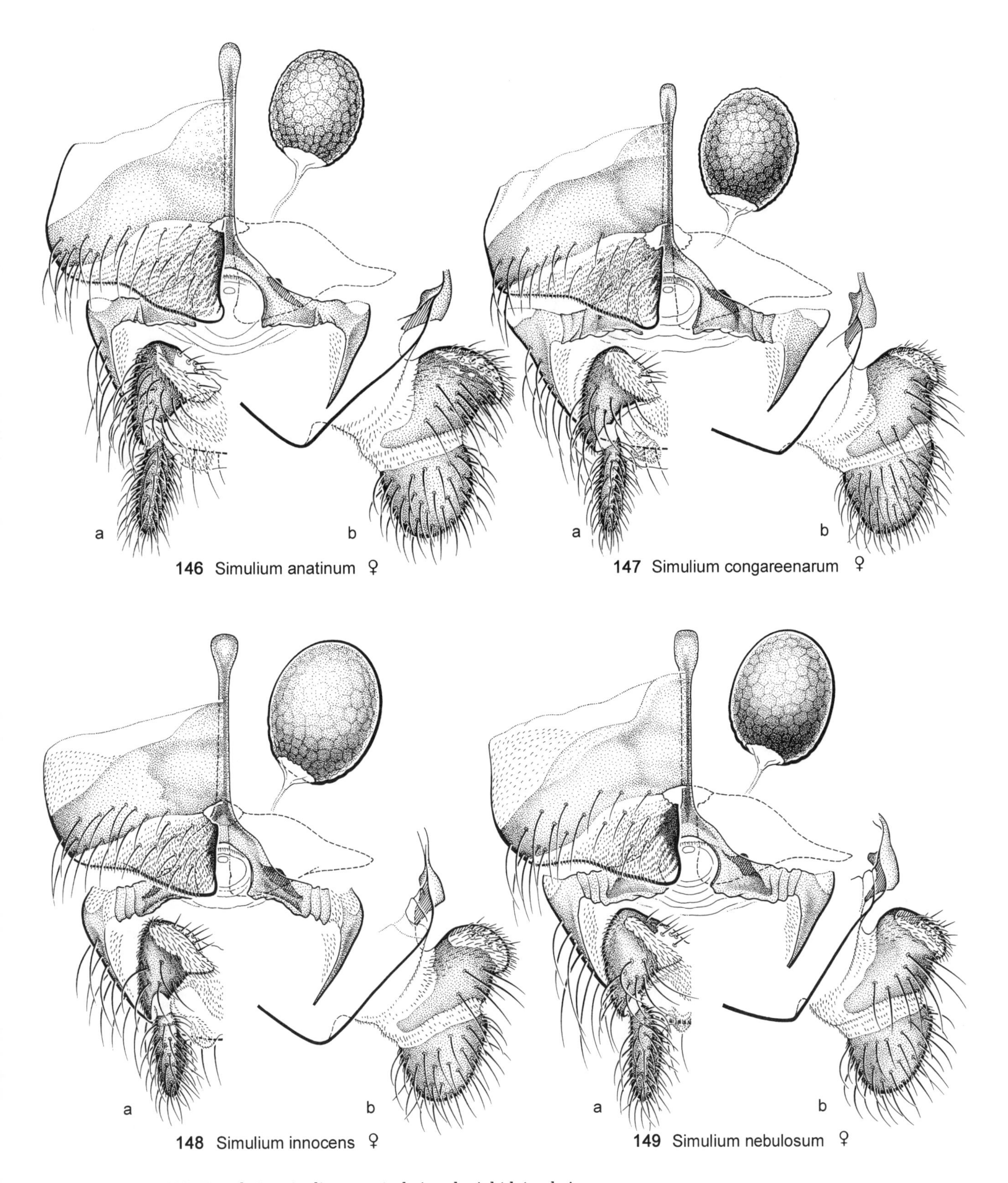


Figs. 10.146–10.149. Female terminalia; a, ventral view; b, right lateral view.

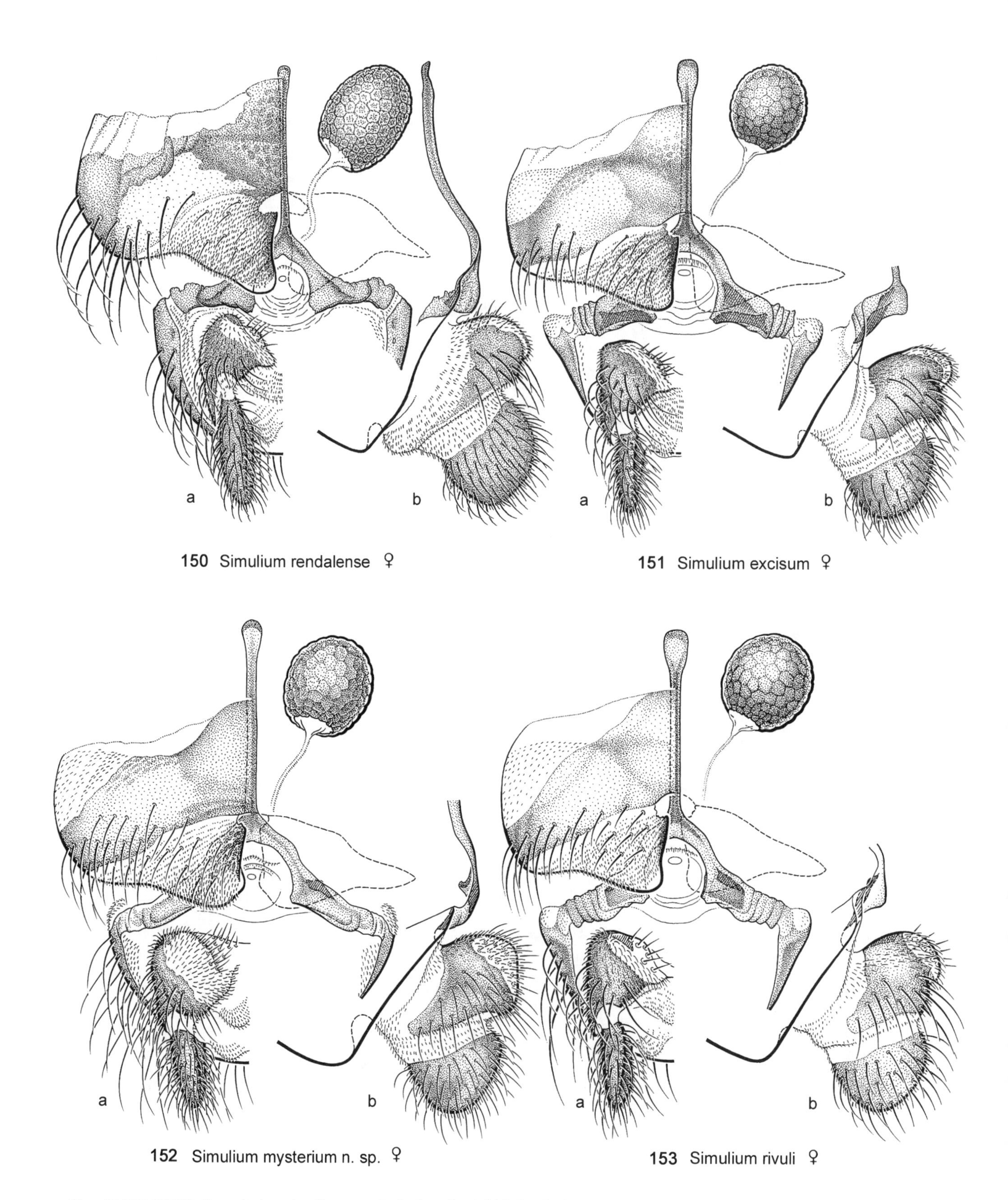


Figs. 10.150–10.153. Female terminalia; a, ventral view; b, right lateral view.

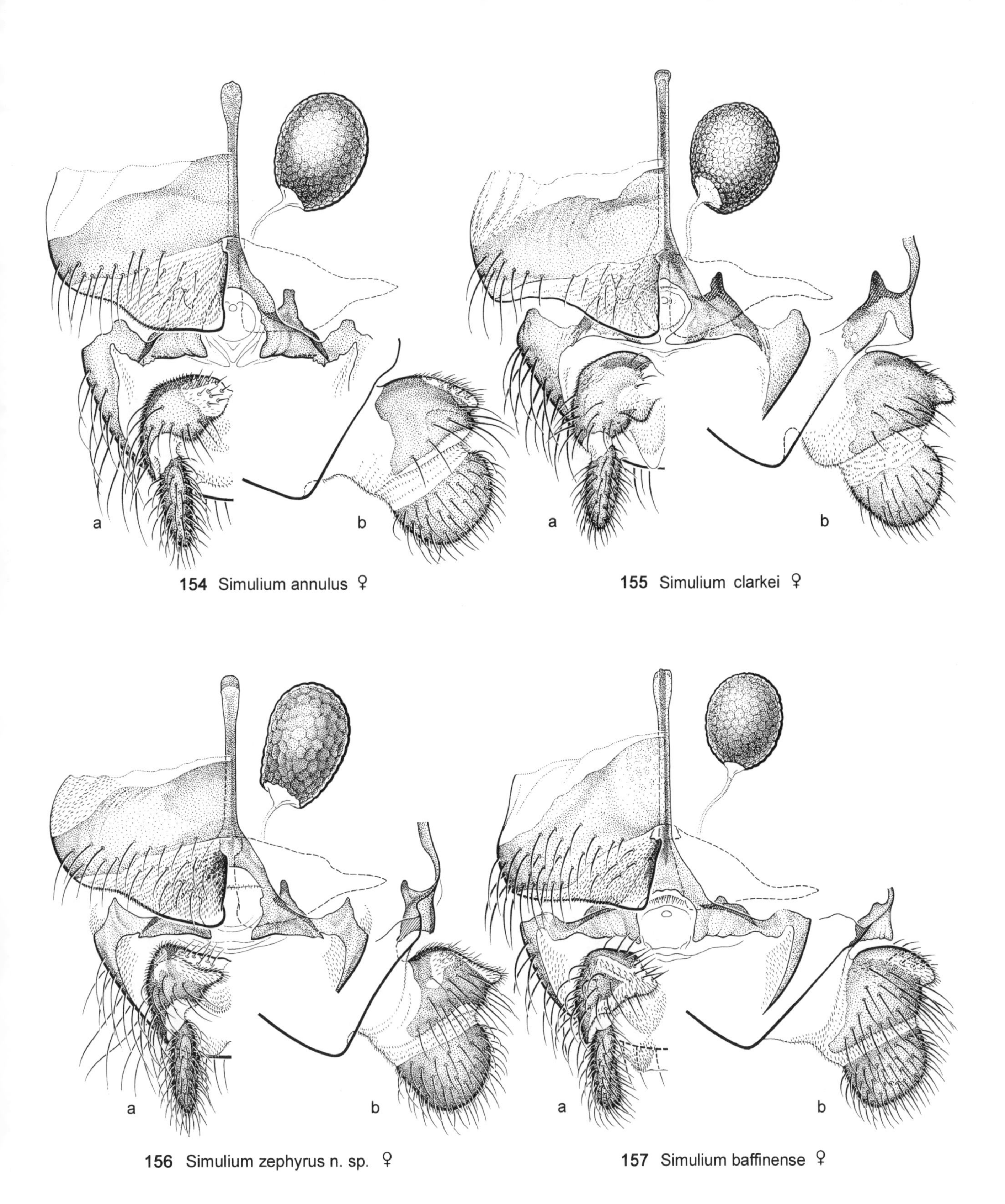


Figs. 10.154–10.157. Female terminalia; a, ventral view; b, right lateral view.

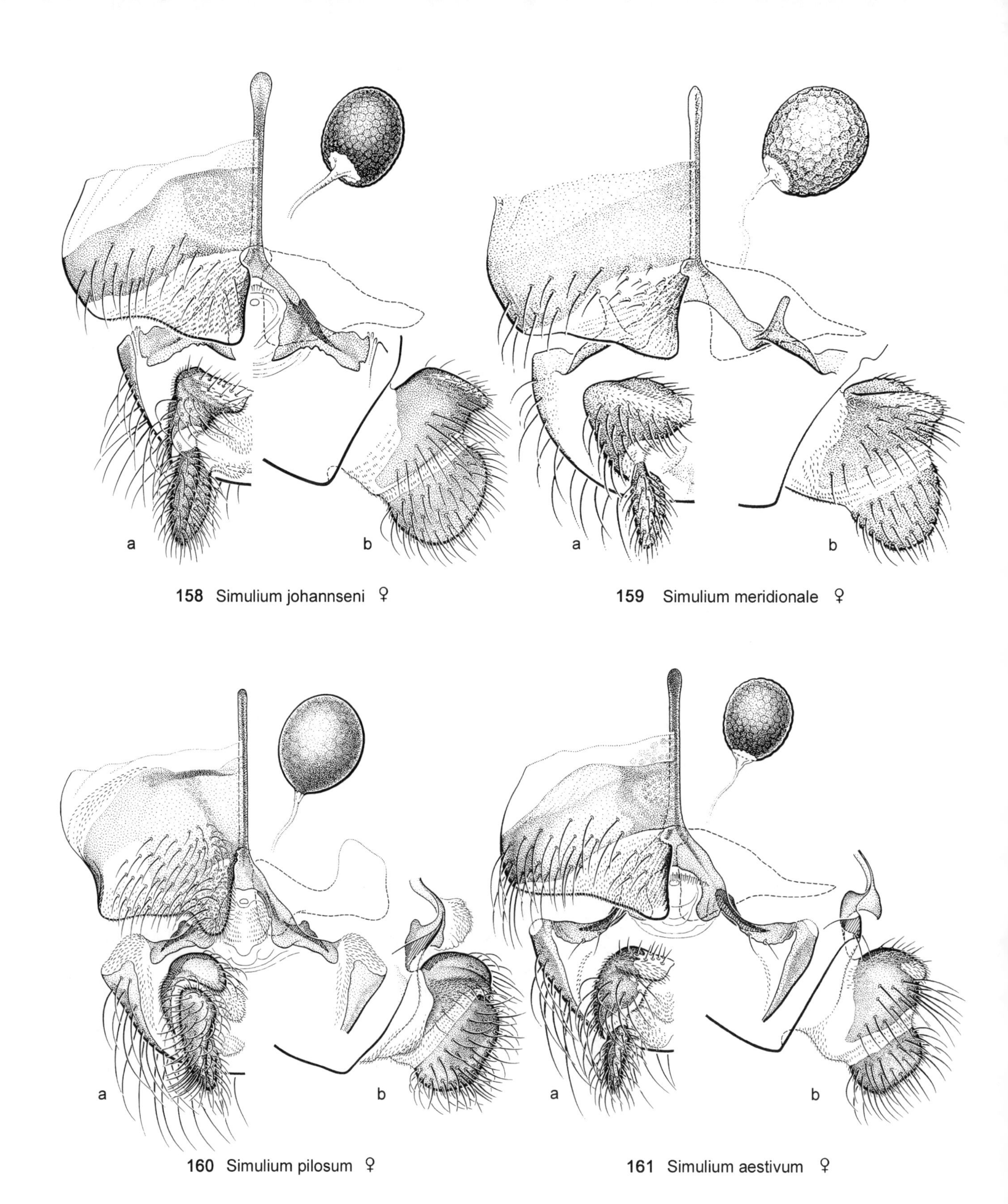


Figs. 10.158–10.161. Female terminalia; a, ventral view; b, right lateral view. (159) From Peterson (1981) by permission.

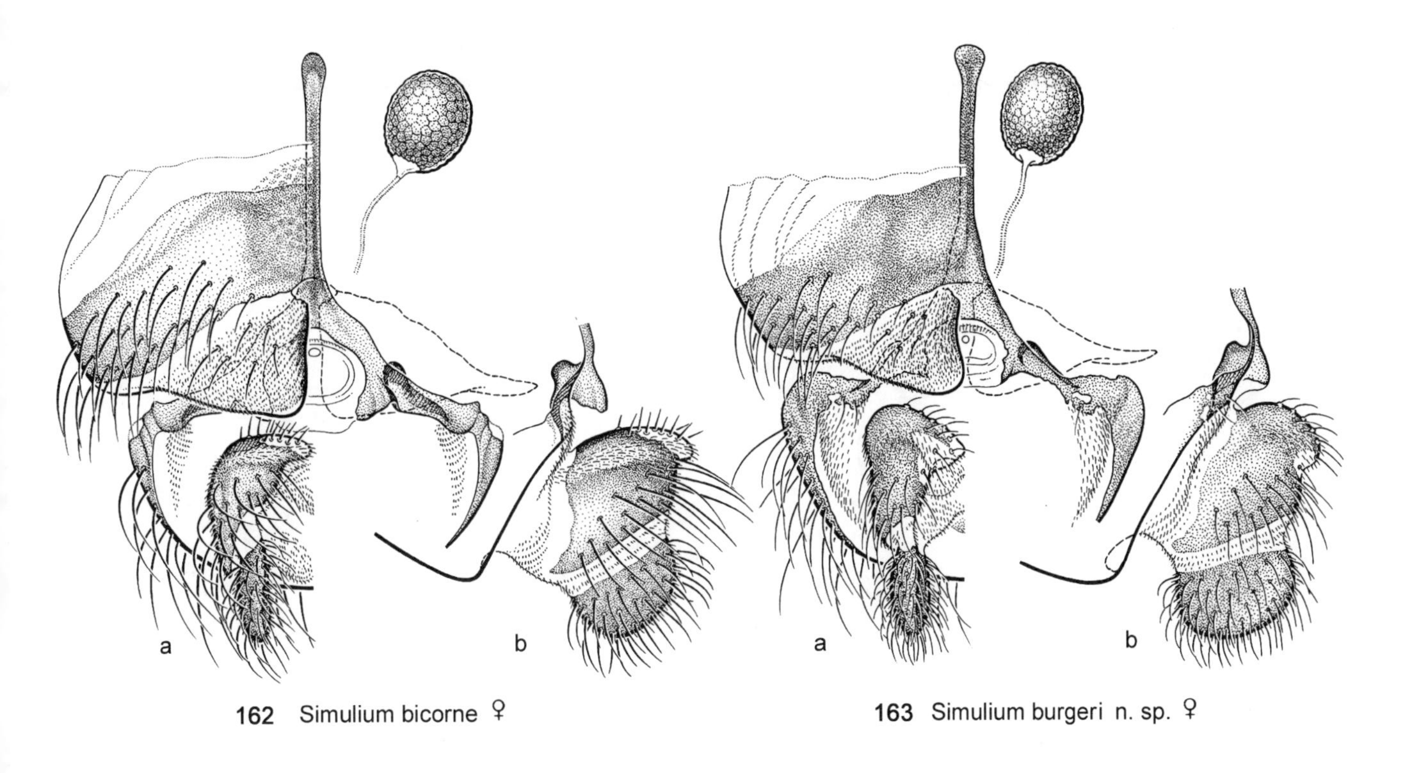


**162** Simulium bicorne ♀

**163** Simulium burgeri n. sp. ♀

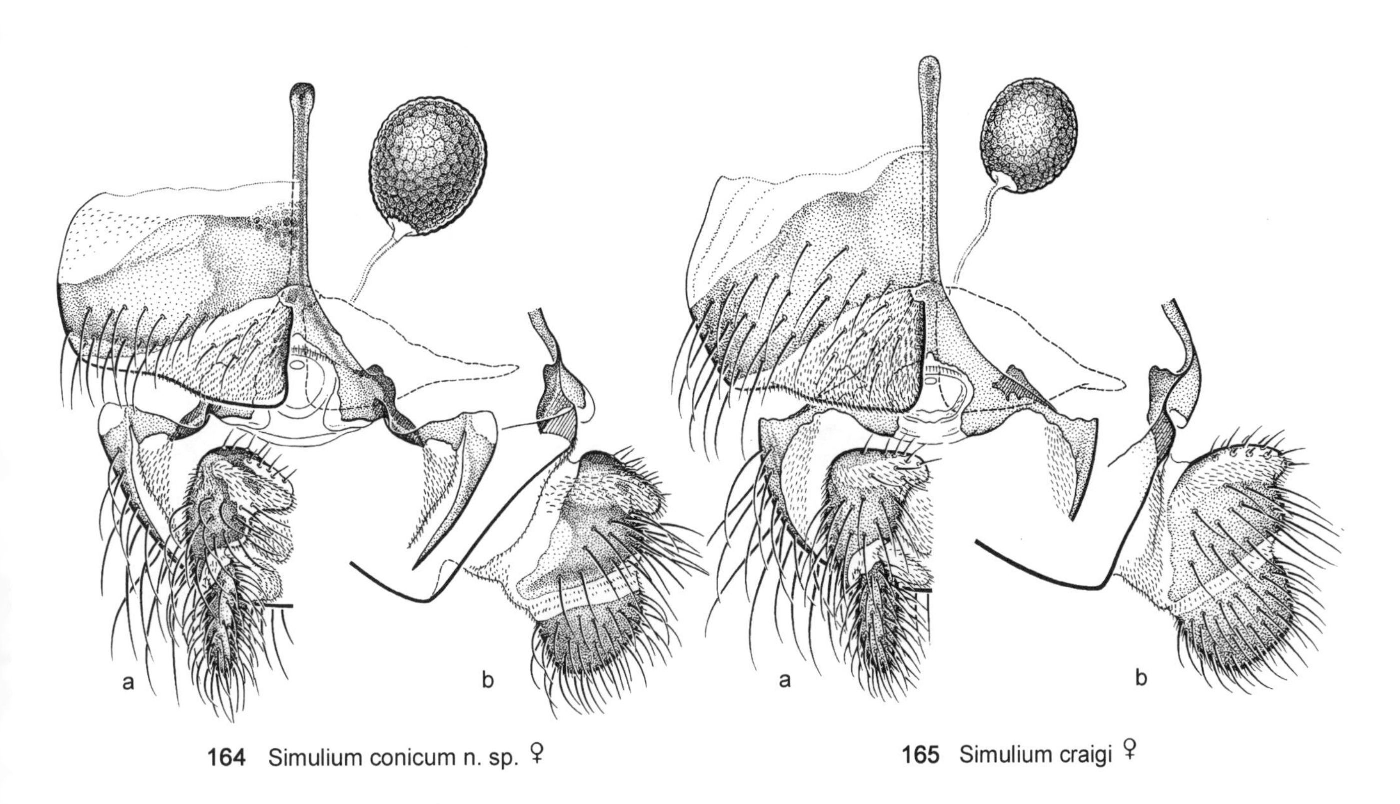


**164** Simulium conicum n. sp. ♀

**165** Simulium craigi ♀

Figs. 10.162–10.165. Female terminalia; a, ventral view; b, right lateral view.

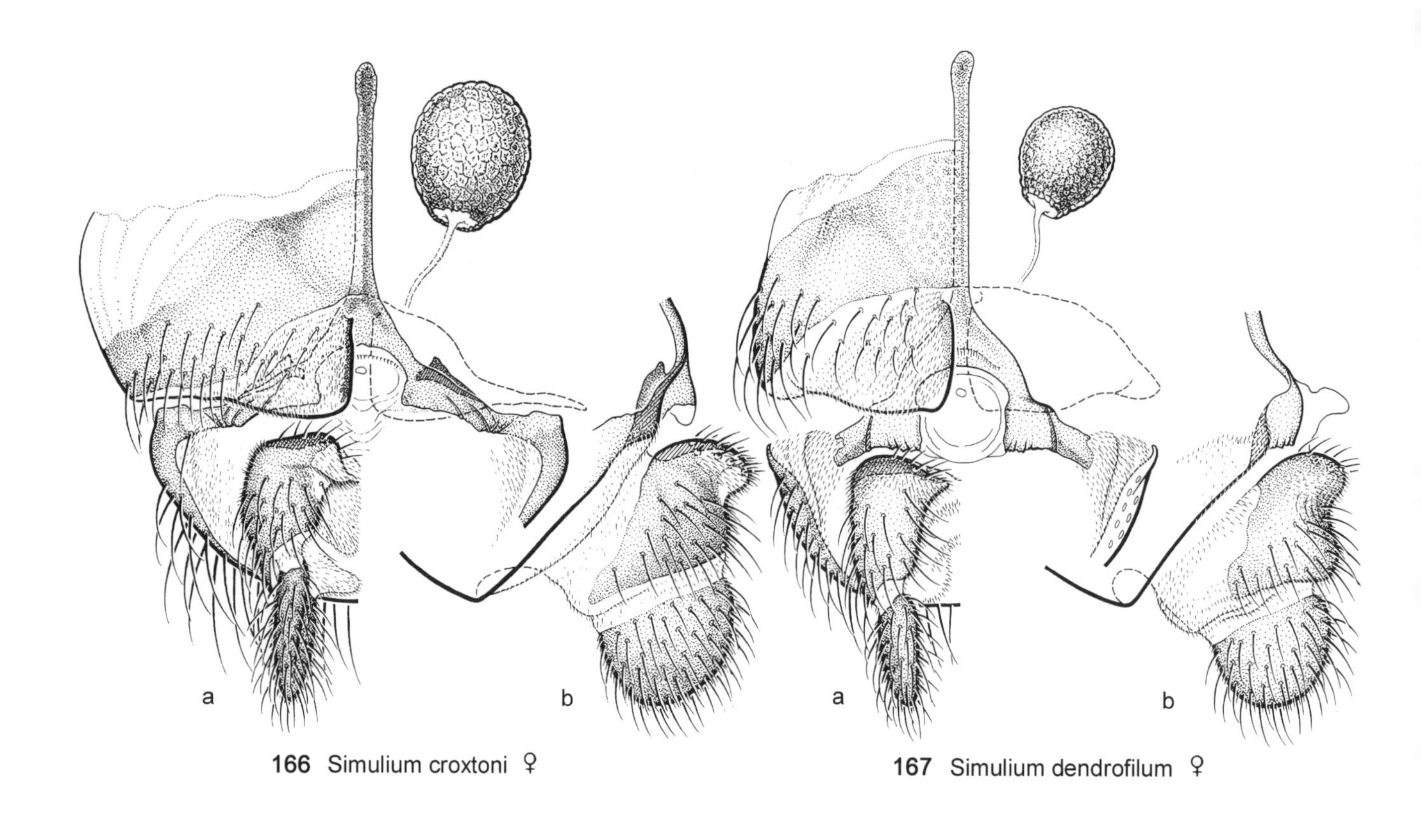


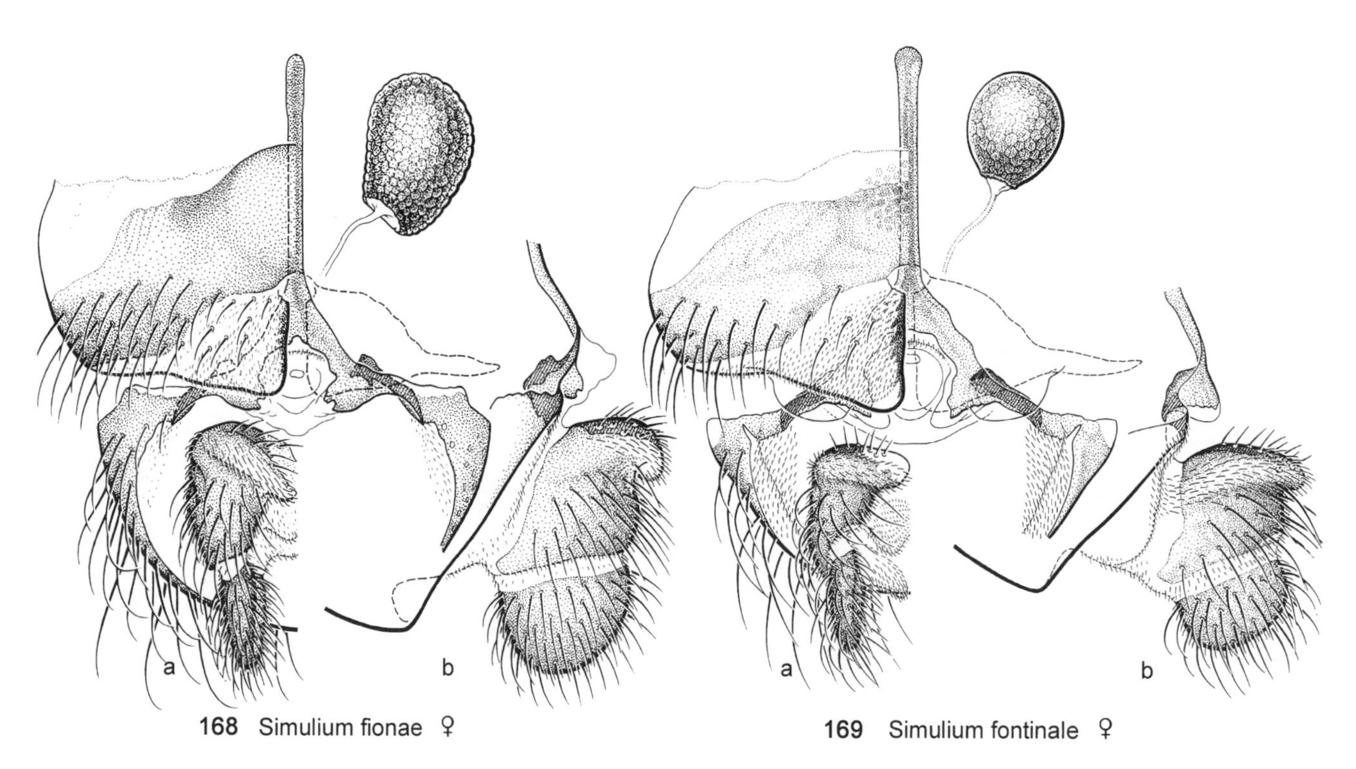


Figs. 10.166–10.169. Female terminalia; a, ventral view; b, right lateral view.

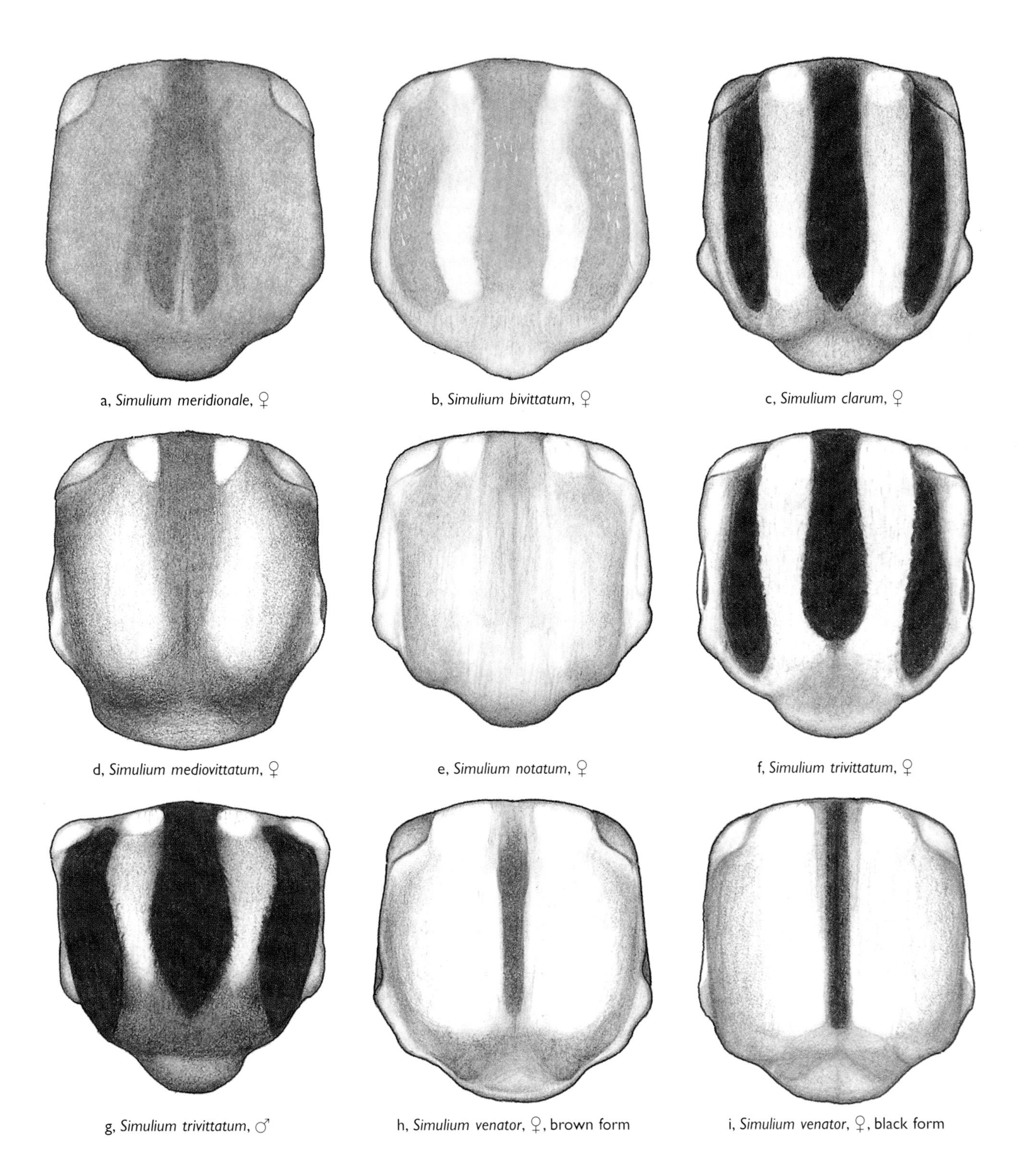


PLATE 1. Adult scutal patterns, dorsal view.

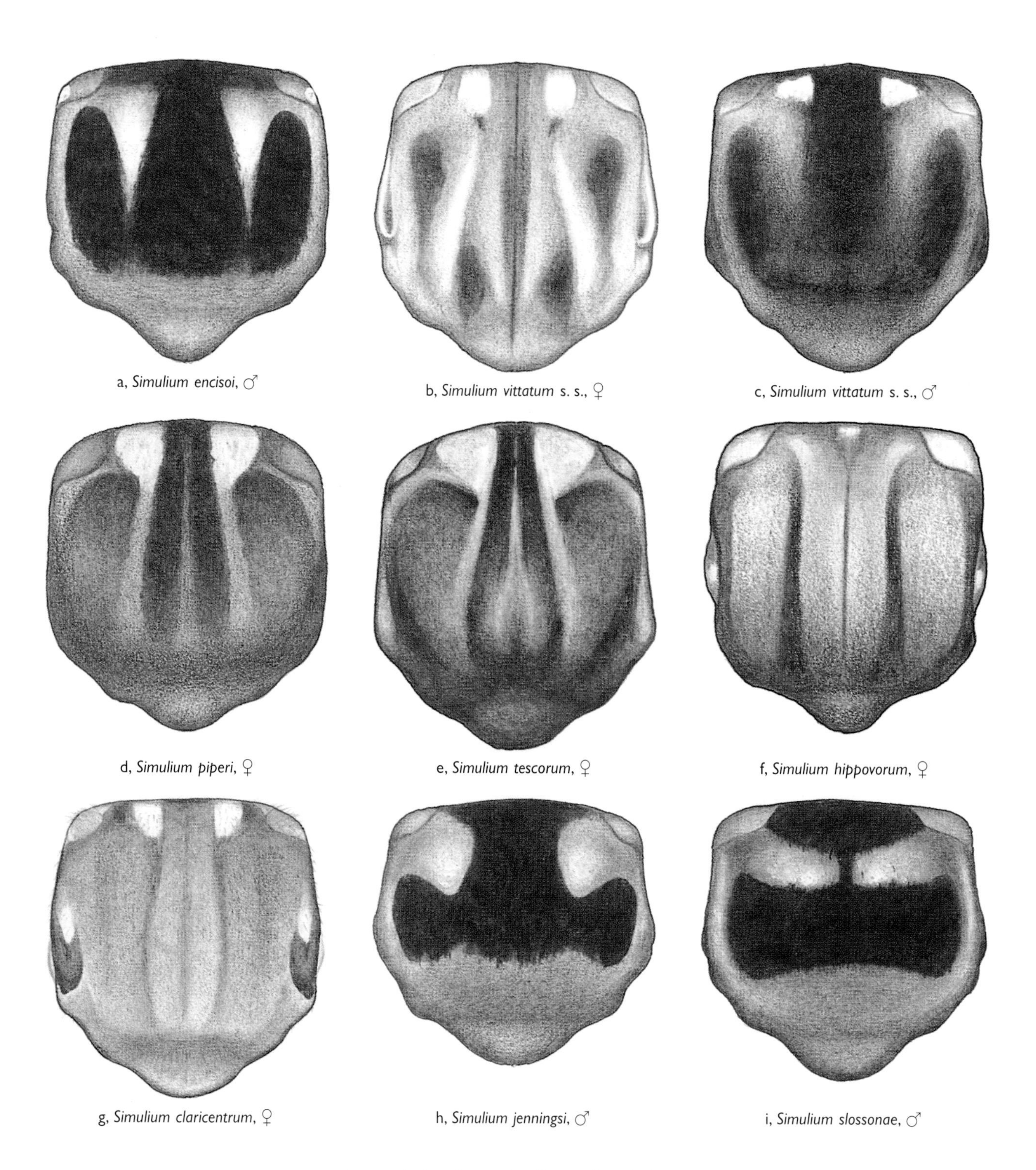


PLATE 2. Adult scutal patterns, dorsal view.

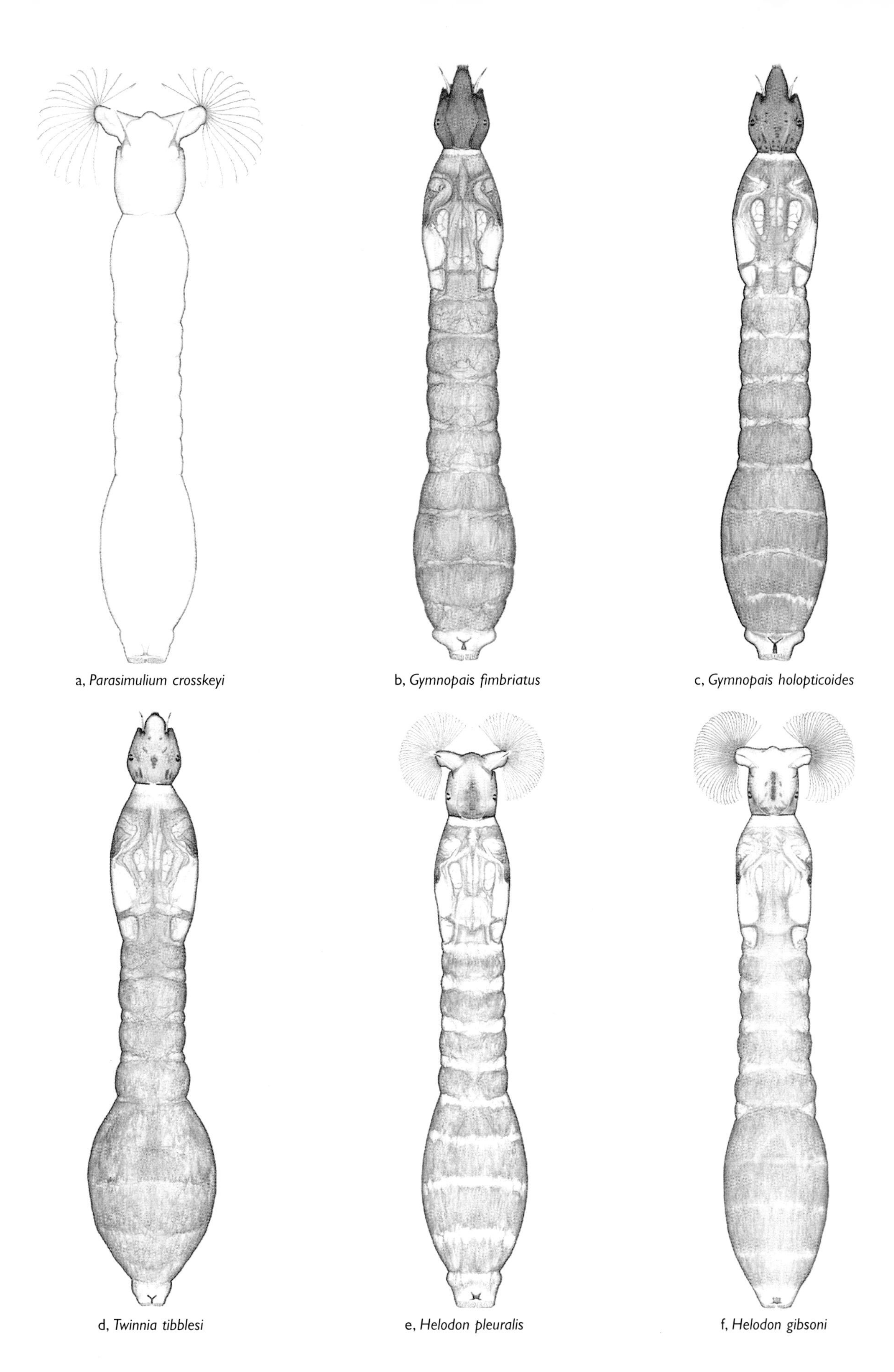

a, *Parasimulium crosskeyi* b, *Gymnopais fimbriatus* c, *Gymnopais holopticoides*

d, *Twinnia tibblesi* e, *Helodon pleuralis* f, *Helodon gibsoni*

PLATE 3. Larvae, dorsal view.

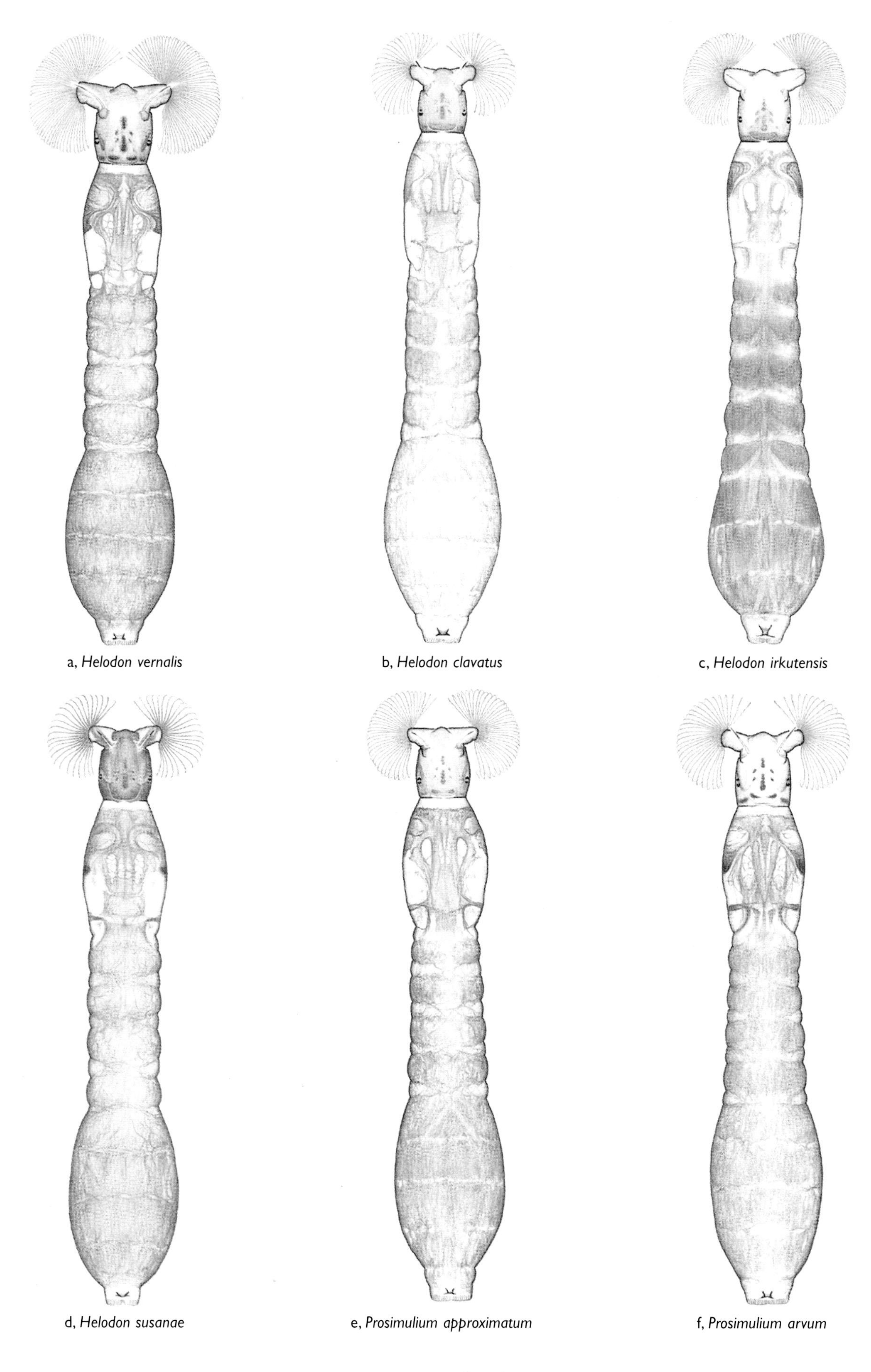


PLATE 4. Larvae, dorsal view.

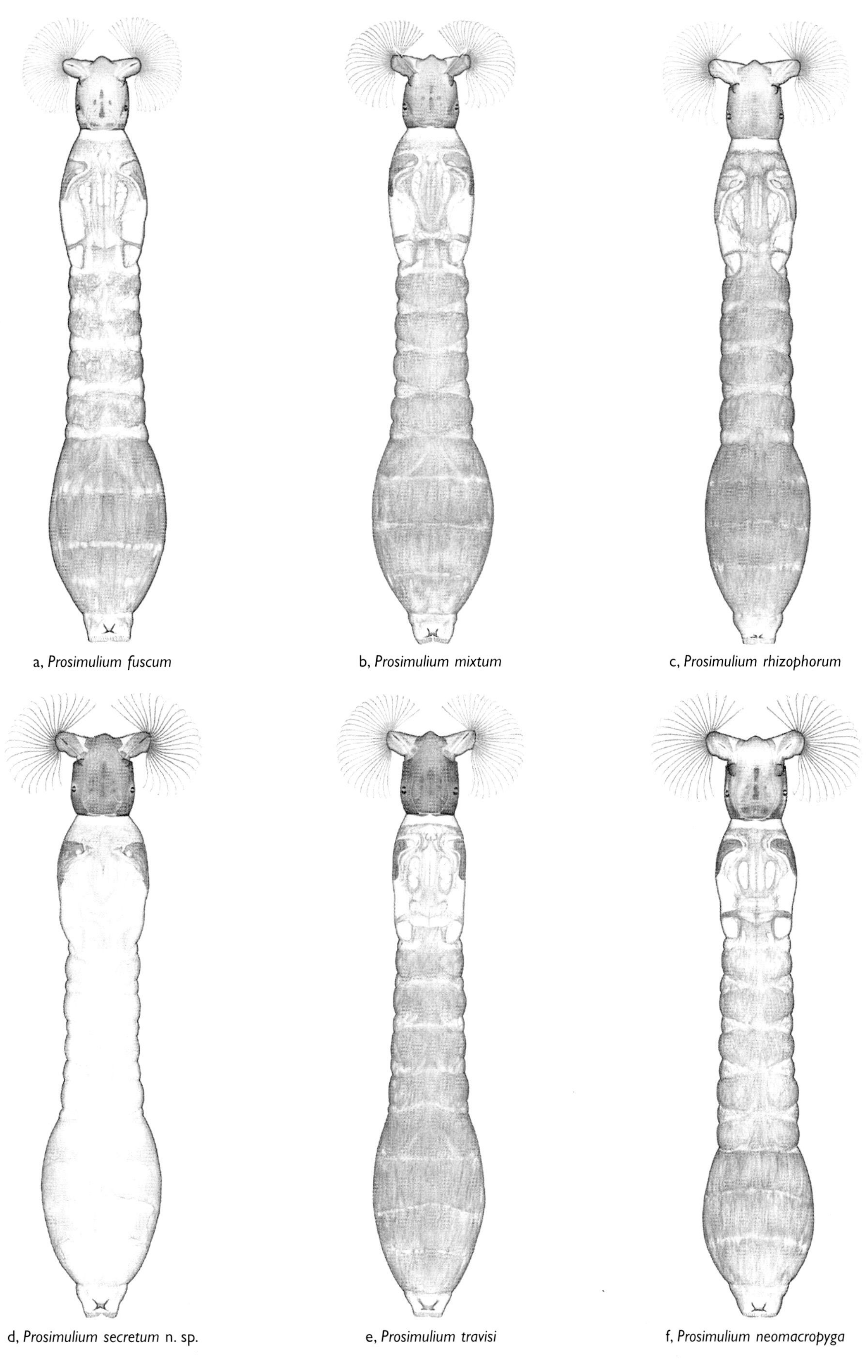

a, *Prosimulium fuscum* b, *Prosimulium mixtum* c, *Prosimulium rhizophorum*

d, *Prosimulium secretum* n. sp. e, *Prosimulium travisi* f, *Prosimulium neomacropyga*

PLATE 5. Larvae, dorsal view.

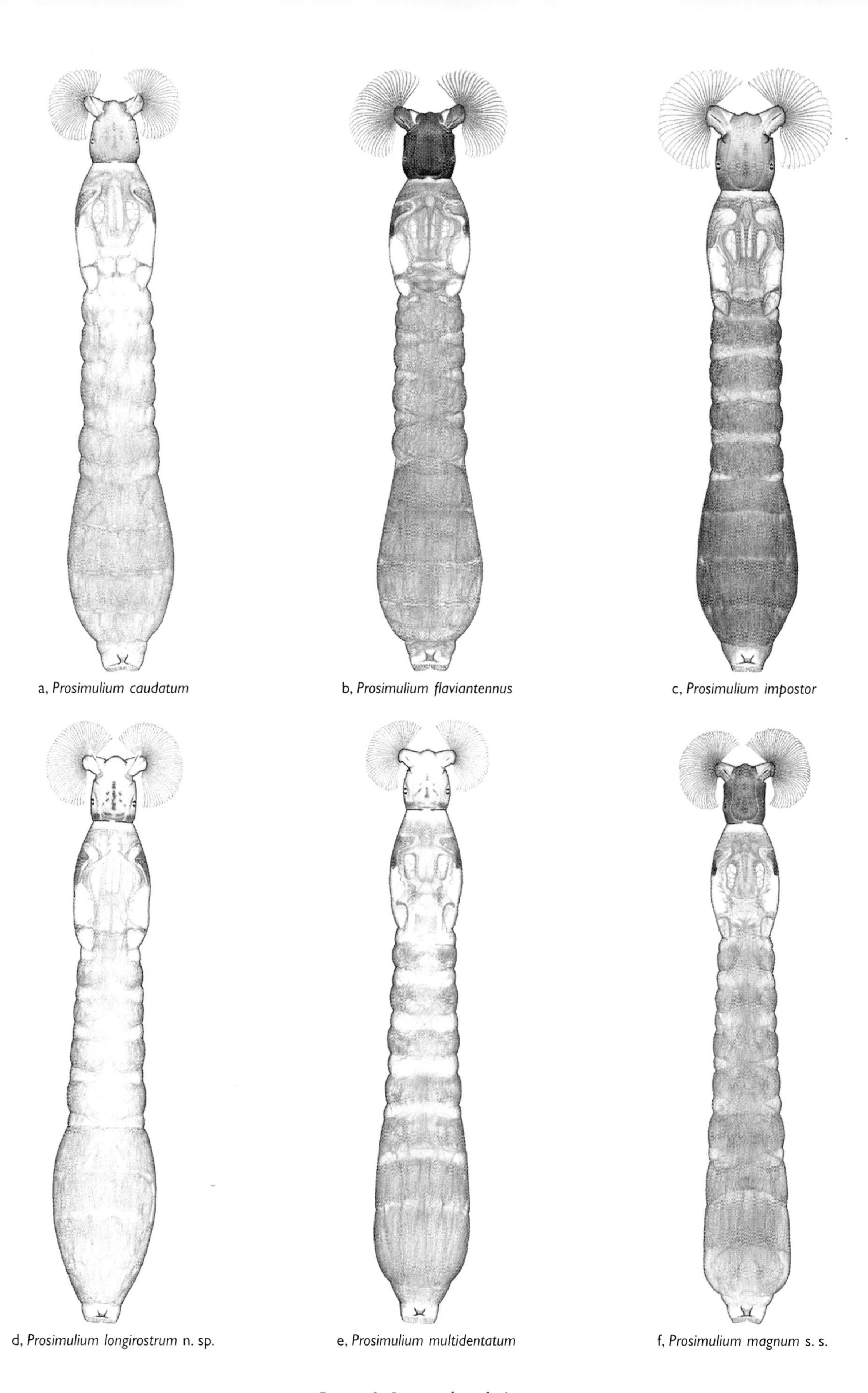


PLATE 6. Larvae, dorsal view.

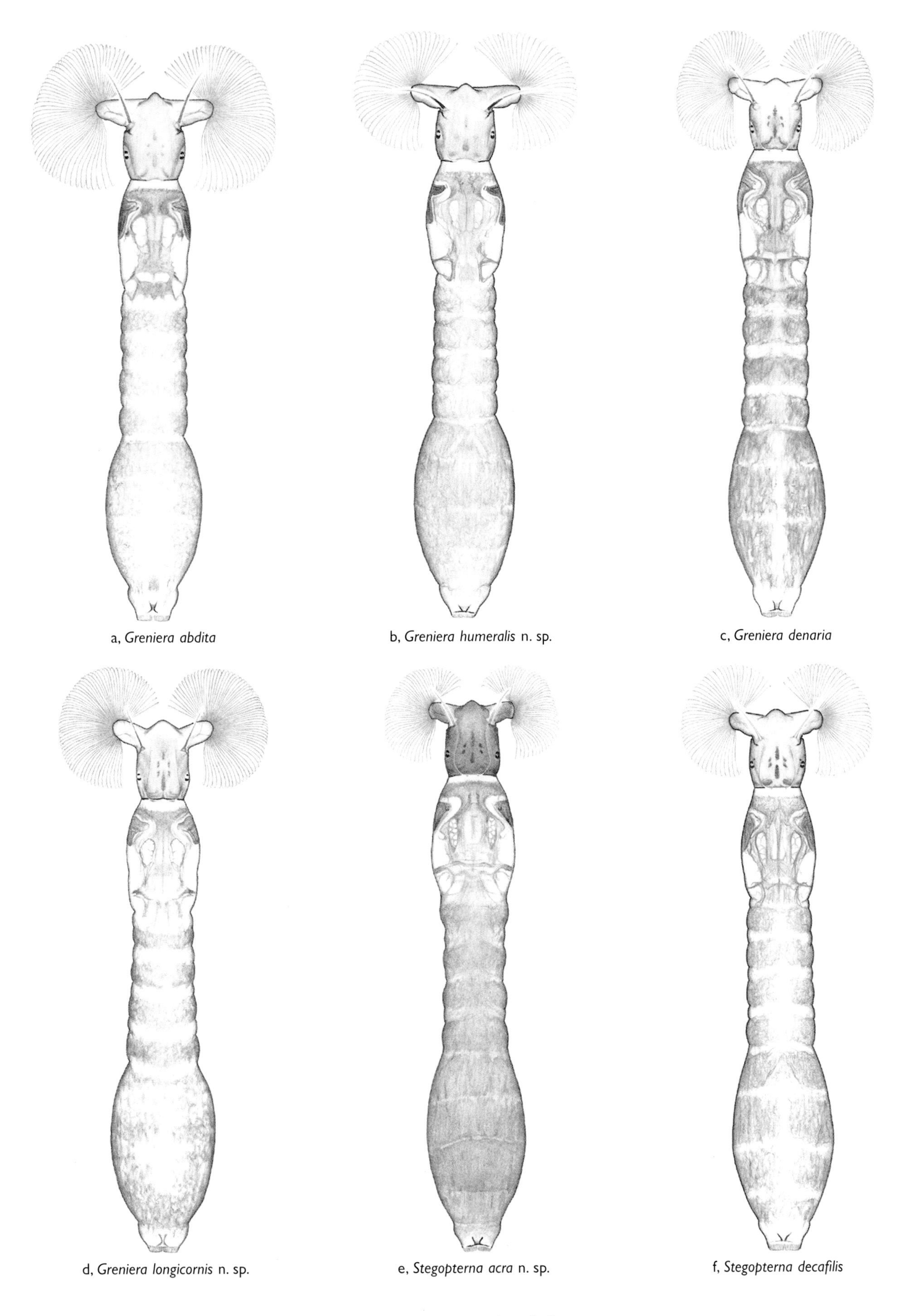

a, *Greniera abdita* b, *Greniera humeralis* n. sp. c, *Greniera denaria*

d, *Greniera longicornis* n. sp. e, *Stegopterna acra* n. sp. f, *Stegopterna decafilis*

PLATE 7. Larvae, dorsal view.

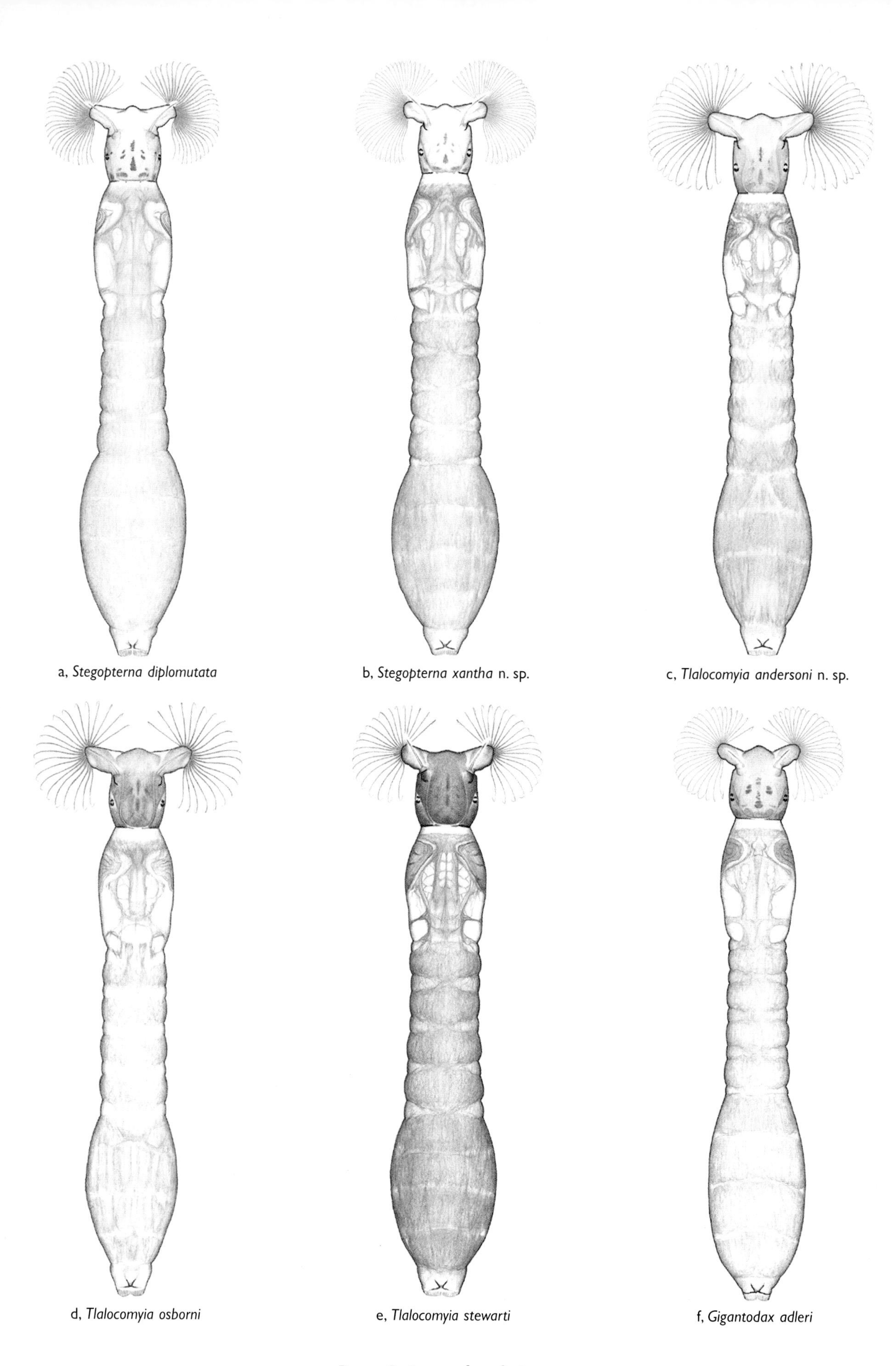

a, *Stegopterna diplomutata*

b, *Stegopterna xantha* n. sp.

c, *Tlalocomyia andersoni* n. sp.

d, *Tlalocomyia osborni*

e, *Tlalocomyia stewarti*

f, *Gigantodax adleri*

PLATE 8. Larvae, dorsal view.

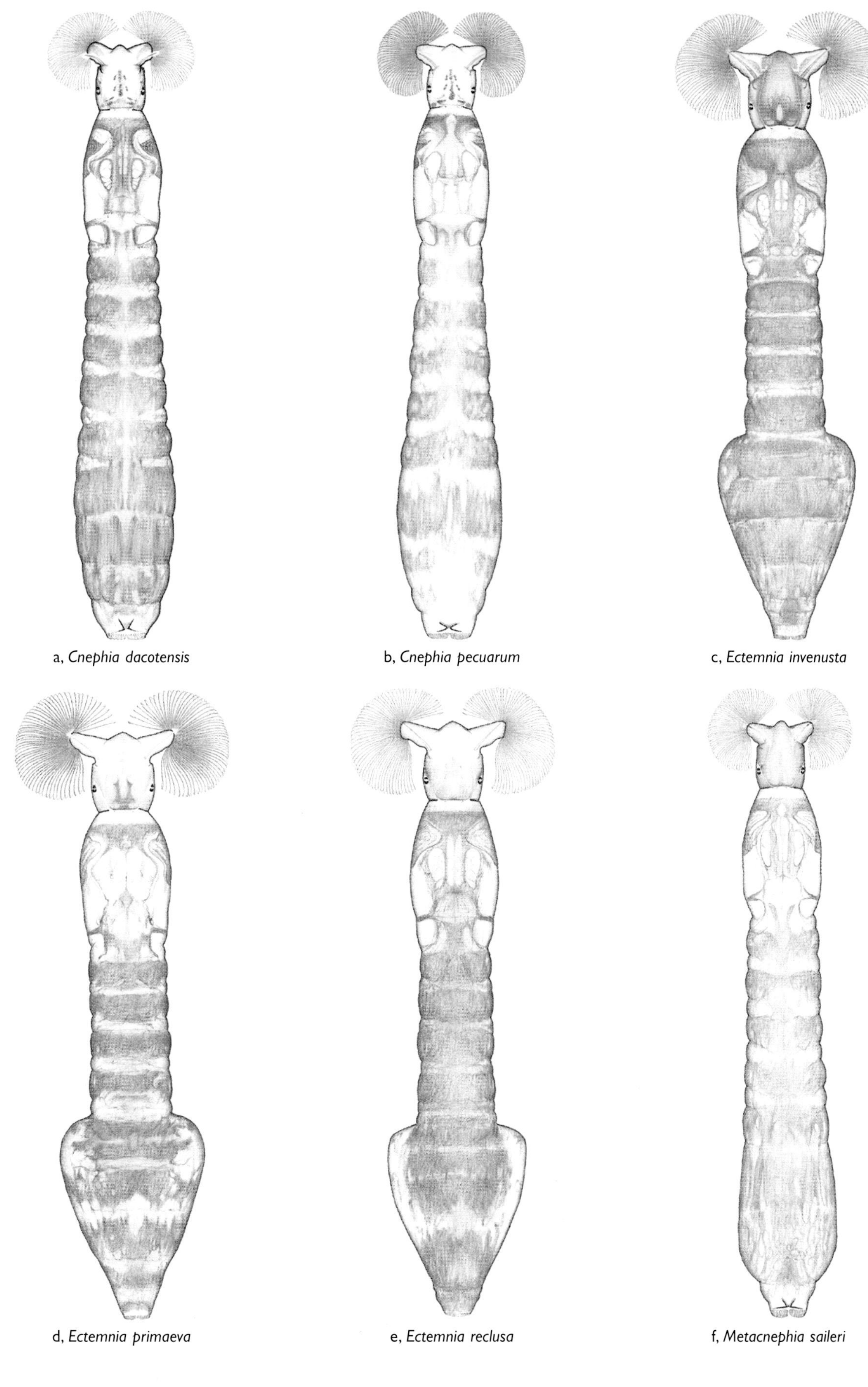

a, *Cnephia dacotensis* b, *Cnephia pecuarum* c, *Ectemnia invenusta*

d, *Ectemnia primaeva* e, *Ectemnia reclusa* f, *Metacnephia saileri*

PLATE 9. Larvae, dorsal view.

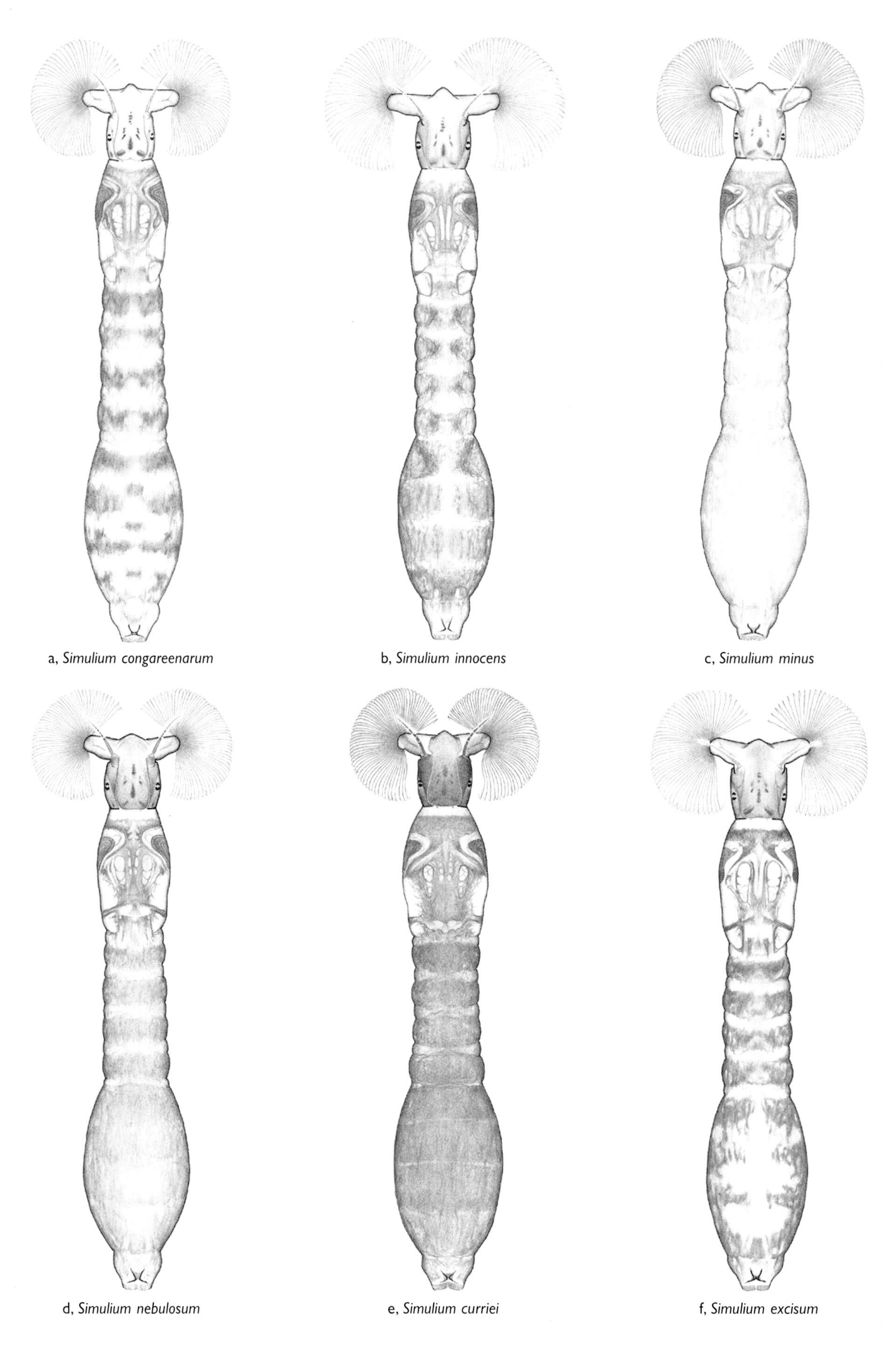


PLATE 10. Larvae, dorsal view.

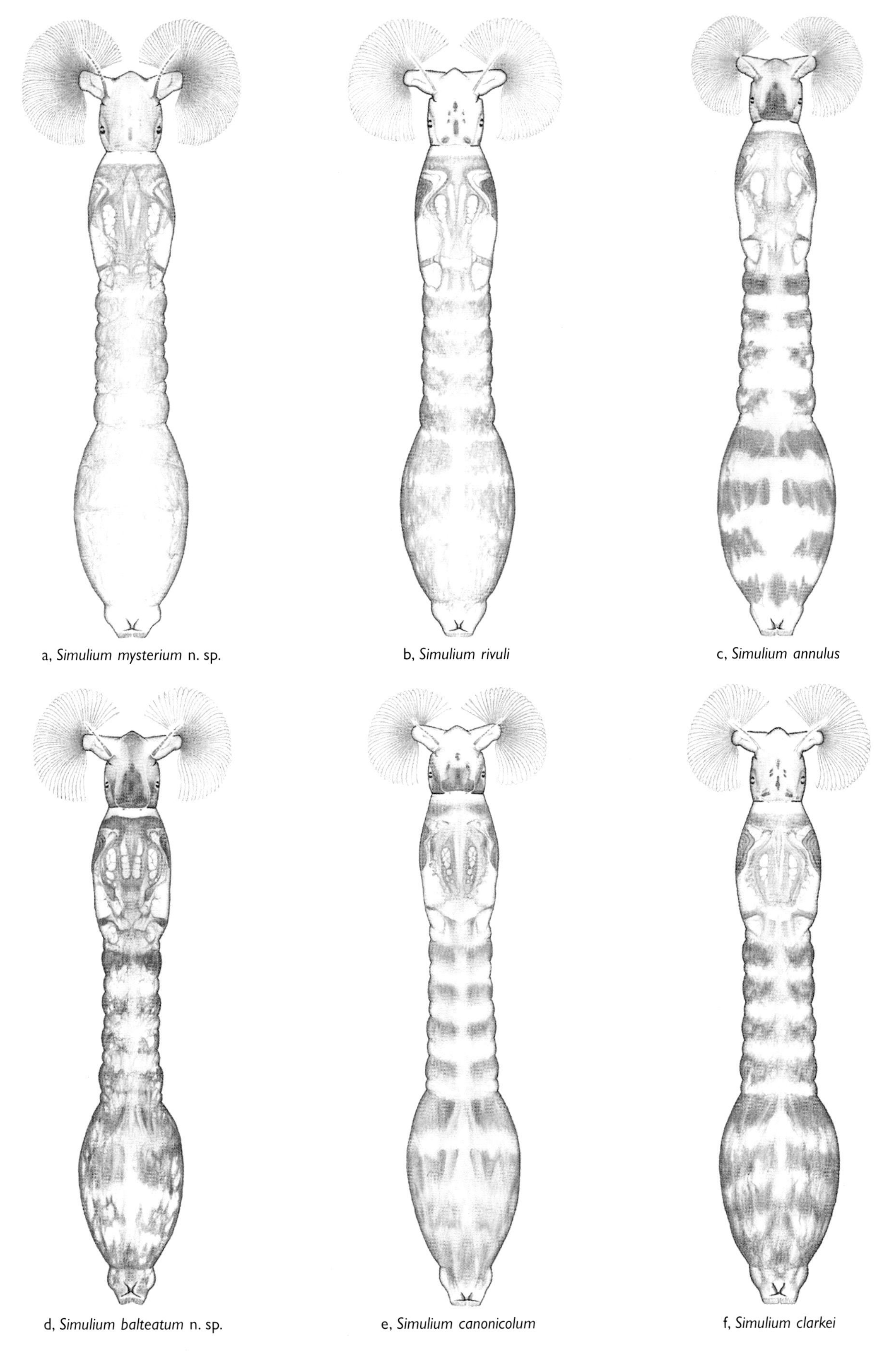


PLATE 11. Larvae, dorsal view.

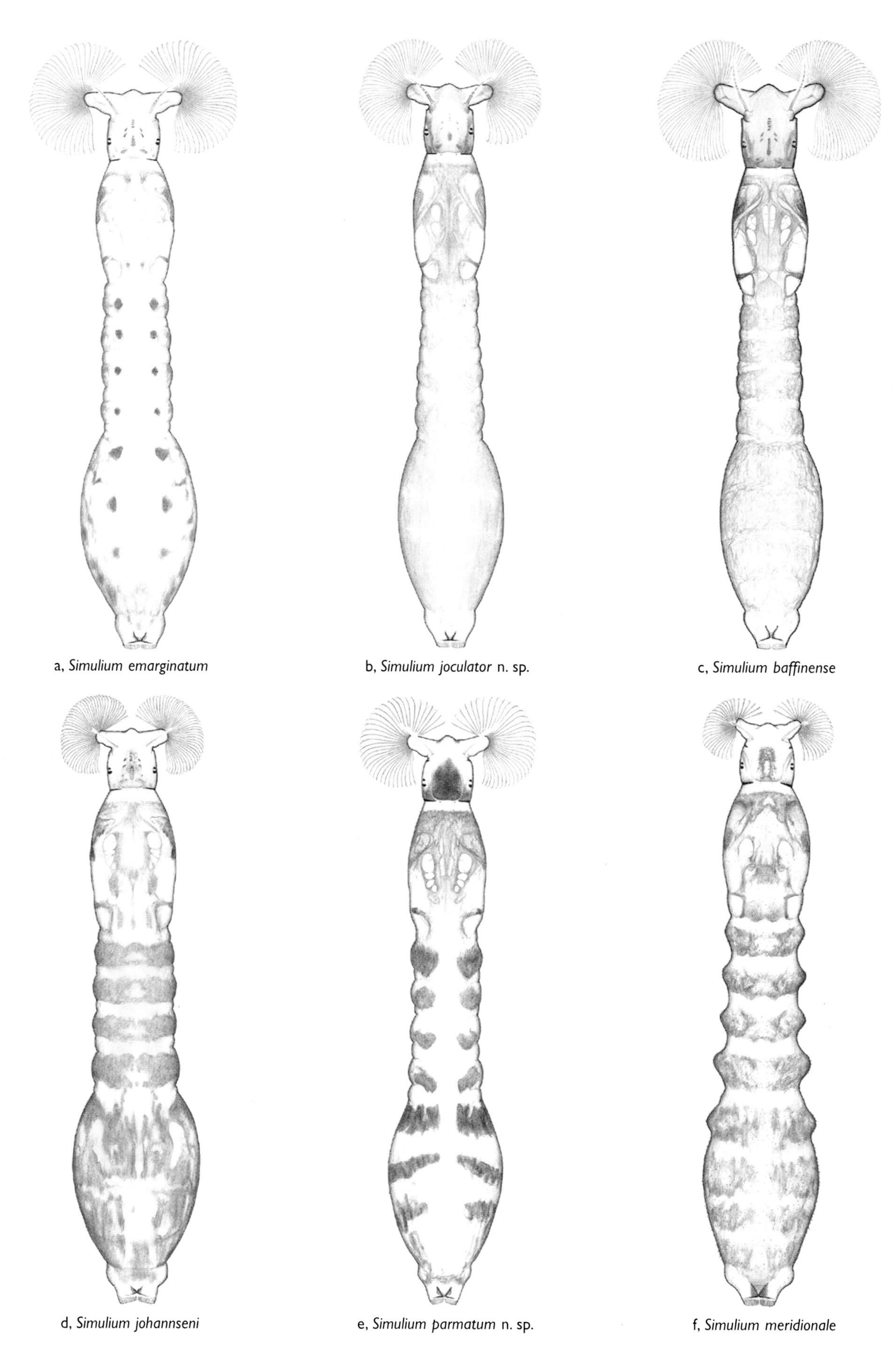

a, *Simulium emarginatum*

b, *Simulium joculator* n. sp.

c, *Simulium baffinense*

d, *Simulium johannseni*

e, *Simulium parmatum* n. sp.

f, *Simulium meridionale*

PLATE 12. Larvae, dorsal view.

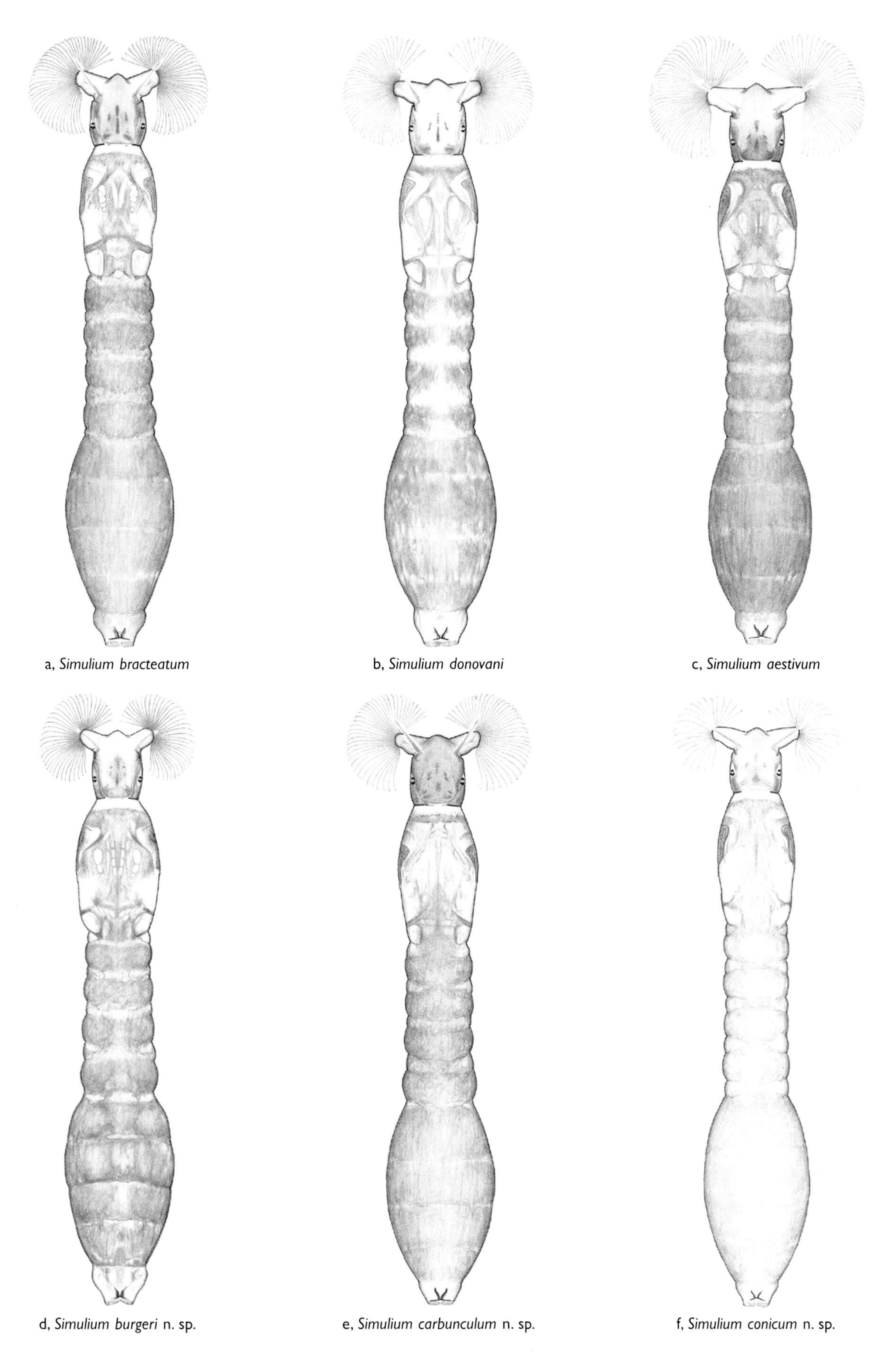
a, *Simulium bracteatum* b, *Simulium donovani* c, *Simulium aestivum*

d, *Simulium burgeri* n. sp. e, *Simulium carbunculum* n. sp. f, *Simulium conicum* n. sp.

PLATE 13. Larvae, dorsal view.

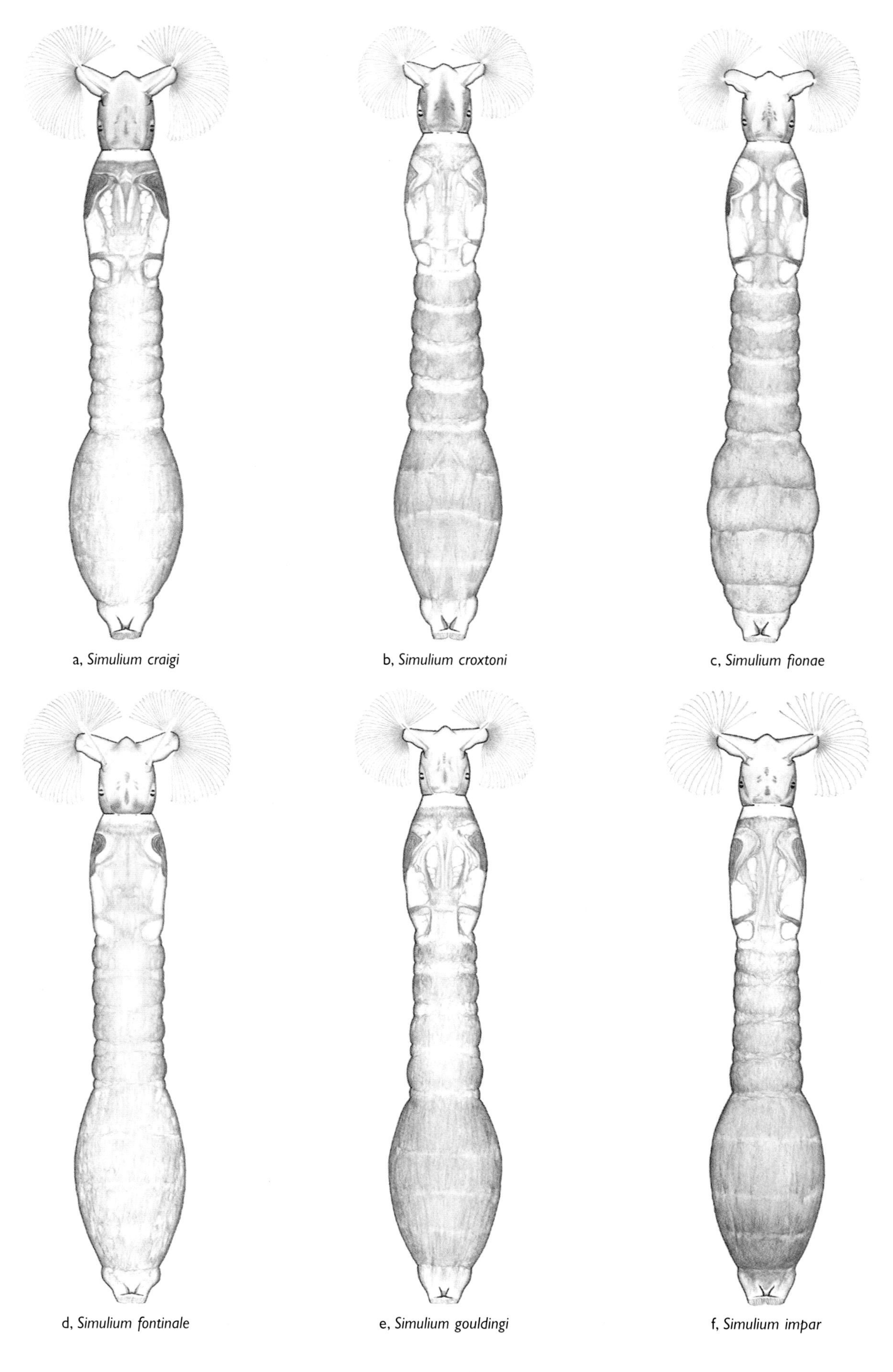


PLATE 14. Larvae, dorsal view.

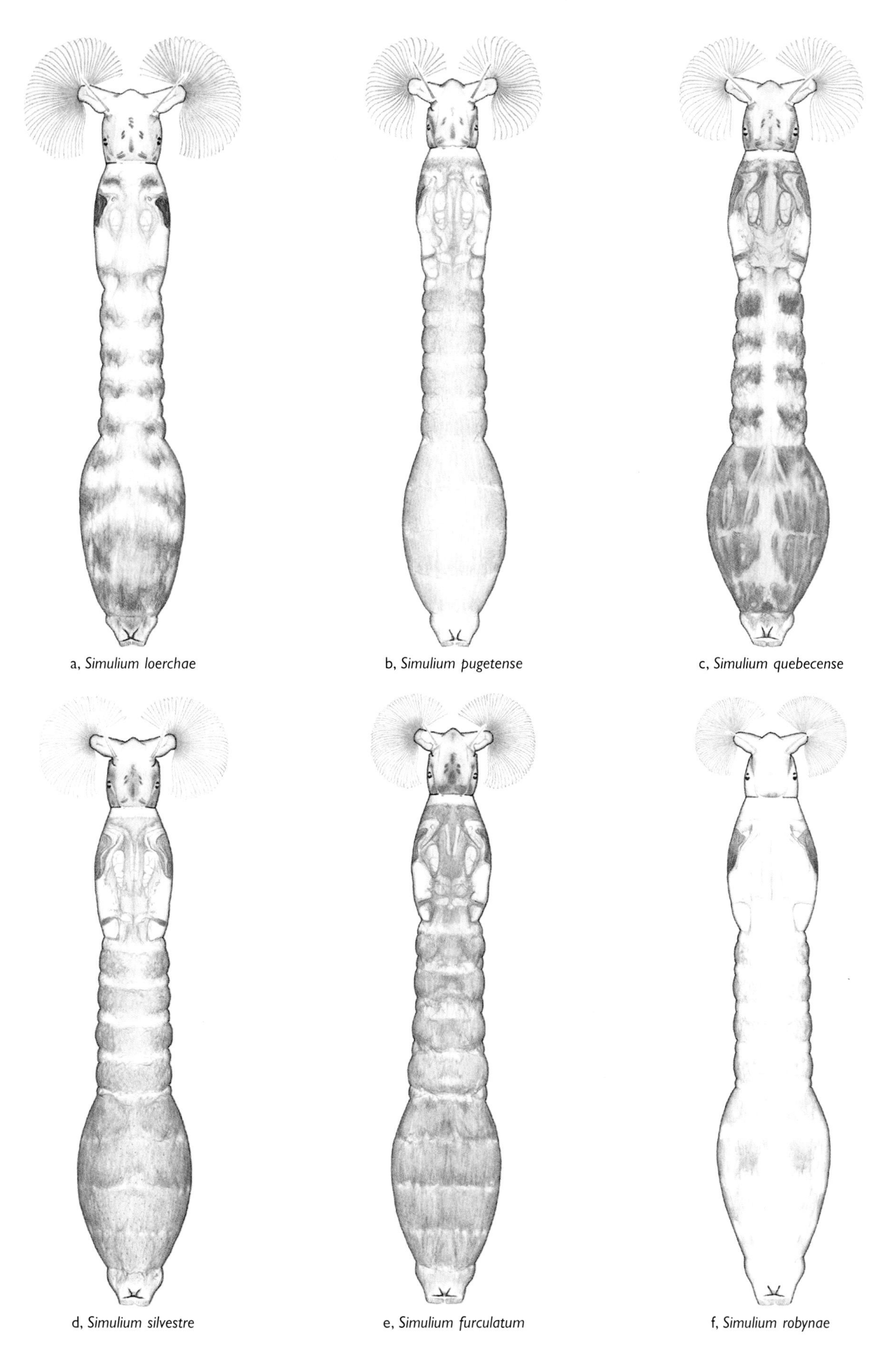
a, *Simulium loerchae* b, *Simulium pugetense* c, *Simulium quebecense*

d, *Simulium silvestre* e, *Simulium furculatum* f, *Simulium robynae*

PLATE 15. Larvae, dorsal view.

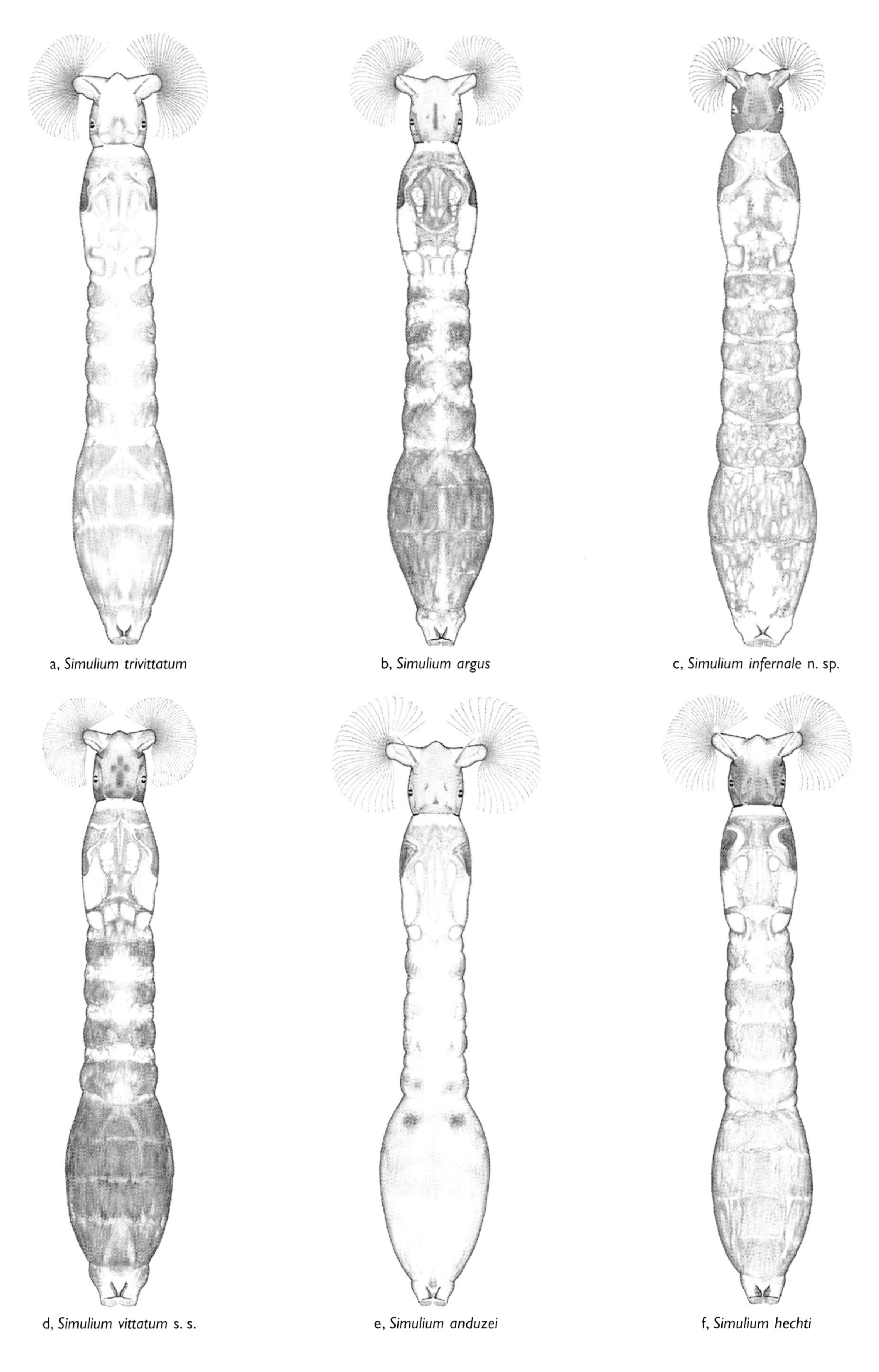
a, *Simulium trivittatum*
b, *Simulium argus*
c, *Simulium infernale* n. sp.
d, *Simulium vittatum* s. s.
e, *Simulium anduzei*
f, *Simulium hechti*

PLATE 16. Larvae, dorsal view.

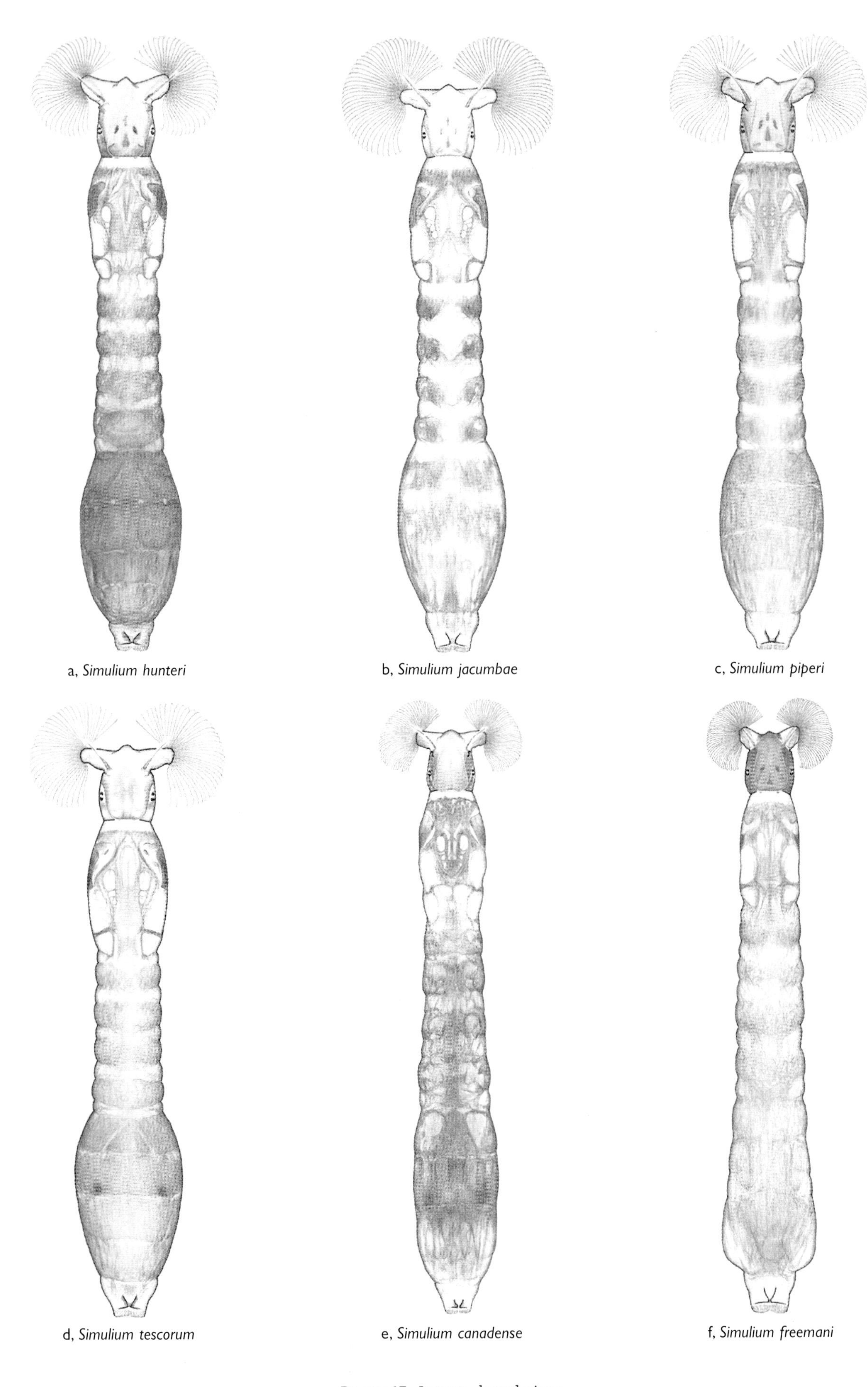

a, *Simulium hunteri*  b, *Simulium jacumbae*  c, *Simulium piperi*

d, *Simulium tescorum*  e, *Simulium canadense*  f, *Simulium freemani*

PLATE 17. Larvae, dorsal view.

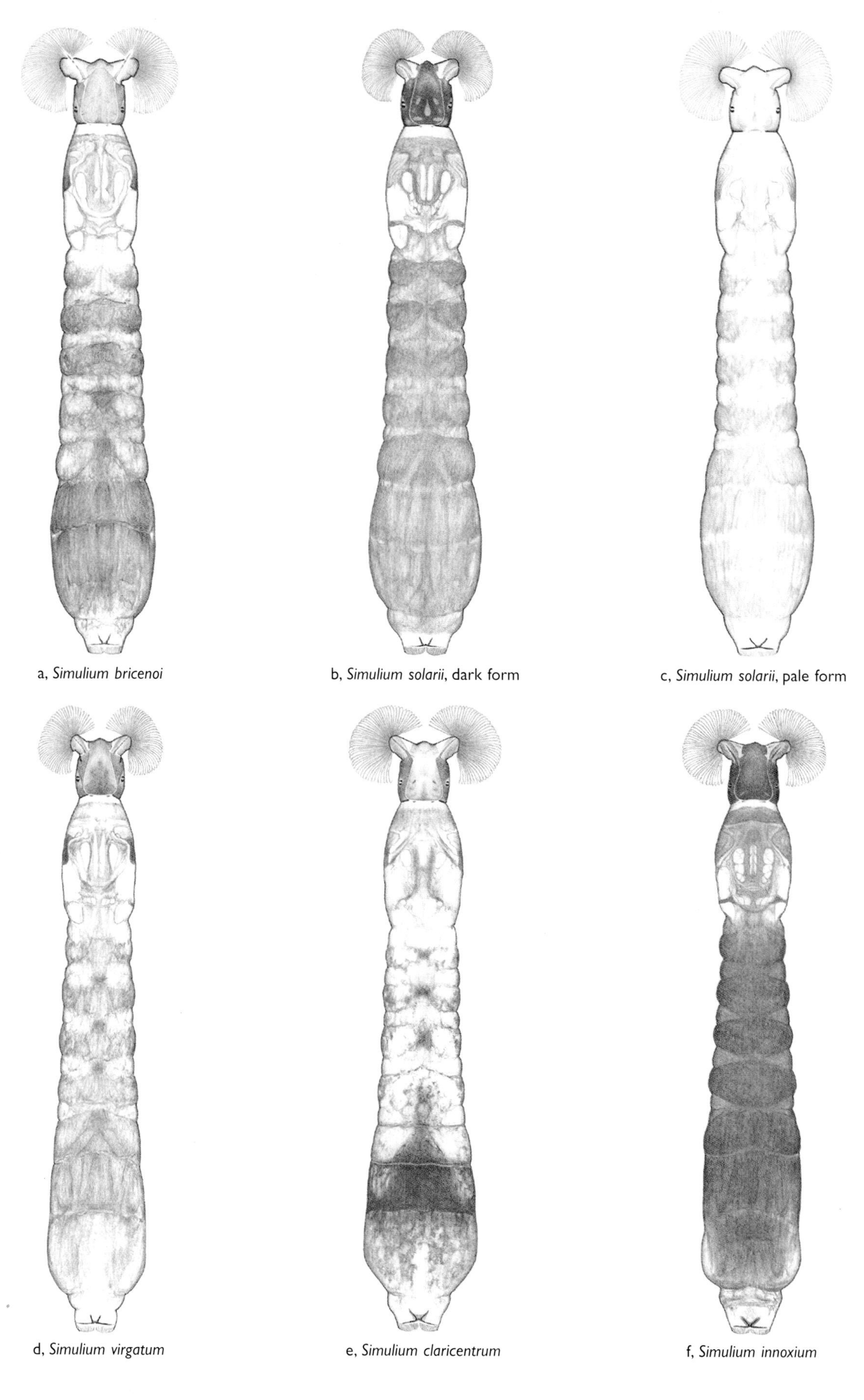


PLATE 18. Larvae, dorsal view.

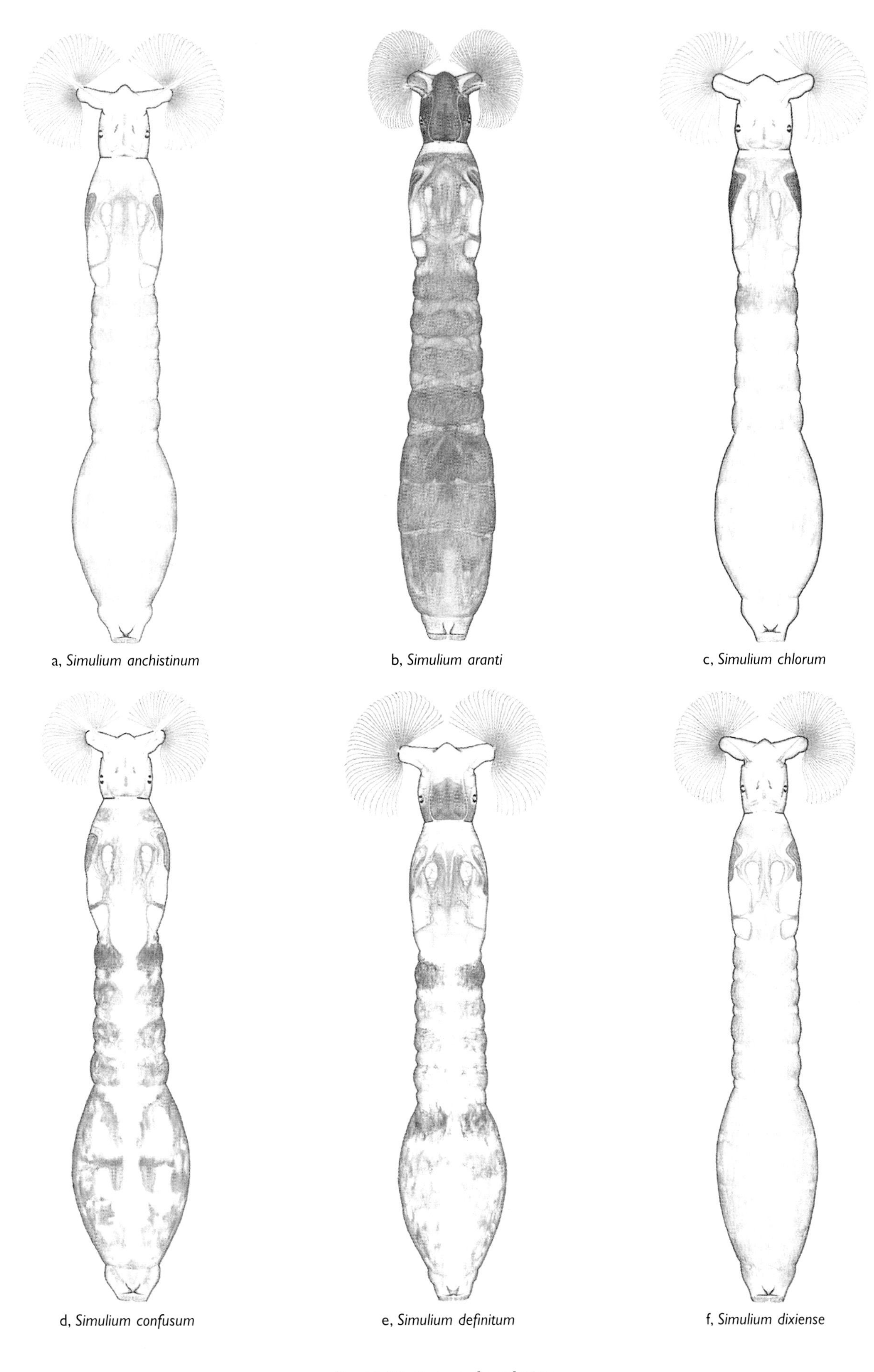

a, *Simulium anchistinum* b, *Simulium aranti* c, *Simulium chlorum*

d, *Simulium confusum* e, *Simulium definitum* f, *Simulium dixiense*

PLATE 19. Larvae, dorsal view.

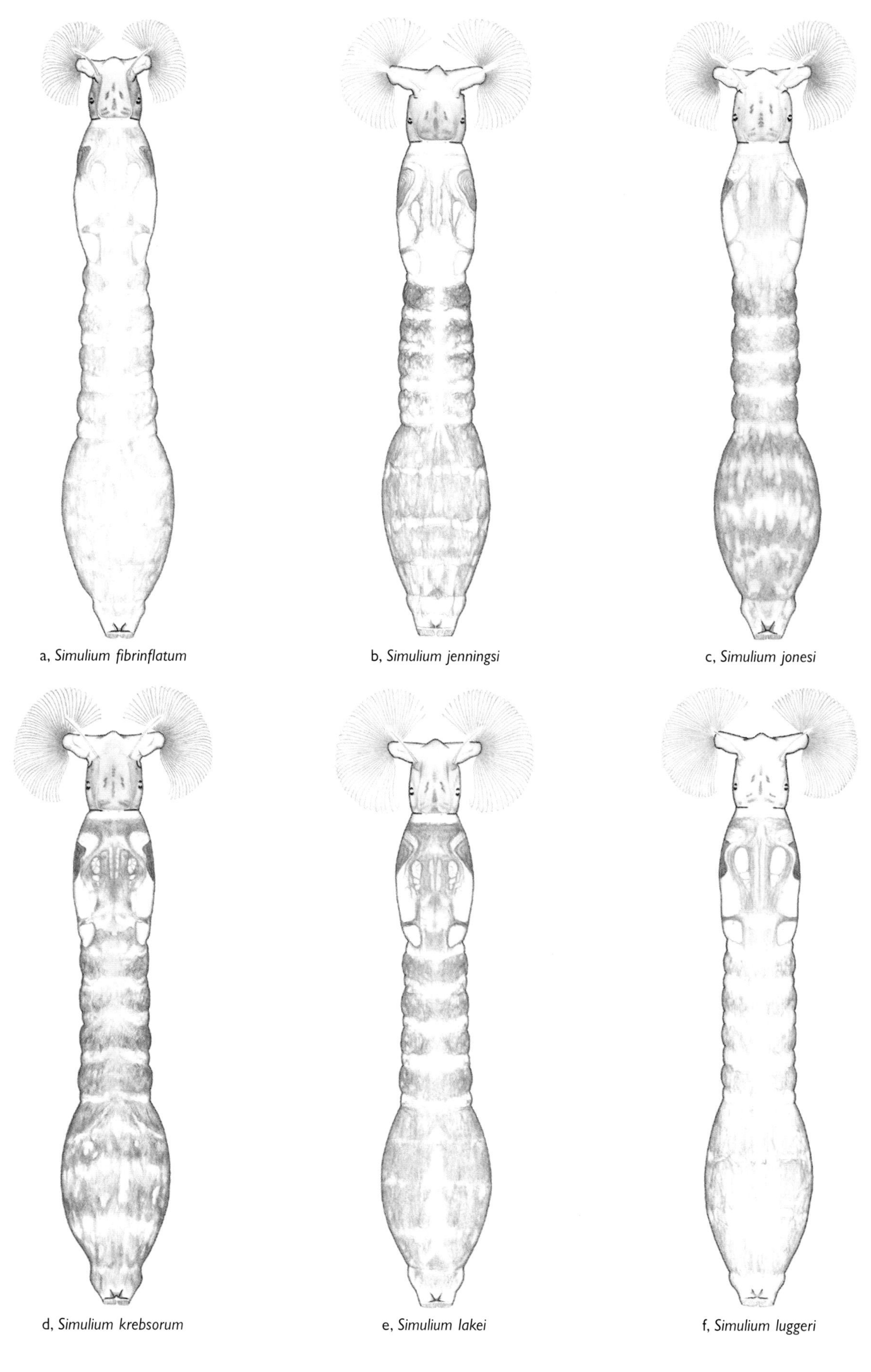

a, *Simulium fibrinflatum* b, *Simulium jenningsi* c, *Simulium jonesi*

d, *Simulium krebsorum* e, *Simulium lakei* f, *Simulium luggeri*

PLATE 20. Larvae, dorsal view.

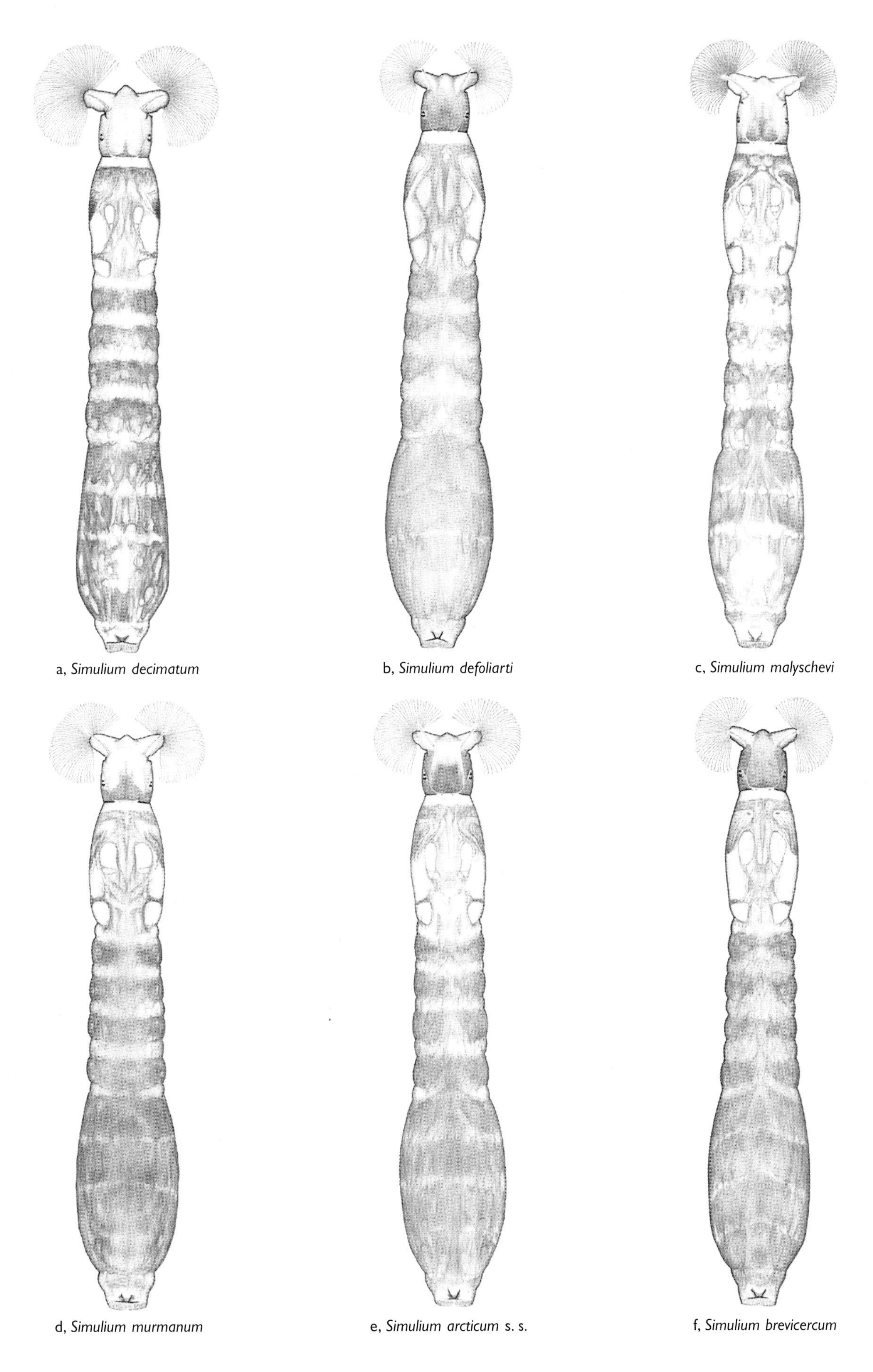


PLATE 21. Larvae, dorsal view.

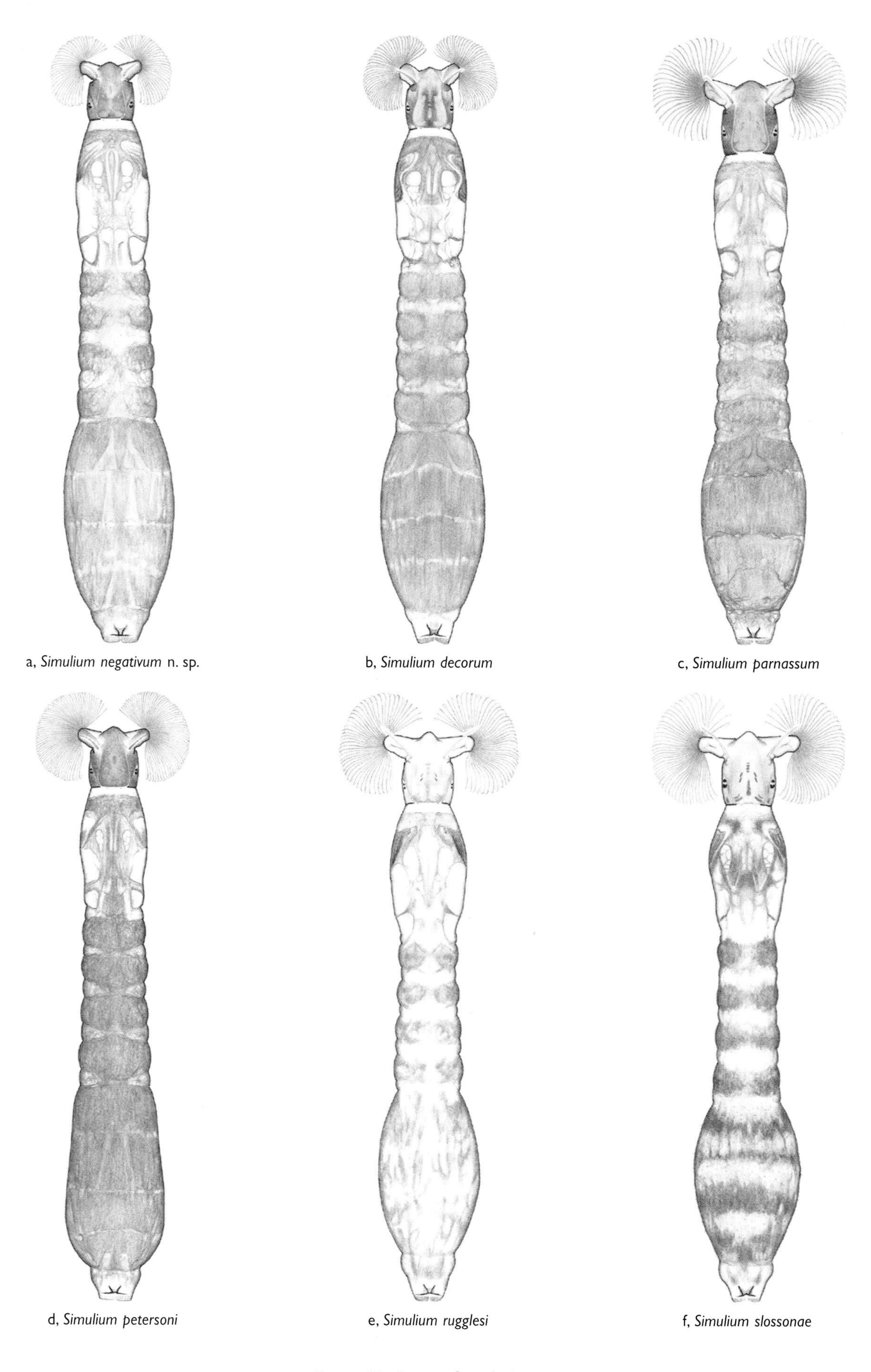


Plate 22. Larvae, dorsal view.

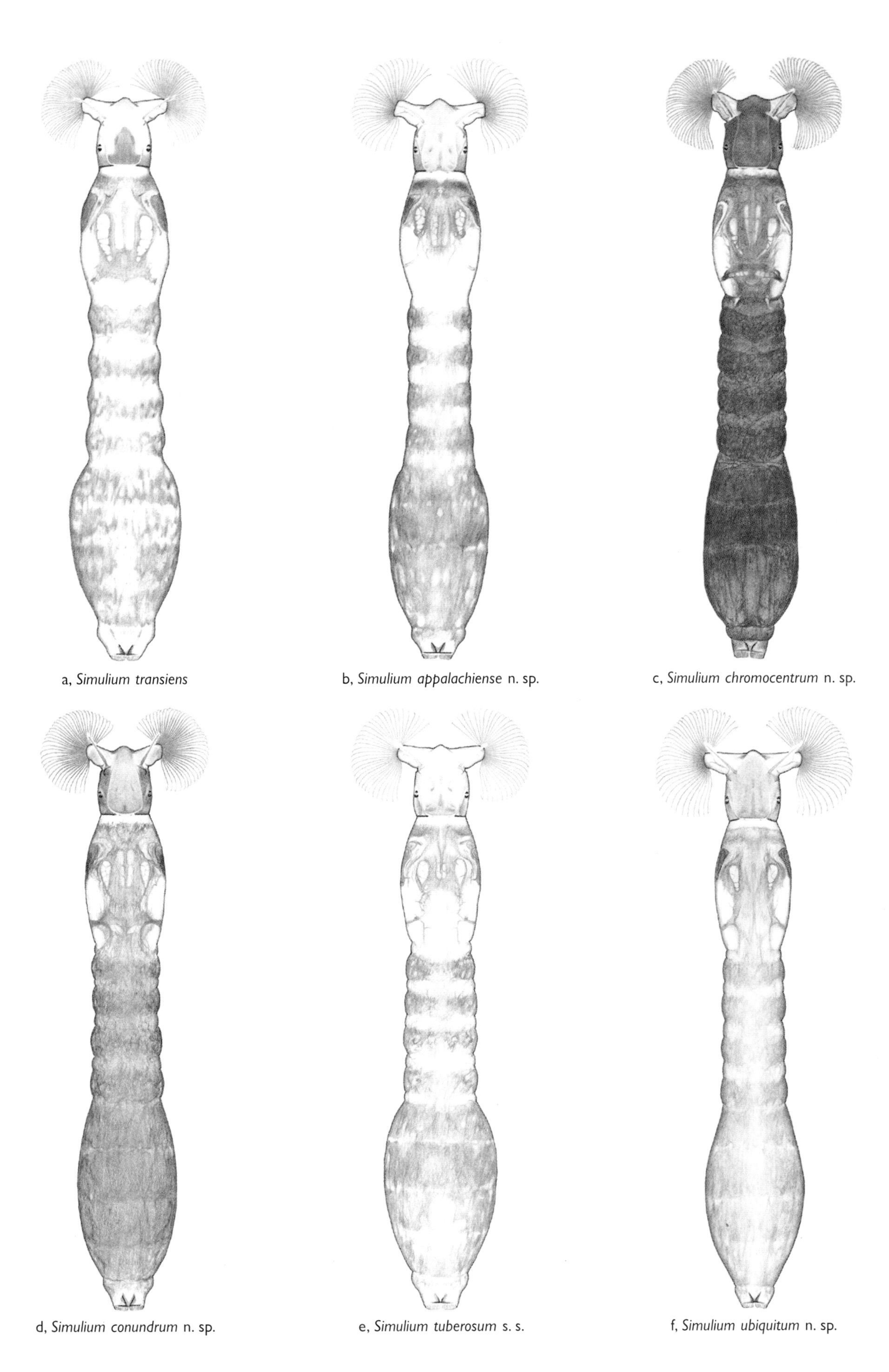


PLATE 23. Larvae, dorsal view.

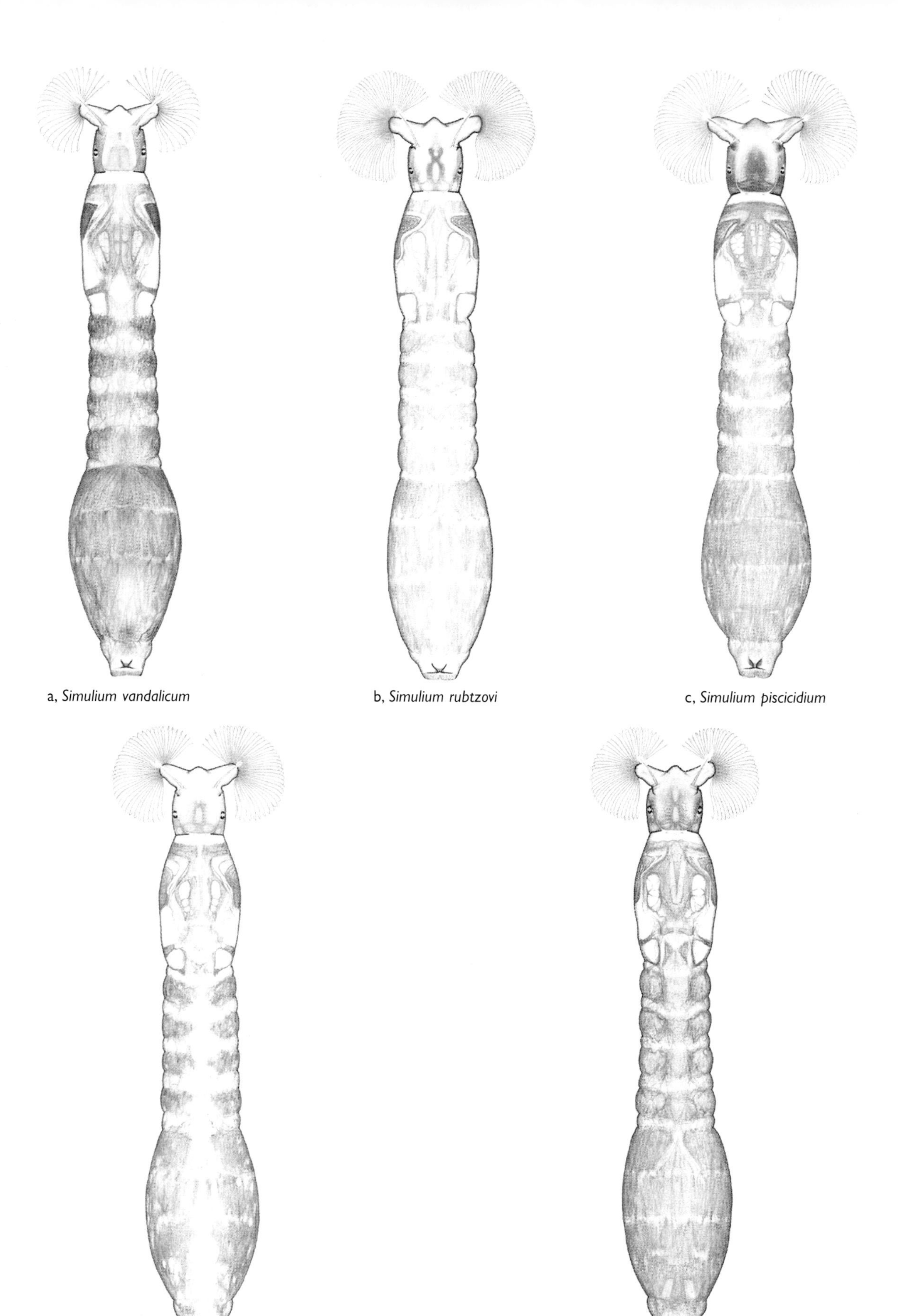


PLATE 24. Larvae, dorsal view.

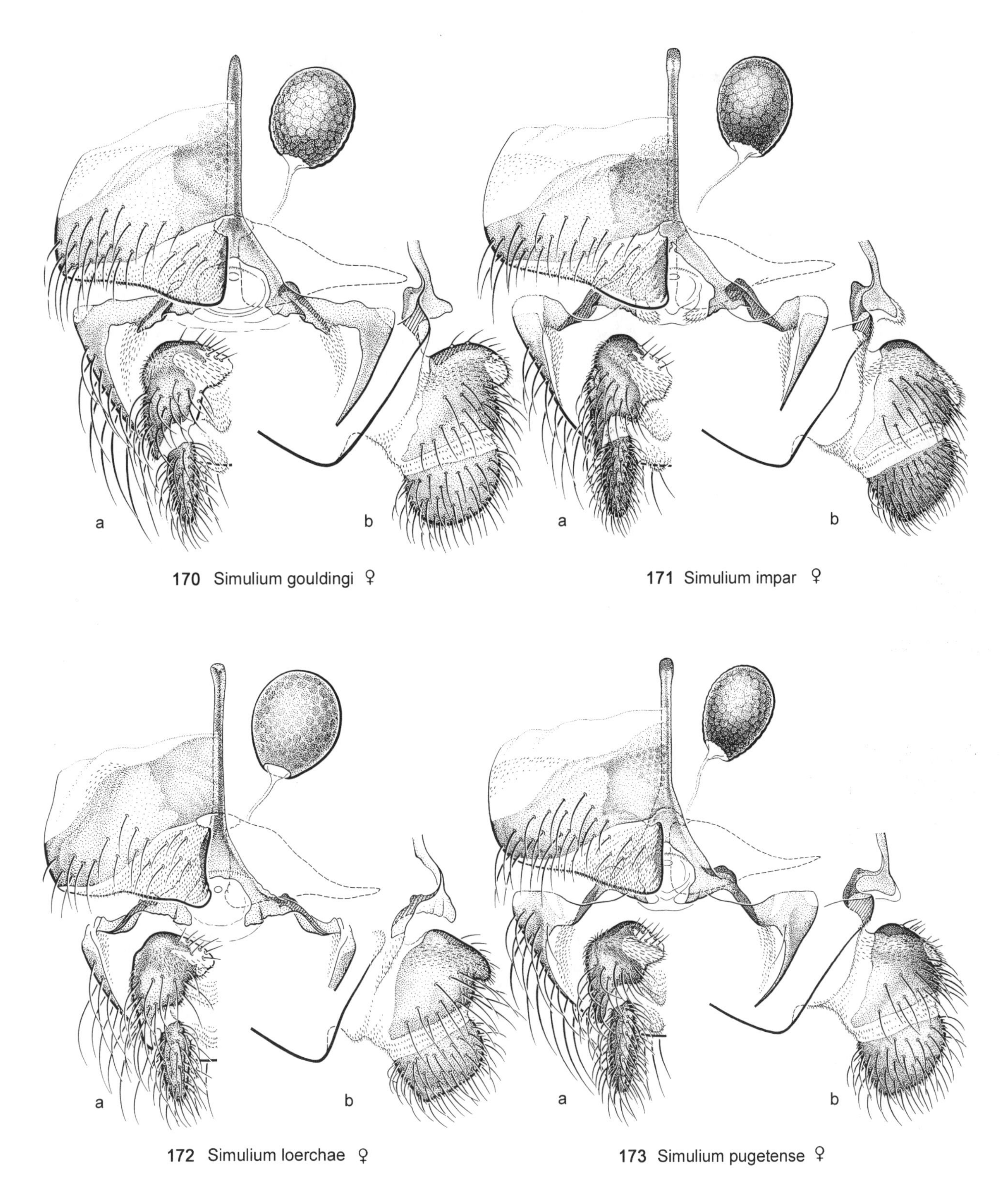


Figs. 10.170–10.173. Female terminalia; a, ventral view; b, right lateral view.

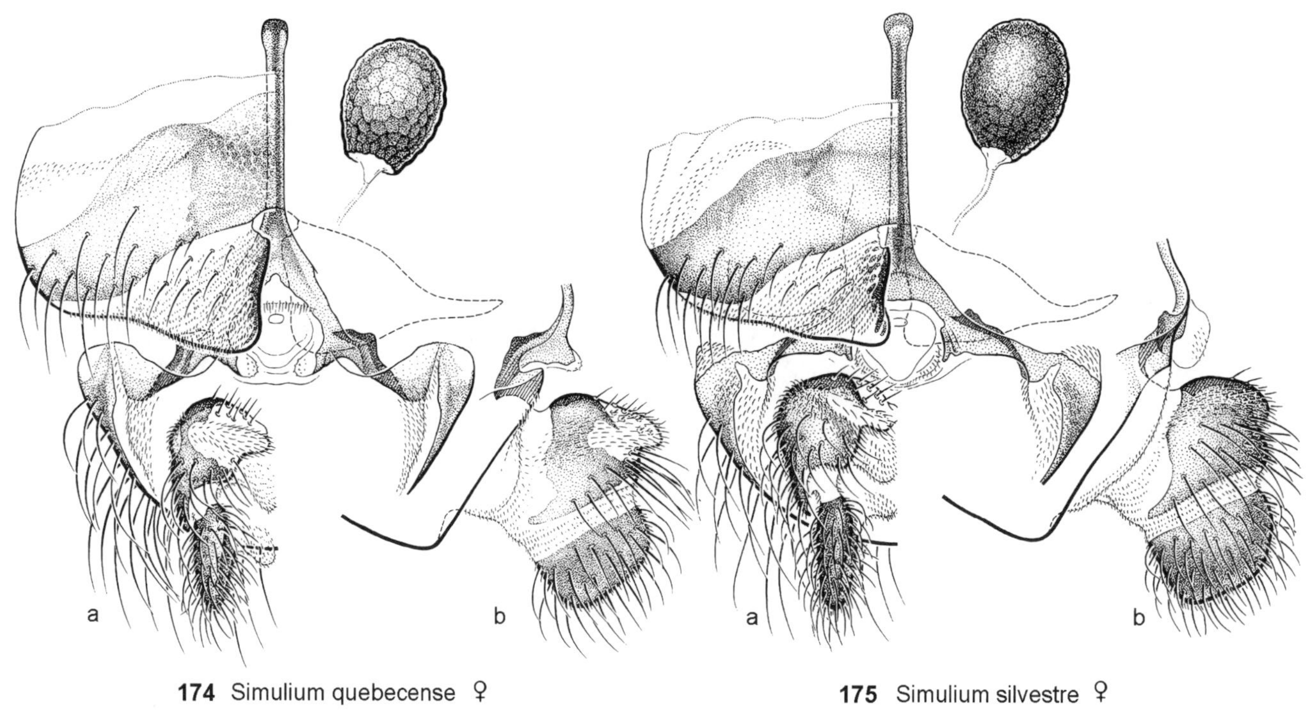


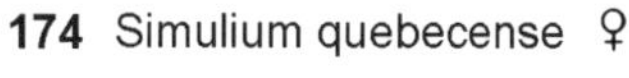

**174** Simulium quebecense ♀

**175** Simulium silvestre ♀

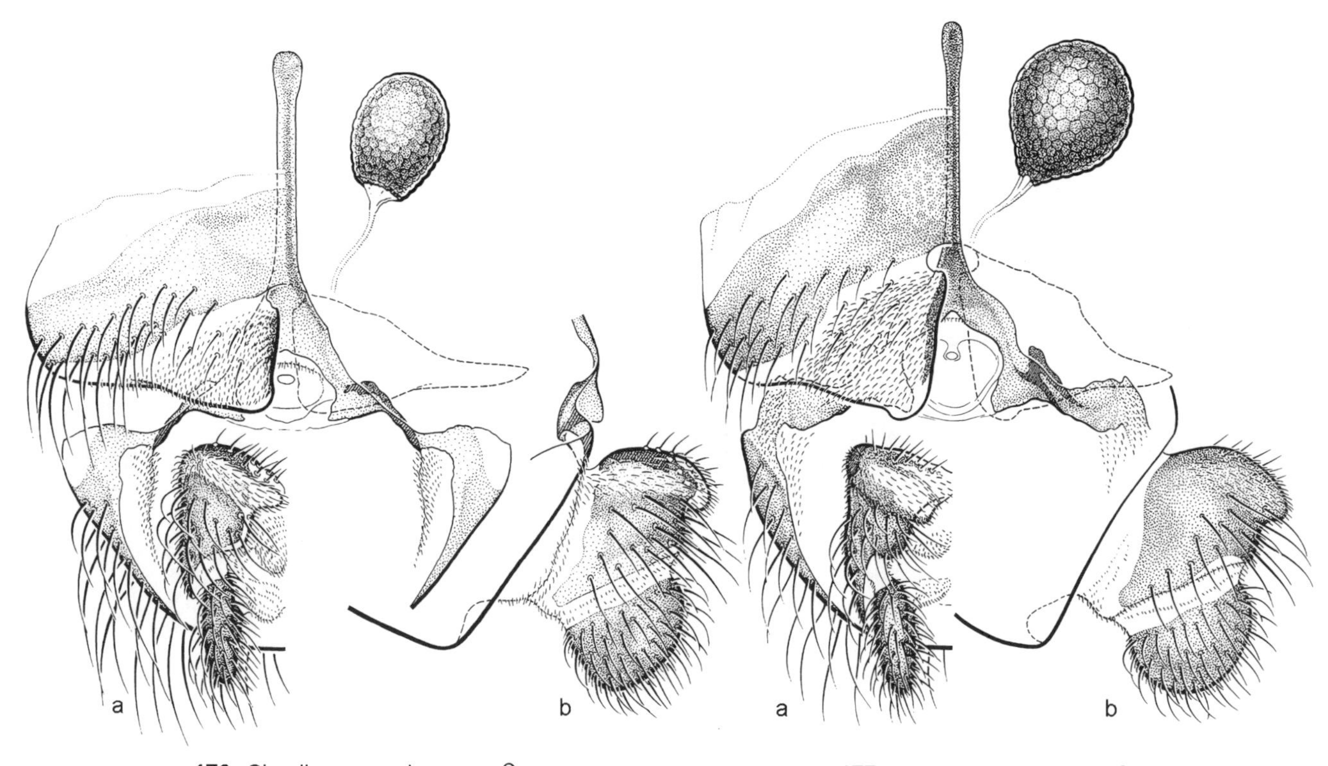


**176** Simulium wyomingense ♀

**177** Simulium furculatum ♀

Figs. 10.174–10.177. Female terminalia; a, ventral view; b, right lateral view.

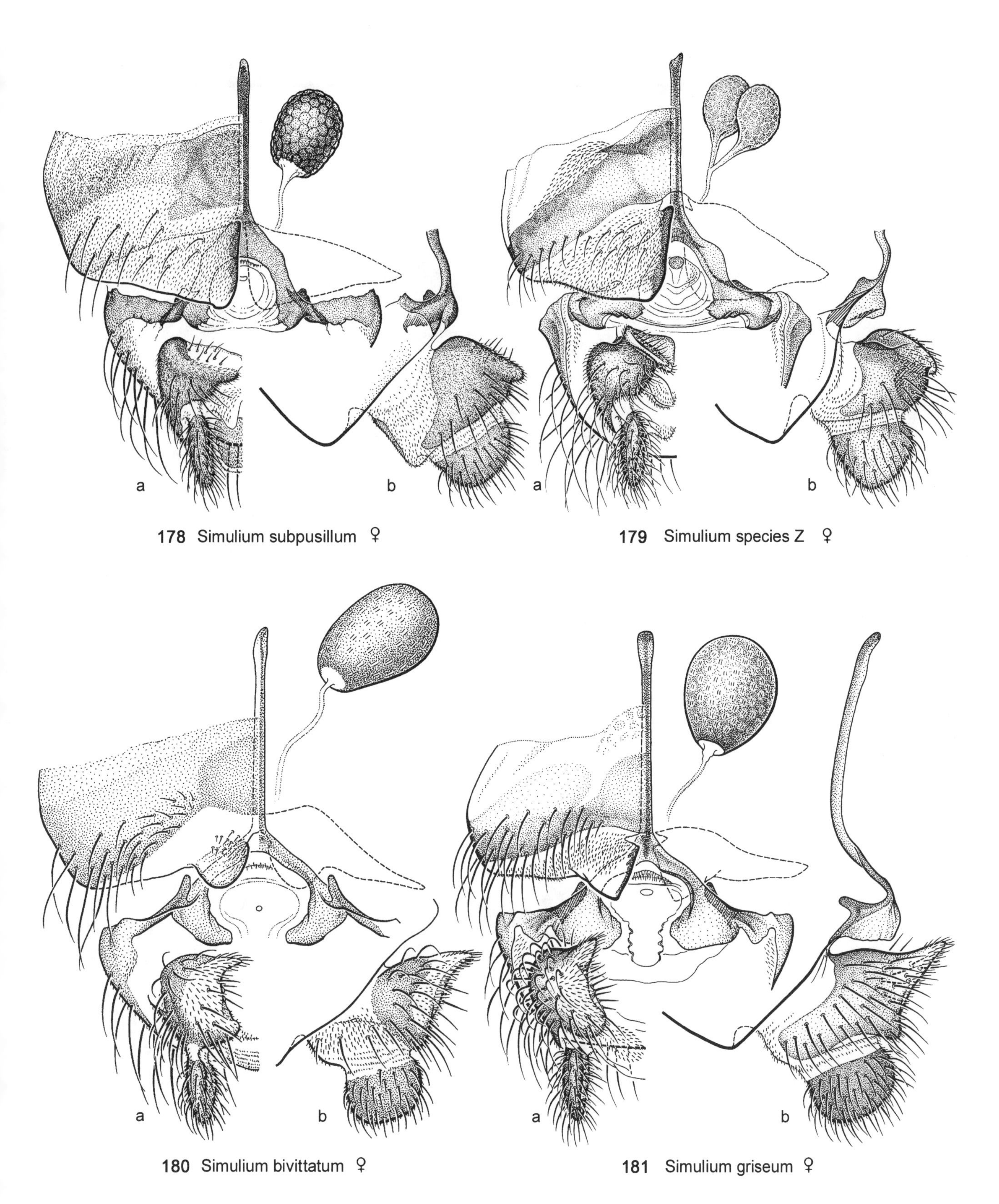


Figs. 10.178–10.181. Female terminalia; a, ventral view; b, right lateral view. (180) From Peterson (1981) by permission.

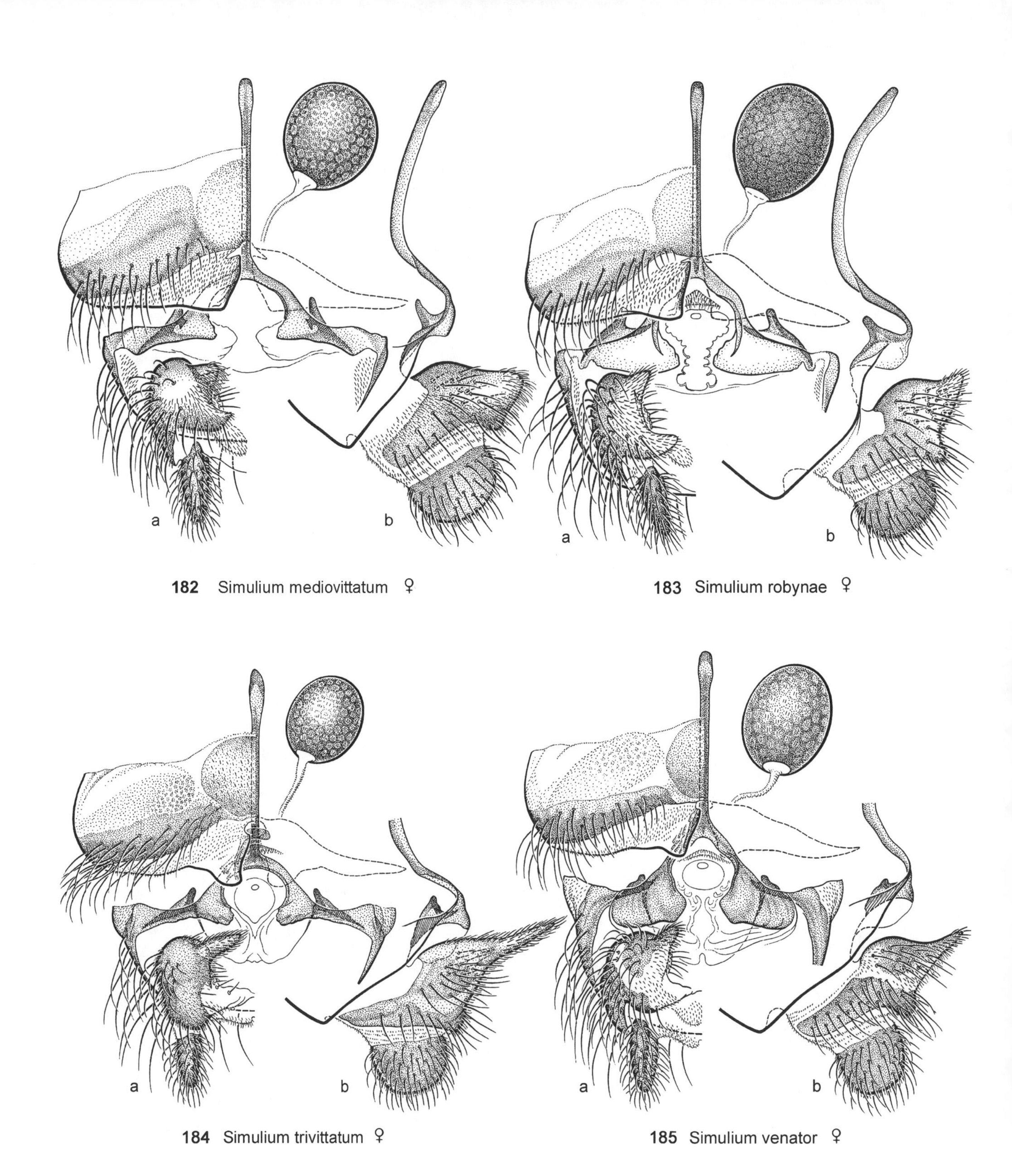


Figs. 10.182–10.185. Female terminalia; a, ventral view; b, right lateral view.

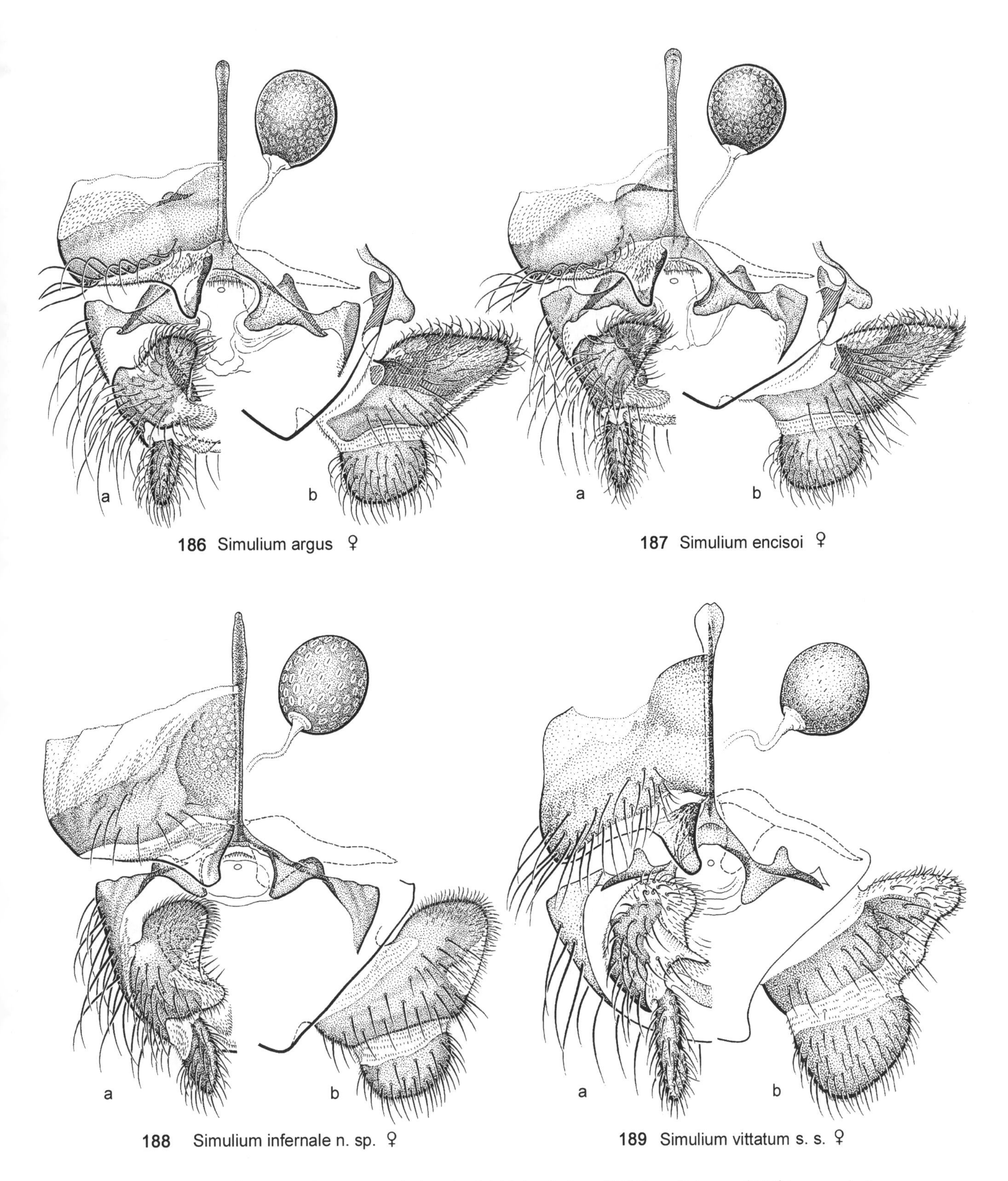


Figs. 10.186–10.189. Female terminalia; a, ventral view; b, right lateral view. (189) From Peterson (1981) by permission.

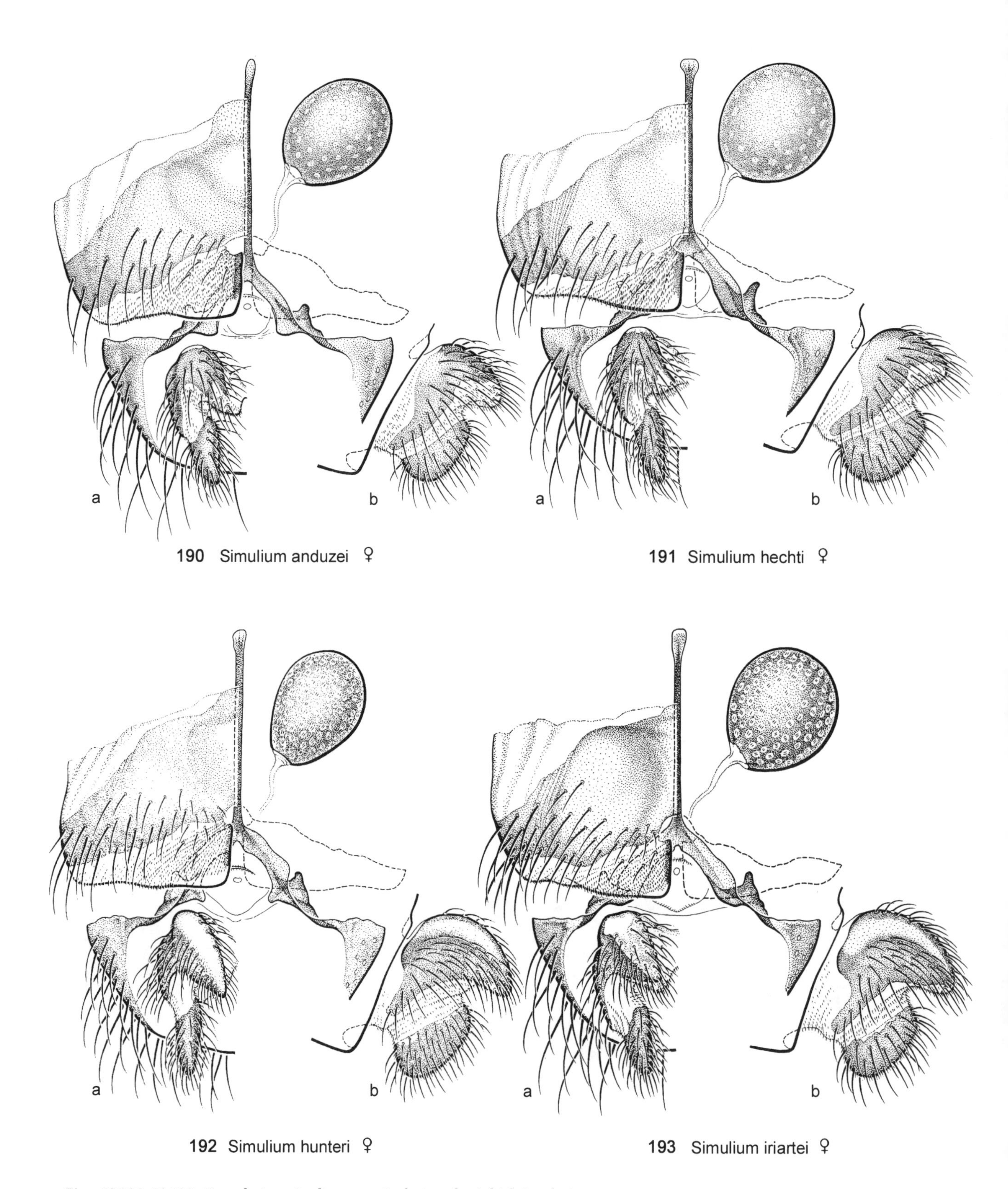


Figs. 10.190–10.193. Female terminalia; a, ventral view; b, right lateral view.

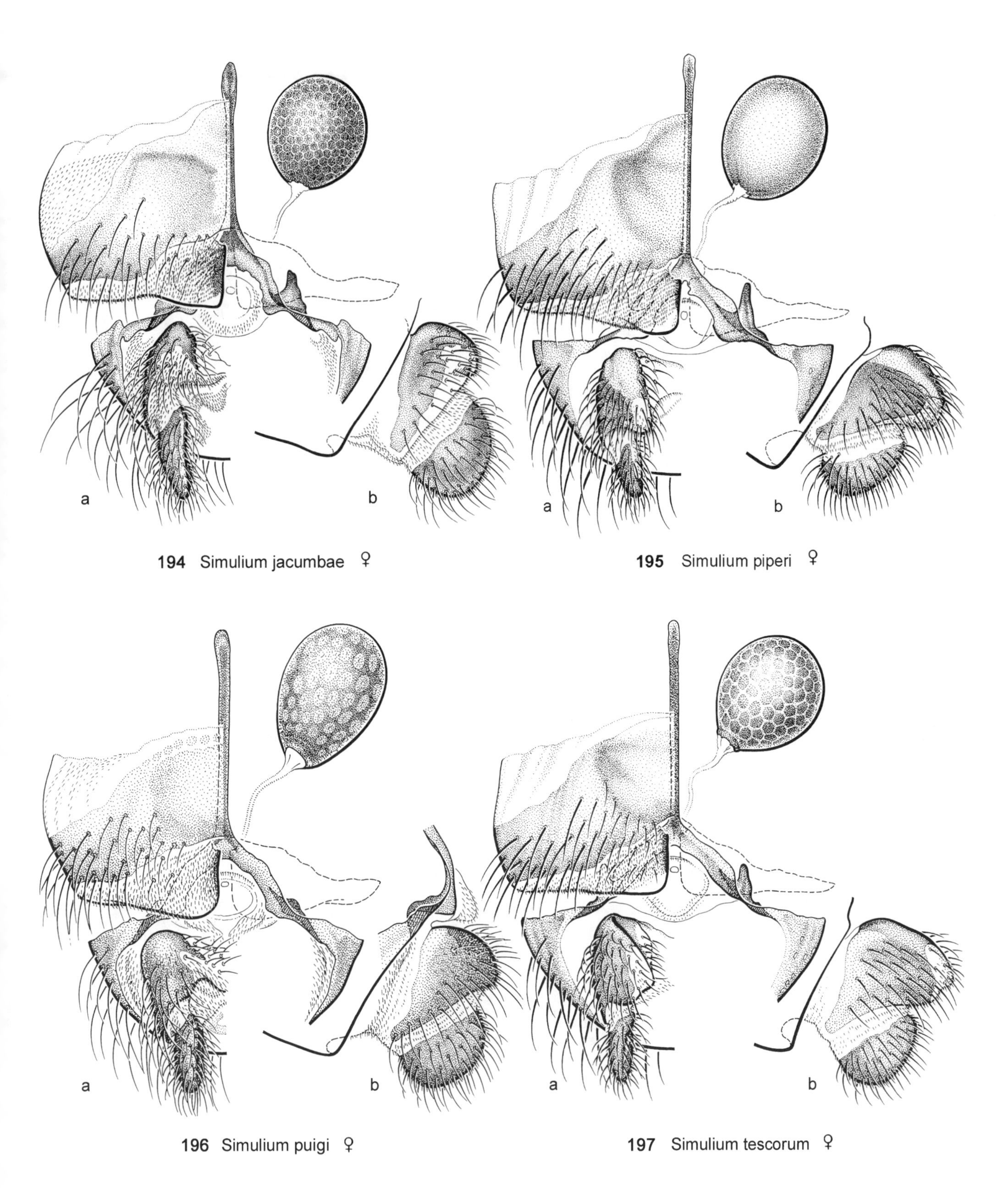


Figs. 10.194–10.197. Female terminalia; a, ventral view; b, right lateral view.

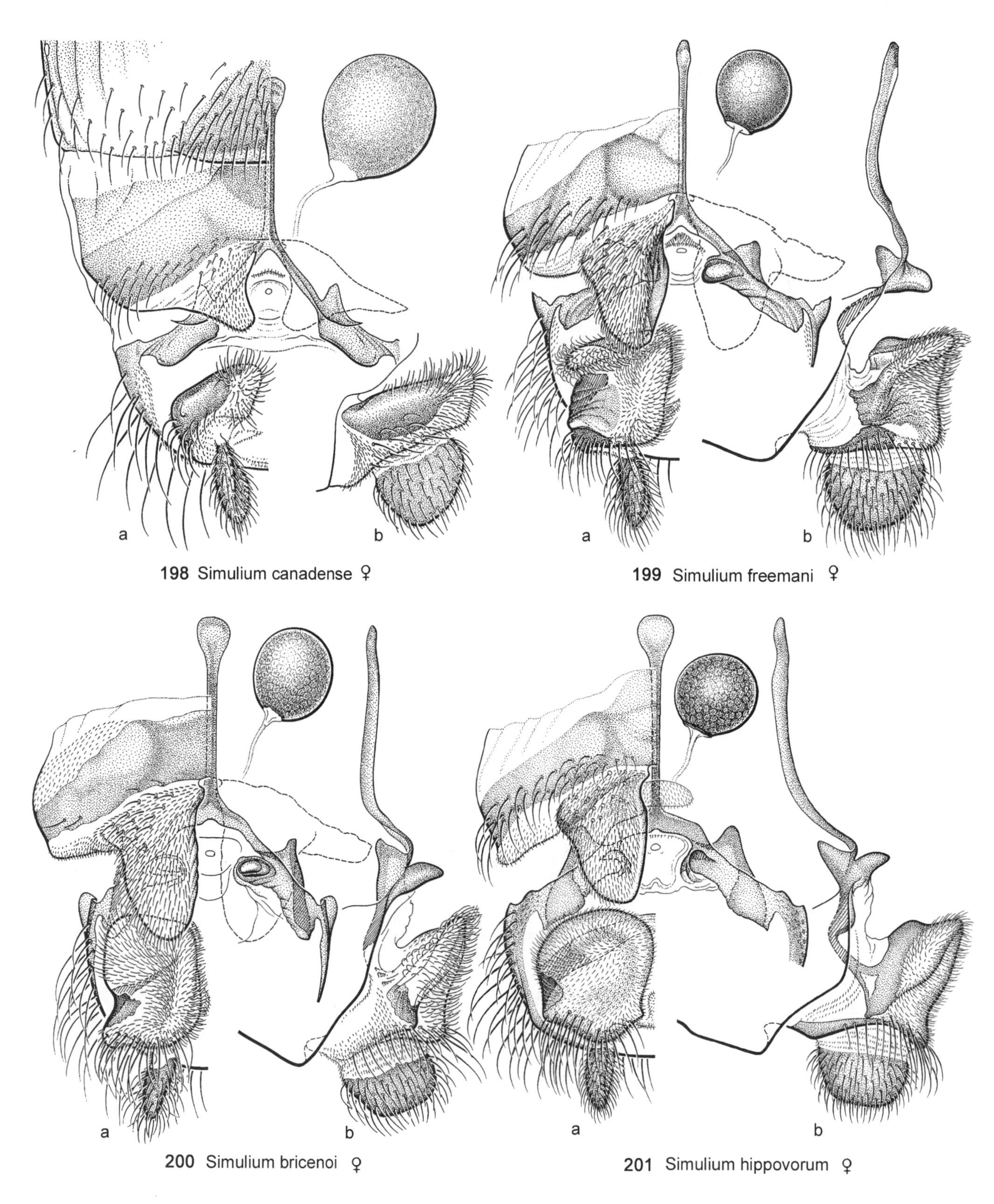


Figs. 10.198–10.201. Female terminalia; a, ventral view; b, right lateral view. (198) From Peterson (1981) by permission.

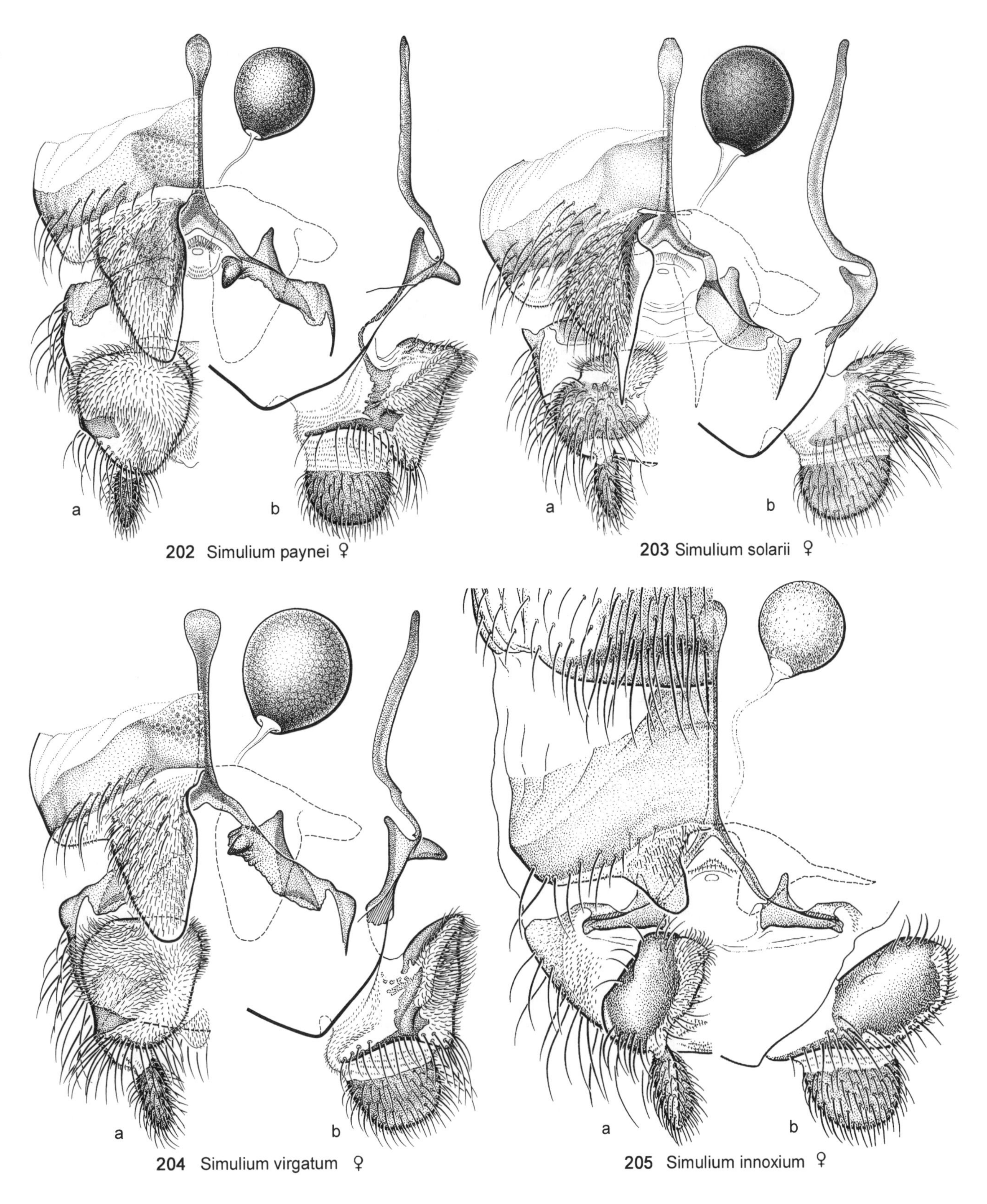


Figs. 10.202–10.205. Female terminalia; a, ventral view; b, right lateral view. (205) From Peterson (1981) by permission.

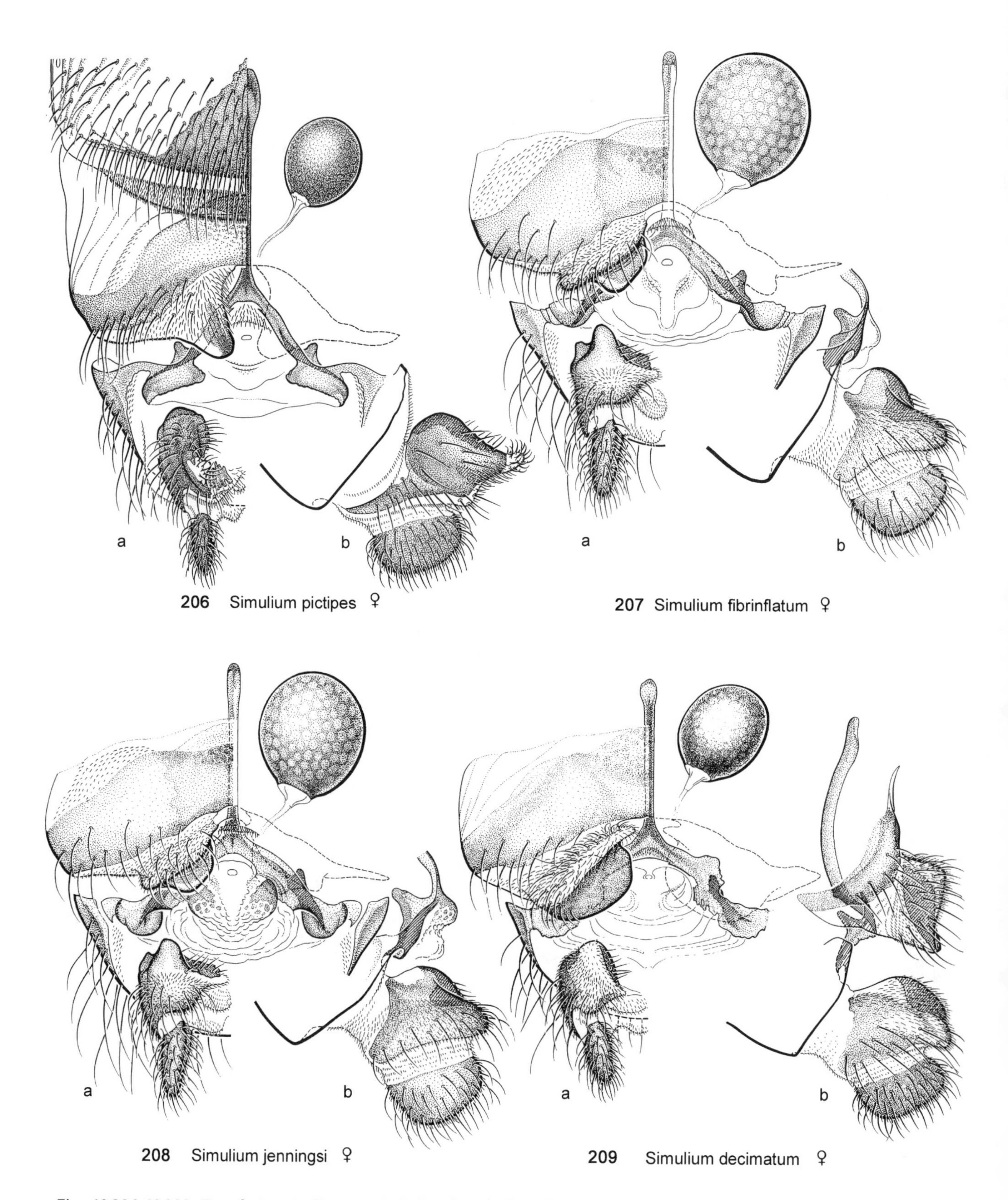


Figs. 10.206–10.209. Female terminalia; a, ventral view; b, right lateral view.

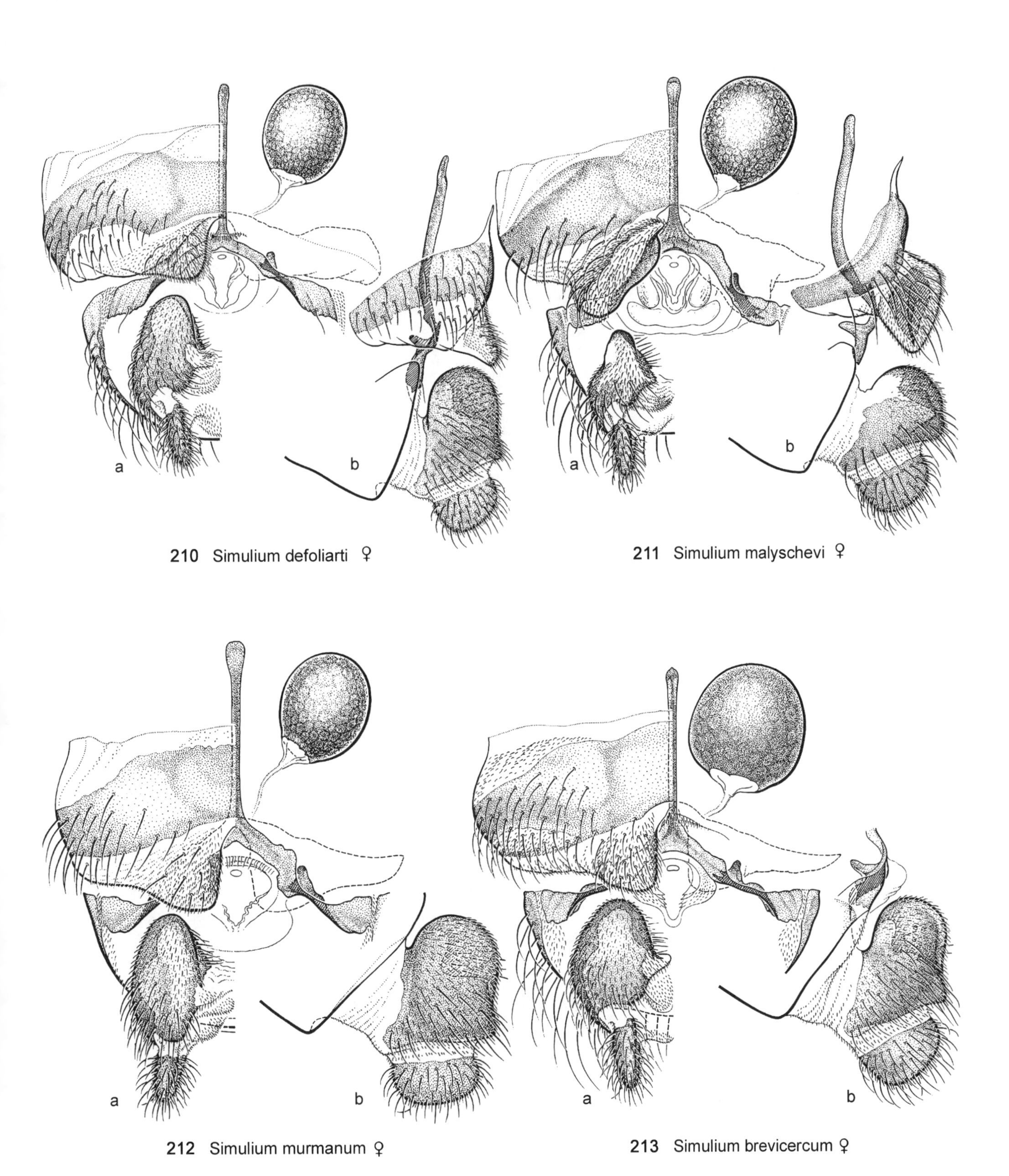


Figs. 10.210–10.213. Female terminalia; a, ventral view; b, right lateral view.

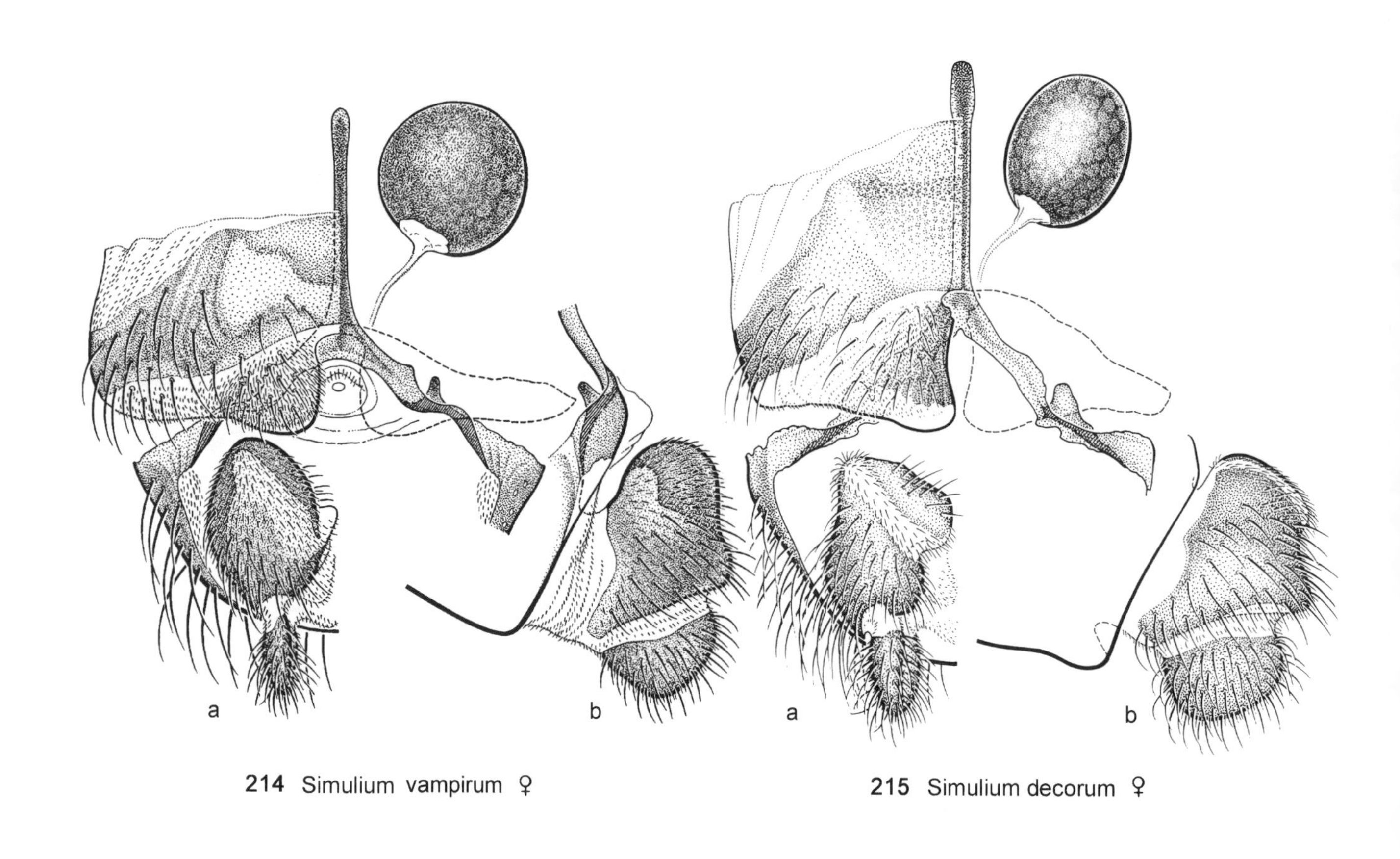


**214** Simulium vampirum ♀

**215** Simulium decorum ♀

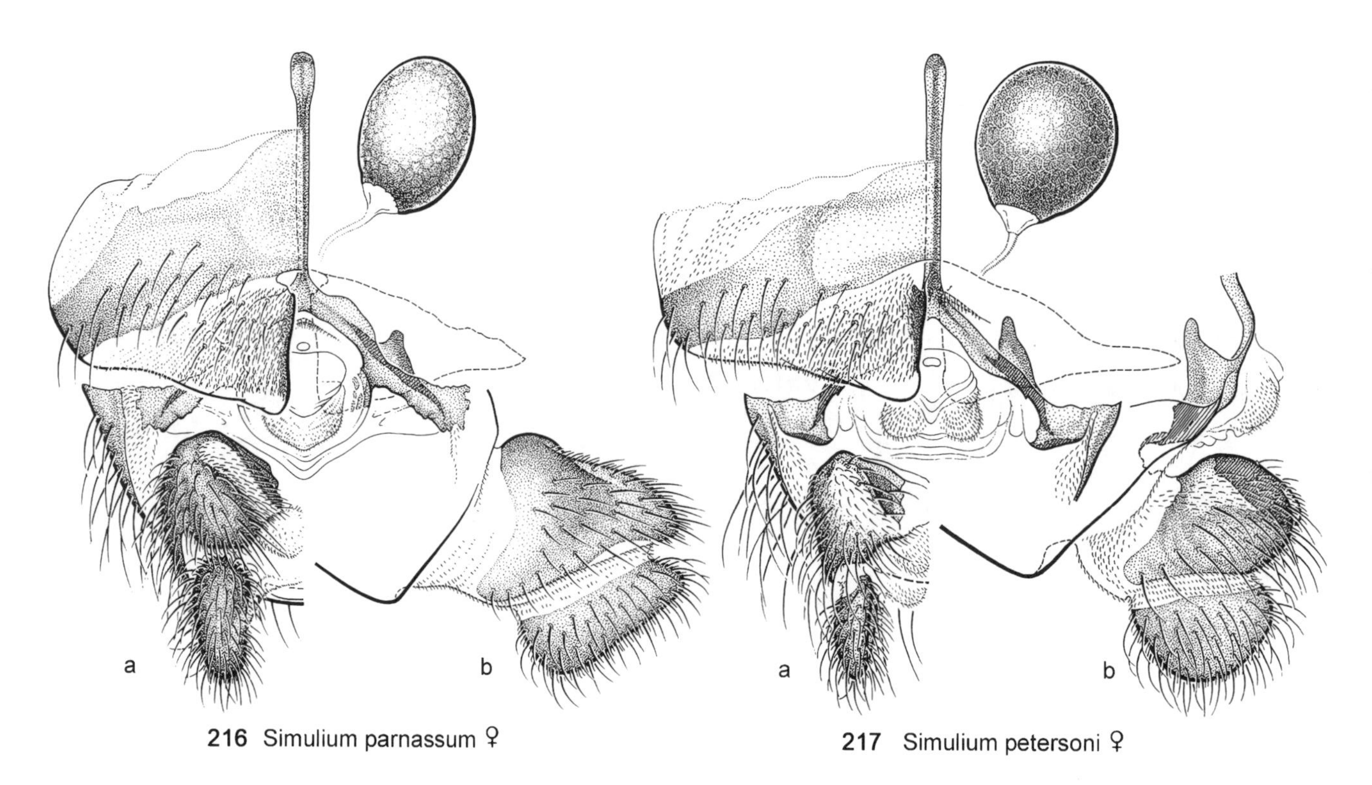


**216** Simulium parnassum ♀

**217** Simulium petersoni ♀

Figs. 10.214–10.217. Female terminalia; a, ventral view; b, right lateral view.

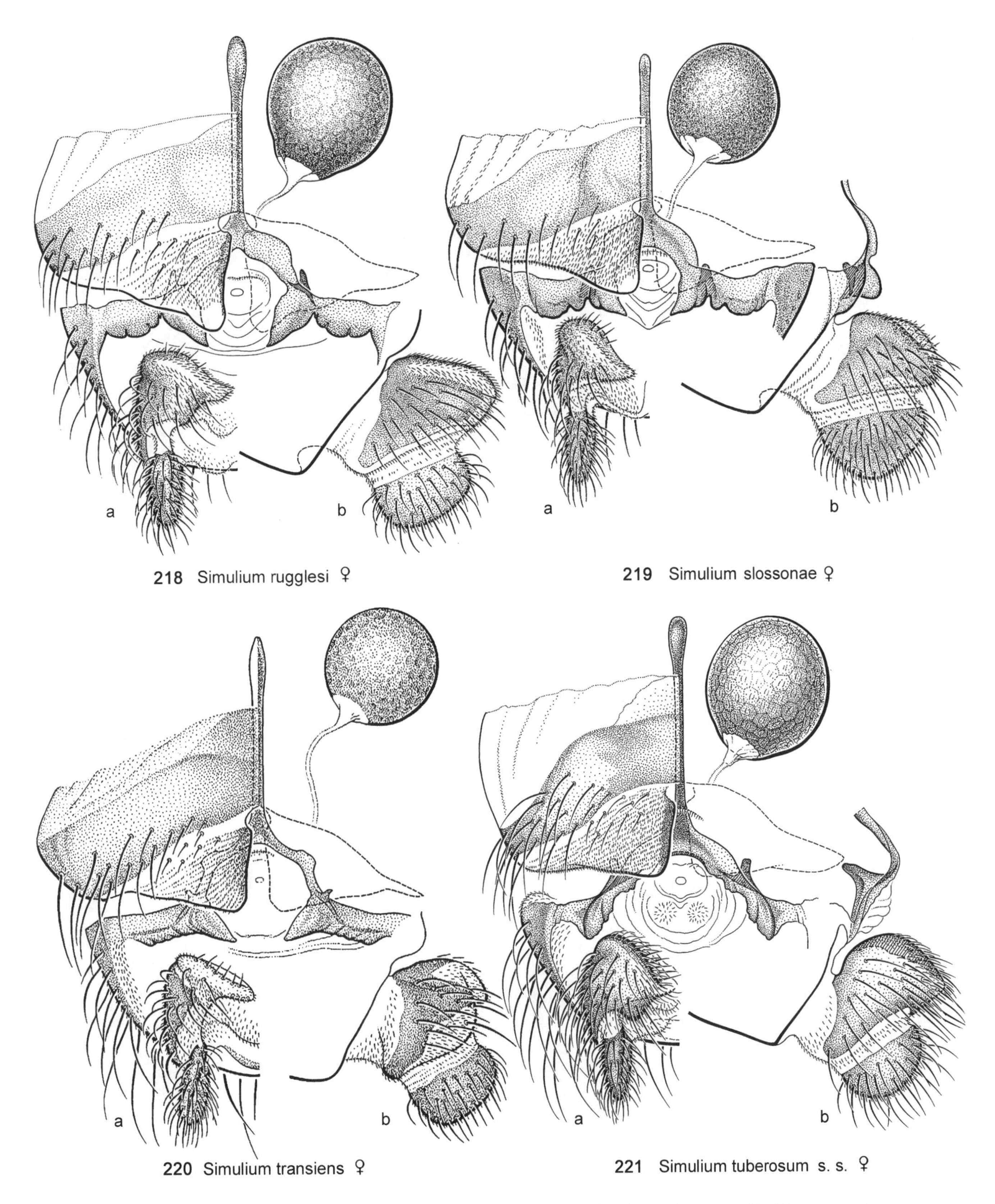


Figs. 10.218–10.221. Female terminalia; a, ventral view; b, right lateral view. (220) From Peterson (1981) by permission.

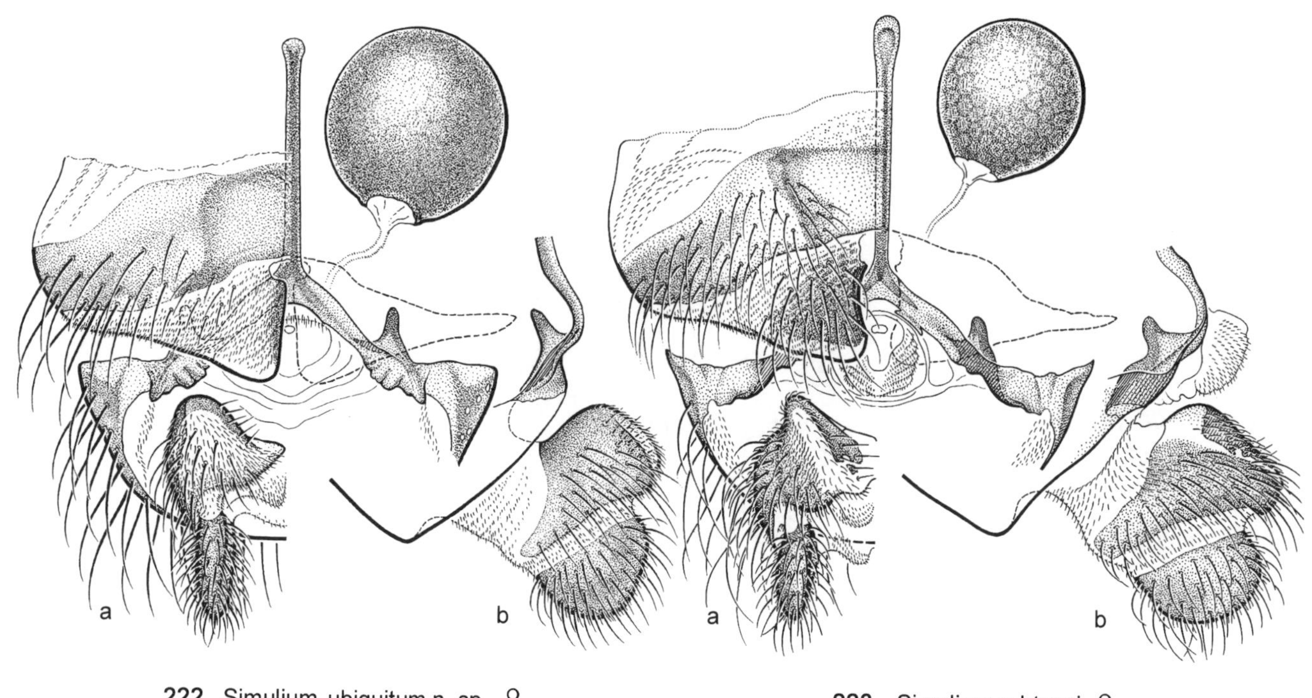


**222** Simulium ubiquitum n. sp. ♀

**223** Simulium rubtzovi ♀

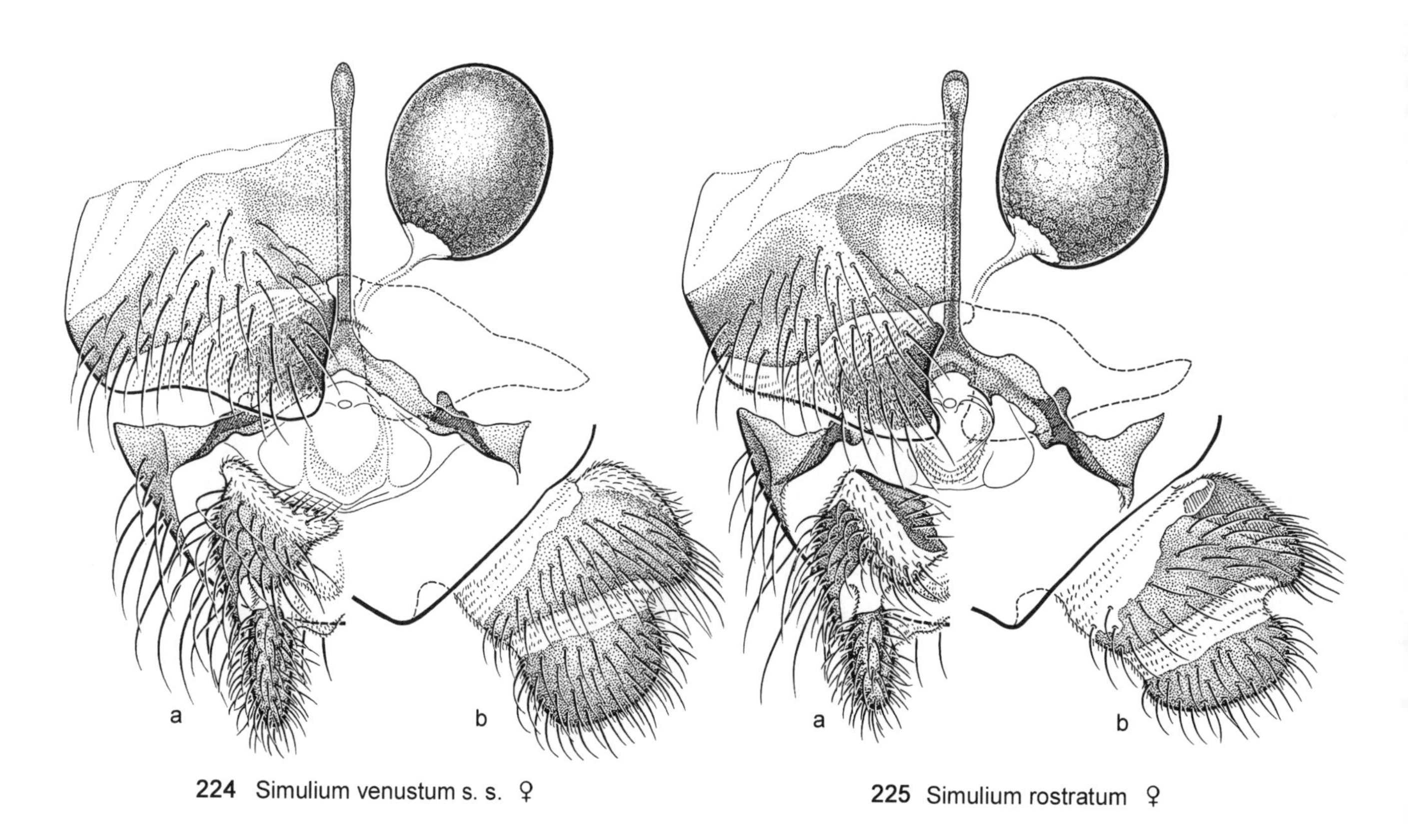


**224** Simulium venustum s. s. ♀

**225** Simulium rostratum ♀

Figs. 10.222–10.225. Female terminalia; a, ventral view; b, right lateral view.

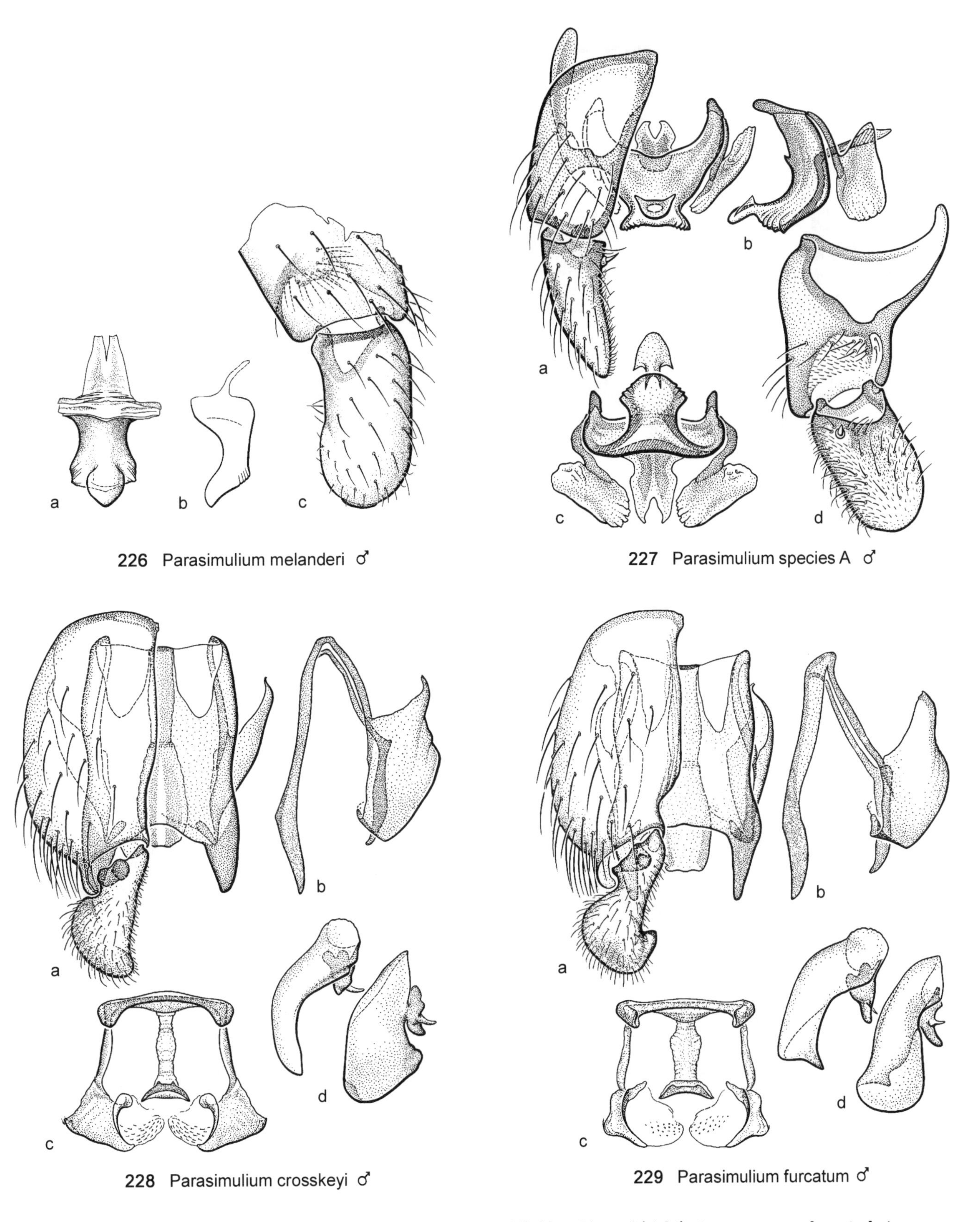


Figs. 10.226–10.229. Male genitalia. (226) a, b, Ventral plate, ventral (left) and lateral (right) views; c, gonopod, ventral view. (227–229) a, Ventral view; b, left lateral view; c, terminal view; d, right gonopod ([227d] dorsal view; [228d, 229d] narrowest and broadest views of gonostylus, setae omitted). (226, 229a) From Peterson (1981) by permission.

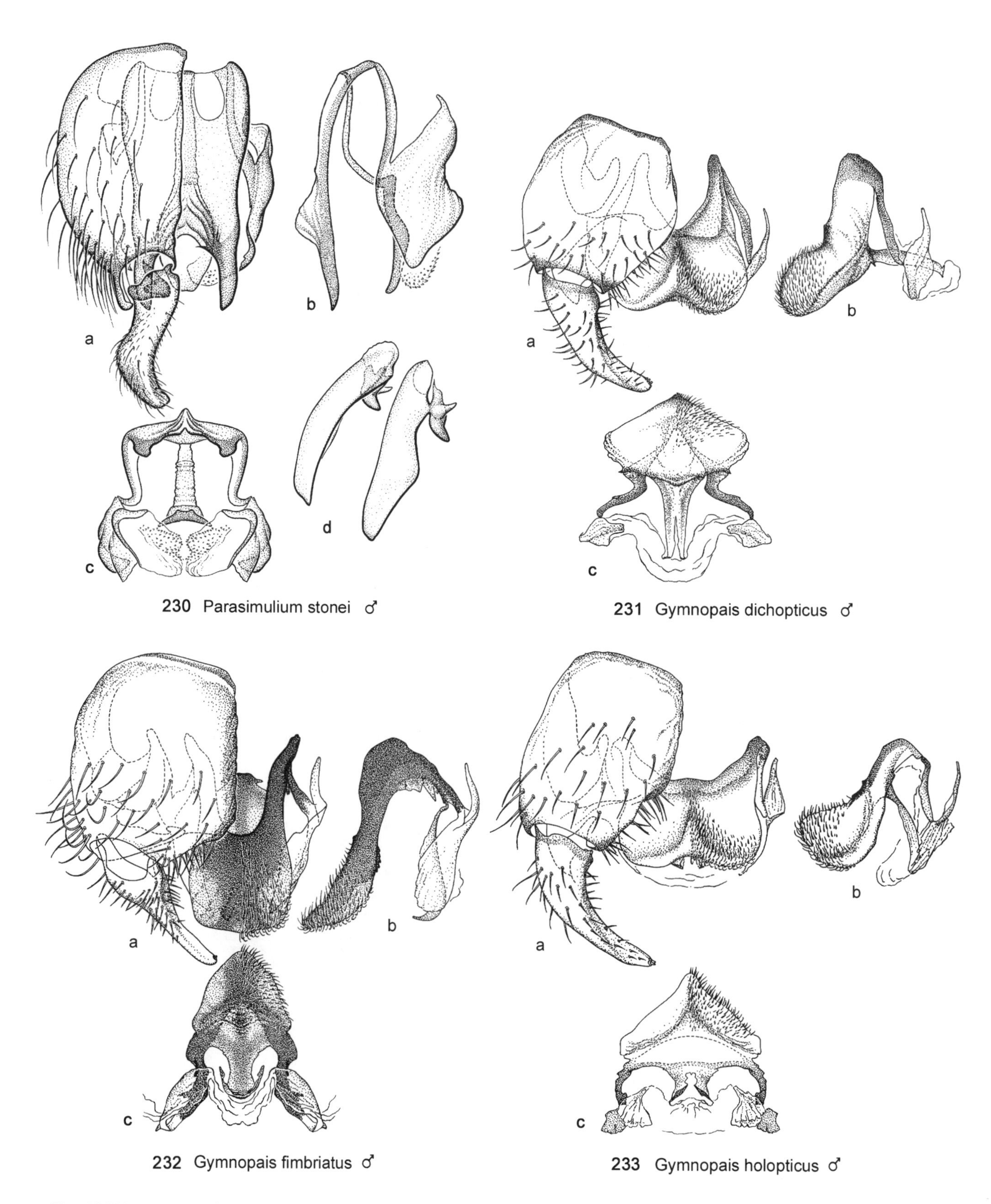


Figs. 10.230–10.233. Male genitalia; a, ventral view; b, left lateral view; c, terminal view; d, right gonostylus, narrowest and broadest views, setae omitted.

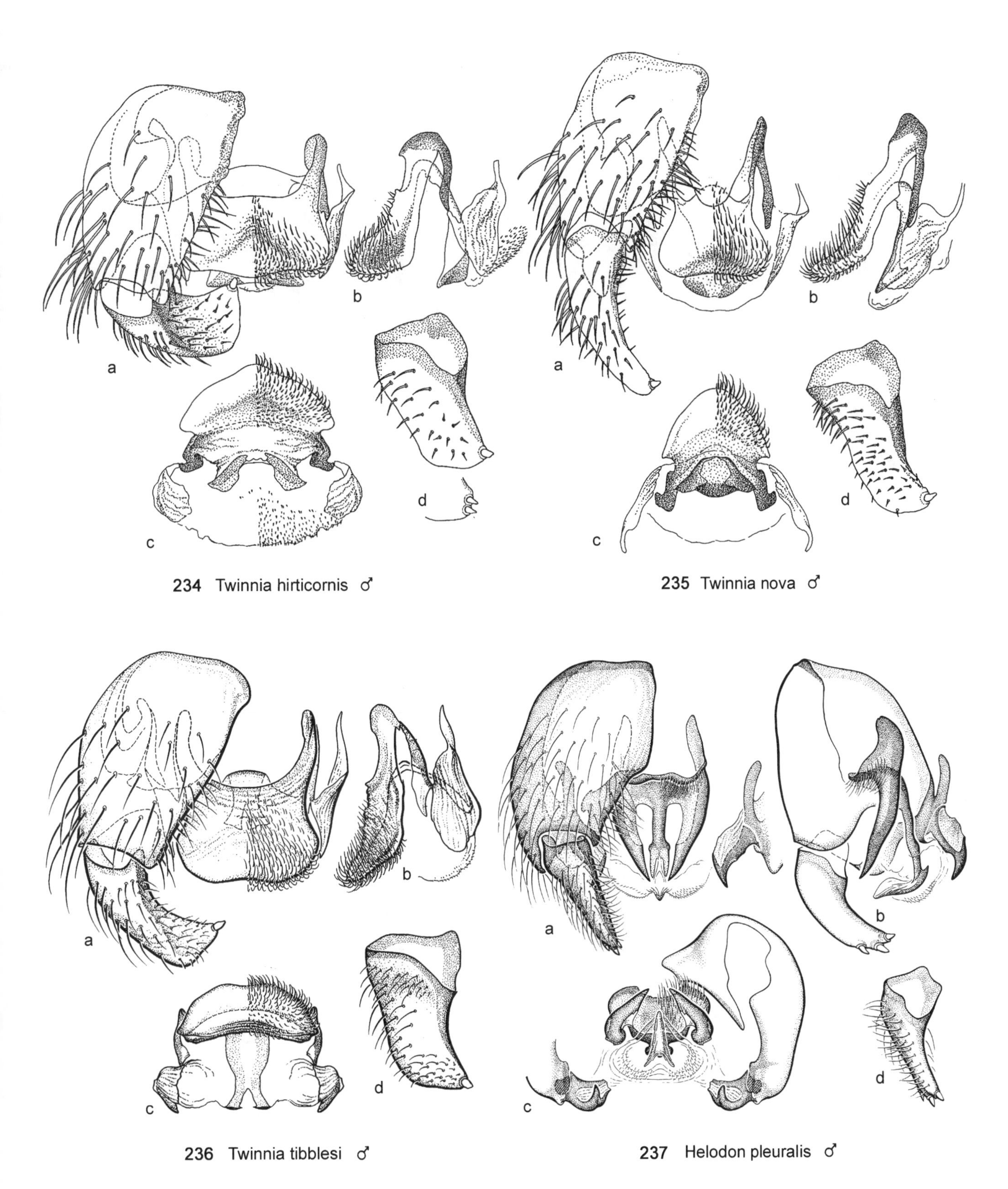


Figs. 10.234–10.237. Male genitalia; a, ventral view; b, left lateral view; c, terminal view; d, right gonostylus, inner lateral view. (236) From Peterson (1981) by permission.

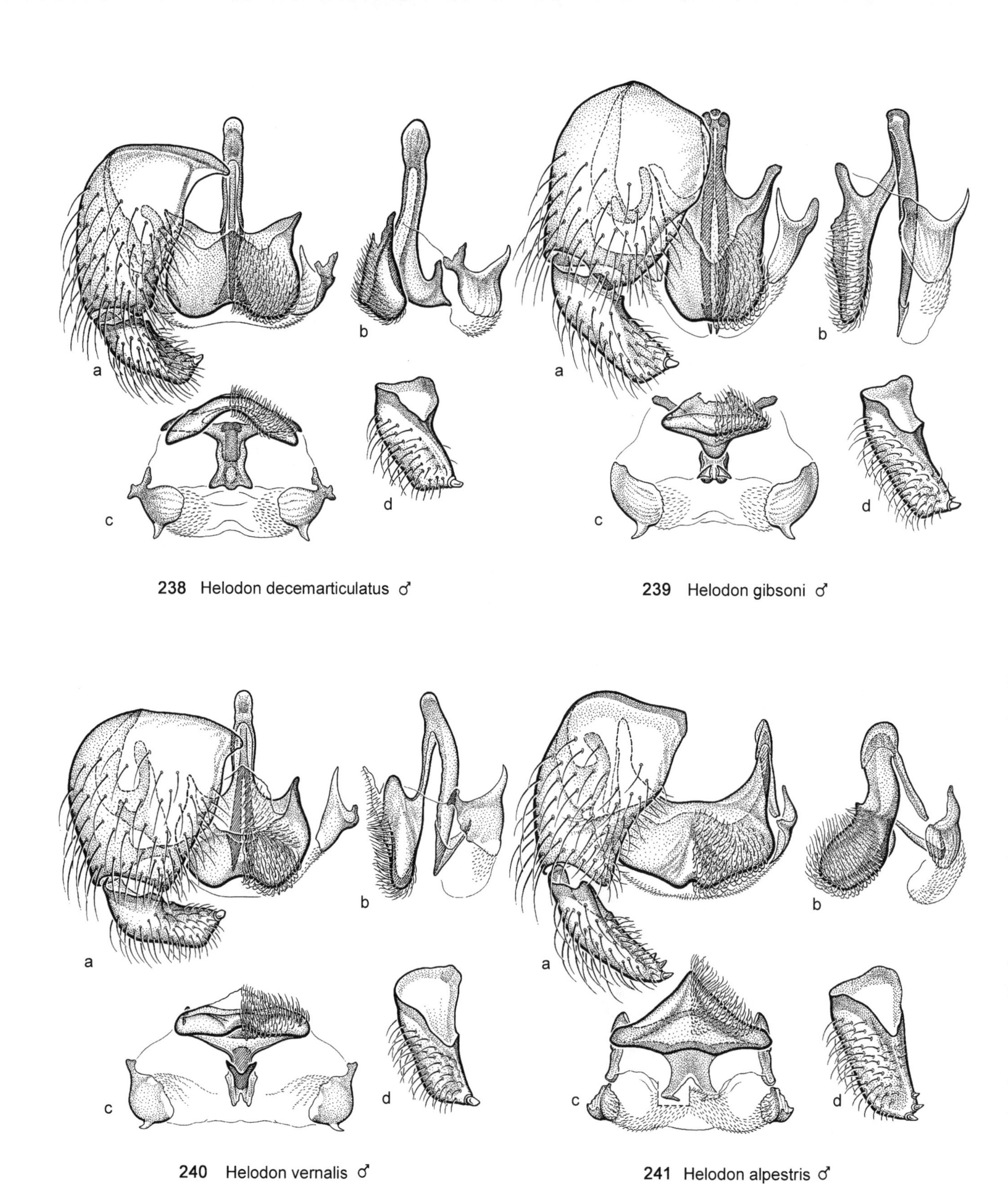


Figs. 10.238–10.241. Male genitalia; a, ventral view; b, left lateral view; c, terminal view; d, right gonostylus, inner lateral view.

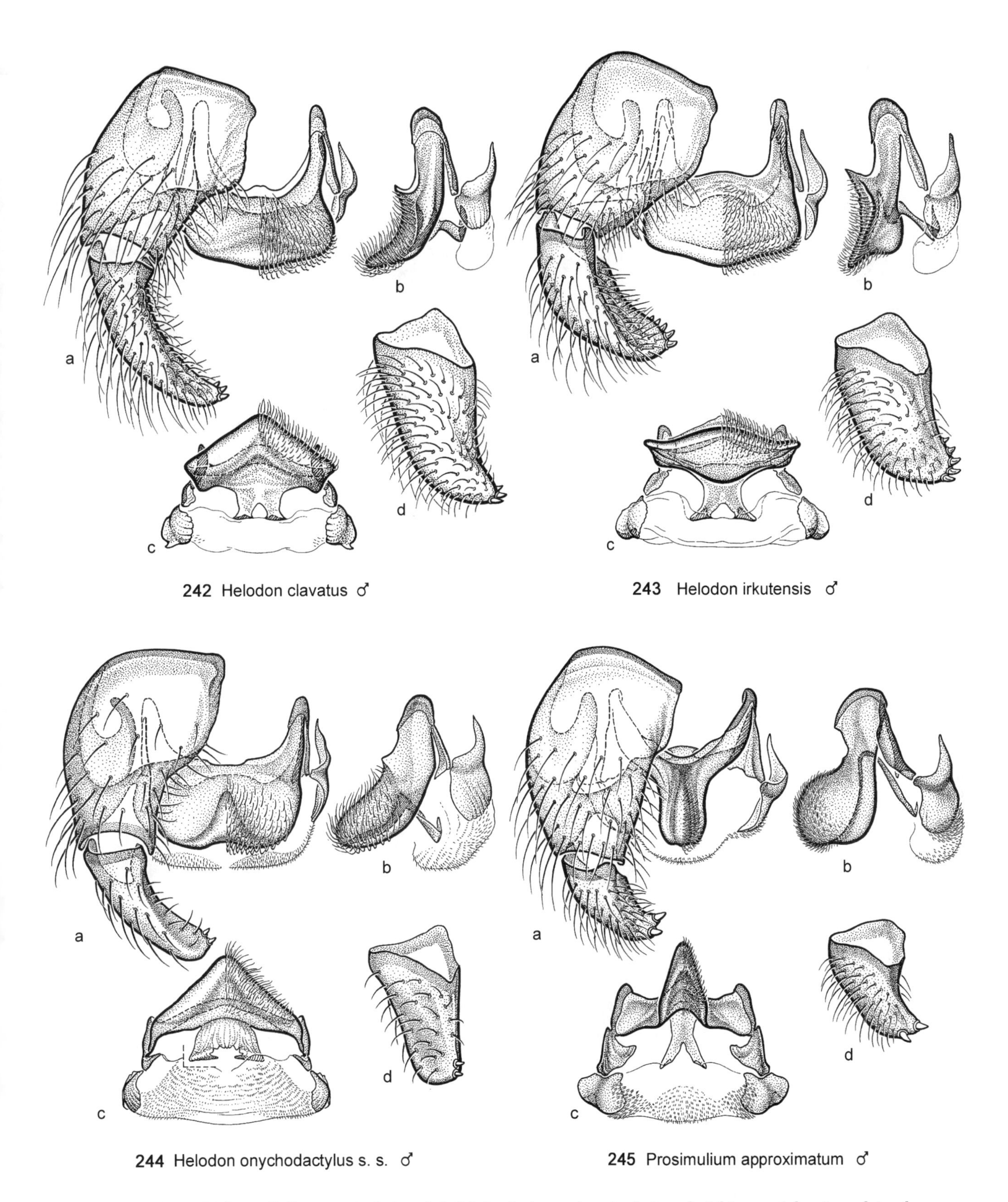


Figs. 10.242–10.245. Male genitalia; a, ventral view; b, left lateral view; c, terminal view; d, right gonostylus, inner lateral view.

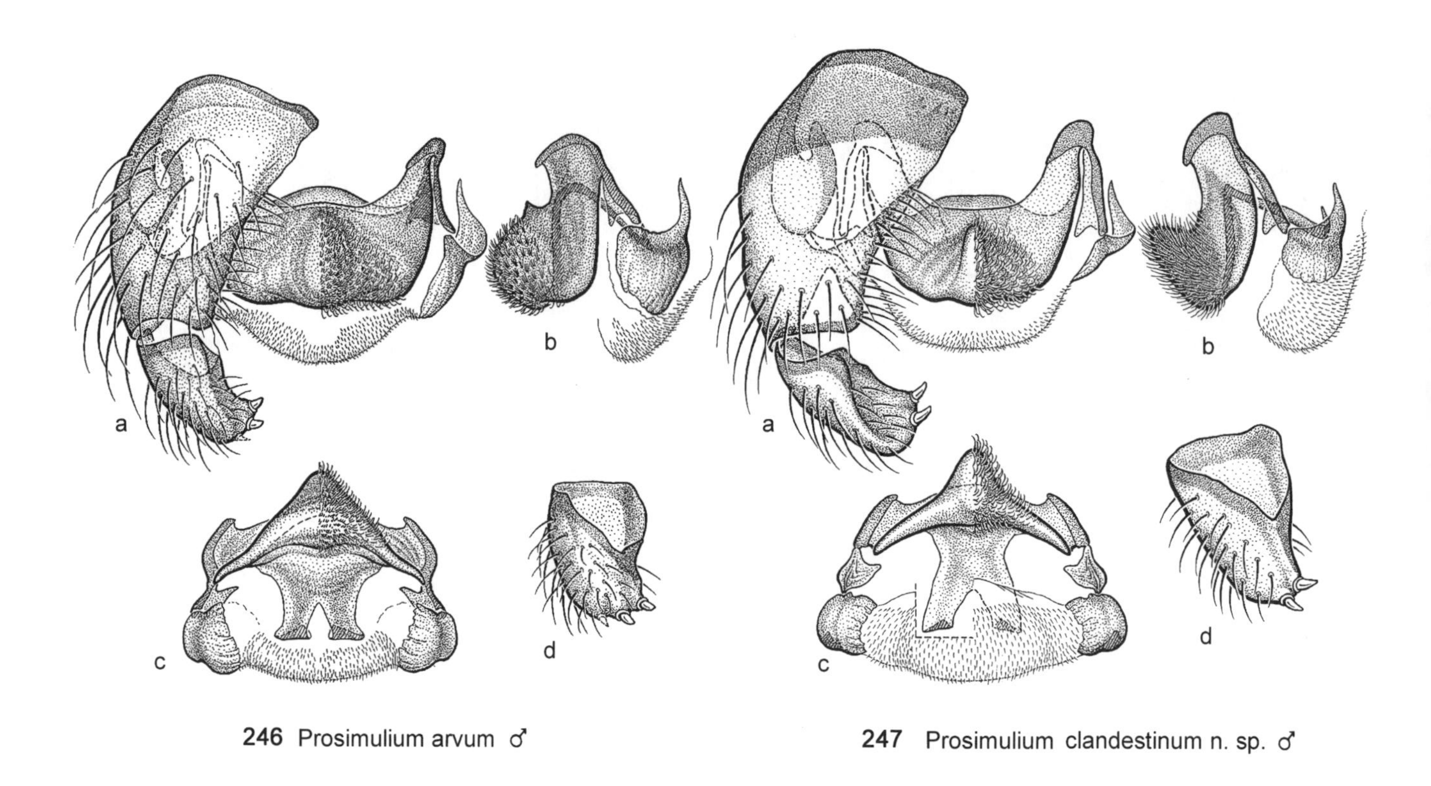


**246** Prosimulium arvum ♂

**247** Prosimulium clandestinum n. sp. ♂

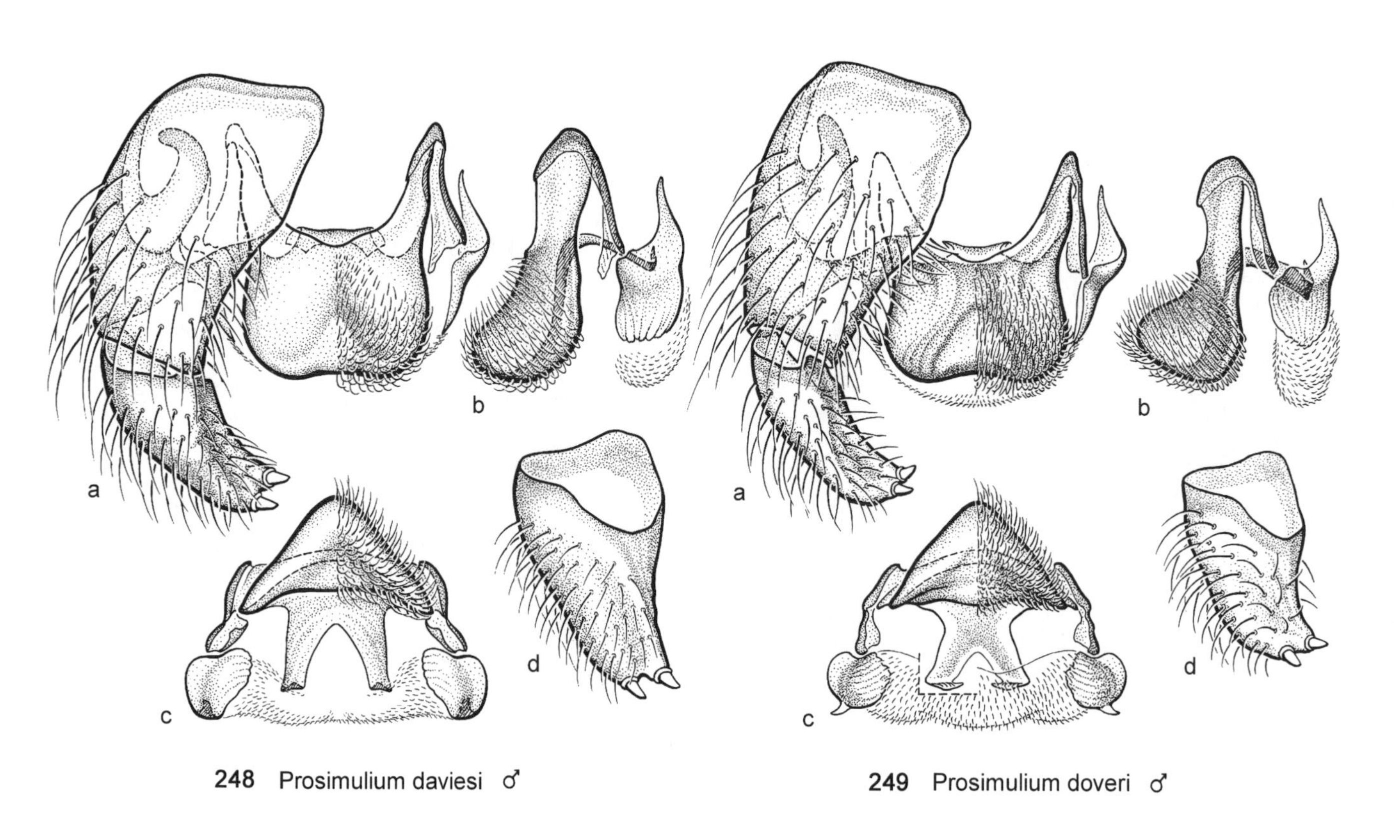


**248** Prosimulium daviesi ♂

**249** Prosimulium doveri ♂

Figs. 10.246–10.249. Male genitalia; a, ventral view; b, left lateral view; c, terminal view; d, right gonostylus, inner lateral view.

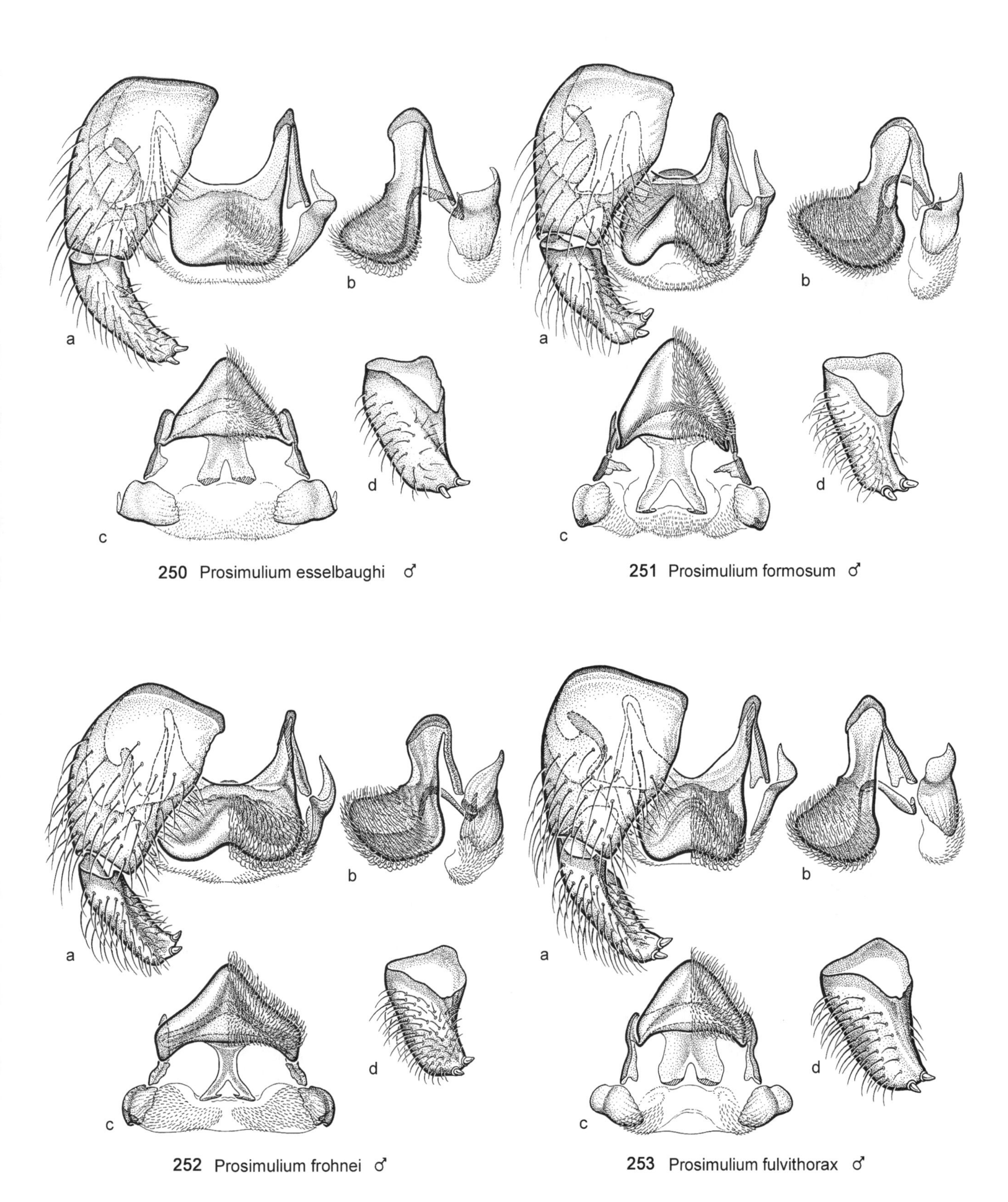


Figs. 10.250–10.253. Male genitalia; a, ventral view; b, left lateral view; c, terminal view; d, right gonostylus, inner lateral view.

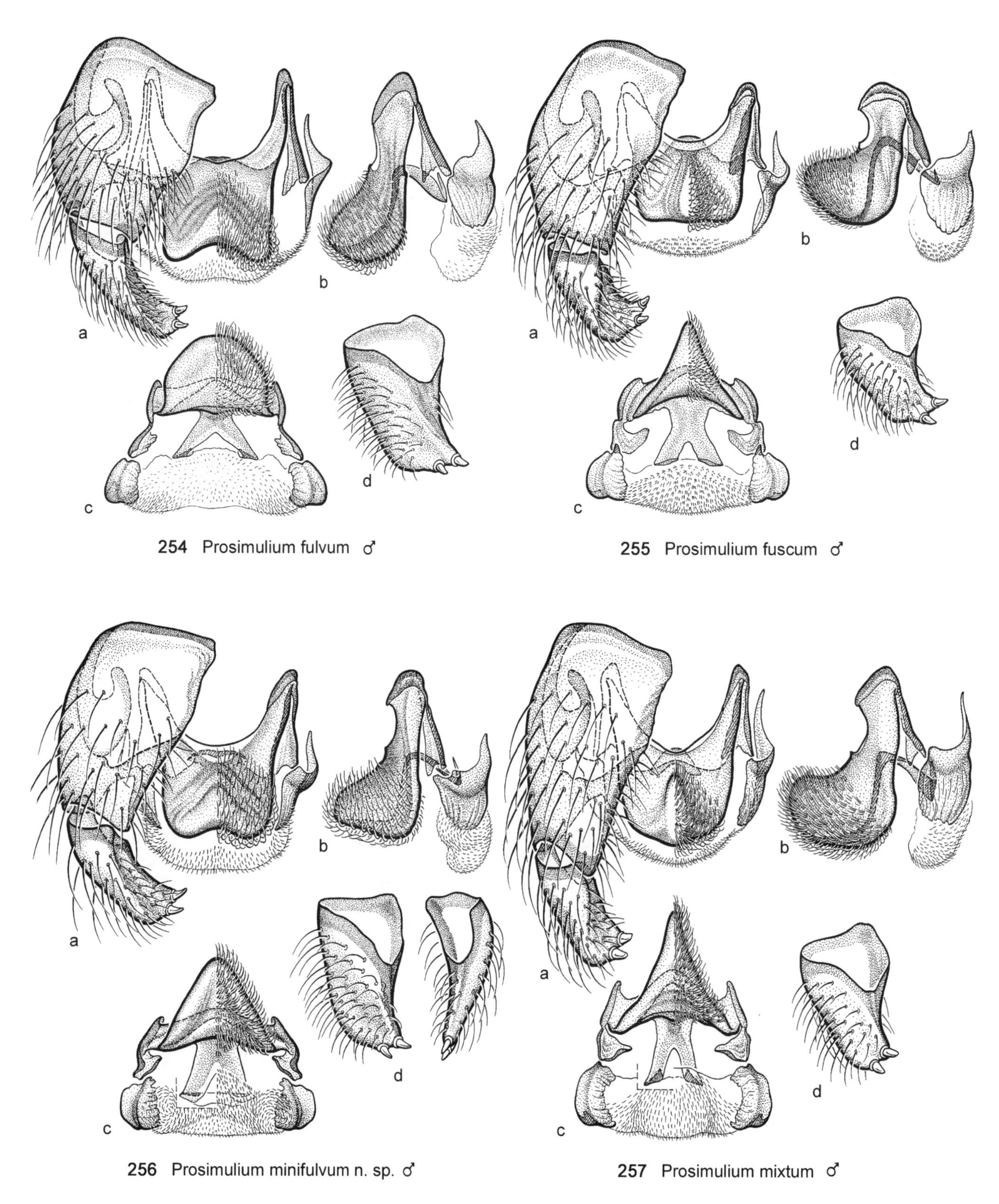


Figs. 10.254–10.257. Male genitalia; a, ventral view; b, left lateral view; c, terminal view; d, right gonostylus ([254d, 255d, 257d] inner lateral view; [256d] inner lateral view [left], inner dorsolateral view [right]).

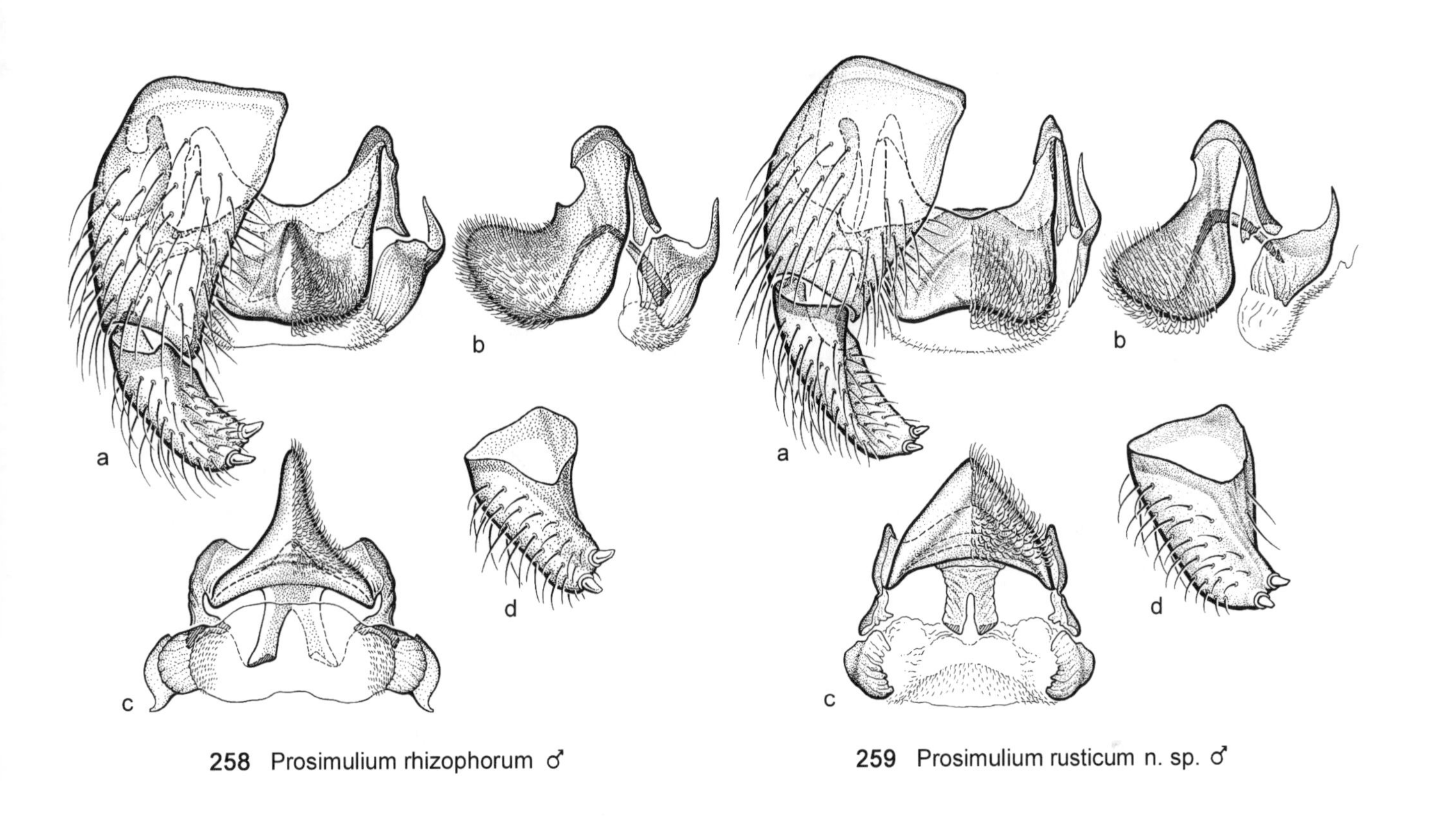


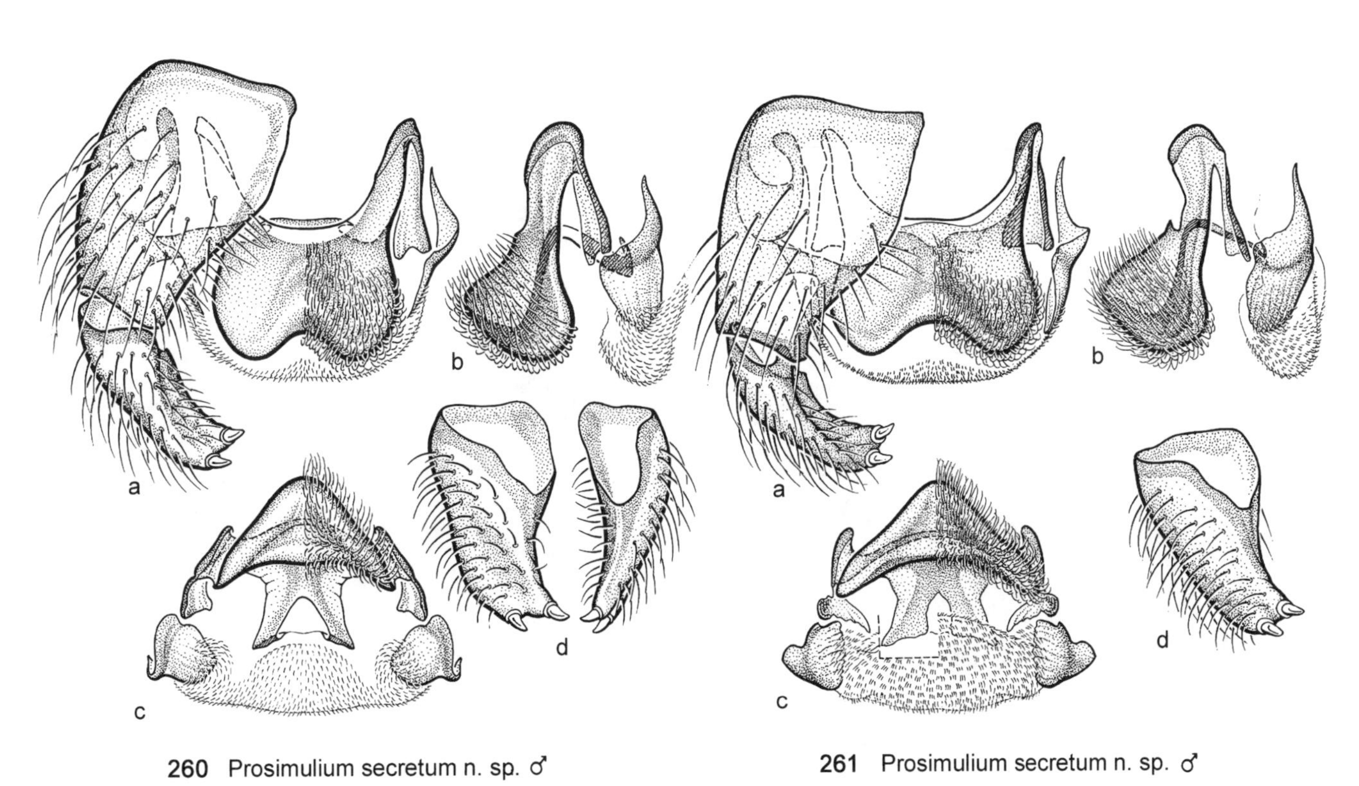


Figs. 10.258–10.261. Male genitalia; a, ventral view; b, left lateral view; c, terminal view; d, right gonostylus ([258d, 259d, 261d] inner lateral view; [260d] inner lateral view [left], inner dorsolateral view [right]).

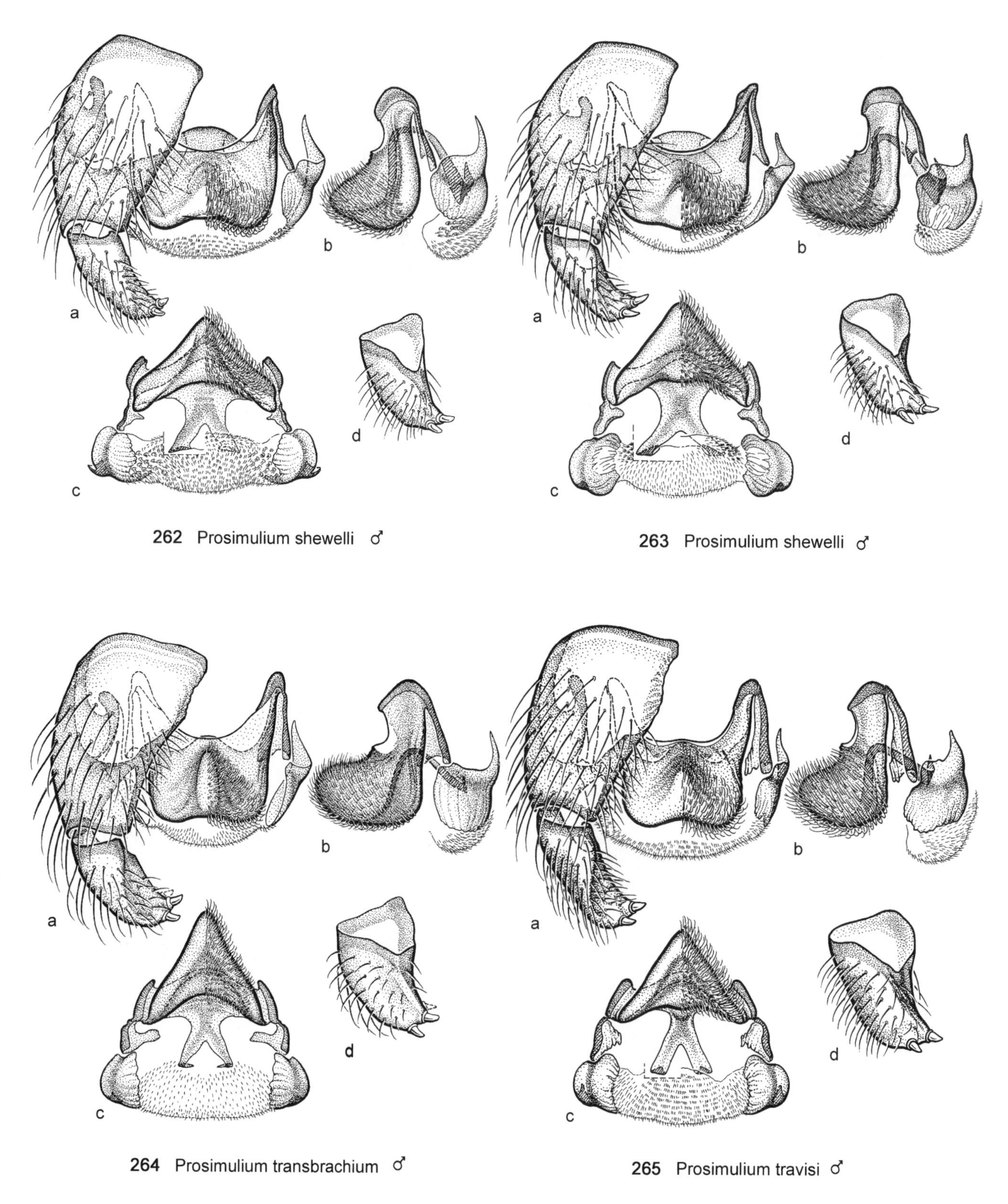


Figs. 10.262–10.265. Male genitalia; a, ventral view; b, left lateral view; c, terminal view; d, right gonostylus, inner lateral view.

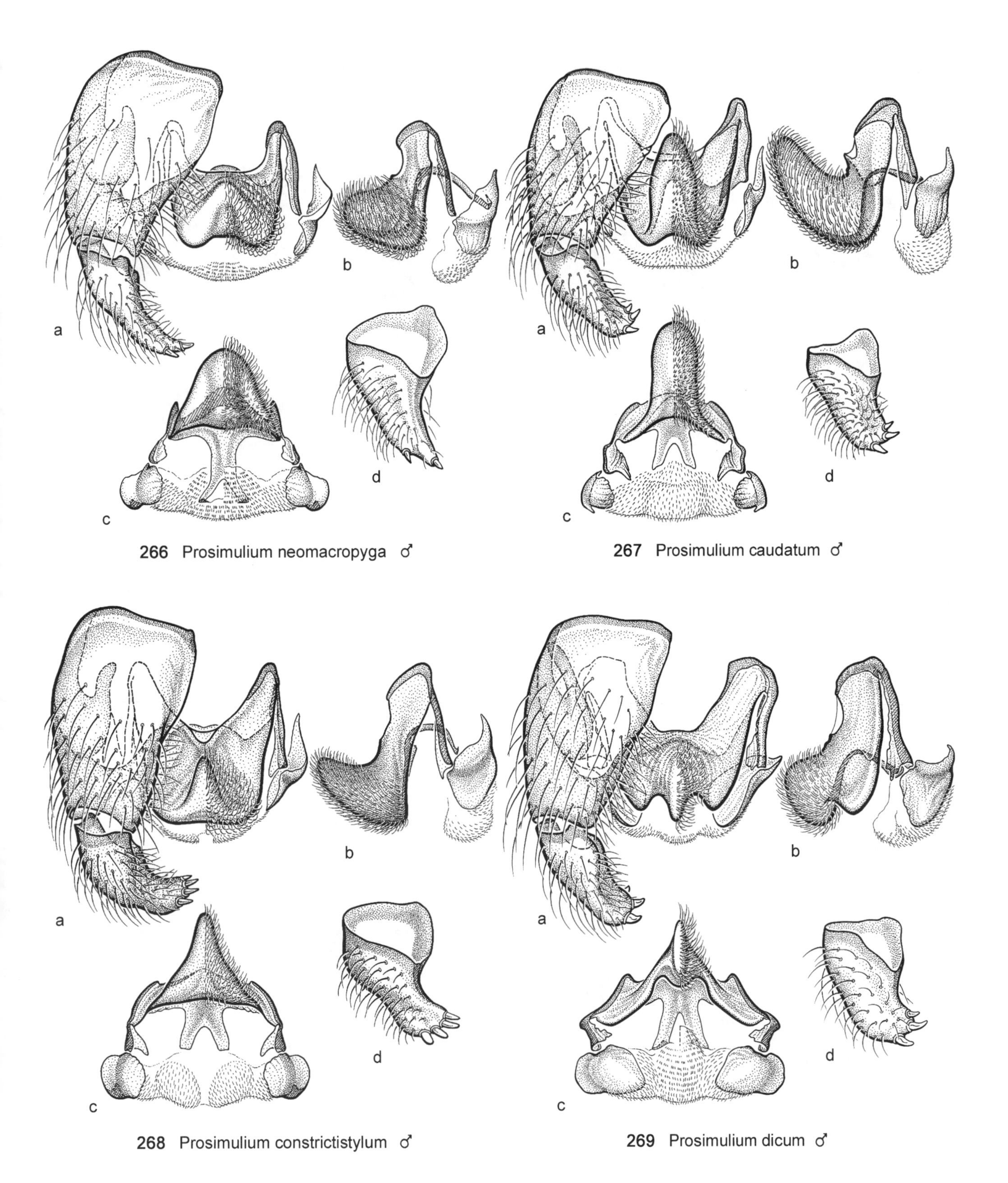


Figs. 10.266–10.269. Male genitalia; a, ventral view; b, left lateral view; c, terminal view; d, right gonostylus, inner lateral view.

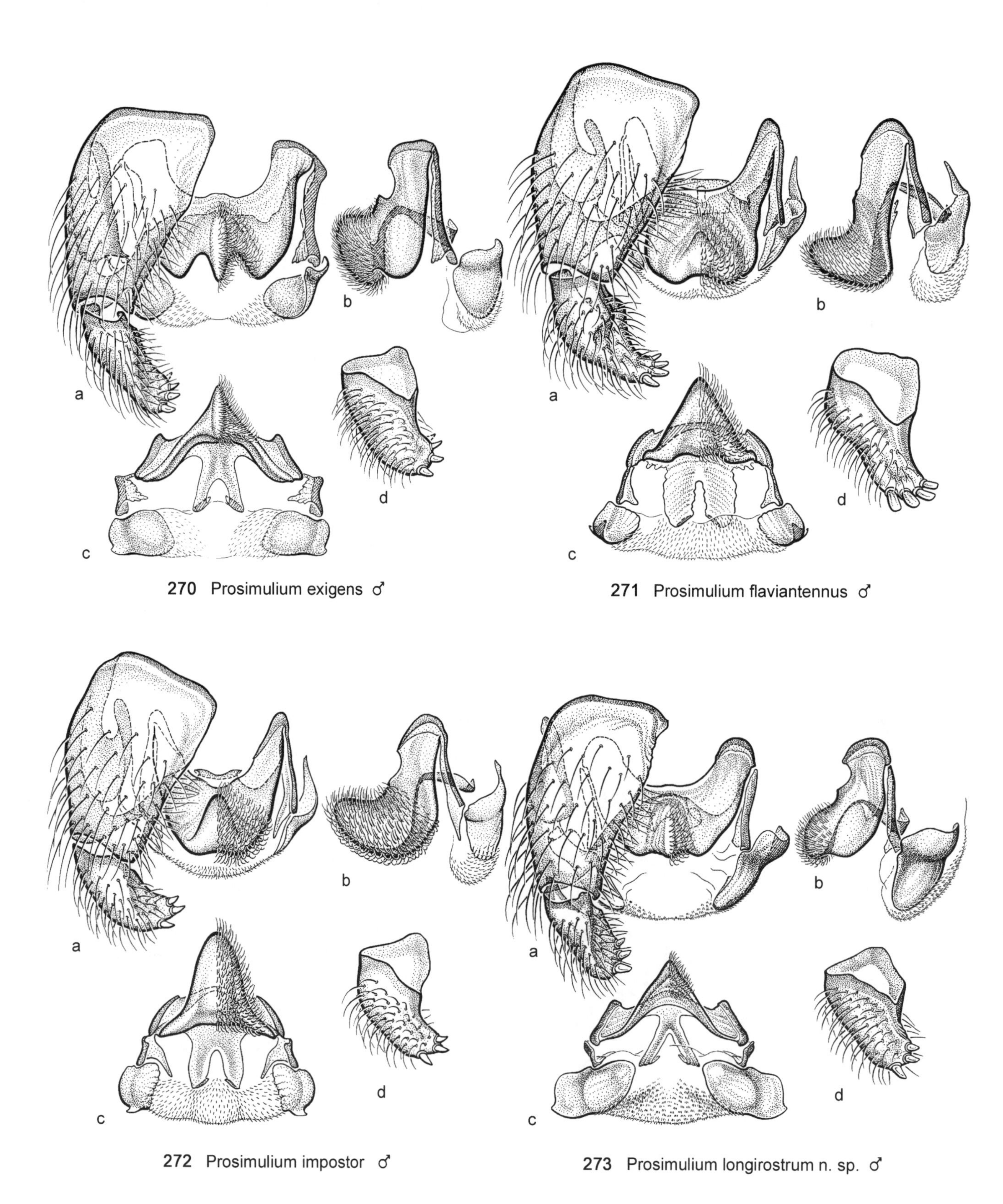


Figs. 10.270–10.273. Male genitalia; a, ventral view; b, left lateral view; c, terminal view; d, right gonostylus, inner lateral view.

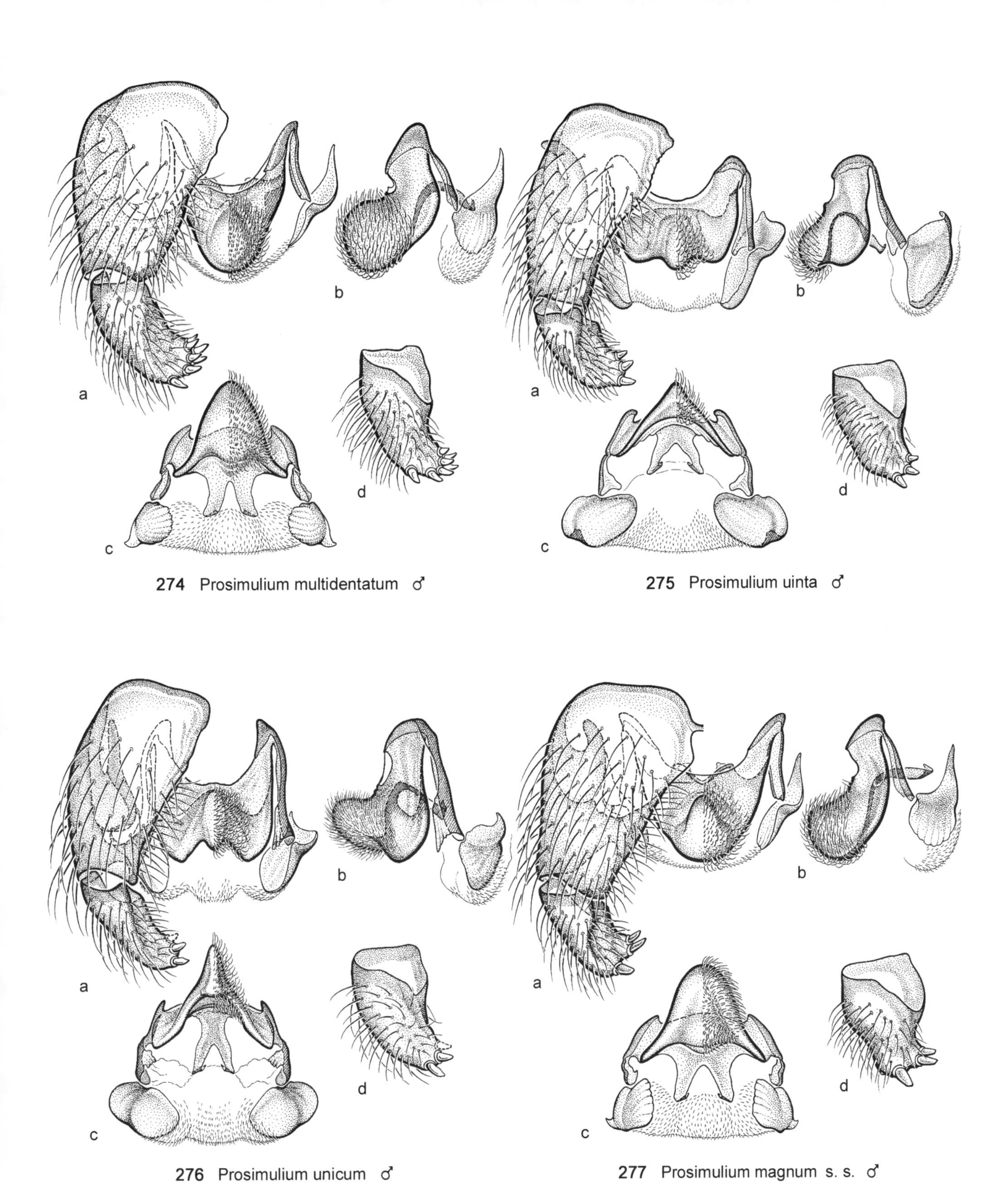


Figs. 10.274–10.277. Male genitalia; a, ventral view; b, left lateral view; c, terminal view; d, right gonostylus, inner lateral view.

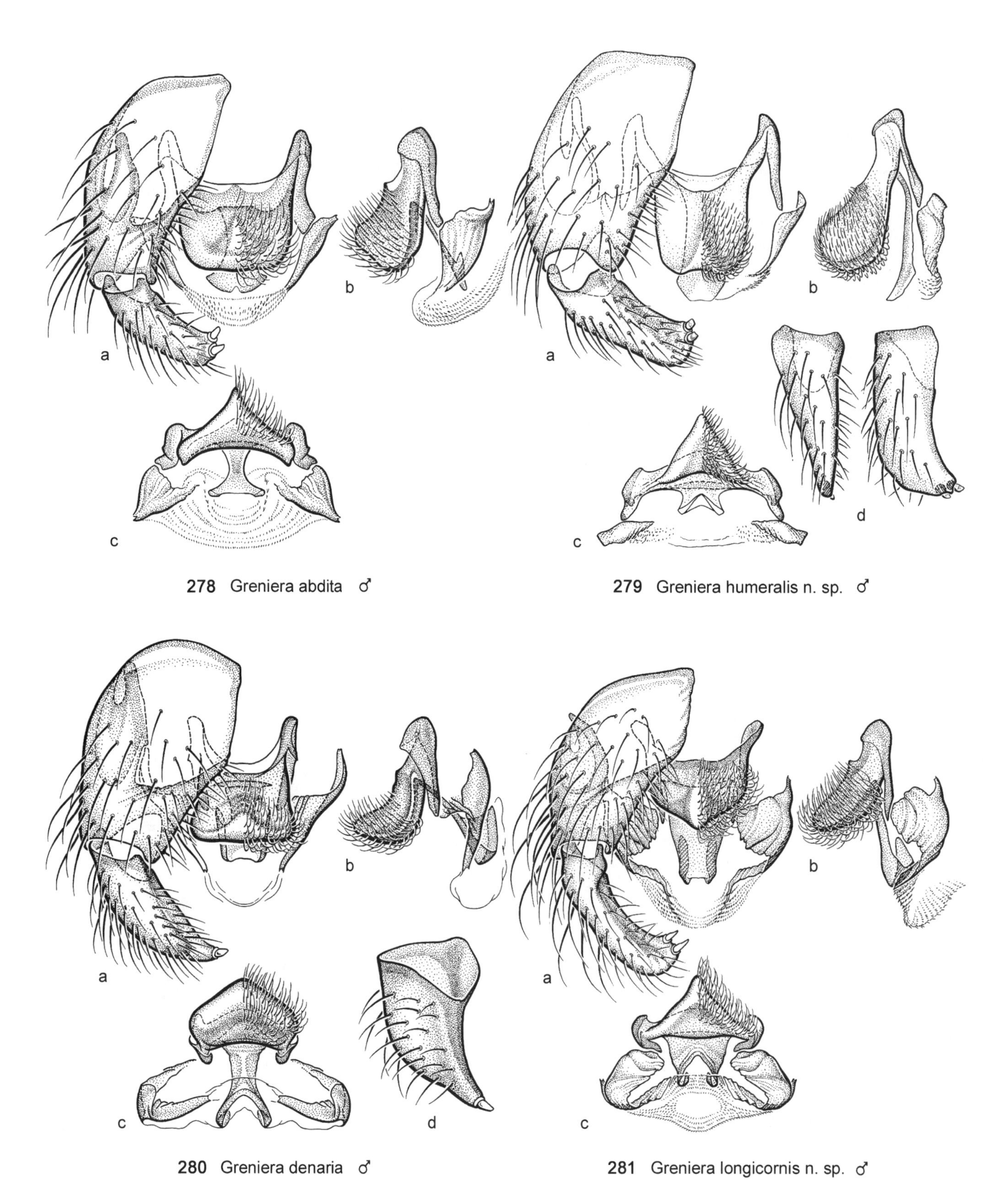


Figs. 10.278–10.281. Male genitalia; a, ventral view; b, left lateral view; c, terminal view; d, right gonostylus ([279d] outer ventrolateral view [left], inner lateral view [right]; [280d] inner lateral view).

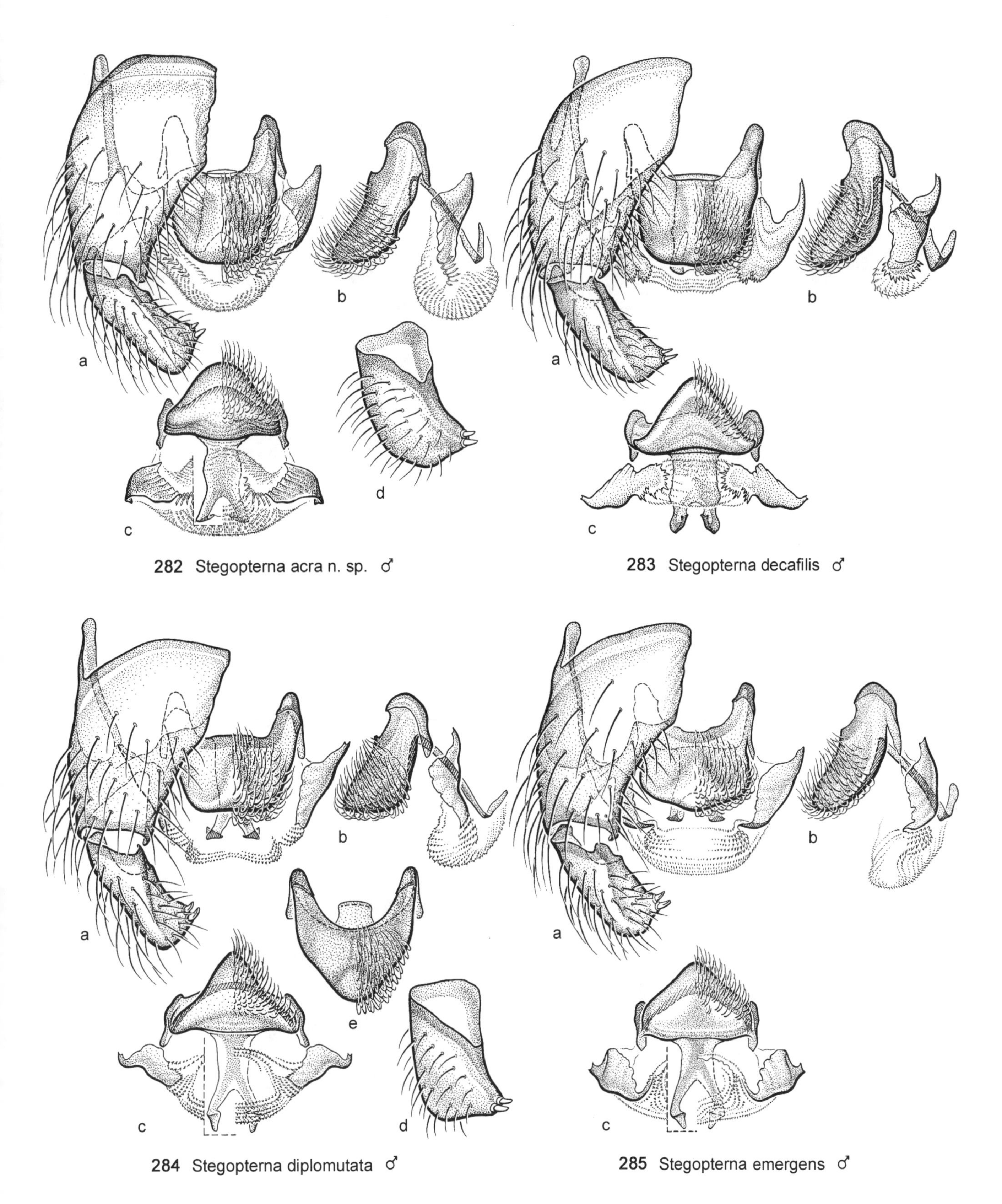


Figs. 10.282–10.285. Male genitalia; a, ventral view; b, left lateral view; c, terminal view; d, right gonostylus, inner lateral view; (284e) ventral plate in ventral view tilted dorsally.

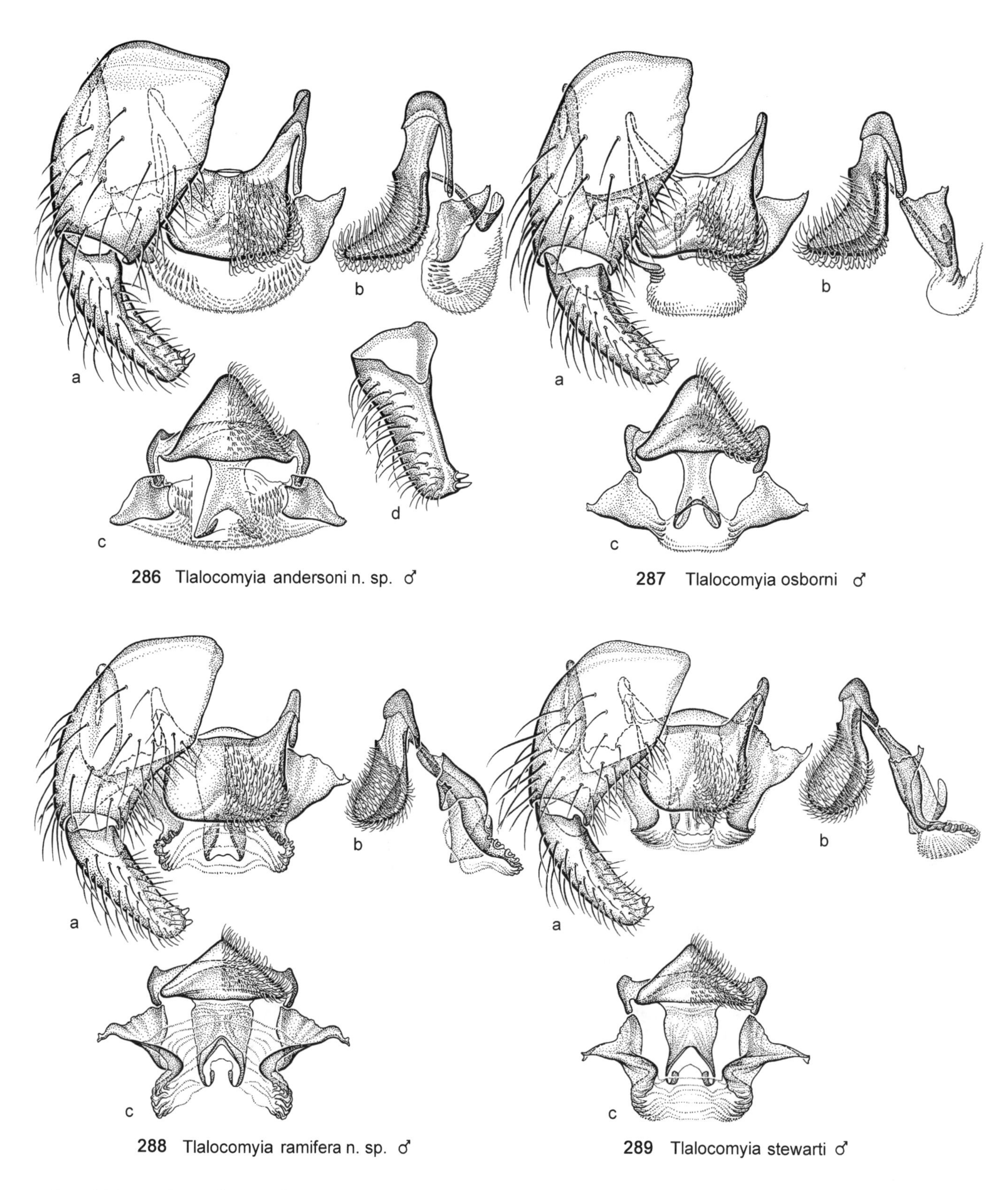


Figs. 10.286–10.289. Male genitalia; a, ventral view; b, left lateral view; c, terminal view; d, right gonostylus, inner lateral view.

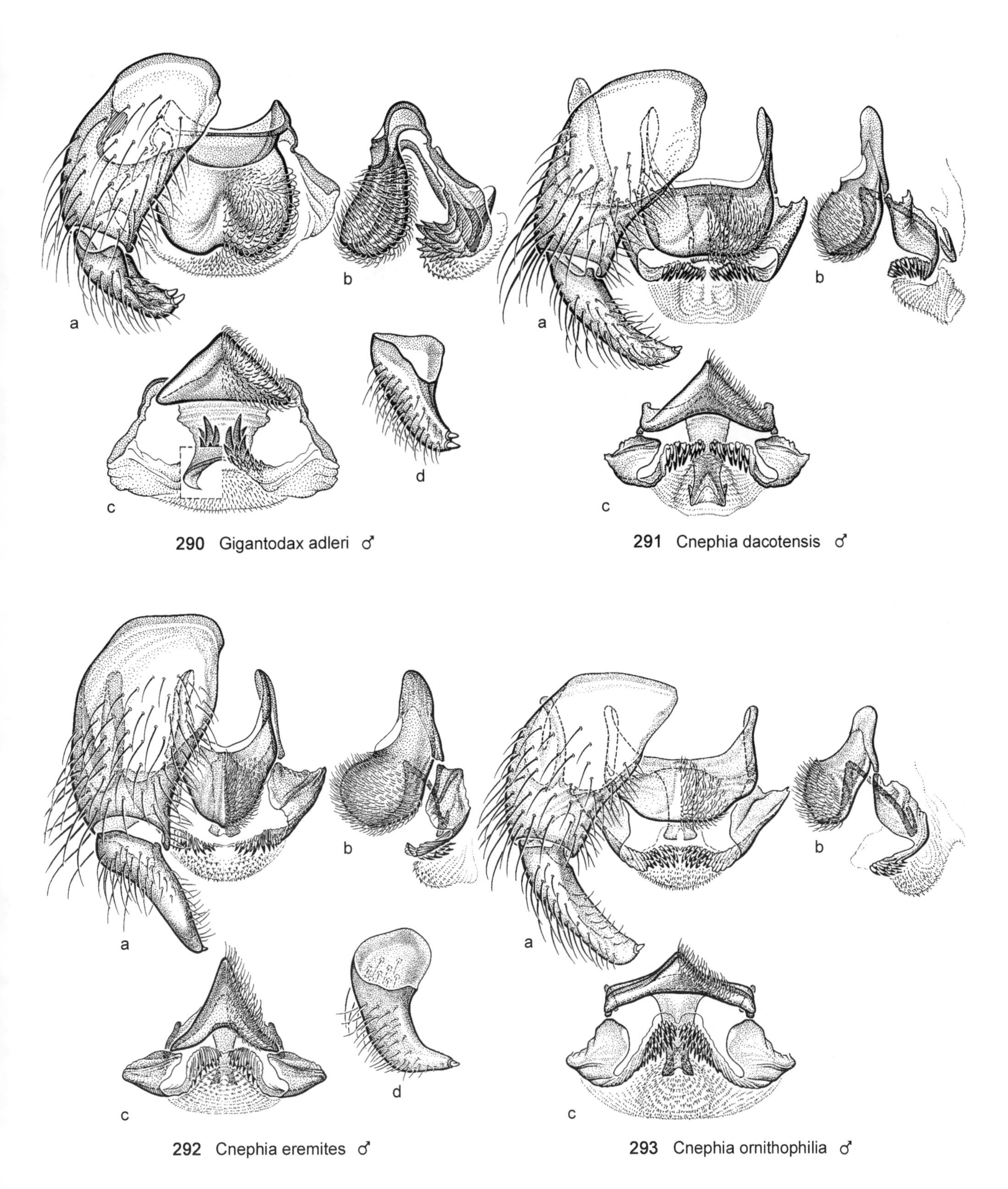


Figs. 10.290–10.293. Male genitalia; a, ventral view; b, left lateral view; c, terminal view; d, right gonostylus, inner lateral view.

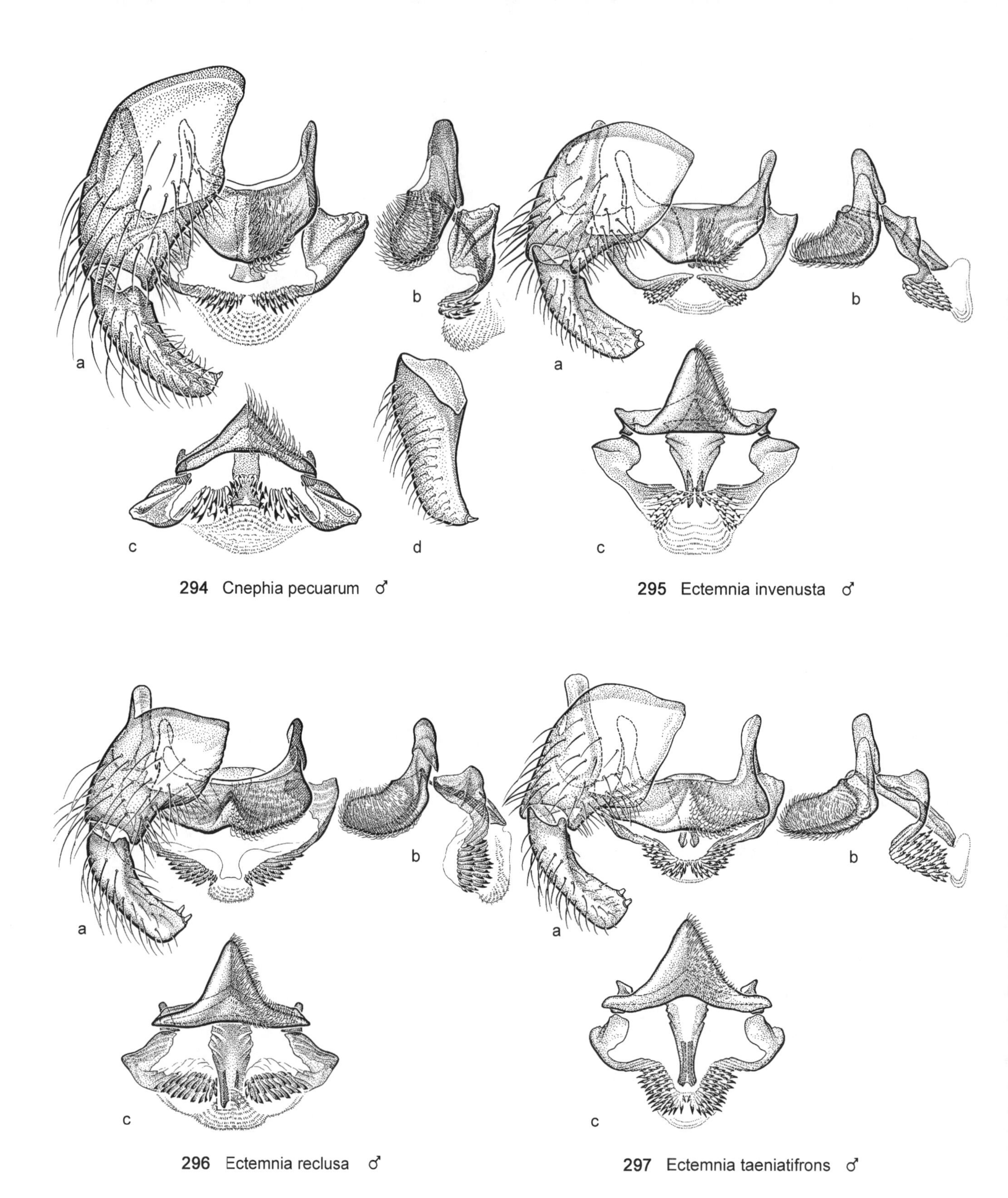


Figs. 10.294–10.297. Male genitalia; a, ventral view; b, left lateral view; c, terminal view; d, right gonostylus, inner lateral view.

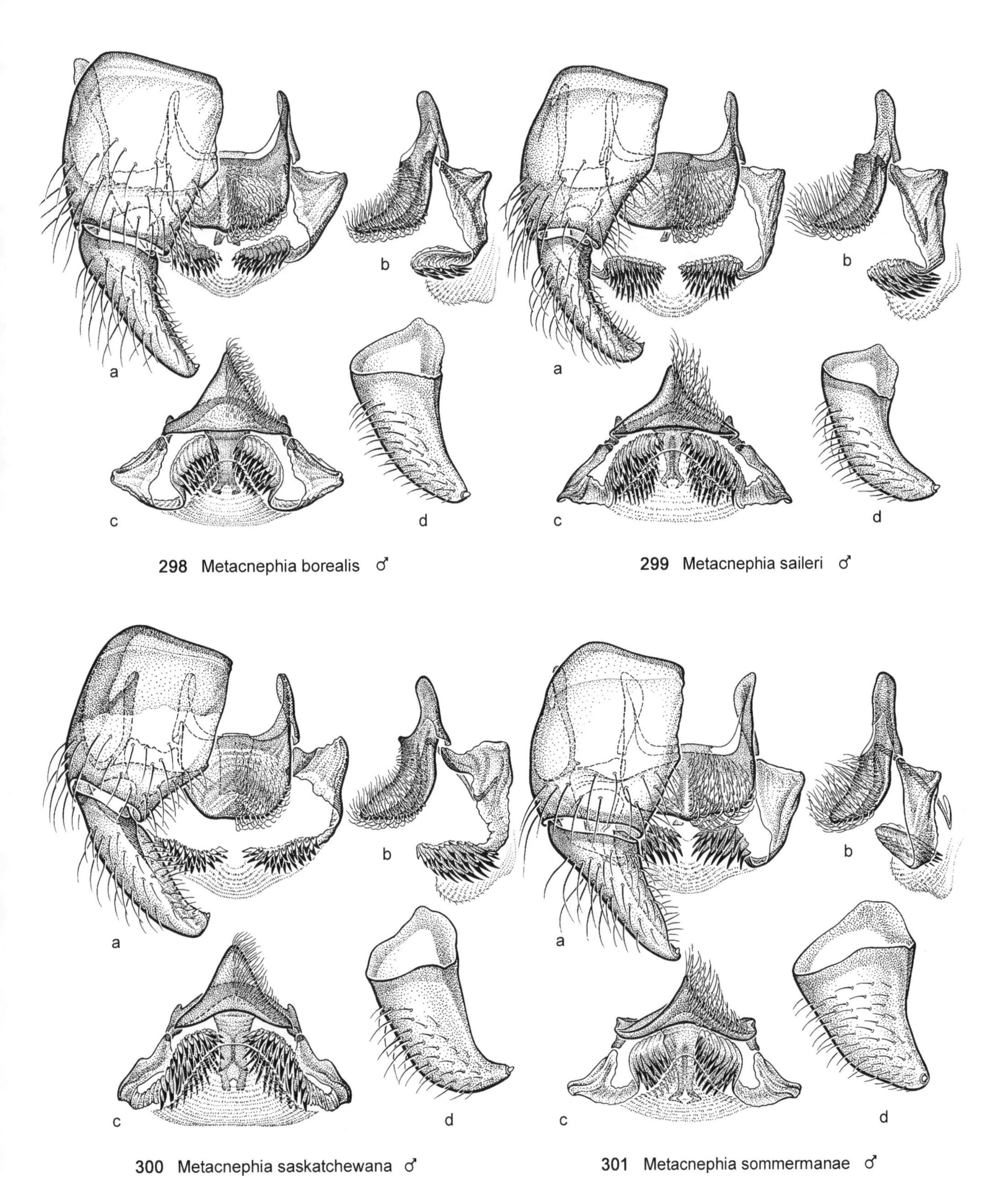


Figs. 10.298–10.301. Male genitalia; a, ventral view; b, left lateral view; c, terminal view; d, right gonostylus, inner lateral view.

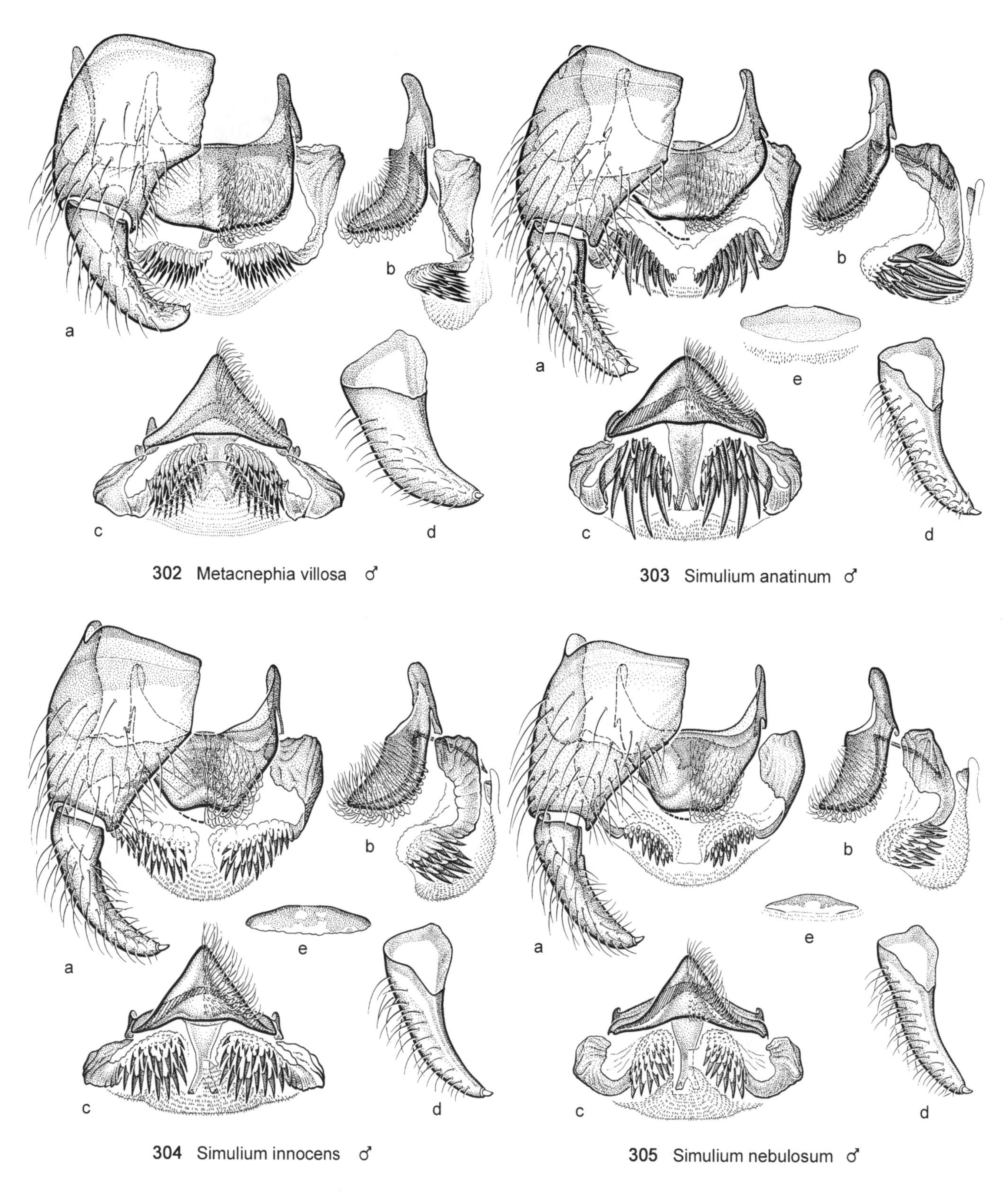


Figs. 10.302–10.305. Male genitalia; a, ventral view; b, left lateral view; c, terminal view; d, right gonostylus, inner lateral view; e, dorsal plate.

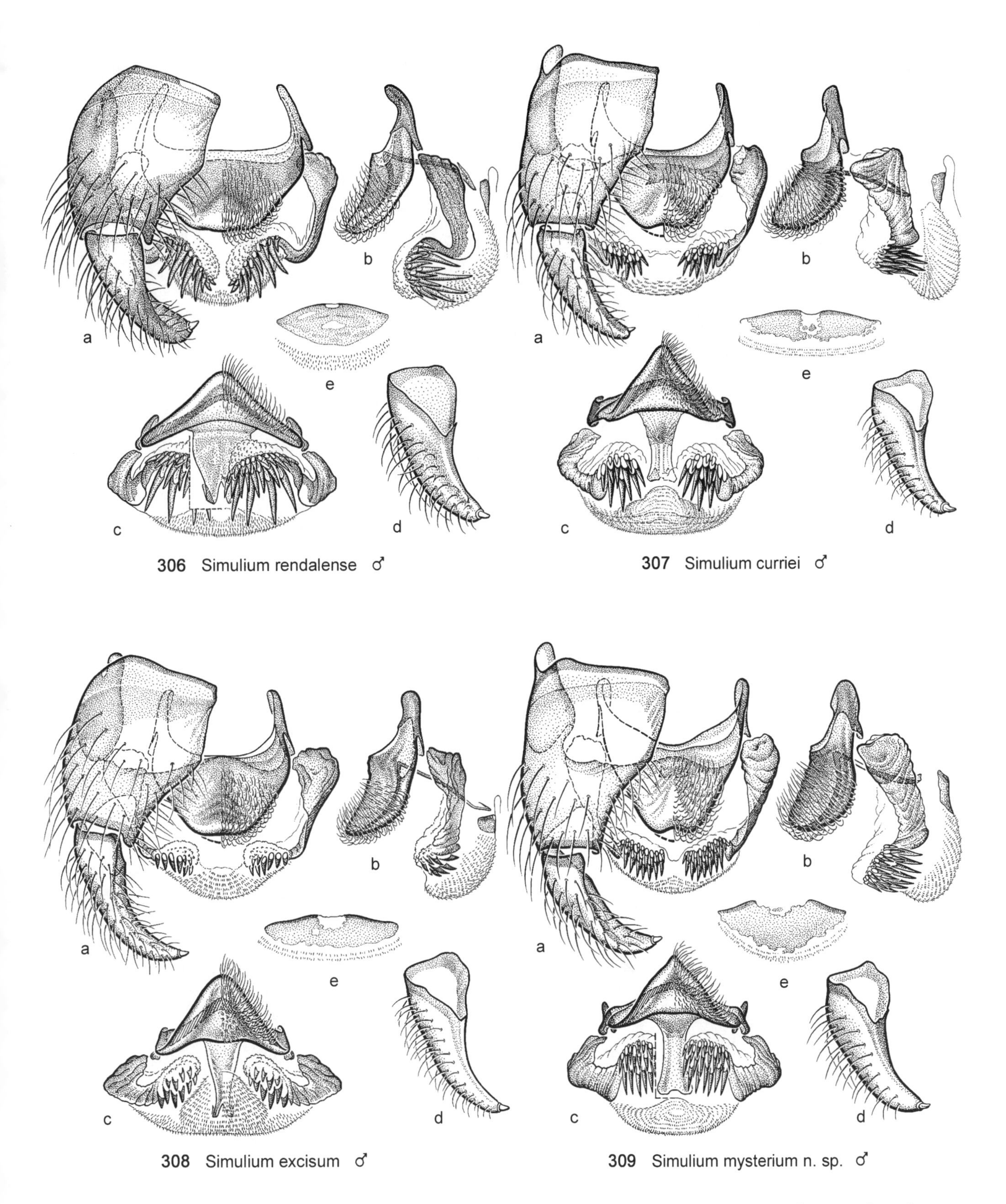


Figs. 10.306–10.309. Male genitalia; a, ventral view; b, left lateral view; c, terminal view; d, right gonostylus, inner lateral view; e, dorsal plate.

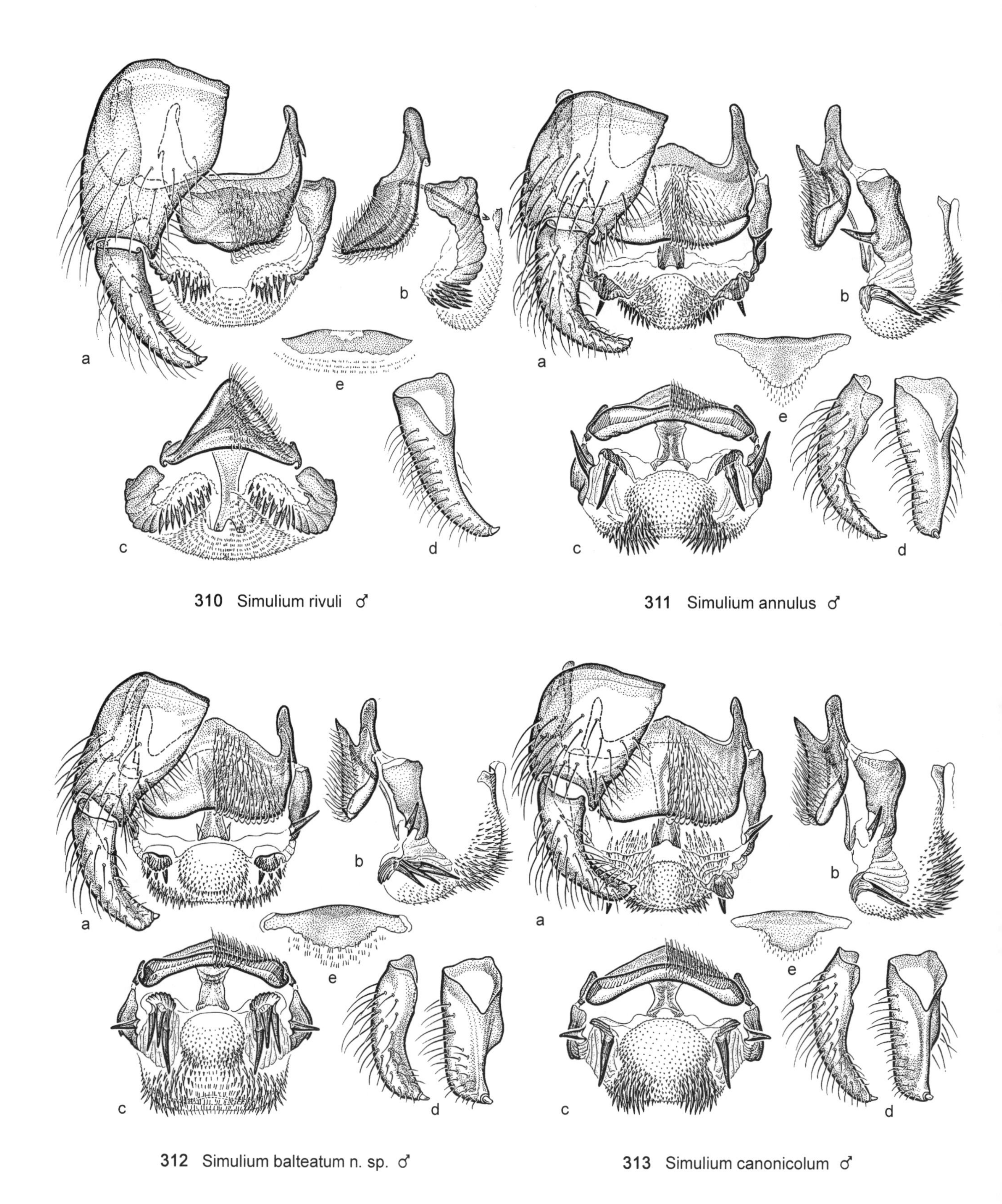


Figs. 10.310–10.313. Male genitalia; a, ventral view; b, left lateral view; c, terminal view; d, right gonostylus ([310d] inner lateral view; [311d, 312d, 313d] inner ventrolateral view [left], inner lateral view [right]); e, dorsal plate.

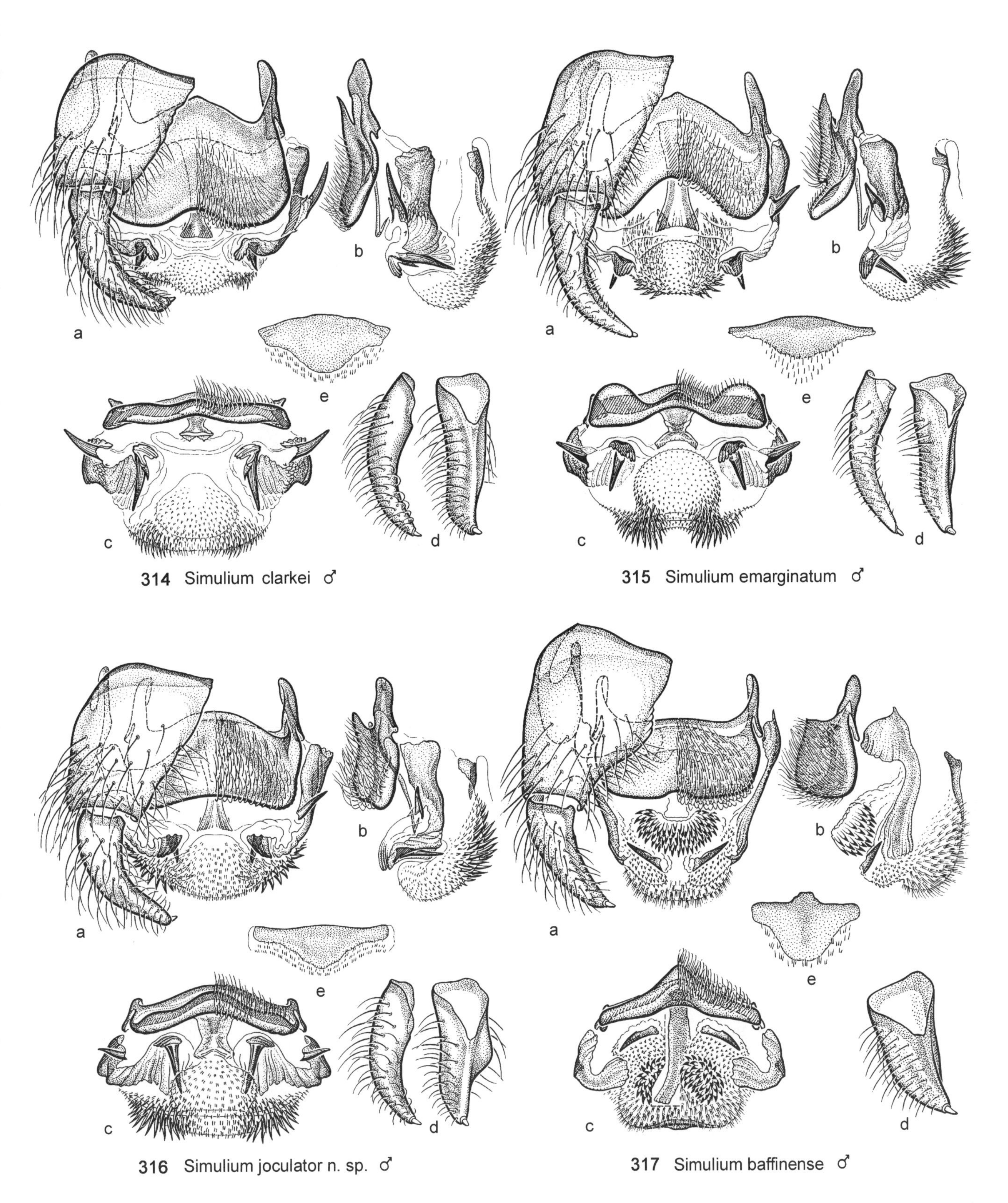


Figs. 10.314–10.317. Male genitalia; a, ventral view; b, left lateral view; c, terminal view; d, right gonostylus ([314d, 315d, 316d] inner ventrolateral view [left], inner lateral view [right]; [317d] inner lateral view); e, dorsal plate. (314a, b, e) From Moulton and Adler (2002b) by permission.

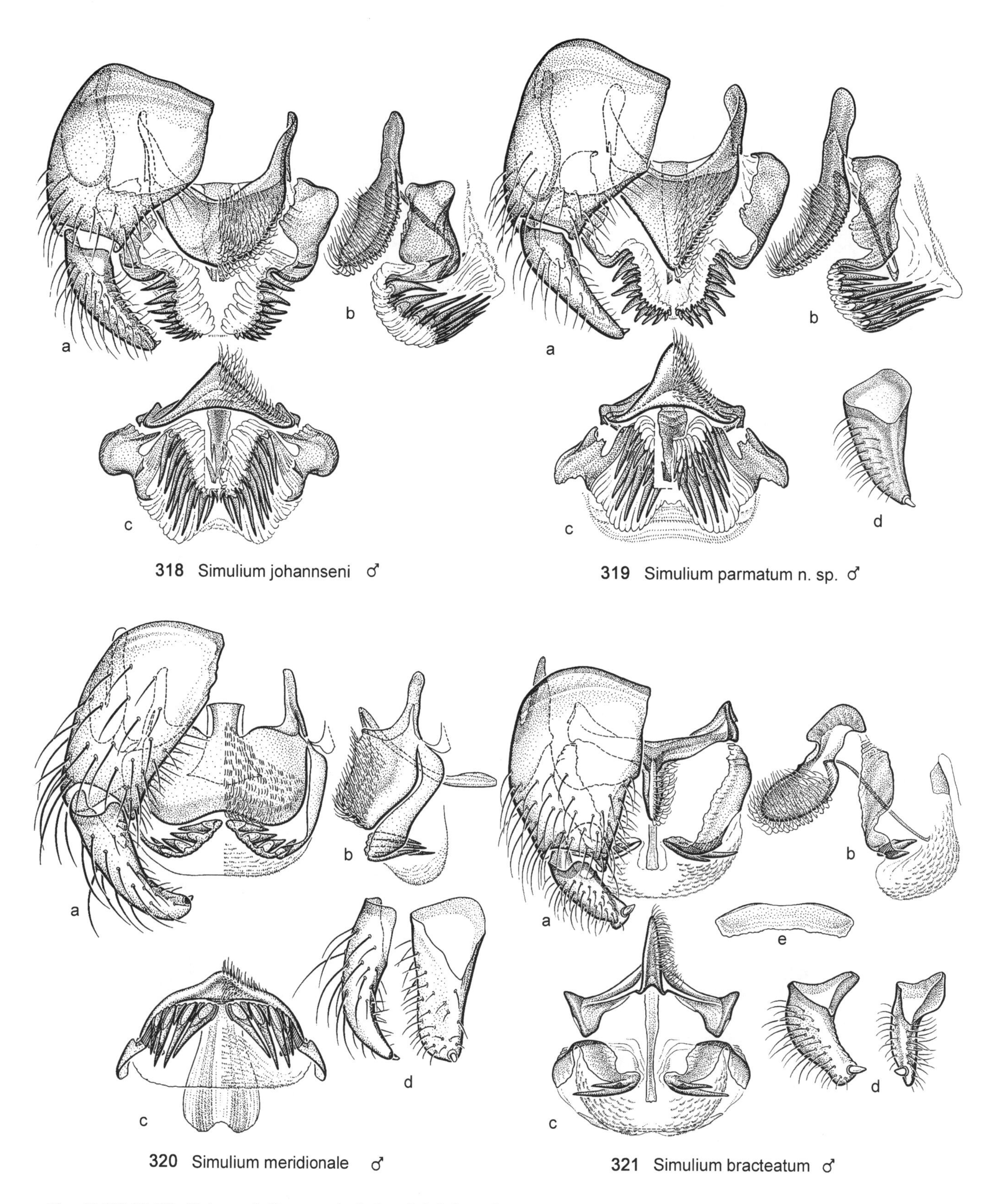


Figs. 10.318–10.321. Male genitalia; a, ventral view; b, left lateral view; c, terminal view; d, right gonostylus ([319d] inner lateral view; [320d] inner ventrolateral view [left], inner lateral view [right]; [321d] inner lateral view [left], dorsal view [right]); e, dorsal plate. (320) From Peterson (1981) by permission.

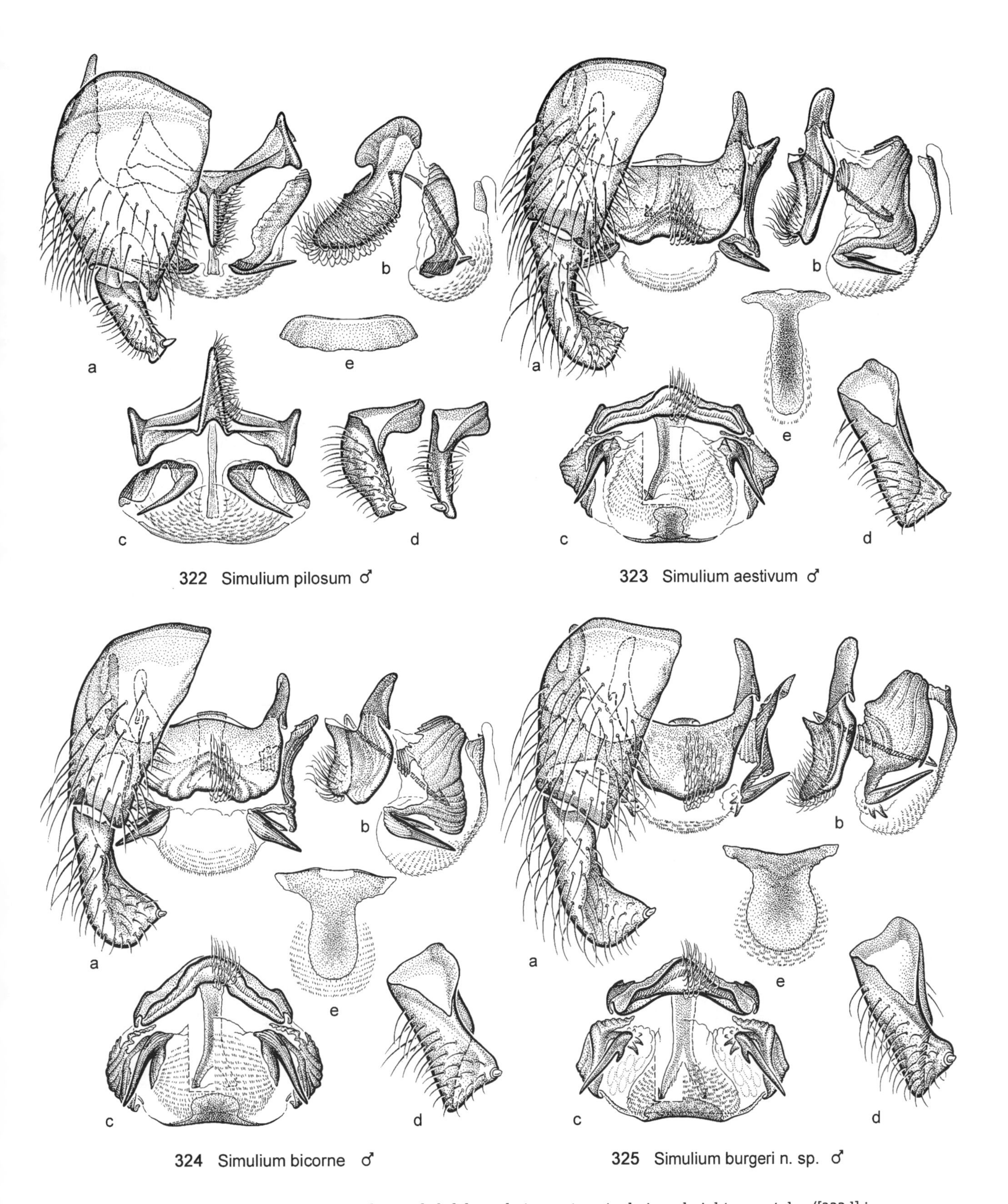


Figs. 10.322–10.325. Male genitalia; a, ventral view; b, left lateral view; c, terminal view; d, right gonostylus ([322d] inner lateral view [left], dorsal view [right]; [323d, 324d, 325d] inner lateral view); e, dorsal plate.

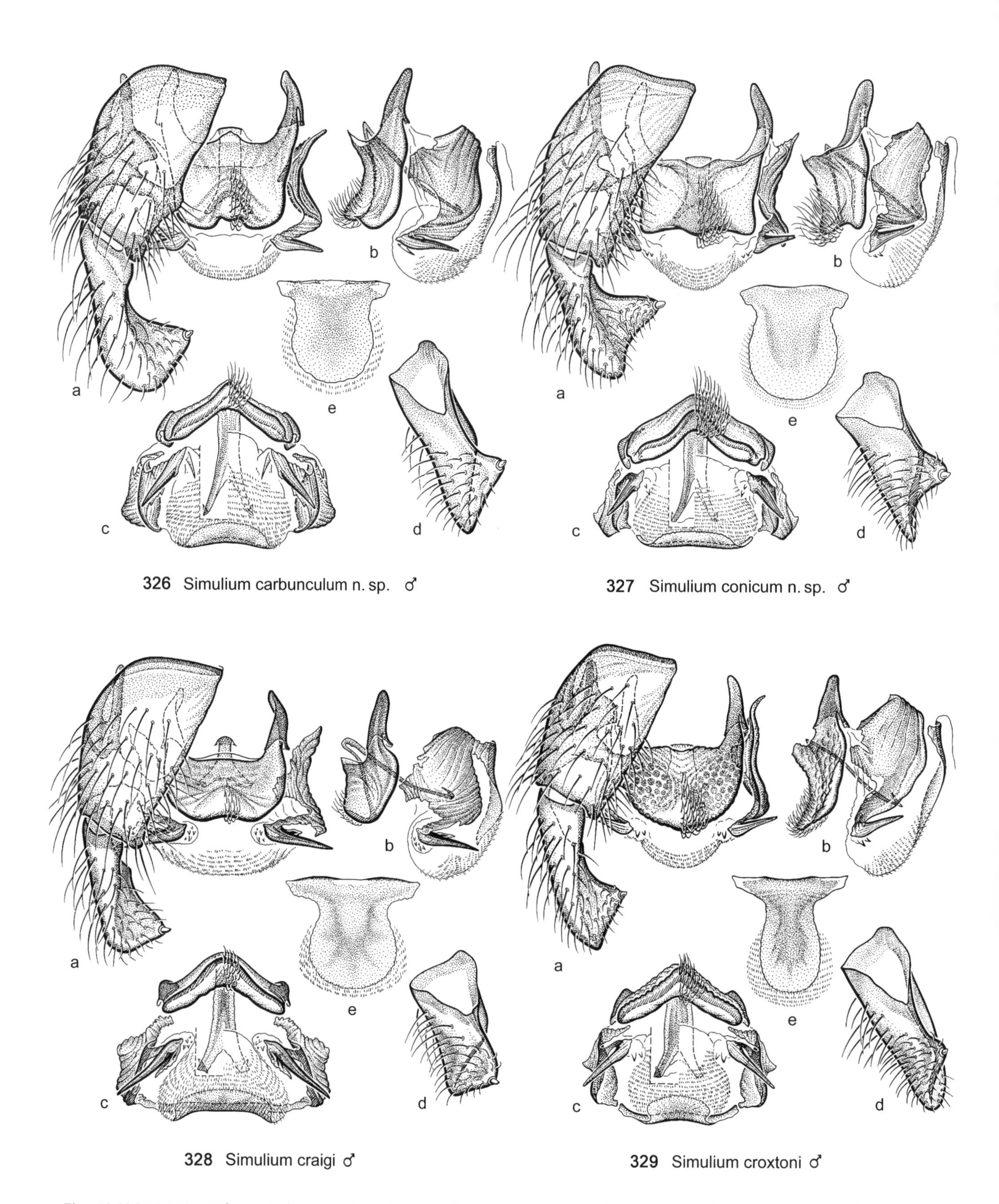


Figs. 10.326–10.329. Male genitalia; a, ventral view; b, left lateral view; c, terminal view; d, right gonostylus, inner lateral view; e, dorsal plate.

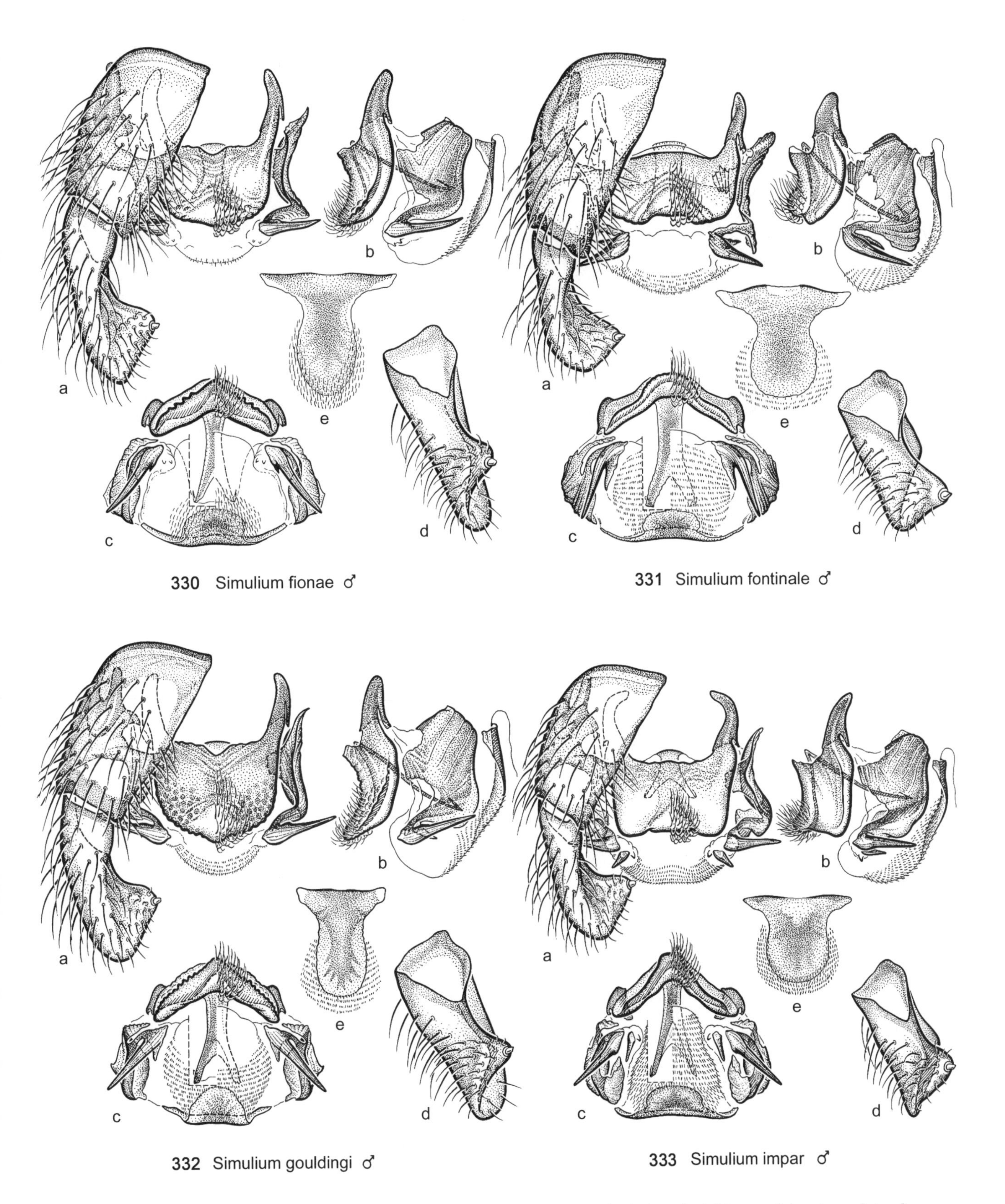


Figs. 10.330–10.333. Male genitalia; a, ventral view; b, left lateral view; c, terminal view; d, right gonostylus, inner lateral view; e, dorsal plate.

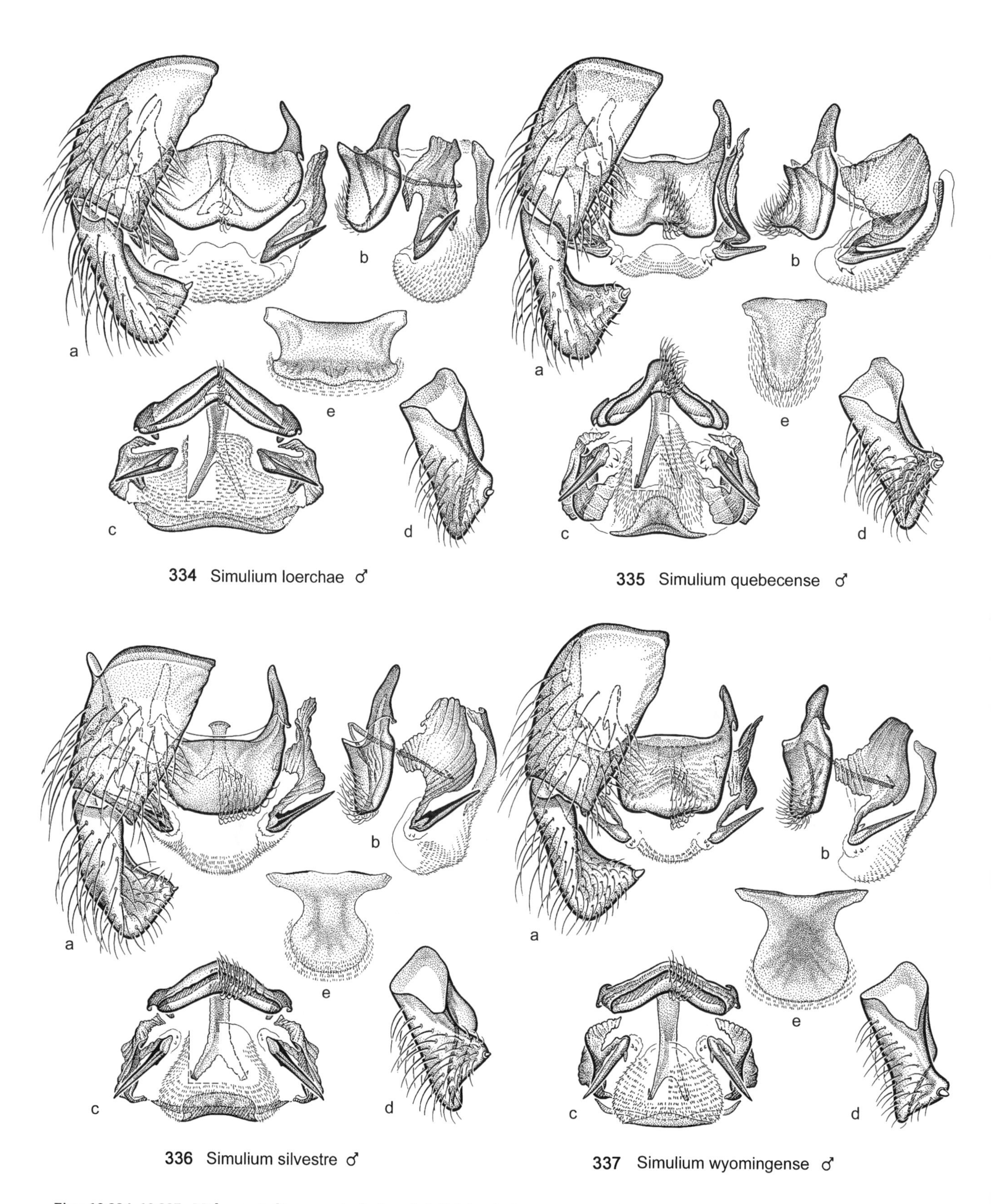


Figs. 10.334–10.337. Male genitalia; a, ventral view; b, left lateral view; c, terminal view; d, right gonostylus, inner lateral view; e, dorsal plate.

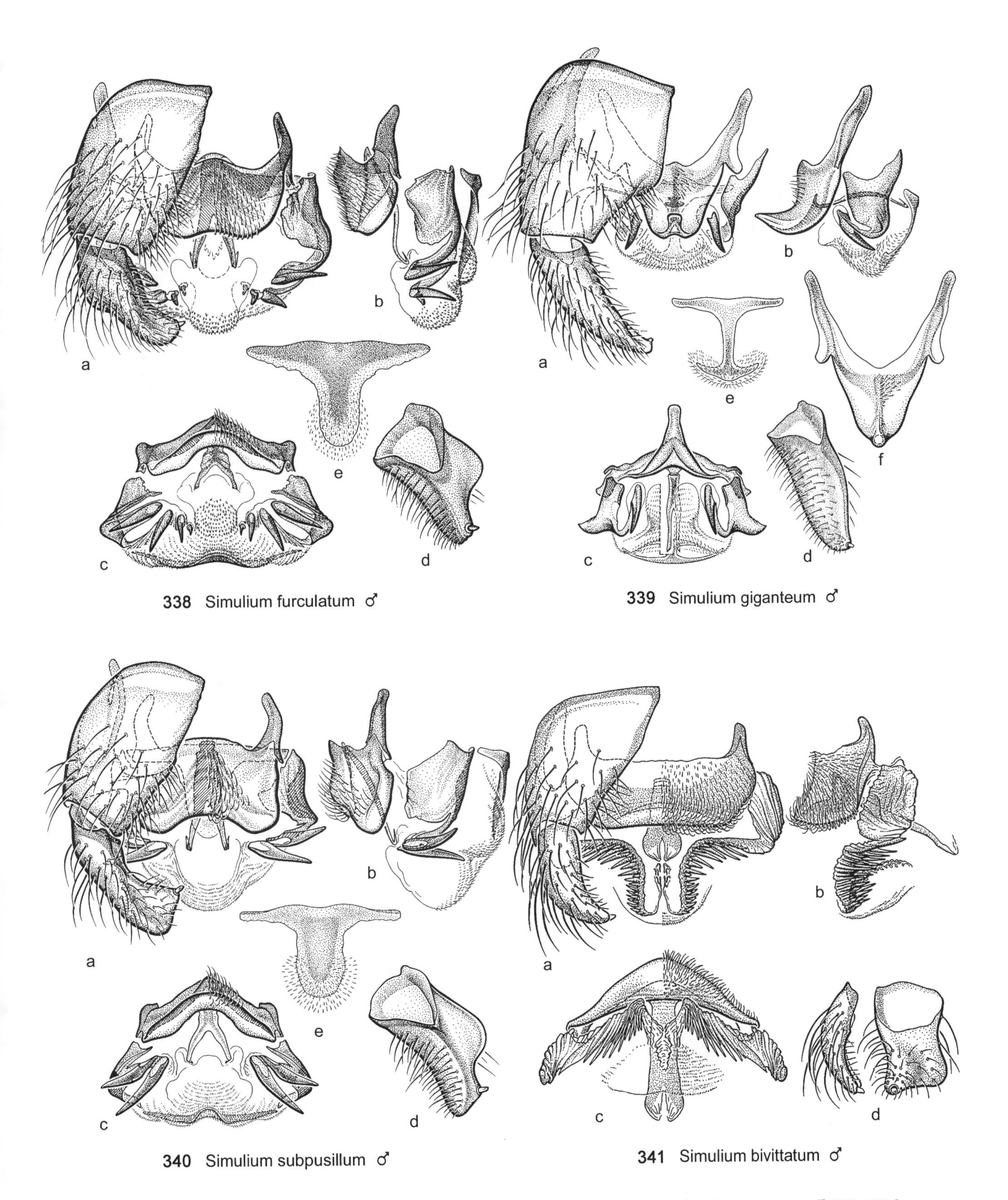


Figs. 10.338–10.341. Male genitalia; a, ventral view; b, left lateral view; c, terminal view; d, right gonostylus ([338d, 339d, 340d] inner lateral view; [341d] inner ventrolateral view [left], inner lateral view [right]); e, dorsal plate; (339f) ventral plate in ventral view tilted dorsally. (341) From Peterson (1981) by permission.

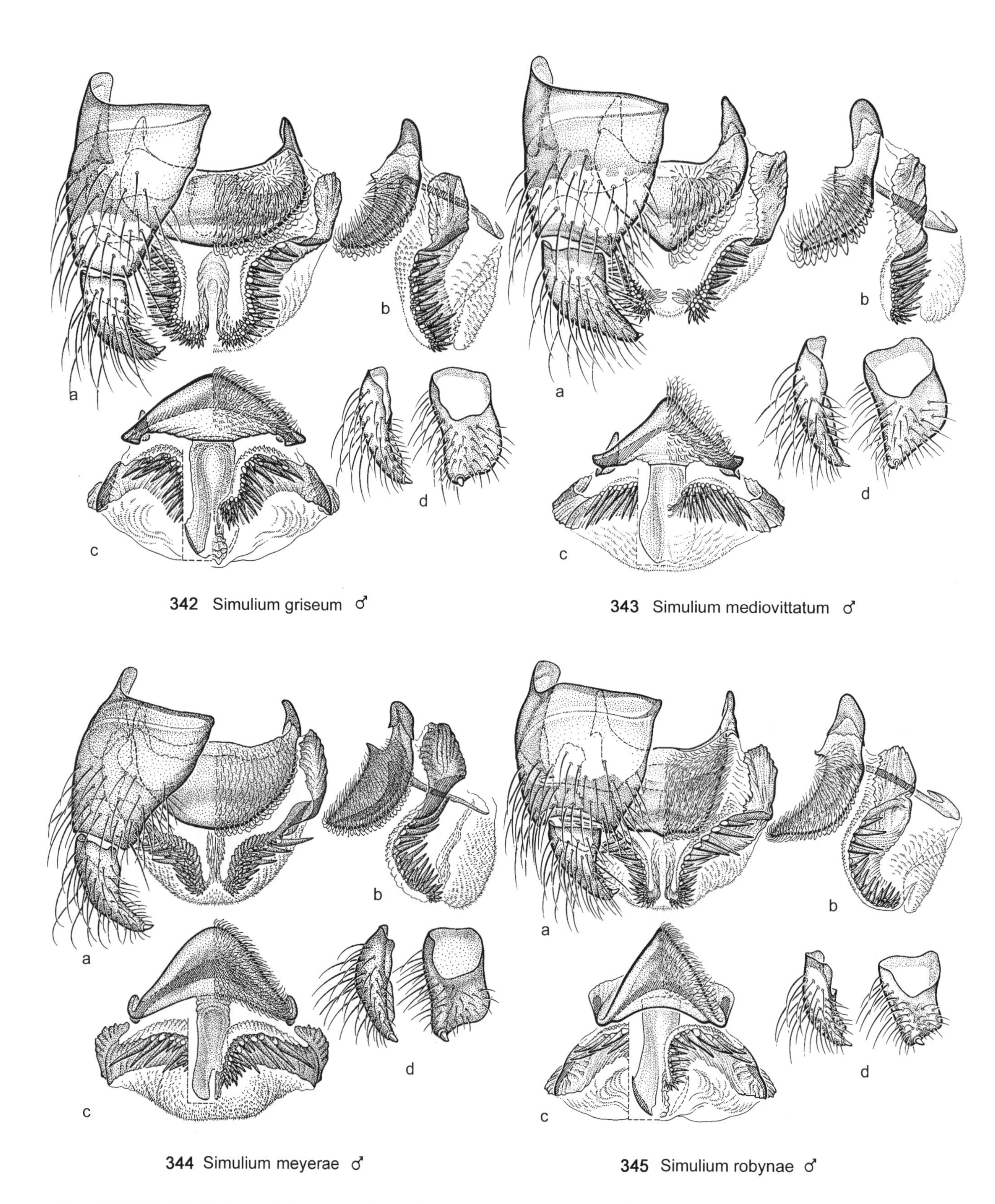


Figs. 10.342–10.345. Male genitalia; a, ventral view; b, left lateral view; c, terminal view; d, right gonostylus, inner ventrolateral view (left), inner lateral view (right). (344) From Moulton and Adler (2002a) by permission.

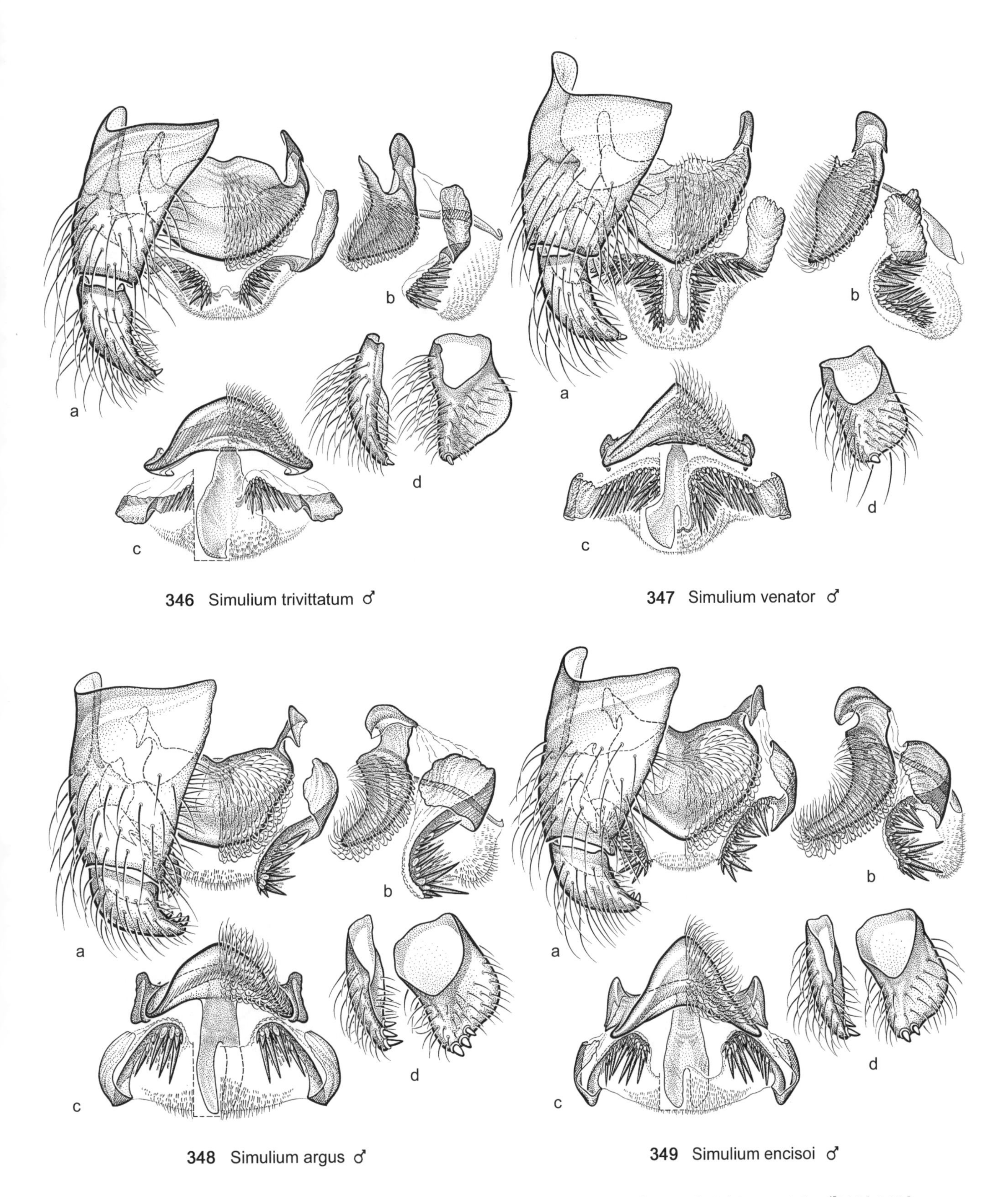


Figs. 10.346–10.349. Male genitalia; a, ventral view; b, left lateral view; c, terminal view; d, right gonostylus ([346d, 348d, 349d] inner ventrolateral view [left], inner lateral view [right]; [347d] inner lateral view).

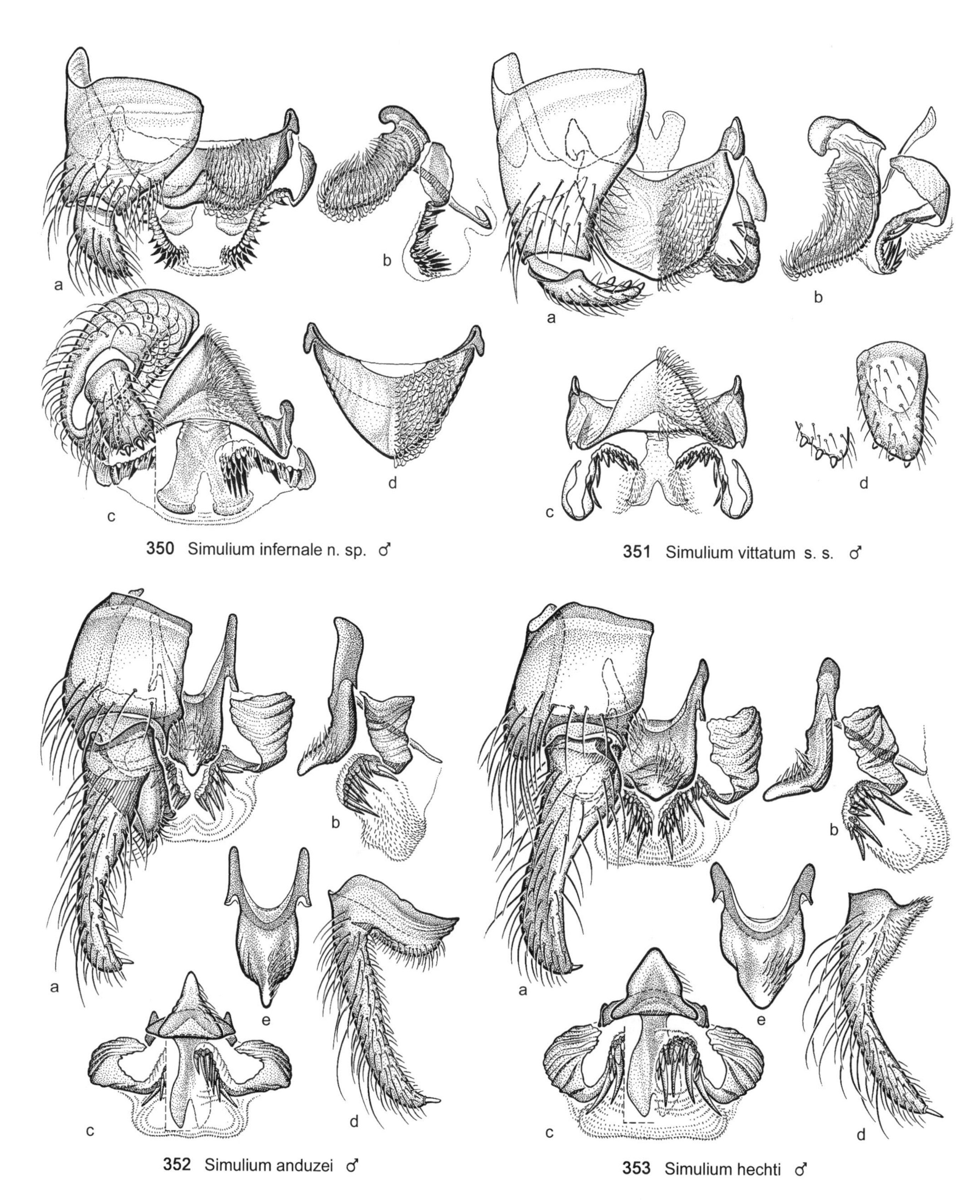


Figs. 10.350–10.353. Male genitalia; a, ventral view; b, left lateral view; c, terminal view; (350d) ventral plate in ventral view tilted dorsally; (351d) right gonostylus, apex (left), terminal view (right); (352d, 353d) right gonostylus, inner lateral view; e, ventral plate in ventral view tilted dorsally. (351) From Peterson (1981) by permission.

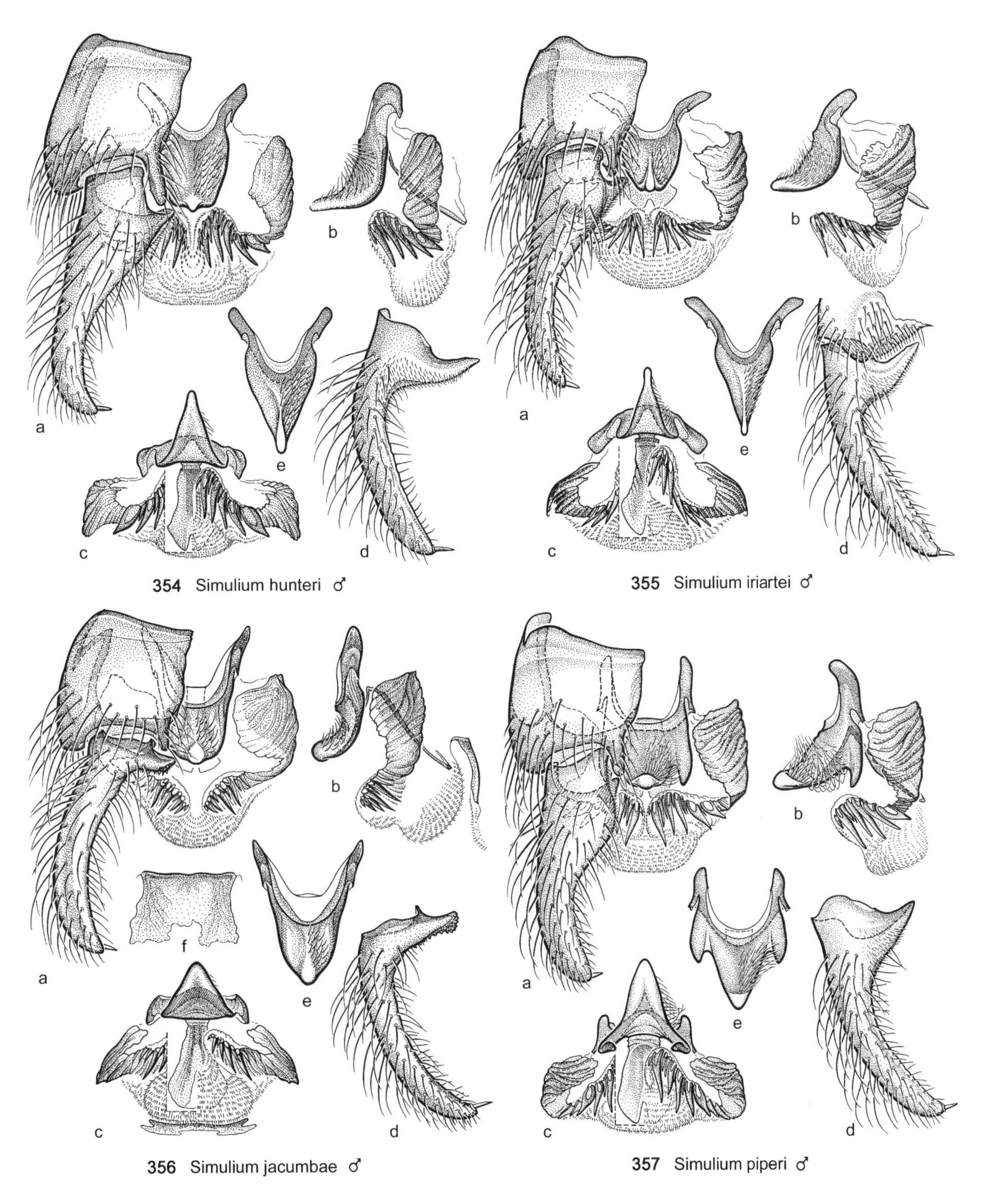


Figs. 10.354–10.357. Male genitalia; a, ventral view; b, left lateral view; c, terminal view; d, right gonostylus, inner lateral view; e, ventral plate in ventral view tilted dorsally; f, dorsal plate.

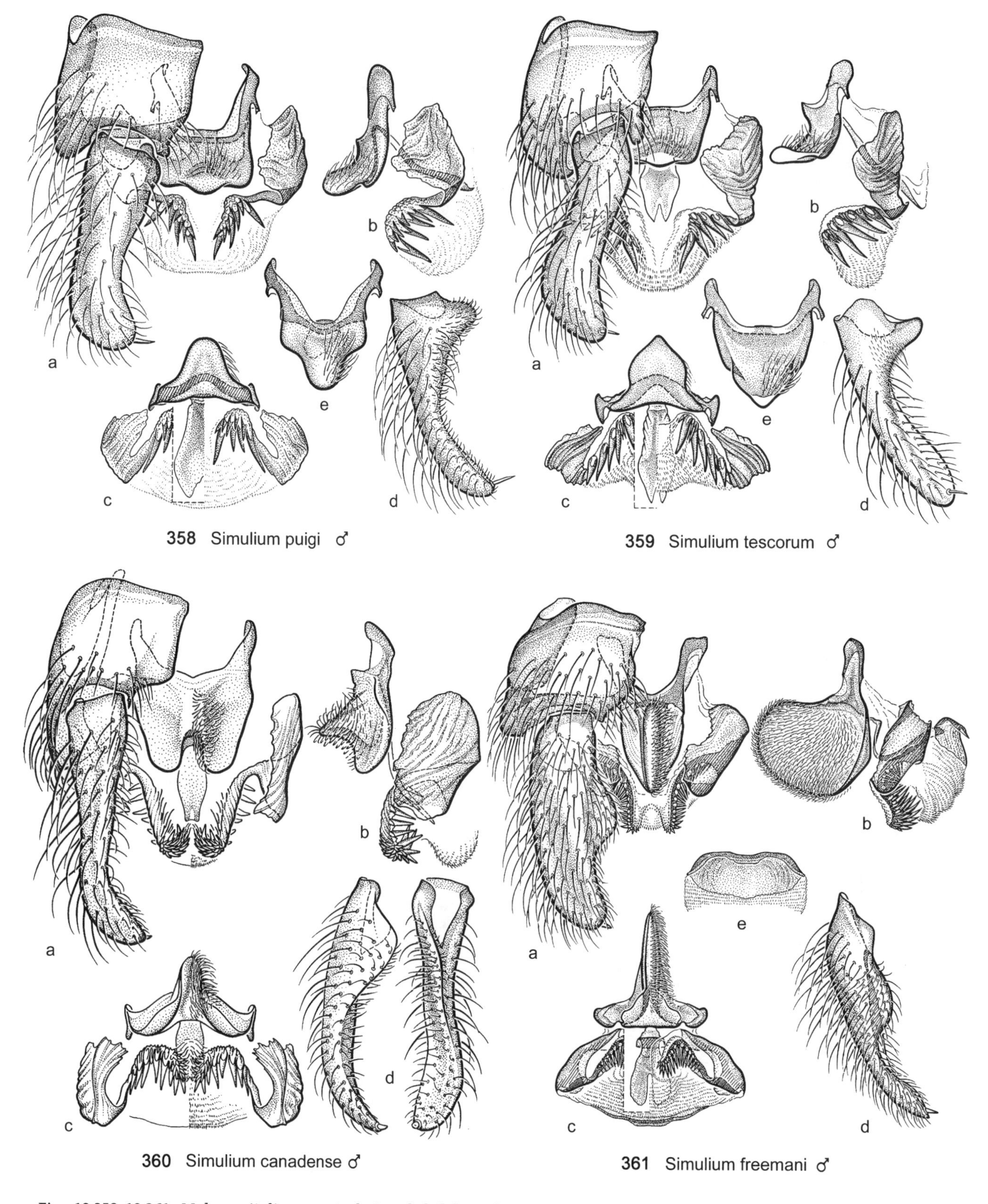


Figs. 10.358–10.361. Male genitalia; a, ventral view; b, left lateral view; c, terminal view; d, right gonostylus ([358d, 359d] inner lateral view; [360d] inner lateral view [left], dorsal view [right]; [361d] inner lateral view); (358e, 359e) ventral plate in ventral view tilted dorsally; (361e) dorsal plate. (360) From Peterson (1981) by permission.

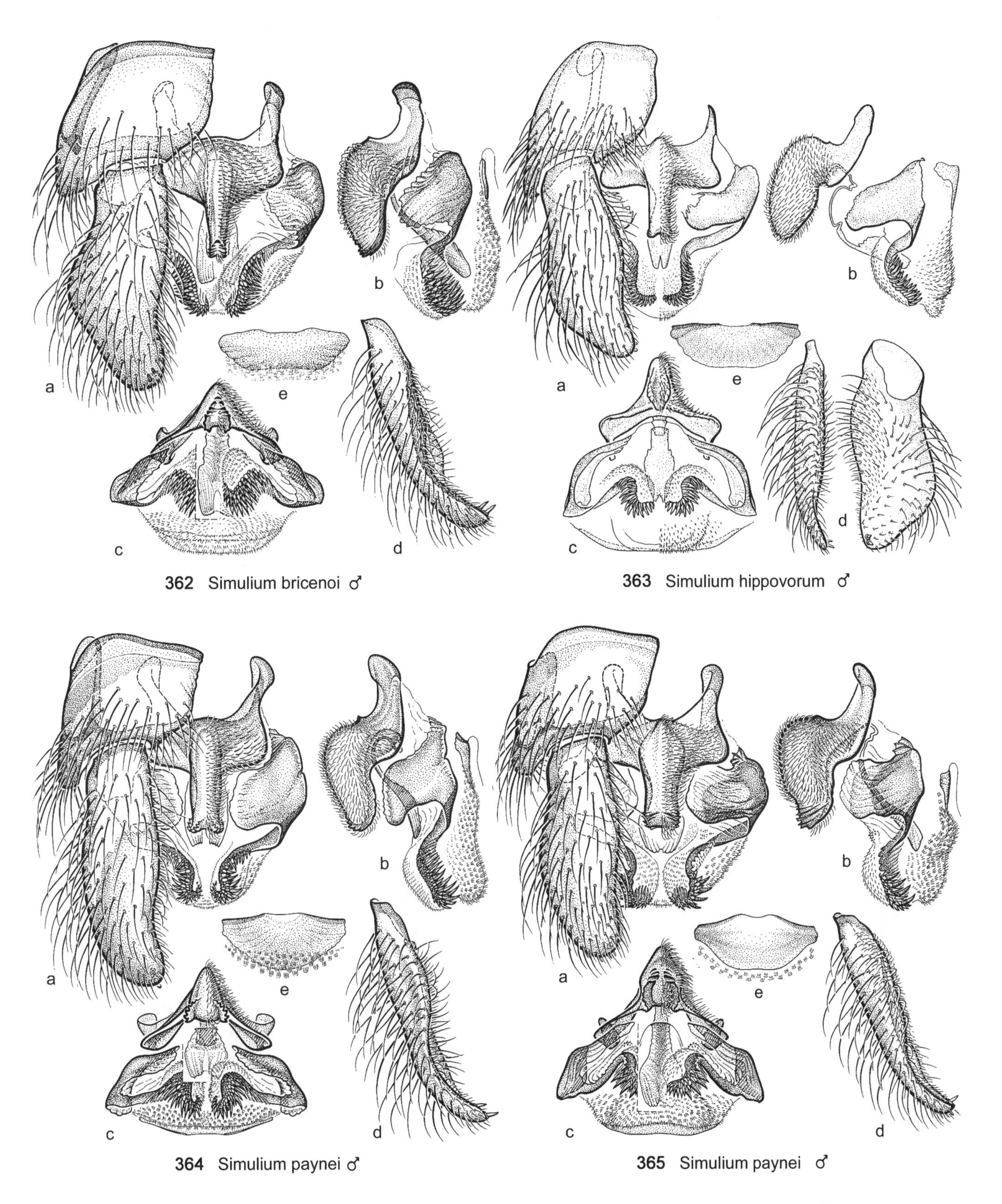


Figs. 10.362–10.365. Male genitalia; a, ventral view; b, left lateral view; c, terminal view; d, right gonostylus ([362d, 364d, 365d] inner lateral view; [363d] ventrolateral view [left], dorsal view [right]); e, dorsal plate. (363) From Peterson (1981) by permission.

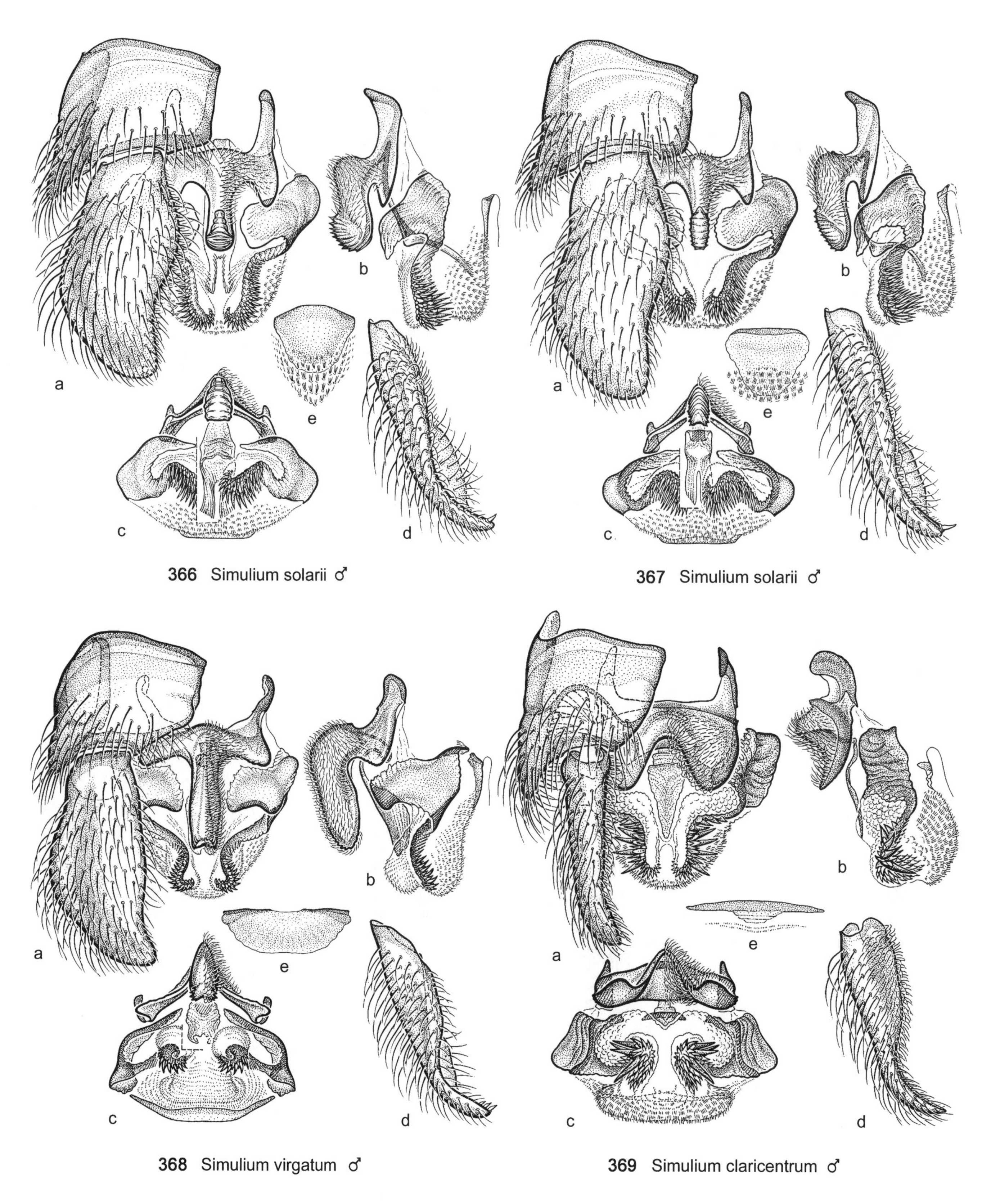


Figs. 10.366–10.369. Male genitalia; a, ventral view; b, left lateral view; c, terminal view; d, right gonostylus, inner lateral view; e, dorsal plate.

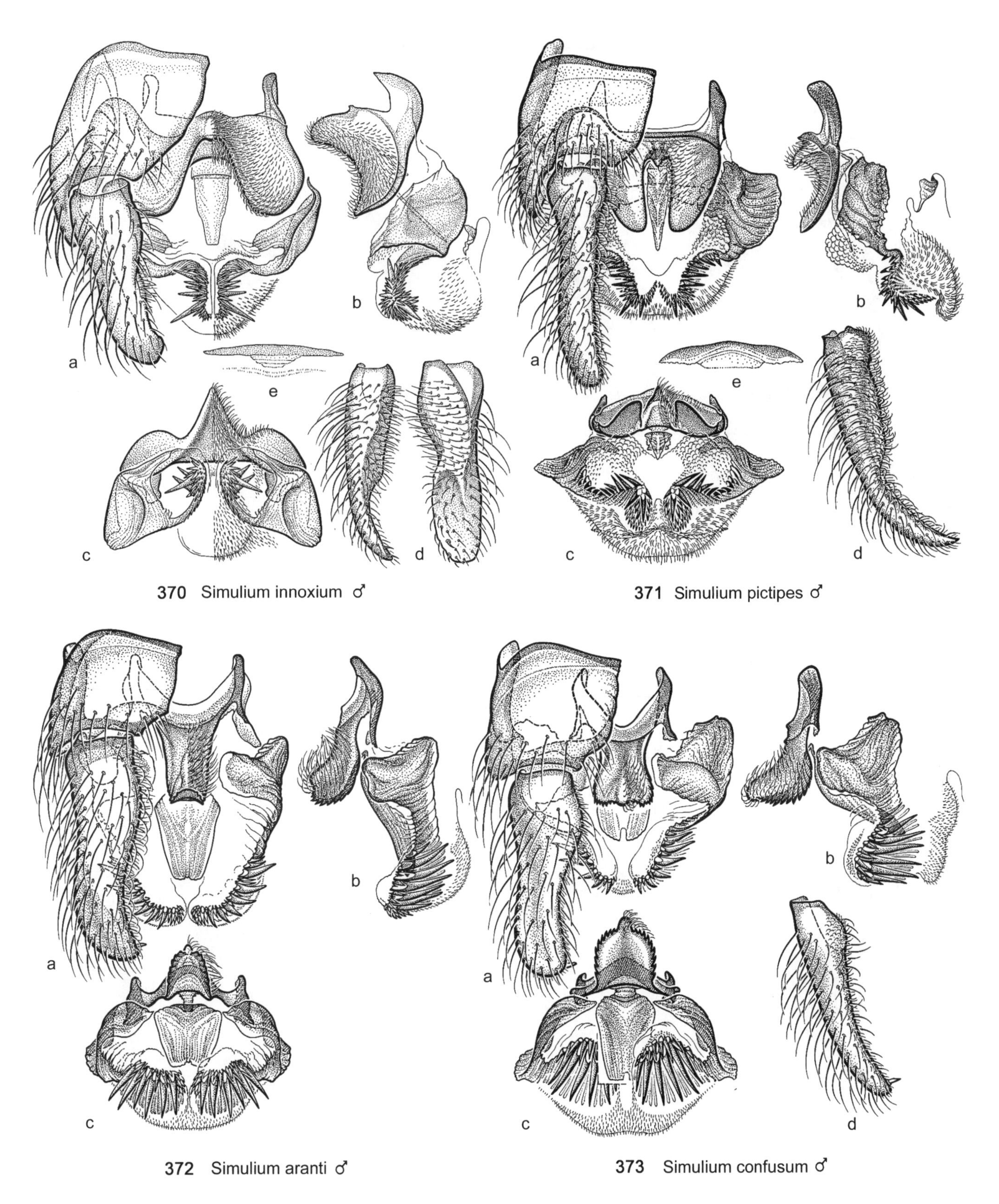


Figs. 10.370–10.373. Male genitalia; a, ventral view; b, left lateral view; c, terminal view; d, right gonostylus ([370d] inner lateral view [left], dorsal view [right]; [371d, 373d] inner lateral view); e, dorsal plate. (370) From Peterson (1981) by permission.

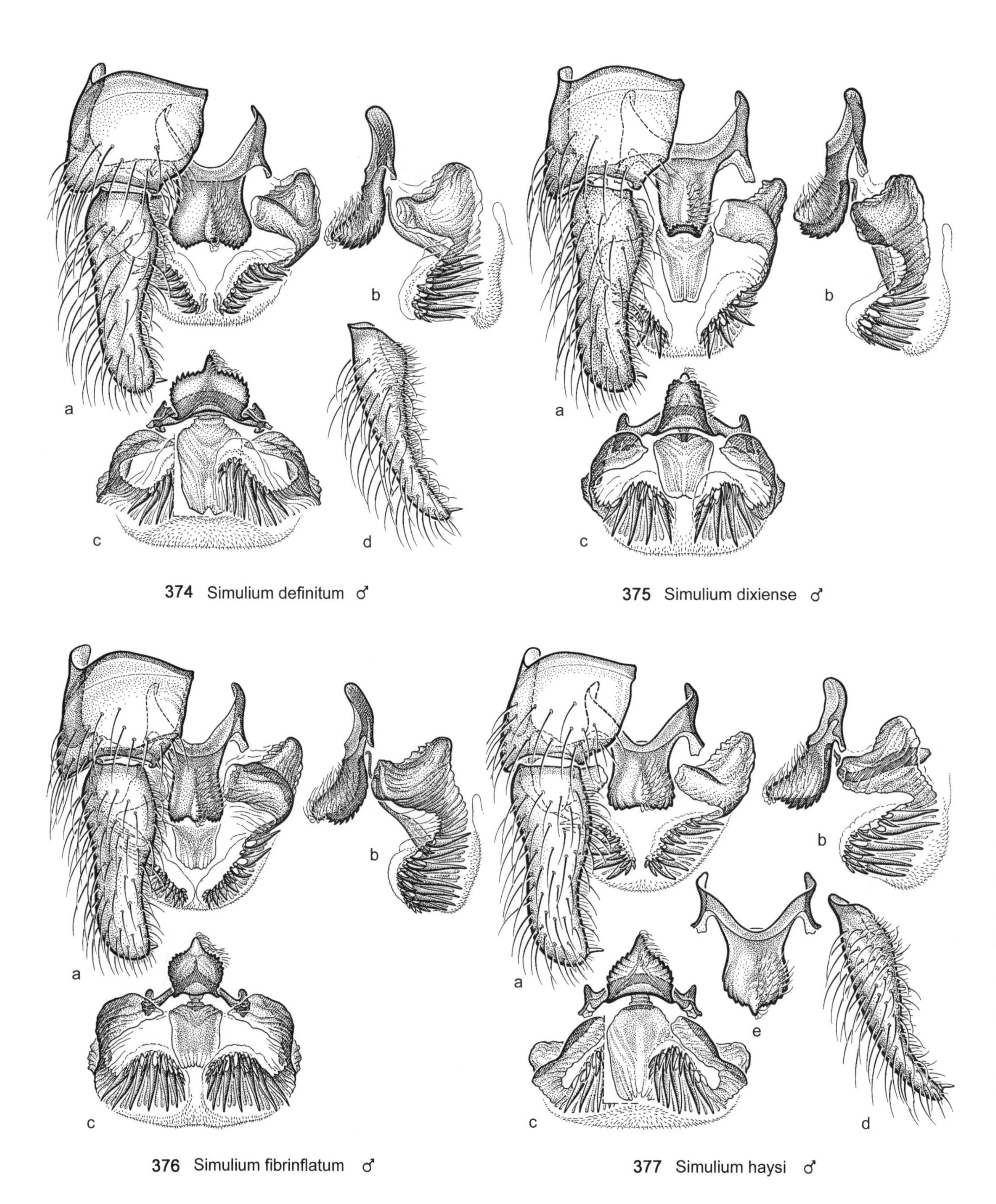


Figs. 10.374–10.377. Male genitalia; a, ventral view; b, left lateral view; c, terminal view; d, right gonostylus, inner lateral view; e, ventral plate in ventral view tilted dorsally.

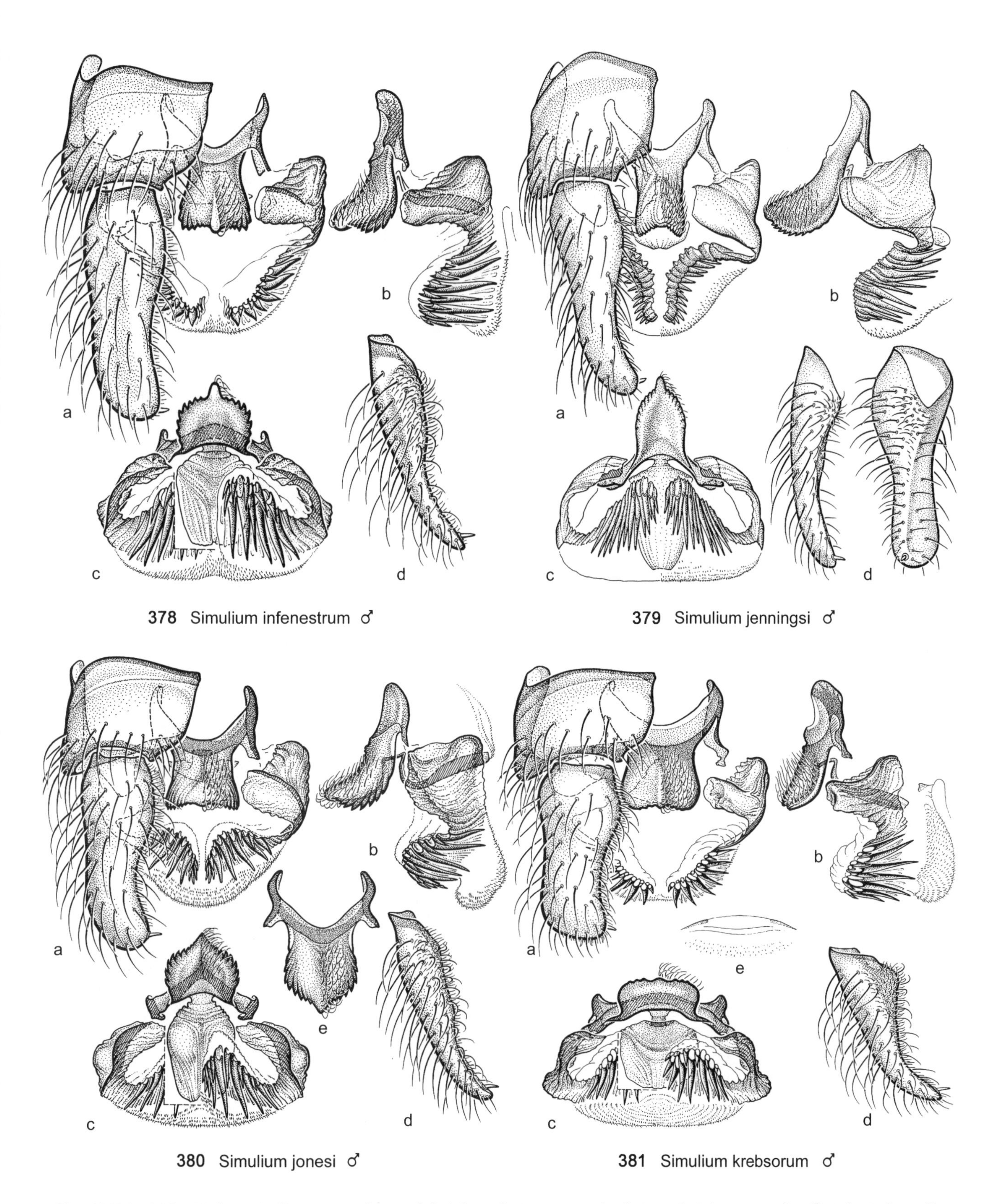


Figs. 10.378–10.381. Male genitalia; a, ventral view; b, left lateral view; c, terminal view; d, right gonostylus ([378d, 380d, 381d] inner lateral view; [379d] inner lateral view [left], dorsal view [right]); (380e) ventral plate in ventral view tilted dorsally; (381e) dorsal plate. (379) From Peterson (1981) by permission.

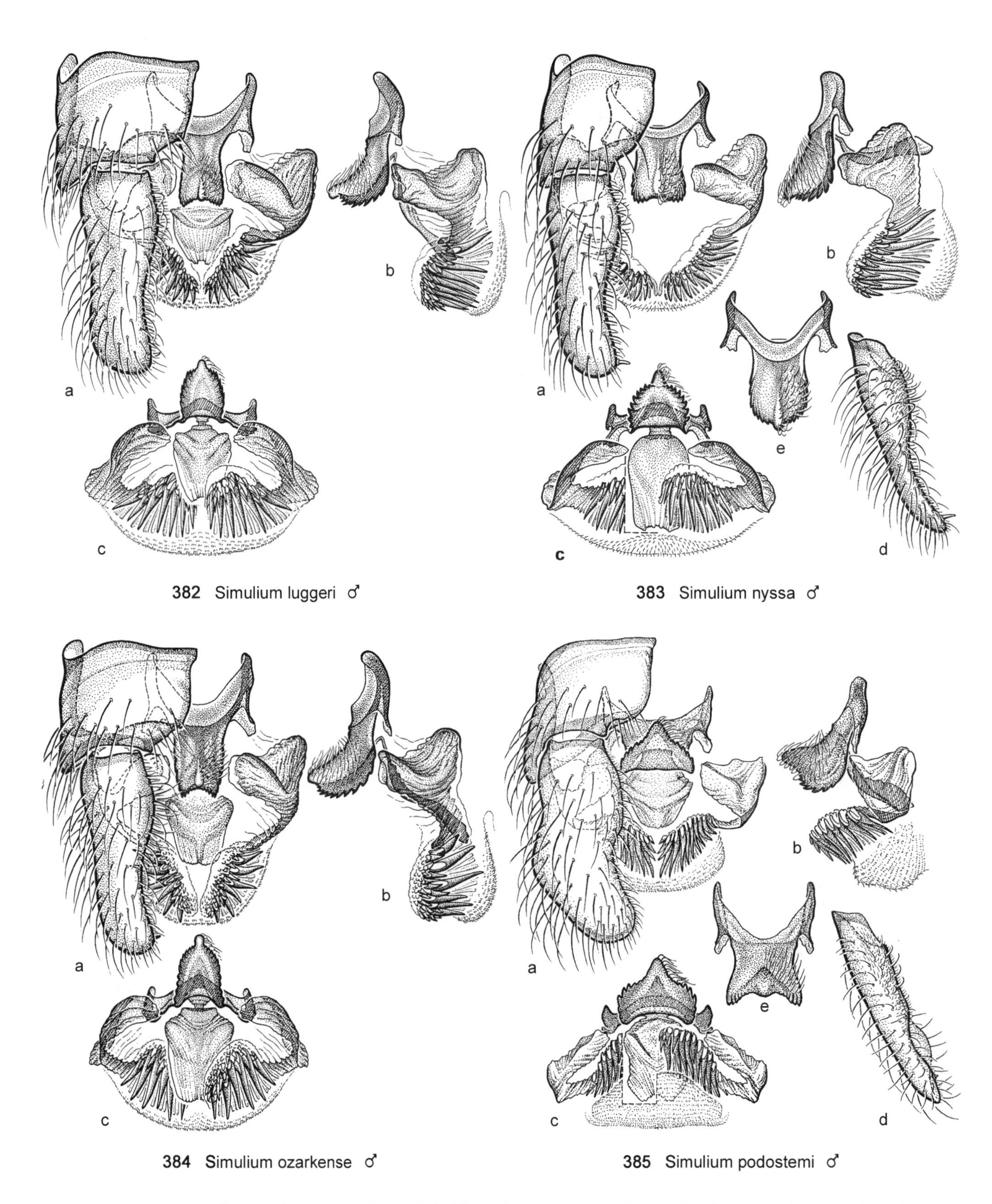


Figs. 10.382–10.385. Male genitalia; a, ventral view; b, left lateral view; c, terminal view; d, right gonostylus, inner lateral view; e, ventral plate in ventral view tilted dorsally.

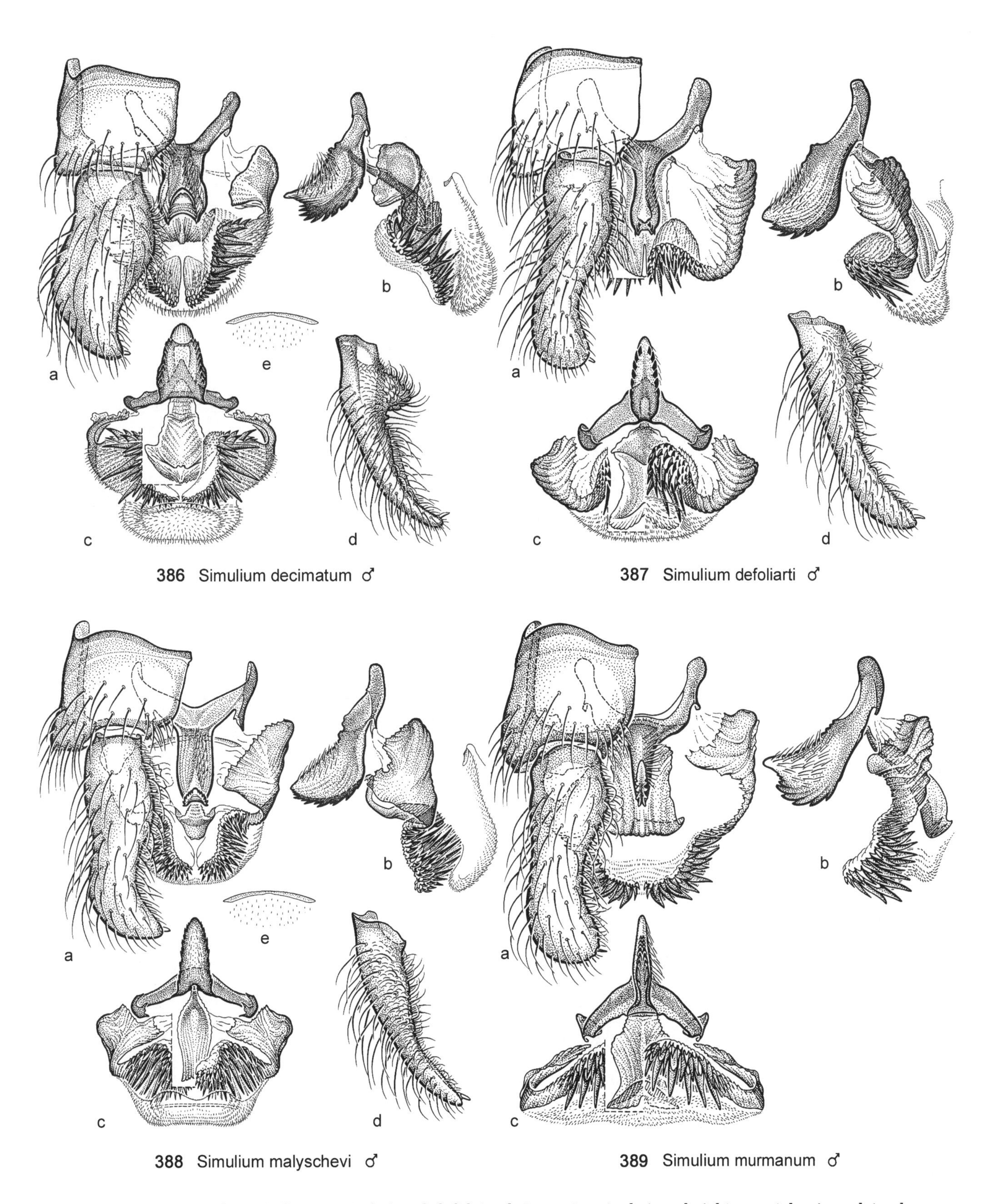


Figs. 10.386–10.389. Male genitalia; a, ventral view; b, left lateral view; c, terminal view; d, right gonostylus, inner lateral view; e, dorsal plate.

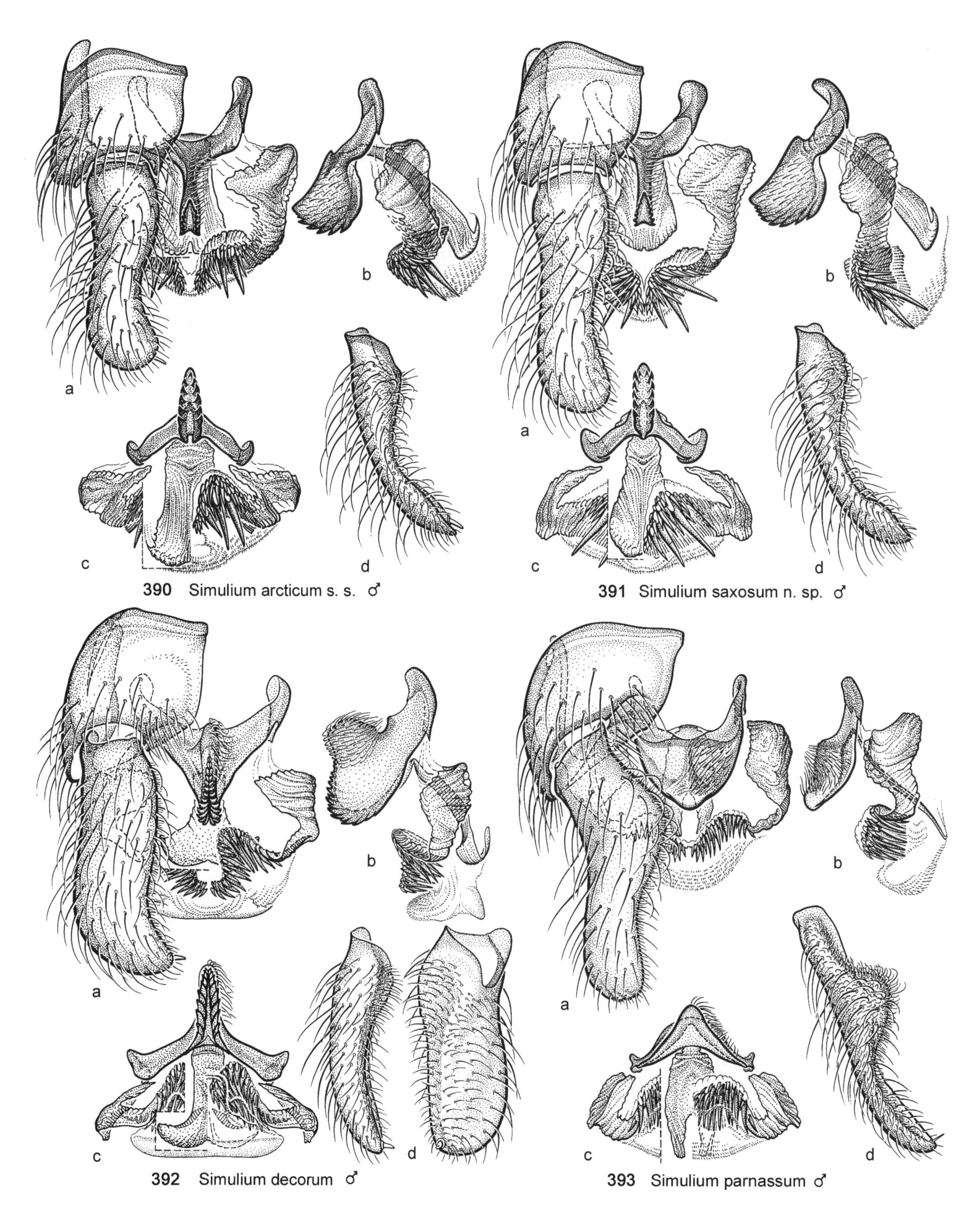


Figs. 10.390–10.393. Male genitalia; a, ventral view; b, left lateral view; c, terminal view; d, right gonostylus ([390d, 391d, 393d] inner lateral view; [392d] inner lateral view [left], dorsal view [right]).

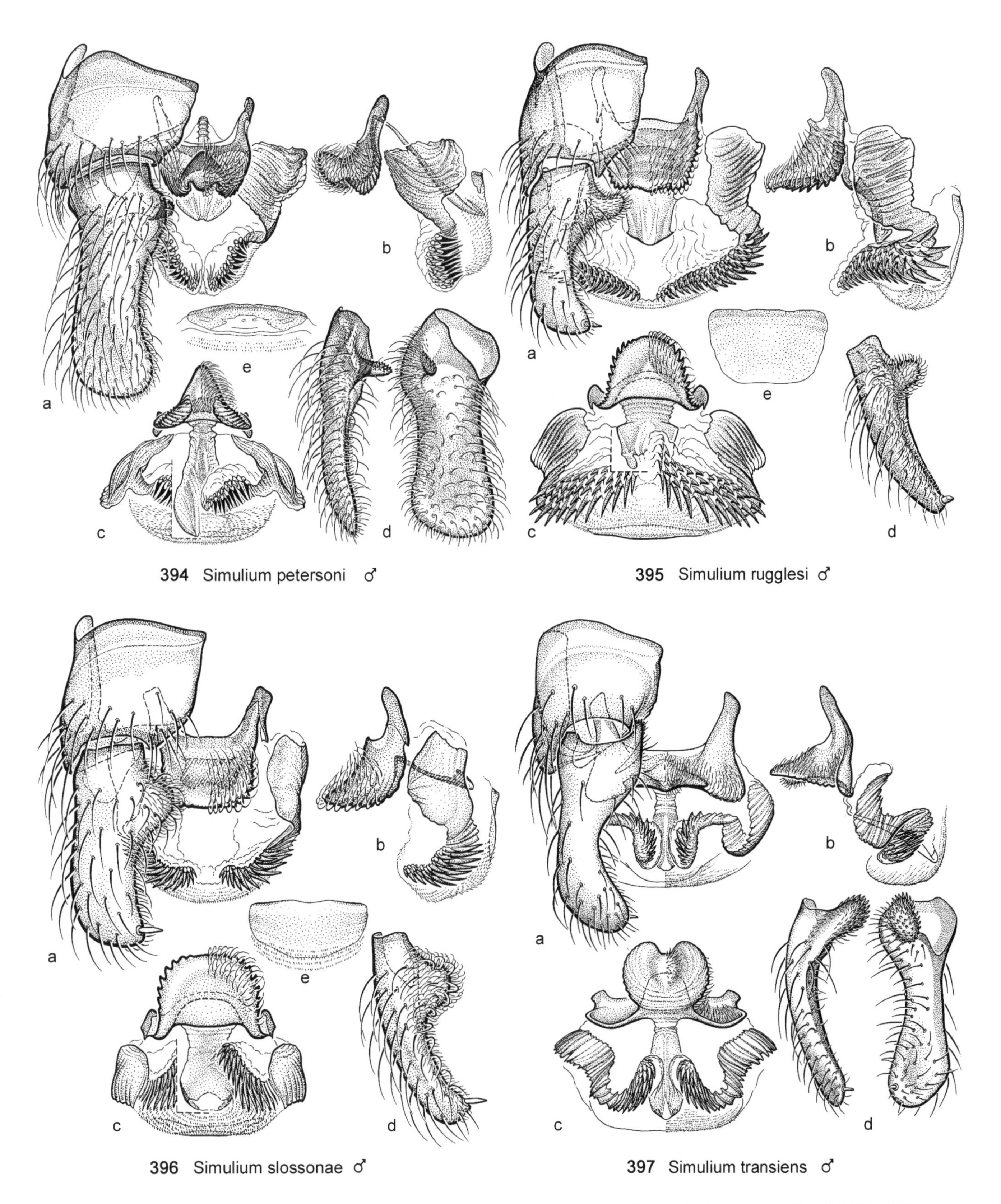


Figs. 10.394–10.397. Male genitalia; a, ventral view; b, left lateral view; c, terminal view; d, right gonostylus ([394d, 397d] inner lateral view [left], dorsal view [right]; [395d, 396d] inner lateral view); e, dorsal plate. (397) From Peterson (1981) by permission.

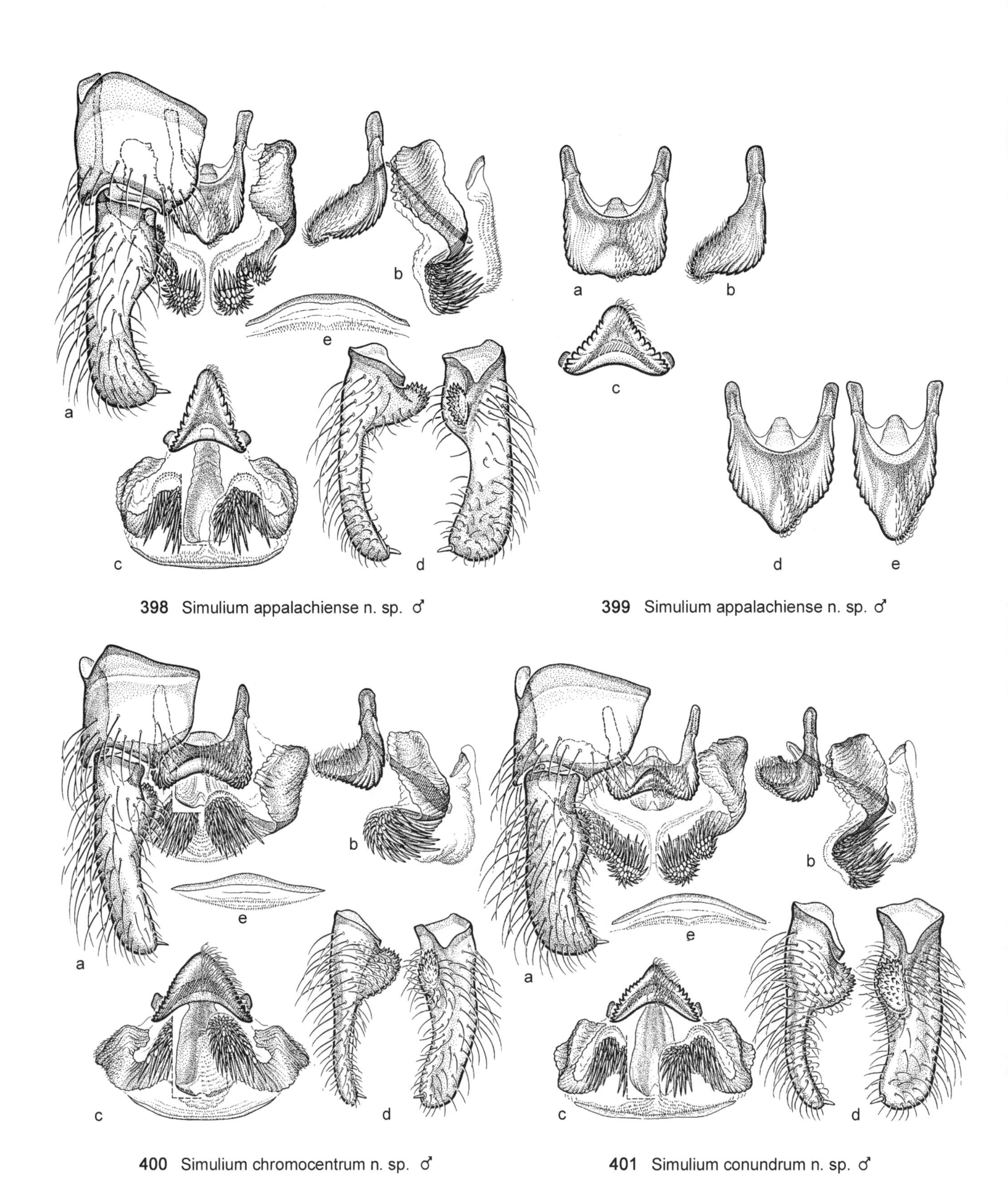


Figs. 10.398–10.401. Male genitalia; (398, 400, 401) a, ventral view; b, left lateral view; c, terminal view; d, right gonostylus, inner lateral view (left), dorsal view (right); e, dorsal plate. (399) Ventral plates; a, d, e, ventral view; b, left lateral view; c, terminal view.

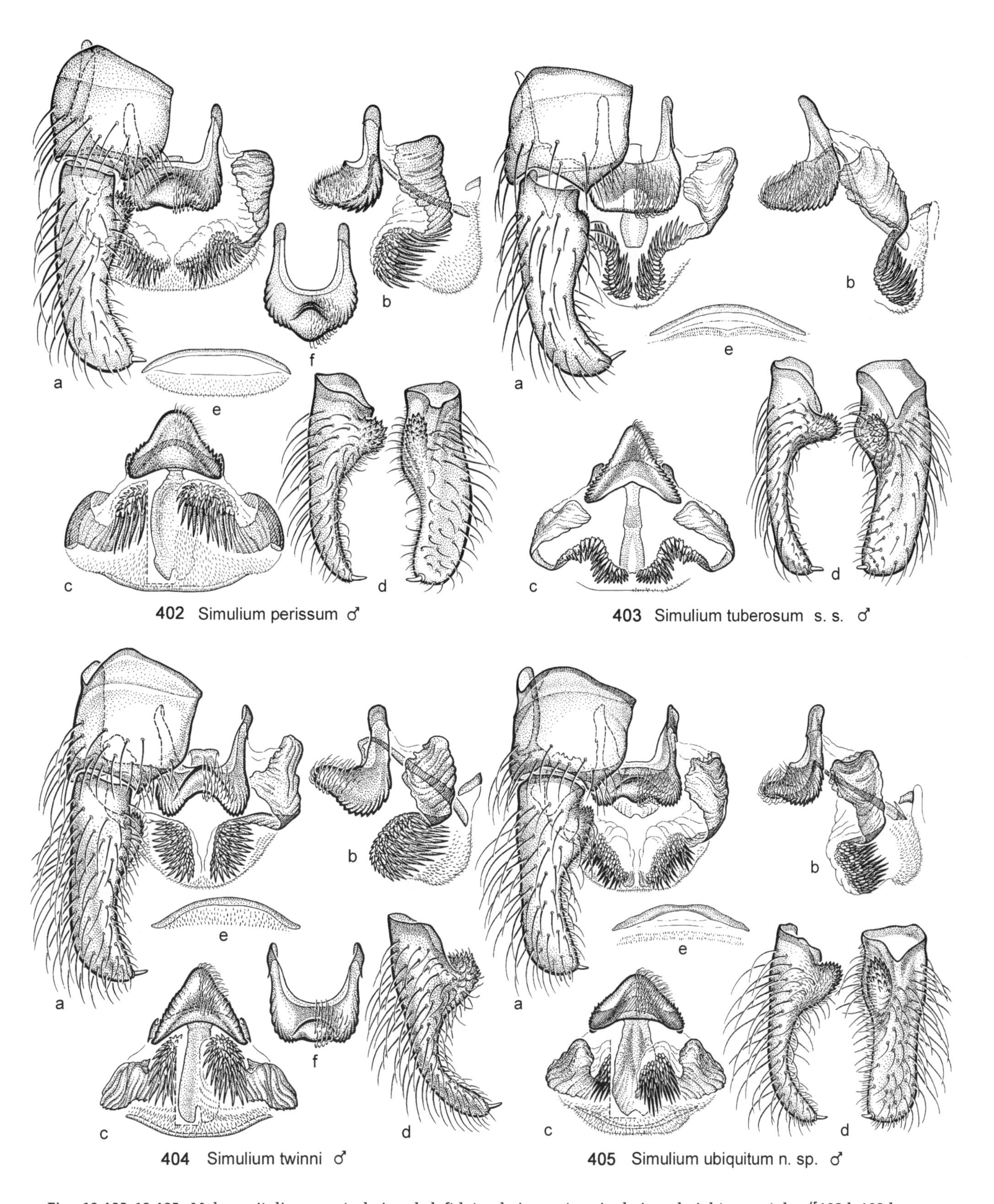


Figs. 10.402–10.405. Male genitalia; a, ventral view; b, left lateral view; c, terminal view; d, right gonostylus ([402d, 403d, 405d] inner lateral view [left], dorsal view [right]; [404d] inner lateral view); e, dorsal plate; f, ventral plate in ventral view tilted dorsally. (403) From Peterson (1981) by permission.

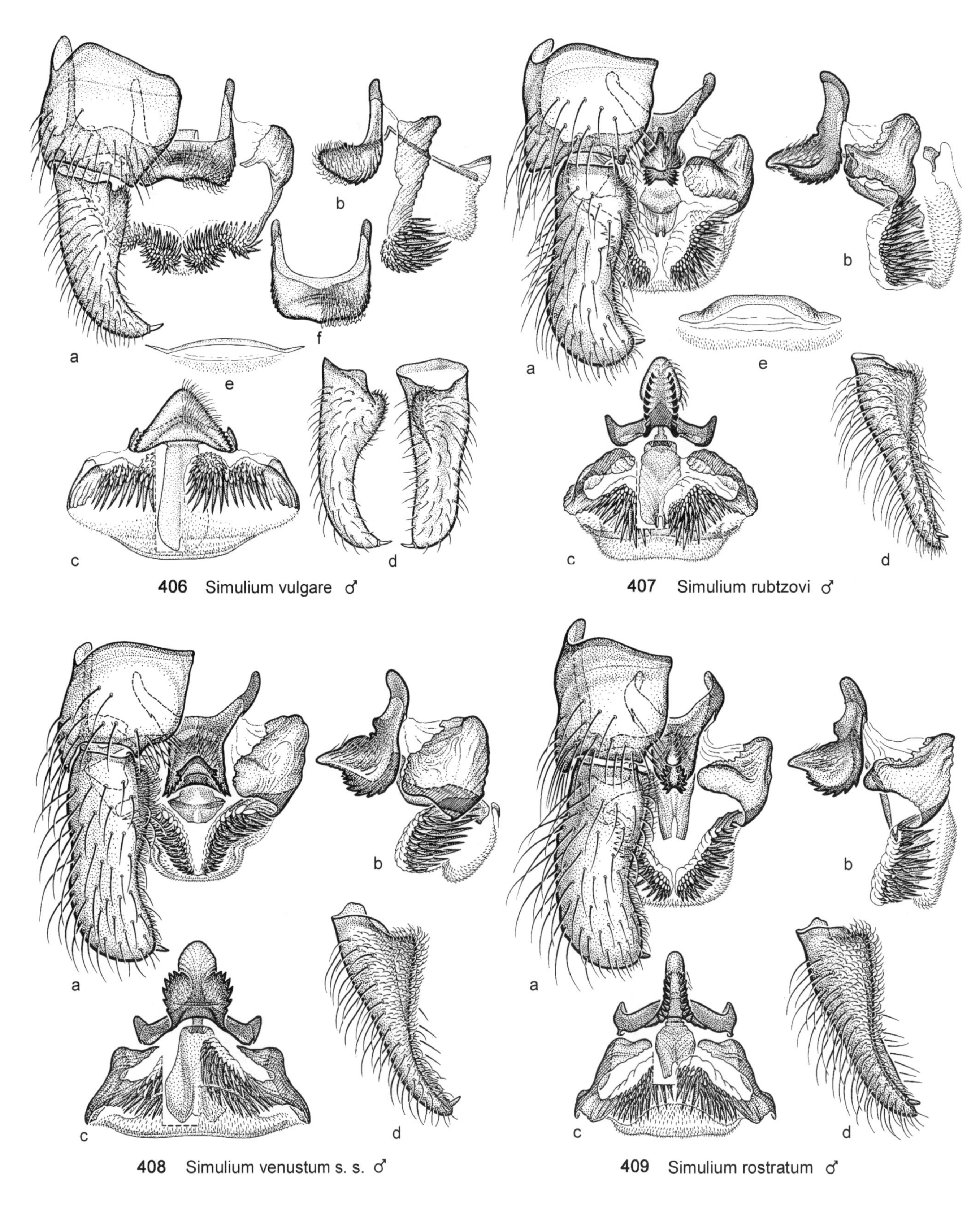


Figs. 10.406–10.409. Male genitalia; a, ventral view; b, left lateral view; c, terminal view; d, right gonostylus ([406d] inner lateral view [left], dorsal view [right]; [407d, 408d, 409d] inner lateral view); e, dorsal plate; f, ventral view of ventral plate tilted dorsally.

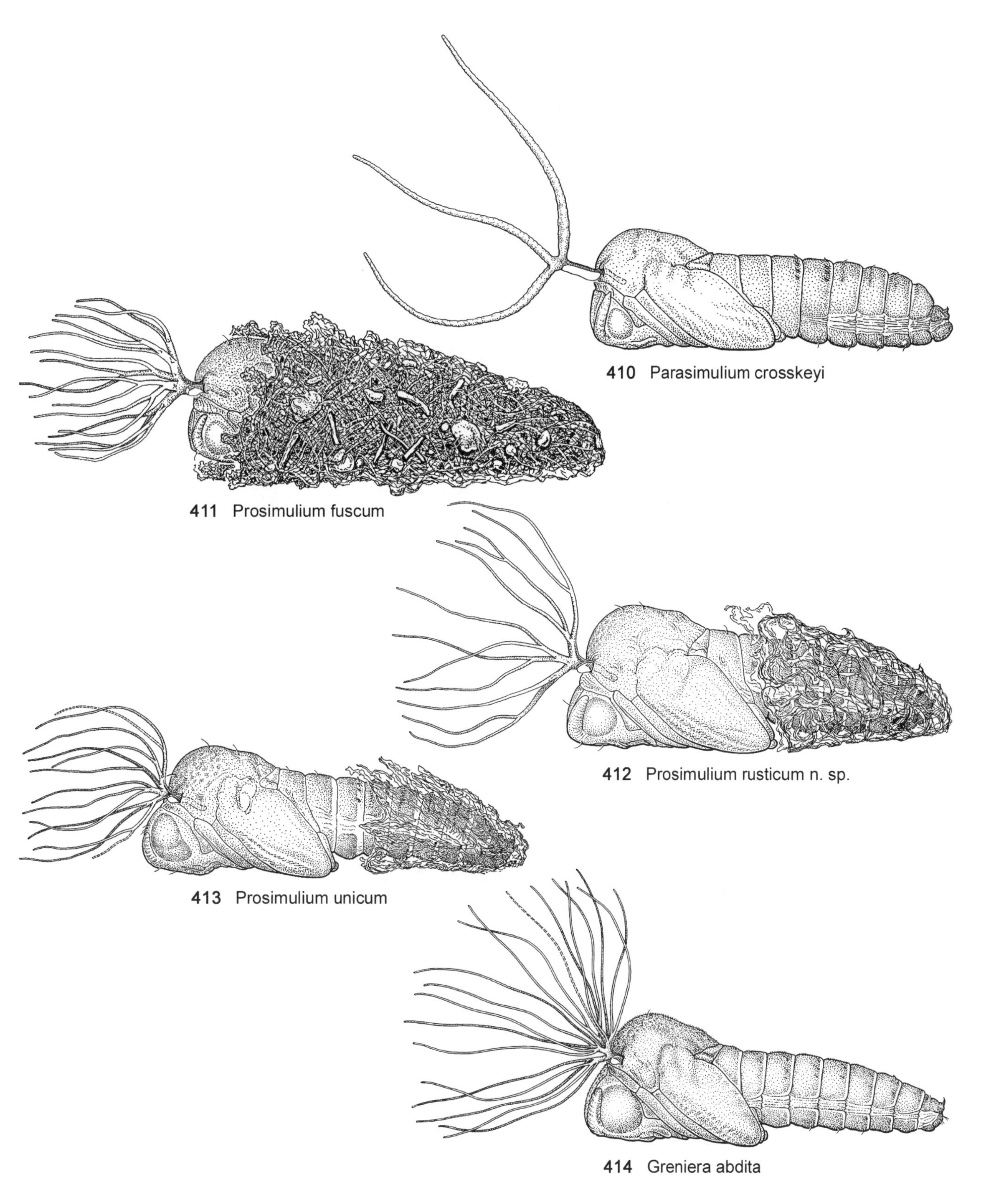


Figs. 10.410–10.414. Pupae, lateral view; (410, 414) cocoon omitted.

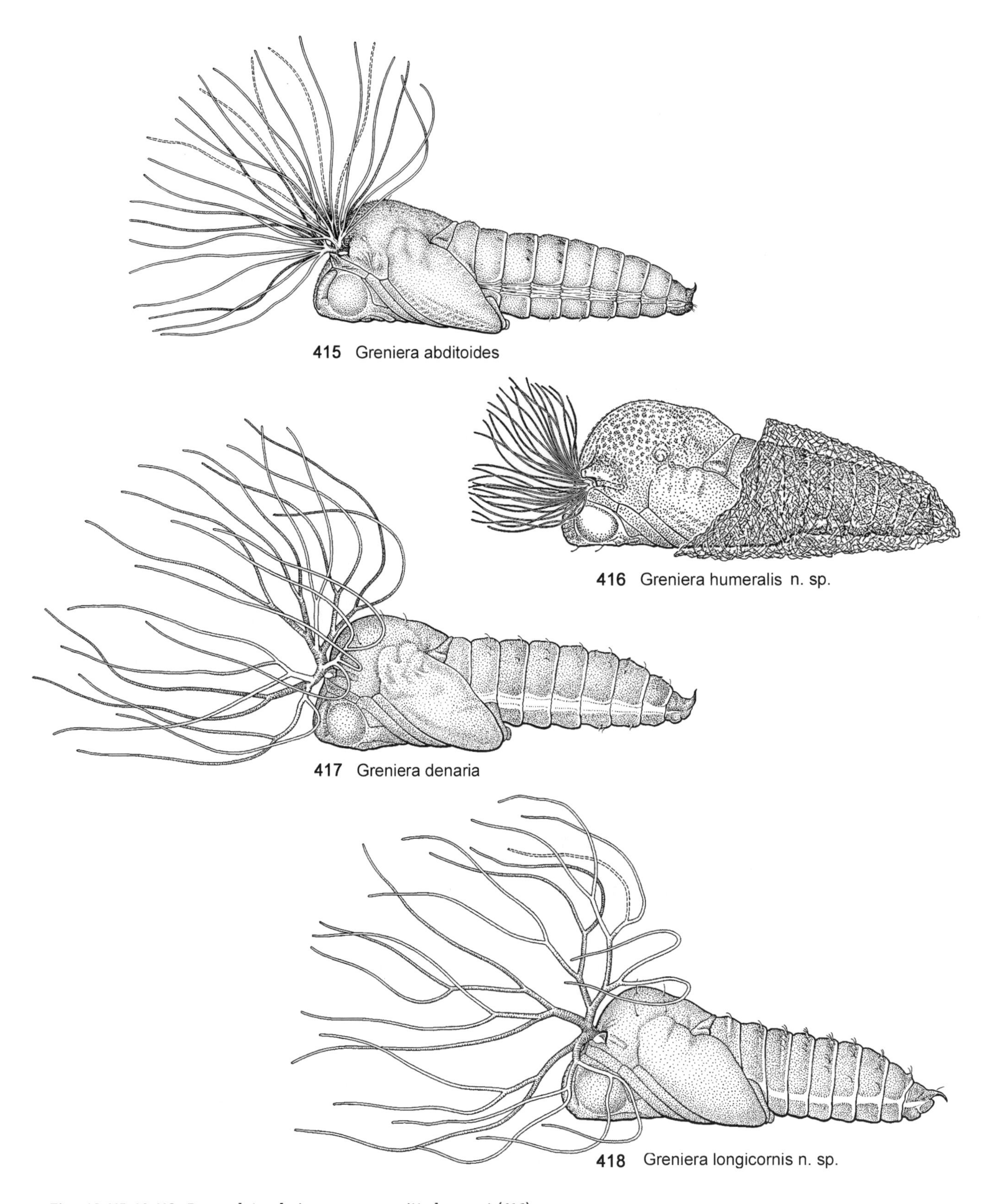


Figs. 10.415–10.418. Pupae, lateral view; cocoon omitted, except (416).

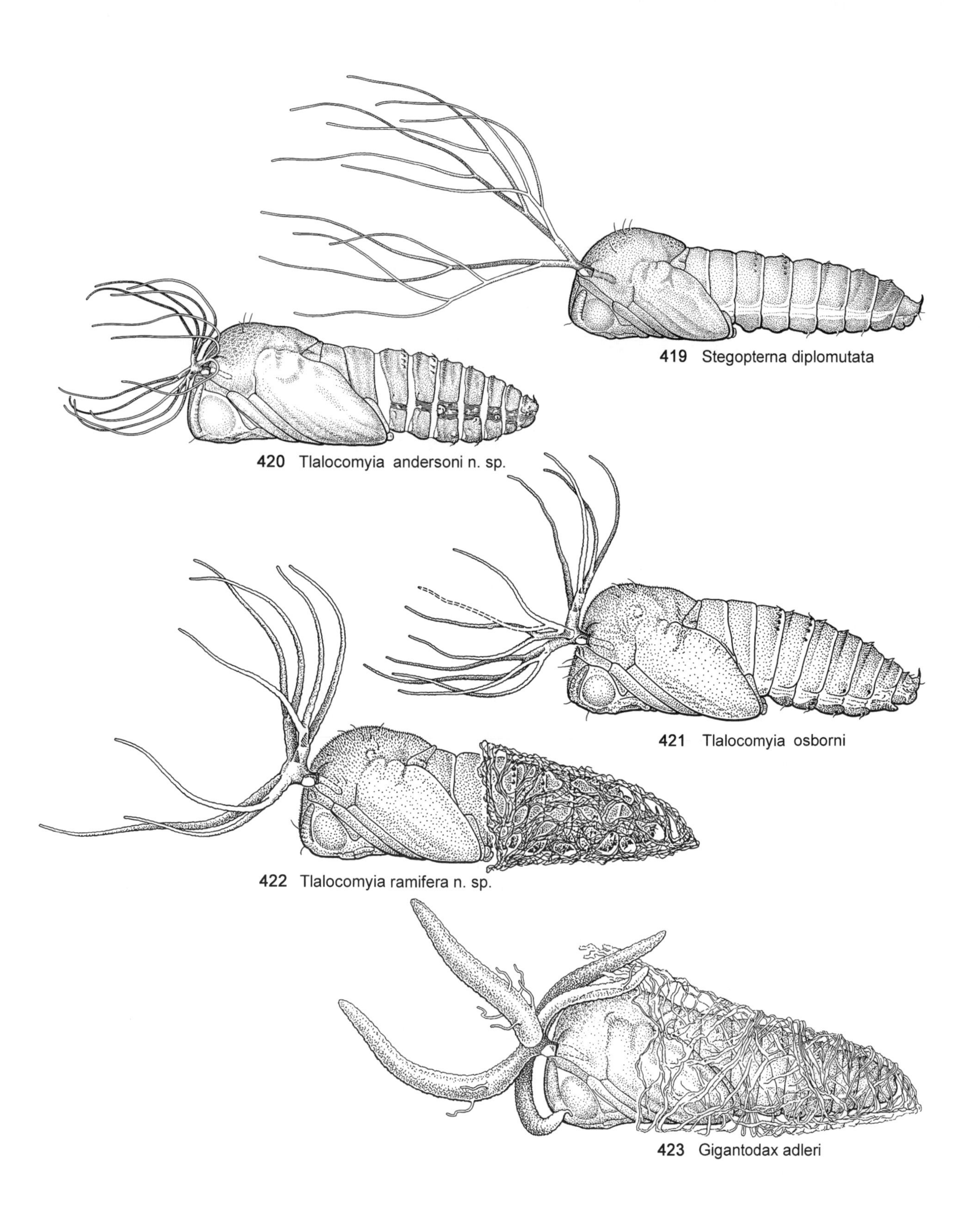


Figs. 10.419–10.423. Pupae, lateral view; (419–421) cocoon omitted.

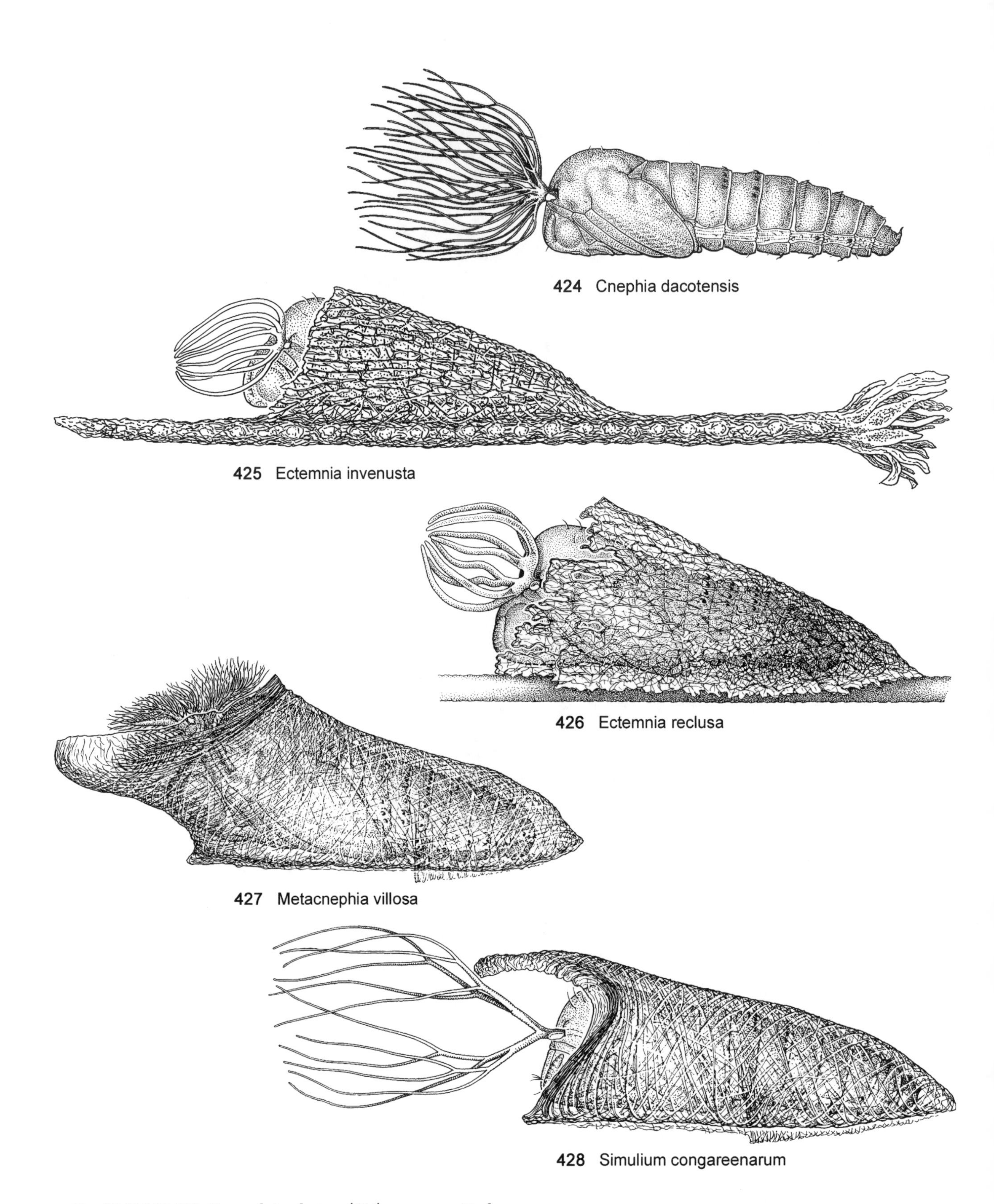


Figs. 10.424–10.428. Pupae, lateral view; (424) cocoon omitted.

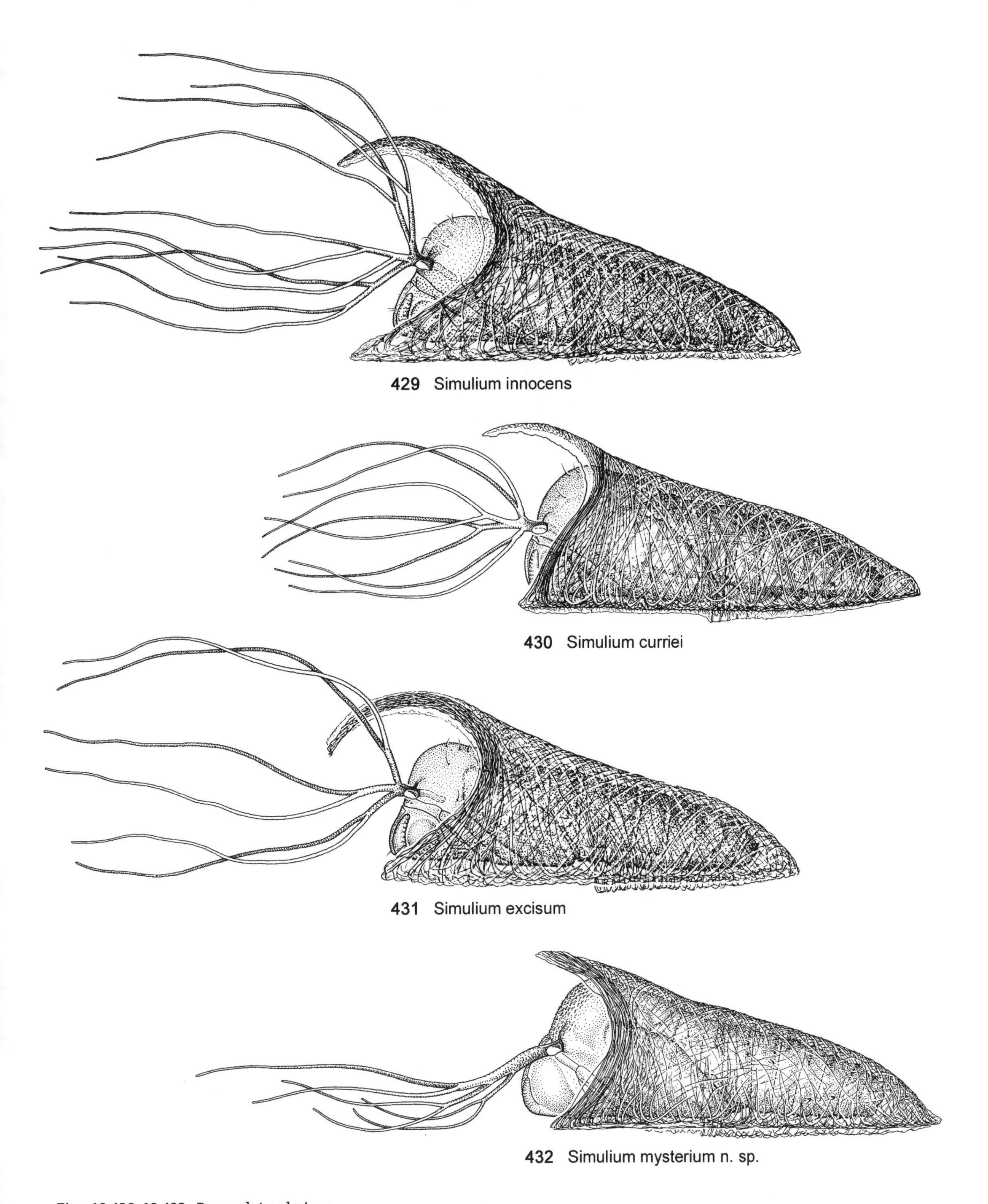


Figs. 10.429–10.432. Pupae, lateral view.

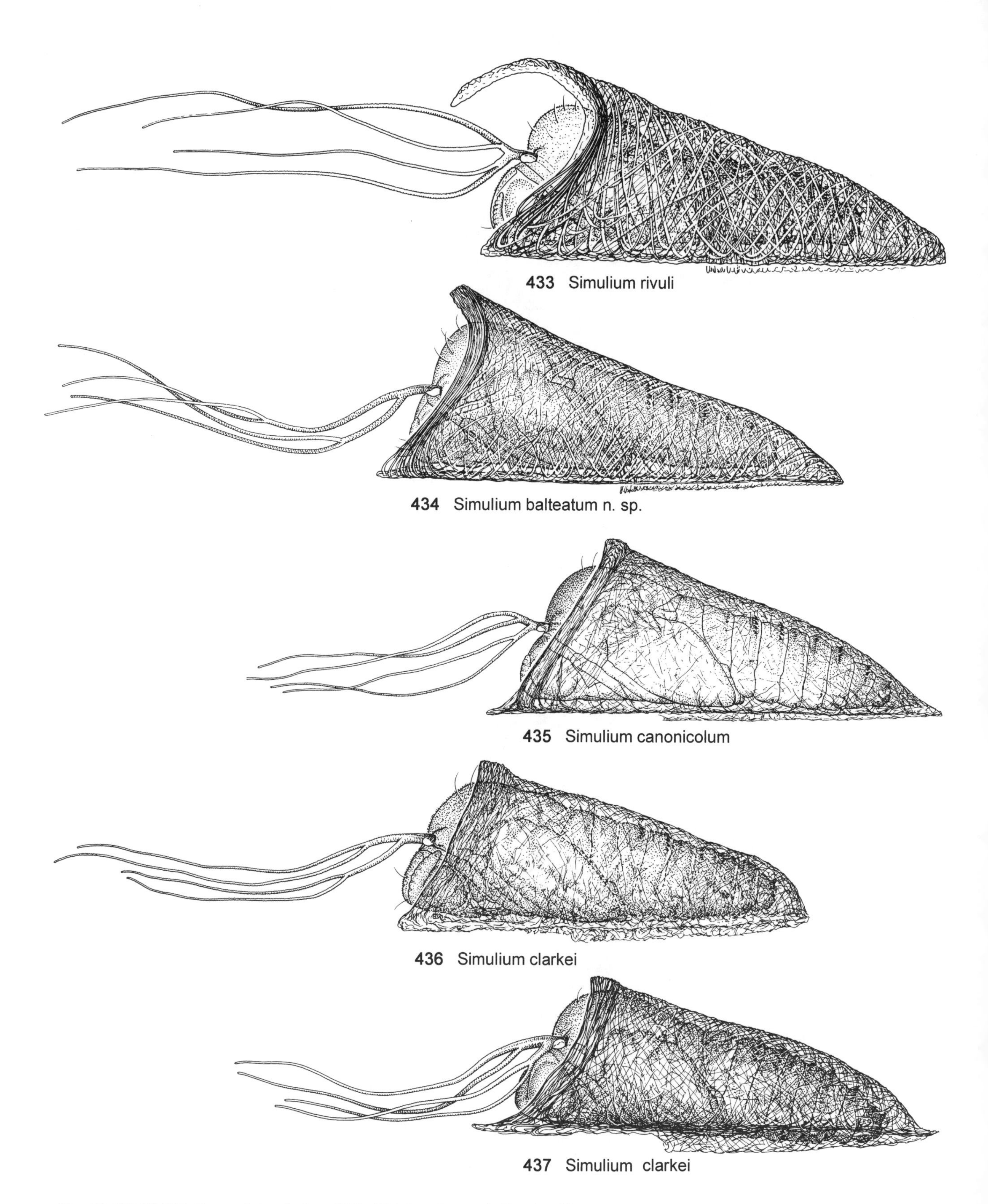


Figs. 10.433–10.437. Pupae, lateral view; (436, 437) showing variation in gill.

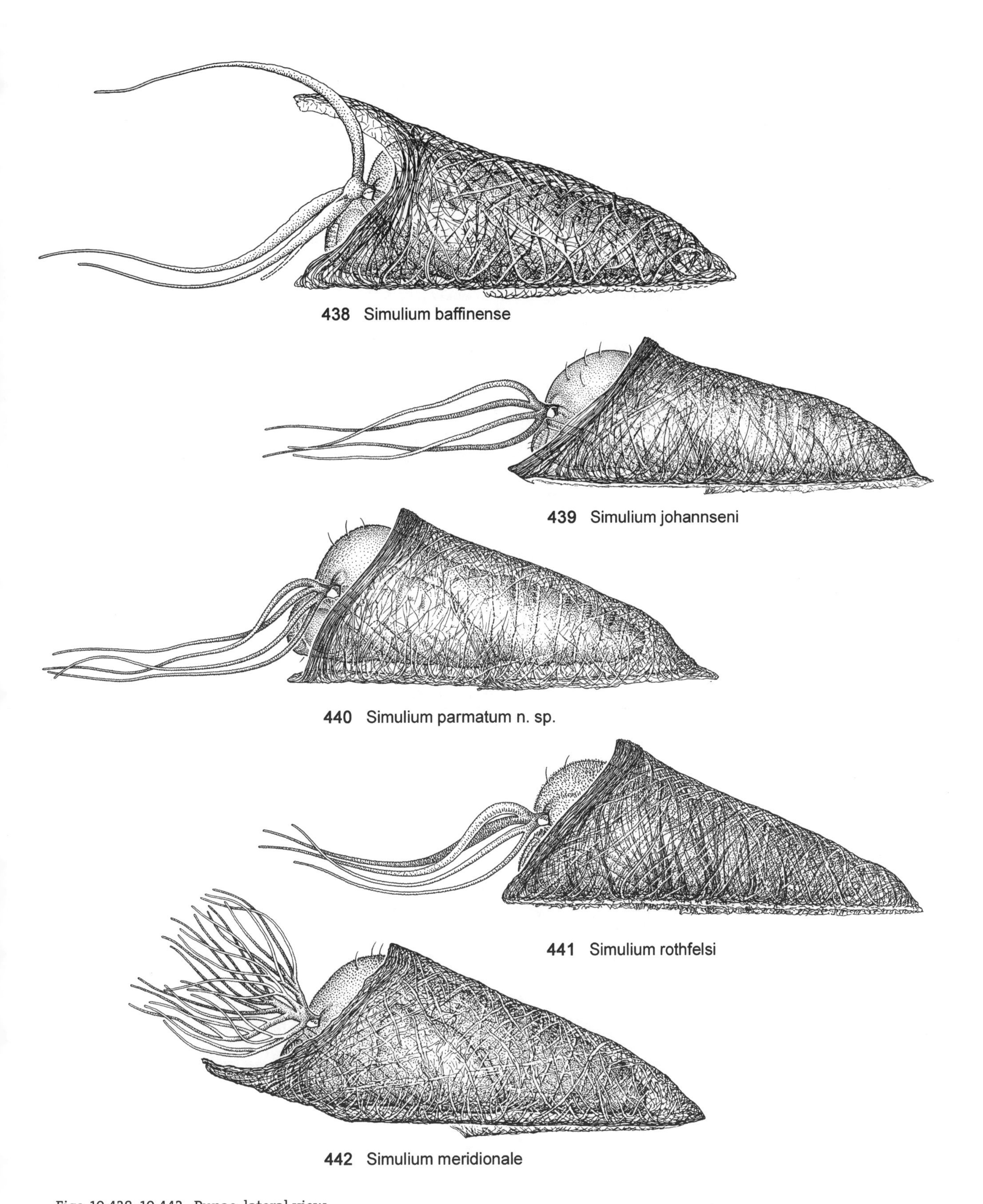


Figs. 10.438–10.442. Pupae, lateral view.

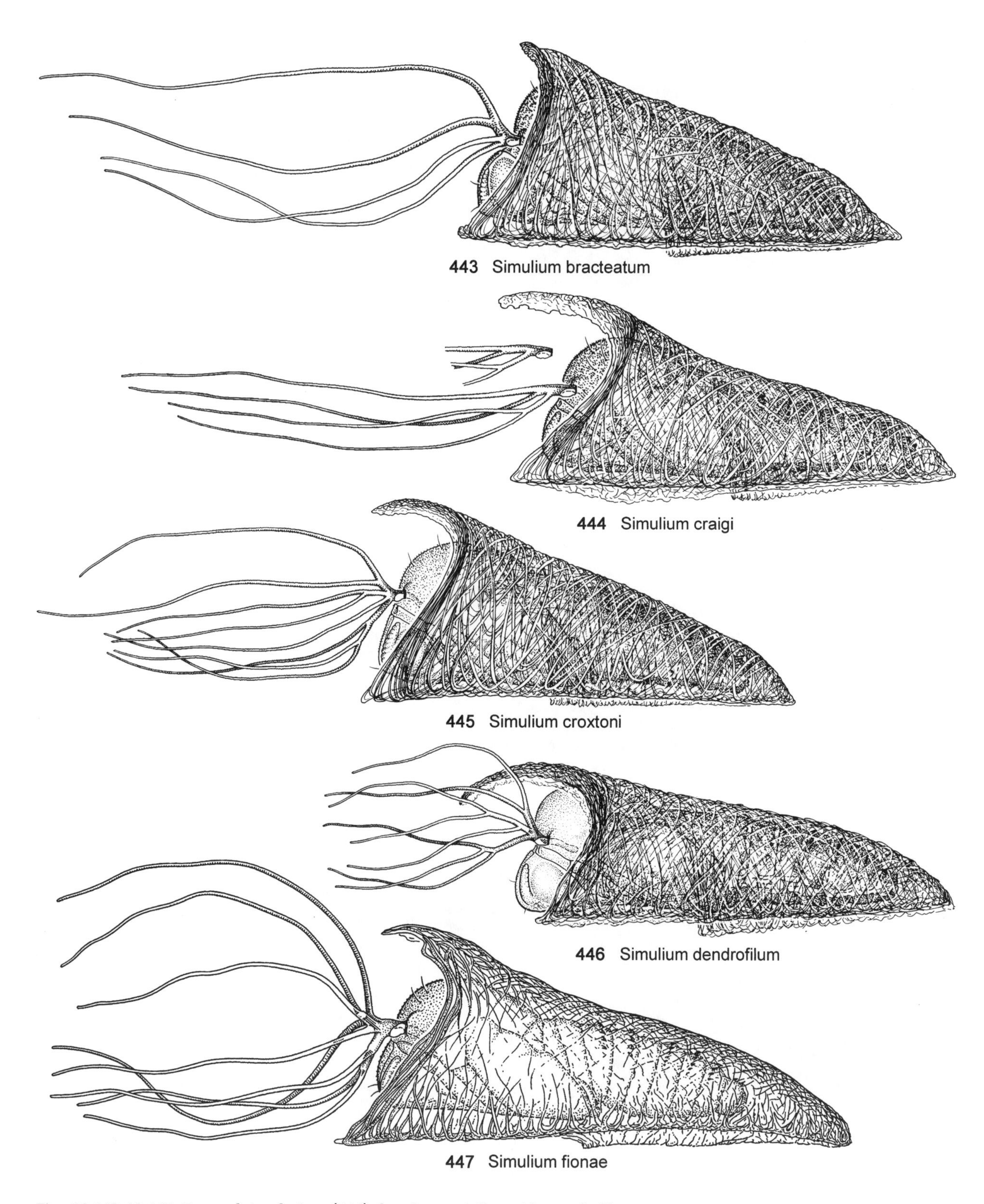


Figs. 10.443–10.447. Pupae, lateral view; (444) showing variation at base of gill.

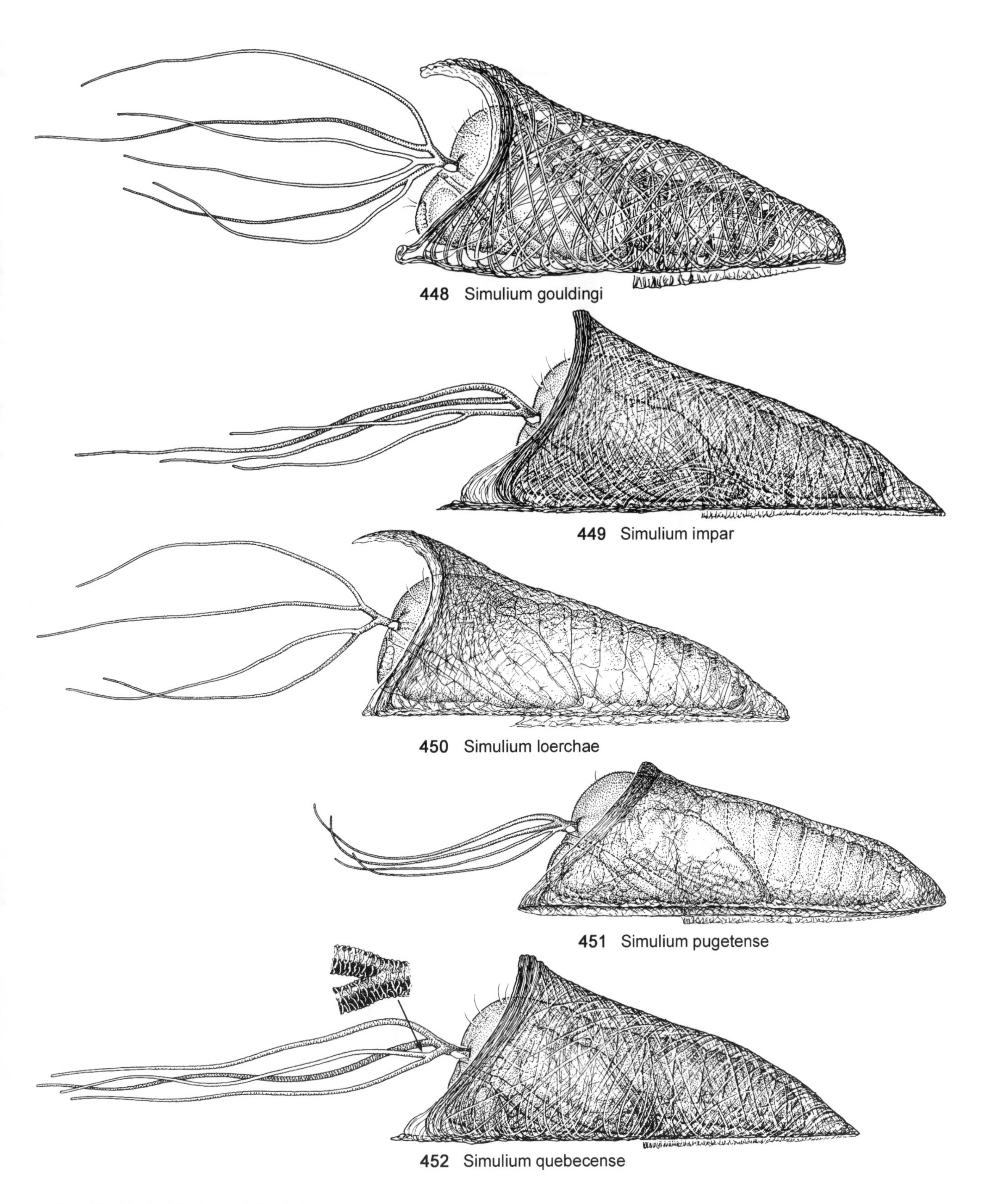


Figs. 10.448–10.452. Pupae, lateral view.

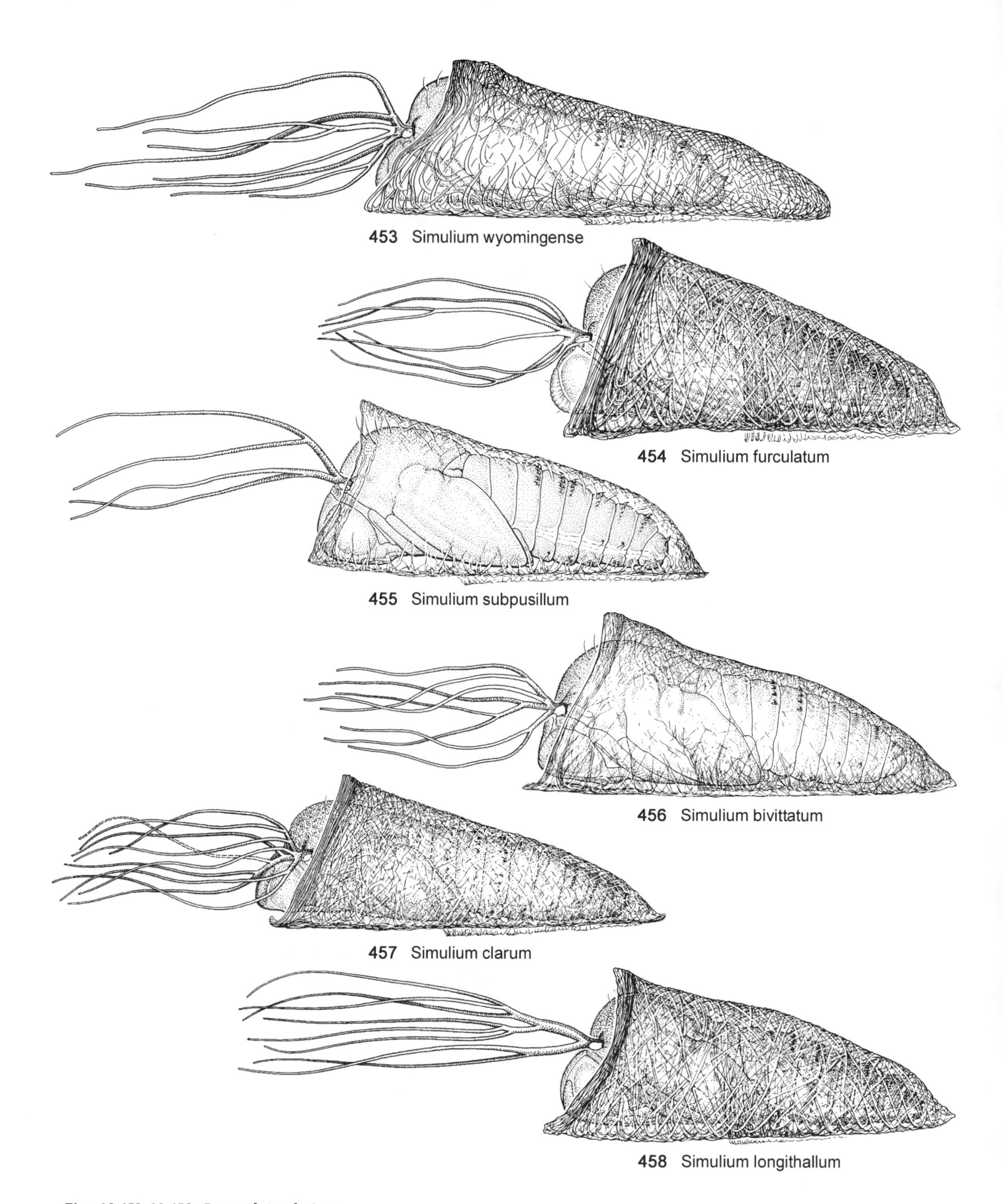


Figs. 10.453–10.458. Pupae, lateral view.

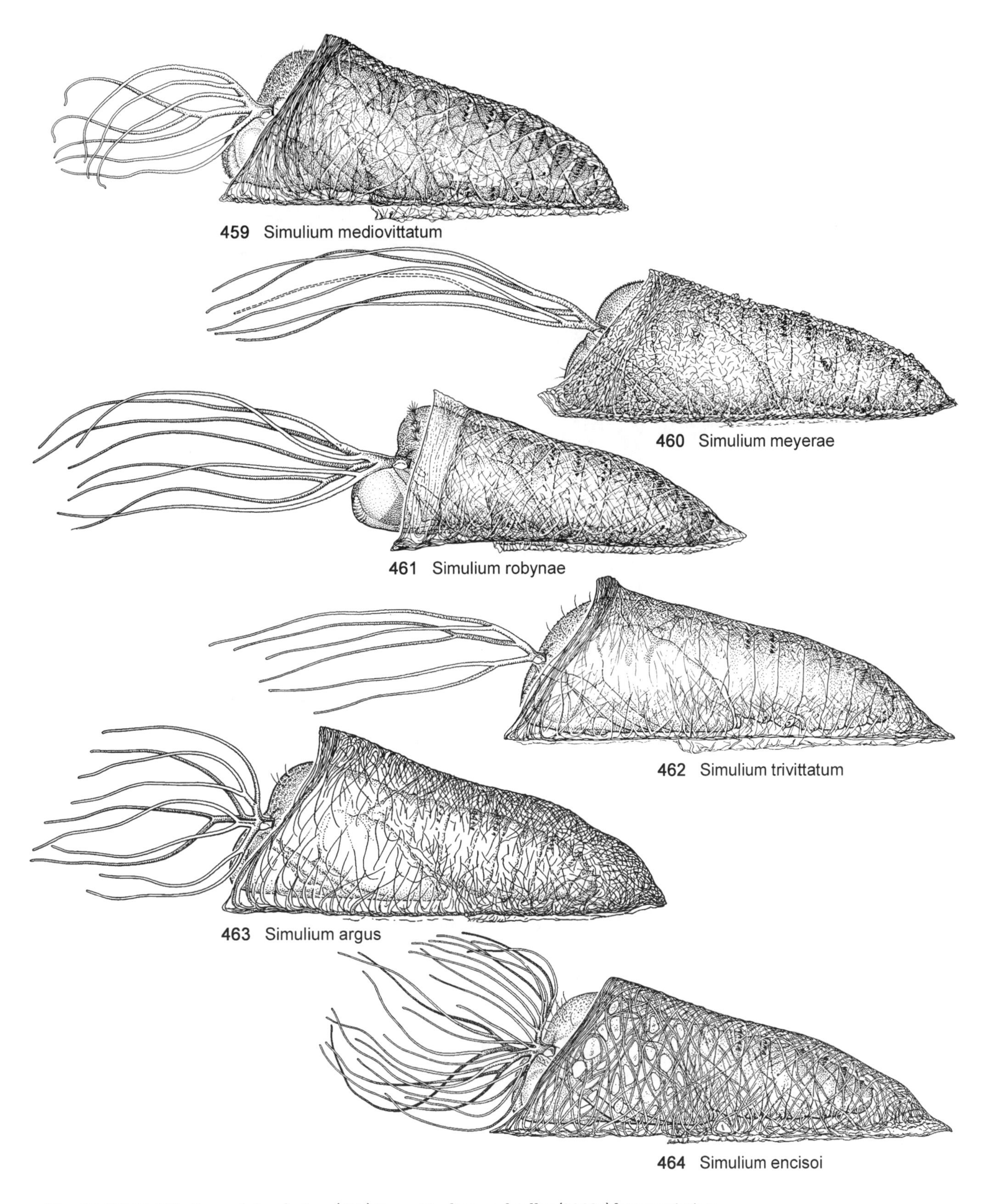


Figs. 10.459–10.464. Pupae, lateral view. (460) From Moulton and Adler (2002a) by permission.

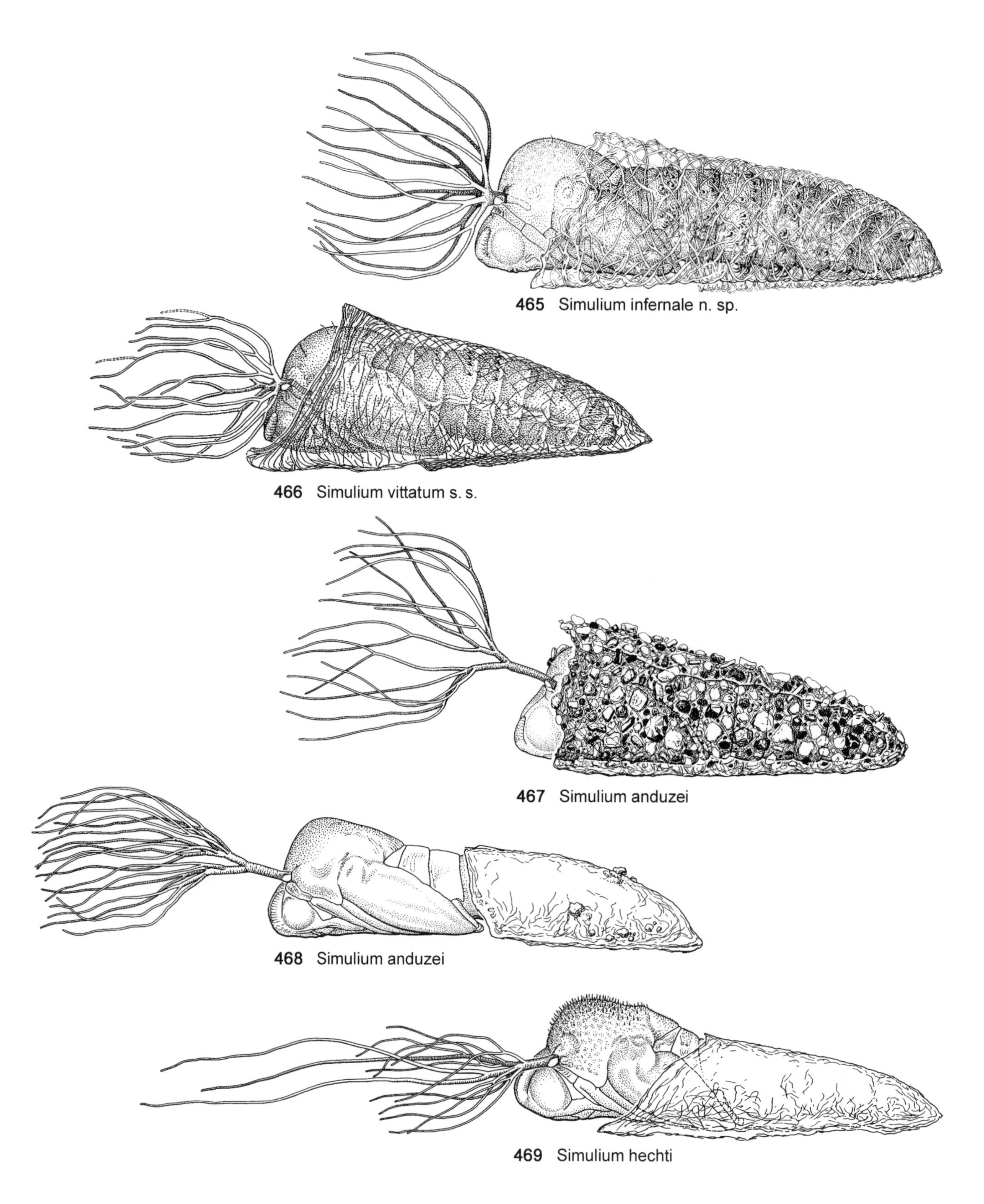


Figs. 10.465–10.469. Pupae, lateral view; (469) pupa slightly extricated from cocoon. (466) From Peterson (1981) by permission.

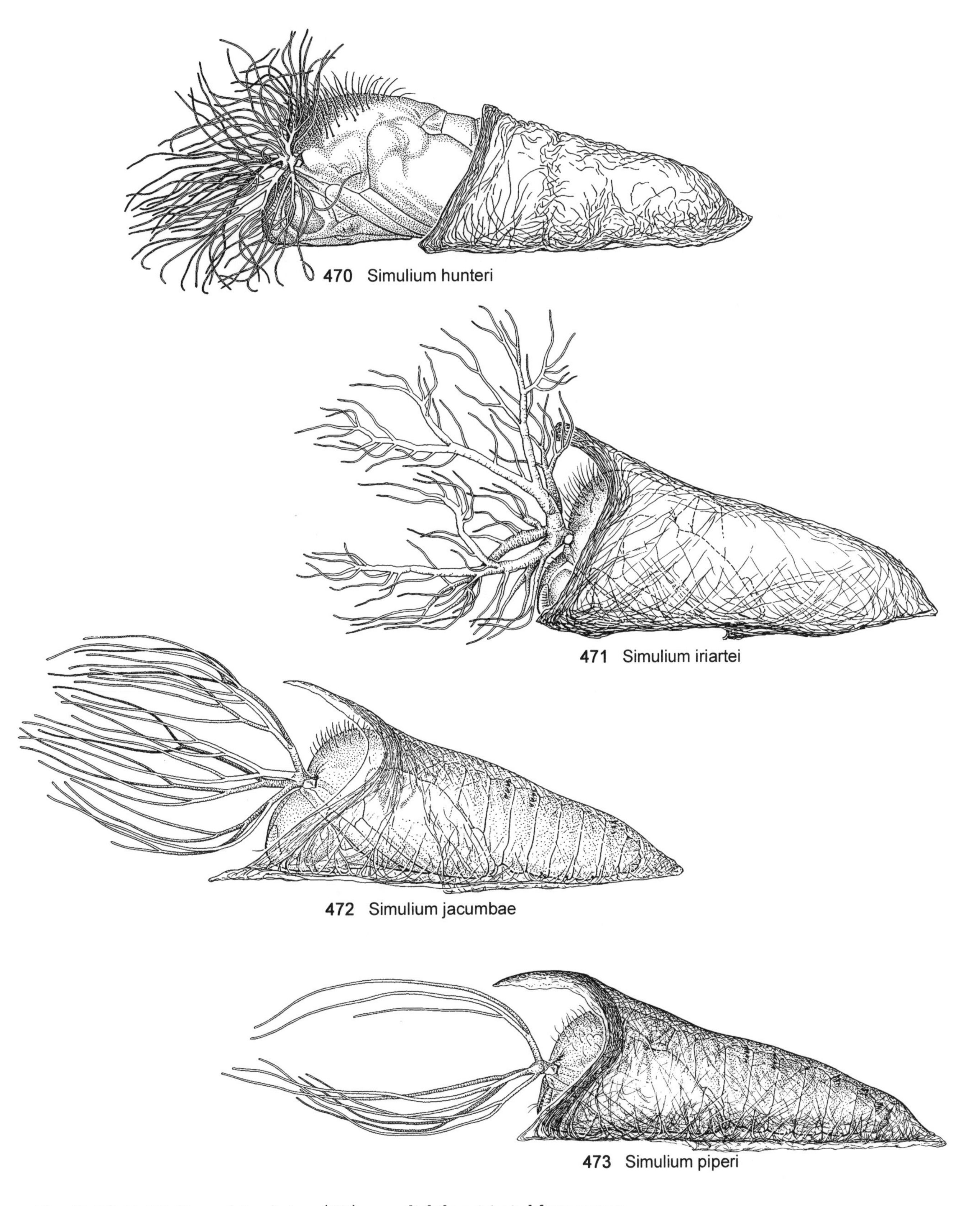


Figs. 10.470–10.473. Pupae, lateral view; (470) pupa slightly extricated from cocoon.

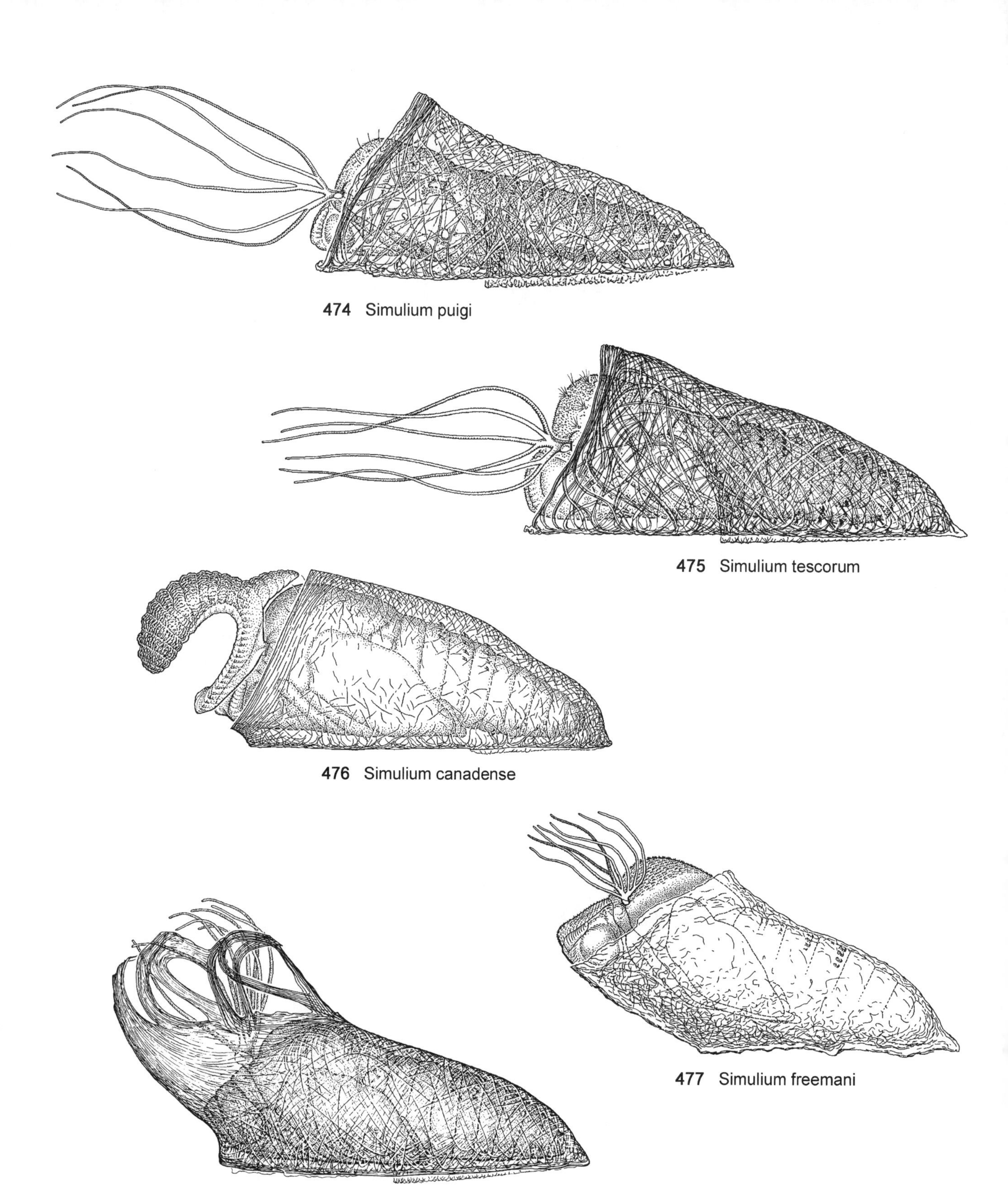


Figs. 10.474–10.478. Pupae, lateral view.

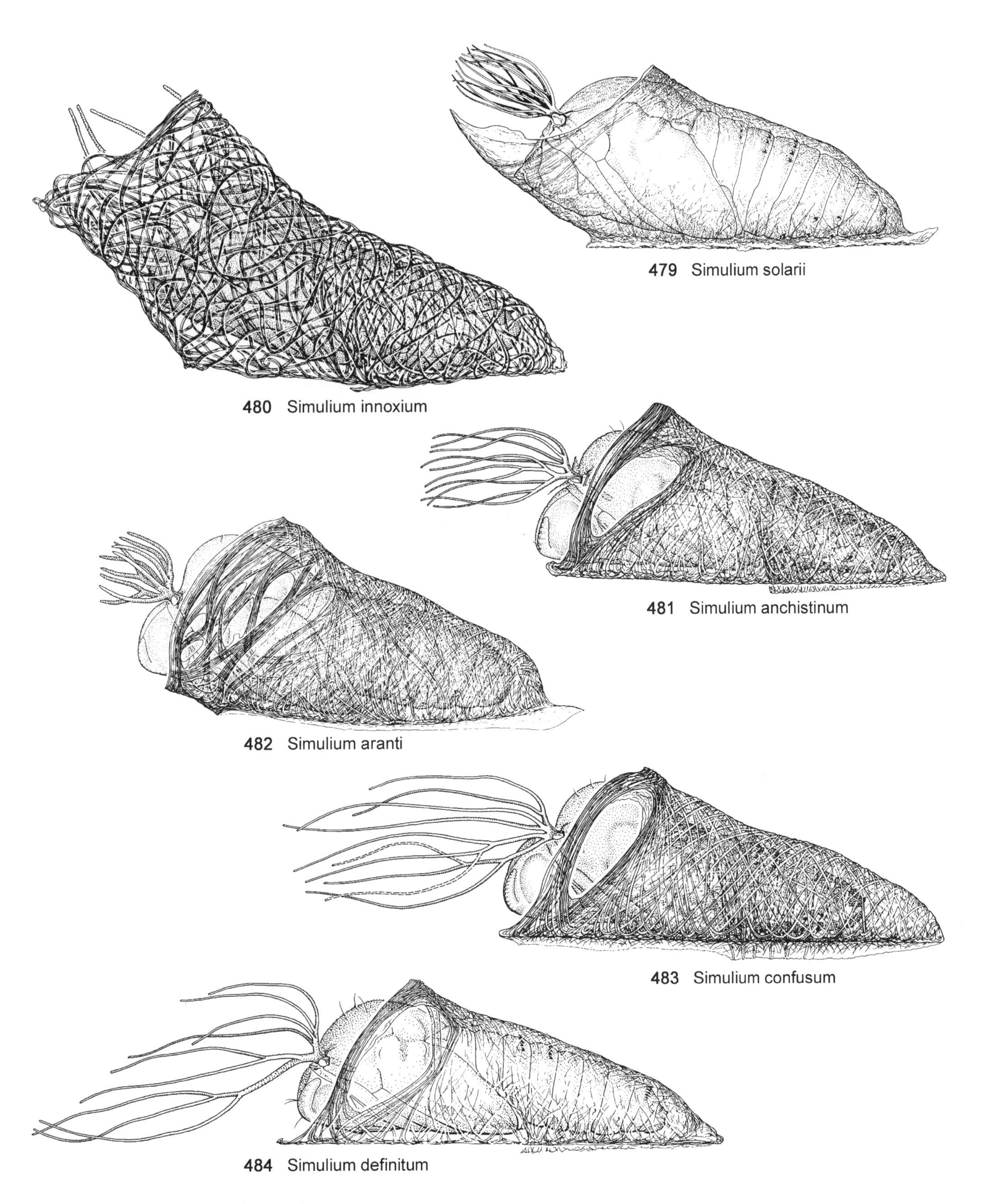


Figs. 10.479–10.484. Pupae, lateral view.

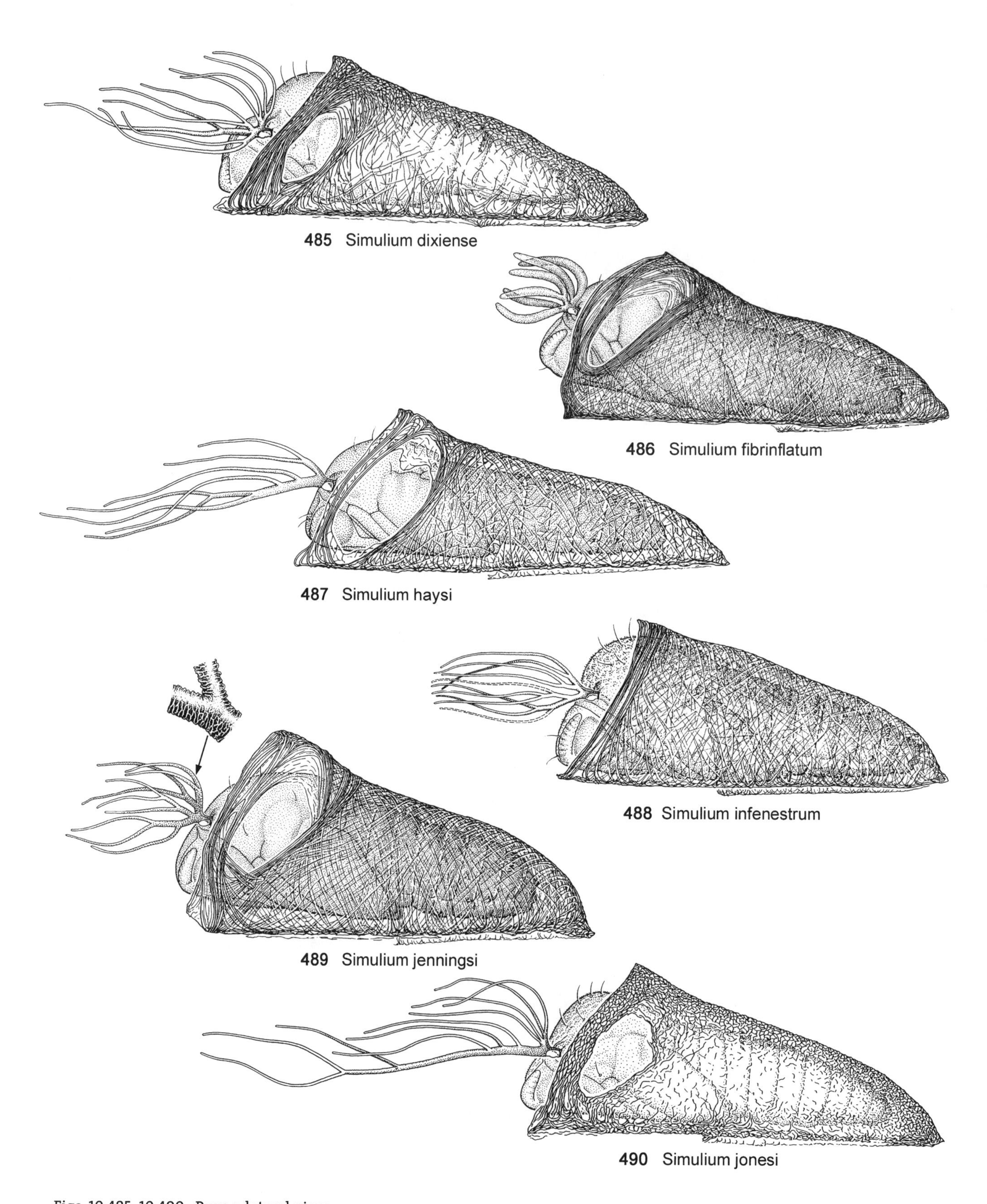


Figs. 10.485–10.490. Pupae, lateral view.

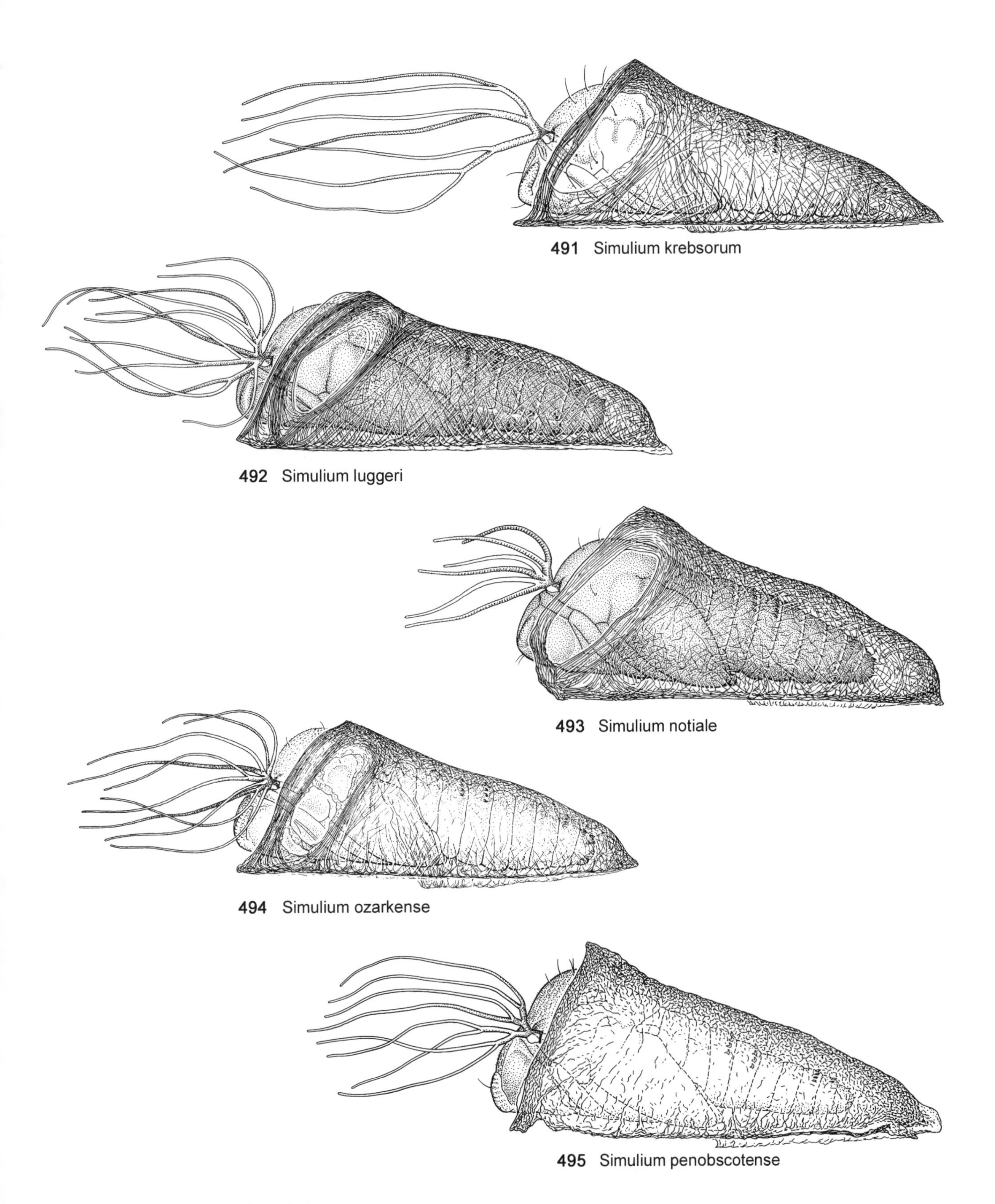


Figs. 10.491–10.495. Pupae, lateral view.

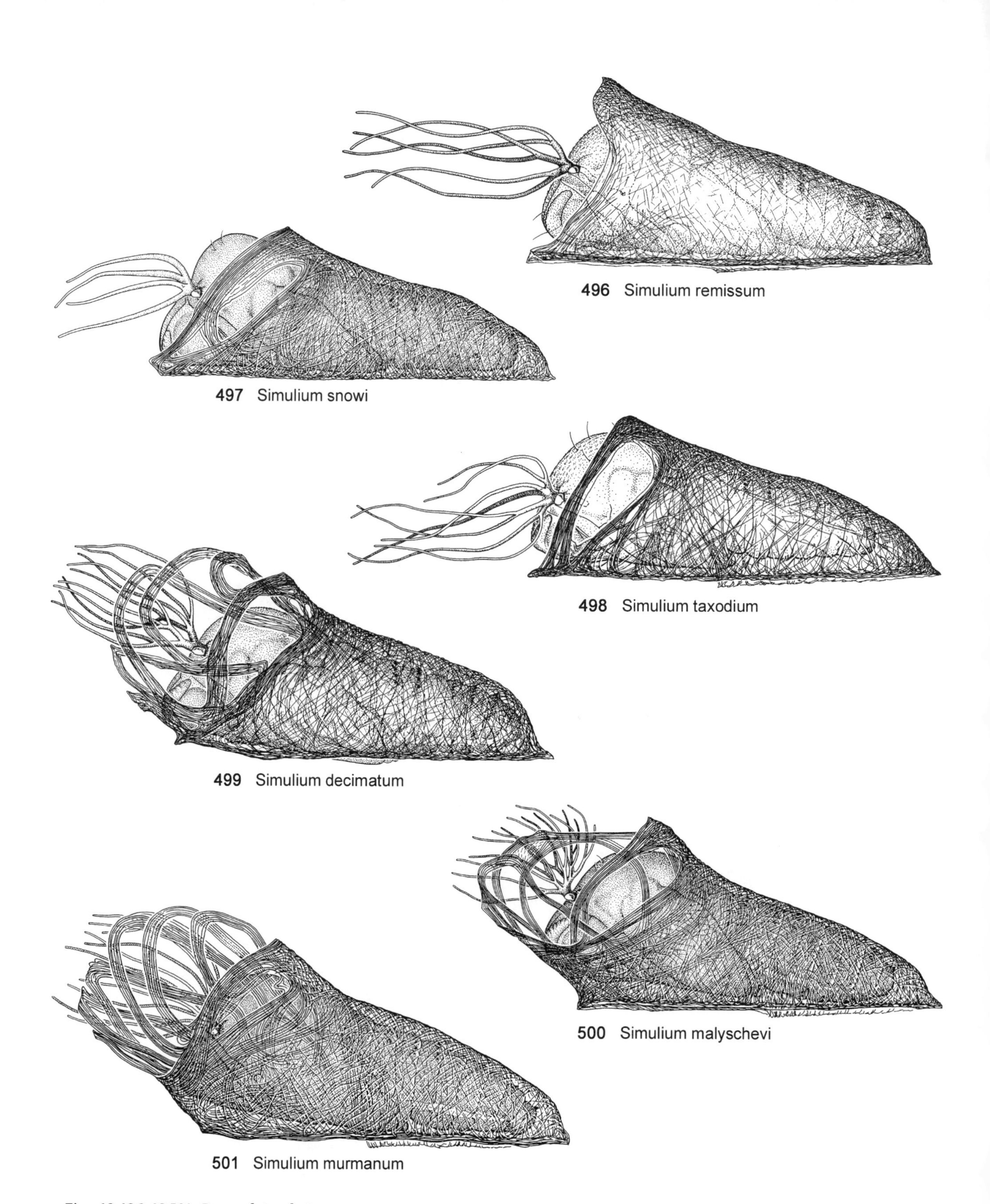


Figs. 10.496–10.501. Pupae, lateral view.

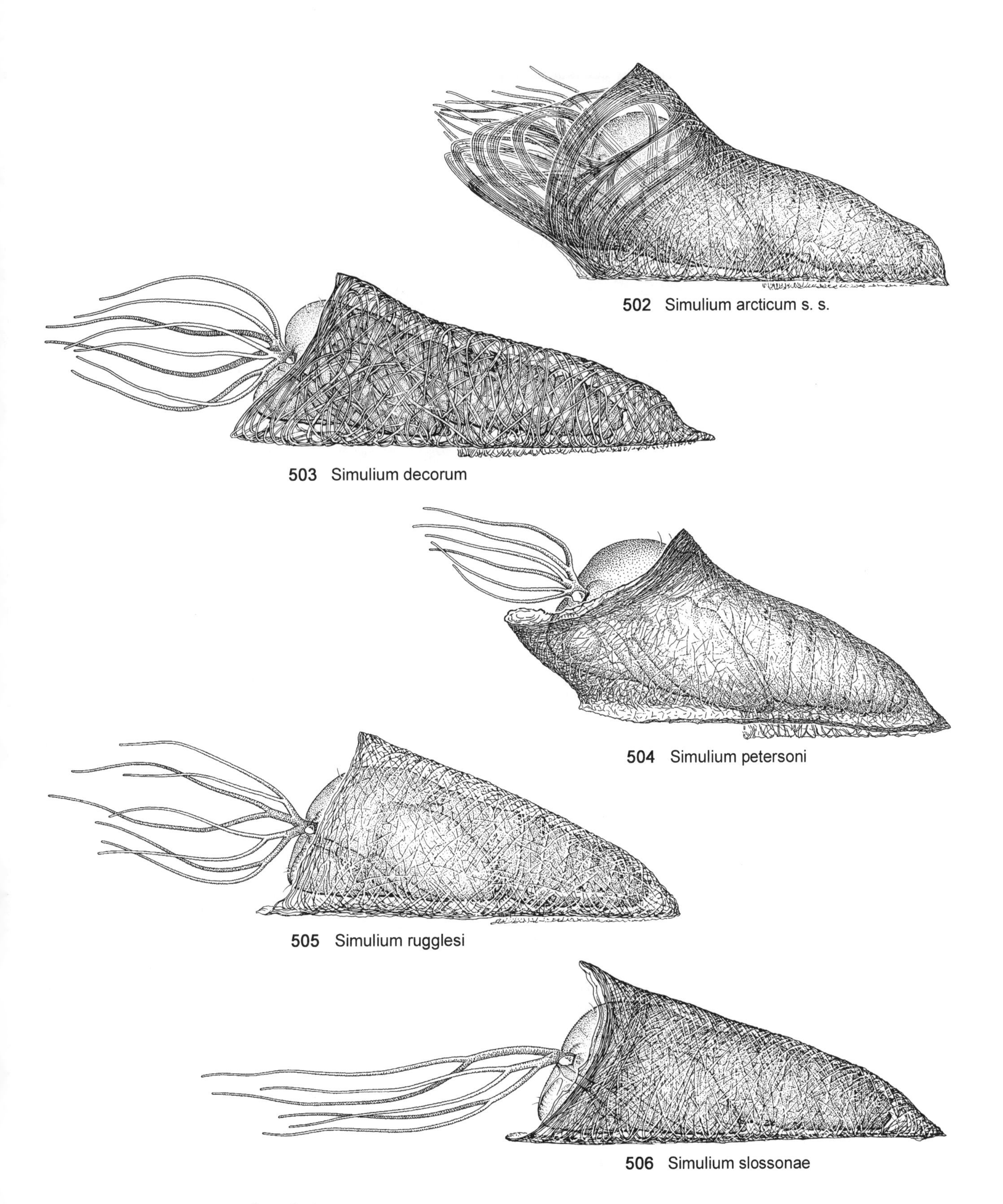


Figs. 10.502–10.506. Pupae, lateral view.

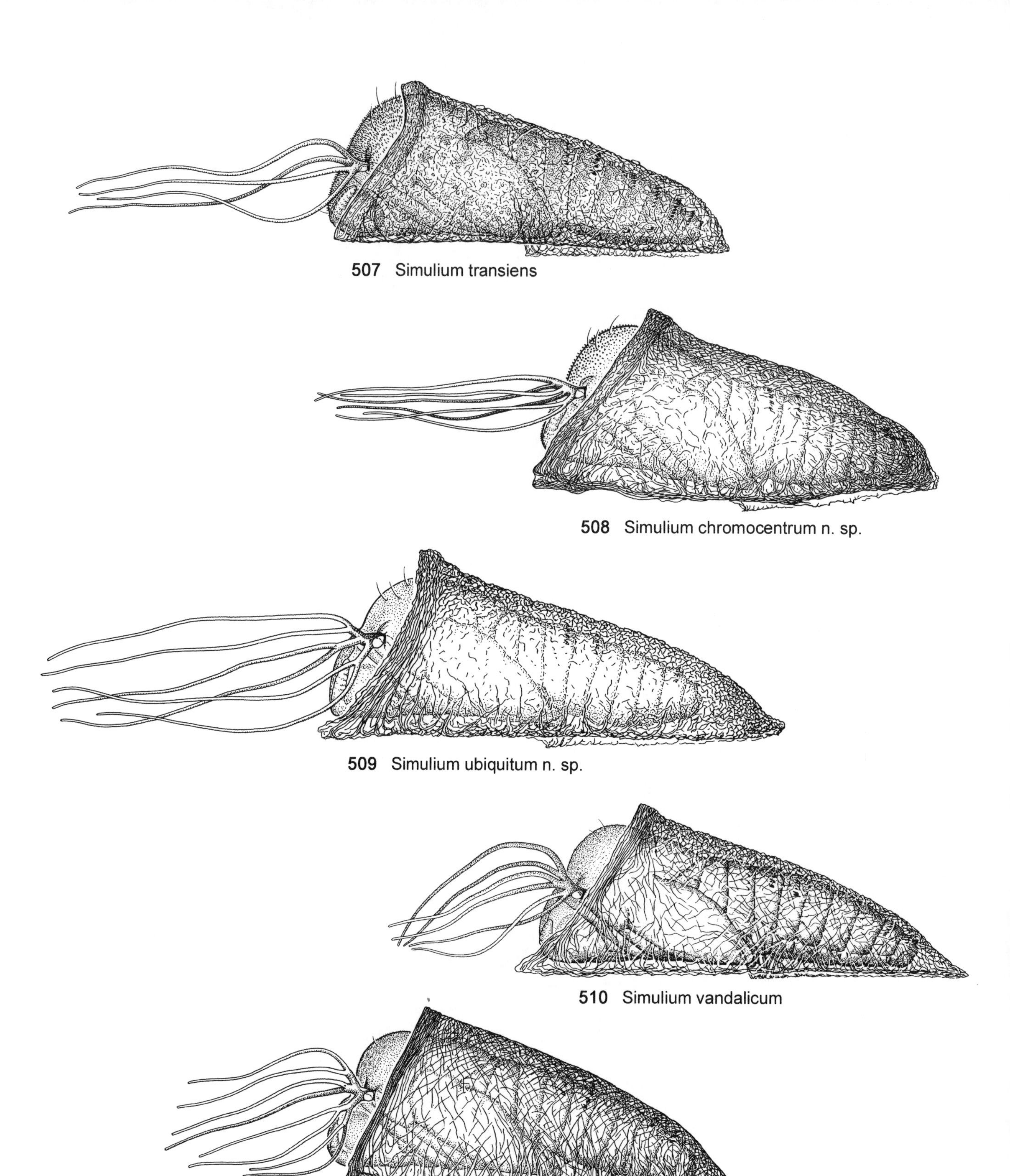


Figs. 10.507–10.511. Pupae, lateral view.

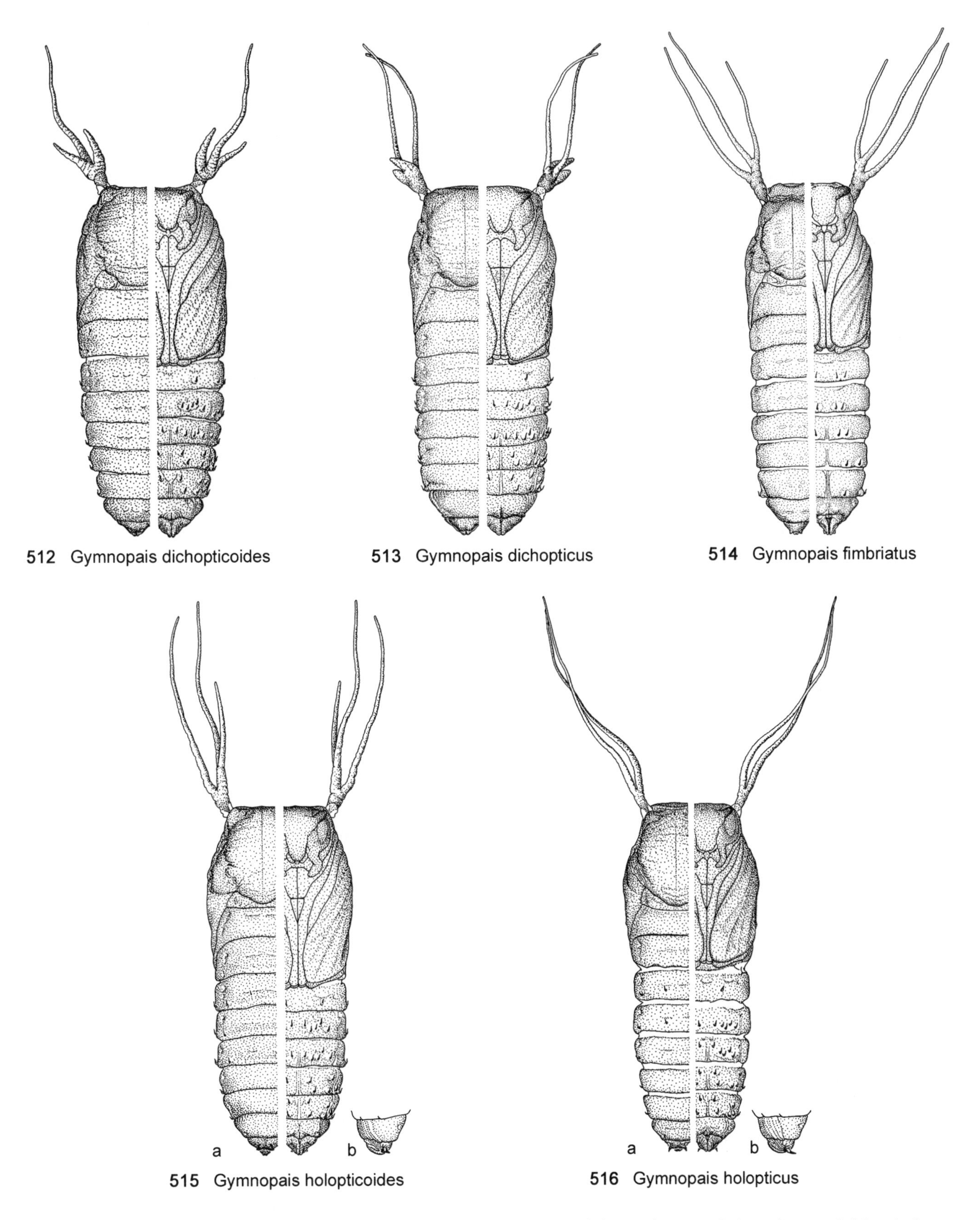


Figs. 10.512–10.516. Pupae, dorsal three fifths (left); ventral three fifths (right); (515, 516) a, pupa; b, lateral view of abdominal segments VIII and IX.

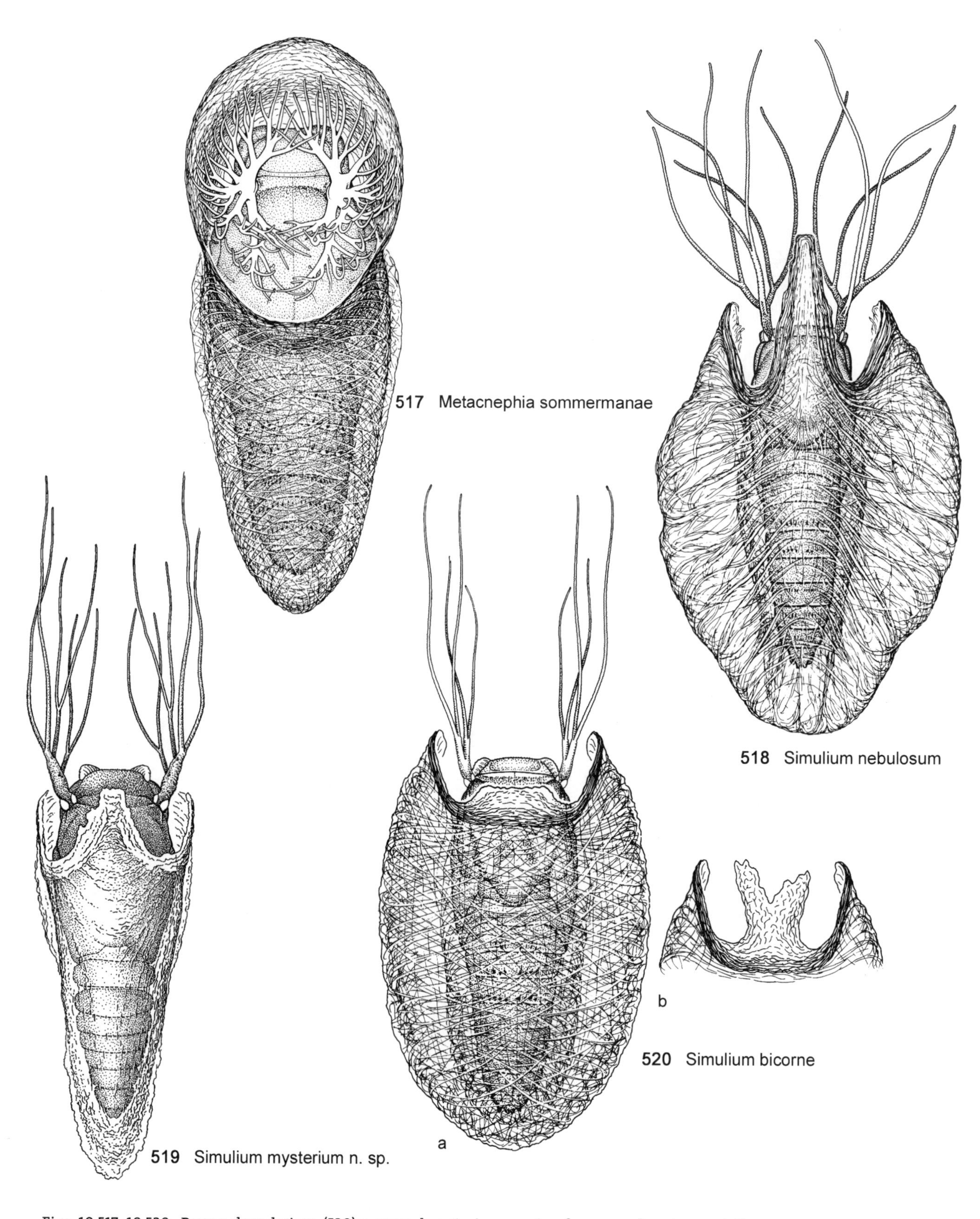


Figs. 10.517–10.520. Pupae, dorsal view; (520) a, pupa; b, anterior margin of cocoon, showing variation.

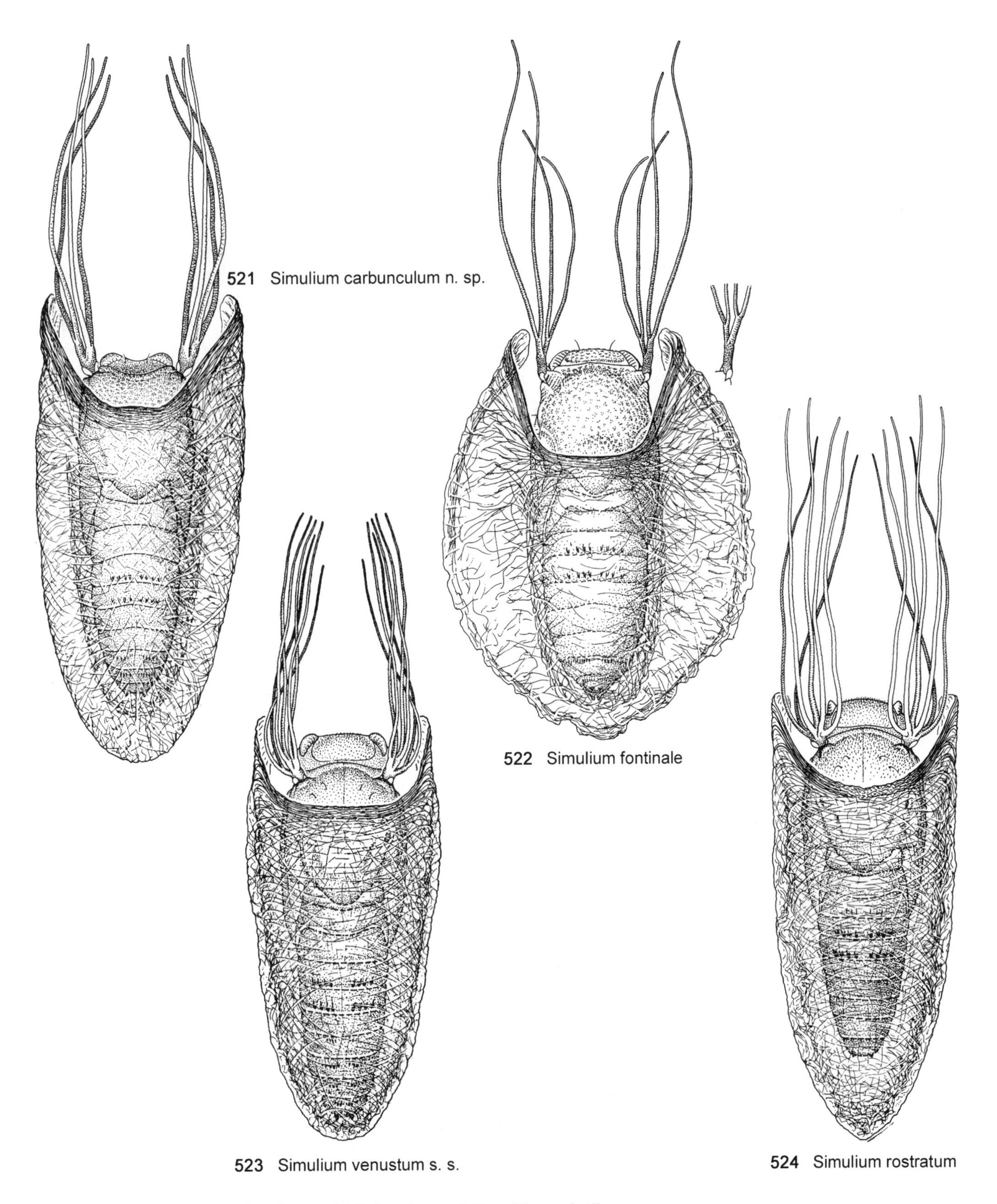


Figs. 10.521–10.524. Pupae, dorsal view; (522) showing variation at base of gill.

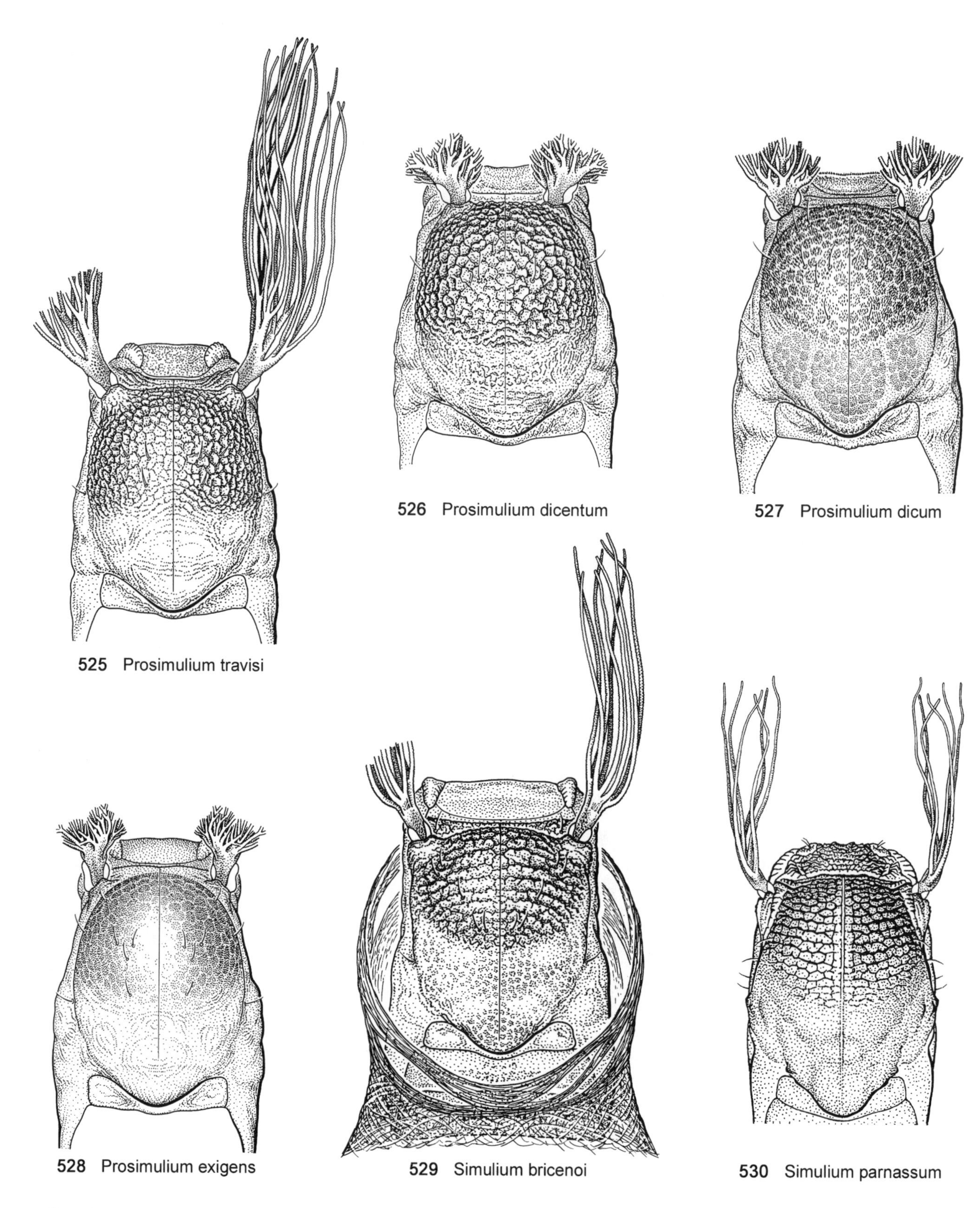


Figs. 10.525–10.530. Female pupal heads and thoraces, dorsal view; apices of gill filaments omitted on one or both sides, except (530); (529) anterior portion of cocoon included.

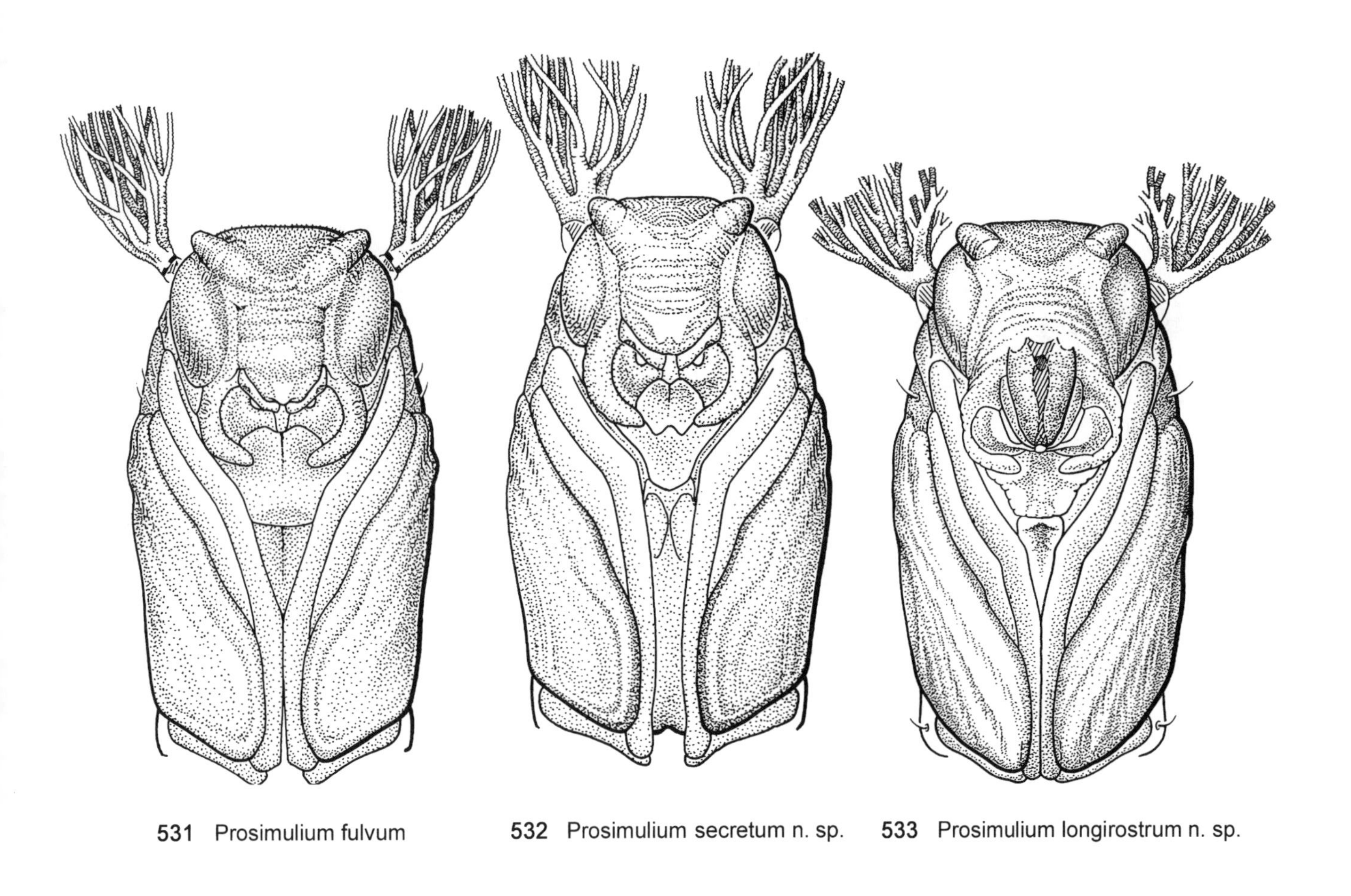


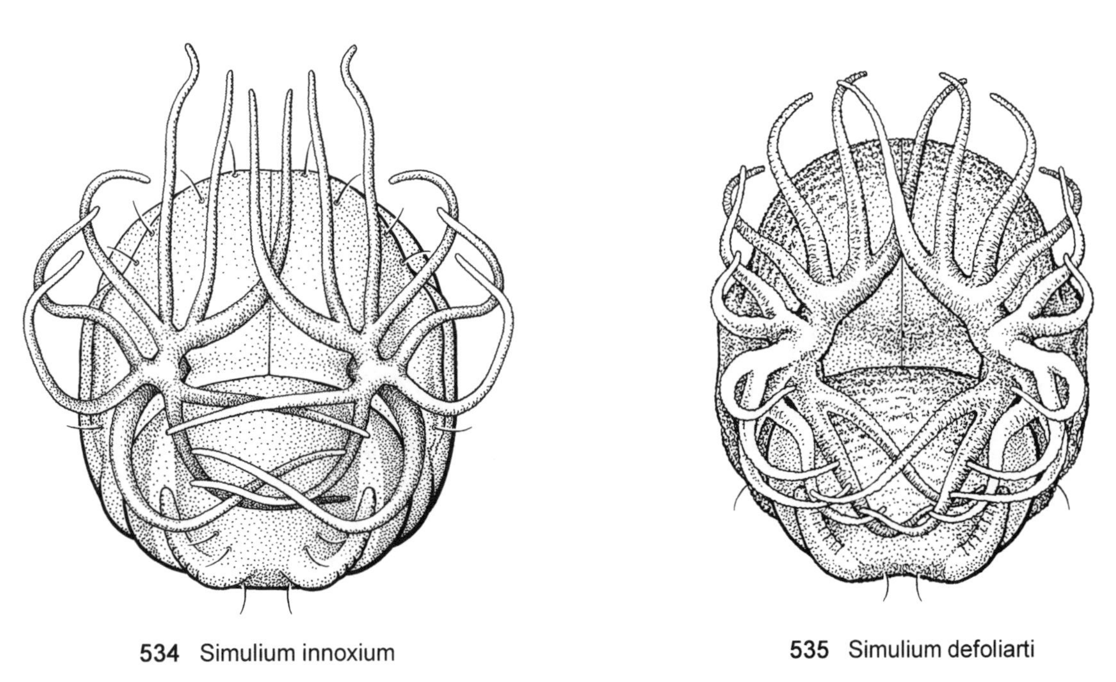


Figs. 10.531–10.535. Male pupal heads and thoraces; (531–533) ventral view, with apices of gill filaments omitted; (534, 535) anterior view.

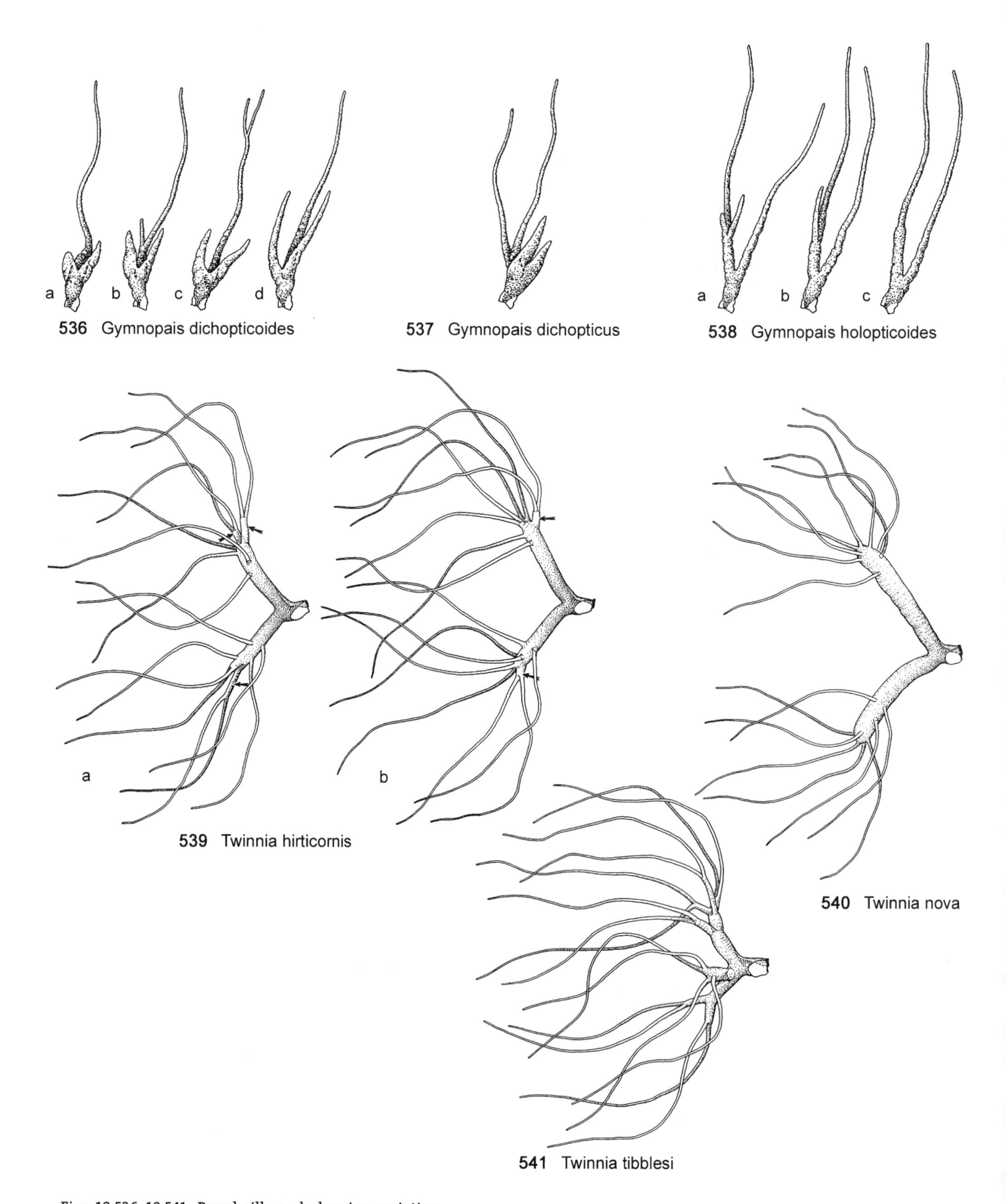


Figs. 10.536–10.541. Pupal gills; a–d, showing variation.

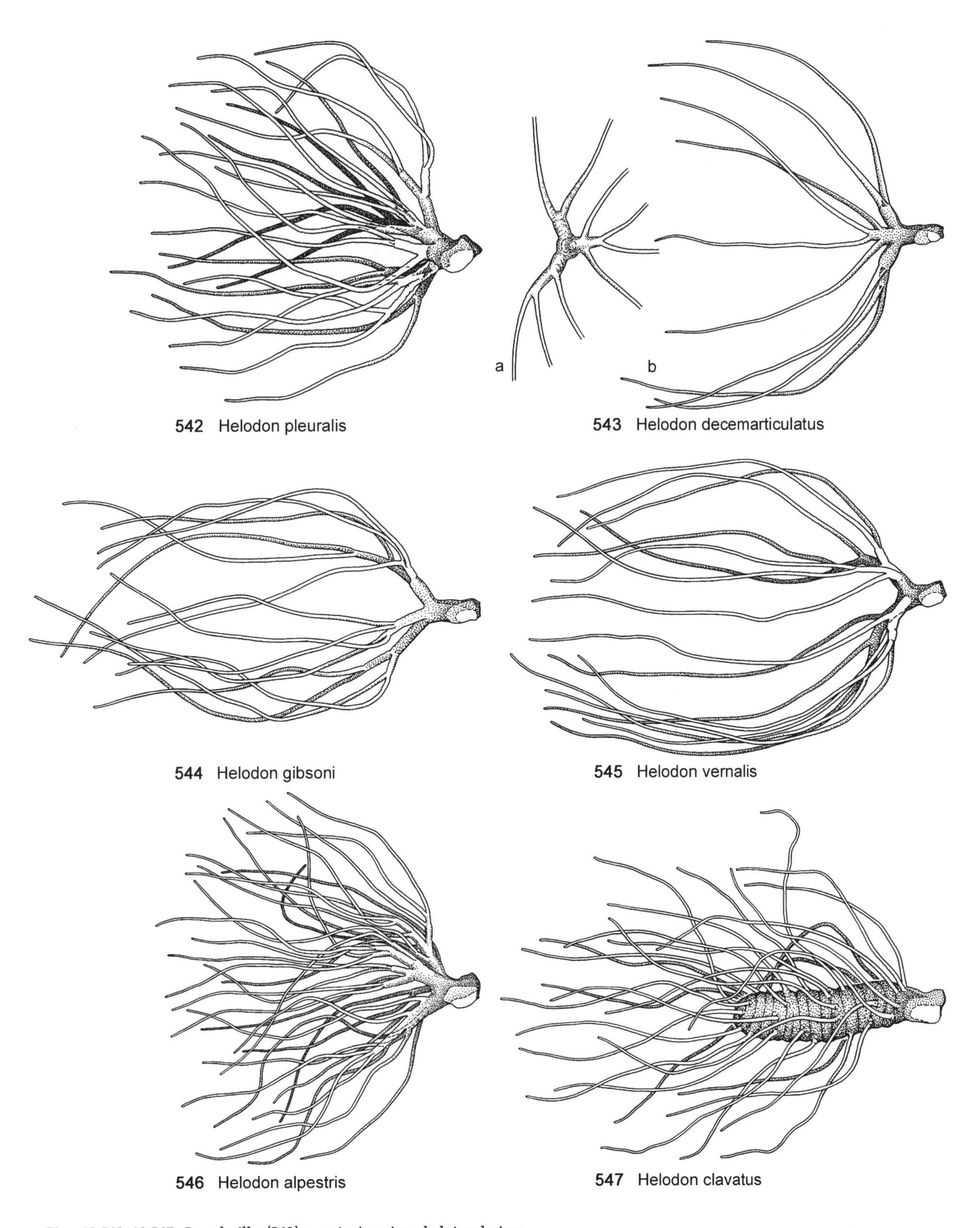


Figs. 10.542–10.547. Pupal gills; (543) a, anterior view; b, lateral view.

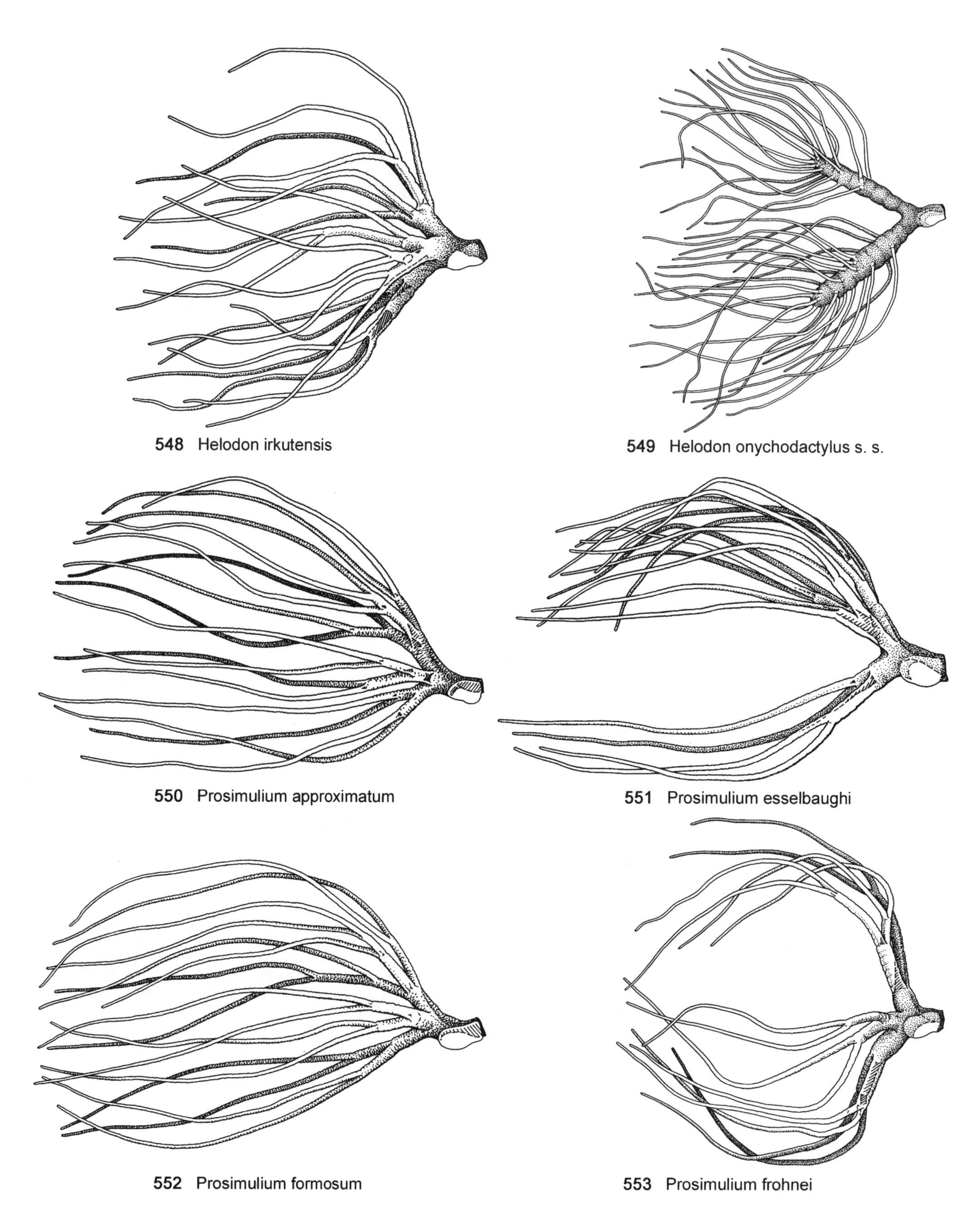


Figs. 10.548–10.553. Pupal gills.

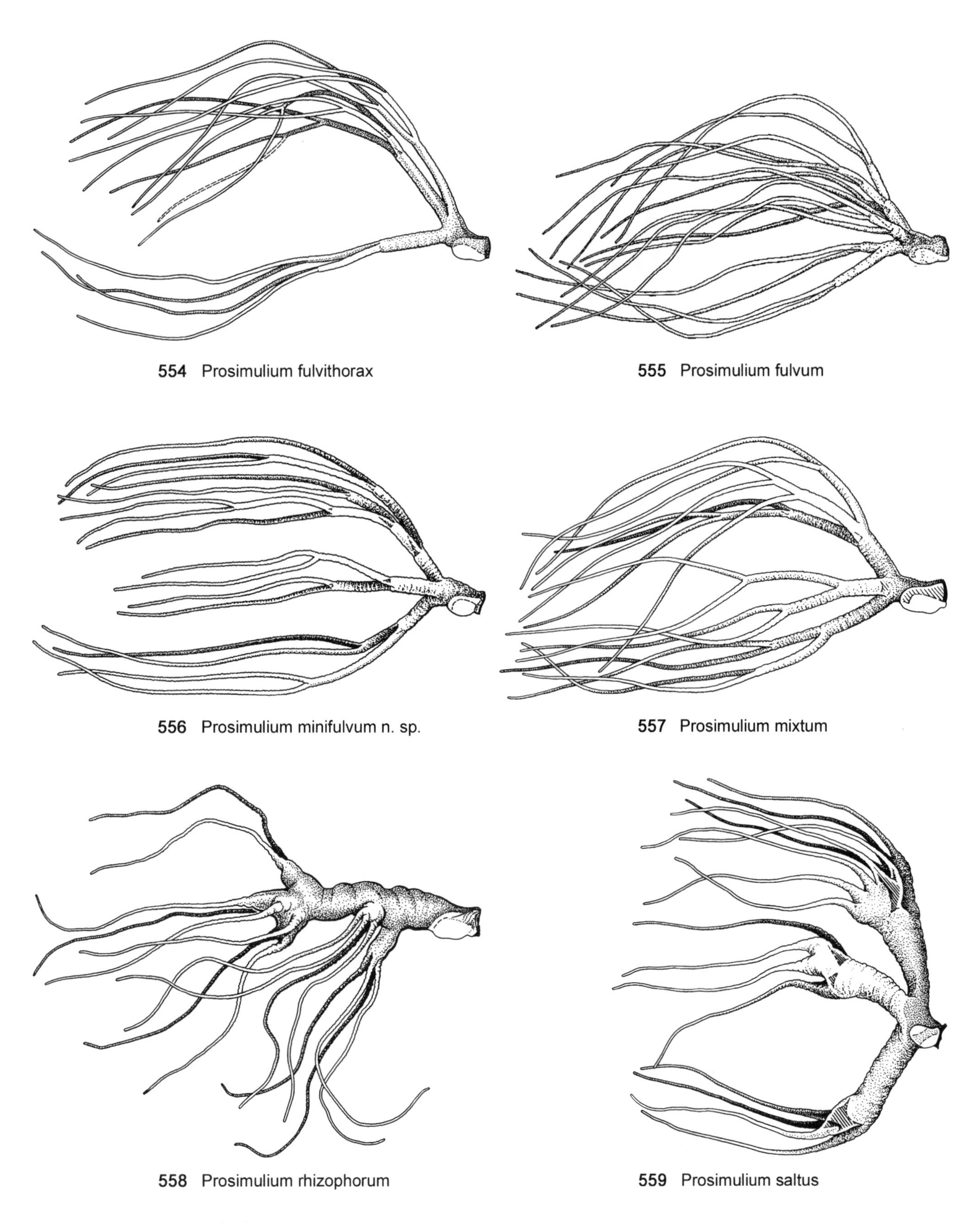


Figs. 10.554–10.559. Pupal gills.

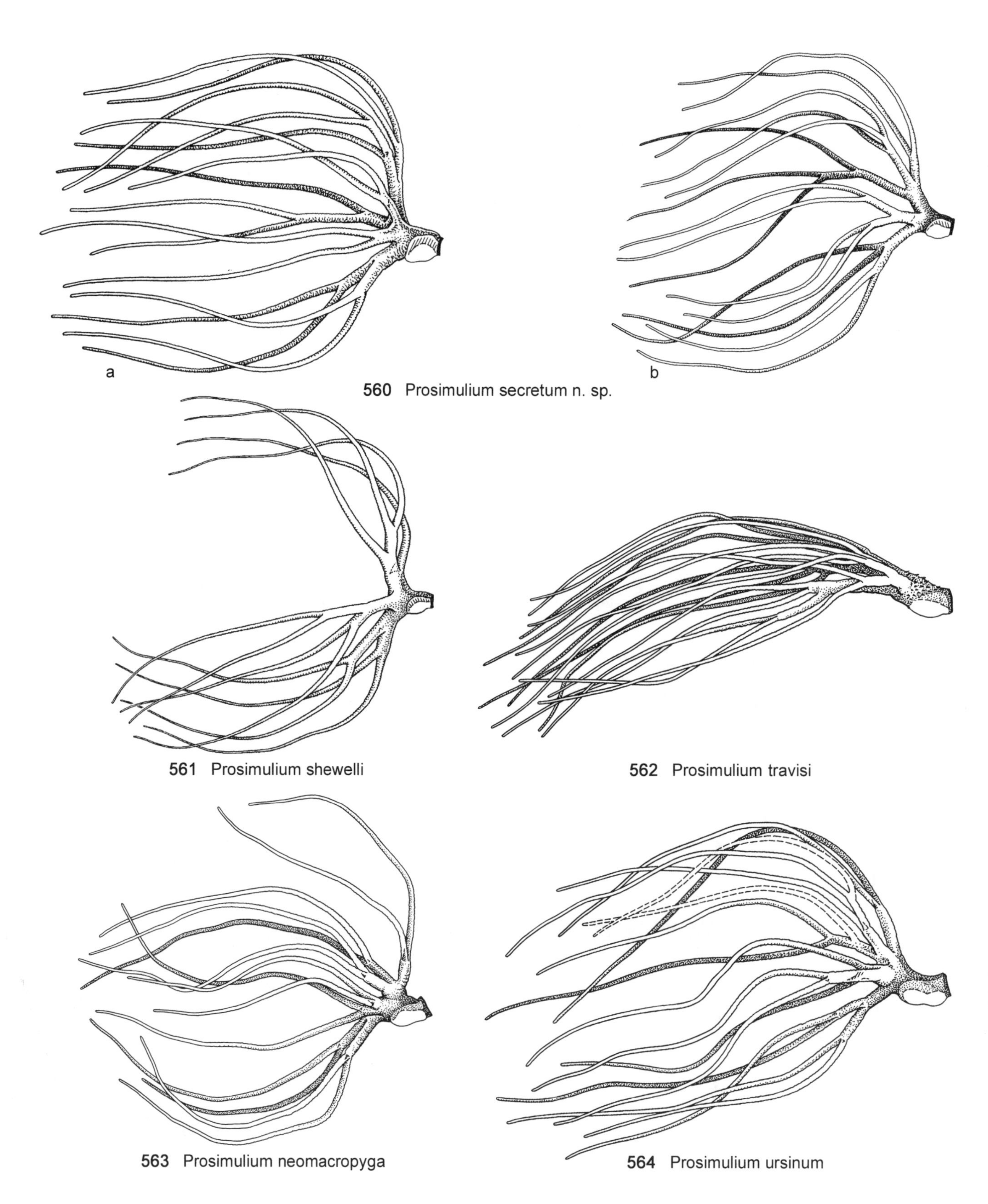


Figs. 10.560–10.564. Pupal gills; (560) showing variation.

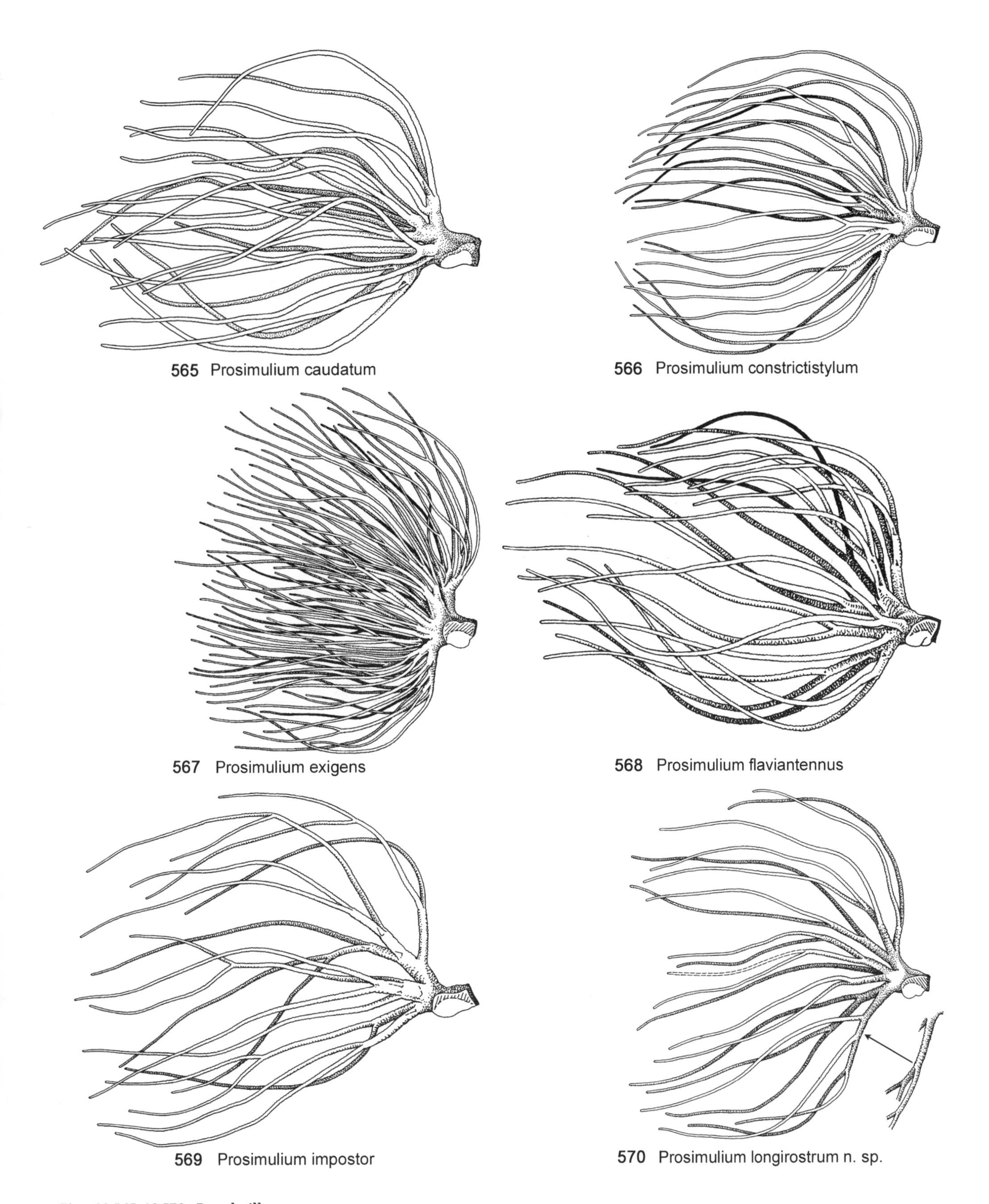


Figs. 10.565–10.570. Pupal gills.

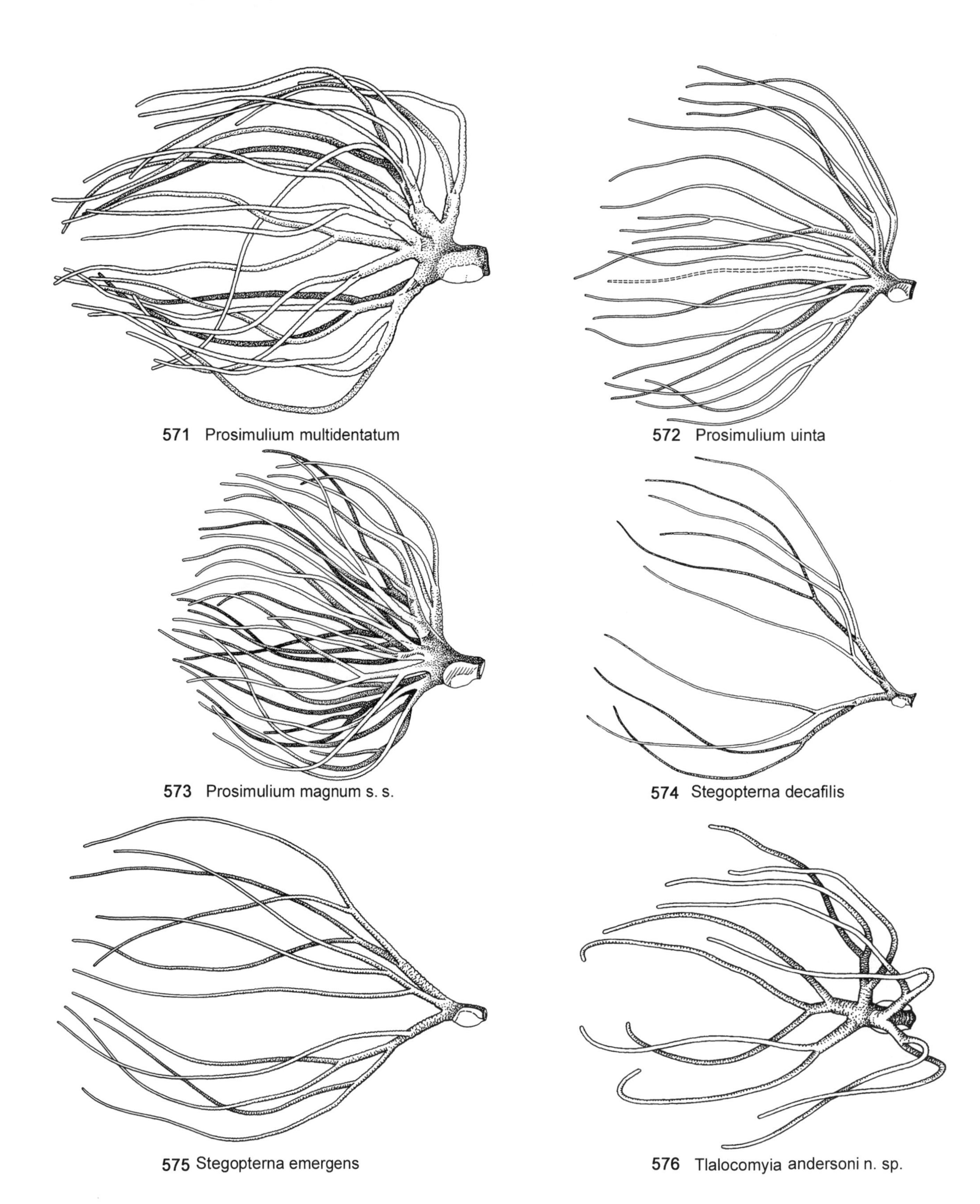


Figs. 10.571–10.576. Pupal gills.

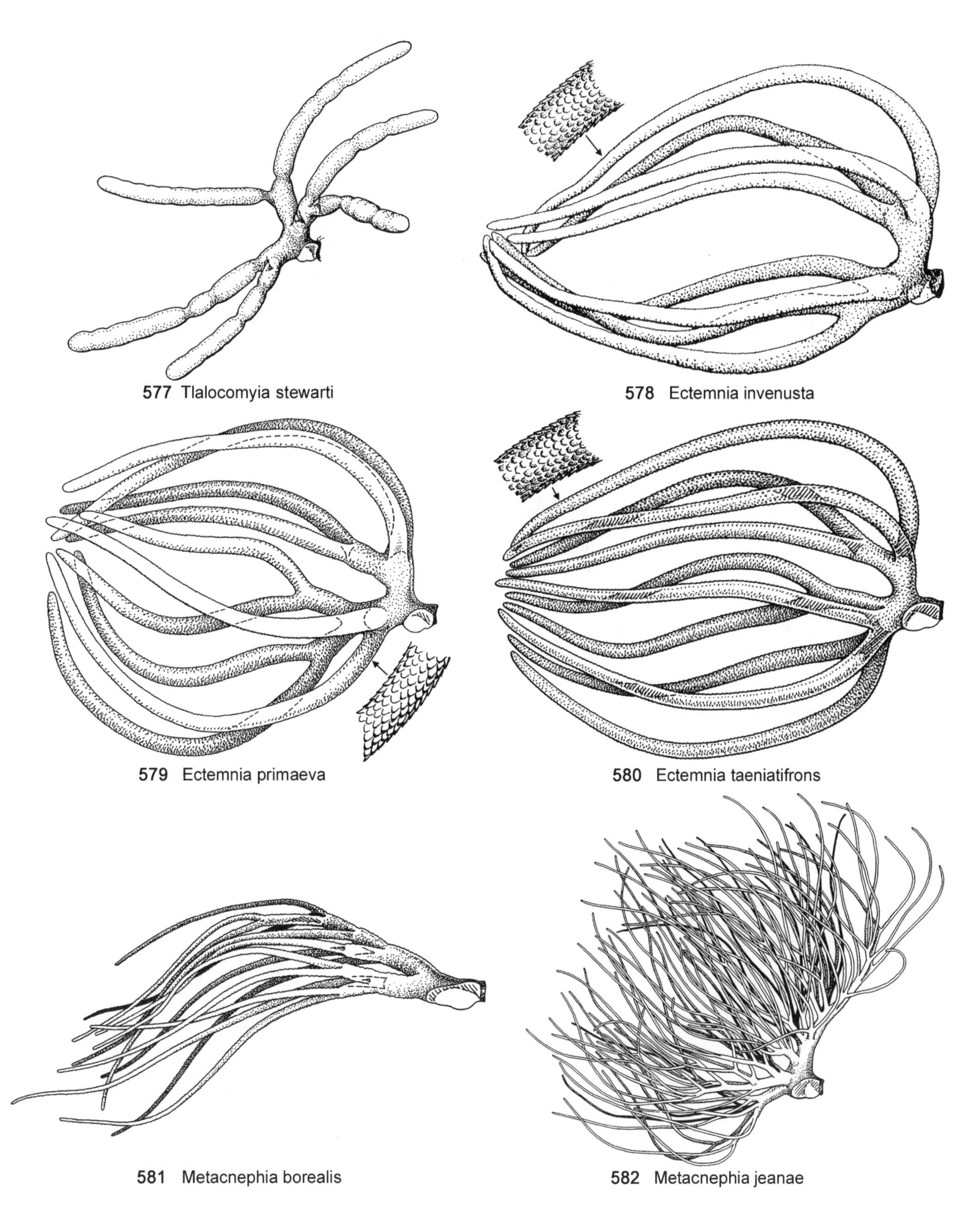


Figs. 10.577–10.582. Pupal gills.

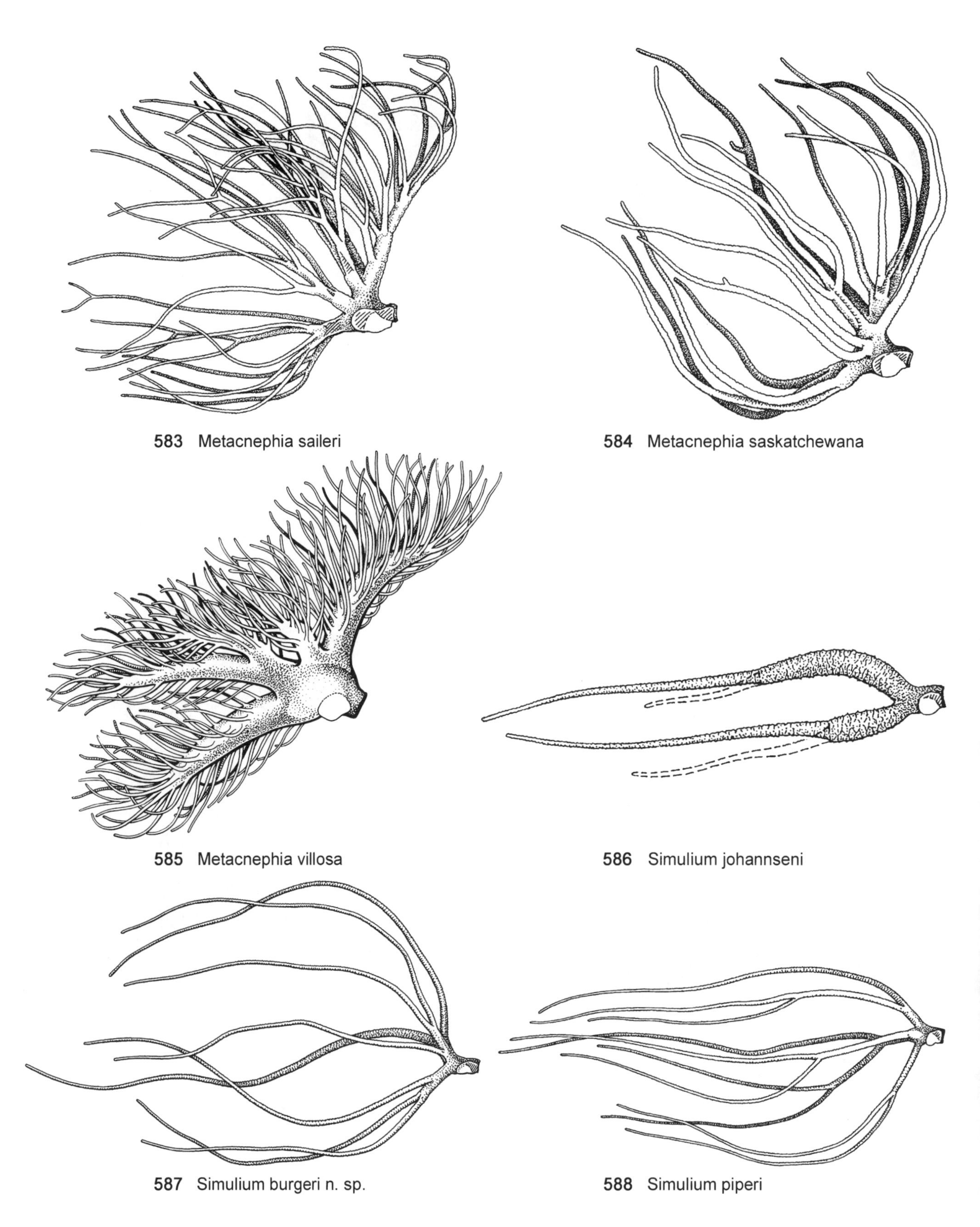


Figs. 10.583–10.588. Pupal gills.

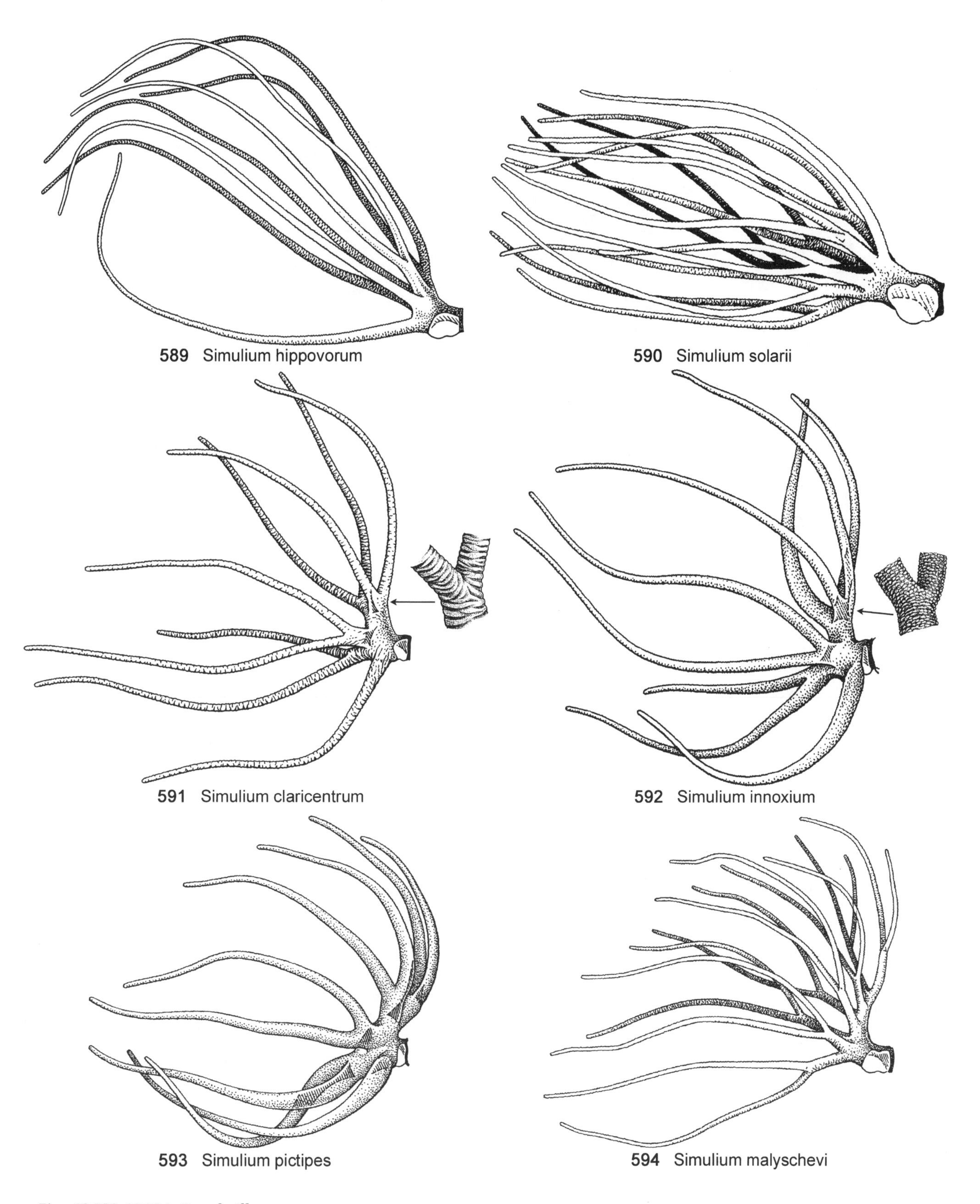


Figs. 10.589–10.594. Pupal gills.

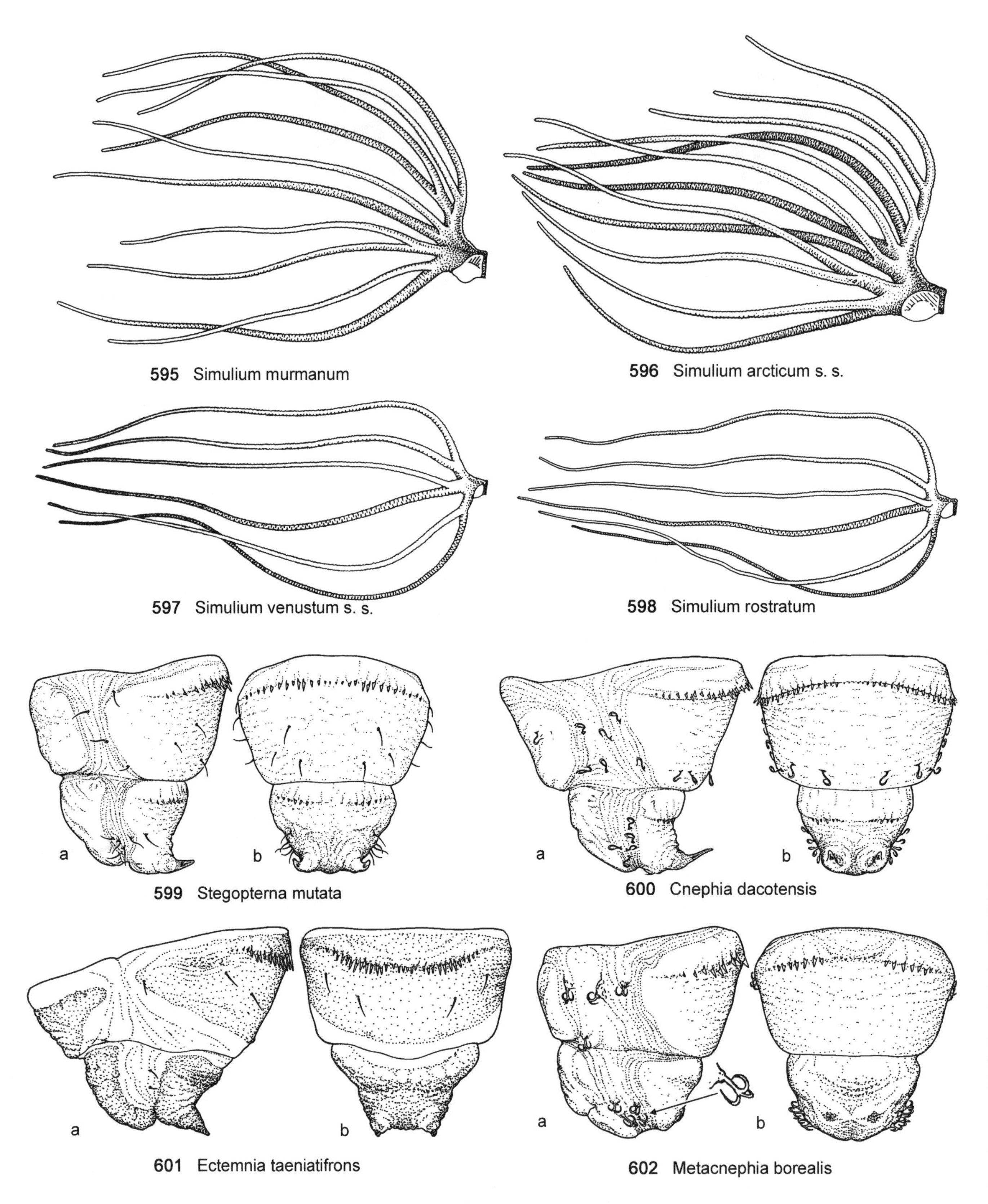


Figs. 10.595–10.602. Pupal features; (595–598) gills; (599–602) abdominal segments VIII–IX; a, left lateral view; b, dorsal view.

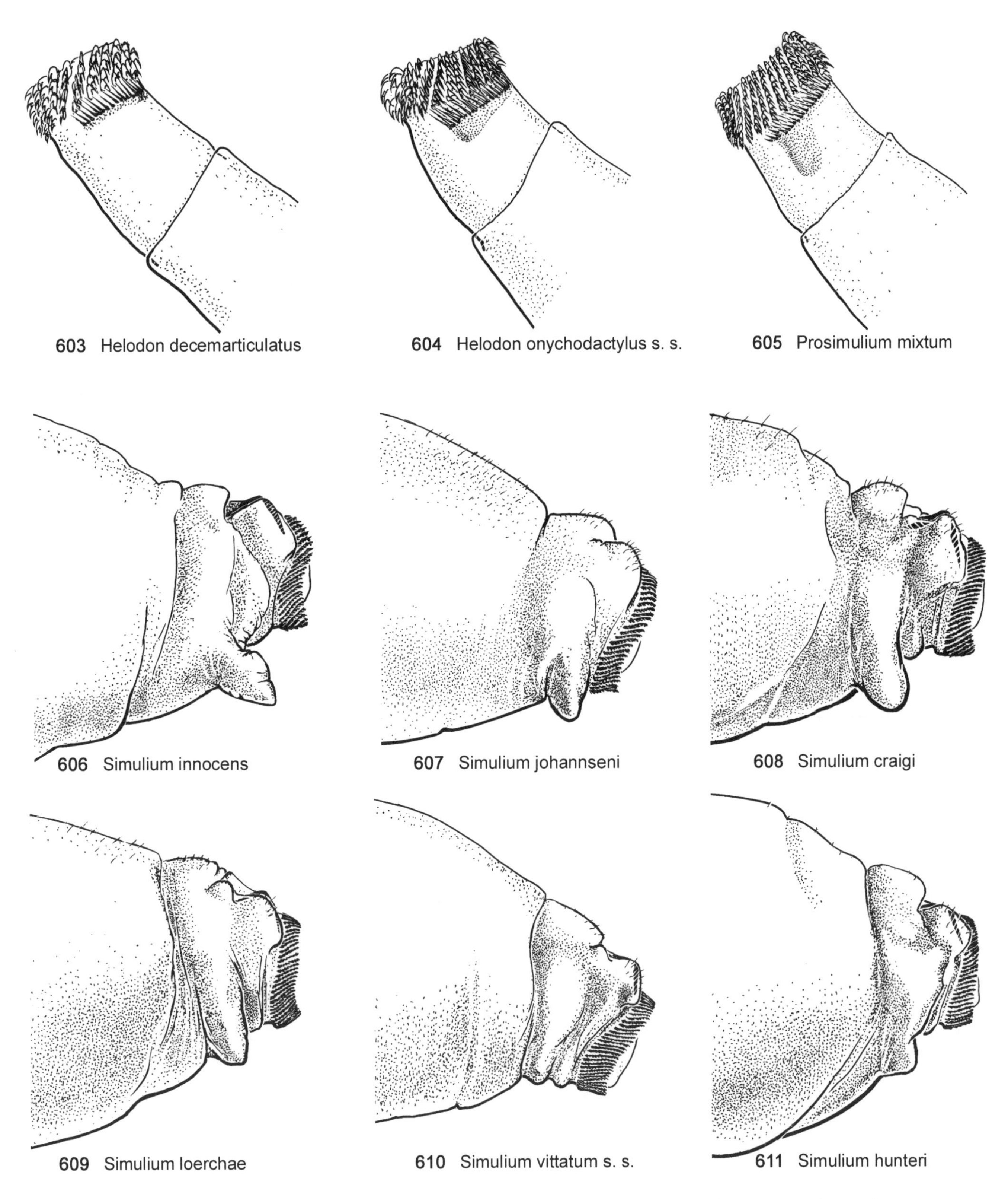


Figs. 10.603–10.611. Larval features; (603–605) prothoracic proleg, left lateral view; (606–611) posterior abdominal segments, left lateral view.

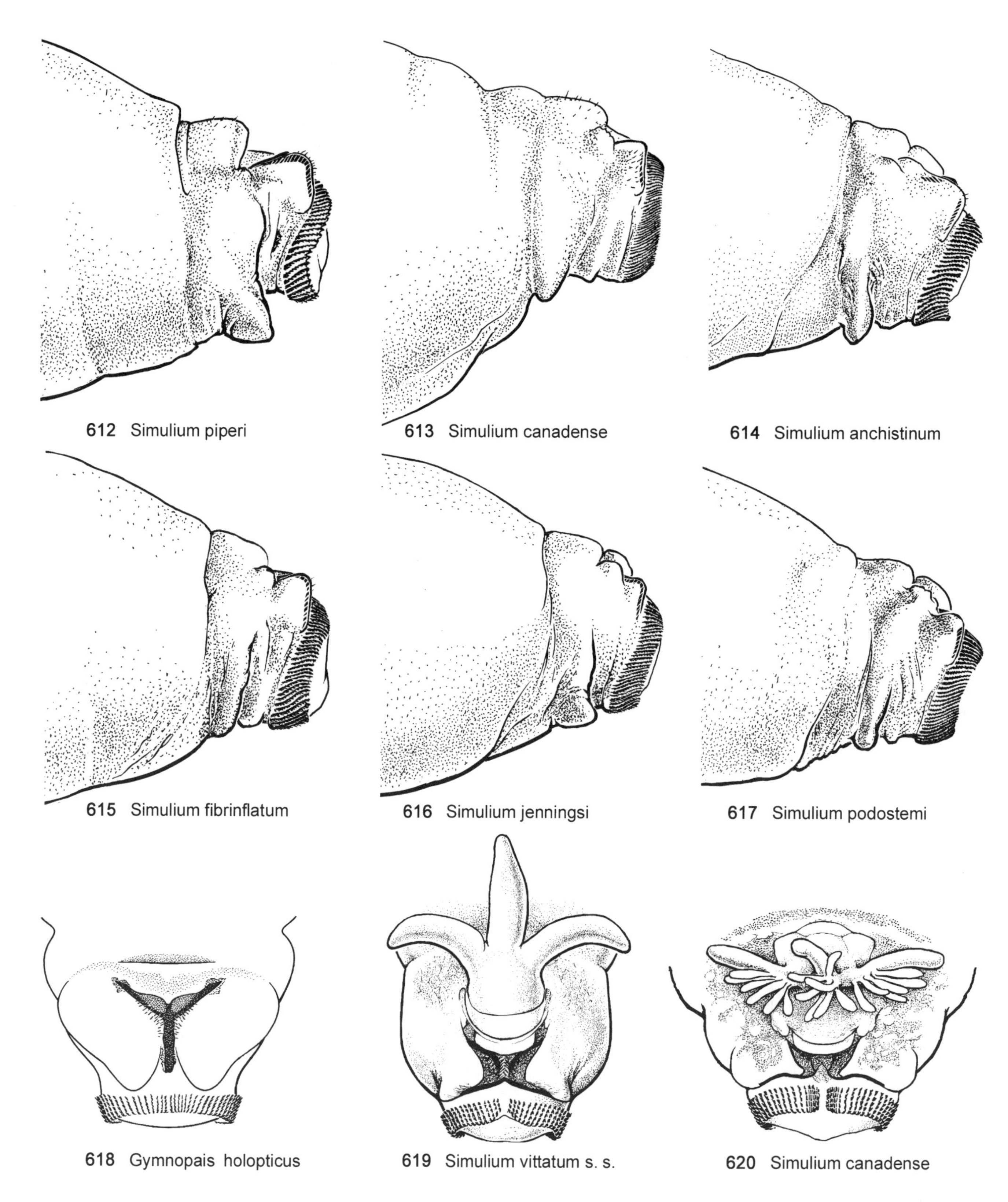


Figs. 10.612–10.620. Larval abdomen, posterior portion; (612–617) left lateral view; (618–620) dorsal view. (618) From Peterson (1981); (619, 620) redrawn and modified from Peterson (1996).

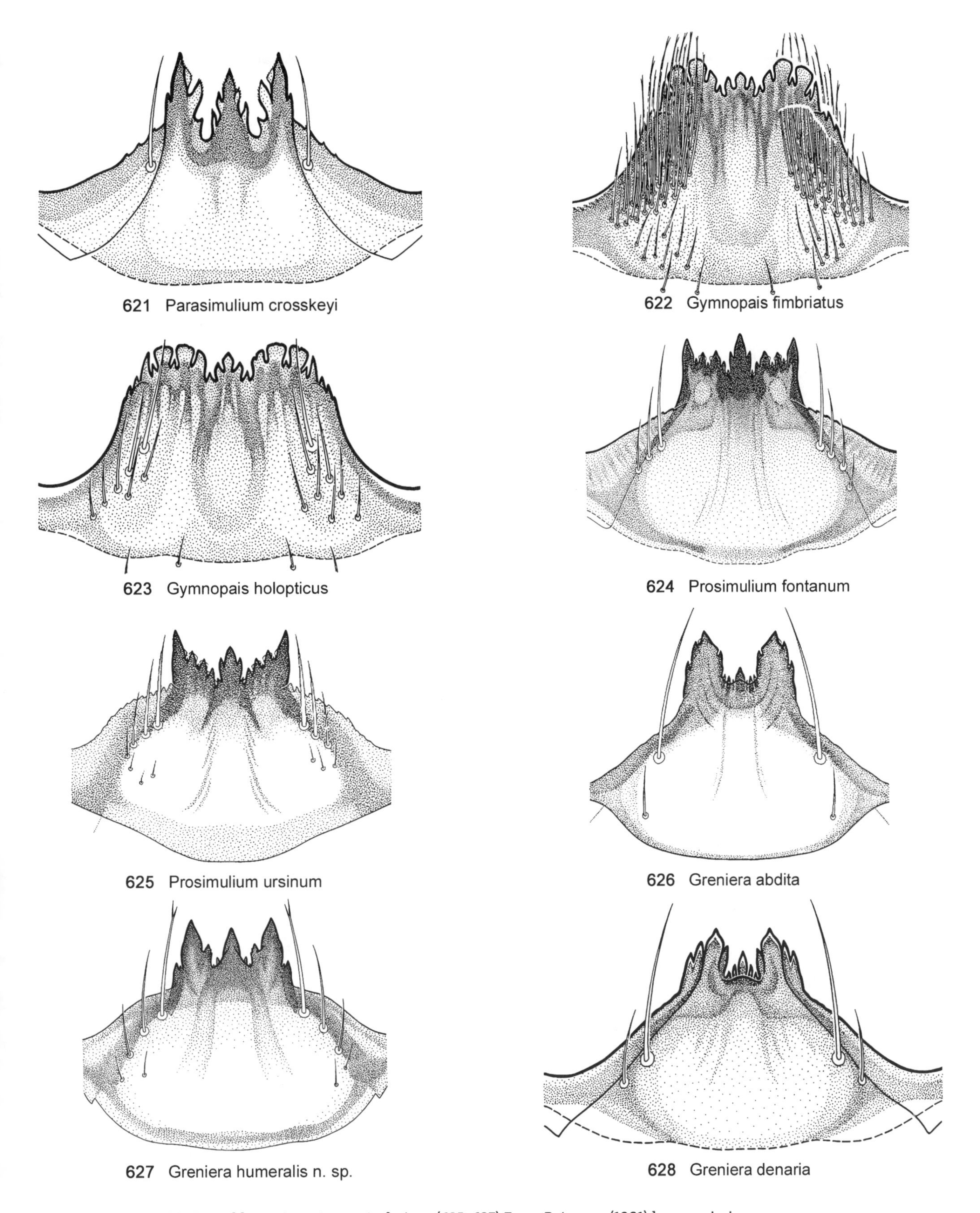


Figs. 10.621–10.628. Larval hypostomata, ventral view. (625–627) From Peterson (1981) by permission.

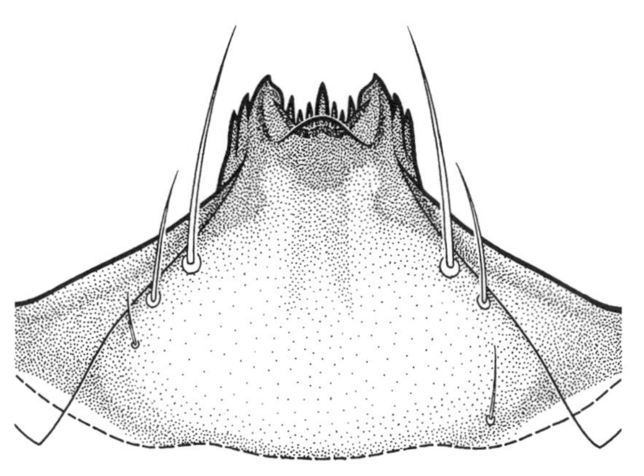

**629** Greniera longicornis n. sp.

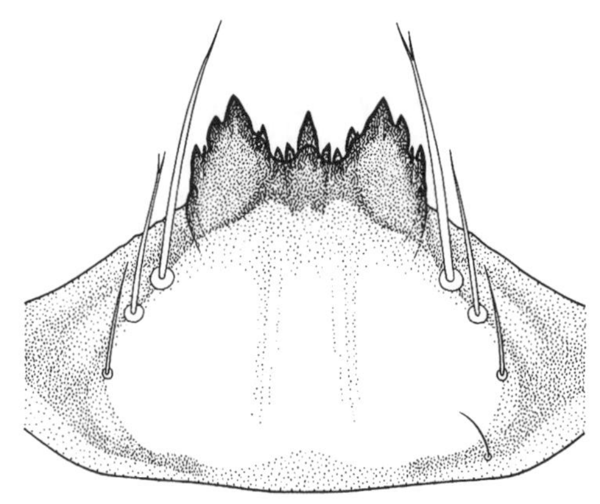

**630** Stegopterna mutata

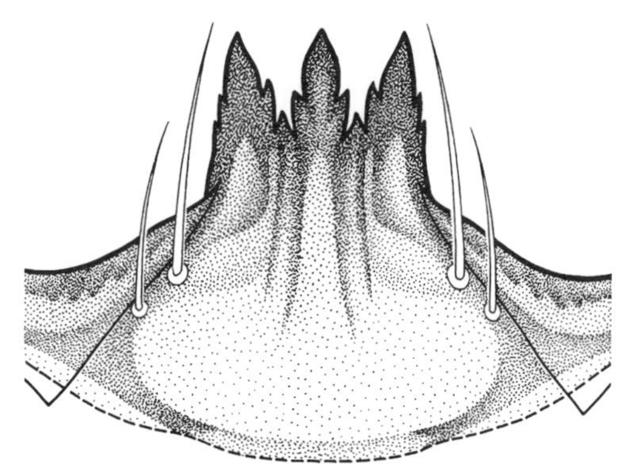

**631** Tlalocomyia andersoni n. sp.

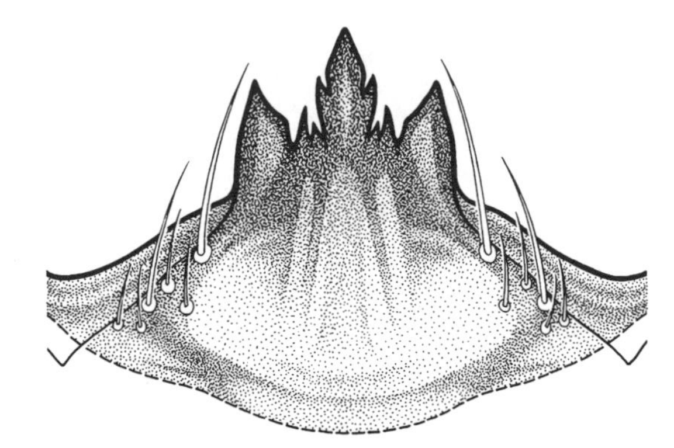

**632** Tlalocomyia osborni

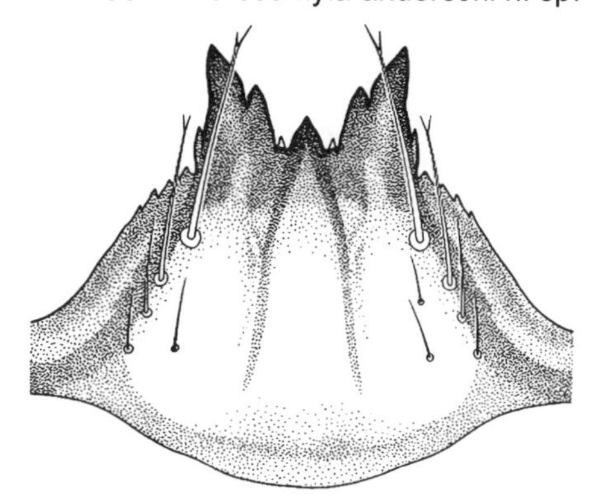

**633** Tlalocomyia ramifera n. sp.

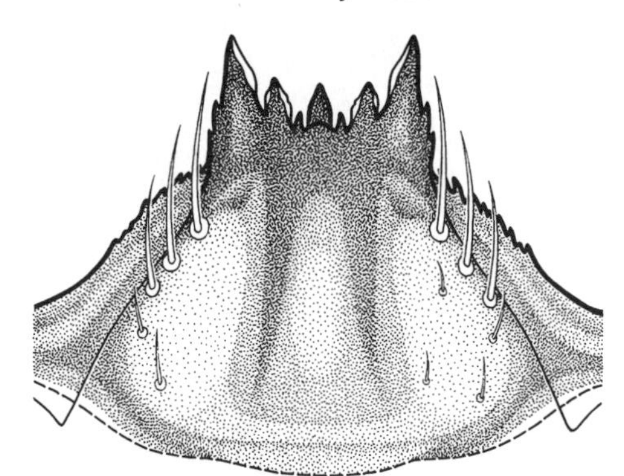

**634** Tlalocomyia stewarti

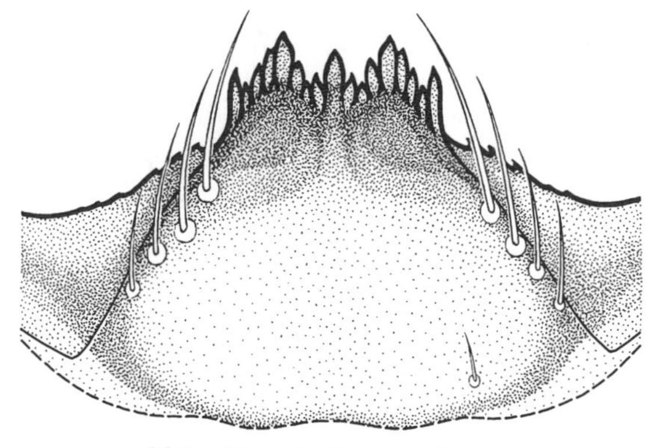

**635** Gigantodax adleri

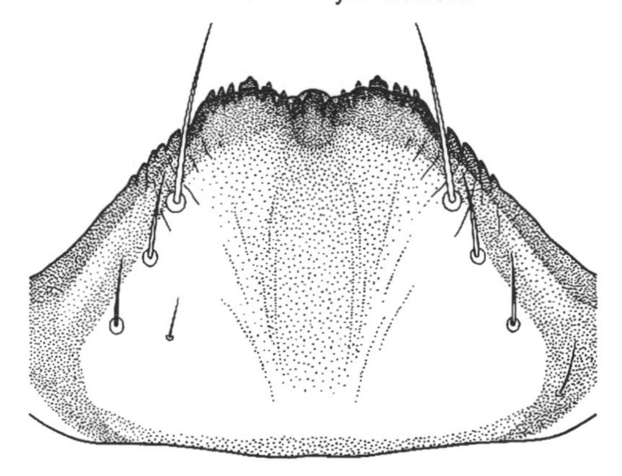

**636** Cnephia dacotensis

Figs. 10.629–10.636. Larval hypostomata, ventral view. (630, 633, 636) From Peterson (1981) by permission.

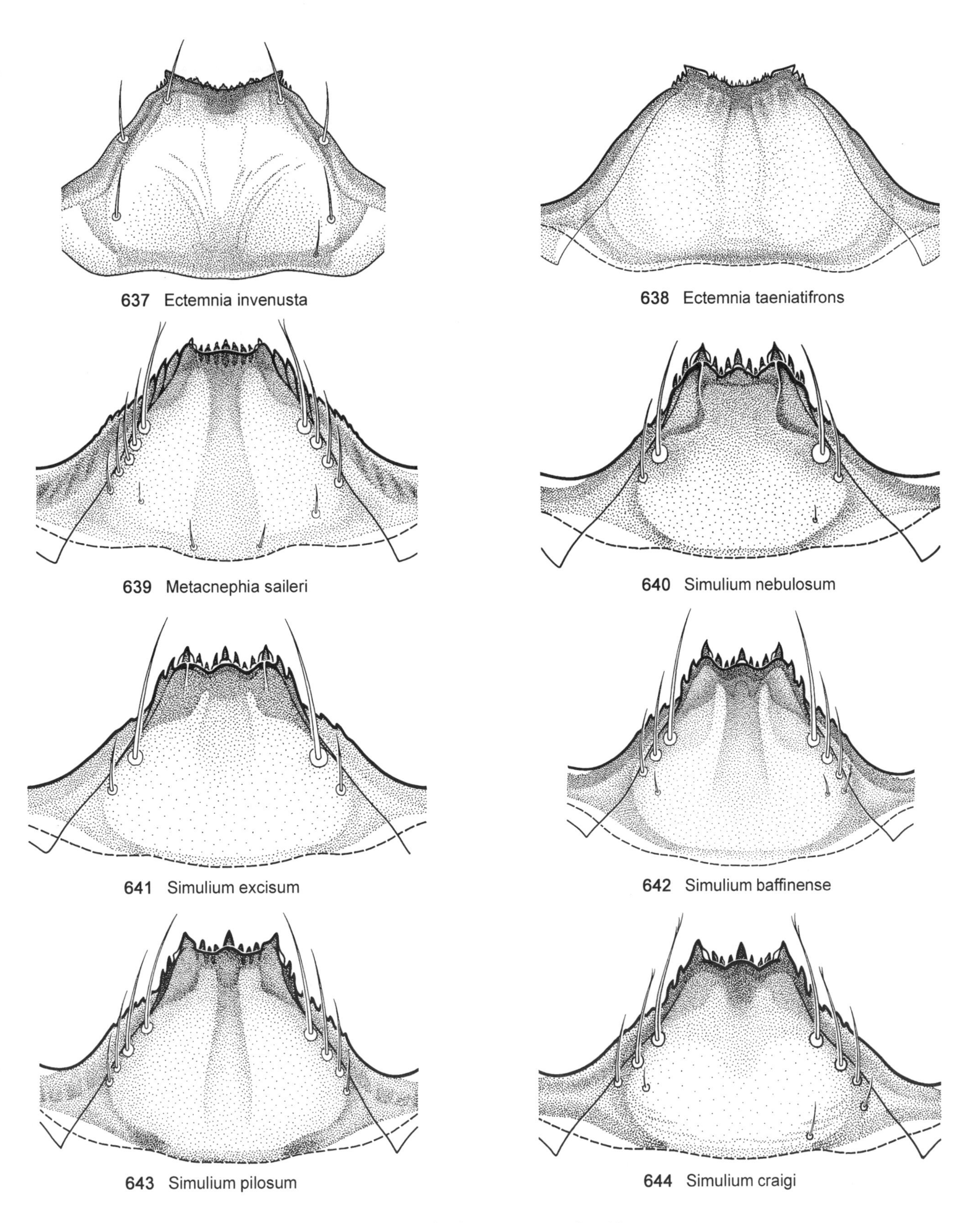


Figs. 10.637–10.644. Larval hypostomata, ventral view. (637) From Peterson (1981) by permission.

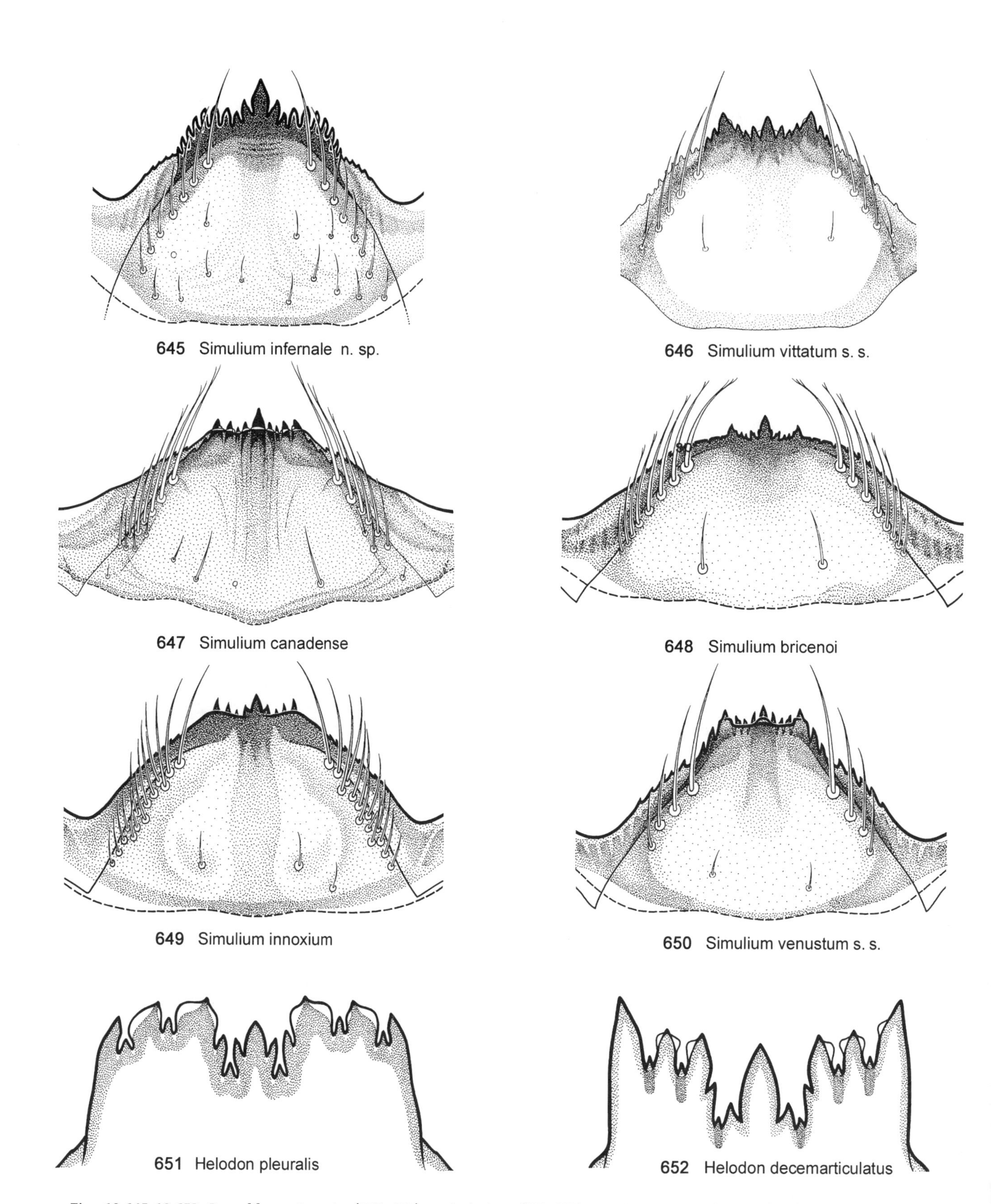


Figs. 10.645–10.652. Larval hypostomata; (645–650) ventral view; (651, 652) anterior margin of teeth. (646) From Peterson (1981) by permission.

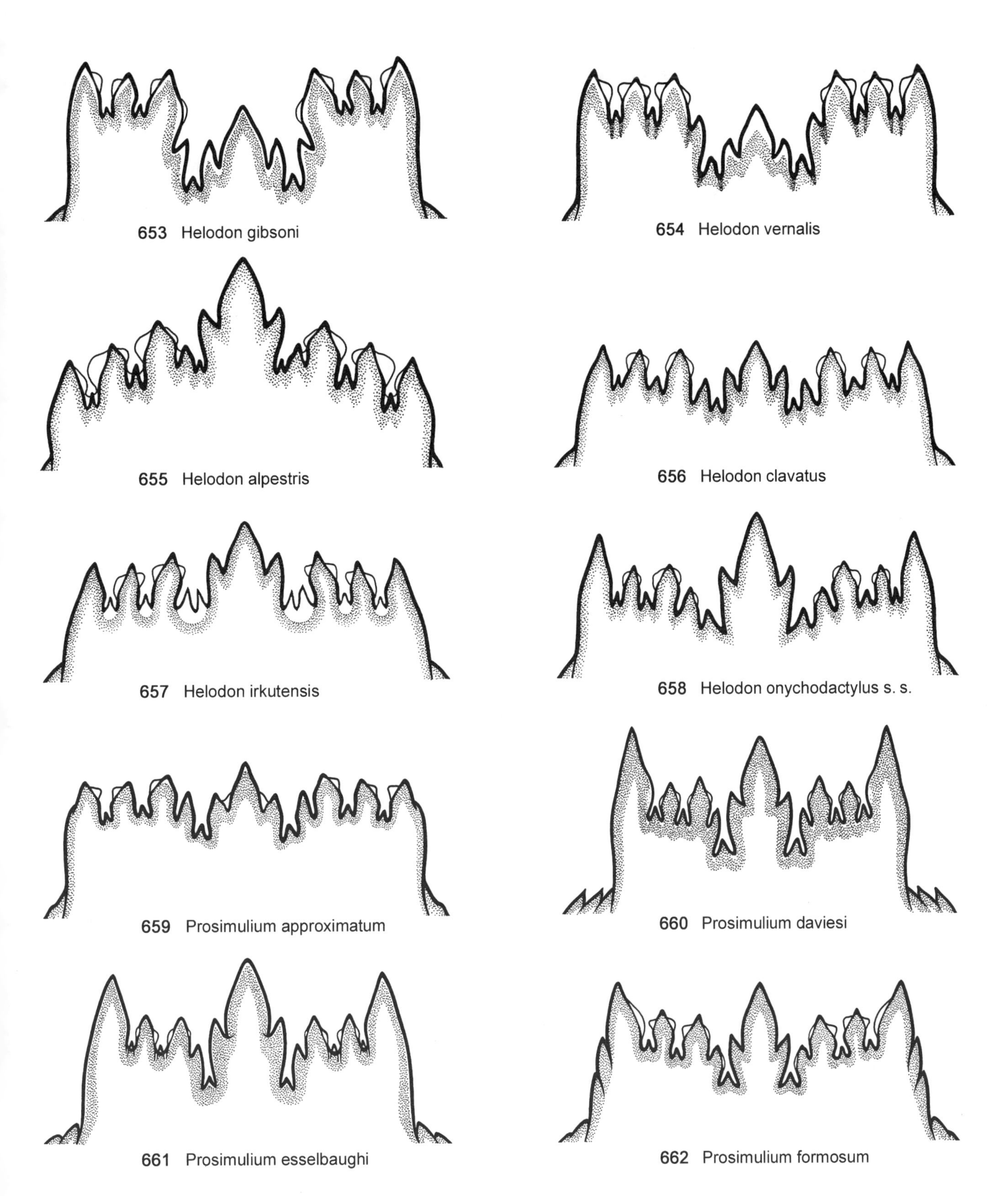


Figs. 10.653–10.662. Larval hypostomal teeth, anterior margin.

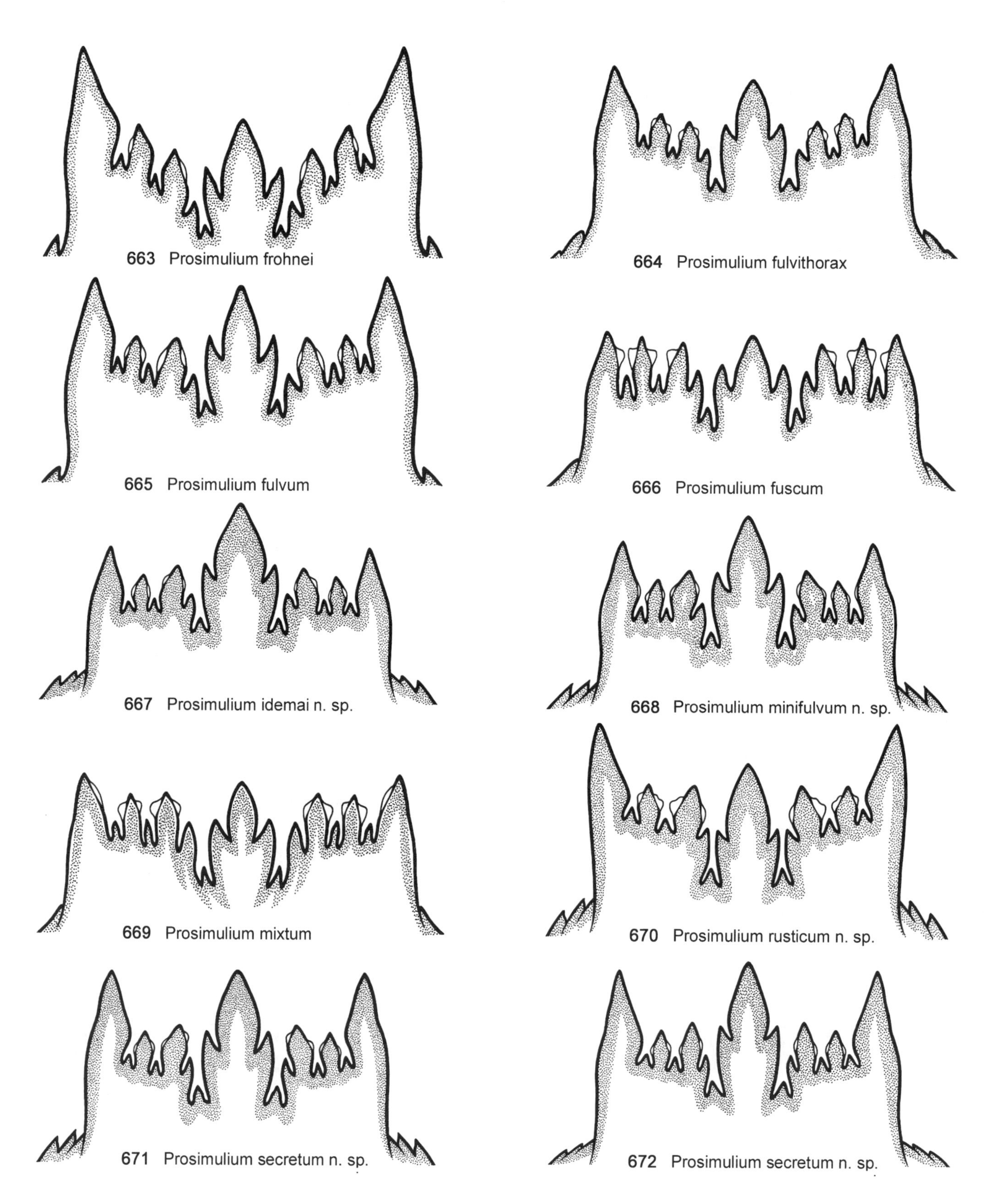


Figs. 10.663–10.672. Larval hypostomal teeth, anterior margin; (671, 672) showing variation.

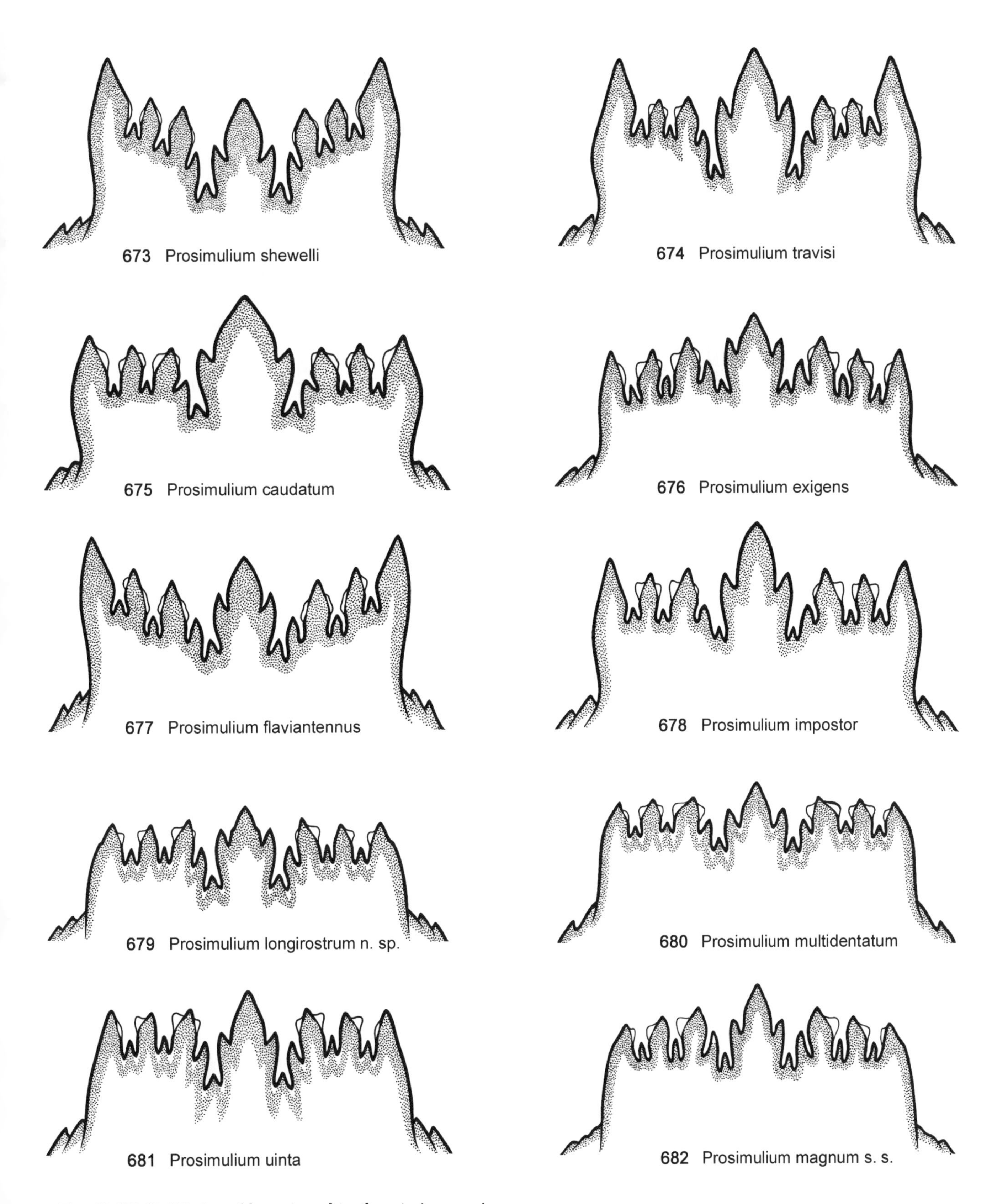


Figs. 10.673–10.682. Larval hypostomal teeth, anterior margin.

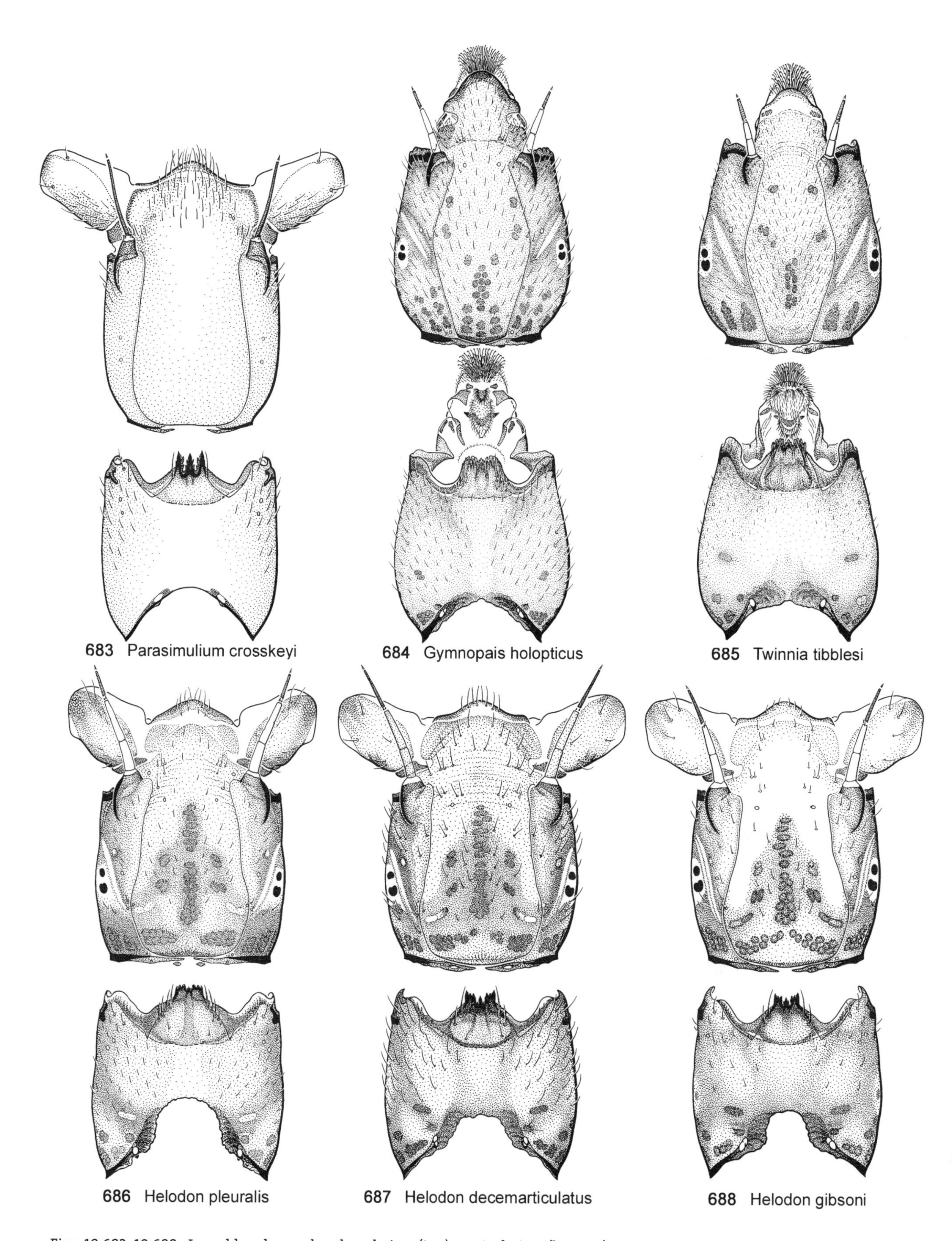


Figs. 10.683–10.688. Larval head capsules, dorsal view (top), ventral view (bottom).

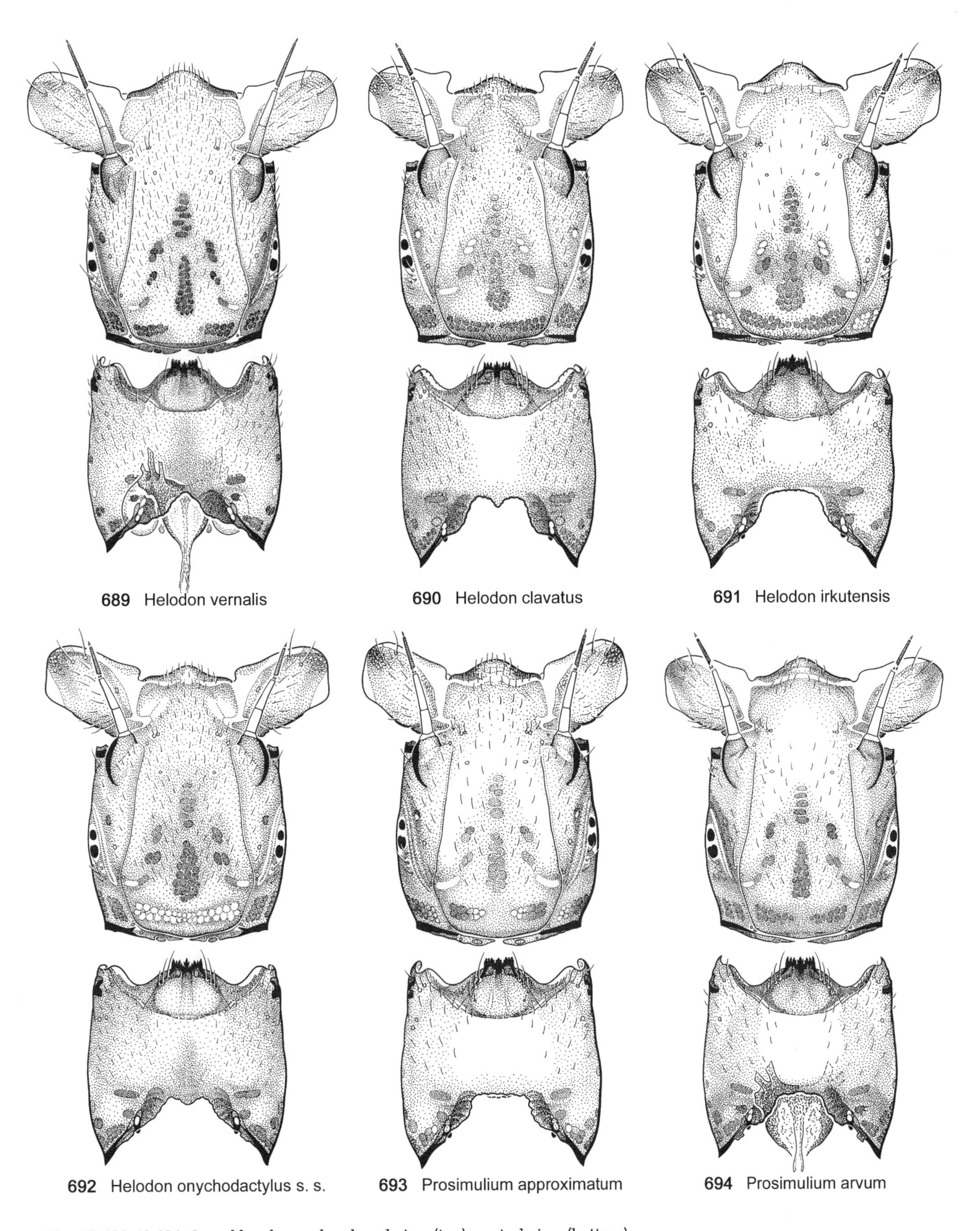


Figs. 10.689–10.694. Larval head capsules, dorsal view (top), ventral view (bottom).

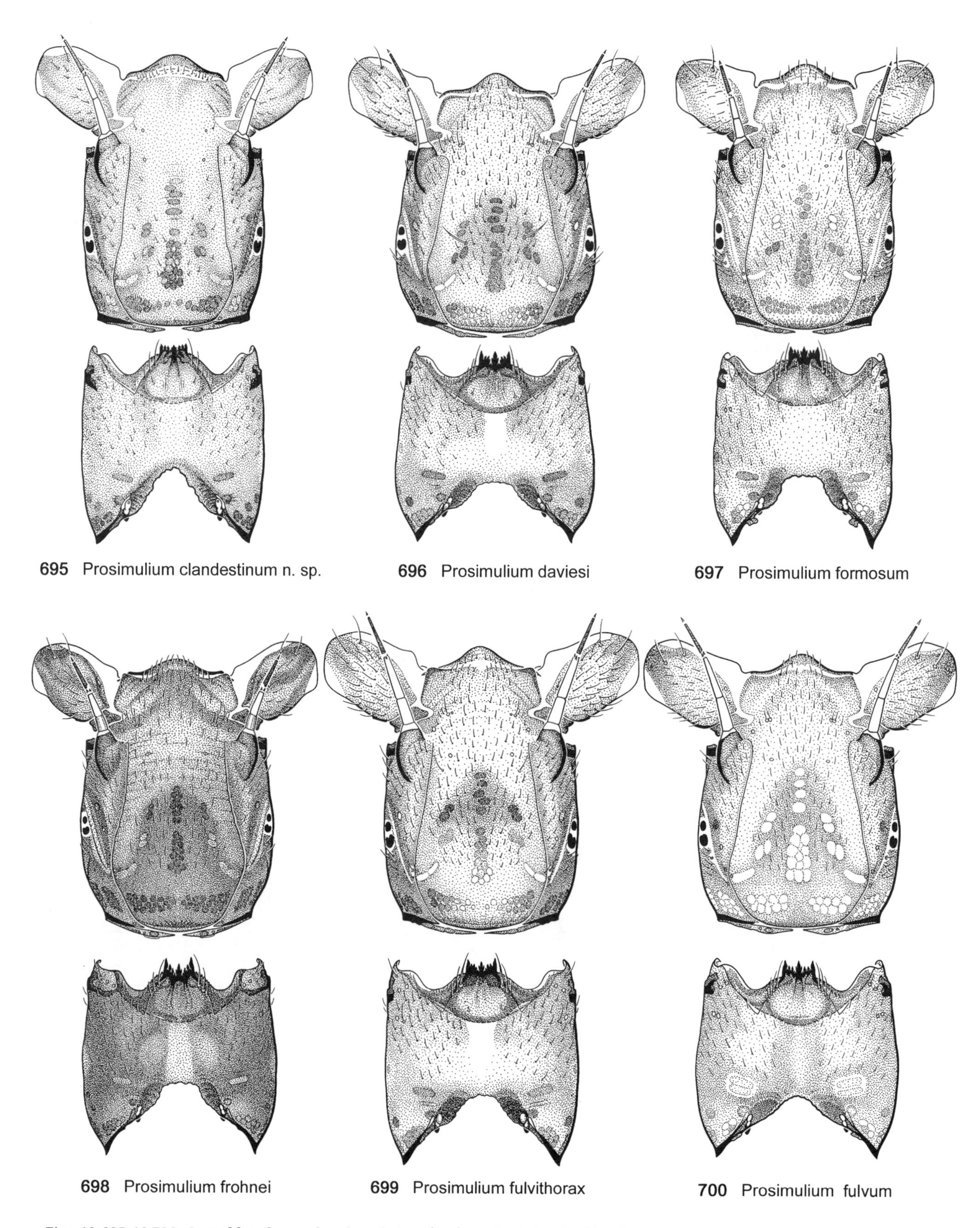


Figs. 10.695–10.700. Larval head capsules, dorsal view (top), ventral view (bottom).

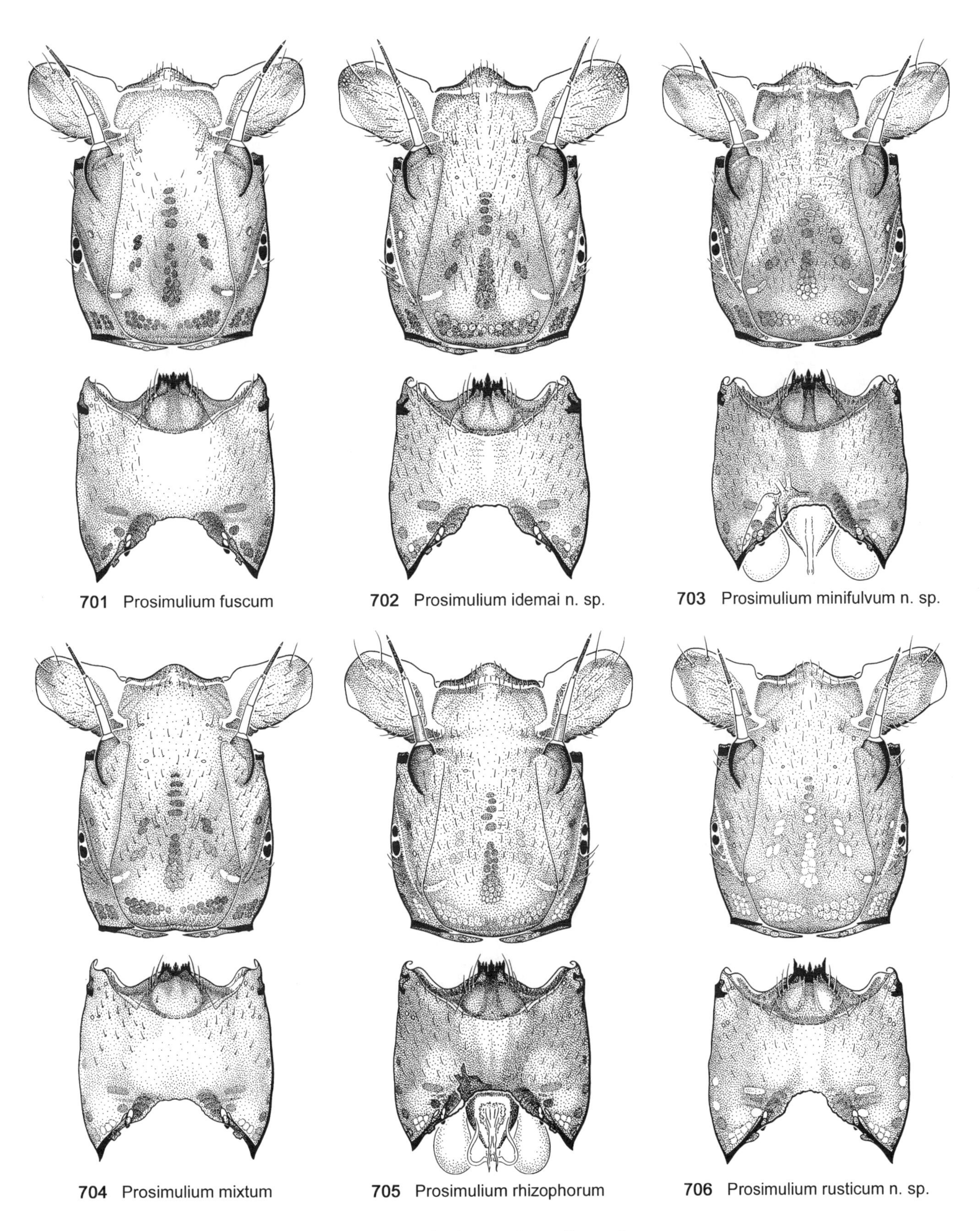


Figs. 10.701–10.706. Larval head capsules, dorsal view (top), ventral view (bottom).

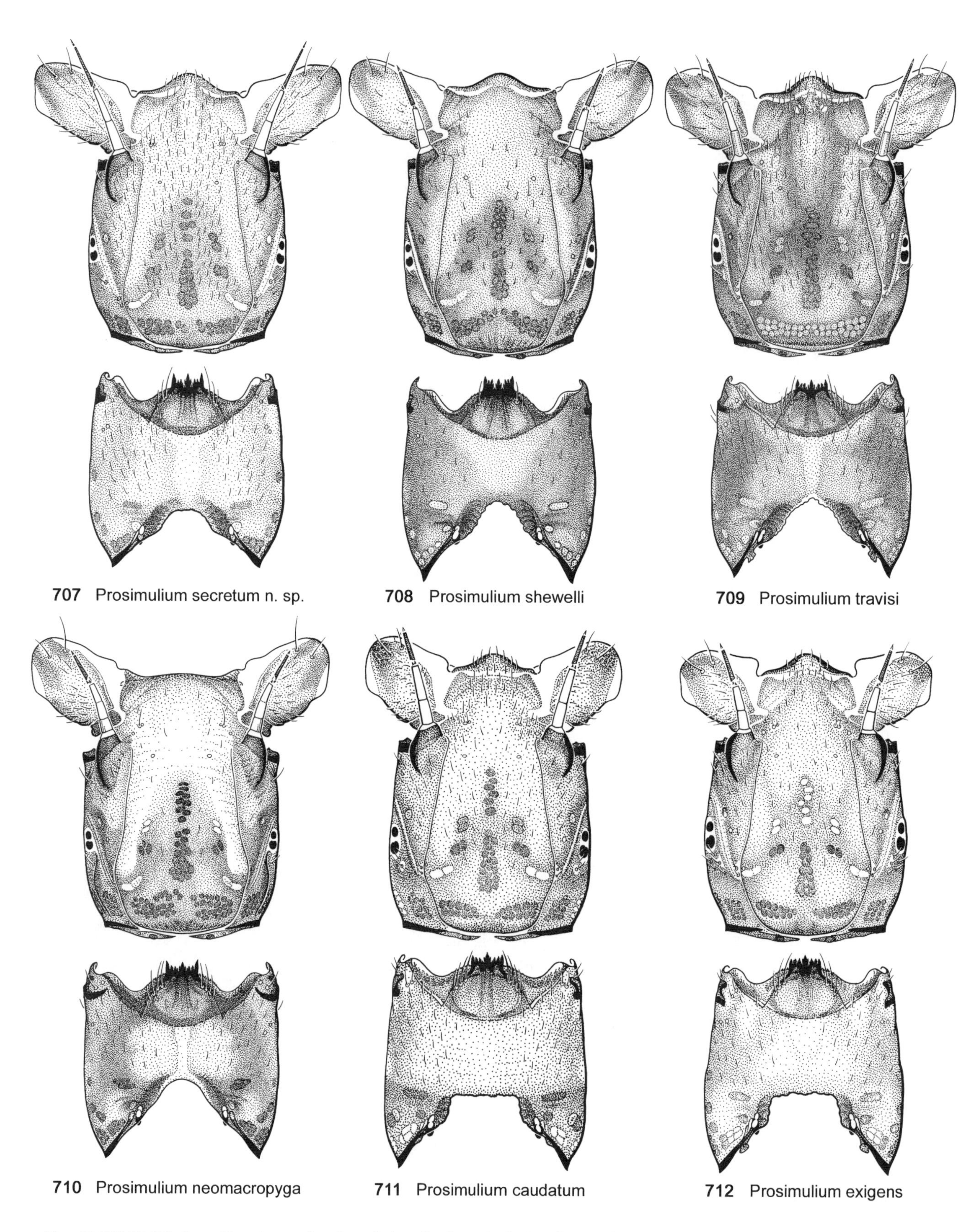


Figs. 10.707–10.712. Larval head capsules, dorsal view (top), ventral view (bottom).

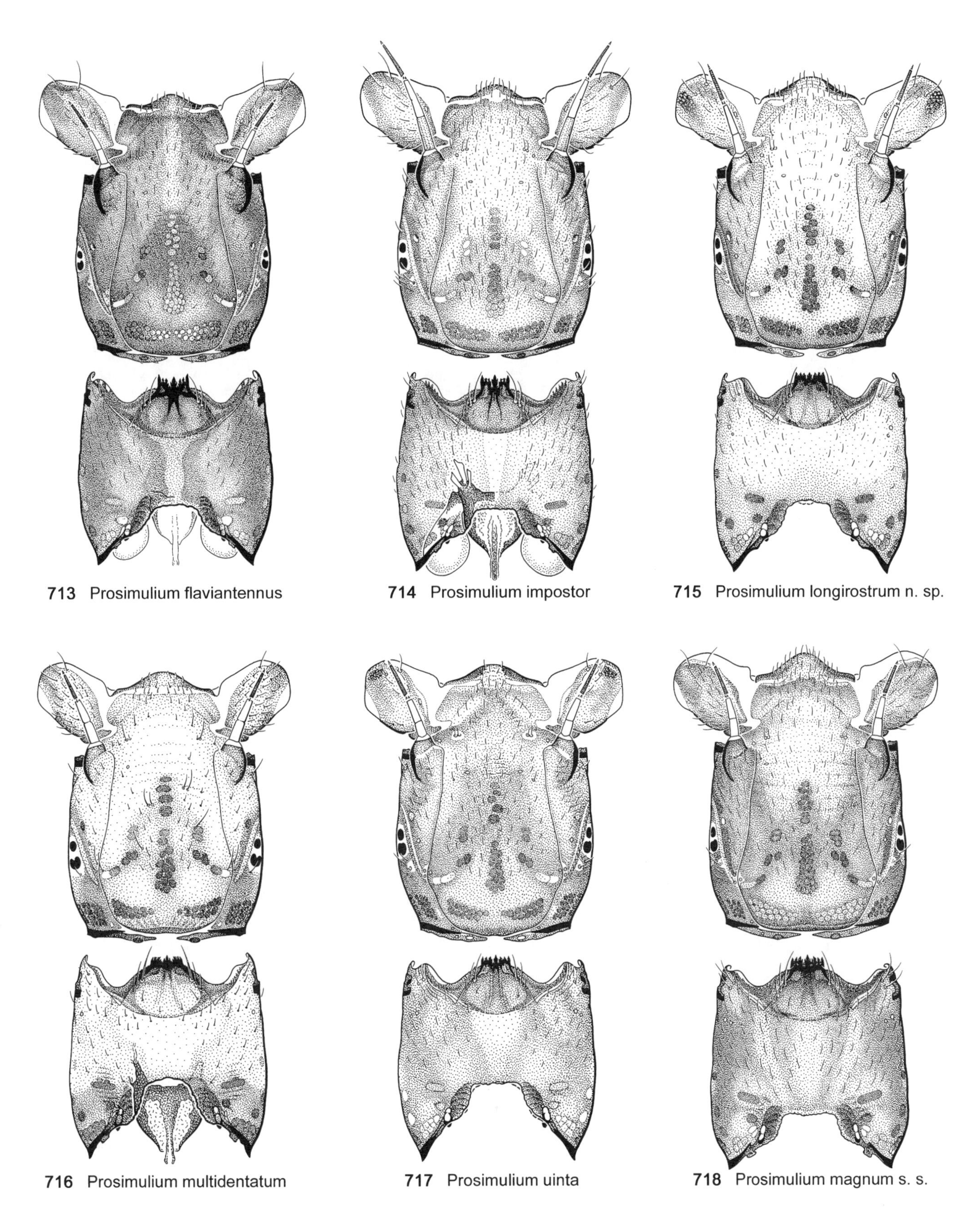


Figs. 10.713–10.718. Larval head capsules, dorsal view (top), ventral view (bottom).

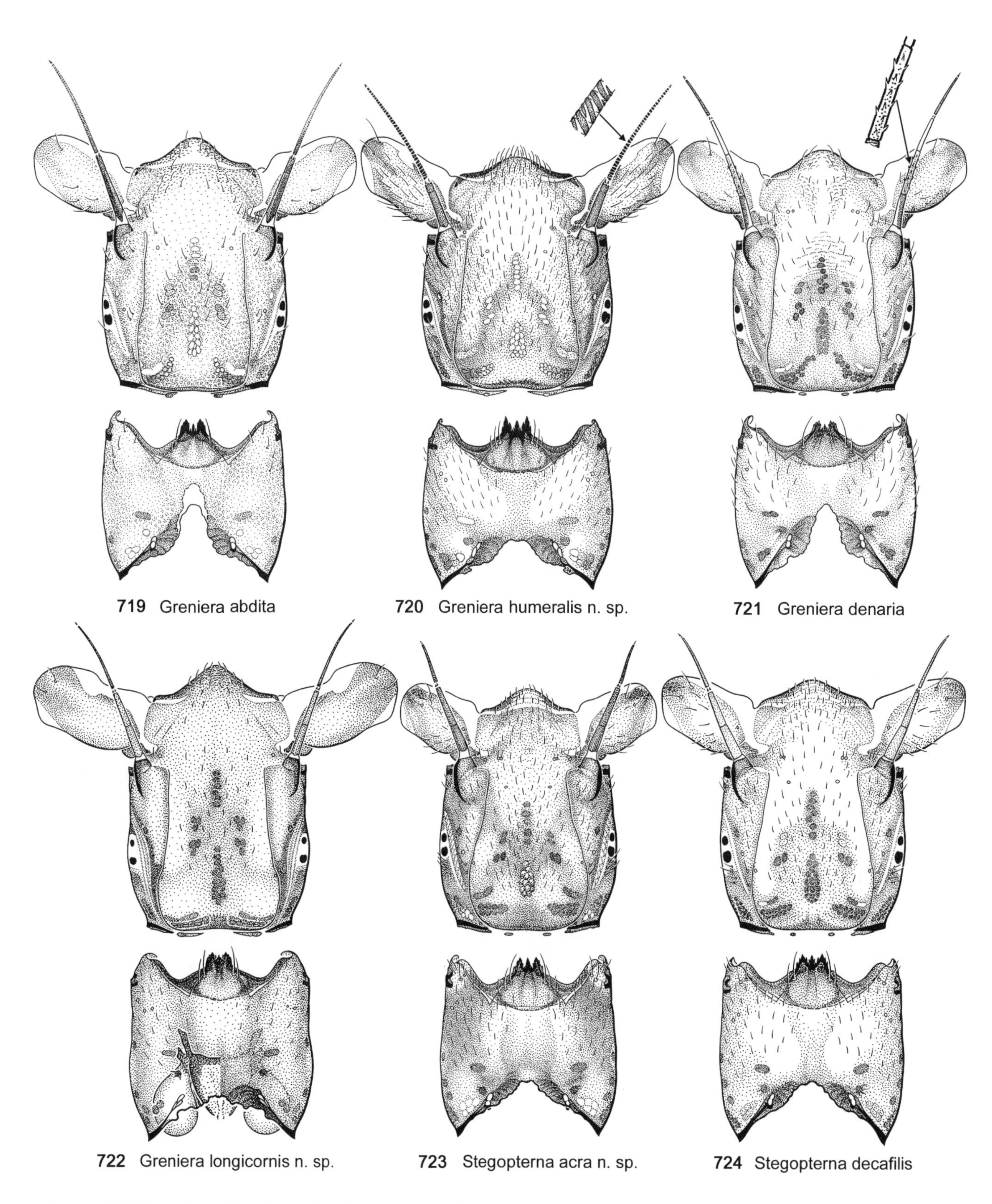


Figs. 10.719–10.724. Larval head capsules, dorsal view (top), ventral view (bottom).

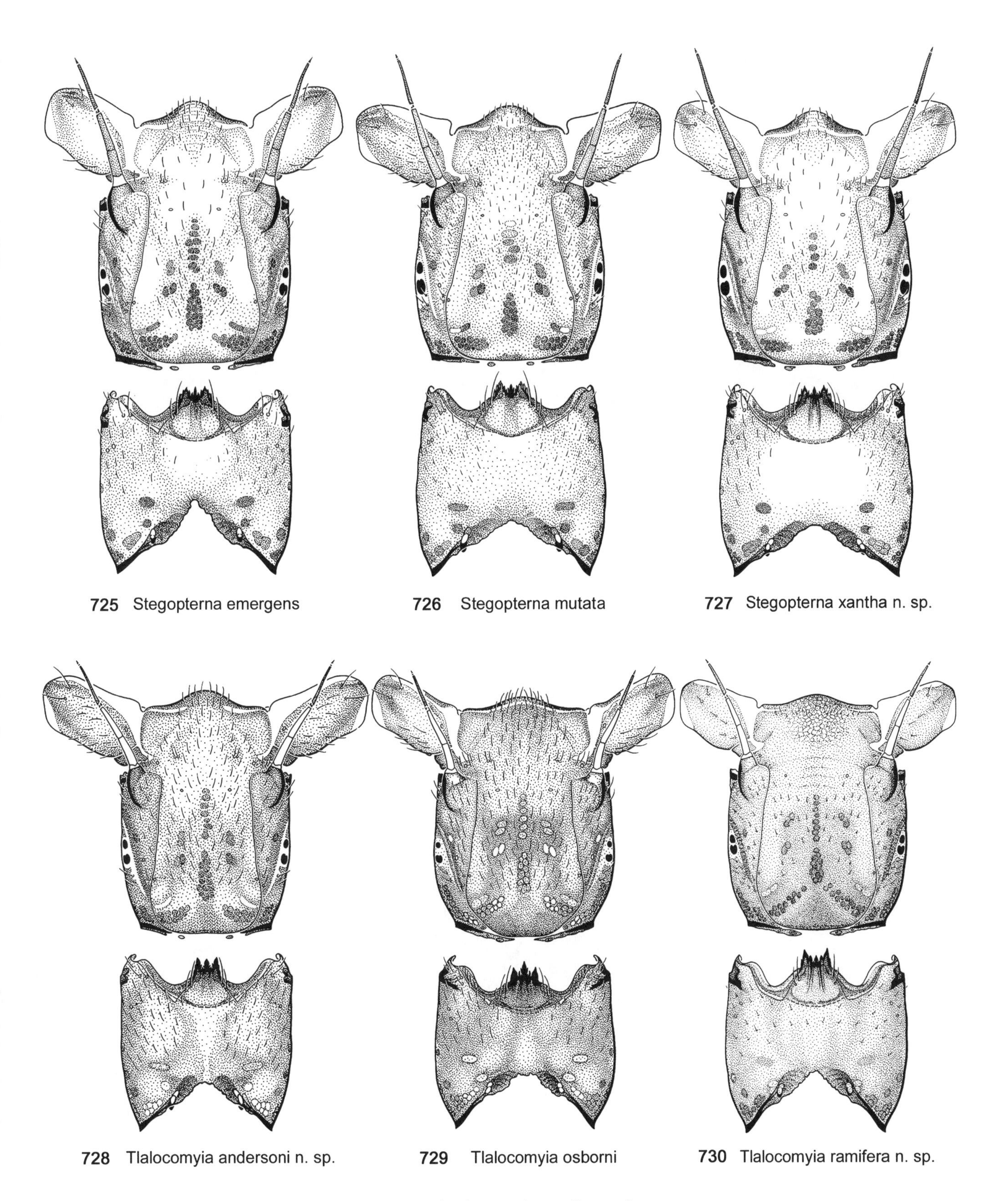


Figs. 10.725–10.730. Larval head capsules, dorsal view (top), ventral view (bottom).

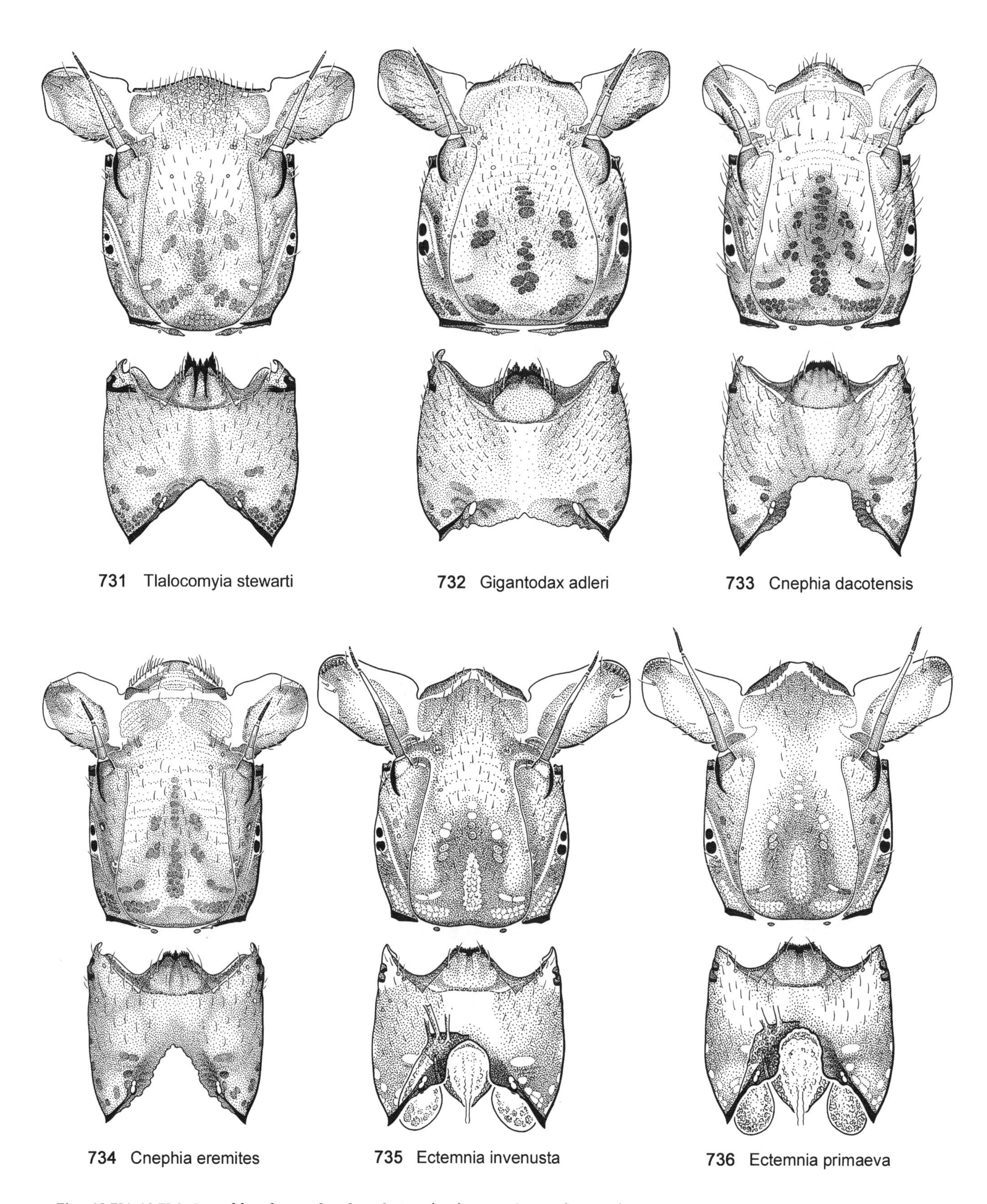


Figs. 10.731–10.736. Larval head capsules, dorsal view (top), ventral view (bottom).

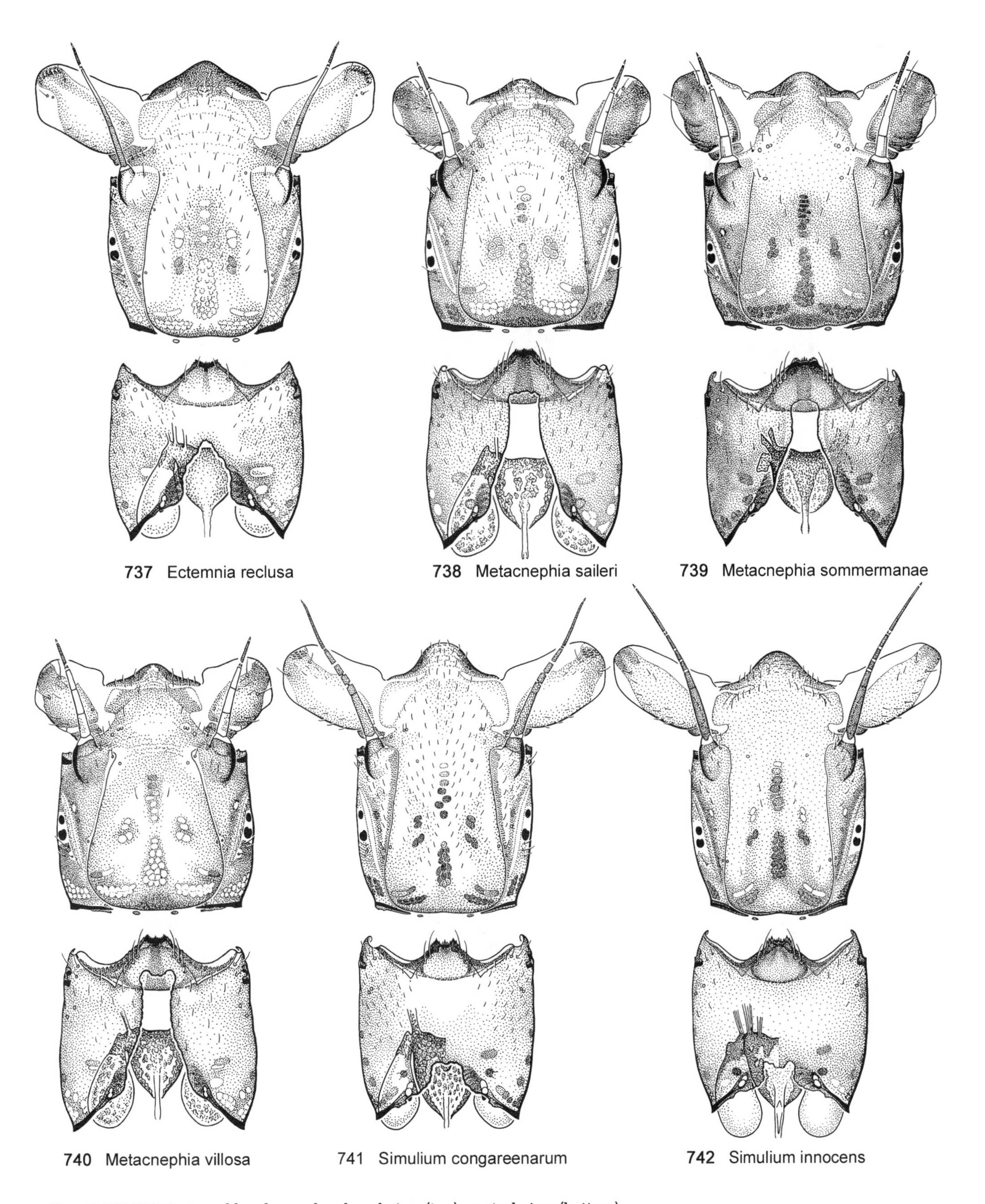


Figs. 10.737–10.742. Larval head capsules, dorsal view (top), ventral view (bottom).

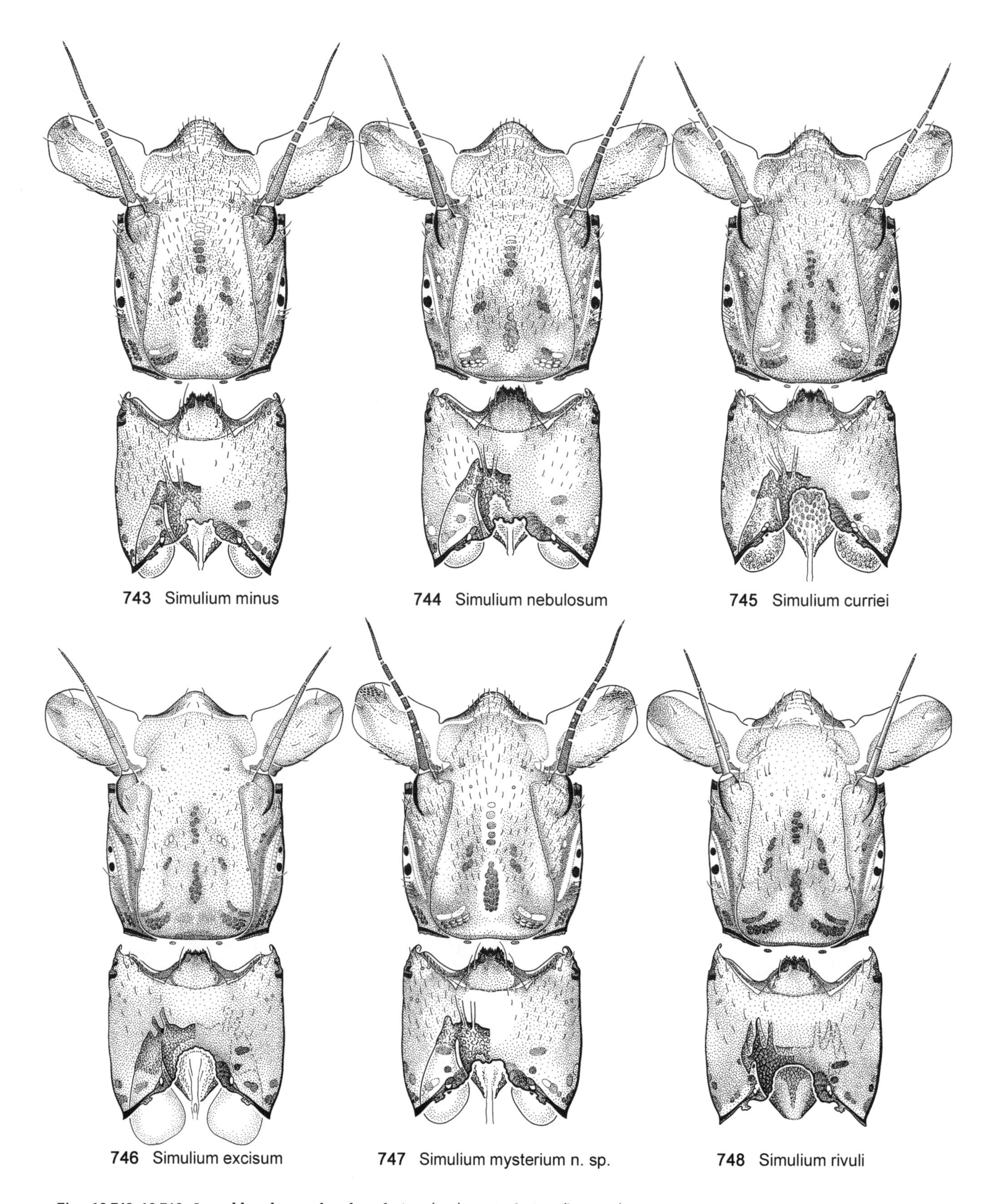


Figs. 10.743–10.748. Larval head capsules, dorsal view (top), ventral view (bottom).

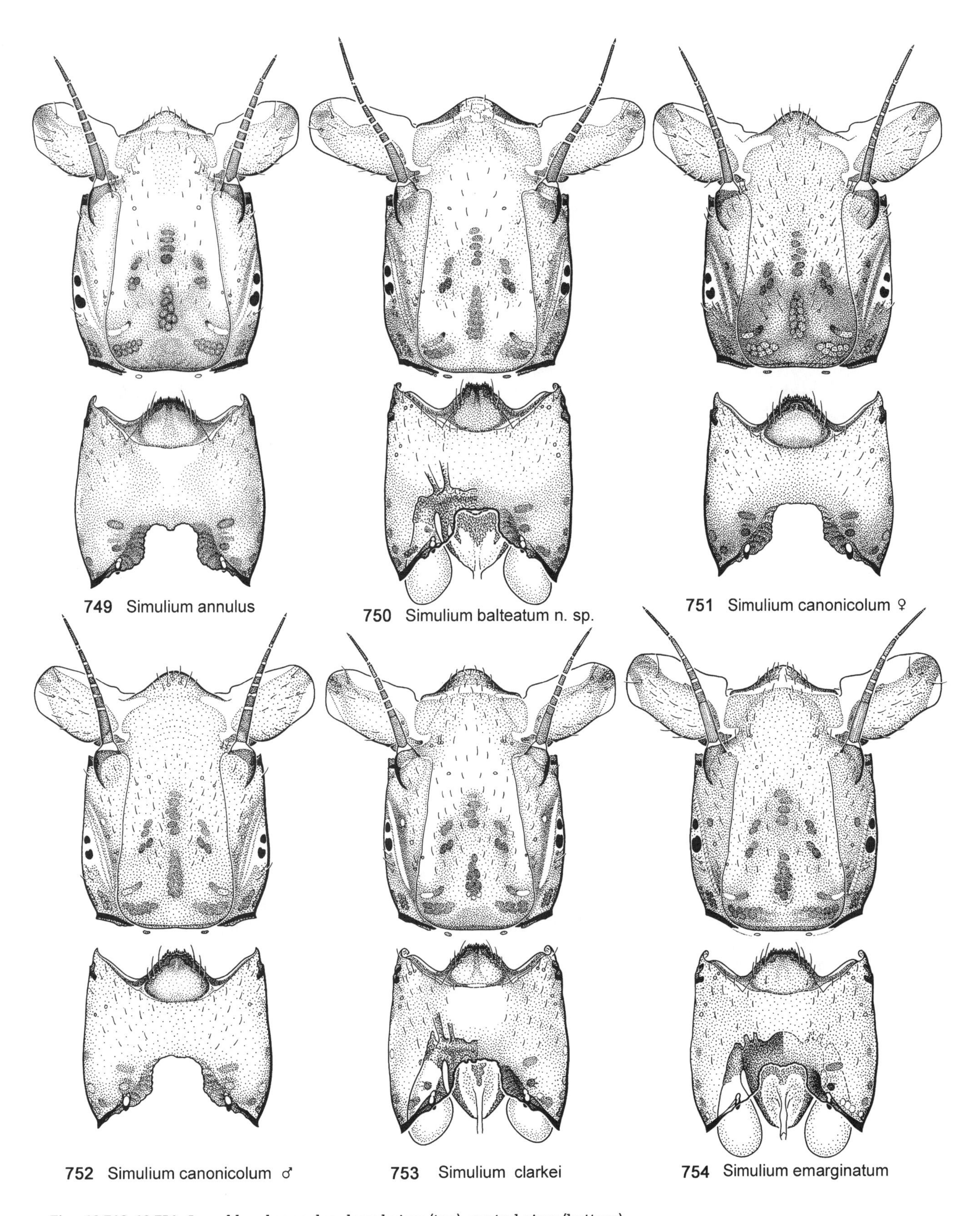


Figs. 10.749–10.754. Larval head capsules, dorsal view (top), ventral view (bottom).

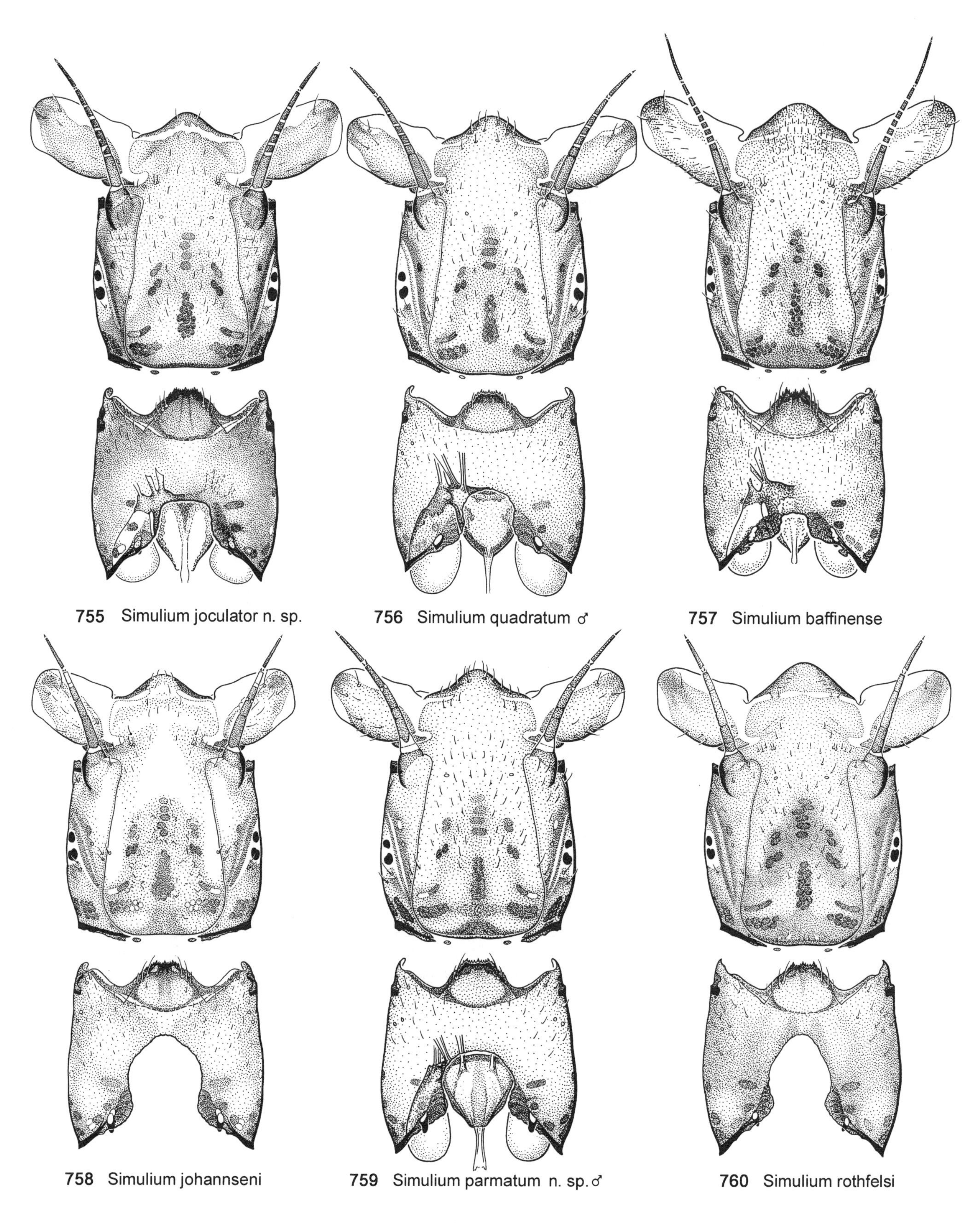


Figs. 10.755–10.760. Larval head capsules, dorsal view (top), ventral view (bottom).

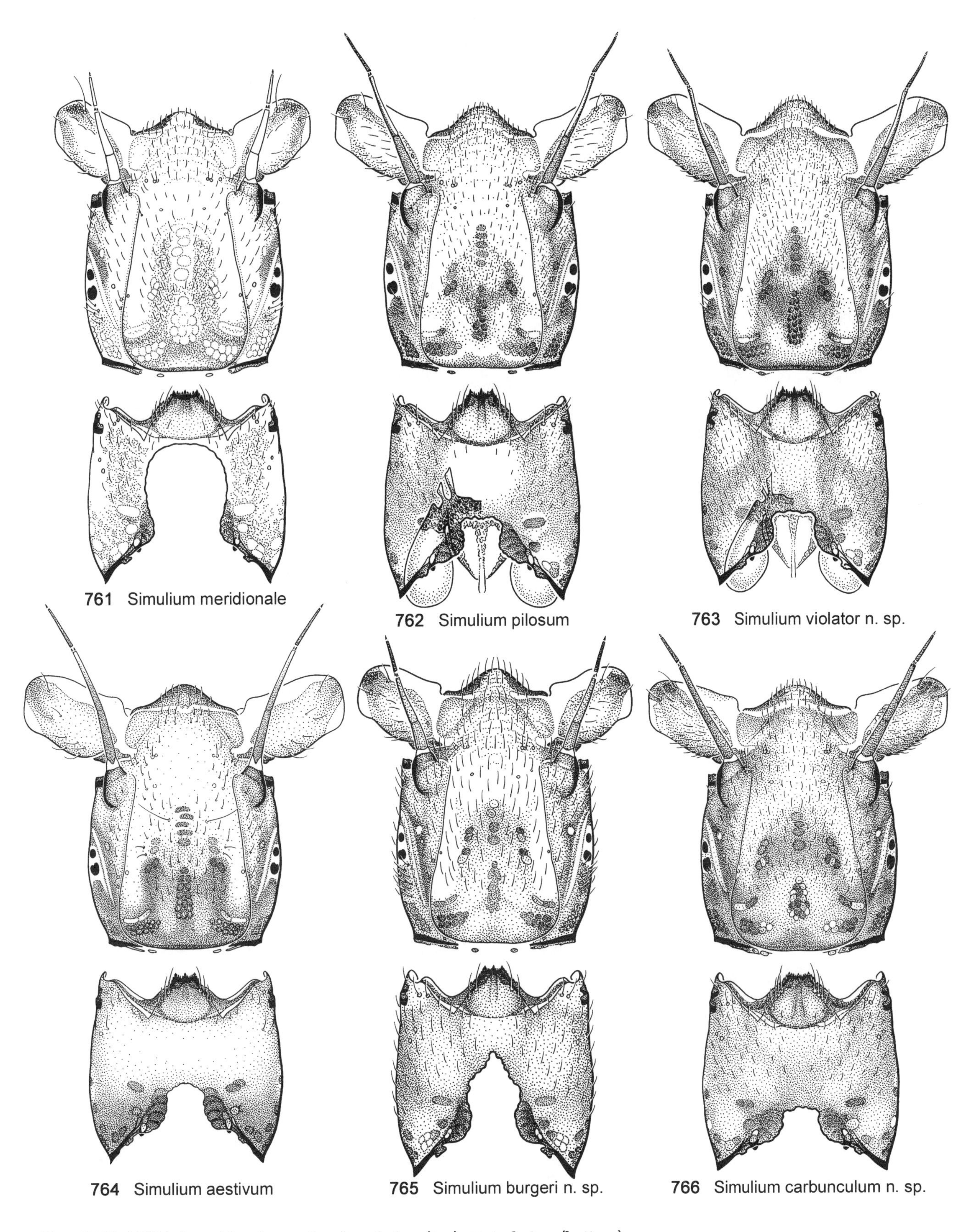


Figs. 10.761–10.766. Larval head capsules, dorsal view (top), ventral view (bottom).

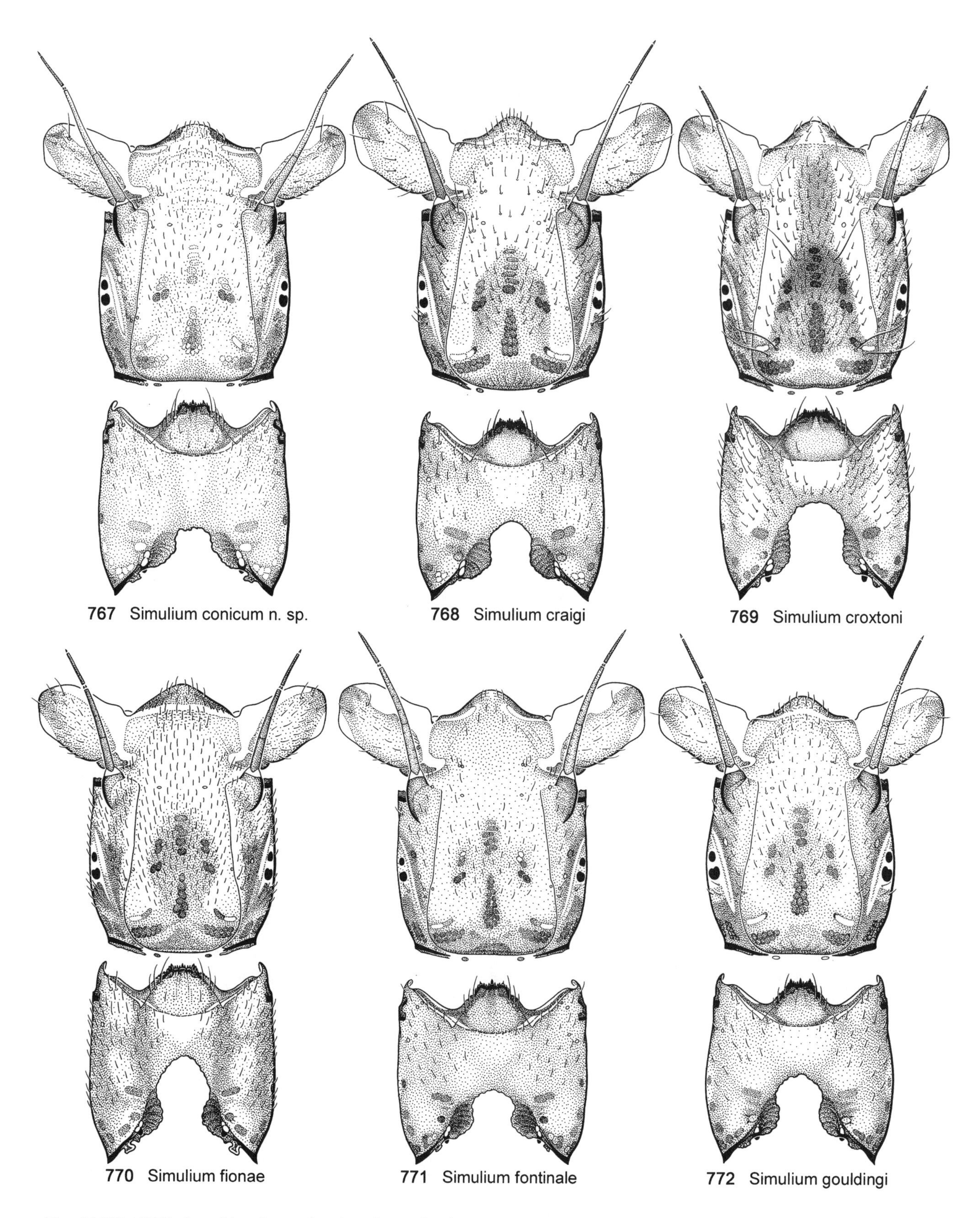


Figs. 10.767–10.772. Larval head capsules, dorsal view (top), ventral view (bottom).

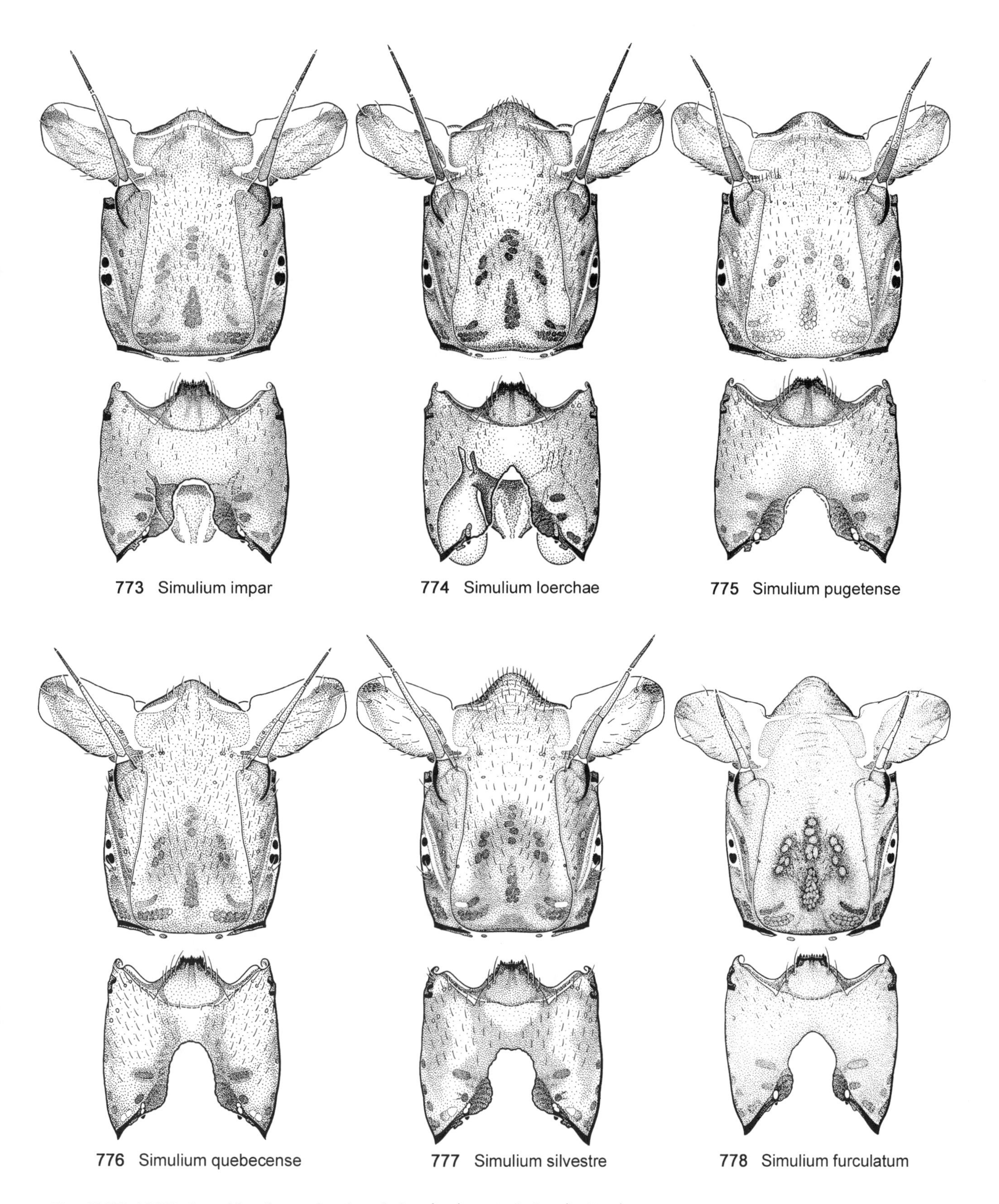


Figs. 10.773–10.778. Larval head capsules, dorsal view (top), ventral view (bottom).

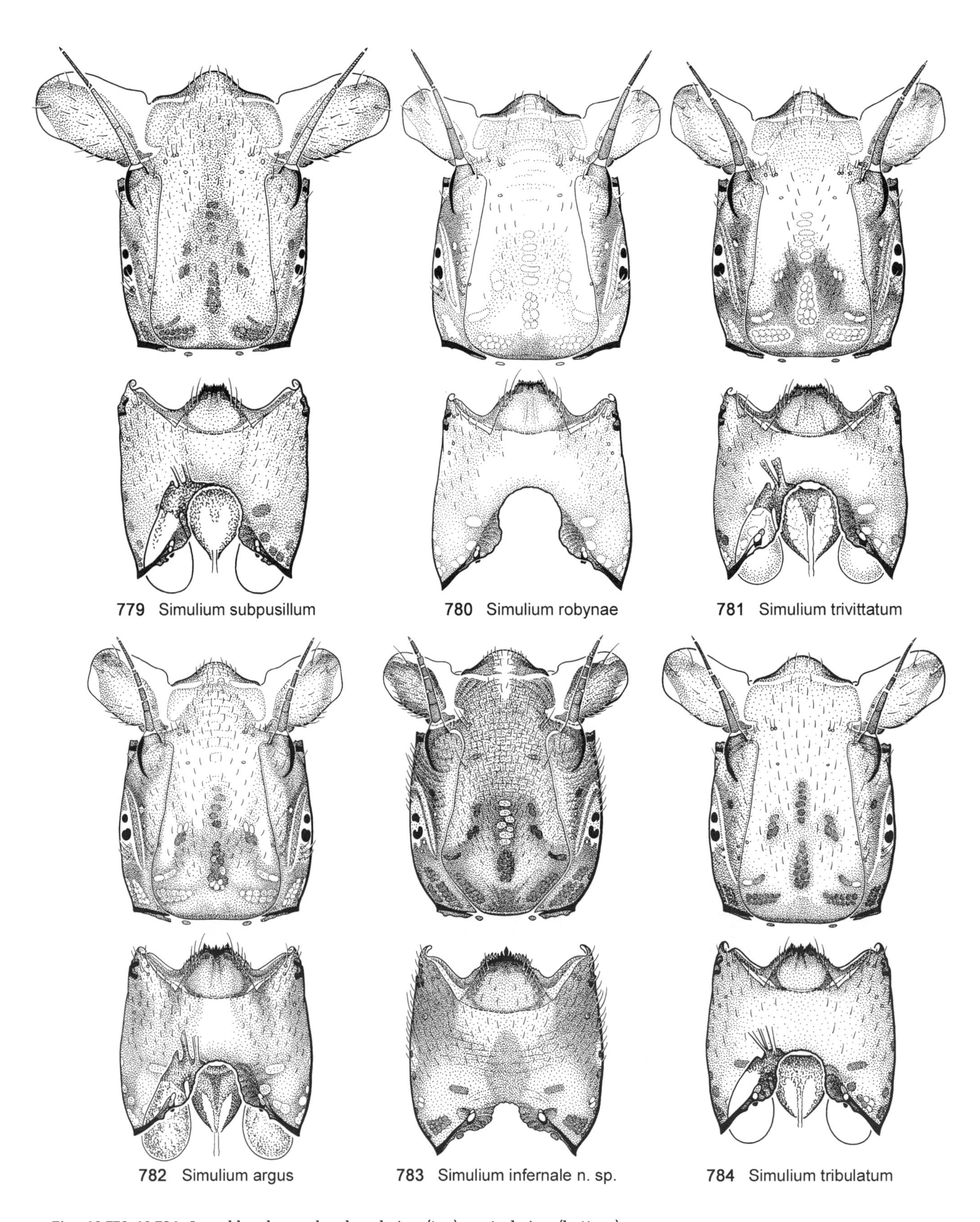


Figs. 10.779–10.784. Larval head capsules, dorsal view (top), ventral view (bottom).

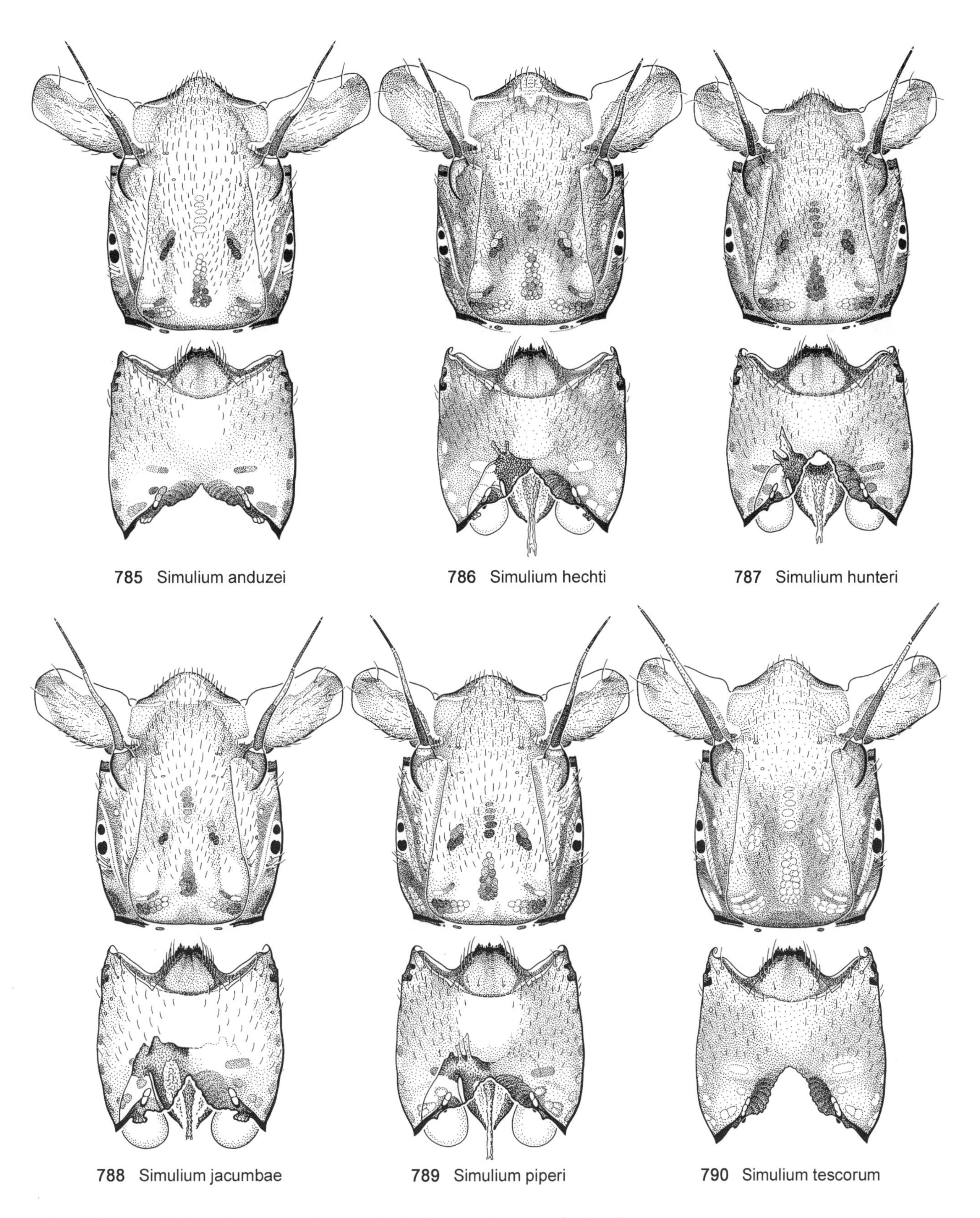


Figs. 10.785–10.790. Larval head capsules, dorsal view (top), ventral view (bottom).

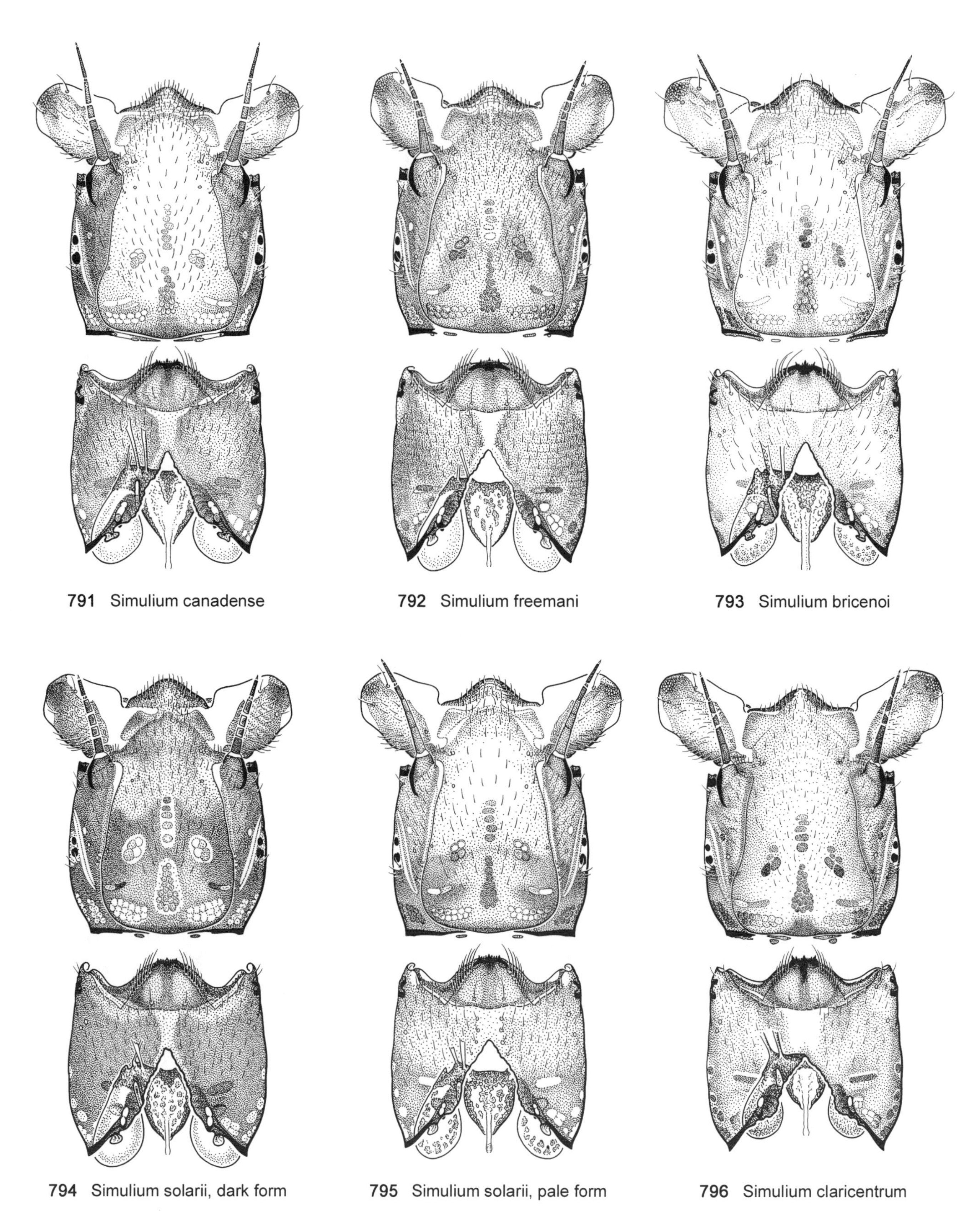


Figs. 10.791–10.796. Larval head capsules, dorsal view (top), ventral view (bottom).

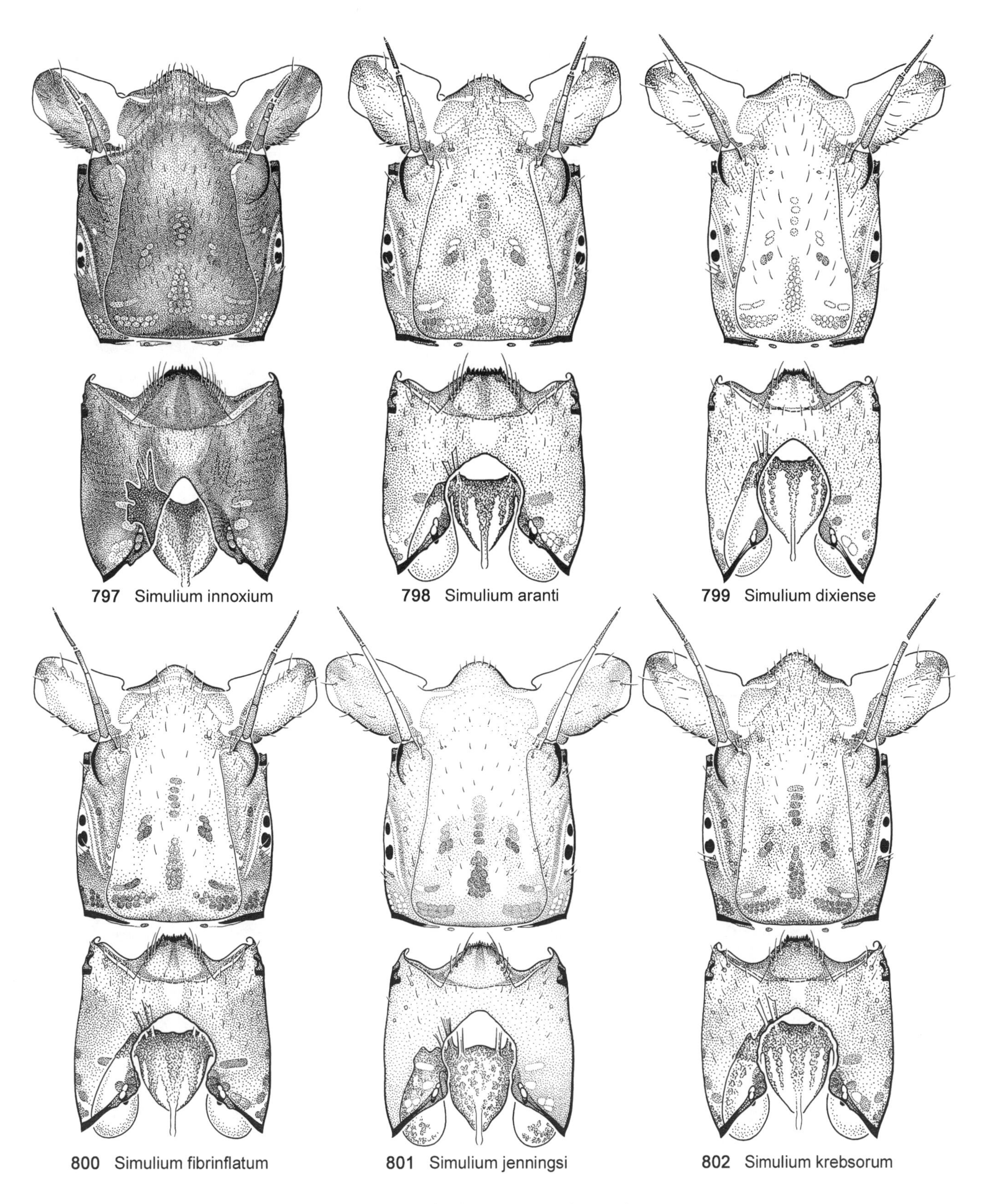


Figs. 10.797–10.802. Larval head capsules, dorsal view (top), ventral view (bottom).

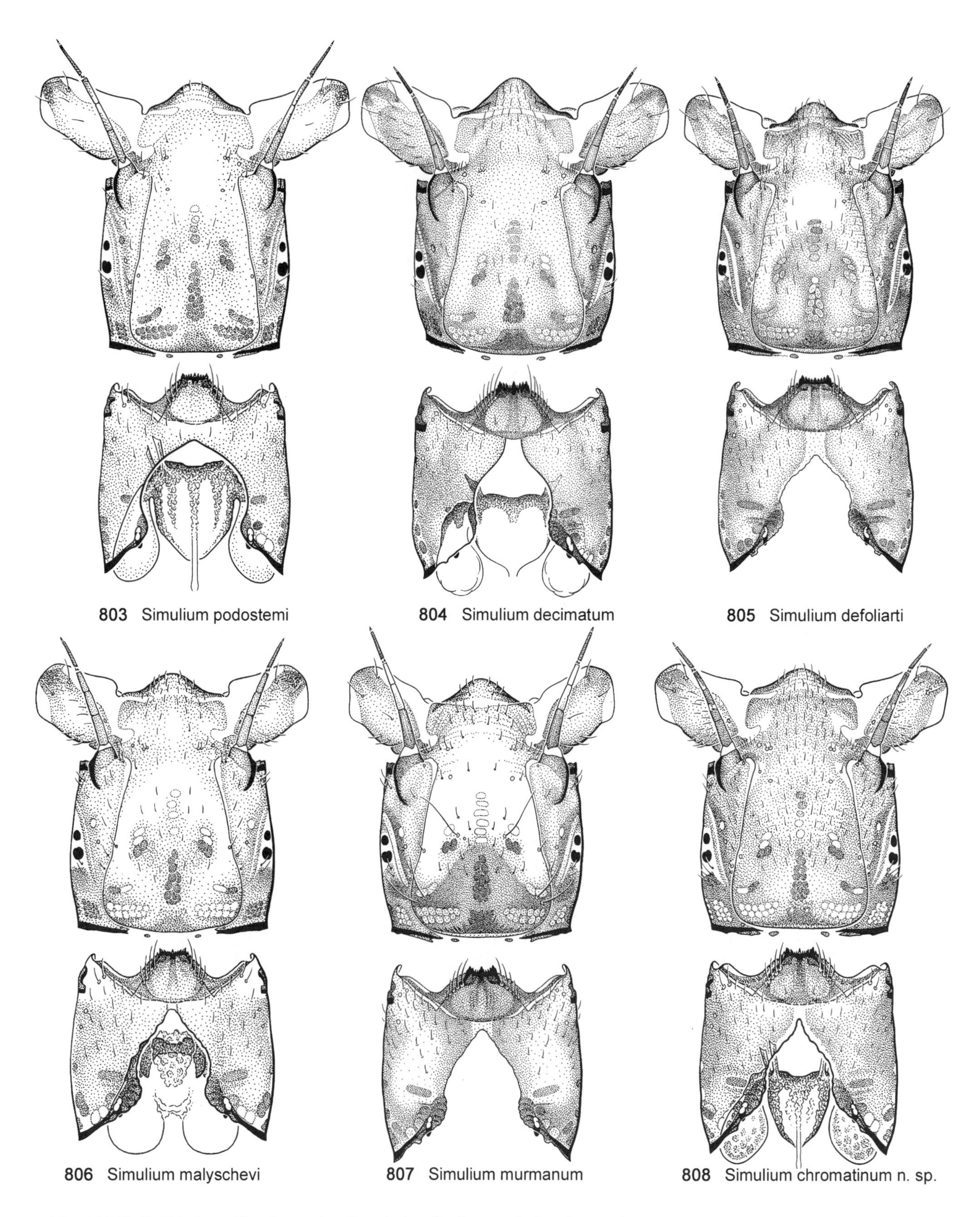


Figs. 10.803–10.808. Larval head capsules, dorsal view (top), ventral view (bottom).

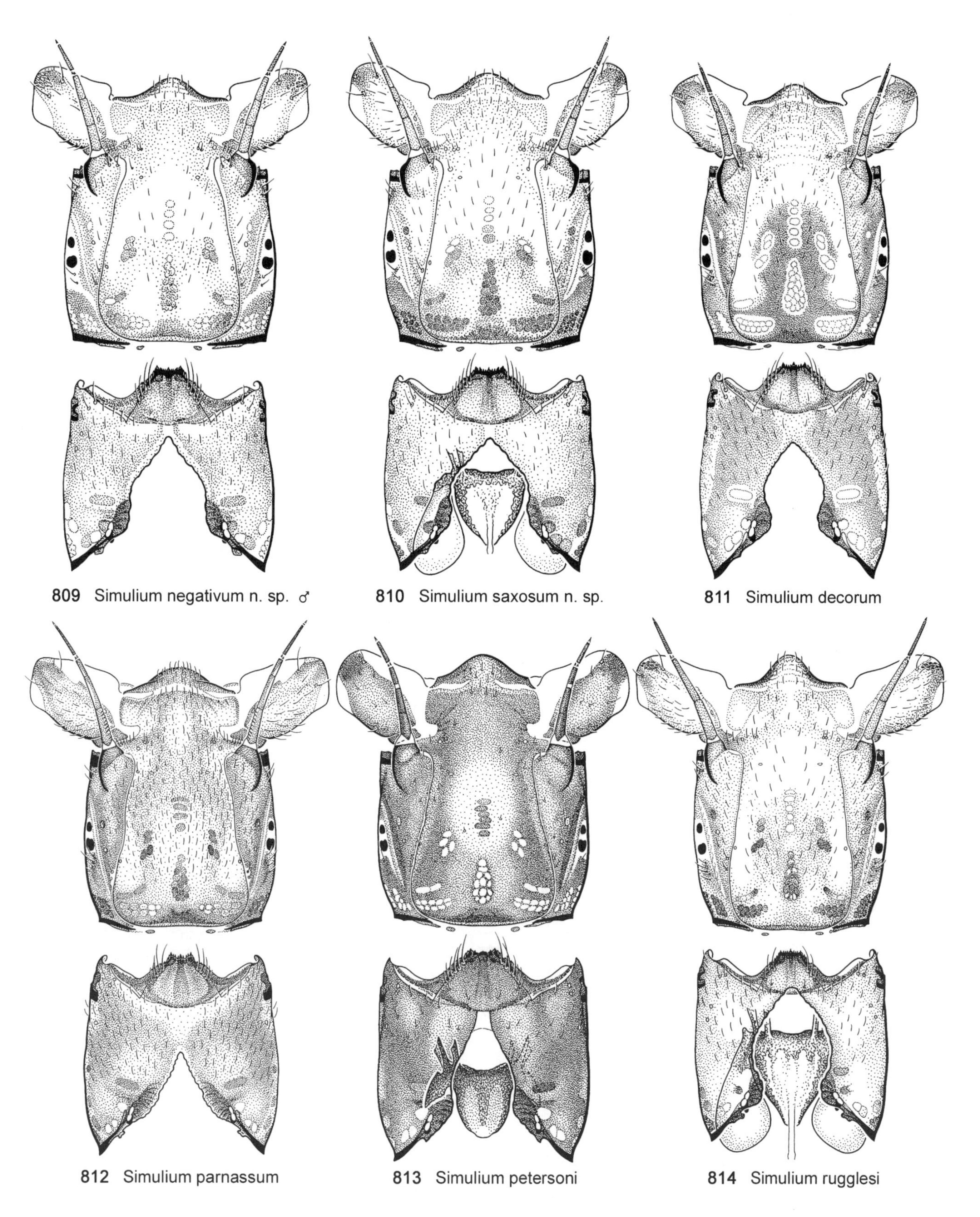


Figs. 10.809–10.814. Larval head capsules, dorsal view (top), ventral view (bottom).

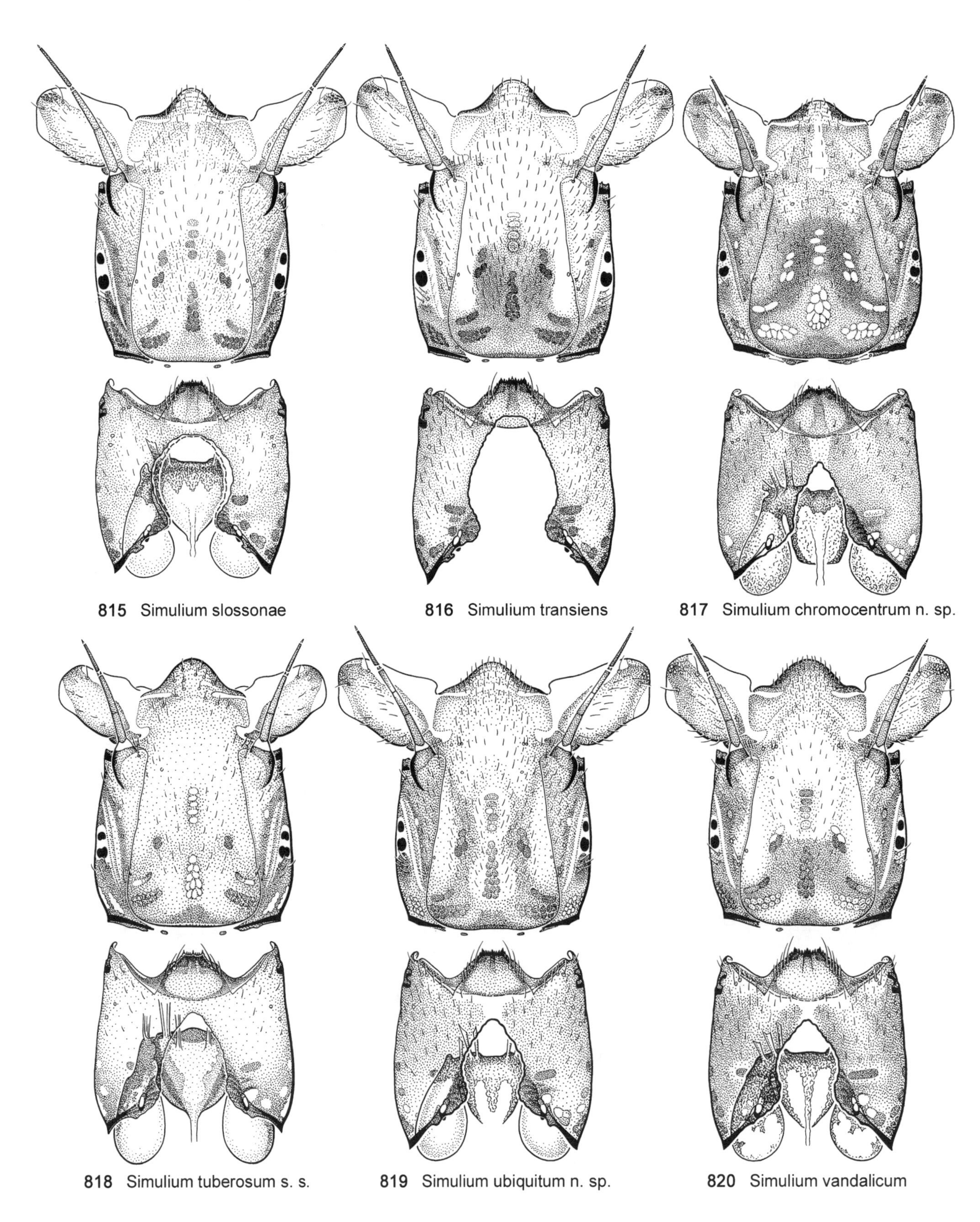


Figs. 10.815–10.820. Larval head capsules, dorsal view (top), ventral view (bottom).

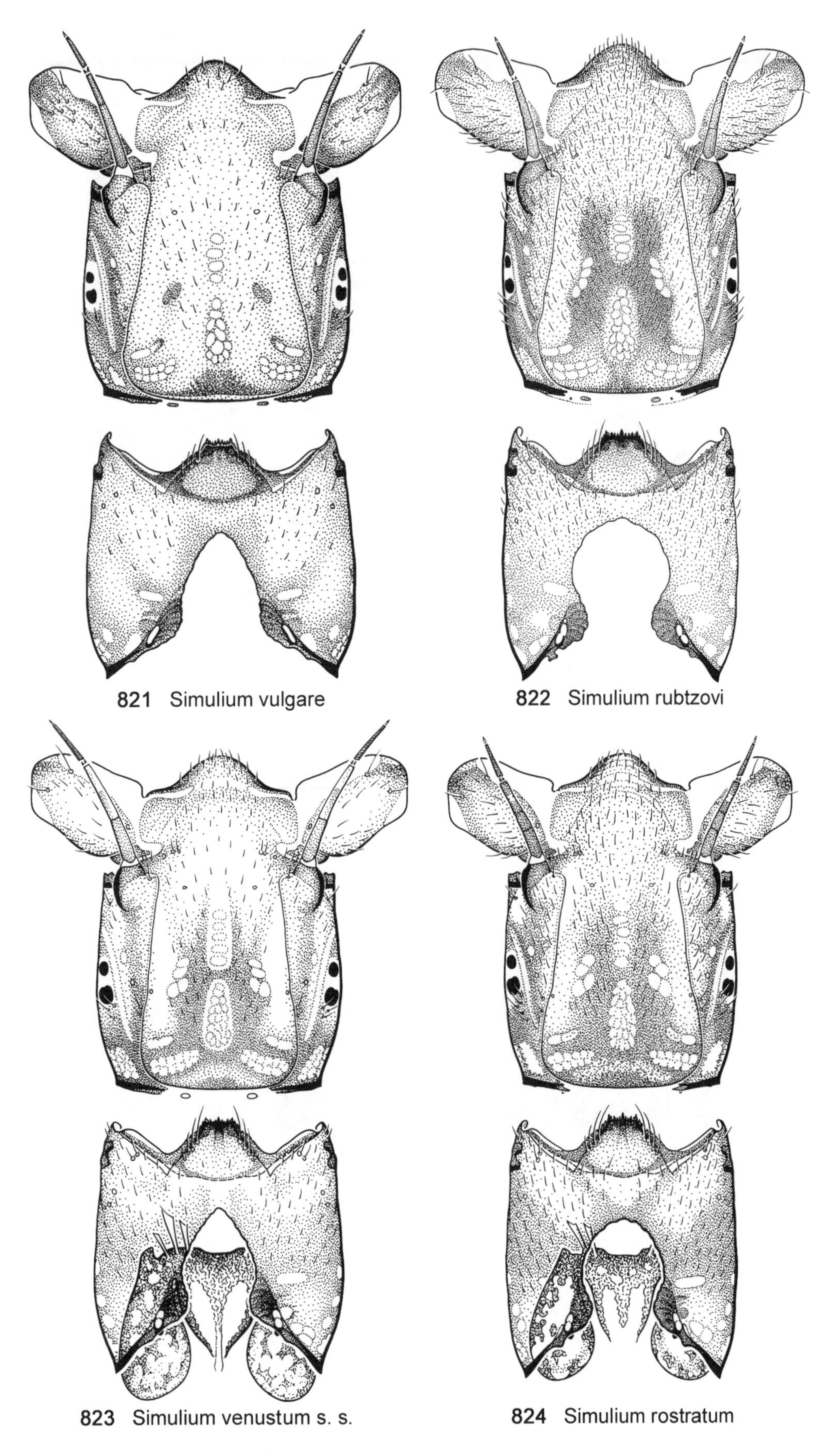


Figs. 10.821–10.824. Larval head capsules, dorsal view (top), ventral view (bottom).

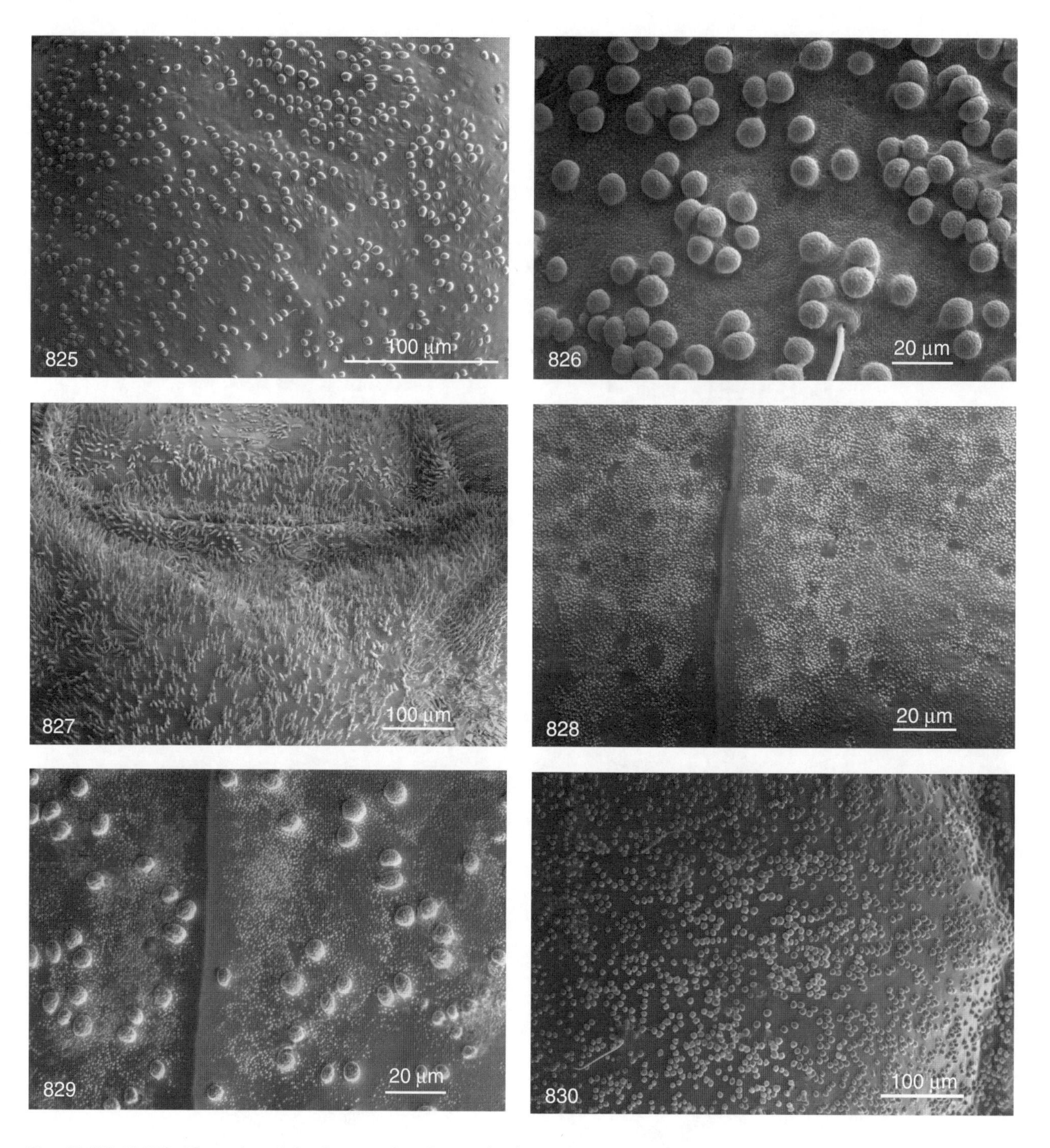


Figs. 10.825–10.830. Thoracic cuticle of pupae, dorsal view. (825) *Simulium mysterium* n. sp. (826) *S. carbunculum* n. sp. (827) *S. conicum* n. sp. (828) *S. merritti* n. sp. (829) *S. pugetense.* (830) *S. chromocentrum* n. sp.

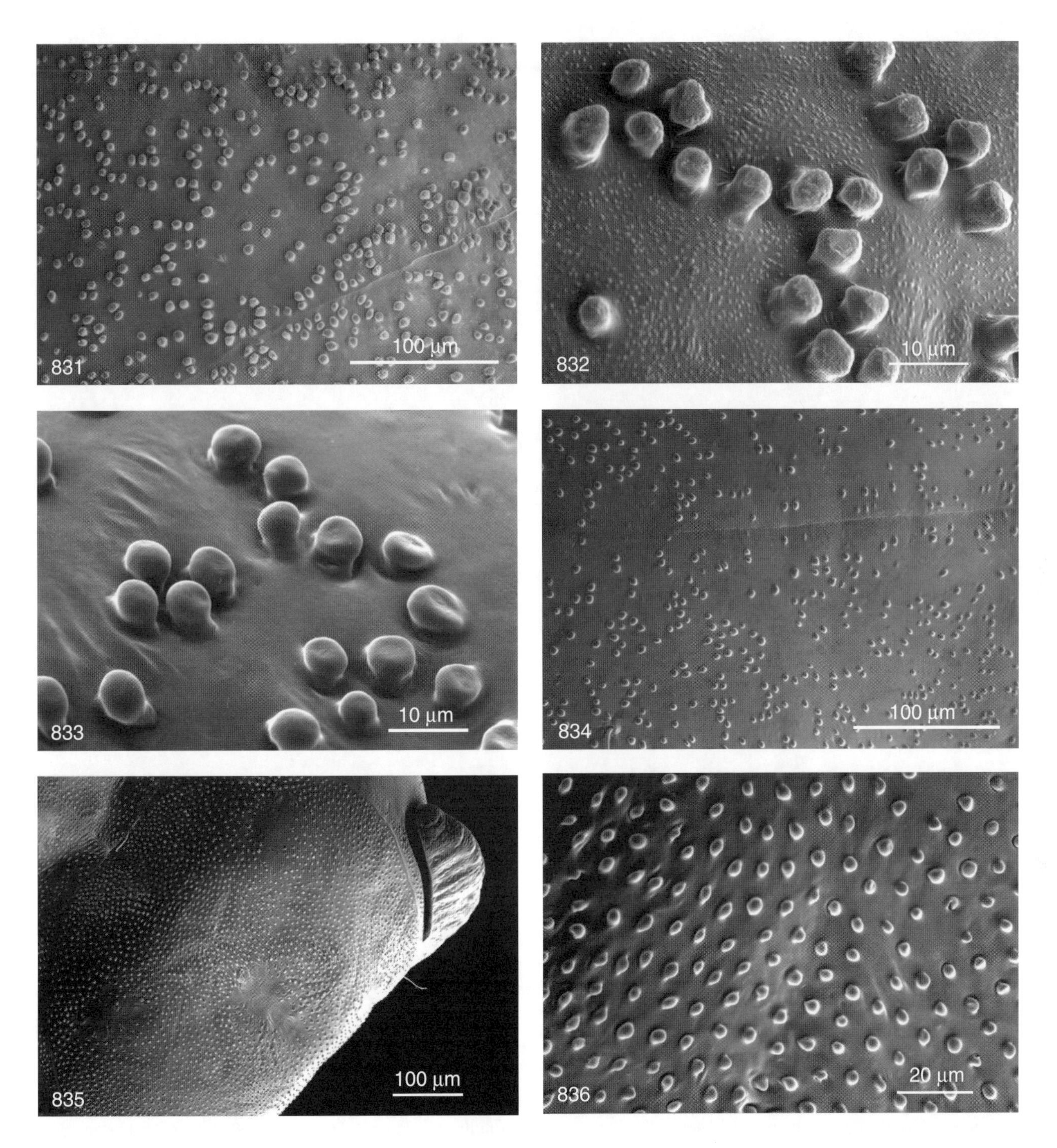


Figs. 10.831–10.836. Thoracic and cephalic cuticle of pupae, dorsal view. (831, 832) *Simulium conundrum* n. sp.; Tumbler Ridge, British Columbia. (833) *S. conundrum* n. sp.; type locality. (834) *S. tuberosum* s. s., cytotype AB. (835, 836) *S. ubiquitum* n. sp., cephalic cuticle.

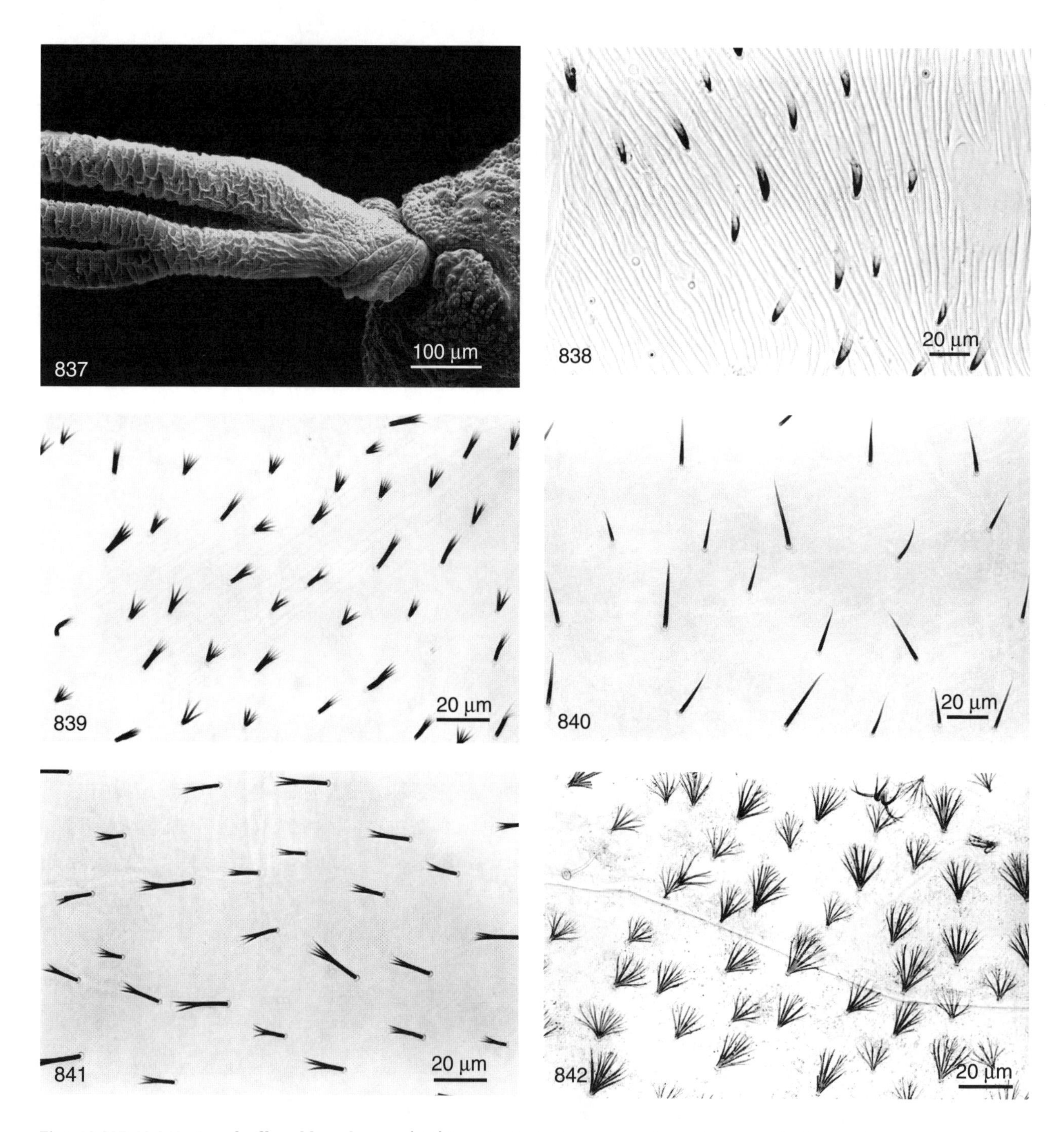


Figs. 10.837–10.842. Pupal gill and larval setae. (837) Pupal gill of *Simulium carbunculum* n. sp. (838–842) Larval setae on posterodorsal cuticle of abdomen. (838) *S. meridionale.* (839) *S. fionae.* (840) *S. silvestre.* (841) *S. slossonae.* (842) *S. transiens.*

North America north of Mexico

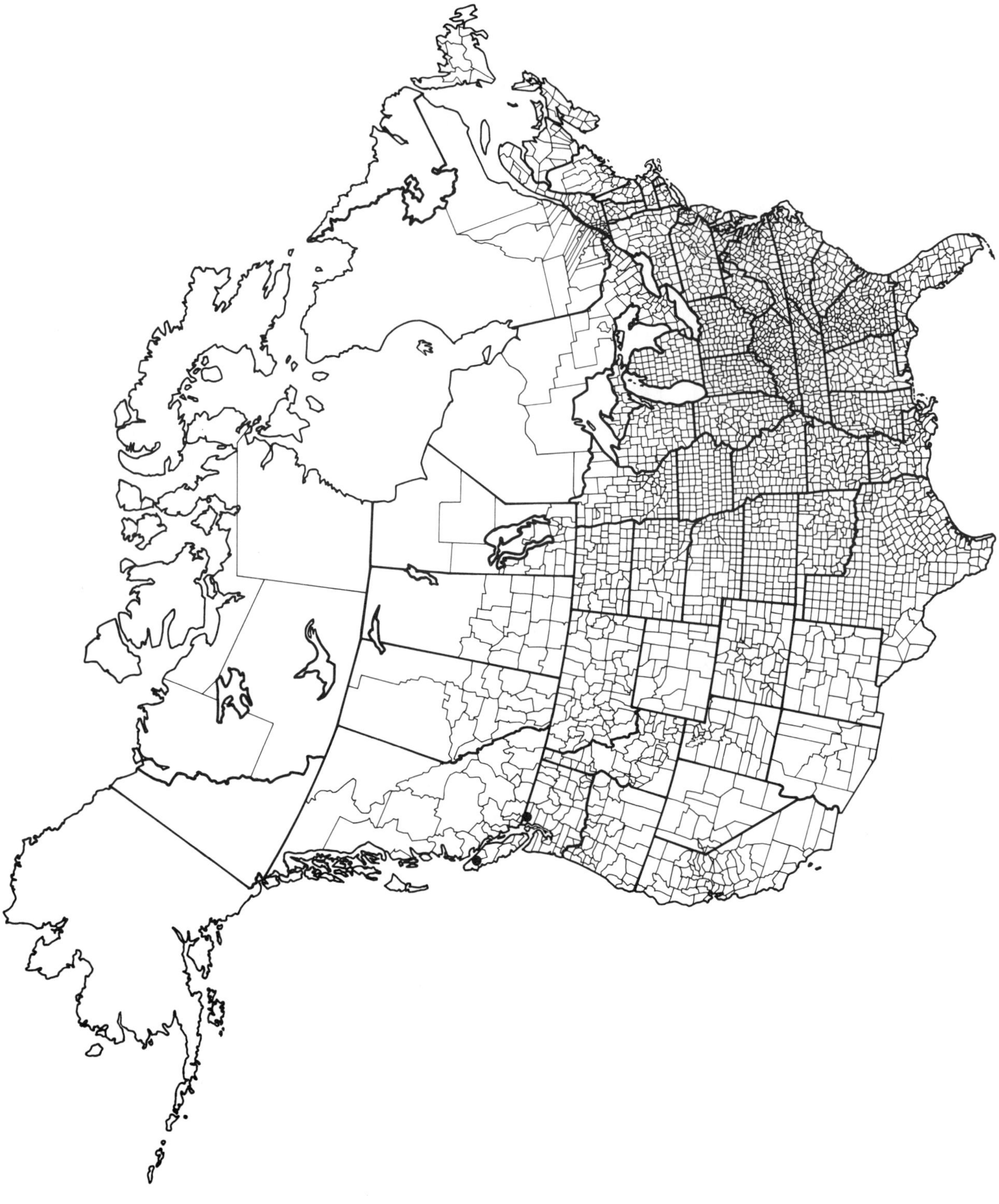

Map 1 *Parasimulium melanderi*

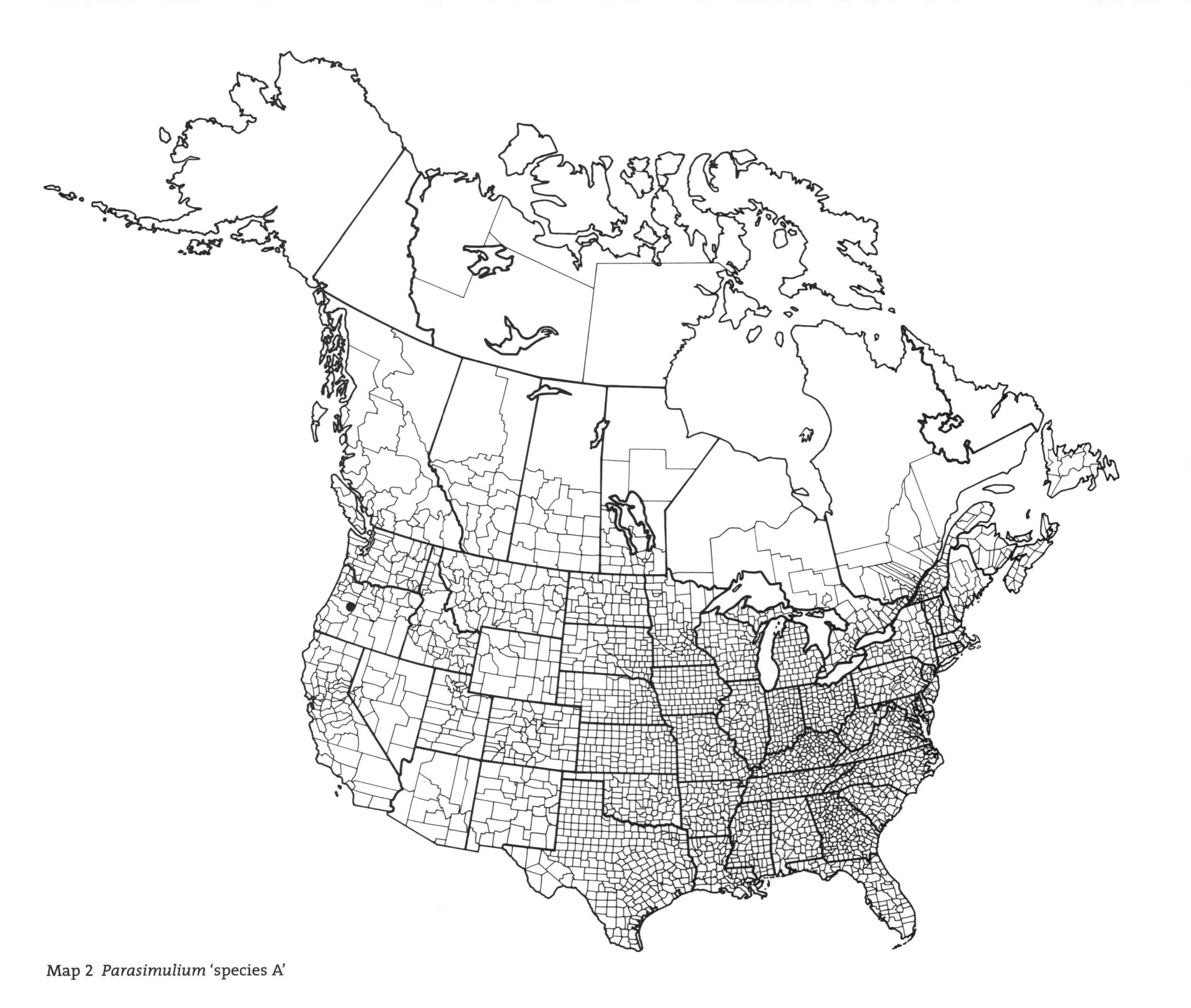

Map 2 *Parasimulium* 'species A'

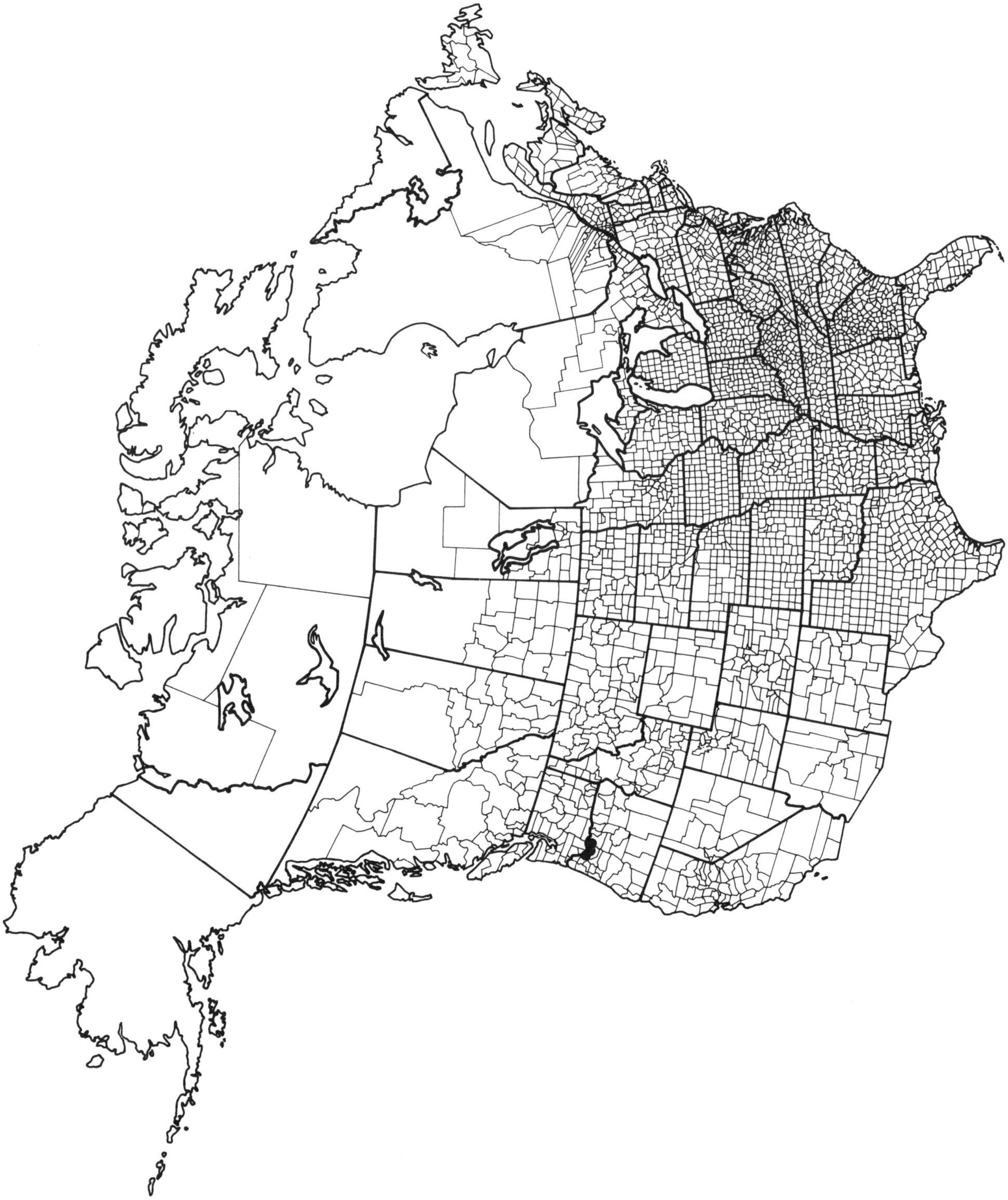

Map 3 *Parasimulium crosskeyi*

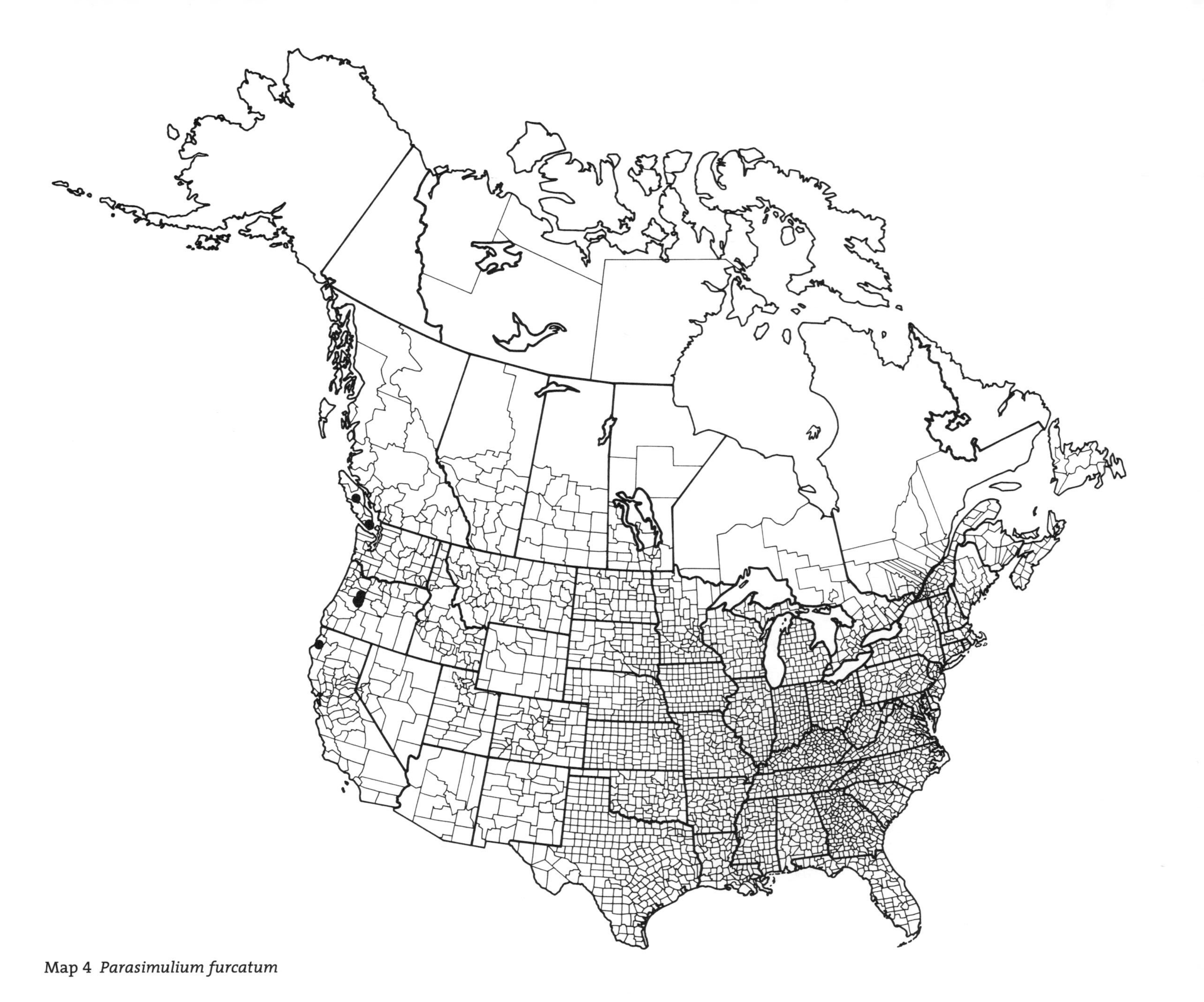

Map 4 *Parasimulium furcatum*

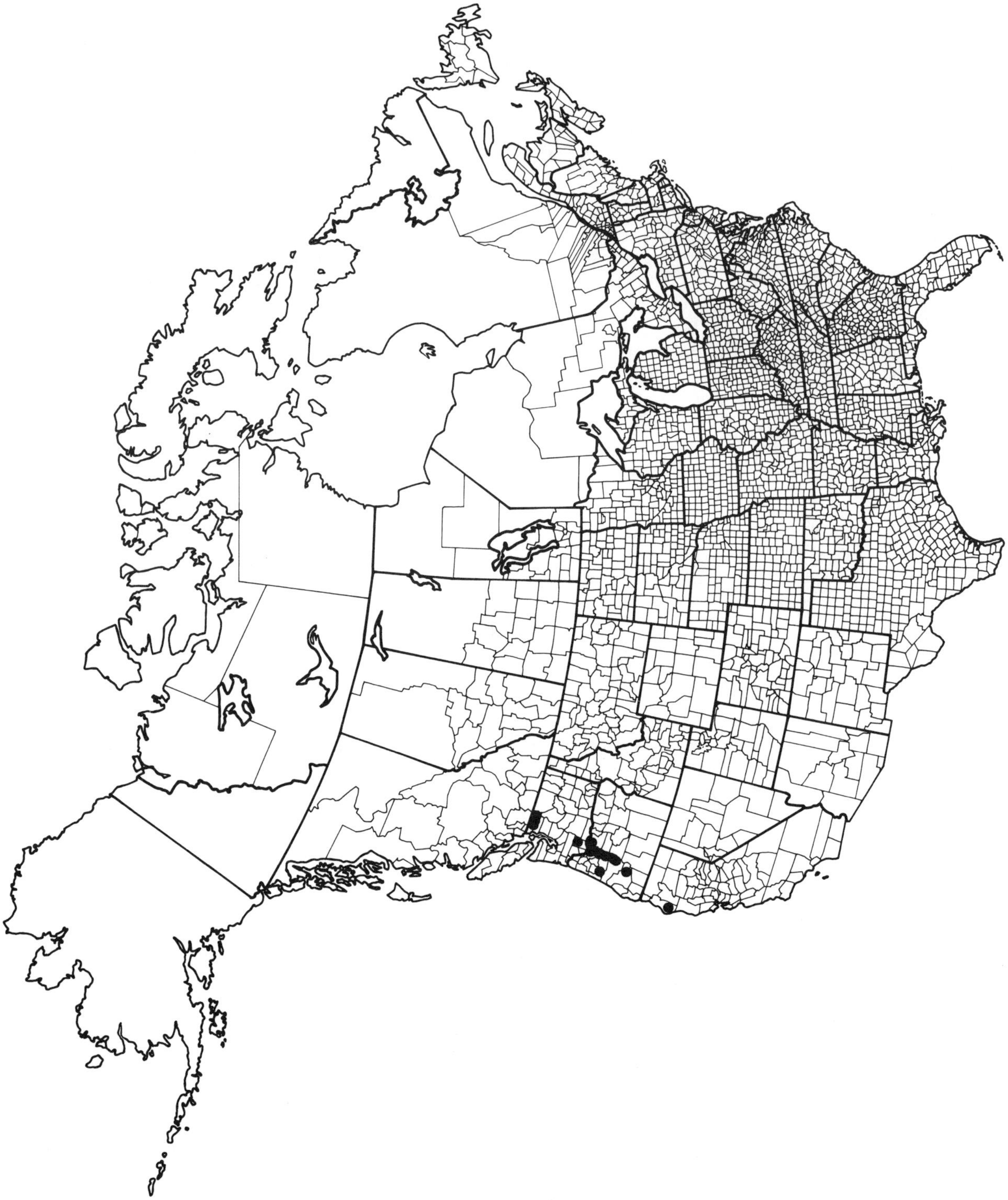

Map 5 *Parasimulium stonei*

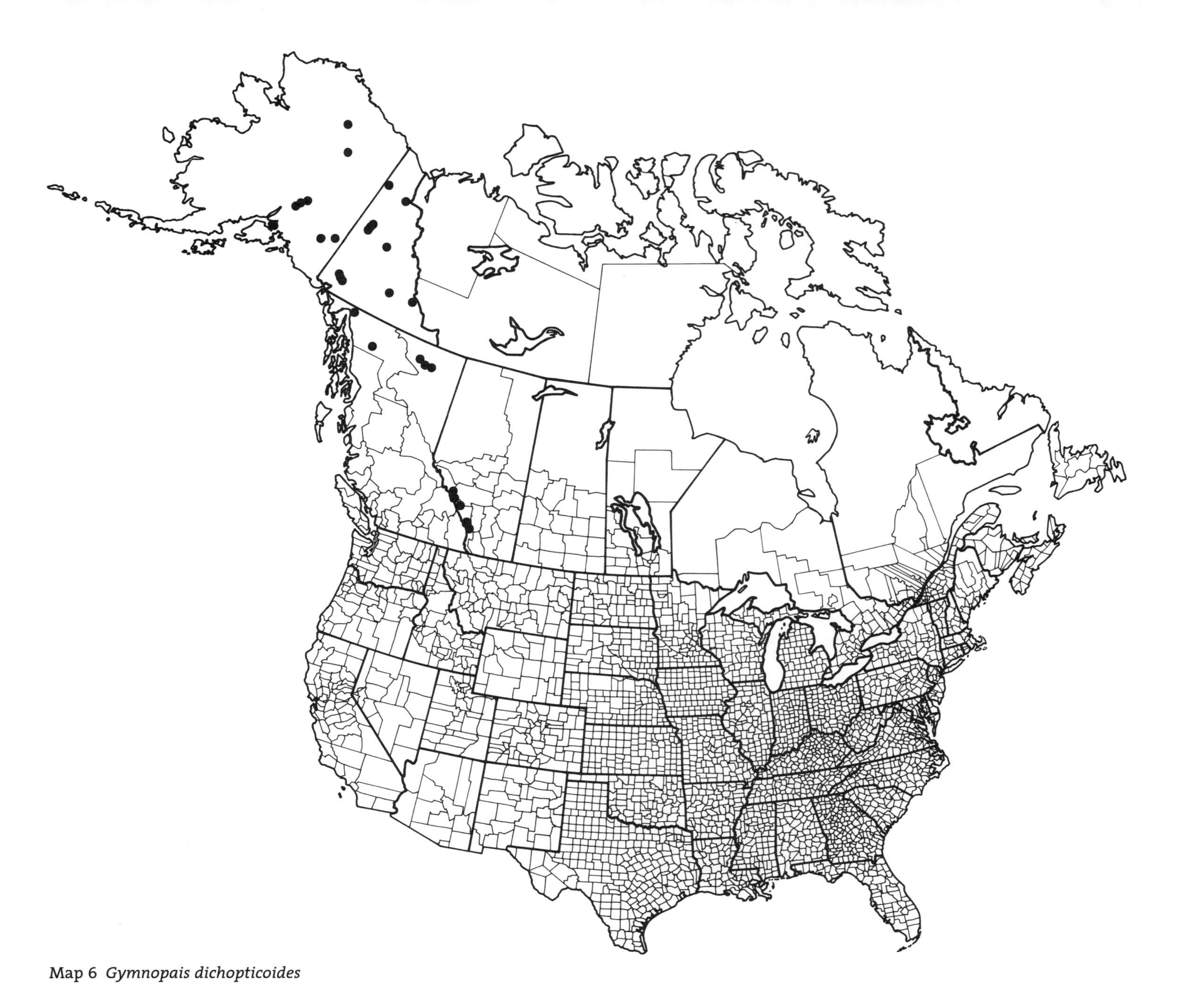

Map 6 *Gymnopais dichopticoides*

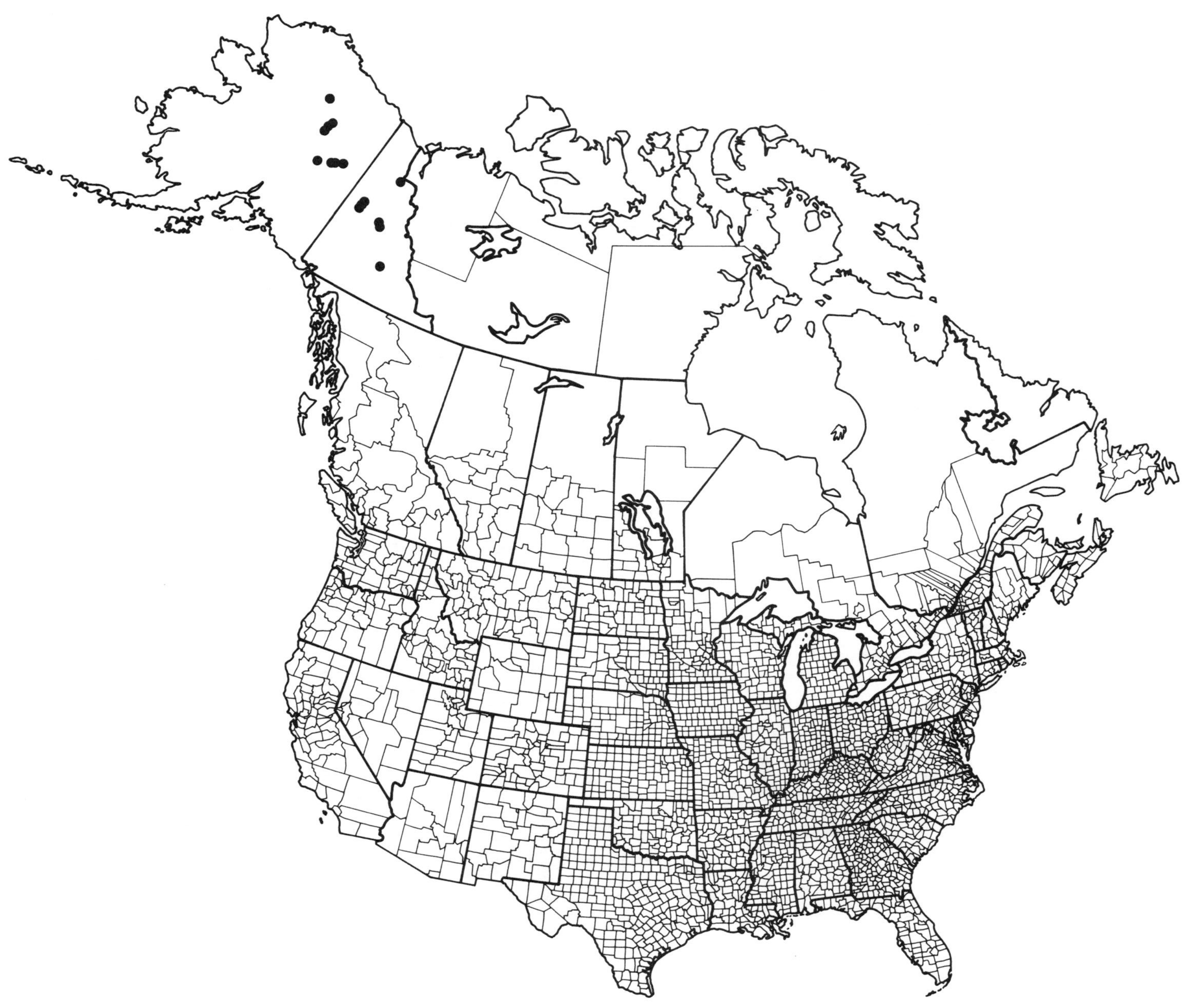

Map 7 *Gymnopais dichopticus*

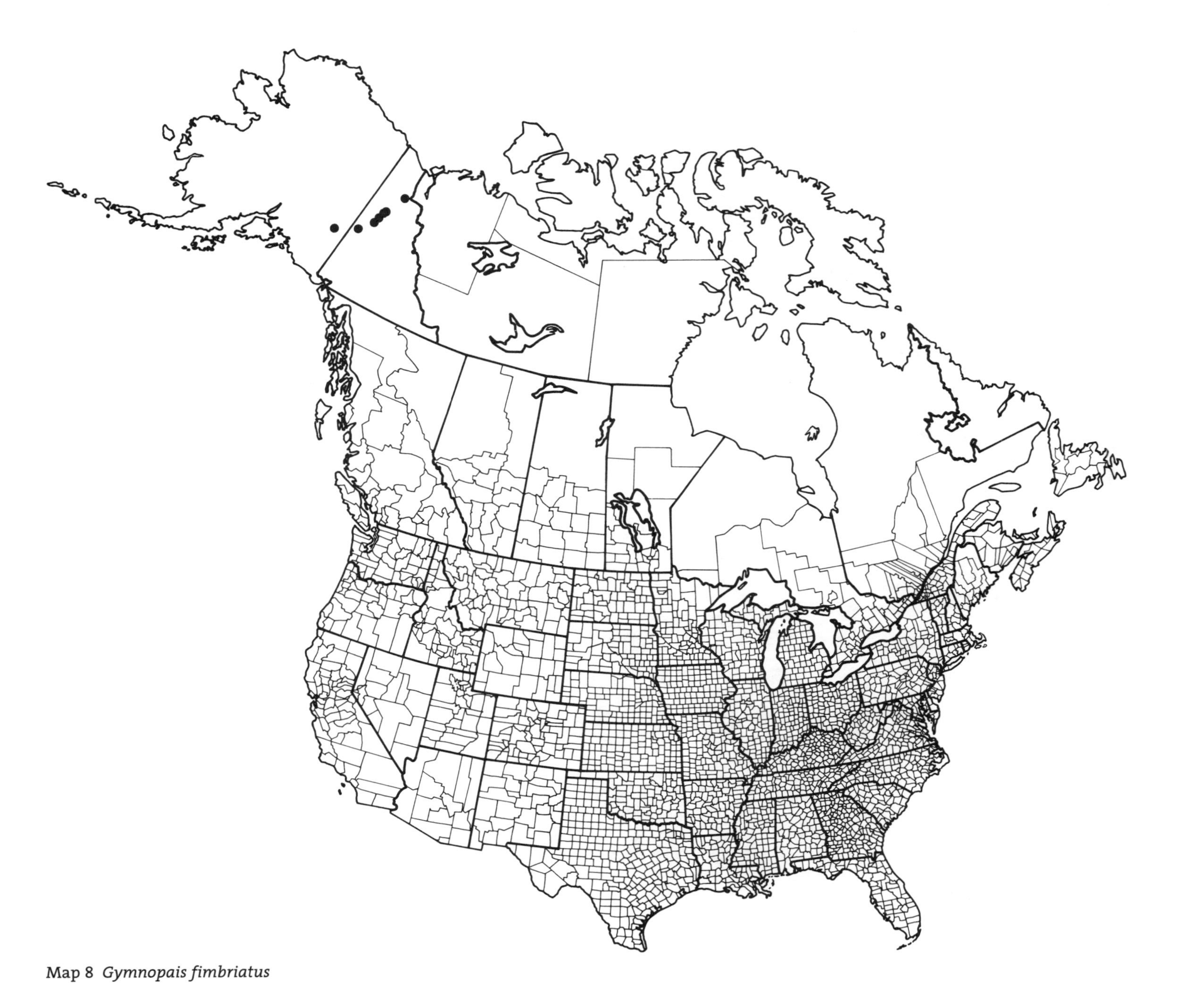

Map 8 *Gymnopais fimbriatus*

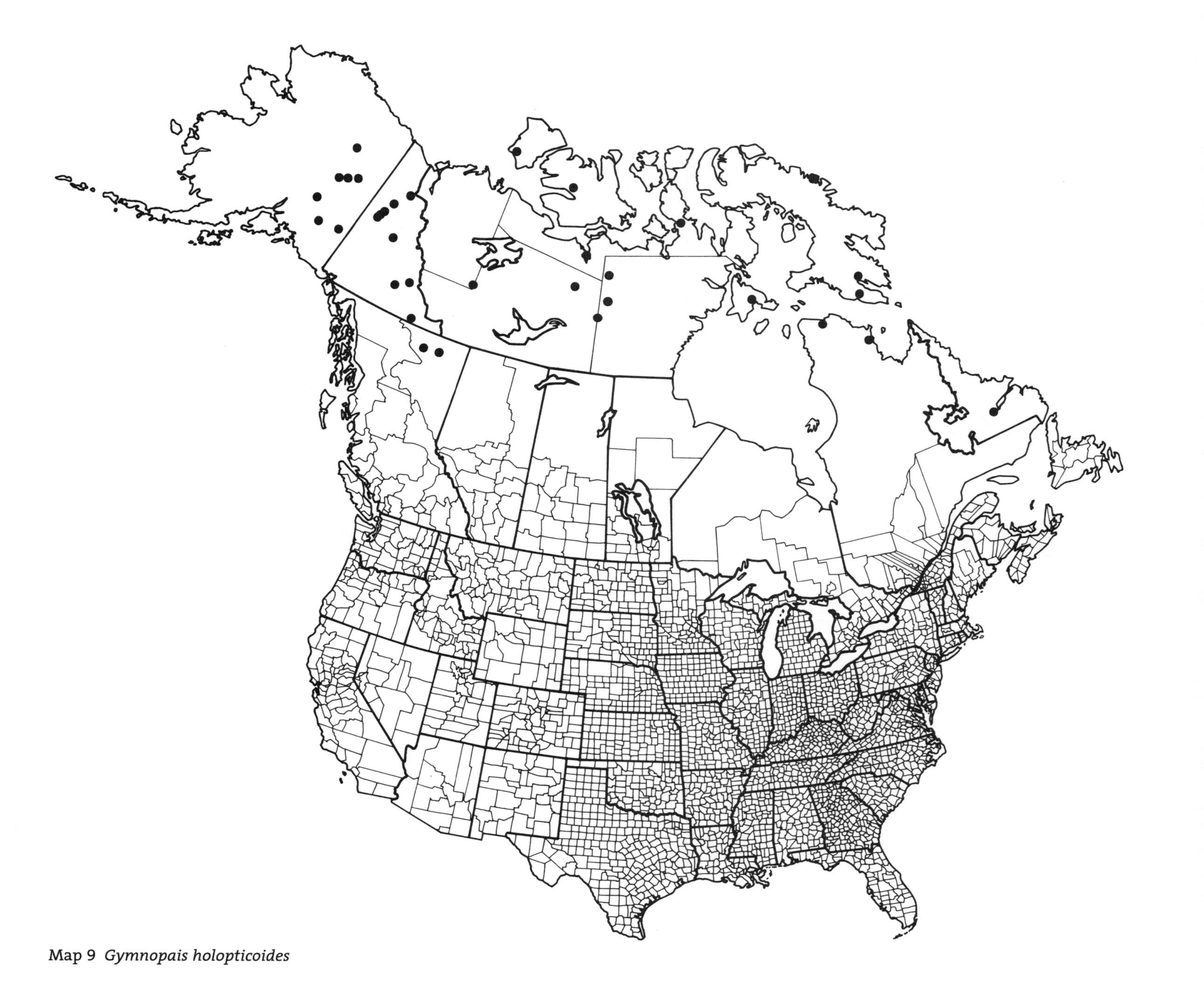

Map 9 *Gymnopais holopticoides*

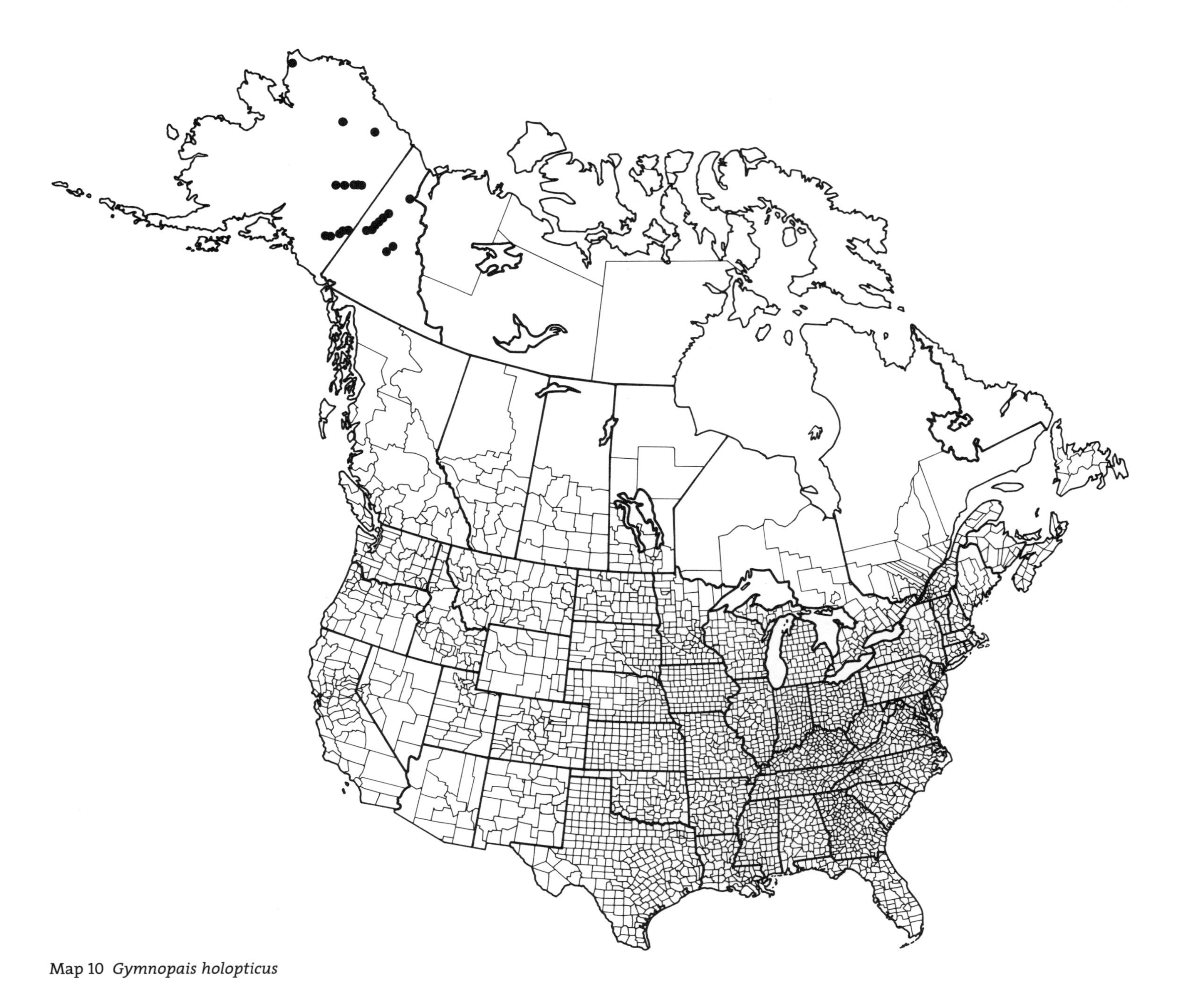

Map 10 *Gymnopais holopticus*

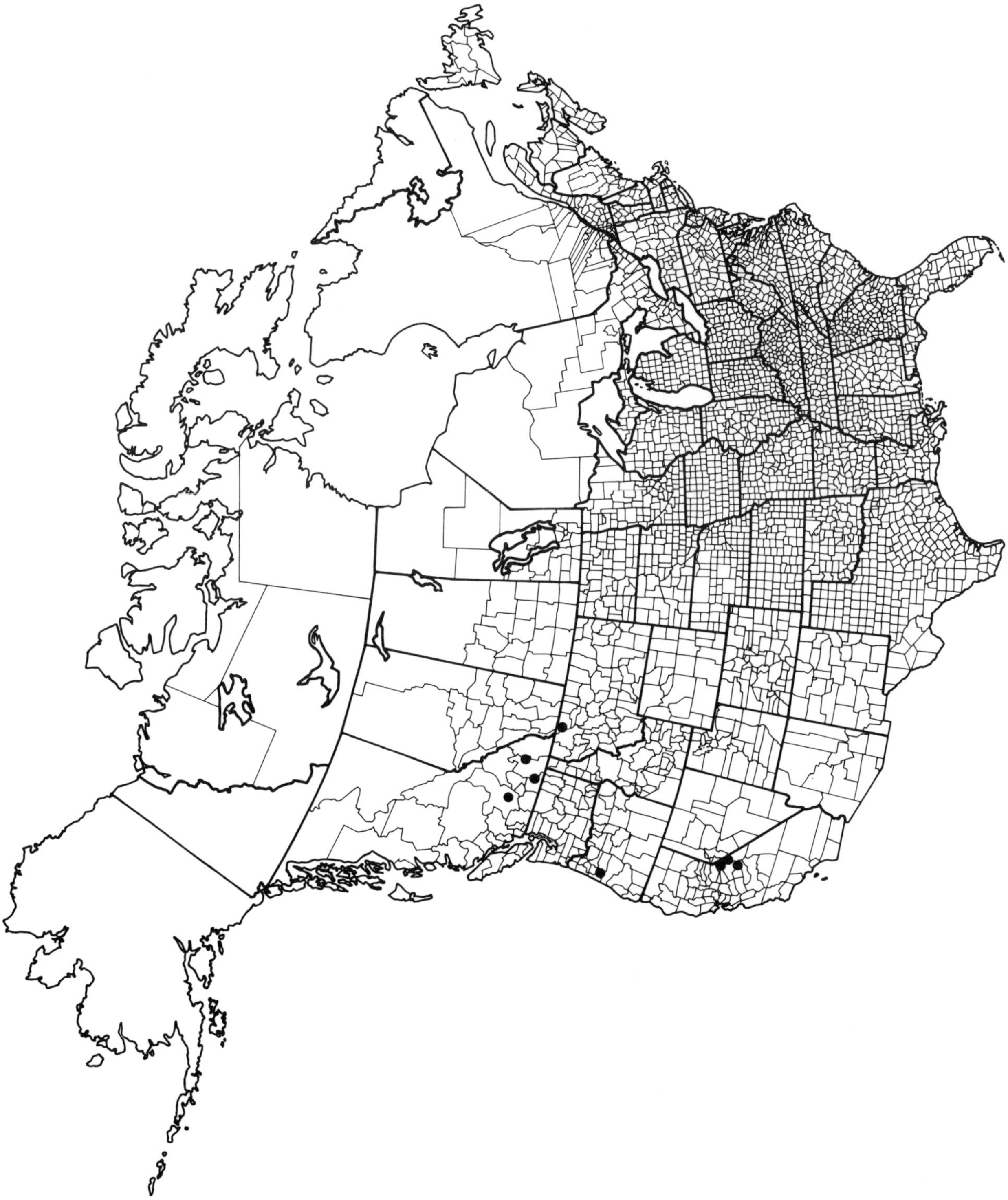

Map 11 *Twinnia hirticornis*

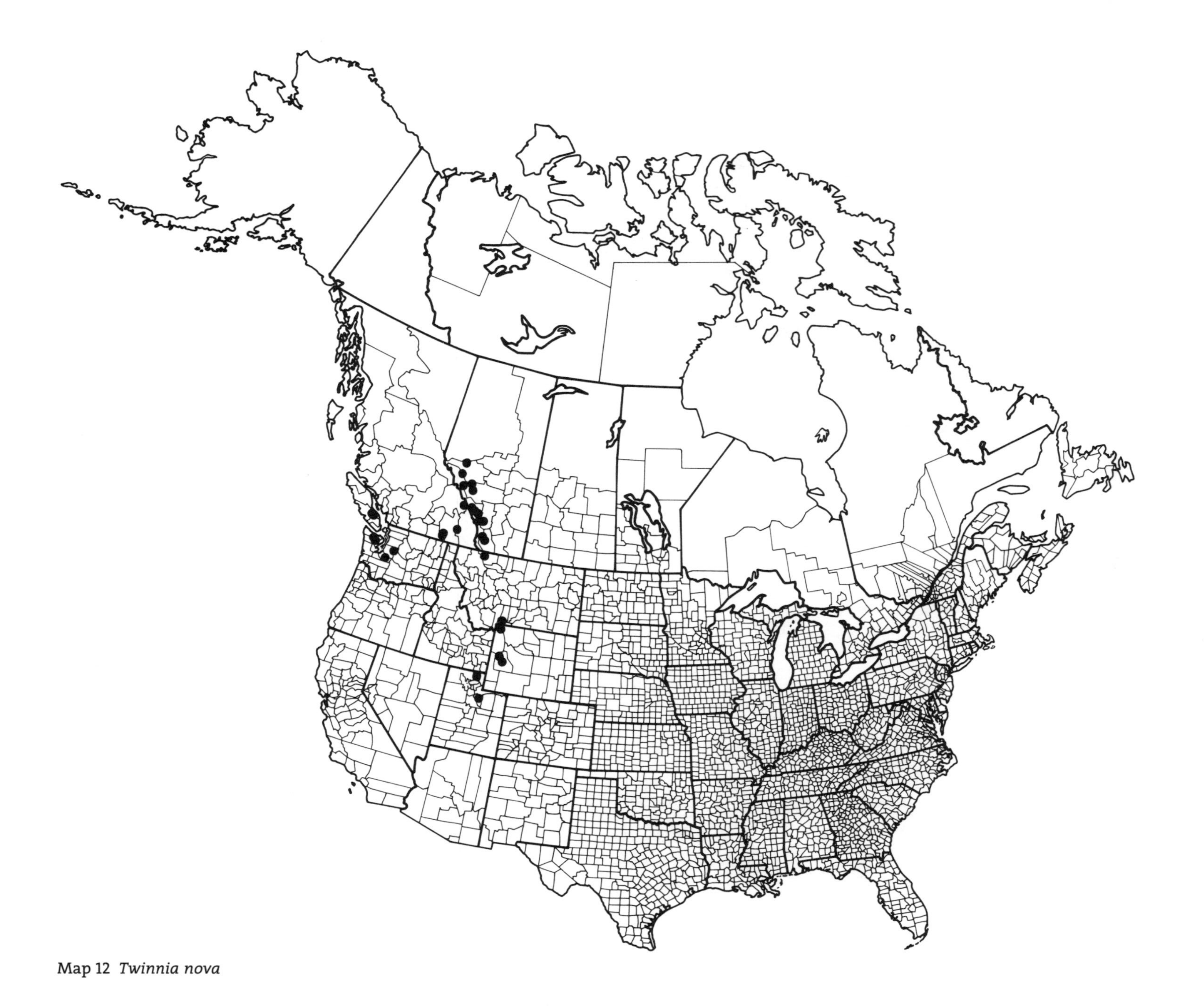

Map 12 *Twinnia nova*

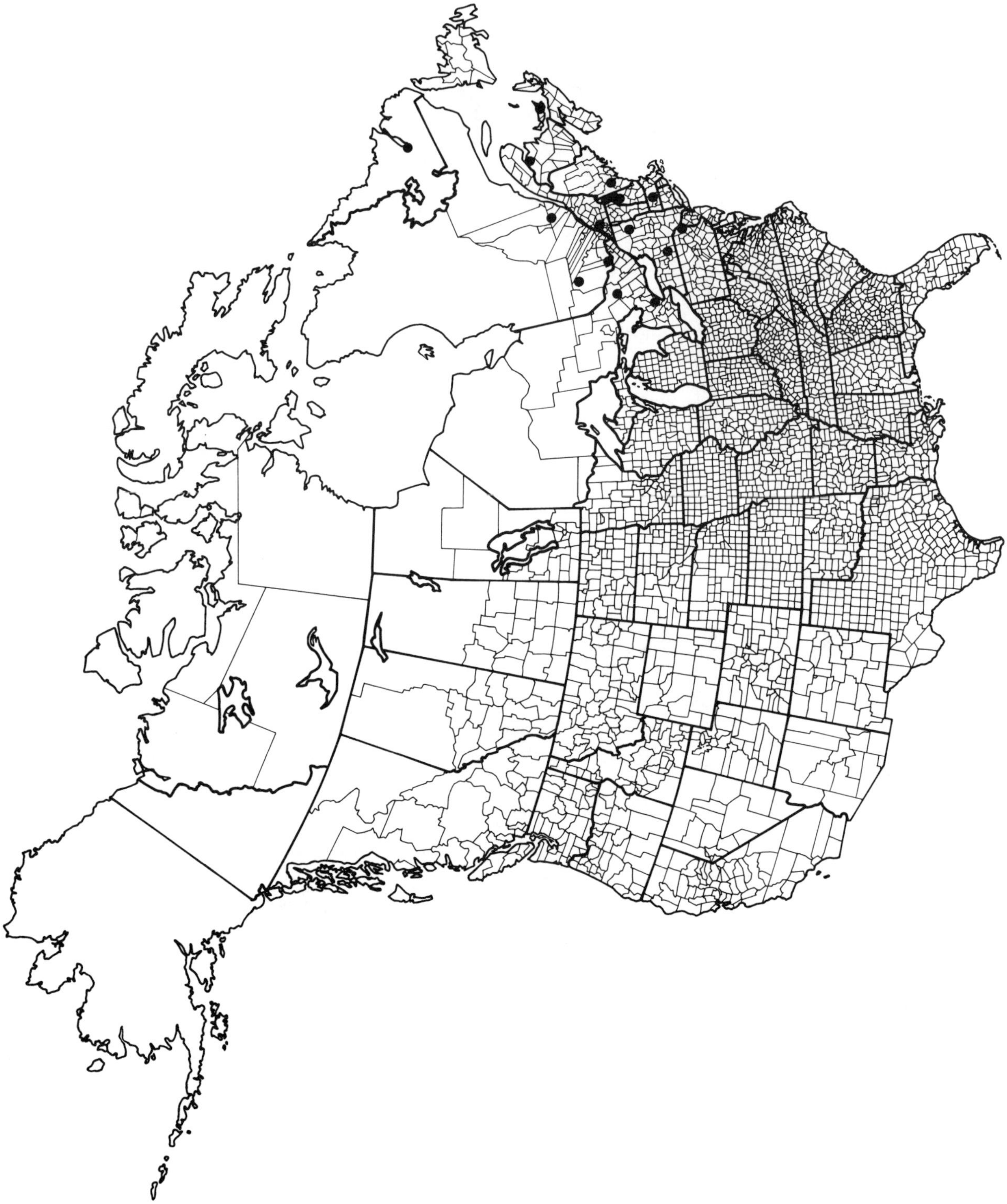

Map 13 *Twinnia tibblesi*

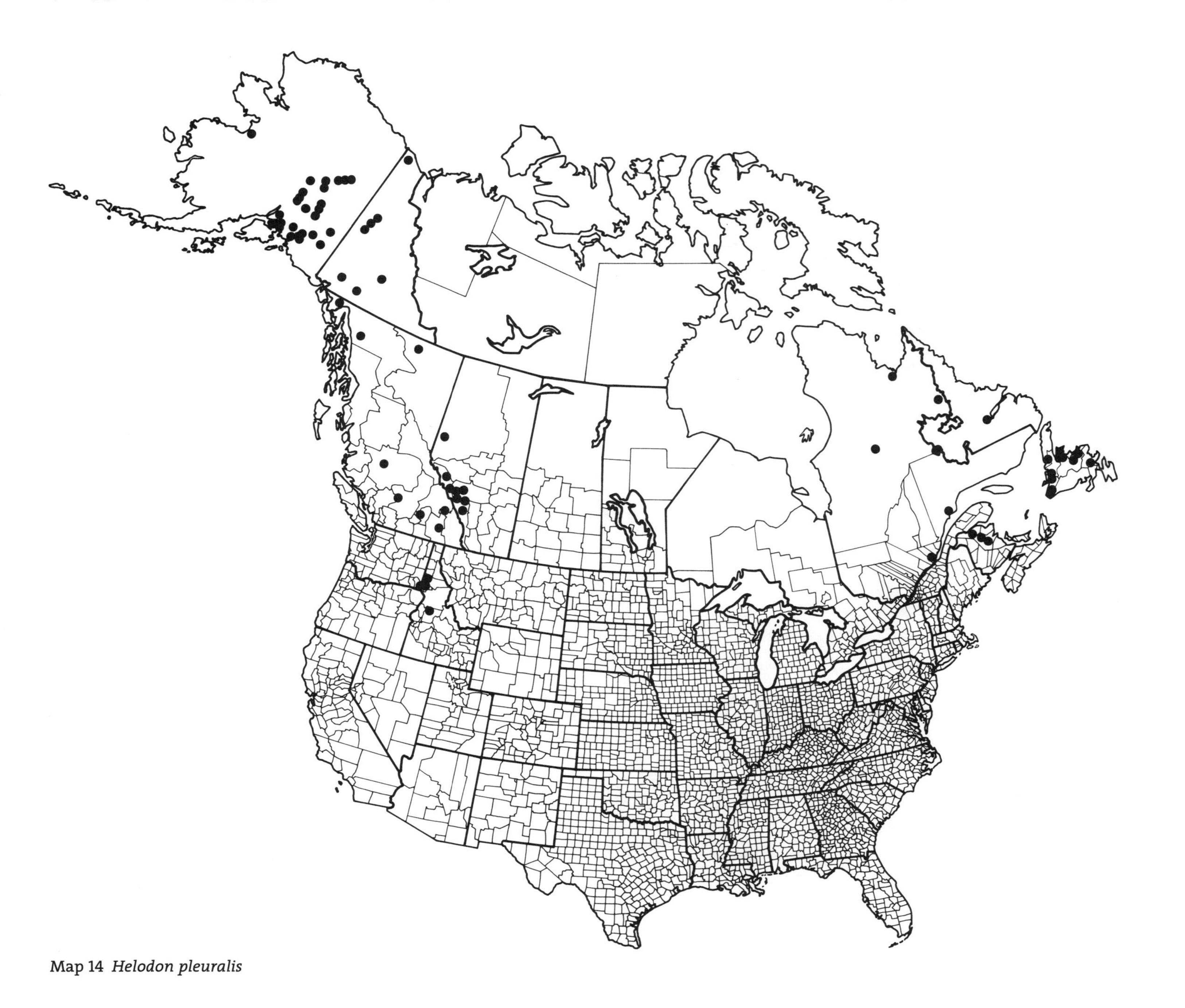

Map 14 *Helodon pleuralis*

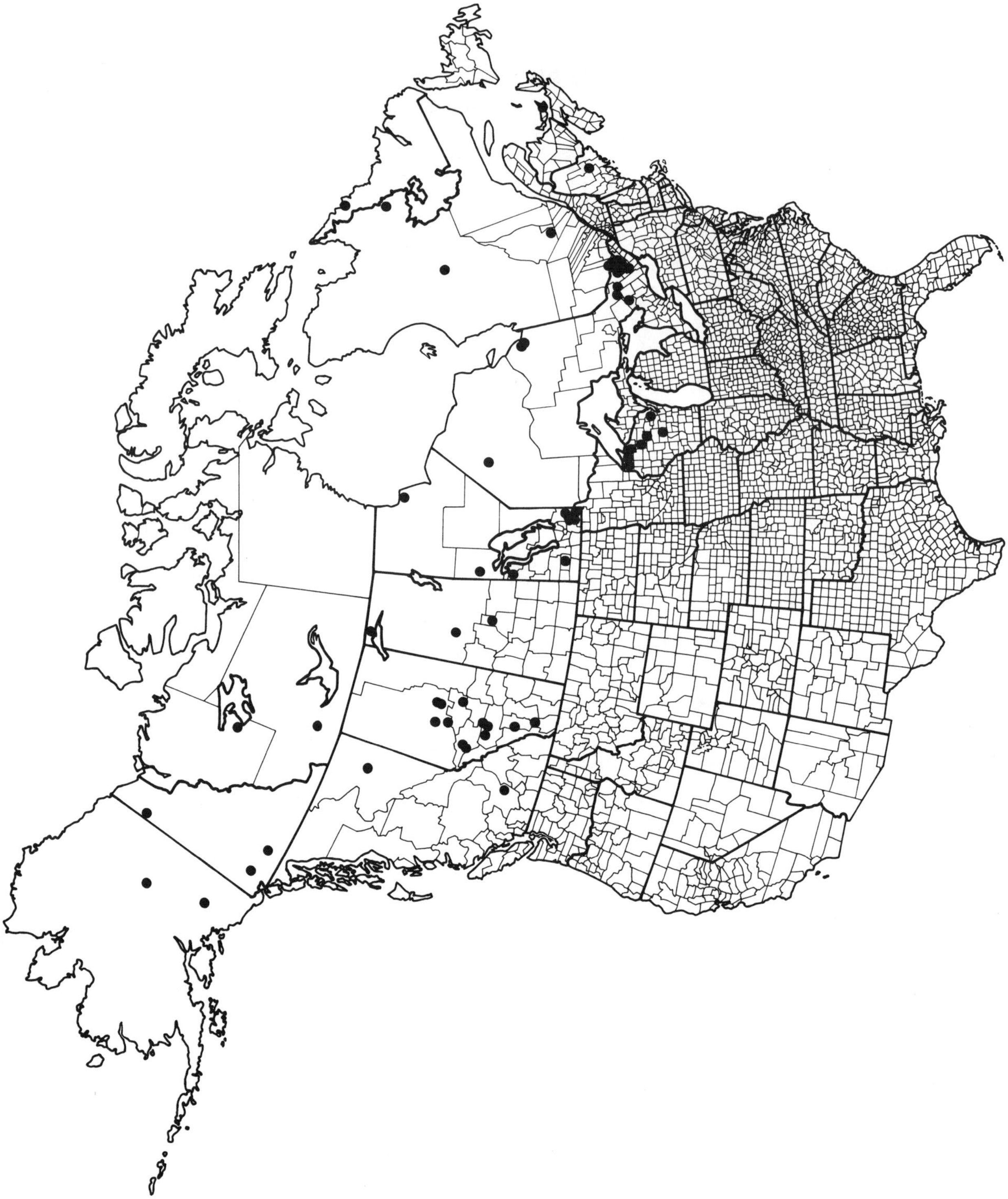

Map 15 *Helodon decemarticulatus*

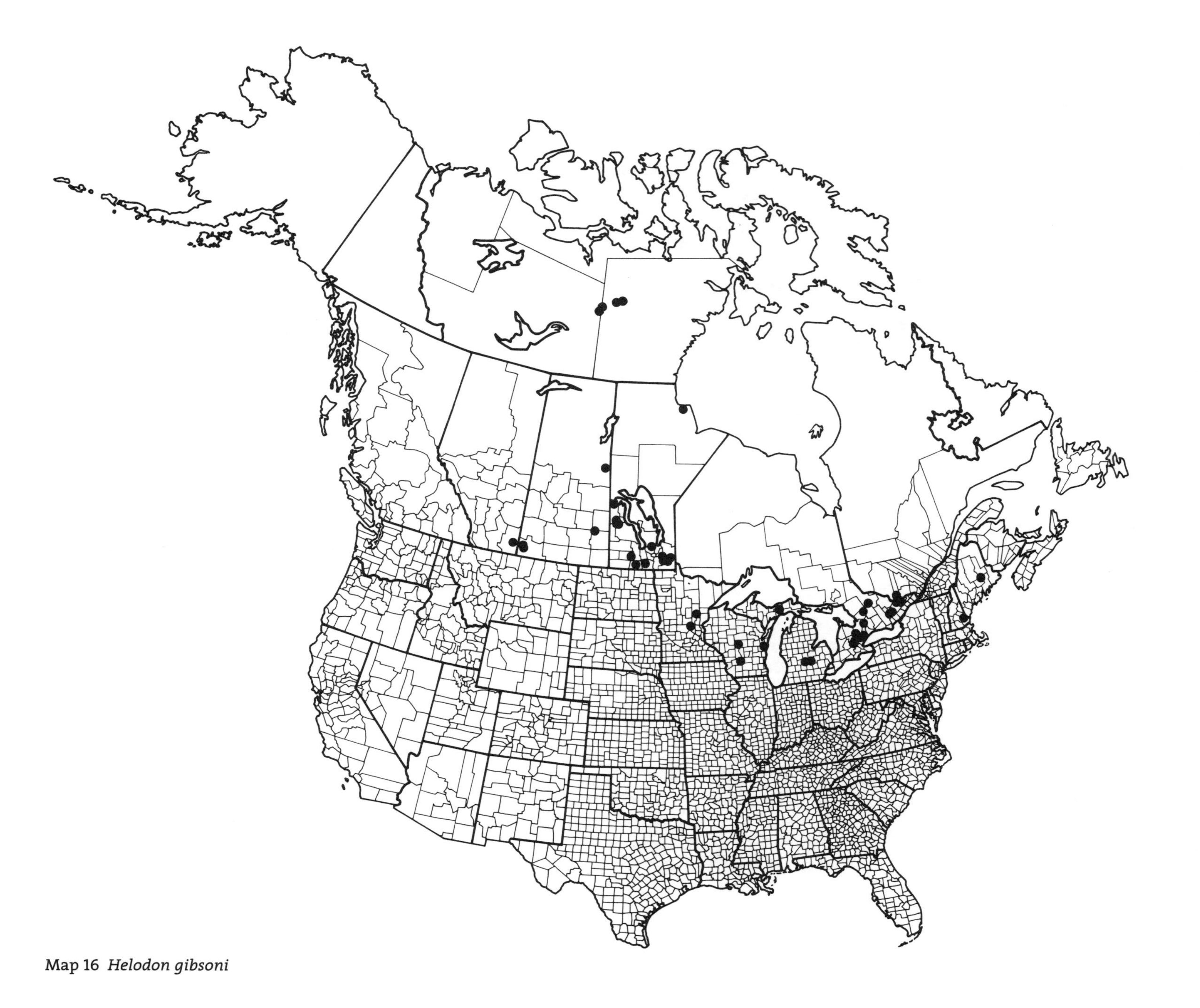

Map 16 *Helodon gibsoni*

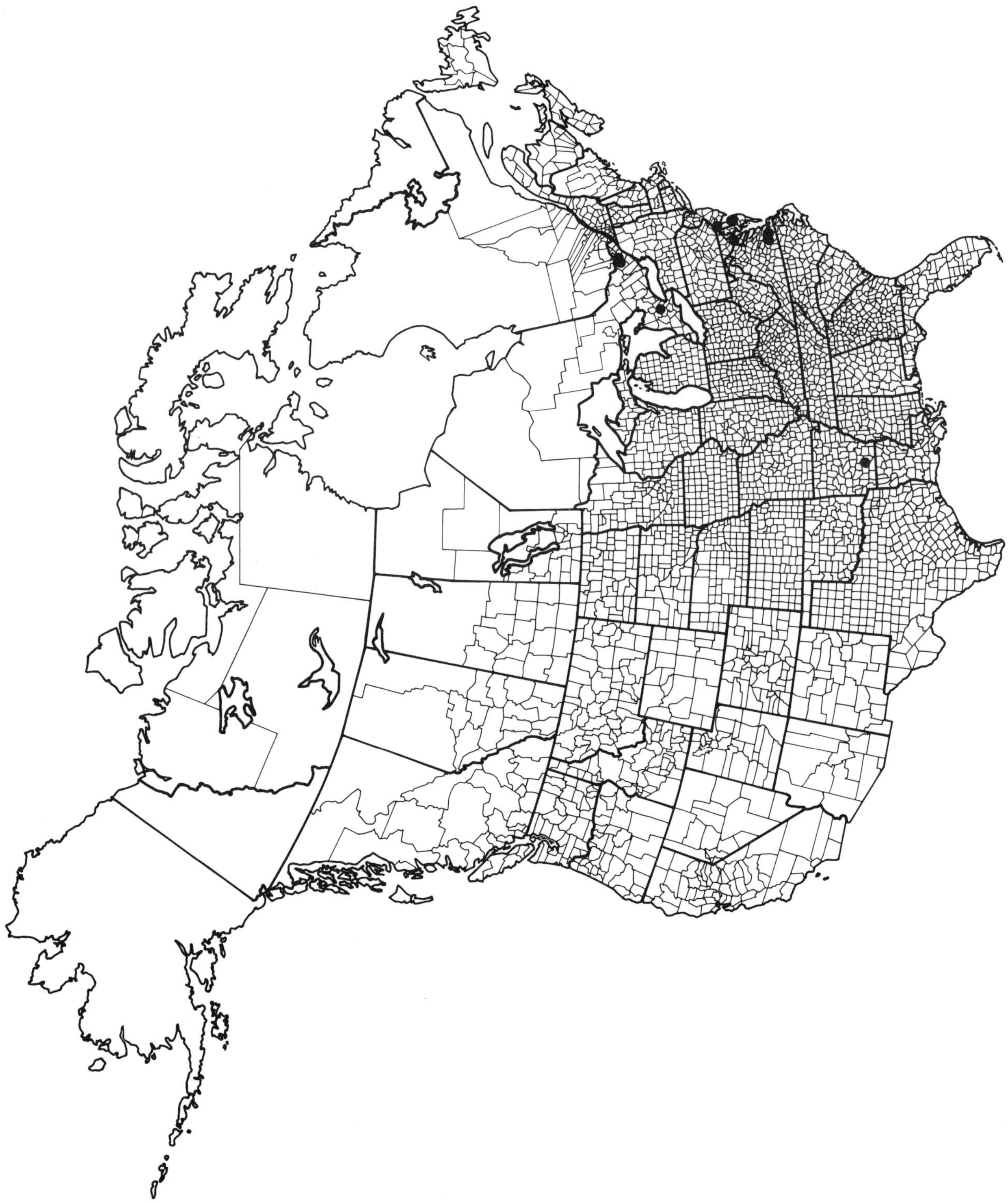

Map 17 *Helodon vernalis*

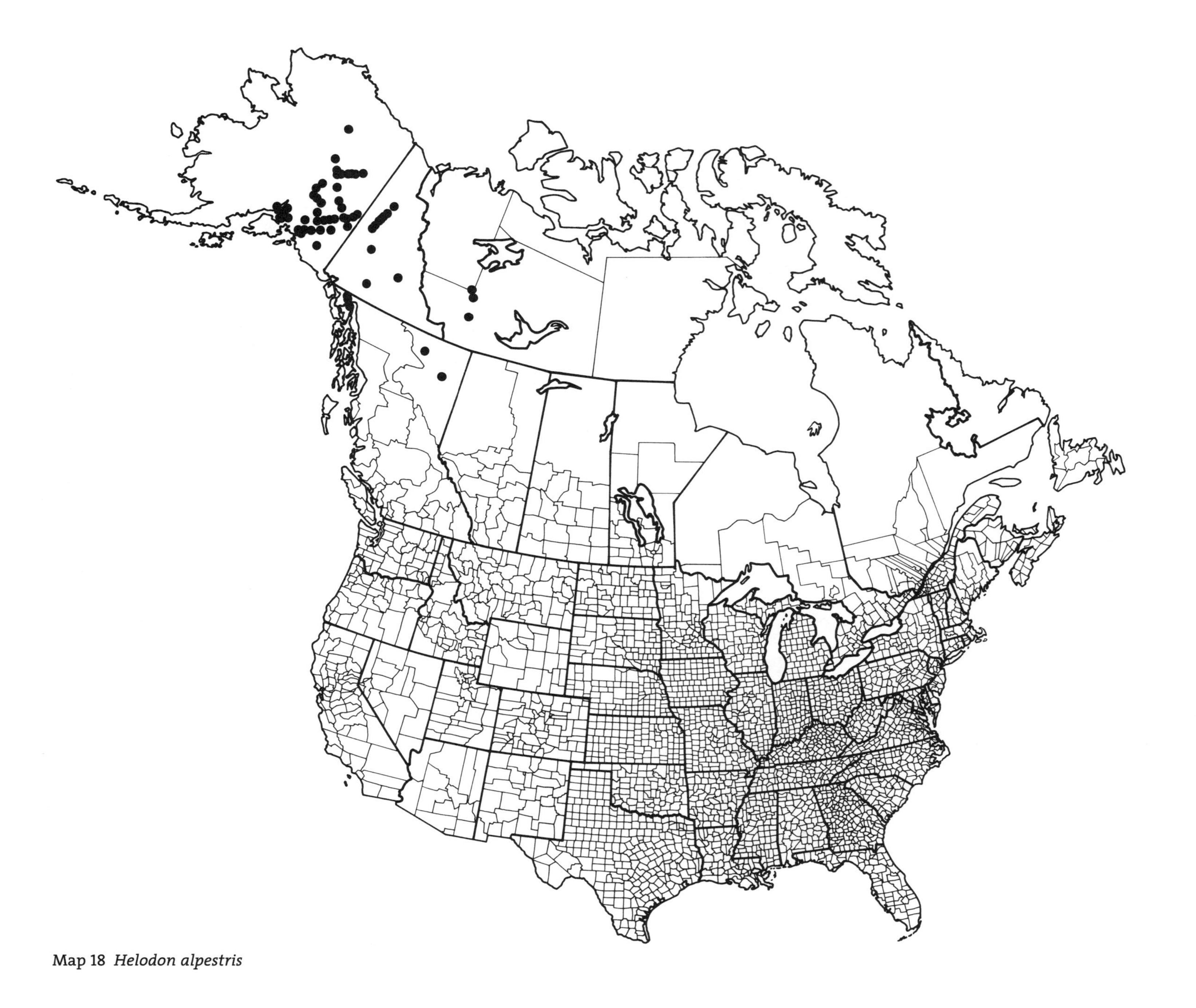

Map 18 *Helodon alpestris*

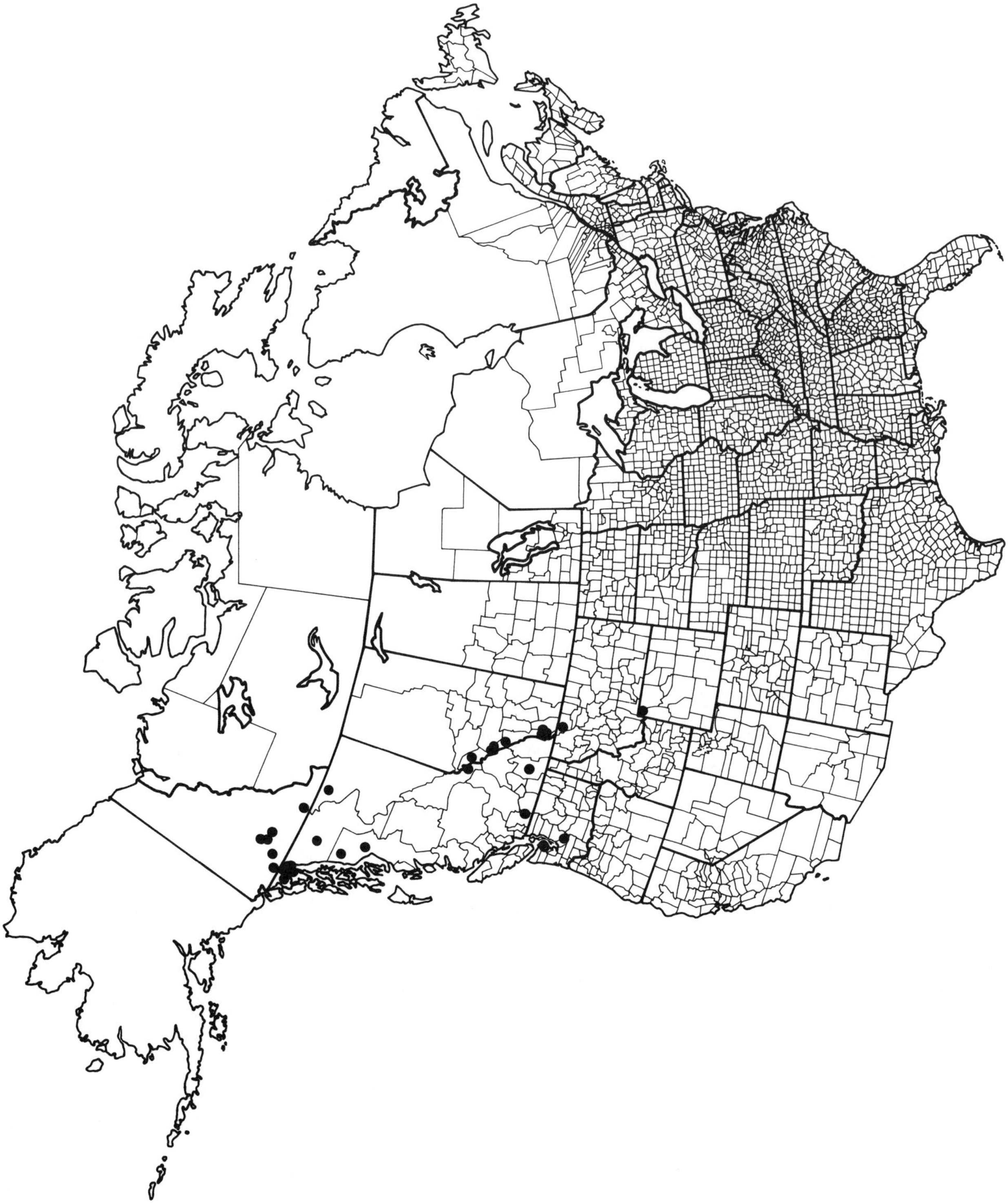

Map 19 *Helodon clavatus*

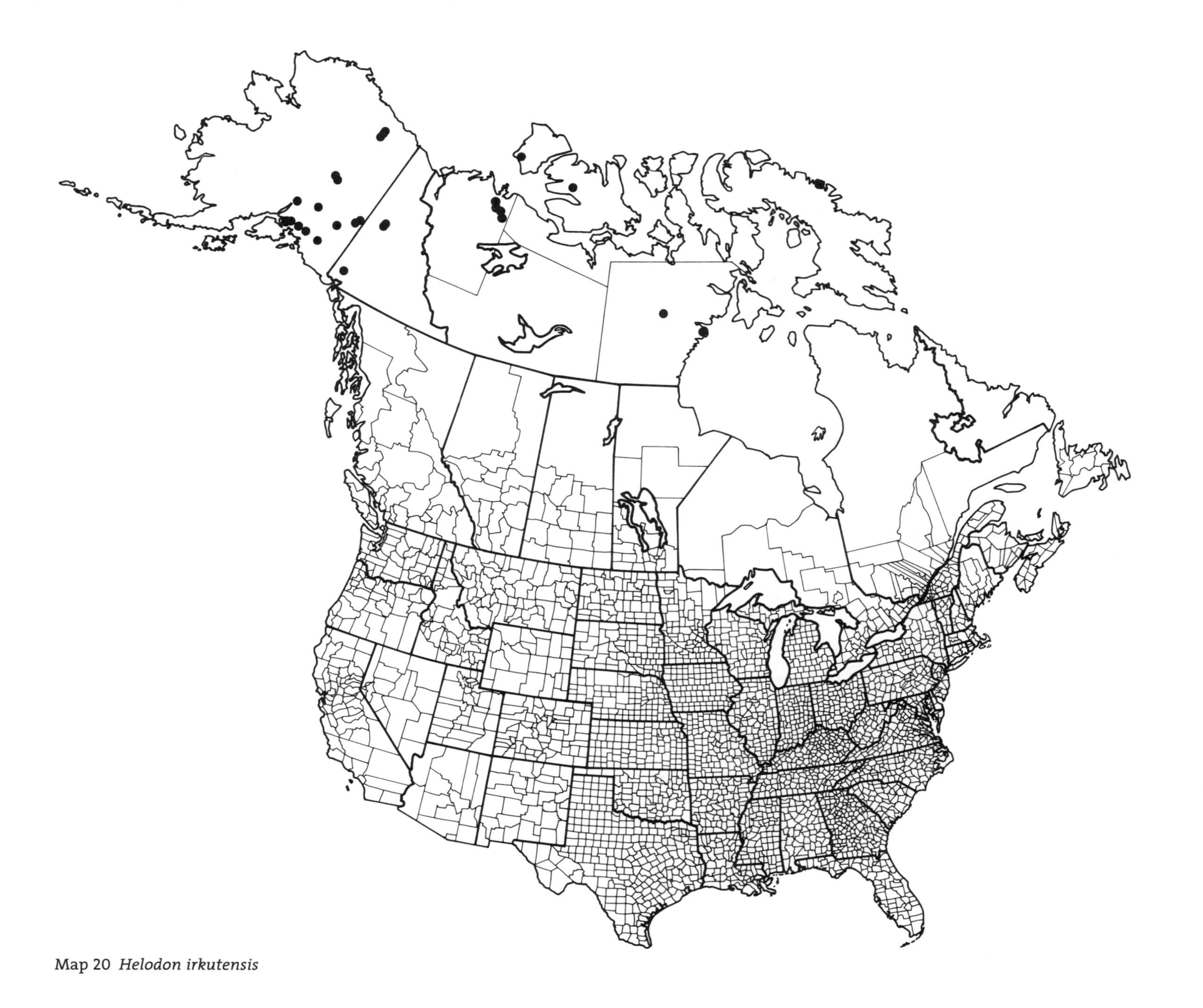

Map 20 *Helodon irkutensis*

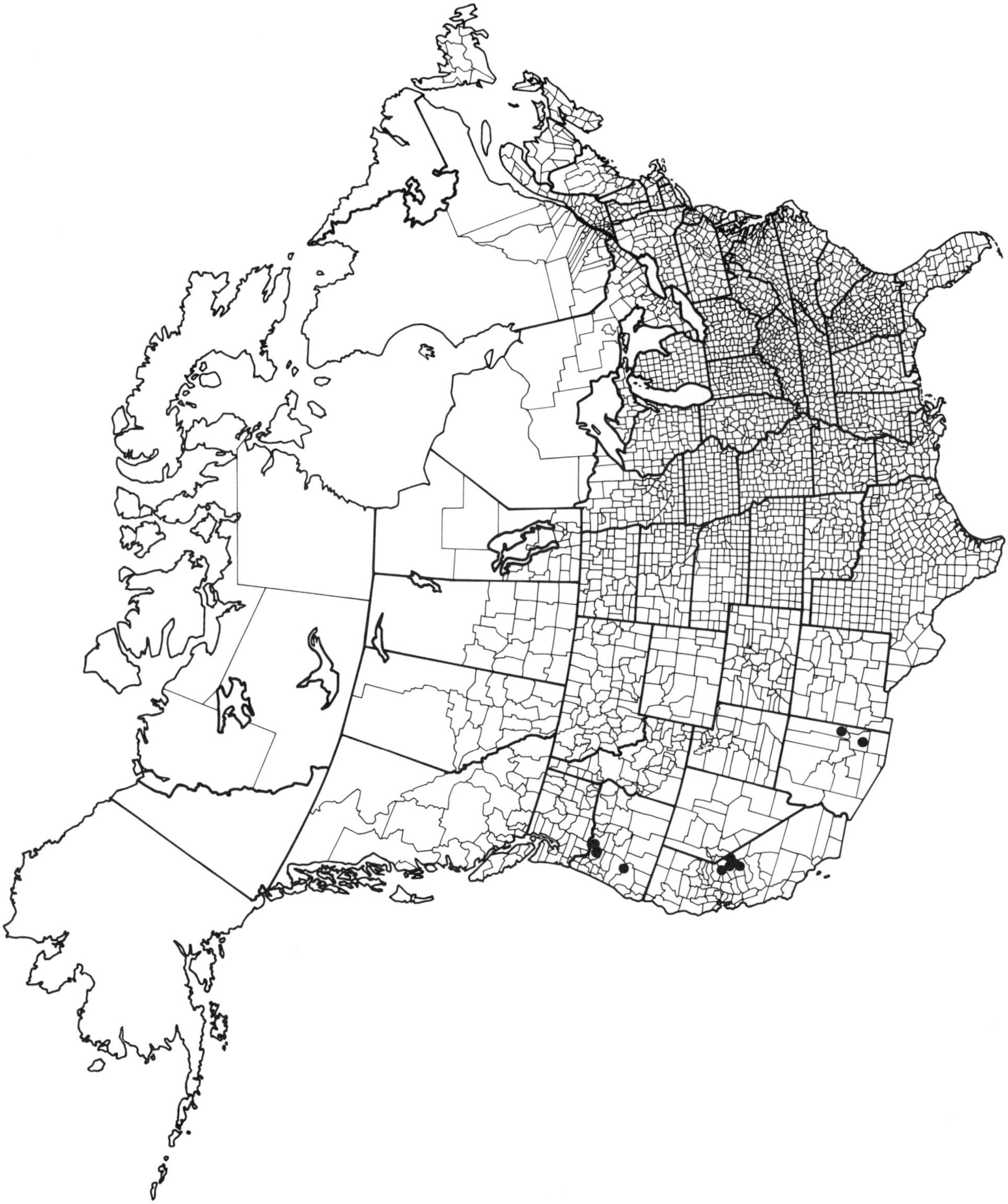

Map 21 *Helodon beardi* n. sp.

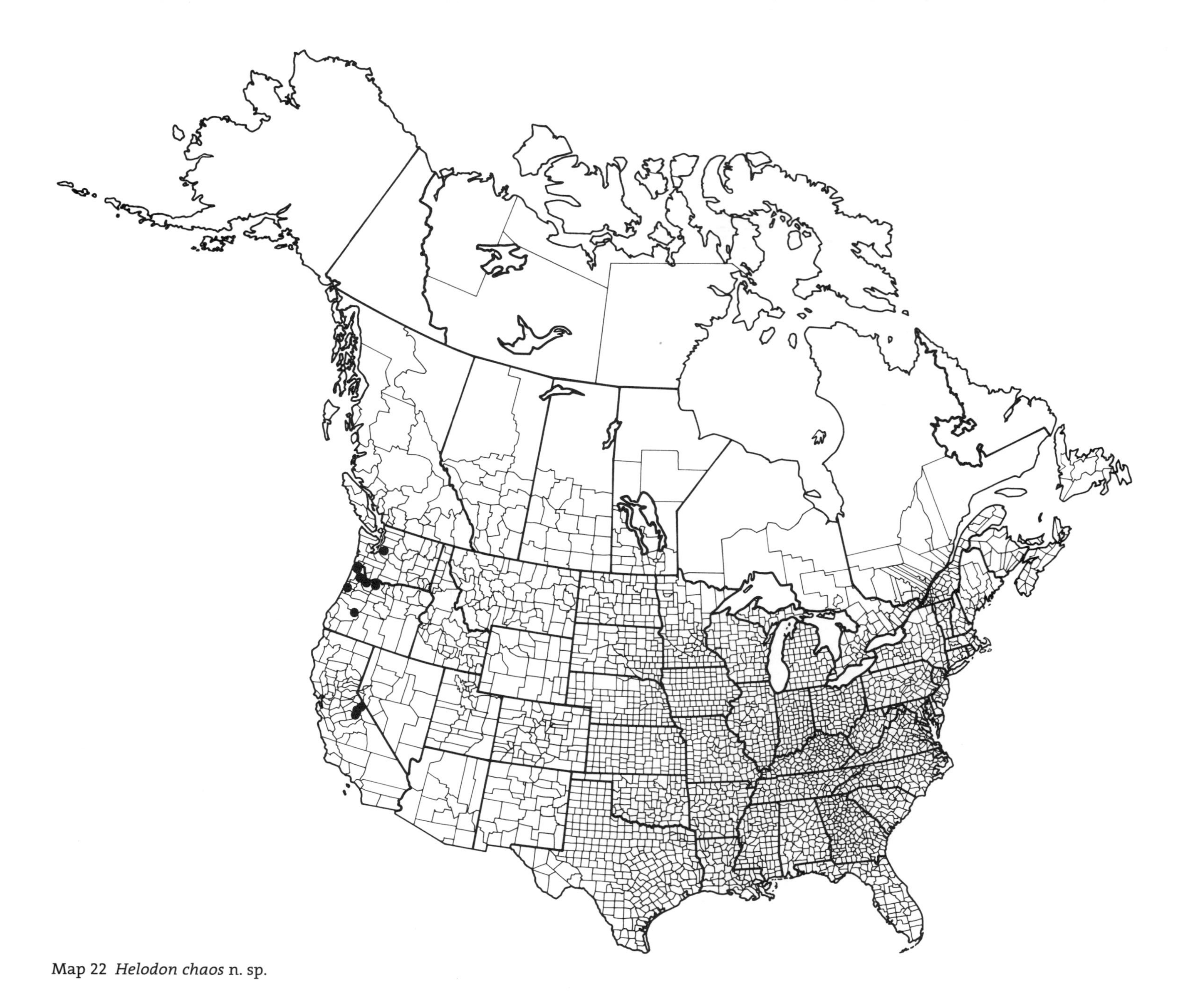

Map 22 *Helodon chaos* n. sp.

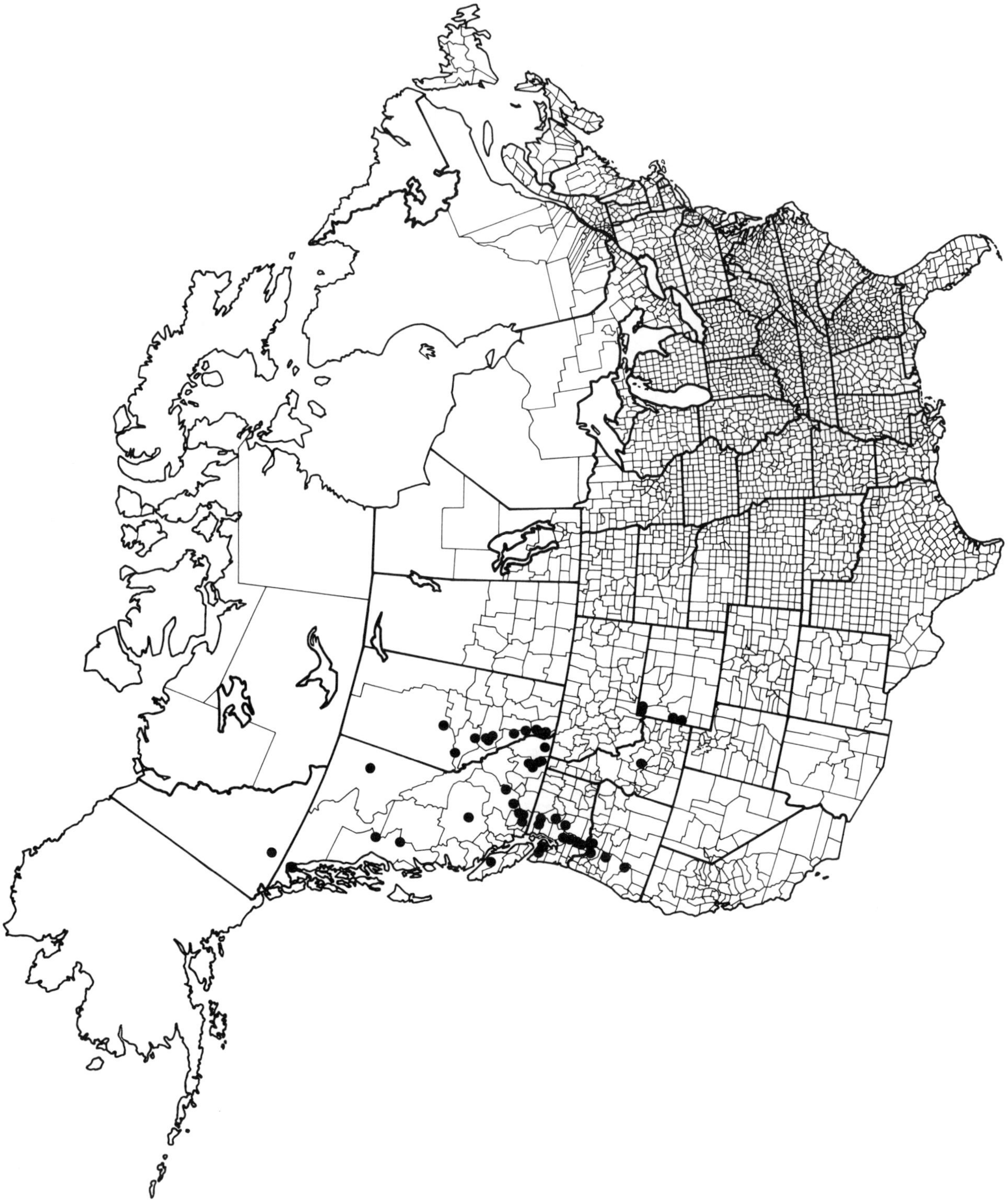

Map 23 *Helodon diadelphus* n. sp.

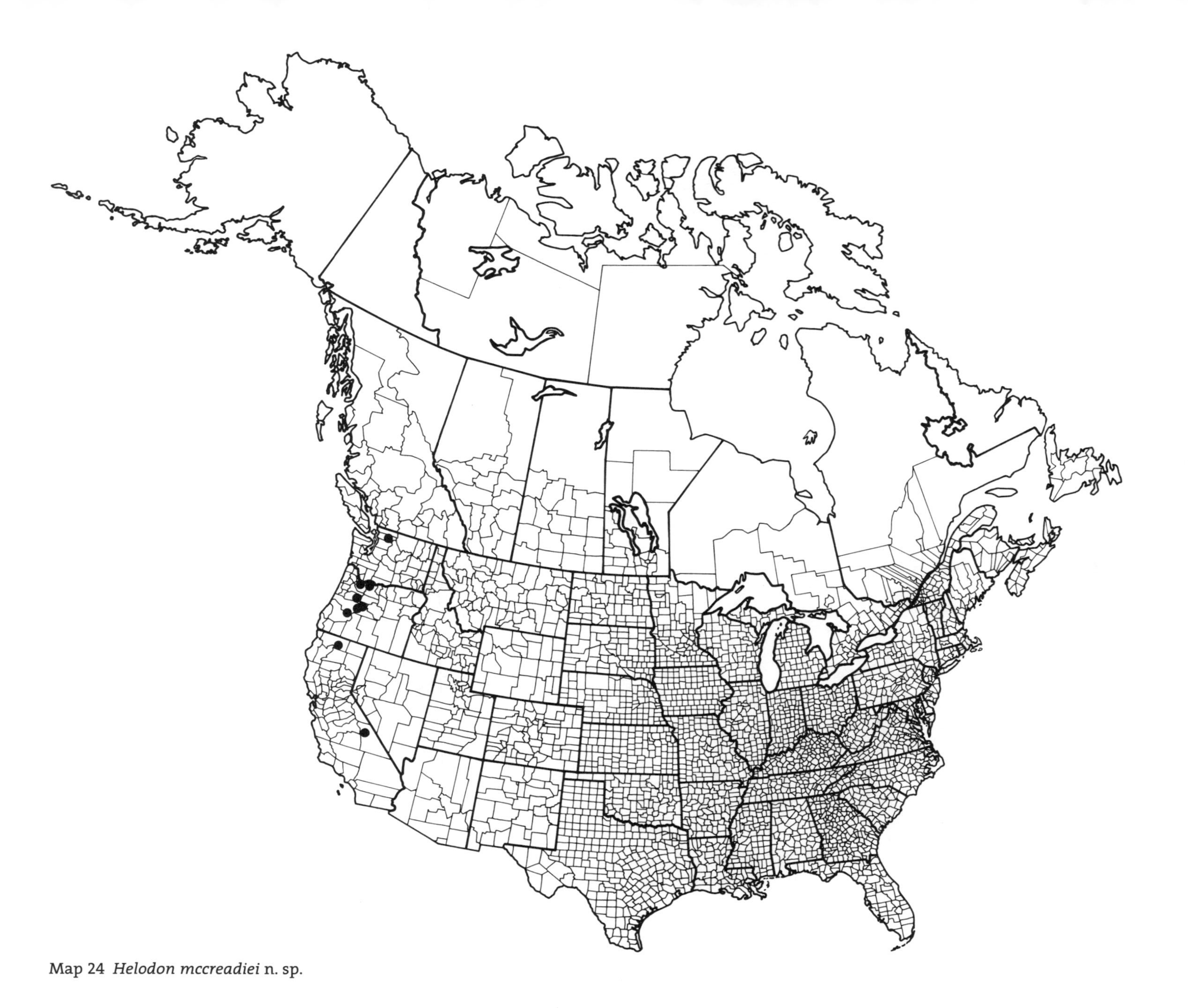

Map 24 *Helodon mccreadiei* n. sp.

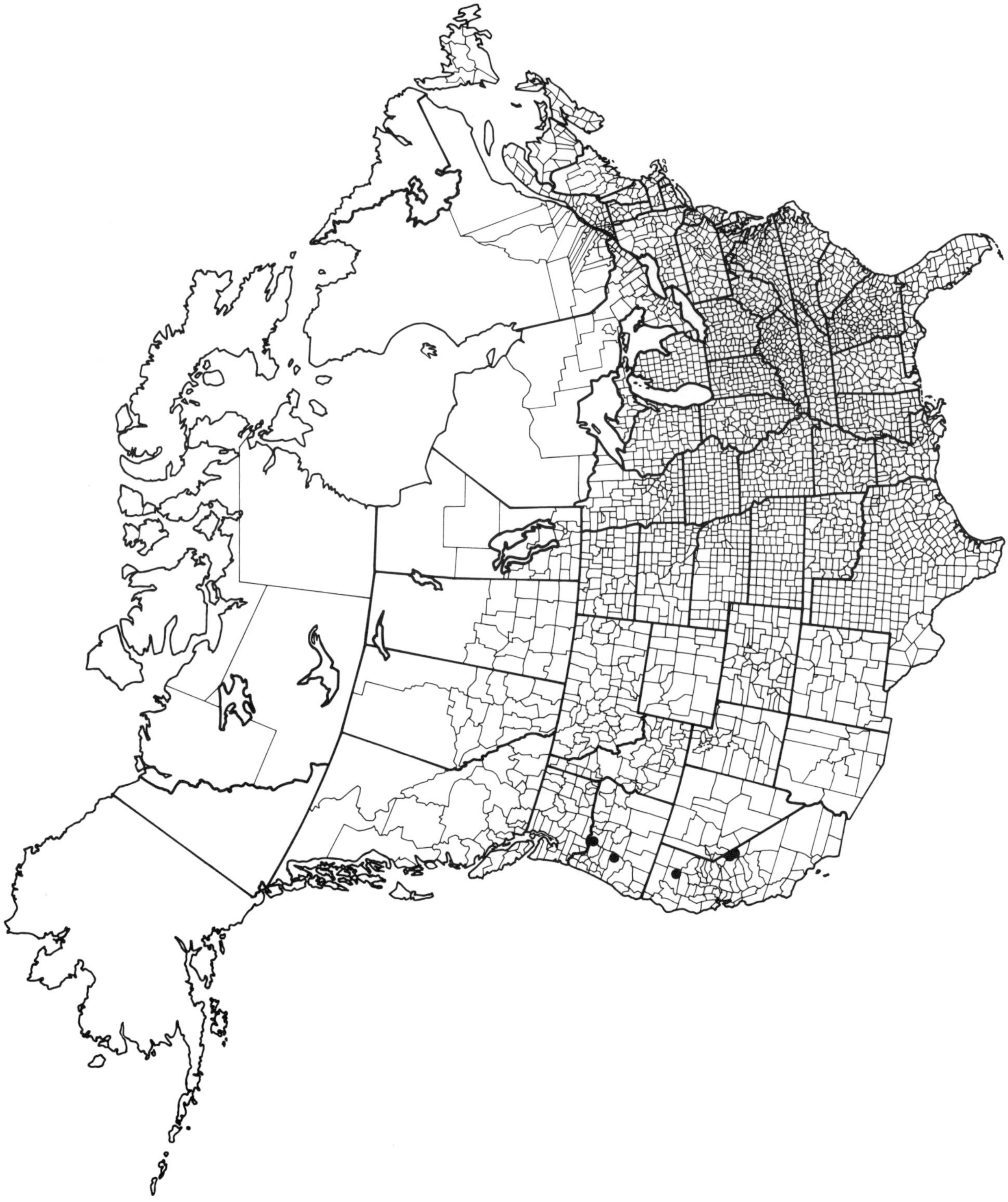

Map 25 *Helodon newmani* n. sp.

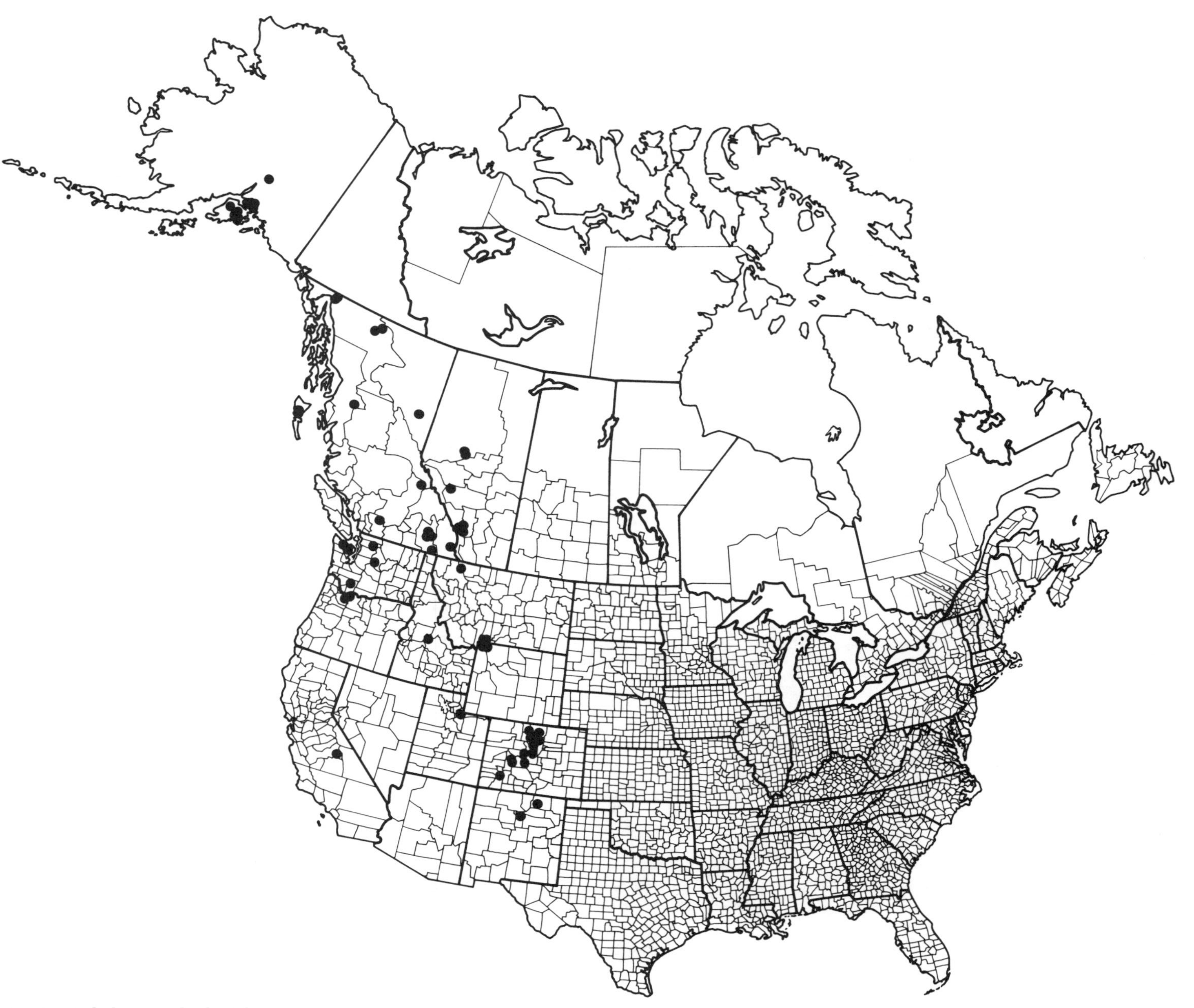

Map 26 *Helodon onychodactylus* s. s.

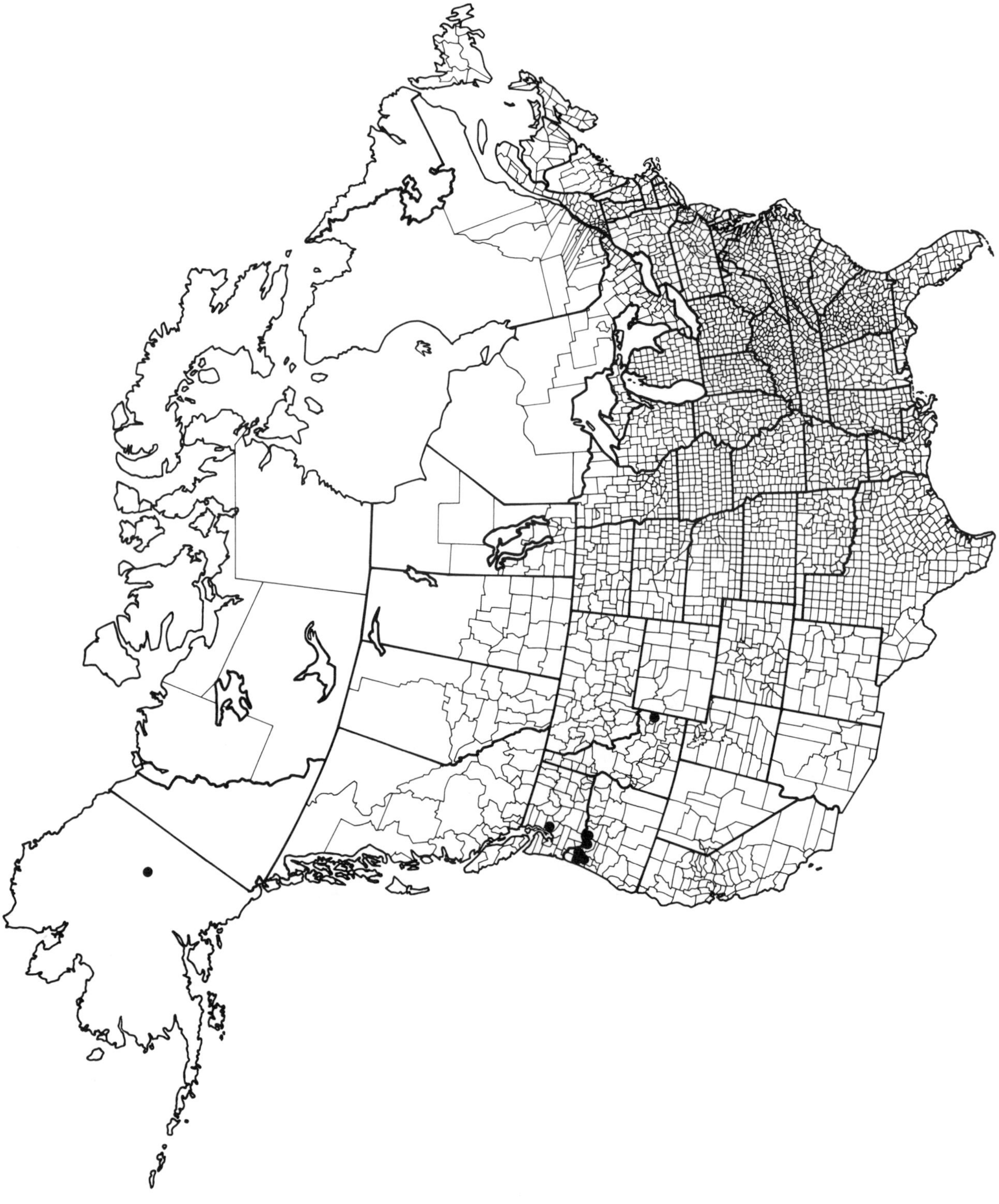

Map 27 *Helodon protus* n. sp.

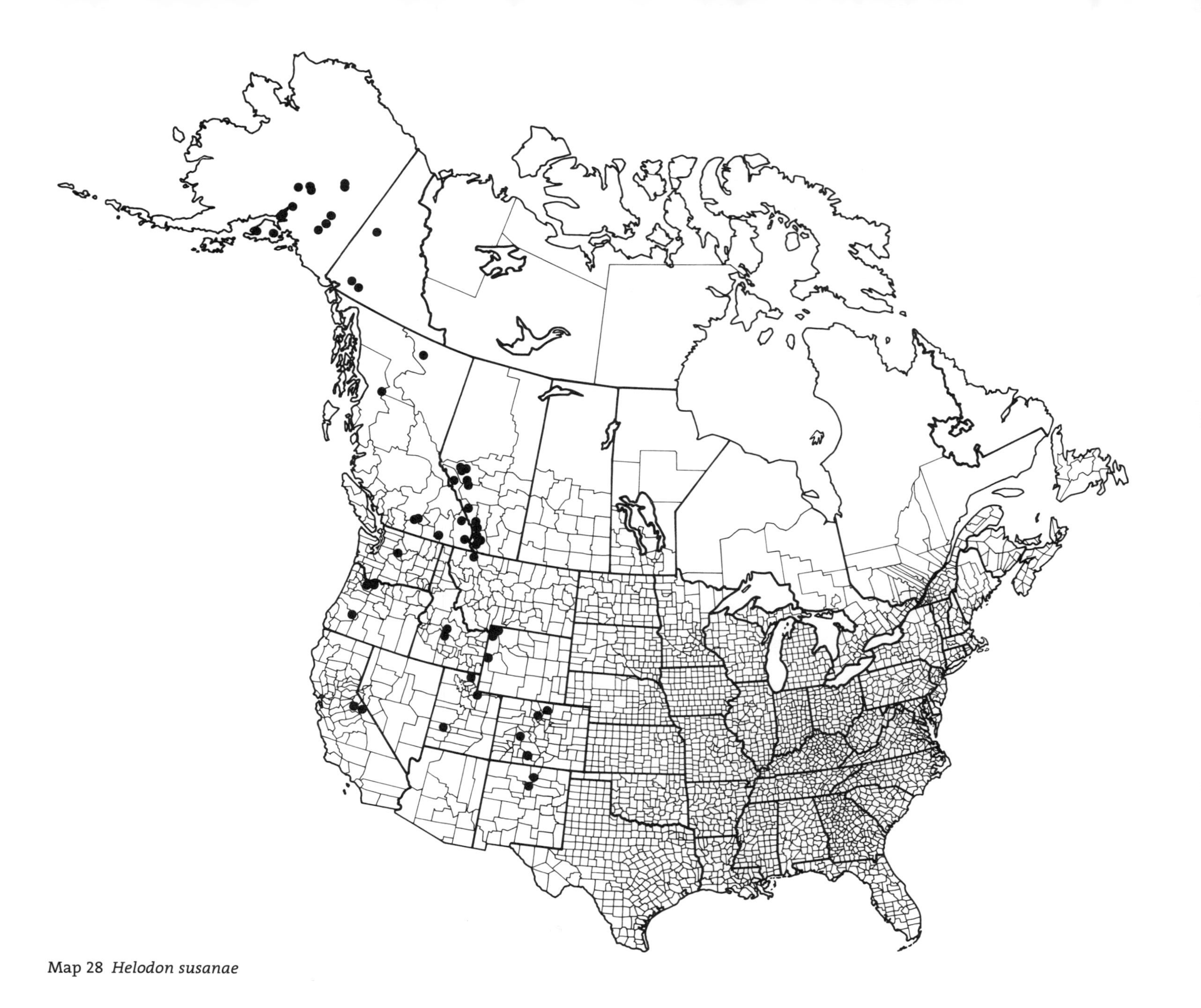

Map 28 *Helodon susanae*

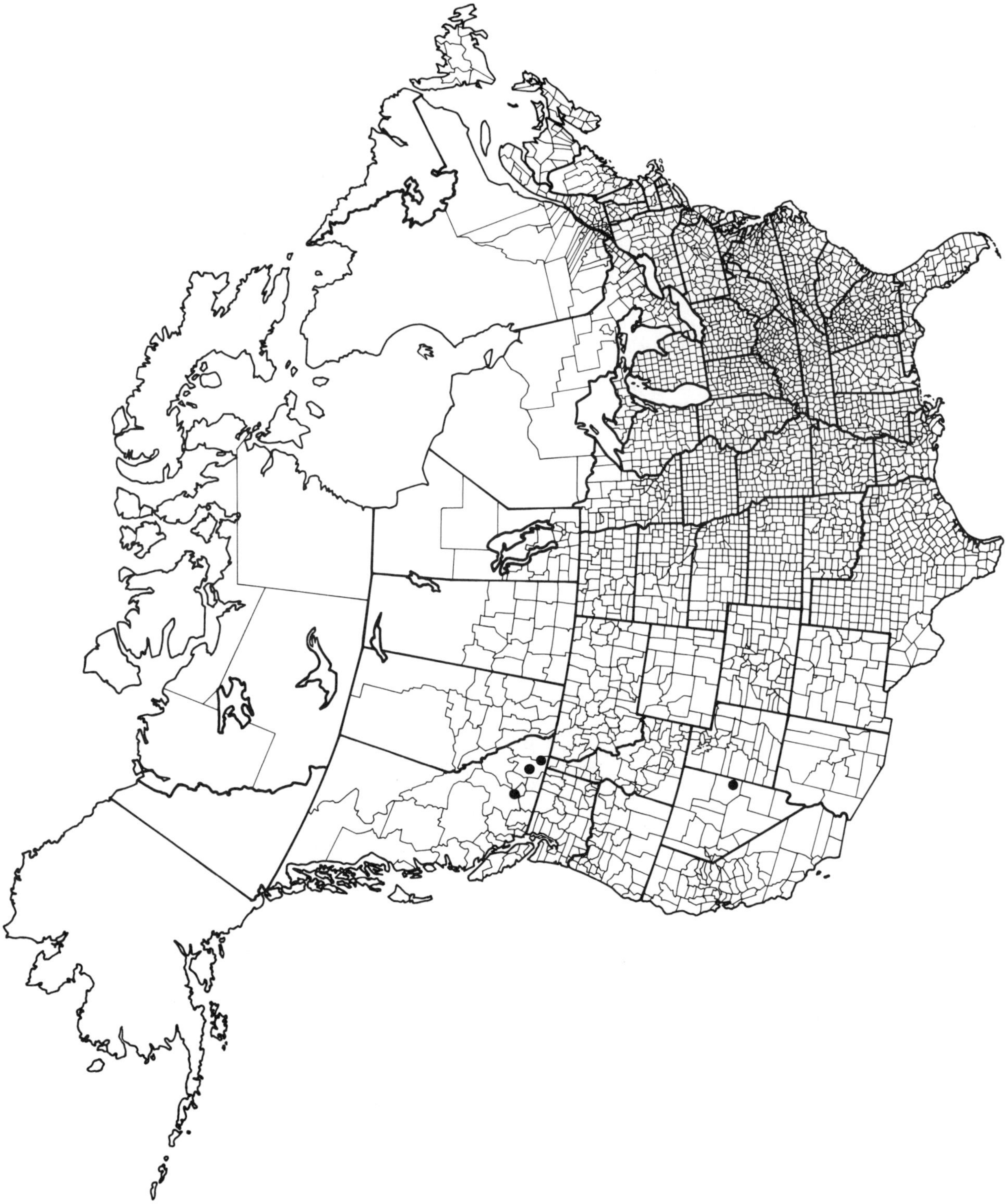

Map 29 *Helodon trochus* n. sp.

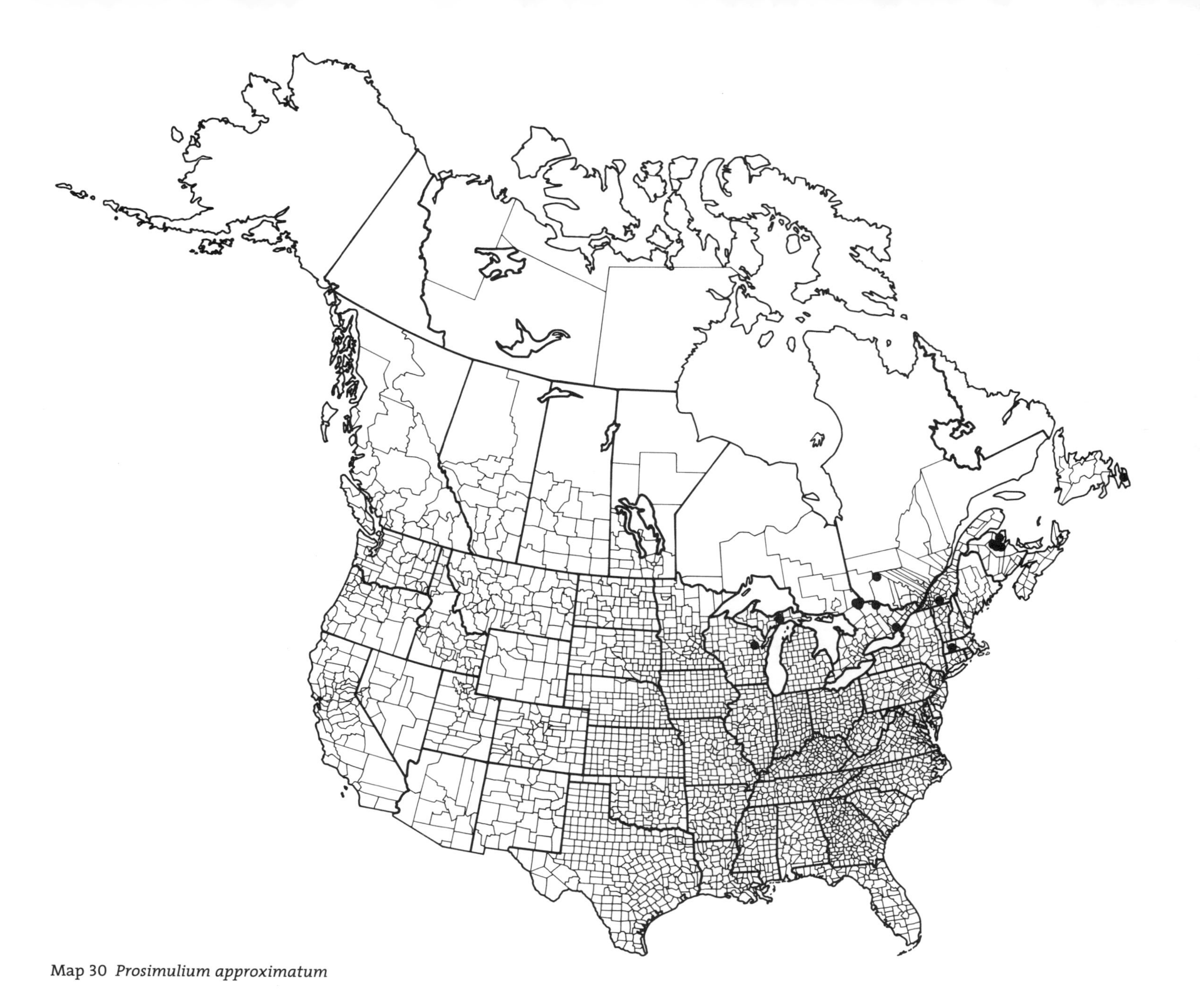

Map 30 *Prosimulium approximatum*

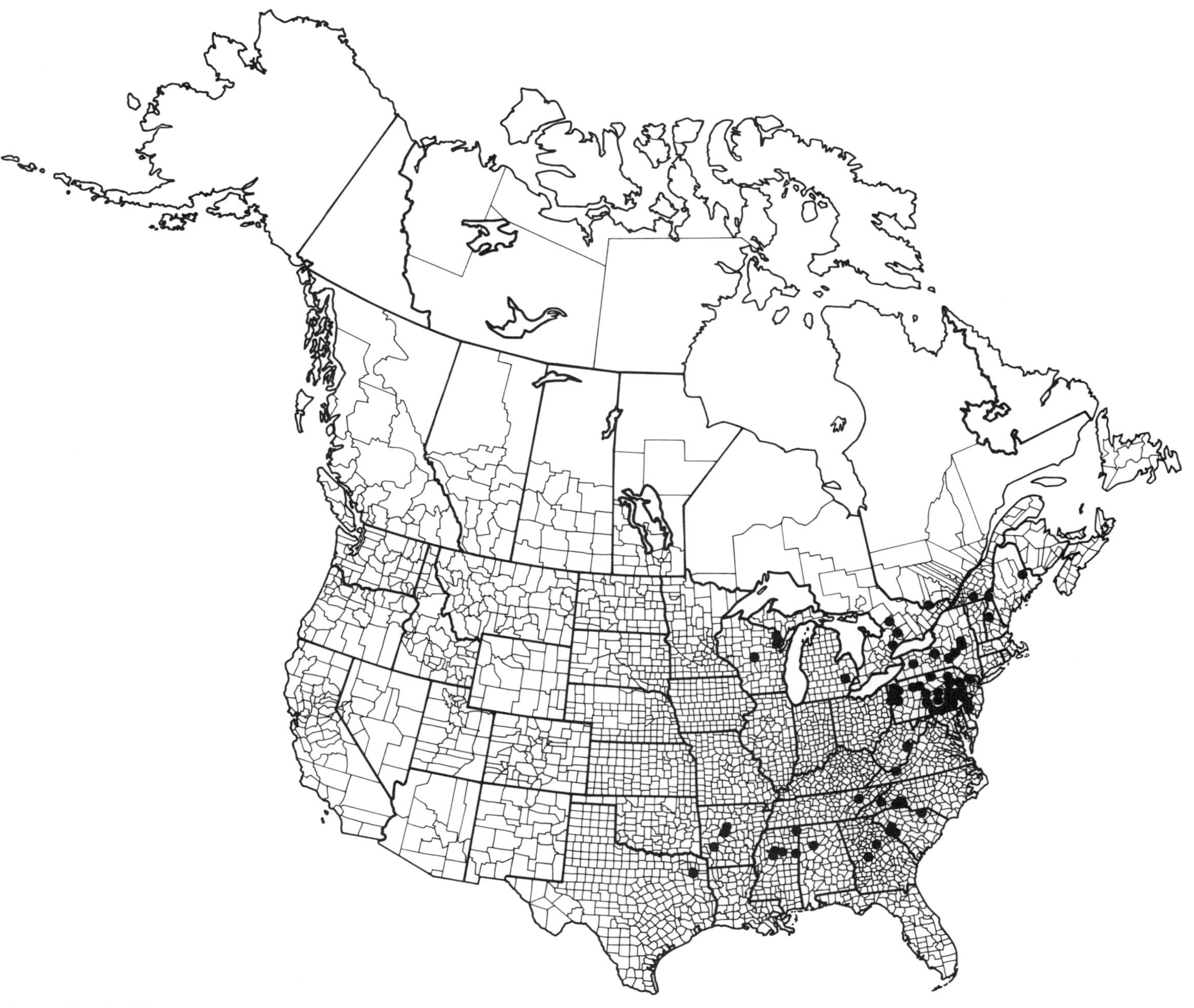

Map 31 *Prosimulium arvum*

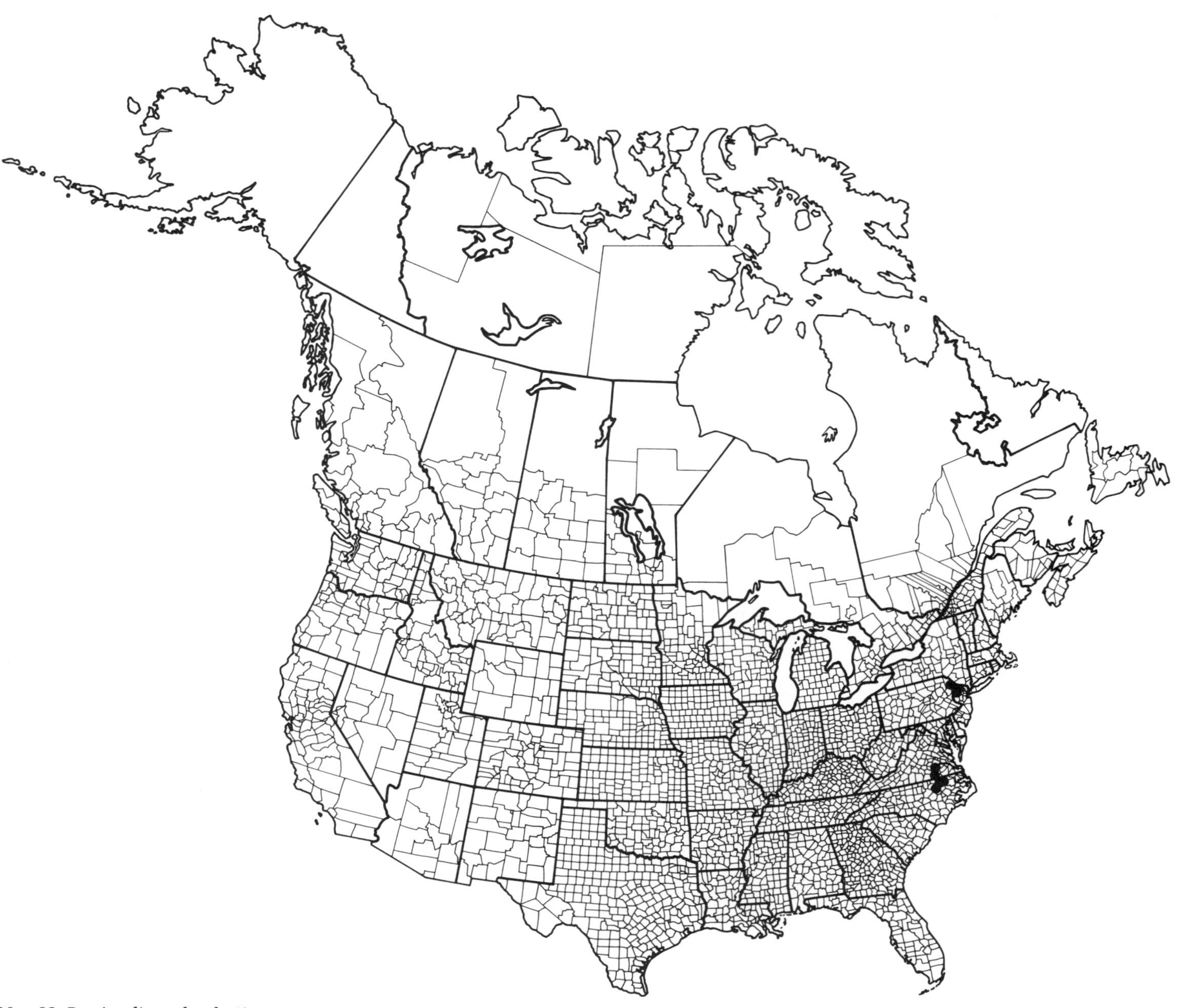

Map 32 *Prosimulium clandestinum* n. sp.

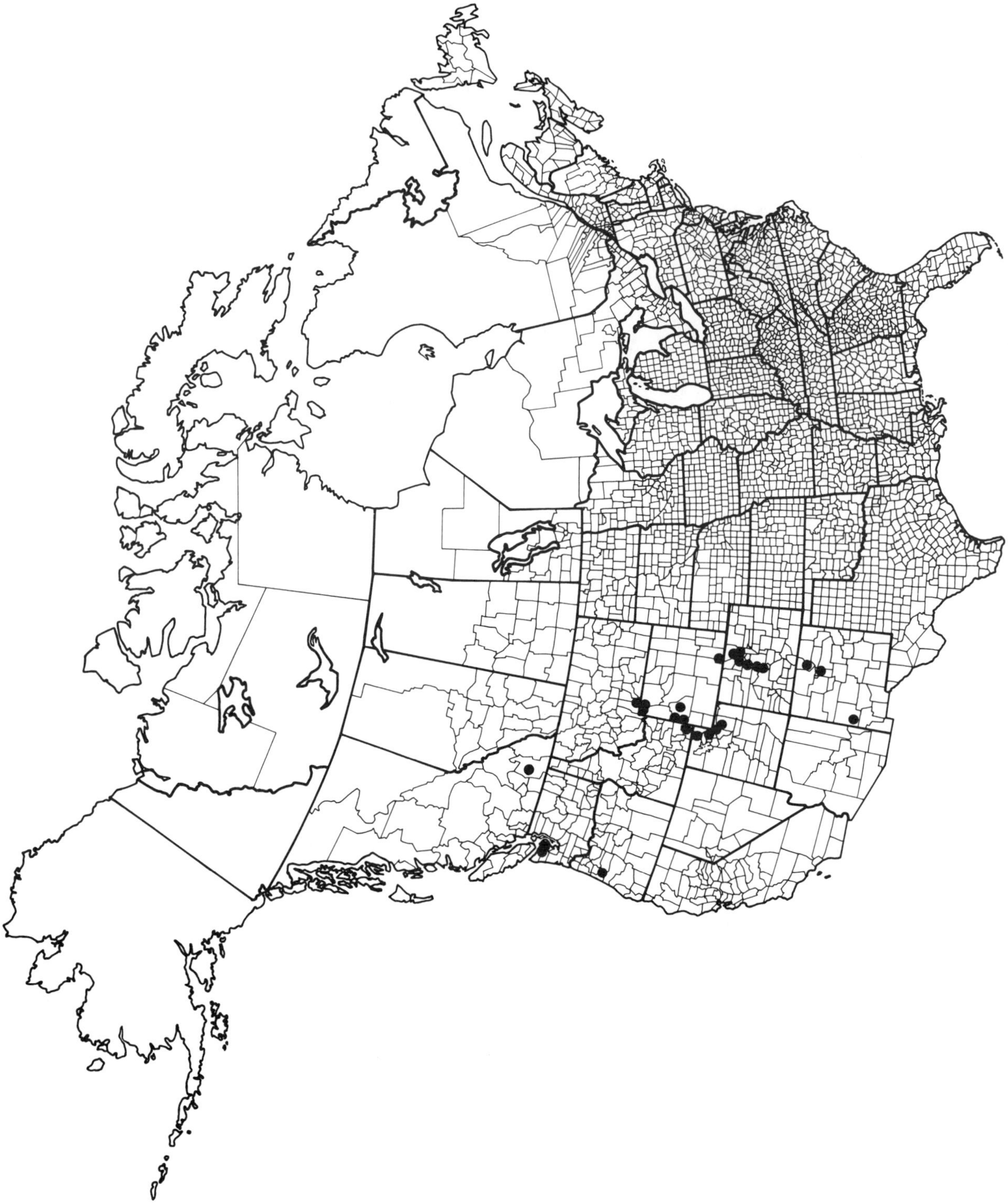

Map 33 *Prosimulium daviesi*

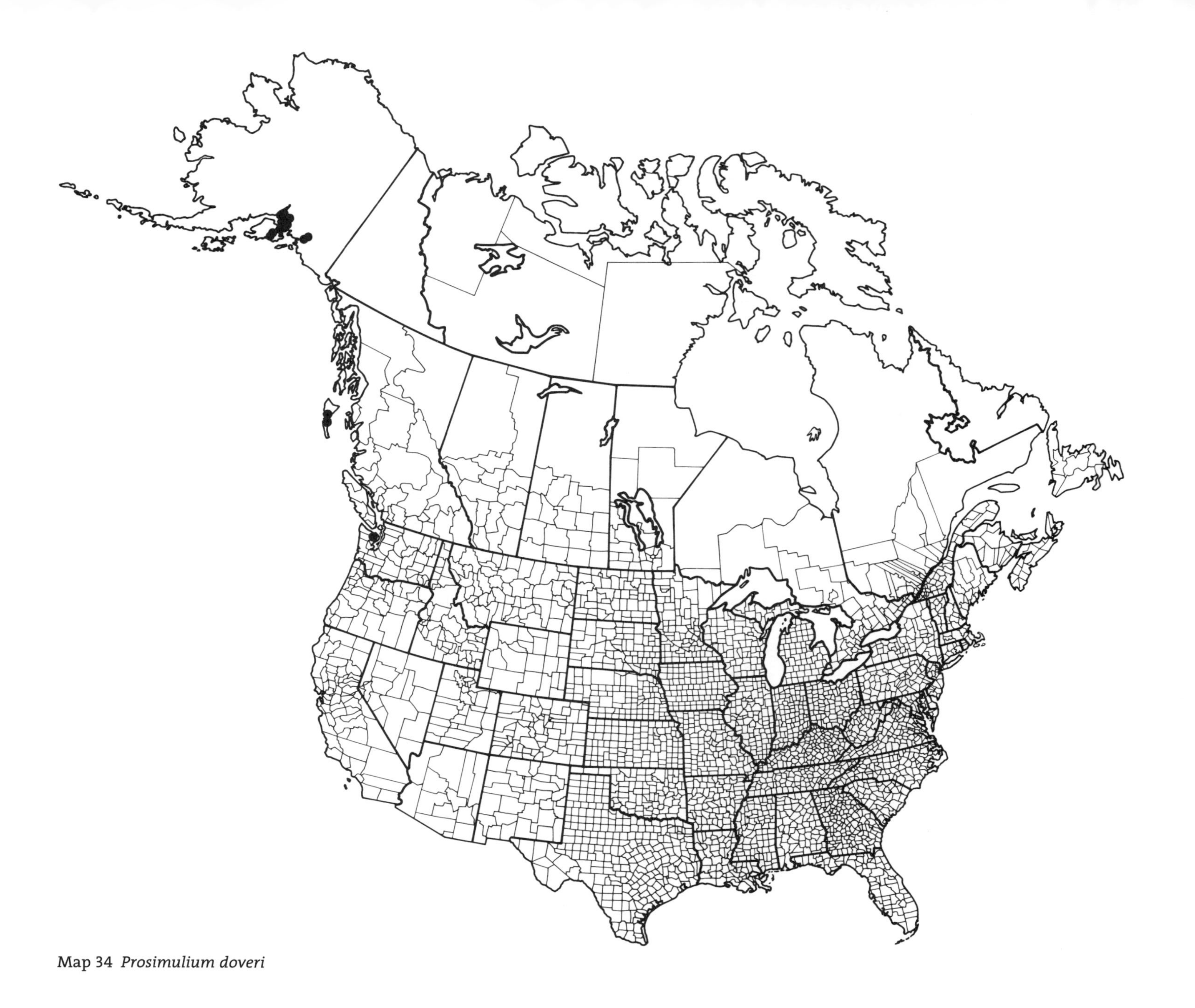

Map 34 *Prosimulium doveri*

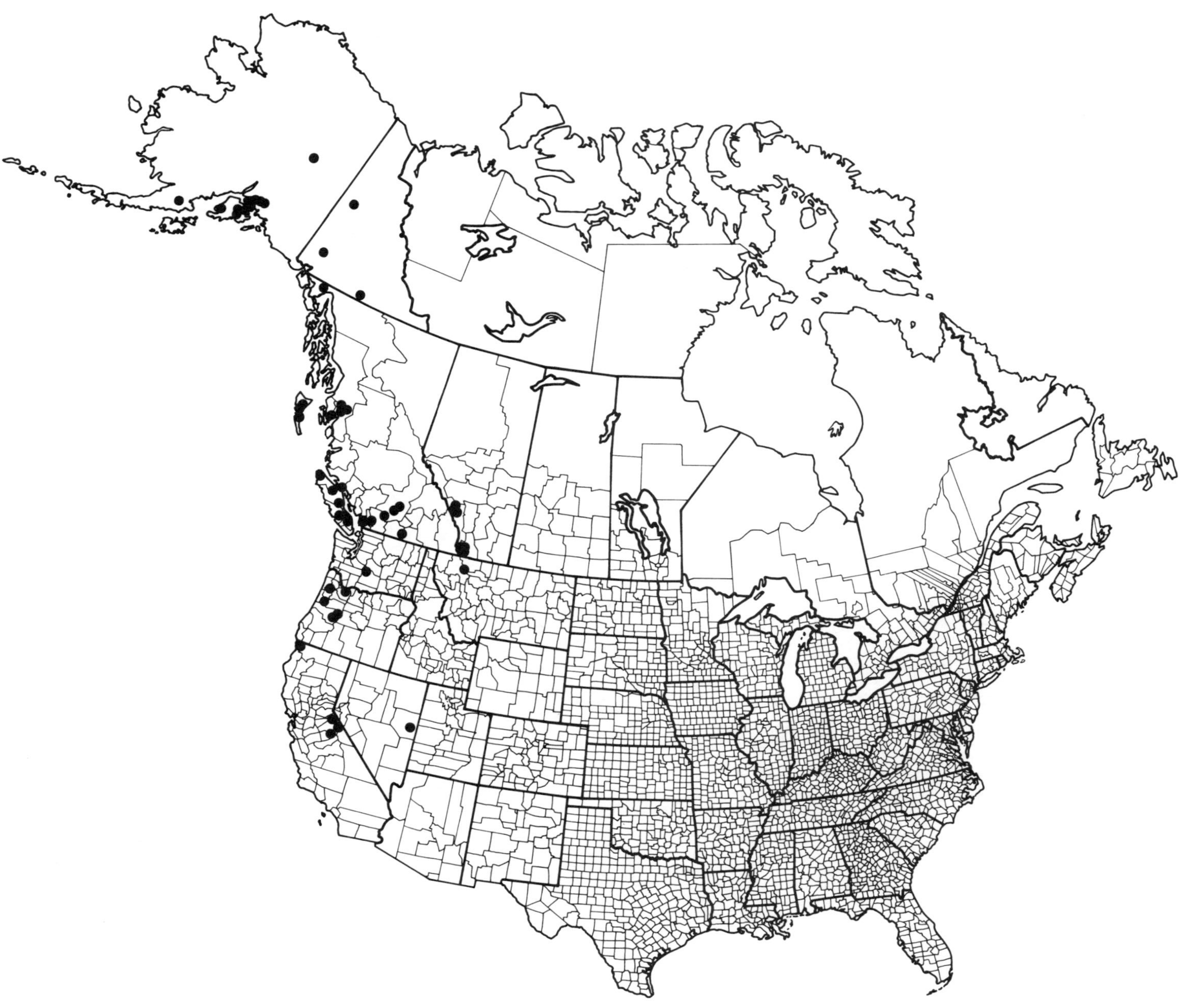

Map 35 *Prosimulium esselbaughi*

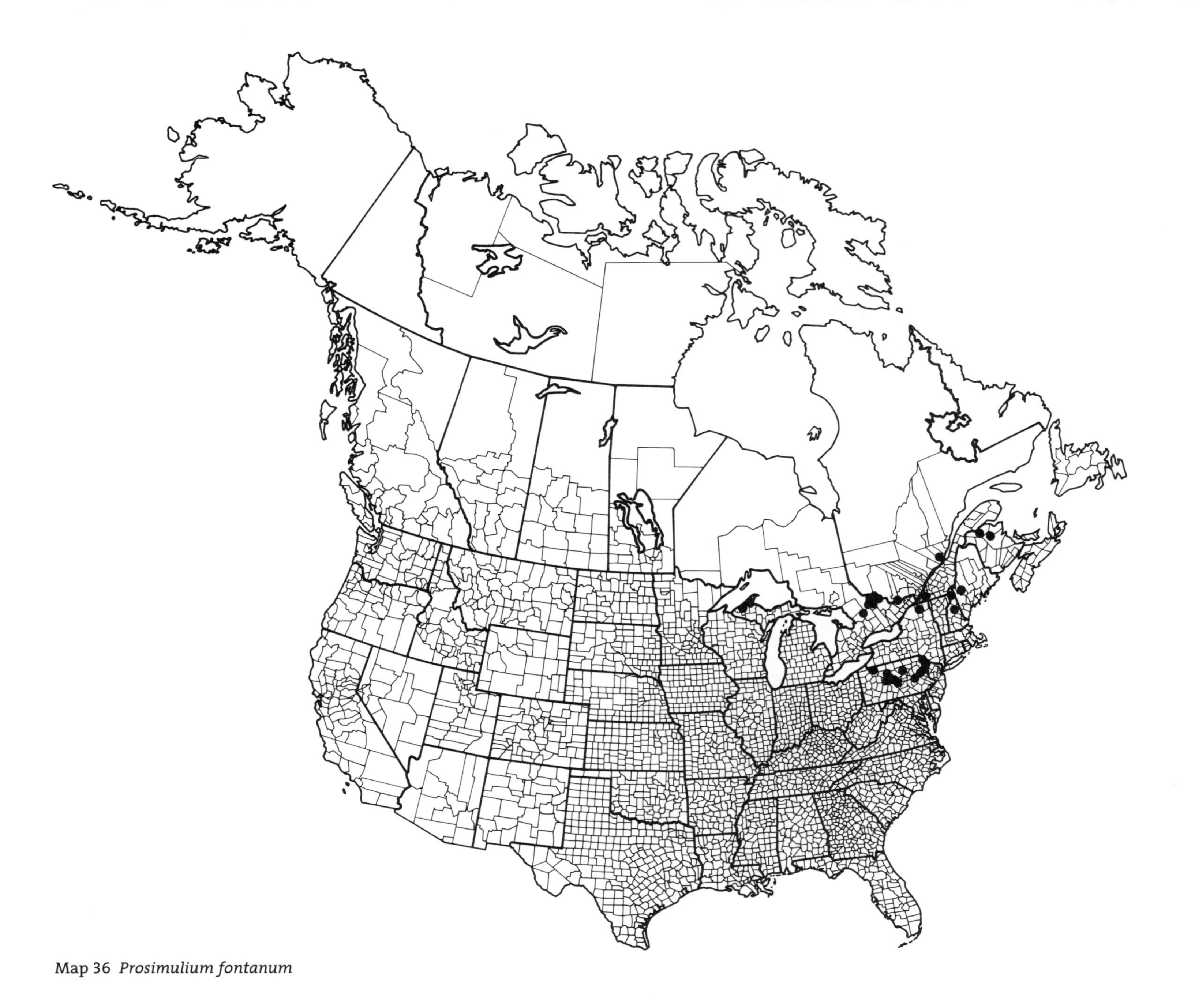

Map 36 *Prosimulium fontanum*

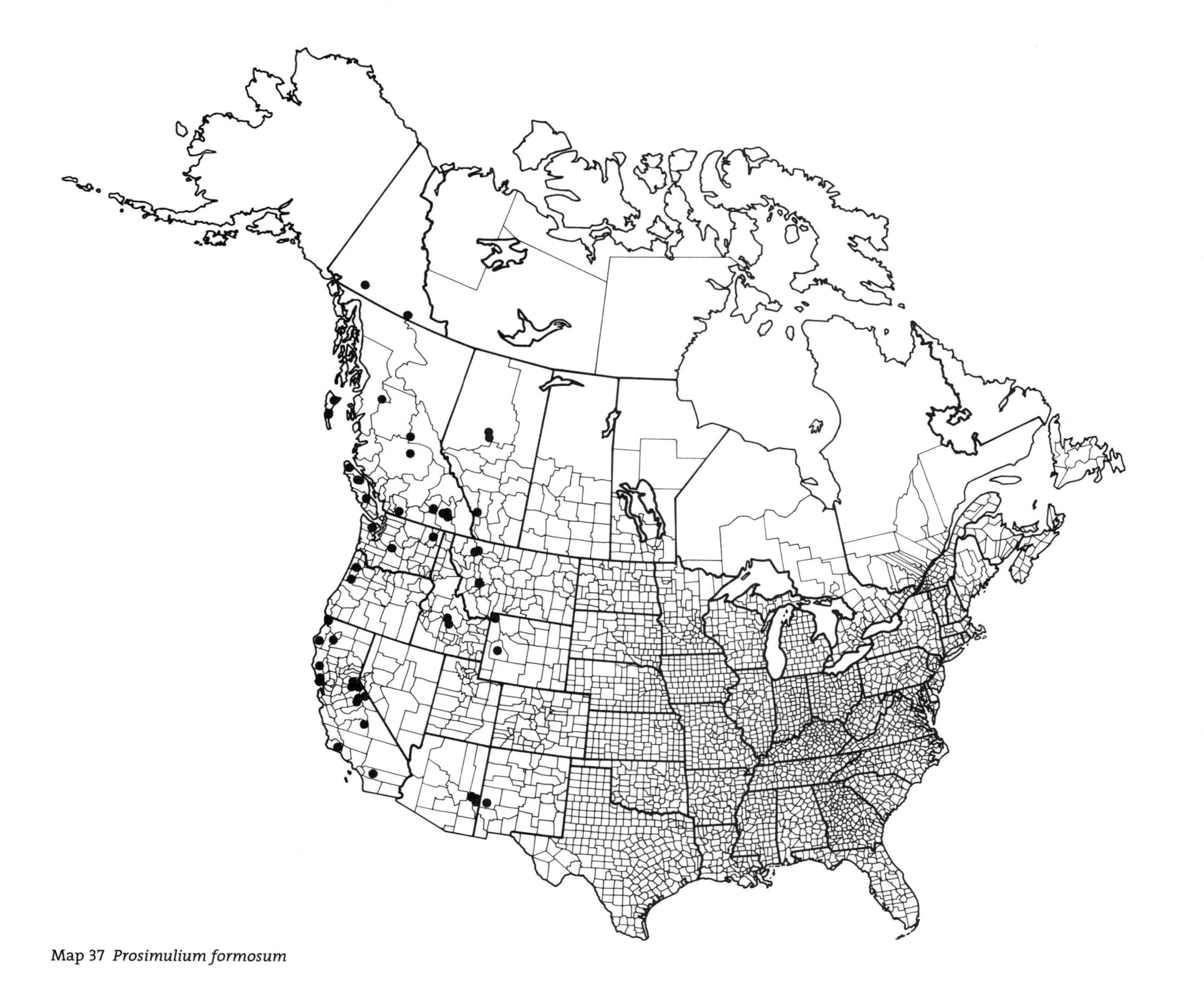

Map 37 *Prosimulium formosum*

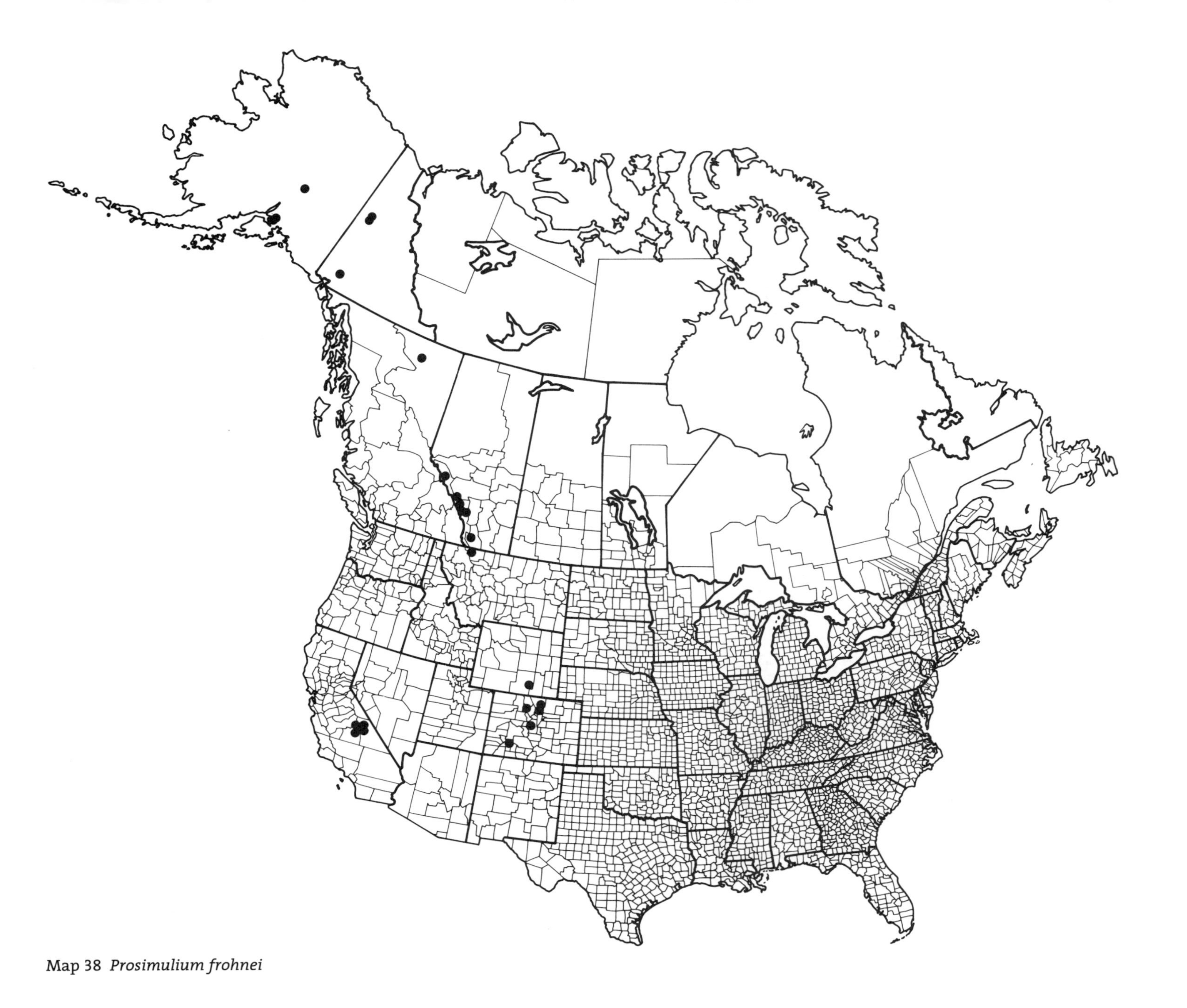

Map 38 *Prosimulium frohnei*

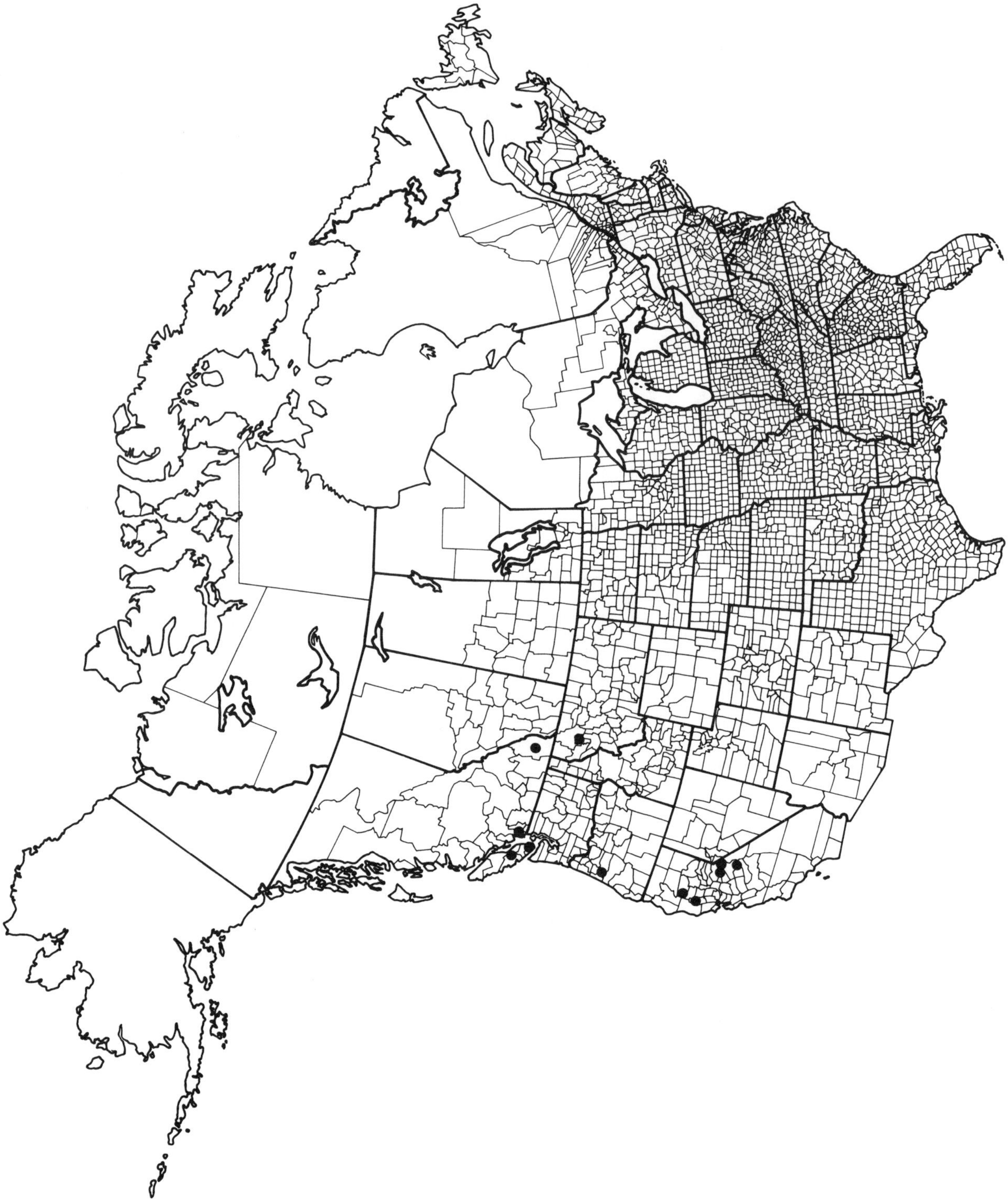

Map 39 *Prosimulium fulvithorax*

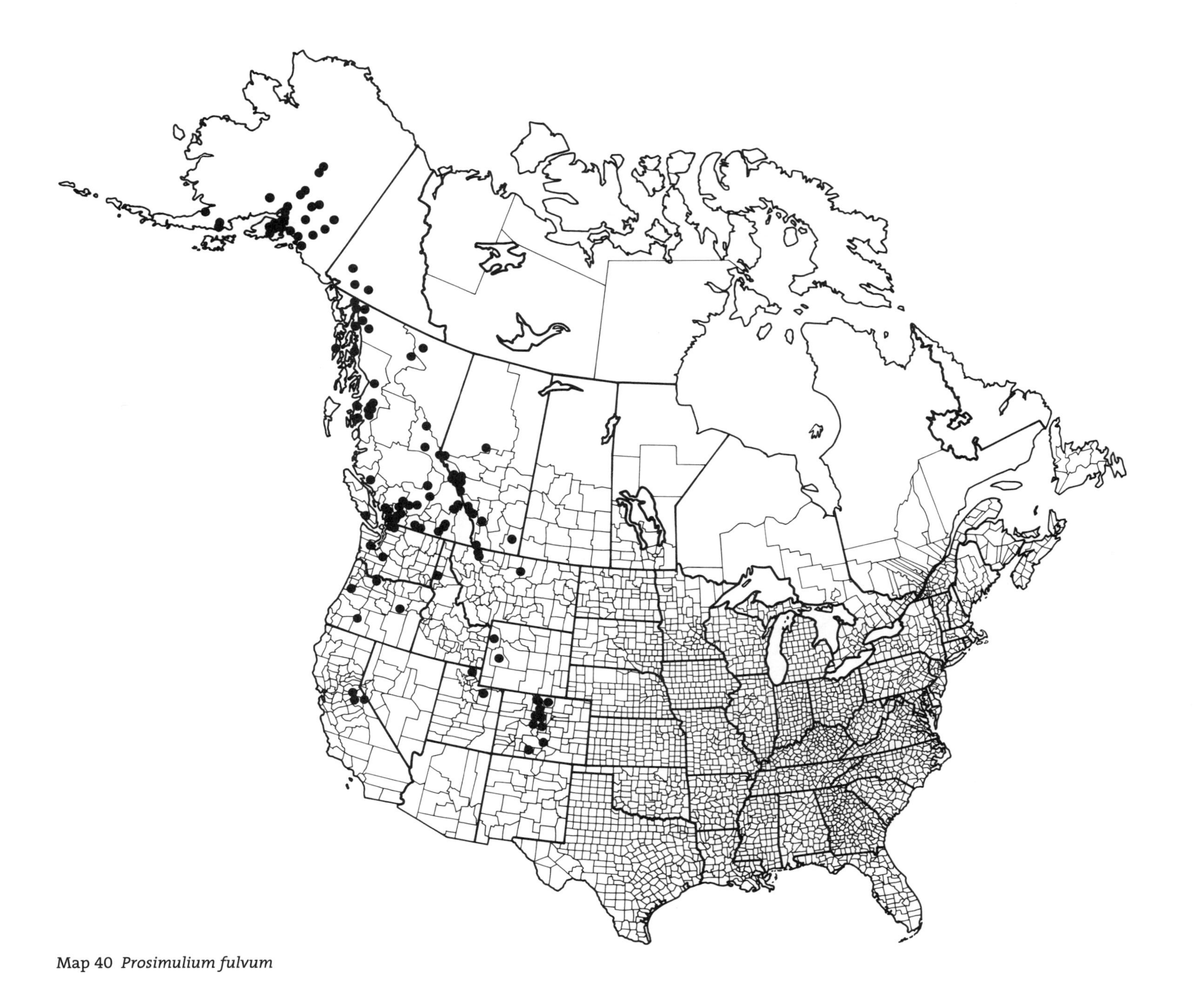

Map 40 *Prosimulium fulvum*

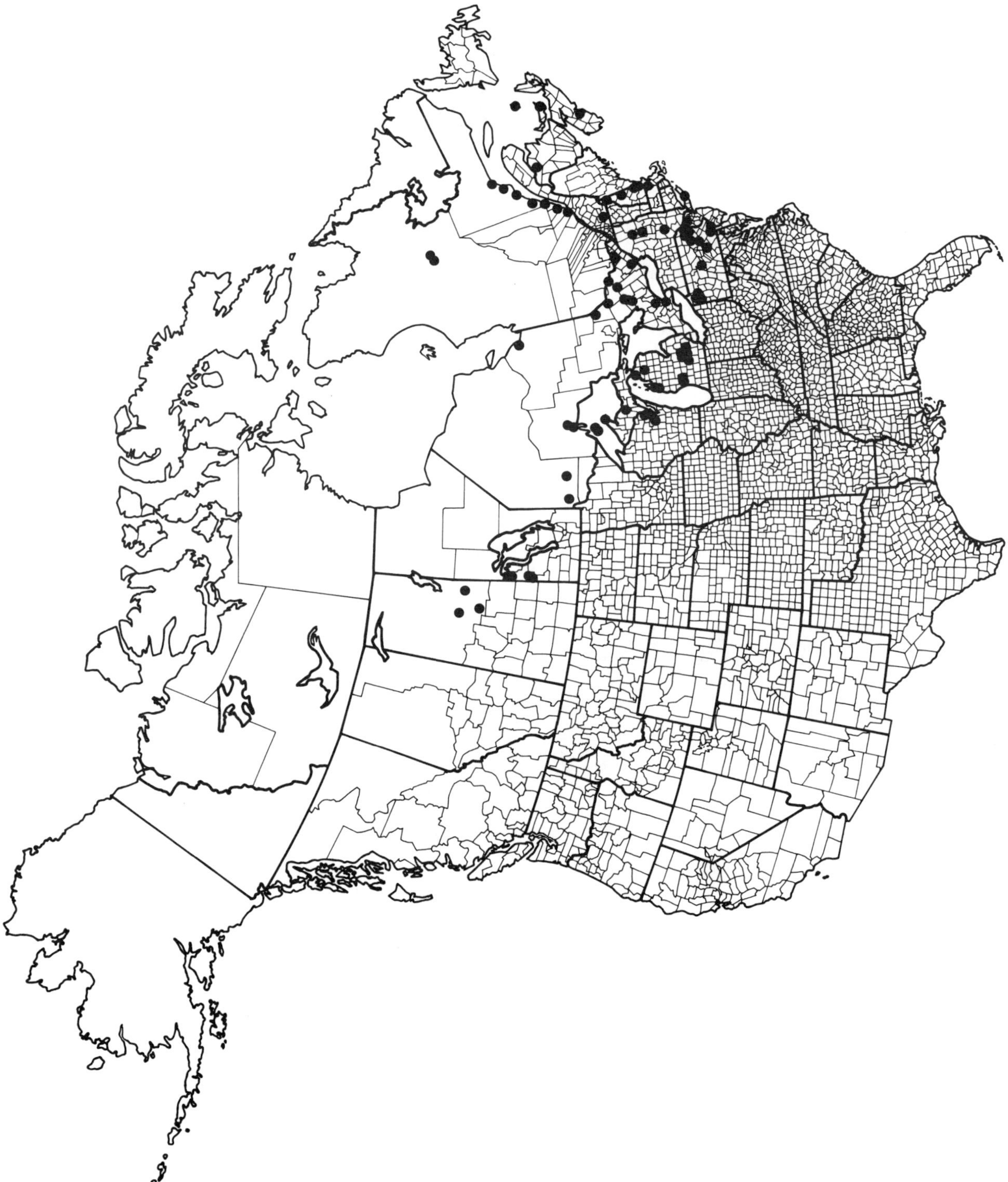

Map 41 *Prosimulium fuscum*

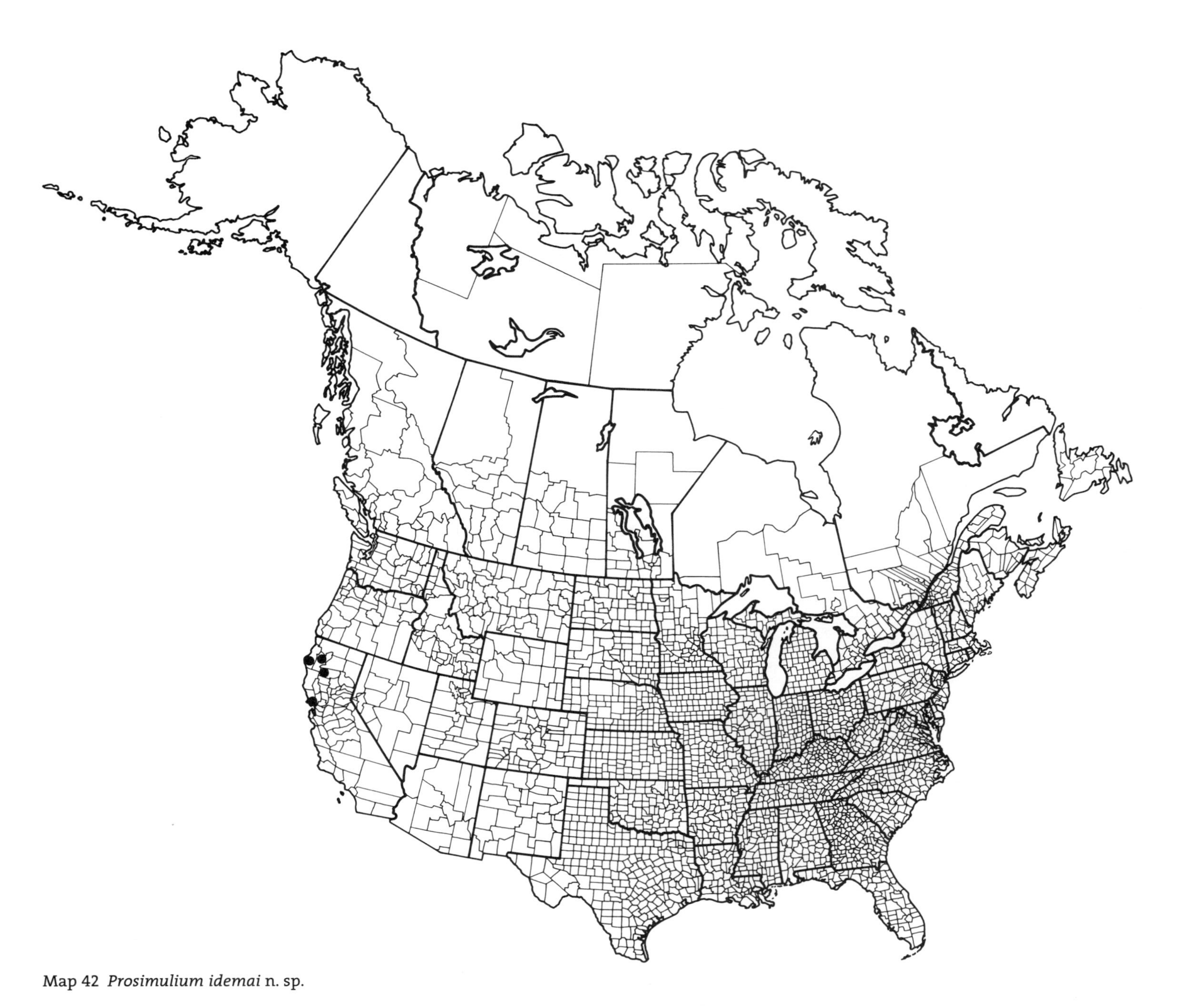

Map 42 *Prosimulium idemai* n. sp.

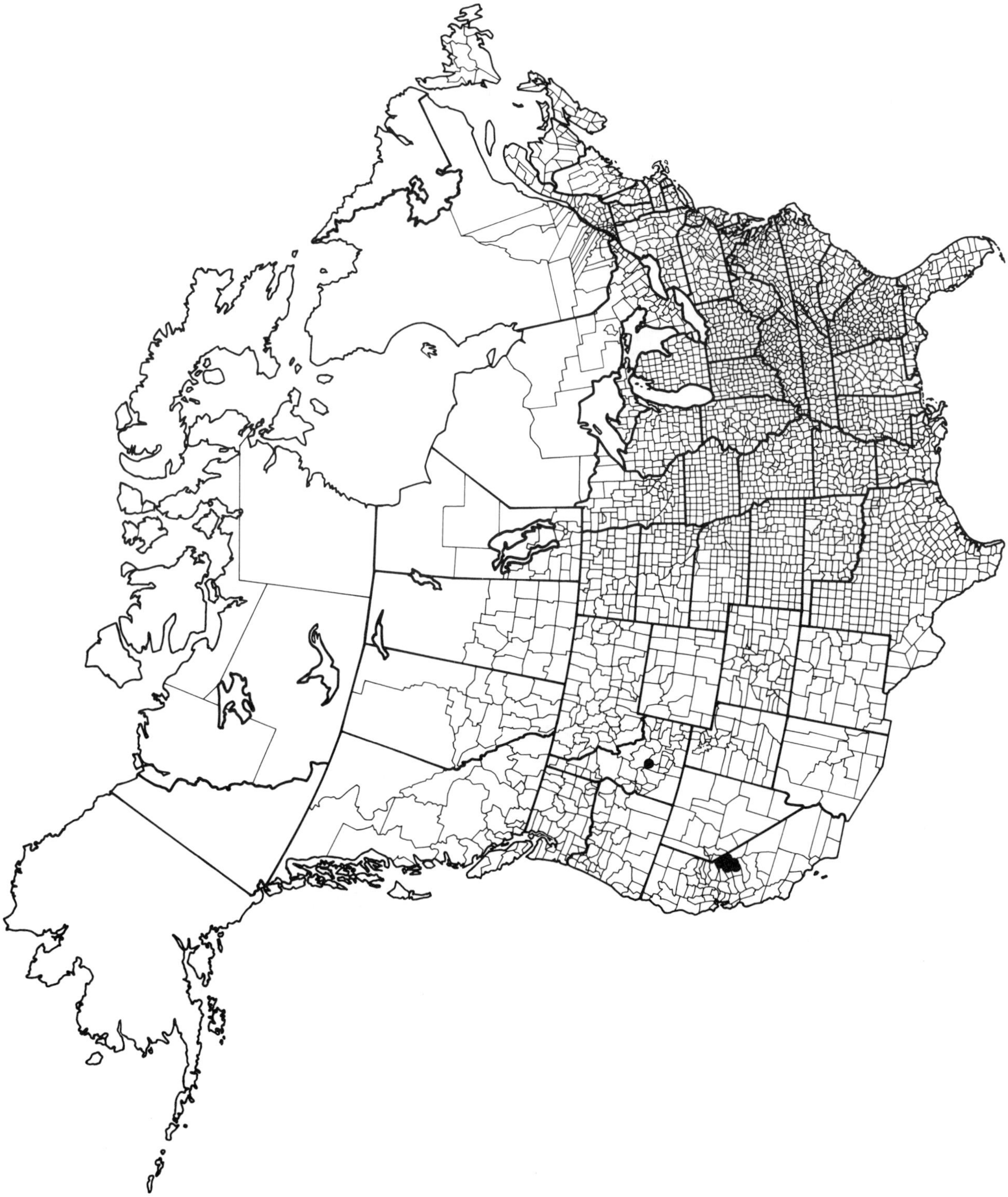

Map 43 *Prosimulium minifulvum* n. sp.

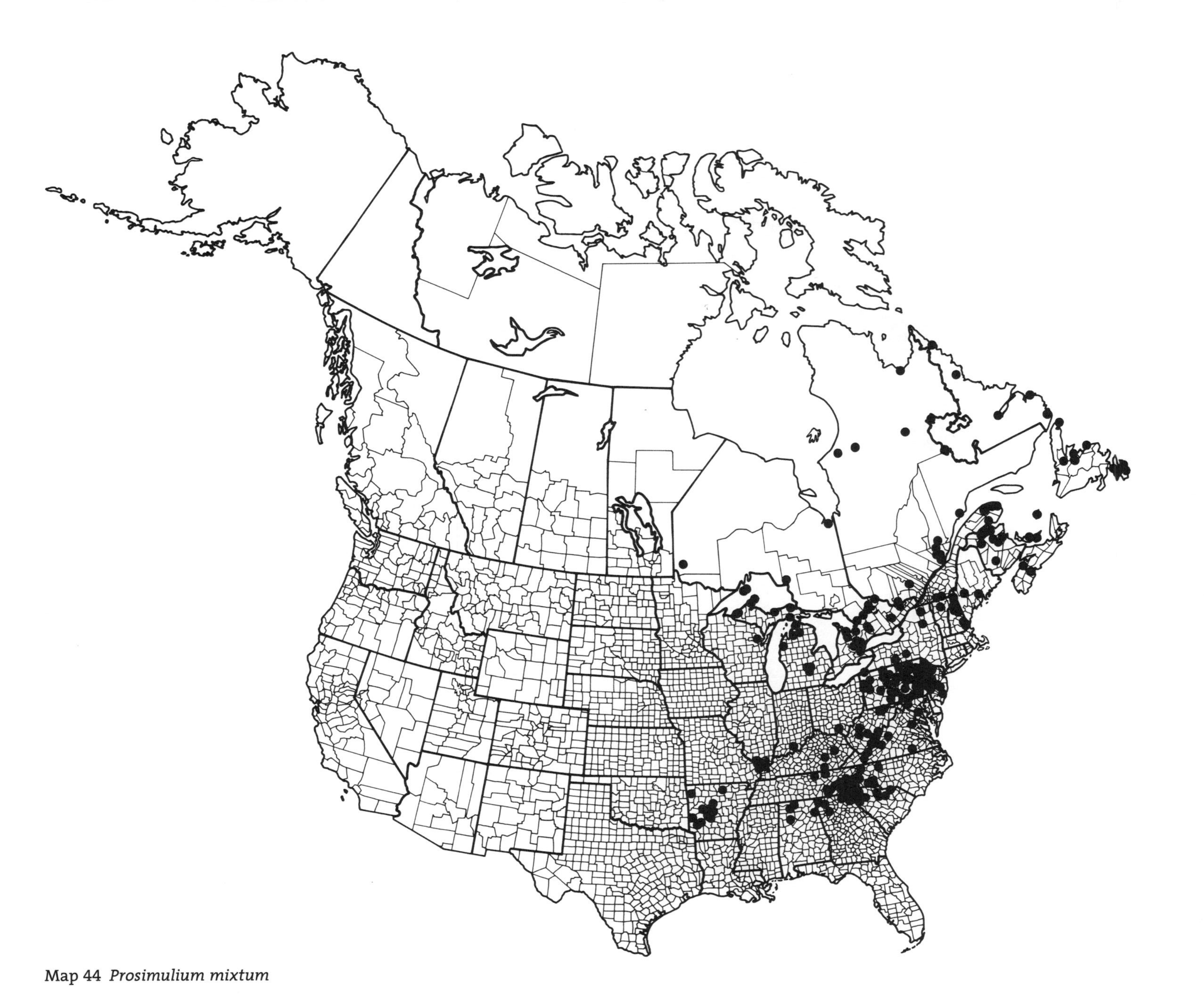

Map 44 *Prosimulium mixtum*

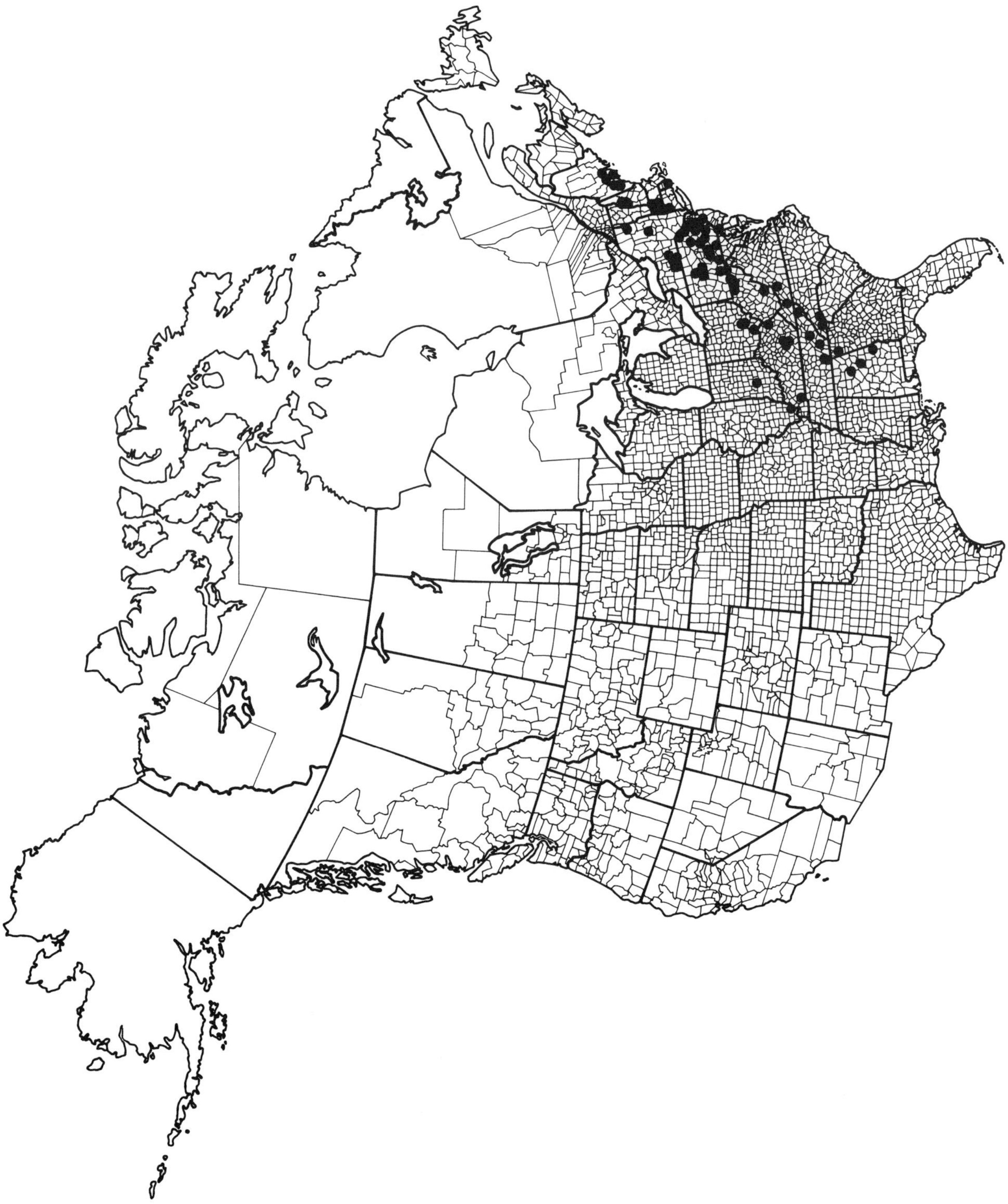

Map 45 *Prosimulium rhizophorum*

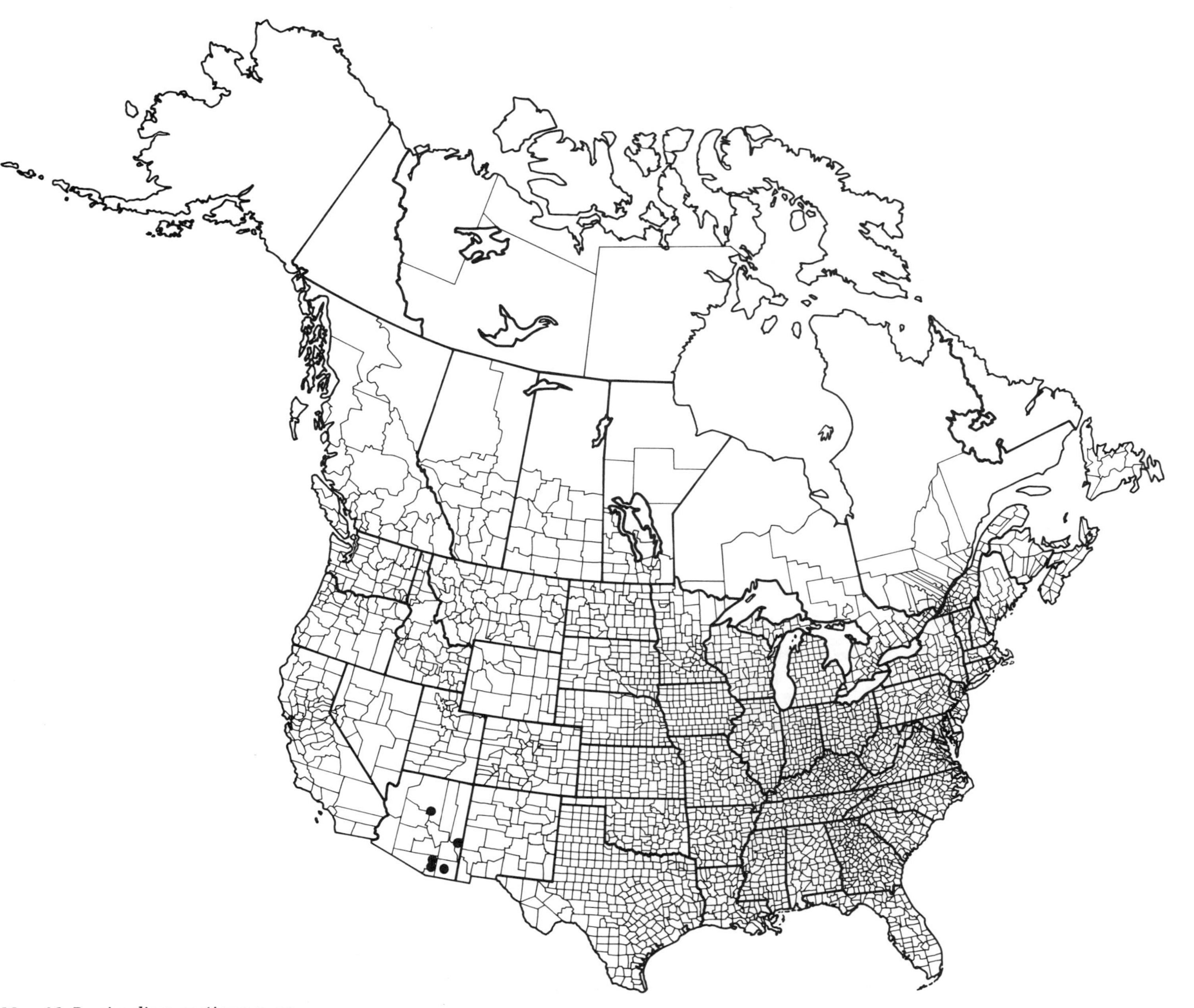

Map 46 *Prosimulium rusticum* n. sp.

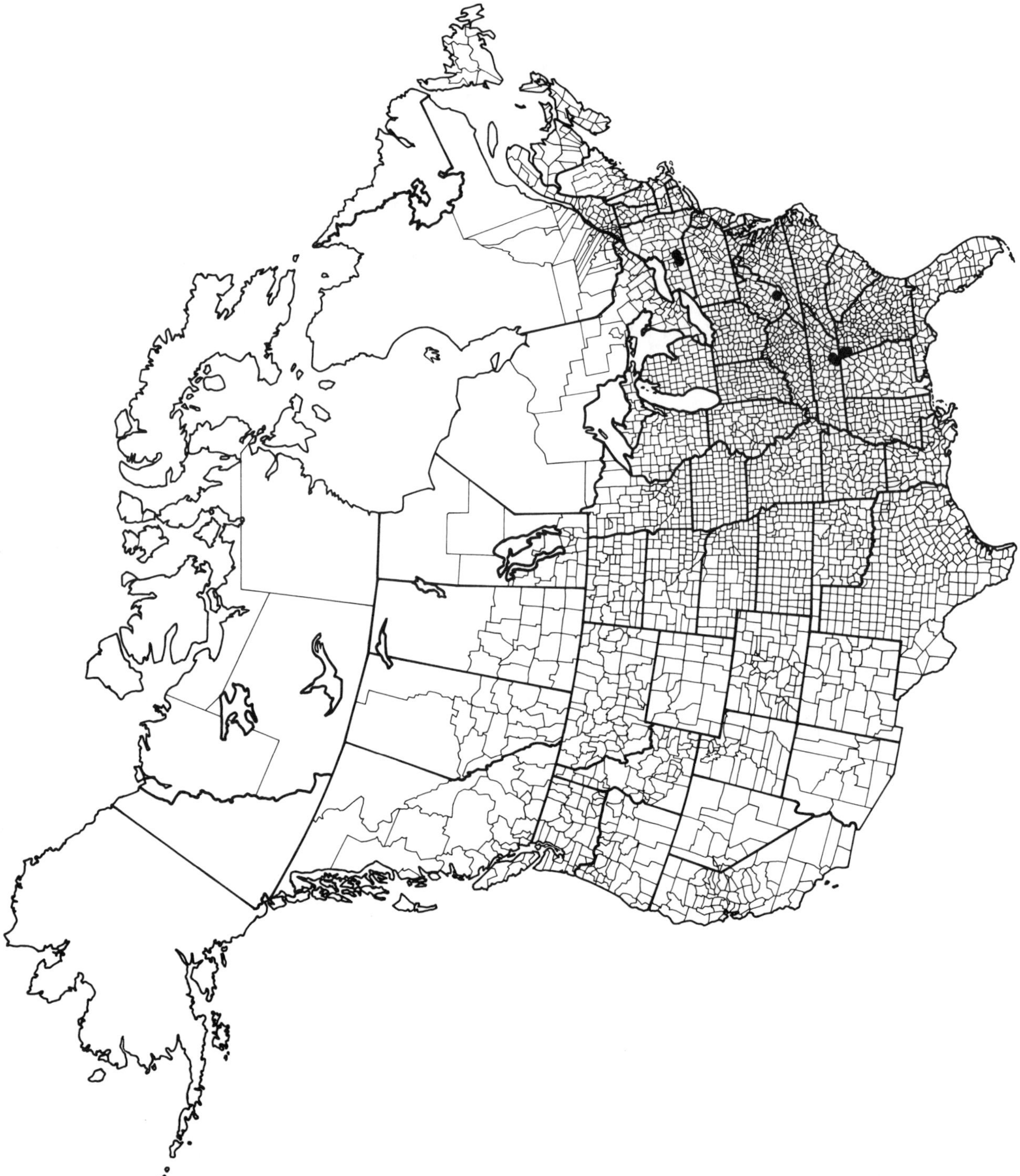

Map 47 *Prosimulium saltus*

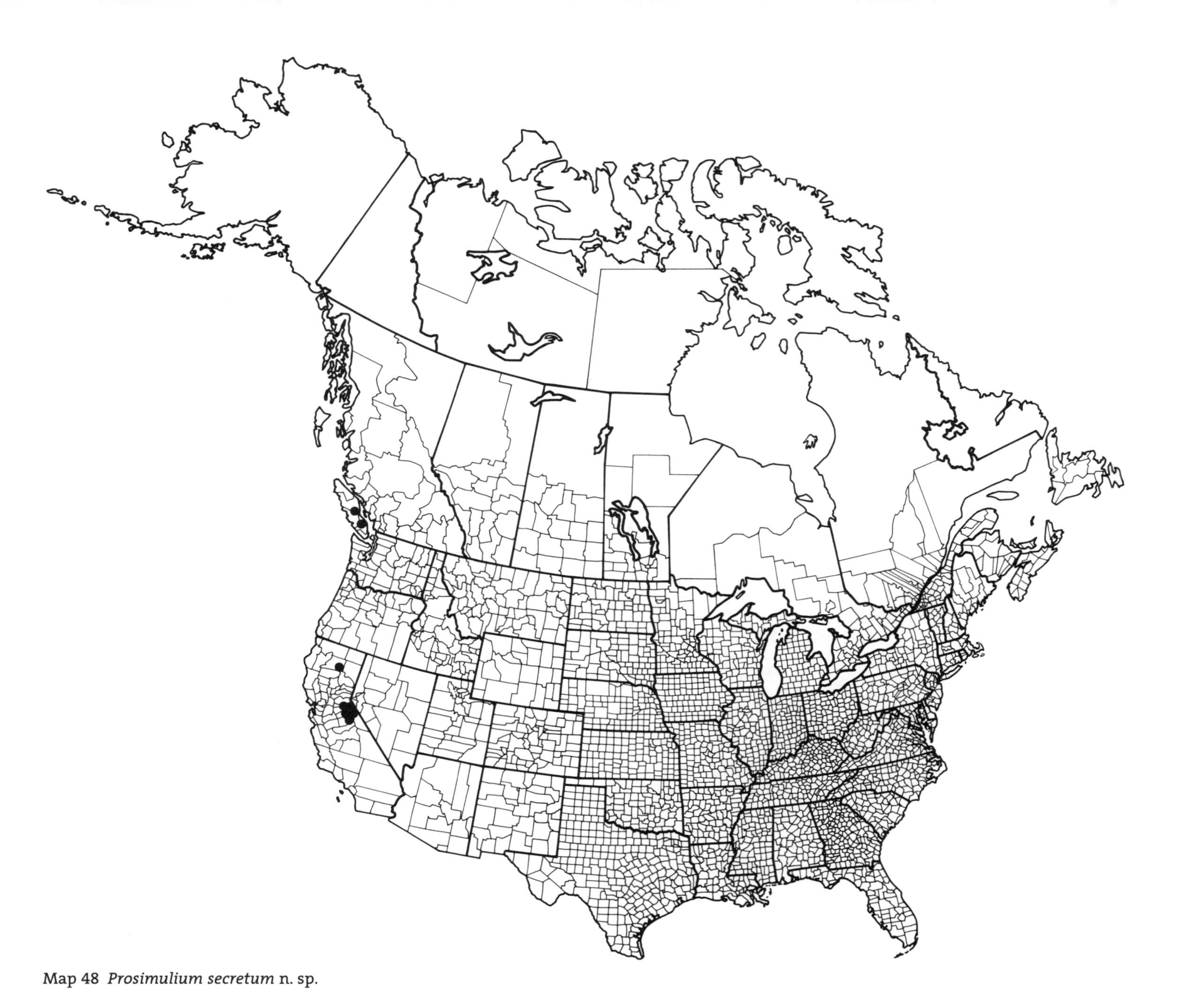

Map 48 *Prosimulium secretum* n. sp.

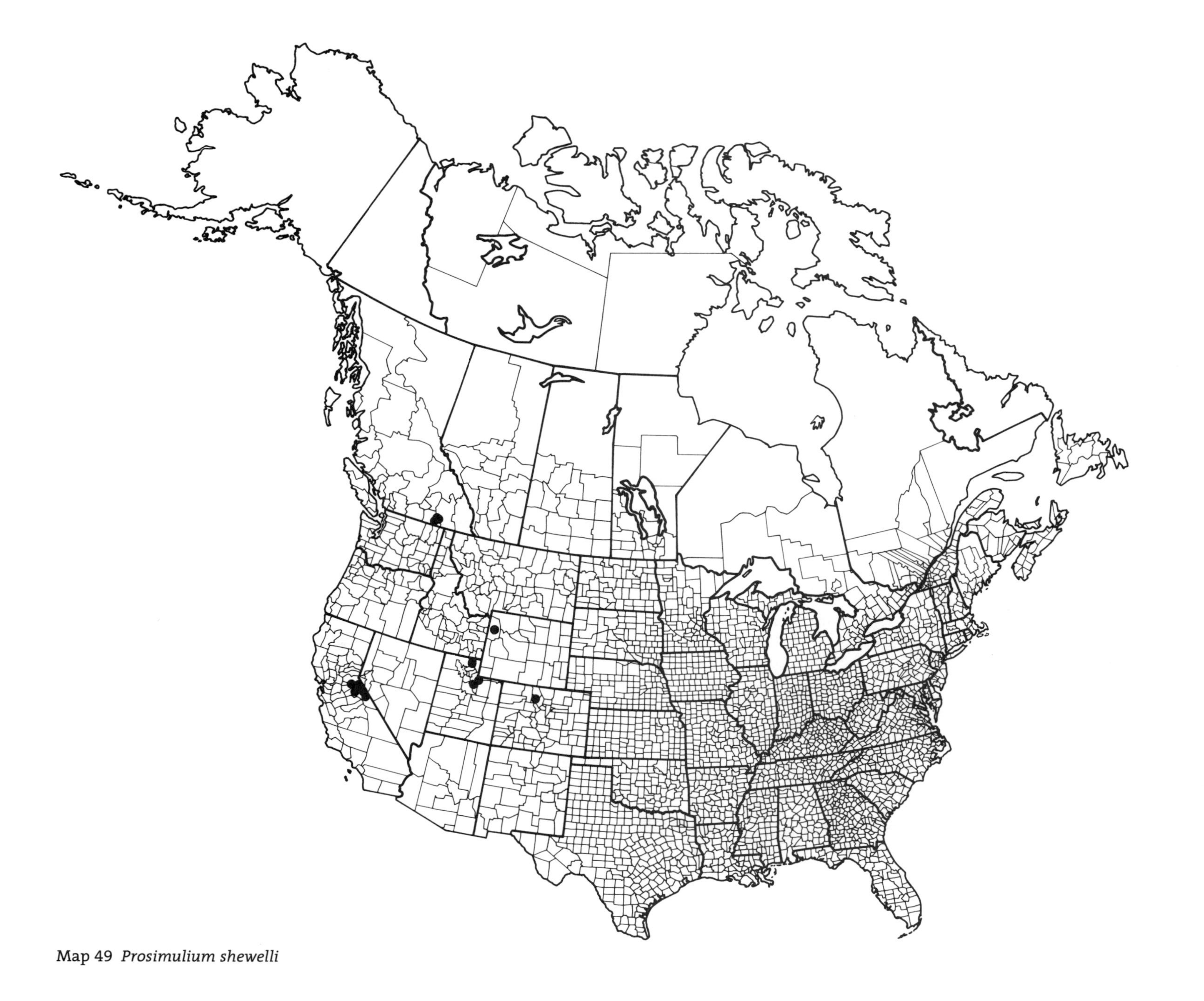

Map 49 *Prosimulium shewelli*

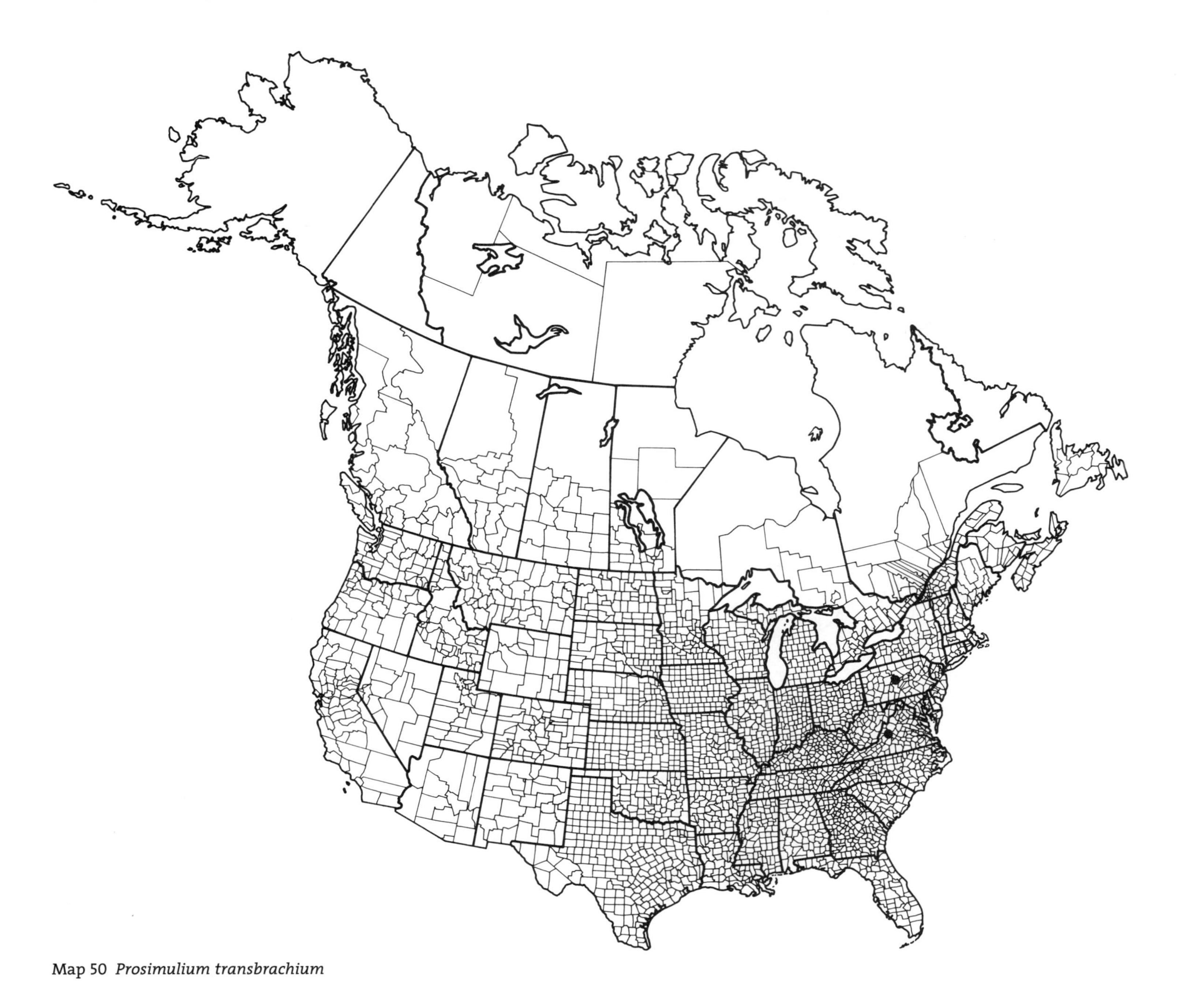

Map 50 *Prosimulium transbrachium*

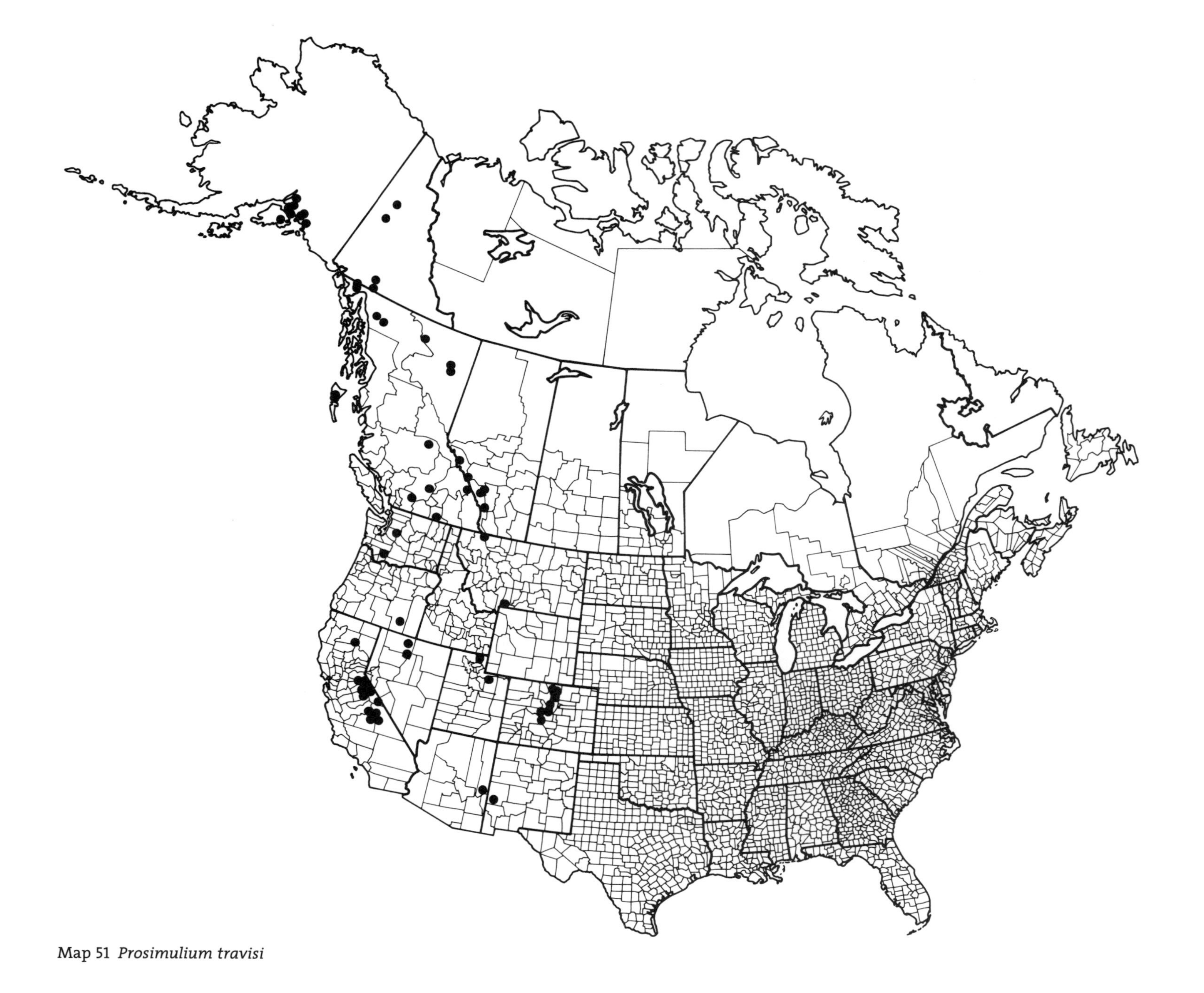

Map 51 *Prosimulium travisi*

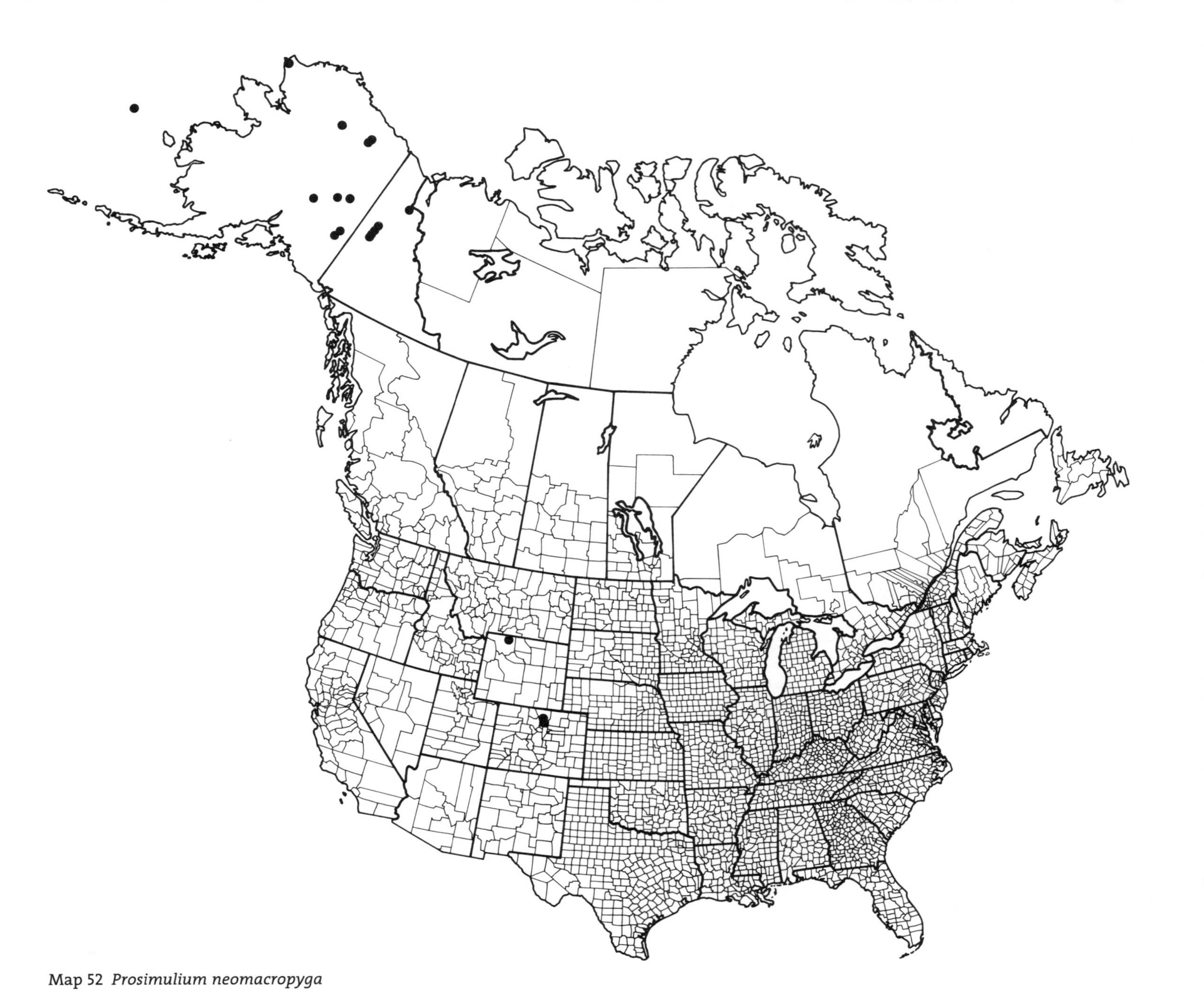

Map 52 *Prosimulium neomacropyga*

Map 53 *Prosimulium ursinum*

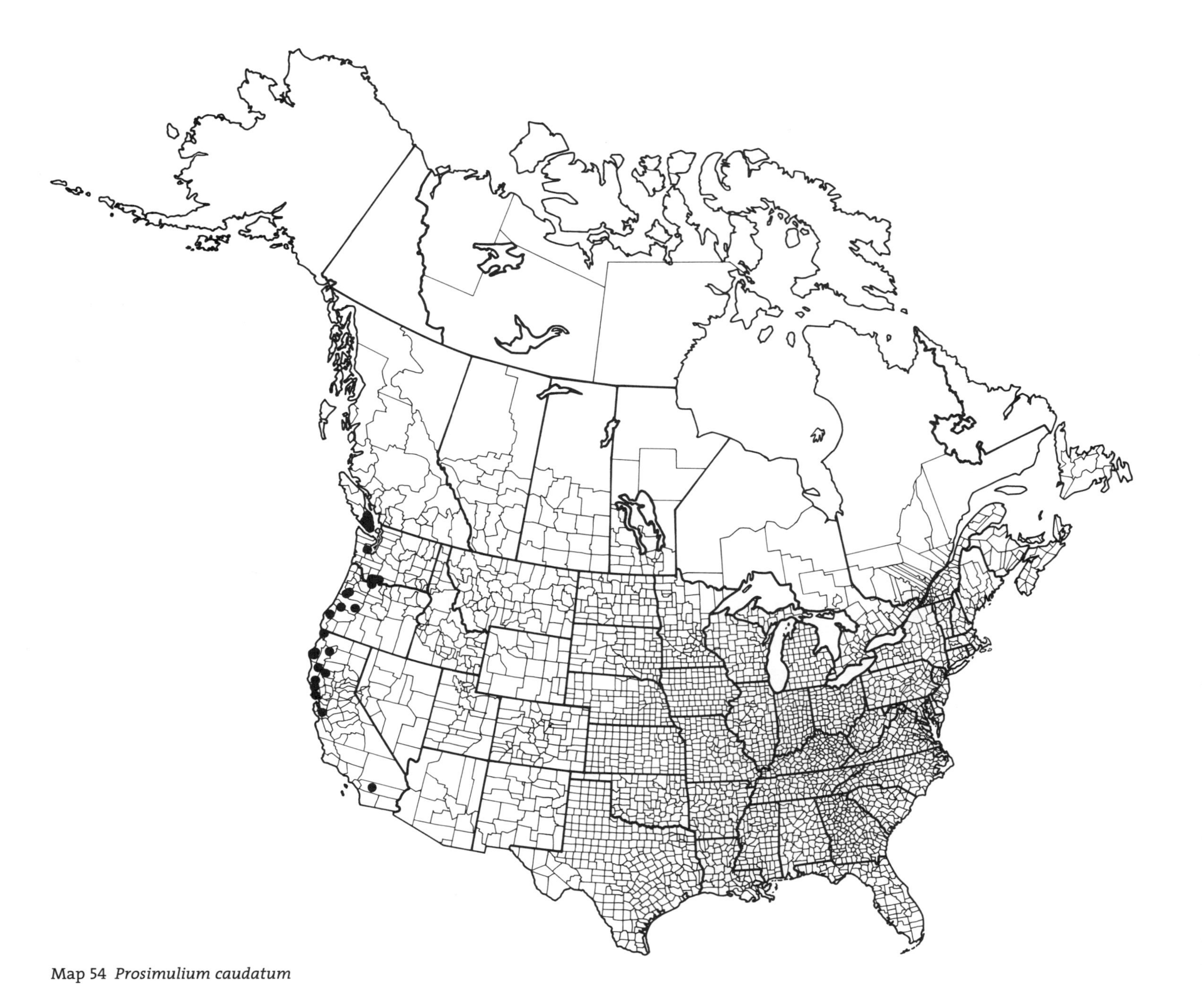

Map 54 *Prosimulium caudatum*

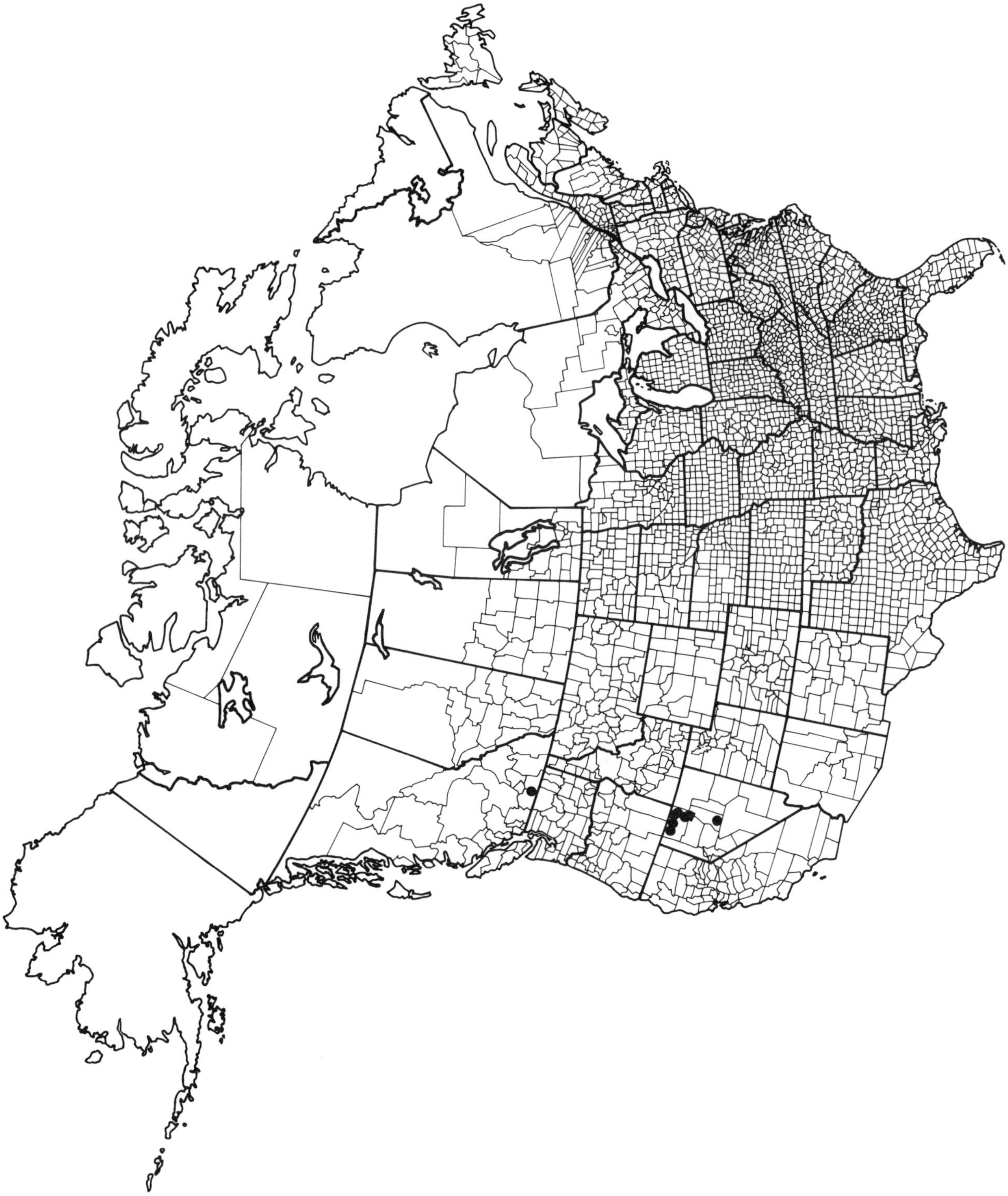

Map 55 *Prosimulium constrictistylum*

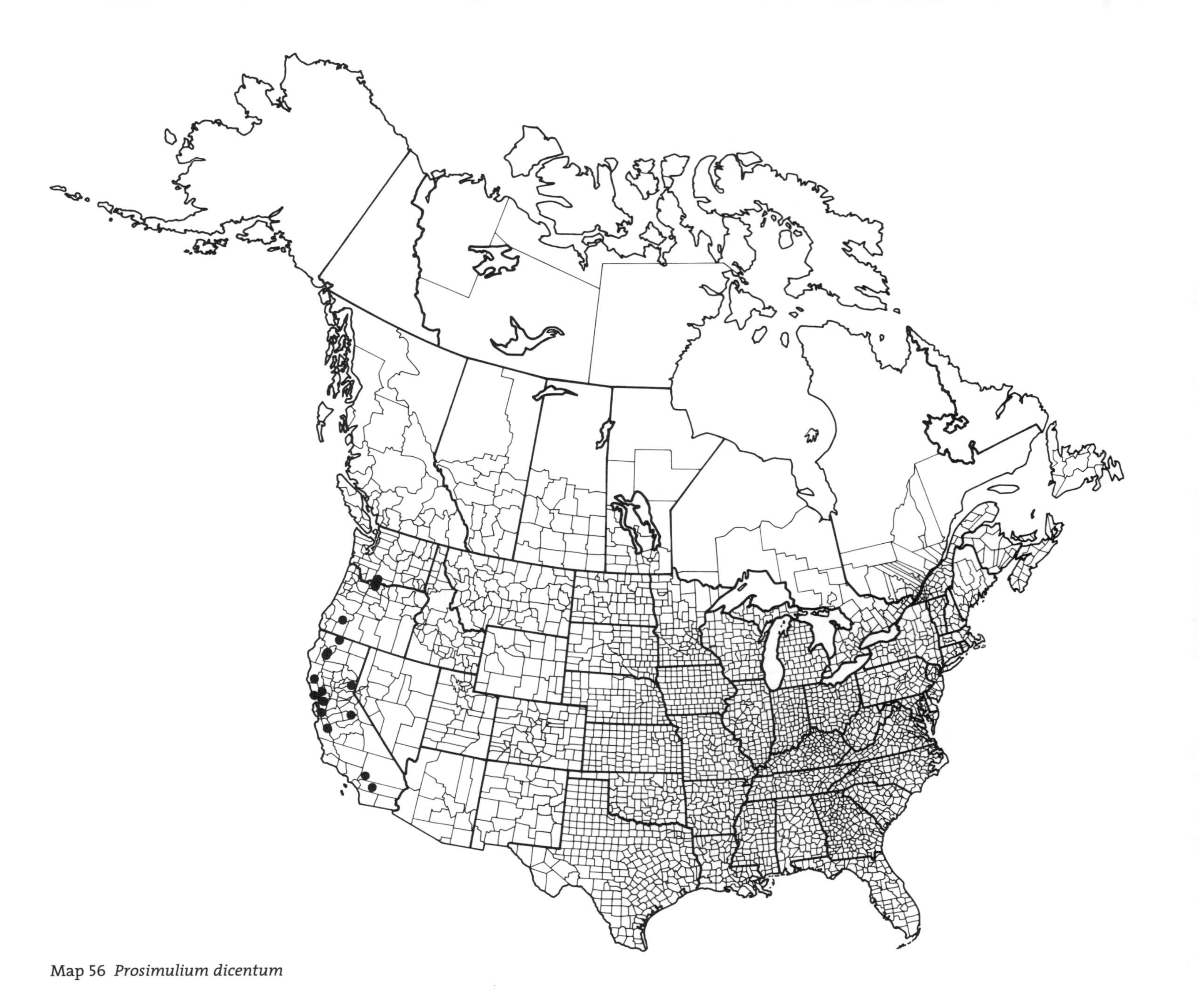

Map 56 *Prosimulium dicentum*

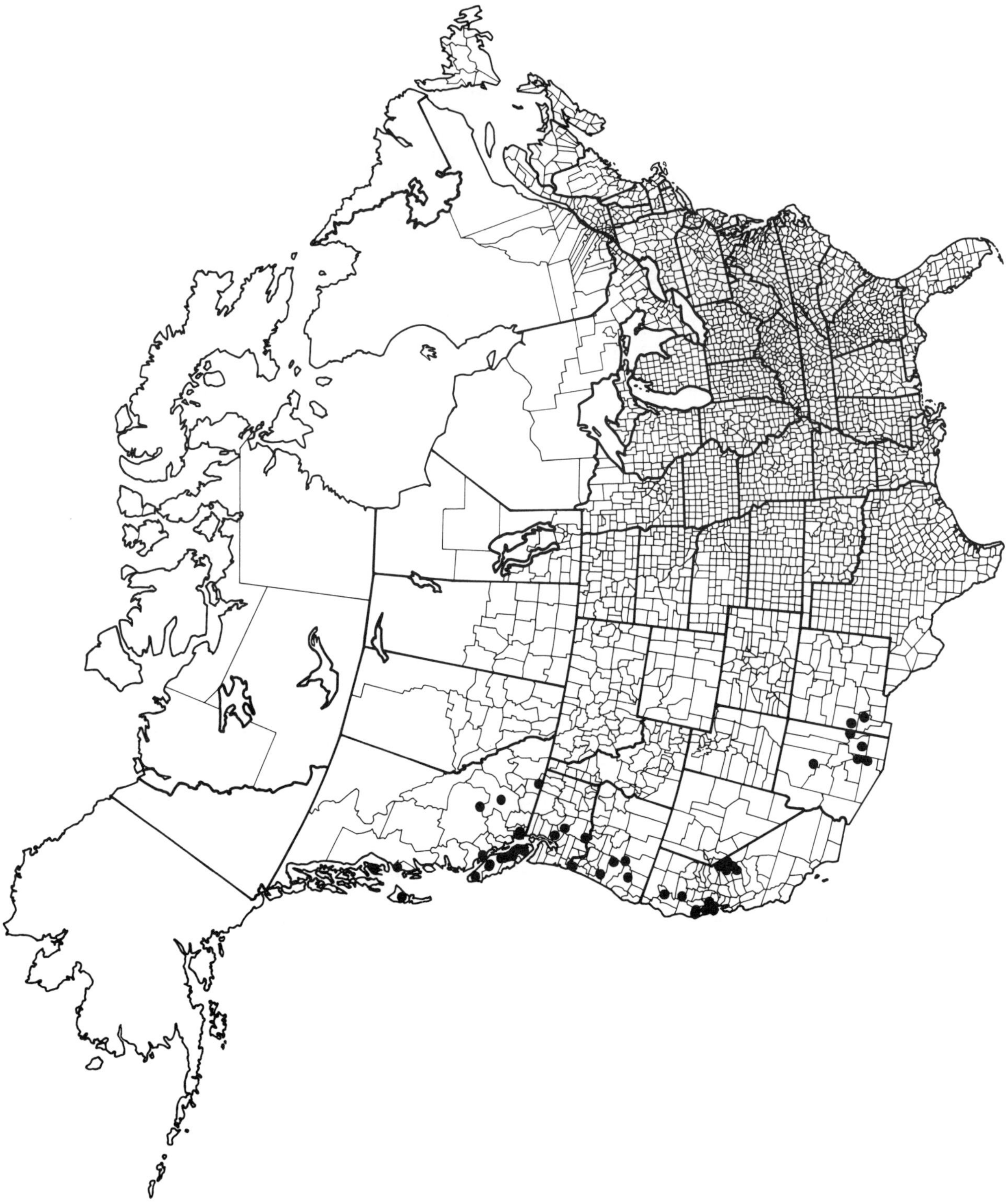

Map 57 *Prosimulium dicum*

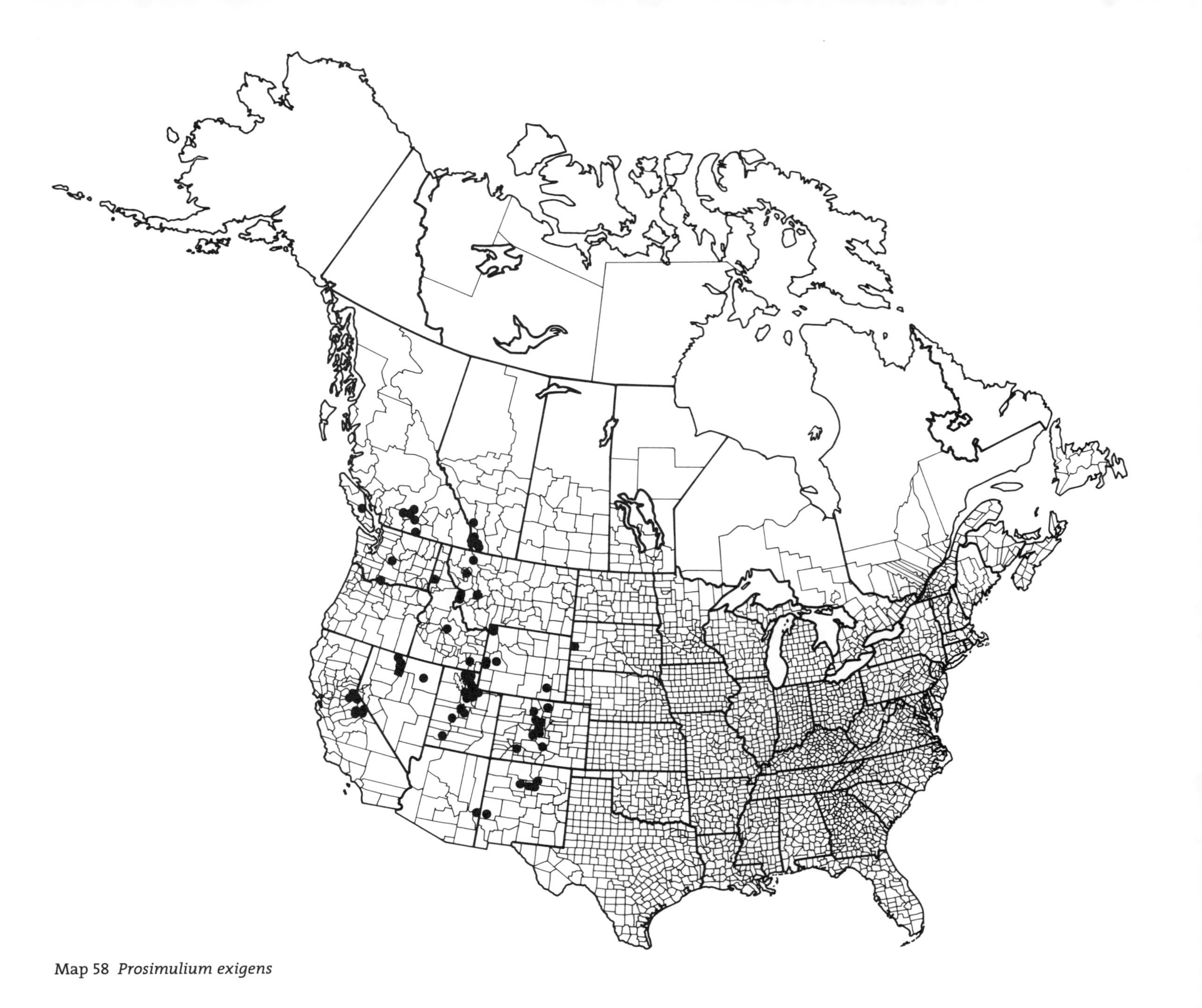

Map 58 *Prosimulium exigens*

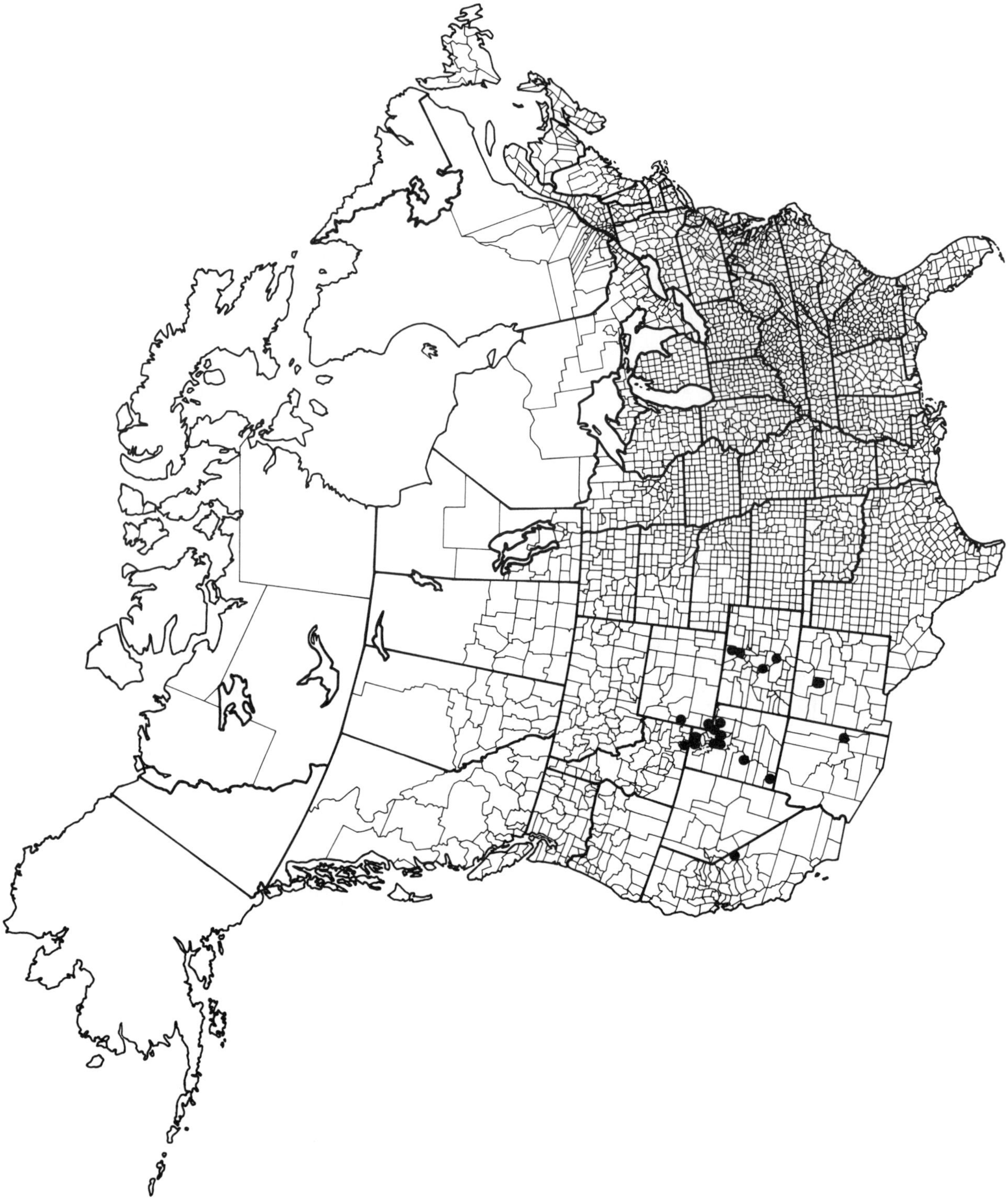

Map 59 *Prosimulium flaviantennus*

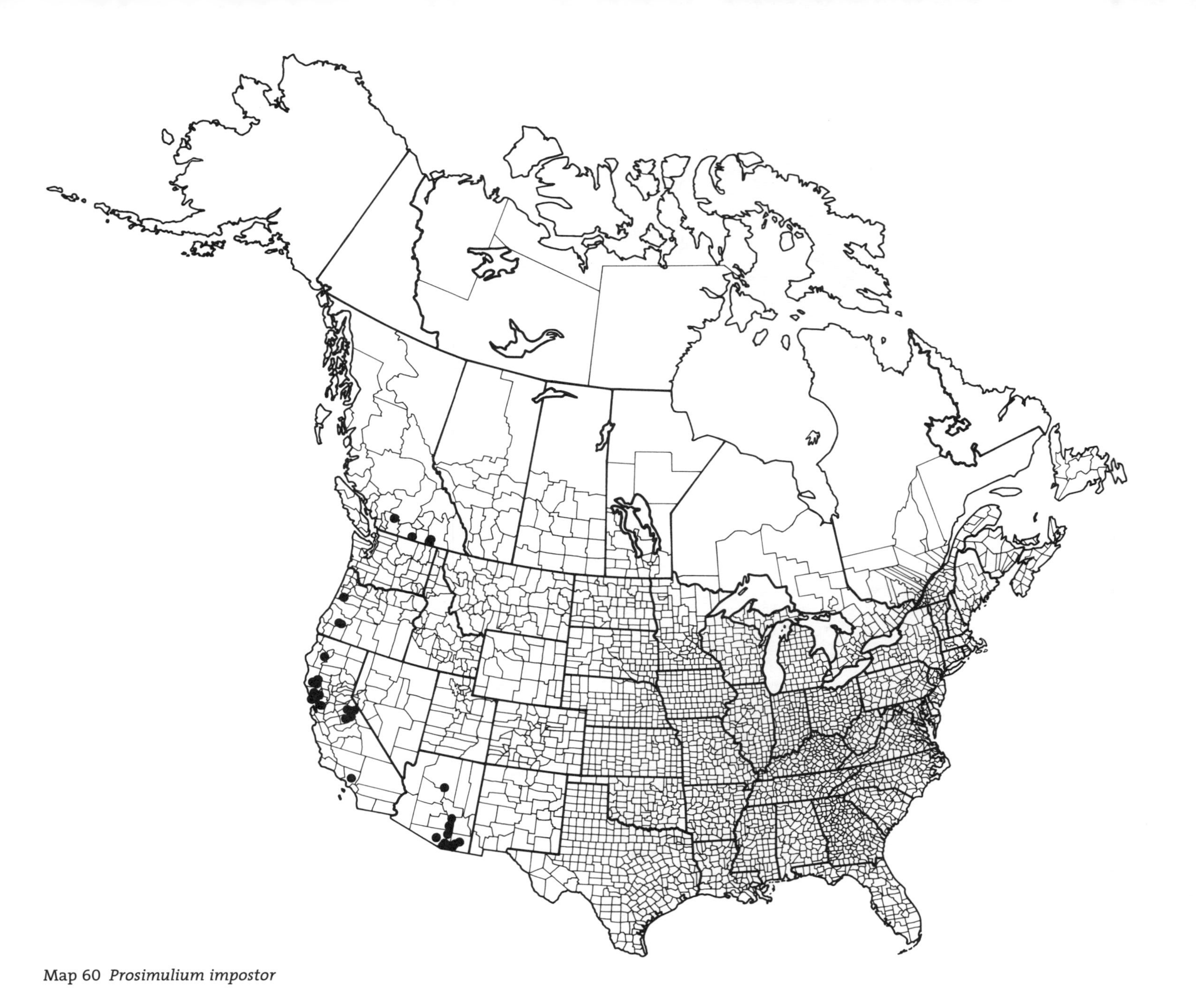

Map 60 *Prosimulium impostor*

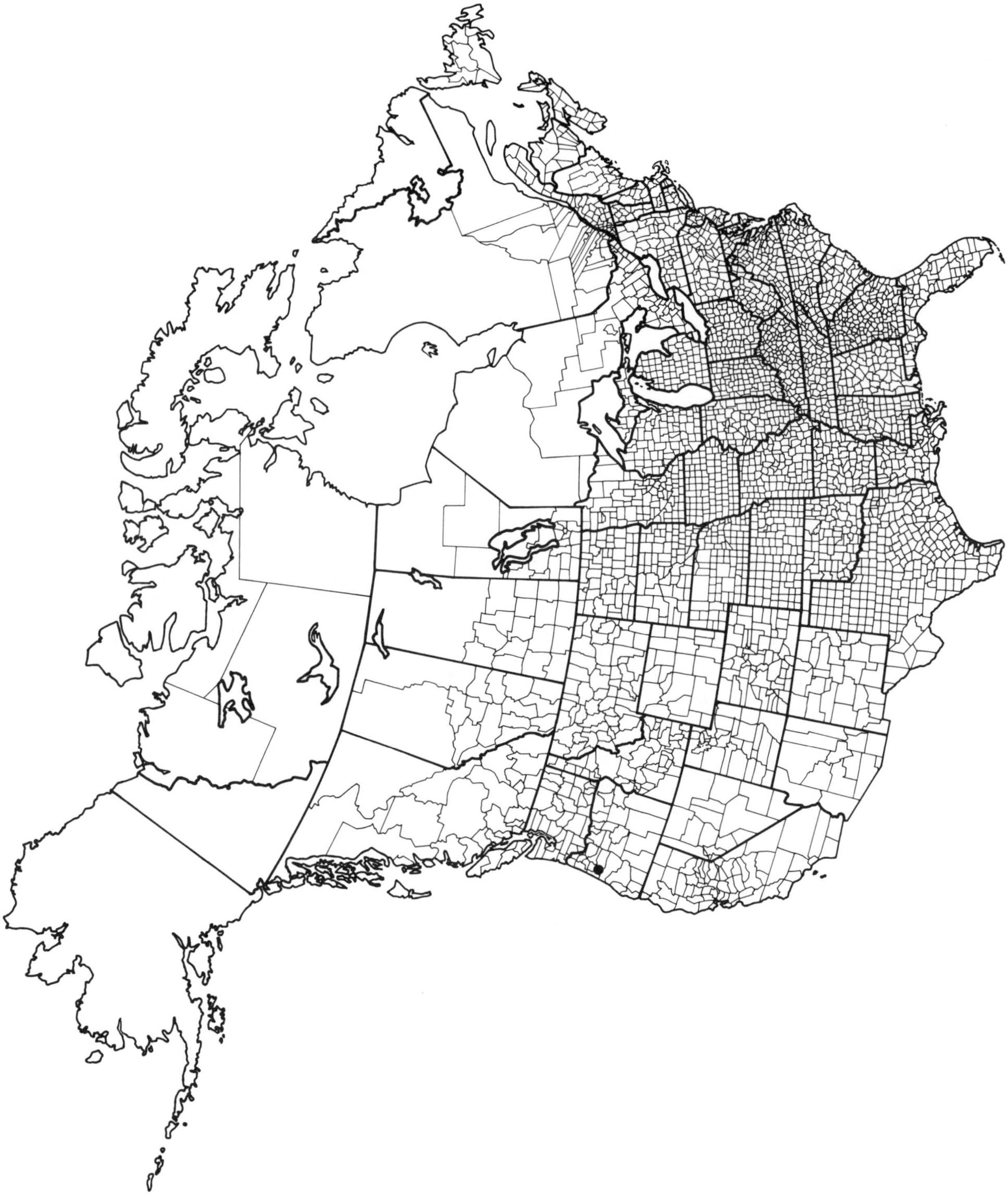

Map 61 *Prosimulium longirostrum* n. sp.

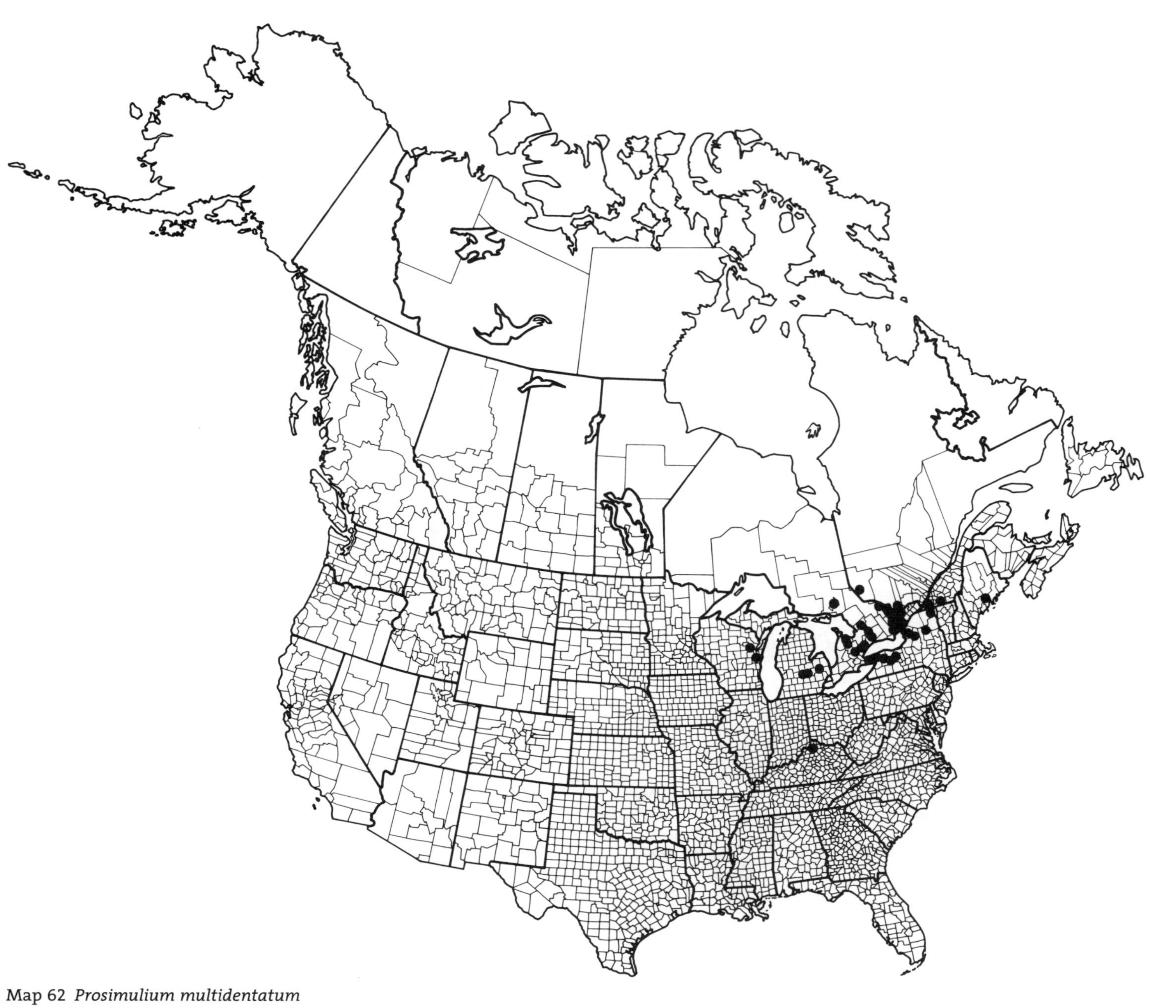

Map 62 *Prosimulium multidentatum*

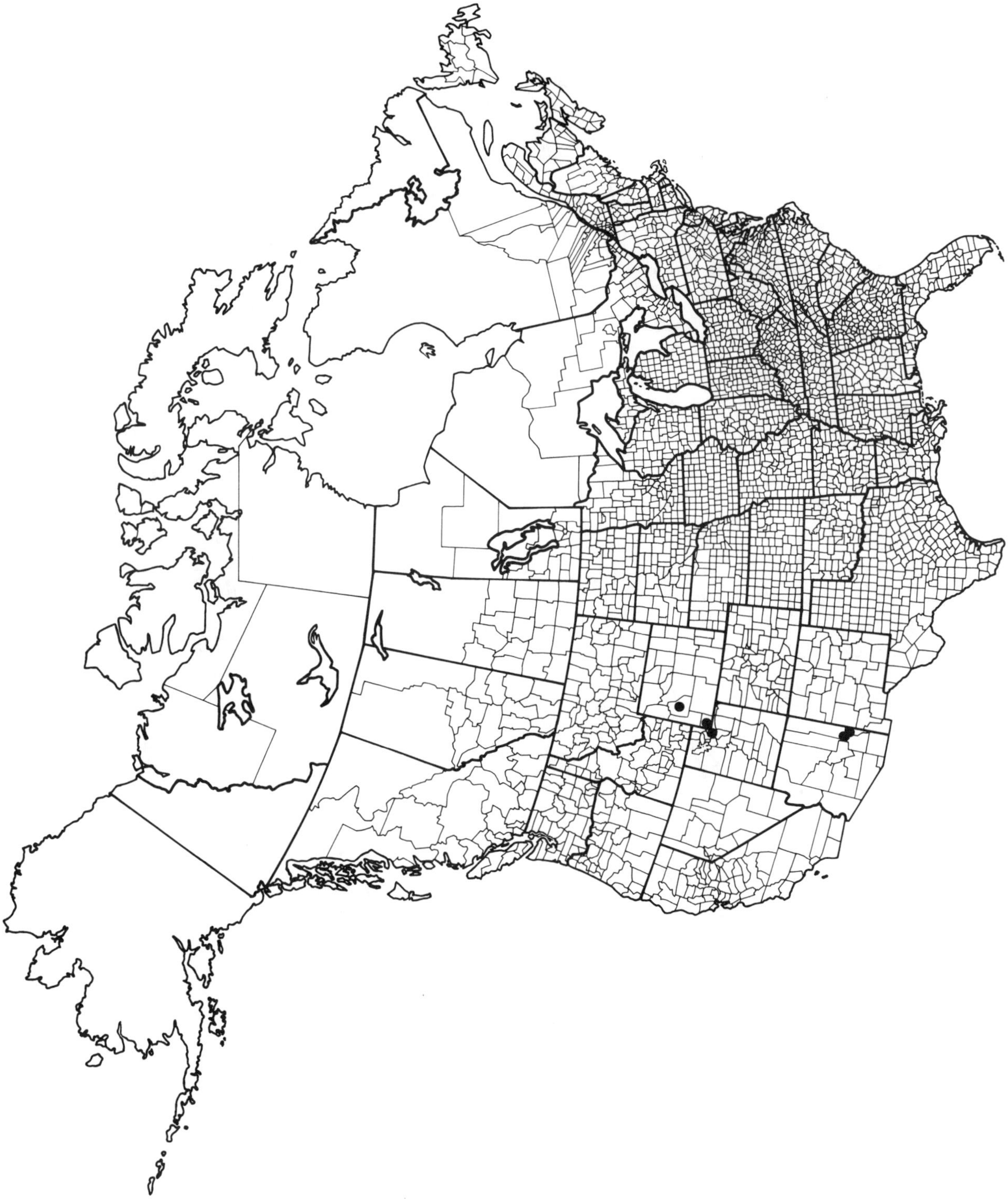

Map 63 *Prosimulium uinta*

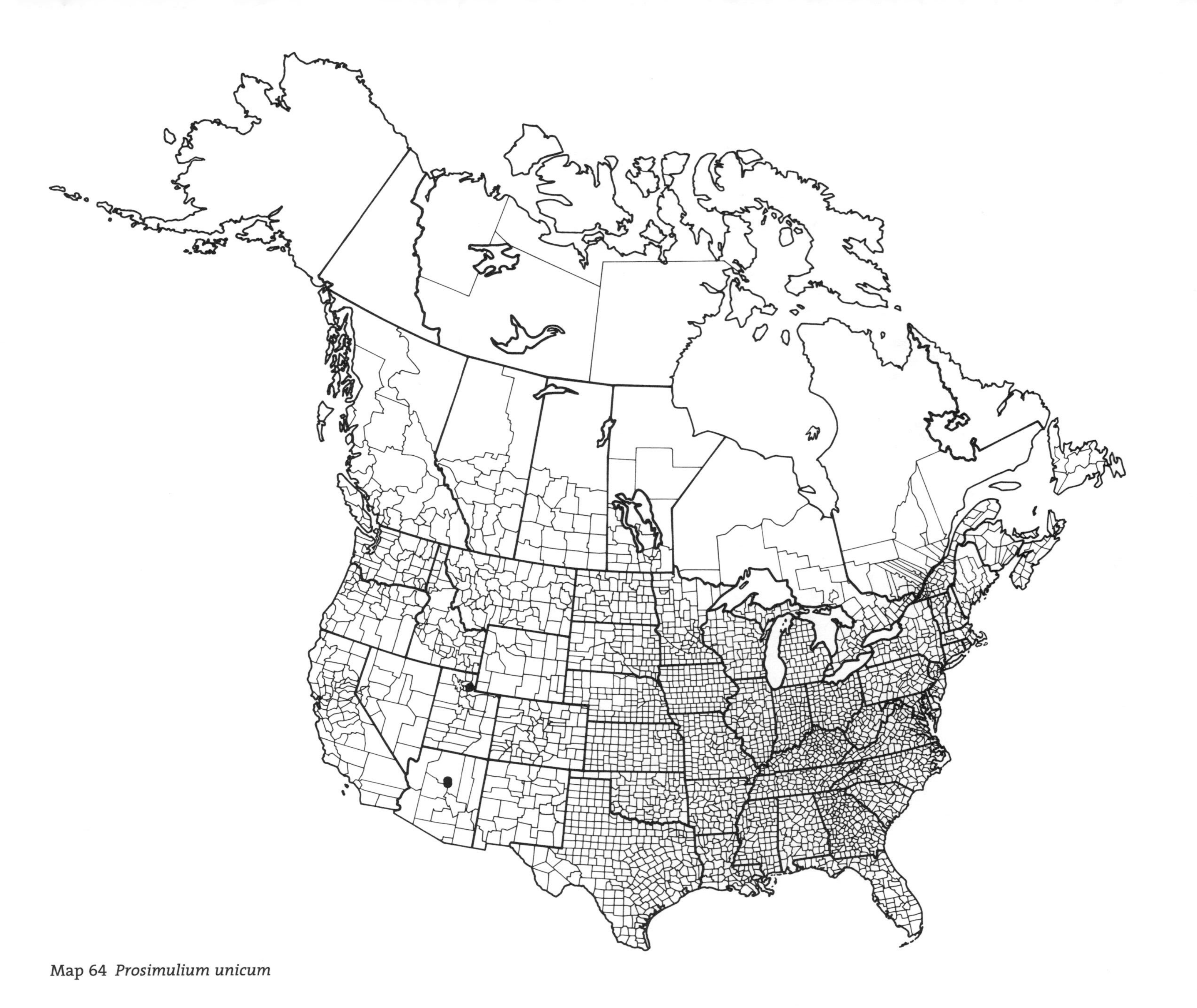

Map 64 *Prosimulium unicum*

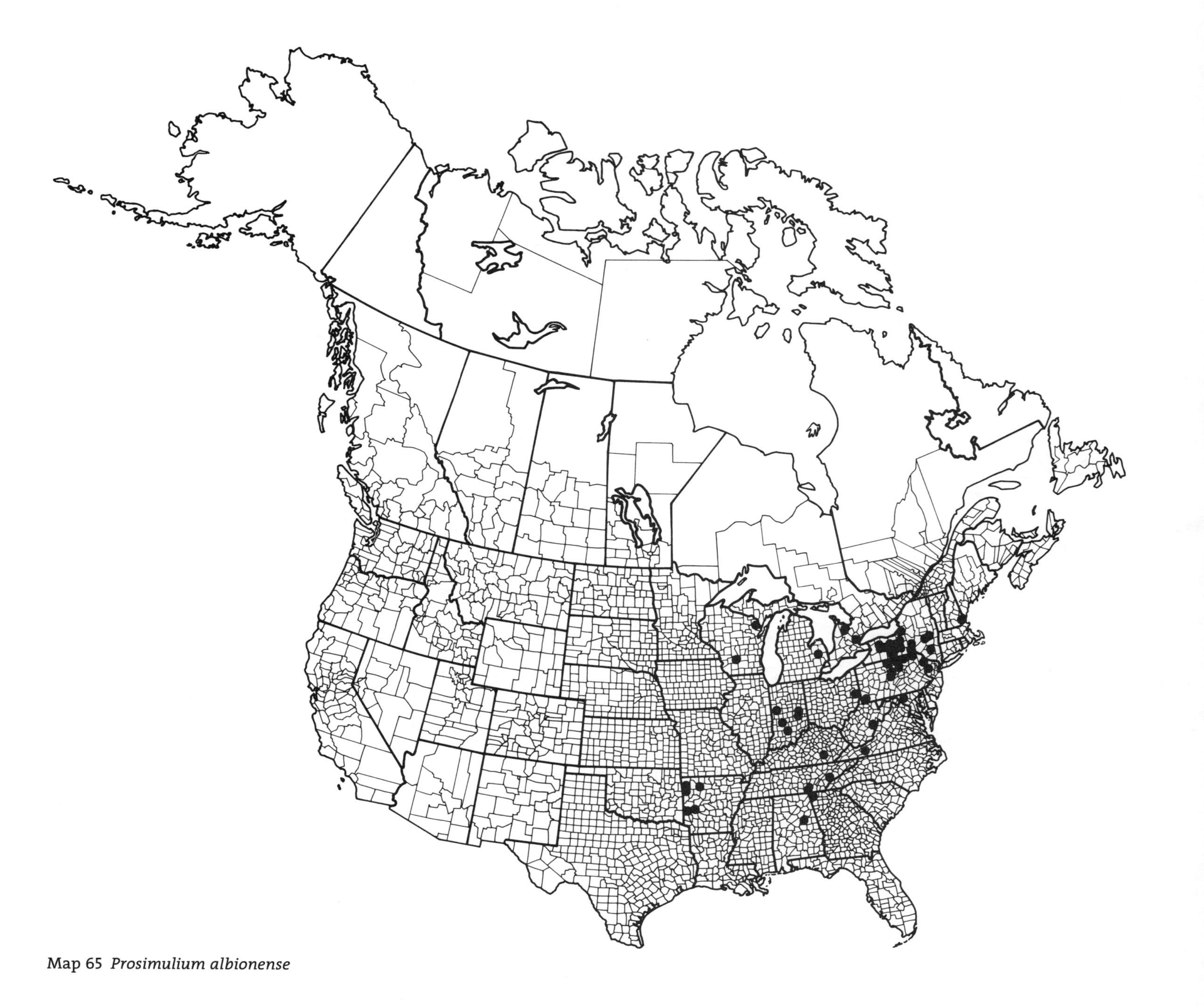

Map 65 *Prosimulium albionense*

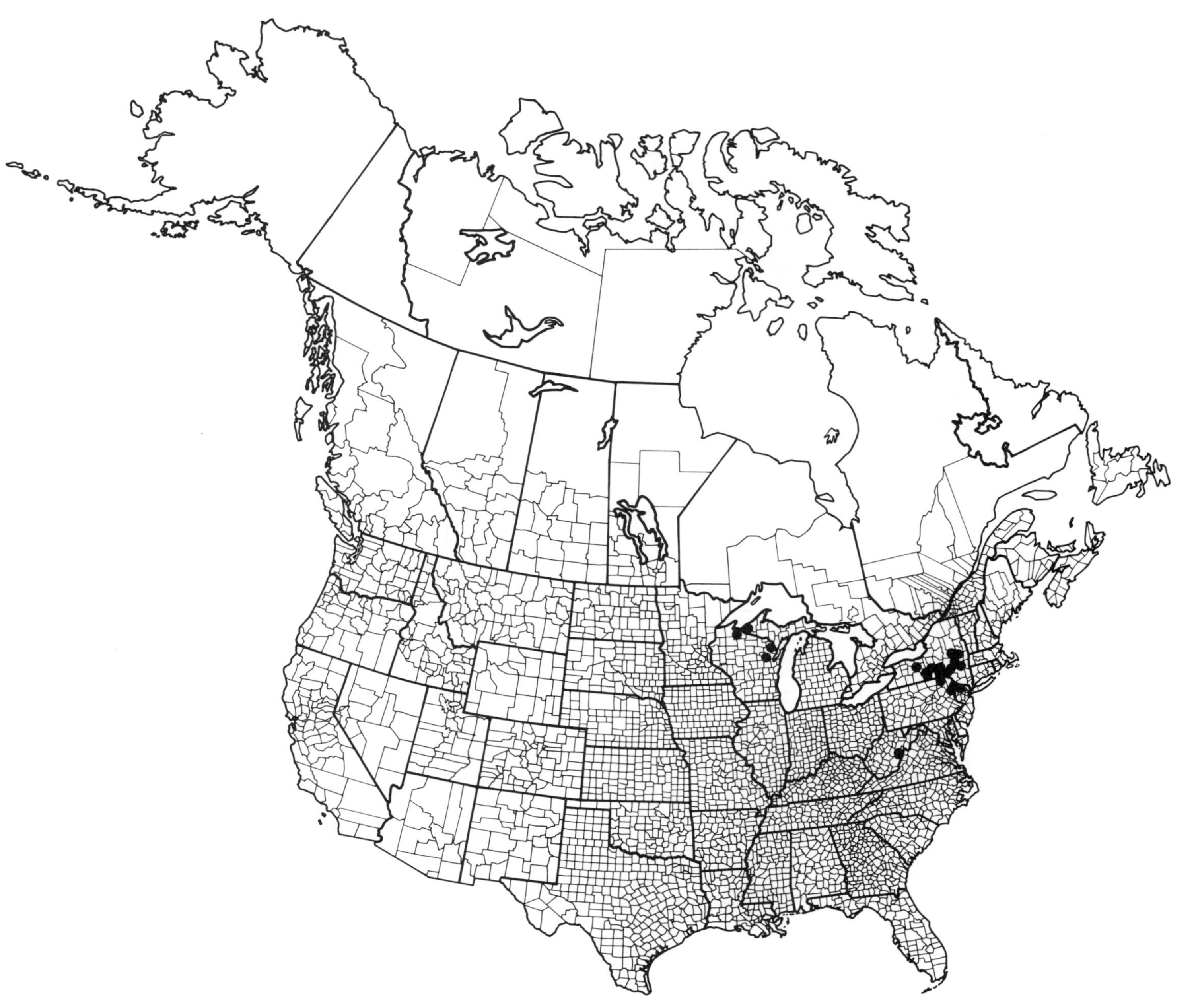

Map 66 *Prosimulium canutum* n. sp.

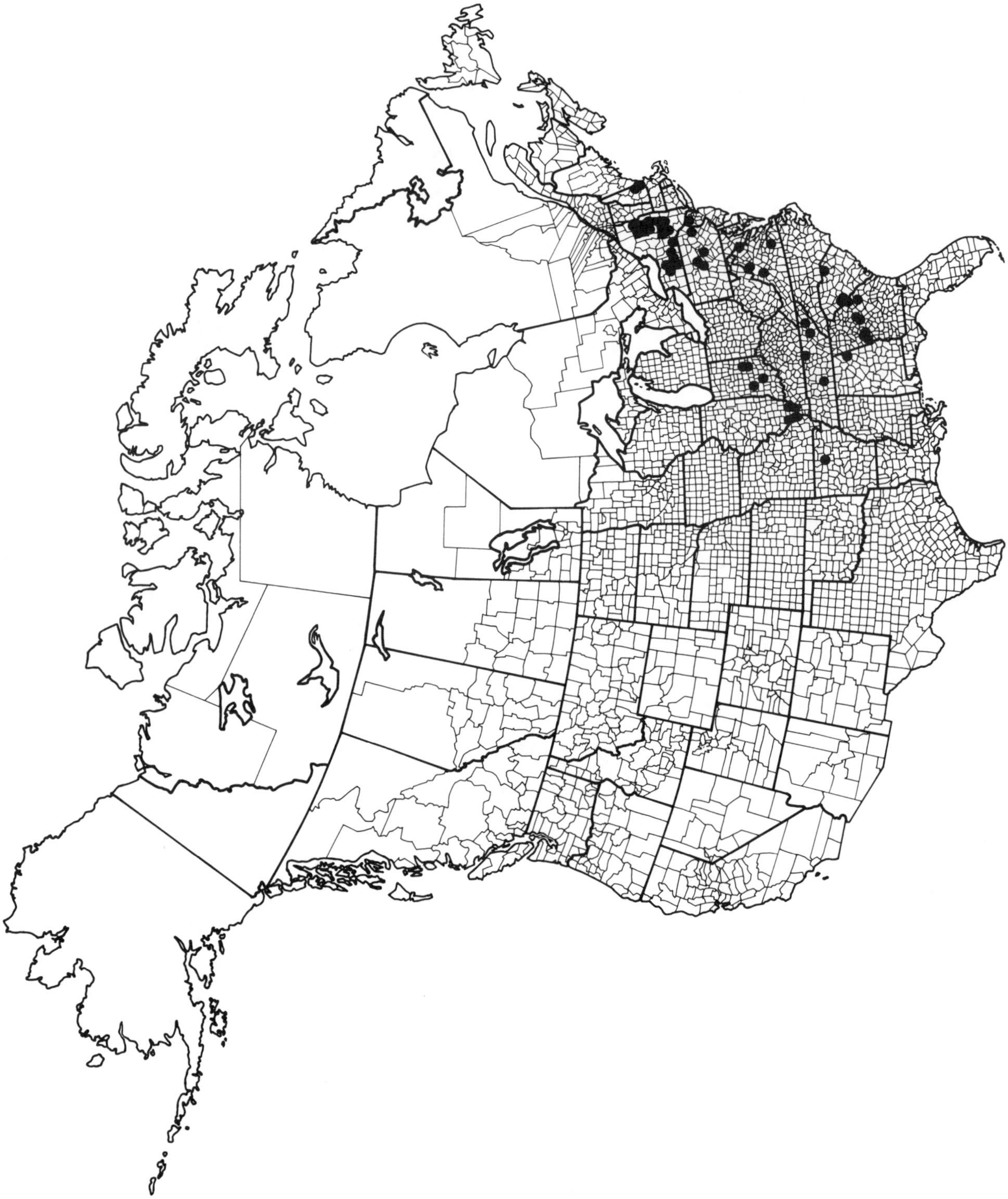

Map 67 *Prosimulium magnum* s. s.

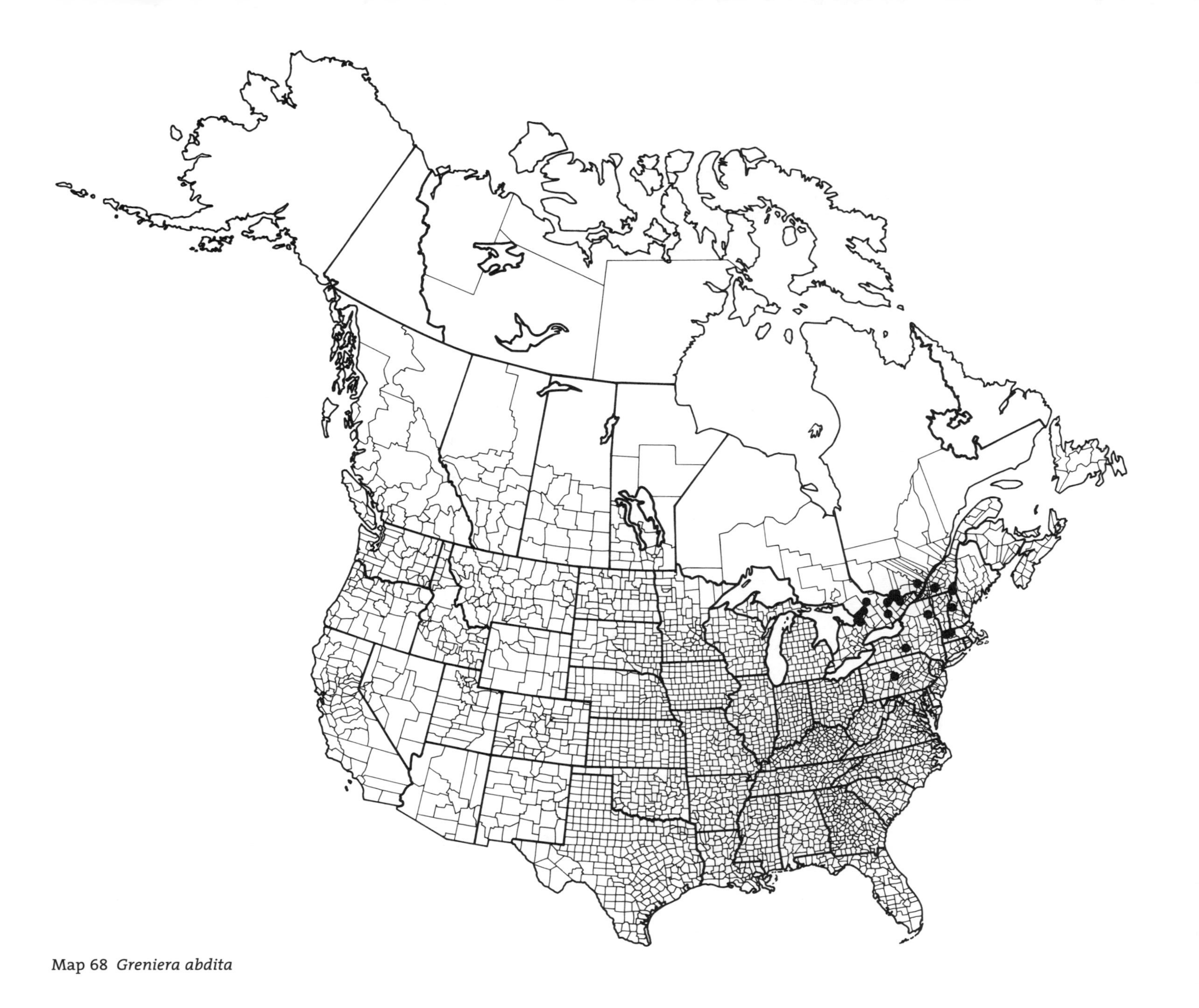

Map 68 *Greniera abdita*

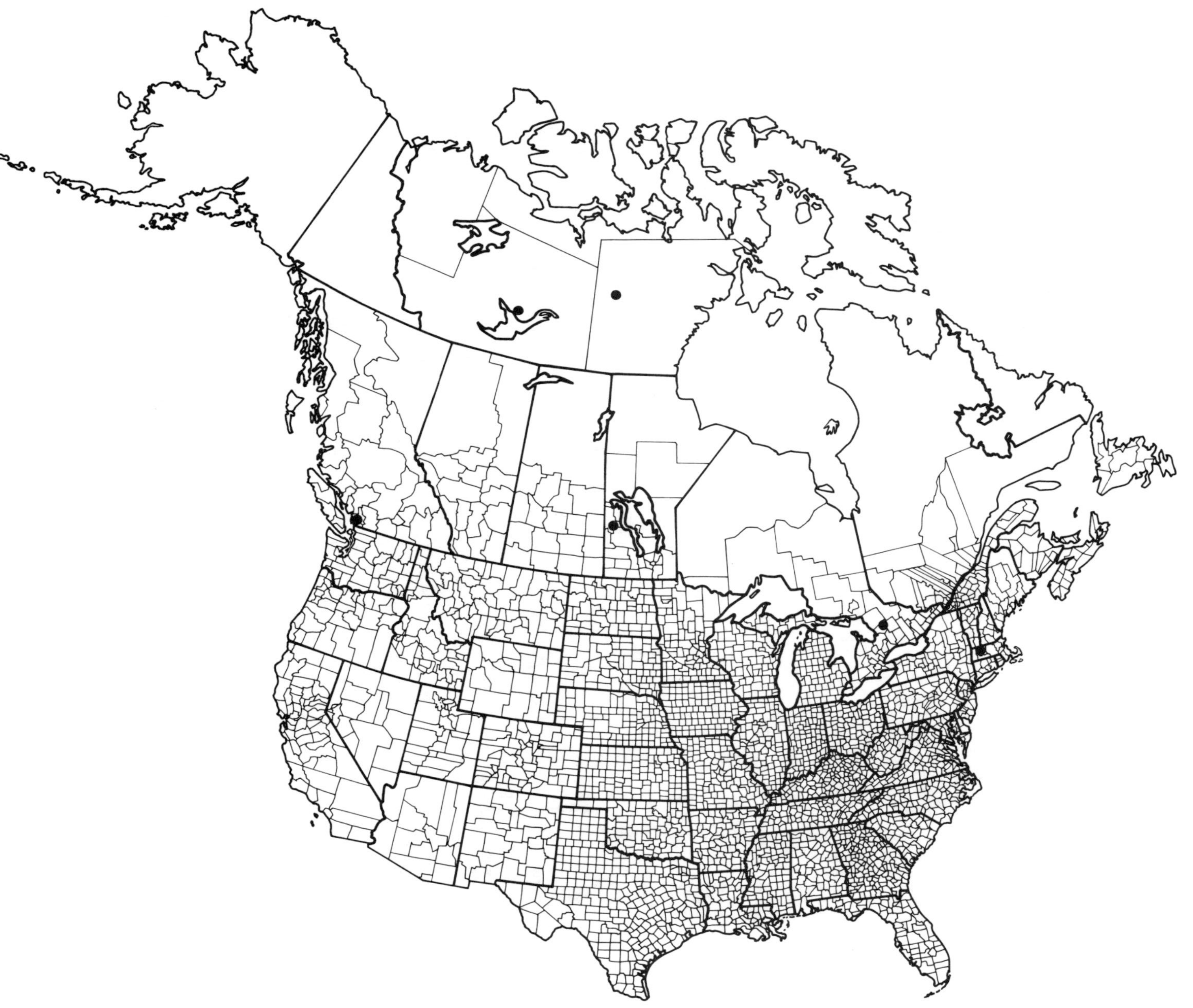

Map 69 *Greniera abditoides*

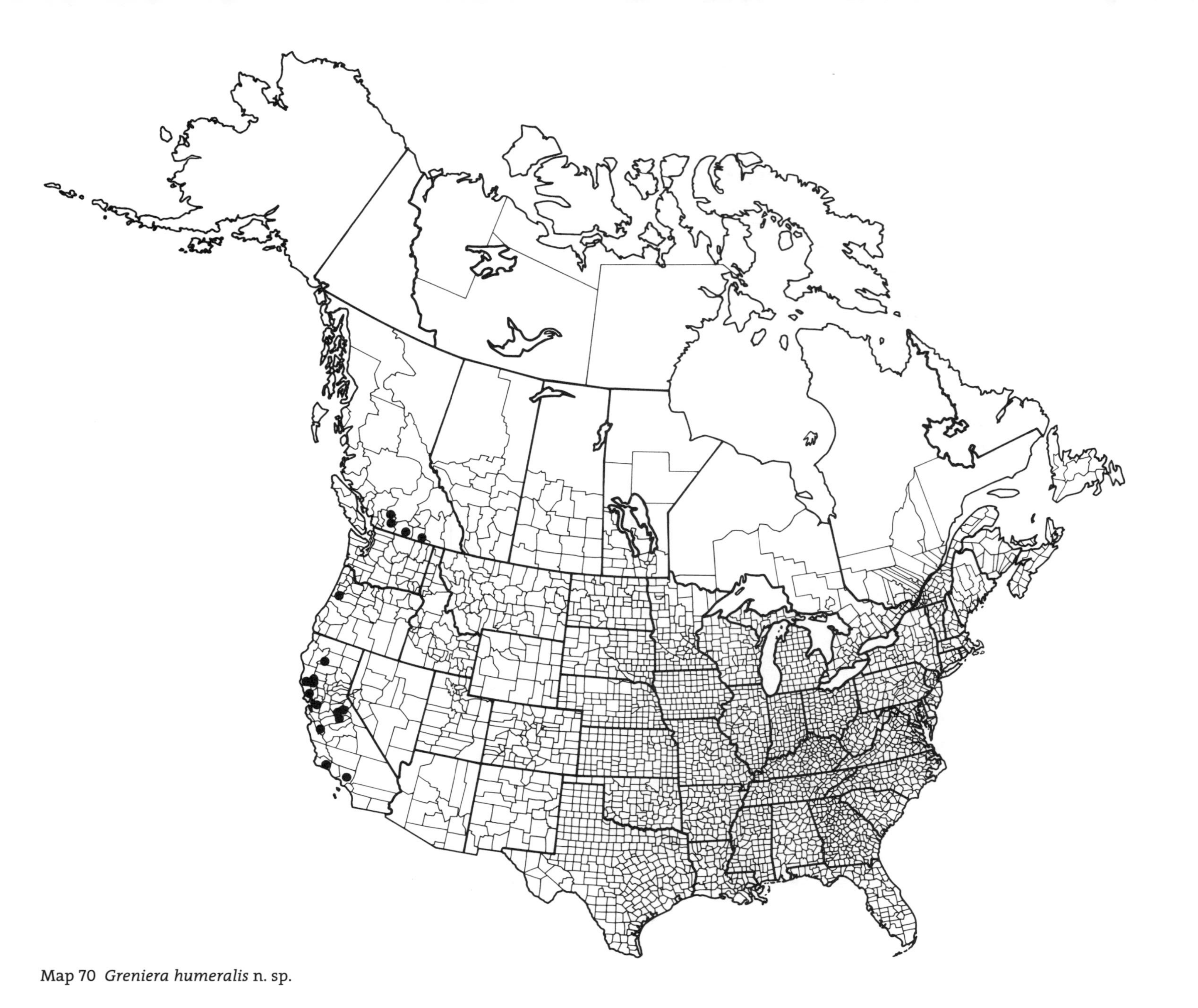

Map 70 *Greniera humeralis* n. sp.

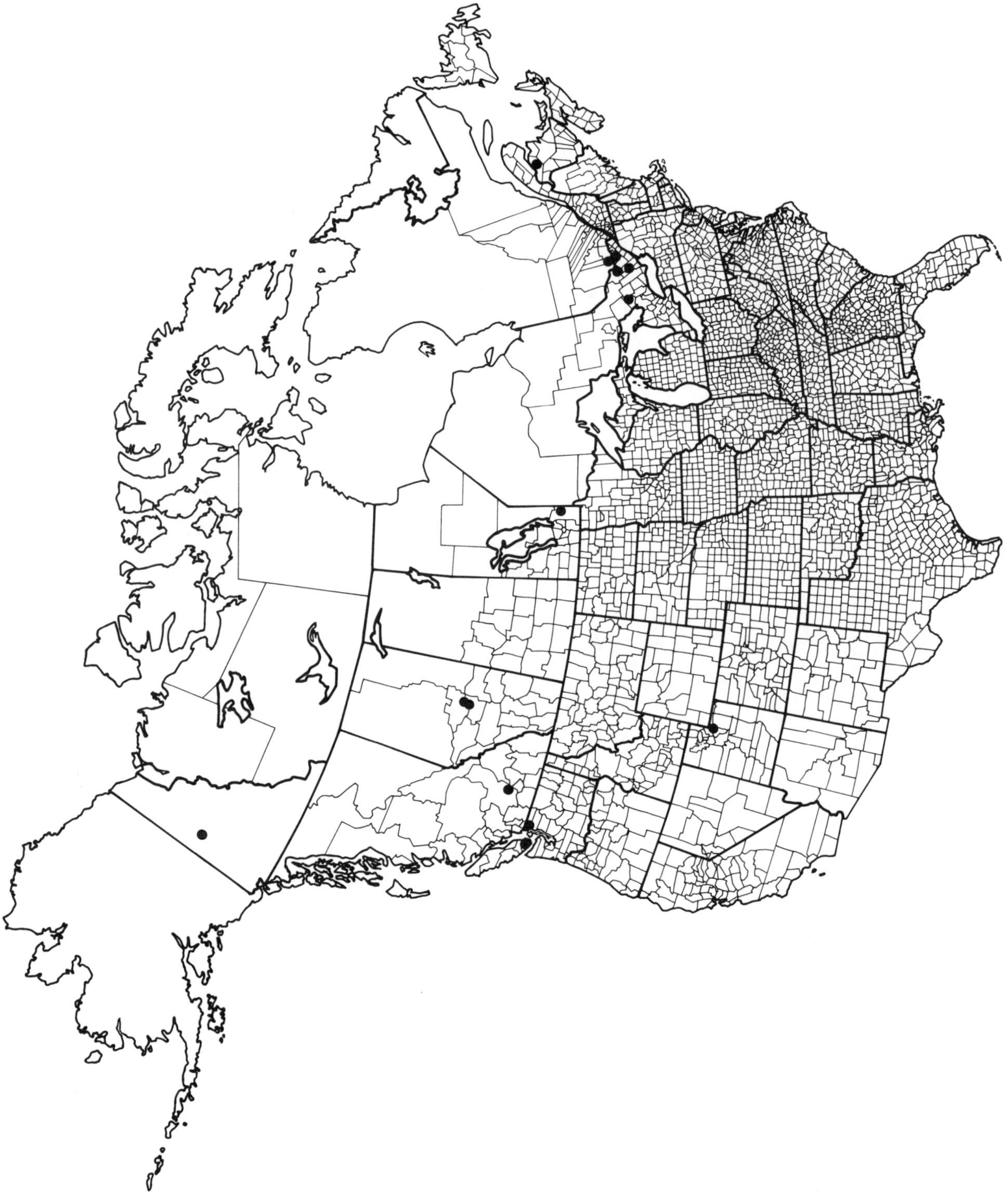

Map 71 *Greniera denaria*

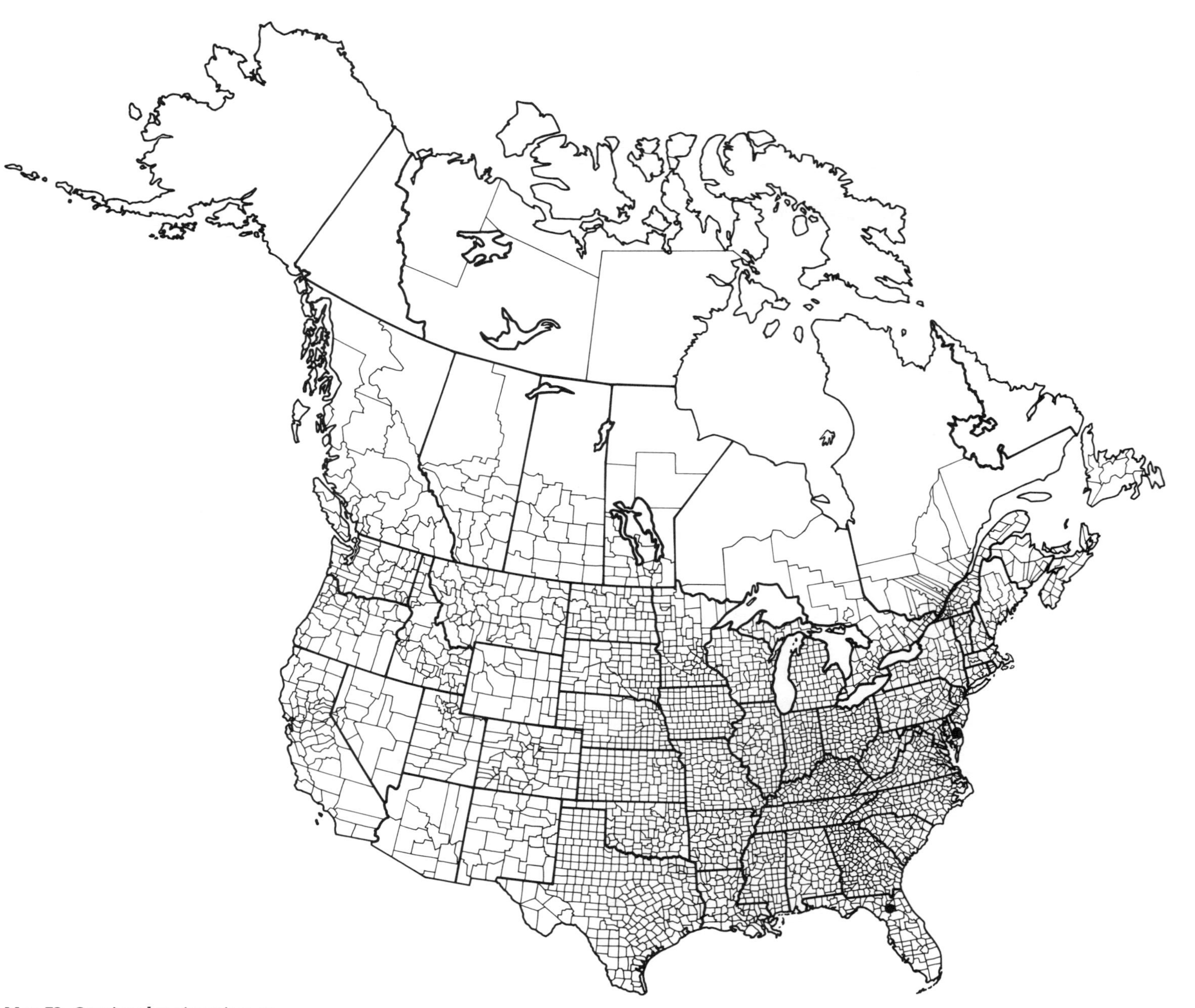

Map 72 *Greniera longicornis* n. sp.

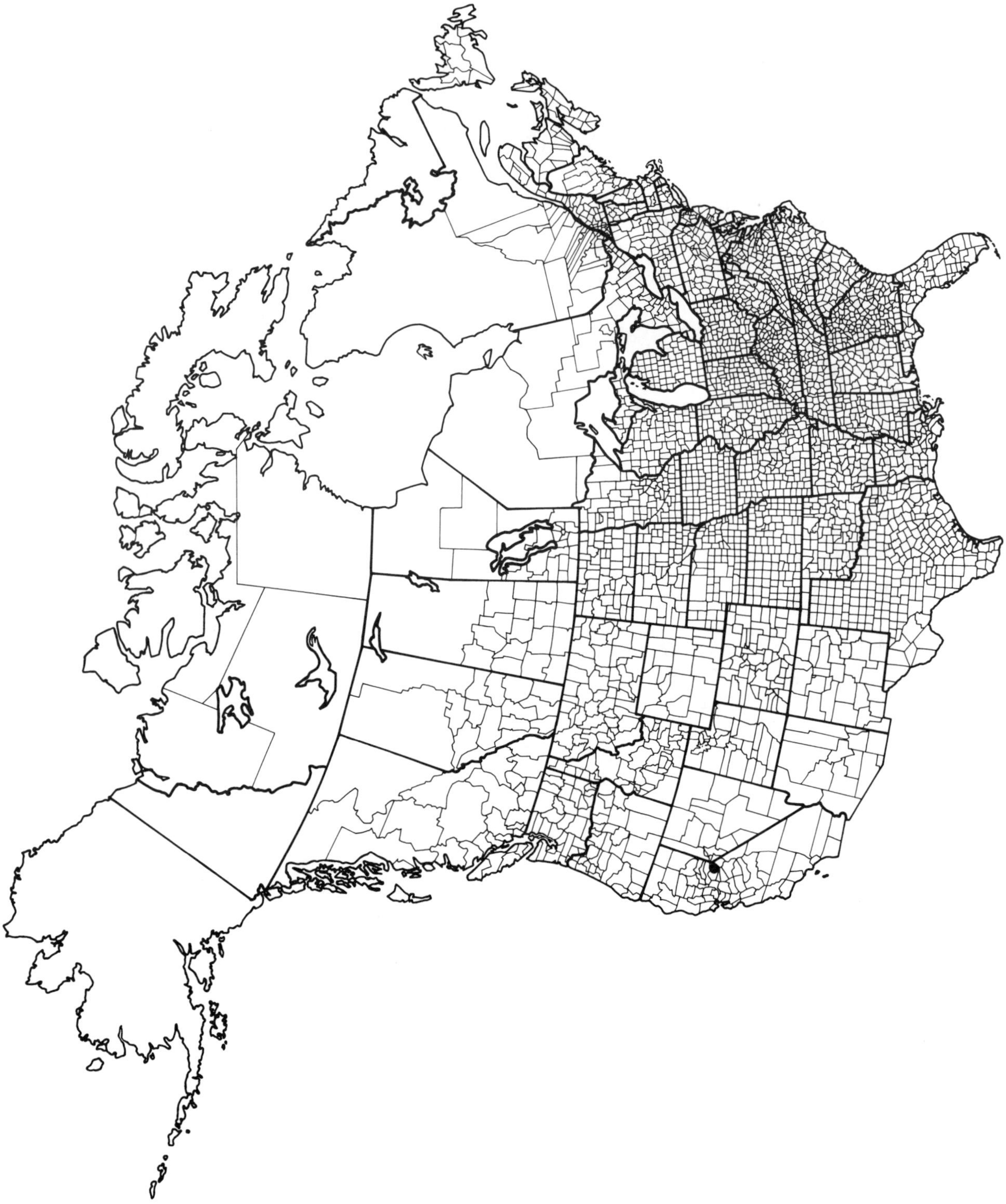

Map 73 *Greniera* 'species F'

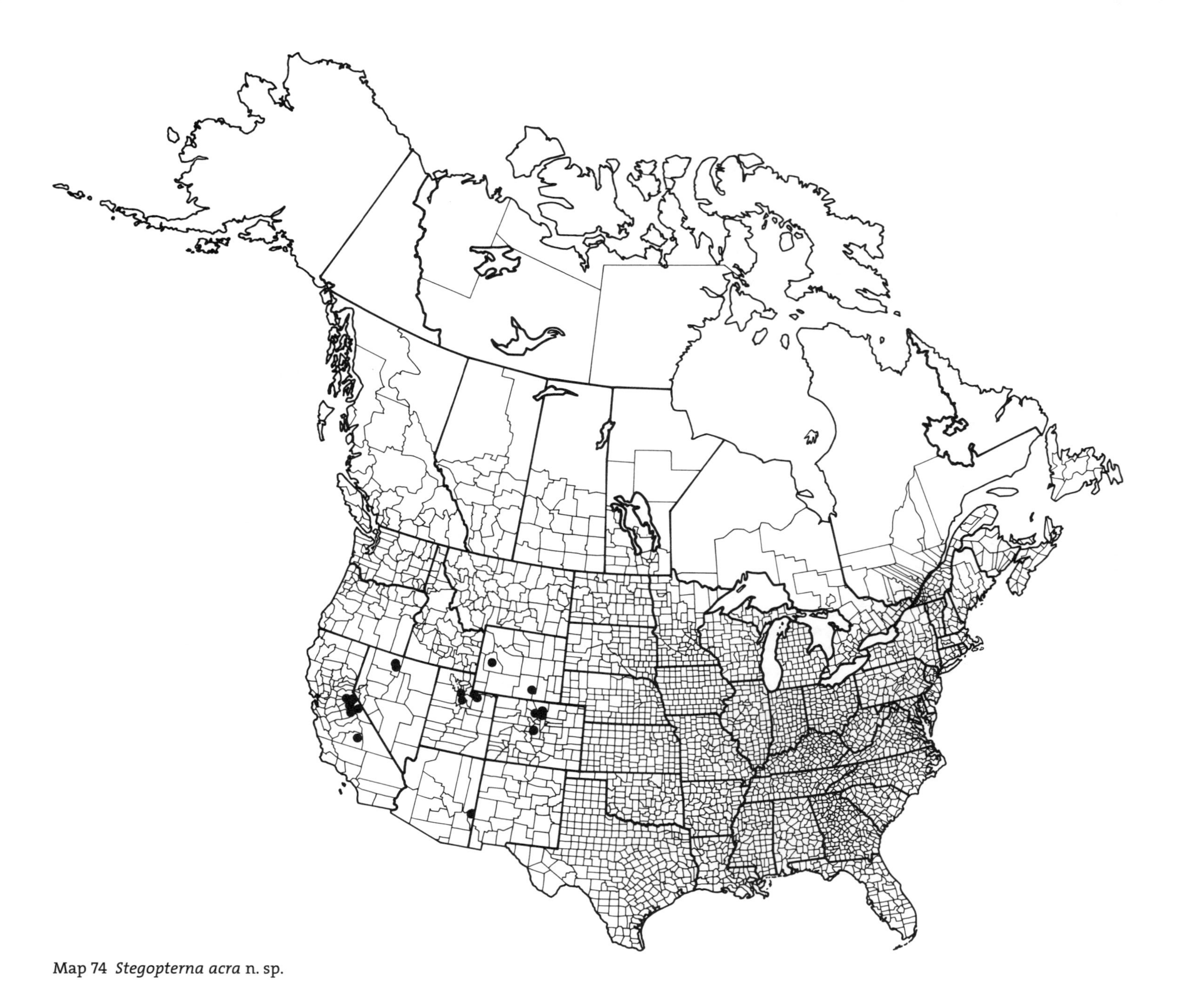

Map 74 *Stegopterna acra* n. sp.

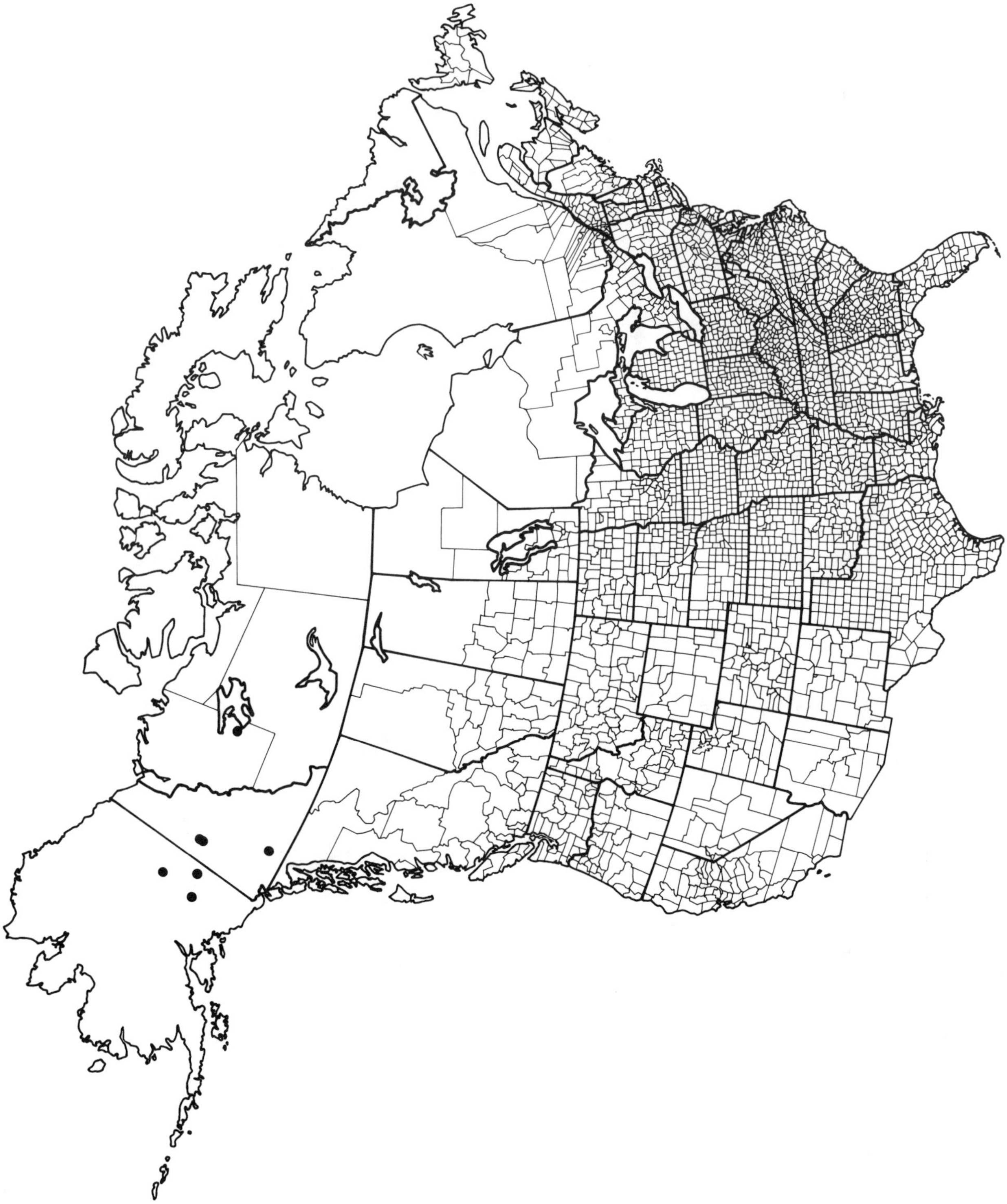

Map 75 *Stegopterna decafilis*

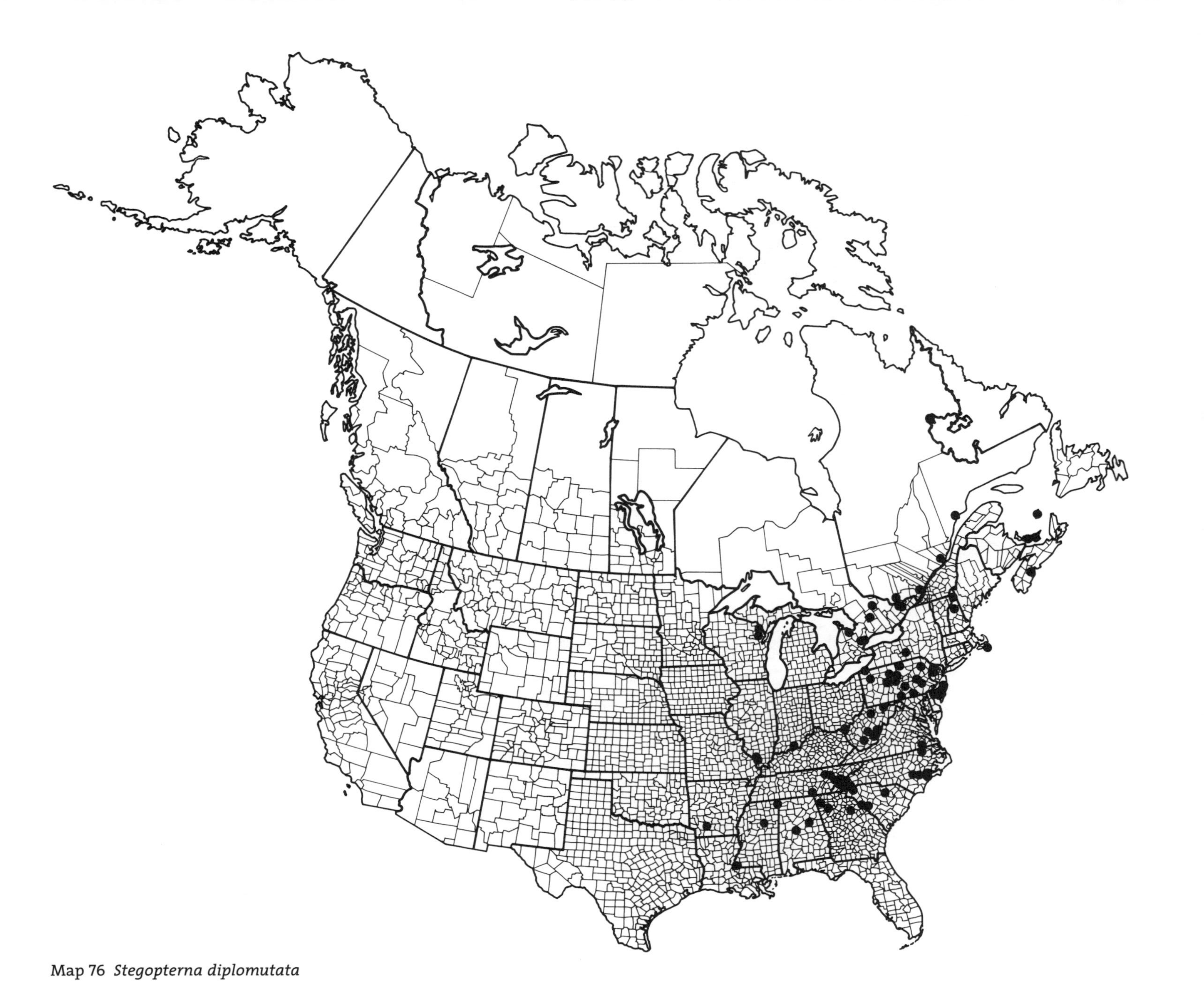

Map 76 *Stegopterna diplomutata*

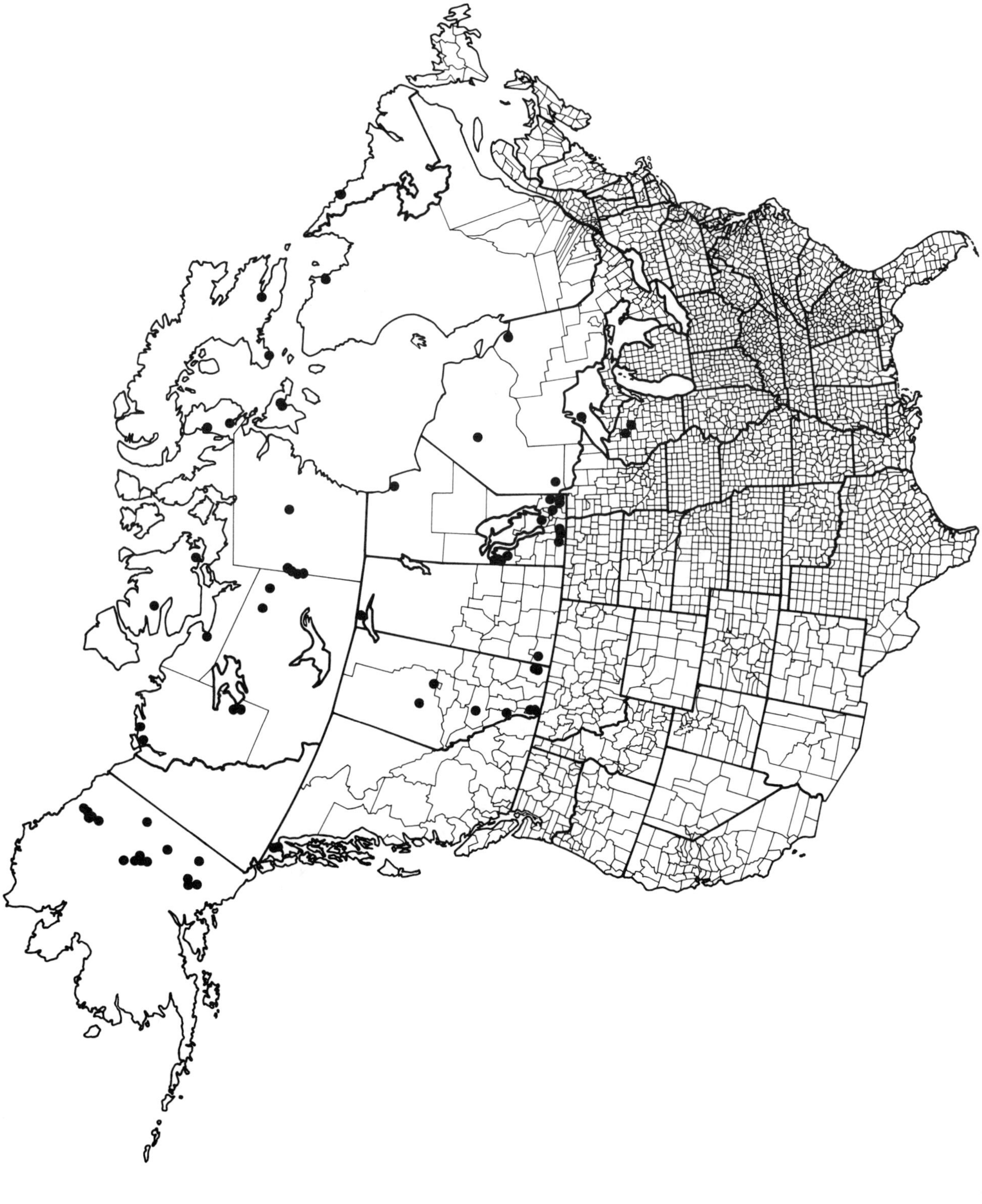

Map 77 *Stegopterna emergens*

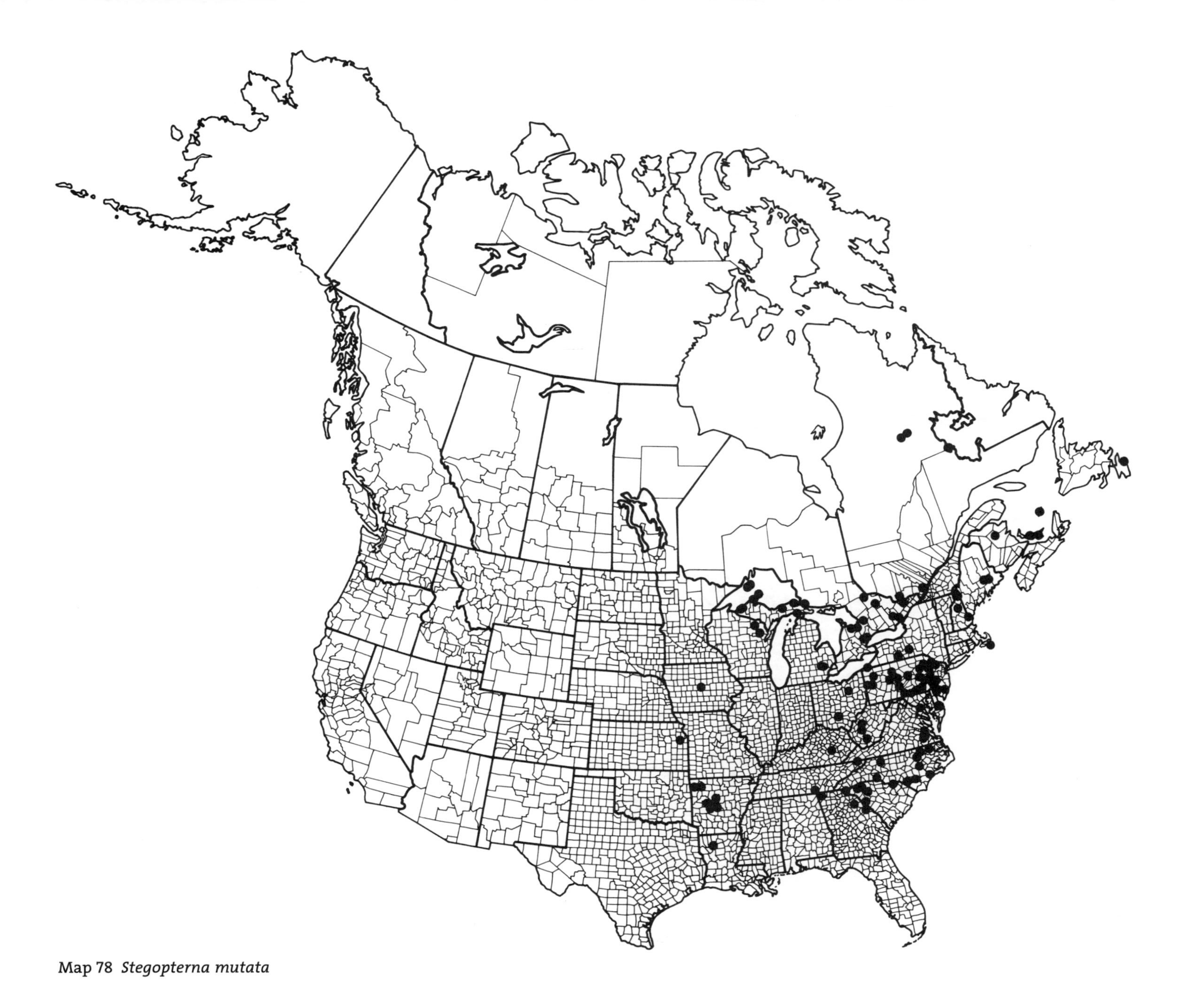

Map 78 *Stegopterna mutata*

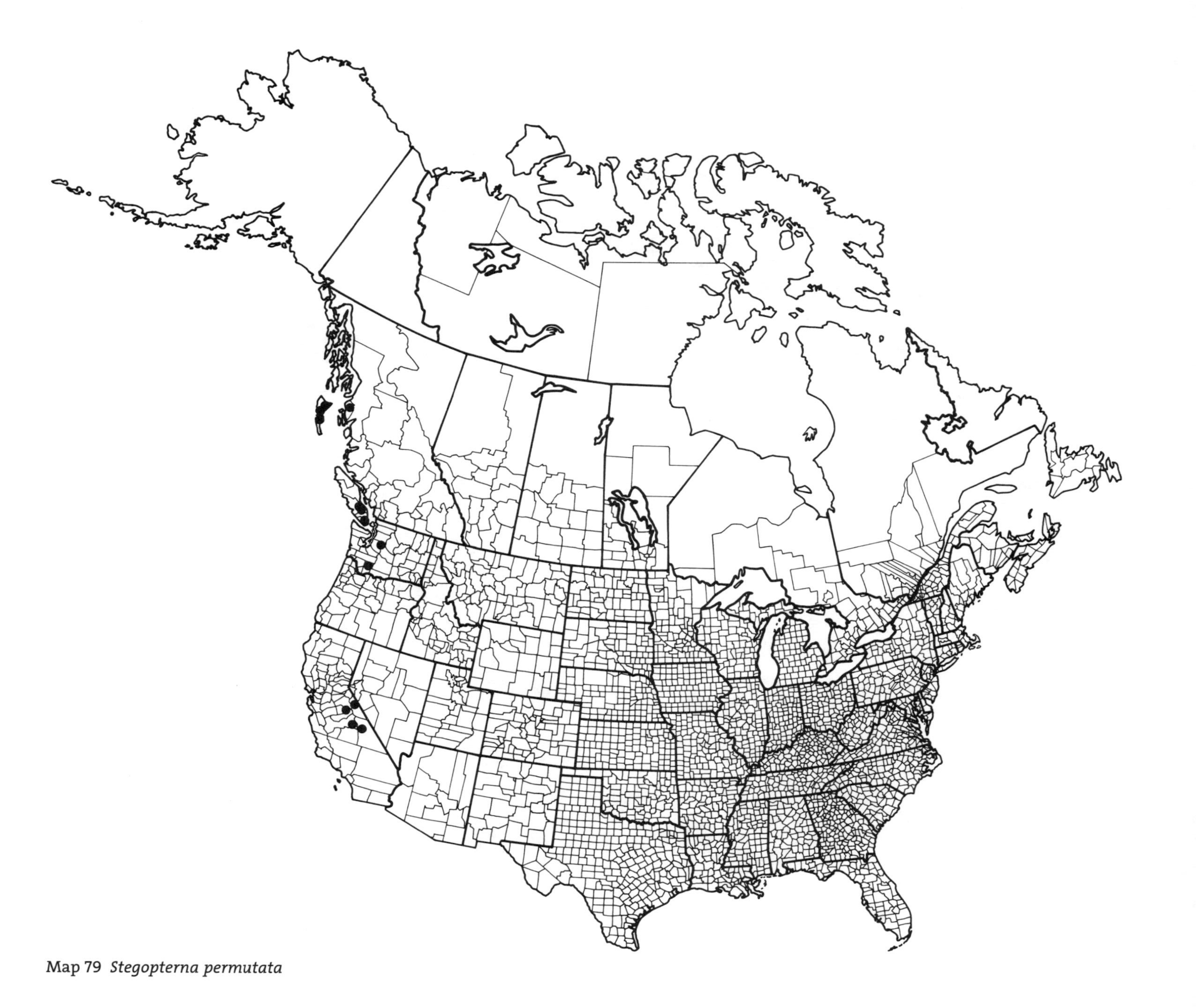

Map 79 *Stegopterna permutata*

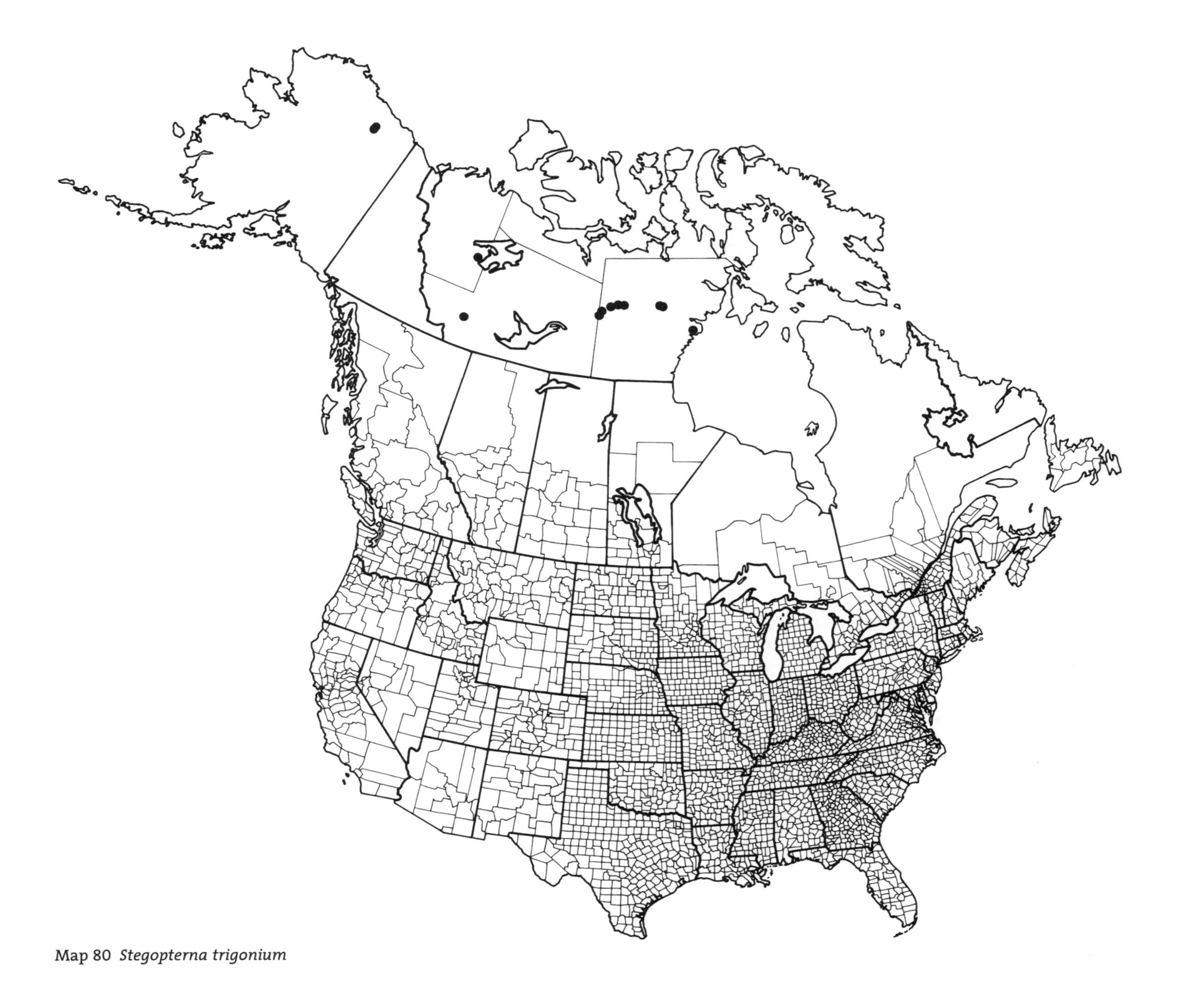

Map 80 *Stegopterna trigonium*

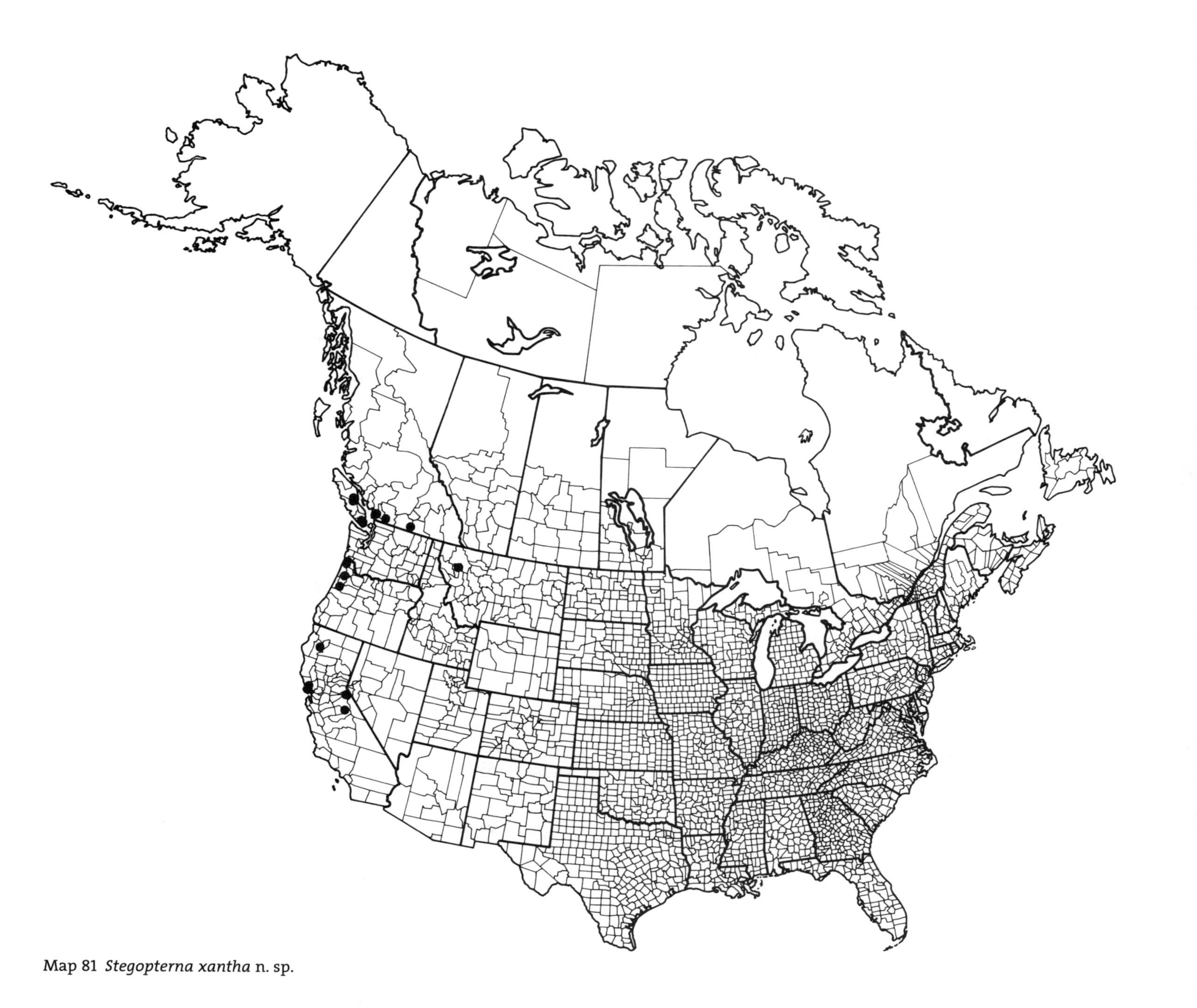

Map 81 *Stegopterna xantha* n. sp.

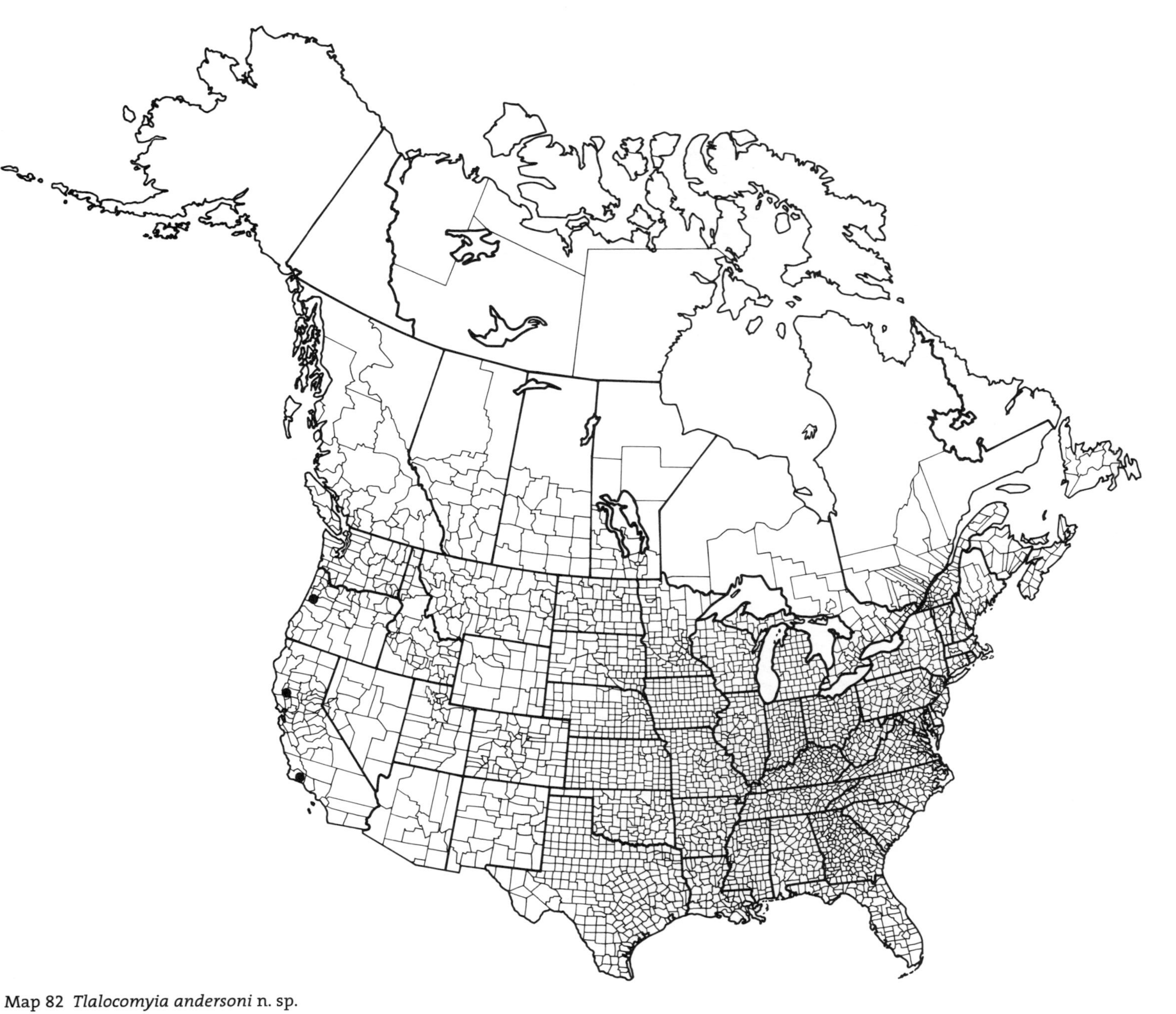

Map 82 *Tlalocomyia andersoni* n. sp.

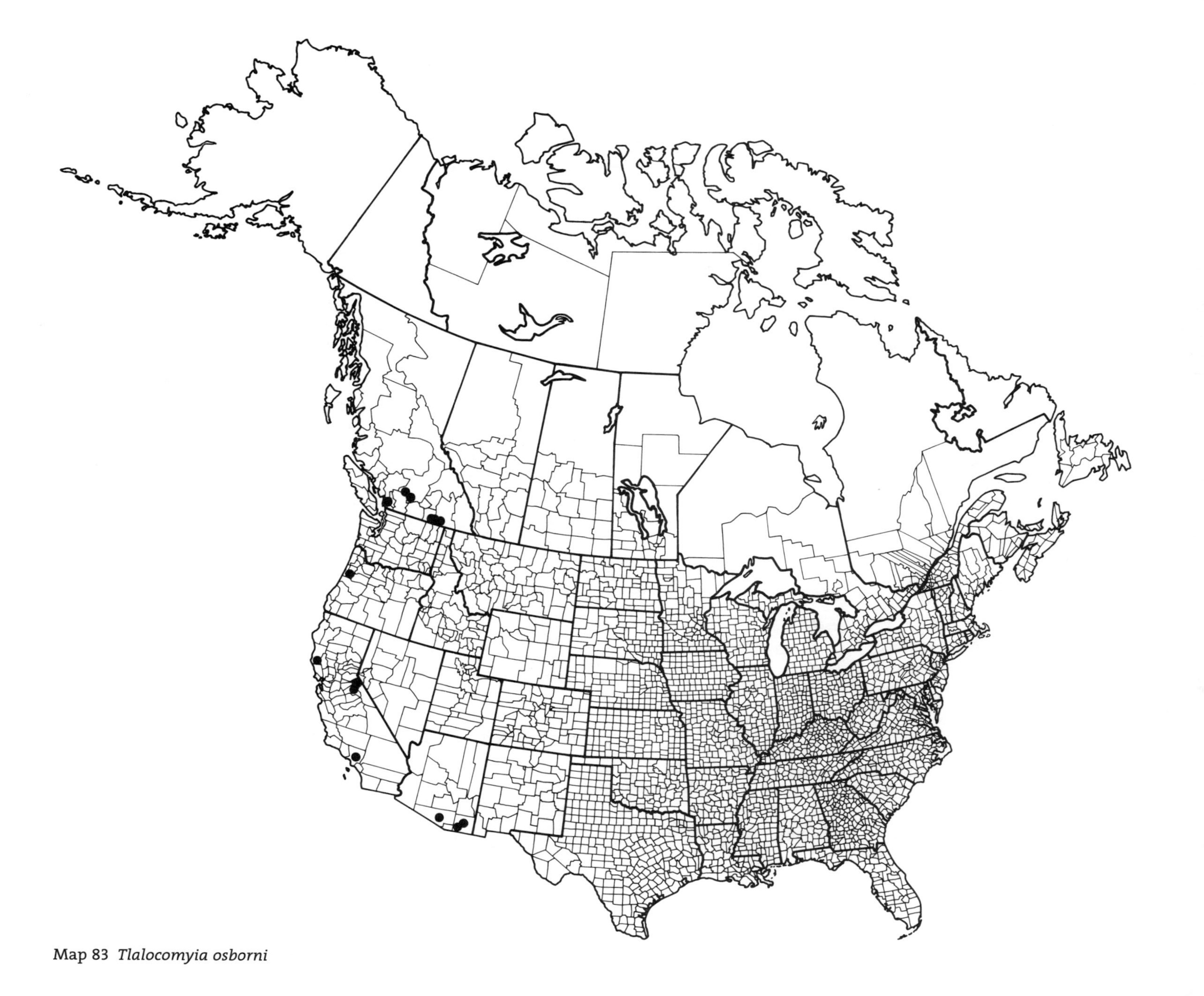

Map 83 *Tlalocomyia osborni*

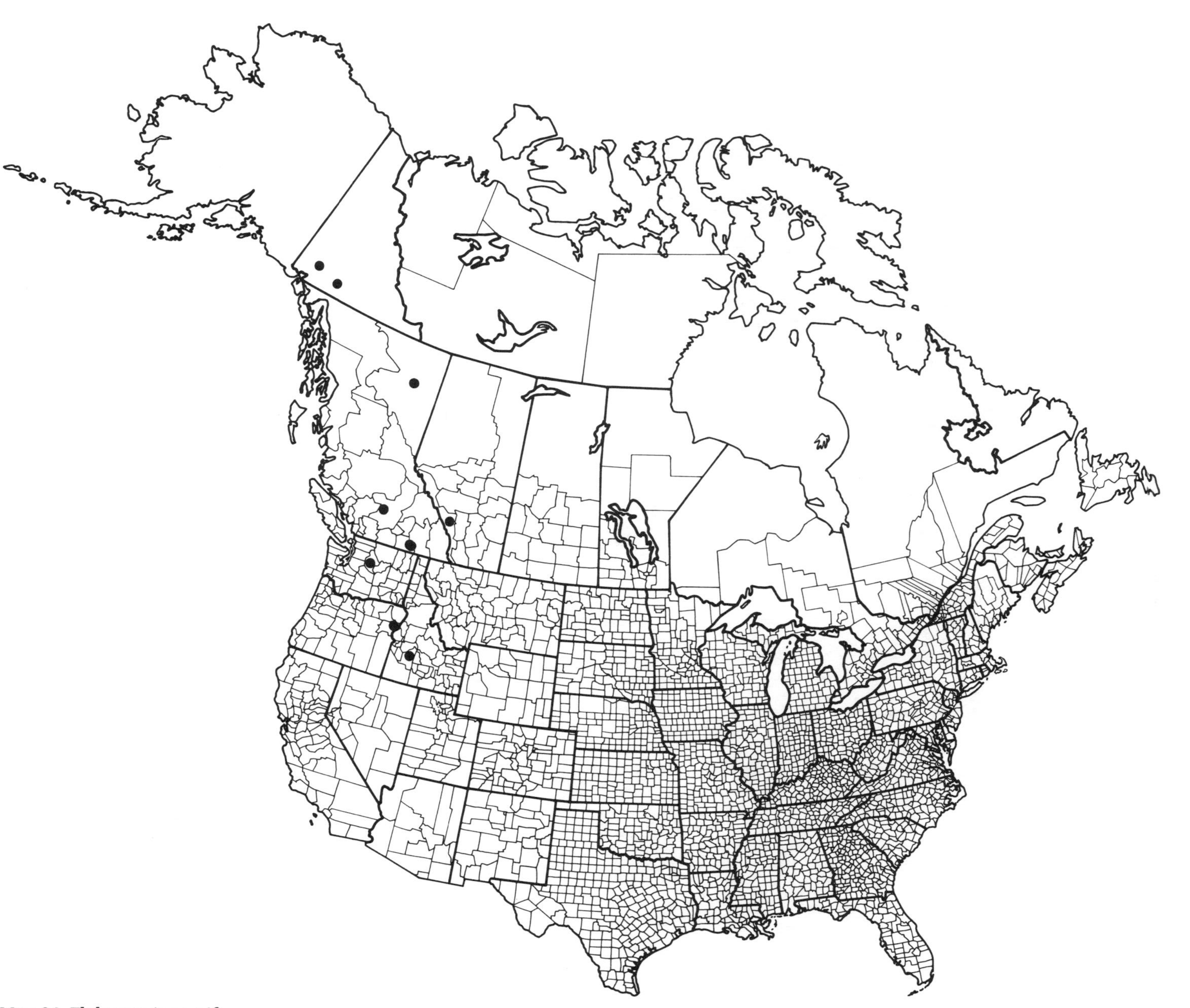

Map 84 *Tlalocomyia ramifera* n. sp.

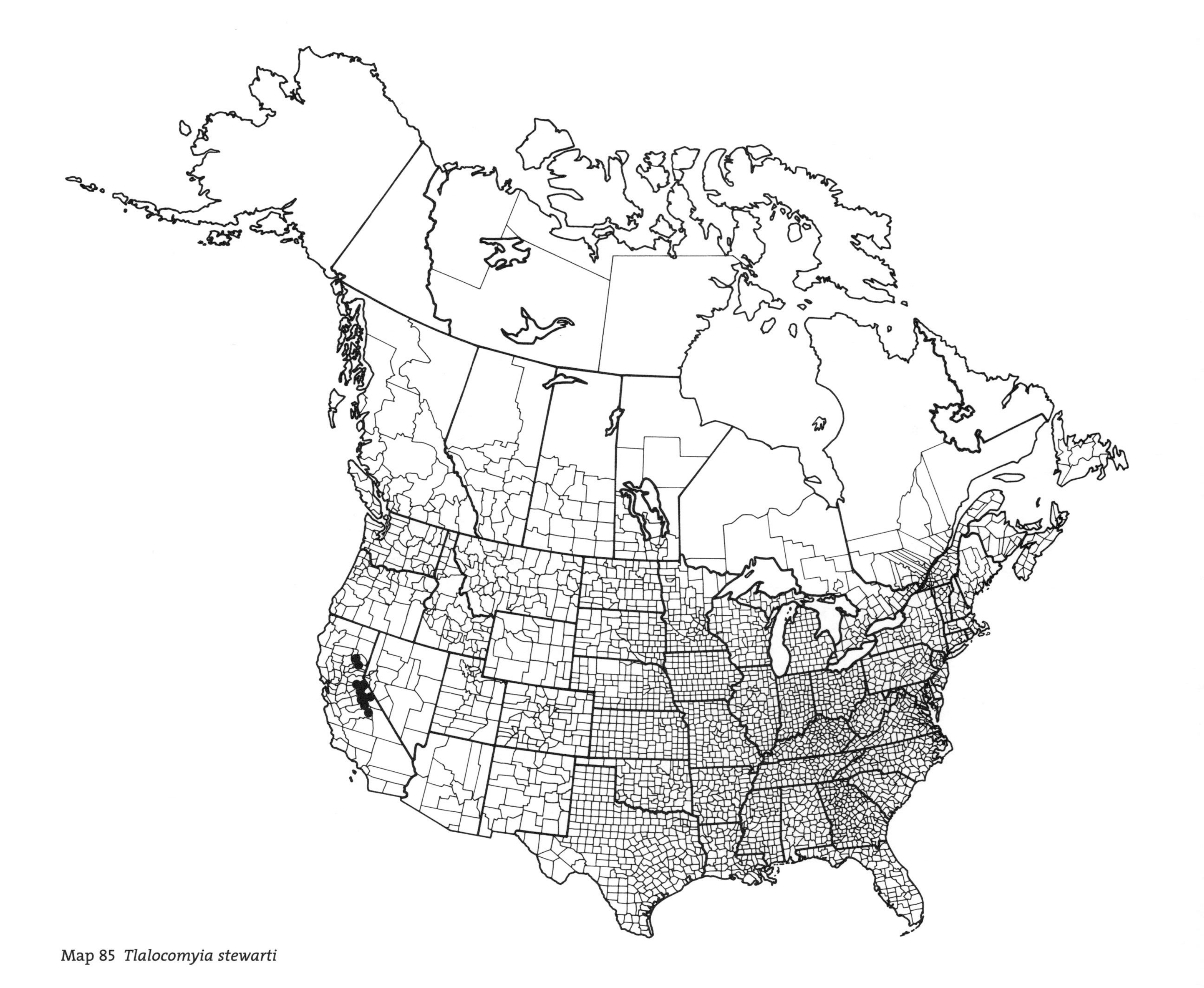

Map 85 *Tlalocomyia stewarti*

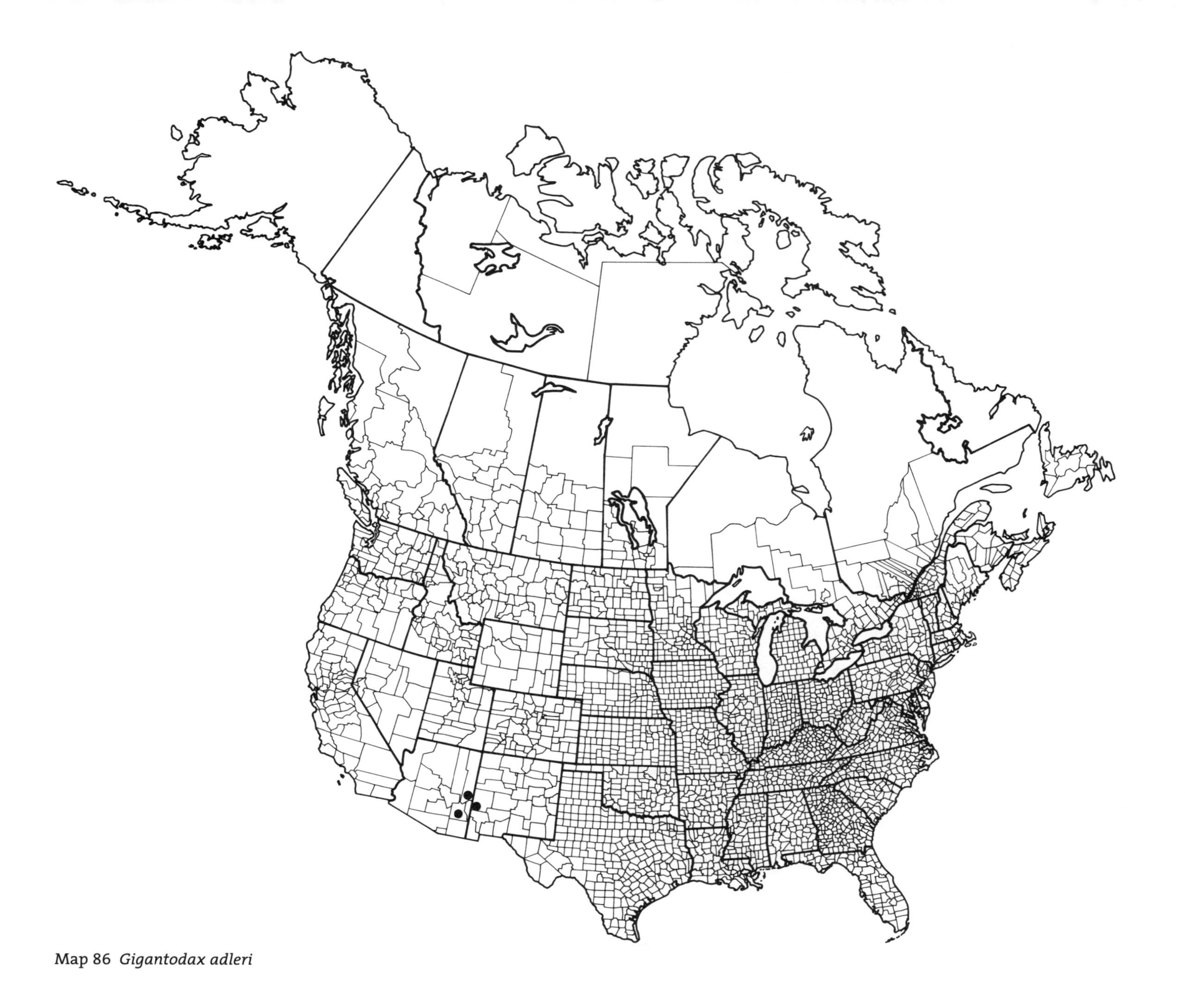

Map 86 *Gigantodax adleri*

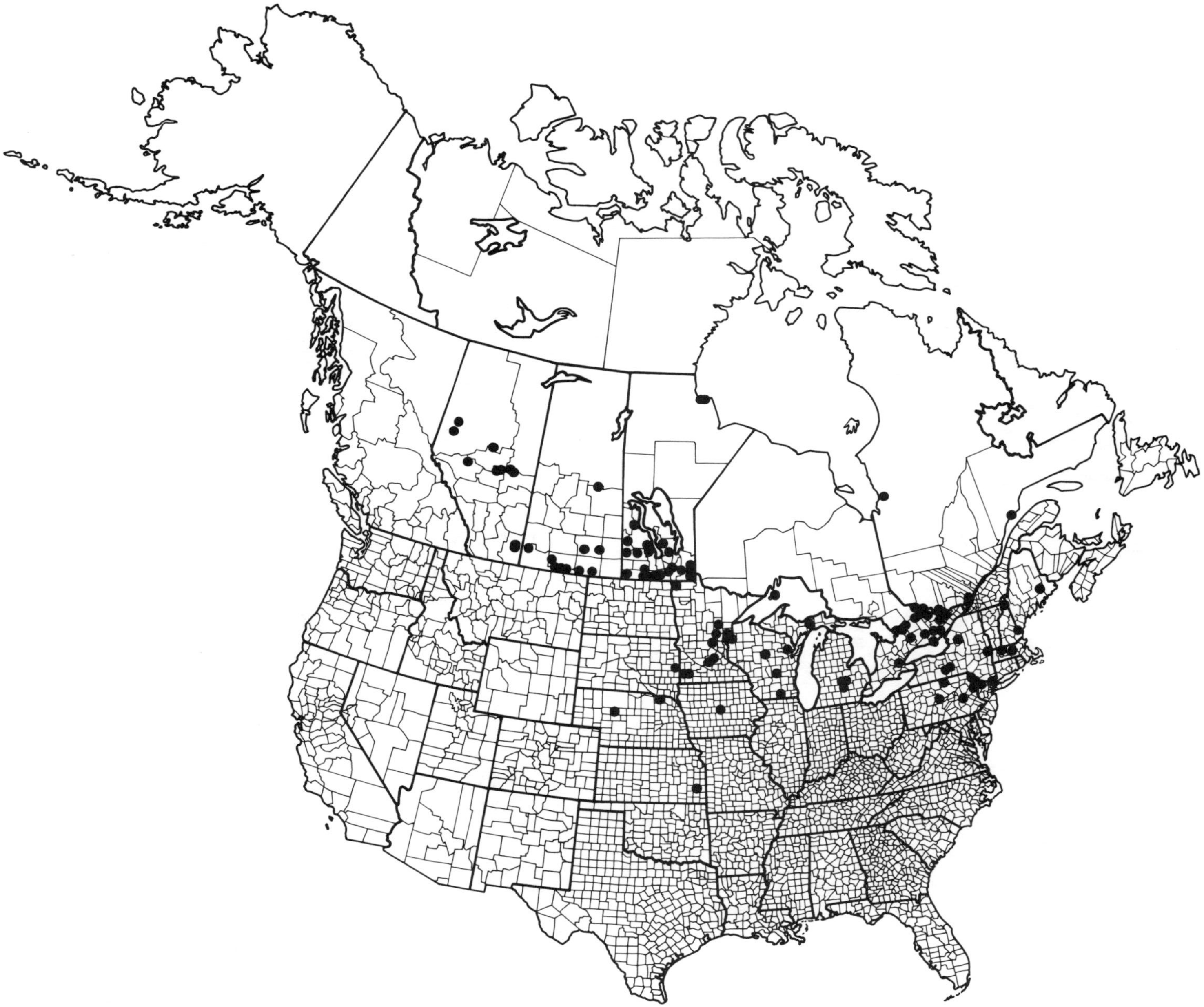

Map 87 *Cnephia dacotensis*

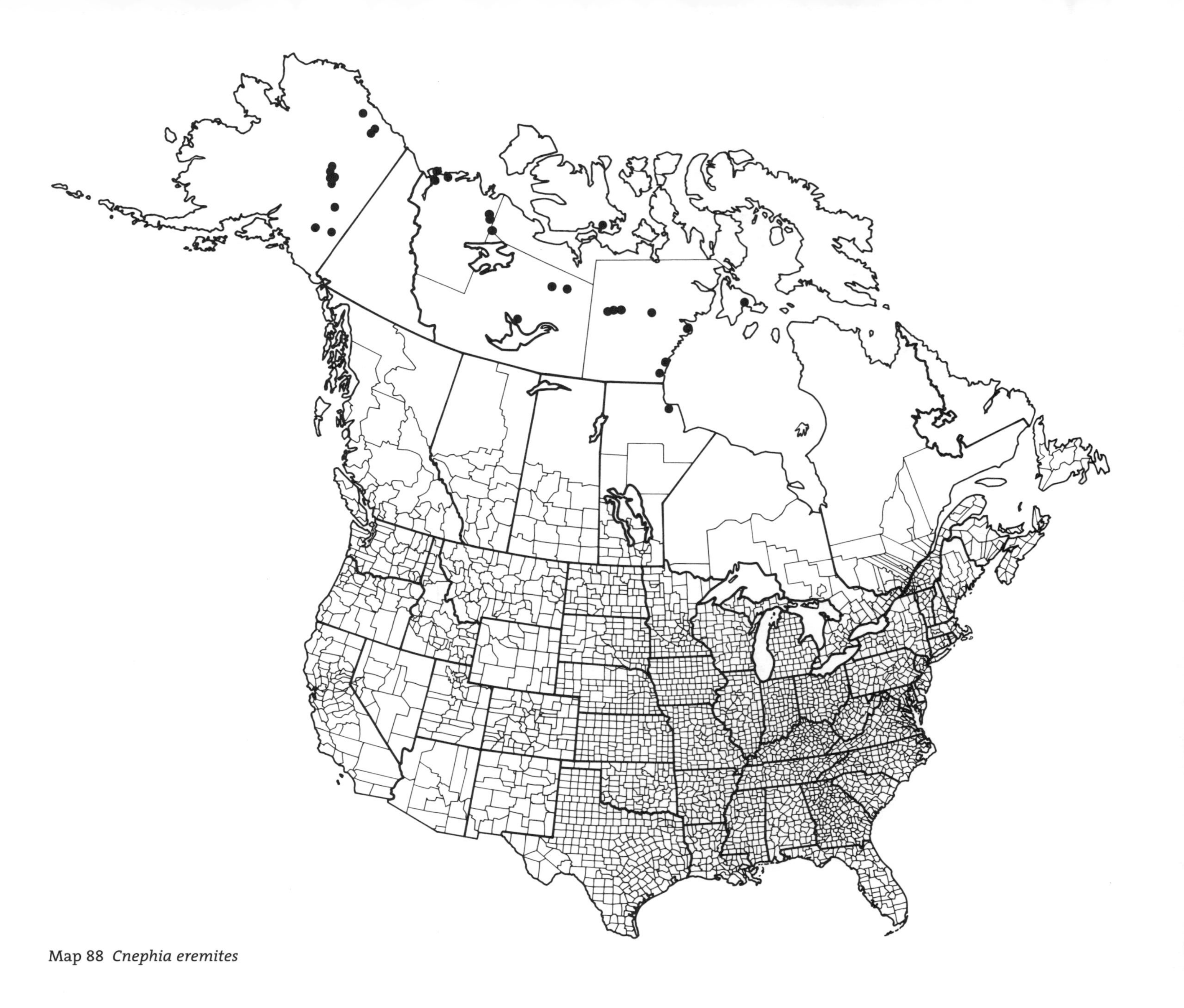

Map 88 *Cnephia eremites*

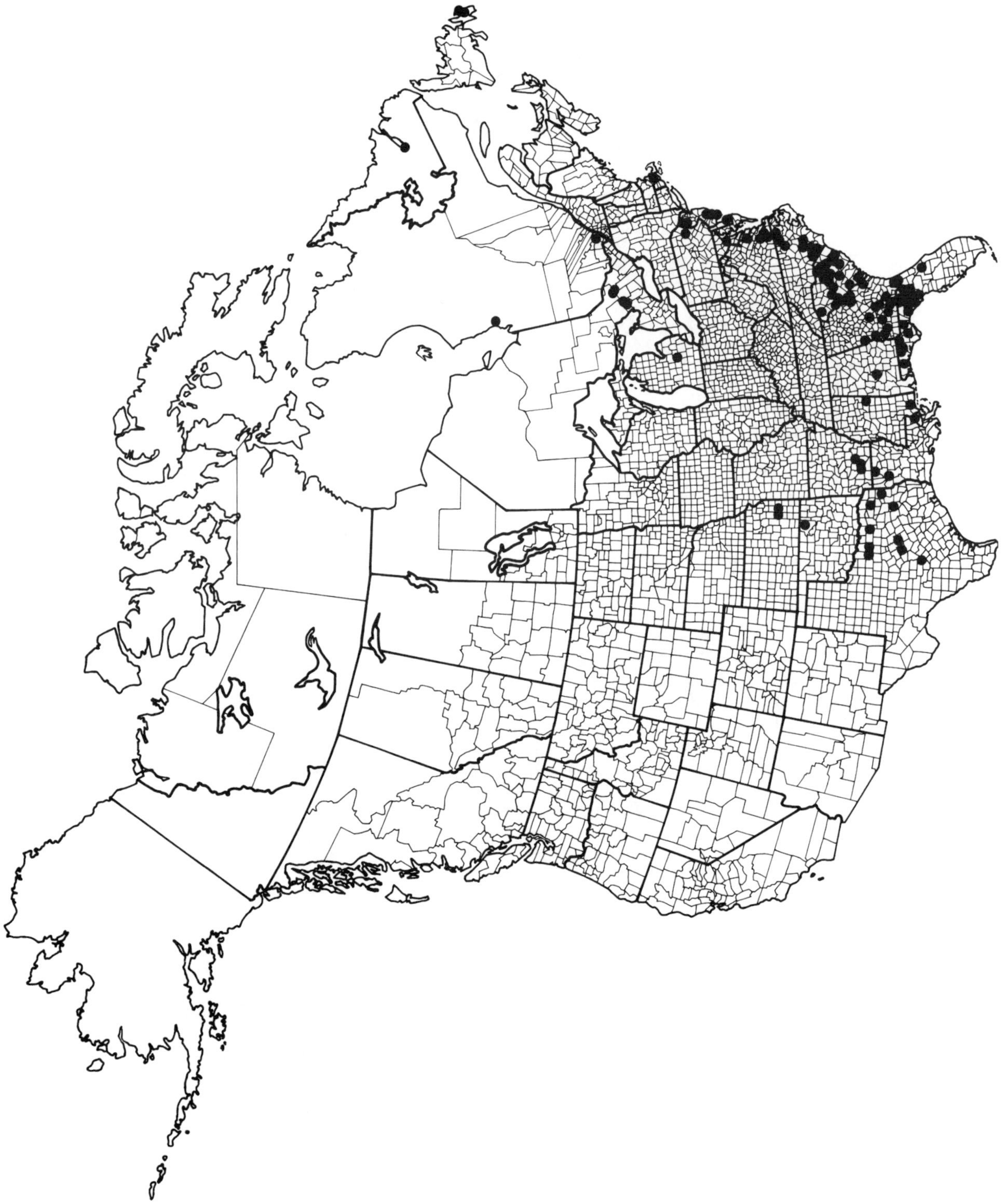

Map 89 *Cnephia ornithophilia*

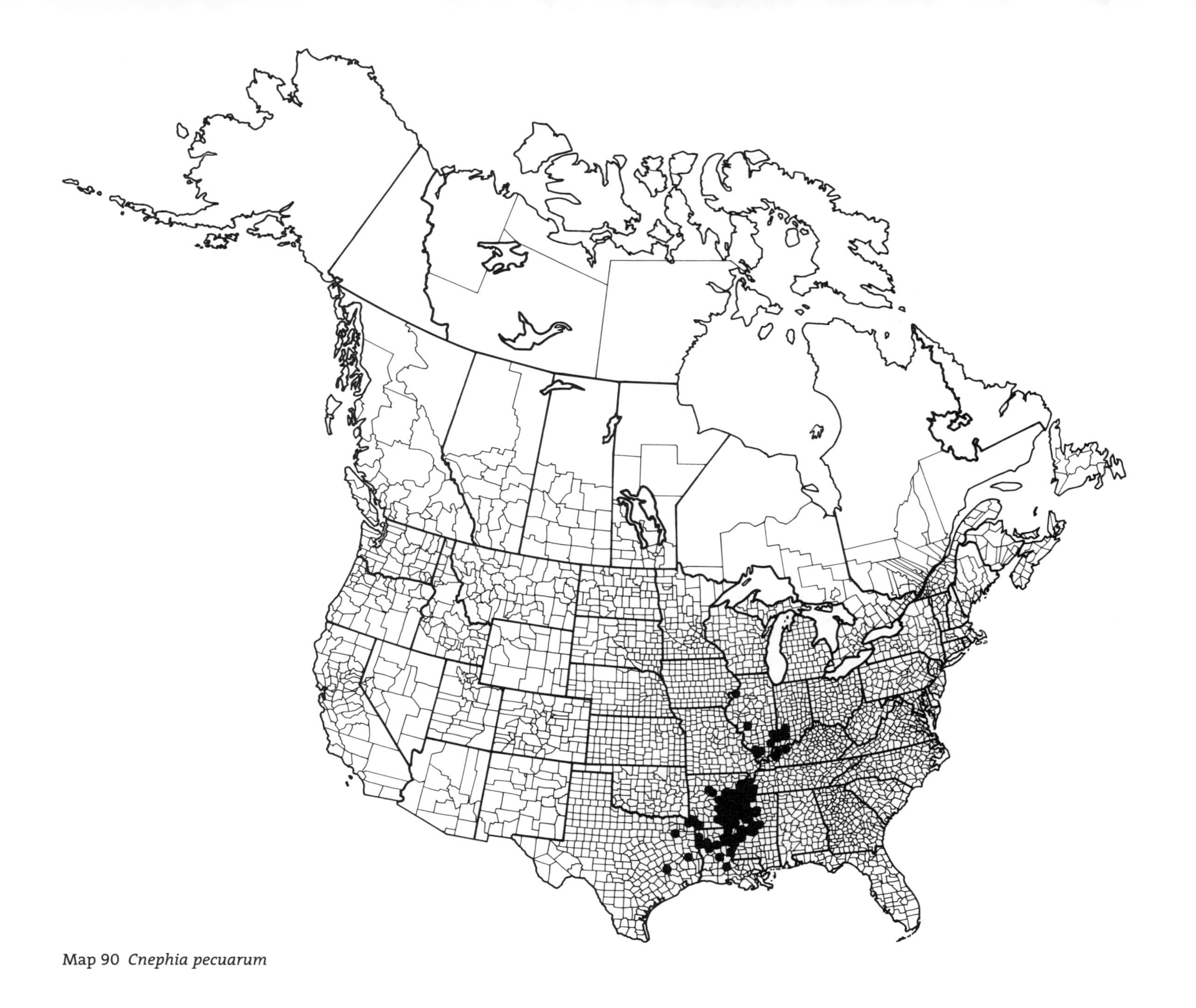

Map 90 *Cnephia pecuarum*

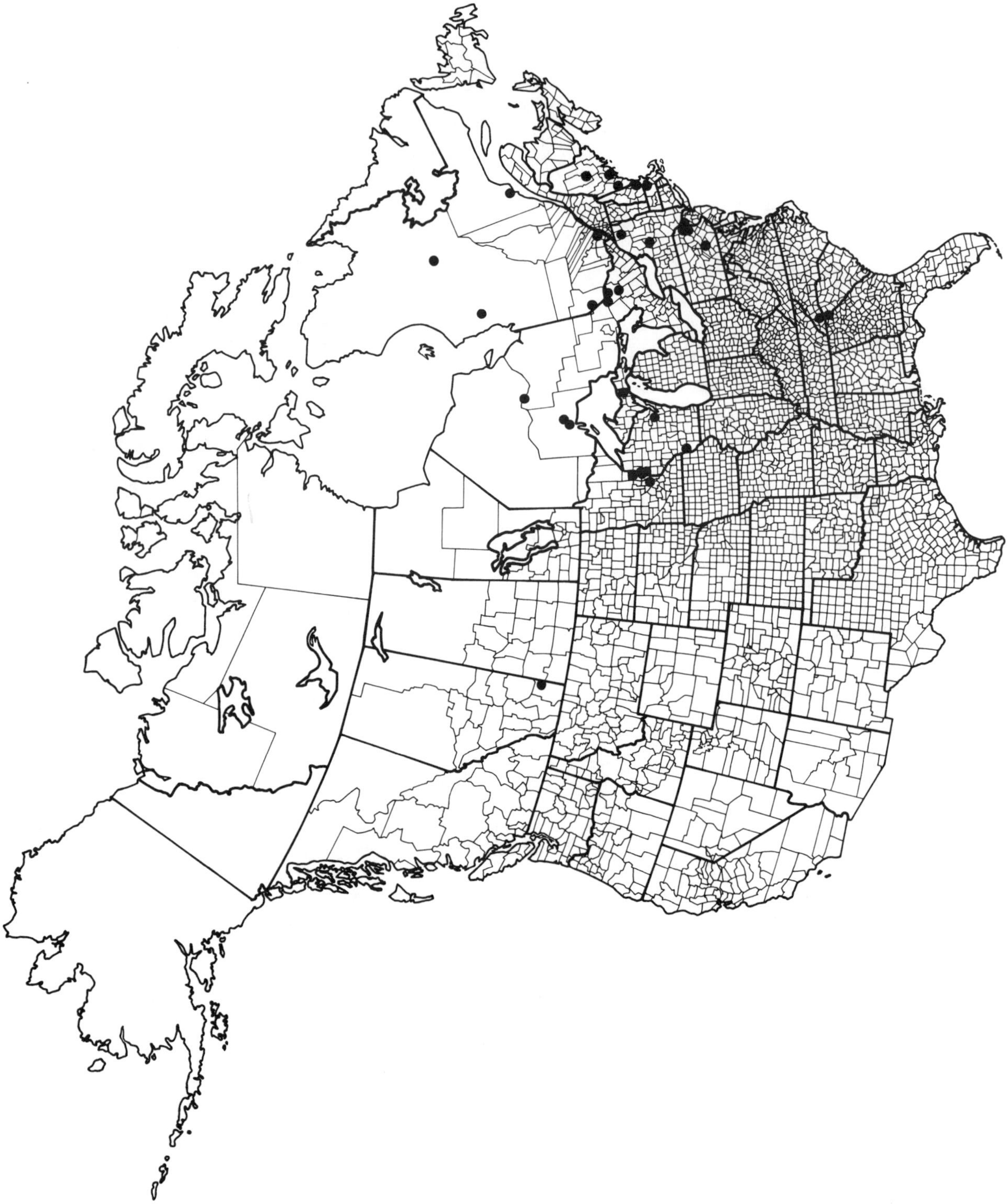

Map 91 *Ectemnia invenusta*

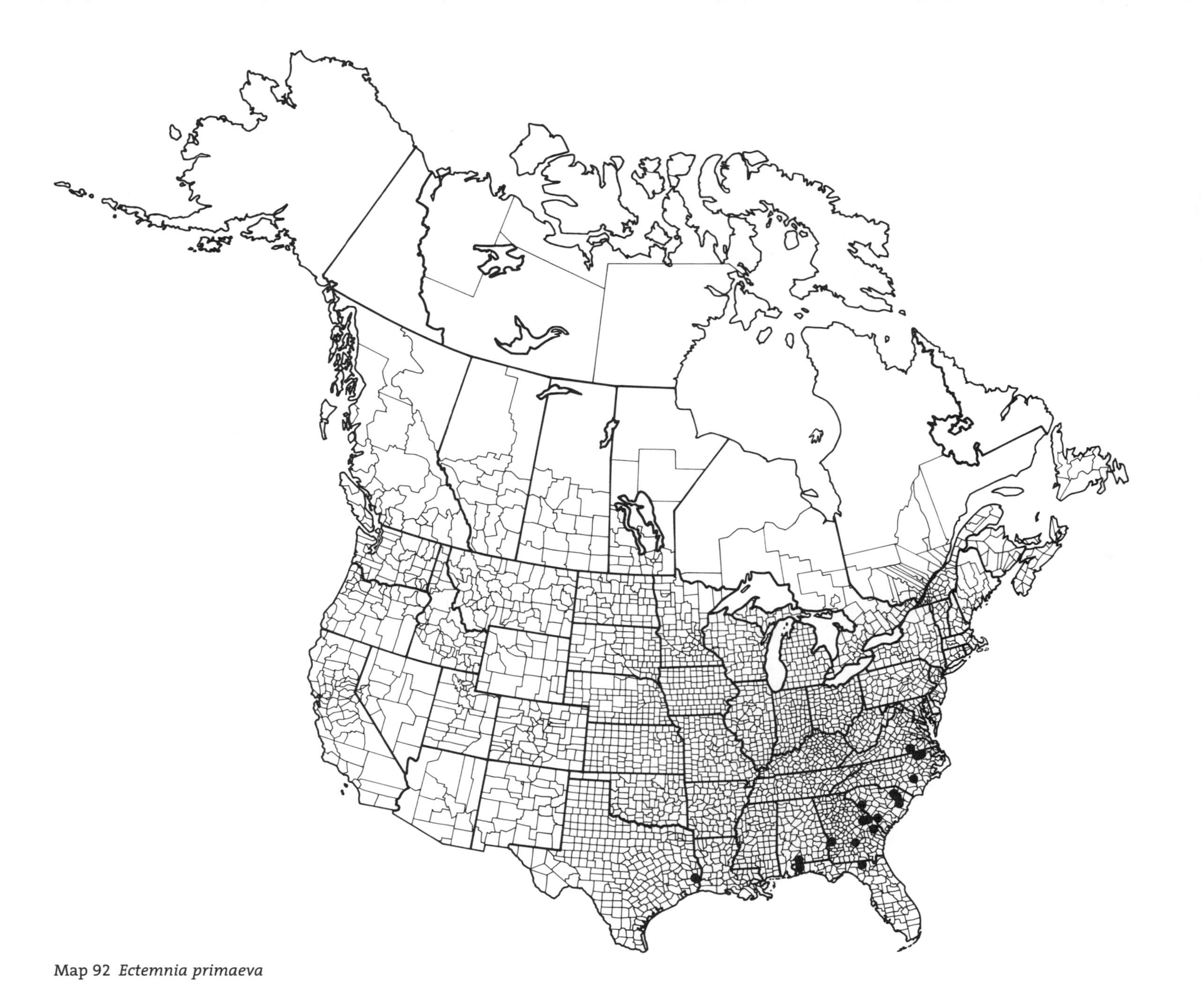

Map 92 *Ectemnia primaeva*

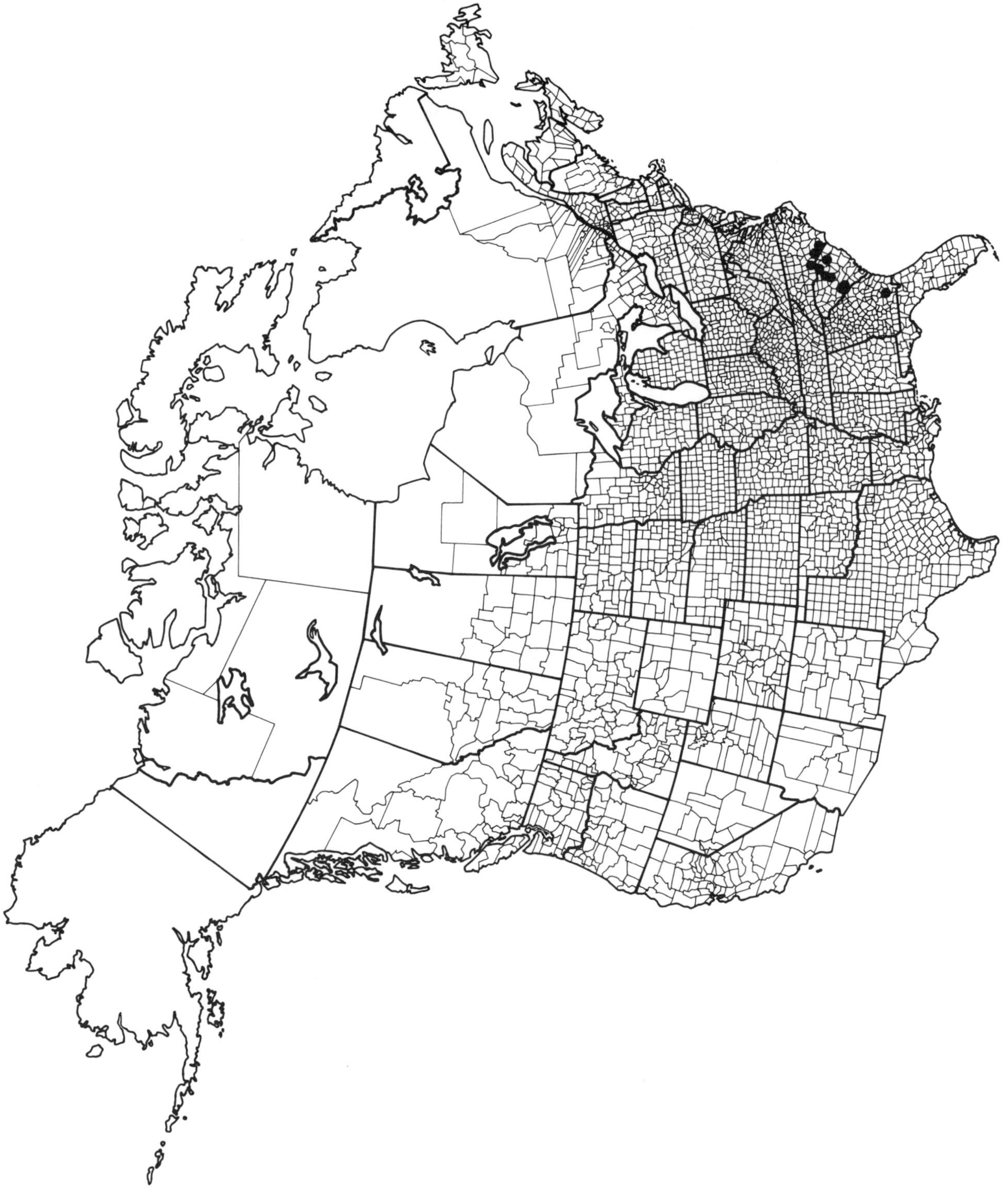

Map 93 *Ectemnia reclusa*

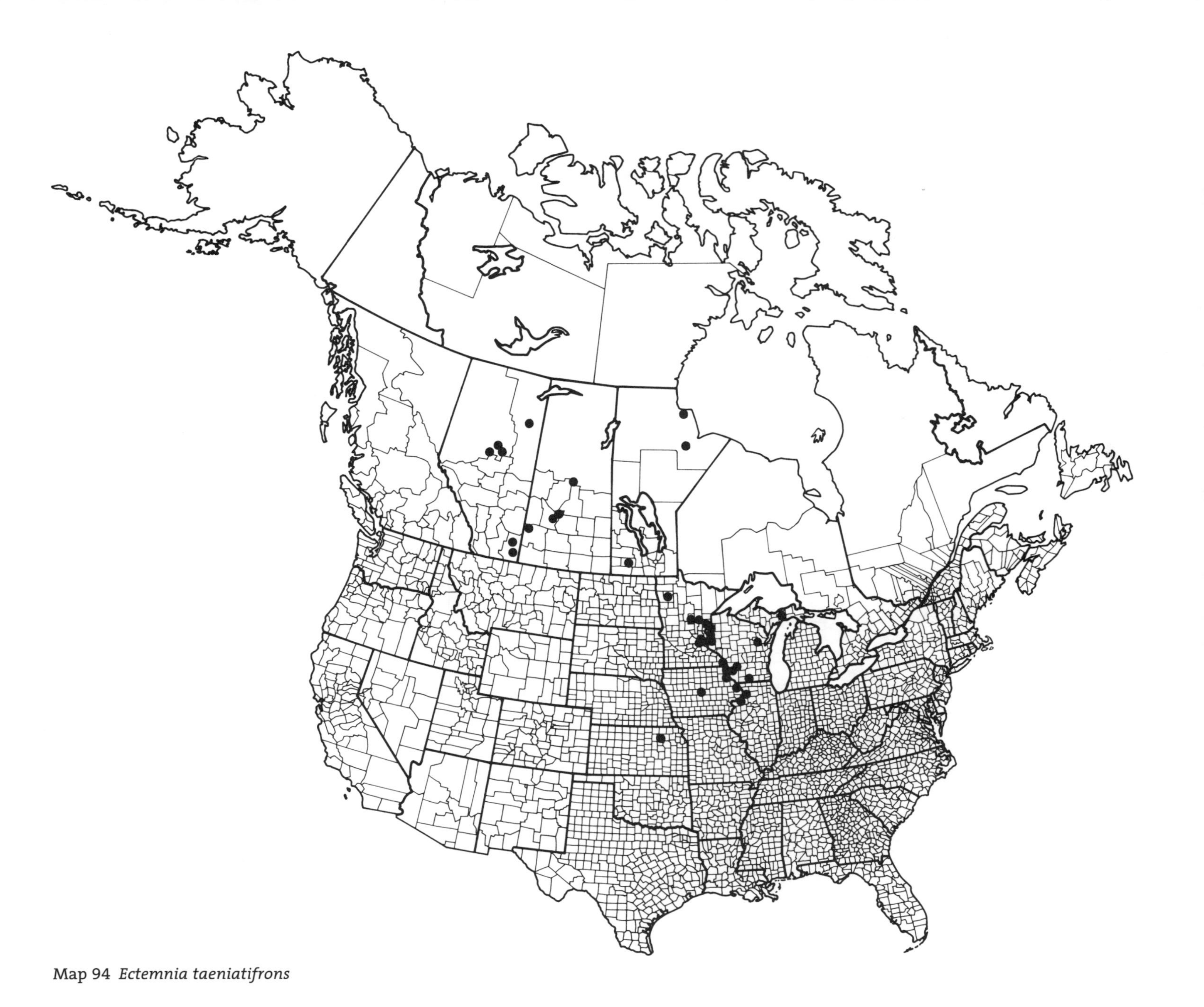

Map 94 *Ectemnia taeniatifrons*

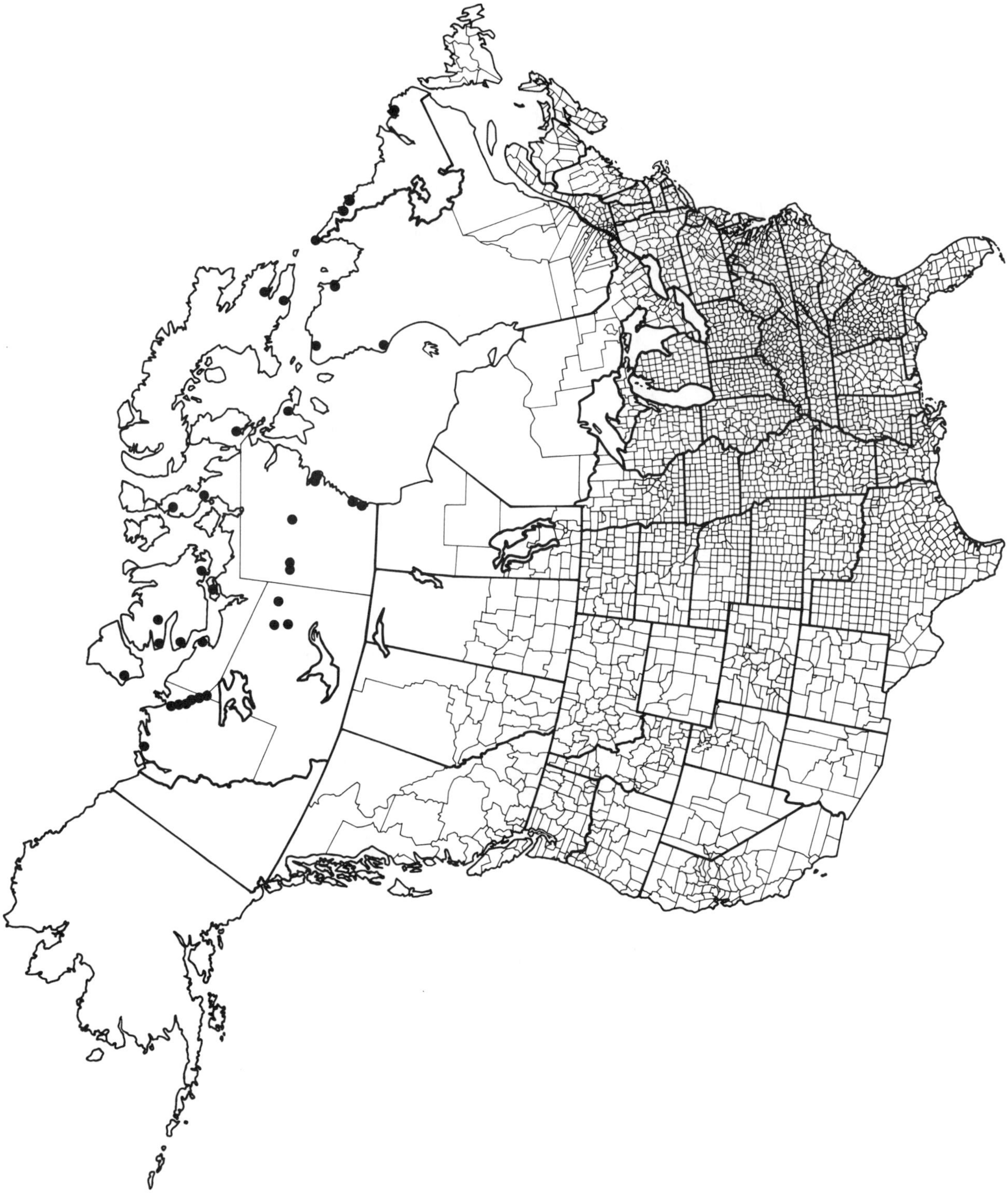

Map 95 *Metacnephia borealis*

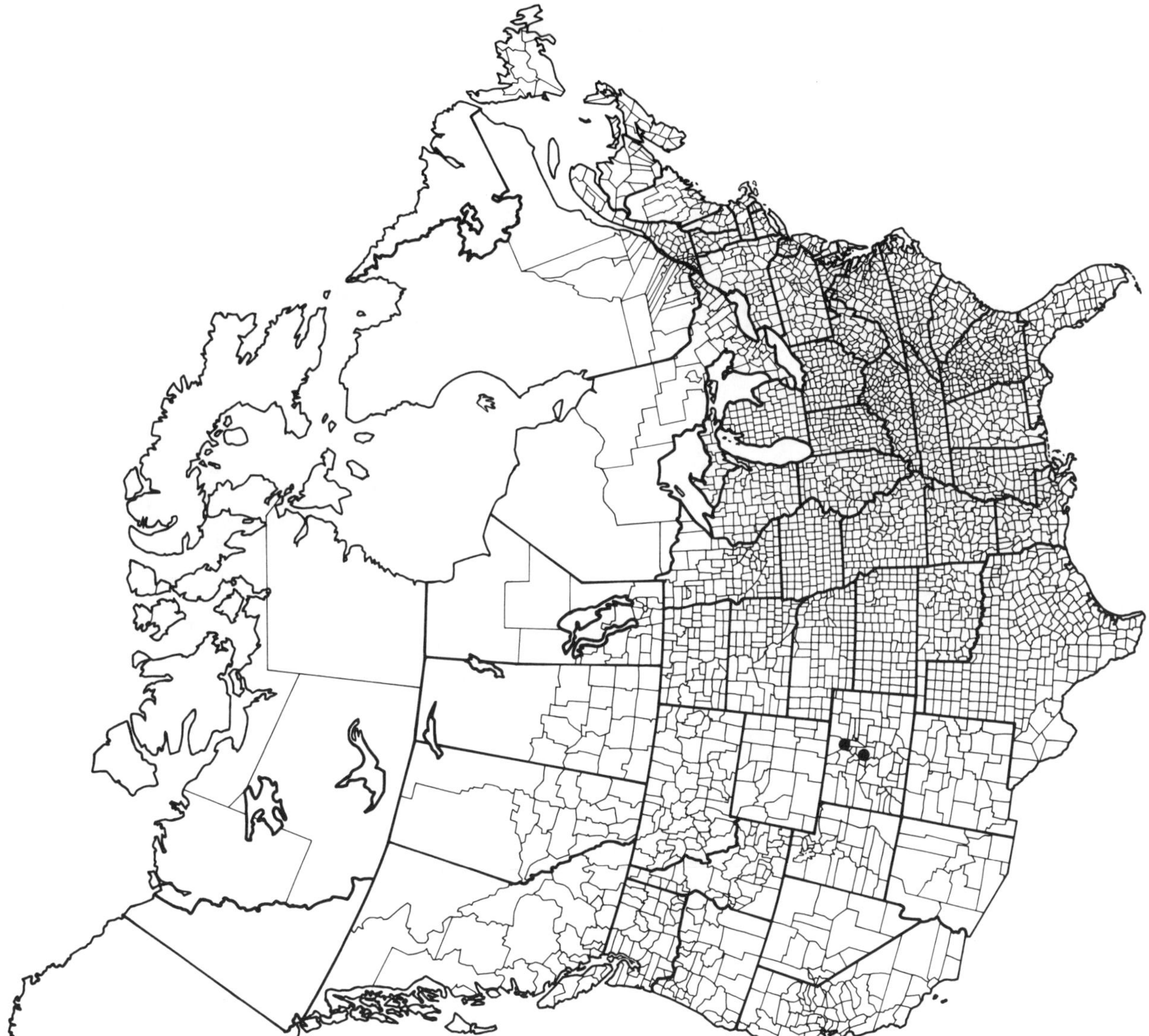

Map 96 *Metacnephia coloradensis*

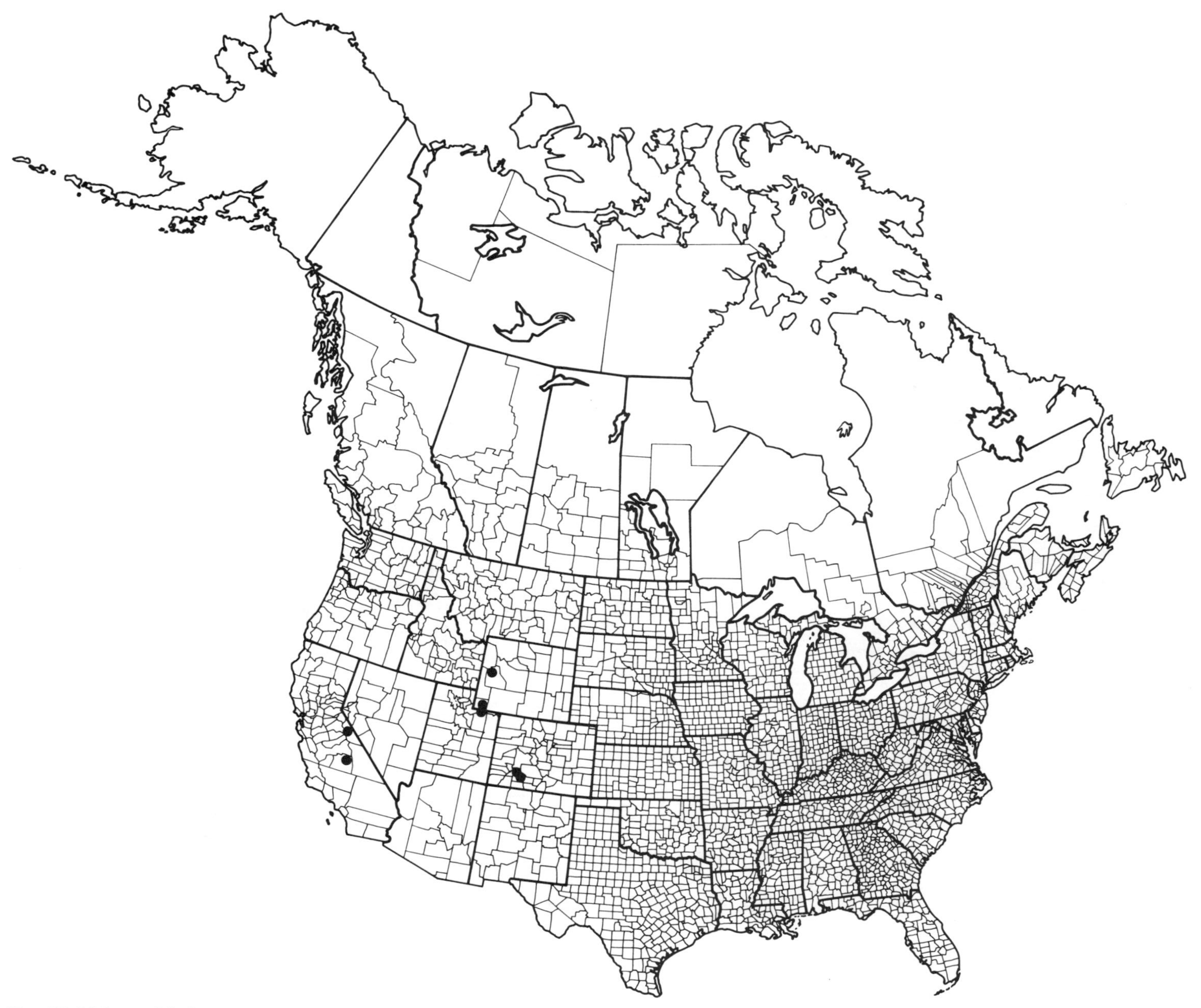

Map 97 *Metacnephia jeanae*

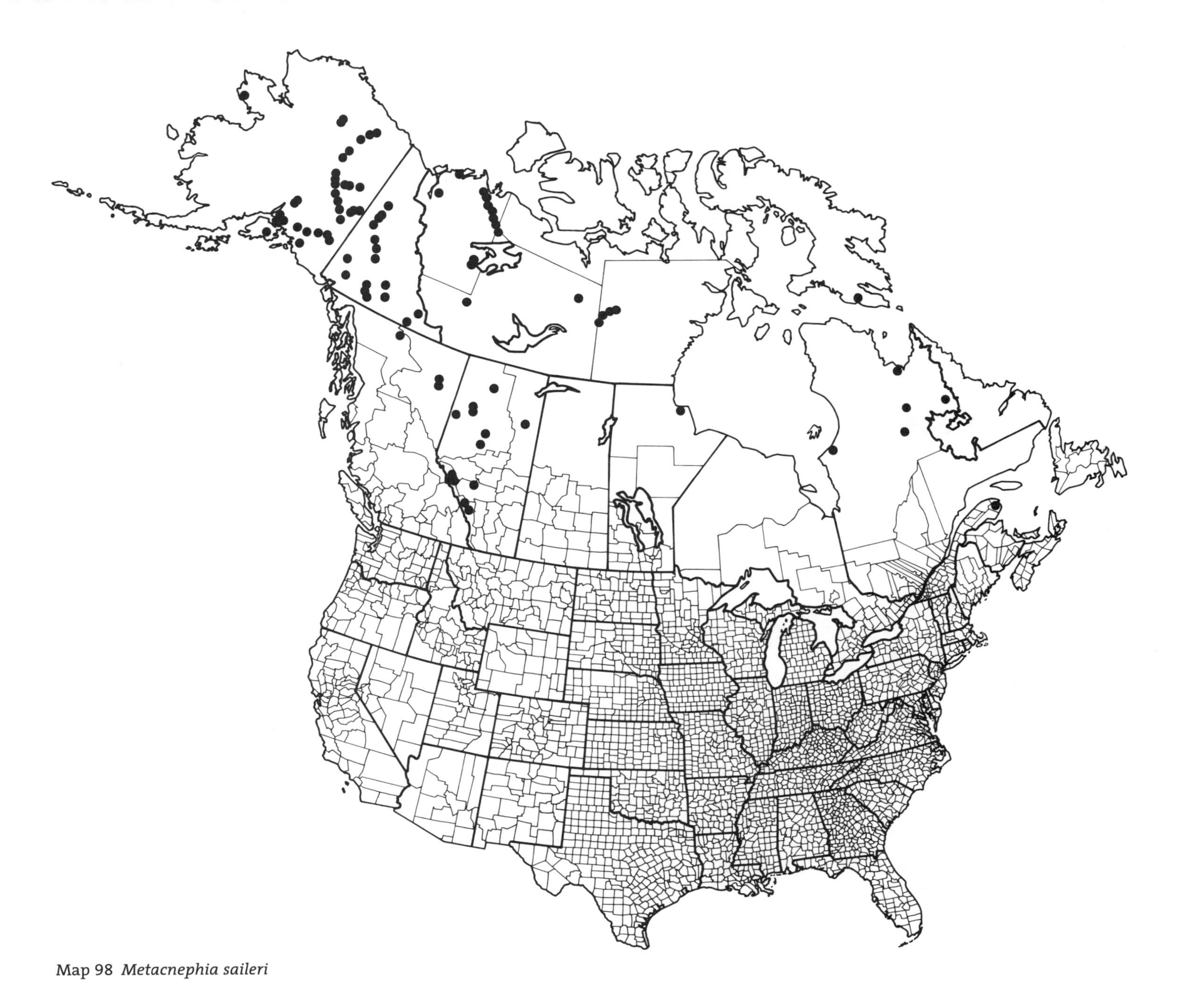

Map 98 *Metacnephia saileri*

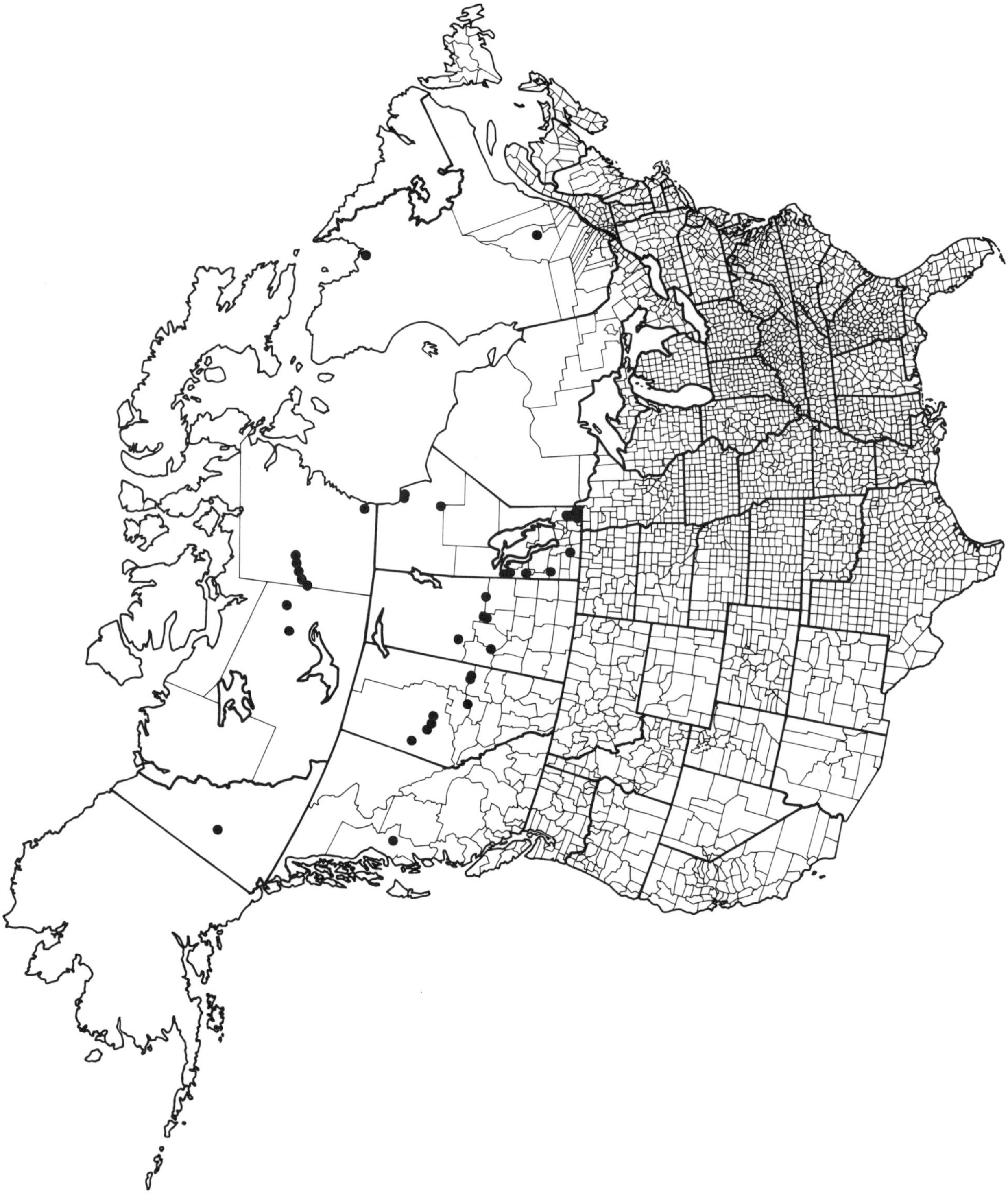

Map 99 *Metacnephia saskatchewana*

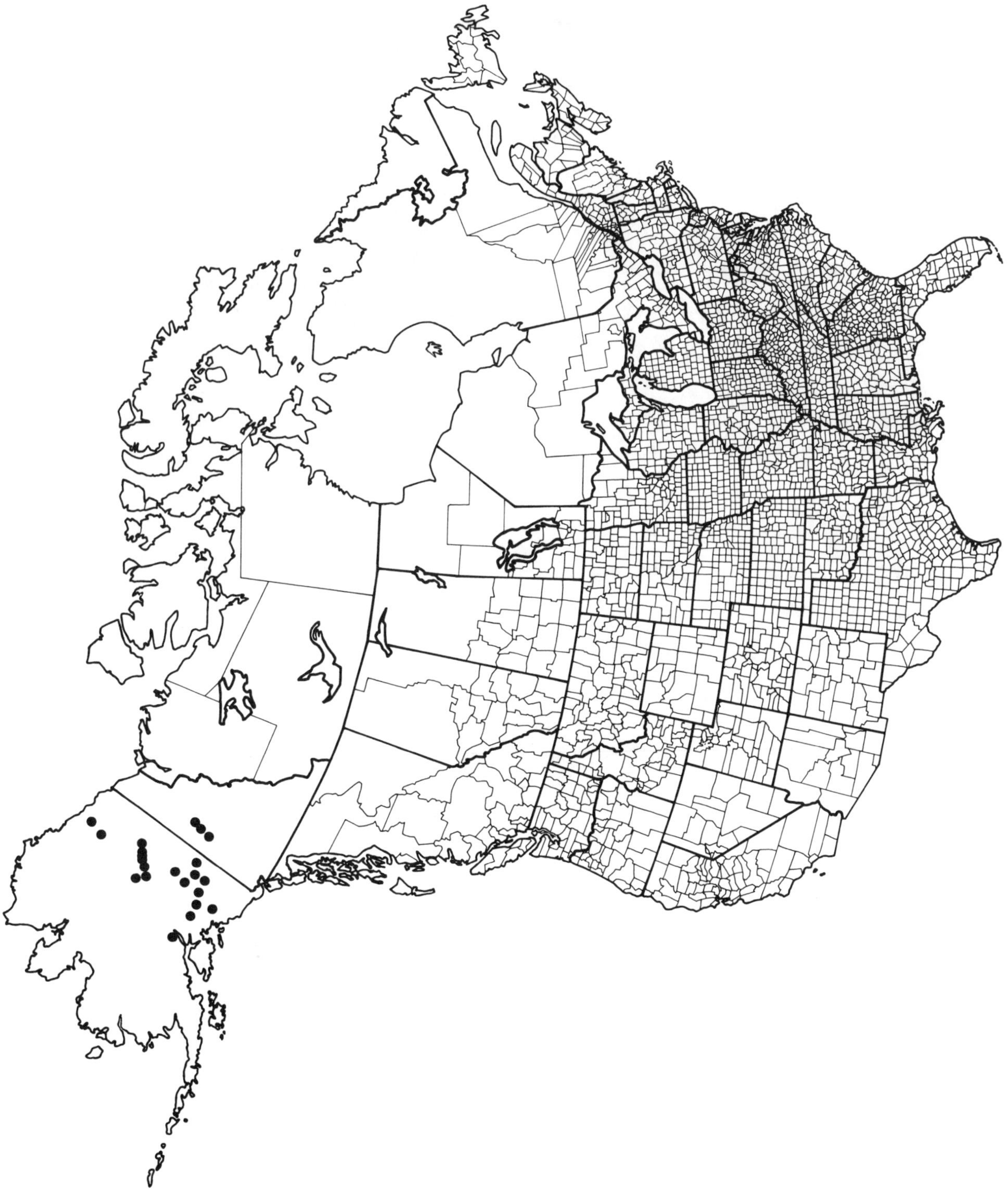

Map 100 *Metacnephia sommermanae*

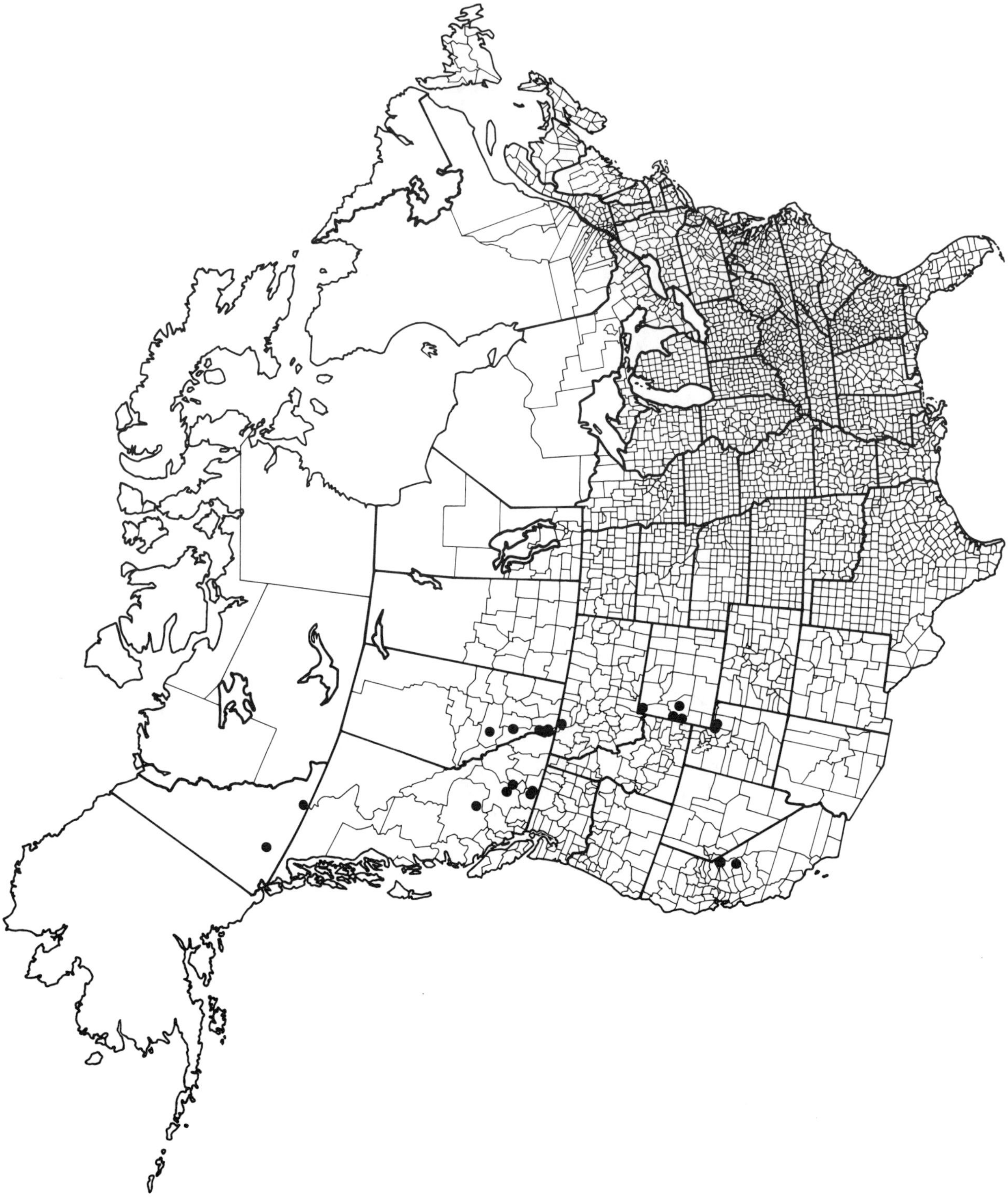

Map 101 *Metacnephia villosa*

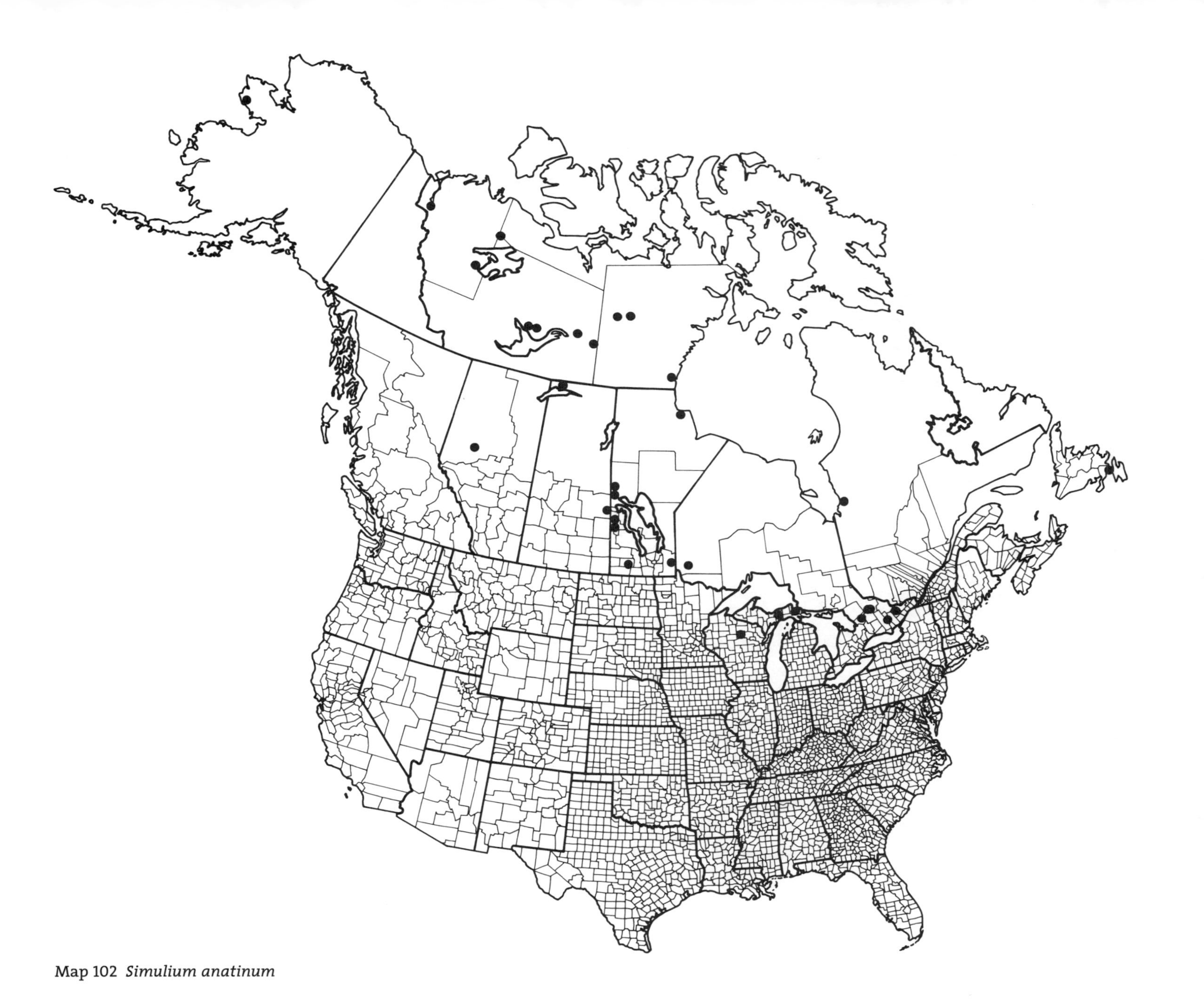

Map 102 *Simulium anatinum*

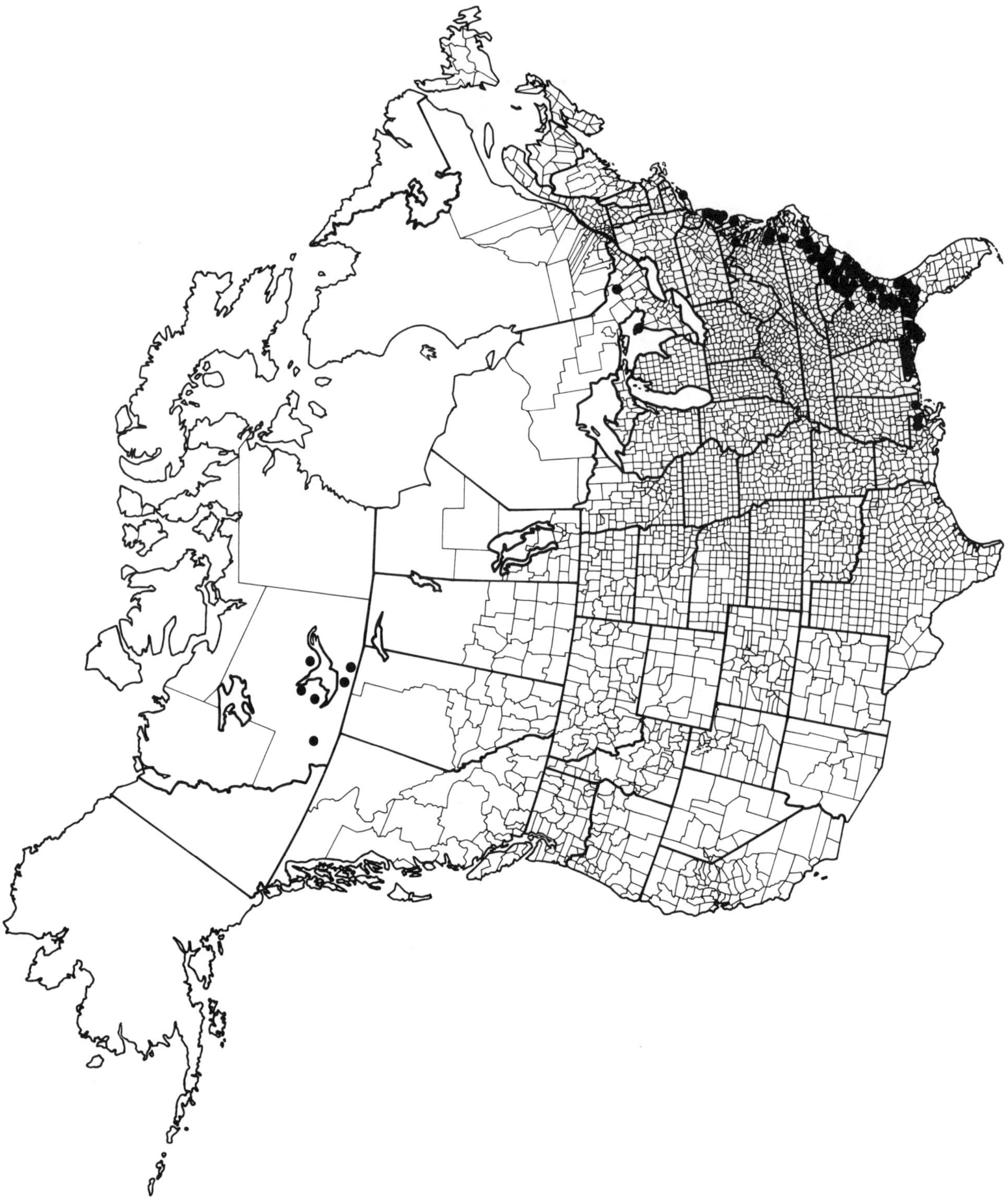

Map 103 *Simulium congareenarum*

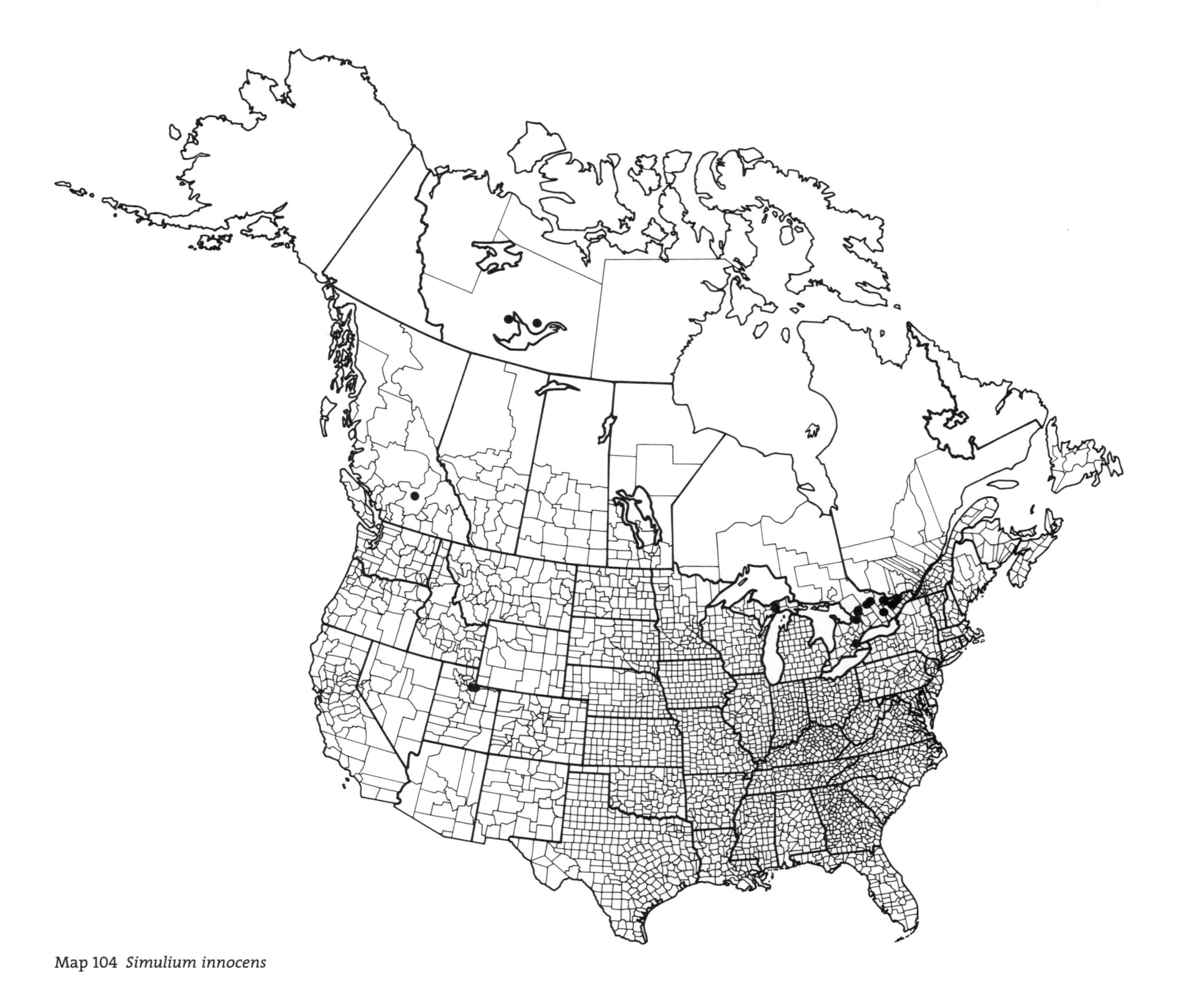

Map 104 *Simulium innocens*

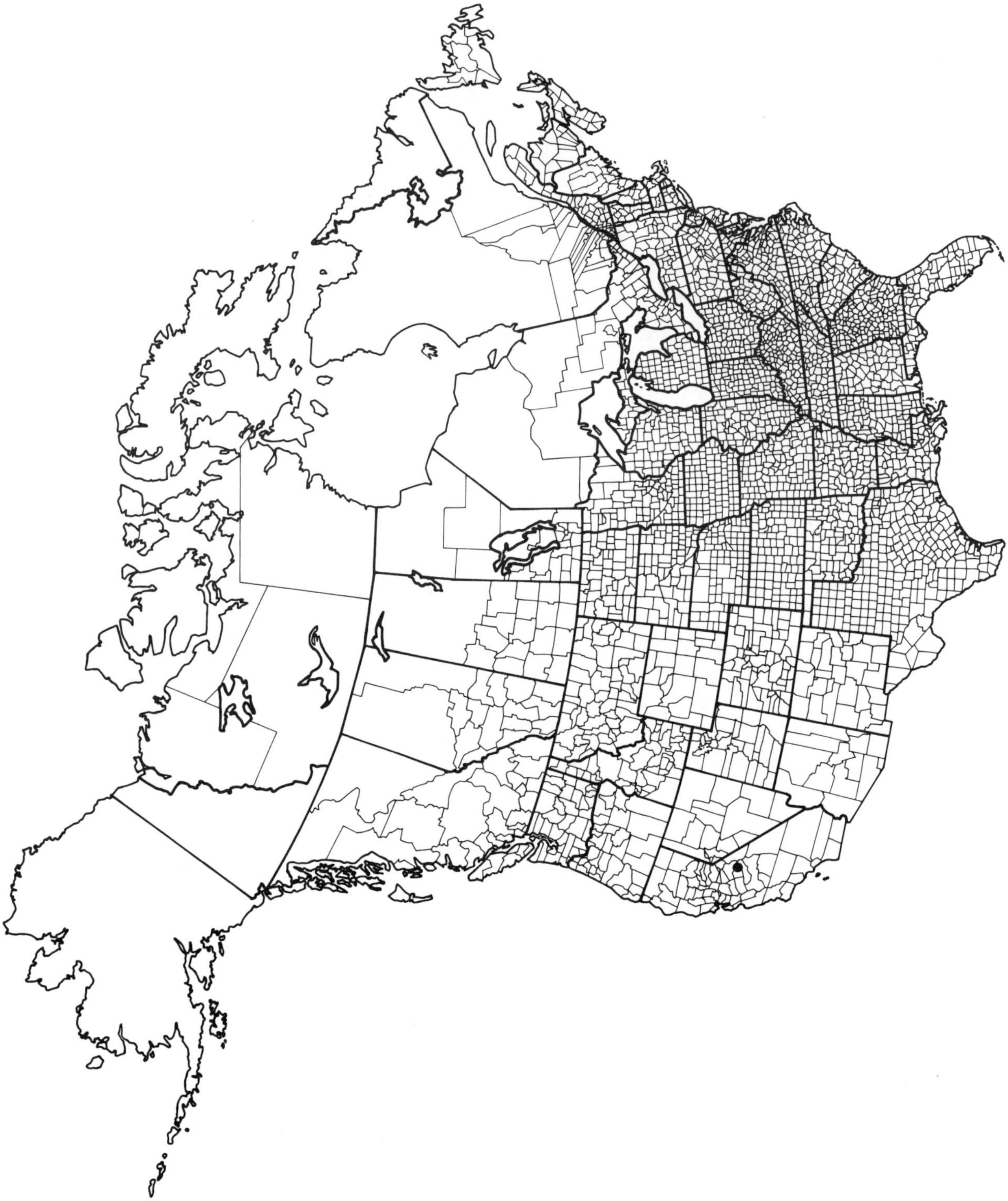

Map 105 *Simulium minus*

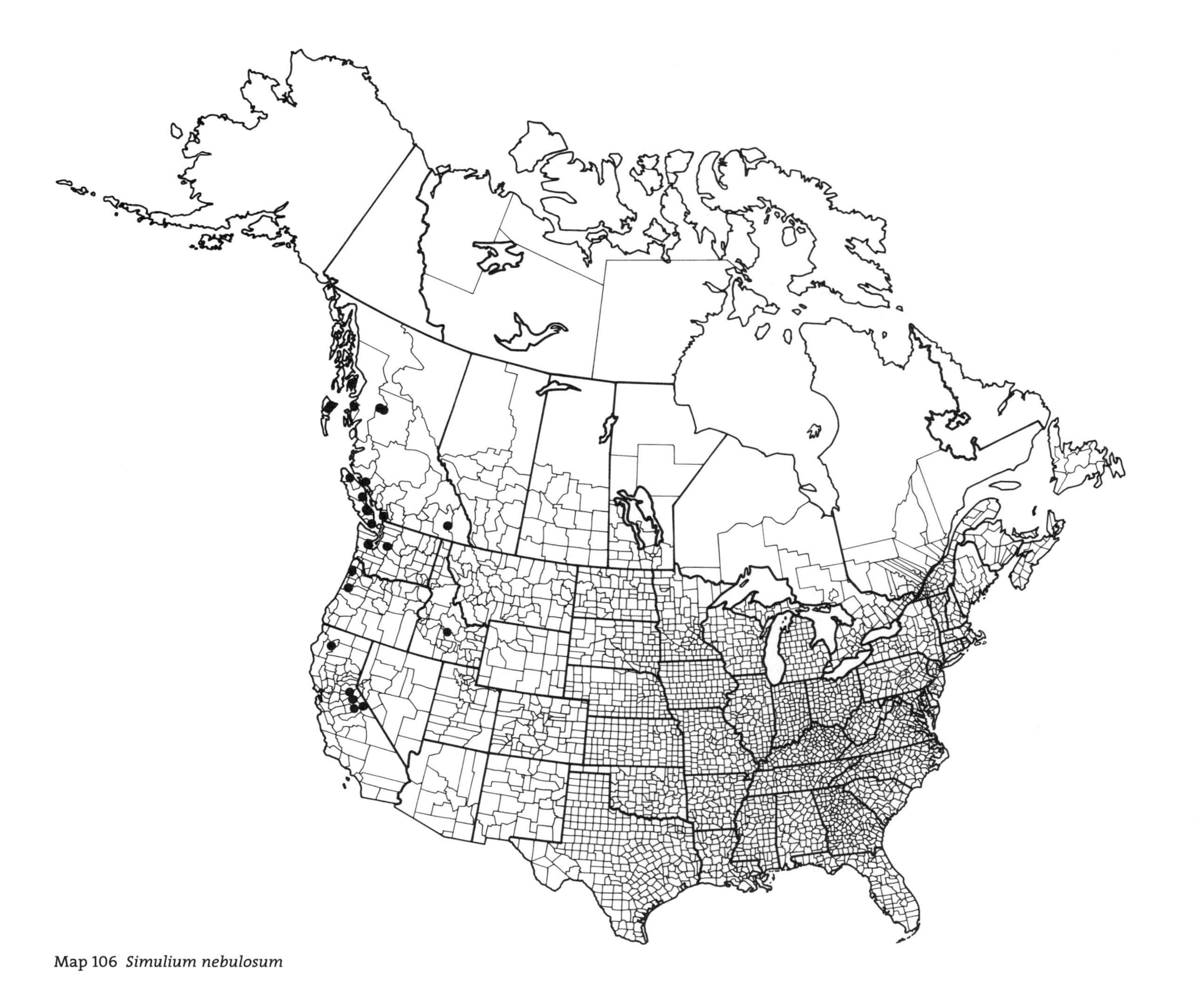

Map 106 *Simulium nebulosum*

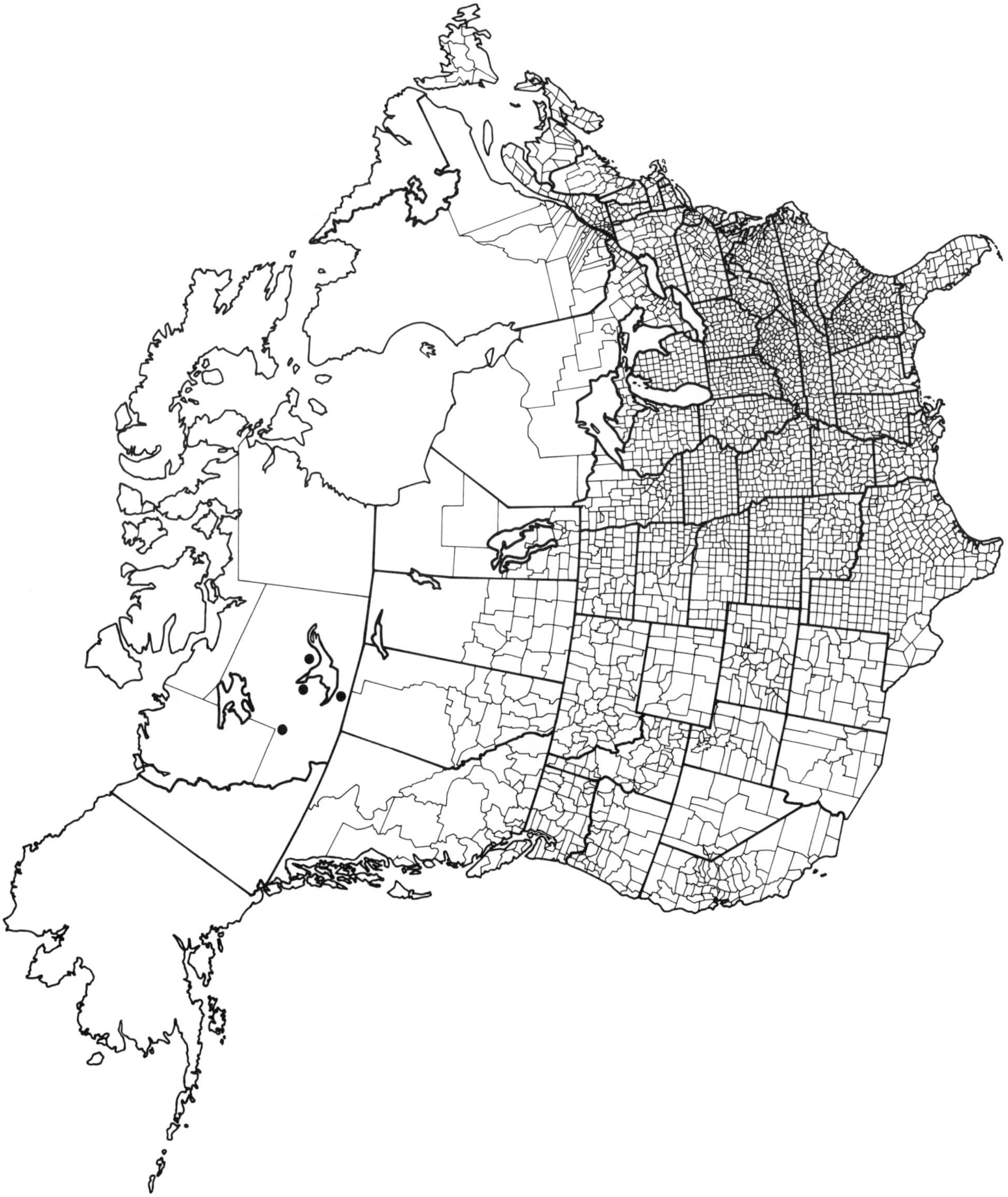

Map 107 *Simulium rendalense*

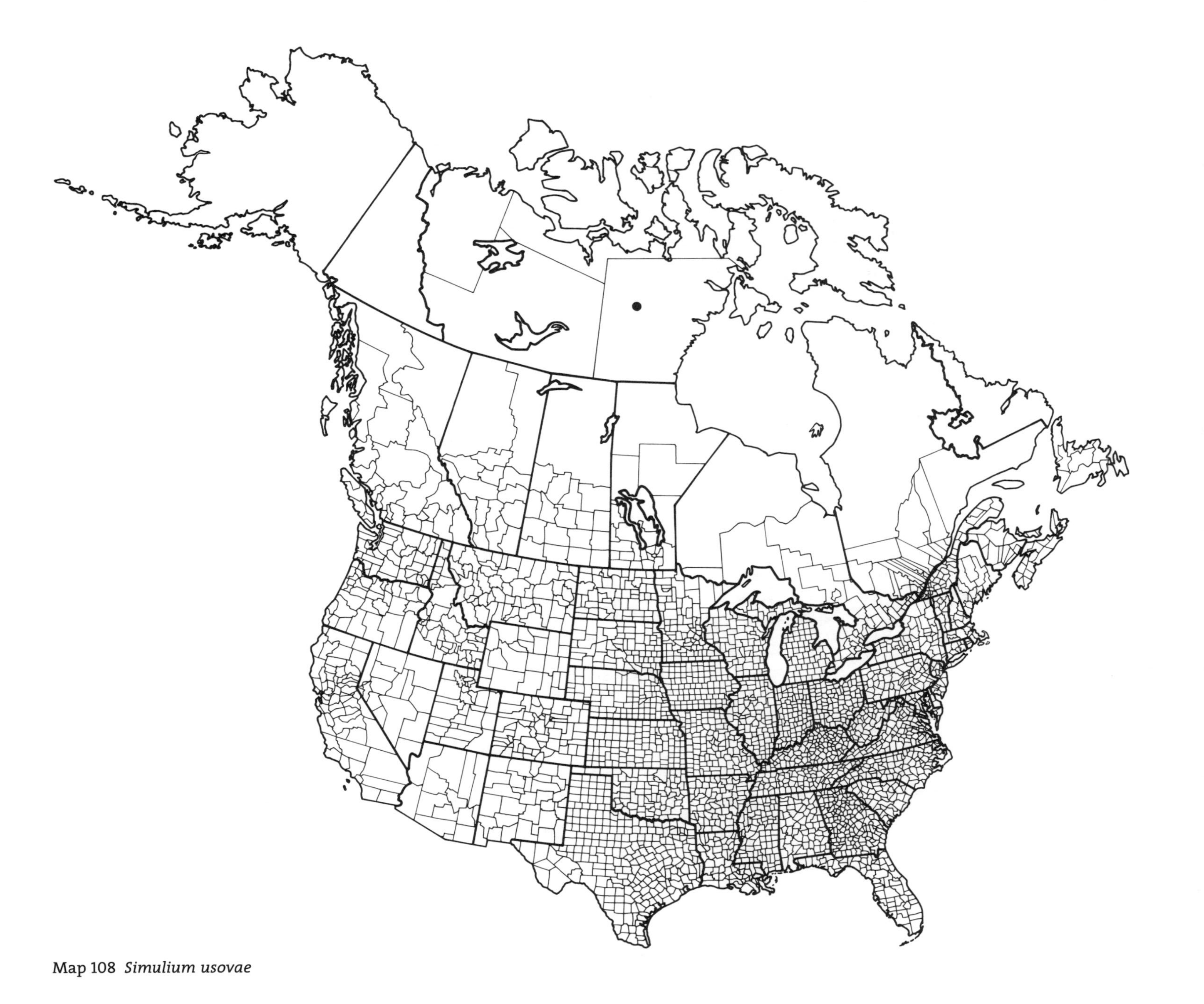

Map 108 *Simulium usovae*

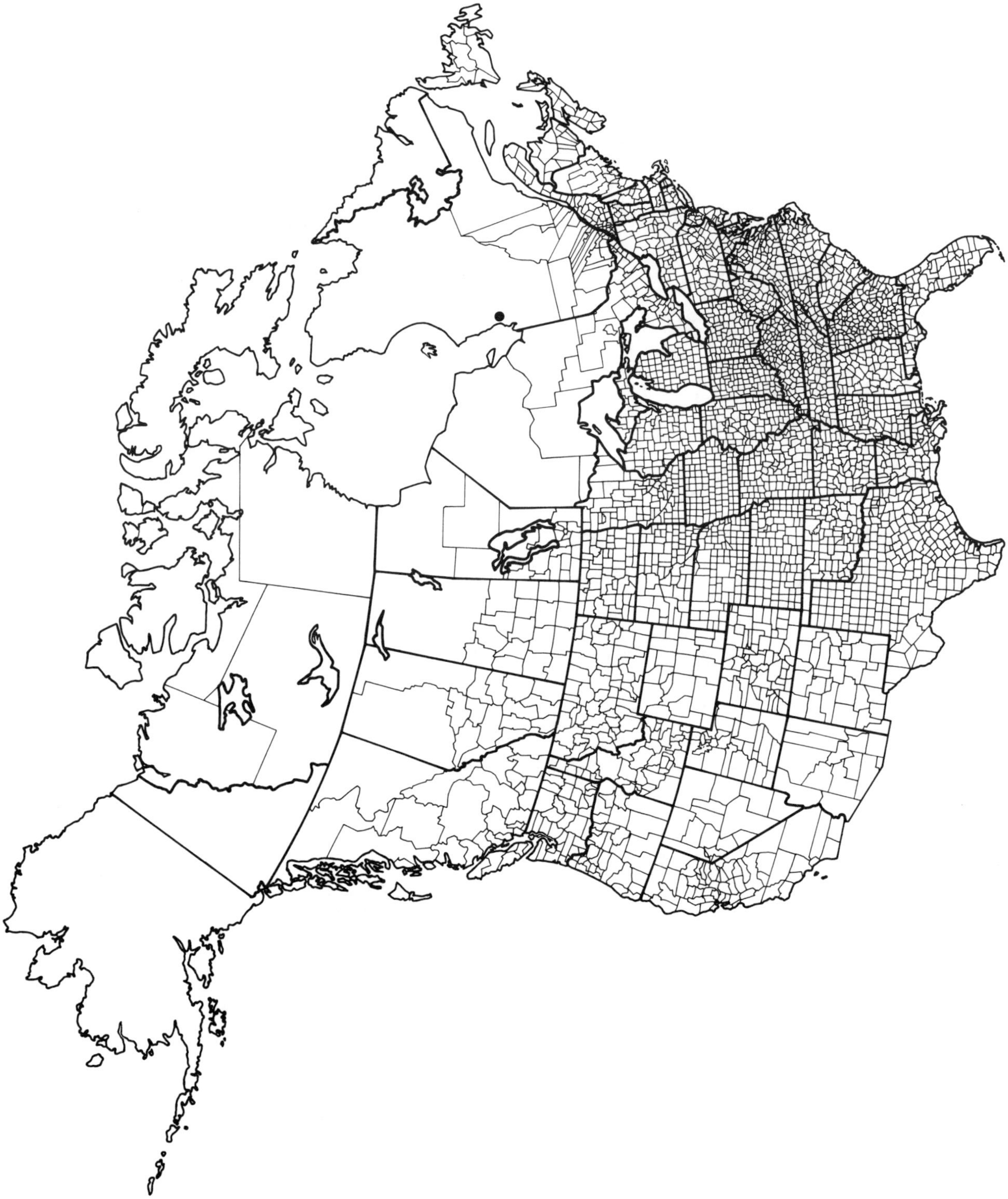

Map 109 *Simulium* 'species O'

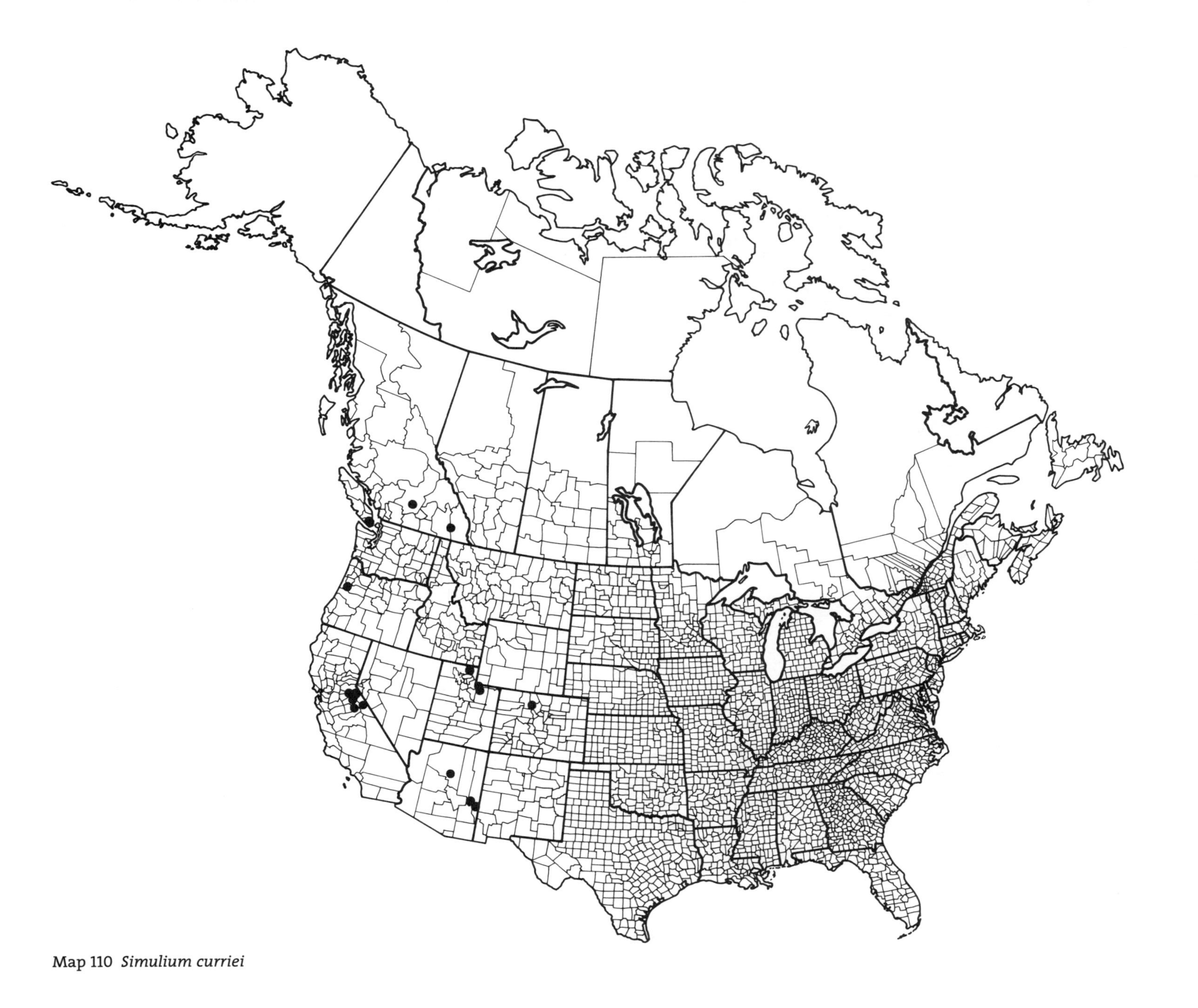

Map 110 *Simulium curriei*

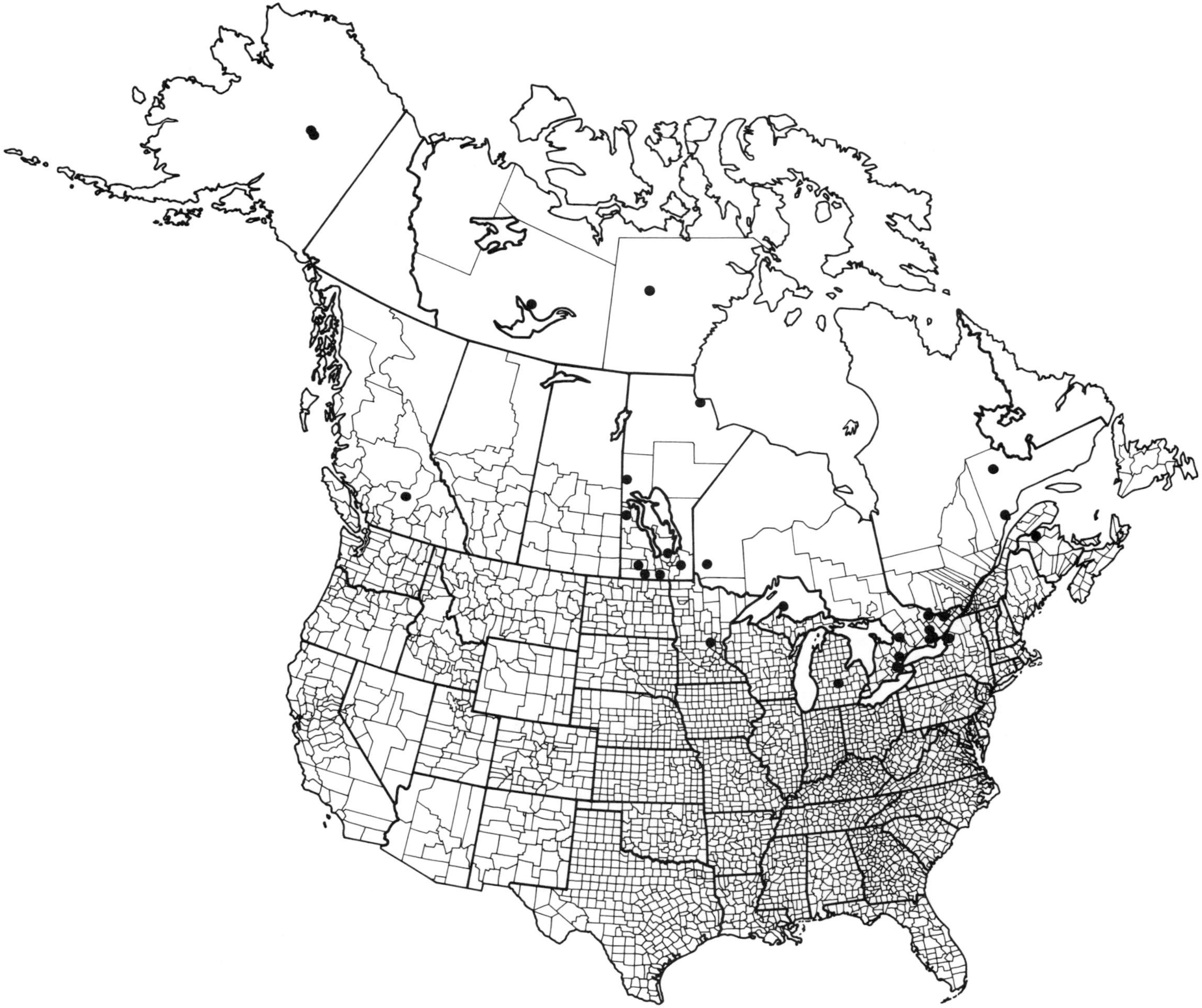

Map 111 *Simulium excisum*

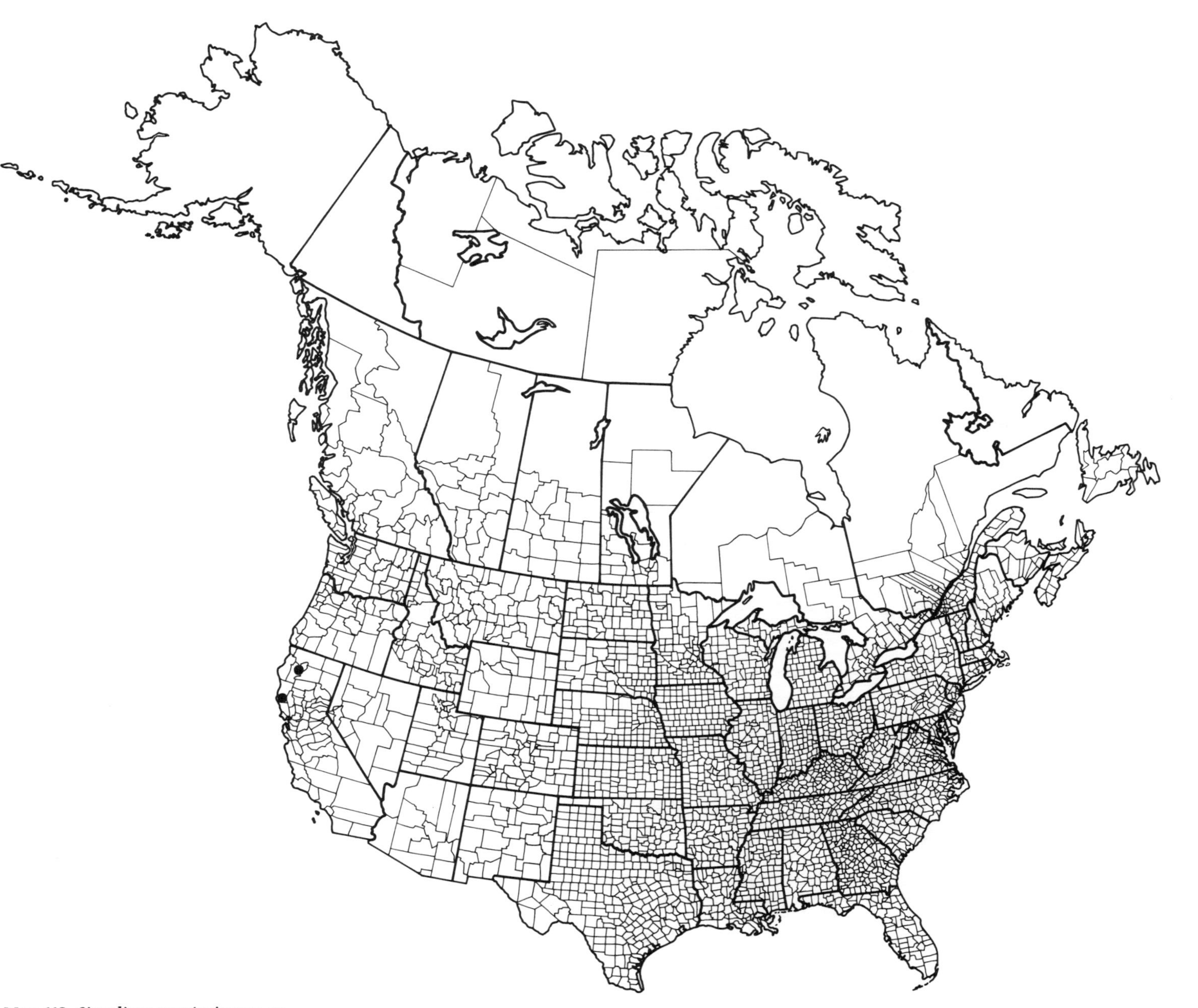

Map 112 *Simulium mysterium* n. sp.

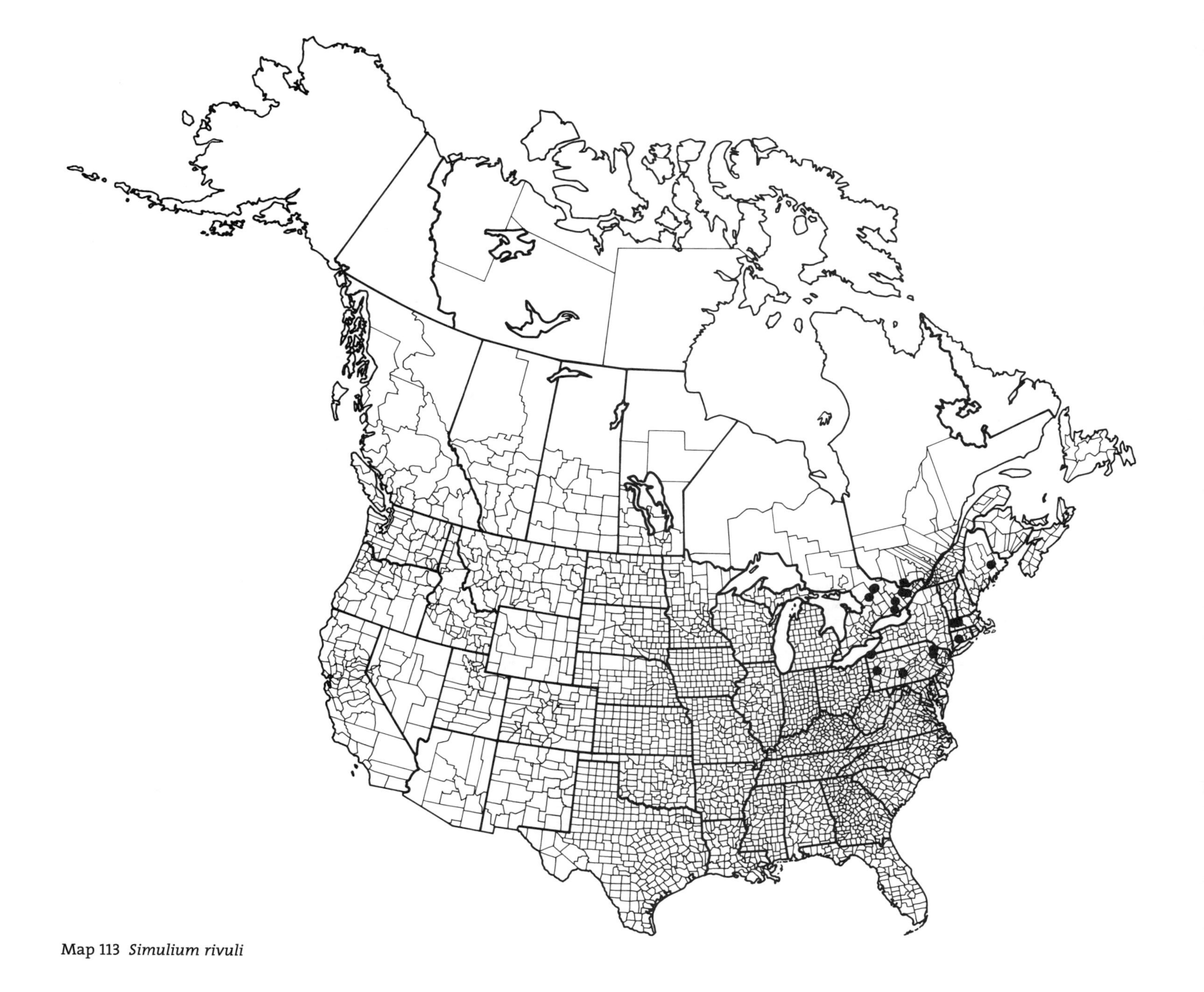

Map 113 *Simulium rivuli*

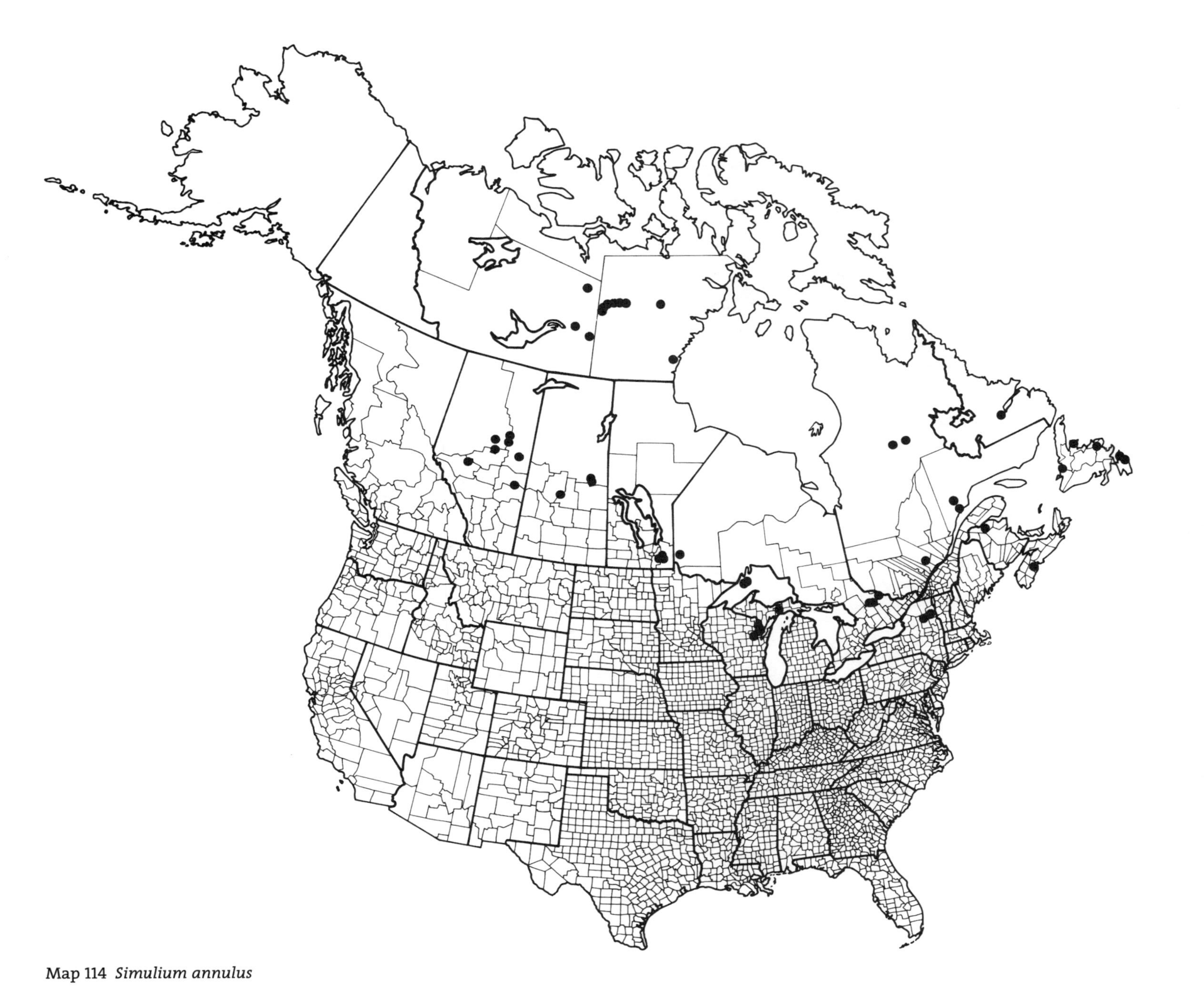

Map 114 *Simulium annulus*

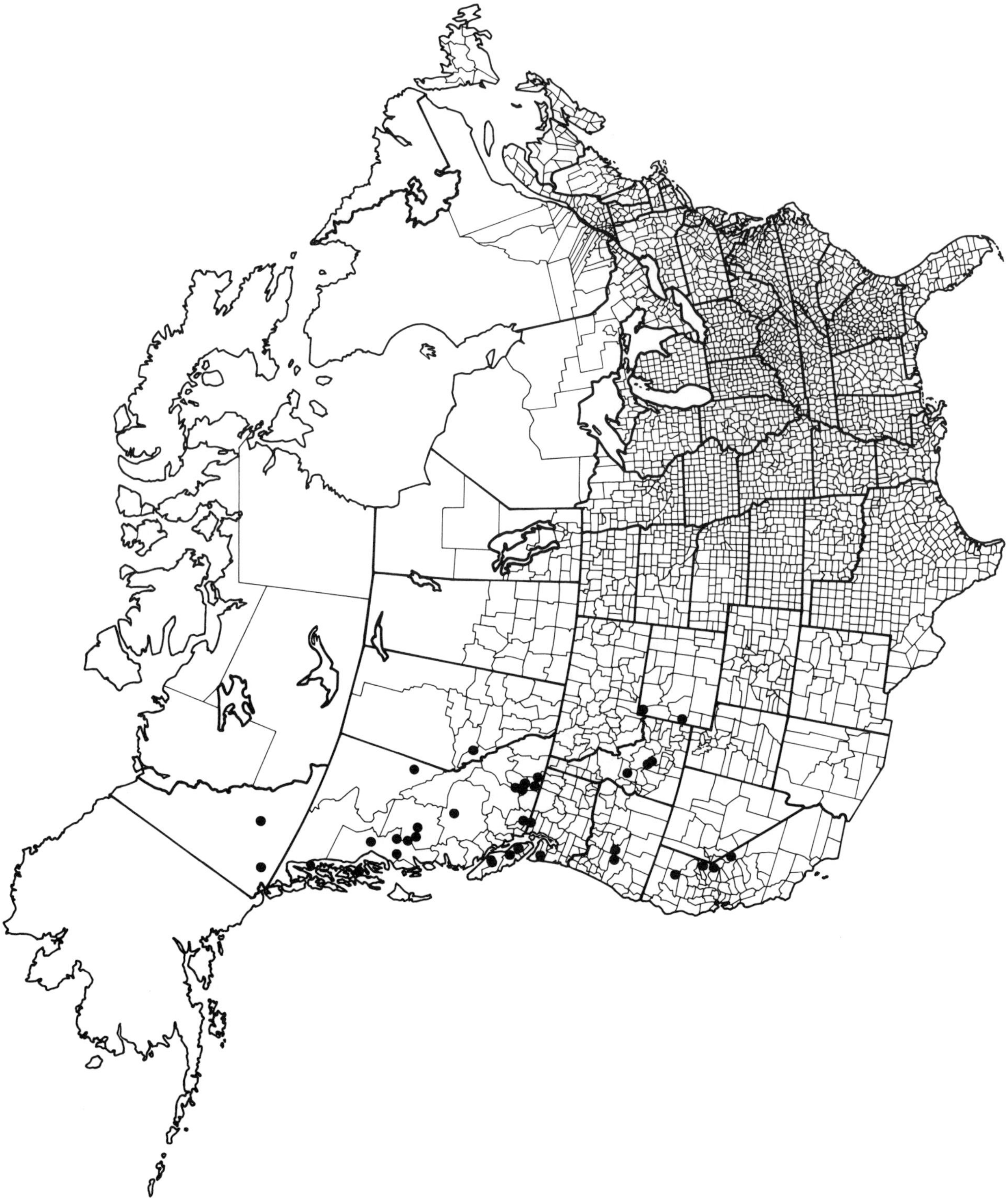

Map 115 *Simulium balteatum* n. sp.

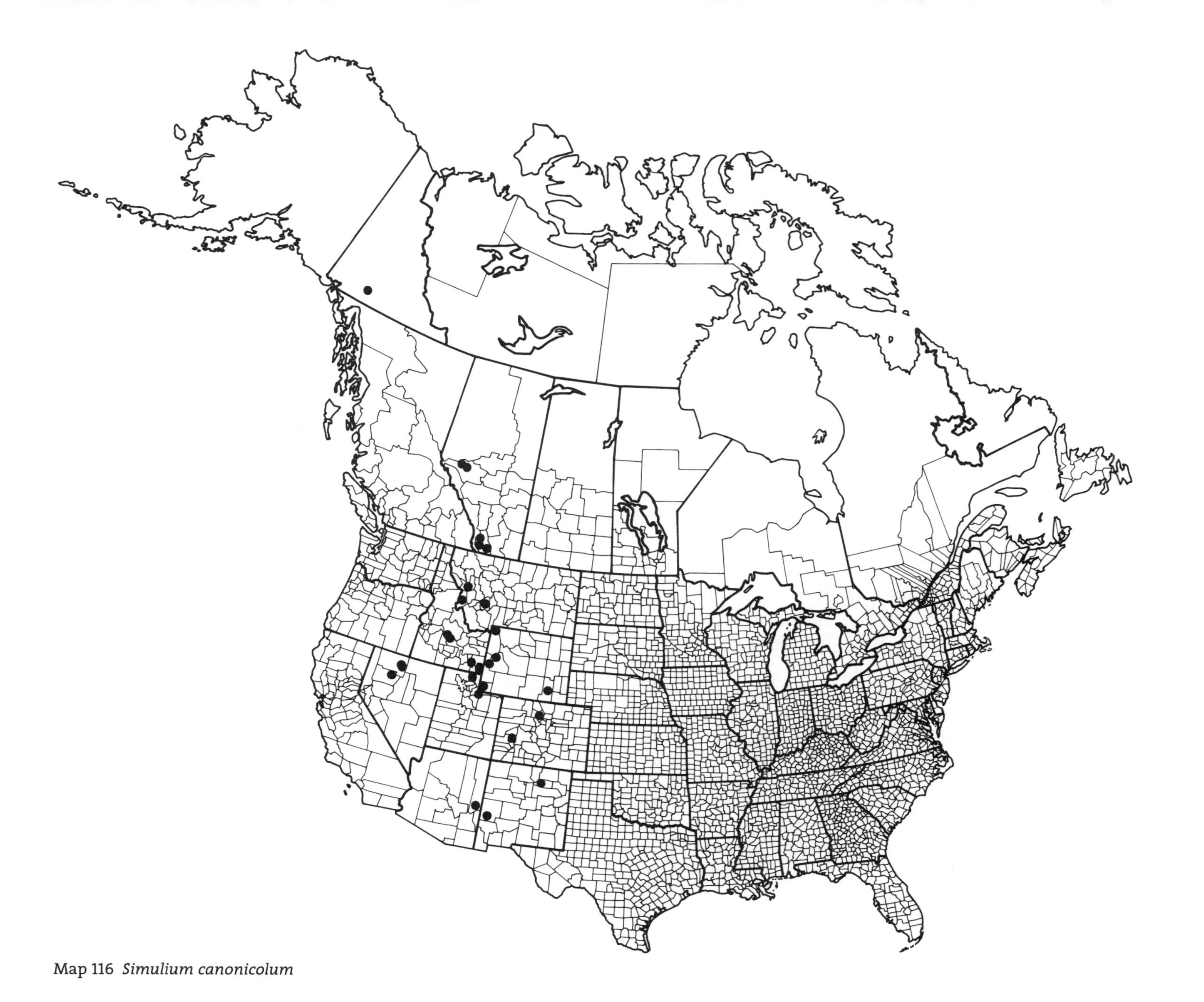

Map 116 *Simulium canonicolum*

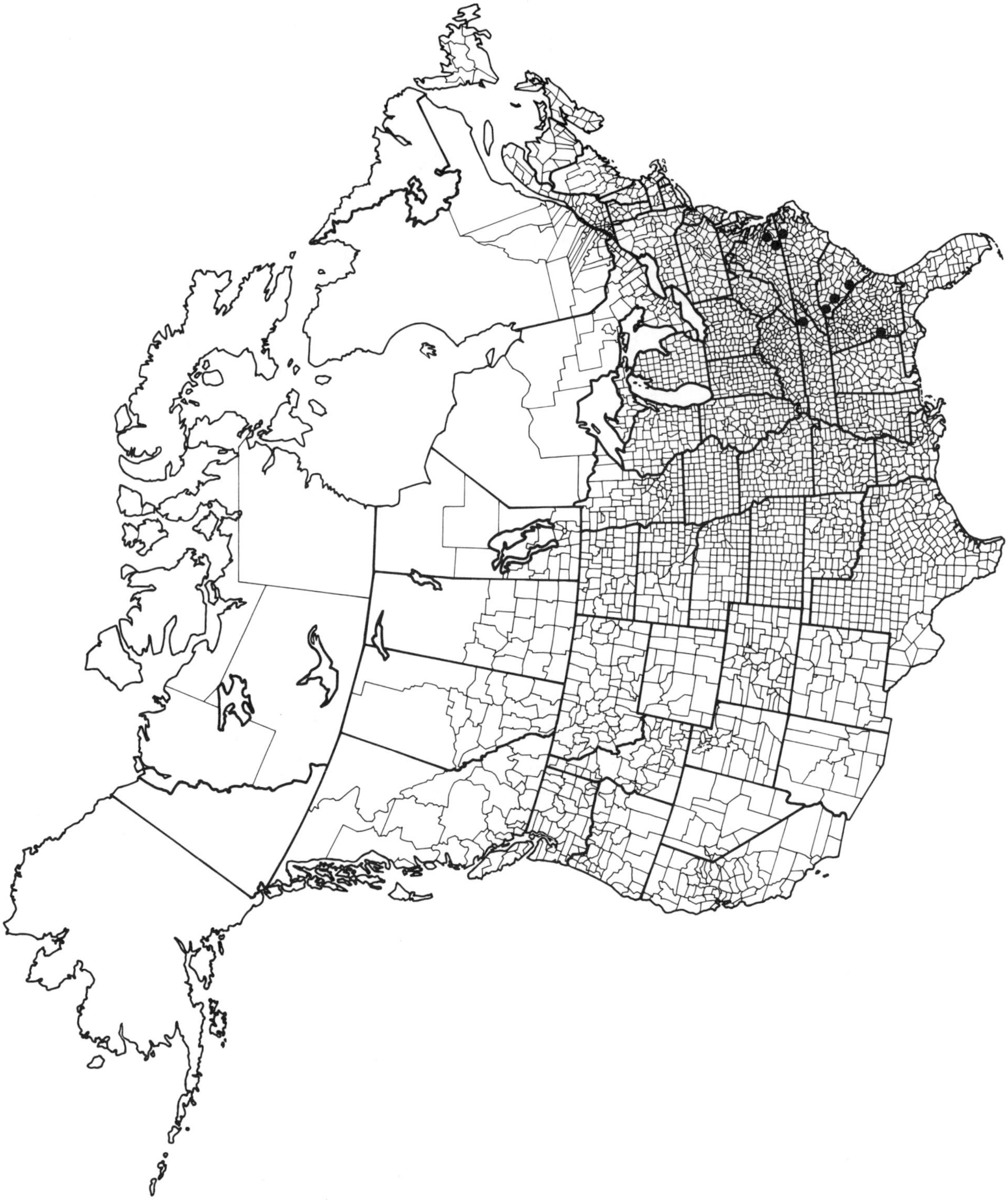

Map 117 *Simulium clarkei*

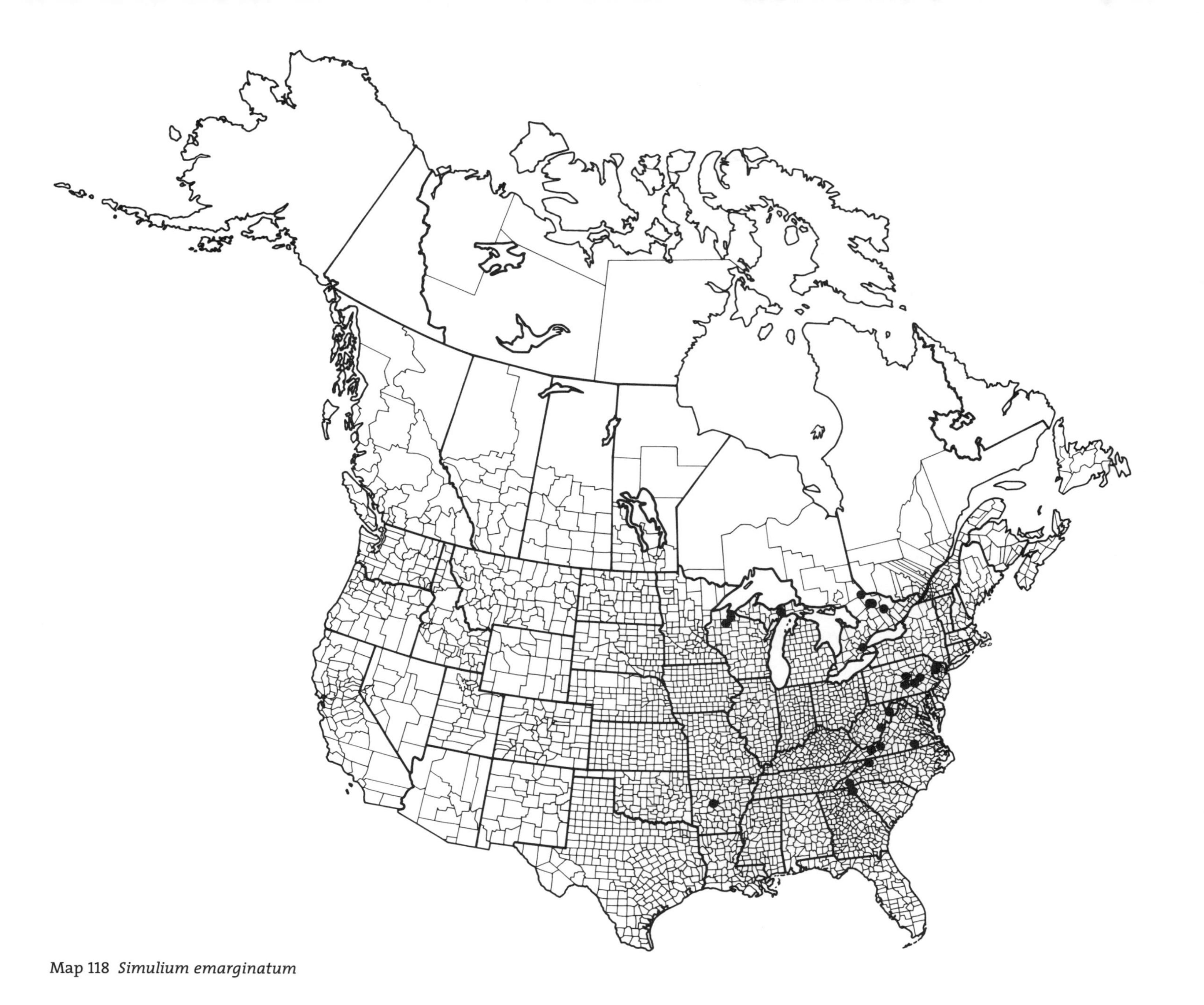

Map 118 *Simulium emarginatum*

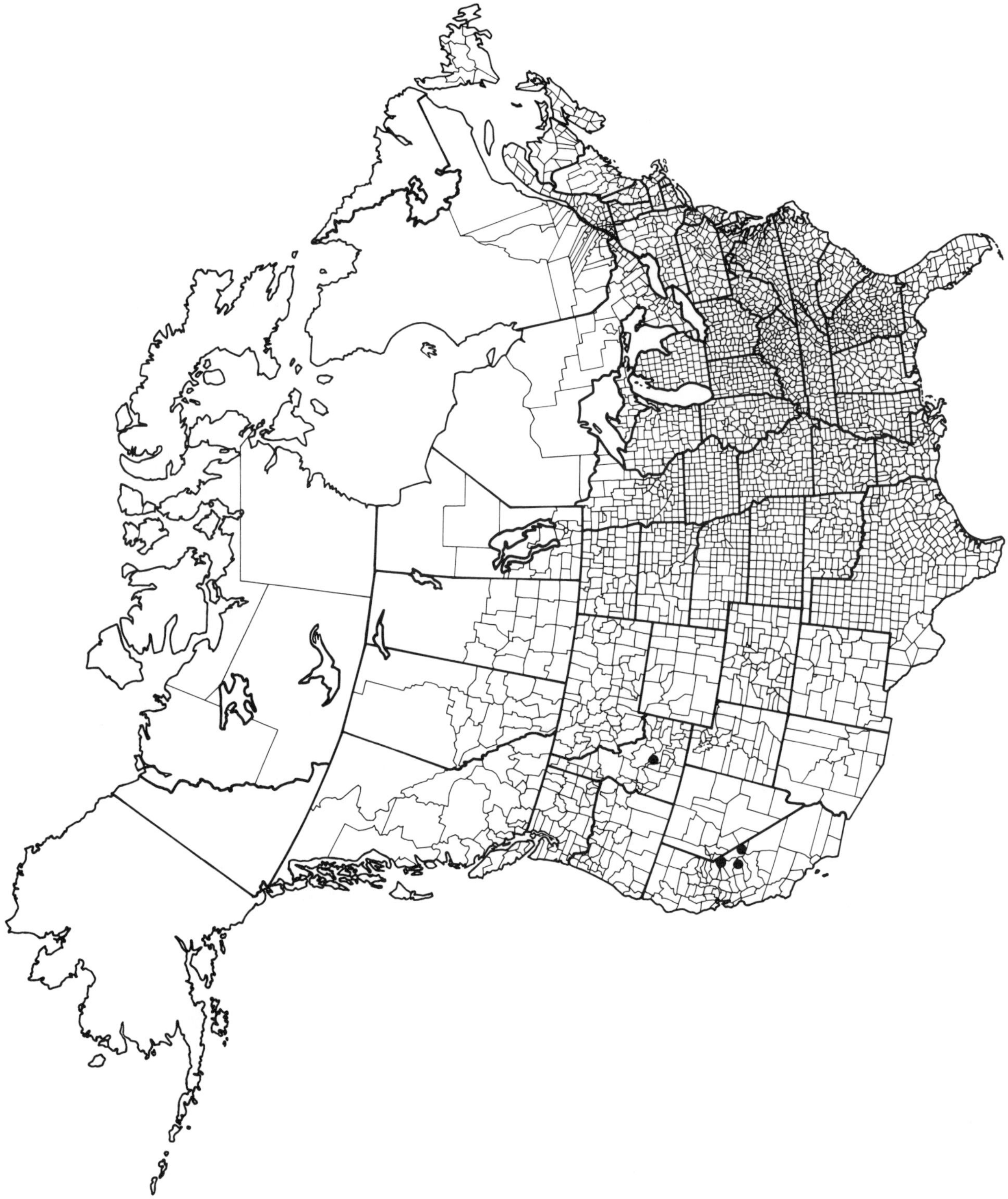

Map 119 *Simulium joculator* n. sp.

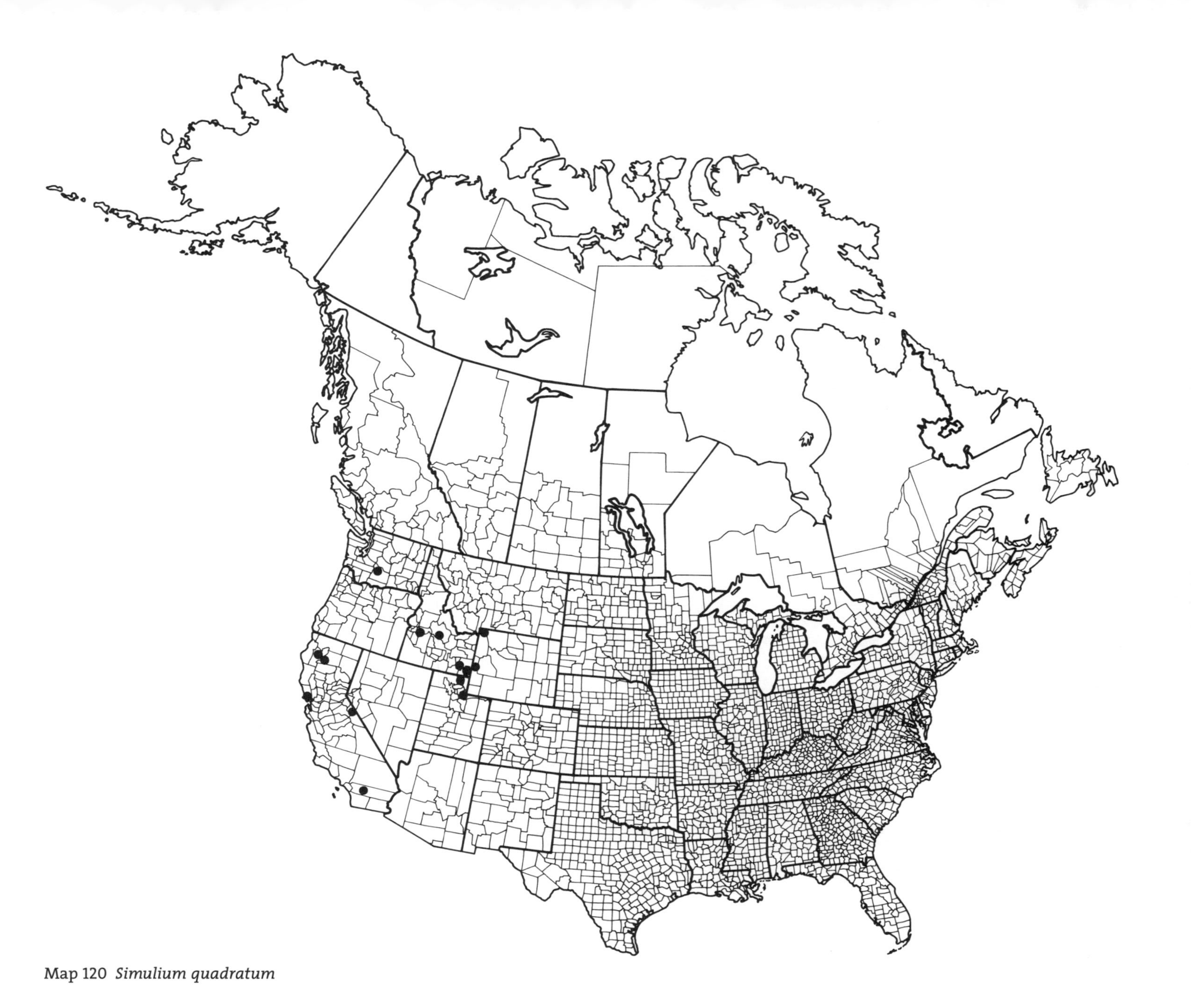

Map 120 *Simulium quadratum*

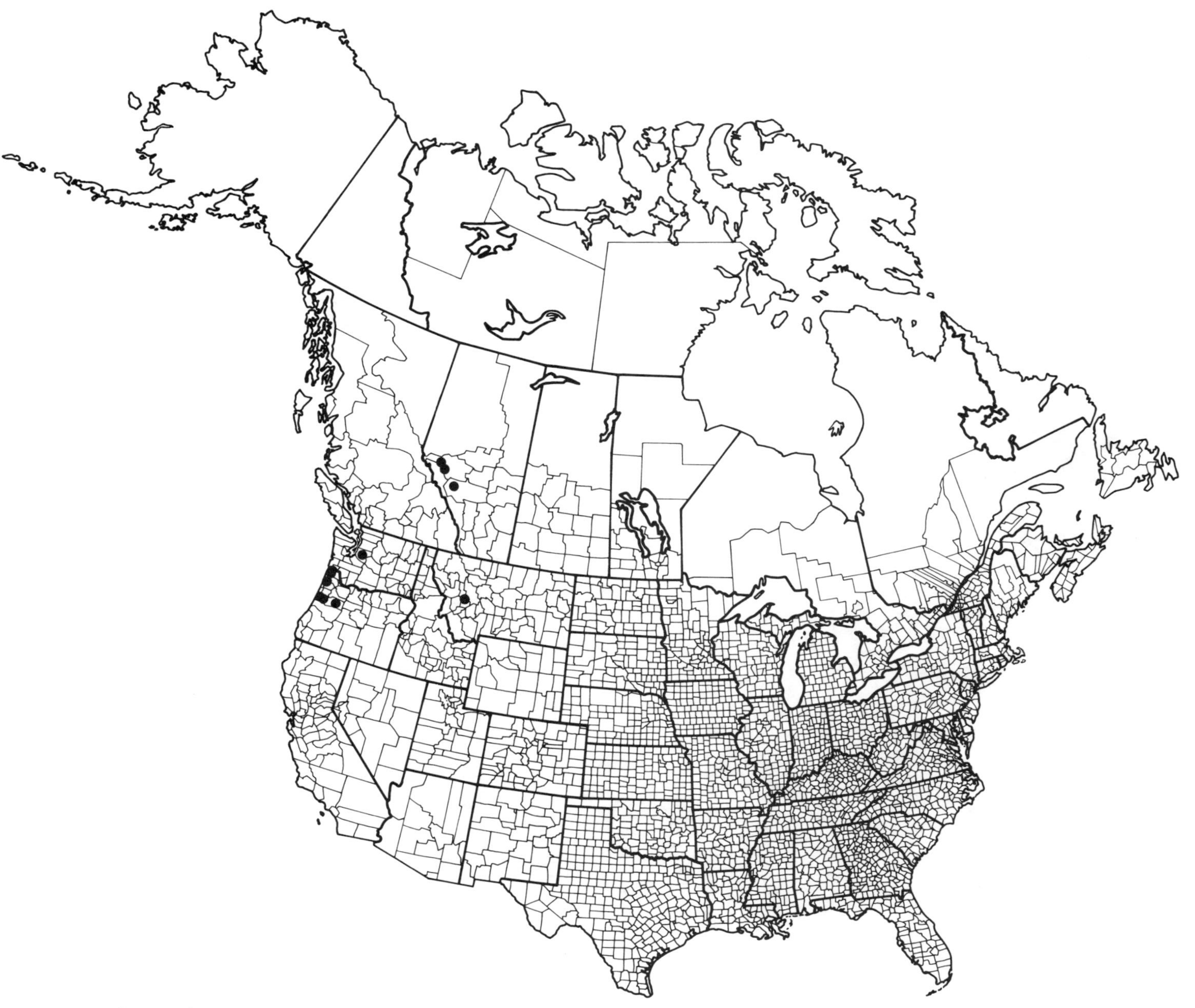

Map 121 *Simulium zephyrus* n. sp.

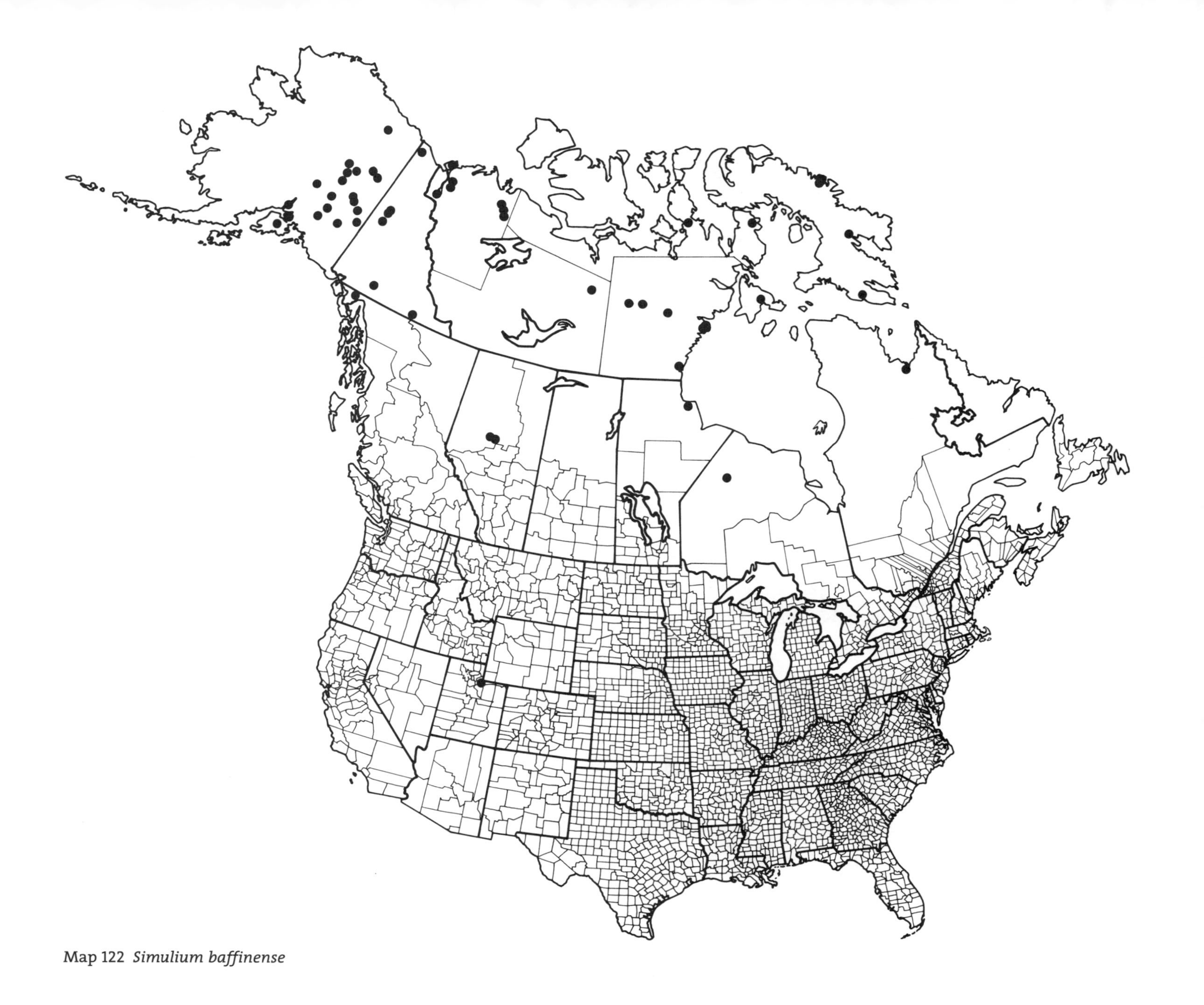

Map 122 *Simulium baffinense*

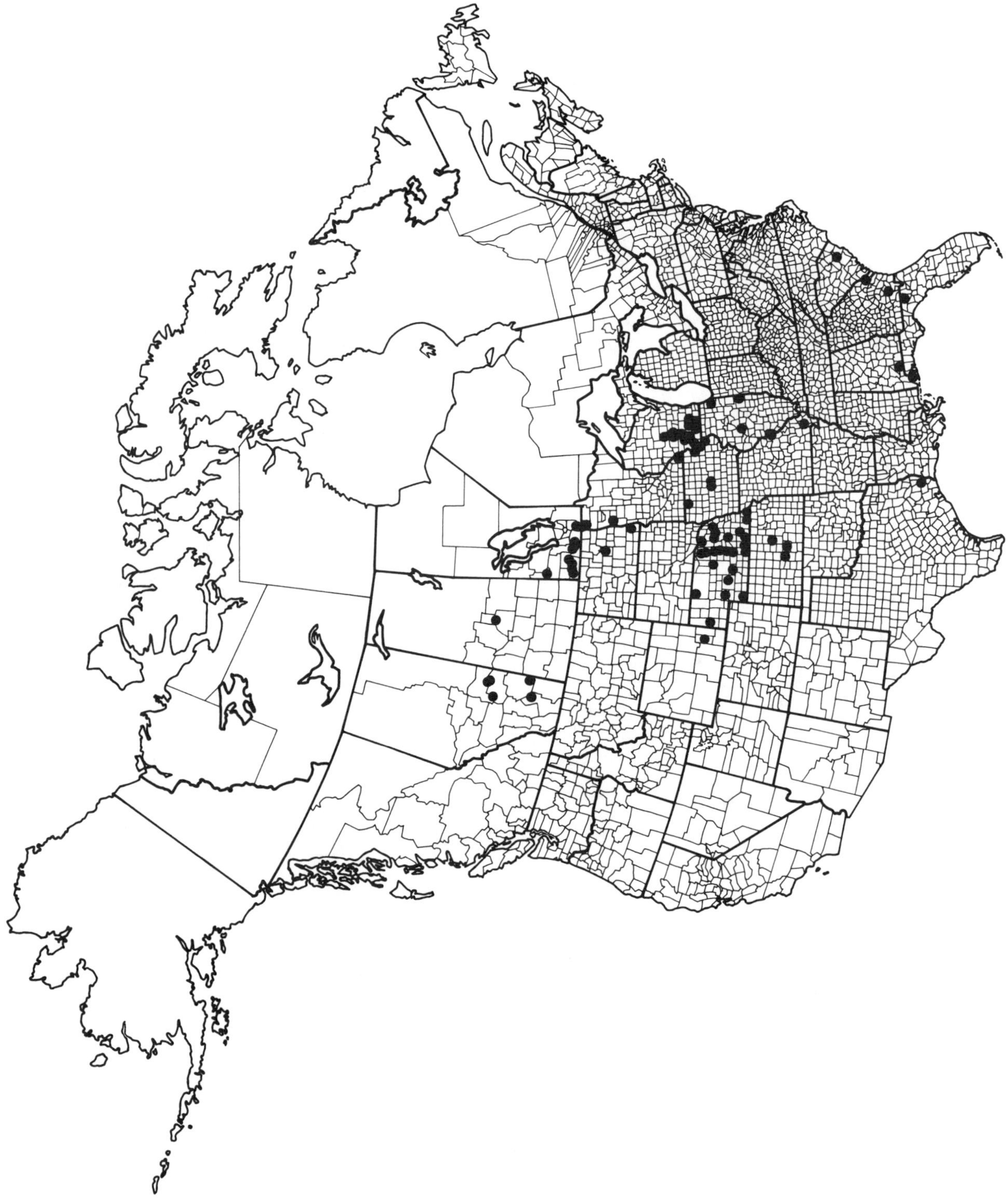

Map 123 *Simulium johannseni*

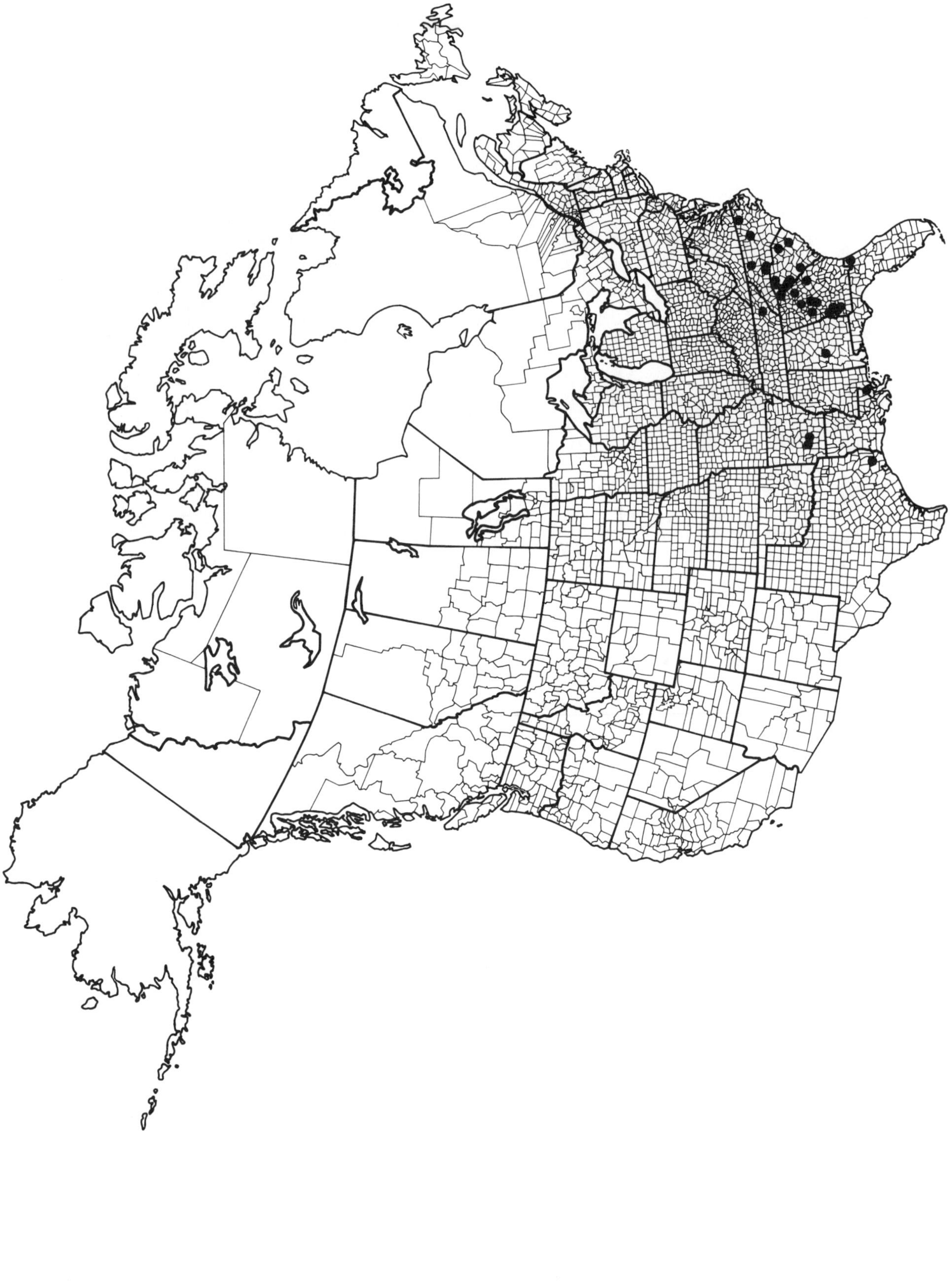

Map 124 *Simulium parmatum* n. sp.

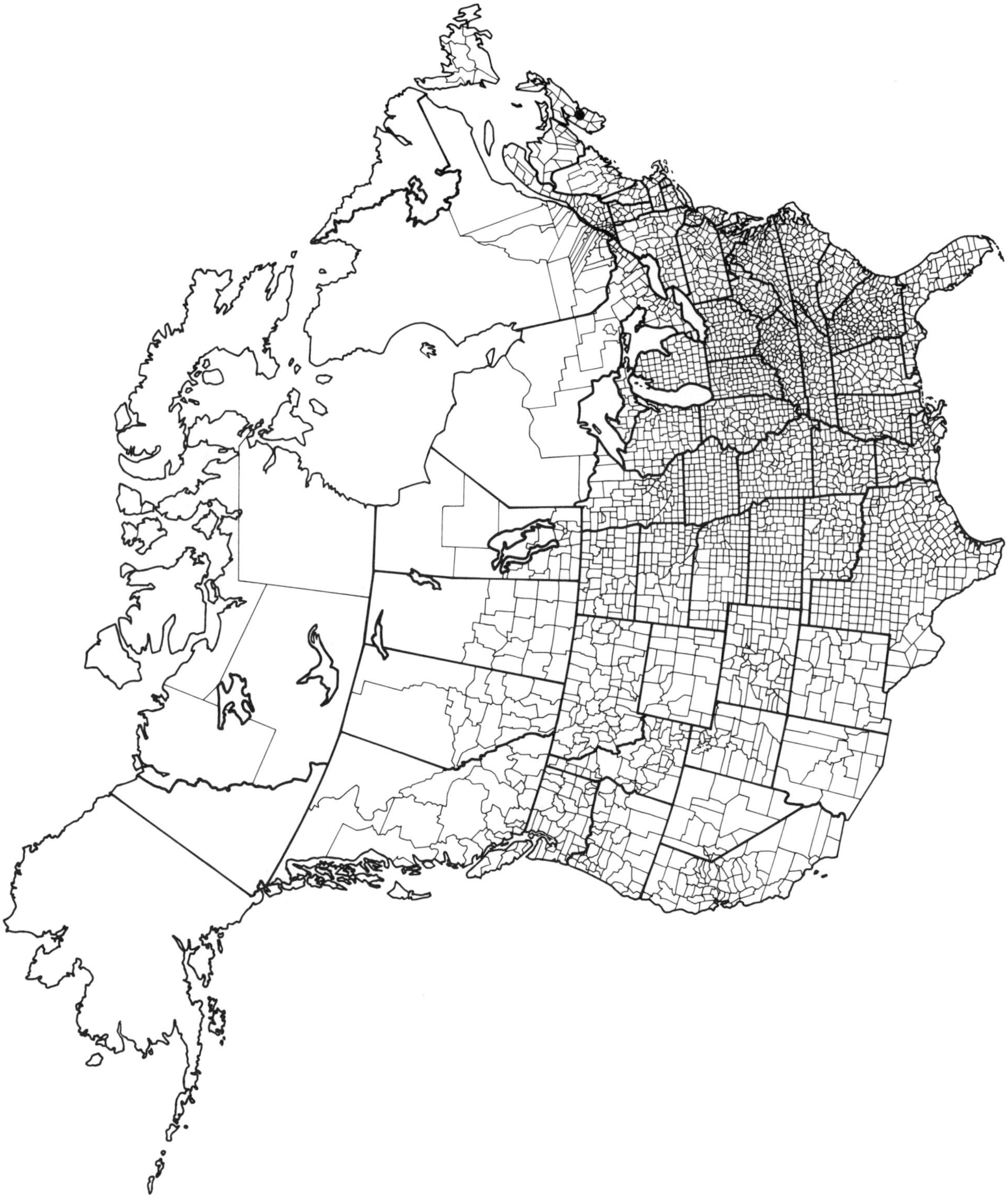

Map 125 *Simulium rothfelsi*

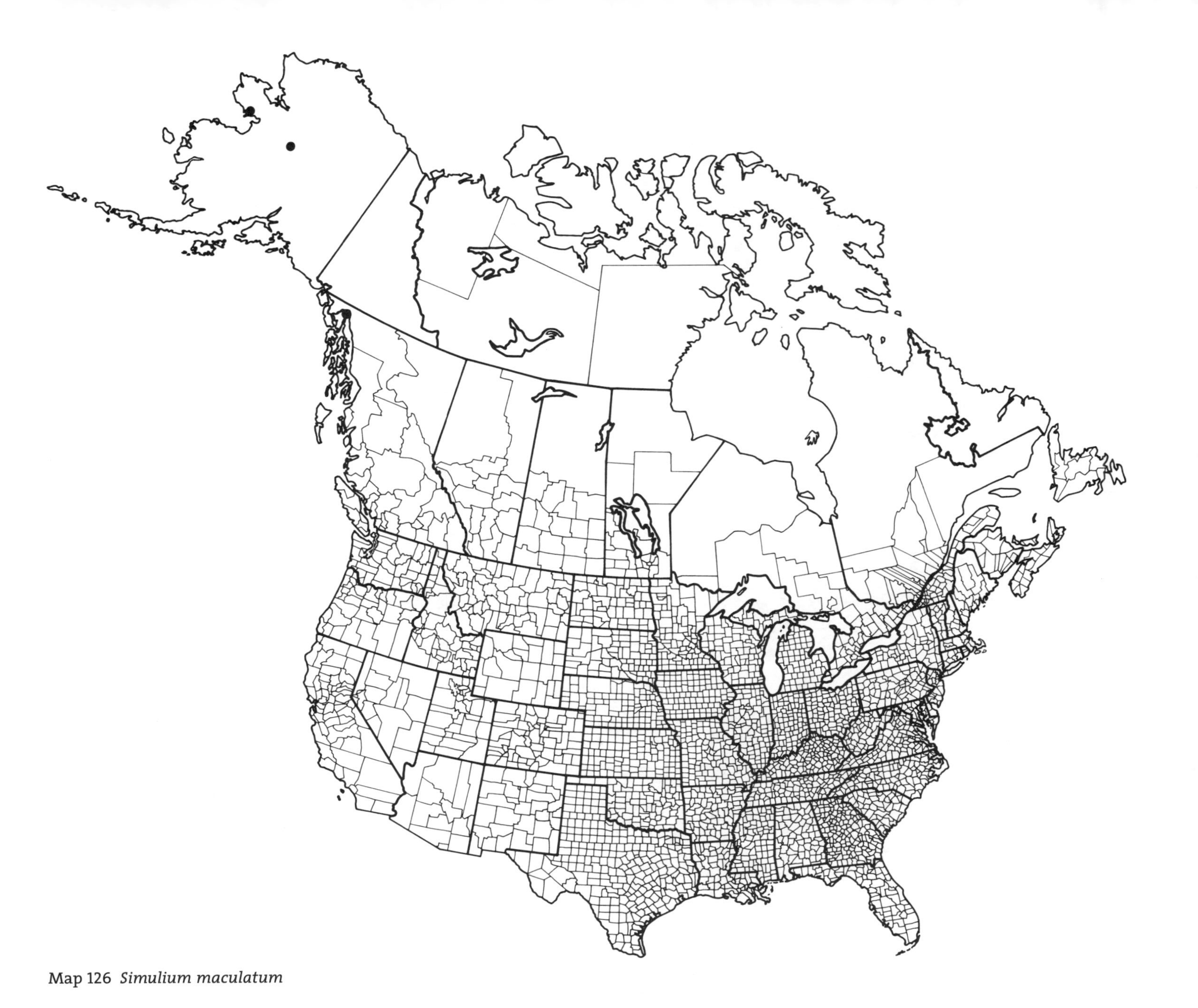

Map 126 *Simulium maculatum*

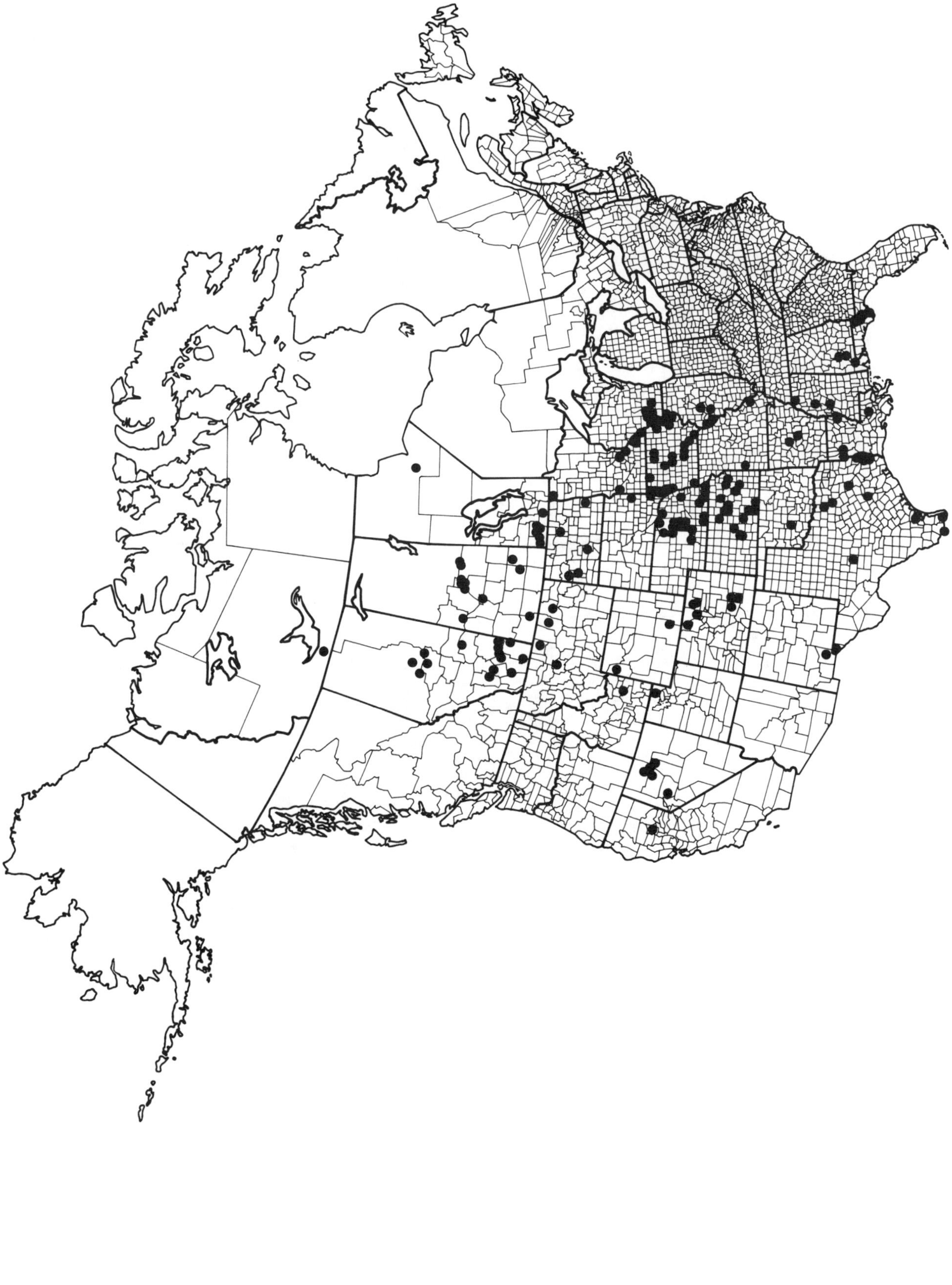

Map 127 *Simulium meridionale*

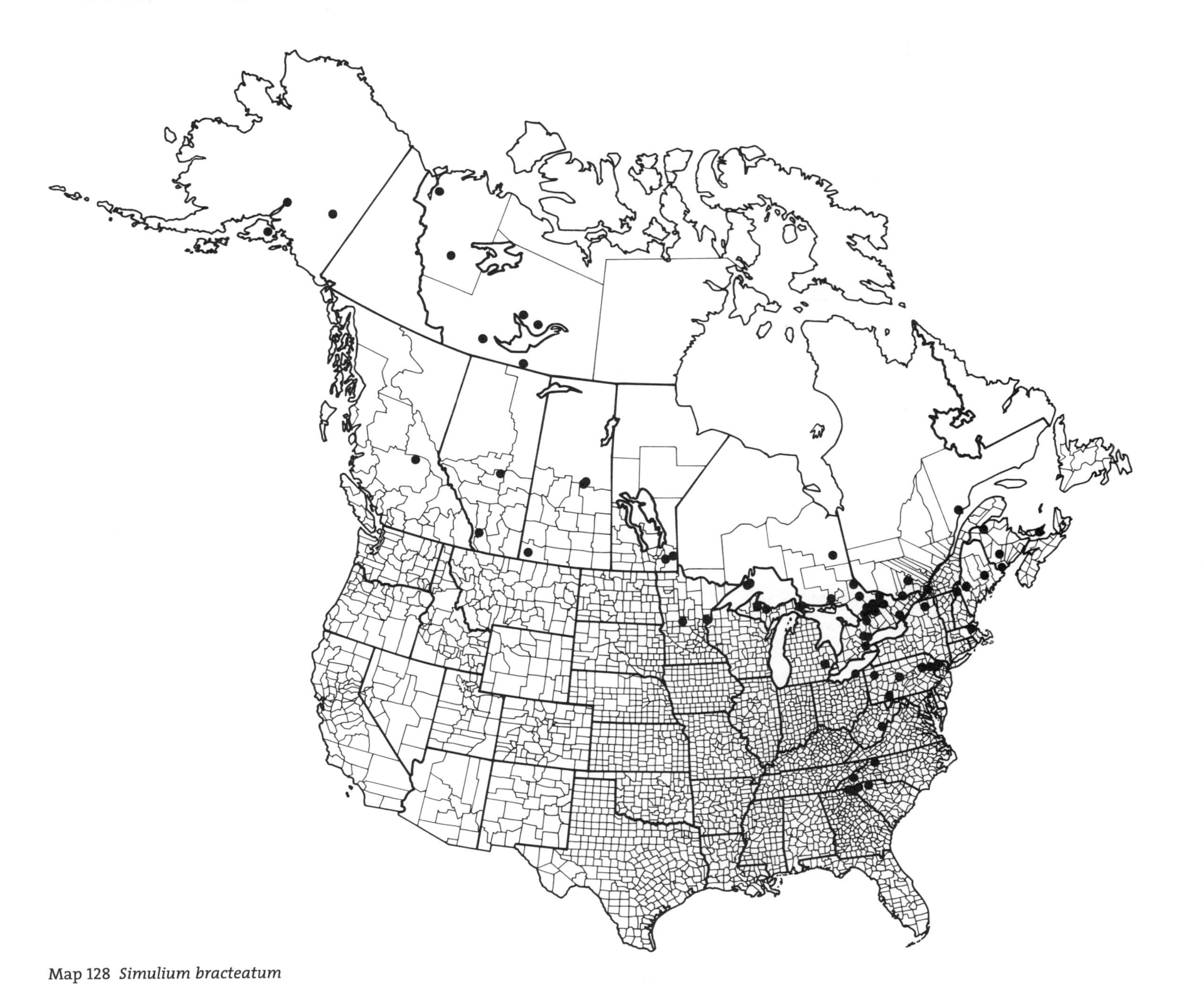

Map 128 *Simulium bracteatum*

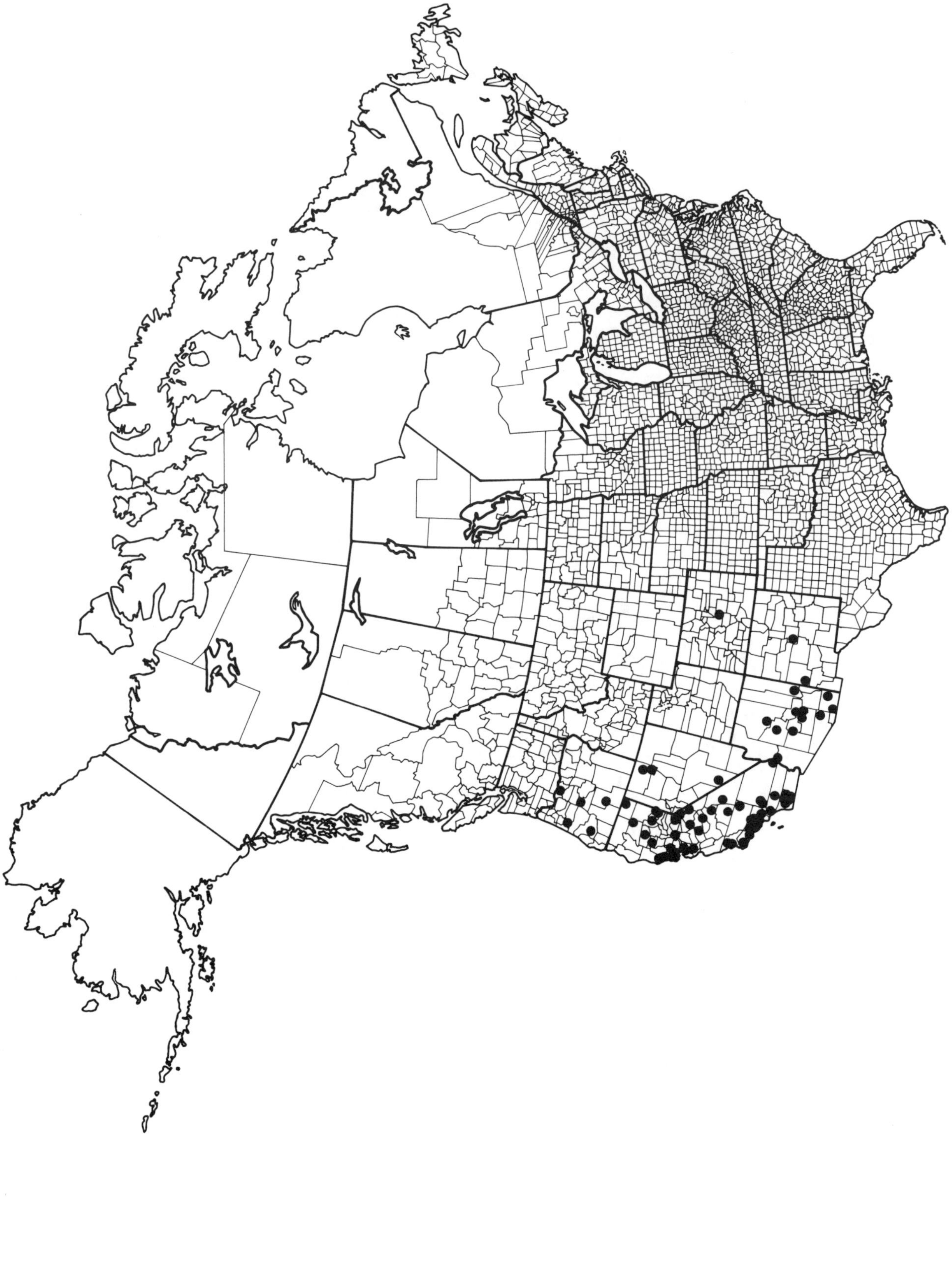

Map 129 *Simulium donovani*

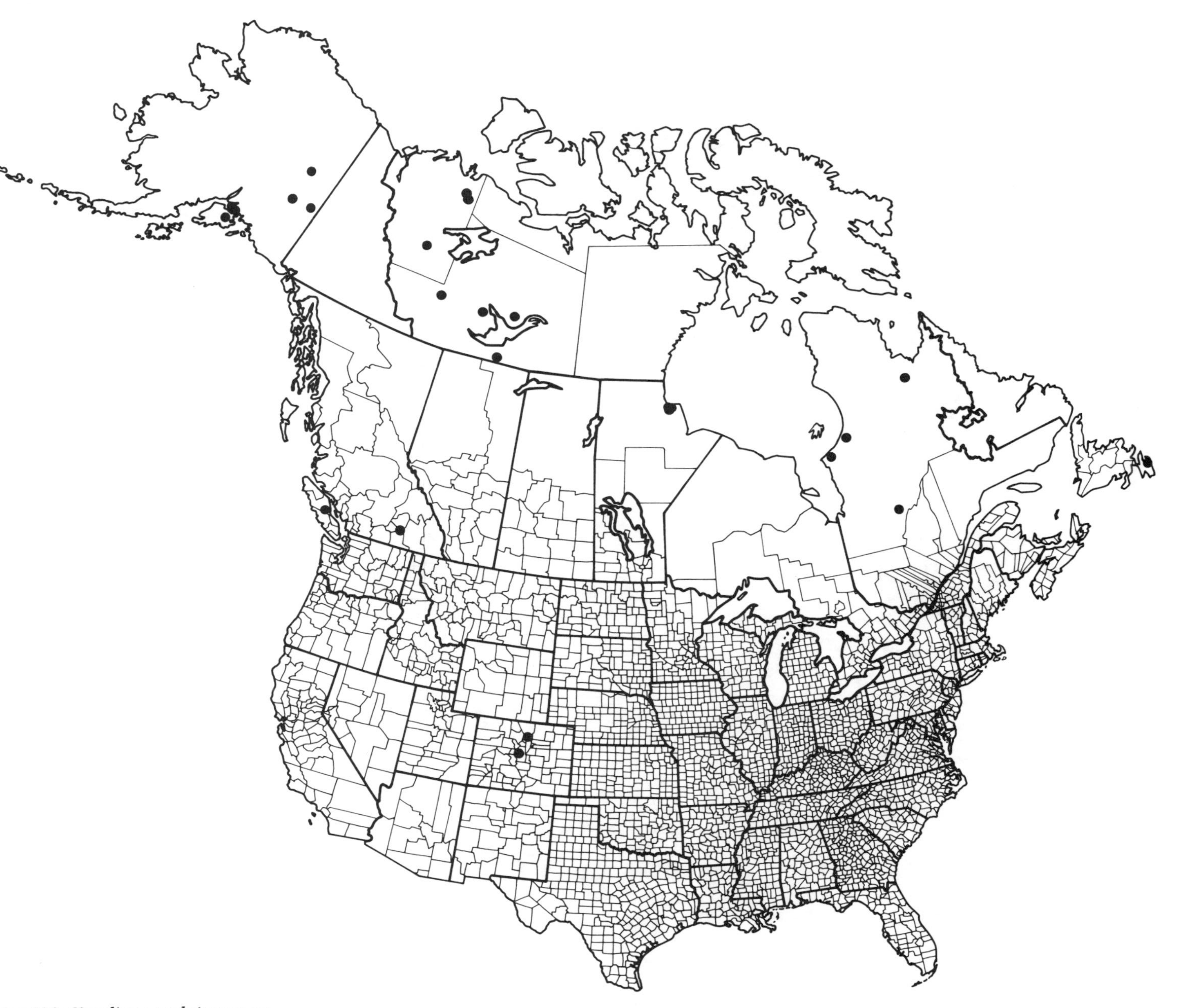

Map 130 *Simulium exulatum* n. sp.

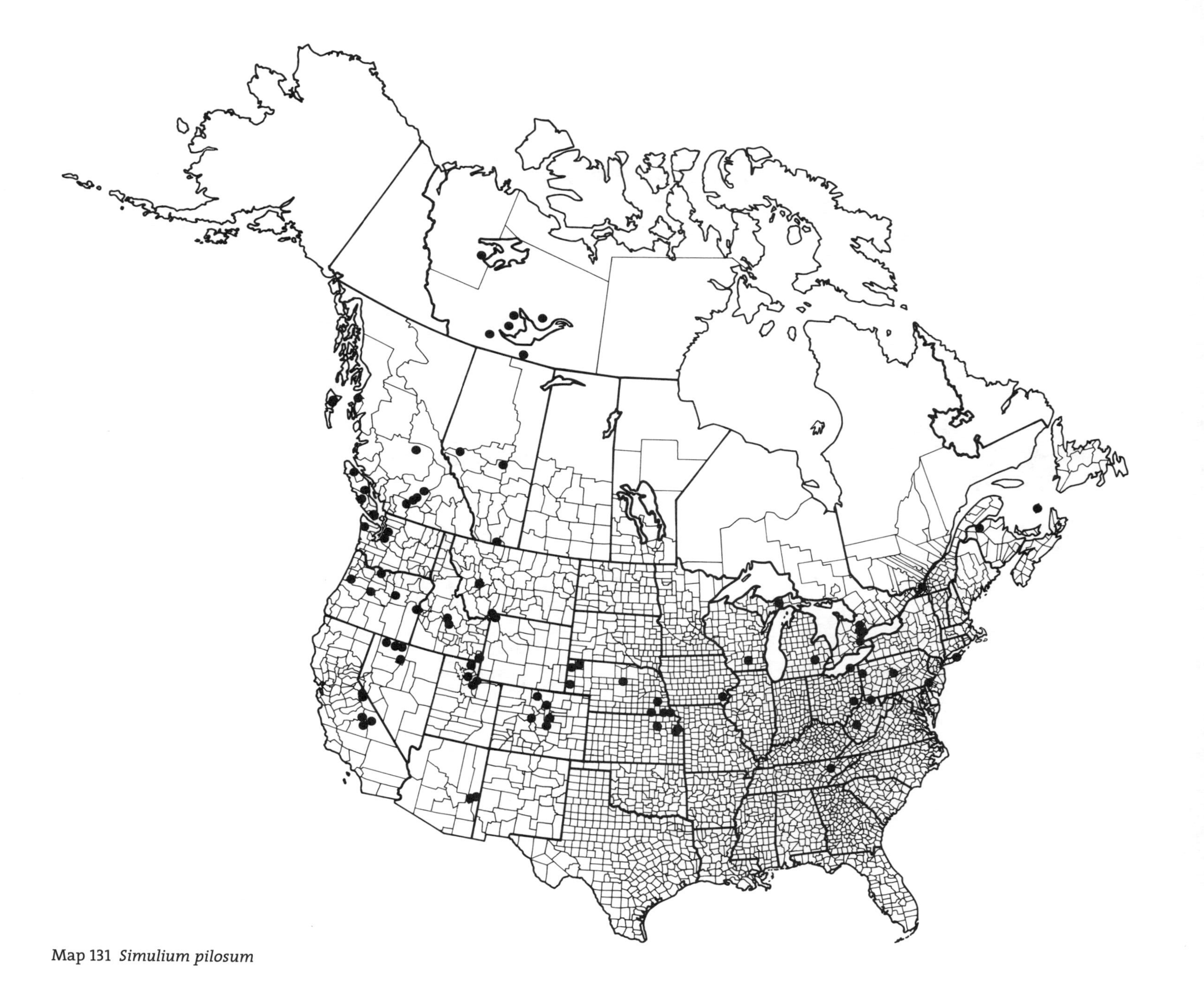

Map 131 *Simulium pilosum*

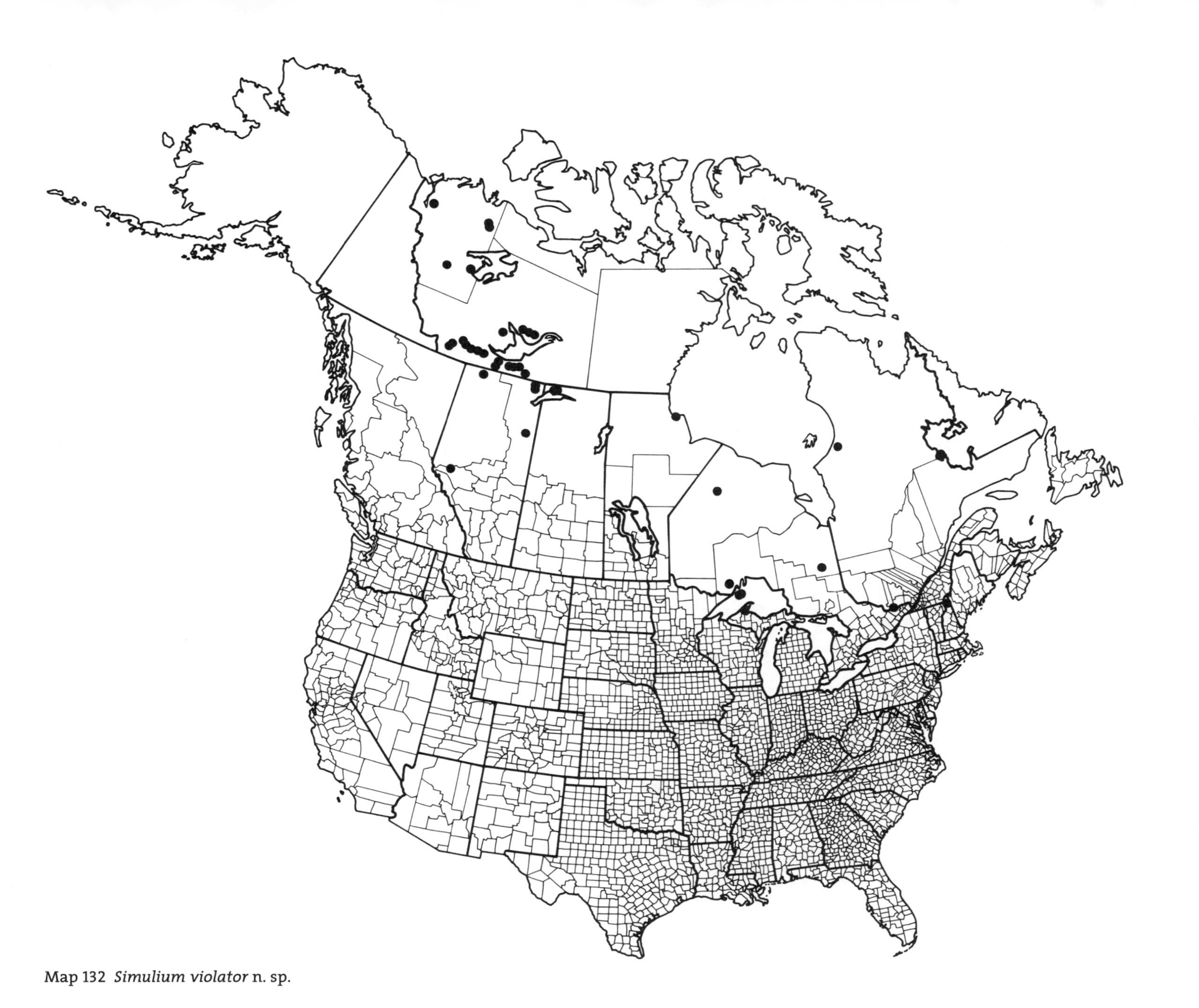

Map 132 *Simulium violator* n. sp.

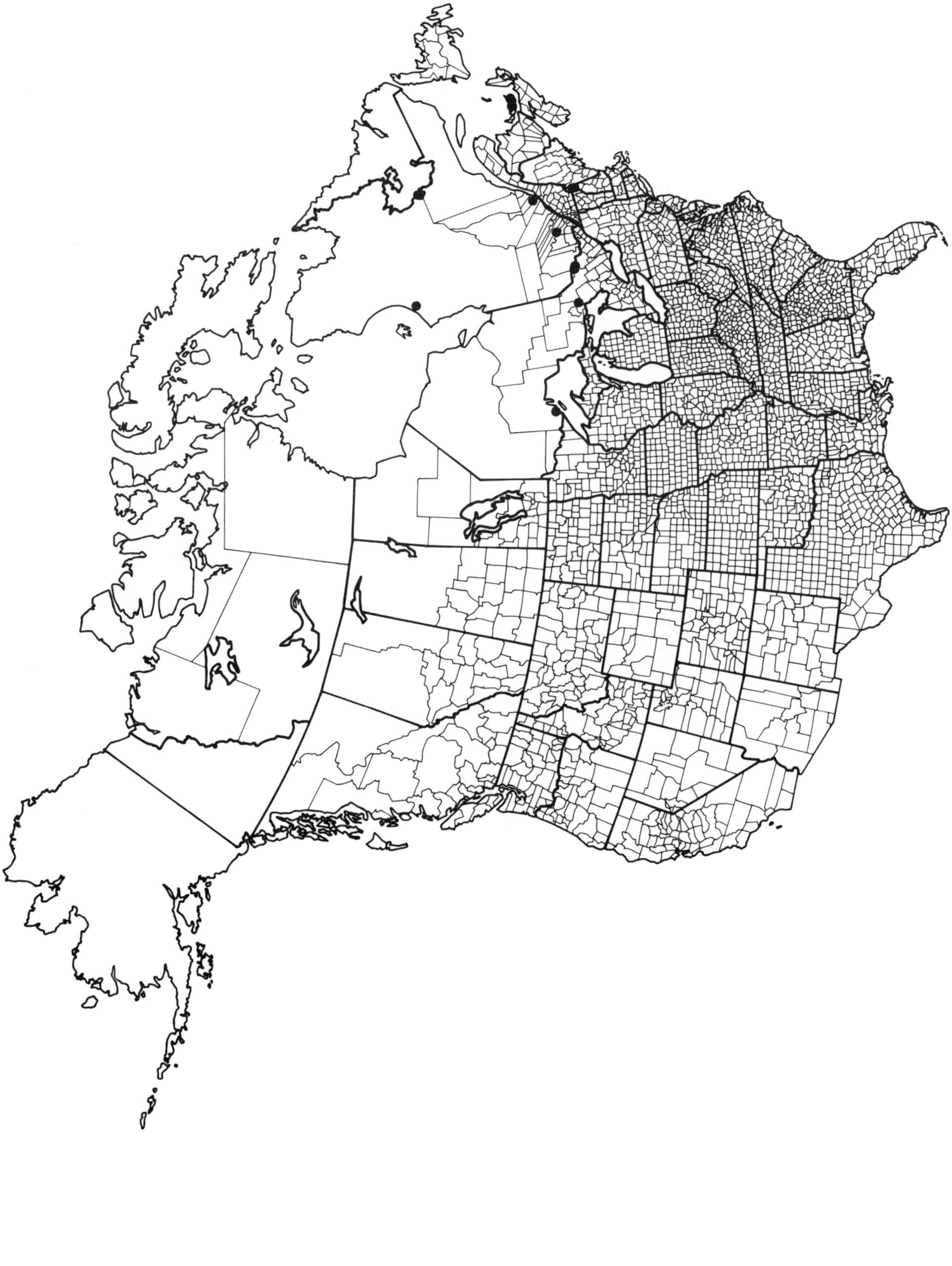

Map 133 *Simulium aestivum*

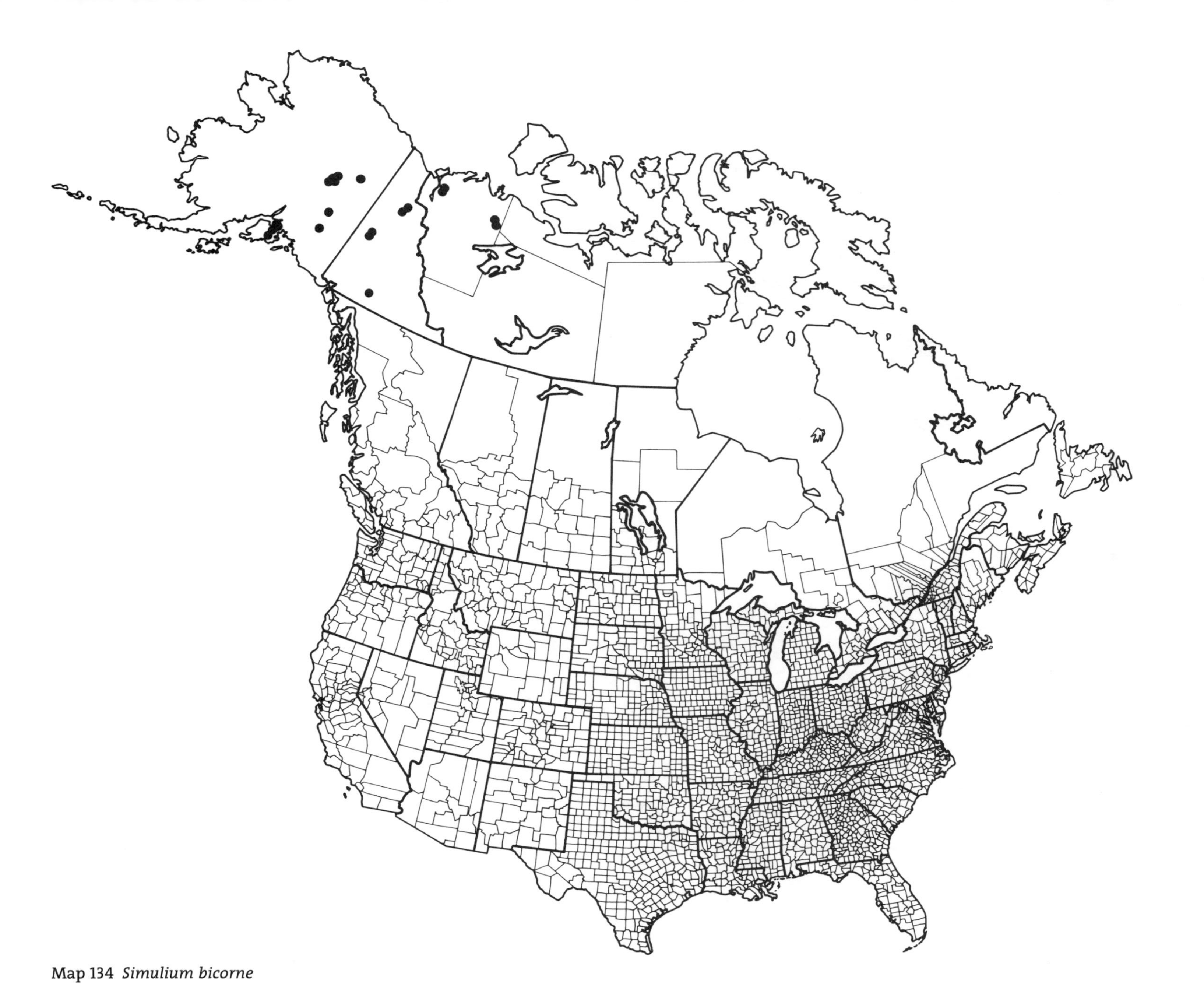

Map 134 *Simulium bicorne*

Map 135 *Simulium burgeri* n. sp.

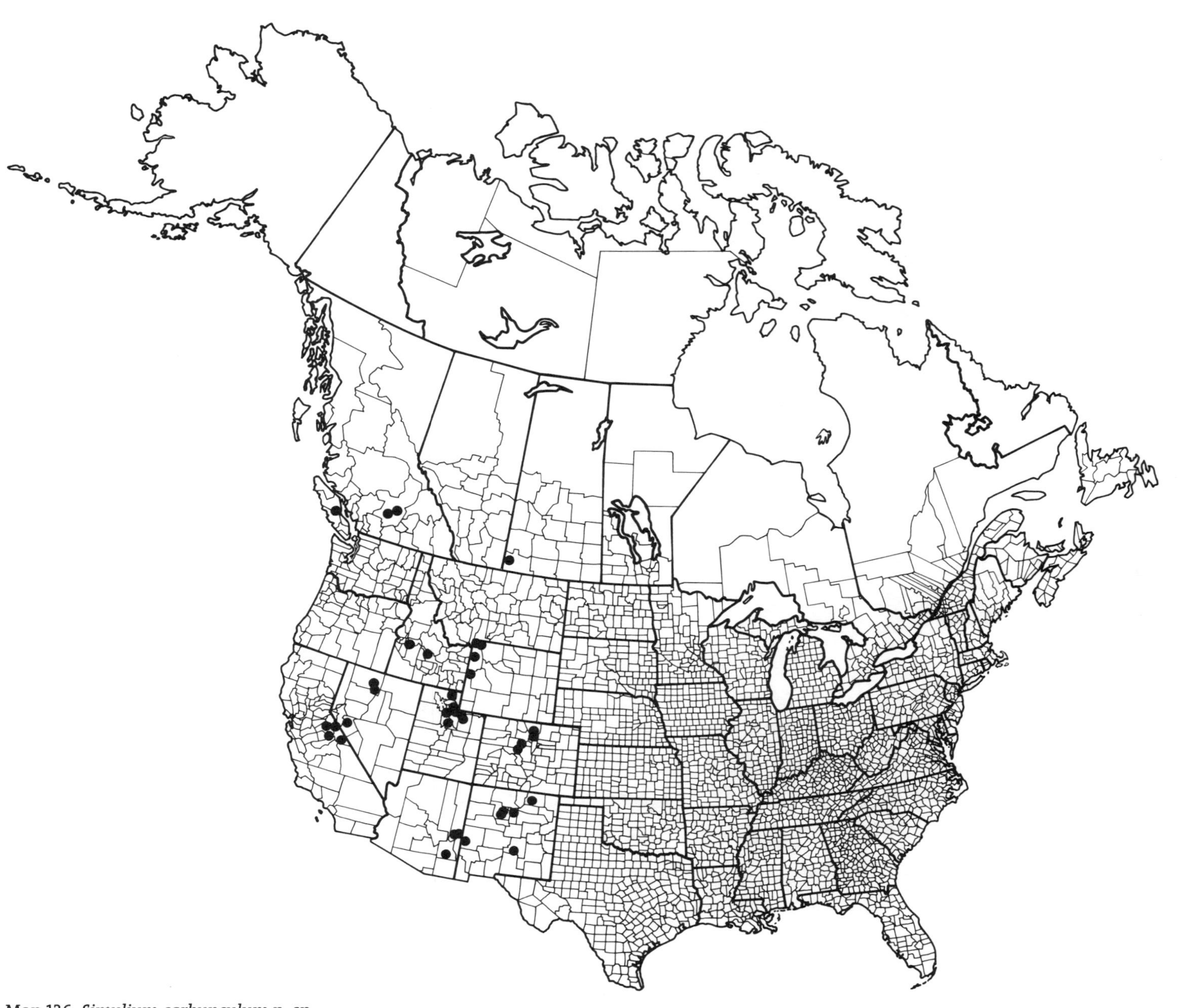

Map 136 *Simulium carbunculum* n. sp.

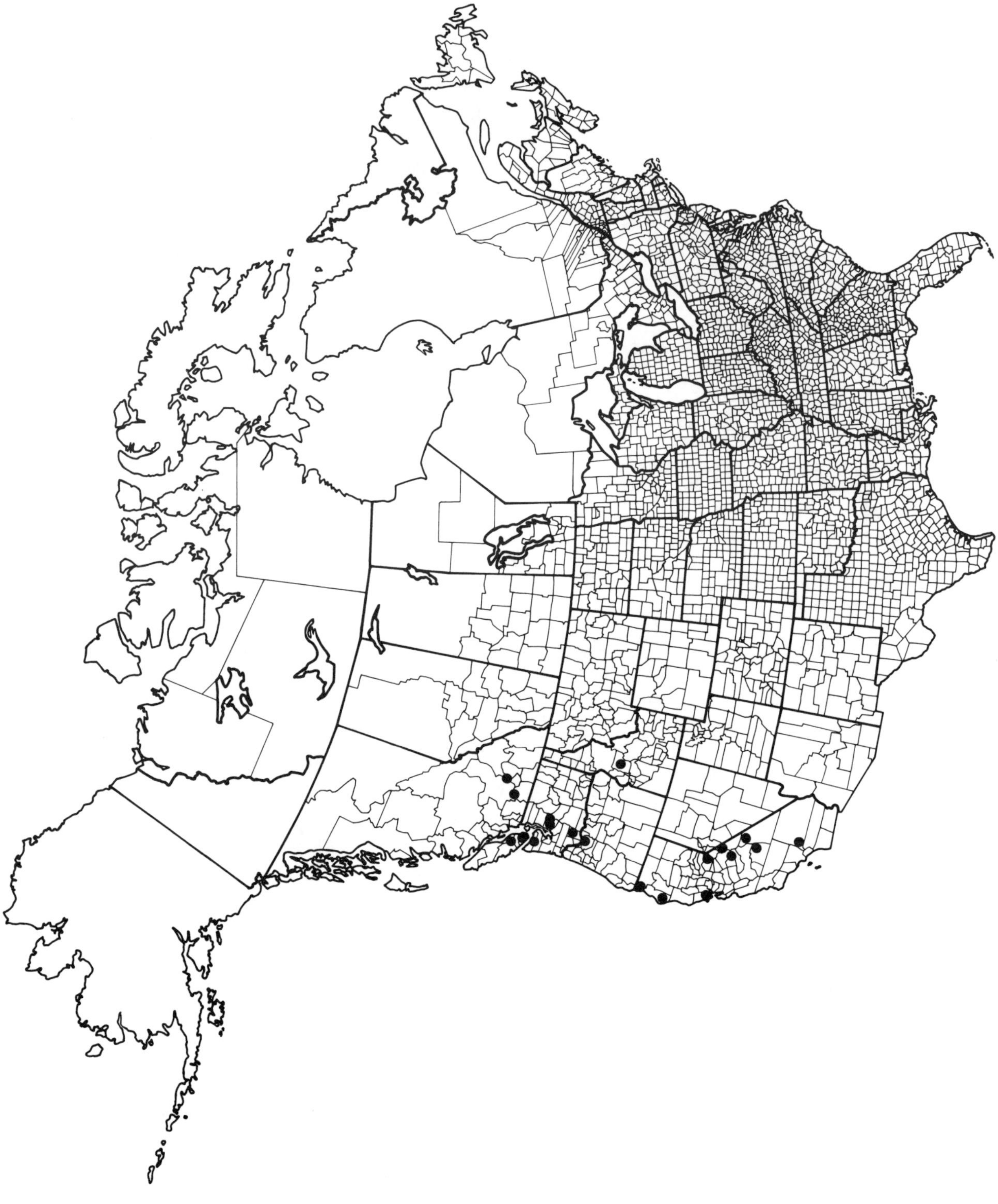

Map 137 *Simulium conicum* n. sp.

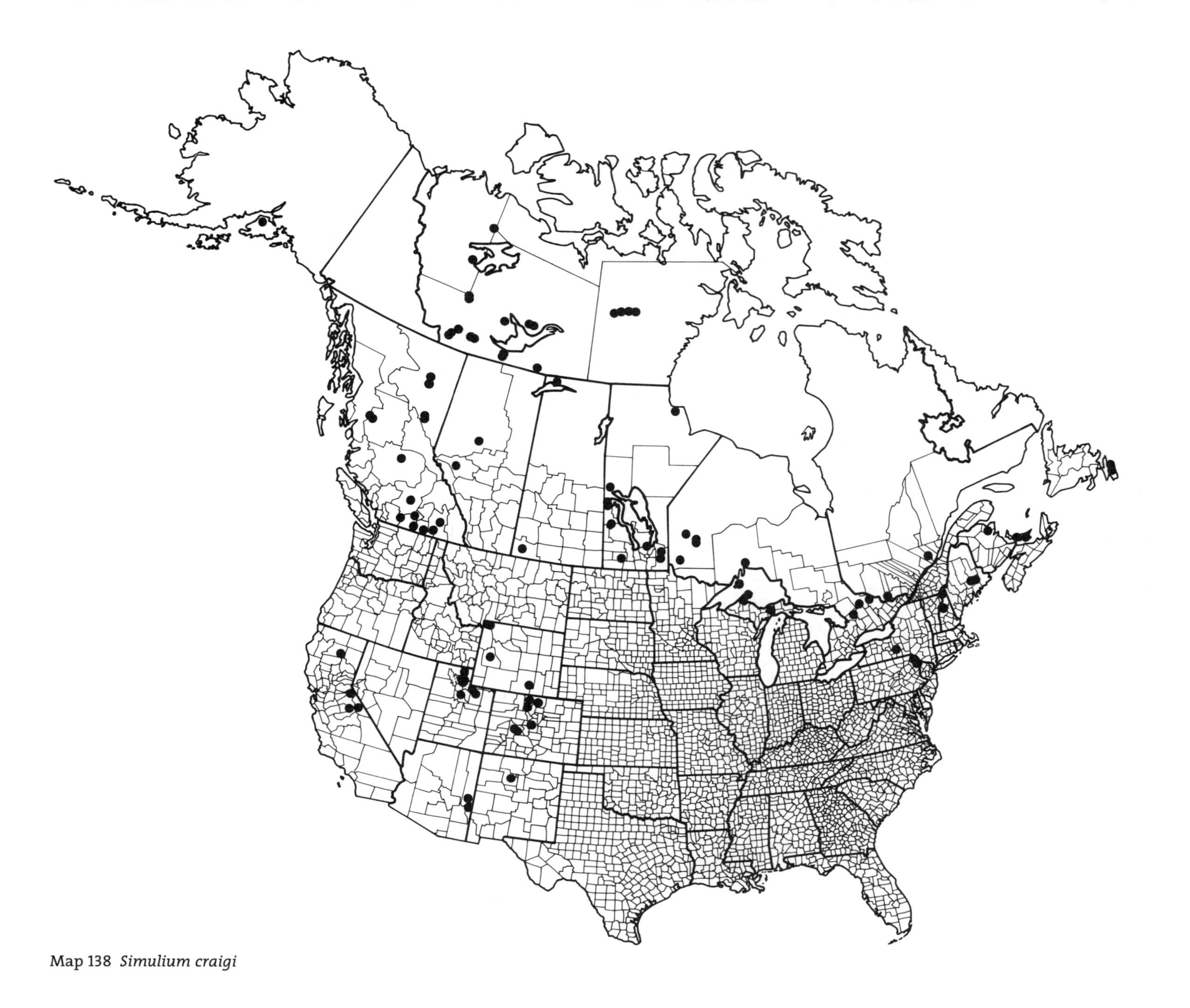

Map 138 *Simulium craigi*

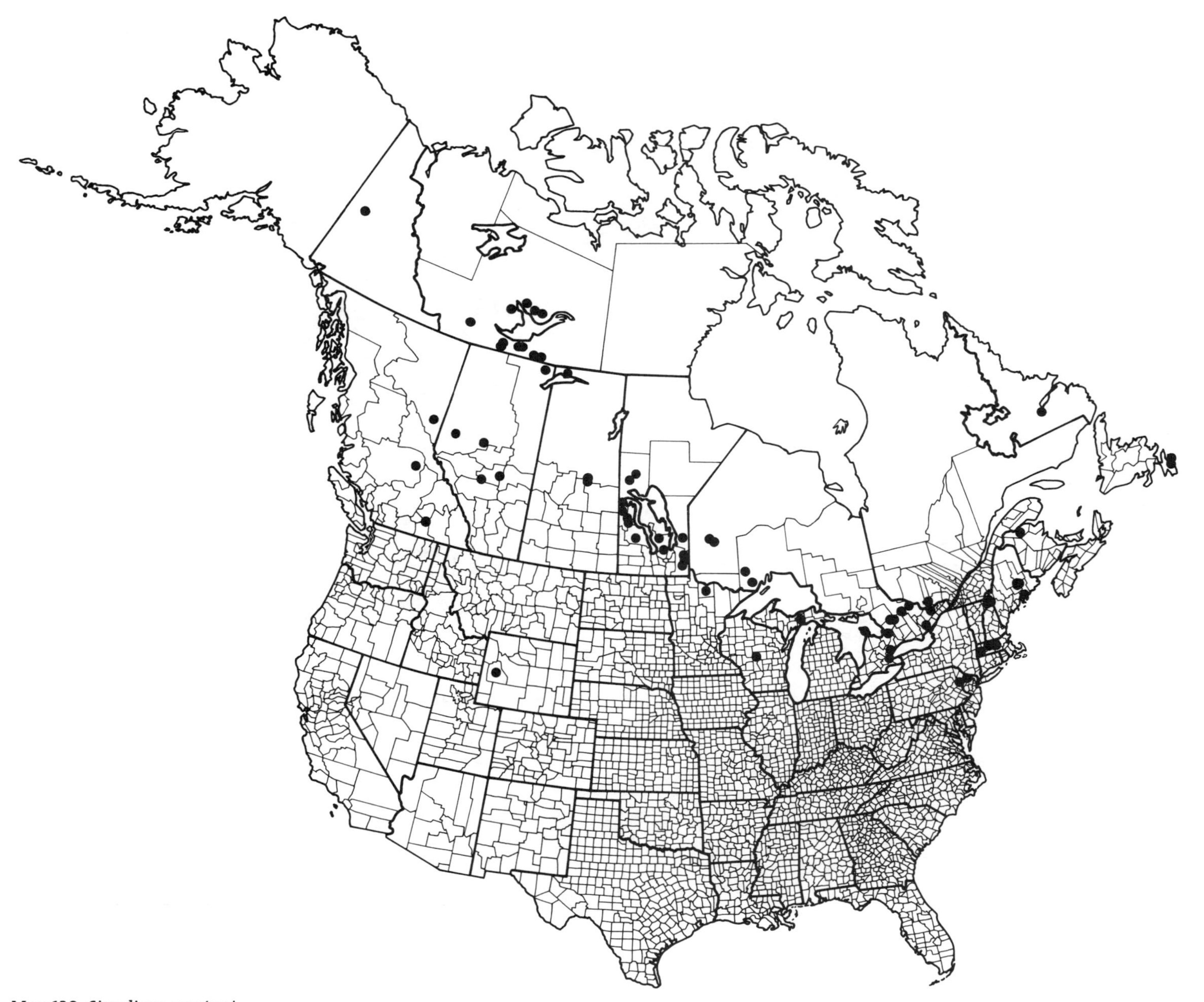

Map 139 *Simulium croxtoni*

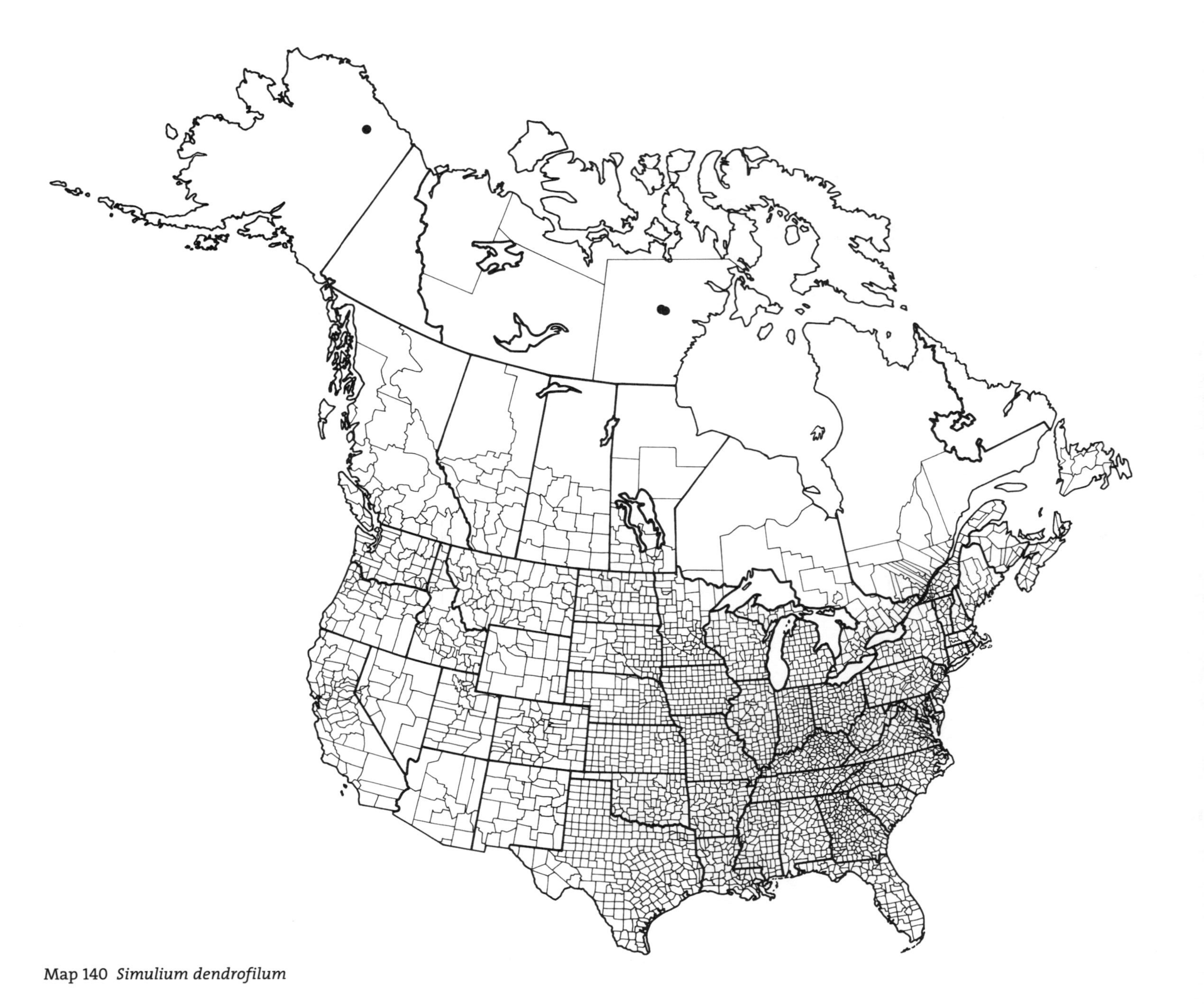

Map 140 *Simulium dendrofilum*

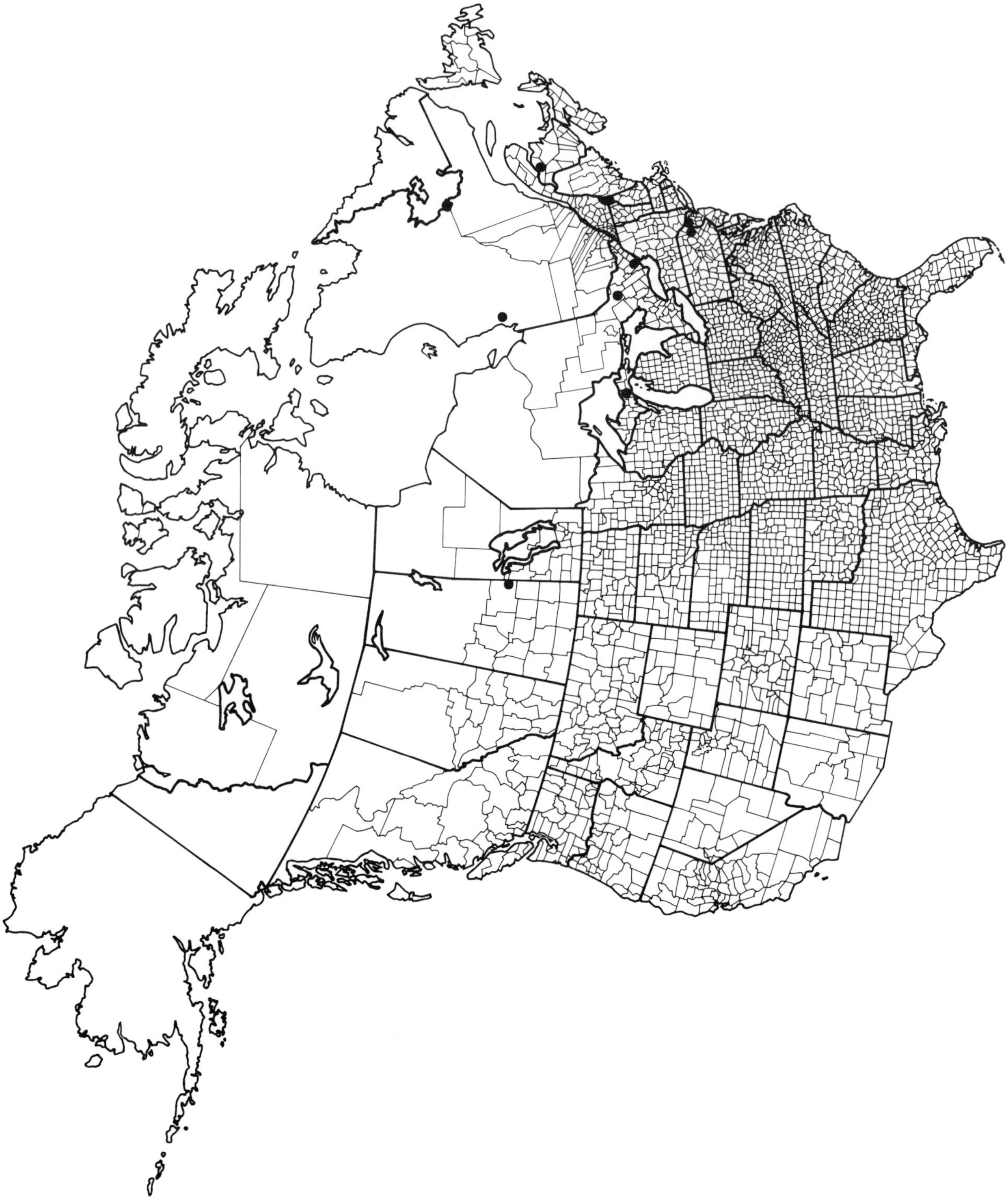

Map 141 *Simulium fionae*

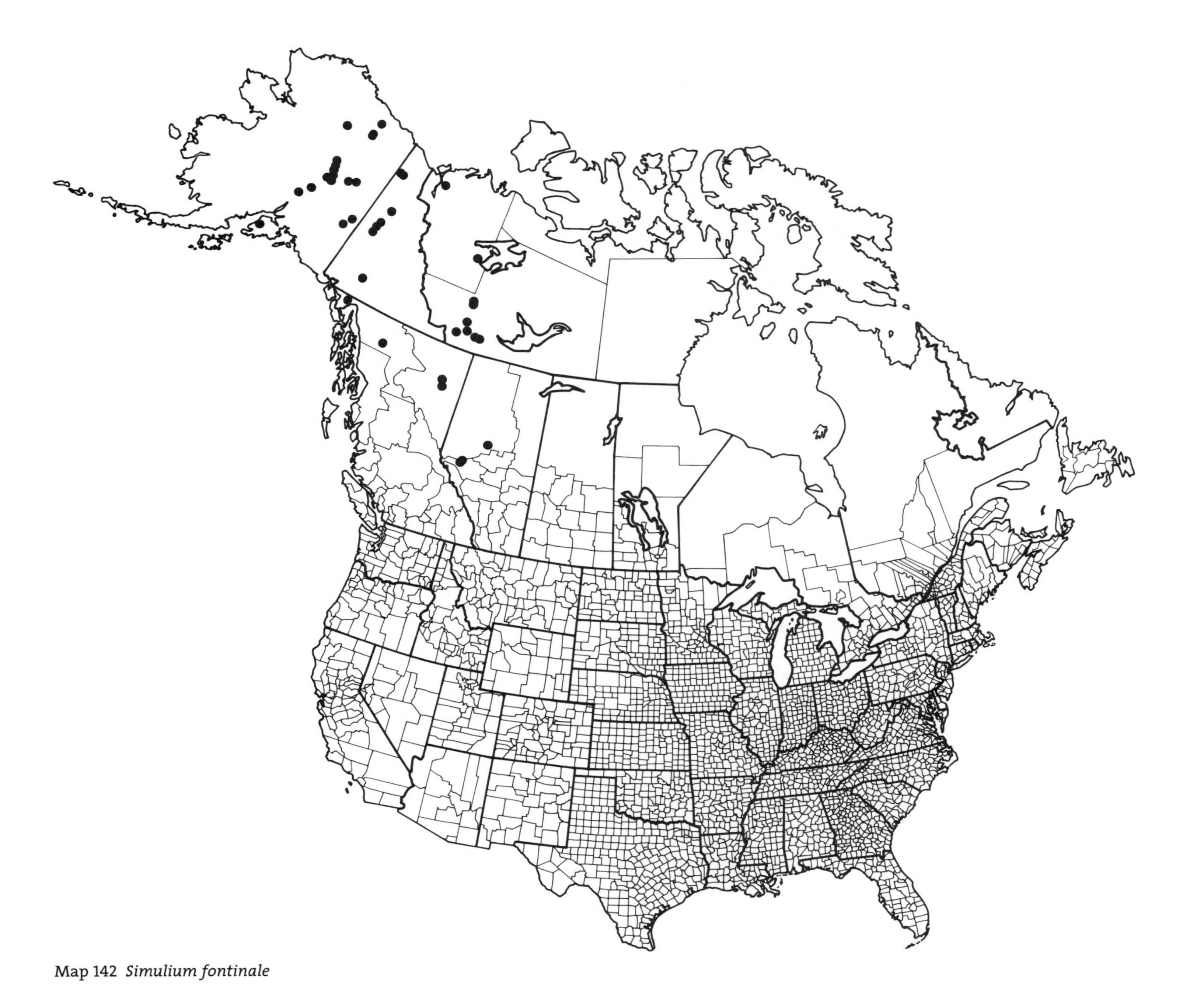

Map 142 *Simulium fontinale*

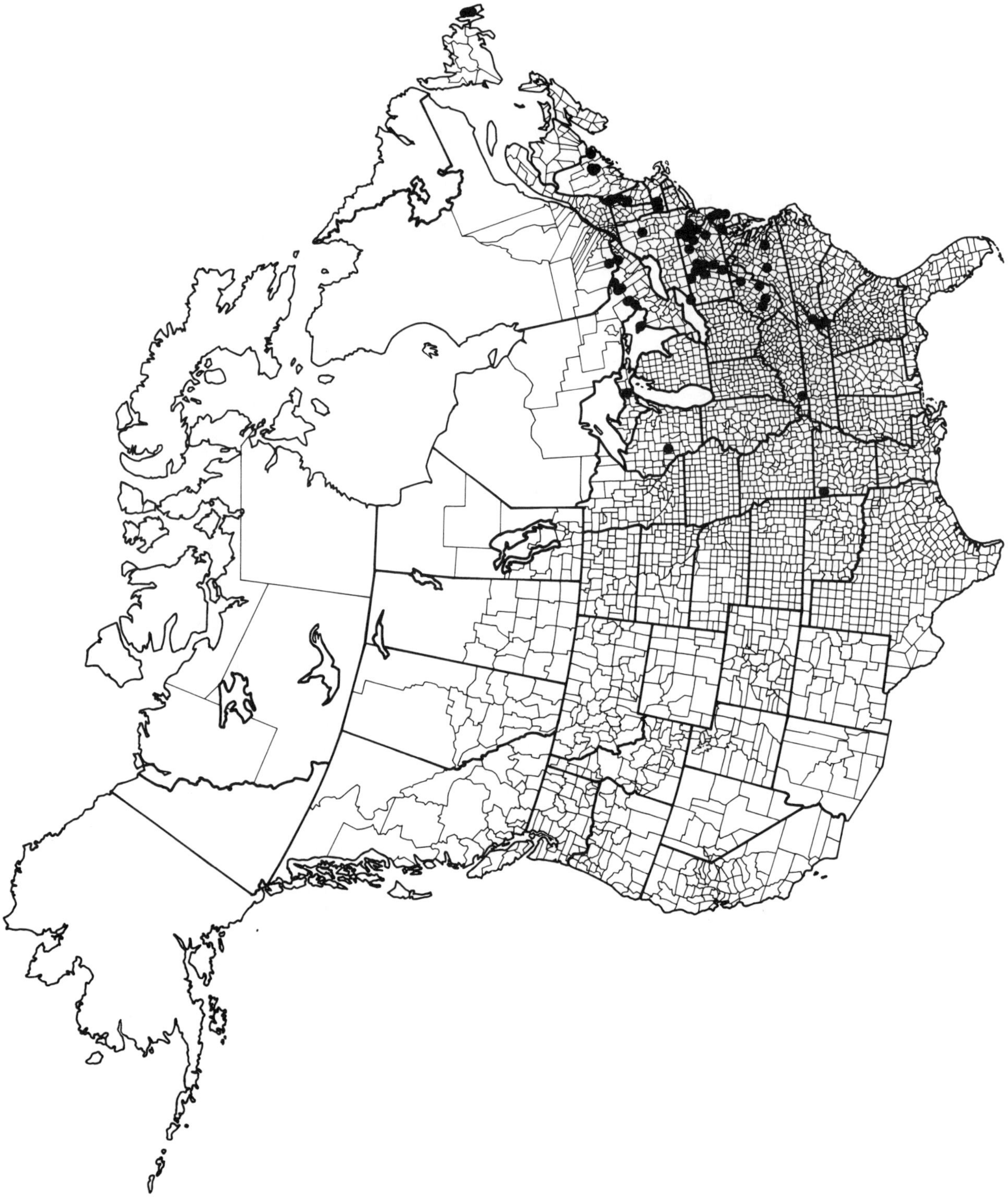

Map 143 *Simulium gouldingi*

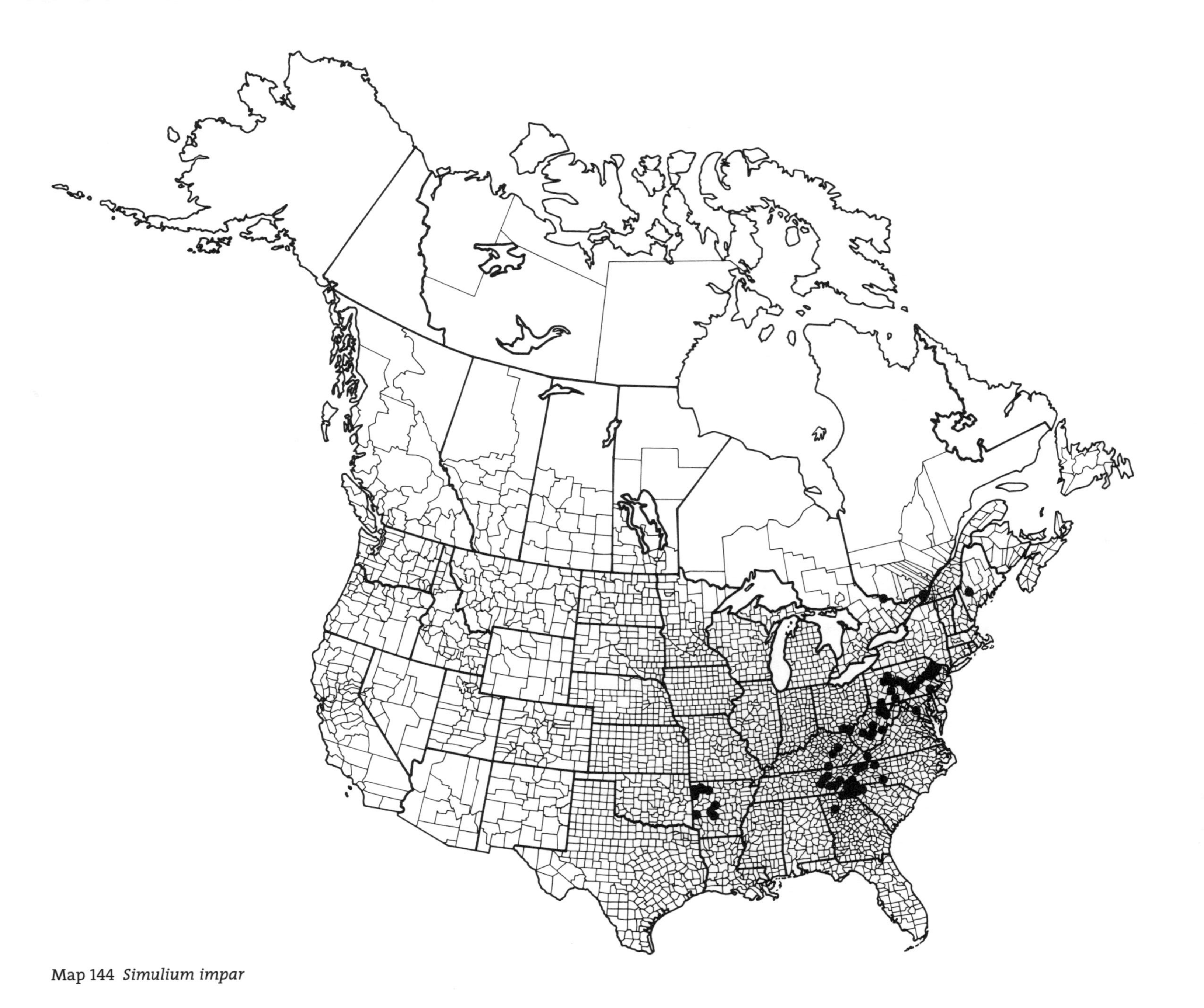

Map 144 *Simulium impar*

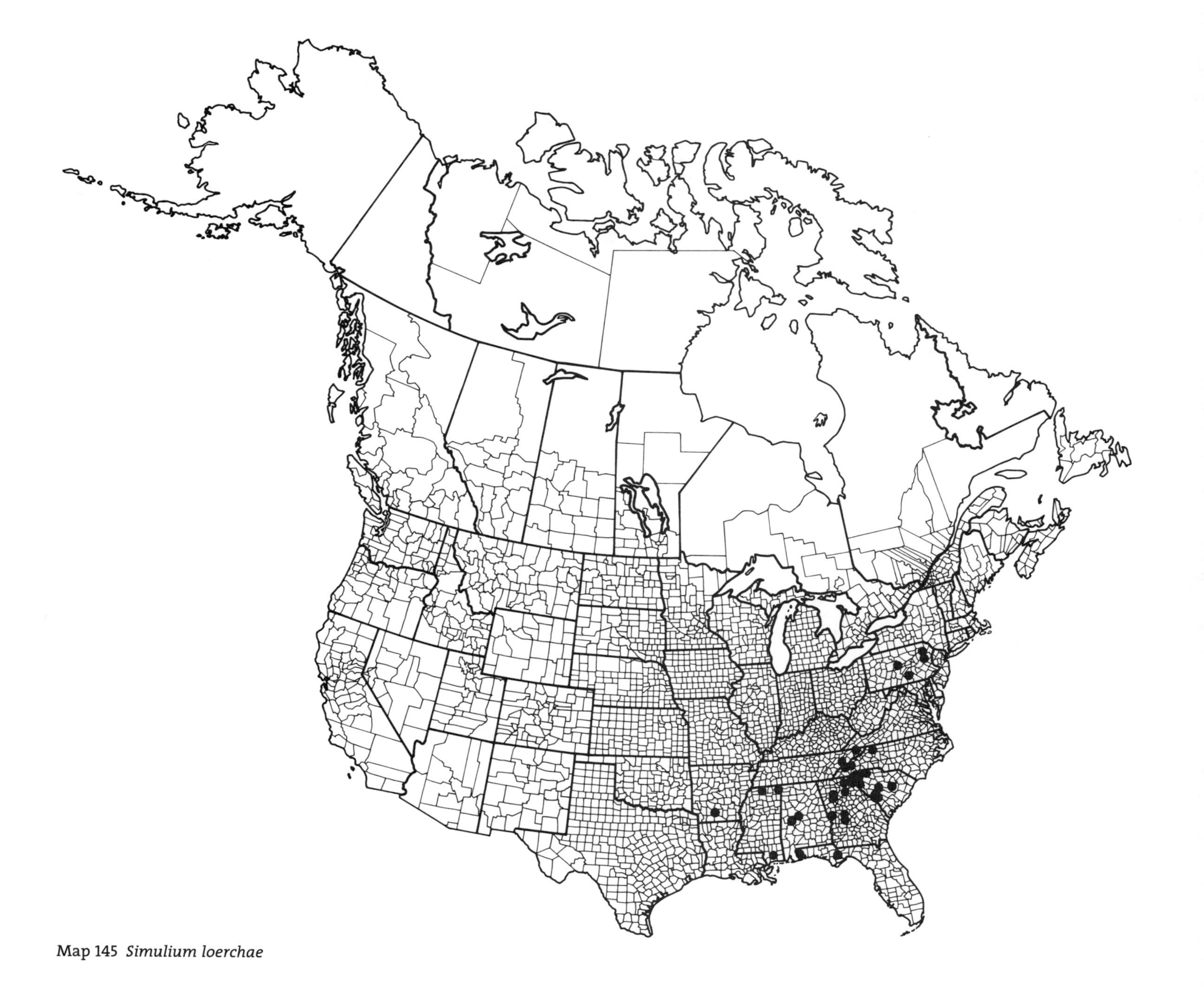

Map 145 *Simulium loerchae*

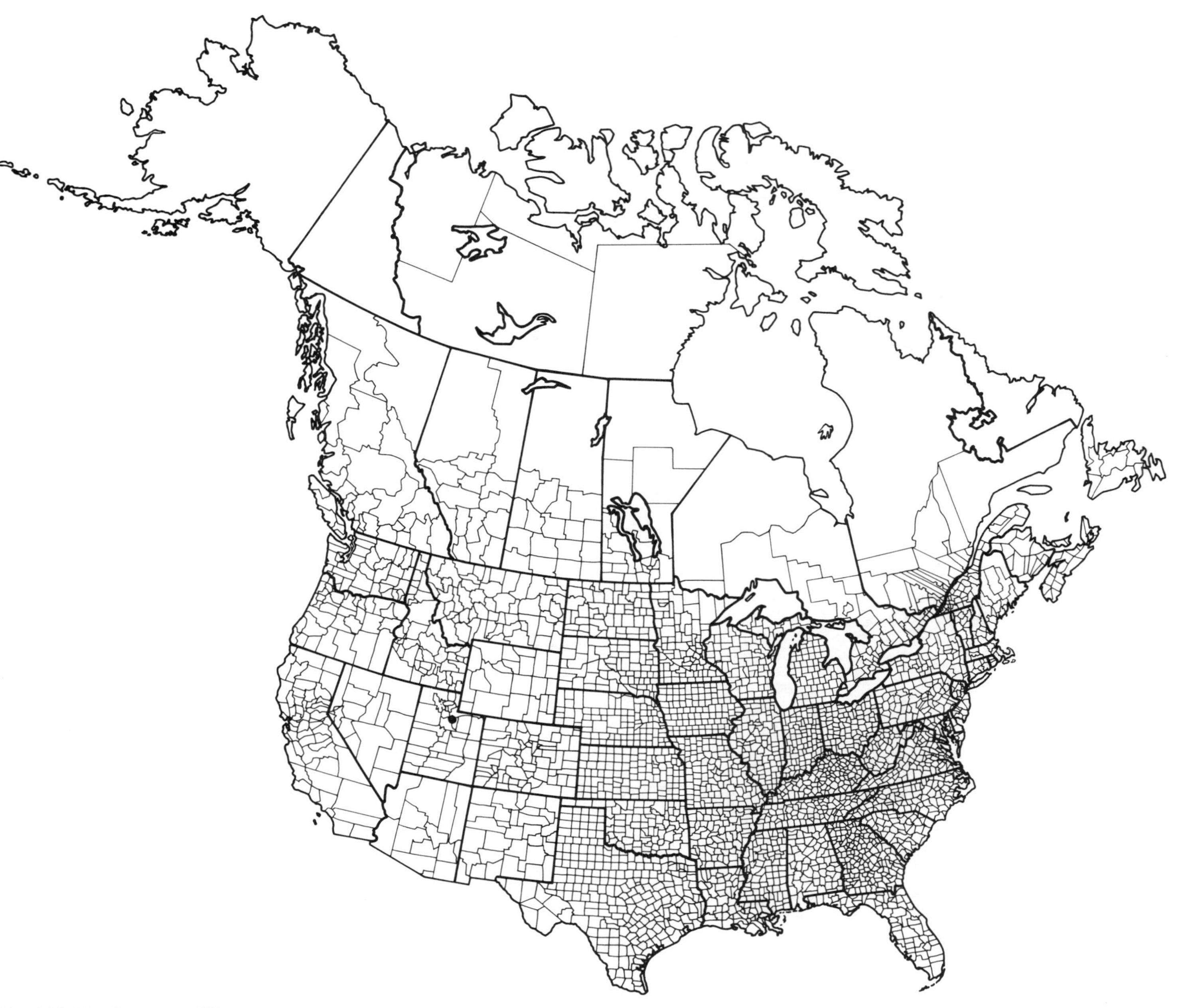

Map 146 *Simulium merritti* n. sp.

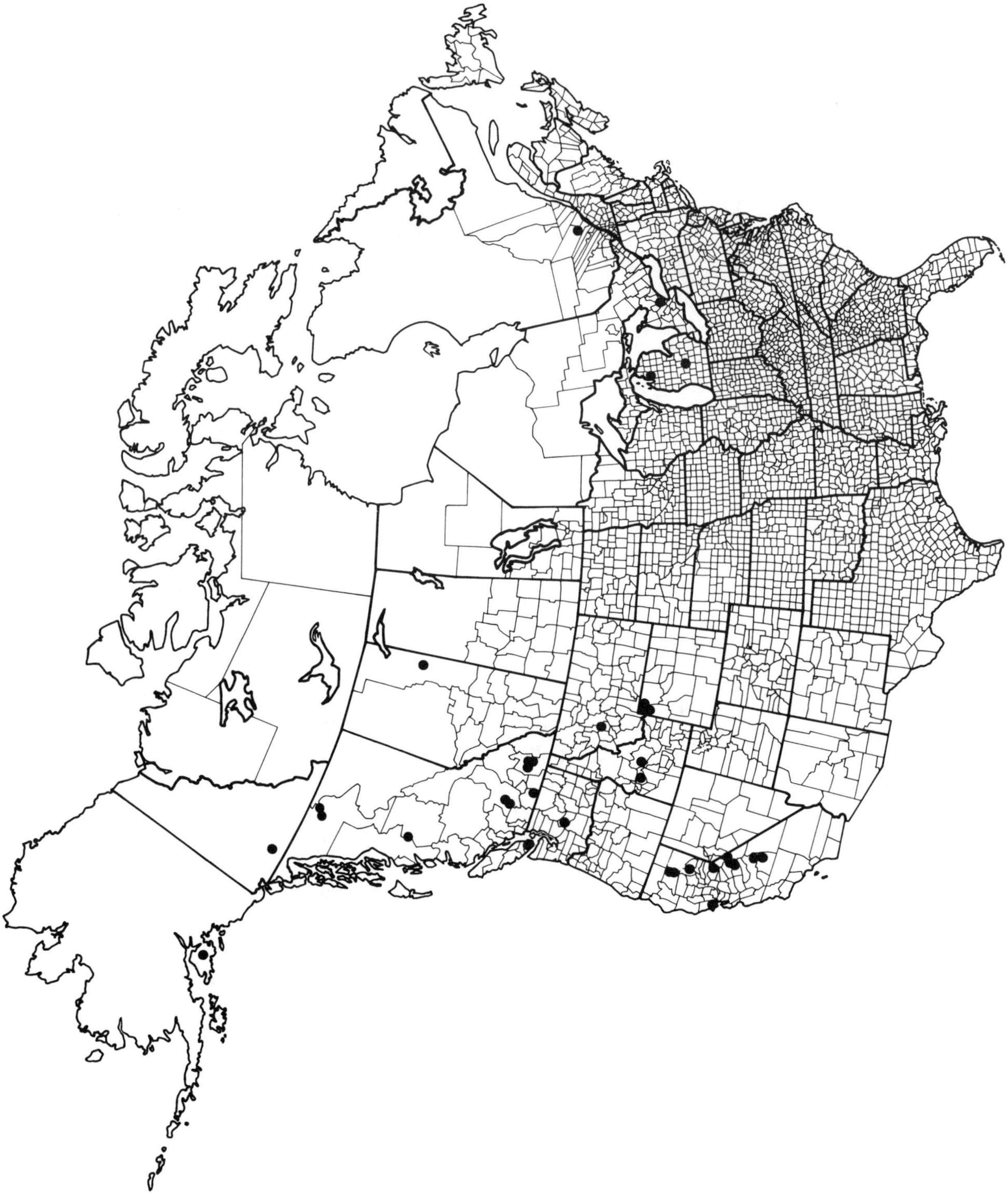

Map 147 *Simulium modicum* n. sp.

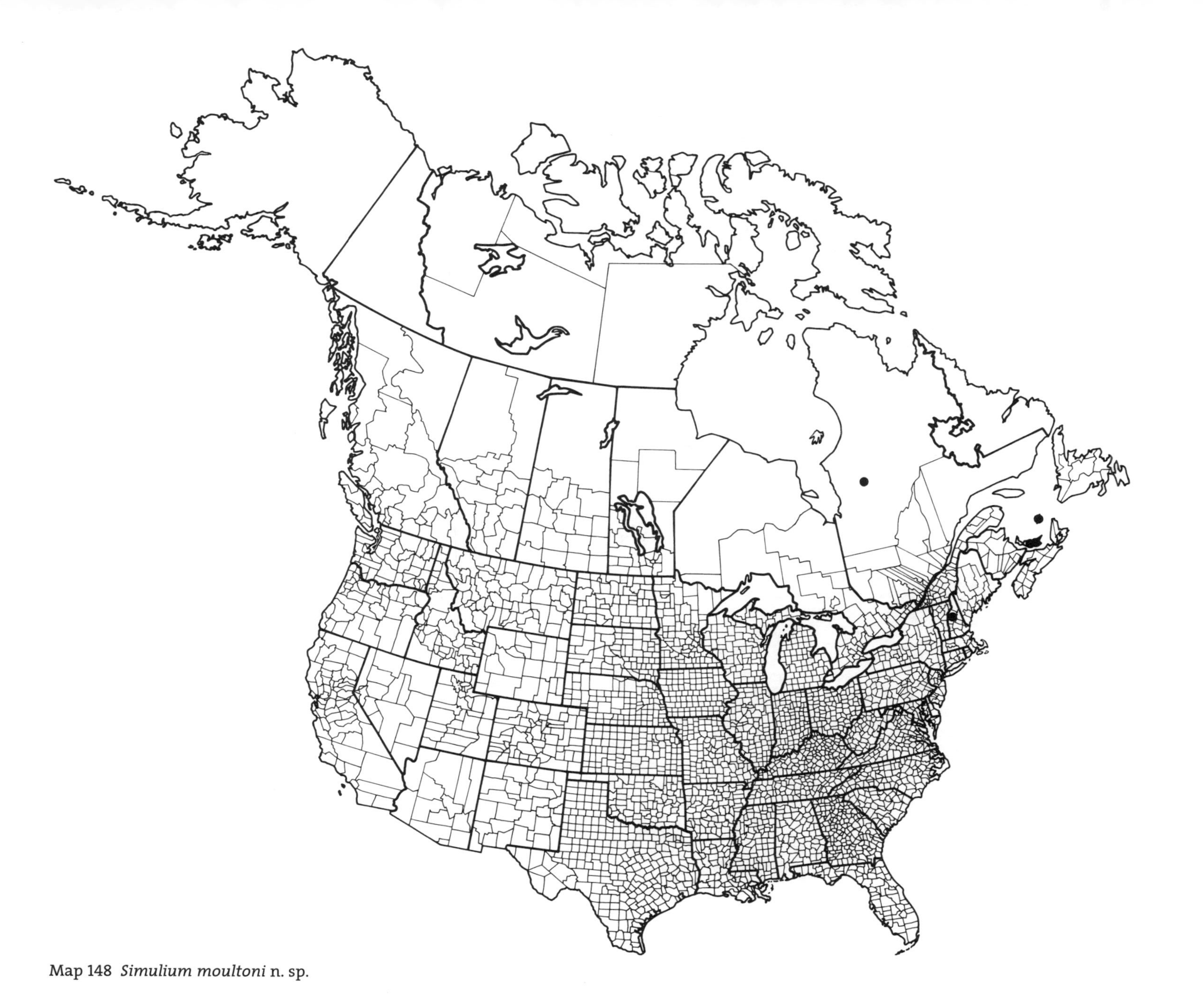

Map 148 *Simulium moultoni* n. sp.

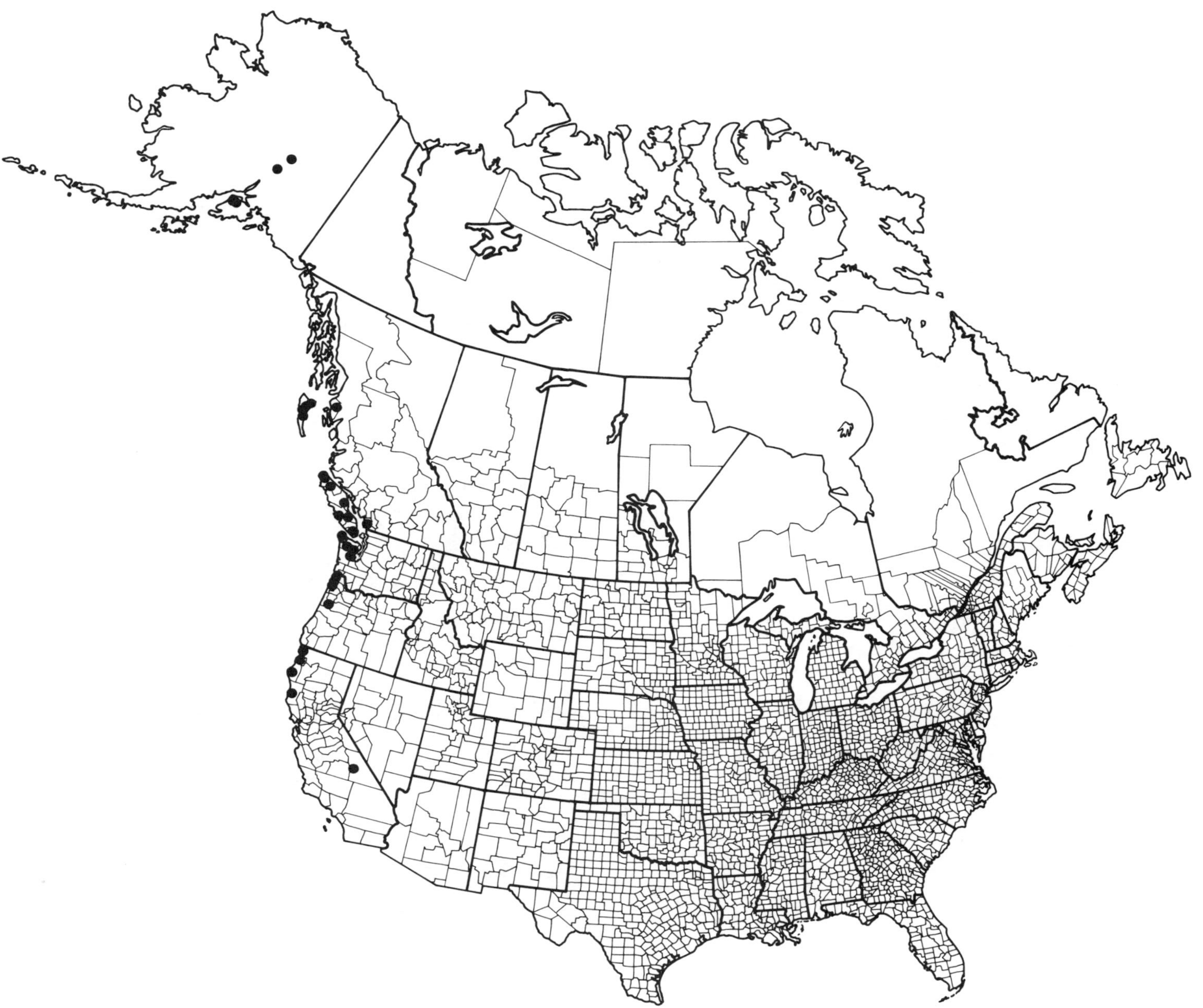

Map 149 *Simulium pugetense*

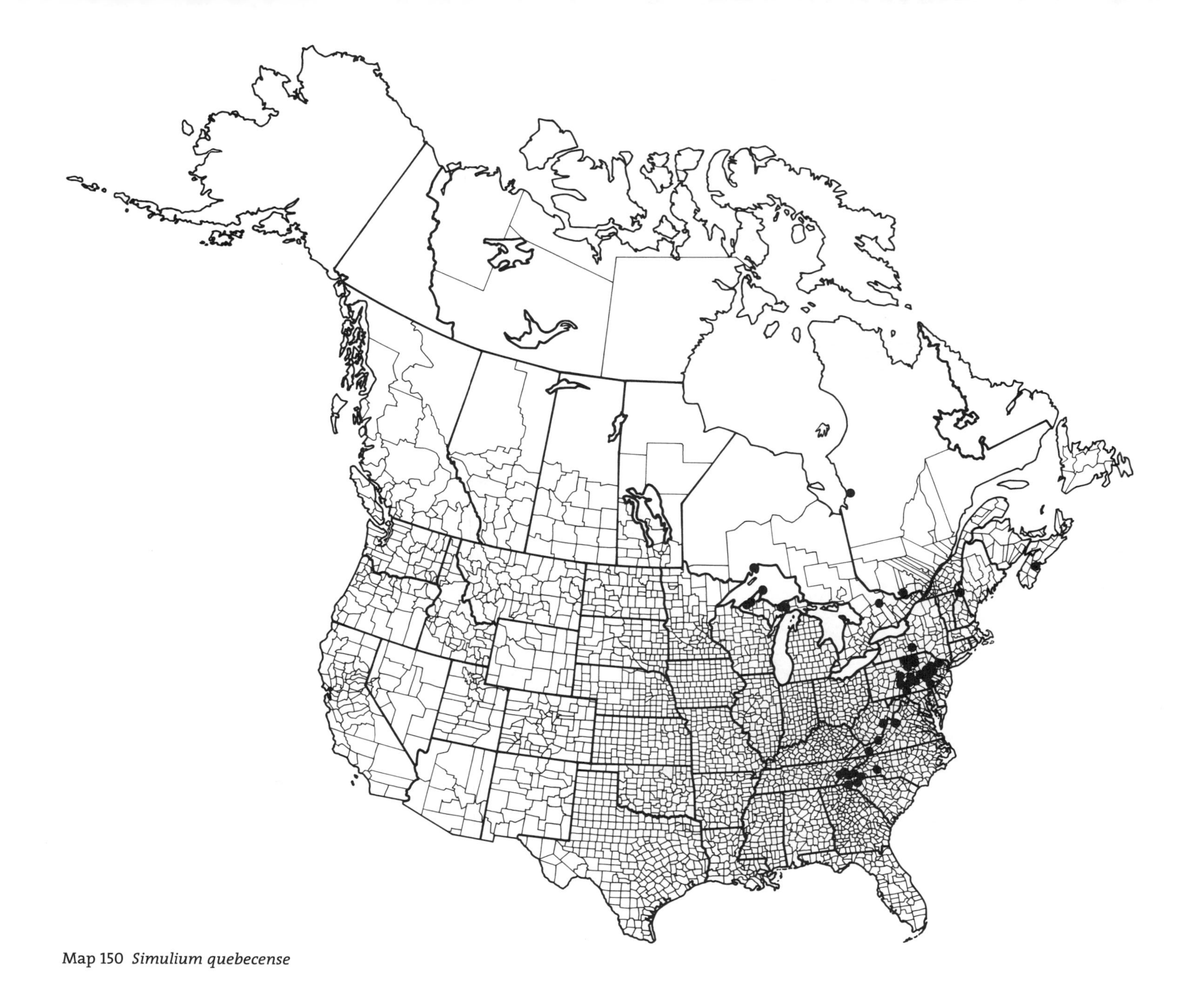

Map 150 *Simulium quebecense*

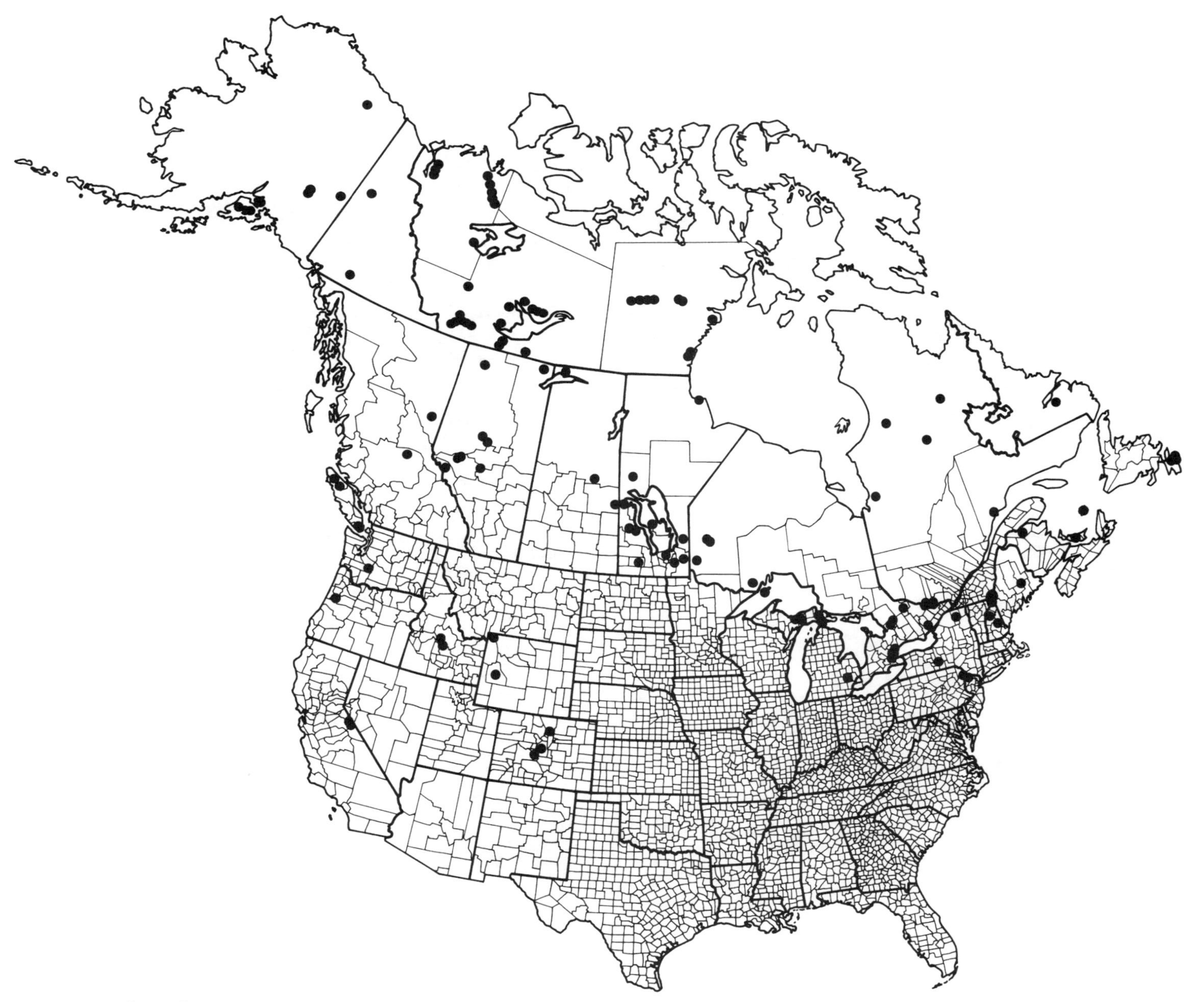

Map 151 *Simulium silvestre*

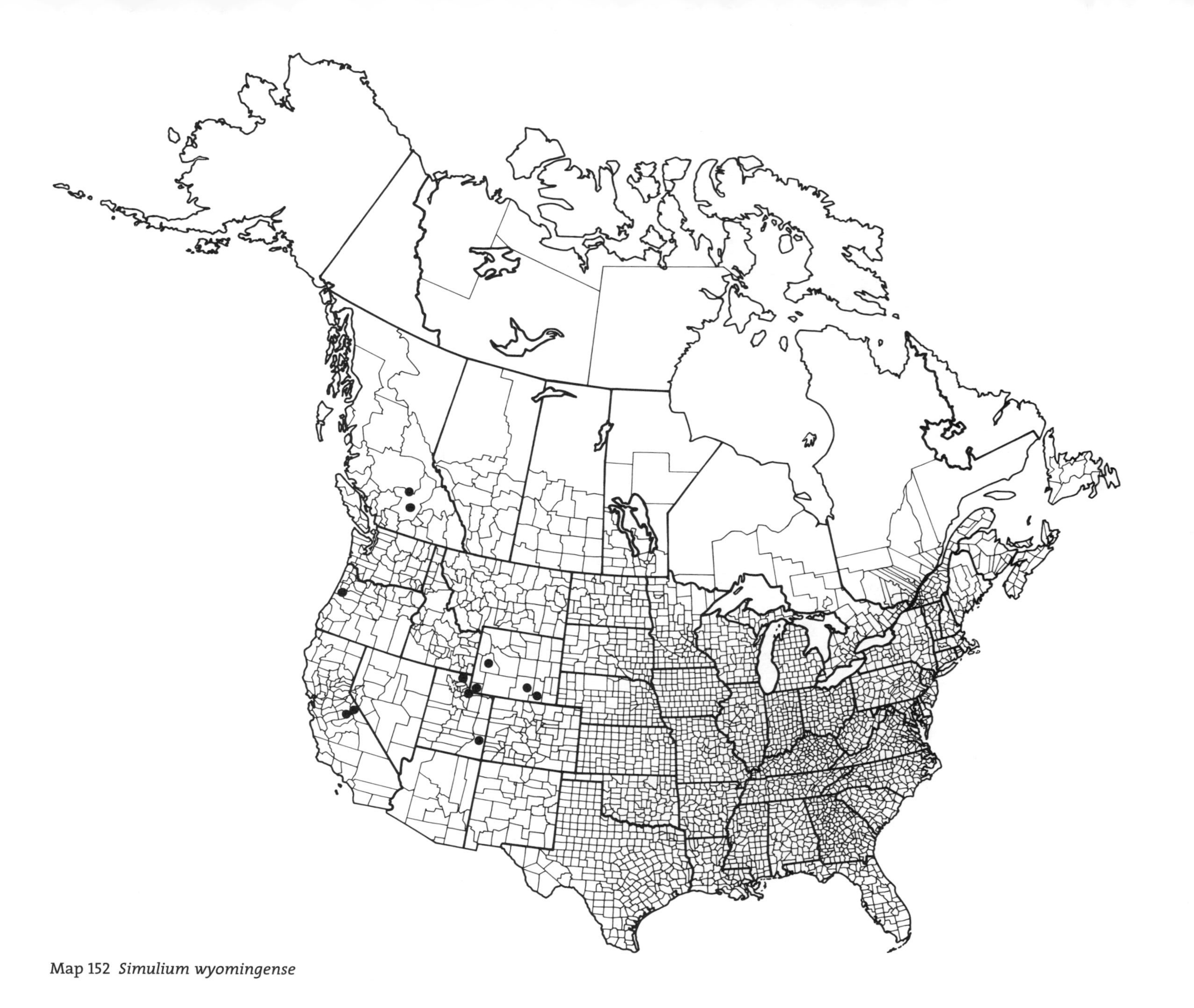

Map 152 *Simulium wyomingense*

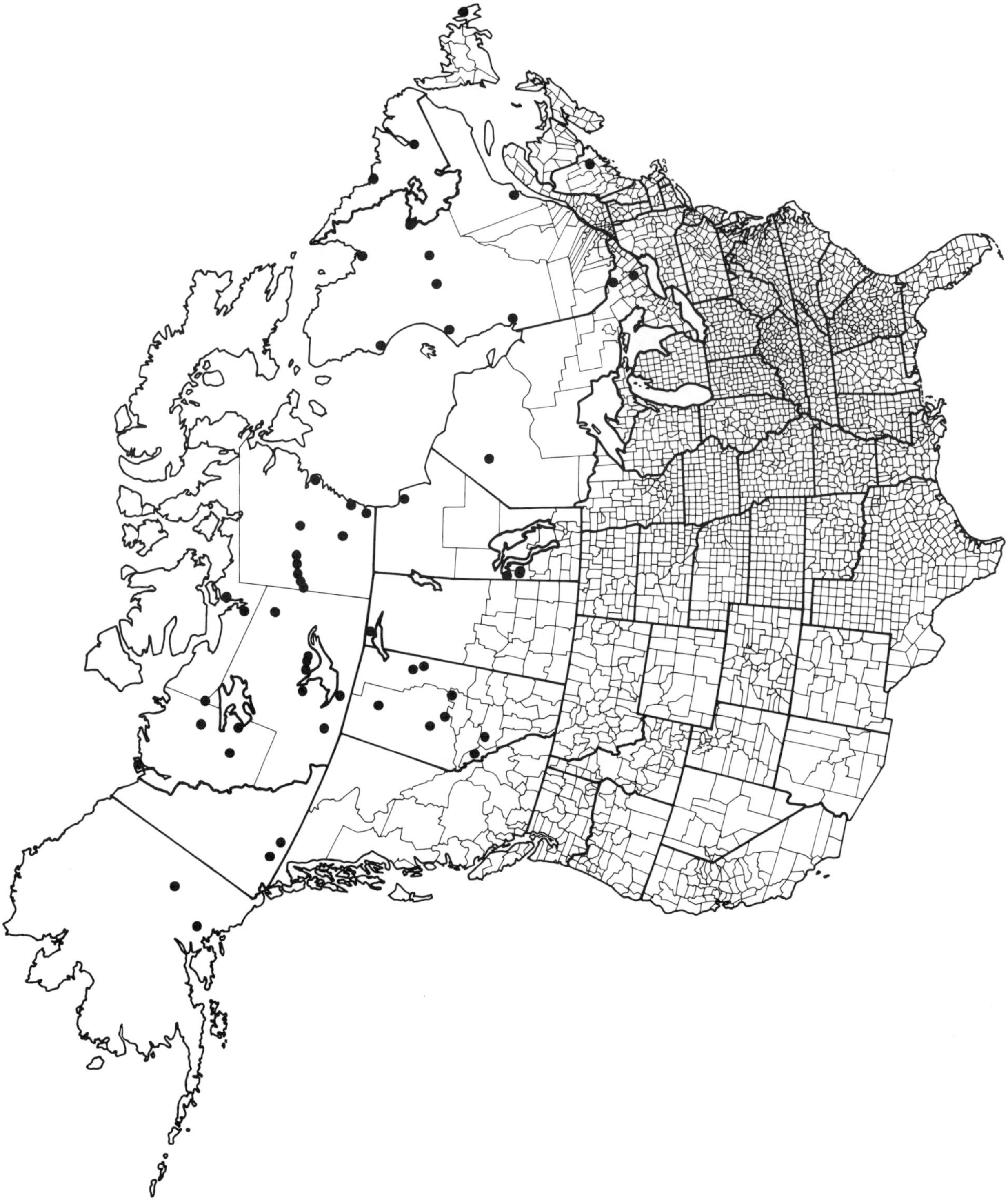

Map 153 *Simulium furculatum*

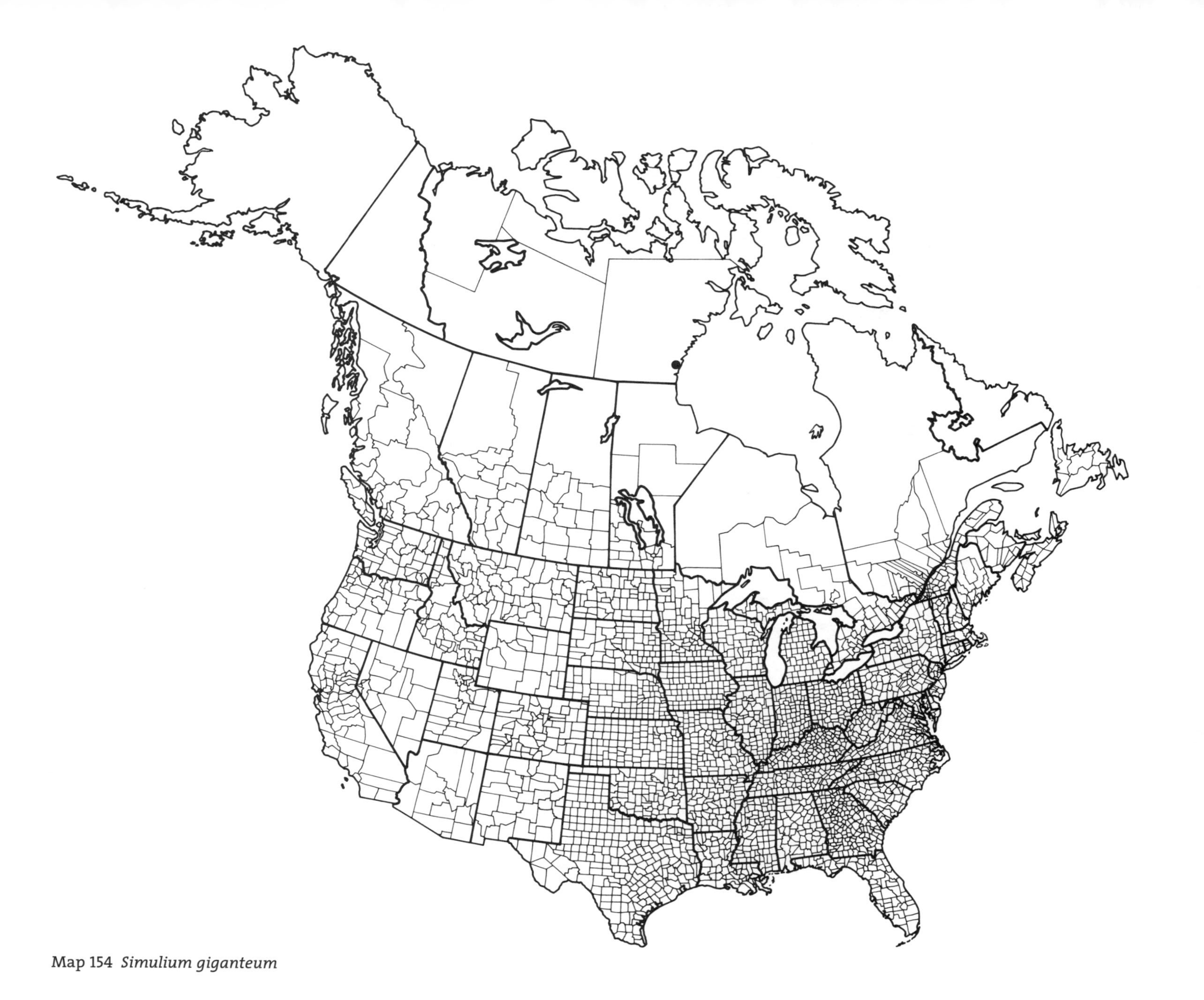

Map 154 *Simulium giganteum*

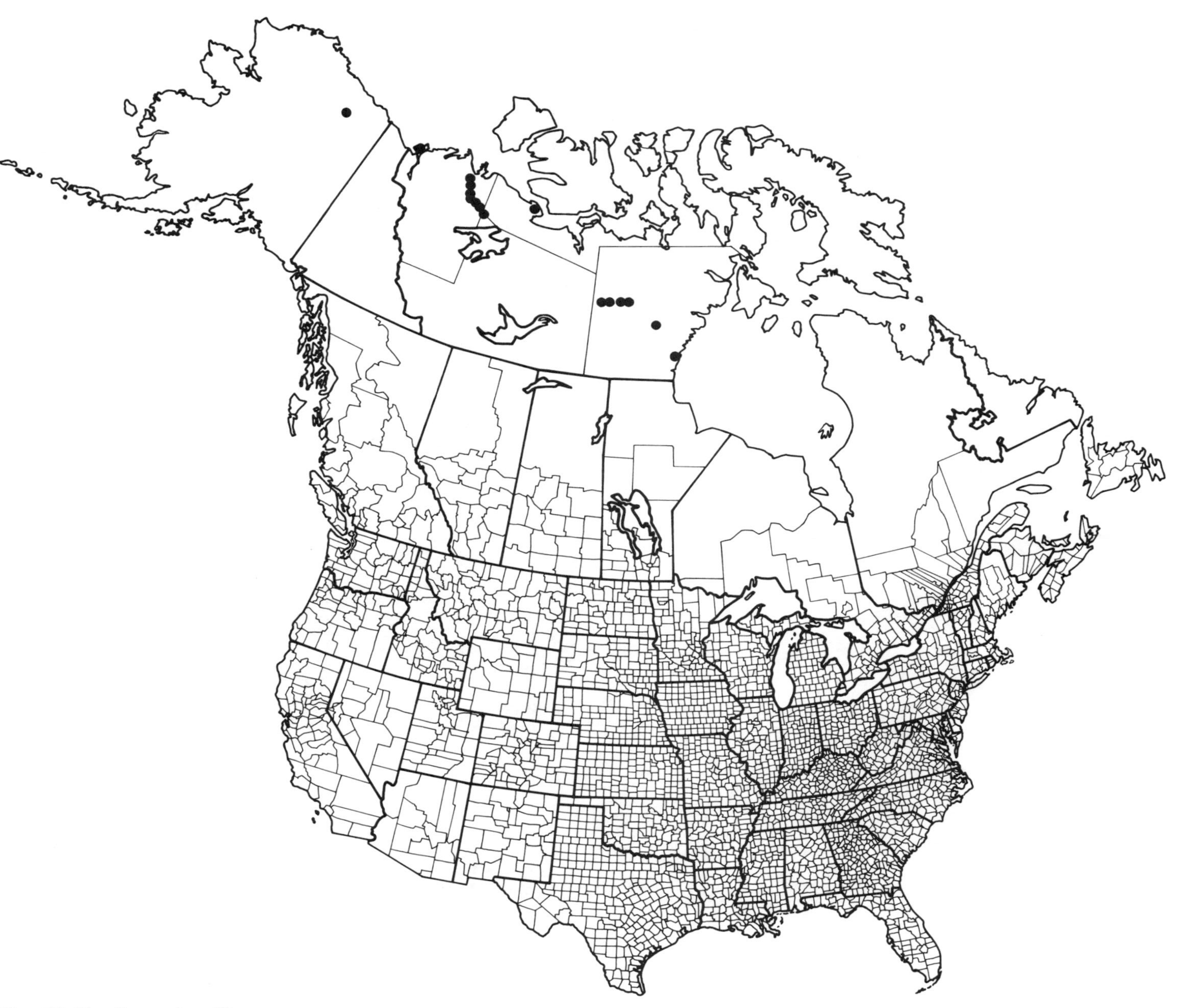

Map 155 *Simulium subpusillum*

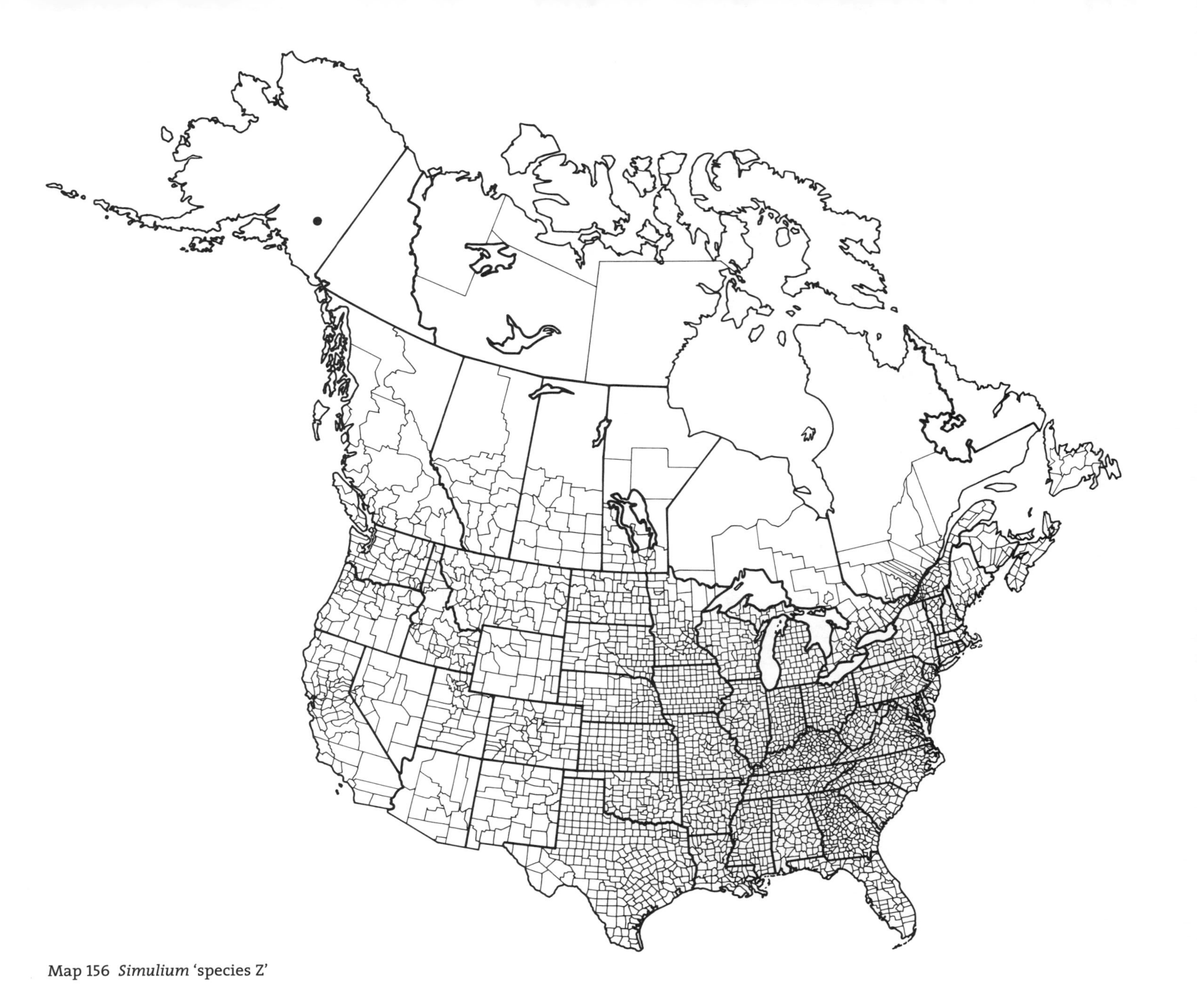

Map 156 *Simulium* 'species Z'

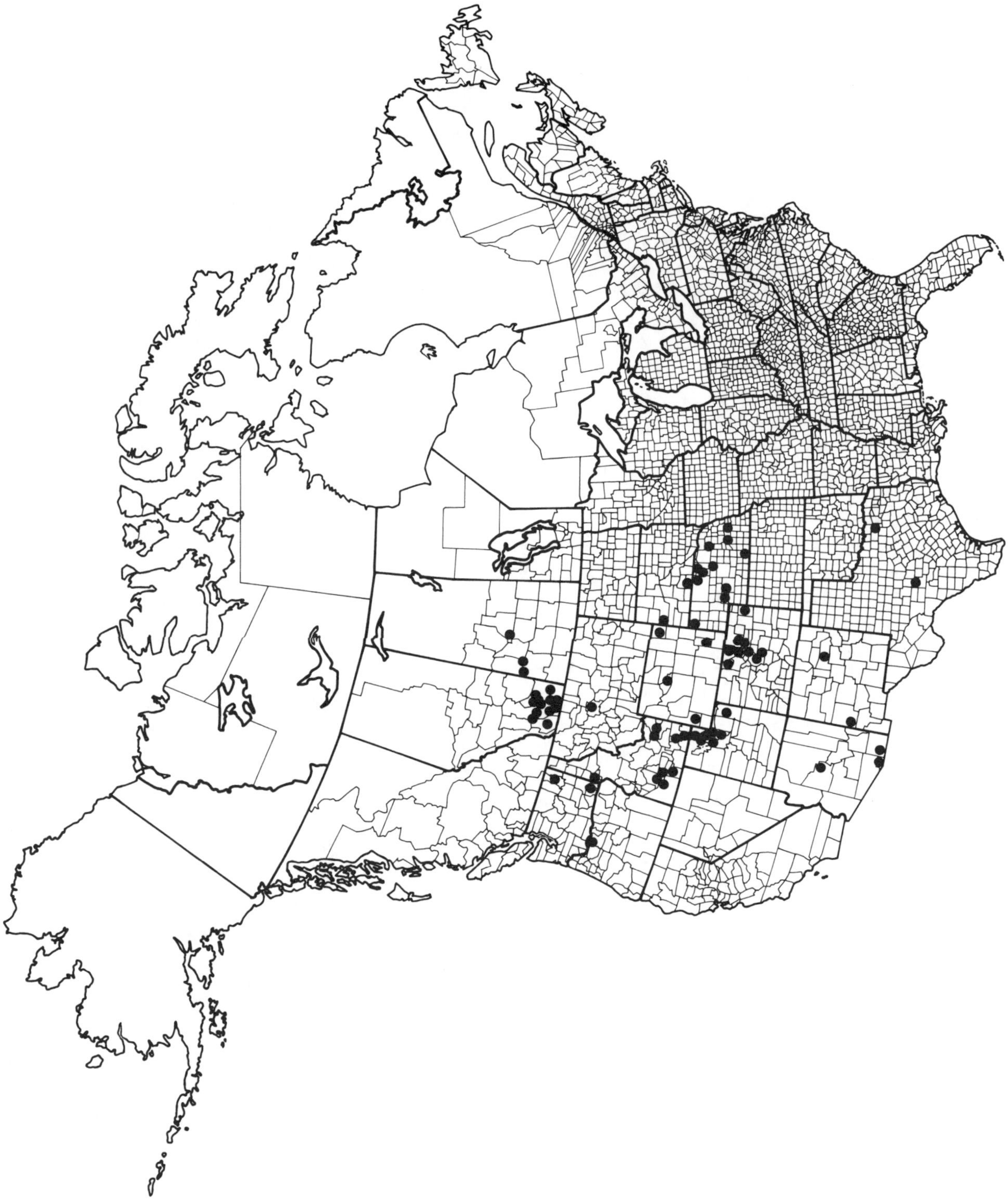

Map 157 *Simulium bivittatum*

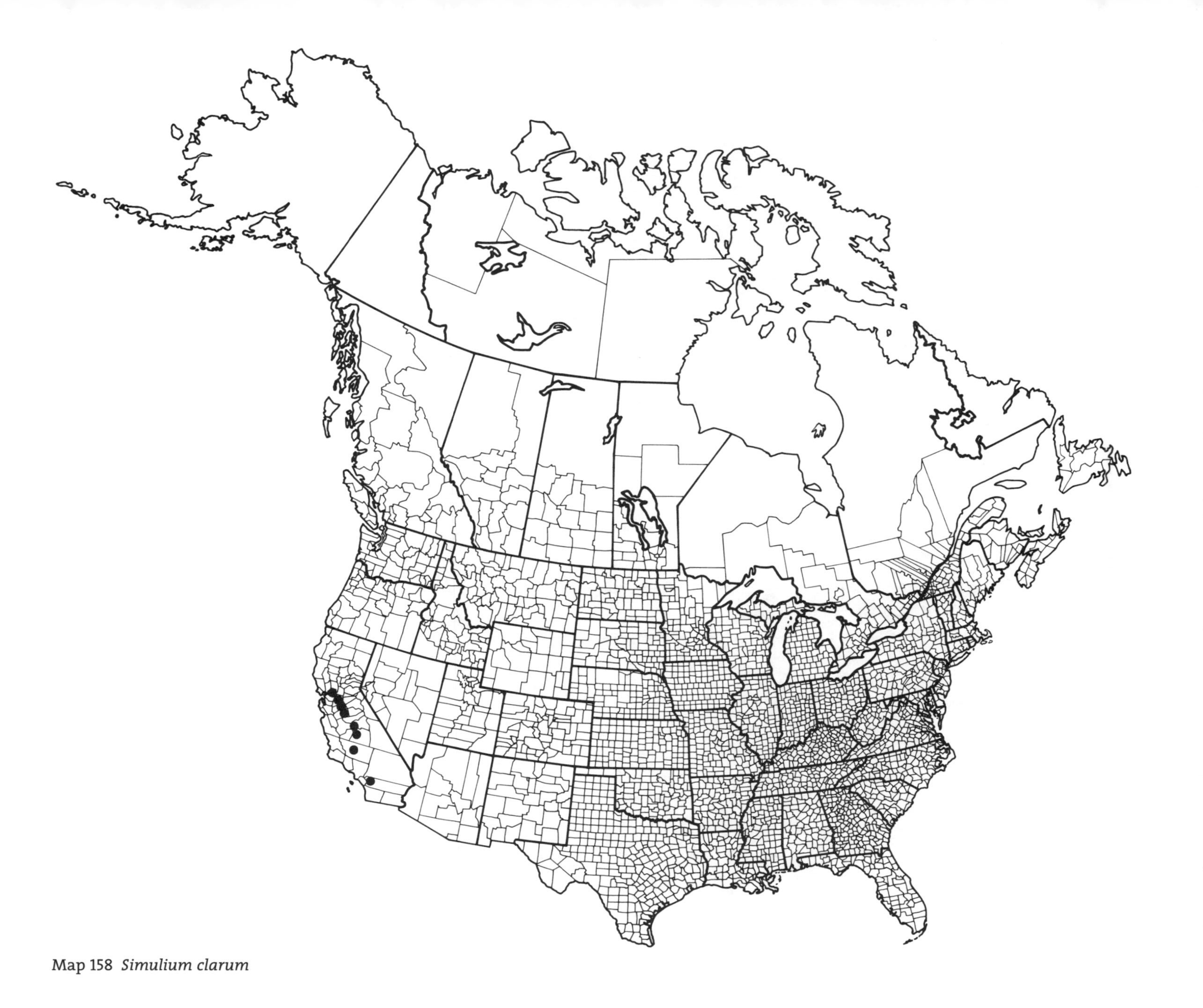

Map 158 *Simulium clarum*

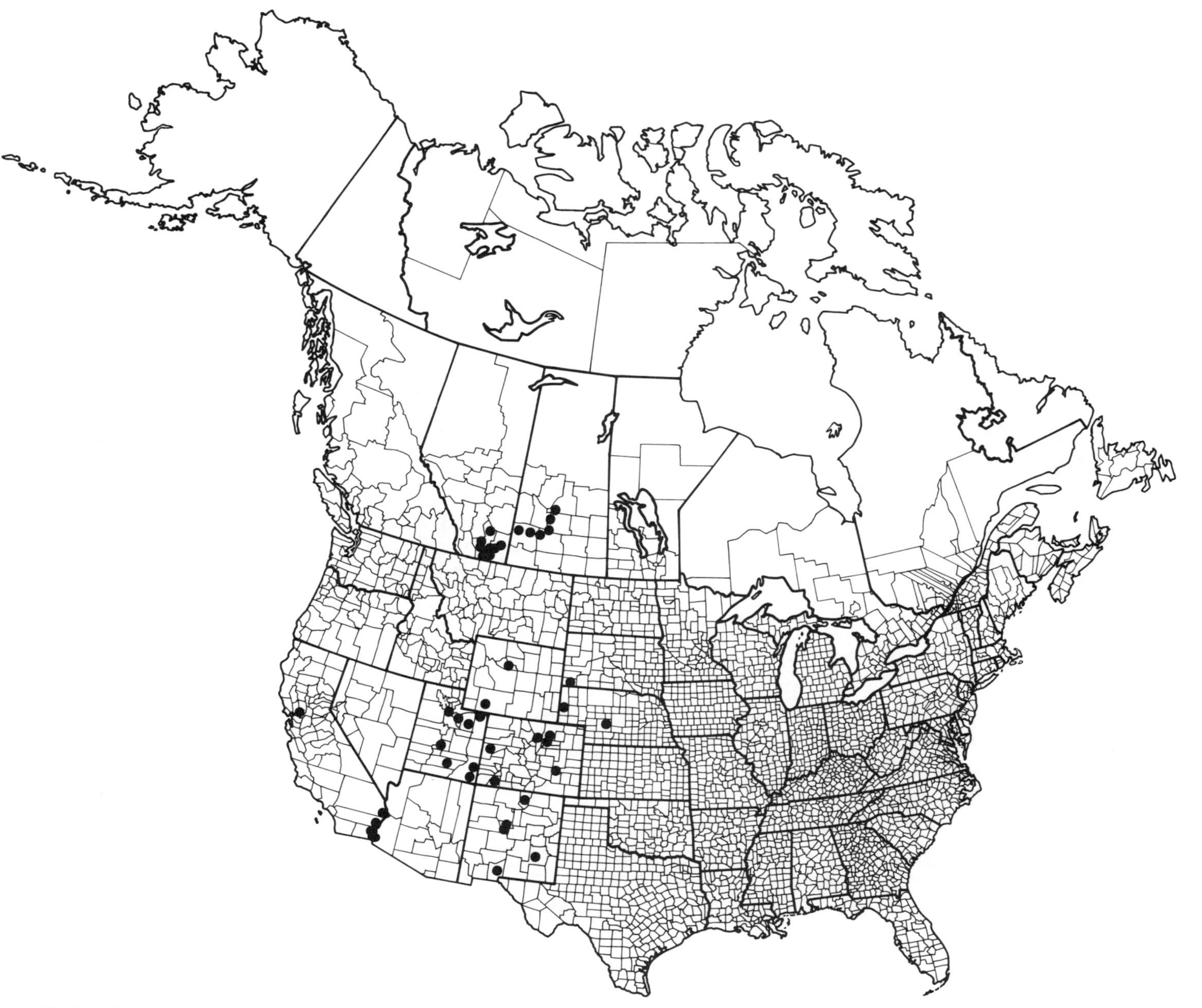

Map 159 *Simulium griseum*

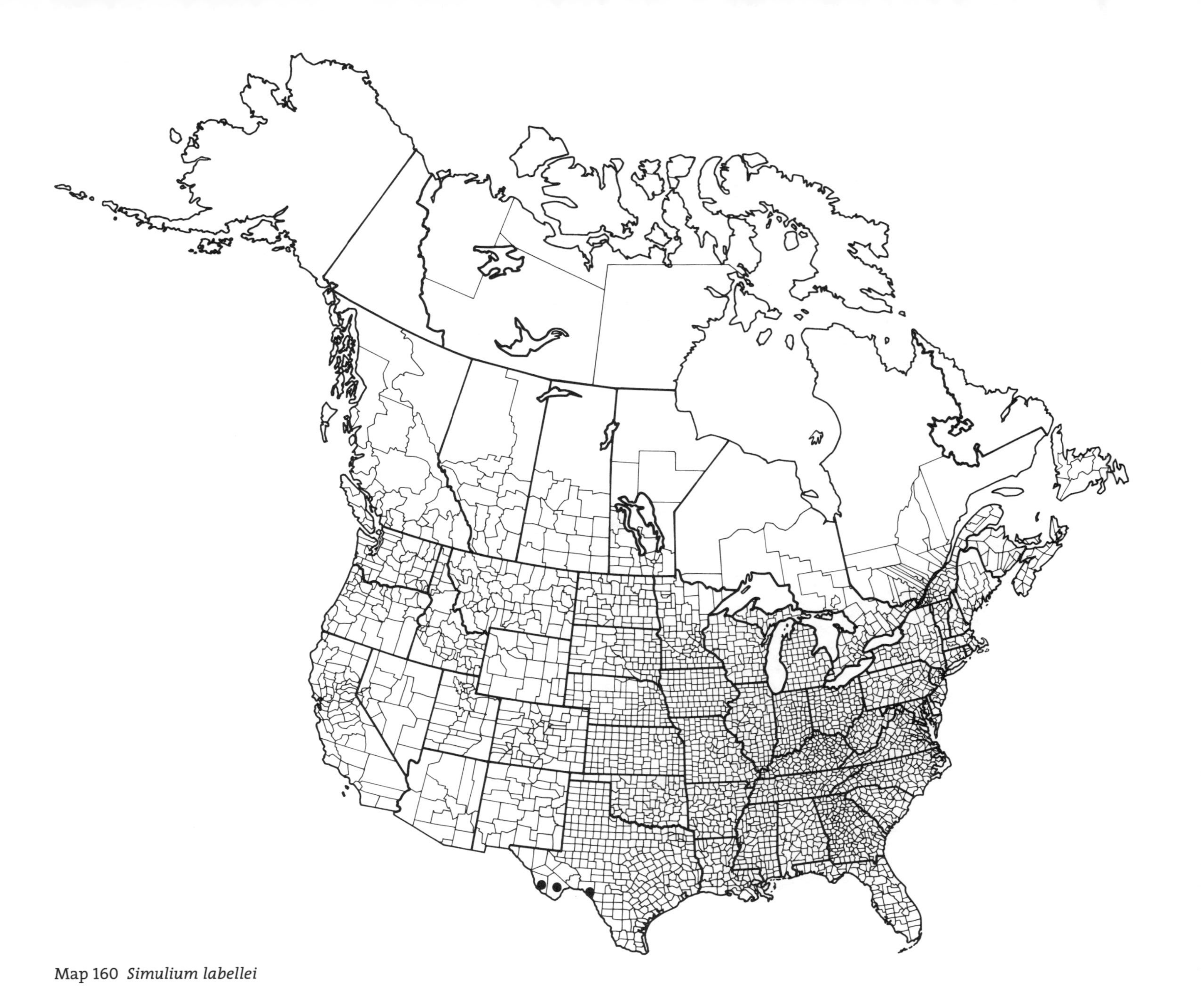

Map 160 *Simulium labellei*

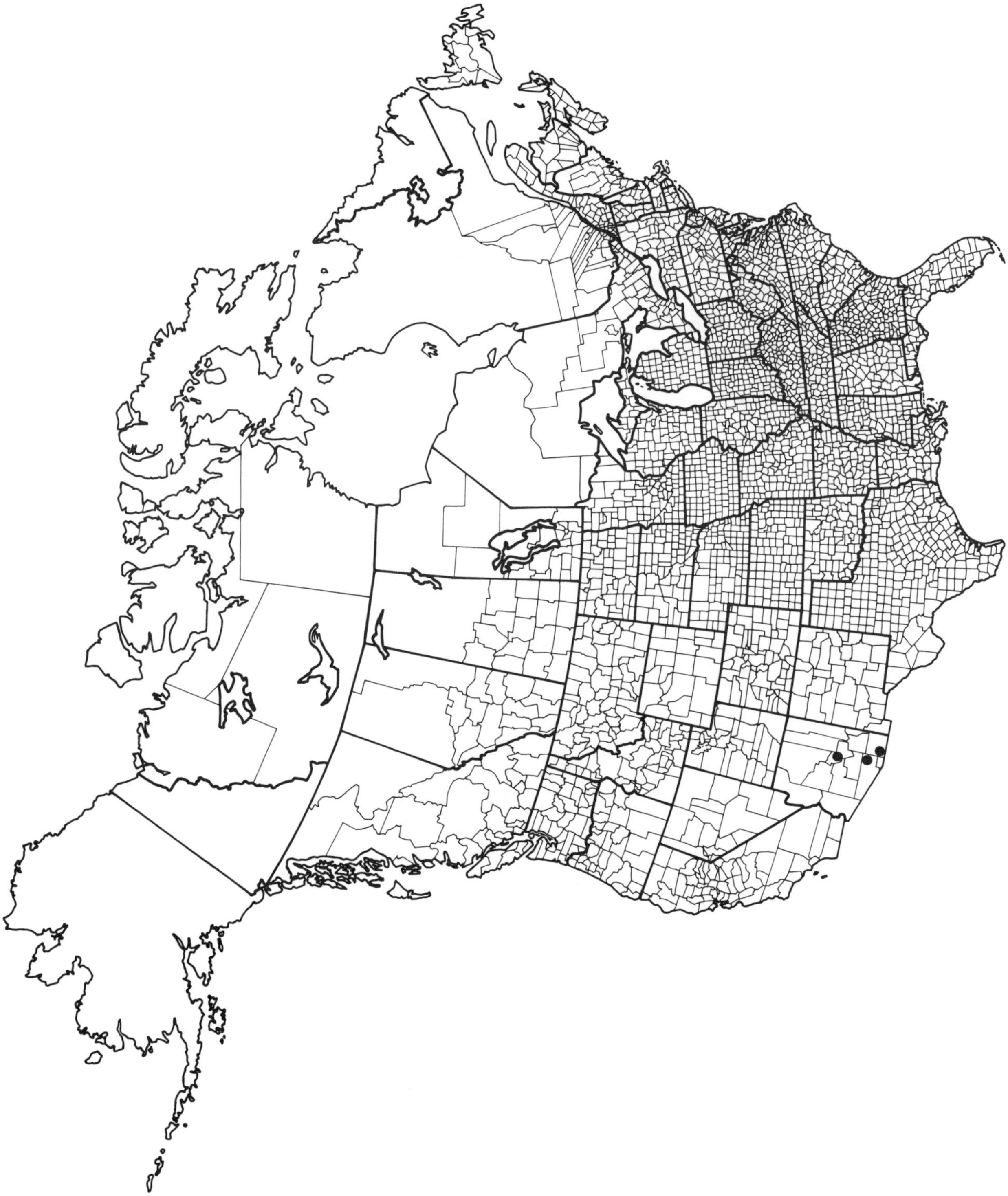

Map 161 *Simulium longithallum*

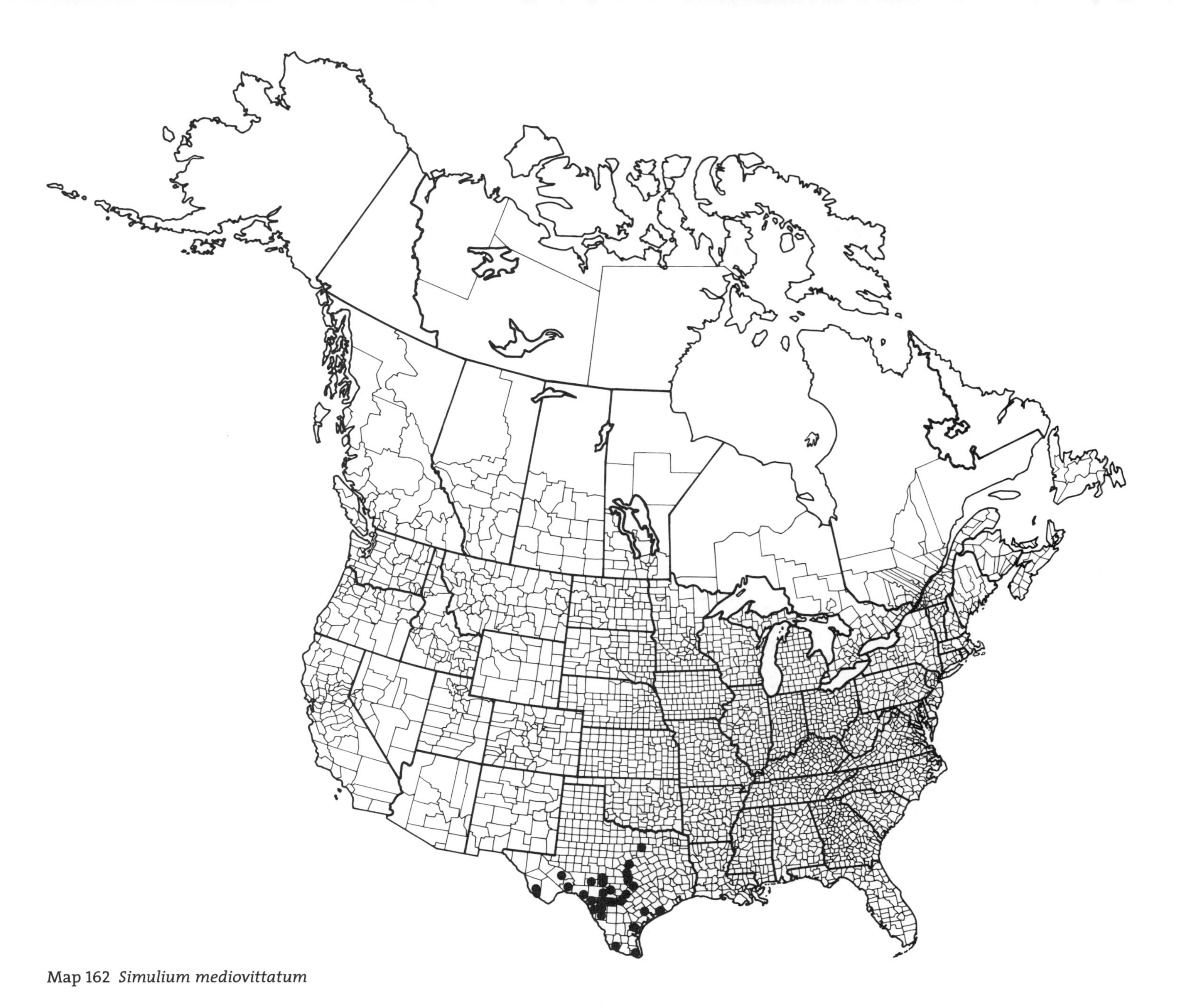

Map 162 *Simulium mediovittatum*

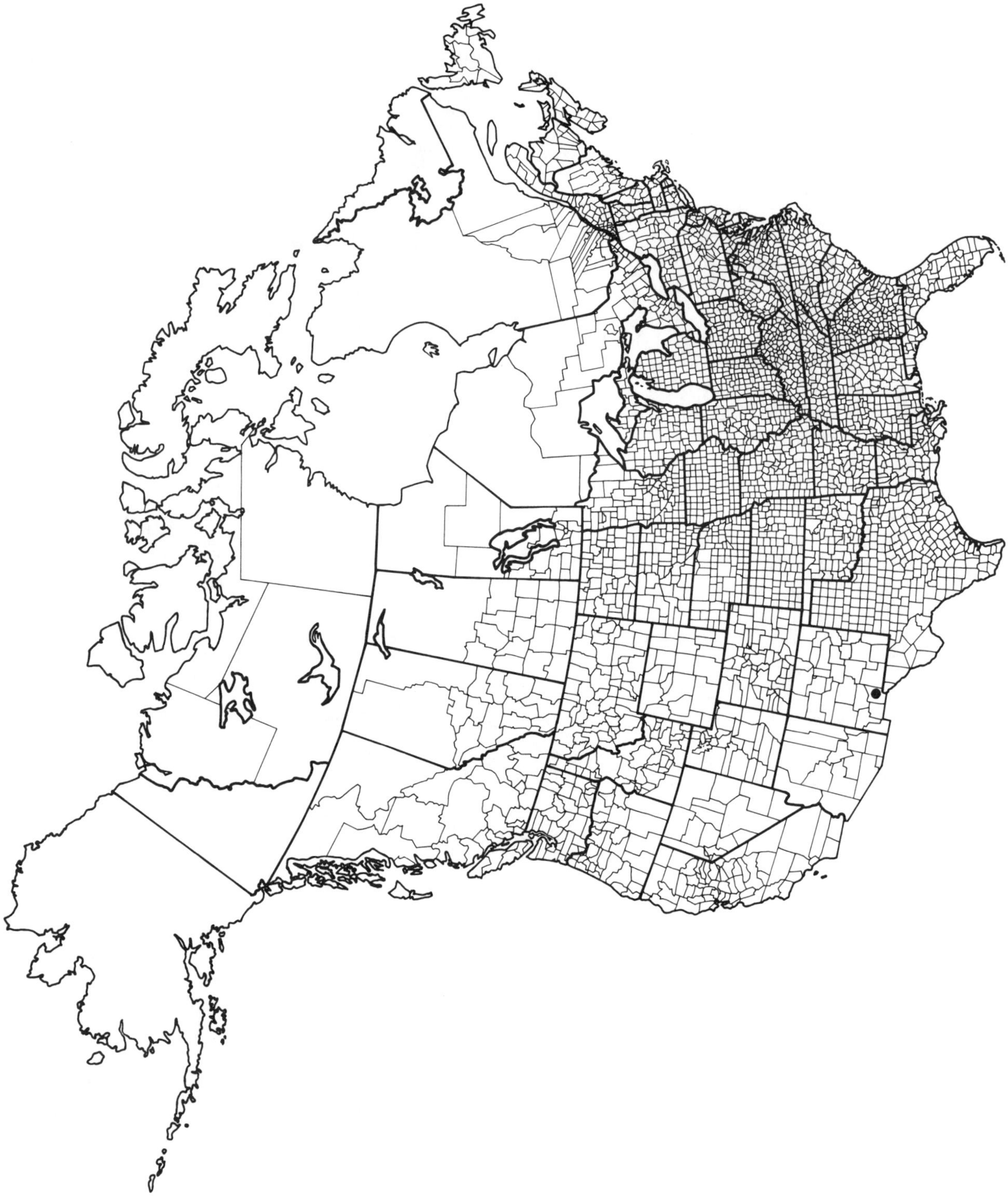

Map 163 *Simulium meyerae*

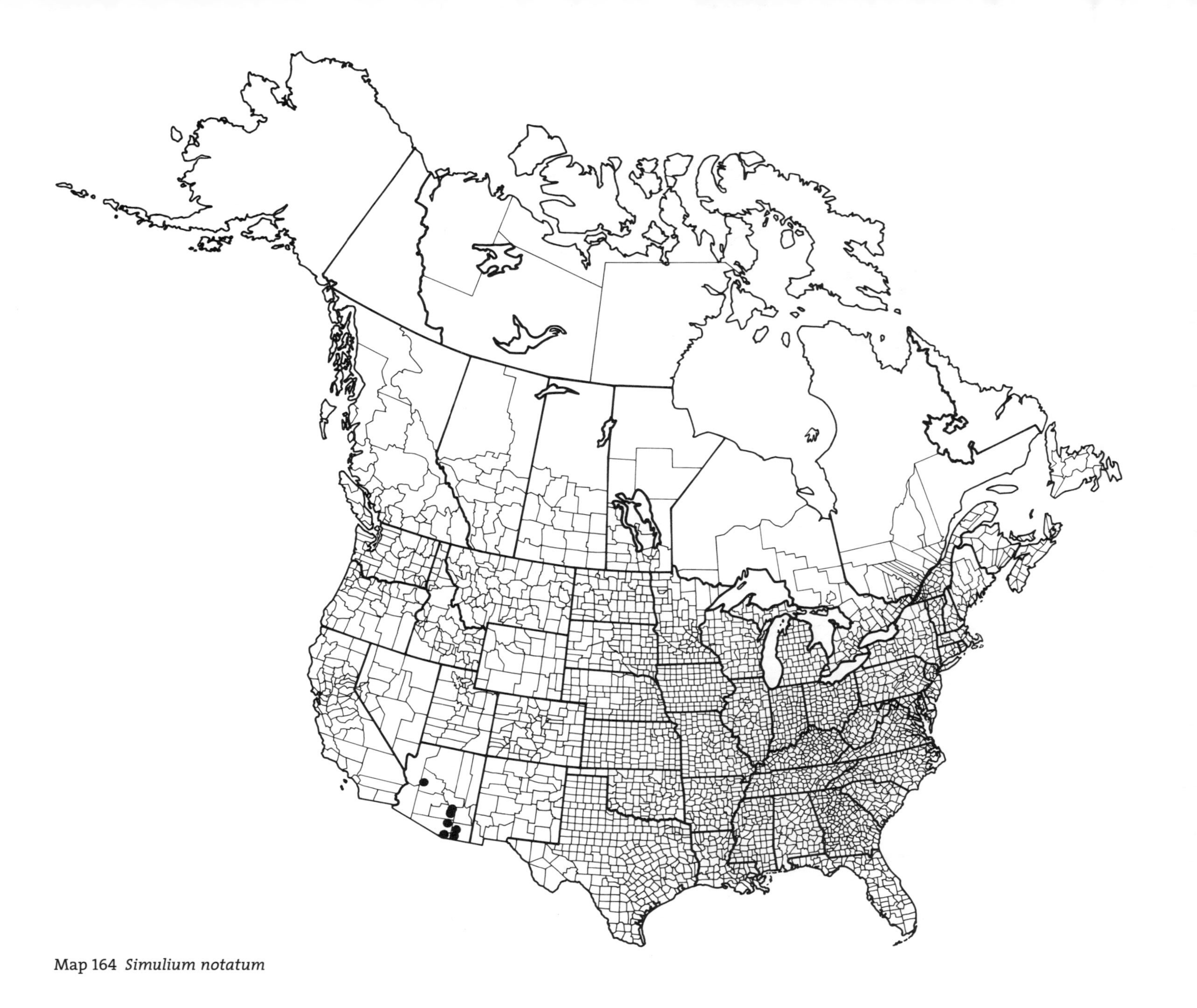

Map 164 *Simulium notatum*

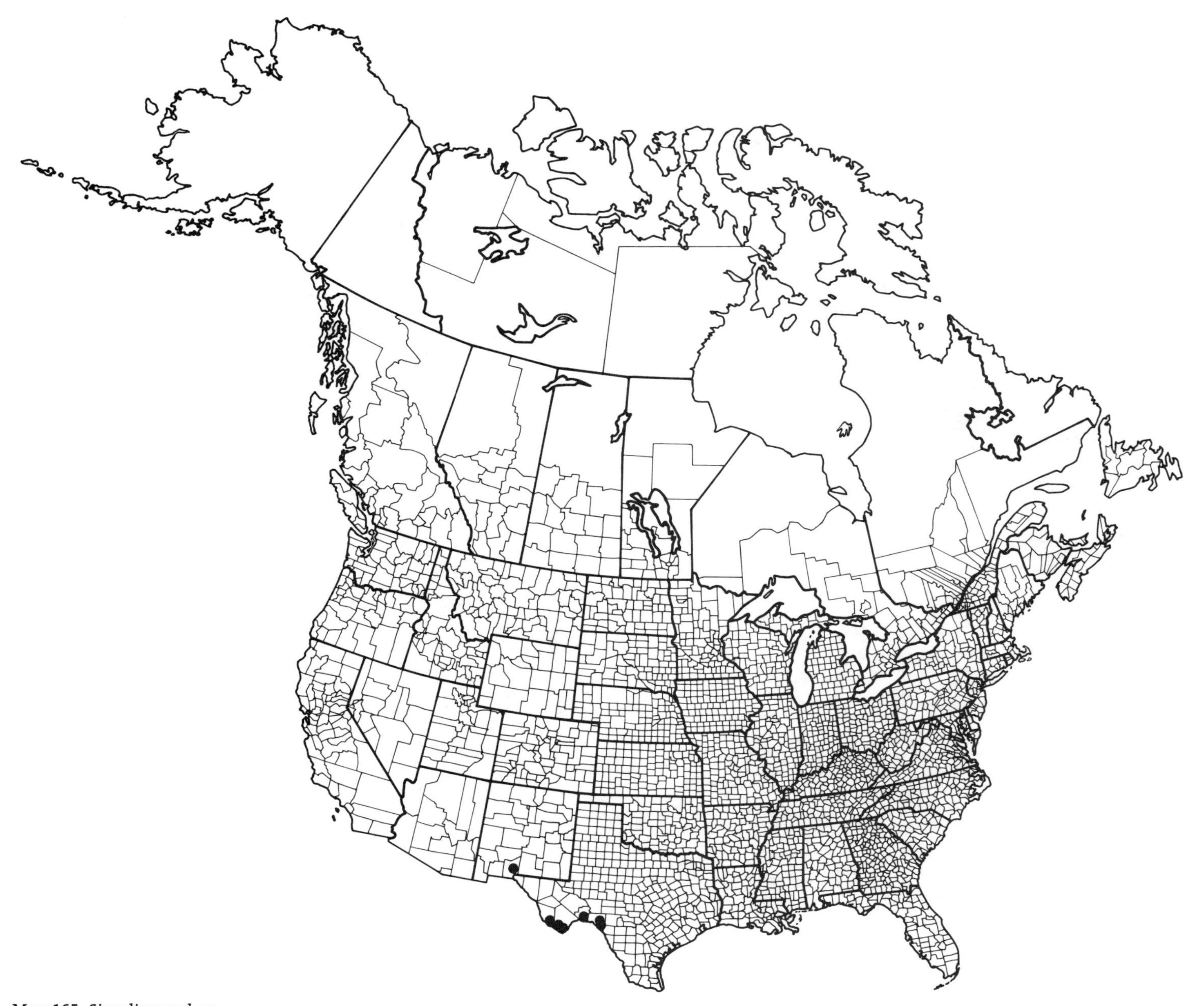

Map 165 *Simulium robynae*

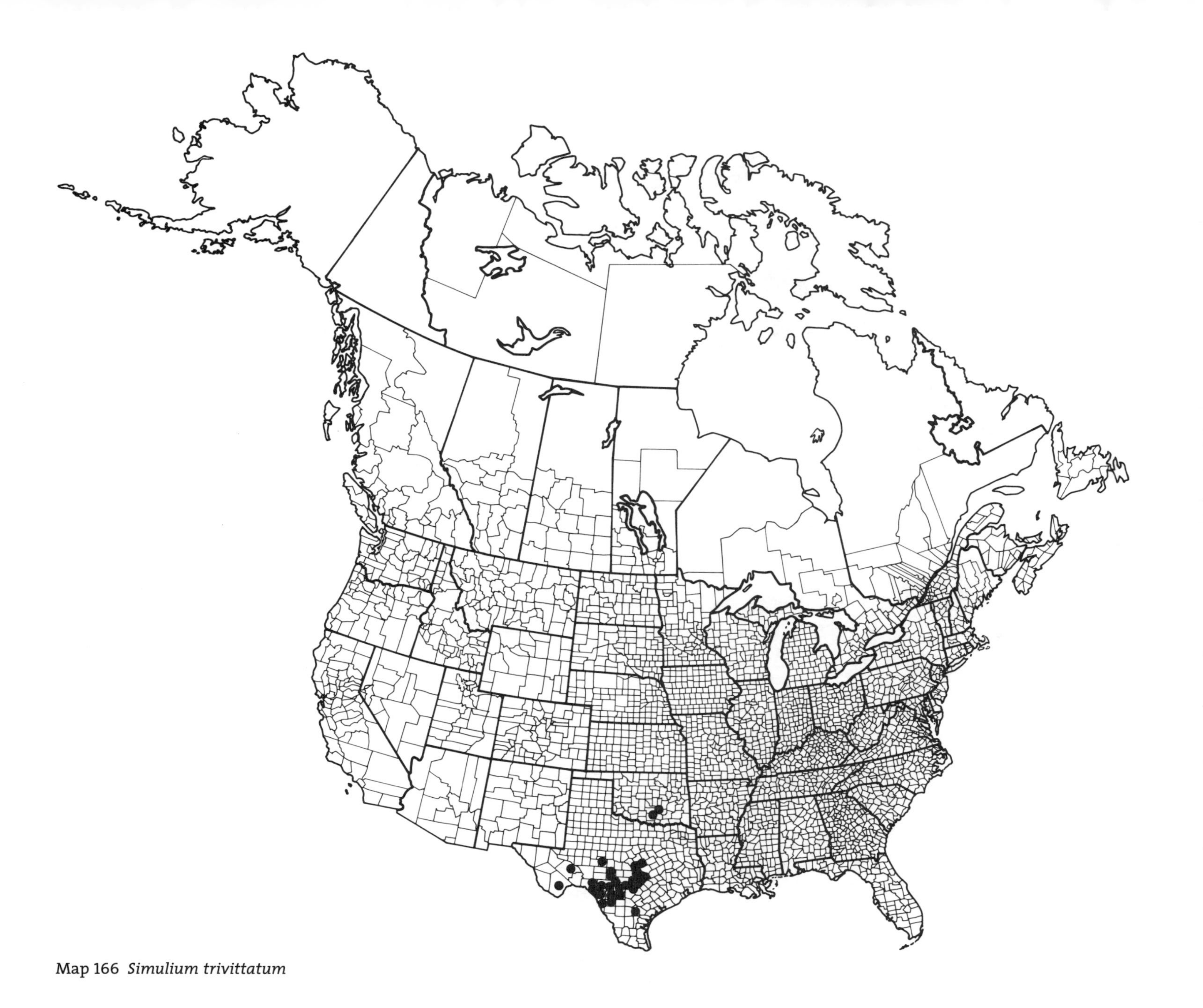

Map 166 *Simulium trivittatum*

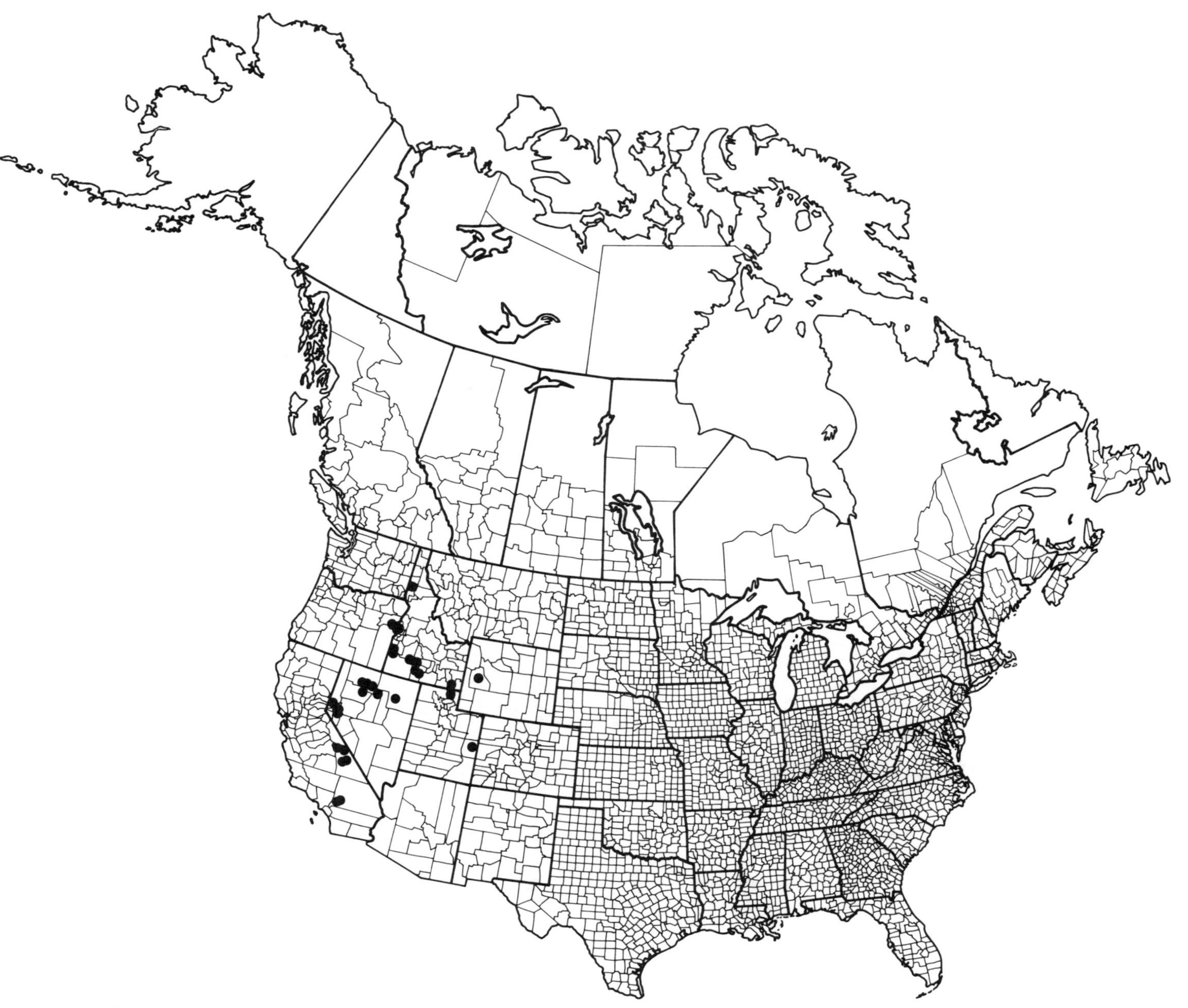

Map 167 *Simulium venator*

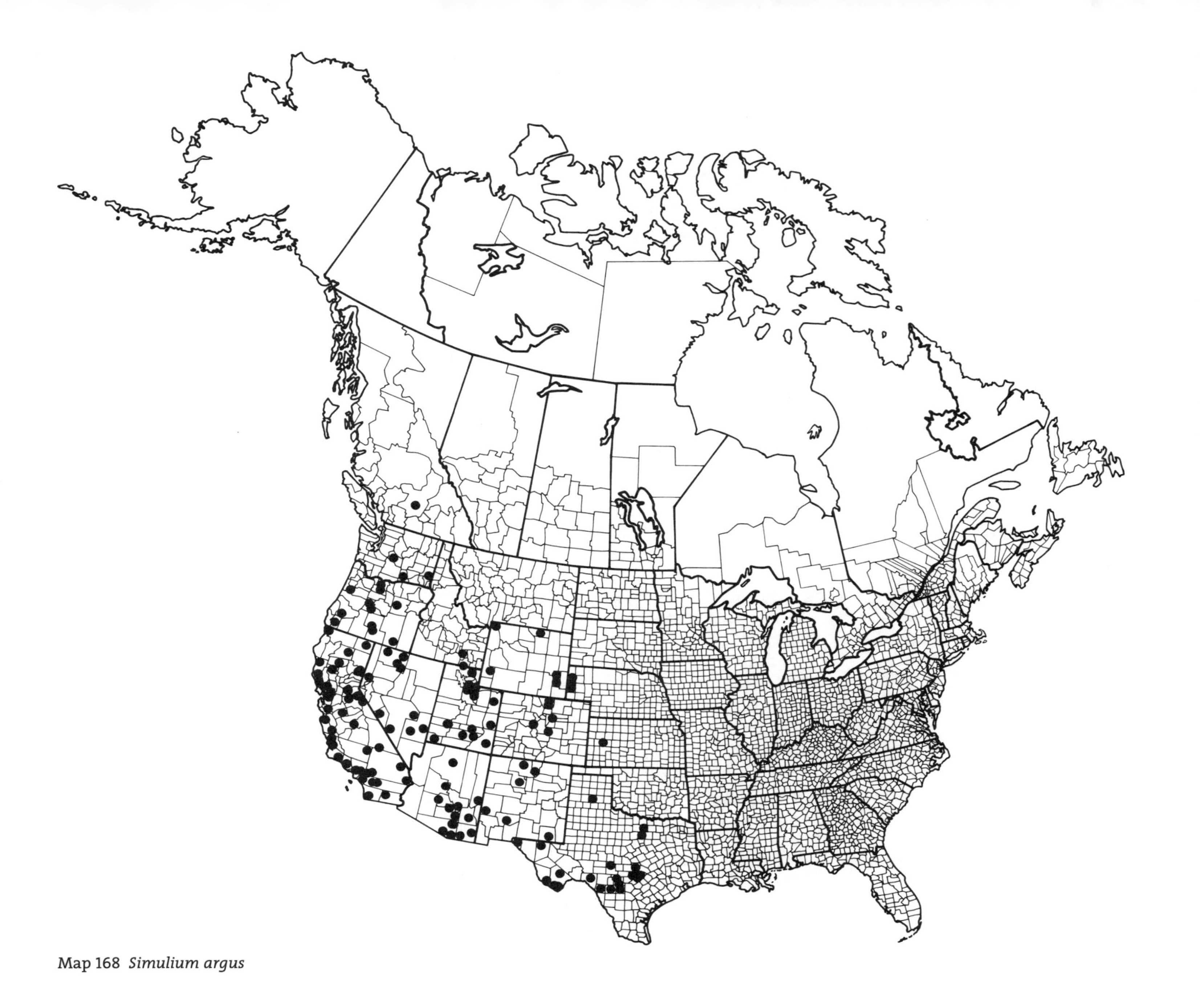

Map 168 *Simulium argus*

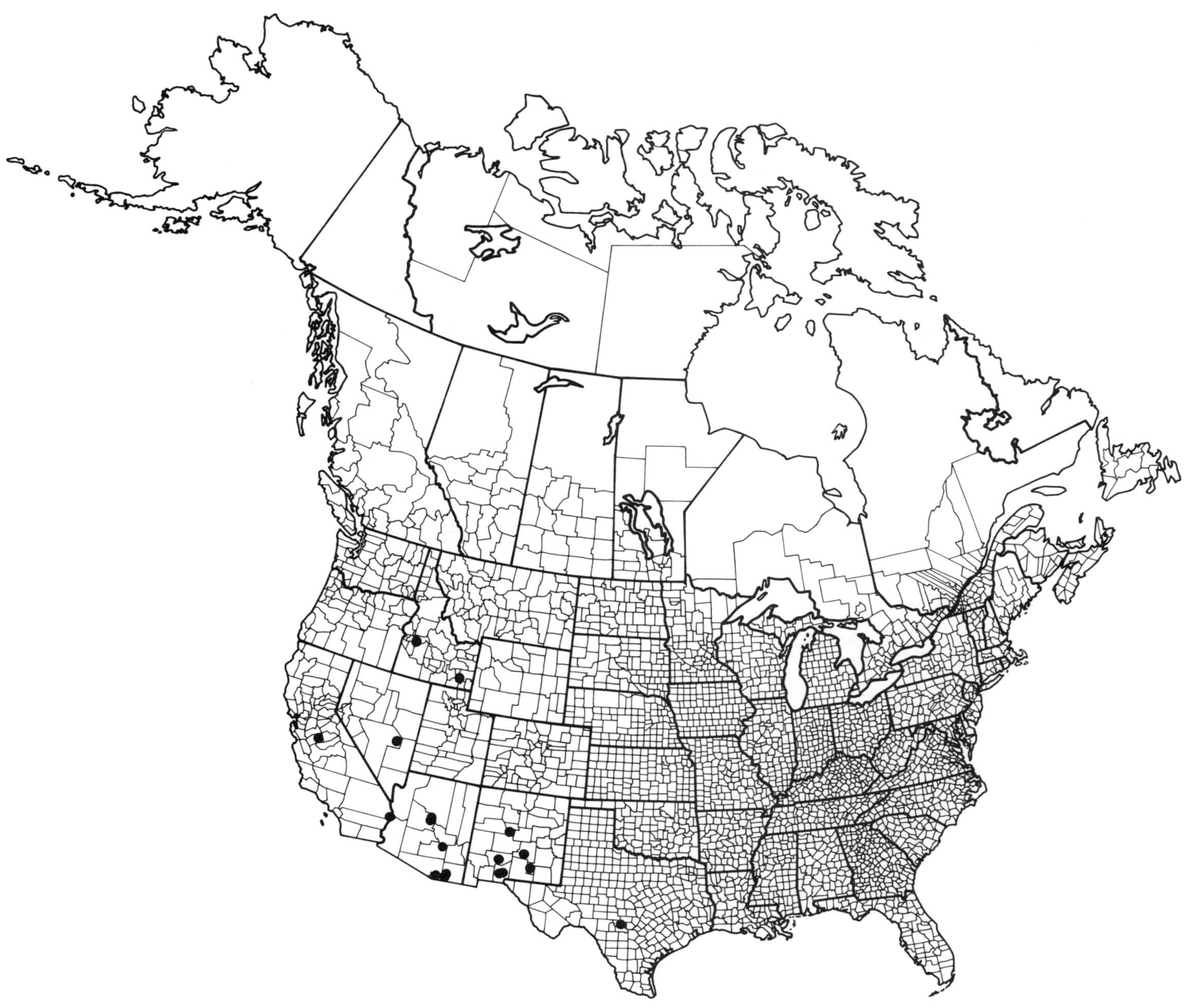

Map 169 *Simulium encisoi*

Map 170 *Simulium infernale* n. sp.

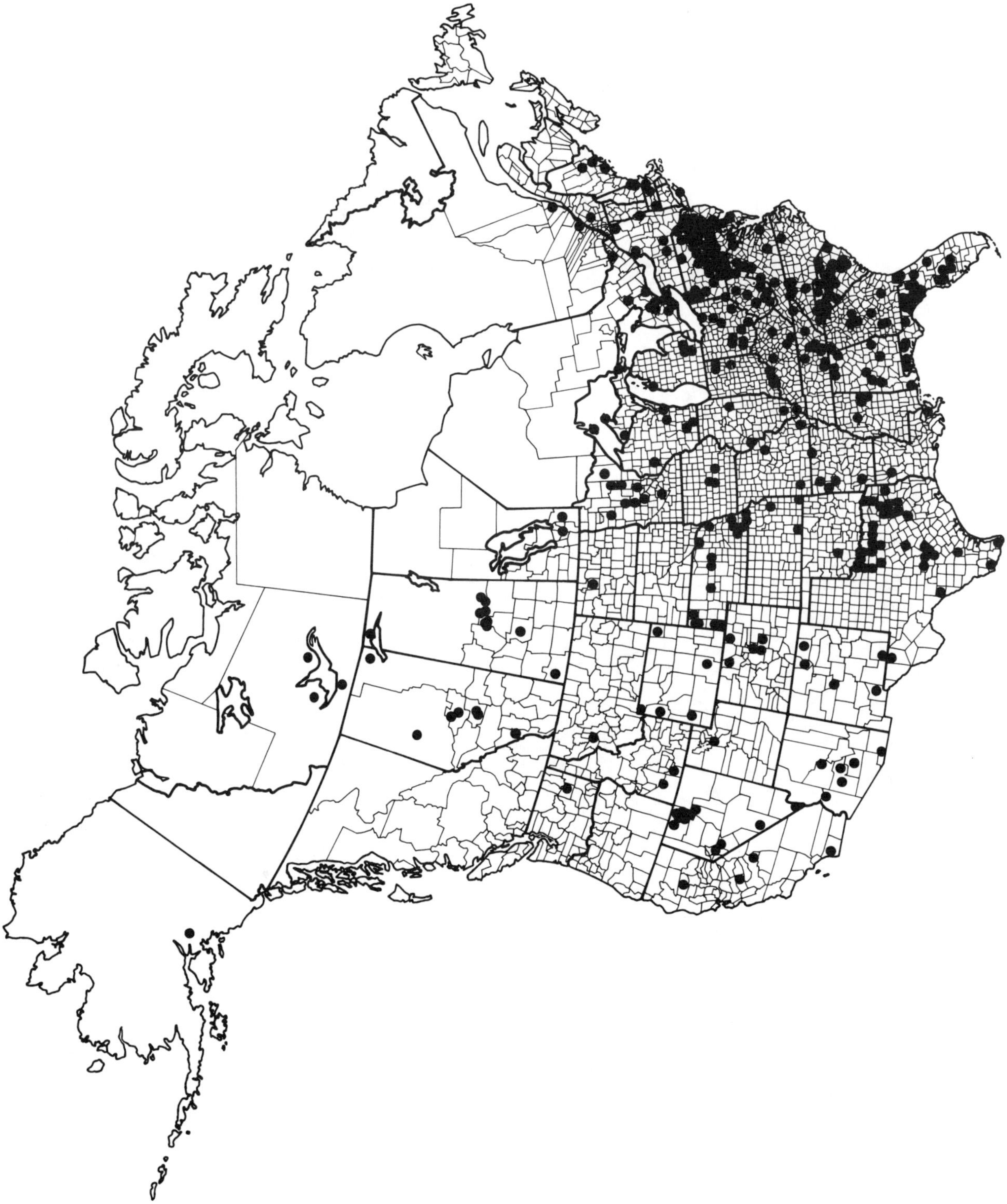

Map 171 *Simulium tribulatum*

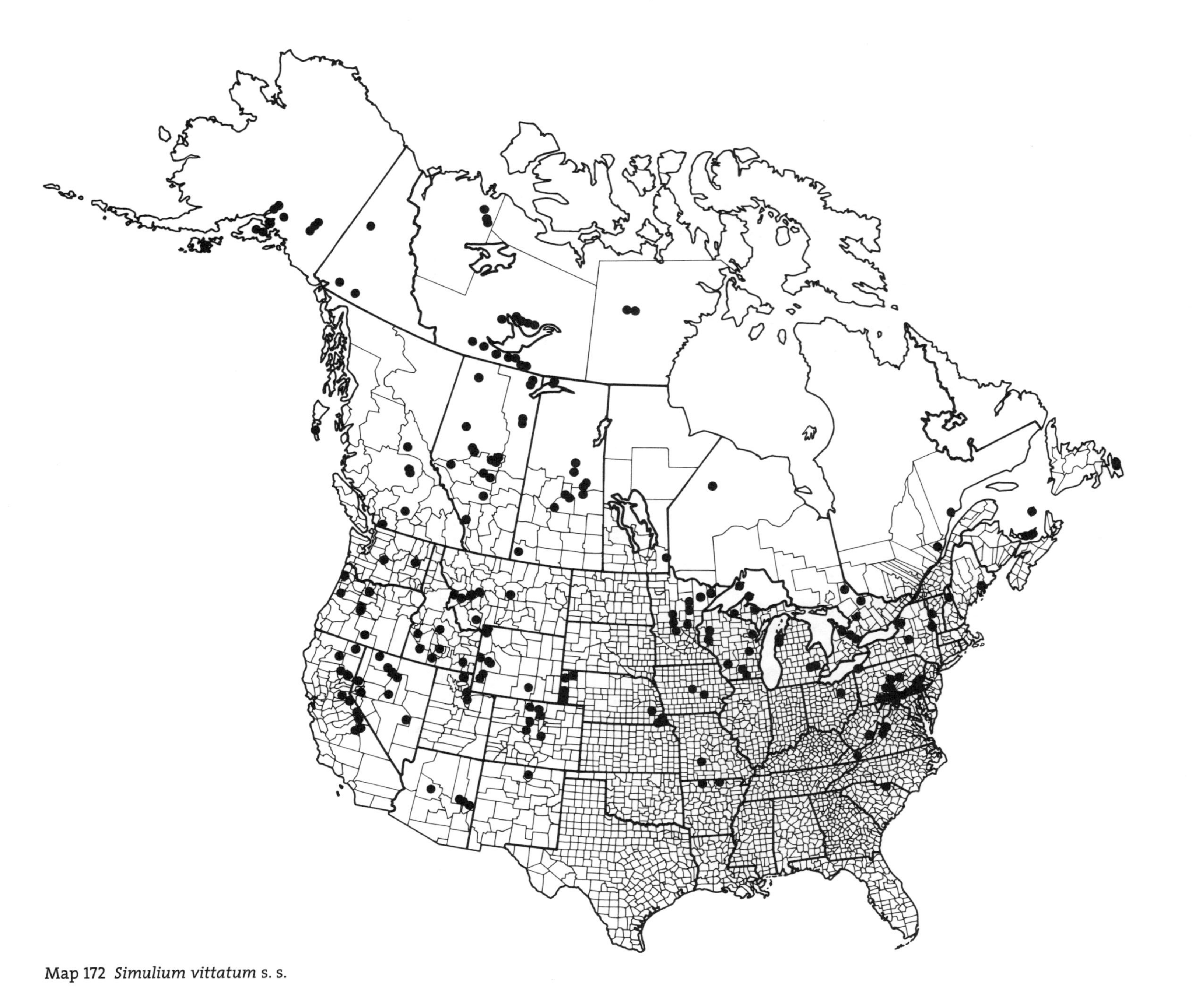

Map 172 *Simulium vittatum* s. s.

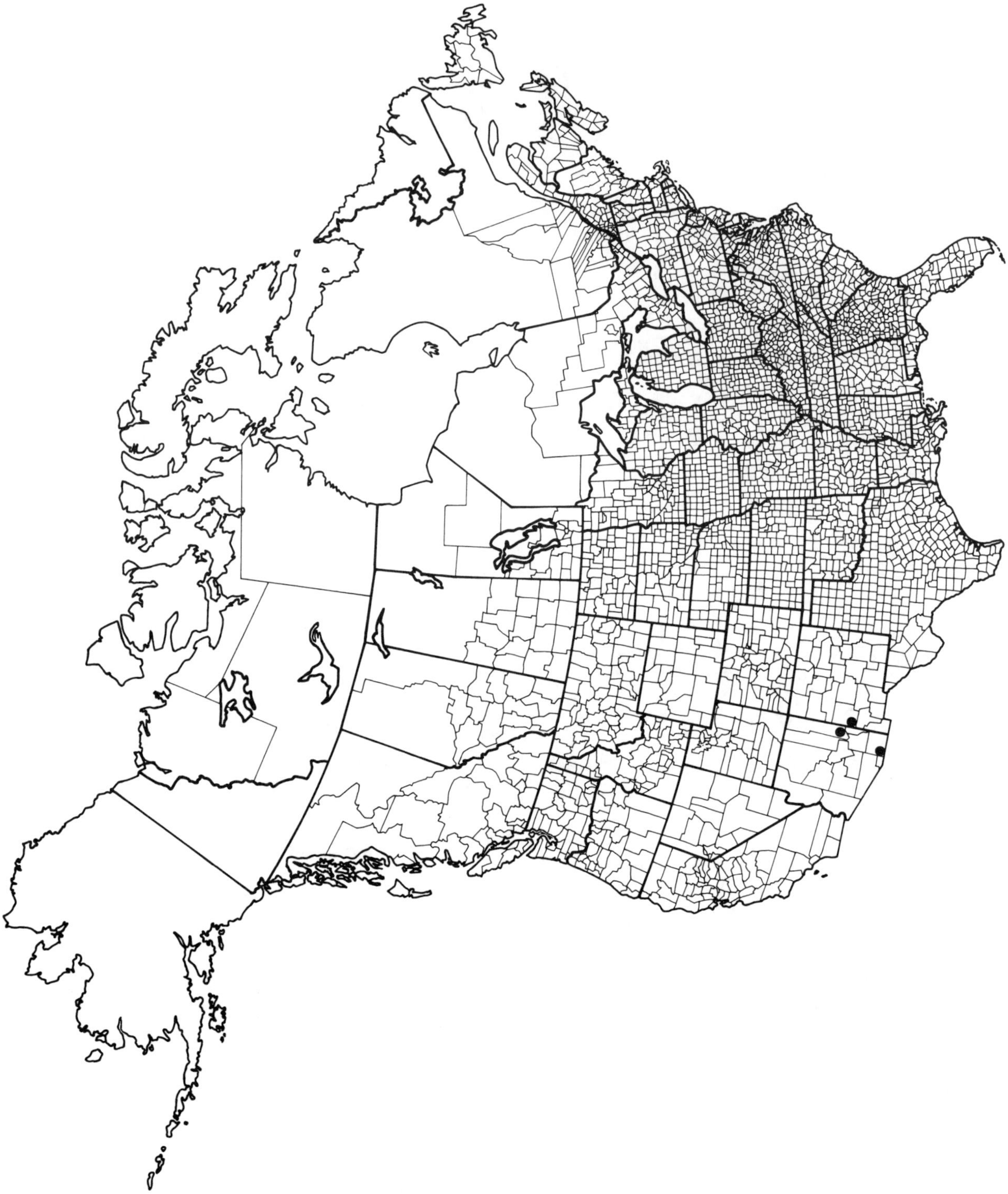

Map 173 *Simulium anduzei*

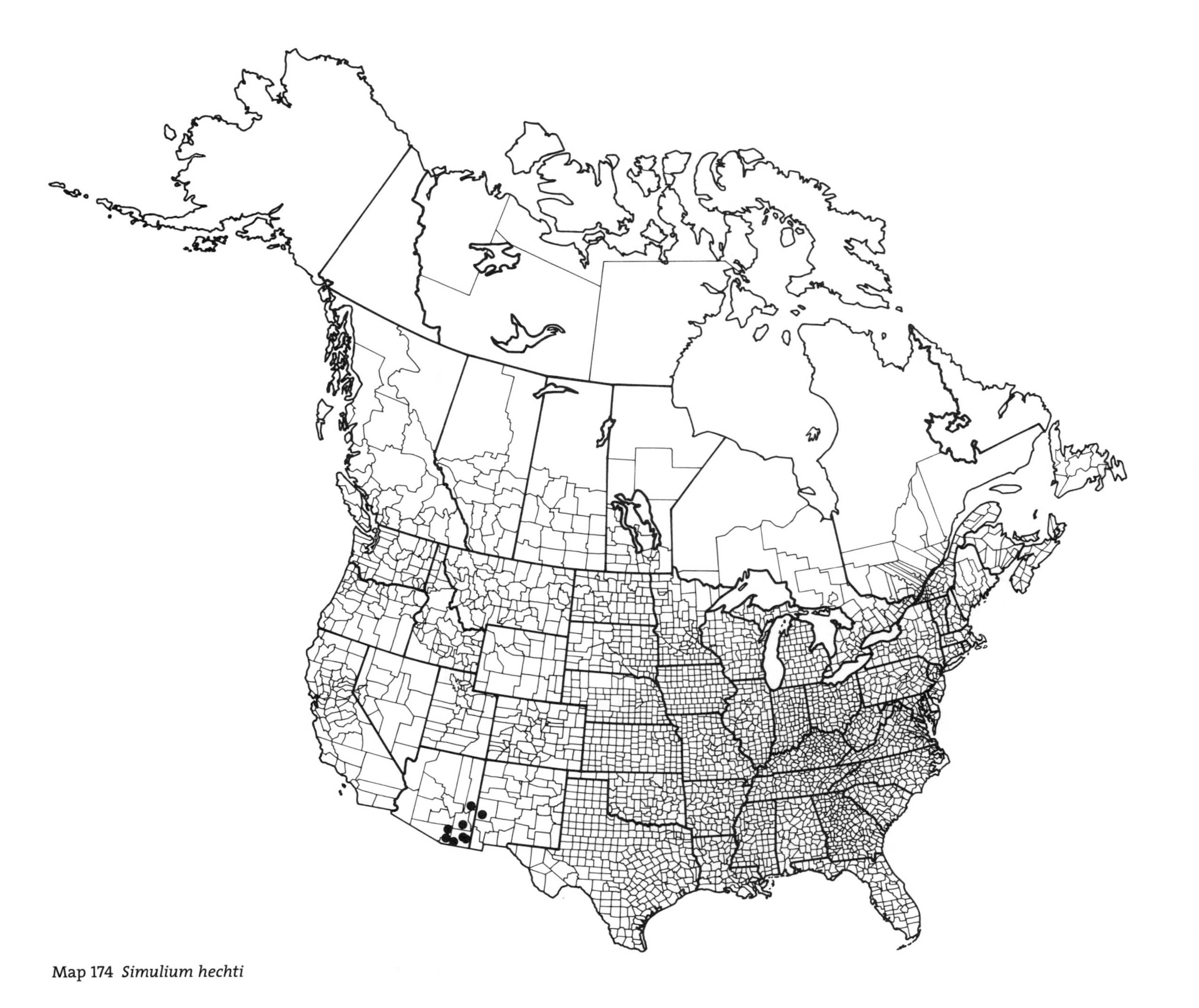

Map 174 *Simulium hechti*

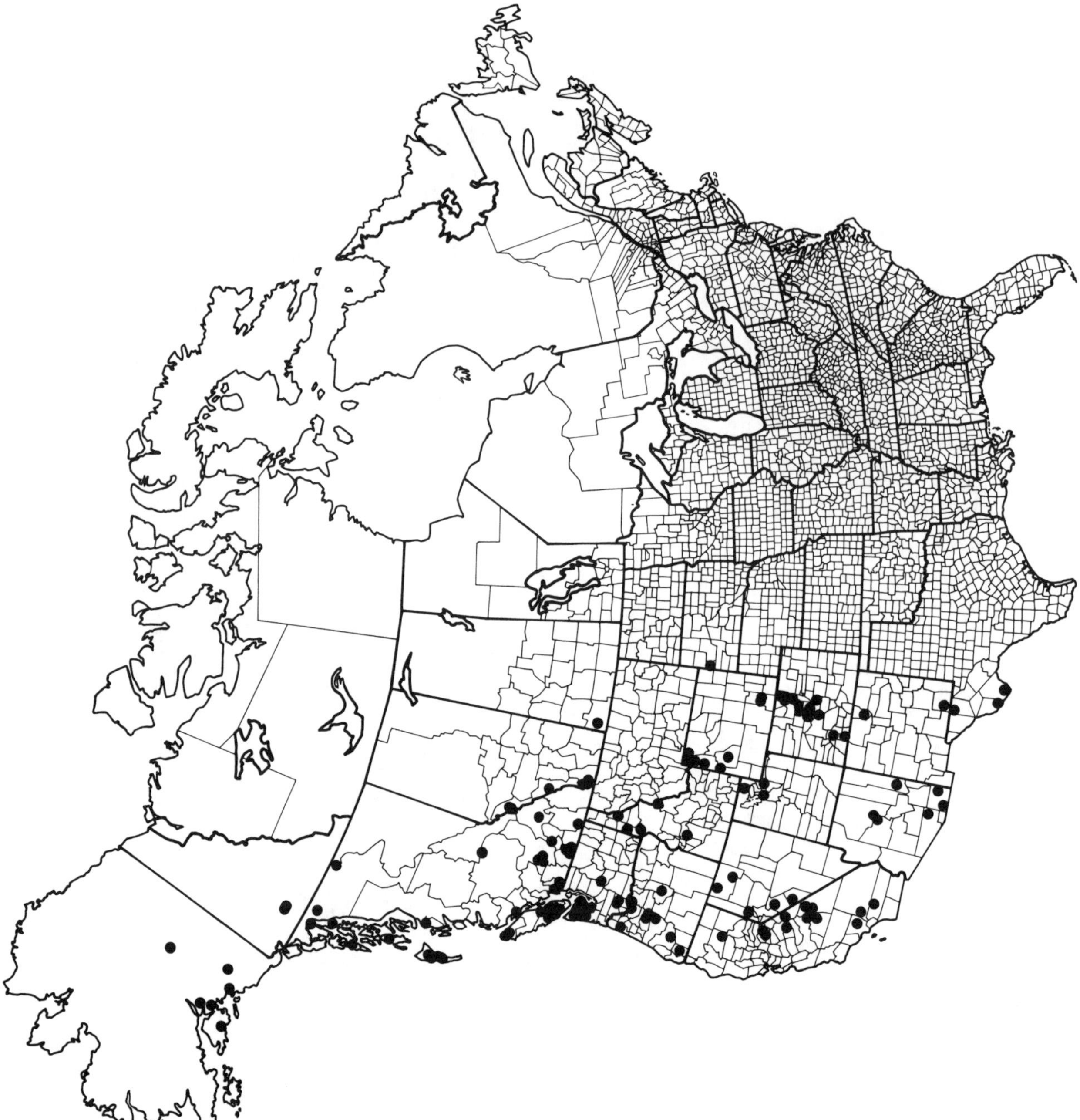

Map 175 *Simulium hunteri*

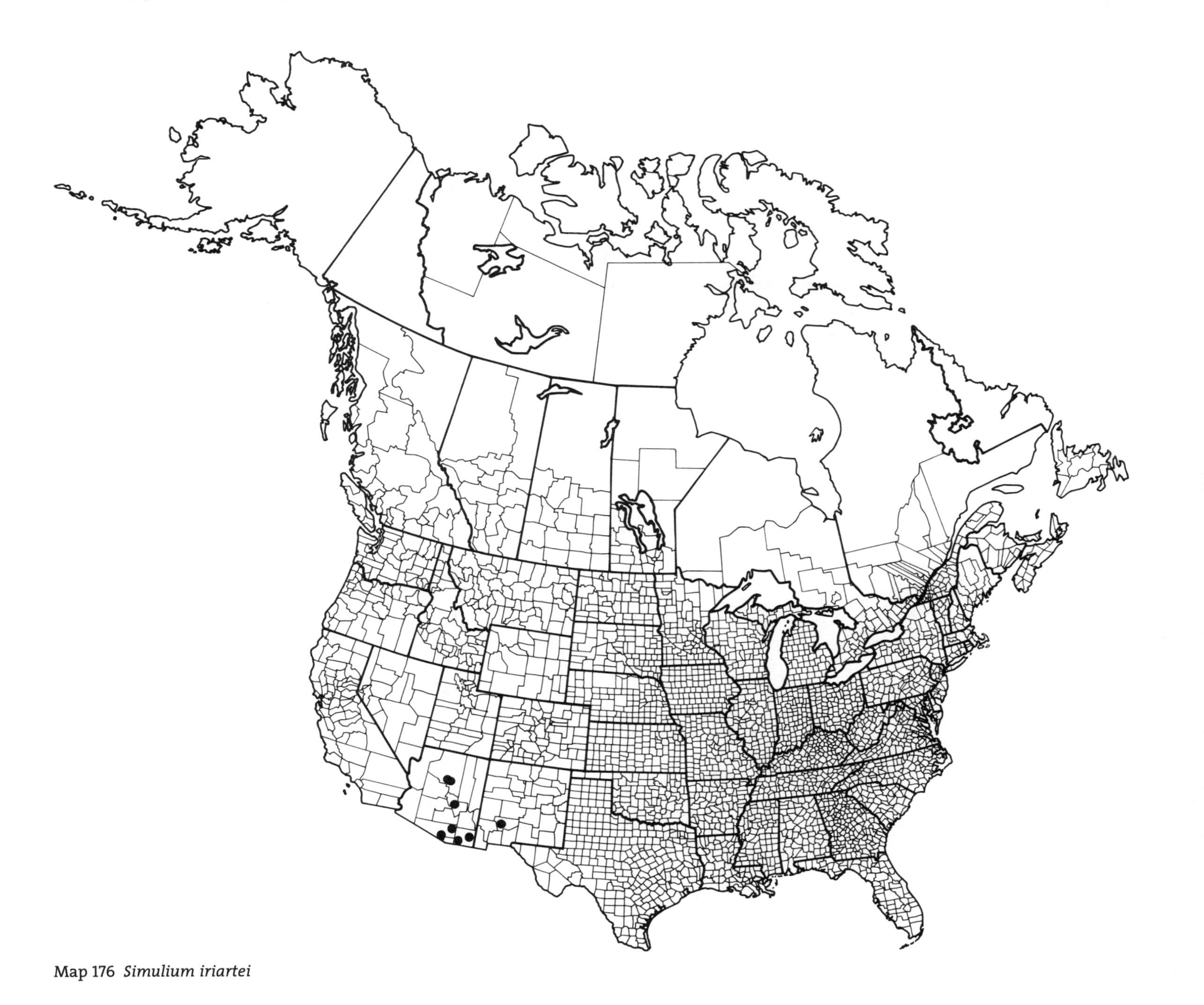

Map 176 *Simulium iriartei*

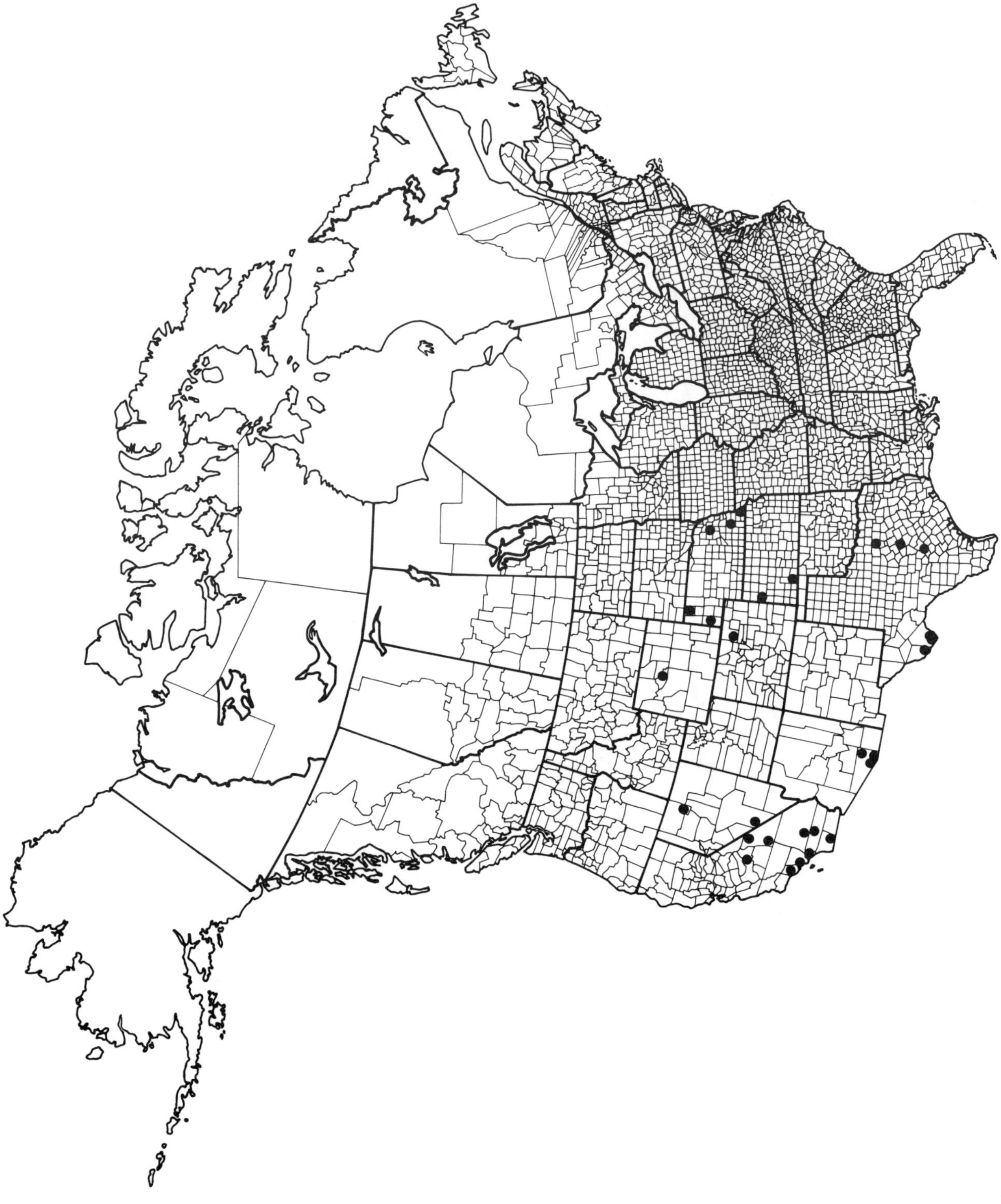

Map 177 *Simulium jacumbae*

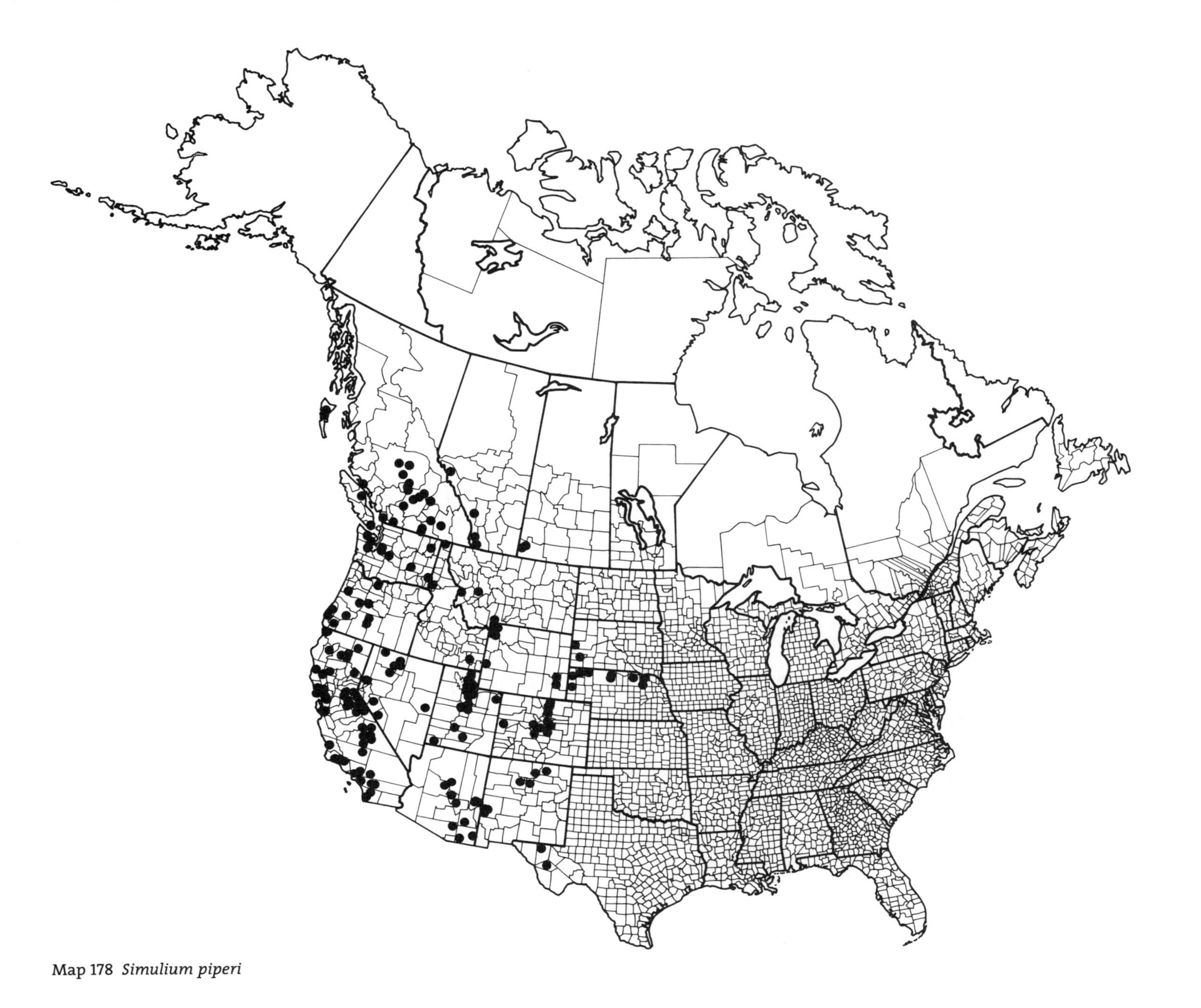

Map 178 *Simulium piperi*

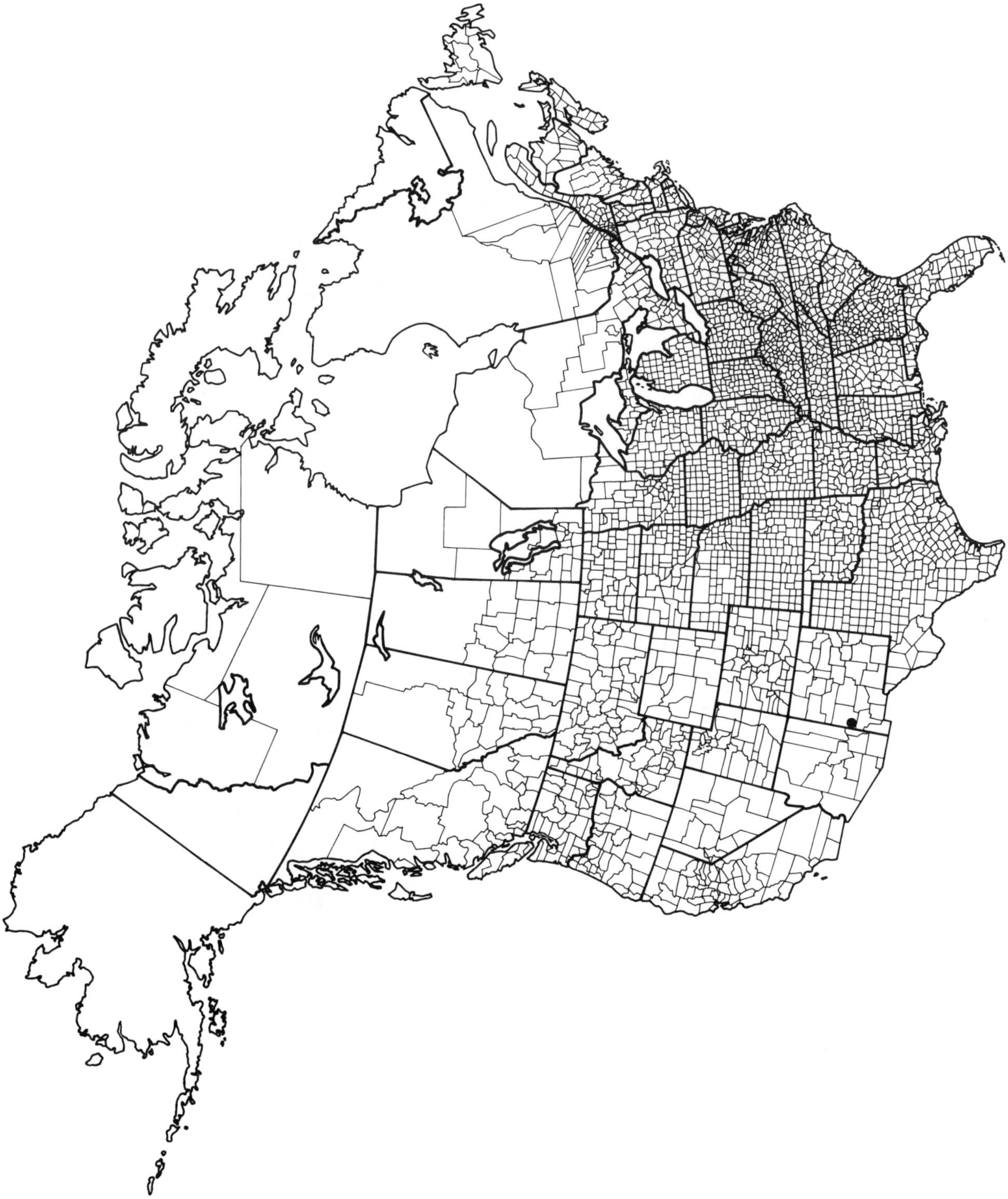

Map 179 *Simulium puigi*

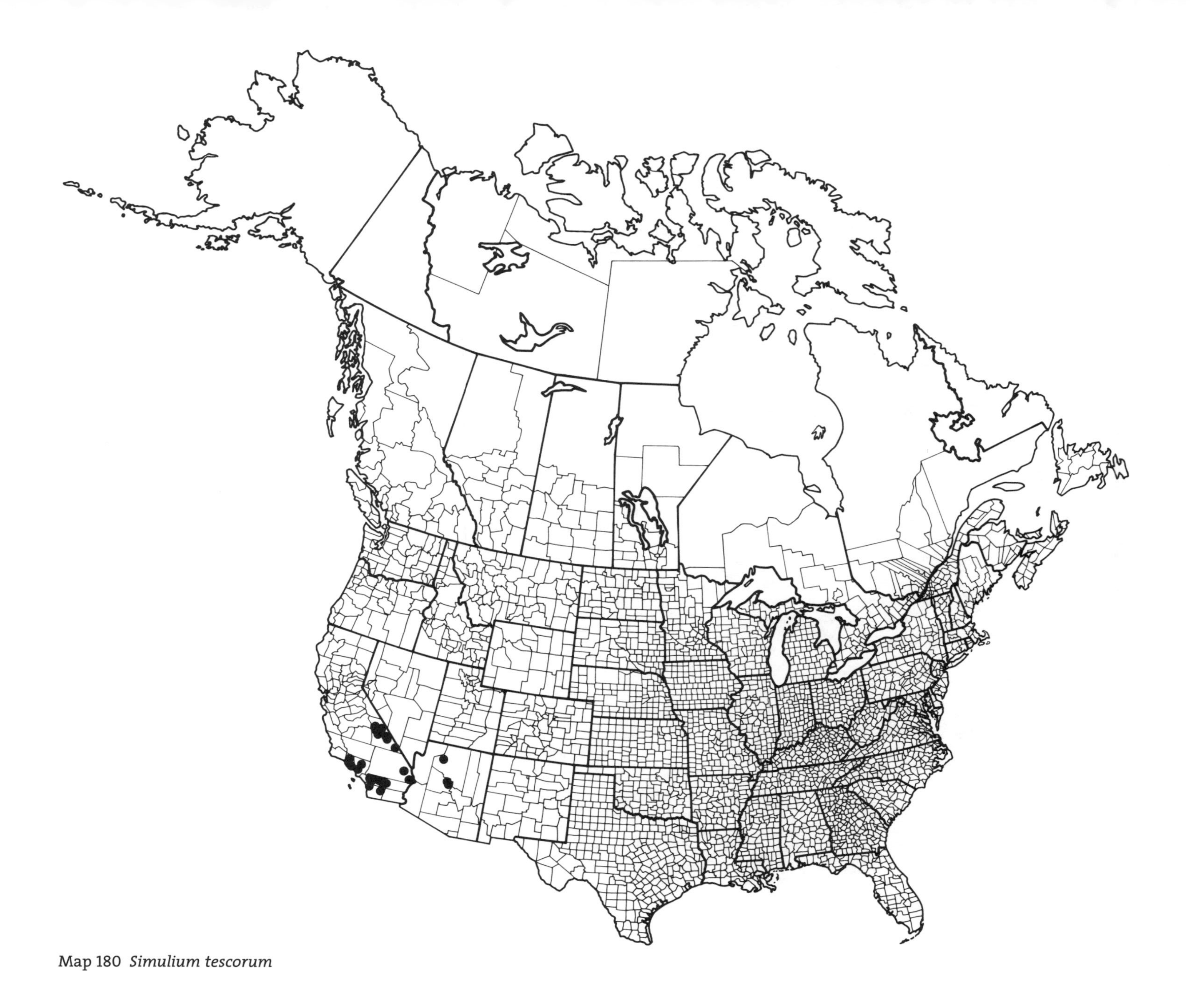

Map 180 *Simulium tescorum*

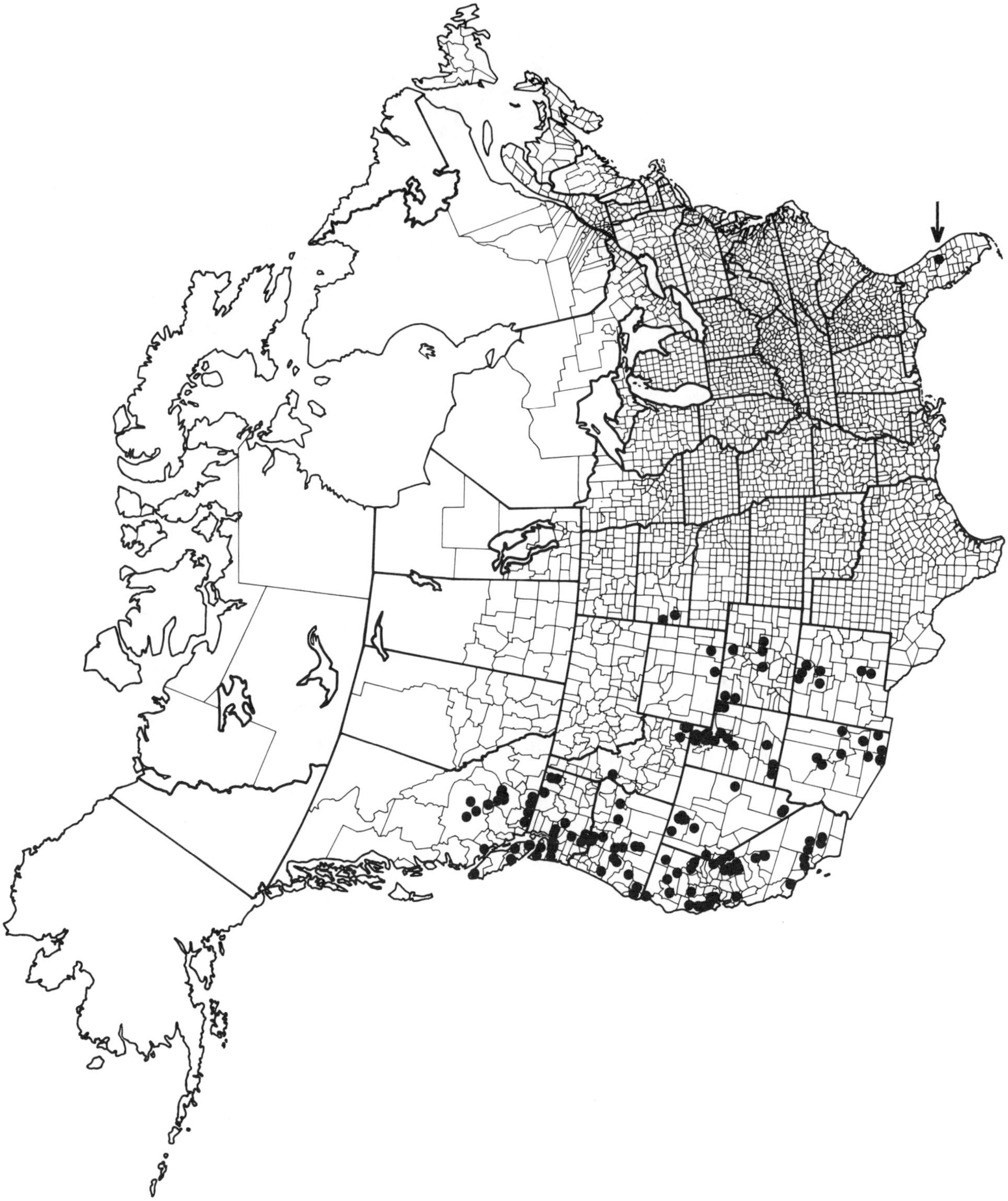

Map 181 *Simulium canadense*

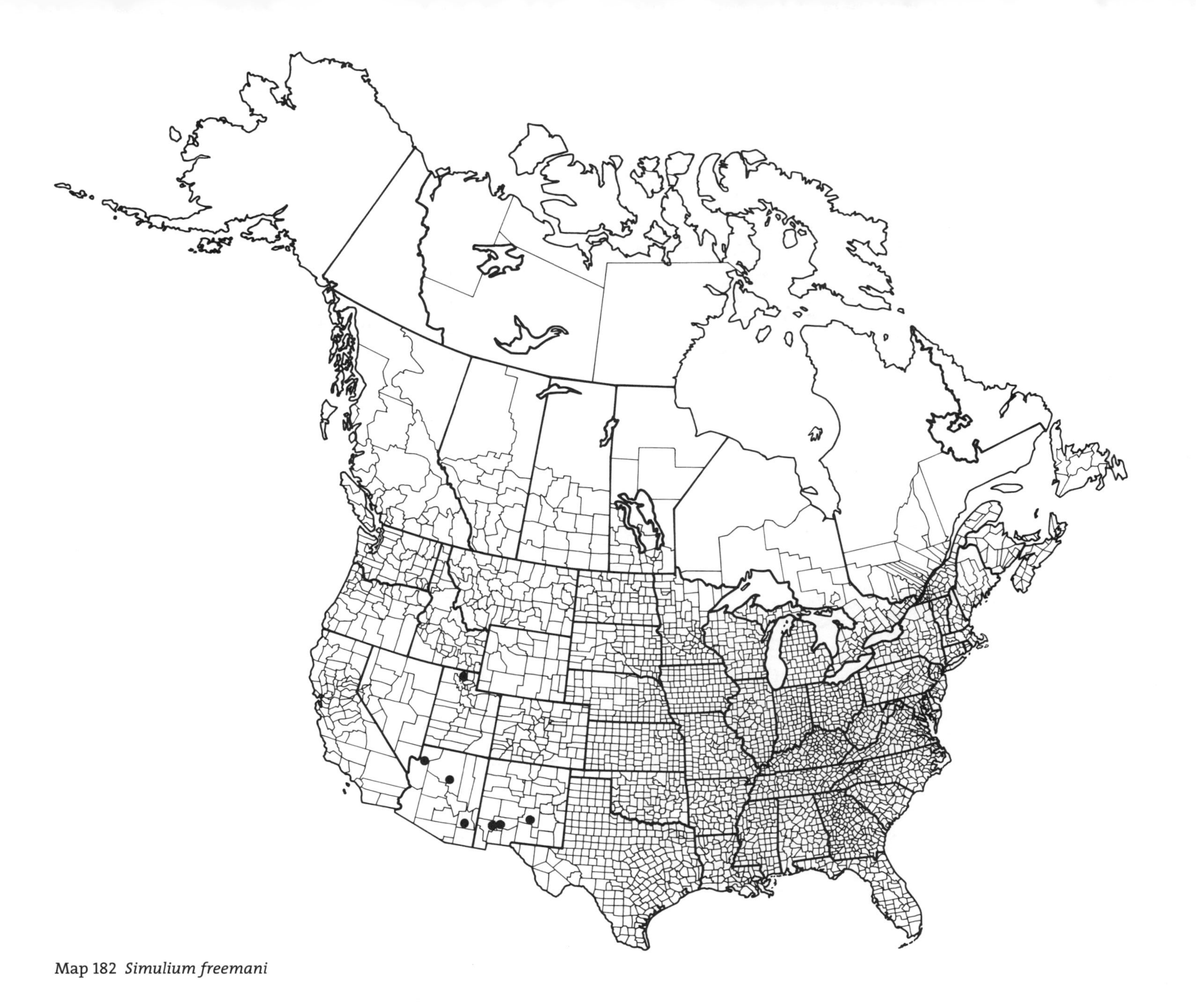

Map 182 *Simulium freemani*

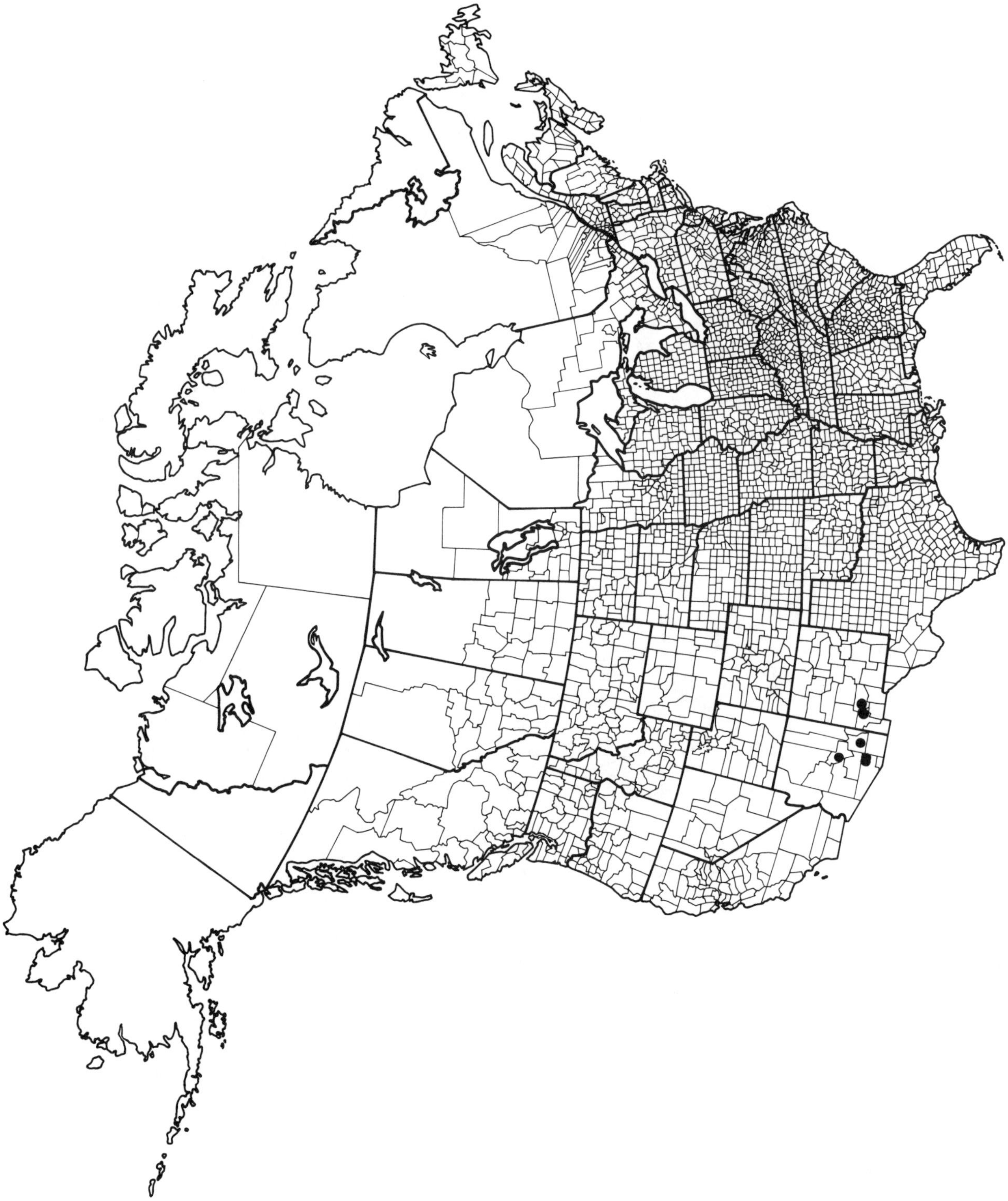

Map 183 *Simulium bricenoi*

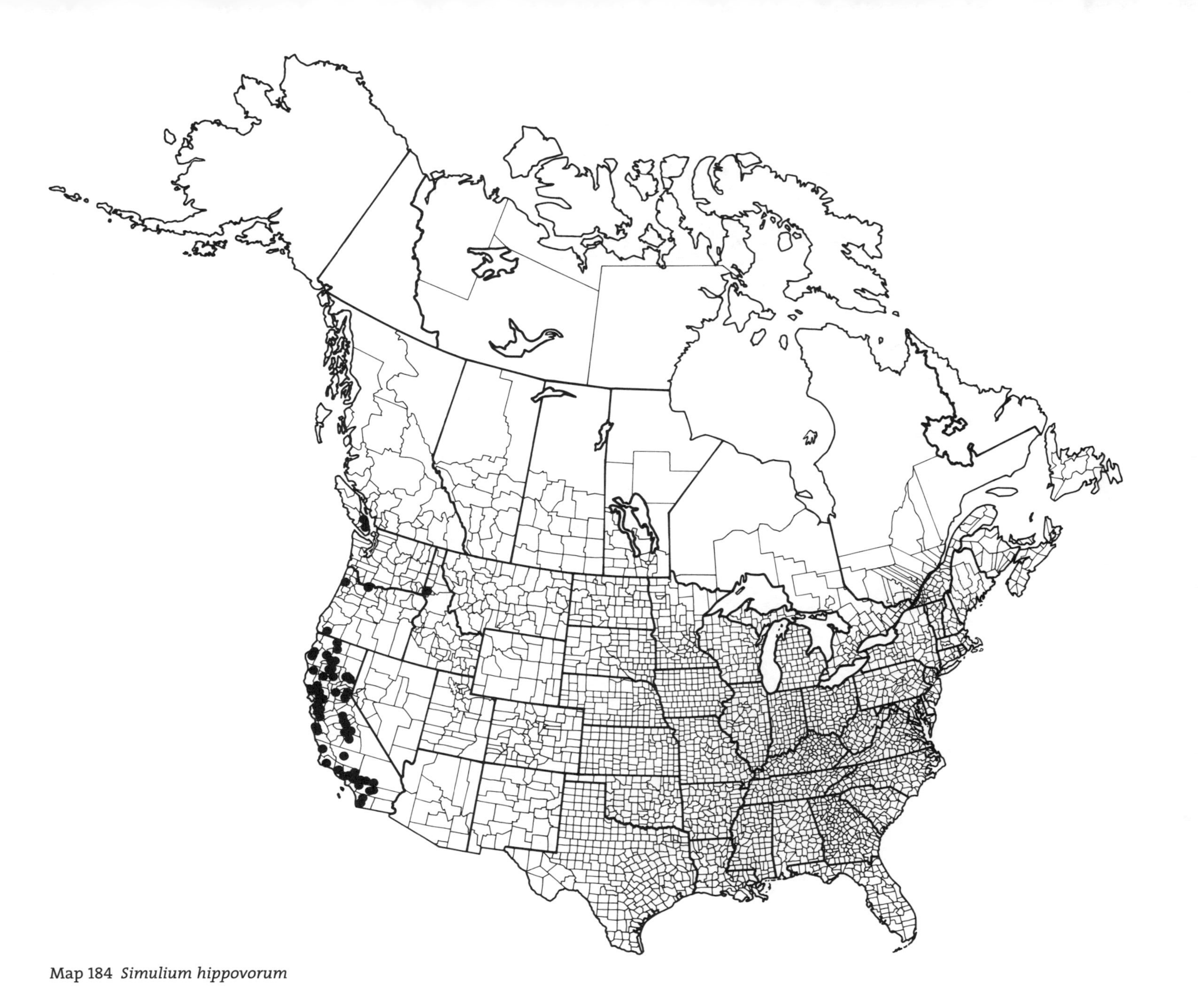

Map 184 *Simulium hippovorum*

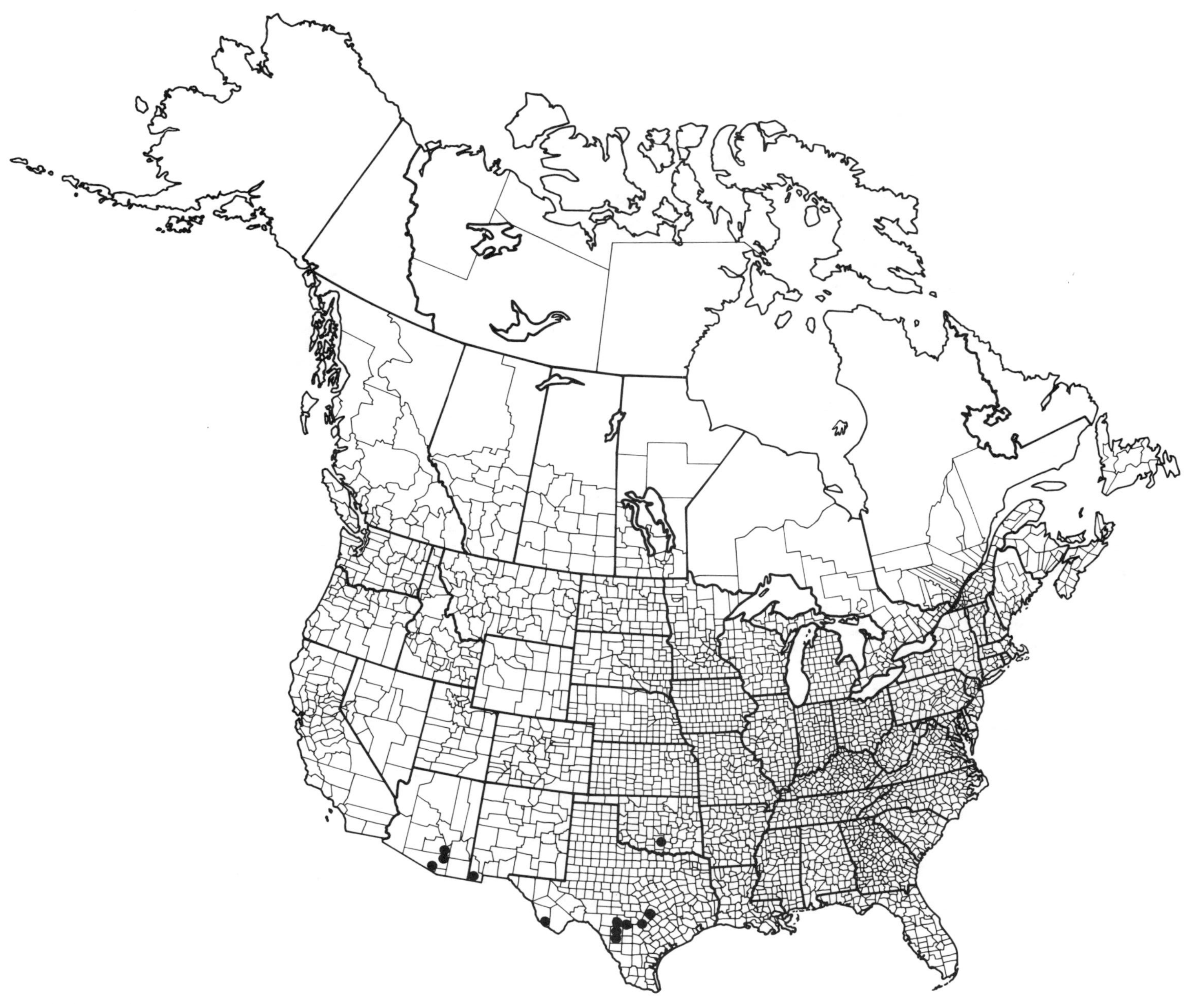

Map 185 *Simulium paynei*

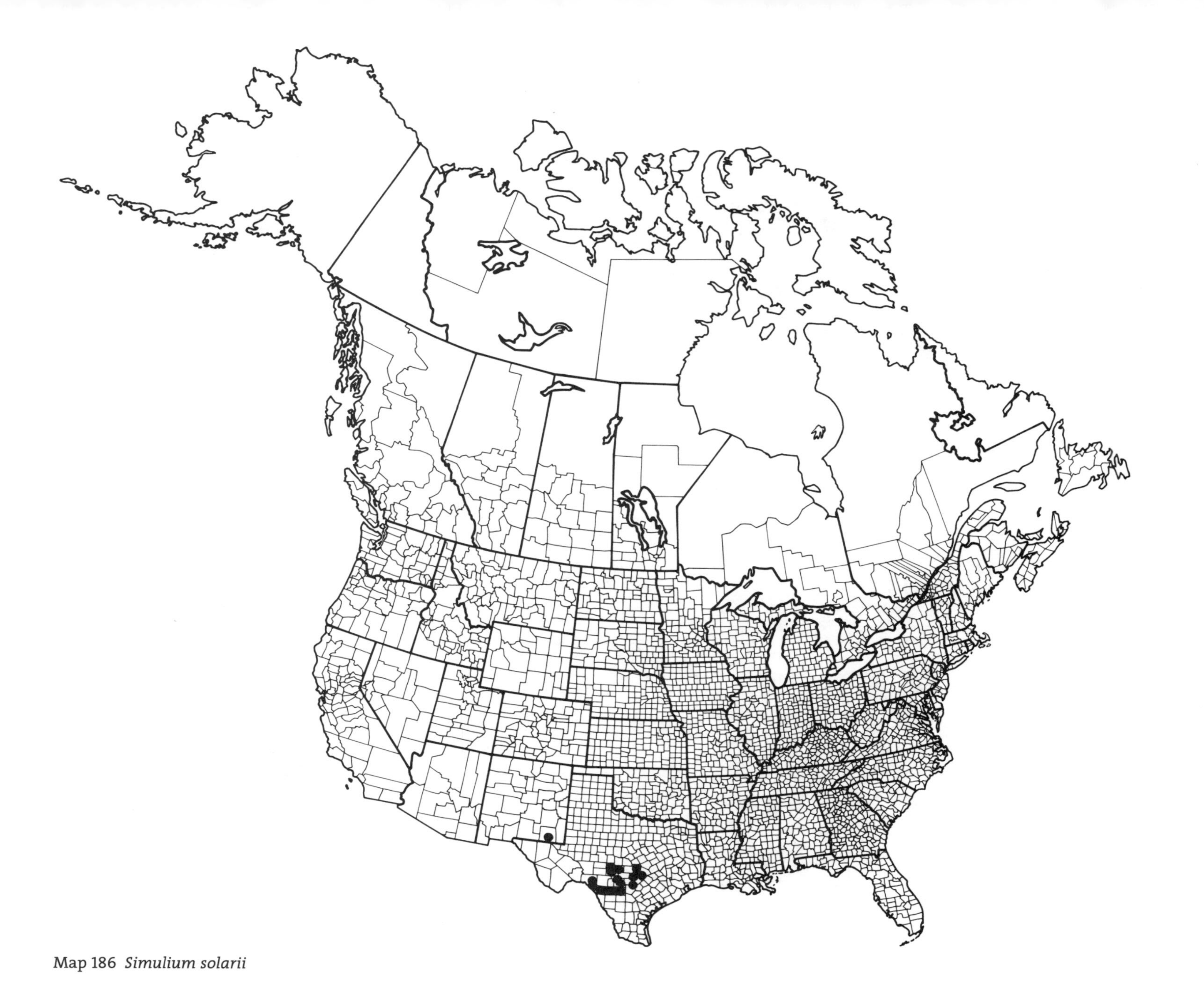

Map 186 *Simulium solarii*

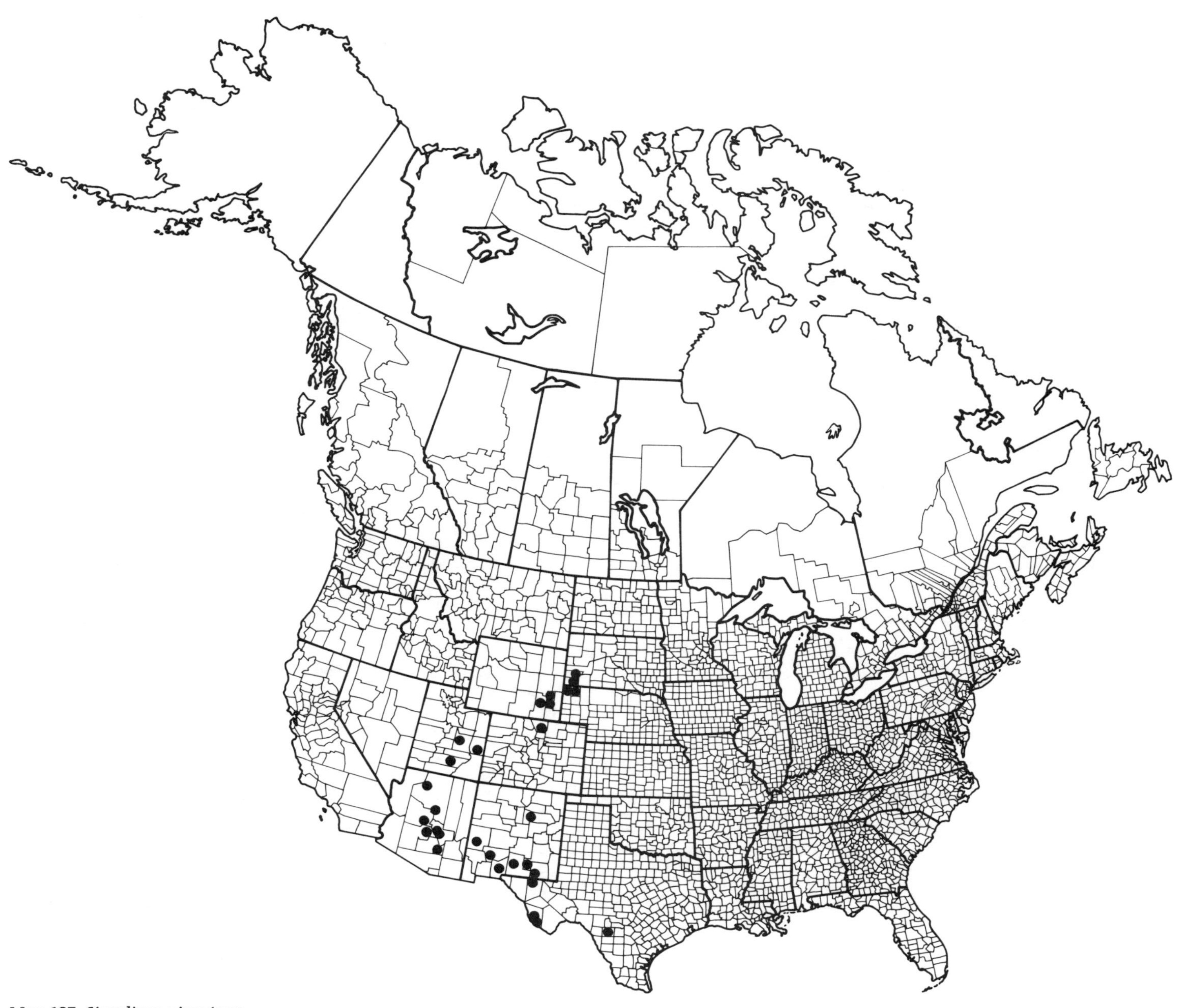

Map 187 *Simulium virgatum*

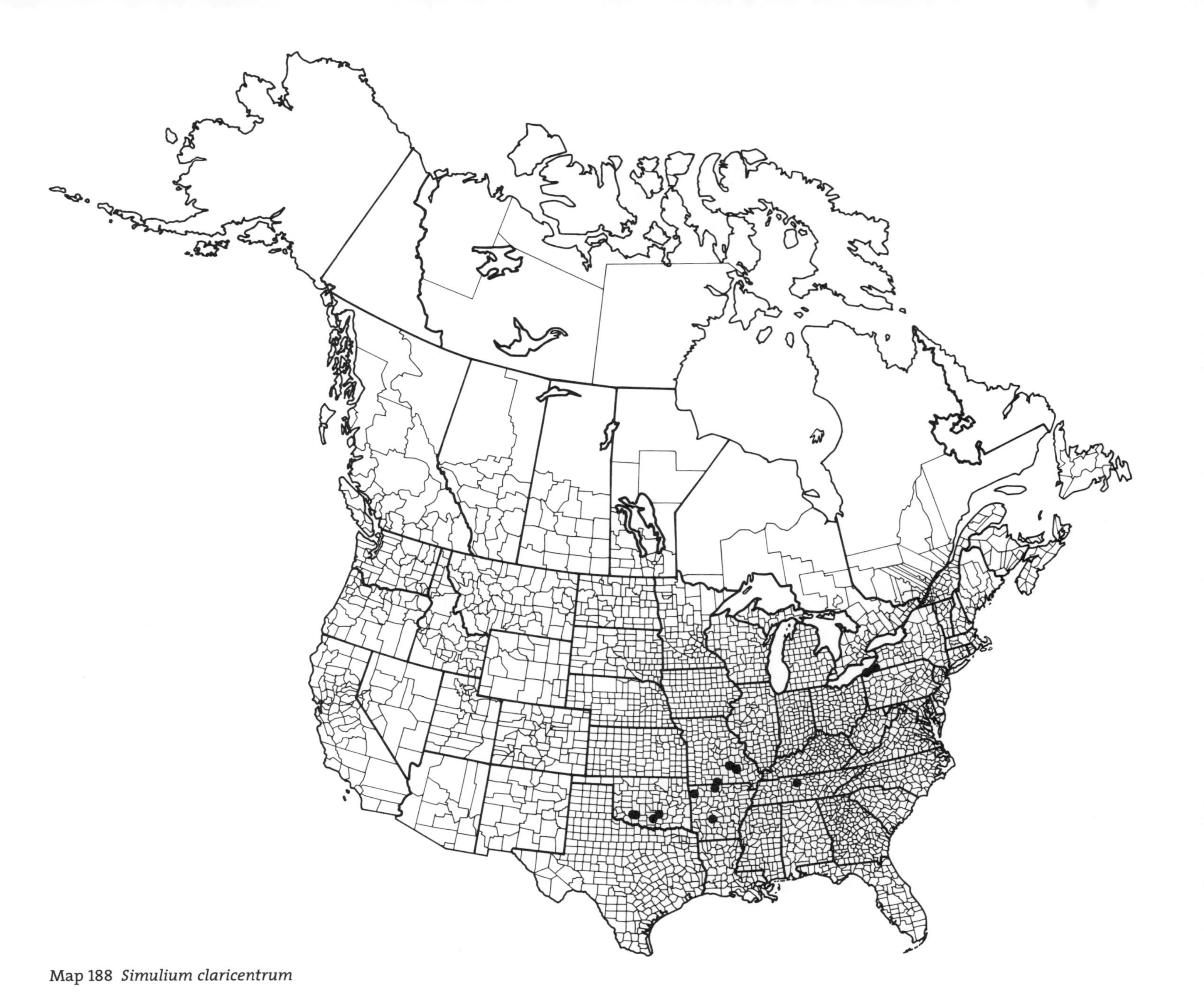

Map 188 *Simulium claricentrum*

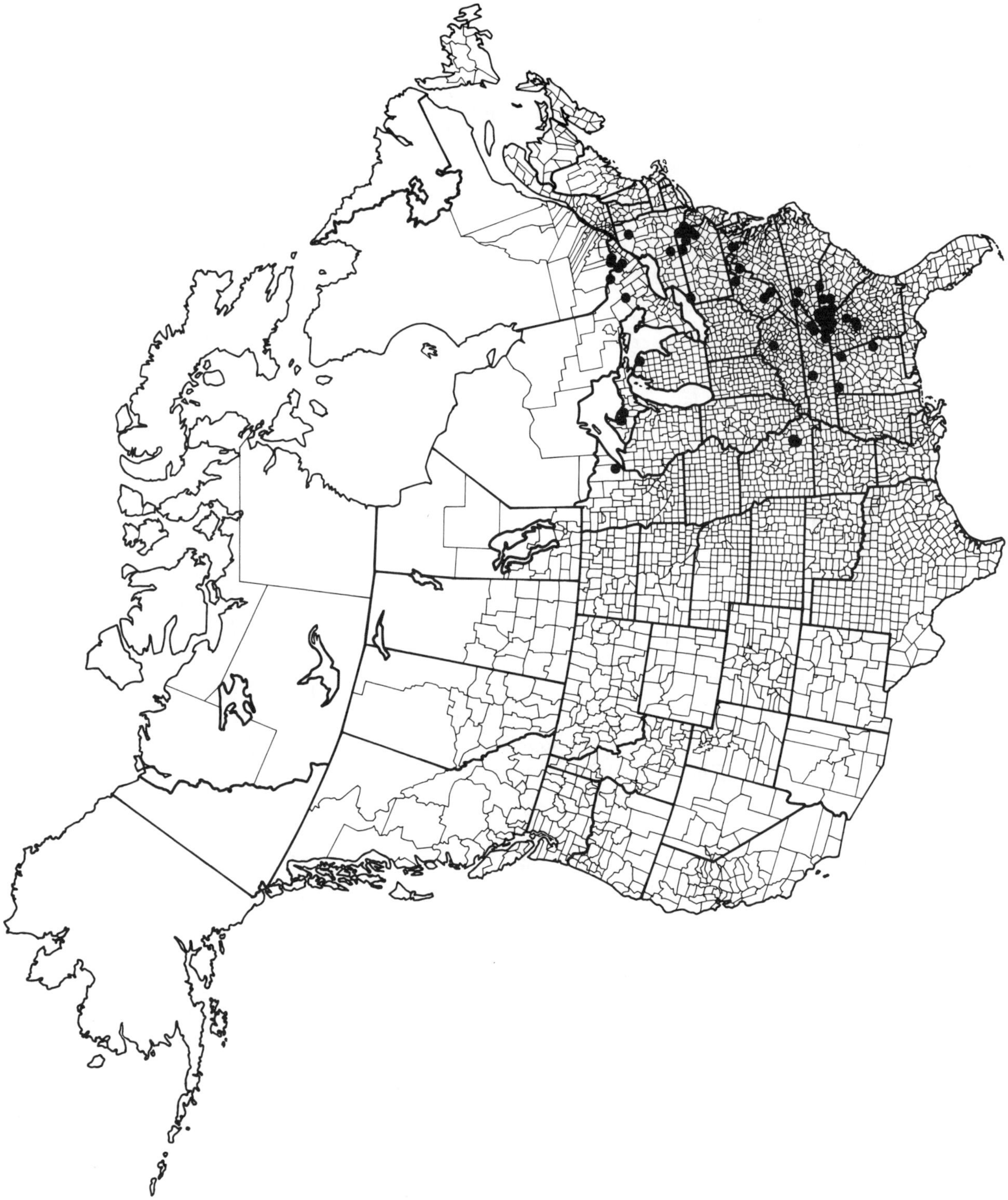

Map 189 *Simulium innoxium*

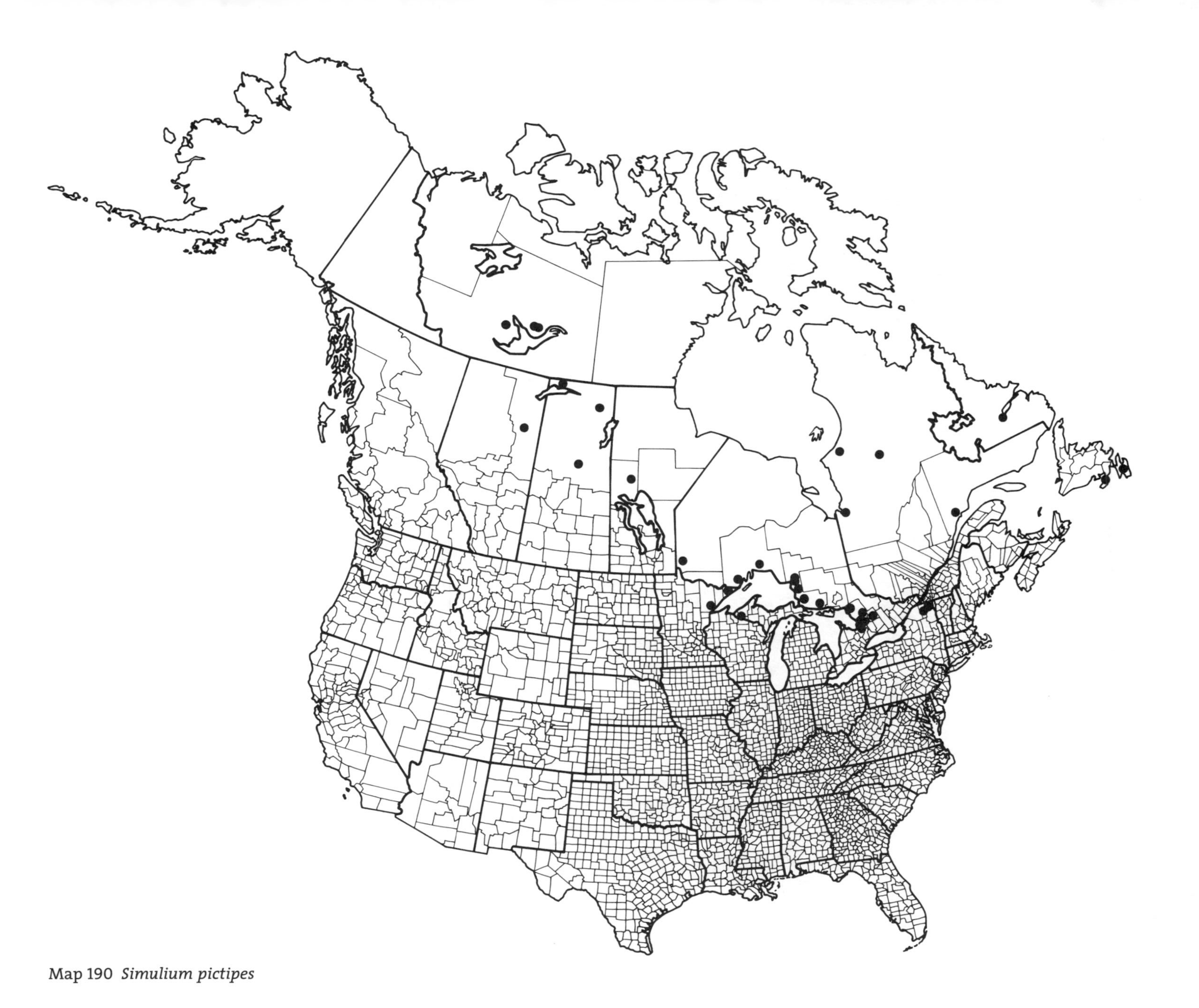

Map 190 *Simulium pictipes*

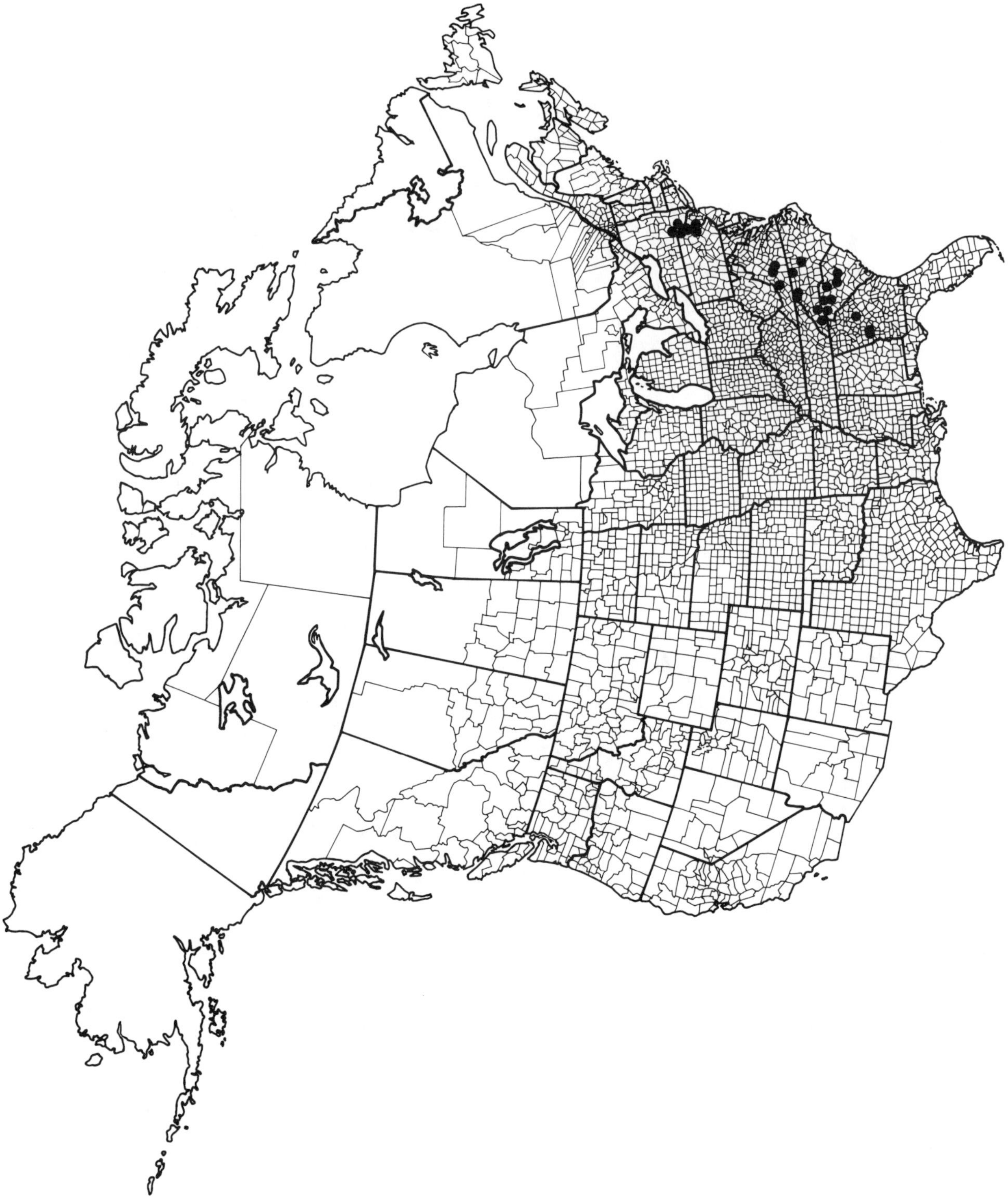

Map 191 *Simulium anchistinum*

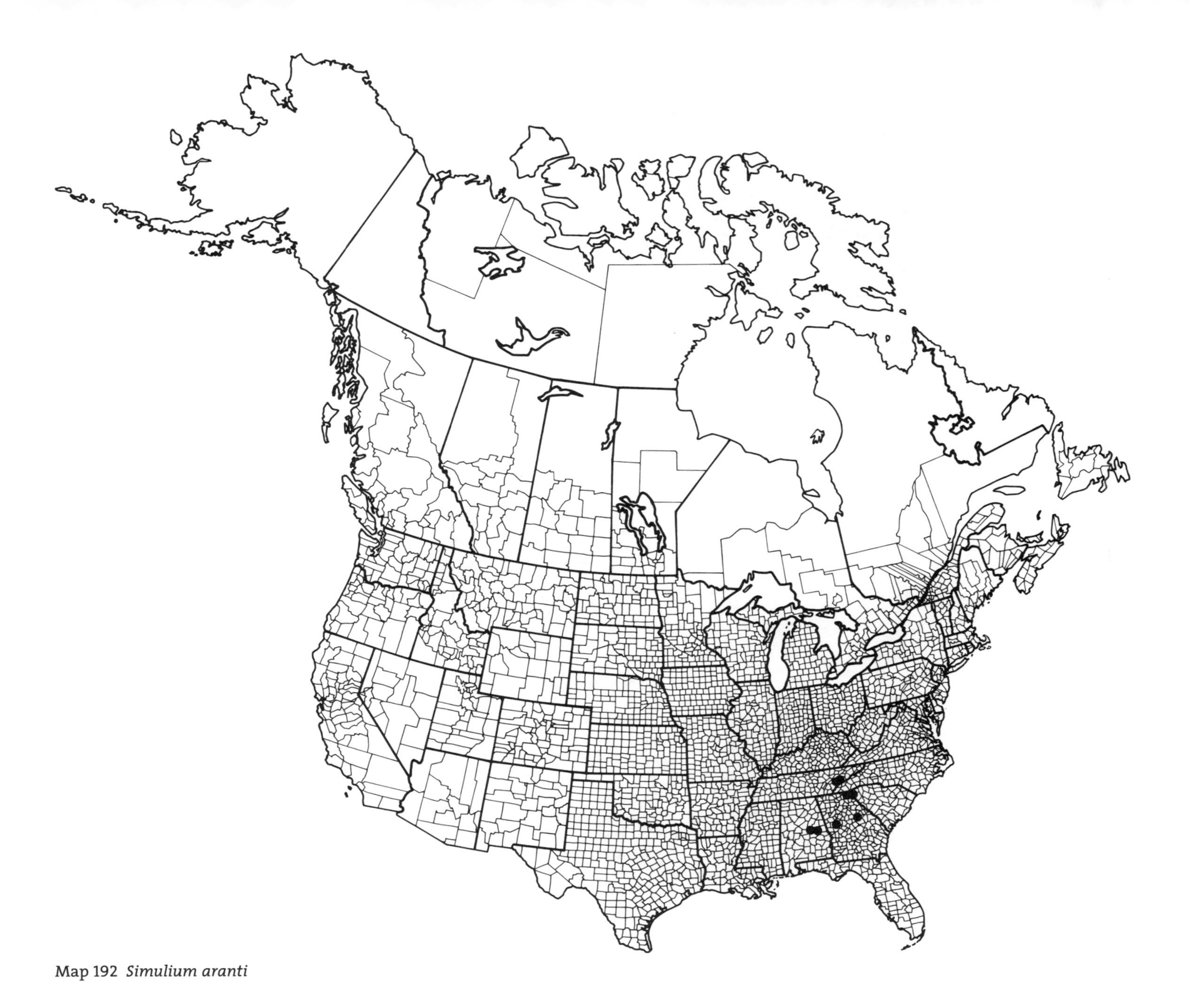

Map 192 *Simulium aranti*

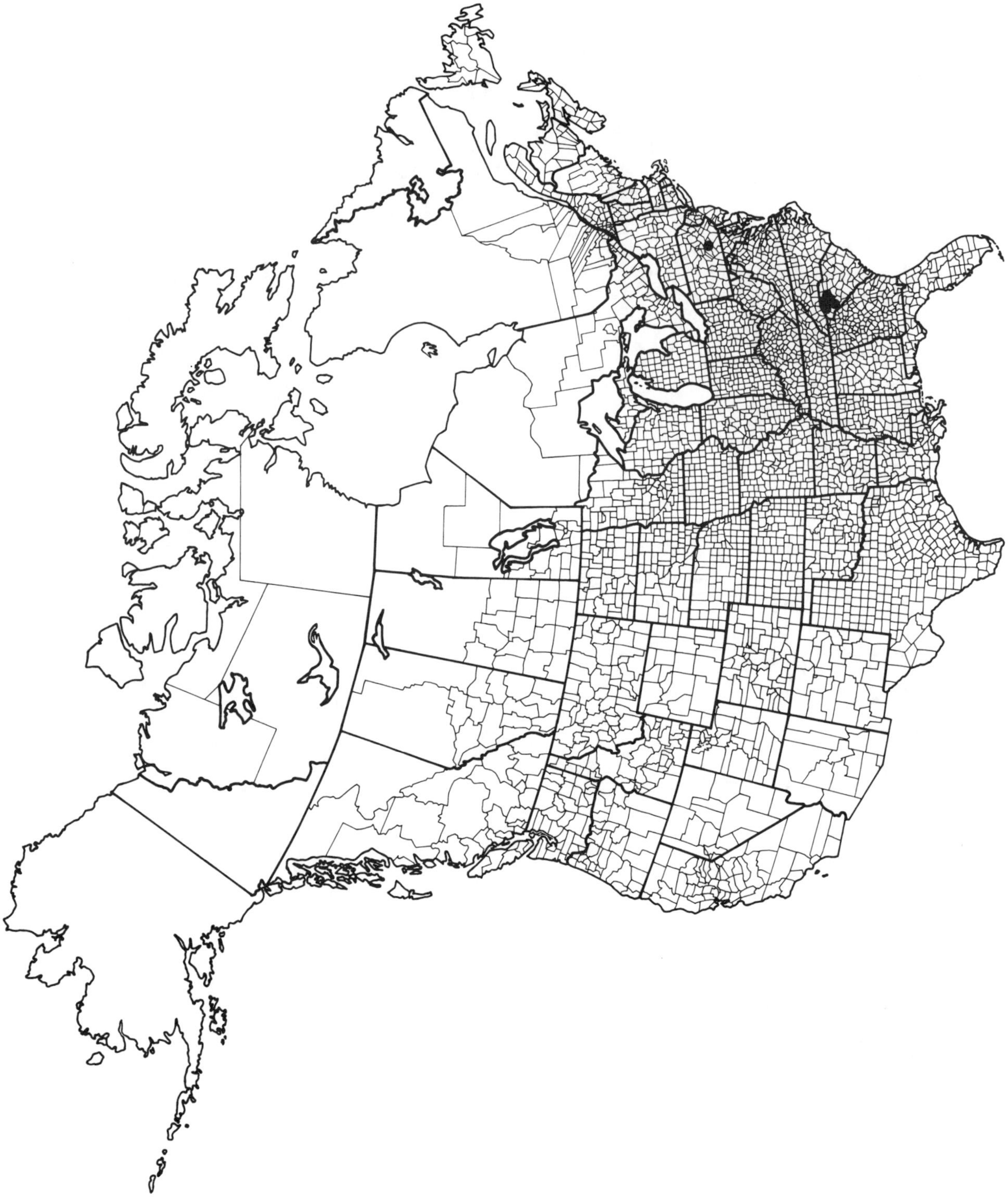

Map 193 *Simulium chlorum*

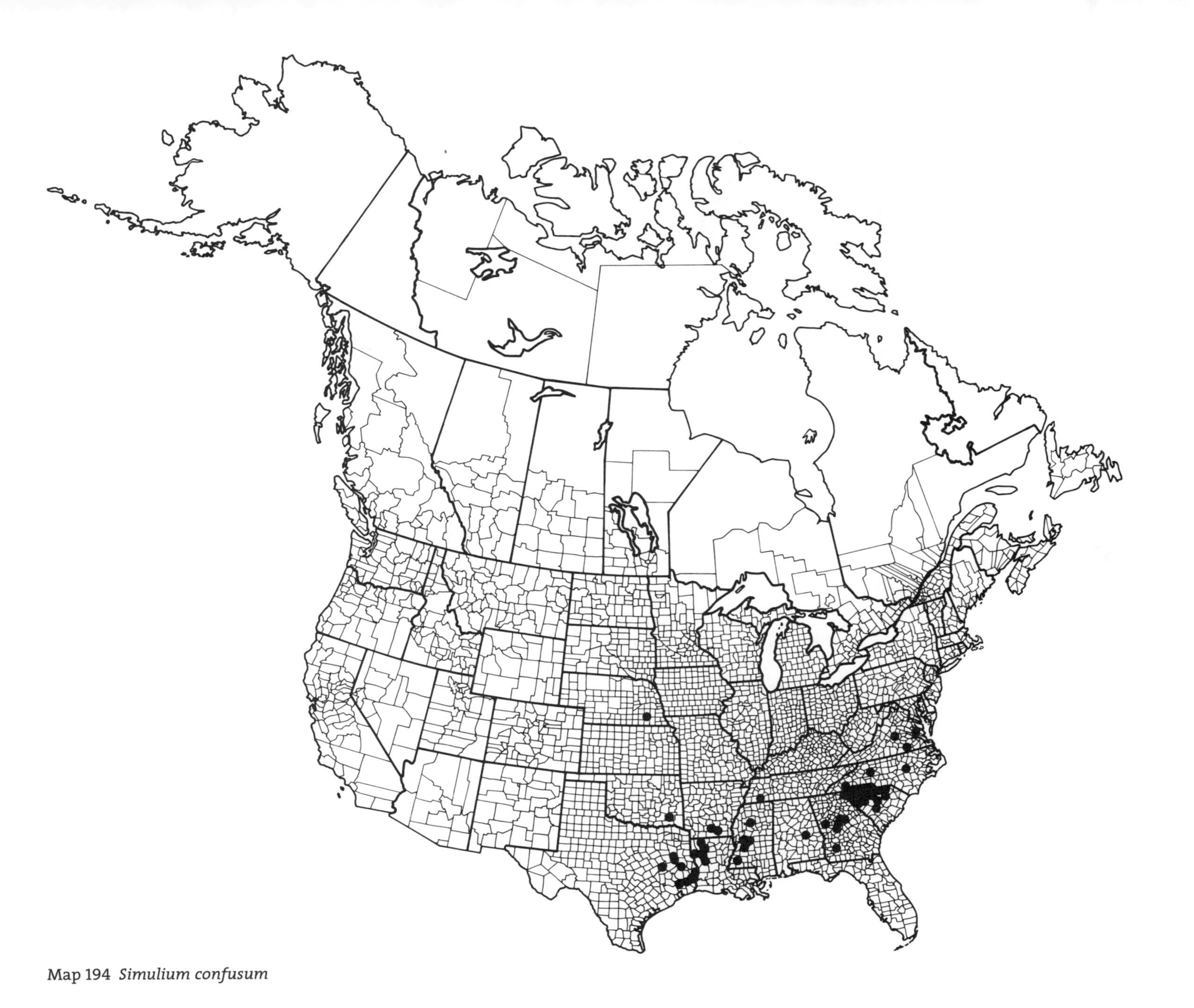

Map 194 *Simulium confusum*

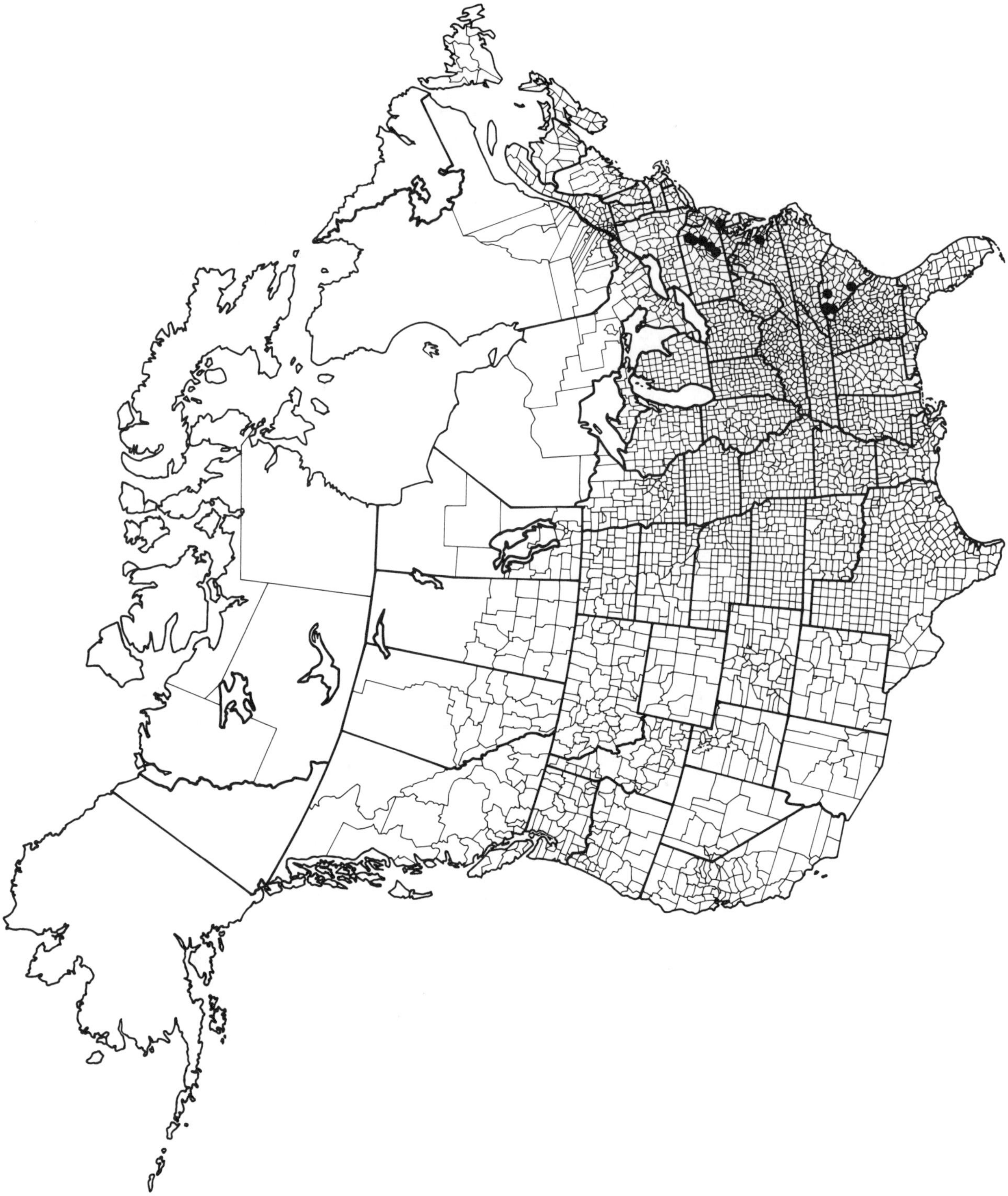

Map 195 *Simulium definitum*

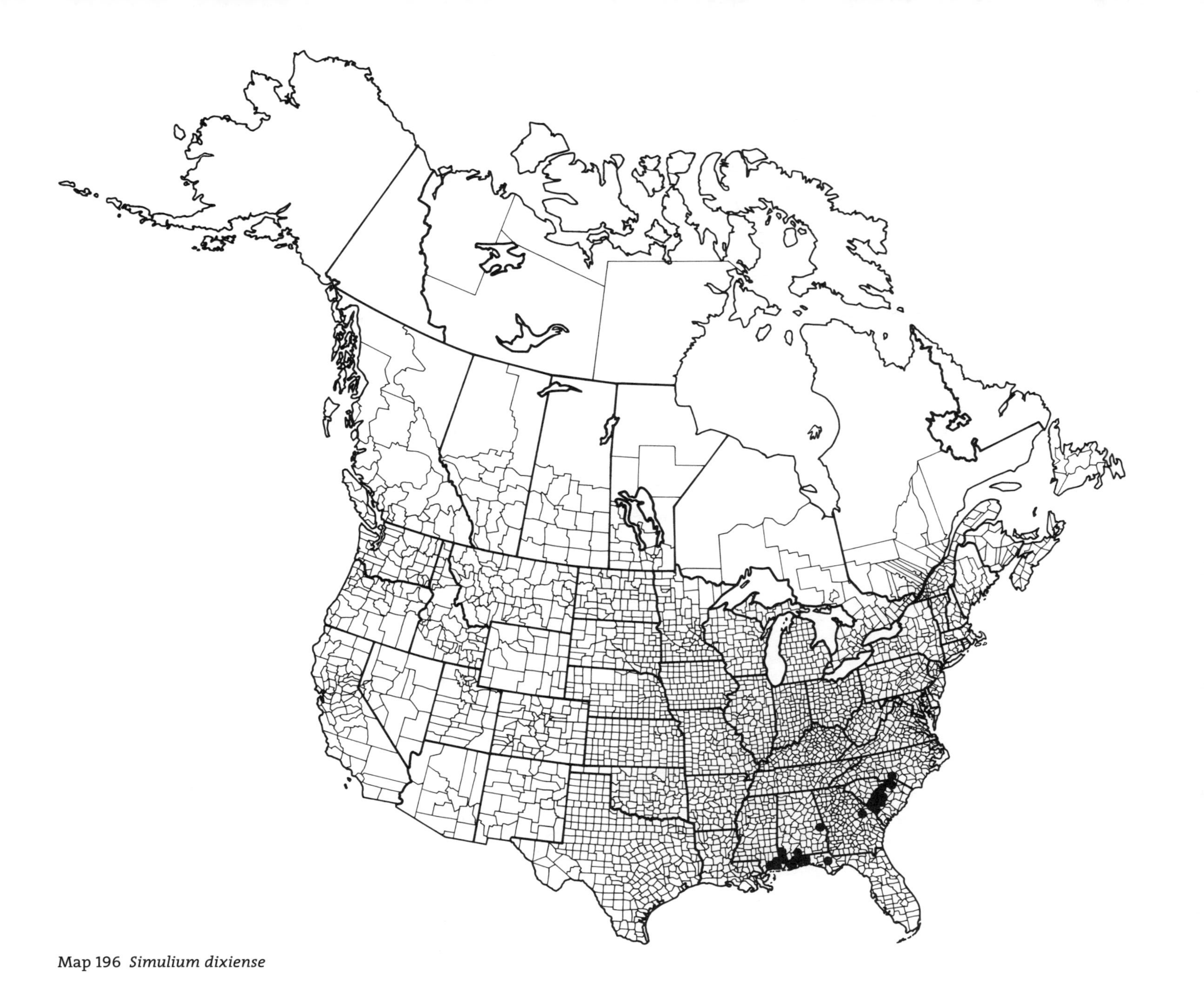

Map 196 *Simulium dixiense*

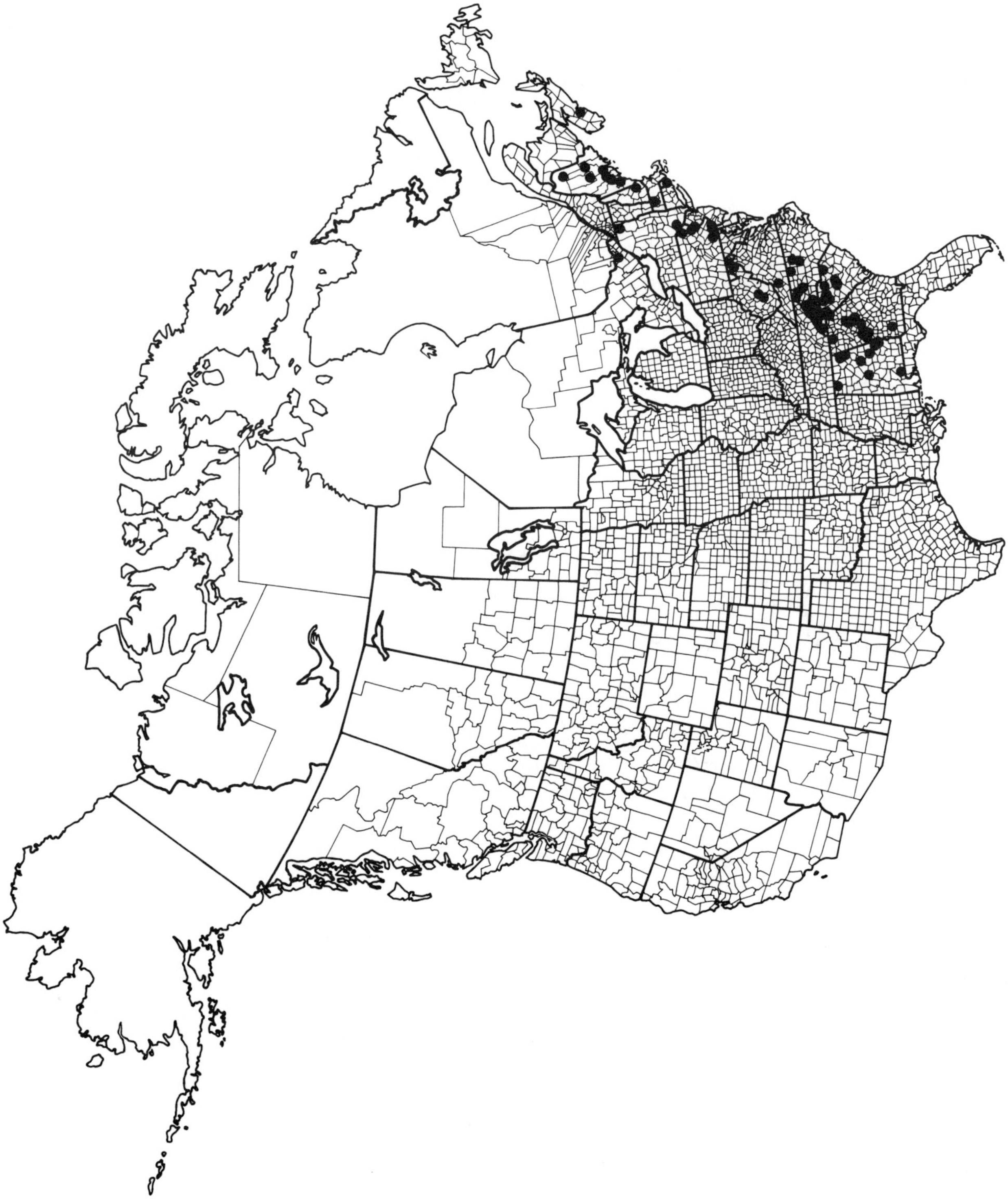

Map 197 *Simulium fibrinflatum*

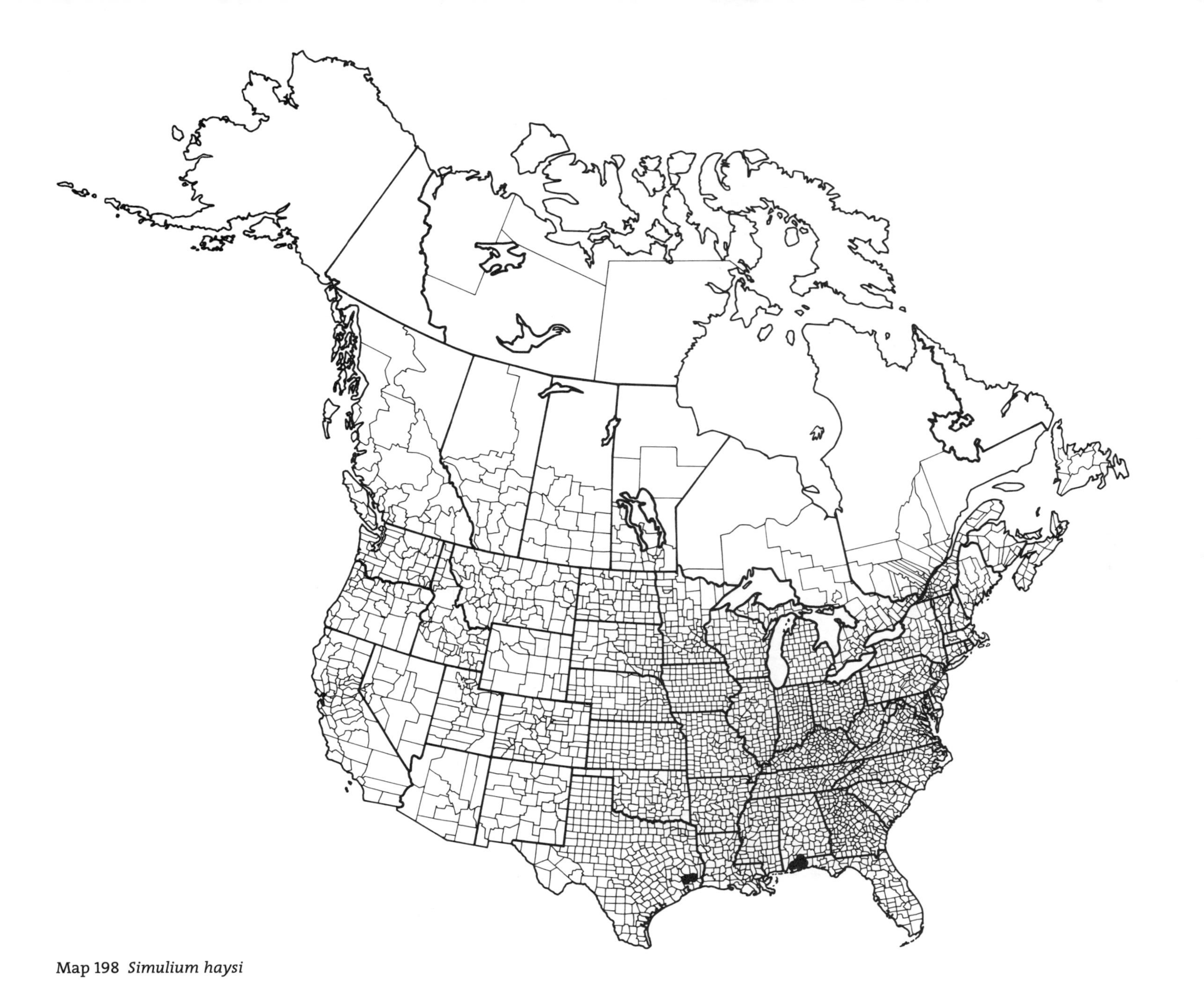

Map 198 *Simulium haysi*

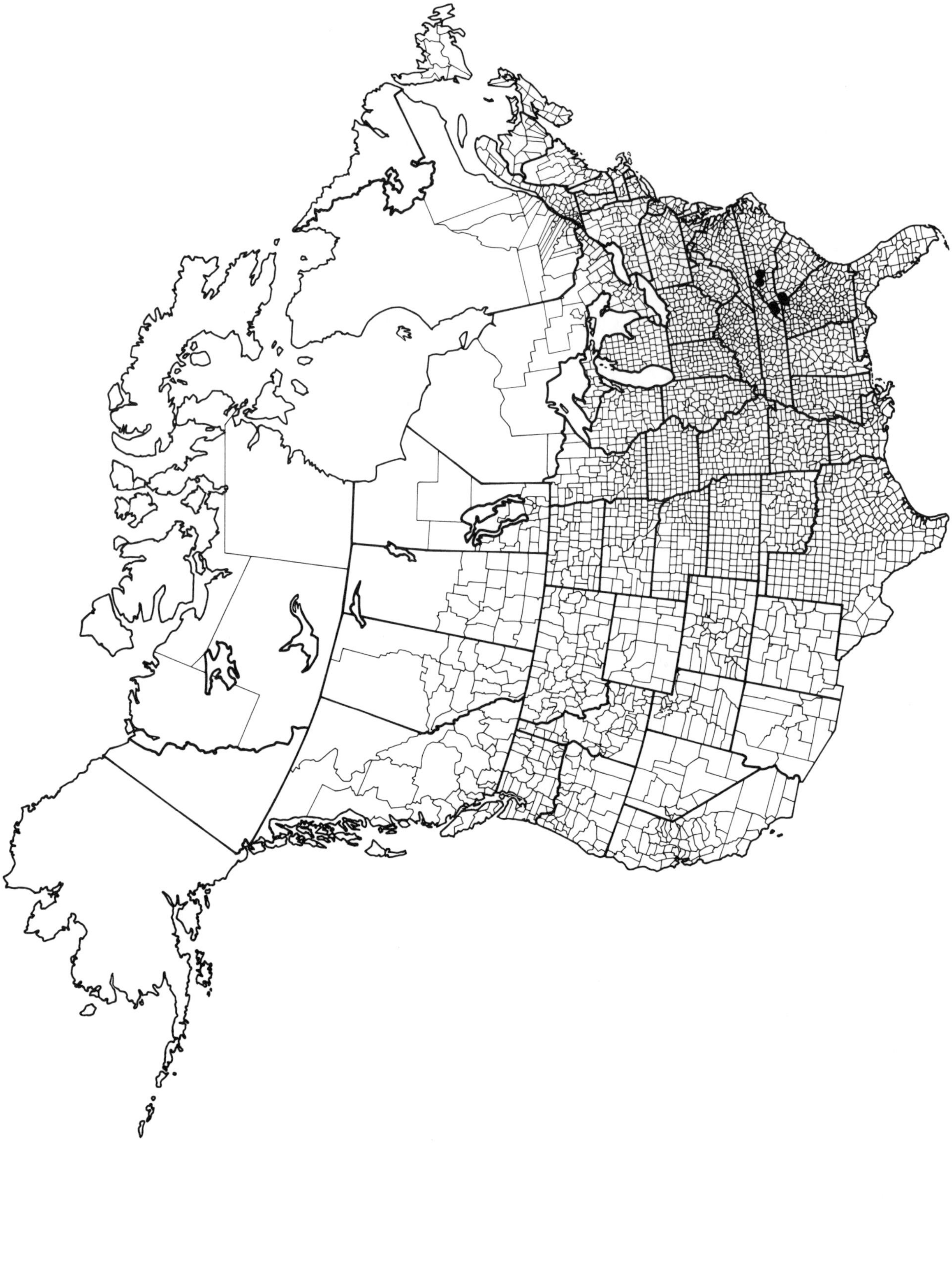

Map 199 *Simulium infenestrum*

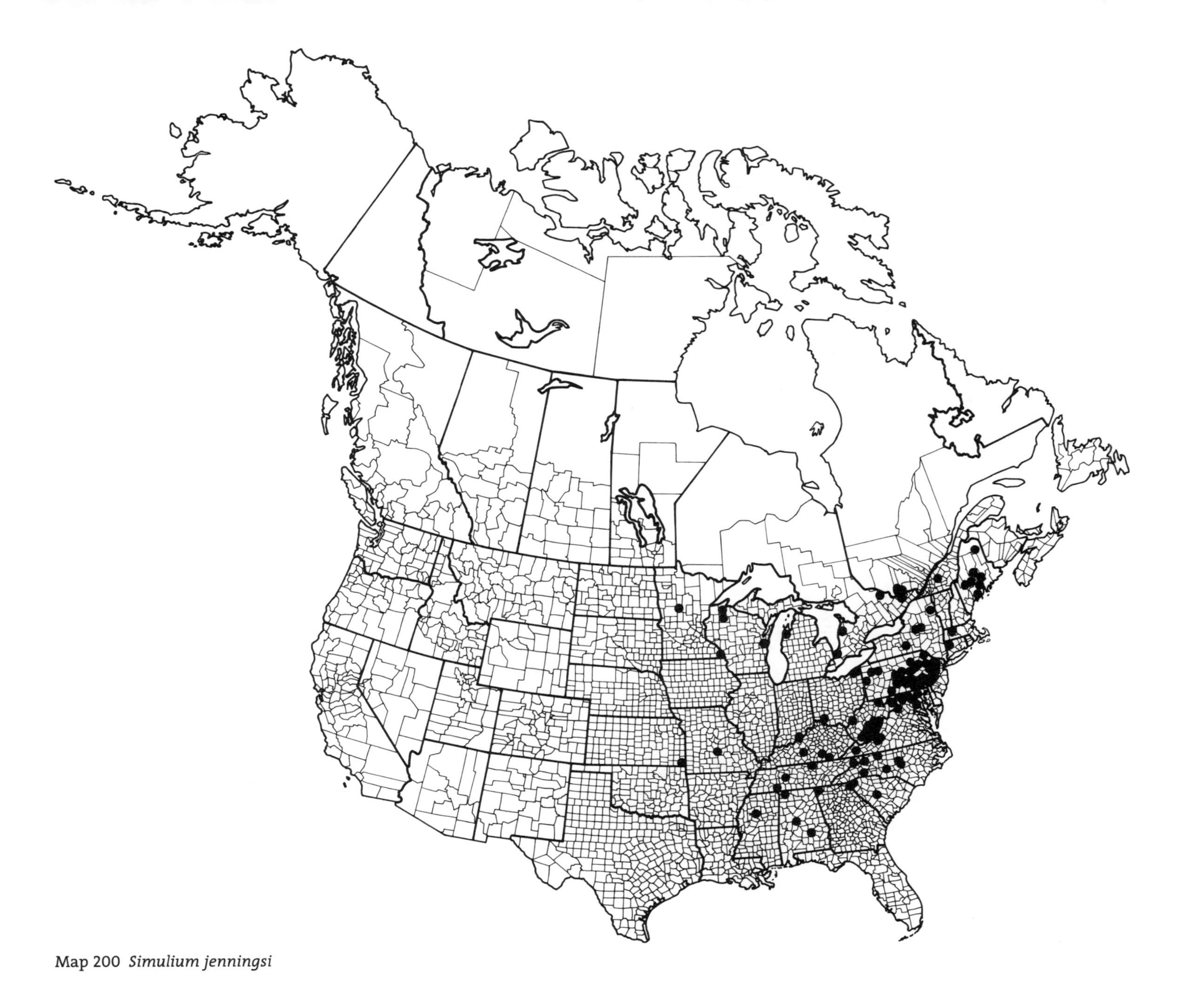

Map 200 *Simulium jenningsi*

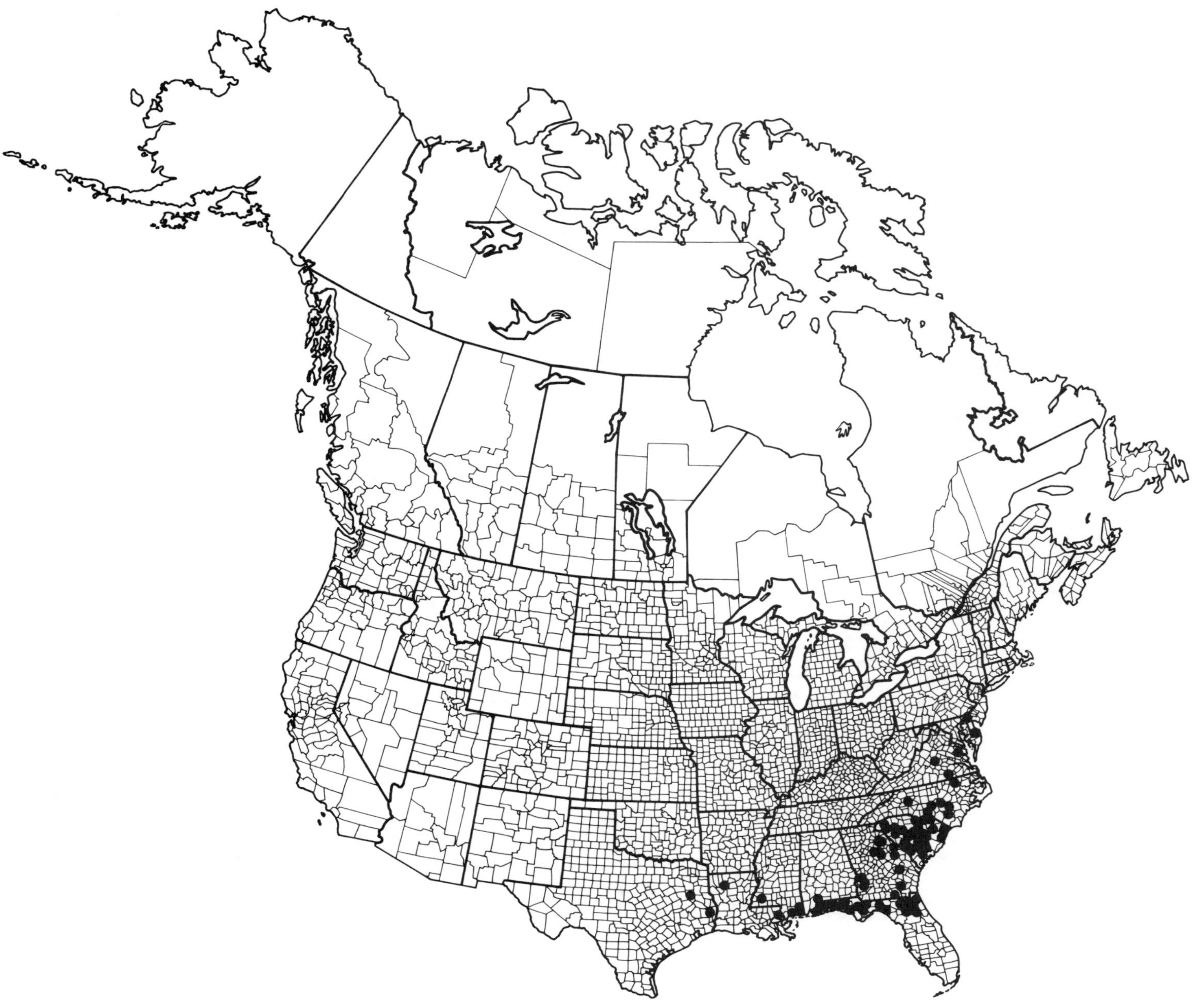

Map 201 *Simulium jonesi*

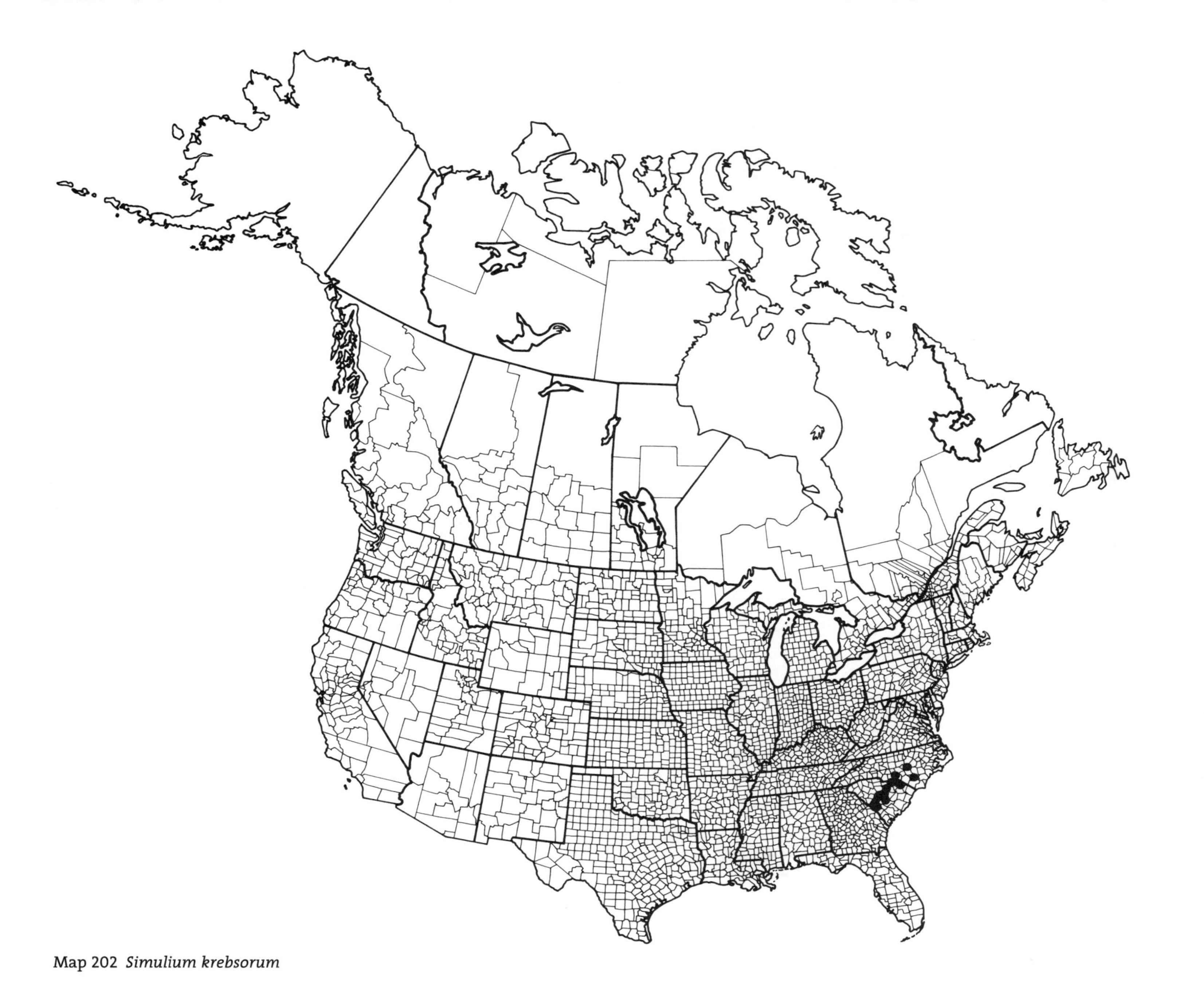

Map 202 *Simulium krebsorum*

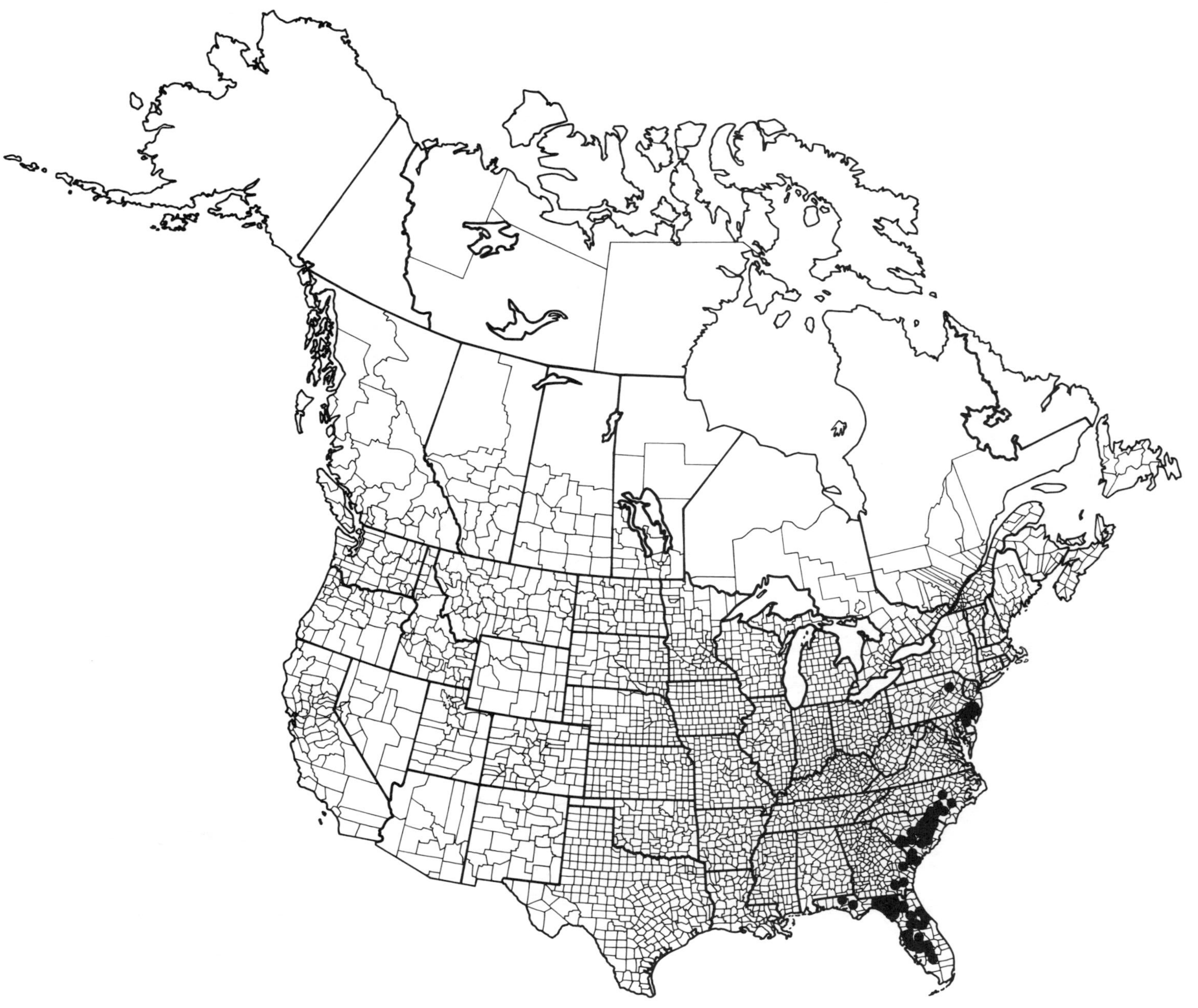

Map 203 *Simulium lakei*

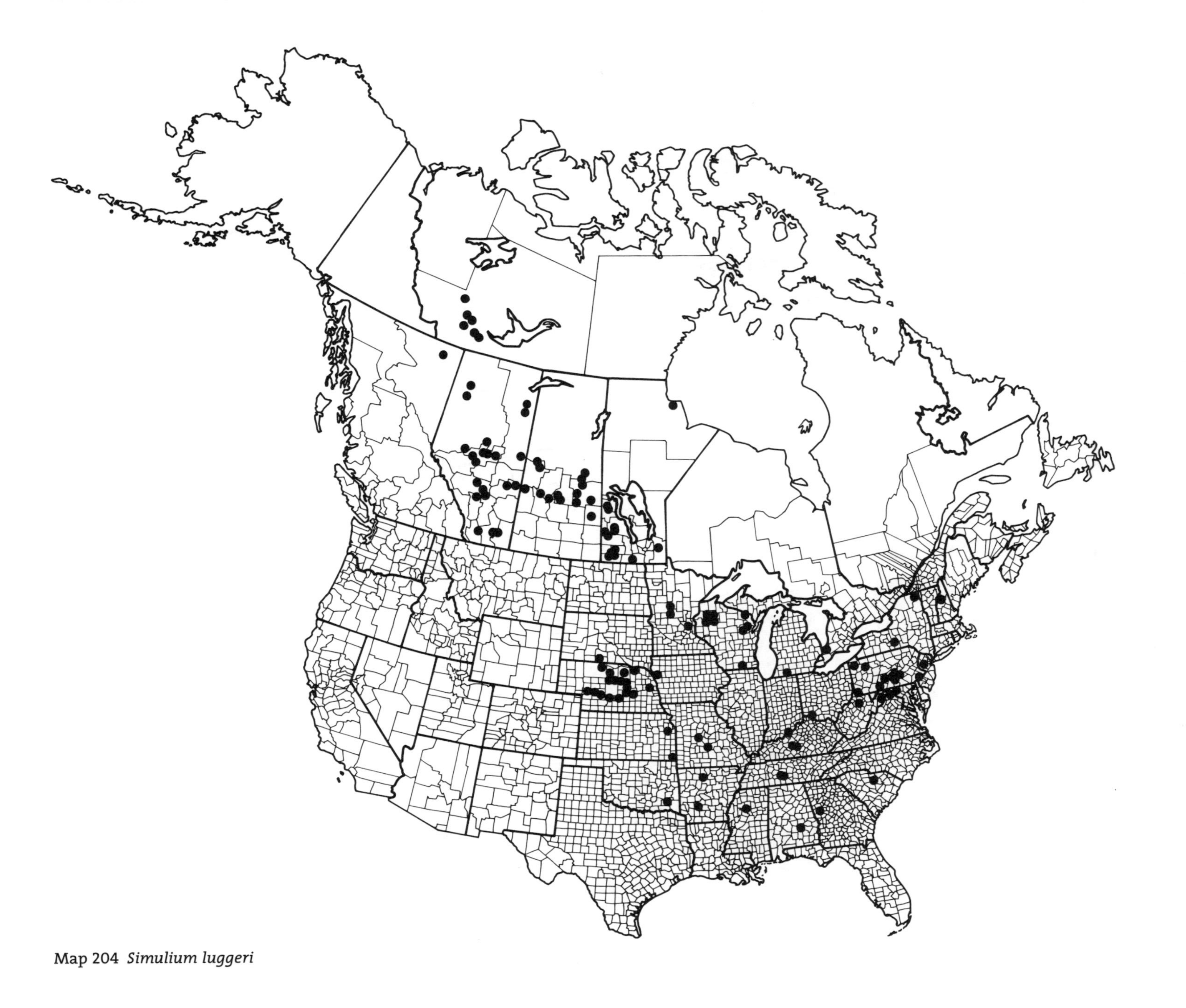

Map 204 *Simulium luggeri*

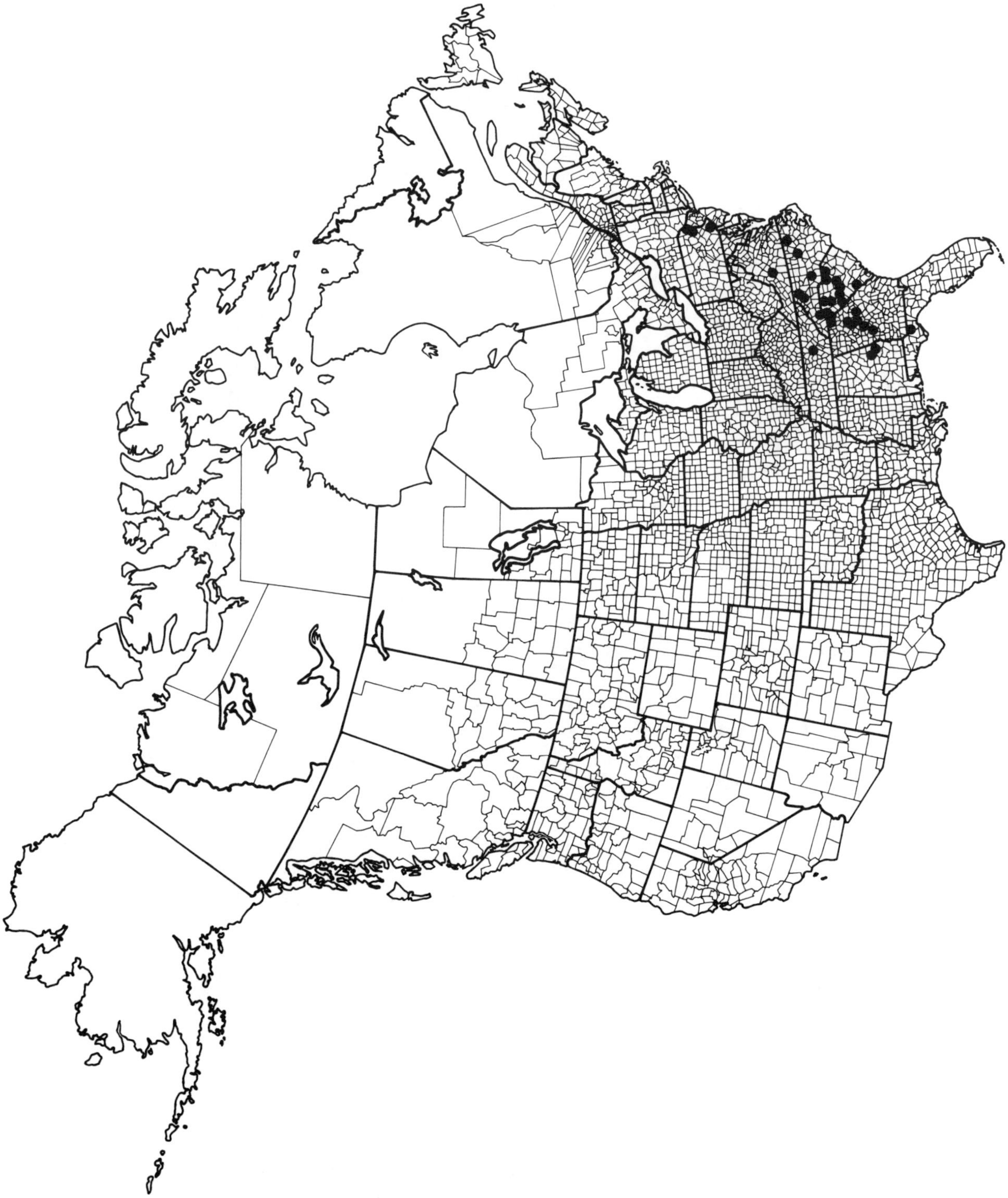

Map 205 *Simulium notiale*

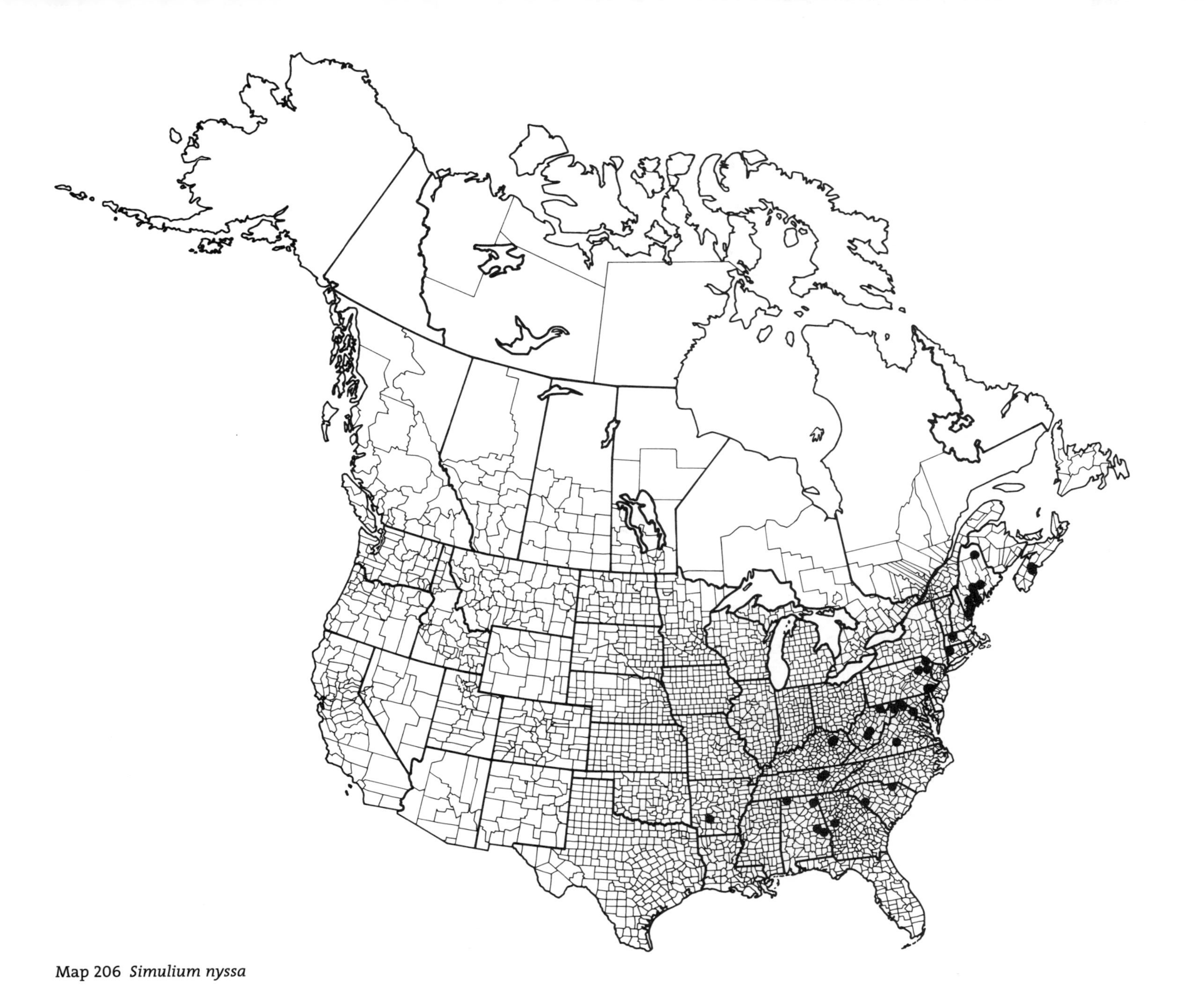

Map 206 *Simulium nyssa*

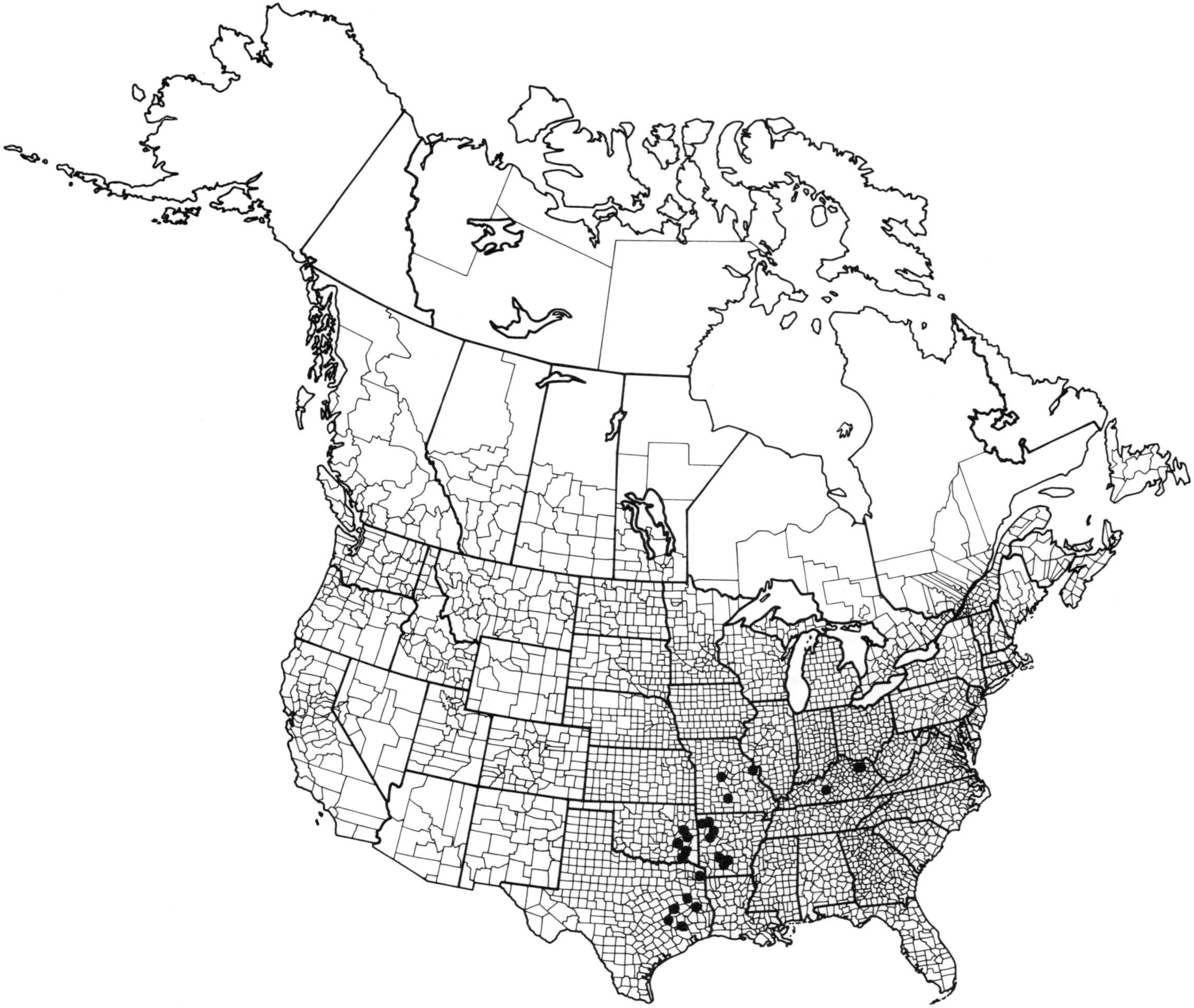

Map 207 *Simulium ozarkense*

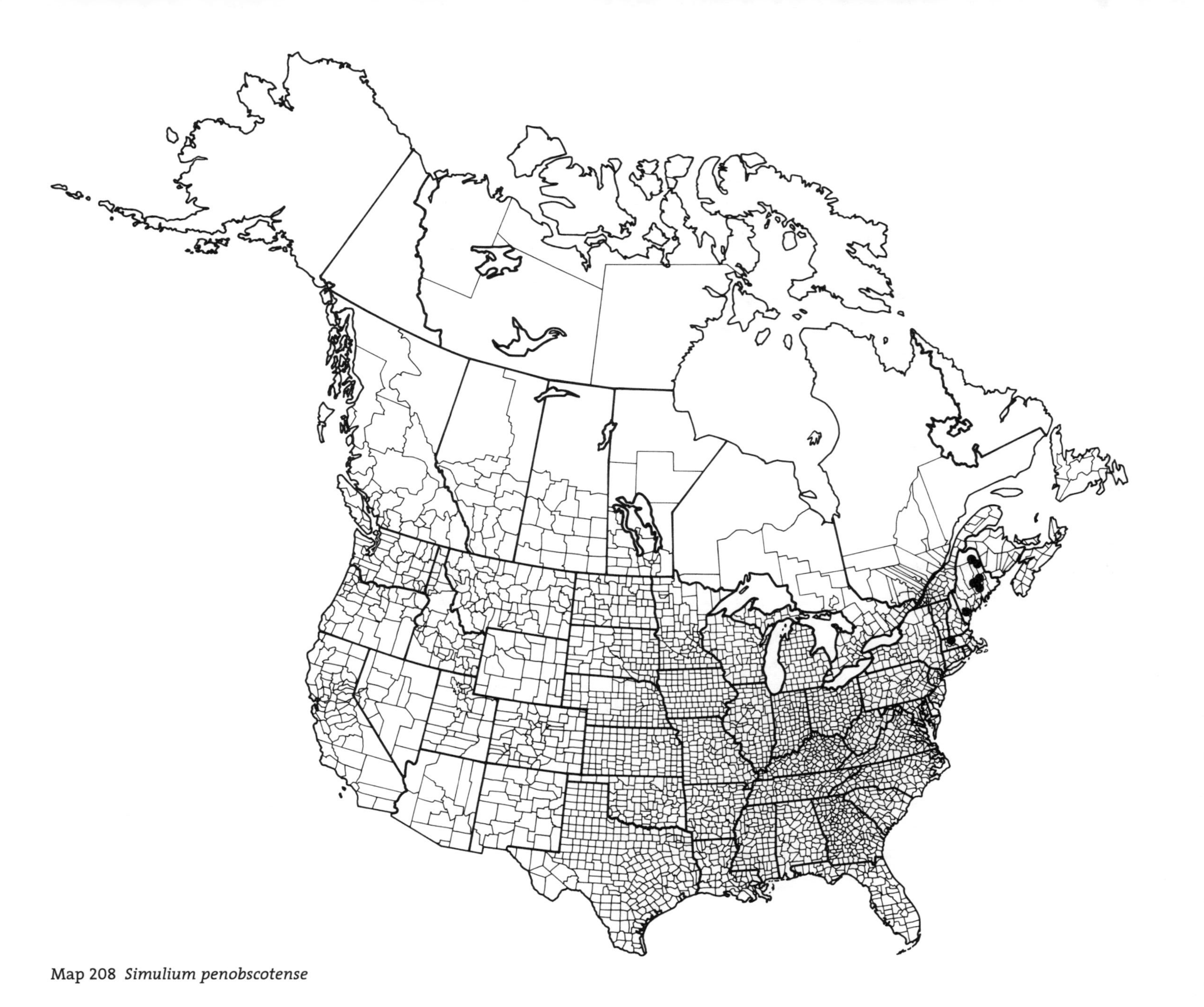

Map 208 *Simulium penobscotense*

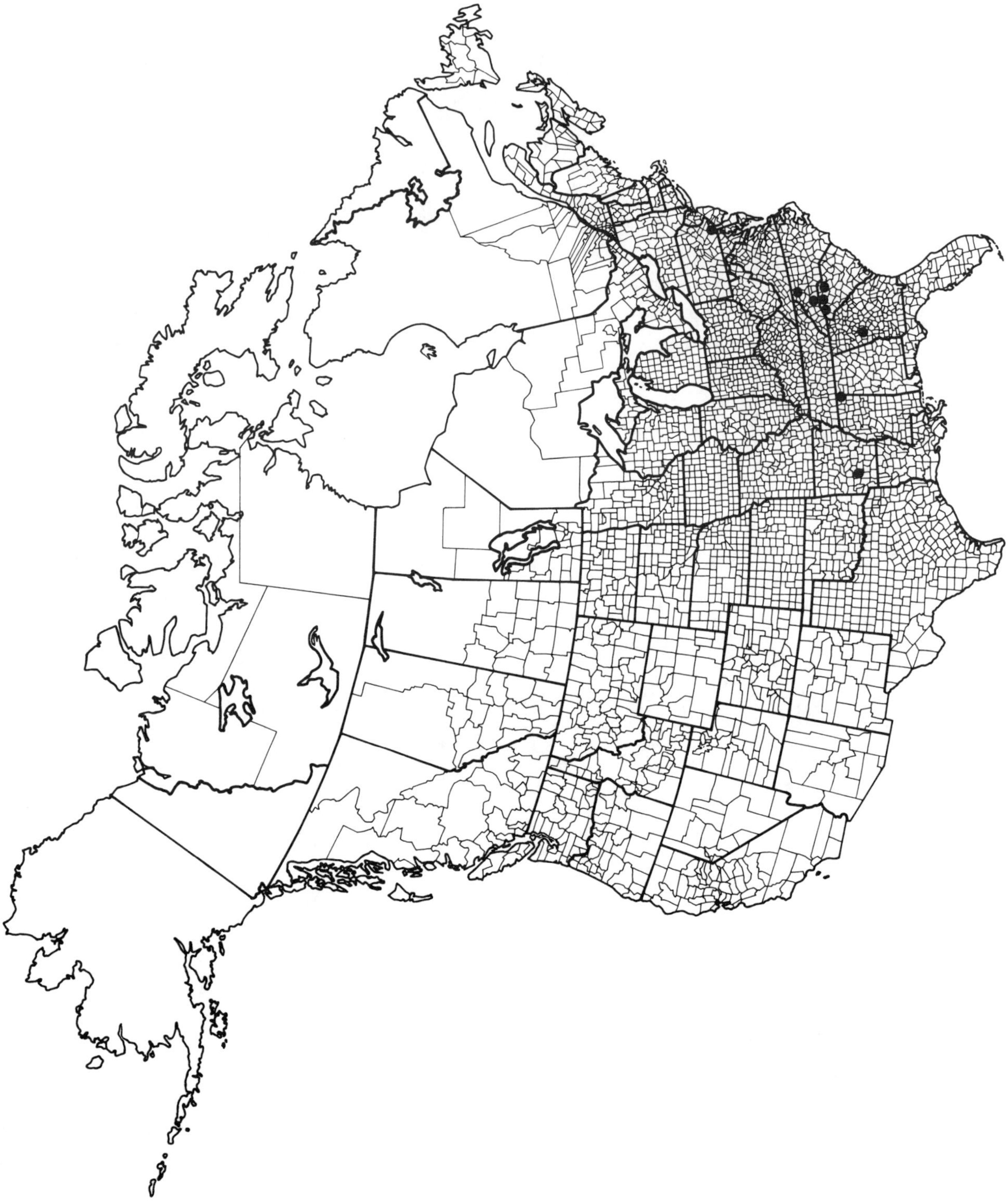

Map 209 *Simulium podostemi*

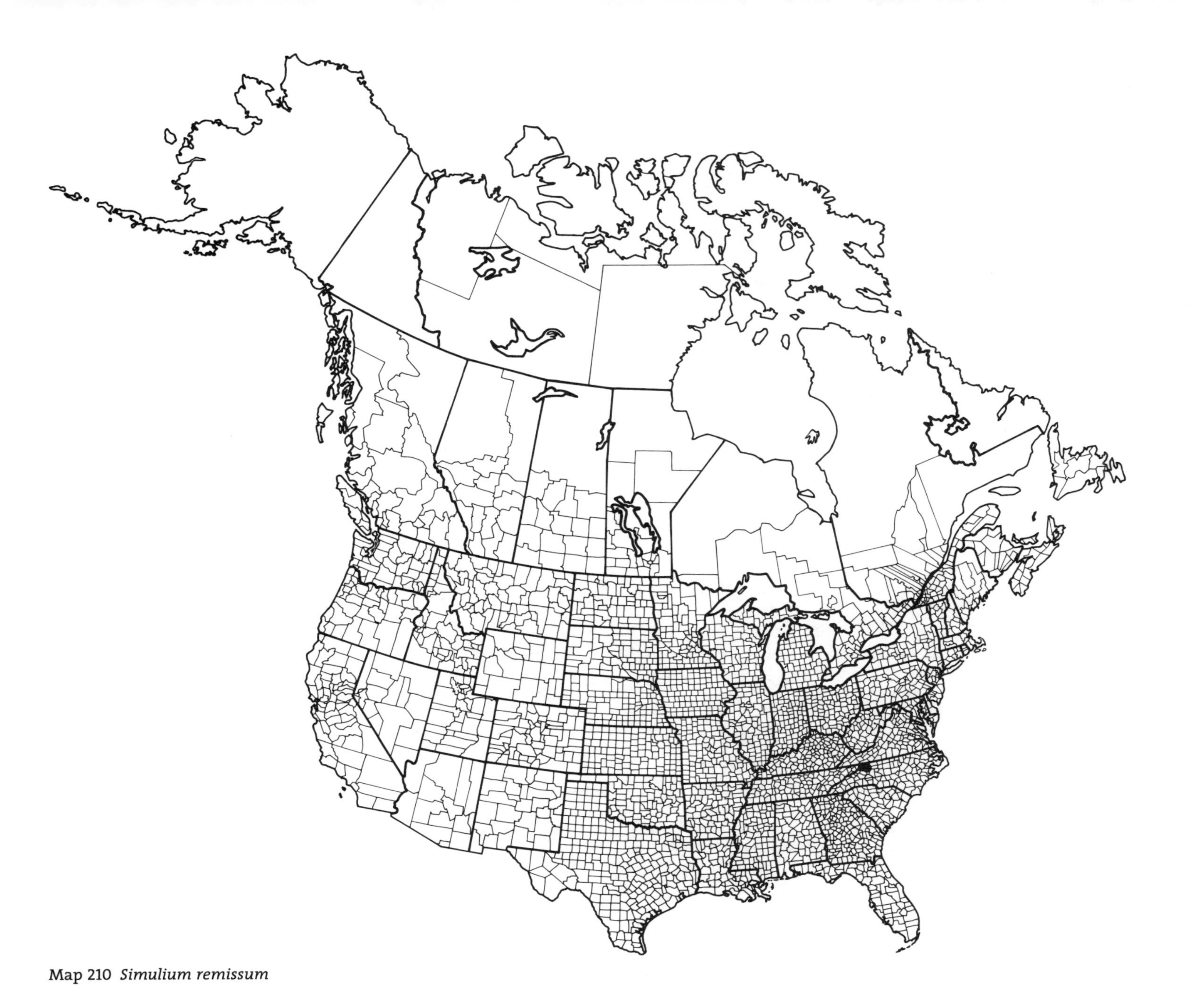

Map 210 *Simulium remissum*

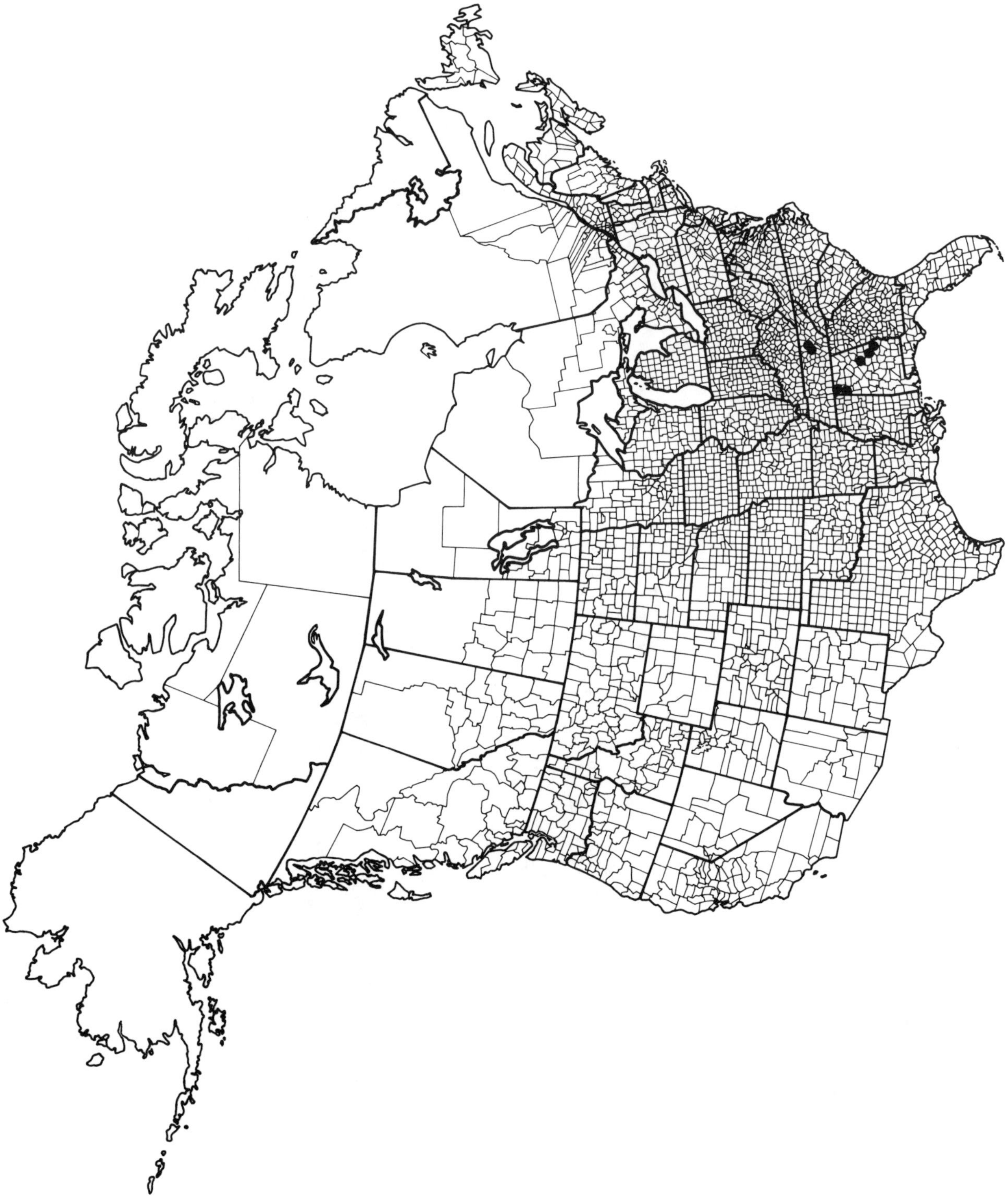

Map 211 *Simulium snowi*

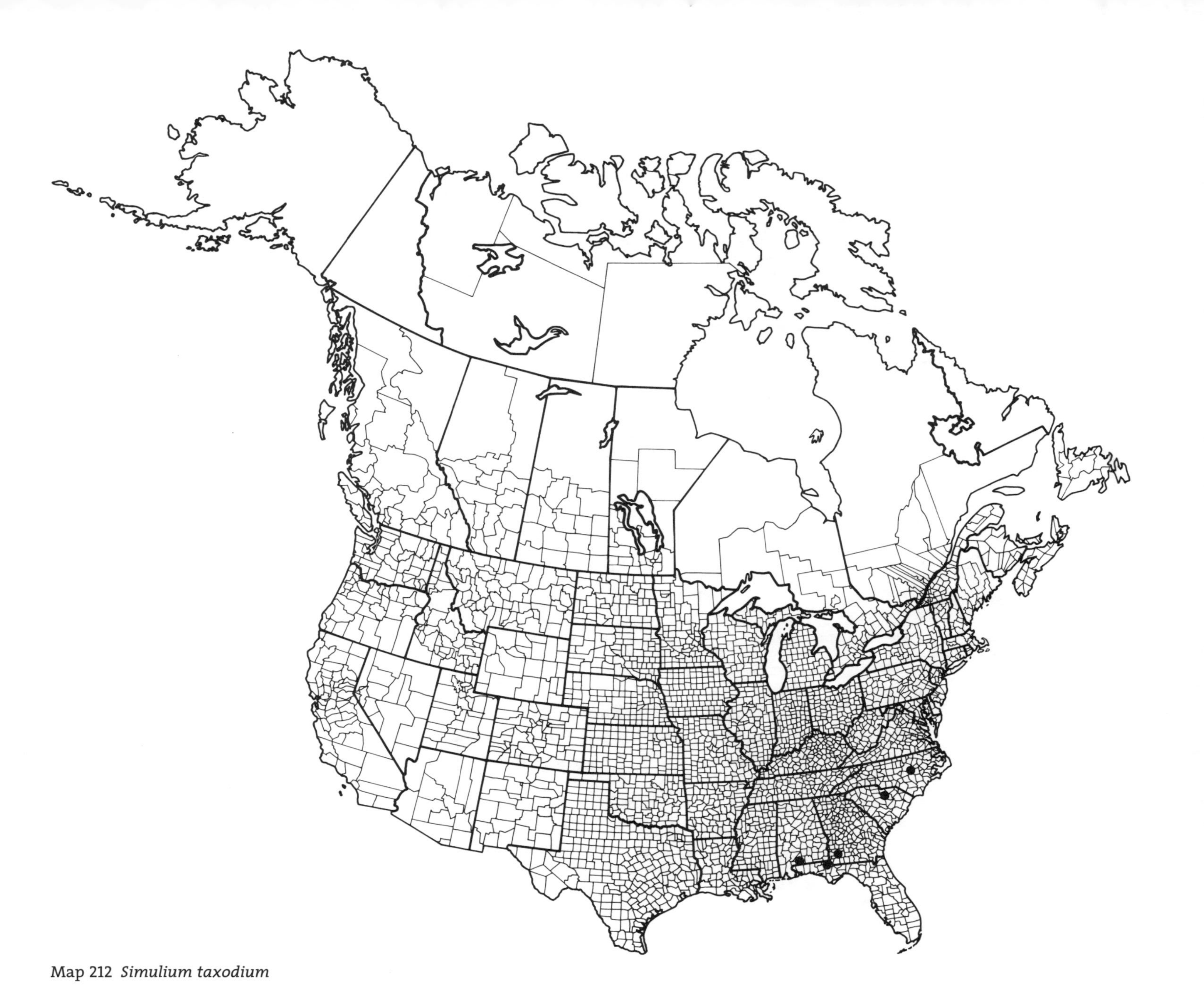

Map 212 *Simulium taxodium*

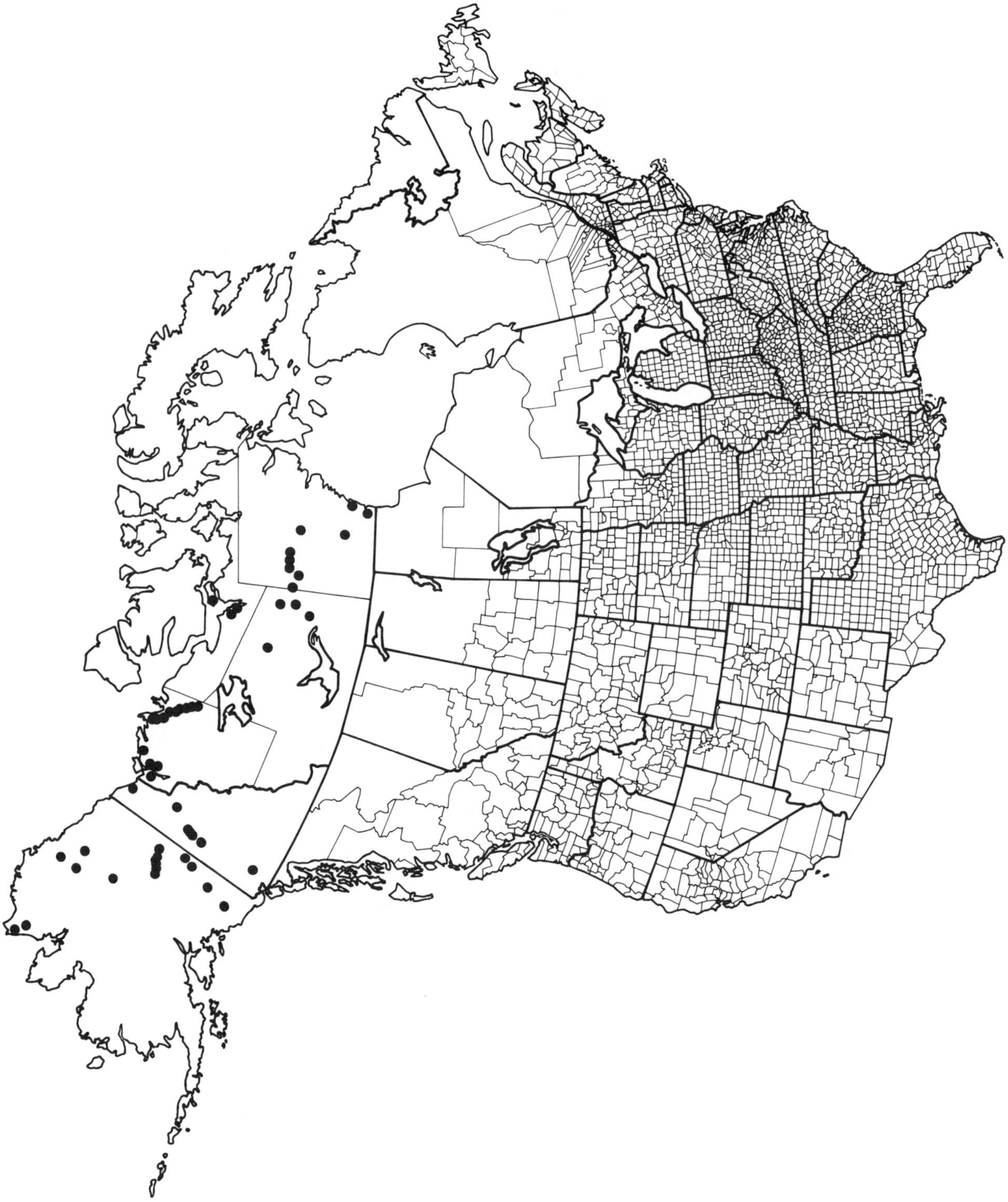

Map 213 *Simulium decimatum*

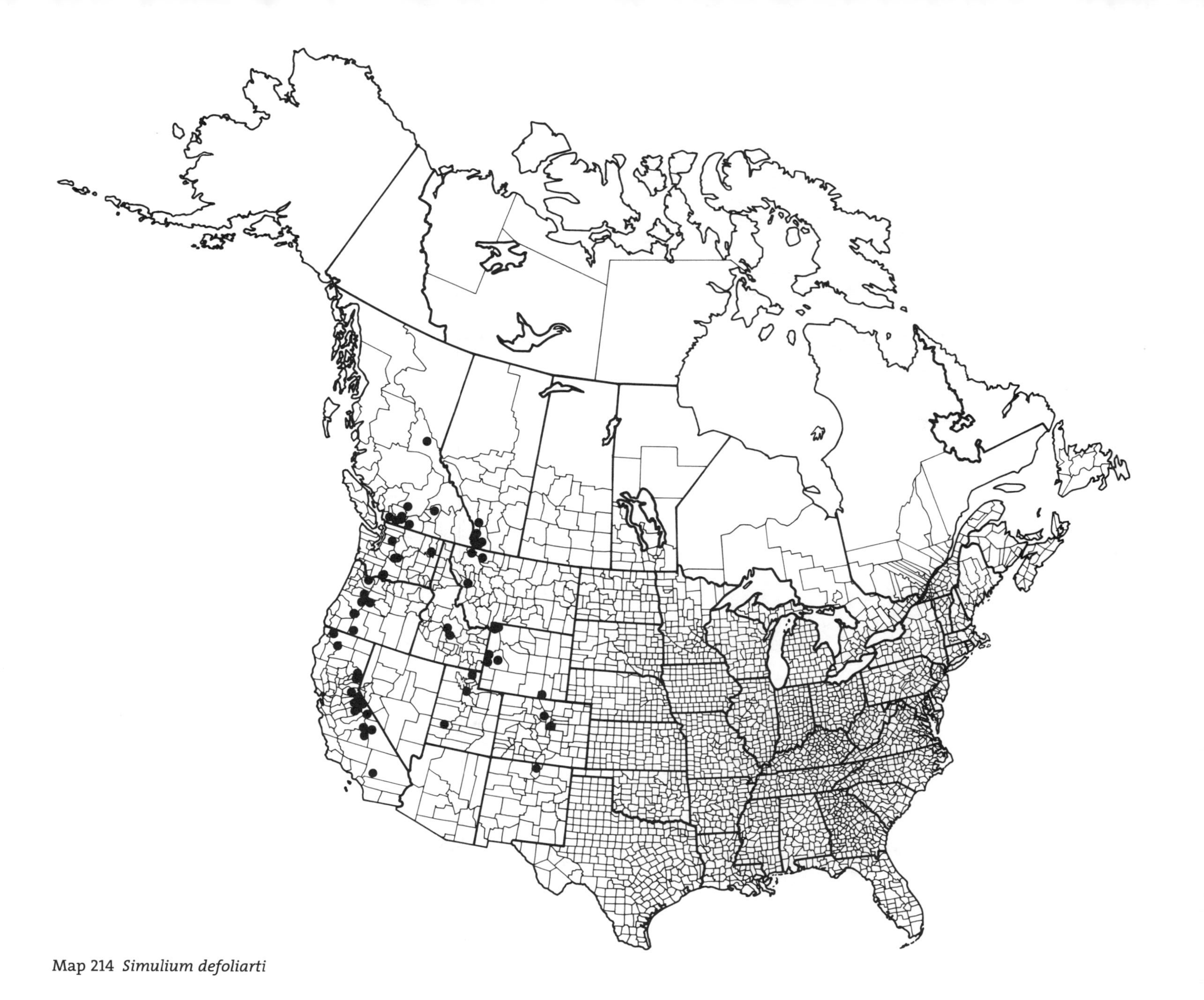

Map 214 *Simulium defoliarti*

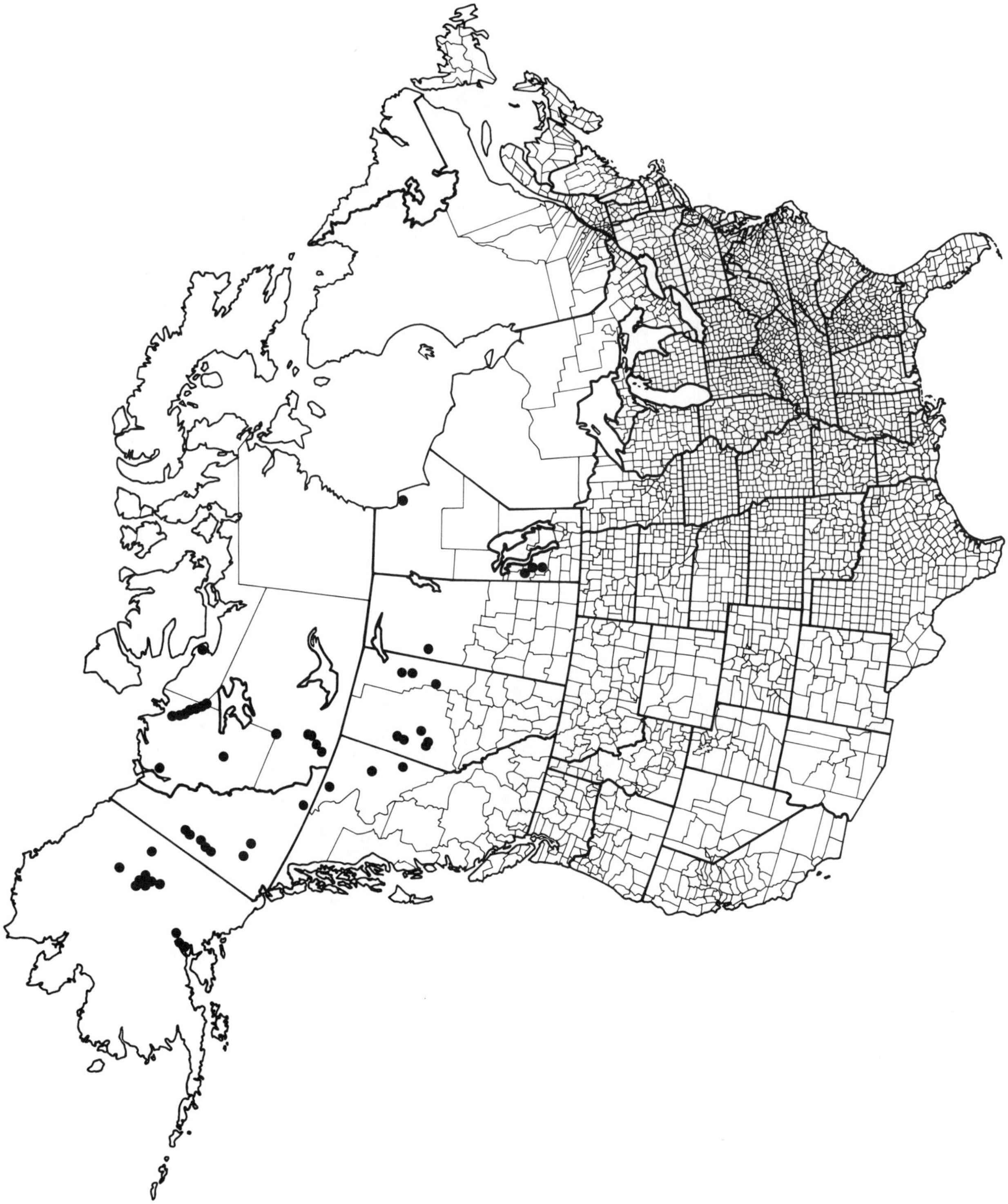

Map 215 *Simulium malyschevi*

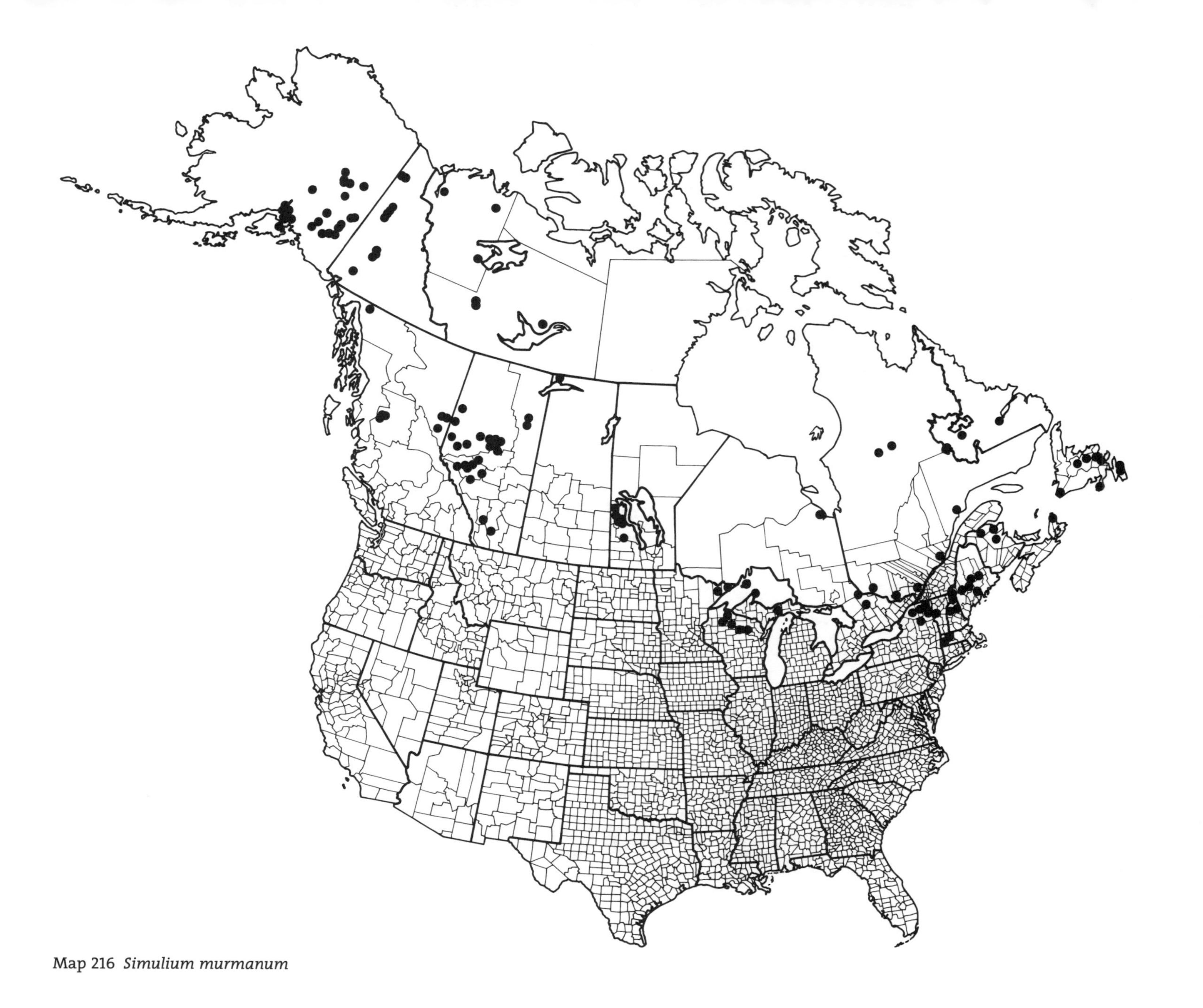

Map 216 *Simulium murmanum*

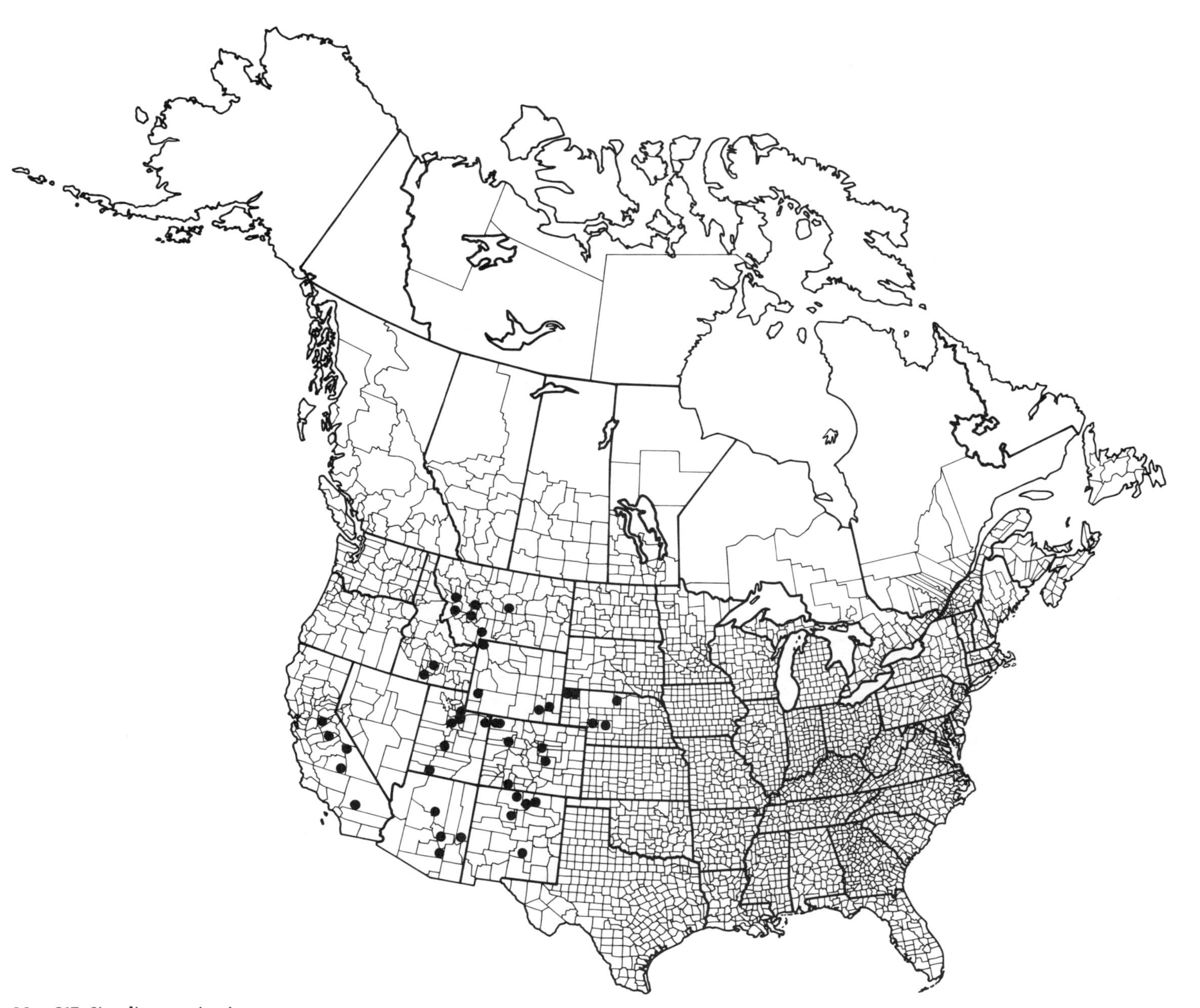

Map 217 *Simulium apricarium* n. sp.

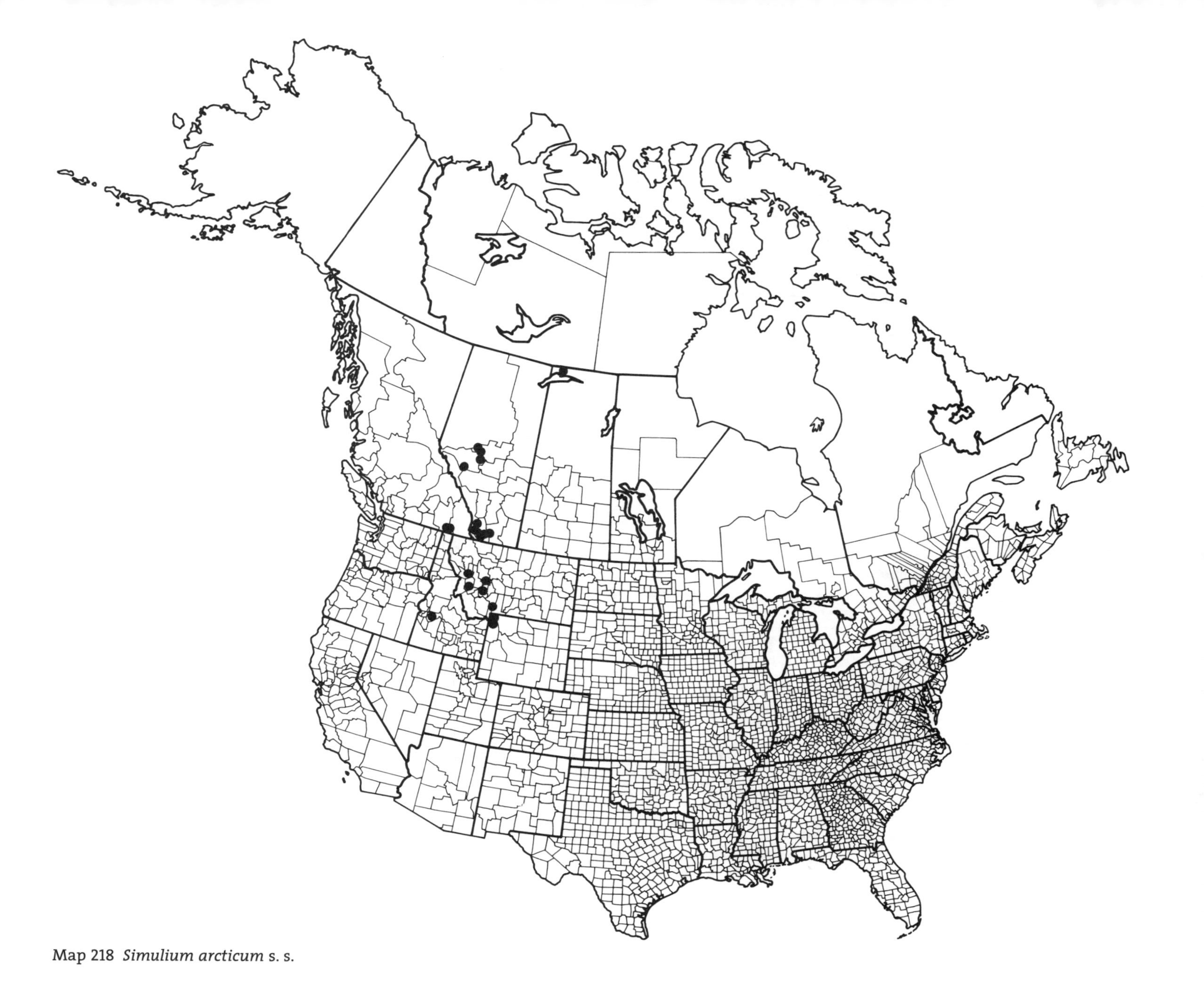

Map 218 *Simulium arcticum* s. s.

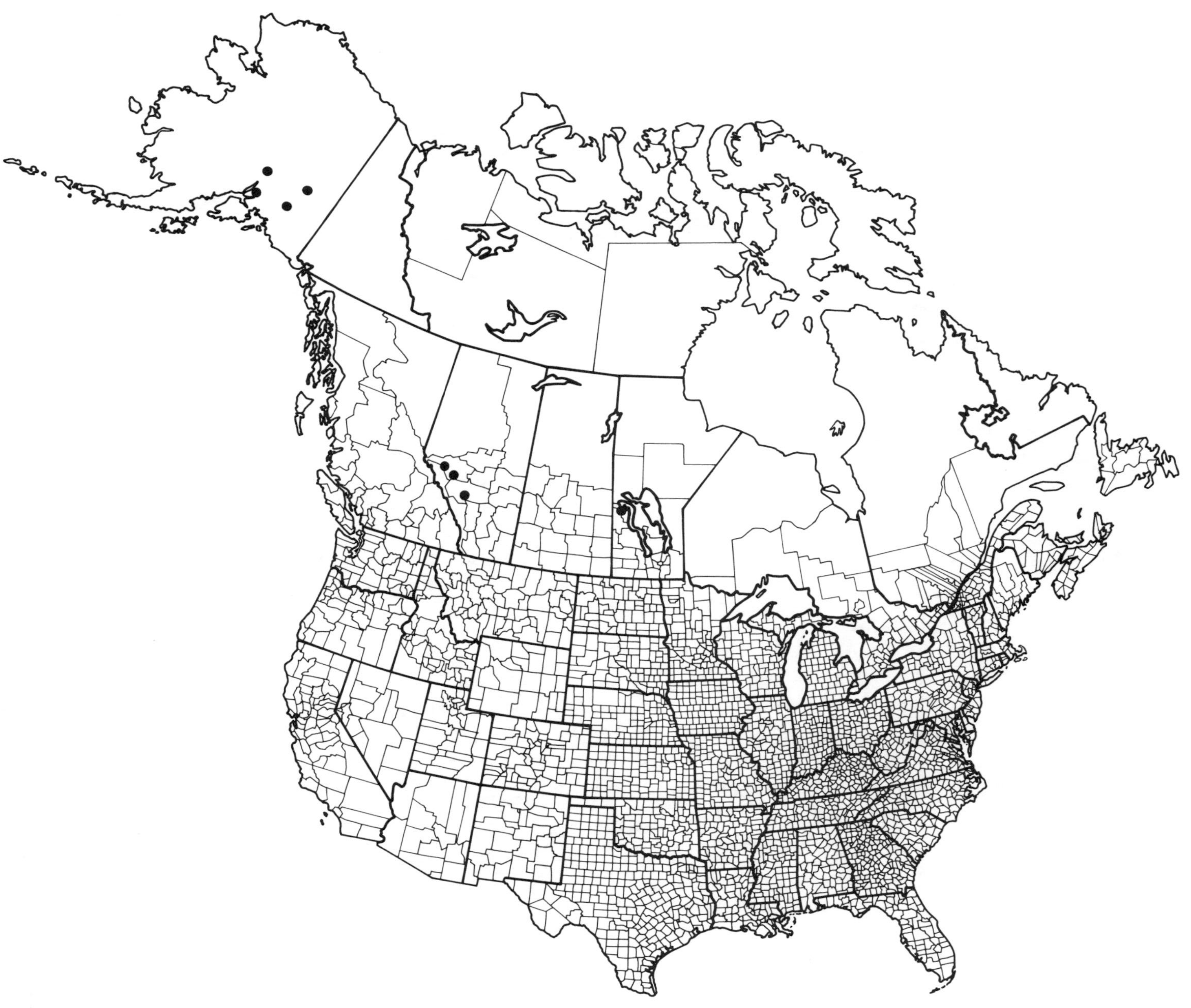

Map 219 *Simulium arcticum* 'cytospecies IIL-1'

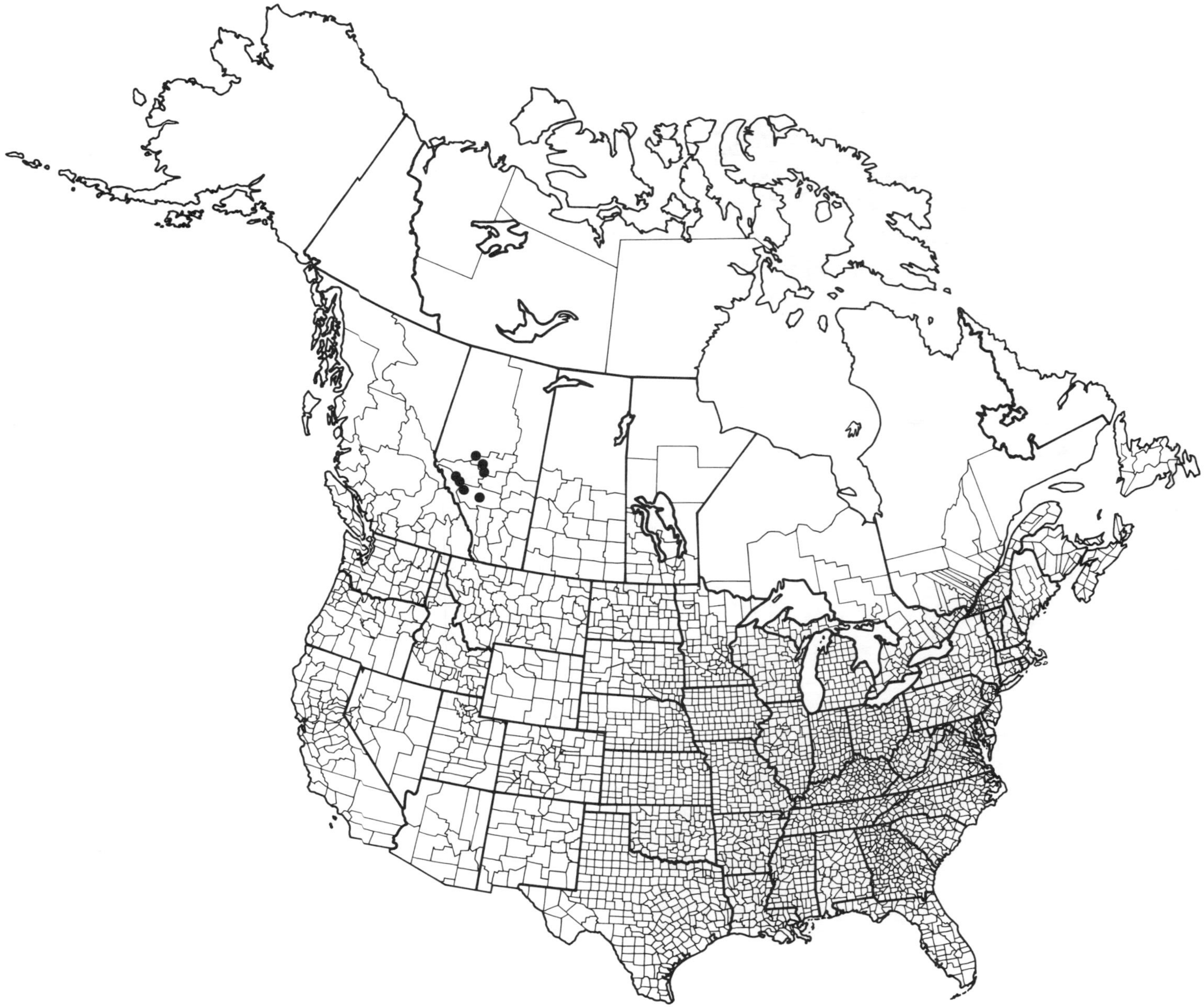

Map 220 *Simulium arcticum* 'cytospecies IIS-4'

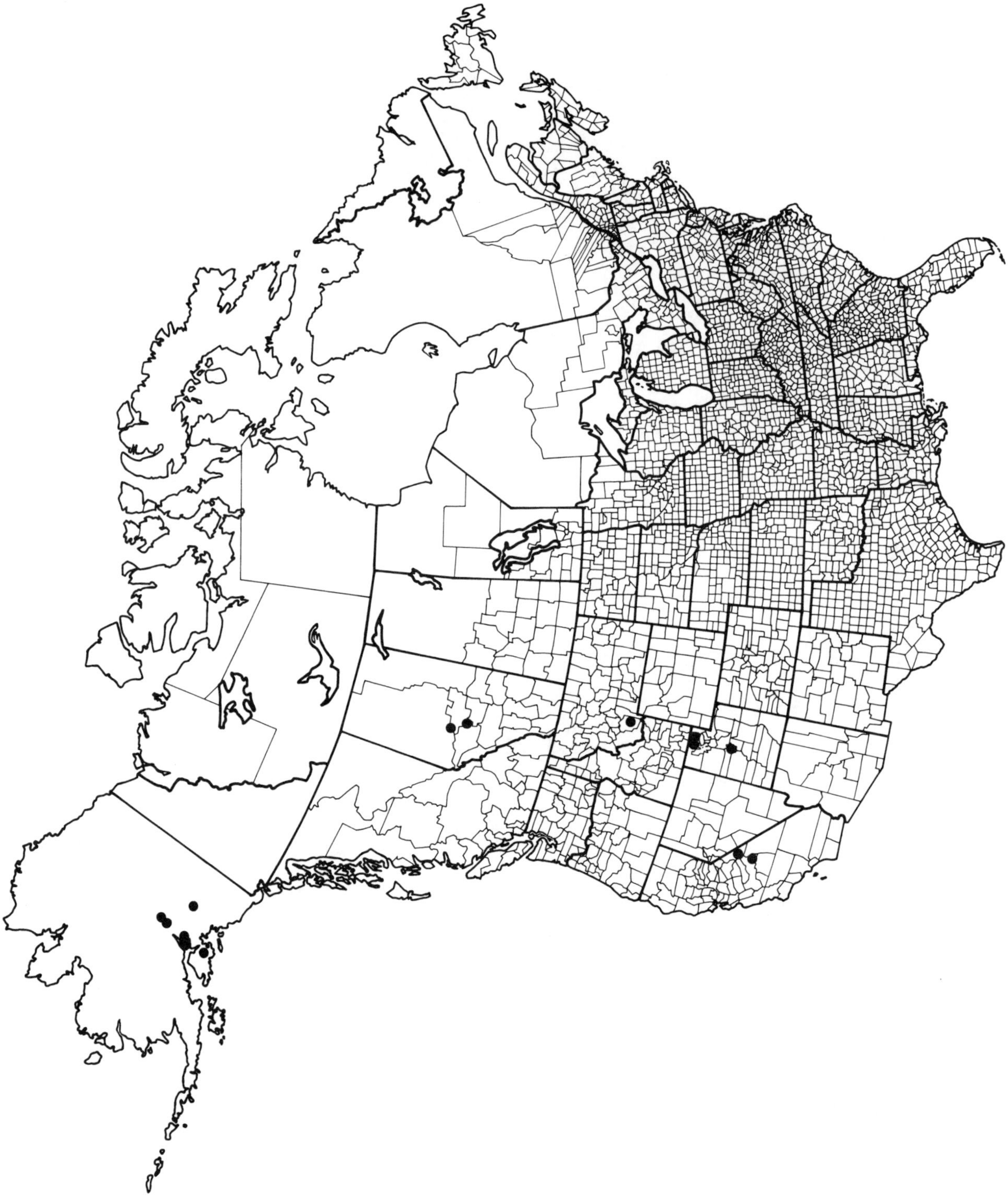

Map 221 *Simulium brevicercum*

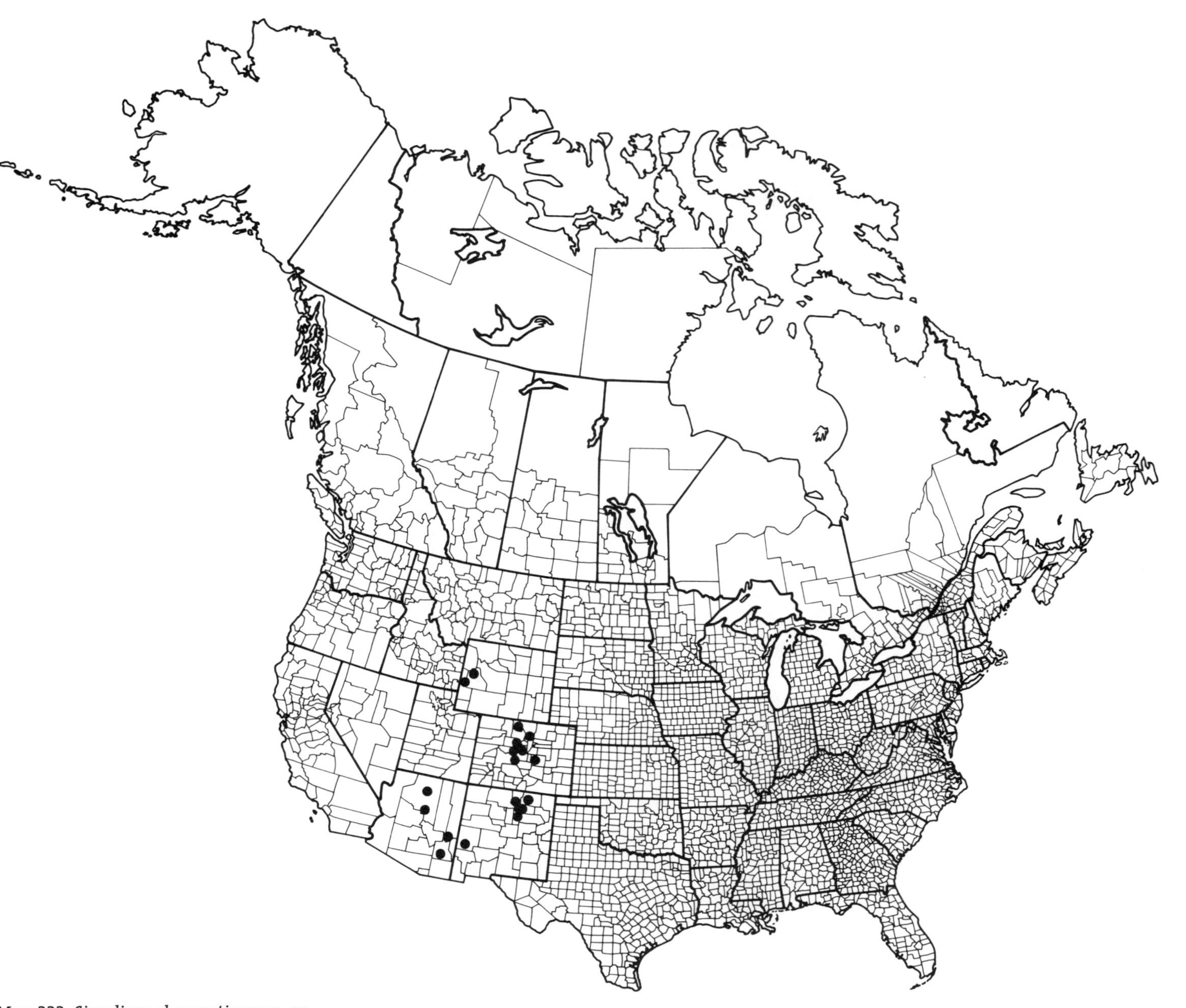

Map 222 *Simulium chromatinum* n. sp.

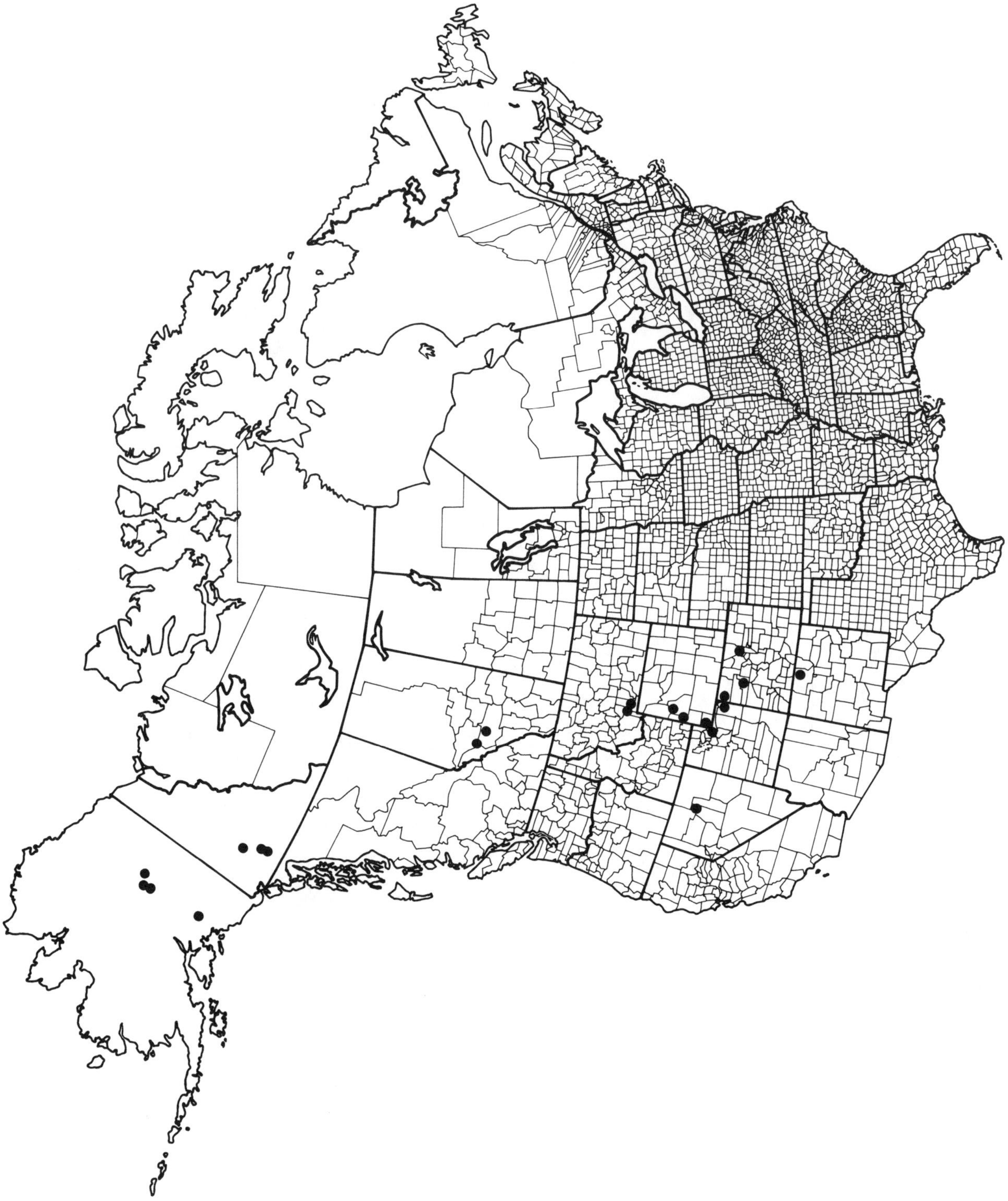

Map 223 *Simulium negativum* n. sp.

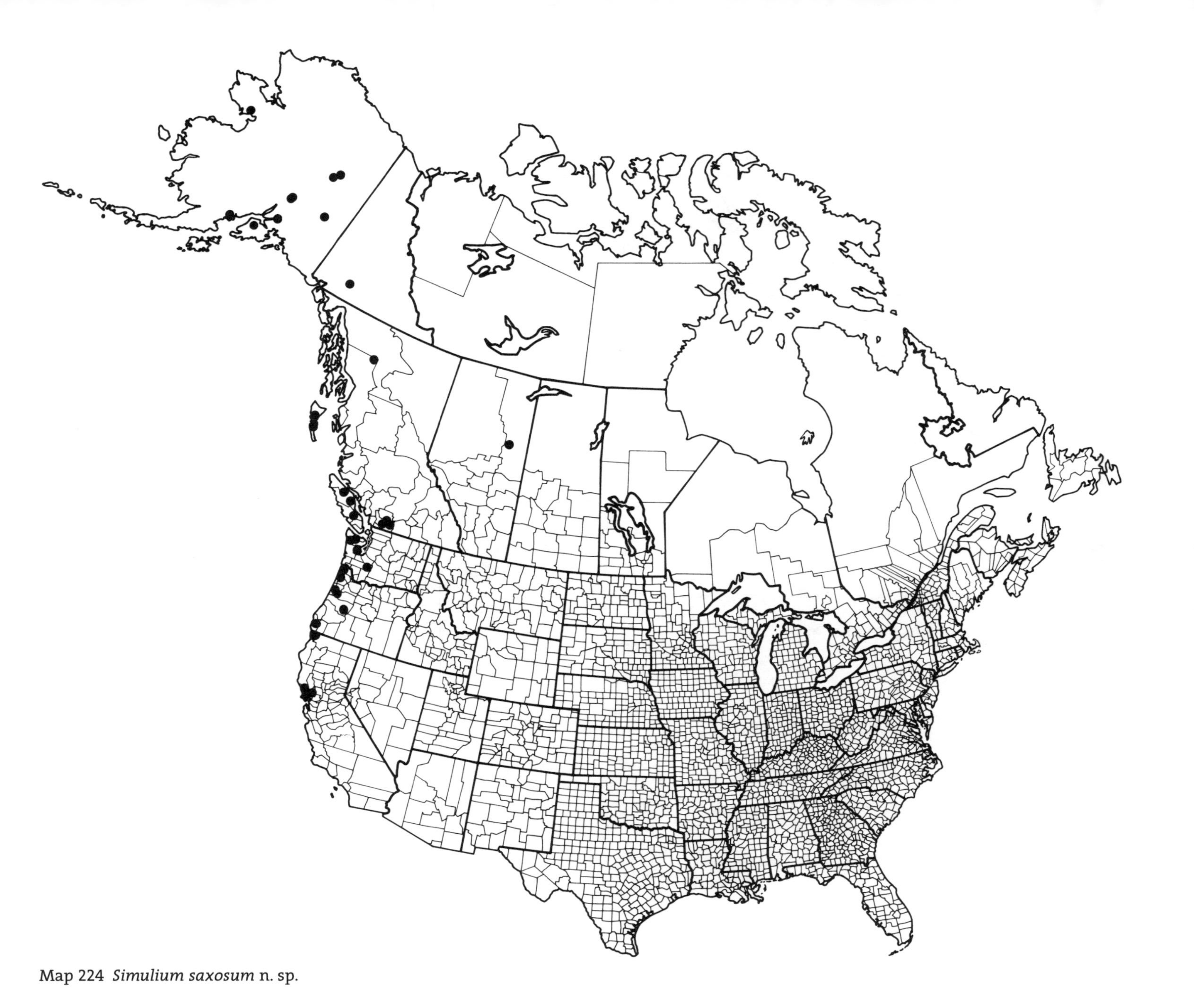

Map 224 *Simulium saxosum* n. sp.

Map 225 *Simulium vampirum*

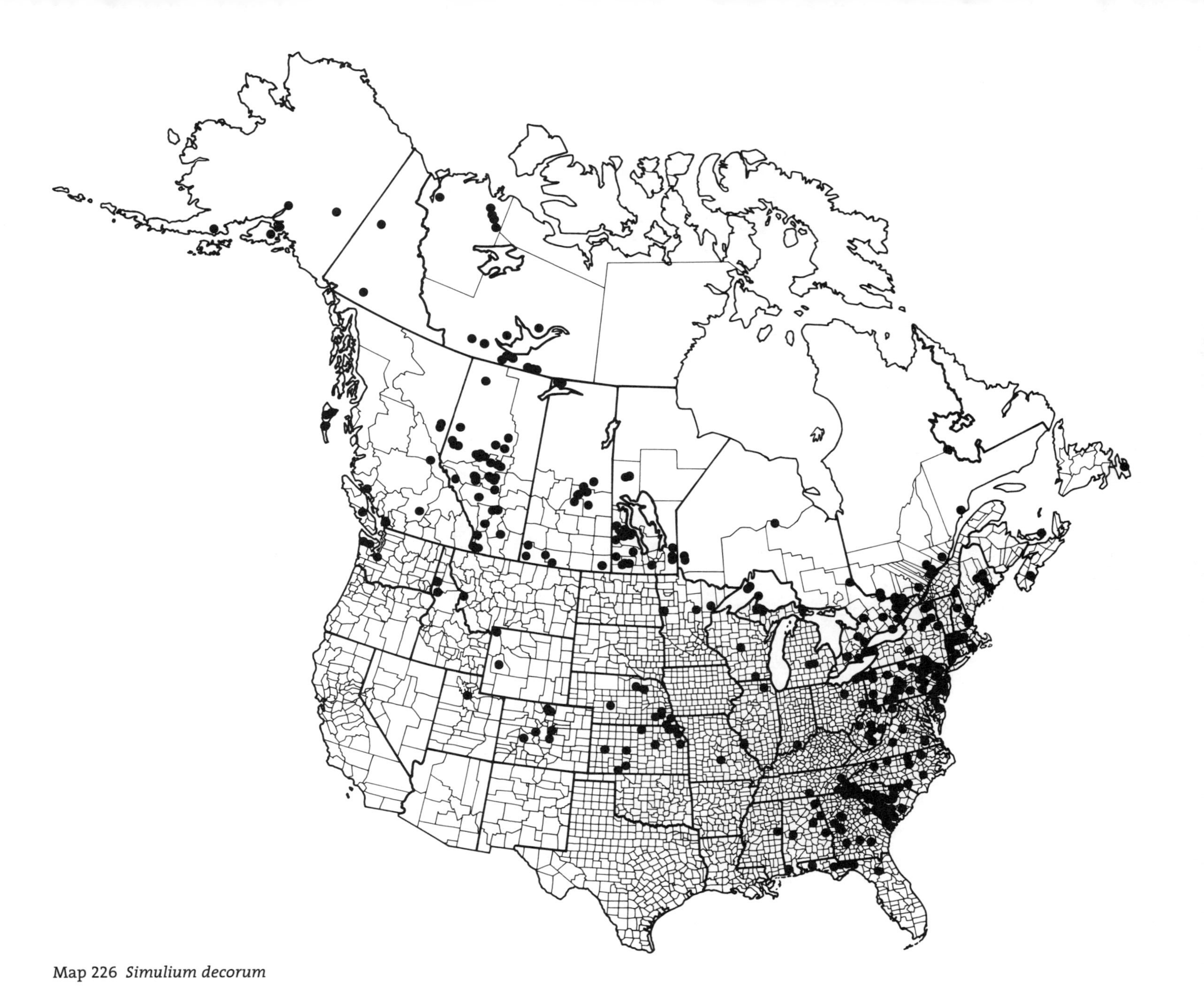

Map 226 *Simulium decorum*

Map 227 *Simulium noelleri*

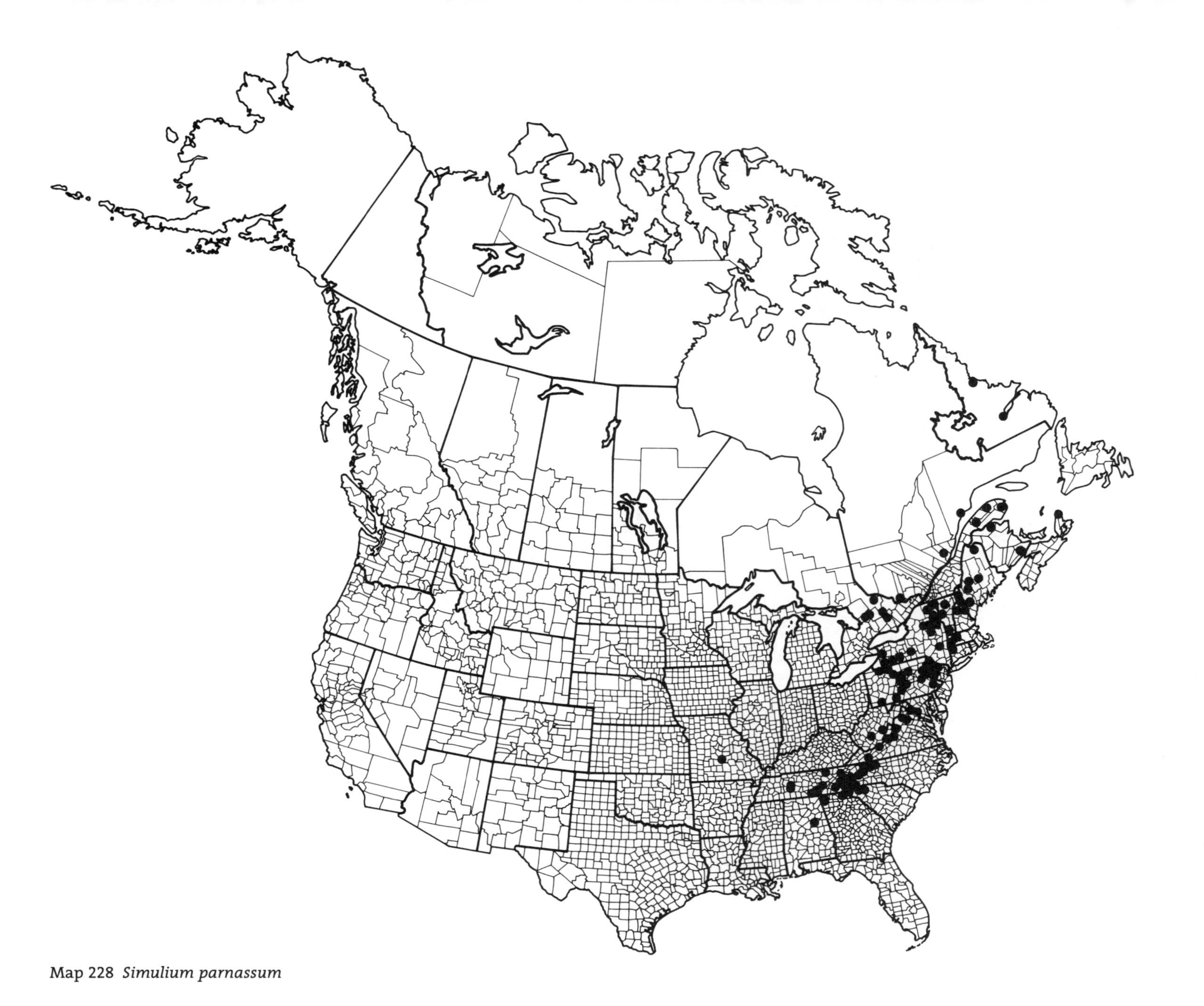

Map 228 *Simulium parnassum*

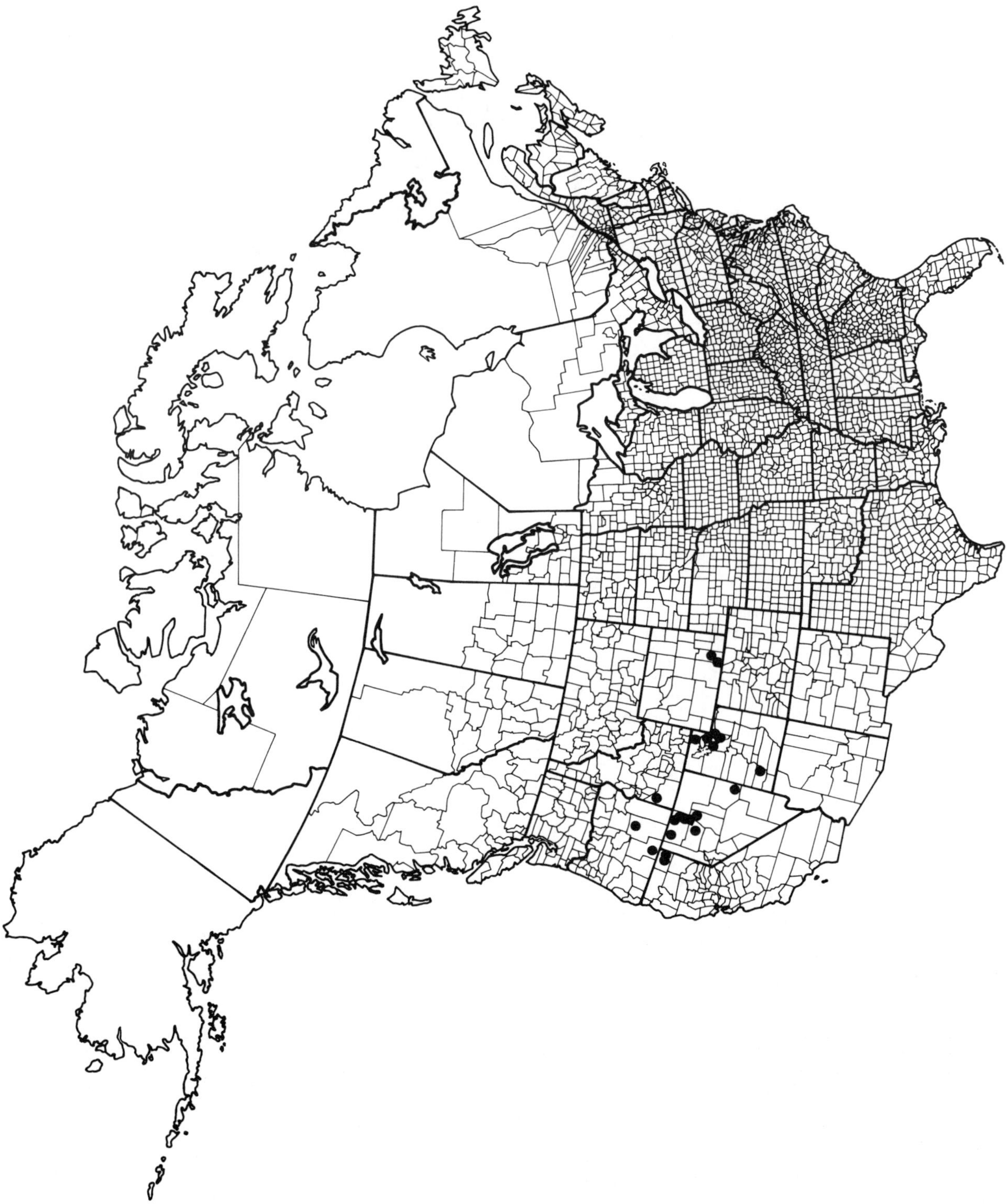

Map 229 *Simulium petersoni*

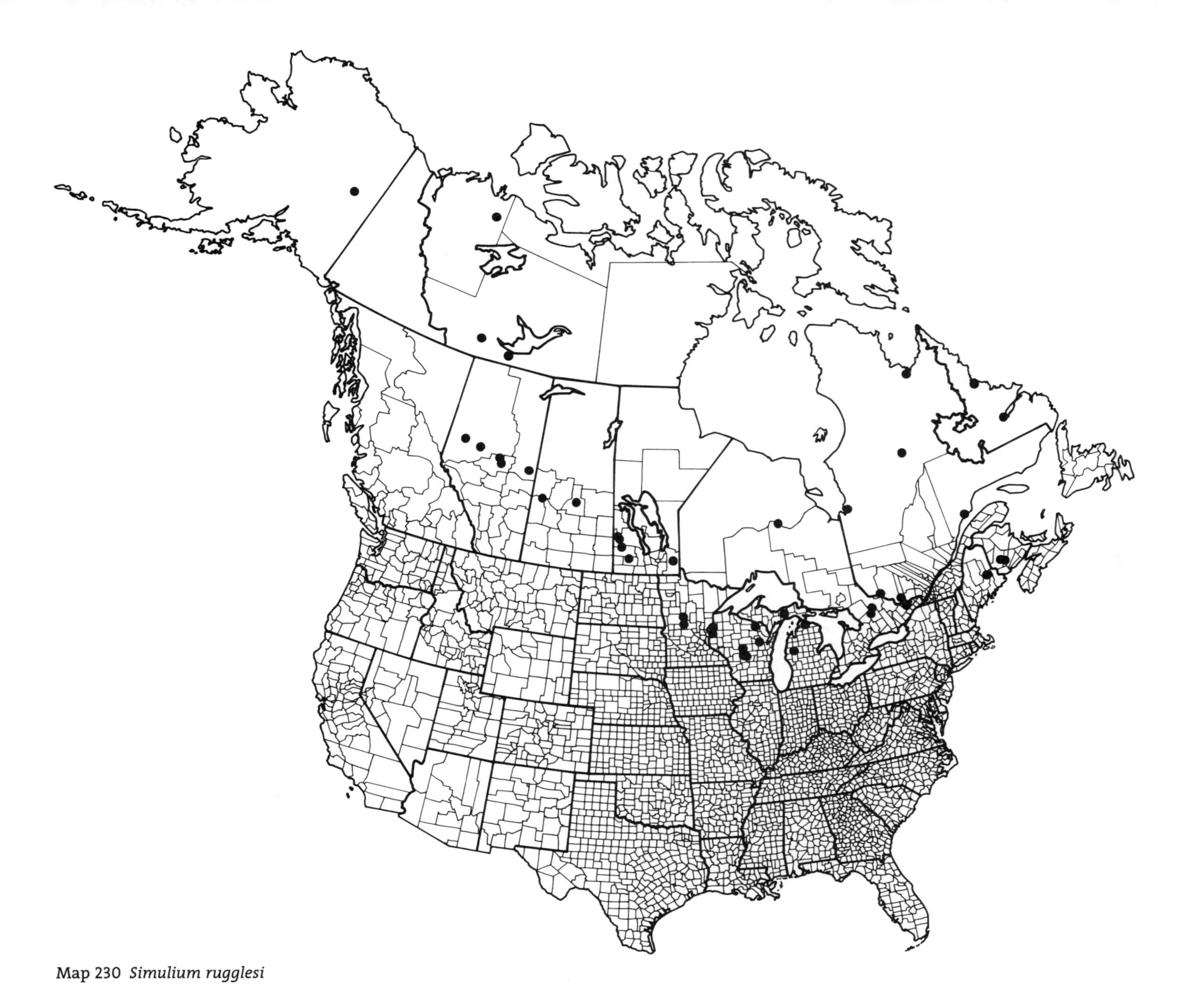

Map 230 *Simulium rugglesi*

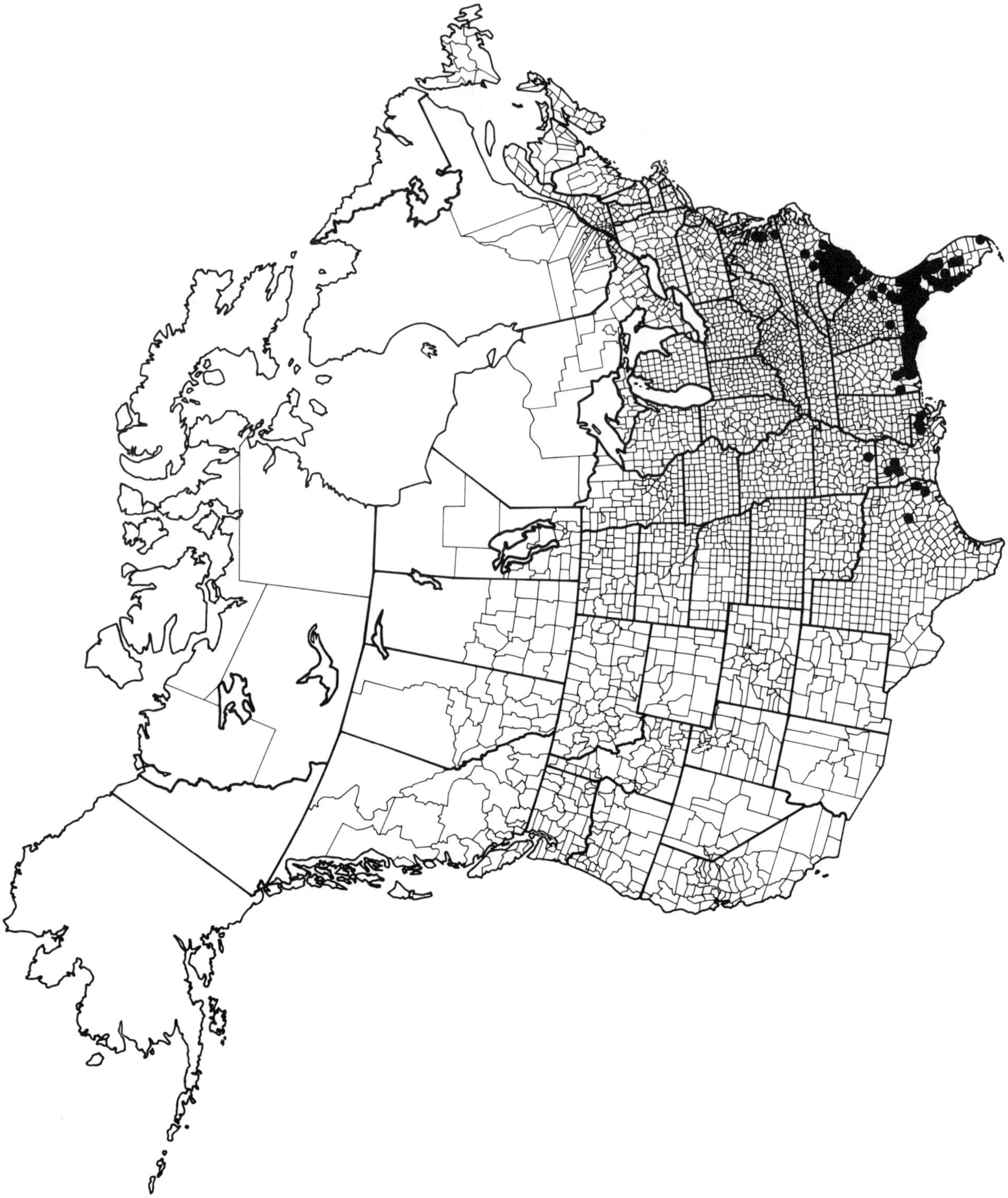

Map 231 *Simulium slossonae*

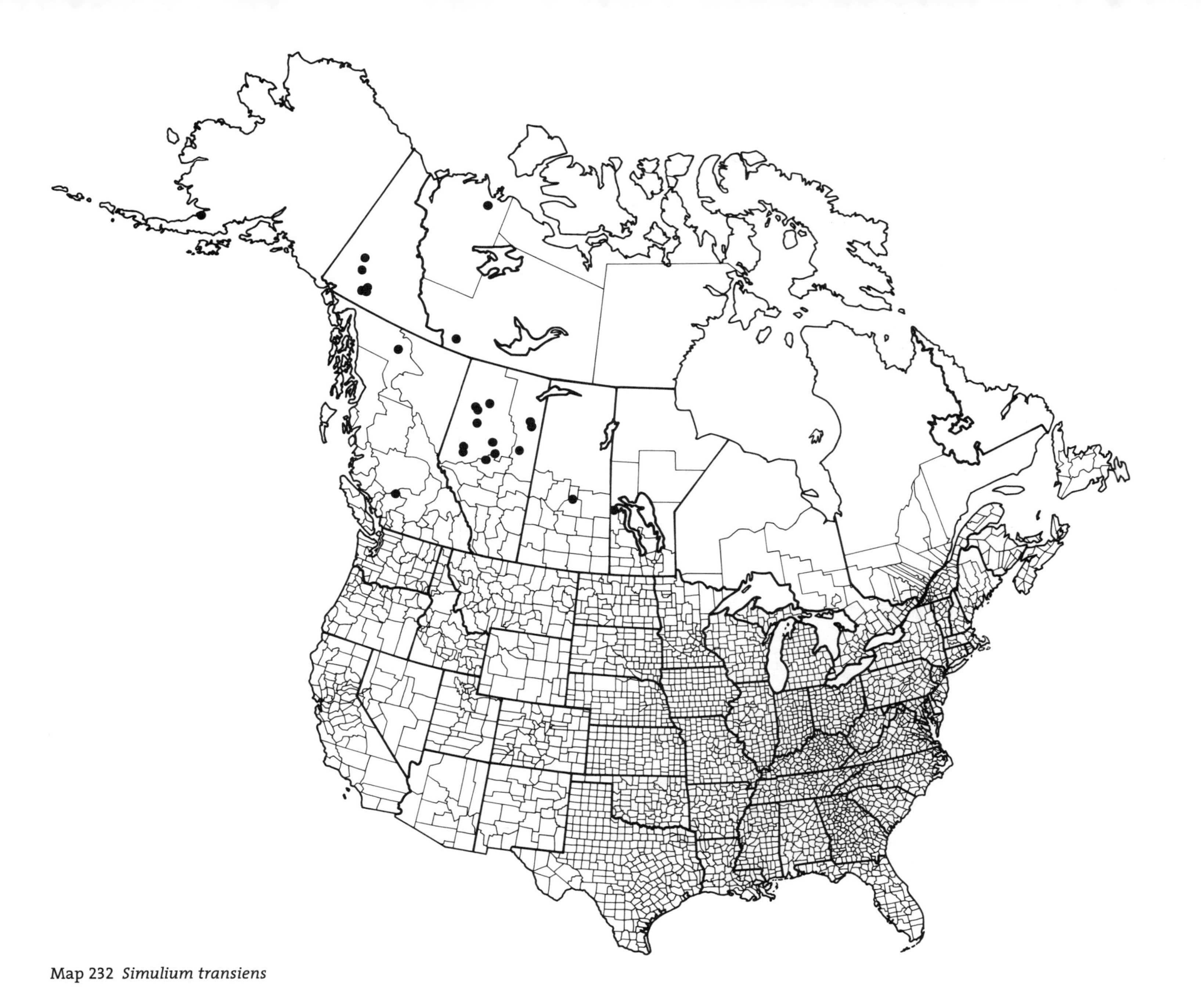

Map 232 *Simulium transiens*

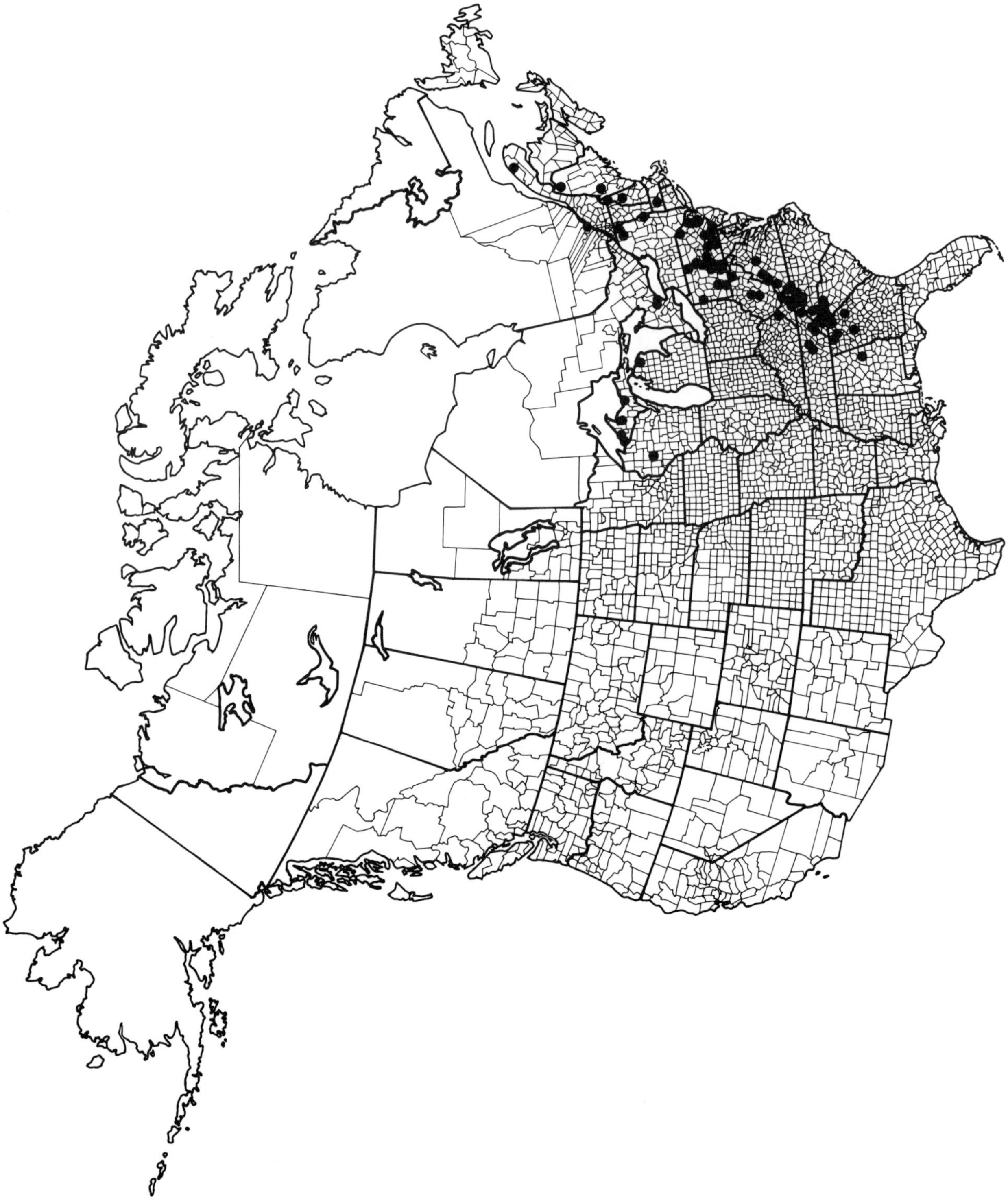

Map 233 *Simulium appalachiense* n. sp.

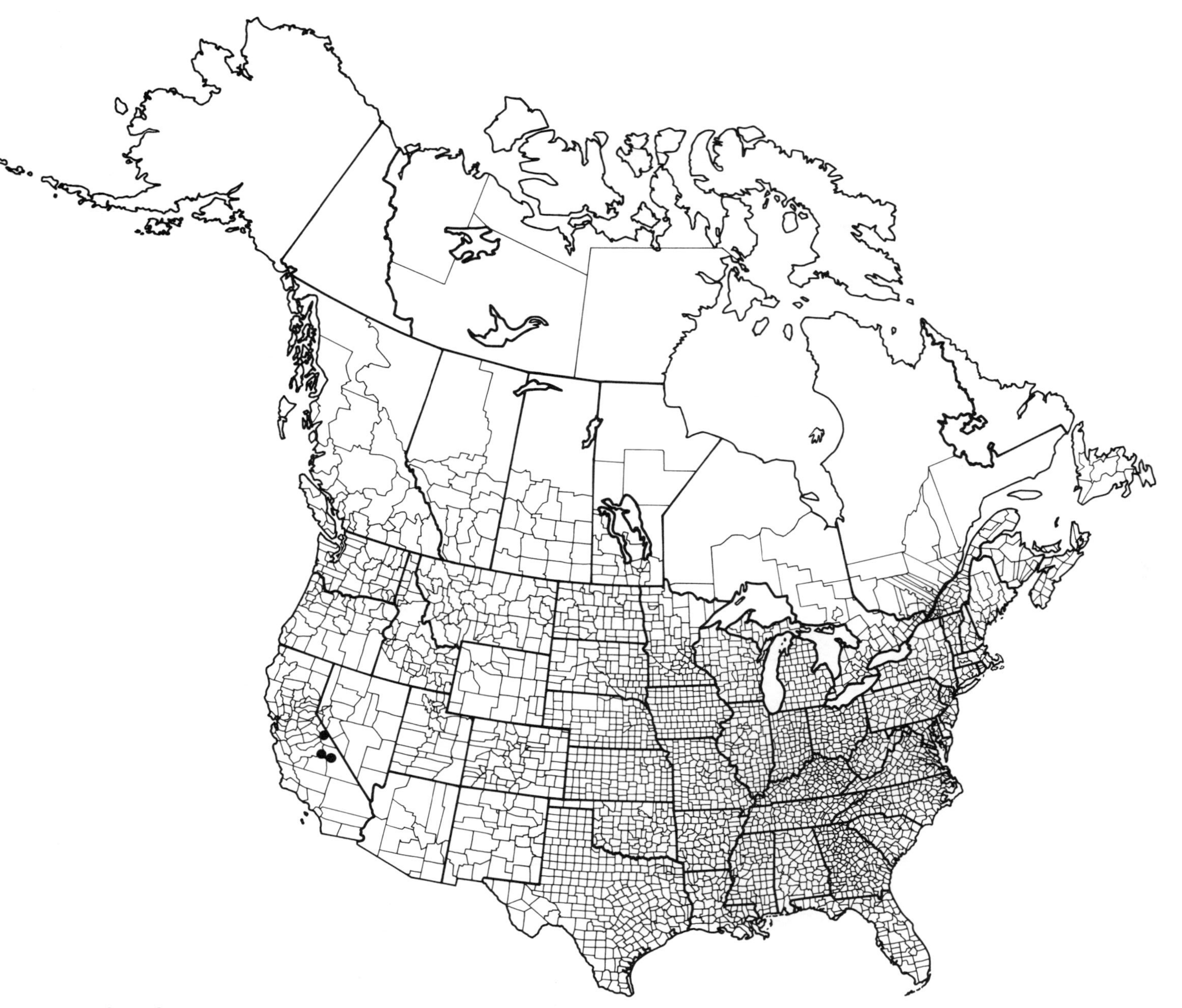

Map 234 *Simulium chromocentrum* n. sp.

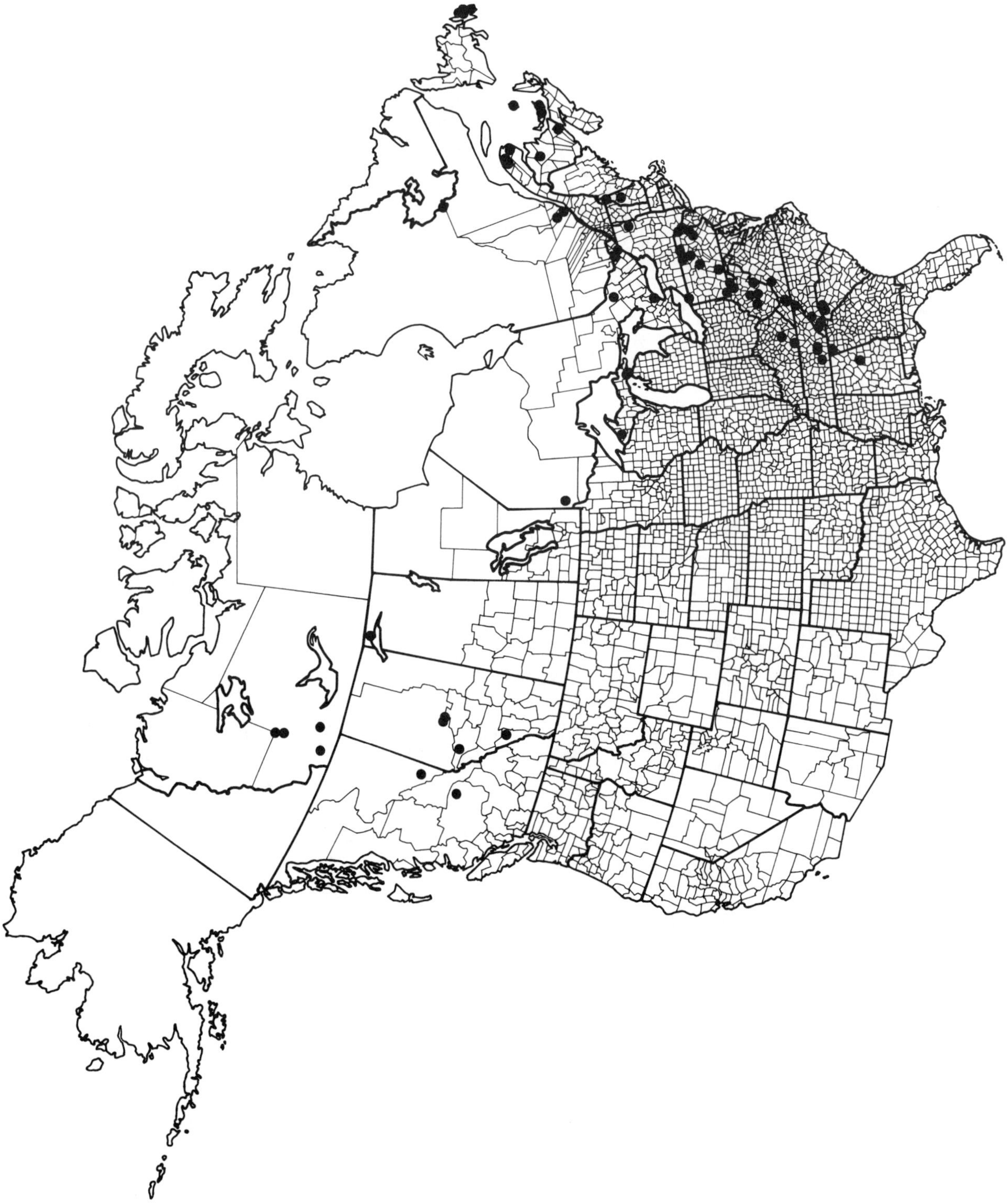

Map 235 *Simulium conundrum* n. sp.

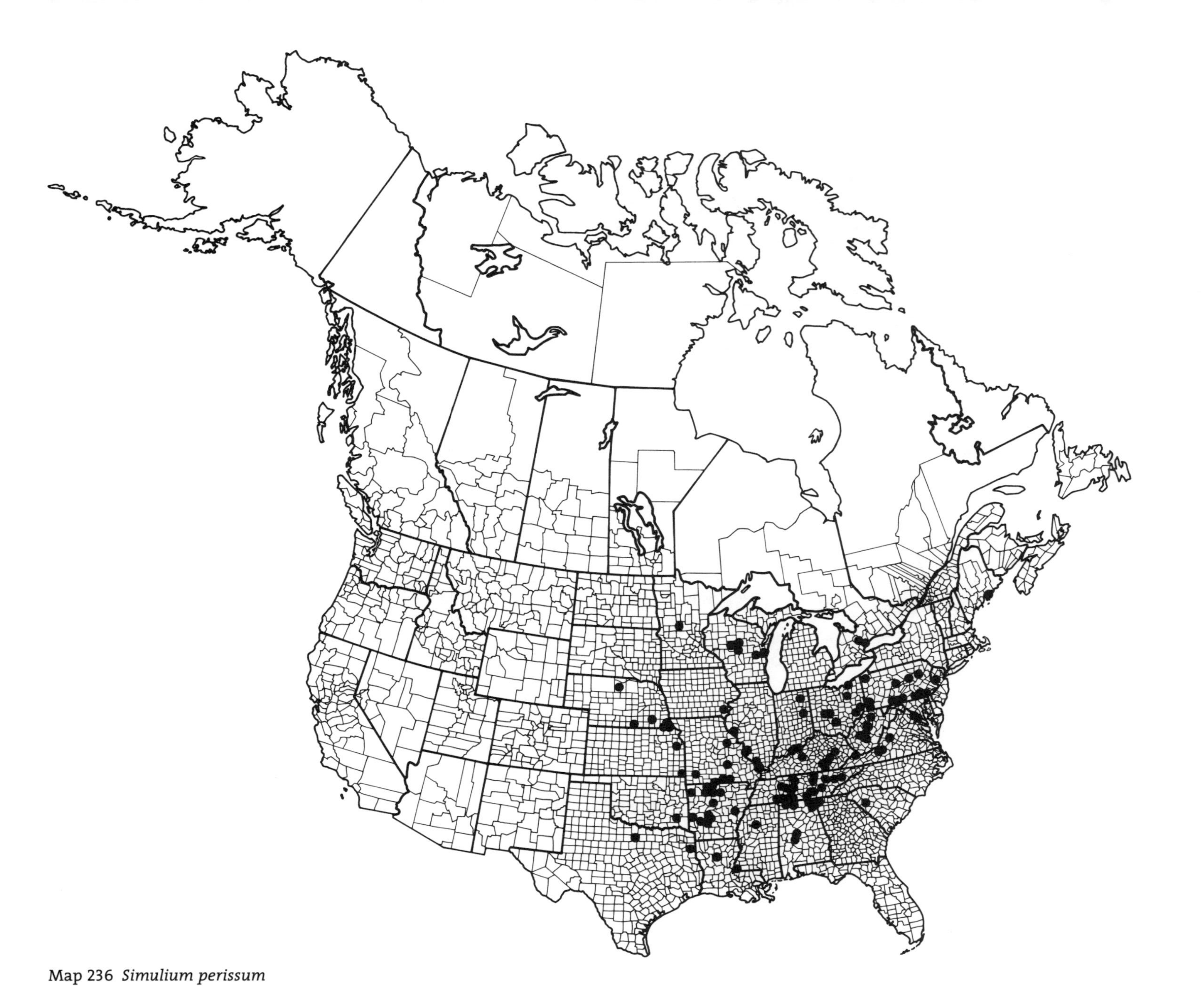

Map 236 *Simulium perissum*

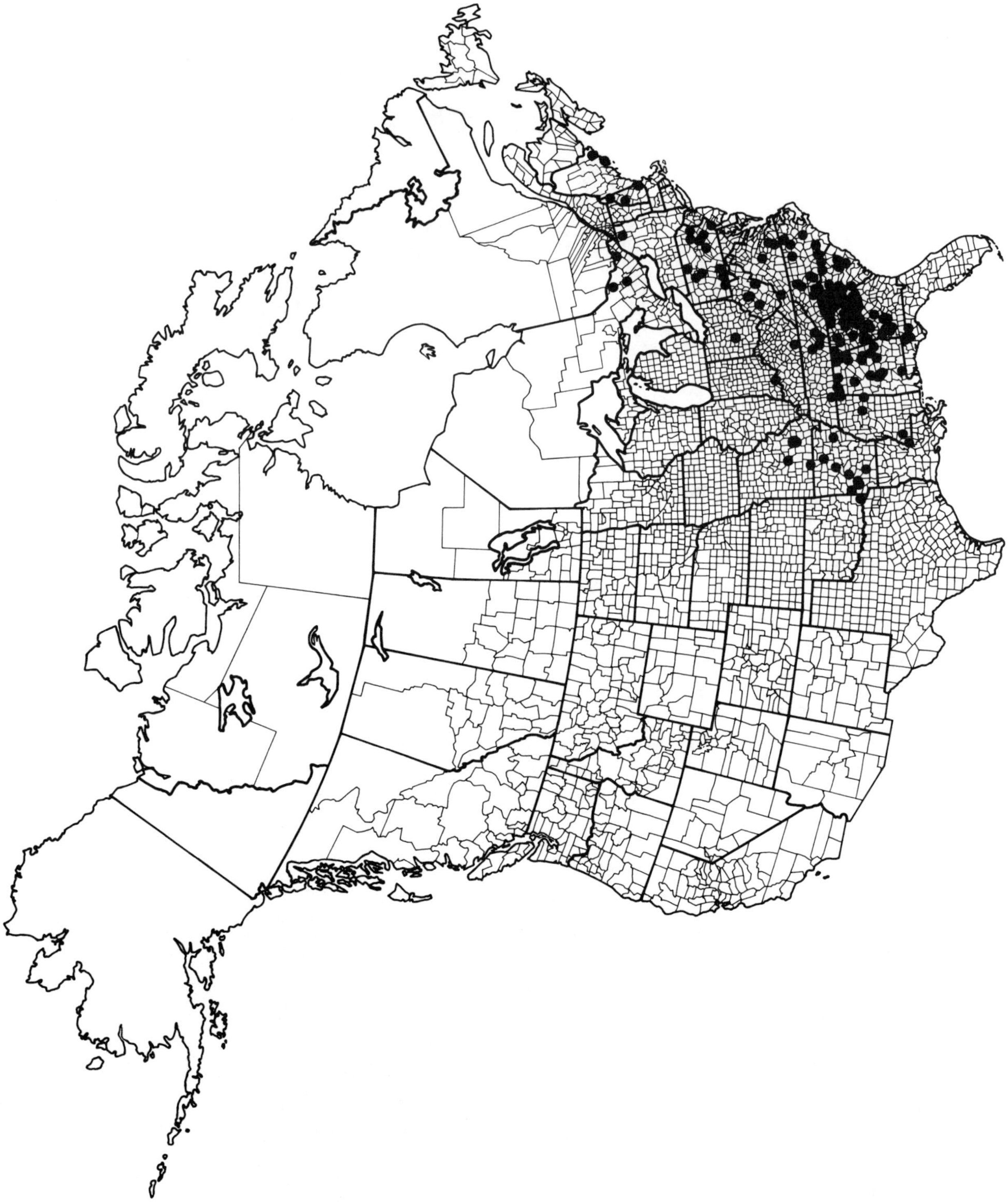

Map 237a *Simulium tuberosum* s. s. 'cytotype A'

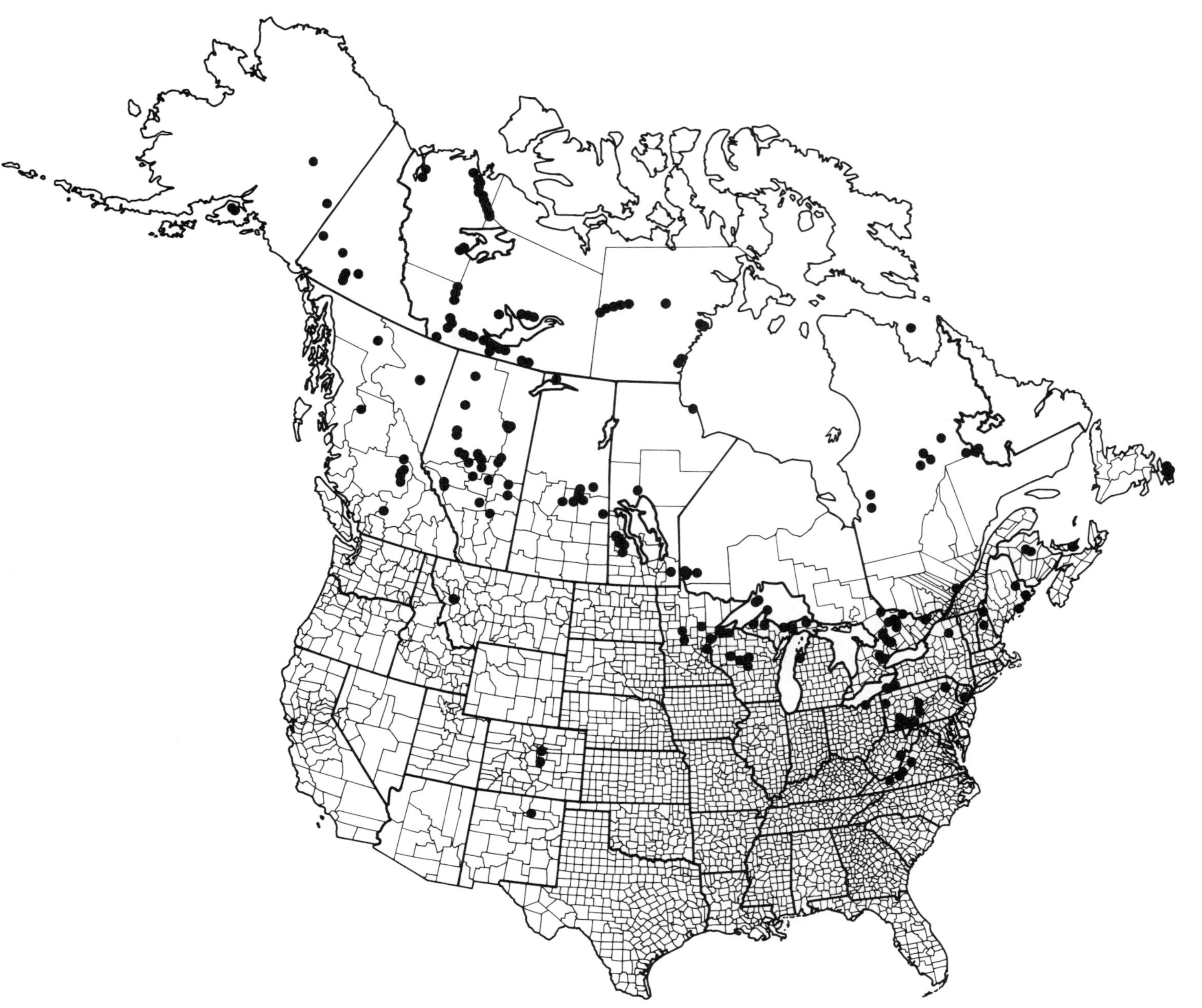

Map 237b *Simulium tuberosum* s. s. 'cytotype AB'

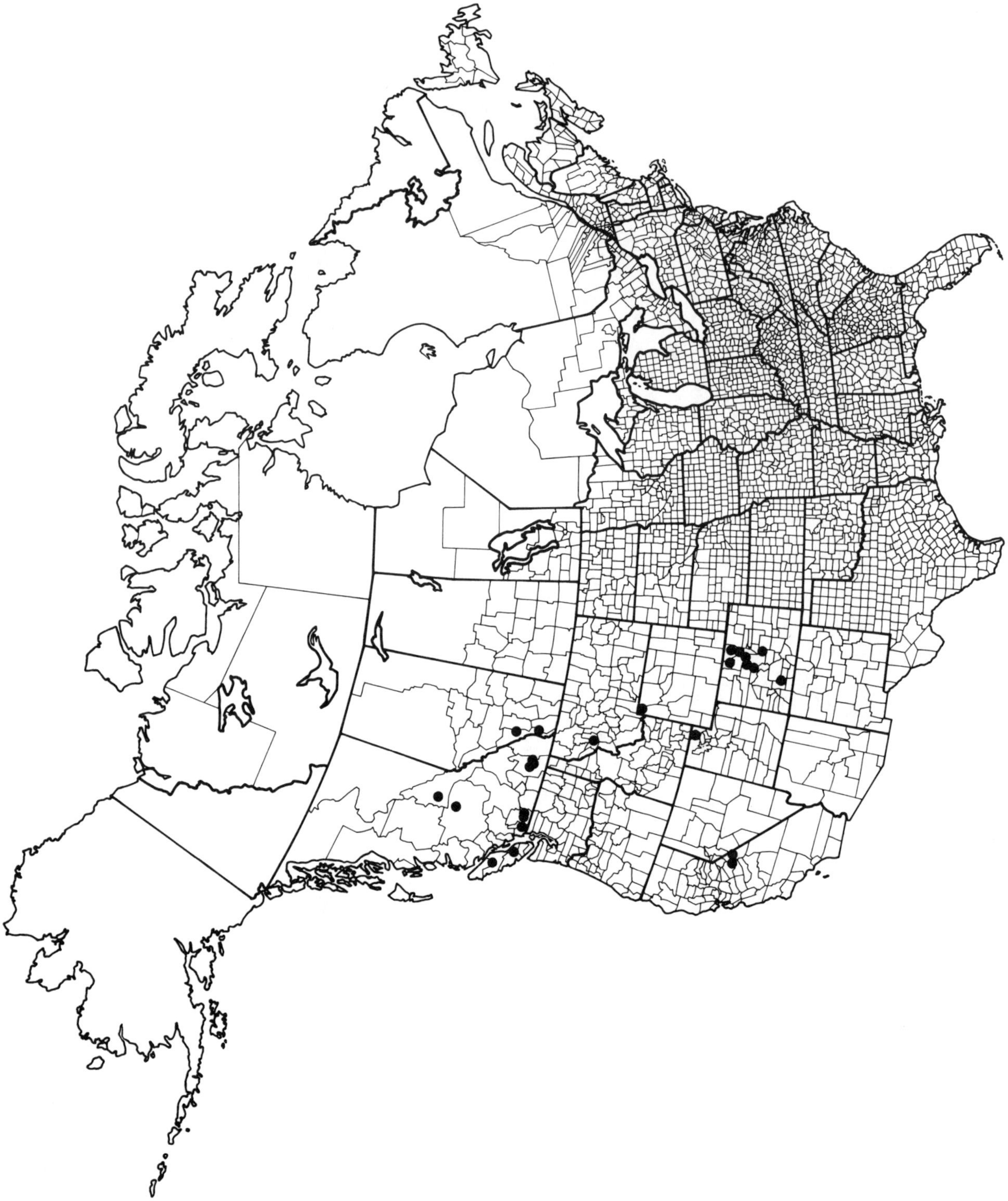

Map 238 *Simulium twinni*

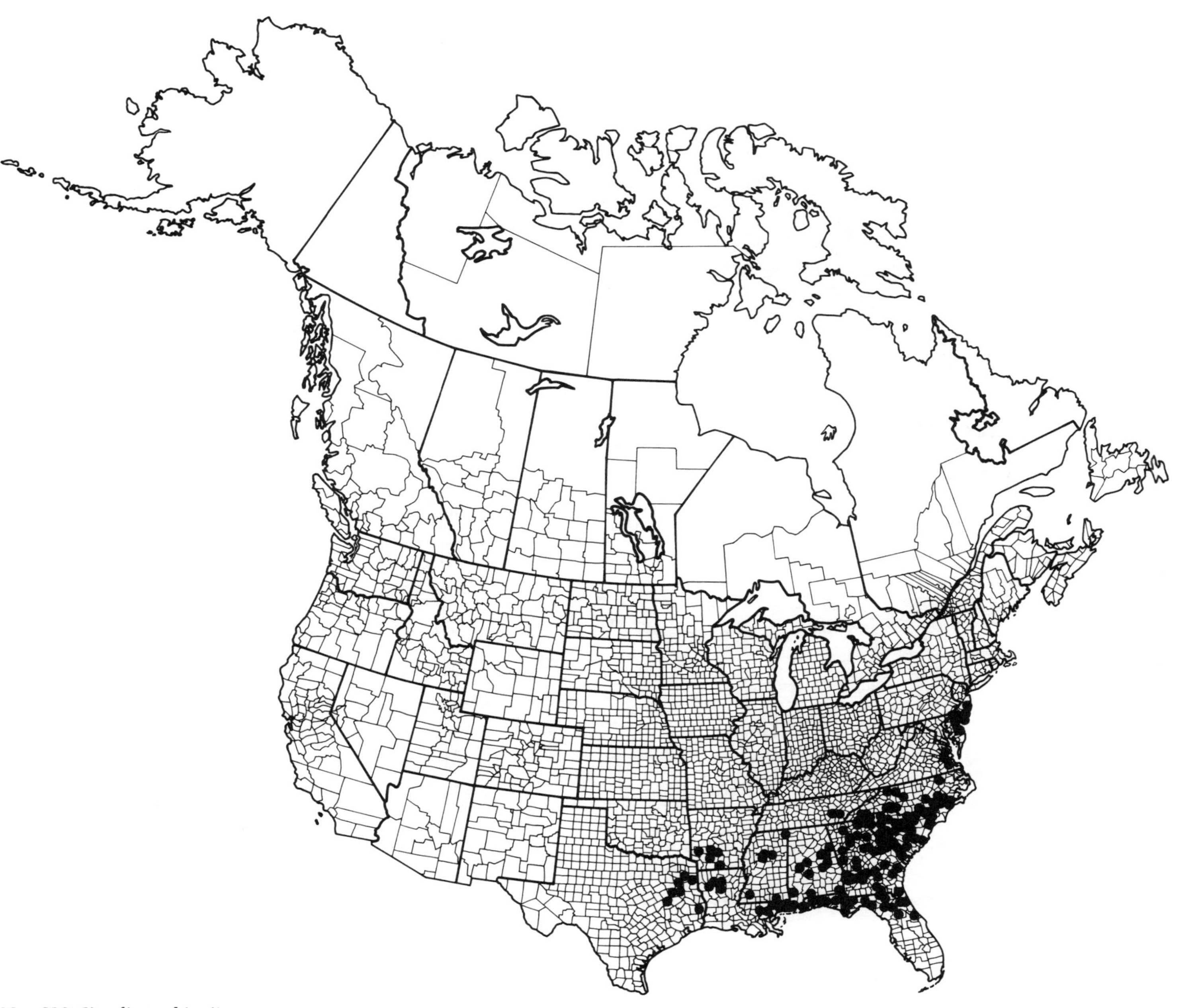

Map 239 *Simulium ubiquitum* n. sp.

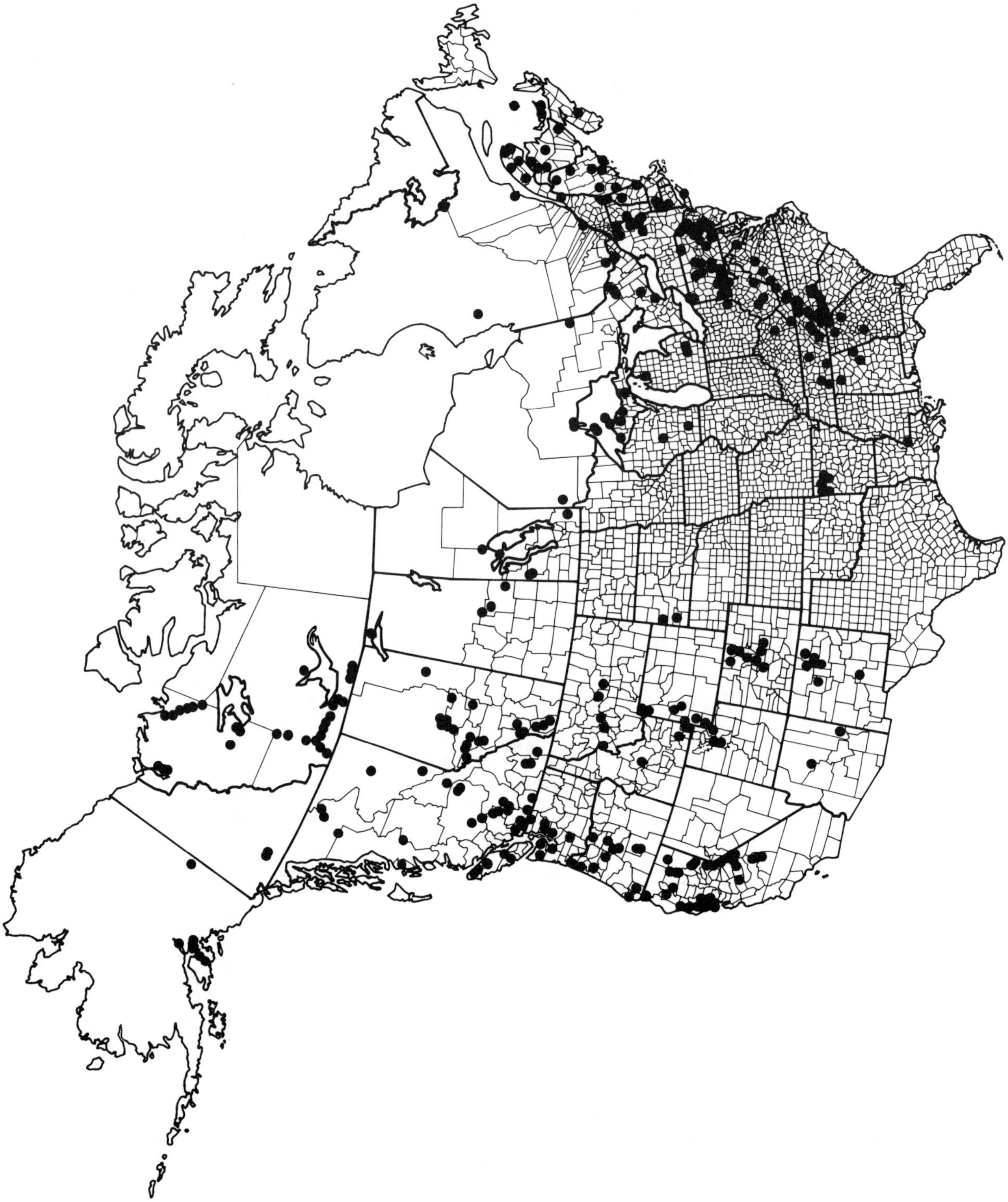

Map 240 *Simulium vandalicum*

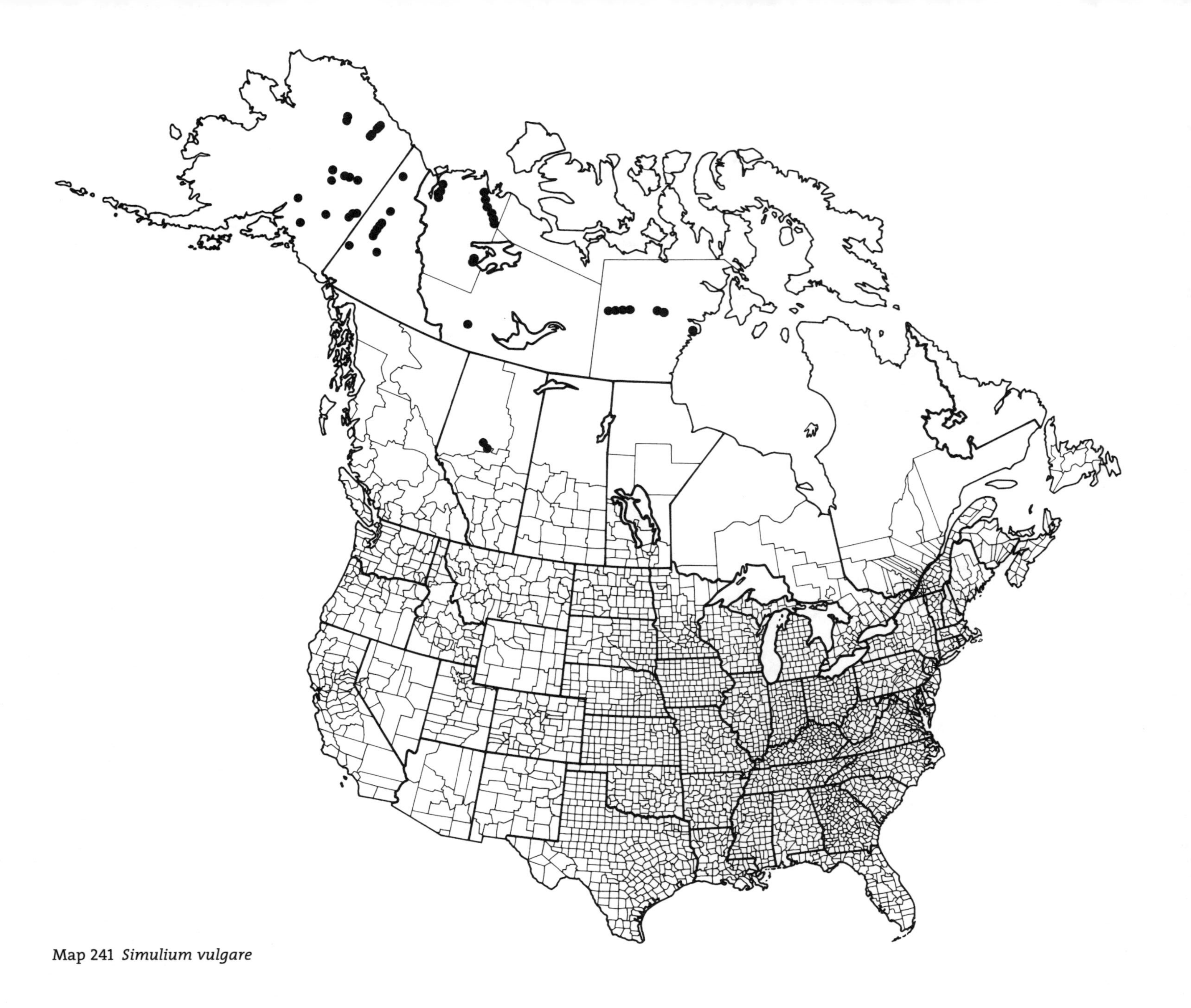

Map 241 *Simulium vulgare*

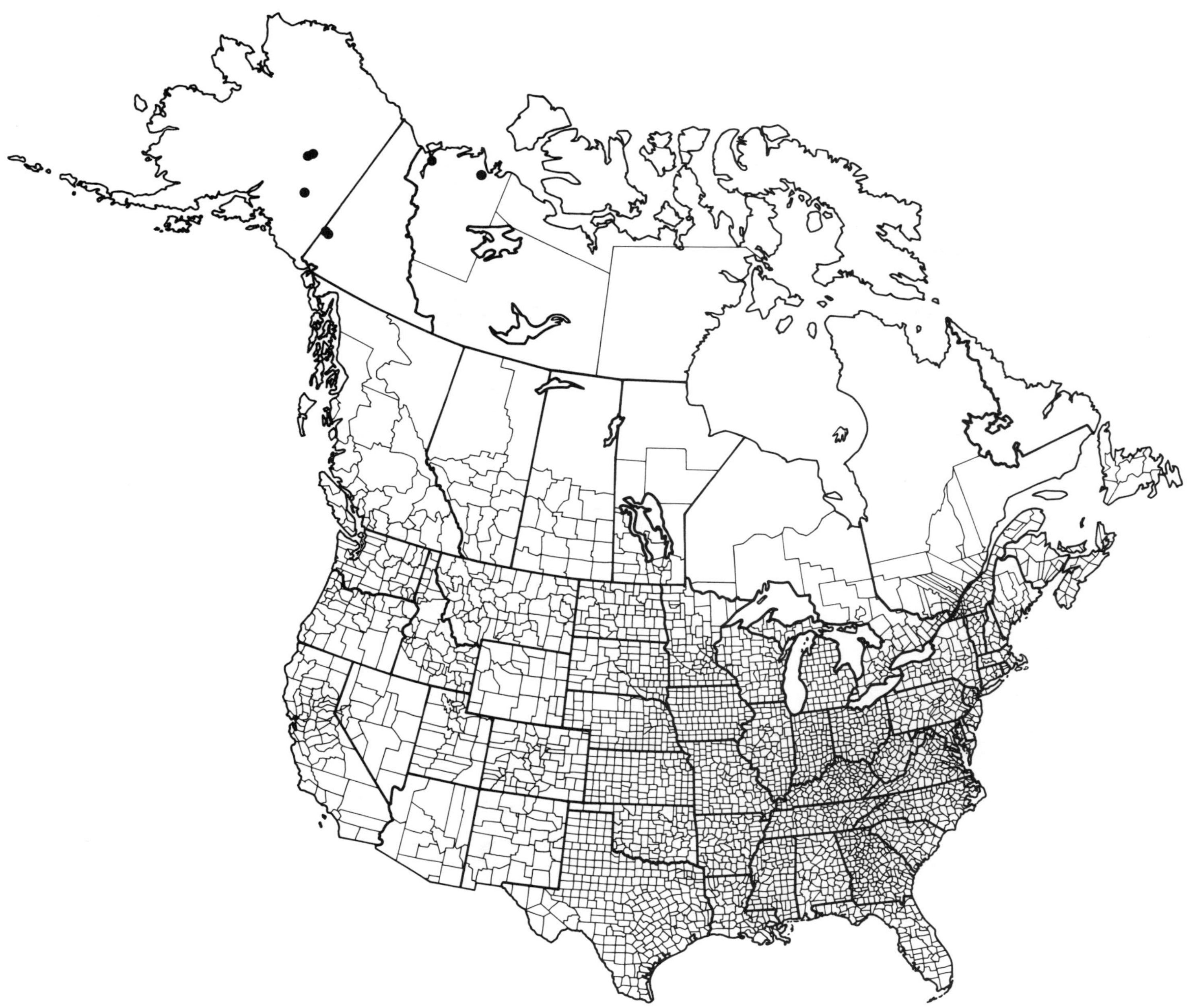

Map 242 *Simulium rubtzovi*

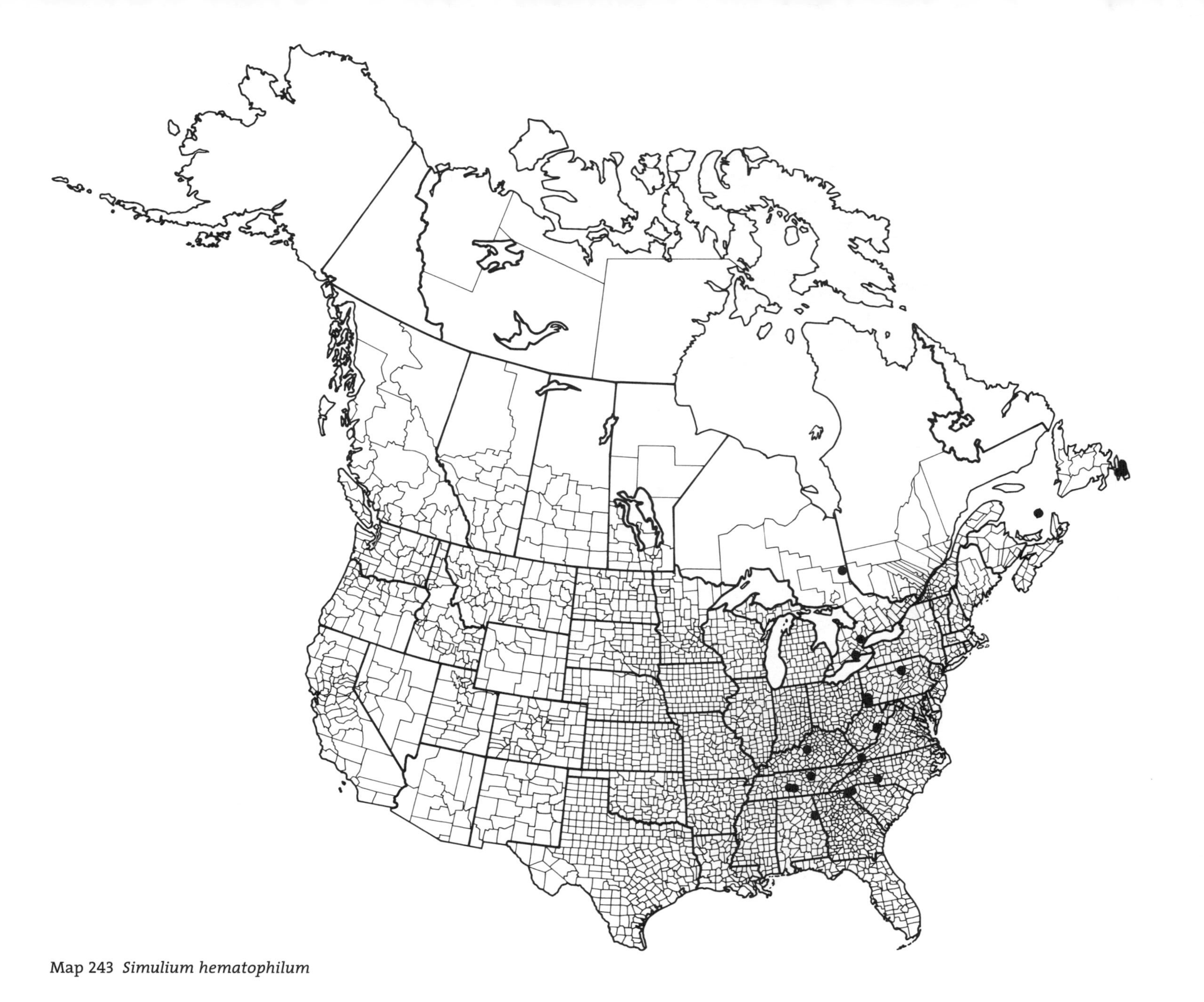

Map 243 *Simulium hematophilum*

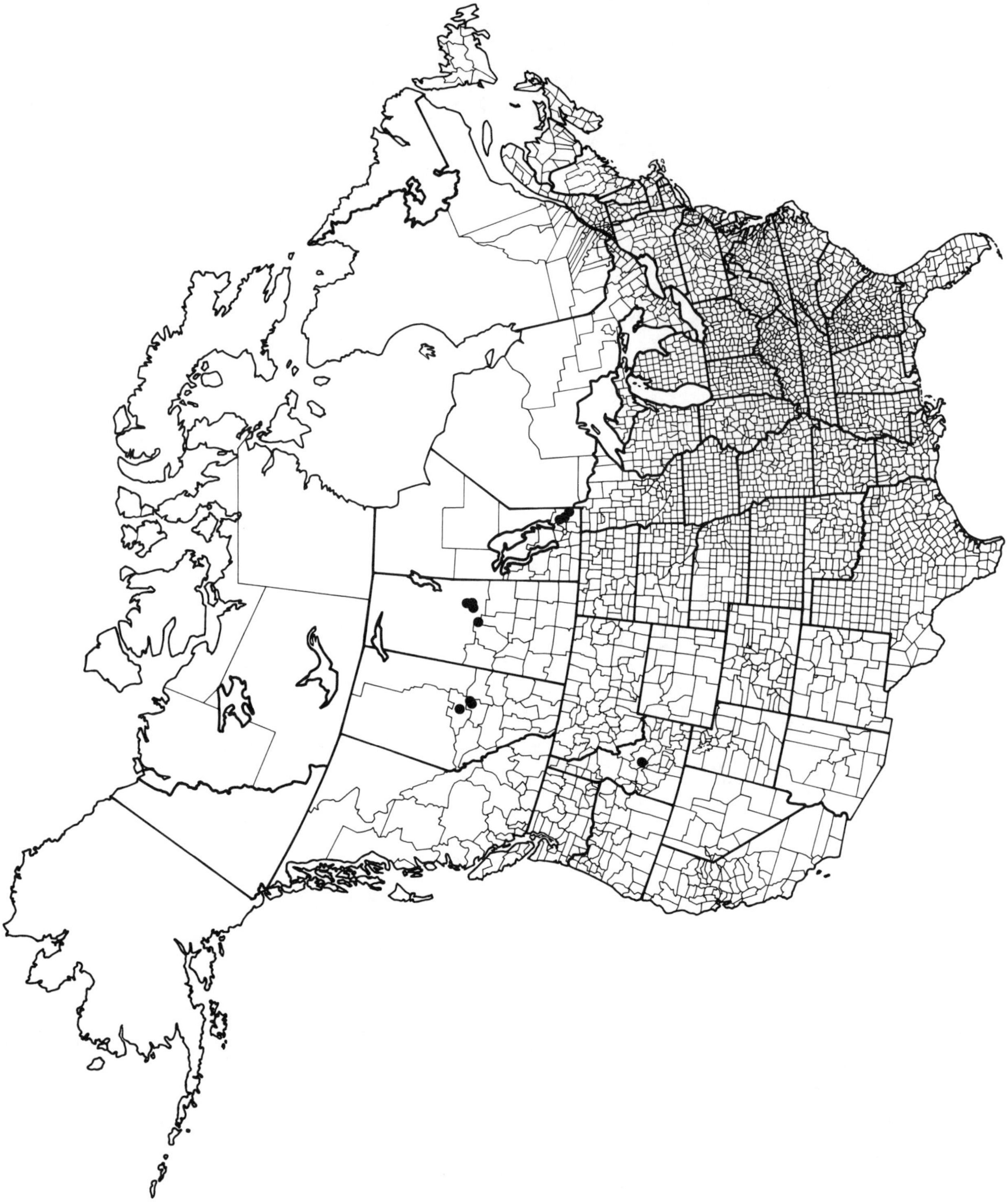

Map 244 *Simulium incognitum*

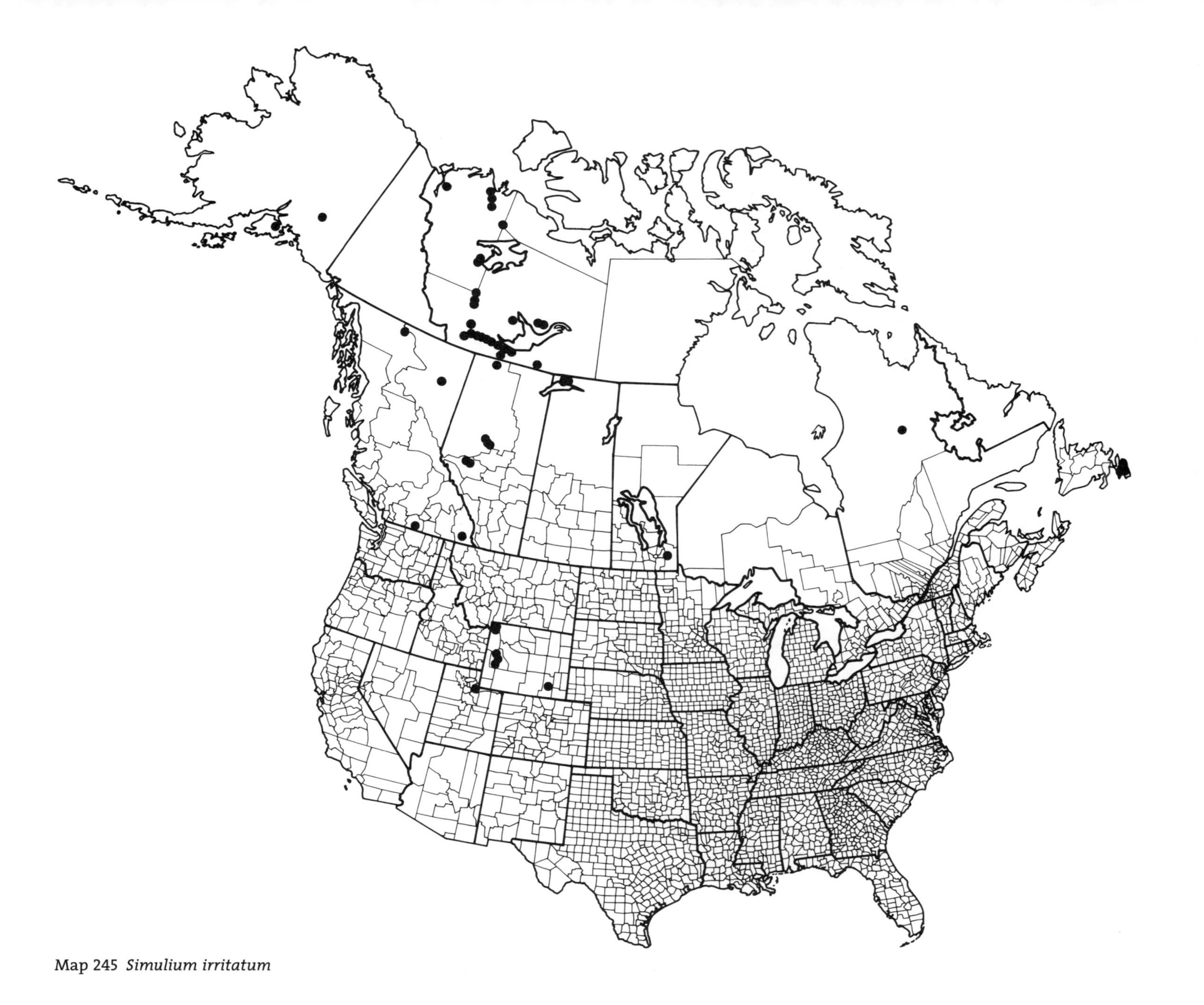

Map 245 *Simulium irritatum*

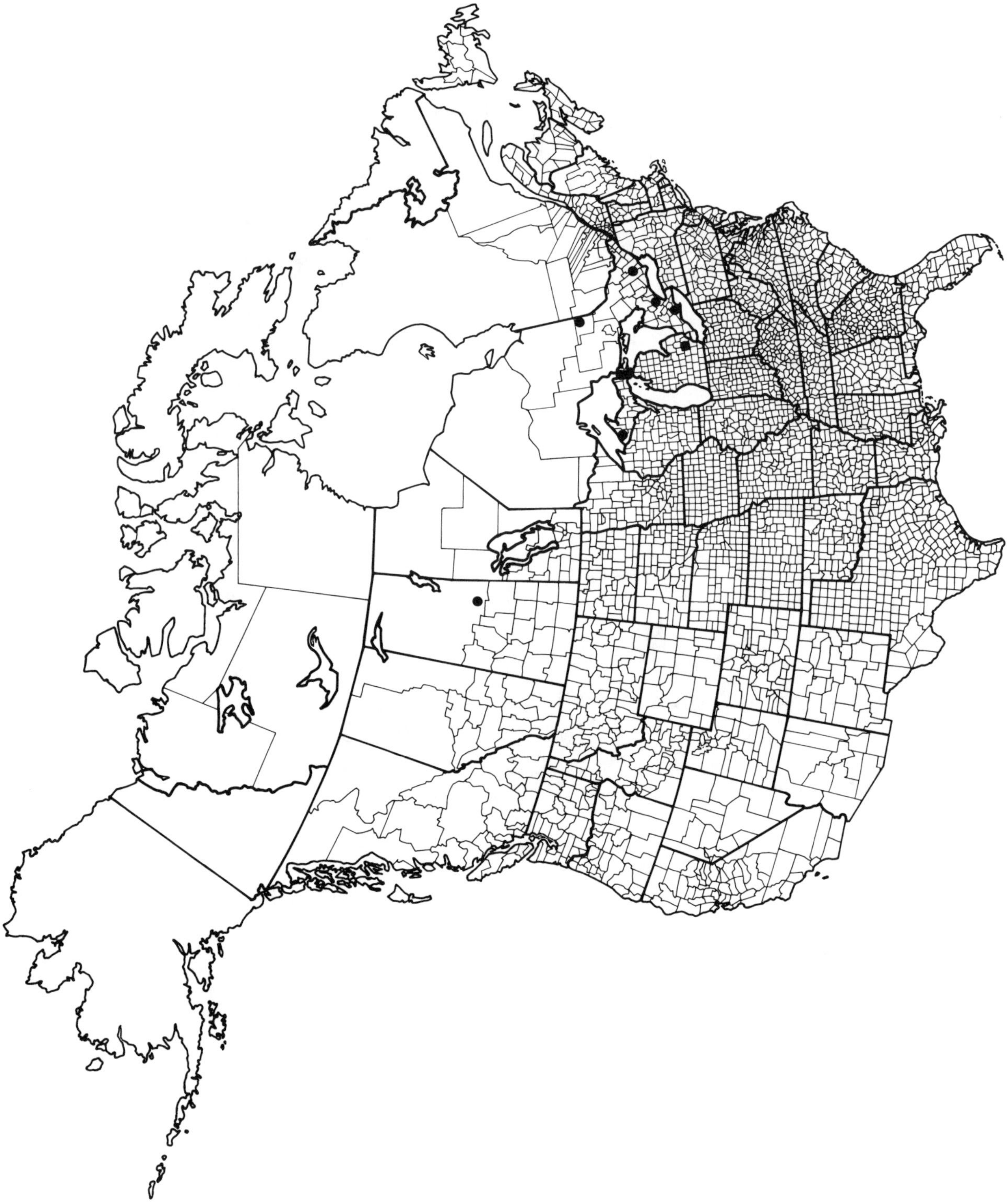

Map 246 *Simulium minutum*

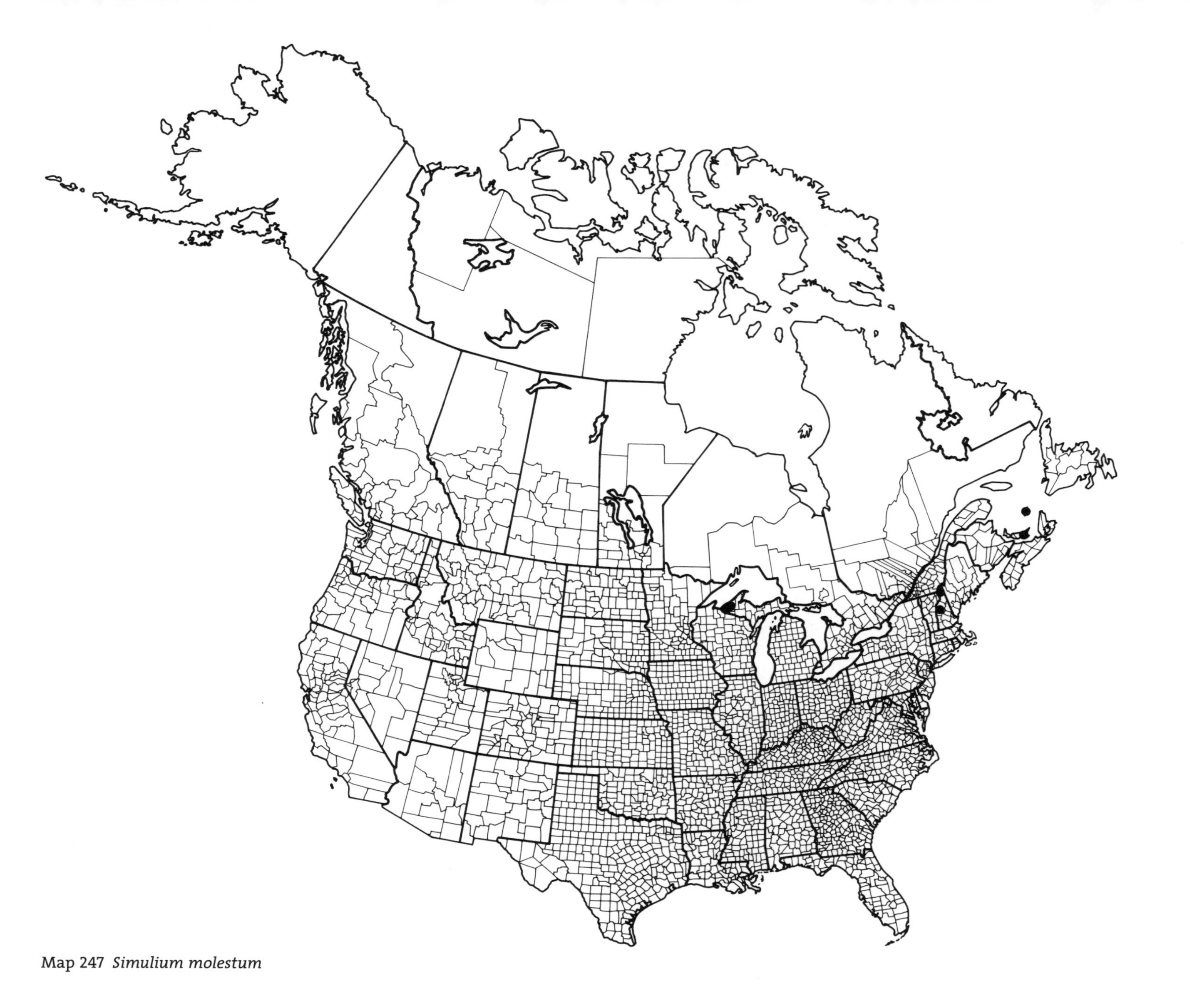

Map 247 *Simulium molestum*

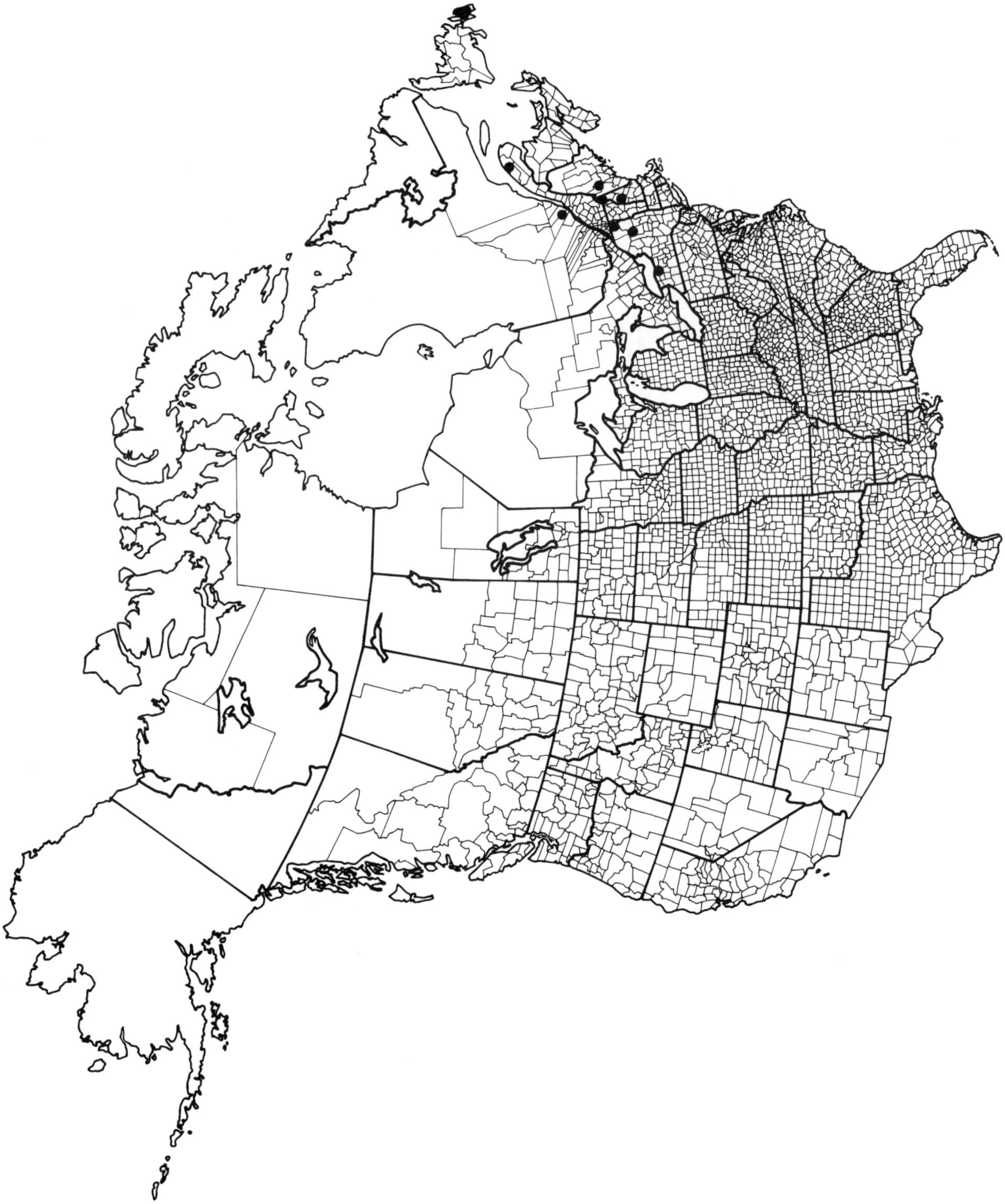

Map 248 *Simulium piscicidium*

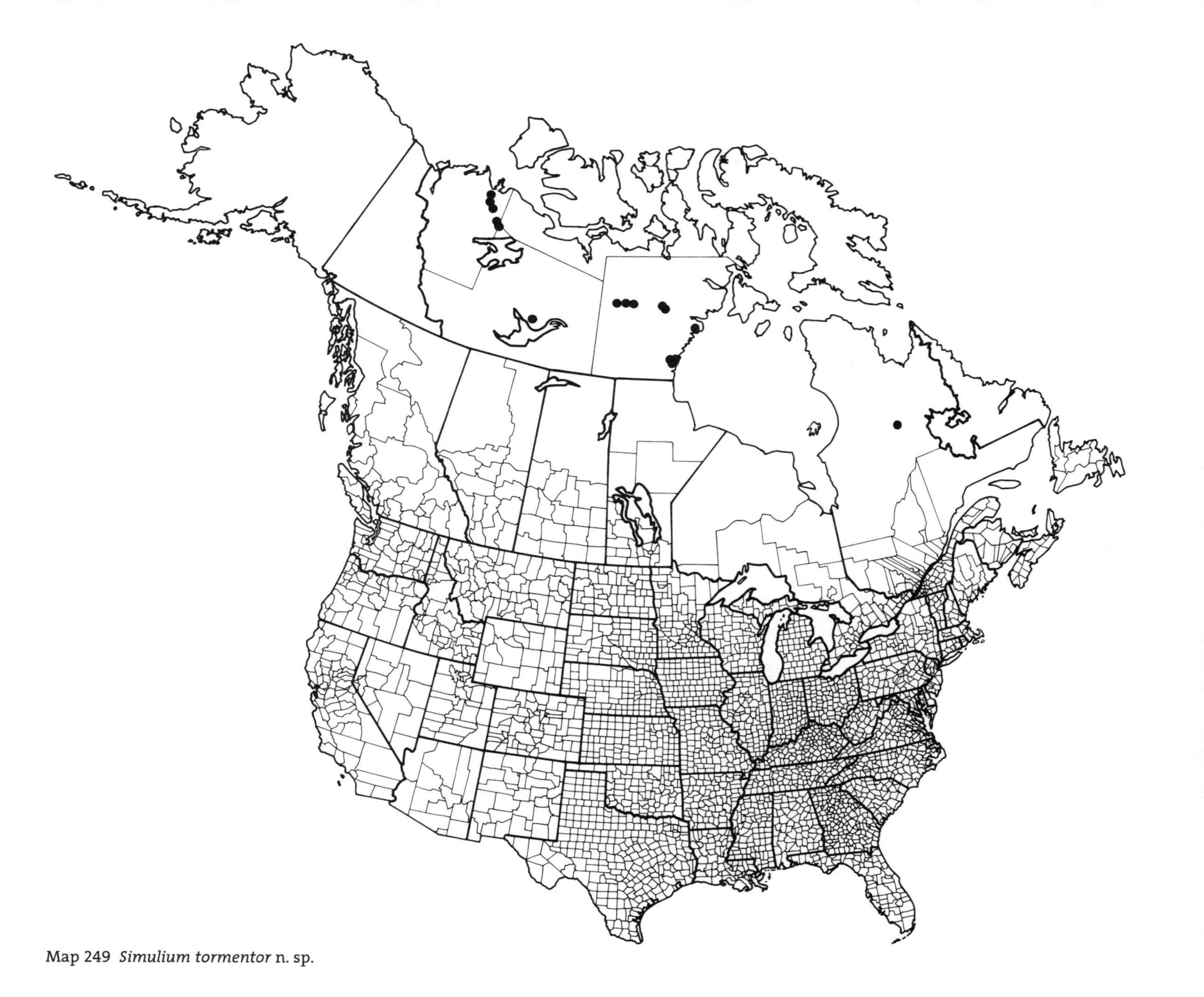

Map 249 *Simulium tormentor* n. sp.

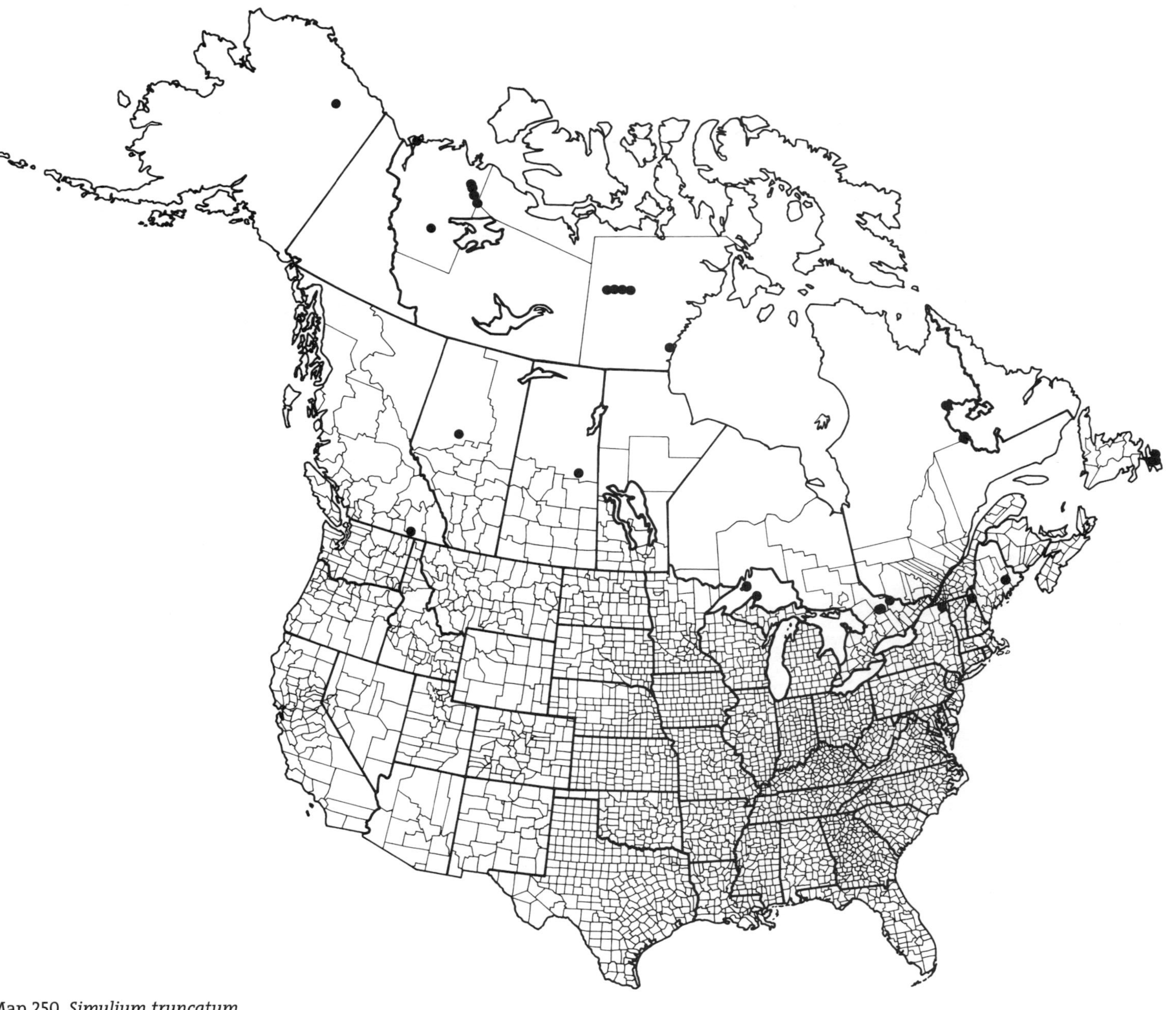

Map 250 *Simulium truncatum*

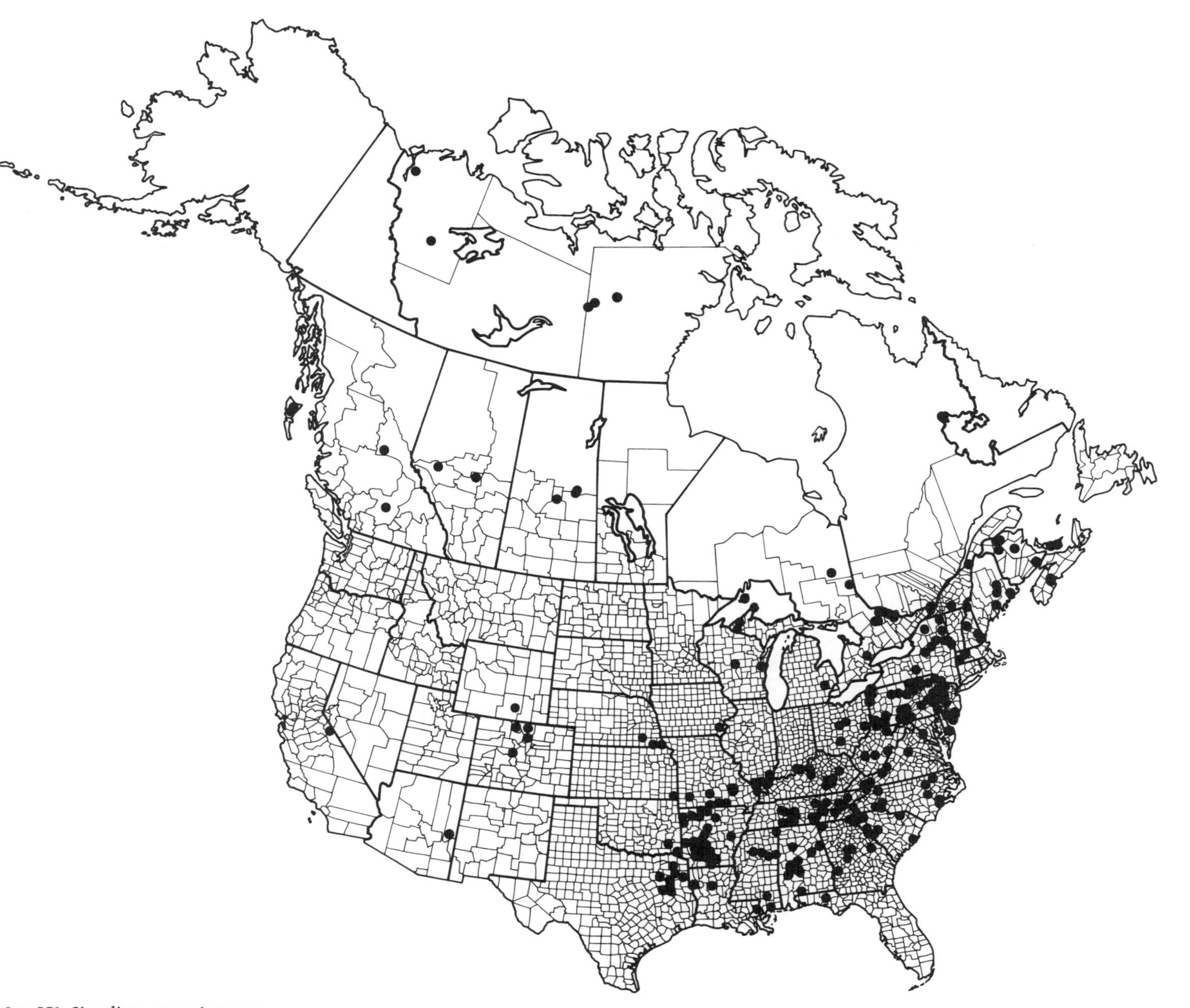

Map 251 *Simulium venustum* s. s.

Map 252 *Simulium venustum* 'cytospecies JJ'

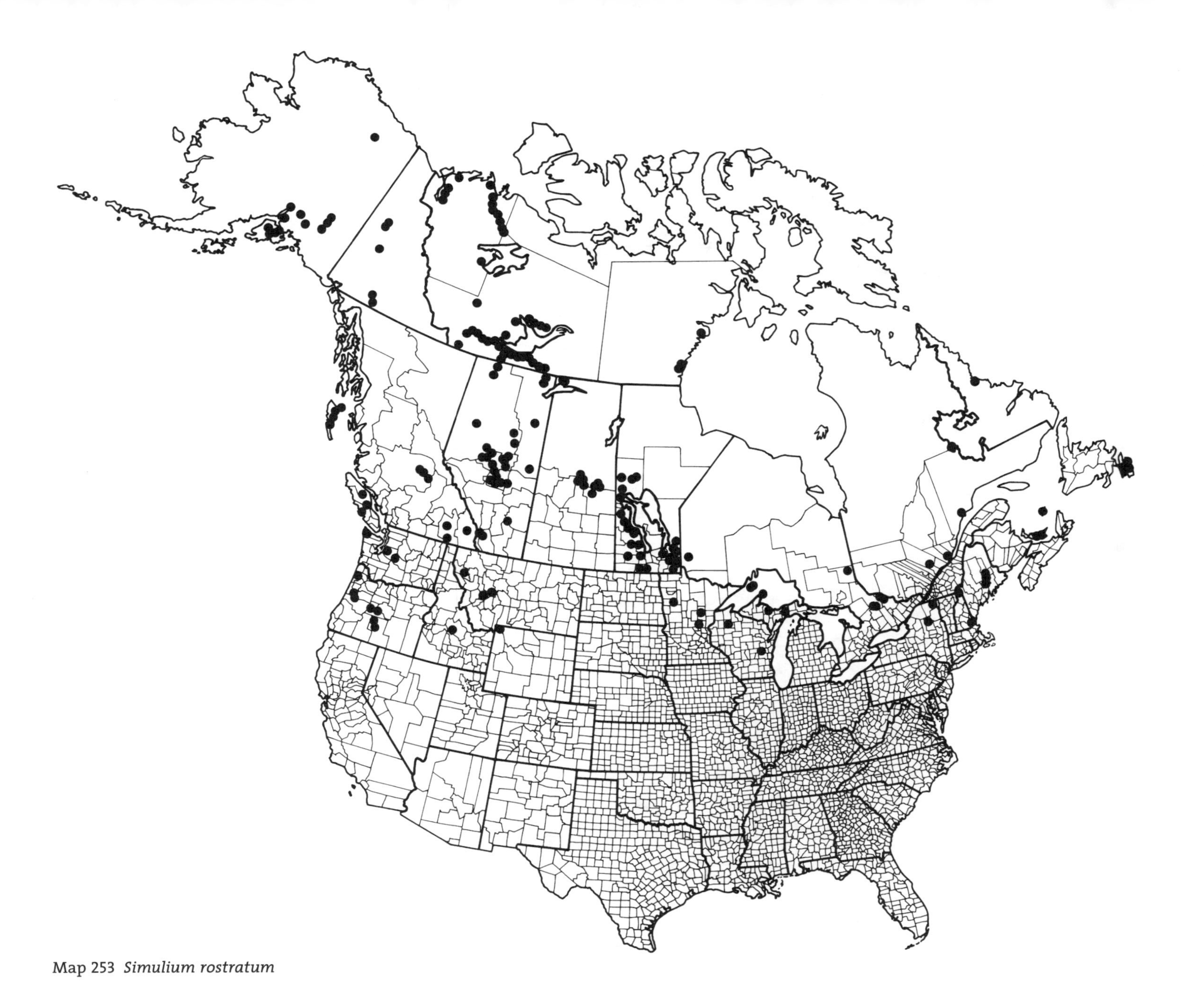

Map 253 *Simulium rostratum*

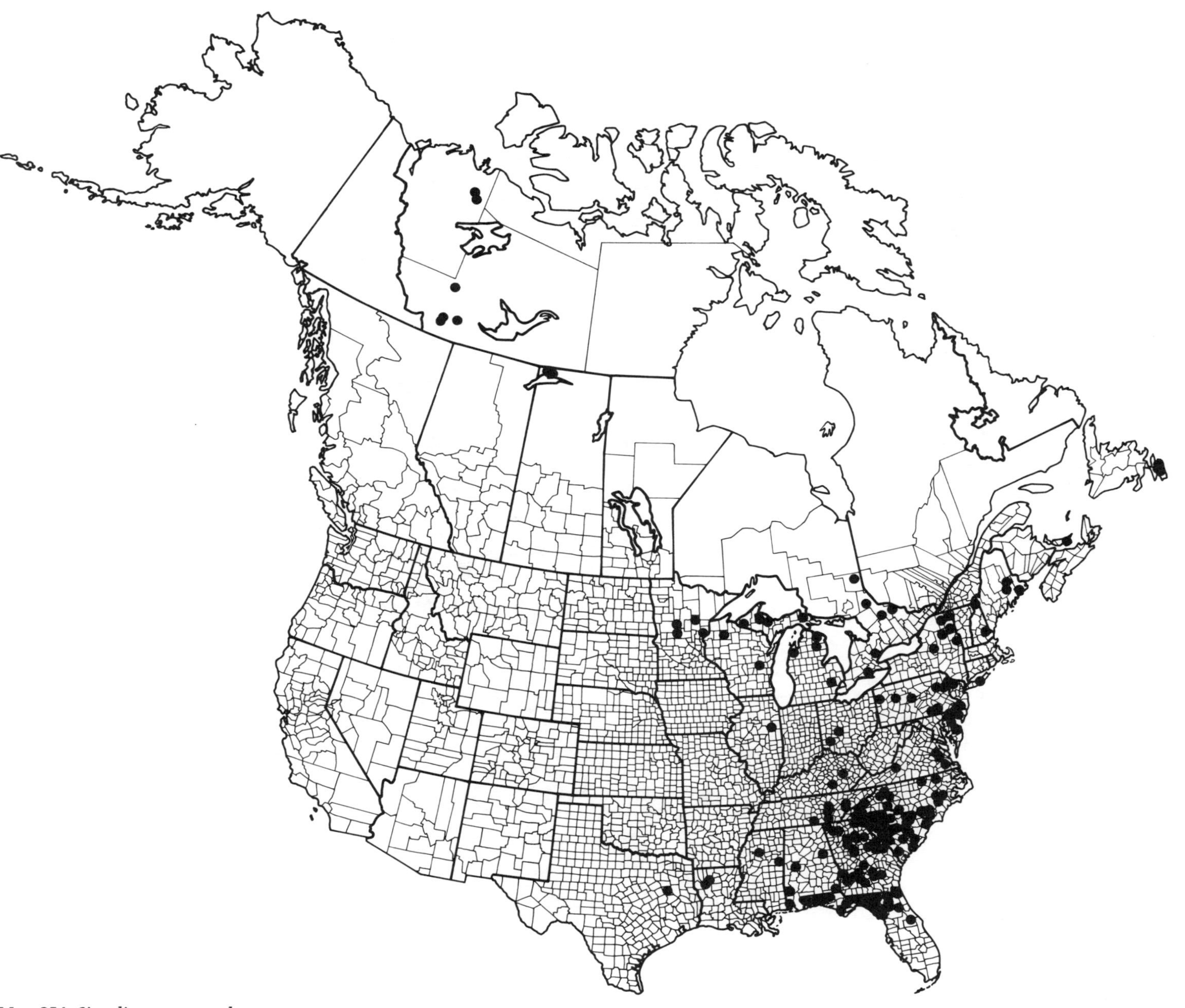

Map 254 *Simulium verecundum* s. s.

# References

References with three or more authors are arranged chronologically and then alphabetically by authors within year. References in Russian follow the format of Crosskey (1999b).


Abdelnur, O. M. 1966. The biology of some black flies (Diptera: Simuliidae) of Alberta. Ph.D. thesis. University of Alberta, Edmonton. 169 pp.

Abdelnur, O. M. 1968. The biology of some black flies (Diptera: Simuliidae) of Alberta. Quaestiones Entomologicae 4:113–174.

Abebe, M. 1994. Some studies on the salivary anticoagulant components of the black fly *Simulium vittatum* (Diptera: Simuliidae). Ph.D. thesis. University of Arizona, Tucson. 137 pp.

Abebe, M., M. S. Cupp, F. B. Ramberg, & E. W. Cupp. 1994. Anticoagulant activity in salivary gland extracts of black flies (Diptera: Simuliidae). Journal of Medical Entomology 31:908–911.

Abebe, M., M. S. Cupp, D. Champagne, & E. W. Cupp. 1995. Simulidin: a black fly (*Simulium vittatum*) salivary gland protein with anti-thrombin activity. Journal of Insect Physiology 41:1001–1006.

Abebe, M., J. M. C. Ribeiro, M. S. Cupp, & E. W. Cupp. 1996. Novel anticoagulant from salivary glands of *Simulium vittatum* (Diptera: Simuliidae) inhibits activity of coagulation factor V. Journal of Medical Entomology 33:173–176.

Abedraabo, S., F. Le Pont, A. J. Shelley, & J. Mouchet. 1993. Introduction et acclimatation d'une simulie anthropophile dans l'ile San Cristobal, archipel des Galapagos (Diptera, Simulidae [*sic*]). Bulletin de la Société Entomologique de France 98:108.

Adams, C. F. 1904. Notes on and descriptions of North American Diptera. Kansas University Science Bulletin 2:433–455.

Addison, E. M. 1980. Transmission of *Dirofilaria ursi* Yamaguti, 1941 (Nematoda: Onchocercidae) of black bears (*Ursus americanus*) by blackflies (Simuliidae). Canadian Journal of Zoology 58:1913–1922.

Adler, P. H. 1983. Ecosystematics of the *Simulium vittatum* complex (Diptera: Simuliidae) and associated black flies. Ph.D. thesis. Pennsylvania State University, University Park. 308 pp.

Adler, P. H. 1986. Ecology and cytology of some Alberta black flies (Diptera: Simuliidae). Quaestiones Entomologicae 22:1–18.

Adler, P. H. 1987. A new North American species in the *Simulium vernum* group (Diptera: Simuliidae) and analysis of its polytene chromosomes. Proceedings of the Entomological Society of Washington 89:673–681.

Adler, P. H. 1988a ["1987"]. Ecology of black fly sibling species. Pp. 63–76. *In* K. C. Kim & R. W. Merritt (eds.), Black flies: ecology, population management, and annotated world list. Pennsylvania State University, University Park. 528 pp. [Actual publication date was 1 March 1988.]

Adler, P. H. 1988b. Black flies are busting out all over. Natural History 97(6):34–40.

Adler, P. H. 1990. Two new species of black flies (Diptera: Simuliidae) from North America. Proceedings of the Entomological Society of Washington 92:431–443.

Adler, P. H., & R. W. Crosskey. 1998. The simuliid collection in the Zoological Institute, St. Petersburg, Russia. British Simuliid Group Bulletin 11:4–7.

Adler, P. H., & D. C. Currie. 1986. Taxonomic resolution of three new species near *Simulium vernum* Macquart (Diptera: Simuliidae). Canadian Entomologist 118:1207–1220.

Adler, P. H., & E. A. Kachvorian. 2001. Cytogenetics of the Holarctic black fly *Simulium noelleri* (Diptera: Simuliidae). Canadian Journal of Zoology 79:1972–1979.

Adler, P. H., & K. C. Kim. 1983. Solving black fly problems starts with larval identity. Science in Agriculture (Pennsylvania State University) 30(3):8.

Adler, P. H., & K. C. Kim. 1984. Ecological characterization of two sibling species, IIIL-1 and IS-7, in the *Simulium vittatum* complex (Diptera: Simuliidae). Canadian Journal of Zoology 62:1308–1315.

Adler, P. H., & K. C. Kim. 1985. Taxonomy of black fly sibling species: two new species in the *Prosimulium mixtum* group (Diptera: Simuliidae). Annals of the Entomological Society of America 78:41–49.

Adler, P. H., & K. C. Kim. 1986. The black flies of Pennsylvania (Simuliidae, Diptera): bionomics, taxonomy, and distribution. Pennsylvania State University Agricultural Experiment Station Bulletin 856:1–88.

Adler, P. H., & K. Kuusela. 1994. Cytological identities of *Simulium tuberosum* and *S. vulgare* (Diptera: Simuliidae), with notes on other Palearctic members of the *S. tuberosum* species-group. Entomologica Scandinavica 25:439–446.

Adler, P. H., & P. G. Mason. 1997. Black flies (Diptera: Simuliidae) of east-central Saskatchewan, with description of a new species and implications for pest management. Canadian Entomologist 129:81–91.

Adler, P. H., & J. W. McCreadie. 1997. The hidden ecology of black flies: sibling species and ecological scale. American Entomologist 43:153–161.

Adler, P. H., & J. W. McCreadie. 2002. Black flies (Simuliidae). Pp. 185–202. *In* G. Mullen & L. Durden (eds.), Medical and veterinary entomology. Academic Press, San Diego. 720 pp.

Adler, P. H., & D. M. Wood. 1991. A new North American species of *Simulium* (*Hellichiella*) (Diptera: Simuliidae). Canadian Journal of Zoology 69:2867–2872.

Adler, P. H., B. L. Travis, K. C. Kim, & E. C. Masteller. 1982. Seasonal emergence patterns of black flies (Diptera: Simuliidae) in northwestern Pennsylvania. Great Lakes Entomologist 15:253–260.

Adler, P. H., K. C. Kim, & R. W. Light. 1983a. Flight patterns of the *Simulium vittatum* (Diptera: Simuliidae) complex over a stream. Environmental Entomology 12:232–236.

Adler, P. H., R. W. Light, & K. C. Kim. 1983b. The aquatic drift patterns of black flies (Diptera: Simuliidae). Hydrobiologia 107:183–191.

Adler, P. H., Z. Wang, & C. E. Beard. 1996. First records of natural enemies from Chinese blackflies (Diptera: Simuliidae). Medical Entomology and Zoology 47:291–292.

Adler, P. H., B. Malmqvist, & Y. Zhang. 1999. Black flies (Diptera: Simuliidae) of northern Sweden: taxonomy, chromosomes, and bionomics. Entomologica Scandinavica 29:361–382.

Adler, P. H., J. J. Becnel, & B. Moser. 2000. Molecular characterization and taxonomy of a new species of Caudosporidae (Microsporidia) from black flies (Diptera: Simuliidae), with host-derived relationships of the North American caudosporids. Journal of Invertebrate Pathology 75:133–143.

Adler, P. H., R. W. Merritt, J. F. Burger, & D. P. Molloy. 2001. A brief history of Northeast Regional Project NE-118 in the USA. British Simuliid Group Bulletin 17:16–20.

Adler, P. H., C. L. Brockhouse, & D. C. Currie. 2003. A new species of black fly (Diptera: Simuliidae) from Nova Scotia. Proceedings of the Entomological Society of Washington 105:9–12.

Agassiz, L. 1850. Lake Superior: its physical character, vegetation and animals, compared with those of other and similar regions. With a narrative of the tour by J. Elliot Cabot. Gould, Kendall & Lincoln, Boston. 428 pp. [Reprint version: 1974 by Robert E. Krieger Publishing, Huntington, N.Y.]

Agricultural Research Service. 1954. Losses in agriculture: a preliminary appraisal for review. U.S. Department of Agriculture, Agricultural Research Service Publication No. 20-1. Washington, D.C. 190 pp.

Ainslie, C. N. 1929. Buffalo gnats (Simuliidae). United States Department of Agriculture Insect Pest Survey Bulletin 9:208.

Al-Amari, O. M. 1988. A cytological study of four species in the *Simulium jenningsi* group (Diptera: Simuliidae) in the state of West Virginia. Ph.D. thesis. West Virginia University, Morgantown. 210 pp.

Ali, S. H., P. B. Burbutis, W. F. Ritter, & R. W. Lake. 1974. Black fly (*Simulium vittatum* Zetterstedt) densities and water quality conditions in Red Clay Creek, Pa.-Del. Environmental Entomology 3:879–881.

Allan, J. D., & A. S. Flecker. 1988. Prey preference in stoneflies: a comparative analysis of prey vulnerability. Oecologia 76:496–503.

Allan, J. D., A. S. Flecker, & N. L. McClintock. 1987. Prey preference of stoneflies: sedentary vs mobile prey. Oikos 49:323–331.

Allan, S. A., J. F. Day, & J. D. Edman. 1987. Visual ecology of biting flies. Annual Review of Entomology 32:297–316.

Allison, L. A., & G. F. Shields. 1989. A cytological description of *Eusimulium vernum*, *E. decolletum*, and the *E. pugetense* complex (Diptera: Simuliidae) in Alaska. Genome 32:550–558.

Alverson, D. R. 1976. Behavioral and simulation models of *Leucocytozoon smithi* epizootiology in South Carolina turkey populations. M.S. thesis. Clemson University, Clemson, S.C. 96 pp.

Alverson, D. R., & R. Noblet. 1976. Response of female black flies to selected meterological [*sic*] factors. Environmental Entomology 5:662–665.

Alverson, D. R., & R. Noblet. 1977. Spring relapse of *Leucocytozoon smithi* (Sporozoa: Leucocytozoidae) in turkeys. Journal of Medical Entomology 14:132–133.

Alverson, D. R., R. Noblet, & J. R. Lambert. 1976. A continuous systems modeling program (CSMP) simulation of *Leucocytozoon smithi* transmission by black flies in South Carolina turkey populations. South Carolina Agricultural Experiment Station Technical Bulletin 1060:1–7.

American Ornithologists' Union. 1998. The A.O.U. check-list of North American birds. 7th ed. Allen Press, Washington, D.C. 829 pp.

Amrine, J. W. 1971. The black flies (Diptera: Simuliidae) of Ohio. M.S. thesis. Ohio State University, Columbus. 278 pp.

Amrine, J. W. 1982. The New River connection to the black fly problem in southern West Virginia. West Virginia University Agricultural and Forestry Experiment Station Bulletin 678:1–30.

Anderson, J. R. 1960. The biology and taxonomy of Wisconsin black flies (Diptera: Simuliidae). Ph.D. thesis. University of Wisconsin, Madison. 186 pp.

Anderson, J. R. 1988 ["1987"]. Reproductive strategies and gonotrophic cycles of black flies. Pp. 276–293. *In* K. C. Kim & R. W. Merritt (eds.), Black flies: ecology, population management, and annotated world list. Pennsylvania State University, University Park. 528 pp. [Actual publication date was 1 March 1988.]

Anderson, J. R., & G. R. DeFoliart. 1961. Feeding behavior and host preferences of some black flies (Diptera: Simuliidae) in Wisconsin. Annals of the Entomological Society of America 54:716–729.

Anderson, J. R., & G. R. DeFoliart. 1962. Nematode parasitism of black fly (Diptera: Simuliidae) larvae in Wisconsin. Annals of the Entomological Society of America 55:542–546.

Anderson, J. R., & R. J. Dicke. 1960. Ecology of the immature stages of some Wisconsin black flies (Simuliidae: Diptera). Annals of the Entomological Society of America 53:386–404.

Anderson, J. R., & J. A. Shemanchuk. 1975. Maintenance and transportation of black fly (Diptera: Simuliidae) larvae in nonagitated water. Proceedings and Papers of the Annual Conference of the California Mosquito Control Association 43:120–122.

Anderson, J. R., & J. A. Shemanchuk. 1987a. The biology of *Simulium arcticum* Malloch in Alberta. Part I. Obligate anautogeny in *S. arcticum* and other black flies (Diptera: Simuliidae). Canadian Entomologist 119:21–27.

Anderson, J. R., & J. A. Shemanchuk. 1987b. The biology of *Simulium arcticum* Malloch in Alberta. Part II. Seasonal parity structure and mermithid parasitism of populations attacking cattle and flying over the Athabasca River. Canadian Entomologist 119:29–44.

Anderson, J. R., & G. H. Voskuil. 1963. A reduction in milk production caused by the feeding of blackflies (Diptera: Simuliidae) on dairy cattle in California, with notes on the feeding activity on other animals. Mosquito News 23:126–131.

Anderson, J. R., & W. L. Yee. 1995. Trapping black flies (Diptera: Simuliidae) in northern California. I. Species composition and seasonal abundance on horses, host models, and in insect flight traps. Journal of Vector Ecology 20:7–25.

Anderson, J. R., V. H. Lee, S. Vadlamudi, R. P. Hanson, & G. R. DeFoliart. 1961. Isolation of eastern encephalitis virus from Diptera in Wisconsin. Mosquito News 21:244–248.

Anderson, J. R., D. O. Trainer, & G. R. DeFoliart. 1962. Natural and experimental transmission of the waterfowl parasite, *Leucocytozoon simondi* M. & L., in Wisconsin. Zoonoses Research 1:155–164.

Anderson, R. C. 1955. Black flies (Simuliidae) as vectors of *Ornithofilaria fallisensis* Anderson, 1954. Journal of Parasitology 41(Suppl.):45.

Anderson, R. C. 1956. The life cycle and seasonal transmission of *Ornithofilaria fallisensis* Anderson, a parasite of domestic and wild ducks. Canadian Journal of Zoology 34:485–525.

Anderson, R. C. 1968. The simuliid vectors of *Splendidofilaria fallisensis* of ducks. Canadian Journal of Zoology 46:610–611.

Anderson, R. C., & M. W. Lankester. 1974. Infectious and parasitic diseases and arthropod pests of moose in North

America. Naturaliste Canadien 101:23–50.
Angioy, A. M., A. Liscia, R. Crnjar, P. Pietra, & J. G. Stoffolano Jr. 1982. Electrophysiological responses of the labellar, tarsal and palpal chemosensilla of the adult female black fly, *Simulium venustum* Say, to NaCl and sucrose stimulation. Bollettino della Società Italiana di Biologia Sperimentale 58:1319–1324.
[Anonymous]. 1875. [no title]. Harper's Weekly 19:666.
[Anonymous]. 1883. Buffalo gnats. American Veterinary Review 7:145–146.
[Anonymous]. 1884. Destructive buffalo gnats. American Veterinary Review 8:153.
[Anonymous]. 1930. Outstanding entomological features in Canada for August, 1930. United States Department of Agriculture Insect Pest Survey Bulletin 10:259–260.
[Anonymous]. 1934. Buffalo gnats kill live stock. Journal of the American Veterinary Medical Association 84(New Series 37):972.
[Anonymous]. 1947. Biting insects of the arctic. Bulletin of the United States Army Medical Department 7:991–993.
[Anonymous]. 1949. Progress report of the Alaska Insect Project for 1948. U.S. Department of Agriculture Bureau of Entomology and Plant Quarantine Interim Report No. 0-137. Agricultural Research Administration, Washington, D.C. 88 pp.
[Anonymous]. 1961 March 24. 15 horses, mules die as result of gnats; buffalo gnats blamed. Bastrop Daily Enterprise (Bastrop, La.), p. 1.
[Anonymous]. 1976. U.S. Department of Agriculture Cooperative Plant Pest Report 1(17):181–196.
[Anonymous]. 1977. Annual report of Memorial University of Newfoundland Research Unit on Vector Pathology. Memorial University of Newfoundland, St. John's. 27 pp.
[Anonymous]. 1982 February 22. Scientists, lawmakers discuss gnat control. Charleston Daily Mail (Charleston, W.V.), p. 7A.
[Anonymous]. 1983 May 28. Black flies called health problem. Charleston Daily Mail (Charleston, W.V.), p. 4B.
[Anonymous]. 1999. Aqueous formulation: VectoBac®. Technical Bulletin. Abbott Laboratories, Agricultural Products, North Chicago, Ill. 8 pp.
Anthony, D. W., & D. J. Richey. 1958. Influence of black fly control on the incidence of *Leucocytozoon* disease in South Carolina turkeys. Journal of Economic Entomology 51:845–847.
Arbegast, D. H. 1994. Blackfly suppression, the Pennsylvania experience. Proceedings of the Annual Meeting of the New Jersey Mosquito Control Association 81:107–113.
Arbegast, D. H. 1996. Survey of warm season blackflies of Bucks and Northampton Counties in Pennsylvania. Proceedings of the Annual Meeting of the New Jersey Mosquito Control Association 83:55–63.
Arnason, A. P., A. W. A. Brown, F. J. H. Fredeen, W. W. Hopewell, & J. G. Rempel. 1949. Experiments in the control of *Simulium arcticum* Malloch by means of DDT in the Saskatchewan River. Scientific Agriculture 29:527–537.
Arnold, D. C. 1974. Black flies (Diptera: Simuliidae) of the coastal plains and sandhills of South Carolina. M.S. thesis. Clemson University, Clemson, S.C. 192 pp.
Arnold, D. C., & R. Noblet. 1975. Distribution of black fly vectors of *Leucocytozoon smithi* of turkeys in the Sandhills and Coastal Plains of South Carolina. South Carolina Agricultural Experiment Station Technical Bulletin 1054:1–9.
Arthur, E. M. 1985. George Barnston (c. 1800–1883). Bulletin of the Entomological Society of Canada 17:20–21.
Asselin, J. A. 1935. Bye, bye, blackfly. Pulp and Paper of Canada 36:261–263.
Atwood, D. W. 1996. Distribution, abundance, control and field observations of the southern buffalo gnat, *Cnephia pecuarum* (Diptera: Simuliidae), in Arkansas and Texas. Ph.D. thesis. University of Arkansas, Fayetteville. 127 pp.
Atwood, D. W., & M. V. Meisch. 1993. Evaluation of 1-octen-3-ol and carbon dioxide as black fly (Diptera: Simuliidae) attractants in Arkansas. Journal of the American Mosquito Control Association 9:143–146.
Atwood, D. W., J. V. Robinson, M. V. Meisch, J. K. Olson, & D. R. Johnson. 1992. Efficacy of *Bacillus thuringiensis* var. *israelensis* against larvae of the southern buffalo gnat, *Cnephia pecuarum* (Diptera: Simuliidae), and the influence of water temperature. Journal of the American Mosquito Control Association 8:126–130.
Avery, S. W., & L. Bauer. 1984. Iridescent virus from *Prosimulium* collected in Maine. Journal of Invertebrate Pathology 43:430–431.
Baba, M., & H. Takaoka. 1991. Oviposition habits of a univoltine blackfly, *Prosimulium kiotoense* (Diptera: Simuliidae), in Kyushu, Japan. Medical and Veterinary Entomology 5:351–357.
Back, C., & P. P. Harper. 1978. Les mouches noires (Diptera: Simuliidae) de deux ruisseaux des Laurentides, Québec. Annales de la Société Entomologique du Québec 23:55–66.
Back, C., & P. P. Harper. 1979. Succession saisonnière, émergence, voltinisme, et répartition de mouches noires des Laurentides (Diptera; Simuliidae). Canadian Journal of Zoology 57:627–639.
Back, C., J.-G. Lanouette, & A. Aubin. 1979. Preliminary tests on the use of temephos for the control of black flies (Diptera: Simuliidae) in northern Quebec. Mosquito News 39:762–767.
Back, C., A. Leblanc, & A. Aubin. 1983. Effets sur la dérive des insectes aquatiques d'un traitment au téméphos contre les larves de moustiques dans le Québec subarctique. Canadian Entomologist 115:703–712.
Back, C., J. Boisvert, J. O. Lacoursière, & G. Charpentier. 1985. High-dosage treatment of a Quebec stream with *Bacillus thuringiensis* serovar. *israelensis*: efficacy against black fly larvae (Diptera: Simuliidae) and impact on non-target insects. Canadian Entomologist 117:1523–1534.
Bacon, M. 1953. A study of the arthropods of medical and veterinary importance in the Columbia Basin. Washington Agricultural Experiment Stations Technical Bulletin 11:1–40.
Bacon, M., & R. H. McCauley Jr. 1959. Black flies (Diptera: Simuliidae) in a newly developed irrigation district (Columbia Basin, Washington). Northwest Science 33:103–110.
Bailey, C. H. 1977. Field and laboratory observations on a cytoplasmic polyhedrosis virus of blackflies (Diptera: Simuliidae). Journal of Invertebrate Pathology 29:69–73.
Bailey, C. H., & R. Gordon. 1977. Observations on the occurrence and collection of mermithid nematodes from blackflies (Diptera: Simuliidae). Canadian Journal of Zoology 55:148–154.
Bailey, C. H., R. Gordon, & J. Mokry. 1974. Procedure for mass collection of mermithid postparasites (Nematoda: Mermithidae) from larval blackflies (Diptera: Simuliidae). Canadian Journal of Zoology 52:660–661.
Bailey, C. H., M. Shapiro, & R. R. Granados. 1975. A cytoplasmic polyhedrosis virus from the larval blackflies *Cnephia mutata* and *Prosimulium mixtum* (Diptera: Simuliidae). Journal of Invertebrate Pathology 25:273–274.
Bailey, C. H., R. Gordon, & C. Mills. 1977. Laboratory culture of the free-living stages of *Neomesomermis flumenalis*, a mermithid nematode parasite of Newfoundland blackflies (Diptera: Simuliidae). Canadian Journal of Zoology 55:391–397.
Bain, O., & A. Renz. 1993. Infective larvae of a new species of Robertdollfusidae (Adenophorea, Nematoda) in the gut of *Simulium damnosum* in Cameroon. Annales de Parasitologie Humaine et Comparée 68:182–184.

Baker, C. F. 1897. *Simulium ochraceum* again. Entomological News 8:172.

Baldwin, W. F., & H. P. Gross. 1972. Fluctuations in numbers of adult black flies (Diptera: Simuliidae) in Deep River, Ontario. Canadian Entomologist 104:1465–1470.

Baldwin, W. F., J. R. Allen, & N. S. Slater. 1966. A practical field method for the recovery of blackflies labelled with phosphorus-32. Nature 212:959–960.

Baldwin, W. F., A. S. West, & J. Gomery. 1975. Dispersal pattern of black flies (Diptera: Simuliidae) tagged with $^{32}$P. Canadian Entomologist 107:113–118.

Baldwin, W. F., H. P. Gross, M. L. Wilson, D. J. Keill, R. J. Stuart, R. J. Sebastien, A. G. Knight, G. D. Chant, P. A. Knight, & A. S. West. 1977. Suppression of black fly populations in Deep River, Ontario. Canadian Entomologist 109:249–254.

Ballard, J. W. O. 1988. A simple technique for sexing blackfly larvae of the taxon *Austrosimulium bancrofti*. Transactions of the Royal Society of Tropical Medicine and Hygiene 82:478.

Banfield, A. W. F. 1954. Preliminary investigation of the barren ground caribou. Part II. Life history, ecology, and utilization. Wildlife Management Bulletin, Ottawa (Series 1) 10B:1–112.

Banks, N. 1911. Notes on African Myrmeleonidae. Annals of the Entomological Society of America 4:1–31.

Baranov ["Baranoff"], N. 1926. Eine neue Simuliiden-Art und einige Bemerkungen über das System der Simuliiden. Neue Beiträge zur systematischen Insektenkunde 3:161–164.

Baranov ["Baranoff"], N. 1935. Neues über die Kolumbatscher-Mücke. Arbeiten über morphologische und taxonomische Entomologie aus Berlin-Dahlem 2:156–158.

Baranov, N. 1938. K poznavanju golubačke mušice VI (Studij golubačke mušice i njezinih sinbiocenonta). Veterinarski Arhiv 8:313–328. [In Croatian; German summary.]

Baranov, N. 1939. *Odagmia ornata barensis* supsp. nova i njen parazit *Megaselia brevissima* Schmitz. Veterinarski Arhiv 9:599–601. [In Croatian; German summary.]

Barley, D. A. 1964. Salivary gland chromosomes of black flies. Turtox News 42:298–300.

Barnard, D. R. 1979. A vehicle-mounted insect trap. Canadian Entomologist 111:851–854.

Barnard, W. S. 1880. Notes on the development of a black-fly (*Simulium*) common in the rapids around Ithaca, N.Y. American Entomologist 3:191–193.

Barnett, B. D. 1977. *Leucocytozoon* disease of turkeys. World's Poultry Science Journal 33:76–87.

Barr, W. B. 1982. Attachment and silk of larvae of *Simulium vittatum* Zetterstedt (Diptera: Simuliidae). M.S. thesis. University of Alberta, Edmonton. 170 pp.

Barr, W. B. 1984. Prolegs and attachment of *Simulium vittatum* (sibling IS-7) (Diptera: Simuliidae) larvae. Canadian Journal of Zoology 62:1355–1362.

Barrow, J. H., Jr., N. Kelker, & H. Miller. 1968. The transmission of *Leucocytozoon simondi* to birds by *Simulium rugglesi* in northern Michigan. American Midland Naturalist 79:197–204.

Barton, W. E., R. Noblet, & D. C. Kurtak. 1991. A simple technique for determining relative toxicities of *Bacillus thuringiensis* var. *israelensis* formulations against larval blackflies (Diptera: Simuliidae). Journal of the American Mosquito Control Association 7:313–315.

Basrur, P. K. 1958. The salivary gland chromosomes of seven segregates of *Prosimulium* (Diptera: Simuliidae) with transformed centromere. Ph.D. thesis. University of Toronto, Toronto. 73 pp.

Basrur, P. K. 1959. The salivary gland chromosomes of seven segregates of *Prosimulium* (Diptera: Simuliidae) with a transformed centromere. Canadian Journal of Zoology 37:527–570.

Basrur, P. K. 1962. The salivary gland chromosomes of seven species of *Prosimulium* (Diptera: Simuliidae) from Alaska and British Columbia. Canadian Journal of Zoology 40:1019–1033.

Basrur, V. R. 1957. Cytotaxonomic studies in black flies and midges. Ph.D. thesis. University of Toronto, Toronto. 49 pp. + 8 plates.

Basrur, V. R., & K. H. Rothfels. 1959. Triploidy in natural populations of the black fly *Cnephia mutata* (Malloch). Canadian Journal of Zoology 37:571–589.

Bass, J. [A. B.]. 1998. Last-instar larvae and pupae of the Simuliidae of Britain and Ireland: a key with brief ecological notes. Freshwater Biological Association Scientific Publication 55:1–102.

Batson, B. S. 1983. *Tetrahymena dimorpha* sp. nov. (Hymenostomatida: Tetrahymenidae), a new ciliate parasite of Simuliidae (Diptera) with potential as a model for the study of ciliate morphogenesis. Philosophical Transactions of the Royal Society of London (B) 301:345–363.

Bauer, L. S. 1977. The black flies (Diptera: Simuliidae) of the Piscataquis River watershed, Maine. M.S. thesis. University of Maine, Orono. 43 pp.

Bauer, L. S., & J. Granett. 1979. The black flies of Maine. Maine Life Sciences and Agriculture Experiment Station Technical Bulletin 95:1–18.

Bazarova, N. D. 1990. New species of black flies in the genus *Metacnephia* from Buryat. Pp. 54–74. *In* L. D. Shestakova (ed.), Fauna and ecology of arthropods in Transbaikalia and Prebaikalia [Fauna i ekologiya chlenistonogikh Zabaikal'ya i Pribaikal'ya]. Sibirskoe Otdelenie, Adakemiya Nauk SSSR, Ulan-Ude, Russia. 157 pp. [In Russian.]

Beard, C. E. 2002. Colonization of black flies (Diptera: Simuliidae) by trichomycete fungi (Zygomycota) in South Carolina, USA. Ph.D. thesis. Clemson University, Clemson, S.C. 126 pp.

Beard, C. E., & P. H. Adler. 2000. Bionomics, axenic culture, and substrate-related variation in trichospores of *Smittium megazygosporum*. Mycologia 92:296–300.

Beard, C. E., & P. H. Adler. 2002. Seasonality of trichomycetes in larval black flies from South Carolina, USA. Mycologia 94:200–209.

Beaucournu-Saguez, F., & Y. Braverman. 1987. A new species of *Levitinia* Chubareva and Petrova (Diptera: Simuliidae) from the Golan Heights, Israel. Annales de Parasitologie Humaine et Comparée 62:59–75.

Beaudoin, R. [L.], & W. Wills. 1965. A description of *Caudospora pennsylvanica* sp. n. (Caudosporidae, Microsporidia), a parasite of the larvae of the black fly, *Prosimulium magnum* Dyar and Shannon. Journal of Invertebrate Pathology 7:152–155.

Beaudoin, R. L., & W. Wills. 1968. *Haplosporidium simulii* sp. n. (Haplosporida: Haplosporidiidae), parasitic in larvae of *Simulium venustum* Say. Journal of Invertebrate Pathology 10:374–378.

Beck, B. E. 1980. Clinical studies of the effect of the black fly *Simulium arcticum* on cattle. Pp. 233–237. *In* W. O. Haufe & G. C. R. Croome (eds.), Control of black flies in the Athabasca River. Alberta Environment Technical Report. Alberta Environment, Pollution Control Division, Edmonton. 241 pp.

Becker, C. D. 1973. Development of *Simulium* (*Psilozia*) *vittatum* Zett. (Diptera: Simuliidae) from larvae to adults at thermal increments from 17.0 to 27.0 C. American Midland Naturalist 89:246–251.

Becker, T. 1903. Ägyptische Dipteren. Mitteilungen aus dem Zoologischen Museum in Berlin 2:67–80 + 4 plates.

Becker, T. 1908. Dipteren der Kanarischen Inseln. Mitteilungen aus dem Zoologischen Museum in Berlin 4:1–180 + 4 plates.

Beckett, S. A. 1992. Microdistribution and colonization of *Simulium vittatum* (Diptera: Simuliidae) larvae on natural and artificial substrates. M.S. thesis. University of Windsor, Windsor, Ontario. 290 pp.

Bedo, D. G. 1973. Sibling species and sex chromosomes in black flies of the

Simulium pictipes group (Diptera: Simuliidae). M.S. thesis. University of Toronto, Toronto. 72 pp.
Bedo, D. G. 1975a. Polytene chromosomes of three species of blackflies in the *Simulium pictipes* group (Diptera: Simuliidae). Canadian Journal of Zoology 53:1147–1164.
Bedo, D. G. 1975b. C banding in polytene chromosomes of *Simulium ornatipes* and *S. melatum* (Diptera: Simuliidae). Chromosoma 51:291–300.
Bedo, D. G. 1976. Polytene chromosomes in pupal and adult blackflies (Diptera: Simuliidae). Chromosoma 57:387–396.
Bedo, D. G. 1977. Cytogenetics and evolution of *Simulium ornatipes* Skuse (Diptera: Simuliidae). I. Sibling speciation. Chromosoma 64:37–65.
Bedo, D. G. 1979. Cytogenetics and evolution of *Simulium ornatipes* Skuse (Diptera: Simuliidae). II. Temporal variation in chromosomal polymorphisms and homosequential sibling species. Evolution 33:296–308.
Bell, R. A. 1960. Observations on the biology and overwintering of certain Simuliidae in Northwest Arkansas. M.S. thesis. University of Arkansas, Fayetteville. 62 pp.
Bellardi, L. 1859. Saggio di ditterologia messicana. Parte I. Memoire della Reale Accademia delle Scienze di Torino (Series 2) 19:1–80 + 2 plates.
Bellardi, L. 1862. Saggio di ditterologia messicana. Appendice. Memoire della Reale Accademia delle Scienze di Torino (Series 2) 21:1–28 + 1 plate.
Bellec, C. 1976. Captures d'adultes de *Simulium damnosum* Theobald, 1903 (Diptera, Simuliidae) á l'aide de plaques d'aluminium, en Afrique de l'Ouest. Cahiers ORSTOM (Entomologie Médicale et Parasitologie) 14:209–217.
Belluck, D., & J. Furnish. 1981. *Psephidonus brunneus* (Say): a stream inhabiting staphylinid beetle (Coleoptera: Staphylinidae) associated with blackfly larvae (Diptera: Simuliidae). Coleopterists Bulletin 35:204.
Benbrook, E. A. 1943. External parasites of poultry. Pp. 599–636. *In* H. E. Biester & L. DeVries (eds.), Diseases of poultry. Collegiate Press, Ames, Iowa. 1005 pp.
Benke, A. C., & K. A. Parsons. 1990. Modelling black fly production dynamics in blackwater streams. Freshwater Biology 24:167–180.
Benke, A. C., T. C. Van Arsdall Jr., & D. M. Gillespie. 1984. Invertebrate productivity in a subtropical blackwater river: the importance of habitat and life history. Ecological Monographs 54:25–63.
Bennet, D., & T. Tiner. 1993. Up north: a guide to Ontario's wilderness from blackflies to the northern lights. McClelland & Stewart, Toronto. 316 pp.
Bennett, G. F. 1960. On some ornithophilic blood-sucking Diptera in Algonquin Park, Ontario, Canada. Canadian Journal of Zoology 38:377–389.
Bennett, G. F. 1961. On the specificity and transmission of some avian trypanosomes. Canadian Journal of Zoology 39:17–33.
Bennett, G. F. 1963a. Use of $P^{32}$ in the study of a population of *Simulium rugglesi* (Diptera: Simuliidae) in Algonquin Park, Ontario. Canadian Journal of Zoology 41:831–840.
Bennett, G. F. 1963b. The salivary gland as an aid in the identification of some simuliids. Canadian Journal of Zoology 41:947–952.
Bennett, G. F., & R. F. Coombs. 1975. Ornithophilic vectors of avian hematozoa in insular Newfoundland. Canadian Journal of Zoology 53:1241–1246.
Bennett, G. F., & A. M. Fallis. 1960. Blood parasites of birds in Algonquin Park, Canada, and a discussion of their transmission. Canadian Journal of Zoology 38:261–273.
Bennett, G. F., & A. M. Fallis. 1971. Flight range, longevity, and habitat preference of female *Simulium euryadminiculum* Davies (Diptera: Simuliidae). Canadian Journal of Zoology 49:1203–1207.
Bennett, G. F., & C. D. MacInnes. 1972. Blood parasites of geese of the McConnell River, N. W. T. Canadian Journal of Zoology 50:1–4.
Bennett, G. F., & M. A. Peirce. 1992. Leucocytozoids of seven Old World passeriform families. Journal of Natural History 26:693–707.
Bennett, G. F., & D. Squires-Parsons. 1992. The leucocytozoids of the avian families Anatidae and Emberizidae s. l., with descriptions of three new *Leucocytozoon* species. Canadian Journal of Zoology 70:2007–2014.
Bennett, G. F., P. C. C. Garnham, & A. M. Fallis. 1965. On the status of the genera *Leucocytozoon* Ziemann, 1898 and *Haemoproteus* Kruse, 1890 (Haemosporidiida: Leucocytozoidae and Haemoproteidae). Canadian Journal of Zoology 43:927–932.
Bennett, G. F., A. M. Fallis, & A. G. Campbell. 1972. The response of *Simulium* (*Eusimulium*) *euryadminiculum* Davies (Diptera: Simuliidae) to some olfactory and visual stimuli. Canadian Journal of Zoology 50:793–800.
Bennett, G. F., R. A. Earlé, M. A. Peirce, F. W. Huchzermeyer, & D. Squires-Parsons. 1991. Avian Leucocytozoidae: the leucocytozoids of the Phasianidae *sensu lato*. Journal of Natural History 25:1407–1428.
Bennett, G. F., R. A. Earlé, & M. A. Peirce. 1993. The Leucocytozoidae of South African birds: the Falconiformes and Strigiformes. Ostrich 64:67–72.
Benny, G. L., & K. O'Donnell. 2000. *Amoebidium parasiticum* is a protozoan, not a Trichomycete. Mycologia 92:1133–1137.
Benoit, P. (ed.). 1975. Noms français d'insectes au Canada. 4th ed. Agriculture Québec Publication No. QA38—R4-30. Ministère de l'Agriculture, Québec. 214 pp.
Bequaert, J. C. 1934. Part III. Notes on the black-flies or Simuliidae, with special reference to those of the *Onchocerca* region of Guatemala. Pp. 175–224. *In* R. P. Strong, J. H. Sandground, J. C. Bequaert, & M. M. Ochoa (eds.), Onchocerciasis with special reference to the Central American form of the disease. Contributions from the Department of Tropical Medicine and the Institute for Tropical Biology and Medicine, Harvard 6:1–234.
Bequaert, J. [C.]. 1945. Dr. Luis Vargas on American black-flies—a review, with critical notes (Diptera). Bulletin of the Brooklyn Entomological Society 40:111–115.
Bernardo, M. J. 1986. Rearing and colonization of black flies (Diptera: Simuliidae) with application to life table statistics and *in vitro* feeding studies. M.S. thesis. Cornell University, Ithaca, N.Y. 134 pp.
Bernardo, M. J., & E. W. Cupp. 1986. Rearing black flies (Diptera: Simuliidae) in the laboratory: mass-scale in vitro membrane-feeding and its application to collection of saliva and to parasitological and repellent studies. Journal of Medical Entomology 23:666–679.
Bernardo, M. J., E. W. Cupp, & A. E. Kiszewski. 1986a. Rearing black flies (Diptera: Simuliidae) in the laboratory: colonization and life table statistics for *Simulium vittatum*. Annals of the Entomological Society of America 79:610–621.
Bernardo, M. J., E. W. Cupp, & A. E. Kiszewski. 1986b. Rearing black flies (Diptera: Simuliidae) in the laboratory: bionomics and life table statistics for *Simulium pictipes*. Journal of Medical Entomology 23:680–684.
Bidlack, D. S. 1997. Factors influencing filter-feeding behavior of *Simulium pictipes* Hagen (Diptera: Simuliidae). Ph.D. thesis. Clemson University, Clemson, S.C. 108 pp.
Bierer, B. W. 1954. Buffalo gnats and *Leucocytozoon* infections of poultry. Veterinary Medicine 49:107–110, 115.
Bishopp, F. C. 1913. Some important insect enemies of live stock in the United States. Pp. 383–396. *In* Yearbook of the United States Department of Agriculture 1912. Government Printing Office, Washington, D.C. 784 pp.
Bishopp, F. C. 1935. Buffalo gnats (*Eusimulium* spp.). United States

Department of Agriculture Insect Pest Survey Bulletin 15:320.
Bishopp, F. C. 1942. Some insect pests of horses and mules. Pp. 492–500. *In* Keeping livestock healthy. Yearbook of Agriculture 1942. U.S. Department of Agriculture, Washington, D.C. 1276 pp.
Blake, B. J., G. S. Cooper, B. V. Helson, C. S. Kirby, C. A. Curry, F. L. McEwen, & D. W. Wilson. 1981. Controlling mosquitoes and blackflies in Ontario. Ministry of the Environment, Toronto. 22 pp.
Boakye, D. A., C. Back, & P. M. Brakefield. 2000. Evidence of multiple mating and hybridization in *Simulium damnosum* s. l. (Diptera: Simuliidae) in nature. Journal of Medical Entomology 37:29–34.
Boardman, E. T. 1939. Field guide to the lower aquarium animals. Cranbrook Institute of Science Bulletin 16:1–186.
Bodrova, Yu. D. 1977. Black flies (Diptera, Simuliidae) of Chukotka. Entomofauna of the Far East [Entomofauna Dal'nevo Vostocka] 46:95–108. [In Russian.]
Bodrova, Yu. D. 1980. Parallelism in the variability of certain morphological features in black flies (Diptera, Simuliidae). Pp. 70–80. *In* V. A. Krasilov (ed.), Parallelism and direction in insect evolution [Parallelizm i napravlennoct' evolutsii nasekomych]. Akademiya Nauk SSSR, Vladivostok, Russia. 128 pp. [In Russian.]
Bodrova, Yu. D. 1988. New and little known species of black flies (Simuliidae) from the Russian Far East. Parazitologiya 22:143–148. [In Russian, English summary.]
Boisvert, M., & J. Boisvert. 2000. Effects of *Bacillus thuringiensis* var. *israelensis* on target and nontarget organisms: a review of laboratory and field experiments. Biocontrol Science and Technology 10:517–561.
Boisvert, M., J. Boisvert, & A. Aubin. 2001a. A new field procedure and method of analysis to evaluate the performance of *Bacillus thuringiensis* subsp. *israelensis* liquid formulations in streams and rivers. Biocontrol Science and Technology 11:261–271.
Boisvert, M., J. Boisvert, & A. Aubin. 2001b. Factors affecting black fly larval mortality and carry of two formulations of *Bacillus thuringiensis* subsp. *israelensis* tested in the same stream during a 3-year experiment. Biocontrol Science and Technology 11:711–725.
Boisvert, M., J. Boisvert, & A. Aubin. 2001c. Factors affecting residual dosages of two formulations of *Bacillus thuringiensis* subsp. *israelensis* tested in the same stream during a 3-year experiment. Biocontrol Science and Technology 11:727–744.
Boisvert, M., J. Boisvert, & A. Aubin. 2002. Influence of stream profile and abiotic factors on the performance and residual dosages of two commercial formulations of *Bacillus thuringiensis* subsp. *israelensis* during a 2-year experiment. Biocontrol Science and Technology 12:19–33.
Boo, K. S., & D. M. Davies. 1980. Johnston's organ of the black fly *Simulium vittatum* Zett. Canadian Journal of Zoology 58:1969–1979.
Boobar, L. R. 1979. Habitat distribution and evaluation of sampling methods for *Simulium penobscotensis* (Diptera: Simuliidae). M.S. thesis. University of Maine, Orono. 39 pp.
Boobar, L. R., & J. Granett. 1978. Evaluation of polyethylene samplers for black fly larvae (Diptera: Simuliidae), with particular reference to Maine species. Canadian Journal of Zoology 56:2245–2248.
Boobar, L. R., & J. Granett. 1980. *Simulium penobscotensis* (Diptera: Simuliidae) habitat characteristics in the Penobscot River, Maine. Environmental Entomology 9:412–415.
Boreham, M. M. 1962. A new record of *Simulium* (*Psilozia*) *encisoi* Vargas and Díaz Nájera, in California. Mosquito News 22:403.
Borkent, A. 1992. *Parasimulium* (Diptera: Simuliidae), a black fly genus new to the extant fauna of Canada. Canadian Entomologist 124:743–744.
Borkent, A., & D. C. Currie. 2001. Discovery of the female of *Parasimulium* (*Astoneomyia*) *melanderi* Stone (Diptera: Simuliidae) in a cave in British Columbia, with a discussion of its phylogenetic position. Proceedings of the Entomological Society of Washington 103:546–553.
Borkent, A., & D. M. Wood. 1986. The first and second larval instars and the egg of *Parasimulium stonei* Peterson (Diptera: Simuliidae). Proceedings of the Entomological Society of Washington 88:287–296.
Boucias, D. G., J. J. Becnel, S. E. White, & M. Bott. 2001. In vivo and in vitro development of the protist *Helicosporidium* sp. Journal of Eukaryotic Microbiology 48:460–470.
Bowles, D. E., & D. D. Pinkovsky. 1993. Occurrence of larval black flies and horse flies in an Ozark headwater stream. Southwestern Naturalist 38:86–87.
Boyer, R. J. 1989. Human nuisance black flies (Diptera: Simuliidae) of the Penobscot River, Maine. M.S. thesis. University of Maine, Orono. 102 pp.
Bradbury, W. C. 1972. Experiments and observations on the near-host orientation and landing behavior of Simuliidae (Diptera). M.S. thesis. Memorial University of Newfoundland, St. John's. 139 pp.
Bradbury, W. C., & G. F. Bennett. 1974a. Behavior of adult Simuliidae (Diptera). I. Responses to color and shape. Canadian Journal of Zoology 52:251–259.
Bradbury, W. C., & G. F. Bennett. 1974b. Behavior of adult Simuliidae (Diptera). II. Vision and olfaction in near-orientation and landing. Canadian Journal of Zoology 52:1355–1364.
Bradley, G. H. 1932. Southern buffalo gnat (*Eusimulium pecuarum* Riley). United States Department of Agriculture Insect Pest Survey Bulletin 12:182.
Bradley, G. H. 1934. Buffalo gnats (*Simulium* spp.). United States Department of Agriculture Insect Pest Survey Bulletin 14:92.
Bradley, G. H. 1935a. Notes on the southern buffalo gnat, *Eusimulium pecuarum* (Riley) (Diptera: Simuliidae). Proceedings of the Entomological Society of Washington 37:60–64.
Bradley, G. H. 1935b. The hatching of eggs of the southern buffalo gnat. Science 82:277–278.
Bradley, G. H., & T. E. McNeel. 1928. Buffalo gnat (*Simulium pecuarum* Riley). United States Department of Agriculture Insect Pest Survey Bulletin 8:79.
Bradt, S. 1932. Notes on Puerto Rican black flies. Puerto Rico Journal of Public Health and Tropical Medicine 8:69–79 + 2 plates.
Braimah, S. A. 1985. Mechanisms and fluid mechanical aspects of filter-feeding in blackfly larvae (Diptera: Simuliidae) and mayfly nymphs (Ephemeroptera: Oligoneuriidae). Ph.D. thesis. University of Alberta, Edmonton. 124 pp.
Braimah, S. A. 1987a. Mechanisms of filter feeding in immature *Simulium bivittatum* Malloch (Diptera: Simuliidae) and *Isonychia campestris* McDunnough (Ephemeroptera: Oligoneuriidae). Canadian Journal of Zoology 65:504–513.
Braimah, S. A. 1987b. Pattern of flow around filter-feeding structures of immature *Simulium bivittatum* Malloch (Diptera: Simuliidae) and *Isonychia campestris* McDunnough (Ephemeroptera: Oligoneuriidae). Canadian Journal of Zoology 65:514–521.
Braimah, S. A. 1987c. The influence of water velocity on particle capture by the labral fans of larvae of *Simulium bivittatum* Malloch (Diptera: Simuliidae). Canadian Journal of Zoology 65:2395–2399.
Brancato, J. C. 1996. Environmental impact of Pennsylvania black fly control programs on fish and aquatic invertebrates. Proceedings of the Annual Meeting of the New Jersey Mosquito Control Association 83:64–71.
Brenner, R. J. 1980. Rearing and colonizing black flies with development of life tables for geographic strains of *Simulium decorum*. Ph.D. thesis. Cornell University, Ithaca, N.Y. 95 pp.

Brenner, R. J., & E. W. Cupp. 1980a. Rearing black flies (Diptera: Simuliidae) in a closed system of water circulation. Tropenmedizin und Parasitologie 31:247–258.

Brenner, R. J., & E. W. Cupp. 1980b. Preliminary observations on parity and nectar feeding in the black fly, *Simulium jenningsi*. Mosquito News 40:390–393.

Brenner, R. J., E. W. Cupp, & M. J. Bernardo. 1980. Laboratory colonization and life table statistics for geographic strains of *Simulium decorum* (Diptera: Simuliidae). Tropenmedizin und Parasitologie 31:487–497.

Brenner, R. J., E. W. Cupp, & M. J. Bernardo. 1981. Growth and development of geographic and crossbred strains of colonized *Simulium decorum* (Diptera: Simuliidae). Canadian Journal of Zoology 59:2072–2079.

Brereton, C., W. A. House, P. D. Armitage, & R. S. Wotton. 1999. Sorption of pesticides to novel materials: snail pedal mucus and blackfly silk. Environmental Pollution 105:55–65.

Bridges, V. E., B. J. McCluskey, M. D. Salman, H. S. Hurd, & J. Dick. 1997. Review of the 1995 vesicular stomatitis outbreak in the western United States. Journal of the American Veterinary Medical Association 211:556–560.

Brimley, C. S. 1938. The insects of North Carolina, being a list of the insects of North Carolina and their close relatives. Division of Entomology, North Carolina Department of Agriculture. Raleigh, N.C. 560 pp.

Brockhouse, C. L. 1984. Sibling species and sex chromosomes in *Eusimulium vernum* (Diptera: Simuliidae). M.S. thesis. University of Toronto, Toronto. 94 pp.

Brockhouse, C. [L.]. 1985. Sibling species and sex chromosomes in *Eusimulium vernum* (Diptera: Simuliidae). Canadian Journal of Zoology 63:2145–2161.

Brockhouse, C. [L.]. 1991. Cytological and molecular studies of sibling species in the Simuliidae (Diptera). Ph.D. thesis. University of Toronto, Toronto. 192 pp.

Brockhouse, C. L., & P. H. Adler. 2002. Cytogenetics of laboratory colonies of *Simulium vittatum* cytospecies IS-7 (Diptera: Simuliidae). Journal of Medical Entomology 39:293–297.

Brockhouse, C. L., & R. M. Tanguay. 1996. A rapid method for the detection of sibling species in samples of black fly larvae (Diptera: Simuliidae). Journal of Agricultural Entomology 13:339–347.

Brockhouse, C. [L.], J. A. B. Bass, & N. A. Straus. 1989. Chromocentre polymorphism in polytene chromosomes of *Simulium costatum* (Diptera: Simuliidae). Genome 32:510–515.

Brockhouse, C. L., C. G. Vajime, R. Marin, & R. M. Tanguay. 1993. Molecular identification of onchocerciasis vector sibling species in black flies (Diptera: Simuliidae). Biochemical and Biophysical Research Communications 194:628–634.

Brown, A. W. A. 1952. Rotary brush units for aerial spraying against mosquitoes and black flies. Journal of Economic Entomology 45:620–625.

Brown, A. W. A., & R. Pal. 1971. Insecticide resistance in arthropods. 2d ed. World Health Organization, Geneva. 491 pp.

Brown, A. W. A., R. P. Thompson, C. R. Twinn, & L. K. Cutkomp. 1951. Control of adult mosquitoes and black flies by DDT sprays applied from aircraft. Mosquito News 11:75–84.

Brown, B. [V.]. 1990. Using Peldri II as an alternative to critical point drying for small flies. Fly Times 4:6.

Brown, B. V. 1993. A further chemical alternative to critical-point-drying for preparing small (or large) flies. Fly Times 11:10.

Brown, H., & H. S. Bernton. 1970. A case of asthma caused by *Simulium jenningsi* (order Diptera) protected by hyposensitization. Journal of Allergy 45:103–104.

Browne, J. H. 1865. Four years in Secessia. O. D. Case, Hartford, Conn. 450 pp.

Browne, S. M., & G. F. Bennett. 1980. Color and shape as mediators of host-seeking responses of simuliids and tabanids (Diptera) in the Tantramar marshes, New Brunswick, Canada. Journal of Medical Entomology 17:58–62.

Bruder, K. W. 1974. The blackflies (Simuliidae: Diptera) of the Stony Brook Watershed of New Jersey with emphasis on parasitism by mermithid nematodes (Mermithidae: Nematoda). Ph.D. thesis. Rutgers University, New Brunswick, N.J. 114 pp.

Bruder, K. W., & W. J. Crans. 1979. The black flies (Simuliidae: Diptera) of the Stony Brook watershed of New Jersey, with emphasis on parasitism by mermithid nematodes (Mermithidae: Nematoda). New Jersey (Rutgers) Agricultural Experiment Station Bulletin 851:1–21.

Brues, C. T. 1932. Further studies on the fauna of North American hot springs. Proceedings of the American Academy of Arts and Sciences 67:184–303.

Bryson, H. R. 1931. Buffalo gnats (Simuliidae). United States Department of Agriculture Insect Pest Survey Bulletin 11:131.

Buerger, G. 1967. Sense organs on the labra of some blood-feeding Diptera. Quaestiones Entomologicae 3:283–290.

Burdick, G. E., H. J. Dean, E. J. Harris, J. Skea, C. Frisa, & C. Sweeney. 1968. Methoxychlor as a blackfly larvicide, persistence of its residues in fish and its effect on stream arthropods. New York Fish and Game Journal 15:121–142.

Burg, J. G., D. G. Powell, & F. W. Knapp. 1991. Arthropod faunal composition on Kentucky equine premises. Journal of Medical Entomology 28:658–662.

Burger, J. F. (ed.). 1977. Proceedings of the First Inter-regional Conference on North American Black Flies. Dixville Notch, New Hampshire, 30 January–2 February 1977. [no publisher]. 181 pp.

Burger, J. F. 1988 ["1987"]. Specialized habitat selection by black flies. Pp. 129–145. *In* K. C. Kim & R. W. Merritt (eds.), Black flies: ecology, population management, and annotated world list. Pennsylvania State University, University Park. 528 pp. [Actual publication date was 1 March 1988.]

Burger, J. F., & L. A. Pistrang. 1987. Is *Simulium tuberosum* (Diptera: Simuliidae) a pest of humans? A problem of interpretation and sibling species. Entomological News 98:53–62.

Burghardt, H. F., J. H. Whitlock, & P. J. McEnerney. 1951. Dermatitis in cattle due to *Simulium* (black flies). Cornell Veterinarian 41:311–313.

Burgin, S. G. 1996. An investigation of sugar feeding by black flies (Diptera: Simuliidae). M.S. thesis. Brock University, St. Catharines, Ontario. 169 pp.

Burgin, S. G., & F. F. Hunter. 1997a. Nectar versus honeydew as sources of sugar for male and female black flies (Diptera: Simuliidae). Journal of Medical Entomology 34:605–608.

Burgin, S. G., & F. F. Hunter. 1997b. Evidence of honeydew feeding in black flies (Diptera: Simuliidae). Canadian Entomologist 129:859–869.

Burgin, S. G., & F. F. Hunter. 1997c. Sugar-meal sources used by female black flies (Diptera: Simuliidae): a four-habitat study. Canadian Journal of Zoology 75:1066–1072.

Burrill, A. C. 1922. Turkey gnat (*Simulium meridionale* Riley). United States Department of Agriculture Insect Pest Survey Bulletin 2:103.

Burton, D. K. 1984. Impact of *Bacillus thuringiensis* var. *israelensis* in dosages used for black fly (Simuliidae) control, against target and non-target organisms in the Torch River, Saskatchewan. M.S. thesis. University of Manitoba, Winnipeg. 129 pp.

Burton, G. J. 1971. Cannibalism among *Simulium damnosum* (Simulidae [*sic*]) larvae. Mosquito News 31:602–603.

Burton, G. J. 1973. Feeding of *Simulium hargreavesi* Gibbins larvae on *Oedegonium* algal filaments in Ghana. Journal of Medical Entomology 10:101–106.

Bushnell, J. H., S. Q. Foster, & B. M. Wahle. 1987. Annotated inventory of invertebrate populations of an alpine lake and

stream chain in Colorado. Great Basin Naturalist 47:500–511.
Bušta, J., & V. Našincová. 1986. Record of *Plagiorchis neomidis* Brendow, 1970 (Trematoda: Plagiorchidae) in Czechoslovakia and studies on its life cycle. Folia Parasitologica 33:123–129 + 2 plates.
Byrd, B. 1984 March 20. Black fly report suggests more studies of problem. Charleston Daily Mail (Charleston, W.V.), p. 3B.
Byrd, M. A. 1959. Observations on *Leucocytozoon* in pen-raised and free-ranging wild turkeys. Journal of Wildlife Management 23:145–156.
Byrtus, G., & R. Jackson. 1988. Monitoring of the black fly (Diptera: Simuliidae) abatement program on the Athabasca River—1986. Alberta Environment, Edmonton. 88 pp.
Cameron, A. E. 1918. Some blood-sucking flies of Saskatchewan. Agricultural Gazette of Canada 5:556–561. [The identical paper, without figures, was published the same year in Journal of the American Veterinary Medical Association 53:632–638.]
Cameron, A. E. 1922. The morphology and biology of a Canadian cattle-infesting black fly, *Simulium simile* Mall. (Diptera, Simuliidae). Canada Department of Agriculture Bulletin (New Series) (Entomological Bulletin 20) 5:1–26.
Canning, E. U., & E. I. Hazard. 1982. Genus *Pleistophora* Gurley, 1893: an assemblage of at least three genera. Journal of Protozoology 29:39–49.
Carestia, R. R., R. L. Frommer, R. W. Vavra Jr., & V. A. Loy. 1974. Field evaluation of blackfly control—aerial applications. Mosquito News 34:330–332.
Carlough, L. A. 1990. Sestonic protists in the foodweb of a southeastern blackwater river. Ph.D. thesis. University of Georgia, Athens. 186 pp.
Carlson, D. A., & J. F. Walsh. 1981. Identification of two West African black flies (Diptera: Simuliidae) of the *Simulium damnosum* species complex by analysis of cuticular paraffins. Acta Tropica 38:235–239.
Carlsson, G. 1962. Studies on Scandinavian black flies (Fam. Simuliidae Latr.). Opuscula Entomologica 21 (Suppl.):1–280.
Carlsson, G. 1966. Cytology of the black fly *Prosimulium frohnei* Sommerman. Hereditas 55:73–78.
Carlsson, G. 1968. Morphology of the black fly *Prosimulium frohnei* Sommerman (Dipt. Simuliidae). Opuscula Entomologica 33:191–196.
Carlsson, M., L. M. Nilsson, B. Svensson, S. Ulfstrand, & R. S. Wotton. 1977. Lacustrine seston and other factors influencing the blackflies (Diptera: Simuliidae) inhabiting lake outlets in Swedish Lapland. Oikos 29:229–238.
Carstens, J. 2001. Black flies. Pest Control 69(7):68.
Castello Branco, A., Jr. 1999. Effects of *Polydispyrenia simulii* (Microspora; Duboscqiidae) on development of the gonads of *Simulium pertinax* (Diptera; Simuliidae). Memórias do Instituto Oswaldo Cruz 94:421–424.
Cawthon, J. 1981 October 4. Entomologist defends black fly eradicator. Gazette-Mail (Charleston, W.V.), p. 4A.
Chambers, L., & J. R. Hill III. 1996. Dr's house calls. Purple Martin Update 7(1):12–17.
Chance, M. M. 1969. Functional morphology of the mouthparts of blackfly larvae (Diptera: Simuliidae). M.S. thesis. University of Alberta, Edmonton. 132 pp.
Chance, M. M. 1970a. The functional morphology of the mouthparts of blackfly larvae (Diptera: Simuliidae). Quaestiones Entomologicae 6:254–284.
Chance, M. M. 1970b. A review of chemical control methods for blackfly larvae (Diptera: Simuliidae). Quaestiones Entomologicae 6:287–292.
Chance, M. M. 1977. Influence of water flow and particle concentration on larvae of the black fly *Simulium vittatum* Zett. (Diptera: Simuliidae), with emphasis on larval filter-feeding. Ph.D. thesis. University of Alberta, Edmonton. 236 pp.
Chance, M. M., & D. A. Craig. 1986. Hydrodynamics and behaviour of Simuliidae larvae (Diptera). Canadian Journal of Zoology 64:1295–1309.
Charnetski, W. A., & W. O. Haufe. 1981. Control of *Simulium arcticum* Malloch in northern Alberta, Canada. Pp. 117–132. *In* M. Laird (ed.), Blackflies: the future for biological methods in integrated control. Academic Press, New York. 399 pp.
Charpentier, B., & A. Morin. 1994. Effect of current velocity on ingestion rates of black fly larvae. Canadian Journal of Fisheries and Aquatic Sciences 51:1615–1619.
Charpentier, G., C. Back, S. Garzon, & G. Strykowski. 1986. Observations on a new intranuclear virus-like particle infecting larvae of a black fly *Simulium vittatum* (Diptera: Simuliidae). Diseases of Aquatic Organisms 1:147–150.
Chatton, E., & E. Roubaud. 1909. Sur un *Amoebidium* du rectum des larves de simulies (*Simulium argyreatum* Meig. et *S. fasciatum* Meig.). Comptes Rendus Hebdomadaires des Séances et Mémoires de la Société de Biologie 66:701–703.
Cheer, A. Y. L., & M. A. R. Koehl. 1987. Fluid flow through filtering appendages of insects. IMA Journal of Mathematics Applied in Medicine and Biology 4:185–199.
Chen, A. W. 1969. Histology and histochemistry of the ovary during oogenesis in the autogenous black-fly *S. vittatum* Zett. M.S. thesis. McMaster University, Hamilton, Ontario. 100 pp.
Chivers-Wilson, V. S. 1952. A preliminary investigation of the pharmacological properties of ustilagic acid and the presence of histamine and an anticoagulant in the black fly *Simulium venustum* Say (Diptera: Simuliidae). M.S. thesis. University of Saskatchewan, Saskatoon. 98 + 51 pp.
Chmielewski, C. M., & R. J. Hall. 1992. Response of immature blackflies (Diptera: Simuliidae) to experimental pulses of acidity. Canadian Journal of Fisheries and Aquatic Sciences 49:833–840.
Chmielewski, C. M., & R. J. Hall. 1993. Changes in the emergence of blackflies (Diptera: Simuliidae) over 50 years from Algonquin Park streams: is acidification the cause? Canadian Journal of Fisheries and Aquatic Sciences 50:1517–1529.
Choate, L. A. 1984. A cytological description of *Eusimulium vernum*, *E.* X, and the *E. pugetense* complex (Diptera: Simuliidae) in Alaska. M.S. thesis. University of Alaska, Fairbanks. 122 pp.
Choe, J. C., P. H. Adler, K. C. Kim, & R. A. J. Taylor. 1984. Flight patterns of *Simulium jenningsi* (Diptera: Simuliidae) in central Pennsylvania, USA. Journal of Medical Entomology 21:474–476.
Chubareva, L. A. 1978. Karyotypical peculiarities of the genus *Prosimulium* Roub. (Simuliidae) and problems of taxonomy. Parazitologiya 12:37–43. [In Russian; English summary.]
Chubareva, L. A., & N. A. Petrova. 1979. Basic characteristics of karyotypes of black flies (Diptera, Simuliidae) of the world fauna. Pp. 58–95. *In* O. A. Skarlato & L. A. Chubareva (eds.), Karyosystematics of the invertebrate animals [Kariosistematika bespozvonochnych jivotnych]. Akademia Nauk SSSR, Leningrad [= St. Petersburg], Russia. 119 pp. + 11 plates. [In Russian.]
Chubareva, L. A., & N. A. Petrova. 1981. A new genus of black flies (Diptera, Simuliidae) from Tadzhikistan. Entomologicheskoe Obozrenie 60:898–900. [In Russian; English translation in Entomological Review, Washington 60:140–144.]
Chubareva, L. A., I. A. Rubtsov, & N. A. Petrova. 1976. Morphological and karyological similarities and differences between Palearctic and Neotropical species of the genus *Hemicnetha* End. (Diptera, Simuliidae). Entomologicheskoe Obozrenie 55:452–457. [In Russian;

English translation in Entomological Review, Washington 55:137–142.]
Chutter, F. M. 1970. A preliminary study of factors influencing the number of oocytes present in newly emerged blackflies (Diptera: Simuliidae) in Ontario. Canadian Journal of Zoology 48:1389–1400.
Ciborowski, J. J. H., & P. H. Adler. 1990. Ecological segregation of larval black flies (Diptera: Simuliidae) in northern Saskatchewan, Canada. Canadian Journal of Zoology 68:2113–2122.
Ciborowski, J. J. H., & D. A. Craig. 1989. Factors influencing dispersion of larval black flies (Diptera: Simuliidae): effects of current velocity and food concentration. Canadian Journal of Fisheries and Aquatic Sciences 46:1329–1341.
Ciborowski, J. J. H., & D. A. Craig. 1991. Factors influencing dispersion of larval black flies (Diptera: Simuliidae): effects of the presence of an invertebrate predator. Canadian Journal of Zoology 69:1120–1123.
Ciborowski, J. J. H., D. A. Craig, & K. M. Fry. 1997. Dissolved organic matter as food for black fly larvae (Diptera: Simuliidae). Journal of the North American Benthological Society 16:771–780.
Cibulsky, R. J., & R. A. Fusco. 1988 ["1987"]. Recent experiences with Vectobac for black fly control: an industrial perspective on future developments. Pp. 419–424. *In* K. C. Kim & R. W. Merritt (eds.), Black flies: ecology, population management, and annotated world list. Pennsylvania State University, University Park. 528 pp. [Actual publication date was 1 March 1988.]
Clifford, H. F., G. M. Wiley, & R. J. Casey. 1993. Macroinvertebrates of a beaver-altered boreal stream of Alberta, Canada, with special reference to the fauna on the dams. Canadian Journal of Zoology 71:1439–1447.
Cockerell, T. D. A. 1895. The buffalo gnat. Insect Life 7:426.
Cockerell, T. D. A. 1897. A buffalo gnat new to the United States. Entomological News 8:100.
Coffman, W. P. 1967. Community structure and trophic relations in a small woodland stream Linesville Creek, Crawford County, Pennsylvania. Ph.D. thesis. University of Pittsburgh, Penn. 573 pp.
Colbo, M. H. 1979. Distribution of winter-developing Simuliidae (Diptera), in eastern Newfoundland. Canadian Journal of Zoology 57:2143–2152.
Colbo, M. H. 1982. Size and fecundity of adult Simuliidae (Diptera) as a function of stream habitat, year, and parasitism. Canadian Journal of Zoology 60:2507–2513.
Colbo, M. H. 1984. Control of black flies (Simuliidae) using *Bacillus thuringiensis* var. *israelensis* (*Bti*) as a larvicide, with emphasis on the northern programs. Proceedings of the Annual Meeting of the Canadian Pest Management Society 31:98–109.
Colbo, M. H. 1985. Variation in larval black fly populations at three sites in a stream system over five years (Diptera: Simuliidae). Hydrobiologia 121:77–82.
Colbo, M. H. 1988 ["1987"]. Problems in estimating black fly populations in their aquatic stages. Pp. 77–89. *In* K. C. Kim & R. W. Merritt (eds.), Black flies: ecology, population management, and annotated world list. Pennsylvania State University, University Park. 528 pp. [Actual publication date was 1 March 1988.]
Colbo, M. H. 1989. *Simulium vittatum* (Simuliidae: Diptera), a black fly with a variable instar number. Canadian Journal of Zoology 67:1730–1732.
Colbo, M. H. 1990. Persistence of Mermithidae (Nematoda) infections in black fly (Diptera: Simuliidae) populations. Journal of the American Mosquito Control Association 6:203–206.
Colbo, M. H., & H. O'Brien. 1984. A pilot black fly (Diptera: Simuliidae) control program using *Bacillus thuringiensis* var. *israelensis* in Newfoundland. Canadian Entomologist 116:1085–1096.
Colbo, M. H., & A. N. Okaeme. 1988. The larval instars of *Cnephia ornithophilia* (Diptera: Simuliidae), a black fly with a variable molting pattern. Canadian Journal of Zoology 66:2084–2089.
Colbo, M. H., & G. N. Porter. 1979. Effects of the food supply on the life history of Simuliidae (Diptera). Canadian Journal of Zoology 57:301–306.
Colbo, M. H., & G. N. Porter. 1980. Distribution and specificity of Mermithidae (Nematoda) infecting Simuliidae (Diptera) in Newfoundland. Canadian Journal of Zoology 58:1483–1490.
Colbo, M. H., & G. N. Porter. 1981. The interaction of rearing temperature and food supply on the life history of two species of Simuliidae (Diptera). Canadian Journal of Zoology 59:158–163.
Colbo, M. H., & B. H. Thompson. 1978. An efficient technique for laboratory rearing of *Simulium verecundum* S. & J. (Diptera: Simuliidae). Canadian Journal of Zoology 56:507–510.
Colbo, M. H., & A. H. Undeen. 1980. Effect of *Bacillus thuringiensis* var. *israelensis* on non-target insects in stream trials for control of Simuliidae. Mosquito News 40:368–371.
Colbo, M. H., & R. S. Wotton. 1981. Preimaginal blackfly bionomics. Pp. 209–226. *In* M. Laird (ed.), Blackflies: the future for biological methods in integrated control. Academic Press, New York. 399 pp.
Colbo, M. H., R. M. K. W. Lee, D. M. Davies, & Y. J. Yang. 1979. Labro-cibarial sensilla and armature in adult black flies (Diptera: Simuliidae). Journal of Medical Entomology 15:166–175.
Cole, F. R. 1969. The flies of western North America. University of California Press, Berkeley. 693 pp.
Cole, F. R., & A. L. Lovett. 1921. An annotated list of the Diptera (flies) of Oregon. Proceedings of the California Academy of Sciences (Fourth Series) 11:197–344.
Cole, J. 1996. Neighbors against gnats. Proceedings of the Annual Meeting of the New Jersey Mosquito Control Association 83:72–74.
Coleman, L. C. 1938. Preparation of leuco basic fuchsin for use in the Feulgen reaction. Stain Technology 13:123–124.
Coleman, R. W. 1951. The Simuliidae of California. Ph.D. thesis. University of California, Berkeley. 173 pp.
Coleman, R. W. 1953. A new blackfly species from California. Proceedings of the Entomological Society of Washington 55:45–46.
Colgan, P. 1995. Interpreting the hierarchy of nature: from systematic patterns to evolutionary process theories [book review]. Global Biodiversity 5:39.
Collins, D. L., & H. Jamnback. 1958. Ten years of blackfly control in New York State. Proceedings of the Tenth International Congress of Entomology, Montreal (1956) 3:813–818.
Collins, D. L., B. V. Travis, & H. Jamnback. 1952. The application of larvicide by airplane for control of blackflies (Simuliidae). Mosquito News 12:75–77.
Comstock, J. H. 1924. An introduction to entomology. First complete edition. Comstock Publishing, Ithaca, N.Y. 1044 pp.
Comstock, J. H., & A. B. Comstock. 1895. A manual for the study of insects. Comstock Publishing, Ithaca, N.Y. 701 pp.
Condon, W. J. 1975. Some aspects of the host/parasite relations of Newfoundland blackflies and their mermithid parasites. M.S. thesis. Memorial University of Newfoundland, St. John's. 87 pp.
Condon, W. J., & R. Gordon. 1977. Some effects of mermithid parasitism on the larval blackflies *Prosimulium mixtum fuscum* and *Simulium venustum*. Journal of Invertebrate Pathology 29:56–62.
Condon, W. J., R. Gordon, & C. H. Bailey. 1976. Morphology of the neuroendocrine systems of two larval blackflies, *Prosimulium mixtum/fuscum* and *Simulium venustum*. Canadian Journal of Zoology 54:1579–1584.
Conn, J., K. H. Rothfels, W. S. Procunier, & H. Hirai. 1989. The *Simulium metallicum*

species complex (Diptera: Simuliidae) in Latin America: a cytological study. Canadian Journal of Zoology 67:1217–1245.

Conradi, A. F. 1905. Black-fly studies. United States Department of Agriculture Bureau of Entomology Bulletin 52:100–101.

Cooper, B. E. 1991. Diptera types in the Canadian National Collection of Insects. Part 1. Nematocera. Publication No. 1845/B. Research Branch, Agriculture Canada, Ottawa. 113 pp.

Cooter, R. J. 1982. Studies on the flight of black-flies (Diptera: Simuliidae). I. Flight performance of *Simulium ornatum* Meigen. Bulletin of Entomological Research 72:303–317.

Cooter, R. J. 1983. Studies on the flight of black-flies (Diptera: Simuliidae). II. Flight performance of three cytospecies in the complex of *Simulium damnosum* Theobald. Bulletin of Entomological Research 73:275–288.

Cope, O. B., C. M. Gjullin, & A. Storm. 1949. Effects of some insecticides on trout and salmon in Alaska, with reference to blackfly control. Transactions of the American Fisheries Society 77:160–177.

Coquillett, D. W. 1898. The buffalo-gnats, or black-flies, of the United States. United States Department of Agriculture Division of Entomology Bulletin (New Series) 10:66–69.

Coquillett, D. W. 1900. Papers from the Harriman Alaska Expedition. IX. Entomological results (3): Diptera. Proceedings of the Washington Academy of Sciences 2:389–464.

Coquillett, D. W. 1902. New Diptera from North America. Proceedings of the United States National Museum 25(1280):83–126.

Corbet, P. S. 1967. The diel oviposition periodicity of the black fly, *Simulium vittatum*. Canadian Journal of Zoology 45:583–584.

Corkum, L. D., & D. C. Currie. 1987. Distributional patterns of immature Simuliidae (Diptera) in northwestern North America. Freshwater Biology 17:201–221.

Cornelius, E. 1818. On the geology, mineralogy, scenery, and curiosities of parts of Virginia, Tennessee, and of the Alabama and Mississippi Territories, etc. with miscellaneous remarks, etc. American Journal of Science 1:317–331. [The identical information was repeated in the following reference: Osborne, H. 1888. An old American account of the buffalo gnat. Insect Life 1:224–225.]

Corredor, D. 1975. The black flies of Washington (Diptera: Simuliidae). M.S. thesis. Washington State University, Pullman. 94 pp.

Corti, E. 1914. Le simulie Italiane. Atti della Società Italiana di Scienze Naturali e del Museo Civico di Storia Naturale di Milano 53:192–206.

Coscarón, S. 1987. El género *Simulium* Latreille en la Región Neotropical: análisis de los grupos supraespecíficos, especies que los integran y distribución geográfica (Simuliidae, Diptera). Museu Paraense Emílio Goeldi. Belém, Brasil. 112 pp.

Coscarón, S. 1990. Taxonomía y distribución del subgénero *Simulium* (*Ectemnaspis*) Enderlein (Simuliidae, Diptera, Insecta). Iheringia (Série Zoologia) 70:109–170.

Coscarón, S., & S. Ibáñez Bernal. 1993. Sobre una nueva especie de Simuliidae del sur de Mexico: *Simulium* (*Hemicnetha*) *biuxinisa* n. sp. (Diptera). Folia Entomologica Mexicana 88:61–68.

Coscarón, S., & D. R. Miranda-Esquivel. 1998. *Pedrowygomyia*, a new Neotropical genus of Prosimuliini (Diptera: Simuliidae): *Gigantodax* s. lat. split into two genera. Entomologica Scandinavica 29:161–167.

Coscarón, S., S. Ibáñez Bernal, & C. L. Coscarón Arias. 1996. Revisión de *Simulium* (*Psilopelmia*) Enderlein en la Región Neotropical y análisis cladístico de sus especies (Diptera: Simuliidae). Acta Zoologica Mexicana (New Series) 69:37–104.

Coscarón, S., S. Ibáñez Bernal, & C. L. Coscarón Arias. 1999. Revision of *Simulium* (*Simulium*) in the Neotropical realm (Insecta: Diptera: Simuliidae). Memoirs on Entomology, International 14:543–604.

Coscarón Arias, C. L. 1998. The polytene chromosomes of *Cnesia dissimilis* (Edwards) and three species of *Gigantodax* Enderlein (Diptera: Simuliidae) from Lanin National Park (Argentina). Memórias do Instituto Oswaldo Cruz 93:445–458.

Cottral, G. E. 1938. Buffalo gnats in the ears of horses. Veterinary Medicine 33:376–377.

Couch, C. A., J. L. Meyer, & R. O. Hall Jr. 1996. Incorporation of bacterial extracellular polysaccharide by black fly larvae (Simuliidae). Journal of the North American Benthological Society 15:289–299.

Coupland, J. B. 1992. Effect of egg mass age on subsequent oviposition by *Simulium reptans* (Diptera: Simuliidae). Journal of Medical Entomology 29:293–295.

Courtney, G. W. 1986. Discovery of the immature stages of *Parasimulium crosskeyi* Peterson (Diptera: Simuliidae), with a discussion of a unique black fly habitat. Proceedings of the Entomological Society of Washington 88:280–286.

Courtney, G. W. 1993. Archaic black flies and ancient forests: conservation of *Parasimulium* habitats in the Pacific Northwest. Aquatic Conservation: Marine and Freshwater Ecosystems 3:361–373.

Cowell, B. C., & W. C. Carew. 1976. Seasonal and diel periodicity in the drift of aquatic insects in a subtropical Florida stream. Freshwater Biology 6:587–594.

Cox, J. A. 1938. Morphology of the digestive tract of the blackfly (*Simulium nigroparvum*). Journal of Agricultural Research 57:443–448.

Craig, D. A. 1968. The clarification of a discrepancy in descriptions of maxillary musculature in larval Simuliidae. Quaestiones Entomologicae 4:31–32.

Craig, D. A. 1969. The embyrogenesis of the larval head of *Simulium venustum* Say (Diptera: Nematocera). Canadian Journal of Zoology 47:495–503.

Craig, D. A. 1972a. Rapid orientation of wax embedded specimens. Quaestiones Entomologicae 8:61–62.

Craig, D. A. 1972b. Observing insect embryos with a scanning electron microscope. Canadian Entomologist 104:761–763.

Craig, D. A. 1972c. Blackflies. University of Alberta Extension Leaflet 921-a:1–3.

Craig, D. A. 1974. The labrum and cephalic fans of larval Simuliidae (Diptera: Nematocera). Canadian Journal of Zoology 52:133–159.

Craig, D. A. 1975. The larvae of Tahitian Simuliidae (Diptera: Nematocera). Journal of Medical Entomology 12:463–476.

Craig, D. A. 1977. Mouthparts and feeding behaviour of Tahitian larval Simuliidae (Diptera: Nematocera). Quaestiones Entomologicae 13:195–218.

Craig, D. A. 1985. Black flies are a drag! University of Alberta Agriculture and Forestry Bulletin 8(2):15–18.

Craig, D. A. 1997. A taxonomic revision of the Pacific black fly subgenus *Inseliellum* (Diptera: Simuliidae). Canadian Journal of Zoology 75:855–904.

Craig, D. A., & H. Batz. 1982. Innervation and fine structure of antennal sensilla of Simuliidae larvae (Diptera: Culicomorpha). Canadian Journal of Zoology 60:696–711.

Craig, D. A., & A. Borkent. 1980. Intra- and inter-familial homologies of maxillary palpal sensilla of larval Simuliidae (Diptera: Culicomorpha). Canadian Journal of Zoology 58:2264–2279.

Craig, D. A., & M. M. Chance. 1982. Filter feeding in larvae of Simuliidae (Diptera: Culicomorpha): aspects of functional morphology and hydrodynamics. Canadian Journal of Zoology 60:712–724.

Craig, D. A., & R. E. G. Craig. 1986. Simuliidae (Diptera: Culicomorpha) of Rarotonga, Cook Islands, South Pacific. New Zealand Journal of Zoology 13:357–366.

Craig, D. A., & M. M. Galloway. 1988 ["1987"]. Hydrodynamics of larval black flies. Pp. 171–185. *In* K. C. Kim & R. W. Merritt (eds.), Black flies: ecology, population management, and annotated world list. Pennsylvania State University, University Park. 528 pp. [Actual publication date was 1 March 1988.]

Craig, D. A., O. Fossati, & Y. Séchan. 1995. Black flies (Diptera: Simuliidae) of the Marquesas Islands, French Polynesia: redescriptions and new species. Canadian Journal of Zoology 73:775–800.

Crans, A., F. Carle, T. Rainey, & C. Musa. 2001. Involvement of county mosquito personnel in New Jersey's black fly supression [*sic*] program. Proceedings of the Annual Meeting of the New Jersey Mosquito Control Association 88:25–31.

Crans, W. J. 1996. The biology of New Jersey's black flies. Proceedings of the Annual Meeting of the New Jersey Mosquito Control Association 83:50–54.

Crans, W. J., & L. G. McCuiston. 1970a. A checklist of the blackflies of New Jersey (Diptera: Simuliidae). Mosquito News 30:654–655.

Crans, W. J., & L. G. McCuiston. 1970b. The current status of black fly investigations in New Jersey. Proceedings of the Annual Meeting of the New Jersey Mosquito Extermination Association 57:103–106.

Cranshaw, W. S. 1995. Notes on black flies. Pest Alert (Cooperative Extension, Colorado State University) 12(14):6–7.

Cressa, C. 1977. Selection of substrate by blackfly larvae. M.S. thesis. University of Waterloo, Waterloo, Ontario. 70 pp.

Cresswell, J. E. 1991. Capture rates and composition of insect prey of the pitcher plant *Sarracenia purpurea*. American Midland Naturalist 125:1–9.

Crnjar, R., A. Liscia, A. M. Angioy, P. Pietra, & J. G. Stoffolano Jr. 1983. Observations on the morphology and electrophysiological responses of labellar, tarsal and palpal chemosensilla of the black fly, *Simulium decorum* Walker (Diptera Simuliidae). Monitore Zoologico Italiano (New Series) 17:133–141.

Cross, M. L., M. S. Cupp, E. W. Cupp, F. B. Ramberg, & F. J. Enriquez. 1993a. Antibody responses of BALB/c mice to salivary antigens of hematophagous black flies (Diptera: Simuliidae). Journal of Medical Entomology 30:725–734.

Cross, M. L., M. S. Cupp, E. W. Cupp, A. L. Galloway, & F. J. Enriquez. 1993b. Modulation of murine immunological responses by salivary gland extract of *Simulium vittatum* (Diptera: Simuliidae). Journal of Medical Entomology 30:928–935.

Cross, M. L., E. W. Cupp, & F. J. Enriquez. 1994. Modulation of murine cellular immune responses and cytokines by salivary gland extract of the black fly *Simulium vittatum*. Tropical Medicine and Parasitology 45:119–124.

Crosskey, R. W. 1969. A re-classification of the Simuliidae (Diptera) of Africa and its islands. Bulletin of the British Museum (Natural History). Entomology Supplement 14:1–195.

Crosskey, R. W. 1981. Simuliid taxonomy—the contemporary scene. Pp. 3–18. *In* M. Laird (ed.), Blackflies: the future for biological methods in integrated control, Academic Press, New York. 399 pp.

Crosskey, R. W. 1985. The blackfly fauna of the London area (Diptera: Simuliidae). Entomologist's Gazette 36:55–75.

Crosskey, R. W. 1988 ["1987"]. An annotated checklist of the world black flies (Diptera: Simuliidae). Pp. 425–520. *In* K. C. Kim & R. W. Merritt (eds.), Black flies: ecology, population management, and annotated world list. Pennsylvania State University, University Park. 528 pp. [Actual publication date was 1 March 1988.]

Crosskey, R. W. 1990. The natural history of blackflies. John Wiley, Chichester, U.K. 711 pp.

Crosskey, R. W. 1993. Blackflies (Simuliidae). Pp. 241–287. *In* R. P. Lane & R. W. Crosskey (eds.), Medical insects and arachnids. Chapman & Hall, London. 723 pp.

Crosskey, R. W. 1994a. The Manitoban blackfly fauna with special reference to prospections made in 1983–1985 in central and southern Manitoba (Diptera: Simuliidae). Journal of Natural History 28:87–107.

Crosskey, R. W. 1994b. The Baltic amber blackfly fossil *Nevermannia cerberus* Enderlein and its reassignment to *Ectemnia* Enderlein (Diptera: Simuliidae). Entomologist's Gazette 45:275–280.

Crosskey, R. W. 1999a. First update to the taxonomic and geographical inventory of world blackflies (Diptera: Simuliidae). Natural History Museum, London. 10 pp.

Crosskey, R. W. 1999b. An annotated bibliography in English of the work of I. A. Rubtsov (1902–1993) on the dipterous family Simuliidae (blackflies). Studia Dipterologica 6:3–32.

Crosskey, R. W. 2002. Second update to the taxonomic and geographical inventory of world blackflies (Diptera: Simuliidae). Natural History Museum, London. 14 pp.

Crosskey, R. W., & L. Davies. 1972. The identities of *Simulium lineatum* (Meigen), *S. latipes* (Meigen) and *S. vernum* Macquart (Diptera: Simuliidae). Entomologist's Gazette 23:249–258.

Crosskey, R. W., & T. M. Howard. 1997. A new taxonomic and geographical inventory of world blackflies (Diptera: Simuliidae). Natural History Museum, London. 144 pp.

Crosskey, R. W., & C. A. Lowry. 1990. Simuliidae. Pp. 201–235. *In* B. C. Townsend (collator), A catalogue of the types of bloodsucking flies in the British Museum (Natural History). Occasional Papers on Systematic Entomology 7:1–371.

Cumming, J. E., & B. McKague. 1973. Preliminary studies of effects of juvenile hormone analogues on adult emergence of black flies (Diptera: Simuliidae). Canadian Entomologist 105:509–511.

Cupp, E. W. 1981. Blackfly physiology. Pp. 199–206. *In* M. Laird (ed.), Blackflies: the future for biological methods in integrated control. Academic Press, New York. 399 pp.

Cupp, E. W. 1988 ["1987"]. The epizootiology of livestock and poultry diseases associated with black flies. Pp. 387–395. *In* K. C. Kim & R. W. Merritt (eds.), Black flies: ecology, population management, and annotated world list, Pennsylvania State University, University Park. 528 pp. [Actual publication date was 1 March 1988.]

Cupp, E. W., & M. S. Cupp. 1997. Black fly (Diptera: Simuliidae) salivary secretions: importance in vector competence and disease. Journal of Medical Entomology 34:87–94.

Cupp, E. W., & A. E. Gordon (eds.). 1983. Notes on the systematics, distribution, and bionomics of black flies (Diptera: Simuliidae) in the northeastern United States. Search: Agriculture. Cornell University Agricultural Experiment Station 25:1–76.

Cupp, E. W., & F. B. Ramberg. 1997. Care and maintenance of blackfly colonies. Pp. 31–40. *In* J. M. Crampton, C. B. Beard, & C. Louis (eds.), The molecular biology of insect disease vectors: a methods manual. Chapman & Hall, London. 578 pp.

Cupp, E. W., C. J. Maré, M. S. Cupp, & F. B. Ramberg. 1992. Biological transmission of vesicular stomatitis virus (New Jersey) by *Simulium vittatum* (Diptera: Simuliidae). Journal of Medical Entomology 29:137–140.

Cupp, M. S., E. W. Cupp, & F. B. Ramberg. 1993. Salivary gland apyrase in black flies (*Simulium vittatum*). Journal of Insect Physiology 39:817–821.

Cupp, M. S., J. M. C. Ribeiro, & E. W. Cupp. 1994. Vasodilative activity in black fly salivary glands. American Journal of Tropical Medicine and Hygiene 50:241–246.

Cupp, M. S., E. W. Cupp, J. O. Ochoa-A, & J. K. Moulton. 1995. Salivary apyrase in New World blackflies (Diptera: Simuliidae) and its relationship to onchocercia-

sis vector status. Medical and Veterinary Entomology 9:325–330.
Cupp, M. S., Y. Chen, & E. W. Cupp. 1997. Cellular hemolymph response of *Simulium vittatum* (Diptera: Simuliidae) to intrathoracic injection of *Onchocerca lienalis* (Filarioidea: Onchocercidae) microfilariae. Journal of Medical Entomology 34:56–63.
Cupp, M. S., J. M. C. Ribeiro, D. E. Champagne, & E. W. Cupp. 1998. Analyses of cDNA and recombinant protein for a potent vasoactive protein in saliva of a blood-feeding black fly, *Simulium vittatum*. Journal of Experimental Biology 201:1553–1561.
Currie, D. C. 1986. An annotated list of and keys to the immature black flies of Alberta (Diptera: Simuliidae). Memoirs of the Entomological Society of Canada 134:1–90.
Currie, D. C. 1988. Morphology and systematics of primitive Simuliidae (Diptera: Culicomorpha). Ph.D. thesis. University of Alberta, Edmonton. 331 pp.
Currie, D. C. 1995. Little demons. Rotunda 28(1):16–25.
Currie, D. C. 1997a. Black flies (Diptera: Simuliidae) of the Yukon, with reference to the black-fly fauna of northwestern North America. Pp. 563–614. *In* H. V. Danks & J. A. Downes (eds.). Insects of the Yukon. Biological Survey of Canada (Terrestrial Arthropods), Ottawa. 1034 pp.
Currie, D. C. 1997b. Feature species: the black fly *Gymnopais holopticoides*. Arctic Insect News 8:15–17.
Currie, D. C., & P. H. Adler. 1986. Blackflies (Diptera: Simuliidae) of the Queen Charlotte Islands, British Columbia, with discussion of their origin and description of *Simulium* (*Hellichiella*) *nebulosum* n. sp. Canadian Journal of Zoology 64:218–227.
Currie, D. C., & P. H. Adler. 2000. Update on a survey of the black flies (Diptera: Simuliidae) from the Northwest Territories and Nunavut Project. Arctic Insect News 11:6–9.
Currie, D. C., & D. A. Craig. 1988 ["1987"]. Feeding strategies of larval black flies. Pp. 155–170. *In* K. C. Kim & R. W. Merritt (eds.), Black flies: ecology, population management, and annotated world list. Pennsylvania State University, University Park. 528 pp. [Actual publication date was 1 March 1988.]
Currie, D. C., & D. Grimaldi. 2000. A new black fly (Diptera: Simuliidae) genus from mid Cretaceous (Turonian) amber of New Jersey. Pp. 473–485. *In* D. Grimaldi (ed.), Studies on fossils in amber, with particular reference to the Cretaceous of New Jersey. Backhuys Publishers, Leiden, The Netherlands. 498 pp.
Currie, D. C., & F. F. Hunter. 2003. A new species of *Stegopterna* Enderlein and its relationship to the allotriploid species *St. mutata* (Malloch 1914) (Diptera: Simuliidae). Zootaxa 214:1–11.
Currie, D. C., & I. R. Walker. 1992. Recognition and palaeohydrologic significance of fossil black fly larvae, with a key to the Nearctic genera (Diptera: Simuliidae). Journal of Paleolimnology 7:37–54.
Currie, D. C., D. Giberson, & P. H. Adler. 2002. Insect biodiversity in the Thelon Wildlife Sanctuary. Newsletter of the Biological Survey of Canada (Terrestrial Arthropods) 21:59–64.
Curry, R. A., & P. M. Powles. 1991. The insect community in an outlet stream of an acidified lake. Naturaliste Canadien 118:27–34.
Curtis, J. 1839. British Entomology; being illustrations and descriptions of the genera of insects found in Great Britain and Ireland: containing coloured figures from nature of the most rare and beautiful species, and in many instances of the plants upon which they are found, vol. 16, pp. 722–769. Richard and John E. Taylor, London.
Curtis, L. C. 1954. Observations on a black fly pest of cattle in British Columbia (Diptera: Simuliidae). Proceedings of the Entomological Society of British Columbia 51:3–6.
Curtis, L. C. 1968. A method for accurate counting of blackfly larvae (Diptera: Simuliidae). Mosquito News 28:238–239.
Dalmat, H. T. 1950. Induced oviposition of *Simulium* flies by exposure to $CO_2$. United States Public Health Reports 65:545–546.
Dalmat, H. T. 1951. Notes on the Simuliidae (Diptera) of Guatemala, including descriptions of three new species. Annals of the Entomological Society of America 44:31–58.
Dalmat, H. T. 1952. Longevity and further flight range studies on the blackflies (Diptera, Simuliidae), with the use of dye markers. Annals of the Entomological Society of America 45:23–37.
Dalmat, H. T. 1955. The black flies (Diptera, Simuliidae) of Guatemala and their role as vectors of onchocerciasis. Smithsonian Miscellaneous Publications 125(1):1–425.
Dang, S., & R. W. Lichtwardt. 1979. Fine structure of *Paramoebidium* (Trichomycetes) and a new species with viruslike particles. American Journal of Botany 66:1093–1104.
David, J.-P., D. Rey, M.-P. Pautou, & J.-C. Meyran. 2000. Differential toxicity of leaf litter to dipteran larvae of mosquito developmental sites. Journal of Invertebrate Pathology 75:9–18.
Davies, D. M. 1949a. Variation in taxonomic characters of some Simuliidae (Diptera). Canadian Entomologist 81:18–21.
Davies, D. M. 1949b. Description of *Simulium euryadminiculum*, a new species of blackfly (Simuliidae: Diptera). Canadian Entomologist 81:45–49.
Davies, D. M. 1949c. The ecology and life history of blackflies (Simuliidae, Diptera) in Ontario with a description of a new species. Ph.D. thesis. University of Toronto, Toronto. 166 pp.
Davies, D. M. 1950. A study of the black fly population of a stream in Algonquin Park, Ontario. Transactions of the Royal Canadian Institute 28:121–159.
Davies, D. M. 1951. Some observations of the number of black flies (Diptera, Simuliidae) landing on colored cloths. Canadian Journal of Zoology 29:65–70.
Davies, D. M. 1952. The population and activity of adult female black flies in the vicinity of a stream in Algonquin Park, Ontario. Canadian Journal of Zoology 30:287–321.
Davies, D. M. 1953. Longevity of black flies in captivity. Canadian Journal of Zoology 31:304–312.
Davies, D. M. 1957 ["1956"]. Microsporidian infections in Ontario black flies. Annual Report of the Entomological Society of Ontario 87:79. [Actual publication date was October 1957.]
Davies, D. M. 1959. The parasitism of black flies (Diptera, Simuliidae) by larval water mites mainly of the genus *Sperchon*. Canadian Journal of Zoology 37:353–369.
Davies, D. M. 1960. Microsporidia in a sperchonid mite, and further notes on Hydracarina and simuliids (Diptera). Proceedings of the Entomological Society of Ontario 90:53.
Davies, D. M. 1961. Colour affects the landing of bloodsucking black flies (Diptera: Simuliidae) on their hosts. Proceedings of the Entomological Society of Ontario 91:267–268.
Davies, D. M. 1972. The landing of blood-seeking female black-flies (Simuliidae: Diptera) on coloured materials. Proceedings of the Entomological Society of Ontario 102:124–155.
Davies, D. M. 1978. Ecology and behaviour of adult black flies (Simuliidae): a review. Quaestiones Entomologicae 14:3–12.
Davies, D. M. 1981. Predators upon blackflies. Pp. 139–158. *In* M. Laird (ed.), Blackflies: the future for biological methods in integrated control. Academic Press, New York. 399 pp.
Davies, D. M. 1989 ["1988"]. Gynandromorphs in Canadian Simuliidae (Diptera). Proceedings of the Entomolog-

ical Society of Ontario 119:87–89. [Actual publication date was February 1989.]
Davies, D. M. 1991. Additional records of predators upon black flies (Simuliidae: Diptera). Bulletin of the Society for Vector Ecology 16:256–268.
Davies, D. M., & H. Györkös. 1990. Autogeny in Canadian Simuliidae (Diptera), with some experiments on rate of oögenesis. Canadian Journal of Zoology 68:2429–2436.
Davies, D. M., & B. V. Peterson. 1956. Observations on the mating, feeding, ovarian development, and oviposition of adult black flies (Simuliidae, Diptera). Canadian Journal of Zoology 34:615–655.
Davies, D. M., & B. V. Peterson. 1957. Black flies over lakes (Simuliidae, Diptera). Annals of the Entomological Society of America 50:512–514.
Davies, D. M., & P. D. Syme. 1958. Three new Ontario black flies of the genus *Prosimulium* (Diptera: Simuliidae). Part II. Ecological observations and experiments. Canadian Entomologist 90:744–759.
Davies, D. M., B. V. Peterson, & D. M. Wood. 1962. The black flies (Diptera: Simuliidae) of Ontario. Part I. Adult identification and distribution with descriptions of six new species. Proceedings of the Entomological Society of Ontario 92:70–154.
Davies, L. 1954. Observations on *Prosimulium ursinum* Edw. at Holandsfjord, Norway. Oikos 5:94–98.
Davies, L. 1957. A new *Prosimulium* species from Britain, and a re-examination of *P. hirtipes* Fries from the Holarctic region (Diptera: Simuliidae). Proceedings of the Royal Entomological Society of London (B) 26:1–10.
Davies, L. 1960. The first-instar larva of a species of *Prosimulium* (Diptera: Simuliidae). Canadian Entomologist 92:81–84.
Davies, L. 1961. Ecology of two *Prosimulium* species (Diptera) with reference to their ovarian cycles. Canadian Entomologist 93:1113–1140.
Davies, L. 1963. Seasonal and diurnal changes in the age-composition of adult *Simulium venustum* Say (Diptera) populations near Ottawa. Canadian Entomologist 95:654–667.
Davies, L. 1965a. On spermatophores in Simuliidae (Diptera). Proceedings of the Royal Entomological Society of London (A) 40:30–34 + 1 plate.
Davies, L. 1965b. The structure of certain atypical Simuliidae (Diptera) in relation to evolution within the family, and the erection of a new genus for the Crozet Island black-fly. Proceedings of the Linnean Society of London 176:159–180.
Davies, L. 1966. The taxonomy of British black-flies (Diptera: Simuliidae). Transactions of the Royal Entomological Society of London 118:413–511.
Davies, L. 1968. A key to the British species of Simuliidae (Diptera) in the larval, pupal and adult stages. Freshwater Biological Association Scientific Publication 24:1–126.
Davies, L. 1974. Evolution of larval headfans in Simuliidae (Diptera) as inferred from the structure and biology of *Crozetia crozetensis* (Womersley) compared with other genera. Zoological Journal of the Linnean Society of London 55:193–224 + 2 plates.
Davies, L., & D. M. Roberts. 1973. A net and a catch-segregating apparatus mounted in a motor vehicle for field studies on flight activity of Simuliidae and other insects. Bulletin of Entomological Research 63:103–112 + 1 plate.
Davis, A. N., J. B. Gahan, J. A. Fluno, & D. W. Anthony. 1957. Larvicide tests against blackflies in slow-moving streams. Mosquito News 17:261–265.
Davis, C. C. 1971. A study of the hatching process in aquatic invertebrates. XXV. Hatching in the blackfly, *Simulium* (probably *venustum*) (Diptera, Simuliidae). Canadian Journal of Zoology 49:333–336.
Davis, H. G., & M. T. James. 1957. Black flies attracted to meat bait. Proceedings of the Entomological Society of Washington 59:243–244.
Davis, J. R., & D. R. Cook. 1985. An unusual new species of *Sperchon* (Acari: Sperchontidae) from Texas, with notes on its biology. Transactions of the American Microscopical Society 104:113–121.
Debaisieux, P. 1919. Une Chytridinée nouvelle: *Coelomycidium simulii*, nov. gen. nov. spec. Comptes Rendus Hebdomadaires des Séances et Mémoires de la Société de Biologie 82:899–900.
Debboun, M., D. Strickman, V. B. Solberg, R. C. Wilkerson, K. R. McPherson, C. Golenda, L. Keep, R. A. Wirtz, R. Burge, & T. A. Klein. 2000. Field evaluation of deet and a piperidine repellent against *Aedes communis* (Diptera: Culicidae) and *Simulium venustum* (Diptera: Simuliidae) in the Adirondack Mountains of New York. Journal of Medical Entomology 37:919–923.
Dechene, R. M. 1973. A study of the chromosomes of three populations of *Simulium pictipes* Hagen (Diptera: Simuliidae) in Tompkins County, New York. M.S. thesis. Cornell University, Ithaca, N.Y. 38 pp.
DeFoliart, G. R. 1951a. The life histories, identification and control of blackflies (Diptera: Simuliidae) in the Adirondack mountains. Ph.D. thesis. Cornell University, Ithaca, N.Y. 98 pp.
DeFoliart, G. R. 1951b. A comparison of several repellents against blackflies. Journal of Economic Entomology 44:265.
DeFoliart, G. R., & B. V. Peterson. 1960. New North American Simuliidae of the genus *Cnephia* Enderlein (Diptera). Annals of the Entomological Society of America 53:213–219.
DeFoliart, G. R., & M. R. Rao. 1965. The ornithophilic black fly *Simulium meridionale* Riley (Diptera: Simuliidae) feeding on man during autumn. Journal of Medical Entomology 2:84–85.
DeFoliart, G. R., M. R. Rao, & C. D. Morris. 1967. Seasonal succession of bloodsucking Diptera in Wisconsin during 1965. Journal of Medical Entomology 4:363–373.
DeFoliart, G. R., R. O. Anslow, R. P. Hanson, C. D. Morris, O. Papadopoulos, & G. E. Sather. 1969. Isolation of Jamestown Canyon serotype of California encephalitis virus from naturally infected *Aedes* mosquitoes and tabanids. American Journal of Tropical Medicine and Hygiene 18:440–447.
De León, J. R. 1943. Preliminares para la descripción de cinco nuevas especies de simúlidos en Guatemala. Boletin Sanitario de Guatemala 51:94–101.
De León, J. R. 1945 ["1944"]. Apuntes para una monografía sobre los simúlidos de Guatemala. Nuevas especies de simúlidos en la region occidental de Guatemala. Boletin Sanitario de Guatemala 52:66–77. [Actual publication date was an unspecified time in 1945.]
De Meijere, J. C. H. 1907. Studien über südostasiatische Dipteren. I. Tijdschrift voor Entomologie 50:196–264 + 2 plates.
De Meillon, B., & F. Hardy. 1951. New records and species of biting insects from the Ethiopian Region—III. Journal of the Entomological Society of Southern Africa 14:30–35.
Depner, K. R. 1971. The distribution of black flies (Diptera: Simuliidae) of the mainstream of the Crowsnest-Oldman River system of southern Alberta. Canadian Entomologist 103:1147–1151.
Depner, K. R. 1978. Application of black fly larvicide to large rivers. Pp. 57–59. *In* Research Highlights—1977. Agriculture Canada Research Station, Lethbridge, Alberta. 79 pp.
Depner, K. R. 1979. Effect of methoxychlor on *Simulium arcticum* and other invertebrates in the Athabasca River. Pp. 88–90. *In* G. C. R. Croome & N. D. Holmes (eds.), Research Highlights—1978. Agriculture Canada Research Station, Lethbridge, Alberta. 94 pp.
Dermott, R. M., & H. J. Spence. 1984. Changes in populations and drift of stream invertebrates following lampricide treatment. Canadian Journal of

Fisheries and Aquatic Sciences 41:1695–1701.
Desser, S. S. 1977. Ultrastructural observations on the epimastigote stages of *Trypanosoma avium* in *Simulium rugglesi*. Canadian Journal of Zoology 55:1359–1367.
Desser, S. S., & A. K. Ryckman. 1976. The development and pathogenesis of *Leucocytozoon simondi* in Canada and domestic geese in Algonquin Park, Ontario. Canadian Journal of Zoology 54:634–643.
Desser, S. S., & Y. J. Yang. 1973. Sporogony of *Leucocytozoon* spp. in mammalophilic simuliids. Canadian Journal of Zoology 51:793.
Desser, S. S., S. B. McIver, & D. Jez. 1975. Observations on the role of simuliids and culicids in the transmission of avian and anuran trypanosomes. International Journal for Parasitology 5:507–509.
Desser, S. S., J. Stuht, & A. M. Fallis. 1978. Leucocytozoonosis in Canada geese in Upper Michigan. I. Strain differences among geese from different localities. Journal of Wildlife Diseases 14:124–131.
Devito, E. 1996. Cascading trophic interactions: views on black flies from the food web. Proceedings of the Annual Meeting of the New Jersey Mosquito Control Association 83:75–76.
Dey, P. D. 1992. Food quality of ultrafine seston: growth of a stream filter feeder *Simulium vittatum*. M.S. thesis. Idaho State University, Pocatello. 67 pp.
Diarrassouba, S. 1977. Parasites and predators of mosquitoes (Diptera: Culicidae) and black flies (Diptera: Simuliidae). An assessment of the possibility of using natural enemies in biting fly control programmes. Master of Pest Management Professional Paper. Simon Fraser University, Burnaby, British Columbia. 92 pp.
Díaz Nájera, A. 1981 ["1979"]. Una nueva especie del género *Simulium* (Diptera: Simuliidae) del Estado de Coahuila, México. Anales del Instituto de Biologia Universidad Nacional Autónoma México 50(Serie Zoología):553–561. [Actual publication date was after 12 June 1981 and before November 1981.]
Díaz Nájera, A., & M. A. Vulcano. 1962a ["1961"]. Descripción de *Simulium (Psilopelmia) longithallum* n. sp. (Diptera, Simuliidae). Revista del Instituto Salubridad y Enfermedades Tropicales (México) 21:221–235. [Actual publication date was between March and October 1962.]
Díaz Nájera, A., & M. A. Vulcano. 1962b. Claves para identificar las larvas de simúlidos del subgénero *Hearlea*, con descripción de dos nuevas especies (Diptera: Simuliidae). Revista del Instituto Salubridad y Enfermedades Tropicales (México) 22:91–144.
Dimond, J. B., & W. G. Hart. 1953. Notes on the blackflies (Simuliidae) of Rhode Island. Mosquito News 13:238–242.
Dinulescu, G. 1966. Diptera Fam. Simuliidae (Muştele Columbace). Fauna Republicii Socialiste România 11(8):1–600. [In Romanian.]
Doby, J.-M., & F. David. 1959. *Greniera*, genre nouveau de Simuliidé (Diptères-Nématocères). Discussion de la position systématique. Comptes Rendus Hebdomadaires des Séances de l'Académie des Sciences 249:763–765.
Doisy, K. E., R. D. Hall, & F. J. Fischer. 1986. The black flies (Diptera: Simuliidae) of an Ozark stream in southern Missouri and associated water quality measurements. Journal of the Kansas Entomological Society 59:133–142.
Donahue, W. F., & D. W. Schindler. 1998. Diel emigration and colonization responses of blackfly larvae (Diptera: Simuliidae) to ultraviolet radiation. Freshwater Biology 40:357–365.
Doran, E. W. 1887. Report on the economic entomology of Tennessee. Pp. 173–267. *In* Biennial Report of the Commissioner of Agriculture, Statistics and Mines of the State of Tennessee (1885–1886). Marshall and Bruce, Nashville, Tenn. ["The buffalo gnat" appears on pp. 239–242.]
Dorogostaisky, V. C., I. A. Rubtsov, & N. M. Vlasenko. 1935. Notes for study of the systematics, distribution and biology of the blackflies (Simuliidae) of eastern Siberia. Parazitologichesky Sbornik 5:107–204. [In Russian; English summary.]
Dosdall, L. M., & D. M. Lehmkuhl. 1989. The impact of methoxychlor treatment of the Saskatchewan River system on artificial substrate populations of aquatic insects. Environmental Pollution 60:209–222.
Dosdall, L. M., M. M. Galloway, J. T. Arnason, & P. Morand. 1991. Field evaluation of the phototoxin, alpha-terthienyl, for reducing larval populations of black flies (Diptera: Simuliidae) and its impact on drift of aquatic invertebrates. Canadian Entomologist 123:439–449.
Dosdall, L. M., M. M. Galloway, & R. M. Gadawski. 1992. New self-marking device for dispersal studies of black flies (Diptera: Simuliidae). Journal of the American Mosquito Control Association 8:187–190.
Douglass, J. R. 1940. A buffalo gnat (*Simulium vittatum* Zett.). United States Department of Agriculture Insect Pest Survey Bulletin 20:199.
Dove, R. F., & A. B. McKague. 1975. Effects of insect developmental inhibitors on adult emergence of black flies (Diptera: Simuliidae). II. Canadian Entomologist 107:1211–1213.
Downe, A. E. R. 1957. Precipitin test studies on rate of digestion of blood meals in black flies (Diptera: Simuliidae). Canadian Journal of Zoology 35:459–462.
Downe, A. E. R., & P. E. Morrison. 1957. Identification of blood meals of blackflies (Diptera: Simuliidae) attacking farm animals. Mosquito News 17:37–40.
Downes, J. A. 1958. Assembly and mating in the biting Nematocera. Proceedings of the Tenth International Congress of Entomology, Montreal (1956) 2:425–434.
Downes, J. A. 1962. What is an arctic insect? Canadian Entomologist 94:143–162.
Downes, J. A. 1964. Arctic insects and their environment. Canadian Entomologist 96:279–307.
Downes, J. A. 1965. Adaptations of insects in the arctic. Annual Review of Entomology 10:257–274.
Downes, J. A. 1969. The swarming and mating flight of Diptera. Annual Review of Entomology 14:271–298.
Drake, C. J. 1933. A black fly (*Simulium occidentale* Towns.). United States Department of Agriculture Insect Pest Survey Bulletin 13:184.
Drake, C. J. 1935. Buffalo gnats (*Eusimulium* spp.). United States Department of Agriculture Insect Pest Survey Bulletin 15:269.
Duckitt, G. S. 1986. *Bacillus thuringiensis* var. *israelensis*: the bacterium, its use in black fly control and effects on nontarget organisms. Master of Pest Management Professional Paper. Simon Fraser University, Burnaby, British Columbia. 72 pp.
Ducros, G., J.-M. Quiot, S. Belloncik, & G. Charpentier. 1992. Établissement de nouvelles lignées cellulaires de diptères piqueurs. Mémoires de la Société Royale Belge d'Entomologie 35:125–128.
Dudley, T. L., S. D. Cooper, & N. Hemphill. 1986. Effects of macroalgae on a stream invertebrate community. Journal of the North American Benthological Society 5:93–106.
Dudley, T. L., C. M. D'Antonio, & S. D. Cooper. 1990. Mechanisms and consequences of interspecific competition between two stream insects. Journal of Animal Ecology 59:849–866.
Dumbleton, L. J. 1963. The classification and distribution of the Simuliidae (Diptera) with particular reference to the genus *Austrosimulium*. New Zealand Journal of Science 6:320–357.
Dunbar, R. W. 1958a. The salivary gland chromosomes of seven forms of black flies included in *Eusimulium aureum* Fries. M.A. thesis. University of Toronto, Toronto. 53 pp.
Dunbar, R. W. 1958b. The salivary gland chromosomes of two sibling species of

black flies included in *Eusimulium aureum* Fries. Canadian Journal of Zoology 36:23–44.
Dunbar, R. W. 1959. The salivary gland chromosomes of seven forms of black flies included in *Eusimulium aureum* Fries. Canadian Journal of Zoology 37:495–525.
Dunbar, R. W. 1962. Cytotaxonomic studies in Simuliidae. Ph.D. thesis. University of Toronto, Toronto. 101 pp.
Dunbar, R. W. 1965. Chromosome inversions as blocks to genetic exchange leading to sympatric speciation in black flies (Simuliidae; Diptera). Proceedings of the Twelfth International Congress of Entomology, London (1964):268–269.
Dunbar, R. W. 1966. Cytotaxonomic studies in black flies (Diptera: Simuliidae). Pp. 179–181. *In* C. D. Darlington & K. R. Lewis (eds.), Chromosomes today, vol. 1. Plenum Press, New York. 274 pp.
Dunbar, R. W. 1967. The salivary gland chromosomes of six closely related black flies near *Eusimulium congareenarum* (Diptera: Simuliidae). Canadian Journal of Zoology 45:377–396.
Duncan, D., & K. Burns. 1997. Lewis and Clark: the journey of the Corps of Discovery. Alfred A. Knopf, New York. 250 pp.
Dunn, M. B. 1925. Methods of protection from mosquitoes, black-flies and similar pests in the forest. Dominion of Canada, Department of Agriculture Pamphlet 55(New Series):1–11.
Dyar, H. G., & R. C. Shannon. 1927. The North American two-winged flies of the family Simuliidae. Proceedings of the United States National Museum 69(10):1–54 + 7 plates.
Ebsary, B. A. 1973. The mermithid (Nematoda) and other endoparasites of Simuliidae (Diptera) in insular Newfoundland. M.S. thesis. Memorial University of Newfoundland, St. John's. 106 pp.
Ebsary, B. A., & G. F. Bennett. 1973. Molting and oviposition of *Neomesomermis flumenalis* (Welch, 1962) Nickle, 1972, a mermithid parasite of blackflies. Canadian Journal of Zoology 51:637–639.
Ebsary, B. A., & G. F. Bennett. 1974. Redescription of *Neomesomermis flumenalis* (Nematoda) from blackflies in Newfoundland. Canadian Journal of Zoology 52:65–68.
Ebsary, B. A., & G. F. Bennett. 1975a. The occurrence of some endoparasites of blackflies (Diptera: Simuliidae) in insular Newfoundland. Canadian Journal of Zoology 53:1058–1062.
Ebsary, B. A., & G. F. Bennett. 1975b. Studies on the bionomics of mermithid nematode parasites of blackflies in Newfoundland. Canadian Journal of Zoology 53:1324–1331.
Eckhart, P. H. 1968. A survey of the black flies (Diptera: Simuliidae) of northeastern Pennsylvania. M.S. thesis. Pennsylvania State University, University Park. 122 pp.
Eckhart, P. [H.], & R. Snetsinger. 1969. Black flies (Diptera: Simuliidae) of northeastern Pennsylvania. Melsheimer Entomological Series 4:1–7.
Edgar, S. A. 1953. A field study of the effect of black fly bites on egg production of laying hens. Poultry Science 32:779–780.
Edman, J. D., & K. R. Simmons. 1985a. Rearing and colonization of black flies (Diptera: Simuliidae). Journal of Medical Entomology 22:1–17.
Edman, J. D., & K. R. Simmons. 1985b. Simuliids (mainly *Simulium decorum* Walker). Pp. 145–152. *In* P. Singh & R. F. Moore (eds.), Handbook of insect rearing, vol. II, Elsevier, Amsterdam. 514 pp.
Edman, J. D., & K. R. Simmons. 1988 ["1987"]. Maintaining black flies in the laboratory. Pp. 305–314. *In* K. C. Kim & R. W. Merritt (eds.), Black flies: ecology, population management, and annotated world list. Pennsylvania State University, University Park. 528 pp. [Actual publication date was 1 March 1988.]
Edmund, A. G. 1952. The relation between black fly activity and meteorological conditions (Simuliidae, Diptera). M.A. thesis. University of Toronto, Toronto. 91 pp.
Edmunds, L. R. 1954. A note on irrigation drop structures as breeding sites of blackflies in western Nebraska (Diptera: Simuliidae). Mosquito News 14:65–66.
Edwards, F. W. 1915. On the British species of *Simulium*.—I. The adults. Bulletin of Entomological Research 6:23–42.
Edwards, F. W. 1920. On the British species of *Simulium*.—II. The early stages; with corrections and additions to Part I. Bulletin of Entomological Research 11:211–246.
Edwards, F. W. 1931a. Oxford University Greenland expedition, 1928.—Diptera Nematocera. Annals and Magazine of Natural History (Series 10) 8:617–618.
Edwards, F. W. 1931b. Simuliidae. Pp. 121–154. *In* Diptera of Patagonia and South Chile. Part II. Nematocera (excluding crane-flies and Mycetophilidae). British Museum (Natural History), London. 331 pp.
Edwards, F. W. 1933. Oxford University Expedition to Hudson's Strait, 1931: Diptera Nematocera. With notes on some other species of the genus *Diamesa*. Annals and Magazine of Natural History (Series 10) 12:611–620.
Edwards, F. W. 1935a. Diptera Nematocera from east Greenland. Annals and Magazine of Natural History (Series 10) 15:467–473.
Edwards, F. W. 1935b. Diptera from Bear Island. Annals and Magazine of Natural History (Series 10) 15:531–543.
Edwards, R. T., & J. L. Meyer. 1987. Bacteria as a food source for black fly larvae in a blackwater river. Journal of the North American Benthological Society 6:241–250.
Eide, A., & A. M. Fallis. 1972. Experimental studies of the life cycle of *Leucocytozoon simondi* in ducks in Norway. Journal of Protozoology 19:414–416.
Einarsson, A. 1988. Distribution and movements of Barrow's goldeneye *Bucephala islandica* young in relation to food. Ibis 130:153–163.
El-Buni, A. M., & R. W. Lichtwardt. 1976. Asexual sporulation and mycelial growth in axenic cultures of *Smittium* spp. (Trichomycetes). Mycologia 68:559–572.
Elgmork, K., & O. A. ["R."] Saether. 1970. Distribution of invertebrates in a high mountain brook in the Colorado Rocky Mountains. University of Colorado Studies, Series in Biology 31:1–55.
Elliott, L. 1983. Biting back. Adirondack Life 14(3):37–40.
Ellsworth, S. D. 2000. Influence of substrate size, *Cladophora*, and caddisfly pupal cases on colonization of macroinvertebrates in Sagehen Creek, California. Western North American Naturalist 60:311–319.
Emery, W. T. 1913. The morphology and biology of *Simulium vittatum* and its distribution in Kansas. Kansas University Science Bulletin 8:323–362 + 5 plates.
Enderlein, G. 1914. Diptera: Scatopsidae, Simuliidae. Transactions of the Linnean Society of London (Second Series), Zoology 16:373–375.
Enderlein, G. 1921a. Das System der Kriebelmücken (Simuliidae). Deutsche Tierärztliche Wochenschrift 29:197–200. [Actual publication date was April 1921.]
Enderlein, G. 1921b. Neue paläarktische Simuliiden. Sitzungsberichte der Gesellschaft naturforschender Freunde zu Berlin 1921:212–224. [Actual publication date was June 1921.]
Enderlein, G. 1922. Weitere Beiträge zur Kenntnis der Simuliiden. Konowia 1:67–76.
Enderlein, G. 1925. Weitere Beiträge zur Kenntnis der Simuliiden und ihrer Verbreitung. Zoologischer Anzeiger 62:201–211.
Enderlein, G. 1929a. Beitrag zur Kenntnis der Prosimuliinen und Hellichiinen. Sitzungsberichte der Gessellschaft naturforschender Freunde zu Berlin 1929:222–224.
Enderlein, G. 1929b. Über einige neotropische Simuliiden des Genus *Friesia*. (Dipt.).

Deutsche Entomologische Zeitschrift 1929:327–328.
Enderlein, G. 1929c. Neue Arten des Simuliidengenus *Cnetha*. (Dipt.). Wiener Entomologische Zeitung 46:73–77.
Enderlein, G. 1930. Der heutige Stand der Klassifikation der Simuliiden. Archiv für Klassifikatorische und Phylogenetische Entomologie 1:77–97.
Enderlein, G. 1934a. Weiterer Ausbau des Systems der Simuliiden. (Dipt.). Deutsche Entomologische Zeitschrift 1933:273–292.
Enderlein, G. 1934b. Außereuropäische Simuliiden aus dem Wiener Museum. Sitzungsberichte der Gessellschaft naturforschender Freunde zu Berlin 1934:190–195.
Enderlein, G. 1935. Neue Simuliiden, besonders aus Afrika. Sitzungsberichte der Gesellschaft naturforschender Freunde zu Berlin 1934:358–364.
Enderlein, G. 1936a. 22. Ordnung: Zweiflügler, Diptera. Section 16. Pp. 1–259. *In* P. Brohmer, P. Ehrmann, & G. Ulmer (eds.), Die Tierwelt Mitteleuropas 6 (3). Quelle & Meyer, Leipzig, Germany.
Enderlein, G. 1936b. Simuliologica I. Sitzungsberichte der Gesellschaft naturforschender Freunde zu Berlin 1936:113–130.
Erlandson, M. A., & P. G. Mason. 1990. An iridescent virus from *Simulium vittatum* (Diptera: Simuliidae) in Saskatchewan. Journal of Invertebrate Pathology 56:8–14.
Essig, E. O. 1928. Some vacation biters. Pan-Pacific Entomologist 4:185–186.
Essig, E. O. 1942. College entomology. Macmillan, New York. 900 pp.
Etheridge, E. W. 1972. The butterfly caste, a social history of pellagra in the South. Greenwood, Westport, Conn. 278 pp.
Evans, C. L. 2001. Cytogenetics and systematics of the *Leucocytozoon* vector *Simulium slossonae* (Diptera: Simuliidae). Ph.D. thesis. Clemson University, Clemson, S.C. 213 pp.
Evans, C. L., & P. H. Adler. 2000. Microsculpture and phylogenetic significance of the spermatheca of black flies (Diptera: Simuliidae). Canadian Journal of Zoology 78:1468–1482.
Expert Committee on Arthropod Pests of Animals. 1993. Black fly control. Alberta Agriculture Agri-Fax (Agdex 651–6), Edmonton. 3 pp.
Eymann, M. 1985. The behaviours of the blackfly larvae *Simulium vittatum* and *S. decorum* (Diptera: Simuliidae) associated with establishing and maintaining dispersion patterns on natural and artificial substrate. M.S. thesis. University of Toronto, Toronto. 110 pp.
Eymann, M. 1988. Drag on single larvae of the black fly *Simulium vittatum* (Diptera: Simuliidae) in a thin, growing boundary layer. Journal of the North American Benthological Society 7:109–116.
Eymann, M. 1991a. Dispersion patterns exhibited by larvae of the black flies *Cnephia dacotensis* and *Simulium rostratum* (Diptera: Simuliidae). Aquatic Insects 13:99–106.
Eymann, M. 1991b. Flow patterns around cocoons and pupae of black flies of the genus *Simulium* (Diptera: Simuliidae). Hydrobiologia 215:223–229.
Eymann, M. 1991c. Fluid flow and the behaviour, ecology, and morphology of subimaginal black flies (Diptera: Simuliidae). Ph.D. thesis. University of Alberta, Edmonton. 169 pp.
Eymann, M. 1993. Some boundary layer characteristics of microhabitats occupied by larval black flies (Diptera: Simuliidae). Hydrobiologia 259:57–67.
Eymann, M., & W. G. Friend. 1986. Avoidance of scouring by larvae of *Simulium vittatum* (Diptera: Simuliidae) during a spring flood. Journal of the American Mosquito Control Association 2:382–383.
Eymann, M., & W. G. Friend. 1988. Behaviors of larvae of the black flies *Simulium vittatum* and *S. decorum* (Diptera: Simuliidae) associated with establishing and maintaining dispersion patterns on natural and artificial substrates. Journal of Insect Behavior 1:169–186.
Eymann, M., J. M. Schmidt, & W. G. Friend. 1987. Computer analysis of polarized spacing patterns with special reference to the larvae of the black fly, *Simulium vittatum*. Canadian Journal of Zoology 65:602–604.
Ezenwa, A. O. 1973. Mermithid and microsporidan parasitism of blackflies (Diptera: Simuliidae) in the vicinity of Churchill Falls, Labrador. Canadian Journal of Zoology 51:1109–1111.
Ezenwa, A. O. 1974a. Ecology of Simuliidae, Mermithidae, and Microsporida in Newfoundland freshwaters. Canadian Journal of Zoology 52:557–565.
Ezenwa, A. O. 1974b. Studies on host-parasite relationships of Simuliidae with mermithids and microsporidans. Journal of Parasitology 60:809–813.
Ezenwa, A. O., & N. E. Carter. 1975. Influence of multiple infections on sex ratios of mermithid parasites of blackflies. Environmental Entomology 4:142–144.
Fairchild, G. B. 1940. Notes on the Simuliidae of Panama (Dipt., Nematocera). Annals of the Entomological Society of America 33:701–719.
Fallis, A. M. 1964. Feeding and related behavior of female Simuliidae (Diptera). Experimental Parasitology 15:439–470.
Fallis, A. M., & G. F. Bennett. 1958. Transmission of *Leucocytozoon bonasae* Clarke to ruffed grouse (*Bonasa umbellus* L.) by the black flies *Simulium latipes* Mg. and *Simulium aureum* Fries. Canadian Journal of Zoology 36:533–539.
Fallis, A. M., & G. F. Bennett. 1961. Sporogony of *Leucocytozoon* and *Haemoproteus* in simuliids and ceratopogonids and a revised classification of the Haemosporidiida. Canadian Journal of Zoology 39:215–228.
Fallis, A. M., & G. F. Bennett. 1962. Observations on the sporogony of *Leucocytozoon mirandae*, *L. bonasae*, and *L. fringillinarum* (Sporozoa: Leucocytozoidae). Canadian Journal of Zoology 40:395–400.
Fallis, A. M., & G. F. Bennett. 1966. On the epizootiology of infections caused by *Leucocytozoon simondi* in Algonquin Park, Canada. Canadian Journal of Zoology 44:101–112.
Fallis, A. M., & S. M. Smith. 1964a. Ether extracts from birds and $CO_2$ as attractants for some ornithophilic simuliids. Canadian Journal of Zoology 42:723–730.
Fallis, A. M., & S. M. Smith. 1964b. Attraction of some simuliids to ether extracts from birds and to carbon dioxide. Proceedings of the First International Congress of Parasitology 2:934–935.
Fallis, A. M., D. M. Davies, & M. A. Vickers. 1951. Life history of *Leucocytozoon simondi* Mathis and Leger in natural and experimental infections and blood changes produced in the avian host. Canadian Journal of Zoology 29:305–328.
Fallis, A. M., R. C. Anderson, & G. F. Bennett. 1956. Further observations on the transmission and development of *Leucocytozoon simondi*. Canadian Journal of Zoology 34:389–404.
Fallis, A. M., G. F. Bennett, G. Griggs, & T. Allen. 1967. Collecting *Simulium venustum* female [*sic*] in fan traps and on silhouettes with the aid of carbon dioxide. Canadian Journal of Zoology 45:1011–1017.
Fantham, H. B., A. Porter, & L. R. Richardson. 1941. Some microsporidia found in certain fishes and insects in eastern Canada. Parasitology 33:186–208.
Featherston, D. W. 1982. The origins of the IS-7 and the IIIL-1 sibling species within the morpho-species *Simulium vittatum* (Zett.). M.S. thesis. University of Toronto, Toronto. 71 pp.
Federici, B. A., & L. A. Lacey. 1976. Densonucleosis virus and cytoplasmic polyhedrosis virus diseases in larvae of the blackfly, *Simulium vittatum*. Proceedings and Papers of the Annual Conference of the California Mosquito Control Association 44:124.
Federici, B. A., & L. A. Lacey. 1987. Intranuclear disease of uncertain etiology in larvae of the blackfly, *Simulium vittatum*. Journal of Invertebrate Pathology 50:184–190.

Federici, B. A., L. A. Lacey, & M. S. Mulla. 1977. *Coelomycidium simulii*: a fungal pathogen in larvae of *Simulium vittatum* from the Colorado River. Proceedings and Papers of the Annual Conference of the California Mosquito Control Association 45:110–113.

Feraday, R. [M.]. 1984. Weak male-determining genes and female heterogamety in *Chironomus tentans*. Canadian Journal of Zoology 26:748–751.

Feraday, R. M., & K. G. Leonhardt. 1989. Absence of population structure in black flies as revealed by enzyme electrophoresis. Genome 32:531–537.

Feraday, R. M., K. G. Leonhardt, & C. L. Brockhouse. 1989. The role of sex chromosomes in black fly evolution. Genome 32:538–542.

Fessenden, T. G. 1826. [no title]. New England Farmer 4(44):347.

Field, G., R. J. Duplessis, & A. P. Breton. 1967. Progress report on laboratory rearing of black flies (Diptera: Simuliidae). Journal of Medical Entomology 4:304–305.

Finelli, C. M., D. D. Hart, & R. A. Merz. 2002. Stream insects as passive suspension feeders: effects of velocity and food concentration on feeding performance. Oecologia 131:145–153.

Finney, J. R. 1976. The in vitro culture of mermithid parasites of blackflies and mosquitoes. Journal of Nematology 8:284.

Finney, J. R. 1981a. Potential of mermithids for control and *in vitro* culture. Pp. 325–333. *In* M. Laird (ed.), Blackflies: the future for biological methods in integrated control. Academic Press, New York. 399 pp.

Finney, J. R. 1981b. Mermithid nematodes: in vitro culture attempts. Journal of Nematology 13:275–280.

Finney, J. R., & J. B. Harding. 1982. The susceptibility of *Simulium verecundum* (Diptera: Simuliidae) to three isolates of *Bacillus thuringiensis* serotype 10 (*darmstadiensis*). Mosquito News 42:434–435.

Finney, J. R., & J. E. Mokry. 1980. *Romanomermis culicivorax* and simuliids. Journal of Invertebrate Pathology 35:211–213.

Finogle, K. 2001. Black flies and mosquitoes . . . what good are they? New Hampshire Wildlife Journal 14(1):4–7.

Fitch, H. S., F. Swenson, & D. F. Tillotson. 1946. Behavior and food habits of the red-tailed hawk. Condor 48:205–237.

Flannagan, J. F., B. E. Townsend, B. G. E. de March, M. K. Friesen, & S. L. Leonhard. 1979. The effects of an experimental injection of methoxychlor on aquatic invertebrates: accumulation, standing crop, and drift. Canadian Entomologist 111:73–89.

Fletcher, M. G. 1987. Determination of the possible role of arthropods as vectors for "Potomac horse fever" in equines. Ph.D. thesis. Virginia Polytechnic Institute and State University, Blacksburg. 111 pp.

Fletcher, M. G., E. C. Turner, J. W. Hansen, & B. D. Perry. 1988. Horse-baited insect trap and mobile insect sorting table used in a disease vector identification study. Journal of the American Mosquito Control Association 4:431–435.

Flook, D. R. 1959. Moose using water as refuge from flies. Journal of Mammalogy 40:455.

Fonseca, D. M. 1996. Fluid-mediated dispersal: effects on the foraging behavior and distribution of stream insects. Ph.D. thesis. University of Pennsylvania, Philadelphia. 163 pp.

Fonseca, D. M. 1999. Fluid-mediated dispersal in streams: models of settlement from the drift. Oecologia 121:212–223.

Fonseca, D. M., & D. D. Hart. 1996. Density-dependent dispersal of black fly neonates is mediated by flow. Oikos 75:49–58.

Fonseca, D. M., & D. D. Hart. 2001. Colonization history masks habitat preferences in local distributions of stream insects. Ecology 82:2897–2910.

Forbes, S. A. 1912. On black-flies and buffalo-gnats (*Simulium*) as possible carriers of pellagra in Illinois. Report of the Illinois State Entomologist 27:21–55.

Forbes, S. A. 1913. The *Simulium*-pellagra problem in Illinois, U. S. A. Science (New Series) 37:86–91.

Fortin, C., D. Lapointe, & G. Charpentier. 1986. Susceptibility of brook trout (*Salvelinus fontinalis*) fry to a liquid formulation of *Bacillus thuringiensis* serovar. *israelensis* (Teknar®) used for blackfly control. Canadian Journal of Fisheries and Aquatic Sciences 43:1667–1670.

Francy, D. B., C. G. Moore, G. C. Smith, W. L. Jakob, S. A. Taylor, & C. H. Calisher. 1988. Epizoötic vesicular stomatitis in Colorado, 1982: isolation of virus from insects collected along the northern Colorado Rocky Mountain Front Range. Journal of Medical Entomology 25:343–347.

Frazier, C. A. 1973. Biting insects. Archives of Dermatology 107:400–402.

Fredeen, F. J. H. 1951. The black flies of Saskatchewan (Diptera: Simuliidae). M.A. thesis. University of Saskatchewan, Saskatoon. 215 pp. + 18 plates.

Fredeen, F. J. H. 1956. Research on black flies, pests of livestock and man on the Canadian prairies. Proceedings of the Entomological Society of Manitoba 12:2–10.

Fredeen, F. J. H. 1958. Black flies (Diptera: Simuliidae) of the agricultural areas of Manitoba, Saskatchewan, and Alberta. Proceedings of the Tenth International Congress of Entomology, Montreal (1956) 3:819–823.

Fredeen, F. J. H. 1959a. Rearing black flies in the laboratory (Diptera: Simuliidae). Canadian Entomologist 91:73–83.

Fredeen, F. J. H. 1959b. Collection, extraction, sterilization and low-temperature storage of black-fly eggs (Diptera: Simuliidae). Canadian Entomologist 91:450–453.

Fredeen, F. J. H. 1960. Bacteria as a source of food for black-fly larvae. Nature 187:963.

Fredeen, F. J. H. 1961. A trap for studying the attacking behaviour of black flies, *Simulium arcticum* Mall. Canadian Entomologist 93:73–78.

Fredeen, F. J. H. 1962. DDT and heptachlor as black-fly larvicides in clear and turbid water. Canadian Entomologist 94:875–880.

Fredeen, F. J. H. 1963. Oviposition in relation to the accumulation of bloodthirsty black flies (*Simulium* (*Gnus*) *arcticum* Mall. (Diptera)) prior to a damaging outbreak. Nature 200:1024.

Fredeen, F. J. H. 1964a. On the determination of the approximate age of a black fly (Diptera: Simuliidae) and its significance. Canadian Entomologist 96:109.

Fredeen, F. J. H. 1964b. Bacteria as food for blackfly larvae (Diptera: Simuliidae) in laboratory cultures and in natural streams. Canadian Journal of Zoology 42:527–548.

Fredeen, F. J. H. 1969. Outbreaks of the black fly *Simulium arcticum* Malloch in Alberta. Quaestiones Entomologicae 5:341–372.

Fredeen, F. J. H. 1970a. Sexual mosaics in the black fly *Simulium arcticum* (Diptera: Simuliidae). Canadian Entomologist 102:1585–1592.

Fredeen, F. J. H. 1970b. A constant-rate liquid dispenser for use in blackfly larviciding. Mosquito News 30:402–405.

Fredeen, F. J. H. 1973. Black flies. Agriculture Canada Publication No. 1499. Agriculture Canada, Ottawa. 19 pp.

Fredeen, F. J. H. 1974. Tests with single injections of methoxychlor black fly (Diptera: Simuliidae) larvicides in large rivers. Canadian Entomologist 106:285–305.

Fredeen, F. J. H. 1975a. Effects of a single injection of methoxychlor black-fly larvicide on insect larvae in a 161-km (100-mile) section of the North Saskatchewan River. Canadian Entomologist 107:807–817.

Fredeen, F. J. H. 1975b. Controlling black fly outbreaks. Canada Agriculture 20:15–17.

Fredeen, F. J. H. 1976. The seven larval instars of *Simulium arcticum* (Diptera:

Simuliidae). Canadian Entomologist 108:591–600.

Fredeen, F. J. ["G."] H. 1977a. Some recent changes in black fly populations in the Saskatchewan River system in western Canada coinciding with the development of reservoirs. Canadian Water Resources Journal 2:90–102.

Fredeen, F. J. H. 1977b. A review of the economic importance of black flies (Simuliidae) in Canada. Quaestiones Entomologicae 13:219–229.

Fredeen, F. J. H. 1977c. Black fly control and environmental quality with reference to chemical larviciding in western Canada. Quaestiones Entomologicae 13:321–325.

Fredeen, F. J. H. 1979. Blackfly species adapt to change. Canada Agriculture 24:15–18.

Fredeen, F. J. H. 1981a. Keys to the black flies (Simuliidae) of the Saskatchewan River in Saskatchewan. Quaestiones Entomologicae 17:189–210.

Fredeen, F. J. H. 1981b. The seven larval instars of *Simulium* (*Phosterodoros*) *luggeri* (Diptera: Simuliidae). Canadian Entomologist 113:161–165.

Fredeen, F. J. H. 1983. Trends in numbers of aquatic invertebrates in a large Canadian river during four years of black fly larviciding with methoxychlor (Diptera: Simuliidae). Quaestiones Entomologicae 19:53–92.

Fredeen, F. J. H. 1984. Effects of outbreaks of the black fly *Simulium luggeri* on livestock in east-central Saskatchewan. Canada Agriculture 30:26–31.

Fredeen, F. J. H. 1985a. The black flies (Diptera: Simuliidae) of Saskatchewan. Saskatchewan Culture and Recreation Museum of Natural History Contribution 8:1–41 + 31 maps.

Fredeen, F. J. H. 1985b. Some economic effects of outbreaks of black flies (*Simulium luggeri* Nicholson & Mickel) in Saskatchewan. Quaestiones Entomologicae 21:175–208.

Fredeen, F. J. H. 1988 ["1987"]. Black flies: approaches to population management in a large temperate-zone river system. Pp. 295–304. *In* K. C. Kim & R. W. Merritt (eds.), Black flies: ecology, population management, and annotated world list. Pennsylvania State University, University Park. 528 pp. [Actual publication date was 1 March 1988.]

Fredeen, F. J. H., & P. G. Mason. 1991. Meteorological factors influencing host-seeking activity of female *Simulium luggeri* (Diptera: Simuliidae). Journal of Medical Entomology 28:831–840.

Fredeen, F. J. H., & J. A. Shemanchuk. 1960. Black flies (Diptera: Simuliidae) of irrigation systems in Saskatchewan and Alberta. Canadian Journal of Zoology 38:723–735.

Fredeen, F. J. H., & D. T. Spurr. 1978. Collecting semi-quantitative samples of black fly larvae (Diptera: Simuliidae) and other aquatic insects from large rivers with the aid of artificial substrates. Quaestiones Entomologicae 14:411–431.

Fredeen, F. J. H., J. G. Rempel, & A. P. Arnason. 1951. Egg-laying habits, overwintering stages, and life-cycle of *Simulium arcticum* Mall. (Diptera: Simuliidae). Canadian Entomologist 83:73–76.

Fredeen, F. J. H., A. P. Arnason, B. Berck, & J. G. Rempel. 1953a. Further experiments with DDT in the control of *Simulium arcticum* Mall. in the North and South Saskatchewan Rivers. Canadian Journal of Agricultural Science 33:379–393.

Fredeen, F. J. H., A. P. Arnason, & B. Berck. 1953b. Adsorption of DDT on suspended solids in river water and its role in blackfly control. Nature 171:700–701.

Fredeen, F. J. H., J. W. T. Spinks, J. R. Anderson, A. P. Arnason, & J. G. Rempel. 1953c. Mass tagging of black flies (Diptera: Simuliidae) with radiophosphorus. Canadian Journal of Zoology 31:1–15.

Fredeen, F. J. H., J. G. Saha, & M. H. Balba. 1975. Residues of methoxychlor and other chlorinated hydrocarbons in water, sand, and selected fauna following injections of methoxychlor black fly larvicide into the Saskatchewan River, 1972. Pesticides Monitoring Journal 8:241–246.

Freeman, M. C., & J. B. Wallace. 1984. Production of net-spinning caddisflies (Hydropsychidae) and black flies (Simuliidae) on rock outcrop substrate in a small southeastern piedmont stream. Hydrobiologia 112:3–15.

Freeman, P., & B. De Meillon. 1953. Simuliidae of the Ethiopian Region. British Museum (Natural History), London. 224 pp.

Frempong-Boadu, J. 1966a. A study of behavior of blackfly larvae (Diptera: Simuliidae) in insecticide screening tests for potential blackfly larvicides. Ph.D. thesis. Yale University, New Haven, Conn. 91 pp.

Frempong-Boadu, J. 1966b. A laboratory study of the effectiveness of methoxychlor, fenthion and carbaryl against blackfly larvae (Diptera: Simuliidae). Mosquito News 26:562–564.

Friederichs, K. 1920. Neues über Kribbelmücken. Berliner und Münchener Tierärztliche Wochenschrift 36:567–569.

Friederichs, K. 1922. Untersuchungen über Simuliiden. Zeitschrift für angewandte Entomologie 8:31–92.

Frierson, G. A. 1889. Buffalo gnats on the Red River. Insect Life 1:313–314.

Frierson, G. A. 1892. Notes on buffalo gnats. Insect Life 4:143–144.

Fries, B. F. 1824. Monographia Simuliarum Sveciae. Pp. 5–20. *In* Observationes Entomologicae, part I. Lundae [= Lund], Sweden. 20 pp.

Frison, T. H. 1927. A list of the insect types in the collections of the Illinois State Natural History Survey and the University of Illinois. Illinois Natural History Survey Bulletin 16(Article IV):137–309. [Simuliidae on p. 181.]

Frizzi, G., A. Lecis, & C. Contini. 1970. Sex determination in *Urosimulium stefanii* Cont. (Diptera: Simuliidae). Zeitschrift für zoologische Systematik und Evolutionsforschung 8:154–159.

Frohne, W. C., & D. A. Sleeper. 1951. Reconnaissance of mosquitoes, punkies, and blackflies in Southeast Alaska. Mosquito News 11:209–213.

Frommer, R. L. 1981. *Bacillus thuringiensis* Berliner var *israelensis* as a potential microbial insecticide for use in black fly (*Simulium* sp.) larval control programs. Ph.D. thesis. University of Minnesota, Minneapolis. 136 pp.

Frommer, R. L., R. R. Carestia, & R. W. Vavra Jr. 1974. A modified CDC trap using carbon dioxide for trapping blackflies (Simuliidae: Diptera). Mosquito News 34:468–469.

Frommer, R. L., R. R. Carestia, & R. W. Vavra Jr. 1975. Field evaluation of deet-treated mesh jacket against black flies (Simuliidae). Journal of Medical Entomology 12:558–561.

Frommer, R. L., B. A. Schiefer, & R. W. Vavra Jr. 1976. Comparative effects of $CO_2$ flow rates using modified CDC light traps on trapping adult black flies (Simuliidae: Diptera). Mosquito News 36:355–358.

Frommer, R. L., S. C. Hembree, J. H. Nelson, M. [P.] Remington, & P. H. Gibbs. 1980. The susceptibility of *Simulium vittatum* larvae (Diptera: Simuliidae) to *Bacillus thuringiensis* var. *israelensis* in the laboratory. Mosquito News 40:577–584.

Frommer, R. L., S. C. Hembree, J. H. Nelson, M. P. Remington, & P. H. Gibbs. 1981a. The distribution of *Bacillus thuringiensis* var. *israelensis* in flowing water with no extensive aquatic vegetative growth. Mosquito News 41:331–338.

Frommer, R. L., S. C. Hembree, J. H. Nelson, M. P. Remington, & P. H. Gibbs. 1981b. The evaluation of *Bacillus thuringiensis* var. *israelensis* in reducing *Simulium vittatum* (Diptera: Simuliidae) larvae in their natural habitat with no extensive aquatic vegetative growth. Mosquito News 41:339–347.

Frommer, R. L., J. H. Nelson, M. P. Remington, & P. H. Gibbs. 1981c. The influence of extensive aquatic vegetative growth on the larvicidal activity of

*Bacillus thuringiensis* var. *israelensis* in reducing *Simulium vittatum* (Diptera: Simuliidae) larvae in their natural habitat. Mosquito News 41:707–712.

Frommer, R. L., J. H. Nelson, M. P. Remington, & P. H. Gibbs. 1981d. The effects of extensive aquatic vegetative growth on the distribution of *Bacillus thuringiensis* var. *israelensis* in flowing water. Mosquito News 41:713–724.

Frommer, R. L., J. H. Nelson, P. H. Gibbs, & J. Vorgetts. 1983. Dose-time response between *Simulium vittatum* (Diptera: Simuliidae) larvae and Abate 200E (temephos). Mosquito News 43:70–71.

Frost, S. 1970. Microsporidia (Protozoa: Microsporidia) in Newfoundland blackfly larvae (Diptera: Simuliidae). Canadian Journal of Zoology 48:890–891.

Frost, S., & J.-F. Manier. 1971. Notes on Trichomycetes (Harpellales: Harpellaceae and Genistellaceae) in larval blackflies (Diptera: Simuliidae) from Newfoundland. Canadian Journal of Zoology 49:776–778.

Frost, S., & R. A. Nolan. 1972. The occurrence and morphology of *Caudospora* spp. (Protozoa: Microsporida) in Newfoundland and Labrador blackfly larvae (Diptera: Simuliidae). Canadian Journal of Zoology 50:1363–1366.

Frost, S. W. 1932. Fishermen of the brook. Nature Magazine 20:267–269.

Frost, S. W. 1949. The Simuliidae of Pennsylvania (Dipt.). Entomological News 60:129–131.

Frost, S. W. 1964. Insects taken in light traps at the Archbold Biological Station, Highlands County, Florida. Florida Entomologist 47:129–161.

Fry, K. M. 1994. Origin, composition, and function of glycoconjugates in the cephalic region in larvae of Culicidae and Simuliidae (Diptera: Culicomorpha). Ph.D. thesis. University of Alberta, Edmonton. 119 pp.

Fry, K. M., & D. A. Craig. 1995. Larval black fly feeding (Diptera: Simuliidae): use of endogenous glycoconjugates. Canadian Journal of Zoology 73:615–622.

Fuller, H. S. 1940. Black-flies bite woodchuck. Bulletin of the Brooklyn Entomological Society 35:155.

Fuller, R. L., & P. A. DeStaffan. 1988. A laboratory study of the vulnerability of prey to predation by three aquatic insects. Canadian Journal of Zoology 66:875–878.

Fuller, R. L., & T. J. Fry. 1991. The influence of temperature and food quality on the growth of *Hydropsyche betteni* (Trichoptera) and *Simulium vittatum* (Diptera). Journal of Freshwater Ecology 6:75–86.

Fuller, R. L., T. J. Fry, & J. A. Roelofs. 1988. Influence of different food types on the growth of *Simulium vittatum* (Diptera) and *Hydropsyche betteni* (Trichoptera). Journal of the North American Benthological Society 7:197–204.

Fussel, E. M. 1970. Semi-annual report of activities for period ending 31 December 1970. U.S. Navy Disease Control Center, Alameda, Calif. 14 pp.

Gaard, G. 2001. June nest box mortality. Wisconsin Bluebird 16(2):1, 4–5.

Gaard, G. 2002. Black fly induced nest mortality—prevention possibilities. Wisconsin Bluebird 17(1):13–16.

Gambrell, L. A. ["F. L."]. 1933. The embryology of the black fly, *Simulium pictipes* Hagen. Annals of the Entomological Society of America 26:641–671.

Gardarsson, A., & A. Einarsson. 1994. Responses of breeding duck populations to changes in food supply. Hydrobiologia 279/280:15–27.

Garman, H. 1893. Silk spinning fly larvae. Science 22:215–217.

Garman, H. 1912. A preliminary study of Kentucky localities in which pellagra is prevalent. Kentucky Agricultural Experiment Station Bulletin 159:1–79 + figures 25–65.

Garman, H. 1917. A few notes from Kentucky. Journal of Economic Entomology 10:413–415.

Garms, R., & J. F. Walsh. 1988 ["1987"]. The migration and dispersal of black flies: *Simulium damnosum* s. l., the main vector of human onchocerciasis. Pp. 201–214. *In* K. C. Kim & R. W. Merritt (eds.), Black flies: ecology, population management, and annotated world list. Pennsylvania State University, University Park. 528 pp. [Actual publication date was 1 March 1988.]

Garris, G. I. 1973. Biological studies involving black flies and transmission of *Leucocytozoon smithi* of turkeys. M.S. thesis. Clemson University, Clemson, S.C. 83 pp.

Garris, G. I., & T. R. Adkins Jr. 1974. The effects of Altosid®, an insect developmental inhibitor, on the last instar larva of *Simulium pictipes*. Mosquito News 34:335–336.

Garris, G. I., & R. Noblet. 1975. Notes on parasitism of black flies (Diptera: Simuliidae) in streams treated with Abate®. Journal of Medical Entomology 12:481–482.

Garris, G. I., & R. Noblet. 1976. Investigations on black flies in Chesterfield County, South Carolina, an area epizootic for *Leucocytozoon smithi* of turkeys. South Carolina Agricultural Experiment Station Technical Bulletin 1056:1–17.

Garris, G. I., R. Noblet, & T. R. Adkins Jr. 1975. Observations on black flies (Diptera: Simuliidae) in Sumter County, South Carolina, an area epizootic for *Leucocytozoon smithi* of turkeys. South Carolina Agricultural Experiment Station Technical Bulletin 1053:1–10.

Gaugler, R., & J. R. Finney. 1982. A review of *Bacillus thuringiensis* var. *israelensis* (serotype 14) as a biological control agent of black flies (Simuliidae). Miscellaneous Publications of the Entomological Society of America 12:1–17.

Gaugler, R., & S. Jaronski. 1983. Assessment of the mosquito-pathogenic fungus *Culicinomyces clavosporus* as a black fly (Diptera: Simuliidae) pathogen. Journal of Medical Entomology 20:575–576.

Gaugler, R., & D. Molloy. 1980. Feeding inhibition in black fly larvae (Diptera: Simuliidae) and its effects on the pathogenicity of *Bacillus thuringiensis* var. *israelensis*. Environmental Entomology 9:704–708.

Gaugler, R., & D. Molloy. 1981a. Instar susceptibility of *Simulium vittatum* (Diptera: Simuliidae) to the entomogenous nematode *Neoaplectana carpocapsae*. Journal of Nematology 13:1–5.

Gaugler, R., & D. Molloy. 1981b. Field evaluation of the entomogenous nematode, *Neoaplectana carpocapsae*, as a biological control agent of black flies (Diptera: Simuliidae). Mosquito News 41:459–464.

Gaugler, R., D. Molloy, T. Haskins, & G. Rider. 1980. A bioassay system for the evaluation of black fly (Diptera: Simuliidae) control agents under simulated stream conditions. Canadian Entomologist 112:1271–1276.

Geitler, L. 1934. Die Schleifenkerne von *Simulium*. Zoologische Jahrbücher (Allgemeine Zoologie und Physiologie der Tiere) 54:237–248.

Georgis, R., H. K. Kaya, & R. Gaugler. 1991. Effect of steinernematid and heterorhabditid nematodes (Rhabditida: Steinernematidae and Heterorhabditidae) on nontarget arthropods. Environmental Entomology 20:815–822.

Gersabeck, E. F., Jr. 1978. The effect of physical factors on colonization of artificial substrates by immature black flies (Diptera: Simuliidae). M.S. thesis. Michigan State University, East Lansing. 48 pp.

Gersabeck, E. F., Jr., & R. W. Merritt. 1979. The effect of physical factors on the colonization and relocation behavior of immature black flies (Diptera: Simuliidae). Environmental Entomology 8:34–39.

Gibbs, K. E., F. C. Brautigam, C. S. Stubbs, & L. M. Zibilske. 1986. Experimental applications of *B.t.i.* for larval black fly control: persistence and downstream carry, efficacy, impact on non-target invertebrates and fish feeding. Maine Agricultural Experiment Station Technical Bulletin 123:1–25.

Gibbs, K. E., R. J. Boyer, B. P. Molloy, & D. A. Hutchins. 1988. Experimental stream applications of *B.t.i.* for human nuisance black fly management in a recreational area. Maine Agricultural Experiment Station Technical Bulletin 133:1–15.

Giberson, D., L. Purcell, C. Brockhouse, & L. Hale. 2002. Black flies in salt: unique salinity tolerance in *Simulium vittatum* Zett. larvae (Diptera: Simuliidae). Parks Canada Atlantic Technical Reports in Ecosystem Science. Prince Edward Island National Park, Charlottetown. 11 pp.

Gibson, A. 1930. Insect and other external parasites of poultry in Canada. Scientific Agriculture 11:208–220.

Gibson, G. G. 1965. The taxonomy and biology of Splendidofilariine nematodes of the Tetraonidae of British Columbia. Ph.D. thesis. University of British Columbia, Vancouver. 241 pp.

Gilbert, P. 1977. A compendium of the biographical literature on deceased entomologists. Publication no. 786. British Museum (Natural History), London. 455 pp.

Gill, G. D., & L. S. West. 1955. Notes on the ecology of certain species of Simuliidae in the Upper Peninsula of Michigan. Papers of the Michigan Academy of Science, Arts, and Letters 40:119–124 + 4 plates.

Gíslason, G. M. 1985. The life cycle and production of *Simulium vittatum* Zett. in the River Laxá, North-east Iceland. Verhandlungen der Internationalen Vereinigung für theoretische und angewandte Limnologie 22:3281–3287.

Gíslason, G. M., & A. Gardarsson. 1988. Long term studies on *Simulium vittatum* Zett. (Diptera: Simuliidae) in the River Laxá, North Iceland, with particular reference to different methods used in assessing population changes. Verhandlungen der Internationalen Vereinigung für theoretische und angewandte Limnologie 23:2179–2188.

Gíslason, G. M., & V. Jóhannsson. 1985. The biology of the blackfly *Simulium vittatum* Zett. (Diptera: Simuliidae) in the River Laxá, northern Iceland. Náttúrufraeðingurinn 55:175–194. [In Icelandic; English summary.]

Gíslason, G. M., & V. Jóhannsson. 1991. Effects of food and temperature on the life cycle of *Simulium vittatum* Zett. (Diptera: Simuliidae) in the River Laxá, N-Iceland. Verhandlungen der Internationalen Vereinigung für theoretische und angewandte Limnologie 24:2912–2916.

Gíslason, G. M., T. Hrafnsdóttir, & A. Gardarsson. 1994. Long-term monitoring of numbers of Chironomidae and Simuliidae in the River Laxá, North Iceland. Verhandlungen der Internationalen Vereinigung für theoretische und angewandte Limnologie 25:1492–1495.

Gistel, J. 1848. Naturgeschichte des Thierreichs. Für höhere Schulen. Stuttgart, Germany. 216 pp.

Gistel, J., & T. Bromme. 1850. Handbuch der Naturgeschichte aller drei Reiche, für Lehrer und Lernende, für Schule und Haus. Stuttgart, Germany. 1037 pp.

Gjullin, C. M., O. B. Cope, B. F. Quisenberry, & F. R. DuChanois. 1949a. The effect of some insecticides on black fly larvae in Alaskan streams. Journal of Economic Entomology 42:100–105.

Gjullin, C. M., D. A. Sleeper, & C. N. Husman. 1949b. Control of black fly larvae in Alaskan streams by aerial applications of DDT. Journal of Economic Entomology 42:392.

Gjullin, C. M., H. F. Cross, & K. H. Applewhite. 1950. Tests with DDT to control black fly larvae in Alaskan streams. Journal of Economic Entomology 43:696–697.

Glasgow, R. D. 1936. Buffalo gnats (*Simulium* spp.). United States Department of Agriculture Insect Pest Survey Bulletin 16:211.

Glasgow, R. D. 1939. Control of blackflies (Simuliidae). Journal of Economic Entomology 32:882–883.

Glasgow, R. D. 1942. New Jersey mosquito larvicide for control of blackflies (Simuliidae). Mosquito News 2:33–37.

Glasgow, R. D. 1948. Blackfly. New York State Conservationist 2(5):8–9.

Glasgow, R. D., & D. L. Collins. 1946. The thermal aerosol fog generator for large scale application of DDT and other insecticides. Journal of Economic Entomology 39:227–235.

Gledhill, T., J. Cowley, & R. J. M. Gunn. 1982. Some aspects of the host: parasite relationships between adult blackflies (Diptera; Simuliidae) and larvae of the water-mite *Sperchon setiger* (Acari; Hydrachnellae) in a small chalk stream in southern England. Freshwater Biology 12:345–357.

Glick, P. A. 1939. The distribution of insects, spiders, and mites in the air. United States Department of Agriculture Technical Bulletin 673:1–151.

Goldie, P. 1982. Progress toward cryopreservation of black fly eggs: *in vitro* experiments and ultrastructure observations. M.S. thesis. Cornell University, Ithaca, N.Y. 198 pp.

Goldsmith, J. B., C. N. Husman, A. W. A. Brown, W. C. McDuffie, & J. F. Sharp. 1949. Exploratory studies on the control of adult mosquitoes and blackflies with DDT under arctic conditions. Mosquito News 9:93–97.

Golini, V. I. 1970. Observations on some factors involved in the host-seeking behaviour of simuliids (Diptera) in Ontario and Norway. M.S. thesis. McMaster University, Hamilton, Ontario. 230 pp.

Golini, V. I. 1974. Relative response to coloured substrates by ovipositing blackflies (Diptera: Simuliidae). III. Oviposition by *Simulium (Psilozia) vittatum* Zetterstedt. Proceedings of the Entomological Society of Ontario 105:48–55.

Golini, V. I. 1975. Simuliidae (Diptera) of Rendalen, Norway. I. *Eusimulium rendalense* n. sp. and *E. fallisi* n. sp. feeding on ducks. Entomologica Scandinavica 6:229–239.

Golini, V. I. 1981. A simple technique for rearing pupae of Simuliidae and other Diptera. Entomologica Scandinavica 12:426–428.

Golini, V. I. 1982. Cytology, taxonomy and ecology of species in the genus *Hellichiella* (Diptera: Simuliidae). Ph.D. thesis. McMaster University, Hamilton, Ontario. 230 pp.

Golini, V. I. 1987. Simuliidae (Diptera) of Rendalen, Norway. Description of a new species of *Eusimulium (Hellichiella)* and previously undescribed stages of *Eusimulium rendalense* and *Eusimulium fallisi*. Canadian Journal of Zoology 65:708–721.

Golini, V. I. 2000. In memory of Dr. Klaus Rothfels (d. 1986). Bulletin of the Entomological Society of Canada 32:4–5.

Golini, V. I., & D. M. Davies. 1971. Upwind orientation of female *Simulium venustum* Say (Diptera) in Algonquin Park, Ontario. Proceedings of the Entomological Society of Ontario 101:49–54.

Golini, V. I., & D. M. Davies. 1975. Relative response to colored substrates by ovipositing blackflies (Diptera: Simuliidae). I. Oviposition by *Simulium (Simulium) verecundum* Stone and Jamnback. Canadian Journal of Zoology 53:521–535.

Golini, V. I., & D. M. Davies. 1988 ["1987"]. Oviposition of black flies. Pp. 261–275. *In* K. C. Kim & R. W. Merritt (eds.), Black flies: ecology, population management, and annotated world list. Pennsylvania State University, University Park. 528 pp. [Actual publication date was 1 March 1988.]

Golini, V. I., & K. Rothfels. 1984. The polytene chromosomes of North American blackflies in the *Eusimulium canonicolum* group (Diptera: Simuliidae). Canadian Journal of Zoology 62:2097–2109.

Golini, V. I., D. M. Davies, & J. E. Raastad. 1976. Simuliidae (Diptera) of Rendalen, Norway. II. Adult females attacking cows and humans. Norwegian Journal of Entomology 23:79–86.

Gordon, A. E. 1984a. Observations on the limnological factors associated with three species of the *Simulium jenningsi* group (Diptera: Simuliidae) in New York

State. Freshwater Invertebrate Biology 3:48–51.

Gordon, A. E. 1984b. The cytotaxonomy of three species in the *jenningsi*-group of the subgenus *Simulium* (Diptera: Simuliidae) in New York State. Canadian Journal of Zoology 62:347–354.

Gordon, A. E., & E. W. Cupp. 1980. The limnological factors associated with cytotypes of the *Simulium* (*Simulium*) *venustum*/*verecundum* complex (Diptera: Simuliidae) in New York State. Canadian Journal of Zoology 58:973–981.

Gordon, R. 1984. Nematode parasites of blackflies. Pp. 821–847. *In* W. R. Nickle (ed.), Plant and insect nematodes. Marcel Dekker, New York. 925 pp.

Gordon, R., & C. H. Bailey. 1974. Free amino acid composition of the hemolymph of the larval blackfly *Simulium venustum* (Diptera: Simuliidae). Experientia 30:902–903.

Gordon, R., & C. H. Bailey. 1976. Free amino acids, ions, and osmotic pressure of the hemolymph of three species of blackflies. Canadian Journal of Zoology 54:399–404.

Gordon, R., B. A. Ebsary, & G. F. Bennett. 1973. Potentialities of mermithid nematodes for the biocontrol of blackflies (Diptera: Simuliidae)—a review. Experimental Parasitology 33:226–238.

Gordon, R., W. J. Condon, W. J. Edgar, & S. J. Babie. 1978. Effects of mermithid parasitism on the haemolymph composition of the larval blackflies *Prosimulium mixtum*/*fuscum* and *Simulium venustum*. Parasitology 77:367–374.

Gordon, R., J. R. Finney, W. J. Condon, & T. N. Rusted. 1979. Lipids in the storage organs of three mermithid nematodes and in the hemolymph of their hosts. Comparative Biochemistry and Physiology, B 64:369–374.

Gordon, R., W. J. Condon, & J. M. Squires. 1980. Sterols in the trophosomes of the mermithid nematodes *Neomesomermis flumenalis* and *Romanomermis culicivorax* relative to sterols in the host hemolymph. Journal of Parasitology 66:585–590.

Gorham, J. R. 1975. Survey of stored-food insects and other Alaskan insect pests. Bulletin of the Entomological Society of America 21:113–117.

Gosbee, J., J. R. Allen, & A. S. West. 1969. The salivary glands of adult blackflies. Canadian Journal of Zoology 47:1341–1344.

Goulding, R. L., Jr., & C. C. Deonier. 1950. Observations on the control and ecology of black flies in Pennsylvania. Journal of Economic Entomology 43:702–704.

Grace, B., & J. L. Shipp. 1988. A laboratory technique for examining the flight activity of insects under controlled environment conditions. International Journal of Biometeorology 32:65–69.

Graham, J. L. 1992. Spatial distribution of female Tabanidae and Simuliidae (Diptera) among different terrestrial habitats in central Newfoundland. M.S. thesis. Memorial University of Newfoundland, St. John's. 116 pp.

Grande, J. 1997. History of New Jersey's first black fly treatment program. Proceedings of the Annual Meeting of the New Jersey Mosquito Control Association 84:49–52.

Granett, J. 1977. Black flies in Maine: biology, damage, and control. University of Maine Life Sciences and Agriculture Experiment Station Miscellaneous Report 188:1–18.

Granett, J. 1979a. Instar frequency and depth distribution of *Simulium penobscotensis* (Diptera: Simuliidae) on aquatic vegetation. Canadian Entomologist 111:161–164.

Granett, J. 1979b. Generation and instar succession of the black fly *Simulium penobscotensis* (Diptera: Simuliidae). Mosquito News 39:792–796.

Granett, J., & L. R. Boobar. 1979. Possible control of the black fly *Simulium penobscotensis* by temporary habitat alteration. Maine Life Sciences and Agriculture Experiment Station Miscellaneous Report 215:1–10.

Grant, D. F. 1991. Evolution of glutathione *S*-transferase subunits in Culicidae and related Nematocera: electrophoretic and immunological evidence for conserved enzyme structure and expression. Insect Biochemistry 21:435–445.

Gray, E. W., & R. Noblet. 1999. Large scale laboratory rearing of black flies. Pp. 85–105. *In* K. Maramorosch & F. Mahmood (eds.), Maintenance of human, animal, and plant pathogen vectors. Oxford & IBH Publishing, New Delhi, India. 328 pp.

Gray, E. W., P. H. Adler, & R. Noblet. 1996. Economic impact of black flies (Diptera: Simuliidae) in South Carolina and development of a localized suppression program. Journal of the American Mosquito Control Association 12:676–678.

Gray, E. W., P. H. Adler, C. Coscarón-Arias, S. Coscarón, & R. Noblet. 1999. Development of the first black fly (Diptera: Simuliidae) management program in Argentina and comparison with other programs. Journal of the American Mosquito Control Association 15:400–406.

Greger, P. D., & J. E. Deacon. 1987. Diel food utilization by woundfin, *Plagopterus argentissimus*, in Virgin River, Arizona. Environmental Biology of Fishes 19:73–77.

Gregg, J. 1844. Commerce of the prairies, vol. 2. H. G. Langley, New York. 318 pp. + glossary + 6 plates.

Greiner, E. C. 1975. Prevalence and potential vectors of *Haemoproteus* in Nebraska mourning doves. Journal of Wildlife Diseases 11:150–156.

Greiner, E. C. 1991. Leucocytozoonosis in waterfowl and wild galliform birds. Bulletin of the Society for Vector Ecology 16:84–93.

Greiner, E. C., & D. J. Forrester. 1979. Prevalence of sporozoites of *Leucocytozoon smithi* in Florida blackflies. Journal of Parasitology 65:324–326.

Greiner, E. C., G. F. Bennett, E. M. White, & R. F. Coombs. 1975. Distribution of the avian hematozoa of North America. Canadian Journal of Zoology 53:1762–1787.

Grenier, P. 1953. Simuliidae de France et d'Afrique du Nord (systématique, biologie, importance médicale). Encyclopédie Entomologique Série A 29:1–170.

Grenier, P., & J. Rageau. 1960. Simulies (Dipt., Simuliidae) de Tahiti. Remarques sur la classification des Simuliidae. Bulletin de la Société de Pathologie Exotique 53:727–742.

Griffen, A. B. 1939. The structure and development of the salivary gland chromosome of *Simulium*. Ph.D. thesis. University of Texas, Austin. 47 pp.

Gross, H. P., W. F. Baldwin, & A. S. West. 1972. Introductory studies on the use of radiation in the control of black flies (Diptera: Simuliidae). Canadian Entomologist 104:1217–1222.

Gryaznov, A. I. 1984a. Morphological adaptations of bloodsucking blackflies to their hosts. Pp. 31–34. *In* E. P. Narchuk & V. V. Zlobin (eds.), Diptera (Insecta) of the fauna of the USSR and their significance in ecosystems [Dvukrylye fauny SSSR i ikh rol' v ekosistemakh]. Akademiya Nauk SSSR, Leningrad [= St. Petersburg], Russia. [In Russian; English translation: 1992. Entomological Review, Washington 71:143–145.]

Gryaznov, A. I. 1984b. Morphology of the abdominal sclerotization of female blood-sucking black flies Simuliidae. Pp. 51–73. *In* A. M. Pankova (ed.), Morphology and ecology of flies—potential carriers of infectious diseases [Morphologiya i ekologiya dvukrylykh—potentsial'nykh perenoschikov zaraznykh zabolevanii]. Sbornik Nauchykh Trudov, Ivanovskii Gosudarstvennyi Meditsinskii Institut, Ivanovo, Russia. [In Russian.]

Gryaznov, A. I. 1989. The signs of adaptation to nutrition on various feeders in the eye structure of blood-sucking black flies (Diptera, Simuliidae). Zoologichesky Zhurnal 68:149–153. [In Russian; English summary.]

Gryaznov, A. I. 1995a. Age-grading in blackflies (Diptera: Simuliidae) by ovariolar morphology. Bulletin of Entomological Research 85:339–344.

Gryaznov, A. I. 1995b. The number of ovarioles in ovaries and its correlation with feeding and reproductive behaviour in bloodsucking black flies (Diptera, Simuliidae). Zoologichesky Zhurnal 74:47–56. [In Russian; English summary.]

Gudgel, E. F., & F. H. Grauer. 1954. Acute and chronic reactions to black fly bites (*Simulium* fly). Archives of Dermatology and Syphilology 70:609–615.

Gutowski, M. J., & J. R. Stauffer Jr. 1993. Selective predation by *Noturus insignis* (Richardson) (Teleostei: Ictaluridae) in the Delaware River. American Midland Naturalist 129:309–318.

Guttman, D. 1967a. Cephalic fan structure as an aid in the identification of black fly larvae (Diptera: Simuliidae). M.S. thesis. Cornell University, Ithaca, N.Y. 107 pp.

Guttman, D. 1967b. MS-222 Sandoz as an anesthetic for black fly larvae (Diptera: Simuliidae). Journal of Medical Entomology 4:477–478.

Guttman, D., B. V. Travis, & R. R. Crafts. 1966. A technique for testing suspensions in simulated stream tests for blackfly larvicides. Mosquito News 26:155–157.

Guyette, J. E. 1994. BTi [*sic*] treatments knock out black flies. Pest Control 62(9):62.

Guyette, J. [E.]. 1999. Battling black flies for profit. Pest Control 67(6):52, 55–56.

Hackman, W., & R. Väisänen. 1985. The evolution and phylogenetic significance of the costal chaetotaxy in the Diptera. Annales Zoologici Fennici 22:169–203.

Hadi, U. K., H. Takaoka, & C. Aoki. 1995. Larval salivary gland chromosomes of the blackfly, *Simulium* (*Gomphostilbia*) *yaeyamaense* (Diptera: Simuliidae) from Ryukyu Islands, Japan. Japanese Journal of Sanitary Zoology 46:235–239.

Hadwen, S. 1923. Insects affecting live stock. Canada Department of Agriculture Bulletin 29 (New Series) [Entomological Bulletin 24]:1–32.

Hafele, R. 1997. Black flies and the fly fisher. American Angler 20(March/April):10–14.

Hagen, H. A. 1880. A new species of *Simulium* with a remarkable nympha case. Proceedings of the Boston Society of Natural History 20:305–307.

Hagen, H. A. 1881. On *Simulium*. Canadian Entomologist 13:150–151.

Hagen, H. A. 1883. *Simulium* feeding upon chrysalids. Entomologist's Monthly Magazine 19:254–255.

Hahn, N. [E.]. 1990. Investigations into the vector competency of arthropods for two ehrlichias: *Ehrlichia risticii* and *Cowdria ruminantium*. Ph.D. thesis. Virginia Polytechnic Institute and State University, Blacksburg. 85 pp.

Hahn, N. E., B. D. Perry, R. M. Rice, J. W. Hansen, & E. C. Turner. 1989. Role of blackflies in the epidemiology of Potomac horse fever. Veterinary Record 125:273–274.

Hall, F. 1972. Observations on black flies of the genus *Simulium* in Los Angeles County, California. California Vector Views 19:53–58.

Hall, F. 1973. The blackflies, genus *Simulium* (Diptera: Simuliidae), of southern California. M.A. thesis. California State University, Long Beach. 109 pp.

Hall, F. 1974. A key to the *Simulium* larvae of southern California (Diptera: Simuliidae). California Vector Views 21:65–71.

Hall, R. J., R. C. Bailey, & J. Findeis. 1988. Factors affecting survival and cation concentration in the blackflies *Prosimulium fuscum/mixtum* and the mayfly *Leptophlebia cupida* during spring snowmelt. Canadian Journal of Fisheries and Aquatic Sciences 45:2123–2132.

Hall, R. O., Jr., C. L. Peredney, & J. L. Meyer. 1996. The effect of invertebrate consumption on bacterial transport in a mountain stream. Limnology and Oceanography 41:1180–1187.

Hallock, H. C. 1922. Studies in the anatomy of the larva of *Simulium pictipes* Hagen. M.S. thesis. Cornell University, Ithaca, N. Y. 49 pp. + 6 plates.

Hamada, N., J. W. McCreadie, & P. H. Adler. 2002. Species richness and spatial distribution of blackflies (Diptera: Simuliidae) in streams of Central Amazonia, Brazil. Freshwater Biology 47:31–40.

Hannay, C. L., & E. F. Bond. 1971a. Blackfly wing surface. Canadian Journal of Zoology 49:543–549 + 4 plates.

Hannay, C. L., & E. F. Bond. 1971b. Blackfly thoracic pollinosity. Canadian Journal of Zoology 49:572–573 + 1 plate.

Hansen, R. A., D. D. Hart, & R. A. Merz. 1991. Flow mediates predator-prey interactions between triclad flatworms and larval black flies. Oikos 60:187–196.

Harding, J., & M. H. Colbo. 1981. Competition for attachment sites between larvae of Simuliidae (Diptera). Canadian Entomologist 113:761–763.

Harkrider, J. R. 1988. A field study of parasitism of larval black flies (Diptera: Simuliidae) by *Neomesomermis flumenalis* (Nematoda: Mermithidae) in southern California. Environmental Entomology 17:391–397.

Harned, R. W. 1927. Buffalo gnat (*Simulium pecuarum* Riley). United States Department of Agriculture Insect Pest Survey Bulletin 7:50–51.

Harned, R. W. 1929. Buffalo gnats (Simuliidae). United States Department of Agriculture Insect Pest Survey Bulletin 9:93–94.

Harned, R. W. 1931. Buffalo gnats (Simuliidae). United States Department of Agriculture Insect Pest Survey Bulletin 11:130–131.

Harper, F. 1955. The barren ground caribou of Keewatin. University of Kansas Museum of Natural History Miscellaneous Publication No. 6. University of Kansas, Lawrence. 163 pp.

Harris, T. W. 1835. Insects. Pp. 553–602. *In* E. Hitchcock (ed.), Report on the geology, mineralogy, botany, and zoology of Massachusetts. 2nd ed. J. S. & C. Adams, Amherst, Mass. 702 pp. + 19 plates + 3 maps.

Harris, T. W. 1841. A report on the insects of Massachusetts, injurious to vegetation. 1st ed. Folsom, Wells & Thurston, Cambridge, Mass. 459 pp.

Hart, D. D. 1986. The adaptive significance of territoriality in filter-feeding larval blackflies (Diptera: Simuliidae). Oikos 46:88–92.

Hart, D. D. 1987. Feeding territoriality in aquatic insects: cost-benefit models and experimental tests. American Zoologist 27:371–386.

Hart, D. D. 1988 ["1987"]. Processes and patterns of competition in larval black flies. Pp. 109–128. *In* K. C. Kim & R. W. Merritt (eds.), Black flies: ecology, population management, and annotated world list. Pennsylvania State University, University Park. 528 pp. [Actual publication date was 1 March 1988.]

Hart, D. D., & S. C. Latta. 1986. Determinants of ingestion rates in filter-feeding larval blackflies (Diptera: Simuliidae). Freshwater Biology 16:1–14.

Hart, D. D., & R. A. Merz. 1998. Predator-prey interactions in a benthic stream community: a field test of flow-mediated refuges. Oecologia 114:263–273.

Hart, D. D., R. A. Merz, S. J. Genovese, & B. D. Clark. 1991. Feeding postures of suspension-feeding larval black flies: the conflicting demands of drag and food acquisition. Oecologia 85:457–463.

Hart, D. D., B. D. Clark, & A. Jasentuliyana. 1996. Fine-scale field measurement of benthic flow environments inhabited by stream invertebrates. Limnology and Oceanography 41:297–308.

Hartley, C. F. 1955. Rearing simuliids in the laboratory from eggs to adults. Proceedings of the Helminthological Society 22:93–95.

Hatfield, C. T. 1969. Effects of DDT larviciding on aquatic fauna of Bobby's Brook, Labrador. Canadian Fish Culturist 40:61–72.

Hauer, F. R., & A. C. Benke. 1987. Influence of temperature and river hydrograph on black fly growth rates in a subtropical blackwater river. Journal of the North

American Benthological Society 6:251–261.
Haufe, W. O., & G. C. R. Croome (eds.). 1980. Control of black flies in the Athabasca River. Alberta Environment Technical Report. Alberta Environment, Pollution Control Division, Edmonton. 241 pp.
Hayton, A. 1979. The age structure and population dynamics of some black-flies in Algonquin Park, Ontario. M.S. thesis. University of Waterloo, Waterloo, Ontario. 191 pp.
Hazard, E. I., & S. W. Oldacre. 1975. Revision of Microsporida (Protozoa) close to *Thelohania*, with descriptions of one new family, eight new genera, and thirteen new species. United States Department of Agriculture Technical Bulletin 1530:1–104.
Headlee, T. J. 1906. Blood gills of *Simulium pictipes*. American Naturalist 40:875–885.
Hearle, E. 1929. A remarkable simuliid pupa. Notes on *Simulium virgatum* in British Columbia. Proceedings of the Entomological Society of British Columbia 26:48–54.
Hearle, E. 1932. The blackflies of British Columbia (Simuliidae, Diptera). Proceedings of the Entomological Society of British Columbia 29:5–19.
Hearle, E. 1935. Notes on *Simulium canadense* Hearle and *Simulium virgatum* Coquillett and its varieties. Canadian Entomologist 67:15–18.
Hearle, E. 1938. Insects and allied parasites injurious to livestock and poultry in Canada. Canada Department of Agriculture Publication No. 604 (Farmers' Bulletin 53). Department of Agriculture, Ottawa. 108 pp.
Helle, T., J. Aspi, K. Lempa, & E. Taskinen. 1992. Strategies to avoid biting flies by reindeer: field experiments with silhouette traps. Annales Zoologici Fennici 29:69–74.
Helson, B. V. 1972. The selective effects of particulate formulations of insecticides on stream fauna when applied as blackfly (Diptera: Simuliidae) larvicides. M.S. thesis. Queen's University, Kingston, Ontario. 187 pp.
Helson, B. V., & A. S. West. 1978. Particulate formulations of Abate® and methoxychlor as black fly larvicides: their selective effects on stream fauna. Canadian Entomologist 110:591–602.
Hembree, S. C., R. L. Frommer, & M. P. Remington. 1980. A bioassay apparatus for evaluating larvicides against black flies. Mosquito News 40:647–650.
Hemphill, N. 1988. Competition between two stream dwelling filter-feeders, *Hydropsyche oslari* and *Simulium virgatum*. Oecologia 77:73–80.
Hemphill, N. 1989. The effects of competition and disturbance on the relative abundances and distribution of *Simulium virgatum* and *Hydropsyche oslari* in a stream. Ph.D. thesis. University of California, Santa Barbara. 175 pp.
Hemphill, N. 1991. Disturbance and variation in competition between two stream insects. Ecology 72:864–872.
Hemphill, N., & S. D. Cooper. 1983. The effect of physical disturbance on the relative abundances of two filter-feeding insects in a small stream. Oecologia 58:378–382.
Henderson, C. A. P. 1985. A cytological study of sibling species in *Prosimulium onychodactylum* (Diptera: Simuliidae). M.S. thesis. University of Toronto, Toronto. 108 pp.
Henderson, C. A. P. 1986a. A cytological study of the *Prosimulium onychodactylum* complex (Diptera, Simuliidae). Canadian Journal of Zoology 64:32–44.
Henderson, C. A. P. 1986b. Homosequential species 2*a* and 2*b* within the *Prosimulium onychodactylum* complex (Diptera): temporal heterogeneity, linkage disequilibrium, and Wahlund effect. Canadian Journal of Zoology 64:859–866.
Henderson, G., P. G. Holland, & G. L. Werren. 1979. The natural history of a subarctic adventive: *Epilobium angustifolium* L. (Onagraceae) at Schefferville, Québec. Naturaliste Canadien 106:425–437.
Hennig, W. 1973. Diptera (Zweiflüger). *In* W. Kükenthal (ed.), Handbuch der Zoologie. Band 4, Hälfte 2, Teil 2. Walter de Gruyter, New York. 337 pp.
Herman, C. M., J. H. Barrow Jr., & I. B. Tarshis. 1975. Leucocytozoonosis in Canada geese at the Seney National Wildlife Refuge. Journal of Wildlife Diseases 11:404–411.
Hernández, O., G. Maldonado, & T. Williams. 2000. An epizootic of patent iridescent virus disease in multiple species of blackflies in Chiapas, Mexico. Medical and Veterinary Entomology 14:458–462.
Herrick, G. W. 1899. Some insects injurious to stock and remedies therefor. Mississippi Agricultural Experiment Station Bulletin 53:1–8.
Herrick, G. W. 1916. Insects injurious to the household and annoying to man. Macmillan, New York. 470 pp.
Hershey, A. E., & A. L. Hiltner. 1988. Effect of a caddisfly on black fly density: interspecific interactions limit black flies in an arctic river. Journal of the North American Benthological Society 7:188–196.
Hershey, A. E., R. W. Merritt, & M. C. Miller. 1995. Insect diversity, life history, and trophic dynamics in arctic streams, with particular emphasis on black flies (Diptera: Simuliidae). Pp. 283–295. *In* F. S. Chapin & C. Körner (eds.), Arctic and alpine biodiversity: patterns, causes, and ecosystem consequences. Springer-Verlag, New York. 332 pp.
Hershey, A. E., R. W. Merritt, M. C. Miller, & J. S. McCrea. 1996. Organic matter processing by larval black flies in a temperate woodland stream. Oikos 75:524–532.
Hershey, A. E., W. B. Bowden, L. A. Deegan, J. E. Hobbie, B. J. Peterson, G. W. Kipphut, G. W. Kling, M. A. Lock, R. W. Merritt, M. C. Miller, J. R. Vestal, & J. A. Schuldt. 1997. The Kuparuk River: a long-term study of biological and chemical processes in an arctic river. Pp. 107–129. *In* A. M. Milner & M. W. Oswood (eds.), Freshwaters of Alaska: ecological syntheses. Springer-Verlag, New York. 369 pp.
Hewitt, C. G. 1910. *Simulium* flies and pellagra. Nature 85:169–170.
Hidalgo Escalante, E. 1959. Simúlidos del Estado de Morelos (Dipt. Simuliidae). Acta Zoologica Mexicana 3:1–63.
Hill, J. R., III. 1994. What's bugging your birds? An introduction to the ectoparasites of purple martins. Purple Martin Update 5(1):1–7.
Hilsenhoff, W. L. 1982. Using a biotic index to evaluate water quality in streams. Wisconsin Department of Natural Resources Technical Bulletin 132:1–22.
Hilsenhoff, W. L. 1987. An improved biotic index of organic stream pollution. Great Lakes Entomologist 20:31–39.
Hiltner, A. L., & A. E. Hershey. 1992. Black fly (Diptera: Simuliidae) response to phosphorus enrichment of an arctic tundra stream. Hydrobiologia 240:259–265.
Hilton, D. F. J. 1970. A method for providing personal protection against simuliids. Mosquito News 30:474.
Hinton, H. E. 1957. Some little known respiratory adaptations. Science Progress 180:692–700.
Hinton, H. E. 1958. The pupa of the fly *Simulium* feeds and spins its own cocoon. Entomologist's Monthly Magazine 94:14–16.
Hinton, H. E. 1959. The function of chromatocytes in the Simuliidae, with notes on their behaviour at the pupal-adult moult. Quarterly Journal of Microscopical Science 100:65–71.
Hinton, H. E. 1964. The respiratory efficiency of the spiracular gill of *Simulium*. Journal of Insect Physiology 10:73–80.
Hinton, H. E. 1965. The spiracular gill of the fly *Orimargula australiensis* and its relation to those of other insects. Australian Journal of Zoology 13:783–800.
Hinton, H. E. 1968. Spiracular gills. Advances in Insect Physiology 5:65–162.
Hinton, H. E. 1976. The fine structure of the pupal plastron of simuliid flies. Journal of Insect Physiology 22:1061–1070.

Hocking, B. 1950. Further tests of insecticides against black flies (Diptera: Simuliidae) and a control procedure. Scientific Agriculture 30:489–508.
Hocking, B. 1952a. Protection from northern biting flies. Mosquito News 12:91–102.
Hocking, B. 1952b. Two predators as prey. Canadian Field-Naturalist 66:107.
Hocking, B. 1953a. On the intrinsic range and speed of flight of insects. Ph.D. thesis. University of Alberta, Edmonton. 240 pp.
Hocking, B. 1953b. On the intrinsic range and speed of flight of insects. Transactions of the Royal Entomological Society of London 104:223–345.
Hocking, B. 1953c. Developments in the chemical control of black flies (Diptera: Simuliidae). Canadian Journal of Agricultural Science 33:572–578.
Hocking, B., & L. R. Pickering. 1954. Observations on the bionomics of some northern species of Simuliidae (Diptera). Canadian Journal of Zoology 32:99–119.
Hocking, B., & W. R. Richards. 1952. Biology and control of Labrador black flies (Diptera: Simuliidae). Bulletin of Entomological Research 43:237–257 + 2 plates.
Hocking, B., C. R. Twinn, & W. C. McDuffie. 1949. A preliminary evaluation of some insecticides against immature stages of blackflies (Diptera: Simuliidae). Scientific Agriculture 29:69–80.
Hoffman, C. C. 1930a. Un *Simulium* nuevo de la Zona Cafetera de Chiapas. Anales del Instituto de Biologia (México) 1:51–53.
Hoffman, C. C. 1930b. Los simúlidos de la región onchocercosa de Chiapas (con descripción de nuevas especies). Anales del Instituto de Biologia (México) 1:293–306.
Hoffman, C. C. 1931. Los simúlidos de la región onchocercosa de Chiapas, segunda parte; los estados larvales. Anales del Instituto de Biologia (México) 2:207–218.
Holbrook, F. R. 1967. The black flies (Diptera: Simuliidae) of western Massachusetts. Ph.D. thesis. University of Massachusetts, Amherst. 265 pp.
Hopla, C. E. 1965. Alaskan hematophagous insects, their feeding habits and potential as vectors of pathogenic organisms. II: The feeding habits and colonization of subarctic mosquitoes. Technical Report of the U.S. Air Force Arctic Aeromedical Laboratory. Publication No. AAL-TR-64-12. Fort Wainwright, Alaska. 90 pp.
Horosko, S., III. 1982. Investigations using *Bacillus thuringiensis* var. *israelensis* for control of black flies and suppression of leucocytozoonosis in turkeys. Ph.D. thesis. Clemson University, Clemson, S.C. 63 pp.
Horosko, S., III, & R. Noblet. 1983. Efficacy of *Bacillus thuringiensis* var. *israelensis* for control of black fly larvae in South Carolina. Journal of the Georgia Entomological Society 18:531–537.
Horosko, S., III, & R. Noblet. 1986a. Black fly control and suppression of leucocytozoonosis in turkeys. Journal of Agricultural Entomology 3:10–24.
Horosko, S., III, & R. Noblet. 1986b. Local area control of black flies in the Southeast with Vectobac®-AS and Vectobac®-12AS. South Carolina Agricultural Experiment Station Bulletin 658:1–9.
Howard, L. O. 1888. Notes on a *Simulium* common at Ithaca, N.Y. Insect Life 1:99–101.
Howard, L. O. 1894. Death web of young trout. Insect Life 7:50.
Howard, L. O. 1895. The buffalo gnat. Insect Life 7:426.
Howard, L. O. 1901. The insect book. Doubleday, Page, Garden City, N.Y. 429 pp.
Hudson, D. K. M., & K. L. Hays. 1975. Some factors affecting the distribution and abundance of black fly larvae in Alabama. Journal of the Georgia Entomological Society 10:110–122.
Hudson, S. B. 1994. Factors affecting the survival of black fly larvae (Diptera: Simuliidae) when exposed to *Bacillus thuringiensis* serovar *israelensis* in the orbital shaker bioassay. M.S. thesis. Clemson University, Clemson, S.C. 55 pp.
Hungerford, H. B. 1913. Anatomy of *Simulium vittatum*. Kansas University Science Bulletin 8:365–382 + 3 plates.
Hunter, D. B., C. Rohner, & D. C. Currie. 1997a. Mortality in fledgling great horned owls from black fly hematophaga and leucocytozoonosis. Journal of Wildlife Diseases 33:486–491.
Hunter, D. B., C. Rohner, & D. C. Currie. 1997b. Black-flies and *Leucocytozoon* spp. as causes of mortality in juvenile great horned owls in the Yukon, Canada. Pp. 243–245. *In* J. R. Duncan, D. H. Johnson, & T. H. Nicholls (eds.), Biology and conservation of owls of the Northern Hemisphere. U.S. Department of Agriculture Forest Service General Technical Report No. NC-190. U.S. Department of Agriculture, North Central Research Station, Forest Service, St. Paul, Minn. 635 pp.
Hunter, F. F. 1987a. A cytotaxonomic study of species in the *Simulium vernum* group (Diptera: Simuliidae). M.S. thesis. University of Toronto, Toronto. 181 pp.
Hunter, F. F. 1987b. Cytotaxonomy of four European species in the *Eusimulium vernum* group (Diptera: Simuliidae). Canadian Journal of Zoology 65:3102–3115.
Hunter, F. F. 1989. The polytene chromosomes of *Simulium furculatum* (Shewell) (Diptera: Simuliidae). Genome 32:522–530.
Hunter, F. F. 1990. Ecological, morphological, and behavioural correlates to cytospecies in the *Simulium venustum/verecundum* complex (Diptera: Simuliidae). Ph.D. thesis. Queen's University, Kingston, Ontario. 185 pp.
Hunter, F. F. 2002. Polytene chromosomes of *Simulium craigi* (Diptera: Simuliidae). Genetica 114:207–215.
Hunter, F. F., & R. Bayly. 1991. ELISA for identification of blood meal source in black flies (Diptera: Simuliidae). Journal of Medical Entomology 28:527–532.
Hunter, F. F., & V. Connolly. 1986. A cytotaxonomic investigation of seven species in the *Eusimulium vernum* group (Diptera: Simuliidae). Canadian Journal of Zoology 64:296–311.
Hunter, F. F., & H. Jain. 2000. Do gravid black flies (Diptera: Simuliidae) oviposit at their natal site? Journal of Insect Behavior 13:585–595.
Hunter, F. F., J. F. Sutcliffe, & A. E. R. Downe. 1993. Blood-feeding host preferences of the isomorphic species *Simulium venustum* and *S. truncatum*. Medical and Veterinary Entomology 7:105–110.
Hunter, F. F., S. G. Burgin, & D. M. Wood. 1994. New techniques for rearing black flies from pupae (Diptera: Simuliidae). Journal of the American Mosquito Control Association 10:456–459.
Hunter, F. F., S. G. Burgin, & A. Woodhouse. 2000. Shattering the folklore: black flies do not pollinate sweet lowbush blueberry. Canadian Journal of Zoology 78:2051–2054.
Hunter, S. J. 1912a. The sand-fly and pellagra. Journal of Economic Entomology 5:61–64.
Hunter, S. J. 1912b. The sand-fly and pellagra. Journal of the American Medical Association 58:547–548.
Hunter, S. J. 1913a. Pellagra and the sand-fly, II. Journal of Economic Entomology 6:96–101.
Hunter, S. J. 1913b. University experiments with sand fly and pellagra. Kansas University Science Bulletin 8:313–320.
Hunter, S. J. 1914. The sandfly and pellagra, III. Journal of Economic Entomology 7:293–294.
Hutcheon, D. E., & V. S. Chivers-Wilson. 1953. The histaminic and anticoagulant activity of extracts of the black fly (*Simulium vittatum* and *Simulium venustum*). Revue Canadienne de Biologie 12:77–85.
Hyder, A. H. 1998. Black flies (Diptera: Simuliidae): bioindicator potential and toxic responses to chloripyrifos and the microbial pesticide Vectobac®. Ph.D. thesis. Clemson University, Clemson, S.C. 122 pp.

Ibáñez Bernal, S. 1992. Las especies Mexicanas de *Simulium* (*Hemicnetha*) y *S.* (*Notolepria*) (Diptera: Simuliidae). M.S. thesis. Universidad Nacional Autonoma de México, Federal District, Mexico. 302 pp.

Ide, F. P. 1942. Availability of aquatic insects as food of the speckled trout, *Salvelinus fontinalis*. Transactions of the North American Wildlife Conference 7:442–450.

Imhof, J. E. 1977. The oviposition behaviour of certain species of black flies (Diptera: Simuliidae) and some aspects of the biology of their eggs. M.S. thesis. University of Waterloo, Waterloo, Ontario. 181 pp.

Imhof, J. E., & S. M. Smith. 1979. Oviposition behaviour, egg-masses and hatching response of the eggs of five Nearctic species of *Simulium* (Diptera: Simuliidae). Bulletin of Entomological Research 69:405–425.

International Commission on Zoological Nomenclature. 1986. Opinion 1416. *Cnetha* Enderlein, 1921 and *Pseudonevermannia* Baranov, 1926 (Insecta, Diptera): type species designated; *Atractocera latipes* Meigen, 1804: confirmation of holotype. Bulletin of Zoological Nomenclature 43:264–266.

Jackson, J. K., B. W. Sweeney, T. L. Bott, J. D. Newbold, & L. A. Kaplan. 1994. Transport of *Bacillus thuringiensis* var. *israelensis* and its effect on drift and benthic densities of nontarget macroinvertebrates in the Susquehanna River, northern Pennsylvania. Canadian Journal of Fisheries and Aquatic Sciences 51:295–314.

Jackson, J. K., R. J. Horwitz, & B. W. Sweeney. 2002. Effects of *Bacillus thuringiensis israelensis* on black flies and nontarget macroinvertebrates and fish in a large river. Transactions of the American Fisheries Society 131:910–930.

Jacobs, J. W., E. W. Cupp, M. Sardana, & P. A. Friedman. 1990. Isolation and characterization of a coagulation factor Xa inhibitor from black fly salivary glands. Thrombosis and Haemostasis 64:235–238.

Jacobs, P. 1991. *Plagiorchis noblei* and blackfly larvae: factors affecting parasite acquisition and the effect of infection on host survival. M.S. thesis. McGill University, Montreal, Quebec. 94 pp.

Jacobs, P., M. E. Rau, & D. J. Lewis. 1993. Factors affecting the acquisition of *Plagiorchis noblei* (Trematoda: Plagiorchiidae) cercariae by black fly (Diptera: Simuliidae) larvae and the effect of metacercariae on host survival. Journal of the American Mosquito Control Association 9:36–45.

Jacobs-Lorena, M., M. Doman, & A. Mahowald. 1988. Identification of species-specific DNA sequences in North American blackflies. Tropical Medicine and Parasitology 39:31–34.

Jahn, L. A. 1932. The embryology of the black fly. Ph.D. thesis. Ohio State University, Columbus. 43 pp. + 12 plates.

James, H. G. 1968. Bird predation on black fly larvae and pupae in Ontario. Canadian Journal of Zoology 46:106–107.

Jamnback, H. 1951. An investigation of certain aspects of blackfly control. M.S. thesis. University of Massachusetts, Amherst. 76 pp.

Jamnback, H. 1952. The importance of correct timing of larval treatments to control specific blackflies (Simuliidae). Mosquito News 12:77–78.

Jamnback, H. 1953. Blackfly control investigations. Ph.D. thesis. University of Massachusetts, Amherst. 370 pp.

Jamnback, H. 1956. An illustrated key to the blackfly larvae commonly collected in New York State. New York State Science Service, Albany, N.Y. 10 pp.

Jamnback, H. 1962. An eclectic method of testing the effectiveness of chemicals in killing blackfly larvae (Simuliidae: Diptera). Mosquito News 22:384–389.

Jamnback, H. 1964. Description of a trough testing technique useful in evaluating the effectiveness of chemicals as blackfly larvicides. World Health Organization mimeographed document WHO/Oncho./28.64 & WHO/VC/96.64. Geneva. 4 pp. + 2 plates.

Jamnback, H. 1969a. Bloodsucking flies and other outdoor nuisance arthropods of New York State. New York State Museum and Science Service Memoir 19:1–90.

Jamnback, H. 1969b. Field tests with larvicides other than DDT for control of blackfly (Diptera: Simuliidae) in New York. Bulletin of the World Health Organization 40:635–638.

Jamnback, H. 1969c. Field tests with larvicides other than DDT for blackfly control (Diptera: Simuliidae) in New York. World Health Organization mimeographed document WHO/VBC/69.120. Geneva. 7 pp.

Jamnback, H. 1970. *Caudospora* and *Weiseria*, two genera of microsporidia parasitic in blackflies. Journal of Invertebrate Pathology 16:3–13.

Jamnback, H. 1973a. Recent developments in control of blackflies. Annual Review of Entomology 18:281–304.

Jamnback, H. 1973b. The blackfly. NAHO (New York State Museum, Albany) 6:6–8.

Jamnback, H. 1981. The origins of blackfly control programmes. Pp. 71–73. *In* M. Laird (ed.), Blackflies: the future for biological methods in integrated control. Academic Press, New York. 399 pp.

Jamnback, H., & D. L. Collins. 1955. The control of blackflies (Diptera: Simuliidae) in New York. New York State Museum Bulletin 350:1–113.

Jamnback, H., & H. S. Eabry. 1962. Effects of DDT, as used in black fly larval control, on stream arthropods. Journal of Economic Entomology 55:636–639.

Jamnback, H., & J. Frempong-Boadu. 1966. Testing blackfly larvicides in the laboratory and in streams. Bulletin of the World Health Organization 34:405–421.

Jamnback, H., & R. Means. 1966. Length of exposure period as a factor influencing the effectiveness of larvicides for blackflies (Diptera: Simuliidae). Mosquito News 26:589–591.

Jamnback, H. A., & R. G. Means. 1968. Formulation as a factor influencing the effectiveness of Abate® in control of blackflies (Diptera: Simuliidae). Proceedings of the Annual Meeting of the New Jersey Mosquito Extermination Association 55:89–94.

Jamnback, H., & A. Stone. 1957. A first record of *Simulium* (*Eusimulium*) *congareenarum* (D. & S.) from New York, with descriptions of the male, female, pupa, and larva. Annals of the Entomological Society of America 50:395–399.

Jamnback, H., & A. S. West. 1970. Decreased susceptibility of blackfly larvae to p,p′-DDT in New York State and eastern Canada. Journal of Economic Entomology 63:218–221.

Jarvis, B. J. 1987. Phenology and drift dynamics of preimaginal Simuliidae (Diptera) in a large temperate river. M.S. thesis. University of Saskatchewan, Saskatoon. 174 pp.

Jefferies, D. 1987. Labrocibarial sensilla in the female of the black fly *Simulium damnosum* s. l. (Diptera: Simuliidae). Canadian Journal of Zoology 65:441–444.

Jell, P. A., & P. M. Duncan. 1986. Invertebrates, mainly insects, from the freshwater, Lower Cretaceous, Koonwarra Fossil Bed (Korumburra Group), South Gippsland, Victoria. Memoirs of the Association of Australasian Palaeontologists 3:111–205.

Jenkins, D. W. 1948. Ecological observations on the blackflies and punkies of central Alaska. Mosquito News 8:148–154.

Jenkins, D. W. 1964. Pathogens, parasites and predators of medically important arthropods—annotated list and bibliography. Bulletin of the World Health Organization 30(Suppl.):1–150.

Jennings, A. H. 1914. Summary of two years' study of insects in relation to pellagra. Journal of Parasitology 1:10–21.

Jennings, A. H., & W. V. King. 1913a. One of the possible factors in the causation of pellagra. Journal of the American Medical Association 60:271–274. [The identical paper was published in 1912 in

Transactions of the National Association for the Study of Pellagra 2:51–60.]
Jennings, A. H., & W. V. King. 1913b. An intensive study of insects as a possible etiologic factor in pellagra. American Journal of the Medical Sciences (New Series) 146:411–440. [The identical paper was published in 1914 (pp. 81–110) in the First Progress Report of the Thompson-McFadden Pellagra Commission of the New York Post-Graduate Medical School and Hospital.]
Jensen, F. 1984. A revision of the taxonomy and distribution of the Danish blackflies (Diptera: Simuliidae), with keys to the larval and pupal stages. Natura Jutlandica 21:69–116.
Jensen, F. 1997. Diptera Simuliidae, blackflies. Pp. 209–241. *In* A. N. Nilsson (ed.), Aquatic insects of North Europe—a taxonomic handbook, vol. 2, Apollo Books, Stenstrup, Denmark. 440 pp.
Jessen, J. I. 1977. Black flies (Diptera: Simuliidae) which affect sheep in southern Idaho. Ph.D. thesis. University of Idaho, Moscow. 154 pp.
Jobbins-Pomeroy, A. W. 1916. Notes on five North American buffalo gnats of the genus *Simulium*. United States Department of Agriculture Bulletin 329:1–48.
Jobling, B. 1987. Anatomical drawings of biting flies. British Museum (Natural History) and Wellcome Trust, London. 119 pp.
Johannsen, O. A. 1902. On aquatic Diptera. M.A. thesis. Cornell University, Ithaca, N.Y. 111 pp. + 16 plates.
Johannsen, O. A. 1903a. Notes on some Adirondack Diptera collected by Messrs. MacGillivray and Houghton. Entomological News 14:14–17.
Johannsen, O. A. 1903b. Part 6. Aquatic nematocerous Diptera. Pp. 328–441, 492–494, plates 32–38. *In* Aquatic insects in New York State. New York State Museum Bulletin 68:199–517 + 52 plates.
Johannsen, O. A. 1912. Insect notes for 1910. Maine Agricultural Experiment Station Bulletin 187(27th Annual Report):1–24 + 8 plates.
Johannsen, O. A. 1934. Aquatic Diptera. Part I. Nemocera exclusive of Chironomidae and Ceratopogonidae. Cornell University Agricultural Experiment Station Memoir 164:1–71 + 24 plates.
Jóhannsson, V. 1984. Seasonal changes in the gut content of *Simulium vittatum* Zett. larvae in River Laxá. Náttúruverndarráð 14:65–72. [In Icelandic; English summary.]
Jóhannsson, V. 1988. The life cycles of *Simulium vittatum* Zett. in Icelandic lake-outlets. Verhandlungen der Internationalen Vereinigung für theoretische und angewandte Limnologie 23:2170–2178.
Johnson, A. F., & D. H. Pengelly. 1966. A cone trap for immature black flies (Diptera: Simuliidae). Proceedings of the Entomological Society of Ontario 96:120.
Johnson, A. F., & D. H. Pengelly. 1969. The larval instars of *Simulium rugglesi* Nicholson and Mickel (Diptera: Simuliidae). Proceedings of the Entomological Society of Ontario 100:182–187.
Johnson, C. G., R. W. Crosskey, & J. B. Davies. 1982. Species composition and cyclical changes in numbers of savanna blackflies (Diptera: Simuliidae) caught by suction traps in the Onchocerciasis Control Programme area of West Africa. Bulletin of Entomological Research 72:39–63 + 1 plate.
Johnson, C. W. 1925. Diptera of the Harris Collection. Proceedings of the Boston Society of Natural History 38:57–99.
Johnson, E. P. 1942. Further observations on a blood protozoan of turkeys transmitted by *Simulium nigroparvum* (Twinn). American Journal of Veterinary Research 3:214–218.
Johnson, E. P., G. W. Underhill, J. A. Cox, & W. L. Threlkeld. 1938. A blood protozoon of turkeys transmitted by *Simulium nigroparvum* (Twinn). American Journal of Hygiene 27:649–665.
Jones, C. M., & D. J. Richey. 1956. Biology of the black flies in Jasper County, South Carolina, and some relationships to a *Leucocytozoon* disease of turkeys. Journal of Economic Entomology 49:121–123.
Jones, M. P. 1934. Buffalo gnats (*Simulium* spp.). United States Department of Agriculture Insect Pest Survey Bulletin 14:92.
Jones, R. H. 1961. Some observations on biting flies attacking sheep. Mosquito News 21:113–115.
Jones, R. H. 1981. Biting flies collected from recumbent bluetongue-infected sheep in Idaho. Mosquito News 41:183.
Jones, R. H., R. O. Hayes, H. W. Potter Jr., & D. B. Francy. 1977. A survey of biting flies attacking equines in three states of the southwestern United States, 1972. Journal of Medical Entomology 14:441–447.
Jónsson, E., A. Gardarsson, & G. M. Gíslason. 1986. A new window trap used in the assessment of the flight periods of Chironomidae and Simuliidae (Diptera). Freshwater Biology 16:711–719.
Judd, W. W. 1957. Studies of the Byron Bog in southwestern Ontario. IV. Seasonal distribution of the black fly, *Simulium vittatum* Zett. (Diptera: Simuliidae). Entomological News 68:263–265.
Kachvoryan, E. A., L. A. Chubareva, N. A. Petrova, & L. S. Mirumyan. 1996. Frequency changes of B chromosomes in synanthropic species of bloodsucking blackflies (Diptera, Simuliidae). Genetika 32:637–640. [In Russian; English translation in Russian Journal of Genetics 32:554–557.]
Kalugina, N. S. 1991. New Mesozoic Simuliidae and Leptoconopidae and blood-sucking origin in lower dipterans. Paleontologichesky Zhurnal 1991:69–80 + 1 plate. [In Russian; English summary.]
Kampani, D. V. 1986. An economic evaluation of available farm level technologies for the control of black flies on Saskatchewan livestock farms. M.S. thesis. University of Saskatchewan, Saskatoon. 168 pp.
Kellogg, V. L. 1901. Food of larvae of *Simulium* and *Blepharocera*. Psyche 9:166–167.
Kelly, D. J., J. C. Clare, & M. L. Bothwell. 2001. Attenuation of solar ultraviolet radiation by dissolved organic matter alters benthic colonization patterns in streams. Journal of the North American Benthological Society 20:96–108.
Kesler, D. H. 1982. Cellulase activity in four species of aquatic insect larvae in Rhode Island, U.S.A. Journal of Freshwater Ecology 1:559–562.
Khachatourians, G. G. 1990. Blackfly larvicide efficacy, bacterial distribution and regeneration of *Bacillus thuringiensis* ser. H14 (BTH-14) in a 10 kilometer segment of the South Saskatchewan River. Final Report No. R-89-01-0447. Agriculture Development Fund, Regina, Saskatchewan. 32 pp.
Khan, M. A. 1978. Protection of cattle from black flies. Pp. 54–56. *In* Research Highlights—1977. Agriculture Canada Research Station, Lethbridge, Alberta. 79 pp.
Khan, M. A. 1980. Protection of cattle from black flies. Pp. 217–232. *In* W. O. Haufe & G. C. R. Croome (eds.), Control of black flies in the Athabasca River. Alberta Environment Technical Report. Alberta Environment, Pollution Control Division, Edmonton. 241 pp.
Khan, M. A. 1981. Protection of pastured cattle from black flies (Diptera: Simuliidae): improved weight gains following a dermal application of phosmet. Veterinary Parasitology 8:327–336.
Khan, M. A., & G. C. Kozub. 1985. Response of Angus, Charolais, and Hereford bulls to black flies (*Simulium* spp.), with and without phosmet treatment. Canadian Journal of Animal Science 65:269–272.
Khan, R. A. 1975. Development of *Leucocytozoon ziemanni* (Laveran). Journal of Parasitology 61:449–457.
Khan, R. A., & A. M. Fallis. 1968. Comparison of infections with *Leucocytozoon simondi* in black ducks (*Anas rubripes*), mallards (*Anas platyrhynchos*), and white Pekins (*Anas bochas*). Canadian Journal of Zoology 46:773–780.

Khan, R. A., & A. M. Fallis. 1970. Life cycles of *Leucocytozoon dubreuili* Mathis and Leger, 1911 and *L. fringillinarum* Woodcock, 1910 (Haemosporidia: Leucocytozoidae). Journal of Protozoology 17:642–658.

Khan, R. A., & A. M. Fallis. 1971. Speciation, transmission, and schizogony of *Leucocytozoon* in corvid birds. Canadian Journal of Zoology 49:1363–1367.

Kiel, E. 1997. Durability of simuliid silk pads (Simuliidae, Diptera). Aquatic Insects 19:15–22.

Kiel, E., & A. Frutiger. 1997. Behavioural responses of different blackfly species to short-term oxygen depletion. Internationale Revue der Gesamten Hydrobiologie 82:107–120.

Kiel, E., F. Böge, & W. Rühm. 1998a. Sustained effects of larval blackfly settlement on further substrate colonisers. Archiv für Hydrobiologie 141:153–166.

Kiel, E., F. Böge, & W. Rühm. 1998b. Do silk pad remnants of larval blackflies (Simuliidae, Diptera) affect further substrate colonisation processes? Limnologica 28:307–312.

Kiel, E., K. Dickmann, & W. Rühm. 1998c. Effects of frequent drift events on blackfly (Simuliidae, Diptera) development: a laboratory study. Limnologica 28:301–305.

Kiffney, P. M., E. E. Little, & W. H. Clements. 1997a. Influence of ultraviolet-B radiation on the drift response of stream invertebrates. Freshwater Biology 37:485–492.

Kiffney, P. M., W. H. Clements, & T. A. Cady. 1997b. Influence of ultraviolet radiation on the colonization dynamics of a Rocky Mountain stream benthic community. Journal of the North American Benthological Society 16:520–530.

Kim, K. C., & R. W. Merritt (eds.). 1988 ["1987"]. Black flies: ecology, population management, and annotated world list. Pennsylvania State University, University Park. 528 pp. [Actual publication date was 1 March 1988.]

Kindler, J. B., & F. R. Regan. 1949. Larvicide tests on blackflies in New Hampshire. Mosquito News 9:108–112.

King, D. G. 1991. The origin of an organ: phylogenetic analysis of evolutionary innovation in the digestive tract of flies (Insecta: Diptera). Evolution 45:568–588.

Kissam, J. B., R. Noblet, & H. S. Moore IV. 1973. *Simulium*: field evaluation of Abate larvacide [*sic*] for control in an area endemic for *Leucocytozoon smithi* of turkeys. Journal of Economic Entomology 66:426–428.

Kissam, J. B., R. Noblet, & G. I. Garris. 1975. Large-scale aerial treatment of an endemic area with Abate® granular larvicide to control black flies (Diptera: Simuliidae) and suppress *Leucocytozoon smithi* of turkeys. Journal of Medical Entomology 12:359–362.

Kiszewski, A. E. 1984. Black fly transmission of *Leucocytozoon smithii* in Tompkins County, New York. M.S. thesis. Cornell University, Ithaca, N.Y. 83 pp.

Kiszewski, A. E., & E. W. Cupp. 1986. Transmission of *Leucocytozoon smithi* (Sporozoa: Leucocytozoidae) by black flies (Diptera: Simuliidae) in New York, USA. Journal of Medical Entomology 23:256–262.

Klowden, M. J., & A. O. Lea. 1979. Oocyte maturation in the black fly, *Simulium underhilli* Stone & Snoddy, resulting from blood enemas. Canadian Journal of Zoology 57:1344–1347.

Klowden, M. J., L. A. Bulla Jr., & R. L. Stoltz. 1985. Susceptibility of larval and adult *Simulium vittatum* (Diptera: Simuliidae) to the solubilized parasporal crystal of *Bacillus thuringiensis israelensis.* Journal of Medical Entomology 22:466–467.

Knab, F. 1915a ["1914"]. New data and species in Simuliidae (Diptera). Insecutor Inscitiae Menstruus 2:177–180. [Actual publication date was 14 January 1915.]

Knab, F. 1915b. A new *Simulium* from Texas (Diptera, Simuliidae). Insecutor Inscitiae Menstruus 3:77–78.

Knap, J. 1969. Blackflies. Imperial Oil Review 53(3):10–13.

Knight, A. L. 1980. Host range and temperature requirements of *Culicinomyces clavosporus.* Journal of Invertebrate Pathology 36:423–425.

Knowlton, G. F. 1935. Simuliids annoy livestock. Journal of Economic Entomology 28:1073.

Knowlton, G. F., & L. E. Fronk. 1950. Some blood sucking Diptera of Utah. Utah Agricultural Experiment Station Mimeograph Series 369:1–9.

Knowlton, G. F., & D. R. Maddock. 1944. Snipe flies in Utah. Journal of Economic Entomology 37:119.

Knowlton, G. F., & J. A. Rowe. 1934a. New blood-sucking flies from Utah (Simuliidae, Diptera). Annals of the Entomological Society of America 27:580–584.

Knowlton, G. F., & J. A. Rowe. 1934b. Preliminary studies of insect transmission of equine encephalomyelitis. Utah Academy of Sciences, Arts and Letters 11:267–270.

Knowlton, G. F., & J. A. Rowe. 1934c. Buffalo gnats. Utah Agricultural Experiment Station Leaflet 27:1–2.

Knowlton, G. F., F. C. Harmston, & D. E. Hardy. 1938. Blood-sucking Utah Diptera. Utah Academy of Sciences, Arts and Letters 15:103–105.

Knoz, J. 1965. To identification of Czechoslovakian black-flies (Diptera, Simuliidae). Folia Facultatis Sbientiarum Naturalium Universitatis Purkynianae Brunensis 6(5):1–52 + 425 figures.

Knutti, H. J., & W. R. Beck. 1988 ["1987"]. The control of black fly larvae with Teknar®. Pp. 409–418. *In* K. C. Kim & R. W. Merritt (eds.), Black flies: ecology, population management, and annotated world list. Pennsylvania State University, University Park. 528 pp. [Actual publication date was 1 March 1988.]

Koch, D. A., G. A. Duncan, T. J. Parsons, K. P. Pruess, & T. O. Powers. 1998. Effects of preservation methods, parasites, and gut contents of black flies (Diptera: Simuliidae) on polymerase chain reaction products. Journal of Medical Entomology 35:314–318.

Koehler, A., C. Zia, & S. S. Desser. 2002. Structural organization of the midgut musculature in black flies (*Simulium* spp.). Canadian Journal of Zoology 80:910–917.

Kohler, S. L. 1992. Competition and the structure of a benthic stream community. Ecological Monographs 62:165–188.

Komnick, H. 1977. Chloride cells and chloride epithelia of aquatic insects. International Review of Cytology 49:285–329.

Kondratieff, B. C., & J. R. Voshell. 1983. A study of the black fly invertebrate predators in the New River below Bluestone Dam. Report to West Virginia Department of Natural Resources. Charleston. 42 pp.

Konurbaev, E. O. 1973. Variability of some of the quantitative characters of simuliid larvae (Diptera, Simuliidae) in the mountains of Soviet Central Asia. Entomologicheskoe Obozrenie 52:915–923. [In Russian; English translation in Entomological Review, Washington 52:590–595.]

Konurbaev, E. O. 1977. The ecological classification of running waters in Soviet Central Asia and the distribution pattern of black flies (Diptera, Simuliidae) in watercourses of different types. Entomologicheskoe Obozrenie 56:736–750. [In Russian; English translation in Entomological Review, Washington 56:17–27.]

Koontz, K. 1992. The exterminator. Backpacker 20(3):16.

Koslucher, D. G., & G. W. Minshall. 1973. Food habits of some benthic invertebrates in a northern cool-desert stream (Deep Creek, Curlew Valley, Idaho-Utah). Transactions of the American Microscopical Society 92:441–452.

Krafchick, B. 1941. The structure of the mouthparts of blackflies with special reference to *Eusimulium lascivum* Twinn. M.S. thesis. Cornell University, Ithaca, N.Y. 20 pp. +4 plates.

Krafchick, B. 1942. The mouthparts of blackflies with special reference to *Eusimulium lascivum* Twinn. Annals of

the Entomological Society of America 35:426–434.
Kramer, J. P. 1983. A mycosis of the black fly *Simulium decorum* (Simuliidae) caused by *Erynia curvispora* (Entomophthoraceae). Mycopathologia 82:39–43.
Kramer, W. L., & M. S. Mulla. 1980. Prevalence and potential problems of nuisance and vector insects in Malibu Creek, Los Angeles County, California. Proceedings and Papers of the Annual Conference of the California Mosquito Control Association 48:121–125.
Kramer, W. L., R. H. Jones, F. R. Holbrook, T. E. Walton, & C. H. Calisher. 1990. Isolation of arboviruses from *Culicoides* midges (Diptera: Ceratopogonidae) in Colorado during an epizootic of vesicular stomatitis New Jersey. Journal of Medical Entomology 27:487–493.
Kreutzweiser, D. P., S. B. Holmes, & D. J. Behmer. 1992. Effects of the herbicides hexazinone and triclopyr ester on aquatic insects. Ecotoxicology and Environmental Safety 23:364–374.
Krombein, K. V. 1960. Biological notes on some Hymenoptera that nest in sumach pith. Entomological News 71:63–69.
Kukalová-Peck, J. 1992. The "Uniramia" do not exist: the ground plan of the Pterygota as revealed by Permian Diaphanopterodea from Russia (Insecta: Paleodictyopteroidea). Canadian Journal of Zoology 70:236–255.
Kurtak, D. C. 1973. Observations on filter feeding by the larvae of black flies (Diptera: Simuliidae). Ph.D. thesis. Cornell University, Ithaca, N.Y. 157 pp.
Kurtak, D. [C.]. 1974. Overwintering of *Simulium pictipes* Hagen (Diptera: Simuliidae) as eggs. Journal of Medical Entomology 11:383–384.
Kurtak, D. C. 1978. Efficiency of filter feeding of black fly larvae (Diptera: Simuliidae). Canadian Journal of Zoology 56:1608–1623.
Kurtak, D. C. 1979. Food of black fly larvae (Diptera: Simuliidae): seasonal changes in gut contents and suspended material at several sites in a single watershed. Quaestiones Entomologicae 15:357–374.
Kurtak, D. C., B. V. Travis, & J. R. Meyer. 1972. Tests (1971) with black fly larvicides. Proceedings of the Annual Meeting of the New Jersey Mosquito Extermination Association 59:169–174.
Labeyrie, E. S., D. P. Molloy, & R. W. Lichtwardt. 1996. An investigation of Harpellales (Trichomycetes) in New York State blackflies (Diptera: Simuliidae). Journal of Invertebrate Pathology 68:293–298.
Laboulbène, A. 1882. Sur l'insecte Diptère nuisible de Terre-Neuve signalé par M. le Docteur Treille. Archives de Médecine Navale 38:222–224.
Lacey, L. A. 1978. Evaluation of *Bacillus thuringiensis* Berliner and diflubenzuron as biocides for blackflies—development of laboratory and field bioassay techniques. Ph.D. thesis. University of California, Riverside. 141 pp.
Lacey, L. A. 1997. Bacteria: laboratory bioassay of bacteria against aquatic insects with emphasis on larvae of mosquitoes and black flies. Pp. 79–90. *In* L. A. Lacey (ed.), Manual of techniques in insect pathology. Academic Press, San Diego. 409 pp.
Lacey, L. A., & B. A. Federici. 1979. Pathogenesis and midgut histopathology of *Bacillus thuringiensis* in *Simulium vittatum* (Diptera: Simuliidae). Journal of Invertebrate Pathology 33:171–182.
Lacey, L. A., & C. M. Heitzman. 1985. Efficacy of flowable concentrate formulations of *Bacillus thuringiensis* var. *israelensis* against black flies (Diptera: Simuliidae). Journal of the American Mosquito Control Association 1:493–497.
Lacey, L. A., & M. S. Mulla. 1977a. Evaluation of *Bacillus thuringiensis* as a biocide of blackfly larvae (Diptera: Simuliidae). Journal of Invertebrate Pathology 30:46–49.
Lacey, L. A., & M. S. Mulla. 1977b. Biting flies in the lower Colorado River Basin. II. Adult activities of the blackfly, *Simulium vittatum* Zetterstedt (Diptera: Simuliidae). Proceedings and Papers of the Annual Conference of the California Mosquito and Vector Control Association 45:214–218.
Lacey, L. A., & M. S. Mulla. 1977c. Larvicidal and ovicidal activity of Dimlin® against *Simulium vittatum*. Journal of Economic Entomology 70:369–373.
Lacey, L. A., & M. S. Mulla. 1977d. A new bioassay unit for evaluating larvicides against blackflies. Journal of Economic Entomology 70:453–456.
Lacey, L. A., & M. S. Mulla. 1978a. Factors affecting the activity of diflubenzuron against *Simulium* larvae (Diptera: Simuliidae). Mosquito News 38:264–268.
Lacey, L. A., & M. S. Mulla. 1978b. Biological activity of diflubenzuron and three new IGRs against *Simulium vittatum* (Diptera: Simuliidae). Mosquito News 38:377–381.
Lacey, L. A., & M. S. Mulla. 1979a. Field evaluation of diflubenzuron against *Simulium* larvae. Mosquito News 39:86–90.
Lacey, L. A., & M. S. Mulla. 1979b. Factors affecting feeding rates of black fly larvae. Mosquito News 39:315–319.
Lacey, L. A., & M. S. Mulla. 1980. Observations on the biology and distribution of *Simulium tescorum* (Diptera: Simuliidae) in California and adjacent areas. Pan-Pacific Entomologist 56:323–331.
Lacey, L. A., & A. H. Undeen. 1984. Effect of formulation, concentration, and application time on the efficacy of *Bacillus thuringiensis* (H-14) against black fly (Diptera: Simuliidae) larvae under natural conditions. Journal of Economic Entomology 77:412–418.
Lacey, L. A., & A. H. Undeen. 1986. Microbial control of black flies and mosquitoes. Annual Review of Entomology 31:265–296.
Lacey, L. A., & A. H. Undeen. 1988 ["1987"]. The biological control potential of pathogens and parasites of black flies. Pp. 327–340. *In* K. C. Kim & R. W. Merritt (eds.), Black flies: ecology, population management, and annotated world list. Pennsylvania State University, University Park. 528 pp. [Actual publication date was 1 March 1988.]
Lacey, L. A., M. S. Mulla, & H. T. Dulmage. 1978. Some factors affecting the pathogenicity of *Bacillus thuringiensis* Berliner against blackflies. Environmental Entomology 7:583–588.
Lacey, L. A., A. H. Undeen, & M. M. Chance. 1982. Laboratory procedures for the bioassay and comparative efficacy evaluation of *Bacillus thuringiensis* var. *israelensis* (serotype 14) against black flies (Simuliidae). Miscellaneous Publications of the Entomological Society of America 12:19–23.
Lacoursière, J. O. 1990. Suspension-feeding behaviour of black fly larvae (Diptera: Simuliidae): hydrodynamical perspectives. Ph.D. thesis. University of Alberta, Edmonton. 194 pp.
Lacoursière, J. O. 1992. A laboratory study of fluid flow and microhabitat selection by larvae of *Simulium vittatum* (Diptera: Simuliidae). Canadian Journal of Zoology 70:582–596.
Lacoursière, J. O., & J. L. Boisvert. 1987. Short-term maintenance system for black fly larvae (Diptera: Simuliidae). Journal of Medical Entomology 24:463–466.
Lacoursière, J. O., & G. Charpentier. 1988. Laboratory study of the influence of water temperature and pH on *Bacillus thuringiensis* var. *israelensis* efficacy against black fly larvae (Diptera: Simuliidae). Journal of the American Mosquito Control Association 4:64–72.
Lacoursière, J. O., & D. A. Craig. 1993. Fluid transmission and filtration efficiency of the labral fans of black fly larvae (Diptera: Simuliidae): hydrodynamic, morphological, and behavioural aspects. Canadian Journal of Zoology 71:148–162.
Ladle, M., J. A. B. Bass, & L. J. Cannicott. 1985. A unique strategy of blackfly oviposition (Diptera: Simuliidae). Entomologist's Gazette 36:147–149.

Laird, M. 1961. A lack of avian and mammalian haematozoa in the Antarctic and Canadian Arctic. Canadian Journal of Zoology 39:209–213.
Laird, M. 1972. A novel attempt to control biting flies with their own diseases. Science Forum 30(December):12–14.
Laird, M. 1978. The status of biocontrol investigations concerning Simuliidae. Environmental Conservation 5:133–142.
Laird, M. 1980a. Biocontrol in veterinary entomology. Advances in Veterinary Science and Comparative Medicine 24:145–177.
Laird, M. 1980b. Blackflies, mosquitoes and tsetse flies. Queen's Quarterly 87:401–410.
Laird, M. (ed.). 1981. Blackflies: the future for biological methods in integrated control. Academic Press, New York. 399 pp.
Laird, M., & G. F. Bennett. 1970. The subarctic epizootiology of *Leucocytozoon simondi*. Journal of Parasitology 56(Sect. 11):198.
Laird, M., A. Aubin, P. Belton, M. M. Chance, F. J. H. Fredeen, W. O. Haufe, H. B. N. Hynes, D. J. Lewis, I. S. Lindsay, D. M. McLean, G. A. Surgeoner, D. M. Wood, & M. D. Sutton. 1982. Biting flies in Canada: health effects and economic consequences. National Research Council of Canada Publication No. 19248. National Research Council of Canada, Ottawa. 157 pp.
Lake, D. J. 1980. Effect of physical and hydrochemical factors on the distribution and succession of outlet-breeding black flies (Diptera: Simuliidae) in New Hampshire. M.S. thesis. University of New Hampshire, Durham. 91 pp.
Lake, D. J., & J. F. Burger. 1983. Larval distribution and succession of outlet-breeding blackflies (Diptera: Simuliidae) in New Hampshire. Canadian Journal of Zoology 61:2519–2533.
Lake, R. W. 1983. Simuliidae. P. 311. *In* J. C. Morse, J. W. Chapin, D. D. Herlong, & R. S. Harvey (eds.), Aquatic insects of Upper Three Runs Creek, Savannah River Plant, South Carolina. Part II: Diptera. Journal of the Georgia Entomological Society 18:303–316.
Landau, R. 1962. Four forms of *Simulium tuberosum* (Lundstr.) in southern Ontario: a salivary gland chromosome study. Canadian Journal of Zoology 40:921–939.
LaPointe, D. A., & P. J. S. Martin. 1990. Mosquitoes, black flies and B.t.i. Massachusetts Wildlife 40(2):26–35.
Larson, D. J., & M. H. Colbo. 1983. The aquatic insects: biogeographic considerations. Pp. 593–677. *In* G. R. South (ed.), Biogeography and ecology of the Island of Newfoundland. Dr. W. Junk, The Hague. 736 pp.
LaScala, P. A. 1979. The bionomics of black flies (Diptera: Simuliidae) in Dixville Notch, New Hampshire with emphasis on species of potential economic importance. M.S. thesis. University of New Hampshire, Durham. 61 pp.
LaScala ["La Scala"], P. A., & J. F. Burger. 1981. A small-scale environmental approach to blackfly control in the USA. Pp. 133–136. *In* M. Laird (ed.), Blackflies: the future for biological methods in integrated control. Academic Press, New York. 399 pp.
Latreille, P. A. 1802. Histoire naturelle, générale et particulière des crustacés et des insectes. 3. Familles naturelles des genres. F. Dufart, Paris. 467 pp.
Lavigne, R. J., S. W. Bullington, & G. Stephens. 1993. Ethology of *Holopogon seniculus* (Diptera: Asilidae). Annals of the Entomological Society of America 86:91–95.
Lavinder, C. H. 1910. The theory of the parasitic origin of pellagra. Public Health Reports 25:735–737.
Ledin, K. E. 1994. Microsporidian pathogens of North American black flies (Diptera: Simuliidae), with notes on the ecology of southeastern microsporidia. M.S. thesis. Clemson University, Clemson, S.C. 83 pp.
Léger, L. 1897. Sur une nouvelle Myxosporidie de la famille des Glugeidées. Comptes Rendus Hebdomadaires des Séances de l'Académie des Sciences 125:260–262.
Léger, L., & O. Duboscq. 1929. *Harpella melusinae* n. g. n. sp. Entophyte eccriniforme parasite des larves de Simulie. Comptes Rendus Hebdomadaires des Séances de l'Académie des Sciences 188:951–954.
Lehmann, T. 1994. *Onchocerca lienalis* in *Simulium vittatum:* navigation of microfilariae and the fly defense response. Ph.D. thesis. University of Arizona, Tucson. 144 pp.
Lehmann, T., M. S. Cupp, & E. W. Cupp. 1994a. *Onchocerca lienalis:* rapid clearance of microfilariae within the black fly, *Simulium vittatum*. Experimental Parasitology 78:183–193.
Lehmann, T., M. S. Cupp, & E. W. Cupp. 1994b. *Onchocerca lienalis:* a comparison of microfilarial loss in *Simulium jenningsi* and *Simulium vittatum*. Experimental Parasitology 79:195–197.
Lehmann, T., M. S. Cupp, & E. W. Cupp. 1995a. Analysis of migration success of *Onchocerca lienalis* microfilariae in the haemocoel of *Simulium vittatum*. Journal of Helminthology 69:47–52.
Lehmann, T., M. S. ["S. M."] Cupp, & E. W. ["W. E."] Cupp. 1995b. Chemical guidance of *Onchocerca lienalis* microfilariae to the thorax of *Simulium vittatum*. Parasitology 110:329–337.
Lehmkuhl, D. M. 1990. Efficacy tests on blackfly larvicide (BTH): a new application method for problem areas of the Saskatchewan River. Final Report No. V-89-01-0242. Agriculture Development Fund, Regina, Saskatchewan. 24 pp.
Leonhardt, K. G. 1985. A cytological study of species in the *Eusimulium aureum* group (Diptera: Simuliidae). Canadian Journal of Zoology 63:2043–2061.
Leonhardt, K. G. 1987. A cytological study of species in the *Eusimulium aureum* group with comparisons to *E. vernum* and *E. ruficorne* (Diptera: Simuliidae). M.S. thesis. University of Toronto, Toronto. 104 pp.
Leonhardt, K. G., & R. M. Feraday. 1989. Sex chromosome evolution and population differentiation in the *Eusimulium aureum* group of black flies. Genome 32:543–549.
LePage, S., G. Charpentier, D. Pecqueur, A. Vey, & J.-M. Quiot. 1992. Utilisation des toxines de champignons entomopathogènes dans la lutte contre les diptères piqueurs. Mémoires de la Société Royale Belge d'Entomologie 35:139–143.
Lewis, D. J. 1973. The Simuliidae of insular Newfoundland and their dynamics in small streams on the Avalon Peninsula. Ph.D. thesis. Memorial University of Newfoundland, St. John's. 216 pp.
Lewis, D. J., & G. F. Bennett. 1973. The blackflies (Diptera: Simuliidae) of insular Newfoundland. I. Distribution and bionomics. Canadian Journal of Zoology 51:1181–1187.
Lewis, D. J., & G. F. Bennett. 1974a. An artificial substrate for the quantitative comparison of the densities of larval simuliid (Diptera) populations. Canadian Journal of Zoology 52:773–775.
Lewis, D. J., & G. F. Bennett. 1974b. The blackflies (Diptera: Simuliidae) of insular Newfoundland. II. Seasonal succession and abundance in a complex of small streams on the Avalon Peninsula. Canadian Journal of Zoology 52:1107–1113.
Lewis, D. J., & G. F. Bennett. 1975. The blackflies (Diptera: Simuliidae) of insular Newfoundland. III. Factors affecting the distribution and migration of larval simuliids in small streams on the Avalon Peninsula. Canadian Journal of Zoology 53:114–123.
Lewis, D. J., & G. F. Bennett. 1979. An annotated list of the black flies (Diptera: Simuliidae) of the Maritime Provinces of Canada. Canadian Entomologist 111:1227–1230.
Leyon, H., & G. Eklund. 1950. Some electron microscopical observations on the structure of the wings of *Simulium*. Arkiv för Zoologi 1:471–476.

Lichtwardt, R. W. 1964. Axenic culture of two new species of branched Trichomycetes. American Journal of Botany 51:836–842.
Lichtwardt, R. W. 1967. Zygospores and spore appendages of *Harpella* (Trichomycetes) from larvae of Simuliidae. Mycologia 59:482–491.
Lichtwardt, R. W. 1972. Undescribed genera and species of Harpellales (Trichomycetes) from the guts of aquatic insects. Mycologia 64:167–197.
Lichtwardt, R. W. 1976. Trichomycetes. Pp. 651–671. *In* E. B. G. Jones (ed.), Recent advances in aquatic mycology. Halsted Press, John Wiley, New York. 749 pp.
Lichtwardt, R. W. 1984. Species of Harpellales living within the guts of aquatic Diptera larvae. Mycotaxon 19:529–550.
Lichtwardt, R. W. 1986. The Trichomycetes, fungal associates of arthropods. Springer-Verlag, New York. 343 pp.
Lichtwardt, R. W. 1996. Trichomycetes and the arthropod gut. Pp. 315–330. *In* D. H. Howard & J. D. Miller (eds.), The Mycota. VI. Human and animal relationships. Springer-Verlag, Berlin. 399 pp.
Lichtwardt, R. W. 1997. Costa Rican gut fungi (Trichomycetes) infecting lotic insect larvae. Revista de Biologia Tropical 45:1349–1383.
Lichtwardt, R. W., & M. C. Williams. 1988. Distribution and species diversity of trichomycete gut fungi in aquatic insect larvae in two Rocky Mountain streams. Canadian Journal of Botany 66:1259–1263.
Lichtwardt, R. W., M. M. White, & M. H. Colbo. 2001. Harpellales in Newfoundland aquatic insect larvae. Mycologia 93:764–773.
Lidz, F. 1981. The blackfly festival competitors really had to stick their necks out. Sports Illustrated 55(6):63.
Light, R. W. 1983. Growth and reproductive ecology of the eastern brook trout, *Salvelinus fontinalis*, in streams of differing vulnerability to acidic atmospheric deposition. Ph.D. thesis. Pennsylvania State University, University Park. 198 pp.
Lindner, E. 1943. Neue Dipteren aus dem Gebiet der Alpen. Mitteilungen Münchener Entomologischen Gesellschaft 33:244–247.
Lindsay, I. S., & J. M. McAndless. 1978. Permethrin-treated jackets versus repellent-treated jackets and hoods for personal protection against black flies and mosquitoes. Mosquito News 38:350–356.
Lindsay, R. C., & R. W. Dimmick. 1983. Mercury residues in wood ducks and wood duck foods in eastern Tennessee. Journal of Wildlife Diseases 19:114–117.
Linley, J. R., & K. R. Simmons. 1981. Sperm motility and spermathecal filling in lower diptera [*sic*]. International Journal of Invertebrate Reproduction 4:137–146.
Linley, J. R., & K. R. Simmons. 1983. Quantitative aspects of sperm transfer in *Simulium decorum* (Diptera: Simuliidae). Journal of Insect Physiology 29:581–584.
Linnaeus, C. 1758. Systema naturae per regna tria naturae, secundum classes, ordines, genera, species, cum caracteribus, differentiis, synonymis, locis. 10th ed., Stockholm, Sweden. 824 pp.
Lintner, J. A. 1884. Report of the State Entomologist to the Regents of the University of the State of New York for the year 1883. Annual Report on the New York State Museum of Natural History 37:45–60.
Linton, M. C. 1982. Population structure and genetic variability in *Prosimulium mixtum* and *P. fuscum* (Diptera: Simuliidae). M.S. thesis. Michigan Technological University, Houghton. 38 pp.
Lipsit, S. W. 2001. Effect of *Bacillus thuringiensis* variety *israelensis* on blackflies (Diptera: Simuliidae) in the North Saskatchewan River. M.S. thesis. University of Saskatchewan, Saskatoon. 125 pp.
Liu, T. P. 1972. Ultrastructural changes in the nuclear envelope of larval fat body cells of *Simulium vittatum* (Diptera) induced by microsporidian infection of *Thelohania bracteata*. Tissue and Cell 4:493–502.
Liu, T. P. 1973. Ultrastructure of the yolk protein granules in the frozen-etched oocyte of an insect. Cytobiologie 7:33–41.
Liu, T. P. 1974. Ultrastructure of the lipid inclusions of the yolk in the freeze-etched oocyte of an insect. Cytobiologie 8:412–420.
Liu, T. P., & D. M. Davies. 1971. Ultrastructural localization of glycogen in the flight muscle of the blackfly, *Simulium vittatum* Zett. Canadian Journal of Zoology 49:219–221.
Liu, T. P., & D. M. Davies. 1972a. Ultrastructural localization of glutamic oxaloacetic transaminase in mitochondria of the flight muscle of Simuliidae. Journal of Insect Physiology 18:1665–1671.
Liu, T. P., & D. M. Davies. 1972b. Ultrastructure of the cytoplasm in fat-body cells of the blackfly, *Simulium vittatum*, with microsporidian infection; a freeze-etching study. Journal of Invertebrate Pathology 19:208–218.
Liu, T. P., & D. M. Davies. 1972c. Ultrastructure of the nuclear envelope from blackfly, fat-body cells with and without microsporidian infection. Journal of Invertebrate Pathology 20:176–182.
Liu, T. P., & D. M. Davies. 1972d. Fine structure of frozen-etched lipid granules in the fat body of an insect. Journal of Lipid Research 13:115–118.
Liu, T. P., & D. M. Davies. 1972e. An autoradiographic and ultrastructural study of glycogen metabolism and function in the adult fat body of a black-fly during oögenesis. Entomologia Experimentalis et Applicata 15:265–273.
Liu, T. P., & D. M. Davies. 1972f. Ultrastructure of protein and lipid inclusions in frozen-etched blackfly oöcytes (Simuliidae, Diptera). Canadian Journal of Zoology 50:59–62.
Liu, T. P., & D. M. Davies. 1973. Intramitochondrial transformations during lipid vitellogenesis in oöcytes of a black fly, *Simulium vittatum* Zetterstedt (Diptera: Simuliidae). International Journal of Insect Morphology and Embryology 2:233–245.
Liu, T. P., J. J. Darley, & D. M. Davies. 1975. Differentiation of ovariolar follicular cells and formation of previtelline-membrane substance in *Simulium vittatum* Zetterstedt (Diptera: Simuliidae). International Journal of Insect Morphology and Embryology 4:331–340.
Lok, J. B. 1981. Surrogate and natural vectors for *Onchocerca* spp. Ph.D. thesis. Cornell University, Ithaca, N.Y. 90 pp.
Lok, J. B., E. W. Cupp, & M. J. Bernardo. 1980. The development of *Onchocerca* spp. in *Simulium decorum* Walker and *Simulium pictipes* Hagen. Tropenmedizin und Parasitologie 31:498–506.
Lok, J. B., E. W. Cupp, M. J. Bernardo, & R. J. Pollack. 1983a. Further studies on the development of *Onchocerca* spp. (Nematoda: Filarioidea) in Nearctic black flies (Diptera: Simuliidae). American Journal of Tropical Medicine and Hygiene 32:1298–1305.
Lok, J. B., E. W. Cupp, & M. J. Bernardo. 1983b. *Simulium jenningsi* Malloch (Diptera: Simuliidae): a vector of *Onchocerca lienalis* Stiles (Nematoda: Filarioidea) in New York. American Journal of Veterinary Research 44:2355–2358.
Long, P. L., W. L. Current, & G. P. Noblet. 1987. Parasites of the Christmas turkey. Parasitology Today 3:360–366.
Longstaff, T. G. 1932. An ecological reconnaissance in West Greenland. Journal of Animal Ecology 1:119–142.
Loomis, E. C., J. P. Hughes, & E. L. Bramhall. 1975. The common parasites of horses. University of California Division of Agricultural Sciences Publication No. 4006. Richmond, California. 35 pp.
Lord, C. C., & W. J. Tabachnick. 2002. Influence of nonsystemic transmission on the epidemiology of insect borne arboviruses: a case study of vesicular stomatitis epidemiology in the western United States. Journal of Medical Entomology 39:417–426.
Love, S. D., & R. C. Bailey. 1992. Community development of epilithic invertebrates in streams: independent and interactive

effects of substratum properties. Canadian Journal of Zoology 70:1976–1983.
Lowry, C. A., & A. J. Shelley. 1990. Studies on the scutal patterns of three South American *Simulium* species (Diptera: Simuliidae). Canadian Journal of Zoology 68:956–961.
Lowther, J. K., & D. M. Wood. 1964. Specificity of a black fly, *Simulium euryadminiculum* Davies, toward its host, the common loon. Canadian Entomologist 96:911–913.
Lucas, P. E. 1995. The effect of feeding regime on larval black fly (Diptera: Simuliidae) primary head fan ray number. M.S. thesis. Brock University, St. Catharines, Ontario. 110 pp.
Lucas, P. [E.], & F. F. Hunter. 1999. Phenotypic plasticity in the labral fan of simuliid larvae (Diptera): effect of seston load on primary-ray number. Canadian Journal of Zoology 77:1843–1849.
Luckhart, S., M. S. Cupp, & E. W. Cupp. 1992. Morphological and functional classification of the hemocytes of adult female *Simulium vittatum* (Diptera: Simuliidae). Journal of Medical Entomology 29:457–466.
Lugger, O. 1897a ["1896"]. Insects injurious in 1896. University of Minnesota Agricultural Experiment Station Bulletin 48:31–270 + 16 plates. [Actual publication date was early 1897.]
Lugger, O. 1897b ["1896"]. Insects injurious in 1896. Annual Report of the Entomologist of the University of Minnesota State Experiment Station 2:1–244 + 16 plates. [Actual publication date was early 1897.]
Lundström, C. 1911. Beiträge zur Kenntnis der Dipteren Finlands. VII. Melusinidae (Simuliidae). Acta Societatis pro Fauna et Flora Fennica 34(Part 12):1–23 + 1 plate.
Luoma, J. R. 1984. Blackflies everywhere. Audubon 86(3):14–17.
Lutz, A. 1910. Segunda contribuição para o conhecimento das especies brazileiras do genero "*Simulium*." Memorias do Instituto Oswaldo Cruz 2:213–267. [In German and Portuguese.]
Lyle, C., & Assistants. 1935. A buffalo gnat (*Eusimulium pecuarum*). United States Department of Agriculture Insect Pest Survey Bulletin 15:98.
Lynn, D. H., D. Molloy, & R. LeBrun. 1981. *Tetrahymena rotunda* n. sp. (Hymenostomatida: Tetrahymenidae), a ciliate parasite of the hemolymph of *Simulium* (Diptera: Simuliidae). Transactions of the American Microscopical Society 100:134–141.
Macgregor, H. C., & J. B. Mackie. 1967. Fine structure of the cytoplasm in salivary glands of *Simulium*. Journal of Cell Science 2:137–144.
MacInnes, K. L. 1973. Reproductive ecology of five arctic species of *Pedicularis* (Scrophulariaceae). Ph.D. thesis. University of Western Ontario, London, Ontario. 235 pp.
Maciolek, J. A., & M. G. Tunzi. 1968. Microseston dynamics in a simple Sierra Nevada lake-stream system. Ecology 49:60–75.
MacNay, C. G. (compiler). 1954. Insects affecting man and domestic animals. Canadian Agricultural Insect Pest Review 32:156–159.
MacNay, C. G. (compiler). 1955. Insects affecting man and domestic animals. Canadian Insect Pest Review 33:201–203.
MacNay, C. G. 1956. Insects of potential economic importance new to certain regions or hosts in Canada, 1954: a review. Canadian Insect Pest Review 34:290–294.
MacNay, C. G. (compiler). 1958. Insects attacking man and other mammals. Canadian Insect Pest Review 36:244–245.
MacNay, C. G. (compiler). 1959a. Insects affecting man and other mammals. Canadian Insect Pest Review 37:166–167.
MacNay, C. G. (compiler). 1959b. Insects affecting man and other mammals. Canadian Insect Pest Review 37:186–187.
Macquart, J. 1826. Insectes diptères du nord de la France. Tipulaires. Recueil des Travaux de la Société d'Amateurs des Sciences, de l'Agriculture et des Arts, de Lille. 1823–1824:59–224 + 4 plates.
Macquart, J. 1834. Histoire naturelle des insectes. Diptères. Vol. 1. N. E. Roret, Paris. 578 pp. + 12 plates.
Macquart, J. 1838. Diptères exotiques nouveaux ou peu connus. Vol. 1. Part 1. Mémoires de la Société Royal des Sciences, de l'Agriculture et des Arts de Lille 1838(2):9–225 + 25 plates.
Madahar, D. P. 1967. Cytological studies on black flies (Simuliidae: Diptera). Ph.D. thesis. University of Toronto, Toronto. 104 pp.
Madahar, D. P. 1969. The salivary gland chromosomes of seven taxa in the subgenus *Stegopterna* (Diptera, Simuliidae, *Cnephia*). Canadian Journal of Zoology 47:115–119.
Madahar, D. P. 1973. Cytogenetic study in triploidy in *Prosimulium ursinum* (Simuliidae: Diptera). Genetics 74(Suppl. 2):170–171.
Magnarelli, L. A., & J. F. Burger. 1984. Caloric reserves in natural populations of a blackfly, *Simulium decorum* (Diptera: Simuliidae), and a deerfly, *Chrysops ater* (Diptera: Tabanidae). Canadian Journal of Zoology 62:2589–2593.
Magnarelli, L. A., & E. W. Cupp. 1977. Physiological age of *Simulium tuberosum* and *Simulium venustum* (Diptera: Simuliidae) in New York State, U.S.A. Journal of Medical Entomology 13:621–624.
Mahrt, J. L. 1982. Black flies (Diptera: Simuliidae) from Hardwicke Island, British Columbia. Canadian Journal of Zoology 60:3364–3369.
Mail, G. A., & J. D. Gregson. 1940. Observations on the possible effects of arthropod parasites on wildlife in western Canada. Convention of the International Association of Game, Fish and Conservation Commissioners 34:126–132.
Mallis, A. 1971. American entomologists. Rutgers University Press, New Brunswick, N.J. 549 pp.
Malloch, J. R. 1913. A new species of *Simulium* from Texas. Proceedings of the Entomological Society of Washington 15:133–134.
Malloch, J. R. 1914. American black flies or buffalo gnats. U.S. Department of Agriculture Bureau of Entomology Technical Series, No. 26. Government Printing Office, Washington, D.C. 83 pp.
Malloch, J. R. 1919. The Diptera collected by the Canadian Expedition, 1913–1918 (excluding the Tipulidae and Culicidae). Report of the Canadian Arctic Expedition 1913–1918, 3:34–90.
Malmqvist, B. 1994. Preimaginal blackflies (Diptera: Simuliidae) and their predators in a central Scandinavian lake outlet stream. Annales Zoologici Fennici 31:245–255.
Malmqvist, B., & P.-O. Hoffsten. 2000. Macroinvertebrate taxonomic richness, community structure and nestedness in Swedish streams. Archiv für Hydrobiologie 150:29–54.
Malmqvist, B., Y. Zhang, & P. H. Adler. 1999. Diversity, distribution and larval habitats of North Swedish blackflies (Diptera: Simuliidae). Freshwater Biology 42:301–314.
Malmqvist, B., R. S. Wotton, & Y. Zhang. 2001. Suspension feeders transform massive amounts of seston in large northern rivers. Oikos 92:35–43.
Malone, K. M., & R. A. Nolan. 1978. Aerobic bacterial flora of the larval gut of the black fly *Prosimulium mixtum* (Diptera: Simuliidae) from Newfoundland, Canada. Journal of Medical Entomology 14:641–645.
Mansingh, A., & R. W. Steele. 1973. Studies on insect dormancy. I. Physiology of hibernation in the larvae of the blackfly *Prosimulium mysticum* Peterson. Canadian Journal of Zoology 51:611–618.
Mansingh, A., R. W. Steele, & B. V. Helson. 1972. Hibernation in the blackfly *Prosimulium mysticum*: quiescence or oligopause? Canadian Journal of Zoology 50:31–34.
Maré, C. J., E. W. Cupp, & M. H. Cupp. 1991. Vesicular stomatitis virus (New Jersey) infection and replication in black flies (*Simulium vittatum*). Proceedings of the United States Animal Health Association 19:179–188.

Marlatt, C. L. 1889. Report of a trip to investigate buffalo gnats. Insect Life 2:7–11.
Martin, F. R. 1987. The spatial distribution of host-seeking mammalophilic black flies (Diptera: Simuliidae) in relation to terrestrial habitats on the Avalon Peninsula, Newfoundland. M.S. thesis. Memorial University of Newfoundland, St. John's. 92 pp.
Martin, F. R., J. W. McCreadie, & M. H. Colbo. 1994. Effect of trapsite, time of day, and meteorological factors on abundance of host-seeking mammalophilic black flies (Diptera: Simuliidae). Canadian Entomologist 126:283–289.
Martin, J. W. 1981. The black flies (Diptera: Simuliidae) in Waterville Valley, New Hampshire. M.S. thesis. University of New Hampshire, Durham. 49 pp.
Martin, M. D., R. S. Brown, D. R. Barton, & G. Power. 2000. Abundance of stream invertebrates in winter: seasonal changes and effects of river ice. Canadian Field-Naturalist 115:68–74.
Martin, M. M., J. J. Kukor, J. S. Martin, & R. W. Merritt. 1985. The digestive enzymes of larvae of the black fly, *Prosimulium fuscum* (Diptera, Simuliidae). Comparative Biochemistry and Physiology, B 82:37–39.
Martin, P. J. S. 1988. Comparison of larval black fly and mosquito feeding. Proceedings of the Annual Meeting of the New Jersey Mosquito Control Association 75:40–41.
Martin, P. J. S. 1989. Assimilation of different foods by larvae of *Simulium verecundum* Stone and Jamnback (Diptera: Simuliidae). M.S. thesis. University of Massachusetts, Amherst. 122 pp.
Martin, P. J. S. 1993. Diflubenzuron (Dimlin®): environmental effects and biochemical mode-of-action. Ph.D. thesis. University of Massachusetts, Amherst. 186 pp.
Martin, P. J. S., & J. D. Edman. 1991. Comparison of water ingestion between black flies (Simuliidae) and a mosquito (*Aedes aegypti*; Culicidae) using radiolabelled glucose. Hydrobiologia 209:227–234.
Martin, P. J. S., & J. D. Edman. 1993. Assimilation rates of different particulate foods for *Simulium verecundum* (Diptera: Simuliidae). Journal of Medical Entomology 30:805–809.
Maser, E. 1973. Disease-spreading blackflies under attack at Memorial. Science Dimension 5(4):4–7. [In English and French.]
Mason, G. F. 1982. Cytological studies of sibling species of *Simulium tuberosum* (Lundström) (Diptera: Simuliidae). Canadian Journal of Zoology 60:292–303.
Mason, G. F. 1984. Sex chromosome polymorphism in the *Simulium tuberosum* complex (Lundström) (Diptera: Simuliidae). Canadian Journal of Zoology 62:647–658.
Mason, P. G. 1986. Evaluation of a "cow-type" silhouette trap with and without $CO_2$ bait for monitoring populations of adult *Simulium luggeri* (Diptera: Simuliidae). Journal of the American Mosquito Control Association 2:482–484.
Mason, P. G., & P. M. Kusters. 1990. Seasonal activity of female black flies (Diptera: Simuliidae) in pastures in northeastern Saskatchewan. Canadian Entomologist 122:825–835.
Mason, P. G., & P. M. Kusters. 1991a. Evaluation of backrubber insecticides for black fly control. Final Report No. R-89-01-0423. Agriculture Development Fund, Regina, Saskatchewan. 21 pp.
Mason, P. G., & P. M. Kusters. 1991b. Procedures manual for the Saskatchewan black fly control program. Agriculture Canada Publication Agriculture Canada, Saskatoon, Saskatchewan. 96 pp.
Mason, P. G., & P. M. Kusters. 1993. Procedures for adult black fly monitoring using silhouette traps. Saskatoon Research Station Technical Bulletin 93–001:1–13.
Mason, P. G., & J. A. Shemanchuk. 1990. Black flies. Agriculture Canada Publication No. 1499/E. Agriculture Canada, Ottawa. 19 pp.
Mason, P. G., M. Boisvert, J. Boisvert, & M. H. Colbo. 2002. *Prosimulium* and *Simulium* spp., black flies (Diptera: Simuliidae). Pp. 230–237. *In* P. G. Mason & J. T. Huber (eds.), Biological control programmes in Canada, 1981–2000. CABI Publishing, Wallingford, U.K. 480 pp.
Matheson, R. 1938. Buffalo gnats (Simuliidae). United States Department of Agriculture Insect Pest Survey Bulletin 18:134.
Mathis, A. 2000. Microsporidia: emerging advances in understanding the basic biology of these unique organisms. International Journal for Parasitology 30:795–804.
Matsumura, S. 1931. 6000 illustrated insects of the Japan-Empire. Toko Shoin, Tokyo. 1497 pp. + 191 pp. separately paginated index. [In Japanese.]
Matsuo, K., & J. O. Ochoa A. 1979. Scanning electron microscopic studies on Guatemalan black flies. I. Abdominal dorsal hairs of larvae of 6 species. Japanese Journal of Sanitary Zoology 30:329–333.
Maurand, J. 1975. Les microsporidies des larves de simulies: systématique, données cytochimiques, pathologiques et écologiques. Annales de Parasitologie Humaine et Comparée 50:371–396.
Maury, D. H. 1894. Recollections of a Virginian in the Mexican, Indian, and Civil Wars. Charles Scribner's Sons, New York. 279 pp.
May, B., L. S. Bauer, R. L. Vadas, & J. Granett. 1977. Biochemical genetic variation in the family Simuliidae: electrophoretic identification of the human biter in the isomorphic *Simulium jenningsi* group. Annals of the Entomological Society of America 70:637–640.
McAlpine, J. F. 1981. Morphology and terminology—adults. Pp. 9–63. *In* J. F. McAlpine, B. V. Peterson, G. E. Shewell, H. J. Teskey, J. R. Vockeroth, & D. M. Wood (eds.), Manual of Nearctic Diptera, vol. 1, Monograph No. 27. Research Branch, Agriculture Canada, Ottawa. 674 pp.
McAlpine, J. F., B. V. Peterson, G. E. Shewell, H. J. Teskey, J. R. Vockeroth, & D. M. Wood (eds.), 1981. Manual of Nearctic Diptera, vol. 1, Monograph No. 27. Research Branch, Agriculture Canada, Ottawa. 674 pp.
McAtee, W. L. 1922. *Prosimulium fulvum* Coquillett a biting species (Dip., Simuliidae). Entomological News 33:79.
McAuliffe, J. R. 1984. Competition for space, disturbance, and the structure of a benthic stream community. Ecology 65:894–908.
McCall, P. J., & P. A. Lemoh. 1997. Evidence for the "invitation effect" during bloodfeeding by blackflies of the *Simulium damnosum* complex (Diptera: Simuliidae). Journal of Insect Behavior 10:299–303.
McCall, P. J., R. R. Heath, B. D. Dueben, & M. D. Wilson. 1997. Oviposition pheromone in the *Simulium damnosum* complex: biological activity of chemical fractions from gravid ovaries. Physiological Entomology 22:224–230.
McComb, C. W., & W. E. Bickley. 1959. Observations on black flies in two Maryland counties. Journal of Economic Entomology 52:629–632.
McCracken, I. R., & S. L. Matthews. 1997. Effects of *Bacillus thuringiensis* subsp. *israelensis* (*B.t.i.*) applications on invertebrates from two streams on Prince Edward Island. Bulletin of Environmental Contamination and Toxicology 58:291–298.
McCreadie, J. W. 1991. Ecological characterization of cytotypes of the *Simulium venustum/verecundum* complex (Diptera: Simuliidae) found on the Avalon Peninsula, Newfoundland. Ph.D. thesis. Memorial University of Newfoundland, St. John's. 347 pp.
McCreadie, J. W., & P. H. Adler. 1998. Scale, time, space, and predictability: species distributions of preimaginal black flies (Diptera: Simuliidae). Oecologia 114:79–92.
McCreadie, J. W., & P. H. Adler. 1999. Parasites of larval black flies (Diptera:

Simuliidae) and environmental factors associated with their distributions. Invertebrate Zoology 118:310–318.
McCreadie, J. W., & M. H. Colbo. 1990. Allometry in the last larval instars of *Simulium truncatum* (Lundström) and *S. rostratum* (Lundström) (Diptera: Simuliidae). Canadian Entomologist 122:1137–1140.
McCreadie, J. W., & M. H. Colbo. 1991a. The influence of temperature on the survival, development, growth, and chromosome preparation quality of the EFG/C, ACD, and AA cytotypes of the *Simulium venustum-verecundum* complex (Diptera: Simuliidae). Canadian Journal of Zoology 69:1356–1365.
McCreadie, J. W., & M. H. Colbo. 1991b. Spatial distribution patterns of larval cytotypes of the *Simulium venustum/verecundum* complex (Diptera: Simuliidae) on the Avalon Peninsula, Newfoundland: factors associated with occurrence. Canadian Journal of Zoology 69:2651–2659.
McCreadie, J. W., & M. H. Colbo. 1991c. A critical examination of four methods of estimating the surface area of stone substrate from streams in relation to sampling Simuliidae (Diptera). Hydrobiologia 220:205–210.
McCreadie, J. W., & M. H. Colbo. 1992. Spatial distribution patterns of larval cytotypes of the *Simulium venustum/verecundum* complex (Diptera: Simuliidae) on the Avalon Peninsula, Newfoundland: factors associated with cytotype abundance and composition. Canadian Journal of Zoology 70:1389–1396.
McCreadie, J. W., & M. H. Colbo. 1993a. Seasonal succession and spatial-temporal distribution patterns of six larval cytospecies of the *Simulium venustum/verecundum* complex (Diptera: Simuliidae). Canadian Journal of Zoology 71:116–124.
McCreadie, J. W., & M. H. Colbo. 1993b. Larval and pupal microhabitat selection by *Simulium truncatum* Lundström, *S. rostratum* Lundström, and *S. verecundum* AA (Diptera: Simuliidae). Canadian Journal of Zoology 71:358–367.
McCreadie, J. [W.], & M. Robertson. 1998. Size of the larval black fly *Simulium truncatum* (Diptera: Simuliidae) in relation to distance from a lake outlet. Journal of Freshwater Ecology 13:21–27.
McCreadie, J. W., M. H. Colbo, & G. F. Bennett. 1984. A trap design for the collection of haematophagous Diptera from cattle. Mosquito News 44:212–216.
McCreadie, J. W., M. H. Colbo, & G. F. Bennett. 1985. The seasonal activity of hematophagous Diptera attacking cattle in insular Newfoundland. Canadian Entomologist 117:995–1006.
McCreadie, J. W., M. H. Colbo, & G. F. Bennett. 1986. The influence of weather on host seeking and blood feeding of *Prosimulium mixtum* and *Simulium venustum/verecundum* complex (Diptera: Simuliidae). Journal of Medical Entomology 23:289–297.
McCreadie, J. W., P. H. Adler, & M. H. Colbo. 1994a. Cytogenetics of the *Simulium venustum/verecundum* complex (Diptera: Simuliidae) on the Avalon Peninsula, Newfoundland. Canadian Entomologist 126:23–30.
McCreadie, J. W., P. H. Adler, & E. C. Masteller. 1994b. Long-term emergence patterns of black flies (Diptera: Simuliidae) in northwestern Pennsylvania. Hydrobiologia 288:39–46.
McCreadie, J. W., M. H. Colbo, & F. F. Hunter. 1994c. Notes on sugar feeding and selected wild mammalian hosts of black flies (Diptera: Simuliidae) in Newfoundland. Journal of Medical Entomology 31:566–570.
McCreadie, J. W., P. H. Adler, & M. H. Colbo. 1995. Community structure of larval black flies (Diptera: Simuliidae) from the Avalon Peninsula, Newfoundland. Annals of the Entomological Society of America 88:51–57.
McCreadie, J. W., P. H. Adler, & J. F. Burger. 1997. Species assemblages of larval black flies (Diptera: Simuliidae): random or predictable? Journal of the North American Benthological Society 16:760–770.
McCullough, D. A. 1975. The bioenergetics of three aquatic insects determined by radioisotopic analyses. M. S. thesis. Idaho State University, Pocatello. 326 pp.
McCullough, D. A., G. W. Minshall, & C. E. Cushing. 1979. Bioenergetics of lotic filter-feeding insects *Simulium* spp. (Diptera) and *Hydropsyche occidentalis* (Trichoptera) and their function in controlling organic transport in streams. Ecology 60:585–596.
McDaniel, I. N. 1971. Research on the black fly problem in Maine: a report to the 105th Maine Legislature. University of Maine (Orono) Life Sciences and Agriculture Experiment Station Miscellaneous Report 131:1–8.
McDaniel, I. N. 1975. Is a black fly survey worthwhile? Journal of the New York Entomological Society 83:245.
McIver, S. B., & G. E. O'Grady. 1987. Possible role of the central retinular cells in the ommatidia of male black flies (Diptera: Simuliidae). Canadian Journal of Zoology 65:3186–3188.
McIver, S. B., & J. F. Sutcliffe. 1988 ["1987"]. Sensory basis of behavior and structural adaptations for feeding in black flies. Pp. 228–249. *In* K. C. Kim & R. W. Merritt (eds.), Black flies: ecology, population management, and annotated world list. Pennsylvania State University, University Park. 528 pp. [Actual publication date was 1 March 1988.]
McIver, S. [B.], R. Siemicki, & J. Sutcliffe. 1980. Bifurcate sensilla on the tarsi of female black flies, *Simulium venustum* (Diptera: Simuliidae): contact chemosensilla adapted for olfaction? Journal of Morphology 165:1–11.
McKague, A. B., & R. B. Pridmore. 1979. Drift response of black fly larvae to Dimlin. Mosquito News 39:678–679.
McKague, [A.] B., & P. M. Wood. 1974. Effects of insect developmental inhibitors on adult emergence of black flies (Diptera: Simuliidae). Canadian Entomologist 106:253–256.
McKague, A. B., R. B. Pridmore, & P. M. Wood. 1978. Effects of Altosid and Dimilin on black flies (Diptera: Simuliidae): laboratory and field tests. Canadian Entomologist 110:1103–1110.
McKibben, B. 1999. Consuming nature. Pp. 87–95. *In* R. Rosenblatt (ed.), Consuming desires: consumption, culture, and the pursuit of happiness. Island Press/Shearwater Books, Washington, D.C. 240 pp.
McNelly, J. R., & W. J. Crans. 1999. An update on New Jersey's black fly control program in the Delaware River. Proceedings of the Annual Meeting of the New Jersey Mosquito Control Association 86:42.
Mead, D. G. 1995. Studies on the ecology of vesicular stomatitis virus in southwestern Arizona. M. S. thesis. University of Arizona, Tucson. 130 pp.
Mead, D. G. 1999. Maintenance and transmission of vesicular stomatitis viruses: new data for an old puzzle. Ph.D. thesis. University of Arizona, Tucson. 141 pp.
Mead, D. G., C. J. Maré, & E. W. Cupp. 1997. Vector competence of select black fly species for vesicular stomatitis virus (New Jersey serotype). American Journal of Tropical Medicine and Hygiene 57:42–48.
Mead, D. G., C. J. Maré, & F. B. Ramberg. 1999. Bite transmission of vesicular stomatitis virus (New Jersey serotype) to laboratory mice by *Simulium vittatum* (Diptera: Simuliidae). Journal of Medical Entomology 36:410–413.
Mead, D. G., F. B. Ramberg, & C. J. Maré. 2000a. Laboratory vector competence of black flies (Diptera: Simuliidae) for the Indiana serotype of vesicular stomatitis virus. Annals of the New York Academy of Sciences 916:437–443.
Mead, D. G., F. B. Ramberg, D. G. Besselsen, & C. J. Maré. 2000b. Transmission of vesicular stomatitis virus from infected to noninfected black flies co-feeding on

nonviremic deer mice. Science 287:485–487.
Meigen, J. W. 1803. Versuch einer neuen Gattungs Eintheilung der europäischen zweiflügligen Insekten. Magazin Insektenk (Illiger) 2:259–281.
Meigen, J. W. 1804. Klassifikazion und Beschreibung der europäischen zweiflügeligen Insekten (Diptera Linn.). Erster Band. K. Reichard, Braunschweig [= Brunswick], Germany. 314 pp. + 15 plates.
Meigen, J. W. 1818. Systematische Beschreibung der bekannten europäischen zweiflügeligen Insekten 1. Forstmann, Aachen, Germany. 332 pp.
Meigen, J. W. 1838. Systematische Beschreibung der bekannten europäischen zweiflügeligen Insekten 7 (supplementary volume). Schulz, Hamm, Germany. 434 pp. + 8 plates.
Mercer, K. L. 1972. Sensory perception in selected Simuliidae (Diptera). M. S. thesis. University of Toronto, Toronto. 154 pp.
Mercer, K. L., & S. B. McIver. 1973a. Studies on the antennal sensilla of selected blackflies (Diptera: Simuliidae). Canadian Journal of Zoology 51:729–734.
Mercer, K. L., & S. B. McIver. 1973b. Sensilla on the palps of selected blackflies (Diptera: Simuliidae). Journal of Medical Entomology 10:236–239.
Meredith, S. E. O., D. Kurtak, & J. H. Adiamah. 1986. Following movements of resistant populations of *Simulium soubrense/sanctipauli* (Diptera: Simuliidae) by means of chromosome inversions. Tropical Medicine and Parasitology 37:290–294.
Merritt, R. W. 1979. Black flies. Michigan Natural Resources Magazine 48(May/June):36–39.
Merritt, R. W., & J. R. Wallace. 2001. The role of aquatic insects in forensic investigations. Pp. 177–222. *In* J. H. Byrd & J. L. Castner (eds.), Forensic entomology: the utility of arthropods in legal investigations, CRC Press, Boca Raton, Fla. 456 pp.
Merritt, R. W., & R. S. Wotton. 1988. The life history and behavior of *Limnophora riparia* (Diptera: Muscidae), a predator of larval black flies. Journal of the North American Benthological Society 7:1–12.
Merritt, R. W., M. M. Mortland, E. F. Gersabeck, & D. H. Ross. 1978a. X-ray diffraction analysis of particles ingested by filter-feeding animals. Entomologia Experimentalis et Applicata 24:27–34.
Merritt, R. W., D. H. Ross, & B. V. Peterson. 1978b. Larval ecology of some lower Michigan black flies (Diptera: Simuliidae) with keys to the immature stages. Great Lakes Entomologist 11:177–208.
Merritt, R. W., D. H. Ross, & G. L. Larson. 1982. Influence of stream temperature and seston on the growth and production of overwintering larval black flies (Diptera: Simuliidae). Ecology 63:1322–1331. [Condensed version: 1983. Stream temperature, food, and black fly production. BioScience 33:51–52.]
Merritt, R. W., E. D. Walker, M. A. Wilzbach, K. W. Cummins, & W. T. Morgan. 1989. A broad evalution of *B.t.i.* for black fly (Diptera: Simuliidae) control in a Michigan river: efficacy, carry and nontarget effects on invertebrates and fish. Journal of the American Mosquito Control Association 5:397–415.
Merritt, R. W., M. S. Wipfli, & R. S. Wotton. 1991. Changes in feeding habits of selected nontarget aquatic insects in response to live and *Bacillus thuringiensis* var. *israelensis* de Barjac-killed black fly larvae (Diptera: Simuliidae). Canadian Entomologist 123:179–185.
Merritt, R. W., D. A. Craig, R. S. Wotton, & E. D. Walker. 1996. Feeding behavior of aquatic insects: case studies on black fly and mosquito larvae. Invertebrate Biology 115:206–217.
Metcalf, C. L. 1932. Black flies and other biting flies of the Adirondacks. New York State Museum Bulletin 289:5–58.
Metcalf, C. L., & W. E. Sanderson. 1931. Black flies, mosquitoes and punkies of the Adirondacks. New York State Museum Circular 5:1–38.
Metcalf, C. L., & W. E. Sanderson. 1932. Control of biting flies in the Adirondacks. New York State Museum Bulletin 289:59–78.
Michelbacher, A. E. 1938. Buffalo gnats (Simuliidae). United States Department of Agriculture Insect Pest Survey Bulletin 18:134–135.
Millar, J. L., & J. G. Rempel. 1944. Live stock losses in Saskatchewan due to blackflies. Canadian Journal of Comparative Medicine 8:334–337.
Miller, B. R., M. B. Crabtree, & H. M. Savage. 1997. Phylogenetic relationships of the Culicomorpha inferred from 18S and 5.8S ribosomal DNA sequences (Diptera: Nematocera). Insect Molecular Biology 6:105–114.
Miller, M. C., M. Kurzhals, A. E. Hershey, & R. W. Merritt. 1998. Feeding behavior of black fly larvae and retention of fine particulate organic matter in a high-gradient blackwater stream. Canadian Journal of Zoology 76:228–235.
Milliron, H. E. 1958. Economic insect and allied pests of Delaware. University of Delaware Agricultural Experiment Station Bulletin 321:1–87.
Minakawa, N. 1997. The dynamics of aquatic insect communities associated with salmon spawning. Ph.D. thesis. University of Washington, Seattle. 105 pp.
Miranda Esquivel, D. R., & P. Muñoz de Hoyos. 1995. *¿Ectemnaspis* o *Psilopelmia?* he ahí el dilema. Revista Colombiana de Entomologia 21:129–144.
Mitrokhin, V. U. 1979. The infection of black flies (Simuliidae) with microsporidians in water bodies of the Ob' and Irtish Rivers. Parazitologiya 13:245–249. [In Russian; English summary.]
Mock, D. E., & P. H. Adler. 2002. Black flies (Diptera: Simuliidae) of Kansas: review, new records, and pest status. Journal of the Kansas Entomological Society 75:203–213.
Mohsen, Z. H. 1981. Ecology and control of blackflies (Diptera: Simuliidae) in southern California streams—impact of chemical control agents on nontarget insects. Ph.D. thesis. University of California, Riverside. 193 pp.
Mohsen, Z. H. 1982. Diel rhythms in susceptibility of *Simulium argus* Williston to Abate®. Proceedings and Papers of the Annual Conference of the California Mosquito and Vector Control Association 50:104–105.
Mohsen, Z. H., & M. S. Mulla. 1981a. Additional records of *Simulium defoliarti* and *Prosimulium onychodactylum* from southern California. Mosquito News 41:379.
Mohsen, Z. H., & M. S. Mulla. 1981b. Toxicity of blackfly larvicidal fomulations to some aquatic insects in the laboratory. Bulletin of Environmental Contamination and Toxicology 26:696–703.
Mohsen, Z. H., & M. S. Mulla. 1982a. The ecology of black flies (Diptera: Simuliidae) in some southern California streams. Journal of Medical Entomology 19:72–85.
Mohsen, Z. H., & M. S. Mulla. 1982b. Field evaluation of *Simulium* larvicides: effects on target and nontarget insects. Environmental Entomology 11:390–398.
Mohsen, Z. H., & M. S. Mulla. 1984. Notes on the occurrence of microsporida and ib [*sic*] *Simulium* (Simuliidae: Diptera) larvae in southern California. Journal of Biological Sciences Research 15:55–59.
Mokry, J. E. 1975. Studies on the ecology and biology of blackfly larvae utilizing an in situ benthobservatory. Verhandlungen der Internationalen Vereinigung für theoretische und angewandte Limnologie 19:1546–1549.
Mokry, J. E. 1976a. A simplified membrane technique for feeding blackflies (Diptera: Simuliidae) on blood in the laboratory. Bulletin of the World Health Organization 53:127–129.
Mokry, J. E. 1976b. Laboratory studies on the larval biology of *Simulium venustum* Say (Diptera: Simuliidae). Canadian Journal of Zoology 54:1657–1663.

Mokry, J. E. 1978. Progress towards the colonization of *Cnephia mutata* (Diptera: Simuliidae). Bulletin of the World Health Organization 56:455–456.
Mokry, J. E. 1979. A study of some factors affecting blood-feeding, autogeny and fecundity of *Simulium vittatum* Zetterstedt and *Prosimulium mixtum* Syme and Davies. M. S. thesis. Memorial University of Newfoundland, St. John's. 89 pp.
Mokry, J. E. 1980a. Laboratory studies on blood-feeding of blackflies. 1. Factors affecting the feeding rate. Tropenmedizin und Parasitologie 31:367–373.
Mokry, J. E. 1980b. Laboratory studies on blood-feeding of blackflies (Diptera: Simuliidae). 2. Factors affecting fecundity. Tropenmedizin und Parasitologie 31:374–380.
Mokry, J. E. 1980c. An *in vitro* technique for the bioassay of repellents against black flies. Mosquito News 40:448–449.
Mokry, J. E., & J. R. Finney. 1977. Notes on mermithid parasitism of Newfoundland blackflies, with the first record of *Neomesomermis flumenalis* from adult hosts. Canadian Journal of Zoology 55:1370–1372.
Mokry, J. E., M. H. Colbo, & B. H. Thompson. 1981. Laboratory colonization of blackflies. Pp. 299–306. *In* M. Laird (ed.), Blackflies: the future for biological methods in integrated control. Academic Press, New York. 399 pp.
Molloy, D. P. 1976. Biological control of blackflies with mermithid parasites. Ph.D. thesis. State University of New York, Syracuse. 70 pp.
Molloy, D. [P.]. 1979. Description and bionomics of *Mesomermis camdenensis* n. sp. (Mermithidae), a parasite of black flies (Simuliidae). Journal of Nematology 11:321–328.
Molloy, D. P. 1981. Mermithid parasitism of black flies (Diptera: Simuliidae). Journal of Nematology 13:250–256.
Molloy, D. [P.] (ed.). 1982. Biological control of black flies (Diptera: Simuliidae) with *Bacillus thuringiensis* var. *israelensis* (serotype 14): a review with recommendations for laboratory and field protocol. Miscellaneous Publications of the Entomological Society of America 12(4):1–30.
Molloy, D. P. 1984. The black fly debate. NAHO (New York State Museum, Albany) 17(1):7–10.
Molloy, D. P. 1988 ["1987"]. The ecology of black fly parasites. Pp. 315–326. *In* K. C. Kim & R. W. Merritt (eds.), Black flies: ecology, population management, and annotated world list. Pennsylvania State University, University Park. 528 pp. [Actual publication date was 1 March 1988.]
Molloy, D. P. 1990. Progress in the biological control of black flies with *Bacillus thuringiensis israelensis*, with emphasis on temperate climates. Pp. 161–186. *In* H. de Barjac & D. J. Sutherland (eds.), Bacterial control of mosquitoes and black flies: biochemistry, genetics and applications of *Bacillus thuringiensis israelensis* and *Bacillus sphaericus*. Rutgers University Press, New Brunswick, N.J. 349 pp.
Molloy, D. P. 1992. Impact of the black fly (Diptera: Simuliidae) control agent *Bacillus thuringiensis* var. *israelensis* on chironomids (Diptera: Chironomidae) and other nontarget insects: results of ten field trials. Journal of the American Mosquito Control Association 8:24–31.
Molloy, D. [P.], & H. Jamnback. 1975. Laboratory transmission of mermithids parasitic in blackflies. Mosquito News 35:337–342.
Molloy, D. [P.], & H. Jamnback. 1977. A larval black fly control field trial using mermithid parasites and its cost implications. Mosquito News 37:104–108.
Molloy, D. [P.], & H. Jamnback. 1981. Field evaluation of *Bacillus thuringiensis* var. *israelensis* as a black fly biocontrol agent and its effect on nontarget stream insects. Journal of Economic Entomology 74:314–318.
Molloy, D. P., & R. H. Struble. 1989. Investigation of the feasibility of the microbial control of black flies (Diptera: Simuliidae) with *Bacillus thuringiensis* var. *israelensis* in the Adirondack Mountains of New York. Bulletin of the Society for Vector Ecology 14:266–276.
Molloy, D. [P.], R. Gaugler, & H. Jamnback. 1980. The pathogenicity of *Neoaplectana carpocapsae* to blackfly larvae. Journal of Invertebrate Pathology 36:302–306.
Molloy, D. [P.], R. Gaugler, & H. Jamnback. 1981. Factors influencing efficacy of *Bacillus thuringiensis* var. *israelensis* as a biological control agent of black fly larvae. Journal of Economic Entomology 74:61–64.
Molloy, D. [P.], S. P. Wraight, B. Kaplan, J. Gerardi, & P. Peterson. 1984. Laboratory evaluation of commercial formulations of *Bacillus thuringiensis* var. *israelensis* against mosquito and black fly larvae. Journal of Agricultural Entomology 1:161–168.
Monaghan, M. T., S. A. Thomas, G. W. Minshall, J. D. Newbold, & C. E. Cushing. 2001. The influence of filter-feeding benthic macroinvertebrates on the transport and deposition of particulate organic matter and diatoms in two streams. Limnology and Oceanography 46:1091–1099.
Monro, H. A. U., L. J. Briand, R. Delisle, & C. C. Smith. 1943. Some experiments with the pyrethrum aerosol under Canadian conditions. Entomological Society of Ontario Report 74:42–45.
Moobola, S. M. 1981. The relationship of gonotrophic cycles to vector potential of black flies (Diptera: Simuliidae) in Tompkins County, New York. Ph.D. thesis. Cornell University, Ithaca, N.Y. 141 pp.
Moore, H. S., IV. 1973. Flight range of black flies and biological studies of *Leucocytozoon smithi* of domestic turkeys. M. S. thesis. Clemson University, Clemson, S.C. 33 pp.
Moore, H. S., IV, & R. Noblet. 1974. Flight range of *Simulium slossonae*, the primary vector of *Leucocytozoon smithi* of turkeys in South Carolina. Environmental Entomology 3:365–369.
Moore, J. W. 1977a. Some factors effecting [*sic*] algal consumption in subarctic Ephemeroptera, Plecoptera and Simuliidae. Oecologia 27:261–273.
Moore, J. W. 1977b. Relative availability and utilization of algae in two subarctic rivers. Hydrobiologia 54:201–208.
Moore, J. W., & I. A. Moore. 1974. Food and growth of arctic char, *Salvelinus alpinus* (L.), in the Cumberland Sound area of Baffin Island. Journal of Fish Biology 6:79–92.
Moreira, G. R. P., & G. Sato. 1996. Blackfly oviposition on riparian vegetation of waterfalls in an Atlantic rain forest stream. Anais da Sociedade Entomológica do Brasil 25:557–562.
Morin, A. 1987a. Estimation and prediction of black fly abundance and productivity. Ph.D. thesis. McGill University, Montreal. 289 pp.
Morin, A. 1987b. Unsuitability of introduced tiles for sampling blackfly larvae (Diptera: Simuliidae). Freshwater Biology 17:143–150.
Morin, A. 1991. Intensity and importance of abiotic control and inferred competition on biomass distribution patterns of Simuliidae and Hydropsychidae in southern Québec streams. Journal of the North American Benthological Society 10:388–403.
Morin, A., & R. H. Peters. 1988. Effect of microhabitat features, seston quality, and periphyton on abundance of overwintering black fly larvae in southern Québec. Limnology and Oceanography 33:431–446.
Morin, A., P.-P. Harper, & R. H. Peters. 1986. Microhabitat-preference curves of blackfly larvae (Diptera: Simuliidae): a comparison of three estimation methods. Canadian Journal of Fisheries and Aquatic Sciences 43:1235–1241.
Morin, A., T. A. Mousseau, & D. A. Roff. 1987. Accuracy and precision of secondary production estimates. Limnology and Oceanography 32:1342–1352.
Morin, A., C. Back, A. Chalifour, J. Boisvert, & R. H. Peters. 1988a. Effect of black fly ingestion and assimilation on seston

transport in a Quebec lake outlet. Canadian Journal of Fisheries and Aquatic Sciences 45:705–714.

Morin, A., C. Back, A. Chalifour, J. Boisvert, & R. H. Peters. 1988b. Empirical models predicting ingestion rates of black fly larvae. Canadian Journal of Fisheries and Aquatic Sciences 45:1711–1719.

Morin, A., M. Constantin, & R. H. Peters. 1988c. Allometric models of simuliid growth rates and their use for estimation of production. Canadian Journal of Fisheries and Aquatic Sciences 45:315–324.

Morin, A., C. Back, J. Boisvert, & R. H. Peters. 1989. A conceptual model for the estimation of the sensitivity of black fly larvae to *Bacillus thuringiensis* var. *israelensis*. Canadian Journal of Fisheries and Aquatic Sciences 46:1785–1792.

Morris, C. D. 1978. A vehicle trap for collecting low-flying mosquitoes and black flies. Proceedings of the Annual Meeting of the New Jersey Mosquito Control Association 65:145–148.

Moss, S. T. 1970. Trichomycetes inhabiting the digestive tract of *Simulium equinum* larvae. Transactions of the British Mycological Society 54:1–13.

Moss, S. T., & E. Descals. 1986. A previously undescribed stage in the life cycle of Harpellales (Trichomycetes). Mycologia 78:213–222.

Moss, S. T., & R. W. Lichtwardt. 1980. *Harpella leptosa*, a new species of Trichomycetes substantiated by electron microscopy. Canadian Journal of Botany 58:1035–1044.

Moulton, J. K. 1992. Revision of the *Simulium jenningsi* group (Insecta: Diptera: Simuliidae). M. S. thesis. Clemson University, Clemson, S.C. 244 pp.

Moulton, J. K. 1996. A new species of *Gigantodax* Enderlein (Diptera: Simuliidae) from the United States. Proceedings of the Entomological Society of Washington 98:741–751.

Moulton, J. K. 1997. Molecular systematics of the Simuliidae (Diptera: Culicomorpha). Ph.D. thesis. University of Arizona, Tucson. 279 pp.

Moulton, J. K. 1998. Reexamination of *Simulium* (*Psilopelmia*) Enderlein (Diptera: Simuliidae) of America north of Mexico. Proceedings of the Entomological Society of Washington 100:50–71.

Moulton, J. K. 2000. Molecular sequence data resolves basal divergences within Simuliidae (Diptera). Systematic Entomology 25:95–113.

Moulton, J. K., & P. H. Adler. 1992. New species of black fly in the *Simulium jenningsi* group (Diptera: Simuliidae) from the southeastern United States. Annals of the Entomological Society of America 85:393–399.

Moulton, J. K., & P. H. Adler. 1995. Revision of the *Simulium jenningsi* species-group (Diptera: Simuliidae). Transactions of the American Entomological Society 121:1–57.

Moulton, J. K., & P. H. Adler. 1997. The genus *Ectemnia* (Diptera: Simuliidae): taxonomy, polytene chromosomes, new species, and phylogeny. Canadian Journal of Zoology 75:1896–1915.

Moulton, J. K., & P. H. Adler. 2002a. A new species of *Simulium* (*Psilopelmia*) Enderlein (Diptera: Simuliidae) from southern New Mexico, USA. Studia Dipterologica 9:213–218.

Moulton, J. K., & P. H. Adler. 2002b. Taxonomy and biology of *Simulium clarkei* Stone & Snoddy (Diptera: Simuliiidae), a poorly known black fly of the southeastern United States. Zootaxa 31:1–7.

Muhammad, A. 1988. A cytological description of *Simulium virgatum* Coquillett and related species. M. S. thesis. Sul Ross State University, Alpine, Tex. 94 pp. + 18 plates.

Muldrow, M. W. 1937. Buffalo gnats (*Eusimulium* spp.). United States Department of Agriculture Insect Pest Survey Bulletin 17:66.

Mulla, M. S., & L. A. Lacey. 1976a. Feeding rates of *Simulium* larvae on particulates in natural streams (Diptera: Simuliidae). Environmental Entomology 5:283–287.

Mulla, M. S., & L. A. Lacey. 1976b. Biting flies in the Lower Colorado River Basin: economic and public health implications of *Simulium* (Diptera—Simuliidae). Proceedings and Papers of the Annual Conference of the California Mosquito Control Association 44:130–133.

Mullens, B. A., & C. E. Dada. 1992. Insects feeding on desert bighorn sheep, domestic rabbits, and Japanese quail in the Santa Rosa Mountains of southern California. Journal of Wildlife Diseases 28:476–480.

Muotka, T. 1993. Microhabitat use by predaceous stream insects in relation to seasonal changes in prey availability. Annales Zoologici Fennici 30:287–297.

Muttkowski, R. A., & G. M. Smith. 1929. The food of trout stream insects in Yellowstone National Park. Roosevelt Wild Life Annals 2:241–263.

Nadeau, M. P. 1995. Physiological ecology of *Erynia conica* and *Erynia curvispora* (Zygomycetes: Entomophthorales) attacking black flies (Diptera: Simuliidae) in Quebec. Ph.D. thesis. McGill University, Montreal. 204 pp.

Nadeau, M. P., & J. L. Boisvert. 1994. Larvicidal activity of the entomopathogenic fungus *Tolypocladium cylindrosporum* (Deuteromycotina: Hyphomycetes) on the mosquito *Aedes triseriatus* and the black fly *Simulium vittatum* (Diptera: Simuliidae). Journal of the American Mosquito Control Association 10:487–491.

Nadeau, M. P., G. B. Dunphy, & J. L. Boisvert. 1994. Entomopathogenic fungi of the order Entomophthorales (Zygomycotina) in adult black fly populations (Diptera: Simuliidae) in Quebec. Canadian Journal of Microbiology 40:682–686.

Nadeau, M. P., G. B. Dunphy, & J. L. Boisvert. 1995. Effects of physical factors on the development of secondary conidia of *Erynia conica* (Zygomycetes: Entomophthorales), a pathogen of adult black flies (Diptera: Simuliidae). Experimental Mycology 19:324–329.

Nadeau, M. P., G. B. Dunphy, & J. L. Boisvert. 1996. Development of *Erynia conica* (Zygomycetes: Entomophthorales) on the cuticle of the adult black flies *Simulium rostratum* and *Simulium decorum* (Diptera: Simuliidae). Journal of Invertebrate Pathology 68:50–58.

Needham, J. G., & C. Betten. 1901. Aquatic insects in the Adirondacks. New York State Museum Bulletin 47:383–612 + 36 plates.

Newell, R. L. 1970. Checklist of some aquatic insects from Montana. Proceedings of the Montana Academy of Sciences 30:45–56.

Newman, E. 1834. Attempted division of British insects into natural orders. Entomological Magazine 2:379–431.

Newman, L. J. 1983. Sibling species of the blackfly *Prosimulium onychodactylum* (Simuliidae, Diptera): a salivary gland chromosome study. Canadian Journal of Zoology 61:2816–2835.

Newson, H. D. 1977. Arthropod problems in recreation areas. Annual Review of Entomology 22:333–353.

Nicholson, H. P. 1941. The morphology of the mouthparts of the non-biting blackfly, *Eusimulium dakotense* [*sic*] D. & S., as compared with those of the biting species, *Simulium venustum* Say (Simuliidae: Diptera). M. S. thesis. University of Minnesota, Minneapolis. 42 pp.

Nicholson, H. P. 1945. The morphology of the mouthparts of the non-biting blackfly, *Eusimulium dacotense* D. & S., as compared with those of the biting species, *Simulium venustum* Say. Annals of the Entomological Society of America 38:281–297.

Nicholson, H. P. 1949. The Simuliidae of Minnesota with reference to their taxonomy and biologies. Ph.D. thesis. University of Minnesota, Minneapolis. 150 pp.

Nicholson, H. P., & C. E. Mickel. 1950. The black flies of Minnesota (Simuliidae). University of Minnesota Agricultural Experiment Station Technical Bulletin 192:1–64.

Nixon, K. E. 1988. The effect of *Simulium vittatum* Zett. (Diptera: Simuliidae) larval feeding behavior on the efficacy of *Bacillus thuringiensis* serotype H-14 (de Barjac). M. S. thesis. University of Manitoba, Winnipeg. 150 pp.

Noble, L. L. 1861. After icebergs with a painter. A summer voyage to Labrador and around Newfoundland. Appleton, New York. 366 pp.

Noblet, R., & D. R. Alverson. 1978. Sampling methods for black flies (Diptera: Simuliidae). South Carolina Agricultural Experiment Station Technical Bulletin 1067:1–6.

Noblet, R., & H. S. Moore IV. 1975. Prevalence and distribution of *Leucocytozoon smithi* and *Haemoproteus meleagridis* in wild turkeys in South Carolina. Journal of Wildlife Diseases 11:516–518.

Noblet, R., T. R. Adkins, & J. B. Kissam. 1972. *Simulium congareenarum* (Diptera: Simuliidae), a new vector of *Leucocytozoon smithi* (Sporozoa: Leucocytozoidae) in domestic turkeys. Journal of Medical Entomology 9:580.

Noblet, R., J. B. Kissam, & T. R. Adkins Jr. 1975. *Leucocytozoon smithi*: incidence of transmission by black flies in South Carolina (Diptera: Simuliidae). Journal of Medical Entomology 12:111–114.

Noblet, R., D. C. Arnold, & E. L. Snoddy. 1976. Distribution of *Simulium* vectors of *Leucocytozoon smithi* in South Carolina. Journal of Economic Entomology 69:481–483.

Noblet, R., W. A. Gardner, D. C. Arnold, R. E. Moore Jr., E. L. Snoddy, & T. R. Adkins Jr. 1978. An annotated list of the Simuliidae (Diptera) of South Carolina. Journal of the Georgia Entomological Society 13:333–338.

Nolan, R. A. 1976. Physiological studies on an isolate of *Saprolegnia ferax* from the larval gut of the blackfly *Simulium vittatum*. Mycologia 68:523–540.

Nolan, R. A. 1981. Mass production of pathogens. Pp. 319–324. *In* M. Laird (ed.), Blackflies: the future for biological methods in integrated control. Academic Press, New York. 399 pp.

Nolan, R. A., & D. J. Lewis. 1974. Studies on *Pythiopsis cymosa* from Newfoundland. Transactions of the British Mycological Society 62:163–179.

Noriega, R., F. B. Ramberg, & H. H. Hagedorn. 2002. Ecdysteroids and oocyte development in the black fly *Simulium vittatum*. BMC Developmental Biology 2:6.

Nübel, E. 1984. Rasterelektronenmikroskopische Untersuchungen an den Filtrierstrukturen von Kriebelmücken-Larven (Simuliidae, Diptera) [SEM-investigations on the filter feeding organs of blackfly larvae (Simuliidae, Diptera). Archiv für Hydrobiologie 66(Suppl.): 223–253.

Nyhof, J. M., & S. B. McIver. 1987. Fine structure of ocelli of the larval black fly *Simulium vittatum* (Diptera: Simuliidae). Canadian Journal of Zoology 65:142–150.

O'Connor, P. F., J. E. Hazelrigg, M. W. Shaw, & M. B. Madon. 2001. Black fly control program at Greater Los Angeles County Vector Control District. Proceedings and Papers of the Annual Conference of the Mosquito and Vector Control Association of California 69:26–31.

Odum, A. 1973. Meet the bloodthirsty blackfly. Reader's Digest (Canadian edition) 103(August):128–133.

O'Grady, G. E. 1986. Fine structure of the compound eye of *Simulium vittatum* Zetterstedt (Diptera: Simuliidae). M. S. thesis. University of Toronto. 205 pp.

O'Grady, G. E., & S. B. McIver. 1987. Fine structure of the compound eye of the black fly *Simulium vittatum* (Diptera: Simuliidae). Canadian Journal of Zoology 65:1454–1469.

Ojanen, U., O. Rätti, P. H. Adler, K. Kuusela, B. Malmqvist, & P. Helle. 2002. Blood feeding by black flies (Diptera: Simuliidae) on the black grouse (*Tetrao tetrix*) in Finland. Entomologica Fennica 13:153–158.

Okaeme, A. N. 1983. On the biology of *Cnephia ornithophilia* and *Simulium vernum* (Diptera: Simuliidae) in insular Newfoundland. M. S. thesis. Memorial University of Newfoundland, St. John's. 77 pp.

O'Kane, W. C. 1926. Black flies in New Hampshire. New Hampshire Agricultural Experiment Station Technical Bulletin 32:1–24.

Olsen, O. W., & R. Fenstermacher. 1942. Parasites of moose in northern Minnesota. American Journal of Veterinary Research 3:403–408.

Onishi, O., T. Okazawa, & J. O. Ochoa A. 1977. Clave gráfica para la identificación de los simulidos del área de San Vicente Pacaya, por los caracteres externos de larvas y pupas. Guatemala-Japan Cooperation Research and Control Program of Onchocerciasis, Malaria Erradication [*sic*] National Service and Adjointed [*sic*] Programs. Serie No. 2. Guatemala. 11 pp.

Ono, H. 1982. Taxonomic study of black flies in Hokkaido, with notes on their veterinary viewpoint (Diptera: Simuliidae). Research Bulletin of Obihiro University 12:277–316.

O'Roke, E. C. 1930. The incidence, pathogenicity and transmission of *Leucocytozoon anatis* of ducks. Journal of Parasitology 17:112.

O'Roke, E. C. 1931. The life history of *Leucocytozoon anatis* Wickware. Journal of Parasitology 18:127.

O'Roke, E. C. 1934. A malaria-like disease of ducks caused by *Leucocytozoon anatis* Wickware. University of Michigan School of Forestry and Conservation Bulletin 4:1–44 + 5 plates.

Osborn, H. 1896. Insects affecting domestic animals: an account of the species of importance in North America, with mention of related forms occurring on other animals. United States Department of Agriculture Division of Entomology Bulletin (New Series) 5:1–302.

Osborne, L. L., E. E. Herricks, & V. Alavian. 1985. Characterization of benthic microhabitat: an experimental system for aquatic insects. Hydrobiologia 123:153–160.

Osten Sacken, R. 1870. On the transformations of *Simulium*. American Entomologist and Botanist 2:229–231.

Ottonen, P. O. 1964. The salivary gland chromosomes of six species in the IIIS-1 group of *Prosimulium* (Diptera: Simuliidae). M. A. thesis. University of Toronto, Toronto. 87 pp. + 9 plates.

Ottonen, P. O. 1966. The salivary gland chromosomes of six species in the IIIS-1 group of *Prosimulium* Roub. (Diptera: Simuliidae). Canadian Journal of Zoology 44:677–701.

Ottonen, P. O., & R. Nambiar. 1969. The salivary gland chromosomes of *Prosimulium multidentatum* Twinn and three forms included in *Prosimulium magnum* (Dyar and Shannon) (Diptera: Simuliidae). Canadian Journal of Zoology 47:943–949.

Pachón, R. T. 1999. Effects of a small-scale fish farm on a desert stream: the impact of enrichment on resources and abundance of aquatic macroinvertebrates with an emphasis on the Simuliidae. M. S. thesis. University of California, Riverside. 145 pp.

Packard, A. S., Jr. 1869. A chapter on flies. American Naturalist 2:586–596 + 1 plate. [Reprint version in Packard, A. S., Jr. 1873. Our common insects: a popular account of the insects of our fields, forests, gardens and houses. Salem Press, Boston, Mass. 225 pp.]

Packard, A. S. [Jr.]. 1886. History of the buffalo gnat. American Naturalist 20:650–651.

Page, P. 1996 August 5. Taking a bite out of clime. Times (Princeton, N.J.), pp. 1, 5.

Paine, G. H., Jr., & A. R. Gaufin. 1956. Aquatic Diptera as indicators of pollution in a midwestern stream. Ohio Journal of Science 56:291–304.

Painter, T. S. 1939. The structure of salivary gland chromosomes. American Naturalist 73:315–330. [The identical paper was published in 1940 in Biological Symposia 1:215–230.]

Painter, T. S., & A. B. Griffen. 1937a. The origin and structure of the salivary gland chromosomes of *Simulium virgatum*. Genetics 22:202–203.

Painter, T. S., & A. B. Griffen. 1937b. The structure and the development of the salivary gland chromosomes of *Simulium*. Genetics 22:612–633.

Palmer, R. W., & D. A. Craig. 2000. An ecological classification of primary labral fans of filter-feeding black fly (Diptera: Simuliidae) larvae. Canadian Journal of Zoology 78:199–218.

Panzer, G. W. F. 1806. Favnae insectorvm germanicae initia oder Devtschlands Insecten. Heft. 105. Nürnberg [= Nuremberg], Germany. 24 pp. [Actual publication date was sometime between 1806 and 1809.]

Parker, R. R. 1934. Recent studies of tick-borne diseases made at the United States Public Health Service Laboratory at Hamilton, Montana. Proceedings of the Pacific Science Conference 5:3367–3374.

Pascuzzo, M. C. 1976. Fecundity, ovarian development, and physiological age in adult black-flies (Simuliidae) with some observations on vertical distribution. M. S. thesis. McMaster University, Hamilton, Ontario. 250 pp.

Pasternak, J. J. 1961. Chromosomal polymorphism in the black fly *Simulium vittatum* Zetterstedt. M. A. thesis. University of Toronto, Toronto. 87 pp.

Pasternak, J. [J.]. 1964. Chromosome polymorphism in the blackfly *Simulium vittatum* (Zett.). Canadian Journal of Zoology 42:135–158.

Patrusheva, V. D. 1962. Blackfly fauna of the central Ob' River Basin. Izvestiya Sibirskogo Otdeleniya Akademii Nauk SSSR (Seriya Biologicheskaya) 3:94–110. [In Russian.]

Patrusheva, V. D. 1971. The ecology and systematics of *Titanopteryx maculata* Mg. and *Schoenbaueria pusilla* Fries (Diptera, Simuliidae) in the Ob' River Basin. Entomologicheskoe Obozrenie 50:770–779. [In Russian; English translation in Entomological Review, Washington 50:438–443.]

Patrusheva, V. D. 1973. The male of *Prosimulium ircutense* [*sic*] Rubz. (Diptera, Simuliidae). Pp. 117–119. *In* A. I. Cherepanov (ed.), New and little-known species in the fauna of Siberia [Novye i maloizvestnye vidy fauny Sibiri] 7. "Nauka," Sibirskoe Otdelenie, Novosibirsk, Russia. 148 pp. [In Russian.]

Patrusheva, V. D. 1975. New species of blackflies from the genus *Prosimulium* Roub. from Taymyr and Kolyma. Pp. 65–72. *In* A. I. Cherepanov (ed.), New and little-known species in the fauna of Siberia [Novye i maloizvestnye vidy fauny Sibiri] 9, "Nauka," Sibirskoe Otdelnie, Novosibirsk, Russia. 125 pp. [In Russian; English summary.]

Patrusheva, V. D. 1982. Black flies of Siberia and the Far East. Izdatel'stvo "Nauka," Sibirskoe Otdelenie, Novosibirsk, Russia. 322 pp. [In Russian.]

Paysen, E. S. 1999. Cytogenetics, ecology, and evolution of the black fly *Simulium parnassum* Malloch (Diptera: Simuliidae). M. S. thesis. Clemson University, Clemson, S.C. 87 pp.

Paysen, E. S., & P. H. Adler. 2000. Taxonomy and polytene chromosomes of *Simulium parnassum* Malloch (Diptera: Simuliidae). Proceedings of the Entomological Society of Washington 102:843–851.

Pearce, J. 1888. An application for buffalo gnat bites. Insect Life 1:15.

Pearson, W. D., & D. R. Franklin. 1968. Some factors affecting drift rates of *Baetis* and Simuliidae in a large river. Ecology 49:75–81.

Pelsue, F. W. 1971. Black flies in the Southeast Mosquito Abatement District. Proceedings and Papers of the Annual Conference of the California Mosquito Control Association 39:50.

Pelsue, F. W., G. C. McFarland, & H. I. Magy. 1970. Buffalo gnat (Simuliidae) control in the Southeast Mosquito Abatement District. Proceedings and Papers of the Annual Conference of the California Mosquito Control Association 38:102–104.

Perez, J. M. 1999. Associations among selected physcio[*sic*]-chemical parameters and Simuliidae (Diptera) from 23 lake-outlet sites in Newfoundland. M.S. thesis. Memorial University of Newfoundland, St. John's. 173 pp.

Peschken, D. P. 1960. Investigations on the sensory behaviour and ecology of black flies Simuliidae: Diptera in the Whiteshell Forest Reserve, Manitoba. M.S. thesis. University of Manitoba, Winnipeg. 72 pp.

Peschken, D. [P.], & A. J. Thorsteinson. 1965. Visual orientation of black flies (Simuliidae: Diptera) to colour, shape and movement of targets. Entomologia Experimentalis et Applicata 8:282–288.

Peters, R. H., & D. J. Womeldorf. 1966. *Simulium* annoying to humans in San Joaquin County, California. California Vector Views 13:41.

Peterson, B. J., J. E. Hobbie, A. E. Hershey, M. A. Lock, T. E. Ford, J. R. Vestal, V. L. McKinley, M. A. J. Hullar, M. C. Miller, R. M. Ventullo, & G. S. Volk. 1985. Transformation of a tundra river from heterotrophy to autotrophy by addition of phosphorus. Science 229:1383–1386.

Peterson, B. J., L. Deegan, J. Helfrich, J. E. Hobbie, M. Hullar, B. Moller, T. E. Ford, A. Hershey, A. Hiltner, G. Kipphut, M. A. Lock, D. M. Fiebig, V. McKinley, M. C. Miller, J. R. Vestal, R. Ventullo, & G. Volk. 1993. Biological responses of a tundra river to fertilization. Ecology 74:653–672.

Peterson, B. V. 1955. A preliminary list of the black flies (Diptera: Simuliidae) of Utah. Proceedings of the Utah Academy of Sciences, Arts and Letters 32:113–115.

Peterson, B. V. 1956. Observations on the biology of Utah black flies (Diptera: Simuliidae). Canadian Entomologist 88:496–507.

Peterson, B. V. 1958a. The taxonomy and biology of Utah species of black flies (Diptera: Simuliidae). Ph.D. thesis. University of Utah, Salt Lake City. 336 pp. + 14 plates.

Peterson, B. V. 1958b. A redescription of the female and first descriptions of the male, pupa and larva of *Prosimulium flaviantennus* (S. and K.) with notes on the biology and distribution. Canadian Entomologist 90:469–473.

Peterson, B. V. 1958c. Simuliidae (family). Buffalo gnats, black flies. Pp. 152–153. *In* A. M. Woodbury (ed.). Preliminary report on biological resources of the Glen Canyon reservoir. University of Utah (Department of Anthropology) Anthropological Paper No. 31. University of Utah, Salt Lake City, 219 pp.

Peterson, B. V. 1959a. Observations on mating, feeding, and oviposition of some Utah species of black flies (Diptera: Simuliidae). Canadian Entomologist 91:147–155.

Peterson, B. V. 1959b. Notes on the biology of some species of Utah blackflies (Diptera: Simuliidae). Mosquito News 19:86–90.

Peterson, B. V. 1959c. Three new black fly records from Utah. Proceedings of the Entomological Society of Washington 61:21.

Peterson, B. V. 1960a. The Simuliidae (Diptera) of Utah, part I. Keys, original citations, types and distribution. Great Basin Naturalist 20:81–104.

Peterson, B. V. 1960b. Notes on some natural enemies of Utah black flies (Diptera: Simuliidae). Canadian Entomologist 92:266–274.

Peterson, B. V. 1962a. *Cnephia abdita*, a new black fly (Diptera: Simuliidae) from eastern North America. Canadian Entomologist 94:96–102.

Peterson, B. V. 1962b. Observations on mating swarms of *Simulium venustum* Say and *Simulium vittatum* Zetterstedt (Diptera: Simuliidae). Proceedings of the Entomological Society of Ontario 92:188–190.

Peterson, B. V. 1965a. Lectotype designation for *Simulium vittatum* Zetterstedt (Dipt. Simuliidae). Opuscula Entomologica 30:231–233.

Peterson, B. V. 1965b. The status of the name *Prosimulium albionense* Rothfels. Proceedings of the Entomological Society of Ontario 95:142.

Peterson, B. V. 1970a. The identities of three closely related western species of *Prosimulium* (Diptera: Simuliidae). Canadian Entomologist 102:118–128.

Peterson, B. V. 1970b. The *Prosimulium* of Canada and Alaska. Memoirs of the Entomological Society of Canada 69:1–216.

Peterson, B. V. 1975. A new name for a subgeneric homonym in the Simuliidae (Diptera). Canadian Entomologist 107:111.

Peterson, B. V. 1977a. A synopsis of the genus *Parasimulium* Malloch (Diptera: Simuliidae), with descriptions of one new subgenus and two new species. Proceedings of the Entomological Society of Washington 79:96–106.

Peterson, B. V. 1977b. The black flies of Iceland (Diptera: Simuliidae). Canadian Entomologist 109:449–472.

Peterson, B. V. 1981. Simuliidae. Pp. 355–391. *In* J. F. McAlpine, B. V. Peterson, G. E. Shewell, H. J. Teskey, J. R. Vockeroth, & D. M. Wood (eds.), Manual of Nearctic Diptera, vol. 1, Monograph No. 27. Research Branch, Agriculture Canada, Ottawa. 674 pp.

Peterson, B. V. 1989. An unusual black fly (Diptera: Simuliidae), representing a new genus and new species. Journal of the New York Entomological Society 97:317–331.

Peterson, B. V. 1993. The black flies of the genus *Simulium*, subgenus *Psilopelmia* (Diptera: Simuliidae), in the contiguous United States. Journal of the New York Entomological Society 101:301–390.

Peterson, B. V. 1996. Simuliidae. Pp. 591–634. *In* R. W. Merritt & K. W. Cummins (eds.), An introduction to the aquatic insects of North America, 3rd ed. Kendall/Hunt, Dubuque, Iowa. 862 pp.

Peterson, B. V., & G. W. Courtney. 1985. First description of the female of *Parasimulium stonei* Peterson (Diptera: Simuliidae), with notes and a discussion on collection sites. Proceedings of the Entomological Society of Washington 87:656–661.

Peterson, B. V., & M. E. Craig. 1997 ["1996"]. *Simulium* (*Hemicnetha*) *wirthi*, a new species of black fly (Diptera: Simuliidae) from New Mexico. Memoirs of the Entomological Society of Washington 18:212–220. [Actual publication date was 3 January 1997.]

Peterson, B. V., & D. M. Davies. 1960. Observations on some insect predators of black flies (Diptera: Simuliidae) of Algonquin Park, Ontario. Canadian Journal of Zoology 38:9–18.

Peterson, B. V., & G. R. DeFoliart. 1960. Four new species of *Prosimulium* (Diptera: Simuliidae) from western United States. Canadian Entomologist 92:85–102.

Peterson, B. V., & K. R. Depner. 1972. A new species of *Prosimulium* from Alberta (Diptera: Simuliidae). Canadian Entomologist 104:289–294.

Peterson, B. V., & B. C. Kondratieff. 1995 ["1994"]. The black flies (Diptera: Simuliidae) of Colorado: an annotated list with keys, illustrations and descriptions of three new species. Memoirs of the American Entomological Society 42:1–121. [Actual publication date was 15 February 1995.]

Peterson, B. V., M. Vargas V., & J. Ramírez-Pérez. 1988. *Simulium* (*Hemicnetha*) *hieroglyphicum* (Diptera: Simuliidae), a new black fly species from Costa Rica. Proceedings of the Entomological Society of Washington 90:76–86.

Peterson, D. G. 1955. Biology and control of biting flies in pulpwood cutting areas. Pulp and Paper Magazine of Canada 56:182–186.

Peterson, D. G., & L. S. Wolfe. 1958. The biology and control of black flies (Diptera: Simuliidae) in Canada. Proceedings of the Tenth International Congress of Entomology, Montreal (1956) 3:551–564.

Peterson, D. G., & A. S. West. 1960. Control of adult black flies (Diptera: Simuliidae) in the forests of eastern Canada by aircraft spraying. Canadian Entomologist 92:714–719.

Peterson, S. W., & R. W. Lichtwardt. 1987. Antigenic variation within and between populations of three genera of Harpellales (Trichomycetes). Transactions of the British Mycological Society 88:189–197.

Petrova, N. A., I. A. Rubtsov, & L. A. Chubareva. 1971. On the position of *Simulium* (*Schönbaueria*) *gigantea* Rubz. in the classification of black flies (morphological and karyological characters in the systematics). Parazitologiya 5:40–50. [In Russian; English translation in Parasitology 1:45–57.]

Pettit, R. H. 1929. Buffalo gnats (Simuliidae). United States Department of Agriculture Insect Pest Survey Bulletin 9:208.

Phelps, R. J. 1962. Nematode parasitism of larval Simuliidae. Ph.D. thesis. University of Wisconsin, Madison. 231 pp.

Phelps, R. J., & G. R. DeFoliart. 1964. Nematode parasitism of Simuliidae. University of Wisconsin Agricultural Experiment Station Research Bulletin 245:1–78.

Philip, C. B., & W. L. Jellison. 1986. Field tests in western Montana of mechanical transmission of tularemia by biting flies (Diptera) between immobilized laboratory animals. Bulletin of the Society of Vector Ecologists 11:197–198.

Phillips, R. O. 1890. The transformations of *Simulium innoxium*, Williston. B.S. thesis. Cornell University, Ithaca, N.Y. 35 pp. + 8 plates.

Philogène, B. J. R., J. T. Arnason, C. W. Berg, F. Duval, D. Champagne, R. G. Taylor, L. C. Leitch, & P. Morand. 1985. Synthesis and evaluation of the naturally occurring phototoxin, alpha-terthienyl, as a control agent for larvae of *Aedes intrudens*, *Aedes atropalpus* (Diptera: Culicidae) and *Simulium verecundum* (Diptera: Simuliidae). Journal of Economic Entomology 78:121–126.

Pickavance, J. R., G. F. Bennett, & J. Phipps. 1970. Some mosquitoes and blackflies from Newfoundland. Canadian Journal of Zoology 48:621–624.

Pilfrey, R. J. 1963. Studies on the life history and ecology of *Simulium rugglesi* N. & M. (Diptera: Simuliidae). M.S. thesis. University of Toronto, Ontario Agricultural College, Guelph. 116 pp. + 3 plates.

Pinkovsky, D. D. 1970. An ecological survey of the black flies (Diptera: Simuliidae) of Tompkins County, New York. M.S. thesis. Cornell University, Ithaca, N.Y. 250 pp.

Pinkovsky, D. D. 1976. The black flies (Diptera: Simuliidae) of Florida and their involvement in the transmission of *Leucocytozoon smithi* to turkeys. Ph.D. thesis. University of Florida, Gainesville. 331 pp.

Pinkovsky, D. D., & J. F. Butler. 1978. Black flies of Florida I. Geographic and seasonal distribution. Florida Entomologist 61:257–267.

Pinkovsky, D. D., D. J. Forrester, & J. F. Butler. 1981. Investigations on black fly vectors (Diptera: Simuliidae) of *Leucocytozoon smithi* (Sporozoa: Leucocytozoidae) in Florida. Journal of Medical Entomology 18:153–157.

Pistrang, L. A. 1983. The *Simulium tuberosum* complex (Diptera: Simuliidae) in Waterville Valley, New Hampshire: a study of sibling species. M.S. thesis. University of New Hampshire, Durham. 70 pp.

Pistrang, L. A., & J. F. Burger. 1984. Effect of *Bacillus thuringiensis* var. *israelensis* on a genetically-defined population of black flies (Diptera: Simuliidae) and associated insects in a montane New Hampshire stream. Canadian Entomologist 116:975–981.

Pistrang, L. A., & J. F. Burger. 1988. The spatial and temporal distribution of four *Simulium tuberosum* (Diptera: Simuliidae) cytospecies in Waterville Valley, New Hampshire, U.S.A. Canadian Journal of Zoology 66:904–911.

Pledger, D. J. 1978. Black flies (Diptera, Simuliidae) of the Swan Hills, Alberta as possible vectors of *Onchocerca cervipedis* Wehr and Dikmans, 1935 (Nematoda, Onchocercidae) in moose (*Alces alces* Linnaeus). M.S. thesis. University of Alberta, Edmonton. 144 pp.

Pledger, D. J., W. M. Samuel, & D. A. Craig. 1980. Black flies (Diptera, Simuliidae) as possible vectors of legworm (*Onchocerca cervipedis*) in moose of central Alberta. Proceedings of the North American Moose Conference Workshop 16:171–202.

Podszuhn, H. 1967. Gattungsbestimmung von europäischen Simuliiden-Larven (Diptera). Gewässer und Abwässer 44/45:87–95.

Poff, N. L., & J. V. Ward. 1991. Drift responses of benthic invertebrates to experimental streamflow variation in a hydrologically stable stream. Canadian Journal of Fisheries and Aquatic Sciences 48:1926–1936.

Poinar, G. O., Jr. 1981. Mermithid nematodes of blackflies. Pp. 159–170. *In* M. Laird (ed.), Blackflies: the future for biological methods in integrated control. Academic Press, New York. 399 pp.

Poinar, G. O., Jr., & R. Hess. 1979. *Mesomermis paradisus* sp. n. (Mermithidae: Nematoda), a parasite of *Prosimulium exigens* D. & S. (Diptera: Simuliidae) in California. Nematologica 25:368–372.

Poinar, G. O., Jr., R. Hess, E. Hansen, & J. W. Hansen. 1979. Laboratory infection of blackflies (Simuliidae) and midges (Chironomidae) by the mosquito mermithid, *Romanomermis culicivorax*. Journal of Parasitology 65:613–615.

Poirier, D. G., & G. A. Surgeoner. 1987. Laboratory flow-through bioassays of four forestry insecticides against stream invertebrates. Canadian Entomologist 119:755–763.

Popov, V. D. 1968. A new species of the genus *Prosimulium* Roub. (Diptera, Simuliidae) from the Far East. Parazitologiya 2:444–447. [In Russian; English summary.]

Post, R. J. 1985. Sex chromosome evolution in *Simulium erythrocephalum* (Diptera: Simuliidae). Heredity 54:149–158.

Post, R. J., P. K. Flook, & A. L. Millest. 1993. Methods for the preservation of insects for DNA studies. Biochemical Systematics and Ecology 21:85–92.

Power, G. 1969. The salmon of Ungava Bay. Arctic Institute of North America Technical Paper 22:1–72.

Prevost, G. 1946. Fourth report of the Biological Bureau. Pp. 40–91. Game and Fisheries Department, Quebec. [In English and French; also published as General report of the Minister of Game and Fisheries concerning the activities of fish and game for the year ending March 31st 1946.]

Prevost, G. 1949 December 29. Eradication of black fly larvae (*Simulium* sp.) for a long term by the use of DDT at a critical time. Paper read at the meeting of the Limnological Society of the American Association for the Advancement of Science, Chicago, Ill. 8 pp.

Price, W. A. 1936. A buffalo gnat (*Simulium* sp.). United States Department of Agriculture Insect Pest Survey Bulletin 16:146.

Procunier, W. S. 1974. A cytological study of two closely related blackfly species *Cnephia dacotensis* and *Cnephia ornithophilia* (Diptera: Simuliidae). M.S. thesis. University of Toronto, Toronto. 52 pp. + 6 plates.

Procunier, W. S. 1975a. A cytological study of two closely related blackfly species: *Cnephia dacotensis* and *Cnephia ornithophilia* (Diptera: Simuliidae). Canadian Journal of Zoology 53:1622–1637.

Procunier, W. S. 1975b. B-chromosomes of *Cnephia dacotensis* and *C. ornithophilia* (Diptera: Simuliidae). Canadian Journal of Zoology 53:1638–1647.

Procunier, W. S. 1980. Cytological studies in *Metacnephia* and *Cnephia s. str.* (Diptera: Simuliidae). Ph.D. thesis. University of Toronto, Toronto. 182 pp.

Procunier, W. S. 1982a. A cytological description of 10 taxa in *Metacnephia* (Diptera: Simuliidae). Canadian Journal of Zoology 60:2852–2865.

Procunier, W. S. 1982b. A cytological study of species in *Cnephia* s. str. (Diptera: Simuliidae). Canadian Journal of Zoology 60:2866–2878.

Procunier, W. S. 1982c. The interdependence of B chromosomes, nucleolar organizer expression, and larval development in the blackfly species *Cnephia dacotensis* and *Cnephia ornithophilia* (Diptera: Simuliidae). Canadian Journal of Zoology 60:2879–2896.

Procunier, W. S. 1984. Cytological identification of pest species of the *Simulium arcticum* complex present in the Athabasca River and associated tributaries. Alberta Research Council Farming for the Future Final Technical Report, No. 82-101. Agriculture Canada Research Station, Lethbridge. 44 pp.

Procunier, W. S. 1989. Cytological approaches to simuliid biosystematics in relation to the epidemiology and control of human onchocerciasis. Genome 32:559–569.

Procunier, W. S., & A. I. Muro. 1994. A midarm interchange as a potential reproductive isolating mechanism in the medically important *Simulium neavei* group (Diptera: Simuliidae). Genome 37:957–969.

Procunier, W. S., & R. J. Post. 1986. Development of a method for the cytological identification of man-biting sibling species within the *Simulium damnosum* complex. Tropical Medicine and Parasitology 37:49–53.

Procunier, W. S., & J. A. Shemanchuk. 1983. Identification of sibling species of black flies in Alberta using polytene chromosome analysis. Pp. 33–36. *In* L. J. L. Sears & T. G. Atkinson (eds.), Research Highlights—1982. Agriculture Canada Research Station, Lethbridge, Alberta. 124 pp.

Procunier, W. S., & J. J. Smith. 1993. Localization of ribosomal DNA in *Rhagoletis pomonella* (Diptera: Tephritidae) by *in situ* hybridization. Insect Molecular Biology 2:163–174.

Procunier, W. S., J. A. Shemanchuk, & W. B. Barr. 1984. Cytological identification of *Simulium* (IIS-10.11) *arcticum* pest populations from the Athabasca area using larval and adult polytene chromosomes. Pp. 37–40. *In* L. J. L. Sears & D. B. Wilson (eds.), Research Highlights—1983. Agriculture Canada Research Station, Lethbridge, Alberta. 127 pp.

Procunier, W. S., A. J. Shelley, & M. Arzube. 1987. Cytological identification of *Simulium oyapockense* manabi form (Diptera: Simuliidae): a potential vector of onchocerciasis in Ecuador. Tropical Medicine and Parasitology 38:71.

Prokofyeva, K. K. 1959. On the types of egg maturation in non-bloodsucking species of black-flies (Diptera, Simuliidae). Entomologicheskoe Obozrenie 38:58–63. [In Russian; English translation in Entomological Review, Washington 38:51–55.]

Pruess, K. P. 1989. Colonization of immature black flies (Diptera: Simuliidae) on artificial substrates in a Nebraska sandy river. Environmental Entomology 18:433–437.

Pruess, K. P., & B. V. Peterson. 1987. The black flies (Diptera: Simuliidae) of Nebraska: an annotated list. Journal of the Kansas Entomological Society 60:528–534.

Pruess, K. P., X. Zhu, & T. O. Powers. 1992. Mitochondrial transfer RNA genes in a black fly, *Simulium vittatum* (Diptera: Simuliidae), indicate long divergence from mosquito (Diptera: Culicidae) and fruit fly (Diptera: Drosophilidae). Journal of Medical Entomology 29:644–651.

Pruess, K. P., B. J. Adams, T. J. Parsons, X. Zhu, & T. O. Powers. 2000. Utility of the mitochondrial cytochrome oxidase II gene for resolving relationships among black flies (Diptera: Simuliidae). Molecular Phylogenetics and Evolution 16:286–295.

Ptáček, V., & J. Knoz. 1971. Über der [die] larvale Entwicklung der Art *Simulium* (*S.*) *argyreatum* Meigen 1838. Scripta Facultatis Scientiarum Naturalium Universitatis Purkynianae Brunensis 1:179–195.

Puri, I. M. 1925. On the life-history and structure of the early stages of Simuliidae (Diptera, Nematocera). Part I. Parasitology 17:295–334 + 2 plates.

Puri, I. M. 1926. On the early stages of some of the Scandinavian species of *Simulium* (Simuliidae, Diptera, Nematocera). Parasitology 18:160–167.

Py-Daniel, V. 1990. Revisão da Tribo Prosimuliini seg. Crosskey (Diptera, Culicomorpha, Simuliidae) nas regiões zoogeográficas Néartica e Neotropical (larvas e pupas). Ph.D. thesis. Instituto Nacional de Pesquisas da Amazônia, Manaus, Brazil. 320 pp.

Py-Daniel, V., & R. T. Moreira Sampaio. 1994. *Jalacingomyia* gen. n. (Culicomorpha); a ressurreição de Gymnopaidinae; a eliminação do nivel tribal; apresentação de novos caracteres e a redescrição dos estágios larval e pupal de *Simulium colombaschense* (Fabricius, 1787) (Diptera: Simuliidae). Memorias del CAICET 4:101–148.

Raastad, J. E. 1979. Fennoscandian blackflies (Diptera, Simuliidae): annotated list of the species and their gross distribution. Rhizocrinus 11:1–28.

Raastad, J. E., & P. H. Adler. 2001. The identity of *Simulium murmanum* Enderlein, 1935 (Diptera: Simuliidae). Aquatic Insects 23:183–186.

Radzivilovskaya, Z. A. 1948. The fauna and ecology of black flies (Simuliidae) of mountainous regions of South Ussirian taiga. Parazitologichesky Sbornik 10:131–150. [In Russian.]

Raminani, L. N., & E. W. Cupp. 1978. The male reproductive system of the black fly, *Simulium pictipes* Hagen. Mosquito News 38:591–594.

Ramírez-Pérez, J. 1983. Los jejenes de Venezuela. Simposio de oncocercosis Americana. CAICET. Puerto Ayacucho, Venezuela, 15–17 October 1983. 156 pp.

Ramírez-Pérez, J., & M. A. Vulcano 1973. Descripción y redescripciones de algunos simúlidos de Venezuela (Diptera: Simuliidae). Archivos Venezolanos de Medicina Tropical y Parasitologia Medica 5:375–399.

Ramírez-Pérez, J., B. V. Peterson, & M. Vargas V. 1988. *Mayacnephia salasi* (Diptera: Simuliidae), a new black fly species from Costa Rica. Proceedings of the Entomological Society of Washington 90:66–75.

Ramos, A. G. 1991. Analysis of the peritrophic membrane and two protease-like genes from the American black fly, *Simulium vittatum*. Ph.D. thesis. Case Western Reserve University, Cleveland, Ohio. 153 pp.

Ramos, A. [G.], A. Mahowald, & M. Jacobs-Lorena. 1993. Gut-specific genes from the black fly *Simulium vittatum* encoding trypsin-like and carboxypeptidase-like proteins. Insect Molecular Biology 1:149–163.

Ramos, A. [G.], A. Mahowald, & M. Jacobs-Lorena. 1994. Peritrophic matrix of the black fly *Simulium vittatum*: formation, structure, and analysis of its protein components. Journal of Experimental Zoology 268:269–281.

Rao, M. R. 1966. Phenology and faunistics of bloodsucking Diptera and arbovirus isolations in Wisconsin with emphasis on the Simuliidae (including a comparative study of larval simuliid cephalic fans). Ph.D. thesis. University of Wisconsin, Madison. 82 pp.

Rees, D. M., & B. V. Peterson. 1953. The black flies (Diptera: Simuliidae) in the canyons near Salt Lake City, Utah. Proceedings of the Utah Academy of Sciences, Arts and Letters 30:57–59.

Reeves, C. D. 1910. A remedy for the black fly pest in certain streams of the southern peninsula of Michigan. Michigan Academy of Science Report 12:77–78.

Reeves, W. C., & M. M. Milby. 1990. Natural infection in arthropod vectors. Pp. 128–144. *In* W. C. Reeves (ed.), Epidemiology and control of mosquito-borne arboviruses in California, 1943–1987. California Mosquito and Vector Control Association. Sacramento. 508 pp.

Reeves, W. K., & D. Nayduch. 2002. Pathogenic *Bacillus* from a larva of the *Simulium tuberosum* species complex (Diptera: Simuliidae). Journal of Invertebrate Pathology 79:126–128.

Reeves, W. K., & E. S. Paysen. 1999. Black flies (Diptera: Simuliidae) and a new species of caddisfly (Trichoptera: Hydropsychidae) in a northwestern Georgia cave stream. Entomological News 110:253–259.

Reid, G. D. F. 1994. Structure and function of the cibarial armature in Simuliidae. Medical and Veterinary Entomology 8:295–301.

Reidelbach, J., & E. Kiel. 1990. Observations on the behavioural sequences of looping and drifting by blackfly larvae (Diptera: Simuliidae). Aquatic Insects 12:49–60.

Reiling, S. D., K. J. Boyle, M. L. Phillips, V. A. Trefts, & M. W. Anderson. 1988. The economic benefits of late-season black fly control. Maine Agricultural Experiment Station Bulletin 822:1–43.

Reiling, S. D., K. J. Boyle, H. Cheng, & M. L. Phillips. 1989. Contingent valuation of a public program to control black flies. Northeastern Journal of Agricultural and Resource Economics 18:126–134.

Reilly, J. J., III. 1975. Some aspects of the biology of *Neomesomermis flumenalis* (Mermithidae: Nematoda) parasitizing New Jersey black flies. M.S. thesis. Rutgers University, New Brunswick, N.J. 47 pp.

Reisen, W. K. 1974a. The ecology of larval blackflies (Diptera: Simuliidae) in a south central Oklahoma stream. Ph.D. thesis. University of Oklahoma, Norman. 170 pp.

Reisen, W. K. 1974b. The ecology of Honey Creek: a preliminary evaluation of the influence of *Simulium* spp. (Diptera: Simuliidae) larval populations on the concentration of total suspended particles. Entomological News 85:275–278.

Reisen, W. K. 1975a. Quantitative aspects of *Simulium virgatum* Coq. and *S.* species life history in a southern Oklahoma stream. Annals of the Entomological Society of America 68:949–954.

Reisen, W. K. 1975b. The ecology of Honey Creek, Oklahoma: spatial and temporal distributions of the macroinvertebrates. Proceedings of the Oklahoma Academy of Science 55:25–31.

Reisen, W. K. 1977. The ecology of Honey Creek, Oklahoma: population dynamics and drifting behavior of three species of *Simulium* (Diptera: Simuliidae). Canadian Journal of Zoology 55:325–337.

Reisen, W. K., & R. C. Fox. 1970. Some ecological notes on lotic dipteran emergence in Prater's Creek, South Carolina. Annals of the Entomological Society of America 63:624–625.

Reisen, W. K., & R. Prins. 1972. Some ecological relationships of the invertebrate drift in Praters Creek, Pickens County, South Carolina. Ecology 53:876–884.

Rempel, J. G., & A. P. Arnason. 1947. An account of three successive outbreaks of the black fly, *Simulium arcticum*, a serious livestock pest in Saskatchewan. Scientific Agriculture 27:428–445.

Reuter, U., & W. Rühm. 1976. Über die zeitliche Verteilung der anfliegenden Weibchen von *Boophthora erythrocephala* De Geer und *Simulium sublacustre* Davies bei der Eiablage (Simuliidae, Dipt.). Zeitschrift für angewandte Zoologie 63:385–391.

Ribeiro, J. M. C., R. Charlab, E. D. Rowton, & E. W. Cupp. 2000. *Simulium vittatum* (Diptera: Simuliidae) and *Lutzomyia longipalpis* (Diptera: Psychodidae) salivary gland hyaluronidase activity. Journal of Medical Entomology 37:743–747.

Richey, D. J., & R. E. Ware. 1955. Schizonts of *Leucocytozoon smithi* in artificially infected turkeys. Cornell Veterinarian 45:642–643.

Riegert, P. W. 1999a. The Northern Biting Fly Project 1947–1955. Entomology Series 8. Rampeck Publishers, Regina, Saskatchewan. 36 pp.

Riegert, P. W. 1999b. A history of outbreaks of blackflies and their control in western

Canada 1886–1980. Entomology Series 14. Rampeck Publishers, Regina, Saskatchewan. 68 pp.
Riley, C. M., & R. Fusco. 1990. Field efficacy of Vectobac®-12AS and Vectobac®-24AS against black fly larvae in New Brunswick streams (Diptera: Simuliidae). Journal of the American Mosquito Control Association 6:43–46.
Riley, C. V. 1870a. The death web of young trout. American Entomologist and Botanist 2:174.
Riley, C. V. 1870b. The death-web of young trout. American Entomologist and Botanist 2:211.
Riley, C. V. 1870c. The death-web of young trout. American Entomologist and Botanist 2:227–228.
Riley, C. V. 1870d. The so-called web-worm of young trout. American Entomologist and Botanist 2:365–367.
Riley, C. V. 1881. Entomology. American Naturalist 15:322–331.
Riley, C. V. 1884. Report of the entomologist. United States Department of Agriculture Report 1884:285–418 + 10 plates.
Riley, C. V. 1887. Report of the entomologist. United States Department of Agriculture Report 1886:459–592 + 11 plates.
Riley, C. V. 1888a. A Virginia *Simulium* called "Cholera Gnat." Insect Life 1:14–15.
Riley, C. V. 1888b. Buffalo-gnats attacking man. Insect Life 1:60–61.
Riley, C. V., & L. O. Howard. 1889. A phase of buffalo gnat injury. Insect Life 2:2.
Riley, C. V., & L. O. Howard. 1892a. Great damage by buffalo gnats. Insect Life 4:406.
Riley, C. V., & L. O. Howard. 1892b. A new *Simulium*. Insect Life 5:61.
Riley, W. A., & O. A. Johannsen. 1915. Handbook of medical entomology. Comstock Publishing, Ithaca, N.Y. 348 pp.
Ritcher, P. O. 1931. An undescribed species of simuliid larva and the corresponding pupa (Diptera: Simuliidae). Entomological News 42:241–246 + plate VI.
Ritter, D. G., & E. T. Feltz. 1974. On the natural occurrence of California encephalitis virus and other arboviruses in Alaska. Canadian Journal of Microbiology 20:1359–1366.
Rivosecchi, L. 1964a. Contributo alla conoscenza dei Simulidi italiani. IX: La collezione dell'Istituto di Zoologia "L. Spallanzani" di Pavia. Rivista di Parassitologia 25:129–143.
Rivosecchi, L. 1964b. The Simuliidae of the Apennine Mountains. (A note on the distribution.) Proceedings of the First International Congress of Parasitology 2:935–937.
Rivosecchi, L. 1967. Contributo alla conoscenza dei Simulidi italiani. XV. Un nuovo *Eusimulium* del gruppo *annulum*. Rivista di Parassitologia 28:63–70.
Rivosecchi, L. 1978. Simuliidae: Diptera Nematocera. Fauna d'Italia 13:1–533.
Rivosecchi, L., & R. Cardinali. 1975. Contributo alla conoscenza dei Simulidi italiani. XXIII. Nuovi dati tassonomici. Rivista di Parassitologia 36:55–78.
Roach, A. G. 1954. Controlling the black fly. Canadian Mining Journal 75:62–64.
Robert, L. L., R. E. Coleman, D. A. LaPointe, P. J. S. Martin, R. Kelly, & J. D. Edman. 1992. Laboratory and field evaluation of five repellents against the black flies *Prosimulium mixtum* and *P. fuscum* (Diptera: Simuliidae). Journal of Medical Entomology 29:267–272.
Robert, M., & L. Cloutier. 2001. Summer food habits of harlequin ducks in eastern North America. Wilson Bulletin 113:78–84.
Roberts, R. H. 1965. A steer-baited trap for sampling insects affecting cattle. Mosquito News 25:281–285.
Roberts, S. R. 1911. Sambon's new theory of pellagra and its application to conditions in Georgia. Journal of the American Medical Association 56:1713–1715.
Robertson, C. 1928. Flowers and insects. Science Press Printing, Lancaster, Penn. 221 pp.
Robertson, L. 1937. The southern buffalo gnat. Arkansas Agriculturist 14(4):5, 8.
Rodrigues, C. S. 1982. Effects of insecticides including insect growth regulators on black fly (Diptera: Simuliidae) larvae and associated nontarget stream invertebrates. Ph.D. thesis. University of Guelph, Guelph, Ontario. 240 pp.
Rodrigues, C. S., & N. K. Kaushik. 1984a. A bioassay apparatus for the evaluation of black fly (Diptera: Simuliidae) larvicides. Canadian Entomologist 116:75–78.
Rodrigues, C. S., & N. K. Kaushik. 1984b. The effect of temperature on the toxicity of temephos to black fly (Diptera: Simuliidae) larvae. Canadian Entomologist 116:451–455.
Rodrigues, C. S., & N. K. Kaushik. 1986. Laboratory evaluation of the insect growth regulator diflubenzuron against black fly (Diptera: Simuliidae) larvae and its effects on nontarget stream invertebrates. Canadian Entomologist 118:549–558.
Rodrigues, C. S., D. Molloy, & N. K. Kaushik. 1983. Laboratory evaluation of microencapsulated formulations of chlorpyrifos-methyl against black fly larvae (Diptera: Simuliidae) and selected nontarget invertebrates. Canadian Entomologist 115:1189–1201.
Rodway, M. S. 1998. Activity patterns, diet, and feeding efficiency of harlequin ducks breeding in northern Labrador. Canadian Journal of Zoology 76:902–909.
Rohner, C., C. J. Krebs, D. B. Hunter, & D. C. Currie. 2000. Roost site selection of great horned owls in relation to black fly activity: an anti-parasite behavior? Condor 102:950–955.
Root, F. A., & W. E. Connelley. 1901. The Overland Stage to California: personal reminiscences and authentic history of the Great Overland Stage Line and Pony Express from the Missouri River to the Pacific Ocean. Published by the authors, Topeka, Kans. 627 pp.
Ross, D. H. 1977. The ecology and distribution of immature black flies (Diptera: Simuliidae) of the Rose Lake Wildlife Research Area, Michigan. M.S. thesis. Michigan State University, East Lansing. 91 pp.
Ross, D. H. 1979. The larval instars of the black flies *Stegopterna mutata* and *Simulium vittatum* (Diptera: Simuliidae). Canadian Entomologist 111:693–697.
Ross, D. H., & D. A. Craig. 1979. The seven larval instars of *Prosimulium mixtum* Syme and Davies and *Prosimulium fuscum* Syme and Davies (Diptera: Simuliidae). Canadian Journal of Zoology 57:290–300.
Ross, D. H., & D. A. Craig. 1980. Mechanisms of fine particle capture by larval black flies (Diptera: Simuliidae). Canadian Journal of Zoology 58:1186–1192.
Ross, D. H., & R. W. Merritt. 1978. The larval instars and population dynamics of five species of black flies (Diptera: Simuliidae) and their responses to selected environmental factors. Canadian Journal of Zoology 56:1633–1642.
Ross, D. H., & R. W. Merritt. 1988 ["1987"]. Factors affecting larval black fly distributions and population dynamics. Pp. 90–108. *In* K. C. Kim & R. W. Merritt (eds.), Black flies: ecology, population management, and annotated world list. Pennsylvania State University, University Park. 528 pp. [Actual publication date was 1 March 1988.]
Ross, H. H. 1940. The Rocky Mountain "black fly," *Symphoromyia atripes* (Diptera: Rhagionidae). Annals of the Entomological Society of America 33:254–257.
Rothfels, K. H. 1956. Black flies: siblings, sex and species groupings. Journal of Heredity 47:113–122.
Rothfels, K. H. 1979. Cytotaxonomy of black flies (Simuliidae). Annual Review of Entomology 24:507–539.
Rothfels, K. H. 1980. Chromosomal variability and speciation in blackflies. Pp. 207–224. *In* R. L. Blackman, G. M. Hewitt, & M. Ashburner (eds.), Insect cytogenetics. Symposium of the Royal Entomological Society of London, No. 10. Blackwell Scientific, Oxford, U.K. 278 pp.
Rothfels, K. [H.]. 1981a. Cytotaxonomy: principles and their application to some northern species-complexes in *Simulium*. Pp. 19–29. *In* M. Laird (ed.),

Blackflies: the future for biological methods in integrated control. Academic Press, New York. 399 pp.
Rothfels, K. [H.]. 1981b. Cytological approaches to the study of blackfly systematics and evolution. Pp. 67–83. *In* M. W. Stock (ed.), Application of genetics and cytology in insect systematics and evolution. Forest, Wildlife and Range Experiment Station, University of Idaho, Moscow. 152 pp.
Rothfels, K. H. 1988 ["1987"]. Cytological approaches to black fly taxonomy. Pp. 39–52. *In* K. C. Kim & R. W. Merritt (eds.), Black flies: ecology, population management, and annotated world list. Pennsylvania State University, University Park. 528 pp. [Actual publication date was 1 March 1988.]
Rothfels, K. [H.]. 1989. Speciation in black flies. Genome 32:500–509.
Rothfels, K. H., & R. W. Dunbar. 1953. The salivary gland chromosomes of the black fly *Simulium vittatum* Zett. Canadian Journal of Zoology 31:226–241.
Rothfels, K. [H.], & D. Featherston. 1981. The population structure of *Simulium vittatum* (Zett.): the IIIL-1 and IS-7 sibling species. Canadian Journal of Zoology 59:1857–1883.
Rothfels, K. [H.], & [D.] M. Freeman. 1966. The salivary gland chromosomes of three North American species of *Twinnia* (Diptera: Simuliidae). Canadian Journal of Zoology 44:937–945.
Rothfels, K. H., & D. M. Freeman. 1977. The salivary gland chromosomes of seven species of *Prosimulium* (Diptera, Simuliidae) in the *mixtum* (*IIIL-1*) group. Canadian Journal of Zoology 55:482–507.
Rothfels, K. [H.], & D. M. Freeman. 1983. A new species of *Prosimulium* (Diptera: Simuliidae): an interchange as a primary reproductive isolating mechanism? Canadian Journal of Zoology 61:2612–2617.
Rothfels, K. [H.], & V. I. Golini. 1983. The polytene chromosomes of species of *Eusimulium* (*Hellichiella*) (Diptera: Simuliidae). Canadian Journal of Zoology 61:1220–1231.
Rothfels, K. H., & G. F. Mason. 1975. Achiasmate meiosis and centromere shift in *Eusimulium aureum* (Diptera-Simuliidae). Chromosoma 51:111–124.
Rothfels, K. [H.], & R. Nambiar. 1975. The origin of meiotic bridges by chiasma formation in heterozygous inversions in *Prosimulium multidentatum* (Diptera: Simuliidae). Chromosoma 52:283–292.
Rothfels, K. [H.], & R. Nambiar. 1981. A cytological study of natural hybrids between *Prosimulium multidentatum* and *P. magnum* with notes on sex determination in the Simuliidae (Diptera). Chromosoma 82:673–691.
Rothfels, K. [H.], R. Feraday, & A. Kaneps. 1978. A cytological description of sibling species of *Simulium venustum* and *S. verecundum* with standard maps for the subgenus *Simulium* Davies [*sic*] (Diptera). Canadian Journal of Zoology 56: 1110–1128.
Roubaud, E. 1906. Aperçus nouveaux, morphologiques et biologiques, sur les Diptères piqueurs du groupe des Simulies. Comptes Rendus Hebdomadaires des Séances de l'Académie des Sciences 143:519–521.
Roubaud, E. 1909. Description d'une Simulie nouvelle du Pérou. Bulletin de la Société de Pathologie exotique et de ses Filiales 2:428–430.
Rubtsov, I. A. 1940a ["1939"]. On the blackfly fauna (Simuliidae) of Transbaikalia. Parazitologichesky Sbornik 7:193–201. [In Russian; English summary and species descriptions.] [Actual publication date was 22 February 1940.]
Rubtsov, I. A. 1940b. Black flies (fam. Simuliidae) [Moshki (sem. Simuliidae)]. Fauna of the USSR. New Series No. 23, Insects, Diptera 6 (6). Akademii Nauk SSSR, Leningrad [= St. Petersburg], Russia. 533 pp. [In Russian; English keys and descriptions of new species.]
Rubtsov, I. A. 1947. Composition, relationships, geographical distribution and probable origins of the black fly fauna of Central Asia. Izvestiya Akademii Nauk SSSR (Seriya Biologicheskaya) 1:87–116. [In Russian; part in English.]
Rubtsov, I. A. 1955. Peculiar new species of black flies (Diptera, Simuliidae) from eastern Siberia. Entomologicheskoe Obozrenie 34:323–339. [In Russian.]
Rubtsov, I. A. 1956. Blackflies (fam. Simuliidae) [Moshki (sem. Simuliidae)]. Fauna of the USSR. New Series No. 64, Insects, Diptera 6 (6). Akademii Nauk SSSR, Leningrad [= St. Petersburg], Russia. 859 pp. [In Russian; English translation: 1990. Blackflies (Simuliidae). 2nd ed. Fauna of the USSR. Diptera, 6 (6). E. J. Brill, Leiden. 1042 pp.]
Rubtsov, I. A. 1960a. The gonotrophic cycle in non-bloodsucking species of blackflies (Diptera, Simuliidae). Entomologicheskoe Obozrenie 39:556–573. [In Russian; English translation in Entomological Review, Washington 39:392–405.]
Rubtsov ["Rubzow"], I. A. 1960b. Simuliidae (Melusinidae). *In* E. Lindner (ed.), Die Fliegen der palaearktischen Region (III) 14:97–160.
Rubtsov ["Rubzow"], I. A. 1961a. Simuliidae (Melusinidae). *In* E. Lindner (ed.), Die Fliegen der palaearktischen Region (III) 14:161–208.
Rubtsov ["Rubzow"], I. A. 1961b. Simuliidae (Melusinidae). *In* E. Lindner (ed.), Die Fliegen der palaearktischen Region (III) 14:209–256.
Rubtsov ["Rubzow"], I. A. 1962a. Simuliidae (Melusinidae). *In* E. Lindner (ed.), Die Fliegen der palaearktischen Region (III) 14:257–304.
Rubtsov ["Rubzow"], I. A. 1962b. Simuliidae (Melusinidae). *In* E. Lindner (ed.), Die Fliegen der palaearktischen Region (III) 14:305–336.
Rubtsov ["Rubzow"], I. A. 1962c. Simuliidae (Melusinidae). *In* E. Lindner (ed.), Die Fliegen der palaearktischen Region (III) 14:337–368.
Rubtsov, I. A. 1962d. Short keys to bloodsucking blackflies in the fauna of the USSR [Kratky opredeliteli' krovososushchikh moshek fauny SSSR]. Akademii Nauk SSSR, Leningrad [= St. Petersburg], Russia. 227 pp. [In Russian; English translation: 1969. Short keys to the bloodsucking Simuliidae of the USSR. Israel Program for Scientific Translations, Jerusalem. 228 pp.]
Rubtsov ["Rubzow"], I. A. 1962e. Simuliidae (Melusinidae). *In* E. Lindner (ed.), Die Fliegen der palaearktischen Region (III) 14:369–400.
Rubtsov, I. A. 1962f. Genera of blackflies (family Simuliidae) in the fauna of the Ethiopian Region. Zoologichesky Zhurnal 41:1488–1502. [In Russian; English summary.]
Rubtsov ["Rubzow"], I. A. 1962g. Simuliidae (Melusinidae). *In* E. Lindner (ed.), Die Fliegen der palaearktischen Region (III) 14:401–432.
Rubtsov ["Rubzow"], I. A. 1962h. Simuliidae (Melusinidae). *In* E. Lindner (ed.), Die Fliegen der palaearktischen Region (III) 14:433–464.
Rubtsov ["Rubzow"], I. A. 1963a. Simuliidae (Melusinidae). *In* E. Lindner (ed.), Die Fliegen der palaearktischen Region (III) 14:497–528.
Rubtsov ["Rubzow"], I. A. 1963b. Simuliidae (Melusinidae). *In* E. Lindner (ed.), Die Fliegen der palaearktischen Region (III) 14:529–560.
Rubtsov ["Rubzow"], I. A. 1963c. Simuliidae (Melusinidae). *In* E. Lindner (ed.), Die Fliegen der palaearktischen Region (III) 14:561–592.
Rubtsov ["Rubzow"], I. A. 1964a. Simuliidae (Melusinidae). *In* E. Lindner (ed.), Die Fliegen der palaearktischen Region (III) 14:593–616.
Rubtsov, I. A. ["Rubzov, J. A."] 1964b. Simuliidae d'Italia: Memoria I. Memorie della Società Entomologica Italiana 43(Fasciculo Suppl.):1–124.
Rubtsov ["Rubzow"], I. A. 1964c. Simuliidae (Melusinidae). *In* E. Lindner (ed.), Die Fliegen der palaearktischen Region (III) 14:617–656.
Rubtsov, I. A. 1964d. On the mode and range of larval blackfly migrations (Diptera, Simuliidae). Entomologicheskoe

Obozrenie 43:52–66. [In Russian; English translation in Entomological Review, Washington 43:27–43.]
Rubtsov, I. A. 1965. Additional data on the black fly fauna (Diptera, Simuliidae) of the Palearctic Region. Entomologicheskoe Obozrenie 44:649–651. [In Russian; English translation in Entomological Review, Washington 44:381–382.]
Rubtsov, I. A. 1966. Mermithids (fam. Mermithidae) parasitizing blackflies (fam. Simuliidae). New species of the genus *Mesomermis* Dad. Pp. 109–147. *In* A. I. Cherepanov (ed.), New species in the fauna of Siberia and adjoining regions [Novye vidy fauny Sibiri y prilegayushchikh regionov]. "Nauka," Sibirskoe Otdelenie, Novosibirsk, Russia. 154 pp. [In Russian; English summary.]
Rubtsov, I. A. 1967. Mermithids (Nematoda, Mermithidae), endoparasites of blackflies (Diptera, Simuliidae). II. New species of the genus *Gastromermis* Micoletzky, 1923. Trudy Zoologicheskogo Instituta 43:59–92. [In Russian.]
Rubtsov, I. A. 1968. A new species of the genus *Isomermis* (Nematoda, Mermithidae)—parasite of blackflies—and its variability. Zoologichesky Zhurnal 47:510–524. [In Russian; English summary.]
Rubtsov, I. A. ["Rubzow, J. A."] 1969. Simuliidae II. Ergebnisse der zoologischen Forschungen von Dr. Z. Kaszab in der Mongolei (Diptera). Reichenbachia 12:113–130.
Rubtsov, I. A. 1971a. New and little known species of black-flies (Diptera, Simuliidae). III. Pp. 167–183. *In* S. Asahina, J. L. Gressitt, Z. Hidaka, T. Nishida, & K. Nomura (eds.), Entomological essays to commemorate the retirement of Professor K. Yasumatsu. Hokuryukan Publishing, Tokyo. 389 pp. [In Russian; English summary.]
Rubtsov, I. A. 1971b. New and little known species of black flies. Pp. 89–108. *In* A. I. Cherepanov (ed.), New and little known species in the fauna of Siberia [Novye i maloizvestnye vidy fauny Sibiri] 5. "Nauka," Sibirskoe Otdelenie, Novosibirsk, Russia. 138 pp. [In Russian; English summary.]
Rubtsov, I. A. 1972a. Phoresy in black flies (Diptera, Simuliidae) and new phoretic species from mayfly larvae. Entomologicheskoe Obozrenie 51:403–411. [In Russian; English translation in Entomological Review, Washington 51:243–247.]
Rubtsov, I. A. 1972b. Aquatic mermithids [Vodnye mermitidy]. Part I. "Nauka," Leningrad [= St. Petersburg], Russia. 254 pp. [In Russian; English translation: 1977. Aquatic Mermithidae of the fauna of the USSR. Amerind Publishing, New Delhi, India. 280 pp.]
Rubtsov, I. A. 1974a. On the evolution, phylogeny and classification of the family of blackflies (Simuliidae, Diptera). Trudy Zoologicheskogo Instituta 53:230–281. [In Russian; English translation: Evolution, phylogeny and classification of the family Simuliidae (Diptera). British Library Lending Division, Boston Spa, U.K. 83 pp.]
Rubtsov, I. A. 1974b. Aquatic mermithids [Vodnye mermitidy]. Part II. "Nauka," Leningrad [= St. Petersburg], Russia. 222 pp. [In Russian; English translation: 1981. Aquatic Mermithidae of the fauna of the USSR. Oxonian Press, New Delhi, India. 274 pp.]
Rubtsov, I. A. 1977. The genus *Ahaimophaga* Rubtsov and Chubareva (Diptera, Simuliidae) and some considerations about phylogenetic connections between the Palearctic and Nearctic fauna of black-flies of the subfamily Prosimuliinae. Pp. 47–49. *In* O. A. Skarlato (ed.), New and little known species of insects of the European part of the USSR [Novye i maloizvestnye vidy nasekomykh Evropeiskoi chasti SSSR]. Akademiya Nauk SSSR, Leningrad [= St. Petersburg], Russia. 116 pp. [In Russian; English summary.]
Rubtsov, I. A. 1980. New species of mermithids in arthropods. Pp. 95–102. *In* M. D. Sonin (ed.), Helminths of insects [Gel'minty nasekomykh]. "Nauka," Moscow. 161 pp. [In Russian; English translation: 1987. Pp. 137–146. *In* M. D. Sonin (ed.), Helminths of insects. Oxonian Press, New Delhi, India. 227 pp.]
Rubtsov, I. A., & G. Carlsson. 1965. On the taxonomy of black flies from Scandinavia and northern USSR. Acta Universitatis Lundensis (II) 18:1–40.
Rubtsov, I. A., & N. A. Petrova. 1977. Black flies of the tribe Cnephiini (Diptera, Simuliidae) and diagnoses of the genera *Cnephia* Enderlein and *Astega* Enderlein. Entomologicheskoe Obozrenie 56:691–697. [In Russian; English translation in Entomological Review, Washington 56:145–149.]
Rubtsov, I. A., & N. A. Violovich. 1965. Black flies of Tuva [Moshki Tuvy]. "Nauka," Sibirskoe Otdelenie, Novosibirsk, Russia. 63 pp. [In Russian.]
Rubtsov, I. A., & A. V. Yankovsky. 1982. New genera and subgenera of black flies (Diptera, Simuliidae). Entomologicheskoe Obozrenie 61:183–187. [In Russian; English translation in Entomological Review, Washington 61:176–181].
Rubtsov, I. A., & A. V. Yankovsky. 1984. Keys to the genera of Palearctic black flies. Keys to the Fauna of the USSR [Opredeliteli po Faune SSSR] 142:1–176. "Nauka," Leningrad [= St. Petersburg], Russia. [In Russian.]
Rubtsov ["Rubzov"], I. A., & A. V. Yankovsky. 1988. Family Simuliidae. Pp. 114–186. *In* A. Soós & L. Papp (eds.), Catalogue of Palaearctic Diptera, vol. 3, Ceratopogonidae-Mycetophilidae. Elsevier, Amsterdam. 448 pp.
Ruggles, A. G. 1933. A buffalo gnat (*Simulium venustum* Say). United States Department of Agriculture Insect Pest Survey Bulletin 13:318.
Rühm, W. 1975. Freilandbeobachtungen zum Funktionskreis der Eiablage verschiedener Simuliidenarten unter besonderer Berücksichtigung von *Simulium argyreatum* Meig. (Dipt. Simuliidae). Zeitschrift für angewandte Entomologie 78:321–334.
Rutley, M. S. 2000. Black fly control, the Adirondack way. Adirondack Journal of Environmental Studies 7(2):17–21.
Rutley, M. S. 2001. Black fly control, the Adirondack experience. Adirondack Journal of Environmental Studies 8(1):9–14.
Ryan, J. K., & G. J. Hilchie. 1982. Black fly problem in Athabasca County and vicinity, Alberta, Canada. Mosquito News 42:614–616.
Ryckman, R. E. 1961. Parasitic Ceratopogonidae and Simuliidae (Diptera) from Imperial County, California. Journal of Parasitology 47:405.
Saether, O. A. 1970. Chironomids and other invertebrates from North Boulder Creek, Colorado. University of Colorado Studies, Series in Biology 31:57–114.
Saether, O. A. 2000. Phylogeny of Culicomorpha (Diptera). Systematic Entomology 25:223–234.
Sailer, R. I. 1953. The blackfly problem in Alaska. Mosquito News 13:232–235.
Sailer, R. I. 1954. Invertebrate research in Alaska. Arctic (Journal of the Arctic Institute of North America) 7:266–274.
Sambon, L. W. 1910. Progress report on the investigation of pellagra. Journal of Tropical Medicine and Hygiene 13:271–282, 287–300, 305–315, 319–321.
Sanderson, C. L., D. R. C. McLachlan, & U. De Boni. 1982. Altered steroid induced puffing by chromatin bound aluminum in a polytene chromosome of the blackfly *Simulium vittatum*. Canadian Journal of Genetics and Cytology 24:27–36.
Sanderson, E. D. 1910. Controlling the black fly in the White Mountains. Journal of Economic Entomology 3:27–29.
Sanford, D., B. Eikenhorst, T. Lamb, J. E. Cates, J. Robinson, J. Olsen, C. Hoelscher, & J. Jeffrey. 1993. Black flies cause costly losses in East Texas ostriches and emus. Texas Agricultural Extension Service Veterinary Quarterly Review 9(2):1–2.
Sanzone, J. F. 1995. Organization and operations of the Metropolitan Mosquito Control District. Proceedings of the

Annual Meeting of the New Jersey Mosquito Control Association 82:78–84.
Sasaki, H. 1988. Morphological and immunological studies on the blood sources of black flies in Hokkaido, Japan (Diptera: Simuliidae). Journal of Rakuno Gakuen University 13:29–81 + 14 plates.
Say, T. 1823. Descriptions of dipterous insects of the United States. Journal of the Academy of Natural Sciences of Philadelphia 3:9–54.
Schaefer, G. W., & G. A. Bent. 1984. An infra-red remote sensing system for the active detection and automatic determination of insect flight trajectories (IRADIT). Bulletin of Entomological Research 74:261–278.
Schiefer, B. A., R. W. Vavra Jr., R. L. Frommer, & E. J. Gerberg. 1976. Field evaluation of several repellents against black flies (Diptera, Simuliidae). Mosquito News 36:242–247.
Schmidtmann, E. T. 1987. A trap/sampling unit for ear-feeding black flies. Canadian Entomologist 119:747–750.
Schmidtmann, E. T., W. J. Tabachnick, G. J. Hunt, L. H. Thompson, & H. S. Hurd. 1999. 1995 epizootic of vesicular stomatitis (New Jersey serotype) in the western United States: an entomologic perspective. Journal of Medical Entomology 36:1–7.
Schmidtmann, E. T., J. E. Lloyd Sr., R. J. Bobian, R. Kumar, J. W. Waggoner Jr., W. J. Tabachnick, & D. Legg. 2001. Suppression of mosquito (Diptera: Culicidae) and black fly (Diptera: Simuliidae) blood feeding from Hereford cattle and ponies treated with permethrin. Journal of Medical Entomology 38:728–734.
Schnitzlein, W. M., & M. E. Reichmann. 1985. Characterization of New Jersey vesicular stomatitis virus isolates from horses and black flies during the 1982 outbreak in Colorado. Virology 142:426–431.
Schofield, S. W. 1994. The bases of human attractiveness and bitability for black flies (Diptera: Simuliidae) in Algonquin Park, Ontario. M.S. thesis. Trent University, Peterborough, Ontario. 133 pp.
Schofield, S. W., & J. F. Sutcliffe. 1996. Human individuals vary in attractiveness for host-seeking black flies (Diptera: Simuliidae) based on exhaled carbon dioxide. Journal of Medical Entomology 33:102–108.
Schofield, S. [W.], & J. F. Sutcliffe. 1997. Humans vary in their ability to elicit biting responses from *Simulium venustum* (Diptera: Simuliidae). Journal of Medical Entomology 34:64–67.
Schönbauer, J. A. 1795. Geschichte der schädlichen Kolumbatczer Mücken im Bannat, als ein Beytrag zur naturgeschichte von Ungarn. Patzowsky, Wien [= Vienna], Austria. 100 pp. + 1 plate.
Schreck, C. E., N. Smith, T. P. McGovern, D. Smith, & K. Posey. 1979. Repellency of selected compounds against black flies (Diptera: Simuliidae). Journal of Medical Entomology 15:526–528.
Schreck, C. E., N. Smith, K. Posey, & D. Smith. 1980. Observations on the biting behavior of *Prosimulium* spp., and *Simulium venustum*. Mosquito News 40:113–115.
Schroeder, H. O. 1939. Black gnats (Simuliidae). United States Department of Agriculture Insect Pest Survey Bulletin 19:347.
Schuh, R. T. 2000. Biological systematics: principles and applications. Cornell University Press, Ithaca, N.Y. 236 pp.
Schütte, G. 1990. Die Anpassung der Populationen der autogenen Kriebelmückenart *Simulium noelleri* Friederichs 1920 (Diptera: Simuliidae) an ein extremes Habitat. Ph.D. thesis. Universität Hamburg, Hamburg, Germany. 106 pp.
Schwardt, H. H. 1935a. Buffalo gnats (*Eusimulium pecuarum* Riley). United States Department of Agriculture Insect Pest Survey Bulletin 15:55.
Schwardt, H. H. 1935b. Lubricating oil emulsion as a buffalo gnat repellent. Journal of the Kansas Entomological Society 8:141.
Schwardt, H. H. 1938. Buffalo gnats (*Eusimulium* spp.). United States Department of Agriculture Insect Pest Survey Bulletin 18:28.
Scopoli, J. A. 1780. Zwölfter Brief. Über eine Gattung schädlicher Insekten, die man die kolombaschischen Mücken nennt, welche unter den Ochsen, Schafen, Ziegen. Schweinen und Pferden in Serviert, dem Temeswarer Bannat, und der abendländischen Wallachen groβen Schaden anrichten. Pp. 123–135. *In* F. Griselini, Versuch einer politischen und natürlichen Geschichte des Temeswarer Bannats in Briefen an Standespersonen und Gelehrte. Zweyter Theil. Johann Paul Krauss, Wien [= Vienna], Austria. 135 pp. + 9 plates.
Sebastien, R. J., & W. L. Lockhart. 1981. The influence of formulation on toxicity and availability of a pesticide (methoxychlor) to black fly larvae (Diptera: Simuliidae), some non-target aquatic insects and fish. Canadian Entomologist 113:281–293.
Séguy, E., & A. Dorier. 1936. Description d'une nouvelle espèce de simulie récoltée en Dauphiné. Annales de l'Université de Grenoble (new series) (Sciences-Médecine) 13:133–147.
Service, M. W. 1981. Sampling methods for adults. Pp. 287–296. *In* M. Laird (ed.), Blackflies: the future for biological methods in integrated control. Academic Press, New York. 399 pp.
Service, M. W. 1988 ["1987"]. Monitoring adult simuliid populations. Pp. 187–200. *In* K. C. Kim & R. W. Merritt (eds.), Black flies: ecology, population management, and annotated world list. Pennsylvania State University, University Park. 528 pp. [Actual publication date was 1 March 1988.]
Shaw, F. R. 1959. New records and distributions of the biting flies of Mt. Desert Island, Maine. Mosquito News 19:189–191.
Shcherbakov, E. S. 1966. The morphology of polytene chromosomes of salivary glands of *Simulium nölleri*. Tsitologiya 8:703–713. [In Russian; English summary.]
Sheldon, A. L., & M. W. Oswood. 1977. Blackfly (Diptera: Simuliidae) abundance in a lake outlet: test of a predictive model. Hydrobiologia 56:113–120.
Shelley, A. J., & A. P. A. Luna Dias. 1989. First report of man eating blackflies (Dipt., Simuliidae). Entomologist's Monthly Magazine 125:44.
Shelley, A. J., M. Arzube, & C. A. Couch. 1989. The Simuliidae (Diptera) of the Santiago onchocerciasis focus of Ecuador. Bulletin of the British Museum of Natural History (Entomology) 58:79–130.
Shemanchuk, J. A. 1978a. Repellents for cattle protection. Pp. 53–54. *In* Research Highlights—1977. Agriculture Canada Research Station, Lethbridge, Alberta. 79 pp.
Shemanchuk, J. A. 1978b. A bait trap for sampling the feeding populations of blood-sucking Diptera on cattle. Quaestiones Entomologicae 14:433–439.
Shemanchuk, J. A. 1980a. Distribution, seasonal incidence and infestation of cattle by *Simulium arcticum* and other black fly adults. Pp. 201–205. *In* W. O. Haufe & G. C. R. Croome (eds.), Control of black flies in the Athabasca River. Alberta Environment Technical Report. Alberta Environment, Pollution Control Division, Edmonton. 241 pp.
Shemanchuk, J. A. 1980b. Protection of cattle on farms. Pp. 215–216. *In* W. O. Haufe & G. C. R. Croome (eds.), Control of black flies in the Athabasca River. Alberta Environment, Technical Report. Alberta Environment, Pollution Control Division, Edmonton. 241 pp.
Shemanchuk, J. A. 1980c. *Entomophthora culicis*: a fungus parasitic on adult black flies. Pp. 87–88. *In* G. C. R. Croome & D. B. Wilson (eds.), Research Highlights—1979. Agriculture Canada Research Station, Lethbridge, Alberta. 93 pp.
Shemanchuk, J. A. 1981. Repellent action of permethrin, cypermethrin and

resmethrin against black flies (*Simulium* spp.) attacking cattle. Pesticide Science 12:412–416.

Shemanchuk, J. A. 1982a. Protecting cattle against blood-sucking flies. Alberta Agriculture Farming for the Future Final Technical Report, No. 79-0070. Alberta Agriculture, Lethbridge, Alberta. 31 pp. + 12 figs.

Shemanchuk, J. A. 1982b. Protecting cattle from black flies with repellents. Pp. 28–30. *In* L. J. L. Sears, K. K. Krogman, & T. G. Atkinson (eds.), Research Highlights—1981. Agriculture Canada Research Station, Lethbridge, Alberta. 86 pp.

Shemanchuk, J. A. 1985. Dual-cloud electrostatic sprayer for application of repellents and insecticides to cattle. Pp. 23–24. *In* L. J. L. Sears & E. E. Swierstra (eds.), Research Highlights—1984. Agriculture Canada Research Station, Lethbridge, Alberta. 85 pp.

Shemanchuk, J. A. 1987. Resistance of cattle breeds to black-fly attack. P. 36. *In* L. J. L. Sears & D. L. Struble (eds.), Research Highlights—1985–1986. Agriculture Canada Research Station, Lethbridge, Alberta. 119 pp.

Shemanchuk, J. A. 1988 ["1987"]. Host-seeking behavior and host preference of *Simulium arcticum*. Pp. 250–260. *In* K. C. Kim & R. W. Merritt (eds.), Black flies: ecology, population management, and annotated world list. Pennsylvania State University, University Park. 528 pp. [Actual publication date was 1 March 1988.]

Shemanchuk, J. A., & J. R. Anderson. 1980. Bionomics of biting flies in the agricultural area of central Alberta. Pp. 207–214. *In* W. O. Haufe & G. C. R. Croome (eds.), Control of black flies in the Athabasca River. Alberta Environment Technical Report. Alberta Environment, Pollution Control Division, Edmonton. 241 pp.

Shemanchuk, J. A., & K. R. Depner. 1971. Seasonal distribution and abundance of females of *Simulium aureum* Fries (Simuliidae: Diptera) in irrigated areas of Alberta. Journal of Medical Entomology 8:29–33.

Shemanchuk, J. A., & R. A. Humber 1978. *Entomophthora culicis* (Phycomycetes: Entomophthorales) parasitizing black fly adults (Diptera: Simuliidae) in Alberta. Canadian Entomologist 110:253–256.

Shemanchuk, J. A., & W. G. Taylor. 1983. Repellents protect cattle from black flies. Canada Agriculture 29(3–4):14–17. [In English and French.]

Shemanchuk, J. A., & W. G. Taylor. 1984. Protective action of fenvalerate, deltamethrin, and four stereoisomers of permethrin against black flies (*Simulium* spp.) attacking cattle. Pesticide Science 15:557–561.

Shewell, G. E. 1952. New Canadian black flies (Diptera: Simuliidae). I. Canadian Entomologist 84:33–42.

Shewell, G. E. 1955. Identity of the black fly that attacks ducklings and goslings in Canada (Diptera: Simuliidae). Canadian Entomologist 87:345–349.

Shewell, G. E. 1957. Interim report on distributions of the black flies (Simuliidae) obtained in the Northern Insect Survey. Defence Research Board Environmental Protection Technical Report No. 7. Canada Department of National Defence, Ottawa. 3 pp. + 47 maps.

Shewell, G. E. 1958. Classification and distribution of arctic and subarctic Simuliidae. Proceedings of the Tenth International Congress of Entomology, Montreal (1956) 1:635–643.

Shewell, G. E. 1959a. New Canadian black flies (Diptera: Simuliidae). II. Canadian Entomologist 91:83–87.

Shewell, G. E. 1959b. New Canadian black flies (Diptera: Simuliidae) III. Canadian Entomologist 91:686–697.

Shewell, G. E., & F. J. H. Fredeen. 1958. Two new black flies from Saskatchewan (Diptera: Simuliidae). Canadian Entomologist 90:733–738.

Shields, G. F. 1990. Interchange chromosomes in *Simulium nigricoxum* Stone Diptera: Simuliidae. Genome 33:683–685.

Shields, G. F., & W. S. Procunier. 1981. The utility of polytene chromosome analysis in identification of biting black flies. Proceedings of the Alaska Science Conference 32:151.

Shields, G. F., & W. S. Procunier. 1982. A cytological description of sibling species of *Simulium* (*Gnus*) *arcticum* (Diptera: Simuliidae). Polar Biology 1:181–192.

Shipitsina, N. K. 1962. Age-groups and comparative ecology of bloodsucking black flies (Diptera, family Simuliidae) near Krasnoyarsk. Meditzinskaya Parazitologiya 31:415–424. [In Russian; English summary.]

Shipitsina, N. K. 1963. Infestation of simuliids (Simuliidae, Diptera) with their parasites and its effect upon ovarian functioning. Zoologichesky Zhurnal 42:291–294. [In Russian; English summary.]

Shipp, J. L. 1983. Effectiveness of different traps for collecting adult black flies. Pp. 37–39. *In* L. J. L. Sears & T. G. Atkinson (eds.), Research Highlights—1982. Agriculture Canada Research Station, Lethbridge, Alberta. 124 pp.

Shipp, J. L. 1985a. Distribution of and notes on blackfly species (Diptera: Simuliidae) found in the major waterways of southern Alberta. Canadian Journal of Zoology 63:1823–1828.

Shipp, J. L. 1985b. Comparison of silhouette, sticky, and suction traps with and without dry-ice bait for sampling black flies (Diptera: Simuliidae) in central Alberta. Canadian Entomologist 117:113–117.

Shipp, J. L. 1985c. Evaluation of a portable $CO_2$ generator for sampling black flies. Journal of the American Mosquito Control Association 1:515–517.

Shipp, J. L. 1987. Diapause induction of eggs of *Simulium arcticum* Malloch (IIS-10.11) (Diptera: Simuliidae). Canadian Entomologist 119:497–499.

Shipp, J. L. 1988. Classification system for embryonic development of *Simulium arcticum* Malloch (IIS-10.11) (Diptera: Simuliidae). Canadian Journal of Zoology 66:274–276.

Shipp, J. L., & G. Byrtus. 1984. Effectiveness of plexiglass [*sic*] as a black fly egg sampler on the Athabasca River. Pp. 47–49. *In* L. J. L. Sears & D. B. Wilson (eds.), Research Highlights—1983. Agriculture Canada Research Station, Lethbridge, Alberta. 127 pp.

Shipp, J. L., & G. Grace. 1988. Microclimate and black-fly flight activity. Pp. 23–25. *In* L. J. L. Sears & P. A. O'Sullivan (eds.), Research Highlights—1987. Agriculture Canada Research Station, Lethbridge, Alberta. 70 pp.

Shipp, J. L., & E. G. Kokko. 1987. Mouthparts of two sibling black-fly species of *Simulium arcticum*. Pp. 39–40. *In* L. J. L. Sears & T. G. Atkinson (eds.), Research Highlights—1985–1986. Agriculture Canada Research Station, Lethbridge, Alberta. 119 pp.

Shipp, J. L., & W. S. Procunier. 1986. Seasonal occurrence of, development of, and the influences of selected environmental factors on the larvae of *Prosimulium* and *Simulium* species of blackflies (Diptera: Simuliidae) found in the rivers of southwestern Alberta. Canadian Journal of Zoology 64:1491–1499.

Shipp, J. L., & G. H. Whitfield. 1987a. Influence of temperature on embryonic development and egg hatching of *Simulium arcticum* Malloch IIS-10.11 (Diptera: Simuliidae). Environmental Entomology 16:683–686.

Shipp, J. L., & G. H. Whitfield. 1987b. Model to predict egg development of a black-fly species. Pp. 37–38. *In* L. J. L. Sears & T. G. Atkinson (eds.), Research Highlights—1985–1986. Agriculture Canada Research Station, Lethbridge, Alberta. 119 pp.

Shipp, J. L., B. W. Grace, & G. B. Schaalje. 1987. Effects of microclimate on daily flight activity of *Simulium arcticum* Malloch (Diptera: Simuliidae). International Journal of Biometeorology 31:9–20.

Shipp, J. L., B. Grace, & H. H. Janzen. 1988a. Influence of temperature and water vapour pressure on the flight activity of *Simulium arcticum* Malloch (Diptera: Simuliidae). International Journal of Biometeorology 32:242–246.

Shipp, J. L., J. F. Sutcliffe, & E. G. Kokko. 1988b. External ultrastructure of sensilla on the antennal flagellum of a female black fly, *Simulium arcticum* (Diptera: Simuliidae). Canadian Journal of Zoology 66:1425–1431.

Sibley, P. K., & N. K. Kaushik. 1991. Toxicity of microencapsulated permethrin to selected nontarget aquatic invertebrates. Archives of Environmental Contamination and Toxicology 20:168–176.

Siebold, C. T. von. 1848. Ueber die Fadenwürmer der Insecten. (Zweiter Nachtrag.). Entomologische Zeitung, Stettin 9:290–300.

Silva Figueroa, C. 1917. Dos nuevos simúlidos de Chile. Boletin del Museo Nacional de Chile 10:28–35.

Simmons, K. R. 1982. Quantitative sampling of preimaginal black flies (Diptera: Simuliidae) and drift ecology of *Simulium tuberosum* Lundstrom complex. M.S. thesis. University of Massachusetts, Amherst. 122 pp.

Simmons, K. R. 1985. Reproductive ecology and host-seeking behavior of the black fly, *Simulium venustum* Say (Diptera: Simuliidae). Ph.D. thesis. University of Massachusetts, Amherst. 204 pp.

Simmons, K. R., & J. D. Edman. 1978. Successful mating, oviposition, and complete generation rearing of the multivoltine black fly *Simulium decorum* (Diptera: Simuliidae) in the laboratory. Canadian Journal of Zoology 56:1223–1225.

Simmons, K. R., & J. D. Edman. 1981. Sustained colonization of the black fly *Simulium decorum* Walker (Diptera: Simuliidae). Canadian Journal of Zoology 59:1–7.

Simmons, K. R., & R. D. Sjogren. 1984. Black fly (Diptera: Simuliidae) problems and their control strategies in Minnesota. Bulletin of the Society of Vector Ecologists 9:21–22.

Simmons, K. R., J. D. Edman, & S. R. Bennett. 1989. Collection of blood-engorged black flies (Diptera: Simuliidae) and identification of their source of blood. Journal of the American Mosquito Control Association 5:541–546.

Simpson, C. F., D. W. Anthony, & F. Young. 1956. Parasitism of adult turkeys in Florida by *Leucocytozoon smithi* (Lavern and Lucet). Journal of the American Veterinary Medical Association 129:573–576.

Simpson, J. A., & E. S. C. Weiner (preparers). 1989. The Oxford English dictionary, vol. II. Clarendon Press, Oxford, U.K. 1078 pp.

Singh, M. P., & S. M. Smith. 1985. Emergence of blackflies (Diptera: Simuliidae) from a small forested stream in Ontario. Hydrobiologia 122:129–135.

Sites, R. W., & B. J. Nichols. 1990. Life history and descriptions of immature stages of *Ambrysus lunatus lunatus* (Hemiptera: Naucoridae). Annals of the Entomological Society of America 83:800–808.

Skidmore, L. V. 1931. *Leucocytozoon smithi* infection in turkeys and its transmission by *Simulium occidentale* Townsend. Journal of Parasitology 18:130.

Skidmore, L. V. 1932. *Leucocytozoon smithi* infection in turkeys and its transmission by *Simulium occidentale* Townsend. Zentralblatt für Bakteriologie, Parasitenkunde und Infektionskrankheiten I (Originale) 125:329–335.

Skipwith, P. H. 1888. Formula for a buffalo gnat application. Insect Life 1:143.

Skuse, F. A. A. 1890 ["1891"]. Diptera of Australia. Proceedings of the Linnean Society of New South Wales (Second Series) 5:595–640 + 1 plate. [Actual publication date was 16 December 1890.]

Slaymaker, A. 1998. Diversity of trichomycete species and prevalence in aquatic hosts in two north-eastern Kansas streams. M.A. thesis. University of Kansas, Lawrence. 93 pp.

Sleeper, F. 1975. Visit from a small monster. Sports Illustrated 43(8):46–49.

Smart, J. 1934. Notes on the biology of *Simulium pictipes* Hagen. Canadian Entomologist 66:62–66.

Smart, J. 1935. The internal anatomy of the black-fly, *Simulium ornatum* Mg. Annals of Tropical Medicine and Parasitology 29:161–170.

Smart, J. 1944a. The British Simuliidae with keys to the species in the adult, pupal and larval stages. Freshwater Biological Association Scientific Publication 9:1–57.

Smart, J. 1944b. Notes on Simuliidae (Diptera). II. Proceedings of the Royal Entomological Society of London (B) 13:131–136.

Smart, J. 1945. The classification of the Simuliidae (Diptera). Transactions of the Royal Entomological Society of London 95:463–528.

Smart, J. 1946. A new name in Simuliidae (Diptera). Entomologist 79:22.

Smart, R. A. 1952. Biting insects in the arctic and sub-arctic. Journal of the Royal Army Medical Corps 98:8–14.

Smith, C. N., W. C. McDuffie, & R. I. Sailer. 1952. Insects affecting man and animals in Alaska—faunal and control studies. Proceedings of the Alaskan Science Conference 3:199–200.

Smith, J. B. 1890. A contribution toward a knowledge of the mouth parts of the Diptera. Transactions of the American Entomological Society 17:319–339.

Smith, J. J. B., & W. G. Friend. 1982. Feeding behaviour in response to blood fractions and chemical phagostimulants in the black-fly, *Simulium venustum*. Physiological Entomology 7:219–226.

Smith, J. P., & W. F. Rapp. 1985. Black flies (Diptera: Simuliidae) in Nebraska. Proceedings of the Annual Meeting of the New Jersey Mosquito Control Association 72:135–136.

Smith, M. L. 2002. Molecular systematics of the Nearctic black fly subgenus *Simulium* s. s. (Diptera: Simuliidae: *Simulium* s. l.). M.S. thesis. University of Toronto, Toronto. 172 pp.

Smith, R. C., E. G. Kelly, G. A. Dean, H. R. Bryson, & R. L. Parker. 1943. Common insects of Kansas. Kansas State Board of Agriculture Report No. 62. Kansas State Printing Plant, Topeka. 440 pp. [Also issued as Gates, D. E., & L. L. Peters. 1962. Insects in Kansas. Kansas State University, Extension Division, Manhattan, Kans. 307 pp.]

Smith, R. N., S. L. Cain, S. H. Anderson, J. R. Dunk, & E. S. Williams. 1998. Blackfly-induced mortality of nestling red-tailed hawks. Auk 115:368–375.

Smith, S. M. 1966. Observations on some mechanisms of host finding and host selection in the Simuliidae and Tabanidae (Diptera). M.S. thesis. McMaster University, Hamilton, Ontario. 144 pp.

Smith, S. M. 1969. The black fly, *Simulium venustum*, attracted to the turtle, *Chelydra serpentina*. Entomological News 80:107–108.

Smith, S. M. 1970. The biting flies of the Baker Lake region, Northwest Territories [Diptera: Culicidae and Simuliidae]. Ph.D. thesis. University of Manitoba, Winnipeg. 280 pp.

Smith, S. M., & A. Hayton. 1995. The gonotrophic-age structure of a population of the *Simulium venustum* complex (Diptera: Simuliidae) in Algonquin Park, Ontario. Great Lakes Entomologist 28:185–198.

Smock, L. A. 1988. Life histories, abundance and distribution of some macroinvertebrates from a South Carolina, USA coastal plain stream. Hydrobiologia 157:193–208.

Smock, L. A., & C. E. Roeding. 1986. The trophic basis of production of the macroinvertebrate community of a southeastern U. S. A. blackwater stream. Holarctic Ecology 9:165–174.

Smock, L. A., E. Gilinsky, & D. L. Stoneburner. 1985. Macroinvertebrate production in a southeastern United States blackwater stream. Ecology 66:1491–1503.

Snider, E. C. 1958. 1957 black fly control at Mont Apica, Que. Pulp and Paper Magazine of Canada 59:93–106.

Snoddy, E. L. 1966. The black flies (Diptera: Simuliidae) of Alabama and some aspects of their ecology. Ph.D. thesis. Auburn University, Auburn, Ala. 129 pp.

Snoddy, E. L. 1967. The common grackle, *Quiscalus quiscula quiscula* L. (Aves: Icteridae), a predator of *Simulium pictipes* Hagen larvae (Diptera: Simuliidae). Journal of the Georgia Entomological Society 2:45–46.

Snoddy, E. L. 1968. Simuliidae, Ceratopogonidae, and Chloropidae as prey of *Oxybelus emarginatum*. Annals of the Entomological Society of America 61:1029–1030.

Snoddy, E. L. 1971. *Simulium* (*Phosterodoros*) *podostemi*, a new species of black fly (Diptera: Simuliidae) from central Georgia. Journal of the Georgia Entomological Society 6:196–199.

Snoddy, E. L. 1976. *Simulium* (*Phosterodoros*) *lakei*, a new species of black fly (Diptera: Simuliidae) from the eastern United States. Journal of the Georgia Entomological Society 11:173–176.

Snoddy, E. L., & L. S. Bauer. 1978. *Simulium* (*Phosterodoros*) *penobscotensis*, a new species of black fly (Diptera: Simuliidae) from Maine, U.S.A. Journal of Medical Entomology 14:579–581.

Snoddy, E. L., & R. J. Beshear. 1968. *Simulium* (*Simulium*) *taxodium*, a new species of black fly (Diptera: Simuliidae) from southwestern Georgia. Journal of the Georgia Entomological Society 3:123–125.

Snoddy, E. L., & J. R. Chipley. 1971. Bacteria from the intestinal tract of *Simulium underhilli* (Diptera: Simuliidae) as a possible index to water pollution. Annals of the Entomological Society of America 64:1467–1468.

Snoddy, E. L., & K. L. Hays. 1966. A carbon dioxide trap for Simuliidae (Diptera). Journal of Economic Entomology 59:242–243.

Snoddy, E. L., & R. Noblet. 1976. Identification of the immature black flies (Diptera: Simuliidae) of the southeastern U. S. with some aspects of the adult role in transmission of *Leucocytozoon smithi* to turkeys. South Carolina Agricultural Experiment Station Technical Bulletin 1057:1–58.

Snodgrass, R. E. 1944. The feeding apparatus of biting and sucking insects affecting man and animals. Smithsonian Miscellaneous Collections 104(7):1–113.

Snow, W. E., E. Pickard, & C. M. Jones. 1958a. Observations on the activity of *Culicoides* and other Diptera in Jasper County, South Carolina. Mosquito News 18:18–21.

Snow, W. E., E. Pickard, & J. B. Moore. 1958b. Observations on blackflies (Simuliidae) in the Tennessee River Basin. Journal of the Tennessee Academy of Science 33:5–23.

Snyder, T. P. 1981. Electrophoretic identification of siblings in the *Simulium venustum/verecundum* species complex. Pp. 95–102. *In* M. W. Stock (ed.), Application of genetics and cytology in insect systematics and evolution. Forest, Wildlife and Range Experiment Station, University of Idaho, Moscow. 152 pp.

Snyder, T. P. 1982. Eletrophoretic [*sic*] characterizations of black flies in the *Simulium venustum* and *verecundum* species complexes (Diptera: Simuliidae). Canadian Entomologist 114:503–507.

Snyder, T. P., & D. G. Huggins. 1980. Kansas black flies (Diptera: Simuliidae) with notes on distribution and ecology. Technical Publication of the State Biological Survey of Kansas 9:30–34.

Snyder, T. P., & M. C. Linton. 1983. Electrophoretic and morphological separation of *Prosimulium fuscum* and *P. mixtum* larvae (Diptera: Simuliidae). Canadian Entomologist 115:81–87.

Snyder, T. P., & M. C. Linton. 1984. Population structure in black flies: allozymic and morphological estimates for *Prosimulium mixtum* and *P. fuscum* (Diptera: Simuliidae). Evolution 38:942–956.

Sohn, U. I. 1973. Molecular hybridization of iodinated blackfly DNA. M.S. thesis. University of Toronto, Toronto. 44 pp.

Sohn, U. I., K. H. Rothfels, & N. A. Straus. 1975. DNA: DNA hybridization studies in black flies. Journal of Molecular Evolution 5:75–85.

Sommerman, K. M. 1953. Identification of Alaskan black fly larvae (Diptera, Simuliidae). Proceedings of the Entomological Society of Washington 55:258–273.

Sommerman, K. M. 1958. Two new species of Alaskan *Prosimulium*, with notes on closely related species (Diptera, Simuliidae). Proceedings of the Entomological Society of Washington 60:193–202.

Sommerman, K. M. 1962a ["1961"]. *Prosimulium doveri*, a new species from Alaska, with keys to related species (Diptera: Simuliidae). Proceedings of the Entomological Society of Washington 63:225–235. [Actual publication date was 8 January 1962.]

Sommerman, K. M. 1962b. Notes on two species of *Oreogeton*, predaceous on black fly larvae, Diptera: Empididae and Simuliidae. Proceedings of the Entomological Society of Washington 64:123–129.

Sommerman, K. M. 1964. *Prosimulium esselbaughi* n. sp., the Alaskan *P. hirtipes* 2 (Diptera: Simuliidae). Proceedings of the Entomological Society of Washington 66:141–145.

Sommerman, K. M. 1977. Biting fly-arbovirus probe in interior Alaska (Culicidae) (Simuliidae)-(SSH: *California* complex) (Northway: Bunyamwera group). Mosquito News 37:90–103.

Sommerman, K. M., R. I. Sailer, & C. O. Esselbaugh. 1955. Biology of Alaskan black flies (Simuliidae, Diptera). Ecological Monographs 25:345–385.

Spalatin, J., L. Karstad, J. R. Anderson, L. Lauerman, & R. P. Hanson. 1961. Natural and experimental infections in Wisconsin turkeys with the virus of eastern encephalitis. Zoonoses Research 1:29–48.

Speir, J. A. 1969. Biological and ecological aspects of the black flies of the Marys River drainage system (Diptera: Simuliidae). M.S. thesis. Oregon State University, Corvallis. 80 pp.

Speir, J. A. 1975. The ecology and production dynamics of four black fly species (Diptera: Simuliidae) in western Oregon streams. Ph.D. thesis. Oregon State University, Corvallis. 297 pp.

Speir, J. A., & N. H. Anderson. 1974. Use of emergence data for estimating annual production of aquatic insects. Limnology and Oceanography 19:154–156.

Speiser, P. 1904. Zur Nomenclatur blutsaugender Dipteren Amerikas. Insekten-Börse 21:148.

Sprague, V., J. J. Becnel, & E. I. Hazard. 1992. Taxonomy of Phylum Microspora. Critical Reviews in Microbiology 18:285–395.

Stains, G. S. 1941. A taxonomic and distributional study of Simuliidae of western United States. M.S. thesis. Utah State Agricultural College [= Utah State University], Logan. 42 pp.

Stains, G. S., & G. F. Knowlton. 1940. Three new western Simuliidae (Diptera). Annals of the Entomological Society of America 33:77–80.

Stains, G. S., & G. F. Knowlton. 1943. A taxonomic and distributional study of Simuliidae of western United States. Annals of the Entomological Society of America 36:259–280.

Stallings, T., M. S. Cupp, & E. W. Cupp. 2002. Orientation of *Onchocerca lienalis* Stiles (Filarioidea: Onchocercidae) microfilariae to black fly saliva. Journal of Medical Entomology 39:908–914.

Stamps, D. 1985. The black fly: airborne scourge of stream and forest. Minnesota Volunteer 48(281):49–53.

State Plant Board of Mississippi. 1933. Buffalo gnats (Simuliidae). United States Department of Agriculture Insect Pest Survey Bulletin 13:95.

Steele, E. J., & G. P. Noblet. 2001. Gametogenesis, fertilization and ookinete differentiation of *Leucocytozoon smithi*. Journal of Eukaryotic Microbiology 48:118–125.

Steingrímsson, S. Ó., & G. M. Gíslason. 2002. Body size, diet and growth of land-locked brown trout, *Salmo trutta*, in the subarctic River Laxá, North-East Iceland. Environmental Biology of Fishes 63:417–426.

Stokes, J. H. 1914. A clinical, pathological and experimental study of the lesions produced by the bite of the "black fly" (*Simulium venustum*). Journal of Cutaneous Diseases 32:751–769, 830–856.

Stoltz, R. L. 1982. Blackfly larval control in irrigation canals with *Bacillus thuringiensis* var *israelensis*, 1981. Insecticide and Acaricide Tests 7:254–255.

Stone, A. 1941. A restudy of *Parasimulium furcatum* Malloch (Diptera, Simuliidae). Proceedings of the Entomological Society of Washington 43:146–149.

Stone, A. 1948. *Simulium virgatum* Coquillett and a new related species (Diptera: Simuliidae). Journal of the Washington Academy of Sciences 38:399–404.

Stone, A. 1949a. The identity of two Nearctic Simuliidae (Diptera). Bulletin of the Brooklyn Entomological Society 44:138–140.

Stone, A. 1949b. A new genus of Simuliidae from Alaska (Diptera). Proceedings of the Entomological Society of Washington 51:260–267.

Stone, A. 1952. The Simuliidae of Alaska (Diptera). Proceedings of the Entomological Society of Washington 54:69–96.

Stone, A. 1962a. Notes on the types of some Simuliidae (Diptera) described by Enderlein. Annals of the Entomological Society of America 55:206–209.

Stone, A. 1962b. A new record for *Parasimulium furcatum* Malloch (Diptera: Simuliidae). Proceedings of the Entomological Society of America 64:174.

Stone, A. 1963a. An annotated list of genus-group names in the family Simuliidae (Diptera). United States Department of Agriculture Technical Bulletin 1284:1–28.

Stone, A. 1963b. A new *Parasimulium* and further records for the type species (Diptera: Simuliidae). Bulletin of the Brooklyn Entomological Society 58:127–129.

Stone, A. 1964. Guide to the insects of Connecticut. Part VI. The Diptera or true flies of Connecticut. Ninth fascicle. Simuliidae and Thaumaleidae. State Geological and Natural History Survey of Connecticut Bulletin 97:1–126.

Stone, A. 1965a. Family Simuliidae. Pp. 181–189. *In* A. Stone, C. W. Sabrosky, W. W. Wirth, R. H. Foote, & J. R. Coulson (eds.), A catalog of the Diptera of America north of Mexico. United States Department of Agriculture Handbook No. 276. Agricultural Research Service, Washington, D.C. 1696 pp.

Stone, A. 1965b. The relationships of Nearctic and Palaearctic Simuliidae (Diptera). Proceedings of the Twelfth International Congress of Entomology, London (1964):762.

Stone, A., & M. M. Boreham. 1965. A new species of *Simulium* from the southwestern United States (Diptera: Simuliidae). Journal of Medical Entomology 2:164–170.

Stone, A., & G. R. DeFoliart. 1959. Two new black flies from the western United States (Diptera, Simuliidae). Annals of the Entomological Society of America 52:394–400.

Stone, A., & H. Jamnback. 1955. The black flies of New York State (Diptera: Simuliidae). New York State Museum Bulletin 349:1–144.

Stone, A., & B. V. Peterson. 1958. *Simulium defoliarti*, a new black fly from the western United States (Diptera, Simuliidae). Bulletin of the Brooklyn Entomological Society 53:1–6.

Stone, A., & E. L. Snoddy. 1969. The black flies of Alabama (Diptera: Simuliidae). Auburn University Agricultural Experiment Station Bulletin 390:1–93.

Stoneburner, D. L., & L. A. Smock. 1979. Seasonal fluctuations of macroinvertebrate drift in a South Carolina piedmont stream. Hydrobiologia 63:49–56.

Strand, M. A., C. H. Bailey, & M. Laird. 1977. Pathogens of Simuliidae (blackflies). Pp. 213–237. *In* D. W. Roberts & M. A. Strand (eds.), Pathogens of medically important arthropods. Bulletin of the World Health Organization 55(Suppl. 1). 419 pp.

Strickland, E. H. 1911. Some parasites of *Simulium* larvae and their effects on the development of the host. Biological Bulletin of the Marine Biological Laboratory, Woods Hole 21:302–339.

Strickland, E. H. 1913a. Further observations on the parasites of *Simulium* larvae. Journal of Morphology 24:43–105.

Strickland, E. H. 1913b. Some parasites of *Simulium* larvae and their possible economic value. Canadian Entomologist 45:405–413.

Strickland, E. H. 1938. An annotated list of the Diptera (flies) of Alberta. Canadian Journal of Research (D) 16:175–219.

Strobl, P. G. 1898. Diptera fauna of Bosnia, Herzegovina and Dalmatia. Glasnik Zemaljskog Muzeja u Bosni i Hercegovini 10:561–616. [In Russian.]

Stuart, A. E. 1995. Behavioural characters in phylogenetics: a case study using black fly (Diptera: Simuliidae) cocoon spinning behaviour. M.S. thesis. Brock University, St. Catharines, Ontario. 157 pp.

Stuart, A. E., & F. F. Hunter. 1995. A redescription of the cocoon-spinning behaviour of *Simulium vittatum* (Diptera Simuliidae). Ethology, Ecology, and Evolution 7:363–377.

Stuart, A. E., & F. F. Hunter. 1998a. Phylogenetic placement of *Ectemnia*, an autapomorphic black fly (Diptera: Simuliidae), using behavioural characters. Canadian Journal of Zoology 76:1942–1948.

Stuart, A. E., & F. F. Hunter. 1998b. End-products of behaviour versus behavioural characters: a phylogenetic investigation of pupal cocoon construction and form in some North American black flies (Diptera: Simuliidae). Systematic Entomology 23:387–398.

Stuart, A. E., F. F. Hunter, & D. C. Currie. 2002. Using behavioural characters in phylogeny reconstruction. Ethology, Ecology & Evolution 14:129–139.

Surcouf, J. M. R., & R. Gonzalez-Rincones. 1911. Essai sur les diptères vulnérants du Venezuela: matériaux pour servir à l'étude des diptères piqueurs et suceurs de sang de l'Amérique intertropicale. Part 1. A. Maloine, Paris. 320 pp.

Sutcliffe, J. F. 1975. Studies on the sensory receptors and behaviour of selected black flies (Simuliidae: Diptera). M.S. thesis. University of Toronto, Toronto. 278 pp.

Sutcliffe, J. F. 1978. Feeding behaviour in *Simulium venustum* (Diptera: Simuliidae), sensory aspects and mouthpart mechanics. Ph.D. thesis. University of Toronto, Toronto. 198 pp.

Sutcliffe, J. F. 1985. Anatomy of membranous mouthpart cuticles and their roles in feeding in black flies (Diptera: Simuliidae). Journal of Morphology 186:53–68.

Sutcliffe, J. F. 1986. Black fly host location: a review. Canadian Journal of Zoology 64:1041–1053.

Sutcliffe, J. F. 1987. Distance orientation of biting flies to their hosts. Insect Science and Its Application 8:611–616.

Sutcliffe, J. F., & S. B. McIver. 1974. Head appendages and wing cleaning in blackflies (Diptera: Simuliidae). Annals of the Entomological Society of America 67:450–452.

Sutcliffe, J. F., & S. B. McIver. 1975. Artificial feeding of simuliids (*Simulium venustum*): factors associated with probing and gorging. Experientia 31:694–695.

Sutcliffe, J. F., & S. B. McIver. 1976. External morphology of sensilla on the legs of selected black fly species (Diptera: Simuliidae). Canadian Journal of Zoology 54:1779–1787.

Sutcliffe, J. F., & S. B. McIver. 1979. Experiments on biting and gorging behaviour in the black fly, *Simulium venustum*. Physiological Entomology 4:393–400.

Sutcliffe, J. F., & S. B. McIver. 1982. Innervation and structure of mouth part sensilla in females of the black fly *Simulium venustum* (Diptera: Simuliidae). Journal of Morphology 171:245–258.

Sutcliffe, J. F., & S. B. McIver. 1984. Mechanics of blood-feeding in black flies (Diptera, Simuliidae). Journal of Morphology 180:125–144.
Sutcliffe, J. F., & S. B. McIver. 1987. Fine structure of tarsal sensilla of male and female *Simulium vittatum* (Diptera: Simuliidae). Journal of Morphology 192:13–26.
Sutcliffe, J. F., & J. A. Shemanchuk. 1993. Investigation of attractants of the cattle-biting black fly of the Athabasca region of Alberta. Alberta Agriculture Farming for the Future Final Report, No. 91-0841. Alberta Agriculture, Edmonton, Alberta. 73 pp.
Sutcliffe, J. F., J. L. Shipp, & E. G. Kokko. 1987. Ultrastructure of the palpal bulb sensilla of the black fly *Simulium arcticum* (Diptera: Simuliidae). Journal of Medical Entomology 24:324–331.
Sutcliffe, J. F., E. G. Kokko, & J. L. Shipp. 1990. Transmission electron microscopic study of antennal sensilla of the female black fly, *Simulium arcticum* (IIL-3; IIS-10.11) (Diptera: Simuliidae). Canadian Journal of Zoology 68:1443–1453.
Sutcliffe, J. F., J. A. Shemanchuk, & D. B. McKeown. 1994. Preliminary survey of odours that attract the black fly, *Simulium arcticum* (Malloch) (IIS-10.11) (Diptera: Simuliidae) to its cattle hosts in the Athabasca region of Alberta. Insect Science and Its Application 15:487–494.
Sutcliffe, J. F., D. J. Steer, & D. Beardsall. 1995. Studies of host location behaviour in the black fly *Simulium arcticum* (IIS-10.11) (Diptera: Simuliidae): aspects of close range trap orientation. Bulletin of Entomological Research 85:415–424.
Sutherland, D. W. S., & R. F. Darsie Jr. 1960a. A report on the blackflies (Simuliidae) of Delaware. Part I. Record of Delaware species and an introduction to a survey of the western branches of the Christiana River, New Castle County. Bulletin of the Brooklyn Entomological Society 55:46–52.
Sutherland, D. W. S., & R. F. Darsie Jr. 1960b. A report on the blackflies (Simuliidae) of Delaware. Part II. Description and discussion of blackfly habitats. Bulletin of the Brooklyn Entomological Society 55:53–61.
Sutton, J. M. 1929. Buffalo gnats (Simuliidae). United States Department of Agriculture Insect Pest Survey Bulletin 9:94.
Swabey, Y. H., C. F. Schenk, & G. L. Parker. 1967. Evaluation of two organophosphorus compounds as blackfly larvicides. Mosquito News 27:149–155.
Swales, W. E. 1936. Two important diseases of ducks in Quebec. Journal of Agriculture and Horticulture 39:13, 40–41, 43.
Sweeney, A. W., & D. W. Roberts. 1983. Laboratory evaluation of the fungus *Culicinomyces clavosporus* for control of blackfly (Diptera: Simuliidae) larvae. Environmental Entomology 12:774–778.
Swenk, M. H. 1922. Turkey gnat (*Simulium meridionale* Riley). United States Department of Agriculture Insect Pest Survey Bulletin 2:103.
Swenk, M. H. 1938. A buffalo gnat (*Simulium vittatum* Zett.). United States Department of Agriculture Insect Pest Survey Bulletin 18:137.
Swenk, M. H., & F. E. Mussehl. 1928. The insects and mites injurious to poultry in Nebraska and their control. University of Nebraska Agricultural Experiment Station Circular 37:1–31.
Syme, P. D. 1957. Three new Ontario black flies of the genus *Prosimulium* (Diptera: Simuliidae) and observations on their ecology. M.S. thesis. McMaster University, Hamilton, Ontario. 94 pp.
Syme, P. D., & D. M. Davies. 1958. Three new Ontario black flies of the genus *Prosimulium* (Diptera: Simuliidae). Part I. Descriptions, morphological comparisons with related species, and distribution. Canadian Entomologist 90:697–719.
Takahasi, H. 1940. Description of five new species of Simuliidae from Manchoukuo (studies on Simuliidae of Manchoukuo, I). Insecta Matsumurana 15 (1/2):63–74.
Takaoka, H. 1980. Pathogens of blackfly larvae in Guatemala and their influence on natural populations of three species of onchocerciasis vectors. American Journal of Tropical Medicine and Hygiene 29:467–472.
Takaoka, H., & P. H. Adler. 1997. A new subgenus, *Simulium* (*Daviesellum*), and a new species, *S.* (*D.*) *courtneyi*, (Diptera: Simuliidae) from Thailand and Peninsular Malaysia. Japanese Journal of Tropical Medicine and Hygiene 25:17–27.
Takaoka, H., & D. M. Davies. 1996. The black flies (Diptera: Simuliidae) of Java, Indonesia. Bishop Museum Bulletin in Entomology 6:1–81.
Takaoka, H., & C. Kuvangkadilok. 1999. Four new species of black flies (Diptera: Simuliidae) from Thailand. Japanese Journal of Tropical Medicine and Hygiene 27:497–509.
Takaoka, H., & K. Saito. 2002. Description of a new species of *Simulium* (*Simulium*) from Hokkaido, Japan (Diptera: Simuliidae). Japanese Journal of Tropical Medicine and Hygiene 30:311–317.
Takaoka, H., & H. Takahasi. 1982. A new species of black fly (Diptera: Simuliidae) from upland areas of Guatemala. Journal of Medical Entomology 19:63–67.
Tanaka, T. 1924. The alimentary canal of the larva of *Simulium pictipes* Hagen. M.S. thesis. Cornell University, Ithaca, N.Y. 77 pp. +7 plates.
Tanaka, T. 1934. The cytological studies of the alimentary canal of *Simulium pictipes* Hagen: its structure and metamorphosis. Bulletin of the Kagoshima Imperial College of Agriculture and Forestry (Papers on the 25th Anniversary of Kagoshima Advanced Agricultural School) 11:813–912 + 23 plates.
Tang, J., L. Toè, C. Back, & T. R. Unnasch. 1995a. Mitochondrial alleles of *Simulium damnosum sensu lato* infected with *Onchocerca volvulus*. International Journal for Parasitology 25:1251–1254.
Tang, J., L. Toè, C. Back, P. A. Zimmerman, K. Pruess, & T. R. Unnasch. 1995b. The *Simulium damnosum* species complex: phylogenetic analysis and molecular identification based upon mitochondrially encoded gene sequences. Insect Molecular Biology 4:79–88.
Tang, J., K. Pruess, E. W. Cupp, & T. R. Unnasch. 1996a. Molecular phylogeny and typing of blackflies (Diptera: Simuliidae) that serve as vectors of human or bovine onchocerciasis. Medical and Veterinary Entomology 10:228–234.
Tang, J., K. Pruess, & T. R. Unnasch. 1996b. Genotyping North American black flies by means of mitochondrial ribosomal RNA sequences. Canadian Journal of Zoology 74:39–46.
Tang, J., J. K. Moulton, K. Pruess, E. W. Cupp, & T. R. Unnasch. 1998. Genetic variation in North American black flies in the subgenus *Psilopelmia* (*Simulium*: Diptera: Simuliidae). Canadian Journal of Zoology 76:205–211.
Tarrant, C. A. 1984. The vertical transmission of black fly (Diptera: Simuliidae) parasites. Ph.D. thesis. Cornell University, Ithaca, N.Y. 120 pp.
Tarrant, C. A., & R. Soper. 1986. Evidence for the vertical transmission of *Coelomycidium simulii* (Myceteae (Fungi): Chytridiomycetes). P. 212. *In* R. A. Samson, J. M. Vlak, & D. Peters (eds.), Fundamental and applied aspects of invertebrate pathology, Foundation of the Fourth International Colloquium of Invertebrate Pathology. Wageningen, The Netherlands. 711 pp.
Tarrant, C. [A.], S. Moobola, G. Scoles, & E. W. Cupp. 1983. Mating and oviposition of laboratory-reared *Simulium vittatum* (Diptera: Simuliidae). Canadian Entomologist 115:319–323.
Tarrant, C. A., G. Scoles, & E. W. Cupp. 1987. Techniques for inducing oviposition in *Simulium vittatum* (Diptera: Simuliidae) and for rearing sibling cohorts of simuliids. Journal of Medical Entomology 24:694–695.
Tarshis, I. B. 1965a. A simple method for the collection of black fly larvae. Bulletin of the Wildlife Disease Association 1:8.

Tarshis, I. B. 1965b. Procurement and shipment of black fly eggs. Bulletin of the Wildlife Disease Association 1:8–9.
Tarshis, I. B. 1966. A method of shipping live larvae of *Simulium vittatum* long distances (Diptera: Simuliidae). Annals of the Entomological Society of America 59:866–867.
Tarshis, I. B. 1968a. Use of fabrics in streams to collect black fly larvae. Annals of the Entomological Society of America 61:960–961.
Tarshis, I. B. 1968b. Collecting and rearing black flies. Annals of the Entomological Society of America 61:1072–1083.
Tarshis, I. B. 1971. Individual black fly rearing cylinders (Diptera: Simuliidae). Annals of the Entomological Society of America 64:1192–1193.
Tarshis, I. B. 1972. The feeding of some ornithophilic black flies (Diptera: Simuliidae) in the laboratory and their role in the transmission of *Leucocytozoon simondi*. Annals of the Entomological Society of America 65:842–848.
Tarshis, I. B. 1973. Studies on the collection, rearing, and biology of the blackfly (*Cnephia ornithophilia*). United States Fish and Wildlife Service Special Scientific Report—Wildlife No. 165. U.S. Department of the Interior, Fish and Wildlife Service, Washington, D.C. 16 pp.
Tarshis, I. B. 1976. Further laboratory studies on the rearing and feeding of *Cnephia ornithophilia* (Diptera: Simuliidae) and the transmission of *Leucocytozoon simondi* by this black fly. Journal of Medical Entomology 13:337–341.
Tarshis, I. B. 1978. Black flies (Family Simuliidae). Pp. 11–18. *In* R. A. Bram (ed.), Surveillance and collection of arthropods of veterinary importance. United States Department of Agriculture Handbook No. 518. Agricultural Research Service, Washington, D.C. 125 pp.
Tarshis, I. B., & T. R. Adkins Jr. 1971. Equipment for transporting live black fly larvae (Diptera: Simuliidae). Annals of the Entomological Society of America 64:1194–1195.
Tarshis, I. B., & C. M. Herman. 1965. Is *Cnephia invenusta* (Walker) a possible important vector of *Leucocytozoon* in Canada geese? Bulletin of the Wildlife Disease Association 1:10–11.
Tarshis, I. B., & W. Neil. 1970. Mass movement of black fly larvae on silken threads (Diptera: Simuliidae). Annals of the Entomological Society of America 63:607–610.
Tarshis, I. B., & J. N. Stuht. 1970. Two species of Simuliidae (Diptera), *Cnephia ornithophilia* and *Prosimulium vernale*, from Maryland. Annals of the Entomological Society of America 63:587–590.
Tartar, A., D. G. Boucias, B. J. Adams, & J. J. Becnel. 2002. Phylogenetic analysis identifies the invertebrate pathogen *Helicosporidium* sp. as a green alga (Chlorophyta). International Journal of Systematic and Evolutionary Microbiology 52:273–279.
Taylor, M. R., S. T. Moss, & M. Ladle. 1996. Temporal changes in the level of infestation of *Simulium ornatum* Meigen (complex) (Simuliidae: Diptera) larvae by the endosymbiotic fungus *Harpella melusinae* Lichtwardt (Harpellales: Trichomycetes). Hydrobiologia 328:117–125.
Taylor, T. H. 1902. On the tracheal system of *Simulium*. Transactions of the Entomological Society of London 1902:701–716.
Teshima, I. E. 1970. A study of conditions for molecular hybridization in black flies. Ph.D. thesis. University of Toronto, Toronto. 118 pp.
Teshima, I. [E.]. 1972. DNA-DNA hybridization in blackflies (Diptera: Simuliidae). Canadian Journal of Zoology 50:931–940.
Teskey, H. J. 1960. Survey of insects affecting livestock in southwestern Ontario. Canadian Entomologist 92:531–544.
Teskey, H. J. 1981. Morphology and terminology—larvae. Pp. 65–88. *In* J. F. McAlpine, B. V. Peterson, G. E. Shewell, H. J. Teskey, J. R. Vockeroth, & D. M. Wood (eds.), Manual of Nearctic Diptera, vol. 1, Monograph No. 27. Research Branch, Agriculture Canada, Ottawa. 674 pp.
Tessler, S. 1991. Structure of a mountain black fly community (Diptera: Simuliidae) in a first-order drainage. Ph.D. thesis. Pennsylvania State University, University Park. 214 pp.
Thaxter, R. 1888. The Entomophthoreae of the United States. Memoirs of the Boston Society of Natural History 4:133–201.
Thomas, A. W. 1967. Effects of age and thermal acclimation on the selected temperature and thermal resistance of culicid and simuliid larvae (Diptera). M.S. thesis. McMaster University, Hamilton, Ontario. 150 pp.
Thomas, L. J. 1946. Black fly incubator-aerator cabinet. Science 103:21.
Thompson, B. H. 1987a. The use of algae as food by larval Simuliidae (Diptera) of Newfoundland streams. I. Feeding selectivity. Archiv für Hydrobiologie 76(Suppl.):425–442.
Thompson, B. H. 1987b. The use of algae as food by larval Simuliidae (Diptera) of Newfoundland streams. II. Digestion of algae, and environmental factors affecting feeding rates and the degree of digestion. Archiv für Hydrobiologie 76(Suppl.):443–457.
Thompson, B. H. 1987c. The use of algae as food by larval Simuliidae (Diptera) of Newfoundland streams. III. Growth of larvae reared on different algal and other foods. Archiv für Hydrobiologie 76(Suppl.):459–466.
Thompson, B. H. 1989. The ingestion and digestion of algal and other foods by larval black flies (Diptera: Simuliidae) of Newfoundland. Ph.D. thesis. Memorial University of Newfoundland, St. John's. 276 pp.
Thompson, B. H., & B. G. Adams. 1979. Laboratory and field trials using Altosid® insect growth regulator against black flies (Diptera: Simuliidae) of Newfoundland, Canada. Journal of Medical Entomology 16:536–546.
Thompson, F. C. 2001. The name of the type species of *Simulium* (Diptera: Simuliidae): an historical footnote. Entomological News 112:125–129.
Tietze, N. S., & M. S. Mulla. 1989. Species composition and distribution of blackflies (Diptera: Simuliidae) in the Santa Monica Mountains, California. Bulletin of the Society for Vector Ecology 14:253–261.
Tonnoir, A. L. 1925. Australasian Simuliidae. Bulletin of Entomological Research 15:213–255.
Townsend, C. H. T. 1891. A new *Simulium* from southern New Mexico. Psyche 6:106–107.
Townsend, C. H. T. 1893. On a species of *Simulium* from the Grand Canon [*sic*] of the Colorado. Transactions of the American Entomological Society 20:45–48.
Townsend, C. H. T. 1895. On the correlation of habit in nemoscerous and brachycerous Diptera between aquatic larvae and blood-sucking adult females. Journal of the New York Entomological Society 33:134–136.
Townsend, C. H. T. 1897. Diptera from the Lower Rio Grande or Tamaulipan region of Texas.—I. Journal of the New York Entomological Society 5:171–178.
Townsend, L. H., Jr. 1975. Feeding activity, a study of control measures, and a survey of black fly pests (Diptera: Simuliidae) of horses in Virginia. M.S. thesis. Virginia Polytechnic Institute and State University of Virginia, Blacksburg. 95 pp.
Townsend, L. H., Jr., & E. C. Turner Jr. 1976. Field evaluation of several chemicals against ear-feeding black fly pests of horses in Virginia. Mosquito News 36:182–186.
Townsend, L. H., Jr., E. C. Turner Jr., & W. A. Allen. 1977. An assessment of feeding damage threshold of *Simulium vittatum*, a black fly pest of horses in Virginia. Mosquito News 37:742–744.
Traoré, S. 1992. The black flies (Simuliidae: Diptera) of the Great Smoky Mountain National Park and vicinity. M.S. thesis. University of Tennessee, Knoxville. 51 pp.

Travis, B. V. 1949. Studies of mosquito and other biting-insect problems in Alaska. Journal of Economic Entomology 42:451–457.

Travis, B. V. 1966. The biology and control of black flies. Proceedings of the Annual Meeting of the New Jersey Mosquito Extermination Association 53:194–197.

Travis, B. V. 1967. Biology and control of biting flies in New York State. Cornell University Extension Bulletin 1186:1–40.

Travis, B. V. 1968. Some problems with simulated stream tests of black-fly larvicides. Proceedings of the Annual Meeting of the New Jersey Mosquito Extermination Association 55:129–134.

Travis, B. V., & D. Guttman. 1966. Additional tests with blackfly larvicides. Mosquito News 26:157–160.

Travis, B. V., & S. M. Schuchman. 1968. Tests (1967) with black fly larvicides. Journal of Economic Entomology 61:843–845.

Travis, B. V., & D. P. Wilton. 1965. A progress report on simulated stream tests of blackfly larvicides. Mosquito News 25:112–118.

Travis, B. V., D. L. Collins, G. DeFoliart, & H. Jamnback. 1951a. Strip spraying by helicopter to control blackfly larvae. Mosquito News 11:95–98.

Travis, B. V., A. L. Smith, & A. H. Madden. 1951b. Effectiveness of insect repellents against black flies. Journal of Economic Entomology 44:813–814.

Travis, B. V., D. Guttman, & R. R. Crafts. 1967. Tests (1966) with black fly larvicides. Proceedings of the Annual Meeting of the New Jersey Mosquito Extermination Association 54:49–53.

Travis, B. V., D. C. Kurtak, & L. C. Rimando. 1970. Tests (1969) with black fly larvicides. Proceedings of the Annual Meeting of the New Jersey Mosquito Extermination Association 57:106–110.

Travis, B. V., M. Vargas V., & J. C. Swartzwelder. 1974. Bionomics of black flies (Diptera: Simuliidae) in Costa Rica. I. Species biting man, with an epidemiological summary for the Western Hemisphere. Revista de Biologia Tropical 22:187–200.

Treille, E. 1882. Quelques considérations sur un insecte Diptère nuisible de Terre-Neuve. Archives de Médecine Navale 38:216–221.

Tucker, E. S. 1918. Occurrences of black flies in Louisiana during recent years. Transactions of the Kansas Academy of Science 29:65–75.

Twinn, C. R. 1933a. The blackfly, *Simulium venustum* Say, and a protozoon disease of ducks. Canadian Entomologist 65:1–3.

Twinn, C. R. 1933b. Blackfly control with larvicides. Scientific Agriculture 13:404.

Twinn, C. R. 1936a. The blackflies of eastern Canada (Simuliidae, Diptera). Part I. Canadian Journal of Research (D) 14:97–130.

Twinn, C. R. 1936b. The blackflies of eastern Canada (Simuliidae, Diptera). Part II. Canadian Journal of Research (D) 14:131–150.

Twinn, C. R. 1938. Blackflies from Utah and Idaho, with descriptions of new species (Simuliidae, Diptera). Canadian Entomologist 70:48–55.

Twinn, C. R. 1939. Notes on some parasites and predators of blackflies (Simuliidae, Diptera). Canadian Entomologist 71:101–105.

Twinn, C. R. 1950. Studies of the biology and control of biting flies in northern Canada. Arctic (Journal of the Arctic Institute of North America) 3:14–26.

Twinn, C. R. 1952. A review of studies of blood-sucking flies in northern Canada. Canadian Entomologist 84:22–28.

Twinn, C. R. 1954. Present trends and future needs of entomological research in northern Canada. Part II. Biology and control of biting flies. Arctic (Journal of the Arctic Institute of North America) 7:279–283.

Twinn, C. R., & D. G. Peterson. 1955. Control of black flies in Canada. Canada Department of Agriculture Publication No. 940. Entomology Division, Canada Department of Agriculture, Ottawa. 9 pp.

Twinn, C. R., B. Hocking, W. C. McDuffie, & H. F. Cross. 1948. A preliminary account of the biting flies at Churchill, Manitoba. Canadian Journal of Research (D) 26:334–357.

Uemoto, K., T. Okazawa, & O. Onishi. 1976. Revision of the genus *Prosimulium* Roubaud (Diptera, Simuliidae) of Japan. II. The subgenus *Distosimulium* Peterson, 1970. Japanese Journal of Sanitary Zoology 27:97–104.

Undeen, A. H. 1979a. Observations on the ovarian phycomycete of Newfoundland blackflies. Pp. 251–255. *In* J. Weiser (ed.), Progress in Invertebrate Pathology (Proceedings of the International Colloquium on Invertebrate Pathology, 1978). Prague, Czechoslovakia. 265 pp.

Undeen, A. H. 1979b. Simuliid larval midgut pH and its implications for control. Mosquito News 39:391–392.

Undeen, A. H. 1980. Control of black flies (Simuliidae) using *Bacillus thuringiensis* var. *israelensis*. Proceedings of the Florida Anti-Mosquito Association 51:55–58.

Undeen, A. H. 1981. Microsporida infections in adult *Simulium vittatum*. Journal of Invertebrate Pathology 38:426–427.

Undeen, A. H., & M. H. Colbo. 1980. The efficacy of *Bacillus thuringiensis* var. *israelensis* against blackfly larvae (Diptera: Simuliidae) in their natural habitat. Mosquito News 40:181–184.

Undeen, A. H., & L. A. Lacey. 1982. Field procedures for the evaluation of *Bacillus thuringiensis* var. *israelensis* (serotype 14) against black flies (Simuliidae) and nontarget organisms in streams. Miscellaneous Publications of the Entomological Society of America 12:25–30.

Undeen, A. H., & W. L. Nagel. 1978. The effect of *Bacillus thuringiensis* ONR-60A strain (Goldberg) on *Simulium* larvae in the laboratory. Mosquito News 38:524–527.

Undeen, A. H., & R. A. Nolan. 1977. Ovarian infection and fungal spore oviposition in the blackfly *Prosimulium mixtum*. Journal of Invertebrate Pathology 30:97–98.

Undeen, A. H., L. A. Lacey, & S. W. Avery. 1984a. A system for recommending dosage of *Bacillus thuringiensis* (H-14) for control of simuliid larvae in small streams based upon stream width. Mosquito News 44:553–559.

Undeen, A. H., J. Vàvra, and K. H. Rothfels. 1984b. The sex of larval simuliids infected with microsporidia. Journal of Invertebrate Pathology 43:126–127.

Underhill, G. W. 1939. Two simuliids found feeding on turkeys in Virginia. Journal of Economic Entomology 32:765–768.

Underhill, G. W. 1940. Some factors influencing feeding activity of simuliids in the field. Journal of Economic Entomology 33:915–917.

Underhill, G. W. 1944. Blackflies found feeding on turkeys in Virginia, (*Simulium nigroparvum* Twinn and *Simulium slossonae* Dyar and Shannon). Virginia Agricultural Experiment Station Technical Bulletin 94:1–32.

Usova, Z. V. 1961. The blackfly fauna of the Karelia and Murmansk Region (Diptera, Simuliidae). Izdatel'stvo Akademii Nauk SSSR, Leningrad [= St. Petersburg], Russia. 286 pp. [In Russian; English translation: 1964. Flies of Karelia and the Murmansk Region (Diptera, Simuliidae). Israel Program for Scientific Translations, Jerusalem. 268 pp.]

Usova, Z. V., & Yu. D. Bodrova. 1979. Discovery of the male of *Stegopterna sibirica decafilis* Rubz. (Diptera, Simuliidae) in Primorskiy Region. Pp. 46–47. *In* A. I. Cherepanov (ed.), Arthropods and helminths (New and little known fauna of Siberia). Izdatel'stvo "Nauka," Sibirskoe Otdelenie, Novosibirsk, Russia. 136 pp. [In Russian.]

Usova ["Ussova"], Z. V., & M. Reva. 2000. New genus of blackfly—*Gallipodus* (Diptera, Simuliidae) and its type species *G. raastadi* sp. n. International Journal of Dipterological Research 11:109–112.

Van Deveire, P. J. 1981. Flight and host seeking behaviour of adult black flies in the Souris River Area, Manitoba. M.S.

thesis. University of Manitoba, Winnipeg. 133 pp.

Vargas, L. 1942. Notas sobre la terminalia de algunos simúlidos de Mexico. *S.* (*E.*) *paynei* n. n. Vargas, 1942. Revista del Instituto de Salubridad y Enfermedades Tropicales 3:229–249 + 4 plates.

Vargas, L. 1943a. Nombres y datos nuevos de simúlidos del Nuevo Mundo. Revista de la Sociedad Mexicana de Historia Natural 4:135–146.

Vargas, L. 1943b. Nuevos datos sobre simúlidos mexicanos (Dipt. Simulidae [*sic*]). Revista del Instituto de Salubridad y Enfermedades Tropicales 4:359–370.

Vargas, L. 1945a. Cuatro nuevas especies y otros datos sobre simúlidos de Mexico. Revista de la Sociedad Mexicana de Historia Natural 6:71–82 + 6 plates.

Vargas, L. 1945b. Tres nuevos nombres propuestos para simúlidos del Nuevo Mundo. Revista Medicina Tropical y Parasitologia (Habana) 11:4–5.

Vargas, L. 1945c. Simúlidos del Nuevo Mundo. Monografia del Instituto de Salubridad y Enfermedades Tropicales 1:1–241.

Vargas, L., & A. Díaz Nájera. 1948a. Nota sobre la identificación de los simúlidos de Mexico. El subgénero *Mallochianella* n. n. Revista del Instituto de Salubridad y Enfermedades Tropicales 9:65–73.

Vargas, L., & A. Díaz Nájera. 1948b. Nuevas especies de simúlidos de Mexico y consideraciones diversas sobre especies ya descritas. Revista del Instituto de Salubridad y Enfermedades Tropicales. 9:321–369.

Vargas, L., & A. Díaz Nájera. 1949. Claves para identificar las pupas de los simúlidos de Mexico. Descripción de *Simulium* (*Dyarella*) *freemani* n. sp. de *Simulium* (*Neosimulium*) *encisoi* n. sp. y referencias adicionales sobre *S. anduzei* y *S. ruizi*. Revista del Instituto de Salubridad y Enfermedades Tropicales 10:283–319.

Vargas, L., & A. Díaz Nájera. 1954. Algunas consideraciones morfológicas y de nomenclatura relativas a simúlidos Americanos (Diptera: Simuliidae). Revista del Instituto de Salubridad y Enfermedades Tropicales 14:57–72 + 1 plate.

Vargas, L., & A. Díaz Nájera. 1957. Simúlidos Mexicanos. Revista del Instituto de Salubridad y Enfermedades Tropicales 17:143–399.

Vargas, L., & A. Díaz Nájera. 1958. Nota sobre *Simulium* (*Psilopelmia*) *bivittatum* Malloch, 1914 (Diptera: Simuliidae). Revista del Instituto de Salubridad y Enfermedades Tropicales 18:13–30.

Vargas, L., A. Díaz Nájera, & A. Martínez Palacios. 1943. Tres simúlidos nuevos para Mexico. Revista del Instituto de Salubridad y Enfermedades Tropicales 4:287–290 + 2 plates.

Vargas, L., A. Martínez Palacios, & A. Díaz Nájera. 1946. Simúlidos de Mexico. Datos sobre sistemática y morfología, descripción de nuevos subgéneros y especies. Revista del Instituto de Salubridad Enfermedades Tropicales 7:101–192 + 25 plates.

Vatne, R. D. 1972. Turkey leucocytozoonosis. Down to Earth 28(3):3–5.

Vávra, J., & A. H. Undeen. 1980. The failure of the attempt to infect blackfly larvae by feeding of microsporidian spores. Journal of Protozoology (Program and Abstracts for the Thirty-third Annual Meeting of the Society of Protozoologists) 27:74A.

Vávra, J., & A. H. Undeen. 1981. Microsporidia (Microspora: Microsporida) from Newfoundland blackflies (Diptera: Simuliidae). Canadian Journal of Zoology 59:1431–1446.

Veit, M. F. 1986. Population dynamics and population control of black flies (Diptera: Simuliidae) at a resort area in northern New Hampshire. M.S. thesis. University of New Hampshire, Durham. 94 pp.

Vinikour, W. S. 1982. Phoresis between the snail *Oxytrema* (= *Elima* [*sic*]) *carinifera* and aquatic insects, especially *Rheotanytarsus* (Diptera: Chironomidae). Entomological News 93:143–151.

Vockeroth, J. R. 1966. A method of mounting insects from alcohol. Canadian Entomologist 98:69–70.

Vogt, W. 1936. Editorial. Bird-Lore 38:287.

von Ahlefeldt, J. P. 1968. The histopathology of the major abdominal tissues of *Prosimulium magnum* Dyar and Shannon (Diptera: Simuliidae) parasitized by *Mesomermis* Dadai (Nematoda: Mermithidae). M.S. thesis. Cornell University, Ithaca, N.Y. 50 pp.

Voshell, J. R., Jr. 1991. Life cycle of *Simulium jenningsi* (Diptera: Simuliidae) in southern West Virginia. Journal of Economic Entomology 84:1220–1226.

Wahlberg, P. F. 1844. Nya Diptera fran Norrbotten och Lulea Lappmark. Öfversigt af Kongliga Vetenskaps-Akademiens Förhandlingar 1:107–110. [In Swedish; description in Latin.]

Waldron, M. 1931. Snow man: John Hornby in the Barren Lands. Houghton Mifflin, Boston. 292 pp. [Republished: 1997. Kodansha America, New York.]

Walker, F. 1848. List of the specimens of dipterous insects in the collection of the British Museum. Part 1. London. 229 pp.

Walker, F. 1861. Characters of undescribed Diptera in the collection of W. W. Saunders, Esq., F. R. S., &c. Transactions of the Entomological Society of London (New Series) 5:268–334.

Walker, G. P. 1927. A black fly, (*Simulium bracteatum*), fatal to goslings. Canadian Entomologist 59:123.

Wallace, J. B., & R. W. Merritt. 1980. Filter-feeding ecology of aquatic insects. Annual Review of Entomology 25:103–132.

Wallace, R. R. 1971. The effects of several insecticides on blackfly larvae (Diptera: Simuliidae) and on other stream-dwelling aquatic invertebrates. M.S. thesis. Queen's University, Kingston, Ontario. 144 pp.

Wallace, R. R., & B. N. Hynes. 1981. The effect of chemical treatments against blackfly larvae on the fauna of running waters. Pp. 237–258. *In* M. Laird (ed.), Blackflies: the future for biological methods in integrated control. Academic Press, New York. 399 pp.

Wallace, R. R., A. S. West, A. E. R. Downe, & H. B. N. Hynes. 1973. The effects of experimental blackfly (Diptera: Simuliidae) larviciding with Abate, Dursban, and methoxychlor on stream invertebrates. Canadian Entomologist 105:817–831.

Wallace, R. R., H. B. N. Hynes, & W. F. Merritt. 1976. Laboratory and field experiments with methoxychlor as a larvicide for Simuliidae (Diptera). Environmental Pollution 10:251–269.

Walsh, D. J., D. Yeboah, & M. H. Colbo. 1981. A spherical sampling device for black fly larvae. Mosquito News 41:18–21.

Walsh, J. F. 1980. Sticky trap studies on *Simulium damnosum* s. l. in northern Ghana. Tropenmedizin und Parasitologie 31:479–486.

Washburn, F. L. 1905. The Diptera of Minnesota: two-winged flies affecting the farm, garden, stock and household. University of Minnesota Agricultural Experiment Station Bulletin 93:18–168.

Waters, L. H. 1969. Bionomics of black flies in Maine. M.S. thesis. University of Maine, Orono. 91 pp.

Watson, D. G., J. J. Davis, & W. C. Hanson. 1966. Terrestrial invertebrates. Pp. 565–584. *In* N. J. Wilimovsky & J. N. Wolfe (eds.), Environment of the Cape Thompson region, Alaska. United States Atomic Energy Commission, Division of Technical Information Extension, Oak Ridge, Tenn. 1250 pp. + 6 plates.

Watts, S. B. 1976. Blackflies (Diptera: Simuliidae): a problem review and evaluation. Pest Management Papers (Simon Fraser University) 5:1–118.

Watts, S. B. 1981a. A characterization of the salivary gland proteins of the blood-sucking blackflies, *Simulium vittatum* and *Simulium decorum* (Diptera: Simuliidae). Ph.D. thesis. University of British Columbia, Vancouver. 117 pp.

Watts, S. B. 1981b. A rapid technique for extracting salivary glands from live adult black flies (Diptera: Simuliidae). Mosquito News 41:380–381.

Webb, D. W., & W. U. Brigham. 1982. Aquatic Diptera. Pp. 11.1–11.111. *In* A. R. Brigham, W. U. Brigham, & A. Gnilka (eds.), Aquatic insects and oligochaetes of North and South Carolina. Midwest Aquatic Enterprises, Mahomet, Ill. 837 pp.

Webster, F. M. 1887. Report on buffalo-gnats. United States Department of Agriculture Division of Entomology Bulletin 14:29–39.

Webster, F. M. 1889. *Simulium*, or buffalo gnats. United States Department of Agriculture Fourth and Fifth Annual Reports of the Bureau of Animal Industry 1887–1888:456–465.

Webster, F. M. 1891. Buffalo gnats (Simuliidae) [*sic*] in Indiana and Illinois. Proceedings of the Indiana Academy of Science 1:155–159.

Webster, F. M. 1902. Winds and storms as agents in the diffusion of insects. American Naturalist 36:795–801.

Webster, F. M. 1904. The suppression and control of the plague of buffalo gnats in the Valley of the Lower Mississippi River, and the relations thereto of the present levee system, irrigation in the arid West and tile drainage in the Middle West. Proceedings of the Annual Meeting of the Society for the Promotion of Agricultural Science 25:53–72.

Webster, F. M. 1914. Natural enemies of *Simulium*: notes. Psyche 21:95–99.

Webster, F. M., & W. Newell. 1902. Insects of the year in Ohio. United States Department of Agriculture Division of Entomology Bulletin 31(New Series): 84–90.

Webster, J. M. 1973. Manipulation of environment to facilitate use of nematodes in biocontrol of insects. Experimental Parasitology 33:197–206.

Weed, C. M. 1904a. An experiment with black flies. United States Department of Agriculture Division of Entomology Bulletin 46:108–109.

Weed, C. M. 1904b. Experiments in destroying black-flies. New Hampshire College Agricultural Experiment Station Bulletin 112:132–136.

Wehr, E. E. 1962. Studies on leucocytozoonosis of turkeys, with notes on schizogony, transmission, and control of *Leucocytozoon smithi*. Avian Diseases 6:195–210.

Weinmann, C. J., J. R. Anderson, W. M. Longhurst, & G. Connolly. 1973. Filarial worms of Columbian black-tailed deer in California. 1. Observations in the vertebrate host. Journal of Wildlife Diseases 9:213–220.

Weiser, J. 1968. Iridescent virus from the blackfly *Simulium ornatum* Meigen in Czechoslovakia. Journal of Invertebrate Pathology 12:36–39.

Weiser, J. 1978. A new host, *Simulium argyreatum* Meig., for the cytoplasmic polyhedrosis virus of blackflies in Czechoslovakia. Folia Parasitologica 25:361–365.

Weiser, J., & A. H. Undeen. 1981. Diseases of blackflies. Pp. 181–196. *In* M. Laird (ed.), Blackflies: the future for biological methods in integrated control. Academic Press, New York. 399 pp.

Weiser, J., & Z. Žižka. 1974a. The ultrastructure of the chytrid *Coelomycidium simulii* Deb. I. Ultrastructure of the thalli. Česká Mykologie 28:159–162.

Weiser, J., & Z. Žižka. 1974b. The ultrastructure of the chytrid *Coelomycidium simulii* Deb. II. Division of the thalius [*sic*] and structures of zoospores. Česká Mykologie 28:227–232.

Welch, H. E. 1962. New species of *Gastromermis*, *Isomermis*, and *Mesomermis* (Nematoda: Mermithidae) from black fly larvae. Annals of the Entomological Society of America 55:535–542.

Wellington, W. G. 1974. Black-fly activity during cumulus-induced pressure fluctuations. Environmental Entomology 3:351–353.

Wenk, P. 1981. Bionomics of adult blackflies. Pp. 259–279. *In* M. Laird (ed.), Blackflies: the future for biological methods in integrated control. Academic Press, New York. 399 pp.

Wenk, P. 1988 ["1987"]. Swarming and mating behavior of black flies. Pp. 215–227. *In* K. C. Kim & R. W. Merritt (eds.), Black flies: ecology, population management, and annotated world list. Pennsylvania State University, University Park. 528 pp. [Actual publication date was 1 March 1988.]

Werner, D. 1996. Die Simuliiden-Typen der Dipteren-Sammlung des Zoologischen Museums Berlin (Diptera: Simuliidae) [The types of Simuliidae in the Diptera Collection of the Zoological Museum, Berlin (Diptera: Simuliidae)]. Mitteilungen aus dem Zoologischen Museum in Berlin 72:221–258. [In German.]

Werner, D. 1998. Studies on bioacoustic behaviour in blackflies (Simuliidae). British Simuliid Group Bulletin 12:13.

West, A. S. 1961. Biting fly control on the Quebec North Shore. Proceedings of the Annual Meeting of the New Jersey Mosquito Extermination Association 48:87–96.

West, A. S. 1964. Canadian experience in handling blackflies under laboratory conditions. Bulletin of the World Health Organization 31:487–489.

West, A. S. 1973. Recent Canadian developments in area chemical control of biting flies. Pp. 15–18. *In* A. Hudson (ed.), Symposium on biting fly control and environmental quality. Publication No. DR 217. Defence Research Board, Ottawa, Canada. 162 pp.

West, A. S., A. W. A. Brown, & D. G. Peterson. 1960. Control of black fly larvae (Diptera: Simuliidae) in the forests of eastern Canada by aircraft spraying. Canadian Entomologist 92:745–754.

Westwood, A. R. 1979. Ecological studies on the black flies of the Souris River, Manitoba. M.S. thesis. University of Manitoba, Winnipeg. 183 pp.

Westwood, A. R., & R. A. Brust. 1981. Ecology of black flies (Diptera: Simuliidae) of the Souris River, Manitoba as a basis for control strategy. Canadian Entomologist 113:223–234.

Wetmore, S. H., R. J. Mackay, & R. W. Newbury. 1990. Characterization of the hydraulic habitat of *Brachycentrus occidentalis*, a filter-feeding caddisfly. Journal of the North American Benthological Society 9:157–169.

White, D. A. 1969. The infection of immature aquatic insects by larval *Paragordius* (Nematomorpha). Great Basin Naturalist 29:44.

White, D. J. 1983. Predation of *Prosimulium mixtum/fuscum* (Diptera: Simuliidae) copulating pairs by *Formica* ants (Hymenoptera: Formicidae). New York Entomological Society 91:90–91.

White, D. J. 1984. Bionomics of the hominoxious Simuliidae and Tabanidae of the New York State Adirondack Mountain region. Ph.D. thesis. State University of New York, Syracuse. 135 pp.

White, D. J., & C. D. Morris. 1985a. Bionomics of anthropophilic Simuliidae (Diptera) from the Adirondack Mountains of New York State, USA. 1. Adult dispersal and longevity. Journal of Medical Entomology 22:190–199.

White, D. J., & C. D. Morris. 1985b. Seasonal abundance of anthropophilic Simuliidae from the Adirondack Mountains of New York State and effectiveness of an experimental treatment program using *Bacillus thuringiensis* var. *israeliensis* [*sic*]. Environmental Entomology 14:464–469.

Whittaker, R. H. 1952. A study of summer foliage insect communities in the Great Smoky Mountains. Ecological Monographs 22:1–44.

Wiley, M. J., & S. L. Kohler. 1981. An assessment of biological interactions in an epilithic stream community using time-lapse cinematography. Hydrobiologia 78:183–188.

Williams, M. C., & R. W. Lichtwardt. 1971. A new *Pennella* (Trichomycetes) from *Simulium* larvae. Mycologia 63:910–914.

Williams, M. C., & R. W. Lichtwardt. 1987. Three new species of *Smittium* (Tri-

chomycetes) with notes on range extensions. Mycologia 79:832–838.
Williams, N. A., J. L. Mahrt, & F. C. Zwickel. 1980. The ecology of blood parasites in blue grouse from Vancouver Island, British Columbia. Canadian Journal of Zoology 58:2175–2186.
Williams, T., & J. Cory. 1991. A new record of hymenopterous parasitism of an immature blackfly (Diptera: Simuliidae). Entomologist's Gazette 42:220–221.
Williams, T. R. 1974. Egg membranes of Simuliidae. Transactions of the Royal Society of Tropical Medicine and Hygiene 68:15–16.
Williams, T. R., C. E. Denley, & H. M. Wain. 1993. Is the simuliid pupal gill a plastron? British Simuliid Group Bulletin 2:12–13.
Williston, S. W. 1893. List of Diptera of the Death Valley Expedition. Pp. 253–259. *In* C. V. Riley (ed.), The Death Valley Expedition. A biological survey of parts of California, Nevada, Arizona, and Utah. Part II. 4. Report on a small collection of insects made during the Death Valley Expedition. North American Fauna 7:235–268.
Williston, S. W. 1908. Manual of North American Diptera. 3rd ed. J. T. Hathaway, New Haven, Conn. 405 pp.
Wilson, C. S., K. H. Applewhite, & L. M. Redlinger. 1949. Heavy ground aerosol generators for the control of adult biting insects in Alaska. Mosquito News 9:97–101.
Wilson, D. E., & D. M. Reeder (eds.). 1993. Mammal species of the world. Smithsonian Institution Press, Washington, D. C. 1206 pp.
Wilton, D. P., & B. V. Travis. 1965. An improved method for simulated stream tests of blackfly larvicides. Mosquito News 25:118–123.
Wipfli, M. S. 1992. Direct and indirect effects of the black fly (Diptera: Simuliidae) larvicide, *Bacillus thuringiensis* var. *israelensis*, on selected non-target aquatic insects and trout. Ph.D. thesis. Michigan State University, East Lansing. 162 pp.
Wipfli, M. S., & R. W. Merritt. 1994a. Disturbance to a stream food web by a bacterial larvicide specific to black flies: feeding responses of predatory macroinvertebrates. Freshwater Biology 32:91–103.
Wipfli, M. S., & R. W. Merritt. 1994b. Effects of *Bacillus thuringiensis* var. *israelensis* on nontarget benthic insects through direct and indirect exposure. Journal of the North American Benthological Society 13:190–205.
Wipfli, M. S., R. W. Merritt, & W. W. Taylor. 1994. Low toxicity of the black fly larvicide *Bacillus thuringiensis* var. *israelensis* to early stages of brook trout (*Salvelinus fontinalis*), brown trout (*Salmo trutta*), and steelhead trout (*Oncorhynchus mykiss*) following direct and indirect exposure. Canadian Journal of Fisheries and Aquatic Sciences 51:1451–1458.
Wirth, W. W., & A. Stone. 1956. Aquatic Diptera. Pp. 372–482. *In* R. L. Usinger (ed.), Aquatic insects of California with keys to North American genera and California species. University of California Press, Berkeley. 508 pp.
Wiseman, J. S., & R. B. Eads. 1960. Texas blackfly records (Diptera: Simuliidae). Mosquito News 20:45–49.
Wolfe, L. S., & D. G. Peterson. 1958. A new method to estimate levels of infestations of black-fly larvae (Diptera: Simuliidae). Canadian Journal of Zoology 36:863–867.
Wolfe, L. S., & D. G. Peterson. 1959. Black flies (Diptera: Simuliidae) of the forests of Quebec. Canadian Journal of Zoology 37:137–159.
Wolfe, L. S., & D. G. Peterson. 1960. Diurnal behavior and biting habits of black flies (Diptera: Simuliidae) in the forests of Quebec. Canadian Journal of Zoology 38:489–497.
Wolkomir, R., & J. Wolkomir. 1992. When scientists become sleuths. National Wildlife 30(2):8–15.
Wong, C. J. L.-k. 1976. Drift of black-fly larvae and the influence of water-velocity, substrate roughness and incident light intensity on their microdistribution (Diptera: Simuliidae). M.S. thesis. McMaster University, Hamilton, Ontario. 289 pp.
Woo, P. T. K. 1964. A study on the blood protozoa of blue grouse on Vancouver Island. M.S. thesis. University of British Columbia, Vancouver. 71 pp.
Wood, D. M. 1963a. An interpretation of the phylogeny of the *Eusimulium*-group (Diptera: Simuliidae) with descriptions of six new species. Ph.D. thesis. McMaster University, Hamilton, Ontario. 234 pp.
Wood, D. M. 1963b. Two new species of Ontario black flies (Diptera: Simuliidae). Proceedings of the Entomological Society of Ontario 93:94–98.
Wood, D. M. 1978. Taxonomy of the Nearctic species of *Twinnia* and *Gymnopais* (Diptera: Simuliidae) and a discussion of the ancestry of the Simuliidae. Canadian Entomologist 110:1297–1337.
Wood, D. M. 1985. Biting flies attacking man and livestock in Canada. Agriculture Canada Publication No. 1781/E. Agriculture Canada, Ottawa. 38 pp.
Wood, D. M. 1991. Homology and phylogenetic implications of male genitalia in Diptera. The ground plan. Pp. 255–284. *In* L. Weismann, I. Országh, & A. C. Pont (eds.), Proceedings of the Second International Congress of Dipterology. SPB Academic Publishing, The Hague. 367 pp.
Wood, D. M., & A. Borkent. 1982. Description of the female of *Parasimulium crosskeyi* Peterson (Diptera: Simuliidae), and a discussion of the phylogenetic position of the genus. Memoirs of the Entomological Society of Washington 10:193–210.
Wood, D. M., & A. Borkent. 1989. Phylogeny and classification of the Nematocera. Pp. 1333–1370. *In* J. F. McAlpine (ed.), Manual of Nearctic Diptera, vol. 3, Monograph No. 32. Research Branch, Agriculture Canada, Ottawa. 249 pp.
Wood, D. M., & D. M. Davies. 1965. The rearing of simuliids (Diptera). Proceedings of the Twelfth International Congress of Entomology, London (1964):821–823.
Wood, D. M., & D. M. Davies. 1966. Some methods of rearing and collecting black flies (Diptera: Simuliidae). Proceedings of the Entomological Society of Ontario 96:81–90.
Wood, D. M., B. V. Peterson, D. M. Davies, & H. Györkos. 1963. The black flies (Diptera: Simuliidae) of Ontario. Part II. Larval identification, with descriptions and illustrations. Proceedings of the Entomological Society of Ontario 93:99–129.
Woods, J. 1822. Two years' residence in the settlement on the English Prairie, in the Illinois country, United States. Longman, Hurst, Rees, Orme, and Brown, London. 310 pp. [Reprint version: 1968 by Lakeside Press, R. R. Donnelley & Sons, Chicago, Illinois. 242 pp.; reference to buffalo gnats appears on p. 200 of reprint version].
Wotton, R. S. 1976. Evidence that blackfly larvae can feed on particles of colloidal size. Nature 261:697.
Wotton, R. S. 1978a. The feeding-rate of *Metacnephia tredecimatum* larvae (Diptera: Simuliidae) in a Swedish lake outlet. Oikos 30:121–125.
Wotton, R. S. 1978b. Growth, respiration, and assimilation of blackfly larvae (Diptera: Simuliidae) in a lake-outlet in Finland. Oecologia 33:279–290.
Wotton, R. S. 1980. Coprophagy as an economic feeding tactic in blackfly larvae. Oikos 34:282–286.
Wotton, R. S. 1982a. Difference in carbon weight of the immature stages of two co-existing species of blackflies (Diptera: Simuliidae) with contrasting reproductive strategies. Hydrobiologia 94:279–283.
Wotton, R. S. 1982b. Does the surface film of lakes provide a source of food for animals living in lake outlets? Limnology and Oceanography 27:959–960.
Wotton, R. S. 1984. The relationship between food particle size and larval size in *Simulium noelleri* Friederichs. Freshwater Biology 14:547–550.

Wotton, R. S. 1985. The reaction of larvae of *Simulium noelleri* (Diptera) to different current velocities. Hydrobiologia 123:215–218.

Wotton, R. S. 1986. The use of silk life-lines by larvae of *Simulium noelleri* (Diptera). Aquatic Insects 8:255–261.

Wotton, R. S. 1987. Lake outlet blackflies—the dynamics of filter feeders at very high population densities. Holarctic Ecology 10:65–72.

Wotton, R. S. 1988a ["1987"]. The ecology of lake-outlet black flies. Pp. 146–154. *In* K. C. Kim & R. W. Merritt (eds.), Black flies: ecology, population management, and annotated world list. Pennsylvania State University, University Park. 528 pp. [Actual publication date was 1 March 1988.]

Wotton, R. S. 1988b. Very high secondary production at a lake outlet. Freshwater Biology 20:341–346.

Wotton, R. S. 1992. Feeding by blackfly larvae (Diptera: Simuliidae) forming dense aggregations at lake outlets. Freshwater Biology 27:139–149.

Wotton, R. S., and R. W. Merritt. 1988. Experiments on predation and substratum choice by larvae of the muscid fly, *Limnophora riparia*. Holarctic Ecology 11:151–159.

Wotton, R. S., M. S. Wipfli, L. Watson, & R. W. Merritt. 1993. Feeding variability among individual aquatic predators in experimental channels. Canadian Journal of Zoology 71:2033–2037.

Wotton, R. S., B. Malmqvist, T. Muotka, & K. Larsson. 1998. Fecal pellets from a dense aggregation of suspension-feeders in a stream: an example of ecosystem engineering. Limnology and Oceanography 43:719–725.

Wright, R. E., & G. R. DeFoliart. 1970. Some hosts fed upon by ceratopogonids and simuliids. Journal of Medical Entomology 7:600.

Wu, Y. F. 1931. A contribution to the biology of *Simulium* (Diptera). Papers of the Michigan Academy of Science, Arts, and Letters 13:543–599.

Wygodzinsky, P. 1973. On a species of *Simulium* (*Ectemnaspis*) from the northeastern United States (Diptera). Journal of the New York Entomological Society 81:10–12.

Wygodzinsky, P., & S. Coscarón. 1973. A review of the Mesoamerican and South American black flies of the tribe Prosimuliini (Simuliinae, Simuliidae). Bulletin of the American Museum of Natural History 151:129–199.

Wygodzinsky, P., & S. Coscarón. 1989. Revision of the black fly genus *Gigantodax* (Diptera: Simuliidae). Bulletin of the American Museum of Natural History 189:1–269.

Wygodzinsky, P., & A. Díaz Nájera. 1970. Un nuevo género de simúlido de la República Mexicana (Diptera: Simuliidae). Revista de Investigación en Salud Pública (México) 30:83–110.

Xiong, B. 1992. Intraspecific variability and phylogeny of sibling species of *Simulium venustum* and *S. verecundum* complexes (Diptera: Simuliidae) revealed by the sequence of the mitochondrial large (16S) ribosomal RNA gene. Ph.D. thesis. University of New Hampshire, Durham. 104 pp.

Xiong, B., & M. Jacobs-Lorena. 1995a. The black fly *Simulium vittatum* trypsin gene: characterization of the 5′-upstream region and induction by the blood meal. Experimental Parasitology 81:363–370.

Xiong, B., & M. Jacobs-Lorena. 1995b. Gut-specific transcriptional regulatory elements of the carboxypeptidase gene are conserved between black flies and *Drosophila*. Proceedings of the National Academy of Sciences of the United States of America 92:9313–9317.

Xiong, B., & T. D. Kocher. 1991. Comparison of mitochondrial DNA sequences of seven morphospecies of black flies (Diptera: Simuliidae). Genome 34:306–311.

Xiong, B., & T. D. Kocher. 1993a. Intraspecific variation in sibling species of *Simulium venustum* and *Simulium verecundum* complexes (Diptera: Simuliidae) revealed by the sequence of the mitochondrial 16S rRNA gene. Canadian Journal of Zoology 71:1202–1206.

Xiong, B., & T. D. Kocher. 1993b. Phylogeny of sibling species of *Simulium venustrum* [*sic*] and *S. verecundum* (Diptera: Simuliidae) based on sequences of the mitochondrial 16S rRNA gene. Molecular Phylogenetics and Evolution 2:293–303.

Yakuba, V. N. 1960. On the number of stages in larvae of black flies (Simuliidae, Diptera). Trudy Vostochno-Sibirskogo Filiala (Biologicheskaya) 22:136–140. [In Russian.]

Yakuba, V. N. 1963. The blood-sucking Diptera of Yakutia and their epidemiological importance. Pp. 431–433. *In* Problems of parasitology. Transactions of the IV Scientific Conference of Parasitologists of the Ukranian SSR [Problemy parazitologee. Trudy IV Nauchnoi Konferentsee Parazitologov USSR.] Izdatel'stvo Akademii Nauk Ukrainskoi SSR, Kiev, Ukraine. 501 pp. [In Russian.]

Yang, Y. J. 1968. A study of ingestion and digestion, emphasizing the peritrophic membrane and digestive enzymes in adult simuliids (Diptera) fed blood, blood-sucrose mixtures and sucrose. Ph.D. thesis. McMaster University, Hamilton, Ontario. 177 pp.

Yang, Y. J., & D. M. Davies. 1968a. Amylase activity in black-flies and mosquitoes (Diptera). Journal of Medical Entomology 5:9–13.

Yang, Y. J., & D. M. Davies. 1968b. Digestion, emphasizing trypsin activity, in adult simuliids (Diptera) fed blood, blood-sucrose mixtures, and sucrose. Journal of Insect Physiology 14:205–222.

Yang, Y. J., & D. M. Davies. 1968c. Occurrence and nature of invertase activity in adult black-flies (Simuliidae). Journal of Insect Physiology 14:1221–1232.

Yang, Y. J., & D. M. Davies. 1974. The saliva of adult female blackflies (Simuliidae: Diptera). Canadian Journal of Zoology 52:749–753.

Yang, Y. J., & D. M. Davies. 1977. The peritrophic membrane in adult simuliids (Diptera) before and after feeding on blood and blood-sucrose mixtures. Entomologia Experimentalis et Applicata 22:132–140.

Yankovsky, A. V. 1977. The taxonomic importance of premandibular structure in black fly larvae (Diptera, Simuliidae). Entomologicheskoe Obozrenie 56:435–440. [In Russian; English translation in Entomological Review, Washington 56:143–146.]

Yankovsky, A. V. 1978. The ecology and systematics of the striped black flies of the group *Byssodon maculata* Meigen (Diptera, Simuliidae). Entomologicheskoe Obozrenie 57:169–179. [In Russian; English translation in Entomological Review, Washington 57:118–124.]

Yankovsky, A. V. 1979. Two new species of the genus *Cnetha* (Diptera: Simuliidae). Entomologicheskoe Obozrenie 58:172–178. [In Russian; English translation in Entomological Review, Washington 58:100–105.]

Yankovsky, A. V. 1982. Two new species of black-flies (Simuliidae), *Hemicnetha tsharae* sp. n. and *Simulium vershininae* sp. n., from Transbaikal region (The Chita District, Charskaya Hollow). Parazitologiya 16:248–253. [In Russian; English summary.]

Yankovsky, A. V. 1992. A review of the system of blackflies (Diptera, Simuliidae) on the suprageneric level. Parazitologichesky Sbornik 37:203–217. [In Russian; English summary.]

Yankovsky, A. V. 1995. Family Simuliidae Newman. Pp. 1–61. *In* A. V. Yankovsky & K. N. Ulyanov (compilers), Catalogue of type specimens in the collection of the Zoological Institute, Russian Academy of Sciences [Katalog tipovykh ekzemplyarov kollektsii Zoologicheckogo Instituta RAN.] Diptera. 5. Simuliidae, Culicidae. Rossiiskaya Akademiya Nauk, Zoologicheskii Institut, St. Petersburg, Russia. 64 pp. [In Russian.]

Yankovsky, A. V. 1996. Additions to the synonymy of generic names, and to the homonymy of species names of Palaearctic blackflies (Diptera: Simuliidae). Parazitologiya 30:113–116. [In Russian; English summary.]

Yankovsky, A. V. 1999. Simuliidae. Pp. 154–182, 564–579. *In* S. J. Tsalolikhin (ed.), Key to freshwater invertebrates of Russia and adjacent lands [Opredelitel' presnovodnykh bespozvonochnykh Rossii i sopredel'nykh territorii]. Rossiiskaya Akademiya Nauk, Zoologicheskii Institut, St. Petersburg, Russia. 1000 pp. [In Russian.]

Yeboah, D. O. 1980. A survey of the prevalence and study of the effects of an ovarian phycomycete in some Newfoundland blackflies. M.S. thesis. Memorial University of Newfoundland, St. John's. 55 pp.

Yeboah, D. O., A. H. Undeen, & M. H. Colbo. 1984. Phycomycetes parasitizing the ovaries of blackflies (Simuliidae). Journal of Invertebrate Pathology 43:363–373.

Yee, W. L., & J. R. Anderson. 1995. Trapping black flies (Diptera: Simuliidae) in northern California. II. Testing visual cues used in attraction to $CO_2$-baited animal head models. Journal of Vector Ecology 20:26–39.

Yoerg, S. I. 1994. Development of foraging behaviour in the Eurasian dipper, *Cinclus cinclus*, from fledging until dispersal. Animal Behavior 47:577–588.

Yoon, I.-B., & M. Y. Song. 1989. A revision of the taxonomy of Korean black-flies (Simuliidae: Diptera). I. The larval and pupal stages of subgenus *Simulium*. Entomological Research Bulletin (Korea) 15:35–64.

Young, W. W. 1958. Biting Diptera of medical importance in Kansas. M.S. thesis. Kansas State College of Agriculture and Applied Science (= Kansas State University), Manhattan. 143 pp.

Zetterstedt, J. W. 1837. Conspectus familiarum, generum et specierum dipterorum, in fauna insectorum Lapponica descriptorum, pp. 28–67. Isis. L. Oken, Leipzig, Germany.

Zetterstedt, J. W. 1838. Sectio tertia, Diptera. Dipterologis scandinaviae amicis et popularibus carissimis. Pp. 477–868. *In* Insecta Lapponica. L. Voss, Lipsiae [= Leipzig], Germany, "1840." 1140 pp.

Zettler, J. A. 1996. Factors influencing larval color in the *Simulium vittatum* complex (Diptera: Simuliidae). M.S. thesis. Clemson University, Clemson, S.C. 85 pp.

Zettler, J. A., P. H. Adler, & J. W. McCreadie. 1998. Factors influencing larval color in the *Simulium vittatum* complex (Diptera: Simuliidae). Invertebrate Biology 117:245–252.

Zettler, L. W., & J. E. Fairey III. 1990. The status of *Platanthera integrilabia*, an endangered terrestrial orchid. Lindleyana 5:212–217.

Zhang, Y. 2000. Effects of fan morphology and habitat on feeding performance of blackfly larvae. Archiv für Hydrobiologie 149:365–386.

Zhang, Y., & B. Malmqvist. 1996. Relationships between labral fan morphology, body size and habitat in North Swedish blackfly larvae (Diptera: Simuliidae). Biological Journal of the Linnean Society 59:261–280.

Zhang, Y., & B. Malmqvist. 1997. Phenotypic plasticity in a suspension-feeding insect, *Simulium lundstromi* (Diptera: Simuliidae), in response to current velocity. Oikos 78:503–510.

Zhu, X. 1990. Mitochondrial DNA polymorphism in black flies (Diptera: Simuliidae). Ph.D. thesis. University of Nebraska, Lincoln. 230 pp.

Zhu, X., K. P. Pruess, & T. O. Powers. 1998. Mitochondrial DNA polymorphism in a black fly, *Simulium vittatum* (Diptera: Simuliidae). Canadian Journal of Zoology 76:440–447.

Zimring, R. 1953. A comparative study of the salivary gland chromosomes of five related black fly species. M.A. thesis. University of Toronto. Toronto. 32 pp. + 13 plates.

Zwick, H. 1974. Faunistisch-ökologische und taxonomische Untersuchungen an Simuliidae (Diptera), unter besonderer Berücksichtigung der Arten des Fulda-Gebietes. Abhandlungen der senckenbergischen naturforschenden Gesellschaft 533:1–116.

Zwick, H. 1986. Lectotype designation for *Simulium noelleri* Friederichs, 1920 (Diptera: Simuliidae). Aquatic Insects 8:140.

Zwick, H. 1987. Identity of *Simulium rostratum* (Diptera: Simuliidae). Aquatic Insects 9:26.

Zwick, H. 1995. Contribution to the European blackfly taxa (Diptera: Simuliidae) named by Enderlein. Aquatic Insects 17:129–173.

Zwick, H., & R. W. Crosskey. 1981 ["1980"]. The taxonomy and nomenclature of the blackflies (Diptera: Simuliidae) described by J. W. Meigen. Aquatic Insects 2:225–247. [Actual publication date was 10 February 1981.]

Zwick, H., & W. Rühm. 1973. Erstnachweis von *Simulium sublacustre* Davies 1966 in Mitteleuropa. Ein Beitrag zur Simuliidenfauna des Aller-Leine-Gebietes. Zeitschrift für angewandte Entomologie 72:429–434.

Zwick, H., & P. Zwick. 1990. Terrestrial mass-oviposition of *Prosimulium*-species (Diptera: Simuliidae). Aquatic Insects 12:33–46.

Zwolski, W. 1956. Mustyki (Simuliidae) Lubelszczyzny. Annales Universitatis Mariae Curie-Skłodowska 13:231–259. [In Polish; English translation: 1963. Simuliidae of the Lublin district. Centralny Instytut Informacji Naukowo-Technicznej i Ekonomicznej. 25 pp.]

# Figure Credits

The Smithsonian Institution Archives, Washington, D.C., granted permission to publish photographs of J. H. Comstock, H. G. Dyar, and C. V. Riley (Record Unit 95, Photograph Collection, 1850s–; negatives #89-3785, #SA-1278, and #10583, respectively); G. F. Knowlton, J. R. Malloch, and R. C. Shannon (Record Unit 7298, Charles P. Alexander Papers, circa 1870–1979; negatives #99-42639, #99-1581, and #83-14174, respectively); and O. Lugger and C. H. T. Townsend (Record Unit 7323, Systematic Entomology Laboratory, USDA, 1797–1988 and undated, Photographs and Biographical Information; negatives #99-1580 and 99-1579, respectively). Pictures of the following individuals were obtained from the L. O. Howard photographic collection of American and foreign scientists (Record Group 7H) in the National Archives, College Park, Maryland: W. S. Barnard, A. B. Comstock, D. W. Coquillett, S. A. Forbes, H. W. Garman, H. A. Hagen, T. W. Harris, C. A. Hart, T. J. Headlee, O. A. Johannsen, F. W. Knab, T. Say, F. M. Webster, and S. W. Williston. The picture of G. S. Stains was provided courtesy of the U. S. Navy Disease Vector Ecology and Control Center, Bangor, Washington.

Figures 10.344 and 10.460 are reproduced from Moulton and Adler (2002a) with permission from Andreas Stark. Figures 10.314a, b, and e are reproduced from Moulton and Adler (2002b) with permission from Magnolia Press. The following figures are reproduced from Peterson (1981), Agriculture and Agri-Food Canada, with permission of the Minister of Public Works and Government Services, 2002: 4.3–4.7, 4.9–4.12, 4.14, 4.15, 4.17–4.23, 4.30, 4.31, 4.44–4.49, 10.18, 10.19, 10.34, 10.38, 10.42, 10.43, 10.75, 10.77–10.81, 10.90, 10.93, 10.101, 10.159, 10.180, 10.189, 10.198, 10.205, 10.220, 10.226, 10.229a, 10.236, 10.320, 10.341, 10.351, 10.360, 10.363, 10.370, 10.379, 10.403, 10.466, 10.618, 10.625–10.627, 10.630, 10.633, 10.636, 10.637, and 10.646. Permission to reproduce Figure 4.64 was granted by Peter Goldie, and permission to reproduce the frontispiece was given by the Canadian Forest Service, Ottawa. Neil K. Dawe of the Canadian Wildlife Service gave permission to reproduce Fig. 7.3. The following people provided their original photographic prints or slides (with the respective figure numbers in parentheses): Stephen A. Marshall (6.1), Jacques L. Boisvert and Martin P. Nadeau (6.3), Robert W. Lichtwardt (6.6), J. Robert Harkrider (6.7), Gayle P. Noblet (6.15), James F. Sutcliffe (7.2), Raymond Noblet (7.4), Daniel P. Molloy (8.5), Robert A. Fusco (8.6), and John Smink (8.7). Figure 2.28 was provided by Carl C. Reading. Figures 10.542–10.549, 10.551, 10.553–10.555, 10.557, 10.562–10.565, 10.567, 10.569, and 10.571 were redrawn and modified from Peterson (1970) and 10.228a, d; 10.229d; and 10.230a, d from Peterson (1977a). Figure 3.1 was redrawn and modified from Hunter et al. (1994). Figures 4.33, 4.50, 4.51, 10.619, and 10.620 were redrawn and modified from Peterson (1996).

# Index to Names of Black Flies

This index lists all formal, vernacular, and misspelled names of North American black flies, as well as many names of black flies in other areas of the world. Vernacular names are given in single quotes. Misspelled names are followed by the word "[*sic*]." Misidentifications typically are not indexed. Authors and generic names are given only for vernacular names that are not associated with a formal species name or for taxa of the same name but described by different authors. The endings of all formal, adjectival species names (except misspellings) in the index conform with the gender of the genus in which they are currently recognized. Names in boldface represent valid North American species. Page numbers in boldface indicate the location of the principal taxonomic entry for each valid North American species and higher taxonomic category. Page numbers in italics indicate the locations of figures.

# Index to Names of Organisms Other Than Black Flies

This index includes common and scientific names of organisms other than black flies. Names appearing in the text that are derived informally from scientific category names are given under the relevant Latin names; for example, empidid fly is listed as Empididae and mermithid nematode as Mermithidae. Common or general names are provided parenthetically for each scientific name.

# Subject Index

Names of people are given only if mentioned in a biographical or historical context. Morphological terms in the taxonomic descriptions and diagnoses are not indexed. Names of all rivers mentioned in the text are listed (under "rivers"), along with the country, state, or region corresponding to the sites indicated in the text; names of smaller flows are not indexed. States and provinces are indexed only for Chapters 1–9; state and provincial references in Chapter 10 can be obtained from the distribution maps (pages 597–852).

# About the Authors

Peter H. Adler was born in South Charleston, West Virginia, in 1954. He received his bachelor of science degree in biology from Washington and Lee University and his master's degree and doctorate from Pennsylvania State University. He conducted postdoctoral work at the University of Alberta before moving to Clemson University in 1984, where he is currently Professor of Entomology.

Douglas C. Currie was born in Edmonton, Alberta, in 1955. He received his bachelor of science degree and doctorate from the University of Alberta. His postdoctoral work was carried out at the University of British Columbia and with the Biological Resources Program in Ottawa. He has been a curator at the Royal Ontario Museum, with a joint appointment at the University of Toronto, since 1993.

D. Monty Wood was born in London, Ontario, in 1933. He received his bachelor's and master's degrees from the University of Toronto and his doctorate from McMaster University. He was employed with Agriculture Canada in Ottawa as a specialist on Diptera from 1964 to 1990 and is currently an honorary research associate with that institution.

# About the Artists

Ralph Idema has been drawing insects for more than 40 years and is recognized as one of the world's premier biological illustrators, having produced illustrations for many publications such as the *Manual of Nearctic Diptera*. He continues to work as a biological illustrator in Ottawa, Ontario.

Lawrence W. Zettler completed the color illustrations of black flies while he was a doctoral student at Clemson University. He has illustrated a book on dragonflies and is currently a professor of biology and a specialist on orchids at Illinois College.